Lecture Notes in Computer Science 3195

Commenced Publication in 1973
Founding and Former Series Editors:
Gerhard Goos, Juris Hartmanis, and Jan van Leeuwen

Editorial Board

David Hutchison
  Lancaster University, UK
Takeo Kanade
  Carnegie Mellon University, Pittsburgh, PA, USA
Josef Kittler
  University of Surrey, Guildford, UK
Jon M. Kleinberg
  Cornell University, Ithaca, NY, USA
Friedemann Mattern
  ETH Zurich, Switzerland
John C. Mitchell
  Stanford University, CA, USA
Moni Naor
  Weizmann Institute of Science, Rehovot, Israel
Oscar Nierstrasz
  University of Bern, Switzerland
C. Pandu Rangan
  Indian Institute of Technology, Madras, India
Bernhard Steffen
  University of Dortmund, Germany
Madhu Sudan
  Massachusetts Institute of Technology, MA, USA
Demetri Terzopoulos
  New York University, NY, USA
Doug Tygar
  University of California, Berkeley, CA, USA
Moshe Y. Vardi
  Rice University, Houston, TX, USA
Gerhard Weikum
  Max-Planck Institute of Computer Science, Saarbruecken, Germany

Carlos G. Puntonet   Alberto Prieto (Eds.)

Independent Component Analysis and Blind Signal Separation

Fifth International Conference, ICA 2004
Granada, Spain, September 22-24, 2004
Proceedings

 Springer

Volume Editors

Carlos G. Puntonet
Alberto Prieto
Universidad de Granada, E.T.S. de Ingeniería Informática
Departamento de Arquitectura y Tecnología de Computadores
C/Periodista Daniel Saucedo, sn, 18071 Granada, Spain
E-mail: carlos@atc.ugr.es, aprieto@ugr.es

Library of Congress Control Number: 2004112253

CR Subject Classification (1998): C.3, F.1.1, E.4, F.2.1, G.3, H.1.1, H.5.1, I.2.7

ISSN 0302-9743
ISBN 3-540-23056-4 Springer Berlin Heidelberg New York

This work is subject to copyright. All rights are reserved, whether the whole or part of the material is concerned, specifically the rights of translation, reprinting, re-use of illustrations, recitation, broadcasting, reproduction on microfilms or in any other way, and storage in data banks. Duplication of this publication or parts thereof is permitted only under the provisions of the German Copyright Law of September 9, 1965, in its current version, and permission for use must always be obtained from Springer. Violations are liable to prosecution under the German Copyright Law.

Springer is a part of Springer Science+Business Media

springeronline.com

© Springer-Verlag Berlin Heidelberg 2004
Printed in Germany

Typesetting: Camera-ready by author, data conversion by Olgun Computergrafik
Printed on acid-free paper   SPIN: 11320678   06/3142   5 4 3 2 1 0

Preface

In many situations found both in Nature and in human-built systems, a set of mixed signals is observed (frequently also with noise), and it is of great scientific and technological relevance to be able to isolate or separate them so that the information in each of the signals can be utilized.

Blind source separation (BSS) research is one of the more interesting emerging fields nowadays in the field of signal processing. It deals with the algorithms that allow the recovery of the original sources from a set of mixtures only. The adjective "blind" is applied because the purpose is to estimate the original sources without any a priori knowledge about either the sources or the mixing system. Most of the models employed in BSS assume the hypothesis about the independence of the original sources. Under this hypothesis, a BSS problem can be considered as a particular case of independent component analysis (ICA), a linear transformation technique that, starting from a multivariate representation of the data, minimizes the statistical dependence between the components of the representation. It can be claimed that most of the advances in ICA have been motivated by the search for solutions to the BSS problem and, the other way around, advances in ICA have been immediately applied to BSS.

ICA and BSS algorithms start from a mixture model, whose parameters are estimated from the observed mixtures. Separation is achieved by applying the inverse mixture model to the observed signals (separating or unmixing model). Mixture models usually fall into three broad categories: instantaneous linear models, convolutive models and nonlinear models, the first one being the simplest but, in general, not near realistic applications. The development and test of the algorithms can be accomplished through synthetic data or with real-world data. Obviously, the most important aim (and most difficult) is the separation of real-world mixtures. BSS and ICA have strong relations also, apart from signal processing, with other fields such as statistics and artificial neural networks.

As long as we can find a system that emits signals propagated through a mean, and those signals are received by a set of sensors and there is an interest in recovering the original sources, we have a potential field of application for BSS and ICA. Inside that wide range of applications we can find, for instance: noise reduction applications, biomedical applications, audio systems, telecommunications, and many others.

This volume comes out just 20 years after the first contributions in ICA and BSS appeared[1]. Thereinafter, the number of research groups working in ICA and BSS has been constantly growing, so that nowadays we can estimate that far more than 100 groups are researching in these fields.

As proof of the recognition among the scientific community of ICA and BSS developments there have been numerous special sessions and special issues in several well-

---

[1] J. Herault, B. Ans, "Circuits neuronaux à synapses modifiables: décodage de messages composites para apprentissage non supervise", *C.R. de l'Académie des Sciences*, vol. 299, no. III-13, pp. 525–528, 1984.

Independently, two years before, although without the scientific impact of the previous reference, the following contribution was published:

J.W. Barness, Y. Carlin, and M.L. Steinberger, "Bootstrapping adaptive interference cancelers: some practical limitations", *Proc. of the Globecom Conference*, pp. 1251–1255, 1982.

known conferences and journals, such as GRETSI, NOLTA, ISCAS, EUSIPCO, NIPS, ESANN, IWANN, etc. An important landmark in the development of ICA was the organization of the first conference dedicated to ICA and BSS: the 1st International Workshop on Blind Source Separation and Independent Component Analysis (ICA 1999), which occurred in Aussois (France) in 1999. This workshop has been taking place since then about every 18 months, in Helsinki, Finland (ICA 2000), San Diego, California, USA (ICA 2001), and Nara, Japan (ICA 2003).

This volume of Lecture Notes in Computer Science encompasses the contributions to ICA 2004 which was held in Granada (Spain), presenting new ideas and challenges for ICA and BSS. This volume is composed of 156 contributions, selected after a peer-review process from 203 original contributions. The manuscripts were organized into the following eight sections:

1. Theory and fundamentals
2. Linear models
3. Convolutive models
4. Nonlinear models
5. Speech processing applications
6. Image processing applications
7. Biomedical applications
8. Other applications

We would like to thank all members of the International Advisory and Program Committees, who evaluated many contributions in a short time schedule. Other volunteers participated in the reviewing process, to whom we express our gratitude.

We are also very grateful to the University of Granada, the MEC (*Ministerio de Educación y Cultura*), *Consejería de Universidades* of the Andalusian Regional Government, and the Spanish CICYT (*Comisión Interministerial de Ciencia y Tecnología*) for their generous help.

The editors would like to thank Springer-Verlag, in particular Mr. Alfred Hofmann, for their collaboration and attention.

Finally, many thanks to all the authors who submitted their contributions and who made this volume possible.

September 2004   Alberto Prieto and Carlos G. Puntonet

ICA 2004 Committee Listings

Local Organizing Committee

General Chairman

Carlos G. Puntonet, University of Granada, Spain

Organizing Chairman

Alberto Prieto Espinosa, University of Granada, Spain

Members

Julio Ortega Lopera, University of Granada, Spain
Francisco Pelayo Valle, University of Granada, Spain
Manuel Rodriguez Álvarez, University of Granada, Spain
Fernando Rojas Ruiz, University of Granada, Spain
Ignacio Rojas Ruiz, University of Granada, Spain
Moisés Salmeron Campos, University of Granada, Spain

Web Administrator

Francisco M. Illeras García, University of Granada, Spain

International Program Committee

Members

Aapo Hyvarinen, Helsinki University of Technology, Finland
Adel Belouchrani, National Technical School, Algeria
Adriana Dapena, University of La Coruña, Spain
Ali Mansour, IEEE, France
Allan K. Barros, UFMA, Brazil
Ana María Tomé, University of Aveiro, Portugal
Antonio Rubio Ayuso, University of Granada, Spain
Arie Yeredor, Tel Aviv University, Israel
Asoke K. Nandi, University of Liverpool, UK
Deniz Erdogmus, University of Florida, USA
Diego Ruiz Padillo, University of Granada, Spain
Dinh-Tuam Pham, LMC, France
Elmar Lang, University of Regensburg, Germany
Enric Monte, Polytechnic University of Catalunya, Spain
Eric Moreau, Université de Toulon et du Var, France
Fabian Theis, University of Regensburg, Germany
Francisco J. González Serrano, Carlos III University, Spain

Harri Valpola, Helsinki University of Technology, Finland
Jagath C. Rajapakse, Nanyang Technological University, Singapore
Jean-Francois Cardoso, CNRS/LTCI, France
Jean M. Vesin, Federal Institute of Technology, Lausanne, Switzerland
Jonathon Chambers, Cardiff University, UK
Jordi Solé, University of Vic, Spain
Jorge Igual, University of Valencia, Spain
Jose Principe, University of Florida, Gainesville, USA
Juan Manuel Górriz Sáez, University of Cadiz, Spain
Juan José Murillo, University of Seville, Spain
Juha Karhunen, Helsinki University of Technology, Finland
Justinian Rosca, Siemens, USA
Kari Torkkola, Motorola Labs, USA
Kiyotoshi Matsuoka, Kyushu Institute of Technology, Japan
Klaus Obermayer, Technical University of Berlin, Germany
Konstantinos I. Diamantaras, TEI, Thessaloniki, Greece
Lieven De Lathauwer, CNRS Cergy-Pontoise, France
Lucas Parra, City College of New York, USA
Maria del Carmen Carrión, University of Granada, Spain
Mark Girolami, University of Glasgow, UK
Masashi Ohata, RIKEN, BMC Research Center, Japan
Michael Zibulevsky, Technion – Israel Institute of Technology, Israel
Miguel Ángel Lagunas, Technical University of Catalunya, Spain
Mitsuru Kawamoto, Shimane University, Japan
Noboru Ohnishi, Nagoya University, Japan
Pedro Gómez Vilda, Technical University of Madrid, Spain
Philippe Loubaton, University of Marne la Vallée, France
Pierre Comon, Université de Nice, France
Ricardo Vigário, Helsinki University of Technology, Finland
Rubén Martín Clemente, University of Seville, Spain
Santiago Zazo, Technical University of Madrid, Spain
Scott Makeig, University of California, San Diego, USA
Sergio Cruces, University of Seville, Spain
Simone Fiori, University of Perugia, Italy
Soo Young Lee, Advanced Institute of Science and Technology, Korea
Susana Hornillo, University of Seville, Spain
Tom Heskes, Neurocomputing
Tony Bell, Salk Institute of San Diego, USA
Toshihisa Tanaka, RIKEN BSI, Japan
Vincent Vigneron, University of Paris, France
Vicente Zarzoso, University of Liverpool, UK
Visa Koivunen, Helsinki University of Technology, Finland
Yannick Deville, Paul Sabatier University, France
Yolanda Blanco Archilla, European University of Madrid, Spain
Yujiro Inouye, Shimane University, Japan
Yuanqing Li, RIKEN BSI, Japan

International Advisory Committee

Members

Shun-Ichi Amari, RIKEN BSI, Japan
Christian Jutten, INPG, France
Andrew Cichocki, RIKEN BSI, Japan
Erkki Oja, Helsinki University of Technology, Finland
Simon Haykin, McMaster University, Canada
Kart Muller, Frauenhofer FIRST, Germany
Liqing Zhang, Shangai Jiatong University, China
Noboru Murata, Waseda University, Japan
Scout Douglas, Southern Methodist University, USA
Shoji Makino, NTT CS Labs, Japan
Te-Won Lee, UCSD, USA
Carlos G. Puntonet, University of Granada, Spain

Table of Contents

Theory and Fundamentals

A FastICA Algorithm for Non-negative Independent Component Analysis 1
Zhijian Yuan and Erkki Oja

Blind Source Separation by Adaptive Estimation
of Score Function Difference 9
Samareh Samadi, Massoud Babaie-Zadeh, Christian Jutten, and Kambiz Nayebi

Exploiting Spatiotemporal Information
for Blind Atrial Activity Extraction in Atrial Arrhythmias 18
Francisco Castells, Jorge Igual, Vicente Zarzoso, José Joaquín Rieta, and José Millet

Gaussianizing Transformations for ICA 26
Deniz Erdogmus, Yadunandana N. Rao, and José Carlos Príncipe

New Eigensystem-Based Method for Blind Source Separation 33
Rubén Martín-Clemente, Susana Hornillo-Mellado, Carlos G. Puntonet, and José I. Acha

Optimization Issues in Noisy Gaussian ICA 41
Jean-François Cardoso and Dinh-Tuan Pham

Optimization Using Fourier Expansion over a Geodesic for Non-negative ICA ... 49
Mark D. Plumbley

The Minimum Support Criterion for Blind Signal Extraction:
A Limiting Case of the Strengthened Young's Inequality 57
Sergio Cruces and Iván Durán

Accurate, Fast and Stable Denoising Source Separation Algorithms 65
Harri Valpola and Jaakko Särelä

An Overview of BSS Techniques Based on Order Statistics:
Formulation and Implementation Issues 73
Yolanda Blanco and Santiago Zazo

Analytical Solution of the Blind Source Separation Problem Using Derivatives ... 81
Sebastien Lagrange, Luc Jaulin, Vincent Vigneron, and Christian Jutten

Approximate Joint Diagonalization Using a Natural Gradient Approach 89
Arie Yeredor, Andreas Ziehe, and Klaus-Robert Müller

BSS, Classification and Pixel Demixing 97
  Albert Bijaoui, Danielle Nuzillard, and Tanusree Deb Barma

Blind Identification of Complex Under-Determined Mixtures 105
  Pierre Comon and Myriam Rajih

Blind Separation of Heavy-Tailed Signals Using Normalized Statistics 113
  Mohamed Sahmoudi, Karim Abed-Meraim, and Messaoud Benidir

Blind Source Separation of Linear Mixtures with Singular Matrices 121
  Pando Georgiev and Fabian J. Theis

Closely Arranged Directional Microphone for Source Separation 129
  *Yusuke Katayama, Masanori Ito, Allan Kardec Barros, Yoshinori Takeuchi,
  Tetsuya Matsumoto, Hiroaki Kudo, Noboru Ohnishi, and Toshiharu Mukai*

Estimating Functions for Blind Separation
when Sources Have Variance-Dependencies 136
  Motoaki Kawanabe and Klaus-Robert Müller

Framework of Constrained Matrix Gradient Flows 144
  Gen Hori

Identifiability, Subspace Selection and Noisy ICA 152
  Mike Davies

Improving GRNNs in CAD Systems 160
  *Fulgencio S. Buendía Buendía, J. Miguel Barrón-Adame,
  Antonio Vega-Corona, and Diego Andina*

Fisher Information in Source Separation Problems 168
  Vincent Vigneron and Christian Jutten

Localization of P300 Sources in Schizophrenia Patients Using Constrained BSS .. 177
  Saeid Sanei, Loukianos Spyrou, Wenwu Wang, and Jonathon A. Chambers

On the Estimation of the Mixing Matrix for Underdetermined Blind Source
Separation in an Arbitrary Number of Dimensions 185
  *Luis Vielva, Ignacio Santamaría, Jesús Ibáñez, Deniz Erdogmus,
  and José Carlos Príncipe*

On the Minimum ℓ_1-Norm Signal Recovery
in Underdetermined Source Separation 193
  Ichigaku Takigawa, Mineichi Kudo, Atsuyoshi Nakamura, and Jun Toyama

On the Strong Uniqueness of Highly Sparse Representations
from Redundant Dictionaries 201
  Rémi Gribonval and Morten Nielsen

Reliability of ICA Estimates with Mutual Information 209
  Harald Stögbauer, Ralph G. Andrzejak, Alexander Kraskov,
  and Peter Grassberger

Robust ICA for Super-Gaussian Sources 217
  Frank C. Meinecke, Stefan Harmeling, and Klaus-Robert Müller

Robustness of Prewhitening Against Heavy-Tailed Sources 225
  Aiyou Chen and Peter J. Bickel

Simultaneous Extraction of Signal Using Algorithms
Based on the Nonstationarity 233
  Juan Charneco Fernández

Space-Time Variant Blind Source Separation with Additive Noise 240
  Ivica Kopriva and Harold Szu

The Use of ICA in Speckle Noise 248
  D. Blanco, B. Mulgrew, S. McLaughlin, D.P. Ruiz, and M.C. Carrion

Theoretical Method for Solving BSS-ICA Using SVM 256
  Carlos G. Puntonet, Juan Manuel Górriz, Moisés Salmerón,
  and Susana Hornillo-Mellado

Wavelet De-noising for Blind Source Separation in Noisy Mixtures 263
  Bertrand Rivet, Vincent Vigneron, Anisoara Paraschiv-Ionescu,
  and Christian Jutten

Linear Mixture Models

A Gaussian Mixture Based Maximization of Mutual Information
for Supervised Feature Extraction 271
  José M. Leiva-Murillo and Antonio Artés-Rodríguez

Blind Separation of Nonstationary Sources by Spectral Decorrelation 279
  Shahram Hosseini and Yannick Deville

Delayed AMUSE – A Tool for Blind Source Separation and Denoising 287
  Ana R. Teixeira, Ana Maria Tomé, Elmar W. Lang, and Kurt Stadlthanner

Dimensionality Reduction in ICA and Rank-$(R_1, R_2, \ldots, R_N)$ Reduction
in Multilinear Algebra 295
  Lieven De Lathauwer and Joos Vandewalle

Linear Multilayer Independent Component Analysis
Using Stochastic Gradient Algorithm 303
  Yoshitatsu Matsuda and Kazunori Yamaguchi

Minimax Mutual Information Approach for ICA
of Complex-Valued Linear Mixtures 311
  Jian-Wu Xu, Deniz Erdogmus, Yadunandana N. Rao, and José Carlos Príncipe

Signal Reconstruction in Sensor Arrays
Using Temporal-Spatial Sparsity Regularization 319
  Dmitri Model and Michael Zibulevsky

Underdetermined Source Separation with Structured Source Priors 327
  Emmanuel Vincent and Xavier Rodet

A Grassmann-Rayleigh Quotient Iteration for Dimensionality Reduction in ICA . 335
  Lieven De Lathauwer, Luc Hoegaerts, and Joos Vandewalle

An Approach of Moment-Based Algorithm for Noisy ICA Models 343
  Daisuke Ito and Noboru Murata

Geometrical ICA-Based Method for Blind Separation
of Super-Gaussian Signals 350
  *Manuel Rodríguez-Álvarez, Fernando Rojas Ruiz,
  Rubén Martín-Clemente, Ignacio Rojas Ruiz, and Carlos G. Puntonet*

A Novel Method to Recover N Sources from N-1 Observations
and Its Application to Digital Communications 358
  Adriana Dapena

A Sufficient Condition for Separation of Deterministic Signals
Based on Spatial Time-Frequency Representations 366
  Nadège Thirion-Moreau, El Mostafa Fadaili, and Eric Moreau

Adaptive Robust Super-exponential Algorithms
for Deflationary Blind Equalization of Instantaneous Mixtures 374
  *Masanori Ito, Masashi Ohata, Mitsuru Kawamoto, Toshiharu Mukai,
  Yujiro Inouye, and Noboru Ohnishi*

Application of Gaussian Mixture Models for Blind Separation
of Independent Sources 382
  Koby Todros and Joseph Tabrikian

Asymptotically Optimal Blind Separation of Parametric Gaussian Sources 390
  Eran Doron and Arie Yeredor

Bayesian Approach for Blind Separation of Underdetermined Mixtures
of Sparse Sources 398
  Cédric Févotte, Simon J. Godsill, and Patrick J. Wolfe

Blind Source Separation
Using the Block-Coordinate Relative Newton Method 406
  Alexander M. Bronstein, Michael M. Bronstein, and Michael Zibulevsky

Hybridizing Genetic Algorithms with ICA in Higher Dimension 414
  *Juan Manuel Górriz, Carlos G. Puntonet, Moisés Salmerón,
  and Fernando Rojas Ruiz*

ICA Using Kernel Entropy Estimation with NlogN Complexity 422
  Sarit Shwartz, Michael Zibulevsky, and Yoav Y. Schechner

Soft-LOST: EM on a Mixture of Oriented Lines 430
  Paul D. O'Grady and Barak A. Pearlmutter

Some Gradient Based Joint Diagonalization Methods for ICA 437
  Bijan Afsari and Perinkulam S. Krishnaprasad

Underdetermined Independent Component Analysis by Data Generation 445
  Sang Gyun Kim and Chang D. Yoo

Convolutive Models

Batch Mutually Referenced Separation Algorithm
for MIMO Convolutive Mixtures 453
  Ali Mansour

Frequency Domain Blind Source Separation for Many Speech Signals 461
  Ryo Mukai, Hiroshi Sawada, Shoko Araki, and Shoji Makino

ICA Model Applied to Multichannel Non-destructive Evaluation
by Impact-Echo 470
  *Addisson Salazar, Luis Vergara, Jorge Igual, Jorge Gosálbez,
  and Ramón Miralles*

Monaural Source Separation Using Spectral Cues 478
  Barak A. Pearlmutter and Anthony M. Zador

Multichannel Speech Separation Using Adaptive Parameterization
of Source PDFs 486
  Kostas Kokkinakis and Asoke K. Nandi

Non-negative Matrix Factor Deconvolution;
Extraction of Multiple Sound Sources from Monophonic Inputs 494
  Paris Smaragdis

Optimal Sparse Representations for Blind Deconvolution of Images 500
  *Alexander M. Bronstein, Michael M. Bronstein,
  Michael Zibulevsky, and Yehoshua Y. Zeevi*

Separation of Convolutive Mixtures of Cyclostationary Sources:
A Contrast Function Based Approach 508
  Pierre Jallon, Antoine Chevreuil, Philippe Loubaton, and Pascal Chevalier

A Continuous Time Balanced Parametrization Approach
to Multichannel Blind Deconvolution 516
   Liang Suo Ma and Ah Chung Tsoi

A Frequency-Domain Normalized Multichannel Blind Deconvolution
Algorithm for Acoustical Signals 524
   Seung H. Nam and Seungkwon Beack

A Novel Hybrid Approach to the Permutation Problem
of Frequency Domain Blind Source Separation 532
   Wenwu Wang, Jonathon A. Chambers, and Saeid Sanei

Application of Geometric Dependency Analysis to the Separation
of Convolved Mixtures 540
   Samer Abdallah and Mark D. Plumbley

Blind Deconvolution of SISO Systems with Binary Source
Based on Recursive Channel Shortening 548
   Konstantinos I. Diamantaras and Theophilos Papadimitriou

Blind Deconvolution Using the Relative Newton Method 554
   Alexander M. Bronstein, Michael M. Bronstein, and Michael Zibulevsky

Blind Equalization Using Direct Channel Estimation 562
   Hyung-Min Park, Sang-Hoon Oh, and Soo-Young Lee

Blind MIMO Identification Using the Second Characteristic Function 570
   Eran Eidinger and Arie Yeredor

Blind Signal Separation of Convolutive Mixtures:
A Time-Domain Joint-Diagonalization Approach 578
   Marcel Joho

Characterization of the Sources in Convolutive Mixtures:
A Cumulant-Based Approach 586
   Susana Hornillo-Mellado, Carlos G. Puntonet, Rubén Martín-Clemente,
   Manuel Rodríguez-Álvarez, and Juan Manuel Górriz

CICAAR: Convolutive ICA with an Auto-regressive Inverse Model 594
   Mads Dyrholm and Lars Kai Hansen

Detection by SNR Maximization:
Application to the Blind Source Separation Problem 602
   Bernard Xerri and Bruno Borloz

Estimating the Number of Sources
for Frequency-Domain Blind Source Separation 610
   Hiroshi Sawada, Stefan Winter, Ryo Mukai, Shoko Araki, and Shoji Makino

Estimating the Number of Sources in a Noisy Convolutive Mixture Using BIC ... 618
Rasmus Kongsgaard Olsson and Lars Kai Hansen

Evaluation of Multistage SIMO-Model-Based Blind Source Separation
Combining Frequency-Domain ICA and Time-Domain ICA 626
Satoshi Ukai, Hiroshi Saruwatari, Tomoya Takatani, Kiyohiro Shikano, Ryo Mukai, and Hiroshi Sawada

On Coefficient Delay in Natural Gradient Blind Deconvolution
and Source Separation Algorithms 634
Scott C. Douglas, Hiroshi Sawada, and Shoji Makino

On the FIR Inversion of an Acoustical Convolutive Mixing System:
Properties and Limitations 643
Markus Hofbauer

Overcomplete BSS for Convolutive Mixtures Based on Hierarchical Clustering .. 652
Stefan Winter, Hiroshi Sawada, Shoko Araki, and Shoji Makino

Penalty Function Approach
for Constrained Convolutive Blind Source Separation 661
Wenwu Wang, Jonathon A. Chambers, and Saeid Sanei

Permutation Alignment for Frequency Domain ICA
Using Subspace Beamforming Methods.. 669
Nikolaos Mitianoudis and Mike Davies

QML Blind Deconvolution: Asymptotic Analysis 677
Alexander M. Bronstein, Michael M. Bronstein, Michael Zibulevsky, and Yehoshua Y. Zeevi

Super-exponential Methods Incorporated with Higher-Order Correlations
for Deflationary Blind Equalization of MIMO Linear Systems 685
Kiyotaka Kohno, Yujiro Inouye, and Mitsuru Kawamoto

Nonlinear ICA and BSS

Blind Maximum Likelihood Separation of a Linear-Quadratic Mixture 694
Shahram Hosseini and Yannick Deville

Markovian Source Separation in Post-nonlinear Mixtures 702
Anthony Larue, Christian Jutten, and Shahram Hosseini

Non-linear ICA by Using Isometric Dimensionality Reduction................. 710
John A. Lee, Christian Jutten, and Michel Verleysen

Postnonlinear Overcomplete Blind Source Separation Using Sparse Sources 718
Fabian J. Theis and Shun-ichi Amari

Second-Order Blind Source Separation
Based on Multi-dimensional Autocovariances 726
Fabian J. Theis, Anke Meyer-Bäse, and Elmar W. Lang

Separating a Real-Life Nonlinear Mixture of Images 734
Luís B. Almeida and Miguel Faria

Independent Slow Feature Analysis and Nonlinear Blind Source Separation 742
Tobias Blaschke and Laurenz Wiskott

Nonlinear PCA/ICA for the Structure from Motion Problem 750
Jun Fujiki, Shotaro Akaho, and Noboru Murata

Plugging an Histogram-Based Contrast Function on a Genetic Algorithm
for Solving PostNonLinear-BSS 758
*Fernando Rojas Ruiz, Carlos G. Puntonet, Ignacio Rojas Ruiz,
Manuel Rodríguez-Álvarez, and Juan Manuel Górriz*

Post-nonlinear Independent Component Analysis
by Variational Bayesian Learning 766
Alexander Ilin and Antti Honkela

Temporal Decorrelation as Preprocessing for Linear and Post-nonlinear ICA 774
Juha Karvanen and Toshihisa Tanaka

Tree-Dependent and Topographic Independent Component Analysis
for fMRI Analysis 782
Anke Meyer-Bäse, Fabian J. Theis, Oliver Lange, and Carlos G. Puntonet

Using Kernel PCA for Initialisation
of Variational Bayesian Nonlinear Blind Source Separation Method 790
Antti Honkela, Stefan Harmeling, Leo Lundqvist, and Harri Valpola

Speech Processing Applications

A Geometric Approach for Separating Several Speech Signals 798
Massoud Babaie-Zadeh, Ali Mansour, Christian Jutten, and Farrokh Marvasti

A Novel Method for Permutation Correction in Frequency-Domain
in Blind Separation of Speech Mixtures 807
Christine Serviere and Dinh-Tuan Pham

Convolutive Acoustic Mixtures Approximation to an Instantaneous Model
Using a Stereo Boundary Microphone Configuration 816
Juan Manuel Sanchis, Francisco Castells, and José Joaquín Rieta

DOA Detection from HOS by FOD Beamforming and Joint-Process Estimation .. 824
*Pedro Gómez Vilda, R. Martínez, Agustín Álvarez Marquina,
Victor Nieto Lluis, María Victoria Rodellar Biarge, F. Díaz, and F. Rodríguez*

Nonlinear Postprocessing for Blind Speech Separation 832
   Dorothea Kolossa and Reinhold Orglmeister

Real-Time Convolutive Blind Source Separation
Based on a Broadband Approach 840
   Robert Aichner, Herbert Buchner, Fei Yan, and Walter Kellermann

A New Approach to the Permutation Problem
in Frequency Domain Blind Source Separation 849
   Koutaro Kamata, Xuebin Hu, and Hidefumi Kobatake

Adaptive Cross-Channel Interference Cancellation
on Blind Source Separation Outputs................................... 857
   Changkyu Choi, Gil-Jin Jang, Yongbeom Lee, and Sang Ryong Kim

Application of the Mutual Information Minimization
to Speaker Recognition / Verification Improvement...................... 865
   Jordi Solé-Casals and Marcos Faúndez-Zanuy

Single Channel Speech Enhancement:
MAP Estimation Using GGD Prior Under Blind Setup 873
   Rajkishore Prasad, Hiroshi Saruwatari, and Kiyohiro Shikano

Stable and Low-Distortion Algorithm Based on Overdetermined
Blind Separation for Convolutive Mixtures of Speech 881
   Tsuyoki Nishikawa, Hiroshi Saruwatari, Kiyohiro Shikano,
   and Atsunobu Kaminuma

Two Channel, Block Adaptive Audio Separation
Using the Cross Correlation of Time Frequency Information 889
   Daniel Smith, Jason Lukasiak, and Ian Burnett

Underdetermined Blind Separation of Convolutive Mixtures
of Speech with Directivity Pattern Based Mask and ICA 898
   Shoko Araki, Shoji Makino, Hiroshi Sawada, and Ryo Mukai

Image Processing Applications

A Digital Watermarking Technique Based on ICA Image Features............. 906
   Wei Lu, Jian Zhang, Xiaobing Sun, and Kanzo Okada

A Model for Analyzing Dependencies Between Two ICA Features
in Natural Images 914
   Mika Inki

An Iterative Blind Source Separation Method for Convolutive Mixtures
of Images... 922
   Marc Castella and Jean-Christophe Pesquet

Astrophysical Source Separation Using Particle Filters 930
  Mauro Costagli, Ercan E. Kuruoğlu, and Alijah Ahmed

Independent Component Analysis in the Watermarking of Digital Images 938
  Juan José Murillo-Fuentes

Spatio-chromatic ICA of a Mosaiced Color Image 946
  David Alleysson and Sabine Süsstrunk

An Extended Maximum Likelihood Approach for the Robust Blind
Separation of Autocorrelated Images from Noisy Mixtures 954
  Ivan Gerace, Francesco Cricco, and Anna Tonazzini

Blind Separation of Spatio-temporal Data Sources 962
  Hilit Unger and Yehoshua Y. Zeevi

Data Hiding in Independent Components of Video 970
  Jiande Sun, Ju Liu, and Huibo Hu

Biomedical Applications

3D Spatial Analysis of fMRI Data on a Word Perception Task 977
  *Ingo R. Keck, Fabian J. Theis, Peter Gruber, Elmar W. Lang,
  Karsten Specht, and Carlos G. Puntonet*

Decomposition of Synthetic Multi-channel Surface-Electromyogram
Using Independent Component Analysis 985
  Gonzalo A. García, Kazuya Maekawa, and Kenzo Akazawa

Denoising Using Local ICA and a Generalized Eigendecomposition
with Time-Delayed Signals 993
  *Peter Gruber, Kurt Stadlthanner, Ana Maria Tomé, Ana R. Teixeira,
  Fabian J. Theis, Carlos G. Puntonet, and Elmar W. Lang*

MEG/EEG Source Localization Using Spatio-temporal Sparse Representations . 1001
  Alexey Polonsky and Michael Zibulevsky

Reliable Measurement of Cortical Flow Patterns Using Complex
Independent Component Analysis of Electroencephalographic Signals 1009
  Jörn Anemüller, Terrence J. Sejnowski, and Scott Makeig

Sensor Array and Electrode Selection for Non-invasive Fetal
Electrocardiogram Extraction by Independent Component Analysis 1017
  Frédéric Vrins, Christian Jutten, and Michel Verleysen

A Comparison of Time Structure and Statistically Based BSS Methods
in the Context of Long-Term Epileptiform EEG Recordings 1025
  Christopher J. James and Christian W. Hesse

A Framework for Evaluating ICA Methods of Artifact Removal
from Multichannel EEG 1033
    Kevin A. Glass, Gwen A. Frishkoff, Robert M. Frank, Colin Davey,
    Joseph Dien, Allen D. Malony, and Don M. Tucker

A New Method for Eliminating Stimulus Artifact
in Transient Evoked Otoacoustic Emission Using ICA 1041
    Ju Liu, Yu Du, Jing Li, and Kaibao Nie

An Efficient Time-Frequency Approach to Blind Source Separation
Based on Wavelets 1048
    Christian W. Hesse and Christopher J. James

Blind Deconvolution of Close-to-Orthogonal Pulse Sources Applied
to Surface Electromyograms 1056
    Ales Holobar and Damjan Zazula

Denoising Mammographic Images Using ICA 1064
    P. Mayo, Francisco Rodenas Escriba, and Gumersindo Verdú Martín

Independent Component Analysis of Pulse Oximetry Signals
Based on Derivative Skew 1072
    Paul F. Stetson

Mixing Matrix Pseudostationarity and ECG Preprocessing Impact
on ICA-Based Atrial Fibrillation Analysis 1079
    José Joaquín Rieta, César Sánchez, Juan Manuel Sanchis, Francisco Castells,
    and José Millet

'Signal Subspace' Blind Source Separation Applied
to Fetal Magnetocardiographic Signals Extraction 1087
    Giulia Barbati, Camillo Porcaro, and Carlo Salustri

Suppression of Ventricular Activity in the Surface Electrocardiogram
of Atrial Fibrillation 1095
    Mathieu Lemay, Jean-Marc Vesin, Zenichi Ihara, and Lukas Kappenberger

Unraveling Spatio-temporal Dynamics in fMRI Recordings
Using Complex ICA 1103
    Jörn Anemüller, Jeng-Ren Duann, Terrence J. Sejnowski, and Scott Makeig

Wavelet Domain Blind Signal Separation
to Analyze Supraventricular Arrhythmias from Holter Registers 1111
    César Sánchez, José Joaquín Rieta, Francisco Castells, Raúl Alcaraz,
    and José Millet

Other Applications

A New Auditory-Based Index to Evaluate the Blind Separation Performance
of Acoustic Mixtures 1118
    Juan Manuel Sanchis, José Joaquín Rieta, Francisco Castells, and José Millet

An Application of ICA to Identify Vibratory Low-Level Signals Generated
by Termites 1126
  *Juan Jose G. de la Rosa, Carlos G. Puntonet, Juan Manuel Górriz,
  and Isidro Lloret*

Application of Blind Source Separation to a Novel Passive Location 1134
  Gaoming Huang, Luxi Yang, and Zhenya He

Blind Source Separation in the Adaptive Reduction
of Inter-channel Interference for OFDM 1142
  Rafael Boloix-Tortosa and Juan José Murillo-Fuentes

BSS for Series of Electron Energy Loss Spectra 1150
  Danielle Nuzillard and Noël Bonnet

HOS Based Distinctive Features for Preliminary Signal Classification 1158
  Maciej Pędzisz and Ali Mansour

ICA as a Preprocessing Technique for Classification 1165
  V. Sanchez-Poblador, Enric Monte-Moreno, and Jordi Solé-Casals

Joint Delay Tracking and Interference Cancellation
in DS-CDMA Systems Using Successive ICA for Oversaturated Data 1173
  Tapani Ristaniemi and Toni Huovinen

Layered Space Frequency Equalisation for MIMO-MC-CDMA Systems
in Frequency Selective Fading Channels 1181
  Sonu Punnoose, Xu Zhu, and Asoke K. Nandi

Multiuser Detection and Channel Estimation
in MIMO OFDM Systems via Blind Source Separation 1189
  Luciano Sarperi, Asoke K. Nandi, and Xu Zhu

Music Transcription with ISA and HMM 1197
  Emmanuel Vincent and Xavier Rodet

On Shift-Invariant Sparse Coding 1205
  Thomas Blumensath and Mike Davies

Reliability in ICA-Based Text Classification 1213
  Xavier Sevillano, Francesc Alías, and Joan Claudi Socoró

Source Separation on Astrophysical Data Sets from the WMAP Satellite 1221
  Guillaume Patanchon, Jacques Delabrouille, and Jean-François Cardoso

Multidimensional ICA for the Separation of Atrial and Ventricular Activities
from Single Lead ECGs in Paroxysmal Atrial Fibrillation Episodes 1229
  *Francisco Castells, Cibeles Mora, José Millet, José Joaquín Rieta,
  César Sánchez, and Juan Manuel Sanchis*

Music Indexing Using Independent Component Analysis
with Pseudo-generated Sources 1237
   E.S. Gopi, R. Lakshmi, N. Ramya, and S.M. Shereen Farzana

Invited Contributions

Lie Group Methods for Optimization with Orthogonality Constraints 1245
   Mark D. Plumbley

A Hierarchical ICA Method for Unsupervised Learning
of Nonlinear Dependencies in Natural Images 1253
   Hyun-Jin Park and Te-Won Lee

Author Index 1263

A FastICA Algorithm for Non-negative Independent Component Analysis

Zhijian Yuan and Erkki Oja*

Helsinki University of Technology
Neural Networks Research Centre
P.O.Box 5400, 02015 HUT, Finland
{zhijian.yuan,erkki.oja}@hut.fi

Abstract. The non-negative ICA problem is here defined by the constraint that the sources are non-negative with probability one. This case occurs in many practical applications like spectral or image analysis. It has then been shown by [10] that there is a straightforward way to find the sources: if one whitens the non-zero-mean observations and makes a rotation to positive factors, then these must be the original sources. A fast algorithm, resembling the FastICA method, is suggested here, rigorously analyzed, and experimented with in a simple image separation example.

1 The Non-negative ICA Problem

The basic linear instantaneous ICA mixing model $\mathbf{x} = \mathbf{As}$ can be considered to be solved, with a multitude of practical algorithms and software; for reviews, see [1, 3]. However, if one makes some further assumptions which restrict or extend the model, then there is still ground for new analysis and solution methods. One such assumption is *positivity or non-negativity* of the sources and perhaps the mixing coefficients; for applications, see [9, 5, 13, 2]. Such a constraint is usually called *positive matrix factorization* [8] or *non-negative matrix factorization* [4]. We refer to the combination of non-negativity and independence assumptions on the sources as *non-negative independent component analysis*.

Recently, Plumbley [10, 11] considered the non-negativity assumption on the sources and introduced an alternative way of approaching the ICA problem, as follows. He calls a source s_i *non-negative* if $\Pr(s_i < 0) = 0$, and *well-grounded* if $\Pr(s_i < \delta) > 0$ for any $\delta > 0$; i.e. s_i has non-zero pdf all the way down to zero. The following key result was proven [10]:

Theorem 1. *Suppose that* $\mathbf{s}$ *is a vector of non-negative well-grounded independent unit-variance sources* s_i, $i = 1, ..., n$, *and* $\mathbf{y} = \mathbf{Qs}$ *where* $\mathbf{Q}$ *is a square orthonormal rotation, i.e.* $\mathbf{Q}^T\mathbf{Q} = \mathbf{I}$. *Then* $\mathbf{Q}$ *is a permutation matrix, i.e. the elements* y_j *of* $\mathbf{y}$ *are a permutation of the sources* s_i, *if and only if all* y_j *are non-negative.*

---

* This work was supported by the Academy of Finland as part of its Center of Excellence project "New Information Processing Principles".

The result of Theorem 1 can be used for a simple solution of the non-negative ICA problem. The sources of course are unknown, and $\mathbf{Q}$ cannot be found directly. However, it is a simple fact that an arbitrary rotation of $\mathbf{s}$ can also be expressed as a rotation of a pre-whitened observation vector. Denote it by $\mathbf{z} = \mathbf{Vx}$ with $\mathbf{V}$ the whitening matrix. Assume that the dimensionality of $\mathbf{z}$ has been reduced to that of $\mathbf{s}$ in the whitening, which is always possible in the overdetermined case (number of sensors is not smaller than number of sources).

It holds now $\mathbf{z} = \mathbf{VAs}$. Because both $\mathbf{z}$ and $\mathbf{s}$ have unit covariance matrices (for $\mathbf{s}$, this is assumed in Theorem 1), the matrix $\mathbf{VA}$ must be square orthogonal. This holds even in the case when $\mathbf{s}$ and $\mathbf{z}$ have non-zero means. We can write

$$\mathbf{y} = \mathbf{Qs} = \mathbf{Q}(\mathbf{VA})^T \mathbf{z} = \mathbf{Wz}$$

where the matrix $\mathbf{W}$ is a new parametrization of the problem. The key fact is that $\mathbf{W}$ is orthogonal, because it is the product of two orthogonal matrices $\mathbf{Q}$ and $(\mathbf{VA})^T$. By Theorem 1, to find the sources, it now suffices to *find an orthogonal matrix $\mathbf{W}$ for which $\mathbf{y} = \mathbf{Wz}$ is non-negative*. The elements of $\mathbf{y}$ are then the sources.

It was further suggested by [10] that a suitable cost function for actually finding the rotation could be constructed as follows: suppose we have an output truncated at zero, $\mathbf{y}^+ = (y_1^+, ..., y_n^+)$ with $y_i^+ = \max(0, y_i)$, and we construct a reestimate of $\mathbf{z} = \mathbf{W}^T \mathbf{y}$ given by $\hat{\mathbf{z}} = \mathbf{W}^T \mathbf{y}^+$. Then a suitable cost function would be given by

$$J(\mathbf{W}) = E\{\|\mathbf{z} - \hat{\mathbf{z}}\|^2\} = E\{\|\mathbf{z} - \mathbf{W}^T \mathbf{y}^+\|^2\}. \tag{1}$$

Due to the orthogonality of matrix $\mathbf{W}$, this is in fact equal to

$$J(\mathbf{W}) = E\{\|\mathbf{y} - \mathbf{y}^+\|^2\} = \sum_{i=1}^{n} E\{\min(0, y_i)^2\}. \tag{2}$$

Obviously, the value will be zero if $\mathbf{W}$ is such that all the y_i are positive.

The minimization of this cost function by various numerical algorithms was suggested in [11, 12, 7]. In [11], explicit axis rotations as well as geodesic search over the Stiefel manifold of orthogonal matrices were used. In [12], the cost function (1) was taken as a special case of "nonlinear PCA" for which an algorithm was earlier suggested by one of the authors [6]. Finally, in [7], it was shown that the cost function (2) is a Liapunov funtion for a certain matrix flow in the Stiefel manifold, providing global convergence.

However, the problem with the gradient type of learning rules is slow speed of convergence. It would be tempting therefore to develop a "fast" numerical algorithm for this problem, perhaps along the lines of the well-known FastICA method [3]. In this paper, such an algorithm is suggested and its convergence is theoretically analyzed.

2 The Classical FastICA Algorithm

Under the whitened zero-mean demixing model $\mathbf{y} = \mathbf{Wz}$, the classical FastICA algorithm finds the extrema of a generic cost function $E\{G(\mathbf{w}^T \mathbf{z})\}$, where $\mathbf{w}^T$

is one of the rows of the demixing matrix $\mathbf{W}$. The cost function can be e.g. a normalized cumulant or an approximation of the marginal entropy which is minimized in ICA in order to find maximally nongaussian projections $\mathbf{w}^T\mathbf{z}$. Under fairly weak assumptions, the true independent sources are among the extrema of $E\{G(\mathbf{w}^T\mathbf{z})\}$ [3]. FastICA updates $\mathbf{w}$ according to the following rule:

$$\mathbf{w} \leftarrow E\{\mathbf{z}g(\mathbf{w}^T\mathbf{z})\} - E\{g'(\mathbf{w}^T\mathbf{z})\}\mathbf{w}. \tag{3}$$

Here g is the derivative of G, and g' is the derivative of g. After (3), the vectors $\mathbf{w}$ are orthogonalized either in a deflation mode or symmetrically. The algorithm typically converges in a small number of steps to a demixing matrix $\mathbf{W}$, and $\mathbf{y}$ becomes a permutation of the source vector $\mathbf{s}$ with arbitrary signs.

3 The Non-negative FastICA Algorithm

For the non-negative independent components, our task becomes to find an orthogonal matrix $\mathbf{W}$ such that $\mathbf{y} = \mathbf{W}\mathbf{z}$ is nonnegative with the pre-whitened vector $\mathbf{z}$.

The classical FastICA is now facing two problems. First, the non-negative sources cannot have zero means. The mean values must be explicitly included in the analysis. Second, in FastICA, the function g in equation (3) is assumed to be an odd fuction, the derivative of the even function G. If this condition fails to be satisfied, the FastICA as such may not work. Applying FastICA to minimizing the cost function (2), we see that $G(y) = min(0, y)^2$ whose negative derivative (dropping the 2) is

$$g_-(y) = -min(0, y) = \begin{cases} -y, & y < 0 \\ 0, & y \geq 0. \end{cases} \tag{4}$$

We see that it does not satisfy the condition for FastICA.

In order to correct these problems, first, we use non-centered but whitened data $\mathbf{z}$, which satisfies $E\{(\mathbf{z} - E\{\mathbf{z}\})(\mathbf{z} - E\{\mathbf{z}\})^T\} = \mathbf{I}$. Second, we add a control parameter μ on the FastICA update rule (3), giving the following update rule:

$$\mathbf{w} \leftarrow E\{(\mathbf{z} - E\{\mathbf{z}\})g_-(\mathbf{w}^T\mathbf{z})\} - \mu E\{g'_-(\mathbf{w}^T\mathbf{z})\}\mathbf{w}, \tag{5}$$

where g'_- is the derivative of g_-. This formulation shows the similarity to the classical FastICA algorithm. Substituting function g_- from (4) simplifies the terms; for example, $E\{g'_-(\mathbf{w}^T\mathbf{z})\} = -E\{1|\mathbf{w}^T\mathbf{z} < 0\}P\{\mathbf{w}^T\mathbf{z} < 0\}$. The scalar $P\{\mathbf{w}^T\mathbf{z} < 0\}$, appearing in both terms in (5), can be dropped because the vector $\mathbf{w}$ will be normalized anyway. In practice, expectations are replaced by sample averages.

In (5), μ is a parameter determined by:

$$\mu = \min_{\{\mathbf{z}:\mathbf{z}\in\Delta)\}} \frac{E\{(\mathbf{z} - E\{\mathbf{z}\})\mathbf{w}^T\mathbf{z}|\mathbf{w}^T\mathbf{z} < 0\}^T\mathbf{z}}{E\{1|\mathbf{w}^T\mathbf{z} < 0\}\mathbf{w}^T\mathbf{z}}. \tag{6}$$

There the set $\Delta = \{\mathbf{z} : \mathbf{z}^T\mathbf{z}(0) = 0\}$, with $\mathbf{z}(0)$ the vector satisfying $||\mathbf{z}(0)|| = 1$ and $\mathbf{w}^T\mathbf{z}(0) = \max(\mathbf{w}^T\mathbf{z})$. Computing this parameter is computationally somewhat heavy, but on the other hand, now the algorithm converges in a fixed number of steps.

The nonnegative FastICA algorithm is shown in Table 1.

Table 1. The Non-negative FastICA algorithm for estimating several ICs.

---

1. Whiten the data to get vector $\mathbf{z}$.
2. Set counter $p \leftarrow 1$.
3. Choose an initial vector $\mathbf{w}_p$ of unit norm, and orthogonalize it as

$$\mathbf{w}_p \leftarrow \mathbf{w}_p - \sum_{j=1}^{p-1}(\mathbf{w}_p^T\mathbf{w}_j)\mathbf{w}_j$$

   and then normalize by $\mathbf{w}_p \leftarrow \mathbf{w}_p/||\mathbf{w}_p||$.
4. If $\max_{\mathbf{z}\neq 0}(\mathbf{w}_p^T\mathbf{z}) \leq 0$, update $\mathbf{w}_p$ by $-\mathbf{w}_p^T$.
5. If $\min_{\mathbf{z}\neq 0}(\mathbf{w}_p^T\mathbf{z}) \geq 0$, update $\mathbf{w}_p$ by the vector orthogonal to $\mathbf{w}_p$ and the source vectors that are orthogonal to $\mathbf{w}_p$. (See equation (11)).
6. Update $\mathbf{w}_p$ by the equation (5), replacing expectations by sample averages.
7. Let $\mathbf{w}_p \leftarrow \mathbf{w}_p/||\mathbf{w}_p||$.
8. If $\mathbf{w}_p$ has not converged, go back to step (4).
9. Set $p \leftarrow p+1$. If $p < n$ where n is the number of independent components, go back to step (3).

---

4 Analysis of the Algorithm

To make use of the properties of the non-negative independent sources, we perform the following orthogonal variable change:

$$\mathbf{q} = \mathbf{A}^T\mathbf{V}^T\mathbf{w} \tag{7}$$

where $\mathbf{A}$ is the mixture matrix and $\mathbf{V}$ is the whitening matrix. Then

$$\mathbf{w}^T\mathbf{z} = \mathbf{q}^T(\mathbf{VA})^T(\mathbf{VAs}) = \mathbf{q}^T\mathbf{s}. \tag{8}$$

Remember that matrix $\mathbf{VA}$ is orthogonal.

Our goal is to find the orthogonal matrix $\mathbf{W}$ such that $\mathbf{Wz}$ is non-negative. This is equivalent to finding a permutation matrix $\mathbf{Q}$, whose rows will be denoted by vectors $\mathbf{q}^T$, such that $\mathbf{Qs}$ is non-negative. In the space of the $\mathbf{q}$ vectors, the convergence result of the non-negative FastICA algorithm must be a unit vector $\mathbf{q}$ with exactly one entry nonzero and equal to one.

4.1 The Proof of the Convergence

Using the above transformation of eq. (7), the definition of the function g_-, and the parameter μ, the update rule (5) for the variable $\mathbf{q}$ becomes

$$\mathbf{q} \leftarrow \mu E\{1|\mathbf{q}^T\mathbf{s} < 0\}\mathbf{q} - E\{(\mathbf{s} - E\{\mathbf{s}\})(\mathbf{q}^T\mathbf{s})|\mathbf{q}^T\mathbf{s} < 0\}. \tag{9}$$

Before each iteration, there are three cases for $\mathbf{q}^T\mathbf{s}$. If $\mathbf{q}^T\mathbf{s} \leq 0$ for all the sources $\mathbf{s}$, that is, $\mathbf{q} \leq 0$, we simply update it by $\mathbf{q} \leftarrow -\mathbf{q}$ as shown in step 4 in the algorithm. So we only need to consider the other two cases, $\mathbf{q} \geq 0$, and $\min(\mathbf{q}^T\mathbf{s}) < 0$ with $\max(\mathbf{q}^T\mathbf{s}) > 0$.

A. *Consider the case that* $\min(\mathbf{q}^T\mathbf{s}) < 0$ *and* $\max(\mathbf{q}^T\mathbf{s}) > 0$. Since the sources are positive, then at least one element of $\mathbf{q}$ is negative, and one element is positive. Let $\mathbf{q}(k)$ be the vector after kth iteration. It is easy to see that the update equation (9) keeps zero elements unvariable, that is, $\mathbf{q}(k+1)_i = 0$ if $\mathbf{q}(k)_i = 0$. This can be shown from equation (9)

$$\mathbf{q}(k+1)_i = \mu E\{\mathbf{q}(k)_i | \mathbf{q}(k)^T\mathbf{s} < 0\} - E\{(\mathbf{s}_i - E\{\mathbf{s}_i\})(\mathbf{q}(k)^T\mathbf{s}) | \mathbf{q}(k)^T\mathbf{s} < 0\}$$
$$= 0 - E\{\mathbf{s}_i - E\{\mathbf{s}_i\}\} E\{(\mathbf{q}(k)^T\mathbf{s}) | \mathbf{q}(k)^T\mathbf{s} < 0\} = 0$$

by noticing that $\mathbf{s}_i$ is independent to $\mathbf{q}(k)^T\mathbf{s} = \sum_{j \neq i} \mathbf{q}(k)_j \mathbf{s}_j$.

Let I and J be the index sets such that $\mathbf{q}(k)_i < 0$ for all $i \in I$ and $\mathbf{q}(k)_j > 0$ for all $j \in J$. Let $\mathbf{s}(0)$ be the source vector such that $\mathbf{q}(k)^T\mathbf{s}(0) = \max(\mathbf{q}(k)^T\mathbf{s})$ and $\|\mathbf{s}(0)\| = 1$. The vector $\mathbf{s}(0)$ exists and $\mathbf{s}(0)_i = 0$ for $i \in I$. Further, let the source set $\Delta' := \{\mathbf{s} : \mathbf{s}^T\mathbf{s}(0) = 0\}$, which is not empty; we have for all $\mathbf{s} \in \Delta'$, $\mathbf{s}_j = 0$ for $j \in J$.

By the equation (6) and the transformation equation (7), we have the parameter estimation with variable $\mathbf{q}$

$$\mu = \min_{\{\mathbf{s} : \mathbf{s} \in \Delta'\}} \frac{E\{(\mathbf{s} - E\{\mathbf{s}\}) \mathbf{q}(k)^T\mathbf{s} | \mathbf{q}(k)^T\mathbf{s} < 0\}^T \mathbf{s}}{E\{1 | \mathbf{q}(k)^T\mathbf{s} < 0\} \mathbf{q}(k)^T\mathbf{s}}. \tag{10}$$

Then for $\mathbf{s} \in \Delta'$, $\mathbf{q}^T\mathbf{s} < 0$ and

$$\mu \leq \frac{E\{(\mathbf{s} - E\{\mathbf{s}\}) \mathbf{q}(k)^T\mathbf{s} | \mathbf{q}(k)^T\mathbf{s} < 0\}^T \mathbf{s}}{E\{1 | \mathbf{q}(k)^T\mathbf{s} < 0\} \mathbf{q}(k)^T\mathbf{s}}.$$

Therefore, $\mathbf{q}(k+1)^T\mathbf{s} \geq 0$.

Since $\mathbf{e}_i s_i$ belongs to the set Δ' if $i \in I$, where $\mathbf{e}_i$ is the unit vector with the ith entry one and the others zero, it holds $\mathbf{q}(k+1)^T \mathbf{e}_i s_i \geq 0$. This implies that $\mathbf{q}(k+1)_i \geq 0$ for $i \in I$. According to the choice of parameter μ, there exists at least one source $\mathbf{s} \in \Delta'$ such that $\mathbf{q}(k+1)^T\mathbf{s} = 0$, that is $\sum_{\{i \in I\}} \mathbf{q}(k+1)_i s_i = 0$. Since the sources are nonnegative, and also for $i \in I$, $\mathbf{q}(k+1)_i$ is nonnegative, there is at least one index $i_0 \in I$, such that $\mathbf{q}(k+1)_{i_0} = 0$. Therefore, after this iteration, the number of zero elements of vector $\mathbf{q}$ increases.

B. *Consider the case that* $\mathbf{q} \geq 0$. Then, the algorithm updates $\mathbf{q}$ by the orthogonal vector of $\mathbf{q}$ which keeps the zero elements of $\mathbf{q}$ zero. Since $\mathbf{q} \geq 0$, its orthogonal vector will not be nonnegative or negative, and the iteration goes back to the case A we just discussed.

To find this update vector, consider the sources $\hat{\mathbf{S}} := \{\mathbf{s} \neq 0 : \mathbf{q}(k)^T\mathbf{s} = 0\}$. The updated vector $\mathbf{q}(k+1)$ can be chosen as the vector, which is orthogonal to $\mathbf{q}(k)$ and $\hat{\mathbf{S}}$. To do this, let all the vectors in the sources space $\hat{\mathbf{S}}$ be column vectors forming matrix $\mathbf{B}$. Then the null space null($\mathbf{B}$) is orthogonal to the sources space $\hat{\mathbf{S}}$. If null($\mathbf{B}$) contains only one column, then this column vector is what we want, and the iteration goes to next step. Otherwise, take any column $\mathbf{q}(r)$ from null($\mathbf{B}$) which is different from $\mathbf{q}(k)$, and the update rule is

$$\mathbf{q}(k+1) = \mathbf{q}(r) - (\mathbf{q}(r)^T \mathbf{q}(k))\mathbf{q}(k). \tag{11}$$

Therefore, after each iteration, the updated vector $\mathbf{q}$ keeps the old zero entries zero and gains one more zero entry. So within $n-1$ iteration steps, the vector $\mathbf{q}$ is updated to be a unit vector $\mathbf{e}_i$ for certain i. With total iterative steps $\sum_{i=1}^{n-1} i = n(n-1)/2$, the permutation matrix $\mathbf{Q}$ is formed.

4.2 Complexity of the Computation

As the analysis in the above section shows, the total iteration steps of our algorithm are less than or equal to $n(n-1)/2$. During each iteration, the computational differences compared to classic FastICA come from step 4, 5 and 6 as shown in Table 1. The step 4 does not increase the computation much, so we can almost omit it. In step 5, we need to calculate the value of $\mathbf{w}_p^T \mathbf{z}$ once, and solve a $m \times n$ line equation (m is the number of vectors in the source space $\{\mathbf{s} \neq 0 : \mathbf{q}(k)^T \mathbf{s} = 0\}$). This can be solved by Matlab command null() immediately. Step 6 is the main update rule, just as in FastICA, and we need to calculate the expectation $E\{\mathbf{z} - E\{\mathbf{z}\}\}$. Furthermore, in our algorithm, to calculate the parameter μ, we need to go through the data $\mathbf{z}$ once more.

5 Experiments

In this section we present some numerical simulations, run in Matlab, to demonstrate the behaviour of the algorithm. The demixing matrix $\mathbf{W}$ is initialized to the identity matrix, ensuring initial orthogonality of $\mathbf{W}$ and hence of $\mathbf{H} = \mathbf{WVA}$.

The algorithm was applied to an image unmixing task. 4 image patches of size 252×252 were selected from a set of images of natural scenes, and downsampled by a factor of 4 in both directions to yield 63×63 images (see [7]). Each of the $n = 4$ images was treated as one source, with its pixel values representing the $63 \times 63 = 3969$ samples. The source image values were shifted to have a minimum of zero, to ensure they were *well-grounded* as required by the algorithm, and the images were scaled to ensure they were all unit variance.

After scaling, the source covariance matrix $\overline{\mathbf{ss}^T} - \bar{\mathbf{s}}\bar{\mathbf{s}}^T$ was computed and the largest off-diagonal element turned out to be 0.16. This is an acceptably small covariance between the images: as with any ICA method based on prewhitening, any covariance between sources would prevent accurate identification of the sources. A mixing matrix $\mathbf{A}$ was generated randomly and used to construct the four mixture images.

After iteration, the source-to-output matrix $\mathbf{H} = \mathbf{WVA}$ was

$$\mathbf{H} = \begin{pmatrix} 0.058 & \mathbf{1.010} & -0.106 & 0.062 \\ -0.106 & 0.042 & -0.078 & \mathbf{1.002} \\ -0.003 & -0.017 & \mathbf{1.014} & 0.076 \\ \mathbf{0.997} & -0.105 & -0.102 & -0.086 \end{pmatrix} \tag{12}$$

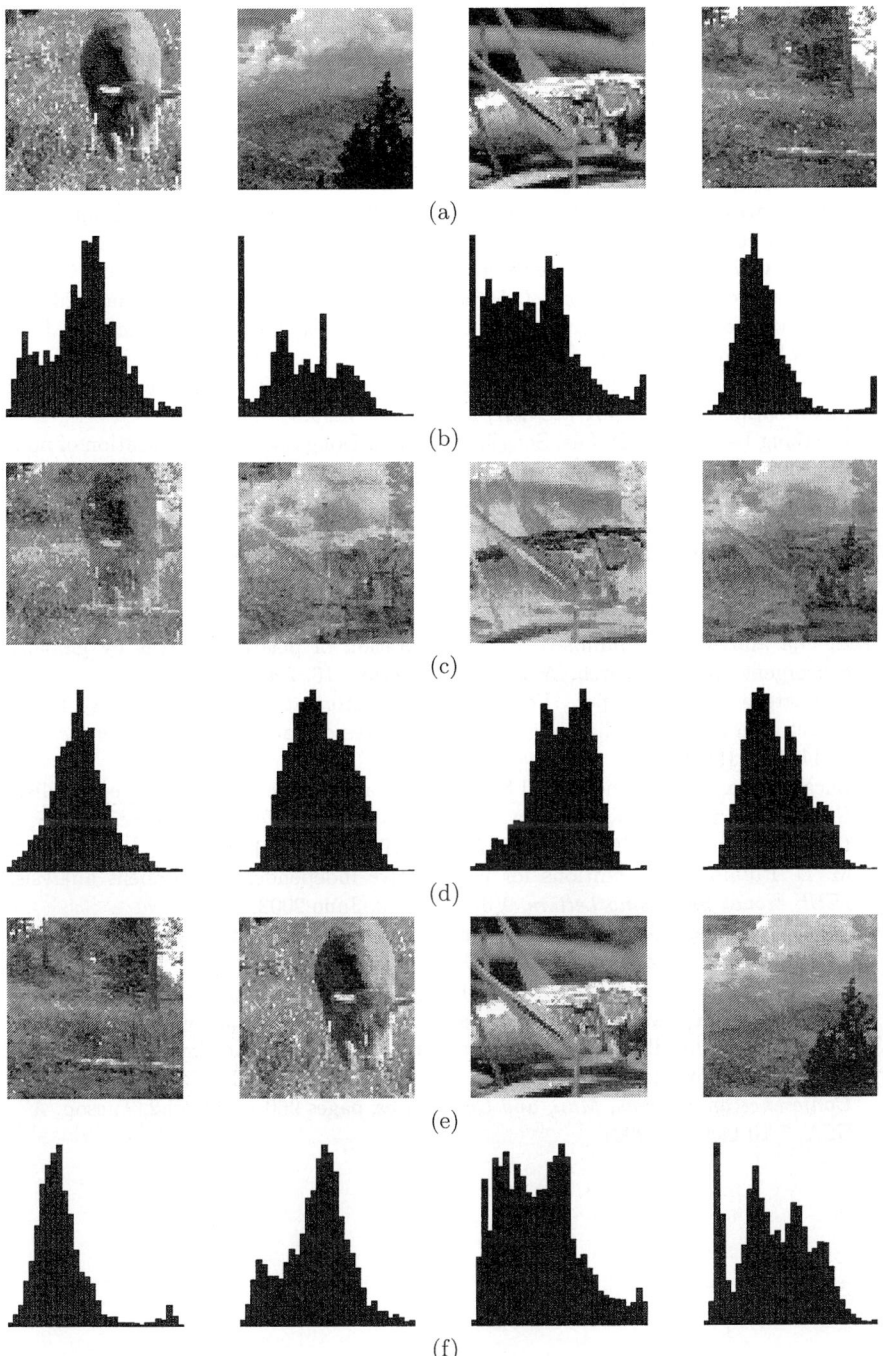

Fig. 1. Images and histograms for the image separation using the non-negative FastICA algorithm, showing (a) source images and (b) their histograms, (c), (d) the mixed images and their histograms, and (e), (f) the separated images and their histograms.

Figure 1 shows the original, mixed and separated images and their histograms. The algorithm converges in 6 steps and is able to separate the images reasonably well.

References

1. S-I. Amari and A. Cichocki. *Adaptive Blind Signal and Image Processing.* John Wiley & sons, Inc., 2002.
2. Ronald C. Henry. Multivariate receptor models–current practice and future trends. *Chemometrics and Intelligent Laboratory Systems*, 60(1-2):43–48, January 2002.
3. A. Hyvärinen, J. Karhunen, and E. Oja. *Independent Component Analysis.* John Wiley & Sons, 2001.
4. D. D. Lee and H. S. Seung. Learning the parts of objects by non-negative matrix factorization. *Nature*, 401:788–791, 21 October 1999.
5. Jae Sung Lee, Daniel D. Lee, Seugjin Choi, and Dong Soo Lee. Application of non-negative matrix factorization to dynamic positron emission tomography. In T.-W. Lee, T.-P. Jung, S. Makeig, and T. J. Sejnowski, editors, *Proceedings of the International Conference on Independent Component Analysis and Signal Separation (ICA2001), San Diego, California*, pages 629–632, December 9-13 2001.
6. E. Oja. The nonlinear PCA learning rule in independent component analysis. *Neurocomputing*, 17(1):25–46, 1997.
7. E. Oja and M. D. Plumbley. Blind separation of positive sources by globally convergent gradient search. *Neural Computation*, 16, 2004.
8. P. Paatero and U. Tapper. Positive matrix factorization: A non-negative factor model with optimal utilization of error estimates of data values. *Environmetrics*, 5:111–126, 1994.
9. Lucas Parra, Clay Spence, Paul Sajda, Andreas Ziehe, and Klaus-Robert Müller. Unmixing hyperspectral data. In *Advances in Neural Information Processing Systems 12 (Proc. NIPS*99)*, pages 942–948. MIT Press, 2000.
10. M. D. Plumbley. Conditions for nonnegative independent component analysis. *IEEE Signal Processing Letters*, 9(6):177 –180, June 2002.
11. M. D. Plumbley. Algorithms for non-negative independent component analysis. *IEEE Transaction on Neural Networks*, 14(3):534 –543, 2003.
12. M. D. Plumbley and E. Oja. A 'non-negative PCA' algorithm for independent component analysis. *IEEE Transactions on Neural Networks*, 15(1):66 – 76, 2004.
13. S. Tsuge, M. Shishibori, S. Kuroiwa, and K. Kita. Dimensionality reduction using non-negative matrix factorization for information retrieval. In *IEEE International Conference on Systems, Man, and Cybernetics*, pages 960 – 965 vol.2, Tucson, AZ, USA, 7-10 October 2001.

Blind Source Separation by Adaptive Estimation of Score Function Difference

Samareh Samadi[1], Massoud Babaie-Zadeh[1,2],
Christian Jutten[3], and Kambiz Nayebi[1,*]

[1] Electrical Engineering Department
Sharif University of Technology, Tehran, Iran
{samarehsamadi,mbzadeh}@yahoo.com, knayebi@sina.sharif.edu
[2] Multimedia Lab. Iran Telecom Research Center(ITRC), Tehran, Iran
[3] Institut National Polytechnique de Grenoble (INPG)
Laboratoire des images et des signaux (LIS), Grenoble, France
Christian.Jutten@infg.fr

Abstract. In this paper, an adaptive algorithm for blind source separation in linear instantaneous mixtures is proposed, and it is shown to be the optimum version of the EASI algorithm. The algorithm is based on minimization of mutual information of outputs. This minimization is done using adaptive estimation of a recently proposed non-parametric "gradient" for mutual information.

1 Introduction

Blind Source Separation (BSS) is a relatively new subject in signal processing, which has been considered extensively since mid 80's [1]. It consists in retrieving unobserved independent mixed signals from mixtures of them, assuming there is information neither about the original sources, nor about the mixing system. The simplest BSS model is the linear instantaneous model. In this case, the mixture is supposed to be of the form $\mathbf{x} = \mathbf{As}$, where $\mathbf{s}$ is the source vector, $\mathbf{x}$ is the observation vector, and $\mathbf{A}$ is the (constant) mixing matrix which is supposed to be an unknown matrix of full rank. The separating system, $\mathbf{B}$, tries to estimate the sources via $\mathbf{y} = \mathbf{Bx}$. For linear mixtures, it can be shown that the independence of the components of $\mathbf{y}$, is a necessary and sufficient condition for achieving the separation up to a scale and a permutation indeterminacy, provided that there is at most one Gaussian source [2].

The early works on BSS were concerned linear instantaneous mixture and by now a lot of algorithms are available for separating them (see [1,3] for a review and extensive references). These methods can not be easily generalized to more complicated models. Source separation can be obtained by optimizing a "contrast function" *i.e.* a scalar measure of some "distributional property" of the outputs [4]. One of the most general contrast functions is mutual information,

---

[*] This work has been partially funded by the European project Blind Source Separation and applications (BLISS, IST 1999-13077), Sharif University of Technology, and by Iran Telecom Research Center (ITRC).

which has been shown [4] to be an asymptotically Maximum-Likelihood (ML) estimation of source signals. Recently a non-parametric "gradient" for mutual information has been proposed [5], which has been used successfully in separating different mixing models [6]. The proposed algorithms based on this gradient are all batch algorithms, which makes them unsuitable for being used in real-time applications.

In this paper, we propose an adaptive method for estimating this "gradient" of mutual information, and we use it to construct a new adaptive algorithm for separating linear instantaneous mixtures. This approach not only leads to good separation results, but also constructs a framework that can also be generalized to more complicated models. More interestingly, we will show that, for linear instantaneous mixtures, this new approach has a close relation to the famous EASI algorithm [7], and is, in fact, an optimal version of EASI. The paper is organized as follows. Section 2 reviews the essential materials to express the "gradient" of mutual information. The iterative equations of the algorithm are developed in Section 3, and its relation to EASI is considered in Section 4. The proposed algorithm is introduced as an optimum verion of EASI in section 5. This algorithm can be adaptively implemented using the adaptive estimation method of Section 6. In Section 7, the normalization method of the output energies is explained. Finally, Section 8 presents some experimental results.

2 Preliminary Issues

The objective of this section is to review mutual information definition, as the independence criterion, and its "gradient". Expressing this gradient, requires reviewing the definition of multivariate score functions of a random vector, which have been first introduced in [8].

2.1 Multivariate Score Functions

In statistics, the score function of a random variable y is defined as $-p'_y(y)/p_y(y)$, where $p_y(y)$ is the probability density function (PDF) of y. For an N-dimensional random vector $\mathbf{y} = (y_1, \ldots, y_N)^T$, two forms of score function are defined in [8]:

Definition 1 (MSF) *The marginal score function (MSF) of* $\mathbf{y}$*, is the vector of score functions of its components, i.e.:*

$$\boldsymbol{\psi}_\mathbf{y}(\mathbf{y}) = (\psi_1(y_1), \ldots, \psi_N(y_N))^T \qquad (1)$$

where:

$$\psi_i(\mathbf{y}) = -\frac{d}{dy_i} \ln p_{y_i}(y_i) = -\frac{p'_{y_i}(y_i)}{p_{y_i}(y_i)} \qquad (2)$$

and $p_{y_i}(y_i)$ *is the marginal PDF of* y_i.

Definition 2 (JSF) *The joint score function (JSF) of* $\mathbf{y}$*, is the vector function* $\boldsymbol{\varphi}_\mathbf{y}(\mathbf{y})$*, such that its i-th component is:*

$$\varphi_i(\mathbf{y}) = -\frac{\partial}{\partial y_i} \ln p_\mathbf{y}(\mathbf{y}) = -\frac{\frac{\partial}{\partial y_i} p_\mathbf{y}(\mathbf{y})}{p_\mathbf{y}(\mathbf{y})} \qquad (3)$$

where $p_\mathbf{y}(\mathbf{y})$ *is the joint PDF of* $\mathbf{y}$.

Definition 3 (SFD) *The score function difference (SFD) of* $\mathbf{y}$, *is the difference between its JSF and MSF, i.e.:*

$$\beta_{\mathbf{y}}(\mathbf{y}) = \psi_{\mathbf{y}}(\mathbf{y}) - \varphi_{\mathbf{y}}(\mathbf{y}) \tag{4}$$

2.2 Mutual Information and Its Gradient

For measuring the statistical independence of random variables $y_1, \ldots, y_N$, one can use their mutual information, defined by:

$$\begin{aligned} I(\mathbf{y}) &= D\left(p_{\mathbf{y}}(\mathbf{y}) \parallel \prod_i p_{y_i}(y_i)\right) \\ &= \int_{\mathbf{y}} p_{\mathbf{y}}(\mathbf{y}) \ln \frac{p_{\mathbf{y}}(\mathbf{y})}{\prod_i p_{y_i}(y_i)} d\mathbf{y} \\ &= E\left\{\ln \frac{p_{\mathbf{y}}(\mathbf{y})}{\prod_i p_{y_i}(y_i)}\right\} \end{aligned} \tag{5}$$

where $\mathbf{y} = (y_1, \ldots, y_N)^T$, and D denotes the Kullback-Leibler divergence. This function is always positive, and is zero if and only if the y_i's are independent.

For designing a source separation algorithm, one can use mutual information as a criterion for measuring output independence. In other words, the parameters of the separating system must be computed in such a way that the mutual information of the outputs be minimized. For doing this, the gradient based algorithms may be used. To calculate the gradient of the output mutual information with respect to the parameters of the separating system, the following theorem [5] will be quite helpful.

Theorem 1 *Let $\boldsymbol{\Delta}$ be a 'small' random vector, with the same dimension as $\mathbf{x}$. Then:*

$$I(\mathbf{x} + \boldsymbol{\Delta}) - I(\mathbf{x}) = E\left\{\boldsymbol{\Delta}^T \beta_{\mathbf{x}}(\mathbf{x})\right\} + o(\boldsymbol{\Delta}) \tag{6}$$

where $o(\boldsymbol{\Delta})$ denotes higher order terms in $\boldsymbol{\Delta}$.

This theorem points out that SFD can be called the "stochastic gradient" of mutual information.

Remark. Equation (6) may be stated in the following form (which is similar to what is done in [9]):

$$I(\mathbf{x} + \mathcal{E}\mathbf{y}) - I(\mathbf{x}) = E\left\{(\mathcal{E}\mathbf{y})^T \beta_{\mathbf{x}}(\mathbf{x})\right\} + o(\mathcal{E}) \tag{7}$$

where $\mathbf{x}$ and $\mathbf{y}$ are bounded random vectors, $\mathcal{E}$ is a matrix with small entries, and $o(\mathcal{E})$ stands for a term that converges to zero faster than $\|\mathcal{E}\|$. This equation is mathematically more sophisticated, because in (6) the term 'small random vector' is somewhat ad-hoc. Conversely, (6) is simpler, and easier to be used in developing gradient based algorithms for optimizing a mutual information.

3 Estimating Equations

In linear instantaneous mixture, the separating system is:

$$\mathbf{y} = \mathbf{B}\mathbf{x} \tag{8}$$

and $\mathbf{B}$ must be computed to minimize $I(\mathbf{y})$, where I stands for mutual information. For calculating $\mathbf{B}$, the steepest descent algorithm may be applied:

$$\mathbf{B}_{n+1} = \mathbf{B}_n - \mu \left.\frac{\partial I}{\partial \mathbf{B}}\right|_{\mathbf{B}=\mathbf{B}_n} \tag{9}$$

where μ is a small positive constant. However, to design an equivariant algorithm [7], that is, an algorithm whose separation performance does not depend on the conditioning of the mixing matrix, one must use the serial (multiplicative) updating rule:

$$\mathbf{B}_{n+1} = \left(\mathbf{I} - \mu \left[\nabla_\mathbf{B} I\right]_{\mathbf{B}=\mathbf{B}_n}\right)\mathbf{B}_n \tag{10}$$

where $\mathbf{I}$ denotes the identity matrix, and $\nabla_\mathbf{B} I \triangleq \frac{\partial I}{\partial \mathbf{B}}\mathbf{B}^T$ is the relative (or natural) gradient [7, 10] of $I(\mathbf{y})$ with respect to $\mathbf{B}$.

Using theorem 1, $\nabla_\mathbf{B} I$ can be easily obtained [5] (although, for this simple linear instantaneous case, this gradient may be directly calculated):

$$\nabla_\mathbf{B} I = E\left\{\boldsymbol{\beta}_\mathbf{y}(\mathbf{y})\mathbf{y}^T\right\} \tag{11}$$

By dropping the expectation operation, the stochastic version of (10) is obtained:

$$\mathbf{B}_{n+1} = \left(\mathbf{I} - \mu \boldsymbol{\beta}_\mathbf{y}(\mathbf{y})\mathbf{y}^T\right)\mathbf{B}_n \tag{12}$$

For developing the above algorithm in adaptive form, adaptive estimation of SFD is required, which will be discussed in Section 6.

4 Relation to EASI

The EASI algorithm has been proposed by Cardoso and Laheld [7]. In developing this algorithm, they showed that if the separation is achieved by minimizing a contrast function $\phi(\mathbf{B}) = E\{f(\mathbf{y})\}$ with respect to $\mathbf{B}$, the performance of the following serial updating algorithm, is independent of the mixing matrix:

$$\mathbf{B}_{n+1} = \left(\mathbf{I} - \mu \nabla \phi(\mathbf{B}_n)\right)\mathbf{B}_n \tag{13}$$

where the relative gradient $\nabla \phi(\mathbf{B})$ is:

$$\nabla \phi(\mathbf{B}) = \nabla E\left\{f(\mathbf{y})\right\} = E\left\{f'(\mathbf{y})\mathbf{y}^T\right\} \tag{14}$$

Consequently, the stochastic version of (13) becomes:

$$\mathbf{B}_{n+1} = \left(\mathbf{I} - \mu g(\mathbf{y})\mathbf{y}^T\right)\mathbf{B}_n \tag{15}$$

where $g \triangleq f'$. Developing EASI is then continued by choosing a "component-wise" g, and implementing a pre-whitening stage in the above algorithm, which is required by some contrast functions. This makes the final EASI equation more complicated than (15).

Now, let the contrast function be the mutual information:

$$\phi(\mathbf{B}) = I(\mathbf{y}) = E\left\{\ln \frac{p_\mathbf{y}(\mathbf{y})}{\prod_i p_{y_i}(y_i)}\right\} \tag{16}$$

Comparing with (14), we have:

$$f(\mathbf{y}) = \ln \frac{p_\mathbf{y}(\mathbf{y})}{\prod_i p_{y_i}(y_i)} \tag{17}$$

Then, the relative gradient (14) becomes (11), and the algorithm updating rule is (12). In fact, it is a special case of (15), where the contrast function is the mutual information of the outputs, and g is the SFD of $\mathbf{y}$. However, contrary to the "standard" EASI, where g is a "component-wise" and fixed function, here $g(\mathbf{y}) = \boldsymbol{\beta}_\mathbf{y}(\mathbf{y})$ is a multi-variate function and depends on the distribution of $\mathbf{y}$.

5 Optimum EASI

As mentioned in sectin 4, $\phi(\mathbf{B})$ is a contrast function in EASI. Recall now that minimizing mutual information of outputs for source separation tends asymptotically towards a Maximum Likelihood (ML) estimation of sources [4]. Consequently, the optimal (in ML sense) contrast function in the EASI algorithm is mutual information of outputs and hence, the algorithm (12) can be considered as an *optimal* version of EASI (in ML sense). In other words, we have shown that the optimum choice of the non-linearity ($g(\mathbf{y})$) in the EASI algorithm is not a fixed and component-wise non-linearity, it is a multi-variate function which depends on the output statistics.

Moreover, in the "standard" EASI, one must take into account the necessity of existence of a pre-whitening stage, and implementing it in the algorithm. This makes the final equation of EASI [7] more complicated than (15). However, when using mutual information contrast, no pre-whitening is required.

Finally, besides its performance (see Section 8), one great advantage of this new algorithm is that it can be generalized to more complicated mixtures. In fact, it is based on SFD, which has been successfully used in separating other mixtures (especially, post-nonlinear and convolutive) in batch algorithms [6].

We recall that these advantages are obtained at the expense of higher computational load: a multi-variate nonlinear function (SFD) has to be estimated, based on the output statistics.

6 Adaptive SFD Estimation

For estimating the MSF, one must simply estimate the score functions of its components. It can be seen that for a function f with continuous first derivative and bounded sources we have [11]:

$$E\{f(x)\psi_x(x)\} = E\{f'(x)\} \tag{18}$$

where ψ_x is the MSF of the random variable x. Now, let the score function ψ_x be modeled as a linear combination of some basis functions $k_1(x), k_2(x), \ldots, k_L(x)$:

$$\hat{\psi}_x(x) = \sum_{i=1}^{L} w_i k_i(x) = \mathbf{k}(x)^T \mathbf{w} \tag{19}$$

where $\mathbf{k}(x) \triangleq (k_1(x), \ldots, k_L(x))^T$ and $\mathbf{w} \triangleq (w_1, \ldots, w_L)^T$. For calculating $\mathbf{w}$, we minimize the mean square error:

$$\mathcal{E} \triangleq E\left\{\left(\psi_x(x) - \hat{\psi}_x(x)\right)^2\right\} \tag{20}$$

Expanding the above expression and using (18), it is seen that the minimizer of $\mathcal{E}$ minimizes also:

$$\xi \triangleq \frac{1}{2} E\left\{\hat{\psi}_x(x)^2\right\} - E\left\{\frac{\partial}{\partial x}\hat{\psi}_x(x)\right\} \tag{21}$$

For minimizing ξ with respect to $\mathbf{w}$, the Newton method can be used:

$$\mathbf{w} \leftarrow \mathbf{w} - \mu E\left\{\left(\frac{\partial^2 \xi}{\partial \mathbf{w}^2}\right)\right\}^{-1} E\left\{\left(\frac{\partial \xi}{\partial \mathbf{w}}\right)\right\} \tag{22}$$

where:

$$\frac{\partial \xi}{\partial \mathbf{w}} = \mathbf{k}(x)\mathbf{k}(x)^T \mathbf{w} - \frac{\partial \mathbf{k}(x)}{\partial x} \tag{23}$$

and:

$$\frac{\partial^2 \xi}{\partial \mathbf{w}^2} = \mathbf{k}(x)\mathbf{k}(x)^T \tag{24}$$

This method can be easily generalized for estimating JSF. It has been shown [8] that for bounded sources and an arbitrary multivariate function $f(\mathbf{x})$ with continuous derivative with respect to x_i:

$$E\{f(\mathbf{x})\varphi_i(\mathbf{x})\} = E\left\{\frac{\partial}{\partial x_i} f(\mathbf{x})\right\} \tag{25}$$

Let now $\varphi_i(\mathbf{x})$, the i-th component of JSF, be estimated as the linear combination of the (multivariate) basis functions $k_1(\mathbf{x}), \ldots, k_L(\mathbf{x})$, that is:

$$\hat{\varphi}_i(\mathbf{x}) = \sum_{i=1}^{L} w_i k_i(\mathbf{x}) = \mathbf{k}(\mathbf{x})^T \mathbf{w} \tag{26}$$

and $\mathbf{w}$ is the minimizer of $E\{(\varphi_i(\mathbf{x}) - \hat{\varphi}_i(\mathbf{x}))^2\}$.

Following similar calculation as above, we obtain the same algorithm given by equations (22), (23) and (24), where in this case:

$$\xi = \frac{1}{2} E\{\hat{\varphi}_i(\mathbf{x})^2\} - E\left\{\frac{\partial}{\partial x_i}\hat{\varphi}_i(\mathbf{x})\right\} \tag{27}$$

Finally, SFD is estimated by calculating the difference of the estimated MSF and JSF.

7 Normalization of Output Energies

From the scale indeterminacy it is deducted that the algorithm (12) has no restriction on the energy of outputs. Consequently, this algorithm does not converge to a unique solution. To overcome this indeterminacy, and making the algorithm to converge to unit energy outputs, we replace the i-th diagonal element of $\boldsymbol{\beta}_{\mathbf{y}}(\mathbf{y})\mathbf{y}^T$ by $1-y_i^2$ to force the separating system to create unit variance outputs. This is similar to what is done in [11].

8 Experimental Result

As an experiment, two independent sources with normal and uniform distributions and with zero means and unit variances are mixed by:

$$\mathbf{A} = \begin{bmatrix} 1 & 0.7 \\ 0.5 & 1 \end{bmatrix} \tag{28}$$

Basis functions for estimating $\psi_i(y_i)$ are:

$$k_1(y) = 1,\ k_2(y) = y,\ k_3(y) = y^2,\ k_4(y) = y^3 \tag{29}$$

and basis functions for estimating $\varphi_i(\mathbf{y})$ are:

$$k_1(y_1, y_2) = 1,$$
$$k_2(y_1, y_2) = y_1,\ k_3(y_1, y_2) = y_1^2,\ k_4(y_1, y_2) = y_1^3$$
$$k_5(y_1, y_2) = y_2,\ k_6(y_1, y_2) = y_2^2,\ k_7(y_1, y_2) = y_2^3$$

To compare the separation result of the proposed algorithm with EASI, we have separated this mixture, using both algorithms. In our method, the adaptation rate of the Newton algorithm is 0.1 and the adaptation rate of the separation algorithm is 0.001. In EASI, the component-wise nonlinear function $g(y_i) = y_i|y_i|^2$ has been used, with the same adaptation rate (0.001). Figure 1 shows the averaged output signal to noise ratios (SNR) taken over 50 runs of the algorithms. SNR is defined as:

$$\text{SNR} = 10 \log_{10} \frac{E\{s^2\}}{E\{(y-s)^2\}} \tag{30}$$

where y is the output corresponding to the source s. The figure shows that the proposed algorithm has better separation performance than EASI, as was expected because this algorithm is an optimal version of EASI (see Section 4). However the cost of this better performance is a higher complexity (which increases with the source number), since a multivariate non-linear function must be estimated at each iteration.

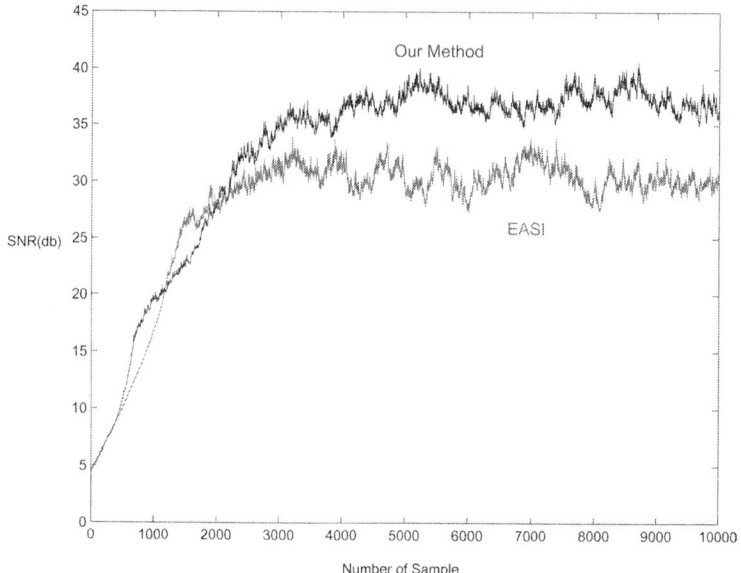

Fig. 1. Output SNRs versus iteration for EASI and our method.

9 Conclusion

In this paper an adaptive algorithm for blind separating linear instantaneous mixtures has been proposed, which is based on adaptive estimation of SFD. It has been shown that this algorithm can be seen as an optimum version of the EASI algorithm. Moreover, it is conjectured that this method can be generalized to separating more complicated (than linear instantaneous) mixing models, such as convolutive and non-linear mixtures. This is because SFD has been successfully used in separating these models [6]. Such a generalization is currently under study. The drawback of this method is that, despite of EASI, this algorithm requires the estimation of multivariate score functions (which are related to joint PDFs). This estimation becomes too difficult, and requires a lot of data, when the dimension (i.e. number of sources) grows. Practically, this method is suitable only up to 3 or 4 sources.

References

1. A. Hyvärinen, J. Karhunen, and E. Oja, *Independent Component Analysis*, John Wiely & Sons, 2001.
2. P. Comon, "Independent component analysis, a new concept?," *Signal Processing*, vol. 36, no. 3, pp. 287–314, 1994.
3. A. Cichocki and S.I. Amari, *Adaptive Blind Signal and Image Processing*, Wiley, 2002.
4. J.-F. Cardoso, "Blind signal separation: statistical principles," *Proceedings of IEEE*, vol. 9, pp. 2009–2025, 1998.

5. M. Babaie-Zadeh, C. Jutten, and K. Nayebi, "Differential of mutual information function," *IEEE Signal Processing Letters*, vol. 11, no. 1, pp. 48–51, January 2004.
6. M. Babaie-Zadeh, *On blind source separation in convolutive and nonlinear mixtures*, Ph.D. thesis, INP Grenoble, 2002.
7. J.-F. Cardoso and B. Laheld, "Equivariant adaptive source separation," *IEEE Trans. on SP*, vol. 44, no. 12, pp. 3017–3030, December 1996.
8. M. Babaie-Zadeh, C. Jutten, and K. Nayebi, "Separating convolutive mixtures by mutual information minimization," in *Proceedings of IWANN'2001*, Granada, Spain, Juin 2001, pp. 834–842.
9. D. T. Pham, "Mutual information approach to blind separation of stationary sources," *IEEE Transactions on Information Theory*, vol. 48, no. 7, pp. 1–12, July 2002.
10. S. I. Amari, "Natural gradient works efficiently in learning," *Neural Computation*, vol. 10, pp. 251–276, 1998.
11. A. Taleb and C. Jutten, "Entropy optimization, application to blind source separation," in *ICANN*, Lausanne, Switzeland, October 1997, pp. 529–534.

Exploiting Spatiotemporal Information for Blind Atrial Activity Extraction in Atrial Arrhythmias

Francisco Castells[1], Jorge Igual[2], Vicente Zarzoso[3],
José Joaquín Rieta[1], and José Millet[1]

[1] Universidad Politécnica de Valencia
46730 Gandia, Spain
{fcastells,jjrieta,jmillet}@eln.upv.es
[2] Universidad Politécnica de Valencia
03800 Alcoi, Spain
jigual@dcom.upv.es
[3] Department of Electrical Engineering and Electronics, The University of Liverpool
Brownlow Hill, Liverpool L69 3GJ, UK
vicente@liv.ac.uk

Abstract. The analysis and characterization of atrial tachyarrhythmias requires the previous estimation of the atrial activity (AA) free from any ventricular activity and other artefacts. This contribution considers a blind source separation (BSS) model to separate the AA from multilead electrocardiograms (ECGs). Previously proposed BSS methods for AA extraction exploit only the spatial diversity introduced by the multiple electrodes. However, AA typically shows certain degree of temporal correlation, featuring a narrowband spectrum. Taking advantage of this observation, we put forward a novel two-step BSS-based technique which exploits both spatial and temporal information. The spatiotemporal BSS algorithm is validated on real ECGs from a significant number of patients, and proves consistently superior to a spatial-only ICA method. In real ECG recordings, performance can be measured by the main frequency peak and the spectral concentration. The spatiotemporal algorithm outperforms the ICA method, obtaining a spectral concentration of 58.8% and 44.7%, respectively.

1 Introduction

Biomedical engineering is one of the research areas where the statistical tool of independent component analysis (ICA) has demonstrated a remarkable success. Indeed, ICA techniques are suitable to solve a large number of biomedical problems in electroencephalography (EEG) [1], magnetoencephalography (MEG), electrocardiography (ECG) [2], functional magnetic resonance imaging (fMRI) [3], etc. In the area of cardiac signal analysis, ICA methods can be employed for the separation of the ventricular activity (VA) and the atrial activity (AA). This separation is particularly useful in the study of atrial arrhythmias, e.g., atrial fibrillation (AF) or atrial flutter (AFL) [4], where AA and VA are temporally and spectrally overlapped.

The analysis and characterization of atrial arrhythmias from the ECG requires the previous estimation of AA. The main difficulty is that AA and VA appear mixed at the electrode outputs. The separation of these cardiac activities from the 12-lead standard ECG has already been modelled as a blind source separation (BSS) problem [5] although only spatial information has been utilized for imposing statistical independence in the estimated sources. Indeed, any prior information about the temporal structure of the sources is disregarded in most ICA algorithms applied to this problem.

Motivated by the observation that VA and AA present specific spatiotemporal statistical properties, the present study presents a novel separation method more adapted to the biomedical problem in hand. In effect, VA can be modelled as a supergaussian distribution [6] whereas AA signals are typically quasi-gaussian and exhibit a narrowband spectrum with a main frequency of between 3.5-9 Hz [7][8]. Taking into account these considerations, a new BSS-based algorithm aiming to utilize more fully the spatiotemporal information of the ECG recordings is developed in a bid to enhance the quality of the estimated AA. The proposed approach combines ICA based on spatial-only higher-order statistics (HOS) with spatiotemporal second-order processing. The first stage is implemented with the FastICA method, whereas the second is carried out via the second order blind identification (SOBI) algorithm.

2 Methods

2.1 Statistical Source Analysis

The sources contained in an ECG recording can be divided into three types. VA sources are the ECG components with the highest energy. These components have a high amplitude during ventricular depolarization and repolarization (QRS complex and T wave respectively), but the rest of the time they present values close to zero due to the inactive period. Accordingly, VA sources possess supergaussian random distributions, even with kurtosis values above those of Laplacian distributions, as will be confirmed in the results section.

In the second place, AA consists of small and continuous wavelets with a cycle between 125 ms and 300 ms. A statistical analysis of the sources shows that AA has kurtosis values very close to zero (as will be discussed later on), typical of quasi-gaussian distributions. AA waves have a characteristic spectrum, with a main peak due to the refractory period. This fact, which is neglected by practically all AA extraction methods to date, is exploited by the algorithm proposed in this paper. Fig. 1 shows an example of VA and AA waveforms, and their corresponding distribution estimates.

Finally, noise and other artefacts are the contributions with the lowest energy, although in more than a few leads they could show an amplitude of the same order of

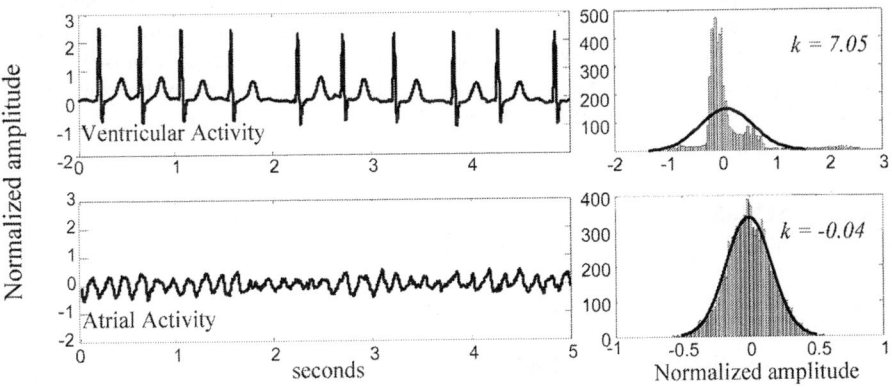

Fig. 1. Examples of VA and AA, and their histogram including kurtosis values.

magnitude as the atrial sources, or even higher. The statistical behaviour of the noise may be different for each recording; even several noise sources with different statistical behaviour may be found in a single ECG. Hence, no assumption about the noise pdf or correlation can generally be made. The only noise assumption in the separation model we propose is that the noise and the AA source have different spectra. This hypothesis is verified in practically all cases.

2.2 Two Stage Methodology

The body-surface potentials as a result of cardiac electrical activity can be modelled as a blind source separation (BSS) problem:

$$\mathbf{x}(t) = \mathbf{A}\mathbf{s}(t) \tag{1}$$

where $\mathbf{x}(t)$ is a length-m vector which represents the electrode outputs at time instant t, $s(t)$ is a length-n ($n \leq m$) vector that represents the bioelectric sources, and $\mathbf{A}$ is the $m \times n$ mixing matrix which models the propagation from sources to electrodes. For the standard ECG, we have $m = 12$. Neither the original sources nor the transfer coefficients from the epicardial surface towards the body surface are known.

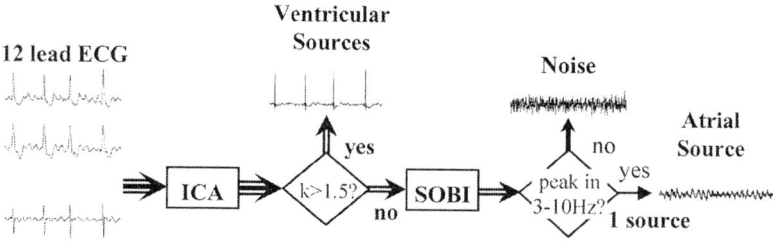

Fig. 2. Block diagram of the proposed hybrid approach to AA signal extraction.

In order to separate the AA free from VA and other interference, we propose a two-stage methodology, as illustrated in Fig. 2. The first stage exploits the super-gaussian character of the VA to remove the ventricular contributions, and is implemented with spatial-only HOS-based ICA. Since ventricular components show the highest amplitude, this stage, if successful, eliminates the major source of interference. The remaining non-ventricular components (AA, artefacts and noise) are the inputs of the second stage. This step, which is implemented with spatiotemporal SOS-based ICA, takes advantage of the characteristic spectrum of AA in order to enhance the AA estimation quality.

2.2.1 First Stage: Higher-Order Spatial Information (HOS-Based ICA)

In general, ICA methods estimate a separation matrix $\mathbf{B}$ such that the estimated sources

$$\hat{\mathbf{s}}(t) = \mathbf{B}\mathbf{x}(t) \tag{2}$$

fulfil certain statistical independence criterion. HOS-based ICA techniques are most suitable to separate independent nongaussian sources. These techniques are able to estimate the independent sources by using certain measures of independence provided by the HOS of the multilead signal [9][10]. In this study we have chosen an algorithm that estimates nongaussianity as a function of the following approximation of negentropy $J(\cdot)$ [11]:

$$J(y) \propto \left[\mathrm{E}[G(y)] - \mathrm{E}[G(v)] \right]$$
$$G(y) = \log \cosh y$$
(3)

where y is the output signal and v is a unit variance Gaussian variable. The maximisation of the contrast function can carried out, after pre-whitening, by means of a robust fixed point algorithm known as FastICA [12]. Note that the aim of this paper is not to emphasize the convenience of a specific ICA algorithm but to demonstrate the suitability of HOS-based ICA as a general concept for this first processing stage.

HOS-based ICA algorithms are especially equipped to extract all nongaussian sources, but are unable to separate gaussian sources since their HOS are null. Hence, all gaussian sources will appear mixed at the ICA output. The practical consequence over AF recordings is that VA sources will be correctly extracted, but the AA source can appear combined with gaussian-like sources of interference such as thermal noise and other artefacts. Due to the very low amplitude of the AA signal, the separation of AA from these sources of interference becomes an important necessary task. This task will be carried out in the second stage, which is described next.

2.2.2 Second Stage: Second-Order Spatiotemporal Information (SOBI)

The inputs to the second processing stage are the non-ventricular source components estimated by the first stage. The decision as to which components belong to the ventricular subspace and which components belong to the non-ventricular subspace can be done automatically. Due to the existence of the QRS complex, the ventricular sources are highly kurtic; by contrast, AA usually displays kurtosis values marginally different from zero. Consequently, a kurtosis-based threshold can be employed to distinguish between ventricular and non-ventricular sources. Preliminary experiments show that a conservative normalized-kurtosis threshold of around 1.5 allows us to retain the AA information in the non-ventricular subspace (the signal subspace which lies orthogonal to that spanned by the mixing-matrix columns associated to the ventricular sources) and reject all other sources that contain QRS complexes.

The so-called second-order blind identification (SOBI) is designed to separate a mixture of uncorrelated sources with different spectral content through a second-order statistical analysis which capitalizes on the source temporal information [13]. For this purpose, SOBI aims to find a transformation that simultaneously diagonalizes several correlation matrices at different lags. Since, in general, no transformation may exist that accomplishes such a stringent condition, a function that objectively measures the degree of joint approximate diagonalization (JD) at different lags is employed instead.

Let $\mathbf{z}$ denote the non-ventricular sources inaccurately estimated at the first processing stage, and $\mathbf{s}$ the associated actual sources, among which the desired AA source appears. In the simplified two-signal case, the real sources $\mathbf{s}$ and the whitened observations $\mathbf{z}$ are related through a Givens transformation:

$$\mathbf{z} = \mathbf{Q}\mathbf{s}, \quad \mathbf{Q} = \begin{bmatrix} \cos\theta & -\sin\theta \\ \sin\theta & \cos\theta \end{bmatrix}$$
(4)

where θ is an unknown rotation angle. The rotation angle that maximizes the JD criterion allows the recovery of the original sources. The extension of this procedure to more than two signals is easily carried out through a Jacobi-like iteration. Full details are given in [13], and are omitted here due to the lack of space.

The SOBI algorithm is appropriate for extracting sources with a narrowband spectrum; hence its suitability for AA estimation. The number of matrices for joint diagonalization and their respective time lags must be properly selected. Since the autocorrelation of the AA source in AF episodes is quasi-periodic with a period around 160 ms – i.e., 160 samples at a sampling rate of 1 Khz –, correlation matrices with time lags comprising two cycles (that is, 320 ms) are chosen. This choice guarantees that even for AF signals with larger AA cycle the lag range spans at least one complete cycle period. Choosing correlation matrices at evenly spaced lags of 20 ms (i.e., a total of 17 correlation matrices) guarantees a high proportion of significant (non-zero) autocorrelation values among the selected lags with an affordable computational complexity.

We refer to the proposed two-stage hybrid method as ICA-SOBI.

3 Results

28 ECGs digitised during 30 s at a constant sampling rate of $f_s = 1$ Khz with 16-bit amplitude resolution were employed in our study. All patients were suffering from atrial arrhythmias, including 17 AF and 11 AFL episodes. HOS-based ICA only (without the SOBI step) and ICA-SOBI were applied to this database. The estimation of the AA source was successful in all cases.

A spectral analysis was carried out in order to detect the main frequency f_p. The AA source estimated with ICA provided the same main frequency as the AA source estimated with ICA-SOBI, being of 6.19±0.73Hz for AF and 4.06±0.65 Hz for AFL. As an objective measure of AA extraction quality, the spectral concentration of the AA source around its main peak was computed according to the following expression:

$$SC = \frac{\sum_{0.82 f_p}^{1.17 f_p} P_{AA}(f_i)}{\sum_{0}^{f_s/2} P_{AA}(f_i)} \quad (5)$$

where P_{AA} is the power spectrum of the AA signal, computed using the Welch's method over a 8192-point FFT with a 50%-overlap 4096-sample Hamming window, and f_i denote the FFT discrete frequency values.

The AA source obtained with ICA-SOBI had a higher spectral concentration around the main frequency peak in all cases. In average, ICA obtained a spectral concentration of 37.1% for AF and 54.5% for AFL. The spectral concentration was increased with ICA-SOBI up to 53.7% and 65.2% for AF and AFL, respectively. The higher spectral concentration of the AA signal obtained after SOBI processing indicates that part of the noise present in the AA signal after ICA is effectively removed by SOBI. Fig. 3 compares the spectral concentration levels of the estimated AA using both methodologies. A typical example of the estimated AA and its spectrum where ICA-SOBI outperforms ICA is shown in Fig. 4.

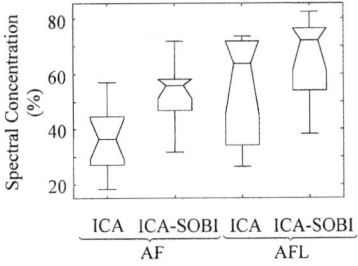

Fig. 3. Box-and-whiskers plot of the spectral concentration of the estimated AA.

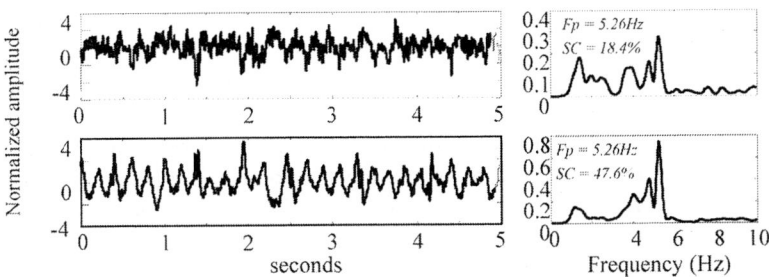

Fig. 4. An example where ICA-SOBI outperforms ICA.

Regarding the kurtosis values of the VA and the AA, the results confirm the hypotheses employed in the separation model. With a kurtosis value of 16.5±5.9 for the ECGs under test, VA is indeed supergaussian. In contrast, AA cannot be assumed nongaussian, with a kurtosis value of -0.21±0.45 for this database. The fact that the estimated ventricular and atrial sources fulfilled the hypothesis assumed in the problem formulation regarding their statistical behaviour and spectral characteristics endorses the proposed approach for the enhanced estimation of AA in patients with atrial arrhythmias.

The improvement in the quality of the estimated AA after the latter stage appears closely correlated with the gaussianity of the sources. In the cases where the AA source presented a nongaussian character (i.e., kurtosis values significantly different from zero) the improvement in the spectral concentration was more important than in those cases with a higher gaussianity degree (kurtosis near zero). Fig. 5 illustrates the improvement in the spectral concentration as a function of the AA kurtosis.

4 Discussion and Conclusions

This paper has demonstrated that the source temporal information is indeed relevant in the estimation of AA from multi-lead ECG recordings of atrial arrhythmias episodes. A spatiotemporal BSS algorithm adapted to this specific problem has been designed and implemented. The algorithm consists of an initial spatial-HOS based separation stage (ICA) aiming to remove nongaussian interference (mainly VA), followed by a time-SOS based separation stage (SOBI) aiming to cancel gaussian-like noise. In this manner, the AA can be separated not only from VA, but also from other independent sources of noise and interference regardless of their distribution (gaussian or otherwise). An experimental study with real AF and AFL signals has validated the appropriateness of the proposed method.

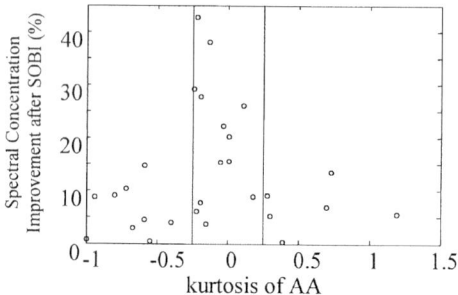

Fig. 5. Improvement in the spectral concentration as a function of AA source kurtosis.

In the experimental results, AA estimation has always improved with the application of the second separation stage exploiting temporal information. Even in ECGs where ICA had already estimated the AA accurately (because the existing AA was sufficiently nongaussian), the second step has been able to maintain the separation quality. Since the statistical behaviour of the AA source is not known a priori, but it may well change across patients, or even evolve in the same patient, it seems sensible to make use of the full two-step approach in all cases.

This contribution improves the existing solutions for AF analysis. Once the AA has been extracted, it can be further analyzed for spectral characterization, pattern recognition, time-frequency parameter extraction, etc. The proposed methodology thus emerges as a helpful tool in clinical diagnosis.

Acknowledgements

This study has been partly funded by the research grant TIC2002-00957 and the Universidad Politecnica de Valencia. V. Zarzoso is in receipt of a Post-doctoral Research Fellowship awarded by the Royal Academy of Engineering of the UK. This work was partially carried out while on leave at Laboratoire I3S, Sophia Antipolis, France. He gratefully acknowledges Pierre Comon's kind hospitality.

References

1. Makeig S., Bell A.J., Jung T.P., Sejnowski T.J., "Independent component analysis of electroencephalographic data", *Advances in Neural Information Processing Systems*, Vol. 8, 1996, pp. 145-151.
2. Barros A.K., Mansour A., Ohnishi N., "Adaptive blind elimination of artifacts in ECG signals", *I&ANN'98*, Tenerife, Spain, Feb. 1998, pp. 1380-1386.
3. McKeown M.J., Makeig S., Brown G.G., Jung T.P., Kindermann S.S., Sejnowski T.J., "Analysis of fMRI data by blind separation into independent spatial components", *Human Brain Mapping*, Vol. 6, No 3, 1998, pp. 160-188.
4. Rieta J.J., Millet-Roig J., Zarzoso V., Castells F., Sánchez C., García-Civera G., Morell S., "Atrial fibrillation, atrial flutter and normal sinus rhythm discrimination by means of blind source separation and spectral parameters extraction", *IEEE Computers in Cardiology*, Memphis, Sep. 2002, pp. 25-28.

5. Rieta J.J., Castells F., Sanchez C., Zarzoso V., Millet J., "Atrial activity extraction for atrial fibrillation analysis using blind source separation", *IEEE Trans. Biomed. Eng.*, Vol. 51, 2004, pp. 1176-86.
6. Castells F., Igual J., Rieta J.J., Sánchez C., Millet J., "Atrial fibrillation analisis based on ICA including statistical and temporal source information", *ICASSP-2003*, Hong Kong, Apr. 2003, Vol. V, pp. 94-96.
7. Bollmann A., Kanuru N.K., McTeague K.K., Walter P.F., DeLurgio D.B., Langberg J.J., "Frequency analysis of human atrial fibrillation using the surface electrocardiogram and its response to ibutilide", *Am. J. Cardiol.* Vol. 81, 1998, pp. 1439-45.
8. Stridh M., Sörnmo L., Meurling C., Olsson B., "Characterization of atrial fibrillation using the surface ECG: Spectral analysis and timedependent properties", *IEEE Trans. Biomed. Eng.*, Vol. 48, 2001, pp. 19-27.
9. Comon P., "Independent component analysis – a new concept?", *Signal Processing*, Vol. 36, 1994, pp. 287-314.
10. Cardoso J.-F., Souloumiac A., "Blind beamforming for non Gaussian signals", *IEE Proceedings-F*, Vol. 140, 1993, pp. 362-370.
11. Hyvärinen A., Karhunen J., Oja E., *Independent Component Analysis*, John Willey & Sons, New York, 2001.
12. Hyvärinen A., "Fast and robust fixed-point algorithms for independent component analysis", *IEEE Trans. on Neural Networks*, Vol. 10, 1999, pp. 626-634.
13. Belouchrani A., Abed-Meraim K., Cardoso J.-F., Moulines E., "A blind source separation technique using second-order statistics", *IEEE Trans. Sig. Proc.*, Vol. 45, 1997, pp. 434-444.

Gaussianizing Transformations for ICA

Deniz Erdogmus, Yadunandana N. Rao, and José Carlos Príncipe

CNEL, Electrical and Computer Engineering Department,
University of Florida, Gainesville, Florida 32611, USA
{deniz,yadu,principe}@cnel.ufl.edu
http://www.cnel.ufl.edu

Abstract. Nonlinear principal components analysis is shown to generate some of the most common criteria for solving the linear independent components analysis problem. These include minimum kurtosis, maximum likelihood and the contrast score functions. In this paper, a topology that can separate the independent sources from a linear mixture by specifically utilizing a Gaussianizing nonlinearity is demonstrated. The link between the proposed topology and nonlinear principal components is established. Possible extensions to nonlinear mixtures and several implementation issues are also discussed.

1 Introduction

Independent components analysis (ICA) is now a mature field with numerous approaches and algorithms to solve the basic instantaneous linear mixture case as well as a variety of extensions of these basic principles to solve the more complicated problems involving convolutive or nonlinear mixtures [1-3]. Due to the existence of a wide literature and excellent survey papers [4,5], in addition to the books listed above, we shall not go into a detailed literature survey. Interested readers are referred to the references mentioned above and the references therein.

In this paper, we will focus on a special type of homomorphic transformation, called the Gaussianizing function. Several interesting observations about this transformation and its utility in ICA will be addressed in this paper. Especially, we will establish a link between a Gaussianizing function based topology for solving linear instantaneous mixture problems and the established technique of nonlinear principal components analysis (NPCA) [6], which has already been shown to encompass a number of linear ICA optimization criteria as special cases [1] corresponding to certain choices of the *nonlinear functions of projection*. Nevertheless, the selection of these nonlinear projection functions stemming from the principal of mutual independence has not been yet addressed. Determining such a function is intellectually appealing since "mutual information is a canonical contrast for ICA" [7]. Finally, we would like to stress that the goal of this paper is *not* to present yet another linear ICA algorithm, but to demonstrate an interesting selection of the nonlinearity in NPCA as this method is applied to solving the ICA problem.

2 Gaussianizing Transformations

Given an n-dimensional random vector $\mathbf{Y}$ with joint probability density function (pdf) $p_\mathbf{Y}(\mathbf{y})$, there exist many functions $\mathbf{g}:\Re^n \to \Re^n$ such that $\mathbf{Z}=\mathbf{g}(\mathbf{Y})$ is jointly Gaussian. In particular we are interested in the elementwise Gaussianization of $\mathbf{Y}$. Suppose Y_i has marginal pdf $p_i(y_i)$, whose corresponding cumulative distribution function (cdf) is $P_i(y_i)$. Let $\phi(.)$ denote the cdf of a zero-mean unit-variance single dimensional Gaussian variable, i.e.,

$$\phi(\xi) = \int_{-\infty}^{\xi} \frac{1}{\sqrt{2\pi}} e^{-\alpha^2/2} d\alpha \tag{1}$$

Then, according to the fundamental theorem of probability [8], $Z_i=\phi^{-1}(P_i(Y_i))$ is a zero-mean and unit-variance Gaussian random variable.

We define $g_i(\xi)=\phi^{-1}(P_i(\xi))$ and call this the Gaussianizing transformation for Y_i. Combining $g_i(.)$ into a vector valued function, we get the elementwise Gaussianizing transformation for $\mathbf{Y}$ as $\mathbf{Z}=\mathbf{g}(\mathbf{Y})$. Since this $\mathbf{g}:\Re^n \to \Re^n$ is acting on each argument separately, its Jacobian matrix is *diagonal* at every point in its domain. Furthermore, since every Z_i is zero mean and unit-variance Gaussian, the vector $\mathbf{Z}$ is jointly Gaussian denoted by $\mathbf{G}(\mathbf{z},\Sigma)$ with zero mean and covariance

$$\Sigma = E[\mathbf{Z}\mathbf{Z}^T] = \begin{bmatrix} 1 & & \rho_{ij} \\ & \ddots & \\ \rho_{ji} & & 1 \end{bmatrix} \tag{2}$$

The utility of this Gaussianizing transformation was pointed out earlier for multidimensional pdf estimation [9]. Clearly, if one estimates the marginal pdfs of $\mathbf{Y}$ and the covariance of $\mathbf{Z}$ after Gaussianizing $\mathbf{Y}$ as described above, then an estimate of the joint pdf of $\mathbf{Y}$ can be obtained using the fundamental theorem of probability [8].

$$p_\mathbf{Y}(\mathbf{y}) = \frac{\mathbf{G}(\mathbf{g}(\mathbf{y}),\Sigma)}{|\nabla \mathbf{g}^{-1}(\mathbf{g}(\mathbf{y}))|} = \mathbf{G}(\mathbf{g}(\mathbf{y}),\Sigma)|\nabla \mathbf{g}(\mathbf{y})|$$

$$= \mathbf{G}(\mathbf{g}(\mathbf{y}),\Sigma) \cdot \prod_{i=1}^{n} g'_i(y_i) = \mathbf{G}(\mathbf{g}(\mathbf{y}),\Sigma) \cdot \prod_{i=1}^{n} \frac{p_i(y_i)}{G(g_i(y_i),1)} \tag{3}$$

3 Homomorphic Linear ICA Topology

The linear ICA problem is described by a generative signal model that assumes the observed signals, denoted by $\mathbf{x}$, and the sources, denoted by $\mathbf{s}$, are obtained by a *square* linear system of equations. The sources are assumed to be statistically independent. In summary, assuming an unknown mixing matrix $\mathbf{H}$, we have

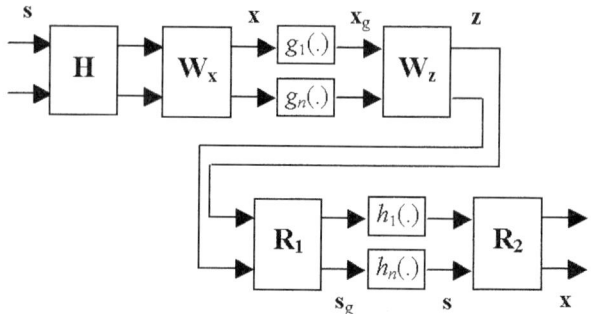

Fig. 1. A schematic diagram of the proposed homomorphic ICA topology.

$$\mathbf{x}_k = \mathbf{H}\mathbf{s}_k \qquad (4)$$

where the subscript k is the sample/time index. The linear ICA problem exhibits the following uncertainties, which cannot be resolved by the independence assumption alone: permutation of separated source estimates and scaling factors (including sign changes).

The goal is to recover the sources from the observed mixtures. For the sake of simplicity in the following arguments, we will assume that the marginal pdfs of the sources and the mixtures are known and all are strictly positive valued (to guarantee the invertibility of Gaussinizing transformations). It is assumed without loss of generality that the sources are already zero-mean.

Consider the topology shown in Fig. 1 as a solution to linear ICA. The observed mixtures are first spatially whitened by $\mathbf{W}_x$ to generate the whitened mixture vector $\mathbf{x}$. Since whitening reduces the mixing matrix to only a coordinate rotation, without loss of generality, we can always focus on mixing matrices that are orthonormal. In this case, we assume that the mixing matrix is $\mathbf{R}_2 = \mathbf{W}_x\mathbf{H}$. Since the marginal pdfs of the mixtures are known, one can construct the Gaussianizing functions $g_i(.)$ according to the previous section to obtain the Gaussianized mixtures $\mathbf{x}_g$. Whitening the Gaussianized mixtures will yield zero-mean unit-variance and uncorrelated signals $\mathbf{z}$. Since $\mathbf{z}$ is jointly Gaussian, uncorrelatedness corresponds to mutual independence. However, considering the function from the sources ($\mathbf{s}$) to the Gaussianized mixtures ($\mathbf{x}_g$) as a post-nonlinear mixture, we notice that although by obtaining $\mathbf{z}$ we have obtained independent components, due to the inherent rotation ambiguity of nonlinear mixtures in the ICA framework [10], we have not yet achieved source separation. Consequently, there is still an unknown orthonormal matrix $\mathbf{R}_1$ that will transform $\mathbf{z}$ into Gaussianized versions of the original sources. If the marginal source pdfs are known, the inverse of the Gaussianizing transformations for the sources could be obtained in accordance with the previous section (denoted by $h_i(.)$ in the figure), which would transform $\mathbf{s}_g$ to the original source distribution, thus yield the separated source signals (at least their estimates).

In summary, given the whitened mixtures, their marginal pdfs and the marginal pdfs of the sources (up to permutation and scaling ambiguities in accordance with the theory of linear ICA), it is possible to obtain an estimate of the orthonormal mixing matrix $\mathbf{R}_2$ and the sources $\mathbf{s}$ by training a constrained multilayer perceptron (MLP) topology with first layer weights given by $\mathbf{R}_1$ and second layer weights given by $\mathbf{R}_2$. The nonlinear functions of the hidden layer processing elements (PE) are determined by the inverse Gaussianizing transformations of the source signals. This MLP with square first and second layer weight matrices would be trained according to the following constrained optimization problem:

$$\min_{\mathbf{R}_1,\mathbf{R}_2} E\left[\left\|\mathbf{x}-\mathbf{R}_2\mathbf{h}(\mathbf{R}_1\mathbf{z})\right\|^2\right] \quad \text{subject to} \quad \mathbf{R}_1\mathbf{R}_1^T = \mathbf{I}, \mathbf{R}_2\mathbf{R}_2^T = \mathbf{I}, \tag{5}$$

Constrained neural structures of this type have been considered previously by Fiori [11]. Interested readers are referred to his work and the references therein to gain a detailed understanding of this subject.

4 Relationship with Nonlinear PCA

NPCA is known to solve the linear (and nonlinear) ICA problem when the nonlinear projection functions are properly selected. Various choices of these functions correspond to different ICA criteria ranging from kurtosis to maximum likelihood (ML) [1]. In the most general sense, the NPCA problem is compactly defined by the following optimization problem:

$$\min_{\mathbf{W}} E\left[\left\|\mathbf{x}-\mathbf{W}\mathbf{f}(\mathbf{W}^T\mathbf{x})\right\|^2\right] \tag{6}$$

where $\mathbf{f}(.)$ is an elementwise function (i.e. with a diagonal Jacobian at every point) that is selected *a priori*. For the special case of $\mathbf{f}(\mathbf{z})=\mathbf{z}$, this optimization problem reduces to the linear bottleneck topology, which is utilized by Xu to obtain the LMSER algorithm for linear PCA [12].

Returning to the topology in Fig. 1, under the assumptions of invertibility (which is satisfied if and only if the source pdfs are strictly greater than zero[1]) we observe that $\mathbf{z}=\mathbf{W}_z\mathbf{g}(\mathbf{x})$ and $\mathbf{x}=\mathbf{R}_2\mathbf{s}$, therefore, the cost function in (5) is $E[\|\mathbf{R}_2\mathbf{s}-\mathbf{R}_2\mathbf{h}(\mathbf{R}_1\mathbf{W}_z\mathbf{g}(\mathbf{R}_2\mathbf{s}))\|^2]$. Being orthonormal, $\mathbf{R}_2$ does not affect the Euclidean norm, and the cost becomes $E[\|\mathbf{s}-\mathbf{h}(\mathbf{R}_1\mathbf{W}_z\mathbf{g}(\mathbf{R}_2\mathbf{s}))\|^2]$. In the ICA setting, $\mathbf{s}$ is approximated by its estimate, the separated outputs $\mathbf{y}$, which is the output of the $\mathbf{h}(.)$ stage of Fig. 1. In the same setting, assuming whitened mixtures, NPCA would optimize

---

[1] In the case of zero probability densities, the Gaussianizing functions will not be invertible in general, since locally at these points the Jacobian might become singular. However, since the probability of occurrence of such points is also zero for the same reason, for the given signal-mixture case global invertibility is not necessary. However, it is assumed for simplicity.

$$\min_{\mathbf{W}} E\left[\|\mathbf{y}-\mathbf{f}(\mathbf{y})\|^2\right] \qquad (7)$$

where $\mathbf{y}=\mathbf{Wx}$, in accordance with (6) [1]. A direct comparison of (7) and the expression given above that is equivalent to (5) yields $\mathbf{f}(\mathbf{y})=\mathbf{h}(\mathbf{R}_1\mathbf{W}_z\mathbf{g}(\mathbf{R}_2\mathbf{y}))$.

In summary, the homomorphic ICA approach described in the previous section and formulated in (5) tries to determine a nonlinear subspace projection of the separated outputs such that the projections become independent. While an arbitrary selection of the nonlinear projection functions would not necessarily imply independence of the separated outputs, the proposed approach specifically exploits homomorphic Gaussianizing transformations of the signals such that orthogonality (uncorelatedness of zero-mean signals) is equivalent to mutual independence.

5 Alternative Approaches

The Gaussianizing transformations could be utilized in alternative linear ICA solution strategies. Here, we will briefly discuss a few. The obvious approach would be to utilize the Gaussianizing transformation to estimate the joint density of the mixtures or the separated outputs. This leads to two possible approaches.

Estimating the joint density of the mixtures: Suppose the whitened mixtures are related to the sources by $\mathbf{x}=\mathbf{Rs}$ and the marginal source distributions are known. Since the sources are independent, the joint source distribution, denoted by $p_S(\mathbf{s})$, is simply the product of the marginals. Due to the fundamental theorem of probability, the joint pdf of the mixtures could be determined as $p_X(\mathbf{x})=p_S(\mathbf{R}^T\mathbf{x})$. At the same time, from (3), we have $p_X(\mathbf{x})=G(\mathbf{g}(\mathbf{x}),\Sigma)|\nabla\mathbf{g}(\mathbf{x})|$. These two joint distributions must be identical, therefore one can determine $\mathbf{R}$ by minimizing any suitable divergence measure between the two representations of the mixture pdf. If the appropriate definition of Kullback-Leibler (KL) divergence is utilized as the measure, then the estimate would also be asymptotically maximum likelihood, due to the well-known relationships between ML and KL divergence.

Estimating the joint density of the separated outputs: Suppose that $\mathbf{x}=\mathbf{Hs}$ and $\mathbf{y}=\mathbf{Wx}$. Suppose that an estimate of the marginal pdfs of $\mathbf{y}$ is available at every step of learning iterations (nonparametric density estimations could be utilized at this stage). Then, one could construct the elementwise Gaussianizing functions of $\mathbf{y}$ to estimate its joint density using (3). The separation matrix $\mathbf{W}$ can be optimized to minimize the mutual information in $\mathbf{y}$ estimating Shannon's definition using the nonparametric marginal and joint distribution estimates of $\mathbf{y}$.

6 Extension to Nonlinear Mixtures

With some modifications, the topology shown in Fig. 1 could also be utilized to obtain independent components from mixtures generated by invertible nonlinear functions of the sources. In fact, given any n dimensional random vector $\mathbf{x}$ (regardless of it being generated from independent sources or not) one can determine n independent components. A proof of existence is provided in [10]. A much simpler proof of existence is as follows: Given $\mathbf{x}$, $\mathbf{z}=\mathbf{W}_z\mathbf{g}(\mathbf{x})$ are independent components, where $\mathbf{W}_z$ and $\mathbf{g}(.)$ are obtained as described above and in Fig. 1. In [10], the rotation ambiguity of nonlinear ICA is also addressed. This ambiguity is also readily observed in Fig. 1. Since $\mathbf{z}$ are independent components, $\mathbf{R}_1\mathbf{z}$ for any orthonormal matrix $\mathbf{R}_1$ also yields independent components for $\mathbf{x}$. Nevertheless, if one is not concerned about these ambiguities, nonlinear ICA is reduced to estimating the marginal pdfs of the mixture and applying whitening to the Gaussianized mixtures.

Actual separation of sources in the nonlinear mixture case requires additional constraints. For example if the mixture is post-nonlinear and the source distributions are known, the structure in Fig. 1 can be used as described in (5) with some modifications to solve the problem. Since the nonlinearities would be absorbed by the initial Gaussianizing transformation $\mathbf{g}(.)$, similar Gaussianizing functions must be employed at the output stage and the desired output should be $\mathbf{x}_g$. The latter Gaussianizing functions will be required to change at every learning iteration as they include the most current estimate of the nonlinearities of the post-nonlinear mixture and the following Gaussianizing function $\mathbf{g}(.)$. An approach along these lines was also proposed by Ziehe et al. [13].

7 Conclusions

In this paper, we have presented a topology based on using Gaussianizing homomorphic transformations that allows handling higher order statistics by considering only second order statistics in the ICA problem setup. The proposed topology is extremely interesting in that it lies at the intersection of nonlinear principal component analysis and learning in neural networks with orthonormality constraints on weight matrices.

Some alternative approaches that basically correspond to directly minimizing an estimate of the mutual information between the separated outputs are also sketched based on the density estimates obtained through the Gaussianizing transformations.

Extensions of the proposed topology to solve nonlinear ICA problems is discussed with special emphasis on post-nonlinear mixtures. The proposed topology also points out much simpler proofs for the existence of nonlinear ICA and its rotation ambiguity.

Acknowledgments

This work is supported by NSF grant ECS-0300340. The authors would like to thank K.E. Hild for useful discussions.

References

1. Hyvarinen, A., Karhunen, J., Oja, E.: Independent Component Analysis. Wiley, New York (2001)

2. Cichocki, A., Amari, S.I.: Adaptive Blind Signal and Image Processing: Learning Algorithms and Applications. Wiley, New York (2002)
3. Lee, T.W.: Independent Component Analysis: Theory and Applications. Kluwer, New York (1998)
4. Hyvarinen, A.: Survey on Independent Component Analysis. Neural Computing Surveys. 2 (1999) 94-128
5. Jutten, C., Karhunen, J.: Advances in Nonlinear Blind Source Separation. Proceedings of ICA'03, Nara, Japan. (2003) 245-256
6. Karhunen, J., Joutsensalo, J.: Representation ans Separation of Signals Using Nonlinear PCA Type Learning. Neural Networks. 7 (1994) 113-127
7. Cardoso, J.F., Souloumiac, A.: Blind Beamforming for Non-Gaussian Signals. IEE Proceedings F: Radar and Signal Processing. 140 (1993) 362-370
8. Papoulis, A.: Probability, Random Variables, and Stochastic Processes. 3rd edn. McGraw-Hill, New York (1991)
9. Chen, S., Gopinath, R.A.: Gaussianization. Proceedings of NIPS'01, Denver, Colorado. (2001) 423-429
10. Hyvarinen, A., Pajunen, P.: Nonlinear Independent Component Analysis: Existence and Uniqueness Results. Neural Networks. 12 (1999) 429-439
11. Fiori, S.: A Theory for Learning by Weight Flow on Stiefel-Grassman Manifold. Neural Computation. 13 (2001) 1625-1647
12. Xu, L.: Least Mean Square Error Reconstruction Principle for Self- Organizing Neural Nets. Neural Networks. 6 (1993) 627-648
13. Ziehe, A., Kawanabe, M., Harmeling, S., Muller, K.R.: Blind Separation of Post-nonlinear Mixtures Using Linearizing Transformations and Temporal Decorrelation. Journal of Machine Learning Research. 4 (2003) 1319-1338

New Eigensystem-Based Method for Blind Source Separation

Rubén Martín-Clemente[1], Susana Hornillo-Mellado[1],
Carlos G. Puntonet[2], and José I. Acha[1]

[1] Área de Teoría de la Señal y Comunicaciones, Universidad de Sevilla
Avda. de los Descubrimientos s/n., 41092-Sevilla, Spain
{ruben,susanah}@us.es

[2] Departamento de Arquitectura y Tecnología de Computadores
Universidad de Granada, E-18071, Granada, Spain
{carlos,mrodriguez}@atc.ugr.es

Abstract. In this paper, it is presented an algorithm to construct a cumulant matrix that has a well-separated extremal eigenvalue. The corresponding eigenvector is well-conditioned and could be used to develop robust algorithms for blind source extraction. Simulations demonstrate the effectiveness of the proposed approach.

1 Introduction

Blind Source Separation (BSS) is a challenging problem in Signal Processing. It consists in extracting source signals from sensor measurements. Here, the 'blind' qualification emphasizes that neither the sources nor the mapping between the sources and the sensor measurements are known *a priori*. Applications arise in numerous fields: e.g., array processing, speech enhancement, noise cancellation, data communications, biomedical signal processing *et cetera*.

Consider the linear instantaneous BSS model:

$$\mathbf{x}(t) = \mathbf{A}\,\mathbf{s}(t) \tag{1}$$

where $\mathbf{s}(t)$ denotes the $N \times 1$ vector whose components $s_i(t)$ are the sources, $\mathbf{x}(t)$ is the $N \times 1$ sensor measurement and $\mathbf{A}$ denotes an unknown mixing matrix. Starting from the seminal work [11], the problem has been studied by a large number of researchers (see [6, 10] and the references therein). In recent times, independence criteria which are based on Information-Theoretic models have attracted a great deal of attention. The algebraic structure of the so-called 'quadricovariance' has been exploited as well: roughly speaking, the quadricovariance is a fourth-order tensor whose coordinates are the cumulants of the whitened sensor measurements; the matrix $\mathbf{N}$ formed by the contraction of the quadricovariance with any arbitrary matrix $\mathbf{M}$ is always diagonalized by the mixing matrix [3] – consequently, the eigenvectors of $\mathbf{N}$ give the columns of the mixing matrix. *The problem arises when matrix $\mathbf{N}$ has close eigenvalues since, in this case, its eigenvectors are very sensitive to errors in the computation of the* *statistics of the data.* To obtain more robust estimates, the joint diagonalization of several matrices $\mathbf{N}_i$ has been proposed [3,5], whereby each matrix $\mathbf{N}_i$ is the contraction of the cumulant tensor with a different matrix $\mathbf{M}_i$; however, this approach is computationally demanding.

The purpose of this paper is to propose a simple algorithm that produces a cumulant matrix $\mathbf{N}$ that has a well-separated extremal eigenvalue. Consequently, the corresponding eigenvector is expected to be numerically stable, in the sense that small changes in $\mathbf{N}$ do not induce large changes in the eigenvector. The new method could be used to develop fast and robust algorithms for blind source extraction.

2 Problem Statement and Notation

The aim of BSS is to determine an $N \times N$ matrix $\mathbf{B}$ from the sole observation of the data $\mathbf{x}(t)$ such that:

$$\mathbf{y}(t) = \mathbf{B}\,\mathbf{x}(t) = \mathbf{G}\,\mathbf{s}(t) \qquad (2)$$

is an estimate of the source vector (up to permutation and scaling). The following hypotheses are assumed:

(H1) The sources $s_i(t)$ are statistically independent.
(H2) Each source $s_i(t)$ is a stationary zero-mean unity-variance process.
(H3) At most one source is gaussian distributed.
(H4) The mixing matrix $\mathbf{A}$ is nonsingular.
(H5) The observed vector $\mathbf{x}(t)$ is spatially white at order 2, i.e.:

$$E\left[\mathbf{x}(t)\,\mathbf{x}(t)^H\right] = \mathbf{I}$$

Hypothesis (H5) is not restrictive: one can always *whiten* the observations.

It follows that matrix $\mathbf{A}$ is *unitary*, i.e., $\mathbf{A}\mathbf{A}^H = \mathbf{I}$. This can be seen from:

$$\mathbf{I} \stackrel{<1>}{=} E\left[\mathbf{x}(t)\,\mathbf{x}(t)^H\right] = \mathbf{A}\,E\left[\mathbf{s}(t)\,\mathbf{s}(t)^H\right]\mathbf{A}^H \stackrel{<2>}{=} \mathbf{A}\mathbf{A}^H$$

where eq. $<1>$ follows from (H5) and eq. $<2>$ follows from (H1)–(H2). Consequently, the search for the inverse of $\mathbf{A}$ can be restricted to the space of the unitary matrices. Therefore, matrix $\mathbf{B}$ is supposed to be unitary. Similarly, it follows that $\mathbf{G} = \mathbf{B}\mathbf{A}$ is unitary as well.

2.1 Quadricovariance: Definition and Properties

Under the term "quadricovariance" [2], we understand the fourth-order tensor with coordinates:

$$q_{il}^{jk} = cum(x_i, x_j^*, x_k^*, x_l) \qquad (3)$$

where:

$$cum(x_i, x_j^*, x_k^*, x_l) = E\{x_i\,x_j^*\,x_k^*\,x_l\} - E\{x_i\,x_j^*\}\,E\{x_k^*\,x_l\} - \\ - E\{x_i\,x_k^*\}\,E\{x_j^*\,x_l\} - E\{x_i\,x_l\}\,E\{x_j^*\,x_k^*\}$$

Let **N** be the matrix with entries defined as:

$$n_{ij} \stackrel{def}{=} \sum_{1 \leq k,l \leq N} q_{il}^{jk} m_{kl} \qquad (4)$$

where m_{kl} are arbitrary constants[1]. It can be shown that [3, 5]

$$\mathbf{N}\mathbf{A} = \mathbf{A}\mathbf{\Lambda}_M, \qquad (5)$$

where $\mathbf{\Lambda}_M$ is a diagonal matrix whose diagonal elements depend on the statistics of $\mathbf{x}(t)$ as well as the particular constants m_{kl}. Eqn. (5) is the usual definition of eigenvalues and eigenvectors: in view of (5), it is inferred that the eigenvectors of **N** are the columns of the unitary mixing matrix **A** (up to complex constants of unit norm). Hence, a true separating matrix **B** is just obtained by transferring the columns from (the complex conjugate) of matrix **A** to the rows in matrix **B**.

This approach is very elegant. However, if **N** has close eigenvalues, the method is very sensitive to errors in the estimation of the cumulants: small changes in **N** produce large changes in its eigenvectors. Several ideas have been proposed to overcome this serious drawback [2, 3, 15]. In particular, our own approach is presented in the next Section.

3 Extraction of a Single Source

Our idea is to produce a matrix **N** *that has one eigenvalue that is well-separated from the others*. The corresponding eigenvector is hence expected to be numerically stable, in the sense that small changes in **N** do not induce large changes in the vector (see [8], Theorem 8.1.12 and Example 8.1.6).

Let $\mathbf{m} = (m_1, \ldots, m_N)^T$ be a unit-norm vector, i.e. $\sum_{k=1}^{N} |m_k|^2 = 1$ and **N** be the $N \times N$ matrix defined entrywise by

$$(\mathbf{N})_{ij} \stackrel{def}{=} \sum_{1 \leq k,l \leq N} q_{il}^{jk} m_k^* m_l \qquad (6)$$

– observe that (6) is nothing but a particular instance of (4). It is obtained that the eigenvalues of **N** are

$$\boxed{|h_1|^2 \kappa_{s_1}, \ldots, |h_N|^2 \kappa_{s_N}} \qquad (7)$$

where

$$h_n \stackrel{def}{=} \sum_{k=1}^{N} m_k a_{kn}, \qquad (8)$$

and $\kappa_{s_n} = cum(s_n, s_n^*, s_n, s_n^*)$ is the kurtosis of the nth source signal; since **A** is unitary, we have that $\sum_{n=1}^{N} |h_n|^2 = 1$.

---
[1] Technically speaking, **N** is said to be the contraction of the quadricovariance with matrix $\mathbf{M} = (m_{ij})$.

We propose to make zero all the eigenvalues of **N** *excepting one.* The rationale is as follows: the separation between one eigenvalue, e.g., $\mid h_i \mid^2 \kappa_{s_i}$, and the closest other eigenvalue, say $\mid h_j \mid^2 \kappa_{s_j}$, is

$$\mid h_i \mid^2 \kappa_{s_i} - \mid h_j \mid^2 \kappa_{s_j}$$

Suppose that $\mid \kappa_{s_i} \mid > \mid \kappa_{s_j} \mid$ (no loss of generality). Then, the distance between the two eigenvalues is maximized when $\mid h_i \mid = 1$, which implies that $\mid h_k \mid = 0$ for $k \neq i$ since $\sum_{\forall n} \mid h_n \mid^2 = 1$. This is the situation in which *there is only one nonzero eigenvalue*; the separation between this particular eigenvalue and the others will be, hence, maximum, as desired. As a further note, it is known that the sensitivity of the corresponding eigenvector is upper bounded by the inverse of that separation [8].

3.1 Choice of the Coefficients m_k

The question arises, how do we compute the coefficients m_k? In this paper, vector $\mathbf{m} = [m_1 \ldots m_N]^T$ *is computed* as the solution to the optimization problem:

$$\max_{\mathbf{m}} \mathbf{m}^H {\mathbf{N}'_i}^T \mathbf{m}, \text{ subject to } \|\mathbf{m}\|^2 = 1 \qquad (9)$$

where $\mathbf{N}'_i$ is the $N \times N$ matrix defined entrywise by

$$(\mathbf{N}'_i)_{mn} = cum(x_m, x_n^*, y_i^*, y_i) \qquad (10)$$

Using basic algebra [8], the optimum $\mathbf{m}$ is immediately found to be the conjugate of the principal eigenvector (the one associated with the largest eigenvalue) of $\mathbf{N}'_i$.

The rationale behind this choice of $\mathbf{m}$ is the following: the change of variables $\mathbf{G} = \mathbf{B} \mathbf{A}$ allows us to rewrite[12]

$$\mathbf{m}^H {\mathbf{N}'_i}^T \mathbf{m} \equiv \sum_{n=1}^{N} \mid h_n \mid^2 \mid g_{in} \mid^2 \kappa_{s_n} \qquad (11)$$

where g_{in} is the (i,n)th coordinate of the global matrix $\mathbf{G} = \mathbf{B} \mathbf{A}$ and $h_1, \ldots, h_N$ were defined in (8). Suppose that

$$\mid g_{i1} \mid^2 \kappa_{s_1} > \mid g_{i2} \mid^2 \kappa_{s_2} > \ldots > \mid g_{iN} \mid^2 \kappa_{s_N} \qquad (12)$$

For instance, (12) holds (no loss of generality) when the coordinates b_{ij} have been randomly chosen, which allows us to establish that the numbers $\mid g_{in} \mid^2 \kappa_{s_n}$ are distinct with probability one. Then, (11) is maximized by making $\mid h_1 \mid$ as large as possible and *this occurs with* $\mid h_1 \mid = 1$ and $\mid h_n \mid = 0$ for $n \neq 1$ (as $\sum_{\forall n} \mid h_n \mid^2 = 1$). In other words: the optimum $\mathbf{m}$ makes zero all the coefficients $h_1, \ldots, h_N$ excepting one.

Returning to the main problem, since the eigenvalues of $\mathbf{N}$ are precisely

$$\mid h_1 \mid^2 \kappa_{s_1}, \ldots, \mid h_N \mid^2 \kappa_{s_N},$$

it readily follows that the matrix $\mathbf{N}$ obtained by substituting $\mathbf{m}$ in definition (6) *possesses only one nonzero eigenvalue,* as desired.

3.2 Algorithm

The computation of **m**, **N** and its principal eigenvector, collectively and in that order, constitute the basis of our method for extracting a single source. The corresponding algorithm may take the following simple form:

```
(0)  Apply the whitening transformation to the data.
(1)  Start with unit-norm vector  b_i = (b_i1,...,b_iN)^T  (initial guess).
(2)  for k = 1,2,...,k_max
     (2.1) Set  y_i = Σ_n b_in x_n.
     (2.2) Estimate matrix  N'_i.
     (2.3) Set m to the conjugate of the principal eigenvector of N'_i.
     (2.4) Estimate matrix  N.
     (2.5) Set b_i to the conjugate of the principal eigenvector of N.
(3)  end for
(4)  return  y_i = Σ_n b_in x_n,  the estimated source.
```

Regarding the **for** step, this is just a mechanism for the iterative refinement of the solutions. It is necessary since the true cumulant matrices $\mathbf{N}$ and $\mathbf{N}'_i$ cannot be perfectly estimated in practice, due to the finite sample size.

4 Extraction of Several Sources

The rows of **B** are decoupled from each other by virtue of their orthogonality. This decoupling property makes it possible for us to accomplish the global problem as a sequence of local optimizations. That is, in order to estimate $M \leq N$ sources, we may compute at each step the principal eigenvectors of different matrices $\mathbf{N}'_i$ ($i = 1, \ldots, M$) and then apply Gram-Schmidt to orthonormalize them. The procedure is repeated until convergence. A similar approach is used in [9].

5 Numerical Experiments

In this Section we explore the algorithm through a simulation example. The performance is measured by the signal to noise ratio (SNR) of each source at the separator output. It is defined for source k by the following expression

$$SNR_k = 10 \log \frac{E\{|s_k|^2\}}{E\{|s_k - \hat{s}_k|^2\}} = -10 \log E\{|s_k - \hat{s}_k|^2\}$$

where $\hat{s}_k$ is the estimate of the kth source.

The source signals $s_1(t)$ and $s_2(t)$ were 16-PSK digitally modulated signals, whereas $s_3(t)$ and $s_4(t)$ were 16-QASK baseband signals. All of them are often used in communication systems. The complex baseband equivalent waveform of each source signal was used in the simulations. The coefficients of the mixing were random numbers whose real and imaginary parts were drawn from the

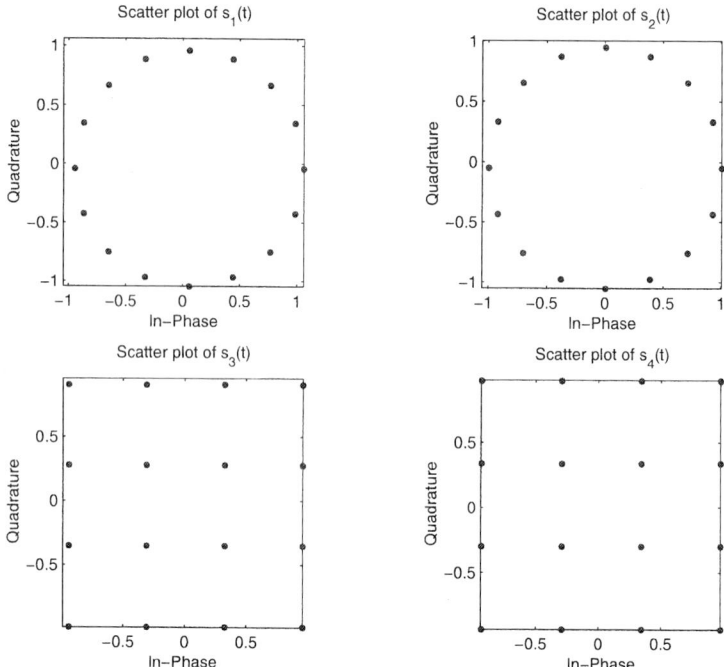

Fig. 1. Scatter Plots of the four sources $s_1(n), s_2(n), s_3(n), s_4(n)$.

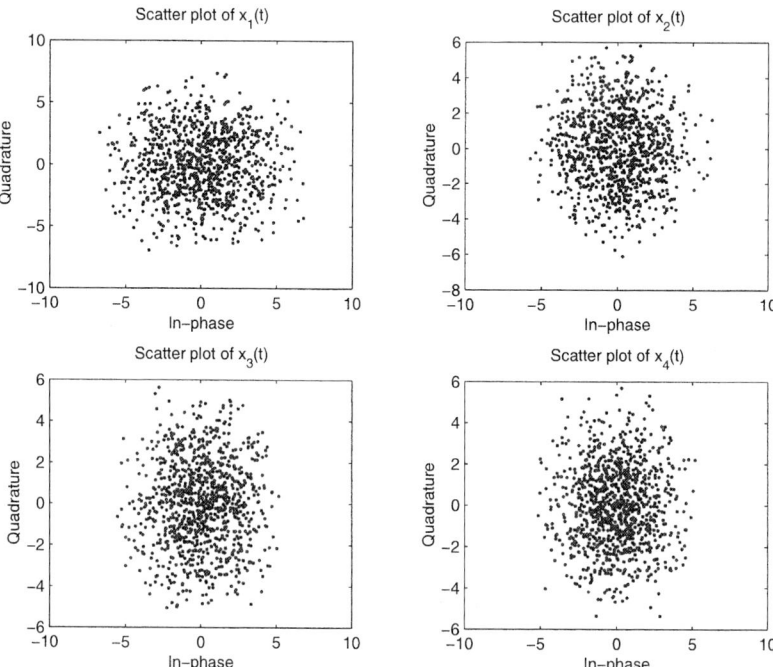

Fig. 2. Scatter Plots of the four measured signals $x_1(n), x_2(n), x_3(n), x_4(n)$.

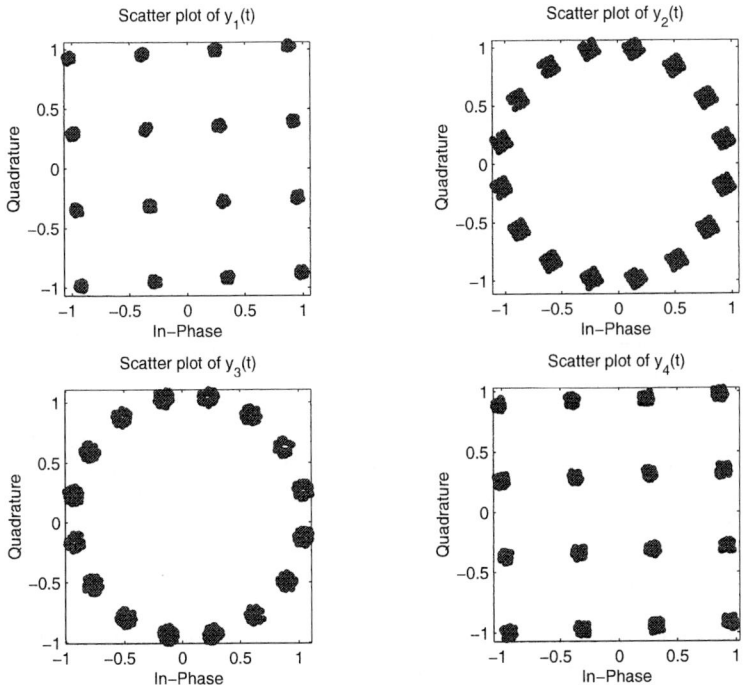

Fig. 3. Scatter Plots of the four estimated sources $y_1(n), y_2(n), y_3(n), y_4(n)$.

normal distribution with zero-mean and unit-variance. Figures 1, 2 and 3 depict, respectively, the scatter plots of the sources, the measured signals and the estimated sources[2].

The constellation of each estimated source appears clearly in Figure 3, showing that the separation is successful. In fact, the mean signal to noise ratio equals 31.64 dB after the separation (averaged over 100 independent experiments).

6 Conclusions

This paper introduces a cumulant matrix which is defined as the contraction of the fourth-order cumulant tensor with $\mathbf{m}^*\mathbf{m}^T$, where $\mathbf{m}$ is a unit-norm vector. The specific structure of this definition allows us to make zero all the eigenvalues of $\mathbf{N}$ excepting one. This makes the computation of the associated eigenvector more robust. The method is then used to develop a new algorithm for BSS.

References

1. A. Belouchrani, K. Abed Meraim, J.-F. Cardoso and E. Moulines, "A Blind Source Separation Technique based on Second Order Statistics", in *IEEE Transactions on Signal Processing*, vol. 45(2), pp. 434-444, 1997.

---

[2] The scatter plot represents the imaginary part (which is termed 'Quadrature Component') versus the real part (which is termed 'In-Phase Component') of the signal.

2. J.-F. Cardoso, "Eigenstructure of the Fourth-Order Cumulant Tensor with Application to the Blind Source Separation Problem", in *Proceedings ICASSP'90*, pp. 2655-2658, Albuquerque, 1990.
3. J.-F. Cardoso and A. Souloumiac, "Blind Beamforming for non-Gaussian Signals", in *Proceedings of the IEE*, vol. 140 (F6), pp. 362-370, 1993.
4. available at: ftp://tsi.enst.fr/pub/jfc/Algo/Jade/jade.m
5. J.-F. Cardoso, "High-Order Contrasts for Independent Component Analysis", in *Neural Computation*, vol. 11, pp. 157-192, 1999.
6. A. Cichocki and S. I. Amari, "Adaptive Blind Signal and Image Processing", *John Willey and Sons*, 2002.
7. N. Delfosse, P. Loubaton, "Adaptive Blind Separation of Independent Sources: A Deflation Approach", in *Signal Processing*, vol. 45, pp. 59-83, 1995.
8. G. Golub and C. van Loan, "Matrix Computations", *The John Hopkins University Press*, 1996.
9. A. Hyvärinen and E. Oja, "A Fast Fixed-Point Algorithm for Independent Component Analysis", in *Neural Computation*, vol. 6, pp. 1484-1492, 1997.
10. A. Hyvärinen, J. Karhunen and E. Oja, "Independent Component Analysis", *John Willey and Sons*, 2001.
11. C. Jutten and J. Herault, "Blind Separation of Sources, Part I: an adaptive algorithm based on neuromimetic architecture", in *Signal Processing*, vol. 24, pp. 1-10, 1991.
12. R. Martín-Clemente and J. I. Acha, "Eigendecomposition of Self-Tuned Cumulant Matrices for Blind Source Separation", *submitted*.
13. E. Moulines and J.-F. Cardoso, "Second-order versus fourth-order MUSIC algorithms. An asymptotical statistical performance analysis", in *Proceedings Workshop on Higher-Order Statistics*, pp. 121-130, Chamrousse, France, 1991.
14. C. Nikias and A. Petropulu "Higher-order spectra analysis", *Prentice-Hall*, 1993.
15. L. Tong, Y. Inouye and R. Liu, "Waveform preserving blind estimation of multiple independent sources", in *IEEE Transactions on Signal Processing*, vol. 41, pp. 2461-2470, 1993.

Optimization Issues in Noisy Gaussian ICA

Jean-François Cardoso[1] and Dinh-Tuan Pham[2]

[1] CNRS/LTCI, UMR 5141, Paris France
cardoso@tsi.enst.fr
[2] CNRS/LMC-IMAG, Grenoble, France
Dinh-Tuan.Pham@imag.fr

Abstract. This paper addresses the blind separation of *noisy* mixtures of independent sources. It discusses issues and techniques related to computing maximum likelihood estimates in Gaussian models.

1 Introduction

This paper is concerned with the blind separation of *noisy* mixtures of independent sources. A sequence $\{\mathbf{X}(t)\}$ of J-dimensional observations is modeled as an instantaneous mixture of K independent source sequences $\{\mathbf{S}_1(t)\}, \ldots \{\mathbf{S}_K(t)\}$, contaminated by an additive independent noise sequence $\{\mathbf{N}(t)\}$:

$$\mathbf{X}(t) = \mathbf{AS}(t) + \mathbf{N}(t)$$

where $\mathbf{A}$ is an unknown $J \times K$ matrix.

The noisy case has not received much attention in the ICA literature, maybe because it was felt that dealing explicitly with noise is useless in a high SNR context and hopeless in a low SNR context while in not-so-bad SNR situation, processing noisy data using noise-free models yields 'good enough' results. Another reason may be that, in the standard approach to ICA, the sources are modeled as non Gaussian i.i.d. sequences, with the effect that including noise in the model very much changes the structure of the estimation problem and makes it significantly more difficult to tackle.

In this paper, we consider noisy *Gaussian* ICA models which ignore the possible non Gaussianity of the sources but instead exploit the time structure of the source sequences. In these models, it is significantly easier to deal explicitly with additive noise. The paper discusses several techniques for (and issues in) computing maximum likelihood solutions in noisy Gaussian models. It is organized as follows. Section 2 recalls basic ideas regarding maximum likelihood estimation in Gaussian ICA. It emphasizes the difficulties encountered in low SNR. Section 3 describes two incarnations (depending on the noise model) of the EM algorithm and discusses its benefits and limitations at low SNR. Section 4 shows how the main limitation of the EM approach (namely: a relatively slow convergence) can be overcome by resorting to Newton algorithms, specifically tailored for ICA.

2 Noisy Gaussian Models for ICA

Gaussian ICA models assume non i.i.d. source sequences and build estimates of the mixing matrix from second-order statistics, assuming stationary correlated sources (early papers include [1–3]) or simple non stationary models (e.g.[4,5]). Among all possible approaches, we focus on those which can be derived from the maximum likelihood principle under simple working assumptions such that the likelihood depends on the data set via a set of Q sample covariance matrices $\hat{\mathbf{R}}_1, \ldots, \hat{\mathbf{R}}_Q$. For instance, when the source sequences are non stationary [6], the observation interval is split into Q subintervals $I_1, \ldots, I_Q$ and the qth matrix $\hat{\mathbf{R}}_q$ is computed as $\hat{\mathbf{R}}_q = \frac{1}{n_q} \sum_{t \in I_q} \mathbf{X}(t)\mathbf{X}(t)^\dagger$ where n_q is the number of samples in I_q. Another example is the separation of stationary colored sources; in this case, one splits the frequency domain into Q bands, i.e. $I_1, \ldots, I_Q$ are frequency intervals, and matrix $\hat{\mathbf{R}}_q$ is an average over the qth band after the data have been Fourier transformed and n_q is the number of DFT points in this band. A more flexible option is to compute sample covariance matrix over appropriate time-frequency or time-scale domains. In all these cases, the assumption of statistical independence between sources implies that each $\hat{\mathbf{R}}_q$ is an estimate of a 'true covariance matrix' $\mathbf{R}_q$ with structure:

$$\mathbf{R}_q = \mathbf{A}\mathbf{D}_q\mathbf{A}^\dagger + \mathbf{N}_q \qquad \mathbf{D}_q = \text{vdiag}\{d_{1q}, \ldots, d_{Kq}\} \qquad 1 \leq q \leq Q$$

where $\mathbf{N}_q$ is the $J \times J$ noise covariance matrix over the the qth domain, where d_{iq} is the variance of the ith source in the qth domain, where $\text{vdiag}(\cdots)$ denotes the diagonal matrix with the arguments on the diagonal and 0 elsewhere.

In the following, we denote by $\mathcal{D}$ (resp. $\mathcal{N}$) all the parameters which collectively describe the source (resp. noise) variances. Even more concisely, the whole parameter set is denoted by $\theta = (\mathbf{A}, \mathcal{D}, \mathcal{N})$ so that $\mathbf{R}_q = \mathbf{R}_q(\theta)$. Maximum likelihood estimates of the parameters are obtained by minimizing the cost

$$\phi(\mathbf{A}, \mathcal{D}, \mathcal{N}) = \phi(\theta) = \sum_{q=1}^{Q} n_q D(\hat{\mathbf{R}}_q, \mathbf{R}_q(\theta)) \qquad (1)$$

where $D(\cdot, \cdot)$ measures the mismatch between two $J \times J$ positive matrices as

$$D(\mathbf{R}_1, \mathbf{R}_2) = \frac{1}{2}\left(\text{tr}(\mathbf{R}_1\mathbf{R}_2^{-1}) - \log\det(\mathbf{R}_1\mathbf{R}_2^{-1}) - J\right).$$

Thus, a set of sample covariance matrices forms a sufficient statistic and the log-likelihood appears as a measure of mismatch between covariance matrices and their sample estimates.

No noise. In the square ($J = K$) and noise free case ($\mathbf{N}_q = 0$), the objective $\phi = \phi(\mathbf{A}, \mathcal{D}, 0)$ can be minimized explicitly with respect to the source variances d_{kq} for any value of $\mathbf{A}$. The reduced criterion $\min_{\mathcal{D}} \phi(\mathbf{A}, \mathcal{D}, 0)$ only depends on $\mathbf{A}$ and is found to be a measure of the joint diagonality of the matrix set $\{\mathbf{A}^{-1}\hat{\mathbf{R}}_q\mathbf{A}^{-\dagger}\}$; a fast and simple algorithm is available for its minimization [6]. If we cannot or do not want to assume $\mathbf{N}_q = 0$, we are facing the noisy case and the minimization of ϕ is a more difficult task.

White noise and free noise. This paper presents some ideas for maximizing the likelihood (minimizing ϕ) under two extreme noise scenarios: white noise and free noise. By 'white noise case', we mean that $\mathbf{N}_q$ does not depend on q so $\mathbf{N}_q = \mathbf{N} = \text{vdiag}(\sigma_1^2, \ldots, \sigma_J^2)$ and $\mathcal{N}$ has J free parameters. By 'free noise case', we mean that each noise covariance $\mathbf{N}_q$ can be any positive diagonal matrix: $\mathbf{N}_q = \text{vdiag}(\sigma_{1q}^2, \ldots, \sigma_{Jq}^2)$; there are $J \times Q$ free parameters in $\mathcal{N}$. Intermediate cases where $\mathbf{N}_q$ would depend smoothly on q are not considered in this paper.

Source recovery. We do not give details regarding source separation itself but only note that, once available, the parameters $(\mathbf{A}, \mathcal{D}, \mathcal{N})$, can be used to compute the Gaussian Wiener filter for source recovery.

Trouble at low SNR. In practical situations, one may have a good SNR on average over all domains but a poor SNR in a particular domain (*i.e* for some value of q). Think for instance of a stationary white noise corrupting speech: even at good global SNR, due to nature of speech, one expects to find time frames and spectral windows where the noise dominate the speech signal.

Estimating the variance of a signal buried in noise is not easy, even in the simplest setting: consider a model where a scalar observation x is modeled as $x = s + v$ where s (signal, say) and v (noise, say) are independent zero-mean normal variables with variances σ_s^2 and σ_v^2. Assume the noise variance σ_v^2 is known and the signal variance σ_s^2 is to be estimated. For a sample of independent observations, its most likely value is $\hat{\sigma}_s^2 = \max(0, \hat{\sigma}_x^2 - \sigma_v^2)$ where $\hat{\sigma}_x^2$ is the variance of the sample. The mean value of $\hat{\sigma}_x^2 - \sigma_v^2$ is σ_s^2 but its variance, for n i.i.d. samples of x, is $2n^{-1}(\sigma_s^2 + \sigma_v^2)^2$. Hence, unless the number n of available samples is significantly larger than $(\sigma_v^2/\sigma_s^2 + 1)^2$, there is a high probability that $\hat{\sigma}_x^2 < \sigma_v^2$, so that $\hat{\sigma}_s^2 = 0$. In other words, in presence of weak sources, the likelihood may be maximum on the boundary of the parameter domain. Such a situation may create difficulties.

3 The EM Algorithm for Noisy Gaussian Models

Because the cost function (1) actually is a likelihood in disguise, it may be minimized using the iterative EM algorithm. This method depends on defining 'hidden data' which, in our case, are taken to be the source signals. The algorithm itself is an updating rule for the parameters which is guaranteed to increase the likelihood, *i.e.* to decrease the cost (1). In the white noise case, the derivation of the EM updating rule has been reported elsewhere [7]; in the free noise case, they are similar and, due to lack of space, not included herein. We only give the final result.

When applied to the noisy Gaussian ICA model, each EM update of the current parameters starts with the computation, for all q, of the matrices

$$\mathbf{C}_q = (\mathbf{A}^\dagger \mathbf{N}_q^{-1} \mathbf{A} + \mathbf{D}_q^{-1})^{-1}, \qquad \mathbf{W}_q = \mathbf{C}_q \mathbf{A}^\dagger \mathbf{N}_q^{-1},$$

using the old values of the parameters and then the computation of

$$\mathbf{R}_q^{xs} = \hat{\mathbf{R}}_q \mathbf{W}_q^\dagger, \qquad \mathbf{R}_q^{ss} = \mathbf{W}_q \hat{\mathbf{R}}_q \mathbf{W}_q^\dagger + \mathbf{C}_q.$$

The re-estimates of the source variances are the given by

$$\mathbf{D}_q = \mathrm{diag}(\mathbf{R}_q^{ss}) \qquad (2)$$

where diag() is the vector made of the diagonal entries of its argument. The re-estimation of $\mathbf{A}$ and $\mathbf{N}_q$, however, depends on the noise model.

White noise. In the white noise case, where $\mathbf{N}_q = \mathbf{N}$, the re-estimates of $\mathbf{A}$ and $\mathbf{N}$ are, with $\tilde{n}_q = n_q / \sum_l n_l$, given by:

$$\mathbf{A} = (\sum_q n_q \mathbf{R}_q^{xs})(\sum_q n_q \mathbf{R}_q^{ss})^{-1}$$
$$\mathbf{N} = (\sum_q \tilde{n}_q \hat{\mathbf{R}}_q) - (\sum_q \tilde{n}_q \mathbf{R}_q^{xs})(\sum_q \tilde{n}_q \mathbf{R}_q^{ss})^{-1}(\sum_q \tilde{n}_q \mathbf{R}_q^{xs})$$

Free noise. In the free noise case, we have not found an explicit form for re-estimating all the parameters simultaneously. However, it is possible to alternate re-estimations of $(\{\mathbf{D}_q\}, \mathbf{A})$ with re-estimations of $(\{\mathbf{D}_q\}, \{\mathbf{N}_q\})$. When re-estimating $(\{\mathbf{D}_q\}, \mathbf{A})$, the new $\mathbf{A}$ is

$$\mathbf{A} = \left(\sum_q n_q\, \mathbf{R}_q^{ss} \otimes \mathbf{N}_q^{-1}\right)^{-1} \left(\sum_q n_q\, \mathbf{N}_q^{-1} \mathbf{R}_q^{xs}\right) \qquad (3)$$

and when re-estimating $(\{\mathbf{D}_q\}, \{\mathbf{N}_q\})$, the new noise variances are

$$\mathbf{N}_q = \mathrm{diag}\left((I - \mathbf{AW}_q)\hat{\mathbf{R}}_q(I - \mathbf{AW}_q)^\dagger + \mathbf{AC}_q \mathbf{A}^\dagger\right) \qquad (4)$$

In both cases, re-estimation of the source variances $\{\mathbf{D}_q\}$ is by equation (2). It would be wrong to use both equations (3) and (4) simultaneously.

Slowness of EM in low SNR. The EM updates are simple, but eventually very slow when it comes to finding the variance of a signal buried in noise. To see this, we return to the scalar example at the end of section 2 and we look at the EM solution for re-estimating σ_s^2. Basic computations show that, upon EM re-estimation of σ_s^2, the quantity $\sigma_s^2 - (\hat{\sigma}_x^2 - \sigma_v^2)$ is multiplied by a factor

$$\left[1 - \left(\frac{\sigma_s^2}{\sigma_s^2 + \sigma_n^2}\right)^2\right].$$

This factor is close to 1 in low SNR, showing that EM eventually becomes very slow at re-estimating the variances of signals buried in noise.

4 Newton-Like Optimization

4.1 Introduction

Newton-like techniques for minimizing an objective $\phi(\theta)$ are based on the second-order expansion of the objective. The quadratic approximation of $\phi(\theta + \delta\theta)$ being

$$\phi(\theta) + \sum_i \frac{\partial \phi}{\partial \theta_i} \delta\theta_i + \frac{1}{2} \sum_{ij} \frac{\partial^2 \phi(\theta)}{\partial \theta_i \partial \theta_j} \delta\theta_i \delta\theta_j$$

is minimized for
$$\delta\theta = -\left(\frac{\partial^2\phi(\theta)}{\partial\theta^2}\right)^{-1}\frac{\partial\phi(\theta)}{\partial\theta} \tag{5}$$

Thus, in raw form, the iterative Newton descent consists in updating θ into $\theta+\delta\theta$ with $\delta\theta$ given as above. Thanks to the gradient 'rectification' provided by the inverted Hessian matrix, Newton algorithms can achieve *quadratic* convergence speed near the solution, unlike simple gradient based techniques which only have *linear* convergence speed. There are however some caveats and also room for simplifications in the specific case of minimizing the spectral mismatch (1). This is the topic of this section.

We will need the first and second derivatives of the spectral mismatch:

$$\frac{\partial\phi}{\partial\theta_i} = \frac{1}{2}\sum_q n_q \operatorname{tr}\left(\mathbf{R}_q^{-1}(\mathbf{R}_q - \hat{\mathbf{R}}_q)\mathbf{R}_q^{-1}\frac{\partial \mathbf{R}_q}{\partial\theta_i}\right) \tag{6}$$

$$\frac{\partial^2\phi}{\partial\theta_i\partial\theta_j} = \frac{1}{2}\sum_q n_q \operatorname{tr}\left(\frac{\partial \mathbf{R}_q}{\partial\theta_i}\mathbf{R}_q^{-1}\frac{\partial \mathbf{R}_q}{\partial\theta_j}\mathbf{R}_q^{-1} - (\mathbf{R}_q - \hat{\mathbf{R}}_q)\frac{\partial^2 \mathbf{R}_q^{-1}}{\partial\theta_i\partial\theta_j}\right) \tag{7}$$

where the dependence of $\mathbf{R}_q$ on θ is not explicitly denoted.

A problem with the "raw" Newton algorithm is that the Hessian matrix may not be *positive definite*. This often occurs at the beginning of the algorithm where the iterate is still far from the solution (at the solution, the Hessian matrix must be positive definite as the solution must be a local minimum of the objective function). The non positive definiteness of the Hessian makes it impossible to ensure that the objective is decreased at each step of the algorithm and likely causes its divergence. In the following variants of the Newton algorithm, the Hessian is actually approximated in such a way that is is always positive.

4.2 Quasi Newton Algorithm

A popular variant of the Newton algorithm (called quasi Newton) consists in approximating the Hessian by its expectation, which is no other than the Fisher information matrix in the case where the objective function is the negative of the log likelihood. The advantage is that this approximate Hessian is (i) much simpler to compute and (ii) guaranteed to be positive. In the present case, the approximate Hessian is simply obtained by dropping the last term in (7):

$$H_{ij}(\theta) = \frac{1}{2}\sum_q n_q \operatorname{tr}\left(\frac{\partial \mathbf{R}_q}{\partial\theta_i}\mathbf{R}_q^{-1}\frac{\partial \mathbf{R}_q}{\partial\theta_j}\mathbf{R}_q^{-1}\right). \tag{8}$$

This is a very reasonable approximation since it is exact when $\hat{\mathbf{R}}_q = \mathbf{R}_q(\theta)$.

4.3 The Broyden-Fletcher-Goldfarb-Shanno (BFGS) Algorithm

One problem in implementing a Newton algorithm may be the cost and complexity of computing the Hessian matrix (or its inverse). This can be alleviated by using methods like BFGS [8] which, as they walk down the criterion, are able to build an approximation to the inverse of the Hessian, which should converge to the true inverse as the iterate converges to the solution. This approach was used in [7]. While perfectly acceptable, this approach does not exploit the specificities of criterion (1). We can however use the inverse of the approximate Hessian (8) to initialize the algorithm.

4.4 Fixing the Scale Indetermination

In the natural parameterization where θ consists of the elements of $\mathbf{A}$ and the diagonal elements of $\mathbf{D}_q$, $\mathbf{N}_q$, there is a scale indetermination: multiplying the matrices $\mathbf{D}_q$ by a diagonal matrix and post multiplying $\mathbf{A}$ by the inverse of its square root doesn't change the $\mathbf{R}_q$. To avoid this problem, one can always reparameterize non-redundantly the model, but this is awkward as the parameterization is not natural. We propose instead the following methods.

1: Changing the objective function. One minimizes, instead of $\phi(\theta)$,

$$\tilde{\phi}(\theta) = \phi(\theta) + p(\mathbf{A}) \qquad p(\mathbf{A}) = \sum_{k=1}^{K} l(|\mathbf{a}_k|^2)$$

where $\mathbf{a}_k$ denotes the kth column of $\mathbf{A}$ and $l(\cdot)$ is a penalty function like, for instance, $l(u) = (u-1)^2$ or $l(u) = u - \log u$. It is easy to see that the minimum of $\tilde{\phi}(\theta)$ is attained by a point minimizing $\phi(\theta)$ and such that the matrix $\mathbf{A}$ has columns of unit norm, that is a point minimizing both $\phi(\theta)$ and $p(\mathbf{A})$. This method is well fitted to the BFGS approach.

2: Changing the Hessian. The scale indeterminations make the Hessian and its approximation $H(\theta)$ is *singular*. Therefore, it is incorrect to write $\delta\theta = -H^{-1}(\theta)\partial\phi(\theta)/\partial\theta$. Instead, one should take $\delta\theta$ as a solution of[1]

$$H(\theta)\delta\theta = -\frac{\partial\phi(\theta)}{\partial\theta}. \tag{9}$$

This linear system always admits a solution since it can be shown that the image space of $H(\theta)$ contains the gradient vector $\partial\phi(\theta)/\partial\theta$. Actually, there is an infinite number of solutions, differing only by a vector in the null space of $H(\theta)$. This is a consequence of the scale indetermination. To fix this indetermination, our method consists in imposing a constraint $C(\theta)\delta\theta = 0$ where $C(\theta)$ is a symmetric positive semi-definite matrix with a null space and an image space complementing respectively the image space and the null space of $H(\theta)$. Then it may be shown that under this constraint, the system (9) always admits an unique solution and the matrix $H(\theta) + C(\theta)$ is positive definite. This unique solution can thus be simply obtained by solving

$$(H(\theta) + C(\theta))\delta\theta = \partial\phi(\theta)/\partial\theta.$$

---

[1] Actually, this is the standard way to compute $-H^{-1}(\theta)\partial\phi(\theta)/\partial\theta$ in numerical computations even in cases where $H(\theta)$ is not singular.

Method 2 is very similar to method 1 by taking $C(\theta)$ to be the Hessian of $p(\mathbf{A})$. Then $C(\theta)\delta\theta = 0$ is equivalent to $\mathbf{a}_k^\dagger \delta\mathbf{a}_k = 0$. Thus, while method 1 tries to find a solution for which the columns of $\mathbf{A}$ have unit norm, method 2 operates in such a way that the norms of these columns remain unchanged up to the first order at each step of the algorithm. Method 2 is however not suitable for the BFGS as this algorithm computes its own approximate inverse of the Hessian.

4.5 Exploiting the Block Diagonal Property of the Hessian

Advantage can be taken of the block structure of the Hessian to speed up solving eq. (5) or rather eq. (9). Indeed, if a linear system is partitioned as

$$\begin{bmatrix} H_{11} & H_{12} \\ H_{21} & H_{22} \end{bmatrix} \begin{bmatrix} \delta_1 \\ \delta_2 \end{bmatrix} = \begin{bmatrix} g_1 \\ g_2 \end{bmatrix},$$

one can compute δ_2 and then δ_1 by solving successively the equations:

$$(H_{22} - H_{21}H_{11}^{-1}H_{12})\delta_2 = g_2 - H_{21}H_{11}^{-1}g_1 \qquad (10)$$
$$H_{11}\delta_1 = g_1 - H_{12}\delta_2 \qquad (11)$$

This route offers large computational savings when H_{11} is large but diagonal or block-diagonal.

Indeed, in the free noise case, for $q \neq q'$, we find the decoupling:

$$\frac{\partial^2 \phi}{\partial \mathbf{N}_q \partial \mathbf{N}_{q'}} = 0 \qquad \frac{\partial^2 \phi}{\partial \mathbf{N}_q \partial \mathbf{D}_{q'}} = 0 \qquad \frac{\partial^2 \phi}{\partial \mathbf{D}_q \partial \mathbf{D}_{q'}} = 0$$

so we should put in δ_1 the diagonal elements of $\mathbf{D}_q$ and $\mathbf{N}_q$ and in δ_2 the elements of $\mathbf{A}$. In the white noise case ($\mathbf{N}_q \equiv \mathbf{N}$), the smart partitioning is for δ_1 to contain the diagonal elements of $\mathbf{D}_q$ and for δ_2 to contain the diagonal elements of $\mathbf{N}$ and the elements of $\mathbf{A}$.

5 Conclusion

We have seen that many options are available for computing the maximum likelihood estimates of the parameters of a noisy Gaussian ICA model.

The EM algorithm is straightforward to implement: there is no parameter to tune and each iteration does increase the likelihood of the parameters. However, after some quick progress, EM will eventually slow down if it has to estimate variances of components which are locally (*i.e.* for some q) buried under the other components. Note that slow EM re-estimation of noise variances is also likely to happen in the free noise if the SNR is high and the noise is 'buried under the signals' (but at least as many sources as sensors are needed to 'bury' the noise). In summary, EM is very good at doing 'most of the job' but, if *accurate* estimation is required, one has to resort to faster, Newton-like methods, to complete likelihood maximization.

There are many possible variants to the basic Newton-like update. We note that the issue of initialization is readily solved by using EM to provide a good starting point. The issue of scale indetermination *has* to be taken into account because it makes the Hessian non invertible and this would be fatal to Newton-like technique; solutions for this problem were proposed at sec. 4.4. Regarding optimization itself, one may use a plain BFGS algorithm (which builds up its own approximation to the inverse Hessian) or, as sketched at sec. 4.5, take advantage of the structure of the likelihood to invert at low cost an approximate Hessian.

Further approximations to the Hessian are also possible, like assuming that $\frac{\partial^2 \phi}{\partial \mathbf{A} \partial \mathbf{D}_q} = 0$ and $\frac{\partial^2 \phi}{\partial \mathbf{A} \partial \mathbf{N}_q} = 0$. This seems to be a reasonable approximation in the free noise case and amounts to complete decoupling between $\mathbf{A}$ and the variance parameters. If decoupling is assumed, one may optimize independently over the 'small' matrix parameter $\mathbf{A}$ and the noise and signal variances for each q; in this approximation, all subproblems have small size.

All these ideas are still subject to improvement; finding the right trade-offs between varying degrees of simplicity and efficiency requires more experiments and probably is problem-dependent.

We note that an important issue should also be addressed: the design of a statistically significant stopping criterion for the minimization algorithms. This is particular relevant for the slow EM algorithm. Even if it is only used to intitialize a Newton-like iteration, we have no objective rule to decide when to stop the sequence of EM iterations.

References

1. Tong, L., Soon, V., Huang, Y., Liu, R.: AMUSE: a new blind identification algorithm. In: Proc. ISCAS. (1990)
2. Molgedey, L., Schuster, H.G.: Separation of a mixture of independent signals using time delayed correlations. Physical Review Letters **72** (1994) 3634–3637
3. Belouchrani, A., Abed Meraim, K., Cardoso, J.F., Éric Moulines: A blind source separation technique based on second order statistics. IEEE Trans. on Sig. Proc. **45** (1997) 434–44
4. Matsuoka, K., Ohya, M., Kawamoto, M.: A neural net for blind separation of nonstationary signals. Neural networks **8** (1995) 411–419
5. Parra, L., Spence, C.: Convolutive blind source separation of non-stationary sources. IEEE Trans. on Speech and Audio Processing (2000) 320–327
6. Pham, D.T., Cardoso, J.F.: Blind separation of instantaneous mixtures of non stationary sources. IEEE Trans. on Sig. Proc. **49** (2001) 1837–1848
7. Cardoso, J.F., Snoussi, H., Delabrouille, J., Patanchon, G.: Blind separation of noisy Gaussian stationary sources. Application to cosmic microwave background imaging. In: Proc. EUSIPCO. Volume 1. (2002) 561–564
8. Luenberger, D.: Linear and Nonlinear Programming. Addison-Wesley (1984)

Optimization Using Fourier Expansion over a Geodesic for Non-negative ICA

Mark D. Plumbley

Department of Electronic Engineering, Queen Mary, University of London
Mile End Road, London E1 4NS, UK
mark.plumbley@elec.qmul.ac.uk

Abstract. We propose a new algorithm for the non-negative ICA problem, based on the rotational nature of optimization over a set of square orthogonal (orthonormal) matrices $\mathbf{W}$, i.e. where $\mathbf{W}^T\mathbf{W} = \mathbf{W}\mathbf{W}^T = \mathbf{I}_n$. Using a truncated Fourier expansion of $J(t)$, we obtain a Newton-like update step along the steepest-descent geodesic, which automatically approximates to a usual (Taylor expansion) Newton update step near to a minimum. Experiments confirm that this algorithm is effective, and it compares favourably with existing non-negative ICA algorithms. We suggest that this approach could modified for other algorithms, such as the normal ICA task.

1 Introduction

The task of non-negative independent component analysis (*non-negative ICA*) is to estimate the source vectors $\mathbf{s} = (s_1, \ldots, s_n)$ and mixing matrix $\mathbf{A}$ in the linear generative model $\mathbf{x} = \mathbf{A}\mathbf{s}$ given a observation vectors $\mathbf{x} = (x_1, \ldots, x_n)$, where the sources are *non-negative*, i.e. $\Pr(s_i < 0) = 0$, and *independent*, i.e. $p(s_i s_j) = p(s_i)p(s_j)$ if $i \neq j$. We can also write this in matrix form as $\mathbf{X} = \mathbf{A}\mathbf{S}$ where each column of $\mathbf{X}$ and $\mathbf{S}$ represent a sample of $\mathbf{x}$ and $\mathbf{s}$ respectively.

There are two particular reasons why the non-negative ICA problem is interesting. Firstly, many real-world problems such as the analysis of images, text or musical signals, contain mixtures of sources which are non-negative [1]. Secondly, the non-negativity constraint introduces new approaches which are not available to the more general ICA problem [2–4]. Specifically, in previous work, we showed that for sources for which $\Pr(s < \delta) > 0$ for any $\delta > 0$, which we term *well-grounded*, the sources will be identified by finding a rotation of prewhitened observations which is non-negative [5]. We also introduced a number of algorithms to perform this rotation [6, 7]. Some of these algorithms use the concept of a *geodesic search*, analogous to a line search, but on the manifold of orthogonal rotation matrices (see e.g. [8–12]). We previously used the tangent gradient at a point to determine the geodesic direction, and then perform a line search along that geodesic [6].

In this paper we explore the use of second order information to assist this line search, deriving a convenient form for the second derivative of mean squared non-negative reconstruction error on the geodesic.

Then, since we are on a rotation-like geodesic, we propose the use of a first order Fourier expansion around the optimum point to find the zero derivative point along the line, rather than using the usual second order Taylor expansion, leading to the normal Newton method. This will allow us to take large steps towards the bottom of the solution, even if we are near to a 'peak' where the Newton method would converge to the maximum instead of a minimum.

2 Non-negative ICA System

The non-negative ICA system we consider is similar to that in [6]. Given a sequence of observed n-dimensional data vectors $\mathbf{x}$, we first carry out a pre-whitening step [13], although being careful not to zero-mean the data in the process, since this would lose any information about the non-negativity of the sources [5]. Let $\mathbf{\Sigma_x} \equiv E((\mathbf{x} - \mu_\mathbf{x})(\mathbf{x} - \mu_\mathbf{x})^T)$, where $\mu_\mathbf{x} = E(\mathbf{x})$ is the mean of $\mathbf{x}$. We form the eigenvector-eigenvalue decomposition $\mathbf{\Sigma_x} = \mathbf{EDE}^T$ where $\mathbf{D} = \mathrm{diag}(d_1, \ldots, d_n)$ is a diagonal matrix containing the eigenvalues of $\mathbf{\Sigma_x}$, and $\mathbf{E} = (\mathbf{e}_1, \ldots, \mathbf{e}_n)$ is a square orthonormal matrix whose columns are the corresponding eigenvectors. Then the pre-whitened data is given by the sequence of vectors $\mathbf{z} = \mathbf{Vx}$ where $\mathbf{V} = \mathbf{MD}^{-1/2}\mathbf{E}^T$ for some square orthonormal matrix $\mathbf{M}$. For example, we can choose $\mathbf{M} = \mathbf{I}_n$ so we have simply $\mathbf{V} = \mathbf{D}^{-1/2}\mathbf{E}^T$. It is easy to verify that $\mathbf{\Sigma_z} \equiv E((\mathbf{x} - \bar{\mathbf{x}})(\mathbf{x} - \bar{\mathbf{x}})^T) = \mathbf{I}_n$.

Given an $n \times n$ orthonormal weight matrix $\mathbf{W}$, i.e. $\mathbf{W}^T\mathbf{W} = \mathbf{WW}^T = \mathbf{I}_n$, we calculate $\mathbf{y} = \mathbf{Wx}$, together with positive rectified version $\mathbf{y}_+ = (y_1^+, \ldots, y_n^+)$ where $y_i^+ = g_+(y_i) \equiv \max(y_i, 0)$. We often regard $\mathbf{y}_+$ as the 'output' of the system, together with a complementary error vector $\mathbf{y}_- = \mathbf{y} - \mathbf{y}_+$.

Typically a sequence of p samples of n-dimensional input vectors $\mathbf{x}$ is represented as the columns of a $n \times p$ matrix $\mathbf{X}$, and $\mathbf{\Sigma_x}$ is estimated from these data samples. We then have corresponding matrices $\mathbf{Z} = \mathbf{VX}$, $\mathbf{Y} = \mathbf{WZ}$, $\mathbf{Y}_+ = g_+(\mathbf{Y})$ and $\mathbf{Y}_- = \mathbf{Y} - \mathbf{Y}_+$ where $g_+(\cdot)$ is applied element-wise to the matrix $\mathbf{Y}$.

Now, at each update step we shall multiplicatively update $\mathbf{W}$ by some square orthonormal "rotation" matrix $\mathbf{R} \in SO(n)$, i.e. $\mathbf{R}^T\mathbf{R} = \mathbf{RR}^T = \mathbf{I}_n$ and $\det \mathbf{R} = +1$. We can easily see that the update $\mathbf{W}_{new} \leftarrow \mathbf{RW}_{old}$ will retain the orthonormality of $\mathbf{W}$, since e.g. $\mathbf{RW}(\mathbf{RW})^T = \mathbf{RWW}^T\mathbf{R}^T = \mathbf{I}_n$ [10]. We often find it convenient to express $\mathbf{Y}_{new}$ in terms of $\mathbf{W}$ and $\mathbf{R}$ directly, i.e. $\mathbf{Y}_{new} = \mathbf{RW}_{old}\mathbf{Z}$, or simply $\mathbf{Y} = \mathbf{RWZ}$. Thus both $\mathbf{R}$ and $\mathbf{W}$ are always elements of the set of $n \times n$ orthogonal (orthonormal) rotation matrices.

Now we can write an orthonormal rotation matrix $\mathbf{R} \in SO(n)$ as the exponential of a skew-symmetric matrix, i.e. $\mathbf{R} = e^\mathbf{B}$ where $\mathbf{B}^T = -\mathbf{B}$ is skew-symmetric, $\mathbf{B} \in so(n)$ [14]. We use the non-zero elements $\{\phi_{ij} \mid i < j\}$ of $\Phi = \mathrm{UT}_+(\mathbf{B})$ to be the coordinates of an $(n(n-1)/2)$-dimensional parameter space, where the strict upper triangle operator $\mathrm{UT}_+(\cdot)$ sets elements on or below the diagonal to zero. For example, in the special case of $n = 2$, we have

$$\mathbf{B} = \begin{pmatrix} 0 & \phi \\ -\phi & 0 \end{pmatrix} \quad \text{giving} \quad \mathbf{R} = \begin{pmatrix} \cos\phi & \sin\phi \\ -\sin\phi & \cos\phi \end{pmatrix}.$$

3 Optimization by Rotation in Steepest-Descent Geodesic

For our non-negative ICA algorithm, it is sufficient to minimize the distortion measure

$$J = \frac{1}{2}\|\mathbf{Y}_-\|_F^2 = \frac{1}{2}\sqrt{\sum_{ij} y_{ij}^-} \qquad (1)$$

where $\|\cdot\|_F$ is the Frobenius norm, which will be zero if and only if the sequence of output vectors $\mathbf{y}_+$ ($=\mathbf{y}$ if $J=0$) is some positive scaling and/or permutation of the non-negative sources $\mathbf{s}$ [5]. Calculating the derivative of J with respect to Φ, we find [6]

$$\nabla_\Phi J = \mathbf{U} \mathbf{T}_+ (\mathbf{Y}_- \mathbf{Y}^T - \mathbf{Y}\mathbf{Y}_-^T) \qquad (2)$$

where $[\nabla_\Phi J]_{ij} \equiv dJ/d\phi_{ij}$. Let us define an inner product in Φ-space of $\langle \Phi, \Theta \rangle = \sum_{ij} \phi_{ij}\theta_{ij}$ and let the corresponding norm $|\Phi| = \sqrt{\langle \Phi, \Phi \rangle} = \|\Phi\|_F$ be the distance measure. Then the matrix gradient $-\nabla_\Phi J$ in (2) is the *steepest descent* gradient for J in Φ-space,

$$\theta = |\nabla_\Phi J| = \frac{1}{2}\|\mathbf{Y}_-\mathbf{Y}^T - \mathbf{Y}\mathbf{Y}_-^T\|_F \qquad (3)$$

is the norm of this gradient, and the matrix $\mathbf{H}_\Phi = -\nabla_\Phi J/\theta = -\mathbf{U}\mathbf{T}_+(\mathbf{Y}_-\mathbf{Y}^T - \mathbf{Y}\mathbf{Y}_-^T)/\theta$ is the unit-norm steepest descent direction. For the zero gradient case $\theta = 0$ the matrix $\mathbf{H}_\Phi$ is undefined.

Starting from $\Phi(0) = \mathbf{0}$ and hence $\mathbf{R}(0) = \mathbf{I}_n$, using $\Phi(t) = t\mathbf{H}_\Phi$ defines a geodesic $\mathbf{R}(t) = e^{\mathbf{B}}(t)$ parametrized by t in $\mathbf{R}$-space, where $\mathbf{B}(t) = \Phi(t) - \Phi(t)^T$ [10]. If we define

$$\mathbf{H} = \mathbf{H}_\Phi - \mathbf{H}_\Phi^T = -(\mathbf{Y}_-\mathbf{Y}^T - \mathbf{Y}\mathbf{Y}_-^T)/\theta \qquad (4)$$

as the equivalent steepest descent direction in $\mathbf{B}$-space, we can therefore also write $\mathbf{B}(t) = t\mathbf{H}$. Thus we can reduce J by performing a "line search" to minimize $J(t)$ along this steepest-descent geodesic and then repeat in a new direction. There are a number of algorithms which can be used, from a small update step leading to gradient flow [10,12], or larger steps based on approximating the $J(t)$ by a quadratic function [6]. However, we shall next consider a modification of this approach, based on the rotational nature of $\mathbf{R}(t)$.

3.1 Rotational Geometry of R for $n \leq 3$

Let us now consider the case of rotations $\mathbf{R}$ for $n = 2$ and $n = 3$ (for $n = 1$ we have trivially $\mathbf{R} = \mathbf{I}_1 = 1$ and a single point search space $\{\mathbf{B}\} = \{\Phi\} = \{0\}$). In the general case, we use the matrix exponential $\mathbf{R}(t) = e^{\mathbf{B}}(t) = \exp(t\mathbf{H})$ but for $n \leq 3$ this matrix exponential takes a convenient and easy to calculate form. For $n = 2$ we get a simple Givens rotation [15]

$$\mathbf{H} = \pm \begin{pmatrix} 0 & 1 \\ -1 & 0 \end{pmatrix} \qquad \mathbf{B}(t) = \pm \begin{pmatrix} 0 & t \\ -t & 0 \end{pmatrix} \qquad \mathbf{R} = \begin{pmatrix} \cos t & \pm \sin t \\ \mp \sin t & \cos t \end{pmatrix} \qquad (5)$$

while for $n = 3$ we can use the normalized Rodrigues formula [16, 14]

$$\mathbf{R}(t) = \mathbf{I}_n + \sin t \mathbf{H} + (1 - \cos(t))\mathbf{H}^2 \qquad (6)$$

where $\mathbf{H}$ is the normalized skew-symmetric matrix as constructed above. We can easily see this also applies in the $n = 2$ case in (5) since $\mathbf{H}^2 = -\mathbf{I}_2$ for $n = 2$. For both of these cases, we can see that $\mathbf{R}(t) = \mathbf{R}(t + 2k\pi)$ and hence we are looking for a minimum of a function $J(t)$ which repeats every 2π.

Of course, it is possible that $J(t)$ repeats at some smaller interval $2\pi/l$ for integer l. In fact, for the usual (not non-negative) ICA problem, the solutions $\mathbf{Y}$ and $-\mathbf{Y}$ are considered to be equivalent, so $\mathbf{R}(t)$ and $\mathbf{R}(t + \pi)$ would also be equivalent, yielding a $J(t)$ which repeats every π (or even $\pi/2$ for $n = 2$, since a quarter-turn will also align with a solution). However, this is not the case for non-negative ICA, where the solution $-\mathbf{Y}$ is not equivalent to $\mathbf{Y}$.

3.2 Fourier Expansion of $J(t)$

If we are close to the minimum of $J(t)$ along the line, we could use a Taylor expansion about the minimum t^* to write

$$J(t) \approx a_0 + a_1(t - t^*) + a_2(t - t^*)^2$$

for which we have $J'(t) \approx a_1 + 2a_2(t - t^*)$ and $J''(t) \approx 2a_2$. Since $J'(t) = 0$ at the minimum point $t = t^*$, we must have $a_1 = 0$, leading to an estimate of distance from t^* of $(t - t^*) \approx J'(t)/J''(t)$. We therefore estimate that the minimum is at $\hat{t} = t - J'(t)/J''(t)$, i.e. a Newton update step. However, the Newton update method can suffer from problems away from the minimum. In particular, if we are close to the maximum rather than the minimum, the Newton method will converge to the maximum (where $J'(t)$ is also zero) instead of the minimum.

In the present system, we can use an alternative approach. Since we know that $J(t)$ repeats every $t = 2k\pi$, we can use a Fourier expansion instead. In its simplest form, we get

$$J(t) \approx a_0 - a_1 \cos(t - t^*) \qquad (7)$$

where we use the minus cosine so that the minimum of $J(t)$ is at $t = t^*$. Proceeding as for the Newton method from the Taylor expansion, we get $J'(t) \approx a_1 \sin(t - t^*)$ and $J''(t) \approx a_1 \cos(t - t^*)$, leading to an estimate for the minimum of $\hat{t} = t - \arctan(J'(t), J''(t))$ where $\arctan(\cdot, \cdot)$ is a four-quadrant arc tan function defined such that $\psi = \arctan(\sin \psi, \cos \psi)$ for all $-\pi < \psi \leq \pi$ (see e.g. the Matlab function `atan2`). We notice that $\arctan(J'(t), J''(t)) \approx J'(t)/J''(t)$ for small $t - t^*$, leading to the Newton method as $t \to t^*$.

For $n > 3$ the situation is more complex [10, 14], with multiple orthogonal rotations so that $\mathbf{R}(t)$ does not repeat every $t = 2\pi$. Nevertheless, we have found experimentally that one rotation direction often dominates a long way from the solution, when a large change to t is required. For small t multiple rotations do emerge, but in this range we approximate the Newton method, so the non-repeating distant behaviour does not seem to be a concern.

3.3 Calculating the Line Derivatives

Once the geodesic (line) is defined by θ and $\mathbf{H}$ as in (3) and (4), to move to the estimated minimum along this geodesic using either the Fourier expansion or the Newton (Taylor expansion) method, we need to calculate the line derivatives $J'(t)$ and $J''(t)$. First differentiating (1) with respect to y_{ij} yields $dJ/dy_{ij} = y_{ij}^-(dy_{ij}^-/dy_{ij})$. We notice that the slope dy_{ij}^-/dy_{ij} is discontinuous at $y_{ij} = 0$, changing between 1 and 0. However, the product $y_{ij}^-(dy_{ij}^-/dy_{ij})$ is well defined, since the discontinuity in dy_{ij}^-/dy_{ij} is 'hidden' by the zero in y_{ij}^-. Letting $\mathbf{K}_- = [k_{ij}^-]$ be an indicator matrix for $\mathbf{Y}_-$, such that $k_{ij}^- = 1$ if $y_{ij} < 0$, and zero otherwise, so $y_{ij}^- = k_{ij}^- y_{ij}$, we get $dJ/dy_{ij} = k_{ij}^- y_{ij} = y_{ij}^-$ so $J'(t) = \langle \mathbf{Y}_-, \mathbf{Y}'(t) \rangle$. Differentiating $\mathbf{Y}(t) = \mathbf{RWX} = e^{t\mathbf{H}}\mathbf{WX}$ with respect to t we get $\mathbf{Y}'(t) = \mathbf{H}e^{t\mathbf{H}}\mathbf{WX} = \mathbf{HY}$ so $J'(t) = \langle \mathbf{Y}_-, \mathbf{HY} \rangle = \text{trace}(\mathbf{Y}_-^T \mathbf{HY})$ which, substituting for $\mathbf{H}$ from (4), eventually gives

$$J'(t) = -2\theta = -\|\mathbf{Y}_-\mathbf{Y}^T - \mathbf{Y}\mathbf{Y}_-^T\|_F. \tag{8}$$

For the second derivative, we would like to differentiate $J'(t) = \langle \mathbf{Y}_-, \mathbf{HY} \rangle$. This is no longer strictly differentiable, since the slope dy_{ij}^-/dy_{ij} jumps from 1 for $y_{ij} < 0$ to 0 for $y_{ij} > 0$ as we remarked above. However, for the purposes of optimization of J we will define $dy_{ij}^-/dy_{ij} = 0$ for $y_{ij} = 0$, giving $dy_{ij}^-/dy_{ij} = k_{ij}^-$. Writing $\mathbf{Y}_- = \mathbf{K}_- \circ \mathbf{Y}$ where $\circ$ represents element-wise multiplication, and noting that $\langle \mathbf{K}_- \circ \mathbf{Y}, \mathbf{HY} \rangle = \langle \mathbf{K}_- \circ \mathbf{Y}, \mathbf{K}_- \circ (\mathbf{HY}) \rangle$, we get

$$\begin{aligned} J''(t) &= \langle \mathbf{K}_- \circ \mathbf{Y}'(t), \mathbf{K}_- \circ (\mathbf{HY}) \rangle + \langle \mathbf{Y}_-, \mathbf{HY}'(t) \rangle \\ &= \|\mathbf{K}_- \circ (\mathbf{HY})\|_F^2 + \langle \mathbf{Y}_-, \mathbf{H}^2\mathbf{Y} \rangle. \end{aligned} \tag{9}$$

4 Proposed Algorithm

Given an input data matrix $\mathbf{X}$, whitened to give $\mathbf{Z} = \mathbf{VX}$ as described above, the algorithm is as follows:

1. Initialize $\mathbf{W} = \mathbf{I}_n$
2. Calculate $\mathbf{Y} = \mathbf{WZ}$, $\mathbf{Y}_- = g_-(\mathbf{Y})$ and θ as in (3).
3. If $\theta = 0$, finish.
4. Calculate $\mathbf{H}$ as in (4), $J'(t)$ as in (8) and $J''(t)$ as in (9), and set $t_1 = -\arctan(J'(t), J''(t))$.
5. Calculate $\mathbf{B} = t_1\mathbf{H}$ and $\mathbf{R} = e^{\mathbf{B}}$ (using e.g. the Rodrigues formula for $n \leq 3$).
6. Update $\mathbf{W} \leftarrow \mathbf{RW}$.
7. Repeat from step 2 until $\theta = 0$.

To use the Newton method instead of a Fourier expansion update, set $t_1 = -J'(t)/J''(t)$ in step 4. While this algorithm has been derived specifically for the non-negative ICA problem, it should be possible to modify it to other tasks for optimization over orthonormal matrices, modifying as necessary if $J(t)$ repeats at $t = \pi$, as for normal ICA, instead of $t = 2\pi$ as for non-negative ICA.

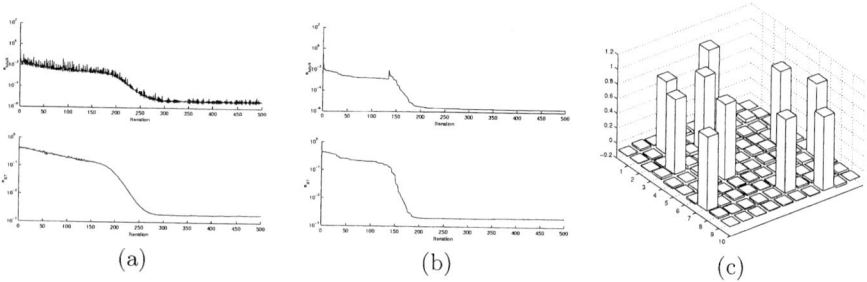

Fig. 1. Results on artificial data for $n = 10$, showing (a) learning curve for the geodesic 'single step' algorithm in [6], (b) learning curve for the Fourier step algorithm described in the current paper, and (c) the values of **WVA** for (b) after 250 iterations, showing this approximates a positive permutation matrix.

5 Experiments

Experiments were carried out in Matlab to confirm the operation of the algorithm. For artificial data, we generated $p = 1000$ unit variance random source data vectors $\mathbf{s}_p$, mixed using a random matrix $\mathbf{A}$ as for the non-negative PCA experiments described in [7]. To measure the separation performance of the algorithm, we use two performance measures: a nonnegative reconstruction error

$$e_{NNR} = \frac{1}{np}\|\mathbf{Z} - \mathbf{W}^T\mathbf{Y}_+\|_F^2 \tag{10}$$

which, for orthonormal $\mathbf{W}$, is a scaled version of J; and a cross-talk error

$$e_{XT} = \frac{1}{n^2}\|\mathrm{abs}(\mathbf{WVA})^T\mathrm{abs}(\mathbf{WVA}) - \mathbf{I}_n\|_F^2 \tag{11}$$

where $\mathrm{abs}(\mathbf{M})$ is the matrix of absolute values of the elements of $\mathbf{M}$, which is zero only if $\mathbf{y} = \mathbf{WVAs}$ is a positive permutation of the sources, i.e. only if the sources have been successfully separated.

For $n = 10$, after 250 iterations (18.8s of CPU time on an 850MHz Pentium 3), we had $e_{NNR} = 2.20 \times 10^{-6}$ and $e_{XT} = 2.00 \times 10^{-3}$, bettering the non-negative PCA algorithm [7] which took 10^5 iterations to obtain $e_{NNR} = 5.02 \times 10^{-4}$ and $e_{XT} = 0.0553$ (called e_{MSE} and e_{Perm} respectively in [7]). We also compared this with the geodesic 'single step' algorithm introduced in [6] which is almost as fast as the current algorithm, but exhibits a very noisy learning curve (Fig. 1(a)), indicating the assumption in that algorithm of a quadratic bowl with $J = 0$ at the bottom is not completely justified. For the current algorithm, the Fourier expansion leads to a particularly fast initial one or two iterations, with an unexpected increase in reconstruction error e_{NNR} around iteration 140 coinciding with a corresponding step decrease in crosstalk error e_{XT}. For details see Fig. 1. Similar results were observed in an image separation task (not shown).

We also applied this algorithm to a music analysis problem. Here a small segment of a Liszt 'Etude' has been played on a MIDI synthesized piano, preprocessed into power spectrogram with $p = 467$ frames and reduced to 10 dimensions using PCA (for details of this task, see [6]). The geodesic 'single step' algorithm becomes unstable very quickly on this task, since $J(t)$ does not reach zero at the best solution. We therefore compared this to the geodesic flow algorithm [10,6], using $\mathbf{B} = -\mu(\mathbf{Y}_-\mathbf{Y}^T - \mathbf{Y}\mathbf{Y}_-^T)$ in the update algorithm with the update factor $\mu = 0.001$ chosen experimentally to yield fastest convergence without instability (Fig. 2). The Fourier update algorithm is similar in speed to the fastest geodesic flow algorithm (actually slightly faster in this case), but does not require any update parameter μ to be selected.

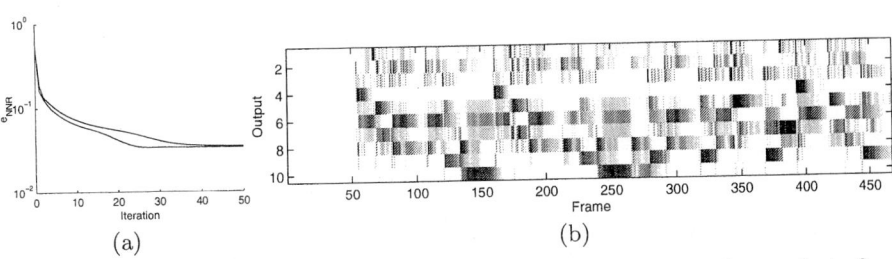

Fig. 2. Results on music analysis task, showing (a) learning curves for geodesic flow (upper curve) and Fourier update algorithm (lower curve), and (b) output for the Fourier update algorithm showing identified notes. See [6] for task details.

6 Conclusions

The non-negative ICA problem can be tackled using algorithms which minimize a cost function J over the space of orthogonal (orthonormal) rotation matrices $\mathbf{W}$, i.e. where $\mathbf{W}\mathbf{W}^T = \mathbf{W}^T\mathbf{W} = \mathbf{I}_n$. For the special case of $n \leq 3$, a geodesic $\mathbf{R}(t) = e^{t\mathbf{H}}$ where $\mathbf{H}$ is skew-symmetric yields a single plane of rotation, giving a cost function $J(t)$ that repeats every $t = 2k\pi$.

We proposed an algorithm that takes advantage of the cyclical nature of $J(t)$, using a truncated Fourier expansion of $J(t)$ to yield a Newton-like update step along the steepest-descent geodesic. This automatically approximates to a usual Newton update step near to a minimum. Experiments confirm that this algorithm is effective, even for $n > 3$, and it compares favourably with existing non-negative ICA algorithms.

We suggest that this approach could be modified for other tasks requiring optimization over orthogonal matrices, such as the standard prewhitened ICA problem, but using a rotation period of π (or $\pi/2$) instead of 2π.

Acknowledgements

This work was partially supported by EPSRC grant GR/R54620, and by the EU-FP6-IST-507142 project SIMAC (Semantic Interaction with Music Audio Contents: www.semanticaudio.org). The music sequence is used by permission of the Classical Piano Midi Page www.piano-midi.de, copyright Bernd Krueger.

References

1. Lee, D.D., Seung, H.S.: Algorithms for non-negative matrix factorization. In Leen, T.K., Dietterich, T.G., Tresp, V., eds.: Advances in Neural Information Processing Systems 13, MIT Press (2001) 556–562 Proceedings of NIPS*2000
2. Paatero, P., Tapper, U.: Positive matrix factorization: A non-negative factor model with optimal utilization of error estimates of data values. Environmetrics **5** (1994) 111–126
3. Harpur, G.F.: Low Entropy Coding with Unsupervised Neural Networks. PhD thesis, Department of Engineering, University of Cambridge (1997)
4. Cichocki, A., Georgiev, P.: Blind source separation algorithms with matrix constraints. IEICE Transactions on Fundamentals of Electronics, Communications and Computer Sciences **E86–A** (2003) 522–531
5. Plumbley, M.D.: Conditions for nonnegative independent component analysis. IEEE Signal Processing Letters **9** (2002) 177–180
6. Plumbley, M.D.: Algorithms for nonnegative independent component analysis. IEEE Transactions on Neural Networks **14** (2003) 534–543
7. Plumbley, M.D., Oja, E.: A "nonnegative PCA" algorithm for independent component analysis. IEEE Transactions on Neural Networks **15** (2004) 66–76
8. Brockett, R.W.: Dynamical systems that sort lists, diagonalize matrices, and solve linear programming problems. Linear Algebra and its Applications **146** (1991) 79–91
9. Edelman, A., Arias, T.A., Smith, S.T.: The geometry of algorithms with orthogonality constraints. SIAM J. Matrix Anal. Appl. **20** (1998) 303–353
10. Nishimori, Y.: Learning algorithm for ICA by geodesic flows on orthogonal group. In: Proceedings of the International Joint Conference on Neural Networks (IJCNN'99). Volume 2., Washington, DC (1999) 933–938
11. Douglas, S.C.: Self-stabilized gradient algorithms for blind source separation with orthogonality constraints. IEEE Transactions on Neural Networks **11** (2000) 1490–1497
12. Fiori, S.: A theory for learning by weight flow on Stiefel-Grassman manifold. Neural Computation **13** (2001) 1625–1647
13. Hyvärinen, A., Karhunen, J., Oja, E.: Independent Component Analysis. John Wiley & Sons (2001)
14. Gallier, J., Xu, D.: Computing exponentials of skew-symmetric matrices and logarithms of orthogonal matrices. International Journal of Robotics and Automation **18** (2003) 10–20
15. Comon, P.: Independent component analysis - a new concept? Signal Processing **36** (1994) 287–314
16. Fiori, S., Rossi, R.: Stiefel-manifold learning by improved rigid-body theory applied to ICA. International Journal of Neural Systems **13** (2003) 273–290

The Minimum Support Criterion for Blind Signal Extraction: A Limiting Case of the Strengthened Young's Inequality

Sergio Cruces and Iván Durán

Área de Teoría de la Señal y Comunicaciones
Camino Descubrimientos s/n, 41092-Seville, Spain
{sergio,iduran}@us.es
http://viento.us.es/~sergio

Abstract. In this paper, we address the problem of the blind extraction of a subset of "interesting" independent signals from a linear mixture. We present a novel criterion for the extraction of the sources whose density has the minimum support measure. By extending the definition of the Renyi's entropies to include the zero-order case, this criterion can be regarded as part of a more general entropy minimization principle. It is known that Renyi's entropies provide contrast functions for the blind extraction of independent and identically distributed sources under an ∞-norm constraint on the global transfer system. The proposed approach gives sharper lower-bounds for the zero-order Renyi's entropy case and, contrary to the existing results, it allows the extraction even when the sources are non identically distributed. Another interesting feature is that it is robust to the presence of certain kinds of additive noise and outliers in the observations.

1 Introduction

The criteria to solve ICA problems are usually mathematically expressed in the form of the optimization of a contrast function with some specific properties. Several mathematician and geophysicist proposed them to solve the problem of blind deconvolution [1]-[2]. Since its origins, although much later, the field of ICA have also stressed the importance of the *information theoretic contrasts* as driven criteria to solve the problem of blind signal extraction (see [3]-[5] and references therein).

In this paper, we present a novel criterion for extraction in ICA, which is based on the minimization of the measure of the support set (or of its convex hull) of the probability density function of the output.

Let us consider the standard linear mixing model. The vector process of observations $\mathbf{X}(t) = [X_1(t), \cdots, X_N(t)]^T$ obeys the following equation

$$\mathbf{X}(t) = \mathbf{A}\mathbf{S}(t), \qquad (1)$$

where $\mathbf{S}(t) = [S_1(t), \cdots, S_N(t)]^T$ is the signal vector process of N independent components, and $\mathbf{A} = [\mathbf{a}_1, \ldots, \mathbf{a}_N] \in \mathbf{R}^{N \times N}$ is the mixing matrix. In order to extract one non-Gaussian source from the mixture, one computes the inner product of the observations with the vector $\mathbf{u}$, to obtain the output random variable or estimated source

$$Y = \mathbf{u}^T \mathbf{X} = \mathbf{g}^T \mathbf{S}, \qquad (2)$$

where $\mathbf{g}^T = \mathbf{u}^T \mathbf{A}$ denotes the global mixture from the sources to the output.

2 A Review of Existing Results

In the late 1970's several mathematics improved the classical Young's (convolution) inequality with sharp constants. From these works, the following strengthened inequalities resulted [9].

Theorem 1 (Strengthened Young's inequality). *Let $0 < p, q, r$ satisfy $\frac{1}{r} = \frac{1}{p} + \frac{1}{q} - 1$, and let $f \in L^p(\mathbf{R})$ and $g \in L^p(\mathbf{R})$ be non-negative functions. Then*

Young's inequality   Reverse Young inequality

$$\|f * g\|_r \leq C^{1/2} \|f\|_p \|g\|_q \qquad \|f * g\|_r \geq C^{1/2} \|f\|_p \|g\|_q \qquad (3)$$

$$\text{for } p, q, r \geq 1, \qquad \text{for } 0 < p, q, r \leq 1.$$

where $C = C_p C_q / C_r$ and $C_\alpha = \frac{|\alpha|^{1/\alpha}}{|\alpha'|^{1/\alpha'}}$, for $1/\alpha + 1/\alpha' = 1$.

Let us denote the density of the output Y as $f(y)$. One of the possible generalization of Shannon's entropy is given by Renyi's entropies of order r [8]

$$h_r(Y) = \tfrac{1}{1-r} \log \left[\int f^r(y) dy \right] \text{ for } r \in \{(0,1) \cup (1, \infty)\}. \qquad (4)$$

Renyi's entropies can be extended to consider the two limiting cases

$$h_r(Y) = \begin{cases} \log(\mu\{y : f(y) > 0\}) & \text{for } r = 0, \\ -\int f(y) \log f(y) dy & \text{for } r = 1, \end{cases} \qquad (5)$$

where $\mu\{\cdot\}$ denotes the Lebesgue measure of the support set of the density. For certain orders, Renyi's entropies may overcome the difficulties that arise in the estimation of Shannon's entropy. The quadratic Renyi's entropy ($r = 2$) is one of such cases, since it can be easily optimized when it is combined with kernel based estimators of the density [5]. Two recent papers [6,7] propose the the minimization of Renyi's quadratic entropy to solve the problem of blind deconvolution. They independently arrived to the same inequality for the blind extraction of i.i.d. sources, which is summarized in the following lemma.

Lemma 1 (Existing lower-bound). *Let $S_1, \ldots, S_N$ be N independent and non-Gaussian sources, identically distributed to S. Then, the r-th order Renyi entropy of any linear combination ($Y = \mathbf{g}^T \mathbf{S}$) of them, is lower-bounded by*

$$h_r(Y) \geq h_r(\|\mathbf{g}\|_\infty S) \qquad (6)$$

The equality occurs if and only if $g_j = \|\mathbf{g}\|_\infty \delta_{ij}$, $j = 1, \ldots, N$, and for any given $i \in [1, N]$ (where δ denotes the Kronecker delta).

In [7] the definition of the Renyi entropy power and the Jensen's inequality are used to prove the lemma. Whereas in [6], the authors' proof was based on the strengthened Young's inequality for the following specific choice of parameters: $q = 1$ and $r = p$. In this paper, we focus our attention on those cases of Young's Inequality for which r, p and q coincide, i.e., when $r = p = q = 1$ and also (after taking limits) when $r = p = q = 0$.

3   Minimum Entropy and Minimum Support Criteria

For two independent random variables A and B, the super-additivity of the function

$$e^{(1+r)h_r(A+B)} \geq e^{(1+r)h_r(A)} + e^{(1+r)h_r(B)}, \qquad r = 0, 1. \qquad (7)$$

follows from the strengthened Young's inequality [8]. For $r = 1$, this equation is the entropy power inequality. Excluding the trivial non-mixing case, the equality holds true if and only if A and B are Gaussian random variables. For $r = 0$, the function $e^{(1+r)h_r(A)}$ coincides with the measure of the support set of the density of A. In this later case, equation (7) is the Brunn-Minkowski inequality. Since the equation is only meaningful when the support sets of the densities of A and B have finite Lebesgue measure, hereinafter, when referring to the zero-order entropy we will implicitly assume sources whose densities have measurable and non-zero support. Under these conditions, the equality in (7) is only obtained when the support sets of the densities $f_A(a)$ and $f_B(b)$ are both convex and homothetic, i.e., equal under translation and dilatation.

The next theorem is used to lower-bound the entropies of the output (of orders 0 and 1) in terms of the respective entropies of the sources.

Theorem 2. *Let $Y = \mathbf{g}^T \mathbf{S}$, then for $r = 0, 1$, $\forall\, m \in \mathbf{N}^+$, $k = (1+r)m$, a lower bound of the entropy of the output is given by*

$$h_r(Y) \geq \sum_{j=1}^{N} \left|\frac{g_j}{\|\mathbf{g}\|_k}\right|^k h_r(\|\mathbf{g}\|_k S_j) \qquad (8)$$

Proof: The proof of this result is based on the following chain of inequalities

$$h_r(Y) \overset{(a)}{\geq} \frac{1}{(1+r)m} \log \left(\sum_{j=1}^{N} |g_j|^{1+r} e^{(1+r)h_r(S_j)} \right)^m , \quad r \in \{0,1\}, \, m \in \mathbf{N}^+. \quad (9)$$

$$\overset{(b)}{\geq} \frac{1}{k} \log \sum_{j=1}^{N} |g_j|^k e^{k h_r(S_j)}, \quad k = (1+r)m. \qquad (10)$$

$$\overset{(c)}{\geq} \sum_{j=1}^{N} \left|\frac{g_j}{\|\mathbf{g}\|_k}\right|^k h_r(\|\mathbf{g}\|_k S_j) \qquad (11)$$

The inequality (a) follows from the super-additivity of the function $e^{(1+r)h_r(Y)}$ for the considered orders, as it has been shown in equation (7). Between (9) and (10) there is equality for $m = 1$ and strict inequality for $m > 1$. The greater m is, the looser is this inequality. Finally, inequality (c) follows from the strict concavity of the logarithm. □

Let us denote by $\hat{\mathbf{g}}^{(i)}$ any solution of the vector $\mathbf{g} = \mathbf{A}^T\mathbf{u}$ that extracts the i-th source, i.e, for any $i \in [1, N]$

$$\hat{\mathbf{g}}^{(i)} = [0, \ldots, 0, \underbrace{\mathbf{a}_i^T\mathbf{u}}_{i\text{-th position}}, 0, \ldots, 0]^T . \qquad (12)$$

where $|\mathbf{a}_i^T\mathbf{u}| = \|\hat{\mathbf{g}}^{(i)}\|_k \neq 0$ for any k-norm.

The minimization of the right-hand-size of (8), with respect to the indices of the sources, yields the following two corollaries.

Corollary 1 (Minimum entropy). *Let Ω_1 be the set of indices of the sources with minimum Shannon's entropy*

$$\Omega_1 = \{i \ : \ i = \arg\min_{j=1,\ldots,N} h_1(S_j)\}.$$

For $m \in \mathbf{N}^+$ and $k = 2m$, the following inequality holds true

$$h_1(Y) \geq h_1(\|\mathbf{g}\|_k S_i) \quad i \in \Omega_1, \qquad (13)$$

and the minimum value of $h_1(Y)$ is only reached at the extraction of one of the sources with minimum entropy, i.e., for $\mathbf{g} = \hat{\mathbf{g}}^{(i)}$, $\forall i \in \Omega_1$.

Corollary 2 (Minimum support). *Let Ω_0 be the set of indices of the sources whose densities have the support sets of minimum measure*

$$\Omega_0 = \{i \ : \ i = \arg\min_{j=1,\ldots,N} h_0(S_j)\}.$$

For $k \in \mathbf{N}^+$, the following inequality holds true

$$h_0(Y) \geq h_0(\|\mathbf{g}\|_k S_i) \quad i \in \Omega_0, \qquad (14)$$

where the tighter lower-bound is obtained for $k = 1$, and the loosest bound for $k = \infty$. For $k > 1$ the minimum of $h_0(Y)$ is only reached at the extraction of one of the sources with the minimum support, i.e., for $\mathbf{g} = \hat{\mathbf{g}}^{(i)}$, $\forall i \in \Omega_0$.

The tighter lower-bound for corollary 1 is obtained for the 2-norm, which leads to the well known minimum entropy criterion [2, 4] under the unit 2-norm constraint on $\mathbf{g}$. Other norms give looser bounds, but some of them are easier to enforce when there is additive Gaussian noise in the mixture. In corollary 2, the tighter bound is obtained for the 1-norm. Although, for the 1-norm, the equality in (14) may hold true for non-extracting solutions, this can only happen when all the non-vanishing contributions $g_j S_j \neq 0$ to the output Y have convex support and a common value of $h_0(g_j S_j)$.

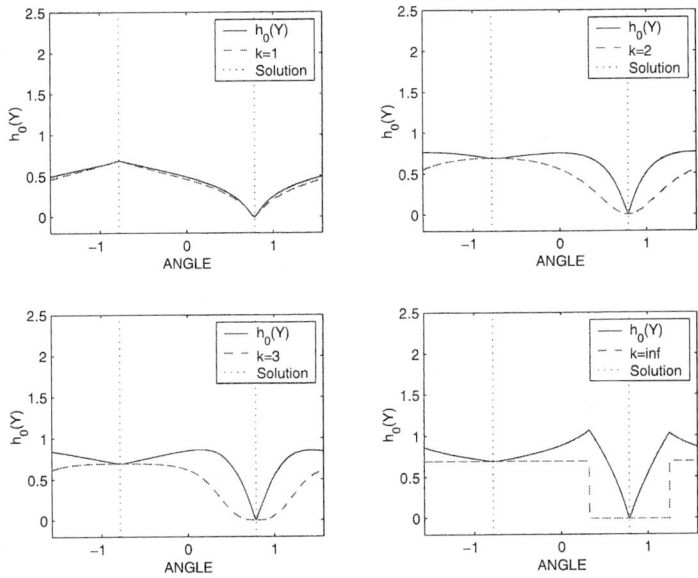

Fig. 1. Zero-order Renyi entropy (continuous line) and lower bound (dashed line) under different k-norm equality constraints The mixing matrix was $A = [2, 1; -2, 1]$ and, in the figures, the axis of abscissas represents the angle $\text{atan}\left(\frac{u_2}{u_1}\right)$ of the vector $\mathbf{u}^T = [u_1, u_2]$.

It is evident from both corollaries that criteria for blind signal extraction are obtained by minimizing $h_r(Y)$, $r = 0, 1$, under the k-norm equality constraint $\|\mathbf{g}\|_k = 1$. Figure 1 illustrates this situation for $k \in \{1, 2, 3, \infty\}$ and with sources of different support. Although this approach seems to work, it is difficult to carry out, because, in practice, we don't know the vector $\mathbf{g}$. Another alternative, consists in enforcing the following k-norm inequality constraints

$$\|\mathbf{g}\|_k \geq 1, \text{ with } \|\mathbf{g}\|_k = 1 \text{ for those } \mathbf{g} = \hat{\mathbf{g}}^{(i)} \; \forall i \in \Omega_r. \tag{15}$$

These are more suited for the practical implementation, since they can be enforced by normalizing the kth-order cumulant of the output, as figure 2 and the next lemma illustrate.

Lemma 2 (Normalization). *Let the location (or the scaling) of each source be defined in such a way that the modulo of the k-th order cumulant ($k \in \mathbf{N}^+$) is upper-bounded by 1, and equal to 1 for the sources belonging to the set Ω_r*

$$|Cum_k(S_j)| \leq 1, \forall j = 1, \ldots, N; \text{ with equality iff } i \in \Omega_r. \tag{16}$$

Then, the normalization of the modulo of the k-th order cumulant of the output ($|Cum_k(Y)| = 1$) automatically enforces the constraints of equation (15).

For the minimum support criterion, an implicit normalization of $\|\mathbf{g}\|_1$ can be obtained by just constraining the extraction vector $\mathbf{u}$ to have unit 2-norm. Theorem 3 summarizes this result.

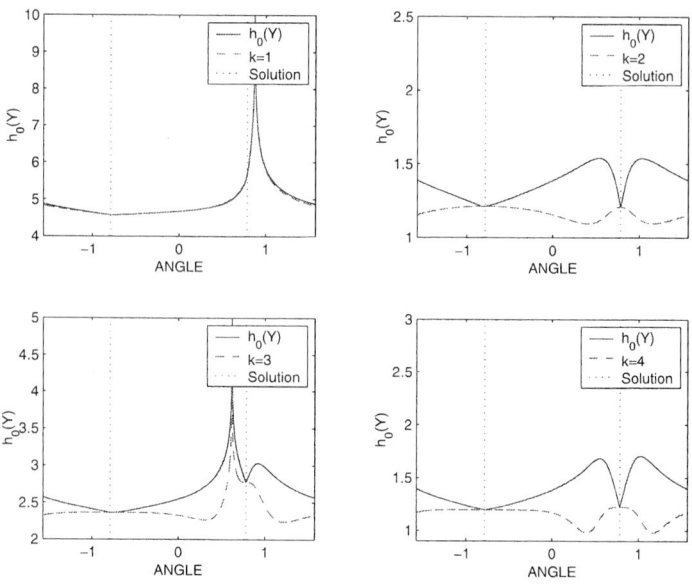

Fig. 2. Zero-order Renyi entropy of the output and lower bound, under the normalization described in lemma 2. The observations were formed from a mixture of two sources through the matrix $A = [2, 1; -2, 1]$. From the figures one can observe that the minimum of the zero-order Renyi entropy (i.e, the minimum support of the output), coincides with the extraction of one of the sources.

Theorem 3 (Implicit normalization). *Under the constraint $\|\mathbf{u}\|_2 = 1$, the zero-order entropy of the output random variable $Y = \mathbf{u}^T \mathbf{X}$ is lower bounded by*

$$h_0(Y) \geq \min_i h_0(\|\hat{\mathbf{g}}^{(i)}\|_1 S_i) \quad s.t. \ \|\mathbf{u}\|_2 = 1, \tag{17}$$

where $\hat{\mathbf{g}}^{(i)}$ was defined in (12). The minimum is only reached at the extraction solutions $\mathbf{g} = \hat{\mathbf{g}}^{(i)}$ whose indices belong to the set

$$\Omega' = \{i \ : \ i = \arg\min_{j=1,\ldots,N} h_0(\hat{g}_j S_j)\}. \tag{18}$$

Due to the limited space we skip the proof. The next lemma presents an extension of this result based on the convex hull. Note that the convex hull of a set of points denotes the smallest convex set that contains them.

Lemma 3. *Corollary 2 and Theorem 3 still hold true when the Renyi's zero order entropy (the logarithm of the support set) is replaced by the logarithm of the convex hull of the support set*

$$h_0(Y) = \log \mu(\{y : f(y) > 0\}) \dashrightarrow \log \mu(conv\{y : f(y) > 0\}) . \tag{19}$$

Proof: The proof of the lemma is straightforward, because both quantities coincide under the hypothetical assumption of sources with compact and convex support, thus, Corollary 2 and Theorem 3 also apply to them. □

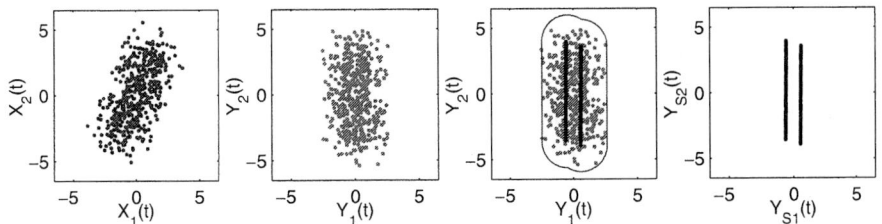

Fig. 3. Extraction of a binary and uniform sources in presence of isotropic and bounded noise. The first and the second figure show, respectively, the scatter plot of the observations and of the recovered sources. The third figure shows how the noise increases equally, in all directions, the support of the observations. From the fourth figure, which presents the scatter plot of the signal components at the output, one can observe that the minimum support projection leads to the extraction of the binary source.

The main advantage of using the convex hull is that the contrast function becomes more regular and easier to optimize. In absence of noise, the convex hull of the observation is a convex polytope whose edges are coincident with the columns of the mixing system.

The presence of additive noise may destroy the connection between the columns of the mixing system and the edges, but the original edge's directions are preserved when the noise has a p.d.f. of compact support and an isotropic behavior in all directions, i.e., the p.d.f. is invariant under rotations around his mean. This kind of noise is commonly found in practical problems, therefore, the immunity to it is an interesting feature of the proposed contrast function. Moreover, additional robustness against outliers can be obtained by defining certain thresholds of minimum contiguity, or of minimum density, for the inclusion of an observation point into the support set.

4 Simulations

In this section we report the results of two different simulations. The first one illustrates the potential of the minimum support criterion for the extraction of sources in presence of noise. The second one shows one practical implementation of the extraction algorithm for mixtures of more than two sources.

In the first simulation we mixed 500 samples of a binary and a uniform signal through a random mixing matrix. The observations $\mathbf{X} = \mathbf{AS} + \mathbf{N}$ were obtained in presence of a strong additive and bounded noise $\mathbf{N}$, and whose joint density was isotropic in all directions. The signal-to-noise ratio (SNR) in the observations was about 0 dB. Then, we chose the vector $\mathbf{u}$ for which the convex-hull of the support of the output $Y = \mathbf{u}^T \mathbf{X}$ was minimum. The minimum led to the extraction of the binary source with a global mixing vector of

$$\mathbf{g} = [1.00,\ 0.03]^T,$$

The results of this simulation are shown in figure 3.

In the second simulation we consider only 150 samples of ten binary signals, and mixed then in presence of additive white Gaussian noise with a maximum SNR of 10 dB. Then, we solved the problem in an iterative fashion. By using a cyclic sequence of planar orthogonal Jacobi rotations of the observations, we minimized the support of the convex hull of the output. After 60 iterations, the algorithm converged to a global mixing vector of

$$\mathbf{g} = [0.08,\ 0.04,\ -0.01,\ -0.04,\ 0.03,\ 0.02,\ 0.01,\ \underline{-0.99},\ 0.10,\ 0.07]^T,$$

and one of the binary signals was extracted.

5 Conclusions

We have presented a new criterion for the blind extraction of sources whose densities have compact support. The new criterion, with roots in information theory and geometry, consists in the extraction of the source that minimizes the measure of support of the output density (or of its convex hull). The criterion has been shown to be robust with respect to the existence of strong levels of isotropic and compact additive noise in the observations, a novel and interesting feature.

Acknowledgements

This research was supported by the CICYT Spanish project TIC2001-0751-C04-04.

References

1. B. Godfrey, "An information theory approach to deconvolution," Stanford exploration project, Report No. 15, pp. 157–182, 1978.
2. D. Donoho, *On Minimum Entropy Deconvolution*, Applied Time Series Analysis II, D. F. Findley Editor, Academic Press, New York, pp. 565–608, 1981.
3. P. Comon, "Independent component analysis, a new concept?," *Signal Processing*, vol. 3, no. 36, pp. 287–314, 1994.
4. S. Cruces, A. Cichocki, and S-i. Amari, "From blind signal extraction to blind instantaneous signal separation: criteria, algorithms and stability", *IEEE Trans. on Neural Networks*, scheduled print: July, 2004.
5. J. Principe, D. Xu and J. Fisher, *Information Theoretic Learning*, in "Unsupervised Adaptive Filtering" Volume I, Simon Haykin Editor, pp. 265–319, Wiley, 2000.
6. J.-F. Bercher, C. Vignat, "A Renyi entropy convolution inequality with application", in *Proc. of EUSIPCO*, Toulouse, France, 2002.
7. D. Erdogmus, J. C. Principe, L. Vielva, "Blind deconvolution with minimum Renyi's entropy", in *Proc. of EUSIPCO*, vol. 2, pp. 71-74, Toulouse, France, 2002.
8. T. M. Cover, J. A. Thomas, *Elements of Information Theory*, Wiley series in telecommunications. John Wiley, 1991.
9. R. J. Gardner, "The Brunn-Minkowski Inequality", *Bulletin of the American Mathematical Society*, vol. 39(3), pp. 355–405, 2002.

Accurate, Fast and Stable Denoising Source Separation Algorithms

Harri Valpola[1,2,*] and Jaakko Särelä[2,**]

[1] Artificial Intelligence Laboratory, University of Zurich
Andreasstrasse 15, 8050 Zurich, Switzerland
[2] Neural Networks Research Centre, Helsinki University of Technology
P.O.Box 5400, FI-02015 HUT, Espoo, Finland
{harri.valpola,jaakko.sarela}@hut.fi

Abstract. Denoising source separation is a recently introduced framework for building source separation algorithms around denoising procedures. Two developments are reported here. First, a new scheme for accelerating and stabilising convergence by controlling step sizes is introduced. Second, a novel signal-variance based denoising function is proposed. Estimates of variances of different source are whitened which actively promotes separation of sources. Experiments with artificial data and real magnetoencephalograms demonstrate that the developed algorithms are accurate, fast and stable.

1 Introduction

In denoising source separation (DSS) framework [1], separation algorithms are built around a denoising function. This makes it easy to tailor source separation algorithms for the task at hand. Good denoisings usually result in fast and accurate algorithms. Furthermore, explicit objective function is not needed, in contrast to most existing source separation algorithms.

Here we report further developments of two aspects. First, we introduce a new method for stabilising and accelerating convergence which is inspired by predictive controllers. Second, we develop further the signal-variance-based denoising principles. The resulting algorithms yield good results in terms of signal-to-noise ratio (SNR) and exhibit fast and stable convergence.

2 Source Separation by Denoising

Consider a linear instantaneous mixing of sources:

$$\mathbf{X} = \mathbf{AS} + \boldsymbol{\nu}, \tag{1}$$

---

[*] Funded by the European Commission, under the project ADAPT (IST-2001-37173) and by the Academy of Finland, under the project New information processing principles.
[**] Funded by the Academy of Finland.

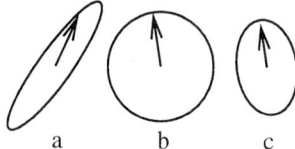

Fig. 1. a) Original data set, b) after sphering and c) after denoising. After these steps, the projection yielding the best signal-to-noise ratio, denoted by arrow, can be obtained by simple correlation-based learning.

where the $N \times T$ matrix $\mathbf{S}$ are the sources, the $M \times T$ matrix $\mathbf{X}$ are the observations and there is noise $\boldsymbol{\nu}$. If the sources are assumed Gaussian, this is a general, linear factor analysis model with rotational invariance.

DSS, as many other computationally efficient ICA algorithms, resorts to sphering. In the case of DSS, the main reason is that after sphering, denoising combined with simple correlation based estimation akin to Hebbian learning (on-line) or power method (batch) is able to retrieve the signal with the highest SNR. Here SNR is implicitly defined by the denoising. The effect of sphering and subsequent denoising is depicted in Fig. 1.

Assuming that $\mathbf{X}$ is already sphered and $\mathbf{f}(\mathbf{s})$ is the denoising procedure, a simple DSS algorithm can be written as follows:

$$\mathbf{s} = \mathbf{w}^T \mathbf{X} \qquad (2)$$
$$\mathbf{s}^+ = \mathbf{f}(\mathbf{s}) \qquad (3)$$
$$\mathbf{w}^+ = \mathbf{X}\mathbf{s}^{+T} \qquad (4)$$
$$\mathbf{w}_{\text{new}} = \text{orth}(\mathbf{w}^+), \qquad (5)$$

where $\mathbf{s}$ is the source estimate (a row vector), $\mathbf{s}^+$ is the denoised source estimate, $\mathbf{w}$ is the previous weight vector (a column vector), $\mathbf{w}^+$ is the new weight vector before and $\mathbf{w}_{\text{new}}$ after orthonormalisation (e.g., deflatory or symmetric orthogonalisation as in FastICA [2]).

Note that if $\mathbf{X}$ were not sphered and no denoising were applied, i.e., $\mathbf{f}(\mathbf{s}) = \mathbf{s}$, the above equations would describe the power method for computing the principal eigenvector. When $\mathbf{X}$ is sphered, all eigenvalues are equal to one and without denoising the solution is degenerate, i.e., any unit vector $\mathbf{w}$ is a fixed point of the iterations. This shows that for sphered $\mathbf{X}$, even the slightest denoising $\mathbf{f}(\mathbf{s})$ can determine the convergence point.

If, for instance, $\mathbf{f}(\mathbf{s})$ is chosen to be low-pass filtering, implicitly signals are assumed to have relatively more low frequencies than noise and the above iteration converges to the signal which has the most low-frequency components. On the other hand, if $\mathbf{f}(\mathbf{s})$ is a shrinkage function, suppressing small components of $\mathbf{s}$ while leaving large components relatively untouched, signals are implicitly assumed to have heavy tails and thus super-Gaussian distributions.

It is possible to begin with an objective function $g(\mathbf{s})$ in which case the denoising can be chosen[1] to be the gradient: $\mathbf{f}(\mathbf{s}) = \nabla g(\mathbf{s})$. In practice, denoising functions can easily be designed without explicitly starting from objective functions. They often work exceedingly well and good denoisings result in fast and accurate algorithms.

3 Accelerating and Stabilising Convergence by Spectral Shift and Adaptation of Learning Rate

If the denoising function is not able to reduce noise significantly more than signal, the basic DSS iterations (2)–(5) may converge slowly. This is closely related to the fact that power method converges slowly if the largest eigenvalue is only slightly larger than the next largest. Consequently, convergence in DSS can be accelerated in a very similar manner as in power method.

A well-known speedup for power method is spectral shift. It is based on modifying an iteration of the form $\mathbf{w}^+ = \mathbf{A}\mathbf{w}$ into $\mathbf{w}^+ = \mathbf{A}\mathbf{w} + \beta\mathbf{w}$. In the original iteration, it holds $\mathbf{w}^+ = \lambda\mathbf{w}$ at the fixed points and consequently $\mathbf{w}^+ = (\lambda + \beta)\mathbf{w}$ after the modification. The fixed points remain the same but the eigenvalues λ are shifted by β, hence the name spectral shift.

If all eigenvalues are large and their differences are small, convergence can be greatly accelerated by using β which is negative and whose absolute value is close to the second largest eigenvalue. On the other hand, power method converges to the eigenvector that corresponds to the eigenvalue having the largest absolute value. This means that instead of finding the principal component, the minor component is obtained with negative enough β.

In DSS, (3) can be modified into

$$\mathbf{s}^+ = \alpha(\mathbf{s})\mathbf{f}(\mathbf{s}) + \beta(\mathbf{s})\mathbf{s} \qquad (6)$$

without changing the fixed points as long as $\alpha(\mathbf{s})$ and $\beta(\mathbf{s})$ are scalar functions. Since $\alpha(\mathbf{s})$ only scales the source estimate, from now on we assume $\alpha(\mathbf{s}) = 1$.

In DSS, $\mathbf{s}^+ \mathbf{s}^T/T$ plays the role of the eigenvalue [1]. Since Gaussian signals are the least desirable ones in source separation, a reasonable choice for β is the one that shifts the eigenvalue of Gaussian signals to zero:

$$\beta = E\{\mathbf{f}(\boldsymbol{\nu})\boldsymbol{\nu}^T/T\}, \qquad (7)$$

where $\boldsymbol{\nu}$ is a normally distributed signal.

It is interesting to note that the fixed-point equation of FastICA [2] can be interpreted within this framework although normally the speedup used in FastICA is justified as an approximation to Netwon's method. In [1], it was shown that if $\beta(\mathbf{s})$ is based on a linearisation of $\mathbf{f}(\mathbf{s})$ around the current source estimate $\mathbf{s}$, the spectral shift (7) will be

$$\beta(\mathbf{s}) = -\operatorname{tr}\mathbf{J}(\mathbf{s})/T, \qquad (8)$$

---

[1] There is some freedom in this choice because there are several denoising functions which have the same convergence points. They are given in (6).

which is identical to the one used in FastICA. Here $\mathbf{J}(\mathbf{s})$ is the Jacobian of $\mathbf{f}(\mathbf{s})$. Interpreting the speedup as a spectral shift corresponding to Gaussian noise gives an intuitive explanation to why FastICA is able to extract both super- and sub-Gaussian signals with the same nonlinearity: power-method-like iterations converge to the eigenvector whose eigenvalue has the largest magnitude. The sign of the eigenvalue is different depending on whether the component is super- or sub-Gaussian but the magnitude increases when moving away from Gaussian signal whose eigenvalue has been shifted to zero.

In general, iterations converge faster with the FastICA-type spectral shift (8) than with the global Gaussian approximation (7) but the latter has the benefit that no gradients need to be computed. This is important when the denoising is defined by a complex nonlinear procedure such as median filtering.

Neither of the spectral shifts, (7) or (8), always results in stable or fast convergence. Sometimes the spectral shift is too large, which due to the nonlinearity of denoising typically leads to oscillatory behaviour: the iteration oscillates between two weight values. Some other times the spectral shift is too modest leading to slow convergence characterised by small changes of $\mathbf{w}$ in the same direction during several iterations.

For this reason, we have suggested a simple stabilisation rule [1]: instead of updating $\mathbf{w}$ into $\mathbf{w}_{\text{new}}$ defined by (5), it is updated into

$$\mathbf{w}_{\text{adapted}} = \text{orth}(\mathbf{w} + \gamma \Delta \mathbf{w}) \qquad (9)$$

$$\Delta \mathbf{w} = \mathbf{w}_{\text{new}} - \mathbf{w}, \qquad (10)$$

where γ is the step size. Originally $\gamma = 1$, but if the consecutive steps are taken in nearly opposite directions, i.e., the angle between $\Delta \mathbf{w}$ and $\Delta \mathbf{w}_{\text{old}}$ is greater than $179°$, then $\gamma = 0.5$ for the rest of the iterations. There exist a stabilised version of FastICA as well [2] and a similar procedure has been used in practice.

The above modification is able to stabilise convergence in case of oscillations but sometimes the spectral shift is too small and then an increase in step size would be appropriate, i.e., $\gamma > 1$. We propose a simple rule for adapting γ which is inspired by predictive controllers used in robotics: a simple but slow and possibly unstable reactive controller is used for teaching a new, predictive controller. Usually stable and rapid convergence are difficult to achieve simultaneously, but in this setup the new controller can be both faster and stabler.

Translated in our problem, the old slow and unstable controller is the weight modification rule which proposes a modification of weight according to (10). The new controller is implemented by (9), i.e., it modifies the step size. The new controller tries to do immediately what the old controller would do in the future. The step at the previous time instant was apparently optimal if the step proposed at this time instant is orthogonal with it. If not, γ should have been different and, assuming that the optimal γ is constant, the gamma used at this time step should be

$$\gamma_{\text{new}} = \gamma_{\text{old}} + \Delta \mathbf{w}_{\text{old}}^T \Delta \mathbf{w}/||\Delta \mathbf{w}_{\text{old}}||^2. \qquad (11)$$

As it does not seem productive to take steps in the direction opposite from what is suggested by $\Delta \mathbf{w}$ or to take extremely short steps, we require that $\gamma \geq 0.5$.

The above adaptation of γ has turned out to be very useful and it can both stabilise and accelerate convergence. According to (11), γ keeps increasing as long as the steps are taken to the same direction and decreases if they are taken backwards.

4 Denoising Based on Estimated Signal Variance

Several denoising procedures based on masking the source estimate were proposed in [1]. The basic idea is to multiply the source estimate by a positive envelope, a mask which has low values when SNR is low and vice versa. Depending on how the mask is computed, several types of prior information about the sources can be used for separation.

A simple and well-founded mask can be obtained from the maximum-a-posteriori (MAP) estimate. Assuming that the signals are Gaussian with changing variance $\sigma_s^2(t)$ (for related methods, see, e.g., [3]) and additive Gaussian noise σ_n^2, the MAP estimate of the signal is

$$s^+(t) = s(t)\frac{\sigma_s^2(t)}{\sigma_{\text{tot}}^2(t)}, \qquad (12)$$

where $\sigma_{\text{tot}}^2(t) = \sigma_s^2(t) + \sigma_n^2(t)$ is the total variance of the observation. Masking then boils down to estimating $\sigma_s^2(t)$ and $\sigma_{\text{tot}}^2(t)$ from the observations.

A naïve estimate of the signal variance is $\sigma_s^2(t) \approx s^2(t)$. It can be improved by low-pass filtering in time, e.g., by convolving with a Gaussian kernel. Simple estimation of the baseline noise-level σ_n^2 was suggested in [1] resulting in a simple DSS algorithm. However, from the viewpoint of the estimated signal, other signals should be treated as noise. DSS algorithm using the above approximation separates easily the signal subspace from noise but the separation in the signal subspace is slow and may even fail. In [1], this was solved by using $\sigma_s^{2\mu}(t)$ with $\mu > 1$ in (12). This way the mask does not saturate so quickly for large signal variances, giving competitive edge to the source which is strongest. A close connection to the familiar tanh-nonlinearity was shown: $\mathbf{f}(\mathbf{s}) = \mathbf{s} - \tanh \mathbf{s}$ has the same fixed points as $\mathbf{f}(\mathbf{s}) = \tanh \mathbf{s}$ but the former can be interpreted as $\mathbf{s}$ masked by a slowly saturating envelope.

In this paper, we propose a new and better founded solution to the separation problem. One can simply whiten the estimated total variance $\sigma_{\text{tot}}(t)$ by a symmetric whitening matrix. This bares resemblance to proposals of the role of divisive normalisation on cortex [4] and to the classical ICA-method called JADE [5]. Whitening naturally requires that all sources are estimated simultaneously and deflation approach is thus not applicable. The total variance is obtained by smoothing $s^2(t)$ as described above. We obtain $\sigma_s^2(t)$ by taking the positive part of the whitened $\sigma_{\text{tot}}^2(t)$. Whitening here includes removing the mean. Separation by (12) is accelerated significantly because the differences between the envelopes of source estimates are actively emphasised.

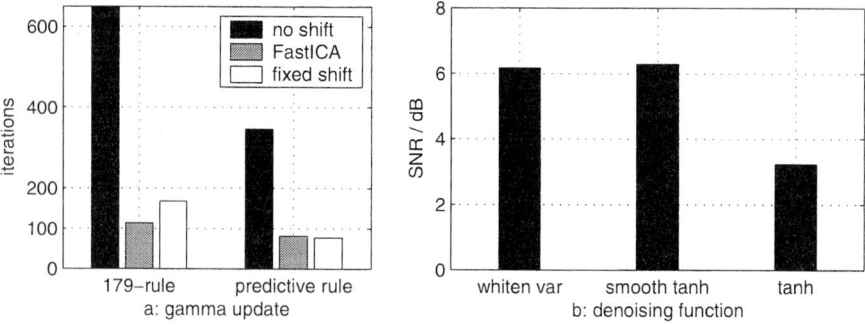

Fig. 2. Speedup tests. a) Effects of spectral shift and step-size adaptation on convergence speed. The leftmost bar not fully shown. b) Average SNRs for different denoising functions: variance whitening and tanh with and without smoothing.

5 Experiments

In this section, we show that the developed algorithms are fast, stable, accurate and produce meaningful results. First, in Sec. 5.1, we demonstrate the different spectral shifts and step-size adaptation. Then the accuracy of different denoising algorithms is tested with artificial data (Sec. 5.2). Finally, we demonstrate the separation capability and convergence speed of the variance-based-denoising in real MEG data (Sec. 5.3).

5.1 Speedup Comparison

In Sec. 3, we reviewed two spectral shifts that can accelerate convergence in DSS algorithms. Later in the section, we proposed two additional methods to adapt these spectral shifts to increase stability. In this section, we compare these adaptive-spectral-shift methods together with the stability improvements in deflatory separation. The data consists of $M = 50$ channels and $T = 8192$ time samples of rhythmic magnetoencephalograms (MEG) [6, 1]. The data was preprocessed as in [1] to enhance weak phenomena. Simple $\mathbf{f}(\mathbf{s}) = \mathbf{s} - \tanh \mathbf{s}$ was used as the denoising function. DSS was run to extract 30 components from this data and average number of iterations was calculated. To be fair for all the methods, each of them was run until convergence, where the angle between old and new projection vectors ($\mathbf{w}$ and $\mathbf{w}_{\text{new}}$) was less than $0.0001°$. We then measured the number of iterations that had taken $\mathbf{w}$ within $0.1°$ of the final solution.

The results are shown in Fig. 2a. Both types of spectral shift and γ adaptation always accelerated convergence. Convergence without any speedups took on average more than 1500 iterations. Without γ adaptation, the FastICA-type scheme (8) converged faster on average than the fixed-shift scheme (7), but γ adaptation reversed the situation. Standard FastICA used about 50% more iterations than the best method.

5.2 Comparison of Denoising Functions

We next compare DSS schemes based on source-variance estimates to the classical tanh-based approach in symmetrical separation of artificial signals. The signals were generated as follows. First, six signals were generated by modulating Gaussian noise with slowly changing envelope. Then the signals were divided into two subspaces, three signals in each. In each of the subspaces, the signals were modulated by another envelope common to all the signals in the subspace. The common envelopes of the subspaces were stronger than the individual envelopes of the sources. Finally, the unit-variance signals were mixed linearly (with $M = N$). Mixing coefficients were sampled from normal distribution and Gaussian noise with variance $\sigma_\nu^2 = 0.09$ was added.

One hundred different data sets were generated and DSS was used to separate the sources with three different denoising functions. Two methods were based on smoothed estimate of source variance. Either the whitening scheme described in Sec. 4 or tanh-based scheme were used in order to promote separation, the tanh-mask being $1 - \tanh[\sigma_{\text{tot}}(t)]/\sigma_{\text{tot}}(t)$. If $\sigma_{\text{tot}}^2(t) = s^2(t)$, this reduces to the popular tanh-nonlinearity. With these methods, spectral shift was computed by assuming that the mask does not significantly depend on any individual source value, i.e. $-\beta$ equals to the average of elements of the mask. The third method was the popular tanh-nonlinearity with FastICA-type spectral shift. The step size was adapted by the 179-rule.

As before, the algorithms were run until convergence. The average SNRs of the separation over the one hundred runs are shown in Fig. 2b. Smoothing the variance estimate clearly improves the SNR with tanh-nonlinearity. Variance whitening achieved comparable SNR but used significantly less iterations.

5.3 MEG Signal Separation

Finally, we used the DSS algorithms and acceleration methods studied in the previous sections to separate sources from rhythmic MEG data. The whole data set ($M = 122$ and $T = 65536$) was used and 30 components were extracted using the same denoising functions as in the previous section. Both the 179-rule and the predictive rule (11) were tested. The number of iterations was taken to be the limit where the projection vector $\mathbf{w}$ of the slowest converging component reaches $0.1°$ of the final projection. Enhanced spectrograms of some interesting components extracted by the variance-whitening DSS are depicted in Fig. 3a.

Tanh-nonlinearity with smoothed variance estimate extracted similar components, but the usual tanh-nonlinearity without smoothing seemed to have trouble in finding the weak steady frequencies shown in the bottom row of Fig. 3a. The processing times of different denoising functions and different step size adaptations are shown in Fig. 3b. Since the computational complexity of one iteration depends on the denoising function, the total CPU-time is reported. Compared to the variance-whitening DSS, the tanh-nonlinearities used more than two times more processing time, independent of the step-size adaptation. Compared to the 179-rule, the adaptive γ reduced the total processing time by 20–50 %, depending

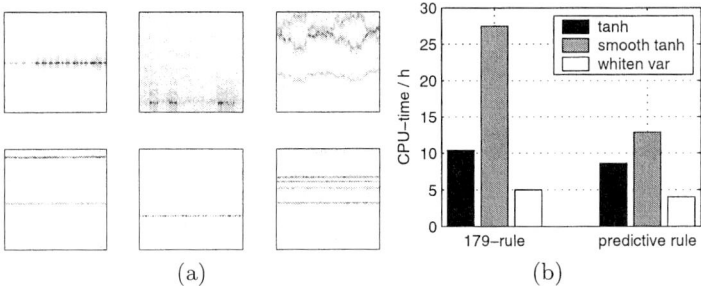

Fig. 3. a) Spectrograms of some of the sources separated using variance whitening. Time on the horizontal and frequency on the vertical axis. b) Used processing time for different denoising functions and step sizes.

on the denoising function. Tanh-nonlinearity with smoothed variance estimate used a fixed spectral shift and benefitted more from adaptation of γ than the plain tanh-nonlinearity with FastICA-type spectral shift.

6 Conclusion

DSS framework offers a sound basis for developing simple but efficient and accurate source separation algorithms. We proposed a method for stabilising and accelerating convergence and showed that convergence is faster than with FastICA. Additional benefit is that gradient of the nonlinearity is not needed. We also proposed a new denoising procedure which was justified as the MAP-estimate of signals with changing variance. Denoising which makes use of non-stationarity of variance was shown to yield better results than the popular tanh-nonlinearity as measured by SNR in the artificially generated data. The variance-whitening DSS also extracted cleaner signals in MEG data, while the tanh-nonlinearity had difficulties with some weak but clear phenomena. Whitening the estimated variances of different sources significantly improved convergence.

References

1. Särelä, J., Valpola, H.: Denoising source separation. Submitted to a journal (2004) Available at Cogprints http://cogprints.ecs.soton.ac.uk/archive/00003493/.
2. Hyvärinen, A.: Fast and robust fixed-point algorithms for independent component analysis. IEEE Trans. on Neural Networks **10** (1999) 626–634
3. Matsuoka, K., Ohya, M., Kawamoto, M.: A neural net for blind separation of nonstationary signals. Neural Networks **8** (1995) 411–419
4. Schwartz, O., Simoncelli, E.P.: Natural signal statistics and sensory gain control. Nature Neuroscience **4** (2001) 819 – 825
5. Cardoso, J.F.: High-order contrasts for independent component analysis. Neural computation **11** (1999) 157 – 192
6. Särelä, J., Valpola, H., Vigário, R., Oja, E.: Dynamical factor analysis of rhythmic magnetoencephalographic activity. In: Proc. Int. Conf. on Independent Component Analysis and Signal Separation (ICA2001), San Diego, USA (2001) 451–456

An Overview of BSS Techniques Based on Order Statistics: Formulation and Implementation Issues

Yolanda Blanco and Santiago Zazo

Departamento de Ingenieria Eléctrica y Electrónica, Universidad Europea de Madrid
Villaviciosa de Odón, 28670 Madrid, Spain
ETS. Ingenieros Telecomunicación, Universidad Politécnica de Madrid
28040 Madrid, Spain
myolanda.blanco@tel.uem.es

Abstract. The main goal of this paper is to review the fundamental ideas of the method called "ICA with OS". We review a set of alternative statistical distances between distributions based on the Cumulative Density Function (cdf). In particular, these gaussianity distances provide new cost functions whose maximization perform the extraction of one independent component at each successive stage of a new proposed deflation ICA procedure. These measures are estimated through Order Statistics (OS) that are consistent estimators of the inverse cdf. The new Gaussianity measures improve the ICA performance and also increase the robustness against outliers compared with the traditional ones.

1 Introduction

The goal of BSS (Blind Source Separation) is to extract N unknown independent sources ($\mathbf{s} = [s_1...s_N]^H$) from a set of linear mixtures ($\mathbf{y} = [y_1...y_M]^H$). Most of the methods perform a spatial decorrelation preprocessing over $\mathbf{y}$ to obtain the decorrelated observable $\mathbf{z} = [z_1...z_M]^H$. Thus, the global mixture is expressed by $\mathbf{z}(n) = \mathbf{V}\mathbf{s}(n)$, where $\mathbf{V}$ is an unknown orthogonal matrix. Afterwards, the Independent Component Analysis (ICA) is applied to find out a linear unitary transformation $\mathbf{w}(n) = \mathbf{B}\mathbf{z}(n)$, in such a way to have w_i components as independent as possible.

Most of the ICA methods consist of extracting all the independent components simultaneously by means of a maximization of an independence measure between the output signals. The original independence measure is the Mutual Information (MI) [8] as the Kullback - Leibler divergence between the joint pdf (probabilty density function) and the product of marginal pdf's. Many methods are deduced from MI: let us mention "Infomax" [2] and fourth cross cumulants methods [8],[7]. Also the authors in [12] proposed the estimation of the MI by means of the Entropy expressed in terms of the Order Statistics.

More recently a new ICA approach has appeared which is based on the following consequence of the Central Limit Theorem (CLT): the linear combination of independent random variables always increases the Gaussianity of the resultant distribution, therefore, ICA must simply decrease the Gaussianity of one single output analysis channel to extract one of the original non-Gaussian independent components; this idea was formalized by Hyvärien through his "*ICA estimation principle 2: Maximum non-gaussianity*" [9]. Therefore, the IC's could be extracted one by one through a multistage procedure [9],[10],[5]. One of the main advantages of this new method compared to traditional ICA is that it decreases computational cost since the degrees of freedom are reduced. Besides this method has a great potential when the number of sensors and sources are not equal, even when unknown.

In the next two sections we deduce a set of Gaussianity measures in terms of inverse cdf's instead of widespread pdf's, whose main advantage is the direct implementation by means of the Order Statistics. The estimation of the Order Statistics is easily performed ordering the samples. *Section 4* will treat the adaptive processing of the new cost functions and the new deflation procedure. Some successful comparative results conclude the paper.

2 Gaussianity Measures Based on Inverse cdf's

Let's remark that a Gaussianity measure is an appropriate distance between the analyzed output signal w_i and the equivalent Gaussian distribution g with the same power.

Negentropy is the original Gaussianity measure and it was defined as the Kullback Leibler distance between both pdf's: f_{w_i} and f_g. Kurtosis arose like an estimation of Negentropy. More recently a family of Gaussianity measures (built with non-linearities) has been proposed like an extension of the Kurtosis measure (see [9]).

Also the distance between two distributions can be equally evaluated using cdf's instead of pdf's. Specifically, let us define a family of distribution distances through the norm concept applied to the difference between both cdf's. For ICA proposal the distance between the analyzed output signal w_i and its equivalent Gaussian distribution would be:

$$d_\infty(F_{w_i}, F_g) = \max_x |F_{w_i}(x) - F_g(x)| \quad (L^\infty norm) \tag{1}$$

$$d_p(F_{w_i}, F_g) = \left(\int_{-\infty}^{\infty} |F_{w_i}(x) - F_g(x)|^p dx \right)^{\frac{1}{p}} \quad (L^p norm) \tag{2}$$

Furthermore, it can be appreciated that the Infinite norm in *eq.*(1) is exactly the well known Kormogorov - Smirnov evaluation test [11] to determine weather two distributions are coincident.

On the other hand, the analitycal expressions of F_w and F_g were obtained in [6] in order to prove that d_p and d_∞ are non-Gaussianity measures since they offer a global minimum when the distribution is Gaussian and the functions grow monotonically according to the distribution which shifts far away from the Gaussian, as seen in figure 1. Let us remark that c is the "Gaussianity parameter"

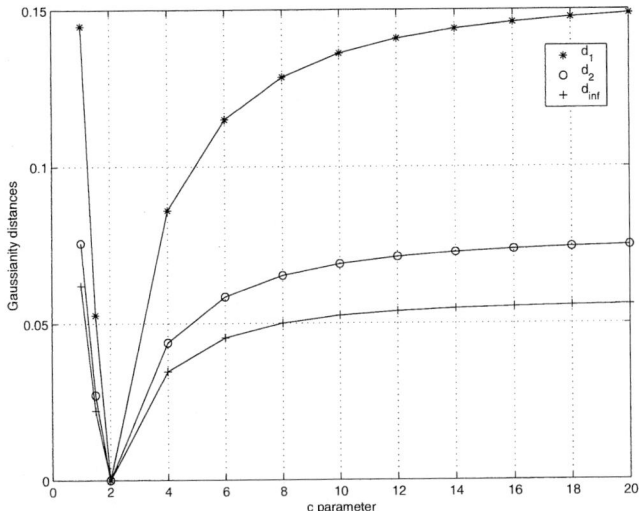

Fig. 1. Gaussianity Measures based on cdf's expressions of the Generalized Family Gaussian.

in the well known "Generalized Gaussian Family" ([11]); $c=2$ implies that w_i is Gaussian, $c<2$ corresponds to a super Gaussian signal and $c>2$ to a sub Gaussian one.

In practice the estimation $\widehat{F}_{w_i}(w)$ would be necessary in the evaluation of previous distance, but would require a high computational cost if it is performed through accumulated histograms. Fortunately, an equivalent distance, easily estimated, can be established in terms of inverse cdf's: $Q_{w_i} = F_{w_i}^{-1}, Q_g = F_g^{-1}$:

$$D_\infty(Q_{w_i}, Q_g) = \max_x |Q_{w_i}(x) - Q_g(x)| \qquad (3)$$

$$D_p(Q_{w_i}, Q_g) = \left(\int_0^1 |Q_{w_i}(x) - Q_g(x)|^p \, dx \right)^{\frac{1}{p}} \qquad (4)$$

Since Q is the inverse of F, both functions present the same characteristic (growing or decreasing); consequently the distances d and its correspondent D present the same behavior, in other words curves in figure 1 for Q have the same properties as curves for F. Therefore, eq.(3) and eq.(4) are appropriate non-Gaussianity measures.

3 Practical Implementation: Cost Functions Based on Order Statistics

The estimation of Q_{w_i} can be performed very robustly in a simple and practical way ordering a set large enough of n temporal discrete samples from

$\{w_i[1],....w_i[n]\}$ and obtaining the set of Order Statistics (OS) $w_{i(1)} < w_{i(2)} < ... < w_{i(n)}$; The OS are a consistent estimation of the quantiles of the distribution:

$$\widehat{Q}_{w_i}(\frac{k}{n}) = w_{i(k)} \Leftrightarrow \widehat{F}(w_{i(k)}) = \frac{k}{n} \tag{5}$$

Consequently, the estimation of eq. 3 and eq. 4 can be expressed using OS notation:

$$\widehat{D}_\infty(Q_{w_i}, Q_g) = \max_k |w_{i(k)} - Q_g(\frac{k}{n})| \quad (L^\infty \text{Norm}) \tag{6}$$

$$\widehat{D}_p(Q_{w_i}, Q_g) = \left(\frac{1}{n}\sum_{k=1}^{n}|w_{i(k)} - Q_g(\frac{k}{n})|^p\right)^{\frac{1}{p}} \quad (L^p \text{ Norm}) \tag{7}$$

where $Q_g(\frac{k}{n})$ is the known $\frac{k}{n}$ quantile of the equivalent Gaussian distribution.

Previous equations imply the estimation of a finite set of discrete values of the inverse cdf Q in the non-Gaussianity measures in eqs. (3) and (4). If n is large enough (around 1000 samples) the estimation is really robust. At this point it is necessary to clarify that the modified Kolmogorov - Smirnov (KS) estimator in eq. 6 is using only one ordered sample and, therefore, can be applied if one wants to reduce the complexity of the algorithm. When using any measure with the whole set of Order Statistics more infomation is obtained since cdf is estimated through a large set of discrete values.

In [6], we evaluated the Gaussianity measure $\widehat{D}_\infty(Q_{w_i}, Q_g)$ in order to establish a prefixed value of k which maximizes the absolute value in eq. 6, thus it was found that:

1. The extreme (1%, 100%) are the k/n values presenting the global maxima of $|w_{i(k)} - Q_g(\frac{k}{n})|$, but also a local maxima appear at (20%,80%); therefore rewriting the first Gaussian measure (eq.6) associated with the KS test, we obtain:

$$\widehat{D}_\infty(Q_{w_i}, Q_g) = |w_{i(k)} - Q_g(\frac{k}{n})| \text{ with } k = \begin{Bmatrix} 1 \\ n \\ 0.20\,n \\ 0.80\,n \end{Bmatrix} \tag{8}$$

2. For symmetric distributions the couple of symmetric Order Statistics: $w_{i(k)}$, $w_{i(l)}$ (with $l=n-k$) provides the same information, and after some simple operations we get:

$$\widehat{D}_\infty(Q_{w_i}, Q_g) = |w_{i(k)} - w_{i(l)} + 2Q_g(\frac{l}{n})| \text{ with } \begin{matrix} k,l = 80\%n, 20\%n \\ \text{either} \\ k,l = n,1 \end{matrix} \tag{9}$$

Let us mention that cost function in eq. (9) was presented in [3],[4] and [5], where the capability of the Order Statistics in order to perform ICA was introduced.

In conclusion, eq. 7 (with $p = 1$ either $p = 2$) and eq. 9 are both possible cost functions to maximize in an ICA multistage procedure.

4 Multistage Deflation Procedure Plus Gradient Rule

We propose an ICA multistage deflation algorithm that is presented in the scheme of the figure 2. Its main difference compared to other multistage approaches [5, 9] is that it decreases the dimension of the problem and doesn't need to realize any orthogonal projection of the separation vector inside the adaptive processing, (for more detail see [6]).

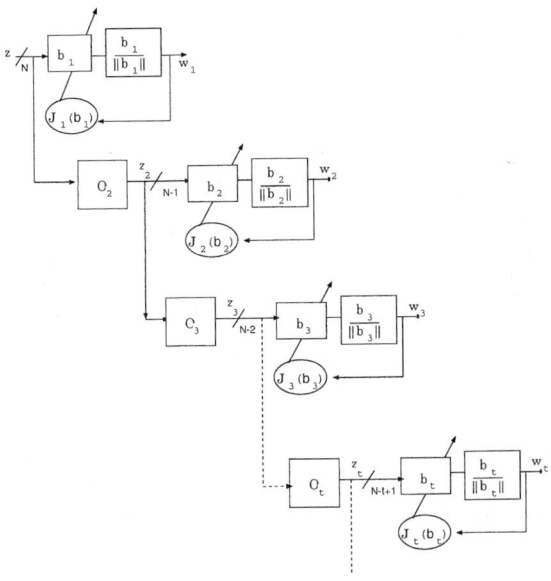

Fig. 2. Multistage ICA deflation algorithm for any t ($t = 1...N$) stage: adaptive rule plus orthogonalization constraint.

Taking into account that $w_i = \mathbf{b}_i^T \mathbf{z}$, where $\mathbf{b}_i^T$ is an unitary separation vector, the goal is to update $\mathbf{b}_i$ properly at each i stage by means of the maximization of an ICA cost function $J(w_i(\mathbf{b}_i))$. Let us take $J(w_i(\mathbf{b}_i)) = \widehat{D}_\infty$. It will be optimized by means of a stochastic gradient rule (plus a normalization):

$$\mathbf{b}_i[t+1] = \mathbf{b}_i[t] + \mu \nabla J|_{\mathbf{b}_i[t]} \tag{10}$$

The gradient of $J(w_i(\mathbf{b}_i))$ in eq.9 is:

$$\nabla J|_{\mathbf{b}_i[t]} = S \left. \frac{d(w_{i(k)} - w_{i(l)})}{d\mathbf{b}_i} \right|_{\mathbf{b}_i[t]} \tag{11}$$

where $S = sign(w_{i(k)} - w_{i(l)} + 2Q_g(\frac{l}{n}))_{\mathbf{b}_i[t]}$

Applying the chain rule we obtain

$$\left. \frac{d(w_{i(k)} - w_{i(l)})}{d\mathbf{b}_i} \right|_{\mathbf{b}_i[t]} = \mathbf{z} \left(\left. \frac{dw_{i(k)}}{dw_i} \right|_{\mathbf{b}_i[t]} - \left. \frac{dw_{i(l)}}{dw_i} \right|_{\mathbf{b}_i[t]} \right) \tag{12}$$

The derivative of any r-order statistic with respect to w_i was obtained by means of taking a derivative of a vector in [4] which gives:

$$\left.\frac{dw_{i(r)}}{dw_i}\right|_{\mathbf{b}_i[t]} = \mathbf{e}_r = [0, 0, ..0, 1, 0..0]^T;$$

where

$$\mathbf{e}_r[j] = \left\{\begin{array}{c} 1 \; if \; w_i[j] = w_{i(r)} \\ 0 \; rest \end{array}\right\}_{j=1...n} \tag{13}$$

The generalization of the previous adaptive algorithm to $J = \widehat{D}_p$ is straightforward, for example for $\widehat{D}_1$:

$$\begin{array}{c} \nabla J_t|_{\mathbf{b}_i[t]} = \frac{1}{n}\mathbf{z}\sum_{k=1}^n S(k)\mathbf{e}_k|_{\mathbf{b}_i[t]} \\ \text{where } S(k) = sign(w_{t(k)} - Q_g(\frac{k}{n}))|_{\mathbf{b}_i[t]} \end{array} \tag{14}$$

Additionally, this optimization must be accomplished by using the proper constraint: $\mathbf{b}_i$ must be orthonormal to the vectors obtained at each one of previous stages in order to guarantee the independence among all the extracted sources. On the other side, it is expected that when one source is extracted, the problem could also be reduced to one dimension. Thus, the multistage procedure in figure 2 is performed constraining the search for any separation vector $\mathbf{b}_i$ (of decreasing dimension $(N-i+1)$) to the orthonormal subspace $\mathbf{O}_i$ (of dimension $(N-i)\text{x}(N-i+1)$). $\mathbf{O}_i$ is spanned by the previous separation vectors, in other words, the matrix $\mathbf{O}_i$ is blocking previous separation vectors $\mathbf{b}_1, \mathbf{b}_2...\mathbf{b}_{i-1}$. Therefore, the vector $\mathbf{b}_i$ is previously forced to belong to the orthonormal subspace $\mathbf{O}_i$.

The calculus of matrix $\mathbf{O}_i$ is based on the Gram-Schmidt (GS) procedure:

$$\{\mathbf{b}_{i-1}, \mathbf{e}_1, ..., \mathbf{e}_{N-i}\} \Rightarrow \{\mathbf{u}_1, ..., \mathbf{u}_{N-i}\} \tag{15}$$

where $\{\mathbf{e}_k\}$ are the unitary vectors to complete an $N-i$ dimension basis and $\{\mathbf{u}_k\}$ are the orthonormal-basis components through the GS procedure. Therefore, the blocking matrix is $\mathbf{O}_i = [\mathbf{u}_2 \; \mathbf{u}_3 \; ... \; \mathbf{u}_{N-i}]$. Let us observe that $\mathbf{O}_i$ is orthonormal to guarantee that vector $\mathbf{z}_i$ maintains the spatial decorrelation feature.

5 Results

In order to compare the ICA performance of the new cost functions and its corresponding algorithm convergence with other representative ICA methods (Infomax [2] and Fast ICA [9]) we have considered three representative examples of mixtures. For each case, one hundred of arbitrary mixtures have been generated and separated averaging the Amari's performance index in dB ([1]); one thousand samples per output channel have been processed; $\mu = 0.5$ and $\mu = 0.01$ have been chosen as the adaptation step for $\widehat{D}_1$ and $\widehat{D}_\infty$ respectively. On the other hand, the computational cost has been evaluated averaging the number of needed iterations for the algorithm to converge; however, let's remark that any iteration requieres ordering samples. The results are shown in table 1.

Table 1. Quality/cost results for several ICA methods. *PI*: Amari performance Index. *Cost*: average number of convergence iterations (U: Uniform, L: Laplacian, G: Gaussian).

	ICA-OS D_1	ICA-OS D_∞	INFOMAX	FAST-ICA
3U, *PI*	-14.5	-21.14	-13.8	-12.59
Cost	10	20 (Extreme OS)	50	8
3L, *PI*	-12.52	-8.3	-12.45	-12.45
Cost	10	200 (Central OS)	50	8
4U+3L+G, *PI*	-8.02	-7.12	-6.12	-7.69
Cost	30	40 (Extreme OS)	70	15

Next, in order to compare the robustness against outliers of the new 'Gaussianity measure based on OS' with other representative measures (Kurtosis [10] and a non-linearity proposed in [9]), several kind of sources have been mixed through a fixed orthogonal mixture matrix $\mathbf{V}(2x2)$. After that, the corresponding ICA algorithm was applied to the decorrelated observable $\mathbf{z}=\mathbf{Vs}$. In this scenario, every original source s_o has been contaminated with uniform outliers of power $\sigma_u^2 = 4\sigma_{s_o}^2$ according to the following pdf's relation:

$$f_s(x) = (1-\lambda)f_{s_o}(x) + \lambda f_u(x) \tag{16}$$

In table 2 we present an example of obtained comparative results expressed by means of the Amari's index [1]; the number of processed samples is 1000. In presence of outliers he processed Order Statistics must be (80%, 20%) because the contaminated samples are put close to the extreme OS (around 100%) and therefore they are filtered by the $D_{\infty(20\%,80\%)}$ function.

Table 2. Comparative ICA performance Index for TITO mixtures in presence of outliers.

$\lambda = 10^{-2}$	D_1	D_2	D_∞	KURTOSIS	log(cosh(.))
UU	-15.97	-17.49	-14.72	-10.29	-8.93
UG	-18.22	-19.37	-11.98	-8.03	-8.00
LU	-16.52	-16.42	-14.45	-7.92	-10.11
LG	-16.05	-12.17	-11.52	-8.69	-10.86
LL	-15.86	-15.56	-14.24	-12.40	-14.58

6 Conclusions

From observation of tables 1 and 2 some conclusions are drawn: Non-Gaussianity measures D_1, D_2 and $D_{\infty(20\%,80\%)}$ are more robust against outliers than the others. ICA with OS using D_1 or D_2 improves quality and its convergence level is comparable to Fast ICA for any kind of distribution. The use of $D_{\infty(1\%,100\%)}$

is highly recommended when most of the distributions are sub Gaussian, but it has to be guaranteed that no outliers are present. The advantage of using D_∞ is the reduced complexity of the algorithms. Another advantage is the easy and efficient implementation of a gradient rule, since the optimization is based only on simple sample reordering. The implementation is performed very efficiently by constraining the search of any separation vector to the subspace orthonormal to the previous extracted vectors.

References

1. S. Amari, A. Cichocki, H. H. Yang. A New Learning Algorithm for Blind Signal Separation. *Proc. of Neural Information Processing Systems, NIPS 96*, vol. 8, pp 757-763.
2. Te-Won Lee. *Independent Component Analysis: Theory and Applications.* Kluwer Academic Publishers, 1998
3. Y. Blanco, S. Zazo, JM Páez-Borrallo. Adaptive Processing of Blind Source Separation through 'ICA with OS'. *Proceedings of the International Conference On Acoustics And Signal Processing ICASSP'00.* Vol I, 233-236 S.
4. Y.Blanco, S. Zazo, J.C. Principe. Alternative Statistical Gaussianity Measure using the Cumulative Density Function. *Proceedings of the Second Workshop on Independent Component Analysis and Blind Signal Separation: ICA'00.* pp 537-542.
5. Y. Blanco, S. Zazo, JM Páez-Borrallo. Adaptive ICA with Order Statistics: An efficient approach based on serial orthogonal projections. *Proceedings of the 6th International Work Conference in artificial and natural neural networks*: IWANN'01. Vol II, pp 770-778
6. Y. Blanco, S. Zazo. New Gaussianity Measures based on Order Statistics. Application to ICA. *Elsevier. Neurocomputing* 51. pp 303-320. 2003.
7. JF Cardoso, A. Soulimiac. Blind beamforming for Non Gaussian Signals. *IEE Proceedings-F*, vol 140, n 6, pp 362-370, December 1993.
8. P.Common. Independent Component Analysis, A New Concept. *Signal Processing*, n 36, pp 287-314. 1992.
9. A. Hyvärien. J. Karhunen, E. Oja. Independent Component Analysis. Ed. John Wiley & sons. 2001
10. S.Y. Kung, C.Mejuto. Extraction of Independent Components from Hybrid Mixture: Knicnet Learning Algorithm and Applications. In *proc. International Conference On Acoustics And Signal Processing ICASSP'98.* vol II, 1209,1211.
11. A. Papoulis. Probability and Statistics. Prentice Hall International, INC. 99
12. E. G. Learned-Miller. ICA Using Spacing Estimates of Entropy. *Journal of Machine Learning Research 4 (2003)* 1271-1295

Analytical Solution of the Blind Source Separation Problem Using Derivatives

Sebastien Lagrange[1,2], Luc Jaulin[2], Vincent Vigneron[1], and Christian Jutten[1]

[1] Laboratoire Images et Signaux, Institut National Polytechnique de Grenoble
46 avenue Félix Viallet, 38031 Grenoble, France
Vincent.Vigneron@lis.inpg.fr, Christian.Jutten@inpg.fr
http://www.lis.inpg.fr

[2] Laboratoire LISA, Institut des Sciences et Techniques de l'Ingénieur
62 avenue notre dame du lac, 49000 Angers
{sebastien.lagrange,luc.jaulin}@isita.univ-angers.fr
http://www.istia.univ-angers.fr

Abstract. In this paper, we consider independence property between a random process and its first derivative. Then, for linear mixtures, we show that cross-correlations between mixtures and their derivatives provide a sufficient number of equations for analytically computing the unknown mixing matrix. In addition to its simplicity, the method is able to separate Gaussian sources, since it only requires second order statistics. For two mixtures of two sources, the analytical solution is given, and a few experiments show the efficiency of the method for the blind separation of two Gaussian sources.

1 Introduction

Blind source separation (BSS) consists in finding unknown sources $s_i(t)$, $i = 1, ..., n$ supposed statistically independent, knowing a mixture of these sources, called observed signals $x_j(t)$, $j = 1, ..., p$. In the literature, various mixtures have been studied : linear instantaneous [1–3] or convolutive mixtures [4–6], nonlinear and especially post-nonlinear mixtures [7,8]. In this paper, we assume (i) the number of sources and observations are equal, $n = p$, (ii) the observed signals are linear instantaneous mixtures of the sources, *i.e*,

$$x_j(t) = \sum_{i=1}^{n} a_{ij} s_i(t), \quad j = 1, \ldots n. \tag{1}$$

In vector form, denoting the source vector $\mathbf{s}(t) = [s_1(t), ..., s_n(t)]^T \in \mathbb{R}^n$, and the observation vector $\mathbf{x}(t) = [x_1(t), ..., x_n(t)]^T \in \mathbb{R}^n$, the observed signal is

$$\mathbf{x}(t) = \mathbf{A}\mathbf{s}(t), \tag{2}$$

where $\mathbf{A} = [a_{ij}]$ is the $n \times n$ mixing matrix, assumed regular.

Without prior knowledge, the BSS problem can be solved by using independent component analysis (ICA) [9], which involves higher (than 2) order statistics, and requires that at most one source is Gaussian. With weak priors, like source coloration [10–12] or non-stationarity [13, 14], it is well known that BSS can be solved by jointly diagonalizing variance-covariance matrices, *i.e.* using only second order statistics, and thus allowing separation of Gaussian sources.

For square $(n \times n)$ mixtures, the unknown sources can be indirectly estimated by estimating a separating matrix denoted $\mathbf{B}$, which provides a signal $\mathbf{y}(t) = \mathbf{B}\mathbf{x}(t)$ with independent components. However, it is well known that independence of the components of $\mathbf{y}(t)$ is not sufficient for estimating exactly $\mathbf{B} = \mathbf{A}^{-1}$, but only $\mathbf{BA} = \mathbf{DP}$, pointing out a scale (diagonal matrix $\mathbf{D}$) and permutation (permutation matrix $\mathbf{DP}$) indeterminacies [9]. It means that source power cannot be estimated. Thus, in the following, we will assumed unit power sources.

In this paper, we propose a new method based on second order statistics between the signals and their first derivatives. In Section 2, a few properties concerning statistical independence are derived. The main result is presented in Section 3, with the proof in Section 4, and a few experiments in Section 5, before the conclusion.

2 Statistical Independence

In this section, we will introduce the main properties used below. For p random variables $x_1, \ldots x_p$, a simple definition of independence is based on the factorisation of the joint density as the product of the marginal densities:

$$p_{x_1,\ldots,x_p}(u_1, \ldots, u_p) = \prod_{i=1}^{p} p_{x_1}(u_i). \qquad (3)$$

We can also define the independence of random processes.

Definition 1. *Two random processes $x_1(t)$ and $x_2(t)$ are independent if and only if any random vectors, $x_1(t_1), \ldots, x_1(t_1 + k_1)$ and $x_2(t_2), \ldots, x_2(t_2 + k_2)$, $\forall t_i$, and k_j, $(i, j = 1, 2)$, extracted from them, are independent.*

Consequently, if two random signals (processes) $x_1(t)$ and $x_2(t)$ are statistically independent, then $\forall t_1, t_2$, $x_1(t_1)$ and $x_2(t_2)$ are statistically independent random variables, too [16].

Notation 1 *In the following, the independence between two random signals $x_1(t)$ and $x_2(t)$ will be denoted $x_1(t) \; \mathbb{I} \; x_2(t)$.*

Proposition 1. *Let $x_1(t), x_2(t), \ldots, x_n(t)$ and $u(t)$ be random signals. We have*

$$\begin{array}{rl} x_1(t)\mathbb{I}u(t) & \Longrightarrow \quad u(t)\mathbb{I}x_1(t) \\ x_1(t)\mathbb{I}u(t),\ldots,x_n(t)\mathbb{I}u(t) & \Longrightarrow \quad (x_1(t) + \ldots + x_n(t))\,\mathbb{I}u(t)\,. \\ \forall \alpha \in \mathbb{R}, x_1(t)\mathbb{I}u(t) & \Longrightarrow \quad \alpha x_1(t)\mathbb{I}u(t) \end{array} \qquad (4)$$

We now consider independence properties involving signals and their derivatives.

Lemma 1. *Let $x_1(t)$ and $x_2(t)$ be differentiable (with respect to t) signals. Then,*

$$x_1(t) \mathbb{I} x_2(t) \implies \begin{cases} x_1(t) \; \mathbb{I} \; \dot{x}_2(t) \\ \dot{x}_1(t) \; \mathbb{I} \; x_2(t) \\ \dot{x}_1(t) \; \mathbb{I} \; \dot{x}_2(t) \end{cases} . \tag{5}$$

As a direct consequence, if $x_1(t)$ and $x_2(t)$ are sufficiently differentiable, for all $m, n \in \mathbb{N}$,

$$x_1(t) \mathbb{I} x_2(t) \implies x_1^{(n)}(t) \mathbb{I} x_2^{(m)}(t). \tag{6}$$

Proof. If $x_1(t)$ and $x_2(t)$ are statistically independent then

$$x_1(t) \mathbb{I} x_2(t) \implies \forall t_1, \forall t_2, \; x_1(t_1) \mathbb{I} x_2(t_2). \tag{7}$$

According to (4), $\forall t_1, \forall t_2$,

$$\left. \begin{array}{c} x_1(t_1) \mathbb{I} x_2(t_2) \\ x_1(t_1) \mathbb{I} x_2(t_2 + \tau) \end{array} \right\} \implies x_1(t_1) \mathbb{I} \frac{x_2(t_2) - x_2(t_2 + \tau)}{\tau}. \tag{8}$$

Hence, since x_2 is differentiable with respect to t

$$\lim_{\tau \to 0} \frac{x_2(t_2) - x_2(t_2 + \tau)}{\tau} < \infty, \tag{9}$$

and we have $\forall t_1, t_2, \; x_1(t_1) \mathbb{I} \dot{x}_2(t_2)$ where $\dot{x}(t)$ denotes the derivative of $x(t)$ with respect to t. Similar proof can be done for showing $\forall t_1, t_2, \dot{x}_1(t_1) \mathbb{I} x_2(t_2)$, and more generally $\forall t_1 \in \mathbb{R}, t_2 \in \mathbb{R}, \; x_1^{(n)}(t_1) \mathbb{I} x_2^{(m)}(t_2)$.

Lemma 2. *Let $x(t)$ be a differentiable signal with the auto-correlation function $\gamma_{xx}(\tau) = E(x(t)x(t - \tau))$, then $E(x\dot{x}) = 0$.*

Proof. Since $x(t)$ is derivable, its autocorrelation function is derivable in zero:

$$\dot{\gamma}_{xx}(0) = \lim_{\tau \to 0} \frac{\gamma_{xx}(0) - \gamma_{xx}(\tau)}{-\tau} \tag{10}$$

$$= \lim_{\tau \to 0} \frac{E(x(t)x(t)) - E(x(t)x(t - \tau))}{-\tau} \tag{11}$$

$$= \lim_{\tau \to 0} E(x(t)(\frac{x(t) - x(t - \tau)}{-\tau})) \tag{12}$$

$$= E(x(t). \lim_{\tau \to 0} (\frac{x(t) - x(t - \tau)}{-\tau})) = -E(x\dot{x}). \tag{13}$$

Finally, since γ_{xx} is even, $\dot{\gamma}_{xx}(0) = 0$, and consequently $E(x\dot{x}) = 0$.

Lemma 3. *If $\mathbf{x} = \mathbf{As}$, where component s_i of $\mathbf{s}$ are mutually independent, then $E(\mathbf{x}\dot{\mathbf{x}}^T) = 0$.*

Proof. Since $\mathbf{x} = \mathbf{As}$, we have $E(\mathbf{x}\dot{\mathbf{x}}^T) = \mathbf{A} E(\mathbf{s}\dot{\mathbf{s}}^T) \mathbf{A}^T$

Using Lemmas 2 and 1, one has $E(s_i \dot{s}_i) = 0$ and $E(s_i \dot{s}_j) = 0$, respectively. Consequently, $E(\mathbf{x}\dot{\mathbf{x}}^T) = 0$.

3 Theorem

In this section, we present the main result of the paper. The proof will be shown in the next section (4).

First, let us define the set $\mathcal{T}$ of trivial linear mixings, *i.e.* linear mappings which preserve independence for any distributions. One can show that $\mathcal{T}$ is set of square regular matrices which are the product of one diagonal regular matrix and one permutation matrix. In other words, $\mathbf{B}$ is a separating matrix if $\mathbf{BA} = \mathbf{DP} \in \mathcal{T}$.

Theorem 1. *Let* $\mathbf{x}(t) = \mathbf{A}\mathbf{s}(t)$, *be an unknown regular mixture of sources* $\mathbf{s}(t)$, *whose components* $s_i(t)$ *are ergodic, stationary, derivable and mutually independent signals, the separating matrices* $\mathbf{B}$, *such that* $\mathbf{y}(t) = \mathbf{B}\mathbf{x}(t)$ *has mutually independent components, are the solutions of the equation set:*

$$\mathbf{B}E(\mathbf{x}\mathbf{x}^T)\mathbf{B}^T = E(\mathbf{y}\mathbf{y}^T)$$
$$\mathbf{B}E(\dot{\mathbf{x}}\dot{\mathbf{x}}^T)\mathbf{B}^T = E(\dot{\mathbf{y}}\dot{\mathbf{y}}^T)$$

where $E(\mathbf{y}\mathbf{y}^T)$ *and* $E(\dot{\mathbf{y}}\dot{\mathbf{y}}^T)$ *are diagonal matrices.*

4 Proof of Theorem

The proof is given for 2 mixtures of 2 sources. It will be admitted in the general case.

The estimated sources are $\mathbf{y} = \mathbf{B}\mathbf{x}$ where $\mathbf{B}$ is a separating matrix of $\mathbf{A}$. After derivation, one has a second equation: $\dot{\mathbf{y}} = \mathbf{B}\dot{\mathbf{x}}$.

The independence assumption of the estimated sources $\mathbf{y}$ implies that the following matrix is diagonal:

$$E\left(\mathbf{y}\mathbf{y}^T\right) = E\left(\mathbf{B}\mathbf{x}\mathbf{x}^T\mathbf{B}^T\right) = \mathbf{B}E\left(\mathbf{x}\mathbf{x}^T\right)\mathbf{B}^T. \tag{14}$$

Moreover, using Lemma 1 and 2, the following matrix is diagonal, too:

$$E\left(\dot{\mathbf{y}}\dot{\mathbf{y}}^T\right) = E\left(\mathbf{B}\dot{\mathbf{x}}\dot{\mathbf{x}}^T\mathbf{B}^T\right) = \mathbf{B}E\left(\dot{\mathbf{x}}\dot{\mathbf{x}}^T\right)\mathbf{B}^T. \tag{15}$$

By developing the vectorial equations (14) and (15), one gets four scalar equations:

$$\begin{cases} b_{11}^2 E\left(x_1^2\right) + 2b_{11}b_{12}E\left(x_1 x_2\right) + b_{12}^2 E\left(x_2^2\right) & = E\left(y_1^2\right) \\ b_{21}^2 E\left(x_1^2\right) + 2b_{21}b_{22}E\left(x_1 x_2\right) + b_{22}^2 E\left(x_2^2\right) & = E\left(y_2^2\right) \\ b_{11}b_{21}E\left(x_1^2\right) + b_{11}b_{22}E\left(x_1 x_2\right) + b_{12}b_{21}E\left(x_1 x_2\right) + b_{12}b_{22}E\left(x_2^2\right) & = 0 \\ b_{11}b_{21}E\left(\dot{x}_1^2\right) + b_{11}b_{22}E\left(\dot{x}_1 \dot{x}_2\right) + b_{12}b_{21}E\left(\dot{x}_1 \dot{x}_2\right) + b_{12}b_{22}E\left(\dot{x}_2^2\right) & = 0. \end{cases} \tag{16}$$

This system is a set of polynomials with respect to the b_{ij}. It has six unknowns for only four equations. In fact, the two unknowns, $E(y_1^2)$ and $E(y_2^2)$ are not relevant, due to the scale indeterminacies of source separation. Since the source power cannot be estimated, we can then consider this unknowns as parameters, or even constraint these parameter to be equal to a constant (e.g. 1). Here, we parameterize the set of solutions (two dimensional manifold) by two real parameters λ_1 and λ_2 such that $|\lambda_1|=E\left(y_1^2\right)$ and $|\lambda_2|=E\left(y_2^2\right)$ in Eq. (16).

Groëbner Basis decomposition [15] give the solutions

$$\mathbf{B} = \begin{pmatrix} \lambda_1 & 0 \\ 0 & \lambda_2 \end{pmatrix} \begin{pmatrix} \phi_1 & \phi_1\eta_1 \\ \phi_2 & \phi_2\eta_2 \end{pmatrix} \quad \text{or} \quad \mathbf{B} = \begin{pmatrix} 0 & \lambda_1 \\ \lambda_2 & 0 \end{pmatrix} \begin{pmatrix} \phi_1 & \phi_1\eta_1 \\ \phi_2 & \phi_2\eta_2 \end{pmatrix}, \qquad (17)$$

where

$$\phi_1 = \left(E(x_1^2) + 2\eta_1 E(x_1 x_2) + \eta_1^2 E(x_2^2)\right)^{-\frac{1}{2}}, \qquad (18)$$

$$\phi_2 = \left(E(x_1^2) + 2\eta_2 E(x_1 x_2) + \eta_2^2 E(x_2^2)\right)^{-\frac{1}{2}}, \qquad (19)$$

$$\eta_1 = -\beta \left(1 + \sqrt{1 - \frac{\alpha}{\beta^2}}\right), \qquad (20)$$

$$\eta_2 = -\beta \left(1 - \sqrt{1 - \frac{\alpha}{\beta^2}}\right), \qquad (21)$$

$$\alpha = \frac{E(x_1^2)E(\dot{x}_1\dot{x}_2) - E(x_1 x_2)E(\dot{x}_1^2)}{E(x_1 x_2)E(\dot{x}_2^2) - E(x_2^2)E(\dot{x}_1\dot{x}_2)}, \qquad (22)$$

$$\beta = \frac{1}{2}\frac{E(x_1^2)E(\dot{x}_2^2) - E(x_2^2)E(\dot{x}_1^2)}{E(x_1 x_2)E(\dot{x}_2^2) - E(x_2^2)E(\dot{x}_1\dot{x}_2)}. \qquad (23)$$

Then, let

$$\tilde{\mathbf{B}} = \begin{pmatrix} \phi_1 & \phi_1\eta_1 \\ \phi_2 & \phi_2\eta_2 \end{pmatrix}, \qquad (24)$$

any matrix $\mathbf{T}\tilde{\mathbf{B}}$ where $\mathbf{T} \in \mathcal{T}$, is still a solution of (16). Especially, it exists a particular matrix $\tilde{\mathbf{T}} \in \mathcal{T}$ with $\lambda_1 = E(y_1^2)$ and $\lambda_2 = E(y_2^2)$ such that:

$$\mathbf{A}^{-1} = \tilde{\mathbf{T}}\tilde{\mathbf{B}}. \qquad (25)$$

Thus, all the possible separating matrices are solutions of (16), and the Theorem 1 is proved.

5 Experiments

Consider two independent Gaussian sources obtained by Gaussian white noises filtered by low-pass second-order filters. Filtering ensures to obtain differentiable sources by preserving the Gaussian distribution of the sources.

The two sources are depicted on Figures 1 and 2 (90,000 samples, sampling period $Te = 0.1$). The corresponding joint distribution is represented on Figure 3.

The derivative joint distribution (Fig. 4) shows the two signals $\dot{s}_1(t)$ and $\dot{s}_2(t)$ are independent (as predicted by Lemma 2).

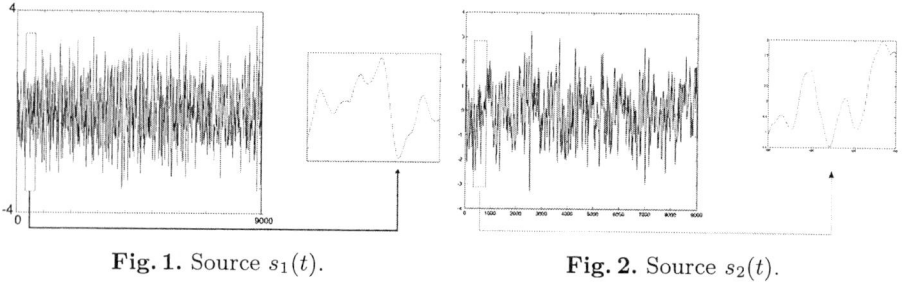

Fig. 1. Source $s_1(t)$. **Fig. 2.** Source $s_2(t)$.

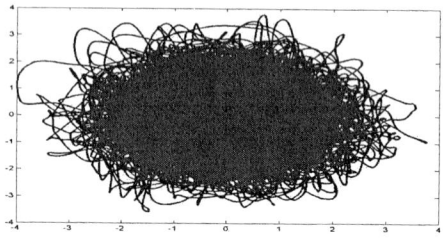
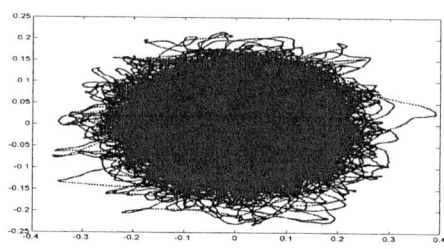

Fig. 3. Distribution of the sources $(s_1(t), s_2(t))$.

Fig. 4. Distribution of the signals $(\dot{s}_2(t), \dot{s}_1(t))$.

The mixing matrix

$$\mathbf{A} = \begin{pmatrix} 1.5 & 0.8 \\ 0.6 & 1.2 \end{pmatrix}, \qquad (26)$$

provides the observation signals (mixtures) $x_1(t)$ and $x_2(t)$. The joint distribution of the mixtures (Fig. 4) and of their derivatives (Fig. 5) points out the statistical dependence of $x_1(t)$ and $x_2(t)$ as well as $\dot{x}_1(t)$ and $\dot{x}_2(t)$.

From the Theorem 1, the separation matrix $\tilde{\mathbf{B}}$ is analytically computed:

$$\tilde{\mathbf{B}} = \begin{pmatrix} 0.456 & -.606 \\ -0.454 & 1.136 \end{pmatrix}. \qquad (27)$$

The estimated sources $\tilde{\mathbf{B}}\mathbf{x}(t)$ are independent and unit power. We also can check that there exists $\tilde{\mathbf{T}} \in \mathcal{T}$ such as $\tilde{\mathbf{T}}\tilde{\mathbf{B}}\mathbf{x}(t) = \mathbf{s}(t)$, i.e. $\tilde{\mathbf{T}}\tilde{\mathbf{B}} = \mathbf{A}^{-1}$. In this example, let

$$\tilde{\mathbf{T}} = \begin{pmatrix} 2 & 0 \\ 0 & 2 \end{pmatrix} \in \mathcal{T}, \; (\lambda_1 = \lambda_2 = 2), \qquad (28)$$

we have

$$\tilde{\mathbf{T}}\tilde{\mathbf{B}} = \begin{pmatrix} 0.912 & -.613 \\ -0.453 & 1.135 \end{pmatrix}, \qquad (29)$$

and

$$\mathbf{A}^{-1} = \begin{pmatrix} 0.909 & -.606 \\ -0.454 & 1.136 \end{pmatrix}. \qquad (30)$$

In order to study the robustness of the solution for Gaussian mixtures, Fig. 7 shows the separation performance (using the index: $E(norm(\mathbf{s} - \hat{\mathbf{s}})))$ versus

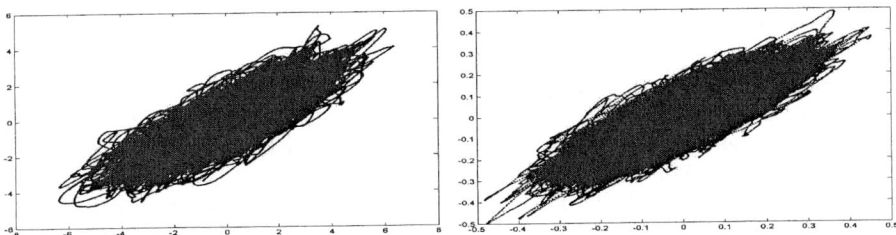

Fig. 5. Joint distribution of the mixtures $(x_1(t), x_2(t))$.

Fig. 6. Joint distribution of the mixture derivatives $(\dot{x}_1(t), \dot{x}_2(t))$.

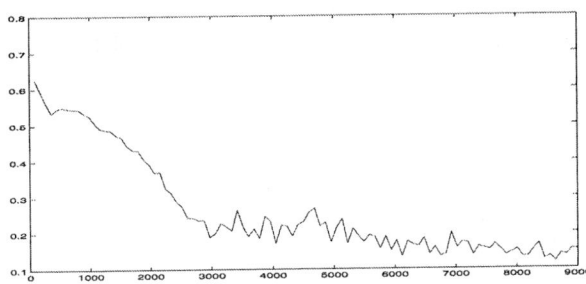

Fig. 7. The error estimation of the sources according to the number of samples.

the sample number. Over 2600 samples, the analytical solution provides good performance, with an error less than -20 dB.

6 Conclusion

In this article, we proposed a new source separation criterion based on variance-covariance matrices of observations and of their derivatives. Since the method only uses second-order statistics, source separation of Gaussians remains possible. The main (and weak) assumption is the differentiability of the unknown sources.

We derived the analytical solution for two mixtures of two sources. In the general case (n mixtures of n sources), the analytical solution seems tricky to compute. Moreover, for an ill-conditioned set of equations (16), the analytical solution can be very sensitive to the statistical moment estimations. For overcoming this problem, we could estimate the solution, by using approximate joint diagonalization algorithm of $E(\mathbf{xx}^T)$ and $E(\dot{\mathbf{x}}\dot{\mathbf{x}}^T)$. Moreover, other variance-covariance matrices, based on higher-order derivatives or using different delays (assuming sources are colored) or on different temporal windows (assuming sources are non stationary), could be used for estimating the solution by joint diagonalization.

Further investigations include implementation of joint diagonalization, and extension to more complex signals mixtures, *e.g.* based on state variable models.

References

1. Jutten, C., Hérault, J.: Blind Separation of Sources, Part I: an Adaptive Algorithm Based on a Neuromimetic Architecture. Signal Processing, Vol. 24 (1991) 1–10
2. Cardoso, J.-F., Souloumiac, A.: Blind beamforming for non Gaussian signals. IEE Proceedings-F, Vol. 140 (1993) 362–370
3. Bell, T., Sejnowski, T.: An Information-Maximization Approach to Blind Separation and Blind Deconvolution. Neural Comutation, Vol. 7 (1995) 1004-1034
4. Yellin, D., Weinstein, E.: Criteria for multichannel signal separation. IEEE Trans. on Signal Processing, Vol. 42 (1994) 2158–2168
5. Nguyen Thi, H. L., Jutten, C.: Blind Sources Separation For Convolutive mixtures. Signal Processing, Vol. 45 (1995) 209–229
6. Babaie-Zadeh, M.: On blind source separation in convolutive and nonlinear mixtures. PhD thesis, INP Grenoble (2002)
7. Pajunen, P., Hyvärinen, A., Karhunen, J.: Non linear source separation by self-organizing maps. Proceedings of ICONIP 96, Hong-Kong (1996)
8. Taleb, A., Jutten, C.: Source separation in post nonlinear mixtures. IEEE Transactions on Signal Processing, Vol. 47 (1999) 2807–2820
9. Comon, P.: Independent Component Analysis, a new concept? Signal Processing, Vol. 36 (1994) 287–314
10. Tong, L., Soon, V., Liu, R., Huang, Y.: AMUSE: a new blind identification algorithm. Proceedings ISCAS, New Orleans, USA (1990)
11. Belouchrani, A., Abed Meraim, K., Cardoso, J.-F., Moulines, E.: A blind source separation technique based on second order statistics. IEEE Trans. on Signal Processing, Vol. 45 (1997) 434–444
12. Molgedey, L., Schuster, H. G.: Separation of a mixture of independent signals using time delayed correlation. Physical Review Letters, Vol. 72 (1994) 3634–3636
13. Matsuoka, K., Ohya, M., Kawamoto, M.: A neural net for blind separation of nonstationary signals. Neural Networks, Vol. 8 (1995) 411–419
14. Pham, D. T., Cardoso, J.-F.: Blind Separation of Instantaneous Mixtures of Non Stationary Sources. IEEE Transaction on Signal Processing, Vol. 49 (2001) 1837–1848
15. Adams, W. W., Loustaunau, P.: An Introduction to Gröbner Bases. Providence, RI: Amer. Math. Soc., 1994
16. Papoulis, A.: Probability, Random Variables, and Stochastic Processes. Third Edition, McGraw-Hill, International Editions. Electronical and Electronic Engineering Series

Approximate Joint Diagonalization Using a Natural Gradient Approach

Arie Yeredor[1], Andreas Ziehe[2], and Klaus-Robert Müller[2]

[1] School of Electrical Engineering, Tel-Aviv University, Israel
arie@eng.tau.ac.il
[2] Fraunhofer FIRST, Germany
{ziehe,klaus}@first.fhg.de

Abstract. We present a new algorithm for non-unitary approximate joint diagonalization (AJD), based on a "natural gradient"-type multiplicative update of the diagonalizing matrix, complemented by step-size optimization at each iteration. The advantages of the new algorithm over existing non-unitary AJD algorithms are in the ability to accommodate non-positive-definite matrices (compared to Pham's algorithm), in the low computational load per iteration (compared to Yeredor's AC-DC algorithm), and in the theoretically guaranteed convergence to a true (possibly local) minimum (compared to Ziehe et al.'s FFDiag algorithm).

1 Introduction

The approximate joint diagonalization (AJD) of a set of matrices constitutes a fundamental stage in many batch-type algorithms for Independent Components Analysis (ICA) or Blind Source Separation (BSS). Usually, in this context, a set of unknown "target matrices" exists, which, assuming a linear static noiseless BSS model, admits exact joint diagonalization. The diagonalizing matrix (or the mixing matrix), can thus be theoretically extracted by jointly diagonalizing these matrices, which usually amounts to applying a generalized eigenvalue decomposition to any couple of matrices from the set. However, in practice the "target set" is unknown, and has to be estimated from the available data. In the presence of estimation errors, the estimated set usually no longer admits exact joint diagonalization. In such cases, one must resort to *approximate* joint diagonalization of the entire set in order to estimate the mixing matrix (or its inverse), as the matrix which diagonalizes the estimated set "as closely as possible".

To formulate the problem, let $\check{M}_1, \check{M}_2, ... \check{M}_K \in \mathbb{C}^{N \times N}$ denote the set of K true (usually unavailable) "target matrices" satisfying the exact joint diagonalization model

$$\check{M}_k = \check{A}\check{\Lambda}_k\check{A}^T \quad \text{or} \quad \check{\Lambda}_k = \check{B}\check{M}_k\check{B}^T \quad , \quad k=1,2,...,K \qquad (1)$$

where $\check{A}$ is the true mixing matrix (assumed non-singular), $\check{B}$ is its inverse (the true "demixing" matrix) and $\{\check{\Lambda}_k\}_{k=1}^{K}$ is a set of diagonal matrices, usually associated with the sources' statistical or structural properties, so that their diagonality dwells on the statistical independence of the sources. Some examples of such sets as used in BSS algorithms are:

- Cumulant matrices (in JADE, [1]);
- Correlation matrices of differently time-lagged or filtered signals (in SOBI [2] or OFI [3]);
- Joint time-frequency distributions at selected times and frequencies (in [4]);
- Hessians of the joint characteristic function (in CHESS, [5]);

and many more, extending also to the context of convolutive BSS, e.g., when working on separate frequency bins, such as in [6,7].

As mentioned earlier, only estimates $\{M_k\}_{k=1}^{K}$ (of $\{\check{M}_k\}_{k=1}^{K}$) are available in practice, and the AJD problem consists of seeking the implied estimate A of $\check{A}$ (or B of $\check{B}$), along with "nuisance estimates" $\{\Lambda_k\}_{k=1}^{K}$ of $\{\check{\Lambda}_k\}_{k=1}^{K}$, such that the respective relation in (1) is most closely satisfied.

Thus, AJD is essentially a non-convex (possibly constrained) optimization problem, whose solution depends on the precise formulation of the target criterion (which has to reflect the proximity of the solution to the state of exact diagonalization). Numerous approaches have been proposed in recent years both for the formulation of the diagonalization criterion and for the iterative solution taken in its minimization:

- One of the most popular and computationally appealing approaches is the unitary AJD (Cardoso and Souloumiac, [8]), which minimizes the criterion

$$C_1(B) = \sum_{k=1}^{K} \text{off}_1(BM_k B^T) \qquad (2)$$

with respect to (w.r.t.) B, subject to the unitarity constraint $B^T B = I$, where

$$\text{off}_1(P) \triangleq \sum_{i \neq j} |P_{ij}|^2. \qquad (3)$$

The unitarity constraint avoids the trivial minimizer $B = 0$, but implies the assumption of a unitary mixing matrix. Hence, in the general case a pre-processing "spatial hard-whitening" stage is required, in which the non-unitary factor of the overall demixing matrix is found and applied to the data. In turn, this "hard whitening" stage implies exact joint diagonalization of the (spatial) correlation matrix, possibly at the expense of poor diagonalization of other matrices in the set. This implied unbalanced weighting has been observed [9] to limit the performance in the context of a general BSS problem.

- In order to avoid the unitarity constraint, an approach for non-unitary AJD has been proposed (the "AC-DC" algorithm, Yeredor [10]), which minimizes

$$C_2(A) = \sum_{k=1}^{K} \|M_k - A\Lambda_k A^T\|_F^2 \qquad (4)$$

(where $\| \bullet \|_F$ denotes the Frobenius norm) w.r.t. A and $\{\Lambda_k\}_{k=1}^{K}$, without constraining A. While computationally efficient in small-scale problems, this algorithm has been observed [11] to exhibit extremely slow convergence in large-scale problems.

- A computationally efficient unconstrained minimization algorithm w.r.t. B was proposed as well (Pham, [12]), whose target criterion is given by

$$C_3(B) = \sum_{k=1}^{K} \text{off}_3(BM_kB^T), \tag{5}$$

where in this case $\text{off}_3(\bullet)$ measures the Kullback-Leibler divergence between the $N \times N$ operand and the diagonal matrix with the same diagonal as the operand. This approach requires all the target matrices to be positive-definite, which poses a limit on its applicability as a generic BSS tool.

- A recently proposed approach (Ziehe et al., [11]) offers another computationally efficient algorithm, which avoids both the unitarity constraint and the positive-definiteness requirement. It aims at minimizing $C_1(B)$ with a different constraint on B: Rather than impose unitarity, it inherently requires (by construction) that B be representable as a product of matrices of the following form:

$$B = \left[\prod_{m=1}^{M}(I + W^{(m)})\right] B^{(0)}, \tag{6}$$

where $B^{(0)}$ is some initial guess, M is the number of iterations, and $W^{(m)}$ are small "update matrices" with imposed zero diagonals, calculated along the iterations. Thus, if $B^{(0)}$ is nonsingular and the norms of all $W^{(m)}$ are maintained sufficiently small, it can be shown that the resulting B must be invertible, hence the trivial minimizer is avoided. Moreover, this constraint does not limit the generality of the solution, since any two nonsingular matrices, say B_1 and B_2, maintain the relationship $B_2 = D(I+W)B_1$, where D is some nonsingular diagonal matrix and W has a null diagonal. Thus, considering the inherent scale-ambiguity in BSS, the structural constraint (6) does not pose any practical restriction on the attainable solutions.

While computationally attractive, this algorithm has a few weak points from a theoretical point of view. It dwells on an approximation that may not always be valid in the presence of large errors in estimating the target matrices, and it involves some heuristics which are justified more on the practical-empirical side than on the theoretical side. Consequently, although its fast convergence has been verified empirically, it is not theoretically guaranteed to converge, and even upon convergence, B is not always guaranteed to be a true (even local) minimizer of $C_1(B)$.

In this paper we propose a novel AJD algorithm, also aimed at the minimization of $C_1(B)$ subject to the same non-restrictive structural constraint (6) as in [11]. Similarly, our algorithm is computationally attractive, and does not require positive-definiteness of the set. Moreover, $C_1(B)$ is guaranteed to decrease in each iteration, so that its convergence is guaranteed. Also, since no approximations or heuristics are involved, upon convergence B is guaranteed to be a true (possibly local) minimizer of $C_1(B)$.

The algorithm is based on the notion of a multiplicative "natural-gradient" (e.g., [13]), as opposed to the "standard" gradient (used, e.g., in [14]). The "natural gradient" is often applied in the context of "on-line" BSS algorithms, but also suits the AJD problem with the structural constraint (6). Our algorithm was named DOMUNG[1] (Diagonalization Of Matrices Using Natural Gradient).

2 Algorithm Derivation

Throughout the derivation we shall frequently use the operation of nullifying the diagonal of a matrix. We shall denote this operation by using an upper bar. More specifically, for any square matrix P we define the notation $\overline{P}$ as

$$\overline{P} \triangleq P - \tilde{P} = P - P \odot I. \tag{7}$$

The $\text{off}_1(\bullet)$ operator (3) can then be expressed based on the trace of a matrix:

$$\text{off}_1(P) = ||\overline{P}||_F^2 = \text{tr}\{\overline{P}^T \overline{P}\} = \text{tr}\{P^T \overline{P}\}. \tag{8}$$

For simplicity of the derivations we shall assume that the target matrices are all real-valued and symmetric, which is often (but not always) the case in BSS applications. Extension to the more general case along similar guidelines is possible, but would extend beyond the scope of this limited-length paper.

We propose the following iterative process. Denote $B^{(m)}$ the estimated diagonalizing (demixing) matrix after the m-th iteration, updated using $B^{(m)} = (I + W^{(m)})B^{(m-1)}$ $m = 1, 2, ...$, where $B^{(0)}$ is some initial guess and $W^{(m)}$ is a matrix with a null main diagonal, which we shall eventually specify. Denoting

$$M_k^{(m)} = B^{(m-1)} M_k B^{(m-1)T} \quad k = 1, 2, ..., K \ \ m = 1, 2, ... \tag{9}$$

as the "transformed" target set after the $(m-1)$-th iteration, it is readily seen that at the m-th iteration the criterion function is given by

$$C_1(B^{(m)}) = \sum_{k=1}^{K} \text{off}_1(B^{(m)} M_k B^{(m)T}) = \sum_{k=1}^{K} \text{off}_1((I+W^{(m)}) M_k^{(m)} (I+W^{(m)})^T). \tag{10}$$

We may therefore define, for each iteration m,

$$C_1^{(m)}(W) \triangleq \sum_{k=1}^{K} \text{off}_1((I+W) M_k^{(m)} (I+W)^T), \tag{11}$$

as a criterion function which we seek to minimize (w.r.t. W) at that iteration, subject to the constraint on the structure of W, namely that W should have a null main diagonal. To this end, we now seek the gradient $\partial C_1^{(m)}(W)/\partial W$, which is a matrix whose (i,j)-th element is the derivative of $C_1^{(m)}(W)$ w.r.t.

---

[1] DOMUNG is a language spoken in Papua New Guinea.

W_{ij} (W_{ij} denoting the (i,j)-th element of $\boldsymbol{W}$). To find this gradient matrix, let us first find the gradient of each summand in (11). We do so by expressing the $\text{off}_1(\bullet)$ function in (11) in the vicinity of $\boldsymbol{W} = \boldsymbol{0}$ up to first-order terms in $\boldsymbol{W} = \boldsymbol{\mathcal{E}}$, where $\boldsymbol{\mathcal{E}}$ is a sufficiently small matrix (for shorthand we shall use, in the following expressions, $\boldsymbol{M}$ instead of $\boldsymbol{M}_k^{(m)}$):

$$\begin{aligned}
\text{off}_1((\boldsymbol{I}+\boldsymbol{\mathcal{E}})\boldsymbol{M}(\boldsymbol{I}+\boldsymbol{\mathcal{E}})^T) &= \text{tr}\{[(\boldsymbol{I}+\boldsymbol{\mathcal{E}})\boldsymbol{M}(\boldsymbol{I}+\boldsymbol{\mathcal{E}})^T]^T \overline{(\boldsymbol{I}+\boldsymbol{\mathcal{E}})\boldsymbol{M}(\boldsymbol{I}+\boldsymbol{\mathcal{E}})^T} \\
&= \text{tr}\{(\boldsymbol{I}+\boldsymbol{\mathcal{E}})\boldsymbol{M}(\boldsymbol{I}+\boldsymbol{\mathcal{E}})^T \overline{(\boldsymbol{I}+\boldsymbol{\mathcal{E}})\boldsymbol{M}(\boldsymbol{I}+\boldsymbol{\mathcal{E}})^T}\} \\
&\approx \text{tr}\{(\boldsymbol{M}+\boldsymbol{\mathcal{E}}\boldsymbol{M}+\boldsymbol{M}\boldsymbol{\mathcal{E}}^T)\overline{(\boldsymbol{M}+\boldsymbol{\mathcal{E}}\boldsymbol{M}+\boldsymbol{M}\boldsymbol{\mathcal{E}}^T)}\} \\
&\approx \text{tr}\{\boldsymbol{M}\overline{\boldsymbol{M}}+\boldsymbol{M}\overline{\boldsymbol{\mathcal{E}}\boldsymbol{M}}+\boldsymbol{M}\overline{\boldsymbol{M}\boldsymbol{\mathcal{E}}^T}+\boldsymbol{\mathcal{E}}\boldsymbol{M}\overline{\boldsymbol{M}}+\boldsymbol{M}\boldsymbol{\mathcal{E}}^T\overline{\boldsymbol{M}}\} \\
&= \text{tr}\{\boldsymbol{M}\overline{\boldsymbol{M}}+\boldsymbol{M}\overline{\boldsymbol{M}}\boldsymbol{\mathcal{E}}+\boldsymbol{M}\overline{\boldsymbol{M}}\boldsymbol{\mathcal{E}}+\boldsymbol{M}\overline{\boldsymbol{M}}\boldsymbol{\mathcal{E}}+\boldsymbol{M}\overline{\boldsymbol{M}}\boldsymbol{\mathcal{E}}\} \\
&= \text{tr}\{\boldsymbol{M}\overline{\boldsymbol{M}}\}+2\text{tr}\{(\boldsymbol{M}\overline{\boldsymbol{M}}+\overline{\boldsymbol{M}\overline{\boldsymbol{M}}})\boldsymbol{\mathcal{E}}\}. \quad (12)
\end{aligned}$$

We used (8) in the first line, and the identities $\text{tr}\{\boldsymbol{P}\} = \text{tr}\{\boldsymbol{P}^T\}$, $\text{tr}\{\boldsymbol{PQ}\} = \text{tr}\{\boldsymbol{QP}\}$ and $\text{tr}\{\boldsymbol{P}\overline{\boldsymbol{Q}}\} = \text{tr}\{\overline{\boldsymbol{P}}\boldsymbol{Q}\}$ in the transition from the fourth line to the fifth. The $\approx$ symbol on the third and fourth lines indicates the elimination of terms of second or higher order in $\boldsymbol{\mathcal{E}}$ in the respective transitions.

Noting that $\partial \text{tr}\{\boldsymbol{P}\boldsymbol{\mathcal{E}}\}/\partial \boldsymbol{\mathcal{E}} = \boldsymbol{P}^T$, we obtain that the gradient of the $\text{off}_1(\bullet)$ function w.r.t. $\boldsymbol{W}$ is $4(\overline{\boldsymbol{M}}\boldsymbol{M})$. Reinstating the full notation we obtain the gradient of $C_1^{(m)}$ w.r.t. $\boldsymbol{W}$ at the m-th iteration:

$$\boldsymbol{G}^{(m)} \triangleq \frac{\partial C_1^{(m)}(\boldsymbol{W})}{\partial \boldsymbol{W}} = 4 \sum_{k=1}^{K} \overline{\boldsymbol{M}_k^{(m)}} \boldsymbol{M}_k^{(m)}. \quad (13)$$

Since we wish to decrease $C_1^{(m)}$ in each iteration, we shall apply a "steepest descent" step, by setting $\boldsymbol{W}$ to $\mu \boldsymbol{D}^{(m)}$, where μ is some positive constant (whose optimal value will be discussed shortly), and $\boldsymbol{D}^{(m)} \triangleq -\boldsymbol{G}^{(m)}$ is an "antigradient" matrix. The use of $\overline{\boldsymbol{G}^{(m)}}$ (rather than $\boldsymbol{G}^{(m)}$) as the gradient direction is due to the null-diagonal constraint on $\boldsymbol{W}$, which implies that its diagonal elements must remain zero, so that the only elements participating in the descent are the off-diagonal ones.

We now wish to ensure that the step-size in the anti-gradient direction yields the largest decrease in the criterion $C_1^{(m)}(\boldsymbol{W})$. Since this step-size is controlled by the parameter μ, we may now minimize $C_1^{(m)}(\boldsymbol{W}) = C_1^{(m)}(\mu \boldsymbol{D}^{(m)})$ w.r.t. μ. More specifically, substituting into (11) we obtain

$$\begin{aligned}
C_1^{(m)}(\mu \boldsymbol{D}^{(m)}) &= \sum_{k=1}^{K} \text{off}_1((\boldsymbol{I}+\mu \boldsymbol{D}^{(m)})\boldsymbol{M}_k^{(m)}(\boldsymbol{I}+\mu \boldsymbol{D}^{(m)})^T) \\
&= \sum_{k=1}^{K} \text{tr}\{(\boldsymbol{I}+\mu \boldsymbol{D}^{(m)})\boldsymbol{M}_k^{(m)}(\boldsymbol{I}+\mu \boldsymbol{D}^{(m)})^T \overline{(\boldsymbol{I}+\mu \boldsymbol{D}^{(m)})\boldsymbol{M}_k^{(m)}(\boldsymbol{I}+\mu \boldsymbol{D}^{(m)})^T}\} \\
&\triangleq a_0^{(m)} + a_1^{(m)}\mu + a_2^{(m)}\mu^2 + a_3^{(m)}\mu^3 + a_4^{(m)}\mu^4 \quad (14)
\end{aligned}$$

where the coefficients $\{a_l^{(m)}\}_{l=0}^2$ are given[4] by $a_l^{(m)} = \sum_{k=1}^K \text{tr}\{F_{l,k}^{(m)}\}$, with $F_{l,k}^{(m)}$ summarized in Table 1:

Table 1.

$F_{0,k}^{(m)}$	$M_k^{(m)}\overline{M_k^{(m)}}$
$F_{1,k}^{(m)}$	$4M_k^{(m)}\overline{M_k^{(m)}}D^{(m)}$
$F_{2,k}^{(m)}$	$2\left[(D^{(m)}M_k^{(m)} + M_k^{(m)}D^{(m)T})\overline{D^{(m)}M_k^{(m)}} + D^{(m)}M_k^{(m)}D^{(m)T}\overline{M_k^{(m)}}\right]$
$F_{3,k}^{(m)}$	$4D^{(m)}M_k^{(m)}D^{(m)T}\overline{D^{(m)}M_k^{(m)}}$
$F_{4,k}^{(m)}$	$4D^{(m)}M_k^{(m)}D^{(m)T}\overline{D^{(m)}M_k^{(m)}D^{(m)T}}$

Thus, since $C_1^{(m)}(\mu D^{(m)})$ is evidently a fourth-order polynomial in μ, the optimal μ for the m-th iteration can be found by polynomial rooting of the derivative third-order polynomial, namely by solving (w.r.t. μ)

$$4a_4^{(m)}\mu^3 + 3a_3^{(m)}\mu^2 + 2a_2^{(m)}\mu + a_1^{(m)} = 0, \qquad (15)$$

To which there is at least one real-valued solution. In the case of three real-valued solutions, the true minimum can be found by substituting each solution back into the polynomial (14) and selecting the solution that yields the smallest value. The algorithm is summarized below.

DOMUNG - Diagonalization Of Matrices Using Natural Gradient

– Denote the original "target set" as $M_1^{(0)}, M_2^{(0)}, ..., M_K^{(0)}$, and let $W^{(0)} = 0$ and $B^{(0)} = I$.
– For $m = 1, 2, ...$ until convergence
  • Compute the updated target set
    $M_k^{(m)} = (I + W^{(m-1)})M_k^{(m-1)}(I + W^{(m-1)})^T$ for $k = 1, 2, ..., K$;
  • Compute $G^{(m)} = 4\sum_{k=1}^K \overline{M_k^{(m)}}M_k^{(m)}$ and set $D^{(m)} = -\overline{G^{(m)}}$;
  • Compute the coefficients $a_0^{(m)}, a_1^{(m)}, ..., a_4^{(m)}$ using Table 1, and compute the real-valued root / three roots of the polynomial (15);
  • Set μ to the root that yields the smallest value in (14);
  • Set $W^{(m)} = \mu D^{(m)}$, $B^{(m)} = (I + W^{(m)})B^{(m-1)}$.
– Upon convergence ($m = M$), the unmixing matrix is given by $B^{(M)}$.

We did not specify a convergence criterion - but since the target criterion $C_1(B^{(m)})$ is guaranteed to decrease (or at least not to increase) in each iteration, and it is bounded below, the sequence of its values over iterations must converge. Thus a stopping criterion that halts when the decrease in $C_1(B^{(m)})$ falls below any (arbitrarily small) specified positive value, is guaranteed to be met.

---

[2] After using similar algebraic manipulations as in (12).

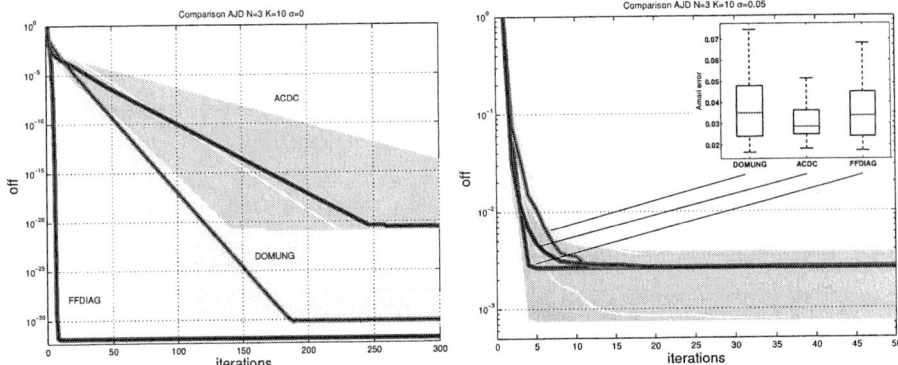

Fig. 1. Diagonalization errors on perfectly diagonalizable matrices

Fig. 2. Diagonalization errors and Amari errors [15] on non-diagonalizable matrices

3 Simulations

Here we provide a comparison of DOMUNG with two previously established algorithms: ACDC [10] and FFDIAG [11].

Noise free case The test data for the experiments is generated as follows. We use $K = 10$ diagonal matrices Λ_k of size 3×3 where the elements on the diagonal are drawn from a uniform distribution in the range $[-1 \ldots 1]$. These matrices are 'mixed' using the fixed matrix $A = \begin{bmatrix} 8 & 1 & 6 \\ 3 & 5 & 7 \\ 4 & 9 & 2 \end{bmatrix}$ according to the model $A\Lambda_k A^T$ to obtain the set of matrices $\{M_k\}$ to be diagonalized.

The convergence behavior of the 3 algorithms in 10 runs is shown in Fig. 1. The diagonalization error is measured by the $\mathrm{off}_1(\cdot)$ function. The shaded area denotes the minima and maxima, while the bold line indicates the median over the 10 runs. In all cases the algorithms converged to the correct solution within the numerical computing precision. The differences in the final levels are only due to the use of slightly different stopping criteria.

Noisy case of non-diagonalizable matrices We also investigated robustness of the three algorithms against non-diagonalizability of the set of matrices.

Non-diagonalizability is modeled by adding random "noise" matrices to the input matrices:

$$M_k = A\Lambda_k A^T + \sigma R_k,$$

where R_k are symmetric matrices, whose free elements are independently drawn from a standard normal distribution. The parameter σ determines the noise level, i.e. impact of the non-diagonalizable component.

Fig. 2 shows the error curves of 10 trials for a noise level of $\sigma = 0.05$, as well as distances from the true solution as measured by the Amari error [15] for 10 trials. One can see that all algorithms converge to the same level of the (normalized) cost function.

4 Conclusions

We proposed a new algorithm for simultaneous diagonalization of a set of symmetric matrices, where we combined: (i) a structural constraint to prevent the trivial solutions (ii) optimal (exact) line search procedure (iii) multiplicative updates based on natural gradient.

Extensions for further research would be to develop other "direction set methods", e.g. conjugate gradient, using the new optimal line search procedure. Additionally, a scale-invariant target criterion would better reflect the BSS-related optimization requirement. Such a modification to the criterion, along with the implied adaptation of the algorithm, are also subject of our future research.

Acknowledgement AZ and KRM acknowledge partly funding by the EU PASCAL network (IST-2002506778).

References

1. Cardoso, J.F., Souloumiac, A.: Blind beamforming for non Gaussian signals. IEE - Proceedings -F **140** (1993) 362–370
2. Belouchrani, A., Abed-Meraim, K., Cardoso, J.F., Moulines, E.: A blind source separation technique using second-order statistics. IEEE Trans. Signal Processing **45** (1997) 434–444
3. Ziehe, A., Nolte, G., Curio, G., Müller, K.R.: OFI: Optimal filtering algorithms for source separation. In: Proc. ICA2000, Helsinki, Finland (2000) 127–132
4. Belouchrani, A., Amin, M.G.: Blind source separation based on time-frequency signal representations. IEEE Trans. Signal Processing **46** (1998) 2888–2897
5. Yeredor, A.: Blind source separation via the second characteristic function. Signal Processing **80** (2000) 897–902
6. Murata, N., Ikeda, S., Ziehe, A.: An approach to blind source separation based on temporal structure of speech signals. Neurocomputing **41** (2001) 1–24
7. Rahbar, K., Reilly, J.P., Manton, J.H.: Blind identification of MIMO FIR systems driven by quasistationary sources using second-order statistics: A frequency domain approach. IEEE Trans. Signal Processing **52** (2004) 406–417
8. Cardoso, J.F., Souloumiac, A.: Jacobi angles for simultaneous diagonalization. SIAM Journal on Matrix Analysis and Applications **17** (1996) 161–164
9. Cardoso, J.F.: On the performance of orthogonal source separation algorithms. Proceedings of EUSIPCO'94 (1994) 776–779
10. Yeredor, A.: Non-orthogonal joint diagonalization in the least-squares sense with application in blind source separation. IEEE Trans. Signal Processing **50** (2002) 1545–1553
11. Ziehe, A., Laskov, P., Müller, K.R., Nolte, G.: A linear least-squares algorithm for joint diagonalization. Proceedings ICA2003 (2003) 469–474
12. Pham, D.T.: Joint approximate diagonalization of positive definite matrices. SIAM J. on Matrix Anal. and Appl. **22** (2001) 1136–1152
13. Amari, S.I., Douglas, S.: Why natural gradient. ICASSP'98 **2** (1998) 1213–1216
14. Joho, M., Mathis, H.: Joint diagonalization of correlation matrices by using gradient methods with application to blind signal separation. In: Proc. of IEEE Sensor Array and Multichannel Signal Processing Workshop SAM. (2002) 273–277
15. Amari, S., Cichocki, A., Yang, H.H.: A new learning algorithm for blind source separation. In Touretzky, D.S., Mozer, M.C., Hasselmo, M.E., eds.: Advances in Neural Information Processing Systems. Volume 8. MIT Press (1996) 757–763

BSS, Classification and Pixel Demixing

Albert Bijaoui[1], Danielle Nuzillard[2], and Tanusree Deb Barma[1]

[1] Cassiopée laboratory, UMR 6202, Côte d'Azur Observatory
BP 4229, 06304 Nice Cedex 4, France
bijaoui@obs-nice.fr

[2] LAM, UFR Sciences, Moulin de la Housse
51687 Reims cedex 2, France
Danielle.nuzillard@univ-reims.fr

Abstract. In the framework of the analysis of remote sensing images, the pixel mixture is a difficult task to solve. As it is considered that a mixture of pure elements is observed, it is necessary to identify them and to determine their proportions. Thus we associate statistical methods of Blind Source Separation (BSS) to complementary techniques of classification. Our purpose is developed and illustrated through an application on images for which a ground analysis was carried out. A comparison between a statistical approach and a clustering one is performed. Even if the BSS approach does not provide the classes associated to the ground analysis, it allows us to refind these classes from a simple learning.

1 Blind Sources Separation and Multispectral Images

The emissivity of a light source depends on the wavelength according to a law variable with their nature. The multispectral imaging makes possible to identify its physical nature if the number of channels is sufficient. So space observations of the Earth are carried out with a growing number of spectral channels, leading to more and more accurate maps of the physical components. Nevertheless the multispectral analysis is limited by the pixel mixture. Indeed the pixel value can result from the mixture of several physical components [6]. This fact is as true as the ground resolution is low.

This situation is close to the astronomical one for which the intensity of a pixel can result from a mixture coming from various physical sources. An example is the case of the observation of the Cosmological Microwave Background (CMB). Several physical, galactical and extragalactical sources are superimposed on this background whose statistical properties contain an essential information over the first moments of the Universe. Important work on the separations, blind or not, of the corresponding sources were carried out [3].

In previous work, we showed [7] that the exploitation of blind methods on Hubble Space Telescope (HST) images of the 3C120 radiosource allow one to enhance the various physical components. This led us to examine the case of remote sensing images, for which an independent ground analysis can be established, contrary to astronomical images.

Thus, one supposes that it exists pure elements which can be differentiated by their spectral distribution. Each of them has a specific map with different statistical properties. We expect from this separation that it determines the spectral distribution of each element and their proportion map. Pure elements are designed in remote sensing literature as *endmember*.

Then, one admits a linear model of mixture, it means that the presence of a source does not influence the emissivity of the other sources even so strictly speaking, this assumption is not exact. For example the presence of water modifies the spectral reflectance of close elements, a humid element having not a reflectance equal to a combination of the water and the pure element ones, taking into account their proportions.

The application of blind methods is based on statistical properties of the distribution of the pure elements. The considered property depends on the method selected to separate the sources. In the case of *Independent Component Analysis* (ICA), it deals with the probability densities of the proportions. For the *Second Order Blind Identification* (SOBI), cross-correlations between the shifted sources are taken into account [2].

The following equation is considered:

$$x_k = \sum_{i=1}^{n} a_{ik} s_i + n_k \quad k = 1, ..., p \qquad (1)$$

where x_k indicate the images, a_{ik} the spectral distributions of the unknown proportions, called also the sources s_i, and n_k the noise associated to x_k images. Solving this system consists in determining the set of the spectral distributions and the unknown sources. It is clearly an ill-posed mathematical problem.

The Karhunen-Loève expansion, associated to the *Principal Component Analysis* (PCA), constitutes the first approach. This one is largely used for the multispectral analysis of astronomical or remote sensing images and its limits are also well known. The resulting sources are not generally associated to physical pure elements.

Many methods of BSS were proposed, specific ones were applied to remote sensing images, especially for identifying the pure elements [1]. Within the framework of the analysis of the radiosource 3C120 [7], it appeared that estimated sources are nearly always identical either from the methods based on high order properties or from the other ones which take into account spatial cross-correlations. So, in this communication, a comparison between different statistical approaches is not described. Our goal, as shown in Fig. 1, is the analysis of the relation between BSS and clustering algorithms, through an application on a set of images for which a ground analysis was given in Fig. 2.

2 A BSS Application to Remote Sensing Images

2.1 Sources, Classes and Pure Elements

A set of nine images of CASI aerial observations were used (Fig. 2). These observations were done by the GSTB (*Groupement Scientifique de Télédétection*

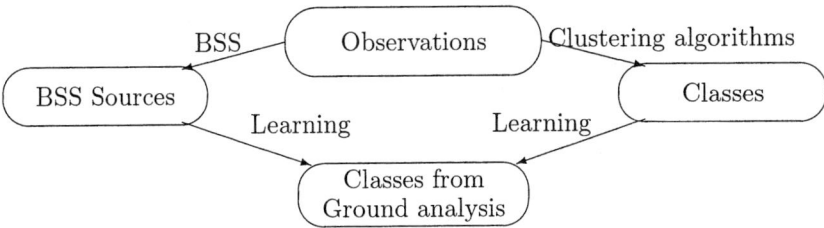

Fig. 1. Relation between BSS and classification.

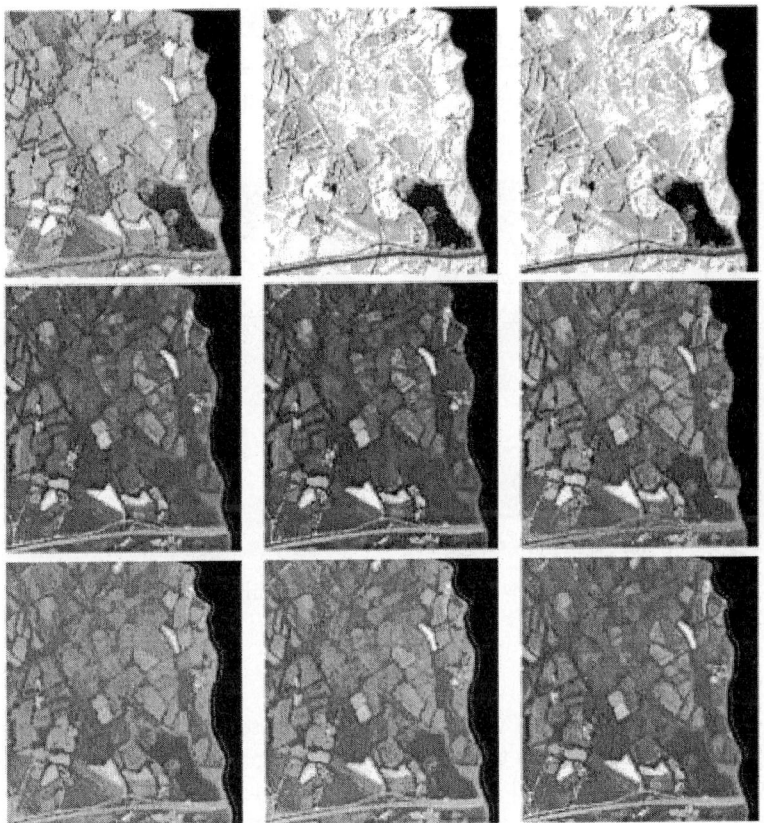

Fig. 2. Processed CASI images. These images were provided by the GSTB thanks to Pr. K. Chehdi.

de Bretagne) with a high ground resolution (0.5 m) and they were distributed thanks to Pr. K. Chehdi. A ground analysis of this area was provided with the images in Fig. 3 from which seventeen classes were identified.

A first BSS analysis was carried out on the raw images providing very noisy sources. Then since BSS algorithms are generally sensitive to the noise, the observations were denoised before the separation as shown in Fig. 4.

Fig. 3. The ground analysis provided with the images.

Fig. 4. The first CASI image without and with denoising. The denoising algorithm is based on a wavelet transform.

The algorithm JADE [4] was applied to separate the sources. Compared to other ICA algorithms, it has the advantage of being much faster in convergence. Clearly, the sources obtained from the separation in Fig. 5 do not correspond to the classes indicated by the ground analysis in Fig. 3. It is essential to couple several sources in order to be able to retrieve an identified class.

The ground analysis used as reference was done by specialists taking into account a knowledge which was not contained in the pixel statistics. It is not surprising that the relation between the sources and the classes is not dual.

Let us come back to the pixel demixing, a BSS technique proposes a set of proportions s_k and a set of spectral distributions (a_{ik}). It is possible to constraint their positivity in order to get an available physical solution. The spectral distribution is obtained without any information on the nature of the sources. Thus the results are clearly not related to any pure physical element. It is possible to interpret the BSS as a pixel demixing only with an extended meaning.

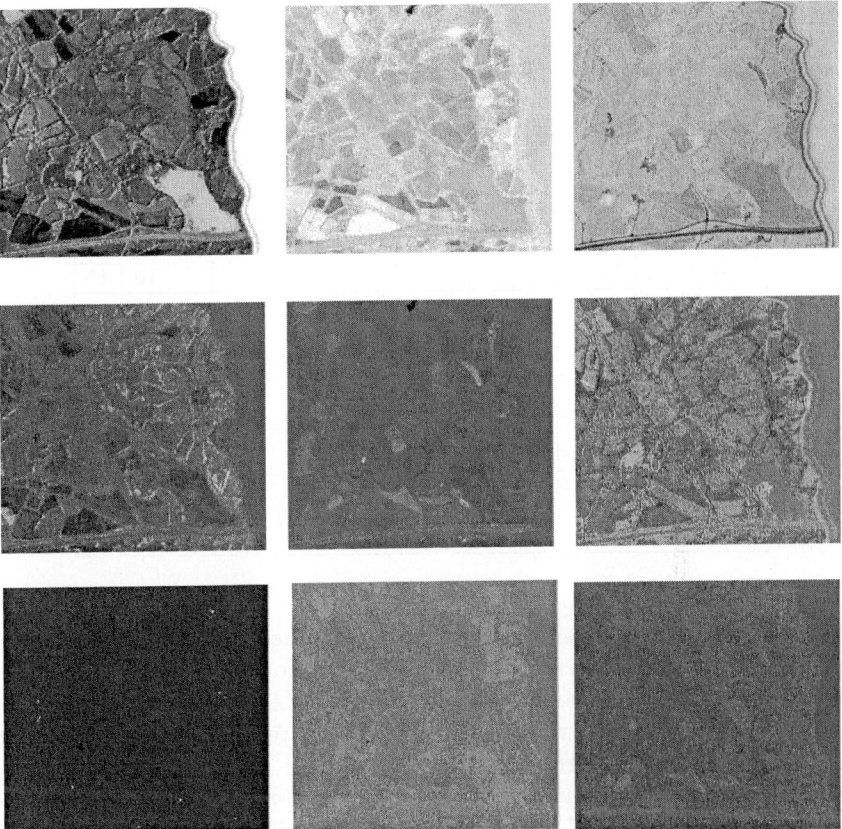

Fig. 5. The sources obtained by applying JADE algorithm on the denoised images.

2.2 Classes Derived from the BSS

The BSS sources are examined by a decreasing energy. For each one, its histogram is determined, allowing one to estimate two thresholds V1 and V2 such that 30% of values are smaller than V1 and 30% greater than V2. If the pixels of a class identified by the interpretation have mostly a value little than V1 the symbol (-) is introduced in a table. This symbol is (+) if the value is the most often greater then V2. If a symbol is placed, it means that the source is discriminant for the given class, but a same source may be associated to several classes (Table 1). A decision tree is deduced by taking into account each source by a decreasing energy order.

Then all pixels are converted into class indexes using this decision tree. The resulting images in Fig. 6 show well the various classes providing by the ground analysis. The residual noise is due to the fact that a discrete classification is done.

It is noticeable that an available classification can be determine from the sources (obtained with BSS) after a simple learning. We failed to retrieve the

Table 1. One determines the pixel values $V1$ and $V2$ corresponding to values 0.3 and 0.7 of the corresponding distribution function. This makes it possible to distribute the pixels in three categories according to whether the intensity is lower than $V1$ (-), higher than $V2$ (+) or lies between the two thresholds. Each column corresponds to a class identified from the ground analysis. Each line is a source obtained from the JADE algorithm. It can be noted that source 7 carries out any information on the classes, while class 10 is not recognized from any source.

	Classes from the ground analysis										
Classes from JADE	1	2	3	4	5	6	7	8	9	10	11
Class 1	-		-	+	+						
Class 2	+	-				-		-			+
Class 3	-			-	-						
Class 4		-	+	-		+		+	+		
Class 5					+						
Class 6					+			+			
Class 7											
Class 8			+			-	+	-			
Class 9								+			

defined classes by a similar method applied on the original images. Each of them is not sufficiently discriminant and the confusion cannot be clarified using other ones. Similarly, the sources obtained from PCA, *i.e.* the Karhunen-Loève expansion, are not also discriminant. The resulting tree is more complicated and the class noise is too important.

The spectral distributions obtained by a blind algorithm cannot be considered as the ones of real physical elements. Nevertheless BSS allows one to carry out sources which correspond to a mixing a few elements. So that, by a supervised method, it is possible to get an available classification. At this step, BSS methods can be seen as an auxiliary tool which simplifies a further classification.

Our conclusion is in agreement with Farah et al.'s one [5] who applied also JADE on remote sensing images and claimed that the resulting sources were more pertinent for a further data fusion.

2.3 Sources Derived from a Classification

In an usual way, pixel classification is made from clustering algorithms. Many of them were proposed for remote sensing images. Here, the classical k-means algorithm was applied by using its implementation in the ENVI software of RSI. The best results were obtained with only six classes (see in Fig. 6).

Rather similar results are displayed in Fig. 6, but while 11 classes were deduced from the BSS and the learning, only 6 classes are displayed leading to a more uniform image.

Each class can be considered as a pure element. It is natural to search for sources associated to them. Taking into account equation (1) the inverse problem has to be solved knowing the mixing matrix, $[a_{ik}]$. Many algorithms can be

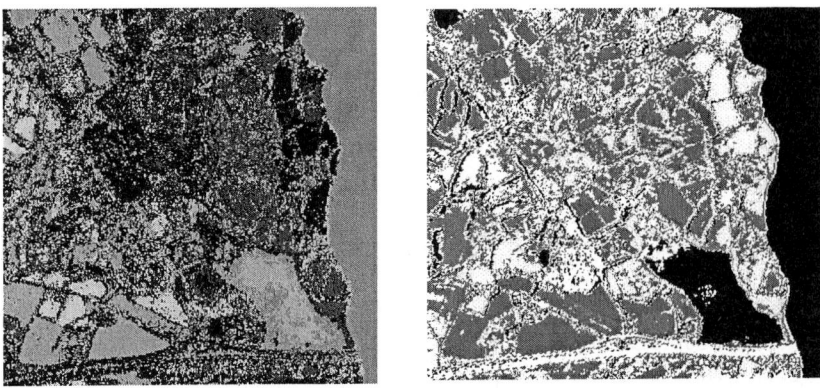

Fig. 6. On the left, classification obtained by using the rules deduced from JADE sources. On the right, classification deduced from the application of k-means.

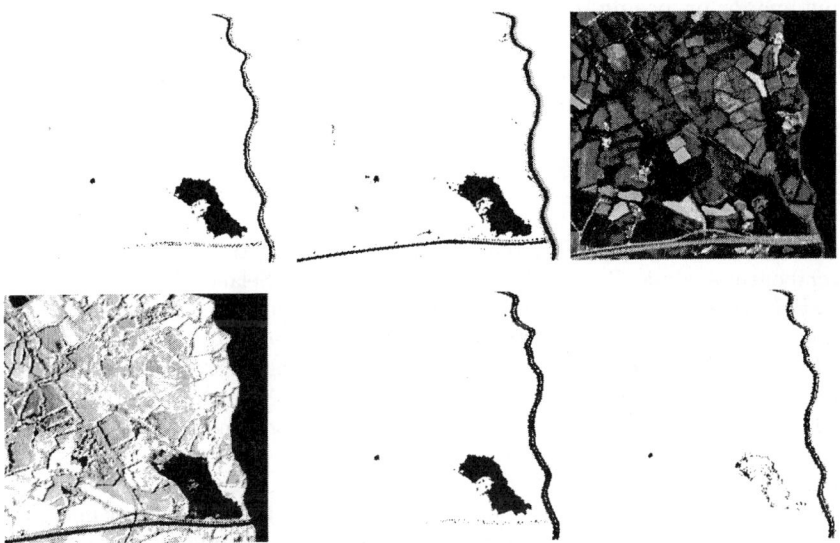

Fig. 7. The sources obtained from the k-means algorithm.

proposed to solve this classical problem. Here, we applied an inversion using the steepest descent to obtain the sources plotted in Fig. 7.

It can be noted that the resulting sources from k-means algorithm are very different from the ones obtained by JADE. The information seems to be only distributed in two sources. At this level, classes determined from the ground analysis cannot be retrieved by learning.

3 Conclusion

The application of BSS methods does not make possible to directly retrieve the images of the classes provided by the ground analysis. It appears that the statistical hypotheses are not sufficient for obtaining an available solution for a physical point of view. Different BSS approaches were tested, even if the source maps can vary from one algorithm to another one, the sources never correspond to pure element maps.

However, we pointed out that it was possible to find a good classification with a raw algorithm of pixel categorization.

As it was showed, an inverse way allows us to search for pure elements by a clustering method (k-means for example). Therefore, the images of percentages can be calculated by inversion. Thus, the blind separation from sources is obtained by means of a classification, whose the advantage is to separate whatever the number of images p for the number of sources n. In the current case where n is greater than p, this inverse problem is ill-posed and a linear inversion, with a demixing matrix, is not the only one. We have now under examination non linear solutions, using matching pursuit algorithms.

Nevertheless, we demonstrated that a BSS algorithm is better than a clustering one for proposing sources allowing one to retrieve classes provided from a ground analysis.

Further development may involve the implementation of a classification rule from various image data and ground analysis.

References

1. J. Bayliss, J.A. Gualtieri, R. Cromp. Analyzing hyperspectral data with independent component analysis. Proc. SPIE AIPR workshop, J.M. Selander ed., 9, SPIE (1997).
2. A. Belouchrani, K. Abed-Meraim, J.F. Cardoso, E. Moulines. A blind source separation technique using second-order statistics. IEEE Signal Proc., 45, (1997), 434-444.
3. J.F. Cardoso, J. Delabrouille, G. Patanchon: Independent Component Analysis of the Cosmic Microwave Background. Proc. of ICA'03, Nara, Japan (2003).
4. J.F. Cardoso, A. Souloumiac. Blind beamforming for non-Gaussian signals. IEE proceedings-F, 40, (1993), 362-370.
5. I.R. Farah, M.S. Naceur, M. Ben Ahmed, M.R. Boussema. Blind separation of sources based on Independent Component Analysis for the extraction of information from satellite images. IEEE conf. on Signal Processing and Information Technology, Le Caire, (2001).
6. N. Keshava, J.F. Mustard. Spectral unmixing, IEEE Signal Processing Magazine, 19 (2002), 44-57.
7. D. Nuzillard and A. Bijaoui. Blind Source Separation and Analysis of multispectral astronomical images, Astronomy and Astrophysics, suppl. Ser., vol 147 (2000), 129-138.

Blind Identification of Complex Under-Determined Mixtures*

Pierre Comon and Myriam Rajih

Lab. I3S, Algorithms/Euclide/B, BP.121, 2000 route des Lucioles
F-06903, Sophia-Antipolis cedex, France
{comon,rajih}@i3s.unice.fr
http://www.i3s.unice.fr

Abstract. Linear Mixtures of independent random variables (the so-called sources) are sometimes referred to as Under-Determined Mixtures (UDM) when the number of sources exceeds the dimension of the observation space. The algorithm proposed is able to identify algebraically a complex mixture of complex sources. It improves an algorithm proposed by the authors for mixtures received on a single sensor, also based on characteristic functions. Computer simulations demonstrate the ability of the algorithm to identify mixtures with typically 3 complex sources received on 2 sensors.

1 Introduction

In the present framework, a $P \times N$ linear mixture of N independent sources is observed:

$$\boldsymbol{x} = \boldsymbol{A}\,\boldsymbol{s} \tag{1}$$

where $\boldsymbol{x}$ is the column vector formed of $x_p = \sum_n A_{pn} s_n$, $\boldsymbol{A}$ denotes the $P \times N$ mixing matrix, and $\boldsymbol{s}$ the source column vector. When $P \geq N$, the mixture is said to be over-determined, whereas in the case we are interested in, namely $P < N$, the mixture is referred to as *under-determined*. If there exists now a large literature on Over-Determined Mixtures (ODM), much less attention has been drawn on Under-Determined Mixtures (UDM).

Under particular hypotheses, UDM can be sometimes deflated to ODM, for instance with the help of sparse decompositions in overcomplete bases [1]. On the contrary, we shall focus our attention to UDM that cannot be deflated. In the Statistics community, the first basic theorems can be traced back to the fifties, and can be found in [2] for instance. The Blind Identification of UDM's can be viewed as a problem of Factor Analysis, in which the number of factors exceeds the dimension [3]; as such, it has been addressed in the seventies, but under restricting assumptions [4] [5]; the decomposition is then known as PARAFAC. In the Signal Processing community, the problem has been addressed only ten years ago [7] [8]. Several approaches are possible, under various assumptions [9] [3] [10] [11] [12]. This will be briefly surveyed in the next section.

---

* This work has been supported in part by the European Network of Excellence PASCAL no.506778 (www.pascal-network.org).

2 Identifiability

We are interested in the Blind Identification (BI) of mixing matrix A; if the solution is unique, then A is *identifiable*. But we may also want to uniquely determine source distributions [13] [14]. It turns out that a unique solution for A indeed does not always yield a unique set of source distributions. Under hypothesis **H1** for instance, this holds true only for ODM. Uniqueness should be understood throughout this paper up to a permutation among the sources, and up to a scale factor; because of this inherent indeterminacy, we shall rather talk about *essentially unique* solutions. It is then useful to introduce the following hypotheses:

H1 the columns of A are pairwise linearly independent.
H2 source distributions are unknown and non Gaussian
H3 the number N of sources is known
H4 the characteristic function of x does not vanish
H5 for a given order $r > 2$, all source marginal cumulants of order r are unknown but finite, and it is known that at most one of them is null
H6 source cumulants are all known up to some order r.
H7 source distributions are known, discrete, and indecomposable

Assumption **H1** is not restrictive; in fact, if two columns i and j of A are proportional, then we can add sources s_i and s_j to form a new source, still independent from the others, and model (1) holds with merely $N-1$ sources instead of N.

Assumptions **H1** and **H2** together yield the unicity of A if it is known to be invertible [2, pp.89-90]. But this cannot be the case when $N > P$. Different instances of the problem can be obtained, depending of the hypotheses assumed. For instance:

P1 under **H1**, **H2**, and **H3**, A can be shown to be essentially unique [2, pp.311-313].
P2 under **H1**, **H2**, and **H4**, A and the N source distributions p_{s_n} are essentially unique, provided [2, pp.470-471]:

$$N \leq P(P+1)/2 \qquad (2)$$

P3 under assumptions **H1**, **H3**, and **H6**, A and the N source distributions p_{s_n} are essentially unique, provided the condition below holds true [13]

$$N \leq \binom{P+r}{r+1} \qquad (3)$$

If in addition sources are complex and non circular at order r, then the bound in the right hand side can be made larger [9].

P5 if sources and mixture are real, then under assumptions **H1**, **H3**, and **H5**, A is essentially unique if the number of sources is small enough [4] [5], viz:

$$2N \leq r(P-1) + 1 \qquad (4)$$

P6 under **H1** and **H7**, then A is essentially unique, and for any finite P, there is no upper bound on N, except for rare ambiguous mixtures [15].

For instance, if $(N, P) = (3, 2)$, then (2) holds true, as well as (3) for $r = 4$, but not (4). This is why PARAFAC methods are considered to be restrictive, even if they are recognized to be useful for large P.

3 Mixture Received on a Single Complex Sensor

Our concern is to solve the BI problem **P1**, and we suppose that the corresponding identifiability conditions are verified. Taleb proposed in [11] an algorithm for the BI of real mixtures of N independent real source signals received on $P = 2$ sensors. The algorithm uses the joint second characteristic function of the 2 sensors, $\psi_x(u, v)$. More precisely, (1) can be rewritten as:

$$\begin{aligned} x_1 &= a_1 s_1 + a_2 s_2 + ... + a_N s_N \\ x_2 &= b_1 s_1 + b_2 s_2 + ... + b_N s_N \end{aligned} \quad (5)$$

and the joint second characteristic function of x_1 and x_2 can be written as: $\psi_x(u,v) = \log E[\exp(iux_1 + ivx_2)]$, $(u,v) \in \Omega$ where Ω is the largest subset of $\mathbb{R}^2$ containing the origin and where the characteristic function of the pair (x_1, x_2) does not vanish. As sources are independent, one may write [11] [6]:

$$\psi_x(u,v) = \sum_{n=1}^{N} \psi_{s_n}(a_n u + b_n v) \quad (6)$$

A complex mixture of N complex sources received on P sensors can be viewed as a particular real mixture of $2N$ real sources received on $2P$ sensors, provided sources have independent real and imaginary parts. This will be subsequently assumed. Thus, for $1 \times N$ complex mixtures, it is possible to use this algorithm, but an appropriate association procedure is necessary in order to group the relevant real and imaginary parts together; this will not be described here for reasons of space, but further details may be found in [6].

4 Two Complex Sensors

When the complex mixture is observed in dimension 2, one may still separate real and imaginary parts to carry out the computations, but by doing so, we form 4 real measurements, which makes the problem much more complicated, because a homogeneous polynomial in more than 2 variables cannot be rooted as a polynomial in 1 variable.

For a complex variable x, denote by $\bar{x}$ and $\tilde{x}$ the real and imaginary parts of x, respectively. Then, (5) becomes:

$$\begin{aligned} \bar{x}_1 &= \sum_{n=1}^{N}(\bar{a}_n \bar{s}_n - \tilde{a}_n \tilde{s}_n), \; \tilde{x}_1 = \sum_{n=1}^{N}(\tilde{a}_n \bar{s}_n + \bar{a}_n \tilde{s}_n) \\ \bar{x}_2 &= \sum_{n=1}^{N}(\bar{b}_n \bar{s}_n - \tilde{b}_n \tilde{s}_n), \; \tilde{x}_2 = \sum_{n=1}^{N}(\tilde{b}_n \bar{s}_n + \bar{b}_n \tilde{s}_n) \end{aligned} \quad (7)$$

Suppose again that the real and imaginary parts of the sources are independent, which is satisfied for numerous basic modulations, as QPSK. Then, the joint second characteristic function of $(\bar{x}_1, \tilde{x}_1, \bar{x}_2, \tilde{x}_2)$ can be written, for $(u_1, v_1, u_2, v_2) \in \Omega$:

$$\psi_{\bar{x}_1,\tilde{x}_1,\bar{x}_2,\tilde{x}_2}(u_1, v_1, u_2, v_2) = \sum_{n=1}^{N} \psi_{\bar{s}_n}(\bar{a}_n u_1 + \tilde{a}_n v_1 + \bar{b}_n u_2 + \tilde{b}_n v_2) \\ + \psi_{\tilde{s}_n}(\bar{a}_n v_1 - \tilde{a}_n u_1 + \bar{b}_n v_2 - \tilde{b}_n u_2) \tag{8}$$

Define differential operator D_n as:

$$D_n = -(\bar{b}_n^2 + \tilde{b}_n^2)\partial v_1 + (\tilde{a}_n \bar{b}_n - \bar{a}_n \tilde{b}_n)\partial u_2 + (\bar{a}_n \bar{b}_n + \tilde{a}_n \tilde{b}_n)\partial v_2 \tag{9}$$

By applying D_n we remove the nth term of the sum in (8). When applying all the D_n's, $n = 1, ..N$, we obtain:

$$\sum_{n=0}^{N}\sum_{k=0}^{n} d_{nk} \frac{\partial^N \psi_{\bar{x}_1,\tilde{x}_1,\bar{x}_2,\tilde{x}_2}(u_1, v_1, u_2, v_2)}{\partial v_1^{N-n} \partial u_2^{n-k} \partial v_2^k} = 0 \tag{10}$$

By replacing $\psi_{\bar{x}_1,\tilde{x}_1,\bar{x}_2,\tilde{x}_2}(u_1, v_1, u_2, v_2)$ by its expression (8) we get:

$$\sum_{j=1}^{N}\Big(\sum_{n=0}^{N}\sum_{k=0}^{n}\big(d_{nk}\tilde{a}_j^{N-n}\bar{b}_j^{n-k}\tilde{b}_j^k \psi_{\bar{s}_j}^{(n)}(\bar{a}_j u_1 + \tilde{a}_j v_1 + \bar{b}_j u_2 + \tilde{b}_j v_2) \\ + d_{nk}\bar{a}_j^{N-n}(-\tilde{b}_j)^{n-k}\bar{b}_j^k \psi_{\tilde{s}_j}^{(n)}(\bar{a}_j v_1 - \tilde{a}_j u_1 + \bar{b}_j v_2 - \tilde{b}_j u_2)\big)\Big) = 0$$

which implies:

$$\sum_{n=0}^{N}\sum_{k=0}^{n} d_{nk}\tilde{a}_j^{N-n}\bar{b}_j^{n-k}\tilde{b}_j^k = 0 \tag{11}$$

$$\sum_{n=0}^{N}\sum_{k=0}^{n} d_{nk}\bar{a}_j^{N-n}(-\tilde{b}_j)^{n-k}\bar{b}_j^k = 0 \tag{12}$$

$\forall j = 1, ..N$. In a similar manner, one can define 3 other differential operators Q_n, R_n and T_n as follows:

$$Q_n = -(\bar{a}_n \bar{b}_n + \tilde{a}_n \tilde{b}_n)\partial u_1 + (\bar{a}_n \tilde{b}_n - \tilde{a}_n \bar{b}_n)\partial v_1 + (\bar{a}_n^2 + \tilde{a}_n^2)\partial u_2$$

$$R_n = -(\bar{b}_n^2 + \tilde{b}_n^2)\partial u_1 + (\bar{a}_n \bar{b}_n + \tilde{a}_n \tilde{b}_n)\partial u_2 + (\bar{a}_n \tilde{b}_n - \tilde{a}_n \bar{b}_n)\partial v_2$$

$$T_n = (\tilde{a}_n \bar{b}_n - \bar{a}_n \tilde{b}_n)\partial u_1 - (\bar{a}_n \bar{b}_n + \tilde{a}_n \tilde{b}_n)\partial v_1 + (\bar{a}_n^2 + \tilde{a}_n^2)\partial v_2$$

When applying all the Q_n's, R_n's, and T_n's, $n = 1, ..N$ we obtain:

$$\sum_{n=0}^{N}\sum_{k=0}^{n} q_{nk}\frac{\partial^N \psi_{\bar{x}_1,\tilde{x}_1,\bar{x}_2,\tilde{x}_2}(u_1, v_1, u_2, v_2)}{\partial u_1^{N-n} \partial v_1^{n-k} \partial u_2^k} = 0 \tag{13}$$

$$\sum_{n=0}^{N}\sum_{k=0}^{n} r_{nk}\frac{\partial^N \psi_{\bar{x}_1,\tilde{x}_1,\bar{x}_2,\tilde{x}_2}(u_1, v_1, u_2, v_2)}{\partial u_1^{N-n} \partial u_2^{n-k} \partial v_2^k} = 0 \tag{14}$$

$$\sum_{n=0}^{N}\sum_{k=0}^{n} t_{nk} \frac{\partial^N \psi_{\bar{x}_1,\tilde{x}_1,\bar{x}_2,\tilde{x}_2}(u_1,v_1,u_2,v_2)}{\partial u_1^{N-n} \partial v_1^{n-k} \partial v_2^k} = 0 \qquad (15)$$

As for D_n, each of the previous operators gives 2 equations in 3 unknowns, $\forall j = 1,..,n$:

$$\sum_{n=0}^{N}\sum_{k=0}^{n} q_{nk} \bar{a}_j^{N-n} \tilde{a}_j^{n-k} \bar{b}_j^k = 0 \qquad (16)$$

$$\sum_{n=0}^{N}\sum_{k=0}^{n} q_{nk} (-\tilde{a}_j)^{N-n} \bar{a}_j^{n-k} (-\tilde{b}_j)^k = 0 \qquad (17)$$

$$\sum_{n=0}^{N}\sum_{k=0}^{n} r_{nk} \bar{a}_j^{N-n} \bar{b}_j^{n-k} \tilde{b}_j^k = 0 \qquad (18)$$

$$\sum_{n=0}^{N}\sum_{k=0}^{n} r_{nk} (-\tilde{a}_j)^{N-n} (-\tilde{b}_j)^{n-k} \bar{b}_j^k = 0 \qquad (19)$$

$$\sum_{n=0}^{N}\sum_{k=0}^{n} t_{nk} \bar{a}_j^{N-n} \tilde{a}_j^{n-k} \tilde{b}_j^k = 0 \qquad (20)$$

$$\sum_{n=0}^{N}\sum_{k=0}^{n} t_{nk} (-\tilde{a}_j)^{N-n} \bar{a}_j^{n-k} \bar{b}_j^k = 0 \qquad (21)$$

As a consequence, by computing directly the joint second characteristic function of $(\bar{x}_1, \tilde{x}_1, \bar{x}_2, \tilde{x}_2)$, we end up with an over-determined system of 8 homogeneous equations of the form: $\sum_n \sum_k \alpha_{nk} x^{N-n} y^{n-k} z^k = 0$. Our contribution in this section was to show that it was possible to obtain $4N$ pairs of polynomial equations in which only 3 unknowns are involved. A solution of such a system is studied in the next section.

5 An Algebraic Solution

In order to solve the previous system of 8 equations, we solve separately the four systems of two equations in 3 unknowns each. Equations (16) and (21) constitute system I, (20) and (17) system II, (18) and (12) system III, (11) and (19) system IV. The first step consists of estimating the coefficients $\hat{\mathbf{d}} = [d_{00}, d_{10}, d_{11}, ...d_{N0}, ..., d_{NN}]^T$, $\hat{\mathbf{q}}$, $\hat{\mathbf{r}}$, and $\hat{\mathbf{t}}$ (defined the same way as $\hat{\mathbf{d}}$) from (10), (13), (14), and (15). To do so, we proceed like in section 3 by selecting K points $(u_1^k, v_1^k, u_2^k, v_2^k) \in \Omega$ and estimating for each of these points all the Nth order derivatives. This allows to form four $K \times (N+1)(N+2)/2$ matrices $\mathbf{H_d}$, $\mathbf{H_q}$, $\mathbf{H_r}$, and $\mathbf{H_t}$. Due to the lack of space we only define $\mathbf{H_d}$, matrices $\mathbf{H_q}$, $\mathbf{H_r}$, and $\mathbf{H_t}$ being defined similarly. Also denote by ψ_i the value $\psi_x(u_1^i, v_1^i, u_2^i, v_2^i)$. Then, $\mathbf{H_d}$ is defined as:

$$\mathbf{H_d} = \begin{pmatrix} \frac{\partial^N \psi_1}{\partial v_1^N} & \cdots & \frac{\partial^N \psi_1}{\partial v_1^{N-n} \partial u_2^{n-k} \partial v_2^k} & \cdots & \frac{\partial^N \psi_1}{\partial v_2^N} \\ \frac{\partial^N \psi_2}{\partial v_1^N} & \cdots & \frac{\partial^N \psi_2}{\partial v_1^{N-n} \partial u_2^{n-k} \partial v_2^k} & \cdots & \frac{\partial^N \psi_2}{\partial v_2^N} \\ \vdots & \vdots & \vdots & \vdots & \vdots \\ \frac{\partial^N \psi_K}{\partial v_1^N} & \cdots & \frac{\partial^N \psi_K}{\partial v_1^{N-n} \partial u_2^{n-k} \partial v_2^k} & \cdots & \frac{\partial^N \psi_K}{\partial v_2^N} \end{pmatrix}$$

with $k = 0, .., n$ and $n = 0, .., N$. In order to solve $\mathbf{H_d d} = 0$, we compute the right singular vector $\hat{\mathbf{d}}$, associated with the smallest singular value of $\mathbf{H_d}$.

Once we have $\hat{\mathbf{d}}, \hat{\mathbf{q}}, \hat{\mathbf{r}},$ and $\hat{\mathbf{t}}$, the second step consists of writing the 8 homogeneous equations in three unknowns as equations of two unknowns. System I for example is equivalent to:

$$\sum_{n=0}^{N} \sum_{k=0}^{n} q_{nk} \frac{\tilde{a}_j}{\bar{a}_j}^{n-k} \frac{\bar{b}_j}{\bar{a}_j}^{k} = 0, \quad \sum_{n=0}^{N} \sum_{k=0}^{n} r_{nk} (-1)^{N-n} \frac{\tilde{a}_j}{\bar{a}_j}^{N-n} \frac{\bar{b}_j}{\bar{a}_j}^{k} = 0$$

so that $\frac{\tilde{a}_j}{\bar{a}_j}$ and $\frac{\bar{b}_j}{\bar{a}_j}$ (for $j = 1, .., N$) are solutions of the system:

$$\sum_{n=0}^{N} \sum_{k=0}^{n} q_{nk} x^{n-k} y^k = 0, \quad \sum_{n=0}^{N} \sum_{k=0}^{n} r_{nk} (-1)^{N-n} x^{N-n} y^k = 0$$

which can be solved by using a resultant method; more precisely, two polynomials in a single variable are rooted in order to get the set of solutions. We end up with N pairs $(\frac{\tilde{a}_j}{\bar{a}_j}, \frac{\bar{b}_j}{\bar{a}_j})$, $j = 1, .., N$. Solving the system II the same way, we end up with N pairs $(\frac{\tilde{a}_j}{\bar{a}_j}, \frac{\tilde{b}_j}{\bar{a}_j})$, but in an order different from the order of the first N pairs obtained from system I. To restore this order and associate the two groups of pairs we use the common coefficient $\frac{\tilde{a}_j}{\bar{a}_j}$. We get $(\frac{\tilde{a}_j}{\bar{a}_j}, \frac{\bar{b}_j}{\bar{a}_j}, \frac{\tilde{b}_j}{\bar{a}_j})$, and by taking $\bar{a}_j = 1, \forall j = 1, .., N$ we obtain the coefficients $(1, \tilde{a}_j, \bar{b}_j, \tilde{b}_j)$, for $j = 1, .., N$, and then an estimate of the complex $2 \times N$ channel matrix.

In the latter approach, we divided by $\bar{a}_j$. To preserve symmetry among the four coefficients, we repeat the same steps by dividing by $\tilde{a}_j$ in systems I and II, by $\bar{b}_j$ in systems III and IV, and then by $\tilde{b}_j$ again in systems III and IV. We obtain respectively $(\bar{a}_j, 1, \bar{b}_j, \tilde{b}_j)$, $(\bar{a}_j, \tilde{a}_j, 1, \tilde{b}_j)$, and $(\bar{a}_j, \tilde{a}_j, \bar{b}_j, 1)$, for $j = 1, .., N$.

Now, it remains to select the best solution. Every time a solution is computed, it is always possible to search for a perturbation in the coefficients of the original polynomial system for which the solution is exact. To do this, it suffices to solve a linear system in the LS sense; in fact, the original polynomial system is linear in its coefficients. The chosen perturbation vector is that of minimum norm, yielded by an SVD. The solution eventually retained is the one yielding the perturbation of smallest norm.

6 Computer Results

Sources that have been generated are i.i.d QPSK, and have therefore independent real and imaginary parts. The theoretical expressions of the requested derivatives of ψ_x have been computed as a function of successive derivatives of ϕ_x. This rather cumbersome calculation has been carried out by Maple once for all. Then, one can eventually replace the terms involved in the latter expressions by their sample estimates. As an example, in the real case, $\frac{\partial \psi}{\partial u} = \frac{\partial \phi}{\partial u}\frac{1}{\phi}$, and a sample estimate of $\partial^{k+\ell+m}\phi/\partial v_1^k \partial u_2^\ell \partial v_2^m$ is:

$$i^{k+\ell+m} \frac{1}{T} \sum_{t=1}^{T} \tilde{x}_1(t)^k \bar{x}_2(t)^\ell \tilde{x}_2(t)^m \exp\{i(\tilde{x}_1(t)v_1 + \bar{x}_2(t)u_2 + \tilde{x}_2(t)v_2)\} \quad (22)$$

6.1 1×3 Complex Mixture of 3 Complex Sources

A numerical algorithm dedicated to mixtures received on a single complex sensor may be found in [6] and performs quite well; results are not reported here for reasons of space. Nevertheless, for known reasons, it performs poorly with two sensors, hence the present contribution.

6.2 2×3 Complex Mixture of 3 Complex Sources

The mixture used for these simulations was:

$$\boldsymbol{A} = \begin{bmatrix} \cos\pi/12 + i\sin\pi/12, & \cos\pi/6 + i\sin\pi/6, & \cos\pi/3 + i\sin\pi/3 \\ \cos 2\pi/5 + i\sin 2\pi/5, & \cos\pi/5 + i\sin\pi/5, & \cos\pi/10 + i\sin\pi/10 \end{bmatrix}$$

The performance criterion used in this section is that described in [16] for $P \times N$ mixtures (computing the optimal scale and permutation ambiguities would be too computationally costly). Results are reported in figure 1 and point out a satisfactory behavior for high SNR.

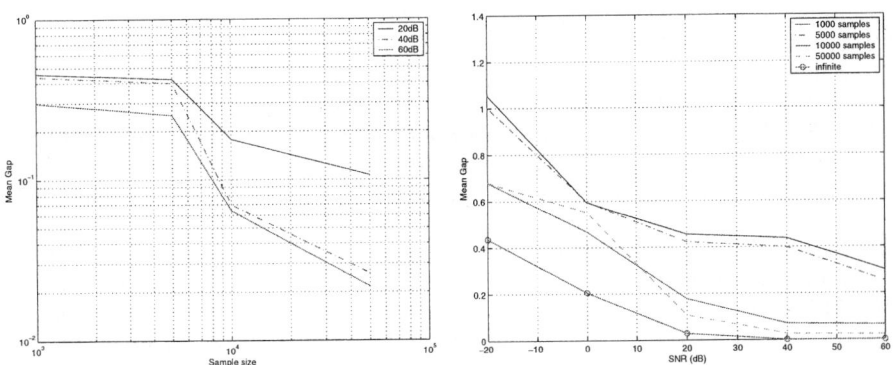

Fig. 1. Average gap obtained as a function of SNR (right) or sample size (left), for 2×3 complex mixtures.

7 Concluding Remarks

Our BI algorithm based on the joint characteristic function of observations is attractive because (i) theoretically not limited by the number of sources, and (ii) of algebraic nature, which means that it terminates after a finite number of operations without convergence problems. However, this method is not very robust to Gaussian noise nor short samples.

References

1. LEWICKI, M.S., SEJNOWSKI, T.J.: Learning overcomplete representations. Neural Computation **12** (2000) 337–365
2. KAGAN, A.M., LINNIK, Y.V., RAO, C.R.: Characterization Problems in Mathematical Statistics. Prob. Math. Stat. Wiley, New York (1973)
3. COMON, P.: Tensor decompositions. In McWhirter, J.G., Proudler, I.K., eds.: Mathematics in Signal Processing V. Clarendon Press, Oxford, UK (2002) 1–24
4. KRUSKAL, J.B.: Three-way arrays: Rank and uniqueness of trilinear decompositions. Linear Algebra and Applications **18** (1977) 95–138
5. SIDIROPOULOS, N.D., BRO, R.: On the uniqueness of multilinear decomposition of N-way arrays. Jour. Chemo. **14** (2000) 229–239
6. COMON, P., RAJIH, M.: Blind Identification of Under-Determined Complex Mixtures of Independent Sources. IEEE SAM Conf., Barcelona (2004)
7. CARDOSO, J.F.: Super-symmetric decomposition of the fourth-order cumulant tensor. Blind identification of more sources than sensors. In: ICASSP, Toronto (1991) 3109–3112
8. CAO, X.R., LIU, R.W.: General approach to blind source separation. IEEE Trans. Sig. Proc. **44** (1996) 562–570
9. COMON, P.: Blind identification and source separation in 2x3 under-determined mixtures. IEEE Trans. Signal Processing (2004) 11–22
10. TALEB, A., JUTTEN, C.: On underdetermined source separation. In: ICASSP99, Phoenix, Arizona (1999)
11. TALEB, A.: An algorithm for the blind identification of N independent signals with 2 sensors. In: Sixth International Symposium on Signal Processing and its Applications (ISSPA'01), Kuala-Lumpur, Malaysia, IEEE (2001)
12. de LATHAUWER, L., de MOOR, B., VANDEWALLE, J.: ICA techniques for more sources than sensors. In: Sixth Sig. Proc. Workshop on Higher Order Statistics, Caesarea, Israel (1999)
13. SZEKELY, G.J.: Identifiability of distributions of independent random variables by linear combinations and moments. Sankhya, Indian Jour. Stat. **62** (2000) 193–202
14. ERIKSSON, J., KOIVUNEN, V.: Identifiability and separability of linear ICA models revisited. In: 4th Int. Symp. ICA, Nara, Japan (2003)
15. GRELLIER, O., COMON, P.: Performance of blind discrete source separation. In: Eusipco. Volume IV., Rhodes, Greece (1998) 2061–2064 invited session.
16. ALBERA, L., FERREOL, A., et al.: Sixth order blind identification of under-determined mixtures - BIRTH. In: 4th Int. Symp. ICA, Nara, Japan (2003)

Blind Separation of Heavy-Tailed Signals Using Normalized Statistics

Mohamed Sahmoudi[1], Karim Abed-Meraim[2], and Messaoud Benidir[1]

[1] LSS, SUPELEC, 91192, Gif-sur-Yvette, France
sahmoudi@lss.supelec.fr
[2] TSI Dept., Telecom-Paris, 75634, Paris cedex, France
abed@tsi.enst.fr

Abstract. This paper introduces a new approach for the blind separation (BS) of heavy tailed signals that can be modeled by real-valued symmetric α-stable (SαS) processes. As the second and higher order moments of the latter are infinite, we propose to use normalized statistics of the observation to achieve the BS of the sources. More precisely, we show that the considered normalized statistics are convergent (i.e., take finite values) and have the appropriate structure that allows for the use of standard BS techniques based on second and higher order cumulants.

1 Introduction

By the generalized central limit theorem, the α-stable laws are the *only* class of distributions that can be the limiting distribution for sums of i.i.d. random variables [12]. Therefore, many signals are impulsive in nature or after certain pre-processing, e.g. using for example wavelet transform, and can be modeled as α-stable processes [8, 1]. Unlike most statistical models, the α-stable distributions except the Gaussian have infinite second and higher order moments. Consequently, standard blind source separation (BSS) methods would be inadequate in this case as most of them are based on second or higher order statistics [4]. An alternative solution consists in achieving the BSS using fractional lower order moments (FLOM) [9]. Other solutions exist in the literature based on the maximum likelihood principle [7], the spectral measure [7], the signal truncature [10] and the order statistics [10], respectively.

In this paper we propose a new approach for the BS of heavy tailed sources using normalized statistics (NS). It is first shown that suitably normalized second- and fourth-order cumulants exist and have the appropriate structure for the BSS. This is a similar result to those of [11] in the ARMA stable context. Then, for extracting α-*stable* source signals from their observed mixtures one can use any standard procedure based on second- or forth-order cumulants. This BSS method has several advantages over the existing ones that are discussed in the sequel. Simulation-based comparisons with the minimum dispersion (MD) criterion based method in [9] are also provided.

2 Stable Distributions

We introduces briefly the stable distribution family and some of its statistical properties. In the literature, stable distributions are defined in several equivalent ways and *stable* refers to the fact that the density function is a closed class under addition. Continuing this property, the following definition is used [12]:

Definition 1. *A random variable X has a stable distribution if for any positive number a and b, there is a positive number c_1 and a real nomber c_2 such that*

$$aX_1 + bX_2 \stackrel{d}{=} c_1 X + c_2 \tag{1}$$

where X_1 and X_2 are independent copies of X and $\stackrel{d}{=}$ means equality in distribution. If $c_2 = 0$ the random variable is strictly stable.

Definition 1 defines the stable distribution based on the stability property, but it does not give a concrete way to parameterize it. The most convenient parameterization of stable distributions is through the characteristic function.

Proposition 1. *A univariate distribution function is stable if and only if its characteristic function is of the form*

$$\varphi(t) = \exp\{j\mu t - \gamma \mid t \mid^\alpha [1 + j\beta sign(t)\omega(t,\alpha)]\} \tag{2}$$

where

$$\omega(t,\alpha) = \begin{cases} \tan \frac{\alpha\pi}{2} & , \text{ if } \alpha \neq 1 \\ 2/\pi \log \mid t \mid, & \text{ if } \alpha = 1 \end{cases} \tag{3}$$

and $-\infty < \mu < \infty$, $\gamma > 0$, $0 < \alpha \leq 2$, $-1 \leq \beta \leq 1$.

We will denote the stable distributions by $S_\alpha(\gamma, \beta, \mu)$. Thus, a stable distribution is completely determined by four parameters. α, the *characteristic exponent*, is a measure of the thickness of the tails of the distribution; β is the *symmetry parameter*: $\beta = 0$ corresponds to a distribution that is symmetric around μ, in which case the distribution is called Symmetric $\alpha-$ Stable ($S\alpha S$), μ is the *location parameter* and for $S\alpha S$ distributions it is the symmetry axis; γ is the *dispersion* and is similar to the variance of the Gaussian distribution in the sense that it is a measure of the deviation around the mean. A $S\alpha S$ is called standard distribution if $\gamma = 1$.

The $S\alpha S$ pdfs present several similarities to the Gaussian pdf: They are smooth and bell-shaped, satisfy the stability property, and naturally arise via a generalized from of the limit theorem. However, they also differ from the Gaussian pdf in several significant ways. For example, the $S\alpha S$ pdfs have sharper maxima than the Gaussian pdf and algebraic (inverse power) tails in contrast to the exponential tails of the Gaussian pdf.

Property 1. If $X \sim S\alpha S(\gamma, 0, \mu)$ and $\alpha \neq 2$, then

$$\lim_{t \to \infty} t^\alpha Pr(\mid X \mid > t) = \gamma C_\alpha \tag{4}$$

where C_α is a constant that depends on α only.

For this reason, the pth-order moments of the $S\alpha S$ pdfs are finite only for $0 < p < \alpha$ (except for the limiting case of $\alpha = 2$). As a result, for α strictly less than 2 (i.e., $0 < \alpha < 2$), α−stable random variables have infinite variance and more generally infinite moments for orders larger than α.

3   Problem Formulation

Consider m mutually independent signals whose $n \geq m$ linear combinations are observed : $\mathbf{x}(t) = \mathbf{A}\mathbf{s}(t)$, where $\mathbf{s}(t) = [s_1(t), \cdots, s_m(t)]^T$ is the $m \times 1$ real valued *impulsive source vector* and $\mathbf{A}$ is a $n \times m$ full rank *mixing matrix*. The source signals $s_i(t), i = 1, \cdots, m$ are assumed to be mutually independent, zero-mean, symmetric α−stable processes. The purpose of blind source separation is to find a separating matrix, i.e., an $m \times n$ matrix $\mathbf{B}$ such that $\mathbf{z}(t) = \mathbf{B}\mathbf{x}(t)$ is an estimate of the source signals. Before proceeding, note that there are two inherent ambiguities in the problem. First there is no way of knowing the original labeling of the sources, and second, exchanging a fixed scalar factor between a source signal and the corresponding column of $\mathbf{A}$ does not affect the observations. It follows that the best that one can do is to determine $\mathbf{B}$ (or equivalently the matrix $\mathbf{A}$) up to a permutation and scaling of its columns. Therefore, $\mathbf{B}$ is said to be a separating matrix if $\mathbf{B}\mathbf{x}(t) = \mathbf{P}\mathbf{\Lambda}\mathbf{s}(t)$ where $\mathbf{P}$ is a permutation matrix and $\mathbf{\Lambda}$ a non-singular diagonal matrix. Similarly, blind identification of $\mathbf{A}$ is understood as the determination of a matrix equal to $\mathbf{A}$ up to a permutation matrix and a non-singular diagonal matrix.

4   Normalized Statistics

4.1   Normalized Moments

Thanks to the algebraic tail-behavior (property 1), we demonstrate here that the ratio of the k-th moments of two random $S\alpha S$ variables with $\alpha \neq 2$ converges to a finite value (even though the moments themselves are infinite). More precisely, we have the following theorem:

Theorem 1. *Let X_1 and X_2 be two $S\alpha S$ variables of dispersions γ_1 and γ_2 and pdfs $f_1(.)$ and $f_2(.)$, respectively. Then, for $k \geq \alpha$, we have*

$$\frac{E(|X_1|^k)}{E(|X_2|^k)} \triangleq \lim_{T \to \infty} \frac{\int_{-T}^{T} |x|^k f_1(x) dx}{\int_{-T}^{T} |u|^k f_2(u) du} = \frac{\gamma_1}{\gamma_2} \qquad (5)$$

Proof. Let R_k represents the above ratio, then due to the symmetric pdf of X_1 and X_2, we have

$$R_k \triangleq \frac{\int_{-T}^{T} |x|^k f_1(x) dx}{\int_{-T}^{T} |u|^k f_2(u) du} = \frac{\int_{0}^{T} x^k f_1(x) dx}{\int_{0}^{T} u^k f_2(u) du} \qquad (6)$$

Using integration by parts, we get

$$R_k = \frac{[-x^k(1-\Phi_1(x))]_0^T + k\int_0^T x^{k-1}(1-\Phi_1(x))dx}{[-u^k(1-\Phi_2(u))]_0^T + k\int_0^T u^{k-1}(1-\Phi_2(u))du} \qquad (7)$$

where $\Phi(.)$ denotes the cumulative function of the considered pdf. From property 1, we can observe that for any $S\alpha S$ cumulative function Φ, we have $(1-\Phi(x)) \sim \frac{C_\alpha}{2}\gamma x^{-\alpha}$ as $x \to \infty$. Then, as $T \to \infty$, R_k is equivalent to:

$$R_k \sim \frac{C_\alpha\gamma_1}{C_\alpha\gamma_2} \frac{[-x^{k-\alpha}]_0^T + k\int_0^T x^{k-1-\alpha}dx}{[-u^{k-\alpha}]_0^T + k\int_0^T u^{k-1-\alpha}du} \to \frac{\gamma_1}{\gamma_2} \qquad \diamond$$

Using a similar proof, one can demonstrate that the ratio of the square of the k-th moment to the 2k-th moment of a random $S\alpha S$ variable ($\alpha \neq 2$) converges to zero for $k > \alpha$. More precisely, we have the following theorem:

Theorem 2. *Let X be a $S\alpha S$ variable of dispersion γ and pdf $f(.)$. Then, for $k > \alpha$, we have:*

$$\frac{(E|X|^k)^2}{E|X|^{2k}} \triangleq \lim_{T\to\infty} \frac{(\int_{-T}^T |x|^k f(x)dx)^2}{\int_{-T}^T |x|^{2k} f(x)dx} = 0 \qquad (8)$$

4.2 Normalized Second and Forth Order Cumulants

Using above results, we can establish now that the normalized covariance matrix of the mixture signal converges to a finite valued matrix with the desired algebraic structure. We have established the following result:

Theorem 3. *Let $\mathbf{x}$ be an $S\alpha S$ vector given by $\mathbf{x} = \mathbf{As}$ ($\mathbf{s}$ being a vector of $S\alpha S$ independent random variables). Then the normalized covariance matrix of $\mathbf{x}$ satisfies:*

$$R(i,j) \triangleq \frac{Cum[x(i),x(j)]}{\sum_{k=1}^n Cum[x(k),x(k)]} = \sum_{k=1}^m d_k a_k(i) a_k(j)$$

or equivalently: $\quad \mathbf{R} = \mathbf{ADA}^T$ *where* $\mathbf{D} = diag(d_1,\cdots,d_m)$ *and*

$$d_i = \frac{\gamma_i}{\sum_{j=1}^m \gamma_j \parallel \mathbf{a}_j \parallel^2}$$

$\mathbf{a}_j$ *being the j-th column vector of $\mathbf{A}$.*

Similarly, the normalized quadri-covariance tensor [2] of the mixture signal converges to a finite valued tensor with the desired algebraic structure. We have established the following result:

Theorem 4. Let $\mathbf{x}$ be an $S\alpha S$ vector given by $\mathbf{x} = \mathbf{As}$ ($\mathbf{s}$ being a vector of $S\alpha S$ independent random variables). Then the normalized quadri-covariance tensor of $\mathbf{x}$ satisfies:

$$Q(i,j,k,l) \triangleq \frac{Cum[x(i),x(j),x(k),x(l)]}{\sum_{r=1}^{n} Cum[x(r),x(r),x(r),x(r)]}$$

$$= \sum_{r=1}^{m} \kappa_r a_r(i) a_r(j) a_r(k) a_r(l)$$

where

$$\kappa_i = \frac{\gamma_i}{\sum_{j=1}^{m} \gamma_j \parallel \mathbf{a}_j \parallel^4}$$

5 Blind Source Separation

Thanks to theorems 3 and 4, we can now use existing BSS methods based on 2nd and 4th order cumulants, e.g. [5, 2]. In this work, we have applied the JADE algorithm [3] to the normalized 2nd and 4th order cumulants of the observations. In summary, we describe the proposed algorithm which is referred to as the Robust-JADE by the following steps:

Step 1. Compute a whitening matrix $\hat{\mathbf{W}}$ from the normalized sample covariance $\hat{R}_x$ (that is estimated as the standard sample covariance matrix divided by its trace value).

Step 2. Compute the most significant eigenpairs (see [3] for more details) $\{\hat{\lambda}_r, \hat{\mathbf{M}}_r; 1 \leq r \leq m\}$ from the normalized sample 4th-order cumulants of the whitened process $\mathbf{z}(t) \triangleq \hat{\mathbf{W}} \mathbf{x}(t)$.

Step 3. Diagonalize jointly the set $\hat{\lambda}_r \hat{\mathbf{M}}_r; 1 \leq r \leq m$ by a unitary matrix $\hat{\mathbf{U}}$.

Step 4. Estimate $\mathbf{A}$ by $\hat{\mathbf{A}} = \hat{\mathbf{W}}^{\#} \hat{\mathbf{U}}$.

We provide here some remarks about the above separation method and discuss certain advantages of the use of normalized statistics.

- Based on theorem 2, the normalized 4-th order cumulants are equal to the normalized 4-th order moments of the $S\alpha S$ source mixture (recall here that for a real valued zero-mean random variable x, we have $cum(x,x,x,x) = E(x^4) - 3(E(x^2))^2$). In other words, for $S\alpha S$ sources, one can replace the 4-th order cumulants by the 4-th order moments of the mixture signal.
- One major advantage of the proposed method compared to the FLOM based methods is that no a priori knowledge or pre-estimation of source pdf parameters (in particular, the characteristic exponent α) is required. Consequently, the normalized-statistics based method is robust to modelization errors with respect the source's pdf.
- In the case where the sources are non-impulsive, the proposed method coincides with the standard one (in our case, with the JADE method). Indeed, because of the scaling indeterminacy, the normalization would have no effect in this case.

– Another advantage of the NS-based method is that it can easily be extended to the case where the sources are of different types: i.e., sources with different characteristic exponents or non-impulsive sources in presence of other impulsive ones. That can be done for example by using the above NS-based method in conjunction with a deflation technique [6]. Indeed, in that case, one can prove that the normalized statistics coincide with those of the mixture of the 'most impulsive' sources only (i.e the ones with the smallest characteristic exponent) which can be estimated first then removed (by deflation) to allow the estimation and separation of the other sources. This point is still under investigation and will be presented in details in future work.
– In this paper, we have established only the convergence of the 'exact' normalized statistics (expressed by the mathematical expectation). In fact, one can prove along the same lines of [11] that the sample estimates of the 2-nd and 4-th order cumulants converge in probability to the exact normalized statistics given by theorems 3 and 4.

6 Simulation Results

This section examines the statistical performances of the separation procedure. The numerical results presented below have been obtained in the following setting. The source signals are i.i.d. impulsive symmetric standard α-stable ($\beta = 0$, $\mu = 0$ and $\gamma = 1$). The number of sources is $m = 3$ and the number of observations is $n = 4$. The statistics are evaluated over 100 Monte-Carlo runs and the mixing matrix is generated randomly at each run. To measure the quality of source separation, we did use the generalized rejection level criterion defined as follows: If source k is the desired signal, the related generalized rejection level would be:

$$I_k \stackrel{\text{def}}{=} \frac{\gamma(\sum_{l \neq k} C_{kl} s_l)}{\gamma(C_{kk} s_k)} = \frac{\sum_{l \neq k} |C_{kl}|^\alpha \gamma_l}{|C_{kk}|^\alpha \gamma_k} \quad (9)$$

where $\gamma(x)$ (resp. γ_l) denotes the dispersion of an $S\alpha S$ random variable x (resp. source s_l) and $\mathbf{C} \stackrel{\text{def}}{=} \hat{\mathbf{A}}^{\#} \mathbf{A}$. Therefore, the averaged rejection level is given by

$$I_{perf} = \frac{1}{m} \sum_{i=1}^{m} I_i = \frac{1}{m} \sum_{i=1}^{m} \sum_{j \neq i} \frac{|C_{ij}|^\alpha \gamma_j}{|C_{ii}|^\alpha \gamma_i}.$$

The performances of the NS-based Robust-JADE method are compared with those of the MD (Minimum Dispersion) method in [9]. Figures 1, 2 and 3 present the mean rejection level versus the characteristic exponent α (the sample size is set to $N = 1000$ and the mixture is noise-free), versus the additive Gaussian noise power ($N = 1000$ and $\alpha = 1.5$), and versus the sample size ($\alpha = 1.5$ and the mixture is noise-free), respectively. We can observe a certain performance gain in favor of the Robust-JADE except for the noisy mixture case (note, however, that for the MD method the characteristic exponent α is not estimated but assumed to be exactly known).

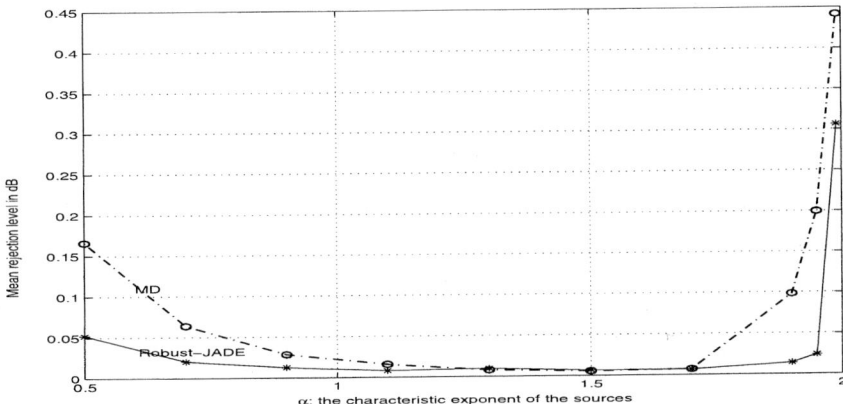

Fig. 1. Generalized mean rejection level versus the characteristic exponent α.

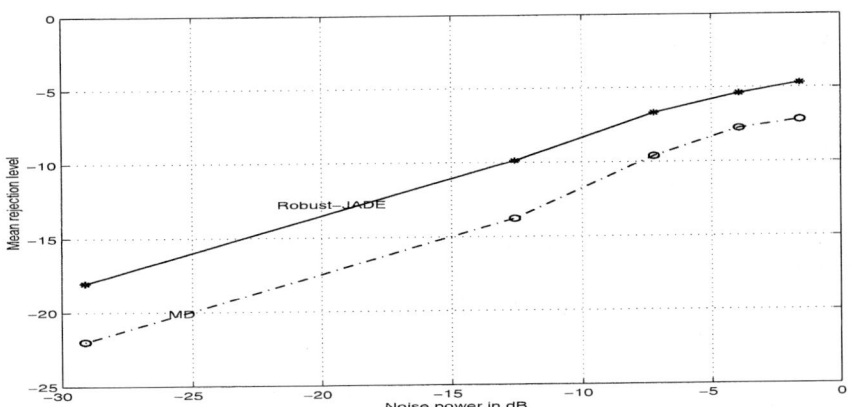

Fig. 2. Generalized mean rejection level versus the additive noise power.

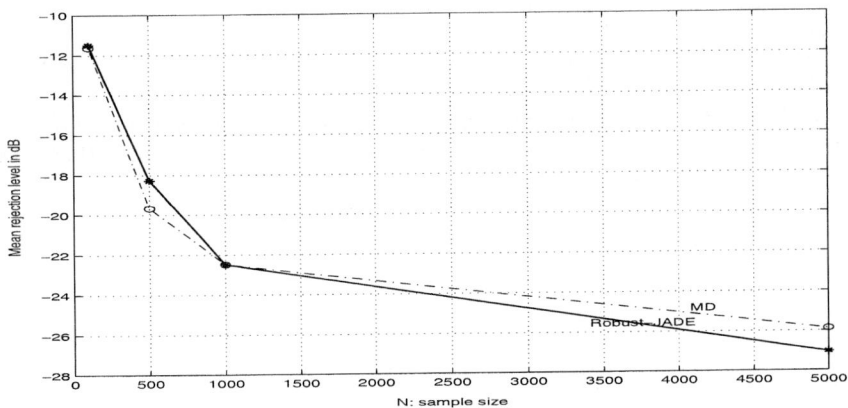

Fig. 3. Generalized mean rejection level versus the sample size.

7 Conclusions

A new NS-based blind separation method for impulsive source signals is introduced. The normalized 2-nd and 4-th order cumulants of the mixture signal are shown to be convergent to finite-valued matrices with the appropriate algebraic structure that is traditionally used in many 2-nd and higher order statistics based BSS methods. The advantages of the proposed method are discussed and a simulation based comparison with the MD method is provided to illustrate and assess its performances.

References

1. A. Achim, A. Bezerianos, P. Tsakalides, "Novel Bayesian Multiscale Method for Speckle Removal in Medical Ultrasound Images", *IEEE Tr. Med. Imag.*, Aug. 2001.
2. J. -F. Cardoso, "Super-symmetric Decomposition of the Fourth-Order Cumulant Tensor. Blind Identification of More Sources Than Sensors", *ICASSP-91*, Toronto, Canada, Apr. 14-17, 1991.
3. J.F. Cardoso and A. Souloumiac, "Blind beamforming for non-Gaussian signals", *Radar and Signal Processing, IEE Proceedings F*, pp. 362–370, Dec. 1993.
4. A. Cichocki and S. Amari, *Adaptive Blind Signal and Image Processing*, John Wiley & Sons, Ltd, Singapore 2002.
5. P. Comon, "Independent component analysis, a new concept?", *Signal Processing*, vol. 36, pp. 287–314, 1994.
6. N. Delfosse and Ph. Loubaton, "Adaptive separation of independent sources: a deflation approach", *Proc. ICASSP*, vol. IV , pp. 41–44, 1994.
7. P. Kidmose, "Blind Separation of Heavy Tail Signals", *Ph. D. Thesis, Technical University of Denmark*, Lyngby 2001.
8. C. L. Nikias and M. Shao, *Signal Processing with Alpha-Stable Distributions and Applications*, New York: John Wiley & Sons, 1995.
9. M. Sahmoudi, K. Abed-Meraim and M. Benidir, "Blind Separation of Instantaneous Mixtures of Impulsive alpha-stable sources", *3rd International Symposium on Image and Signal Processing and Analysis (ISISPA 2003)*, Rome, Italy, 2003.
10. Y. Shereshevski, "Blind Signal Separation of Heavy Tailed Sources", *M.Sc. thesis, Tel Aviv Uni.*, Mar. 2002.
11. A. Swami and B.M. Sadler, "On some detection and estimation problems in heavy-tailed noise", *Signal Processing*, pp. 1829–1846, 2002.
12. G. Samorodnitsky and M. S. Taqqu, *Stable Non-Gaussian Random Processes: Stochastic Models with Infinite Variance*, New York, NY: Chapman & Hall, 2000.

Blind Source Separation of Linear Mixtures with Singular Matrices

Pando Georgiev[1] and Fabian J. Theis[2]

[1] Laboratory for Advanced Brain Signal Processing, Brain Science Institute
The Institute for Physical and Chemical Research (RIKEN)
2-1, Hirosawa, Wako-shi, Saitama, 351-0198, Japan
[2] Institute of Biophysics, University of Regensburg
D-93040 Regensburg, Germany

Abstract. We consider the Blind Source Separation problem of linear mixtures with singular matrices and show that it can be solved if the sources are sufficiently sparse. More generally, we consider the problem of identifying the source matrix $\mathbf{S} \in \mathbb{R}^{n \times N}$ if a linear mixture $\mathbf{X} = \mathbf{AS}$ is known only, where $\mathbf{A} \in \mathbb{R}^{m \times n}, m \leqslant n$ and the rank of $\mathbf{A}$ is less than m. A sufficient condition for solving this problem is that the level of sparsity of $\mathbf{S}$ is bigger than $m - rank(\mathbf{A})$ in sense that the number of zeros in each column of $\mathbf{S}$ is bigger than $m - rank(\mathbf{A})$. We present algorithms for such identification and illustrate them by examples.

1 Introduction

One goal of the Blind Signal Separation (BSS) is the recovering of underlying source signals of some given set of observations obtained by an unknown linear mixture of the sources. BSS has potential applications in many different fields such as medical and biological data analysis, communications, audio and image processing, etc. In order to decompose the data set, different assumptions on the sources have to be made. The most common assumption nowadays is statistical independence of the sources, which leads to the field of *Independent Component Analysis* (ICA), see for instance [1], [5] and references therein. ICA is very successful in the linear *complete* case, when as many signals as underlying sources are observed, and the mixing matrix is non-singular. In [2] it is shown that the mixing matrix and the sources are identifiable except for permutation and scaling. In the *overcomplete* or *underdetermined* case, less observations than sources are given. It can be seen that still the mixing matrix can be recovered [3], but source identifiability does not hold. In order to approximatively detect the sources, additional requirements have to be made, usually sparsity of the sources. We refer to [6–9] and reference therein for some recent papers on sparsity and underdetermined ICA (   ).

Recently, we have shown in [4] that, based on sparsity alone, we can still detect both mixing matrix and sources uniquely (except for trivial indeterminacies) given sufficiently high sparsity of the sources (*Sparse Component Analysis, SCA*). We also proposed algorithms for reconstructing the mixing matrix and the sources.

2 Blind Source Separation Using Sparseness

Definition 1. *A vector* $\mathbf{v} \in \mathbf{R}^m$ *is said to be k-sparse if $\mathbf{v}$ has at least k zero entries. A matrix* $\mathbf{S} \in \mathbf{R}^{m \times n}$ *is said to be k-sparse if each column of it is k-sparse.*

The goal of *Blind Signal Separation* of level k (k-BSS) is to decompose a given m-dimensional random vector $\mathbf{X}$ into

$$\mathbf{X} = \mathbf{AS} \tag{1}$$

with a real $m \times n$-matrix $\mathbf{A}$ and an $n \times N$-dimensional k-sparse matrix $\mathbf{S}$. $\mathbf{S}$ is called the *source matrix*, $\mathbf{X}$ the *mixtures* and $\mathbf{A}$ the *mixing matrix*. We speak of *complete*, *overcomplete* or *undercomplete* k-BSS if $m = n$, $m < n$ or $m > n$ respectively.

Note that in contrast to the ICA model, the above problem is not translation invariant. However it is easy to see that if instead of $\mathbf{A}$ we choose an affine linear transformation, the translation constant can be determined from $\mathbf{X}$ only, as long as the sources are non-determined. Termed differently, this means that instead of assuming k-sparseness of the sources we could also assume that in any column of $\mathbf{S}$ only $n - k$ components are allowed to vary from a previously fixed constant (which can be different for each source).

In the following without loss of generality we will assume $m \leqslant n$: the undercomplete case can be reduced to the complete case by projection of $\mathbf{X}$.

The following theorem is a generalization of a similar one from [4]. Here, for illustrative purposes, we formulate the theorem for the case when the rank of $\mathbf{A}$ is $m - 1$, but its formulation in full generality is straightforward.

Theorem 1 (Matrix identifiability 1). *Assume that $\mathbf{X}$ satisfies (1) and*

1) every $m - 1$ columns of the matrix $\mathbf{A}$ are linearly independent; the indexes $\{1, ..., n\}$ are divided in two groups $\mathcal{N}_1$ and $\mathcal{N}_2$ such that

2) vectors from the group $\{\mathbf{S}_1 = \mathbf{S}(:, i) : i \in \mathcal{N}_1\}$ are sufficiently rich represented in the sense that for any index set of $n - m + 2$ elements $\subset \{1, ..., n\}$ there exist at least $m - 1$ vectors $\mathbf{s}_1, ..., \mathbf{s}_{m-1}$ from $\mathbf{S}_1$ (depending on i) such that each of them has zero elements in places with indexes in i[1] and there exists at least one subgroup of $\{\mathbf{s}_1, ..., \mathbf{s}_{m-1}\}$ consisting of $m - 2$ linearly independent elements;

3) the vectors from the group $\{\mathbf{X}(:, i), i \in \mathcal{N}_2\}$ have the property that no subset of $m - 1$ elements from them lie on a 2-codimensional subspace[2].

Then $\mathbf{A}$ is uniquely determined by $\mathbf{X}$ except for right-multiplication with permutation and scaling matrices, i.e. if $\mathbf{X} = \mathbf{AS} = \hat{\mathbf{A}}\hat{\mathbf{S}}$, then $\mathbf{A} = \hat{\mathbf{A}}\mathbf{PL}$ with a permutation matrix $\mathbf{P}$ and a nonsingular diagonal scaling matrix $\mathbf{L}$.

Proof. It is clear that any column $\mathbf{a}_j$ of the mixing matrix lies in the intersection of all $\binom{n-1}{m-3}$ 2-codimensional subspaces generated by those groups of columns of $\mathbf{A}$, in which $\mathbf{a}_j$ participates.

---

[1] i.e. the vectors $\mathbf{s}_1, ..., \mathbf{s}_{m-1}$ are $(n - m + 2)$-sparse.
[2] Subspace in $\mathbb{R}^m$ with dimension $m - 2$.

We will show that these 2-codimensional subspaces can be obtained by the columns $\{\mathbf{X}(:,\ ),\ \in \mathcal{N}_1\}$ under the condition of the theorem. Let $\mathcal{J}$ be the set of all subsets of $\{1,...,\ \}$ containing -2 elements and let $\in \mathcal{J}$. Note that $\mathcal{J}$ consists of $\binom{n}{m-2}$ elements. We will show that the 2-codimensional subspace (denoted by $_J$) spanned by the columns of $\mathbf{A}$ with indexes from can be obtained by some elements from $\{\mathbf{X}(:,\ ),\ \in \mathcal{N}_1\}$. By 2), there exist -1 indexes $\{\ _k\}_{k=1}^{m-1} \subset \mathcal{N}_1$ and -2 vectors from the group $\{\mathbf{S}(:,\ _k)\}_{k=1}^{m-1}$, which form a basis of the $(\ -2)$-dimensional coordinate subspace of $\mathbb{R}^n$ with zero coordinates given by the indexes $\{1,...,\ \} \setminus$. Because of the mixing model, vectors of the form

$$\mathbf{v}_k = \sum_{j \in J} (\ ,\ _k)\mathbf{a}_j, \quad = 1,...,\ -1,$$

belong to the group $\{\mathbf{X}(:,\ ):\ \in \mathcal{N}_1\}$. Now, applying condition 1) we obtain that there exists a subgroup of -2 vectors from $\{\mathbf{v}_k\}_{k=1}^{m-1}$ which are linearly independent. This implies that the vectors $\{\mathbf{v}_k\}_{k=1}^{m-1}$ will span the same 2-codimensional subspace $_J$. By 1) it follows that the 2-codimensional subspaces $_{J_1}$ and $_{J_2}$ are different, if the indexes $_1 \in \mathcal{J}$ and $_2 \in \mathcal{J}$ are different. By the above reasonings and by 3) it follows that if we cluster the columns of $\mathbf{X}$ in 2-codimensional subspaces containing more than -2 elements from the columns of $\mathbf{X}$, we will obtain $\binom{n}{m-2}$ unique 2-codimensional subspaces, containing all elements of $\{\mathbf{X}(:,\ ),\ \in \mathcal{N}_1\}$ and no elements from $\{\mathbf{X}(:,\ ),\ \in \mathcal{N}_2\}$. Now we cluster the 2-codimensional subspaces obtained in such a way in the smallest number of groups such that the intersection of all 2-codimensional subspaces in one group gives a single one-dimensional subspace. It is clear that such one-dimensional subspace will contain one column of the mixing matrix, the number of these groups is and each group consists of $\binom{n-1}{m-3}$ 2-codimensional subspaces.

In such a way we can identify the columns of the mixing matrix up to scaling and permutation. In other words, if $\mathbf{X} = \mathbf{AS} = \hat{\mathbf{A}}\hat{\mathbf{S}}$, then $\mathbf{A} = \hat{\mathbf{A}}\mathbf{PL}$ with a permutation matrix $\mathbf{P}$ and a nonsingular diagonal scaling matrix $\mathbf{L}$. ∎

In a similar way we can prove the following generalization of the above theorem.

Theorem 2 (Matrix identifiability 2). *Assume that $\mathbf{X}$ satisfies (1) and*

1) every -1 columns of the matrix $\mathbf{A}$ are linearly independent;
the indexes $\{1,...,\ \}$ are divided in two groups $\mathcal{N}_1$ and $\mathcal{N}_2$ such that

2) vectors from the group $\mathbf{S}_1 = \{\mathbf{S}(:,\ )\},\ \in \mathcal{N}_1$ are sufficiently rich represented in the sense that for any index set of $-\ +2$ elements $\subset \{1,...,\ \}$ there exist $_I \geqslant$ vectors $\mathbf{s}_1,...,\mathbf{s}_{N_I}$ from $\mathbf{S}_1$ (depending on) such that each of them has zero elements in places with indexes in and there exists a subset of $\{\mathbf{s}_1,...,\mathbf{s}_{N_I}\}$ containing -2 linearly independent elements;

3) the vectors from the group $\{\mathbf{X}(:,\ ),\ \in \mathcal{N}_2\}$ have the property that at most $\min\{\ _{I_1},...,\ _{I_p}\} - 1$ of them lie on a common 2-codimensional subspace, where

$\{\mathcal{I}_1, ..., \mathcal{I}_p\}$ is the set of all subsets of $\{1, ..., m\}$ with $m - n + 2$ elements and $p = \binom{n}{m-2}$.

Then $\mathbf{A}$ is uniquely determined by $\mathbf{X}$ except for right-multiplication with permutation and scaling matrices, i.e. if $\mathbf{X} = \mathbf{A}\mathbf{S} = \hat{\mathbf{A}}\hat{\mathbf{S}}$, then $\mathbf{A} = \hat{\mathbf{A}}\mathbf{P}\mathbf{L}$ with a permutation matrix $\mathbf{P}$ and a nonsingular diagonal scaling matrix $\mathbf{L}$.

The proof of Theorem 1 gives the idea for the matrix identification algorithm.

Algorithm for identification of the mixing matrix (under assumption of Theorems 1 or 2)

1) Cluster the columns $\{\mathbf{X}(:, i) : i \in \mathcal{N}_1\}$ in $\binom{n}{m-2}$ groups $\mathcal{H}_k$, $k = 1, ..., \binom{n}{m-2}$ such that the span of the elements of each group $\mathcal{H}_k$ produces one 2-codimensional subspace and these 2-codimensional subspaces are different.

2) Calculate any basis of the orthogonal complement of each of these 2-codimensional subspaces.

3) Cluster these bases in the smallest number of groups $\mathcal{G}_j$, $j = 1, ..., n$ (which gives the number of sources n) such that the bases of the 2-codimensional subspaces in each group $\mathcal{G}_j$ are orthogonal to a common (unit) vector, say $\mathbf{a}_j$. The vectors $\mathbf{a}_j$, $j = 1, ..., n$ are estimations of the columns of the mixing matrix (up to permutation and scaling).

Remark 1. The above algorithm is quite general and allows different realizations. Below we propose another method for matrix identification, based on PCA.

The above theorems shows that we can recover the mixing matrix from the mixtures uniquely, up to permutation and scaling of the columns. The next theorem shows that in this case also the sources $\{\mathbf{S}(:, i) : i \in \mathcal{N}_1\}$ can be recovered uniquely (up to a measure zero of the "bad" data points with respect to the "good" data points).

3 Identification of Sources

The following theorem is generalization of those in [4] and the proof is the same.

Theorem 3. (Uniqueness) *Let $\mathcal{H}$ be the set of all $\mathbf{x} \in \mathbb{R}^m$ such that the linear system $\mathbf{A}\mathbf{s} = \mathbf{x}$ has a solution with at least $n - k + 1$ zero components ($k \geq 1$). If any $m - k$ columns of $\mathbf{A}$ are linearly independent, then there exists a subset $\mathcal{H}_0 \subset \mathcal{H}$ with measure zero with respect to $\mathcal{H}$, such that for every $\mathbf{x} \in \mathcal{H} \setminus \mathcal{H}_0$ this system has no other solution with this property.*

From Theorem 3 it follows that the sources are uniquely identifiable generically, i.e. up to a set with a measure zero, if they compose a matrix which is ($n - k + 1$)-sparse, and the mixing matrix is known. Below we present an algorithm based on the observation in Theorem 3.

Source Recovery Algorithm:

1. Identify the the set of -codimensional subspaces $\mathcal{H}$ produced by taking the linear hull of every subsets of the columns of $\mathbf{A}$ with − elements;
2. Repeat for = 1 to :
   2.1. Identify the space ∈ $\mathcal{H}$ containing $\mathbf{x}_i := \mathbf{X}(:,)$, or, in practical situation with presence of noise, identify the one to which the distance from $\mathbf{x}_i$ is minimal and project $\mathbf{x}_i$ onto to $\tilde{\mathbf{x}}_i$;
   2.2. if is produced by the linear hull of column vectors $\mathbf{a}_{i_1}, ..., \mathbf{a}_{i_{m-k}}$, then find coefficients $\mathbf{L}_{i,j}$ such that

$$\tilde{\mathbf{x}}_i = \sum_{j=1}^{m-k} \mathbf{L}_{i,j} \mathbf{a}_{i_j}.$$

These coefficients are uniquely determined if $\tilde{\mathbf{x}}_i$ doesn't belong to the set $\mathcal{H}_0$ with measure zero with respect to to $\mathcal{H}$ (see Theorem 3);
   2.3. Construct the solution $\mathbf{s}_i = \mathbf{S}(:,)$: it contains $\mathbf{L}_{i,j}$ in the place $_j$ for = 1, ..., − , the other its components are zero.

4 Computer Simulation Example

We created artificially four source signals, sparse of level 2, i.e. each column of the source matrix contains at least 2 zeros (shown in Figure 1). They are mixed with a square singular matrix $\mathbf{A}$ such that any 3 columns of it are linearly independent:

$$\mathbf{A} = \begin{pmatrix} -0.4326 & -1.1465 & 0.3273 & -1.2517 \\ -1.6656 & 1.1909 & 0.1746 & -0.3000 \\ 0.1253 & 1.1892 & -0.1867 & 1.1278 \\ 0.2877 & -0.0376 & 0.7258 & 0.9758 \end{pmatrix}.$$

Since the mixing matrix is singular, the data lie on a hyperplane in $\mathbb{R}^4$, i.e. in a 3-dimensional subspace. We apply PCA: $\mathbf{X}_1 = \mathbf{V}\mathbf{X}$, where $\mathbf{V} = \mathbf{L}_3^{-\frac{1}{2}}\mathbf{U}_3^T$, $\mathbf{U}_3$ is the matrix of those eigenvectors of $\mathbf{X}\mathbf{X}^T$, which correspond to the positive eigenvalues of it, $\mathbf{X}\mathbf{X}^T = \mathbf{U}\mathbf{L}\mathbf{U}^T$ (by Singular Value Decomposition) and $\mathbf{L}_3$ is a diagonal matrix which diagonal elements are the positive diagonal elements of $\mathbf{L}$. So we obtain an overcomplete BSS problem $\mathbf{X}_1 = \mathbf{A}_1\mathbf{S}$ with a (3 × 4) mixing matrix $\mathbf{A}_1 = \mathbf{V}\mathbf{A}$. After that we apply the matrix identification algorithm and the source recovery algorithm from [4] (described in this paper in a more general case). The mixed sources are shown in Fig.2, the recovered sources by our algorithm are shown in Fig. 3. For a comparison we show the result of applying ICA and BSS algorithms: Fast ICA algorithm, JADE and SOBI (see Figures 4, 5 and 6 respectively). For a numerical evaluation of our algorithm, we compare the matrix $\mathbf{B}$, estimation of $\mathbf{A}_1$, produced by our algorithms (which has normalized columns - with norm 1) and the matrix $\mathbf{A}_1 = \mathbf{VA}$ after normalization of the columns:

$$\mathbf{V} = \begin{pmatrix} 0.0036 & 0.0049 & -0.0011 & 0.0128 \\ -0.0031 & -0.0069 & 0.0019 & 0.0037 \\ -0.0027 & 0.0021 & 0.0027 & 0.0002 \end{pmatrix},$$

$$\mathbf{B} = \begin{pmatrix} -0.3995 & 0.3667 & -0.9971 & -0.0059 \\ 0.9088 & 0.8309 & -0.0099 & -0.2779 \\ -0.1201 & 0.4184 & 0.0760 & 0.9606 \end{pmatrix},$$

$$\mathbf{A}_2 = \begin{pmatrix} -0.3995 & -0.0059 & 0.9971 & 0.3667 \\ 0.9088 & -0.2779 & 0.0099 & 0.8309 \\ -0.1201 & 0.9606 & -0.0760 & 0.4184 \end{pmatrix}.$$

The normalized matrix $\mathbf{A}_2$=normalized($\mathbf{A}_1$) is different only by permutation and sign of columns from $\mathbf{B}$, which shows the good performance of our method.

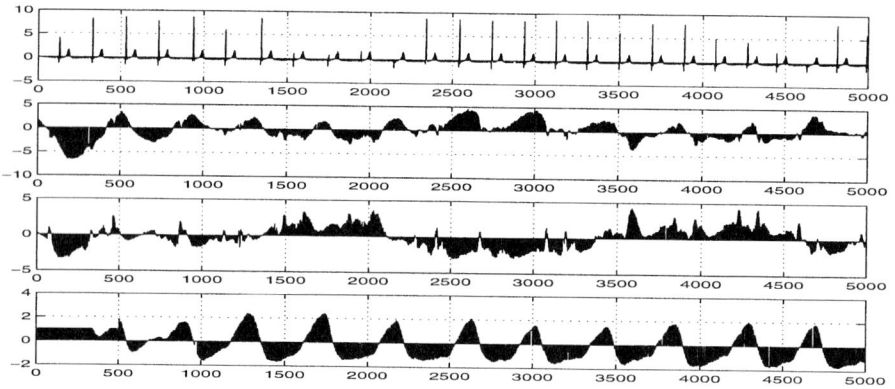

Fig. 1. Original source signals.

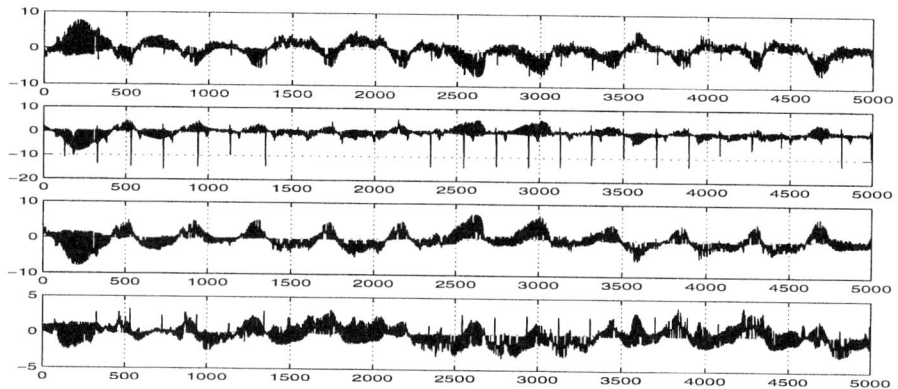

Fig. 2. Mixed signals.

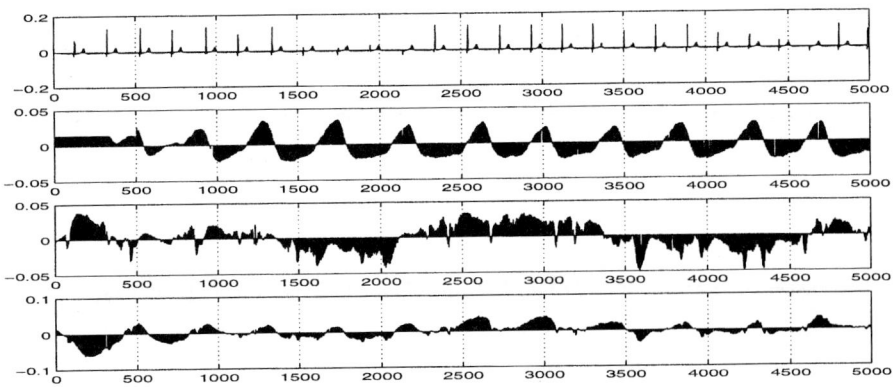

Fig. 3. Estimated sources by our algorithms.

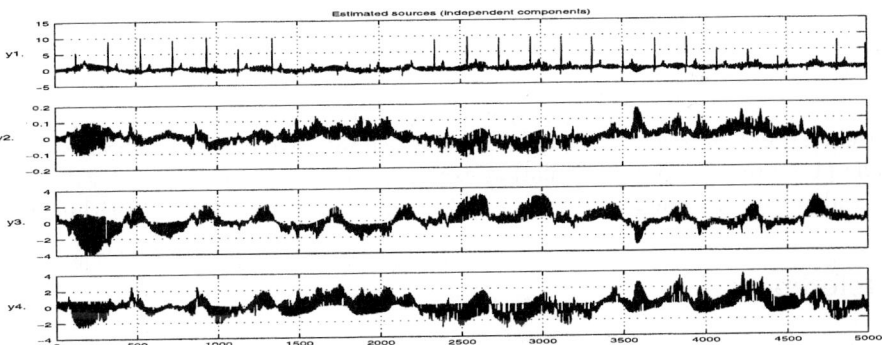

Fig. 4. Estimated sources by the Fast ICA algorithm.

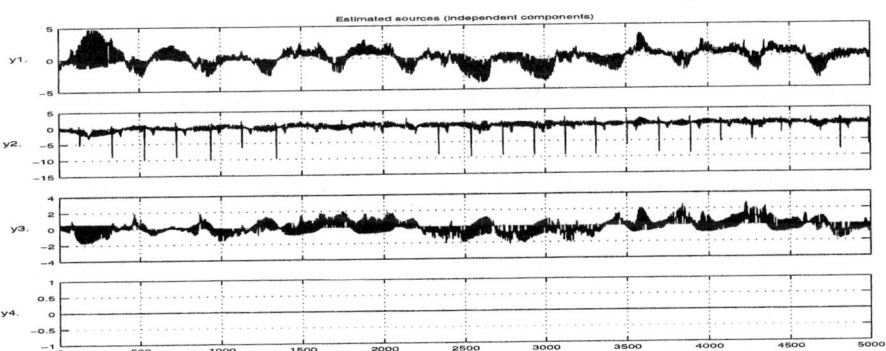

Fig. 5. Results by applying of JADE.

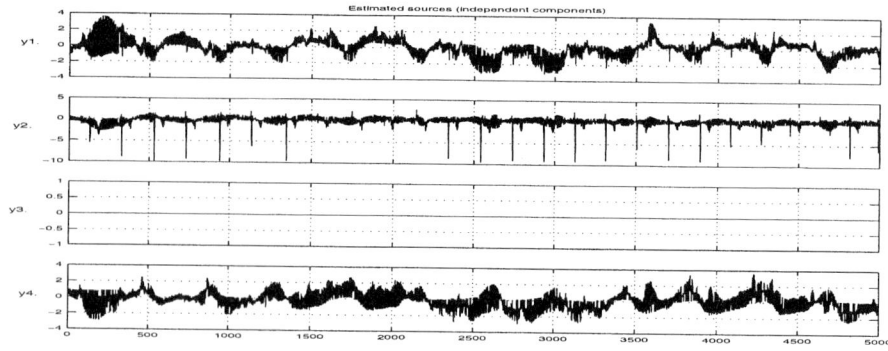

Fig. 6. Results by applying of SOBI.

5 Conclusion

We showed how to solve BSS problems of linear mixtures with singular matrices using sparsity of the source signals and presented sufficient conditions for their solvability. We presented two methods for that: 1) a general one (see matrix identification algorithm and source recovery algorithm) and 2) using reduction of the original problem to an overcomplete one, which we solve by sparse BSS methods. The presented computer simulation example shows the excellent separation by our algorithms, while the Fast ICA algorithm, JADE and SOBI algorithms fail.

References

1. A. Cichocki and S. Amari. *Adaptive Blind Signal and Image Processing.* John Wiley, Chichester, 2002.
2. P. Comon. *Independent component analysis - a new concept?* Signal Processing, 36: 287–314, 1994.
3. J. Eriksson and V. Koivunen. *Identifiability and separability of linear ica models revisited.* In Proc. of ICA 2003, pages 23–27, 2003.
4. P. Georgiev, F.J. Theis, and A. Cichocki. *Blind source separation and sparse component analysis of overcomplete mixtures.* In Proc. of ICASSP 2004, Montreal, Canada, 2004.
5. A. Hyvärinen, J. Karhunen and E. Oja, *Independent Component Analysis*, John Wiley & Sons, 2001.
6. T.-W. Lee, M.S. Lewicki, M. Girolami, T.J. Sejnowski, "Blind sourse separation of more sources than mixtures using overcomplete representaitons", *IEEE Signal Process. Lett.*, Vol. 6, no. 4, pp. 87–90, 1999.
7. F.J. Theis, E.W. Lang, and C.G. Puntonet, A geometric algorithm for overcomplete linear ICA. *Neurocomputing*, in print, 2003.
8. K. Waheed, F. Salem, "Algebraic Overcomplete Independent Component Analysis", in *Proc. Int. Conf. ICA2003*, Nara, Japan, pp. 1077–1082.
9. M. Zibulevsky, and B. A. Pearlmutter, "Blind source separation by sparse decomposition in a signal dictionary", *Neural Comput.*, Vol. 13, no. 4, pp. 863–882, 2001.

Closely Arranged Directional Microphone for Source Separation
Effectiveness in Reduction of the Number of Taps and Preventing Factors

Yusuke Katayama[1], Masanori Ito[2], Allan Kardec Barros[3],
Yoshinori Takeuchi[2,4], Tetsuya Matsumoto[2], Hiroaki Kudo[2],
Noboru Ohnishi[2], and Toshiharu Mukai[4]

[1] Graduate School of Engineering, Nagoya University, Japan
katayama@ohnishi.nuie.nagoya-u.ac.jp
[2] Graduate School of Information Science, Nagoya University, Japan
{ito,takeuchi,matumoto,kudo,ohnishi}@ohnishi.nuie.nagoya-u.ac.jp
[3] Universidade Federal do Maranhao, Brazil
allan@ufma.br
[4] Bio-Mimetic Control Research Center of RIKEN, Japan
tosh@bmc.riken.jp

Abstract. In this work, we work on the problem of sound source separation in convolutive mixtures. Particularly, we propose a method for reducing the number of filter taps while guaranteeing adequate separation performance. We recorded the mixed signals using directional microphones placed close to each other. As a result, we demonstrate that the proposed method successfully separates sources with fewer taps and better separation than conventional methods. In order to enhance the performance, we consider three main factors to prevent the reduction of number of taps: Echoes; frequency property of a directional microphone; and size of sound source. In experimental results, we found that echoes have little influence in preventing the reduction of the number of taps, but the other two factors affect the number of taps.

1 Introduction

In actual reverberant environment, the transfer functions from a certain source to each sensor differ depending upon many factors, such as room temperature, furniture distribution and distance. To construct a filter to separate such signals, we propose to use finite impulse response (FIR) filters because they yield an intrinsic stability[2].

We therefore propose a method to reduce the number of taps required for separating mixed signals in real environment. In the proposed method, we observe mixed signals by directional microphones placed close together. However, there are various factors that may affect the performance of the method[3]. Based on experiments, we consider how much three main factors prevent from reducing the number of taps.

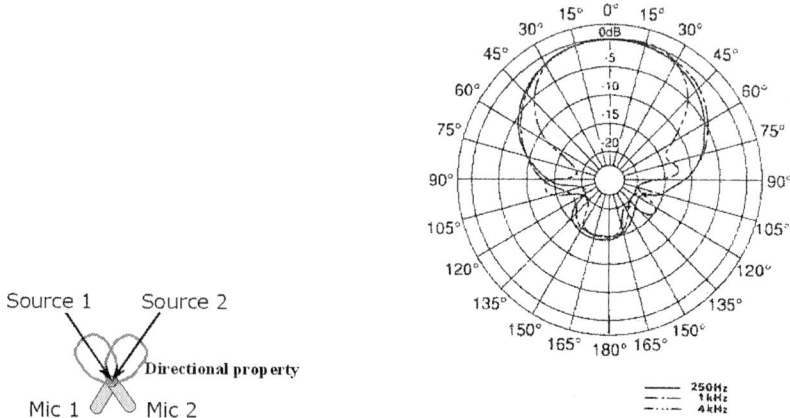

Fig. 1. Arrangement for measuring mixed signals using two directional microphones. **Fig. 2.** The directivity pattern of the microphone we used.

In addition, we quantitatively evaluate the separation performance by using the noise reduction rate (NRR) and signal to noise ratio (SNR) in experiments both in a reverberant chamber and in an anechoic chamber.

2 Converting to Instantaneous Mixture

We deal with two-input two-output model. Two directional microphones are placed as shown in Fig. 1. The tip of each microphone is placed at almost the same point. Fig. 2 shows curves that represent the directional properties of a microphone, that is, one microphone is roughly directed toward one source and the other microphone toward the other source. Because of these directional properties, the contribution of each sound source to the two microphones differs. The idea is that we observe two kinds of mixed signals from Mic 1 and Mic 2 even though Mic 1 and Mic 2 observe the sound sources at the same point.

Let $S = [S_1(z), S_2(z)]^T$ denote the Z-transform of two sound source signals, $X = [X_1(z), X_2(z)]^T$ that of two observed signals, and $H_{ij}(z)$ the transfer function from sound source i to microphone j ($i, j = 1, 2$). For the time being, we neglect the contribution due to echoes. The validity of this hypothesis will be analyzed by the experimental result in section 4.2.

Let us suppose that the observed delay at the microphone does not depend on the angle of incidence and that it is constant. Let us also define a gain function, $g(\theta)$, and a delay when the angle of incidence is zero, $F_0(z)$. Thus, the mixing process can be modeled as:

$$X = \begin{bmatrix} H_{11}(z) \cdot g(\theta_{11}) F_0(z) & H_{21}(z) \cdot g(\theta_{21}) F_0(z) \\ H_{12}(z) \cdot g(\theta_{12}) F_0(z) & H_{22}(z) \cdot g(\theta_{22}) F_0(z) \end{bmatrix} \cdot S \qquad (1)$$

where θ_{ij} denotes the angle of incidence from sound source i to microphone j.

When two directional microphones are placed at the same point, we assume that the two transfer functions from a specific output of sound source to each microphone become the same. We can then write:

$$H_1(z) \equiv H_{11}(z) = H_{12}(z) \\ H_2(z) \equiv H_{21}(z) = H_{22}(z) \qquad (2)$$

In this case, Eq. (1) can be rewritten as:

$$X = \begin{bmatrix} H_1(z) \cdot g(\theta_{11}) F_0(z) & H_2(z) \cdot g(\theta_{21}) F_0(z) \\ H_1(z) \cdot g(\theta_{12}) F_0(z) & H_2(z) \cdot g(\theta_{22}) F_0(z) \end{bmatrix} \cdot s \\ = \begin{bmatrix} g(\theta_{11}) & g(\theta_{21}) \\ g(\theta_{12}) & g(\theta_{22}) \end{bmatrix} \cdot \begin{bmatrix} H_1(z) F_0(z) S_1(z) \\ H_2(z) F_0(z) S_2(z) \end{bmatrix} \qquad (3)$$

We can consider $H_i(z) F_0(z) S_i(z)$ as a new source. Then Eq. (3) means that, under that assumption, the observed mixed signals become an instantaneous mixture.

Remark: In practical situations, however, it was difficult to realize an instantaneous mixture due to several factors. The number of taps needed to separate mixed signals depends on these factors, about which we discuss about in the next section. In addition, we achieve adequate separation by considering fewer taps.

3 Reducing the Number of Taps

There are three factors that affects adversely the need of reducing the number of taps. Firstly, there is the reverberation effect, which can be understood as a number of echoes occurring due to reflection of the sound source on walls, furniture and other objects, until it arrives at the measuring unit, which, in our case, are microphones. The second factor to be considered is the change which occurs in the microphone frequency responses according to the angle of incidence. Indeed, this fact turns the fact of assuming an instantaneous mixture much difficult. The third factor is that the sound source is not a point source. For example, if the diameter of loud-speaker is 0.20m, there is a small time difference, about five taps, between signals arriving at a microphone from one side of the loud-speaker and those arriving from the other side. Similarly, the width of the sound source can also turn the problem more complex, as the angle turns out to be more spread. We confirm this influence using two outputs of sound source which have different width in section4.4.

4 Separation Experiment

We conducted experiments to verify the effectiveness of our method and to investigate how the mentioned three factors may prevent the reduction of the number of taps. In each experiment, we changed either the number of taps, the microphone arrangement, or the sound source. Through the experiments, we used a SONY ECM-670 microphone. The directivity pattern of the microphone is shown in Fig. 2.

We used the method developed by Kawamoto et al. [4,5] to iteratively estimate the parameters of the separation filter. The update formula of this method is written as:

$$\Delta b_{12}(k) = -\alpha \frac{y_1(t-L)y_2(t-k)}{\phi_1(t)}$$
$$\Delta b_{21}(k) = -\alpha \frac{y_2(t-L)y_1(t-k)}{\phi_2(t)}, \quad (4)$$

where $\Delta b_{ij}(k)$ ($k = 0, \ldots, M$) are components of the separation filter matrix $B(z)$, M is the number of taps, L is the time lag, $y_i(t)$ is the output of separation filter, and $\phi_i(t)$ is a moving average estimate of the output energy. $\phi_i(t)$ is calculated as:

$$\phi_i(t) = \beta \phi_i(t-1) + (1-\beta) y_i^2(t-L), \quad (5)$$

where parameters α and β were set 5×10^{-6} and 0.9, respectively. In the experiments, parameter updating was repeated through 20 sweeps.

The separation performance was quantified by the noise reduction rate (NRR). The output of separation filter is the sum of the signal to be extracted and the one to be suppressed. Letting $s_1(t)$ be the desired signal and $s_2(t)$ be the suppressed signal as noise, we calculate signal-to-noise ratio (SNR) as:

$$SNR[dB] = 10 \log_{10} \frac{\sum_{t=0}^{N-1} s_1^2(t)}{\sum_{t=0}^{N-1} s_2^2(t)}, \quad (6)$$

where N is the number of samples. And NRR is defined as output SNR minus input SNR. We used NRR either to evaluate how much the undesired signal is suppressed by separation or to compare SNR before and after separation.

4.1 Confirmation of the Proposed Method

Firstly, we analyzed our method in reverberant chamber (conference room). The reverberation time of the chamber is around 0.5 sec. Two loud-speakers and two microphones are placed in two different arrangements as shown in Figs. 3(a)(b). Figure 3(a) represents the proposed method, and Fig. 3(b), Shows the conventional method in which two microphones are separated from each other.

We used six combinations from four kinds of speech as sound source (Source 1, Source 2). Sound sources were emitted by two loud-speakers, and two mixed signals (10 sec) were observed by two microphones.

Figures 4(a) and (b) show the transition of NRR of Source 1 and Source 2, respectively, caused by changing the number of taps. Figure 4 indicates the following: 1) Stable separation is achieved in the proposed method for five taps, while it is realized over 30 taps in the separated microphone arrangement. 2) This property does not depend on the kind of source. 3) The proposed method exhibits adequate separation performance equal or superior to the conventional method. In particular, when fewer taps were used in the separation filter, the sources are well separated in our method, while they were not in the conventional method.

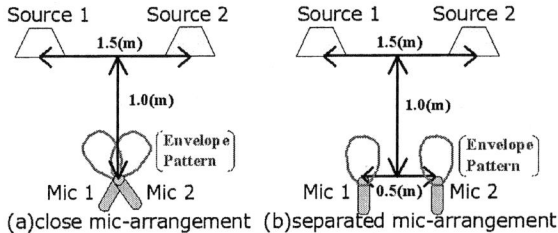

Fig. 3. Arrangement of microphones and sources.

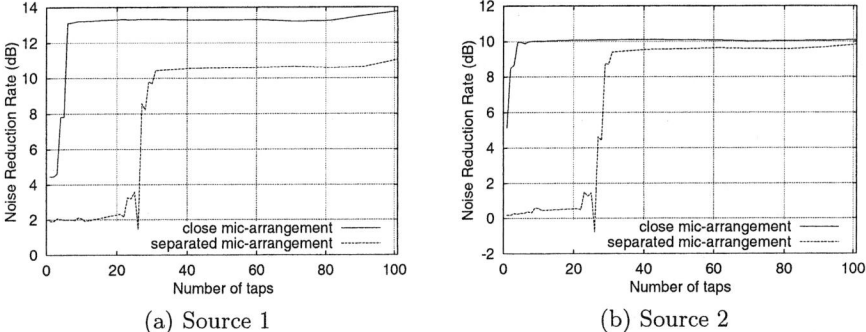

(a) Source 1   (b) Source 2

Fig. 4. The transition of NRR in a reverberant chamber.

4.2 Influence of Echoes

Next, to discuss whether the influence of echoes prevents the effectiveness of our method, we conducted similar experiments in an anechoic chamber. The arrangement of microphones and loud-speakers as source is shown in Fig. 3(a). If the echoes have some influence on our method, mixed signals observed in an anechoic chamber could be separated with less number of taps than that observed in a reverberant chamber.

Figure 5 shows the transition of NRR of Source 1 and Source 2 along with the result in a reverberant chamber for comparison. In each plot of Figs. 5(a) and (b), there is no difference between the number of taps that realize stable separation in a reverberant chamber and in an anechoic chamber. From the point of view of separation performance, the value of NRR in an anechoic chamber exhibits much better separation than that in a reverberant chamber. Therefore, we can say that the influence of echoes does not affect the effectiveness of reduction the number of taps although it affects the degradation of the separation performance.

4.3 Influence of Directional Property of Microphone

Since the microphone we used has asymmetrical directivity pattern as shown in Fig. 2, it may lead to different spectral distribution depending on whether a source is located at the right or left-side of the microphone. By using this asymmetry, we verify the influence of directional property of microphone.

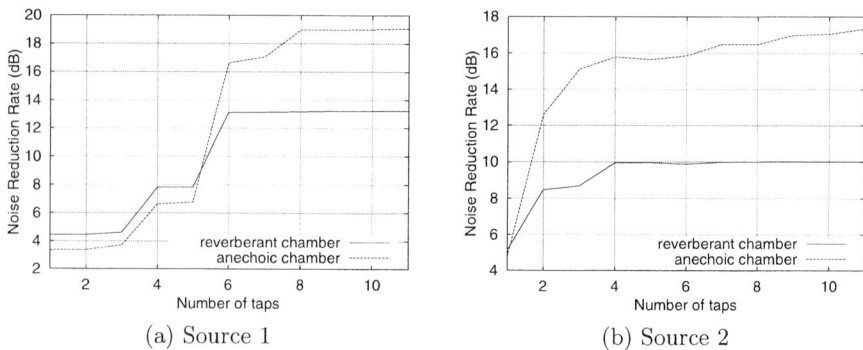

Fig. 5. The transition of NRR comparing result in a reverberant chamber to that in an anechoic chamber.

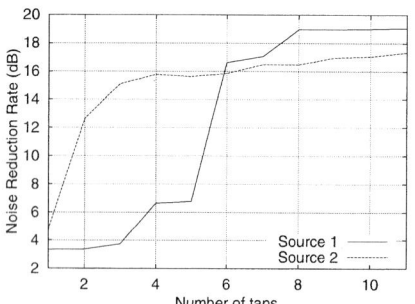

Fig. 6. The transition of NRR comparing Source 1 to Source 2 in an anechoic chamber.

Fig. 7. Arrangement of microphones and sources for verification of output.

Figure 6 shows the NRR transition both Source 1 and Source 2 for the anechoic chamber in Fig. 5. From Fig. 6, we find that the number of taps which realizes stable separation performance differ between Source 1 and Source 2. Therefore, we can say that the directional property of microphone has a influence on the proposed arrangement.

4.4 Influence of Size of Sound Emitting Object

Finally, we conducted experiments in an anechoic chamber to verify the influence of the width of sound source. Two microphones and two sources were placed as shown in Fig. 7. We used either loud-speakers or humans as sound sources. It is important to say that the width of a loud-speaker and human mouth is about 0.12m and 0.04m, respectively.

Figure 8 shows the transition of NRR for Source 1 and Source 2. The solid line is the result of human mouth and broken line that of loud-speaker. The number of taps at stable separation with human mouth is reduced around two taps for that with loud-speaker. Figure 8 indicates that smaller output of sound source can be separated by fewer taps.

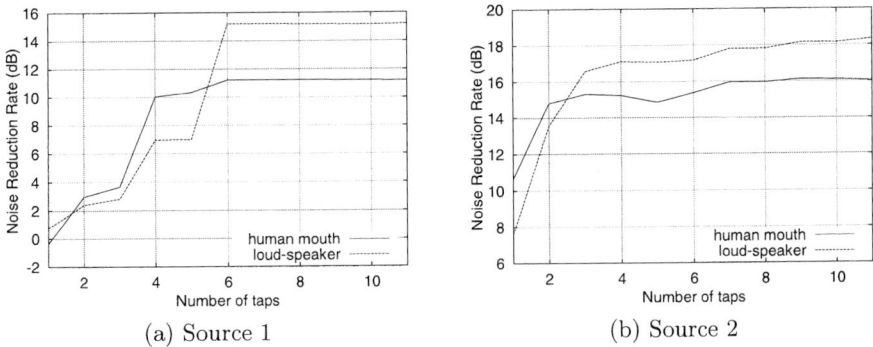

Fig. 8. The transition of NRR comparing result using human mouth to that using loud-speaker.

5 Conclusion

We proposed a method for blind source separation of two sound sources. In order to shorten the computation time required for separation, we proposed a microphone structure so that one reduces the number of taps in the recovery filter. This was carried out by placing the two directional microphones close to each other.

When our method was applied, the effectiveness to reduce the number of tap depend on different factors. Although echo has no influence on the efficacy, we confirmed the influence of directional property of a microphone and size of sound source.

References

1. T. W. LEE: Independent Component Analysis. Kuluwer Academic Publishers (1998)
2. A. Mansour and A. K. Barros and N. Ohnishi: Blind Separation of Sources: Methods, Assumptions and Applications. IEICE Trans. Fundamentals Vol.E83-A No.8 (2000) pp.1498–1511
3. Y. Katayama and Y. Takeuchi and T. Matsumoto and H. Kudo and N. Ohnishi and T. Mukai: Reduction of the Number of Taps for Blind Source Separation in the Real Environment. Proceeding of General Conference of IEICE (2002) p.157 [in Japanese]
4. M. Kawamoto and K. Matsuoka and N. Ohnishi: A method of blind separation for convolved non stationary signals. Neurocomputing 22 (1998) pp.157-171
5. K. Matsuoka and M. Ohya and M. Kawamoto: A Neural Net for Blind Separation of Nonstationary Signals. Neural Networks Vol.8 No.3 (1995) pp.411–419

Estimating Functions for Blind Separation when Sources Have Variance-Dependencies

Motoaki Kawanabe[1] and Klaus-Robert Müller[1,2]

[1] Fraunhofer FIRST.IDA, Kekuléstr. 7, 12489 Berlin, Germany
[2] University of Potsdam, August-Bebel-Strasse 89, 14482 Potsdam, Germany
{nabe,klaus}@first.fhg.de

Abstract. The blind separation problem where the sources are not independent, but have variance-dependencies is discussed. Hyvärinen and Hurri[1] proposed an algorithm which requires no assumption on distributions of sources and no parametric model of dependencies between components. In this paper, we extend the semiparametric statistical approach of Amari and Cardoso[2] under variance-dependencies and study estimating functions for blind separation of such dependent sources. In particular, we show that many of ICA algorithms are applicable to the variance-dependent model as well. Our theoretical consequences were confirmed by artificial and realistic examples.

1 Introduction

Independent component analysis (ICA) is based on the assumption that the observed signals are linear superpositions of mutually independent source signals. Let us denote the n source signals by $s(t) = (s_1(t), \ldots, s_n(t))^\top$ in a vector formula, and the observed signals by $x(t) = (x_1(t), \ldots, x_m(t))^\top$. The mixing process can be expressed as the equation

$$x(t) = As(t), \tag{1}$$

provided that it is not contaminated by any noise, where $A = (a_{ij})$ denotes the mixing matrix. For simplicity, we consider the case where the number of source signals equals that of observed signals ($n = m$).

Among many extensions of the basic ICA models, several researchers have studied the case where the source signals are not independent [3–6]. The dependencies either need to be exactly known beforehand, or they are simultaneously estimated by the algorithms. Recently, a novel idea called double-blind approach was introduced by Hyvärinen and Hurri[1]. In contrast to previous research, their method requires no assumption on the distributions of the sources and no parametric model of dependencies between the components. It is only assumed that the sources are dependent solely through their variances and that the sources have temporal dependencies.

A statistical basis of ICA was established by Amari and Cardoso[2]. They pointed out that the ICA model is an example of semiparametric statistical models[7,8] and studied estimating functions for it. In particular, they showed that the quasi maximum likelihood (QML) estimation and the natural gradient learning give a correct solution regardless of the true source densities. In this paper, we extend their approach to the blind source separation (BSS) problem considered in [1]. Investigating estimating functions for the model, we show that many of ICA algorithms based on the independence assumption work properly, even if there exist variance-dependencies.

This paper is organized as follows. After explaining our framework in Section 2 and Section 3, estimating functions for the variance-dependent BSS model are studied in Section 4. There, the quasi maximum likelihood estimation is taken as an example, while properties of other ICA algorithms are summarized in Section 5. We carried out numerical experiments with artificial and realistic examples (Section 6). Although only the double-blind algorithm gave correct solutions in the example described in [1], many ICA algorithms also worked for the other datasets.

2 Variance-Dependent BSS Model

Hyvärinen and Hurri[1] introduced the following framework. Let us assume that each source signal $s_i(t)$ is a product of non-negative activity level $v_i(t)$ and underlying i.i.d. signal $z_i(t)$, i.e.

$$s_i(t) = v_i(t) z_i(t). \qquad (2)$$

In practice, the activity levels v_i's are often dependent among different signals. In their formulation, each observed signal is expressed as

$$x_i(t) = \sum_{j=1}^{n} a_{ij} v_j(t) z_j(t), \qquad i = 1, \ldots, n, \qquad (3)$$

where $v_i(t)$ and $z_i(t)$ satisfy:

(i) v_i's and z_j's are independent,
(ii) each $z_i(t)$ is i.i.d. in time, z_i and z_j are mutually independent,
(iii) $z_i(t)$ have zero mean and unit variance.

No assumption on the distribution of z_i is made except (iii). Regarding the general activity levels v_i's, $v_i(t)$ and $v_j(t)$ are allowed to be statistically dependent, and furthermore, no particular assumption on these dependencies are made (double-blind situation). We refer to this framework as the variance-dependent BSS model in this paper.

They also proposed an algorithm which can separate the sources under the variance-dependent BSS model. Let $u(t)$ be the preprocessed signal of $x(t)$ by spatial whitening. Their method maximizes the objective function

$$J(W) = \sum_{i,j} [\widehat{\text{cov}}([w_i^\top u(t)]^2, [w_j^\top u(t - \Delta t)]^2)]^2$$

over an orthogonal matrix $W = (w_1, \ldots, w_n)^\top$, where $\widehat{\text{cov}}$ denotes the sample covariance. It was proved that the objective function J is maximized when WA equals a signed permutation matrix, if $K_{ij} = \text{cov}(s_i^2(t), s_j^2(t-\Delta t))$ is of full rank. This method works quite well, provided that there exist temporal variance-dependencies and the data is not spoiled by outliers.

3 Semiparametric Statistical Models and Estimating Functions

Amari and Cardoso[2] established a statistical basis of the ICA problem. They pointed out that the standard ICA model

$$p(X|B, \kappa_s) = |\det B|^T \prod_{t=1}^{T} \prod_{i=1}^{n} \kappa_{s_i} \{b_i^\top x(t)\} \tag{4}$$

is an example of semiparametric statistical models [7, 8], where $X = (x(1), \ldots, x(T))$ is the whole data sequence, $B = (b_1, \ldots, b_n)^\top = A^{-1}$ is the demixing matrix to be estimated and $\kappa_s(s) = \prod_{i=1}^{n} \kappa_{s_i}(s_i)$ is the density of the sources s. As the function κ_s in (4), semiparametric models contain infinite dimensional or functional nuisance parameters which are difficult to estimate. Moreover, they even disturb inference on parameters of interest.

In the variance-dependent BSS model, the sources $s(t)$ are decomposed of two components, the normalized signals $z(t) = (z_1(t), \ldots, z_n(t))^\top$ and the general activity levels $v(t) = (v_1(t), \ldots, v_n(t))^\top$. Since the former have mutual independence in the origin of ICA model, the density of the data X is factorized as

$$p(X|V; B, \kappa) = |\det B|^T \prod_{t=1}^{T} \prod_{i=1}^{n} \frac{1}{v_i(t)} \kappa_i \left\{ \frac{b_i^\top x(t)}{v_i(t)} \right\}, \tag{5}$$

when $V = (v(1), \ldots, v(T))$ is fixed. Therefore, the marginal distribution can be expressed as

$$p(X|B, \kappa, \nu) = \int p(X|V; B, \kappa) \nu(V) dV, \tag{6}$$

where the density ν of V becomes an extra nuisance function.

Estimating functions are a tool for constructing valid estimators in such semiparametric models. Let us consider a general semiparametric model $p(x|\theta, \kappa)$, where θ is an r-dimensional parameter of interest and κ is a nuisance parameter. An r-dimensional vector valued function $f(x, \theta)$ is called an estimating function, when it satisfies the following conditions for any θ and κ,

$$\mathrm{E}[\,f(x, \theta) \,|\, \theta, \kappa] = 0, \tag{7}$$

$$|\det Q| \neq 0, \quad \text{where } Q = \mathrm{E}\left[\frac{\partial}{\partial \theta} f(x, \theta) \,\Big|\, \theta, \kappa\right], \tag{8}$$

$$\mathrm{E}\left[\,\|f(x, \theta)\|^2 \,\big|\, \theta, \kappa\right] < \infty, \tag{9}$$

where $\mathrm{E}[\cdot|\theta, \kappa]$ means the expectation over x with the density $p(x|\theta, \kappa)$ and $\|\cdot\|$ denotes Euclidean norm [9]. Suppose that i.i.d. samples $x(1), \ldots, x(T)$ are obtained from the model $p(x|\theta^*, \kappa^*)$. If such a function exists, an M-estimator is obtained by solving the estimating equation

$$\sum_{t=1}^{T} f(x(t), \hat{\theta}) = 0. \tag{10}$$

The estimator $\hat{\theta}$ is consistent regardless of the true nuisance parameter κ^*, when the sample size T goes to infinity.

4 Estimating Functions for Blind Separation

Estimating functions for the ICA model (4) were discussed by Amari and Cardoso[2] and Cardoso[10]. In this case, the parameter of interest is the $n \times n$ matrix $B = A^{-1}$ and hence it is convenient to write estimating functions in $n \times n$ matrix form $F(x, B)$. Amari and Cardoso[2] showed that the quasi maximum likelihood method is a semiparametric algorithm based on estimating functions.

In the variance-dependent BSS model, in contrast to the ICA model studied by Amari and Cardoso[2], the data sequence $X = (x(1), \ldots, x(T))$ is not i.i.d. in time, but might have temporal dependencies. Therefore, we have to consider more general functions $\bar{F}(X, B)$ of the whole sequence X. General estimating functions $\bar{F}(X, B)$ must satisfy

$$\mathrm{E}[\bar{F}(X, B) | B, \kappa, \nu] = 0, \tag{11}$$

$$|\det Q| \neq 0, \quad \text{where } Q = \mathrm{E}\left[\frac{\partial \mathrm{vec}\{\bar{F}(X, B)\}}{\partial \mathrm{vec}(B)} \bigg| B, \kappa, \nu\right], \tag{12}$$

$$\mathrm{E}\left[\|\bar{F}(X, B)\|_F^2 \big| B, \kappa, \nu\right] < \infty, \tag{13}$$

for all (B, κ, ν). An M-estimator $\hat{B}$ is derived from the estimating equation

$$\bar{F}(X, \hat{B}) = 0. \tag{14}$$

Suppose that the data X is subject to $p(X|B^*, \kappa^*, \nu^*)$ defined by (5) and (6). It is known that the M-estimator $\hat{B}$ is consistent and asymptotically normal.

Theorem 1. *If the function $\bar{F}(X, B)$ satisfies the conditions (11)~(13) and appropriate regularity conditions, the M-estimator $\hat{B}$ derived from the equation (14) is asymptotically Gaussian distributed, i.e. $\mathrm{vec}(\hat{B}) \sim N(\mathrm{vec}(B^*), \mathrm{Av})$, where*

$$\begin{aligned}
\mathrm{Av} = \mathrm{Av}(B^*, \kappa^*, \nu^*) &= Q^{-1} \Sigma (Q^{-1})^\top, \\
\Sigma = \Sigma(B^*, \kappa^*, \nu^*) &= E\left[\mathrm{vec}\{\bar{F}(X, B^*)\} \mathrm{vec}\{\bar{F}(X, B^*)\}^\top \big| B^*, \kappa^*, \nu^*\right], \\
Q = Q(B^*, \kappa^*, \nu^*) &= \mathrm{E}\left[\frac{\partial \mathrm{vec}\{\bar{F}(X, B^*)\}}{\partial \mathrm{vec}(B)} \bigg| B^*, \kappa^*, \nu^*\right].
\end{aligned} \tag{15}$$

Now let us describe our main result. We can show that the function

$$\bar{F}(X, B) = \sum_{t=1}^{T} F(x(t), B) \tag{16}$$

constructed from an estimating function $F(x, B)$ for the ICA model becomes a candidate of estimating functions for the variance-dependent BSS model.

Theorem 2. *The function $\bar{F}(X, B)$ defined in (16) satisfies the two conditions (11) and (13), provided that $F(x, B)$ is an estimating function for the ICA model (4).*

Because it is difficult to check the other condition (12) in the general form, let us consider the quasi maximum likelihood estimation

$$\bar{F}^{\mathrm{QML}}(X,B) = \sum_{t=1}^{T} \left[I - \varphi\{\boldsymbol{y}(t)\}\,\boldsymbol{y}^\top(t) \right], \qquad (17)$$

as an example in the class (16), where $\varphi(\boldsymbol{y}) = (\varphi_1(y_1),\ldots,\varphi_n(y_n))^\top$ is a vector of nonlinear functions.

Theorem 3. *Suppose that the conditions*

$$\sum_{t=1}^{T} \mathrm{E}\left[m_i\{v_i(t)\}\right] + T \neq 0, \qquad \forall i, \qquad (18)$$

$$\det \begin{pmatrix} \sum_{t=1}^{T} \mathrm{E}[\,k_i\{v_i(t)\}v_j^2(t)\,] & T \\ T & \sum_{t=1}^{T} \mathrm{E}[\,k_j\{v_j(t)\}v_i^2(t)\,] \end{pmatrix} \neq 0, \qquad \forall i \neq j, \qquad (19)$$

hold, where

$$k_i\{v_i(t)\} = \mathrm{E}\left[\,\dot{\varphi}_i\{v_i(t)z_i(t)\} \mid V; B, \kappa\,\right], \qquad (20)$$
$$m_i\{v_i(t)\} = v_i^2(t)\,\mathrm{E}\left[\,\dot{\varphi}_i\{v_i(t)z_i(t)\}\,z_i^2(t) \mid V; B, \kappa\,\right], \qquad (21)$$

and $\dot{\varphi}_i$ is the derivative of φ_i. Then, the function $\bar{F}^{\mathrm{QML}}(X,B)$ satisfies the conditions (11)~(13) and becomes an estimating function. Under appropriate regularity conditions, the quasi maximum likelihood estimator $\hat{B}^{\mathrm{QML}}$ derived from the equation $\bar{F}^{QML}(X, \hat{B}^{\mathrm{QML}}) = 0$ is consistent regardless of the true nuisance functions (κ^*, ν^*).

5 Statistical Properties of ICA Algorithms

Although we concentrated on estimating functions of the form (16) in the previous section, we can deal with more general functions and investigate other ICA algorithms within the framework of estimating functions or asymptotic estimating functions[10] as well. Here we examined the unbiasedness condition (11) under the variance-dependent BSS model (Results are summarized in Table 1). In fact, this condition holds at least asymptotically in many algorithms. If the other conditions are satisfied, these algorithms give valid solutions regardless of the nuisance densities (κ^*, ν^*). We remark that our extension also enables us to analyze algorithms based on temporal structure such as TDSEP/SOBI[11, 12].

6 Numerical Experiments

We carried out at first experiments with several artificial datasets. We applied the quasi maximal likelihood methods QML-t and QML-3 (-t and -3 denote tanh and cubic non-linearity, resp.), the double-blind algorithm 'DB' [1], JADE, FastICA-t and FastICA-3,

Table 1. Unbiasedness condition of other ICA algorithms.

algorithm	unbiasedness	inapplicable cases
FastICA[13]	yes	Gaussian sources.
double-blind[1]	asymptotically	same variance-structures or no temporal variance-dependency
JADE[14]	asymptotically	Gaussian sources
TDSEP/SOBI[11, 12]	yes	always
nonstationary[15]	yes	unclear

TDSEP/SOBI [11, 12] and the 'sepagaus' algorithm for nonstationary signals[15], For evaluating the results, we used the index defined in Amari et al.[16]

$$\text{AmariIndex}(B, A^*) = \sum_{i=1}^{n} \left\{ \frac{\sum_{j=1}^{n} C_{ij}}{\max_k C_{ik}} - 1 \right\} + \sum_{j=1}^{n} \left\{ \frac{\sum_{i=1}^{n} C_{ij}}{\max_k C_{kj}} - 1 \right\}, \quad (22)$$

where A^* is the true mixing matrix and $C = BA^*$. If $B = PD(A^*)^{-1}$ with a permutation matrix P and a diagonal matrix D, then $\text{AmariIndex}(B, A^*) = 0$.

In all artificial datasets, five source signals of various types were generated and the data were observed after mixing with multiplying a random 5×5 matrix. We prepared eight artificial datasets ar_subG, ar_uni, sin_supG, sin_subG, com_supG, com_subG, exp_supG, and uni_subG. For the activity levels, the abbreviation 'ar' means that the random vector $v(t)$ was the absolute value of a multivariate AR(1) process. The activity levels of 'sin' datasets were sinusoidal functions with different frequencies, while those of 'com' were ones with same frequency. In the case of 'exp' and 'uni' datasets they were linear transformations of i.i.d. Laplace and uniform random vectors, respectively. For the normalized signals, 'uni' and 'supG' denote uniform and Laplace random variables, while 'subG' sequences were signed fourth roots of uniform random variables.

Table 2. AmariIndex of the estimators. The values are the medians of 100 replications.

	QML-t	QML-3	DB	JADE	FastICA-t	FastICA-3	TDSEP	sepagaus
ar_subG	8.25	11.32	0.52	10.79	9.25	12.52	15.07	1.19
ar_uni	0.30	27.77	0.70	0.66	0.38	0.73	14.92	0.85
sin_supG	0.17	29.97	0.79	0.43	0.23	0.41	15.31	0.08
sin_subG	19.21	0.32	0.27	0.31	0.68	0.33	15.70	0.08
com_supG	0.39	28.37	6.45	0.84	0.48	0.87	16.02	1.28
com_subG	26.53	0.14	22.05	26.49	27.04	26.65	16.23	27.08
exp_supG	0.35	28.43	7.63	1.24	0.44	1.20	16.47	1.28
uni_subG	27.38	0.13	18.56	0.17	0.18	0.18	16.20	27.08
sss	0.03	3.82	0.02	0.02	0.19	0.09	0.01	0.01
v12	0.01	3.73	0.21	0.19	0.17	0.08	0.14	0.01

As Hyvärinen and Hurri[1] showed, almost all algorithms except DB did not give a proper solution in ar_subG. However, DB showed poor performance, when (i) the variance-structures are the same or (ii) there is no temporal dependency. As expected, TDSEP did not work for any data, because there are no temporal correlations. QML-t is applicable to supergaussian cases, while QML-3 can be used for subgaussian data. The other algorithm returned acceptable results except in the difficult case com_subG.

Then, we also studied speech signals as more realistic examples. In the first example 'sss'[1], speakers counts from 1 to 10 in English and in Spanish, respectively (see the left panels of Figure 1). In the second experiment 'v12'[2], we took two speech signals from Japanese text, and modified the second so that the two sequences have large variance-dependency(see the right panels of Figure 1). The correlation of the variances in each example is substantially positive, i.e. 0.65 and 0.74, respectively. The results are shown in Table 2, too. All algorithm except QML-3 gave a proper answer. On these realistic examples, TDSEP also worked, because the statistical model (5) and (6) did not hold perfectly.

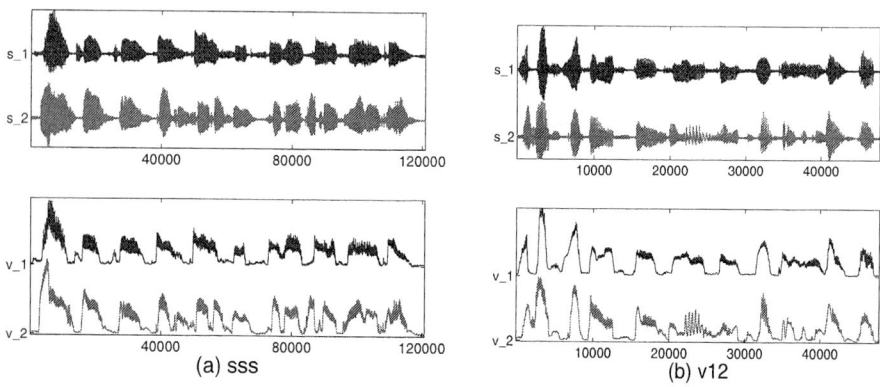

Fig. 1. The sources $s(t)$ (upper panels) and the estimators $v(t)$ of their activity levels with an appropriate smoother (lower panels).

7 Conclusions

In this paper, we discussed semiparametric estimation for blind separation, when sources have variance-dependencies. Extending the semiparametric statistical approach [2] under variance-dependencies, we investigated estimating functions for the variance-dependent BSS model. In particular, we proved that the quasi maximum likelihood estimator is derived from such an estimating function, and hence consistent regardless of the true nuisance densities. Although we omitted details in this paper, we also analyzed other ICA algorithms within the framework of (asymptotic) estimating functions and showed that many of them can separate sources with coherent variances. The theoretical results were confirmed by artificial and realistic examples with speech signals. Further research aims necessary to find good applications of the current framework.

---

[1] http://inc2.ucsd.edu/~tewon/

[2] http://www.islab.brain.riken.go.jp/~mura/ica/v1.wav and v2.wav

Acknowledgement

The authors acknowledge A. Ziehe, S. Harmeling, F. Meinecke and N. Murata for valuable discussion and EU PASCAL for partial funding.

References

1. Hyvärinen, A., Hurri, J.: Blind separation of sources that have spatiotemporal variance dependencies. Signal Processing (2004) to appear.
2. Amari, S., Cardoso, J.F.: Blind source separation – semiparametric statistical approach. IEEE Trans. on Signal Processing **45** (1997) 2692–2700
3. Cardoso, J.F.: Multidimensional independent component analysis. In: Proc. IEEE Int. Conf. on Acoustics, Speech and Signal Processing (ICASSP'98), Seattle, WA (1998)
4. Hyvärinen, A., Hoyer, P.O., Inki, M.: Topographic independent component analysis. Neural Computation **13** (2001)
5. Bach, F.R., Jordan, M.I.: Tree-dependent component analysis. In: Uncertainty in Artificial Intelligence: Proceedings of the Eighteenth Conference (UAI-2002). (2002)
6. Valpola, H., Harva, M., Karhunen, J.: Hierachical models of variance sources. In: Proc. of ICA2003, Nara, Japan (2003)
7. Bickel, P., Klaassen, C., Ritov, Y., Wellner, J.: Efficient and Adaptive Estimation for Semi-paramtric Models. John Hopkins Univ. Press, Baltimore, MD (1993)
8. Amari, S., Kawanabe, M.: Information geometry of estimating functions in semiparametric statistical models. Bernoulli **3** (1997) 29–54
9. Godambe, V., ed.: Estimating Functions. Oxford Univ. Press, New York (1991)
10. Cardoso, J.F.: Estimating equations for source separation. In: Proc. IEEE Int. Conf. on Acoustics, Speech and Signal Processing (ICASSP'97). Volume 5., Munich, Germany (1997) 3449–3452
11. Ziehe, A., Müller, K.R.: TDSEP – an efficient algorithm for blind separation using time structure. In Niklasson, L., Bodén, M., Ziemke, T., eds.: Proc. of the 8th Int. Conf. on Artificial Neural Networks (ICANN '98), Berlin, Springer Verlag (1998) 675 – 680
12. A. Belouchrani, K. Abed Meraim, J.F.C., Moulines, E.: A blind source separation technique based on second order statistics. IEEE Trans. on Signal Processing (1996) 1009–1020
13. Hyvärinen, A., Oja, E.: A fast fixed-point algorithm for independent component analysis. Neural Computation **9** (1997) 1483–1492
14. Cardoso, J.F., Souloumiac, A.: Blind beamforming for non gaussian signals. IEE Proceedings-F (1993) 362–370
15. Pham, D.T., Cardoso, J.F.: Blind separation of instantaneous mixtures of non-stationary sources. In: Proc. of ICA2000, Helsinki, Finland (2000) 187–193
16. Amari, S., Cichocki, A., Yang, H.: A new learning algorithm for blind source separation. In: Advances in Neural Information Processing Systems 8. MIT Press (1996) 757–763

Framework of Constrained Matrix Gradient Flows

Gen Hori

Brain Science Institute, RIKEN, Saitama 351-0198, Japan
hori@brain.riken.jp

Abstract. The paper presents general formulas of constrained matrix gradient flows which can be used to derive algorithms for specific problems appeared in the aspects of ICA including joint diagonalization and joint SVD problems. Some previous and novel examples of constrained matrix gradient flows are derived using the general formulas.

1 Introduction

Taking advantage of the theory of stochastic gradient in neural network learning, many of the ICA algorithms are implemented in the form of matrix gradient flows. Due to their suitability to parallel online computation, matrix gradient flows are among active research fields in adaptive signal processing. The purpose of the paper is to present ready-to-use general formulas of constrained matrix gradient flows which can be used to derive algorithms for various problems appeared in ICA including joint diagonalization and joint SVD problems. Following preliminaries in Section 2, Section 3, 4, 5 and 6 introduce general formulas of matrix gradient flows constrained to four different types of submanifolds, among which the formulas introduced in Section 3, 4 and 5 are for EVD(eigenvalue decomposition) type problems while the one introduced in Section 6 is for SVD(singular value decomposition) type problems. Some previous and novel examples of matrix flows are derived using the general formulas.

2 Preliminaries

We use the following notations in the rest of the paper. $\operatorname{Re} x$ and $\operatorname{Im} x$ denote the real and the imaginary parts of a complex number x respectively. $M(m \times n, R)$ and $M(m \times n, C)$ denote the sets of real and complex $m \times n$ matrices respectively. I_n denotes the $n \times n$ identity matrix. A^* denotes the Hermitian transpose of A and $\operatorname{Tr} A$ the trace of A. $[A, B] = AB - BA$ is a commutator product. $\mathfrak{U}(n)$ and $\mathfrak{u}(n)$ denote the Lie group of $n \times n$ unitary matrices and its Lie algebra of $n \times n$ skew-Hermitian matrices respectively,

$$\mathfrak{U}(n) = \{\, U \in M(n \times n, C) \mid U^*U = I_n \,\},$$
$$\mathfrak{u}(n) = \{\, X \in M(n \times n, C) \mid X^* = -X \,\},$$

and $GL(n,C)$ and $\mathfrak{gl}(n,C)$ the Lie group of $n \times n$ non-degenerate matrices and its Lie algebra,

$$GL(n,C) = \{ T \in M(n \times n, C) \mid \det T \neq 0 \},$$
$$\mathfrak{gl}(n,C) = \{ X \in M(n \times n, C) \}.$$

In the following sections, we derive matrix gradient flows of an arbitrary real-valued potential function defined on the set of $m \times n$ complex matrices,

$$\phi : M(m \times n, C) \to R,$$

where the flows are constrained to respective submanifolds of $M(m \times n, C)$. In the derivation, we regard $M(m \times n, C)$ not as an mn-dimensional complex space C^{mn} but as a $2mn$-dimensional real space R^{2mn}. Accordingly, we introduce the following notation,

$$\frac{d\phi}{dA} = \begin{pmatrix} \frac{\partial \phi(A)}{\partial (\text{Re} a_{11})} & \cdots & \frac{\partial \phi(A)}{\partial (\text{Re} a_{1n})} \\ \vdots & \ddots & \vdots \\ \frac{\partial \phi(A)}{\partial (\text{Re} a_{m1})} & \cdots & \frac{\partial \phi(A)}{\partial (\text{Re} a_{mn})} \end{pmatrix} + \begin{pmatrix} \frac{\partial \phi(A)}{\partial (\text{Im} a_{11})} & \cdots & \frac{\partial \phi(A)}{\partial (\text{Im} a_{1n})} \\ \vdots & \ddots & \vdots \\ \frac{\partial \phi(A)}{\partial (\text{Im} a_{m1})} & \cdots & \frac{\partial \phi(A)}{\partial (\text{Im} a_{mn})} \end{pmatrix} i.$$

Using the notation and the relation

$$\text{Re}\,\text{Tr}\, X^*Y = \sum_{i,j} (\text{Re}\, x_{ij} \text{Re}\, y_{ij} + \text{Im}\, x_{ij} \text{Im}\, y_{ij}),$$

the chain rule can be written briefly as

$$\frac{d}{dt}\phi(A(t)) = \sum_{i,j} \left(\frac{\partial \phi(A)}{\partial (\text{Re}\, a_{ij})} \frac{d(\text{Re}\, a_{ij})}{dt} + \frac{\partial \phi(A)}{\partial (\text{Im}\, a_{ij})} \frac{d(\text{Im}\, a_{ij})}{dt} \right) = \text{Re}\,\text{Tr}(\frac{d\phi}{dA})^* \frac{dA}{dt}.$$

There are continuous correspondences between the Lie groups and the submanifolds introduced in the following sections and consequently the flows on the submanifolds can be regarded as the flows on the Lie groups $GL(n,C)$ or $\mathfrak{u}(n)$ as well. We introduce metrics on the Lie groups and derive the gradient flows on the groups with respect the metrics, and then map the flows onto the submanifolds to define the constrained matrix gradient flows. We introduce the metric on $\mathfrak{gl}(n,C)$ by

$$< X_1, X_2 > = \text{Re}\,\text{Tr}\, X_1^* X_2,$$

and one on $\mathfrak{u}(n)$ by the same equation with the restriction of the domain. The metric on the tangent space at the identity is extended to the tangent space at each point of the Lie group by the left action and determines a Riemannian structure.

The orthogonal projection from $\mathfrak{gl}(n,C)$ to its subspace $\mathfrak{u}(n)$ with respect to the above-defined metric is given by

$$\pi_{\mathfrak{u}(n)} X = \frac{1}{2}(X - X^*).$$

3 Self-similar Gradient Flows (Unitary)

Self-similar flows are matrix flows which evolve with the eigenvalues preserved and are also called "isospectral flows". This section presents the general formula of self-similar gradient flows introduced by Brockett[2] and some previous examples of the flows.

3.1 General Gradient Ascent Equation

Let $\mathfrak{S}_1$ denote the set of all the $n \times n$ complex matrices which are unitarily similar to $A_0 \in M(n \times n, C)$,

$$\mathfrak{S}_1 = \{A = U^* A_0 U \mid U \in \mathfrak{U}(n)\},$$

and consider a continuous dynamical system on $\mathfrak{S}_1$,

$$A(t) = U(t)^* A_0 U(t), \quad U(t) \in \mathfrak{U}(n).$$

When $U(t)$ evolves as

$$\dot{U}(t) = U(t) X(t), \quad U(0) = I_n, \quad X(t) \in \mathfrak{u}(n), \tag{1}$$

we have

$$\dot{A} = (UX)^* A_0 U + U^* A_0 (UX) = X^* A + AX = -XA + AX = [A, X],$$

therefore $A(t)$ evolves as

$$\dot{A}(t) = [A(t), X(t)], \quad A(0) = A_0, \quad X(t) \in \mathfrak{u}(n). \tag{2}$$

Using the chain rule and the relations $\mathrm{Tr}AB = \mathrm{Tr}BA$, $(AB)^* = B^* A^*$ and $\mathrm{Tr}A^* = \overline{\mathrm{Tr}A}$, we have

$$\frac{d}{dt}\phi(A(t)) = \mathrm{Re}\,\mathrm{Tr}\,(\frac{d\phi}{dA})^* \frac{dA}{dt} = \mathrm{Re}\,\mathrm{Tr}\,(\frac{d\phi}{dA})^* (-XA + AX)$$

$$= \mathrm{Re}\,\mathrm{Tr}\,(A^* \frac{d\phi}{dA} - \frac{d\phi}{dA} A^*)^* X = \mathrm{Re}\,\mathrm{Tr}\,[A^*, \frac{d\phi}{dA}]^* X = <[A^*, \frac{d\phi}{dA}], X>$$

which gives the steepest ascent direction of $\phi(A)$ in terms of X as

$$X = \pi_{\mathfrak{u}(n)}[A^*, \frac{d\phi}{dA}] = \frac{1}{2}([A^*, \frac{d\phi}{dA}] - [A^*, \frac{d\phi}{dA}]^*) = \frac{1}{2}([A^*, \frac{d\phi}{dA}] + [A, (\frac{d\phi}{dA})^*]).$$

Substituting this in (1) and (2), we obtain the following general gradient ascent equations in terms of A and U,

$$\frac{dA}{dt} = \frac{1}{2}[A, [A^*, \frac{d\phi}{dA}] + [A, (\frac{d\phi}{dA})^*]], \quad \frac{dU}{dt} = \frac{1}{2}U([A^*, \frac{d\phi}{dA}] + [A, (\frac{d\phi}{dA})^*]).$$

If A_0 is Hermitian then $A(t)$ is always Hermitian which reduces the equations to

$$\frac{dA}{dt} = \frac{1}{2}[A, [A, \frac{d\phi}{dA} + (\frac{d\phi}{dA})^*]], \quad \frac{dU}{dt} = \frac{1}{2}U[A, \frac{d\phi}{dA} + (\frac{d\phi}{dA})^*].$$

Furthermore, if $\frac{d\phi}{dA}$ is Hermitian for all A then the equations reduce to

$$\boxed{\frac{dA}{dt} = [A, [A, \frac{d\phi}{dA}]], \quad A(0) = A_0, \quad \frac{dU}{dt} = U[A, \frac{d\phi}{dA}], \quad U(0) = I_n}. \tag{3}$$

3.2 Previous Examples

Brockett[1] introduced a self-similar flow

$$\dot{A} = [A, [A, C]]$$

where $C \in M(n \times n, C)$ is a constant matrix, which can be derived from (3) by substituting $\phi(A) = \operatorname{Re}\operatorname{Tr} C^* A$. He proved that the flow globally converges to a diagonal matrix as $t \to \pm\infty$ for almost all real symmetric initial matrices where C is a real diagonal matrix with distinct diagonal elements. Hori[5] extended the flow to

$$\dot{A} = \underbrace{[A, [A, \cdots, [A, C]\cdots]]}_{m\text{-fold}}$$

and proved that, when m is an even number, it converges to a diagonal matrix under the same conditions.

Chu and Driessel[3] introduced a self-similar flow

$$\dot{A} = [A, [A, \operatorname{diag}(A)]]$$

where $\operatorname{diag}(A) \in M(n \times n, R)$ is a diagonal matrix whose diagonals are the same as A, which can be derived using (3) as the gradient descent equation of the sum of squares of the off-diagonal elements of A. They proved that the flow globally converges to a diagonal matrix as $t \to \infty$ for almost all real symmetric initial matrices.

Hori[7] introduced a flow on the unitary group $\mathfrak{U}(n)$ for solving joint diagonalization problems utilizing the general formula (3).

4 Self-similar Gradient Flows (Non-unitary)

There are two possibilities of non-unitary extension of the self-similar flows introduced in the previous section, depending on wether the flowing matrix expresses a $(1,1)$-tensor or a $(0,2)$-tensor. This section discusses the former and the next section the latter.

4.1 General Gradient Ascent Equation

Let $\mathfrak{S}_2$ denote the set of all the $n \times n$ complex matrices which are similar to $A_0 \in M(n \times n, C)$,

$$\mathfrak{S}_2 = \{A = T^{-1} A_0 T \mid T \in GL(n, C)\},$$

and consider a continuous dynamical system on $\mathfrak{S}_2$,

$$A(t) = T(t)^{-1} A_0 T(t), \quad T(t) \in GL(n, C).$$

When $T(t)$ evolves as

$$\dot{T}(t) = T(t) X(t), \quad T(0) = I_n, \quad X(t) \in \mathfrak{gl}(n, C), \tag{4}$$

using $\frac{d}{dt}(T^{-1}) = -T^{-1} \dot{T} T^{-1}$, we have

$$\dot{A} = -T^{-1}(TX)T^{-1}A_0T + T^{-1}A_0(TX) = -XA + AX = [A, X],$$

therefore $A(t)$ evolves as

$$\dot{A}(t) = [A(t), X(t)], \quad A(0) = A_0, \quad X(t) \in \mathfrak{gl}(n, C). \tag{5}$$

Using the chain rule, we have

$$\frac{d}{dt}\phi(A(t)) = \operatorname{Re}\operatorname{Tr}(\frac{d\phi}{dA})^* \frac{dA}{dt} = \operatorname{Re}\operatorname{Tr}(\frac{d\phi}{dA})^*(-XA + AX)$$
$$= \operatorname{Re}\operatorname{Tr}(A^*\frac{d\phi}{dA} - \frac{d\phi}{dA}A^*)^* X = \operatorname{Re}\operatorname{Tr}[A^*, \frac{d\phi}{dA}]^* X = <[A^*, \frac{d\phi}{dA}], X>$$

which gives the steepest ascent direction of $\phi(A)$ in terms of X as

$$X = [A^*, \frac{d\phi}{dA}].$$

Substituting this in (4) and (5), we obtain the following general gradient ascent equations in terms of A and T,

$$\boxed{\frac{dA}{dt} = [A, [A^*, \frac{d\phi}{dA}]], \quad A(0) = A_0, \quad \frac{dT}{dt} = T[A^*, \frac{d\phi}{dA}], \quad T(0) = I_n.} \tag{6}$$

4.2 Previous Example

Hori[6] introduced a self-similar flow

$$\dot{A} = -[A, [A^*, L(A)]]$$

where $L(A) \in M(n \times n, C)$ is a strictly lower triangular matrix whose lower triangular elements are the same as A, which can be derived using (6) as the gradient descent equation of the sum of squares of the lower triangular elements of A. He proved that all the fixed points of the flow are upper triangular matrices whose diagonal elements are the eigenvalues of the initial matrix.

5 Self-congruent Gradient Flows

This section introduces the non-unitary extension of the self-similar flows introduced in Section 3 for the matrices expressing $(0, 2)$-tensors. The general formula of self-congruent gradient flows is newly introduced and used to derive a flow for solving joint diagonalization problems.

5.1 General Gradient Ascent Equation

Let $\mathfrak{C}$ denote the set of all the $n \times n$ complex matrices which are congruent with $A_0 \in M(n \times n, C)$,

$$\mathfrak{C} = \{A = T^*A_0T \mid T \in GL(n, C)\},$$

and consider a continuous dynamical system on $\mathfrak{C}$,
$$A(t) = T(t)^* A_0 T(t), \quad T(t) \in GL(n, C).$$

When $T(t)$ evolves as
$$\dot{T}(t) = T(t) X(t), \quad T(0) = I_n, \quad X(t) \in \mathfrak{gl}(n, C), \tag{7}$$

we have
$$\dot{A} = (TX)^* A_0 T + T^* A_0 (TX) = X^* A + AX,$$

therefore $A(t)$ evolves as
$$\dot{A}(t) = X(t)^* A(t) + A(t) X(t), \quad A(0) = A_0, \quad X(t) \in \mathfrak{gl}(n, C). \tag{8}$$

Using the chain rule, we have
$$\frac{d}{dt}\phi(A(t)) = \operatorname{Re} \operatorname{Tr}\left(\frac{d\phi}{dA}\right)^* \frac{dA}{dt} = \operatorname{Re} \operatorname{Tr}\left(\frac{d\phi}{dA}\right)^* (X^* A + AX)$$
$$= \operatorname{Re} \operatorname{Tr}\left(A^* \frac{d\phi}{dA} + A\left(\frac{d\phi}{dA}\right)^*\right)^* X = \left\langle A^* \frac{d\phi}{dA} + A\left(\frac{d\phi}{dA}\right)^*, X \right\rangle$$

which gives the steepest ascent direction of $\phi(A)$ in terms of X as
$$X = A^* \frac{d\phi}{dA} + A\left(\frac{d\phi}{dA}\right)^*.$$

Substituting this in (7) and (8), we obtain the following general gradient ascent equations in terms of A and T,
$$\frac{dA}{dt} = A\left(A^* \frac{d\phi}{dA} + A\left(\frac{d\phi}{dA}\right)^*\right) + \left(\frac{d\phi}{dA} A^* + \left(\frac{d\phi}{dA}\right)^* A\right) A, \quad \frac{dT}{dt} = T\left(A^* \frac{d\phi}{dA} + A\left(\frac{d\phi}{dA}\right)^*\right).$$

If A_0 is Hermitian then $A(t)$ is always Hermitian which reduces the equations to
$$\frac{dA}{dt} = A^2 \left(\frac{d\phi}{dA} + \left(\frac{d\phi}{dA}\right)^*\right) + \left(\frac{d\phi}{dA} + \left(\frac{d\phi}{dA}\right)^*\right) A^2, \quad \frac{dT}{dt} = TA\left(\frac{d\phi}{dA} + \left(\frac{d\phi}{dA}\right)^*\right).$$

Furthermore, if $\frac{d\phi}{dA}$ is Hermitian for all A then the equations reduce to

$$\boxed{\frac{dA}{dt} = 2\left(A^2 \frac{d\phi}{dA} + \frac{d\phi}{dA} A^2\right), \quad A(0) = A_0, \quad \frac{dT}{dt} = 2TA\frac{d\phi}{dA}, \quad T(0) = I_n.} \tag{9}$$

5.2 Example

To derive an algorithm for the joint diagonalization problem of Hermitian matrices $A_1, A_2, \ldots, A_K$, we substitute $\phi(A_k) = \frac{1}{4} \sum_i |a_{ii}^{(k)}|^2$ in (9) and superpose the right hand sides for $k = 1, \ldots, K$ to obtain
$$\dot{T} = T \sum_{k=1}^{K} T^* A_k T \operatorname{diag}(T^* A_k T).$$

It is observed through simulations that the flow converges to the joint diagonalizer of the given Hermitian matrices.

6 Self-equivalent Gradient Flows

Self-equivalent flows are matrix flows which evolve with the singular values preserved and can perform SVD of non-square matrices. This section presents the general formula of self-equivalent gradient flows introduced by Hori[8] and some previous examples of the flows.

6.1 General Gradient Ascent Equation

Let $\mathfrak{E}$ denote the set of all the $m \times n$ complex matrices which share the singular values with $A_0 \in M(m \times n, C)$,

$$\mathfrak{E} = \{A = U^* A_0 V \mid U \in \mathfrak{U}(m), V \in \mathfrak{U}(n)\},$$

where $m \geq n$, and consider a continuous dynamical system on $\mathfrak{E}$,

$$A(t) = U(t)^* A_0 V(t), \quad U(t) \in \mathfrak{U}(m), V(t) \in \mathfrak{U}(n).$$

When $U(t)$ and $V(t)$ evolve as

$$\begin{aligned} \dot{U}(t) &= U(t) X(t), \quad U(0) = I_m, \quad X(t) \in \mathfrak{u}(m), \\ \dot{V}(t) &= V(t) Y(t), \quad V(0) = I_n, \quad Y(t) \in \mathfrak{u}(n), \end{aligned} \qquad (10)$$

we have

$$\dot{A} = (UX)^* A_0 V + U^* A_0 (VY) = X^* A + AY = -XA + AY,$$

therefore $A(t)$ evolves as

$$\dot{A}(t) = -X(t) A(t) + A(t) Y(t), \quad A(0) = A_0, \quad X(t) \in \mathfrak{u}(m), Y(t) \in \mathfrak{u}(n). \qquad (11)$$

Using the chain rule, we have

$$\begin{aligned} \frac{d}{dt} \phi(A(t)) &= \operatorname{Re} \operatorname{Tr} \left(\frac{d\phi}{dA}\right)^* \frac{dA}{dt} = \operatorname{Re} \operatorname{Tr} \left(\frac{d\phi}{dA}\right)^* (-XA + AY) \\ &= -\operatorname{Re} \operatorname{Tr} \left(\frac{d\phi}{dA} A^*\right)^* X + \operatorname{Re} \operatorname{Tr} \left(A^* \frac{d\phi}{dA}\right)^* Y \\ &= < -\frac{d\phi}{dA} A^*, X > + < A^* \frac{d\phi}{dA}, Y >, \end{aligned}$$

which gives the steepest ascent direction of $\phi(A)$ in terms of X and Y as

$$\begin{aligned} X &= -\pi_{\mathfrak{u}(m)} \frac{d\phi}{dA} A^* = \frac{1}{2} \left(A \left(\frac{d\phi}{dA}\right)^* - \frac{d\phi}{dA} A^* \right), \\ Y &= \pi_{\mathfrak{u}(n)} A^* \frac{d\phi}{dA} = \frac{1}{2} \left(A^* \frac{d\phi}{dA} - \left(\frac{d\phi}{dA}\right)^* A \right). \end{aligned}$$

Substituting these in (10) and (11), we obtain the following general gradient ascent equations in terms of A and (U, V),

$$\begin{cases} \dfrac{dA}{dt} = \dfrac{1}{2}A(A^*\dfrac{d\phi}{dA} - (\dfrac{d\phi}{dA})^*A) - \dfrac{1}{2}(A(\dfrac{d\phi}{dA})^* - \dfrac{d\phi}{dA}A^*)A, \quad A(0) = A_0, \\ \dfrac{dU}{dt} = \dfrac{1}{2}U(A(\dfrac{d\phi}{dA})^* - \dfrac{d\phi}{dA}A^*), \quad U(0) = I_m, \\ \dfrac{dV}{dt} = \dfrac{1}{2}V(A^*\dfrac{d\phi}{dA} - (\dfrac{d\phi}{dA})^*A), \quad V(0) = I_n. \end{cases} \quad (12)$$

6.2 Previous Examples

Helmke and Moore[4] and Smith[9] introduced a self-equivalent flow

$$\dot{A} = A(A^*C - C^*A) - (AC^* - CA^*)A$$

where $C \in M(m \times n, C)$ is a constant matrix, which can be derived from (12) by substituting $\phi(A) = 2\operatorname{Re}\operatorname{Tr} C^*A$. They proved that the flow converges to the SVD of the initial matrix as $t \to \pm\infty$ where C is an extended diagonal matrix with distinct real diagonal elements and the initial matrix meets suitable conditions.

Chu and Driessel[3] introduced a self-equivalent flow

$$\dot{A} = A(A^*\operatorname{diag}(A) - \operatorname{diag}(A)^*A) - (A\operatorname{diag}(A)^* - \operatorname{diag}(A)A^*)A$$

where $\operatorname{diag}(A) \in M(m \times n, C)$ is an extended diagonal matrix whose diagonals are the same as A, which can be derived from (12) by substituting $\phi(A) = \sum_i |a_{ii}|^2$. They proved that the flow converges to the SVD of the initial matrix as $t \to \infty$ if the initial matrix meets suitable conditions.

Hori[8] introduced a flow on the unitary groups for solving joint SVD problems utilizing the general formula (12).

References

1. Brockett, R.W. : Linear Algebra Appl., vol. 146, pp. 79–91 (1991)
2. Brockett, R.W. : In *Differential Geometry: Partial Differential Equations on Manifolds* (eds. R.Green and S-T Yau), Amer. Math. Soc., Providence, pp. 69–92 (1993)
3. Chu, M.T. and Driessel, K.R. : SIAM J. Numer. Anal, vol. 27, pp. 1050–1060 (1990)
4. Helmke, U. and Moore, J.B. : Linear Algebra Appl., vol. 169, pp. 223–248 (1992)
5. Hori, G. : Japan J. Indust. Appl. Math., vol. 14, pp. 315–327 (1997)
6. Hori, G. : Japan J. Indust. Appl. Math., vol. 17, pp. 27–42 (2000)
7. Hori, G. : Proc. ICA2000, pp. 151–155 (2000)
8. Hori, G. : Proc. IEEE ICASSP2003, vol. 2, pp. 693–696 (2003)
9. Smith, S.T. : System and Control Letters, vol. 16, pp. 319–328 (1991)

Identifiability, Subspace Selection and Noisy ICA

Mike Davies

DSP & Multimedia Group
Queen Mary University of London
Mile End Road, London E1 4NS, UK
michael.davies@elec.qmul.ac.uk

Abstract. We consider identifiability and subspace selection for the ICA model with additive Gaussian noise. We discuss a canonical decomposition that allows us to decompose the system into a signal and a noise subspace and show that an unbiased estimate of these can be obtained using a standard ICA algorithm. This can also be used to estimate the relevant subspace dimensions and may often be preferable to PCA dimension reduction. Finally we discuss the identifiability issues for the subsequent 'square' noisy ICA model after projection.

1 Introduction

Most work on ICA to date has ignored the presence of noise even when it is known to exist. While there has been some debate (see for example [12]) on the advantages and disadvantages of including the noise term, there are applications where background noise is known to be present (e.g audio source separation [3], or telecommunications [15]) and where noise reduction is important.

There have been a number of attempts to tackle the problem of noisy ICA. Recently solutions based on Higher Order Statistics (HOS) only [5], bias removal (quasi-whitening) [10] and density estimation, e.g. using a Mixture of Gaussians (MoG) model, [14, 1], have been proposed. Each of these comes with its own drawbacks: bias removal requires prior knowledge of the noise covariance matrix; HOS only methods typically require more data and are sensitive to outliers; and density estimation, while on the face of it, the most general solution, is typically computationally expensive and difficult to train. We will also see that in fact the noisy ICA model is not fully identifiable.

It is well known that neglecting the presence of noise in the ICA formalism introduces bias into the subsequent estimators. Furthermore consideration of the noise is actually fundamental in tackling the case of more sensors than sources ($M > N$), since, if there is no noise, we can arbitrarily throw away all but N of the observations (this is due to the famous equivariance property of ICA, [6]). Most papers, e.g. [7], currently advocate projecting down to an equal number of sources and sensors using Principal Component Analysis (PCA) to improve the Signal-to-Noise Ratio (SNR). This can be justified if the noise term is 'small' in comparison with the Independent Components (ICs). However PCA does not always provide the appropriate projection for all noise models. Instead we consider the optimal subspace selection, which turns out to be well defined, and subsequently show that this can be achieved using the outputs of a standard FastICA algorithm *without* the need for bias correction. We also show that negentropy approximations used in algorithms such as FastICA provide a much more apropriate measure for order selection than the eigenvalues from PCA. In fact we are using ICA here much more in the spirit of projection pursuit since component independence is not strictly required. We give an example that illustrates how dramatically different to PCA projection this can be.

Finally we conclude with a discussion of the identifiability issues of the subsequent 'square' noisy ICA model, indicating a weakness in the overall system when both the source densities and noise covariance are unknown *a priori*.

2 The Noisy ICA Model

Let $x \in \mathbb{R}^M$ be an observed random vector. We will assume that x has the following decomposition:

$$x = As + z \tag{1}$$

where A is a constant unknown $M \times N$ mixing matrix, s is a zero mean non-Gaussian random vector with independent components and z is a zero mean Gaussian random vector with unknown covariance C_z. We furthermore assume that A has full column rank, the covariance matrices, C_x and C_z are positive definite and that $M \geq N$ (for $M < N$ the issues become significantly more complicated *even* in the absence of noise - see [8]).

Our statistical model consists of the following unknowns: $\{A, C_z, p_i(s_i)\}$. As with standard ICA, [6], we can consider different degrees of blindness. For example it is quite common to assume knowledge of the noise covariance and/or the individual source densities, $p_i(s_i)$, e.g. [10], in which case the problem is well posed. Here we consider the fully blind case where we assume only that the source components are mutually independent and non-Gaussian.

2.1 Identifiability of A

The identifiability of non-Gaussian components was comprehensively tackled in the seminal work of Kagan *et al.*, [13] chapter 10, in the early seventies and has recently been rediscovered to shed light on the identifiability conditions for both overcomplete ICA, [8] and noisy ICA, [4]. We thus have the following result, taken from [4] which is directly deducible from [13], theorem 10.3.1.

Theorem 1. *Let $x = As + z$ be the random vector as defined above. Then A is unique up to an arbitrary permutation and scaling.*

The only difference between identifiability with and without noise is that here all the sources must be non-Gaussian so as to be able to distinguish them from the Gaussian noise.

2.2 A Canonical Decomposition for Noisy ICA with $M > N$

The noisy ICA problem admits the following canonical form that splits the observation space into a signal subspace and a noise subspace (we first reported this in [4] but have subsequently found an equivalent decomposition was given in [2]).

$$\begin{bmatrix} y_1 \\ y_2 \end{bmatrix} = \begin{bmatrix} s + z_1 \\ z_2 \end{bmatrix} \qquad (2)$$

where z_1 is an N dimensional Gaussian random vector, z_2 is an $M - N$ dimensional Gaussian random vector and s, z_1 and z_2 are all mutually independent. Furthermore these subspaces are unique. That is: it is always possible to reduce the $M \times N$ problem into the $N \times N$ dimensional problem $y_1 = s + z_1$ where the other $M - N$ directions can be completely removed (If $M = N$ then clearly z_2 is degenerate).

One way to construct such a transform is to use the linear Minimum Mean Squared Error (MMSE) estimator for s given A and x which takes the form: $\hat{s} = C_s A^T C_x^{-1} x$ (see, for example, [9]). Although we do not know the covariance C_S this simply rescales the source estimates and does not change the subspace thus we arbitrarily set $C_s = I$. Being linear, this estimator serves to decorrelate the source and noise subspaces (as required in our canonical form). We can therefore construct the canonical form as follows:

$$y = \begin{bmatrix} y_1 \\ y_2 \end{bmatrix} = \begin{bmatrix} (A^T C_x^{-1} A)^{-1} A^T C_x^{-1} \\ U_{\bar{A}}^T \end{bmatrix} x \qquad (3)$$

where $U_{\bar{A}}$ is the $(M - N) \times N$ dimensional matrix of orthogonal vectors that span null(A^T) (obtainable from the singular value decomposition of A. It is easy to show that this also meets the requirement that z_1 and z_2 are mutually independent.

Clearly we can also spatially whiten this decomposition while retaining the subspace decomposition:

$$u = \begin{bmatrix} u_1 \\ u_2 \end{bmatrix} = \begin{bmatrix} C_{y_1}^{-\frac{1}{2}} & 0 \\ 0 & C_{y_2}^{-\frac{1}{2}} \end{bmatrix} y \qquad (4)$$

Since the spatially whitened data is unique up to an arbitrary rotation this shows that, once we have spatially whitened data, the signal subspace, U_s, and the noise subspace, U_n, are orthogonal. In contrast, the IC directions are not necessarily orthogonal.

3 Order and Subspace Selection

As mentioned in the introduction, the noiseless ICA framework offers no indications of how best to deal with more sensors than sources. Indeed choosing any N sensors (generically) should be equivalent. The usual advice (though it has its critics, [11]) is to use PCA to project the observation data onto a vector of the same dimension as the number of sources (assuming we know this number!) down to the square ICA problem. However, ironically, we will see that direct application of noiseless ICA deals with noise in a much more principled manner.

PCA projection can be justified when the noise covariance is isotropic ($C_z = \sigma_z^2 I$) in which case:

$$C_x = AA^T + \sigma_z^2 I = U_A(\Sigma_A^2 + \sigma_z^2 I)U_A^T \qquad (5)$$

where $A = U_A \Sigma_A V_A^T$ is the singular value decomposition of A. Similarly, if the noise is significantly smaller than the signal components, [16], PCA can distinguish between the noise floor (the smallest $N - M$ principal values) and the subspace spanning A. However this argument no longer holds for a general noise covariance (If we know C_z we can always transform x such that this is the case). Indeed when the noise is directional and not insignificant with respect to the component size the PCA projection can result in an extremely poor transformation.

In comparison the canonical decomposition in section 2.2, identifys an optimal projection, assuming that we know A. That is: choosing y_1 in equation (2) is optimal, projecting out the noise subspace z_2. This alone can have a significant de-noising effect.

To see the difference between the PCA and the optimal projection we considered a pair of independent binary sources observed through a 4×2 dimensional mixing matrix, A, chosen randomly, with additive noise that was strongly anisotropic. Figure 1 shows scatter plots of 10000 samples projected first onto the optimal 2-D subspace (left) and the same data projected onto the first two principal components (middle). It is clear that the PCA projection has not only failed to reduce the noise but that the four clusters associated with the binary data are completely indistinguishable. In constrast to this the optimal projection nicely separates out the clusters indicating the potential for significant noise reduction.

Of course, the optimal subspace projection currently requires knowledge of the mixing matrix *a priori*. We now show that we can identify this using standard FastICA, without restorting to bias correction, HOS only techniques or complicated density modelling.

We concentrate on ICA methods that search for orthogonal directions within pre-whitened data, u, (either gradient or fixed point based) and that have as their aim the maximization of some approximation of negentropy (e.g. see [11])

$$J(u) \propto (E\{G(u)\} - E\{G(v)\})^2 \qquad (6)$$

where $G(\cdot)$ is some nonlinear function and v is a zero mean unit variance Gaussian random variable.

For simplicity let us assume that $G(\cdot) = u^4$ so that we are dealing with a kurtosis based ICA algorithm (though the conclusions should be more generally applicable). If $u \in U_n$ we have $J(u) = 0$ while if $u \in U_s$ we know that $J(u) \geq 0$. Finally suppose that u has the form:

$$u = \alpha u_s + \sqrt{(1-\alpha^2)} u_n, u_s \in U_s, u_n \in U_n \qquad (7)$$

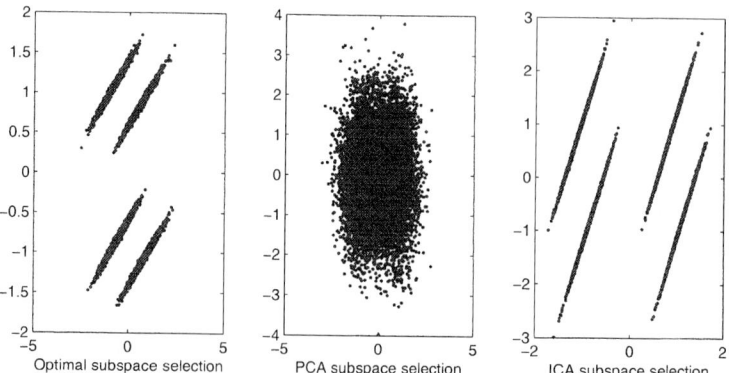

Fig. 1. Scatter plots of 2-dimensional projections using: the optimal canonical form (left), the first two Principal Components (middle) and the first to Independent Components using fastICA without bias correction (right).

with $0 \leq \alpha \leq 1$. Then from the mutual independence of u_s and u_n and the linearity property of kurtosis, $J(u)$ is maximum for $\alpha = 1$. Thus the ICA algorithm will select $M - N$ directions spaning U_s with the remaining directions spanning U_n due to the orthogonality constraint.

In summary, while standard ICA produces biased estimates of the individual components, which do not necessarily form an orthogonal set of directions in the pre-whitened data, it does produce unbiased estimates of the signal and noise subspaces and therefore the first N IC directions provide an estimate for the optimal projection.

To illustrate this the righthand plot in figure 1 shows the subspace associated with the first two ICs estimated for the noisy data using $G(u) = \ln \cosh(u)$. We stress that these estimates were made using the FastICA algorithm, [11], *without* any bias correction. It is clear that we have achieved a similar level of de-noising to the optimal projection.

This still leaves us with the problem of order selection. Here, since the ICs are assumed to be non-Gaussian, $J(u)$, itself, provides us with a good indicator of model order. We thus propose using the negentropy estimates in a similar manner to the eigenvalue spectrum used in PCA.

Figure 2 shows plots of the negentropy spectrum for the FastICA and the eigenvalue spectrum for the PCA. We can clearly identify from the negentropy spectrum that there are 2 non-Gaussian components followed by a 2 dimensional noise subspace. In contrast to this the eigenvalue spectrum tells us nothing about the order of the noisy ICA.

4 Identifiability Issues for 'Square' Noisy ICA

We conclude by looking at the identifiability of the 'square' noisy ICA problem, which we can now easily obtain using standard ICA. Although, in theory, we can

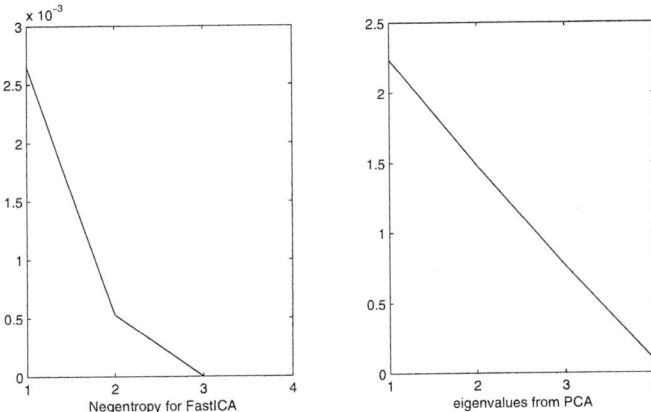

Fig. 2. A plot of the negentropy values for Scatter plots of the 2-dimensional projections using: the canonical form (left); and the pseudo-inverse (right).

identify the mixing matrix A, due to the presence of noise, inverting A does not give us direct access to the independent sources. Furthermore optimal estimation of the sources requires the knowledge of the source distributions, $p_i(s_i)$, and the noise covariance, C_Z. In [4] the following identifiability result was derived for the canonical form:

Theorem 2. *Let y_1, and z_1 be defined as above. Then only the off-diagonal elements of the C_{z_1} are uniquely identifiable. The diagonal elements of C_{z_1} and the source distributions, $p(s_i)$, are not.*

This is not really surprising as it is generally possible to incorporate some of the noise into the sources and vice versa. The ambiguity is related to the notion of non-separability in [8].

What is perhaps surprising is that the amount of noise allowed to be incorporated into one source is dependent on the noise being incorporated into the other sources. This is best illustrated with a simple example.

4.1 A Simple Example

Consider the 4×2 example with binary sources used above. We will assume that we have projected out the noise subspace and have removed the remaining bias in the estimates by one means or other. Let us denote the (observed) covariance matrix of y_1 by:

$$C_{y_1} = \begin{bmatrix} r_{11} & r_{12} \\ r_{12} & r_{22} \end{bmatrix} = I + C_{z_1}$$

We can then write the noise covariance C_{z_1} as:

$$C_{z_1} = \begin{bmatrix} c_1 & r_{12} \\ r_{12} & c_2 \end{bmatrix} \quad (8)$$

The off-diagonal terms are immediately observable from C_{y_1}. In contrast c_1 and c_2 are ambiguous. The extent of the ambiguity region is shown as the shaded area in figure 3, below.

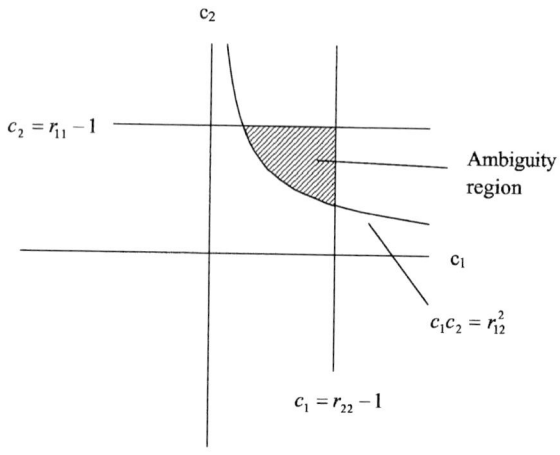

Fig. 3. The ambiguity region for the simple binary source example.

The upper bounds for c_1 and c_2 are defined by, in this case, the true source distributions (we have set the scale ambiguity to one). This is because $p_i(s_i)$ cannot be further de-convolved by a Gaussian of finite variance. In general the top right hand corner of the ambiguity region gives the *minimum variance source estimates*.

The lower bound on c_1 and c_2 depends on the degree of correlation between the components of z_1 and corresponds to the noise covariance becoming rank deficient. Note it is a joint function of c_1 and c_2. Thus there is no unique solution that 'leaves' as much *independent* noise as possible in each of the individual sources.

4.2 Noisy ICA and ML Estimation

Given this ambiguity, ML estimates, such as those in [1] based upon MoG models, are not neccessarily optimal since the selection of the diagonal terms in the noise covariance, C_{z_1} will be essentially arbitrary. One solution is to include additional knowledge of the nature of the source densities. For example, the minimum variance MoG source density will be characterized by the presence of at least one degenerate Gaussian (variance = 0). Another alternative that makes the problem well-posed again is to assume that the noise covariance is isotropic, $C_z \propto I$. In this case PCA can also be justified for the dimension reduction stage.

5 Conclusion

We have shown that a standard ICA algorithm using pre-whitening, in the spirit of projection pursuit, provides unbiased estimates of the signal and noise subspace for the undercomplete noisy ICA model and thus offers optimal de-noising when projecting onto the signal subspace. This is in contrast to PCA which, while more popular for dimension reduction, is only meaningful in a restricted set of circumstances.

Once a 'square' noisy ICA model has been obtained a number of techniques can be used to avoid the bias induced by standard ICA. However further noise reduction is made more difficult since the model is not fully identifiable.

Acknowledgment

I would like to thanks Nikolaos Mitianoudis for provision of the FastICA code.

References

1. H. Attias. Independent Factor Analysis. *Neural Comp.*, 11, 803-851, 1998.
2. O. Bermond and J-F. Cardoso, Approximate Likelihood for noisy mixtures, *Proc. ICA '99*, 1999.
3. M.E. Davies. Audio Source Separation. In *Mathematics in Signal Processing V*, edited by J. McWhirter and I. Proudler, 2002.
4. M. E. Davies, Identifiability Issues in noisy ICA. To appear in *IEEE Sig. Proc. Lett.*, May, 2004.
5. L. De Lathauwer, B. De Moor and J. Vandewalle. A technique for higher-order-only blind source separation. In *Proc. ICONIP*, Hong Kong 1996.
6. J-F. Cardoso. Blind signal separation: statistical principles. Proceedings of the IEEE, 9(10), 2009-2025, 1998.
7. P. Comon. Independent Component Analysis: a new concept? *Signal Processing*, 36(3), 287-314, 1994.
8. J. Eriksson and V. Koivunen. Identifiability and separability of linear ICA models revisited. *4th International Symposium on Independent Component Analysis and Blind Signal Separation (ICA2003)*, Nara, Japan, 2003.
9. M.H. Hayes, *Statistical Digital Signal Processing and Modeling*, John Wiley & Sons, 1996.
10. A. Hyvarinen. Gaussian Moments for noisy Independent Component Analysis. *IEEE Signal Processing Letters*, 6(6):145–147, 1999.
11. A. Hyvarinen, J. Karhunen and E. Oja, *Independent Component Analysis*, John Wiley & Sons, inc, 2001.
12. ICA mailing list maintained by J-F. Cardoso, at: www.tsi.enst.fr/icacentral.
13. A.M. Kagan, Y.V. Linnik and C.R. Rao. *Characterization Problems in Mathematical Statistics.* Wiley, New York, 1973.
14. E. Moulines, J.-F. Cardoso, E. Gassiat. Maximum Likelihood for blind signal separation and deconvolution of noisy signals using mixture models. *ICASSP-97*, 1997.
15. C.B. Papadias. Globally Convergent Blind Source Separation Based on a Mulituser Kurtosis maximization criterion. *IEEE Trans. Signal Processing*, 48(12), 2000.
16. J-P. Nadal, E. Korutcheva and F. Aires, Blind source processing in the presence of weak sources. *Neural Networks*, 13(6):589-596. 2000.

Improving GRNNs in CAD Systems

Fulgencio S. Buendía Buendía[1], J. Miguel Barrón-Adame[2],
Antonio Vega-Corona[2], and Diego Andina[1]

[1] Universidad Politécnica de Madrid
Departamento de Señales, Sistemas y Radiocomunicaciones, E.T.S.I.
Telecomunicación, Madrid, Spain
{wac,diego}@gc.ssr.upm.es
[2] Universidad de Guanajuato
F.I.M.E.E., Guanajuato, México
tono@salamanca.ugto.mx, miguel@gc.ssr.upm.es

Abstract. Different Computer Aided Diagnosis (CAD) systems have been recently developed to detect microcalcifications (MCs) in digitalized mammography, among other techniques, applying General Regression Neural Networks (GRNNs), or Blind Signal Separation techniques. The main problem of GRNNs to achieve an optimal classification performance, is fitting the kernel parameters (KPs). In this paper we present two novel algorithms to fit the KPs, that have been successfully applied in our CAD system achieving an improvement in the classification rates. Important remarks about the application of Gradient Algorithms (GRDAs) are assessed. We make a brief introduction to our CAD system comparing it to other architectures designed to detect MCs.

1 Introduction

Breast cancer is a major cause of death among women; several researches have been presented to develop CAD systems capable to detect MCs in digitalized mammographies [6–9], that is an early symptom of breast cancer. Next figure shows the overall architecture of our CAD system:

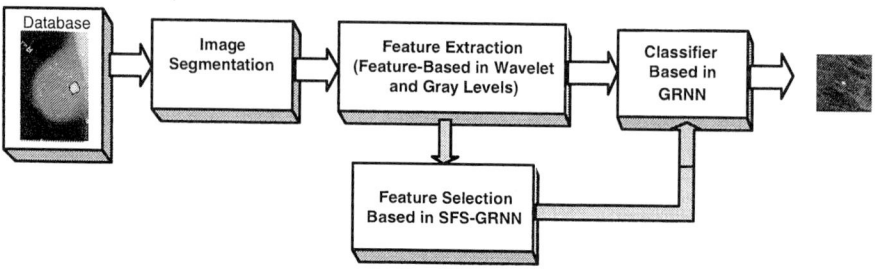

Fig. 1. Block diagram of our CAD system.

Suspicious area of each mammography is selected, avoiding to process the whole image. This part of the image is known as Region of Interest (ROI), and contains 16384 pixels. Different feature extraction strategies can be followed, for instance, Christoyianni in [7] applies Blind Signal Separation (BSS) technics obtaining the set of features assuming that the mammographies are made of a set of independent sources. We apply Wavelet filters, aiming to detect high frequency components in the mammographies [9], that characterizes MCs. Other alternatives to find high frequency components, as Histograms (that presented worse performance) or Independent Component Analysis (ICA) as described in [10] could be applied.

In [6], the proposed CAD system applies GRNNs in the selection of the best features and FFNN networks in the final classifier. Our CAD system applies GRNN structures in both stages, the selection of the best features and the final classification. Different algorithms to fit the Kernel Parameters (KPs) of GRNN structures have been proposed by many researchers [1-4]. Nowadays the strategy proposed by Specht in [1], is the most applied one, although some recent researches applies Genetic Algorithms (GAs) [4]. GRNNs classification performance depends on the KPs fit and the number of samples in the Training Data Set (TDS).In this paper we present two new algorithms to fit the KPs and analyze the application of Gradient Algorithms (GRDAs). The proposed algorithms allow us to apply GRNNs in the final classifier, if the KPs weren't properly fit, the classification would become very noisy and inaccurate. In order to test this structure a comparison to FFNN networks was performed; although similar results were obtained with both classifiers, the training time was significantly lesser with GRNNs.

In Section 2, a brief introduction about the GRNN structures is performed. In Section 3, we explain the new algorithms. In Section 4 explains the results obtained with the proposed strategies. Conclusions are summarized in Section 5.

2 General Regression Neural Networks

GRNN structures are a regression method proposed by Nadaraya-Watson and introduced by Specht [1]. The principal advantages of GRNN are fast learning and convergence to the optimal regression surface as the number of samples becomes very large. These structures just need to be trained once to achieve optimal performance in classification. Being $\mathbf{x}$ a pattern vector to be classified, y a scalar value to be estimated and $f_{\mathbf{x}y}(\mathbf{x}, y)$, the joint probability density function (**pdf**) of $\mathbf{x}$ and y. Expected value of y, given $\mathbf{x}$, is defined in Eq.(1) as

$$E[y|\mathbf{x}] = \frac{\int_{-\alpha}^{\alpha} y f_{\mathbf{x}y}(\mathbf{x}, y) dy}{\int_{-\alpha}^{\alpha} f_{\mathbf{x}y}(\mathbf{x}, y) dy} \qquad (1)$$

Probability distribution function is unknown, it must be estimated from the samples set $\{\mathbf{X}, \mathbf{Y}\}$, where $\mathbf{X} = \{\mathbf{x}^{(q)} : q = 1,..,M\}$ is the n-dimensional vectors set and $\mathbf{Y} = \{y^{(q)} : q = 1,..,M\}$ is the target set. Applying a gaussian parametric estimation the Eq.(1) becomes as follows

$$\hat{y}(\mathbf{x}) = \frac{\sum_{q=1}^{M} y^{(q)} \exp(-\frac{D_q^2}{2\sigma_i^2})}{\sum_{q=1}^{M} \exp(-\frac{D_q^2}{2\sigma_i^2})} \quad (2)$$

where M is the feature vector number, σ_i (i can be equal to q) is the width of the i-th class gaussian kernel, and $D_q^2 = (\mathbf{x} - \mathbf{x}^{(q)})^T(\mathbf{x} - \mathbf{x}^{(q)})$ is the Euclidean distance among the q-th sample and the input vector. Clustering samples and obtaining the centres allow to rewrite Eq.(2) as

$$\hat{y}(x) = \frac{\sum_{i=1}^{m} A_i \exp(-\frac{D_{c_i}^2}{2\sigma^2})}{\sum_{i=1}^{m} B_i \exp(-\frac{D_{c_i}^2}{2\sigma^2})} \quad (3)$$

- A_i is the samples number of the i-th class multiplied by the output value of the class.
- B_i is the samples number of the i-th class.
- D_{c_i} is the Euclidean distance to the c_i center.

To build up the decision regions, it is necessary to fit the $\boldsymbol{\sigma}$ minimizing the Mean Square Error (MSE), that is a n-dimensional function without local minimums [1]. It can be seen that this estimator is the likelihood ratio in the Bayes sense [11].

3 Fitting Algorithms in Clustering Problems

When a TDS given, the A_i, B_i and D_i parameters of Eq. 3 are directly obtained, the only parameter to be fit is the vector of sigmas $\boldsymbol{\sigma}$. First, we have normalized all the input vectors the [0,1] interval, it assures the output to be bounded. Fit $\boldsymbol{\sigma}$ is an optimization problem that consists on finding the vector that satisfies $\nabla(E) = \bar{0}$, being E the error function, $E(\bar{\sigma}) = \sum_{i=1}^{m}(\bar{y} - \hat{y}(\bar{\sigma}))^2$, that is $E(\bar{\sigma})$: $[0, 1] \times [0, 1] \times ...n... \times [0, 1] \rightarrow [0, 1]$. n is the number of clusters and m the number of elements in the TDS. We have developed the fitting algorithms with a problem of two clusters to be able to represent them into 3D graphs.

The strategy proposed by Specht in [1], consists on leaving all the $\boldsymbol{\sigma}$'s components fixed but one, that is optimized minimizing the error. Once the first sigma has been fit, it becomes fixed. It is repeated with all the sigmas. This solution doesn't ensure that the minimum error is reached, since it can be easily seen that the MSE obtained depends on the initial values of the weights. Figure 3(a), shows the minimization of the error varying σ_1 fixed σ_2. Figure 3(b), σ_2 is being fit. Figure 3 shows that the obtained error is not the minimum of the error surface.

We propose to apply this strategy iteratively, as Algorithm 1; this solution doesn't depend on the initial point achieving impressive results. Figure 4 shows the behavior of this algorithm.

The first point of Algorithm 1, initializes all the $\boldsymbol{\sigma}$ components to 0.5. The main loop is performed between the points two to seven, each iteration of this loop performs an optimization of $\boldsymbol{\sigma}$. The loop between the points three to five recovers the n components of $\boldsymbol{\sigma}$. To optimize each $\boldsymbol{\sigma}$'s component leaving the

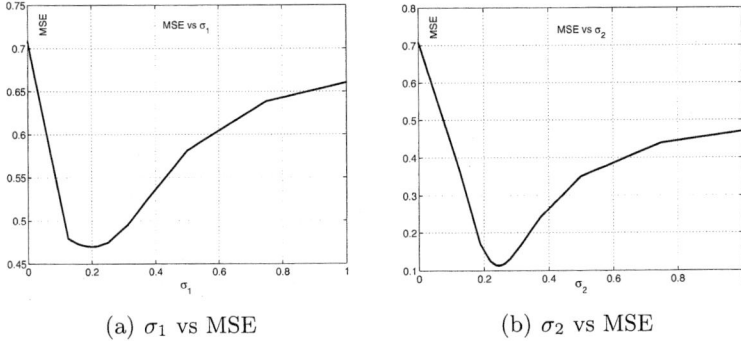

Fig. 2. 2-D KP fit representation applying the algorithm proposed by Specht in [1].

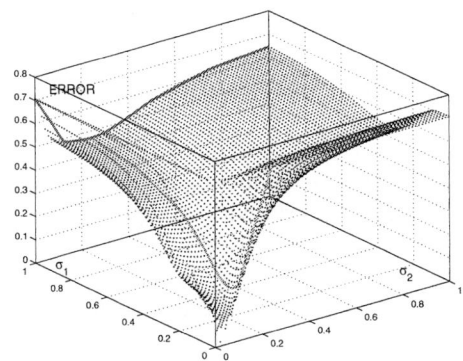

Fig. 3. 3-D representation of the algorithm proposed by Specht in [1].

Algorithm 1 Applying iteratively the strategy proposed by Specht in [1], the error of the TDS descends drastically.

❶ $\sigma_i = 0.5 \ \forall \ i \in 1..n$, ires$= \frac{1}{\text{resolution}}$
❷   for$(t = 0 : t < \text{iters} : t = t + 1)$
❸   for$(i = 0 : i < \text{n} : i = i + 1)$
        $len = 0.5$, $E = E(\bar{\sigma})$
❹   for $(j = 0 : j < \text{ires} : j = j + 1)$
           $len = len/2$, $a = \sigma_i - len$, $b = \sigma_i + len$
           $E_a = E(\bar{\sigma})|_{\sigma_i = a}$ $E_b = E(\bar{\sigma})|_{\sigma_i = b}$
           if $(E_a < E)$
              $E = E_a$, $\sigma_i = a$
           else if $(E_b < E)$
              $E = E_b$, $\sigma_i = b$
           end if
❺   end for
❻   end for
❼   end for

rest fixed, we have applied a variation of the Newton's Successive Approximation Algorithm, [12], between the points 4 and 5.

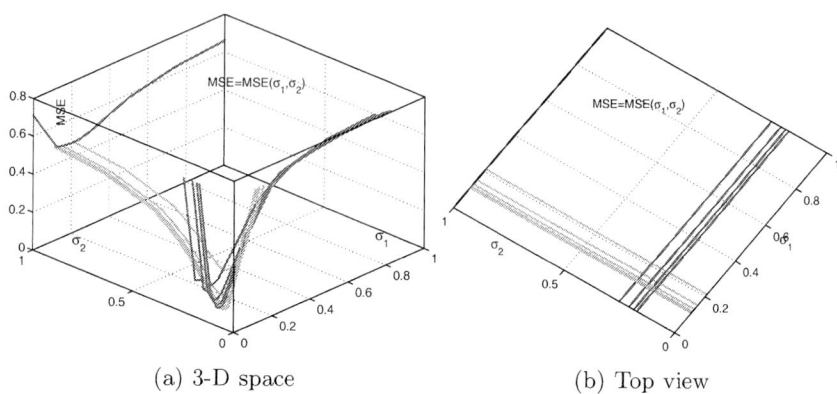

(a) 3-D space (b) Top view

Fig. 4. KP evolution with **Algorithm 1**.

The second Algorithm proposed consists on discretizing the $\boldsymbol{\sigma}$ space $[0,1] \times [0,1] \times ...n... \times [0,1]$, obtaining a grid with M^n points, where M is the number of divisions in every $[0,1]$ interval, Figure 3(a) shows a grid example where $n = 2$ and $M = 6$. Fitting $\boldsymbol{\sigma}$, just consists on calculate the error in every point of the obtained grid and select the one with the smallest error.

Algorithm 2 Swapping the classes space with a $n \times n$ grid. The bi-dimensional case is outlined.

❶ $error_{min} = 1$
❷ for($\sigma_1 = 0.0 : \sigma_1 = 1.0 : \sigma_1 = \sigma_1 + 1/n$)
　　for($\sigma_2 = 0.0 : \sigma_2 = 1.0 : \sigma_1 = \sigma_1 + 1/n$)
　　　(a)error = MSE(σ)
　　　(b)if ($error < error_{min}$)
　　　　$error_{min} = error$
　　　　$\boldsymbol{\sigma}_{min} = \boldsymbol{\sigma}$
　　end for
　end for
❸ end

However this algorithm can only be applied when the number of clusters is low, it assures to reach a perfect fit of the KP. In the detection stage of the CAD system proposed in [9] a surprisingly improvement of the system's performance have been achieved. This is a perfect solution to fit two classes problems varying σ_1 and σ_2 in the interval $[0,1] \times [0,1]$, Figure 3b shows the error surface obtained with a 70×70 grid. In order to reduce computing time, it can be obtained a

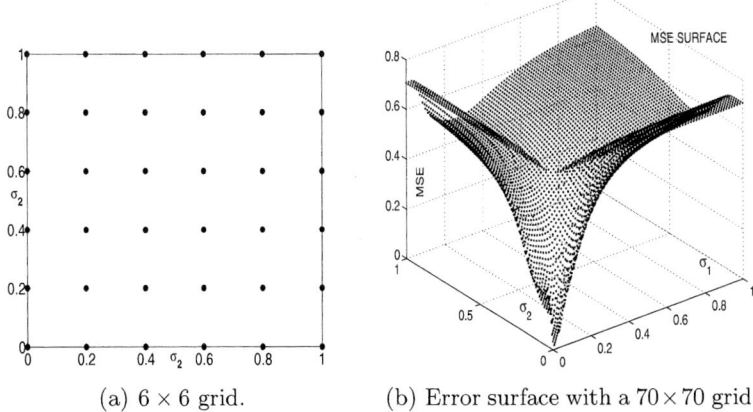

(a) 6 × 6 grid. (b) Error surface with a 70×70 grid.

Fig. 5. Example grid and error surface obtained for **Algorithm 2**.

first approximation of the sigmas vector with another strategy, and then apply Algorithm 2 in a reduced area, ensuring to fit the sigmas vector.

3.1 Gradient Algorithms

Some authors have studied the application of GDRA to fit the kernel parameters of GRNN structures [2]. Nowadays GRDA are being applied in the training of RBF Networks, that have the same kernel [5].

Since the error surface doesn't have local minima, GRDA are a ideal solution to solve the problem, obtaining the σ that satisfies $\nabla(E) = \bar{0}$, this condition can be applied to verify that a obtained σ is a properly solution. The general equation of GRDAs is:

$$\sigma_{n+1} = \sigma_n + K\bar{\nabla}E \qquad (4)$$

Where $\bar{\nabla}E$ is the gradient vector and K is the step length. GRDA have been broadly studied, with several strategies to find the optimal step length [12,13]. In this work we have successfully applied the Step Descend Gradient Algorithm, [12] in order to compare the performance of all algorithms.

4 Results

Data have been split into the Training Data Set (TDS) and the verification one (VDS). Figures. 4, 5 and 6, show the training processes in different algorithms. Figure 3(a) shows the outputs of the Network, introducing a TDS with the classical classification strategy, while Figure 3(b) shows the same samples classified with the other algorithms. Next table shows the results obtained with all algorithms and the number of calls to measure the error.

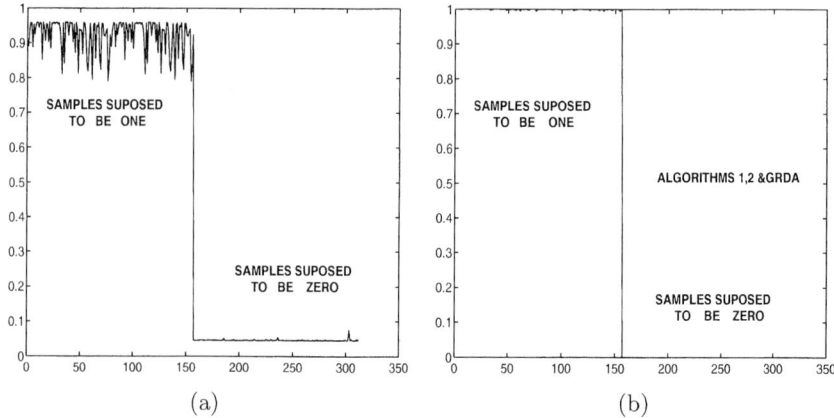

Fig. 6. (a) Noise fitting of the TDS and (b)Perfect fitting of the TDS .

Table 1. Comparative among the different fitting algorithms.

	MSE in TDS	MSE in VDS	Calls to obtain MSE
Classical strategy	0.1037	0.573	30
Algorithm 1	$2.5640 * 10^{-7}$	0.052	300
Algorithm 2	0	$2.2845 * 10^{-4}$	4900
Step Descend GDR	$1.4885 * 10^{-5}$	0.154	383

As Figures 4 and Table 1 show there is a big difference between a properly fit and a noisy one. The TDS had was 180 samples and the VDS 49152 samples. Best performance was achieved by Algorithm 2, that is the one that is currently being used in our CAD system [9].

5 Conclusions

In this paper we briefly describe a CAD system designed to detect MCs and the improvement in its performance when the kernel parameters are properly obtained. Two Algorithms to fit the KPs have been proposed. Presently, Algorithm 2 is the one that is being applied to fit σ. This allows to apply GRNNs in the classification stage of the system. If it weren't properly fit, the final classification would be very poor and noisy. Algorithm 2 might be the best solution to fit σ in detection problems. When the number of cluster grows, it is not suitable; in this case we propose to apply either Algorithm 1 or GRDAs.
Nevertheless our CAD system achieves an excellent classification performance, other CAD architectures have been considered, like BSS techniques, to perform feature extraction in the classification stage.

References

1. Specht D.F. A General Regression Neural Network. *IEEE Transactions on Neural Networks*, **2**(6) (1991) 568–576
2. Masters T. and Land W. A New Training Algorithm for the General Regression Neural Network. *IEEE Transactions on Neural Networks*, **2** (6) (1997) 1990–1994
3. Zhang J., Li Z., Sun J. and Wei Z. Computing Models Based on GRNNS. *IEEE Trancsations on Neural Networks*,(2003) 1853–1856
4. Ling S.H., Frank H.F., Lam H.K. and Peter K.S. Tuning of the Structure and Parameters of a Neural Network Using an Improved Genetic Algorithm. *IEEE Trancsations on Neural Networks*, **14**(1) (2003) 1853–1856
5. Karayiannis N.B. and Randolph-Gips M.M. On the construction and Training of Reformulated Radial Basis Function Neural Networks. *IEEE Trancsations on Neural Networks*, **14**(4) (2003) 835–846
6. Songyang Y. and Ling G. A CAD system for automatic detection of clustered microcalcifications in digitized mammogram films. *IEEE Transactions on Medical Imaging*,**2**(2) (2000) 115–126
7. Christoyianni I., Koutras A. and Dermatas E. Computer Aided Classification of Mammographic Tissue Using Independent Component Analysis *Digital Signal Processing, 2002. 14th International Conference on ,Volume: 1 , 1-3 July* , **14**(4) (2002) 163–166
8. Abe H., Ashizawa K., Katsuragawa S., MacMahon H. and Kiuno D. Use of an Artificial Neural Network to Determine the Diagnostic Value of Specific Clinical and Radiologic Parameters in the Diagnosis of Interstitial Lung Disease on Chest Radiographs. *Journal Academic Radiology*, **9**(1) (2002) 13–17
9. A. Vega-Corona, A. Álvarez, and D. Andina. Feature Vectors Generation for Detection of Microcalcifications in Digitized Mammography Using Neural Networks. *Artificial Neural Nets Problem and Solving Methods, Springer Verlag, LNCS2687*, **2**(2687) (2003) 583–590
10. S. Hornillo-Mellado, R. Martín-Clemente, J Acha C. Puntonet. Application of Independent Component Analysis to Edge Detection and Watermarking. *Artificial Neural Nets Problem and Solving Methods, Springer Verlag, LNCS2687*, **2**(2687) (2003) 273–280
11. Kupinski M., Edwards D., Giger M. and Metz C. Ideal Observer Approximation Using Bayesian Classification Neural Networks. *IEEE Transaction on Medical Imaging*, 20(9) (2001) 886–889
12. Tood K.M. and Wynn C.S. *Mathematical Methods and Algorithms*. Prentice Hall Inc., Upper Saddle River, New Jersey (2000). ISBN 0-201-36186-8
13. Tood C. M. *Neural Networks for Pattern Recognition*. Clarendon Press,Oxford (1996). ISBN 0-19-853849-9196

Fisher Information in Source Separation Problems

Vincent Vigneron and Christian Jutten

INPG-LIS CNRS UMR 5083
46, avenue Félix Viallet
38031 Grenoble cedex, France
vvigne@lis.inpg.fr, Christian.Jutten@inpg.fr

Abstract. The ability to estimate a specific set of parameters, without regard to an unknown set of other parameters that influence the measured data, or nuisance parameters, is described by the Fisher Information matrix (FIM), and its inverse the Cramer-Rao bound. In many adaptive gradient algorithm, the effect of multiplication by the latter is to make the update larger in directions in which the variations of the parameter θ have less statistical *significance*. In this paper, we examine the relationship between the Fisher information and the covariance of the estimation error under the scope of the source separation problem.

1 Introduction

In the Blind Source Separation (BSS) problem, we observe mixed zero-mean signals from independent sources. Ref. [4] contains an extensive bibliography on the subject. In the simplest noiseless case, n primary source signals, represented by the vector[1] $S(t) = [S_1(t), \ldots, S_n(t)]^T$, are observed through $m(\geq n)$ instantaneous mixtures of these signals $X(t) = [X_1(t), \ldots, X_m(t)]^T$, given by (for each time instant t)

$$X(t) = AS(t), \tag{1}$$

where $A = (a_{ij})$ forms the unknown non singular $m \times n$ matrix which does not depend on time t. For sake of readability, $X_t \equiv X(t)$.

This problem is closely related to *independent component analysis* (ICA) introduced by Comon [2]. The i-th component of S is denoted S_i (and similarly for the other vectors) and has the probability density function (pdf) p_{S_i}. We observe T realizations $x(t)$ of $X(t)$ such that $x(t) = As(t)$. The following assumptions hold throughout:

1. the components of $S(t)$ are mutually independent random variables (r.v.) with zero mean white random variables, with non gaussian marginal distributions and such that $\forall i, \mathbb{E}[S_i] = 0$ ($\mathbb{E}[.]$ denotes the expectation operator).
2. the matrix A exists and is a square full rank matrix ($n = m$).

---
[1] In the following, ^{T} is the transpose operator.

3. $\hat{Y} = BX$ is an estimator of the source signals, which is achieved as soon as BA is a $n \times n$ matrix with exactly one non-zero entry in each row and each column.
4. $p_S, p_{S_1}, p_{S_2}, \ldots, p_{S_n}$ are the (unknown) joint distribution of the sources and the marginal distributions of $S_1, \ldots, S_n$ respectively.

Mathematically, we wish to recover the sources through the demixing process $Y = BX = BAS$, that is, $BA = \Lambda$, where Λ is the product of a diagonal (scaling) matrix and a permutation matrix $(\delta_{i,\sigma(j)})$ accounting for not properly ordering the elements of Y [2].

Let's define a *pseudo log-likelihood* (scalar):

$$U_T(B) = \frac{1}{T} \sum_{t=1}^{T} \log\left(|\det B| g(BX_t)\right), \qquad (2)$$

where the vectors $X_t, t = 1, \ldots, T$ are the observations and g is a model for the density of the random variable BX. Statistically, $U_T(B)$ is a *contrast process* if it converges in probability toward a *contrast function* whose maximum is our solution. It is well known (see [5]) that B is the solution of

$$I_n + \mathbb{E}[\phi(BX)(BX)^T] = 0, \qquad (3)$$

where I_n is the identity matrix. In the following, $\lambda_{i,j}$ is the set of solutions of the integral equations $1 + \mathbb{E}[\phi_i(\lambda_{i,j})\lambda_{j,i}S_j] = 0, \forall i,j \in \{1,\ldots,n\}$, with $\phi(u_1, \ldots, u_n) = -\left[\frac{g'(u_1)}{g(u_1)}, \ldots, \frac{g'(u_n)}{g(u_n)}\right]^T$ defined as a score function. For any permutation σ of $\{1,\ldots,n\}$, we define by Λ_σ the matrix whose components are $\lambda_{i,\sigma(i)}\delta_{\sigma(i),j}$. In the following, the permutation operator σ will be omitted.

The main contributions of this article are threefold. First we review some properties of the likelihood function and characterize the estimation in the source separation problem. Second, we give an original parametrization of the FIM and examine the case of equalities of the unknown and the guessed distribution of the sources. Third we examine the case of a 3 sources problem and give an interpretation that highlights the concept of extractable source from a mixture of signals and its relationship with the Hessian matrix.

2 Matrix Fisher Information

In this section, we follow a standard framework to provide convergence results upon the law of the estimator $\hat{B}$ [3, p. 101-102]. As we don't need to assume that B is a matrix, the matrix B is rewritten for sake of readability as a vector $\theta(B)$ or $\theta = (\theta_1, \ldots, \theta_k), k = n^2$. The convention for reordering the elements b_{ij} of B in the vector θ is not relevant to this section. Furthermore, we introduce the notations $\theta_0 = \arg\max_\theta \mathbb{E}[U_T(\theta)]$ and $\hat{\theta} = \arg\max_\theta \mathbb{E}[\hat{U}_T(\theta)]$ is the solution of $U_T(\theta) = 0$, to match, respectively, the inverse of the mixing matrix, and the computed estimation.

2.1 Preliminary Lemmas

Let ϱ_T be the gradient vector of $U_T(\theta)$ defined as $\frac{\partial U_T(\theta)}{\partial \theta} = (\varrho_T^i)_{1 \le i \le k}$, where $\frac{\partial}{\partial \theta}$ is the vector gradient operator and $\varrho_T^i = \frac{\partial}{\partial \theta_i} U_T(\theta)$.

Let us define the cost function $\ell_T(\theta) = -TU_T(\theta) = \sum_{t=1}^{T} \log f(\theta, X_t)$ with $f(\theta, x) = |\det B| g(Bx)$. For $1 \le i, j \le k$, let $I_{ijT}(\theta) = \mathbb{E}[\frac{\partial}{\partial \theta_i} \ell_T(\theta) \frac{\partial}{\partial \theta_j} \ell_T(\theta)^T]$. The matrix $I_T(\theta) = \{I_{ijT}(\theta)\}_{1 \le i, j \le k}$ is called the *Fisher information matrix* w.r.t. θ at time T, abusively written $I(\theta)$.

As ϱ_T is a sequence of *iid* centered vectors, then, by the central limit theorem, the cdf of the sequence of random variables $(\frac{\varrho_T}{\sqrt{T}}, T \ge 1)$ converges in law to the the centered gaussian distribution $\mathcal{N}(0, I(\theta))$ [1]. This is denoted by

$$\frac{\varrho_T}{\sqrt{T}} \stackrel{\mathcal{L}(P_\theta)}{\to} \mathcal{N}(0, I(\theta)). \tag{4}$$

Recall that $X_t \stackrel{Pr}{\to} X$ means that $\lim_{t \to \infty} P(|X_t - X| \ge \epsilon) = 0$ for all $\epsilon > 0$, and it reads $(X_t, t \ge 1)$ converges in probability to X. If the *almost sure* limit of a sequence $(X_t, t \ge 1)$ exists, it is *essentially unique*, that is $X_t \stackrel{as}{\to} X$ (read 'X_t converges to X almost surely when t goes to infinity') and $X_t \stackrel{as}{\to} X'$, then $X = X'$, probably almost sure (P-as) [1]. We will also need the following theorem (see [3], page 9):

Theorem 1. *Let (Q_T) and (ϱ_T) be random sequences, such that $Q_T \stackrel{\mathcal{L}(P)}{\to} X$ and $\varrho_T \stackrel{Pr}{\to} \alpha$, where X is a random variable, and α is a real number, then the concatenate variable $(Q_T, \varrho_T) \stackrel{\mathcal{L}(P)}{\to} (X, \alpha)$.*

This theorem means that a vector of r.v. converges in law toward the limits vector of these r.v.

Then, suppose there exists a maximum likelihood estimator $\hat{\theta}$ which converges in probability P_θ to θ_0. Let $\mathcal{V}$ be a convex neighbourhood of θ_0 in which the function $f(\theta, x)$ is twice differentiable for any x. If $\hat{\theta} \in \mathcal{V}$, then the gradient $\nabla \ell_T(\hat{\theta}) = 0$ and the integral form of the Taylor expansion about some value θ_0 near to $\hat{\theta}$ is given below [3]:

$$0 = \frac{1}{\sqrt{T}} \varrho_T^i(\theta_0) + \sum_{j=1}^{k} \sqrt{T}(\hat{\theta}_j - \theta_{j0}) \frac{1}{T} \sum_{t=1}^{T} \int_0^1 \psi_{ij}(\theta_0 + u(\hat{\theta} - \theta_0), X_t) du, \quad 1 \le i \le k, \tag{5}$$

where we define $\psi_{ij}(\theta, x) = \frac{\partial^2}{\partial \theta_i \partial \theta_j} f(\theta, x), \forall i, j$. From [3], $\mathbb{E}_\theta[\psi_{ij}(\theta_0, X_T)] = -H_{ij}$, and $H = (H_{ij})_{1 \le i, j \le k}$ is the Hessian matrix. From the large number theorem, we know:

$$\frac{1}{T} \sum_{t=1}^{T} \psi_{ij}(\theta_0, X_t) \stackrel{P_\theta - as}{\to} -H_{ij}, 1 \le i, j \le k. \tag{6}$$

Suppose now that an algorithm provides $\hat{\theta}$ that converges in P_θ-probability, inside $\mathcal{V}$. By permuting the sum integral in (5) and using (6), we can write with vector-matrix notations:

$$\frac{1}{\sqrt{T}}\varrho_T(\theta_0) - \sqrt{T}(\hat{\theta} - \theta_0)H \xrightarrow{P_\theta} 0. \tag{7}$$

Assuming that the norm $|H_{ij}| \leq \infty$ and that H is inversible (this is generally true in the neighbourhood of $\mathcal{V}$), then by multiplying by H^{-1}, we have:

$$\frac{1}{\sqrt{T}}H^{-1}\varrho_T(\theta_0) \xrightarrow{\mathcal{L}(P_\theta)} \mathcal{N}(0, H^{-1}I(\theta)H^{-1}). \tag{8}$$

Hence[2],

$$-TH^{-1}\sum_{t=1}^{T}\psi_{ij}(\theta_0, X_t) \xrightarrow{P_\theta} I_n. \tag{9}$$

From theorem 1, it is straightforward to write:

$$\frac{1}{\sqrt{T}}(\hat{\theta} - \theta_0) \xrightarrow{\mathcal{L}(P_\theta)} \mathcal{N}(0, H^{-1}I(\theta)H^{-1}). \tag{10}$$

In (10), the value of interest is the covariance of the estimation error $H^{-1}I(\theta)H^{-1}$. If $\hat{\theta}$ is a maximum likelihood estimator which converges in P_θ-probability toward θ_0, we have when $g = p$ [3]:

$$I(\theta) = -H \Rightarrow \sqrt{T}(\hat{\theta} - \theta_0) \xrightarrow{\mathcal{L}(P_\theta)} \mathcal{N}(0, I(\theta)^{-1}). \tag{11}$$

The next sections intend to compute formally this expression for BSS estimators.

2.2 Reparametrization of the Fisher Information Matrix

θ-paramatrization was useful to evaluate the law of the error, but not for the calculation of the Fisher matrix. The simplest way to compute it is to keep B. Fisher information can be written as a fourth order tensor where each component

$$I_{ab,cd}(B) = \mathbb{E}[G_{ab}(B), G_{cd}(B)], \tag{12}$$

with $G_{ab}(B) = \frac{\partial U_T(B)}{\partial \beta_{ab}} = -\sum_k (\delta_{ak} + \phi_a(BX)(BX)_k^T)B_{kb}^{-T}$, where B_{kb}^{-T} refers to the component k, b of B^{-T}, $B_{kb}^{-T} = \{\alpha_{kb}\}$, and $Y = BX$ is the solution of the equation $I_n + \mathbb{E}[\phi(Y)Y^T] = 0$. Then, the Fisher information tensor is given by:

$$I_{ab,cd}(B) = \sum_{p,\ell} \mathbb{E}[(\delta_{ap} + \phi_a(Y_a)Y_p)\alpha_{pb}(\delta_{c\ell} + \phi_c(Y_c)Y_\ell)\alpha_{\ell d}],$$
$$= \sum_{p,\ell} d_{ap,c\ell}\alpha_{pb}\alpha_{\ell d}. \tag{13}$$

In (13), α_{ij} are unknown parameters, so $I(B)$ is not tractable. Nicer properties can be found if $I(B)$ is projected in the base of positive definite tensors. In this

---
[2] I_n is not a random variable, hence the convergence is in probability.

space, we can prove that $I(B)$ has a block-diagonal structure. This "controlled" change of base is obtained using left and right multiplication with matrices.

Consider the following the *quadratic form* $Q(CB) = \sum_{ab,cd} I_{ab,cd}(B)(CB)_{ab}$ $(CB)_{cd}$ where C is a matrix. We can write:

$$Q(CB) = \sum_{abcdijp\ell} d_{ap,c\ell}\alpha_{pb}\alpha_{\ell d}C_{ci}\beta_{id}C_{aj}\beta_{jb}$$

$$= \sum_{acijp\ell} d_{ap,c\ell}C_{ci}C_{aj}\delta_{i\ell}\delta_{jp} = \sum_{acp\ell} d_{ap,c\ell}C_{c\ell}C_{ap},$$

in which the Fisher matrix components depends only of B. Then, we can go on studying the term in (13):

$$d_{ab,cd} = \delta_{ab}\delta_{cd} + \mathbb{E}[\phi_a(Y_a)\phi_c(Y_c)Y_bY_d]. \tag{14}$$

Table 1 details the expression of the Fisher tensor components $D = \{d_{ab,cd}\}$ in two cases: the first case consider that the true distribution is known ($g = p$), the second case consider that g is chosen as a *symetric* distribution but $g \neq p$. In this table, $\{1, 2, 3, 4\}$ stands for tensor indices and, in general, $a \leq b \leq c \leq d$.

Table 1. Values of the Fisher tensor components when p is known or not.

indices				Fisher components	
a	b	c	d	$g \neq p$ and g symetric	$g = p$
1	2	3	4	0	0
1	1	2	3	0	0
1	2	1	3	0	0
1	2	3	1	0	0
1	2	2	3	0	0
1	2	3	2	$\mathbb{E}[Y^2]\mathbb{E}[\phi_a(Y_a)]\mathbb{E}[\phi_c(Y_c)]$	0
1	2	3	3	0	0
1	1	2	2	0	0
1	2	1	2	$\mathbb{E}[\phi_a^2(Y_a)]\mathbb{E}[Y_b^2]$	$\mathbb{E}[\phi_a^2(Y_a)]\mathbb{E}[Y_b^2]$
1	2	2	1	1	1
1	2	2	2	$\mathbb{E}[\phi_a(Y_a)]\mathbb{E}[\phi_b(Y_b)Y_b^2]$	0
2	1	2	2	0	0
2	2	1	2	$\mathbb{E}[\phi_c(Y_c)]\mathbb{E}[\phi_a(Y_a)Y_a^2]$	0
2	2	2	1	0	0
1	1	1	1	$-1 + \mathbb{E}[\phi_a^2(Y_a)Y_a^2]$	$-1 + \mathbb{E}[\phi_a^2(Y_a)Y_a^2]$

Table 1 shows that the Fisher matrix is more regular when the true distribution of the sources p is known. If the score function ϕ is odd (*i.e.* if g is symetric) and p is symetric, then we have a *block diagonal matrix* similar to the Hessian matrix, as demonstrated in [6]. If ϕ is derived from p, that is $\phi = -\frac{p'}{p}$, then we have a more simple form (see table 1). This is due to the fact that when $g = p$, the Fisher matrix entries $\mathbb{E}[\phi_b(Y_b)Y_b^2]$ will cancel as $\mathbb{E}[\phi(y)y^2] = -\int_y y^2 \frac{p'(y)}{p(y)}p(y)dy = -[y^2] + \int_y 2yp(y)dy = 0.$

Let us organize the Fisher matrix as we did in [6] for the Hessian matrix. We have 4 kind of blocks, located by the relation orders between the indices $abcd$:

$$D = \begin{bmatrix} (a,b) = (c,d) & & (a,b) > (c,d) & & & \\ & \ddots & & & (a,a,c,d) & \\ (a,b) < (c,d) & & (a,b) = (c,d) & & & \\ & & & (a,a,a,a) & & (a,a) > (c,c) \\ & (a,b,c,c) & & & \ddots & \\ & & & (a,a) < (c,c) & & (a,a,a,a) \end{bmatrix},$$

in which, using (3):

- if $(a,b) = (c,d)$:

$$D_{(a,b),(a,b)} = \begin{bmatrix} \mathbb{E}[\phi_a^2(Y_a)]\mathbb{E}[Y_b^2] & \mathbb{E}[\phi_a(Y_a)Y_a]\mathbb{E}[\phi_b(Y_b)Y_b] \\ \mathbb{E}[\phi_a(Y_a)Y_a]\mathbb{E}[\phi_b(Y_b)Y_b] & \mathbb{E}[\phi_b^2(Y_b)]\mathbb{E}[Y_a^2] \end{bmatrix}$$

$$= \begin{bmatrix} \mathbb{E}[\phi_a^2(Y_a)]\mathbb{E}[Y_b^2] & 1 \\ 1 & \mathbb{E}[\phi_b^2(Y_b)]\mathbb{E}[Y_a^2] \end{bmatrix} \quad (15)$$

- if $(a,b) > (c,d)$, that is, in the general case $b > a \geq c$ and $d > c$:

$$D_{(a,b),(c,d)} = \begin{bmatrix} \mathbb{E}[\phi_a(Y_a)\phi_c(Y_c)Y_b Y_d] & \mathbb{E}[\phi_b(Y_b)\phi_c(Y)Y_a Y_d] \\ \mathbb{E}[\phi_a(Y_a)\phi_d(Y_d)Y_b Y_c] & \mathbb{E}[\phi_b(Y)\phi_d(Y)Y_a Y_c] \end{bmatrix} \quad (16)$$

if $a \neq c$, then either

$$D_{(a \neq c, a \neq d)} = \begin{bmatrix} \mathbb{E}[\phi_c(Y_c)]\mathbb{E}[\phi_a(Y_a)]\mathbb{E}[Y_b Y_d] & 0 \\ 0 & 0 \end{bmatrix}, \quad \text{if } a \neq d \quad (17)$$

$$D_{(a \neq c, b \neq d)} = \begin{bmatrix} 0 & \mathbb{E}[\phi_c(Y_c)]\mathbb{E}[\phi_b(Y_b)]\mathbb{E}[Y_a Y_d] \\ 0 & 0 \end{bmatrix}, \quad \text{if } b \neq d \quad (18)$$

if $a = c$, then

$$D_{(a=c)} = \begin{bmatrix} 0 & 0 \\ 0 & \mathbb{E}[\phi_d(Y_d)]\mathbb{E}[\phi_b(Y_b)]\mathbb{E}[Y_a^2] \end{bmatrix}. \quad (19)$$

3 Covariance of the Estimation Error

Our goal is now to find out a more general representation for $H^{-1}I(\theta)H^{-1}$ such that the matrice H has a more tractable form. Let Ψ be some regular matrix. Then, Ψ^{-1} exists and we have, from previous considerations:

$$H^{-1}I(\theta)H^{-1} = (\Psi^{-T}\Psi^T H \Psi \Psi^{-1})^{-1}(\Psi^{-T}\Psi^T I(\theta)\Psi \Psi^{-1})(\Psi^{-T}\Psi^T H \Psi \Psi^{-1})^{-1}$$

$$= \Psi(\Psi^T H \Psi)^{-1} \cdot \Psi^{-T}\Psi^T I(\theta)\Psi \Psi^{-1} \cdot \Psi \left(\Psi^T H \Psi\right)^{-1}\Psi^T$$

$$= \Psi \Gamma^{-1} D \Gamma^{-1} \Psi^T, \quad (20)$$

with positive definite matrices $D = \Psi^T I(\theta)\Psi$ and $\Gamma = \Psi^T H\Psi$. Let $\Psi_{ij,k\ell} = \delta_{i,k}B_{\ell,j}$, then $\Psi_{ij,k\ell}^T = \delta_{i,k}B_{j\ell}$. The $\Psi_{ij,k\ell}$ notation is for matrix representation of the Fisher and Hessian tensors as in section 2.2. From (20), H is transformed in the new basis defined by the matrix Ψ as a block diagonal matrix Γ, similar to the Fisher matrix (see [6] for details on the notations):

$$\Gamma = \begin{bmatrix} \ddots & & & & \\ & \begin{matrix} \Gamma_{ijij} & \Gamma_{jiij} \\ \Gamma_{ijji} & \Gamma_{jiji} \end{matrix} & & 0 & \\ & & \ddots & & 0 \\ & 0 & & \Gamma_{iiii} & \\ & & 0 & & \ddots \end{bmatrix}, \qquad (21)$$

In equation (21),

$$\Gamma_{ijk\ell} = \delta_{jk}\delta_{i\ell}\frac{1}{\lambda_i \lambda_k} - \mathbb{E}[\phi'_i(\lambda_i S_i)S_j^2]\delta_{j\ell}\delta_{ik}, \qquad (22)$$

with the notations $\Gamma_{ij} = \Gamma_{ijij} = \begin{bmatrix} \kappa_{ij} & \eta_{ij} \\ \eta_{ij} & \kappa_{ji} \end{bmatrix}$, $\Gamma_i = \Gamma_{iiii} = \eta_{ij} + \kappa_{ij}$, $\eta_{ij} = \frac{1}{\lambda_i \lambda_j}$, $\kappa_{ij} = -\mathbb{E}[\phi'_i(\lambda_i S_i)S_j^2]$ and $\sigma_i^2 = \mathbb{E}[S_i^2]$. The inverse Γ^{-1} is also block diagonal, with blocks:

$$U_{i,j}^{-1} = -\frac{1}{\Delta}\begin{bmatrix} \kappa_{ji} & -\eta_{ij} \\ -\eta_{ij} & \kappa_{ij} \end{bmatrix} = -\frac{1}{\Delta}\begin{bmatrix} -b_j\sigma_i^2 & -\eta_{ij} \\ -\eta_{ij} & b_i\sigma_j^2 \end{bmatrix}, \qquad (23)$$

with $\Delta = \kappa_{ij}\kappa_{ji} - \eta_{ij}^2$, $b_i = \mathbb{E}[\phi'(\lambda_i S_i)]$ and $U_i^{-1} = \frac{1}{\alpha_{ii}+\kappa_{ii}}$. Let us illustrate this on a more explicit problem.

3.1 Estimation Error for a 3 Sources Separation Problem

In this case, the true Fisher information (estimated with the true cdf, $g = p$) can be written as a tri-diagonal matrix:

$$\hat{I}(\theta) = \begin{bmatrix} \gamma_1\sigma_2^2 & 1 & 0 & 0 & 0 & 0 & 0 & 0 & 0 \\ 1 & \gamma_2\sigma_1^2 & 0 & 0 & 0 & 0 & 0 & 0 & 0 \\ 0 & 0 & \gamma_1\sigma_3^2 & 1 & 0 & 0 & 0 & 0 & 0 \\ 0 & 0 & 1 & \gamma_3\sigma_1^2 & 0 & 0 & 0 & 0 & 0 \\ 0 & 0 & 0 & 0 & \gamma_2\sigma_3^2 & 1 & 0 & 0 & 0 \\ 0 & 0 & 0 & 0 & 1 & \gamma_1\sigma_2^2 & 0 & 0 & 0 \\ 0 & 0 & 0 & 0 & 0 & 0 & \ell_1 & 0 & 0 \\ 0 & 0 & 0 & 0 & 0 & 0 & 0 & \ell_2 & 0 \\ 0 & 0 & 0 & 0 & 0 & 0 & 0 & 0 & \ell_3 \end{bmatrix}. \qquad (24)$$

In (24), $\hat{I}(\theta)$ is block diagonal and easily invertible, with $\gamma_i = \mathbb{E}[\phi_i(Y_i)]$, $\phi_i = -\frac{p'_i}{p_i}$; ℓ_1, ℓ_2, ℓ_3 are non zero (non relevant) diagonal terms. For instance:

$$\begin{bmatrix} \gamma_1\sigma_2^2 & 1 \\ 1 & \gamma_2\sigma_1^2 \end{bmatrix}^{-1} = \begin{bmatrix} \gamma_2\frac{\sigma_1^2}{\gamma_1\sigma_2^2\gamma_2\sigma_1^2-1} & -\frac{1}{\gamma_1\sigma_2^2\gamma_2\sigma_1^2-1} \\ -\frac{1}{\gamma_1\sigma_2^2\gamma_2\sigma_1^2-1} & \gamma_1\frac{\sigma_2^2}{\gamma_1\sigma_2^2\gamma_2\sigma_1^2-1} \end{bmatrix}.$$

If p and g are symetric, then the covariance matrix computed in section 2.2 has the same shape as the Hessian matrix with the bloc diagonal components:

$$D_{ij} = \begin{bmatrix} a_i \sigma_j^2 & 1 \\ 1 & a_j \sigma_i^2 \end{bmatrix}, \quad D_{ii} = -1 + \mathbb{E}[\phi_i^2(Y_i) Y_i^2], \tag{25}$$

where $a_i = \mathbb{E}[\phi_i^2(Y_i)]$. Thus computing $\Gamma^{-1} D \Gamma^{-1}$ gives for each diagonal block (ij, ij):

$$U_{ij}^{-1} D_{ij} U_{ij}^{-1} = \frac{1}{\Delta^2} \begin{bmatrix} -b_j \sigma_i^2 & -\eta_{ij} \\ -\eta_{ij} & b_i \sigma_j^2 \end{bmatrix} \begin{bmatrix} a_i \sigma_j^2 & 1 \\ 1 & a_j \sigma_i^2 \end{bmatrix} \begin{bmatrix} -b_j \sigma_i^2 & -\eta_{ij} \\ -\eta_{ij} & b_i \sigma_j^2 \end{bmatrix} \tag{26}$$

$$= \frac{1}{\Delta^2} \begin{bmatrix} w_{11} & w_{12} \\ w_{12} & w_{22} \end{bmatrix} \tag{27}$$

where

$$\begin{cases} w_{11} = \sigma_i^2 (b_j^2 \sigma_i^2 a_i \sigma_j^2 + 2 b_j \eta_{ij} + \eta_{ij}^2 a_j) \\ w_{12} = \eta_{ij}^2 + (\eta_{ij} b_j a_i + b_i (b_j + \eta_{ij} a_j)) \sigma_j^2 \sigma_i^2 \\ w_{22} = (-\eta_{ij}^2 a_i + 2 \eta_{ij} b_i + b_i^2 \sigma_j^2 a_j \sigma_i^2) \sigma_j^2 \end{cases} \tag{28}$$

In the simplified case where the normalisation constants are all units, i.e. $\eta_{ij} = 1$ (which means physically that the powers of sources i and j are inversely proportional $\lambda_i = \frac{1}{\lambda_j}$), then:

$$U_{ij}^{-1} D_{ij} U_{ij}^{-1} = \frac{\sigma_j^2 \sigma_i^2}{\Delta^2} \left(\begin{bmatrix} 2\frac{b_j}{\sigma_j^2} & 1 \\ 1 & \frac{(a_i + 2b_i)}{\sigma_i^2} \end{bmatrix} + \begin{bmatrix} b_j \sigma_i^2 a_i & b_j a_i + b_i b_j + a_j b_i \\ b_j a_i + b_i b_j + a_j b_i & b_i \sigma_j^2 a_i \end{bmatrix} \right) \tag{29}$$

According (11), $U_{ij}^{-1} D_{ij} U_{ij}^{-1}$ in (29) should match the corresponding block of the Fisher information matrix

$$I_{ij}^{-1} = \frac{1}{\gamma_1 \sigma_2^2 \gamma_2 \sigma_1^2 - 1} \begin{bmatrix} \sigma_1^2 \gamma_2 & -1 \\ -1 & \sigma_2^2 \gamma_1 \end{bmatrix}, \tag{30}$$

if we replace a_i and b_i by the values obtained using the true distribution. Hence we can write $a_i = \gamma_i, b_i = 0$ and

$$U_{ij}^{-1} D_{ij} U_{ij}^{-1} = \begin{bmatrix} \sigma_i^2 \gamma_j & -1 \\ -1 & \sigma_j^2 \gamma_i \end{bmatrix}.$$

Equation (30) means that optimal convergence in the sense of Cramer-Rao is obtained when the Fisher information matrix follows a simple structural form, which depends mainly on the choice of g.

4 Conclusion

The fisher information matrix fundamentaly related to the precision of maximum likelihood estimators. Such information criterion is important to know for at least 3 reasons:

- the understanding of the theory puts one in a much better position to accept use of a BSS algorithm and understand its strengths and weaknesses. Equation (20) show that even when $g \neq p$, good solutions can be found if g is *symetric*.
- we must used *expected* Fisher information as the quatity of interest when model parameters must be estimated,
- the use of Fisher information criteria in the analysis of real data is not based on the existence of a "true" model. Model selection could an important byproduct of the approach.

References

1. P. Brémaud. *An introduction to probabilistic modeling.* Undergraduate texts in mathematics. Springer-Verlag, 1987.
2. P. Comon. Independent component analysis, a new concept. *Signal Processing*, 36(3):287–314, 1994.
3. D. Dacunha-Castelle and M. Duflo. *Probabilités et statistiques*, volume 2 of *Collection Mathématiques Appliquées pour la Maîtrise*. Masson, 2e edition, 1993.
4. A. Hyvärinen, K. Karhunen, and E. Oja. *Independent Component Analysis*. Wiley, 2001.
5. D.T. Pham, Ph. Garat, and C. Jutten. Separation of mixture of independent sources through a maximum likelihood approach. *Signal Processing VI, Proceeding EUSIPCO '92, Bruxelles, Belgium, Eds. J. Vandewalle, R. Boite, M. Moonen and A. Oosterlinck. Amsterdam: Elsevier*, 36(3):771–774, 1992.
6. V. Vigneron, L. Aubry, and C. Jutten. General conditions of stability in blind source separation models and score function selection. *Neurocomputing*, 2004.

Localization of P300 Sources in Schizophrenia Patients Using Constrained BSS

Saeid Sanei[1], Loukianos Spyrou[1], Wenwu Wang[2], and Jonathon A. Chambers[2]

[1] Centre for Digital Signal Processing Research, King's College London, WC2R 2LS, UK
saeid.sanei@kcl.ac.uk
[2] Communications and Information Technologies Research Group
Cardiff School of Engineering, Cardiff University, Cardiff, CF24 0YF, UK

Abstract. A robust constrained blind source separation (CBSS) algorithm has been proposed for separation and localization of the P300 sources in schizophrenia patients. The algorithm is an extension of the Infomax algorithm, based on minimization of mutual information, for which a reference P300 signal is used as a constraint. The reference signal forces the unmixing matrix to separate the sources of both auditory and visual P300 resulted from the corresponding stimulations. The constrained problem is then converted to an unconstrained problem by means of a set of nonlinear penalty functions. This leads to the modification of the overall cost function, based on the natural gradient algorithm (NGA). The P300 sources are then localized based on electrode − source correlations.

1 Introduction

Based on clinical investigations, P300 is a positive event-related potential (ERP), which occurs with a latency of about 300 ms after rare or task relevant stimuli [1]. This is nicely relevant to psychological aspects such as cognition or attention. There are two P300 sub-components that overlap at the scalp; P3b has a more centro-parietal distribution and corresponds to the classical P300 recorded within an oddball paradigm after rare and task relevant events. P3a occurs after novel events independently of task relevance and is characterized by a more frontal distribution, a shorter latency and fast habituation. These sub-components reflect functionally different processes. P3a has been interpreted as an orienting response. P3b has been related to many different psychological constructs such as control information processing, the information content of the events, memory processes, the reorganization of an internal expectancy model. The parietal and temporal cortex are involved in the generation of the auditory P3b. Concerning P3a, the superior temporal plane, the association cortices, limbic structure and frontal as well as pre-frontal cortices appear to play a major role. With local recording the hippocampus shows the largest P300. It has been proved clinically that P300 potentials recorded at the scalp result from intracortical currents induced by post-synaptic potential. The P300 activity of the temporo-basal dipoles corresponds mainly to the classical P3b and that of the frontally oriented temporo-superior dipoles to P3a. An increase of P300 latency with age is found for the temporo-basal but not for the temporo-superior dipoles. Figure 1 illustrates some typical P3a and P3b waves from temporo-basal and temporo-superior dipoles [1].

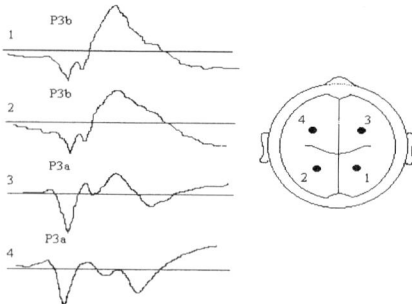

Fig. 1. Typical subcomponents of the P300 signals; a. P3a and b. P3b signals.

Attenuation of auditory and visual P300 signal can be a sign of schizophrenia. Detection and localization of the P300 source have been the objectives of psychiatry clinicians. Although in the healthy person P300 can be seen about 300 millisecond after applying the stimuli for the patients suffering from schizophrenia, the shape, amplitude, and even the position of the P300 may change. Furthermore, children of schizophrenic parents and other subjects with an enhanced risk of developing schizophrenia showed P300 abnormalities. Also small P300 amplitudes have been found in healthy sibling of schizophrenics. However, the reduction in the amplitude of P300 is also found in patients with dementia and with affective disorders. Furthermore the P300 is reduced only in a subgroup of schizophrenic patients. This indicates that the amplitude reduction of P300 is neither a sufficient nor a necessary marker of schizophrenic disorders. But it has been clinically observed that the patients with a P300 amplitude reduction are a sub-group with neuro-developmental disorders. Moreover, the amplitude reduction is more consistent in P3b subcomponents.

A reference signal can be modelled by averaging a number of (electroencephalogram (EEG)) segments obtained after applying a periodic stimulus.

Blind separation of the EEG signals on the other hand, has been followed by a number of researchers [2] [3] [4]. Infomax algorithm [5] has been reported to be robust for separation of EEG signals. Some source separation problems such as signal detection and noise cancellation often expect to estimate a desired single source or a subset of sources from the mixtures. In such cases a separate objective function, as a constraint, has to be minimized (or maximized) in parallel with minimization of the original cost function. Exploitation of Lagrange multipliers [6] and nonlinear penalty functions [7] incorporate the constraint terms into the original cost functions thereby convert the constrained problems to unconstrained algorithms.

The BSS criterion (or equivalently ICA) for instantaneous mixtures such as EEGs, is formulated as follows. Denote the time varying observed signals by $\mathbf{x} = [x_1(t), x_2(t), \ldots, x_n(t)]^T$ where $\mathbf{x} \in R^n$ and the unknown independent sources $\mathbf{s} = [s_1(t), s_2(t), \ldots, s_m(t)]^T$ where $\mathbf{s} \in R^m$.

$$\mathbf{x} = A\mathbf{s} + \mathbf{v} \qquad (1)$$

and

$$\mathbf{y} = W\mathbf{x} \qquad (2)$$

Here $\mathbf{v} \in R^n$ is assumed to be a white Gaussian noise vector, $A \in R^{m \times n}$ and $W \in R^{n \times m}$ are unknown constant mixing and unmixing matrices respectively, and $(.)^T$ is vector transpose. The mixture is assumed to be over-determined (valid for usual cases), i.e. m<n. $\mathbf{y} = [y_1(t), y_2(t), \ldots, y_m(t)]^T$, where $\mathbf{y} \in R^m$ is the output vector. The unconstrained separation matrix can be found by finding the global minima(or maxima) of a cost function $J_M(W)$, which provides a measure of independency of the estimated sources.

Incorporation of the constraint requires another cost function such as $J_C(W)$ to be minimized together with $J_M(W)$. The constraint term is then joined to the main objective function by using either a Lagrange multiplier or a set of penalty functions. Application of the Lagrange multiplier however, ignores nonlinearity of the system whereby the nonstationarity of the mixtures is not exploited. A general overall cost function is best defined as follows:

$$J(\mathbf{W}) = J_M(\mathbf{W}) + kG(J_C(\mathbf{W})) \qquad (3)$$

where G(.) is the penalty function and k is the matrix of penalty coefficients. In the following sections a new constrained BSS method based on the original Infomax BSS system and incorporating of a reference signal as a constraint is introduced for detection and localization of the P300 sources within the brain from the EEG signals.

2 Constrained Infomax Algorithm

In an undetermined BSS system the estimated ICs do not necessarily represent the actual sources. This happens when EEGs are to be separated. In the development of this project we aim at separation of the scalp EEG mixtures in such away that the desired P300 signal is one of the estimated ICs. The Infomax BSS algorithm is based on minimization of the mutual information or maximization of the entropy. The Infomax cost function $J_M(\mathbf{w})$, can be found in the literature [5]. The unmixing matrix is recursively updated by finding a solution to the minimization of such unconstrained overall cost function i.e.

$$\arg\min_{\mathbf{W}} J(\mathbf{w}) = \arg\min_{\mathbf{W}} [J_M(\mathbf{w}) + kG(J_C(\mathbf{w}))] \qquad (4)$$

where $J_M(\mathbf{w})$ is the main objective function of the Infomax BSS algorithm and

$$J_C(\mathbf{W}) = \|\mathbf{p} - \mathbf{y}\|^2 = \|\mathbf{p} - \mathbf{wx}\|^2 \qquad (5)$$

where **p** is a matrix whose rows are the P300 reference signal. The reference signal is obtained by averaging several segments of the same electrode signal after a visual or auditory periodic stimulation. The update equation is generally denoted as

$$\mathbf{w}(t+1) = \mathbf{w}(t) + \Delta \mathbf{w}(t) \tag{6}$$

where by considering the extension to the NGA proposed by Amari [6] we have

$$\Delta \mathbf{w}(t) = \mu \frac{\partial J(\mathbf{w})}{\partial \mathbf{w}} \mathbf{w}^T \mathbf{w}$$
$$= \mu \left(\gamma \mathbf{I} + (1 - \frac{2}{1 + \exp(\mathbf{wx})})(\mathbf{wx})^T + 2\mathbf{q}(\mathbf{x}(\mathbf{wx} - \mathbf{p})^T)^T \mathbf{w}^T \right) \mathbf{w} \tag{7}$$

Here μ is the learning rate, γ is a constant, and **I** is a unitary matrix. **w** is initialised to $\mathbf{w}_{init} = \mathbf{I}$ and μ is calculated empirically via the following adaptive criterion:

$$\mu(t) = \mu_0 \left(\frac{\alpha}{\|\mathbf{Rx}\|_F^2} + \frac{\beta}{\zeta + \|\Delta J_C\|} \right) \tag{8}$$

where μ_0, α, β, and ζ are constants adjusted for adaptation. In the above formulation **q** is updated iteratively based on the new **w** in the direction of minimizing the distance between the output ICs and the P300 reference signal i.e.

$$\mathbf{q} = k \cdot diag\left[(\mathbf{wx} - \mathbf{p})(\mathbf{wx} - \mathbf{p})^T \right] / p^2 \tag{9}$$

where k is the penalty parameter. In the above analysis we ignored the effect of noise, which is inherently contained in **x**. However, incorporation of the constraint into the original Infomax update equation does not change the performance of the system in terms of noise effect.

3 Localization Criterion

Localization of the EEG sources has been investigated recently [3] [8] [9]. With some indeterminacy in the result we can approximate the location of the sources within the brain. Unlike the methods in [8] and [9], which consider the sources as magnetic dipoles, in our approach we simply consider them as the sources of isotropic propagating signals. Therefore the head (mixing media) model only mixes and attenuates the signals. Therefore based on Figure 2 we have

$$\|\mathbf{f}_k - \mathbf{a}_j\|_2 = d_j \tag{10}$$

where **f** and $\mathbf{a}_j$ refer to the source and the electrode coordinates respectively, and d_j are nonlinearly proportional to the inverse of the correlations between the estimated source k and the electrode signals. $j = 1,2,3$ represents the electrode involved in calculation of the correlation values, and $k = 1, 2, \ldots, M$, shows the source number. In this equation all the variables except **f** is known. Incorporating more than three mixtures does not affect the result whereas it makes the computation more intensive. The nonlinearity stated above comes from the fact that head is not a homogenous region.

In a spherical model of the head we may consider three main layers; brain, skull, and scalp for which the thickness is known. The conductivity of the skull is about 100 times less than those of the brain and the scalp i.e. $\sigma_{brain} = \sigma_{scalp} = 80\sigma_{skull}$. In order to incorporate the non-homogeneity and ensure that there will be a solution to equation (10) within the brain region, these values have to be nonlinearly mapped and normalized such that all the estimated sources fall within the brain region. A more accurate method requires the information about non-homogeneity of the media between the sources and the electrodes including the brain (white and gray tissues), the skull and the scalp. Utilizing the described method, it has been shown that the P300 sources within the brain can be localized by a high degree of accuracy.

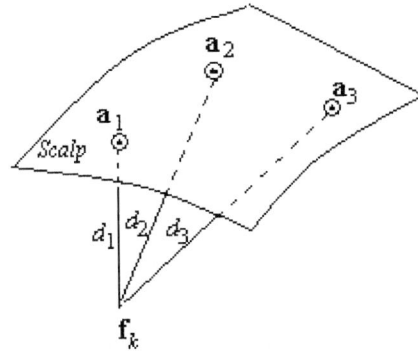

Fig. 2. Part of the scalp including three electrodes, and the location of the source to be identified (assuming the head is homogenous).

4 Experimental Results

In this part the proposed CBSS algorithm is applied to the simulated as well as real EEG data from both a healthy person and a schizophrenia patient. The sampling frequency is 200 sample/sec. The data is pre-whitened and **w**, the separation matrix is initialized to **I**. The P300 reference signal for natural mixtures is achieved by averaging a number of segments of each electrode signal. This signal is then used as a reference for updating the unmixing matrix.

The algorithm maximizes the P300 component as an output of the unmixing matrix. In all the cases the convergence of the algorithm is very fast. In the first experiment only three mixtures including the P300 have been modelled and used. Figure 3 compares the results of separation using traditional Infomax BSS algorithm and those of the proposed constrained technique. Figure 4 shows the convergence of the system with respect to that of the well-known NGA-based joint diagonalization criterion. It is very obvious that the new system converges much faster than the conventional NGA. In the next two experiments the reference P300 signal is estimated from the mixtures by taking an average of a number of segments for each electrode. To have a better reference signal a periodic visual (or audio) stimulation is necessary. The period of

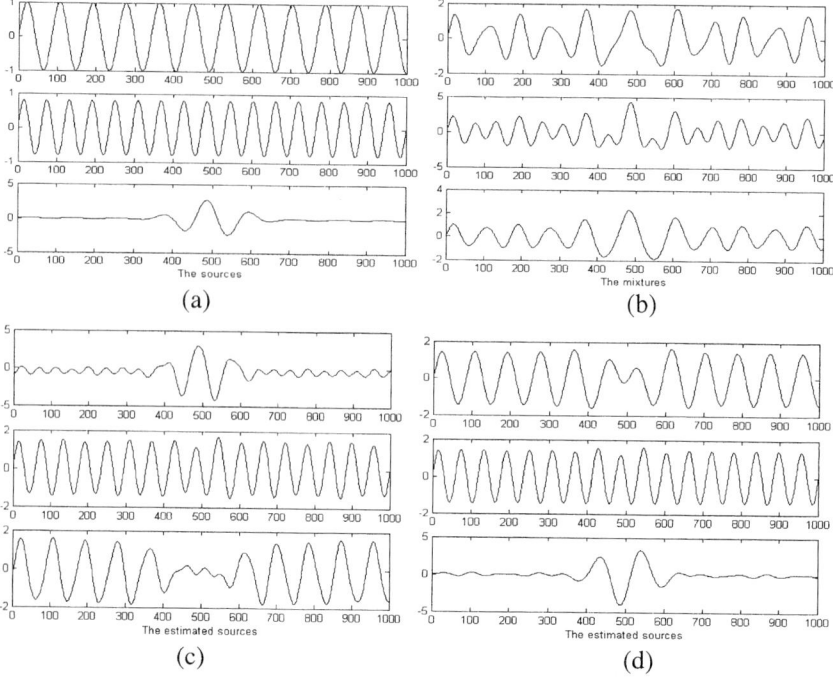

Fig. 3. The results of separation of simulated P300 signal; (a) the original sources, (b) the mixtures, (c) the estimated sources using the traditional and (d) the estimated sources using the constrained Infomax algorithms.

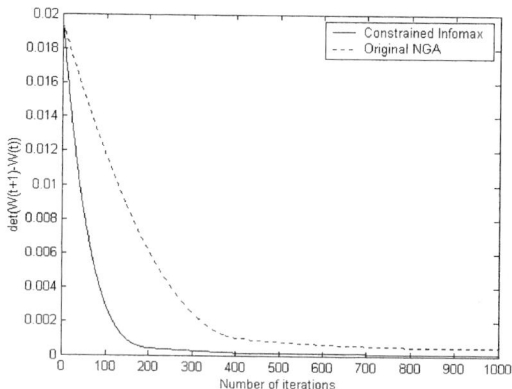

Fig. 4. The convergence comparison between the CBSS and the traditional NGA.

the stimulation is ten seconds and it lasts for one second. We avoid any eye blinking or other artifacts. Figure 5 illustrates the results from separation of the P300 signal for a schizophrenia patient. In this case the P300 signal is not clear in the EEG signals.

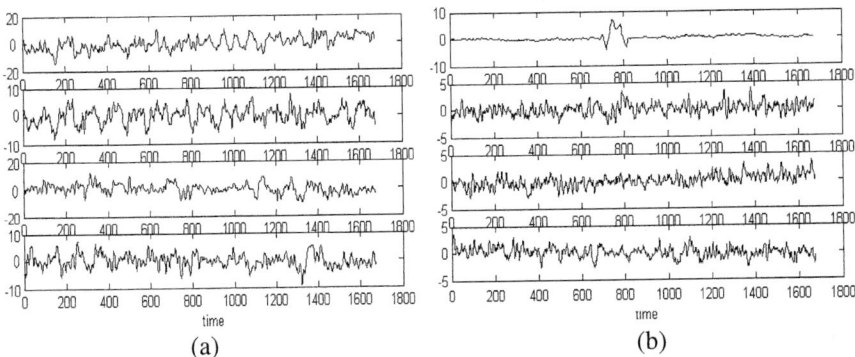

Fig. 5. The results of separation of a real P300 signal in a schizophrenia patient, using the constrained Infomax algorithm; (a) the EEGs and (b) the separated signals using CBSS, the top signal is the separated P300 (only 4 out of 10 signals have been displayed here).

After separation, the estimated highlighted component is localized. In order to guarantee a solution to equation (10) the following steps were followed: (a) The correlation values are inversed and normalized into between 0 and 1, (b) A nonlinear μ-law function, with $100 > \mu > 25$, was used for transformation of the correlation values, (c) The transformed values were scaled into between 12 mm and the radius of the head, as d_j, and (d) The solution to the following least square problem was obtained [10]:

$$\min S(\mathbf{f}_k), \quad \mathbf{f}_k \in R^n, \text{ where } S(\mathbf{f}_k) = \sum_{j=1}^{3} \left\{ \left\| \mathbf{f}_k - \mathbf{a}_j \right\|_2 - d_j \right\}^2 \quad (11)$$

Following the above steps for a head phantom, in more than 95% of the cases an exact localization of the sources has been achieved. Both P3a and P3b signals are localized around the temporo-superior dipoles and temporo-basal dipoles respectively. Based on the information achieved in this project an informative display platform has been provided which greatly assist the clinicians in diagnosis of neurological disorders.

5 Conlusions

An effective and robust CBSS method has been developed and used for separation and localization of the P300 sources in healthy individuals and the schizophrenic patients. The algorithm is an extension of original Infomax algorithm, based on minimization of mutual information, for which a reference P300 signal is used as a constraint. The constrained problem is then converted to an unconstrained problem by means of nonlinear penalty functions weighted by the penalty terms. The algorithm has been examined on both simulated and natural EEG signals of both healthy and schizophrenic patients. Both auditory and visual P300 can be separated and well localized. The reference signal forces the unmixing matrix to separate the sources of

P300 resulted from the corresponding stimulations. The results are compared with those of the traditional Infomax algorithm and the traditional NGA method in terms of the convergence speed. The method is an effective tool in investigation of the schizophrenia disease (as well as some other neurological disorders such as Alzheimer's) in neurophysiology and psychiatry departments.

References

1. E. Niedermeyer and F. L. Da Silva, Electroencephalography; basic principles, clinical applications, and related fields, Ed. 4, LW&W, (1999).
2. R. N. Vigario, "Extraction of ocular artefacts from EEG using independent component analysis," Electroencephalography and Clinical Neurophysiology, 103, pp. 395-404, (1997).
3. A. Cichocki et al., "Neural networks for blind separation with unknown number of sources," Neurocomputing, 24(1-3): pp. 55-93, February (1999).
4. S. Makeig, et al. "Independent component analysis of electroencephalographic data," Advances in neural information processing systems 8, pp. 145-151 MIT Press, Cambridge, MA, (1996).
5. J. F. Cardoso, "Infomax and maximum likelihood for blind source separation," IEEE Signal Processing Letter, 4, pp. 109-111, April (1997).
6. A. Cichocki and S. I. Amari, Adaptive blind signal and image processing, J. Wiley, (2002).
7. W. Wang, S. Sanei, and J. A. Chambers, "Penalty function based joint diagonalization approach for convolutive blind separation of nonstationary sources," to be published in IEEE Transactions on Signal Processing.
8. J. C. Mosher & R. M. Leahy, "Source localization using recursively applied and projected (RAP) MUSIC," IEEE Trans. on SP, 47(2), pp. 332-340, Feb. (1999).
9. J. C. Mosher & R. M. Leahy, and P. S. Lewis, "EEG and MEG: Forward solutions for inverse methods," IEEE Trans. on Biomedical Engineering, 46(3), pp. 245-259, March (1999).
10. I. D. Coope, "Reliable computation of the points of intersection of n spheres in R^n," ANZIAM J., 42(E), pp. C461-C477, (2000).

On the Estimation of the Mixing Matrix for Underdetermined Blind Source Separation in an Arbitrary Number of Dimensions*

Luis Vielva[1], Ignacio Santamaría[1], Jesús Ibáñez[1],
Deniz Erdogmus[2], and José Carlos Príncipe[2]

[1] Dpt. Ingeniería de Comunicaciones, Universidad de Cantabria, España
{luis,nacho,jesus}@gtas.dicom.unican.es
[2] CNEL, University of Florida, Gainesville, USA
{deniz,principe}@cnel.ufl.edu

Abstract. Blind Source Separation consists of estimating n sources from the measurements provided by m sensors. In this paper we deal with the underdetermined case, $m < n$, where the solution can be implemented in two stages: first estimate the mixing matrix from the measurements and then estimate the best solution to the underdetermined linear problem. Instead of being restricted to the conventional two-measurements scenario, in this paper we propose a technique that is able to deal with this underdetermined linear problem at an arbitrary number of dimensions. The key points of our procedure are: to parametrize the mixing matrix in spherical coordinates, to estimate the projections of the maxima of the multidimensional PDF that describes the mixing angles through the marginals, and to reconstruct the maxima in the multidimensional space from the projections. The results presented compare the proposed approach with estimation using multidimensional ESPRIT.

1 Introduction

The blind source separation (BSS) problem consists of estimating n sources from the measurements provided by m sensors. In the noise-free linear model, the measurements are related to the sources through an unknown linear combination

$$\mathbf{As} = \mathbf{x}, \tag{1}$$

where $\mathbf{s} \in R^n$ is the source random vector, $\mathbf{x} \in R^m$ is the measurement random vector, and $\mathbf{A} \in R^{m \times n}$ is the unknown mixing matrix. Depending on the relation between m and n, we are faced with three different scenarios. The square ($m = n$) and the strictly overdetermined ($m > n$) cases have been extensively studied in the literature [1,2], and all we need to separate the sources is to estimate the

---

* This work has been partially supported by Spanish Ministry of Science and Technology under project TIC2001-0751-C04-03.

mixing matrix **A**, since the inverse solves the square problem, and the pseudo-inverse provides the solution with minimum-norm error in the overdetermined case [3].

The last scenario, in which we are interested in this paper, arises when the number of sensors is smaller than the number of sources ($m < n$). In this underdetermined case, the solution process can be divided in two stages: first estimate the mixing matrix from the measurements and then estimate the sources that "best" solve the underdetermined linear problem [4,5]. This procedure relies on the premise that the sources are sparse or that a suitable linear transformation is applied to convert the non-sparse sources into a sparse representation [6]. To parametrically model sources with different degrees of sparsity, the following model for the source densities is used

$$p_{S_j}(s_j) = p_j \, \delta(s_j) + (1 - p_j) f_{S_j}(s_j), \quad j = 1, \ldots, n, \tag{2}$$

where s_j is the j-th source, p_j is the sparsity factor for s_j, and $f_{S_j}(s_j)$ is the PDF when the source j—that is assumed to be zero-mean—is active. The performance of this two-stage procedure strongly depends on the sparsity of the sources, both for the estimation of the mixing matrix and for [7] the estimation of the sources [8]: the higher the sparsity factor the better the estimation of mixing matrix and the recovery of the sources.

Most of the results on underdetermined BSS [6,8] consider the case with two sensors ($m = 2$), in which the mixing matrix can be obtained, from a geometrical point of view [9], by finding the maxima of a unidimensional probability density function (PDF). However, the direct extension of this method to scenarios with more than two sensors requires finding the maxima of a multidimensional PDF [10], that, in addition to be computationally more complex, requires a number of samples that depends exponentially on the number of dimensions.

In this paper, we extend our previous work on underdetermined BSS [4] to deal with an arbitrary number of sensors (more than one) and an arbitrary number of sources. The organization of the paper is as follows: In Section 2, we present the problem of estimating the mixing matrix as the problem of finding the maxima of an $(m-1)$-dimensional PDF. In Section 3, we introduce the projection procedure that reduces the peak estimation problem from a multidimensional PDF to $m-1$ decoupled unidimensional PDFs, and show how to elucidate the spurious combinations of peaks from those that are true maxima of the $(m-1)$-dimensional PDF. In Section 4, we validate the proposed method with a series of Montecarlo simulations. In Section 5 we present the conclusions of this work.

2 Estimation of the Mixing Matrix

Equation (1) can be interpreted from a geometrical point of view as the projection of the source vectors **s** from R^n into the vector space R^m of the measurement vectors **x**. If we denote by $\mathbf{a}_j$ the j-th column of the mixing matrix, so that $\mathbf{A} = [\mathbf{a}_1, \mathbf{a}_2, \cdots, \mathbf{a}_n]$, (1) can be rewritten as

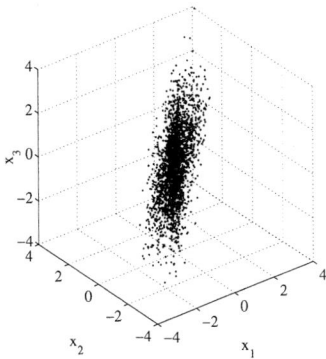
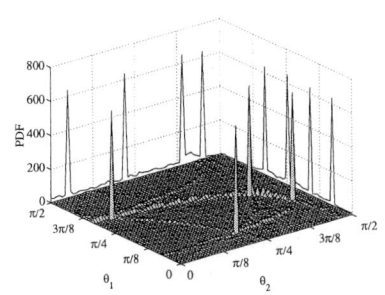

Fig. 1. Scatter plot of measurements for a scenario with three sensors ($m = 3$) and four sources ($n = 4$) of sparsity factor: 0.5.

Fig. 2. Histogram of angles for the measurements of Figure 1. The $(m - 1)$-unidimensional projections onto the plane of angle θ_i, $i = 1, \ldots, m - 1$ are also shown.

$$\mathbf{x} = \sum_{j=1}^{n} s_j \mathbf{a}_j, \qquad (3)$$

that explicitly shows that the measurement vector is a linear combination of the columns of the mixing matrix. According to this interpretation, if at a given time only the j-th source is non-zero, the measurement vector will be collinear with $\mathbf{a}_j$. When more than one source is active at the same time, the measurement will be a linear combination of the corresponding columns of the mixing matrix. In Figure 1 we show a scatter plot for a scenario with four sources and three sensors that is simulated for sources with sparsity factors of 0.5. For higher sparsity factors, the measurements are more concentrated along the directions of the columns of the mixing matrix [4].

The first step in our recovery procedure is to convert all the points of the m-dimensional vector space of the measurements and the columns of the mixing matrix from a Cartesian representation to a spherical coordinate system, where every point $\mathbf{x}$ of Cartesian coordinates $(x_1, \ldots, x_m)$ is represented by its modulus r and by $m - 1$ angles θ_i. According to this definition, the angles can be determined from the rectangular coordinates as

$$\theta_i = \arctan \frac{x_{i+1}}{\sqrt{\sum_{l=1}^{i} x_l^2}}, \quad i = 1, \ldots, m - 1. \qquad (4)$$

If we apply (4) to the measurements of Figure 1, and represent an histogram taking as independent variables the $m - 1$ angles, we obtain the results shown in Figure 2.

3 Dimension Reduction by Projection

Up to this point, we have reduced the problem of estimating the mixing matrix **A** to the problem of estimating the n peaks of an $(m-1)$-dimensional PDF, since those peaks define the spherical angles that parametrize the n columns of the mixing matrix. It is well known that the problem of estimating the peaks of a multidimensional PDF requires much more data samples as the dimensionality of the problem grows [11]. However, the idiosyncrasy of the underdetermined BSS problem will help us to circumvent this problem. The sparsity of the sources, which is a prerequisite for the proposed underdetermined BSS procedures to work, will be determinant to the ability of estimating a multidimensional PDF by means of unidimensional estimations. In Figure 2 it can be observed that the $(m-1)$-dimensional PDF is composed of a set of n peaks that, even for an sparsity factor of 0.5, are quite narrow. In Figure 3, a top view of the $(m-1)$-dimensional PDF is shown. The black spots correspond to the locations of the maxima from Figure 2.

Since we are interested in determining only the position of the peaks, and not the complete shape of the PDF, all the information we are looking for can be extracted from the $m-1$ projections onto the unidimensional vector spaces corresponding to conserving only one spherical coordinate and making zero all the other angles. These projections are shown in Figure 2 for the case of three sensors and four sources, which we are using as an example. They can be considered as the set of $m-1$ unidimensional PDFs of the $m-1$ spherical angles that are shown as projections in Figure 2.

To each of these $m-1$ unidimensional PDFs of the angles that parametrize the measurements, a method has to be applied to find up to n maxima, whose locations correspond to the estimates $\hat{\theta}_{ij}$, $i = 1,\ldots,m-1$, $j = 1,\ldots,n$. A number of methods could be applied, from the simpler one of calculating the histogram and finding the maxima, to the use of nonparametric estimation by means of Parzen windowing [7], or to the use of spectral estimation techniques suitable for the estimation of sinusoids in noise [12].

Once the estimations of the individual spherical angles are obtained, it is necesary to reconstruct the position of the maxima of the multidimensional PDF from the unidimensional projections. The problem arises from the loss of information inherent to the projection process, and can be visualized by reconsidering Figure 3. We are interested on the $(m-1)$-dimensional position of the maxima indicated by the black spots, but all we have access to from the unidimensional estimations is the projections of these spots onto each of the coordinate axes. From these projections, all the combinations of angles could be constructed, as it is shown with dotted lines in Figure 3, and a method has to be implemented that allows to distinguish the correct combinations from the spurious solutions.

Fortunately, there exist an easy way for the correct combinations to stand out: all that we need to do is to define a small area around each combination of angles, that constitutes a tentative solution, and count how many measurements fall into that area. The correct combinations will have a high number of occurrences, but a point falling into the region associated to a spurious combination will be an improbable event. Since the number of combinations of angles is n^{m-1}, the procedure to elucidate which are the correct combinations of angles is to construct an $(m-1)$-dimensional count array $\mathbf{C}$ of length n in each of the dimensions and find the maxima for each intersection of the $m-1$ dimensions. In our example of four sources and three sensors, the $(m-1)$-dimensional array is a 4×4 matrix. In equation (5) the calculated matrix for a simulation with sparsity factor 0.5 is shown. The matrix is shown upside-down to facilitate comparison with Figure 3. The higher the sparsity factor, the more concentrated the measurements along the columns of the mixing matrix. As an example, for a sparsity factor of 0.9, almost all the measurements fall into the regions associated with the correct combinations.

$$\mathbf{C}(0.5) = \begin{pmatrix} 669 & 0 & 1 & 2 \\ 2 & 3 & 705 & 1 \\ 0 & 1 & 3 & 632 \\ 1 & 674 & 3 & 0 \end{pmatrix}. \tag{5}$$

Since the method of estimating the peaks on the multidimensional space is based on the information obtained by projecting, a potential problem could appear when more than one peak is projected along any direction into the same point. In this situation, we would not detect the limit of up to n peaks in each coordinate, but a smaller number of peaks in some angles. However, this is not really a problem, because with the help of the count vector $\mathbf{C}$ we would detect the situation (there would be high count numbers for multiple combinations of the same angle, instead of a single maximum per row and column of $\mathbf{C}$) and we could estimate the position of all the peaks.

4 Numerical Results

To characterize the performance of our method, Montecarlo simulations have been performed to estimate the mixing matrix from scenarios with different numbers of sources and sensors. In all the cases, the source realizations have been generated according to the model in (2), using as $f_{S_j}(s_j)$, $j = 1, \ldots, m$, Gaussian densities with zero mean and unit variance. The simulations have been performed as follows: for each scenario, twenty thousand samples of sources with sparsity factors from 0.05 to 0.95 have been produced. For each scenario and sparsity factor, four hundred mixing matrices have been randomly generated, the spherical angles have been estimated from the unidimensional projected PDFs, and the criterion to select the correct combination of angles has been applied. The different scenarios considered are those associated with a number of sensors ranging from two to five, and a number of sources ranging from one to ten. As the figure of merit we have selected the number of errors in the estimation of the angles (defining a tolerance on the basis of the bin length used on the histograms). We define the mean error rate as the mean number of errors for all the mixing matrices divided by the total number of angles to estimate.

Figure 4 shows the results from scenarios with five sensors ($m = 5$) and a number of sources from six to twelve ($6 \leq n \leq 12$) for all the sparsity factors

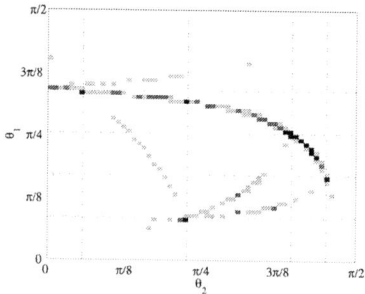
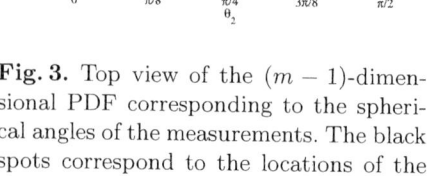
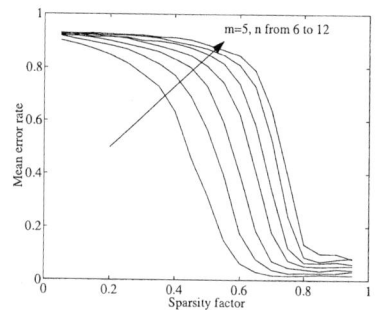

Fig. 3. Top view of the $(m-1)$-dimensional PDF corresponding to the spherical angles of the measurements. The black spots correspond to the locations of the maxima from Figure 2.

Fig. 4. Mean error rate for scenarios with a fixed number of five sensors and a number of sources ranging from six to twelve, as a function of the sparsity factor of the sources.

considered. It can be observed that the number of errors grows with the number of sources (more peaks have to be estimated from the same data, and the mean distance between peaks decreases), and diminish with the sparsity factor (the measurements tend to be more concentrated along the columns of the mixing matrix, reducing the spreading that confuses the estimation).

Figure 5 shows the results from scenarios with seven sources ($n = 7$) and a number of sensors ranging from two to six ($2 \leq m \leq 6$) for all the sparsity factors considered. It can be observed that the number of errors diminish as the number of available measurements increases.

Figure 6 shows the mean squared error (MSE) for the estimation of the angles of the mixing matrix for an scenario with four sources and two sensors obtained with the proposed reconstruction by projection method. In the same figure, the results obtained by applying two-dimensional ESPRIT to the direct estimation of the angles from the bidimensional PDF of Figure 2 are also shown. It is remarkable that the estimation from the projections, that is much easier and faster than the bidimensional ESPRIT, provides even better results.

5 Conclusions

In this paper we have presented a procedure to estimate the mixing matrix for underdetermined BSS problems in an arbitrary number of dimensions. The approach is based on parametrizing both the measurements and the columns of the mixing matrix in spherical coordinates and on estimating the peaks of the multidimensional PDF associated with the angles of the measurements. Since the estimation of multidimensional PDFs is a complex problem, we propose to project onto as many unidimensional PDFs as the number of spherical angles (the number of sensors minus one). Once the individual angles are estimated from the projections, the location of the peaks on the original multidimensional

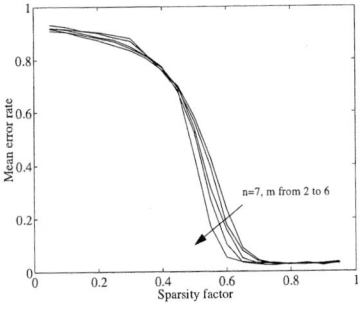
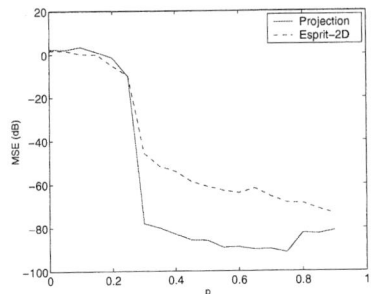

Fig. 5. Mean error rate for scenarios with a fixed number of seven sources and a number of sensors ranging from two to six, as a function of the sparsity factor of the sources.

Fig. 6. MSE of the estimated angles $\hat{\theta}_{ij}$, $j = 1,\ldots,4$ as a function of the sparsity factor (p) of the sources using both Esprit-2D (solid-line) and $m - 1$ projections (dashed-line) for an scenario with two sensors and four sources.

measurement space can be reconstructed. Since there exist different multidimensional PDFs compatible with the given projections, we propose a method to distinguish the spurious combinations of angles and to elucidate the correct combinations. We would like to point out that the procedure presented in this paper is not exclusive for underdetermined cases, since nothing prevents us from using this method in scenarios with less or equal sources than sensors. The reason why we focus on the underdetermined case is twofold: on the one hand, there exist other excellent approaches for the overdetermined and squared scenarios; on the other hand, the performance of our method increases with the sparsity factor of the sources, that is a prerequisite only for the underdetermined scenario. The Montecarlo simulations have shown that our method provides excellent results for an arbitrary number of sources and sensors provided that the sparsity factor is high enough (around 0.75). The intuitive result that the performance improves with the number of measurements and the sparsity factor, and degrades with the number of sources has also been corroborated.

References

1. A. Hyvärinen, Juha Karhunen, and Erkki Oja, *Independent Component Analysis*, John Wiley & Sons, New York, 2001.
2. S. Haykin, Ed., *Unsupervised Adaptive Filtering, Vol I: Blind Source Separation*, John Wiley & Sons, New York, 2000.
3. Gene H. Golub and Charles F. Van Loan, *Matrix Computations*, Johns Hopkins University Press, 3rd edition, 1996.
4. L. Vielva, D. Erdoğmuş, and J. C. Príncipe, "Underdetermined blind source separation in a time-varying environment," in *Proceedings ICASSP-02 (IEEE International Conference on Acoustics, Speech and Signal Processing)*", Orlando, FL, May 2002, pp. 3049–3052.

5. M. Zibulevsky, B. Pearlmutter, P. Bofill, and P. Kisilev, *Independent Components Analysis: Principles and Practice*, chapter Blind source separation by sparse decomposition in a signal dictionary, Cambridge University Press, 2000.
6. P. Bofill and M. Zibulevsky, "Underdetermined blind source separation using sparse representations," *Signal Processing*, vol. 81, no. 11, pp. 2353–2362, 2001.
7. D. Erdoğmuş, L. Vielva, and J. C. Príncipe, "Nonparametric estimation and tracking of the mixing matrix for underdetermined blind source separation," in *Proceedings of ICA-2001, Independent Component Analysis*, San Diego. CA, Dec. 2001, pp. 189–193.
8. L. Vielva, D. Erdoğmuş, and J. C. Príncipe, "Underdetermined blind source separation using a probabilistic source sparsity model," in *Proceedings of ICA-2001, Independent Component Analysis*, San Diego. CA, Dec. 2001, pp. 675–679.
9. C. G. Puntonet, A. Prieto, C. Jutten, M. Rodriguez-Alvarez, and J.Ortega, "Separation of sources: A geometry-based procedure for reconstruction of n-valued signals," *IEEE Trans. on Acoustics, Speech, and Signal Processing*, vol. 46, no. 3, pp. 267–284, 1995.
10. A. Jung, F. Theis, C. Puntonet, and E. Lang, "Fastgeo - a histogram based approach to linear geometric ICA," in *Independent Component Analysis*, San Diego. CA, 2001, pp. 349–354.
11. Sergios Theodoridis and Konstantinos Koutroumbas, *Pattern Recognition*, Academic Press, San Diego, California, 1999.
12. L. Vielva, I. Santamaría, C. Pantaleón, J. Ibáñez, D. Erdoğmuş, and J. C. Príncipe, "Estimation of the mixing matrix for underdetermined blind source separation using spectral estimation techniques," in *Proceedings Eusipco-2002 (XI European Signal Processing Conference)*, Toulouse, France, September 3–6 2002, vol. I, pp. 557–560.

On the Minimum ℓ_1-Norm Signal Recovery in Underdetermined Source Separation

Ichigaku Takigawa, Mineichi Kudo, Atsuyoshi Nakamura, and Jun Toyama

Graduate School of Information Science and Technology, Hokkaido University
Kita 13, Nishi 8, Kita-ku, Sapporo 060-8014, Japan
{1gac,mine,atsu,jun}@main.ist.hokudai.ac.jp

Abstract. This paper studied the minimum ℓ_1-norm signal recovery in underdetermined source separation, which is a problem of separating n sources blindly from m linear mixtures for $n > m$. Based on our previous result of submatrix representation and decision regions, we describe the property of the minimum ℓ_1-norm sequence from the viewpoint of source separation, and discuss how to construct it geometrically from the observed sequence and the mixing matrix, and the unstability for a perturbation of mixing matrix.

1 Introduction

In blind separation of n sources from m linear mixtures, there are sometimes more sources than sensors, i.e., $n > m$. In such cases, the mixing linear system is called *underdetermined*. Let $\boldsymbol{x}(t) = (x_1, \ldots, x_m)^\top$ be an m-dimensional vector of the sensor output at a time point t and $\|\boldsymbol{x}\|_p := (\sum_{i=1}^m |x_i|^p)^{1/p}$ (ℓ_p-norm of $\boldsymbol{x}$). Assuming that the vector $\boldsymbol{x}(t)$ is a linear superposition of an unkown n sources

$$\boldsymbol{x}(t) = \boldsymbol{A}\boldsymbol{s}(t), \quad t = 1, \ldots, T,$$

our goal is to estimate the unknown mixing matrix $\boldsymbol{A}$ and the unknown source sequence $\{\boldsymbol{s}(t)\}_{t=1}^T$ from the given data $\{\boldsymbol{x}(t)\}_{t=1}^T$. Thus, in underdetermined cases of $n > m$, the problem can be divided into two steps, estimation of $\boldsymbol{A}$ (blind identification) and estimation of $\{\boldsymbol{s}(t)\}_{t=1}^T$ (signal recovery). It should be noted that source cancellation instead of source recovery is also proposed [1].

This paper focuses on the latter problem of estimating the sequence of realization $\{\boldsymbol{s}(t)\}_{t=1}^T$ for underdetermined cases assuming that $\boldsymbol{A}$ is given. The discussion including the former problem may be found in [2–5]. In general, it requires *a priori* knowledge on sources to solve an underdetermined problem. As the assumption which can treat continuous-valued signals in application such as to speech signal separation, we consider the sparse signals or signals which have a linear sparse representation. Here, the word "sparse" means that only few components will be nonzero. When we can use such a *sparsity*, the practical signal recovery is possible. Indeed, for sparse signals, the minimum ℓ_1-norm sequence

$$\hat{\boldsymbol{s}}(t) = \arg \min_{\boldsymbol{z} \in \Re^n} \{\, \|\boldsymbol{z}\|_1 \mid \boldsymbol{A}\boldsymbol{z} = \boldsymbol{x}(t) \}, \quad t = 1, \ldots, T \quad (1)$$

is often used and some good experimental results in speech signal separation is reported [2–5]. When signals themself are not sparse but they have a sparse linear representation, this sequence also work well [6, 4, 5]. For example, Fig. 1 shows the case in which three periodical signals are mixed into two signals. For this case, using discrete cosine transform (DCT), we can recover the original signal well since each signal has a different pitch and it means the sparsity of these signals in frequency domain. Speech signals are also often sparse in frequency domain [4, 5].

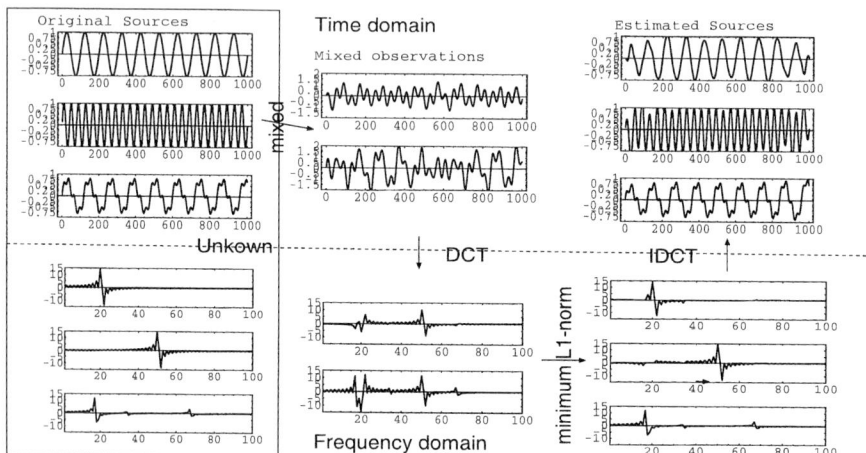

Fig. 1. Examples of minimum ℓ_1-norm sequence for periodic signal separation. DCT plot is shown only for the first 100 points since higher points are almost zero.

One of the motivation to use the minimum ℓ_1 sequence is that it becomes the maximum likelihood estimator when $s(t)$ has a Laplace distribution [2, 7]. However, each subproblem is solved separately from only one sample $x(t)$ and it is insufficient to explain the reported effectiveness in practical applications. Moreover, it is pointed out that even if $s(t)$ is Laplacian, the performance of $\hat{s}(t)$ is not always successful and sparsity or disjointness of s are more important [4, 8]. Another question is on the relation between the solution of each subproblem and the whole sequence.

We studied the properties of the minimum ℓ_1-norm sequence through the concept of submatrix representation and decision regions in our previous work [8] and gave an answer for the question why this simple sequence works well for practical problems such as speech signal separation, and for the question when it will work well.

2 Minimum ℓ_1-Norm Sequence in Signal Recovery

Hereafter, for blind identification step before signal recovery, we assume that the mixing matrix A is given and it has pairwise linearly independent column vectors, therefore, $\boldsymbol{A}$ is full-row rank. Minimum ℓ_1-norm sequence (1) is a sequence of the minimum ℓ_1-norm solution for each $\boldsymbol{x}(t)$. Other minimum ℓ_p type of solutions are also possible:

$$\hat{\boldsymbol{s}}_0 = \arg \min_{\boldsymbol{z} \in \Re^n} \{ \|\boldsymbol{z}\|_0 \mid \boldsymbol{A}\boldsymbol{z} = \boldsymbol{x}(t) \}, \tag{2}$$

$$\hat{\boldsymbol{s}}_p = \arg \min_{\boldsymbol{z} \in \Re^n} \{ \|\boldsymbol{z}\|_p^p \mid \boldsymbol{A}\boldsymbol{z} = \boldsymbol{x}(t), 0 < p < 1 \}, \tag{3}$$

$$\hat{\boldsymbol{s}}_2 = \arg \min_{\boldsymbol{z} \in \Re^n} \{ \|\boldsymbol{z}\|_2 \mid \boldsymbol{A}\boldsymbol{z} = \boldsymbol{x}(t) \} = \boldsymbol{A}^\top (\boldsymbol{A}\boldsymbol{A}^\top)^{-1} \boldsymbol{x}(t), \tag{4}$$

where $\|\boldsymbol{z}\|_0$ is the ℓ_0 quasi-norm of the vector $\boldsymbol{z}$ defined by

$$\|\boldsymbol{z}\|_0 := \lim_{p \to 0} \|\boldsymbol{z}\|_p^p = \#\{i \in [1,n] \mid z_i \neq 0\} \quad \text{for } \boldsymbol{z} = (z_1, \ldots, z_n)^\top$$

which denotes the number of nonzero entries.

These solutions including the minimum ℓ_1-norm case (1) were studied in context of sparse linear representation of data such as the time-frequency linear decomposition in computational harmonic analysis [9–15]. From this viewpoint, the matrix $\boldsymbol{A}$ is a signal dictionary (often two-orthobasis). For such problems, the minimization of the ℓ_0-norm (2) is called *Matching Pursuit* [15], the minimization of the ℓ_1-norm (1) is called *Basis Pursuit* [9]. Recently, the solution (3) is also studied [12]. It should be noted that there are an example of application to blind identification [10] but other works are based on two-orthobases and one of main issues are the condition for the equivalence of (1) and (2) and for the uniqueness of solution.

We discuss them from the viewpoint of sourse separation. First, the minimum ℓ_2-norm solution becomes a linear inverse solution (by the Moore-Penrose psudoinverse), therefore, in underdetermined cases, it becomes just linear embedding of an m-dimensional space into an higher n-dimensional space. Then, it cannot remove the correlation between the samples and it cannot separate the observed mixtures.

If $\boldsymbol{s}$ are very sparse, we can use the minimum ℓ_0 quasi norm solution (2). However, for the cases such as separation problem in Fig. 1, it cannot be unique generally. For the uniqueness, we require the following sparsity.

Theorem 1 ([5, 12]). *Assume that $\boldsymbol{A}$ has pairwise linearly independent column vectors. Also, suppose that for some $\boldsymbol{s}^*$, $\boldsymbol{x} = \boldsymbol{A}\boldsymbol{s}^*$, and $\|\boldsymbol{s}^*\|_0 = L$. Then (2) has a unique solution $\hat{\boldsymbol{s}}_0 = \boldsymbol{s}^*$ if and only if $L < (m+1)/2$.*

Moreover, the minimum ℓ_0 quasi norm solution includes a combinatorial examination and it is difficult to solve (2). Unlike the non-unique solution (2), the minimum ℓ_1-norm solution is unique without the assumption of sparsity. It is also reported that it successfully finds the sparsest representation in many cases when $\boldsymbol{A}$ is pairs of bases [9, 11, 13, 14]. Thus we can expect that the solution (2) is sparse enough, then (1) solver finds it exactly. The minimum ℓ_1-norm solution (1) is easier to solve than (2).

If we can assume the sufficient sparsity for the sources, (2) will find the exact solution successfully. In sourse separation, however, there are low-dimensional problem such as the separation from 2 mixtures (2 channel recordings) and we cannot sometimes expect the sufficient sparsity of Theorem 1 for sources s. Thus we investigate what happens in such cases if we use (1).

3 Submatrix Representation and Decision Regions

In order to investigate the properties of the minimum ℓ_1-norm solution $\hat{s}$ in (1) for arbitrary $x(t)$, we introduce two important concepts, submatrix representation and decision regions.

First, we describe the existence of the submatrix representation of the minimum ℓ_1-norm solution $\hat{s}$. In order to compute $\hat{s}$, the first thing that springs to mind is an optimization of ℓ_1-norm over the null space of A by using a particular solution like (4). However, it requires complicated procedure for numerical stability, therefore it is more popular to use the linear programming. It is well known [9, 4, 5] that $\hat{s}$ can be obtained using the optimal solution in a linear program as follows:

$$\hat{s} = u^* - v^*, \qquad \begin{pmatrix} u^* \\ v^* \end{pmatrix} = \arg \min_{\binom{u}{v} \in \mathcal{P}(A,x)} \left\langle 1, \begin{pmatrix} u \\ v \end{pmatrix} \right\rangle, \qquad (5)$$

where $\mathcal{P}(A, x) \subset \Re^{2n}$ is a polyhedron defined as

$$\mathcal{P}(A, x) = \left\{ \begin{pmatrix} u \\ v \end{pmatrix} \Big| \tilde{A} \begin{pmatrix} u \\ v \end{pmatrix} = x, u \geq 0, v \geq 0 \right\} \quad \text{for } \tilde{A} = (A, -A)$$

where and $u \geq 0$ means that all components of u are zero or positive.

In the linear programming, we can find the optimal solution $z^* = \binom{u^*}{v^*} \in \Re^{2n}$ that gives the minimum ℓ_1-norm solution $\hat{s} = u^* - v^* \in \Re^n$. We know that this z^* must be the feasible basic solution, and its basic part z_B^* and nonbasic part z_N^* are represented as $z_B^* = (\tilde{A}_B^*)^{-1} x$, $\tilde{A}_B^* \subset \tilde{A}$ and $z_N^* = 0$ respectively. Thus, in order to represent the minimum ℓ_1-norm solution $\hat{s}$, we have to connect z_B^*, z_N^* to u^*, v^*. Indeed, we can obtain the following result.

Theorem 2 ([8]). *There exists an $m \times m$ submatrix B^* of A and $\hat{s}$ has $n - m$ zero entries as $\hat{s}_N$ and nonzero part $\hat{s}_B$ are given by the inverse of B^*, i.e.,*

$$\hat{s} = u^* - v^* \Leftrightarrow \hat{s}_B = (B^*)^{-1} x \text{ and } \hat{s}_N = 0,$$

where the indices of nonzero components are determined as follows:

$$A = (a_1, \ldots, a_n), B^* = (a_{i_1}, \ldots, a_{i_m}), \hat{s} = (\hat{s}_1, \ldots, \hat{s}_n)^\top \Rightarrow \hat{s}_B = (\hat{s}_{i_1}, \ldots, \hat{s}_{i_m})^\top$$

From this result, we can understand that the minimization of ℓ_1-norm implies an adaptive method to transform the underdetermined equation $x = As$ to the invertible equation $x = B^* \hat{s}_B$ by assuming that some $n - m$ components are zero, i.e., by setting $\hat{s}_N = 0$ for each time point t.

Hence the remaining problem is how to choose the indices of zero components. In other words, the problem is how to choose the optimal submatrix for each $x(t)$. This problem is resolved by the concept of decision region.

Theorem 3 ([8]). Let $\widetilde{\boldsymbol{A}} = (\boldsymbol{A}, -\boldsymbol{A})$ and let $\mathcal{B}$ be a set of $m \times m$ submatrices of the $\widetilde{\boldsymbol{A}}$ defined by

$$\mathcal{B} := \{\widetilde{\boldsymbol{B}} \mid \boldsymbol{1}^\top - \boldsymbol{1}_B^\top \widetilde{\boldsymbol{B}}^{-1} \widetilde{\boldsymbol{A}} \geqslant \boldsymbol{0}^\top, \det(\widetilde{\boldsymbol{B}}) \neq 0\},$$

where $\boldsymbol{1}$ and $\boldsymbol{1}_B$ denote the vectors with appropriate dimension whose components are all one. Then the observation space $\Re^m$ is decomposed into disjoint polyhedral cones

$$S(\widetilde{\boldsymbol{B}}) = \{\boldsymbol{x} \in \Re^m \mid \widetilde{\boldsymbol{B}}^{-1}\boldsymbol{x} \geqslant \boldsymbol{0}\}, \quad \widetilde{\boldsymbol{B}} \in \mathcal{B}$$

which satisfy

$$S(\widetilde{\boldsymbol{B}}_1)^\circ \cap S(\widetilde{\boldsymbol{B}}_2)^\circ = \varnothing \text{ for } \widetilde{\boldsymbol{B}}_1, \widetilde{\boldsymbol{B}}_2 \in \mathcal{B}, \widetilde{\boldsymbol{B}}_1 \neq \widetilde{\boldsymbol{B}}_2, \quad \bigcup_{\widetilde{\boldsymbol{B}} \in \mathcal{B}} S(\widetilde{\boldsymbol{B}}) = \Re^m,$$

where S° denotes the interior of the set S.

Theorem 4 ([8]). Let $\Psi(\widetilde{\boldsymbol{B}})$ be the matrix which changes the sign of column vector of $\widetilde{\boldsymbol{B}} = (\boldsymbol{b}_1, \ldots, \boldsymbol{b}_m)$ so that $\widetilde{\boldsymbol{B}}$ becomes a submatrix of $\boldsymbol{A}$, which defined as $\Psi(\widetilde{\boldsymbol{B}}) := \mathrm{diag}\{\mathrm{sign}(\tilde{\boldsymbol{b}}_1), \ldots, \mathrm{sign}(\tilde{\boldsymbol{b}}_m)\}$ for $\mathrm{sign}(\tilde{\boldsymbol{b}}_i) = 1$ ($\tilde{\boldsymbol{b}}_i \in \boldsymbol{A}$), $\mathrm{sign}(\tilde{\boldsymbol{b}}_i) = -1$ ($\tilde{\boldsymbol{b}}_i \in -\boldsymbol{A}$). If $\boldsymbol{x}(t) \in S(\widetilde{\boldsymbol{B}})$, i.e., $\widetilde{\boldsymbol{B}}^{-1}\boldsymbol{x}(t) \geqslant \boldsymbol{0}$, then

$$\boldsymbol{B}^* = \widetilde{\boldsymbol{B}}\Psi(\widetilde{\boldsymbol{B}}) \subset \boldsymbol{A}$$

is the optimal submatrix of $\boldsymbol{A}$ for that $\boldsymbol{x}(t)$.

From these three results, we can obtain $\hat{\boldsymbol{s}}(t)$ from $\boldsymbol{x}(t)$. For each data point $\boldsymbol{x}(t)$, first, find $\widetilde{\boldsymbol{B}} \in \mathcal{B}$ such that $\widetilde{\boldsymbol{B}}^{-1}\boldsymbol{x}(t) \geq \boldsymbol{0}$, then using $\boldsymbol{B}^* = \widetilde{\boldsymbol{B}}\Psi(\widetilde{\boldsymbol{B}})$, we can obatain $\hat{\boldsymbol{s}}(t)$ by $\hat{\boldsymbol{s}}_B(t) = (\boldsymbol{B}^*)^{-1}\boldsymbol{x}(t)$ and $\hat{\boldsymbol{s}}_N(t) = \boldsymbol{0}$. This procedure becomes an generalization to $\Re^m$ of the shortest path algorithm [4] which is proposed for $\Re^2$ mixtures and it can reduce the computational cost [16]. An example of decision regions are shown in Fig.2. Note that $\mathcal{B}$ in this example does not contain the submatrices such as $(-\boldsymbol{a}_2, \boldsymbol{a}_3)$ and it contains only submatrix which can define the decision region. For the cases of $m > 2$, this decomposition is based on a triangularion of convex hull of basis $\boldsymbol{A}$, and not much intuitive than the cases of $m = 2$ in which we can obtain it just by sorting in terms of angular parameter.

4 Piecewise Linearity of Minimum ℓ_1-Norm Solution

Decision regions are disjoint and all of decision regions cover the whole observation space. Thus for any data point $\boldsymbol{x}(t)$, there exists a corresponding decision region uniquely. Therefore, we can show that $\boldsymbol{B}^*$ or $\hat{\boldsymbol{s}}(t)$ is also unique. Moreover, the adjacency between decision regions is preserved in the original signal space (Fig. 2). Then, the estimated sequence has piecewise linearlity and it is continuous when the original sources are continuous (with respect to time). This indicates the relation between each $\hat{\boldsymbol{s}}(t)$ and the whole sequence $\{\hat{\boldsymbol{s}}(t)\}_{t=1}^T$.

In general, even if $\boldsymbol{s}(t)$ has many zero entries, it cannot be guaranteed $\boldsymbol{s}(t) = \hat{\boldsymbol{s}}(t)$. Moreover, even if there exists a minimum ℓ_0-norm solution uniquely

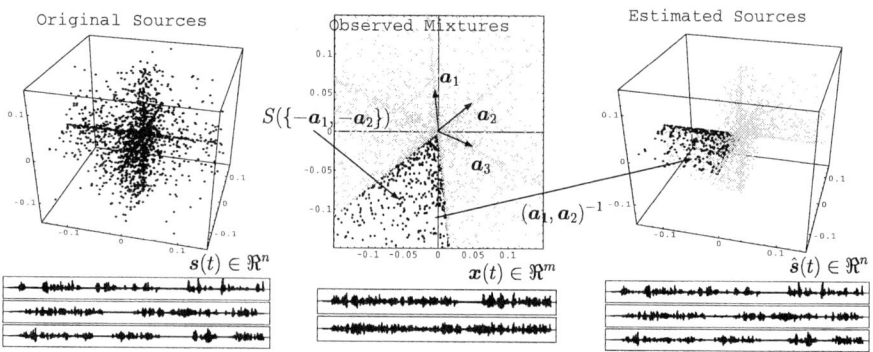

Fig. 2. Example of decision regions in a case with 2 sensors from 3 sources. Each decision regions are mapped linearly into $\Re^3$.

(like Theorem 1), we cannot obtain it as the minimum ℓ_1-norm solution. The zero part of $\hat{s}(t)$ is not always assigned for the recovery of original zero components and the result depends on which decision regions $As(t)$ falls in. The true value for assigned non-basic part can determine the degree of the error [8, 16]. However, for example, when only one component s_i is nonzero, we can obtain a perfect recovery because $As(t)$ always falls in the decision regions including a basis a_i. When this condition is almost satisfied, we can also obtain almost perfect recovery. In practical application, this is a key point for the explanation of successful performance of minimum ℓ_1-norm sequence.

5 Unstability for Perturbation of Mixing Matrix

The matrix A is assumed so far to be given but in actual it must be estimated in the blind identification step before signal recovery step. Then there is always a possibility that the estimated matrix A has a small error.

We should examine whether the minimum ℓ_1-norm solution will still be close to the solution that is estimated using the true matrix A when the estimated matrix has a small perturbation. Thus, we need a kind of perturbation analysis of the minimum ℓ_1-norm solution for mis-estimated matrix.

Unfortunately the minimum ℓ_1-norm solution is generally unstable for perturbation of the matrix. We can see this again from the decision region. For example, consider the 2×3 case that the column vector a_2 of A is mis-estimated as a'_2 by small perturbation (see Fig. 3).

Let $S(a_i, a_j)$ be the decision regions defined by the basis vectors a_i and a_j. The behavior of perturbated solutions is classified into the following three types:

- for samples in the area such as $S(a_1, a_3)$ (I in Fig. 3) which is unrelated to both a_2 and a'_2, a perturbation of a_2 does not affect the original minimum ℓ_1-norm solution.
- for samples in the area such as $S(a_1, a_2)$ or $S(a'_2, -a_3)$ (II in Fig. 3) including a_2 or a'_2 but not both, we use a'_2 instead of a_2 as basis vector of optimal submatrix. Then we can obtain the same perturbation result for linear cases.

- for samples in the area such as $S(\boldsymbol{a}_2', \boldsymbol{a}_2) = S(\boldsymbol{a}_1, \boldsymbol{a}_2') \cap S(\boldsymbol{a}_2, -\boldsymbol{a}_3)$ (III in Fig. 3) including both $\boldsymbol{a}_2$ or $\boldsymbol{a}_2'$, we obtain the different indices for the non-basic part (zero part).

The third case is most problematic for successfull results. In this case, we will obtain a different zero part. The minimum ℓ_1-norm sequence has generally this unstable property for perturbation of mixing matrix $\boldsymbol{A}$. However, if a matrix $\boldsymbol{A}$ is not ill-conditioned, we can see that the distance in $\Re^n$ between the perturbed one and the true one is still small.

The figure on the right in Fig. 3 shows this case (III). If $\boldsymbol{A}$ is not perturbated, the length of segment OA will be assigned as s_2, that of segment OC as s_1, and $s_3 = 0$. But if $\boldsymbol{a}_2$ changes to $\boldsymbol{a}_2'$, the length of segment OB will be assigned as s_2, that of segment OD as s_3, and $s_2 = 0$. The zero part is the third component $s_3 = 0$ in the former case and the second component $s_2 = 0$ in the latter case, and they are different. However, if $\boldsymbol{A}$ is not ill-conditioned and a perturbation is small, the lengths of OA and OB will be similar and the lengths of OC and OD are small. As a result, for many practical cases, permutation of components occurs just for the nearly zero component and zero component (Only these two are interchanged). Thus, for samples in III, important (comparatively large) value such as s_2 in this example will be recoverd by minimum ℓ_1-norm solution, then the effect of perturvation for recovery is little. Existing good experimantal reports in underdetermined separation is based on this fact.

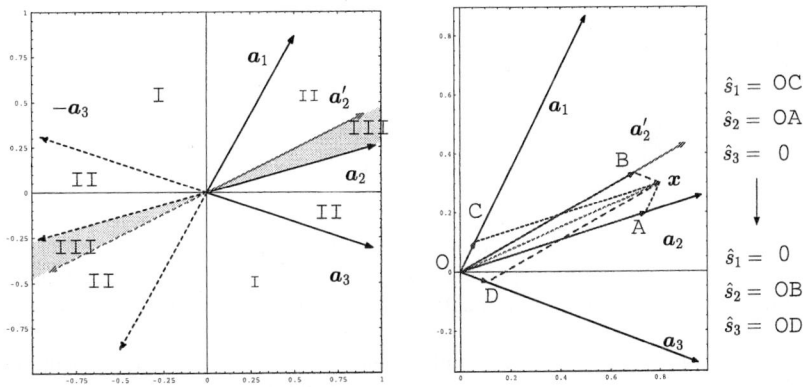

Fig. 3. Pertubation in mixing matrix.

6 Conclusion

In this study, the minimum ℓ_1-norm solution is analyzed. It is shown that there exists a submatrix representation of minimum ℓ_1-norm solution and an optimal submatrix for each data point can be selected using disjoint partitioning of mixture space by decision regions. From the viewpoint of the above submatrix representation and optimal submatrix selection by decision regions, the following properites of ℓ_1-norm solutions are clarified.

- Decision regions are disjoint and all of decision regions cover the whole observation space. The adjacency of decision regions is preserved after recovering into the original signal space.
- Estimated sequence has piecewise linearlity and continuous when the original sources is continuous (with respect to time).
- Non-basic part (zero part) of the estimated solution is not always assigned for the recovery of originally zero components even if an original source have sufficiently many zeros.
- The minimum ℓ_1-norm solution is generally unstable for a perturbation of the estimated matrix A.

References

1. Abrard, F., Deville, Y., White, P.: From blind source separation to blind source cancellation in the underdetermined case: a new approach based on time-frequency analysis. Proceedings of ICA2001 (2001)
2. Lewicki, M., T.J.Sejnowski: Learning overcomplete representations. Neural Computation **12** (2000) 337–365
3. Lee, T.W., Lewicki, M.S., Girolami, M., Sejnowski, T.J.: Blind source separation of more sources than mixtures using overcomplete representations. IEEE Signal Processing Letters **4** (1999)
4. Bofill, P., Zibulevsky, M.: Underdetermined blind source separation using sparse representations. Signal Processing **81** (2001) 2353–2362
5. Li, Y., Cichocki, A., Amari, S.: Sparse component analysis for blind source separation with less sensors than sources. Proceedings of ICA2003 (2003)
6. Zibulevsky, M., Pearlmutter, B.A.: Blind source separation by sparse decomposition. Neural Computations **13** (2001)
7. Olshausen, B.A., Field, D.J.: Sparse coding with an overcomplete basis set: A strategy employed by v1? Vision Research **37** (1997) 3311–3325
8. Takigawa, I., Kudo, M., Toyama, J.: Performance analysis of minimum ℓ_1-norm solutions for underdetermined source separation. IEEE Transactions on Signal Processing **52** (2004) 582–591
9. Chen, S.S., Donoho, D.L., Saunders, M.A.: Atomic decomposition by basis pursuit. SIAM Journal on Scientific Computing **20** (1998) 33–61
10. Donoho, D.L., Elad, M.: Optimally sparse representation in general (nonorthogonal) dictionaries via ℓ_1 minimization. Proceedings of the National Academy of Sciences **100** (2004)
11. Donoho, D.L., Huo, X.: Uncertainty principles and ideal atomic decomposition. IEEE Transactions on Information Theory **47** (2001)
12. Malioutov, M., Çetin, M., Willsky, A.S.: Optimal sparse representations in general overcomplete bases. accepted to IEEE International Conference on Acoustics, Speech, and Signal Processing (2004)
13. Elad, M., Bruckstein, A.M.: A generalized uncertainty principle and sparse representation in pairs of bases. IEEE Transactions on Information Theory **48** (2002)
14. Freuer, A., Nemirovski, A.: On sparse representation in pairs of bases. IEEE Transactions on Information Theory **49** (2003)
15. Mallat, S., Zhang, Z.: Matching pursuits with time-frequency dictionaries. IEEE Transactions on Signal Processing **41** (1993)
16. Takigawa, I.: Analysis of Solutions in Underdetermined Source Separation. PhD thesis, Graduate School of Engineering, Hokkaido University (2004)

On the Strong Uniqueness of Highly Sparse Representations from Redundant Dictionaries

Rémi Gribonval[1] and Morten Nielsen[2,*]

[1] IRISA-INRIA, Campus de Beaulieu, F-35042 Rennes CEDEX, France
remi.gribonval@inria.fr
[2] Department of Mathematical Sciences, Aalborg University
Fredrik Bajers Vej 7G, DK-9220 Aalborg East, Denmark
mnielsen@math.aau.dk

Abstract. A series of recent results shows that if a signal admits a sufficiently sparse representation (in terms of the number of nonzero coefficients) in an "incoherent" dictionary, this solution is unique and can be recovered as the unique solution of a linear programming problem. We generalize these results to a large class of sparsity measures which includes the ℓ^p-sparsity measures for $0 \le p \le 1$. We give sufficient conditions on a signal such that the simple solution of a linear programming problem simultaneously solves all the non-convex (and generally hard combinatorial) problems of sparsest representation w.r.t. arbitrary admissible sparsity measures. Our results should have a practical impact on source separation methods based on sparse decompositions, since they indicate that a large class of sparse priors can be efficiently replaced with a Laplacian prior without changing the resulting solution.

1 Introduction

Sparse decompositions of signals in redundant dictionaries provide quite a succesfull practical tool for blind source separation (BSS), including the degenerate case where there are more sources than sensors [18]. In this paper, we prove that estimators based on sparse decompositions are relatively robust to the choice of the sparse prior within a fairly large class. Our results directly apply to some noise-free single sensor BSS problems [1], but further work is needed to extend them to the case of multiple sensors and noisy measurements.

Given a redundant signal (or image) dictionary, every signal y has infinitely many possible representations, and it is common to choose one according to some *sparsity measure*. When the dictionary is indeed a basis, each signal has a unique representation and it does not matter which sparsity measure is used. However, in the redundant case, it is not clear when the sparsest representation is unique and how it is influenced by the choice of the sparsity measure.

---

* This work was supported in part by the European Union's Human Potential Programme, under contract HPRN-CT-2002-00285 (HASSIP) and in part by the Danish Technical Science Foundation, Grant no. 9701481.

A **dictionary** in a real or complex Hilbert space $\mathbb{R}^N$ or $\mathbb{C}^N$ is a family of $K \geq N$ unit vectors $\{g_k\}_{k \in K}$ which spans the entire space. One can think of g_k as the k-th column of a $N \times K$ matrix $\mathbf{D} = [g_1, \ldots, g_K]$, and any vector $y \in \mathbb{R}^N$ (resp. $\mathbb{C}^N$) has at least one representation $y = \sum_k x_k g_k = \mathbf{D}x$ with coefficient vector $x \in \mathbb{R}^K$ (resp. $x \in \mathbb{C}^K$). When $\mathbf{D}$ is **redundant** $(K > N)$, among the infinite number of representations of a vector y, it is often desirable to choose a *sparse* one. However, sparsity can be measured with diverse quantities such as the ℓ^p measures $\|x\|_p := \sum_k |x_k|^p$ for $0 \leq p \leq 1$ (with the convention $t^0 = 1$ if $t > 0$ and $0^0 = 0$). In this paper, we consider a large class $\mathcal{M}$ of **admissible sparsity measures** $\|x\|_f := \sum_k f(|x_k|)$ where $f : [0, \infty) \to [0, \infty)$ is non-decreasing, not identically zero, $f(0) = 0$ and $t \mapsto f(t)/t$ is non-increasing on $(0, \infty)$. To each sparsity measure corresponds the "f-**sparsest representation**" optimization problem

$$\text{minimize } \|x\|_f \text{ subject to } y = \sum_k x_k g_k. \qquad (1)$$

We address two rather natural questions related to this class of problems:

1/ when is the f-sparsest representation of y unique?
2/ if it is unique, does it depend on the choice of the sparsity measure f?

Our main result[1] is that when a signal y has a very sparse representation (in terms of the total number $\|x\|_0$ of nonzero coefficients), this representation is the *simultaneous and unique* sparsest representation with any admissible sparsity measure. More precisely we have the following theorem.

Theorem 1. *Let $\mathbf{D}$ be a dictionary. Assume m is an integer such that for any x and y with $y = \mathbf{D}x$ and $\|x\|_0 \leq m$, x is the unique ℓ^1-sparsest representation of y. Then, for any x and y such that $y = \mathbf{D}x$ and $\|x\|_0 \leq m$, x is indeed the unique f-sparsest representation of y for any admissible sparsity measure. In particular it is the ℓ^p-sparsest representation for $0 \leq p \leq 1$.*

The interesting consequence is that if y has a *highly sparse representation* x (with at most m elements) from the dictionary, then the combinatorial/highly nonlinear search for the f-sparsest representation of y can be replaced with a polynomial time computation based on linear programming [2, 15], which solves the ℓ^1-optimization problem.

This extends a series of recent results about recovery of sparse expansions from dictionaries by Pursuit algorithms. In the early 1990's, the Matching Pursuit and Basis Pursuit strategies were introduced with the purpose of getting good representations of signals with redundant dictionaries. Soon, it was experimentally noticed that, when y has a sufficiently sparse expansions (in the sense of $\|x\|_0$) in the Dirac/Fourier dictionary, Basis Pursuit can exactly recover it. The experimental observation was turned into a theorem and extended to unions of "incoherent" bases as well as to more general "incoherent" dictionaries [8, 5–7, 13, 3, 11]. Theorems in the same spirit were also recently proved, under slightly stronger assumptions, for exact recovery with Matching Pursuits [9, 10, 16, 14].

---

[1] We refer the reader to our technical report [12] for the proofs.

Previous Basis Pursuit results stated that if y has an expansion x with $\|x\|_0$ sufficiently small, then x is *simultaneously* the *unique* ℓ^0-sparsest and ℓ^1-sparsest representation of y [2, 15]. In between the ℓ^0 and the ℓ^1 sparsity measures lie the ℓ^p ones, and it seemed only natural that by some sort of "interpolation", the Basis Pursuit results should extend to *simultaneous uniqueness* of the ℓ^p-sparsest representations. It turns out that the interpolation can be done and our result show that it extends to the much larger class of admissible sparsity measures.

The structure of the paper is as follows. In Section 2 we give general conditions on an index set $I \subset K$ such that any expansion $y = \mathbf{D}_I x$ from the sub-dictionary $\mathbf{D}_I := \{g_k, k \in I\}$ is the unique f-sparsest representation of y in the whole dictionary. An example shows that the admissible sparsity measures f for which the conditions are satisfied may depend on the considered index set I. In Section 3 we give our main theorems and obtain necessary and sufficient conditions card$(I) \leq m_f(\mathbf{D})$ which ensure that for all admissible sparsity measures $g \in \mathcal{M}$ "between" a given $f \in \mathcal{M}$ and the ℓ^0 sparsity measure, the highly sparse representation is unique and independent of g. To conclude this paper, we briefly discuss how the numbers $m_f(\mathbf{D})$ which appear in the "highly sparse" conditions can be estimated from the *coherence* of the dictionary.

2 Sufficient Uniqueness Conditions

In this section we provide sufficient conditions on a representation $y = \mathbf{D}x$ which ensure that x is the unique f-sparsest representation of y, where f is an arbitrary admissible sparsity measure. A crucial property of $f \in \mathcal{M}$ is the quasi-triangle inequality

$$f(|u+v|) \leq f(|u|+|v|) \leq f(|u|) + f(|v|) \tag{2}$$

which is an easy consequence of the fact that $t \mapsto f(t)/t$ is non-increasing, see [12]. The sufficient uniqueness conditions are expressed in terms of the **support** $I(x) := \{k, x_k \neq 0\}$ of the coefficient vector $x = (x_k) \in \mathbb{R}^K$ (resp. $\mathbb{C}^K$), i.e. they depend on the set of elements of the dictionary which are used in the representation. The **kernel** Ker$(\mathbf{D}) := \{z, \mathbf{D}z = 0\}$ of the dictionary will play a special role. For $f \in \mathcal{M}$, $\mathbf{D}$ a dictionary and $I \subset K$ a set of indices, we define

$$\theta_f(I, z) := \frac{\sum_{k \in I} f(|z_k|)}{\|z\|_f} \quad \text{and} \quad \Theta_f(I, \mathbf{D}) := \sup_{z \in \text{Ker}(\mathbf{D}), z \neq 0} \theta_f(I, z) \tag{3}$$

The value of $\Theta_f(I, \mathbf{D})$ (almost) completely characterizes the uniqueness of f-sparsest expansions from $\mathbf{D}_I$, as expressed in the following lemma.

Lemma 1. *Let $\mathbf{D}$ be a dictionary, f an admissible sparsity measure, and $I \subset K$ an index set.*

1. *Assume that for all $z \in$ Ker$(\mathbf{D})$ $(z \neq 0)$, $\theta_f(I, z) < 1/2$, and let x, y such that $y = \mathbf{D}x$. If $I(x) \subset I$, x is the unique f-sparsest representation of y.*
2. *Assume that for some $z \in$ Ker$(\mathbf{D})$, $\theta_f(I, z) \geq 1/2$. Then, there exists $x \neq x'$ such that $\mathbf{D}x = \mathbf{D}x'$, $I(x) \subset I$ and $\|x'\|_f \leq \|x\|_f$.*

The proof is a slight refinement of ideas from [5, 13], see [12].

Even though the value of $\Theta_f(I, \mathbf{D})$ essentially characterizes the uniqueness of the f-sparsest representation of expansions from the sub-dictionary $\mathbf{D}_I = \{g_k\}_{k \in I}$, its evaluation for a given index set I is not trivial in general. In particular, it is not clear when the condition $\Theta_f(I, \mathbf{D}) < 1/2$ is simultaneously satisfied for all $f \in \mathcal{M}$, i.e., when the unique f-sparsest representation is the same for all sparsity measures f. The following example shows that f-sparsest representations do not necessarily coincide for different f, and that estimating $\Theta_f(I, \mathbf{D})$ for some admissible sparsity measure $f \in \mathcal{M}$ does not tell much about $\Theta_g(I, \mathbf{D})$ for other ones $g \in \mathcal{M}$.

Example 1. Let $\mathbf{B} = [g_1, \ldots, g_N]$ be an orthonormal basis in dimension N, $g_{N+1} := \sum_{k=1}^{N} \frac{1}{\sqrt{N}} g_k$ and $\mathbf{D} = [\mathbf{B}, g_{N+1}]$. Clearly, the kernel of $\mathbf{D}$ is the line generated by the vector $z = (1, \ldots, 1, \sqrt{N})$. Let us consider $I = \{1 \leq k \leq L\}$ an index set where $L \leq N$ and denote Θ_p for Θ_{f_p} where $f_p(t) = t^p$, $0 \leq p \leq 1$. Since

$$\Theta_1(I, \mathbf{D}) = \frac{L}{N + \sqrt{N}} < \frac{L}{N+1} = \Theta_0(I, \mathbf{D})$$

we have $\Theta_1(I, \mathbf{D}) < 1/2 < \Theta_0(I, \mathbf{D})$ whenever $(N+1)/2 < L < (N + \sqrt{N})/2$. On the other hand, let us now consider $J = \{1 \leq k \leq L\} \cup \{N+1\}$. As

$$\Theta_1(J, \mathbf{D}) = \frac{L + \sqrt{N}}{N + \sqrt{N}} \quad \text{and} \quad \Theta_0(I, \mathbf{D}) = \frac{L+1}{N+1}$$

we obtain $\Theta_0(J, \mathbf{D}) < 1/2 < \Theta_1(J, \mathbf{D})$ whenever $(N - \sqrt{N})/2 < L < (N-1)/2$.

3 Uniqueness of Highly Sparse Expansions

In the previous section, Example 1 illustrated the fact that, for arbitrary index sets I, not much can be said about the simultaneity of the f-sparsest representation for different admissible sparsity measures. In this section, we will show that the picture completely changes when we look for conditions on the *cardinal* of I so that $\Theta_f(I, \mathbf{D}) < 1/2$. Let us immediately state the main results of this section. The first result gives the theorem advertised in the introduction, which is the natural generalization to a series of recent results [8, 5–7, 13, 3, 11].

Theorem 2. *Let $\mathbf{D}$ be a dictionary, and f an admissible sparsity measure. Let m be an integer and assume that whenever $y = \mathbf{D}x$ with $\|x\|_0 \leq m$, x is the ℓ^1-sparsest representation of y. Then, whenever $y = \mathbf{D}x$ with $\|x\|_0 \leq m$, x is the simultaneous unique f-sparsest representation of y for any $f \in \mathcal{M}$.*

Theorem 2 is indeed only a special case of the following more general result.

Theorem 3. *Let $\mathbf{D}$ be a dictionary, and f an admissible sparsity measure. Let m be an integer and assume that whenever $y = \mathbf{D}x$ with $\|x\|_0 \leq m$, x is the f-sparsest representation of y. Then, whenever $y = \mathbf{D}x$ with $\|x\|_0 \leq m$, x is the simultaneous unique $(g \circ f)$-sparsest representation of y for any $g \in \mathcal{M}$.*

Note that one can easily check that if $f, g \in \mathcal{M}$ then $g \circ f \in \mathcal{M}$.

3.1 Sketch of the Proof of Theorem 3

We will study in more details in the next section which integers m satisfy the assumptions of Theorem 3, but let us first sketch the proof. For any sequence $z = \{z_k\}_{k \in K}$, denote $|z|^\star$ a decreasing rearrangement of $|z|$, i.e., $|z|^\star_k = |z_{\phi(k)}|$ where ϕ is one to one and $|z|^\star_k \geq |z|^\star_{k+1}$. With a slight abuse of notation, consider the "growth function"

$$\theta_f(m, z) := \max_{\text{card}(I) \leq m} \theta_f(I, z) = \frac{\sum_{k=1}^m f(|z|^\star_k)}{\|z\|_f} = \theta_f(m, |z|^\star) \qquad (4)$$

defined for any $f \in \mathcal{M}$, $m \geq 0$ and $z \neq 0$. We have the following lemma [12].

Lemma 2. *For any $f, g \in \mathcal{M}$, $m \geq 0$ and $z \neq 0$ we have*

$$\theta_0(m, z) \leq \theta_{g \circ f}(m, z) \leq \theta_f(m, z) \leq \theta_1(m, z). \qquad (5)$$

Let us just mention that the result relies crucially on the property that $t \mapsto f(t)/t$ is non-increasing, since the fact that $\theta_f(m, z) \leq \theta_1(m, z)$ for all m and z implies in particular that for any $a < b$ we must have $f(b)/(f(a) + f(b)) \leq b/(a + b)$, i.e., $1 + f(a)/f(b) \geq 1 + a/b$.

Theorem 3 is proved as follows: from Lemma 1, the assumption on m implies that, for all I with $\text{card}(I) \leq m$ and $z \in \text{Ker}(\mathbf{D})$ ($z \neq 0$), $\theta_f(m, z) < 1/2$. It follows from Lemma 2 that for all such I and z, and any $g \in \mathcal{M}$, $\theta_{g \circ f}(m, z) < 1/2$, which gives the desired result using again Lemma 1.

3.2 Explicit Sparsity Conditions

For any dictionary $\mathbf{D}$ and sparsity measure f, one can consider the largest integer $m_f(\mathbf{D})$ that satisfies the assumption of Theorem 3, i.e., such that for any x and y such that $y = \mathbf{D}x$ and $\|x\|_0 \leq m$, x is indeed the unique f-sparsest representation of y. Another formulation of Theorem 3 is simply that for any $f, g \in \mathcal{M}$, $m_{g \circ f}(\mathbf{D}) \geq m_f(\mathbf{D})$. Indeed, it follows from Lemma 2 that

$$m_0(\mathbf{D}) \geq m_{g \circ f}(\mathbf{D}) \geq m_f(\mathbf{D}) \geq m_1(\mathbf{D}) \qquad (6)$$

where m_p, $0 \leq p \leq 1$ is a shorthand for m_{f_p} with $f_p(t) := t^p$.

It is a challenge to compute the numbers $m_f(\mathbf{D})$ for an arbitrary dictionary (the computation of $m_0(\mathbf{D})$ is generally NP-hard). Let us however give a few examples of dictionaries where it is possible to get some non trivial bounds on the **strong sparsity number** $m_1(\mathbf{D})$ and the **weak sparsity number** $m_0(\mathbf{D})$ based on easily computable characteristics of the dictionary. Denoting $\lfloor t \rfloor$ the largest integer such that $\lfloor t \rfloor < t \leq \lfloor t \rfloor + 1$, we have the following lemma [12].

Lemma 3. *For any admissible sparsity measure $f \in \mathcal{M}$ and any dictionary $\mathbf{D}$,*

$$m_f(\mathbf{D}) \geq m_1(\mathbf{D}) \geq \lfloor Z_1(\mathbf{D})/2 \rfloor \qquad \text{and} \qquad m_0(\mathbf{D}) = \lfloor Z_0(\mathbf{D})/2 \rfloor \qquad (7)$$

where

$$Z_0(\mathbf{D}) := \inf_{z \in Ker(\mathbf{D}), z \neq 0} \|z\|_0 \qquad \text{and} \qquad Z_1(\mathbf{D}) := \inf_{z \in Ker(\mathbf{D}), \|z\|_\infty = 1} \|z\|_1. \qquad (8)$$

*are respectively called the **spark** and the **spread** of the dictionary.*

The spark was introduced in [3] and its numerical computation is generally combinatorial. The spread was introduced by the authors in [11]. The above estimates are not quite explicit, but the next one is easily computable.

Lemma 4. *The* **coherence** *of a dictionary* $\mathbf{D} = \{g_k\}$ *is defined [5] as*

$$M(\mathbf{D}) := \sup_{k \neq k'} |\langle g_k, g_{k'} \rangle|. \tag{9}$$

For any admissible sparseness measure $f \in \mathcal{M}$ *we have the lower estimate*

$$m_f(\mathbf{D}) \geq m_1(\mathbf{D}) \geq \lfloor (1 + 1/M(\mathbf{D}))/2 \rfloor. \tag{10}$$

Proof. Consider $x \in \text{Ker}(\mathbf{D})$. For every k we have $x_k g_k = -\sum_{k' \neq k} x_{k'} g_{k'}$ hence, taking the inner product of both hand sides with g_k, $|x_k| \leq M(\mathbf{D}) \cdot \sum_{k' \neq k} |x_{k'}|$. It follows that $(1 + M) \cdot |x_k| \leq M \cdot \|x\|_1$. Taking the supremum over k we get $(1 + M)\|x\|_\infty \leq M \cdot \|x\|_1$ or equivalently $Z_1(\mathbf{D}) \geq 1 + 1/M$, and the result follows using Lemma 3.

When $\mathbf{D}$ contains an orthonormal basis $\mathbf{B}$ in dimension N, the coherence satisfies $M(\mathbf{D}) \geq 1/\sqrt{N}$, and it is possible to find up to $N+1$ orthonormal bases $\{\mathbf{B}_j\}_{j=1}^{N}$ such that their union $\mathbf{D} := [\mathbf{B}_1 \ldots \mathbf{B}_{N+1}]$ is a dictionary of coherence $m(\mathbf{D}) = 1/\sqrt{N}$. For such highly redundant dictionaries, the lemma shows that $m_1(\mathbf{D}) \geq \lfloor (1 + \sqrt{N})/2 \rfloor$. Lemma 4 was in germ in Donoho and Huo's early paper [5] on exact recovery of sparse expansion through Basis Pursuit, where it was only used for $\mathbf{D}$ a union of two orthonormal bases and $f(t) = t^p, p \in \{0, 1\}$. In [13] and [3] it was extended to arbitrary dictionaries, and in [11] to $f(t) = t^p, p \in [0, 1]$. Finer estimates of $m_1(\mathbf{D})$ can be obtained from the properties of the Gram matrix of $\mathbf{D}$, see [12].

4 Conclusion and Statistical Perspectives

We have studied sparse representation of signals using an arbitrary dictionary and a very general admissible sparsity measure $\| \cdot \|_f$. Given a dictionary and a signal y, we provided sufficient conditions for the minimization problem

$$\text{minimize } \|x\|_f \text{ subject to } y = \sum_k x_k g_k, \tag{11}$$

to have the same unique solution as the problem

$$\text{minimize } \|x\|_1 \text{ subject to } y = \sum_k x_k g_k, \tag{12}$$

and the conditions are *independent* of the particular admissible sparsity measure f. The latter minimization problem (12) can be solved using a linear programming technique, *i.e.*, by a polynomial time algorithm. For a dictionary in a Hilbert space we proved that the condition $\|x\|_0 \leq \lfloor 1/2(1 + 1/M) \rfloor$, where M is the coherence of the dictionary, is sufficient for (11) to have the same solution as (12) for any sparsity measure f. The results generalize previous results by Donoho and Elad [3] and by the authors [13], where only two types of sparsity measures were considered: the ℓ^0-norm and the ℓ^1-norm.

The f-sparsest representation problems (11) that we have considered in this paper are related to the statistical problem of Bayesian estimation of unknown parameters (x_k) given the noise-free observation $y = \mathbf{D}x$ and the prior probability density function $P_{f,h}(x) = \frac{1}{Z_{f,h}} \exp(-h(\|x\|_f))$, where $h : [0, \infty) \to [0, \infty)$ is an increasing function and $Z_{f,h}$ a normalizing constant such that $P_{f,h}(x)$ is a probability density on $\mathbb{R}^K$ (resp. $\mathbb{C}^K$).

In this Bayesian estimation setting, our results have an interpretation in terms of *robust estimation* with respect to modeling error. Assume that the prior on x has the above structure, where $\|\cdot\|_f$ is an admissible sparsity measure. Then, for any noise-free observation y that admits a sufficiently sparse representation (with $\|x\|_0 \leq m_1(\mathbf{D})$), it does not matter which admissible sparse prior we use to model the data and search for the sparsest representation: each admissible sparse model yields the same estimate, which is indeed the MAP estimate under the true prior. In particular, we can as well model the parameters with a Laplacian prior $P_1(x) \propto \exp(-\|x\|_1)$, and this relaxed model will recover the "good" parameters (x_k).

To see how strong is the robustness to modeling error, let us simply give an example. First, notice that the Laplacian prior is a model where we assume the independence of the x_k, since $P_1(x) = \prod_k P_1(x_k)$. However, as shown in [12], since the class $\mathcal{M}$ of admissible sparsity measures is stable by $\min(\cdot)$ and $\max(\cdot)$, it contains some nontrivial measures such as $\|\cdot\|_f$ with

$$f(t) := \max(t/2, \min(t^{1/2}, t^0)) = \begin{cases} \sqrt{t}, & 0 \leq t \leq 1 \\ 1, & 1 \leq t \leq 2 \\ t/2, & 2 \leq t < \infty \end{cases}. \qquad (13)$$

Moreover, the use of a "sufficiently increasing" function h to define $P_{f,h}$ can introduce a dependence between the coefficients x_k, since $P_{f,h}$ will no longer be the product of its marginals. Yet, if the solution to the true Bayesian estimation problem is sparse enough, it will be recovered with the Laplacian model, where the parameters are assumed independent!

The main limitation to the theory developed in this paper is certainly that the sparsity condition $\|x\|_0 \leq m_1(\mathbf{D})$ is quite restrictive, since the set of observations y that admits such a sparse representation is of Lebesgue measure zero in $\mathbb{R}^K$ and probability zero under the sparse prior. A second limitation comes from the fact that the results do not apply to noisy data $y = \mathbf{D}x + n$. Recent results [17, 4] indicate that similar robustness properties can be proved even with approximate and noisy sparse representations, and the authors are also investigating the problem of simultaneous sparse representation/approximation of several observations in a single dictionary, which is a widely-spread tool to perform blind source separation [18].

References

1. L. Benaroya, R. Gribonval, and F. Bimbot. Représentations parcimonieuses pour la séparation de sources avec un seul capteur. In *GRETSI 2001*, Toulouse, France, 2001. Article ♯ 434.
2. D. Bertsekas. *Non-Linear Programming*. Athena Scientific, Belmont, MA, 2nd edition, 1995.
3. D. Donoho and M. Elad. Optimally sparse representation in general (non-orthogonal) dictionaries via ℓ^1 minimization. *Proc. Nat. Aca. Sci.*, 100(5):2197–2202, Mar. 2003.
4. D. Donoho, M. Elad, and V. Temlyakov. Stable recovery of sparse overcomplete representations in the presence of noise. Working draft, Feb. 2004.
5. D. Donoho and X. Huo. Uncertainty principles and ideal atomic decompositions. *IEEE Trans. Inform. Theory*, 47(7):2845–2862, Nov. 2001.
6. M. Elad and A. Bruckstein. A generalized uncertainty principle and sparse representations in pairs of bases. *IEEE Trans. Inform. Theory*, 48(9):2558–2567, Sept. 2002.
7. A. Feuer and A. Nemirovsky. On sparse representations in pairs of bases. *IEEE Trans. Inform. Theory*, 49(6):1579–1581, June 2003.
8. J.-J. Fuchs. On sparse representations in arbitrary redundant bases. Technical report, IRISA, Dec. 2003. to appear in IEEE Trans. Inform. Theory.
9. A. Gilbert, S. Muthukrishnan, and M. Strauss. Approximation of functions over redundant dictionaries using coherence. In *The 14th ACM-SIAM Symposium on Discrete Algorithms (SODA'03)*, Jan. 2003.
10. A. Gilbert, S. Muthukrishnan, M. Strauss, and J. Tropp. Improved sparse approximation over quasi-incoherent dictionaries. In *Int. Conf. on Image Proc. (ICIP'03)*, Barcelona, Spain, Sept. 2003.
11. R. Gribonval and M. Nielsen. Approximation with highly redundant dictionaries. In M. Unser, A. Aldroubi, and A. F. Laine, editors, *Proc. SPIE '03*, volume 5207 Wavelets: Applications in Signal and Image Processing X, pages pp. 216–227, San Diego, CA, Aug. 2003.
12. R. Gribonval and M. Nielsen. Highly sparse representations from dictionaries are unique and independent of the sparseness measure. Technical Report R-2003-16, Dept of Math. Sciences, Aalborg University, Oct. 2003.
13. R. Gribonval and M. Nielsen. Sparse decompositions in unions of bases. *IEEE Trans. Inform. Theory*, 49(12):3320–3325, Dec. 2003.
14. R. Gribonval and P. Vandergheynst. Exponential convergence of Matching Pursuit in quasi-incoherent dictionaries. Technical report 1619, IRISA, 2004.
15. A. Shrijver. *Theory of Linear and Integer Programming*. John Wiley, 1998.
16. J. Tropp. Greed is good : Algorithmic results for sparse approximation. Technical report, Texas Institute for Computational Engineering and Sciences, 2003.
17. J. Tropp. Just relax: Convex programming methods for subset selection and sparse approximation. Technical Report ICES Report 04-04, UT-Austin, Feb. 2004.
18. M. Zibulevsky and B. Pearlmutter. Blind source separation by sparse decomposition in a signal dictionary. *Neural Computations*, 13(4):863–882, 2001.

Reliability of ICA Estimates with Mutual Information

Harald Stögbauer[1], Ralph G. Andrzejak[1],
Alexander Kraskov[1], and Peter Grassberger[1]

John-von-Neumann Institute for Computing, Forschungszentrum Jülich
D-52425 Jülich, Germany
{h.stoegbauer,r.g.andrzejak,a.kraskov,p.grassberger}@fz-juelich.de
http://www.fz-juelich.de/nic/cs/

Abstract. Obtaining the most independent components from a mixture (under a chosen model) is only the first part of an ICA analysis. After that, it is necessary to measure the actual dependency between the components and the reliability of the decomposition. We have to identify one- and multidimensional components (i.e., clusters of mutually dependent components) or channels which are too close to Gaussians to be reliably separated. For the determination of the dependencies we use a new highly accurate mutual information (MI) estimator. The variability of the MI under remixing provides us a measure for the stability. A rapid growth of the MI under mixing identifies stable components. On the other hand a low variability identifies unreliable components. The method is illustrated on artificial datasets. The usefulness in real-world data is shown on biomedical data.

1 Introduction

Independent component analysis (ICA) is a statistical method for transforming an observed multivariate data set $\mathbf{X}(t) = (X_1(t), X_2(t), ..., X_n(t))$ into components that are statistically as independent from each other as possible [1]. By construction ICA always finds some decomposition, even if the data do not satisfy the assumption that they are a superposition of independent non-Gaussian sources. Therefore, in applications to real-world data it is not clear which components of the output can be interpreted meaningfully and how reliable they are. To address this problem one should first check the appropriateness of these assumptions which define the model underlying the ICA. The better the model fits the reality the more reliable are the components derived from the ICA. On the other hand, even if the data are consistent with the model, this does not ensure that all components found are meaningful because the solution might not be unique. In consequence, it is necessary to test these two aspects prior to the interpretation of the output of ICA. We here propose a simple scheme for this testing.

While recently proposed reliability tests [2],[3], [4] are based on bootstrap methods or noise injection we here present an alternative two step procedure:

1. Model test - actual dependencies between the ICA components
2. Uniqueness test - variability of the dependencies under remixing

In the following, we restrict ourselves to the class of instantaneous linear ICA algorithms. In order to test its appropriateness we determine the actual independencies of the ICA components. Here components are assumed to be truly independent if their MI is below a certain threshold. The efficient MI estimator proposed in Ref. [5] serves as a dependency measure. Components which are truly independent from all other components will be called one-dimensional components whereas groups of components which are mutually dependent but independent from the rest constitute multi-dimensional components [6]. If we find only one multi-dimensional component, i.e. all obtained channels are dependent among each other, the instantaneous linear ICA model is not appropriate.

Provided that we pass this first step of our procedure, we have to test for the uniqueness of the components. For this purpose, we have to check whether the solution of the ICA indeed corresponds to distinct minima of the dependencies or whether other linear combinations exist which show approximately the same overall dependencies. An example for the latter case is given by two white uncorrelated Gaussian signals which remain independent under rotation. A good estimation for the uniqueness of the ICA output is the variability of the pairwise MI under remixing, i.e. under rotations in the two-dimensional plane. While for unique solutions the MI will change significantly it will stay almost constant for ambiguous outputs of the ICA such as two white Gaussian signals.

The paper is organized as follows. In Section 2 we will give a brief introduction to the MI estimator, followed by the formal development of the reliability test. In Section 3 we apply our method to artificial data to illustrate its behavior. In Section 4 the usefulness of our method in real-world data is shown on the ECG of a pregnant woman. Conclusions are drawn in the last Section.

2 Methods

2.1 Efficient MI Estimator

In contrast to other estimators based on cumulant expansions, entropy maximalization, parameterizations of the densities, kernel density estimators or binnings, the used algorithm is based on entropy estimates from k-nearest neighbor distances [5]. This means that it is data efficient (with $k = 1$ we resolve structures down to the smallest possible scales), adaptive (the resolution is higher where data are more numerous), and has minimal bias. Numerically, it seems to become *exact* for independent distributions, i.e. the estimator $\hat{M}(X, Y)$ vanishes (up to statistical fluctuations) if $\mu(x, y) = \mu(x)\mu(y)$. This holds for all tested marginal distributions and for all dimensions of x and y.

Typically, one has a set of N bivariate measurements, $z_i = (x_i, y_i)$, $i = 1, \ldots N$, which are assumed to be iid (independent identically distributed) realizations of a random variable $Z = (X, Y)$ with density $\mu(x, y)$. The marginal densities of X and Y are $\mu_x(x) = \int dy\, \mu(x, y)$ and $\mu_y(y) = \int dx\, \mu(x, y)$. The MI is defined as

$$I(X,Y) = \iint dxdy\, \mu(x,y) \log \frac{\mu(x,y)}{\mu_x(x)\mu_y(y)}\,. \tag{1}$$

The aim is to estimate $I(X,Y)$ from the set $\{z_i\}$ alone, without knowing the densities μ, μ_x, and μ_y. We will start from the Kozachenko-Leonenko estimate for Shannon entropy [7]:

$$\hat{H}(X) = -\psi(k) + \psi(N) + \log c_d + \frac{d}{N}\sum_{i=1}^{N} \log \epsilon(i) \tag{2}$$

where $\psi(x)$ is the digamma function, $\epsilon(i)$ is twice the distance from x_i to its k-th neighbor, d is the dimension of x and c_d is the volume of the d-dimensional unit ball. The mutual information could be obtained by estimating $H(X)$, $H(Y)$ and $H(X,Y)$ separately and using

$$I(X,Y) = H(X) + H(Y) - H(X,Y)\,. \tag{3}$$

For any fixed k, the distance to the k-th neighbor in the joint space will be larger than the distances to the neighbors in the marginal spaces. Since the bias from the non-uniformity of the density depends of course on these distances, the biases in $\hat{H}(X)$, $\hat{H}(Y)$, and in $\hat{H}(X,Y)$ would not cancel.

To avoid this, we notice that Eq.(2) holds for *any* value of k, and that we do not have to choose a fixed k when estimating the marginal entropies.

So let us use $n_x(i)$ and $n_y(i)$ (the number of points with $||x_i - x_j|| \leq \epsilon_x(i)/2$ and $||y_i - y_j|| \leq \epsilon_y(i)/2$) as the numbers of neighbors in the marginal spaces. The estimate for MI is then (a detailed derivation can be found in [5]):

$$\hat{I}(X,Y) = \psi(k) - 1/k - \langle \psi(n_x) + \psi(n_y) \rangle + \psi(N)\,. \tag{4}$$

We denote by $\langle \ldots \rangle$ averages both over all $i \in [1, \ldots N]$ and over all realizations of the random samples.

The most conspicuous feature that was found in the numerical experiments of [5] is that the systematic error of the estimate for uncorrelated Gaussian signals is compatible with zero, independent from their sample size. This is a property which makes the estimator particularly interesting for ICA because there we are searching for uncorrelated signals. For non-Gaussian signals, we can assume that our estimator still has a small systematic error.

Any estimator has to find a compromise between statistical versus systematical errors. The only parameter which must be chosen in our estimator is the number k of nearest neighbors. It provides the user the possibility to control these two errors. The higher k, the lower the statistical error of $\hat{I}$, since we calculate the average of a higher number of points. The systematical error shows exactly the opposite behavior.

2.2 Dependency Matrix – Model Test

In the simplest case a multivariate signal represents an instantaneous linear mixture of independent signals. In real-world data, however, we are confronted with deviations from this simple model. To test the consistency of this model with a given dataset we calculate the pairwise MIs between all ICA components, $\hat{I}(X_i, X_j)$ with $i, j = 1, ..., n$ and $i \neq j$. One exemplary dependency matrix is

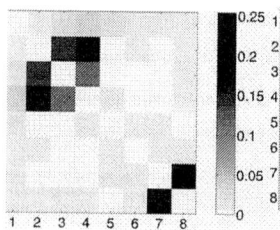

Fig. 1. Exemplary dependency matrix: $\hat{I}(X_i, X_j)$ (k=6) between all pairwise channel combinations (the diagonal is set to zero). Channels 1, 5 and 6 are one-dimensional components. Channels 2,3,4 form a three-dimensional component and channels 7, 8 a two-dimensional component.

shown in Fig. 1. A pair of components whose MI falls below a defined threshold $(\hat{I}(X_i, X_j) < D_{MAX})$ is considered as truly independent. Components which are truly independent from all other components are regarded as one-dimensional components[1]. A group of components which is defined by dependencies between pairs of individual components $(\hat{I}(X_p, X_q) > D_{MAX})$ is regarded as a multi-dimensional component. Notice that within such a group not *all* possible pairs of components are required to be dependent. In this way the dependency matrix allows to identify one- and multi-dimensional components in the output of any ICA.

2.3 Uniqueness Test

Suppose we have identified two one-dimensional components X and Y using the dependency matrix. Because of their independence from all other components the assumption of an instantaneous linear model is fulfilled for these components. To test for the uniqueness of these components we introduce rotations in the two-dimensional sub-space:

$$\mathbf{R}(\phi)(X, Y) = (X', Y') \quad (5)$$

with

$$X' = \cos\phi \, X + \sin\phi \, Y, \quad Y' = -\sin\phi \, X + \cos\phi \, Y \quad (6)$$

and measure $\hat{I}(\mathbf{R}(\phi)(X, Y))$ versus the rotation angle ϕ. In Fig. 2 we show three cases. They correspond to X being given by uniformly distributed noise,

---

[1] Note however it can happen that a group of m components is pairwise independent, although the components are not globally independent. For a rigorous discussion of this, see [8].

uniformly distributed noise superposed with Gaussian noise, and purely Gaussian noise, respectively, while Y is always given by white Gaussian noise. The depth of the minimum in $\hat{I}(\mathbf{R}(\phi)(X,Y))$ reflects the uniqueness of the solution for X and Y. A flat curve reveals ambiguity for X and Y, whereas a distinct minimum shows the uniqueness of the solution. We measure the variability of

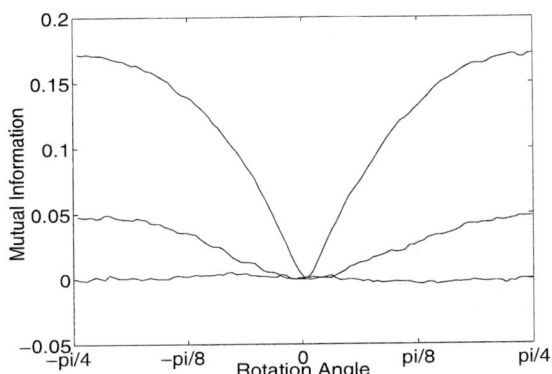

Fig. 2. $\hat{I}(\mathbf{R}(\phi)(X,Y))$ (k=6) versus the rotation angle. From top to bottom: X= uniform, uniform+Gaussian, Gaussian ; Y= always Gaussian.

$\hat{I}(\mathbf{R}(\phi)(X,Y))$ between all channels of the ICA output X_i $i = 1,...,n$:

$$\sigma_{ij} = \overline{I(X_i, X_j)} - I(\mathbf{R}(0)(X_i, X_j)) \tag{7}$$

where the global minimum of $\hat{I}$ is at $\phi = 0$, and

$$\overline{I(X_i, X_j)} = \frac{1}{2\pi} \int_0^{2\pi} d\phi I(\mathbf{R}(\phi)(X_i, X_j)) . \tag{8}$$

Fig. 2 suggests that it would be enough to estimate $\overline{I(X_i, X_j)}$ from the values at the two angles $\phi = (0, \pi/4)$ but this might be dangerous because of statistical fluctuations. Moreover, the shape of the $\hat{I}(\mathbf{R}(\phi)(X,Y))$ curve can be more complicated than the one shown in Fig. 2 (e.g. for signals with multi-modal distributions).

3 Application I: Artificial Data

To illustrate the behavior of our methods we apply them first to artificial data. We start with eight sources: Two channels of uniformly distributed noise, two channels of uniformly distributed noise superposed with Gaussian noise, and two channels of purely Gaussian noise. The last two channels are derived from a nonlinear mixing of two uniformly distributed signals ($x^* = (x+y)^2$ and $y^* =$

$(x-y)^2$). All eight channels were mixed using a non-singular random matrix. Subsequently, we applied the ICA-algorithm of [5] to this mixture which directly minimizes the MI. As shown in Fig. 3(left) we obtained one two-dimensional component and six one-dimensional components using a value of $D_{MAX} = 0.05$. The two-dimensional component corresponds to the nonlinearly mixed signals which could not be decomposed by the instantaneous linear ICA model. We can correctly conclude from this plot that we are dealing with a linear mixture of two dependent (beyond linear) and six independent signals.

We can see in Fig. 3(right) that $\sigma_{78} \approx 0$. This means that channels 7 and 8 can not be decomposed uniquely. Hence, these two channels correspond to the two purely Gaussian signals which are not separable. The relatively low values for σ_{5i} with $i = 6, 7, 8$ and σ_{6j} with $j = 7, 8$ reflects the fact that uniform noise superimposed with Gaussian noise is very similar to pure Gaussian noise. Since $\sigma_{3i} > 0$ and $\sigma_{4i} > 0$ for all i, channels 3 and 4 are stable and represent the two uniformly distributed noise signals. In general, it depends on the specific application whether one should attribute any meaning to σ_{ij} when components i and j are not independent.

In many applications it is useful to project multi-dimensional components back in the measurement space since there they carry a physical meaning. In summary, we can reliably separate the dependent signals from the others.

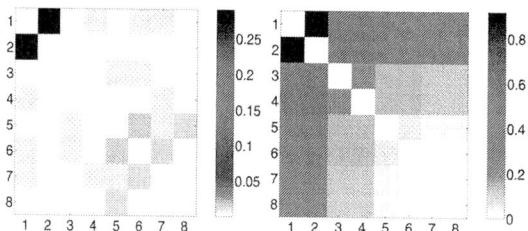

Fig. 3. Right panel: $\hat{I}(X_i, X_j)$ (k=6) between the all pairwise channel combinations. Left panel: the variability of $\hat{I}(R(X_i, X_j))$ depicted be σ_{ij}. In both panels the diagonal is set to zero. Channels 1 and 2 are nonlinearly mixed and thus mutually dependent signals, 3 and 4 are uniform noise, 5 and 6 are uniform plus Gaussian noise, 7 and 8 are pure Gaussians.

4 Application II: ECG

We applied ICA to an ECG recording from the abdomen and thorax of a pregnant woman (8 electrodes, 500 Hz, 5s). The data was also analyzed in [2,6,9] and is available in the public domain [10]. The resulting ICA components are shown in Fig. 4. For such an application it is possible to estimate the success of the decomposition by visual inspection. Obviously channels 1-2 are dominated by the heartbeat of the mother and channel 5 by that of the child. Channels 3, 4 and 6 also contain heartbeat components (of the mother and child, respectively). But these channels look much more noisy. Channels 7-8 seem to be dominated by noise but with rather different spectral content.

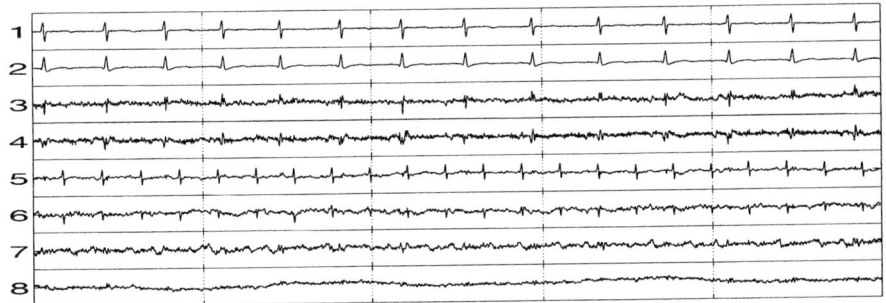

Fig. 4. ICA output of the ECG of a pregnant woman.

In Fig. 5(left) we can see that most values of $\hat{I}$ are indeed small. However, the first two components are still strongly dependent. Using a value of $D_{MAX} = 0.15$ we obtain one two-dimensional (channels 1 and 2) and six one-dimensional components. This value of D_{MAX} is considerably higher than the one used for the artificial signals in the previous section. This indicates that the assumption of an instantaneous linear mixing is not exactly fulfilled for this example of real-world data. Following the arguments used in the last Section we can conclude

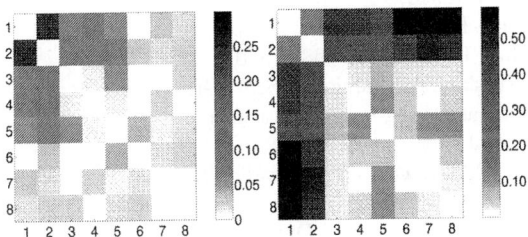

Fig. 5. Right panel: $\hat{I}(X_i, X_j)$ (k=6) between the all pairwise channel combinations (signal shown in Fig. 4). Left panel: the variability of $\hat{I}(R(X_i, X_j))$ depicted be σ_{ij}. In both panels the diagonal is set to zero.

from Fig. 5(right) that channel 5 and the pair (1,2) are reliable one- and two-dimensional components, respectively. This test reveals that these components should be considered for further interpretation. Indeed, Fig. 4 shows that they carry a physiological meaning. Channels 1 and 2 represent the heartbeat of the mother, while the fetal heartbeat is expressed in channel 5. The remaining one-dimensional channels contain mostly Gaussian noise which is reflected in low σ values.

5 Conclusions

We have discussed a reliability test for the output of a linear ICA algorithm. First we check the validity of the linear model. For components which follow the model a uniqueness test is performed afterwards. ICA components can carry only physical meaning when they pass both tests. On biomedical data, the ECG of a pregnant woman, we showed the usefulness of this procedure. The reliability test can be extended to different ICA models by simply changing the remixing operation.

References

1. Hyvärinen, A., Karhunen, J., Oja, E.: Independent Component Analysis. Wiley, New York (2001)
2. Meinecke, F., Ziehe A., Kawanabe, M., Müller, K.-R.: A resampling approach to estimate the stability of one-dimensional or multidimensional independent components. IEEE Trans. Biomed. Eng. **49** (2002) 1514-25
3. Harmeling, S., Meinecke F., Müller, K.-R.: Analysing ICA component by injection noise. In: Proc. Int. Workshop on Independent Component Analysis (2003)
4. Himberg, J., Hyvärinen, A.: Icasso: software for investigating the reliability of ICA estimates by clustering and visualization. In: Processing of the Workshop on Neural Networks and Signal Processing. Toulouse, France (2003)
5. Kraskov, A., Stögbauer H., Grassberger, P.: Estimating mutual information. Phys. Rev. E, in press.
6. Cardoso, J.-F.: Multidimensional independent component analysis. In: Processing of ICASSP. Seattle (1998)
7. Kozachenko, L.F., Leonenko, N.N.: Sample estimate of the entropy of a random vector. Probl. Inf. Transm. **23** (1987) 9-16
8. Stögbauer H., Kraskov, A., Astakhov, S.A., Grassberger, P.: Least Dependent Component Analysis Based on Mutual Information. submitted. http://arXiv.org/abs/physics/0405044
9. Lathauwer, L.D., Moor, B.D., Vandewalle, J.: Fetal electrocardiogram extraction by source subspace separation. In: Processing of HOS, Aiguabla, Spain (1995)
10. De Moor, B.L.R.(ed.): Daisy: Database for the identification of systems. www.esat.kuleuven.ac.be/sista/daisy (1997)

Robust ICA for Super-Gaussian Sources

Frank C. Meinecke[1], Stefan Harmeling[1], and Klaus-Robert Müller[1,2]

[1] Fraunhofer FIRST, IDA group, Kekuléstr. 7, 12489 Berlin, Germany
{meinecke,harmeli,klaus}@first.fhg.de
[2] University of Potsdam, Department of Computer Science
August-Bebel-Strasse 89, 14482 Potsdam, Germany

Abstract. Most ICA algorithms are sensitive to outliers. Instead of robustifying existing algorithms by outlier rejection techniques, we show how a simple outlier index can be used directly to solve the ICA problem for super-Gaussian source signals. This ICA method is outlier-robust by construction and can be used for standard ICA as well as for overcomplete ICA (i.e. more source signals than observed signals (mixtures)).

1 Introduction

ICA models multi-variate time-series $x_n(t)$ with $n = 1 \ldots N$ as a linear combination of statistically independent source signals $s_m(t)$ with $m = 1 \ldots M$:

$$x_n(t) = \sum_m A_{nm} s_m(t). \tag{1}$$

The task of an ICA algorithm is to estimate the *mixing matrix* A given only the observations $x(t)$. Typically, it is assumed that $M \leq N$ and that the columns of A are linearly independent. In this case, Eq. (1) is invertible and the source signals $s(t)$ can be recovered[1].

In the over-complete[2] case, where the number of sources exceeds the number of mixtures (i.e. $M > N$), it is often (if the sources are supergaussian or sparse) still possible to identify the mixing matrix A. However, in general the source signals cannot be recovered, since the model Eq. (1) is not invertible. For very sparse signals (or signals that can be represented sparsely, [1–3]) the underdetermined blind source separation problem is solvable, because each data point can be uniquely assigned to one source (at least approximately).

There exists many of algorihms that can solve the task of estimating the mixing matrix A. Most of them make use of statistical properties of the projections (i.e. kurtosis, negentropy, time lagged covariance matrices...). However, most existing ICA algorithms are highly sensitive to outliers (especially algorithms that employ higher-order statistics).

---

[1] Of course, the source signals can be recovered only up to scaling and permutation, since a scalar factor can be exchanged between each source and the corresponding column of A without changing $x(t)$. The numbering of the sources (and the columns of A) has no physical interpretation and is nothing but a notational device.
[2] also called under-determined.

Recently, Harmeling et al. [4] proposed an outlier detection method based on indices that sort data from very typical points (inliers) to very untypical points (outliers). A simple strategy to robustify existing algorithms is to use these indices for outlier rejection. This is indeed possible, as shown in section 3.2. Moreover, we show that an appropriately defined outlier index can be used *directly* to solve the ICA problem for super-Gaussian source signals. The idea is to look for 'inliers' rather than outliers and use them as estimators for the ICA directions (i.e. columns of the mixing matrix A). Figure 1 shows a scatter

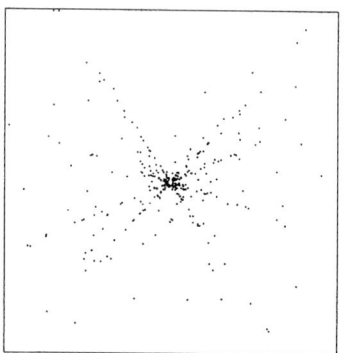 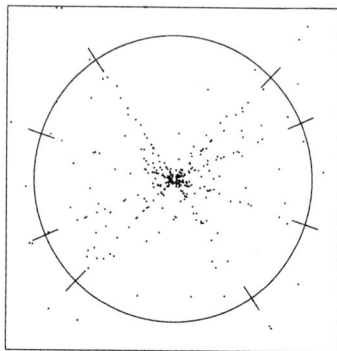

Fig. 1. The left panel shows a scatterplot of a two dimensional mixture of four super-Gaussian source signals. The right panel shows additionally the directions of the points of highest density on the unit circle. Those directions correspond to the columns of A.

plot of a two-dimensional mixture of four super-gaussian source signals (left and right panel). The colums of the mixing matrix are clearly visible as directions in the data space with higher density (right panel). To find these directions, we define a variation of the outlier index γ ([4]) that sorts the data points from very dense ('inlier') to very sparse ('outlier'). The inlier points are estimators for the columns of A. Since the scaling of the columns is arbitrary, also γ must ignore the scaling. This implies two requirements for the index:

- γ must be invariant under rescaling, i.e. $\gamma(\alpha v) = \gamma(v)$ for $\alpha > 0$.
- γ must be invariant under inversion, i.e. $\gamma(-v) = \gamma(v)$.

In other words, 'dense' or 'sparse' should be defined with respect to a distance measure between the *directions* of the data points (i.e. angle distance).

2 The Algorithm

We now describe the Inlier-Based ICA algorithm (abbr. IBICA). Note that some of the presented ideas appeared earlier in other geometrical algorithms (see [2, 5]). The main difference of IBICA is its usage of the inlier index which makes it particularly robust and allows it to be used even in high dimensions.

Step 1: Project the Data on the Unit Sphere

Project all data points $x(1), \ldots, x(T)$ onto the unit sphere by normalizing to length one,

$$z(t) = \frac{x(t)}{\sqrt{x(t)^\top x(t)}} = \frac{x(t)}{|x(t)|}.$$

This step ensures the needed scaling invariance; the distances between the points $z(t)$ on the unit sphere do not depend on the scaling of the original points $x(t)$ but only on the directions. The ICA directions are now given by the dense regions on the sphere. Note that some fraction of the points at the disc around zero have to be removed, because in noisy settings these points do not contain much information about the correct signal directions (and we avoid division by zero for points exactly from the origin).

Step 2: Calculate γ for an Inversion Invariant Distance

The natural distance measure (angle distance) between two normalized points a and b is the geodesic distance on the unit sphere, but we will use a distance measure based on the Euclidean distance since it is easier to calculate and yields similar results[3],

$$d(a,b) = \min(|a-b|, |a+b|).$$

This distance is invariant under the inversion operation (which maps a vector v onto $-v$). This is the natural distance measure to use for our problem, since we are not interested in the orientation of a vector.

Let now $\mathrm{nn}_1(z), \ldots, \mathrm{nn}_k(z)$ be the k nearest neighbors of z according to the distance d. We call the average distance of z to its k nearest neighbors the γ index of z, i.e.

$$\gamma(z) = \frac{1}{k} \sum_{j=1}^{k} d(z, \mathrm{nn}_j(z)).$$

Intuitively speaking, $\gamma(z)$ is large if z lies in a sparse region (z is probably an outlier), and $\gamma(z)$ is small if z lies in a dense region. The data points with the smallest z are good candidates for the directions of the signals, i.e. for the columns of A. We call these points inliers.

Step 3: Pick the Signal Directions Among the Points with Small γ

In order to obtain an estimate for the mixing matrix A, the first idea that comes to mind is to pick the M directions with the smallest values of γ and stack them

---

[3] For two points a and b on the unit sphere ($|a| = |b| = 1$) the geodesic distance is the angle between those vectors, i.e. $\arccos(a^\top b)$. However, for small angles this distance is proportional to the Euclidean distance, $|a-b| = \sqrt{(a-b)^\top(a-b)}$, and in general the relationship is monotonic, i.e. $\arccos(a^\top b) < \arccos(a^\top c) \Leftrightarrow |a-b| < |a-c|$ for another unit vector c.

together. The problem with this approach is that those M columns of A might originate all from the same direction, which by chance happened to be denser than the other directions. To be able to deal with such situations, we need a heuristic that avoids to pick a direction that is similar to a direction that has already been chosen.

Step 3a: Deflational. In the standard ICA setting (i.e. square mixing matrix), this is no problem, since it is possible to find the columns of A one after another in a deflation style: After whitening of the data, the γ values are calculated. The data point with the smallest γ is the first column of the estimated mixing matrix $\hat{A}$. The data set is projected onto the orthogonal subspace and the γ values are re-calculated. The next column of $\hat{A}$ is again given by the smallest gamma and so on. This ensures, that each column of A captures a different source signal since the search is always restricted to a subspace that is orthogonal to the one spanned by the directions found before.

Step 3b: Symmetric. If there are more source signals than mixtures, or if one would like to avoid the whitening step, the deflation procedure is not applicable.

The point density on the sphere (and therefore the distribution of γ values) peaks around the directions of interest. Our task is to find exactly one representative for each of the peaks with (locally) minimal γ. Therefore, after chosing a direction (i.e. a data point) with (globally) minimal γ, the data points forming the corresponding peak should be removed. These are all the data points that can be reached from the γ-minimum along the k-nearest-neighbor graph in a monotonically increasing sequence of γ. This idea is implemented in the following algorithm:

GREEDY PEAK SEARCH
· start with an empty matrix A
· put all points in the pool
· WHILE the pool is not empty
 · pick the point p from the pool with the smallest γ
 · store p as a new column of A
 · color p
 · WHILE there exist colored points in the pool
  · pick a colored point q from the pool
  · remove q from the pool
  · color the k nearest neighbors of q that have a larger γ than q
    and that are still in the pool
 · END
· END

Figure 2 shows the γ-landscape over the angle in the region $[-\frac{\pi}{2}, \frac{\pi}{2}]$ for an example of a two-dimensional mixture of four super-Gaussian sources for 10 and 50 nearest neighbors. Both figures show four pronounced peaks, but in the left panel the landscape is less smooth and and it has additional local minima. Using the heuristic with $k = 10$, more than four directions (shown as circles) are chosen. The choice of k influences, how many columns the estimated mixing matrix A has. Taking into account more neighbors (see the middle panel with $k = 50$), the γ-landscape is smoother and less components are chosen.

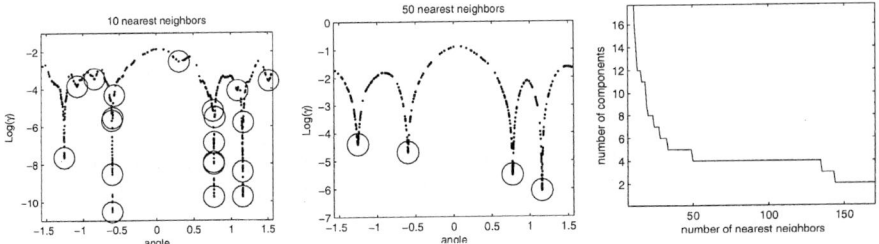

Fig. 2. The γ-Landscape of a two dimensional mixture of four source signals using 10 nearest neighbors (left) or 50 nearest neighbors (middle). In the first case, 21 directions (circles, see text for explanation) have been found, in the second only 4. In the first case the algorithm will therefore return a 2×21 mixing matrix, in the second the (correct) 2×4 Matrix. If the number of sources is not known in advance, one could try several k and look for a plateau (right).

If the number of components M in the mixture is known in advance, we can search for the smallest k that leads to M directions. This can be done very efficiently since the distance matrix has to be calculated and sorted only once. The choice of k influences only the calculation of γ. On the other hand, if the number of components is not known, the algorithm can be repeated efficiently (see previous paragraph) for several choices of k. By looking for a plateau in the number of chosen directions (i.e. by looking for a longer range of values of k that yield the same number of sources) a meaningful k can be found (see Fig. 2).

2.1 Speeding up IBICA

Since the computational costs of calculating the distance matrix grows quadratically with the number of data points, it is appropriate to divide big data sets into smaller subsets, calculate the γ on each of them, and keep only the best data points (i.e. those with the smallest γ) from each subset. Depending on the size of the data set and its subsets, the speed of IBICA can thus be significantly improved. Another side-effect is that this procedure makes IBICA more noise robust. When it comes to the final ICA step, the worst outliers are already removed. This reduces particularly the error, that is made by the whitening in the deflation mode of the algorithm. In the following experiments, we divide the data sets such that we have to deal with distance matrices of size of at most 1000×1000.

3 Experiments

3.1 Performance Measures

To compare our algorithm with other standard ICA algorithms, we will use the following performance measure: Assume, that both the mixing matrix A and its estimator $\hat{A}$ are column normalized (i.e. the norm of the columns of these matrices is one). We then define:

$$pm(A, \hat{A}) = 1 - \left(\frac{1}{2M} \sum_{i=1}^{M} \max_{j} |A^\top \hat{A}|_{ij} + \frac{1}{2M} \sum_{j=1}^{M} \max_{i} |A^\top \hat{A}|_{ij} \right)$$

This performance measure is symmetrical ($pm(A, \hat{A}) = pm(\hat{A}, A)$), smaller or equal to 1 and zero only if $\hat{A} = AP$ with P being a permutation matrix (i.e. perfect solution).

3.2 Robustness Against Outliers

In the following experiments, we produc super-Gaussian source signals by taking gaussian noise to the power of three. The data sets contain 7000 data points each. We compare our algorithm (IBICA) with JADE [6] and FastICA [7].

First, we test the robustness against outliers. We mix two-dimensional super-Gaussian source signals with randomly chosen mixing matrices. Without outliers, the performances of IBICA, JADE and FastICA are all excellent (performance index ≈ 0.01). To test for outlier-robustness, we replace 50 data points with outliers, i.e. uniformly distributed data points within a disc of radius 500 around the origin (the norm of the original data points is roughly within the range from zero to 100).

As expected, IBICA still works fine. In fact, typically it does not even change its solution, because it simply ignores the outliers. JADE and FastICA however, produce arbitrary results because outliers can create directions of high kurtosis, which are attractive for algorithms that use higher order statistics.

3.3 Robustness Against Super-Gaussian Noise

In the next experiment we add noise to the mixtures according to

$$x(t) = As(t) + \sigma \eta(t)$$

with $\eta(t)$ being a N-dimensional noise source of unit variance. We track the evolution of the perfomance index as a function of the noise level σ for kurtotic noise (we used multi-dimensional Gaussian noise, where we change the absolute value to the power of 5) in two dimensions and in 10 dimensions. Figure 3 shows that JADE and FastICA start to fail at a certain noise level, whereas IBICA continues to produce good ICA solutions. In the low dimensional case this difference is more pronounced than in higher dimensions, but even in 10 dimensions

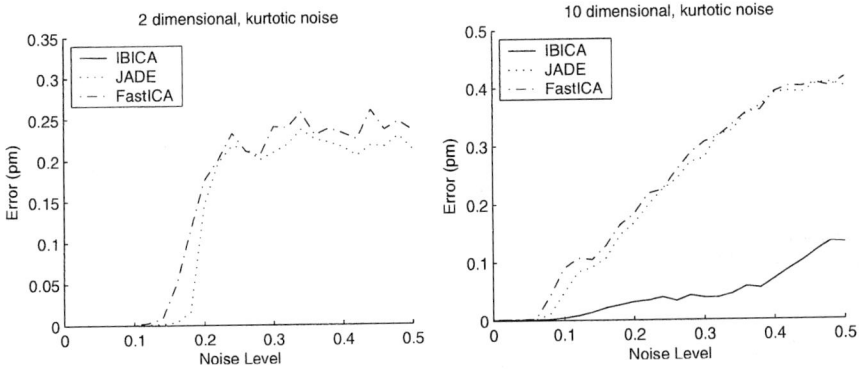

Fig. 3. Performance-index vs. noise level for kurtotic noise in two-dimensional (left) and 10-dimensional mixtures (shown is the median of 50 runs).

IBICA is still clearly superior. Note, that we have chosen the median over 50 runs because the separation performance of the algorithms depend strongly on the actual realization of the noise. However, the signals presented to the different algorithms are of course always the same.

3.4 Overcomplete ICA

As a last experiment, we will now test the ability of IBICA to solve overcomplete ICA problems. We will use the same data sets as before (super-Gaussian signals, 7000 data points). In order to reconstruct the source signals in an overcomplete setting, it is not enough to estimate the mixing matrix. In principle, a source signal reconstruction is only possible if the signals are sparse. There exists a number of techniques than can sparsify certain signals (see, e.g. [1, 8, 3]). However, here we simply assume, that the data can be sparsified by a suitable preprocessing step and focus only on the estimation of A.

We start with a two-dimensional mixture of four source signals (again 7000 data points) (see Figs. 1 and 2). The error is typically at $pm \approx 10^{-5}$, which is a perfect reconstruction of the mixing matrix.

The next example is a five-dimensional mixture containing 20 signals. Here, the error is at $pm \approx 0.01$. The largest angle deviation between one of the 20 source directions and their respective estimators is only about 1.5 degree.

4 Conclusion

Obtaining robust meaningful decompositions is essential when applying blind source separation techniques to data from the real world (see e.g. [9]). In most applications the data is strongly contaminated with measurement noise and outliers where unusual events not belonging to the probability distribution of interest or non-standard noise are measured. Such outlier events pose a severe problem to most existing ICA algorithms, especially the ones that optimize kurtosis-based indices. Our contribution – besides pointing out this fundamental issue – is to use 'inlier' data points only, for performing the decomposition. As this novel framework for ICA does not depend on the dimensionality of the problem, it can be readily used also in overcomplete/underdetermined scenarios. Simulations underline these insights.

Future research will continue the quest for more robust blind source separation algorithms that can have a wider practical applicability.

Acknowledgement

The authors would like to thank Andreas Ziehe, Motoaki Kawanabe and Christin Schäfer for valuable discussions. This research has been partly supported by the PASCAL network of excellence (IST-2002-506778).

References

1. Zibulevsky, M., Pearlmutter, B.A.: Blind source separation by sparse decomposition in a signal dictionary. Neural Computation **13** (2001) 863–882
2. Bofill, P., Zibulevsky, M.: Underdetermined blind source separation using sparse representations. Signal Processing **81** (2001) 2353–2362
3. Lee, T.W., Lewicki, M., Girolami, M., Sejnowski, T.: Blind source separation of more sources than mixtures using overcomplete representations. IEEE Signal Process. Lett. **6** (1999) 78–90
4. Harmeling, S., Dornhege, G., Tax, D., Meinecke, F., Müller, K.R.: From outliers to prototypes: ordering data. Technical report (2004)
5. Puntonet, C.G., Prieto, A., Jutten, C., Rodriguez-Alvarez, M., Ortega, J.: Separation of sources: A geometry-based procedure for reconstruction of n-valued signals. Signal Processing **46** (1995) 267–284
6. Cardoso, J.F., Souloumiac, A.: Blind beamforming for non Gaussian signals. IEEE Proceedings-F **140** (1993) 362–370
7. Hyvärinen, A., Oja, E.: A fast fixed-point algorithm for independent component analysis. Neural Computation **9** (1997) 1483–1492
8. Chen, S., Donoho, D., Saunders, M.: Atomic decomposition by basis pursuit. SIAM Journal on Scientific Computing **20** (1998) 33–61
9. Meinecke, F., Ziehe, A., Kawanabe, M., Müller, K.R.: A resampling approach to estimate the stability of one-dimensional or multidimensional independent components. IEEE Transactions on Biomedical Engineering **49** (2002) 1514–1525

Robustness of Prewhitening Against Heavy-Tailed Sources

Aiyou Chen and Peter J. Bickel

Department of Statistics, University of California
Berkeley, CA, USA 94720
{aychen,bickel}@stat.berkeley.edu

Abstract. Many ICA algorithms use prewhitening (second order decorrelation) as a preprocessing tool. This preprocessing can be shown to be valid when all hidden sources have fintie second moments, which is not required for the identifiability issue[9]. One would conjecture that if one or more sources do not have finite second moments then prewhitening would cause a breakdown. But we discover that this conjecture is not right. We provide some theories for this phenomenon as well as some simulation studies.

1 Introduction

Independent component analysis (ICA) has been used as a standard statistical tool for blind source separation in many application fields [14], e.g., in brain imaging analysis [17]. Formally the classical ICA model is of the form [14]:

$$\mathbf{X} = \mathbf{AS}, \tag{1}$$

where $\mathbf{X} = [X_1, \cdots, X_m]^T$ is a random vector of observations, $\mathbf{S} = [S_1, \cdots, S_m]^T$ is a random vector of hidden sources with mutually independent components (at most one Gaussian component), and $\mathbf{A}$ is a $m \times m$ mixing matrix (nonsingular). Define $\mathbf{W} = \mathbf{A}^{-1}$, which is usually called the demixing matrix. It is well known that $\mathbf{A}$ (thus $\mathbf{W}$) is identifiable up to ambiguity of order, sign and scaling [9]. Without loss of generality we consider a demixing matrix $\mathbf{W}_P$ whose rows are normalized and permuted such that $\mathbf{W}_P \in \Omega$, where Ω defined by

$$\Omega = \{ \mathbf{W}\ m \times m\ \text{matrix} : W_1 \prec \cdots \prec W_m,$$
$$||W_k|| = 1,\ \max_{1 \le j \le m}(W_{kj}) = \max_{1 \le j \le m}(|W_{kj}|),\ \text{for}\ 1 \le k \le m \}.$$

(Note: for $\forall a, b \in R^m$, $a \prec b$ iff there exists $k \in \{1, \cdots, m\}$ such that $a_k < b_k$ and $a_j = b_j$ for $1 \le j < k$) Here and hereafter, we use $||\cdot||$ to denote the l_2 norm for a vector and the Frobenius norm for a matrix, and W_k and W_{kj} denote the kth row and the (k,j)th entry of $\mathbf{W}$. Corresponding to the matrix $\mathbf{W}_P$, the hidden sources can be uniquely identified. Thus we call the model (1) satisfies *the identifiability conditions* if the model holds and $\mathbf{W}_P = \mathbf{A}^{-1} \in \Omega$.

Having n independently and identically distributed (i.i.d.) samples of $\mathbf{X}$, say $\{\mathbf{X}(j) : 1 \leq j \leq n\}$, an ICA algorithm aims to estimate a demixing matrix $\mathbf{W} = \mathbf{A}^{-1}$ and thus to recover each hidden source using the relation $S_k = W_k \mathbf{X}$. This type of unmixing problems is also called blind source separation (BSS) in engineering.

A typical ICA algorithm is primarily composed of a constrast function f and an optimization procedure (let $\mathbf{X}(1:n)$ denote $\{\mathbf{X}(1), \cdots, \mathbf{X}(n)\}$)

$$\hat{W} = \min_{W \in \Omega} f(W; \{\mathbf{X}(1:n)\}), \quad (2)$$

(or equivalently estimating equations), see [1, 5, 14] and references therein (the definition of Ω may be different). There have been many proposals for the contrast function f, derived from the likelihood principle, information theory and measurements of independence among components of $W\mathbf{X}$, see [18] with a comprehensive study.

Prewhitening is a popularly used preprocessing technique in the ICA literature, most due to computational conveniences. When all components of $\mathbf{S}$ have second moments, an ICA algorithm can be realized by first prewhitening, i.e., to obtain

$$\tilde{\mathbf{X}} = \Sigma_{\mathbf{x}}^{-1/2} \mathbf{X},$$

where $\Sigma_{\mathbf{x}}$ is the variance-covariance matrix of $\mathbf{X}$ and $\Sigma_{\mathbf{x}}^{-1/2} = V diag(\{\lambda_i^{-\frac{1}{2}}\}_{i=1}^m) V^T$ if $\Sigma_{\mathbf{x}} = V diag(\{\lambda_i\}_{i=1}^m) V^T$ is the spectrum decomposition, and then rotating, i.e., to obtain an orthogonal matrix $\mathbf{O}$ such that

$$\tilde{\mathbf{S}} = \mathbf{O}\tilde{\mathbf{X}}$$

has mutually independent components. $W = \mathbf{O}\Sigma_{\mathbf{x}}^{-1/2}$ is then a demixing matrix. As a result, the variance-covariance matrix of $\tilde{\mathbf{S}}$, say $\Sigma_{\tilde{\mathbf{s}}}$, must be an identity matrix. The first step can be done by estimating $\Sigma_{\mathbf{x}}$ with its sample covariance matrix $\hat{\Sigma}_{\mathbf{x}}$ and the spectrum decomposition. The rotation matrix O can be estimated by procedures like (2), denoted as $\hat{O}$, i.e.,

$$\hat{O} = \min_{O \in \mathcal{O}(m)} f(O; \{\tilde{\mathbf{X}}(1:n)\}), \quad (3)$$

where $\mathcal{O}(m)$ is the set of $m \times m$ orthogonal matrices, and for simplicity we still use $\tilde{\mathbf{X}}(1:n) = \{\hat{\Sigma}_{\mathbf{x}}^{-1/2} \mathbf{X}(i) : 1 \leq i \leq n\}$. An estimate of $\mathbf{W}_P$ can be obtained by

$$\hat{W} = \hat{O}\hat{\Sigma}_{\mathbf{x}}^{-1/2}, \quad (4)$$

(note: appropriate normalization and row permutation may be needed such that $\hat{W} \in \Omega$). Thus one can say that *prewhitening solves half of the problem of ICA* [14]. In this paper, an algorithm which estimates $\mathbf{W}_P$ following this procedure is called a prewhitened ICA algorithm.

The validity of prewhitening can be expected when all hidden sources have finite second moments. Under second moment constraints Cardoso [4] obtained a lower bound of estimation errors of prewhitened ICA algorithms. We found in simulations that even with heavy-tailed data, some prewhitened ICA algorithms still work very well. This was hard to understand since $\hat{\Sigma}_\mathbf{x}$ diverges in such situations.

In this work, we address this problem. The paper is organized as follows. In Section II, we study the consistency of prewhitening in terms of the estimating parameter space of the demixing matrix. In Section III, we show that prewhitening for the characteristic-function (c.f.) based ICA method [7, 11] can be consistent even when some hidden sources do not have finite second moments. In Section IV, an algorithm based on incomplete Cholesky decomposition is provided to implement PCFICA.

For a $m \times m$ matrix M, we use $[M]_\Omega$ to denote the row-normalized and row-permuted transformation on M such that $[M]_\Omega \in \Omega$.

2 Consistent Prewhitening

From (4), it is clear that the estimating parameter space of the demixing matrix by a prewhitened ICA algorithm is

$$\Omega_n = \{[O\hat{\Sigma}_\mathbf{x}^{-\frac{1}{2}}]_\Omega : O \in \mathcal{O}(m)\},$$

Since $\hat{\Sigma}_\mathbf{x}$ is random, Ω_n is a random set. Obviously Ω_n is a strict subset of Ω.

When $E||\mathbf{S}||^2 < \infty$, by the law of large numbers, $\hat{\Sigma}_\mathbf{x} \to \Sigma_\mathbf{x}$ almost surely and then it is not hard to imagine that we can approximate $\mathbf{W}_P$ in Ω_n. However, when $E||\mathbf{S}||^2 = \infty$, i.e., some components of $\mathbf{S}$ have heavy tails, $\hat{\Sigma}_\mathbf{x}$ diverges. A basic problem is whether it is still possible to approximate $\mathbf{W}_P$ well in Ω_n. If the answer were negative, prewhitening would cause a breakdown of those prewhitened ICA algorithms. To our surprise, simulations of Kernel ICA [2] and CHFICA [7] with prewhitening with heavy tailed distributions (for example, one is uniform on [0,1] and the other is Cauchy distributioned) gave excellent results. This is different from the super-efficiency phenomena for i.i.d. heavy-tailed sources studied in [19]. This phenomenon can be partly explained by the following theorem.

Theorem 1: Under the identifiability conditions,

$$d(\mathbf{W}_P, \Omega_n) \to_P 0,$$

where $d(\mathbf{W}_P, \Omega_n) = inf_{W \in \Omega_n} ||\mathbf{W}_P - W||$.

This theorem tells that for all kinds of hidden sources, there exists a sequence of points in the estimating parameter space for prewhitened ICA algorithms which converges to $\mathbf{W}_P$ in probability. This result is independent of the particular ICA algorithms. The reader is referred to [8] for a complete proof of Theorem 1. Figure 1 shows some simulation results in case of three hidden

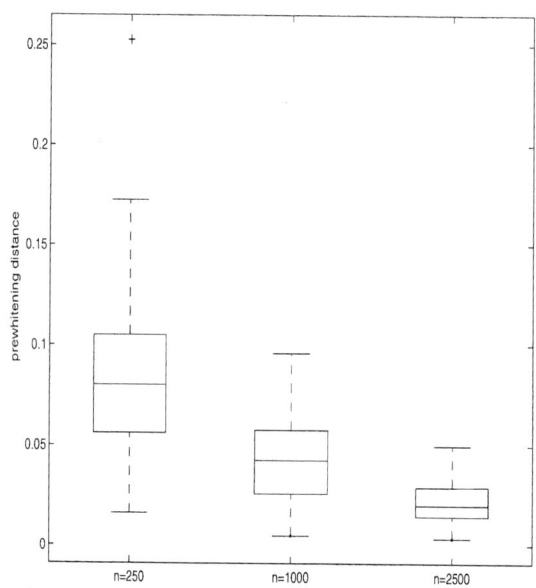

Fig. 1. Consistent Prewhitening (Boxplots with different sample size n).

sources having distributions of $U(0,1)$, Cauchy$(0,1)$ and $\mathcal{N}(0,1)$, separately. A mixing matrix $\mathbf{A}$ was randomly generated such that $\mathbf{W}_P = \mathbf{A}^{-1} \in \Omega$. The prewhitening distance in the vertical axis is defined as $||\mathbf{W}_P - [\hat{\Sigma}_\mathbf{s}^{-\frac{1}{2}} \mathbf{W}_P]_\Omega||$ ($\geq d(\mathbf{W}_P, \Omega_n)$). The sample sizes used were 250, 1000 and 4000, separately and the experiments were replicated 100 times to obtain the boxplots. The decreasing trend is consistent with the claim of Theorem 1.

3 A Consistent Prewhitened ICA Method

Theorem 1 provides the possibility that some prewhitened ICA algorithm may be able to obtain consistent estimates of the demixing matrix with existence of heavy-tail sources. This begs the question whether an implemented algorithm can give good estimates. Our goal in this section is to study the prewhitened c.f.-based ICA method (PCFICA).

The contrast function of the c.f.-based ICA method [16] is given by (let $\mathbf{S} = \mathbf{W}\mathbf{X}$)

$$f(\mathbf{W}; \mathbf{X}) = \int_{\mathbf{t} \in \mathbf{R}^m} |c_\mathbf{S}(\mathbf{t}) - \prod_{j=1}^m c_{S_j}(t_j)|^2 \lambda(\mathbf{t}) d\mathbf{t}, \qquad (5)$$

where $c_\mathbf{S}$ and c_{S_j} stands for the characteristic functions of $\mathbf{S}$, S_j, separately, and λ can be chosen as the density function of $\mathcal{N}(0, I_{m \times m})$. It is clear that $f(\mathbf{W}; \mathbf{X}) \geq 0$ and that the equality holds if and only if the components of $\mathbf{S} = \mathbf{W}\mathbf{X}$ are mutually independent. The c.f.-based ICA method thus estimates

$\mathbf{W}_P$ by minimizing this constrast function but replacing all the characteristic functions with corresponding empirical characteristic functions. This estimator has been shown to be consistent under general conditions and has other nice properties such as $\sqrt{n}$-consistency [8]. A prewhitened version of this estimator can be obtained easily following (3) and (4), and we call it the Prewhitened c.f.-based ICA method (PCFICA). An implementing algorithm of PCFICA is given in Section 4 and was used for simulations here.

Figure 2 shows some simulation results in case of two sources, one has a uniform distribution on $[0, 1]$ and the other has a Cauchy distribution (heavy tail). To detect whether ICA algorithms can obtain consistent estimates in such situation, the sample size was increased from I: $n = 1000$ to II: $n = 8000$. We compare PCFICA and two other famous ICA algorithms, FastICA [15] and JADE [6]. From the boxplots, we can see that as the sample size increases, the estimation error measured by the Amari error [2] for PCFICA decreases more significantly toward zero than for FastICA and JADE. But the simulation also suggests that the convergence rate of the PCFICA estimator is slower than $n^{-1/2}$.

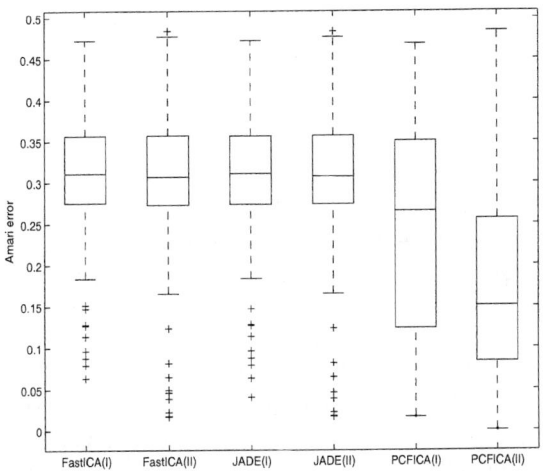

Fig. 2. Consistency of different ICA algorithms with prewhitening when $m = 2$, one is uniform on $[0, 1]$ and the other is Cauchy: 100 replications were used to obtain the boxplots based on quartiles, where the sample sizes were 1000 for case I and 8000 for case II.

The main result of this study is given in the following Theorem 2.

Theorem 2: Suppose that the identifiability conditions hold in model (1). Let $W_P \in \Omega$ be a demixing matrix. The estimator of the demixing matrix defined by PCFICA is consistent, i.e., $||[\hat{W}]_\Omega - \mathbf{W}_P|| \to_P 0$, in either of the two cases:

(i). At most one component of $\mathbf{S}$ has infinite second moment;
(ii). $m = 2$ and both components of $\mathbf{S}$ have heavy tails and stable distributions.

From Theorem 2, when there is at most one heavy-tailed source, PCFICA can always give consistent estimates of the demixing matrix. In cases of more than one heavy-tail source, it is verified for situations with two hidden sources which both have stable distributions. The proof is tedious and the reader is refered to [8] for the details. Theorem 2 does not cover the situation when there are two or more heavy sources which do not have stable distributions, and thus further study of PCFICA may be of interest.

4 Computational Issues for PCFICA

Applying (3) with the contrast function defined in (5), with some algebraic calculation, leads to

$$f(\mathbf{O}; \tilde{\mathbf{X}}(1:n))$$
$$= \frac{1}{n^2} \sum_{i,j=1}^{n} e^{-||\tilde{\mathbf{X}}(i)-\tilde{\mathbf{X}}(j)||^2/2}$$
$$- \frac{2}{n^{m+1}} \sum_{i=1}^{n} \prod_{k=1}^{m} \{\sum_{j=1}^{n} e^{-|\mathbf{O}_k[\tilde{\mathbf{X}}(j)-\tilde{\mathbf{X}}(i)]|^2/2}\}$$
$$+ \frac{1}{n^{2m}} \prod_{k=1}^{m} \{\sum_{i,j=1}^{n} e^{-|\mathbf{O}_k[\tilde{\mathbf{X}}(i)-\tilde{\mathbf{X}}(j)]|^2/2}\}. \tag{6}$$

The first term does not depend on the parameter O and thus can be ignored. Evaluation of the remaining part still requires $O(m^2 n^2)$ operations, which is impossible with a large sample size n (usually $m \ll n$). We provide an algorithm to approximate this contrast function by using the Gaussian kernel and incomplete Cholesky decomposition which makes this doable.

Define $\mathbf{G}^k$ to be a $n \times n$ matrix such that its $(i,j)th$ entry $\mathbf{G}^k(i,j) = e^{-(\mathbf{O}_k[\tilde{\mathbf{X}}(j)-\tilde{\mathbf{X}}(i)])^2/2}$. This is a Gram matrix generated by the 1-dim Gaussian kernel, which is known to be positive semidefinite. Let $\mathbf{G}_+^k(i)$ be the sum of the ith colum of $\mathbf{G}^k$ and $\mathbf{G}_{++}^k = \sum_{i=1}^{n} \mathbf{G}_+^k(i)$ (sum of all entries of $\mathbf{G}^k$). Thus the contrast function consisting of the second term and the third term on the right hand side of (13) becomes

$$f_n(\mathbf{O}) = -2\frac{1}{n^{m+1}} \sum_{i=1}^{n} \prod_{k=1}^{m} \mathbf{G}_+^k(i) + \frac{1}{n^{2m}} \prod_{k=1}^{m} \mathbf{G}_{++}^k.$$

Empirically ([2]) the eigenvalues of such Gram matrices as $\mathbf{G}^k$ are nonnegative and decay quickly, depending on the tail distribution of the associated points. Thus $\mathbf{G}^k \approx UU^T$ for an $h_n \times n$ upper-triangle matrix U, $h_n \ll n$, where h_n is the number of nonignorable eigenvalues. The calculation of U can be done efficiently by using the incomplete Cholesky decomposition [12]. The approximation error $||\mathbf{G}^k - U^T U||$ can be controlled by choosing a threshold value in implementing the incomplete Cholesky decomposition, and we refer readers to [2, 12] for the details. Then $\{\mathbf{G}_+^k(i)\}_{i=1}^n$ can be approximated by taking the advantage of $\mathbf{G}^k \approx U^{1:h_n(k)}[U^{1:h_n(k)}]^T$. The threshold value for incomplete Cholesky decomposition is a constant less than 1 (in simulations we used 0.01), which can make sure that the approximation error for the contrast function is of order $O(\frac{1}{n^m})$, ignorable in comparison with the value of the contrast function of order $O(\frac{1}{\sqrt{n}})$. The partial derivative $\frac{\partial f_n}{\partial \mathbf{O}}$ can be approximated in $O(m^2 n h_n^2)$ operations by using $\mathbf{G}^k \approx U^{1:h_n(k)}[U^{1:h_n(k)}]^T$, where $h_n = \max\{h_n(k) : 1 \leq k \leq m\}$. Theoretical studies [20] show that $h_n = \log(n)$ if the tail distributions of hidden sources decay exponentially fast, in which cases the total cost is $O(m^2 n (\log n)^2)$, which is significantly smaller than $O(m^2 n^2)$.

Once we can evaluate the contrast function $f_n(\mathbf{O})$ and its partial derivative $\frac{\partial f_n}{\partial \mathbf{O}}$, the minimizer of $f_n(\mathbf{O})$ in the domain of orthogonal matrices can be done efficiently using the gradient algorithm described in [10]. Our matlab code which implements PCFICA is downloadable from http://www.stat.berkley.edu/users/aychen/effica.html.

5 Conclusion

In this paper we have studied the statistical properties of prewhitening as a preprocessing technique for ICA algorithms. We have discovered and showed that it is consistent without moment constraints in terms of the estimating parameter space for prewhitened ICA algorithms, and have obtained some consistency results for heavy-tailed sources for a new prewhitened ICA algorithm PCFICA.

References

1. Amari, S. (2002). Independent component analysis and method of estimating functions. *IEICE Trans. Fundamentals* **E85-A**(3) 540-547.
2. Bach, F. and Jordan, M. (2002). Kernel independent component analysis. *Journal of Machine Learning Research* **3** 1-48.
3. Bickel, P., Klaassen, C. , Ritov, Y. and Wellner, J. (1993). *Efficient and Adaptive Estimation for Semiparametric Models*. Springer Verlag, New York, NY.
4. Cardoso, J.F. (1994). On the performance of orthogonal source separation algorithms. *Proc. EUSIPCO*, 776-779. Edinburgh.
5. Cardoso, J. F. (1998). Blind signal separation: statistical principles. *Proceedings of the IEEE* **86**(10) 2009-2025.
6. Cardoso, J.F. (1999). High-order contrasts for independent component analysis. *Neural Computation* **11**(1) 157-192.
7. Chen, A. and Bickel, P.J. (2003). Efficient independent component analysis - based on e.c.f. and one-step MLE. *Technical report #634, Department of Statistics, University of California, Berkeley*.
8. Chen, A. and Bickel, P.J. (2004). Supplement to "Consistent Independent Component Analysis and Prewhitening". *Technical report #656, Department of Statistics, University of California, Berkeley*.

9. Comon, P. (1994). Independent component analysis, a new concept? *Signal Processing* 36(3):287-314.
10. Edelman, A., Arias, T. and Smith, S. (1999). The geometry of algorithms with orghogonality constriants. *SIAM journal on Matrix Analysis and Applications*, **20**(2): 303-353.
11. Jan Eriksson, Visa Koivanen (2003). Characteristic-function based independent component analysis. *Signal Processing*, Vol. 83, pp2195-2208.
12. Golub, G. (1996). *Matrix computation*. Johns Hopkins University Press.
13. Hyvarinen, A. (1999). Fast and robust fixed-point algorithms for independent component analysis. *IEEE Trans. on Neural Networks* **10**(3) 626-634.
14. Hyvarinen, A., Karhunen, J. and Oja, E. (2001). *Independent Component Analysis*. John Wiley & Sons, New York, NY.
15. Hyvarinen, A. and Oja, E. (1997). A fast fixed point algorithm for independent component analysis. *Neural Computation,* **9**(7) 1483-1492.
16. Kagan, A., Linnik, Y. and Rao, C. (1973). *Characterization Problems in Mathematical Statistics*. John Wiley & Sons, USA.
17. Makeig, S., Westerfield, M., Jung, T.-P., Enghoff, S., Townsend, J., Courchesne, E., Sejnowski, T.J. (2002) Dynamic brain sources of visual evoked responses. *Science* 295: 690-694.
18. Pham, D.T. (2001). Contrast functions for ICA and sources separation. *Technical report, BLISS project, France*.
19. Shereshevsk, Y., Yeredor, A. and Messer H. (2001). Super-efficiency in blind signal separation of symmetric heavy-tailed sources. *Proceedings of The 2001 IEEE Workshop on Statistical Signal Processing (SSP2001), pp. 78-81, Singapore, August, 2001*.
20. Widom, H. (1964). Asymptotic behavior of the eigenvalues of certain integral equations. *Transactions of the American Mathematical Society,* **109** 278-295.

Simultaneous Extraction of Signal Using Algorithms Based on the Nonstationarity

Juan Charneco Fernández

Área de Teoría de la señal y Comunicaciones, Escuela Superior de Ingenieros,
Universidad de Sevilla, Camino de los descubrimientos s/n,
41092-Sevilla, Spain
jcharneco@us.es

Abstract. This article is based on the blind extraction of signals from the observations of a linear and instantaneous mixture. Criteria that use the nonstationarity of the sources are analyzed. One use the algorithm of natural gradient to optimize one of these criteria and its result compares the one of the algorithm of fixed point. In the experiments included, the gradient algorithm allows the simultaneous extraction of the desired number of sources.

1 Introduction

ICA is a technique of signal processing that allows the separation of signals in different real applications [5]. Algorithms are based on different concepts. Some authors have demonstrated, Jutten and Herault [6], Cardoso [1], [2], Cichocki and Unbehauen [7], and Delfosse [3], that the signals can be separated using a basic property of the signals. This property is the non-Gaussian and independence between the different signals source. If the sources are non-Gaussian, imposing the statistical independence of the recovered signals yields the separation.

Another innovating hypothesis that allows to separate Gaussian signals, consists of assuming the non-stationarity of the sources [5]. In this article, we study this criterion and the algorithms that allow to optimize it. The new algorithm to be presented is based on the well-known algorithm of the FastICA [8], also known as the fixed point algorithm. This algorithm uses a special method to optimize cost functions in the blind separation of sources (BSS). In this paper, we propose to apply these ideas to a new blind algorithm for the extraction of sources (BES). This method allows the user to extract the desired number of signals. This characteristic it is useful in several environments. Besides, if we have a large number of sources, the computacional cost to separate all of them makes the BSS algorithms unuseful and BES algorithms as the one presented here may be more convenient.

This article is structured as follows. Section 2 presents the model of the system whereas Section 3 presents the fixed point algorithm and its results. These concepts are the key to present a new algorithm in which the user can select the number of signals that wishes to simultaneously extract (Section 4). The simulations will be included in Section 5. In Section 6, a comparative study between BSS and BSE techniques is presented, focusing on the convergence. Finally, Section 7 is devoted to conclusions.

2 System Model

In the problem of the blind separation and extraction of linear instantaneous mixtures it is assumed that N independent signals $s(k) = [s_1(k), \ldots, s_n(k)]^T$ are mixed in a system without memory characterized by the $M \times N$ mixing matrix A where white noise $e(k)$ may be present:

$$x(k) = As(k) + e(k) \tag{1}$$

The recovery of the wished signals $s(k)$ from the observations $x(k)$ can be developed in two steps. In the fiist step the observations are prewhiten, whereas in the second one the whitened signals are extracted. If we define the prewhitening matrix as $B = (\Lambda)^{\frac{-1}{2}} Q^T$, being Q a matrix composed by the autovectors of the R_{xx}, matrix and Λ a diagonal matrix whose entries are the eigenvalues of R_{xx}, the whitened data yields:

$$z(k) = Bx(k) + Be(k) \tag{2}$$

Hence $E\lfloor z(k)z^T(k) \rfloor = I_M + E\lfloor v_N(k)v_N^T(k) \rfloor$, where v_N, is the prewhitened noise.

To obtain the desired source we multiply $z(k)$ by a unitary matrix W obtaining the output signal or estimated source

$$y(k) = Wz(k) \tag{3}$$

where $G = W(\Lambda)^{\frac{-1}{2}} Q^T A$ is the global transfer matrix from the source to the output. The system follows the model of figure 1:

$$\begin{bmatrix} s_1 \\ \vdots \\ s_n \end{bmatrix} = s \rightarrow \boxed{A} \xrightarrow{I} \boxed{B} \rightarrow \boxed{W} \rightarrow y \begin{bmatrix} y_1 \\ \vdots \\ y_n \end{bmatrix}$$

Fig. 1. Signal model

3 Contrast Function Used in the Non-stationary Based Algorithms

Up to our knowledge, the first cost function (also regarded as contrast) based on the non-stationarity of the sources was presented by Matsuoka et al. in [10]. This criterion was based on second order statistics and forced the decorrelation of the sources at different samples along time. Another criterion proposed by Hyvarinen in [4], [5] and is based on the non-stationarity of the energy of the sources. It considers the extraction of a single source $y = w^T x$, where w it is the extraction vector, as follows

$$\psi(k) = |Cum[y(k), y(k), y(k-\tau), y(k-\tau)]| \tag{4}$$

that it consists of the absolute value of the cross fourth-order cumulant between the exit and its delayed versions.

It is possible to notice that if we took $\tau = 0$, the criterion is reduced to take as contract function the kurtosis of the sources. Defining $V = WBA$, $y = Vs$, and

$$S_y = signo(Cum[y_i(k), y_i(k), y_i(k-\tau), y_i(k-\tau)] \qquad (6)$$

we can rewrite the function contrasts like

$$\psi(w) = S_y \sum_{j=1}^{N} v_{ji}^4 Cum(s_j(k), s_j(k), s_j(k-\tau), s_j(k-\tau)), \qquad (7)$$

where v_{ii} are the entrances of matrix V. Using the properties of the cumulantes [5] the gradient of this function with respect to W is represented by the following expression:

$$\nabla_w \psi(w) = 2S_y Cum(y(k), y(k-\tau), y(k-\tau)\mathbf{z}(k))$$
$$+ 2S_y Cum(y(k), y(k), y(k-\tau), \mathbf{z}(k-\tau)) \qquad (8)$$

Hyvarinen in [5] proposes the technique of the fixed point to obtain the adaptation rule and proposes an algorithm based on the following steps in order to obtain 1 signal, that is the first row of of the matrix W. To obtain the rest of signals we will have to use deflaction.

1. Take $w(0) = [1\ 0\ldots 0]$,
2. $w(n) = \nabla_w \psi^T$ \hfill (9)
   taking a large number of piece of information to evaluate the E[],
3. the condition of shutdown will be when $|w(n)w^T(n-1)|$ is near to 1.
4. If it is wanted to extract another source it applies deflaction: $w(n) = w(n) - MM^T w(n)$. Where M it is an orthogonal matrix, since from the operation $\mathbf{z} = \mathbf{B}\mathbf{x} = \mathbf{B}\mathbf{A}\mathbf{s} = \mathbf{M}\mathbf{S}$, we have $E\lfloor \mathbf{z}\mathbf{z}^T \rfloor = \mathbf{M}E\lfloor \mathbf{s}\mathbf{s}^T \rfloor \mathbf{M}^T = \mathbf{M}\mathbf{M}^T = \mathbf{I}$. With this operation, we assured that the vector is orthogonal with the matrix M. We divided $w(n)$ using its norm. We return to step 2 until extracting each one of the original sources.

It can be shown that the convergence rate is cubic, [11] that is to say:

$$\frac{|v_i(n)|}{|v_j(n)|} = \frac{\sqrt{|Cum[s_j(k), s_j(k), s_j(k-\tau), s_j(k-\tau)]|}}{\sqrt{|Cum[s_i(k), s_i(k), s_i(k-\tau), s_i(k-\tau)]|}} \frac{\left(\sqrt{|Cum[s_i(k), s_i(k), s_i(k-\tau), s_i(k-\tau)]|}|v_i(0)|\right)^{3n}}{\left(\sqrt{|Cum[s_j(k), s_j(k), s_j(k-\tau), s_j(k-\tau)]|}\right)^{3n}} \qquad (10)$$

4 Blind Extraction of a Group of Sources

In this part of the article we are able to demonstrate that we can extract P signals from a group of mixed signals, being P a number between 1 and M. Once extracted this set of signals the algorithm will continue processing to obtain by means of deflaction the rest of components.

The contract function we are going to used is the following:

$$\Psi(y) = \sum_{i=1}^{P} |Cum(y_i(n), y_i(n), y_i(n-\tau), y_i(n-\tau))| \quad (11)$$

under the hypothesis that the matrix is W orthogonal, $(WW^T) = \mathbf{I}_p$ that it is analogous to that we have optimized in the previous section but considering that in this occasion E sources are going to be extracted.

The gradient in the space of orthogonal matrices can be expressed based on the ordinary gradient like [9].

$$\hat{\nabla}_\mathbf{w}\Psi = \nabla_\mathbf{w}\Psi - \mathbf{w}(\nabla_\mathbf{w}\Psi)^T\mathbf{w} \quad (12)$$

we know the expression of the regular gradient, equation (8), so in this case we obtain the following algorithm:

1. Take $W(0) = \lfloor I_{p^*p} 0_{p^*(N-P)} \rfloor$
2. $W(i) = W(i-1) + \mu\Big(\nabla_w\psi(k) - W(i-1)(\nabla_\mathbf{w}\Psi)^T W(i-1)\Big) \quad (13)$
   that for taking $\tau = 0$ the form presented in [7].
3. We modified the condition of shutdown to consider that W is a matrix; we were based on the matrix of correlation crossed of the exits.
4. Method of deflaction to continue extracting the rest of the N-P signals.

We divided W(i) using its norm. We return to step 2 until extracting the rest of components of the original sources.

5 Simulation

In this section, we are going to present the main results of the algorithms presented in the preceding section. We left from three independent signals of voice like sources that are mixed through a random matrix to give rise to the observations. The signals of voice on great windows of time reveal a clear nonstationary characteristic in their located energy and therefore they fulfill the hypotheses to begin with necessary to apply the criterion that is analyzed in this article. We present next the obtained results using the first analyzed algorithm, algorithm with the rule of the fixed point.

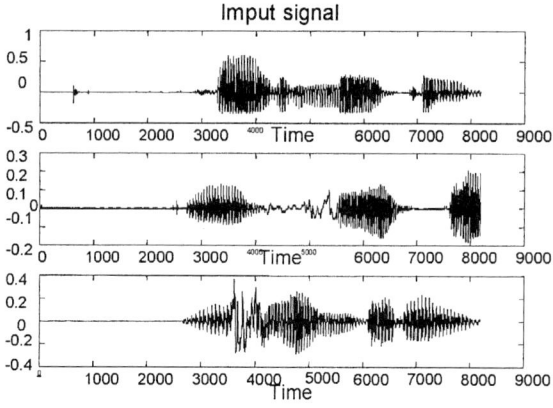

Fig. 2. Signals of voice of entrance used in the different algorithms from extraction

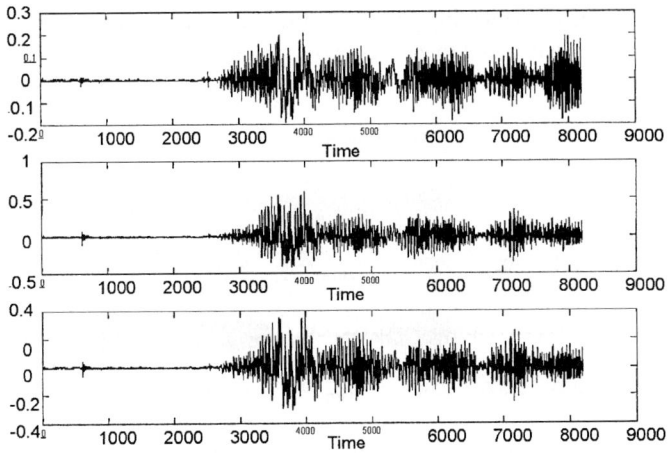

Fig. 3. Signals result of the process of mixture in the channel

After being mixing the signal in the channel, we will apply prewhitening.
Once we apply the prewhitening process, we used the extraction matrix to obtain the desired signal.

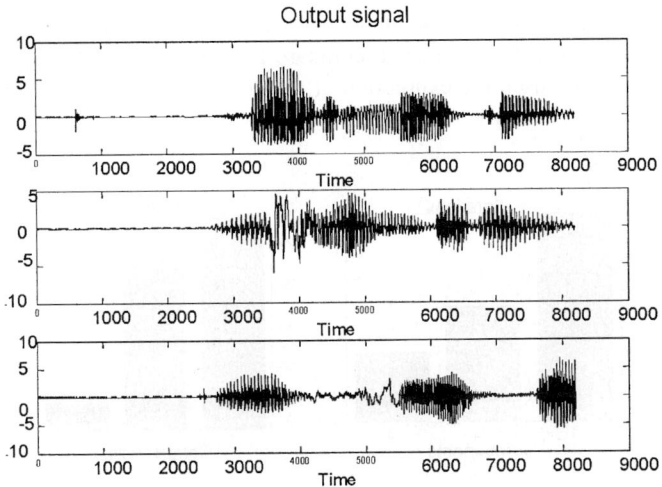

Fig. 4. Signals recovered in the algorithm of extraction of a component. This graphical one shows the final result after extracting each one of the voice signals

The extraction matrix is trained with the algorithm until extracting the wished number. If we presented the same example for the case of the extraction algorithm synchronizes of more than a source, we must consider such steps, generation of signal, mixed of signal, process of prewhitening and the output filter. It is going to be presented the output signal when we decided to extract two sources simultaneously:

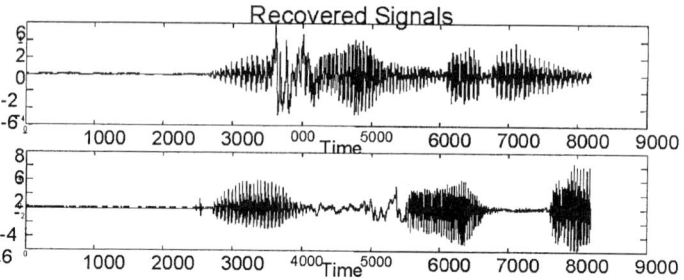

Fig. 5. Signals recovered in the extraction algorithm. It synchronizes of more than a source. In this graph is the case in which two signals of voice are extracted simultaneously

In the following iteration we will recover the last source that we have left to extract.

6 Comparative Between Both Techniques

The most important part of the article that has been introduced is the possibility of extracting more than a source simultaneously using the particular expression of the gradient. In this part, we are going to compare the algorithm of fixed point and the new alternative being used the expression of the gradient (12) to know the number of iterations that are necessary in each case:

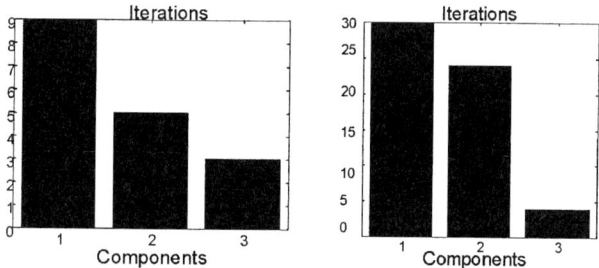

Fig. 6. Convergence of the extraction algorithms. These graphical ones show the number of iterations necessary to extract each one of the components of the voicesignals. The method of the fixed point has been used to reach the solution in the graph of the left and the method of the natural gradient in the right graph

In these two figures can be shown that using the fixed-point the convergence rate is better than the alternative gradient, because the convergence is cubic in the first case and lineal in the second case. However, in the second case it is possible to extract the number of sources you desired. This possibility is a good improvement in the ICA algorithm technique.

7 Conclusion

We have studied how to solve the problem of the blind extraction of sources being used criteria based on the non-stationarity of these. One of these criteria, the based one on the stationarity of the energy have obvious applications in the mixtures of voice signals. In the article the optimization of this criterion sets out according to an algorithm of natural gradient, in contrast with the optimization based on an algorithm of fixed point that sets out in [5].

Acknowledgements

This research was supported by the CYCIT Spanish Project TIC-2003-03781. We also would thanks Dr. Sergio Antonio Cruces for his support.

References

1. J. Cardoso, B. Laheld. Equivariant adaptive source separation. IEEE Transactions on Signal Processing, vol 44, núm 12,págs. 3017-3030, dic. 1996.
2. J.F. Cardoso. Blind source separation: Statistical Principles. Proceeding of the IEEE, vol. 86, núm 10, págs. 2009-2025, 1998.
3. N. Delfosse, P. Loubaton. Adaptive blind separation of independent sources: A deflation approach. Signal Processing, vol. 45, págs. 59-83, 1995.
4. A. Hyvärinen. Fast and robust fixed-point algorithm for independent component Analysis
5. A. Hyvärinen. Independent Component Analysis. A. Hyvärinen, J. Karhunen, E. Oja. Wiley&Sons 2001.
6. C.Jutten, J. Herault. Blind separation of sources, part I:an adaptive algorithm based on neuromimetic architecture. Signal processing, vol. 24, Pág. 1-10, 1991.
7. A. Cichocki, R. Unbehauen. Robust neural networks with online learning for blind identification and separation of sources. IEEE Transactions on circuits and Systems-I, vol. 43, núm. 11, págs. 894-906, 1996.
8. A. Hyvärinen, E. Oja. A fast fixed-point algorithm for independent component analysis. Neural Computing, vol. 9, no. 7, págs. 301-313.
9. A. Edelman, T. Arias, y S.T. Smith. The geometry of algorithm woth ortogonality constraints. SIAM Journal of matrix analysis and application, vol. 20, pp. 303-353, 1998.
10. O. Matsuoka. A neural net for blind separation of nonstationay signal. Neural Networks, vol. 8, núm. 3 411-419, 1995.
11. J. Charneco Fernández. Study of algorithms of separation and blind extraction of sources, págs. 92-94, 2001.

Space-Time Variant Blind Source Separation with Additive Noise

Ivica Kopriva and Harold Szu

Digital Media RF Laboratory
Department of Electrical and Computer Engineering
The George Washington University
725 23rd Street NW, Washington DC 20052, USA
ikopriva@gwu.edu, SZUH@onr.navy.mil

Abstract. We propose a method for solving linear space-time variant blind source separation (BSS) problem with additive noise, **x=As+n,** on the "pixel-by-pixel" basis i.e. assuming that unknown mixing matrix is different for every space or time location. Solution corresponds with the isothermal-T_0 equilibrium of the free energy H =U-T_0S contrast function where U represents the input/output energy exchange and S represents the Shannon entropy. Solution of the inhomogeneous equation (data model with additive noise) is obtained by augmenting inhomogeneous equation into homogeneous "noise free" equation. Consequently, data model with additive noise can be solved by algorithm for the noise free space-time variant BSS problems, [1],[2]. We demonstrate the algorithm capability to perfectly recover images from the space variant mixture of two images with additive noise.

1 Introduction

The BSS problem with additive noise and positivity constraints is defined with

$$\mathbf{x}(r) = \mathbf{A}(r)\mathbf{s}(r) + \mathbf{n}(r) \tag{1}$$

where r is generalized coordinate and $\mathbf{x}, \mathbf{s}, \mathbf{n} \in \mathbf{R}^N$, $\mathbf{A} \in \mathbf{R}^{N \times N}$ represent data vector, source vector, additive noise vector and mixing matrix respectively, N represents problem dimension and R is a set of real numbers. We have presented in [1], [2], [3], [4] algorithm that solves the BSS problem without additive noise on the "pixel-by-pixel" basis. Hence, we may assume unknown mixing matrix to be space variant. In this paper we formulate an extension of the algorithm presented in [1],[2] to treat the BSS problems with additive noise. Because we have focused our attention on imaging applications the positivity constraints were imposed on the data vector, source vector, noise vector and mixing matrix as $\mathbf{x}, \mathbf{s}, \mathbf{n} \in \mathbf{R}_0^{+N}$, $\mathbf{A} \in \mathbf{R}_0^{+N \times N}$ where $\mathbf{R}_0^+$ is a set of positive real numbers including zero. In real world applications such as telescope images in astronomy or remotely sensed images the pixel values correspond to intensities and must be positive, [1],[2],[3],[9],[10],[11]. Also mixing matrix itself must be positive if it for example represents point spread function of an imaging system, [13] [16], or spectral reflectance matrix in remote sensing, [3],[11]. Standard BSS approaches, [5],[6],[7],[8], do not take into account these positivity constraints and that can lead to reconstructed images that have areas of negative intensity. The so-called non-negative ICA methods that explicitly take into account positivity constraints are described in [9],[10]. Like other ICA methods they are probabilistic methods and rely on the priors for the source pixels to be mixture of Laplacians with high probability for positive values around zero and zero probability for the negative values. These probabilistic assumptions implicitly assume that unknown mixing matrix is space invariant. We will show how it is possible to apply the same BSS "single-pixel" deterministic method developed for noise free data model, [1],[2] to treat the model with additive noise (1) by augmenting dimensionality of the data model twice. Consequently deterministic algorithm can be used for solving blind space-time variant linear imaging problem with additive noise by selecting among multiple possible solutions the one at the isothermal-T_o equilibrium of the free energy $H = U - T_o S$ where U represents the input/output energy exchange and S represents the Shannon entropy. Derivation of the algorithm is given in Section 2. We demonstrate the algorithm capability to perfectly recover images from the synthetic space variant linear mixture of two images with additive noise in Section 3. Conclusion is given in Section 4.

2 The Algorithm

The inhomogeneous BSS problem is defined by (1). Note that such formulation allows the mixing matrix $\mathbf{A}(r)$ to be space-time variant. The generalized coordinate r can for example represent pixel location $r(p,q)$ in the case of multispectral image [2],[4] or image sequence [16]. We shall keep argument r in the subsequent derivations in order to indicate that BSS problem is formulated on the "pixel-by-pixel" basis. In order to illustrate how to treat space (time)-variant BSS problem with additive noise (1) we shall assume that $\mathbf{n}(r)$ varies extremely rapidly compared with the variations of both $\mathbf{A}(r)$ and $\mathbf{s}(r)$ i.e.

$$\mathbf{A}(r,t) \cong \mathbf{A}(r,t+\Delta t)$$
$$\mathbf{s}(r,t) \cong \mathbf{s}(r,t+\Delta t) \qquad (2)$$
$$\mathbf{n}(r,t) \neq \mathbf{n}(r,t+\Delta t)$$

Eq.(2) is usual assumption in solving Langevin's equation that describes the Brownian motion of a free particle, [17]. Under assumptions (2) data model (1) can be written in the augmented form that assumes two time measurements as

$$\begin{bmatrix} \mathbf{x}(r,t) \\ \mathbf{x}(r,t+\Delta t) \end{bmatrix} = \begin{bmatrix} \mathbf{A}(r,t) & \mathbf{I} \\ \mathbf{A}(r,t) & \mathbf{D}(r,t,\Delta t) \end{bmatrix} \begin{bmatrix} \mathbf{s}(r,t) \\ \mathbf{n}(r,t) \end{bmatrix} \qquad (3)$$

where $\mathbf{I}$ is N-dimensional identity matrix and $\mathbf{D}(r,t,\Delta t)$ is a diagonal matrix defined with

$$\mathbf{D}(r,t,\Delta t) = \text{diag}\left\{ \frac{\mathbf{n}_i(r,t+\Delta t)}{\mathbf{n}_i(r,t)} \right\}_{i=1}^{N} \qquad (4)$$

and t and $t+\Delta t$ denote two time points at which the measurements are taken. In order to ensure that two sets of measurements are linearly independent the following must hold

$$rank\left(\begin{bmatrix} \mathbf{A}(r,t) & \mathbf{I} \\ \mathbf{A}(r,t) & \mathbf{D}(r,t,\Delta t) \end{bmatrix}\right) = 2N \tag{5a}$$

which is fulfilled when

$$n_i(r, t+\Delta t) \neq n_i(r,t) \quad i=1, ..., N \tag{5b}$$

i.e. noise realizations must be different which is consistent with assumptions (2). In order to fulfill conditions (5) the second measurement at each data vector component at the time point $t+\Delta t$ must be repeated until the following condition is satisfied

$$x_i(r,t) \neq x_i(r, t+\Delta t) \quad i=1,..,N \tag{5c}$$

because by assumption both mixing matrix and source vector remain constant during measurements and according to the augmented data model (3) the only contribution that can change data vector component $x_i(r, t+\Delta t)$ can come from the corresponding noise component $n_i(t+\Delta t)$. If due to the positivity reasons the mixing matrix is parameterized in terms of the mixing angles [1],[2],[3],[4] the augmented data model (3) can be rewritten on the component level for the 2-dimensional case as

$$\begin{bmatrix} x_1(r,t) \\ x_2(r,t) \\ x_1(r,t+\Delta t) \\ x_2(r,t+\Delta t) \end{bmatrix} = \begin{bmatrix} \cos\theta_{11}(r,t) & \cos\theta_{12}(r,t) & 1 & 0 \\ \sin\theta_{11}(r,t) & \sin\theta_{12}(r,t) & 0 & 1 \\ \cos\theta_{11}(r,t) & \cos\theta_{12}(r,t) & \tan\theta_{13}(r,t,\Delta t) & 0 \\ \sin\theta_{11}(r,t) & \sin\theta_{12}(r,t) & 0 & \tan\theta_{14}(r,t,\Delta t) \end{bmatrix} \begin{bmatrix} s_1(r,t) \\ s_2(r,t) \\ n_1(r,t) \\ n_2(r,t) \end{bmatrix} \tag{6}$$

In order to be consistent with data model (1)/(3) the following must hold

$$\tan\theta_{13}(r,t,\Delta t) = \frac{n_1(r,t+\Delta t)}{n_1(r,t)} \quad \tan\theta_{14}(r,t,\Delta t) = \frac{n_2(r,t+\Delta t)}{n_2(r,t)} \tag{7}$$

The augmented data model (3)/(6) can now be solved using the algorithm developed for the noise free model [1],[2]. The price that has to be paid to solve the problem with additive noise is the increased number of unknowns. We show on Fig. 1 the vector diagram representation of the data model with additive noise (1) where the mixing matrix column vectors are $\mathbf{a}_1 = [\cos\theta_{11}(r,t) \quad \sin\theta_{11}(r,t)]^T$ and $\mathbf{a}_2 = [\cos\theta_{12}(r,t) \quad \sin\theta_{12}(r,t)]^T$ and $\tilde{\mathbf{x}}$ represents the noise free part of the data vector (1).

It has been shown in [1],[2] that solution of the noise free blind space-variant imaging problem can be found from the minimum of the Helmholtz free energy contrast function

$$H(\mathbf{W},\mathbf{s}) = U - T_0 S = \boldsymbol{\mu}^T \left[\mathbf{W}\mathbf{x} - |\mathbf{s}|\mathbf{s}'\right] + K_B T_0 |\mathbf{s}| \sum_{i=1}^{N} s'_i \ln s'_i + |\mathbf{s}|(\mu_0 - K_B T_0)\left(\sum_{i=1}^{N} s'_i - 1\right) \tag{9}$$

where S in (9) represents Shannon entropy approximated by

$$S = -K_B T_0 \sum_{i=1}^{N} s'_i \ln s'_i + const \times (\sum_{i=1}^{N} s'_i - 1) \tag{10}$$

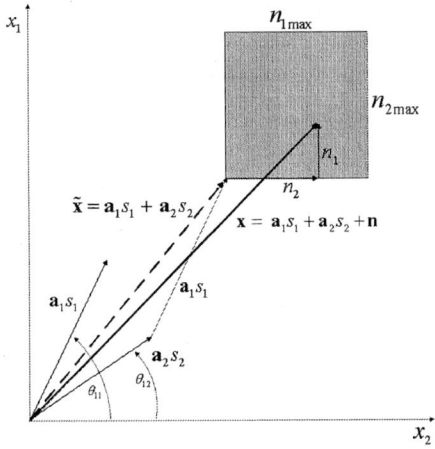

Fig. 1. Vector diagram representation of the 2-D data model (1).

where K_B represents Boltzmann's constant and T_0 represents temperature. They are introduced in (9) due to the dimensionality reasons. Also in (9) $|\mathbf{s}|$ represents L_1 norm of the source vector $\mathbf{s}$, $s'_i = s_i/|\mathbf{s}|$ is the i-th component of the normalized source vector and $\mathbf{W}$ is $N \times N$ matrix that approximates inverse of the mixing matrix i.e. $\mathbf{W} \cong \mathbf{A}^{-1}$ and $\boldsymbol{\mu}$ is the vector of Lagrange multipliers. $U = \boldsymbol{\mu}^T \left[\mathbf{Wx} - |\mathbf{s}| \mathbf{s'} \right]$ in (9) represents a linear error energy term and enables generalization of the Shannon maximum entropy S of the closed system to an open system having non-zero input-output energy exchange U. To solve the BSS imaging problem with the positivity constraints we formulate an algorithm, [14],[1],[15], that looks for the global minimum of the error energy function

$$\left(\mathbf{W}^*, |\mathbf{s}|^* \right) = \arg \min \left(\mathbf{Wx} - |\mathbf{s}| \mathbf{s'} \right)^T \left(\mathbf{Wx} - |\mathbf{s}| \mathbf{s'} \right) \qquad (11)$$

Either deterministic search or stochastic simulated annealing based search, [1],[14],[15], over the phase space could be used in solving optimization problem (11). For a given doublet $\left(\mathbf{W}^{(l)}, |\mathbf{s}|^{(l)} \right)$, where l denotes iteration index in a solution of problem (11), the MaxEnt-like algorithm, [1],[2], computes the most probable solution for the vector of source probabilities, $\mathbf{s'}^{(l)}$

$$s'_j = \frac{1}{1 + \sum_{\substack{i=1 \\ i \neq j}}^{N} \exp\left(\frac{1}{K_B T_0} (\mu_i - \mu_j) \right)} = \sigma(\boldsymbol{\mu}) \qquad (12)$$

with the Lagrange multipliers $\boldsymbol{\mu}$ learning rule given with [2] as

$$\mu_j^{(k+1)} = \mu_j^{(k)} + \left(\frac{K_B T_0}{s'^{(k)}_j} + \mu_j^{(k)} \right) \left(\mathbf{w}_j^{(l)} \mathbf{x} - |\mathbf{s}|^{(l)} s'^{(k)}_j \right) + \sum_{\substack{i=1 \\ i \neq j}}^{N} \mu_i^{(k)} \left(\mathbf{w}_i^{(l)} \mathbf{x} - |\mathbf{s}|^{(l)} s'^{(k)}_i \right) \qquad (13)$$

where k stands for the iteration index related to the Lagrange multipliers learning rule, l stands for the iteration index related to the iterative solution of the optimization problem (11) and $\mathbf{w}_i$ represents the i-th row of the de-mixing matrix $\mathbf{W}$.

3 Simulation Results

To model positivity constraints we have parameterized mixing matrix in terms of the mixing angles as in (6), [1],[2],[3],[4]. Such parameterization reduces a search in higher dimensional parameter space to the first quadrant only and in that sense is an economical representation from the computational complexity standpoint. We illustrate deterministic BSS algorithm on the $N=2$ example of (3). If according to (6) we choose for the particular single pixel case the mixing angles to be $\theta_{11}=5^0$, $\theta_{12}=1^0$, $\theta_{13}=69^0$, $\theta_{14}=60^0$ the model (6) becomes

$$\begin{bmatrix} 350.7711 \\ 100.3941 \\ 268.4710 \\ 107.1244 \end{bmatrix} = \begin{bmatrix} 0.9962 & 0.9998 & 1 & 0 \\ 0.0872 & 0.0175 & 0 & 1 \\ 0.9962 & 0.9998 & 2.6051 & 0 \\ 0.0872 & 0.0175 & 0 & 1.7321 \end{bmatrix} \begin{bmatrix} 54 \\ 154 \\ 143 \\ 43 \end{bmatrix} \quad (14)$$

Fig. 2 shows logarithm of the inverse of the error energy function (11) as a function of angles θ_{11}, θ_{14} for the given model (14) when the mixing angles θ_{12} and θ_{13} were kept at the true values. Note the very sharp peak that correspond with the true solution $\theta_{11}=5^0, \theta_{14}=60^0$.

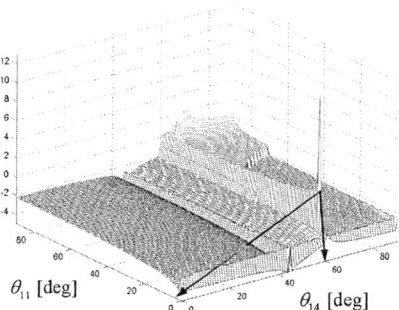

Fig. 2. 2-D plot of the logarithm of the inverse of the error energy (11) in the θ_{11} - θ_{14} domain for data model (14). The other two mixing angles θ_{12} and θ_{13} were assumed to be known.

We now mix two images by mixing matrix that has been changed from pixel to pixel in order to simulate the space variant imaging problem with additive noise, (1). Angles θ_{11} and θ_{12} are changed column wise according to Fig. 3 i.e. for every column index angles were changed for 1^0 and mutual distance between them was 4^0. According to the augmented data model (3)/(6) two measurements per each data channel were assumed to be performed. The angles θ_{13} and θ_{14}, that model the additive noise contribution, were generated randomly. On that way realization of the noise vector at time $t + \Delta t$ was independent from the realization at time t.

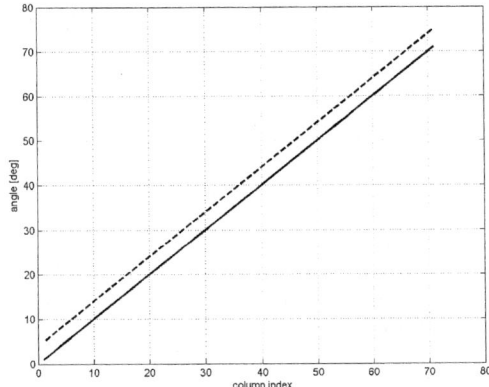

Fig. 3. Change of the mixing angles vs. column index. Solid line – the angle θ_{11}; dashed line – the angle θ_{12} for the mixture given on Fig. 4.

Fig. 4 shows from left to right: two source images, two mixed images without additive noise, two mixed images with additive noise at the time point t, two mixed images with additive noise at the time point $t + \Delta t$ and two separated images obtained by using the deterministic BSS algorithm (9)-(13) and the augmented data model (3)/(6). Thanks to the fact that presented algorithm solves the augmented BSS problem on the "pixel-by-pixel" basis the recovery was perfect although the mixing matrix was space variant and the additive noise was present in the model. Results shown on Fig. 4 are obtained by employing exhaustive search in the mixing angle parameter domain. However, another computationally more efficient strategy would be to employ simulated annealing optimization, [1],[14],[15], to look for global minimum of the error energy function (11). We compare our result with two representative ICA methods that were applied on the same mixture shown on Fig. 4. Fig. 5 shows from left to right separation results obtained by the Infomax algorithm, [6], and by the fourth-order cumulant based JADE algorithm. Due to the space variant nature of the mixing matrix both algorithms fail to recover the original images.

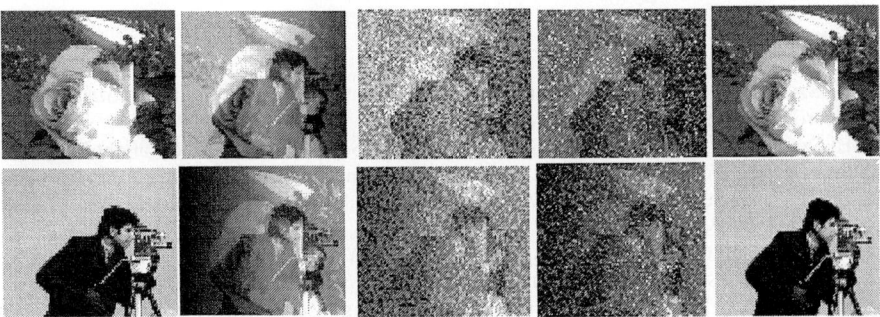

Fig. 4. From left to right: two source images, two mixed images without additive noise, two mixed images with additive noise at the time point t, two mixed images with additive noise at the time point $t + \Delta t$ and two separated images obtained by using the deterministic BSS algorithm (9)-(13) and the augmented data model (3)/(6).

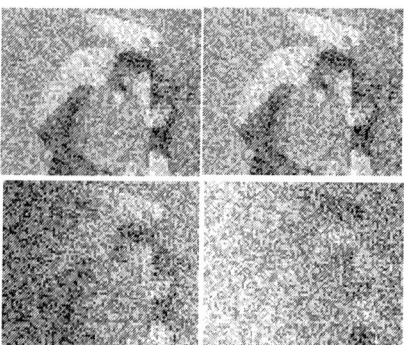

Fig. 5. Source images recovered from the space variant mixture shown on Fig. 4. by the Infomax algorithm, [6], (left) and by the JADE algorithm, [8], (right).

4 Conclusion

The algorithm capable of solving blind linear space-time variant imaging problem with additive noise on the "pixel-by-pixel" basis has been presented. This is accomplished by seeking the global minimum of the free energy contrast function and computing for each pixel the most probable value of the source vector under given macroscopic constraints defined by the data vector. In order to cope with additive noise standard N-dimensional data model has been augmented by one additional measurement per each dimension of the data vector generating the $2N$-dimenisonal "noise free" data model where the additive noise is treated as a source in the extended source vector. It is shown how multiple measurements can be made linearly independent by repeating measurement per each data channel until data channel has different values at the two corresponding time points. The algorithm performance has been demonstrated on the perfect recovery of images from synthetic space variant linear mixture of two images with additive noise. Due to the space variant nature of the mixing matrix the standard ICA algorithms failed to recover the unknown source images.

References

1. H. Szu and I. Kopriva, "Unsupervised Learning with Stochastic Gradient," *submitted* to Neurocomputing.
2. H. Szu and I. Kopriva, "Deterministic Blind Source Separation for Space Variant Imaging," Proc. of the Fourth International Symposium on Independent Component Analysis and Blind Signal Separation, ed. S.I. Amari, A.Cichocki, S. Makino, N. Murata, Nara, Japan, April 1-4, 2003, pp. 669-674.
3. H. H. Szu and C. Hsu, "Landsat spectral Unmixing à la superresolution of blind matrix inversion by constraint MaxEnt neural nets," *Proc. SPIE 3078,* 1997, pp.147-160.
4. H. Szu and I. Kopriva, "Comparison of the Lagrange Constrained Neural Network with Traditional ICA Methods," *Proc. of the IEEE 2002 World Congress on Computational Intelligence-International Joint Conference on Neural Networks*, Hawaii, USA, May 17-22, 2002, pp. 466-471.
5. S. Amari, A. Cihocki, H. H. Yang, "A new learning algorithm for blind signal separation," *Advances in Neural Information Processing Systems,* **8**, MIT Press, 1996, pp. 757-763.

6. A. J. Bell and T. J. Sejnowski, "An information-maximization approach to blind separation and blind deconvolution," *Neural Comp.* **7**, 1995, pp. 1129-1159.
7. A. Hyvärinen and E. Oja, "A fast fixed-point algorithm for independent component analysis," *Neural Computation,* vol. 9, 1997, pp. 1483-1492.
8. J. F. Cardoso, A. Soulomniac, "Blind beamforming for non-Gaussian signals," Proc. IEE F, vol. 140, 1993, pp. 362-370.
9. M. Plumbley, *"Algorithms for Nonnegative Independent Component Analysis",* IEEE Transaction on Neural Networks, Vol.14, No.3, May, 2003, pp.534-543.
10. D. D. Lee and H. S. Seung, "Learning the parts of objects by non-negative matrix factorization," *Nature,* vol. 401, No. 21, 1999, pp.788-791.
11. L. Parra, C. Spence, P. Sajda, A. Ziehe, K-R. Müller, "Unmixing Hyperspectral Data," in *Advances in Neural Information Processing Systems (NIPS12),* S. A. Jolla, T. K. Leen and K-R. Müller (eds), MIT Press, 2000.
12. K. Huang, "Statistical Mechanics," J. Wiley, 1963.
13. H. H. Szu, I. Kopriva, "Artificial Neural Networks for Noisy Image Super-resolution," Optics Communications, Vol. 198 (1-3), 2001, pp. 71-81.
14. Szu H. and Hartley R., "Fast Simulated Annealing," *Physical Letters A,* **122**, No. 3, 1987, pp. 157-162.
15. H. Szu and I. Kopriva, "Cauchy Machine for Blind Inversion in Linear Space-Variant Imaging," Proc. of the 2003 IEEE International Joint Conference on Neural Networks, Portland, OR, Vol. I, July 20-24, 2003, pp. 747-752.
16. I. Kopriva, Q. Du, H. Szu and W. Wasylkiwskyj, "Independent Component Analysis Approach to Image Sharpening in the Presence of Atmospheric Turbulence," *Optics Communications,* vol. 233 (1-3) pp.7-14, 2004.
17. S. Chandrasekhar, "Stochastic Problems in Physics and Astronomy," *Review of Modern Physics,* vol. 15, No. 1, January, 1943, pp. 1-87.

The Use of ICA in Speckle Noise[*]

D. Blanco[1,2], B. Mulgrew[1], S. McLaughlin[1], D.P. Ruiz[2], and M.C. Carrion[2]

[1] Institute of digital communication, University of Edinburgh, UK
[2] Dpt. Física Aplicada, Universidad de Granada, Spain

Abstract. When a linear mixture of independent sources is contaminated by multiplicative noise, also called *speckle noise*, the statistic of the outputs of a linear transformation of the noise data is very different from the statistic that appears when the speckle noise is not present. Specifically, it is not possible find a linear transformation that provides independent outputs and it is necessary study the statistical structure that appears in this case. In this paper, a general approach to obtain the mixture when there exists speckle noise is developed. In order to do this, the linear transformation is searches as the one that reproduces the this theoretical statistic structure.

1 Introduction

The resolution of Blind Source Separation (BSS) by Independent Component Analysis (ICA) has been applied to lots of problems in recent years. An important part of this success resides in the few restrictions of the model of instantaneous linear mixture of independent sources, what it is called in this paper the *ICA model*. Although this problem is widely fulfilled in a lot of applications, there are others where it is not realistic and a more precise model is needed. In this respect, there have been several modifications to the general model, such as the existence of a bias term, the non-instantaneous and non-linear mixtures, additive noise, etc., and the ICA methods have been adapted to deal with these modified models.

In a similar way, this paper studies how the ICA ideas can be applied when the linear mixture of independent sources is contaminated by multiplicative noise. This noise is also called *speckle noise* and it appears in different fields, although all the ICA applications to signals with this kind of noise have been done in image processing. The speckle noise appears, for example, in coherent images, such as synthetic aperture radar images (SAR), ultrasound images, laser images, etc.

The uses of ICA in these images have followed two main directions. On one hand, the ICA methods have been used to build basis functions, looking for the underlying structure. This is done for SAR images in [6], where the basis functions of different classes of sea ice are built. Once with the basis functions

---

[*] Acknowledgements: This work was partially supported by the "Ministerio de Ciencia y Tecnología" of Spain under Project TIC 2001-2902.

for the different types of ice, a heterogeneous sea ice SAR image is studied so that a thematic map showing the position of the different kinds of ice is built, matching each patch in the image with one class of ice. A similar scheme is following in ultrasound breast images in [8], but in this case, the whole image is classified in one class, using previously found basis functions for each one of the different classes. This scheme was first used in [1] to search for the underlying structure in natural black and white images, and it has been extended to colour images, stereo images, hyperspectral images, motion images, face recognition, unsupervised image classification, etc.

On other hand, the ICA methods have been used in [9], and [5] to project the information in different frames in multifrequency polarimetric SAR images, which are different SAR images of a same scene recorded using different frequencies and different polarizations. The ICA method is directly applied to the set of images such that, if the images are the result of independent emissions, the result will be a minor number of images where the information of the independent emitters appears in each one of the different images. This idea has been principally used in hyperspectral image using directly ICA in [7] or using the related approach of projection pursuit in [3].

All of the above techniques for coherent images do the analysis as if there is not speckle noise. This approach is an approximation to the real situation and it is expected to be more accurate when the noise is small and to lose validity when the noise increases. In this paper, the existence of the speckle noise is taken into account and it is studied how it effects the statistic of the mixture. A general approach to deal with a mixture of independent sources in speckle noise is developed. A specific method that follows this general approach was presented in [2]. In Section II, the model used is specified and how the preprocessing step of whitening is affected by the noise. In Section III the statistic properties of this model are used to develop an approach to extract the unmixing matrix in this noisy environment. The paper finishes with the principal conclusions.

2 Whitening

The signals of interest are the contaminated by speckle noise mixture of independent sources, what it is called *speckle ICA model*. This means:

$$z_i = v_i x_i, \text{ for } i = 1, \ldots, N \text{ with } \mathbf{x} = \mathbf{As} , \qquad (1)$$

where $\mathbf{z} = [z_1, \ldots, z_N]^T$ are the N speckle signals recorded in some sensor, $\mathbf{s} = [s_1, \ldots, s_N]^T$ is the vector of N independent sources, the speckle noises $\mathbf{v} = [v_i, \ldots, v_N]^T$ are random variables with one mean and mutually independent each other and with the signals $\mathbf{x}$ [4] and $\mathbf{A}$ is the $N \times N$ mixing matrix. Real signals and same number of sources and signals is assumed for simplicity, but these assumptions can be relaxed without losing generality.

The ICA methods find the inverse $\mathbf{B}$ of the mixing matrix, that is called *unmixing matrix*, as the one whose outputs are as independent as possible. One of the basic difference between the distinct ICA methods resides in how this independence is measured. The first step to produces the separation is the whitening which, although it is not a necessary step, facilitates the convergence and usually appears in the applications. The principal advantage of this whitening is that, after it, the search for the unmixing matrix is limited to unitary matrices. This is based on the fact that these are the unique matrices that preserve the whitening, so the solution of this process is a unitary-shifted version of the unmixing matrix. This always can be done if the data follow the ICA model.

For data following the model (1), the covariance between whatever two components of $\mathbf{z}$, z_i and z_j for $i,j = 1,\ldots,N$, is:

$$\sigma^z_{ij} = \sigma^x_{ij} + \sigma^v_i \delta_{ij} \left(\sigma^x_{ij} + \mu^x_i \mu^x_j\right) , \qquad (2)$$

where σ^x_{ij} is the covariance between x_i and x_j, σ^v_i is the variance of v_i and μ^x_i is the mean of x_i. As the signals $\mathbf{v}$ have one mean, it implies that $\mu^z_i = \mu^x_i$.

A matrix $\mathbf{U}$ that produces white outputs fulfils:

$$\sum_{k=1}^{N} U_{ik} U_{jk} \lambda_k + \sum_{k,l=1}^{N} U_{ik} U_{jl} \sigma^x_{kl} = \delta_{ij} , \qquad (3)$$

where $\lambda_i = \sigma^v_i(\sigma^x_{ii} + (\mu^x_i)^2)$. The unmixing matrix is a solution of this equation only when speckle noise is not present, since in this case $\lambda_i = 0$. A way to resolve this equation assumes that the variance σ^v_i is known is developed in [5].

However, the matrix $\mathbf{U}$ in (3) is not the unmixing matrix or an unitary transformation of this, since if the unmixing matrix, $\mathbf{B} = \mathbf{A}^{-1}$, is applied over the data $\mathbf{z}$, the covariances of whatever two components y_i, y_j, for $i,j = 1\ldots,N$, at its output $\mathbf{y} = \mathbf{Bz}$ is:

$$\sigma^y_{ij} = \delta_{ij} + \sum_{r=1}^{N} B_{ir} B_{jr} \lambda_r . \qquad (4)$$

Then, the matrix provided by the ICA methods, that produces a white output, can not be the unmixing matrix, and only can tend to it when the noise tends to zero. The standard scale indetermination that appears in all the ICA models has been removed setting all the variances of the sources equal to one.

An unitary transformation of the unmixing matrix can be found if the variances of the noises $\{\sigma^v_i\}_{i=1,\ldots,N}$ are known. In speckle images these can be estimated computing the variance in an uniform patch of the image. With these variances, the covariance matrix of $\mathbf{x}$ can be easily computed form (2) and with it the whitening matrix is obtained as always. However, it needs the existence of uniform regions and it will depend on the accuracy of the estimation of the variance of noises and of the theoretical identity $\mu^z_i = \mu^x_i$.

Another option for finding a unitary-shifted version of the mixing matrix is to treat the parameters $\{\lambda_i\}_{i=1,\ldots,N}$ as unknowns of the problem and find a matrix $\mathbf{U}$ such as its outputs reproduce the structure (4) in their covariances. If the outputs of the matrix $\mathbf{U}$ follow the wanted structure with a set of parameter

$\{\lambda_i\}_{i=1,...,N}$, the outputs of a linear transformation $\mathbf{M}$ of the matrix $\mathbf{U}$ will also follow the structure (4) for a set of parameters $\{\lambda'_i\}_{i=1,...,N}$ if it is true:

$$\mathbf{MM}^T + \mathbf{MU\Lambda U}^T \mathbf{M}^T = \mathbf{I} + \mathbf{MU\Lambda' U}^T \mathbf{M}^T, \qquad (5)$$

with $\mathbf{\Lambda} = \text{diag}\{\lambda_1,\ldots,\lambda_N\}$ and $\mathbf{\Lambda}' = \text{diag}\{\lambda'_1,\ldots,\lambda'_N\}$. It is easy to see that if $\mathbf{M}$ is unitary, the above equation is fulfilled for $\mathbf{\Lambda}' = \mathbf{\Lambda}$, but if both sets of parameters are not equal the matrix $\mathbf{M}$ will not be unitary. Then, the whitening matrix $\mathbf{U}$ and the unmixing matrix will be related for a unitary transformation just as if the parameters $\{\lambda_i\}_{i=1,...,N}$ are the correct. If the unmixing matrix is searched only between the unitary matrices using the transformed data $\mathbf{w} = \mathbf{Uz}$, it will not be possible to obtain it if there is any error in the parameters $\{\lambda_i\}_{i=1,...,N}$, since the unitary transformation can not change their values. Without any extra information, is not possible to determine the values of $\{\lambda_i\}_{i=1,...,N}$ by just imposing the structure (4). They only can be estimated through the variance of the noises, and the accuracy of the posterior separation will depend on the accuracy of estimation of the parameters $\{\lambda_i\}_{i=1,...,N}$. This fact produces that in the practice the whitening is not an useful tool, so in the approach of this paper it will not be imposed as a preprocessing step but it will be done jointly with the independence.

The other use of the second order statistic in most ICA methods is find the number of sources when it is not known but it is smaller than the number of signals. This is usually calculated as the dimension of the covariance matrix. However, in the case of speckle ICA model, the covariance (2) will be full-rank even when the number of signals is greater than the number of sources. As with the whitening matrix, the number of the sources could be determined if the variances $\{\sigma^v_i\}_{i=1,...,N}$ are known or can be estimated, but, in the practice, it will not be possible due to the errors in the estimation and in the computation of the mean of the data without noise.

3 Independence

In the ICA model, whitening is only a necessary condition of independence and it is not strong enough to obtain the unmixing matrix. In order to do this, it is necessary to resort to the higher order statistic. If the signal $\mathbf{x}$ follows the ICA model, the joint probability density function (PDF) of the outputs $\mathbf{u} = \mathbf{Wx}$ is

$$f^u(u_1,\cdots,u_N) = \frac{1}{|\mathbf{A}||\mathbf{W}|} \prod_{i=1}^{N} f^s_i \left(\sum_{kl=1}^{N} A^{-1}_{il} W^{-1}_{kl} u_l \right), \qquad (6)$$

where $f^s_i(s_i)$ are the PDFs of the random variables s_i, the elements in the ith row and jth column of the matrices $\mathbf{A}^{-1}$ and $\mathbf{W}^{-1}$ are A^{-1}_{ij} and W^{-1}_{ij}, respectively, and for whatever square matrix $\mathbf{C}$, $|\mathbf{C}|$ means the determinant of this matrix.

It can be seen that, in the case $\mathbf{W} = \mathbf{A}^{-1}$ the joint PDF of $\mathbf{u}$ factorize, i.e. $f(u_1,\ldots,u_N) = f(u_1)\cdots f(u_N)$. As the inverse of a matrix is unique, this transformation is unique, except for scale or permutation. The search is done in the practice by many ICA methods finding a matrix whose outputs follow:

$$\mathcal{E}\{(u_i - \mu_i)g(u_j)\} = 0, \text{ for } i,j = 1,\ldots,N \text{ and } i \neq j , \quad (7)$$

where $\mathcal{E}\{\cdot\}$ is the expectation operator and the function $g(\cdot)$ is a non-linear function such that (7) involves all the higher-order statistic of the random variables.

In the case of speckle ICA model, it is also true that the imposition of the structure (4) is not enough to determine the unmixing matrix, and it will be also necessary to resort to the higher-order statistic. However, the structure of the joint PDF of the outputs after a linear transformation of the speckle data is very different from (6). It can be seen that it is not possible to find a linear transformation such that the PDF in this case factorize.

Specifically, if it is fulfilled that $\mathbf{W} = \mathbf{A}^{-1}$, the joint PDF of the outputs $\mathbf{y} = \mathbf{W}\mathbf{z}$ is:

$$f^y(y_1,\ldots,y_N) = \prod_{i=1}^N \int_{-\infty}^{\infty} \frac{dt_i}{|t_i|} f_i^v(t_i) \prod_{j=1}^N f_i^s \left(\sum_{k,l=1}^N \frac{A_{jk}^{-1} A_{kl}}{t_k} y_l \right) , \quad (8)$$

where $f_i^v(v_i)$ is the PDFs of the random variable v_i.

Although this joint PDF has a special structure, due to the concrete linear transformation used, this is very different from a factorized one and it is difficult to impose on the matrix $\mathbf{W}$, since it is not possible to find an equality as (7) to do that. This special structure is more obvious in the higher-order moments and cumulants of the outputs $\mathbf{y} = \mathbf{A}^{-1}\mathbf{z}$ and the unmixing matrix can be found easier using these statistical functions. The structure in the second-order statistic has been shown in (4), where the covariance of whatever two components of the output $\mathbf{y}$ appears. These covariances depend on some parameters $\{\lambda_i\}_{i=1,\ldots,N}$ that are function of the variance of the speckle noises and the first and second statistic of the data without noise. As these data are not accessible, the parameters need to be obtained jointly with the unmixing matrix.

A similar behaviour occurs in the higher-order statistic. Due to the form of the joint PDF (8), if the linear transformation applied over the speckle data is the unmixing matrix, the third-order cumulants of the outputs have a special structure, although it will not be a diagonal one as in the case of the ICA model. In this structure, some parameters that depend on the first, second and third statistic of the noise and of the non-noisy data, as well as on the mixing matrix, appear and they have to be determined joint with the unmixing matrix. Specifically, if the linear transformation $\mathbf{B}$ is the inverse of the mixing matrix, the third order cumulants of whatever three components y_i, y_j, y_k of the outputs $\mathbf{y} = \mathbf{B}\mathbf{z}$, for $i,j,k = 1,\ldots,N$, is:

$$\gamma_{ijk}^y = \gamma_i^s \delta_{ijk} + \sum_{r=1}^N B_{ir} B_{jr} B_{kr} \beta_r + \sum_{r=1}^N (B_{ir} B_{jr} \alpha_{rk} + B_{jr} B_{kr} \alpha_{ri} + B_{kr} B_{ir} \alpha_{rj}) , \quad (9)$$

where γ_i^s is the skewness of the source s_i and:

$$\beta_i = \gamma_r^v(\gamma_{iii}^x + 3\mu_i^x \sigma_{ii}^x + (\mu_i^x)^3) \\ \alpha_{ij} = \sigma_i^v((A_{ij})^2 \gamma_j^s + 2A_{ij}\mu_i^x) \ , \quad (10)$$

with γ_i^v the skewness of the noise v_i and γ_{ijk}^x is the third-order cumulants of x_i, x_j and x_k. This parameters depend on the statistical properties of the sources and the noises and on the mixing matrix, so they can not be obtain directly from noisy data **z**.

On other hand, the fourth order cumulants of whatever four components y_i, y_j, y_k and y_l, for $i, j, k, l = 1, \ldots, N$, of the output of the unmixing matrix, is:

$$\kappa_{ijkl}^y = \kappa_i^s \delta_{ijkl} + \left\{ \sum_{r=1}^N B_{ir}B_{jr}(\chi_{rkl} + \sum_{r=1}^N \frac{B_{kt}B_{lt}\xi_{rt}}{2}) \right\}_c + \left\{ \sum_{r=1}^N B_{ir}B_{jr}B_{kr}\psi_{rl} \right\}_p , \quad (11)$$

with κ_i^s is the kurtosis of the source s_i and, for whatever function f_{ikjl} depending on the indexes i, j, k and l, it is defined:

$$\{f_{ijkl}\}_m = f_{ijkl} + f_{ikjl} + f_{iljk} \\ \{f_{ijkl}\}_p = f_{ijkl} + f_{jkli} + f_{klij} + f_{lijk} \\ \{f_{ijkl}\}_c = f_{ijkl} + f_{ikjl} + f_{iljk} + f_{jkil} + f_{jlik} + f_{klij} \ . \quad (12)$$

The parameters that appear in the expression (11) are:

$$\xi_{ij} = \left(\frac{\kappa_i^v \delta_{ij}}{3} + \sigma_i^v \sigma_j^v\right)\left(\kappa_{iijj}^x + 2\mu_i^x \gamma_{ijj}^x + 2\mu_j^x \gamma_{iij}^x + 2(\sigma_{ij}^x)^2 + 4\mu_i^x \mu_j^x \sigma_{ij}^x\right)$$

$$+ \frac{\kappa_i^v}{3}\delta_{ij}\left((\sigma_{ii}^x)^2 + 2(\mu_i^x)^2 \sigma_{ii}^x + (\mu_i^x)^4\right) \quad (13)$$

$$\psi_{ij} = \gamma_i^v\left((A_{ij})^3 \kappa_j^s + 3\mu_i^x(A_{ij})^2 \gamma_j^s + 3(\mu_i^x)^2 A_{ij} + 3A_{ij}\sigma_{ii}^x\right)$$

$$\chi_{ijk} = \sigma_i^v\left(\delta_{jk}\left(\kappa_j^s(A_{ij})^2 + 2\mu_i^x A_{ij}\gamma_i^s\right) + 2A_{ij}A_{ik}\right) \ ,$$

where κ_{ijkl}^x is the fourth order cumulant of the signals x_i, x_j, x_k and x_l. As in (10), this parameter can no be estimated from **z**.

Similar structures can be obtained for the cumulants of greater order of the outputs of the unmixing matrix.

Since the structure of the joint PDF (8) is difficult to impose directly in order to find the unmixing matrix, the idea is to search for the matrix whose structure in the output cumulants follows the required theoretical one that has been shown for orders up fourth in (4), (9) and (11).

In order to impose the theoretical structure, the covariances and the higher order cumulants of the outputs are estimated from the speckle data. Specifically, the estimated covariances, third order and fourth order cumulants of the components of **y**, are obtained as:

$$\hat{\sigma}^y_{ij} = \sum_{n,m=1}^{N} B_{in}B_{jm}\hat{\sigma}^z_{mn} \;, \qquad \hat{\gamma}^y_{ijk} = \sum_{n,m,p=1}^{N} B_{in}B_{jm}B_{kp}\hat{\gamma}^z_{mnp} \;,$$

$$\hat{\kappa}^y_{ijkl} = \sum_{n,m,p,q=1}^{N} B_{in}B_{jm}B_{kp}B_{lq}\hat{\kappa}^z_{mnpq} \;, \tag{14}$$

where the functions $\hat{\sigma}^z_{mn}$, $\hat{\gamma}^z_{mnp}$ and $\hat{\kappa}^z_{mnpq}$ are the covariances, third order and fourth order cumulants directly estimated from the components of the speckle data **z**.

The difference between the estimated and theoretical functions can be measured using a cost function as:

$$J = \sum_{1 \le i \le j \le N} \left(\sigma^y_{ij} - \hat{\sigma}^y_{ij}\right)^2 + \sum_{1 \le i \le j \le k \le N} \left(\gamma^y_{ijk} - \hat{\gamma}^y_{ijk}\right)^2$$

$$+ \sum_{1 \le i \le j \le k \le l \le N} \left(\kappa^y_{ijkl} - \hat{\kappa}^y_{ijkl}\right)^2 + \dots \tag{15}$$

This cost function is function of the unmixing matrix but also of the other unknown parameters, and all of them have to be found to resolve the problem. If the matrix **W** is equal to the unmixing matrix **B** and the rest of the parameters take their theoretical values, then $J = 0$. This occurs if and only if all the term in the sums involved in the definition of J are zero, that means a non-linear system of equation, where the unknowns are the components of the matrix **W** and the parameters $\{\lambda_i\}_{i=1,\dots,N}$ and the ones shown in (10) and (13), and the equations are each one of the terms of the sums in (15) equal to zero. The solution of this system of equation will give us the unmixing matrix.

To sum up, the general approach proposed in this paper is to search for the unmixing matrix as the one that reproduces the theoretical structures in the cumulants of its outputs. Different orders in the cumulants can be involved in the cost function that measure the distance between the theoretical structure and the estimated cumulants. Also different set of extra-parameters have to be estimated jointly with the unmixing matrix. In [2], a specific method is developed following this scheme. This method used only the third order cumulants in the cost function, it resolves the non-linear system of equation using the steepest descendant method and it is capable of improving significantly the result shown by the standard ICA method when it is applied to sources with at most one symmetrical. These result are really promising, although more robust and applicable methods can be developed using the fourth order cumulants.

The obtaining of the unmixing matrix is the goal of a lot of ICA applications, as feature extraction, but in others the sources or some kind of classification of the signals is what is wanted. Even with the unmixing matrix, the recovery of the original sources is not as straightforward as in the case of standard ICA model. If the data are not contaminated by speckle noise, the outputs $\mathbf{u} = \mathbf{Bx}$ are the original sources, except with a possible scale or permutation change. However, in the case of the speckle ICA model, the outputs $\mathbf{y} = \mathbf{B}\mathbf{z}$ are just a noisy version of the original sources. Specifically,

$$y_i = s_i + \sum_{k,l=1}^{N} B_{ik} A_{kl} (1 - v_k) s_l \ . \tag{16}$$

The noise term is a zero mean term dependent on all the sources, the speckle noises and the mixing matrix. This estimation of the sources might not be good enough in some applications, but the knowledge of the unmixing matrix allows the use of other tools in the reconstruction of the sources, such as maximum likelihood methods. Also, it allows the classification of the signals and the creation of thematic maps, but, as in the determination of the unmixing matrix, the existence of the speckle noise has to be taken into account in the reconstruction.

4 Conclusions

In this it has been studied how the existence of speckle noise in the linear mixture of independent sources affects to the use of ICA. It has been shown that the ICA solution is not the unmixing matrix, because of the statistic of the output of a linear transformation of the data is very different from the case without speckle noise. The approach of this paper is to find the unmixing matrix that reproduces the theoretical statistical structure. This structure and how it can be imposed on the linear transformation has been shown.

References

1. A. J. Bell and T. J. Sejnowski. Edges are the independent components of natural scenes. *Advances in Neural Information Processing Systems*, 9:831–837, 1997.
2. D. Blanco, B. Mulgrew, and S. McLaughlin. ICA method for speckle signals. In *Proc. of ICASSP*, 2004.
3. S. S. Chiang, C. I. CHang, and I. W. Ginsberg. Unsupervised target detection in hyperspectral images using projection pursuit. *IEEE Trans. on Geos. and Tem. Sens.*, 39:1380–1391, 2001.
4. S. Chitroub and B. Sansal. Statistical characterisation and modelling of SAR images. *Signal Processing*, 82:69–92, 2002.
5. S. Chitroub and B. Sansal. Unsupervised learning rules for POLISAR images analysis. In *Proc. of NNSP*, pages 567–576, 2002.
6. J. Karvonen and M. Similä. Independent component analysis for ice SAR image classification. In *Proc. of IGARSS*, pages 1255–1257, 2001.
7. M. Lennon, G. Mercier, M. C. Mouchot, and L. Hubert-Moy. Independent component analysis as a tool for the dimensionality reduction and the representation of hyperspectral images. In *Proc. of IGARSS*, pages 2893–2895, 2001.
8. H. Neemuchwala, A. Hero, and P. Carson. Image registration using entropic graph-mathching criteria. In *Proc. of Asilomar Conf. on Signal and Sistem*, 2002.
9. X. Zhang and C. H. Chen. A new independent component analyis (ICA) method and its application to SAR images. In *Proc. of NNSP*, pages 283–292, 2001.

Theoretical Method for Solving BSS-ICA Using SVM

Carlos G. Puntonet[2], Juan Manuel Górriz[1],
Moisés Salmerón[2], and Susana Hornillo-Mellado[3]

[1] E.P.S. Algeciras, Universidad de Cádiz,
Avda. Ramón Puyol s/n, 11202 Algeciras Cádiz, Spain
juanmanuel.gorriz@uca.de
[2] E.S.I., Informática, Universidad de Granada
C/ Periodista Daniel Saucedo, 18071 Granada, Spain
{carlos,moises}@atc.ugr.es
[3] Escuela Superior de Ingenieros, Universidad de Sevilla
Avda. de los Descubrimientos s/n 41092 Sevilla , Spain
susanah@us.es

Abstract. In this work we propose a new method for solving the blind source separation (BSS) problem using a support vector machine (SVM) workbench. Thus, we provide an introduction to SVM-ICA, a theoretical approach to unsupervised learning based on learning machines, which has frequently been proposed for classification and regression tasks. The key idea is to construct a Lagrange function from both the objective function and the corresponding constraints, by introducing a dual set of variables and solving the optimization problem. For this purpose we define a specific cost function and its derivative in terms of independence, i.e. inner products between the output and the objective function, transforming an unsupervised learning problem into a supervised learning machine task where optimization theory can be applied to develop effective algorithms.

1 Introduction

Independent Component Analysis (ICA) is a recently developed method in which the goal is to find a suitable representation of non-gaussian sources so that the components are as independent as possible [1]. ICA has been applied successfully to fields such as biomedicine, speech, sonar and radar, signal processing and, more recently, to time series forecasting [2].

There exists a wide range of ICA algorithms for solving blind source separation (BSS) problems, consisting of the minimization (or maximization) of a contrast function [3–7]. In practice, thus ICA, is an algorithm for maximizing the selected statistical principle, i.e. the stochastic gradient descent method can be used to minimize mutual information. The heuristics (learning rates, starting parameters) used in this kind of methods, however, damage the convergence rates. The gradient-based method fails to obtain the correct parameters of the separation system from different initializations due to its limited local search ability and to the complex nonlinear characteristics of the problem (nonlinear or high dimensional ICA)[8].

Optimization Theory is the branch of mathematics concerned with characterizing the solutions to such problems and with developing efficient algorithms for finding such solutions. Any optimization problem can be described using an objective function and equality or inequality constraints (functions defined in a domain $\Omega \subset \mathcal{R}^n$). Depending on the nature of these functions, the problem is called a linear, quadratic, etc. programme.

In this paper, support vector machine (SVM) methodology is applied to ICA in the search for the separation matrix, in order to make use of feature space learning and the numerous regression algorithms developed in this context. The paper is organized as follows; in Section 2 we give a brief overview of basic ICA theory and introduce the notation used in the rest of the paper. The new method is presented in Sections 3 and 4 and some conclusions are drawn in section 5.

2 Definition of ICA

We define ICA using a statistical latent variables model (Jutten & Herault, 1991). Assuming the number of sources n is equal to the number of mixtures, the linear model can be expressed as:

$$x_j(t) = b_{j1}s_1 + b_{j2}s_2 + \ldots + b_{jn}s_n \quad \forall j = 1 \ldots n , \tag{1}$$

where we explicitly emphasize the time dependence of the samples of the random variables and assume that both the mixture variables and the original sources have zero mean without loss of generality. Using matrix notation instead of sums and including additive noise, the latter mixing model can be written as:

$$\mathbf{x}(t) = \mathbf{B} \cdot \mathbf{s}(t) + \mathbf{b}(t) , \text{ or} \tag{2}$$

$$\mathbf{s}(t) = \mathbf{A} \cdot \mathbf{x}(t) + \mathbf{c}(t) , \text{ where } \mathbf{A} = \mathbf{B}^{-1}, \ \mathbf{c}(t) = -\mathbf{B}^{-1} \cdot \mathbf{b}(t) . \tag{3}$$

The conditions that must be satisfied to guarantee the separation are given by Darmois' Theorem in [9]. In brief, the components s_i must be non-gaussian statistically independent. For simplicity, we assume that the unknown matrix is square and that the mixing can be characterized by a linear scenario. Noise is included in the model for two reasons: because the classical statistical linear model is used and because in many applications there is some noise in the measurements (the 'cocktail party' effect).

3 ICA and Convex Optimization Under Discrepancy Constraints

In order to solve ICA problems using the SVM paradigm, we use an approach based on reformulating the determination of the unknown demixing matrix $\mathbf{A} = \mathbf{B}^{-1}$ in the model (3) as a convex optimization problem. The optimization program we formulate is solved using the Lagrange multiplier method combined with an approximation to a given derivative of a convenient discrepancy function based on cumulants or on the characteristic function of the original sources. Note that our approach could easily be modified to take into account other paradigms in ICA research such as density estimation-based approximation methods.

We first restrict the range of possible solutions to the problem, by what is usually a reasonable normalizing constraint: that the Frobenius norm of the matrix $\mathbf{A}$ that we wish to find is minimum. We take the following, however, to be our explicit objective function:

$$minimize \quad \frac{1}{2} \cdot \|\mathbf{A}\|_2^2, \tag{4}$$

because this makes our program a convex one (at least with the Frobenius norm). The discrepancy between the model and what is iteratively observed is contained in the restrictions:

$$-\epsilon < \tilde{L}(\mathbf{a}_i) < \epsilon \ , (i = 1, 2, \ldots, n) \ . \tag{5}$$

where, for each time instant t, we have $\tilde{L}(\mathbf{a}_i) \approx <\mathbf{a}_i, \mathbf{x}> -c_i - s_i$, with $\mathbf{a}_i$ denoting the i-th row of the demixing matrix $\mathbf{A}$, and c_i being the i-th component on vector $\mathbf{c}$. Note that for simplicity we have not written the dependency on the time instant t, but of course this must be taken into account when implementing.

We define the Lagrangian corresponding to (5) as (introducing a soft margin in equation 5)

$$\begin{aligned}\mathcal{L}_i = &\tfrac{1}{2} \cdot \|\mathbf{a}_i\|_2^2 + C \cdot \sum_{j=1}^{l}(\xi_j + \xi_j^*) - \sum_{j=1}^{l} \alpha_j(\epsilon + \xi_j + \tilde{L}(\mathbf{a}_i)) \\ &- \sum_{j=1}^{l} \alpha_j^*(\epsilon + \xi_j^* - \tilde{L}(\mathbf{a}_i)) - \sum_{j=1}^{l}(\eta_j \xi_j + \eta_j^* \xi_j^*) \ .\end{aligned} \tag{6}$$

where l is the number of samples and $\xi_j, \xi_j^*, \alpha_j, \alpha_j^*, \eta_j, \eta_j^*$ are the slack variables introduced in Lagrangian optimization problems.

Now we take the corresponding partial derivatives (according to the Lagrangian method) and equal them to 0, as follows

$$\partial_{c_i} \mathcal{L}_i = \sum_{j=1}^{l}(\alpha_j^* + \alpha_j) = 0 \ . \tag{7}$$

$$\partial_{\xi_i^{(*)}} \mathcal{L}_i = C - \alpha_j^{(*)} - \eta_j^{(*)} = 0 \ . \tag{8}$$

$$\partial_{\mathbf{a}_i} \mathcal{L}_i = \mathbf{a}_i - \sum_{j=1}^{l}(\alpha_j - \alpha_j^*) \cdot \partial_{\mathbf{a}_i} \tilde{L}(\mathbf{a}_i) = 0 \ . \tag{9}$$

From equation 9 we see how the algorithm is able to extract independent components one by one, just working with the maximization of the selected Lagrangian function $\mathcal{L}_i$. The selection of a suitable function $\tilde{L}(\mathbf{a}_i, \mathbf{x})$ determines the current algorithm or strategy used in the process, i.e. if we describe it in terms of neg-entropy we obtain a generalization of FastICA [7]. After some algebraic manipulation, we obtain

$$\mathcal{L}_i = \frac{1}{2} \cdot \left\| \sum_{j=1}^{l}(\alpha_j - \alpha_j^*)\partial_{\mathbf{a}_i}\tilde{L}(\mathbf{a}_i) \right\|^2 - \epsilon \sum_{j=1}^{l}(\alpha_j + \alpha_j^*) - \sum_{j=1}^{l}(\alpha_j - \alpha_j^*)\tilde{L}(\mathbf{a}_i). \quad (10)$$

Finally, ICA is transformed into a multidimensional maximization of the Lagrangian function defined as:

$$\mathcal{L} = \begin{pmatrix} \mathcal{L}_1 \\ \mathcal{L}_2 \\ \vdots \\ \mathcal{L}_n \end{pmatrix}. \quad (11)$$

4 Statistical Independence Criterion

The Statistical Independence of a set of random variables can be described in terms of their joint and individual probability distribution. The independence condition for the independent components of the output vector $\mathbf{y}$ is given by the following definition of independence random variables:

$$p_{\mathbf{y}}(\mathbf{y}) = \prod_{i=1}^{n} p_{\mathbf{y}_i}(y_i) \quad (12)$$

where $p_{\mathbf{y}}$ is the joint pdf of the random vector (observed signals) $\mathbf{y}$ and $p_{\mathbf{y}_i}$ is the marginal PDF of y_i. In order to measure the independence of the outputs, equation 12 is expressed in terms of higher order statistics (cumulants) using the characteristic function (or moment generating function) $\phi(\mathbf{k})$, where $\mathbf{k}$ is a vector of variables in the Fourier transform domain, and considering its natural logarithm $\Phi = log(\phi(\mathbf{k}))$. We first evaluate the difference between the terms in equation 12 to obtain:

$$\pi(\mathbf{y}) = \left\| p_{\mathbf{y}}(\mathbf{y}) - \prod_{i=1}^{n} p_{\mathbf{y}}(y_i) \right\|^2 \quad (13)$$

where the norm $\|\ldots\|^2$ can be defined using the convolution operator with different window functions according to the specific application [8] as follows:

$$\|F(y)\|^2 = \int \{F(\mathbf{y}) * v(\mathbf{y})\}^2 dy \quad (14)$$

and $v(\mathbf{y}) = \prod_{i=1}^{n} w(y_i)$. In the Fourier domain and taking natural log (in order to use higher order statistics, i.e. cumulants) this equation is transformed into:

$$\Pi(\mathbf{k}) = \int \left\| \Psi_{\mathbf{y}}(\mathbf{k}) - \sum_{i=1}^{n} \Psi_{\mathbf{y}_i}(k_i) \right\|^2 \mathbf{V}(\mathbf{k}) d\mathbf{k} \quad (15)$$

where Ψ is the cumulant generating or characteristic function (the natural log of the moment generating function) and V is the Fourier transform of the selected window function $v(\mathbf{y})$. If we take the Taylor expansion around the origin of the characteristic function, we obtain:

$$\Psi_\mathbf{y}(\mathbf{k}) = \sum_\lambda \frac{1}{\lambda!}\frac{\partial^{|\lambda|}\Psi_\mathbf{y}}{\partial k_1^{\lambda_1}\ldots \partial k_n^{\lambda_n}}(\mathbf{0}) k_1^{\lambda_1}\ldots k_n^{\lambda_n} \quad (16)$$

where we define $|\lambda| \equiv \lambda_1 + \ldots + \lambda_n$, $\lambda \equiv \{\lambda_1 \ldots \lambda_n\}$, $\lambda! \equiv \lambda_1!\ldots\lambda_n!$ and:

$$\Psi_{\mathbf{y_i}}(k_i) = \sum_{\lambda_i} \frac{1}{\lambda_i!}\frac{\partial^{\lambda_i}\Psi_{\mathbf{y_i}}}{\partial k_i^{\lambda_i}}(0) k_i^{\lambda_i} \quad (17)$$

where the factors in the latter expansions are the cumulants of the outputs (cross and non-cross cumulants):

$$C_{y_1\ldots y_n}^{\lambda_1\ldots\lambda_n} = (-j)^{|\lambda|}\frac{\partial^{\lambda_1+\ldots+\lambda_n}\Psi_\mathbf{y}}{\partial k_1^{\lambda_1}\ldots\partial k_n^{\lambda_n}}(\mathbf{0}) \qquad C_{y_i}^{\lambda_i} = (-j)^{\lambda_i}\frac{\partial^{\lambda_i}\Psi_{\mathbf{y_i}}}{\partial k_i^{\lambda_i}}(0) \quad (18)$$

Thus, we define the difference between the terms in equation 15 as

$$\beta_\lambda = \frac{1}{\lambda!}(j)^{|\lambda|}C_\mathbf{y}^\lambda \quad (19)$$

which contains the infinite set of cumulants of the output vector $\mathbf{y}$. By substituting 19 into 15 we obtain

$$\Pi(\mathbf{k}) = \int \left\|\sum_\lambda \beta_\lambda k_1^{\lambda_1}\ldots k_n^{\lambda_n}\right\|^2 \mathbf{V}(\mathbf{k})d\mathbf{k} \quad (20)$$

Hence, vanishing cross-cumulants are a necessary condition for $y_1,\ldots,y_n$ to be independent[1]. Equation 20 can be transformed into:

$$\Pi(\mathbf{k}) = \int \sum_{\lambda,\lambda^*}\beta_\lambda\beta_{\lambda^*}^* k_1^{\lambda_1+\lambda_1^*}\ldots k_n^{\lambda_n+\lambda_n^*}\mathbf{V}(\mathbf{k})d\mathbf{k} \quad (21)$$

Finally, by interchanging the sequence of summation and integral equation 21 can be rewritten as:

$$\Pi = \sum_{\lambda,\lambda^*}\beta_\lambda\beta_{\lambda^*}^*\cdot\boldsymbol{\Gamma}_{\lambda,\lambda^*} \quad (22)$$

where $\boldsymbol{\Gamma} = \int k_1^{\lambda_1+\lambda_1^*}\ldots k_n^{\lambda_n+\lambda_n^*}\mathbf{V}(\mathbf{k})d\mathbf{k}$. In this way, we describe the generic function $\tilde{\mathbf{L}}$ in the Lagrangian function $\mathcal{L}$. We must impose some additional restrictions on $\tilde{\mathbf{L}}$, which is a version of the previous one but limiting the set λ. That is, we only consider a finite set of cumulants $\{\lambda,\lambda^*\}$ such as $|\lambda|+|\lambda^*| < \tilde{\lambda}$

---

[1] In practice, we need independence between sources two against two.

and include only the cumulants affecting the current Lagrangian component. Mathematically, these two restrictions are expressed as:

$$\tilde{L}_i \equiv \Pi = \sum_{\{\lambda,\lambda^*\}} \beta_\lambda \beta_{\lambda^*}^* \Gamma_{\lambda,\lambda^*} \quad \setminus \left\{ \begin{array}{c} \{\lambda,\lambda^*\} \cap \{\lambda_i\} \neq 0 \\ |\lambda| + |\lambda^*| < \tilde{\lambda} \end{array} \right\} \quad (23)$$

In order to evaluate the most relevant term in the Lagrangian $\frac{\partial \tilde{L}}{\partial a_i}$ the above equations must be rewritten in terms of the output vector as $y_i = a_i x$, and we must use the connection between cumulants and moments shown in [10]:

$$\frac{\partial \tilde{L}_i}{\partial a_i} \propto \frac{\partial C_y^\lambda}{\partial a_i} \propto \frac{\partial a_i \cdot x}{\partial a_i} \quad (24)$$

4.1 Using the Connection Between Moments and Cumulants

The connection between moments and cumulants can be expressed as:

$$C_y^\lambda = \sum_{p_1,\ldots,p_m} (-1)^{m-1}(m-1)! \cdot E[\prod_{j \in p_1} Y_j] \ldots E[\prod_{j \in p_m} Y_j] \quad (25)$$

where $\{p_1, \ldots, p_m\}$ are all the possible partitions with $m = 1, \ldots, \lambda$ included in the set of integers $\{1, \ldots, \lambda\}$. In SVM methodology, we work with instantaneous values (sample by sample) and thus we have to approximate expected values to instantaneous ones. Finally, by evaluating the derivative term in equation 25 and using the above-mentioned approximations, we obtain

$$\frac{\partial C_y^\lambda}{\partial a_i} = \sum_{p_1,\ldots,p_m} (-1)^{m-1}(m-1)! \cdot \sum_{k=1}^{m} \left(\frac{s_k (A^{-1} \cdot y)^{s_k-1}}{y_i^{s_k}} \prod_{j \in p_1} y_j \cdots \prod_{j \in p_m} y_j \right) \quad (26)$$

where λ satisfies the conditions shown in equation 23 and s_k is an integer in the set $\{1, \ldots, \tilde{\lambda}\}$. In practice, the order of the statistics used never exceeds four or five, and so the latter expression can be simplified significantly, rewriting the cumulants in terms of dot products between the output signals y_i. Expressions of cumulants in terms of moments are well-known and thus equations 26 and 9 allow us to iteratively obtain the coefficients α_j, α_j^* and then the support vector parameters a_i of the separation matrix A:

$$\begin{aligned} a_i &= \sum_{j=1}^{l}(\alpha_j - \alpha_j^*) \cdot \partial_{a_i}\tilde{L}(a_i) = \sum_{j=1}^{l}(\alpha_j - \alpha_j^*) \cdot \sum_{\{\lambda,\lambda^*\}} \partial_{a_i}\left(\beta_\lambda \beta_{\lambda^*}^*\right) \Gamma_{\lambda,\lambda^*} \\ &= \sum_{j=1}^{l}(\alpha_j - \alpha_j^*) \cdot \sum_{\{\lambda,\lambda^*\}} \frac{(j)^{|\lambda|+|\lambda^*|}}{\lambda! \lambda^*!} \partial_{a_i}\left(C_y^\lambda C_y^{\lambda^*}\right) \Gamma_{\lambda,\lambda^*} \end{aligned} \quad (27)$$

5 Conclusions

A support vector-based BSS-ICA method has been developed to solve the BSS problem from linear mixtures of independent sources. The generalization to nonlinear ICA is straightforward considering nonlinear maps to feature spaces. The proposed method obtains a good performance (this statement is back up by the extensive work in the workbench of SVM algorithms), and benefits from the Theoretical Optimization Theory, which consists of solving a uniquely solvable (with order n) optimization problem instead of Newton or gradient descent methods, which require suitable nonlinear optimization, with the consequent risk of getting stuck in local minima.

The tacit assumption in equation 5 avoids cases such as in noisy environments where the separation matrix does not actually exist as a linear function between independent components and observed signals, i.e. the convex optimization problem is not feasible. That is, in cases where the separation is not possible, we use a "soft margin" by introducing slack variables to cope with the otherwise unfeasible constraints of the optimization problem [11]. The main disadvantage of this kind of methods is that Quadratic programs are computationally quite expensive as they scale between quadratic and cubic in the number of patterns although there exists a unique solution, but this is also true for algebraic algorithms like e.g. Cardoso's JADE [12].

References

1. Hyvarynen, A., Oja, E., Independent Component Analysis: Algorithms and Applications Neural Networks Vol 13 411-430 Elsevier (2000)
2. Górriz, J.M., Puntonet, C.G., Salmerón, M., Ortega, J., New method for filtered ICA signals applied to volatile time series 7th International Work Conference on Artificial and Natural Neural Networks IWANN 2003 Lecture Notes in Computer Science Vol 2687 / 2003, Springer pp. 433-440 ISSN: 0302-9743. Menorca, Balearic Islands, Spain. Jun. 2003.
3. Barlow, H.B, Possible principles underlying transformation of Sensory messages. Sensory Communication, W.A. Rosenblith, MIT Press, New York, U.S.A. (1961).
4. Bell,A.J., Sejnowski, T.J. An Information-Maximization Approach to Blind Separation and Blind Deconvolution. Neural Computation, vol 7, 1129-1159 (1995).
5. Cardoso, J.F., Infomax and maximum likelihood for source separation. IEEE Letters on signal processing, 4, 112-114 (1997).
6. Cichoki, A., Unbehauen, R., Robust neural networks with on-line learning for blind identification and blind separation of sources. IEEE Transactions on Circuits and Systems, 43 (11), 894-906 (1996).
7. Hyvärinen, A., Oja, E., A fast fixed point algorithm for independent component analysis. Neural Computation, 9: 1483-1492
8. Tan, Y., Wang, J., Nonlinear Blind Source Separation Using Higher order Statistics and a Genetic Algorithm. IEEE Transactions on Evolutionary Computation, vol. 5, num 6 (2001)
9. Darmois, G., Analyse Générale des Liaisons Stochastiques Rev. Inst. Internat. Stat 21, 2-8 (1953)
10. Nikias, C.L., Mendel, J.M., Signal Processing with Higher order Spectra IEEE Signal Processing Magazine pp 10–37 Jul (1993)
11. Smola, A.J., Schölkopf, B.: A tutorial on Support Vector Regression. NeuroCOLT2. Technical Report Series. NC2-TR-1998-030, October (1998)
12. High-order Contrasts for Independent Component Analysis. Jean-François Cardoso. Neural Computation, vol.11, no1, pp.157-192, Jan 1999

Wavelet De-noising for Blind Source Separation in Noisy Mixtures

Bertrand Rivet[1], Vincent Vigneron[1],
Anisoara Paraschiv-Ionescu[2], and Christian Jutten[1]

[1] Institut National Polytechnique de Grenoble
Laboratoire des Images et des Signaux
Grenoble, France
[2] Swiss Federal Institute of Technology
Lausanne, Switzerland

Abstract. Blind source separation, which supposes that the sources are independent, is a well known domain in signal processing. However, in a noisy environment the estimation of the criterion is harder due to the noise. In strong noisy mixtures, we propose two new principles based on the combination of wavelet de-noising processing and blind source separation. We compare them in the cases of white/correlated Gaussian noise.

1 Introduction

Blind source separation (BSS) is a well known domain in signal processing. Introduced by J. Hérault, C. Jutten and B. Ans [1], its goal is to recover unknown source signals of which only mixtures are observed with only assumptions that the source signals are mutually statistically independent. A lot of BSS models such as instantaneous linear mixtures, convolutive mixtures are presented in recent publications [2–4]. The success of the BSS is its wide range of applications whether it is in telecommunication, speech or medical signal processing. However, the best performances of these methods are obtained for the ideal BSS model and their effectiveness is definitely decreased with observations corrupted by additive noise.

The aim of this paper is to present how to associate wavelet de-noising processing and BSS in order to improve the estimated sources. This paper is organized as follows. Section 2 introduces the BSS problem in noisy mixtures. Section 3 explains the wavelet de-noising principles and proposes two new principles for associating wavelet de-noising and BSS. Section 4 proposes numerical experiments before conclusion and perspectives in section 5.

2 Modelization of the Problem

In an instantaneous linear problem of source separation, the unknown source signals and the observed data are related by (Fig. 1):

$$\mathbf{y}(k) = \mathcal{A}\mathbf{s}(k) + \mathbf{n}(k) = \mathbf{x}(k) + \mathbf{n}(k) \qquad (1)$$

where $\mathcal{A}$ is an unknown full rank $p \times q$ mixing matrix ($p \geq q$), $\mathbf{s}(k)$ is a column vector of q source signals assumed mutually statistically independent, $\mathbf{y}(k)$ a column vector of p mixtures and $\mathbf{n}(k)$ an additive noise. By estimating a $q \times p$ full rank matrix $\mathcal{B}$ one provides estimated sources which are the components (as independent as possible) of the output signal vector $\hat{\mathbf{s}}(k)$ defined as (Fig. 1):

$$\hat{\mathbf{s}}(k) = \mathcal{B}\,\mathbf{y}(k) = \mathcal{B}\,\mathcal{A}\,\mathbf{s}(k) + \mathcal{B}\,\mathbf{n}(k) \qquad (2)$$

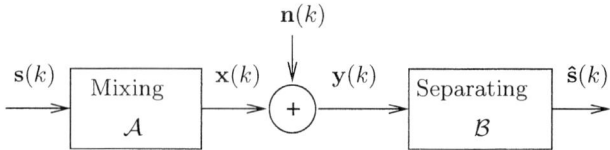

Fig. 1. Blind source separation model in noisy mixtures.

This equation shows that the estimated sources $\hat{\mathbf{s}}$ are affected by the additive noise. Let us illustrate this phenomenon with the figure 2. c) which shows that the estimated separating matrix $\hat{\mathcal{B}}$ is not well estimated ($\hat{\mathcal{B}} \neq \mathcal{B}$) since the ideal sources (see definition 1 below) are different from the original sources. Moreover, even if the separating matrix is well estimated ($\hat{\mathcal{B}} = \mathcal{B}$), the noise affects the estimated sources as shown in d).

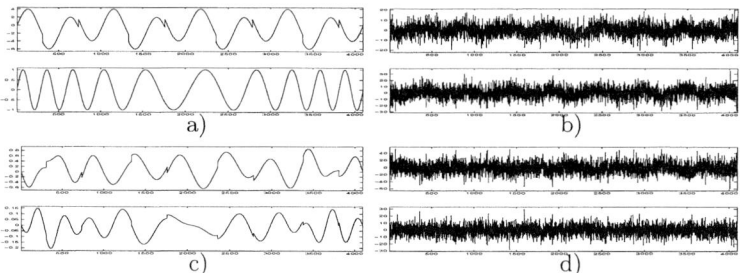

Fig. 2. Illustration of the harmful presence of the noise. a) the original sources $\mathbf{s}$, b) the noisy mixtures $\mathbf{y}$, c) the "ideal" sources (see definition 1) and d) the noisy estimated sources $\hat{\mathbf{s}}$.

Definition 1. *The ideal source signal $\mathbf{s}_{ideal}(k)$ is defined as the product of the separating matrix $\mathcal{B}$ which is estimated from the noisy mixtures, by the noisy-free mixtures $\mathbf{x}(k)$:*

$$\mathbf{s}_{\text{ideal}}(k) = \mathcal{B}_{\text{noisy}}\,\mathbf{x}(k) \qquad (3)$$

3 Wavelet De-noising for BSS

As we said, in BSS the estimated separating matrix is affected by the additive noise. Thus, a powerful de-noising processing before separation seems to be a good solution. In this section we first recall the bases of wavelet de-noising (3.1). Then we propose three methods of wavelet de-noising for BSS: the method proposed by Paraschiv-Ionescu et al. [5] (3.2) and two new methods (3.3 and 3.4).

3.1 Fundamental of Wavelet De-noising

The discret wavelet transform (DWT) is a batch processing, which analyses a finite length time domain signal by breaking up the initial domain in two parts: the detail and approximation information [6]. The approximation domain is successively decomposed into detail and approximation domains.

We use two properties of the discret wavelet transform (DWT):
- the DWT is scattered[1]: a few number of large coefficients dominates the representation,
- the wavelet coefficients are less correlated than the temporal ones.

As a result, we use a nonlinear thresholding function and we treat the coefficients independently to each other. Practicaly, the wavelet de-noising processing consists in applying the DWT to the original noisy signal, chosing the value of the threshold, thresholding the detail coefficients, then inversing the DWT.

Denote $\mathcal{W}(\cdot)$ and $\mathcal{W}^{-1}(\cdot)$ the forward and reverse DWT operators, $d(\cdot)$ the operator which selects the value of the threshold and $\mathcal{T}(\cdot, \lambda)$ the thresholding operator with the threshold λ. Considering the i−th noisy observed signal $\mathbf{y}_i$ from (1), the wavelet de-noising processsing is defined as

$$\begin{cases} \mathbf{w}_i = \mathcal{W}(\mathbf{y}_i) = \theta_i + \mathbf{b}_i \\ \lambda = d(\mathbf{w}_i) \\ \hat{\theta}_i = \mathcal{T}(\mathbf{w}_i, \lambda) \\ \hat{\mathbf{x}}_i = \mathcal{W}^{-1}(\hat{\theta}_i) \end{cases} \quad (4)$$

where $\mathbf{x}_i = (\mathcal{A}\mathbf{s})_i$ is the i−th noisy free mixture. $\theta_i = \mathcal{W}(\mathbf{x}_i)$ and $\mathbf{b}_i = \mathcal{W}(\mathbf{n}_i)$ are respectively the DWT coefficients of the noisy free mixture and the noise. Let denote by $\hat{\mathbf{x}}_i = \mathcal{D}(\mathbf{w}_i)$ the de-noising processing summarizing the four previous stages. Note that the choice of the wavelet function used in the transform is based on one's needs. The choice of a DWT operator can have significant effects on the scheme's performance in terms of noise/signal ratio. Generally, the best choice will resemble "theoretically" the desired feature in the profile, the counterpart being that the analysis fails in direct comparison of the profiles' wavelet transform and identify not-so-similar features. In our analysis, we apply a wavelet transform with the data-adaptive threshold selection rule Sureshrink of ©MATLAB wavelet toolbox to identify sharp gradients.

3.2 P.S. Method

The wavelet de-noising Pre-Separating processing (P.S.) [5] consists in introducing a wavelet de-noising processing before the separating algorithm (Fig. 3). Thus, the separating matrix $\mathcal{B}$ is estimated from de-noised mixtures $\hat{\mathbf{x}}(k)$. The

---

[1] This property is based on the fact that the noise is broad band and is present over all coefficients while deterministic signal is narrow band.

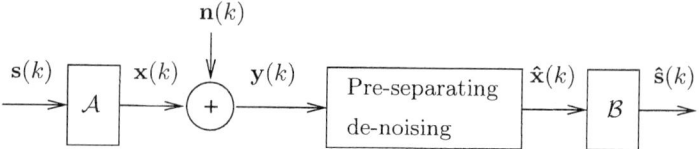

Fig. 3. Principle of the P.S. method.

estimated sources and the noisy mixtures are related by

$$\hat{\mathbf{s}} = \mathcal{B}_{\mathcal{D}(\mathbf{y})}\,\hat{\mathbf{x}} \quad \text{with} \quad \hat{\mathbf{x}} = \mathcal{D}(\mathbf{y}) \tag{5}$$

where the index $\mathcal{D}(\mathbf{y})$ recalls that the separating matrix $\mathcal{B}_{\mathcal{D}(\mathbf{y})}$ is estimated from the de-noised mixtures.

3.3 Serial P.S.P. Method

However, the P.S. method is definitely not efficient. Indeed, the frequency bands or the scales occupied by the mixtures correspond to the union of those occupied by the sources since the mixtures are linear combinations of the sources.

Consequently, we propose the following Serial wavelet Pre-Separating and Post-separating de-noising processing (Serial P.S.P). This method (Fig. 4) allows

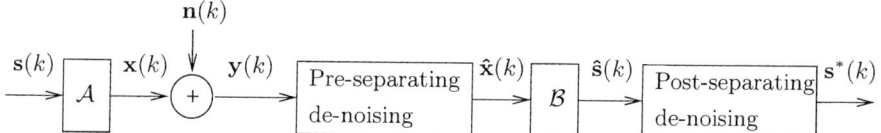

Fig. 4. Principle of the Serial P.S.P. method.

us to adapt the pre-separating de-noising processing to the mixtures and the post-separating de-noising processing to the sources. However, a classical de-noising processing using the variance of the noise $\hat{\sigma}^2$ estimated from the wavelet coefficients at scale 1 cannot succeed. Indeed, the de-noising pre-processing changes the white nature of the noise. In order to overcome this difficulty, we propose the following stages:

1. estimate the variance of the noise $\hat{\sigma}^2_{\mathbf{y}} = (\sigma^2_{y_1}, \cdots, \sigma^2_{y_q})^T$ which corrupts the observed data (cf [6] page 447),
2. calculate $\hat{\sigma}^2_{\hat{\mathbf{s}}} = \mathcal{B}^{*2}\,\hat{\sigma}^2_{\mathbf{y}}$ which is an estimation of the variance of the noise[2] present in the estimated sources $\hat{\mathbf{s}}(k)$, since the noise is white and Gaussian,
3. use a de-noising processing on $\hat{\mathbf{s}}(k)$ using $\hat{\sigma}^2_{\mathbf{y}}$ for determining the value of the threshold.

---

[2] Let denote $\mathcal{B}^{*2}$ the operator which means $(\mathcal{B}^{*2})_{i,j} = (\mathcal{B}_{i,j})^2$.

Using Serial P.S.P., we have to choose carefully the pre-denoising scale. If this scale is overestimated, it provides a distortion of the mixtures $\mathbf{x} = \mathcal{A}\mathbf{s}$, which becomes $\tilde{\mathbf{x}} = \tilde{\mathcal{A}}\tilde{\mathbf{s}}$, where both $\tilde{\mathcal{A}} \neq \mathcal{A}$ and $\tilde{\mathbf{s}} \neq \mathbf{s}$. Thus both estimation of $\tilde{\mathcal{A}}$, and restitution of $\tilde{\mathbf{s}}$, even perfect, do not lead to the good solutions.

3.4 Parallel P.S.P. Method

One of the major problems of the previous methods (P.S. or Serial P.S.P.) lies in the pre-separating de-noising processing: it could remove signal and especially the details (*i.e.* differences between the used wavelet and the signal). Even if this can provide a good estimate of the separating matrix, this may be disastrous for estimating the source signals since the details can contain low power sources (ECG fetal sources for instance). To overcome this problem we propose the following principle (Fig. 5): the Parallel Pre-Separating de-noising and Post-separating de-noising processing (Parallel P.S.P.). The algorithm consists in

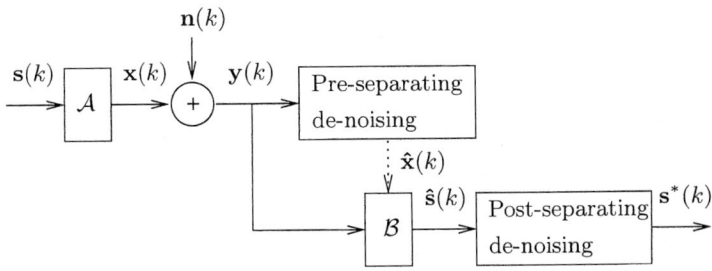

Fig. 5. Principle of Parallel P.S.P. denoising.

1. de-noising the noisy observed data $\mathbf{y}(k) = \mathbf{x}(k) + \mathbf{b}(k)$ using an *ad-hoc* principle to obtain estimated mixed signals $\hat{\mathbf{x}}(k) = \mathcal{D}(\mathbf{y})$,
2. using these estimated mixed signals $\hat{\mathbf{x}}(k)$ in order to estimate the separating matrix $\mathcal{B}_{\mathcal{D}(\mathbf{y})}$,
3. estimating noisy source signals defined as $\hat{\mathbf{s}}(k) = \mathcal{B}_{\mathcal{D}(\mathbf{y})}\,\mathbf{y}(k)$,
4. de-noising the noisy estimated source signal thanks to a post-separating de-noising processing $\mathbf{s}^*(k) = \mathcal{D}(\hat{\mathbf{s}})$.

Thus, noisy estimated source signals and observed data are related by

$$\hat{\mathbf{s}}(k) = \mathcal{B}_{\mathcal{D}(\mathbf{y})}\,\mathbf{y}(k) = \mathcal{B}_{\mathcal{D}(\mathbf{y})}\,\mathcal{A}\,\mathbf{s}(k) + \mathcal{B}_{\mathcal{D}(\mathbf{y})}\,\mathbf{b}(k) \qquad (6)$$

This principle allows us to distinguish the estimation of the separating matrix $\mathcal{B}$ and the restitution of the denoised sources $\mathbf{s}^*$.

4 Simulated Experiments

In the following, we will consider the case of two sources mixed by a 2×2 matrix. We suppose that the mixed signals are corrupted by an additive noise.

In order to compare the principles, we need two indexes: the performance index which quantifies the separating accuracy and the decay index which quantifies the remaining signal after de-noising processing.

Definition 2. *The performance index (PI) [7] which quantifies the separation accuracy is defined as*

$$PI = \sum_{i=1}^{q} \left\{ \left(\sum_{j=1}^{q} \frac{|c_{i,j}|^2}{\max_l |c_{i,l}|^2} - 1 \right) + \left(\sum_{j=1}^{q} \frac{|c_{j,i}|^2}{\max_l |c_{l,i}|^2} - 1 \right) \right\} \quad (7)$$

where $c_{i,j}$ is the (i,j)-th element of the global system $\mathcal{C} = \mathcal{B}\mathcal{A}$.

Definition 3. *The remaining signal $\mathbf{s}_{remaining}$ after de-noising processing is defined as the inverse wavelet transform of the coefficients of the noisy-free signal $\mathbf{s}$ from index where the noisy coefficients $\mathcal{O}(\mathbf{x})$ are larger than the value of the threshold.*

Definition 4. *The decay index (DI) which quantifies the removed signal by the de-noising processing is defined as*

$$DI = \frac{\mathcal{P}_{original}}{\mathcal{P}_{remaining}} \quad (8)$$

where $\mathcal{P}_{original}$ is the power of the noisy-free signal and $\mathcal{P}_{remaining}$ the power of the remaining signal after de-noising processing.

We compare the different principles for two separating algorithms (JADE and EASI) and for different signal to noise ratios (SNR) for the observed mixtures. Each simulation run is repeated 50 times, holding all factors constant except the noise samples. We use the hard shrinkage of the stationary wavelet transform [8] as de-noising processing.

Table 1. Performance for various scales.

Scale	0	1	2	3	4	5	6	7	8	9	10
DI_1	0	0,001	0,003	0,005	0,011	0,027	0,360	1,182	1,515	1,788	1,795
DI_2	0	0	0,001	0,002	0,004	0,023	0,763	1,541	2,042	2,225	2,244
SNR_1	-5,0	-2,0	1,0	4,0	6,7	9,5	10,2	10,1	10,0	10,0	10,1
SNR_2	-5,0	-2,0	1,0	3,9	6,8	9,4	10,3	10,1	9,9	9,9	10,0
PI_{JADE}	-5	-16	-17	-19	-20	-21	-23	-22	-21	-20	-21

4.1 Case of a White Gaussian Noise

Let us begin with a white Gaussian additive noise:

$$\mathbf{y}(k) = \mathcal{A}\,\mathbf{s}(k) + \mathbf{n}(k) \;\; with \;\; \mathbf{n}(k) \stackrel{iid}{\sim} \mathcal{N}(0, \Gamma_\mathbf{n}) \; and \; \Gamma_\mathbf{n} \; diagonal. \quad (9)$$

The table 1 regroups the DI (dB), the SNR (dB) of the denoised mixtures and the PI (dB) versus the scale used for the wavelet de-noising for a SNR of the mixtures egals to -5dB. We note that even if the SNR of the de-noised mixtures is better on scale 6 than on scale 5, the decay index underlines that the removed signal by the de-noising processing is definitively larger at scale 6 than at scale 5.

The table 2 summarizes the results in order to compare the different methods. The numbers between brackets represent the scales used for the de-noising processing pre-separation and post-separation. The performance index PI (dB) and the SNR (dB) for the two estimated sources are reported versus the SNR (dB) of the observed mixtures.

Table 2. Performance for a white Gaussian noise.

	SNR	-10	-5	0	5		-10	-5	0	5
without denoising	PI_{JADE}	0	-5	-22	-24	SNR_1	-9,2	-5,0	-0,3	4,7
	PI_{EASI}	1	-2	-15	-24	SNR_2	-9,1	-4,5	0,3	5,3
P.S. (6,0)	PI_{JADE}	-18	-23	-27	-29	SNR_1	5,9	9,9	13,8	17,3
	PI_{EASI}	-14	-19	-22	-24	SNR_2	5,6	10,1	14,8	19,0
Serial P.S.P. (5,6)	PI_{JADE}	-17	-21	-23	-24	SNR_1	7,4	11,9	15,6	17,8
	PI_{EASI}	-13	-18	-21	-23	SNR_2	7,0	11,6	16,6	21,4
Parallel P.S.P. (6,6)	PI_{JADE}	-18	-23	-27	-29	SNR_1	8,1	12,5	16,1	18,1
	PI_{EASI}	-14	-19	-22	-24	SNR_2	7,8	11.3	16,4	21,4

The denoised principles provide more accurate estimating of the separating matrix $\hat{B}$ and improve the quality of the estimated sources. Serial P.S.P. and Parallel P.S.P. methods have similar performance which is better than the performance obtained with P.S. method.

4.2 Case of a Colored Gaussian Noise

Now, let study the methods with an additive colored Gaussian noise. The simulations were performed with short time dependence noise, modeled by a 2nd-order auto-regressive process AR(2):

$$\mathbf{n}(k) = 1.33\,\mathbf{n}(k-1) - 0.88\,\mathbf{n}(k-2) + \mathbf{w}(k) \qquad (10)$$

with $\mathbf{w}(k)$ an iid Gaussian noise. Since the noise $\mathbf{n}$ has scale-dependent wavelet coefficients, we used a scale-dependent threshold.

In this case we only report the PI (dB) and the SNR of the estimated sources versus the SNR (dB) of the mixtures. The table 3 illustrates the interest of a post-separation de-noising. As in the case of the white noise case, the principles Serial P.S.P. and Parallel P.S.P. improve the performances.

5 Conclusion

The noise strongly limits the separation performance, encouraging us to use wavelet de-noising processing. In this paper, we propose two new principles, Serial P.S.P. and Parallel P.S.P., which associate wavelet de-noising and blind

Table 3. Performance for colored Gaussian noise.

	SNR	-10	-5	0	5		-10	-5	0	5
without de-noising	PI_{JADE}	0	-1	-15	-23	SNR_1	-9,3	-4,7	-0,3	4,7
	PI_{EASI}	0	-2	-12	-21	SNR_2	-9	-4,4	0,3	5,3
P.S. (6,0)	PI_{JADE}	-10	-22	-24	-24	SNR_1	7,2	11,6	16,0	19,0
	PI_{EASI}	-14	-19	-23	-24	SNR_2	7,9	12,8	17,4	21,5
Serial P.S.P. (5,6)	PI_{JADE}	-21	-23	-24	-24	SNR_1	11,2	14,9	17,6	19,6
	PI_{EASI}	-18	-20	-23	-24	SNR_2	10,6	14,8	20,0	25,0
Parallel P.S.P. (6,6 or 5)	PI_{JADE}	-22	-24	-24	-25	SNR_1	11,8	15,3	17,5	19,7
	PI_{EASI}	-19	-21	-23	-24	SNR_2	10,3	14,9	20,0	24,9

source separation. In noisy mixtures, these new principles improve the separation performance and give comparable results, but Parallel P.S.P. method is more robust to a bad pre-separation de-noising with white as well as colored Gaussian noise. Moreover its implementation is easier since there is no trade-off for determining the pre-separation scale.

Finaly, we addressed the fetal ECG extraction from sensors located on the mother's skin [9]. Preliminary experiments [10], performed with success from strong noisy signals, confirm the efficacy of these methods.

References

1. J. Hérault, C. Jutten, and B. Ans. Détection de grandeurs primitives dans un message composite par une architecture de calcul neuromimétrique en apprentissage non supervisé. In *Gretsi*, volume 2, pages 1017–1020, Nice, France, May 1985.
2. C. Jutten and A. Taleb. Source separation: from dusk till dawn. In *Independent compoment analysis 2000*, pages 15–26, Helsinki, Finlande, June 2000.
3. S.I. Amari and A. Cichocki. *Adaptive Blind Signal and Image Processing, Learning Algorithms and Applications*. Wiley, 2002.
4. A. Hyvärinen, J. Karhunen, and E. Oja. *Independent Component Analysis*. Wiley, 2001.
5. A. Paraschiv-Ionescu, C. Jutten, K. Aminian, and al. Source separation in strong noisy mixtures: a study of wavelet de-noising pre-processing. In *ICASSP'2002*, Orlando, Floride, 2002.
6. S. Mallat. *A wavelet tour of signal processing*. Academic Press, second edition, 1999.
7. H.H. Yang, S.I. Amari, and A. Cichocki. Information-theoric approach to blind separation of sources in non-linear mixture. *Signal Processing*, 64(3):291–300, February 1998.
8. R.R. Coifman and D.L. Donoho. Translation-invariant de-noising. In *Wavelets and statistics, Springer lecture notes in Statistics 103*, pages 125–150. New York: Springer-Verlag.
9. L. De Lathauwer, D. Callaerts, B. De Moor, and al. Fetal electrocardiogram extraction by source subspace separation. In *Proc. IEEE Workshop on HOS*, pages 134–138, Girona, Spain, June 12–14 1995.
10. B. Rivet, C. Jutten, and V. Vigneron. Wavelet de-noising for blind source separation in noisy mixtures. Technical report, Technical report for the BLInd Source Separation project (BLISS IST 1999-14190), 2003.

A Gaussian Mixture Based Maximization of Mutual Information for Supervised Feature Extraction

José M. Leiva-Murillo and Antonio Artés-Rodríguez

Department of Signal Theory and Communications
Universidad Carlos III de Madrid
Avda. de la Universidad 30, 28911 Leganés-Madrid, Spain
{jose,antonio}@tsc.uc3m.es

Abstract. In this paper, we propose a new method for linear feature extraction and dimensionality reduction for classification problems. The method is based on the maximization of the Mutual Information (MI) between the resulting features and the classes. A Gaussian Mixture is used for modelling the distribution of the data. By means of this model, the entropy of the data is then estimated, and so the MI at the output. A gradient descent algorithm is provided for its optimization. Some experiments are provided in which the method is compared with other popular linear feature extractors.

1 Introduction

Dimensionality reduction has been paid much attention by researchers involved in statistical learning and data exploration. In the first case, the motivation arises from the "curse of dimensionality" that appears in classification and regression problems when a low number of samples and a high dimensionality are present in the dataset. A dataset with few dimensions requires in general a simpler classifier or regressor that is likely to provide a higher generalization capacity. Data exploration is rather interested in finding a visually useful representation of a dataset in order to extract information about inner relationships in the data. Such a manageable representation is only possible when no more than two or three dimensions are used for the visualization. This is the case of the Self Organizing Maps, Multidimensional Scaling or the Projection Pursuit. In statistical learning, both unsupervised and supervised criteria have been applied to the feature extraction preprocessing. Popular unsupervised linear methods are Principal Component Analysis (PCA) and Independent Component Analysis (ICA). PCA is a common tool for finding the most powerful directions in the data and so provides a set of uncorrelated projections along them. ICA extends this criterion to higher order statistics, in order to achieve statistical independence in addition to un-correlation.

Supervised methods are used as a preprocessing step in classification and regression applications, in which datasets are commonly composed of samples of two variables $\{\mathbf{x}, y\}$ and the objective is then to find the transformation on the primary multidimensional variable $\mathbf{x}$ that more efficiently preserves the ability of regression on the auxiliary variable y. The only difference concerning y between regression and classification is the fact that in the former case y is a continuous variable so that $y \in \Re$. In classification, y is discrete and has a limited number of possible values: $y \in \mathcal{Y} \equiv \{c_1, \cdots, c_{|\mathcal{Y}|}\}$.

Some popular methods for supervised feature extraction are Linear Discriminant Analysis (LDA), Sliced Inverse Regression (SIR), Partial Least Square Regression (PLS) and Canonical Correlation Analysis (CCA). LDA finds the projections along which the Fischer Linear Discriminant of the resulting features are maximized. SIR partitions the range of y and carries out a type of PCA on each of the resulting *slices*. PLS searches the projection that minimizes the square error when doing the regression. The aim of CCA is to obtain a pair of linear transformations in such a way that being applied to the two input signals (in this case one of them is one-dimensional), the correlation between the outputs is maximized. Although widely used in regression, it is not difficult to apply some of these methods to classification problems. In the case of SIR, the slices can be simply determined by the classes. PLS and CCA may require a preprocessing on y, transforming it into a multidimensional signal with $|\mathcal{Y}|$ dimensions, which is the number of classes.

In this paper, we will describe a new method called GM MMI for linear feature extraction that uses an information theoretical criterion for obtaining the components. The number of components that the algorithm is able to yield is not constrained by the number of classes. The novel procedure for entropy estimation provided here may extend its use to a wide range of applications concerning unsupervised and supervised learning.

In Section 2 some properties from Information Theory are provided as well as some previous attempts in feature extraction with Mutual Information as a cost function. Section 3 describes the estimation of the Mutual Information by the use of Gaussian Mixture models. A gradient descent based algorithm for its optimization, called GM MMI, is then provided. Some experiments that prove the validity of the algorithm are displayed in Section 4. In Section 5, some conclusions are stressed.

2 Mutual Information in Feature Extraction

2.1 Information Theory and Feature Extraction

According with Shannon's Information Theory (IT), the Mutual Information (MI) between two signals can be described as the quantity of information that each of the signals carries about each other. It can be seen as a generalization of the concept of entropy: the uncertainty of a variable is now measured with respect to another one:

$$I(\mathbf{x}, y) = h(\mathbf{x}) - h(\mathbf{x}|y) \tag{1}$$

where $\mathbf{x}$ is the primary multidimensional variable and y the auxiliary one-dimensional one. As our work is mainly focused on classification problems, we will consider y as a discrete variable with a finite number of elements. This fact can be very useful since it allows us to decompose the second term in Eq. 1 as:

$$I(\mathbf{x}, y) = h(\mathbf{x}) - \sum_i p(c_i) h(\mathbf{x}|c_i) \qquad (2)$$

This expression is very comfortable to work with, since the MI is defined by the entropy of the subsets of samples of $\mathbf{x}$ belonging to each class. This intuitive criterion has a mathematical justification by means of the Fano's bound, an important inequality from IT [6]:

$$p_e \geq \frac{h(y) - I(\mathbf{x}, y) - 1}{\log(|\mathcal{Y}|)}$$

where $|\mathcal{Y}|$ is the cardinality of the variable y, i.e. the number of classes, and p_e is the expected classification error. This equation reveals how the bound on the error, although being loose, decreases as the MI increases.

Another result from IT is the so called Data Processing inequality [6]:

$$I(T(\mathbf{x}), y) \leq I(\mathbf{x}, y)$$

This intuitive property tells us that we can not get, by means of a transformation on the data, more (mutual) information than we originally got. In fact, the inequality turns into an equality only when the transformation $T(\cdot)$ is invertible. In terms of linear transformations, in which we assume $T(\mathbf{x}) = \mathbf{W}\mathbf{x}$, invertibility takes place only when $\mathbf{W}$ is a full-rank matrix. Due to this fact, in a non-singular situation, a non-square transformation matrix $\mathbf{W}$ will always produce a loss of information.

2.2 Previous Results in Information Theoretical Feature Extraction

Several efforts have been carried out in order to incorporate the MI as a cost function to learning and feature extraction. The study of the flow of information through neural networks and linear systems gave birth to the first ICA algorithms [4].

Considering first the simpler problem of feature selection, in which no transformation is applied to the original features, Battiti [3] proposed a method based on MI estimation. In this case, the entropies are approximated for each feature or component by integrating a histogram of the data. The work provides an important bound on the error committed on the estimation of the MI. A non-parametric estimation was used in a later work by Kwak et al. [10], in which the MI is approximated for each component and so provides a good criterion for feature selection.

When working with continuous signals or variables, it is not easy to compute the integrals present in the expression of the entropy. As an alternative to Shannon's entropy and its drawbacks, Renyi's entropy provides a more manageable definition that can be easily computed when the probability density function (*pdf*) of the signals involved is available. As a result, Principe et al. [8] propose an information theoretical framework that combines the use of Renyi's entropy with non-parametric estimation of the *pdf* of the signals. A related work is described in [12], in which a non-parametric estimation of the *pdf*s is combined with alternative definitions of MI in order to obtain a computationally feasible divergence to be computed.

A generalization of LDA is proposed in the Informative Discriminant Analysis [9], in which the cost function is a likelihood that measures the quality of the prediction ability after the transformation. A non-parametric modelling is used, and an asymptotical equivalence to MI criterion is suggested. Although not directly related to IT, another interesting work is presented in [5], in which a Gaussian Mixture (GM) model is used in order to find a common methodology for unsupervised learning (ICA) as well as pattern classification.

Apart from feature extraction, the Information Bottleneck [11] has found a parallelism between the tradeoff compression vs. preservation from IT in communications, and the tradeoff accuracy vs. simplicity from learning theory.

In the next section, we will describe a method for the estimation of the MI based on a previous modelling of the densities $p(\mathbf{x})$ and $\{p(\mathbf{x}|c_i)\}_{i=1,...,|\mathcal{Y}|}$. A gradient descent based algorithm for the MI optimization, that we call GM MMI, is also provided.

3 Gaussian Mixture Modelling for Maximization of Mutual Information

The main problem of the entropy (and so the MI) of a multidimensional signal $\mathbf{x}$ is the fact that its value is known only for few, analytically defined *pdf*s. In this section we propose an estimation based on semi-parametric modelling by GM models. Even for GM models, the problem is still analytically intractable, but some simplifications can be made by assuming the components of the model not to be very overlapped.

3.1 MI Estimation from GM Models

A GM Model has the form:

$$p(\mathbf{x}) = \sum_{i=1}^{L} \alpha_i p(\mathbf{x}|\Theta_i)$$

where $p(\mathbf{x}|\Theta_i)$ is a Gaussian with parameters $\Theta_i = \{\boldsymbol{\mu}_i, \mathbf{C}_i\}$, being $\boldsymbol{\mu}_i$ its mean and $\mathbf{C}_i$ its covariance matrix, and being the α_i the priors, such that $\sum_i \alpha_i = 1$. The model can be obtained via the Expectation-Maximization (EM) algorithm by Dempster et al. [7]. In our case, mixtures of only two Gaussians are used, since an increment in the number of Gaussians does not raise the performance of the method.

The calculation of the entropy is analytically hard to solve since it implies the logarithm of a sum. The entropy of $\mathbf{x}$ is given by:

$$h(\mathbf{x}) = -\int p(\mathbf{x}) \log p(\mathbf{x}) d\mathbf{x}$$
$$= -\sum_i \alpha_i \int p(\mathbf{x}|\Theta_i) \log \left(\sum_l \alpha_l p(\mathbf{x}|\Theta_l) \right) d\mathbf{x} \qquad (3)$$

We now rewrite the argument inside the logarithm as:

$$\sum_l \alpha_l p(\mathbf{x}|\Theta_l) = \alpha_i p(\mathbf{x}|\Theta_i) + \sum_{l \neq i} \alpha_l p(\mathbf{x}|\Theta_l)$$
$$= \alpha_i p(\mathbf{x}|\Theta_i) \left(1 + \frac{\sum_{l \neq i} \alpha_l p(\mathbf{x}|\Theta_l)}{\alpha_i p(\mathbf{x}|\Theta_i)} \right)$$
$$= \alpha_i p(\mathbf{x}|\Theta_i)(1 + \epsilon_i(\mathbf{x}))$$

Now, we assume $\epsilon_i(\mathbf{x}) \ll 1$, that is equivalent to assume that, in the significant rank of each integral, the contributions of the clusters corresponding to $l \neq i$ are negligible. This assumption may help us simplify things by means of the approximation $\log(1 + \epsilon) \approx \epsilon$ if $\epsilon \ll 1$. The Eq. 3 can be unfolded as:

$$h(\mathbf{x}) \approx -\sum_i \alpha_i \int p(\mathbf{x}|\Theta_i) \left(\log(\alpha_i p(\mathbf{x}|\Theta_i)) + \epsilon_i(\mathbf{x}) \right) d\mathbf{x}$$

The first term defines the entropy of a Gaussian, that is known. The second one has the value:

$$\alpha_i \int p(\mathbf{x}|\Theta_i) \epsilon_i(\mathbf{x}) d\mathbf{x} = \sum_{l \neq i} \alpha_l$$

After grouping and simplifying terms, we obtain:

$$h(\mathbf{x}) \approx \sum_{i=1}^{L} \alpha_i \log \left(\frac{(2\pi e)^{d/2} |\mathbf{C}_i|^{1/2}}{\alpha_i} \right) - (L-1) \qquad (4)$$

For a linear transformation $\mathbf{z} = \mathbf{W}\mathbf{x}$, being $\mathbf{W}$ a reducing matrix, we will use a GM with parameters $\Theta'_i = \{\boldsymbol{\mu}', \mathbf{C}'\}$, being $\boldsymbol{\mu}' = \mathbf{W}\boldsymbol{\mu}_i$ and $\mathbf{C}'_i = \mathbf{W}\mathbf{C}_i\mathbf{W}^t$.

3.2 A Gradient Descent Algorithm

The expression in Eq. 4 can be easily derived in order to make a gradient descent feasible.

$$\frac{\partial}{\partial \mathbf{W}} h(\mathbf{z}) = \sum_{i=1}^{L} \alpha_i \mathbf{C}'^{-1}_i \mathbf{W} \mathbf{C}_i$$

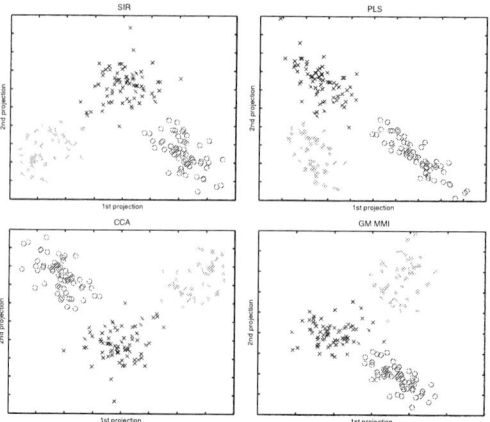

Fig. 1. The two principal projections of the wine data according with SIR, PLS, CCA, and GM MMI.

The gradient is computed for $h(\mathbf{z})$ as well as each of the subsets $\{h(\mathbf{z}|c_i)\}$ and so we dispose of the elements for the gradient of $I(\mathbf{z}, y)$, according with Eq. 2. In each step, the following rule updates the matrix of feature extraction:

$$\mathbf{W}_{n+1} = \mathbf{W}_n + \lambda \frac{\partial}{\partial \mathbf{W}} I(\mathbf{W}_n \mathbf{x}, y)$$

being λ the step factor. A later orthogonization may be applied to the vectors of $\mathbf{W}_n$ in each iteration.

4 Experiments and Results

In this section, we compare GM MMI with other state-of-the-art supervised dimensionality reduction methods in both data exploration and classification applications.

4.1 Data Exploration

We apply our method together with SIR, PLS and CCA to the public Wine dataset, from the UCI repository [1]. This is a set with 178 samples and 13 dimensions, each of them representing a biochemical property of wines from three different locations indicated by the labels. In Fig. 1, the mapping obtained for the data is displayed, projected along the two principal features obtained by each of the methods. It can be seen how our method is able to shatter the samples belonging to each of the classes as perfectly as the other methods.

4.2 Classification: A Face Detection Application

A public database from [2] has been used for measuring the performance of a classifier on the resulting features for CCA, SIR and GM MMI. The data consist

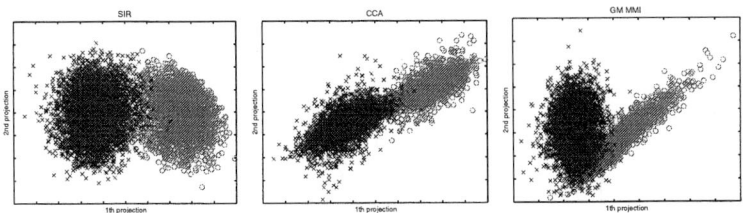

Fig. 2. Two principal projections from SIR, CCA and GM MMI for the CBCL Face Dataset.

of pictures of 19 × 19 pixels and so the dimension of the original dataset is 361. The training set is compounded by 2429 positive pictures (there is a face in the image) and 4548 negative ones. The test set consists of 472 positive samples and 23573 negative ones.

In Fig. 2, the distribution of the samples is displayed along the two main components obtained by each method. Now, PLS projections are not provided since the method is only able to provide $|\mathcal{Y}| - 1$ components, since it leads to a eigen-decomposition that provides less positive eigenvalues than classes. In SIR and PLS, the high difference between the relevance of the first component and the following ones' is high, due to the difference between the first eigenvalue and the other ones obtained by these methods. On the other hand, GM MMI is able to find two projections each of which is significant. A Support Vector Machine (SVM) has been used for classification. As the test data are very unbalanced (there are many more negative samples than positive ones), a set of Receiver Operating Characteristic (ROC) curves has been obtained. We have displayed these curves in Fig. 3 for each method and each reduction degree. The tradeoff between missed detections and false alarms is plotted as the bias parameter of the classifier is swept. When only one component is obtained, SIR and CCA find the same projection as they reach the same eigen-decomposition problem. In each of the experiments, GM MMI outperforms the other methods in the whole range of values for the tradeoff between false alarm and detection rate.

5 Conclusions

We have developed an algorithm for linear feature extraction by maximizing the Mutual Information of the features extracted with respect to an auxiliary variable, the one that describes the class of the samples. The method is based on an entropy estimation according to which the MI is approximated. The promising results suggest that the method is able to find non-linear inner relationship in data that may be not discovered by other methods typically focused on second order statistics. The application of the method of multidimensional entropy estimation proposed in this paper to ICA problems is considered as future work.

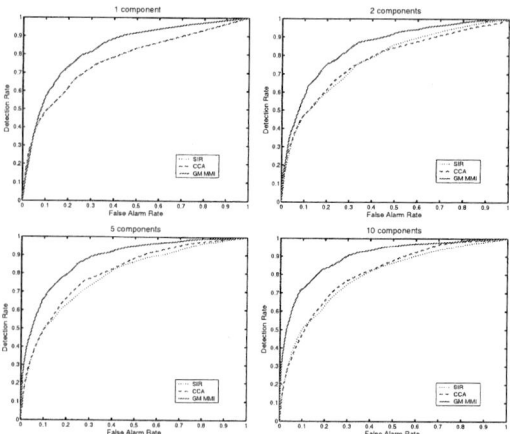

Fig. 3. ROC curves for SIR, CCA and GM MMI methods for several degrees of dimensionality reduction: 1, 2, 5 and 10 features extracted.

References

1. *UCI Repository of Machine Learning Databases*, 1998. http://www.ics.uci.edu/~mlearn/MLRepository.html.
2. *CBCL Software and Datasets, MIT*, Face Images database, 2000. http://www.ai.mit.edu/projects/cbcl/software-datasets/index.html.
3. R. Battiti. Using mutual information for selecting features in supervised neural net learning. *Neural Networks*, 5:537–550, 1994.
4. A. J. Bell and T. Sejnowski. An information maximisation approach to blind separation and blind deconvolution. *Neural Computation*, 7(6):1004–1034, 1995.
5. J. L. Center. Blind source separation, independent component analysis, and pattern classification - connections and synergies. In *Proceedings MaxEnt 23*, Jackson Hole, WY, 2003.
6. T.M. Cover and J.A. Thomas. *Elements of Information Theory*. John Wiley &Sons, 1991.
7. A. P. Dempster, N. M. Laird, and D. B. Rubin. Maximum likelihood from incomplete data via EM algorithm (with discussion). *Journal of the Royal Statistical Society*, B(39):1–38, 1977.
8. D. Xu J. Principe and J. W. Fischer III. *Information-Theoretic Learning*, volume 1. Wiley, 2000.
9. S. Kaski and J. Peltonen. Informative discriminant analysis. In *Proceeding of the ICML*, volume 5, pages 329–336, Washington DC, 2003.
10. N. Kwak and C. Choi. Input feature selection by mutual information based on parzen window. *IEEE Transactions on Pattern Analysis and Machine Intelligence*, 24(12):1667–1671, 2002.
11. F.C. Pereira N. Tishby and W. Bialek. The information bottleneck method. In *37th Annual Allerton International Conference on Communications, Control and Computing*, 1999.
12. K. Torkkola. Feature extraction by non-parametric mutual information maximization. *Journal on Machine Learning Research*, 3:1415–1438, 2003.

Blind Separation of Nonstationary Sources by Spectral Decorrelation

Shahram Hosseini and Yannick Deville

Université Paul Sabatier
Laboratoire d'Acoustique, Métrologie, Instrumentation
Bat. 3R1B2, 118 route de Narbonne, 31062 Toulouse Cedex, France
{hosseini,ydeville}@cict.fr

Abstract. This paper demonstrates and exploits some interesting frequency-domain properties of nonstationary signals. Considering these properties, two new methods for blind separation of linear instantaneous mixtures of mutually uncorrelated, nonstationary sources are proposed. These methods are based on spectral decorrelation of the sources. The second method is particularly important because it allows the existing time-domain algorithms developed for stationary, temporally correlated sources to be applied to nonstationary, temporally uncorrelated sources just by mapping the mixtures in the frequency domain. Moreover, it sets no constraint on the variance profile, unlike previously reported methods.

1 Introduction

Blind source separation can be achieved by exploiting nonGaussianity, time correlation or nonstationarity [1]. In this paper, our goal is to propose new approaches using the nonstationarity of the sources. A few authors have studied this problem [2]-[9]. In many of these works, the nonstationarity of the variance of the sources is used. In [2], separation of nonstationary signals is achieved by computing output components which are uncorrelated *at every time point*. The method requires the joint diagonalization of N covariance matrices, where N represents the number of samples. In [3], the signals are divided in only two subintervals. Then, the joint diagonalization of two covariance matrices, estimated on the two subintervals, allows one to separate the sources. Another approach, presented in [4], is based on the maximization of the nonstationarity, measured by the cross-cumulant, of a linear combination of the observed mixtures. Several methods use the time-frequency diversity of the sources. Some of them [5] are based on a time-frequency version of joint-diagonalization source separation techniques. Others [6]-[8] assume that each source occurs alone in a small time-frequency area and identify the corresponding columns of the scaled mixing matrix in these areas. Pham and Cardoso have developed novel approaches based on the principles of maximum likelihood and minimum mutual information [9].

The methods proposed in the present paper are based on spectral decorrelation of the signals. They result from some interesting frequency-domain properties of nonstationary signals, and may be used for separating linear instantaneous mixtures of Gaussian or nonGaussian nonstationary, mutually uncorrelated signals. For the sake of simplicity, in this paper we only study the case of two mixtures of two sources. However, the method may be extended to more sources and mixtures.

2 Some Mathematical Preliminaries

We here introduce some interesting statistical properties of the Fourier transforms of real random signals. Their proofs are given in Appendix A.

1. Let $u_1(t)$ and $u_2(t)$ be two zero-mean, mutually uncorrelated real signals, i.e. such that $E[u_1(t)u_2(t)] = 0$. Then, denoting their Fourier transforms[1] by $U_1(\omega)$ and $U_2(\omega)$, we have $E[U_1(\omega)U_2(\omega)] = E[U_1(\omega)U_2^*(\omega)] = 0$.
2. Let $u(t)$ be a real stationary signal with Fourier transform $U(\omega)$. Then, $E[U^2(\omega)] = 0$ for $\omega \neq 0$.
3. If $u_1(t)$ and $u_2(t)$ are two stationary, mutually uncorrelated, real, zero-mean signals with Fourier transforms $U_1(\omega)$ and $U_2(\omega)$, and if $V_1(\omega)$ and $V_2(\omega)$ are two linear combinations of $U_1(\omega)$ and $U_2(\omega)$, then $E[V_1^2(\omega)] = E[V_2^2(\omega)] = E[V_1(\omega)V_2(\omega)] = 0$ for $\omega \neq 0$.
4. If $u(t)$ is a temporally uncorrelated, real, zero-mean signal with a nonstationary variance $q(t)$, i.e. if $E[u(t_1)u(t_2)] = q(t_1)\delta(t_1 - t_2)$, then its Fourier transform, $U(\omega)$ is a stationary[2], correlated process with autocorrelation $Q(\omega)$, the Fourier transform of $q(t)$.

3 Source Separation in the Frequency Domain

Given N samples of two linear instantaneous mixtures $x_1(t)$ and $x_2(t)$ of two mutually uncorrelated, nonstationary, real, zero-mean sources $s_1(t)$ and $s_2(t)$, our objective is to estimate $s_1(t)$ and $s_2(t)$ up to a scaling factor and a permutation. Let's denote $\mathbf{s}(t) = [s_1(t), s_2(t)]^T$ and $\mathbf{x}(t) = [x_1(t), x_2(t)]^T$ so that $\mathbf{x}(t) = \mathbf{As}(t)$ where $\mathbf{A}$ is the mixing matrix. Taking the Fourier transform of $\mathbf{x}(t)$, we obtain:

$$\mathbf{X}(\omega) = \mathbf{AS}(\omega) \qquad (1)$$

where $\mathbf{S}(\omega) = [S_1(\omega), S_2(\omega)]^T$, $\mathbf{X}(\omega) = [X_1(\omega), X_2(\omega)]^T$, and $S_1(\omega)$, $S_2(\omega)$, $X_1(\omega)$ and $X_1(\omega)$ are respectively the Fourier transforms of $s_1(t)$, $s_2(t)$, $x_1(t)$ and $x_2(t)$. The spectra $\mathbf{Y}(\omega) = [Y_1(\omega), Y_2(\omega)]^T$ of the estimated sources $\mathbf{y}(t) = [y_1(t), y_2(t)]^T$ may be obtained by multiplying $\mathbf{X}(\omega)$ by a real separating matrix $\mathbf{B}$, i.e. $\mathbf{Y}(\omega) = \mathbf{BX}(\omega)$. It is well known that because of the indeterminacies involved in the problem, this matrix has only two degrees of freedom. Hence, we need at least two equations for estimating it. In the following, we propose two

---
[1] The Fourier transform of a stochastic process $u(t)$ is a stochastic process $U(\omega)$ given by [10] $U(\omega) = \int_{-\infty}^{\infty} u(t)e^{-j\omega t}dt$. The integral is interpreted as a Mean Square limit.
[2] In the sense that $E[U(\Omega+\omega)U^*(\Omega)] = Q(\omega)$, i.e. its autocorrelation depends only on ω, not on Ω.

alternative ideas for obtaining such equations in the frequency domain, using the properties mentioned in Section 2, and knowing that the estimated sources $y_1(t)$ and $y_2(t)$ must be mutually uncorrelated.

3.1 First Source Separation Method, Using Property 1

To avoid the indeterminacy due to the scaling factor, let's fix the entries of the second column of the separating matrix **B** to one, so that $\mathbf{Y}(\omega) = \begin{pmatrix} b_1 & 1 \\ b_2 & 1 \end{pmatrix} \mathbf{X}(\omega)$. Following Property 1, the uncorrelatedness of $y_1(t)$ and $y_2(t)$ implies that

$$E[Y_1(\omega)Y_2^*(\omega)] = E[(b_1 X_1(\omega) + X_2(\omega))(b_2 X_1^*(\omega) + X_2^*(\omega))] = 0$$
$$E[Y_1(\omega)Y_2(\omega)] = E[(b_1 X_1(\omega) + X_2(\omega))(b_2 X_1(\omega) + X_2(\omega))] = 0 \quad (2)$$

Solving these two equations with respect to b_1 and b_2, it can be shown (see Appendix B) that b_1 and b_2 are the two real solutions of the following second-order equation:
$$Az^2 + Bz + C = 0 \quad (3)$$

where

$$A = -E[X_1(\omega)X_2(\omega)]E[X_1(\omega)X_1^*(\omega)] + E[X_1^2(\omega)]E[X_1(\omega)X_2^*(\omega)]$$
$$B = -E[X_1(\omega)X_1^*(\omega)]E[X_2^2(\omega)] + E[X_1^2(\omega)]E[X_2(\omega)X_2^*(\omega)]$$
$$C = -E[X_1(\omega)X_2^*(\omega)]E[X_2^2(\omega)] + E[X_1(\omega)X_2(\omega)]E[X_2(\omega)X_2^*(\omega)] \quad (4)$$

These equations are of interest only if $s_1(t)$ and/or $s_2(t)$ are nonstationary, because, from Property 3, if $s_1(t)$ and $s_2(t)$ are stationary, $E[X_1^2(\omega)] = E[X_2^2(\omega)] = E[X_1(\omega)X_2(\omega)] = 0$ for $\omega \neq 0$ so that the coefficients A, B and C are equal to zero for $\omega \neq 0$. Moreover, since the Fourier transform of a real signal is real at $\omega = 0$, we can write $E[X_1(0)X_2(0)] = E[X_1(0)X_2^*(0)]$, $E[X_1^2(0)] = E[X_1(0)X_1^*(0)]$, and $E[X_2^2(0)] = E[X_2(0)X_2^*(0)]$, so that at $\omega = 0$, $A = B = C = 0$ too, and the sources cannot be separated. This result is not surprising because it is well known that the mutual decorrelation of two sources (which is a second-order statistical parameter) is not a strong enough hypothesis for separating stationary sources[3]. It is therefore necessary to suppose that at least one of the sources is nonstationary for achieving source separation only using mutual decorrelation.

Discussion. From (4), the implementation of the above method requires the computation of the expected values of some spectral functions. Three different cases may be considered.

a) Several realizations of the mixtures $x_1(t)$ and $x_2(t)$ are available. In this case, the expected values may be approximated by averaging the spectral functions on these realizations (for a particular frequency).

b) Only one realization of the mixtures is available but the spectra are ergodic so that the expected values in (4) can be estimated by frequency averages. A

---
[3] Except for temporally correlated sources by exploiting the time correlation.

necessary condition for the ergodicity is the stationarity of the spectral functions, i.e., the expected values in (4) must be independent from ω. However, it seems difficult to find signals satisfying this condition.

c) Only one realization of the mixtures is available but each mixture has nearly the same spectral shape in different time frames (for example, the mixtures are cyclostationary). In this case, the expected values may be estimated by dividing the mixtures in several time frames, computing the Fourier transforms and the spectral functions over each frame, and averaging the results on different frames (for a particular frequency).

3.2 Second Source Separation Method, Using Property 4

If we also suppose that $s_1(t)$ and $s_2(t)$ are temporally uncorrelated, from Property 4, $S_1(\omega)$ and $S_2(\omega)$ are stationary and correlated processes. Moreover, from (1), $\mathbf{X}(\omega)$ is a linear mixture of these two processes. Many algorithms have been proposed for separating such mixtures [11]-[16]. Although these algorithms were originally developed for time-domain stationary, time-correlated processes, nothing prohibits us from applying them to frequency-domain stationary, frequency-correlated processes. Thus, only by mapping the nonstationary temporally uncorrelated mixtures in the frequency domain, they can be separated using one of the numerous methods developed previously for time-correlated stationary mixtures.

4 Simulation Results

In the first experiment, we consider the example used in [2]. The following stationary and nonstationary Gaussian signals are used: $s_1(t) = n_1(t)$, $s_2(t) = \mu_2(t)n_2(t)$, where $n_1(t)$ and $n_2(t)$ are mutually independent Gaussian i.i.d. signals with zero mean and unity variance, and $\mu_2(t) = 2\sin(\omega_0 t)$. The mixing matrix is $\mathbf{A} = \begin{pmatrix} 1 & 0.5 \\ 0.5 & 1 \end{pmatrix}$. It can be easily shown that $E[s_1(t_1)s_1(t_2)] = \delta(t_1 - t_2)$ and $E[s_2(t_1)s_2(t_2)] = 4\sin^2(\omega_0 t_1)\delta(t_1 - t_2)$. Thus, using the same notations as in Property 4, $q_1(t) = 1$ and $q_2(t) = 4\sin^2(\omega_0 t)$, so that $Q_1(\omega) = 2\pi\delta(\omega)$ and $Q_2(\omega) = 2\pi[2\delta(\omega) - \delta(\omega - 2\omega_0) - \delta(\omega + 2\omega_0)]$.

In the first step, we want to separate the sources using the method proposed in Subsection 3.1. The coefficients A, B and C in (4) depend on $E[S_1^2(\omega)]$, $E[S_2^2(\omega)]$, $E[S_1(\omega)S_1^*(\omega)]$ and $E[S_2(\omega)S_2^*(\omega)]$. Using the method employed in the proof of Property 2, it can be shown that $E[S_1^2(\omega)] = Q_1(2\omega)$, $E[S_2^2(\omega)] = Q_2(2\omega)$, $E[S_1(\omega)S_1^*(\omega)] = Q_1(0)$ and $E[S_2(\omega)S_2^*(\omega)] = Q_2(0)$. Since $E[S_1^2(\omega)]$ and $E[S_2^2(\omega)]$ depend on ω, they cannot be considered as ergodic processes so that the coefficients A, B and C in (4) cannot be estimated by frequency averages. However, as $s_1(t)$ is stationary and $s_2(t)$ is cyclostationary, we can estimate the expected values in (4) using the method proposed in part (c) of the discussion of Subsection 3.1.

The experiment was done using 1 second of the sources $s_1(t)$ and $s_2(t)$ containing 8192 samples. The frequency $\omega_0 = 2\pi.256$ of $\mu_2(t)$ was chosen so that each period of $\mu_2(t)$ contains 32 points. Hence, the signal $s_2(t)$ includes 256 periods of $\mu_2(t)$. Then, the 32-point Discrete Fourier Transforms of the mixtures $x_1(t)$ and $x_2(t)$ were computed on each period and the expected values in (4) were estimated by averaging the spectral functions (at $\omega = \omega_0$) on 256 periods. The experiment was repeated 100 times corresponding to 100 different seed values of the random variable generator. For each experiment, the output Signal to Noise Ratio (in dB) was computed by $SNR = 0.5 \sum_{i=1}^{2} 10 \log_{10} \frac{E[s_i^2]}{E[(y_i-s_i)^2]}$, after normalizing the estimated sources, $y_i(t)$, so that they have the same variances as the source signals, $s_i(t)$. The mean and the standard deviation of SNR on the 100 experiments were 27.0 dB and 8.9 dB.

In the second step, we want to separate the sources using the method proposed in Subsection 3.2. This time, we compute the Fourier transforms of $x_1(t)$ and $x_2(t)$ on the whole signals. The autocorrelation function of $X_1(\omega)$ is shown in Figure 1 which presents three peaks at $\omega = 0$ and $\omega = \pm 2\omega_0$, and confirms the theoretical calculus mentioned above (see the expression of $Q_2(\omega)$). The separating matrix may be estimated using the following equations: $E[Y_1(\omega)Y_2^*(\omega)] = 0$, and $E[Y_1(\omega + 2\omega_0)Y_2^*(\omega)] = 0$. We used a modified version of the AMUSE

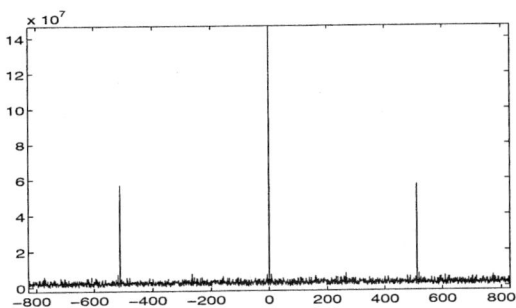

Fig. 1. Autocorrelation function of $X_1(\omega)$.

algorithm [11] for this purpose. This simple and fast algorithm, originally developed for separating time-correlated stationary sources in the time domain, here works as follows. (a) Spatially whiten the data $\mathbf{X}(\omega)$ to obtain $\mathbf{Z}(\omega)$. (b) Compute the eigenvalue decomposition of $\overline{\mathbf{C}_{2\omega_0}^Z} = \frac{1}{2}[\mathbf{C}_{2\omega_0} + \mathbf{C}_{2\omega_0}^T]$, where $\mathbf{C}_{2\omega_0} = E[\mathbf{Z}(\omega + 2\omega_0)\mathbf{Z}^*(\omega)]$ is the covariance matrix corresponding to lag $2\omega_0$. (c) The rows of the separating matrix $\mathbf{B}$ are given by the eigenvectors of $\overline{\mathbf{C}_{2\omega_0}^Z}$. Using the same signals as in the first step, the mean and the standard deviation of SNR were 41.6 dB and 7.2 dB. Other experiments with different profiles of nonstationary variance for the sources $s_1(t)$ and $s_2(t)$ led to similar results.

In the second experiment, the above algorithm based on AMUSE was used for separating mixtures of speech signals. Three tests using three couples of 44100-sample speech signals led to an average SNR of 40.6 dB. This experiment shows that although Property 4 is derived for temporally uncorrelated signals, the proposed method works well also for temporally correlated signals.

5 Conclusion

A major objective of this paper was to demonstrate and exploit some theoretically interesting frequency-domain properties of signals which are nonstationary in the time domain. These properties provide sufficient second-order constraints in the frequency domain for separating instantaneous linear mixtures of nonstationary sources.

Two separating methods were proposed based on Properties 1 and 4. The first method is theoretically interesting but its implementation is difficult unless either many realizations of the mixtures are available or the sources are cyclostationary. The second method is very simple and powerful because it allows the time-domain algorithms developed for stationary time-correlated signals to be applied to temporally uncorrelated sources which are nonstationary in the time domain, just by mapping them in the frequency domain. It should be remarked that this algorithm does not require the variance of the sources to be constant over subintervals, while this hypothesis is necessary in the majority of the source separation algorithms based on the nonstationarity of variance which have been reported in the literature.

A Proofs of the Properties of Section 2

Proof of Property 1: Consider two mutually uncorrelated zero-mean real signals $u_1(t)$ and $u_2(t)$, with Fourier transforms $U_1(\omega)$ and $U_2(\omega)$. We can write:

$$E[U_1(\omega)U_2(\omega)] = \int_{-\infty}^{\infty}\int_{-\infty}^{\infty} E[u_1(t_1)u_2(t_2)]e^{-j\omega(t_1+t_2)}dt_1dt_2 = 0$$

$$E[U_1(\omega)U_2^*(\omega)] = \int_{-\infty}^{\infty}\int_{-\infty}^{\infty} E[u_1(t_1)u_2(t_2)]e^{-j\omega(t_1-t_2)}dt_1dt_2 = 0$$

because $E[u_1(t)u_2(t)] = 0$.

Proof of Property 2: Let $u(t)$ be a real stationary signal with Fourier transform $U(\omega)$. We want to show that $E[U^2(\omega)] = 0$, for $\omega \neq 0$. Using the definition of the Fourier transform, we can write

$$E[U^2(\omega)] = \int_{-\infty}^{\infty}\int_{-\infty}^{\infty} E[u(t_1)u(t_2)]e^{-j\omega(t_1+t_2)}dt_1dt_2$$

Since $u(t)$ is stationary, its autocorrelation function depends only on $t_1 - t_2$: $E[u(t_1)u(t_2)] = R(t_1 - t_2)$. Denoting the auxiliary variable $\tau = t_1 - t_2$,

$$E[U^2(\omega)] = \int_{-\infty}^{\infty}\int_{-\infty}^{\infty} R(\tau)e^{-j\omega(2t_2+\tau)}d\tau dt_2$$

$$= \int_{-\infty}^{\infty} e^{-j2\omega t_2} \int_{-\infty}^{\infty} R(\tau)e^{-j\omega\tau}d\tau dt_2$$

The inner integral represents the power spectral density of $u(t)$, denoted by $\Gamma(\omega)$. Thus $E[U^2(\omega)] = \Gamma(\omega) \int_{-\infty}^{\infty} e^{-j2\omega t_2} dt_2 = 2\pi \Gamma(\omega) \delta(2\omega)$, which yields $E[U^2(\omega)] = 0$ for $\omega \neq 0$.

Proof of Property 3: Suppose $V_1(\omega) = a_{11} U_1(\omega) + a_{12} U_2(\omega)$ and $V_2(\omega) = a_{21} U_1(\omega) + a_{22} U_2(\omega)$. We can write

$$E[V_1^2(\omega)] = a_{11}^2 E[U_1^2(\omega)] + a_{12}^2 E[U_2^2(\omega)] + 2a_{11}a_{12} E[U_1(\omega)U_2(\omega)]$$
$$E[V_2^2(\omega)] = a_{21}^2 E[U_1^2(\omega)] + a_{22}^2 E[U_2^2(\omega)] + 2a_{21}a_{22} E[U_1(\omega)U_2(\omega)]$$
$$E[V_1(\omega)V_2(\omega)] = a_{11}a_{21} E[U_1^2(\omega)] + a_{12}a_{22} E[U_2^2(\omega)] + (a_{11}a_{22} + a_{12}a_{21})$$
$$E[U_1(\omega)U_2(\omega)]$$

Since $u_1(t)$ and $u_2(t)$ are real, zero-mean, uncorrelated and stationary, the first two terms of the right side of all the above equations vanish for $\omega \neq 0$ following Property 2, and the third term of all the equations vanishes whatever ω following Property 1.

Proof of Property 4: (see also [10]) If $E[u(t_1)u(t_2)] = q(t_1)\delta(t_1 - t_2)$, where $\delta(t_1 - t_2)$ is a Dirac distribution, then

$$E[U(\Omega + \omega)U^*(\Omega)] = \int_{-\infty}^{\infty} \int_{-\infty}^{\infty} E[u(t_1)u(t_2)] e^{-j(\Omega+\omega)t_1} e^{j\Omega t_2} dt_1 dt_2$$
$$= \int_{-\infty}^{\infty} \int_{-\infty}^{\infty} q(t_1) \delta(t_1 - t_2) e^{-j\Omega(t_1-t_2)} e^{-j\omega t_1} dt_1 dt_2 \quad (5)$$

Since $\delta(t_1 - t_2) e^{-j\Omega(t_1-t_2)} = \delta(t_1 - t_2)$,

$$E[U(\Omega + \omega)U^*(\Omega)] = \int_{-\infty}^{\infty} q(t_1) e^{-j\omega t_1} \int_{-\infty}^{\infty} \delta(t_1 - t_2) dt_2 dt_1$$
$$= \int_{-\infty}^{\infty} q(t_1) e^{-j\omega t_1} dt_1 = Q(\omega) \quad (6)$$

B Derivation of Equation (3)

For the sake of simplicity, we omit the parameter ω in the following notations. Developing Equations (2), we obtain[4]:

$$b_1 b_2 E[X_1 X_1^*] + (b_1 + b_2) E[X_1 X_2^*] + E[X_2 X_2^*] = 0 \quad (7)$$

$$b_1 b_2 E[X_1^2] + (b_1 + b_2) E[X_1 X_2] + E[X_2^2] = 0 \quad (8)$$

From (8), $b_2 = \frac{-b_1 E[X_1 X_2] - E[X_2^2]}{b_1 E[X_1^2] + E[X_1 X_2]}$. Replacing b_2 in (7), we obtain:

$$\frac{-b_1 E[X_1 X_2] - E[X_2^2]}{b_1 E[X_1^2] + E[X_1 X_2]} (b_1 E[X_1 X_1^*] + E[X_1 X_2^*]) + (b_1 E[X_1 X_2^*] + E[X_2 X_2^*]) = 0$$

---

[4] Note that $E[X_1 X_2^*] = E[X_2 X_1^*]$, because X_1 and X_2 are linear combinations of two spectra S_1 and S_2, and $E[S_1 S_2^*] = E[S_1^* S_2] = 0$, following Property 1.

which yields:

$$(-b_1 E[X_1 X_2] - E[X_2^2])(b_1 E[X_1 X_1^*] + E[X_1 X_2^*]) + (b_1 E[X_1 X_2^*] + E[X_2 X_2^*])$$
$$(b_1 E[X_1^2] + E[X_1 X_2]) = 0$$

Developing the above equation leads to the second-order equation (3), for which b_1 is a real solution. Note that the two equations (7) and (8) are symmetrical with respect to b_1 and b_2. This implies that b_2 is also a real solution of (3). This result is not surprising because the sources may be estimated only up to a permutation.

References

1. J.-F. Cardoso, The three easy routes to independent component analysis: contrast and geometry, in *Proc. ICA2001*, San Diego, 2001, pp. 1-6.
2. K. Matsuoka, M. Ohya, and Mitsuru Kawamoto, A neural net for blind separation of nonstationary signals, *Neural Networks*, vol. 8, no. 3, pp. 411-419, 1995.
3. A. Souloumiac, Blind source detection and separation using second-order nonstationarity, in *Proc. ICASSP*, 1995, pp. 1912-1915.
4. A. Hyvarinen, Blind source separation by nonstationarity of variance: a cumulant based approach, *IEEE Trans. on Neural Networks*, 12(6), pp. 1471-1474, 2001.
5. A. Belouchrani, and M. G. Amin, Blind source separation based on time-frequency signal representation, *IEEE Trans. on Signal Processing*, 46(11), November 1998.
6. F. Abrard, Y. Deville, P. White, From blind source separation to blind source cancellation in the underdetermined case: a new approach based on time-frequency analysis, in *Proc. ICA2001*, pp. 734-739, San Diego, USA, 2001.
7. Y. Deville, Temporal and time-frequency correlation-based blind source separation methods, in *Proc. ICA2003*, pp. 1059-1064, Nara, Japan, 2003.
8. B. Albouy, Y. Deville, A time-frequency blind source separation method based on segmented coherence function, in *Proc. IWANN2003*, vol. 2, pp. 289-296, Mao, Menorca, Spain, 2003.
9. D.-T. Pham, and J.-F. Cardoso, Blind separation of independent mixtures of nonstationary sources, *IEEE Trans. on Signal Processing*, 49(9), 2001.
10. A. Papoulis, and S. U. Pillai, *Probability, random variables and stochastic processes*, 4th Ed., McGraw-Hill, 2002.
11. L. Tong, and V. Soon, Indeterminacy and identifiability of blind identification, *IEEE Trans. Circuits Syst.*, vol. 38, pp. 499-509, May 1991.
12. L. Molgedey, and H. G. Schuster, Separation of a mixture of independent signals using time delayed correlation, *Physical Review Letters*, vol. 72, pp. 3634-3636, 1994.
13. A. Belouchrani, K. Abed Meraim, J.-F. Cardoso, and E. Moulines, A blind source separation technique based on second order statistics, *IEEE Trans. on Signal Processing*, vol. 45, pp. 434-444, Feb. 1997.
14. A. Ziehe, and K. R. Muller, TDSEP - an efficient algorithm for blind separation using time structure, in *Proceedings of Int. Conf. on Artificial Neural Networks*, Skovde, Sweden, pp. 675-680, 1998.
15. S. Degerine, and R. Malki, Second order blind separation of sources based on canonical partial innovations, *IEEE Trans. on Signal Processing*, vol. 48, pp. 629-641, 2000.
16. S. Hosseini, C. Jutten, and D.-T. Pham, Markovian source separation, *IEEE Trans. on Signal Processing*, vol. 51, pp. 3009-3019, 2003.

Delayed AMUSE – A Tool for Blind Source Separation and Denoising

Ana R. Teixeira[1], Ana Maria Tomé[1], Elmar W. Lang[2], and Kurt Stadlthanner[2]

[1] Departamento de Electrónica e Telecomunicações/IEETA
Universidade de Aveiro, P-3810 Aveiro, Portugal
ana@ieeta.pt

[2] Institute of Biophysics, Neuro- and Bioinformatics Group
University of Regensburg, D-93040 Regensburg, Germany
elmar.lang@biologie.uni-regensburg.de

Abstract. In this work we propose a generalized eigendecomposition (GEVD) of a matrix pencil computed after embedding the data into a high-dim feature space of delayed coordinates. The matrix pencil is computed like in AMUSE but in the feature space of delayed coordinates. Its GEVD yields filtered versions of the source signals as output signals. The algorithm is implemented in two EVD steps. Numerical simulations study the influence of the number of delays and the noise level on the performance.

1 Introduction

Blind Source Separation (BSS) methods consider the separation of observed sensor signals into their underlying independent source signals knowing neither these source signals nor the mixing process. BSS methods using second order statistics only can be based on a generalized eigendecomposition (GEVD) of a matrix pencil. They are exact and efficient but sensitive to noise [1].

There are several proposals to improve efficiency and robustness of these algorithms when noise is present[1], [2] which mostly rely on an approximative *joint* diagonalization of a set of correlation or cumulant matrices. Also there exist local projective de-noising techniques which in a first step increase the dimension of the data by joining delayed versions of the signals [3], [4], [5] hence projecting them into a high-dimensional feature space. A similar strategy is used in Singular Spectrum Analysis (SSA) [6] where a matrix composed of the data and their time-delayed versions is considered. Then, a Singular Value Decomposition(SVD) of the data matrix or a Principal Component Analysis (PCA) of the related correlation matrix is computed. The data are then projected onto the principal directions of the eigenvectors of the SVD or PCA analysis. The SSA was used to extract information from short and noisy time series and then provide insight into the underlying system that generates the series [7].

In this work we combine the ideas of solving BSS problems algebraically using a GEVD with local projective denoising techniques. We propose, like in AMUSE, a GEVD of two correlation matrices i.e, the *simultaneous* diagonalization of a matrix pencil formed with a correlation matrix and a matrix of time-delayed correlations. But the proposed algorithm, called dAMUSE, computes the pencil in a high-dimensional feature space of time-delayed coordinates.

In the following section we show, starting from a noise-free model of linearly mixed sensor signals, that the estimated independent signals correspond to filtered versions of the underlying source signals. We also present an algorithm to compute the eigenvector matrix of the pencil which involves a two step procedure based on the standard eigendecomposition (EVD) approach. The advantage of this procedure is concerned with a dimension reduction between the two steps as well as a reduction in the number of independent signals, thus performing a denoising of the estimated underlying source signals. Finally, simulations with artificially mixed signals are discussed to illustrate the proposed method.

2 Generalized Eigendecomposition Using Time-Delayed Signals

Considering the sensor signals x_i, the trajectory matrix [6] of the sensor signals computed for a set of L samples is given by

$$X_i = \begin{bmatrix} x_i[M-1] & x_i[M] & x_i[M+1] & \cdots & x_i[L-1] \\ x_i[M-2] & x_i[M-1] & x_i[M] & \cdots & x_i[L-2] \\ \vdots & \vdots & \ddots & \cdots & \vdots \\ x_i[0] & x_i[1] & x_i[2] & \cdots & x_i[L-M] \end{bmatrix} \quad (1)$$

and encompasses M delayed versions of the signal x_i. Given a group of N sensor signals, x_i, $i = 1 \ldots N$, the trajectory matrix of the set will be a concatenation of the component trajectory matrices X_i computed for each sensor, i.e

$$X = \begin{bmatrix} X_1, X_2, \ldots X_N \end{bmatrix}^T \quad (2)$$

Assuming that each sensor signal is a linear combination $X = HS$ of N underlying but unknown source signals (s_i), a source signal trajectory matrix S can be written in analogy to eqn(1) and eqn(2). Then the mixing matrix (H) is a block matrix with a diagonal matrix in each block

$$H = \begin{bmatrix} h_{11}I_{M \times M} & h_{12}I_{M \times M} & \cdots & h_{1N}I_{M \times M} \\ h_{21}I_{M \times M} & h_{22}I_{M \times M} & \cdots & \\ \vdots & \vdots & \ddots & \vdots \\ h_{N1}I_{M \times M} & \cdots & \cdots & h_{NN}I_{M \times M} \end{bmatrix} \quad (3)$$

The matrix I_{MxM} represents the identity matrix and in accord with an instantaneous mixing model the mixing coefficient h_{ij} relates the sensor signal i with the source signal j.

The time-delayed correlation matrices of the matrix pencil are computed with one matrix (X_r) obtained by eliminating the first k_i columns of X and another matrix, (X_l), obtained by eliminating the last k_i columns. Then, the time-delayed correlation matrix $R_x(k_i) = X_r X_l^T$ will be an $NM \times NM$ matrix. Each of these two matrices can be related with a corresponding matrix in the source signal domain

$$R_x(k_i) = H R_s(k_i) H^T = H S_r S_l^T H^T \tag{4}$$

Then the two pairs of matrices $(R_x(k_1), R_x(k_2))$ and $(R_s(k_1), R_s(k_2))$ represent congruent pencils [8] with the following properties:

- Their eigenvalues are the same, i.e., the eigenvalue matrices of both pencils are identical: $D_x = D_s$.
- If the eigenvalues are non-degenerate (distinct values in the diagonal of the matrix $D_x = D_s$), the corresponding eigenvectors are related by the transformation $E_s = H^T E_x$.

Assuming that all sources are uncorrelated, the matrices $R_s(k_i)$ are block diagonal, having block matrices $R_{mm}(k_i) = S_{ri} S_{li}^T$ along the diagonal

$$R_s(k_i) = \begin{bmatrix} R_{11}(k_i) & 0 & \cdots & 0 \\ 0 & R_{22}(k_i) & \cdots & 0 \\ \vdots & \vdots & \ddots & \vdots \\ 0 & 0 & \cdots & R_{NN}(k_i) \end{bmatrix}$$

The eigenvector matrix of the GEVD of the pencil $(R_s(k_1), R_s(k_2))$ can be written as

$$E_s = \begin{bmatrix} E_{11} & 0 & \cdots & 0 \\ 0 & E_{22} & \cdots & 0 \\ \vdots & \vdots & \ddots & \vdots \\ 0 & 0 & \cdots & E_{NN} \end{bmatrix} \tag{5}$$

where E_{mm} is the $M \times M$ eigenvector matrix of the GEVD of the pencil $(R_{mm}(k_1), R_{mm}(k_2))$. The independent components can be estimated from linearly transformed sensor signals via

$$Y = E_x^T X = E_x^T H S = E_s^T S$$

hence turn out to be filtered versions of the underlying source signals. As the eigenvector matrix E_s (eqn. 5) is a block diagonal matrix, there are M signals in each column of Y which are a linear combination of one of the source signals and its delayed versions. For instance, the block m depends on the source signal m via

$$\sum_{k=1}^{M} E_{mm}(k,j) s_m(n-k+1) \tag{6}$$

Equation (6) defines a convolution operation between column j of E_{mm} and source signal s_m. Then the columns of the matrix E_{mm} represent impulse responses of finite impulse response (FIR) filters. Considering that all the columns of E_{mm} are different, their frequency response might provide different spectral densities of the source signal spectra. Then NM output signals y encompass M filtered versions of each of the N estimated source signals.

3 Implementation of the Algorithm

There are several ways to compute the generalized eigendecomposition. We resume a procedure valid if one of the matrices of the pencil is symmetric positive definite. Thus, we consider the pencil $(R_x(0), R_x(k_2))$ and perform the following steps:

Step 1: Compute a standard eigendecomposition of $R_x(0) = \vee \wedge \vee^T$, i.e, compute the eigenvectors (ν_i) and eigenvalues (λ_i). As the matrix is symmetric positive definite, the eigenvalues can be arranged in descending order ($\lambda_1 > \lambda_2 > ... > \lambda_{NM}$). In AMUSE (and other algorithms) this procedure is used to estimate the number of sources. But it can also be considered a strategy to reduce noise. Dropping small eigenvalues amounts to a projection from a high-dim signal plus noise feature space onto a lower dimensional manifold representing the signal+noise subspace. Thereby it is tacitly assumed that small eigenvalues are related with noise components only. In this work, we consider a variance criterium to choose the most significant eigenvalues, those related with the embedded deterministic signal, according to

$$\frac{\lambda_1 + \lambda_2 + ... + \lambda_l}{\lambda_1 + \lambda_2 + ... \lambda_{NM}} \geq TH \qquad (7)$$

If we are interested in the eigenvectors corresponding to directions of high variance of the signals, the threshold (TH) should be chosen such that their maximum energy is preserved. The transformation matrix can thus be computed using either the l most significant eigenvalues, in which case denoising is achieved, or all eigenvalues and respective eigenvectors. Similar to the whitening phase in many BSS algorithms the data matrix X can be transformed using

$$Q = \wedge^{-\frac{1}{2}} \vee^T \qquad (8)$$

to calculate a transformed matrix of time-delayed correlations $C(k_2)$ to be used in the next step. Also note, that Q represents a $l \times NM$ matrix if denoising is considered.

Step 2: Using the transformed time-delayed correlation matrix $C(k_2) = QR_x(k_2)Q^T$ and its standard eigendecomposition: the eigenvector matrix (U) and eigenvalue matrix (D_x).

The eigenvectors of the pencil $(R_x(0), R_x(k_2))$, which are not normalized, form the columns of the eigenvector matrix $E_x = Q^T U = \vee \wedge^{-\frac{1}{2}} U$. The independent components of the time-delayed sensor signals can then be estimated via the transformation given below, yielding l (or NM) signals, one signal per row of Y

$$Y = E_x^T X = U^T Q X = U^T \wedge^{-\frac{1}{2}} \vee^T X \qquad (9)$$

The first step of this algorithm is thus equivalent to a PCA in a high-dimensional feature space [4], [7] where a matrix similar to Q is used to project the data onto the signal manifold.

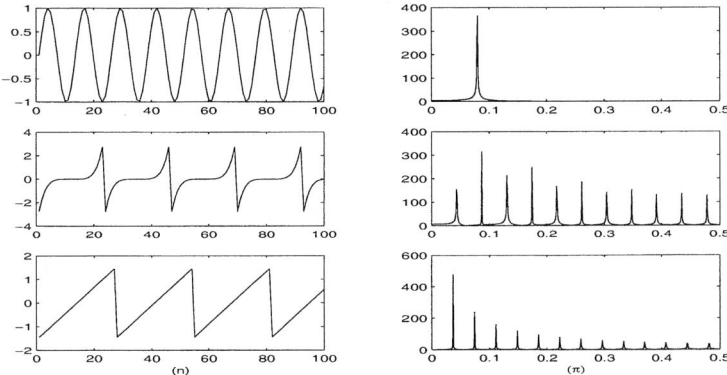

Fig. 1. Artificial signals (left column) and their frequency contents (right column).

4 Numerical Simulations

A group of three source signals with different frequency contents was chosen: one member of the group represents a narrow-band signal, a sinusoid; the second signal encompasses a wide frequency range; and the last one represents a sawtooth wave whose spectral density is concentrated in the low frequency band (see Fig. 1). The simulations were designed to illustrate the method and to study the influence of its parameters (especially M and TH) on the respective performance. As the separation process also involves a linear filtering operation (eqn. (6), each computed independent component has its maximum correlation with one of the source signals for a non-zero delay, besides the usual indeterminacy in order and amplitude. Two experiments will be discussed: a) one changing the number of delays M and the other b) adding noise with different levels. In what concerns noise we also try to find out if there is any advantage of using a GEVD instead of a PCA analysis. Hence the signals at the output of the first step of the algorithm (using the matrix Q to project the data) are also compared with the output signals.

After randomly mixing the source signals, the algorithm was applied for different values of M, with $TH = 0.95$, $k_1 = 0$ and $k_2 = 1$ fixed. In that case for any value of M the number of signals after the first step (or the dimension of matrix C) is $l = 6 < NM$, because the threshold TH eliminates the very low eigenvalues. Even though, the number of output signals is higher than the number of source signals, thus only 3 output signals which have the highest correlations (in the frequency domain) with any of the source signals will be considered in the following. It can be verified easily that upon increasing the number M of delays the estimated independent signals decrease their bandwidth (except for the sinusoid). Fig. 2 shows that source 2 has less components when M increases. The effect is also visible in source 3 but here the time domain characteristics of the wave are less affected as is to be expected.

The second experiment is related with the influence of the threshold parameter TH when noisy signals considered. First random noise was added to the

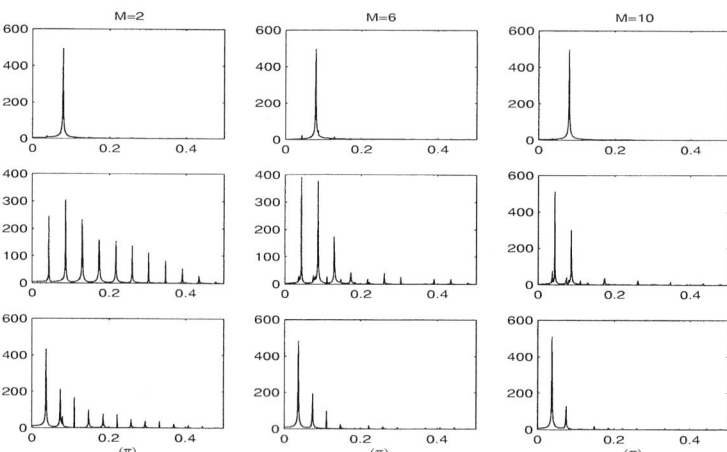

Fig. 2. Frequency contents of the output signals considering different time-delays M to form the input data matrix, hence the embedding dimension of the feature space.

Table 1. Number of output signals correlated with noise or source signals in the steps of the algoritm dAMUSE.

		1st step		2nd step		
SNR	NM	Sources	Noise	Sources	Noise	Total
20dB	12	6	0	6	0	6
15dB	12	5	2	6	1	7
10dB	12	6	2	7	1	8
5dB	12	6	3	7	2	9
0dB	12	7	4	8	3	11

sensor signals yielding a SNR in the range of $[0, 20]dB$. The parameters $M = 4$ and $TH = 0.95$ were kept fixed. As the noise level increases the number of significant eigenvalues also increases, hence at the output of the first step more signals need to be considered. Thus as the noise energy increases, the number of signals (l) or the dimension of matrix C after the application of the first step also increases (last column of table 1). As the noise increases an increasing number of independent components will be available at the output of the two steps. Computing, in the frequency domain, the correlation coefficients between the output signals of each step of the algorithm and noise or source signals we confirm that some are related with the sources and others with noise. Table 1 (columns 3-6) shows that the maximal correlation coefficients are distributed between noise and source signals to a varying degree. We can see that the number of signals correlated with noise (table 1) is always higher in the first level. Results show that for low noise levels the first step (which is mainly a principal component analysis in a space of dimension NM) achieves good solutions already. However, we can also see (for narrow-band signals and/or M low) that the time domain characteristics of the signals resemble the original source signals only after a

GEVD, i.e. at the output of the second step rather than with a PCA, i.e. at the output of first step. Fig. 3 shows examples of signals that have been obtained in the two steps of the algorithm for $SNR = 10dB$. At the output of first level the 3 signals with highest frequency correlation were chosen among the 8 output signals. Using a similar criterium to choose 3 signals at the output of the 2nd step (last column of Fig. 3), we can see that their time course is more similar to the source signals than after the first step (middle column of Fig. 3)

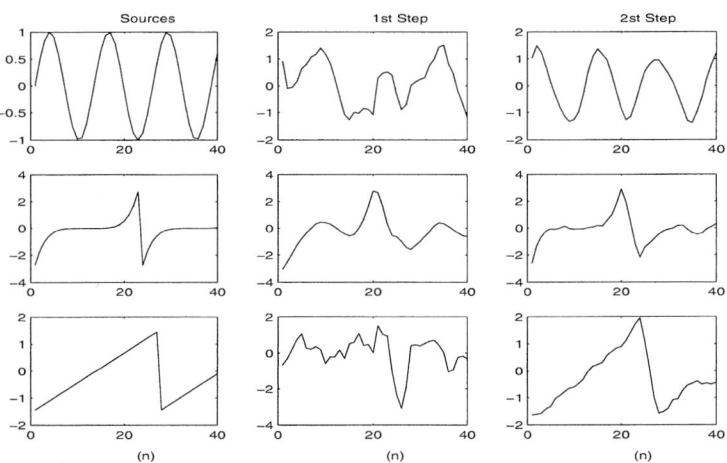

Fig. 3. Comparison of output signals resulting after the first step (second column) and the second step (last column) of dAMUSE.

5 Conclusions

In this work we propose dAMUSE, a modified version of the algorithm AMUSE. The new algorithm is based on a GEVD of a matrix pencil computed with time-delayed sensor signals. It was shown that the independent components estimated with dAMUSE represent filtered versions of the underlying source signals. The algorithm has a set of parameters whose proper choice must be further studied, particularly the number of time-delays M used to build the embedding feature space. Its choice naturally constrains the linear filtering operation that characterizes the method as it was shown in the simulations. The simulations also reveal that denoising cannot completely be achieved by PCA alone. Having at the output a high number of estimated independent components it is also important to find an automatic procedure to choose the relevant components related to the signal as some components might be related with noise only. Another aspect is to find out how the different signals can be joined together if a wide-band signal needs to be reconstructed. The choice of time-delayed matrices to compute the pencil was done according to the algorithm AMUSE as well as the values for time-delays (k_1 and k_2)[1]. Nevertheless other second-order statistics algorithms [9] [10] might be considered as well as they achieved better results when compared with AMUSE [9]. The algorithm was used to study Electrocardiograms and some of the independent signals could be related with the LF (low frequency) and HF (high frequency) fluctuations used to characterize HRV (heart rate variability) studies [11].

References

1. A. Cichocki and S.-I. Amari, *Adaptive Blind Signal and Image Processing*. Wiley, 2002.
2. R. Gharieb and A. Cichocki, "Second-order statistics based blind source separation using a bank of subband filters," *Digital Signal Processing*, vol. 13, pp. 252–274, 2003.
3. T. Schreiber and H. Kantz, "Nonlinear projective filtering II: Application to real time series," in *Proceedings of Nolta*, 1998.
4. R. Vetter, J. Vesin, P. Celka, P. Renevey, and J. Krauss, "Automatic nonlinear noise reduction using local principal component analysis and MDL parameter selection," in *IASTED Internation Conference on Signal Processing, Pattern Recognition&Applications*, (Crete, Greece), pp. 290–294, 2002.
5. P. Gruber, F. J. Theis, A. M. Tomé, and E. W. Lang, "Automatic denoising using local independent component analysis," in *Proceed. EIS'2004*, (Madeira, Portugal), 2004.
6. V. Moskvina and K. M. Schmidt, "Approximate projectors in singular spectrum analysis," *SIAM Journal Mat. Anal. Appl.*, vol. 24, no. 4, pp. 932–942, 2003.
7. M. Ghil, M. Allen, M. D. Dettinger, K. Ide, and e. al, "Advanced spectral methods for climatic time series," *Reviews of Geophysics*, vol. 40, no. 1, pp. 3.1–3.41, 2002.
8. A. M. Tomé and N. Ferreira, "On-line source separation of temporally correlated signals," in *European Signal Processing Conference, EUSIPCO2002*, (Toulouse, France), 2002.
9. A. M. Tomé, "Separation of a mixture of signals using linear filtering and second order statistics," in *10th European Symposium on Artificial Neural Networks*, (Brugges), pp. 307–312, 2002.
10. J. V. Stone, "Blind source separation using temporal predictability," *Neural Computation*, vol. 13, no. 7, pp. 1557–1574, 2001.
11. A. R. Teixeira, A. P. Rocha, R. Almeida, and A. M. Tomé, "The analysis of heart rate variability using independent component signals," in *Second International Conference on Biomedical Engineering*, (Innsbruck, Austria), pp. 240–243, IASTED, 2004.

Dimensionality Reduction in ICA and Rank-$(R_1, R_2, \ldots, R_N)$ Reduction in Multilinear Algebra*

Lieven De Lathauwer[1,2] and Joos Vandewalle[2]

[1] ETIS (CNRS, ENSEA, UCP), UMR 8051, Cergy-Pontoise, France
delathau@ensea.fr
[2] E.E. Dept. (ESAT) - SCD, K.U.Leuven, Leuven, Belgium
vdwalle@esat.kuleuven.ac.be

Abstract. We show that the best rank-$(R_1, R_2, \ldots, R_N)$ approximation in multilinear algebra is a powerful tool for dimensionality reduction in ICA schemes without prewhitening. We consider the application to different classes of ICA algorithms.

1 Introduction

In this paper, the ICA-model is denoted as follows:

$$X = \mathbf{M}S + E, \qquad (1)$$

in which the observations $X \in \mathbb{R}^I$, the sources $S \in \mathbb{R}^R$ and the noise $E \in \mathbb{R}^I$ are zero-mean random vectors. The components of S are assumed to be mutually statistically independent, as well as statistically independent from the noise components. The mixing matrix $\mathbf{M}$ is full column rank. The goal of ICA consists of the estimation of the mixing matrix and/or the corresponding realizations of S, given only realizations of X. For notational convenience we only consider real-valued signals. The generalization to complex signals is straightforward.

Many ICA applications involve high-dimensional data in which however only few sources have significant contributions. Examples are electro-encephalography (EEG), magneto-encephalography (MEG), nuclear magnetic resonance (NMR), hyper-spectral image processing, data analysis, etc. To reduce the computational complexity and to decrease the variance of the results [1], one may wish to reduce the dimensionality of the problem from the number of observation channels, I, to the number of sources, R. In this paper we will discuss algebraic means to perform the dimensionality reduction, other than a classical prewhitening. The

---

* L. De Lathauwer holds a permanent research position with the French C.N.R.S.; he also holds a honorary position with the K.U.Leuven. J. Vandewalle is a Full Professor with the K.U.Leuven. Part of this research was supported by the Research Council K.U.Leuven (GOA-MEFISTO-666), the Flemish Government (F.W.O. project G.0240.99, F.W.O. Research Communities ICCoS and ANMMM, Tournesol project T2004.13) and the Belgian Federal Government (IUAP V-22).

approach makes use of multilinear algebra, which is the algebra of higher-order tensors. Higher-order tensors are the higher-order equivalents of vectors (first order) and matrices (second order), i.e., quantities of which the elements are addressed by more than two indices.

Algebraic prerequisites are introduced in Sect. 2. The application to dimensionality reduction in ICA forms the subject of Sect. 3. Section 4 illustrates the approach by means of two simulations. Section 5 is the conclusion. The algebraic tools are extensively described in [7, 8, 14, 18]. The application to ICA is discussed in [9].

Notation. Scalars are denoted by lower-case letters (a, b, ...), vectors by capitals (A, B, ...), matrices by bold-face capitals (**A**, **B**, ...) and tensors by calligraphic letters ($\mathcal{A}$, $\mathcal{B}$, ...). In this way, the entry with row index i and column index j in a matrix **A**, i.e., $(\mathbf{A})_{ij}$, is symbolized by a_{ij}. However, I, J, N, R denote the upper bounds of indices i, j, n, r. A_j stands for the jth column of **A**. St(R, I) is standard notation for the Stiefel manifold of column-wise orthonormal ($I \times R$) matrices.

2 Basics of Multilinear Algebra

2.1 Elementary Definitions

Consider a tensor $\mathcal{A} \in \mathbb{R}^{I_1 \times I_2 \cdots \times I_N}$ and matrices $\mathbf{U}^{(n)} \in \mathbb{R}^{J_n \times I_n}$, $1 \leqslant n \leqslant N$. Then $\mathcal{B} = \mathcal{A} \times_1 \mathbf{U}^{(1)} \times_2 \mathbf{U}^{(2)} \ldots \times_N \mathbf{U}^{(N)}$ is a ($J_1 \times J_2 \ldots \times J_N$)-tensor of which the entries are given by [18]

$$b_{j_1 j_2 \ldots j_N} = \sum_{i_1 i_2 \ldots i_N} a_{i_1 i_2 \ldots i_N} u^{(1)}_{j_1 i_1} u^{(2)}_{j_2 i_2} \ldots u^{(N)}_{j_N i_N}.$$

In this notation we have $\mathbf{B} = \mathbf{U}^{(1)} \cdot \mathbf{A} \cdot \mathbf{U}^{(2)T} = \mathbf{A} \times_1 \mathbf{U}^{(1)} \times_2 \mathbf{U}^{(2)}$. The *Frobenius-norm* of $\mathcal{A}$ is defined as $\|\mathcal{A}\| = (\sum_{i_1 \ldots i_N} a^2_{i_1 \ldots i_N})^{1/2}$. There are major differences between matrices and tensors when rank properties are concerned. A *rank-1 tensor* is a tensor that consists of the outer product of a number of vectors. For an Nth-order tensor $\mathcal{A}$ and N vectors $U^{(1)}$, $U^{(2)}$, ..., $U^{(N)}$, this means that $a_{i_1 i_2 \ldots i_N} = u^{(1)}_{i_1} u^{(2)}_{i_2} \ldots u^{(N)}_{i_N}$ for all index values, which will be concisely written as $\mathcal{A} = U^{(1)} \circ U^{(2)} \circ \ldots \circ U^{(N)}$. An *n-mode vector* of $\mathcal{A}$ is an I_n-dimensional vector obtained from $\mathcal{A}$ by varying the index i_n and keeping the other indices fixed. In the matrix case, 1-mode vectors are columns and 2-mode vectors are rows. The *n-rank* of a tensor is the obvious generalization of the column (row) rank of matrices: it is defined as the dimension of the vector space spanned by the n-mode vectors. If all the n-mode vectors of a tensor $\mathcal{A}$ are stacked in a matrix $\mathbf{A}_{(n)}$, then the n-rank of $\mathcal{A}$ is equal to the rank of $\mathbf{A}_{(n)}$. In contrast to the matrix case, the different n-ranks of a higher-order tensor are not necessarily the same. A tensor of which the n-ranks are equal to R_n ($1 \leqslant n \leqslant N$) is called a *rank-($R_1, R_2, \ldots, R_N$) tensor*. A *rank-R tensor* is defined in yet an other way: it is a tensor that can be decomposed in a sum of R, but not less than R, rank-1 terms.

2.2 Best Rank-$(R_1, R_2, \ldots, R_N)$ Approximation

We consider the minimization of the least-squares cost function

$$f(\hat{\mathcal{A}}) = \|\mathcal{A} - \hat{\mathcal{A}}\|^2 \tag{2}$$

under the constraint that $\hat{\mathcal{A}}$ is rank-$(R_1, R_2, \ldots, R_N)$ [8, 14]. The n-rank conditions imply that $\hat{\mathcal{A}}$ can be decomposed as

$$\hat{\mathcal{A}} = \mathcal{B} \times_1 \mathbf{U}^{(1)} \times_2 \mathbf{U}^{(2)} \ldots \times_N \mathbf{U}^{(N)}, \tag{3}$$

in which $\mathbf{U}^{(n)} \in \mathrm{St}(R_n, I_n)$, $1 \leqslant n \leqslant N$, and $\mathcal{B} \in \mathbb{R}^{R_1 \times R_2 \times \ldots \times R_N}$.

Similarly to the second-order case, where the best approximation of a given matrix $\mathbf{A} \in \mathbb{R}^{I_1 \times I_2}$ by a matrix $\hat{\mathbf{A}} = \mathbf{U}^{(1)} \cdot \mathbf{B} \cdot \mathbf{U}^{(2)^T}$, with $\mathbf{U}^{(1)} \in \mathrm{St}(R, I_1)$ and $\mathbf{U}^{(2)} \in \mathrm{St}(R, I_2)$, is equivalent to the maximization of $\|\mathbf{U}^{(1)^T} \cdot \mathbf{A} \cdot \mathbf{U}^{(2)}\|$, the minimization of f is equivalent to the maximization of

$$g(\mathbf{U}^{(1)}, \mathbf{U}^{(2)}, \ldots, \mathbf{U}^{(N)}) = \|\mathcal{A} \times_1 \mathbf{U}^{(1)^T} \times_2 \mathbf{U}^{(2)^T} \ldots \times_N \mathbf{U}^{(N)^T}\|^2. \tag{4}$$

The optimal tensor $\mathcal{B}$ follows from

$$\mathcal{B} = \mathcal{A} \times_1 \mathbf{U}^{(1)^T} \times_2 \mathbf{U}^{(2)^T} \ldots \times_N \mathbf{U}^{(N)^T}. \tag{5}$$

It is natural to ask whether, in analogy with the matrix case, the best rank-$(R_1, R_2, \ldots, R_N)$ approximation of a higher-order tensor can be obtained by truncation of a multilinear generalization of the SVD. The situation turns out to be quite different for tensors. Truncation of the Higher-Order Singular Value Decomposition (HOSVD) discussed in [7, 18] generally yields a good but not the optimal approximation. The latter has to be computed by means of tensor generalizations of algorithms for the computation of the dominant subspace of a given matrix. In [8, 14] a higher-order generalization of the orthogonal iteration method is discussed. In [12] we present a higher-order Grassmann-Rayleigh quotient iteration. These algorithms can be initialized by means of the truncated HOSVD components. This means that the columns of a first estimate of $\mathbf{U}^{(n)}$ are determined as an orthonormal basis for the R_n-dimensional dominant subspace of the column space of $\mathbf{A}_{(n)}$, defined in Sect. 2.1 ($1 \leqslant n \leqslant N$). The subsequent optimization can be quite efficient too: in the higher-order Grassmann-Rayleigh quotient iteration, an iteration step consists of solving a square set of linear equations and the convergence is quadratic.

3 Dimensionality Reduction in ICA

3.1 Higher-Order-Only ICA

Many ICA algorithms are prewhitening-based. In the prewhitening step the covariance matrix $\mathbf{C}_X$ of the data is diagonalized. This step has a three-fold goal: (a) determination of the number of source signals R, (b) standardization of the sources to mutually uncorrelated unit-variance signals, and (c) reduction of the $(I \times R)$ mixing matrix to an unknown $(R \times R)$ orthogonal matrix. In a second step the remaining unknown orthogonal factor is determined from the other statistics of the data. This approach has the disadvantage that the results are affected by additive (coloured) Gaussian noise. Errors made in the prewhitening step cannot be compensated in the second step and yield a bound on the overall performance [4, 11]. However, if the sources are non-Gaussian it is possible to identify the mixing matrix by using only higher-order statistics and not the covariance matrix of the observations. Higher-order-only methods (see [10] and the references therein) have the interesting feature that they allow to boost Signal to Noise Ratios (SNR) when the noise is Gaussian.

Let $\mathcal{C}_X^{(N)}$ denote the Nth order cumulant of the observations and κ_r^S the marginal Nth order cumulant of the rth source. We suppose that all values κ_r^S are different from zero. In the absence of noise we have

$$\mathcal{C}_X^{(N)} = \sum_{r=1}^R \kappa_r^S \, M_r \circ M_r \circ \ldots \circ M_r. \tag{6}$$

This is a decomposition of $\mathcal{C}_X^{(N)}$ in a minimal number of rank-1 terms, as the columns of $\mathbf{M}$ are assumed to be linearly independent. As a consequence, the aim of higher-order-only ICA can be formulated as the computation of a rank-revealing decomposition of $\mathcal{C}_X^{(N)}$, taking into account that the sample cumulant equivalent of Eq. (6) may be perturbated by non-Gaussian noise components, finite datalength effects, model misfit, etc.

Tensor $\mathcal{C}_X^{(N)}$ is not only rank-R but also rank-$(R, R, \ldots, R)$. The reason is that every n-mode vector can be written as a linear combination of the R vectors M_r. So, to deal with the situation in which $I > R$, we can first project the sample cumulant on the manifold of rank-$(R, R, \ldots, R)$ tensors, using the techniques mentioned in Sect. 2.2. A subsequent step consists of the further projection on the submanifold of rank-R tensors and the actual computation of decomposition (6). The latter problem can then be solved in a lower-dimensional space.

3.2 ICA Based on Soft Whitening

In prewhitening-based ICA, on one hand, more than half of the parameters of the unknown mixing matrix are calculated from an exact decomposition of the covariance matrix $\mathbf{C}_X$. On the other hand, the complete set of other statistics is used for the estimation of less than half of the unknown parameters; here decompositions are only approximately satisfied. The ratio of the former ($R(2I - R + 1)/2$) over the latter ($R(R - 1)/2$) number of parameters becomes bigger as I/R increases. Contrarily, in higher-order-only schemes the matrix $\mathbf{C}_X$ is not used at all. However, one may also deal with the different statistics in a more balanced way. We call this principle "soft whitening". The idea was first proposed and tested in [19]. In this section we will discuss dimensionality reduction in the context of soft whitening.

Assume that one wants to use the sample estimates of the covariance matrix and the Nth order cumulant tensor, $\hat{\mathbf{C}}_X$ and $\hat{\mathcal{C}}_X^{(N)}$. The ICA problem now amounts to the determination of a matrix $\mathbf{M} \in \mathbb{R}^{I \times R}$ that minimizes the cost function

$$\tilde{f}(\mathbf{M}) = w_1^2 \|\hat{\mathbf{C}}_X - \hat{\mathbf{C}}_S \times_1 \mathbf{M} \times_2 \mathbf{M}\|^2 + w_2^2 \|\hat{\mathcal{C}}_X^{(N)} - \hat{\mathcal{C}}_S^{(N)} \times_1 \mathbf{M} \times_2 \mathbf{M} \ldots \times_n \mathbf{M}\|^2, \quad (7)$$

in which $\hat{\mathbf{C}}_S \in \mathbb{R}^{R \times R}$ is an unknown diagonal matrix and $\hat{\mathcal{C}}_S^{(N)} \in \mathbb{R}^{R \times R \times \ldots \times R}$ an unknown diagonal tensor; w_1 and w_2 are positive weights.

Both the column space of $w_1 \hat{\mathbf{C}}_X$ and the 1-mode vector space of $w_2 \hat{\mathcal{C}}_X^{(N)}$ are theoretically equal to the column space of $\mathbf{M}$. Hence, in comparison with Sect. 2.2, it is natural to replace the truncation of the HOSVD by the computation of the dominant subspace of the column space of a matrix in which all the columns of $w_1 \hat{\mathbf{C}}_X$ and the 1-mode vectors of $w_2 (\hat{\mathbf{C}}_X^{(N)})$ are stacked. The higher-order orthogonal iteration and the higher-order Grassmann-Rayleigh quotient iteration can be easily adapted as well [9, 12].

3.3 ICA Based on Simultaneous Matrix Diagonalization

Many ICA algorithms are based on diagonalization of a set of matrices by means of a simultaneous congruence transformation [2, 5, 16]. Given $\mathbf{A}_1, \ldots, \mathbf{A}_J \in \mathbb{R}^{I \times I}$, the aim is to find a nonsingular matrix $\mathbf{M} \in \mathbb{R}^{I \times R}$ such that, in theory,

$$\mathbf{A}_1 = \mathbf{M} \cdot \mathbf{D}_1 \cdot \mathbf{M}^T$$
$$\vdots$$
$$\mathbf{A}_J = \mathbf{M} \cdot \mathbf{D}_J \cdot \mathbf{M}^T, \quad (8)$$

with $\mathbf{D}_1, \ldots, \mathbf{D}_J \in \mathbb{R}^{R \times R}$ diagonal. In the presence of noise, the difference between the left- and right-hand side of Eqs. (8) has to be minimized.

The original algorithms to solve (8) are prewhitening-based. In the prewhitening step, one picks a positive (semi-)definite matrix from $\{\mathbf{A}_j\}$, say $\mathbf{A}_1$, and computes its EVD. This allows to reduce the other equations to a simultaneous orthogonal diagonalization in a possibly lower-dimensional space. On the other hand, one can also follow the soft whitening approach and solve the different equations in (8) in a more balanced way. In this section we will explain how a dimensionality reduction can be realized here, when $R < I$. For the matrix diagonalization in the lower-dimensional space, one can resort to the techniques presented and referred to in [10, 17, 19].

To see the link with Subsect. 3.1, let us stack the matrices $\mathbf{A}_1, \ldots, \mathbf{A}_J$ in Eq. (8) in a tensor $\mathcal{A} \in \mathbb{R}^{I \times I \times J}$. Define a matrix $\mathbf{D} \in \mathbb{R}^{J \times R}$ of which the subsequent rows are given by the diagonals of $\mathbf{D}_1, \ldots, \mathbf{D}_J$. Then we have

$$\mathcal{A} = \sum_{r=1}^{R} M_r \circ M_r \circ D_r. \quad (9)$$

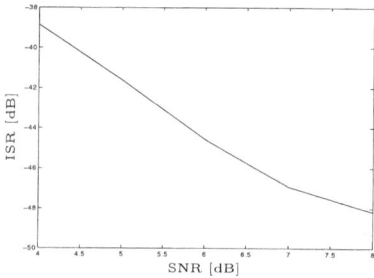 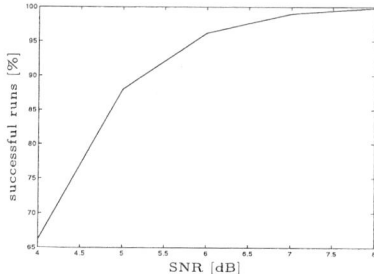

Fig. 1. Average ISR (left) and number of succesful runs (right) in the first simulation.

Let the rank of $\mathbf{D}$ be equal to R_3. Equation (9) is a decomposition of $\mathcal{A}$ in a minimal sum of rank-1 terms (if no columns of $\mathbf{D}$ are collinear [10]). This problem and the simultaneous diagonalization problem are equivalent. Hence, we can proceed in the same way as in Sect. 3.1. The dimensionality reduction can be realized by a rank-(R, R, R_3) reduction of $\mathcal{A}$. The remaining problem is the decomposition of an $(R \times R \times R_3)$-tensor in rank-1 terms.

4 Simulation Results

Our first simulation illustrates the technique described in Subsect. 3.1. Data are generated according to the following model:

$$X = \mathbf{M}_1 S + \mathbf{M}_2 \, \sigma_{\tilde{E}} \, \tilde{E},$$

in which the entries of $S \in \mathbb{R}^4$ are ± 1, with equal probability, and in which $\tilde{E} \in \mathbb{R}^{12}$ is zero-mean unit-variance Gaussian noise. $\mathbf{M}_1 \in \mathbb{R}^{12 \times 4}$ and $\mathbf{M}_2 \in \mathbb{R}^{12 \times 12}$ are random matrices of which the columns have been normalized to unit length. The data length is 500. A Monte Carlo experiment consisting of 500 runs is carried out for different SNR values (controlled by $\sigma_{\tilde{E}}$). Because the observations are corrupted by noise with an unknown colour, the independent components are estimated from the fourth-order cumulant of X. First the dimensionality of the problem is reduced from 12 to 4, and subsequently both sides of (6) are matched in the least-squares sense by means of the technique described in [3]. This algorithm was initialized with the starting value proposed in [15] and with 9 random starting values. The best result was retained. The dimensionality reduction was achieved by means of the algorithm described in [8].

Let the estimate of $\mathbf{M}_1$ be represented by $\hat{\mathbf{M}}_1$ and let the columns of $\hat{\mathbf{M}}_1$ be normalized to unit length and optimally ordered. Then our error measure is defined as the mean of the squared off-diagonal entries of $\hat{\mathbf{M}}_1^{\dagger} \cdot \mathbf{M}_1$. This error measure can be interpreted as an approximate average Interference to Signal Ratio (ISR). Only the results for which the ISR is smaller than 0.04 are retained; the other results are considered as failures. A failure means that 10 initializations were not enough or that the estimate of the low-dimensional cumulant was simply too bad to get sufficiently accurate results. The results are shown in Fig. 1. The

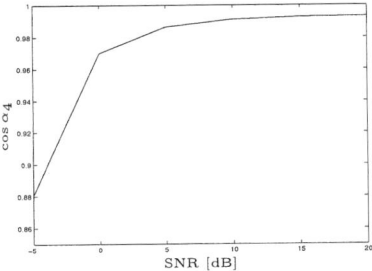

Fig. 2. Average cosine of the largest principal angle in the second simulation.

plots show that after an inexpensive dimensionality reduction, a very accurate source separation is possible for moderate SNR values.

In a second simulation we consider a simultaneous matrix diagonalization. The experiment illustrates that is sometimes possible to extract the subspace associated with a particular class of sources [6]. In this simulation, the sources of interest are the signals obtained by passing 4 mutually independent zero-mean unit-variance Gaussian i.i.d. sequences through the filters $h_1(z^{-1}) = (1+0.9z^{-1})^{-1}$, $h_2(z^{-1}) = (1-0.9z^{-1})^{-1}$, $h_3(z^{-1}) = (1+0.8z^{-1})^{-1}$, $h_4(z^{-1}) = (1-0.8z^{-1})^{-1}$. A second set of 4 independent sources is uniform over $[-\sqrt{3}, \sqrt{3}]$ and i.i.d. The rest of the set-up is the same as in the first simulation. By considering only covariance matrices for nonzero time-lag $\tau = 1, 2, 3, 4$, we are able to estimate the subspace associated to the first class of source signals. The estimate is obtained by computing the best rank-$(4, 4, 4)$ approximation of the $(12 \times 12 \times 4)$ tensor $\mathcal{A}$ in which the covariance matrices are stacked, as explained in Subsect. 3.3. Here we used the higher-order Grassmann-Rayleigh quotient iteration algorithm. After projection, the sources of interest can be separated in a vector space of dimension 4 instead of 12.

The performance is evaluated in terms of the cosine of the largest principal angle [13] between the subspace generated by the first 4 mixing vectors and the 1-mode vector space of the best rank-$(4, 4, 4)$ approximation of $\mathcal{A}$. The results are shown in Fig. 2. The figure shows that even for low SNR the estimate of the signal subspace is very accurate. For high SNR, the accuracy is only bounded by the precision with which the covariance matrices are estimated, given the fact that only 500 snapshots are available.

5 Conclusion

In several ICA applications the number of observation channels exceeds the number of source signals. It is not always appropriate to reduce the dimensionality of the problem by means of a classical prewhitening, for instance because the noise has an unknown colour. In such a case, the dimensionality may be reduced by computing the best rank-$(R_1, R_2, \ldots, R_N)$ approximation of a higher-order tensor.

References

1. André, T.F., Nowak, R.D., Van Veen, B.D.: Low-rank estimation of higher order statistics. IEEE Trans. Signal Processing. **45** (1997) 673–685.
2. Belouchrani, A., Abed-Meraim, K., Cardoso, J.-F., Moulines, E.: A blind source separation technique using second order statistics. IEEE Trans. Signal Processing. **45** (1997) 434–444.
3. Bro, R.: PARAFAC. Tutorial & applications. Chemom. Intell. Lab. Syst. **38** (1997) 149–171.
4. Cardoso, J.-F.: On the performance of orthogonal source separation algorithms. Proc. EUSIPCO-94, Edinburgh, Scotland, U.K. 776–779.
5. Cardoso, J.-F., Souloumiac, A.: Blind beamforming for non-Gaussian signals. IEE Proc.-F **140** (1994) 362–370.
6. Cruces, S., Cichocki, A., De Lathauwer, L.: Thin QR and SVD factorizations for simultaneous blind signal extraction. Proc. EUSIPCO 2004, Vienna, Austria.
7. De Lathauwer, L., De Moor, B., Vandewalle, J.: A multilinear singular value decomposition. SIAM J. Matrix Anal. Appl. **21** (2000) 1253–1278.
8. De Lathauwer, L., De Moor, B., Vandewalle, J.: On the best rank-1 and rank-$(R_1, R_2, \ldots, R_N)$ approximation of higher-order tensors. SIAM J. Matrix Anal. Appl. **21** (2000) 1324–1342.
9. De Lathauwer, L., Vandewalle, J.: Dimensionality reduction in higher-order signal processing and rank-$(R_1, R_2, \ldots, R_N)$ reduction in multilinear algebra. Lin. Alg. Appl. (to appear).
10. De Lathauwer, L., De Moor, B., Vandewalle, J.: Computation of the canonical decomposition by means of a simultaneous generalized Schur decompositition. SIAM J. Matrix Anal. Appl. (to appear).
11. De Lathauwer, L., De Moor, B., Vandewalle, J.: A prewhitening-induced bound on the identification error in independent component analysis, Tech. report.
12. De Lathauwer, L., Hoegaerts, L., Vandewalle, J.: A Grassmann-Rayleigh quotient iteration for dimensionality reduction in ICA. Proc. ICA 2004, Granada, Spain.
13. Golub, G.H., Van Loan, C.F.: Matrix Computations. Johns Hopkins University Press, Baltimore, Maryland (1996).
14. Kroonenberg, P.M.: Three-mode principal component analysis. DSWO Press, Leiden (1983).
15. Leurgans, S.E., Ross, R.T., Abel, R.B.: A decomposition for three-way arrays. SIAM J. Matrix Anal. Appl. **14** (1993) 1064–1083.
16. Pham, D.-T., Cardoso, J.-F.: Blind separation of instantaneous mixtures of non-stationary sources. IEEE Trans. Signal Processing. **49** (2001) 1837–1848.
17. Pham, D.-T. 2001. Joint approximate diagonalization of positive definite Hermitian matrices. SIAM J. Matrix Anal. Appl. **22** (2001) 1136–1152.
18. Tucker, L.R.: Some mathematical notes on three-mode factor analysis. Psychometrika, **31** (1966) 279–311.
19. Yeredor, A.: Non-orthogonal joint diagonalization in the least-squares sense with application in blind source separation. IEEE Trans. Signal Processing **50** (2002) 1545–1553.

Linear Multilayer Independent Component Analysis Using Stochastic Gradient Algorithm

Yoshitatsu Matsuda and Kazunori Yamaguchi

Kazunori Yamaguchi Laboratory, Department of General Systems Studies
Graduate School of Arts and Sciences, The University of Tokyo
3-8-1, Komaba, Meguro-ku, Tokyo, 153-8902, Japan
{matsuda,yamaguch}@graco.c.u-tokyo.ac.jp
http://www.graco.u-tokyo.ac.jp/~matsuda

Abstract. In this paper, the linear (feed-forward) multilayer ICA algorithm is proposed for the blind separation of high-dimensional mixed signals. There are two main phases in each layer. One is the mapping phase, where a one-dimensional mapping is formed by stochastic gradient algorithm which makes the higher-correlated signals be nearer incrementally. Another is the local-ICA phase, where each neighbor pair of signals in the mapping is separated by MaxKurt algorithm. By repetition of these two phase, this algorithm can reduce an ICA criterion monotonically. Some numerical experiments show that this algorithm is quite efficient in natural image processing.

1 Introduction

Independent component analysis (ICA) is a recently-developed method in the fields of signal processing and artificial neural networks, and has been shown to be quite useful for the blind separation problem [1][2][3] [4]. The linear ICA is formalized as follows. Let s and A are N-dimensional original signals and $N \times N$ mixing matrix. Then, the observed signals x are defined as

$$x = As. \qquad (1)$$

The purpose is to find out A (or the inverse W) when the observed (mixed) signals only are given. In other words, ICA blindly extracts the original signals from M samples of the observed signals as follows:

$$\hat{S} = WX, \qquad (2)$$

where X is an $N \times M$ matrix of the observed signals and $\hat{S}$ is the estimate of the original signals. This is a typical ill-conditioned problem, but ICA can solve it by assuming that the original signals are generated according to independent and non-gaussian probability distributions. In general, the ICA algorithms find out the W maximizing a criterion (called the contrast function) such as the higher order statistics (e.g the kurtosis) of every component of $\hat{S}$. That is, the

ICA algorithms can be regarded as an optimization method of such criteria. Some efficient algorithms for this optimization problem have been proposed, for example, the fast ICA algorithm [5][6], the relative gradient algorithm [4], and JADE [7][8].

Now, suppose that quite high-dimensional observed signals (namely, N is quite large) are given such as video images. In this case, even the efficient algorithms are not much useful because they have to find out all the N^2 components of $\boldsymbol{W}$. Recently, we proposed a new algorithm for this problem, which can find out global independent components by integrating the local ICA modules. Developing this approach in this paper, we propose a new efficient ICA algorithm named "the linear multilayer ICA algorithm (LMICA)." LMICA can extract all the independent components approximately by repetition of the following two phases. One is the mapping phase, which forms a one-dimensional mapping by global mapping analysis (GMA) [9] and brings the higher-correlated signals nearer. Another is local-ICA phase, where each neighbor pair of signals in the mapping is separated by MaxKurt algorithm [8]. LMICA is quite efficient ($O(N)$) than other standard ICA algorithms. Fig. 1 illustrates the most ideal case of LMICA.

This paper is organized as follows. In Section 2, the algorithm is described in detail. Section 3 discusses the stability of LMICA and the relation to the MaxKurt algorithm. In Section 4, numerical experiments verify that LMICA is quite efficient in image processing. Lastly, this paper is concluded in Section 5.

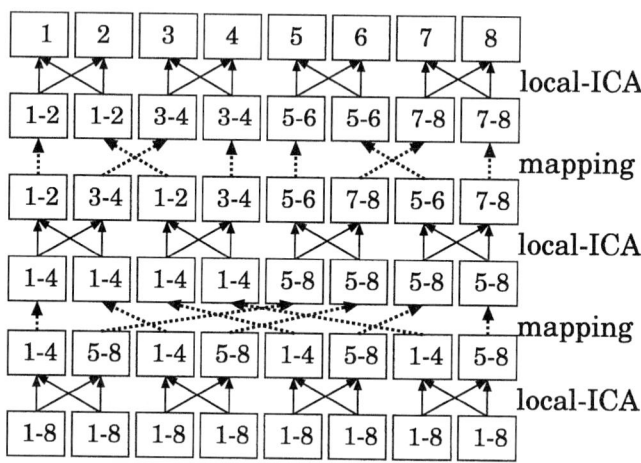

Fig. 1. The illustration of an ideal case of LMICA: Here, the number of (original and mixed) signals is 8. In the first local-ICA phase, the mixed signals are separated into the partially-separated signals of four original ones. Then, the signals of the same four signals are arranged nearer in the following mapping phase. Then, all the original signals are extracted in the last layer.

2 Algorithm

2.1 Contrast Function

In LMICA, the following contrast function (the objective function in ICA) is used:

$$\phi(\boldsymbol{X}, \boldsymbol{W}) = \sum_{i,j>i} \sum_{k=1}^{M} \hat{s}_{ik}^2 \hat{s}_{jk}^2, \tag{3}$$

where $(\hat{s}_{ik}) = \hat{\boldsymbol{S}}$ is given by Eq. (1) and we assume that $\frac{\sum_k \hat{s}_{ik}}{M} = 0$. Then, the following equation also holds for $i \neq j$:

$$\sum_{k=1}^{M} \hat{s}_{ik}^2 \hat{s}_{jk}^2 = \kappa_{iijj} + 2\kappa_{ij}^2 + \kappa_{ii}\kappa_{jj}, \tag{4}$$

where κ_{iijj} and κ_{ij} are the 4th and the 2nd cumulants of $\hat{s}_i$ and $\hat{s}_j$, respectively. κ_{iijj} is always non-negative when all the original independent components are super-gaussian (all the kurtoses are positive). Therefore, if this super-gaussian conditions hold, ϕ is actually a contrast function measuring the independency among $\hat{s}_i$'s. In addition, the separating matrix $\boldsymbol{W}$ is bound to be orthogonal. This limitation guarantees the stability of this algorithm (see Section 3). Note that the pre-whitening is not required.

2.2 Mapping Phase

In the mapping phase, given signals $\boldsymbol{X}$ are arranged in a one-dimensional array so that pairs (i, j) taking high $\sum_k x_{ik}^2 x_{jk}^2$ are placed nearer. Letting $\boldsymbol{Y} = (y_i)$ be the coordinate of the i-th signal x_{ik}, the following objective function μ is defined:

$$\mu(\boldsymbol{Y}) = \sum_{i,j} \sum_k x_{ik}^2 x_{jk}^2 (y_i - y_j)^2. \tag{5}$$

The optimal mapping is found out by minimizing μ with respect to $\boldsymbol{Y}$ under the constraints that $\sum y_i = 1$ and $\sum y_i^2 = 1$. Such an optimization is easily carried out by stochastic gradient algorithm [10][9].

In order to utilize stochastic gradient algorithm, we introduce random variables $(z_i) = \boldsymbol{Z}$ generated from the following probability distribution:

$$P(\boldsymbol{Z}) = \frac{1}{M} \sum_k \delta\left(x_{1k}^2 - z_1, \cdots, x_{Nk}^2 - z_N\right). \tag{6}$$

Then, Eq. (5) is rewritten as

$$\mu(\boldsymbol{Y}) = \sum_{i,j} E(z_i z_j)(y_i - y_j)^2. \tag{7}$$

Here, δ is the N-dimensional Dirac delta function, and $E()$ is the expectation operator. Thus,

$$\frac{\partial \mu}{\partial y_i} = E\left(2M z_i y_i \sum_j z_j - 2z_i \sum_j z_j y_j\right). \tag{8}$$

By applying stochastic gradient algorithm to Eq. (8), the following update equation is derived:

$$y_i(T+1) := y_i(T) - \lambda_T \left(M z_i y_i \zeta + z_i \eta\right), \tag{9}$$

where Z is randomly generated from $P(Z)$ at each step (in other word, $z_i = x_{ik}^2$ for randomly selected k from $\{1, \ldots, M\}$), λ_T is the step size at the T-th step,

$$\zeta = \sum_i z_i, \tag{10}$$

and

$$\eta = \sum_i z_i y_i. \tag{11}$$

Calculating α and *beta* before the updates for each i, each update requires just $O(N)$ computation.

Because the Y in the above method is continuous, the following discretization is applied in the last of the mapping phase:

$$y_i := \text{the ranking of } y_i \text{ in } Y. \tag{12}$$

That is, $y_i := 1$ for the largest y_i, $y_j := N$ for the smallest one, and so on. The corresponding permutation σ is given as $\sigma(i) = y_i$.

The total procedure of the mapping phase for given X as follows:

---

mapping phase

1. $x_{ik} := x_{ik} - \bar{x}_i$ for each i, k, where $\bar{x}_i$ is the mean $\frac{\sum_k x_{ik}}{M}$.
2. $y_i = i$, and $\sigma(i) = i$ for each i.
3. Until the convergence, repeat the following steps:
    (a) Select k randomly from $\{1, \ldots, M\}$, and let $z_i = x_{ik}^2$ for each i.
    (b) Update each y_i by Eq. (9).
    (c) Normalize Y to satisfy $\sum_i y_i = 0$ and $\sum_i y_i^2 = 1$.
4. Discretize y_i by Eq. (12).
5. Update X by $x_{\sigma(i)k} := x_{ik}$ for each i and k.

2.3 Local-ICA Phase

In the local-ICA phase, the contrast function is minimized by "rotating" the neighboring pairs of signals. For each neighboring pair $(i, i+1)$, a rotation matrix $\boldsymbol{R}_i(\theta)$ is given as

$$\boldsymbol{R}_i(\theta) = \begin{pmatrix} \boldsymbol{I}_{i-1} & 0 & 0 & 0 \\ 0 & \cos\theta & \sin\theta & 0 \\ 0 & -\sin\theta & \cos\theta & 0 \\ 0 & 0 & 0 & \boldsymbol{I}_{N-i-2} \end{pmatrix}, \quad (13)$$

where $\boldsymbol{I}_n$ is the $n \times n$ identity matrix. Then, the optimal angle $\hat{\theta}$ is given as

$$\hat{\theta} = \mathrm{argmin}_\theta \phi\left(\boldsymbol{X}'(\theta)\right), \quad (14)$$

where $\boldsymbol{X}'(\theta) = \boldsymbol{R}_i(\theta)\boldsymbol{X}$ and ϕ is given by Eq. (3). After some tedious transformation of the equations, it is shown that $\hat{\theta}$ is determined analytically by the following equations:

$$\sin 4\hat{\theta} = \frac{\alpha_{ij}}{\sqrt{\alpha_{ij}^2 + \beta_{ij}^2}}, \quad \cos 4\hat{\theta} = \frac{\beta_{ij}}{\sqrt{\alpha_{ij}^2 + \beta_{ij}^2}}, \quad (15)$$

where

$$\alpha_{ij} = \sum_k \left(x_{ik}^3 x_{jk} - x_{ik} x_{jk}^3\right), \quad \beta_{ij} = \frac{\sum_k \left(x_{ik}^4 + x_{jk}^4 - 6x_{ik}^2 x_{jk}^2\right)}{4}, \quad (16)$$

and $j = i + 1$.

Now, the procedure of the local-ICA phase for given $\boldsymbol{X}$ as follows:

---

local-ICA phase

1. Let $\boldsymbol{W}_{\mathrm{local}} = \boldsymbol{I}_N$, $\boldsymbol{A}_{\mathrm{local}} = \boldsymbol{I}_N$
2. For each $i = \{1, \ldots, N-1\}$,
   (a) Find out the optimal angle $\hat{\theta}$ by Eq. (15). The second-order statistics
   (b) $\boldsymbol{X} := \boldsymbol{R}_i(\hat{\theta})\boldsymbol{X}$, $\boldsymbol{W}_{\mathrm{local}} := \boldsymbol{R}_i \boldsymbol{W}_{\mathrm{local}}$, and $\boldsymbol{A}_{\mathrm{local}} := \boldsymbol{A}_{\mathrm{local}} \boldsymbol{R}_i^t$.

2.4 Complete Algorithm

The complete algorithm of LMICA for given observed signals $\boldsymbol{X}$ is given by repeating the mapping phase and the local-ICA phase. Here, $\boldsymbol{P}_\sigma$ is the permutation matrix corresponding to σ.

linear multilayer ICA algorithm

1. Initially Settings: Let X be the observed signal matrix, and W and A be I_N.
2. Mapping Phase: Find out optimal permutation matrix P_σ and the optimally-arranged signals X by the mapping phase procedure, and $W := P_\sigma W$ and $A := A P_\sigma^t$.
3. local-ICA Phase: Find out optimal matrices W_{local}, A_{local}, and X. Then, $W := W_{\text{local}} W$ and $A := A A_{\text{local}}$.
4. Repetition: Do the above two phases alternately over L times.

3 Discussions

3.1 Stability

Here, the stability of LMICA is discussed. In the mapping phase, Eq. (9) is guaranteed to converge to a local minimum of the objective function $\mu(Y)$ if λ_T decreases sufficiently slowly ($\lim_{T \to \infty} \lambda_T = 0$ and $\sum \lambda_T = \infty$) [9]. In the local-ICA phase, the following equation plays a crucial role:

$$\phi(X) - \phi(X') = \sum_{k=1}^{T} x_{ik}^2 x_{jk}^2 - \sum_{k=1}^{T} x_{ik}'^2 x_{jk}'^2, \qquad (17)$$

where $X' = R_i(\theta) X$, and i and θ are arbitrary. This equation is straightforwardly derived from the definition of ϕ (see Eq. (3)). Eq. (17) means that $\phi(X)$ is invariant with respect to any pair rotation $R_i(\theta)$ except for the term depending on the pair (i, j) itself. In other words, each local optimization of a neighboring pair reduces ϕ monotonically. So, ϕ always converges to a local minimum in the local-ICA phase. From the above discussions, LMICA is guaranteed to converge to a local minimum.

3.2 Relation to MaxKurt Algorithm

The local-ICA phase is quite similar to the Jacobi algorithms in ICA. Particularly, Eq. (15) is just the same as the MaxKurt algorithm (see [8]) in spite of using the different contrast function. But, our local-ICA phase clearly differs from the MaxKurt algorithm. In LMICA, just the neighboring pairs are optimized in our local-ICA phase. MaxKurt (and almost all Jacobi algorithms in ICA) has to optimized all the $\frac{N(N-1)}{2}$ pairs (i, j). On the other hand, in LMICA, the pairs of higher costs are brought nearer in the mapping phase. So, ϕ can be reduced effectively by optimized just the neighboring pairs.

4 Results

It has been well-known that various local edge detectors can be extracted from natural scenes by the standard ICA algorithm [11][12]. Here, LMICA was applied to the same problem. Here, 30000 samples of 12×12 image patches of natural scenes were given as the observed signals X. That is, N and M was 144 and 30000. The number of layers L was set 720. For comparison, the experiments without the mapping phase were carried out, where the mapping Y was randomly generated. In addition, the standard MaxKurt algorithm [8] was used for 10 steps. The criterion ϕ (Eq. (3)) was calculated at each layer, and it was averaged over 10 independently generated Xs.

Fig. 2 shows the decreasing curves of ϕ of normal LMICA and the one with random mapping. The cross points show the results at each step of MaxKurt. Because one step of MaxKurt is equivalent to 72 layers of LMICA with respect to the times of local optimizations, a scaling ($\times 72$) is applied. Surprisingly, LMICA converges to the optimal point within just 10 layers. The number of parameters within 10 layers is 143×10, which is much fewer than the degree of freedom of A ($\frac{144 \times 143}{2}$). It suggests that LMICA gives a quite suitable model for natural scenes and it is quite effective in large-size image processing. The calculation time is shown in Table. 1. It shows that the costs of the mapping phase are not much higher than those of the local-ICA phase. Because 10 layers of LMICA (22sec.) require much less time than one step of MaxKurt (94sec.), it verifies the efficiency of LMICA.

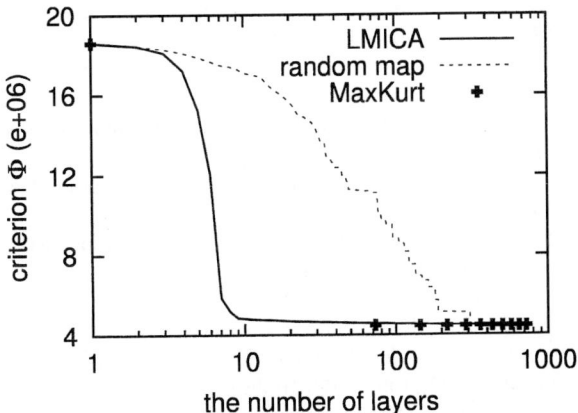

Fig. 2. Decreasing curve of the criterion ϕ along the number of layers (log-scale): The normal and dotted curves are the decreases of the criterion by LMICA and the one without the mapping phase (random mapping), respectively. The cross points show the results of MaxKurt, where one step corresponds to 72 layers.

Table 1. Calculation time: They are the averages over 10 runs of 720 layers (or 10 steps in MaxKurt) in Intel 2.8GHz CPU.

LMICA ($L = 720$)	random mapping ($L = 720$)	MaxKurt ($L = 10$)
1600sec.	670sec.	940sec.

5 Conclusion

In this paper, we proposed the linear multilayer ICA algorithm (LMICA). It can find out the independent components approximately but quite efficiently. In addition, we carried out some numerical experiments of LMICA. Now, we are trying applying this algorithm to quite large-scale signals such as 256×256 images. In addition, we want to utilize LMICA in the data mining. Particularly, in the text mining, LMICA is expected to be effective because the word vectors are generally in quite high-dimensional space.

References

1. Jutten, C., Herault, J.: Blind separation of sources (part I): An adaptive algorithm based on neuromimetic architecture. Signal Processing **24** (1991) 1–10
2. Comon, P.: Independent component analysis - a new concept? Signal Processing **36** (1994) 287–314
3. Bell, A.J., Sejnowski, T.J.: An information-maximization approach to blind separation and blind deconvolution. Neural Computation **7** (1995) 1129–1159
4. Cardoso, J.F., Laheld, B.: Equivariant adaptive source separation. IEEE Transactions on Signal Processing **44** (1996) 3017–3030
5. Hyvärinen, A., Oja, E.: A fast fixed-point algorithm for independent component analysis. Neural Computation **9** (1997) 1483–1492
6. Hyvärinen, A.: Fast and robust fixed-point algorithms for independent component analysis. IEEE Transactions on Neural Networks **10** (1999) 626–634
7. Cardoso, J.F., Souloumiac, A.: Blind beamforming for non Gaussian signals. IEE Proceedings-F **140** (1993) 362–370
8. Cardoso, J.F.: High-order contrasts for independent component analysis. Neural Computation **11** (1999) 157–192
9. Matsuda, Y., Yamaguchi, K.: Global mapping analysis: stochastic approximation for multidimensional scaling. International Journal of Neural Systems **11** (2001) 419–426
10. Matsuda, Y., Yamaguchi, K.: Computer simulation of the formation of global topographic mapping in the visual system. Transactions of Information Processing Society of Japan **40** (1999) 1091–1105 In Japanese.
11. Bell, A.J., Sejnowski, T.J.: The "independent components" of natural scenes are edge filters. Vision Research **37** (1997) 3327–3338
12. van Hateren, J.H., van der Schaaf, A.: Independent component filters of natural images compared with simple cells in primary visual cortex. Proceedings of the Royal Society of London: B **265** (1998) 359–366

Minimax Mutual Information Approach for ICA of Complex-Valued Linear Mixtures

Jian-Wu Xu, Deniz Erdogmus, Yadunandana N. Rao, and José Carlos Príncipe

CNEL, Electrical and Computer Engineering Department,
University of Florida, Gainesville, Florida 32611, USA
{jianwu,deniz,principe}@cnel.ufl.edu
http://www.cnel.ufl.edu

Abstract. Recently, the authors developed the Minimax Mutual Information algorithm for linear ICA of real-valued mixtures, which is based on a density estimate stemming from Jaynes' maximum entropy principle. Since the entropy estimates result in an approximate upper bound for the actual mutual information of the separated outputs, minimizing this upper bound results in a robust performance and good generalization. In this paper, we extend the mentioned algorithm to complex-valued mixtures. Simulations with artificial data demonstrate that the proposed algorithm outperforms FastICA.

1 Introduction

Independent Component Analysis (ICA), which may be viewed as an extension of Principle Component Analysis (PCA), is a method of finding a set of directions to minimize the statistical dependence of the projections of input random vector x on these directions. As a measure of independence between random variables, mutual information is considered as the natural criterion for ICA since minimizing mutual information would make the components of output as independent as possible. One commonly used definition of mutual information is Shannon's mutual information. Given n random variables $Y_1,......,Y_n$ whose joint probability density function (pdf) is $f_\mathbf{Y}(\mathbf{y})$ and marginal probability density functions (pdfs) are defined as $f_1(y^1),..., f_n(y^n)$ respectively, then Shannon's mutual information [1] is defined as follows

$$I(\mathbf{Y}) = \int_{-\infty}^{+\infty} f_\mathbf{Y}(\mathbf{y}) \log \left(f_\mathbf{Y}(\mathbf{y}) \bigg/ \prod_{o=1}^{n} f_o(y^o) \right) d\mathbf{y} \tag{1}$$

where the components y^i, $i=1,...,n$ constitute the vector $\mathbf{y}$. Meanwhile, we can also write Shannon's mutual information as the sum of marginal and joint entropies [1] of these random variables as,

$$I(\mathbf{Y}) = \sum_{o=1}^{n} H(Y_o) - H(\mathbf{Y}) \tag{2}$$

where Shannon's marginal and joint entropies [1] are given by

$$H(Y_o) = \int_{-\infty}^{+\infty} f_o(y^o) \log f_o(y^o) dy^o \qquad H(\mathbf{Y}) = \int_{-\infty}^{+\infty} f_\mathbf{Y}(\mathbf{y}) \log f_\mathbf{Y}(\mathbf{y}) d\mathbf{y} \tag{3}$$

respectively. Three of most widely known algorithms for ICA, namely JADE [2], Infomax [3], and FastICA [4], use the diagonalization of cumulant matrices, maximization of output entropy, and fourth order cumulants separately, instead of using minimization of output mutual information. The difficulties associated with minimum mutual information are the lack of robust pdf estimators; most of them suffer from sensitivity to the underlying data samples.

A common method in developing information theoretic ICA algorithms is to use polynomial expansions to approximate the pdf of the signals, e.g. Gram-Charlier, Edgeworth, and Legendre polynomial expansions. In order to estimate the signal pdf, a truncated polynomial is taken, evaluated in the vicinity of a maximum entropy density [5]. Alternative techniques include Parzen windowing [6], and orthogonal basis functions [7]. Other researchers also use kernel estimates in ICA [8,9,10].

Recently, we used the minimum output mutual information method to develop an efficient and robust ICA algorithm, which is based on a density estimate stemming from Jaynes' maximum entropy principle, where estimated pdfs belong to the exponential family [11, 12]. This approach approximates the solution to a constrained entropy maximization problem and provides an approximate upper bound for the actual mutual information of the output signals, and hence the name Minimax Mutual Information. In addition, this method is related to ICA methods using higher order cumulants when a specific set of constraint functions are selected in the maximum entropy density estimation step.

In this paper, we extend this Minimax Mutual Information algorithm to complex-valued mixtures. The algorithm is compared with the complex-valued FastICA method. The simulations demonstrate that complex-valued Minimax ICA exhibits better performance.

2 The Problem Statement

Suppose that there are n mutual independent sources s, whose components are zero-mean complex-valued signals. We also assume the independence between real and imaginary parts of source signal. The source signal s is mixed by an unknown linear mixture of the form z = Hs to generate n observed random vector z, where the square matrix H is invertible. In this case, the original independent sources can be obtained from z by a two-stage process: spatial whitening to generate uncorrelated but not necessarily independent mixture x = Wz, and a coordinate system rotation in the n-dimensional mixture space to determine the independent components y = Rx [5,8,13]. The whitening matrix W is obtained from the eigendecomposition of the measurement of covariance matrix. Namely, $\mathbf{W} = \mathbf{\Lambda}^{-1/2}\mathbf{\Phi}^T$, where Λ denotes the diagonal eigenvalue matrix and Φ denotes the corresponding orthonormal eigenvector matrix of the mixture covariance matrix $\Sigma=E[zz^T]$ provided that the observations are zero mean. The coordinate rotation is determined by an orthonormal matrix R parameterized by Givens angles [15]. Specifically the procedure involves the minimization of the mutual information between the output signals [5]. Considering the fact that the joint entropy is invariant under rotations, the definition of mutual information in (2) reduces to the summation of marginal output entropies for this case. Namely,

$$J(\boldsymbol{\Theta}) = \sum_{o=1}^{n} H(Y_o) \tag{5}$$

where the vector $\boldsymbol{\Theta}$ is composed of Givens angles $\theta_{ij}, i = 1,....,n-1, j = i+1,....,n$. The Givens parameterization of a rotation matrix involves the multiplication of in-plane rotation matrices. Each of the matrices $R^{ij}(\theta_{ij}^{1,2})$ for the complex-valued signal is constructed by starting with an $n \times n$ identity matrix and replacing the entries $(i,i)^{th}, (i,j)^{th}, (j,i)^{th}, (j,j)^{th}$ by $\cos\theta_{ij}^1 \exp(j\theta_{ij}^2)$, $\sin\theta_{ij}^1$, $-\sin\theta_{ij}^1$, and $\cos\theta_{ij}^1 \exp(-j\theta_{ij}^2)$, respectively, where θ^1 is the angle for the real part and θ^2 is for the imaginary part. The total rotation matrix is then defined as the product of these 2-dimensional rotations parameterized by $n(n-1)$ Givens angles to be optimized:

$$\mathbf{R}(\boldsymbol{\Theta}) = \prod_{i=1}^{n-1} \prod_{j=i+1}^{n} R^{ij}(\theta_{ij}^{1,2}) \tag{6}$$

The described whitening-rotation procedure through Givens angles parameterization of the rotation matrix is widely used in ICA algorithm, and many studies have been done on the efficient ways of dealing with the optimization of these parameters.

3 The Maximum Entropy Principle

Jaynes' maximum entropy principle states that one must maximize the entropy of the estimated distribution under certain constraints so that the estimated pdf fits the known data best without committing extensively to the unknown data because the entropy of a pdf is related with the uncertainty of the associated random variables.

Given the nonlinear moments $\alpha_k = E_X[f_k(X)]$, the maximum entropy pdf estimate for X is obtained by solving the following constrained optimization problem.

$$\max_{p_{\overline{X}}(\cdot)} H = -\int_C p_{\overline{X}}(x) \log p_{\overline{X}}(x) dx \quad s.t. \; E_{\overline{X}}[f_k(\overline{X})] = \alpha_k \; k = 1,...,m \tag{7}$$

where $p_{\overline{X}} : C \to R$ is the pdf of a complex-valued variable, and $f_k : C \to R$ are the constraint functions defined *a priori*. Using calculus of variations and the Lagrange multipliers method, we can get the optimal pdf for the complex-valued signal [1]

$$p_{\overline{X}}(x) = C(\lambda) \exp\left(\sum_{l=1}^{m} \lambda_l f_l(x)\right) \tag{8}$$

where $\lambda = [\lambda_1,....\lambda_m]^T$ is the Lagrange multiplier vector and $C(\lambda)$ denotes the normalization constant. It is not easy to solve the Lagrange multipliers simultaneously from the constraints in case of continuous random variables due to the infinite range of the definite integrals involved. We use the integration by parts method under the assumption that the actual distribution is close to the maximum entropy distribution. Consider the kth constraint equation,

$$\alpha_k = \int_c f_k(x)p(x)dx = \int_{-\infty}^{\infty}\int_{-\infty}^{\infty} f_k(x_r,x_i)p(x_r,x_i)dx_r dx_i \qquad (9)$$

where $f_k(x_r,x_i)$ is the nonlinear moment of the real and imaginary parts of the signal, denoted by x_r, x_i. The integrand covers the whole real and imaginary ranges.

We first give the following definitions:

$$F_k^{(0,1)}(x_r,x_i) = \int_{-\infty}^{+\infty} f_k(x_r,x_i)dx_i, \quad F_k^{(1,0)}(x_r,x_i) = \int_{-\infty}^{+\infty} f_k(x_r,x_i)dx_r$$

$$f_l^{(0,1)}(x_r,x_i) = \frac{\partial}{\partial x_i} f_l(x_r,x_i), \quad f_l^{(1,0)}(x_r,x_i) = \frac{\partial}{\partial x_r} f_l(x_r,x_i) \qquad (10)$$

Integrating by parts over the real part the double integral in (9), we obtain

$$\alpha_k = \int_{-\infty}^{+\infty}\left[p(x_r,x_i) F_k^{(0,1)}(x_r,x_i)\Big|_{-\infty}^{\infty} - \int_{-\infty}^{+\infty} F_k^{(0,1)}(x_r,x_i) \left(\sum_{l=1}^{m} \lambda_l f_l^{(0,1)}(x_r,x_i)\right) p(x_r,x_i)dx_r \right] dx_i \qquad (11)$$

Meanwhile we can also do partial integration over the imaginary part such that

$$\alpha_k = \int_{-\infty}^{+\infty}\left[p(x_r,x_i) F_k^{(1,0)}(x_r,x_i)\Big|_{-\infty}^{\infty} - \int_{-\infty}^{+\infty} F_k^{(1,0)}(x_r,x_i) \left(\sum_{l=1}^{m} \lambda_l f_l^{(1,0)}(x_r,x_i)\right) p(x_r,x_i)dx_i \right] dx_r \qquad (12)$$

If the functions $f_l(x_r,x_i)$ are selected such that their integrals $F_l(x_r,x_i)$ do not diverge faster than the decay rate of the exponential pdf $p_{\overline{X}}(x)$, then the first terms on the right hand sides of (11) and (12) go to zero. For example, this condition would be satisfied if moments of the random variable were defined as the constraint functions since $F_l(x_r,x_i)$ will be a polynomial function and $p_{\overline{X}}(x)$ decays exponentially. Then adding (11) and (12) yields the expression for α_k

$$\alpha_k = -\frac{1}{2}\int_{-\infty}^{+\infty}\int_{-\infty}^{+\infty}\left[F_k^{(0,1)}(x_r,x_i)\left(\sum_{l=1}^{m}\lambda_l f_l^{(0,1)}(x_r,x_i)\right)p(x_r,x_i) \right.$$
$$\left. + F_k^{(1,0)}(x_r,x_i)\left(\sum_{l=1}^{m}\lambda_l f_l^{(1,0)}(x_r,x_i)\right)p(x_r,x_i) \right] dx_i dx_r$$

$$= -\frac{1}{2}\sum_{l=1}^{m}\lambda_l E\left[F_k^{(1,0)}(x_r,x_i)f_l^{(0,1)}(x_r,x_i) + F_k^{(0,1)}(x_r,x_i)f_l^{(1,0)}(x_r,x_i) \right]$$

$$\stackrel{\Delta}{=} -\frac{1}{2}\sum_{l=1}^{m}\lambda_l \beta_{kl}$$

(13)

Note that the coefficients β_{kl} can be estimated using the sample mean. Finally, introducing the vector $\mathbf{a} = [\alpha_1 .... \alpha_m]^T$ and the matrix $\boldsymbol{\beta} = [\beta_{kl}]$, the Lagrange multipliers are given by

$$\boldsymbol{\lambda} = -\frac{1}{2} \boldsymbol{\beta}^{-1} \mathbf{a} \tag{14}$$

This method provides a simple way of finding the coefficients of the estimated pdf directly from the samples when $\mathbf{a}$ and $\boldsymbol{\beta}$ are estimated using sample means.

4 Gradient Update Rule for the Givens Angles

Minimax ICA minimizes the cost function in (5) using the entropy estimate corresponding to the maximum entropy distribution described in the previous section. A gradient descent update rule for the Givens angles is employed to adapt the rotation matrix. The derivative of marginal entropy with respect to a Givens angle is

$$\frac{\partial H(Y_o)}{\partial \theta_{pq}^{1,2}} = -\sum_{k=1}^{m} \lambda_k^o \frac{\partial \alpha_k^o}{\partial \theta_{pq}^{1,2}} \tag{15}$$

where λ^o is the Lagrange multiplier parameter vector for the pdf of oth output signal and α_k^o is the value of the kth constraint for the pdf of the oth output. Using (13) to get the solution for λ^o and the sample mean estimate

$$\alpha_k^o = \frac{1}{N} \sum_{l=1}^{N} f_k(y_{o,l}) \tag{16}$$

where $y_{o,l}$ is the lth sample at the oth output for the current angles, the derivative of α_k^o with respect to $\theta_{pq}^{1,2}$ is obtained as,

$$\begin{aligned}
\frac{\partial \alpha_k^o}{\partial \theta_{pq}^{1,2}} &= \frac{1}{N} \sum_{l=1}^{N} f_k'(y_{o,l}) \frac{\partial y_{o,l}}{\partial \theta_{pq}^{1,2}} = \frac{1}{N} \sum_{l=1}^{N} f_k'(y_{o,l}) \left(\partial y_{o,l} / \partial \mathbf{R}_{o:} \right)^T \left(\partial \mathbf{R}_{o:} / \partial \theta_{pq}^{1,2} \right)^T \\
&= \frac{1}{N} \sum_{l=1}^{N} f_k'(y_{o,l}) \mathbf{x}_l^T \left(\partial \mathbf{R} / \partial \theta_{pq}^{1,2} \right)_{o:}^T
\end{aligned} \tag{17}$$

where the subscripts in $\mathbf{R}_{o:}$ and $\left(\partial \mathbf{R} / \partial \theta_{pq}^{1,2} \right)_{o:}$ denote the oth row of the corresponding matrix. By the definition, the derivative of $\mathbf{R}$ with respect to an angle is

$$\begin{aligned}
\partial \mathbf{R} / \partial \theta_{pq}^{1,2} = &\left(\prod_{i=1}^{p-1} \prod_{j=i+1}^{n} \mathbf{R}^{ij}(\theta_{ij}) \right) \left(\prod_{j=o+1}^{q-1} \mathbf{R}^{pj}(\theta_{pj}) \right) \left(\partial \mathbf{R}^{pq}(\theta_{pq}) / \partial \theta_{pq}^{1,2} \right) \\
&\left(\prod_{j=q+1}^{n} \mathbf{R}^{pj}(\theta_{pj}) \right) \left(\prod_{i=p+1}^{n-1} \prod_{j=i+1}^{n} \mathbf{R}^{ij}(\theta_{ij}) \right)
\end{aligned} \tag{18}$$

Thus, the overall update rule for the Givens angles summing the contributions from each output is

$$\Theta_{t+1} = \Theta_t - \eta \sum_{o=1}^{n} \frac{\partial H(Y_o)}{\partial \Theta} \tag{19}$$

where η is a small step size.

5 Discussion on the Algorithm

In the previous sections, we proposed an approximate numerical solution which replaces the expectation operator over the maximum entropy by a sample mean over the data distribution due to the difficulties associated with solving for the Lagrange multipliers analytically. In this section, we provide how to choose the constraint functions $f_k(\cdot)$ in the formulation. Here we consider the moment constraints for both real and imaginary parts of the output $y_{o,l}$, namely

$$\alpha_k^o = E\left[y_r^{o\,u_k} y_i^{o\,v_k} \right] = \frac{1}{N} \sum_{l=1}^{N} y_{r,l}^{o\,u_k} y_{i,l}^{o\,v_k} \tag{20}$$

where y_r^o and y_i^o are the real and imaginary parts of oth output, u_k, v_k are the moment order. Our brief investigation on the effect of other constraint functions suggests that the simple moment constraint yields significantly better solutions. One motivation to use moment constraint is the asymptotic properties of the exponential pdf estimates in (8).

Besides the desirable asymptotic convergence properties of the exponential family of density estimates, the moment constraint function gives simple gradient updates. Let $y_o = y_r + jy_i = (\mathbf{R}_r + j\mathbf{R}_i) \times (x_r + jx_i) = (\mathbf{R}_r x_r - \mathbf{R}_i x_i) + j(\mathbf{R}_r x_i + \mathbf{R}_i x_r)$. Here $\mathbf{R}_r$ and $\mathbf{R}_i$ are the real and imaginary parts of the rotation matrix R. Then, we can find the derivative of (17) with respect to the Givens angle $\theta_{pq}^{1,2}$ as

$$\frac{\partial \alpha_k^o}{\partial \theta_{pq}^{1,2}} = \frac{1}{N} \sum_{l=1}^{N} y_{r,l}^{o\,(u_k-1)} y_{i,l}^{o\,(v_k-1)} \left(u_k y_{i,l}^o \frac{\partial y_{r,l}^o}{\partial \theta_{pq}^{1,2}} + v_k y_{r,l}^o \frac{\partial y_{i,l}^o}{\partial \theta_{pq}^{1,2}} \right) \tag{22}$$

where the derivative of output with respect to angle is

$$\frac{\partial y_r^o}{\partial \theta_{pq}^{1,2}} = \frac{\partial \mathbf{R}_r^o}{\partial \theta_{pq}^{1,2}} x_r - \frac{\partial \mathbf{R}_i^o}{\partial \theta_{pq}^{1,2}} x_i \qquad \frac{\partial y_i^o}{\partial \theta_{pq}^{1,2}} = \frac{\partial \mathbf{R}_r^o}{\partial \theta_{pq}^{1,2}} x_i + \frac{\partial \mathbf{R}_i^o}{\partial \theta_{pq}^{1,2}} x_r \tag{23}$$

Furthermore, in the computation of (18), we can express $\partial \mathbf{R}^{pq}(\theta_{pq})/\partial \theta_{pq}^{1,2}$ as

$$\frac{\partial \mathbf{R}^{pq}(\theta_{pq})}{\partial \theta_{pq}^1} = \begin{bmatrix} -\sin\theta_{pq}^1 e^{j\theta_{pq}^2} & \cos\theta_{pq}^1 \\ -\cos\theta_{pq}^1 & -\sin\theta_{pq}^1 e^{-j\theta_{pq}^2} \end{bmatrix} \qquad \frac{\partial \mathbf{R}^{pq}(\theta_{pq})}{\partial \theta_{pq}^2} = diag\begin{bmatrix} j\cos\theta_{pq}^1 e^{j\theta_{pq}^2} \\ -j\cos\theta_{pq}^1 e^{-j\theta_{pq}^2} \end{bmatrix} \tag{24}$$

Here $\partial \mathbf{R}_r^o / \partial \theta_{pq}^{1,2}$ and $\partial \mathbf{R}_i^o / \partial \theta_{pq}^{1,2}$ are the real and imaginary parts of $\left(\partial \mathbf{R} / \partial \theta_{pq}^{1,2}\right)^o$.

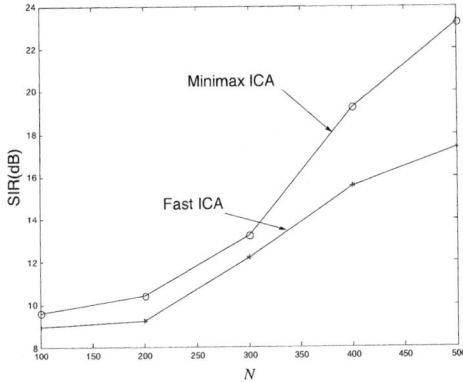

Fig. 1. Average SIR (dB) obtained by complex Minimax ICA and FastICA versus sample size.

6 Simulations

In this section, we present a simple comparison of the proposed complex Minimax ICA algorithm and the popular complex FastICA method [16]. In this controlled environment, the signal-to-interference ratio (SIR) is used as the performance measure:

$$SIR(dB) = \frac{1}{n}\sum_{o=1}^{n} 10\log_{10}\left(\max_{k}(\mathbf{O}^o_{ok})\Big/\left(\mathbf{O}_{o:}\mathbf{O}^T_{o:} - \max_{k}(\mathbf{O}^o_{ok})\right)\right) \quad (25)$$

where O is the overall matrix after separation, i.e. O=RWH. This measure is the average ratio in decibels (dB) of the main signal power in the output channel to the total power of the interfering signals. Minimax ICA uses all complex moments up to order 4 as constraints, thus it considers kurtosis information as FastICA does.

For training set sample sizes (N) ranging from 100 to 500, a set of 100 Monte Carlo simulations are run for each sample size. In each run, N complex samples are generated artificially according to $s_j = r_j(\cos\phi_j + i\sin\phi_j)$, where r_1 is Gaussian, and r_2 and the phases ϕ_j are uniform. In this setup, the sources have independent real and imaginary parts with equal variance. The 2x2 mixing matrix is also complex-valued whose real and imaginary parts of entries are uniformly random in [-1,1].

Fig. 1 shows the SIR for both methods. While Minimax ICA is always better than FastICA, the difference in performance increasingly becomes significant as the sample size is increased. On the other hand, the computational requirement of Minimax ICA is much larger than that of FastICA, as one can assess from the previous sections.

7 Conclusions

In this paper, we extended the Minimax ICA algorithm to complex-valued signals. This algorithm is based on a density estimate stemming from Jaynes' maximum entropy principle. Thus, an approximate upper bound for the mutual information between the separated outputs is obtained from the samples and minimized through the optimization procedure. The density estimation stage utilizes integration by parts in a novel way to arrive at a set of linear equations that uniquely determine the Lagrange multipliers of the constrained maximum entropy density estimation problem.

Numerical simulations conducted using artificial mixtures suggest that the proposed complex Minimax ICA algorithm yields better separation performance compared to complex FastICA at the cost of additional computational burden.

Acknowledgments

This work is supported by NSF grant ECS-0300340.

References

1. Cover, T.M., Thomas, J.A.: Elements of Information Theory. Wiley, New York (1991)
2. Cardoso, J.F., Souloumiac, A.: Blind Beamforming for Non-Gaussian Signals. IEE Proc. F Radar and Signal Processing. 140 (1993) 362-370
3. Bell, A., Sejnowski, T.: An Information-Maximization Approach to Blind Separation and Blind Deconvolution. Neural Computation. 7 (1995) 1129-1159
4. Hyvarinen, A.: Fast and Robust Fixed-Point Algorithms for Independent Component Analysis. IEEE Transactions on Neural Networks. 10 (1999) 626-636
5. Comon, P.: Independent Component Analysis, A New Concept? Signal Processing. 36 (1994) 284-314
6. Parzen, E.: On Estimation of a Probability Density Function and Mode. Annals of Mathematical Statistics. 33 (1962) 1065-176
7. Girolami, M.: Orthogonal Series Density Estimation and the Kernel Eigenvalue Problem. Neural Computation. 14 (2002) 1065-1076
8. Hild II, K.E., Erdogmus, D., Principe, J.C.: Blind Source Separation Using Renyi's Mutual Information. IEEE Signal Processing Letters. 8 (2001) 174-176
9. Xu, D., Principe, J.C., Fisher, J., Wu, H.C.: "A Novel Measure for Independent Component Analysis. Proc. ICASSP'98, Seattle, Washington. (1998) 1161-1164
10. Pham, D.T.: Blind Separation of Instantaneous Mixture of Sources via the Gaussian Mutual Information Criterion. Signal Processing. 81 (2991) 855-870
11. Erdogmus, D., Hild II, K.E., Rao, Y.N., Principe, J.C.: Independent Component Analysis Using Jaynes' Maximum Entropy Principle. Proc. ICA'03, Nara, Japan. (2003) 385-390
12. Erdogmus, D. Hild II, K.E., Rao, Y.N., Principe, J.C.: Minimax Mutual Information Approach for Independent Component Analysis. Neural Computation (2004) to appear
13. Cardoso, J.F.: High-Order Contrasts for Independent Component Analysis. Neural Computation. 11 (1999) 157-192
14. Hild, K.E., Erdogmus, D., Principe, J.C.: Blind Source Separation Using Renyi's Mutual Information. IEEE Signal Processing Letters. 8 (2001) 174-176
15. Golub, G., van Loan, C.: Matrix Computation. John Hopkins Univ. Press, Baltimore (1993)
16. Bingham, E., Hyvärinen, A.: A Fast Fixed-point Algorithm for Independent Component Analysis of Complex-Valued Signals. Int. J. of Neural Systems. 10 (2000) 1-8

Signal Reconstruction in Sensor Arrays Using Temporal-Spatial Sparsity Regularization

Dmitri Model and Michael Zibulevsky

Technion - Israel Institute of Technology
Electrical Engineering Department
Haifa, Israel
dmm@tx.technion.ac.il, mzib@ee.technion.ac.il

Abstract. We propose a technique of multisensor signal reconstruction based on the assumption, that source signals are spatially sparse, as well as have sparse [wavelet-type] representation in time domain. This leads to a large scale convex optimization problem, which involves l_1 norm minimization. The optimization is carried by the Truncated Newton method, using preconditioned Conjugate Gradients in inner iterations. The byproduct of reconstruction is the estimation of source locations.

1 Introduction

The solution of the "Cocktail Party" problem is the active research field. However none of the developed techniques provides an ideal solution. Yet another active research area is source localization. In this paper we propose to benefit from both fields in order to receive a more precise and stable solution.

Our technique is based on the assumption, that incoming signals can be sparsely represented in an appropriate basis or frame (e.g., via the short time Fourier transform, Wavelet transform, Wavelet Packets, etc.). This idea is exploited, for example, in [1],[2]. We also assume that there are few stationary sources, and that they are sparsely located in space. The last assumption is used in [3] and [4]. The combination of both assumptions can lead to an improved performance, as demonstrated by our simulations. An additional advantage of our method, is that it deals with the sensor array model in time domain, and thus is applicable for both narrowband and wideband signals.

The solution of our problem is the restored signals in each location. Only the locations, from which the signals have actually arrived, will contain signals with relatively large energy, others will contain only noise, suppressed by our method and, hence, relatively low energy. Thus, the byproduct of our solution is an estimate of the source locations.

2 Observation Model

Consider several source signals impinging upon an array of n sensors. The arriving signals are sampled and represented in discrete time by T time samples. Let

$\{\theta_1, \ldots, \theta_m\}$ be a discrete grid of all source locations. Hence, the sources can be represented by an $m \times T$ matrix S, whose i-th row represents the signal from the i-th direction. In the same manner, we can introduce the sensor measurement matrix, Y.

Signals from different source positions arrive to each sensor with different delays and, possibly, different attenuations. This leads to the following observation model:

$$Y = \mathcal{A}S + N \qquad (1)$$

where N stands for the measurement noise matrix; $\mathcal{A}$ denotes 'mixing operator', which *shifts*, *attenuates* and *sums* incoming signals modelling the real environment. Note, that the operator $\mathcal{A}$ written in an explicit matrix form will have the huge dimensions of $n \times mT$, hence, for optimization, it is more convenient to implement the product $Y = \mathcal{A}S$ by a series of *shifts*, *multiplications* and *sums* actions:

$$y_i = \sum_{j=1}^{m} \alpha_{ji} U_{\Delta_{ji}}(s_j) \qquad (2)$$

where y_i is the i-th row of the sensor measurement matrix, Y; s_j is the j-th row of sources' matrix S; α_{ji} represents attenuation of the j-th source toward the i-th sensor; $U_{\Delta_{ji}}$ is a shifting operator and Δ_{ji} is the delay of the j-th source toward the i-th sensor.

In the same manner we can implement the application of the adjoint operator $X = \mathcal{A}^* Y$ by a series of *shifts*, *multiplications* and *sums* actions:

$$x_j = \sum_{i=1}^{n} \alpha_{ji} y_i (1 + \Delta_{ji} : T + \Delta_{ji}) \qquad (3)$$

x_j and y_i refer to j-th and i-th rows of X and Y respectively. Matlab-like $y_i(1 + \Delta_{ji} : T + \Delta_{ji})$ stands for the T-length subvector of y_i, starting at $1 + \Delta_{ji}$ position.

As mentioned above, we work with the discrete-time signals. Therefore, a problem arises when Δ_{ji} is not integer. A straightforward solution is to replace the fractional delays with the rounded ones. However, this approach significantly limits the spatial resolution. A better approach suggests upsampling of signals prior to applying the $\mathcal{A}$ operator. The upsampling may be produced using some interpolation kernel.

Let $\mathcal{I}_{N_{up}}$ denote upsampling by factor N_{up} operator, then if S is an $m \times T$ matrix, $S_{up} = \mathcal{I}_{N_{up}} S$ is $m \times T N_{up}$ matrix. We also define the adjoint operator $\mathcal{I}^*_{N_{up}}$, which translates an $m \times T N_{up}$ matrix S_{up} into $m \times T$ matrix $S_r = \mathcal{I}^*_{N_{up}} S_{up}$. Note, that in general $S \neq S_r$.

Now, in our model we will use the modified operators

$$\hat{\mathcal{A}} = \mathcal{A} \cdot \mathcal{I}_{N_{up}} \qquad \hat{\mathcal{A}}^* = \mathcal{I}^*_{N_{up}} \cdot \mathcal{A}^* \qquad (4)$$

instead of $\mathcal{A}$ and $\mathcal{A}^*$, but for simplicity, we will continue to denote the modified operators as $\mathcal{A}$ and $\mathcal{A}^*$ (for more details see [5]). Note, that after upsampling, we should adjust Δ_{ji} to be $\Delta_{ji} * N_{up}$. We will still need to round $\Delta_{ji} * N_{up}$ to the closest integer, but now the rounding error is N_{up} times less. In our simulations we used $N_{up} = 10$ and the 'sinc' interpolation kernel.

3 Method Description

We assume that the sources S are sparsely representable in some basis or overcomplete system of functions [6] (e.g. Gabor, wavelet, wavelet packet, etc.). In other words, there exists some operator Φ, such that $S = C\Phi$, and the matrix of coefficients, C, is sparse. We use the objective function of the following form:

$$F(C) = F_1(C) + F_2(C) + F_3(C) \qquad (5)$$

where $F_1(C)$ is the l_2-norm-based data fidelity term; $F_2(C)$ is the temporal sparsity regularizing term, which is intended to prefer sparsely representable signals; $F_3(C)$ is the spatial sparsity regularizing term, which is intended to prefer solutions with the source signals concentrated in a small number of locations. $F_2(C)$ is based on the l_1-norm, which is proved to be effective in forcing sparsity [6]. Then, the objective function can be written as:

$$F(C) = \frac{1}{2}\|Y - \mathcal{A}(C\Phi)\|_F^2 + \mu_1 \sum_{i,j} |c_{ij}| + \mu_2 \sum_{i=1}^{m} \|c_i\|_2 \qquad (6)$$

where c_i denotes the i-th row of the matrix C (the i-th source' coefficients), and c_{ij} is the j-th element in c_i. The scalars μ_1 and μ_2 are used to regulate the weight of each term. And $\|X\|_F = \sqrt{\sum_{ij} X_{ij}^2}$ denotes a Frobenius norm of matrix X.

In order to minimize the objective (6) numerically, we use a smooth approximation of the l_2-norm, having the following form:

$$\psi(x) = \sqrt{\sum_i x_i^2 + \epsilon} \approx \|x\|_2 \qquad (7)$$

the approximation becomes more precise as $\epsilon \to 0$. It can be easily seen, that if ψ is applied to a single element of x - it becomes the smooth approximation of absolute value:

$$\psi(x_i) = \sqrt{x_i^2 + \epsilon} \approx |x| \qquad (8)$$

Using (7) and (8), we obtain the following objective function:

$$F(C) = \frac{1}{2}\|Y - \mathcal{A}(C\Phi)\|_F^2 + \mu_1 \sum_{i,j} \psi(c_{ij}) + \mu_2 \sum_{i=1}^{m} \psi(c_i) \qquad (9)$$

We can efficiently calculate both the $\mathcal{A}S$ and the $\mathcal{A}^*Y$ products, which enables us to calculate the gradient matrix G and the product of the Hessian operator $\mathcal{H}$ with an arbitrary matrix X (see Appendix A). Hence, the objective (9) can be minimized by one of the numerical optimization methods, for example the *Quasi Newton* method. A problem arises when the dimension of the problem growths. The memory consumption and iteration cost grow as $(mT)^2$. This circumstance leads us to the usage of the *Truncated Newton* method [7],[8]. In the *Truncated Newton* method the Newton direction d is found by the approximate solution of

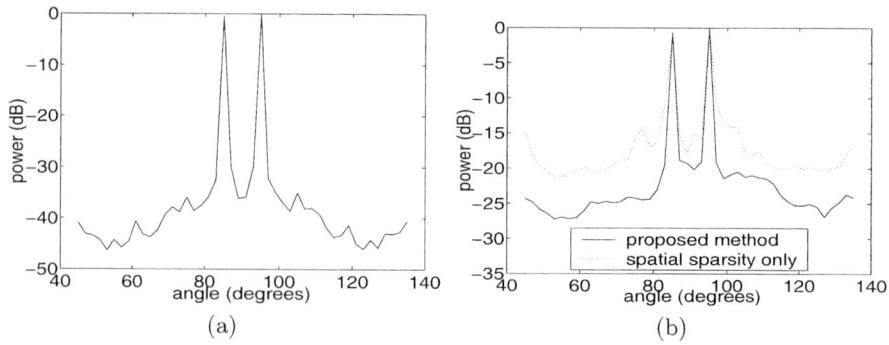

Fig. 1. DOA estimation: (a) - no noise; (b) - SNR=5 db.

the system of linear equations $Hd = -g$. This is done by the *linear Conjugate-Gradients* method. We use diagonal preconditioning in order to further speed up the optimization [9]. Note that in *Truncated Newton* method, the memory consumption growth linearly with the number of variables. This enables us to solve large problems with fair performance.

4 Computational Experiments

Our simulations were restricted to 2D model, far field and sensors lined up with constant distances. The delay of the j-th source location toward the i-th sensor is easy to calculate, given the geometrical position of each sensor and assuming that the source is far enough, so that signal arrives as a planar wave (far field assumption). Note that it is straightforward to extend our simulations to the general case. It only requires to recalculate the delay from each location to each sensor.

The experiment setup is as following: 8 sensors are lined up with $\frac{\lambda_{min}}{2} = \frac{1}{2} \frac{c}{f_{max}}$ distance (we assume our signal to be band limited, and f_{max} denoting the highest frequency). Signals arrive from 45 possible directions, and they are 64 time samples-long. The environment is noisy, with $SNR = 5dB$. There are only 2 active sources, located very close to each other - $10°$. In these conditions conventional methods, such as beamforming and MUSIC fail to superresolve them (as shown in [3],[4]).

We have generated the sensors' measurement matrix Y in the following way: at first, we have generated the sparse coefficients matrix C. Next, the source signals were created $S = C\Phi$ and finally $Y = \mathcal{A}S$ ($\mathcal{A}$ defined in (4) and $N_{up} = 10$).

In the first experiment we have checked that our algorithm can reconstruct signals in noise-free environment. The experiment was successful, and the algorithm has correctly determined the source positions (Fig. 1(a)) and has produced reconstruction with less than $5 * 10^{-3}$ reconstruction error .The error was calculated according to $\frac{\|s_{init} - s_{rec}\|_2}{\|s_{init}\|_2}$.

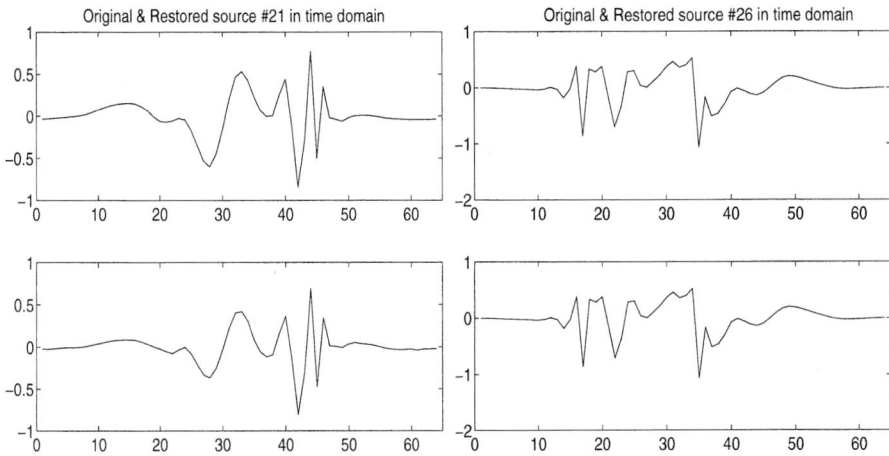

Fig. 2. Source reconstruction (SNR = 5bB). Top: sources from 2 active directions, bottom: restored sources.

In the second experiment, we have also added white Gaussian noise to the matrix Y. The contaminated by the noise matrix Y was used as an input to our algorithm. After successful optimization, we have checked the signals (original vs. reconstructed) from the active directions. As one can see in Figure 2 the active signals were restored rather accurately.

In addition, we have checked our method for DOA estimation, by computing the energy of the restored signal at each direction. We have compared our technique with the method based on spatial sparsity only, in spirit of ([3],[4]), by setting $\mu_1 = 0$ in (9). It can be seen from Figure 1(b) that both methods correctly identify the active directions, however sidelobes are about $5dB$ lower when temporal sparsity is enforced along with spatial sparsity.

5 Conclusions

We have presented a method for reconstruction of multiple source signals from multi-sensor observations, based on temporal-spatial sparsity. We derive the expressions for efficient computation of the gradient, multiplication by Hessian and diagonal preconditioning, necessary for *Truncated Newton* programming.

Computational experiments showed the feasibility of our method. The use of temporal sparsity along with spatial sparsity further lowers the sidelobes. However, more simulations and comparison to other methods should be completed before we can judge the method's performance.

We are planning to test our method in the case of near field sources. As well we wish to further speed up the optimization.

Appendix A.
Gradient and Hessian of the Objective Function

In order to use the *Truncated Newton* method, we need to calculate the gradient G of the objective (9), as well as to implement the product of the Hessian $\mathcal{H}$ with an arbitrary matrix X. Note that $\mathcal{H}$ is a tensor, but if we parse the matrix variable C into a long vector, then a Hessian will be represented by a matrix H. We will use these notations throughout this appendix. We also derive multiplication by the diagonal of H, required for preconditioned *Conjugate-Gradients*.

Let us start with the first term in (9). We will define a new operator $\mathcal{B}$ in the following way:

$$\mathcal{B}C = \mathcal{A}(C\Phi) \qquad \mathcal{B}^*X = (\mathcal{A}^*X)\Phi^* \qquad (10)$$

This enables us to write the first term in (9) as: $F_1 = \frac{1}{2}\|\mathcal{B}C - Y\|_F^2$. If we introduce new variable $U = \mathcal{B}C - Y$, then $F_1 = \frac{1}{2}\|U\|_F^2 = \frac{1}{2}Tr(U^T U)$. Hence, $dF_1 = \frac{1}{2}\left(Tr(U^T dU) + Tr(dU^T U)\right) = Tr(U^T dU)$. Substituting U and $dU = \mathcal{B}dC$ yields $dF_1 = Tr\left((\mathcal{B}C - Y)^T \mathcal{B}dC\right) = \langle \mathcal{B}C - Y, \mathcal{B}dC \rangle = \langle \mathcal{B}^*(\mathcal{B}C - Y), dC \rangle$. Recall that $dF = \langle G, dC \rangle$, and we get the gradient

$$G_1(C) = \mathcal{B}^*(\mathcal{B}C - Y) \qquad (11)$$

Now we can substitute the expressions for $\mathcal{B}$ and $\mathcal{B}^*$ from (10) and we will receive:

$$G_1(C) = \left(\mathcal{A}^*\left(\mathcal{A}(C\Phi) - Y\right)\right)\Phi^* \qquad (12)$$

In order to calculate the multiplication of the Hessian operator $\mathcal{H}$ by an arbitrary matrix X we need to recall that $dG(C) = \mathcal{H}dC$. By (11) $dG_1(C) = \mathcal{B}^*(\mathcal{B}dC)$, and thus for an arbitrary X

$$\mathcal{H}_1 X = \mathcal{B}^*(\mathcal{B}X) \qquad (13)$$

which gives after substituting $\mathcal{B}$ and $\mathcal{B}^*$ from (10):

$$\mathcal{H}_1 X = \left(\mathcal{A}^*\left(\mathcal{A}(X\Phi)\right)\right)\Phi^* \qquad (14)$$

Parentheses are used to ensure correct order of multiplications, $\mathcal{A}X$ and $\mathcal{A}^*X$ are defined in (2),(3),(4).

In order to proceed with the second and the third terms in (9), we need to use the gradient and Hessian of (7):

$$\nabla \psi(x) = \frac{1}{\psi(x)} x \qquad (15)$$

$$\left(\nabla^2 \psi(x)\right)_{ii} = -\frac{1}{\psi^3(x)} x_i^2 + \frac{1}{\psi(x)} \qquad (16)$$

$$\left(\nabla^2 \psi(x)\right)_{ij} = -\frac{1}{\psi^3(x)} x_i x_j \qquad (i \neq j)$$

where $(\nabla^2 \psi(x))_{ii}$ and $(\nabla^2 \psi(x))_{ij}$ are diagonal and off diagonal elements elements of $\nabla^2 \psi(x)$ respectively. Now, by straightforward calculations we can write down the gradients of the second and the third term in (9):

$$(G_2)_{ij} = \mu_1 \frac{1}{\psi(c_{ij})} c_{ij} \tag{17}$$

$$(G_3)_{ij} = \mu_2 \frac{1}{\psi(c_i)} c_{ij} \tag{18}$$

note, that the gradient of (9) is a matrix, because our variable C is also a matrix (hence G_1, G_2 and G_3 are also matrices). It can be noticed in (17), that all elements of G_2 are independent, and thus the $\mathcal{H}_2$ matrix will be diagonal. It is convenient to 'pack' the diagonal of $\mathcal{H}_2$ into a matrix with the same size as C row by row. Let us denote the packed matrix as $\tilde{H}_2$:

$$\tilde{H}_{2_{ij}} = \mu_1 \left(-\frac{1}{\psi^3(c_{ij})} c_{ij}^2 + \frac{1}{\psi(c_{ij})} \right) \tag{19}$$

it is obvious, that

$$\mathcal{H}_2 X = \tilde{H}_2 \odot X \tag{20}$$

where $\odot$ is element-wise multiplication.

In order to define the multiplication $\mathcal{H}_3 X$ we need to rewrite the equation (16):

$$\nabla^2 \psi(c_i^T) = \frac{1}{\psi^3(c_i^T)} c_i^T c_i + \frac{1}{\psi(c_i^T)} I \tag{21}$$

where I represents the identity matrix. Now it is easy to define the i-th row of $\mathcal{H}_3 X$:

$$(\mathcal{H}_3 X)_i = \mu_2 \left(-\frac{1}{\psi^3(c_i^T)} c_i (c_i x_i^T) + \frac{1}{\psi(c_i^T)} x_i \right) \tag{22}$$

where x_i is the i-th row of matrix X.

This calculus is sufficient for the *Truncated Newton* method. However, in order to use *Preconditioned* Conjugate Gradients method for inner iterations, we need to define the diagonal of the Hessian of (9).

We will calculate the elements in the diagonal of $\mathcal{H}_1$ in the following manner: let E be a zero matrix with only one non-zero element equal to 1 at an arbitrary location - i-th row and j-th column. Then:

$$\left(\tilde{H}_1 \right)_{ij} = \langle E, \mathcal{H}_1 E \rangle \tag{23}$$

where $\tilde{H}_1$ is a diagonal of $\mathcal{H}_1$ packed in the same manner as a diagonal of H_2 in (19).

It follows from (13) that $\langle E, \mathcal{H}_1 E \rangle = \langle E, \mathcal{B}^*(\mathcal{B}E) \rangle = \langle \mathcal{B}E, \mathcal{B}E \rangle = \|\mathcal{B}E\|_F^2$, and if we substitute the expression for $\mathcal{B}$ from (10) we will receive $\langle E^T, \mathcal{H}_1 E \rangle = \|\mathcal{A}(E\Phi)\|_F^2$. The elements of $E\Phi$ will be all zeros, except for the i-th row which will be equal to the j-th row of Φ. After applying the operator $\mathcal{A}$ as described in (2),(3),(4), we will receive a *shifted, attenuated* and *upsampled* copy of j-th row of Φ in each row of $\mathcal{A}(E\Phi)$. And, finally, after taking the norm and using (2) and (23), we will receive:

$$\left(\tilde{H}_1\right)_{ij} = \left\|\left(\mathcal{I}_{N_{up}}\Phi\right)_j\right\|_2^2 \sum_{j=1}^{n} \alpha_{ij}^2 \qquad (24)$$

where $\left(\mathcal{I}_{N_{up}}\Phi\right)_j$ is the j-th row of upsampled Φ.

The diagonal of $\mathcal{H}_2$ is already defined in (19). Finally, the diagonal of $\mathcal{H}_3$, packed in the same manner as a diagonal of $\mathcal{H}_2$, is given by:

$$(\tilde{H}_3)_{ij} = \mu_2 \left(-\frac{1}{\psi^3(c_i)}c_{ij}^2 + \frac{1}{\psi^3(c_i)}\right) \qquad (25)$$

References

1. M. Zibulevsky and B. A. Pearlmutter, "Blind source separation by sparse decomposition in a signal dictionary," *Neural Computations*, vol. 13, no. 4, pp. 863–882, 2001.
2. M. Zibulevsky, B. A. Pearlmutter, P. Bofill, and P. Kisilev, "Blind source separation by sparse decomposition," in *Independent Components Analysis: Princeiples and Practice* (S. J. Roberts and R. M. Everson, eds.), Cambridge University Press, 2001.
3. M. Çetin, D. M. Malioutov, and A. S. Willsky, "A variational technique for source localization based on a sparse signal reconstruction perspective," in *IEEE International Conference on Acoustics, Speech, and Signal Processing*, vol. 3, pp. 2965–2968, May 2002.
4. D. M. Malioutov, M. Çetin, J. W. FisherIII, and A. S. Willsky, "Superresolution source localization through data-adaptive regularization," in *IEEE Sensor Array and Multichannel Signal Processing Workshop*, pp. 194–198, Aug. 2002.
5. D. Model and M. Zibulevsky, "Sparse multisensor signal reconstruction," CCIT report #467, EE department, Technion - Israel Institute of Technology, Feb. 2004.
6. S. S. Chen, D. L. Donoho, and M. A. Saunders, "Atomic decomposition by basis pursuit," *SIAM J. Sci. Comput.*, vol. 20, no. 1, pp. 33–61, 1998.
7. R. S. Dembo, S. C. Eisenstat, and T. Steihaug, "Inexact newton methods," *SIAM Journal on Numerical Analysis*, vol. 19, pp. 400–408, 1982.
8. S. Nash, "A survey of truncated-newton methods," *Journal of Computational and Applied Mathematics*, vol. 124, pp. 45–59, 2000.
9. P. E. Gill, W. Murray, and M. H. Wright, *Practical Optimization*. New York: Academic Press, 1981.

Underdetermined Source Separation with Structured Source Priors

Emmanuel Vincent and Xavier Rodet

IRCAM, Analysis-Synthesis Group 1
place Igor Stravinsky, F-75004 Paris
emmanuel.vincent@ircam.fr

Abstract. We consider the source extraction problem for stereo instantaneous musical mixtures with more than two sources. We prove that usual separation methods based only on spatial diversity have performance limitations when the sources overlap in the time-frequency plane. We propose a new separation scheme combining spatial diversity and structured source priors. We present possible priors based on nonlinear Independent Subspace Analysis (ISA) and Hidden Markov Models (HMM), whose parameters are learnt on solo musical excerpts. We show with an example that they actually improve the separation performance.

1 Introduction

In this article we consider the source extraction problem for stereo instantaneous musical mixtures with more than two sources. The goal is to recover for each sample u the $n \times 1$ vector of source signals $\mathbf{s}_u$ satisfying $\mathbf{x}_u = \mathbf{A}\mathbf{s}_u$, where $\mathbf{A}$ is the $2 \times n$ mixing matrix and $\mathbf{x}_u$ the 2×1 mixture vector. It has been shown that this can be solved in two steps [1]: first estimating the (normalized) columns of $\mathbf{A}$ and then estimating $\mathbf{s}_u$ knowing $\mathbf{A}$. We focus here on this second step.

When little information about the sources is available, the usual hypothesis is that in most time-frequency points only one source is present [2–4]. This source is determined exploiting the spatial diversity of the mixture, that is comparing locally the two observed channels. In practice this leads to good results for speech mixtures but not for musical mixtures. Due to western music harmony rules, musical instruments often play notes with overlapping harmonic partials, so that several sources are active in many time-frequency points.

In this article, we investigate the use of structured source priors to improve separation of musical mixtures. We propose a family of priors adapted to instrumental sounds and we show how to use both spatial diversity and source priors into a single separation scheme.

The structure of the article is as follows. In Section 2 we derive a general framework for source extraction and we introduce the three-source example used in the following. In Section 3 we describe some usual separation methods based on spatial diversity and we point their limitations. In Section 4 we propose a family of structured priors adapted to musical sounds and evaluate their performance. We conclude by discussing possible improvements to the proposed method.

2 Source Extraction Framework

In the rest of the article we suppose that **A** has been retrieved from the mixture and has L_2-normalized columns. This is realistic since the spatial directions of the sources can be estimated very precisely when each source is alone in at least one time-frequency point [5]. In this Section we derive a particular piecewise linear separation method and we show that it can potentially recover the sources with very high quality.

2.1 Three-Step Extraction Procedure

Piecewise linear separation methods are three-step procedures [2]: first decompose the mixture channels as weighted sums of time-frequency atoms, then perform a linear separation on each atom, and finally build the estimated sources by summation.

We choose to pass the mixture **x** through a bank of filters regularly spaced on the auditory-motivated ERB frequency scale $f_{ERB} = 9.26 \log(0.00437 f_{Hz} + 1)$ to obtain sub-band signals $(\mathbf{x}_f)$. Then we multiply $(\mathbf{x}_f)$ by disjoint 11 ms rectangular windows to compute short-time sub-band signals $(\mathbf{x}_{ft})$. The ERB frequency scale gives more importance to low frequencies which usually contain more energy. This results in a better separation performance than usual linear frequency scales. Note that as a general notation in the following we use bold letter for vectors or matrices, regular letters for scalars and parentheses for sequences.

Because of the linearity of the time-frequency transform, the relationship $\mathbf{x} = \mathbf{A}\mathbf{s}$ becomes $\mathbf{x}_{ft} = \mathbf{A}\mathbf{s}_{ft}$ for each (f, t). A unique solution $\mathbf{s}_{ft}$ can be estimated for each (f, t) by setting some probabilistic priors on the sources. Here we suppose that the source signals (s_{jft}), $1 \leq j \leq n$, are independent and that (s_{jft}) follows a Gaussian prior with known variance m_{jft}. Then the optimal estimated sources are given by $\widehat{\mathbf{s}_{ft}} = \mathbf{\Sigma}_{ft}^{1/2}(\mathbf{A}\mathbf{\Sigma}_{ft}^{1/2})^+ \mathbf{x}_{ft}$, where $^+$ denotes Moore-Penrose pseudo-inversion [1] and $\mathbf{\Sigma}_{ft}$ is the diagonal matrix containing the source variances (m_{jft}). Note that if at least two sources have nonzero variance then perfect reconstruction of the mixture is verified: $\mathbf{x}_{ft} = \mathbf{A}\widehat{\mathbf{s}_{ft}}$.

Finally the waveforms of the estimated sources are obtained by $\widehat{\mathbf{s}} = \sum_{ft} \widehat{\mathbf{s}_{ft}}$.

2.2 Three-Source Example - Oracle Performance

To compare the source extraction methods proposed hereafter, we build an artificial five-second mixture of s_1 = cello, s_2 = clarinet and s_3 = violin, mixed with relative log-powers $\theta_j = \log(A_{2j}^2/A_{1j}^2)$ equal to 4.8 dB, -4.8 dB and 0 dB respectively. In the rest of the article, we separate this mixture with various methods and evaluate the results by computing Source-to-Interference Ratios (SIR) and Source-to-Artifacts Ratios (SAR) [6]. The sources and the mixture are plotted in Fig. 1 and the results are shown in Table 1. All the corresponding sound files can be listened to on the web page http://www.ircam.fr/anasyn/vincent/ICA04/.

The first test we make is separation of **x** with an oracle estimator of the source power spectrograms $(\mathbf{m}_j)$ (*i.e.* the (m_{jft}) matrices). Performance measures (in

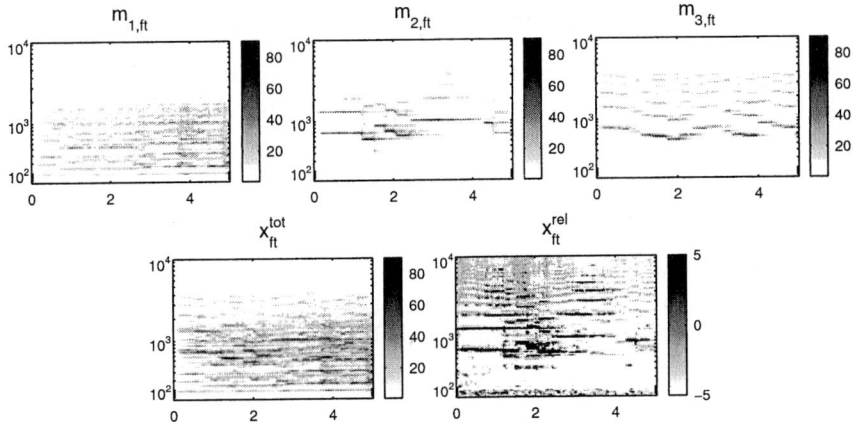

Fig. 1. Power spectrograms of the true sources (top), of the total mixture power and of the relative mixture power (bottom). The horizontal axis is time in seconds, the vertical axis is frequency in Hertz and the color range is in Decibels.

Table 1. Separation of a stereo mixture of three musical sources using several separation methods.

Cues	Method	SIR (dB)			SAR (dB)		
		$\hat{s}_1$	$\hat{s}_2$	$\hat{s}_3$	$\hat{s}_1$	$\hat{s}_2$	$\hat{s}_3$
	All sources	10	8	-5	$+\infty$	$+\infty$	$+\infty$
Spatial diversity	Closest source	36	26	18	11.6	10.3	5.6
	1 or 2 closest sources	27	25	15	13.8	13.9	5.9
Source priors	Bernoulli state priors	13	12	6	11.8	10.6	-3.0
Spatial diversity	Bernoulli state priors	23	22	34	17.1	16.8	7.0
+ Source priors	Markov state priors	30	31	23	17.2	16.8	8.4
	Oracle state sequence	31	35	23	18.7	18.6	10.5
Oracle	Oracle	49	49	44	24.4	30.0	21.9

the last line of Table 1) are higher than 20 dB for all sources. This proves that knowing $(\mathbf{m}_j)$ is enough to recover the sources with high quality.

This test mixture is not completely realistic, however it contains instruments sometimes playing in harmony. This results in notes from different instruments overlapping in the time-frequency plane, either partially (during a limited time or on a limited frequency range) or totally. In practice the oracle separation performance cannot be achieved with blind separation methods, because notes that are totally masked cannot be heard and cannot be recovered except with a musical score. However, notes that are partially masked can generally be heard and should be separated accurately.

3 Separation Methods Based on Spatial Diversity

Now that we have explained how to extract the sources given their power spectrograms $(\mathbf{m}_j)$, the problem becomes: how to estimate $(\mathbf{m}_j)$? In this Section we discuss a few heuristic methods based on spatial diversity inspired from [2–4,1].

3.1 Some Blind Separation Methods and Their Performance

Two quantities of interest are computed from the mixture channels $\mathbf{x}_{1,ft}$ and $\mathbf{x}_{2,ft}$: the total log-power $x_{ft}^{tot} = \log(\|x_{1,ft}\|^2 + \|x_{2,ft}\|^2)$ and the relative log-power $x_{ft}^{rel} = \log(\|x_{2,ft}\|^2) - \log(\|x_{1,ft}\|^2)$, where we use as synonyms the words "power" and "variance". Heuristic separation methods are based on the following remark: if source j_0 has higher power than the other sources in a given time-frequency point (f,t), then the observed direction x_{ft}^{rel} is close to the direction obtained when only source j_0 is present, that is $\theta_{j_0} = \log(A_{2j_0}^2/A_{1j_0}^2)$.

Suppose without loss of generality that the θ_j are sorted in ascending order. The simplest separation method consists in finding the source j_0 that minimizes $|x_{ft}^{rel} - \theta_{j_0}|$ and in setting $\widehat{m_{j_0 ft}} = 1$ and $\widehat{m_{jft}} = 0$ for $j \neq j_0$: we call this the "closest source" method. A derivation is the "1 or 2 closest sources" method, which is to set $\widehat{m_{1,ft}} = 1$ if $x_{ft}^{rel} < \theta_1$, $\widehat{m_{n,ft}} = 1$ if $x_{ft}^{rel} > \theta_n$, and $\widehat{m_{j_0 ft}} = 1$ and $\widehat{m_{j_0+1,ft}} = 1$ if $\theta_{j_0} \leq x_{ft}^{rel} \leq \theta_{j_0+1}$ (and set all other $\widehat{m_{jft}}$ to zero). Finally the "all sources" method consists in setting $\widehat{m_{jft}} = 1$ for all j.

Results for these three separation methods are shown in the first lines of Table 1. Performance is rather good for $\hat{s}_1$ and $\hat{s}_2$ and lower for $\hat{s}_3$, but even for the best method ("1 or 2 closest sources") it remains about 14 dB lower than the oracle performance. There is a compromise between methods that provide high SAR but low SIR ("all sources") and methods that provide high SIR but low SAR ("closest source"). Note that the original "closest source" method described in [3] gave lower performance since it uses only one mixture channel to recover the sources [6]. Computation of mixture sub-bands on a linear frequency scale also yielded lower performance.

3.2 Intrinsic Limitation of Spatial Diversity Cues

We generalize these experimental results by showing that spatial diversity cues have intrinsic ambiguities when the sources overlap in the time-frequency plane. When a source $\mathbf{s}_2$ coming from the left ($\theta_{j_2} < 0$) and a source $\mathbf{s}_1$ from the right ($\theta_{j_1} > 0$) are both present in (f,t) with similar powers, then $x_{ft}^{rel} \approx 0$ so that the source power estimates with the "closest source" method are $\widehat{m_{j_2 ft}} = 0$, $\widehat{m_{j_1 ft}} = 0$ and $\widehat{m_{j_3 ft}} = 1$ for a third source $\mathbf{s}_3$ coming from the center ($\theta_{j_3} \approx 0$). This results in some parts lacking in the "periphery" estimated sources and some excess parts in the "center" estimated sources. This explains why separation performance is generally lower for the "center" source ($\mathbf{s}_3$ here) in a three-source mixture. Note that this limitation generalizes to other estimation methods that use only the single spatial diversity cue x_{ft}^{rel} to determine (m_{jft}). More complex strategies such as [7] suffer from this problem as well in a lesser way.

4 Structured Time-Frequency Source Priors

A way to circumvent this limitation is to use the time-frequency structure of the considered sources. Suppose that $\mathbf{s}_1$ and $\mathbf{s}_2$ play notes with harmonic partials.

Since instruments play in harmony it is very probable that there exists a time-frequency point (f,t) where s_1 and s_2 have similar power. But if they play different notes at that time or the same note with different spectral envelopes, then it is unprobable that s_1 and s_2 have similar power on all time-frequency points (f',t), $1 \leq f' \leq F$. Using the frequency structure of the sources we can remove the ambiguity in x_{ft}^{rel} using information at all frequencies $\mathbf{x}_t^{\text{rel}} = [x_{1,t}^{\text{rel}}, \ldots, x_{F,t}^{\text{rel}}]^T$. Similarly using the time-structure of the sources we can remove ambiguities when sources are masked locally in time (by percussions for example). A problem remains if s_1 and s_2 have the same power on all frequency range for a large time, since all $\mathbf{x}_t^{\text{rel}}$ provide ambiguous information. This problem may also be circumvented using $\mathbf{x}_t^{\text{tot}}$ in conjunction with $\mathbf{x}_t^{\text{rel}}$. For example if $\mathbf{x}_t^{\text{tot}}$ has energy in high frequency bands only, then it is unprobable that instruments playing only low frequency notes are present at that time.

There are two possibilities to use the time-frequency structure of the sources: either decomposing the mixture on structured time-frequency atoms with priors about the decomposition weights and then using estimation laws of Section 3 to derive $(\widehat{\mathbf{m}_j})$, or keeping the same time-frequency decomposition as in Section 2 and then deriving $(\widehat{\mathbf{m}_j})$ with structured priors about $(\mathbf{m}_j)$. We choose here the second solution because musical sources are better described in the time-frequency power domain than in the waveform domain. Relative phases of harmonic partials are rather irrelevant, so that a very large number of atoms would be needed to describe the harmonic structure of most instrumental sounds.

4.1 Structured Priors for Instrumental Sounds

The structured priors we propose here have been used first for single-channel polyphonic music transcription. More details and justifications about our assumptions are available in our companion article [8].

We suppose that each instrument j, $1 \leq j \leq n$, can play a finite number of notes h, $1 \leq h \leq H_j$. At a given time t the presence/absence of note h from instrument j is described with a state variable $E_{jht} \in \{0,1\}$, and its parameters (instantaneous power, instantaneous frequency, instantaneous spectral envelope, etc) are given by a vector of descriptors $\mathbf{p}_{jht} \in \mathbb{R}^{K+1}$. We assume a three-layer generative model, where high-level states (E_{jht}) generate middle-level descriptors $(\mathbf{p}_{jht})$ which in turn generate low-level spectra $(\mathbf{m}_{jt})$. These three layers are termed respectively state layer, descriptor layer and spectral layer.

The spectral layer model is a nonlinear Independent Subspace Analysis (ISA). We write the note descriptors as $\mathbf{p}_{jht} = [e_{jht}, v_{jht}^1, \ldots, v_{jht}^K]$, where e_{jht} is the log-energy of note h from instrument j at time t and (v_{jht}^k) are other variables related to the local spectral shape of this note. Denoting $\mathbf{\Phi'}_{jht}$ the log-power spectrum of note h from instrument j at time t, we assume

$$\mathbf{m}_{jt} = \sum_{h=1}^{H_j} \exp(\mathbf{\Phi'}_{jht}) \exp(e_{jht}) + \mathbf{n}_j, \tag{1}$$

$$\Phi'_{jht} = \Phi_{jh} + \sum_{k=1}^{K} v_{jht}^k \mathbf{U}_{jh}^k, \qquad (2)$$

where exp(.) and log(.) are the exponential and logarithm functions applied to each coordinate. The vector Φ_{jh} is the total-power-normalized mean log-power spectrum of note h from instrument j and $(\mathbf{U}_{jh}^k)$ are L_2-normalized "variation spectra" that model local variations of the spectral shape of this note. The vector $\mathbf{n}_j$ is the power spectrum of the background noise in source j.

The descriptor layer is defined by setting conditional priors on $\mathbf{p}_{jht}$ given E_{jht}. We assume that e_{jht} is constrained to $-\infty$ and v_{jht}^k to 0 given $E_{jht} = 0$, and that e_{jht} and v_{jht}^k follow independent Gaussian laws given $E_{jht} = 1$.

Finally we consider two models for the state layer in order to study the relative importance of frequential and temporal structure for source separation. A product of Bernoulli priors with constant sparsity factor $P_Z = P(E_{ht} = 0)$ results in frequential structure alone, while a factorial Markov chain prior adds some temporal structure by modeling the typical durations of notes and silences.

4.2 Relationship with the Observed Mixture

This model for $(\mathbf{m}_j)$ is completed with a model relating $(\mathbf{m}_j)$ to $\mathbf{x}^{\text{tot}}$ and $\mathbf{x}^{\text{rel}}$:

$$\mathbf{x}_t^{\text{tot}} = \log\left[\sum_{j=1}^{n} \mathbf{m}_{jt}\right] + \epsilon_t^{\text{tot}}, \qquad (3)$$

$$\mathbf{x}_t^{\text{rel}} = \log\left[\sum_{j=1}^{n} A_{2j}^2 \mathbf{m}_{jt}\right] - \log\left[\sum_{j=1}^{n} A_{1j}^2 \mathbf{m}_{jt}\right] + \epsilon_t^{\text{rel}}. \qquad (4)$$

Experiments show that ϵ_t^{tot} and ϵ_t^{rel} can generally be modeled as independent white generalized exponential noises with sparsity parameters $R^{\text{tot}} \simeq 2$ and $R^{\text{rel}} \simeq 0.7$ (i.e. ϵ_t^{tot} is Gaussian and ϵ_t^{rel} is sparser than a Laplacian noise).

4.3 Model Learning and Source Power Spectra Estimation

The probability of $(\mathbf{m}_j)$ given $\mathbf{x}^{\text{tot}}$ and $\mathbf{x}^{\text{rel}}$ is written as the weighted Bayes law

$$P((\mathbf{m}_j)|\mathbf{x}^{\text{tot}}, \mathbf{x}^{\text{rel}}) \propto (P_{\text{spec}})^{w_{\text{spec}}} (P_{\text{desc}})^{w_{\text{desc}}} P_{\text{state}}, \qquad (5)$$

involving probability terms $P_{\text{spec}} = \prod_t P(\epsilon_t^{\text{tot}}) P(\epsilon_t^{\text{rel}})$, $P_{\text{desc}} = \prod_{jht} P(\mathbf{p}_{jht}|E_{jht})$ and $P_{\text{state}} = \prod_{jh} P(E_{jh,1}, \ldots, E_{jh,T})$ and correcting exponents w_{spec} and w_{desc}. Weighting by w_{spec} with $0 < w_{\text{spec}} < 1$ mimics the existence of dependencies between values of ϵ_t^{tot} and ϵ_t^{rel} at adjacent time-frequency points and makes the model distribution closer to the true data distribution.

We learn the model parameters (mean and "variation" spectra, means and variances, initial and transition probabilities) on single-channel solo excerpts of each instrument using a probabilistic model similar to (3) [8].

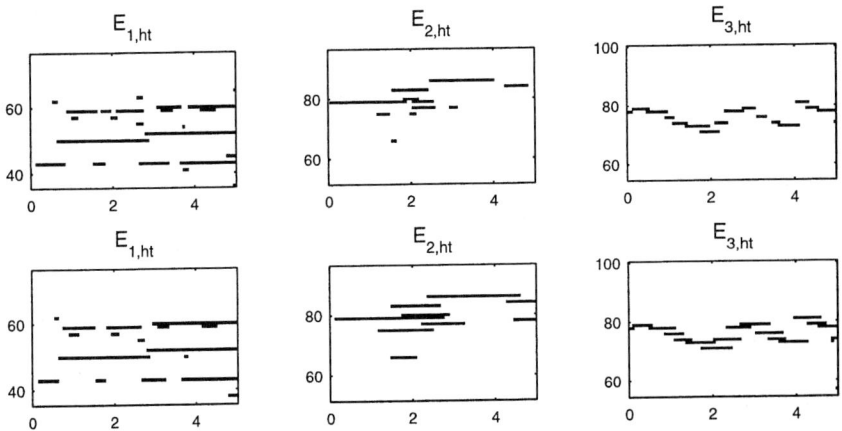

Fig. 2. State sequences obtained with Markov temporal priors (top) compared with oracle state sequences (bottom). The horizontal axis is time in seconds and the vertical axis is note pitch on the MIDI scale.

Then we estimate $(\mathbf{m}_j)$ given $\mathbf{x}^{\text{tot}}$ and $\mathbf{x}^{\text{rel}}$ by finding the states $(\widehat{E_{jht}})$ and the descriptors $(\widehat{e_{jht}})$ and $(\widehat{v^k_{jht}})$ that maximize the posterior (5). Maximization over (E_{jht}) involves a jump procedure with Bernoulli state priors and Viterbi decoding with Markov state priors. Maximization over (e_{jht}) and (v^k_{jht}) is carried out with an approximate second order Newton method. The background noise spectra $(\mathbf{n}_j)$ are re-estimated during transcription to maximize the posterior.

4.4 Performance

The performance of this method was tested using the two defined state models and with an oracle estimator of the state sequence. We also tested separation using only source priors and discarding the spatial likelihood terms $P(\epsilon_t^{\text{rel}})$ in (5). Instrument models were learnt on one-minute solo excerpts taken from other CDs than the test mixture. Results are shown in Table 1.

The combination of spatial diversity and structured source priors provides an average increase of the separation performance of 2.7 dB over spatial diversity alone and 9.7 dB over source priors alone. This proves that our method actually combined the two kinds of information. Results were not significantly improved using larger learning sets.

Moreover results with Markov state priors are a bit better than with Bernoulli state priors, but are still 1.8 dB inferior to results knowing the true state sequence. The main reason for this is not that our method badly estimated the notes played by the instruments, but that some notes were estimated as absent in some zones where they are masked, particularly during reverberation as can be seen in Fig. 2. A way to improve this could be to use more complex state models involving the typical segments "attack, sustain, release" (and reverberation) of musical notes and imposing minimal durations for each of these segments.

5 Conclusion

We considered the source separation problem for underdetermined stereo instantaneous musical mixtures. We proposed a family of probabilistic priors modeling the typical time-frequency structure of musical sources. We showed that combining these priors with spatial diversity leads to a better separation performance than using source priors or spatial diversity alone. This is an important difference with previous works using structured source priors in single-channel [9] and in overdetermined mixtures [10] which did not consider spatial diversity.

A first direction to extend this work is to use simpler source priors involving spectral and temporal continuity but no instrument specific parameters. This could provide faster computations and be useful for the separation of speech mixtures. A second direction we are currently considering is to complexify the source priors with other state models, for example forcing instruments to play monophonic phrases, favoring *legato* note transitions or taking into account the "attack, sustain, release" behavior. We are also studying extension of the method to underdetermined stereo convolutive mixtures using other spatial cues.

References

1. Theis, F., Lang, E.: Formalization of the two-step approach to overcomplete BSS. In: Proc. SIP. (2002) 207–212
2. Gribonval, R.: Piecewise linear separation. In: Wavelets: Applications in Signal and Image Processing, Proc. SPIE. (2003)
3. Yilmaz, O., Rickard, S.: Blind separation of speech mixtures via time-frequency masking. IEEE Transactions on Signal Processing (2002) Submitted.
4. Zibulevsky, M., Pearlmutter, B.: Blind source separation by sparse decomposition in a signal dictionnary. Neural Computation **13** (2001)
5. Deville, Y.: Temporal and time-frequency correlation-based blind source separation methods. In: Proc. ICA. (2003) 1059–1064
6. Gribonval, R., Benaroya, L., Vincent, E., Févotte, C.: Proposals for performance measurement in source separation. In: Proc. ICA. (2003)
7. Vielva, L., Erdoğmuş, D., Príncipe, J.: Underdetermined blind source separation using a probabilistic source sparsity model. In: Proc. ICA. (2001) 675–679
8. Vincent, E., Rodet, X.: Music transcription with ISA and HMM. In: Proc. ICA. (2004)
9. Benaroya, L., Bimbot, F.: Wiener based source separation with HMM/GMM using a single sensor. Proc. ICA (2003) 957–961
10. Reyes-Gomez, M., Raj, B., Ellis, D.: Multi-channel source separation by factorial HMMs. In: Proc. ICASSP. (2003)

A Grassmann-Rayleigh Quotient Iteration for Dimensionality Reduction in ICA*

Lieven De Lathauwer[1,2], Luc Hoegaerts[2], and Joos Vandewalle[2]

[1] ETIS (CNRS, ENSEA, UCP), UMR 8051, Cergy-Pontoise, France
lieven.delathauwer@ensea.fr
[2] E.E. Dept. (ESAT) - SCD, K.U.Leuven, Leuven, Belgium
{luc.hoegaerts,joos.vandewalle}@esat.kuleuven.ac.be

Abstract. We derive a Grassmann-Rayleigh Quotient Iteration for the computation of the best rank-(R_1, R_2, R_3) approximation of higher-order tensors. We present some variants that allow for a very efficient estimation of the signal subspace in ICA schemes without prewhitening.

1 Introduction

Many ICA applications involve high-dimensional data in which however only a few sources have significant contributions. Examples are nuclear magnetic resonance (NMR), electro-encephalography (EEG), magneto-encephalography (MEG), hyper-spectral image processing, data analysis, etc. To reduce the computational complexity and to decrease the variance of the results, one may wish to reduce the dimensionality of the problem from the number of observation channels, which will be denoted by I, to the number of sources, denoted by R. If one wishes to avoid a classical prewhitening, for the reasons given in [7], then the solution can be obtained by means of a so-called best rank-(R_1, R_2, R_3) approximation of a higher-order tensor [4,5]. (Higher-order tensors are the higher-order equivalents of vectors (first order) and matrices (second order), i.e., quantities of which the elements are addressed by more than two indices.) Consequently, in this paper we will derive a numerical algorithm to compute this approximation. It consists of a generalization to tensors of the Rayleigh Quotient Iteration (RQI) for the computation of an invariant subspace of a given matrix [1]. It also generalizes the RQI for the best rank-1 approximation of higher-order tensors [8].

This paper primarily concerns the derivation of the numerical algorithm. Due to space limitations, the relevance of this problem in the context of ICA

---

* L. De Lathauwer holds a permanent research position with the French CNRS; he also holds a honorary position with the K.U.Leuven. L. Hoegaerts is a Ph.D. student supported by the Flemish Institute for the Promotion of Scientific and Technological Research in the Industry (IWT). J. Vandewalle is a Full Professor with the K.U.Leuven. Part of this research was supported by the Research Council K.U.Leuven (GOA-MEFISTO-666), the Flemish Government (F.W.O. project G.0240.99, F.W.O. Research Communities ICCoS and ANMMM, Tournesol project T2004.13) and the Belgian Federal Government (IUAP V-22).

and the link with the best rank-(R_1, R_2, R_3) approximation are discussed in the companion paper [5].

With respect to the numerical aspects of Principal Component Analysis (PCA) world-wide scientific efforts are made. This has led to powerful routines for the computation of the Eigenvalue Decomposition (EVD), Singular Value Decomposition (SVD), dominant subspaces, etc. of high-dimensional matrices. So far, no clear ICA equivalent has emerged. This paper aims to be a first step in this direction.

In Sect. 2 we introduce some basic concepts of multilinear algebra. In Sect. 3 we present our basic algorithm. The formulation is in terms of arbitrary third-order tensors because (i) this allows for the easy derivation of different variants applicable in the context of ICA (Sect. 4), and because (ii) the algorithm has important applications, apart from ICA [4]. Section 5 is the conclusion.

For notational convenience we mainly focus on real-valued third-order tensors. The generalization to complex-valued tensors and tensors of order higher than three is straightforward.

Notation. Scalars are denoted by lower-case letters (a, b, ...), vectors are written as capitals (A, B, ...), matrices correspond to bold-face capitals ($\mathbf{A}$, $\mathbf{B}$, ...) and tensors are written as calligraphic letters ($\mathcal{A}$, $\mathcal{B}$, ...). In this way, the entry with row index i and column index j in a matrix $\mathbf{A}$, i.e., $(\mathbf{A})_{ij}$, is symbolized by a_{ij}. There is one exception: as we use the characters i, j and r in the meaning of indices (counters), I, J and R will be reserved to denote the index upper bounds. $\otimes$ denotes the Kronecker product. $\mathbf{I}$ is the identity matrix. $O(R)$ and $\text{St}(R, I)$ are standard notation for the manifold of $(R \times R)$ orthogonal matrices and the Stiefel manifold of column-wise orthonormal $(I \times R)$ matrices $(I \geqslant R)$, respectively. qf($\mathbf{X}$) denotes the orthogonal factor in a QR-decomposition of a matrix $\mathbf{X}$.

2 Basic Definitions

For a tensor $\mathcal{A} \in \mathbb{R}^{I_1 \times I_2 \times I_3}$, the *matrix unfoldings* $\mathbf{A}_{(1)} \in \mathbb{R}^{I_1 \times I_3 I_2}$, $\mathbf{A}_{(2)} \in \mathbb{R}^{I_2 \times I_1 I_3}$ and $\mathbf{A}_{(3)} \in \mathbb{R}^{I_3 \times I_2 I_1}$ are defined by

$$(\mathbf{A}_{(1)})_{i_1,(i_3-1)I_3+i_2} = (\mathbf{A}_{(2)})_{i_2,(i_1-1)I_1+i_3} = (\mathbf{A}_{(3)})_{i_3,(i_2-1)I_2+i_1} = a_{i_1 i_2 i_3}$$

for all index values. Straightforward generalizations apply to tensors of order higher than three. Consider $\mathbf{U}^{(1)} \in \mathbb{R}^{J_1 \times I_1}$, $\mathbf{U}^{(2)} \in \mathbb{R}^{J_2 \times I_2}$, $\mathbf{U}^{(3)} \in \mathbb{R}^{J_3 \times I_3}$. Then $\mathcal{B} = \mathcal{A} \times_1 \mathbf{U}^{(1)} \times_2 \mathbf{U}^{(2)} \times_3 \mathbf{U}^{(3)}$ is a $(J_1 \times J_2 \times J_3)$-tensor of which the entries are given by

$$b_{j_1 j_2 j_3} = \sum_{i_1 i_2 i_3} a_{i_1 i_2 i_3} u^{(1)}_{j_1 i_1} u^{(2)}_{j_2 i_2} u^{(3)}_{j_3 i_3} .$$

In terms of the matrix unfoldings, we have, for instance,

$$\mathbf{B}_{(1)} = \mathbf{U}^{(1)} \cdot \mathbf{A}_{(1)} \cdot (\mathbf{U}^{(2)} \otimes \mathbf{U}^{(3)})^T.$$

An *n-mode vector* of $\mathcal{A}$ is an I_n-dimensional vector obtained from $\mathcal{A}$ by varying the index i_n and keeping the other indices fixed. It is a column of $\mathbf{A}_{(n)}$. The *n-rank* of a tensor is the obvious generalization of the column (row) rank of matrices: it is defined as the dimension of the vector space spanned by the n-mode vectors and is equal to the rank of $\mathbf{A}_{(n)}$. An important difference with the rank of matrices, is that the different n-ranks of a higher-order tensor are not necessarily the same. A tensor of which the n-ranks are equal to R_n ($1 \leqslant n \leqslant 3$) is called a *rank-$(R_1, R_2, R_3)$ tensor*. A rank-$(1,1,1)$ tensor is briefly called a *rank-1 tensor*. Real-valued tensors are called *supersymmetric* when they are invariant under arbitrary index permutations. Finally, the *Frobenius-norm* of $\mathcal{A}$ is defined as $\|\mathcal{A}\| = (\sum_{i_1 i_2 i_3} a_{i_1 i_2 i_3}^2)^{1/2}$.

Now consider the minimization of the least-squares cost function

$$f(\hat{\mathcal{A}}) = \|\mathcal{A} - \hat{\mathcal{A}}\|^2 \tag{1}$$

under the constraint that $\hat{\mathcal{A}}$ is rank-(R_1, R_2, R_3). This constraint implies that $\hat{\mathcal{A}}$ can be decomposed as

$$\hat{\mathcal{A}} = \mathcal{B} \times_1 \mathbf{X}^{(1)} \times_2 \mathbf{X}^{(2)} \times_3 \mathbf{X}^{(3)}, \tag{2}$$

in which $\mathbf{X}^{(n)} \in \text{St}(R_n, I_n)$, $n = 1, 2, 3$, and $\mathcal{B} \in \mathbb{R}^{R_1 \times R_2 \times R_3}$. The minimization of f can be shown [3] to be equivalent to the maximization of

$$\begin{aligned} g(\mathbf{X}^{(1)}, \mathbf{X}^{(2)}, \mathbf{X}^{(3)}) &= \|\mathcal{A} \times_1 \mathbf{X}^{(1)^T} \times_2 \mathbf{X}^{(2)^T} \times_3 \mathbf{X}^{(3)^T}\|^2 \\ &= \|\mathbf{X}^{(1)^T} \cdot \mathbf{A}_{(1)} \cdot (\mathbf{X}^{(2)} \otimes \mathbf{X}^{(3)})\|^2 \ . \end{aligned} \tag{3}$$

For given $\mathbf{X}^{(1)}, \mathbf{X}^{(2)}, \mathbf{X}^{(3)}$, the optimal $\mathcal{B}$ follows from the linear equation (2).

Now assume that $\mathbf{X}^{(2)}$ and $\mathbf{X}^{(3)}$ are fixed. From (3) we see that $\mathbf{X}^{(1)}$ can only be optimal if its columns span the same subspace as the R_1 dominant left singular vectors of $\tilde{\mathbf{A}}_{(1)} = \mathbf{A}_{(1)} \cdot (\mathbf{X}^{(2)} \otimes \mathbf{X}^{(3)})$. A necessary condition is that the column space of $\mathbf{X}^{(1)}$ is an invariant subspace of $\tilde{\mathbf{A}}_{(1)} \cdot \tilde{\mathbf{A}}_{(1)}^T$. Similar conditions can be derived for the other modes. We obtain:

$$\mathbf{X}^{(1)} \cdot \mathbf{W}_1 = \mathbf{A}_{(1)} \cdot (\mathbf{X}^{(2)} \otimes \mathbf{X}^{(3)}) \cdot (\mathbf{X}^{(2)} \otimes \mathbf{X}^{(3)})^T \cdot \mathbf{A}_{(1)}^T \cdot \mathbf{X}^{(1)} \tag{4}$$

$$\mathbf{X}^{(2)} \cdot \mathbf{W}_2 = \mathbf{A}_{(2)} \cdot (\mathbf{X}^{(3)} \otimes \mathbf{X}^{(1)}) \cdot (\mathbf{X}^{(3)} \otimes \mathbf{X}^{(1)})^T \cdot \mathbf{A}_{(2)}^T \cdot \mathbf{X}^{(2)} \tag{5}$$

$$\mathbf{X}^{(3)} \cdot \mathbf{W}_3 = \mathbf{A}_{(3)} \cdot (\mathbf{X}^{(1)} \otimes \mathbf{X}^{(2)}) \cdot (\mathbf{X}^{(1)} \otimes \mathbf{X}^{(2)})^T \cdot \mathbf{A}_{(3)}^T \cdot \mathbf{X}^{(3)} \tag{6}$$

for some $\mathbf{W}_1 \in \mathbb{R}^{R_1 \times R_1}$, $\mathbf{W}_2 \in \mathbb{R}^{R_2 \times R_2}$, $\mathbf{W}_3 \in \mathbb{R}^{R_3 \times R_3}$.

This set of equations forms the starting point for the derivation of our new algorithm. Note that only the column spaces of $\mathbf{X}^{(1)}$, $\mathbf{X}^{(2)}$ and $\mathbf{X}^{(3)}$ are of importance, and not their individual columns. This means that we are actually working on Grassmann manifolds [6].

3 Higher-Order Grassmann-Rayleigh Quotient Iteration

For $\mathbf{X}^{(1)} \in St(R_1, I_1)$, $\mathbf{X}^{(2)} \in St(R_2, I_2)$, $\mathbf{X}^{(3)} \in St(R_3, I_3)$ and $\mathcal{A} \in \mathbb{R}^{I_1 \times I_2 \times I_3}$ we define *n-mode Rayleigh quotient matrices* as follows:

$$\mathbf{R}_1(\mathbf{X}) = \mathbf{X}^{(1)^T} \cdot \mathbf{A}_{(1)} \cdot (\mathbf{X}^{(2)} \otimes \mathbf{X}^{(3)}) \tag{7}$$

$$\mathbf{R}_2(\mathbf{X}) = \mathbf{X}^{(2)^T} \cdot \mathbf{A}_{(2)} \cdot (\mathbf{X}^{(3)} \otimes \mathbf{X}^{(1)}) \tag{8}$$

$$\mathbf{R}_3(\mathbf{X}) = \mathbf{X}^{(3)^T} \cdot \mathbf{A}_{(3)} \cdot (\mathbf{X}^{(1)} \otimes \mathbf{X}^{(2)}) \ . \tag{9}$$

This definition properly generalizes the existing definitions of Rayleigh quotients associated with an eigenvector, invariant subspace or tensor rank-1 approximation [1, 8]. The cornerstone of our algorithm is the following theorem.

Theorem 1. *Let* $\mathbf{X}^{(1)} \in St(R_1, I_1)$, $\mathbf{X}^{(2)} \in St(R_2, I_2)$, $\mathbf{X}^{(3)} \in St(R_3, I_3)$ *be solutions to (4–6). For small perturbations* $\Delta\mathbf{X}^{(1)}$, $\Delta\mathbf{X}^{(2)}$, $\Delta\mathbf{X}^{(3)}$ *satisfying*

$$\mathbf{X}^{(1)^T} \Delta\mathbf{X}^{(1)} = \mathbf{0}, \quad \mathbf{X}^{(2)^T} \Delta\mathbf{X}^{(2)} = \mathbf{0}, \quad \mathbf{X}^{(3)^T} \Delta\mathbf{X}^{(3)} = \mathbf{0}, \tag{10}$$

we have

$$\|\mathbf{R}_n(\mathbf{X})\mathbf{R}_n(\mathbf{X})^T - \mathbf{R}_n(\mathbf{X}+\Delta\mathbf{X})\mathbf{R}_n(\mathbf{X}+\Delta\mathbf{X})^T\| = O(\|\Delta\mathbf{X}\|^2) \quad n = 1, 2, 3 \ .$$

Proof. Let us consider the case $n = 1$. The cases $n = 2, 3$ are completely similar.
By definition, we have

$$\mathbf{R}_1(\mathbf{X}+\Delta\mathbf{X})\mathbf{R}_1(\mathbf{X}+\Delta\mathbf{X})^T =$$
$$\mathbf{X}^{(1)^T} \cdot \mathbf{A}_{(1)} \cdot (\mathbf{X}^{(2)} \otimes \mathbf{X}^{(3)}) \cdot (\mathbf{X}^{(2)} \otimes \mathbf{X}^{(3)})^T \cdot \mathbf{A}_{(1)}^T \cdot \mathbf{X}^{(1)}$$
$$+(\Delta\mathbf{X}^{(1)})^T \cdot \mathbf{A}_{(1)} \cdot (\mathbf{X}^{(2)} \otimes \mathbf{X}^{(3)}) \cdot (\mathbf{X}^{(2)} \otimes \mathbf{X}^{(3)})^T \cdot \mathbf{A}_{(1)}^T \cdot \mathbf{X}^{(1)}$$
$$+\mathbf{X}^{(1)^T} \cdot \mathbf{A}_{(1)} \cdot (\Delta\mathbf{X}^{(2)} \otimes \mathbf{X}^{(3)}) \cdot (\mathbf{X}^{(2)} \otimes \mathbf{X}^{(3)})^T \cdot \mathbf{A}_{(1)}^T \cdot \mathbf{X}^{(1)}$$
$$+\mathbf{X}^{(1)^T} \cdot \mathbf{A}_{(1)} \cdot (\mathbf{X}^{(2)} \otimes \Delta\mathbf{X}^{(3)}) \cdot (\mathbf{X}^{(2)} \otimes \mathbf{X}^{(3)})^T \cdot \mathbf{A}_{(1)}^T \cdot \mathbf{X}^{(1)}$$
$$+\mathbf{X}^{(1)^T} \cdot \mathbf{A}_{(1)} \cdot (\mathbf{X}^{(2)} \otimes \mathbf{X}^{(3)}) \cdot (\Delta\mathbf{X}^{(2)} \otimes \mathbf{X}^{(3)})^T \cdot \mathbf{A}_{(1)}^T \cdot \mathbf{X}^{(1)}$$
$$+\mathbf{X}^{(1)^T} \cdot \mathbf{A}_{(1)} \cdot (\mathbf{X}^{(2)} \otimes \mathbf{X}^{(3)}) \cdot (\mathbf{X}^{(2)} \otimes \Delta\mathbf{X}^{(3)})^T \cdot \mathbf{A}_{(1)}^T \cdot \mathbf{X}^{(1)}$$
$$+\mathbf{X}^{(1)^T} \cdot \mathbf{A}_{(1)} \cdot (\mathbf{X}^{(2)} \otimes \mathbf{X}^{(3)}) \cdot (\mathbf{X}^{(2)} \otimes \mathbf{X}^{(3)})^T \cdot \mathbf{A}_{(1)}^T \cdot \Delta\mathbf{X}^{(1)} + O(\|\Delta\mathbf{X}\|^2) \ .$$

In this expansion the first term equals $\mathbf{R}_1(\mathbf{X})\mathbf{R}_1(\mathbf{X})^T$. The first-order terms vanish, because of (4–6) and (10). This proves the theorem. □

Consider perturbations $\Delta\mathbf{X}^{(1)}$, $\Delta\mathbf{X}^{(2)}$, $\Delta\mathbf{X}^{(3)}$ satisfying (10). Using Theorem 1, saying that $\mathbf{W}_1 = \mathbf{R}_1(\mathbf{X}) \cdot \mathbf{R}_1(\mathbf{X})^T$ is only subject to second-order perturbations, we have the following linear expansion of (4):

$$(\mathbf{X}^{(1)} + \Delta\mathbf{X}^{(1)}) \cdot \mathbf{R}_1(\mathbf{X}) \cdot \mathbf{R}_1(\mathbf{X})^T =$$
$$\mathbf{A}_{(1)} \cdot \left[(\mathbf{X}^{(2)} \cdot \mathbf{X}^{(2)^T}) \otimes (\mathbf{X}^{(3)} \cdot \mathbf{X}^{(3)^T})\right] \cdot \mathbf{A}_{(1)}^T \cdot (\mathbf{X}^{(1)} + \Delta\mathbf{X}^{(1)}) +$$
$$\mathbf{A}_{(1)} \cdot \left[(\Delta\mathbf{X}^{(2)} \cdot \mathbf{X}^{(2)^T}) \otimes (\mathbf{X}^{(3)} \cdot \mathbf{X}^{(3)^T}) + (\mathbf{X}^{(2)} \cdot \Delta\mathbf{X}^{(2)^T}) \otimes (\mathbf{X}^{(3)} \cdot \mathbf{X}^{(3)^T}) + \right.$$
$$\left. (\mathbf{X}^{(2)}\mathbf{X}^{(2)^T}) \otimes (\Delta\mathbf{X}^{(3)}\mathbf{X}^{(3)^T}) + (\mathbf{X}^{(2)}\mathbf{X}^{(2)^T}) \otimes (\mathbf{X}^{(3)}\Delta\mathbf{X}^{(3)^T})\right] \cdot \mathbf{A}_{(1)}^T \cdot \mathbf{X}^{(1)}. \tag{11}$$

Now, let the (approximate) true solution be given by $\overline{\mathbf{X}}^{(n)} = \mathbf{X}^{(n)} + \Delta \mathbf{X}^{(n)}$, $n = 1, 2, 3$. First we will justify conditions (10). It is well-known [6] that, for $\overline{\mathbf{X}}^{(n)}$ to be on the Stiefel manifold, the perturbation can up to first order terms be decomposed as in

$$\overline{\mathbf{X}}^{(n)} = \mathbf{X}^{(n)}(\mathbf{I} + \Delta \mathbf{E}_1^{(n)}) + (\mathbf{X}^\perp)^{(n)} \Delta \mathbf{E}_2^{(n)},$$

in which $\Delta \mathbf{E}_1^{(n)} \in \mathbb{R}^{R_n \times R_n}$ is skew-symmetric and $(\mathbf{X}^\perp)^{(n)} \in \mathrm{St}(I_n - R_n, I_n)$ perpendicular to $\mathbf{X}^{(n)}$. As a first order approximation we have now

$$\overline{\mathbf{X}}^{(n)} \cdot (\mathbf{I} - \Delta \mathbf{E}_1^{(n)}) = \mathbf{X}^{(n)} + (\mathbf{X}^\perp)^{(n)} \Delta \mathbf{E}_2^{(n)}. \qquad (12)$$

Because of the skew symmetry of $\Delta \mathbf{E}_1^{(n)}$, the matrix $\overline{\mathbf{X}}^{(n)} \cdot (\mathbf{I} - \Delta \mathbf{E}_1^{(n)})$ is in first order column-wise orthonormal, and it has the same column space as $\overline{\mathbf{X}}^{(n)}$. Because only this column space is of importance (and not the individual columns), (12) implies that we can limit ourselves to perturbations satisfying (10).

From (11) we have

$$\overline{\mathbf{X}}^{(1)} \cdot \mathbf{R}_1(\mathbf{X}) \cdot \mathbf{R}_1(\mathbf{X})^T =$$
$$\mathbf{A}_{(1)} \cdot (\mathbf{X}^{(2)} \cdot \mathbf{X}^{(2)T}) \otimes (\mathbf{X}^{(3)} \cdot \mathbf{X}^{(3)T}) \cdot \mathbf{A}_{(1)}^T \cdot (\overline{\mathbf{X}}^{(1)} - 4\mathbf{X}^{(1)}) +$$
$$\mathbf{A}_{(1)} \cdot \left[(\overline{\mathbf{X}}^{(2)} \cdot \mathbf{X}^{(2)T}) \otimes (\mathbf{X}^{(3)} \cdot \mathbf{X}^{(3)T}) + (\mathbf{X}^{(2)} \cdot \overline{\mathbf{X}}^{(2)T}) \otimes (\mathbf{X}^{(3)} \cdot \mathbf{X}^{(3)T}) + \right.$$
$$\left. (\mathbf{X}^{(2)} \cdot \mathbf{X}^{(2)T}) \otimes (\overline{\mathbf{X}}^{(3)} \cdot \mathbf{X}^{(3)T}) + (\mathbf{X}^{(2)} \cdot \mathbf{X}^{(2)T}) \otimes (\mathbf{X}^{(3)} \cdot \overline{\mathbf{X}}^{(3)T}) \right] \cdot \mathbf{A}_{(1)}^T \cdot \mathbf{X}^{(1)} \quad (13)$$

Exploiting the symmetry of the problem, we obtain similar expressions for the 2-mode and 3-mode Rayleigh quotient matrices. The global set consists of linear equations in $\overline{\mathbf{X}}^{(1)}, \overline{\mathbf{X}}^{(2)}, \overline{\mathbf{X}}^{(3)}$. This means that it can be written in the form

$$\mathbf{M}_{\mathcal{A},\mathbf{X}} \overline{X} = B_{\mathcal{A},\mathbf{X}}, \qquad (14)$$

in which the coefficients of $\mathbf{M}_{\mathcal{A},\mathbf{X}} \in \mathbb{R}^{(I_1 R_1 + I_2 R_2 + I_3 R_3) \times (I_1 R_1 + I_2 R_2 + I_3 R_3)}$ and $B_{\mathcal{A},\mathbf{X}} \in \mathbb{R}^{I_1 R_1 + I_2 R_2 + I_3 R_3}$ depend on $\mathcal{A}$ and $\mathbf{X}^{(1)}, \mathbf{X}^{(2)}, \mathbf{X}^{(3)}$ and in which the coefficients of $\overline{\mathbf{X}}^{(1)}, \overline{\mathbf{X}}^{(2)}, \overline{\mathbf{X}}^{(3)}$ are stacked in $\overline{X}$. (Explicit expressions for $\mathbf{M}_{\mathcal{A},\mathbf{X}}$ and $B_{\mathcal{A},\mathbf{X}}$ are not given due to space limitations.) Hence, given $\mathbf{X}^{(1)}, \mathbf{X}^{(2)}, \mathbf{X}^{(3)}$ and the associated n-mode Rayleigh quotient matrices, $\overline{\mathbf{X}}^{(1)}, \overline{\mathbf{X}}^{(2)}, \overline{\mathbf{X}}^{(3)}$ can be estimated by solving a square linear set of equations in $I_1 R_1 + I_2 R_2 + I_3 R_3$ unknowns.

The resulting algorithm is summarized in Table 1. The algorithm can be initialized with the truncated components of the Higher-Order Singular Value Decomposition [2]. This means that the columns of $\overline{\mathbf{X}}_0^{(n)}$ are taken equal to the dominant left singular vectors of $\mathbf{A}_{(n)}$, $n = 1, 2, 3$. See [3, 4] for more details.

The convergence of Alg. 1 is quadratic:

Table 1. GRQI for the computation of the best rank-(R_1, R_2, R_3) approximation of $\mathcal{A} \in \mathbb{R}^{I_1 \times I_2 \times I_3}$.

Given initial estimates $\overline{\mathbf{X}}_0^{(1)} \in \mathbb{R}^{I_1 \times R_1}$, $\overline{\mathbf{X}}_0^{(2)} \in \mathbb{R}^{I_2 \times R_2}$, $\overline{\mathbf{X}}_0^{(3)} \in \mathbb{R}^{I_3 \times R_3}$
Iterate until convergence:
  1. Normalize to matrices on Stiefel manifold:

$$\mathbf{X}_k^{(1)} = \text{qf}(\overline{\mathbf{X}}_k^{(1)}) \qquad \mathbf{X}_k^{(2)} = \text{qf}(\overline{\mathbf{X}}_k^{(2)}) \qquad \mathbf{X}_k^{(3)} = \text{qf}(\overline{\mathbf{X}}_k^{(3)})$$

  2. Compute n-mode Rayleigh quotient matrices:

$$\mathbf{R}_1(\mathbf{X}_k) = \mathbf{X}_k^{(1)^T} \cdot \mathbf{A}_{(1)} \cdot (\mathbf{X}_k^{(2)} \otimes \mathbf{X}_k^{(3)})$$
$$\mathbf{R}_2(\mathbf{X}_k) = \mathbf{X}_k^{(2)^T} \cdot \mathbf{A}_{(2)} \cdot (\mathbf{X}_k^{(3)} \otimes \mathbf{X}_k^{(1)})$$
$$\mathbf{R}_3(\mathbf{X}_k) = \mathbf{X}_k^{(3)^T} \cdot \mathbf{A}_{(3)} \cdot (\mathbf{X}_k^{(1)} \otimes \mathbf{X}_k^{(2)})$$

  3. Solve the linear set of equations

$$\mathbf{M}_{\mathcal{A}, \mathbf{X}_k} \overline{X}_{k+1} = B_{\mathcal{A}, \mathbf{X}_k}$$

Theorem 2. *Let $\overline{\mathbf{X}}^{(1)}, \overline{\mathbf{X}}^{(2)}, \overline{\mathbf{X}}^{(3)}, \mathbf{R}_1(\overline{\mathbf{X}}), \mathbf{R}_2(\overline{\mathbf{X}}), \mathbf{R}_3(\overline{\mathbf{X}})$ correspond to a nonzero solution to (4–6). If $\mathbf{M}_{\mathcal{A}, \overline{\mathbf{X}}}$ is nonsingular, then Alg. 1 converges to $(\overline{\mathbf{X}}^{(1)} \mathbf{Q}_1, \overline{\mathbf{X}}^{(2)} \mathbf{Q}_2, \overline{\mathbf{X}}^{(3)} \mathbf{Q}_3)$, with $\mathbf{Q}_1 \in \mathbf{O}(R_1)$, $\mathbf{Q}_2 \in \mathbf{O}(R_2)$, $\mathbf{Q}_3 \in \mathbf{O}(R_3)$, quadratically in a neighbourhood of $(\overline{\mathbf{X}}^{(1)}, \overline{\mathbf{X}}^{(2)}, \overline{\mathbf{X}}^{(3)})$.*

Proof. Because $\overline{\mathbf{X}}^{(1)}, \overline{\mathbf{X}}^{(2)}, \overline{\mathbf{X}}^{(3)}, \mathbf{R}_1(\overline{\mathbf{X}}), \mathbf{R}_2(\overline{\mathbf{X}}), \mathbf{R}_3(\overline{\mathbf{X}})$ give a solution to (4–6), we have

$$\mathbf{M}_{\mathcal{A}, \overline{\mathbf{X}}} \overline{X} - B_{\mathcal{A}, \overline{\mathbf{X}}} = 0 \ .$$

Consider $\mathbf{X}^{(1)} = \overline{\mathbf{X}}^{(1)} - \Delta \mathbf{X}^{(1)}$, $\mathbf{X}^{(2)} = \overline{\mathbf{X}}^{(2)} - \Delta \mathbf{X}^{(2)}$, $\mathbf{X}^{(3)} = \overline{\mathbf{X}}^{(3)} - \Delta \mathbf{X}^{(3)}$, with $\Delta \mathbf{X}^{(1)}$, $\Delta \mathbf{X}^{(2)}$, $\Delta \mathbf{X}^{(3)}$ satisfying (10). Because of Theorem 1 and (13) we have

$$\mathbf{M}_{\mathcal{A}, \mathbf{X}} \overline{X} - B_{\mathcal{A}, \mathbf{X}} = O(\|\Delta \mathbf{X}\|^2) \ .$$

Because $\mathbf{M}_{\mathcal{A}, \overline{\mathbf{X}}}$ is nonsingular, we can write:

$$(\|\Delta \overline{\mathbf{X}}_{k+1}^{(1)}\|^2 + \|\Delta \overline{\mathbf{X}}_{k+1}^{(2)}\|^2 + \|\Delta \overline{\mathbf{X}}_{k+1}^{(3)}\|^2)^{1/2} = \|\overline{X} - \overline{X}_{k+1}\|$$
$$= \|\overline{X} - \mathbf{M}_{\mathcal{A}, \mathbf{X}_k}^{-1} B_{\mathcal{A}, \mathbf{X}_k}\| = O(\|\mathbf{M}_{\mathcal{A}, \mathbf{X}_k} B_{\mathcal{A}, \mathbf{X}_k} - \overline{X}\|) = O(\|\Delta \mathbf{X}_k\|^2) \ . \quad (15)$$

This equation indicates that the convergence is quadratic. Finally, we verify that

$$\|\Delta \overline{\mathbf{X}}_{k+1}^{(n)}\|^2 = O(\min_{\mathbf{Q} \in \mathbf{O}(R_n)} \|\text{qf}(\overline{\mathbf{X}}_{k+1}^{(n)}) - \mathbf{X}^{(n)} \mathbf{Q}\|^2), \quad n = 1, 2, 3 \ .$$

This means that the normalization in step 1 of Alg. 1 does not decrease the convergence rate. □

4 Variants for Dimensionality Reduction in ICA

Variant 1. Several ICA-methods are based on the joint diagonalization of a set of matrices $\mathbf{A}_1, \ldots, \mathbf{A}_J \in \mathbb{R}^{I \times I}$. In the absence of noise, these matrices satisfy

$$\mathbf{A}_j = \mathbf{M} \cdot \mathbf{D}_j \cdot \mathbf{M}^T, \qquad j = 1, \ldots, J$$

in which $\mathbf{M}$ is the mixing matrix and $\mathbf{D}_j \in \mathbb{R}^{R \times R}$ are diagonal. These matrices can be stacked in a tensor $\mathcal{A} \in \mathbb{R}^{I \times I \times J}$. Because the columns of all $\mathbf{A}_j$ are linear combinations of the columns of $\mathbf{M}$, the 1-mode vector space of $\mathcal{A}$ is the column space of $\mathbf{M}$ and its 1-mode rank equals R. Because of the symmetry, the 2-mode vector space also coincides with the column space of $\mathbf{M}$ and the 2-mode rank is also equal to R. It can be verified that the 3-mode vectors are linear combinations of the vectors $(\mathbf{D}_1(r,r), \ldots, \mathbf{D}_J(r,r))^T$, $r = 1, \ldots, R$. This is shown in detail in [5]. Hence the 3-mode rank is bounded by R.

A dimensionality reduction can thus be achieved by computing the best rank-(R,R,R) approximation of $\mathcal{A}$. A difference with Sect. 3 is that now $\mathbf{X}^{(1)} = \mathbf{X}^{(2)}$, $\overline{\mathbf{X}}^{(1)} = \overline{\mathbf{X}}^{(2)}$, $\mathbf{R}_1(\mathbf{X}) = \mathbf{R}_2(\mathbf{X})$, because of the symmetry. When $R < J$, this can simply be inserted in (13). Equation (14) then becomes a square set in $(I + J)R$ unknowns.

Variant 2. When $R \geqslant J$, the computation can further be simplified. In this case, no dimensionality reduction in the third mode is needed, and $\mathbf{X}^{(3)}$ can be fixed to the identity matrix. Equation (13) reduces to

$$\overline{\mathbf{X}}^{(1)} \cdot \mathbf{R}_1(\mathbf{X}) \cdot \mathbf{R}_1(\mathbf{X})^T =$$
$$\mathbf{A}_{(1)} \cdot \left[(\mathbf{X}^{(1)} \cdot \mathbf{X}^{(1)T}) \otimes \mathbf{I} \right] \cdot \mathbf{A}_{(1)}^T \cdot (\overline{\mathbf{X}}^{(1)} - 2\mathbf{X}^{(1)})$$
$$+ \mathbf{A}_{(1)} \cdot \left[(\overline{\mathbf{X}}^{(1)} \cdot \mathbf{X}^{(1)T} + \mathbf{X}^{(1)T} \cdot \overline{\mathbf{X}}^{(1)}) \otimes \mathbf{I} \right] \cdot \mathbf{A}_{(1)}^T \cdot \mathbf{X}^{(1)} \ . \qquad (16)$$

(Note that the factor 4 in (13) has been replaced by a factor 2, because two of the terms in (11) vanish.) Equation (14) now becomes a square set in IR unknowns.

Variant 3. Now assume that one wants to avoid the use of second-order statistics (e.g. because the observations are corrupted by additive coloured Gaussian noise). We consider the case where the dimensionality reduction is based on the observed fourth-order cumulant $\mathcal{K}^Y \in \mathbb{R}^{I \times I \times I \times I}$ instead. In the absence of noise we have

$$\mathcal{K}^Y = \mathcal{K}^S \times_1 \mathbf{M} \times_2 \mathbf{M} \times_3 \mathbf{M} \times_4 \mathbf{M},$$

in which $\mathcal{K}^S \in \mathbb{R}^{R \times R \times R \times R}$ is the source cumulant. This equation implies that all n-mode vectors, for arbitrary n, are linear combinations of the R mixing vectors. In other words, $\mathcal{K}^Y$ is a supersymmetric rank-(R,R,R,R) tensor. Hence it is natural to look for a matrix $\mathbf{X}^{(1)} \in \operatorname{St}(R,I)$ that maximizes

$$g(\mathbf{X}^{(1)}) = \| \mathcal{K}^Y \times_1 \mathbf{X}^{(1)T} \times_2 \mathbf{X}^{(1)T} \times_3 \mathbf{X}^{(1)T} \times_4 \mathbf{X}^{(1)T} \|^2 \ .$$

A necessary condition is that $\mathbf{X}^{(1)}$ maximizes

$$h(\mathbf{U}) = \| \mathbf{U}^T \cdot \mathbf{K}^Y_{(1)} \cdot (\mathbf{X}^{(1)} \otimes \mathbf{X}^{(1)} \otimes \mathbf{X}^{(1)}) \|^2, \quad \mathbf{U} \in \operatorname{St}(R,I) \ . \qquad (17)$$

The matrix $\mathbf{K}^Y_{(1)} \in \mathbb{R}^{I \times I^3}$ is a matrix unfolding of $\mathcal{K}^Y$. Given (17), we can proceed as in Sect. 2 and 3.

Variant 4. Finally, we consider the mixed use of second- *and* fourth-order statistics. In this case, it is natural to consider the maximization of the function

$$g(\mathbf{X}^{(1)}) = \|\mathbf{X}^{(1)^T} \cdot \mathbf{C}^Y \cdot \mathbf{X}^{(1)}\|^2 + \|\mathbf{X}^{(1)^T} \cdot \mathbf{K}^Y_{(1)} \cdot (\mathbf{X}^{(1)} \otimes \mathbf{X}^{(1)} \otimes \mathbf{X}^{(1)})\|^2, \quad (18)$$

in which the two terms are possibly weighted. The optimal $\mathbf{X}^{(1)}$ has to maximize $h(\mathbf{U}) = \|\mathbf{U}^T \cdot \mathbf{F}^Y(\mathbf{X}^{(1)})\|^2$, with

$$\mathbf{F}^Y(\mathbf{X}^{(1)}) = \left(\mathbf{C}^Y \cdot \mathbf{X}^{(1)} \quad \mathbf{K}^Y_{(1)} \cdot (\mathbf{X}^{(1)} \otimes \mathbf{X}^{(1)} \otimes \mathbf{X}^{(1)}) \right) .$$

A necessary condition is that

$$\mathbf{X}^{(1)} \cdot \mathbf{W}_1 = \mathbf{F}^Y(\mathbf{X}^{(1)}) \cdot (\mathbf{F}^Y(\mathbf{X}^{(1)}))^T \cdot \mathbf{X}^{(1)} \quad (19)$$

for some $\mathbf{W}_1 \in \mathbb{R}^{R \times R}$. From here, we can proceed as in Sect. 3. The role of $\mathbf{R}_1(\mathbf{X}) \cdot \mathbf{R}_1(\mathbf{X})^T$ is played by $\mathbf{W}_1$.

5 Conclusion

We have derived a higher-order Grassmann-Rayleigh Quotient Iteration, which can be used for dimensionality reduction in ICA without prewhitening. The convergence is quadratic and each iteration step merely involves solving a square set of linear equations. This is a big improvement over the algorithm discussed in [3], of which the convergence is at most linear and of which each iteration involves the partial computation of a number of SVDs. The relevance to ICA is further substantiated in [5], which also contains some simulation results.

References

1. Absil, P.-A., Mahony, R., Sepulchre, R., Van Dooren, P.: A Grassmann-Rayleigh quotient iteration for computing invariant subspaces. SIAM Rev. **44** (2002) 57–73.
2. De Lathauwer, L., De Moor, B., Vandewalle, J.: A multilinear singular value decomposition. SIAM J. Matrix Anal. Appl. **21** (2000) 1253–1278.
3. De Lathauwer, L., De Moor, B., Vandewalle, J.: On the best rank-1 and rank-$(R_1, R_2, \ldots, R_N)$ approximation of higher-order tensors. SIAM J. Matrix Anal. Appl. **21** (2000) 1324–1342.
4. De Lathauwer, L., Vandewalle, J.: Dimensionality reduction in higher-order signal processing and rank-$(R_1, R_2, \ldots, R_N)$ reduction in multilinear algebra. Lin. Alg. Appl. (to appear).
5. De Lathauwer, L., Vandewalle, J.: Dimensionality Reduction in ICA and Rank-$(R_1, R_2, \ldots, R_N)$ Reduction in Multilinear Algebra. Proc. ICA 2004.
6. Edelman, A., Arias, T.A., Smith, S.T.: The geometry of algorithms with orthogonality constraints. SIAM J. Matrix Anal. Appl. **20** (1998) 303–353.
7. Yeredor, A.: Non-orthogonal joint diagonalization in the least-squares sense with application in blind source separation. IEEE Trans. Signal Processing **50** (2002) 1545–1553.
8. Zhang, T., Golub, G.H.: Rank-one approximation to high order tensors. SIAM J. Matrix Anal. Appl. **23** (2001) 534–550.

An Approach of Moment-Based Algorithm for Noisy ICA Models

Daisuke Ito and Noboru Murata

Department of Electrical, Electronics, and Computer Engineering
Waseda University
3-4-1 Ohkubo, Shinjuku, Tokyo, 169-8555, Japan
Daisuke.Ito@murata.elec.waseda.ac.jp
Noboru.Murata@eb.waseda.ac.jp

Abstract. Factor analysis is well known technique to uncorrelate observed signals with Gaussina noises before ICA (Independent Component Analysis) algorithms are applied. However, factor analysis is not applicable when the number of source signals are more than that of Ledermann's bound, and when the observations are contaminated by non-Gaussian noises. In this paper, an approach is proposed based on higher-order moments of signals and noises in order to overcome those constraints.

1 Introduction

Independent component analysis (ICA) has become a powerful tool for analyzing observed signals, which are mixtures of mutually independent components. There are two typical models for analysis, a noiseless model and a noisy model. For a noiseless model, an uncorrelation procedure based on principle component analysis (PCA) is often used before ICA algorithms are applied. A noisy model is constructed from noiseless models with contamination by Gaussian noise. In that case, the uncorrelation procedure is usually achieved by factor analysis (FA), instead of PCA[1, 2].

Even though FA is known as a powerful tool to achieve uncorrelation, FA has two critical limitaions. One is, that the number of dimension of source signals must be smaller than that of observed signals. Because of this limitation, a noisy ICA model, in which the dimension of source signals are equal, or almost equal to that of observed signals, can not be uncorrelated. This limitation comes from the fact that FA uses only covariance structure, i.e. second order statistics.

Another critical limitation is that noises are assumed to be a Gaussian most of the cases. However, in practical case, there are various noises that are not i.i.d. or normally distributed but contains some kind of non-Gaussianity. In that case, we can not use FA.

We propose an approach that can overcome the dimensional constraint and non-Gaussian noise in a noisy ICA model. Our proposing approach will be realized by two step iterations; 1) evaluate the higher order moments of decomposed components and noises from observed signals subject to the estimated mixing matrix, and 2) update the mixing matrix subject to estimated moments of decomposed components and noises.

2 ICA Model

Before explaining our approach, we first review two ICA models. A basic ICA model is called a noiseless ICA model

$$\mathbf{x} = A\mathbf{s},$$

where $\mathbf{x} = [x_1, \cdots, x_n]^\tau$ is an n-dimensional observation vector, $A \in \Re^{n \times m}$ is an unknown mixing matrix, $\mathbf{s}$ is an m-dimensional source signal vector with zero mean and unit variance, whose signals are mutually independent, τ is the transposition, and we assume $n \geq m$.

The aim of ICA is to recover the source signal $\mathbf{s}$ by estimating separating matrix that decomposes $\mathbf{x}$ into independent components. PCA is often used as preprocessing to uncorrelate observed signals. Uncorrelated components are obtained by a whitening matrix Q

$$\mathbf{z} = Q\mathbf{x} = \Lambda^{-\frac{1}{2}} U \mathbf{x}, \qquad\qquad V = U^\tau \Lambda U$$

where V is a covariance matrix of $\mathbf{x}$, Λ is a diagonal matrix whose elements are eigenvalue of V, U is an orthogonal matrix, and $\mathbf{z}$ is an m-dimensional vector. Then, ICA algorithms are applied to estimate separating matrix W that decomposes $\mathbf{z}$ into independent components as

$$\mathbf{y} = W\mathbf{z} = WQ\mathbf{x}.$$

In practical situation, observed signals contain noises

$$\mathbf{x} = A\mathbf{s} + \mathbf{n}, \tag{1}$$

where $\mathbf{n}$ is a noise vector. This is called a noisy ICA model. In this noisy case, conventional FA is often used for uncorrelation procedures[3], assuming $\mathbf{n}$ is subject to Gaussian distribution $N(0, \Sigma)$, where Σ is a covariance matrix of noise $\mathbf{n}$. FA is a method to explain high dimensional observed signals by small number of source signals, based on covariance structure

$$V = AA^\tau + \Sigma.$$

When estimates of A and Σ are obtained, observed signals are uncorrelated with a matrix

$$Q = \left(A^\tau \Sigma^{-1} A\right)^{-1} A^\tau \Sigma^{-1}.$$

When to apply FA, the dimensional constraint is required, that is, the dimension of decomposed signals must be smaller than Ledermann's bound[4],

$$m \leq \frac{1}{2}\left(2n + 1 - \sqrt{8n+1}\right), \tag{2}$$

where A, Σ and V have $nm - m(m-1)/2$, n and $n(n+1)/2$ meaningful free parameters, respectively.

However, in case that the dimension of source signals are equal, or almost equal to that of observed signals, FA can not be used as a preprocessing. Also, in case that the contaminating noise is not i.i.d. nor Gaussian, FA can not be used as a preprocesing as well.

Our aim is to deal with a practical noisy ICA model 1) without dimensional constraints, and 2) with non-Gaussian noises.

3 Proposing Approach

We propose an approach of extending FA to overcome constraints of a conventional noisy ICA model, that is to rely on second and fourth order moments to solve a noisy ICA model as (1), where $\mathbf{x} \in \Re^n$, $A \in \Re^{n \times m}$, $\mathbf{s} \in \Re^m$, $\mathbf{n} \in \Re^n$ and $\mathbf{n}$ obeys non-Gaussian distribution, for estimating the entire mixing matrix.

Let us calculate the bound of our approach by comparing FA. The matrix A, fourth order moment of $\mathbf{s}$, and second and fourth moment of $\mathbf{n}$ has nm, m and $2n$ meaningful free parameters, respectively. On the other hand, fourh order moment of $\mathbf{x}$ and two second order moment pair of $\mathbf{x}$ has n and $n(n-1)/2$ free parameters, respectively, since we assume $E[\mathbf{x}] = 0$. Also we have $n(n-1)/2$ free parameters for third and first order moment pair, that it can not be ignore. Here we assume two first order moment pair, such as $(2, 1, 1)$ order moment is close to 0 and we omit them. Therefore, we have sufficient conditions as

$$m \leq \frac{2n(n-1)}{n+1}. \tag{3}$$

It is clear that our method is able to handle wider bound than (2). Note that our approach can assume the same number of sources with observations if $n \geq 3$.

The procedure of our approach is briefly summarized as follows.

---

step.1 Estimate an initial orthogonal mixing matirix $\hat{A}$, by PCA or some ICA algorithms.
step.2 Estimate moments of decomposed components and noises subject to estimated mixing matrix $\hat{A}$.
step.3 Update mixing matrix $\hat{A}$ subject to estimated moments of decomposed components and noise.
step.4 Repeat step.2 and 3 until a certain condition is fulfilled.

3.1 Moment Estimation

Let X_i be random variables with zero mean and x_i be their observations. Let us define the u-th order moment of empirical estimate

$$m(x_i^u) = \frac{1}{T} \sum x_i^u, \tag{4}$$

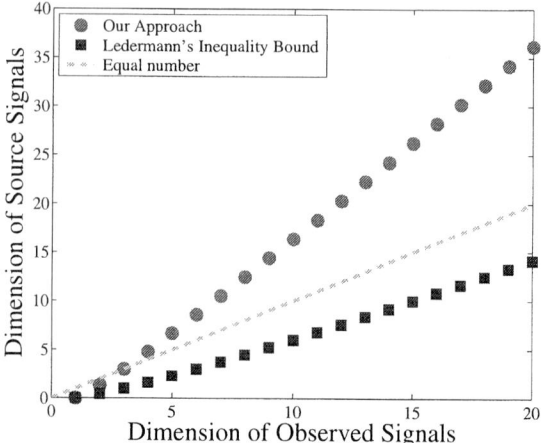

Fig. 1. Bound of number of dimensions.

where u is the order of moment and T is a number of sample. It relates well known statistics, mean and variance, when $u = 1, 2$, respectively. In the following, we assume that odd moments of $\mathbf{s}$ and $\mathbf{n}$, i.e. $u = 1, 3$, vanishes. Also let us define the u-th-v-th cross moment of two distinct variables

$$m(x_i^u x_j^v) = \frac{1}{T} \sum x_i^u x_j^v. \tag{5}$$

Thanks to the linearity of the moment, $m(\mathbf{x}^u)$ can be calculated by using $m(\mathbf{s}^u)$ and $m(\mathbf{n}^u)$, and mixing matrix. We define conditional u-th order moment of observed signals $m(\mathbf{x}^u|\hat{A}, \mathbf{s}, \mathbf{n})$ by mixing matrix $\hat{A}$, source signals $\mathbf{s}$ and noise $\mathbf{n}$, that is to say, estimated moments of x_i under the model (1) where A and moments of $\mathbf{s}$ and $\mathbf{n}$ are given. However, $m(\mathbf{x}^u|\hat{A}, \mathbf{s}, \mathbf{n})$ is determined by u-th and less order moments of $\mathbf{s}$ and $\mathbf{n}$, that is to say, $m(\mathbf{s}^u), \cdots, m(\mathbf{s}), m(\mathbf{n}^u), \cdots, m(\mathbf{n})$, as the linearity of the moment holds.

Loss functions $l_b \in l, (b = 1, \cdots, 5)$ are defined by sum of square difference of higher order moments between observed signals $\mathbf{x}$ and conditional reconstructed observed signals under estimated mixing matrix $\hat{A}$ and moments of $\mathbf{s}$ and $\mathbf{n}$.

$$l_1^{(i)}(\mathbf{x}|\hat{A}, \mathbf{s}, \mathbf{n}) = \left\{ m(x_i^4) - m(x_i^4|\hat{A}, \mathbf{s}, \mathbf{n}) \right\}^2 \tag{6}$$

$$l_2^{(i,j)}(\mathbf{x}|\hat{A}, \mathbf{s}, \mathbf{n}) = \left\{ m(x_i^2 x_j^2) - m(x_i^2 x_j^2|\hat{A}, \mathbf{s}, \mathbf{n}) \right\}^2 \tag{7}$$

$$l_3^{(i,j)}(\mathbf{x}|\hat{A}, \mathbf{s}, \mathbf{n}) = \left\{ m(x_i^3 x_j) - m(x_i^3 x_j|\hat{A}, \mathbf{s}, \mathbf{n}) \right\}^2 \tag{8}$$

$$l_4^{(i)}(\mathbf{x}|\hat{A}, \mathbf{s}, \mathbf{n}) = \left\{ m(x_i^2) - m(x_i^2|\hat{A}, \mathbf{s}, \mathbf{n}) \right\}^2 \tag{9}$$

$$l_5^{(i,j)}(\mathbf{x}|\hat{A}, \mathbf{s}, \mathbf{n}) = \left\{ m(x_i x_j) - m(x_i x_j|\hat{A}, \mathbf{s}, \mathbf{n}) \right\}^2 \tag{10}$$

Proposing method is closely related to the GLS method[5], but fundamental difference is that we use not only higher order moments of signals but higher order moments of noises as well. A total loss function L is a sum of $l_b, (b = 1, \cdots, 5)$ as follow;

$$L(\mathbf{x}|\hat{A}, \mathbf{s}, \mathbf{n}) = \sum_i l_1^{(i)} + \sum_{i \neq j} l_2^{(i,j)} + \sum_{i \neq j} l_3^{(i,j)} \\ + \sum_i l_4^{(i)} + \sum_{i \neq j} l_5^{(i,j)}. \quad (11)$$

Finally, moments of $\mathbf{s}$ and $\mathbf{n}$ are estimated by minimizing $L(\mathbf{x}|\hat{A}, \mathbf{s}, \mathbf{n})$,

$$\underset{m(\mathbf{s}^4), m(\mathbf{n}^4), m(\mathbf{n}^2)}{\text{minimize}} L(\mathbf{x}|\hat{A}, \mathbf{s}, \mathbf{n}), \quad (12)$$

where we assume $m(\mathbf{s}^2) = 1$. However, fouth order moments are not stable than second order moments, (12) will required to minimize first three loss functions subject to latter two.

$$\underset{m(\mathbf{s}^4), m(\mathbf{n}^4), m(\mathbf{n}^2)}{\text{minimize}} \sum_i l_1^{(i)} + \sum_{i \neq j} l_2^{(i,j)} + \sum_{i \neq j} l_3^{(i,j)} \quad (13)$$
$$\text{subject to } l_4^{(i)} = 0, l_5^{(i,j)} = 0.$$

3.2 Mixing Matrix Update

Once moments of $\mathbf{s}^u$ and $\mathbf{n}^u$ are estimated based on the loss functions, we then need to optimize mixing matrix $\hat{A}$ subject to $m(\mathbf{s}^u)$ and $m(\mathbf{n}^u)$. We use an iterative multiplicative update[6] to optimize mixing matrix $\hat{A}$. Let us set

$$\mathcal{M} = \left\{ \Delta \in \Re^{m \times m} | \Delta_{ii} = 0, (1 \leq i \leq m) \right\}. \quad (14)$$

Let an initial mixing matrix be $\hat{A}^{(0)}$, mixing matrix $\hat{A}^{(t)}$ is updated as

$$\hat{A}^{(t)} = e^{\Delta^{(t-1)}} \hat{A}^{(t-1)} \\ = e^{\Delta^{(t-1)}} \hat{A}^{(t-1)} e^{\Delta^{(t-2)}} \hat{A}^{(t-2)} \quad (15) \\ \to e^{\Delta^{(t-1)}} e^{\Delta^{(t-2)}} \cdots e^{\Delta^{(0)}} \hat{A}^{(0)},$$

where $e^{\Delta^{(t)}}$ is a product of three-body interactions

$$e^{\Delta^{(t)}} = \prod_{i<j<k}^{m} e^{\Delta(i,j,k)} \quad (16) \\ = e^{\Delta(1,2,3)} e^{\Delta(1,2,4)} \cdots e^{\Delta(m-2,m-1,m)}$$

where

$$\Delta_{(i,j,k)} = \begin{matrix} & & i & j & k & \\ & \begin{pmatrix} 0 & & \cdots & & & 0 \\ & \ddots & & & & \\ i & & & \Delta_{ij} & \Delta_{ik} & \\ j & \vdots & \Delta_{ji} & \ddots & \Delta_{jk} & \vdots \\ k & & \Delta_{ki} & \Delta_{kj} & \ddots & \\ 0 & & \cdots & & & 0 \end{pmatrix} \end{matrix}. \quad (17)$$

Each $e^{\Delta_{(i,j,k)}}$ is determined by optimizing loss function L,

$$\begin{aligned} &\underset{\Delta_{(i,j,k)}}{\text{minimize}} L(e^{\Delta_{(i,j,k)}} \hat{A} | \mathbf{x}, \mathbf{s}, \mathbf{n}) \\ &\text{subject to } l_4^{(i)} = 0, l_5^{(i,j)} = 0. \end{aligned} \quad (18)$$

However, minimizing (18) is not easy task. We obtain it by calculating each loss function as follow,

$$\begin{aligned} l_b(e^{\Delta_{(i,j,k)}} \hat{A} | \mathbf{x}, \mathbf{s}, \mathbf{n}) &\approx l_b(\hat{A} | \mathbf{x}, \mathbf{s}, \mathbf{n}) + \frac{\partial l_b(\hat{A} | \mathbf{x}, \mathbf{s}, \mathbf{n})}{\partial \Delta_{(i,j,k)}} \Delta_{(i,j,k)} \\ &= l_b(\hat{A} | \mathbf{x}, \mathbf{s}, \mathbf{n}) + \sum_{p,q} \frac{\partial l_b(\hat{A} | \mathbf{x}, \mathbf{s}, \mathbf{n})}{\partial \hat{A}_{pq}} \frac{\partial \hat{A}_{pq}}{\partial \Delta_{(i,j,k)}} \Delta_{(i,j,k)} \quad (19) \\ &= l_b(\hat{A} | \mathbf{x}, \mathbf{s}, \mathbf{n}) + \left(\sum_{p,q} \frac{\partial l_b(\hat{A} | \mathbf{x}, \mathbf{s}, \mathbf{n})}{\partial \hat{A}_{pq}} \frac{\partial \hat{A}_{pq}}{\partial \Delta_{(i,j,k)}} \right) \tilde{\Delta} \end{aligned}$$

where $\tilde{\Delta} = [\Delta_{ij}, \Delta_{ik}, \cdots, \Delta_{kj}]^T \in \Re^{6 \times 1}$ is elements of (17) and $e^{\Delta} \approx I + \Delta$ when Δ is sufficiently small.

Decomposing all elements of all loss functions,

$$l_1^{(i)}(e^{\Delta_{(i,j,k)}} \hat{A} | \mathbf{x}, \mathbf{s}, \mathbf{n}) = l_1^{(i)}(\hat{A} | \mathbf{x}, \mathbf{s}, \mathbf{n}) + \sum_{p,q} \frac{\nabla l_1^{(i)}(\hat{A}, \mathbf{x}, \mathbf{s}, \mathbf{n})}{\partial \hat{A}_{pq}} \frac{\partial \hat{A}_{pq}}{\tilde{\Delta}} \tilde{\Delta}$$

$$\vdots \quad (20)$$

$$l_5^{(j,k)}(e^{\Delta_{(i,j,k)}} \hat{A} | \mathbf{x}, \mathbf{s}, \mathbf{n}) = l_5^{(j,k)}(\hat{A} | \mathbf{x}, \mathbf{s}, \mathbf{n}) + \sum_{p,q} \frac{\nabla l_5^{(j,k)}(\hat{A}, \mathbf{x}, \mathbf{s}, \mathbf{n})}{\partial \hat{A}_{pq}} \frac{\partial \hat{A}_{pq}}{\tilde{\Delta}} \tilde{\Delta},$$

where ∇ denotes the partial derivative with respect to $\tilde{\Delta}$. By minimizing loss functions $l_b(e^{\Delta} \hat{A} | \mathbf{x}, \mathbf{s}, \mathbf{n})$, we can simplify (20) as

$$\mathbf{0} = \mathbf{l} + \mathbf{D}\tilde{\Delta} \quad (21)$$

where

$$\mathbf{l} = \left[l_1^{(i)}(\hat{A}|\mathbf{x},\mathbf{s},\mathbf{n}), \cdots, l_5^{(j,k)}(\hat{A}|\mathbf{x},\mathbf{s},\mathbf{n}) \right]^T \quad (22)$$

$$\mathbf{D} = \left[\sum_{p,q} \frac{\partial \nabla l_1^{(i)}(\hat{A}|\mathbf{x},\mathbf{s},\mathbf{n})}{\partial \hat{A}_{pq}} \frac{\partial \hat{A}_{pq}}{\partial \tilde{\Delta}}, \cdots, \sum_{p,q} \frac{\partial \nabla l_5^{(j,k)}(\hat{A}|\mathbf{x},\mathbf{s},\mathbf{n})}{\partial \hat{A}_{pq}} \frac{\partial \hat{A}_{pq}}{\partial \tilde{\Delta}} \right]^T. \quad (23)$$

We get updating weight $\tilde{\Delta}$ as

$$\tilde{\Delta} = -\mathbf{D}^{-1}\mathbf{l}, \tag{24}$$

and now $\Delta_{(i,j,k)}$ is updated using $\tilde{\Delta}$ followed by (17). Another way to achieve minimization of (20) is using line-search procedure with respect to interaction of $\tilde{\Delta}$,

$$\tilde{\Delta}^{(v+1)} = \tilde{\Delta}^{(v)} - L(e^{\tilde{\Delta}^{(v)}}|\mathbf{x},\mathbf{s},\mathbf{n})\frac{\tilde{\Delta}^{(v)} - \tilde{\Delta}^{(v-1)}}{L(e^{\tilde{\Delta}^{(v)}}|\mathbf{x},\mathbf{s},\mathbf{n}) - L(e^{\tilde{\Delta}^{(v-1)}}|\mathbf{x},\mathbf{s},\mathbf{n})}. \tag{25}$$

Finally updated mixing matrix is obtained by

$$\begin{aligned}\hat{A}^{(t+1)} &= e^{\Delta^{(t)}}\hat{A}^{(t)} \\ &= \prod_{i<j<k} e^{\Delta_{(i,j,k)}}\hat{A}^{(t)}.\end{aligned} \tag{26}$$

4 Conclusion

We have discussed a practical noisy ICA model, that can overcome the dimensional constraint and non-Gaussian noises that FA has. Using second and fourth order moment, we can obtain enough equation for solving the unknowns with less stringent restrictions on the number of sources, m. Procedure of our approach is based on two step iterations; 1) evaluate the higher order moments of $\mathbf{s}$ and $\mathbf{n}$ subject to observed signals and the estimated mixing matrix $\hat{A}$, and 2) update the mixing matrix $\hat{A}$ subject to estimated higher order moments of $\mathbf{x}$ and $\mathbf{n}$.

Numerical simulations and more detailed consideration about adequacy, stability and robustness are needed as a future work.

References

1. Ikeda, S., Toyama, K.: Independent componenta fast fixed-point algorithm for independent component analysi analysis for noisy data - MEG data analysis. Neural Networks **13** (2000) 1063–1074
2. Kawanabe, M., Murata, N.: Independent component analysis in the presence of gaussian noise based on estimating functions. In: Proceedings of East Asian Symposium on Statistics, University of Tokyo, Peking University and Seoul National University, Tokyo (2000) 105–112
3. Kano, Y., Miyamoto, Y., Shimizu, S.: Factor rotation and ica. In: Proceedings of Fourth International Symposium on Independent Component Analysis and Blind Signal Separation (ICA2003). (2003) 101–105
4. Ledermann, W.: On the rank of the reduced correlation matrix in multiple factor analysis. Psychometrika (1937) 85–93
5. Shimizu, S., Kano, Y.: Examination of independence in independent component analysis. Springer-Verlag Tokyo (2003)
6. Akuzawa, T., Murata, N.: Multiplicative nonholonomic/newton -like algorithm. Chaos Solitons & Fractals **12** (2001) 785–793

Geometrical ICA-Based Method for Blind Separation of Super-Gaussian Signals

Manuel Rodríguez-Álvarez[1], Fernando Rojas Ruiz[1], Rubén Martín-Clemente[2], Ignacio Rojas Ruiz[1], and Carlos G. Puntonet[1]

[1] Departamento de Arquitectura y Tecnología de Computadores. Univ. de Granada, Spain
{mrodriguez,frojas,irojas,carlos}@atc.ugr.es
[2] Departamento de Ingeniería Electrónica. Area de Teoría de la Señal. Univ. de Sevilla, Spain
ruben@us.es

Abstract. This work explains a new method for blind separation of a linear mixture of sources, based on geometrical considerations concerning the observation space. This new method is applied to a mixture of several sources and it obtains the estimated coefficients of the unknown mixture matrix A and separates the unknown sources. In this work, the principles of the new method and a description of the algorithm are shown.

1 Introduction

The separation of source signals from mixed observed data is a fundamental and challenging signal processing problem. In many practical situations, one or more desired signals need to be recovered blindly knowing only the observed sensor signals. When p different source signals propagating through a real medium have to be captured by sensors, these sensors are sensitive to all sources $s_i(t)$ and thus the signal $x_k(t)$, observed at the output of sensor k, is a mixture of source signals. With a linear and stationary mixing medium the sensor signals can be described by:

$$\vec{x}(t) = A\,\vec{s}(t) \qquad (1)$$

where $\vec{x}(t) = (x_1(t), ..., x_n(t))^T$ is an experimentally observable $(n \times 1)$-sensor signal vector s(t), with $\vec{s}(t) = (s_1(t), ..., s_p(t))^T$ is a $(p \times 1)$ - unknown source signal vector having stochastic independent and zero-mean non-Gaussian elements $s_i(t)$, and A is a $(n \times p)$ unknown full-rank and non-singular mixing matrix. The solution of the blind signal separation (BSS) problem consists of retrieving the unknown sources $s_i(t)$ from just the observations. To achieve this it is necessary to apply the hypotheses that the sources $s_i(t)$ and the mixture matrix $A = (\vec{a}_1, ..., \vec{a}_n)^T$ are unknown, that the number n of sensors is at least equal to the number p of sources, i.e. $n \geq p$, and that the components of the source vector are statistically independent yielding:

$$p(\vec{s}) = \prod_{i=1}^{n} p(s_i) \qquad (2)$$

In order to solve the BSS problem a separating matrix W is computed whose output is an estimate of the vector $\vec{s}(t)$ of the source signals such that:

$$\vec{y}(t) = W^{-1} \vec{x}(t) \quad (3)$$

Any BSS algorithm can only obtain W subject to:

$$W^{-1} A = DP \quad (4)$$

with a diagonal scaling matrix D modified by a permutation matrix P. Recently, BSS and ICA (Independent Component Analysis) have received much attention because of its potential applications in signal processing. A great diversity of estimation methods have been proposed based on some kind of statistical analysis, neural networks [7], the entropy concept [3], the geometric structure of the signal spaces [1], [6], the fixed-point algorithm FastICA [5], the maximum likelihood stochastic gradient algorithm [2], the Jade algorithm [4], among others. Several geometric procedures have been used to separate either multivalued or analog signals, by analyzing the observed sensor signals in the resulting p-dim space of observations. In this work we will present a geometric ICA algorithm which is based on rough density estimation and its application to separate super-Gaussian mixed signals.

2 Principles of the New Method

With two bounded super-Gaussian signals, the observed signals $(x_1(t), x_2(t))$ takes the form in the $(\vec{x}_1, \vec{x}_2)$ space, like in Figure 1. We have demonstrated [8] that, through a matrix transformation, the coefficients of the matrix coincide with the slopes of the straight lines $\overline{P_1 P_2}$ and $\overline{P_3 P_4}$. The slopes of these segments give the coefficients of the estimated mixture matrix W. In order to obtain these segments, it is necessary to estimate the coordinates of those points P_i, $i = 1, 2, 3, 4$. Assuming super-Gaussian distributed signal as the sources, for example speech signals with an underlying super-Gaussian distribution; the form of the sensor signal distribution in the space of observations is highly super-Gaussian too, as can be seen in Figure 2. In this case it is necessary to detect the directions of high density in the space of observations. These directions are called ICA axes (ICA-1 and ICA-2).

2.1 Description of the Algorithm

First of all, the algorithm computes the kurtosis of each component of the sensor signals and also the correlation coefficients between all observations. This is to detect whether the underlying source signal distributions correspond to sub- or super-Gaussian distributions. According to the Central Limit Theorem, mixtures will tend to be closer to Gaussian than the original ones. Consequently, kurtoses of the mixtures will be closer to zero (Gaussian distribution) than the sources:

$$\left| Kurt(x_i) \right| \leq \max \left\{ \left| Kurt(s_j) \right| \right\} ; \; i, j \in [1, ..., n] \quad (5)$$

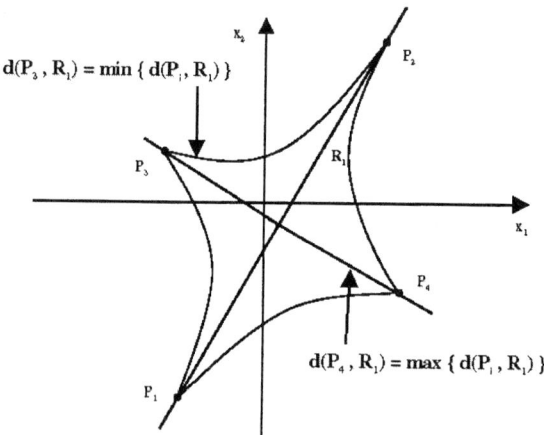

Fig. 1. Space of observations for the mixture of two super-Gaussian signals: Representative points and straight lines.

In any case, for mixtures of two signals, they will tend to preserve the sub- or super-Gaussian nature of the original signals, assuming that both sources have the same sign in the kurtosis. If the kurtoses of all observations are positive, the algorithm searches for high density regions of the sensor signal distribution. The algorithm subdivides the space of observations $(\vec{x}_1, \vec{x}_2)$ into a regular lattice of cells with N-rows and M-columns as shown in Figure 2. Then, the algorithm computes the number of cells in the lattice in which the number of points inside it is greater than a given threshold TH. The distribution of sensor signals within each of these cells then is replaced by a prototype sensor signal vector. The prototype vector mostly does not point towards the centre of the cell because its position is weighted by the density of points (x_{1i}, x_{2i}) in this cell. The next step of the algorithm finds those points which form the high density regions of the sensor signal distribution in the space, by looking for cells that have an empty neighborhood (such cells have fewer points than the threshold TH). Then these cells without a complete neighborhood form the border of the distribution encompassing NR data points in the space of observations. The algorithm then computes the coordinates of $P_1 = (p_{11}, p_{12})$ and $P_2 = (p_{21}, p_{22})$. The space of observations has been reduced to NR data points which, in two dimensions, represent pairs of coordinates (x_{1i}, x_{2i}). In this reduced set of NR data points, there exist data points P_1 and P_2 with largest Euclidean distance between them in the space of observations :

$$d(P_1, P_2) = \max_{i, j \in (1, 2, \ldots NR)} d(P_i, P_j) \qquad (6)$$

Once points P_1 and P_2 have been identified, the algorithm calculates the equation of the straight line R_1 which passes through these points P_1 and P_2 :

$$Ax_1 + Bx_2 + C = 0 \qquad (7)$$

being

$$A = (p_{22} - p_{12}), \quad B = (p_{11} - p_{21}), \quad C = (p_{21} - p_{12}) - (p_{22} - p_{11}) \tag{8}$$

Next, the algorithm estimates the coordinates of the points $P_3 = (p_{31}, p_{32})$ and $P_4 = (p_{41}, p_{42})$ as follows: the straight line R_1 divides the space of observations $(\vec{x}_1, \vec{x}_2)$ into two subspaces, being R_1 the border between them. Data points which lie within one of these subspaces yield a nonzero result in Eq. (7). For example, data points lying above the straight line R_1 yield a negative result in Eq. (7). There is then one data point $P_3 = (p_{31}, p_{32})$ which provides the most negative value of all possible outcomes of Eq. (7), hence which also represents the point with the greatest Euclidean distance from the straight line R_1 in the subspace above R_1. In the same way, points in the other subspace, below the straight line R_1, yield a positive result in Eq. (7). Again, there is one point $P_4 = (p_{41}, p_{42})$ that provides the most positive value of all possible results from Eq. (7), and which is also the point with greatest Euclidean distance from the straight line R_1 in the subspace below R_1.

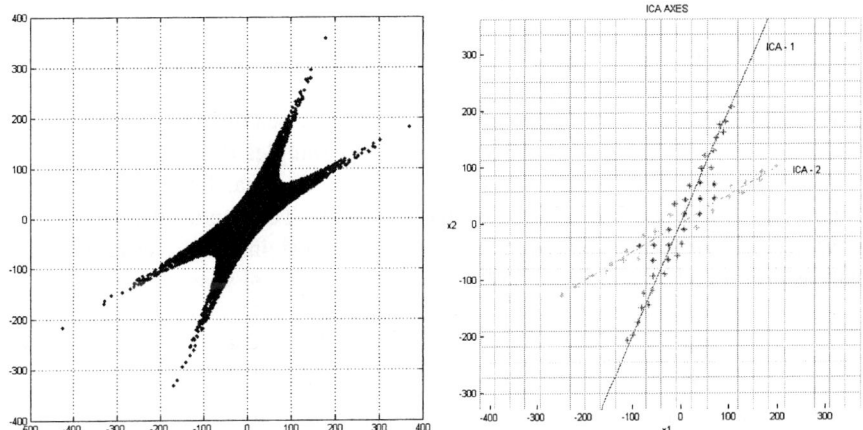

Fig. 2. Linear mixture of two super-Gaussian signals, lattice of the space of observations and ICA axes.

Once the characteristical points have been obtained, the algorithm computes the slopes of the segments diagonals ($\overline{P_1 P_2}$ and $\overline{P_3 P_4}$) for super-Gaussian densities in order to obtain the slopes of the ICA axes and the coefficients of the matrix W as in Eq. (9) (see Figure 2):

$$w_{12}^{-1} = \left(\frac{a_{12}}{a_{22}}\right)^{-1} = \frac{p_{22} - p_{12}}{p_{21} - p_{11}} \quad ; \quad w_{21} = \left(\frac{a_{21}}{a_{11}}\right) = \frac{p_{32} - p_{42}}{p_{31} - p_{41}} \tag{9}$$

Using the coefficients of matrix W, the algorithm computes the inverse matrix W^{-1} and reconstructs the unknown source signals $\vec{s}(t)$ (see Eq. (3)).

2.2 Further Enhancements

The computational order of the algorithm is polynomial:
$$Comput - Order = (DataPoints^2 \cdot XColumns \cdot YRows) \tag{10}$$

As a further improvement, we propose the reduction of the number of points at the beginning of the algorithm with a random elimination through all the space of the joint distribution of the mixtures as long as enough data points are kept to correctly estimate the sources. A more elaborated proposal is eliminating those points of the joint distribution of the mixtures which lay within a calculated radius near the center of the joint distribution, because they are useless for the algorithm, due to its nature of computing using points whose Euclidean distances are the highest. From experimental results, we have derived equation (11) for the calculation of the radius.

For super-Gaussian mixtures (positive kurtosis), the algorithm will search for high density regions of the joint distribution of the mixtures. Thus, the exclusion radius was calculated as:
$$R = 1.5 \cdot \overline{x} \tag{11}$$

3 Simulations and Results

The new algorithm, named as "LatticeICA", has been tested on various ensembles of artificial sensor signals with an arbitrary number of samples drawn at random from sub- and super-Gaussian distributions like uniform, Gamma, Laplacian and Delta distributions, as well as with real world speech signals. To quantify the performance achieved we calculate both a crosstalking error of the original and recovered source signals as proposed by Amari et al. [2] as well as a component wise crosstalk:

$$E(P) = \sum_{i=1}^{n}(\sum_{j=1}^{n}\frac{|p_{ij}|}{\max_k |p_{ik}|} - 1) + \sum_{j=1}^{n}(\sum_{i=1}^{n}\frac{|p_{ij}|}{\max_k |p_{kj}|} - 1) \tag{12}$$

where $P=(p_{ij})= W^{-1} \cdot A$. The parameter MSE (Mean Square Error) measures the similarity of the signals $s_i(t)$ and $y_i(t)$.

3.1 Speech Signals

In this simulation the algorithm separate two super-Gaussian speech voice signals with 10000 samples each. The lattice was automatically computed to be 16 rows and 16 columns, using TH = 29. The original and estimated matrices were:

$$A = \begin{bmatrix} 1 & 0.50 \\ 0.50 & 1 \end{bmatrix}; W = \begin{bmatrix} 1 & 0.53 \\ 0.51 & 1 \end{bmatrix} \tag{13}$$

The joint distribution of the mixtures points out the super-Gaussian nature of the sources (see Figure 3). The matrix performance index for this simulation was $E(W, A) = 0.13$, with Crosstalk1 (E_{s1}) = -27.26 dB and Crosstalk2 (E_{s2}) = -29.11

dB. In Figure 3 it is shown how the algorithm searches for the lines of higher density instead of the contour plot.

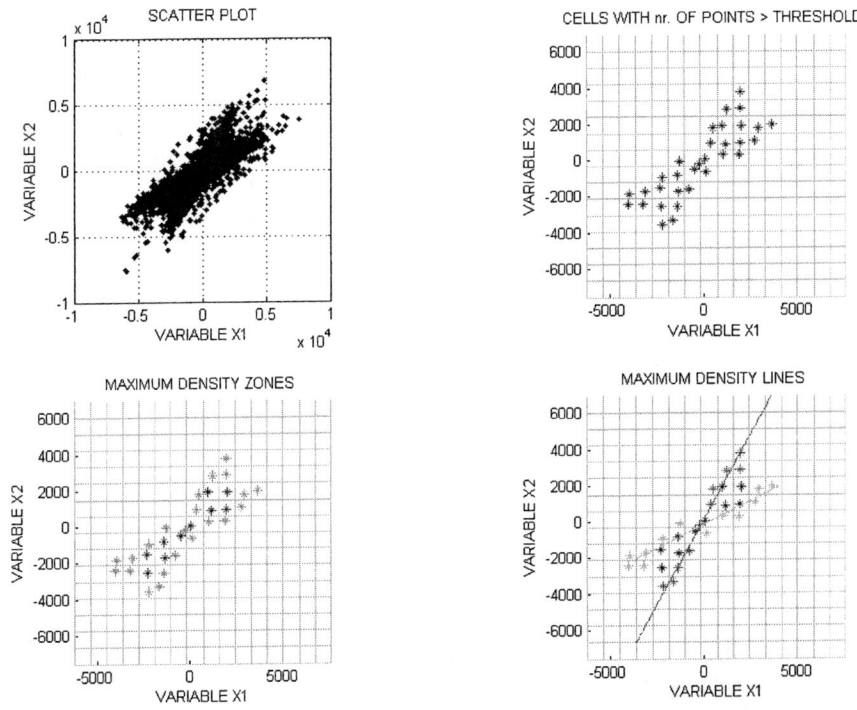

Fig. 3a. Performance of the LatticeICA algorithm for a two real voice signals mixture.

3.2 Comparison with Other Algorithms

In this simulation we started a more systematic exploration of the algorithm and compared the results to those obtained with two other algorithms, the FastICA [5] and Jade algorithms [4]. We tried random mixture matrixes over uniform and Laplacian mixtures of 10000 samples, running 100 simulations each time, with automatic parameters. With FastICA the number of bins has been choosen in all cases to be 180. The NRMS (normalized root mean squared error) in each case and the corresponding average convergence times (Pentium IV 1.5 GHz., 512 MB RAM, under Matlab environment) are summarized in Table 1. Although, both FastICA and Jade algorithms globally get better results than LatticeICA in most of the simulations, LatticeICA shows a great performance especially for super-Gaussian mixtures (speech signals) and it outperforms previous geometric algorithms. As a particular advantage of LatticeICA when compared with FastICA and Jade it remains its easy hardware implementation, due to the fact that it only computes simple arithmetic operations. Future enhancements in fine tuning the radius of exclusion and adjusting the final separation lines will certainly lead to a better performance.

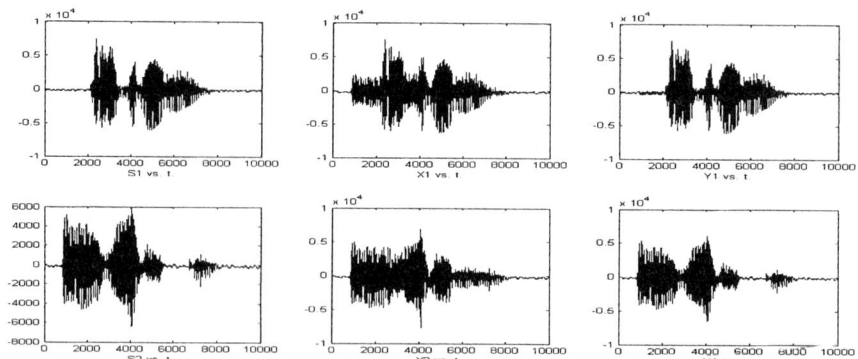

Fig. 3b. Original signals (left), mixed signals (center) and reconstructed signals (right) using LatticeICA algorithm for a two real voice signals mixture.

Table 1. Comparison of performance of the algorithm (LatticeICA) with FastICA and Jade.

Source Type	Procedure	NRMS	Speed of convergence (ms.)
Uniform	Lattice ICA	0.054	808
	FastICA	0.021	501
	Jade	0.028	584
Laplacian	Lattice ICA	0.034	703
	FastICA	0.087	406
	Jade	0.009	273

3.3 Extension to Higher Dimensionality

Finally, we show how this algorithm can be extended to higher dimensionality situations by attempting to separate the projections of p mixed signals from $\mathbb{R}^p$ onto $\mathbb{R}^2$. The signals are shown in figures 4 and 5 (with a Laplacian noise, a music source and a speech signal). The original and obtained matrices are:

$$A = \begin{bmatrix} 1 & 0.5 & 0.5 \\ 0.5 & 1 & 0.5 \\ 0.5 & 0.5 & 1 \end{bmatrix} \quad W = \begin{bmatrix} 1 & 0.64 & 0.73 \\ 0.45 & 1 & 0.63 \\ 0.46 & 0.47 & 1 \end{bmatrix} \quad (14)$$

In Figure 4 can be seen the 3-dimensional mixture and the separated signals of the proposed LatticeICA algorithm.

4 Conclusions

We have developed a new geometry-based method for blind separation of sources which greatly reduces the complexity and computational load inherent in the standard geometric ICA algorithms. This new algorithm is based on a tessellation of the input space where in each cell a code book vector is determined to represent the center of gravity of the local distribution of sample vectors. For signals with super-Gaussian, distribution the slopes of the diagonals of the scatter-plot are determined to obtain the coefficients of the estimated mixing matrix W. The method lends itself for an easy hardware implementation and is also very intuitive in terms of computer applications. Furthermore, this method could be used to detect the perimeter or outlines in simple two-dimensional figures. In the future we will intend to implement this method for more than two signals without using projections but working in the p-dimensional space.

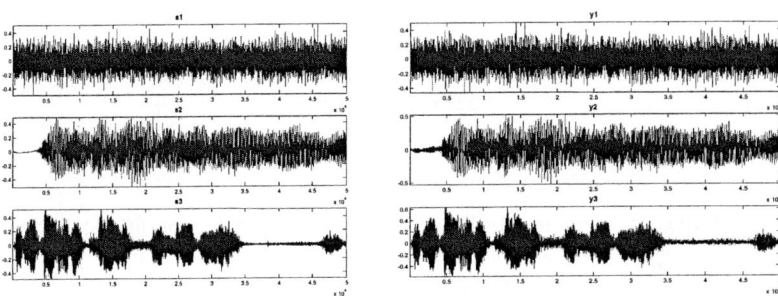

Fig. 4. Left: a laplacian noise, a music and a speech source signals. Right: Estimated signals after reconstruction.

Acknowledgement. This work has been supported by the Spanish CICYT Projects TIC2001-2845 "PROBIOCOM - Procedures for Biomedical and Communications for BSS" and TIC2000-1348 "Hybrid procedures for parallel optimization in clusters".

References

1. Álvarez, M. R., Puntonet, C. G., Rojas, I.: Separation of Sources based on the Partitioning of the Space of Observations. Lecture Notes on Computer Science, Vol. 2085. Springer-Verlag, Berlin-Heidelberg-New York (2001) 762 – 769.
2. Amari, S. I., Cichocki, A., Yang, H. H.: A new learning algorithm for blind signal separation, Proceedings of NIPS'96, (1996) 757-763.
3. Bell A. J., Sejnowski, T. J.: An information-maximisation approach to blind separation and blind deconvolution, Neural Computation, Vol. 7 (1995) 1129-1159.
4. Cardoso, J. F.: High-order contrasts for independent component analysis. Neural Computation, Vol. 11, n° 1 (1999)157-192.
5. Hyvärinen, A., Karhunen, J., Oja, E.: Independent Component Analysis. Wiley & Sons, New York (2001).
6. Jung, A., Theis, F. J., Puntonet, C. G., Lang, E. W.: FASTGEO: A histogram based approach to linear geometric ICA. Proceedings of ICA'01 (2001) 349-354.
7. Jutten, C., Hérault, J., Comon, P., Sorouchiary, E.: Blind separation of sources, Parts I, II, III. Signal Processing, Vol. 24, n° 1 (1991) 1-29.
8. Rodríguez-Álvarez, M.: New geometrical procedure of Blind Sources Separation based on the partitioning of the space of observations. Ph. D. Thesis. University of Granada (2002).

A Novel Method to Recover N Sources from N-1 Observations and Its Application to Digital Communications

Adriana Dapena

Departamento de Electrónica e Sistemas
Universidade da Coruña
Campus de Elviña s/n, 15.071, A Coruña, Spain
Tel: ++34-981-167000, Fax: ++34-981-167160
adriana@udc.es

Abstract. This paper deals with the blind source separation (BSS) problem with fewer sensors than sources. We propose a simple procedure to transform the problem with N sources and $N-1$ observations in a classical BSS problem with N observations which can be solved using many well-known algorithms. We will also show how to apply this idea to digital communications for separating BPSK and QPSK signals.

1 Introduction

Blind Source Separation (BSS) methods aim at recovering a set of N signals (sources) from a set of P observed signals (observations) provided by an array of sensors [6]. Most of BSS algorithms have been developed assuming that the number of observations is equal to the number of sources (see [5] and references therein). Unfortunately, these algorithms cannot be used in those practical applications where there exist less observations than sources due, for instance, to physical limitations in the number of sensors. This problem, known in the literature as *undetermined source separation*, has been considered in few investigations [1, 7–10].

In this paper, we will propose a simple method to solve the undetermined source separation problem assuming that $N-1$ sensors receive instantaneous mixtures of N sources. The basic idea is to use a cancellation or BSS algorithm to recover *one* source. Subsequently, we obtain a set of N observations by combining the original $N-1$ observations and the recovered source. In other words, we transform the undetermined source separation problem in a classical BSS problem which is solved using a well-known BSS algorithm. Unlike previous work [7, 8], the method can be used for both binary and quaternary signals.

This paper is structured as follows. In Section 2, we describe the undetermined source separation problem and in Section 3 we present the new solution. The experimental performance of the new method is presented in Section 4. Finally, Section 5 contains the conclusions.

2 Problem Statement

We will consider the following signal model. Let $\mathbf{s}(n) = [s_1(n), ..., s_N(n)]^T$ be the vector containing N discrete sources modeled as zero-mean complex-valued stationary signals which arrive at an array of P sensors whose output, denoted by $\mathbf{x}(n) = [x_1(n), ..., x_P(n)]^T$, is an instantaneous combination of the N sources. In a compact form, we can write

$$\mathbf{x}(n) = \mathbf{A}\mathbf{s}(n) \tag{1}$$

where

$$\mathbf{A} = \begin{pmatrix} a_{11} & a_{12} & ... & a_{1N} \\ a_{21} & a_{22} & ... & a_{2N} \\ \vdots & & \ddots & \vdots \\ a_{P1} & a_{P2} & ... & a_{PN} \end{pmatrix} \tag{2}$$

is the $P \times N$ mixing matrix. The coefficient a_{ij} represents the contribution of the j-th source in the i-th observation.

The aim in BSS is to recover the sources from the observations assuming that both the sources and the mixing matrix are completely unknown. Towards this end, the observations are processed by a linear system with output

$$\mathbf{y}(n) = \mathbf{W}\mathbf{x}(n) \tag{3}$$

where $\mathbf{W}$ represents the separating matrix. Combining (1) and (3) together, the output can be expressed as a linear combination of the sources

$$\mathbf{y}(n) = \mathbf{G}\mathbf{s}(n) \tag{4}$$

where $\mathbf{G} = \mathbf{W}\mathbf{A}$ is the gain matrix. In those situations where $P = N$ an exact separation of all sources can be done [2]. In particular, it is obvious that the sources can be recovered when $\mathbf{W}$ is made equal to $\mathbf{A}^{-1}$. It is clearly also possible when $P > N$ but it is in general impossible when $P < N$. This latter situation is known in the literature as *undetermined source separation* (see [1, 7–10]).

For discrete sources, the undetermined problem can be solved when the alphabet is known. For instance, BPSK sources can be recovered from only one observation using [8]. Also, assuming that all the sources are MSK signals (or BPSK), the method proposed in [7] allows to recover three sources from two observations with low computational cost.

Unlike previous work [7, 8], our method can be used for both BPSK and QPSK signals. The solution will be proposed considering $P = N-1$ observations.

3 Proposed Solution

We propose to solve the undetermined source separation problem using a two stages separating system: a source extraction stage followed by a BSS stage. The

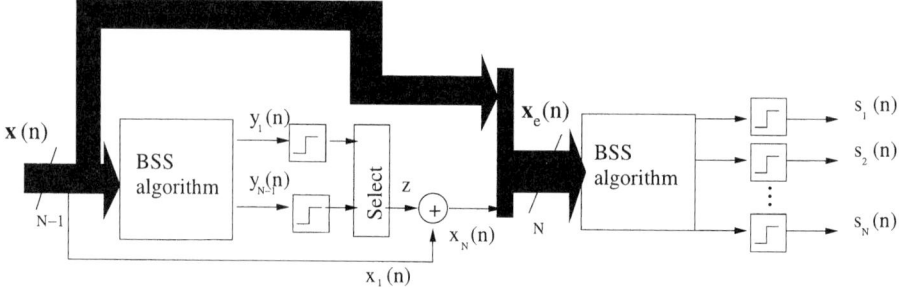

Fig. 1. Separating system for the binary case.

first stage must be designed to recover only one source from the $N-1$ observations taking into account the source properties. This aspect will be considered in Subsection 3.1 for BPSK and QPSK signals.

After recovering one source, e.g. $s_1(n)$, we create a virtual observation by combining this source and the original observations. For instance, the virtual observation $x_N(n) = x_1(n) - s_1(n)$ has the form

$$x_N(n) = (a_{11} - 1)s_1(n) + a_{12}s_2(n) + ... + a_{1N}s_N(n) \tag{5}$$

Denoting by $\mathbf{x}_e(n) = [x_1(n), x_2(n), ..., x_{N-1}(n), x_N(n)]^T$ the extended vector containing the original $N-1$ observations and the virtual one, we can express the new problem as follows

$$\mathbf{x}_e(n) = \mathbf{A}_e \mathbf{s}(n) \tag{6}$$

where $\mathbf{A}_e$ is an $N \times N$ matrix given by

$$\mathbf{A}_e = \begin{pmatrix} a_{11} & a_{12} & ... & a_{1N} \\ a_{21} & a_{22} & ... & a_{2N} \\ \vdots & \ddots & & \vdots \\ a_{(N-1)1} & a_{(N-1)2} & ... & a_{(N-1)N} \\ a_{11} - 1 & a_{12} & ... & a_{1N} \end{pmatrix} \tag{7}$$

As a consequence, the N original sources can be recovered by computing $\mathbf{W}_e = \mathbf{A}_e^{-1}$. Since we are assuming that the mixing system is completely unknown, the second stage of the proposed separating system consists in estimating this matrix by using a BSS algorithm.

3.1 BPSK and QPSK Signals

The procedure presented above takes a simple form for some common digital communication signals like BPSK (with alphabet ± 1) or QPSK (with alphabet $(\pm 1 \pm j)/\sqrt{2}$). In this case, we can use the system shown in Figure 1 where the

$N-1$ observations are processed by a classical BSS algorithm followed by the decision function

$$z_i(n) = \frac{1}{\sqrt{2}}(sign(Re(y_i(n))) + j\; sign(Im(y_i(n))))$$

$$z_i(n) = \frac{z_i(n)}{\sqrt{E[|z_i(n)|^2}}, \; i = 1, ..., N-1 \tag{8}$$

Now, we need to select one of these $N-1$ outputs but note that some outputs are not adequate because some of them are linear combinations of several sources. As decision criterion we propose to evaluate the Mean Square Error (MSE) and select the output corresponding to the small value, i.e.,

$$z(n) = min_{i=0,...,N-1}(E[|z_i(n) - y_i(n)|^2]) \tag{9}$$

Subsequently, a new observation is obtained using this estimated source and the observation $x_1(n)$, i.e., $x_N(n) = x_1(n) - z(n)$. The second stage consists in a BSS algorithm that processes N observations to recover the N sources.

4 Experiment Results

This Section presents some results of computer simulations carried out to validate our method. In all the simulations, the mixing matrix $\mathbf{A}$ has been randomly generated with an uniform distribution in $(-1, 1)$ with $diag(A) = 1$.

The first experiment is devoted to determine if some existing algorithms are suitable for the extraction stage of the proposed system. We have considered the EASI algorithm [4] and the JADE algorithm [3]. The separation achieved in the outputs with the minimum MSE (9) has been measured using the performance index

$$P(\mathbf{g}_j) = \sum_{i=1}^{N} \frac{|g_{ji}|^2}{max_i(|g_{j1}|^2, ..., |g_{jN}|^2)} - 1 \tag{10}$$

where $max_i()$ represents the maximum value. In the simulations we have generated 10,000 samples of BPSK signals for $N = 3, 5, 7$ and 9. Table 1 shows the performance index and the error probability obtained by averaging twenty independent realizations with different sources and mixing matrices. The two algorithms have presented an optimum behavior.

In the second experiment, we have used the JADE algorithm for both the extraction and the BSS stages. For a particular simulation with $N = 4$, Figure 2 (a) shows 50 samples of the sources and part (b) shows the observations obtained using the mixing matrix:

$$\mathbf{A} = \begin{pmatrix} 1.0000 & 0.9784 & -0.0808 & -0.8934 \\ -0.3726 & 1.0000 & 0.3109 & -0.4024 \\ 0.3906 & -0.9984 & 1.0000 & -0.0473 \end{pmatrix} \tag{11}$$

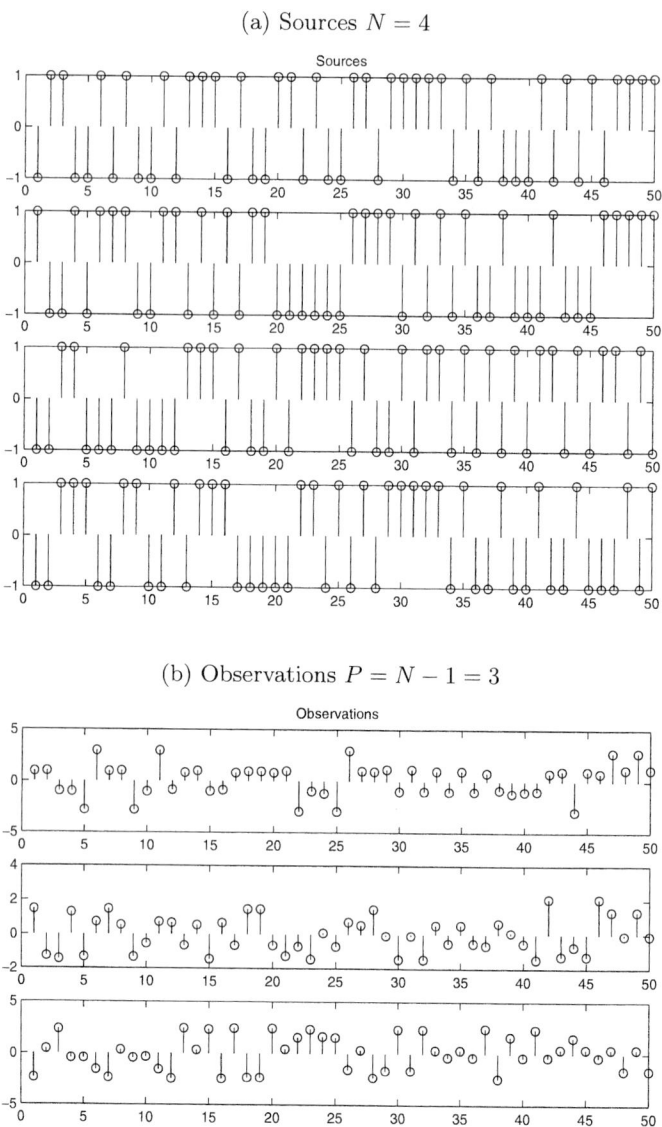

Fig. 2. Simulation with BPSK signals: sources and observations.

After evaluating the MSE criterion (9), we have selected the first output of the JADE algorithm: $y_1(n) = 0.9244 s_1(n) + \epsilon$ where $\epsilon = -0.0855 s_2(n) - 0.0356 s_3(n) - 0.3741 s_4(n)$. Note that the error term ϵ is small in comparison with the gain factor $g_{11} = 0.9244$. As a consequence, the decision step has perfectly recovered the source, $z(n) = s_1(n)$, as is shown in Figure 3 (a). In this case, the virtual observation has the form $x_4(n) = x_1(n) - z(n) = [0, 0.9784, -0.0808, -0.8934]$. In Figure 3 (b) it can be seen that all the sources have been retrieved using the JADE algorithm in the second stage.

Table 1. Simulations with BPSK signals: performance index and error probability obtained in the extraction stage with different algorithms.

JADE algorithm

N	3	5	7	9
$P(\mathbf{g}_i)$ (dB)	-18.7888	-19.0139	-24.9685	-25.3963
Pe	0	0	0	0

EASI algorithm

N	3	5	7	9
$P(\mathbf{g}_i)$ (dB)	-14.8419	-14.3268	-15.8711	-15.6024
Pe	0	0	0	0

Table 2. Simulations with QPSK signals: performance index and error probability.

JADE algorithm

N	3	5	7	9
$P(\mathbf{g}_i)$ (dB)	-20.0382	-24.4333	-22.3954	-27.8593
Pe	0	0	0	0

In order to illustrate the performance of our method for QPSK signals, Table 2 shows the performance index and the error probability obtained using the JADE algorithm in the two stages. We have also averaged twenty independent realizations with different randomly generated sources and mixing matrices. Again, the performance has been optimum.

5 Conclusions

We have proposed to transform the undetermined source separation problem in a classical BSS problem by using a two stages system: a source extraction stage followed by a BSS stage. The basic idea is to obtain virtual observations by combining the original observations and the sources extracted in the first stage. The set of original and virtual observations are then processed by a classical BSS algorithm to recover all the sources.

Computer simulations show that the strategy exhibits promising performance for BPSK and QPSK signals. The immediate improvement deals with the reduction of the computational cost by designing specific algorithms for the extraction stage and with the consideraton of noisy environments.

Acknowledgements

This work has been supported by Ministerio de Ciencia y Tecnología of Spain and FEDER funds (grant TIC2001-0751-C04-01).

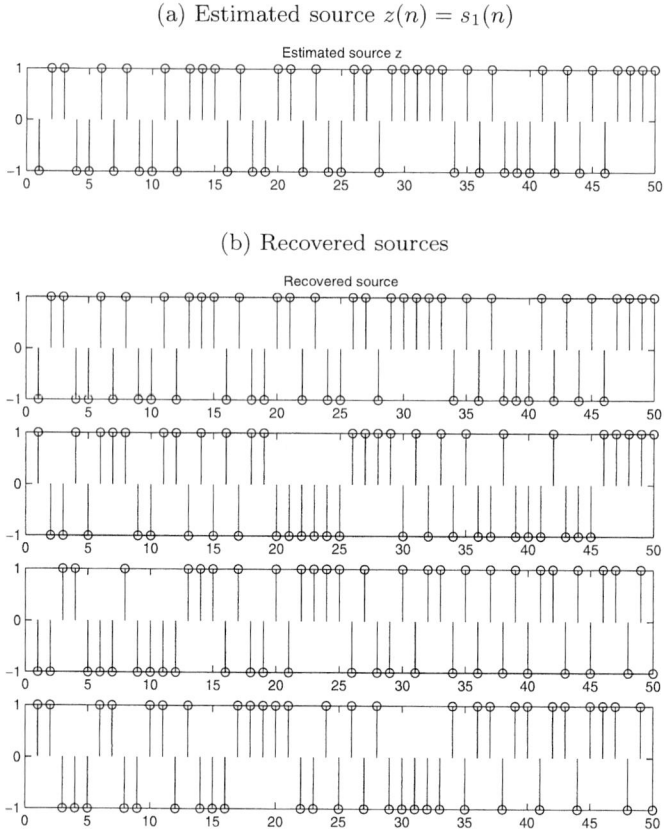

Fig. 3. Simulation with BPSK signals: estimated source $z(n) = s_1(n)$ and recovered sources.

References

1. F. Abrard, Y. Deville, P. White, "From blind source separation to blind source cancellation in the undetermined case: a new approach base on time-frequency analysis", Proc. ICA 2001, San Diego, USA, December 2001, pp. 734–739.
2. X-R Cao and R-W Liu, "General approach to blind source separation", IEEE Trans. Signal Processing, vol. 44, no. 3, March 1996, pp. 562–571.
3. J. F. Cardoso and A. Souloumiac, "Blind Beamforming for non-Gaussian Signals", IEE-Proceedings-F, vol 140, no. 6, pp. 362-370, December 1993.
4. J. F. Cardoso and B. Laheld, "Equivariant adaptive source separation", IEEE Trans. on Signal Processing, vol. 44, no. 2, December 1996, pp. 3017-3030.
5. J. F. Cardoso, "Blind signal separation: statistical principles", Proc. IEEE, vol 86, No. 10, October 1998, pp. 225–254.
6. A. Cichochi, S. I. Amari, Adaptive blind signal and image processing. Learning algorithms and applications, Wiley, England, 2002.
7. P. Comon and O. Grellier, "Non-linear inversion of undetermined mixtures", Proc. ICA'99, Aussois, France, January 1999, pp. 461–465.

8. K. I. Diamantaras, E. Chassioti, "Blind separation of N binary sources from one observation: a deterministic approach", Proc. ICA'2000, Helsinki, Finland, June 2000, pp. 93–98.
9. A. Jourjune, S. Rickard, O. Yilmaz, "Blind separation of disjoint orthogonal signals: demixing N sources from 2 mixtures", Proc. ICASSP 2000, Istanbul, Turkey, June 2000, pp. 2985–2988.
10. A. Taleb, C. Jutten, "On undetermined source separation", Proc. ICASSP'99, Arizona, March 1999, pp. 1445–1448.

A Sufficient Condition for Separation of Deterministic Signals Based on Spatial Time-Frequency Representations

Nadège Thirion-Moreau, El Mostafa Fadaili, and Eric Moreau

STD, ISITV, Université de Toulon
av. G. Pompidou, BP56
F-83162 La Valette du Var, Cedex, France
{thirion,fadaili,moreau}@univ-tln.fr

Abstract. The paper is devoted to blind separation of deterministic source signals based on time-frequency representations. Our main result is to show that the joint-diagonalization of a number of spatial quadratic transform matrices of the observation signals is sufficient for separation and we give the minimum number. A computer simulation illustrate the result.

1 Introduction

Many solutions have been brought to the problem of blind separation of instantaneous mixtures of unknown sources. In this communication, we focus on techniques based on the use of time-frequency representations. Such methods are proved to be particularly interesting when dealing with non stationary signals. It has been recently shown in [1] that the blind sources separation problem can be solved by joint-diagonalizing a combined set of well-chosen "spatial quadratic t-f distributions (SQTFD)" matrices. The selection of time-frequency (t-f) points that correspond only to sources auto-terms (in such t-f points there is no interference between sources (also called sources cross-terms)) with a view to build a set of matrices to be joint-diagonalized, is the main point with this kind of methods. This problem has been discussed in [2][3][4].

The purpose of this communication is to provide a sufficient separation condition when such spatial quadratic time-frequency representations are used by proposing a new contrast function. Contrast functions have been proved to be efficient tools for separation algorithms since their global maximization solves the problem [5][6]. Moreover a contrast function fixes the identifiability condition under the hypotheses of the model.

One of the main originality of the paper is to provide a very first contrast function in the case of deterministic sources. This proposed contrast function takes into consideration some t-f representations. Moreover, it also makes it possible for us to determine the minimum number of t-f points needed to build the set of matrices to be joint-diagonalized.

2 Model, Assumptions and Solutions

We consider the classical instantaneous blind sources separation problem where N sources signals are received on N sensors. In matrix and vector notations, the input/output relationship of the mixing model reads:

$$\mathbf{x}(t) = \mathbf{A}\mathbf{s}(t) \qquad (1)$$

with $\mathbf{A}$ the (N, N) real mixing matrix which is assumed invertible, $\mathbf{x}(t) = [x_1(t), \ldots, x_N(t)]^T$ the $(N, 1)$ observations vector $((\cdot)^T$ denotes the transposition operator) and $\mathbf{s}(t) = [s_1(t), \ldots, s_N(t)]^T$ the $(N, 1)$ deterministic real sources vector. The source signals are assumed to possess different time-frequency representation, see condition (7).

The problem of blind sources separation consists in the estimation of a "separating" matrix, say $\mathbf{B}$, which applied to the observation as

$$\mathbf{y}(t) = \mathbf{B}\mathbf{x}(t) \qquad (2)$$

yields an estimation of the source signals.

Defining $\mathbf{G} = \mathbf{B}\mathbf{A}$ as the matrix of the global system, the source separation problem is solved when one has found a separating matrix $\mathbf{B}$ in such a way that

$$\mathbf{G} = \mathbf{D}\mathbf{P} \qquad (3)$$

where $\mathbf{D}$ is an invertible diagonal matrix which corresponds to arbitrary attenuations for the restored sources and $\mathbf{P}$ is a permutation matrix which corresponds to an arbitrary order of restitution of source signals.

In the following, all the considered matrices are assumed orthogonal. Like in classical sources separation, such an approach can always be considered according to a first normalization stage which is often called "spatial whitening".

3 Backgrounds

3.1 About Contrast Function

The notion of contrast function [5][6] is a key tool for criteria based sources separation. Indeed, the maximization of such a contrast function is a sufficient condition for separation and, then, the question of identifiability is ensured. Moreover if one disposes of such a contrast function then the source separation problem becomes "simply" an optimization one.

Basically, based on the general definition given in [6], a contrast function depends on a characteristic function of the output of the separating matrix, that is, here, $\mathbf{y}(t)$. In the classical statistical framework, it often depends on the probability density function through the very use of high order cumulants. Contrast functions have to satisfy three important properties. The first one states that it is invariant under multiplication of $y_i(t)$, $i = 1, \ldots, N$ by any constant, here of unit absolute value because orthogonal matrices are considered. The second property states that contrast function values have to decrease when a (non trivial) mixture is applied on source signals. Finally, the third property states that contrast function attains its (global) maxima only for matrices of the form given in (3), *i.e.* for separating matrices.

3.2 About Spatial Quadratic Transform

The proposed developments are based on the use of Spatial Quadratic Transforms (SQT) of signals and their properties [10][11][1]. Let us briefly recall the important points related to our utilization.

Considering a real deterministic vectorial signal $\mathbf{z}(t)$, the general form of a SQT is given by a matrix $\mathbf{D}_z(t,\nu) = \left(D_{z_i,z_j}(t,\nu)\right)$ written as

$$\mathbf{D}_z(t,\nu) = \int\int \mathbf{z}(\theta)\mathbf{z}^T(\theta')R(\theta,\theta';t,\nu)d\theta d\theta' \qquad (4)$$

which is defined component-wise by

$$D_{z_i,z_j}(t,\nu) = \int\int z_i(\theta)z_j(\theta')R(\theta,\theta';t,\nu)d\theta d\theta' \qquad (5)$$

for all i and j. The diagonal terms of the SQT $\mathbf{D}_z(t,\nu)$ are called *auto-terms* while the off-diagonal ones are called *inter-terms*. The function $R(\theta,\theta';t,\nu)$ which is generally a complex function is referred to as the *kernel* of the transform. For physical reasons, this kernel is often constrained to satisfy the following property

$$R(\theta,\theta';t,\nu) = R^*(\theta',\theta;t,\nu) \qquad (6)$$

where $(\cdot)^*$ stands for the complex conjugate operator. Then, the SQT satisfies an *hermitian symmetry* as

$$\mathbf{D}_z(t,\nu) = \mathbf{D}_z^H(t,\nu)$$

where $(\cdot)^H$ stands for the complex conjugate and transpose operator.

The auto-terms correspond to the same quadratic transform associated to different scalar deterministic signals. This quadratic transform is said *energetic* if its double integral over t and ν is equal to the energy of the considered signal, *i.e.* for a scalar signal $z(t)$ we have $\int\int D_{z,z}(t,\nu)dtd\nu = \int z^2(t)dt$. Such energetic transforms form the basis of diverse Time-Frequency Representations (TFR).

Let us now give two important examples: the Cohen's class and the Affine class. The Cohen's class is the class of energetic quadratic transforms covariant under time and frequency shifts. Some examples of TFR belonging to the Cohen's class are the Spectrogram, the Wigner-Ville transform, the smoothed Pseudo Wigner-Ville transform the Choï-Williams transform and the Pseudo Wigner-Ville transform. The Affine class is the class of energetic quadratic transforms covariant under time scalings and shifts. Some examples of TFR belonging to the Affine class are the Scalogramme and the Wigner-Ville transform.

In the following, all the derivations are based on any time-frequency representation denoted by $D_{z,z}(t,\nu)$ without, for simplicity, explicit link to the kernel.

4 Sufficient Separation Condition and Joint-Decomposition

Let us consider that we have $N-1$ points in the t-f plane corresponding each of them to a different source signal. In other words, suppose that there exists $N-1$ couples (t_k, ν_k), $k = 1, \ldots, N-1$ such that

$$D_{s_i, s_j}(t_k, \nu_k) = \delta_{i,j,k} D_{s_i} \tag{7}$$

where $D_{s_i} \neq 0$ for all $i = 1, \ldots, N-1$ and $\delta_{i,j,k} = 1$ if $i = j = k$ and 0 otherwise.

Notice that the above assumption for deterministic signals plays the role of the classical statistical independence assumption for random signals. It is clear that a discriminating property for source signals is always required in order to think about separation. Here we consider deterministic signals whose quadratic time-frequency representation do not overlap too much two by two. In other words the signatures of the sources in the time-frequency plane are "sufficiently" different to be able to find time-frequency points satisfying the above assumption.

According to (7), it is easy to see that matrices $\mathbf{D}_s(t_k, \nu_k) = (D_{s_i,s_j}(t_k, \nu_k))$, $k = 1, \ldots, N-1$, are thus diagonal. Then the joint-diagonalization of the SQT of the observations calculated at these t-f points (t_k, ν_k), $k = 1, \ldots, N-1$ is necessary for the identification of the mixing matrix. Our goal, hereafter, is to show the most important fact that it is also sufficient.

4.1 Sufficient Separation Condition

To that aim, let us introduce the following functional

$$\mathrm{D}_N(\mathbf{y}) = \sum_{k=1}^{N-1} \sum_{i=1}^{N} D_{y_i, y_i}^2(t_k, \nu_k) . \tag{8}$$

We now propose the following first result

Proposition 1. *If $\mathbf{y}(t) = \mathbf{G}\mathbf{s}(t)$ where $\mathbf{G}$ is an orthogonal matrix then the following inequality holds*

$$\mathrm{D}_N(\mathbf{y}) \leq \mathrm{D}_N(\mathbf{s}) . \tag{9}$$

Proof. According successively to (5) with $\mathbf{z}(t) = \mathbf{y}(t)$ and to (7), we have for all $k = 1, \ldots, N-1$

$$D_{y_i, y_i}(t_k, \nu_k) = \sum_{\ell_1, \ell_2 = 1}^{N} G_{i,\ell_1} G_{i,\ell_2} D_{s_{\ell_1}, s_{\ell_2}}(t_k, \nu_k)$$

$$= \sum_{\ell_1, \ell_2 = 1}^{N} G_{i,\ell_1} G_{i,\ell_2} \delta_{k,\ell_1,\ell_2} D_{s_k}$$

$$= G_{i,k}^2 D_{s_k} . \tag{10}$$

Hence $D_N(\mathbf{y})$ reads

$$D_N(\mathbf{y}) = \sum_{k=1}^{N-1} \sum_{i=1}^{N} \left(G_{i,k}^2 D_{s_k}\right)^2$$

$$= \sum_{k=1}^{N-1} \left(\sum_{i=1}^{N} G_{i,k}^4\right) D_{s_k}^2 . \qquad (11)$$

Now because $\mathbf{G}$ is an orthogonal matrix then for all $k = 1, \ldots, N-1$ we have, see e.g. [7],

$$\sum_{i=1}^{N} G_{i,k}^4 \leq 1 \qquad (12)$$

and thus

$$D_N(\mathbf{y}) \leq \sum_{k=1}^{N-1} D_{s_k}^2 = D_N(\mathbf{s}) \qquad (13)$$

which is the proposed inequality. □

Hence one can reach a separating state by maximizing $D_N(\mathbf{y})$. An important task still remains to be considered which corresponds to the characterization of all the global maxima of $D_N(\mathbf{y})$. To that aim, we have the following result

Proposition 2. *If* $\mathbf{y}(t) = \mathbf{G}\mathbf{s}(t)$ *where* $\mathbf{G}$ *is an orthogonal matrix then we have*

$$D_N(\mathbf{y}) = D_N(\mathbf{s}) \qquad (14)$$

if and only if $\mathbf{G} = \mathbf{DP}$ *where* $\mathbf{D}$ *is an orthogonal diagonal matrix and* $\mathbf{P}$ *a permutation matrix.*

Proof. We have the equality (14) if and only if for all $k = 1, \ldots, N-1$ we have the equality in (12). Since $\mathbf{G}$ is an orthogonal matrix, it is well-known that this is only possible for orthogonal matrices as specified in the proposition, see e.g. [7]. □

Hence according to the above result, the maximization of $D_N(\mathbf{y})$ in (8) is a (necessary and) sufficient condition for the separation. We now emphasize a link of the above criterion with a joint-diagonalization one.

4.2 Link with Joint-Decomposition

Source separation based on joint-diagonalization of a set of matrices is classically considered through the optimization of a criterion, see e.g. [8],[7]. Hence, given a set $\mathcal{M} = \{\mathbf{M}(m), m = 1, \ldots, N_d\}$ of N_d square matrices, a joint-diagonalizer (JD) of this set is defined as an orthogonal matrix that maximizes:

$$D(\mathbf{U}, \mathcal{M}) = \sum_{m=1}^{N_d} \|\text{diag}\{\mathbf{U}\mathbf{M}(m)\mathbf{U}^T\}\|^2 \qquad (15)$$

where $\|\cdot\|$ is the Euclidean norm of matrices and where the matrix operator diag$\{\cdot\}$ is defined component-wise for a given matrix $\mathbf{A} = (A_{i,j})$ as (diag$\{\mathbf{A}\})_{i,j} = A_{i,j}\delta_{i,j}$ with $\delta_{i,j} = 1$ if $i = j$ and 0 otherwise.

We have the following result

Proposition 3. *For a given observation signal* $\mathbf{x}(t)$, *let us consider the following set of* $N - 1$ *matrices*

$$\mathcal{X} = \{\mathbf{D}_x(t_k, \nu_k) = \left(D_{x_i,x_j}(t_k, \nu_k)\right) \;,\; k = 1, \ldots, N-1\}\;.$$

If $\mathbf{y}(t) = \mathbf{B}\mathbf{x}(t)$ *where* $\mathbf{B}$ *is an* (N, N) *matrix then we have*

$$\mathsf{D}_N(\mathbf{y}) = \mathsf{D}(\mathbf{B}, \mathcal{X})\;.$$

Proof. Considering the $N - 1$ time-frequency points defined according to (7), and using (2), we have

$$D_{y_i,y_i}(t_k, \nu_k) = \sum_{\ell_1,\ell_2=1}^{N} B_{i,\ell_1} B_{i,\ell_2} D_{x_{\ell_1},x_{\ell_2}}(t_k, \nu_k)\;.$$

Then the contrast in (8) can be directly written as

$$\mathsf{D}_N(\mathbf{y}) = \sum_{k=1}^{N-1} \sum_{i=1}^{N} \left(\sum_{\ell_1,\ell_2=1}^{N} B_{i,\ell_1} B_{i,\ell_2} D_{x_{\ell_1},x_{\ell_2}}(t_k, \nu_k) \right)^2 \tag{16}$$

$$= \sum_{k=1}^{N-1} \|\mathrm{diag}\{\mathbf{B}\mathbf{D}_x(t_k, \nu_k)\mathbf{B}^T\}\|^2 \tag{17}$$

$$= \mathsf{D}(\mathbf{B}, \mathcal{X})\;. \tag{18}$$

Which is the proposed equality. □

Hence according to the above three propositions, the joint-diagonalization of $N - 1$ well chosen matrices is sufficient to estimate the mixing matrix and thus to separate the sources. Notice that the question of the automatic determination of the above matrices is considered in [2][3].

5 A Simulation Example

Let us give a computer simulation in the case of $N = 3$ sources. The three considered sources are linearly modulated signals with different modulation law. Their Wigner-Ville transformations are given in figure 1 and is used as their time-frequency representation for that simulation. The mixing matrix is chosen orthogonal in order to avoid the whitening step and is equal to

$$\mathbf{A} = \begin{pmatrix} \cos(\frac{\pi}{6}) & 0 & \sin(\frac{\pi}{6}) \\ 0 & 1 & 0 \\ -\sin(\frac{\pi}{6}) & 0 & \cos(\frac{\pi}{6}) \end{pmatrix} \begin{pmatrix} 1 & 0 & 0 \\ 0 & \cos(\frac{\pi}{6}) & \sin(\frac{\pi}{6}) \\ 0 & -\sin(\frac{\pi}{6}) & \cos(\frac{\pi}{6}) \end{pmatrix} \approx \begin{pmatrix} 0.87 & -0.25 & 0.43 \\ 0 & 0.87 & 0.5 \\ -0.5 & -0.43 & 0.75 \end{pmatrix}\;.$$

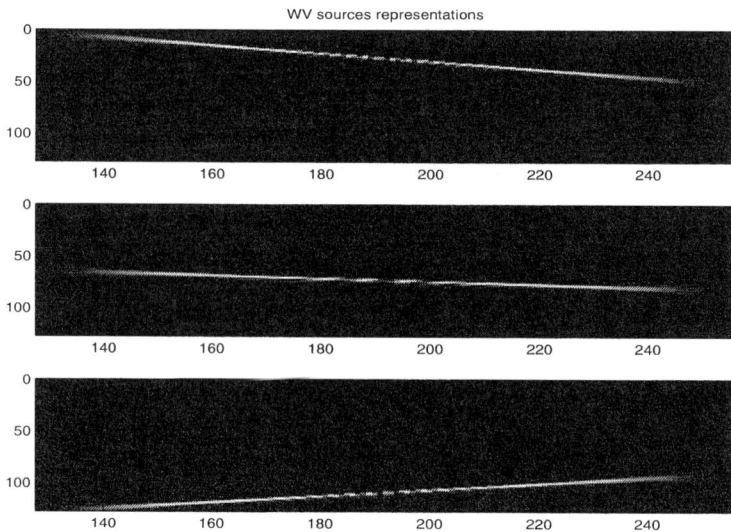

Fig. 1. The Wigner-Ville representations of each of the three linearly modulated signals.

Choosing two t-f points, the first one (resp. the second one) corresponding to an auto term of the first source (resp. the second source), we obtain the two following matrices to be joint-diagonalized:

$$\mathbf{D}_x(t_1,\nu_1) \approx \begin{pmatrix} 10.09 & 0.75 - 0.42i & -4.44 + 0.03i \\ 0.75 + 0.42i & 0.53 & 1.05 + 0.06i \\ -4.44 - 0.03i & 1.05 - 0.06i & 5.18 \end{pmatrix}$$

$$\mathbf{D}_x(t_2,\nu_2) \approx \begin{pmatrix} 12.76 & -0.01 - 0.03i & -7.14 - 0.15i \\ -0.01 + 0.03i & 0.05 & 0.17 + 0.03i \\ -7.14 + 0.15i & 0.17 - 0.03i & 4.21 \end{pmatrix}.$$

Applying a joint-diagonalization algorithm in the complex case [9], we obtain the following estimation of the separating matrix

$$\mathbf{B} \approx \begin{pmatrix} 0.8674 & 0.0003 & -0.4974 \\ -0.2155 & 0.8973 & -0.3745 \\ 0.4454 & 0.4316 & 0.7790 \end{pmatrix} + i \begin{pmatrix} 0.0006 & -0.0101 & -0.0071 \\ -0.0504 & 0.0020 & -0.0746 \\ -0.0111 & -0.0915 & -0.0001 \end{pmatrix}$$

which yields a global matrix whose components modulus read

$$(|G_{i,j}|) \approx \begin{pmatrix} 0.9999 & 0.0060 & 0.0105 \\ 0.0063 & 0.9942 & 0.1070 \\ 0.0103 & 0.1070 & 0.9942 \end{pmatrix}.$$

That is a global matrix near to the identity one which clearly show a "good" separation. Notice that there is no exact identification since the time-frequency representations are "only" estimated and not theoretically calculated.

6 Conclusion

We emphasize in this paper that, after a whitening stage, the separation of deterministic signals having different time-frequency representations can be realized by the simultaneous diagonalization of a number of matrices equal to the number of sources minus one. More importantly this necessary fact is shown here to be sufficient. A computer simulation example illustrates a typical result obtained by this kind of approach using the minimum number of matrices.

References

1. A. Belouchrani, M. G. Amin, "Blind source separation based on time-frequency signal representations," *IEEE Trans. Signal Processing*, Vol. 46, No. 11, pp. 2888–2897, Nov. 1998.
2. A. Belouchrani, K. Abed-Meraim, M. G. Amin, A. M. Zoubir, "Joint anti-diagonalisation for blind source separation", in *Proc. ICASSP 2001*, Salt Lake City, USA, pp. 2789–2792, May 2001.
3. L. Giulieri, N. Thirion-Moreau, P.-Y. Arquès, "Blind sources separation based on bilinear time-frequency representations: a performance analysis", in *Proc. ICASSP 2002*, Orlando, USA, pp. 1649–165, May 2002.
4. L. Giulieri, N. Thirion-Moreau and P.-Y. Arquès, "Blind sources separation based on quadratic time-frequency representations: a method without pre-whitening", in *Proc. ICASSP 2003, Vol. V*, Honk-Kong, pp. 289–292, April 2003.
5. P. Comon, "Independent component analysis, a new concept?", *Signal Processing*, Vol. 36, No. 3, pp. 287–314, Apr. 1994. Special issue on Higher Order Statistics.
6. E. Moreau and N. Thirion-Moreau, "Nonsymmetrical contrasts for source separation", *IEEE Trans. Signal Processing*, Vol. 47, No. 8, pp 2241-2252, August 1999.
7. E. Moreau, "A generalization of joint-diagonalization criteria for source separation", *IEEE Trans. Signal Processing*, Vol. 49, No 3, pp. 530-541, March 2001.
8. J.-F. Cardoso, A. Souloumiac, "Blind beamforming for non-gaussian signals", *IEE proceedings-F*, Vol. 40, pp. 362-370, 1993.
9. J.-F. Cardoso and A. Souloumiac, "Jacobi angles for simultaneous diagonalization", *SIAM Journal of Matrix Analysis and Applications*, Vol. 17, No 1, pp. 161-164, Jan. 1996.
10. P. Flandrin, *Time-Frequency/Time-Scale Analysis*, Academic Press, 1999.
11. L. Cohen, *Time-Frequency Analysis*, Prentice Hall, 1995.

Adaptive Robust Super-exponential Algorithms for Deflationary Blind Equalization of Instantaneous Mixtures

Masanori Ito[1], Masashi Ohata[2], Mitsuru Kawamoto[2,3],
Toshiharu Mukai[2], Yujiro Inouye[3], and Noboru Ohnishi[1]

[1] Graduate School of Information Science, Nagoya University
Furo-cho, Chikusa-ku, Nagoya, 464-8603 Japan
{ito,ohnishi}@ohnishi.nuie.nagoya-u.ac.jp
[2] Bio-mimetic Control Research Center, RIKEN
2271-130 Anagahora, Shimoshidami, Moriyama-ku, Nagoya, 463-0003 Japan
{ohatama,kawa,tosh}@bmc.riken.jp
[3] Department of Electronic and Control Systems Engineering, Shimane University
1060 Nishikawatsu, Matsue, Shimane, 690-8504 Japan
kawa@ecs.shimane-u.ac.jp, inouye@riko.shimane-u.ac.jp

Abstract. The so called "super-exponential" algorithms (SEA's) are attractive algorithms for solving blind signal processing problems. The conventional SEA's, however, have such a drawback that they are very sensitive to Gaussian noise. To overcome this drawback, we propose a new SEA. While the conventional SEA's use the second- and higher-order cumulants of observations, the proposed SEA uses only the higher-order cumulants of observations. Since higher-order cumulants are insensitive to Gaussian noise, the proposed SEA is robust to Gaussian noise, which is referred to as a robust super-exponential algorithm (RSEA). The proposed RSEA is implemented as an adaptive algorithm, which is referred to as an adaptive robust super-exponential algorithm (ARSEA). To show the validity of the ARSEA, some simulation results are presented.

1 Introduction

The present paper deals with the blind equalization problem of a static system driven by (or linear mixtures of) independent source signals. To solve this problem, the ideas of the super-exponential algorithms (SEA's) in [1], [4], and [6] are used. Several researchers (e.g., [1, 4–6, 10]) have proposed some SEA's until now for solving independent component analysis (ICA), blind source separation (BSS), and blind channel equalization (BCE). One of the attractive properties of the SEA's is that they are computationally efficient and they converge to a desired solution at a super-exponential rate. However, almost all the SEA's proposed until now have such a drawback that they are very sensitive to Gaussian noise (this will be shown in Section 4), because they utilize the second-order and the higher-order cumulants of observations.

In the present paper, we propose a new SEA which overcomes the drawback. The proposed SEA utilizes only the higher-order cumulants of observations, and hence the SEA becomes robust to Gaussian noise, which is referred to as a *robust super-exponential algorithm* (RSEA). Moreover, we develop an adaptive algorithm for implementing the RSEA, which is referred to as an adaptive robust super-exponential algorithm (ARSEA). Simulation results show that the proposed ARSEA is robust to Gaussian noise and can successfully implement the equalization of static systems.

2 Problem Formulation

Throughout the present paper, let us consider the following MIMO static system with n inputs and m outputs:

$$\boldsymbol{y}(t) = \boldsymbol{H}\boldsymbol{s}(t) + \boldsymbol{n}(t), \tag{1}$$

where $\boldsymbol{y}(t)$ represents an m-column output vector called the *observed signal*, $\boldsymbol{s}(t)$ represents an n-column input vector called the *source signal*, $\boldsymbol{H}$ is an $m \times n$ matrix, $\boldsymbol{n}(t)$ represents an m-column noise vector. It can be regarded as a linear mixture model with additive noise.

To implement the equalization of the system (1), the following n filters, which are m-input single-output (MISO) static systems driven by the observed signals, are used:

$$z_l(t) = \boldsymbol{w}_l^T \boldsymbol{y}(t), \quad l = 1, 2, \cdots, n, \tag{2}$$

where $z_l(t)$ is the lth output of the filter, $\boldsymbol{w}_l = [w_{l1}, w_{l2}, \cdots, w_{lm}]^T$ is an m-column vector representing the m coefficients of the lth filter. Substituting (1) into (2), we obtain

$$\begin{aligned} z_l(t) &= \boldsymbol{w}_l^T \boldsymbol{H} \boldsymbol{s}(t) + \boldsymbol{w}_l^T \boldsymbol{n}(t) \\ &= \boldsymbol{g}_l^T \boldsymbol{s}(t) + \boldsymbol{w}_l^T \boldsymbol{n}(t), \quad l = 1, 2, \cdots, n, \end{aligned} \tag{3}$$

where $\boldsymbol{g}_l = [g_{l1}, g_{l2}, \cdots, g_{ln}]^T := \boldsymbol{H}^T \boldsymbol{w}_l$ is an n-column vector. The blind equalization problem considered in the present paper can be formulated as follows: Find n filters $\boldsymbol{w}_l$'s denoted by $\tilde{\boldsymbol{w}}_l$'s satisfying the following condition, without the knowledge of $\boldsymbol{H}$, even if the Gaussian noise $\boldsymbol{n}(t)$ is added to the observed signal $\boldsymbol{y}(t)$,

$$\tilde{\boldsymbol{g}}_l = \boldsymbol{H}^T \tilde{\boldsymbol{w}}_l = \tilde{\boldsymbol{\delta}}_l, \quad l = 1, 2, \cdots, n, \tag{4}$$

where $\tilde{\boldsymbol{\delta}}_l$ is an n-column vector whose elements $\tilde{\delta}_{lr}$ ($r = 1, 2, \cdots, n$) are equal to zero expect for the ρ_lth element, that is, $\tilde{\delta}_{lr} = d_l \delta(r - \rho_l)$, $r = 1, 2, \cdots, n$.

Here, $\delta(t)$ is the Kronecker delta function, d_l is a number standing for a scale change, and ρ_l is one of integers $\{1, 2, \cdots, n\}$ such that the set $\{\rho_1, \rho_2, \cdots, \rho_n\}$ is a permutation of the set $\{1, 2, \cdots, n\}$.

To solve the blind equalization problem, we put the following assumptions on the system and the source signals.

A1) The matrix $\boldsymbol{H}$ in (1) is an $m \times n$ ($m \geq n$) matrix and has full column rank.

A2) The input sequence $\{\boldsymbol{s}(t)\}$ is a zero-mean, non-Gaussian vector stationary process whose element processes $\{s_i(t)\}$, $i = 1, 2, \cdots, n$, are mutually statistically independent and have the nonzero $(p+1)$st-order cumulants, γ_i^{p+1} defined as

$$\gamma_i^{p+1} = \mathrm{cum}\{\underbrace{s_i(t), s_i(t), \cdots, s_i(t)}_{p+1}\} \neq 0, \tag{5}$$

where $i = 1, 2, \cdots, n$ and $p \geq 2$ (in this paper we treat the case of $p = 3$).

A3) The noise signal sequence $\{\boldsymbol{n}(t)\}$ is a zero-mean, Gaussian vector stationary process whose element processes $\{n_i(t)\}$, $i = 1, 2, \cdots, m$, are mutually statistically independent.

A4) The two vector sequences $\{\boldsymbol{n}(t)\}$ and $\{\boldsymbol{s}(t)\}$ are mutually statistically independent.

It is assumed for the sake of simplicity in the present paper that all the signals and all the systems are real-valued.

3 Super-exponential Algorithms (SEA's)

3.1 Two-Step Iterative Procedure for Vector $\boldsymbol{w}_l$

To find solutions in (4), the following two-step iterative procedure for $\boldsymbol{w}_l$ is used:

$$\boldsymbol{w}_l^{[1]} := \hat{\boldsymbol{R}}^\dagger \boldsymbol{d}_l, \quad l = 1, 2, \cdots, n, \tag{6}$$

$$\boldsymbol{w}_l^{[2]} := \boldsymbol{w}_l^{[1]} / \sqrt{|\boldsymbol{w}_l^{[1]T} \hat{\boldsymbol{R}} \boldsymbol{w}_l^{[1]}|}, \quad l = 1, 2, \cdots, n, \tag{7}$$

where $\dagger$ denotes the pseudo-inverse operation of a matrix and $\boldsymbol{d}_l = [d_{l1}, d_{l2}, \cdots, d_{lm}]^T$ whose each element d_{lj} is calculated by

$$d_{lj} = \mathrm{cum}\{z_l(t), z_l(t), z_l(t), y_j(t)\}, \tag{8}$$

which means that d_{lj} is given by the forth-order cumulant of $z_l(t)$ and $y_j(t)$.

The above two-step procedure becomes one cycle of iterations in the super-exponential algorithm (SEA) [1, 4–6, 10]. In the conventional SEA's (e.g., [1, 4–6, 10]), $\hat{\boldsymbol{R}}$ in (6) and (7) is calculated by the second-order cumulant of the observed signals $y_j(t)$'s, whereas, in the proposed SEA, the fourth-order cumulants of the observed signals are only used. To be specific, $\hat{\boldsymbol{R}}$ is calculated by

$$\hat{\boldsymbol{R}} = \sum_{i,j=1}^{m} \beta_{ij} \boldsymbol{C}_{\boldsymbol{y}, i, j}^{(4)}, \tag{9}$$

where β_{ij}'s are either 1 or 0, which represent *design parameters* and $\boldsymbol{C}_{\boldsymbol{y}, i, j}^{(4)}$ is defined by

$$\boldsymbol{C}_{\boldsymbol{y}, i, j}^{(4)} = [\mathrm{cum}\{y_q(t), y_r(t), y_i(t), y_j(t)\}]_{q, r}, \tag{10}$$

where $[x]_{q,r}$ denotes the (q,r)th element of the matrix $\boldsymbol{C}_{\boldsymbol{y},i,j}^{(4)}$. Namely, the proposed SEA is implemented by using only higher-order cumulants of the observed signals, from which it becomes insensitive to Gaussian noise. This is a *novel key point* of our proposed SEA, which is referred to as a robust super-exponential algorithm (RSEA).

3.2 Adaptive Robust Super-exponential Algorithm (ARSEA)

We consider that the two-step iterative procedure (6) and (7) is adaptively implemented. To this end, we must specify the dependency of each time t and rewrite (6) and (7) as, respectively,

$$\boldsymbol{w}_l^{[1]}(t) := \boldsymbol{Q}_r(t-1)\boldsymbol{R}_r^{\dagger}(t)\boldsymbol{Q}_r^T(t)\tilde{\boldsymbol{d}}_l(t), \qquad l=1,2,\cdots,n, \quad (11)$$

$$\boldsymbol{w}_l^{[2]}(t) := \boldsymbol{w}_l^{[1]}(t)/\sqrt{|\boldsymbol{w}_l^{[1]T}(t)\boldsymbol{A}(t)\boldsymbol{Q}_r^T(t-1)\boldsymbol{w}_l^{[1]}(t)|}, \qquad l=1,2,\cdots,n, \quad (12)$$

where $\boldsymbol{Q}_r(t)$ and $\boldsymbol{R}_r(t)$ are obtained by the QR decomposition [3] of matrix $\boldsymbol{A}(t)$ defined by $\boldsymbol{A}(t) := \hat{\boldsymbol{R}}\boldsymbol{Q}_r(t-1)$, which is decomposed as

$$\boldsymbol{A}(t) := \boldsymbol{Q}_r(t)\boldsymbol{R}_r(t) \quad (13)$$

and the update of $\boldsymbol{A}(t)$ is

$$\boldsymbol{A}(t) := \mu \boldsymbol{A}(t-1)\boldsymbol{Q}_r^T(t-2)\boldsymbol{Q}_r(t-1) + (1-\mu)\boldsymbol{U}(t)\boldsymbol{V}(t)\boldsymbol{Q}_r(t-1), \quad (14)$$

where μ is a positive constant close to, but less than one.

Here $\boldsymbol{V}(t)$ in (14) is defined by $\boldsymbol{V}(t) := \boldsymbol{y}(t)\boldsymbol{y}^T(t)$, the matrix $\boldsymbol{U}(t)$ in (14) is defined by

$$\boldsymbol{U}(t) := \boldsymbol{V}(t) - 2\tilde{\boldsymbol{V}}(t) - \text{tr}\{\tilde{\boldsymbol{V}}(t)\}\boldsymbol{I}, \quad (15)$$

where $\text{tr}\{X\}$ denotes the trace of the matrix X, and $\tilde{\boldsymbol{V}}(t)$ is a moving average of $\boldsymbol{V}(t)$ which is calculated by

$$\tilde{\boldsymbol{V}}(t) = \lambda \tilde{\boldsymbol{V}}(t-1) + (1-\lambda)\boldsymbol{V}(t), \quad (16)$$

where λ is a positive constant close to, but less than one and $\mu > \lambda$. The update of $\tilde{\boldsymbol{d}}_l(t)$ is

$$\tilde{\boldsymbol{d}}_l(t) = \mu \tilde{\boldsymbol{d}}_l(t-1) + (1-\mu)(z_l^2(t) - 3r_l(t))z_l(t)\boldsymbol{y}(t), \quad (17)$$

where $r_l(t)$ is a moving average of $z_l^2(t)$, which is calculated by

$$r_l(t) = \lambda r_l(t) + (1-\lambda)z_l^2(t). \quad (18)$$

In the present paper, by using the adaptive two-step iterative procedure (11) and (12), We treat the case where only one source is recovered, that is, the case of $l=1$ in (4). A forthcoming paper will treat the case where all the sources are recovered. Therefore, the proposed adaptive RSEA (ARSEA) is summarized as shown in Table 1.

Table 1. The proposed ARSEA.

Step	Contents
1	Set an initial time ($t = 0$) and set l, which denotes the number of the channels equalized, to be 1.
2	Choose randomly an initial value $\boldsymbol{w}_1(0)$, where $\boldsymbol{w}_1(0)$ denotes an initial value of $\boldsymbol{w}_1$.
3	Calculate (2) ($z_l(t) = \boldsymbol{w}_1^T(t)\boldsymbol{y}(t)$).
4	Calculate $\boldsymbol{V}(t) := \boldsymbol{y}(t)\boldsymbol{y}^T(t)$.
5	Calculate (16) ($\tilde{\boldsymbol{V}}(t) = \lambda\tilde{\boldsymbol{V}}(t-1) + (1-\lambda)\boldsymbol{V}(t)$).
6	Calculate (15) ($\boldsymbol{U}(t) = \boldsymbol{V}(t) - 2\tilde{\boldsymbol{V}}(t) - \text{tr}\{\tilde{\boldsymbol{V}}(t)\}\boldsymbol{I}$).
7	Calculate $\boldsymbol{A}(t)$ using (14).
8	Calculate the QR decomposition of $\boldsymbol{A}(t)$.
9	Calculate (18) ($r_l(t) = \lambda r_l(t) + (1-\lambda)z_l^2(t)$).
10	Calculate (17) ($\tilde{\boldsymbol{d}}_l(t) = \mu\tilde{\boldsymbol{d}}_l(t-1) + (1-\mu)(z_l^2(t) - 3r_l(t))z_l(t)\boldsymbol{y}(t)$).
11	Calculate $\boldsymbol{w}_1(t)$ using (11) and (12).
12	If t is less then T (observed data length), then set $t = t+1$, and the procedures (Step 3 to Step 11) are continued until $t = T$.

4 Simulation Results

To demonstrate the validity of the proposed ARSEA, many computer simulations were conducted. Some of the results is shown in this section. We considered a two-input and two-output system, that is, $\boldsymbol{H}$ in (1) was set to be

$$\boldsymbol{H} = \begin{bmatrix} 1.0 & 0.5 \\ 0.4 & 1.0 \end{bmatrix}. \tag{19}$$

Two source signals $s_1(t)$ and $s_2(t)$ were sub-Gaussian and super-Gaussian, respectively, in which $s_1(t)$ takes one of two values, -1 and 1 with equal probability $1/2$, $s_2(t)$ takes one of three values, -2, 0 and 2 with probability $1/8$, $6/8$, and $1/8$, respectively, and they are zero-mean and unit variance. The parameter p in (8) was set to be $p = 3$, that is, γ_j^{p+1} ($j = 1,2$) in (5) were the fourth-order cumulants of the source signals. These values were set to $\gamma_1^4 = -2$ and $\gamma_2^4 = 1$. The parameters λ in (18) and μ in (17) were set to be $\lambda = 0.999$ and $\mu = 0.9999$, respectively. $\beta_{ij} = 1$ for $i = j$ and $\beta_{ij} = 0$ for $i \neq j$ (see (9)). Two independent Gaussian noises (with identical variance σ_n^2) were added to the two outputs $y_i(t)$'s at various SNR levels. The SNR is, for convenience, defined as Input-SNR $:= 10\log_{10}(\sigma_{s_i}^2/\sigma_n^2)$, where $\sigma_{s_i}^2$'s are the variances of $s_i(t)$'s ($\sigma_{s_i}^2 = 1$). As a measure of performance, we used the *signal-to-interference ratio* (SIR) defined in the logarithmic (dB) scale by

$$\text{SIR} = 10\log\frac{E[(g_{1\rho_1}s_{\rho_1})^2]}{E[(\boldsymbol{g}_1^T\boldsymbol{s}(t) - g_{1\rho_1}s_{\rho_1})^2]}, \tag{20}$$

where $E[x]$ denotes the expectation of a random variable x. The value of SIR becomes ∞, if $\tilde{\boldsymbol{g}}_1$ in (4) is obtained, and hence a large value of SIR indicates it

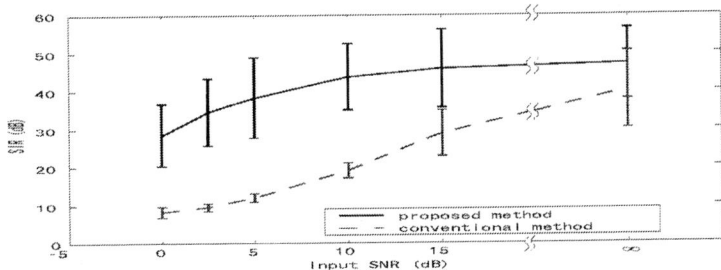

Fig. 1. The performances for the proposed ARSEA and the conventional SEA.

is in close proximity to the desired solution. As a conventional algorithm, the algorithm proposed in [5] was used for comparison. The data length T was set to be 160,000. During first 10,000 data samples, we didn't update $\boldsymbol{w}_l$.

Figure 1 shows the results of performances for the proposed ARSEA (solid line) and the conventional SEA (dashed line) when the Input-SNR levels were respectively taken to be 0[dB] ($\sigma_n^2 = 1$), 2.5[dB], 5[dB], 10[dB], 15[dB] and ∞[dB] (without noise). In each line, the averages of the SIR of the performance results obtained by 100 independent Monte Carlo runs and their standard deviations are shown. It can be seen from Figure 1 that the proposed ARSEA is robust to Gaussian noise and implement successfully finding a desired solution.

Moreover, we investigated the intersymbol-interference (ISI) in each Input-SNR, defined by

$$\text{ISI}(t) = \frac{\left(\sum_i g_{1i}^2(t)\right) - g_{1\rho_1}^2(t)}{g_{1\rho_1}^2(t)}, \qquad (21)$$

The ISI becomes zero, if the $\tilde{g}_1$ in (4) is obtained, and a small value of ISI indicates it is in close proximity to the desired solution. Figures 2 through 6 show the results, which indicate the changes of ISI(t) obtained by using 160,000 data samples. It can be seen from all the figures that the proposed ARSEA makes it possible to be converged to a desired solution by using about 20,000data samples.

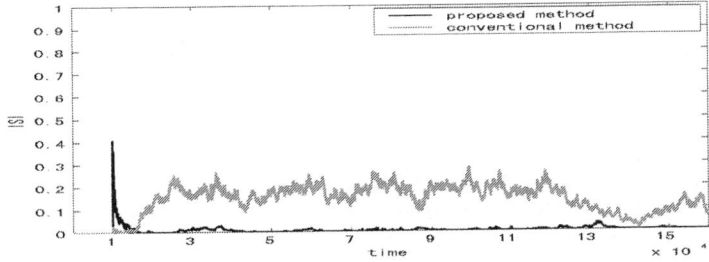

Fig. 2. ISI's for the proposed ARSEA and the conventional SEA (Input-SNR = 0 [dB]).

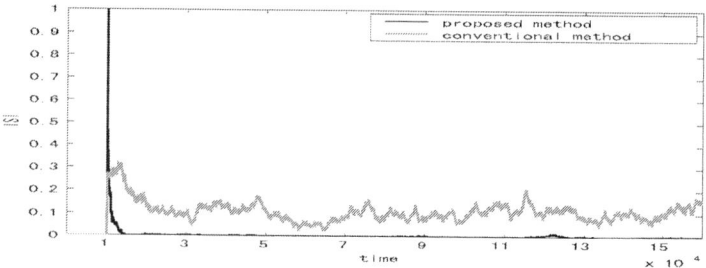

Fig. 3. ISI's for the proposed ARSEA and the conventional SEA (Input-SNR = 2.5 [dB]).

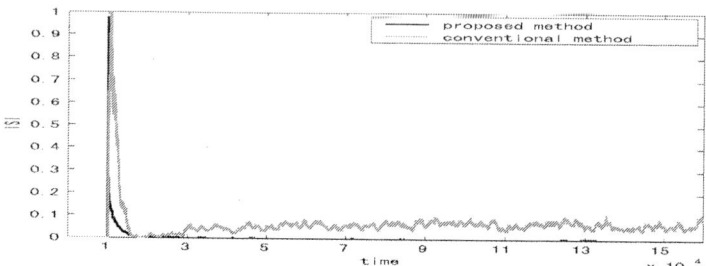

Fig. 4. ISI's for the proposed ARSEA and the conventional SEA (Input-SNR = 5 [dB]).

5 Conclusions

We have proposed an SEA for solving blind equalization problem. The proposed SEA is not sensitive to Gaussian noise, which is referred to as a robust super-exponential algorithm (RSEA). This is a novel property of the proposed algorithm, whereas the conventional algorithms do not posses it. Moreover, we have proposed an algorithm which adaptively implements the proposed RSEA, which is referred to as an adaptive robust super-exponential algorithm (ARSEA). It was shown from the simulation results that the proposed ARSEA is robust to Gaussian noise and could successfully recover one source.

We have not shown a method of recovering all the sources, using the ARSEA. This will be treated in our future study.

References

1. Inouye, Y., Tanebe, K.: Super-exponential algorithms for multichannel blind deconvolution. IEEE Trans. Signal Processing, Vol. 48, No. 3 (2000) 881–888.
2. Kawamoto, M., Inouye, Y.: A deflation algorithm for the blind source-factor separation of MIMO-FIR systems driven by colored sources. IEEE Signal Processing Letters, Vol. 10, No. 11 (2003) 343–346.
3. Lancaster, P., Tismenetsky, M.: The Theory of Matrices, second edition, Academic Press, INC. (1985).

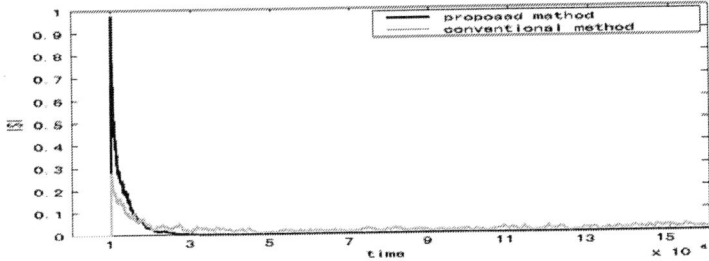

Fig. 5. ISI's for the proposed ARSEA and the conventional SEA (Input-SNR = 10 [dB]).

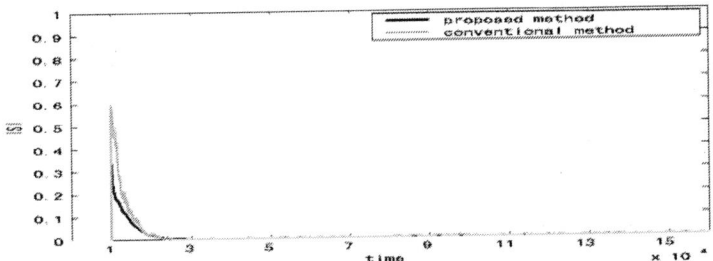

Fig. 6. ISI's for the proposed ARSEA and the conventional SEA (Input-SNR = 15 [dB]).

4. Martone, M.: An adaptive algorithm for antenna array Low-Rank processing in cellular TDMA base stations. IEEE Trans. Communications, Vol. 46, No. 5 (1998) 627–643.
5. Martone, M.: Fast adaptive super-exponential multistage beamforming cellular base-station transceivers with antenna arrays. IEEE Trans. Vehicular Tech., Vol. 48, No. 4 (1999) 1017–1028.
6. Shalvi, O., Weinstein, E.: Super-exponential methods for blind deconvolution. IEEE Trans. Information Theory, Vol. 39, No. 2 (1993) 504–519.
7. Simon, C., Loubaton, P., Jutten, C.: Separation of a class of convolutive mixtures: a contrast function approach. Signal Processing, Vol. 81 (2001) 883–887.
8. Tong, L., Inouye, Y., Liu, R.-w.: Waveform-preserving blind estimation of multiple independent sources. IEEE Trans. Signal Processing, Vol. 41, No. 7 (1993) 2461–2470.
9. Tugnait, J.K.: Identification and deconvolution of multichannel non-Gaussian processes using higher order statistics and inverse filter criteria. IEEE Trans. Signal Processing, Vol. 45, No. 3 (1997) 658–672.
10. Yeung, K.L., Yau, S. F.: A cumulant-based super-exponential algorithm for blind deconvolution of multi-input multi-output systems. Signal Processing, Vol. 67 (1998) 141–162.

Application of Gaussian Mixture Models for Blind Separation of Independent Sources

Koby Todros and Joseph Tabrikian

Department of ECE
Ben-Gurion University of the Negev
Beer-Sheva, 84105 Israel

Abstract. In this paper, we consider the problem of blind separation of an instantaneous mixture of independent sources by exploiting their non-stationarity and/or non-Gaussianity. We show that non-stationarity and non-Gaussianity can be exploited by modeling the distribution of the sources using Gaussian Mixture Model (GMM). The Maximum Likelihood (ML) estimator is utilized in order to derive a new source separation technique. The method is based on estimation of the *sensors* distribution parameters via the Expectation Maximization (EM) algorithm for GMM parameter estimation. The separation matrix is estimated by applying simultaneous joint diagonalization of the estimated GMM covariance matrices. The performance of the proposed method is evaluated and compared to existing blind source separation methods. The results show superior performance.

1 Introduction

Consider the following noiseless instantaneous linear mixture model:

$$\mathbf{x}_t = \mathbf{A}\mathbf{s}_t, \; t=1,2,...,T. \tag{1}$$

The random vector $\mathbf{s}_t = [s_{1,t},...,s_{K,t}]^T$, representing K statistically independent sources, is mixed by a fixed unknown $L \times K$ ($L \geq K$) mixing matrix $\mathbf{A}$ at time instance t. The observations vector $\mathbf{x}_t = [x_{1,t},...,x_{L,t}]^T$ is obtained from an array of L sensors. The problem of BSS (Blind Source Separation) addresses the reconstruction of the source vectors $\{\mathbf{s}_t\}_{t=1}^T$, by estimating a $K \times L$ separation matrix $\mathbf{B}$ for which:

$$\hat{\mathbf{s}}_t = \mathbf{B}\mathbf{x}_t, \; t=1,2,...,T. \tag{2}$$

Blind separation of an instantaneous linear mixture of independent sources can be achieved, up to scaling and permutation of the estimated sources, by exploiting their non-stationarity and or non-Gaussianity [1], [2]. Some existing methods for BSS use restrictive assumptions on the sources distribution, which makes them inapplicable in some cases. For example, cumulant-based methods like JADE [3], which exploit non-Gaussianity of the source signals, assume that the sources have non-zero 4th order cumulant and the probability density function (pdf) of each source is approximated by using only the 2nd and 4th order cumulants (Edgeworth expansion [3]). Other methods, which exploit non-stationarity of the sources, [2], [4], assume that the statistical properties of the sources vary smoothly, so the observed signals can be partitioned into quasi-stationary segments.

Applying Gaussian Mixture Models (GMMs), which are widely used to model highly complex pdf's, can achieve more accurate modeling of the joint pdf of the sources. In this paper, non-Gaussianity and/or non-stationarity of the sources are exploited by modeling the pdf of each source using GMM. The fact that at every time instance $s_{k,t}$ is generated by a different set of Gaussians can be viewed as a non-Gaussian-non-stationary case. By this approach, no prior assumptions on the stationarity of the sources are required, so their statistical properties can vary non- smoothly from sample to sample. An earlier application of GMM for blind separation of noiseless instantaneous linear mixtures was offered by Attaias [5] (Noiseless IFA). According to this method, the pdf of each source is modeled by GMM and an Expectation-Maximization (EM) algorithm, which jointly estimates the *sources* distribution parameters and the separation matrix coefficients, is proposed. The source signals are then reconstructed according to equation (2). This approach has two disadvantages. First, it is very difficult to achieve accurate initialization of the unobserved source distribution parameters, so the EM algorithm may converge into local minima. Second, implementation of this approach is complicated and requires correct empirical choice of a learning rate factor, which is used for updating the estimation of the separation matrix in each step of the algorithm.

In this paper, we show that the maximum likelihood estimator can be utilized for solving the BSS problem in two separate steps. In the first step the *sensors* distribution parameters are estimated via the EM algorithm for GMM parameter estimation [7] and [8]. In the second step, the separation matrix is estimated by applying simultaneous joint diagonalization of the estimated GMM covariance matrices [2]. The source signals are reconstructed according to equation (2). We restrict the discussion to the case of as many sensors as sources ($L = K$), so that **A** is invertible and $\mathbf{B} = \mathbf{A}^{-1}$. The performance of the proposed method is evaluated and compared to existing blind source separation methods. The results show superior performance.

The paper is organized as follows: in Section 2, we derive mathematical models for the pdf's of the source and sensor signals. In Section 3 a technique for the estimation of the separation matrix is developed. In Section 4, the performance of the proposed method is evaluated and compared to other existing methods. Finally, Section 5 summarizes the main points of this contribution.

2 Derivation of Sources and Sensors Distribution Models

In this section, the distributions of the source and sensor signals are modeled by applying GMM.

2.1 Sources Distribution Model

The pdf of each source signal at time instance t is given by:

$$f_s\left(s_{k,t}; \boldsymbol{\theta}_k^{(s)}\right) = \sum_{i=1}^{n_k} \phi_{k,i}\, \mathrm{N}\left(s_{k,t}; \mu_{k,i}; \sigma_{k,i}^2\right), \quad k = 1,\ldots,K . \tag{3}$$

where $N(\cdot,\cdot,\cdot)$ denotes normal distribution function, n_k is the GMM order (i.e. number of Gaussians), $\{\phi_{k,i}\}_{i=1}^{n_k}$, $\{\mu_{k,i}\}_{i=1}^{n_k}$, $\{\sigma_{k,i}^2\}_{i=1}^{n_k}$ are mixing proportions or a-priori probabilities, means and variances of each Gaussian respectively. The vector of unknown distribution parameters of the kth source signal is denoted by $\boldsymbol{\theta}_k^{(s)} = \{\phi_{k,i}, \mu_{k,i}, \sigma_{k,i}^2\}_{i=1}^{n_k}$. By applying the assumption of independent source signals, their joint pdf can be formulated as follows:

$$f_s(\mathbf{s}_t;\boldsymbol{\theta}^{(s)}) = \prod_{k=1}^{K} f_s(s_{k,t};\boldsymbol{\theta}_k^{(s)}) = \sum_{m=1}^{M} w_m N(\mathbf{s}_t;\boldsymbol{\mu}_m;\mathbf{C}_m) . \qquad (4)$$

where $M = \prod_{k=1}^{K} n_k$ is the total number of Guassians in the joint pdf and $w_m = \prod_{k=1}^{K} \phi_{k,l(k)}$; $l(k) = 1,...,n_k$ are the mixing proportions of each gaussian, such that $\sum_{m=1}^{M} w_m = 1$. The mean vectors and covariance matrices of each Gaussian are denoted by $\boldsymbol{\mu}_m = [\mu_{1,l(1)},...,\mu_{K,l(K)}]^T$ and $\mathbf{C}_m = diag(\sigma_{1,l(1)}^2,...,\sigma_{K,l(K)}^2)$ respectively and the vector of unknown parameters of the joint pdf is denoted by $\boldsymbol{\theta}^{(s)} = \{w_m, \boldsymbol{\mu}_m, \mathbf{C}_m\}_{m=1}^{M}$. Equation (4) implies that the joint pdf of the sources is also GMM with diagonal covariance matrices.

2.2 Sensors Distribution Model

In this subsection, the generative model of the observation signals is utilized in order to derive an expression for their joint pdf at time instance t. Figure 1 depicts a graphical model corresponding to the generation process of the observation signals at time instance t.

According to this generative model, at every time instance, a hidden vector $\mathbf{y}_t = [y_{t,1},...,y_{t,M}]^T$, which indicates the generating Gaussian of $\mathbf{s}_t$, is randomized by the following discrete pdf:

$$f_\mathbf{y}(\mathbf{y}_t) = \sum_{m=1}^{M} w_m \delta(y_{t,m} - 1), \qquad (5)$$

where

$$y_{t,m} = \begin{cases} 1, & \text{if } \mathbf{s}_t \text{ was generated by the } m^{th} \text{ Gaussian component} \\ 0, & \text{otherwise} \end{cases} . \qquad (6)$$

Hidden values of $\mathbf{s}_t$ are then mixed by the mixing matrix $\mathbf{A}$ and an observation vector $\mathbf{x}_t$ is formed. The fact that at every time instance $\mathbf{s}_t$ may be associated with a

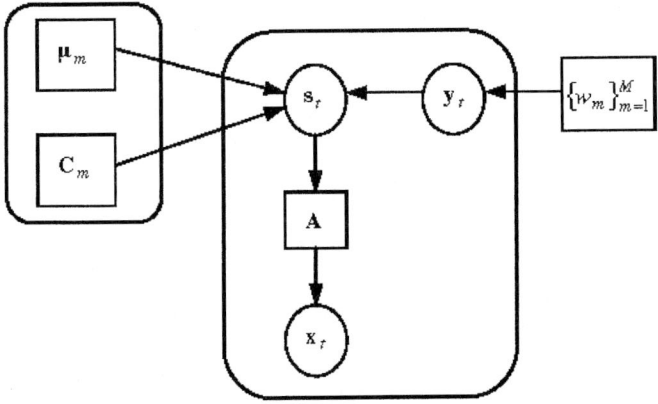

Fig. 1. The generative model of the observation signals at time instance t.

different set of Gaussians can be viewed as a general case of non-Gaussianity-non-stationarity of the observed signals. According to Bayes theorem:

$$f_{\mathbf{x}}(\mathbf{x}_t;\mathbf{\theta}^{(\mathbf{x})}) = E_{\mathbf{y}}\left[f_{\mathbf{x}|\mathbf{y}}(\mathbf{x}_t \mid \mathbf{y}_t;\mathbf{\theta}^{(\mathbf{x})})\right]. \tag{7}$$

Using equations (5) and (7) the distribution of $\mathbf{x}_t$ can be formulated as follows:

$$\begin{aligned} f_{\mathbf{x}}(\mathbf{x}_t;\mathbf{\theta}^{(\mathbf{x})}) &= \sum_{y_{t,m}=0}^{1}\sum_{m=1}^{M} w_m \delta(y_{t,m}-1) f_{\mathbf{x}|\mathbf{y}}(\mathbf{x}_t \mid y_{t,m};\mathbf{\theta}^{(\mathbf{x})}) \\ &= \sum_{m=1}^{M} w_m f_{\mathbf{x}|\mathbf{y}}(\mathbf{x}_t \mid y_{t,m}=1;\mathbf{\theta}^{(\mathbf{x})}), \end{aligned} \tag{8}$$

where:

$$f_{\mathbf{x}|\mathbf{y}}(\mathbf{x}_t \mid y_{t,m}=1;\mathbf{\theta}^{(\mathbf{x})}) = f_{\mathbf{x}}(\mathbf{x}_t;\mathbf{\theta}_m^{\mathbf{x}}) = N(\mathbf{x}_t;\mathbf{A}\mathbf{\mu}_m;\mathbf{A}\mathbf{C}_m\mathbf{A}^T), \tag{9}$$

$\mathbf{\theta}_m^{(\mathbf{x})} = \{w_m, \mathbf{A}\mathbf{\mu}_m, \mathbf{A}\mathbf{C}_m\mathbf{A}^T\}$ is the vector of unknown distribution parameters of the mth Gaussian and $\mathbf{\theta}^{(\mathbf{x})} = \{\mathbf{\theta}_m^{(\mathbf{x})}\}_{m=1}^{M}$. Thus, we conclude that the joint pdf of the sensor signals is also GMM with non-diagonal covariance matrices. The expression of this distribution is given by:

$$f_{\mathbf{x}}(\mathbf{x}_t;\mathbf{\theta}^{(\mathbf{x})}) = \sum_{m=1}^{M} w_m N(\mathbf{x}_t,\mathbf{A}\mathbf{\mu}_m,\mathbf{A}\mathbf{C}_m\mathbf{A}^T). \tag{10}$$

3 Estimation of the Separation Matrix

Using the generative model of the observed signals, described in section 2.2, we derive an objective function, based on the ML estimator, for the estimation of the separation matrix $\mathbf{B}$. According to equations (5) and (9) the joint pdf of $\mathbf{x}_t$ and $\mathbf{y}_t$ is given by:

$$f_{\mathbf{x},\mathbf{y}}(\mathbf{x}_t,\mathbf{y}_t;\boldsymbol{\theta}^{(x)}) = \prod_{m=1}^{M} [w_m N(\mathbf{x}_t;\boldsymbol{\eta}_m;\mathbf{R}_m)]^{y_{t,m}}, \quad (11)$$

where $\boldsymbol{\eta}_m = \mathbf{A}\boldsymbol{\mu}_m$ and $\mathbf{R}_m = \mathbf{A}\mathbf{C}_m\mathbf{A}^T$. We assume that $\mathbf{x}_1,...,\mathbf{x}_T$ and $\mathbf{y}_1,...,\mathbf{y}_T$ are temporally independent[1], so their joint log likelihood is given by:

$$\log(f_{\mathbf{X},\mathbf{Y}}(\mathbf{X},\mathbf{Y};\boldsymbol{\theta}^{(x)})) = \sum_{t=1}^{T}\sum_{m=1}^{M} y_{t,m}[\log(w_m) + \log(N(\mathbf{x}_t,\boldsymbol{\eta}_m,\mathbf{R}_m))], \quad (12)$$

where $\mathbf{X} = [\mathbf{x}_1,...,\mathbf{x}_T]$ and $\mathbf{Y} = [\mathbf{y}_1,...,\mathbf{y}_T]$. The values of the indication vectors $\mathbf{y}_{t,m}$ are unknown, so we shall use their conditional expectation with respect to $\mathbf{x}_t$ in the following manner:

$$E_{\mathbf{Y}|\mathbf{X}}[\log f_{\mathbf{X},\mathbf{Y}}(\mathbf{X},\mathbf{Y};\boldsymbol{\theta}^{(x)})] = \sum_{t=1}^{T}\sum_{m=1}^{M} \gamma_{t,m} \begin{Bmatrix} \log(w_m) - 0.5\log\det(2\pi\mathbf{R}_m) \\ -0.5[(\mathbf{x}_t - \boldsymbol{\eta}_m)^T(\mathbf{R}_m)^{-1}(\mathbf{x}_t - \boldsymbol{\eta}_m)] \end{Bmatrix}, \quad (13)$$

in which

$$\gamma_{t,m} = E_{\mathbf{y}|\mathbf{x}}[y_{t,m}] = 0 \cdot \Pr(y_{t,m} = 0 \mid \mathbf{x}_t;\boldsymbol{\theta}_f^{(x)}) + 1 \cdot \Pr(y_{t,m} = 1 \mid \mathbf{x}_t;\boldsymbol{\theta}_f^{(x)}). \quad (14)$$

Therefore:

$$\gamma_{t,m} = \frac{\Pr(y_{t,m} = 1)f_{\mathbf{x}|\mathbf{y}}(\mathbf{x}_t \mid y_{t,m} = 1;\boldsymbol{\theta}_f^{(x)})}{f_{\mathbf{x}}(\mathbf{x}_t;\boldsymbol{\theta}_f^{(x)})} = \frac{w_m^f N(\mathbf{x}_t,\boldsymbol{\eta}_m^f,\mathbf{R}_m^f)}{\sum_{m=1}^{M} w_m^f N(\mathbf{x}_t,\boldsymbol{\eta}_m^f,\mathbf{R}_m^f)}, \quad (15)$$

where $\boldsymbol{\theta}_f^{(x)} = \{w_m^f, \boldsymbol{\eta}_m^f, \mathbf{R}_m^f\}_{m=1}^{M}$ denotes the distribution parameters of $\mathbf{x}_t$, obtained in the final step of the EM algorithm for GMM parameter estimation [7] and [8]. By normalizing equation (13) by a factor of $-\frac{1}{T}$, the following expression is obtained:

$$Q = \frac{1}{T}\sum_{m=1}^{M}\sum_{t=1}^{T} \gamma_{t,m} \left\{ 0.5tr\left[\mathbf{R}_m^{-1}\left(\frac{\sum_{t'=1}^{T}\gamma_{t',m}(\mathbf{x}_{t'} - \boldsymbol{\eta}_m)(\mathbf{x}_{t'} - \boldsymbol{\eta}_m)^T}{\sum_{t'=1}^{T}\gamma_{t',m}}\right)\right] + 0.5\log\det(2\pi\mathbf{R}_m) - \log(w_m) \right\}. \quad (16)$$

where $Q = -\frac{1}{T}E_{\mathbf{Y}|\mathbf{X}}[\log f_{\mathbf{X},\mathbf{Y}}(\mathbf{X},\mathbf{Y};\boldsymbol{\theta}^{(x)})]$. The EM algorithm for GMM parameter estimation implies that:

---

[1] The performance of the algorithm with signals having temporally dependent samples (speech) is tested in section 5.

$$\hat{\mathbf{R}}_m + \hat{\boldsymbol{\eta}}_m \hat{\boldsymbol{\eta}}_m^T = \frac{\sum_{t=1}^{T} \gamma_{t,m} \mathbf{x}_t \mathbf{x}_t^T}{\sum_{t=1}^{T} \gamma_{t,m}}, \quad \hat{\boldsymbol{\eta}}_m = \frac{\sum_{t=1}^{T} \gamma_{t,m} \mathbf{x}_t}{\sum_{t=1}^{T} \gamma_{t,m}} \quad \text{and} \quad \hat{w}_m = \frac{1}{T} \sum_{t=1}^{T} \gamma_{t,m} .$$

Therefore, applying that $\mathbf{B} = \mathbf{A}^{-1}$, $\boldsymbol{\eta}_m = \mathbf{A}\boldsymbol{\mu}_m$ and $\mathbf{R}_m = \mathbf{A}\mathbf{C}_m\mathbf{A}^T$, equation (16) can be rewritten as an objective function, with respect to $\mathbf{B}$, in the following manner:

$$Q(\mathbf{B}) = \sum_{m=1}^{M} \hat{w}_m \left\{ KL_{norm}\left[\mathbf{B}\hat{\mathbf{R}}_m \mathbf{B}^T \mid \mathbf{C}_m\right] + tr\left[(\boldsymbol{\mu}_m - \mathbf{B}\hat{\boldsymbol{\eta}}_m)\mathbf{C}_m^{-1}(\boldsymbol{\mu}_m - \mathbf{B}\hat{\boldsymbol{\eta}}_m)^T\right] \right\} + const, \quad (17)$$

where $KL_{norm}(\boldsymbol{\Sigma}_1 \mid \boldsymbol{\Sigma}_2)$ denotes the Kullback-Leibler divergence between two zero mean K-variate normal densities with covariance matrices $\boldsymbol{\Sigma}_1$ and $\boldsymbol{\Sigma}_2$.

For any fixed $\mathbf{B}$ we can choose $\boldsymbol{\mu}_m = \mathbf{B}\hat{\boldsymbol{\eta}}_m$, so that equation (17) can be reduced to the following form:

$$Q(\mathbf{B}) = \sum_{m=1}^{M} \hat{w}_m KL_{norm}\left[\underbrace{\mathbf{B}\hat{\mathbf{R}}_m\mathbf{B}^T}_{\text{positive definite}} \mid \underbrace{\mathbf{C}_m}_{\text{diagonal}}\right]. \quad (18)$$

The Pythagorean property of the Kullback-Leibler divergence [2] implies that for any positive definite matrix $\mathbf{R}$ and for any positive definite diagonal matrix $\mathbf{C}$, $KL_{norm}[\mathbf{R} \mid \mathbf{C}]$ can be decomposed as follows:

$$KL_{norm}[\mathbf{R} \mid \mathbf{C}] = KL_{norm}[\mathbf{R} \mid diag(\mathbf{R})] + KL_{norm}[diag(\mathbf{R}) \mid \mathbf{C}]. \quad (19)$$

Thus, equation (18) can be rewritten as follows:

$$Q(\mathbf{B}) = \sum_{m=1}^{M} \hat{w}_m \left\{ KL_{norm}\left[\mathbf{B}\hat{\mathbf{R}}_m\mathbf{B}^T \mid diag(\mathbf{B}\hat{\mathbf{R}}_m\mathbf{B}^T)\right] + KL_{norm}\left[diag(\mathbf{B}\hat{\mathbf{R}}_m\mathbf{B}^T) \mid \mathbf{C}_m\right] \right\}. \quad (20)$$

For any fixed $\mathbf{B}$ we can choose $\mathbf{C}_m = diag(\mathbf{B}\hat{\mathbf{R}}_m\mathbf{B}^T)$ and equation (20) is reduced to:

$$Q(\mathbf{B}) = \sum_{m=1}^{M} \hat{w}_m KL_{norm}\left[\mathbf{B}\hat{\mathbf{R}}_m\mathbf{B}^T \mid diag(\mathbf{B}\hat{\mathbf{R}}_m\mathbf{B}^T)\right]. \quad (21)$$

Therefore, we conclude that the normalized conditional expectation of the joint log-likelihood (16), under the GMM assumption, leads to an objective function, which measures the deviation of $\{\mathbf{B}\hat{\mathbf{R}}_m\mathbf{B}^T\}_{m=1}^{M}$ from diagonality. The minimum of Q is attained for a matrix $\mathbf{B}$ which jointly diagonalizes the estimated GMM covariance matrices. An approximate joint diagonalization algorithm offered by Pham [2] is applied in order to estimate this matrix. This algorithm works similarly to the Jacobi method by applying successive transformations on each pair of distinct rows of $\mathbf{B}$, with the exception that the row vectors of $\mathbf{B}$ are not enforced to be orthogonal. Detailed description of this algorithm appears in [2]. In summary, the proposed algorithm (GMMJD) comprises of the following steps: 1) Estimation of the sensors distribution parameters via the EM algorithm for GMM parameter estimation, 2) Estimation of **B** by applying simultaneous joint diagonalization of the estimated GMM covariance matrices, 3) Sources recovery according to equation (2).

4 Simulations

In the following example, the performance of the proposed method is evaluated. Two 10 seconds long speech signals, sampled at 16 kHz were mixed by the following 2×2 mixing matrix: $\mathbf{A} = \begin{bmatrix} 5 & 3 \\ -7 & 1 \end{bmatrix}$. The GMM parameters $\{\hat{w}_m\}_{m=1}^M$ and $\{\hat{\mathbf{R}}_m\}_{m=1}^M$ were estimated by applying the EM algorithm. The GMM order was found to be 9 according to the BIC (Bayesian information criterion). The separation matrix **B** was estimated by applying the joint diagonalization algorithm and by applying the JADE [3], FastICA [6] and Noiseless IFA [5] algorithms. Separation performances of each method in terms of Interference-to-Signal Ratio (ISR) were: -82.23 dB for the GMMJD algorithm, -38.43 dB for the JADE algorithm, -32.18 dB for the FastICA algorithm and -43.04 dB for the Noiseless IFA algorithm (computation of ISR is described in [1]). Figure 2 depicts the scatter of the mixed sources (left) and the scatter of the reconstructed sources, using the proposed method (right).

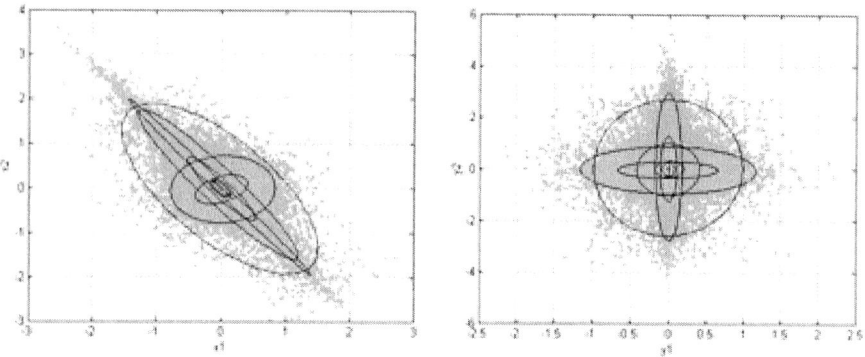

Fig. 2. Left: scatter plot of the observations signals (mixed sources). Right: scatter plot of the reconstructed sources. The ellipses represent the covariance matrices of the Gaussian components.

5 Conclusions

A new algorithm for BSS, based on the ML estimator under the GMM assumption on the sources distribution is proposed. This algorithm exploits both non-stationarity and non-Gaussianity of the mixed sources without using restrictive assumptions on the sources distribution. Finally, results demonstrate superior separation performance achieved by the proposed algorithm in comparison with existing methods.

References

1. J.-F. Cardoso, "Blind Signal Seperation: Statistical Principles," Proceedings of the IEEE, vol. 86, no. 10, pp. 2009-2025, 1998.
2. D.-T. Pham and J.-F. Cardoso, "Blind Seperation of instantaneous mixtures of non-stationary sources," IEEE Trans. on Signal Processing, vol. 49, no. 9, pp. 1837-1848, 2001.
3. J.-F. Cardoso, "High-Order Contrasts for Independent Component Analysis," Neural Computation, vol. 11, pp. 157-192, 1999.
4. A. Hyvarinen, "Blind Source Seperation by Nonstationarity of Variance: A Cumulant-Based Approach," IEEE Trans. on Neural Networks, vol. 12, no. 6, pp. 1471-1474, 2001.
5. H. Attaias, "Independent factor analysis", Neural Computation, vol. 11, pp. 803-851, 1999.
6. A. Hyvarnien and E. Oja, "A fast-fixed-point algorithm for independent component analysis," Neural Computation, vol. 9, pp. 1483-1492, 1997.
7. A.-P. Dempster, N.-M. Laird and D.-B Rubin, "Maximum likelihood from incomplete data via the EM algorithm," Journal of the Royal Statistical Society, vol 39 B, pp. 1-38, 1977.
8. J. Bilmes, "A Gentle Tutorial on the EM Algorithm and its Application to Parameter Estimation for Gaussian Mixture and Hidden Markov Models," Technical Report, University of Berkely, ICSI-TR-97-021, 1997.

Asymptotically Optimal Blind Separation of Parametric Gaussian Sources

Eran Doron and Arie Yeredor

School of Electrical Engineering, Tel-Aviv University
{eran,arie}@eng.tau.ac.il

Abstract. The second-order blind identification (SOBI) algorithm (Belouchrani et al., 1997) is a classical blind source separation (BSS) algorithm for stationary sources. The weights-adjusted SOBI (WASOBI) algorithm (Yeredor 2000) proposed a reformulation of the SOBI algorithm as a weighted nonlinear least squares problem, and showed how to obtain asymptotically optimal weights, under the assumption of Gaussian Moving Average (MA) sources. In this paper, we extend the framework by showing how to obtain the (asymptotically) optimal weight matrix also for the cases of auto-regressive (AR) or ARMA Gaussian sources (of unknown parameters), bypassing the apparent need for estimation of infinitely many correlation matrices. Comparison with other algorithms, with the Cramér Rao bound and with the analytically predicted performance is presented using simulations. In particular, we show that the optimal performance can be attained with fewer estimated correlation matrices than in the Gaussian Mutual Information approach (which is also optimal in this context).

1 Introduction

We address the noiseless static invertible model

$$\mathbf{x}[n] = \mathbf{A}\mathbf{s}[n] \quad n = 1, 2, \ldots, T \tag{1}$$

where $\mathbf{A} \in \mathbb{R}^{N \times N}$ is the unknown mixing matrix, and $\mathbf{x}[n], \mathbf{s}[n] \in \mathbb{R}^N$ are the observations and sources, respectively. In this work, the statistically independent sources are assumed to be different stationary parametric Gaussian auto-regressive (AR) or auto-regressive moving-average (ARMA) processes of known order (or at least with a known upper-bound on their orders).

Considerable research has been directed in recent years (e.g., [1–6]) to the separation of stationary sources with distinct spectra. In [4] Pham and Garat proposed a method for sources whose power spectra are known in advance, based on the maximum-likelihood (ML) approach. However, for cases where the power spectra of the sources are not known in advance[1], they have to be "guessed", rendering the approach "Quasi-ML" (QML), and the algorithm's performance

---
[1] with the exception of the case where all sources are first-order AR processes.

depends on the accuracy of the guess. In another work, Pham suggested to estimate the unknown mixing matrix via the minimization of the Gaussian mutual information (GMI) criterion [5]. The proposed method amounts to jointly approximately diagonalizing a set of estimated spectral density matrices. In [6] Dégerine and Malki proposed a method for colored sources whose underlying AR models are distinct.

The observations' correlation matrices take the structure of

$$\mathbf{R_x}[l] \triangleq \mathbf{A}\mathbf{R_s}[l]\mathbf{A}^T \qquad \forall l \qquad (2)$$

where due to the spatial independence of the sources, their correlation matrices $\mathbf{R_s}[l] \triangleq diag[\lambda_l^{(1)} \lambda_l^{(2)} \cdots \lambda_l^{(K)}]$ are diagonal. Therefore, the second-order blind identification (SOBI) algorithm [1], estimates $\mathbf{A}$ via approximate unitary joint diagonalization (AJD) of some estimated correlation matrices $\hat{\mathbf{R}}_\mathbf{x}[l]$. Both the algorithm proposed in [6] and SOBI involve whitening of $\hat{\mathbf{R}}_\mathbf{x}[0]$. As suggested in [2], this operation implies sub-optimal weighting of the errors in the correlation estimates. It was therefore proposed in [2] to reformulate the AJD as a non-linear WLS problem and to obtain asymptotically optimal weights under the assumption of Gaussian moving average (MA) sources.

That is, since $\mathbf{A}\mathbf{R_s}[l]\mathbf{A}^T$ is a symmetric matrix, (2) can be rewritten in matrix form as: $\mathbf{r}_l \triangleq vec\{\mathbf{R_x}[l]\} = (\mathbf{A} \odot \mathbf{A})\boldsymbol{\lambda}_l$, where $\boldsymbol{\lambda}_l \triangleq [\lambda_l^{(1)} \lambda_l^{(2)} \cdots \lambda_l^{(K)}]^T$, and $\odot$ denotes the Khatri-Rao product (a column-wise Kronecker product).

If $\mathbf{R_x}[l]$ are estimated as $\hat{\mathbf{R}}_\mathbf{x}[l] = \frac{1}{T}\sum_{n=1}^{T} \mathbf{x}[n]\mathbf{x}^T[n+l]$ $0 \leq l < M$ (assuming $T + M - 1$ samples are available), the covariance of the estimates $\{\hat{\mathbf{r}}_l\}_{l=0}^{M-1}$ is given (exploiting the sources' Gaussianity) by:

$$Cov(\hat{\mathbf{r}}_k, \hat{\mathbf{r}}_l) = \frac{1}{T}\sum_{p=-(T-1)}^{(T-1)}(1-\frac{|p|}{T})\left[\mathbf{R_x}[p+l-k]\otimes\mathbf{R_x}[p]+(\mathbf{R_x}[p-k]\otimes\mathbf{R_x}[p+l])\,\mathbf{P}\right] \quad (3)$$

where $\otimes$ denotes Kronecker's product, and $\mathbf{P}$ is the permutation matrix obtained as $\sum_{i,j=1}^{N} \mathbf{E}_{i,j} \otimes \mathbf{E}_{j,i}$, with $\mathbf{E}_{i,j}$ an $N \times N$ all-zeros matrix, with 1 as its (i,j)-th element. Thus, consistent estimates of the correlations can yield consistent estimates of the covariance when substituted into (3). An asymptotically optimal weight matrix for the WLS model can then be obtained as the inverse of the estimated overall covariance matrix.

Apparently, in order to estimate the covariance in (3), we generally need to estimate and sum up $2T$ ($\to \infty$) correlation matrices. In [2], a finite-length correlation assumption is employed in order to obtain a finite number of elements in the calculation of (3). However, this assumption does not hold for AR or ARMA sources. In this paper, based on Stoica et al. [7], we show an alternative, finite summation approach, as we calculate the sum using the Z–transform properties, and extend the WASOBI algorithm to AR and ARMA sources.

We shall focus on the square, real-valued case for simplicity only. Extension to more general cases is rather straightforward, yet notationally more cumbersome.

2 Arma Spectral Estimation

Given that at least one source is not an MA process, generally all the observations are ARMA(p,q) processes, where $p = \sum_{i=1}^{N} p_i$, $q = max\{q_1 + p - p_1, q_2 + p - p_2, \ldots, q_N + p - p_N\}$ and p_i and q_i are i-th source's AR and MA orders, respectively. Since all observations are linear combinations of the same sources, their power spectra and cross-power spectra all have the same denominator and are defined as:

$$S_{\mathbf{x}}^{(i,j)}(z;\boldsymbol{\theta}) = \frac{\sum_{k=-q}^{q} \beta_k^{(i,j)} z^{-k}}{A(z)A^*(\frac{1}{z^*})} \quad i,j = 1,2,\ldots,N \quad, \tag{4}$$

where $S_{\mathbf{x}}^{(i,j)}(z;\boldsymbol{\theta})$ denotes the (i,j)-th element of $\mathbf{S}_{\mathbf{x}}(z;\boldsymbol{\theta})$, the Z–transform of the correlation matrices $\mathbf{R}_{\mathbf{x}}[k]$, and $A(z) \triangleq 1 + \sum_{i=1}^{p} a_i z^{-i}$. The vector

$$\boldsymbol{\theta} \triangleq [a_1 \, a_2 \cdots a_p \, \beta_{-q}^{(1,1)} \, \beta_{-q+1}^{(1,1)} \cdots \beta_q^{(1,1)} \, \beta_{-q}^{(1,2)} \cdots \beta_q^{(1,2)} \cdots \beta_q^{(N,N)}]^T \tag{5}$$

encompasses all the (unknown) parameters of the joint observations' spectrum.

We further assume, without loss of generality, that all the zeros of $A(z)$ lie inside the unit circle.

Although the set $\{\mathbf{R}_{\mathbf{x}}[k]\}_{k=0}^{(p+q)}$ uniquely determines $\boldsymbol{\theta}$ [7], the estimated set $\{\hat{\mathbf{R}}_{\mathbf{x}}[k]\}_{k=0}^{(p+q)}$ does not form a sufficient statistic [8]. Consequently, we begin by computing an augmented set, the estimated correlation matrices up to lag $M-1$, $\{\hat{\mathbf{R}}_{\mathbf{x}}[k]\}_{k=0}^{(M-1)}$, for some $M > p+q$. We then obtain an initial estimate $\hat{\mathbf{a}}$ of the AR parameters $\mathbf{a} \triangleq [a_1 \, a_2 \cdots a_p]^T$, e.g., by the LS solution of the over-determined, extended (for all the observations) modified Yule-Walker (MYW) equations. $\hat{\mathbf{a}}$ is then used in estimating the MA coefficients $\beta_k^{(i,j)}$, $1 \leq i,j \leq N$, via

$$\begin{aligned}\hat{\beta}_k^{(i,j)} &= \sum_{s=0}^{p} \sum_{t=0}^{p} \hat{a}_s \hat{a}_t \hat{R}_{\mathbf{x}}^{(i,j)}[k+s-t] \quad k = 0, 1, \ldots, q \\ \hat{\beta}_k^{(i,j)} &= \hat{\beta}_{-k}^{(i,j)} \quad k = -q, -q+1, \ldots, -1,\end{aligned} \tag{6}$$

where $\hat{R}_{\mathbf{x}}^{(i,j)}[l]$ denotes the (i,j)-th element of $\hat{\mathbf{R}}_{\mathbf{x}}[k]$. In order to exploit the augmented set, so as to refine the estimate of $\{\hat{\mathbf{R}}_{\mathbf{x}}[k]\}_{k=0}^{(p+q)}$, we propose a constrained linear model, to which we shall apply a WLS criterion:

$$\min_{\mathbf{r}_0^{(p+q)}, \mathbf{z}_1^{(M-p-q-1)}} \left\| \begin{bmatrix} \hat{\mathbf{r}}_0^{(p+q)} \\ \hat{\mathbf{z}}_1^{(M-p-q-1)} \end{bmatrix} - \begin{bmatrix} \mathbf{r}_0^{(p+q)} \\ \mathbf{z}_1^{(M-p-q-1)} \end{bmatrix} \right\|_{\mathbf{W}}^2 \quad \text{s.t.} \quad \mathbf{z}_1^{(M-p-q-1)} = \mathbf{0} \tag{7}$$

where the notation $\mathbf{v}_{i_1}^{i_2}$ denotes the concatenation $[\mathbf{v}_{i_1}^T \cdots \mathbf{v}_{i_2}^T]^T$ for any vector $\mathbf{v}$, and the vectors $\mathbf{z}_k$ denote some "constraint-vectors" to be discussed immediately. $\|\cdot\|_{\mathbf{W}}^2$ denotes the $\mathbf{W}$-weighted square norm. We shall denote the minimizing $\mathbf{r}_0^{(p+q)}$ and $\mathbf{z}_1^{(M-p-q-1)}$ as $\tilde{\mathbf{r}}_0^{(p+q)}$ and $\tilde{\mathbf{z}}_1^{(M-p-q-1)}$, respectively.

Note that the unweighted ($\mathbf{W} = \mathbf{I}$) solution to this problem is trivial, $\tilde{\mathbf{r}}_0^{(p+q)} = \hat{\mathbf{r}}_0^{(p+q)}$ and $\tilde{\mathbf{z}}_1^{(M-p-q-1)} = \mathbf{0}$. However, the properly weighted solution modifies, or essentially "corrects" the raw estimates $\hat{\mathbf{r}}_0^{(p+q)}$, by exploiting known correlation between their estimation errors and the estimation errors of $\hat{\mathbf{z}}_1^{(M-p-q-1)}$, whose true values (**0**) are known.

An appealing choice for the "constraint-vectors" $\mathbf{z}_k$ was introduced in [7], $\mathbf{z}_k = \sum_{s,t=0}^{p} a_s a_t \mathbf{r}_{p+q+k-s-t}$, so that $\hat{\mathbf{z}}_k = \sum_{s,t=0}^{p} \hat{a}_s \hat{a}_t \hat{\mathbf{r}}_{p+q+k-s-t}$.

In order to find the optimal weight matrix for (7), we partition $\mathbf{C} \stackrel{\triangle}{=} Cov\left([\hat{\mathbf{r}}_0^{(p+q)T} \; \hat{\mathbf{z}}_1^{(M-p-q-1)T}]^T\right)$ as: $\mathbf{C} = \begin{bmatrix} \mathbf{C}_{11} & \mathbf{C}_{12} \\ \mathbf{C}_{21} & \mathbf{C}_{22} \end{bmatrix}$, where $\mathbf{C}_{11}$ is of dimension $N^2(p+q+1) \times N^2(p+q+1)$, etc. The solution of the constrained minimization (7) with an optimal weight matrix ($\mathbf{W} = \mathbf{C}^{-1}$) is then given by:

$$\tilde{\mathbf{r}}_0^{(p+q)} = \hat{\mathbf{r}}_0^{(p+q)} - \mathbf{C}_{12}\mathbf{C}_{22}^{-1}\hat{\mathbf{z}}_1^{(M-p-q-1)}. \tag{8}$$

The expression for the covariance matrix (**C**) is developed in the next section, and depends only on the unknown parameter vector $\boldsymbol{\theta}$, which may be substituted by an estimate $\hat{\boldsymbol{\theta}}$, leading to an estimate $\hat{\mathbf{C}}$. Moreover, the WLS solutions to the MYW equations and to (7) produces refined estimates of the parameters. It is therefore proposed to employ an iterative scheme, in which the refined estimates are in turn used for attaining a refined estimate of the covariance matrix, leading to a better weight matrix, and so on[2], as follows:

1. Estimate $\hat{\mathbf{r}}_0^{(M-1)}$, initialize $\tilde{\mathbf{r}}_0^{(p+q)[0]} = \hat{\mathbf{r}}_0^{(p+q)}$ and $\hat{\mathbf{C}}^{[0]} = \mathbf{I}$ (the bracketed superscript denotes the iteration index).
2. Repeat steps 3,4 for $l = 1, 2, 3$ (or until the update norm $\|\tilde{\mathbf{r}}_0^{(p+q)[l]} - \tilde{\mathbf{r}}_0^{(p+q)[l-1]}\|$ is small enough):
3. Obtain / refine the estimate of **a** by the WLS solution of the MYW equations: $\hat{\mathbf{a}}^{[l]} = -(\hat{\mathbf{H}}^{[l-1]T}\mathbf{W}\hat{\mathbf{H}}^{[l-1]})^{-1}\hat{\mathbf{H}}^{[l-1]T}\mathbf{W}\hat{\mathbf{h}}^{[l-1]}$, where $\hat{\mathbf{H}}^{[l-1]} \stackrel{\triangle}{=} \begin{bmatrix} \tilde{\mathbf{r}}_q^{(q+p-1)[l-1]} & \cdots & \tilde{\mathbf{r}}_{(q-p+1)}^{(q)[l-1]} \end{bmatrix}$, $\hat{\mathbf{h}}^{[l-1]} \stackrel{\triangle}{=} \tilde{\mathbf{r}}_{(q+1)}^{(q+p)[l-1]}$ and $\mathbf{W} = \left(\hat{\tilde{\mathbf{C}}}^{[l-1]}\right)^{-1}$, where $\hat{\tilde{\mathbf{C}}}^{[l-1]}$ is the estimate of $Cov\left(\hat{\mathbf{h}}^{[l-1]}\right)$, obtained as the $N^2 p \times N^2 p$ lower-right block of the matrix $\hat{\mathbf{C}}_{11}^{[l-1]} - \hat{\mathbf{C}}_{12}^{[l-1]}\hat{\mathbf{C}}_{22}^{[l-1]-1}\hat{\mathbf{C}}_{21}^{[l-1]}$.
4. Obtain / refine the estimates of $\left\{\beta_k^{(n,m)}\right\}$ using (6), estimate $\hat{\mathbf{C}}^{[l]}$ using $\hat{\boldsymbol{\theta}}^{[l]}$, (re)calculate (using $\hat{\mathbf{a}}^{[l]}$) $\hat{\mathbf{z}}_1^{(M-p-q-1)}$ and obtain $\tilde{\mathbf{r}}_0^{(p+q)[l]} = \hat{\mathbf{r}}_0^{(p+q)} - \hat{\mathbf{C}}_{12}^{[l]}\hat{\mathbf{C}}_{22}^{[l]-1}\hat{\mathbf{z}}_1^{(M-p-q-1)}$.

Subsequently, a "standard" WASOBI stage [2] can be applied, using the (final) refined correlations estimates $\tilde{\mathbf{r}}_0^{(p+q)}$ and their refined estimated covariance, given by $\hat{\mathbf{C}}_{11} - \hat{\mathbf{C}}_{12}\hat{\mathbf{C}}_{22}^{-1}\hat{\mathbf{C}}_{21}$. The inverse of this covariance matrix is the (asymptotically) optimal weight matrix, to be used in WASOBI for weighting the AJD of the estimated correlation matrices (extracted from $\tilde{\mathbf{r}}_0^{(p+q)}$).

---

[2] Our empirical experience indicates that asymptotically the process converges within 1-3 iterations; However, there is no general proof (nor claim) of convergence, so we merely propose to apply a limited number of iterations.

3 Expressions for C

In this section we derive expressions for $\mathbf{C}$, required for calculating $\hat{\mathbf{C}}$ from a finite number of correlation estimates, by substituting $\boldsymbol{\theta}$ with $\hat{\boldsymbol{\theta}}$. We assume asymptotic conditions, namely that $T \to \infty$, or at least that T is very large compared to the longest correlation length involved. Note that under non-asymptotic conditions the resulting $\hat{\mathbf{C}}$ may have negative eigenvalues. In such cases it is recommended to apply some *ad-hoc* conditioning, e.g., by replacing any negative eigenvalues with a small positive constant.

The $(k+1, l+1)$-th elements of $\mathbf{C}_{11}$ are of the form:

$$\lim_{T\to\infty} Cov\left(\hat{R}_\mathbf{x}^{(i,j)}[k], \hat{R}_\mathbf{x}^{(m,n)}[l]\right) =$$

$$\lim_{T\to\infty} \frac{1}{T} \sum_{p=-(T-1)}^{(T-1)} (1-\frac{|p|}{T})(R_\mathbf{x}^{(i,m)}[p]R_\mathbf{x}^{(j,n)}[p+l-k] + R_\mathbf{x}^{(i,n)}[p+l]R_\mathbf{x}^{(j,m)}[p-k])$$

$$\approx \frac{1}{T}(\phi^{(i,m)(j,n)}[l-k] + \phi^{(i,n)(j,m)}[l+k]) \quad \begin{smallmatrix} 0\leq k,l\leq p+q \\ 1\leq i,j,n,m\leq N \end{smallmatrix} \quad , \quad (9)$$

where $\phi^{(i,j)(m,n)}[\tau] \triangleq \sum_{p=-\infty}^{\infty} R_\mathbf{x}^{(i,j)}[p]R_\mathbf{x}^{(m,n)}[p+\tau]$. In order to evaluate this infinite sum, we evaluate the inverse parametric Z–transform of $\Phi^{(i,j)(m,n)}(z)$, the Z–transform of $\phi^{(i,j)(m,n)}[\tau]$.

$$\Phi^{(i,j)(m,n)}(z) = \sum_{\tau=-\infty}^{\infty}\sum_{p=-\infty}^{\infty} R_\mathbf{x}^{(i,j)}[p]R_\mathbf{x}^{(m,n)}[p+\tau]z^{-\tau} = S_\mathbf{x}^{(i,j)}(z;\boldsymbol{\theta})S_\mathbf{x}^{(m,n)}(z;\boldsymbol{\theta}) \quad (10)$$

The last equality is justified since $R_\mathbf{x}^{(i,j)}[p] = R_\mathbf{x}^{(j,i)}[p]$, and therefore $R_\mathbf{x}^{(i,j)}[p] = R_\mathbf{x}^{(i,j)}[-p]$, so the convolution translates into a product in the Z–domain.

We evaluate $\phi^{(i,j)(m,n)}[\tau]$ by using the Residue theorem for the evaluation of complex integrals (see [8] for further discussion of computational aspects).

Since $\phi^{(i,j)(m,n)}[\tau]$ is a symmetric function, we only have to evaluate $\phi^{(i,j)(m,n)}[\tau]$ for $\tau = 0, 1, \ldots, 2(p+q)$.

The $(k+1, l)$-th elements of $\mathbf{C}_{12}$ are of the form:

$$\lim_{T\to\infty} Cov\left(\hat{R}_\mathbf{x}^{(i,j)}[k], \sum_{s,t=0}^{p} \hat{a}_s \hat{a}_t \hat{R}_\mathbf{x}^{(m,n)}[p+q+l-s-t]\right)$$

$$\approx \frac{1}{T}(\gamma^{(i,m)(j,n)}[l-k] + \gamma^{(j,m)(i,n)}[l+k]) \quad \begin{smallmatrix} 0\leq k\leq p+q,\ 1\leq l\leq M-p-q-1 \\ 1\leq i,j,n,m\leq N \end{smallmatrix} \quad , \quad (11)$$

where $\gamma^{(i,j)(m,n)}[\tau]$ is the inverse Z–transform of

$$(\sum_{k=-q}^{q} \beta_k^{(i,j)} z^{-k})(\sum_{k=-q}^{q} \beta_k^{(m,n)} z^{-k}) z^{p+q}/(A^*(\frac{1}{z^*}))^2. \quad (12)$$

Since the poles of this expression lie outside the unit circle, the only inner poles in the Z-transform inversion are at $z = 0$. Therefore, $\gamma^{(i,j)(m,n)}[\tau] = 0$ for $\tau > q - p$. That is, we only have to evaluate $\gamma^{(i,j)(m,n)}[\tau]$ for $1 - p - q \leq \tau \leq q - p$.

Similarly, the (k,l)-th elements of $\mathbf{C}_{22}$ are of the form:

$$\lim_{T\to\infty} \text{Cov}\left(\sum_{s,t=0}^{p} \hat{a}_s \hat{a}_t \hat{R}_{\mathbf{x}}^{(i,j)}[p+q+k-s-t], \sum_{s,t=0}^{p} \hat{a}_s \hat{a}_t \hat{R}_{\mathbf{x}}^{(m,n)}[p+q+l-s-t]\right)$$
$$\approx \frac{1}{T}\psi^{(i,m)(j,n)}[l-k] \quad \begin{array}{l} 1\leq k,l\leq M-p-q-1 \\ 1\leq i,j,n,m\leq N \end{array}, \quad (13)$$

where $\psi^{(i,j)(m,n)}[\tau]$ is the inverse Z-transform of $(A(z)A^*(\frac{1}{z^*}))^2 S_x^{(i,j)}(z;\boldsymbol{\theta})S_x^{(m,n)}(z;\boldsymbol{\theta})$ and therefore can be evaluated simply by reading the coefficient of $z^{-\tau}$ from $(\sum_{k=-q}^{q}\beta_k^{(i,j)}z^{-k})(\sum_{k=-q}^{q}\beta_k^{(m,n)}z^{-k})$. Hence, $\psi^{(i,j)(m,n)}[\tau]=0$ for $|\tau|>2q$.

4 Simulations Results

Simulations results are presented in terms of Interference to Signal Ratio (ISR). The global ISR is defined as $ISR \triangleq \sum_{i\neq j} ISR_{ij}$. Under a small errors assumption (See [2] for details), ISR_{ij}, and therefore the global ISR, can be predicted (or bounded) by a linear combination of the elements of the estimated mixing matrix' covariance matrix (or Cramér Rao lower bound (CRLB), respectivey). For the predicted ISR_{ij} we set $\mathbf{W}$ to the inverse of the true covariance, and for the ISR bounds we set the estimated mixing matrix' covariance to the CRLB [8].

The proposed algorithm is given the acronym AOL (Asymptotically OptimaL).

In all the figures, for comparison with Pham's GMI algorithm [5] we estimated (for GMI) the spectral density matrices for 512 frequencies ($L=512$) with the same window that was chosen for the simulations in [5].

In the first experiment we used the following Gaussian sources: $s_1[n]$ - AR(1) with a pole at -0.8 (and its reciprocal), $s_2[n]$ - AR(3) with poles at $0.5e^{\pm j\frac{\pi}{3}}$, and -0.85 (and their reciprocals), and $s_3[n]$ - ARMA(1,2) with a pole at -0.45 and zeros at $0.85e^{\pm j\frac{\pi}{2}}$ (and their reciprocals). We tested the case of $N=2$ with $\mathbf{A} = \begin{bmatrix} 1 & 2 \\ 3 & 4 \end{bmatrix}$, mixing $s_1[n]$ and $s_2[n]$, as well as the case of $N=3$ with $\mathbf{A} = \begin{bmatrix} 1 & 2 & -1 \\ 3 & 4 & 11 \\ -9 & 2 & -4 \end{bmatrix}$, mixing all three.

In Fig. 1 and 2, we compare the empirical ISRs of AOL, SOBI, GMI and QML[3] [4] with the predicted ISRs of AOL and the CRLBs, vs. the number of observations, for the $N=2$ and $N=3$ cases, respectively.

In Fig. 1 for AOL, we set $M=p+q+1=8$. However, only 4 estimated matrices (lags 0 to 3) and their estimated covariance were used for the WASOBI stage, since they form a sufficient statistic in this case [8]. Further, for SOBI we set $M=4$ and for GMI we set $M=16$, since for that number it was observed to attains the CRLBs asymptotically. In Fig. 2 we set $M=p+q+1=12$ for all algorithms. Additionally, in order to demonstrate (empirically) the robustness with respect to the Gaussianity assumption, we also present in both figures the

---

[3] The "guessed" sources' power spectra were $\varphi_1(z) = z^{-1}+2+z$ and $\varphi_2(z) = -z^{-1} + 2 - z$, and in Fig. 2, in addition $\varphi_3(z) = z^{-2} + 2z^{-1} + 3 + 2z + z^2$.

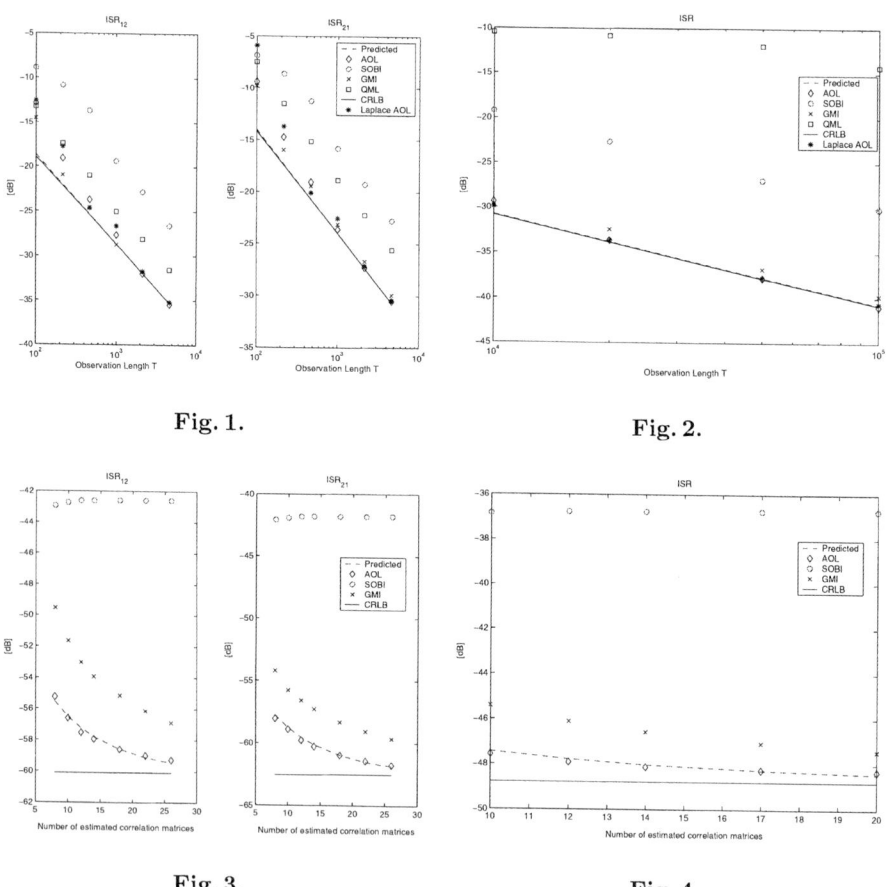

Fig. 1., Fig. 2. Empirical ISRs of AOL, SOBI, GMI [5] and QML [4], predicted ISRs of AOL and the CRLBs, vs. the number of observed samples, for $N = 2$ and $N = 3$, respectively. The AOL ISRs when the sources' driving noises have Laplace distributions are denoted by "*". Each simulation result is an average of 500 trails.

Fig. 3., Fig. 4. Empirical ISRs of AOL, SOBI and GMI [5], predicted ISR of AOL, and the CRLB vs. number of estimated correlation matrices, for $N = 2$ and $N = 3$, respectively. Each simulation result is an average of 500 trails.

performance of AOL when the sources' driving noises have Laplace (rather than Gaussian) distributions.

In the second experiment we used the following Gaussian sources: $s_1[n]$ - ARMA(2,1) with poles at $0.75e^{\pm j\frac{\pi}{3}}$ and a zero at 0.95 (and their reciprocals), $s_2[n]$ - ARMA(2,1) with poles at $0.65e^{\pm j\frac{\pi}{2}}$, and a zero at -0.95 (and their reciprocals), and $s_3[n]$ - AR(1) with a pole at 0.4 (and its reciprocal). The mixing matrices were the same as in the first experiment.

In Fig. 3 and 4 we compare the empirical ISRs of AOL, SOBI and GMI with the predicted ISRs of AOL and the CRLBs, vs. the number of estimated correlation matrices, for $N = 2$ and $N = 3$, respectively, under asymptotic conditions ($T = 100,000$).

These simulation results show that our algorithm is asymptotically superior to other well-known algorithms (SOBI, Pham and Garat's QML, Pham's GMI) in terms of the number of estimated correlation matrices required to attain asymptotically optimal separation. This is a computationally significant advantage, especially under asymptotic conditions, since the estimation of each correlation matrix requires a number of multiplications proportional to the observation length T. Additionally, when using enough estimated correlation matrices, our algorithm is shown to asymptotically attain the CRLB.

5 Conclusion

An asymptotically efficient algorithm for blind separation of parametric Gaussian sources has been introduced, based on a weighted nonlinear least squares formulation. Asymptotically optimal weights are attained by over-parameterization of the processes. The algorithm exploits the parametric information in the kinds and the orders of the sources (previously unexploited in existing algorithms), and can therefore maintain near-optimal performance with fewer estimated correlation matrices.

References

1. A. Belouchrani, K. Abed-Meraim, J.-F. Cardoso, and E. Moulines, "A blind source separation technique using second-order statistics," *IEEE Trans. Signal Processing*, vol. 45, pp. 434–444, Feb. 1997.
2. A. Yeredor, "Blind separation of gaussian sources via second-order statistics with asymptotically optimal weigthting," *IEEE Signal Processing Letters*, vol. 7, pp. 197–200, Jul. 2000.
3. A. Yeredor and E. Doron, "Using farther correlations to further improve the optimally-weighted sobi algorithm," *Proc. EUSIPCO'2002*, Sep. 2002.
4. D.-T. Pham and P. Garat, "Blind separation of mixture of independent sources through a quasi-maximum likelihood approach," *IEEE Trans. Signal Processing*, vol. 45, pp. 1712–1725, Jul. 1997.
5. D.-T. Pham, "Blind separation of instantaneous mixture of sources via the gaussian mutual iformation criterion," *Signal Processing*, vol. 81, pp. 855–870, 2001.
6. S. Dégerine and R. Malki, "Second-order blind separation of sources based on canonical partial innovations," *IEEE Trans. Signal Processing*, vol. 48, pp. 629–641, Mar. 2000.
7. P. Stoica, B. Friendlander, and T. Söderström, "Approximate maximum-likelihood approach to arma spectral estimation," *Int. J. Contr*, vol. 45, no. 4, pp. 1281–1310, 1987.
8. E. Doron, "Asymptotically optimal blind separation of parametric gaussian sources," Master's thesis, Dept. of EE-Systems, Tel-Aviv University, Israel, 2003.

Bayesian Approach for Blind Separation of Underdetermined Mixtures of Sparse Sources

Cédric Févotte, Simon J. Godsill, and Patrick J. Wolfe

Cambridge University Engineering Dept., Cambridge, CB2 1PZ, UK
{cf269,sjg,pjw47}@eng.cam.ac.uk
http://www-sigproc.eng.cam.ac.uk/~cf269/

Abstract. We address in this paper the problem of blind separation of underdetermined mixtures of sparse sources. The sources are given a Student t distribution, in a transformed domain, and we propose a bayesian approach using Gibbs sampling. Results are given on synthetic and audio signals.

1 Introduction

Blind Source Separation (BSS) consists in estimating n signals (the sources) from the sole observation of m mixtures of them (the observations). In this paper we consider linear instantaneous mixtures of time series: at each time index, the observations are a linear combination of the sources at the same time index. Moreover, we are interested in the underdetermined case ($m < n$). This case is very difficult to handle because contrary to (over)determined mixtures ($m \geq n$), estimating the mixing system (a single matrix in the linear instantaneous case) is not sufficient for reconstructing the sources, since for $m < n$ the mixing matrix is not invertible. Then, it appears that separation of underdetermined mixtures requires important prior information on the sources to allow their reconstruction.

In this paper we address the case of sparse sources, meaning that only a few samples are significantly non-zero. The use of sparsity to handle source separation problems has arisen in several papers, see for instance [1,2]. In these papers, source time series are assumed to have a sparse representation on a given or learnt dictionary, possibly overcomplete. The aim of methods then becomes the estimation of the coefficients of the sources on the dictionary and not the time series in themselves. The time series are then reconstructed from the estimated coefficients.

More specifically, in [3,4] the coefficients of the representations of the sources in the dictionary are given a discrete mixture a Gaussian distributions with 2 or 3 states (one Gaussian with very small variance, the other(s) with big variance) and a probabilistic framework is presented for the estimation of the mixing matrix and the sources. In particular, in [4], the authors use EM optimisation and present results with speech signals decomposed on a MDCT orthogonal basis [5]. The use of an orthogonal basis provides equivalence between representations in the time domain and transformed domain, and separation can be simply performed in the transformed domain instead of the time domain. The use of an overcomplete dictionary is very appealing because it allows sparser representations but leads to much more tricky calculations [1].

Motivated by the successful results of Student t modeling for audio restoration in [6], we address in this paper separation of Student t distributed sources, which leads to sparse modeling when the degrees of freedom is low. We emphasize that we will work in the transformed domain: the observations and sources we consider have arisen from the decomposition of some corresponding time series on a dictionary, which is restricted at this point to be an *orthogonal basis* (to satisfy equivalence between time and transformed domains). The method we present is a bayesian approach: a Gibbs sampler is derived to sample from the posterior conditional distribution of the parameters (which include the sources and the mixing matrix).

The paper is organised as follows: section 2 introduces notations and assumptions, in section 3 we derive the posterior distributions of the parameters to be estimated and section 4 presents results on synthetic and audio signals. Conclusions and perspectives are given in section 5.

2 Model and Assumptions

2.1 Model

We consider the following standard linear model, $\forall t = 0, \ldots, N-1$:

$$\mathbf{x}_t = \mathbf{A}\,\mathbf{s}_t + \mathbf{n}_t \tag{1}$$

where $\mathbf{x}_t = [x_{1,t}, \ldots, x_{m,t}]^T$ is vector of size m containing the observations, $\mathbf{s}_t = [s_{1,t}, \ldots, s_{n,t}]^T$ is a vector of size n containing the sources, $\mathbf{n}_t = [n_{1,t}, \ldots, n_{m,t}]^T$ is a vector of size m containing noise. Variables without time index t denote whole sequences of samples, *e.g*, $\mathbf{x} = [\mathbf{x}_0, \ldots, \mathbf{x}_{N-1}]$ and $x_1 = [x_{1,0}, \ldots, x_{1,N-1}]$.

2.2 Assumptions

1) We assume that each source sequence s_i is independently and identically distributed (i.i.d), with Student t distribution $t(\alpha_i, \lambda_i)$:

$$p(s_{i,t}) = K \left(1 + \frac{1}{\alpha_i}\left(\frac{s_{i,t}}{\lambda_i}\right)^2\right)^{-\frac{\alpha_i+1}{2}} \tag{2}$$

α_i is the "degrees of freedom", λ_i is a scale parameter and K is a normalizing constant. With $\lambda = 1$ and $\alpha = 1$, the Student t distribution is equal to the standard Cauchy distribution, and it tends to the standard Gaussian distribution when $\alpha \to +\infty$. Fig. 1 plots Student t densities for several values of α. For small α, the Student t has "fatter tails" than the normal distribution. A nice property of the Student t distribution is the fact that it can be expressed as a Scaled Mixture of Gaussians [7], such that

$$p(s_{i,t}) = \int_0^{+\infty} \mathcal{N}(s_{i,t}|0, v_{i,t})\,\mathcal{IG}\left(v_{i,t}\Big|\frac{\alpha_i}{2}, \frac{2}{\alpha_i \lambda_i^2}\right) dv_{i,t} \tag{3}$$

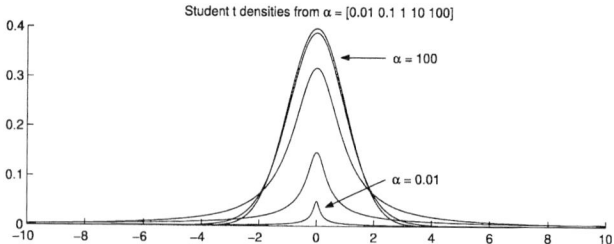

Fig. 1. Student t densities for $\lambda = 1$ and $\alpha = [0.01, 0.1, 1, 10, 100]$ - From $\alpha = 100$ $t(\alpha,1)$ is very close to $\mathcal{N}(0,1)$.

where $\mathcal{N}(x|0,v)$ denotes the normal distribution with mean 0 and variance v and $\mathcal{IG}(x|\gamma,\beta)$ denotes Inverted Gamma distribution, defined by $\mathcal{IG}(x|\gamma,\beta) = (x^{-(\gamma+1)})/(\Gamma(\gamma)\beta^{\gamma})\exp(-1/(\beta x))$, for $x \geq 0$. $p(s_{i,t})$ can thus be interpreted as a marginal of the joint distribution $p(s_{i,t}, v_{i,t})$, defined by:

$$p(s_{i,t}, v_{i,t}) = p(s_{i,t}|v_{i,t})\,p(v_{i,t}|\alpha_i, \lambda_i) \qquad (4)$$

with:

$$p(s_{i,t}|v_{i,t}) = \mathcal{N}(s_{i,t}|0, v_{i,t}) \quad \text{and} \quad p(v_{i,t}|\alpha_i, \lambda_i) = \mathcal{IG}\left(v_{i,t}\Big|\frac{\alpha_i}{2}, \frac{2}{\alpha_i \lambda_i^2}\right) \qquad (5)$$

$v_{i,t}$ of the Gaussian distribution in Eq. (3) is a very convenient property which will help us deriving posterior distributions of the parameters in the implementation of the Gibbs sampler. The Student t can be interpreted as an infinite sum of Gaussians, which contrasts with the finite sums of Gaussians used in [3, 4].

In the following, we note $\mathbf{v}_t = [v_{1,t}, \ldots, v_{n,t}]^T$, $\mathbf{v} = [\mathbf{v}_0, \ldots, \mathbf{v}_{N-1}]$, $\boldsymbol{\alpha} = [\alpha_1, \ldots, \alpha_n]$ and $\boldsymbol{\lambda} = [\lambda_1, \ldots, \lambda_n]$.

2) We assume that the source sequences are mutually independent, such that $p(\mathbf{s}) = \prod_{i=1}^{n} p(s_i)$.

3) We assume that $\mathbf{n}$ is a i.i.d Gaussian noise with covariance $\sigma^2 \mathbf{I}_m$, and σ unknown.

We now present a Markov chain Monte Carlo approach to estimate the set of parameter of interest $\{\mathbf{A}, \mathbf{s}, \sigma\}$ together with the set $\{\mathbf{v}, \boldsymbol{\alpha}, \boldsymbol{\lambda}\}$.

3 Derivations for the Gibbs Sampler

We propose to generate samples from the posterior distribution $p(\mathbf{A}, \mathbf{s}, \sigma, \mathbf{v}, \boldsymbol{\alpha}, \boldsymbol{\lambda}|\mathbf{x})$ of whole set of parameters. We use a Gibbs sampler which only requires to derive the expression of the posterior distribution of each parameter conditionally upon the data $\mathbf{x}$ and the other parameters, see details for instance in [8].

3.1 Likelihood

With the gaussian noise assumption, the likelihood of one sample of the observations is written:

$$p(\mathbf{x}_t|\mathbf{A}, \mathbf{s}_t, \sigma) = \mathcal{N}(\mathbf{x}_t|\mathbf{A}\,\mathbf{s}_t, \sigma^2 \mathbf{I}_m) \qquad (6)$$

where $\mathcal{N}(\mathbf{x}|\boldsymbol{\mu}, \boldsymbol{\Sigma})$ denotes multivariate Gaussian distribution with mean $\boldsymbol{\mu}$ and covariance $\boldsymbol{\Sigma}$. With the i.i.d source assumption, the likelihood of the observations is written:

$$p(\mathbf{x}|\mathbf{A}, \mathbf{s}, \sigma) = \prod_{t=0}^{N-1} \mathcal{N}(\mathbf{x}_t|\mathbf{A}\,\mathbf{s}_t, \sigma^2\,\mathbf{I}_m) \tag{7}$$

$$= \frac{1}{(2\pi\sigma^2)^{\frac{Nm}{2}}} \exp(-\frac{1}{2\sigma^2} \sum_{t=0}^{N-1} \|\mathbf{x}_t - \mathbf{A}\,\mathbf{s}_t\|_F^2) \tag{8}$$

3.2 Expression of $p(\mathbf{s}|\mathbf{A}, \sigma, \mathbf{v}, \boldsymbol{\alpha}, \boldsymbol{\lambda})$

We have:
$$p(\mathbf{s}|\mathbf{A}, \sigma, \mathbf{v}, \boldsymbol{\alpha}, \boldsymbol{\lambda}) \propto p(\mathbf{x}|\mathbf{A}, \mathbf{s}, \sigma)\, p(\mathbf{s}|\mathbf{v}) \tag{9}$$

With $p(\mathbf{s}|\mathbf{v}) = \prod_{t=0}^{N-1} p(\mathbf{s}_t|\mathbf{v}_t) = \prod_{t=0}^{N-1} \mathcal{N}(\mathbf{s}_t|0, \mathrm{diag}\,(\mathbf{v}_t))$ [1] and with Eq. (7), we have

$$p(\mathbf{s}|\mathbf{A}, \sigma, \mathbf{v}, \boldsymbol{\alpha}, \boldsymbol{\lambda}) = \prod_{t=0}^{N-1} \mathcal{N}(\mathbf{s}_t|\boldsymbol{\mu}_{\mathbf{s}_t}, \boldsymbol{\Sigma}_{\mathbf{s}_t}) \tag{10}$$

where $\boldsymbol{\Sigma}_{\mathbf{s}_t} = \left(\frac{1}{\sigma^2}\mathbf{A}^T\mathbf{A} + \mathrm{diag}\,(\mathbf{v}_t)^{-1}\right)^{-1}$ and $\boldsymbol{\mu}_{\mathbf{s}_t} = \frac{1}{\sigma^2}\boldsymbol{\Sigma}_{\mathbf{s}_t}\mathbf{A}^T\mathbf{x}_t$.

3.3 Expression of $p(\mathbf{A}|\mathbf{s}, \sigma, \mathbf{v}, \boldsymbol{\alpha}, \boldsymbol{\lambda})$

Let $\mathbf{r}_1, \ldots, \mathbf{r}_m$ denote the transposed rows of $\mathbf{A}$, such that $\mathbf{A}^T = [\mathbf{r}_1 \ldots \mathbf{r}_m]$. Let $\mathbf{S}_t$ and $\mathbf{a}$ denote the $m \times nm$ matrix and the $nm \times 1$ vector defined by

$$\mathbf{S}_t = \begin{bmatrix} \mathbf{s}_t^T & & 0 \\ & \ddots & \\ 0 & & \mathbf{s}_t^T \end{bmatrix} \quad \text{and} \quad \mathbf{a} = \begin{bmatrix} \mathbf{r}_1 \\ \vdots \\ \mathbf{r}_m \end{bmatrix} \tag{11}$$

By construction, we have:
$$\mathbf{A}\,\mathbf{s}_t = \mathbf{S}_t\,\mathbf{a} \tag{12}$$

Of course, the estimation of $\mathbf{a}$ is equivalent to the estimation of $\mathbf{A}$, and we have:

$$p(\mathbf{a}|\mathbf{s}, \sigma, \mathbf{v}, \boldsymbol{\alpha}, \boldsymbol{\lambda}) \propto p(\mathbf{x}|\mathbf{a}, \mathbf{s}, \sigma)\, p(\mathbf{a}) \tag{13}$$

Without further information on the mixing matrix, we assume uniform prior and set $p(\mathbf{a}) \propto 1$. With Eq. (7) and (12), we have then $p(\mathbf{a}|\mathbf{s}, \sigma, \mathbf{v}, \boldsymbol{\alpha}, \boldsymbol{\lambda}) \propto \prod_{t=0}^{N-1} \mathcal{N}(\mathbf{x}_t|\mathbf{S}_t\,\mathbf{a}, \sigma^2\,\mathbf{I}_m)$ and it follows that:

$$p(\mathbf{a}|\mathbf{s}, \sigma, \mathbf{v}, \boldsymbol{\alpha}, \boldsymbol{\lambda}) = \mathcal{N}(\mathbf{a}|\boldsymbol{\mu}_\mathbf{a}, \boldsymbol{\Sigma}_\mathbf{a}) \tag{14}$$

with $\boldsymbol{\Sigma}_\mathbf{a} = \sigma^2 \left(\sum_{t=0}^{N-1} \mathbf{S}_t^T \mathbf{S}_t\right)^{-1}$ and $\boldsymbol{\mu}_\mathbf{a} = \frac{1}{\sigma^2}\boldsymbol{\Sigma}_\mathbf{a}\sum_{t=0}^{N-1} \mathbf{S}_t^T \mathbf{x}_t$. To fix the well known BSS indeterminacies on gain and permutations, we set in practice the first row of $\mathbf{A}$ to ones and only estimate the other rows.

---

[1] $\mathrm{diag}\,(\mathbf{u})$ is the diagonal matrix whose main diagonal is given by $\mathbf{u}$.

3.4 Expression of $p(\sigma|\mathbf{A},\mathbf{s},\mathbf{v},\boldsymbol{\alpha},\boldsymbol{\lambda})$

As before we have:

$$p(\sigma|\mathbf{A},\mathbf{s},\mathbf{v},\boldsymbol{\alpha},\boldsymbol{\lambda}) \propto p(\mathbf{x}|\mathbf{A},\mathbf{s},\sigma)\,p(\sigma) \quad (15)$$

Using Jeffreys prior $p(\sigma) = 1/\sigma$ and expression (8) of the likelihood, we have:

$$p(\sigma|\mathbf{A},\mathbf{s},\mathbf{v},\boldsymbol{\alpha},\boldsymbol{\lambda}) \propto \sigma^{-(2\gamma_\sigma+1)} \exp\left(-\frac{1}{\beta_\sigma \sigma^2}\right) \quad (16)$$

with $\gamma_\sigma = \frac{mN}{2}$ and $\beta_\sigma = 2/\sum_{t=0}^{N-1}\|\mathbf{x}_t - \mathbf{A}\,\mathbf{s}_t\|_F^2$. It appears that $\sigma|\mathbf{A},\mathbf{s},\mathbf{v},\boldsymbol{\alpha},\boldsymbol{\lambda}$ can be drawn from $\sqrt{\mathcal{IG}(\gamma_\sigma,\beta_\sigma)}$.

3.5 Expression of $p(\mathbf{v}|\mathbf{A},\mathbf{s},\sigma,\boldsymbol{\alpha},\boldsymbol{\lambda})$

Since the data $\mathbf{x}$ does not depend on the parameters $\{\mathbf{v},\boldsymbol{\alpha},\boldsymbol{\lambda}\}$, their posterior distributions only depend on the prior distributions. The posterior distribution of $\mathbf{v}$ is then:

$$p(\mathbf{v}|\mathbf{A},\mathbf{s},\sigma,\boldsymbol{\alpha},\boldsymbol{\lambda}) \propto p(\mathbf{s}|\mathbf{v})\,p(\mathbf{v}|\boldsymbol{\alpha},\boldsymbol{\lambda}) \quad (17)$$

With $p(\mathbf{s}|\mathbf{v})\,p(\mathbf{v}|\boldsymbol{\alpha},\boldsymbol{\lambda}) = \prod_{t=0}^{N-1}\prod_{i=1}^{n} p(s_{i,t}|v_{i,t})\,p(v_{i,t}|\alpha_i,\lambda_i)$ one can show that:

$$p(s_{i,t}|v_{i,t})\,p(v_{i,t}|\alpha_i,\lambda_i) \propto \mathcal{IG}\left(v_{i,t}|\gamma_{v_i},\beta_{v_{i,t}}\right) \quad (18)$$

with $\gamma_{v_i} = (\alpha_i + 1)/2$ and $\beta_{v_{i,t}} = 2/(s_{i,t}^2 + \alpha_i\,\lambda_i^2)$. Thus:

$$p(\mathbf{v}|\mathbf{A},\mathbf{s},\sigma,\boldsymbol{\alpha},\boldsymbol{\lambda}) = \prod_{t=0}^{N-1}\prod_{i=1}^{n}\mathcal{IG}\left(v_{i,t}|\gamma_{v_i},\beta_{v_{i,t}}\right) \quad (19)$$

3.6 Expression of $p(\boldsymbol{\alpha}|\mathbf{A},\mathbf{s},\sigma,\mathbf{v},\boldsymbol{\lambda})$

We have:

$$p(\boldsymbol{\alpha}|\mathbf{A},\mathbf{s},\sigma,\mathbf{v},\boldsymbol{\lambda}) \propto p(\mathbf{v}|\boldsymbol{\alpha},\boldsymbol{\lambda})\,p(\boldsymbol{\alpha}) \quad (20)$$

With $p(\mathbf{v}|\boldsymbol{\alpha},\boldsymbol{\lambda})\,p(\boldsymbol{\alpha}) = \prod_{i=1}^{n}\prod_{t=0}^{N-1} p(v_{i,t}|\alpha_i,\lambda_i)p(\alpha_i)$, one can show that

$$p(\boldsymbol{\alpha}|\mathbf{A},\mathbf{s},\sigma,\mathbf{v},\boldsymbol{\lambda}) \propto \prod_{i=1}^{n}\frac{P_i^{-(\frac{\alpha_i}{2}+1)}}{\Gamma(\frac{\alpha_i}{2})^N}\left(\frac{\alpha_i\,\lambda_i^2}{2}\right)^{\frac{\alpha_i N}{2}}\exp\left(-\frac{\alpha_i\,\lambda_i^2}{2}S_i\right)p(\alpha_i) \quad (21)$$

with $S_i = \sum_{t=0}^{N-1}\frac{1}{v_{i,t}}$ and $P_i = \prod_{t=0}^{N-1} v_{i,t}$. In practice we choose a uniform prior on α_i and set $p(\boldsymbol{\alpha}) \propto 1$. As the distribution of $\boldsymbol{\alpha}|\mathbf{A},\mathbf{s},\sigma,\mathbf{v},\boldsymbol{\lambda}$ is not straightforward to sample from and since the precise value α_i for each source is unlikely to be important, we sample $\boldsymbol{\alpha}$ from a uniform grid of discrete values with probability mass given by Eq. (21).

3.7 Expression of $p(\boldsymbol{\lambda}|\mathbf{A},\mathbf{s},\boldsymbol{\sigma},\mathbf{v},\boldsymbol{\alpha})$

Finally, the posterior distribution of the scale parameters is given by:

$$p(\boldsymbol{\lambda}|\mathbf{A},\mathbf{s},\boldsymbol{\sigma},\mathbf{v},\boldsymbol{\alpha}) \propto p(\mathbf{v}|\boldsymbol{\alpha},\boldsymbol{\lambda})\,p(\boldsymbol{\lambda}) \qquad (22)$$

With $p(\mathbf{v}|\boldsymbol{\alpha},\boldsymbol{\lambda})\,p(\boldsymbol{\lambda}) = \prod_{i=1}^{n}\left(\prod_{t=0}^{N-1} p(v_{i,t}|\alpha_i,\lambda_i)\right) p(\lambda_i)$, one can show that:

$$p(\boldsymbol{\lambda}|\mathbf{A},\mathbf{s},\boldsymbol{\sigma},\mathbf{v},\boldsymbol{\lambda}) \propto \prod_{i=1}^{n} \lambda_i^{\alpha_i N} \exp\left(-\frac{\alpha_i S_i}{2}\lambda_i^2\right) p(\lambda_i) \qquad (23)$$

With Jeffreys prior $p(\lambda_i) = 1/\lambda_i$, it appears that $\lambda_i|\mathbf{A},\mathbf{s},\boldsymbol{\sigma},\mathbf{v},\boldsymbol{\lambda}$ can be drawn from $\sqrt{\mathcal{G}(\gamma_{\lambda_i},\beta_{\lambda_i})}$, with $\gamma_{\lambda_i} = (\alpha_i N)/2$ and $\beta_{\lambda_i} = 2/(\alpha_i S_i)$, and where $\mathcal{G}(\gamma,\beta)$ is the Gamma distribution, whose density is written $\mathcal{G}(x|\gamma,\beta) = x^{\gamma-1}/(\Gamma(\gamma)\,\beta^{\gamma})\exp(-x/\beta)$, for $x \geq 0$.

4 Results

Synthetic Signals. We present results of the method over a mixture of $n = 2$ Student t sources of length $N = 1000$ with $m = 3$ observations. The mixing matrix is arbitrarly chosen as $\mathbf{A} = [1\ 1\ 1; 1\ -0.5\ 0.2]$. The sources are simulated with $\boldsymbol{\alpha} = [0.9\ 0.7\ 0.8]$ and $\boldsymbol{\lambda} = [0.03\ 0.003\ 0.002]$. The values of $\boldsymbol{\alpha}$ are chosen according to a range of values that seem to fit reasonably well MDCT coefficients of several types of audio signals. The values of the scale parameters $\boldsymbol{\lambda}$ are of little importance. Noise was added on the observations with variance $\sigma = 0.1$, which leads to 35dB and 30dB SNR on each observation. We ran 10000 iterations of the Gibbs sampler. The convergence of $\mathbf{r}_2$ [2] (initialized to zeros), σ (initialised with random value between 0 and 1), $\boldsymbol{\alpha}$ (initialised to ones) and $\boldsymbol{\lambda}$ (initialised to [0.01 0.01 0.01]) is shown on Fig. 2.

Estimated sources were computed as mean estimates of the 2000 last sampled values of $\mathbf{s}$ (that is after convergence of all the values of $\mathbf{r}_2$ is obtained). Sources estimates are compared to the original ones by computing the evaluation criteria described in [9]: Source to Distortions Ratio (global criterion), Source to Interference Ratio, Source to Noise Ratio, Source to Artifact Ratio. We obtain (values in dB):

	SDR	SIR	SNR	SAR
$\hat{s}_1$	30.6	40.2	33.7	34.6
$\hat{s}_2$	40.36	47.6	41.6	52.1
$\hat{s}_3$	26.57	44.8	27.0	37.7

With 30 dB corresponding to hearing threshold, the estimates are very good. Furthermore, one can see from Fig. 2 that mixing parameters converge to the exact values of the mixing matrix. The noise variance σ^2 converge to its true value within only a few samples. Besides, the sampled values of $\boldsymbol{\alpha}$ and $\boldsymbol{\lambda}$ show high variance, but considering the quality of the sources estimates, their precise values are of little importance.

---

[2] We recall that $\mathbf{r}_1$ is set to ones.

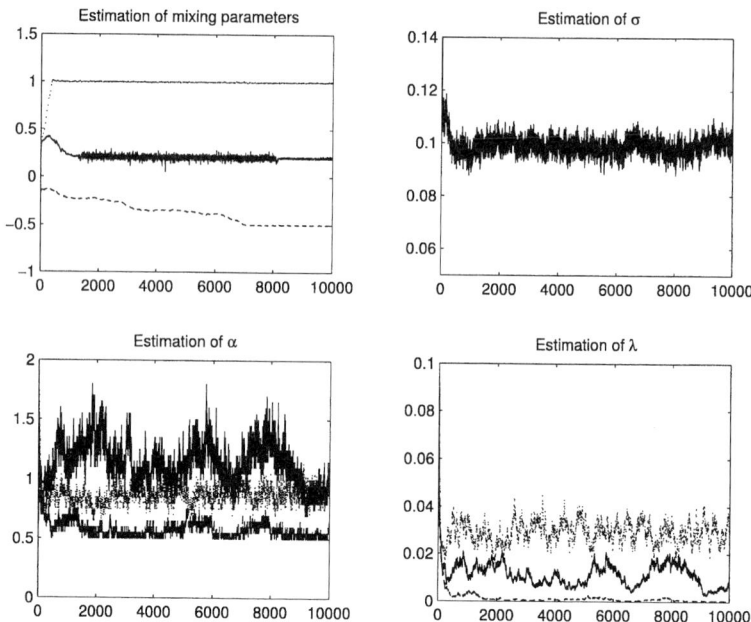

Fig. 2. Estimation of **A**, σ, $\boldsymbol{\alpha}$, $\boldsymbol{\lambda}$ with Gibbs sampler.

Audio Signals. We have applied our model and method a mixture of three musical signals (s_1 = cello, s_2 = percussions, s_3 = piano) with two observations and $\approx$ 20dB SNR on each observation. Separation was performed on MDCT coefficients of the original signals ($\approx$ 3s sampled at 8000Hz) with window length equal to 128 samples (16ms). 5000 iterations of the sampler were run, convergence was obtained after $\approx$ 2000 iterations. Mixing matrix was chosen as in previous section. Audio samples can be listened to at http://www-sigproc.eng.cam.ac.uk/~cf269/ica04/sound_files.html. The obtained performance criteria are:

	SDR	SIR	SNR	SAR
$\hat{s}_1$	11.6	16.6	29.2	13.6
$\hat{s}_2$	1.3	10.8	27.2	2.1
$\hat{s}_3$	4.1	8.7	28.7	6.5

5 Conclusion

Good results of section 4 show the relevance of the bayesian approach to handle separation of underdetermined mixtures of sparse sources. The quality of the audio estimates is average due to a high amount of artifacts, but interference rejection is good. However there is room for improvement. Indeed, the method can be extended to overcomplete dictionaries, and other prior distributions of the coefficients of the decomposition of the sources can be used. For example generalised Gaussian distributions family can be used easily as they can be expressed as scaled mixtures of Gaussians too.

Motivated by these promising results, the next step is to study what kind of prior and dictionary can be used with a particular type of signal.

Acknowledgements

C. Févotte and S. J. Godsill acknowledge the partial support of EU RTN MOUMIR (HP-99-108). P. J. Wolfe and S. J. Godsill acknowledge partial support from EPSRC ROPA Project 67958 "High Level Modelling and Inference for audio signals using Bayesian atomic decompositions". Many thanks to Laurent Daudet for providing us with MDCT code.

References

1. Zibulevsky, M., Pearlmutter, B.A., Bofill, P., Kisilev, P.: Blind source separation by sparse decomposition. In Roberts, S.J., Everson, R.M., eds.: Independent Component Analysis: Principles and Practice. Cambridge University Press (2001)
2. Lewicki, M.S., Sejnowski, T.J.: Learning overcomplete representations. Neural Computations **12** (2000) 337–365
3. Olshausen, B.A., Millman, K.J.: Learning sparse codes with a mixture-of-Gaussians prior. In S. A. Solla, T.K.L., ed.: Advances in Neural Information Processing Systems. MIT press (2000) 841–847
4. Davies, M., Mitianoudis, N.: A simple mixture model for sparse overcomplete ICA. IEE Proceedings on Vision, Image and Signal Processing (2004)
5. Mallat, S.: A wavelet tour of signal processing. Academic Press (1998)
6. Wolfe, P.J., Godsill, S.J., Ng, W.J.: Bayesian variable selection and regularisation for time-frequency surface estimation. J. R. Statist. Soc. **B** (2004)
7. Andrews, D.F., Mallows, C.L.: Scale mixtures of normal distributions. J. R. Statist. Soc. **B** (1974) 99–102
8. Gilks, W.R., Richardson, S., Spiegelhalter, D.J.: Markov Chain Monte Carlo in Practice. Chapman & Hall (1996)
9. Gribonval, R., Benaroya, L., Vincent, E., Févotte, C.: Proposals for performance measurement in source separation. In: Proc. 4th Symposium on Independent Component Analysis and Blind Source Separation (ICA'03), Nara, Japan (2003)

Blind Source Separation
Using the Block-Coordinate Relative Newton Method

Alexander M. Bronstein, Michael M. Bronstein, and Michael Zibulevsky*

Technion – Israel Institute of Technology, Department of Electrical Engineering,
32000 Haifa, Israel
{alexbron,bronstein}@ieee.org, mzib@ee.technion.ac.il

Abstract. Presented here is a generalization of the modified relative Newton method, recently proposed in [1] for quasi-maximum likelihood blind source separation. Special structure of the Hessian matrix allows to perform block-coordinate Newton descent, which significantly reduces the algorithm computational complexity and boosts its performance. Simulations based on artificial and real data show that the separation quality using the proposed algorithm outperforms other accepted blind source separation methods.

1 Introduction

The term *blind source separation* (BSS) refers to a wide class of problems in acoustics, medical signal and image processing, hyperspectral imaging, etc., where one needs to extract the underlying 1D or 2D sources from a set of linear mixtures without any knowledge of the mixing matrix. As a particular case, consider the problem of equal number of sources and mixtures, in which an N-channel sensor signal arises from N unknown scalar source signals, linearly mixed by an unknown $N \times N$ invertible matrix A: $x(t) = As(t)$. When a finite sample $t = 1, .., T$ is given, the latter can be rewritten in matrix notation as $X = AS$, where X and S are $N \times T$ matrices containing $s_i(t)$ and $x_i(t)$ as the rows. In the 2D case, images can be thought of as one-dimensional vectors. Our goal is to estimate the unmixing matrix $W = A^{-1}$, which yields the source estimate $s(t) = Wx(t)$.

Let us assume that the sources $s_i(t)$ are zero-mean i.i.d. and independent on each other. The minus log likelihood of the observed data is given by

$$\ell(X; W) = -\log|W| + \frac{1}{T} \sum_{i,t} h_i(W_i x(t)), \qquad (1)$$

where W_i is the i-th row of W, $h_i(s) = -\log p_i(s)$, and $p_i(s)$ is the PDF of the i-th source. We will henceforth assume for simplicity that $h_i(s) = h(s)$ for all the sources, although the presented method is also valid in the general case. Many times, when h_i are not equal to the exact minus log PDFs of the sources, minimization of (1) leads to a consistent estimator, known as *quasi maximum likelihood* (QML) estimator.

---

* This research has been supported by the HASSIP Research Network Program HPRN-CT-2002-00285, sponsored by the European Commission, and by the Ollendorff Minerva Center.

QML estimation is convenient when the source PDF is unknown, or not well-suited for optimization. For example, when the sources are sparse or sparsely representable, the absolute value function, or its smooth approximation is a good choice for $h(s)$ [2,3]. We use a parametric family of functions

$$h_\lambda(s) = |s| + \frac{1}{|s| + \lambda^{-1}} \quad (2)$$

with a smoothing parameter $\lambda > 0$. Up to an additive constant, $h_\lambda(s) \to |s|$ when $\lambda \to 0^+$. Evaluation of this type of non-linearity and its first- and second-order derivatives has relatively low complexity.

The widely accepted *natural gradient* method shows poor convergence when the approximation of the absolute value becomes too sharp. In order to overcome this obstacle, a relative Newton approach was recently proposed in [1], which is an improvement of the Newton method used in [4]. It was noted that the block-diagonal structure of the Hessian allows its fast approximate inversion, leading to the modified relative Newton step. In current work, we extend this approach by introducing a block-coordinate relative Newton method, which possesses faster convergence in approximately constant number of iterations.

2 Relative Newton Algorithm

The following *relative optimization* (RO) algorithm for minimization of the QML function (1) was used in [5]:

Relative Optimization Algorithm

1. Start with initial estimates of the unmixing matrix $W^{(0)}$ and the sources $X^{(0)} = W^{(0)}X$.
2. For $k = 0, 1, 2, ...$, until convergence
3. Start with $W^{(k+1)} = I$.
4. Using an unconstrained optimization method, find $W^{(k+1)}$ such that $\ell(X^{(k)}; W^{(k+1)}) < \ell(X^{(k)}; I)$.
5. Update source estimate: $X^{(k+1)} = W^{(k+1)} X^{(k)}$.
6. End

The use of a single gradient descent iteration on Step 4 leads to the natural (relative) gradient method [6,7], whereas the use of a Newton iteration leads to the relative Newton method [1].

2.1 Gradient and Hessian of $\ell(X; W)$

The use of the Newton method on Step 4 of the RO algorithm requires the knowledge of the Hessian of $\ell(X; W)$. Since $\ell(X; W)$ is a function of a matrix argument, its gradient w.r.t. W is also a matrix

$$G(W) = \nabla_W \ell(X; W) = -W^{-T} + \frac{1}{T} h'(WX) X^T, \quad (3)$$

where h' is applied element-wise to WX.

The Hessian of $\ell(X;W)$ can be thought as a fourth-order tensor $\mathcal{H}$, which is inconvenient in practice. Alternatively, one can convert the matrix W into an N^2-long column vector $w = \text{vec}(W)$ by row-stacking. Using this notation, the Hessian is an $N^2 \times N^2$ matrix, which can be found from the differential of $g(w)$ (see [1] for derivation). The k-th column of the Hessian of the log-determinant term of $\ell(X;W)$ is given by

$$H^k = \text{vec}\left(A^j A_i\right), \qquad (4)$$

where $A = W^{-1}$, and A_i, A^j are its i-t row and j-th column, respectively, and $k = (i-1)N + j$. The Hessian of the second term of $\ell(X;W)$ containing the sum is a block-diagonal matrix, whose m-th block is an $N \times N$ matrix of the form

$$B^m = \frac{1}{T}\sum_t h''(W_m x(t))x(t)x^T(t). \qquad (5)$$

2.2 The Modified Relative Newton Step

At each relative Newton iteration, the Hessian is evaluated for $W = I$, which simplifies the Hessian of the log-determinant term in (4) to

$$H^k = \text{vec}\left(e_i e_j^T\right), \qquad (6)$$

where e_i is the standard basis vector containing 1 at the i-th coordinate. The second term (5) becomes

$$B^m = \frac{1}{T}\sum_t h''(x_m(t))x(t)x^T(t). \qquad (7)$$

At the solution point, $x(t) = s(t)$, up to scale and permutation. For a sufficiently large sample, the sum approaches the corresponding expected value yielding $B^m \approx \mathbb{E}\left\{h''(x_m)xx^T\right\}$. Invoking the assumption that $s_i(t)$ are mutually-independent zero-mean i.i.d. processes, B^m become approximately diagonal.

Using this approximation of the Hessian, the modified (fast) relative Newton method is obtained. The diagonal approximation significantly simplifies both Hessian evaluation and Newton system solution. Computation of the diagonal approximation requires about $N^2 T$ operations, which is of the same order as the gradient computation. Approximate solution of the Newton system separates to solution of $\frac{1}{2}N(N-1)$ symmetric systems of size 2×2

$$\begin{pmatrix} Q_{ij} & 1 \\ 1 & Q_{ji} \end{pmatrix}\begin{pmatrix} D_{ij} \\ D_{ji} \end{pmatrix} = -\begin{pmatrix} G_{ij} \\ G_{ji} \end{pmatrix}, \qquad (8)$$

for the off-diagonal elements ($i \neq j$), and N additional linear equations

$$Q_{ii}D_{ii} + D_{ii} = -G_{ii} \qquad (9)$$

for the diagonal elements, where D is the $N \times N$ Newton direction matrix, G is the gradient matrix, and Q is an $N \times N$ matrix, in which the Hessian diagonal is packed row-by-row.

In order to guarantee global convergence, the 2×2 systems are modified by forcing positive eigenvalues [1]. Approximate Newton system solution requires about $15N^2$ operations. This implies that the modified Newton step has the asymptotic complexity of a gradient descent step.

3 Block-Coordinate Relative Newton Method

Block-coordinate optimization is based on decomposition of the vector variable into components (blocks of coordinates) and producing optimization steps in the respective block subspaces in a sequential manner. Such algorithms usually have two loops: a step over block (inner iteration), and a pass over all blocks (outer iteration). The main motivation for the use of block-coordinate methods can be that when most variables are fixed, we often obtain subproblems in the remaining variables, which can be solved efficiently. In many cases, block-coordinate approaches require significantly less outer iterations compared to conventional methods [8].

In our problem, the Hessian is approximately separable with respect to the pairs of symmetric elements of W. This brings us to the idea of applying the Newton step block-coordinately on these pairs. As it will appear from the complexity analysis, the relative cost of the nonlinearity computation becomes dominant in this case, therefore, we can do one step further and use pair-wise symmetric blocks of larger size. The matrix W can be considered as consisting of $M = N/K$ blocks of size $K \times K$,

$$W = \begin{pmatrix} W_{11} & W_{12} & \cdots & W_{1M} \\ W_{21} & W_{22} & \cdots & W_{2M} \\ \vdots & \vdots & \ddots & \vdots \\ W_{M1} & W_{M2} & \cdots & W_{MM} \end{pmatrix} \qquad (10)$$

The *block-coordinate* modified relative Newton step (as opposed to the *full* modified relative Newton step described before) is performed by applying the relative Newton algorithm to the subspace of two blocks W_{ij} and W_{ji} at a time, while fixing the rest of the matrix elements. In order to update all the entries of W, $N(N-1)/2K^2$ inner iterations are required. We obtain the following block-coordinate relative Newton algorithm:

Block-Coordinate Relative Newton Algorithm

1. Start with initial estimates of the unmixing matrix $W^{(0)}$ and the sources $X^{(0)} = W^{(0)}X$.
2. For $k = 0, 1, 2, ...$, until convergence
    3. For $i = 1, ..., K$
        4. For $j = 1, ..., K$
            5. Start with $W^{(k+1)} = I$.
            6. Update the blocks W_{ij} and W_{ji} using one block-coordinate relative Newton iteration to find $W^{(k+1)}$ such that $\ell(X^{(k)}; W^{(k+1)}) < \ell(X^{(k)}; I)$.
            7. Efficiently update the source estimate: $X^{(k+1)} = W^{(k+1)}X^{(k)}$.
        8. End
    9. End
10. End

Since only few elements of W are updated at each inner iteration, evaluation of the cost function, its gradient and Hessian can be significantly simplified. In the term $Wx(t)$, only $2K$ elements are updated and consequently, the non-linearity h is applied to a $2K \times T$ stripe to update the sum $\sum h(W_i x(t))$.

Since at each inner step the identity matrix I is substituted as an initial value of W, the updated matrix will have the form

$$W = \begin{pmatrix} I_{K \times K} & & W_{ij} & \\ & I_{K \times K} & & \\ W_{ji} & & \ddots & \\ & & & I_{K \times K} \end{pmatrix} \quad (11)$$

It can be easily shown that the computation of the determinant of W having this form can be reduced to

$$\det W = \det \begin{pmatrix} I & W_{ij} \\ W_{ji} & I \end{pmatrix} \quad (12)$$

and carried out in $2K^3$ operations. Similarly, the computation of the gradient requires applying h' to the updated $2K \times T$ stripe of WX and multiplying the result by the corresponding $2K \times T$ stripe of X^T. In addition, the gradient requires inversion of W. When $i \neq j$, the inverse matrix has the form

$$W^{-1} = \begin{pmatrix} I & & & \\ & A_{ii} & & A_{ij} \\ & & I & \\ & A_{ji} & & A_{jj} \\ & & & & I \end{pmatrix}, \quad (13)$$

where the $K \times K$ blocks A_{ii}, A_{ij}, A_{ji} and A_{jj} are obtained from

$$\begin{pmatrix} A_{ii} & A_{ij} \\ A_{ji} & A_{jj} \end{pmatrix} = \begin{pmatrix} I & W_{ij} \\ W_{ji} & I \end{pmatrix}^{-1}, \quad (14)$$

which also requires $2K^3$ operations. To compute the Hessian, one should update $2K$ elements in $x(t)x^T(t)$ for each $t = 1, ..., T$ and apply h'' to the updated $2K \times T$ stripe of WX.

3.1 Computational Complexity

For convenience, we denote as α, α' and α'' the number of operations required for the computation of the non-linearity h and its derivatives h' and h'', respectively. A reasonable estimate of these constants for h given in (2) is $\alpha = 6, \alpha' = 2, \alpha'' = 2$ [9]. We will also denote $\beta = \alpha + \alpha' + \alpha''$. A single block-coordinate relative Newton inner iteration involves computation of the cost function, its gradient and Hessian, whose respective complexities are $2(K^2T + K^3 + \alpha KT)$, $2(K^2T + K^3 + \alpha' KT)$ and $2(K^2T + (\alpha'' + 1)KT)$. In order to compute the Newton direction, K systems of equations of size 2×2 have to be solved, yielding in total solution of systems per outer iteration, independent of K. Other operations have negligible complexity. Therefore, a single block-coordinate outer Newton iteration will require about $N^2T(3+(\beta+1)/K)$ operations. Substituting $K = N$, the algorithm degenerates to the relative Newton method, with the complexity of about $3N^2T$. Therefore, the block-coordinate approach with $K \times K$ blocks is advantageous, if its runtime is shortened by the factor $\gamma > 1 + (\beta+1)/3K$ compared to the full relative Newton method.

4 Numerical Results

For numerical experiments, three data sets were used: sparse normal signals generated using the MATLAB function sprandn, 50,000 samples from instrumental and vocal music recordings sampled at 11025 Hz, and natural images. In all the experiments, the sources were artificially mixed using an invertible random matrix with uniform i.i.d. elements. The modified relative Newton algorithm with backtracking line search was used, stopped after the gradient norm reached 10^{-10}. Data sets containing audio signals and images were not originally sparse, and thus not the corresponding mixtures. Short time Fourier transform (STFT) and discrete derivative were used to sparsify the audio signals and the images, respectively, as described in [10, 11, 2, 3]. In Table 1, the separation quality (in terms of the signal-to-interference ratio (SIR) in dB units) of the relative Newton method is compared with that of stochastic natural gradient (Infomax) [7, 6, 12], Fast ICA [13, 14] and JADE [15]. We should note that without the sparse representation stage, all algorithms produced very poor separation results. Figure 2 depicts the convergence of the full modified relative Newton algorithm and its block-coordinate version for different block sizes, with audio signals and images. Complete comparison can be found at
http://visl.technion.ac.il/bron/~works/bss/newton.

The block-coordinate algorithm (with block size $K = 1, 3, 5$ and 10) was compared to the full modified relative Newton algorithm ($K = N$) on problems of different size (N from 3 to 50 in integer multiplies of K; $T = 10^3$) with the sparse sources. The total number of the cost function, its gradient and Hessian evaluations were recorded and used for complexity computation. Remarkably, the number of outer iterations is approximately constant with the number of sources N in the block-coordinate method, as opposed to the full relative Newton method (see Figure 1, left). Particularly, for $K = 1$ the number of outer iterations is about 10. Furthermore, the contribution of the non-linearity computation to the overall complexity is decreasing with the block size K. Hence, it explains why in Figure 1 (right) the complexity normalized by the

Table 1. Separation quality (best and worst SIR in dB) of sparse signals, audio signals and images.

SIR	Newton	InfoMax	FastICA	JADE
Sparse	172.98 ÷ 167.99	34.35 ÷ 18.64	23.82 ÷ 21.89	26.78 ÷ 21.89
Audio	46.68 ÷ 25.72	37.34 ÷ 23.35	25.15 ÷ 2.11	25.78 ÷ 9.02
Images	57.35 ÷ 31.74	38.52 ÷ 25.66	30.54 ÷ 19.75	32.35 ÷ 27.85

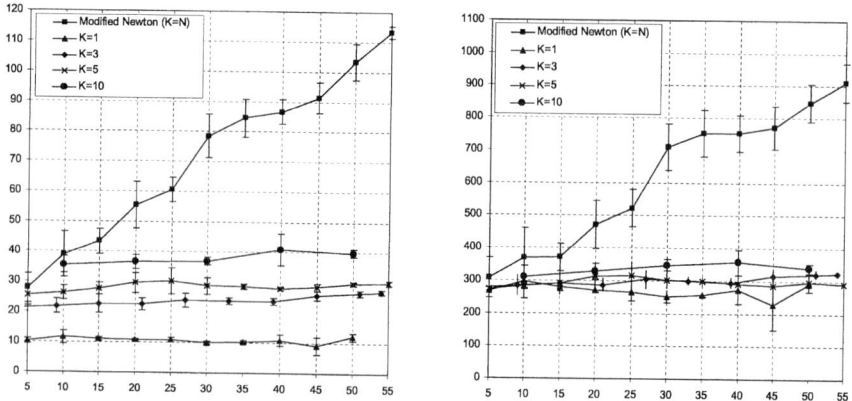

Fig. 1. Average number of outer iterations (left) and the normalized complexity (right) vs. the number of sources N for different block sizes K.

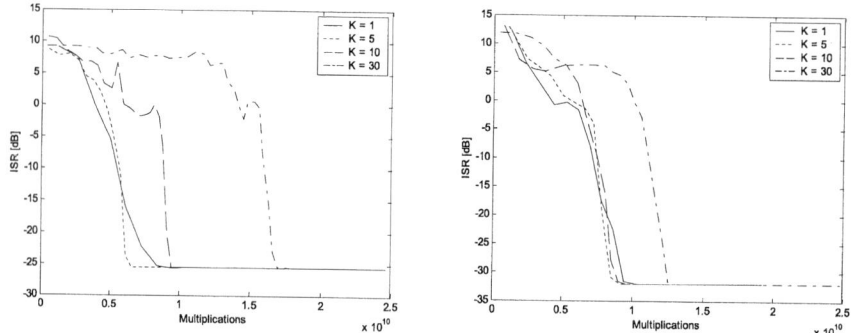

Fig. 2. Convergence of the the block-coordinate relative Newton method for audio sources (left) and images (right) using blocks of different size K ($K = 30$ corresponds to full relative Newton).

factor N^2T is almost the same for blocks of size $K = 1, 3, 5$ and 10. However, CPU architecture considerations may make larger blocks preferable. The block-coordinate algorithm outperformed the relative Newton algorithm by about 3.5 times for $N = 55$.

5 Conclusion

We presented a block-coordinate version of the relative Newton algorithm for QML blind source separation introduced in [1]. In large problems, we observed a nearly three-fold reduction of the computational complexity of the modified Newton step by using the block-coordinate approach. The use of an accurate approximation of the absolute value nonlinearity in the QML function leads to accurate separation of sources, which have sparse representation. Simulations showed that from the point of view of the obtained SIR, such optimization appears to outperform other accepted algorithms for blind source separation. The most intriguing property, demonstrated by computational experiments, is the almost constant number of iterations (independent of the number of sources) of the block-coordinate relative Newton algorithm. Though formal mathematical explanation of this phenomenon is an open question at this point, it is of importance for practical applications.

References

1. Zibulevsky, M.: Sparse source separation with relative Newton method. In: Proc. ICA2003. (2003) 897–902
2. Zibulevsky, M., Pearlmutter, B.A.: Blind source separation by sparse decomposition in a signal dictionary. Neural Comp. **13** (2001) 863–882
3. Zibulevsky, M., Pearlmutter, B.A., Bofill, P., Kisilev, P.: Blind source separation by sparse decomposition. In Roberts, S.J., Everson, R.M., eds.: Independent Components Analysis: Principles and Practice. Cambridge University Press (2001)
4. Pham, D., Garrat, P.: Blind separation of a mixture of independent sources through a quasi-maximum likelihood approach. IEEE Trans. Sig. Proc. **45** (1997) 1712–1725
5. Bell, A.J., Sejnowski, T.J.: An information maximization approach to blind separation and blind deconvolution. Neural Comp. **7** (1995) 1129–1159
6. S. Amari, A.C., Yang, H.H.: A new learning algorithm for blind signal separation. Advances in Neural Information Processing Systems **8** (1996)
7. Cichocki, A., Unbehauen, R., Rummert, E.: Robust learning algorithm for blind separation of signals. Electronics Letters **30** (1994) 1386–1387
8. Grippo, L., Sciandrone, M.: Globally convergent block-coordinate techniques for unconstrained optimization. Optimization Methods and Software **10** (1999) 587–637
9. Bronstein, A.M., Bronstein, M.M., Zibulevsky, M.: Block-coordinate relative Newton method for blind source separation. Technical Report 445, Technion, Israel (2003)
10. Bofill, P., Zibulevsky, M.: Underdetermined blind source separation using sparse representations. Sig. Proc. **81** (2001) 2353–2362
11. Bronstein, A.M., Bronstein, M.M., Zibulevsky, M., Zeevi, Y.Y.: Separation of reflections via sparse ICA. In: Proc. IEEE ICIP. (2003)
12. Makeig, S.: ICA toolbox for psychophysiological research (1998) Online: http://www.cnl.salk.edu/~ica.html.
13. Hyvärinen, A.: The Fast-ICA MATLAB package (1998) Online: http://www.cis.hut.fi/~aapo.
14. Hyvärinen, A.: Fast and robust fixed-point algorithms for independent component analysis. IEEE Trans. Neural Net. **10** (1999) 626–634
15. Cardoso, J.F.: JADE for real-valued data (1999) Online: http://sig.enst.fr:80/~cardoso/guidesepsou.html.

Hybridizing Genetic Algorithms with ICA in Higher Dimension

Juan Manuel Górriz[1], Carlos G. Puntonet[2],
Moisés Salmerón[2], and Fernando Rojas Ruiz[2]

[1] E.P.S. Algeciras, Universidad de Cádiz
Avda. Ramón Puyol s/n, 11202 Algeciras Cádiz, Spain
juanmanuel.gorriz@uca.de
[2] E.S.I., Informática, Universidad de Granada
C/ Periodista Daniel Saucedo, 18071 Granada, Spain
{carlos,moises}@atc.ugr.es

Abstract. In this paper we present a novel method for blindly separating unobservable independent component signals from their linear mixtures, using genetic algorithms (GA) to minimize the nonconvex and nonlinear cost functions. This approach is very useful in many fields such as forecasting indexes in financial stock markets where the search for independent components is the major task to include exogenous information into the learning machine. The GA presented in this work is able to extract independent components with faster rate than the previous independent component analysis algorithms based on Higher Order Statistics (HOS) as input space dimension increases showing significant accuracy and robustness.

1 Introduction

The starting point in the Independent Component Analysis (ICA) research can be found in [1] where a principle of redundancy reduction as a coding strategy in neurons was suggested, i.e. each neural unit was supposed to encode statistically independent features over a set of inputs. But it was in the 90´s when Bell and Sejnowski applied this theoretical concept to the blindly separation of the mixed sources (BSS) using a well known stochastic gradient learning rule [2] and originating a productive period of research in this area [3–6]. In this way ICA algorithms have been applied successfully to several fields such as biomedicine, speech, sonar and radar, signal processing, etc. and more recently also to time series forecasting [7], i.e. using stock data [8]. In the latter application the mixing process of multiple sensors is based on linear transformation making the following assumptions:

1. the original (unobservable) sources are statistically independent which are related to social-economic events.
2. the number of sensors (stock series) is equal to that of sources.
3. the Darmois-Skitovick conditions are satisfied [9].

On the other hand there is a wide class of interesting applications for which no reasonably fast algorithms have been developed, i.e. optimization problems that appear frequently in several applications such as VLSI design or the travelling salesman problem. In general, any abstract task to be accomplished can be viewed as a search through a space of potential solutions and whenever we work with large spaces, GAs are suitable artificial intelligence techniques for developing this optimization [10, 11]. GA are stochastic algorithms whose search methods model some natural phenomena according to genetic inheritance and Darwinian strife for survival. Such search requires balancing two goals: exploiting the best solutions and exploring the whole search space. In order to carry out them GA performs an efficient multi-directional search maintaining a population of potential solutions instead of methods such as simulated annealing or Hill Climbing.

In this work we apply GA to ICA in the search of the separation matrix, in order to improve the performance of endogenous learning machines in real time series forecasting speeding up convergence rates (scenarios with the BSS problem in higher dimension). We organize the essay as follows. In section 2 we give a brief overview of the basic GA theory and introduce a set of new genetic operators in sections 3 and 4. The new search algorithm will be compare to the well-known ICA algorithms and state state some conclusions in section 5.

2 Basis Genetic Algorithms in Higher Dimension

A GA can be modelled by means of a *time inhomogeneous Markov* chain [12] obtaining interesting properties related with weak and strong ergodicity, convergence and the distribution probability of the process [13]. In the latter reference, a canonical GA is constituted by operations of parameter encoding, population initialization, crossover , mutation, mate selection, population replacement, fitness scaling, etc. proving that with these simple operators a GA does not converge to a population containing only optimal members. However, there are GAs that converge to the optimum, *The Elitist GA* [14] and those which introduce *Reduction Operators*[15]. We have borrowed the notation mainly from [13] where the model for GAs is a inhomogeneous Markov chain model on probability distributions (**S**) over the set of all possible populations of a fixed finite size. Let **C** the set of all possible creatures in a given world (number of vectors of genes equal to that of elements of the mixing matrix) and a function $f : \mathbf{C} \to R^+$ (see section 2.1). The task of GAs is to find an element $c \in \mathbf{C}$ for which $f(c)$ is maximal. We encode creatures into genes and chromosomes or individuals as strings of length ℓ of binary digits (size of Alphabet A is $a = 2$) using one-complement representation.

In the Initial Population Generation step (choosing randomly $p \in \wp_N$, where $\wp_N$ is the set of populations, i.e the set of N-tuples of creatures containing $a^{L \equiv N \cdot \ell}$ elements) we assume that creatures lie in a bounded region $[-1, 1]$. After the initial population p has been generated, the fitness of each chromosome $\mathbf{c}_i$ is determined using a contrast function (i.e based on cumulants or neg-entropy) which measures the pair-wise statistical independency between sources in the current individual (see section 2.1).

Table 1. Pseudo-code of GA.

```
Initialize Population
i=0
while not stop do
  do N/2 times
    Select two mates from p_i
    Generate two offspring using crossover operator
    Mutate the two children
    Include children in new generation p_new
  end do
  Build population p̂_i = p_i ∪ p_new
  Apply Reduction Operators (Elitist Strategies) to get p_{i+1}
  i=i+1
end
```

The next step in canonical GA is to define the Selection Operator. New generations for mating are selected depending on their fitness function values using *roulette wheel selection*. Let $p = (c_1, \ldots, c_N) \in \wp_N$, $n \in \mathcal{N}$ and f the fitness function acting in each component of p. Scaled fitness selection of p is a lottery for every position $1 \leq i \leq N$ in population p such that creature c_j is selected with probability proportional to its fitness value. Thus proportional fitness selection can be described by column stochastic matrices $\mathbf{F_n}$, $n \in \mathcal{N}$, with components:

$$\langle q, \mathbf{F_n} p \rangle = \prod_{i=1}^{N} \frac{n(q_i) f_n(p, q_i)}{\sum_{j=1}^{N} f_n(p, j)} \qquad (1)$$

where $p, q \in \wp_N$ so $p_i, q_i \in \mathbf{C}$, $\langle \ldots \rangle$ denotes the standard inner product, and $n(q_i)$ the number of occurrences of q_i in p. Once the two individuals have been selected, an elementary crossover operator $\mathbf{C}(K, P_c)$ is applied (setting the crossover rate at a value, i.e. $P_c \to 0$, which implies children similar to parent individuals) that is given (assuming N even) by:

$$\mathbf{C}(K, P_c) = \prod_{i=1}^{N/2} ((1 - P_c)\mathcal{I} + P_c \mathbf{C}(2i-1, 2i, k_i)) \qquad (2)$$

where $\mathbf{C}(2i-1, 2i, k_i)$ denotes elementary crossover operation of c_i, c_j creatures at position $1 \leq k \leq \ell$ and $\mathcal{I}$ the identity matrix, to generate two offspring (see [13] for further properties of the crossover operator), $K = (k_1, \ldots, k_{N/2})$ a vector of cross over points and P_c the cross over probability.

2.1 Fitness Function Based on Cumulants

The independence condition for the independent components of the output vector $\mathbf{y}$ is given by the definition of independence random variables:

$$p(\mathbf{y}) = \prod_{i=1}^{n} p_{y_i}(y_i); \qquad (3)$$

In order to measure the independence of the outputs we express equation 3 in terms of higher order statistics (cumulants) using the characteristic function (or moment generating function) $\phi(\mathbf{k})$, where $\mathbf{k}$ is a vector of variables in the Fourier transform domain, and considering its natural logarithm $\Phi = log(\phi(\mathbf{k}))$. Thus we get:

$$Cum(\overbrace{y_i, y_j, \ldots}^{stimes}) = \kappa_s^i \delta_{i,j,\ldots} \qquad \forall i, j, \ldots \in [1, \ldots, n] \qquad (4)$$

where $Cum(\overbrace{}^{stimes})$ is the s-th order cross-cumulant and $\kappa_s = Cum(\overbrace{y_i}^{stimes})$ is the auto-cumulant of order s straightforward related to moments [16]. Hence vanishing cross-cumulants are a necessary condition for $y_1, \ldots, y_n$ to be independent[1]. Based on the briefly above discussion, we can define the fitness function for BSS as:

$$f(p_o) = \sum_{i,j,\ldots} \|Cum(\overbrace{y_i, y_j, \ldots}^{stimes})\| \qquad \forall i, j, \ldots \in [1, \ldots, n] \qquad (5)$$

where p_o is the parameter vector (individual) containing the separation matrix and $\|\ldots\|$ denotes the absolute value.

3 Mutation Operator Based on Neighborhood Philosophy

The new Mutation Operator $\mathbf{M_{P_m}}$ is applied (with probability $\mathbf{P}_m$) independently at each bit in a population $p \in \wp_N$, to avoid premature convergence (see [10] for further discussion) and enforcing strong ergodicity. The multi-bit mutation operator with changing probability following a *exponential* law with respect to the position $1 \leq i \leq L$ in $p \in \wp_N$:

$$P_m(i) = \mu \cdot \exp\left(\frac{-mod\{\frac{i-1}{N}\}}{\emptyset}\right) \qquad (6)$$

where $\emptyset$ is a normalization constant and μ the change probability at the beginning of each creature p_i in population p; can be described as a positive stochastic matrix in the form:

$$\langle q, \mathbf{M_{P_m}} p \rangle = \mu^{\Delta(p,q)} \exp\left(-\sum_{dif(i)}^{\Delta(p,q)} \frac{mod\{\frac{i-1}{N}\}}{\emptyset}\right) \cdot \prod_{equ(i)}^{L-\Delta(p,q)} [1 - P_m(i)] \qquad (7)$$

where $\Delta(p,q)$ is the Hamming distance between p and $q \in \wp_N$, $dif(i)$ resp. $equ(i)$ is the set of indexes where p and q are different resp. equal. Following from equation 7 and checking how the matrices act on populations we can write:

$$\mathbf{M_{P_m}} = \prod_{\lambda=1}^{N} \left([1 - P_m(i)]\mathbf{1} + P_m(i)\hat{\mathbf{m}}^1(\lambda)\right) \qquad (8)$$

---

[1] In practice we need independence between sources two against two.

where $\hat{\mathbf{m}}^1(\lambda) = \mathbf{1} \otimes \mathbf{1} \ldots \otimes \overbrace{\hat{m}^1}^{\lambda} \otimes \ldots \otimes \mathbf{1}$ is a linear operator on $V_\wp$, the free vector space over A^L and $\hat{m}^1$ is the linear 1-bit mutation operator on V_1, the free vector space over A. The latter operator is defined acting on Alphabet as:

$$\langle \hat{a}(\tau'), \hat{m}^1 \hat{a}(\tau) \rangle = (a-1)^{-1}, \quad 0 \le \tau' \ne \tau \le a-1 \tag{9}$$

i.e. probability of change a letter in the Alphabet once mutation occurs with probability equal to $L\mu$. The spectrum of $\mathbf{M}_{\mathbf{P}_\mathbf{m}}$ can be evaluated according to the following expression:

$$sp(\mathbf{M}_{\mathbf{P}_\mathbf{m}}) = \left\{ \left(1 - \frac{\mu(\lambda)}{a-1}\right)^\lambda ; \quad \lambda \in [0, L] \right\} \tag{10}$$

where $\mu(\lambda) = \exp\left(\frac{-mod\{\frac{\lambda-1}{N}\}}{\emptyset}\right)$.

The operator presented in equation 8 has similar properties to the Constant Multiple-bit mutation operator $\mathbf{M}_\mu$ presented in [13]. $\mathbf{M}_\mu$ is a contracting map in the sense presented in [13]. It is easy to prove that $\mathbf{M}_{\mathbf{P}_\mathbf{m}}$ is a contracting map too, using the Corollary B.2 in [13] and the eigenvalues of this operator (equation 10). We can also compare the coefficients of ergodicity:

$$\tau_r(\mathbf{M}_{\mathbf{P}_\mathbf{m}}) < \tau_r(\mathbf{M}_\mu) \tag{11}$$

where $\tau_r(\mathbf{X}) = max\{\|\mathbf{X}v\|_r : \quad v \in \mathcal{R}^n, \quad v \perp e \quad and \quad \|v\|_r = 1\}$.

Mutation is more likely at the beginning of the string of binary digits ("small neighborhood philosophy"). In order to improve the speed convergence of the algorithm we have included mechanisms such as elitist strategy (reduction operator [17] consisting of sampling a Boltzmann probability distribution in the extended population) in which the best individual in the current generation always survives into the next (a further discussion about reduction operator, $\mathbf{P}_R$, can be found in [18]).

4 Guided Genetic Algorithm

In order to include statistical information into the algorithm (it would be a nonsense to ignore it!) we define the hybrid statistical genetic operator based on reduction operators as follows (in standard notation acting on populations):

$$\langle q, \mathbf{M}_\mathbf{G}^\mathbf{n} p \rangle = \frac{1}{\aleph(T_n)} \exp\left(-\frac{\|q - \mathbf{S}^\mathbf{n} \cdot p\|^2}{T_n}\right); \quad p, q \in \wp_N \tag{12}$$

where $\aleph(T_n)$ is the normalization constant depending on temperature T_n, n is the iteration and $\mathbf{S}^\mathbf{n}$ is the step matrix which contains statistical properties, i.e based on cumulants it can be expressed using quasi-Newton algorithms as [5]:

$$\mathbf{S}^\mathbf{n} = (\mathbf{I} - \mu^n (\mathbf{C}_{y,y}^{1,\beta} \mathbf{S}_y^\beta - \mathbf{I})); \quad p_i \in \mathbf{C} \tag{13}$$

where $\mathbf{C}_{y,y}^{1,\beta}$ is the cross-cumulant matrix whose elements are $[\mathbf{C}_{y,y}^{\alpha,\beta}]_{ij} = Cum(\underbrace{y_i,\ldots,y_i}_{\alpha},\underbrace{y_j,\ldots,y_j}_{\beta})$ and $\mathbf{S}_y^\beta$ is the sign matrix of the output cumulants.

Such search requires balancing two goals: exploiting the blindly search like a canonical GA and using statistical properties like a standard ICA algorithm. Finally the guided GA (GGA) is modelled, at each step, as the stochastic matrix product acting on probability distributions over populations:

$$\mathbf{G^n} = \mathbf{P}_R^n \cdot \mathbf{F_n} \cdot \mathbf{C}_{\mathbf{P}_c^n}^{\mathbf{k}} \cdot \mathbf{M}_{(\mathbf{P_m},\mathbf{G})^n} \qquad (14)$$

The GA used applies local search (using the selected mutation and crossover operators) around the values (or individuals) found to be optimal (elite) the last time. The computational time depends on the encoding length, number of individuals and genes. Because of the probabilistic nature of the GA-based method, the proposed method almost converges to a global optimal solution on average. In our simulation, however, nonconvergent case was not found. Table 1 shows the GA-pseudocode.

5 Simulations and Conclusions

To check the performance of the proposed hybrid algorithm, 50 computer simulations were conducted to test the GGA vs. the GA method [7] and the most relevant ICA algorithm to date, FastICA [5]. In this paper we neglect the evaluation of the computational complexity of the current methods, described in detail in several references such as [7] or [19]. The main reason lies in the fact that we are using a 8 nodes Cluster Pentium II 332MHz 512Kb Cache, thus the computational requirements of the algorithms (fitness functions, encoding, etc.) are generally negligible compared with the cluster capacity. Logically GA-based BSS approaches suffer from a higher computational complexity.

Consider the mixing cases from 2 to 20 independent random super-gaussian input signals. We focuss our attention on the evolution of the crosstalk vs. the number of iterations using a mixing matrix randomly chosen in the interval $[-1,+1]$. The number of individuals chosen in the GA methods were $N_p = 30$ in the 50 (randomly mixing matrices) simulations for a number of input sources from 2 (standard BSS problem) to 20 (BSS in biomedicine or finances). The standard deviation of the parameters of the separation over the 50 runs never exceeded 1% of their mean values while using the FASTICA method we found large deviations from different mixing matrices due to its limited capacity of local search as dimension increases. The results for the crosstalk are displayed in Table 2. It can be seen from the simulation results that the FASTICA convergence rate decreases as dimension increases whereas GA approaches work efficiently.

A GGA-based BSS method has been developed to solve BSS problem from the linear mixtures of independent sources. The proposed method obtain a good performance overcoming the local minima problem over multidimensional domains. Extensive simulation results prove the ability of the proposed method.

Table 2. Figures: 1) Mean Crosstalk (50 runs) vs. iterations to reach the convergence for num. sources equal to 2 2) Mean Crosstalk (50 runs) vs. iterations to reach the convergence for num. sources equal to 20 3) Evolution of the crosstalk area vs. dimension. 4) Example of independent source used in the simulations.

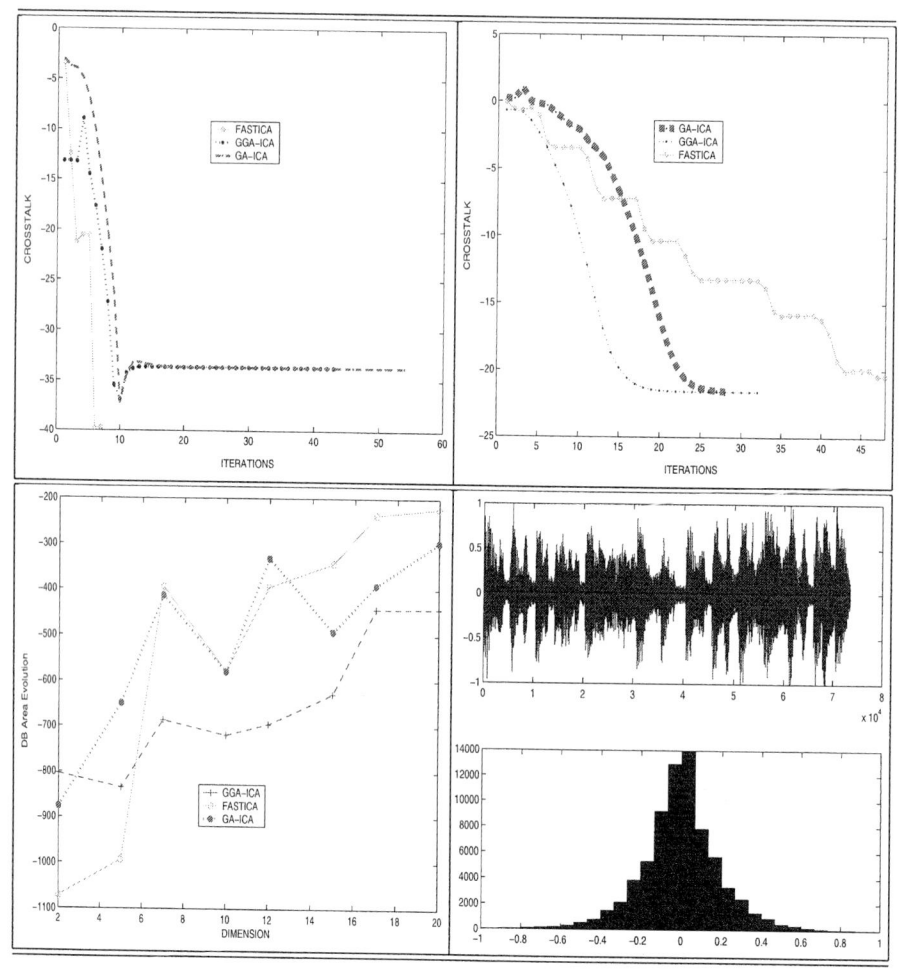

This is particular useful in some medical applications where input space dimension increases and in real time applications where reaching fast convergence rates is the major objective.

References

1. Barlow, H.B, Possible principles underlying transformation of Sensory messages. Sensory Communication, W.A. Rosenblith, MIT Press, New York, U.S.A. (1961).
2. Bell, A.J., Sejnowski, T.J. An Information-Maximization Approach to Blind Separation and Blind Deconvolution. Neural Computation, vol 7, 1129-1159 (1995).

3. Cardoso, J.F., Infomax and maximun likelihood for source separation. IEEE Letters on signal processing, 4, 112-114 (1997).
4. Cichoki, A., Unbehauen, R., Robust neural networks with on-line learning for blind identification and blind separation of sources. IEEE Transactions on Circuits and Systems, 43 (11), 894-906 (1996).
5. Hyvärinen, A., Oja, E., A fast fixed point algorithm for independent component analysis Neural Computation, 9: 1483-1492
6. Puntonet, C.G., Prieto, A. Neural net approach for blind separation of sources based on geometric properties. Neurocomputing 18, 141-164 (1998)
7. Górriz, J.M., Algorítmos Híbridos para la Modelización de Series Temporales con Técnicas AR-ICA. PhD Thesis, University of Cádiz (2003)
8. Back, A.D., Weigend, A.S., A first Application of Independent Component Analysis to Extracting Structure from Stock Returns. International Journal of Neural Systems, vol 8,(5), (1997)
9. Cao, X.R., Liu W.R., General Approach to Blind Source Separation. IEEE Transactions on signal Processing, vol 44, num 3, 562-571 (1996)
10. Michalewicz, Z., *Genetic Algorithms + Data structures = Evolution Programs*, springer Verlag, Berlin 1992.
11. Rojas F., Álvarez M.R., Puntonet C.G., Martin-Clemente R., Applying Neural Networks and Genetic Algorithms to the Separation of Sources Iberama 2002 LNAInteligence 2527,420-429, Sevilla (2002)
12. Haggstrom, O., *Finite Markov Chains and Algorithmic Applications*, Cambridge University,1998.
13. Schmitt, L.M., Nehaniv, C.L., Fujii, R.H., *Linear Analysis of Genetic Algorithms*, Theoretical Computer Science, volume 200, pages 101-134, 1998.
14. Suzuki, J., *A markov Chain Analysis on Simple Genetic Algorithms*, IEEE Transaction on Systems, Man, and Cybernetics, vol 25, 4, 655-659,(1995).
15. Eiben, A.E., Aarts, E.H.L., Van Hee, K.M., Global Convergence of Genetic Algorithms: a Markov Chain Analysis, Parallel Problem Solving from Nature, Lecture Notes in Computer Science, vol 496, (4-12),(1991).
16. Chryssostomos, C., Petropulu, A.P., Higher Order Spectra Analysis: A Non-linear Signal Processing Framework Prentice Hall, London (1993)
17. Lozano, J.A., Larrañaga, P., Graña, M., Albizuri, F.X., *Genetic Algorithms: Bridging the Convergence Gap*, Theoretical Computer Science, vol 229, 11-22, (1999).
18. Rudolph, G., *Convergence Analysis of Canonical Genetic Algorithms*, IEEE Transactions on Neural Networks, vol 5, num 1,(1994) 96-101.
19. Tan, Y., Wang, J., Nonlinear Blind Source Separation Using Higher order Statistics and a Genetic Algorithm. IEEE Transactions on Evolutionary Computation, vol. 5, num 6 (2001)

ICA Using Kernel Entropy Estimation with NlogN Complexity*

Sarit Shwartz, Michael Zibulevsky, and Yoav Y. Schechner

Department of Electrical Engineering
Technion - Israel Institute of Technology, Haifa 32000, Israel
psarit@tx.technion.ac.il, {mzib,yoav}@ee.technion.ac.il

Abstract. Mutual information (MI) is a common criterion in independent component analysis (ICA) optimization. MI is derived from probability density functions (PDF). There are scenarios in which assuming a parametric form for the PDF leads to poor performance. Therefore, the need arises for non-parametric PDF and MI estimation. Existing non-parametric algorithms suffer from high complexity, particularly in high dimensions. To counter this obstacle, we present an ICA algorithm based on accelerated kernel entropy estimation. It achieves both high separation performance and low computational complexity. For K sources with N samples, our ICA algorithm has an iteration complexity of at most $\mathcal{O}(KN \log N + K^2 N)$.

1 Introduction

Mutual information (MI) of signals is a natural criterion for statistical dependency and is thus used in ICA algorithms (see for example in [5,2,7,9,10] and references therein). MI is based on an estimate of the probability density function (PDF) of signals, which is computationally costly. For this reason, existing ICA algorithms have assumed rough models for the PDFs [1,4,8], or used high order cumulants instead of MI [3]. These approximations can sometimes lead to failure, as demonstrated in [2] as well as in our current paper. In contrast, rather robust separation can be achieved with non-parametric kernel-based estimation of PDFs [2]. The drawback of that algorithm is high computational complexity. For K sources, each of which having N samples, that algorithm has a complexity of $\mathcal{O}(K^2 N^2)$. Another existing algorithm [7] has a complexity of $\mathcal{O}(3^K N + K^2 N)$, which may be tolerated for a small K, but has exponential growth in K.

In this study, we develop non-parametric ICA that has $\mathcal{O}(KN \log N + K^2 N)$ complexity by using an approximation of the kernel estimator. The approxima-

---

* This research was supported by the HASSIP Research Network Program HPRN-CT-2002-00285, sponsored by the European Commission. It was also supported by the US-Israel Binational Science Foundation (BSF), and the Ollendorff Minerva Center. Minerva is funded through the BMBF. Yoav Schechner is a Landau Fellow - supported by the Taub Foundation, and an Alon Fellow.

tion is calculated using a fast convolution. The errors caused by the approximation are reasonably small. Therefore, our method makes non-parametric ICA a practical algorithm for large problems.

2 Blind Source Separation and Mutual Information

Let $\{s_1, s_2, ... s_K\}$ be a set of independent sources. Each source is of the form $s_k = [s_k(1), s_k(2), \ldots, s_k(N)]^T$. Let $\{y_1, y_2, ... y_K\}$ be a set of measured signals, each of which being a linear mixture of the sources. Denote $\{\hat{s}_1, \hat{s}_2, ... \hat{s}_K\}$ as the set of the reconstructed sources and W as the separation matrix. Then,

$$[\hat{s}_1, \hat{s}_2, ..., \hat{s}_K]^T = W[y_1, y_2, ... y_K]^T . \quad (1)$$

The MI of the K random variables $\hat{s}_1, \hat{s}_2, ... \hat{s}_K$ is (see for example [5])

$$\mathcal{I}(\hat{s}_1, \hat{s}_2, ... \hat{s}_K) = \mathcal{H}_{\hat{s}_1} + \mathcal{H}_{\hat{s}_2} + ... + \mathcal{H}_{\hat{s}_K} - \log|\det(W)| - \mathcal{H}_{\text{measurements}} , \quad (2)$$

where $\mathcal{H}(\hat{s}_k)$ is the differential entropy (DE) of $\hat{s}_k$. Here, $\mathcal{H}_{\text{measurements}}$ is independent of W and is constant for a given sample set $\{y_1, y_2, ... y_K\}$. Thus, the minimization problem that we solve is

$$\min_{W} \left\{ \sum_{k=1}^{K} \mathcal{H}_{\hat{s}_k} - \log|\det(W)| + \lambda \sum_{k=1}^{K} (\|\hat{s}_k\| - 1)^2 \right\} . \quad (3)$$

The last sum $\sum(\|\hat{s}_k\| - 1)^2$ in Eq. (3) weighted by a constant λ penalizes for un-normalized sources, therefore resolving ambiguities arising from the scale invariance of MI[1]. The gradient of this normalization penalty term is trivial to calculate and efficient to implement [11]. Therefore we do not discuss it further.

For non-parametric estimation of the DEs $\mathcal{H}_{\hat{s}_k}$, we use the Parzen-windows estimator [2, 12]. That estimator has a high computational complexity. Our method bypasses this problem using FFT-based fast convolution.

3 Estimation of MI and Its Gradient

Estimating DE using Parzen-windows [12] enables us to differentiate the estimated entropies, and have a closed form expression for the DE gradients. The Parzen-window estimator for the PDF at a value t is

$$\hat{p}(t|\hat{s}_k) \equiv (1/N) \sum_{n=1}^{N} \varphi[t - \hat{s}_k(n)] , \quad (4)$$

where $\hat{s}_k(n)$ is a sample from $\hat{s}_k$ and $\varphi(t)$ is a smoothing kernel[2]. The Parzen-windows estimator [2, 12] for the DE of $\hat{s}_k$ is

$$\hat{\mathcal{H}}_{\hat{s}_k} = -\frac{1}{N} \sum_{l=1}^{N} \log \hat{p}[\hat{s}_k(l)|\hat{s}_k] , \quad (5)$$

---
[1] This term does not affect the separation quality, but improves convergence of the optimization algorithm [11].
[2] We use a Gaussian kernel with a zero mean and variance σ^2. Following [2], we use $\sigma = 1.06 N^{-1/5}$.

explicitly,

$$\hat{\mathcal{H}}_{\hat{\mathbf{s}}_k} = -\frac{1}{N}\sum_{l=1}^{N}\log\left\{\frac{1}{N}\sum_{n=1}^{N}\varphi[\hat{s}_k(l)-\hat{s}_k(n)]\right\}, \quad (6)$$

The gradient of $\log|\det(\mathbf{W})|$ is $(\mathbf{W}^{-1})^T$ (see for example [5]). Therefore, the MI gradient is

$$\nabla_{\mathbf{W}}\mathcal{I}(\hat{\mathbf{s}}_1,\hat{\mathbf{s}}_2,...\hat{\mathbf{s}}_K) = \nabla_{\mathbf{W}}\sum_{k=1}^{K}\mathcal{H}_{\hat{\mathbf{s}}_k} - (\mathbf{W}^{-1})^T. \quad (7)$$

We calculate the gradients of the sum of DEs $\nabla_{\mathbf{W}}\sum_{k=1}^{K}\mathcal{H}_{\hat{\mathbf{s}}_k}$ in two stages, using a chain rule. First, we calculate the DEs gradients with respect to the estimated sources

$$\nabla_{\hat{\mathbf{s}}_k}\mathcal{H}_{\hat{\mathbf{s}}_k} = \left[\frac{\partial \mathcal{H}_{\hat{\mathbf{s}}_k}}{\partial \hat{s}_k(1)},...,\frac{\partial \mathcal{H}_{\hat{\mathbf{s}}_k}}{\partial \hat{s}_k(N)}\right]^T. \quad (8)$$

Then, we calculate the gradients of the sum of DEs with respect to the separation matrix by

$$\nabla_{\mathbf{W}}\sum_{k=1}^{K}\mathcal{H}_{\hat{\mathbf{s}}_k} = [\nabla_{\hat{\mathbf{s}}_1}\mathcal{H}_{\hat{\mathbf{s}}_1},...,\nabla_{\hat{\mathbf{s}}_K}\mathcal{H}_{\hat{\mathbf{s}}_K}]^T[\mathbf{y}_1,...,\mathbf{y}_K]. \quad (9)$$

The derivatives of Eq. (6) are

$$\frac{\partial \mathcal{H}_{\hat{\mathbf{s}}_k}}{\partial \hat{s}_k(r)} = -\frac{1}{N}\sum_{l=1}^{N}\frac{\frac{1}{N}\sum_{n=1}^{N}\varphi'[\hat{s}_k(l)-\hat{s}_k(n)][\delta_{lr}-\delta_{nr}]}{\frac{1}{N}\sum_{n=1}^{N}\varphi[\hat{s}_k(l)-\hat{s}_k(n)]} =$$

$$= -\frac{1}{N}\frac{\frac{1}{N}\sum_{n=1}^{N}\varphi'[\hat{s}_k(r)-\hat{s}_k(n)]}{\hat{p}[\hat{s}_k(r)|\hat{\mathbf{s}}_k]} + \frac{1}{N}\sum_{l=1}^{N}\frac{\frac{1}{N}\varphi'[\hat{s}_k(l)-\hat{s}_k(r)]}{\hat{p}[\hat{s}_k(l)|\hat{\mathbf{s}}_k]}, \quad (10)$$

where δ_{lr} is the Kroneker delta, φ' is the derivative of φ, and $\hat{p}[\hat{s}_k(l)|\hat{\mathbf{s}}_k]$ is defined in Eq. (4). Define

$$\Phi'[\hat{s}_k(l)|\hat{\mathbf{s}}_k] \equiv \frac{1}{N}\sum_{n=1}^{N}\varphi'[\hat{s}_k(l)-\hat{s}_k(n)], \quad (11)$$

$$F'[\hat{s}_k(l)] \equiv \frac{1}{N}\sum_{n=1}^{N}\frac{\varphi'[\hat{s}_k(n)-\hat{s}_k(l)]}{\hat{p}[\hat{s}_k(n)|\hat{\mathbf{s}}_k]}. \quad (12)$$

Then, Eq. (10) can be written as

$$\frac{\partial \mathcal{H}_{\hat{\mathbf{s}}_k}}{\partial \hat{s}_k(l)} = -\frac{1}{N}\frac{\Phi'[\hat{s}_k(l)|\hat{\mathbf{s}}_k]}{\hat{p}[\hat{s}_k(l)|\hat{\mathbf{s}}_k]} + \frac{1}{N}F'[\hat{s}_k(l)], \quad (13)$$

Calculating the MI gradient explicitly using Eqs. (7-13) has a complexity of $\mathcal{O}(KN^2+K^2N)$, for details see [11]. This complexity is achieved thanks to the exploiting of the chain rule (Eq. 9) and it is lower than the $\mathcal{O}(K^2N^2)$ complexity of gradient calculation presented in [2].

4 Efficient Calculation of the Entropy Estimator

The PDF estimator given by Eq. (4) can be seen as a convolution

$$\hat{p}(t|\hat{s}_k) = f * \varphi , \qquad (14)$$

where

$$f(t) = (1/N) \sum_{n=1}^{N} \delta[t - \hat{s}_k(n)] . \qquad (15)$$

It requires N^2 calculations of φ to compute this convolution in N points as needed in Eq. (5). On the other hand, it is known that fast convolution can be performed in $\mathcal{O}(N \log N)$ operations if done over a *uniform grid*. Therefore we resample (interpolate) the function $f(t)$ to a uniform grid. Then we convolve it with a uniformly sampled version of φ, which we denote φ_{sampled}. Finally, we interpolate the results back to the set of points $\hat{s}_k(l)$ used in entropy calculation Eq. (5). This process is illustrated in Fig. 1. The resampling of f starts by defining a vote function **v** on a uniform grid of length M, with a step size of Δ_v. Let $m^\#$ be the index of the grid node closest to the value of $\hat{s}_k(n)$ that satisfies

$$m^\# \leq \hat{s}_k(n)/\Delta_v \leq m^\# + 1 . \qquad (16)$$

Define the distance of $\hat{s}_k(n)$ from the index $m^\#$ (normalized by Δ_v) by

$$\eta = \frac{\hat{s}_k(n)}{\Delta_v} - m^\#, \quad 0 \leq \eta \leq 1 . \qquad (17)$$

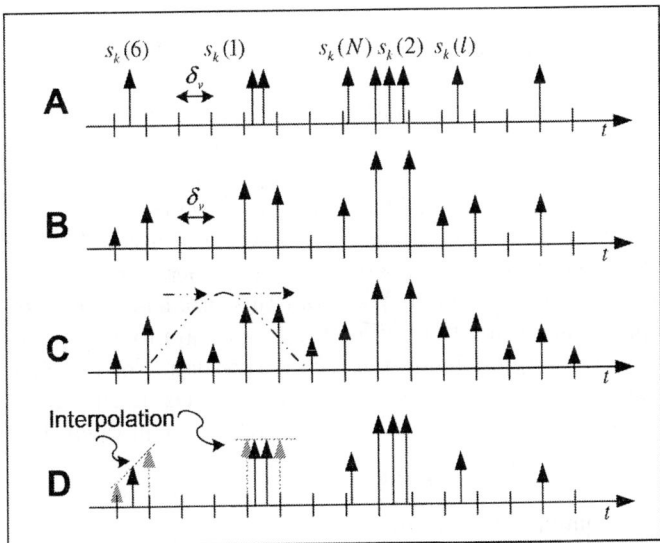

Fig. 1. Efficient calculation of $\hat{p}[\hat{s}_k(l)|\hat{s}_k]$: (A) The function f. (B) The result of voting is a function on a uniform grid. (C) The result of discrete convolution with the sampled kernel. (D) Interpolation to the original $\hat{s}_k(l)$.

Let $h(\eta)$ be a function[3] that satisfies $h(1-\eta) = 1 - h(\eta)$. Then, for each sample $\hat{s}_k(n)$ we update the vote function by

$$v(m) \leftarrow \begin{cases} v(m) + h(\eta) & \text{for } m = m^{\#} \\ v(m) + 1 - h(\eta) & \text{for } m = m^{\#} + 1 \end{cases}. \quad (18)$$

After the voting is over, we associate $\mathbf{v}/N$ with the resampled f. This transfers the function illustrated in Fig. 1A to the function illustrated in Fig. 1B. Then, we convolve[4] $\mathbf{v}/N$ with φ_{sampled} (Fig. 1B, $\rightarrow$ Fig. 1C).

$$\hat{p}_{\mathrm{u}} = (\mathbf{v}/N) * \varphi_{\text{sampled}}. \quad (19)$$

Note that $\hat{p}_{\mathrm{u}}$ resides on the uniform grid. However, the DE (Eq. 5) does not use $\hat{p}_{\mathrm{u}}$, but rather $\hat{p}[\hat{s}_k(l)]$. We obtain an estimation of $\hat{p}[\hat{s}_k(l)]$ by interpolating the values of $\hat{p}_{\mathrm{u}}$ onto the points $\hat{s}_k(l)$, using the same interpolation function $h(\eta)$ as before

$$\hat{p}[\hat{s}_k(l)|\hat{\mathbf{s}}_k] = h(\eta)\hat{p}_{\mathrm{u}}(m^{\#}) + [1 - h(\eta)]\hat{p}_{\mathrm{u}}(m^{\#} + 1), \quad (20)$$

where $m^{\#}$ and η are defined in (16, 17). This step is illustrated in Fig. 1D. Finally, we estimate the DE by Eq. (5).

The voting, the interpolation and the entropy calculation (Eqs. 5,18 and 20) require $\mathcal{O}(N)$ operations. The convolution (Eq. 20) requires $\mathcal{O}(M \log N_{\text{kernel}})$ operations[5], where N_{kernel} is the length of φ_{sampled}. In addition, estimating the sources (Eq. 1) requires $\mathcal{O}(K^2 N)$ operations. Therefore, the overall complexity of calculating the DEs for the K estimated sources is $\mathcal{O}(KM \log N_{\text{kernel}} + K^2 N)$. This is significantly lower than the $\mathcal{O}(KN^2 + K^2 N)$ complexity of the explicit calculation of the K DEs using Eq. (6).

5 Efficient Estimation of the Entropy Gradient

Calculating the DE gradient explicitly using Eq. (11-13) requires $\mathcal{O}(N^2)$ operations. Note we may calculate the gradient of any function in the same complexity of calculating the function itself (see for example [11]). In order to compute the DE gradient with complexity of $\mathcal{O}(M \log N_{\text{kernel}})$ we could have differentiated the DE approximation derived in Sec. 4. However, the resampling is an approximation causing fluctuations in the DE value as a function of $\mathbf{W}$. This may stop MI optimization at local minima. We avoid this problem altogether by taking a different approach. Rather than differentiating an approximation based on resampling, we elect to approximate the DEs derivatives (Eq. 13) directly. We do so in a similar manner to the approximation of the DE itself. In the same way as Eq. (4) is represented by Eq. (14), Eqs. (11,12) are equivalent to

$$\Phi'(t|\hat{\mathbf{s}}_k) = f * \varphi', \quad (21)$$

---
[3] We use a linear interpolation function $h(\eta) = 1 - \eta$.
[4] We used a Matlab code for fast convolution based on FFT, which had been written by Luigi Rosa, `luigi.rosa@tiscali.it`, http://utenti.lycos.it/matlab.
[5] Typically M and N_{kernel} are of the order of N or smaller. Therefore, the complexity needed is at most $\mathcal{O}(N \log N)$.

$$F'(t|\hat{\mathbf{s}}_k) = (f/\hat{p}) * \varphi'^{\text{mirror}} ,\qquad(22)$$

where f is given by Eq. (15), and $\varphi'^{\text{mirror}}(t) = \varphi'(-t)$. We compute the convolution (Eq. 21) in a fast way, using the array $\mathbf{v}$ (which is f, uniformly resampled by Eq. (18))

$$\hat{\Phi}'_{\text{u}} = (\mathbf{v}/N) * \varphi'_{\text{sampled}} \qquad(23)$$

Finally, we interpolate Φ'_{u} to the set of points $\hat{s}_k(l)$. We do so similarly to Eq. (20).

In a somewhat analogous manner, we obtain a fast calculation of Eq. (22), as described next. First, we uniformly resample $f/\hat{p}$: similarly to Eq. (18), we define a weighted vote function $\mathbf{w}$ on the uniform grid. For each sample $\hat{s}_k(n)$ we update the this function,

$$w(m) \leftarrow \begin{cases} w(m) + h(\eta)/\hat{p}[\hat{s}_k(n)|\hat{\mathbf{s}}_k] & \text{for } m = m^\# \\ w(m) + [1 - h(\eta)]/\hat{p}[\hat{s}_k(n)|\hat{\mathbf{s}}_k] & \text{for } m = m^\# + 1 , \end{cases}\qquad(24)$$

where $\hat{p}[\hat{s}_k(n)|\hat{\mathbf{s}}_k]$ has been computed in (20). We associate $\mathbf{w}/N$ with $f/\hat{p}$. In addition, we define a sampled version of φ'^{mirror}, termed $\varphi'^{\text{mirror}}_{\text{sampled}}$. We thus imitate Eq. (22) by

$$\hat{F}'_{\text{u}} = (\mathbf{w}/N) * \varphi'^{\text{mirror}}_{\text{sampled}}. \qquad(25)$$

Finally, we interpolate F'_{u} to the set of points $\hat{s}_k(l)$, similarly to Eq. (20).

Recall from Sec. 4 that the complexity of the voting and the interpolation is $\mathcal{O}(N)$, while the complexity of the discrete convolution is $\mathcal{O}(M \log N_{\text{kernel}})$. Moreover, the complexity of Eqs. (13) is $\mathcal{O}(N)$, while the complexity of Eq. (9) is $\mathcal{O}(K^2 N)$. Thus, the overall complexity of calculating the DEs gradients of the K signals is the same as of calculating the entropy itself, $\mathcal{O}(KM \log N_{\text{kernel}} + K^2 N)$. A pseudo-code for the DE estimator and its gradient is given in [11].

6 Demonstrations

In order to evaluate our method, we performed numerous separation simulations. The first set of simulations dealt with random sources of 3K samples. We simulated six sources: four of the sources were random i.i.d., with an exponential PDF[$\alpha = 2$], an exponential PDF[$\alpha = 0.6$], a normal PDF[0,1] and a Rayleigh PDF[$\beta = 1$] (Here α and β denote the parameters of the respective PDFs [2]). The other two sources were extracted as data vectors from the Lena and Trees standard pictures. The sources were mixed using randomly generated square matrices (condition number≤ 20).

Source separation was attempted using three parametric ICA algorithms [4, 3, 6]: InfoMax, Jade and Fast ICA. In addition, separation was attempted using two non parametric ICA algorithms: the first is based on Sec. 3 and thus does not use fast convolution. The second algorithm is the one we described in Secs. 4 and 5. The software for the prior algorithms [4, 3, 6] was downloaded from the websites of the respective authors.

In order to limit the signals to the grid range we use, we first performed a rough normalization of the raw measurements. We subtracted the mean of each signal and divided it by its standard deviation. The InfoMax and FastICA

Table 1. Simulation results: The accuracy of the separation is measured in terms of the signal to interference ratio (SIR).

Algorithm	SIR [dB]	Time
Non-parametric ICA, based on Sec. 3	18 ± 4	**760** min
Non-parametric ICA with fast kernel convolution, using 1K voting bins	22 ± 3	**1.2** min
Jade	7 ± 4	0.2 sec
InfoMax	1 ± 0.5	1.4 sec
InfoMax with pre-filtering	8 ± 4	1.6 sec
Fast ICA	4 ± 4	1.1 sec
Fast ICA with pre-filtering	5 ± 3	1.9 sec

algorithms are more efficient when the measured signals are sparse. We thus pre-filtered the inputs to these algorithms using the derivative operator $[-1\,0\,1]/2$. Our separation procedure was based on the BFGS Quasi-Newton algorithm as implemented in the MATLAB optimization toolbox (function FMINUNC).

The results of the simulations are presented in Table 1. The separation quality is given by the signal to interference ratio (SIR)[6]. After performing numerous simulations, we report the mean SIR and the standard deviation of the SIR. Clearly, Table 1 shows that practically no degradation of the separation quality is caused by our entropy approximation. On the other hand, the improvement in the run time is huge, compared to the competing non-parametric method. Our method does not compete with the parametric algorithms over run time, but it outperforms them in separation quality. We can separate signals that the parametric methods fail to handle.

Fig. 2. Four samples of a set of 10 pictures involved in a separation simulation. The mixed signals had been filtered by a derivative operator prior to optimization. The separation SIR is 20dB.

---

[6] SIR=$\min_k(\|\mathbf{s}_k\|^2/\|\mathbf{s}_k - \hat{\mathbf{s}}_k\|^2)$. Note that the SIR is based on the signal k having the worst separation quality. As explained in [11], the estimated $\hat{\mathbf{s}}_k$ is prone to permutation and scale ambiguities. Thus, SIR is calculated from separation results which are compensated for these ambiguities.

To visually demonstrate the separation quality, we performed an additional set of separation simulations based on 10 pictures. The pictures were mixed using randomly generated full rank matrices. The results are presented in Fig. 2.

To Conclude: We presented an algorithm that delivers high performance and possesses low computational complexity. The low complexity makes non parametric ICA applicable to high dimensional problems and large sample sizes. We have yet to study the influence of the number of uniform grid nodes on the algorithm performance.

References

1. Anthony J. Bell and Terrence J. Sejnowski. An information-maximization approach to blind separation and blind deconvolution. *Neural Computation*, 7(6):1129–1159, 1995.
2. Riccardo Boscolo, Hong Pan, and Vwani P. Roychowdhury. Non-parametric ICA. In *Proc. ICA*, pages 13–18, 2001.
3. Jean-François Cardoso and Antoine Souloumiac. Blind beamforming for non Gaussian signals. *IEE Proceedings-F*, 140(6):362–370, 1993.
4. A. Hyvärinen. The Fast-ICA MATLAB package. 1998. http://www.cis.hut.fi/~aapo/.
5. Aapo Hyvärinen, Juha Karhunen, and Erkki Oja. *Independent component analysis*. John Wiley and Sons, USA, 2001.
6. S. Makeig, A.J. Bell, T-P Jung, and T.J. Sejnowski. Independent component analysis of electroencephalographic data. *Advances in Neural Inf. Proc. Systems 8*, pages 145–151, 1996.
7. Dinh Tuan Pham. Fast algorithm for estimating mutual information, entropies and score functions. In *Proc. ICA*, pages 17–22, 2003.
8. D.T. Pham and P Garrat. Blind separation of a mixture of independent sources through a quasi-maximum likelihood approach. *IEEE Trans. Sig. Proc.*, 45(7): 1712–1725, 1997.
9. Yoav Y. Schechner, Nahum Kiryati, and Ronen Basri. Separation of transparent layers using focus. *Int. J. Computer Vision*, 89:25–39, 2000.
10. Yoav Y. Schechner, Joseph Shamir, and Nahum Kiryati. Polarization and statistical analysis of scenes containing a semi-reflector. *J. Opt. Soc. America A*, 17:276–284, 2000.
11. Sarit Shwartz, Michael Zibulevsky, and Yoav Y. Schechner. Fast kernel entropy estimation and optimization. Technical report, No. 1431, Dep. Elec. Eng., Technion Israel Inst. Tech., 2004.
12. Paul A. Viola. *Alignment by Maximization of mutual information*. PhD thesis, MIT - Artificial Intelligence Lab., 1995.

Soft-LOST: EM on a Mixture of Oriented Lines

Paul D. O'Grady and Barak A. Pearlmutter

Hamilton Institute,
National University of Ireland Maynooth
Co. Kildare, Ireland
paul.ogrady@may.ie, barak@cs.may.ie

Abstract. Robust clustering of data into overlapping linear subspaces is a common problem. Here we consider one-dimensional subspaces that cross the origin. This problem arises in blind source separation, where the subspaces correspond directly to columns of a mixing matrix. We present an algorithm that identifies these subspaces using an EM procedure, where the E-step calculates posterior probabilities assigning data points to lines and M-step repositions the lines to match the points assigned to them. This method, combined with a transformation into a sparse domain and an L_1-norm optimisation, constitutes a blind source separation algorithm for the under-determined case.

1 Introduction

Mixtures of oriented lines arise in sparse separation when a set of observations from N sensors, $\mathbf{X} = (\mathbf{x}(1)|\cdots|\mathbf{x}(T))$, consist of a linear mixture of M source signals, $\mathbf{S} = (\mathbf{s}(1)|\cdots|\mathbf{s}(T))$, by way of an unknown linear mixing process characterised by the $N \times M$ mixing matrix $\mathbf{A}$ via $\mathbf{x}(t) = \mathbf{A}\,\mathbf{s}(t)$. When $N = M$ the sources can be recovered by an unmixing matrix $\mathbf{W}$ where $\hat{\mathbf{s}}(t) = \mathbf{W}\,\mathbf{x}(t)$ and $\hat{\mathbf{s}}(t)$ holds the estimated sources at time t, $\mathbf{W} = \mathbf{A}^{-1}$ up to permutation and scaling of the rows.

When the sources are sparse the mixtures have special structure corresponding to overlaid lines on a scatter plot. For sources of interest in practice (voice, music) a sparse representation can often be achieved by a transformation into a suitable basis such as such as the Fourier, Gabor or Wavelet basis. The line orientations correspond to the columns of the mixing matrix $\mathbf{A}$, so if the lines can be estimated from the data then an estimate of the mixing matrix can be trivially constructed.

An algorithm for identification of radial line orientation and line separation is presented in Section 2. The application of the algorithm to blind source separation (BSS) of speech signals in both the even-determined and under-determined case, along with experimental results including empirical assessments of robustness to noise, are presented in Section 3.

2 Oriented Lines Separation

2.1 Determining Line Orientation Using Data Covariance

The orientation of a linear cloud of data corresponds to the principal eigenvector of its covariance matrix [1, pages 125-132]. In order to identify multiple lines within a scatter plot, we *soft assigned* data into M classes corresponding to the elements of the mixture, represented by orientation vectors $\mathbf{v}_i$ (eq. 1). This calculation corresponds to the *Expectation* step of an EM algorithm [2]. The covariance matrix is then calculated for the data associated with each class (eq. 2) and the principal eigenvector of the matrix is used as the new line orientation vector estimate (eq. 4), in the *Maximisation* step of our EM algorithm. This process is iterated until convergence, at which point the estimated mixing matrix $\hat{\mathbf{A}}$ is constructed by adjoining the estimated line orientations to form the columns of the matrix (eq. 5). We initialised the line orientation vectors randomly in the unit N-sphere by sampling an N-dimensional zero-mean spherical Gaussian.

2.2 Data Point Separation

For the even-determined case ($N = M$) the estimated mixing matrix $\hat{\mathbf{A}}$ is square and the sensor data can be converted to sources using its inverse. When $N < M$, the under-determined case, $\mathbf{A}$ is not invertible so the sources need to be estimated by some other means. To this end, we assume the source coefficients are sparse. One appropriate technique is the *hard assignment* of coefficients using a mask [3, 4]. Another is partial assignment, in which each coefficient is decomposed into more than one source. This is generally done by minimisation of the L_1-norm, which can be seen as a maximum likelihood reconstruction under the assumption that the coefficients are drawn from a distribution of the form $p(c) \propto \exp -|c|$, *i.e.* a Laplacian [5, 6].

2.3 Algorithm Summary

We present an algorithm called *Soft-LOST*, for *Line Orientation Separation Technique*. The prefix "soft" indicates that data points are assigned to lines using a soft assignment where each data point is weighted by proximity to each line. A discussion of hard and soft assignments is presented by Kearns et al. [7]. The algorithm is composed of a soft line orientation estimation subroutine which is called by the separation algorithm.

Soft Line Orientation Estimation

1. Randomly initialise the M line orientation vectors $\mathbf{v}_i$.
2. Partially assign each data point $\mathbf{d}_j$, where $\mathbf{d}_j = \mathbf{x}(j)$, to each line orientation vector using a soft data assignment

$$z_{ij} = \|\mathbf{d}_j - (\mathbf{v}_i \cdot \mathbf{d}_j)\mathbf{v}_i\|^2$$

$$\hat{z}_{ij} = \frac{e^{-\beta z_{ij}}}{\sum_{i'} e^{-\beta z_{i'j}}} \quad (1)$$

where β controls the softness of the boundaries between the regions attributed to each line and $\hat{z}_{ij}$ are the computed weightings of data point j for each line i.

3. Determine the new line orientation estimate by calculating the principal eigenvector of the covariance matrix. The covariance matrix expression (with zero mean) and assignment weightings are combined as follows:

$$\boldsymbol{\Sigma}_i = \frac{\sum_j \hat{z}_{ij} \mathbf{d}_j \mathbf{d}_j^T}{\sum_j \hat{z}_{ij}} \qquad (2)$$

where $\boldsymbol{\Sigma}_i$ is the covariance of weighted data associated with line i. The eigenvector decomposition of $\boldsymbol{\Sigma}_i$ is expressed as:

$$\boldsymbol{\Sigma}_i = \mathbf{U}_i \boldsymbol{\Lambda}_i \mathbf{U}_i^{-1} \qquad (3)$$

The matrix $\mathbf{U}_i$ contains the eigenvectors of $\boldsymbol{\Sigma}_i$ and the diagonal matrix $\boldsymbol{\Lambda}_i$ contains it's associated eigenvalues $\lambda_i \ldots \lambda_N$. The new line orientation vector estimate is the principal eigenvector of $\boldsymbol{\Sigma}_i$ which is expressed as

$$\mathbf{v}_i = \mathbf{u}_{\max} \qquad (4)$$

where $\mathbf{u}_{\max}$ is the principal eigenvector, the eigenvector whose eigenvalue is $\lambda_{\max}$.

Return to step 2 and repeat until the $\mathbf{v}_i$ converge.

4. After convergence adjoin the line orientations estimates to form the estimated mixing matrix.

$$\hat{\mathbf{A}} = [\mathbf{v}_1 | \cdots | \mathbf{v}_M] \qquad (5)$$

Soft-LOST Line Separation Algorithm

1. Perform *soft line orientation estimation* to calculate $\hat{\mathbf{A}}$.
2. For the even-determined case data points are assigned to line orientations using $\mathbf{s}(t) = \hat{\mathbf{A}}^{-1} \mathbf{x}(t)$. For the under-determined case calculate coefficients $\mathbf{c}_j$ using linear programming for each data point j such that

$$\text{minimise } \|\mathbf{c}_j\|_1 \text{ subject to } \hat{\mathbf{A}} \mathbf{c_j} = \mathbf{d_j}$$

The resultant $\mathbf{c}_j$ coefficients, properly arranged, constitute the estimated linear subspaces, $\hat{\mathbf{S}} = [\mathbf{c}_1 | \cdots | \mathbf{c}_T]$.
3. The final result is a $M \times T$ matrix $\hat{\mathbf{S}}$ that contains the line orientation data sets in each row.

3 Experimental Results

The Soft-LOST algorithm was used for a blind source separation problem, where source attenuation vectors correspond to linear subspaces. The Soft-LOST solution to BSS is presented as follows

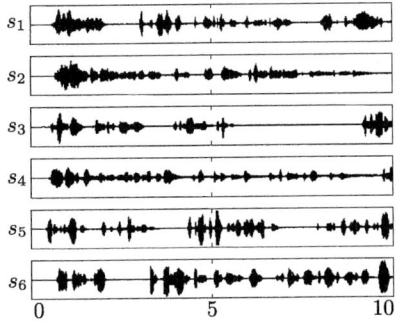

Fig. 1. Ten-second clips of six acoustic sources. Sound wave pressure is plotted against time, in seconds (see Appendix A.).

Soft-LOST for BSS

1. A $N \times T$ data matrix $\mathbf{X}(t)$ is composed of sensor observations of N instantaneous mixtures. The data is transformed into a sparse representation, $\mathbf{X}(t) \mapsto \mathbf{X}(\omega)$.
2. The Soft-LOST algorithm is performed on the data $\mathbf{X}(\omega)$. The algorithm estimates a mixing matrix, which in turn allows sources to be estimated from the mixtures via L_1-norm optimisation.
3. The resultant $M \times T$ matrix $\hat{\mathbf{S}}(\omega)$ contains in its rows the M estimated sources $\hat{\mathbf{s}}_1, \ldots, \hat{\mathbf{s}}_M$. These estimates are then transformed back into the time domain, $\hat{\mathbf{S}}(\omega) \mapsto \hat{\mathbf{S}}(t)$.

3.1 Experimental Method

The Signal-to-Noise Ratios of the estimated sources $\hat{\mathbf{s}}_i$ (in dB) are used to measure the performance of the algorithm, $\mathrm{SNR}_i = 20 \log_{10}(\|\mathbf{s}_i\|/\|\hat{\mathbf{s}}_i - \mathbf{s}_i\|)$.

Speech signals (see Figure 1 and Appendix A) were transformed using a 512-point windowed FFT and the real coefficients were used to create a scatter plot. The experiments were coded for Matlab 6.5.0 and run on a 3.06 GHz Intel Pentium-4 based computer with 768MB of RAM. Experiments for the underdetermined case typically took 35 minutes while the tests for the even-determined case ran for less than six minutes depending on the number of convergence iterations. For comparison the potential performance given a perfect estimate of $\mathbf{A}$ was also evaluated. In these experiments the line orientation estimation phase is skipped and the L_1-norm minimisation phase is tested separately. In general the better defined the line orientations in the scatter plot, the more accurate the source estimates. Experiments were performed for a range of different values of N and M, and the parameter β was varied on an *ad hoc* basis.

3.2 Results

Results are presented for a total of 15 experiments. Data on the number of mixtures, sources used, and the value of the parameter β are contained in the

Table 1. Two Mixtures and Two Sources.

Mixtures	Sources	β	SNR (dB)
2	s_1 s_2	1.5	35.28 43.90
2	s_3 s_4	1.5	41.24 63.32
2	s_5 s_6	1.5	40.30 39.17

Table 2. Five Mixtures and Five Sources.

Mixtures	Sources	β	SNR (dB)
5	s_1 s_2 s_3	6	27.76 24.06 28.31
	s_4 s_5		26.08 28.67
5	s_1 s_2 s_3	6.6	27.95 24.15 28.41
	s_4 s_5		26.2 28.77
5	s_1 s_2 s_3	5.5	27.54 23.96 28.18
	s_4 s_5		25.94 28.54

Table 3. L_1-Norm and True Mixing Matrix.

Mixtures	Sources	SNR (dB)
2	s_1 s_2 s_3	10.41 15.64 7.75
5	s_1 s_2 s_3	20.85 20.62 19.10
	s_4 s_5 s_6	17.08 21.93 48.96

Table 4. Two Mixtures and Three Sources.

Mixtures	Sources	β	SNR (dB)
2	s_1 s_2 s_3	2	10.43 15.58 7.87
2	s_1 s_2 s_3	1.5	10.43 15.58 7.87

Table 5. Five Mixtures and Six Sources.

Mixtures	Sources	β	SNR (dB)
5	s_1 s_2 s_3	6.5	20.17 19.85 18.88
	s_4 s_5 s_6		16.66 21.09 32.19
5	s_1 s_2 s_3	6	20.15 19.83 18.87
	s_4 s_5 s_6		16.65 21.08 32.21

Table 6. Additive Gaussian Noise.

Mixtures	Sources	Noise (dB)	SNR (dB)
2	s_1 s_2	5	34.75 40.05
5	s_1 s_2	15	26.58 23.45
	s_3 s_4		27.02 25.41
	s_5		27.14
5	s_1 s_2	15	17.03 16.78
	s_3 s_4		15.79 13.82
	s_5 s_6		18.27 31.59

tables of results. Results in Tables 1 and 2 demonstrate the effectiveness of the algorithm for the even-determined case. Experiments for testing line separation using L_1-norm minimisation were performed and their results are presented in table 3. These experiments evaluate the effectiveness of the separation phase of the Soft-LOST algorithm in the under-determined case, and provide a benchmark for the subsequent experiments. Results for experiments that test both line orientation estimation and line separation in the under-determined case are presented in Tables 4 and 5. The Soft-LOST algorithm was tested for robustness to noise. Gaussian noise of various intensities was added to the signals of the experiments in Table 6, where the noise introduced to each signal is measured in terms of SNR values. These results, when contrasted with those previously presented, indicate the algorithm's robustness to noise.

The experimental results provided demonstrate that the Soft-LOST algorithm is an effective technique for BSS in both the even-determined and under-

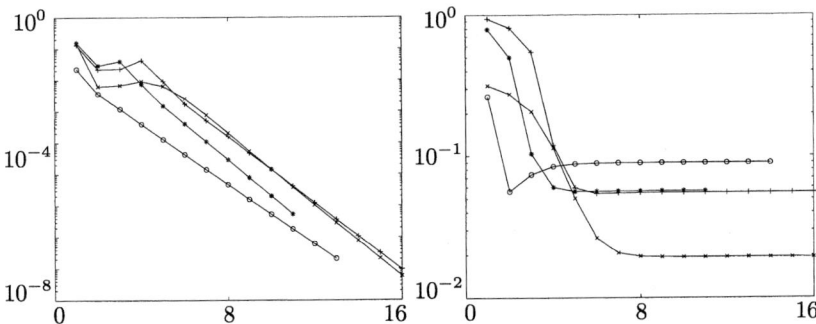

Fig. 2. Convergence plots of estimated mixing matrices for the following experiments; 5 mixtures 6 Sources (∗), 5 mixtures 5 sources (+), 4 mixtures 5 sources (○) and 2 mixtures 3 sources (×). On the left is the difference between consecutive estimates $\|\hat{\mathbf{A}}_l - \hat{\mathbf{A}}_{l-1}\|$, while the right is the difference between the mixing matrix and the current estimate, $\|\mathbf{A}_{\mathrm{orig}} - \hat{\mathbf{A}}_l\|$. The x axis of each plot is in units of algorithm iterations l.

determined case, even in the presence of noise. A plot of convergence illustrating the algorithm convergence properties is provided in Figure 2.

4 Conclusion

The results presented demonstrate that the identification of line orientations using a modified EM procedure is an effective method for determining the mixing matrix of a set of linear mixtures. It has been demonstrated that once the mixing matrix is found, sources can then be separated by minimising the L_1-norm between the data point being considered and the line orientations represented by the columns of the mixing matrix. The Soft-LOST algorithm provides a good solution to blind source separation of instantaneous mixtures even when there are fewer sensors than sources. The experiments presented are concerned with the specific problem of blind source separation of speech signals, however the results can be applied to any situation involving a mixture of oriented lines.

This work follows on from previous research in which we developed a a modified k-means algorithm called *Hard-LOST* [8]. The Soft-LOST results presented here can be contrasted with those of Hard-LOST. In future work, we plan to modify the L_2 norm of the line distance calculation to use the covariance matrix of each line, and to partition the coefficients into classes exhibiting different noise levels to allow optimal combination of evidence using such a noise-sensitive measure.

Acknowledgements

Supported by Higher Education Authority of Ireland (An tÚdarás Um Ard-Oideachas), and Science Foundation Ireland grant 00/PI.1/C067.

References

[1] Juha Karhunen Aapo Hyvärinen and Erkki Oja. *Independent Component Analysis*. John Wiley & Sons, 2001.
[2] A. P. Dempster, N. M. Laird, and D. B. Rubin. Maximum likelihood from incomplete data via the EM algorithm. *Journal of the Royal Statistical Society, Series B*, 39(1):1–38, 1976.
[3] S. T. Rickard and F. Dietrich. DOA estimation of many W-disjoint orthogonal sources from two mixtures using DUET. In *Proceedings of the 10th IEEE Workshop on Statistical Signal and Array Processing (SSAP2000)*, pages 311–314, Pocono Manor, PA, August 2000.
[4] Sam T. Roweis. One microphone source separation. In *Advances in Neural Information Processing Systems 13*, pages 793–799. MIT Press, 2001.
[5] T.-W. Lee, M. S. Lewicki, M. Girolami, and T. J. Sejnowski. Blind source separation of more sources than mixtures using overcomplete representations. *IEEE Signal Processing Letters*, 6(4):87–90, 1999.
[6] Michael Zibulevsky and Barak A. Pearlmutter. Blind source separation by sparse decomposition in a signal dictionary. *Neural Computation*, 13(4):863–882, April 2001.
[7] M. Kearns, Y. Mansour, and A. Y. Ng. An information-theoretic analysis of hard and soft assignment methods for clustering. *Proceedings of the Thirteenth Conference on Uncertainty in Artificial Intelligence*, pages 282–293, 1997.
[8] Paul O'Grady and Barak Pearlmutter. Hard-LOST: Modified k-means for oriented lines. In *Proceedings of the Irish Signals and Systems Conference*, 2004.
[9] Elise Paschen, Charles Osgood, and Rebekah Presson Mosby, editors. *Poetry Speaks: Hear Great Poets Read Their Work from Tennyson to Plath*. Sourcebooks Incorporated, 2001. ISBN 1570717206.

A Source Signals

The source signals were taken from *Poetry Speaks*, a commercial audio CD of poems read by their authors [9]. Audio CD data is recorded as uncompressed 44.1 kHz 16-bit stereo waveforms. Prior to further processing ten-second clips were extracted, the two signal channels were averaged, and the data was downsampled to 8 kHz. The scale of the audio data is arbitrary, leading to the arbitrary units on auditory waveform samples throughout the manuscript.

s_1 *Coole Park and Ballylee*, by William Butler Yeats.
s_2 *The Lake Isle of Innisfree*, by William Butler Yeats.
s_3 *Among Those Killed in the Dawn Raid Was a Man Aged a Hundred*, by Dylan Thomas.
s_4 *Fern Hill*, by Dylan Thomas.
s_5 *Ave Maria*, by Frank O'Hara.
s_6 *Lana Turner Has Collapsed*, by Frank O'Hara.

Some Gradient Based Joint Diagonalization Methods for ICA

Bijan Afsari and Perinkulam S. Krishnaprasad

Institute for Systems Research, University of Maryland
College Park, Maryland 20742, USA
{bijan,krishna}@isr.umd.edu

Abstract. We present a set of gradient based orthogonal and non-orthogonal matrix joint diagonalization algorithms. Our approach is to use the geometry of matrix Lie groups to develop continuous-time flows for joint diagonalization and derive their discretized versions. We employ the developed methods to construct a class of Independent Component Analysis (ICA) algorithms based on non-orthogonal joint diagonalization. These algorithms pre-whiten or sphere the data but do not restrict the subsequent search for the (reduced) un-mixing matrix to orthogonal matrices, hence they make effective use of both second and higher order statistics.

1 Introduction

Simultaneous or Joint Diagonalization (JD) of a set of estimated statistics matrices is a part of many algorithms, especially in the field of ICA and Blind Source Separation (BSS). The early methods developed for JD were those that restrict the joint diagonalizer to belong to the compact Lie group of orthogonal matrices $O(n)$[5]. Accordingly the JD problem is defined as minimization of a function of the form:

$$J_1(\Theta) = \sum_{i=1}^{n} \left\| \Theta C_i \Theta^T - \text{diag}(\Theta C_i \Theta^T) \right\|_F^2 \qquad (1)$$

where $\{C_i\}_{i=1}^{N}$ is the set of symmetric matrices to be diagonalized, $\Theta \in O(n)$ is the joint diagonalizer sought, $\text{diag}(A)$ is the diagonal part of A and $\|A\|_F$ denotes the Frobenius norm of matrix A. We remind the reader that due to compactness of $O(n)$ we know in advance that $J_1(\Theta)$ has a minimum on $O(n)$. Different methods for minimization of this cost function in the context of Jacobi methods [5],[3] and optimization on manifolds [10],[12] have been proposed. Here we shall give a gradient flow expression for this problem, which to our knowledge is referred to in some papers without explicit representation [9].

Non-orthogonal JD is very appealing in the context of noisy ICA. Consider the standard ICA model:

$$\mathbf{x}_{n \times 1} = A_{n \times n} \mathbf{s}_{n \times 1} + \mathbf{n}_{n \times 1} = \mathbf{z} + \mathbf{n} \qquad (2)$$

with **n** a Gaussian noise vector(all random variables are assumed to be of mean zero). We know that if $\{C_i\}_{i=1}^N$ is a collection of matrix slices of cumulant tensor of **x** of order higher than two and B is an un-mixing matrix belonging to the Lie group of non-singular matrices GL(n) then BC_iB^T's are diagonal. The problem of non-orthogonal JD has been addressed by few authors among them: [13] [11] [14]. Defining a suitable cost function for non-orthogonal JD seems to be difficult due to non-compactness of GL(n). In Section (3) we consider extension of J_1 to GL(n) or SL(n) (the group of non-singular matrices with unity determinant) using the scale ambiguity inherent in the ICA problem and we shall derive gradient based continuous flows and their discrete versions for non-orthogonal JD. Although these algorithms are general, they perform much better if the matrix sought is close to orthogonal or a perturbation of the initial condition (the identity matrix in most cases). This is a manifestation of the fact the JD problem is easier to solve on a compact set or locally. Based on this observation, in Section (4) we develop an ICA algorithm based on non-orthogonal JD. These algorithms have the property that although they first sphere the data they do not confine the JD search afterwards to O(n), i.e. they perform non-orthogonal JD after the data is whitened. In Section (5) we present some simulations comparing the performance of the developed ICA algorithms and the celebrated JADE algorithm [5] in noise.

Notation: In the sequel tr(A) is the trace of the matrix A, $\dot{x}$ denotes the time derivative of the variable x, T_pM represents the tangent space to the manifold M at point p and $I_{n\times n}$ is the $n \times n$ identity matrix. All random variables are in boldface small letters and are assumed to be zero mean.

2 Gradient Based Orthogonal JD

Considering O(n) as a Riemannian Lie group with the Riemannian metric defined as $\langle \xi, \eta \rangle_\Theta = \text{tr}((\xi\Theta^T)^T \eta\Theta^T) = \text{tr}(\xi^T\eta)$ for $\xi, \eta \in T_\Theta O(n)$ and, following [8], it is easy to find the gradient flow for minimization of $J_1(\Theta)$ as:

$$\dot{\Theta} = -\Delta\Theta = \sum_{i=1}^{N} \left[\text{diag}(\Theta C_i \Theta^T), \Theta C_i \Theta^T\right]\Theta, \quad \Theta(0) = I_{n\times n} \quad (3)$$

where $[X, Y] = XY - YX$ is the Lie bracket. In [6], a result is given which is essentially the same as (3) but with a different point of view and representation. In discretization of a flow on O(n) it is difficult to ensure that the updates keep the answer always orthogonal. Different methods have been proposed to address this [4], [10], [12]. We mention that in the context of ICA an Euler discretization with small enough fixed step-size, which is equivalent to steepest descent algorithm, is promising.

3 Non-orthogonal JD Based on the Gradient of J_1

Consider a set of symmetric matrices $\{C_i\}_{i=1}^N$ that are assumed to have an exact joint diagonalizer in GL(n). Then the cost function $J_1(B)$ with $B \in$ GL(n) has a minimum of zero. It may seem appropriate to define this as a cost function for JD in the non-orthogonal case. However we can see that this cost function can be reduced by reducing the norm of B. In other words this cost function is not scale-invariant, i.e. $J_1(\Lambda B) \neq J_1(B)$ for non-singular diagonal Λ. By scale-invariance for a JD cost function in terms of un-mixing matrix B, we mean that it does not change under left multiplication of the argument by diagonal matrices in the same manner that mutual information is scale-invariant. In the following we provide some remedies to deal with scale variability of $J_1(B)$.

We consider GL(n) as a Riemannian manifold with the Riemannian metric (also known as Natural Riemannian metric [2]):

$$\langle \xi, \eta \rangle_B = \mathrm{tr}((\xi B^{-1})^T \eta B^{-1}) = \mathrm{tr}(B^{-T} \xi^T \eta B^{-1}) = \mathrm{tr}(\eta (B^T B)^{-1} \xi^T) \quad (4)$$

for $\xi, \eta \in T_B GL(n)$. Again it is easy to see that the gradient flow for minimization of $J_1(B)$ is:

$$\dot{B} = -\Delta B \quad (5)$$

with

$$\Delta = \sum_{i=1}^{N} \left(BC_i B^T - \mathrm{diag}(BC_i B^T) \right) BC_i B^T \quad (6)$$

and $B(0) \in \mathrm{GL}(n)$. An interesting observation is that the equilibria of this flow found by letting $\Delta = 0$ satisfy $BC_i B^T = \mathrm{diag}(BC_i B^T)$ for all $1 \leq i \leq N$. Therefore unless C_i's have an exact joint diagonalizer flow in (5) has no equilibria, which confirms our argument that $J_1(B)$ is not a suitable criterion for non-orthogonal JD. We recall that in [11] a scale invariant cost function is introduced that is applicable only for positive definite C_i's.

One way to ameliorate the problem with non-compactness of GL(n) and scale variability of $J_1(B)$ is to consider minimization of $J_1(B)$ over SL(n). Obviously SL(n) is not a compact group and det(B) = 1 does not put any upper bound on $\|B\|$, but it requires $\|B\|_2 \geq 1$ and this prevents converging to the trivial infimum of $J_1(B)$ at $B = 0$. By restricting B to be in SL(n), we identify all matrices of the form αB for $\alpha \in \mathbb{R} - \{0\}$ with B. It is easy to show that the orthogonal projection of *any* matrix $A_{n \times n}$ on the space of matrices with zero trace is given by: $A^0 = A - \frac{tr(A)}{n} I_{n \times n}$. Accordingly the projection of the gradient flow found in (5) to SL(n) is the gradient flow:

$$\dot{B} = -\Delta^0 B, \quad B(0) = I \quad (7)$$

with Δ as in (6).

A more general way to deal with non-compactness of GL(n) and scale-variability of $J_1(B)$ is to project its gradient on to a subspace such that the projection does not reduce the cost function due to row scaling. This approach maybe considered as equivalent to identifying B and ΛB for all non-singular diagonal Λ. This method leads to a nonholonomic flow [1]. The projected flow is derived from (5) by projecting Δ in (6) to the space of zero diagonal matrices. Letting $\Delta^\perp = \Delta - \mathrm{diag}(\Delta)$, the projected flow is given by:

$$\dot{B} = -\Delta^\perp B, \quad B(0) = I_{n \times n} \quad (8)$$

where Δ is the same as in (6). From the definition of gradient with respect to the Riemannian metric (4) we have, along trajectories of (8):

$$\dot{J}_1 = \text{tr}((\nabla J_1 B^{-1})^T \dot{B} B^{-1}) = -\text{tr}(\Delta^T \Delta^\perp) = -\text{tr}(\Delta^{\perp T} \Delta^\perp) = -\sum_{i \neq j} \Delta_{ij}^2 \leq 0 \tag{9}$$

So, as long as Δ is not diagonal, (8) is a descent flow. Note that the equilibria of the flow (8) is exactly the set $\{B \in GL(n) | \Delta \text{ is diagonal}\}$. On the other hand if $B(0) \in \text{SL}(n)$ then (8) restricts to a flow on $\text{SL}(n)$ and $\|B(t)\|_2 \geq 1$.

By picking small enough step-size we expect to have discretizations of (7) and (8) that decrease the cost function at each step and keep the trajectory on $\text{SL}(n)$ as much as possible. These two flows have the general form:

$$\dot{B} = -XB \tag{10}$$

where X is defined accordingly. The Euler discretization will be:

$$B_{k+1} = (I - \mu_k X_k) B_k, \quad B_0 = I \quad k \geq 0 \tag{11}$$

In practice we can choose a fixed small step-size and change it if we observe instability. A pseudo code for this algorithm is:

Algorithm 1:

1. set μ and ϵ.
2. set $B_0 = I_{n \times n}$ or "*to a good initial guess*".
3. while $\|X_k\|_F > \epsilon$ do
    $B_{k+1} = (I - \mu X_k) B_k$
    if $\|B_{k+1}\|_F$ is "*big*" then "*reduce*" μ and goto 2.
4. end

It is possible to modify the flow (8) such that its discretization yields det $(B_k) = 1$, by construction. Let X^L (X^U) denote a lower (upper) triangular matrix that has the same lower (upper) part as X. Consider the lower-triangular version of (8) $\dot{B} = -\Delta^{\perp L} B$. Note that by the Euler discretization $B_{k+1} = (I - \mu \Delta_k^{\perp L}) B_k$ and if $B_0 = I$ then $\det(B_k) = 1$, by construction. The same is true if we consider the upper triangular version of (8). Therefore based on the LU factorization of the un-mixing matrix we can have an iterative algorithm that alternatively looks for upper and lower triangular factors of the un-mixing matrix and keeps the determinant unity by construction. A pseudo code for this method is:

Algorithm 2:

Consider the set $\{C_i\}_{i=1}^N$ of symmetric matrices. Let (a): $\dot{U} = -\Delta^{\perp U} U$ and (b): $\dot{L} = -\Delta^{\perp L} L$ with $B = U(0) = L(0) = I$ be the corresponding upper and lower triangularized versions of (8).

1. Use Algorithm 1 to find U the solution to (a).
2. set $C_i \leftarrow U C_i U^T$.
3. Use Algorithm 1 to find L the solution to (b)
4. set $C_i \leftarrow L C_i L^T$.
5. set $B \leftarrow L U B$
6. if $\|LU - I\|_F$ is "small" end, else goto 1

4 A Family of ICA Algorithms Based on JD

Here we introduce a general scheme for an ICA algorithm. Consider the data model (2). If we lack information about noise, we use the correlation matrix of **x** instead of that of **z** to find a whitening or sphering matrix W. In this case the sphered signal:

$$\mathbf{y} = W\mathbf{x} = WA\mathbf{s} + W\mathbf{n} = A_1\mathbf{s} + \mathbf{n}_1 \qquad (12)$$

is such that the reduced mixing matrix A_1 can not assumed to be orthogonal as in the noiseless case, however it can assumed to be close to orthogonal where the orthogonality error depends on the signal and noise power and condition number of the matrix A [7]. Note that, by Gaussianity of noise all the higher order cumulant matrix slices of **y** are diagonalizable by A_1. Applicability of the JADE algorithm which jointly diagonalizes a set of fourth order cumulant slices of **y** by an orthogonal matrix will be limited in this case because it reduces the degrees of freedom in the optimization problem involved or in other words leaves the bias introduced in the whitening phase un-compensated. An algorithm such as JADE or mere "sphereing" brings the data (globally) close to independence but we can proceed further by (locally) finding a non-orthogonal un-mixing matrix and reduce mutual information further. This local un-mixing matrix can be incorporated into the whole answer by multiplication due to the multiplicative group structure of the ICA problem. We shall use this idea in developing a new ICA method based on non-orthogonal JD. We emphasize that after whitening although we look for a non-orthogonal joint diagonalizer the fact that it is close to orthogonal makes the search much easier in practice.

Consider the data model (2). The general scheme for ICA based on non-orthogonal JD of fourth (or higher) order cumulant slices is comprised of the following steps:

1. Whiten **x**, let W be a whitening matrix, compute $\mathbf{y} = W\mathbf{x}$ and set $B = W$.
2. Estimate $C = \{C_i\}_{i=1}^{N}$ a subset of the fourth order cumulant matrix slices of **y**.
3. Jointly diagonalize $C = \{C_i\}_{i=1}^{N}$ by an orthogonal matrix Θ and set $C_i \leftarrow \Theta C_i \Theta^T$.
4. Jointly diagonalize $C = \{C_i\}_{i=1}^{N}$ by a non-orthogonal matrix B_{JDN} (using any algorithm such as Algorithms 1 or 2), set $C_i \leftarrow B_{JDN} C_i B_{JDN}^T$ and set $B \leftarrow B_{JDN} \Theta B$.
5. If necessary goto step (3).
6. Compute the recovered signal $\hat{\mathbf{x}} = B\mathbf{x}$.

Steps (1-3) comprise the JADE algorithm. In experiments that the model (2) truly holds, inclusion of step (3) proves to be redundant, but in cases where the model does not hold, compactness of O(n) can be helpful as well as repeating steps (3) and (4). The justification for adopting this scheme is four-fold:

1. Usually by whitening the data the mutual information is reduced so the whitened data is closer to independence.
2. In most cases whitening the data reduces the dynamic range of $\|C_i\|$'s and enables better convergence for numerical methods thereafter.

3. Although estimation of the correlation matrix of **z** in (2) from observation data **x** is biased it has less variance than the estimated higher order cumulant slices (this is pronounced especially in small sample sizes). Therefore it is meaningful to use as much information as possible from this correlation matrix provided we can avoid the harm of the "bias" it introduces.
4. As we mentioned before, solving the ICA or JD problem for **y** is more local than the one for **x**. Also the fact that A_1 in (12) is close to orthogonal makes the non-orthogonal JD of the cumulant slices of **y** instead of those of **x** much more efficient and easier.

In the sequel we consider ICA algorithms that are comprised of steps 1,2,4. An algorithm with its JD part based on the discrete version of flow (7) is referred to as SL(n)-JD, an algorithm with its JD part based on the discrete version of flow (8) is referred to as NH-JD and an algorithm based on the LU factorization (Algorithm 2) is referred to as LU-JD.

5 Simulations

In this section we compare the performance of the developed set of algorithms with the standard JADE in the presence of noise. We consider

$$\mathbf{x} = A\mathbf{s}_{n \times 1} + \sigma \mathbf{n} \tag{13}$$

where **n** is zero mean Gaussian noise with identity correlation matrix then σ^2 indicates the power of noise. We consider $n = 5$ sources. Two of them are uniformly distributed in $[-\frac{1}{2}, \frac{1}{2}]$ and another two are two-side exponentially distributed with parameter $\lambda = 1$ and mean zero and the fifth one is one-side exponential with parameter $\lambda = 1$. The matrix A is randomly generated and to fit in the page the entries are truncated to integers:

$$A = \begin{bmatrix} -4 & 11 & -1 & 1 & 2 \\ -16 & 11 & 7 & 10 & -13 \\ 1 & 0 & -5 & 0 & 7 \\ 2 & 3 & 21 & 0 & 16 \\ -11 & 1 & -1 & -8 & -6 \end{bmatrix}$$

We generate $T = 3500$ samples of data and mix the data through A. Next we run four ICA algorithms. Three algorithms SL(n)-JD, NH-JD and LU-JD in addition to the standard JADE are applied to the data. $N = n^2 = 25$ fourth order cumulant matrix slices are used. For SL(n)-JD and NH-JD $\mu = .01$ and $\epsilon = .01$ are used (see Algorithm 1). For LU-JD $\mu = .05$, $\epsilon = .01$ are used (see Algorithm 2) and the LU iteration is performed five times. These values are not optimal, they were chosen based on few tries. Implementations are in $MATLAB^{\circledR}$ code and the $MATLAB^{\circledR}$ code for JADE was downloaded from: http://tsi.enst.fr/c̃ardoso/icacentral/Algos. The performance measure used is the distance of the product of the estimated un-mixing and the mixing matrix, i.e. $P = BA$, from essential diagonality:

$$\text{Index}(P) = \sum_{i=1}^{n} \left(\sum_{j=1}^{n} \frac{|p_{ij}|}{\max_k |p_{ik}|} - 1 \right) + \sum_{j=1}^{n} \left(\sum_{i=1}^{n} \frac{|p_{ij}|}{\max_k |p_{kj}|} - 1 \right) \tag{14}$$

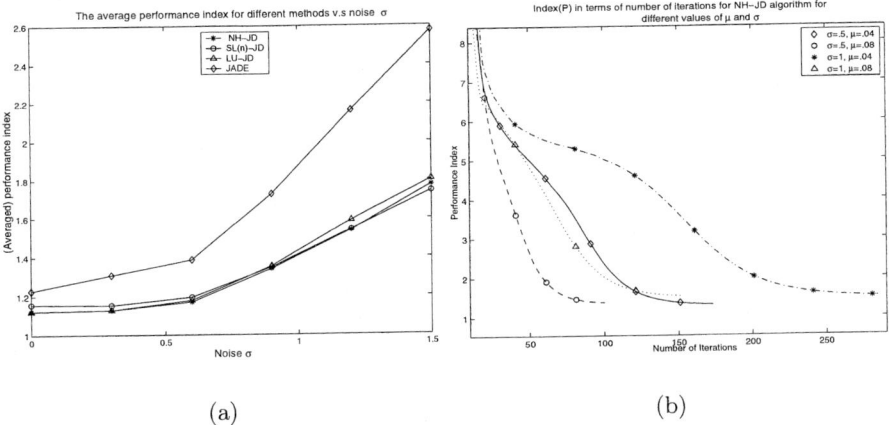

Fig. 1. (a) Average in-noise-performance index (every point is averaged over 100 trials) of different JD based ICA algorithms. The average Index(P) is plotted versus σ. (b) Index(P) in terms of iteration number for NH-JD algorithm for noise $\sigma = 0.5, 1$ and step-size $\mu = 0.04, 0.08$.

For each value of σ the experiment is run $k = 100$ times and the performance measure is averaged over the trials. Figure (1.a) shows the results. We can see that the introduced algorithms all have almost the same performance and outperform the standard JADE especially in high level Gaussian noise. In Figure (1.b) we consider the behavior of Index(P) in terms of number of iterations in the JD part of the NH-JD algorithm for a single realization of data generated as above. Two different values of $\sigma = .5, 1$ and step-size $\mu = .04, .08$ are examined. All the initial conditions in the JD part (i.e. Algorithm 1) are $B_0 = I_{5\times 5}$. As the figure illustrates more iterations are required as noise power increases. By increasing step-size one may combat this, however dramatic increase of μ may result in instability of the algorithm. The run-time for these algorithms (in $MATLAB^\circledR$ code) is higher than JADE's, although we expect faster performance in low-level codes or DSPs. Part of this slower convergence can be attributed to the nature of gradient based methods which have linear convergence. One idea for speed improvement can be to use the result from JADE as the initial condition for the non-orthogonal JD methods introduced here.

6 Conclusion

We introduced gradient based flows for orthogonal and non-orthogonal JD of a set symmetric matrices and developed a family of ICA algorithms upon non-orthogonal JD. The non-orthogonal flows are derived based on defining suitable metrics and the geometry of the groups GL(n) and SL(n). The main drawback of gradient based JD methods is their slow convergence but their implementation requires only addition and multiplication. The developed ICA algorithms have the property that after whitening the data they do not confine the search space to orthogonal matrices. This way we can take advantage of both second order statistics (which has less variance) and higher order statistics which are blind to Gaussian noise. Numerical simulations show better performance for the proposed algorithms than for the standard JADE algorithm in Gaussian noise.

Acknowledgments

This research was supported in part by Army Research Office under ODDR&E MURI01 Program Grant No. DAAD19-01-1-0465 to the Center for Communicating Networked Control Systems (through Boston University).

References

1. S.Amari, T.-P Chen, A. Chichoki, Non-holonomic constraints in learning algorithms for blind source separation, preprint, 1997.
2. S. Amari: Natural Gradient Adaptation. In S. Haykin (ed): Unsupervised Adaptive Filtering, Vlume I, Blind Source Separation, Wiley Interscience, 2000.
3. A. Bunse-Gerstner, R. Byers and V. Mehrmann: Numerical Methods For Simultaneous Diagonalization, SIAM Journal on Matrix Analysis and Applications, vol. 4, pp. 927-949, 1993.
4. M.P. Calvo, A. Iserles and A. Zanna: Runge-Kutta methods for orthogonal and isospectral flows, Appl. Num. Maths 22 (1996)
5. J.F. Cardoso and A. Soulumiac: Blind Beamforming For Non-Gauusian Signals, IEE-Proceedings, Vol.140, No 6, Dec 1993
6. J.F. Cardoso: Perturbation of joint diagonalizers, Technical report, Telecom Paris, 1994.
7. J.F. Cardoso: On the performance of orthogonal source separation algorithms, EUSIPCO-94, Edinburgh.
8. U.Helmke and J.B. Moore: Optimization and Dynamical Systems, Springer-Verlag, 1994
9. G. Hori, J.H. Manton: Critical Point Analysis of Joint Diagonalization Criteria, 4th International Symposium on Independent Component Analysis and Blind Signal Separation (ICA2003), April 2003, Nara, Japan
10. J.H Manton: Optimization Algorithms Exploiting Unitary Constraints, IEEE Transactions on Signal Processing, Vol. 50, No. 3, March 2002
11. D.T. Pham: Joint Approximate Diagonalization of Positive Definite Hermitian Matrices, SIAM Journal of Matrix Analysis and Applications, Vol. 22, No. 4, pp. 136-1152.
12. I. Yamada, T. Ezaki: An Orthogonal Matrix Optimization by Dual Cayley Parametrization Technique, 4th International Symposium on Independent Component Analysis and Blind Signal Separation (ICA2003), April 2003, Nara, Japan
13. A.Yeredor: Non-Orthogonal Joint Diagonalization in the Least-Squares Sense With Application in Blind Source Separation, IEEE Transactions on Signal Processing, Vol 50, No.7.July 2002.
14. A.Ziehe, P. Laskov, K. Muller, G. Nolte: Linear Least-squares Algorithm for Joint Diagonalization, 4th International Symposium on Independent Component Analysis and Blind Signal Separation (ICA2003), Nara, Japan, April 2003.

Underdetermined Independent Component Analysis by Data Generation

Sang Gyun Kim and Chang D. Yoo

Department of Electrical Engineering and Computer Science
Korea Advanced Institute of Science and Technology
Guseong-dong, Yuseong-gu, Daejon, Republic of Korea

Abstract. In independent component analysis (ICA), linear transformation that minimizes the dependence among the components is estimated. Conventional ICA algorithms are applicable when the numbers of sources and observations are equal; however, they are inapplicable to the underdetermined case where the number of sources is larger than that of observations. Most underdetermined ICA algorithms have been developed with an assumption that all sources have sparse distributions. In this paper, a novel method for converting the underdetermined ICA problem to the conventional ICA problem is proposed; by generating hidden observation data, the number of the observations can be made to equal that of the sources. The hidden observation data are generated so that the probability of the estimated sources is maximized. The proposed method can be applied to separate the underdetermined mixtures of sources without the assumption that the sources have sparse distribution. Simulation results show that the proposed method separates the underdetermined mixtures of sources with both sub- and super-Gaussian distributions.

1 Introduction

In independent component analysis (ICA), linear transformation to minimize the statistical dependence of the components of the representation is estimated. Recently, blind source separation by ICA has received great deal of attention because of its potential in speech enhancement, telecommunication, and medical signal processing.

In ICA, the objective is to find an $M \times M$ invertible square matrix $\mathbf{W}$ such that

$$\mathbf{s} = \mathbf{W}\mathbf{x} \qquad (1)$$

where $\mathbf{s}$ and $\mathbf{x}$ are respectively $M \times 1$ source signal and $M \times 1$ observation, and the components of $\mathbf{s} = \{s_1, s_2, \ldots, s_M\}^T$ are as *independent* as possible. In other words, the jth component x_j of $\mathbf{x}$ can be interpreted as a linear combination of the independent sources since

$$\mathbf{x} = \mathbf{W}^{-1}\mathbf{s} = \mathbf{A}\mathbf{s} \qquad (2)$$

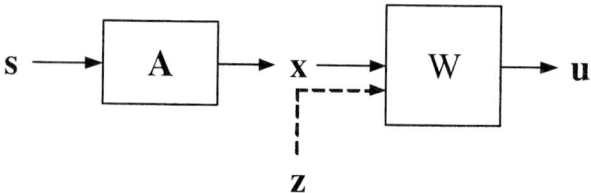

Fig. 1. Block diagram of underdetermined ICA by data generation.

where $\mathbf{A}$ is an $M \times M$ square mixing matrix, and $\mathbf{A} = \mathbf{W}^{-1}$. Therefore, the goal of ICA is to estimate the mixing matrix $\mathbf{A}$ and therefore find the independent sources $\mathbf{s}$ given only the observations $\mathbf{x}$.

Infomax algorithm based on entropy maximization has been developed by Bell and Sejnowski [1]. This algorithm is effective in separating sources with super-Gaussian distribution. However, it fails to separate sources with sub-Gaussian distribution. To separate the mixtures of super- and sub-Gaussian sources, Xu et al. [2] and Attias [3] modelled the underlying probability density function (*pdf*) of sources as a mixture of Gaussians. However, these algorithms are computationally expensive. To simplify the computation and to separate the mixtures of super-Gaussian and sub-Gaussian sources, an extended infomax algorithm was proposed by Lee et al. [4].

Conventional ICA algorithms are inapplicable to an underdetermined case where the number of sources is larger than that of observations, that is, the mixing matrix $\mathbf{A}$ is an $m \times M$ matrix with $m < M$. The underdetermined ICA problem is generally more difficult to tackle than the conventional ICA problem where the number of sources is equal to that of observations, since some of the observation data are hidden in the underdetermined case. Even if the mixing matrix $\mathbf{A}$ is estimated exactly, the sources $\mathbf{s}$ can not be found directly, but have to be inferred [5]. The overcomplete representation and sparse coding were studied by Olshausen and Field [6] and were later developed as learning overcomplete representations for ICA by Lewicki and Sejnowski [5]. This method was applied to blind separation of speech signals in the underdetermined case by Lee et al. [7]. However, methods such as these are based on an assumption that the distribution of the source is sparse. Therefore, if the assumption is not valid, these methods are not effective. When a source does not satisfy this assumption, a method for achieving the sparsity in a sparser transformed domain, such as by short-time Fourier transform [8] or by wavelet packet transform [9], was proposed. However, the method does not guarantee sparsity since achieving sparsity depends on the distributions of the sources.

In this paper, a novel method for converting the underdetermined ICA problem to the conventional ICA problem by generating the hidden observation data $\mathbf{z}$, as shown in Fig. 1, is proposed. The hidden data $\mathbf{z}$ is generated so that the conditional probability of the hidden data $\mathbf{z}$ given the observation $\mathbf{x}$ and the unmixing matrix $\mathbf{W}$ is maximized. The observation data $\mathbf{x}$ and the hidden data $\mathbf{z}$ make up a complete data $\mathbf{y}$ that is defined as

$$\mathbf{y} \equiv \begin{bmatrix} \mathbf{x} \\ \mathbf{z} \end{bmatrix} \qquad (3)$$

where $\mathbf{y} \in \mathbb{R}^M$ and $\mathbf{z} \in \mathbb{R}^{M-m}$. With the complete data $\mathbf{y}$, conventional ICA algorithms can be applied to estimate the sources in the underdetermined case. In order to separate the mixtures of sources that have both sub- and super-Gaussian distribution, the hyperbolic-Cauchy density model in [4], which can describe both super- and sub-Gaussian distribution, is used to model the *pdf* of source. The proposed method does not require the assumption that the source distribution is sparse since the learning of the square unmixing matrix $\mathbf{W}$ is performed based on the extended infomax ICA algorithm [4].

This paper is organized as follows. Section 2 presents the extended infomax algorithm proposed by Lee et al. [4]. Section 3 presents the proposed underdetermined ICA algorithm by data generation. Section 4 shows the simulation results. Section 5 discusses the problem of the proposed method, and Section 6 concludes the paper.

2 The Extended Infomax Algorithm

An unsupervised learning algorithm based on entropy maximization was proposed by Bell and Sejnowski [1]. This algorithm is effective in separating sources that have super-Gaussian distribution. However, it fails to separate sources that have sub-Gaussian distribution. In order to separate the mixtures of super-Gaussian and sub-Gaussian sources, an extended infomax algorithm is proposed preserving the simple architecture of infomax algorithm by Lee et al. [4]. It provides a simple learning rule with a parametric density model that can have various distributions by changing the value of a parameter. One proposed parametric density that may be used to model both sub- and super-Gaussian source data s is given as

$$p_s(s) = \frac{1}{4}\{sech^2(s+b) + sech^2(s-b)\} \qquad (4)$$

where b is a constant. Depending on the value that b takes, $p_s(s)$ can model either sub- or super-Gaussian distribution. For example, when $b=0$, the parametric density is proportional to the hyperbolic-Cauchy distribution and therefore is suited for separating super-Gaussian distributions. When $b=2$, it has a bimodal distribution with negative kurtosis and therefore is suited for separating sub-Gaussian distributions. Switching between the sub- and super-Gaussian is determined according to the sufficient condition that guarantees asymptotic stability [10].

3 Underdetermined ICA by Data Generation

When the number of sources is larger than that of observations, it is difficult to estimate the sources given only the observations. In this section, sparse representation for the underdetermined ICA is briefly reviewed [5], [7], [8], [9], and a novel algorithm for underdetermined ICA when the sources have the super- and sub-Gaussian distributions is proposed.

3.1 Underdetermined ICA Using Sparse Representations

In the underdetermined ICA model, the sources should be inferred even if the mixing matrix, **A**, is known. There are infinitely many solutions to **s**. If the source distribution is sparse, the mixing matrix can be estimated by either external optimization or clustering and, given the mixing matrix, a minimal l_1-norm representation of the sources can be obtained by solving a low-dimensional linear programming problem [5], [7], [8], [9]. In these algorithms, even in the case when the mixing matrix is known, high sparsity is required for good separability. Therefore, these algorithms are not effective in separating the mixtures of the sources, anyone of which has a sub-Gaussian distribution.

3.2 Underdetermined ICA by Data Generation

In this paper, the objective is to separate the underdetermined mixtures of sources that have sub- and super-Gaussian distributions. To achieve this, a novel method for converting the underdetermined ICA problem to the conventional ICA problem by generating hidden observations **z** is proposed. The hidden data **z** is generated by maximizing the conditional probability of the hidden data **z** given the observation, **x**, and the unmixing matrix, **W**. It is given as following

$$\mathbf{z} = arg\ \underset{\mathbf{z}}{max}\ \log p(\mathbf{z}|\mathbf{x},\mathbf{W}) \tag{5}$$

$$= arg\ \underset{\mathbf{z}}{max}\ \log \frac{p(\mathbf{z},\mathbf{x}|\mathbf{W})}{p(\mathbf{x}|\mathbf{W})} \tag{6}$$

$$= arg\ \underset{\mathbf{z}}{max}\ \log p_\mathbf{s}(\mathbf{Wy})|\det \mathbf{W}| \tag{7}$$

$$= arg\ \underset{\mathbf{z}}{max}\ \sum_i^M \log p_{s_i}(\mathbf{w}_i\mathbf{y}) \tag{8}$$

where $\mathbf{w}_i$ is the ith row of the unmixing matrix **W** and **y** is given in (3). From (8), the generation of the hidden data is performed such that the summation of the log-probabilities of the estimated sources is maximized.

After generating the hidden data **z**, as shown in Fig. 1, the sources are estimated as a linear product of **W** and **y** defined in (3) as in the case of conventional ICA algorithms. This is mathematically represented by

$$\mathbf{u} = \mathbf{Wy} \tag{9}$$

where **W** is an $M \times M$ unmixing matrix and **u** are estimated sources.

In order to generate the hidden data well, the probability density of the sources has to be estimated with good precision. In addition, the density estimate of the source plays an important role in the performance of the learning rule of the mixing matrix. To achieve this, the parametric density of (4) is used to model the source distribution. The $M \times 1$ parameter $\mathbf{b}$ representing column vector of constant b in (4) should be updated so as to match the parametric density function to the source density function. Therefore, the learning rule for $\mathbf{b}$ is given as

$$\Delta \mathbf{b} \propto \frac{\partial \log p_s(\mathbf{u}|\mathbf{b})}{\partial \mathbf{b}} \qquad (10)$$

to maximize the log-likelihood of the source.

Next, the learning algorithm for the unmixing matrix $\mathbf{W}$ for sub- and super-Gaussian sources is

$$\Delta \mathbf{W} \propto [\mathbf{I} + 2\tanh(\mathbf{u})\mathbf{u}^T - 2\tanh(\mathbf{u}+\mathbf{b})\mathbf{u}^T - 2\tanh(\mathbf{u}-\mathbf{b})\mathbf{u}^T]\mathbf{W} \qquad (11)$$

which was given in [4].

Therefore, the underdetermined ICA algorithm by data generation is summarized as follows. First, the unmixing matrix $\mathbf{W}$ and the parameter $\mathbf{b}$ of the source density is initialized, respectively. After initialization, the hidden data $\mathbf{z}$ is generated to maximize the summation of the log-probabilities of the estimated sources according to (8) given the observations $\mathbf{x}$ and the unmixing matrix $\mathbf{W}$. After generating the hidden data, the source $\mathbf{s}$ is estimated according to (9) and then the parameter $\mathbf{b}$ and the unmixing matrix $\mathbf{W}$ is updated according to (10) and (11), respectively. Finally, at next iteration, we start again from the data generation using $\mathbf{W}$ and $\mathbf{b}$ of the previous step.

4 Simulation Results

In this section, simulation results are shown to verify the performance of the underdetermined ICA algorithm by data generation for the 2×3 underdetermined case. In Example 1, the performances of two algorithms to separate the underdetermined mixtures of 2 sources of super-Gaussian distributions and 1 source of sub-Gaussian distribution are compared. One algorithm is the proposed underdetermined ICA algorithm by data generation and the other is the underdetermined ICA algorithm based on the minimum l_1-norm solution using the linear programming [5]. In Example 2, it is shown that the proposed algorithm can separate the mixtures of two speech signals and one sub-Gaussian signal.

In all experiments, a same mixing matrix $\mathbf{A}$ is used, which is given as

$$\mathbf{A} = \begin{bmatrix} 1/\sqrt{(2)} & -1/\sqrt{(2)} & 1 \\ 1/\sqrt{(2)} & 1/\sqrt{(2)} & 0 \end{bmatrix}. \qquad (12)$$

The problem of generating the hidden data is solved using a nonlinear optimization subroutine in MATLAB. The hidden data generation based on (8) is

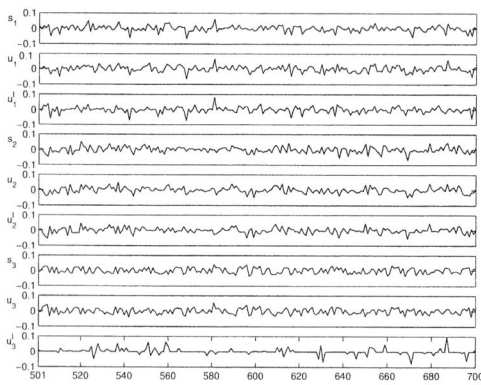

Fig. 2. Separation of the mixtures of the sources that have super- and sub-Gaussian distributions using the proposed and Lewicki's algorithm.

Table 1. Performance comparison between the proposed method and linear programming method.

Source	Original kurtosis	Proposed method		Lewicki's method	
number	κ_i^o	κ_i	$corr_i$	κ_i^l	$corr_i^l$
1	1.16	0.63	0.74	1.65	0.66
2	2.01	0.45	0.76	1.04	0.78
3	-1.34	-0.88	0.92	9.30	0.04

actually performed so that $-\sum_i^M \log p_{s_i}(\mathbf{w}_i \mathbf{y})$ is minimized using the nonlinear optimization (minimization) function in MATLAB.

Example 1: The simulation of separating the 2×3 underdetermined mixtures of the sources that have different distributions is performed. Two sources s_1 and s_2 have the super-Gaussian distributions, and the other s_3 has the sub-Gaussian distribution. The super- and sub-Gaussian sources that are used in the simulation are generated from the hyperbolic-Cauchy density model of (4); $b=0$ for super-Gaussian distribution and $b=2$ for sub-Gaussian, respectively. Some data of length 3000 is used in the learning process that iterates 10 times. The batch size is 100. A batch hidden data are generated one sample at a time using same unmixing matrix for that batch. The unmixing matrix and the density parameter are updated every batch. The learning rates for the unmixing matrix and the density parameter are 0.001 and 0.001, respectively.

In Fig. 2, the simulation result using the proposed and Lewicki's algorithm is shown after reordering and rescaling. In Fig. 2, s_i, u_i, and u_i^l represent the ith original source, the estimate of s_i using the proposed algorithm, and the estimate of s_i using the algorithm proposed by Lewicki et al. for $i=1, 2,$ and 3, respectively. The sources that have super-Gaussian distributions are estimated to some extent in both algorithms, however, the source of sub-Gaussian distribution is estimated well only when using the proposed algorithm as expected.

In Table 1, the simulation results are summarized; κ_i^o, κ_i, and κ_i^l represents the kurtosis of the ith original source, estimated source using the proposed

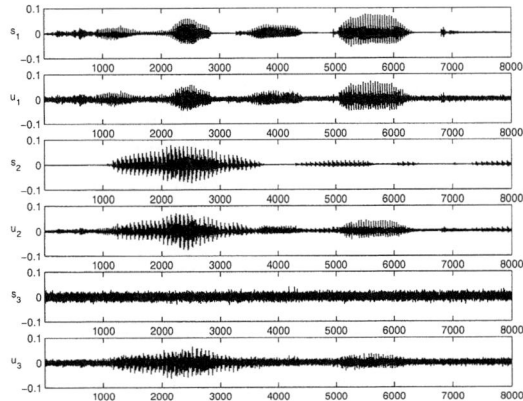

Fig. 3. Separation of the mixtures of two speech signals with super-Gaussian distribution and one noise with sub-Gaussian distribution.

method, and estimated source using Lewicki's method, respectively, and $corr_i$ and $corr_i^l$ represent the correlation coefficient between the original source signal and the estimated source signal using the proposed method and Lewicki's method after reordering, respectively. As shown in Table 1, it is also verified that both methods can estimated the super-Gaussian sources; however, the method proposed by Lewicki et al. fails to separate the sub-Gaussian source. The kurtosis κ_3^l of the estimated source u_3^l is positive, and the correlation coefficient $corr_3^l$ between the sub-Gaussian signal s_3 and the estimated signal u_3^l is very small.

Example 2: Finally, the proposed algorithm is applied to separate the mixtures of two speech signals and one noise that have sub-Gaussian distribution. Fig. 3 shows the separation results. It is shown that the proposed algorithm separates two speech signals and the sub-Gaussian noise to some extent as in Example 2.

5 Discussion

The proposed method is based on a parametric density model. Therefore, when the parametric density model of (4) does not describe the source densities well, e.g., speech density, the performance of the proposed method is degraded. That is also verified in Example 2. In order to solve this problem, nonparametric density estimation method in [11] can be applied to this method. In that case, the generation equation of (8) and the learning rule of (11) should be modified based on the estimated nonparametric density. Further study is need to obtain an underdetermined ICA algorithm using nonparametric density estimation.

6 Conclusion

A novel method for applying the extended infomax algorithm to the underdetermined ICA model is proposed. This is achieved by converting the underdetermined ICA problem to the conventional ICA problem by generating the hidden observation data. The hidden data are generated to maximize the summation of the log-probabilities of the estimated sources. The simulation results show that the proposed algorithm can separate the underdetermined mixtures of the sources that have sub- and super-Gaussian distributions. However, further study is needed to determine until what dimensionality of the hidden data the proposed algorithm is effective and modify this algorithm to be nonparametric.

Acknowledgement

This work was supported in part by grant No. R01-2003-000-10829-0 from the Basic Research Program of the Korea Science and Engineering Foundation and by University IT Research Center Project.

References

1. Bell, A.J. and Sejnowski, T.J.: An information-maximisation approach to blind separation and blind deconvolution. Neural Computation, Vol. 7, No. 6. (1995) 1129–1159
2. Xu, L., Cheung, C., Yang, H., and Amari, S.: Maximum equalization by entropy maximization and mixture of cumulative distribution functions. Proceedings of ICNN. Houston. (1997) 1821–1826
3. Attias, H.: Independent factor analysis. Neural Computation. Vol. 11. (1999) 803–852
4. Lee, T.W., Girolami, M., and Sejnowski, T.J.: Independent component analysis using an extended infomax algorithm for mixed sub-Gaussian and super-Gaussian sources. Neural Computation. Vol. 11, No. 2. (1999) 409–433
5. Lewicki, M.S. and Sejnowski, T.J.: Learning overcomplete representations. Neural Computation. Vol. 12. (2000) 337–365
6. Olshausen, B.A. and Field D.J.: Sparse coding with an overcomplete basis set: A strategy employed by V1?. Vision Research. Vol.11. (1997) 3311–3325
7. Lee, T.W., Lewicki, M.S., Girolami, M., and Sejnowski, T.J.: Blind source separation of more sources than mixtures using overcomplete representation. IEEE Signal Processing Letters. Vol. 6, No. 4. (1999) 87–90
8. Bofill, P. and Zibulevsky, M.: Underdertermined blind source separation using sparse representations. Signal Processing. Vol. 81. (2001) 2353–2362
9. Li, Y., Cichocki, A., and Amari, S.I.: Sparse component analysis for blind source separation with less sensors than sources. 4th International Symposium on Independent Component Analysis and Blind Signal Separation. Nara Japan (2003) 89–94
10. Cardoso, J.F.: Blind signal processing: statistical principles. Proceedings of the IEEE. (1998) 2009–2025
11. Boscolo, R., Pan, H., and Roychowdhury, V. P.: Independent component analysis based on nonparametric density estimation. IEEE transaction on neural networks, Vol. 15, No. 1. (2004) 55–65

Batch Mutually Referenced Separation Algorithm for MIMO Convolutive Mixtures

Ali Mansour

Lab. E^{3}I^2, ENSIETA
29806 Brest cedex 09, France
mansour@ieee.org
http://www.ensieta.fr
http://ali.mansour.free.fr

Abstract. This paper deals with the blind separation problem of Multi-Input Multi-Output (MIMO) convolutive mixtures. Previously, we presented some algorithms based on mutually referenced criterion to separate MIMO convolutive mixtures. However, the proposed algorithms are time consuming and they need a lot of computation efforts. It is obvious that the computation efforts can be reduced as well the convergence time when the adaptive algorithms are well initialized. To choose the best starting point of these algorithms, we propose here a direct and batch minimization of the proposed criteria.

1 Introduction

The blind separation of sources (BSS) (or the Independent Component Analysis "ICA") problem consists of the estimation of the unknown input signals of an unknown channel using only the output signals of that channel (i.e., the observed signals or the mixing signals) [1,2]. The sources are assumed to be statistically independent from each other [3]. Recently, that problem has been addressed and applied in many different situations [4] such as speech enhancement [5], separation of seismic signals [6], sources separation method applied to nuclear reactor monitoring [7], airport surveillance [8], noise removal from biomedical signals [9], and some radar applications have been addressed in [10]. Generally, the transmission channel is considered to be a memoryless channel (i.e., the case of an instantaneous mixture) or a matrix of linear filters (i.e. in the convolutive mixture case). Since 1985, many algorithms have been proposed to solve the ICA problem [11–15]. The criteria of those algorithms have generally been based on high-order statistics [16–18]. Recently, by using only second-order statistics, some subspace methods that blindly separate the sources in the case of convolutive mixtures have been explored [19, 20].

In previous work, we proposed two subspace approaches using LMS [20, 21] or a conjugate gradient algorithm [22] to minimize subspace criteria. Those criteria were derived from the generalization of the method proposed by Gesbert *et*

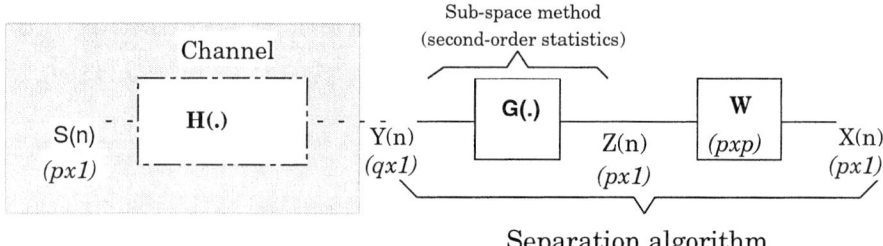

Fig. 1. General Structure.

al. [23] for blind identification[1]. The main advantage of such algorithms is that one can identify almost all parameters of the mixing filters using only second order statistics. The major drawbacks of the previous proposed algorithms are computation efforts and time consuming algorithms. In fact, the convergence of these algorithms is very slow and it can be improved by better choice of algorithm initialization parameters. In the following we discuss some initialization scenarios.

2 Channel Model, Assumptions and Background

Let $S(n)$ denotes the p unknown sources which are statistically independent from each other. $Y(n)$ is the $q \times 1$ observed vector, see Fig. 1. The relationship between $S(n)$ and $Y(n)$ is given by :

$$Y(n) = [\mathcal{H}(z)]S(n) \qquad (1)$$

where $\mathcal{H}(z)$ stands for the channel effect. In the case of convolutive mixture, $\mathbf{H}(z) = (h_{ij}(z))$ becomes a $q \times p$ complex polynomial matrix. In the following, we consider that the channel is a linear and causal one and that the coefficients $h_{ij}(z)$ are RIF filter. Let M denotes the degree of the channel which is the highest degree of $h_{ij}(z)$. The previous equation (1) can be rewritten as:

$$Y(n) = \sum_{i=0}^{M} \mathbf{H}(i)S(n-i) \qquad (2)$$

Here $\mathbf{H}(i)$ denotes the $q \times p$ real constant matrix corresponding to the impulse response of the channel at time i and $S(n-i)$ is the $p \times 1$ source vector at time $(n-i)$. Considering $(N+1)$ observations of the mixture vector $(N > q)$ and using the following notations:

$$Y_N(n) = \begin{pmatrix} Y(n) \\ \vdots \\ Y(n-N) \end{pmatrix} \quad \text{and} \quad S_{M+N}(n) = \begin{pmatrix} S(n) \\ \vdots \\ S(n-M-N) \end{pmatrix}, \qquad (3)$$

---

[1] In the identification problem, the authors generally assume that there is only one source and that the source is an independent and identically distributed (iid) signal.

model (2) can be rewritten as:

$$Y_N(n) = \mathbf{T}_N(\mathbf{H})S_{M+N}(n), \qquad (4)$$

where the $q(N+1) \times p(M+N+1)$ matrix $\mathbf{T}_N(\mathcal{H})$ is the Sylvester matrix corresponding to $\mathcal{H}(z)$. In reference [24], the Sylvester matrix is given by:

$$\mathbf{T}_N(\mathbf{H}) = \begin{bmatrix} \mathbf{H}(0) & \mathbf{H}(1) & \mathbf{H}(2) & \cdots & \mathbf{H}(M) & 0 & 0 & \cdots & 0 \\ 0 & \mathbf{H}(0) & \mathbf{H}(1) & \cdots & \mathbf{H}(M-1) & \mathbf{H}(M) & 0 & \ddots & \vdots \\ \vdots & \ddots & \ddots & \ddots & \ddots & \ddots & \ddots & \ddots & 0 \\ 0 & \cdots & \cdots & \cdots & 0 & \mathbf{H}(0) & \mathbf{H}(1) & \cdots & \mathbf{H}(M) \end{bmatrix}. \qquad (5)$$

Under some mild and realistic assumptions [19] (i.e the number of sensors is larger than the number of sources $q > p$ and H(z) is a column-reduced irreducible matrix), we proposed in [20] a subspace separation algorithm based on the identification algorithm proposed by Gesbert et al. in [23, 25]. That algorithm consists of two steps:

- A blind identification step is done by minimizing a second order criterion

$$\mathcal{C}(\mathbf{G}) = \mathrm{E}\, \|(\mathbf{I}\ \ \mathbf{0})\mathbf{G}Y_N(n) - (\mathbf{0}\ \ \mathbf{I})\mathbf{G}Y_N(n+1)\|^2. \qquad (6)$$

Here, E stands for the expectation, $\mathbf{G}$ denotes a $p(M+N+1) \times q(N+1)$ real matrix and $\mathbf{I}$ is the $(M+N)p \times (M+N)p$ identity matrix. It has been shown in [20] that the above minimization leads to a matrix $\mathbf{G}^\star$ such:

$$\mathbf{Perf} = \mathbf{G}^\star\ \mathbf{T}_N(\mathbf{H}) = \mathrm{diag}(\mathbf{A}, \cdots, \mathbf{A}), \qquad (7)$$

where **Perf** denotes the performance matrix and $\mathbf{A}$ is any $p \times p$ matrix.
- It is clear that the first step does not achieve the separation due to fact that the residual matrix $\mathbf{A}$ isn't a general permutation matrix [3]. Therefore, one should apply any high order statistics BSS algorithm of instantaneous mixtures (we used different algorithms as [13, 26]).

Later on, we derived another algorithm based on the same criterion [27]. In [27], we proved the type and the uniqueness of solution (7). We should mention that equation (7) can lead us to an spurious solution $\mathbf{G}$. To avoid such solution, the minimization of equation (6) has be done with respect to a consistent constraint, please see [27]:

$$\mathbf{G}_1 \mathbf{R}_Y(n) \mathbf{G}_1^T = \mathbf{I}_p, \qquad (8)$$

where $\mathbf{G}_i$ is the ith block row $p \times q(N+1)$ of $\mathbf{G}$, $\mathbf{R}_Y(n) = EY_N(n)Y_N(n)^T$ is the covariance matrix of $Y_N(n)$ and $\mathbf{I}_p$ is a $p \times p$ identity matrix. If the above constraint is satisfied and $\mathbf{G}_1$ is such that $\mathbf{G}_1 \mathbf{Y}_N(n) = \mathbf{A}S(n)$, then:

$$\mathbf{G}_1 \mathbf{R}_Y(n) \mathbf{G}_1^T = \mathbf{A}\mathbf{R}_S(n)\mathbf{A}^T = \mathbf{I}_p, \qquad (9)$$

where $\mathbf{R}_S(n) = ES(n)S(n)^T$ is the source covariance matrix. $\mathbf{R}_S(n)$ is a full rank diagonal matrix as a result of the statistical independence of the p sources from each other. When equation (9) is satisfied, matrix $\mathbf{A}$ becomes invertible.

3 Batch Algorithm

To improve the performance and the convergence speed of the previous proposed algorithms [20, 27], one can at first simplify the constraint. Actually, the proposed constraint (8) is equivalent to p equations. That constraint can be easily change to one equation constraint such as:

$$||\mathbf{G}_1\mathbf{R}_Y(n)\mathbf{G}_1^T|| = 1 \tag{10}$$

Another constraint can also be derived as:

$$\det\{\mathbf{G}_1\mathbf{R}_Y(n)\mathbf{G}_1^T\} = 1 \tag{11}$$

One should mention that the last two equations (10) and (11) can avoid spurious solutions as well as done the first constraint. In addition, the constant "1" used in both equations can be change to any other positive number since that will be reduced to a simple normalization of the residual matrix $\mathbf{A}$.

3.1 Criterion Minimization

Using the criterion derivative found in [20], we can prove that the minimization of the criterion is equivalent to the following matrix equation system:

$$\begin{aligned}
\mathbf{G}_1\mathbf{R}_Y(n) &= \mathbf{G}_2\mathbf{R}_Y^T(n+1) \\
2\mathbf{G}_2\mathbf{R}_Y(n) &= \mathbf{G}_3\mathbf{R}_Y^T(n+1) + \mathbf{G}_1\mathbf{R}_Y(n+1) \\
&\vdots \\
2\mathbf{G}_i\mathbf{R}_Y(n) &= \mathbf{G}_{i+1}\mathbf{R}_Y^T(n+1) + \mathbf{G}_{i-1}\mathbf{R}_Y(n+1) \\
&\vdots \\
\mathbf{G}_{M+N+1}\mathbf{R}_Y(n) &= \mathbf{G}_{M+N}\mathbf{R}_Y^T(n+1)
\end{aligned} \tag{12}$$

Here $\mathbf{R}_Y(n) = EY_N(n)Y_N(n+1)^T$ is the correlation matrix of $Y_N(n)$ and $Y_N(n+1)$.

Under the channel assumptions considered in the previous section, Sylvester matrix $\mathbf{T}_N(\mathbf{H})$ becomes a full rank matrix [27]. Using the previous statement, the fact that the sources are supposed to be persistently exciting and the definition of $\mathbf{R}_Y(n)$ one can easily prove that $\text{Rank}(\mathbf{R}_Y(n)) = (M+N+1)p$. Hence, one can find using a SVD decomposition two matrices $\mathbf{U}$ and $\mathbf{V}$ such that $\mathbf{U}$ is a $q(N+1) \times (M+N+1)p$ left invertible matrix and $\mathbf{V}$ is a $(M+N+1)p \times q(N+1)$ right invertible matrix. Let $\mathbf{V}^\dagger$ be the right pseudo-inverse matrix of $\mathbf{V}$ and let us denote $\mathbf{P} = \mathbf{R}_Y^T(n+1)\mathbf{V}^\dagger$ and $\mathbf{Q} = \mathbf{R}_Y(n+1)\mathbf{V}^\dagger$, then system (12) becomes:

$$\begin{aligned}
\mathbf{G}_1\mathbf{U} &= \mathbf{G}_2\mathbf{P} \\
2\mathbf{G}_2\mathbf{U} &= \mathbf{G}_3\mathbf{P} + \mathbf{G}_1\mathbf{Q} \\
&\vdots
\end{aligned}$$

$$2\mathbf{G}_i\mathbf{U} = \mathbf{G}_{i+1}\mathbf{P} + \mathbf{G}_{i-1}\mathbf{Q} \qquad (13)$$

$$\vdots$$

$$\mathbf{G}_{M+N+1}\mathbf{U} = \mathbf{G}_{M+N}\mathbf{Q}$$

To minimize the criterion, one can solve the above matrix equation system (13).

3.2 Analytical Solution

In this subsection, an analytical solution of system (13) is given. Using the fact that $\mathbf{U}$ is a full rank matrix and a QR decomposition [28], one can find an orthogonal matrix $\mathbf{L}$ and an upper triangular matrix $\mathbf{R}$ such that $\mathbf{U} = \mathbf{LR}$. System (13) can then be rewritten as

$$\tilde{\mathbf{G}}_1\mathbf{R} = \tilde{\mathbf{G}}_2\tilde{\mathbf{P}}$$
$$2\tilde{\mathbf{G}}_2\mathbf{R} = \tilde{\mathbf{G}}_3\tilde{\mathbf{P}} + \tilde{\mathbf{G}}_1\tilde{\mathbf{Q}}$$

$$\vdots$$

$$2\tilde{\mathbf{G}}_i\mathbf{R} = \tilde{\mathbf{G}}_{i+1}\tilde{\mathbf{P}} + \tilde{\mathbf{G}}_{i-1}\tilde{\mathbf{Q}} \qquad (14)$$

$$\vdots$$

$$\tilde{\mathbf{G}}_{M+N+1}\mathbf{R} = \tilde{\mathbf{G}}_{M+N}\tilde{\mathbf{Q}}$$

Where $\tilde{\mathbf{G}}_i = \mathbf{G}_i\mathbf{L}^T$, $\tilde{\mathbf{P}} = \mathbf{P}^T\mathbf{P}$ and $\tilde{\mathbf{Q}} = \mathbf{Q}^T\mathbf{Q}$. Since $\mathbf{R}$ is a full rank upper triangular matrix, then without loss of generality one can write

$$\mathbf{R} = \begin{pmatrix} \bar{\mathbf{R}} \\ \mathbf{0} \end{pmatrix} \qquad (15)$$

where $\mathbf{0}$ is a zero matrix of appropriate dimensions. Let us decompose the three matrices $\tilde{\mathbf{P}}$, $\tilde{\mathbf{Q}}$ and $\tilde{\mathbf{G}}_i$ as following

$$\tilde{\mathbf{P}} = \begin{pmatrix} \tilde{\mathbf{P}}_1 \\ \tilde{\mathbf{P}}_2 \end{pmatrix}$$
$$\tilde{\mathbf{Q}} = \begin{pmatrix} \tilde{\mathbf{Q}}_1 \\ \tilde{\mathbf{Q}}_2 \end{pmatrix} \qquad (16)$$
$$\tilde{\mathbf{G}}_i = \begin{pmatrix} \tilde{\mathbf{g}}_i & \tilde{\mathbf{X}}_i \end{pmatrix}$$

where the different sub-matrices are of appropriate dimensions. Using equations (14) and (16), one can write

$$\tilde{\mathbf{g}}_1 = \tilde{\mathbf{g}}_2\bar{\mathbf{P}}_1 + \tilde{\mathbf{X}}_2\bar{\mathbf{P}}_2$$
$$2\tilde{\mathbf{g}}_2\mathbf{R} = \tilde{\mathbf{g}}_3\bar{\mathbf{P}}_1 + \tilde{\mathbf{X}}_3\bar{\mathbf{P}}_2 + \tilde{\mathbf{g}}_1\bar{\mathbf{Q}}_1 + \tilde{\mathbf{X}}_1\bar{\mathbf{Q}}_2$$

$$\vdots$$

$$2\tilde{\mathbf{g}}_i\mathbf{R} = \tilde{\mathbf{g}}_{i+1}\bar{\mathbf{P}}_1 + \tilde{\mathbf{X}}_{i+1}\bar{\mathbf{P}}_2 + \tilde{\mathbf{g}}_{i-1}\bar{\mathbf{Q}}_1 + \tilde{\mathbf{X}}_{i-1}\bar{\mathbf{Q}}_2 \qquad (17)$$

$$\vdots$$

$$\tilde{\mathbf{g}}_{M+N+1}\mathbf{R} = \tilde{\mathbf{g}}_{M+N}\bar{\mathbf{Q}}_1 + \tilde{\mathbf{g}}_{M+N}\bar{\mathbf{Q}}_2$$

where $\bar{\mathbf{P}}_i\bar{\mathbf{R}} = \tilde{\mathbf{P}}_i$ and $\bar{\mathbf{Q}}_i\bar{\mathbf{Q}} = \tilde{\mathbf{Q}}_i$.

Theoretically, any solution of the previous system (17) can minimize the proposed criterion. We should mention here that system (17) contents $M+N+1$ matrix equations and $2(M+N+1)$ unknown matrices (i.e. $\tilde{\mathbf{g}}_i$ and $\tilde{\mathbf{X}}_i$) which means that we have many solutions. These solutions are natural solutions (up to permutation and scale polynomial filter) and spurious solution as mentioned in the previous section.

3.3 Simplified Approximation

In [29], $\mathbf{R}_Y(n)$ is considered to be a full rank matrix (i.e the channel is noisy one), then system (12) with respect of the constraint can be be solved as following

$$\mathbf{D}_{(i+1)} = \mathbf{B}(2\mathbf{I} - \mathbf{D}_i\mathbf{A})^{-1} \qquad (18)$$
$$\mathbf{G}_{M+N-i-1} = \mathbf{G}_{M+N-i-2}\mathbf{D}_i \qquad (19)$$

where $\mathbf{D}_0 = \mathbf{B} = \mathbf{R}_N(n+1)\mathbf{R}_N^{-1}(n)$ and $\mathbf{A} = \mathbf{R}_N^T(n+1)\mathbf{R}_N^{-1}(n)$. It is proved that with a Signal to Noise Ratio (SNR) over 15dB, the previous system (19) can give satisfactory results.

Due to limitation of page number, Another simplified approximation of he system (17) has been omitted.

4 Conclusion

In this paper, a batch mutually algorithm for MIMO convolutive mixtures is presented. Generally, the batch algorithms have several advantages over adaptive algorithms from computation efforts and time point of view. However, they are very sensitive to noisy channel and to the estimation errors of the different needed parameters. Therefore, they are promising solutions to solve the initialization problems of adaptive algorithms and to improve also their performances.

References

1. A. Hyvärinen and E. Oja, "Independent componenet analysis: algorithms and applications," *Neural Networks*, vol. 13, pp. 411–430, 2000.
2. A. Mansour, A. Kardec Barros, and N. Ohnishi, "Blind separation of sources: Methods, assumptions and applications.," *IEICE Transactions on Fundamentals of Electronics, Communications and Computer Sciences*, vol. E83-A, no. 8, pp. 1498–1512, August 2000.
3. P. Comon, "Independent component analysis, a new concept?," *Signal Processing*, vol. 36, no. 3, pp. 287–314, April 1994.
4. A. Mansour and M. Kawamoto, "Ica papers classified according to their applications & performances.," *IEICE Transactions on Fundamentals of Electronics, Communications and Computer Sciences*, vol. E86-A, no. 3, pp. 620–633, March 2003.
5. L. Nguyen Thi and C. Jutten, "Blind sources separation for convolutive mixtures," *Signal Processing*, vol. 45, no. 2, pp. 209–229, 1995.

6. N. Thirion, J. Mars, and J. L. Boelle, "Separation of seismic signals: A new concept based on a blind algorithm," in *Signal Processing VIII, Theories and Applications (EUSIPCO'96)*, Triest, Italy, September 1996, pp. 85–88, Elsevier.
7. G. D'urso and L. Cai, "Sources separation method applied to reactor monitoring," in *Proc. 3rd Workshop on Higher Order Statistics*, Edinburgh, Scotland, September 1994.
8. E. Chaumette, P. Comon, and D. Muller, "Application of ICA to airport surveillance," in *HOS 93*, South Lake Tahoe-California, 7-9 June 1993, pp. 210–214.
9. A. Kardec Barros, A. Mansour, and N. Ohnishi, "Removing artifacts from ECG signals using independent components analysis," *NeuroComputing*, vol. 22, pp. 173–186, 1999.
10. M. Bouzaien and A. Mansour, "HOS criteria & ICA algorithms applied to radar detection," in *4th International Workshop on Independent Component Analysis and blind Signal Separation, ICA2003*, Nara, Japan, 1-4 April 2003.
11. J. Hérault, C. Jutten, and B. Ans, "Détection de grandeurs primitives dans un message composite par une architecture de calcul neuromimétique en apprentissage non supervisé," in *Actes du Xème colloque GRETSI*, Nice, France, 20-24, May 1985, pp. 1017–1022.
12. J. F. Cardoso and P. Comon, "Tensor-based independent component analysis," in *Signal Processing V, Theories and Applications (EUSIPCO'90)*, L. Torres, E. Masgrau, and M. A. Lagunas, Eds., Barcelona, Espain, 1990, pp. 673–676, Elsevier.
13. C. Jutten and J. Hérault, "Blind separation of sources, Part I: An adaptive algorithm based on a neuromimetic architecture," *Signal Processing*, vol. 24, no. 1, pp. 1–10, 1991.
14. S. I. Amari, A. Cichocki, and H. H. Yang, "A new learning algorithm for blind signal separation," in *Neural Information Processing System 8*, Eds. D.S. Toureyzky et. al., 1995, pp. 757–763.
15. A. Mansour and C. Jutten, "A direct solution for blind separation of sources," *IEEE Trans. on Signal Processing*, vol. 44, no. 3, pp. 746–748, March 1996.
16. M. Gaeta and J. L. Lacoume, "Sources separation without a priori knowledge: the maximum likelihood solution," in *Signal Processing V, Theories and Applications (EUSIPCO'90)*, L. Torres, E. Masgrau, and M. A. Lagunas, Eds., Barcelona, Espain, 1990, pp. 621–624, Elsevier.
17. N. Delfosse and P. Loubaton, "Adaptive blind separation of independent sources: A deflation approach," *Signal Processing*, vol. 45, no. 1, pp. 59–83, July 1995.
18. A. Mansour and C. Jutten, "Fourth order criteria for blind separation of sources," *IEEE Trans. on Signal Processing*, vol. 43, no. 8, pp. 2022–2025, August 1995.
19. A. Gorokhov and Ph. Loubaton, "Subspace based techniques for second order blind separation of convolutive mixtures with temporally correlated sources," *IEEE Trans. on Circuits and Systems*, vol. 44, pp. 813–820, September 1997.
20. A. Mansour, C. Jutten, and Ph. Loubaton, "An adaptive subspace algorithm for blind separation of independent sources in convolutive mixture," *IEEE Trans. on Signal Processing*, vol. 48, no. 2, pp. 583–586, February 2000.
21. A. Mansour, C. Jutten, and Ph. Loubaton, "Subspace method for blind separation of sources and for a convolutive mixture model," in *Signal Processing VIII, Theories and Applications (EUSIPCO'96)*, Triest, Italy, September 1996, pp. 2081–2084, Elsevier.
22. A. Mansour, A. Kardec Barros, and N. Ohnishi, "Subspace adaptive algorithm for blind separation of convolutive mixtures by conjugate gradient method," in *The First International Conference and Exhibition Digital Signal Processing (DSP'98)*, Moscow, Russia, June 30-July 3 1998, pp. I–252–I–260.

23. D. Gesbert, P. Duhamel, and S. Mayrargue, "Subspace-based adaptive algorithms for the blind equalization of multichannel fir filters," in *Signal Processing VII, Theories and Applications (EUSIPCO'94)*, M.J.J. Holt, C.F.N. Cowan, P.M. Grant, and W.A. Sandham, Eds., Edinburgh, Scotland, September 1994, pp. 712–715, Elsevier.
24. T. Kailath, *Linear systems*, Prentice Hall, 1980.
25. D. Gesbert, P. Duhamel, and S. Mayrargue, "On-line blind multichannel equalization based on mutually referenced filters," *IEEE Trans. on Signal Processing*, vol. 45, no. 9, pp. 2307–2317, September 1997.
26. A. Mansour and N. Ohnishi, "Multichannel blind separation of sources algorithm based on cross-cumulant and the levenberg-marquardt method.," *IEEE Trans. on Signal Processing*, vol. 47, no. 11, pp. 3172–3175, November 1999.
27. A. Mansour, "A mutually referenced blind multiuser separation of convolutive mixture algorithm," *Signal Processing*, vol. 81, no. 11, pp. 2253–2266, November 2001.
28. G. H. Golub and C. F. Van Loan, *Matrix computations*, The johns hopkins press-London, 1984.
29. A. Mansour and N. Ohnishi, "A batch subspace ica algorithm.," in *10th IEEE Signal Processing Workshop on Statistical Signal and Array Processing*, Pocono Manor Inn, Pennsylvania, USA, 14 - 16 August 2000, pp. 63–67.

Frequency Domain Blind Source Separation for Many Speech Signals

Ryo Mukai, Hiroshi Sawada, Shoko Araki, and Shoji Makino

NTT Communication Science Laboratories, NTT Corporation
2-4 Hikaridai, Seika-cho, Soraku-gun, Kyoto 619–0237, Japan
{ryo,sawada,shoko,maki}@cslab.kecl.ntt.co.jp

Abstract. This paper presents a method for solving the permutation problem of frequency domain blind source separation (BSS) when the number of source signals is large, and the potential source locations are omnidirectional. We propose a combination of small and large spacing sensor pairs with various axis directions in order to obtain proper geometric information for solving the permutation problem. Experimental results in a room (reverberation time T_R=130 ms) with eight microphones show that the proposed method can separate a mixture of six speech signals that come from various directions, even when two of them come from the same direction.

1 Introduction

Independent component analysis (ICA) is one of the major statistical methods for blind source separation (BSS). It is theoretically possible to solve the BSS problem with a large number of sources by ICA if we assume that the number of observed signals is equal to or greater than the number of source signals. However, there are many practical difficulties, and although a large number of studies have been undertaken on audio BSS in a reverberant environment, only a few studies have dealt with more than two source signals.

In a reverberant environment, the signals are mixed in a convolutive manner with reverberations, and the unmixing system that we have to estimate is a matrix of filters, not just a matrix of scalars. There are two major approaches to solving the convolutive BSS problem. The first is the time domain approach, where ICA is applied directly to the convolutive mixture model. Matsuoka et al. have proved that time domain ICA can solve the convolutive BSS problem of eight sources with eight microphones in a real environment [1]. Unfortunately, the time domain approach incurs considerable computation cost, and it is difficult to obtain a solution in a practical time.

The other approach is frequency domain BSS, where ICA is applied to multiple instantaneous mixtures in the frequency domain. This approach takes much less computation time than time domain BSS. However, it poses another problem in that we need to align the output signal order for every frequency bin so that a separated signal in the time domain contains frequency components from one source signal. This problem is known as the permutation problem.

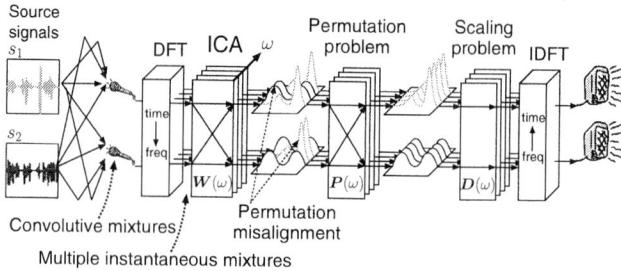

Fig. 1. Flow of frequency domain BSS

Many methods have been proposed for solving the permutation problem, and the use of geometric information, such as beam patterns [2–4], direction of arrival (DOA) and source locations [5], is an effective approach. We have proposed a robust method that combines the DOA based method [2, 3] and the correlation based method [6], which almost completely solves the problem for 2-source cases [7]. However it is insufficient when the number of signals is large or when the signals come from the same or similar direction. In this paper, we propose a method for obtaining proper geometric information for solving the permutation problem in such cases.

2 Frequency Domain BSS Using ICA

When the source signals are $s_i(t)(i = 1, ..., N)$, the signals observed by sensor j are $x_j(t)(j = 1, ..., M)$, and the separated signals are $y_k(t)(k = 1, ..., N)$, the BSS model can be described as: $x_j(t) = \sum_{i=1}^{N}(h_{ji} * s_i)(t)$, $y_k(t) = \sum_{j=1}^{M}(w_{kj} * x_j)(t)$, where h_{ji} is the impulse response from source i to sensor j, w_{kj} are the separating filters, and $*$ denotes the convolution operator. Figure 1 shows the flow of BSS in the frequency domain. A convolutive mixture in the time domain is converted into multiple instantaneous mixtures in the frequency domain. Therefore, we can apply an ordinary independent component analysis (ICA) algorithm [8] in the frequency domain to solve a BSS problem in a reverberant environment. Using a short-time discrete Fourier transform, the model is approximated as: $\mathbf{X}(\omega, m) = \mathbf{H}(\omega)\mathbf{S}(\omega, m)$, where, ω is the angular frequency, and n represents the frame index. The separating process can be formulated in each frequency bin as: $\mathbf{Y}(\omega, m) = \mathbf{W}(\omega)\mathbf{X}(\omega, m)$, where $\mathbf{S}(\omega, m) = [S_1(\omega, m), ..., S_N(\omega, m)]^T$ is the source signal in frequency bin ω, $\mathbf{X}(\omega, m) = [X_1(\omega, m), ..., X_M(\omega, m)]^T$ denotes the observed signals, $\mathbf{Y}(\omega, m) = [Y_1(\omega, m), ..., Y_N(\omega, m)]^T$ is the estimated source signal, and $\mathbf{W}(\omega)$ represents the separating matrix. $\mathbf{W}(\omega)$ is determined so that $Y_i(\omega, m)$ and $Y_j(\omega, m)$ become mutually independent.

The ICA solution suffers permutation and scaling ambiguities. This is due to the fact that if $\mathbf{W}(\omega)$ is a solution, then $\mathbf{D}(\omega)\mathbf{P}(\omega)\mathbf{W}(\omega)$ is also a solution, where $\mathbf{D}(\omega)$ is a diagonal complex valued scaling matrix, and $\mathbf{P}(\omega)$ is an arbitrary permutation matrix. We thus have to solve the permutation and scaling problems to reconstruct separated signals in the time domain.

There is a simple and reasonable solution for the scaling problem: $\mathbf{D}(\omega) = \text{diag}\{[\mathbf{P}(\omega)\mathbf{W}(\omega)]^{-1}\}$, which is obtained by the minimal distortion principle (MDP) [9], and we can use it. On the other hand, the permutation problem is complicated, especially when the number of source signals is large.

3 Geometric Information for Solving Permutation Problem

3.1 Invariant in ICA Solution

If a separating matrix $\mathbf{W}(\omega)$ is calculated successfully and it extracts source signals with scaling ambiguity, $\mathbf{D}(\omega)\mathbf{W}(\omega)\mathbf{H}(\omega) = \mathbf{I}$ holds (except for singular frequency bins). Because of the scaling ambiguity, we cannot obtain $\mathbf{H}(\omega)$ simply from the ICA solution. However, the ratio of elements in the same column $H_{ji}/H_{j'i}$ is invariable in relation to $\mathbf{D}(\omega)$, and given by

$$\frac{H_{ji}}{H_{j'i}} = \frac{[\mathbf{W}^{-1}\mathbf{D}^{-1}]_{ji}}{[\mathbf{W}^{-1}\mathbf{D}^{-1}]_{j'i}} = \frac{[\mathbf{W}^{-1}]_{ji}}{[\mathbf{W}^{-1}]_{j'i}}, \qquad (1)$$

where $[\cdot]_{ji}$ denotes the ji-th element of the matrix. We can estimate several types of geometric information related to source signals by using this invariant. The estimated information is used to solve the permutation problem.

If we have more sensors than sources ($N < M$), principal component analysis (PCA) is performed as a preprocessing of ICA [10] so that the N dimensional subspace spanned by the row vectors of $\mathbf{W}(\omega)$ is almost identical to the signal subspace, and the Moore-Penrose pseudo-inverse $\mathbf{W}^+ \triangleq \mathbf{W}^T(\mathbf{W}\mathbf{W}^T)^{-1}$ is used instead of $\mathbf{W}^{-1}$.

3.2 DOA Estimation with ICA Solution

We can estimate the DOA of source signals by using the above invariant $H_{ji}/H_{j'i}$ [7]. With a farfield model, a frequency response is formulated as:

$$H_{ji}(\omega) = e^{j\omega c^{-1}\mathbf{a}_i^T \mathbf{p}_j}, \qquad (2)$$

where c is the speed of wave propagation, $\mathbf{a}_i$ is a unit vector that points to the direction of source i, and $\mathbf{p}_j$ represents the location of sensor j. According to this model, we have

$$H_{ji}/H_{j'i} = e^{j\omega c^{-1}\mathbf{a}_i^T(\mathbf{p}_j - \mathbf{p}_{j'})} \qquad (3)$$

$$= e^{j\omega c^{-1}\|\mathbf{p}_j - \mathbf{p}_{j'}\|\cos\theta_{i,jj'}}, \qquad (4)$$

where $\theta_{i,jj'}$ is the direction of source i relative to the sensor pair j and j'. By using the argument of (4) and (1), we can estimate:

$$\hat{\theta}_{i,jj'} = \arccos \frac{\arg(H_{ji}/H_{j'i})}{\omega c^{-1}\|(\mathbf{p}_j - \mathbf{p}_{j'})\|}$$

$$= \arccos \frac{\arg([\mathbf{W}^{-1}]_{ji}/[\mathbf{W}^{-1}]_{j'i})}{\omega c^{-1}\|(\mathbf{p}_j - \mathbf{p}_{j'})\|}. \qquad (5)$$

This procedure is valid for sensor pairs with a small spacing.

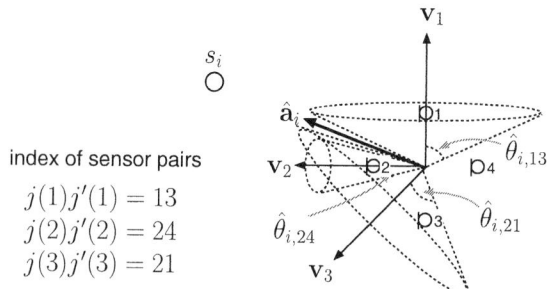

Fig. 2. Solving ambiguity of estimated DOAs

3.3 Ambiguity of DOA Estimation

DOA estimation involves some ambiguities. When we use only one pair of sensors or a linear array, the estimated $\hat{\theta}_{i,jj'}$ determines a cone rather than a direction. If we assume a horizontal plane on which sources exist, the cone is reduced to two half-lines. However, the ambiguity of two directions that are symmetrical with respect to the axis of the sensor pair still remains. This is a fatal problem when the source locations are omnidirectional.

When the spacing between sensors is larger than half a wavelength, spatial aliasing causes another ambiguity, but we do not consider this here.

3.4 Solving Ambiguity of DOA Estimation

The ambiguity can be solved by using multiple sensor pairs. If we use sensor pairs that have different axis directions, we can estimate cones with various vertex angles for one source direction. If the *relative* DOA $\hat{\theta}_{i,jj'}$ is estimated without any error, the *absolute* direction of the source signal $\mathbf{a}_i$ satisfies:

$$\frac{(\mathbf{p}_j - \mathbf{p}_{j'})^T \mathbf{a}_i}{\|\mathbf{p}_j - \mathbf{p}_{j'}\|} = \cos \hat{\theta}_{i,jj'}. \tag{6}$$

When we use L sensor pairs whose indexes are $j(l)j'(l)(1 \leq l \leq L)$, $\mathbf{a}_i$ is given by the solution of the following equation:

$$\mathbf{V}\mathbf{a}_i = \mathbf{c}_i, \tag{7}$$

where $\mathbf{v}_l \triangleq \frac{\mathbf{p}_{j(l)} - \mathbf{p}_{j'(l)}}{\|\mathbf{p}_{j(l)} - \mathbf{p}_{j'(l)}\|}$ is a normalized axis, $\mathbf{V} \triangleq (\mathbf{v}_1, ..., \mathbf{v}_L)^T$, and $\mathbf{c}_i \triangleq [\cos(\hat{\theta}_{i,j(1)j'(1)}), ..., \cos(\hat{\theta}_{i,j(L)j'(L)})]^T$. Sensor pairs should be selected so that $\text{rank}(\mathbf{V}) \geq 3$ if potential source locations are three-dimensional, or $\text{rank}(\mathbf{V}) \geq 2$ if we assume a plane on which sources exist.

Actually, $\hat{\theta}_{i,j(l)j'(l)}$ has an estimation error, and (7) has no solution. Thus we adopt an optimal solution by employing certain criteria such as:

$$\hat{\mathbf{a}}_i = \underset{\mathbf{a}}{\operatorname{argmin}} \|\mathbf{V}\mathbf{a} - \mathbf{c}_i\| \quad \text{(subject to } \|\mathbf{a}\| = 1\text{)} \tag{8}$$

This can be solved approximately by using the Moore-Penrose pseudo-inverse $\mathbf{V}^+ \triangleq (\mathbf{V}^T\mathbf{V})^{-1}\mathbf{V}^T$, and we have:

$$\hat{\mathbf{a}}_i \approx \frac{\mathbf{V}^+\mathbf{c}_i}{\|\mathbf{V}^+\mathbf{c}_i\|}. \tag{9}$$

Accordingly, we can determine a unit vector $\hat{\mathbf{a}}_i$ pointing to the direction of source s_i (Fig. 2).

3.5 Estimation of Sphere with ICA Solution

The interpretation of the ICA solution with a nearfield model yields other geometric information [11]. When we adopt the nearfield model, including the attenuation of the wave, $H_{ji}(\omega)$ is formulated as:

$$H_{ji}(\omega) = \frac{1}{\|\mathbf{q}_i - \mathbf{p}_j\|} e^{\jmath\omega c^{-1}(\|\mathbf{q}_i - \mathbf{p}_j\|)} \tag{10}$$

where $\mathbf{q}_i$ represents the location of source i. By taking the ratio of (10) for a pair of sensors j and j' we obtain:

$$H_{ji}/H_{j'i} = \frac{\|\mathbf{q}_i - \mathbf{p}_{j'}\|}{\|\mathbf{q}_i - \mathbf{p}_j\|} e^{\jmath\omega c^{-1}(\|\mathbf{q}_i - \mathbf{p}_j\| - \|\mathbf{q}_i - \mathbf{p}_{j'}\|)}. \tag{11}$$

By using the modulus of (11) and (1), we have:

$$\frac{\|\mathbf{q}_i - \mathbf{p}_{j'}\|}{\|\mathbf{q}_i - \mathbf{p}_j\|} = \left|\frac{[\mathbf{W}^{-1}]_{ji}}{[\mathbf{W}^{-1}]_{j'i}}\right|. \tag{12}$$

By solving (11) for $\mathbf{q}_i$, we have a sphere whose center $O_{i,jj'}$ and radius $R_{i,jj'}$ are given by:

$$O_{i,jj'} = \mathbf{p}_j - \frac{1}{r_{i,jj'}^2 - 1}(\mathbf{p}_{j'} - \mathbf{p}_j), \tag{13}$$

$$R_{i,jj'} = \|\frac{r_{i,jj'}}{r_{i,jj'}^2 - 1}(\mathbf{p}_{j'} - \mathbf{p}_j)\|, \tag{14}$$

where $r_{i,jj'} \triangleq |[\mathbf{W}^{-1}]_{ji}/[\mathbf{W}^{-1}]_{j'i}|$. Thus, we can estimate a sphere $(\hat{O}_{i,jj'}, \hat{R}_{i,jj'})$ on which $\mathbf{q}_i$ exists by using the result of ICA $\mathbf{W}$ and the locations of the sensors $\mathbf{p}_j$ and $\mathbf{p}_{j'}$. Figure 3 shows an example of the spheres determined by (12) for various ratios $r_{i,jj'}$. This procedure is valid for sensor pairs with a large spacing.

3.6 Solving Permutation Problem

We solve the permutation problem by classification using the geometric information together with a correlation based method. This is similar to our previously reported proposal [7].

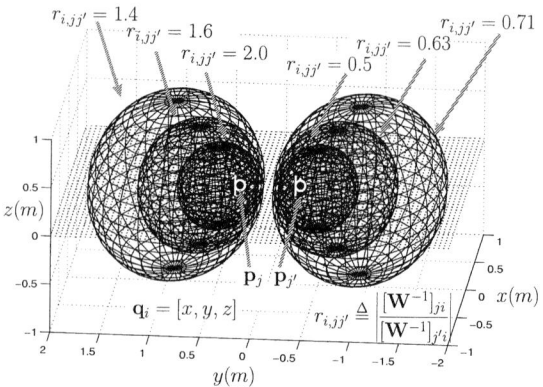

Fig. 3. Example of spheres determined by (12) ($\mathbf{p}_j = [0, 0.3, 0]$, $\mathbf{p}_{j'} = [0, -0.3, 0]$)

The models (2) and (10) are simple approximations without multi-path propagation and reverberation, however we can use them to obtain information for classifying signals. Even when some signals come from the same or a similar direction, we can distinguish between them by using the information obtained by the method described in Sec.3.5. The source locations can be estimated by combining the estimated direction and spheres. Then, we can classify separated signals in the frequency domain according to the estimated source locations.

Unfortunately, classification on the basis of the estimated location tends to be inconsistent especially in a reverberant environment. In many frequency bins, several signals are assigned to the same cluster, and such classification is inconsistent. We solve the permutation only for frequency bins with a consistent classification, and we employ a correlation based method for the rest. The correlation based method solves the permutation so that the inter-frequency correlation for neighboring or harmonic frequency bins is maximized.

4 Experiments

We carried out experiments with 6 sources and 8 microphones using speech signals convolved with impulse responses measured in a room with reverberation time of 130 ms. The room layout and other experimental conditions are shown in Fig. 4. We assume that the number of source signals $N = 6$ is known. The experimental procedure is as follows.

First, we apply ICA to $x_j(t)(j = 1, ..., 8)$, and calculate separating matrix $\mathbf{W}(\omega)$ for each frequency bin. The initial value of $\mathbf{W}(\omega)$ is calculated by PCA. Then we estimate DOAs by using the rows of $\mathbf{W}^+(\omega)$ (pseudo-inverse) corresponding to the small spacing microphone pairs (1-3, 2-4, 1-2 and 2-3). Figure 5 shows a histogram of the estimated DOAs. We can find five clusters in this histogram, and one cluster is twice the size of the others. This implies that two

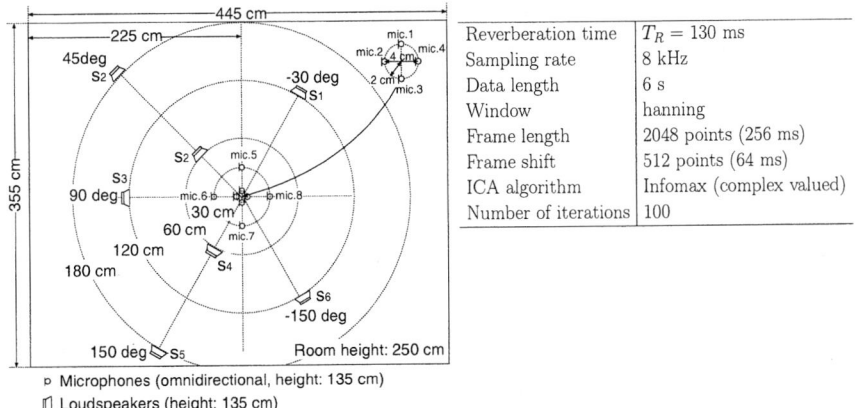

Fig. 4. Room layout and experimental conditions

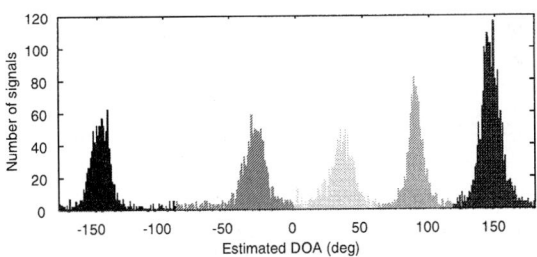

Fig. 5. Histogram of estimated DOAs obtained by using small spacing microphone pairs

signals come from the same direction (about 150°). We can solve the permutation problem for other four sources by using this DOA information (Fig. 6(a)).

Then, we apply the estimation of spheres to the signals that belong to the large cluster by using the rows of $\mathbf{W}^+(\omega)$ corresponding to the large spacing microphone pairs (7-5, 7-8, 6-5 and 6-8). Figure 6(b) shows estimated radiuses for S_4 and S_5 for the microphone pair 7-5. Although the radius estimation includes a large error, it provides sufficient information to distinguish two signals. Finally, we can classify the signals into six clusters. We determine the permutation only for frequency bins with a consistent classification, and we employ a correlation based method for the rest. In addition, we use the spectral smoothing method proposed in [12] to construct separating filters in the time domain from the ICA result in the frequency domain.

The performance is measured from the signal-to-inference ratio (SIR). The portion of $y_k(t)$ that comes from $s_i(t)$ is calculated by $y_{ki}(t) = \sum_{j=1}^{M}(w_{kj} * h_{ji} * s_i)(t)$. If we solve the permutation problem so that $s_i(t)$ is output to $y_i(t)$, the SIR for $y_k(t)$ is defined as:

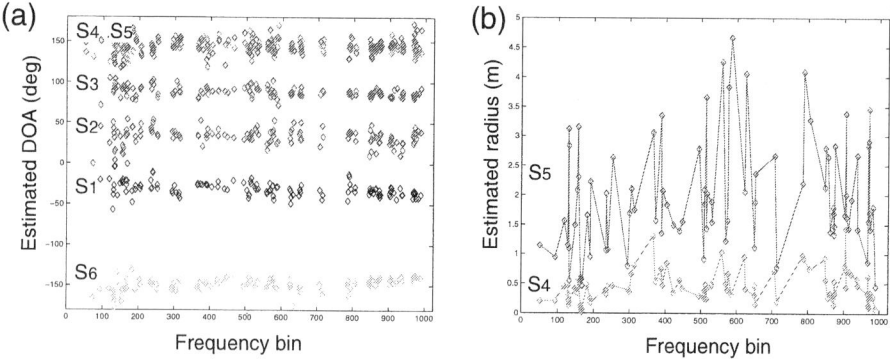

Fig. 6. Permutation solved by using (a) DOAs and (b) estimated radiuses

Table 1. Experimental results (dB), T_R=130 ms

	SIR_1	SIR_2	SIR_3	SIR_4	SIR_5	SIR_6	ave.
Input SIR	-8.3	-6.8	-7.8	-7.7	-6.7	-5.2	-7.1
C	4.4	2.6	4.0	9.2	3.6	-2.0	3.7
D+C	9.6	9.3	14.7	2.7	6.5	14.0	9.4
D+S+C	10.8	10.4	14.5	7.0	11.0	12.2	11.0

$$\mathrm{SIR}_k = 10\log[\sum_t y_{kk}(t)^2 / \sum_t (\sum_{i \neq k} y_{ki}(t))^2] \; (\mathrm{dB}).$$

We measured SIRs for three permutation solving strategies: the correlation based method ("C"), estimated DOAs and correlation ("D+C"), and a combination of estimated DOAs, spheres and correlation ("D+S+C", proposed method). We also measured input SIRs by using the mixture observed by microphone 1 for the reference ("Input SIR"). The results are summarized in Table 1.

Our proposed method succeeded in separating six speech signals. It can be seen that the discrimination obtained by using estimated spheres is effective in improving the separation performance for signals coming from the same direction.

5 Conclusion

We proposed using a combination of small and large spacing microphone pairs with various axis directions to obtain proper geometric information for solving the permutation problem in frequency domain BSS. In experiments (T_R=130 ms), our method succeeded in the separation of six speech signals, even when two came from the same direction. The computation time was about 1 minute for 6 seconds of data. Some sound examples can be found on our web site [13].

References

1. Matsuoka, K., Ohba, Y., Toyota, Y., Nakashima, S.: Blind separation for convolutive mixture of many voices. In: Proc. IWAENC 2003. (2003) 279–282
2. Kurita, S., Saruwatari, H., Kajita, S., Takeda, K., Itakura, F.: Evaluation of blind signal separation method using directivity pattern under reverberant conditions. In: Proc. ICASSP 2000. (2000) 3140–3143
3. Ikram, M.Z., Morgan, D.R.: A beamforming approach to permutation alignment for multichannel frequency-domain blind speech separation. In: Proc. ICASSP 2002. (2002) 881–884
4. Parra, L.C., Alvino, C.V.: Geometric source separation: Merging convolutive source separation with geometric beamforming. IEEE Trans. Speech Audio Processing **10** (2002) 352–362
5. Soon, V.C., Tong, L., Huang, Y.F., Liu, R.: A robust method for wideband signal separation. In: Proc. ISCAS '93. (1993) 703–706
6. Asano, F., Ikeda, S., Ogawa, M., Asoh, H., Kitawaki, N.: A combined approach of array processing and independent component analysis for blind separation of acoustic signals. In: Proc. ICASSP 2001. (2001) 2729–2732
7. Sawada, H., Muaki, R., Araki, S., Makino, S.: A robust and precise method for solving the permutation problem of frequency-domain blind source separation. IEEE Trans. Speech Audio Processing **12** (2004)
8. Hyvärinen, A., Karhunen, J., Oja, E.: Independent Component Analysis. John Wiley & Sons (2001)
9. Matsuoka, K., Nakashima, S.: Minimal distortion principle for blind source separation. In: Proc. ICA 2001. (2001) 722–727
10. Winter, S., Sawada, H., Makino, S.: Geometrical understanding of the PCA subspace method for overdetermined blind source separation. In: Proc. ICASSP 2003. Volume 5. (2003) 769–772
11. Mukai, R., Sawada, H., Araki, S., Makino, S.: Near-field frequency domain blind source separation for convolutive mixtures. In: Proc. ICASSP 2004. (2004)
12. Sawada, H., Mukai, R., de la Kethulle, S., Araki, S., Makino, S.: Spectral smoothing for frequency-domain blind source separation. In: Proc. IWAENC 2003. (2003) 311–314
13. http://www.kecl.ntt.co.jp/icl/signal/mukai/demo/ica2004/

ICA Model Applied to Multichannel Non-destructive Evaluation by Impact-Echo

Addisson Salazar, Luis Vergara, Jorge Igual, Jorge Gosálbez, and Ramón Miralles

Universidad Politécnica de Valencia, Departamento de Comunicaciones,
Camino de Vera s/n, 46022 Valencia, Spain
{asalazar,lvergara,jigual,jorgocas,rmiralle}@dcom.upv.es

Abstract. This article presents an ICA model for applying in Non Destructive Testing by Impact-Echo. The approach consists in considering flaws inside the material as sources for blind separation using ICA. A material is excited by a hammer impact and a convolutive mixture is sensed by a multichannel system. Obtained information is used for classifying in defective or non defective material. Results based on simulation by finite element method are presented, including different defect geometry and location.

1 Introduction

The importance of Non Destructive Evaluation (NDE) of materials is broadly grateful. Its application in different fields of industrial applications of material quality evaluation has been increased recently. Due to new requirements of quality coming from the industry, NDE techniques are in strong period of expansion and development. One of the most effective of those techniques is the impact-echo (I-E).

I-E technique consists in a procedure where a material is excited by a hammer impact, which produces a response from material microstructure. The response can be sensed by a set of transducers located on material surface. Received signals contain backscattering from material grain microstructure and flaw information [1].

In I-E the time domain transient waveforms obtained from the impact of a steel sphere in the material are analysed in frequency domain. This technique has had an extended use in applications of concrete structures in civil engineering. A work type has been the analysis of resonance frequencies in different shape elements, such as, circular, square beams, beams with empty ducts or cement fillings, rectangular columns, post-tensed structure, and tendon ducts.

In addition I-E has been used in determining superficial crack depth, evaluation of early age concrete hardness, evaluation of structural integrity, crack propagation tracing and detection of steel corrosion damages of concrete reinforcement. A displacement of fundamental frequency to a lower value is the key to identify the presence of a crack. Also, spectrum will show a high amplitude pick in frequency corresponding to the crack depth. Recently this technique has been used in testing marble rock blocks for general status classification and discontinuity location inside the blocks [2].

ICA is a powerful statistical technique that has had a successful application in different areas [3] [4]. Principal aim of this paper is to present and test a convolutive mixture ICA model for contributing in defect characterization in noisy means starting from signals captured by multiple sensors on a material inspected by I-E technique. Model defines flaws into the material as the sources for blind separation. Convolutive mixtures suppose that there are contributions of all the sources in each sensor in several time instants; that is to say, each j source contributes to the sensor i as a weighted sum of the values of that source in the different instants of time.

Some works in time-frequency domain applied to convolutive mixtures are: Frequency domain Infomax, cumulant based methods, methods based on beamforming, and multichannel blind deconvolution using FIR matrix algebra. There are not previous references about application of ICA on signals coming from I-E.

2 ICA Model

The impact-echo procedure is illustrated in Figure 1. A solid material is excited by a hammer hit and material response is sensed by a multichannel system.

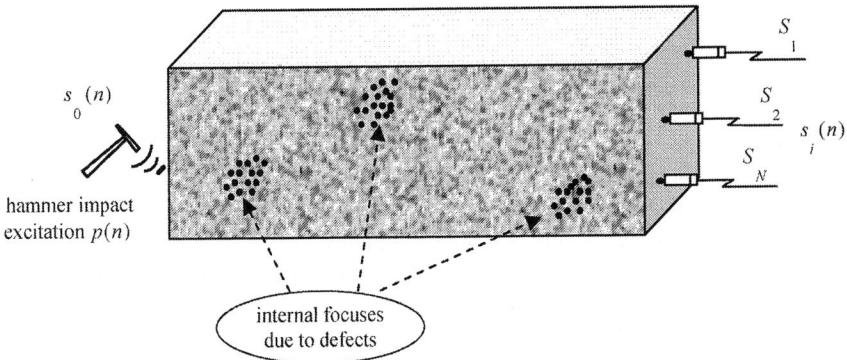

Fig. 1. Rectangular block inspection by impact-echo.

2.1 Signal Modelling

The received signals can be modelled from two points of view, first one based on solid bulk acoustic propagation waves theory and second one based on linear system theory. From this latter approach, signals sensed by transducers can be considered as a convolutive mixture between input signal and inner material defects.

From figure 1, we have

- One attack point (source 0), which generates wave $s_0(n) = p(n)$
- F internal focuses (defects), which generate waves $f_j(n)$ $j = 1, ..., F$
- N sensors, where waves $s_i(n)$ $i = 1, ..., N$ are recorded

Acoustic energy vectors are defined

$$\mathbf{f} = [f_0 f_1 \cdots f_F]^T, f_j(n) \qquad (1)$$

where $f_0(n) = p(n)$

Assuming linear propagation, Fourier transform sensed by i^{th} sensor due to internal focus j, ($i=1,...N$ $j=1,...F$), is given by

$$S_{ij}(\omega) = P(\omega) \cdot H_{j0}(\omega) \cdot H_{ij}(\omega) \qquad (2)$$

where,

- $H_{j0}(\omega)$: Frequency response between impact point and internal focus j, $j=1,...,F$
- $H_{ij}(\omega)$: Frequency response between internal focus and transducer i, $i=1,...,N$
- $P(\omega)$: Impact signal spectrum

The, spectrum of the signal sensed by transducer i, can be calculated as

$$S_i(\omega) = P(\omega) \cdot \left[\left(\sum_{j=1}^{F} H_{j0}(\omega) \cdot H_{ij}(\omega) \right) + H_i(\omega) \right] \qquad (3)$$

where,

$H_i(\omega)$ is the direct path frequency response between excitation and sensor i (in absence of internal defects). It can be obtained from an acoustic model, calculating the frequency response between excitation and transducer in a perfect rectangular solid block. Also, this response can be calculated from non-defect block measurements.

We define "residual deconvolutioned signal" as

$$A_i(\omega) = \frac{S_i(\omega)}{P(\omega)} - H_i(\omega) \qquad (4)$$

Residual signal is used for determining a first material classification as defective or nondefective material. Replacing (3) in (4) becomes,

$$A_i(\omega) = \sum_{j=1}^{F} H_{j0}(\omega) \cdot H_{ij}(\omega) \qquad (5)$$

2.2 Problem Formulation by ICA

In order to get information such as shape and location of defects in a defective material, parameters for a classification process can be extracted from equation (5) or we can undertake a blind separation source process by ICA. Considering a BSS problem by ICA, for finding information of the inner sources or material flaws, the following matrix formulation is outlined.

For each ω, considering all the sensors, from (5) can be written

$$\mathbf{a}(\omega) = \mathbf{M}(\omega) \cdot \mathbf{h}(\omega) \qquad (6)$$

where the matrix and vector elements are defined as follows

$\mathbf{a}[i] = A_i(\omega)$ (sample vector, $i = 1, ..., N$)

$\mathbf{h}[j] = H_{j0}(\omega)$ (source vector, internal focuses, $j = 1, ..., F$)

$\mathbf{M}[i, j] = H_{ij}(\omega)$ (mixture matrix)

2.3 Calculation of the Homogeneous Material Response

In the previous exposition it is fundamental to know $H_i(\omega)$ for each sensor. From the point of view of acoustic wave propagation in solids, stress pulse generated by point impact propagates into the solid as spherical dilatational *(P)* and distortional *(S)* wavefronts. In addition, a surface wave, or Rayleigh wave *(R)* travels throughout a circular wavefront across the material surface. The phenomenon of volumetric wave propagation can be modelled by means of the following two expressions [5],

$$\frac{\partial T_{ij}}{\partial x_j} = \rho_0 \frac{\partial^2 u_i}{\partial t^2} \qquad T_{ij} = c_{ijkl} S_{kl} \qquad (7)$$

where,

ρ_0: Material density. u_i: Length elongation with respect to starting point in force direction. $\dfrac{\partial T_{ij}}{\partial x_j}$: Force variation in i direction due to deformations in j directions.

c_{ijkl}: Elastic constant tensor (Hooke's law). S_{kl}: Strain or relative volume change under deformation in face l in direction k in unitary cube that represents a material element.

That is to say, force variation in the direction due to face contribution of the material elementary cube is equal to the mass per volume (density) times the strain acceleration. To derive an analytical solution to problems that involve stress wave propagation in delimited solids is very difficult, reason why not a very extensive bibliography exists. Numeric models, such as Finite Element Method (FEM), can be used to obtain the material theoretical response. Several studies using FEM have been made, demonstrating a good approximation in the theoretical response calculated by FEM and results obtained in experimental researches [6]. Finite Element models have been made to validate the exposed ICA model; its results are shown in next section.

3 Experiments and Results

Several experiments have been made in the two simulation scenarios shown in figures 2a and 2b (distances in mm.). In both cases the test specimen consists in a parallelepiped shape material of 0.07x0.05x0.22 m. (width, height and length) supported to one third and two thirds of the block length (direction z). Material elastic constants were defined as: density 2300 kg/m3, modulus of elasticity 33100 Mpa and Poisson's ratio 0.2. It corresponds to a concrete rectangular block.

A full transient dynamic analysis by FEM was used for determining the dynamic response of the material structure under the action of an impact transient load. This type of analysis can be used to determine the time-varying displacements in a structure as it responds to a transient load. Elements having dimensions of about 0.01 m were used in the models. These elements can accurately capture the frequency response up to 40 kHz. The impact was simulated by applying a force-time history of a half sine wave with a period of $60\mu s$ and it was applied as a uniform pressure load over two elements at the center of the specimen front face.

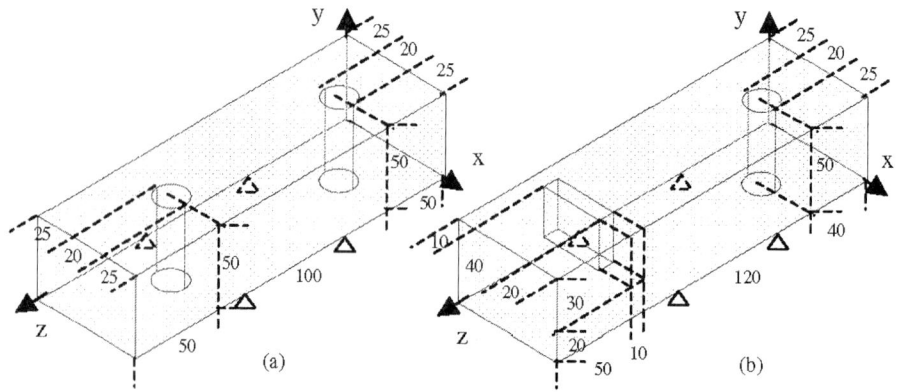

Fig. 2. Simulated test specimens with two internal defects.

Surface displacement waveforms were taken from the simulation results at 6 nodes in different locations over the specimen surface. It would be equivalent to the signals that can be measured by sensors in a real experiment. Signals consisted of 20000 points recorded at a sampling frequency of 100 kHz. Figure 3 shows a sensor configuration set up on a homogeneous specimen.

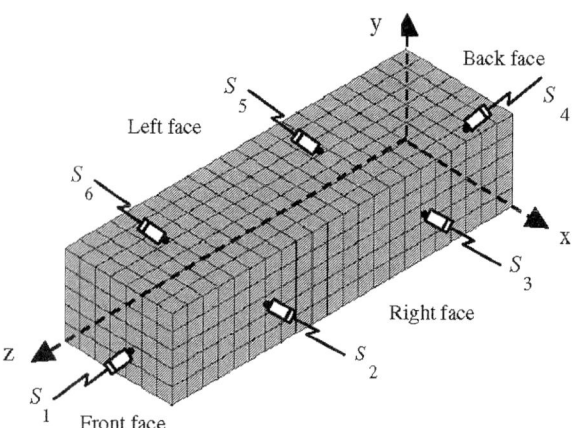

Fig. 3. Meshed homogeneous test specimen and multichannel configuration.

One important condition to confirm the feasibility of applying ICA is mutual statistical independence of the sources. In the presented model sources are defined as the transfer function from the unique impact excitation and one inner location where a defect exists. Firstly, transfer function (source) was calculated in 540 locations in the material throughout 15 depth levels moving in axis z from 0 to material length and different x and y coordinate values. Correlation and higher order statistics between one source with each other of the 540 were calculated using displacement modulus normalized signals. Two different curve types were found, first one was obtained for sources at the extremes of z value coordinates (front and back face) and second one for sources at coordinates no located at the extreme values of z, see Figure 4.

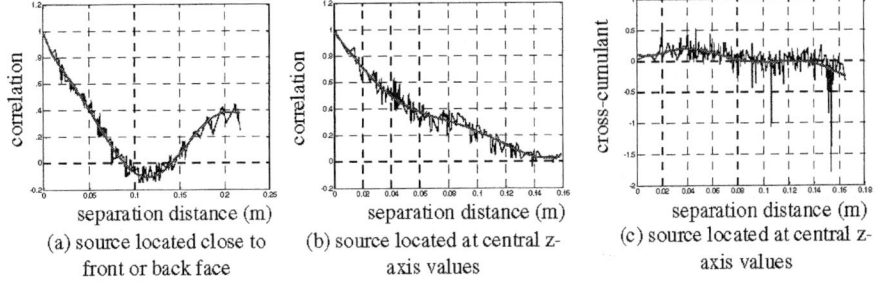

(a) source located close to front or back face

(b) source located at central z-axis values

(c) source located at central z-axis values

Fig. 4. Some statistics for source independence calculation.

Curve in Figure 4a shows that given a first source fixed at one extreme of the material length in axis of load application direction (axis z). Its correlation with a second source moving in axis z decreases the longer is the distance separation between both, until the second one reaches centre of material length. After that, values increase as the second source location is closer to the first one source opposite extreme. Increase of correlation is due to the effects caused by multiple reflections in the close proximity of side boundaries and symmetry factors. Curve in Figure 4b shows that given a first source fixed at intermediate localization on axis z, its correlation with other sources decreases steadies the longer distance separation between both. Calculation of fourth order moments and cumulants has thrown consistent results with correlations. Figure 4c shows cross-cumulant for a source at intermediate localization in axis z. Cross-cumulant has values near to zero over all distances and zero at the source separation distance interval [0.08-0.013 m.].

Final verification is checking source pdf Gaussianity. Figure 5 shows histograms for sources in specimen of Figure 2b. First source is evaluated at coordinates x,y,z (35,25,155 mm.) and second source is evaluated at (35,25,25 mm.). It can be observed that both sources are not Gaussians. From the previous analysis it can be concluded that the sources are independent and can be separated in a source separation distance interval, assuming sources no located at the extreme values at application load axis. In the example explained it is [0.08-0.013 m.].

Source outlined as transfer function between the impact and a flaw inside the material behaves in a way depending on its location inside the material. Such defined source can become distinguishable and discernable according to many factors such as geometry, attenuation, propagation velocity, in-homogeneities and backscattering of the material. In addition, phenomenon of compression, shear and superficial wave propagation is complex which defines a non deterministic problem for source separation by applying ICA.

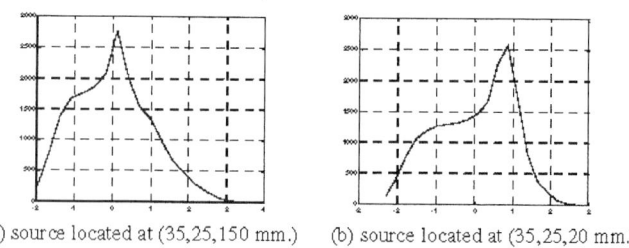

(a) source located at (35,25,150 mm.)   (b) source located at (35,25,20 mm.)

Fig. 5. Pdfs of the sources in Figure 2b.

To test the model, ICA algorithm applied was Equivarient Adaptive Signal Separation Blind Serial Update EASI-BSU [7] using an implementation on FIR polynomial [8]. Figure 6 shows results corresponding to rectangular source at coordinates (35,25,155 mm.) in Figure 2b. Top graphs of Figure 6 show expected waveform for displacements in axis y and spectrum signal, and bottom graphs of Figure 6 show calculated waveform and spectrum signal.

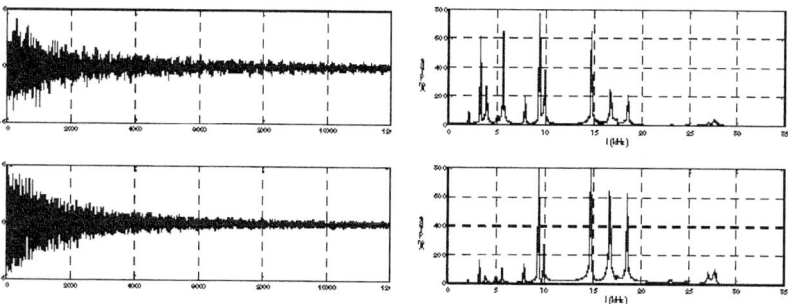

Fig. 6. Signal expected and signal recovered by ICA.

Correlation of expected and calculated waveforms is 0.3 and spectra of both signals are comparables having coincident maximum peaks. In case of cylindrical source located at coordinates (25,25,25 mm.) correlation was 0.25 and spectra were also comparables. Model has had better results with asymmetric defect specimen of Figure 2b than symmetric one in Figure 2a. Additionally, more simulated experiments are being developed for verifying displacement sensitivity in each one of the axes produced by the wavefront propagation to defect shape and location.

4 Conclusions

A novel ICA model applied to multichannel NDE by Impact-Echo has been presented. Model has been validated by means of finite element simulations. It has been found that there exists a material bulk where the model is applicable. This bulk is located surrounding the centre of the material. Results have been presented for detecting two defects in simulated specimens working with modulus of the displacements sensed in a material sensing configuration. Model is being extended to get a 3D material map. In addition it is being tested in real experiments with different materials.

Acknowledgements

This work has been supported by Spanish Administration under grant TIC 2002-04643 and Universidad Politécnica Valencia under interdisciplinary grant 2003-0554.

References

1. Sansalone M. and Streett W. B.: Impact-echo: Non-destructive evaluation of concrete and masonry. Bullbrier Press, USA, June 1997.
2. Vergara,L.,Gosálbez J.,Fuente J.,Miralles R.,Bosch I.,Salazar A.,López A.,Domínguez L.: Ultrasonic Nondestructive testing on Marble Block Rocks, Materials Evaluation. Ed. American Society for Nondestructive Testing, Vol. 62, No. 1, pp. 73-78, Jan 2004.
3. Hyvärinen A.: Independent Component Analysis. John Wiley & Sons, 2001.
4. Cichocki A. and Amari S.: Adaptive Blind Signal and Image Processing: Learning algorithms and applications. Wiley, John & Sons, 2001.
5. Cheeke J. D.: Fundamentals and Applications of Ultrasonic Waves. CRC Press LLC, USA, 2002.
6. Abraham O., Leonard C., Côte P, Piwakowski B.: Time-frequency Analysis of Impact_Echo Signals: Numerical Modeling and Experimental Validation. ACI Materials Journal, v 97 n 6, p 645-657, November-December, 2000.
7. Cardoso J.F. and Laheld B.: Equivariant adaptive source separation. IEEE Trans. On Signal Processing 45(2), 434-444, 1996.
8. Lambert R.: Multichannel blind deconvolution: FIR matrix algebra and separation of multipath mixtures. PhD. thesis at Faculty of Graduate School University of Southern California, 1996.

Monaural Source Separation Using Spectral Cues

Barak A. Pearlmutter[1] and Anthony M. Zador[2]

[1] Hamilton Institute, National University of Ireland Maynooth, Co. Kildare, Ireland
[2] Cold Spring Harbor Laboratory, One Bungtown Rd, Cold Spring Harbor, NY 11724, USA

Abstract. The acoustic environment poses at least two important challenges. First, animals must localise sound sources using a variety of binaural and monaural cues; and second they must separate sources into distinct auditory streams (the "cocktail party problem"). Binaural cues include intra-aural intensity and phase disparity. The primary monaural cue is the spectral filtering introduced by the head and pinnae via the head-related transfer function (HRTF), which imposes different linear filters upon sources arising at different spatial locations.

Here we address the second challenge, source separation. We propose an algorithm for exploiting the monaural HRTF to separate spatially localised acoustic sources in a noisy environment. We assume that each source has a unique position in space, and is therefore subject to preprocessing by a different linear filter. We also assume prior knowledge of weak statistical regularities present in the sources. This framework can incorporate various aspects of acoustic transfer functions (echos, delays, multiple sensors, frequency-dependent attenuation) in a uniform fashion, treating them as cues for, rather than obstacles to, separation. To accomplish this, sources are represented sparsely in an overcomplete basis. This framework can be extended to make predictions about the neural representations required to separate acoustic sources.

1 Introduction

Organisms exploit a variety of binaural and monaural cues to separate acoustic sources, a process sometimes referred to as "stream segregation" [1]. One set of cues that can be used to separate sources is the differential filtering imposed by the head and pinnae (the head-related transfer function, or HRTF) on sources at different positions in space [2]. It is often reasonable to assume that sound arriving from different locations should be treated as arising from distinct sources. While the importance of the HRTF in sound localisation has been studied extensively, its role in source separation *per se* has not received as much scrutiny.

Let us consider a formulation of source separation that includes the HRTF. Suppose there are N acoustic sources $x_i(t)$ located at distinct positions in space. Associated with each position is a distinct spectral filter, given by the corresponding head-related transfer functions $h_i(t)$. The received signal $y(t)$ is then the sum of the filtered signals

$$y(t) = \sum_{i=1}^{N} h_i(t) * x_i(t) \qquad (1)$$

where $*$ indicates convolution. Our goal is to recover the underlying sources $x_i(t)$ from the observed signal $y(t)$, using knowledge of the directional filters[1] $h_i(t)$. Although the HRTF can also be exploited in multi-sensor situations, in the present work we focus only on the more difficult single-sensor case.

2 Monaural Separation Using a Weak Prior

We solve this underdetermined system in a sparse separation framework, with L_1-norm optimisation as a sparseness measure [3–7]. The two-sensor underdetermined case has been addressed in this context [8, 9] but separating multiple sources from a single sensor is harder and requires stronger assumptions [10–13]. In this framework, we model the i-th source $x_i(t)$ as a weighted sum of elements $d_j(t)$ from an overcomplete dictionary,

$$x_i(t) = \sum_j c_{ij} \, d_j(t), \qquad (2)$$

where the weighting associated with dictionary element $d_j(t)$'s contribution to source i is c_{ij}, and the c_{ij} are assumed to be sparse.

In particular, the signals in the dictionary, $d_j(t)$, are chosen with two criteria in mind. First, sources should be sparse when represented in this dictionary, meaning that the coefficients c_{ij} required to represent $x_i(t)$ will have a distribution with more zeros (and more large values) than might be naively expected. A common formalisation of this assumption is that the distribution of coefficients is governed by a Laplacian distribution ($p(c_i) \propto e^{-|c_i|}$); a Laplacian distribution has more elements close to zero (and far from zero) than does a Gaussian with the same variance. Second, dictionary elements should be chosen such that, following transformation by the HRTF, elements differ as much as possible; this is equivalent to minimising the condition number of the matrix $\mathbf{D}$ introduced below.

In what follows, we assume that each source appears at a unique position in space, and that there is only a single source at each position. The components $d_j(t)$ of each source might thus be subject to filtering by any of the HRTFs $h_i(t)$. We therefore construct a new dictionary by applying each possible filter to each original element. We denote the resulting dictionary elements

$$d'_{ij}(t) = h_i(t) * d_j(t). \qquad (3)$$

Note that the number of elements in the new d' dictionary is equal to the number of original dictionary elements times the number of sources N; the original overcomplete basis has now become "more overcomplete" by the factor N.

The source separation problem can now be cast as decomposing $y(t)$ into this overcomplete dictionary by finding appropriate c_{ij} for

$$y(t) = \sum_{ij} c_{ij} \, d'_{ij}(t). \qquad (4)$$

---

[1] The filter terms $h_i(t)$ may be interpreted to include not just the filtering of the head and pinnae, but also the filter function of the acoustic environment, and the audiogram of the ear itself.

Once the coefficients c_{ij} are known, the individual sources can be reconstructed directly from the unfiltered elements $d_j(t)$ using Eq. 2.

Source separation thus requires estimating the coefficients c_{ij}. Let us define **c** as a single column vector containing all the coefficients c_{ij}, with the elements indexed by i, j, and **D** as a matrix whose k-th row holds the elements $d'_{ij}(t_k)$. The columns of **D** are indexed by i and j, and the rows are indexed by k. Finally, let **y** be a column vector whose elements correspond to the discrete-time sampled elements $y(t)$. Thus **y** = **Dc**.

If the dictionary $d'_{ij}(t)$ formed a complete basis, **c** would be given by **c** = $\mathbf{D}^{-1}\mathbf{y}$. However, by assumption the system is now underdetermined – many possible combinations of sources yield the observed sensor data $y(t)$ – so in order to specify a unique solution we must have a way of choosing among them. We therefore introduce a regulariser that incorporates some weak prior information about the problem and renders it well-posed [14]. Here we express the regulariser in terms of an easily stated condition on the norm of the solution vector **c**: Find the **c** that minimises the L_p norm $\|\mathbf{c}\|_p$ subject to **Dc** = **y**, where $\|\mathbf{c}\|_p = (\sum_{ij} |c_{ij}|^p)^{\frac{1}{p}}$.

Different choices for p correspond to different priors and so yield different solutions **c**. A natural choice would seem to be $p = 2$, which corresponds to assuming that the source coefficients c_{ij} were drawn from a Gaussian distribution; this is the solution found by the the pseudo-inverse **c** = $\mathbf{D}^*\mathbf{y}$. However, this choice does not exploit the sparseness assumption about the sources; rather, it seeks a solution in which the power is spread across the sources (Figure 1). With $p = 0$ ($\|\mathbf{c}\|_0$ is the number of nonzero elements of **c**) we would exploit sparseness, but this can be a computationally intractable combinatorial problem, and moreover the solution would not be continuous in **y** and therefore not be robust to noise [15].

Instead, as shown in Figure 1, we use $p = 1$ (the L_1-norm), which is equivalent to a Laplacian prior on the coefficients **c**. That is, we solve

$$\text{minimise} \sum_{ij} |c_{ij}| \text{ subject to } \mathbf{y} = \mathbf{Dc} \tag{5}$$

This has a single global optimum which can be found efficiently using linear programming [3], and is continuous in **y**.

This algorithm can be sensitive to sensor and background noise, as it insists on precisely accounting for the measured signal using some combination of dictionary elements, which can generate large artefacts. However, we can generalise the optimisation problem to include a noise process (simulations not shown) by changing the goal to

$$\text{minimise } \|\mathbf{c}\|_1 \text{ subject to } \|\mathbf{Dc} - \mathbf{y}\|_p \le \beta \tag{6}$$

where β is proportional to the noise level and $p = 1, 2$, or ∞. The Gaussian noise case, $p = 2$, which can also be formulated as unconstrained minimisation, can be solved by Semidefinite Programming [16], or mixed L_1+L_2 optimisation methods used in control theory. Unfortunately these are too computationally burdensome for our purposes. Both $p = 1$ and $p = \infty$ can be solved using linear programming. All these are qualitatively similar, and in them all as $\beta \to 0$ the noise is assumed to be very small, and the solutions converge to that of the zero-noise solution, Eq. 5.

Fig. 1. Minimising the L_1-norm can provide a good and computationally tractable solution to data generated by a sparse prior. In this example, three non-orthogonal basis vectors (*black arrows*) are assumed, and each data point (*black points*) is generated by assuming only one non-zero coefficient c_i (the sparseness assumption), along with a small amount of noise. Since there are three basis vectors in two dimensions, there are many possible solutions, and additional constraints are required to specify the solution. The *red vectors* illustrate solutions found for the *red point* under three different constraints. (*Right*) Minimising the L_0-norm of **c** finds the sparse solution, but is computationally intractable (NP-complete). (*Left*) The L_2-norm can be efficiently minimised by the pseudoinverse, but yields a poor solution because it spreads the power across multiple basis vectors, in violation of the sparseness assumption. (*Centre*) The minimum L_1-norm solution can be found efficiently using linear programming, and under suitable assumptions finds a good approximation of the sparse solution.

Example: Harmonic Comb Prior. We illustrate the algorithm with a simple example. Suppose that the sources can be modelled as simple "musical instruments" playing notes drawn from a 12-tone (Western) scale. Sources are defined by position – there is by definition only a single source at a given position – but each source may play more than one note simultaneously. Each note consists of a "harmonic comb" – a fundamental frequency F and its harmonics nF, $n = 2, 3, \ldots$, with amplitudes $1/n$. Each dictionary element, then, is given by

$$d_i = \sum_{n=1} \frac{1}{n} \sin(2\pi n F_i t). \tag{7}$$

where $F_i = 2^{i/12} F_0$ is the fundamental frequency of the i-th note in the equal-tempered scale, and F_0 is the frequency of the lowest note.

Figure 2 shows that such harmonic comb sources can be readily separated using knowledge of the spectral filtering, provided that one searches for a sparse solution vector **c** by minimising its L_1-norm. In this example three sources were assumed, each playing two "notes" selected from 72. Thus each source is fully described by the values of the two non-zero coefficients.

The top graph of Figure 2 shows the difference between L_1- and L_2-norm minimisation, in the absence of spectral filtering. The L_2-norm solution fits the received signal $y(t)$ using coefficients c_{ij} distributed in a roughly Gaussian fashion, whereas the L_1-norm solution found by linear programming finds a sparse solution in which the only non-zero dictionary coefficients correspond to notes actually present in at least one of the sources. However in the absence of the HRTF, even the L_1-norm solution has no way to assign the notes to the appropriate sources, so it assumes that an equal fraction of each note arises from each source. L_1-norm optimisation thus finds a more interpretable solution than L_2-norm optimisation even without an HRTF, but due to lack of any suitable cues it is equally unable to correctly separate the sources (see Table 1).

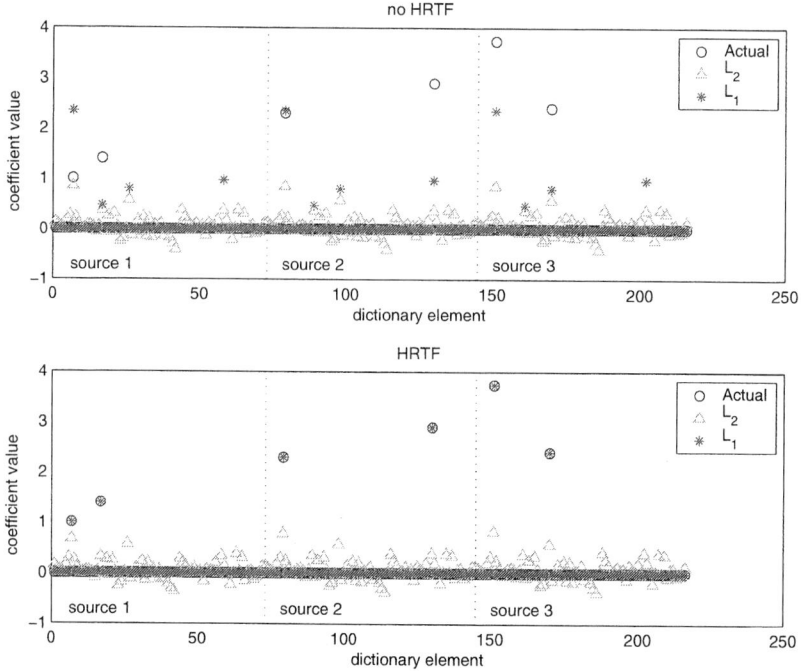

Fig. 2. Spectral cues can be exploited by assuming a sparse prior. The input to the microphone consisted of the sum of three sources (x-axis), each playing two notes but with different amplitudes (y-axis). **(Top)** If no spectral filtering is applied, the algorithm minimising the L_1-norm of the solution **c** accounts for the signal using a small number of coefficients, but cannot assign the correct amplitude to each source. It therefore assumes equal weight among the sources. By contrast, minimising the L_2-norm spreads the energy across many dictionary elements, leading to an uninterpretable solution. **(Bottom)** When a different spectral filter is applied to each source L_1-norm minimisation finds the exact solution, while minimising the L_2-norm yields a solution that remains both uninterpretable and unseparated.

The lower graph of Figure 2 shows how the spectral filtering due to the HRTF can enhance separation. In this case, the L_1-norm constraint is able to separate the sources almost perfectly, while the L_2-norm solution remains poor (see Table 1.) This example, although highly idealised, is intended to capture key features of many real-world problems in which sources have characteristic spectrotemporal signatures. In this framework, more sophisticated models of spectrotemporal structure can be readily accommodated by adding dictionary elements.

3 Discussion

We have described an algorithm for using the head-related transfer function to improve the separation of acoustic sources at different spatial locations. We show how, in certain special cases, the added cues provided by the HRTF permit otherwise unseparable

Table 1. SNR in dB of sources recovered using the proposed algorithm, in a synthetic acoustic environment with versus without an HRTF. Large positive numbers indicate better performance; the best performance is achieved by the algorithm that exploits the HRTF and minimises the L_1-norm of the solution.

norm	SNR without HRTF	SNR with HRTF
L_1	1.78	106.69
L_2	−4.86	−5.19

sources to be separated. We also show how, in the more general case, the cues can be used to improve separation.

The novel contribution of this work is a specific proposal for how the HRTF can be used for source separation, a process related to but distinct from localisation. It has long been known that the HRTF provides important cues for localisation [17–20]. Acoustic sources that bypass the HRTF (*e.g.* those presented with headphones) are typically perceived inside the head, unlike real sounds which are perceived outside the head [20, 21]. The HRTF is not, however, strictly required for localisation; under some conditions, binaural cues are sufficient to localise sounds even in the absence of the HRTF. Conversely, source separation can occur even without spatial cues, for example when selecting out the individual instruments of a concerto presented over a single speaker. Nevertheless, it is clear that the HRTF cues, when present, help in source separation [2].

The present formulation can be readily extended to include binaural information. Each HRTF function is made single-input two-output, and the lengths of the column vectors corresponding to the post-HRTF dictionary elements d'_{ij} and the data vector **y** are doubled. In this way, intra-aural time and level disparity can be used to separate sources. Information from two (or more) sensors can thus be naturally incorporated into the present framework. Similarly, although presented here as a batch algorithm, an online variant which gradually estimates coefficients as the signal becomes available would be straightforward to develop.

3.1 Assumptions About the HRTF

One of the main limitations of the present algorithm is that it requires that the precise HRTF $h_i(t)$ associated with each source be known. This requires knowing both the dependence of HRTF on spatial position, and the spatial position of each source.

The first assumption, that organisms learn their own HRTF, is reasonable and supported by extensive experimental evidence [22–24]. When $h_i(t)$ is interpreted to include not only the HRTF but also the properties of the acoustic environment (reverberations, *etc.*) then this assumption becomes considerably stronger. Animals have, however, been shown to estimate some properties of their acoustic environments quite quickly [25].

The second assumption, that the precise positions of each source are known, is more restrictive. There are, however, several ways in which the source positions might be determined. One possibility is that they might be established by prior or additional knowledge, perhaps using visual information. Indeed, the spatial cues provided by vision can override those inferred from audition, as demonstrated by the "ventriloquist effect." A second possibility is that the positions of the sources could be established through auditory preprocessing, using for example the binaural cues available to the auditory brainstem. Finally, the positions of the sources, as well as the properties of the acoustic environment, could be jointly estimated along with the content of each source; this joint estimation might be made easier by moving the head slightly so as to perturb the HRTFs by some known angle without changing the source positions.

3.2 The Signal Dictionary and Neural Representations

We have not considered the question of how an appropriate signal dictionary might be obtained. Fortunately there is a rich literature on finding a basis matched (in the sense of yielding sparse representations) to an ensemble of signals [6, 26, 27].

The algorithm was developed here in the signal processing framework, with little attention to possible neural implementation. However, overcomplete representations have been suggested for visual areas V1 [27] and IT [28]. Signal dictionaries have been interpreted in terms of models of receptive fields, and receptive field properties have been predicted from the principles of sparse representations [26, 29]. Similarly, the signal elements derived from optimising the matrix D for separating ensembles of natural sounds filtered through the HRTF offers predictions for auditory representations. The extension of such models to auditory cortex is intriguing [30].

Acknowledgements. We thank Didier Depireux, Tomas Hromadka and Mike Deweese for helpful comments. Supported by Higher Education Authority of Ireland and Science Foundation Ireland grant 00/PI.1/C067 (BAP), and grants from the Sloan Foundation, Mathers Foundation, NIH, Packard Foundation and the Redwood Neuroscience Institute (AMZ).

References

[1] Albert S. Bregman. *Auditory Scene Analysis: The Perceptual Organization of Sound*. MIT Press, Cambridge, Massachusetts, 1990. ISBN 0-262-02297-4.

[2] W. A. Yost, Jr. Dye, R. H., and S. Sheft. A simulated "cocktail party" with up to three sound sources. *Percept Psychophys*, 58(7):1026–1036, 1996.

[3] Scott Shaobing Chen, David L. Donoho, and Michael A. Saunders. Atomic decomposition by basis pursuit. *SIAM Journal on Scientific Computing*, 20(1):33–61, 1999.

[4] T.-W. Lee, M. S. Lewicki, M. Girolami, and T. J. Sejnowski. Blind source separation of more sources than mixtures using overcomplete representations. *IEEE Signal Processing Letters*, 4(5):87–90, 1999.

[5] M. Lewicki and B. A. Olshausen. Inferring sparse, overcomplete image codes using an efficient coding framework. In *Advances in Neural Information Processing Systems 10*, pages 815–821. MIT Press, 1998.

[6] M. S. Lewicki and T. J. Sejnowski. Learning overcomplete representations. *Neural Computation*, 12(2):337–365, 2000.

[7] Michael Zibulevsky and Barak A. Pearlmutter. Blind source separation by sparse decomposition in a signal dictionary. *Neural Computation*, 13(4):863–882, April 2001.

[8] P. Bofill and M. Zibulevsky. Underdetermined blind source separation using sparse representations. *Signal Processing*, 81(11):2353–2362, 2001.

[9] S. T. Rickard and F. Dietrich. DOA estimation of many W-disjoint orthogonal sources from two mixtures using DUET. In *Proceedings of the 10th IEEE Workshop on Statistical Signal and Array Processing (SSAP2000)*, pages 311–314, Pocono Manor, PA, August 2000.
[10] G. Cauwenberghs. Monaural separation of independent acoustical components. In *Proc. IEEE Int. Symp. Circuits and Systems (ISCAS'99)*, volume 5, pages 62–65, Orlando FL, 1999.
[11] Sepp Hochreiter and Michael C. Mozer. Monaural separation and classification of mixed signals: A support-vector regression perspective. In Te-Won Lee, Tzyy-Ping Jung, Scott Makeig, and Terrence J. Sejnowski, editors, *3rd International Conference on Independent Component Analysis and Blind Signal Separation*, San Diego, CA, December 9-12 2001.
[12] Gil-Jin Jang and Te-Won Lee. A maximum likelihood approach to single-channel source separation. *Journal of Machine Learning Research*, 4:1365–1392, December 2003.
[13] Sam T. Roweis. One microphone source separation. In *Advances in Neural Information Processing Systems 13*, pages 793–799. MIT Press, 2001.
[14] T. Poggio, V. Torre, and C. Koch. Computational vision and regularization theory. *Nature*, 317(6035):314–319, 1985.
[15] D. L. Donoho and M. Elad. Maximal sparsity representation via l1 minimization. *Proceedings of the National Academy of Sciences*, 100:2197–2202, March 2003.
[16] R. Fletcher. Semidefinite matrix constraints in optimization. *SIAM J. Control and Opt.*, 23: 493–513, 1985.
[17] P. M. Hofman and A. J. Van Opstal. Bayesian reconstruction of sound localization cues from responses to random spectra. *Biol Cybern*, 86(4):305–16, 2002.
[18] E. I. Knudsen and M. Konishi. Mechanisms of sound localization in the barn owl. *Journal of Comparative Physiology*, 133:13–21, 1979.
[19] E. M. Wenzel, M. Arruda, D. J. Kistler, and F. L. Wightman. Localization using nonindividualized head-related transfer functions. *J Acoust Soc Am*, 94(1):111–23, 1993.
[20] F. L. Wightman and D. J. Kistler. Headphone simulation of free-field listening. II: Psychophysical validation. *J Acoust Soc Am*, 85(2):868–78, 1989.
[21] A. Kulkarni and H. S. Colburn. Role of spectral detail in sound-source localization. *Nature*, 396(6713):747–749, 1998.
[22] A. J. King, C. H. Parsons, and D. R. Moore. Plasticity in the neural coding of auditory space in the mammalian brain. *Proc Natl Acad Sci USA*, 97(22):11821–11828, 2000.
[23] B. A. Linkenhoker and E. I. Knudsen. Incremental training increases the plasticity of the auditory space map in adult barn owls. *Nature*, 419(6904):293–296, 2002.
[24] P. M. Hofman, J. G. Van Riswick, and A. J. Van Opstal. Relearning sound localization with new ears. *Nat Neurosci*, 1(5):417–421, 1998.
[25] B. G. Shinn-Cunningham. Models of plasticity in spatial auditory processing. *Audiology and Neuro-Otology*, 6(4):187–191, 2001.
[26] Anthony J. Bell and Terrence J. Sejnowski. The 'independent components' of natural scenes are edge filters. *Vision Research*, 37(23):3327–3338, 1997.
[27] B. A. Olshausen and D. J. Field. Sparse coding with an overcomplete basis set: A strategy employed by V1? *Vision Research*, 37(23):3311–3325, 1997.
[28] M. Riesenhuber and T. Poggio. Models of object recognition. *Nature Neuroscience*, 3 Suppl:1199–1204, 2000.
[29] B. Olshausen and D. J. Field. Emergence of simple-cell receptive field properties by learning a sparse code for natural images. *Nature*, 381:607–609, 1996.
[30] B. A. Olshausen and K. N. O'Connor. A new window on sound. *Nature Neuroscience*, 5: 292–293, 2002.

Multichannel Speech Separation Using Adaptive Parameterization of Source PDFs

Kostas Kokkinakis* and Asoke K. Nandi

Signal Processing and Communications Group
Department of Electrical Engineering and Electronics
The University of Liverpool, Brownlow Hill, Liverpool, L69 3GJ, UK
{kokkinak,a.nandi}@liv.ac.uk

Abstract. Convolutive and temporally correlated mixtures of speech are tackled with an LP-based temporal pre-whitening stage combined with the natural gradient algorithm (NGA), to essentially perform spatial separation by maximizing entropy at the output of a nonlinear function. In the past, speech sources have been parameterized by the generalized Gaussian density (GGD) model, in which the exponent parameter directly relates to the exponent of the corresponding optimal nonlinear function. In this paper, we present an adaptive, source dependent estimation of this parameter, controlled exclusively by the statistics of the output source estimates. Comparative experimental results illustrate the inherent flexibility of the proposed method, as well as an overall increase in convergence speed and separation performance over existing approaches.

1 Introduction

This paper addresses the problem of blind signal separation (BSS), in the general case where any m observed signals $\mathbf{x}(t) = [x_1(t), \ldots, x_m(t)]^T \in \mathbb{R}^m$, are considered to be linear and convolutive mixtures of n unknown and yet statistically independent (at each time instant) sources $\mathbf{s}(t) = [s_1(t), \ldots, s_n(t)]^T \in \mathbb{R}^n$. In this typical scenario for real acoustic environments, the signal observed at the ith microphone is:

$$x_i(t) = \sum_{j=1}^{n} \sum_{k=0}^{l-1} h_{ij}(k)\, s_j(t-k), \quad i = 1, 2, \ldots, m. \tag{1}$$

with t the discrete-time index, $[h_{ij}(k)]$ the room impulse response characterizing the path between the jth source and the ith sensor and $(l-1)$ the order of the FIR filters that model the room acoustic effects. The same model in the z-domain reads:

$$X_i(z) = \sum_{j=1}^{n} H_{ij}(z)\, S_j(z), \quad i = 1, 2, \ldots, m. \tag{2}$$

---

* This work is supported by the Engineering and Physical Sciences Research Council of the UK and the University of Liverpool.

Various authors have used multichannel blind deconvolution (MBD) techniques for speech separation and enhancement, by resorting to the frequency or z-domain [9]–[11]. A typical assumption made here – apart from the spatial independence of the sources – is that each source is also an i.i.d. sequence. In the case of speech sources, this results in extracting whitened (equalized) and therefore audibly unnatural estimates. To resolve this issue, an LP-based MBD method (LP-NGA), capable of retaining the original source spectral characteristics has been recently proposed in [6]–[8]. The same authors in [7], have shown a vast improvement in performance, when LP-NGA is coupled with a parameterized nonlinearity stemming from the GGD model. In [3] and [12], the same GG parametric density is chosen to derive general adaptive activation functions able to tackle both super and sub-Gaussian sources with respect to kurtosis, while in [13] such a task is made simpler with the use of an adaptive threshold parameter.

In BSS via entropy maximization [2], separation efficiency and convergence are closely related to the nonlinear function used to model the source estimates. Dissimilarities are unavoidable, especially when modelling under the assumption of a fixed distribution shape or for a different number of arbitrary sources. To overcome this inefficiency in the case of convolutive mixtures of speech, we introduce a flexible modelling parameter capable of characterizing the unknown source distributions. Through experimental results, we further show that an accurate and adaptive estimation of this parameter, leads to significant increase in separation performance and speed of convergence.

2 MBD in Frequency Domain with the Natural Gradient

In [2], Bell and Sejnowski showed that entropy maximization at the output of a nonlinearity, tuned to the cdf of the sources, can lead to blind extraction of the independent components of a linear instantaneous set of mixtures. An efficient update rule was later proposed by Amari et al. [1], to maximize entropy following its natural gradient. In the framework of convolutive mixtures, the potential of these methods was quickly realized and further explored with the use of FIR polynomials operating in the frequency domain [10]–[11]. Based on [9], any FIR filter mixing matrix may be transformed into an FIR polynomial matrix by performing a Fourier transform on its elements. Thus, for a j-source and i-sensor system configuration, the mixing matrix in (2) can be written as the FIR polynomial matrix $\underline{\mathbf{H}}(z)^{(i \times \ell \times j)}$, with its elements being complex valued FIR polynomials given by:

$$H_{ij}(z) = \sum_{\ell=0}^{k} h_{ij}(\ell) z^{-\ell} \qquad (3)$$

where the indices, $i = [1, 2, \ldots, m]$, $j = [1, 2, \ldots, n]$ and $\ell = [0, 1, \ldots, l-1]$, represent the observations, sources and each filter coefficient, respectively. Consequently, the natural gradient algorithm (NGA) can be shown to adopt the following form:

$$\underline{\mathbf{W}}_{k+1} = \underline{\mathbf{W}}_k + \mu \left[\underline{\mathbf{I}} - \text{FFT}\left[\varphi(\mathbf{u})\right] \underline{\mathbf{u}}^H \right] \underline{\mathbf{W}}_k \qquad (4)$$

where $(\cdot)^H$ is the Hermitian operator, μ the step size and $\mathbf{W}$ is the spatial separation FIR polynomial matrix. In addition, the vector FFT $[\varphi(\mathbf{u})]$ defines the frequency domain representation of the nonlinear monotonic activation function $\varphi(\mathbf{u}) = [\varphi_1(u_1), \ldots, \varphi_m(u_m)]^T$, which in turn operates in the time domain and is equal to:

$$\varphi_i(u_i) = -\frac{\partial \log p_{u_i}(u_i)}{\partial u_i} = -\frac{\frac{\partial p_{u_i}(u_i)}{\partial u_i}}{p_{u_i}(u_i)}, \quad i = 1, 2, \ldots, m. \tag{5}$$

where $p_{u_i}(u_i)$ defines the pdf of each source estimate u_i. Recently in [6]–[8], we have shown that by endorsing a temporal pre-whitening LP-based stage, it is possible to preserve the original spectral characteristics of each source contribution. We follow a previously overlooked avenue and exploit the invariance of the speech temporal model with respect to the mixing model to extract only the contribution of each source by applying the estimated spatial separation filters to the original mixtures. Using the same update rule as in (4), the LP-NGA MBD method yields the spatially separated yet temporally correlated source estimates:

$$\underline{\mathbf{u}}(z) = [U_1(z), \ldots, U_m(z)]^T = \underline{\mathbf{W}}(z) \, \underline{\mathbf{x}}(z) \tag{6}$$

for $i = 1, 2, \ldots, m$.

3 Flexible Nonlinearity Based on the GGD

3.1 Generalized Gaussian Density for Speech

Speech closely follows a Laplacian distribution as long as the samples chosen are restricted in voiced activity intervals, whilst its density has shown to exhibit the characteristics of a Gamma pdf when both voiced and silence regions are taken into account [5]. In general, we may sufficiently approximate the speech distribution by employing the generalized Gaussian density (GGD) model. For any zero-mean speech signal x_i, the generalized Gaussian distribution is defined as:

$$p_{x_i}(x_i) = \frac{\beta_i}{2\alpha_i \Gamma(1/\beta_i)} \, e^{-(|x_i|/\alpha_i)^{\beta_i}} \tag{7}$$

where α_i and β_i are positive real parameters defined for $i = 1, 2, \ldots, m$, while the Gamma function $\Gamma(\cdot)$ is expressed as:

$$\Gamma(y) = \int_0^\infty x_i^{y-1} e^{-x_i} dx_i \tag{8}$$

Each $\alpha_i > 0$ is a generalized measure of the variance of the distribution and is referred to as the dispersion or scale parameter, while $\beta_i > 0$ describes the exponential rate of decay and in general, the shape of the distribution. As special cases of the GGD, a Laplacian distribution is defined for $\beta = 1$, a standard Gaussian distribution for $\beta = 2$ and a Gamma distribution for $\beta = 0.5$. These

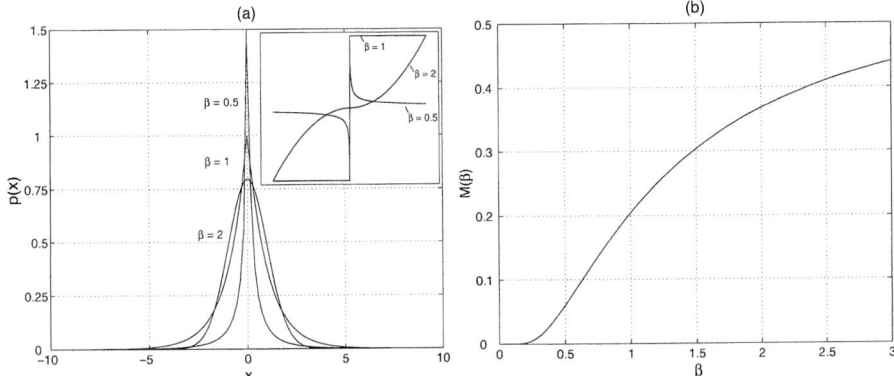

Fig. 1. (a) GGD model for different values of the shape parameter $\beta = 0.5, 1, 2$, with the corresponding nonlinear functions $\varphi_i(\cdot)$ derived for each distribution superimposed in the graph and (b) evolution and behaviour of function $\mathcal{F}_M(\beta)$ in (16) for $\beta \in [0, 3]$.

are all shown in Fig. 1(a). In relation to a Gaussian distribution, essentially a zero-kurtosis distribution referred to as mesokurtic, the above may be also classified as super-Gaussian or leptokurtic distributions, with all having a positive kurtosis. Common nonlinearities proposed for super-Gaussian distributions, generally employ sigmoidal functions, such as sign$(\cdot)$ or tanh$(\cdot)$ [12]. Resorting to the GGD model however, we may define a rather more general expression for such activation functions. As in [3], these can be shown to adopt a parametric structure, based solely on the exponent parameter β of the distribution. The family of the GGD-based nonlinear activation functions is given by:

$$\varphi_i(u_i) = |u_i|^{\beta_i - 1} \operatorname{sign}(u_i) \qquad (9)$$

which by taking into account that $\operatorname{sign}(u_i) = u_i/|u_i|$, further reduces to:

$$\varphi_i(u_i) = \frac{u_i}{|u_i|^{2-\beta_i}}, \quad 0 < \beta_i < 1 \qquad (10)$$

defined for $u_i \neq 0$ with $\varphi_i(u_i)$ acting elementwise on the source estimate components u_i for all $i = 1, 2, \ldots, m$.

3.2 Adaptive Estimation of the Exponent Parameter

For all allowed values of α_i and β_i, it may be further shown that the rth-order absolute central moment for a generalized Gaussian signal is given by:

$$E\left[|x_i|^r\right] = \int_{-\infty}^{\infty} |x_i|^r p_{x_i}(x_i)\, dx_i \qquad (11)$$

where $E[\cdot]$ represents the expectation operator. Substituting (7) into (11), it is simple to show that the rth-order moments are in general defined as:

$$m_r^{(i)} = E\left[|x_i|^r\right] = \alpha_i^r \frac{\Gamma\left(\frac{r+1}{\beta_i}\right)}{\Gamma\left(\frac{1}{\beta_i}\right)}, \quad \beta_i > 0 \tag{12}$$

which for example in the case of $r = 2$, yields the following:

$$m_2^{(i)} = E\left[|x_i|^2\right] = \alpha_i^2 \frac{\Gamma\left(\frac{3}{\beta_i}\right)}{\Gamma\left(\frac{1}{\beta_i}\right)} \tag{13}$$

Assuming a unit variance, the above also provides an expression for the term α_i, defined as:

$$\alpha_i = \sqrt{\frac{\Gamma\left(\frac{1}{\beta_i}\right)}{\Gamma\left(\frac{3}{\beta_i}\right)}} \tag{14}$$

with (11)–(14) all defined for every $i = 1, 2, \ldots, m$. To obtain a complete statistical description for the distributions of the source estimates with respect to the underlying shape parameter, we resort to the method of moments [14]. In particular, we propose the use of the following ratio:

$$\mathcal{F}_{M_i} = \frac{\left|m_1^{(i)}\right|^2}{\sqrt{m_4^{(i)}}} \tag{15}$$

where $m_1^{(i)}$ and $m_4^{(i)}$ are the first and fourth-order moments, respectively, both defined for each x_i. Other consistent moment ratios are also given in [3] and [4]. Combining (15) with (12), we obtain an expression for β_i leading to:

$$\mathcal{F}_M(\beta_i) = \frac{\Gamma^2\left(\frac{2}{\beta_i}\right)}{\sqrt{\Gamma\left(\frac{5}{\beta_i}\right)} \cdot \sqrt{\Gamma^3\left(\frac{1}{\beta_i}\right)}} \tag{16}$$

To estimate β_i, we need to solve (16) above, which cannot be inverted in an explicit form. Alternatively, we use a look-up table, which is simply constructed by projecting the ratio of moments $\mathcal{F}_M(\beta_i)$ on different values of β_i. $\mathcal{F}_M(\beta_i)$, is in fact a steadily increasing function of β_i, as depicted in Fig. 1(b). In addition, several experiments with a large number of different ratios of moments, have shown that in the case of speech – where in the majority of cases β is limited to $[0, 1]$ – the ratio defined in (15) provides a considerably more accurate estimate. As opposed to the flexible ICA algorithm proposed in [3], which focuses only on a limited number of values of β, we propose a continuously adaptive estimation of the shape parameter driven exclusively from the output speech estimates.

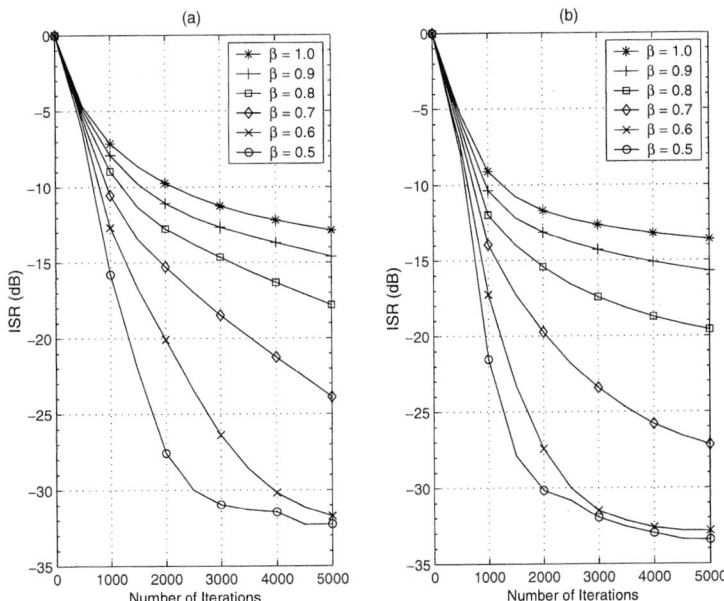

Fig. 2. ISR (dB) versus number of iterations. Performance of LP-NGA for different values of the exponent parameter $\beta \in [0.5, 1.0]$, (a) with a fixed step size $\mu = 0.001$ and (b) an exponentially decaying step size $\mu_{\exp}$.

At the end of each iteration, the vector of the exponents $\boldsymbol{\beta} = \begin{bmatrix} \beta_1, \ldots, \beta_m \end{bmatrix}^T$ is estimated from the available source estimates based on (15)–(16). These values are then used to adaptively estimate the parametric nonlinearity defined in (10) as a function of these parameters. In fact, this method accomplishes two objectives – i) it uses the 'best' estimate of the shape parameter at each iteration, i.e., it evolves, and ii) it uses the 'best' estimate for different sources, i.e., it adapts appropriately to each source distribution.

4 Experimental Results

The data set used as the original sources, employs two female speech signals. The corresponding algorithm parameters are summarized in Table 1. Convolutive mixtures are generated from a non-minimum phase mixing system, while separation performance is measured using the interference-to-signal (ISR) ratio as defined in [6]–[7]. To investigate the separation performance with respect to the exponent parameter, the LP-NGA is first executed using the GGD-based nonlinear function defined in (10) with a fixed step size of $\mu = 0.001$, for different values of β in the range $[0.5, 1.0]$. As Fig. 2(a) reveals, separation performance depends greatly on the chosen nonlinearity and in effect on the assigned exponent value. Next, the same experiment is re-run, with an exponentially decaying learning rate $\mu_{\exp}$ employed in the update of (4). Performance slightly improves

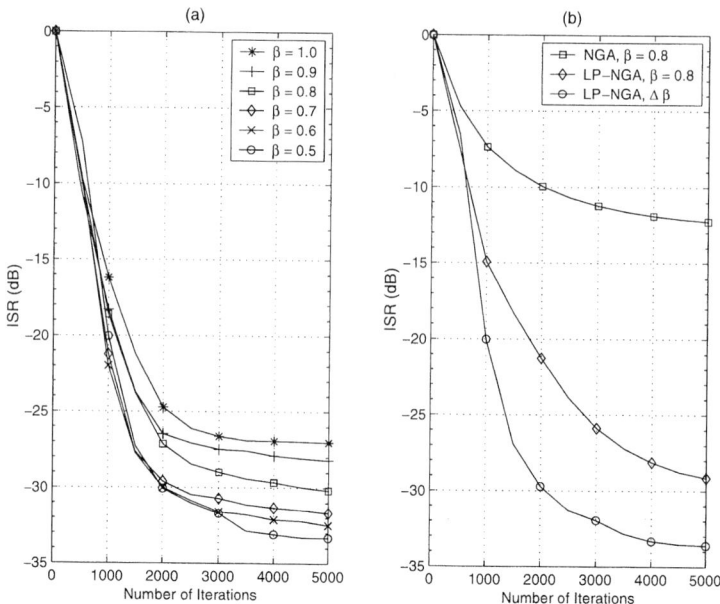

Fig. 3. ISR (dB) versus number of iterations. (a) Performance of LP-NGA for $\beta \in [0.5, 1.0]$ with an individually chosen step size and (b) performance of the NGA and LP-NGA with $\beta = 0.8$ and the LP-NGA with the adaptive parametric nonlinearity.

for each individual β, however as shown in Fig. 2(b), there still exists a substantial deviation in ISR values between $\beta = 0.5$ and $\beta = 1.0$. Fig. 3(a) depicts the ISR achieved, when the learning rate is explicitly tuned for maximum possible performance for each value of the exponent. Although significantly smaller, there is still a considerable difference in performance (mostly by about 5 dB) between $\beta = 0.5$ and $\beta = 1.0$. Finally, the parametric nonlinear function in (10), controlled by continuously adaptive estimates of β is put to use. In this case, the LP-NGA clearly outperforms both the NGA and the LP-NGA when both operate using $\beta = 0.8$ (as suggested in [3]). The increase in separation performance accomplished and shown here in Fig. 3(b), is almost 20 dB when compared with the performance of the former and 5 dB when measured against the latter.

Table 1. Algorithm parameters.

Length of speech signals	5 seconds
Sampling frequency	8 kHz
Blocksize	$M = 128$ points
Order of LP filters	$p = 15$
Separating filters	$\underline{\mathbf{W}} = 2 \times 256 \times 2$
Number of runs	$N = 30$

5 Conclusions

We have introduced source dependent flexibility into the separating parametric nonlinear function, solely controlled by the exponent parameter of the underlying source distributions. Through the LP-NGA BSS method which effectively combines the NGA with entropy maximization, we have demonstrated that the adaptive estimation of the exponent parameters from the current source estimates, increases flexibility, separation performance and convergence speed when compared against approaches, operating only on a fixed-exponent nonlinearity.

References

1. S. Amari, A. Cichocki and H. Yang, "A New Learning Algorithm for Blind Signal Separation" *Advances in Neural Information Processing Systems 8*, MIT Press, Cambridge, 1996, pp. 757–763.
2. A. Bell and T. Sejnowski, "An Information Maximization Approach to Blind Separation and Blind Deconvolution" *Neural Computation*, Vol. 7, No. 6, July 1995, pp. 1129–1159.
3. S. Choi, A. Cichocki and S. Amari, "Flexible Independent Component Analysis" *Journal of VLSI Signal Processing*, Vol. 26, No. 1, August 2000, pp. 25–38.
4. M. N. Do and M. Vetterli, "Wavelet-Based Texture Retrieval Using Generalized Gaussian Density and Kullback-Leibler Distance" *IEEE Trans. on Image Processing*, Vol. 11, No. 2, February 2002, pp. 146–158.
5. S. Gazor and W. Zhang, "Speech Probability Distribution" *IEEE Signal Processing Letters*, Vol. 10, No. 7, July 2003, pp. 204–207.
6. K. Kokkinakis, V. Zarzoso and A. K. Nandi, "Blind Separation of Acoustic Mixtures based on Linear Prediction Analysis" In *Proc. Fourth Int. Symp. on ICA and BSS*, Nara, Japan, April 1–4, 2003, pp. 343–348.
7. K. Kokkinakis and A. K. Nandi, "Optimal Blind Separation of Convolutive Audio Mixtures without Temporal Constraints" In *Proc. IEEE Int. Conf. on Acoustics, Speech and Signal Processing*, Montreal, Canada, May 17–21, 2004, pp. 217–220.
8. K. Kokkinakis and A. K. Nandi, "Multichannel Blind Deconvolution for Source Separation in Convolutive Mixtures of Speech" Submitted to *IEEE Trans. on Speech and Audio Processing*, February 2004.
9. R. H. Lambert, *Multichannel Blind Deconvolution: FIR Matrix Algebra and Separation of Multipath Mixtures*. Ph.D. Thesis, University of Southern California, May 1996.
10. R. H. Lambert and A. J. Bell, "Blind Separation of Multiple Speakers in a Multipath Environment" In *Proc. IEEE Int. Conf. on Acoustics, Speech and Signal Processing*, Munich, Germany, April 21–24, 1997, pp. 423–426.
11. T.-W. Lee, A. J. Bell, and R. Orglmeister, "Blind Source Separation of Real World Signals" In *Proc. ICNN*, Texas, June 9–12, 1997, pp. 2129–2135.
12. T.-W. Lee, M. Girolami and T. Sejnowski, "Independent Component Analysis Using an Extended Infomax Algorithm for Mixed Subgaussian and Supergaussian Sources" *Neural Computation*, Vol. 11, No. 2, February 1999, pp. 417–441.
13. H. Mathis, T. P. von Hoff and M. Joho, "Blind Separation of Signals with Mixed Kurtosis Signs Using Threshold Activation Functions" *IEEE Trans. on Neural Networks*, Vol. 12, No. 3, May 2001, pp. 618–624.
14. M. K. Varanasi and B. Aazhang, "Parametric Generalized Gaussian Density Estimation" *J. Acoust. Soc. America*, Vol. 86, No. 4, October 1989, pp. 1404–1415.

Non-negative Matrix Factor Deconvolution; Extraction of Multiple Sound Sources from Monophonic Inputs

Paris Smaragdis

Mitsubishi Electric Research Laboratories
201 Broadway, Cambridge MA, 02139, USA
paris@merl.com

Abstract. In this paper we present an extension to the Non-Negative Matrix Factorization algorithm which is capable of identifying components with temporal structure. We demonstrate the use of this algorithm in the magnitude spectrum domain, where we employ it to perform extraction of multiple sound objects from a single channel auditory scene.

1 Introduction

Non-Negative Matrix Factorization (NMF), was introduced as a concept independently by Paatero (1997) as the Positive Matrix Factorization, and by Lee and Seung (1999) who also proposed some very efficient algorithms for its computation. Since its inception NMF has been applied successfully to a variety of problems despite a hazy statistical underpinning. In this paper we will introduce an extension of NMF for time series, which is useful for problems akin to source separation for single channel inputs.

2 Non-negative Matrix Factorization

The original formulation of NMF is defined as follows. Starting with a non-negative $M \times N$ matrix $\mathbf{V} \in \mathbb{R}^{\geq 0, M \times N}$ the goal is to approximate it as a product of two non-negative matrices $\mathbf{W} \in \mathbb{R}^{\geq 0, M \times R}$ and $\mathbf{H} \in \mathbb{R}^{\geq 0, R \times N}$ where $R \leq M$, such that we minimize the error of reconstruction of $\mathbf{V}$ by $\mathbf{W} \cdot \mathbf{H}$. The success of the reconstruction can be measured using a variety of cost functions, in this paper we will use a cost function introduced by Lee and Seung (1999):

$$D = \left\| \mathbf{V} \otimes \ln(\frac{\mathbf{V}}{\mathbf{W} \cdot \mathbf{H}}) - \mathbf{V} + \mathbf{W} \cdot \mathbf{H} \right\|_F \quad (1)$$

where $\|\cdot\|_F$ is the Frobenius norm and $\otimes$ is the Hadamard product (an element-wise multiplication); the division is also element-wise. Lee and Seung (2000) also introduced an efficient multiplicative update algorithm to optimize this function without the need for constraints to enforce non-negativity:

$$\mathbf{H} = \mathbf{H} \otimes \frac{\mathbf{W}^\top \cdot \frac{\mathbf{V}}{\mathbf{W} \cdot \mathbf{H}}}{\mathbf{W}^\top \cdot \mathbf{1}}, \quad \mathbf{W} = \mathbf{W} \otimes \frac{\frac{\mathbf{V}}{\mathbf{W} \cdot \mathbf{H}} \cdot \mathbf{H}^\top}{\mathbf{1} \cdot \mathbf{H}^\top} \quad (2)$$

where **1** is a $M \times N$ matrix with all its elements set to unity, and the divisions are again element-wise. The variable R corresponds to the number of basis functions to extract. It is usually set to a small number so that NMF results into a low-rank approximation.

2.1 NMF for Sound Object Extraction

It has been shown (Casey and Westner 2000, Smaragdis 2001) that sequentially applying PCA and ICA on magnitude short-time spectra results in decompositions which permits extraction of multiple simple sounds from single-channel inputs. A similar NMF formulation is developed here. Consider a sound scene $s(t)$, and its short-time Fourier transform packed into a $M \times N$ matrix:

$$\mathbf{F} = DFT \begin{bmatrix} s(t_1) & s(t_2) & & s(t_N) \\ \vdots & \vdots & \cdots & \vdots \\ s(t_1 + M - 1) & s(t_2 + M - 1) & & s(t_N + M - 1) \end{bmatrix} \quad (3)$$

where M is the DFT size and N the overall number of frames computed[1]. From the matrix $\mathbf{F} \in \mathbb{R}^{M \times N}$ we can extract the magnitude of the transform $\mathbf{V} = |\mathbf{F}|, \mathbf{V} \in \mathbb{R}^{\geq 0, M \times N}$ and then apply NMF on it. To better understand the point of this operation consider the spectrogram in figure 1.

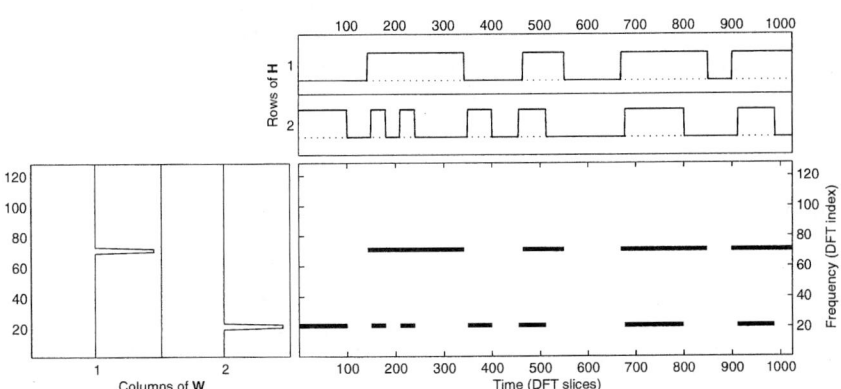

Fig. 1. NMF on spectrograms. The lower right plot is the input magnitude spectrogram, it represents two sinusoids with randomly gated amplitudes. The two columns of **W**, interpreted as spectral bases, are shown in the leftmost plot. The rows of **H**, depicted at the top plot, are the time weights corresponding to the two spectral bases.

It it easily seen that this spectrogram defines a scene that it composed out of sinusoids of two frequencies beeping in and out in some random manner.

---

[1] Ideally we would also apply a window function to the input sound to improve the spectral estimation. Since it this isn't a crucial addition to the process, we omit it for notational simplicity.

Applying a two-component NMF on this signal we obtain the two factors **W** and **H** also shown in figure 1. If we examine the two columns of **W**, shown at the leftmost plots of the figure, we notice that they have energy only at the two frequencies that are present in the input spectrogram. We can interpret these two columns as basis functions for the spectra contained in the spectrogram. Likewise the rows of **H**, shown at the top of the figure, only have energy at the time points where the two sinusoids do. We can interpret the rows of **H** as the weights of the spectral bases at each time. The bases and the weights have a one-to-one correspondence. The first basis describes the spectrum of one of the sinusoids and the first weight vector describes its time envelope. Likewise the other sinusoid is described in both time and frequency by the set of the second basis and second weight vector. In effect we can say that we have performed a rudimentary sound scene description. Although we presented a simplistic scenario this method is powerful enough to dissect even a piece of complex piano music to a set of weights and spectral bases describing each note played and its position in time, effectively performing musical transcription (Smaragdis 2003).

3 Non-negative Matrix Factor Deconvolution

The process we described above works well for many audio tasks. It is however a weak model since it does not take into account the relative positions of each spectrum thereby discarding temporal information. In this section we will introduce an extended version of NMF which deals with this issue. In the previous section we used the model $\mathbf{V} \approx \mathbf{W} \cdot \mathbf{H}$. In this section we will extend it to:

$$\mathbf{V} \approx \sum_{t=0}^{T-1} \mathbf{W}_t \cdot \overset{t \rightarrow}{\mathbf{H}} \qquad (4)$$

where $\mathbf{V} \in \mathbb{R}^{\geq 0, M \times N}$ is the input we wish to decompose, and $\mathbf{W}_t \in \mathbb{R}^{\geq 0, M \times R}$ and $\mathbf{H} \in \mathbb{R}^{\geq 0, R \times N}$ are the bases and weights matrices. The $\overset{i \rightarrow}{(\cdot)}$ operator shifts the columns of its argument by i spots to the right. So that:

$$\mathbf{A} = \begin{bmatrix} 1 & 2 & 3 & 4 \\ 5 & 6 & 7 & 8 \end{bmatrix}, \overset{0 \rightarrow}{\mathbf{A}} = \begin{bmatrix} 1 & 2 & 3 & 4 \\ 5 & 6 & 7 & 8 \end{bmatrix}, \overset{1 \rightarrow}{\mathbf{A}} = \begin{bmatrix} 0 & 1 & 2 & 3 \\ 0 & 5 & 6 & 7 \end{bmatrix}, \overset{2 \rightarrow}{\mathbf{A}} = \begin{bmatrix} 0 & 0 & 1 & 2 \\ 0 & 0 & 5 & 6 \end{bmatrix}, \ldots \qquad (5)$$

The leftmost columns of the matrix are appropriately set to zero so as to maintain the original size of the input. Likewise we define the inverse operation $\overset{\leftarrow i}{(\cdot)}$, which shifts columns to the left.

Just as before our objective is to find a set of $\mathbf{W}_t$ and a $\mathbf{H}$ to approximate $\mathbf{V}$ as best as possible. We set $\mathbf{\Lambda} = \sum_{t=0}^{T-1} \mathbf{W}_t \cdot \overset{t \rightarrow}{\mathbf{H}}$ and define the cost function:

$$D = \left\| \mathbf{V} \otimes \ln(\frac{\mathbf{V}}{\mathbf{\Lambda}}) - \mathbf{V} + \mathbf{\Lambda} \right\|_F \qquad (6)$$

To optimize this model we can use a strategy akin to the one presented above, only this time we will have to optimize more than just two matrices. The update rules for this case will be the same as when performing NMF for each iteration of t, plus some shifting to appropriately line up the arguments:

$$\mathbf{H} = \mathbf{H} \otimes \frac{\mathbf{W}_t^\top \cdot \overset{\leftarrow t}{[\frac{\mathbf{V}}{\mathbf{\Lambda}}]}}{\mathbf{W}_t^\top \cdot \mathbf{1}} \quad \text{and} \quad \mathbf{W}_t = \mathbf{W}_t \otimes \frac{\frac{\mathbf{V}}{\mathbf{\Lambda}} \cdot \overset{t \to}{\mathbf{H}}^\top}{\mathbf{1} \cdot \overset{t \to}{\mathbf{H}}^\top}, \quad \forall t \in [0...T-1] \quad (7)$$

In every training iteration for each t we update $\mathbf{H}$ and each $\mathbf{W}_t$. That way we can optimize the factors in parallel and account for their interplay. In complex cases it is often useful to average the updates of $\mathbf{H}$ over all t's. Due to the rapid convergence properties of the multiplicative rules there is the danger that $\mathbf{H}$ has been more influenced by the last $\mathbf{W}_t$ used for its update, rather than the entire ensemble of $\mathbf{W}_t$. To gain some intuition on the form of the factors $\mathbf{W}_t$ and $\mathbf{H}$, consider the data in figure 2.

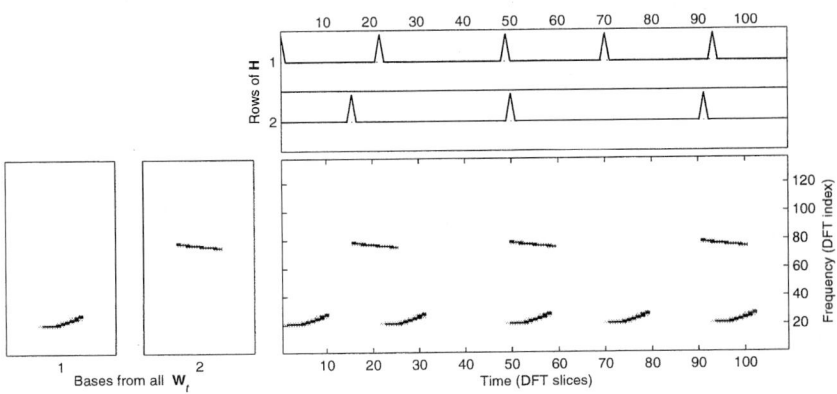

Fig. 2. A spectrogram and the extracted NMFD bases and weights. The lower right plot is the magnitude spectrogram that we used as an input to NMFD. The two leftmost plots are derived from $\mathbf{W}$, and are interpreted as temporal-spectral bases. The rows of $\mathbf{H}$, depicted at the top plot, are the time weights corresponding to the two bases. Note that the leftmost plots have been zero-padded in these figures from left and right so as to appear in the same scale as the input plot.

Just like the previous example the scene contains two randomly repeating elements, however they exhibit a temporal structure which cannot be expressed by spectral bases spanning a single time unit. We perform a two-component NMFD with $T = 10$. This results into a $\mathbf{H}$ and T $\mathbf{W}_t$ matrices of size $M \times 2$. The nth column of the tth $\mathbf{W}_t$ matrix is the nth basis offset by t spots in the left-right dimension (time in our case). In other words the $\mathbf{W}_t$ matrices contain bases that extend in both dimensions of the input. $\mathbf{H}$, like in regular NMF, holds the weights of these functions. Examining figure 2 we see that the bases in $\mathbf{W}_t$ contain the finer temporal information in the present patterns, while $\mathbf{H}$ localizes them in time.

3.1 NMFD for Sound Object Extraction

Using the above formulation of NMFD we analyze a sound snippet which contains a set of drum sounds. In this example the drum sounds exhibit some overlap at both time and frequency. The input was sampled at 11.025 Hz and analyzed with 256-point DFTs which were overlapping by 128-points. A hanning window was applied to the input to improve the spectral estimate. NMFD was performed for 3 basis functions each with a time extend of 10 DFT frames ($R = 3$ and $T = 10$). The results are shown in figure 3. There are three types of drum sounds present into the scece; four instances of the bass drum (the low frequency element), two instances of a snare drum (the two loud wideband bursts), and the hi-hat the repeating high-band burst. Upon analysis we extract a set of spectral/temporal basis functions from $\mathbf{W}_t$. The weights from $\mathbf{H}$ show us how these bases are placed in time. Examining the bases we see that they have encapsulated the short-time spectral evolution of each drum. For example the second basis has adapted to the bass drum structure. Note how the main frequency of the basis drops with time and is preceded from a wide-band element just like the bass drum sound. Likewise the snare drum basis is wide-band with denser energy at the mid-frequencies, and the hi-hat basis is mostly high-band.

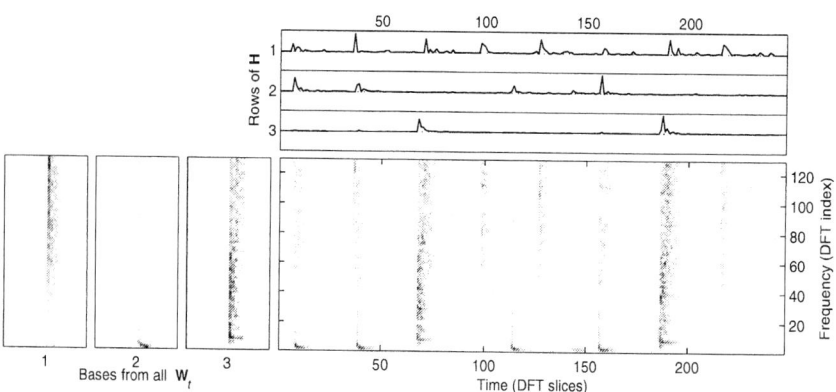

Fig. 3. NMFD bases and weights for drum example. The lower right plot is the magnitude spectrogram that we used as an input. The three leftmost plots are the temporal-spectral bases from $\mathbf{W}_t$. Their corresponding weights and rows of $\mathbf{H}$ are depicted at the top plot. Note how the extracted bases encapsulate the temporal/spectral structure of the three drum sounds in the spectrogram.

Having this description is a valuable guide to perform separation. We can do partial reconstructions of the input spectrogram using one basis function at a time. For example to extract the bass drum which was mapped to the jth basis we do:

$$\hat{\mathbf{V}}_j = \sum_{t=0}^{T-1} \mathbf{W}_t^{(j)} \cdot \overset{t\rightarrow}{\mathbf{H}} \quad (8)$$

where the $(\cdot)^{(j)}$ operator selects the jth column of the argument. This gives us the magnitude spectrogram of one component. We apply this to the original phase of the spectrogram and invert the result to obtain a time series. Subjectively we have found that the extracted elements consistently sound like the elements of the input sound scene. Unfortunately it is very hard to come up with a useful and intuitive measure that otherwise describes the quality of separation due to various non-linear distortions and lost information, problems inherent in the mixing and the analysis processes.

4 Conclusions

In this paper we presented an convolutional version of NMF. We have pinpointed some of the shortcomings of conventional NMF when analyzing temporal patterns and presented an extension which results in the extraction of more expressive basis functions. We have also shown how these basis functions can be used in the same way spectral bases have been used on spectrograms to extract sound objects from single channel sound scenes.

References

Casey, M.A. and A. Westner (2000) "Separation of Mixed Audio Sources by Independent Subspace Analysis", in *Proceedings of the International Computer Music Conference*, Berlin, Germany, August, 2000.

Lee, D.D. and H.S. Seung. (1999) "Learning the parts of objects with nonnegative matrix factorization". In *Nature*, 401:788 791, 1999.

Lee, D.D. and H.S. Seung (2000) "Algorithms for Non-Negative Matrix Factorization". In *Neural Information Processing Systems* 2000, pp. 556-562.

Paatero, P. (1997) "Least Squares Formulation of Robust Non-Negative Factor Analysis", in *Chemometrics and Intelligent Laboratory Systems* **37**, pp. 23-35, 1997.

Smaragdis, P. (2001) "Redundancy Reduction for Computational Audition, a Unifying Approach", *Doctoral Dissertation*, MAS Dept. Massachusetts Institute of Technology, Cambridge MA, USA.

Smaragdis, P. and J.C. Brown. (2003) "Non-Negative Matrix Factorization for Polyphonic Music Transcription", in *IEEE Workshop on Applications of Signal Processing to Audio and Acoustics*. New Paltz, NY, October 2003.

Optimal Sparse Representations
for Blind Deconvolution of Images

Alexander M. Bronstein, Michael M. Bronstein,
Michael Zibulevsky, and Yehoshua Y. Zeevi*

Technion - Israel Institute of Technology, Department of Electrical Engineering
32000 Haifa, Israel
{alexbron,bronstein}@ieee.org
{mzib,zeevi}@ee.technion.ac.il

Abstract. The relative Newton algorithm, previously proposed for quasi maximum likelihood blind source separation and blind deconvolution of one-dimensional signals is generalized for blind deconvolution of images. Smooth approximation of the absolute value is used in modelling the log probability density function, which is suitable for sparse sources. We propose a method of sparsification, which allows blind deconvolution of sources with arbitrary distribution, and show how to find optimal sparsifying transformations by training.

1 Introduction

Two-dimensional *blind deconvolution* (BD) is a special case of a more general problem of *image restoration*. The goal of BD is to reconstruct the original scene from an observation degraded by the action of a linear shift invariant (LSI) system, when no or very little *a priori* information about the scene and the degradation process is available, hence the term "blind". BD is critical in many fields, including astronomy, remote sensing, biological and medical imaging and microscopy.

According to the convolution model, the observed sensor image X is created from the *source image* S passing through an LSI system characterized by the point spread function A, $X = A * S$. We assume that the action of A is invertible (at least approximately), i.e. there exists some other kernel W such that $A * W \approx \delta$. This assumption holds well especially in the case of blurring kernels resulting from scattering (such kernels are usually Lorenzian-shaped and their inverse can be approximated by small FIR kernels). The aim of BD is to find such a *deconvolution* (*restoration*) kernel W that produces an estimate $\tilde{S}$ of S up to integer shift and scaling factor:
$\hat{S}_{mn} = (W * X)_{mn} \approx c \cdot S_{m-\Delta_M, n-\Delta_N}$.

Unlike approaches estimating the image and the blurring kernel [1, 2], we estimate the restoration kernel only, which results in a lower dimensionality of the problem. Here we present a *quasi maximum likelihood* (QML) BD algorithm, which generalizes the fast relative Newton algorithm previously proposed for blind source separation [3] and 1D BD [4]. We also propose optimal distribution-shaping approach (e.g. sparsification), which allows to use simple and convenient sparsity prior for a wide class of images.

---
* This research has been supported by the HASSIP Research Network Program HPRN-CT-2002-00285, sponsored by the European Commission, and by the Ollendorff Minerva Center.

2 QML Blind Deconvolution

Denote by $Y = W * X$ the source estimate and let us assume that S is zero-mean i.i.d. In the zero-noise case, the normalized minus-log-likelihood function of the observed signal X, given the restoration kernel W, is

$$\ell(X;W) = -\frac{1}{4\pi^2} \int_{-\pi}^{\pi} \int_{-\pi}^{\pi} \log|\mathcal{F}W(\xi,\eta)| \, d\xi d\eta + \frac{1}{M_X N_X} \sum_{m,n} \varphi(Y_{mn}), \quad (1)$$

where $\varphi(s) = -\log p_s(s)$, $p_s(s)$ stands for the source probability density function (PDF), $M_X \times N_X$ is the observation sample size, and $\mathcal{F}W(\xi,\eta)$ denotes the Fourier transform of W_{mn}. We will henceforth assume that W is an FIR kernel, supported on $[-M, ..., M] \times [-N, ..., N]$. Cost functions similar to (1) were also obtained in the 1D case using negative joint entropy and information maximization considerations [5]. In practice, it is difficult to evaluate the first term of $\ell(X;W)$ containing the integral. However, it can be approximated with any desired accuracy using FFT.

Source images arising in most applications have usually multi-modal non-log-concave distributions. These are difficult to model and are not suitable for optimization. However, consistent estimator of S can be obtained by minimizing $\ell(X;W)$ even when $\varphi(s)$ is not exactly equal to $-\log p_S(\cdot)$. Such *quasi-ML estimation* has been shown to be practical in instantaneous blind source separation [6, 3, 7] and blind deconvolution of time signals [4]. For example, when the source is super-Gaussian (sparse), a smooth approximation of the absolute value function is a good choice for $\varphi(s)$ [8, 9]. Although natural images are usually far from being sparse, they can be transformed into a space of a sparse representation. We will therefore focus our attention on modelling super-Gaussian distributions using a family of convex smooth functions

$$\varphi_\lambda(s) = |s| - \lambda \log\left(1 + \frac{|s|}{\lambda}\right) \quad (2)$$

with λ being a positive smoothing parameter; $\varphi_\lambda(s) \to |s|$ as $\lambda \to 0^+$.

The gradient of $\ell(X;W)$ w.r.t W_{ij} is given by (for derivation see [10]):

$$\frac{\partial \ell}{\partial W_{ij}} = -Q_{-i,-j} + \frac{1}{M_X N_X} \sum_{m,n} \varphi'(Y_{mn}) X_{m-i,n-j}, \quad (3)$$

where Q_{mn} is the inverse DFT of $\mathcal{F}W_{kl}^{-1}$. The Hessian of $\ell(X;W)$ is:

$$\frac{\partial^2 \ell}{\partial W_{ij} \partial W_{kl}} = \frac{1}{M_X N_X} \sum_{m,n} \varphi''(Y_{mn}) x_{m-i,n-j} x_{m-k,n-l} + R_{-(i+j),-(k+l)}, \quad (4)$$

where R_{mn} is the inverse DFT of $\mathcal{F}W_{kl}^{-2}$. Both the gradient and the Hessian can be evaluated efficiently using FFT.

3 The Fast Relative Newton Method

A fast relative optimization algorithm for blind source separation, based on the Newton method was introduced in [3]. In [4] it was used for BD of 1D signals. Here we use the relative optimization framework for BD of images. The main idea of relative optimization is to iteratively produce an estimate of the source signal and use it as the observed signal at the subsequent iteration:

Relative optimization algorithm

1. Start with initial estimates of the restoration kernel $W^{(0)}$ and the source $X^{(0)} = W^{(0)} * X$.
2. For $k = 0, 1, 2, ...$, until convergence
3. Start with $W^{(k+1)} = \delta$.
4. Using an unconstrained optimization method, find $W^{(k+1)}$ such that $\ell(X^{(k)}; W^{(k+1)}) < \ell(X^{(k)}; \delta)$.
5. Update source estimate: $X^{(k+1)} = W^{(k+1)} * X^{(k)}$.
6. End

The restoration kernel estimate at k-th iteration is $\hat{W} = W^{(0)} * ... * W^{(k)}$, and the source estimate is $\hat{S} = X^{(k)}$. This method allows to construct large restoration kernels growing at each iteration, using a set of relatively low-order factors. It can be seen easily that the relative optimization algorithm has uniform performance, i.e. its step at iteration k depends only on $A * W^{(0)} * ... * W^{(k-1)}$.

Step 4 can be carried out using any unconstrained optimization algorithm. Particulary, it was found that a single Newton step can be used, yielding very fast convergence. However, its use is limited to small values of M, N and M_X, N_X due to the complexity of Hessian construction, and solution of the Newton system. This complexity can be significantly reduced if special Hessian structure is exploited. Near the solution point, $X^{(k)} \approx cS$, hence $\ell(X; \delta)$ evaluated at each relative Newton iteration becomes approximately $\ell(cS; \delta)$. For a zero-mean i.i.d. source and sufficiently large sample size (in practice, $M_X N_X > 10^2$), the Hessian has an approximately diagonal-anti-diagonal form with ones on the anti-diagonal [10]. Using this approximation, only the main diagonal of the Hessian matrix has to be evaluated at each iteration, and the solution of the Newton system $\nabla^2 \ell d = -\nabla \ell$ separates into the set of 2×2 systems of the form

$$\begin{pmatrix} \nabla \ell_{-i,-j} \\ \nabla \ell_{ij} \end{pmatrix} = - \begin{pmatrix} \nabla^2 \ell_{-i,-j,-i,-j} & 1 \\ 1 & \nabla^2 \ell_{ijij} \end{pmatrix} \begin{pmatrix} d_{-i,-j} \\ d_{ij} \end{pmatrix}$$

for $(i, j) \neq \mathbf{0}$, and an additional equation $\nabla \ell_{00} = -\nabla^2 \ell_{0000} d_{00}$. We will henceforth refer to this approximate Newton step as to the *fast relative Newton method*, since its complexity is of the same order as that of the gradient-based methods.

4 Optimal Sparse Representations of Images

The QML framework presented in Section 2 is valid for sparse sources; this type of a prior of source distribution is especially convenient since the prior term in the underlying optimization problem is convex. In addition, deconvolution of sparse sources is reported to be very accurate. However, natural images arising in the majority of BD applications can by no means be considered to be sparse in their native space of representation (usually, they are sub-Gaussian), and thus such a prior is not valid for "real-life" sources. On the other hand, it is very difficult to model actual distributions of natural images, which are often multi-modal and non-log-concave. This apparent gap between a simple model and the real world calls for an alternative approach. In this section, we show how to overcome this problem using sparse representation.

While it is difficult to derive a prior suitable for natural images, it is much easier to transform an image in such a way that it fits some universal prior. In this study, we limit our attention to the sparsity prior, and thus discuss sparsifying transformations, though the idea is general and is suitable for other priors as well. The idea of *sparsification* was successfully exploited in BSS [8, 7, 11, 10]. It was shown in [10] that even such simple transformation as a discrete derivative can make the image sparse. However, most of these transformations were derived from empirical considerations. Here we present a criterion for finding optimal sparsifying transformations.

Let assume that there exists a *sparsifying transformation* $\mathcal{T}_S$, which makes the source S sparse (wherever possible, the subscript S in $\mathcal{T}_S$ will be omitted for brevity). In this case, our algorithm is likely to produce a good estimate of the restoration kernel W since the source properties are in accord with the sparsity prior. The problem is, however, that in the BD setting, S is not available, and $\mathcal{T}$ can be applied only to the observation X. Hence, it is necessary that the sparsifying transformation commute with the convolution operation, i.e. $(\mathcal{T}S) * A = \mathcal{T}(S * A) = \mathcal{T}X$, such that applying $\mathcal{T}$ to X is equivalent to applying it to S. Obviously, $\mathcal{T}$ must be a shift-invariant (SI) transformation[1].

Using the most general nonlinear form of $\mathcal{T}$, we have a wide class of sparsifying transformations. An important example is a family of SI transformations of the form $(\mathcal{T}S)_{mn} = \sqrt{(T_1 * S)^2_{mn} + (T_2 * S)^2_{mn}}$, where T_1, T_2 are some convolution kernels. After sparsification with $\mathcal{T}$, the prior term of the likelihood function becomes

$$\sum_{m,n} |(\mathcal{T}Y)_{mn}| = \sum_n \sqrt{(T_1 * Y)^2_{mn} + (T_2 * Y)^2_{mn}}, \qquad (5)$$

which is a generalization of the 2D *total-variation* (TV) norm. The TV norm, which has been found to be a successful prior in numerous studies related to signal restoration and denoising [12–14], and was also used by Chan and Wong as a regularization in BD [1], is obtained when T_1, T_2 are chosen to be discrete x- and y-directional derivatives.

For simplicity, we limit our attention in this study to linear shift-invariant (LSI) transformations, i.e. $\mathcal{T}$ that can be represented by convolution with a *sparsifying kernel* $\mathcal{T}S = T * S$. Thus, we obtain a general BD algorithm, which is not limited to sparse sources. We first sparsify the observation data X by convolving it with T (which has to be found in a way described in Section 4.1), and then apply the sparse BD algorithm on the result $X * T$. The obtained restoration kernel W is then applied to X to produce the source estimate.

An important practical issue is how to find the kernel T. By definition T must produce a sparse representation of the source; it is obvious that T would usually depend on S, and also, T does not necessarily have to be stable, since we use it as a pre-processing of the data and hence never need its inverse. Let assume that the source S is given

---

[1] In BSS problems, the sparsifying transformation needs to be linear and not necessarily shift-invariant, e.g. wavelet packets were used for sparsification in [8, 7].

(this is, of course, impossible in reality; the issue of what to use instead of S will be addressed in Section 4.1). It is desired that the unity restoration kernel δ_{mn} be a local minimizer of the QML function, given the transformed source $S * T$ as an observation, i.e.: $\nabla \ell(\delta_{mn}; S * T) = 0$. Informally, this means that $S * T$ optimally fits the sparsity prior (at least in local sense). Due to the equivariance property, $\nabla \ell(\delta_{mn}; S * T) = 0$ is equivalent to $\nabla \ell(T; S) = 0$. In other words, we can define the following optimization problem:

$$\min_T \ell(T; S), \qquad (6)$$

whose solution is the optimal sparsifying kernel for S. This problem is equivalent to the problem solved for deconvolution itself. The log-spectrum term in $\ell(T; S)$ eliminates the trivial solution $T = 0$.

4.1 Finding the Sparsifying Kernel by Training

Since the source image S is not available, computation of the sparsifying kernel by the procedure described before is possible only theoretically. However, empirical results indicate that for images belonging to the same class, the proper sparsifying kernels are sufficiently similar.

Let $\mathcal{C}_1$ denote a class of images, e.g. human faces, and assume that the unknown source S belongs to $\mathcal{C}_1$. We can find find images $S^{(1)}, S^{(2)}, ..., S^{(N_T)} \in \mathcal{C}_1$ and use them to find the optimal sparsifying kernel of S. Optimization problem (6) becomes in this case

$$\min_T \left\{ -\frac{1}{4\pi^2} \int_{-\pi}^{\pi} \int_{-\pi}^{\pi} \log |\mathcal{F}T(\xi, \eta)| \, d\xi d\eta + \frac{1}{M_X N_X} \sum_{i=1}^{N_T} \sum_{m,n} \varphi((T * S^{(i)})_{mn}) \right\},$$

i.e. T is required to be the optimal sparsifying kernel for all $S^{(1)}, S^{(2)}, ..., S^{(N_T)}$ simultaneously. The images $S^{(1)}, S^{(2)}, ..., S^{(N_T)}$ constitute a *training set*, and the process of finding such T as *training*. Given that the images in the training set are "sufficiently similar" to S, the optimal sparsifying kernel obtained from training is similar enough to T_S.

5 Simulation Results

The QML-based deconvolution approach was tested on simulated data under zero-noise conditions. As a criterion for evaluation of the reconstruction quality, we used the signal-to-interference-ratio (SIR) in sense of the L_2, L_∞ norms, and the peak SIR (PSIR) in dB units [10]. In the first test, a real aerial photo of a factory was used as the source image, and a synthetic one (drawn using PhotoShop) as the training image (Figure 1). A 3×3 sparsifying kernel is found by training on a single image, then the same kernel is used as a pre-processing for BD applied to a different blurred source image from the same class of images. The source image was convolved with a symmetric FIR 31×31 Lorenzian-shaped blurring kernel. Deconvolution kernel was of size

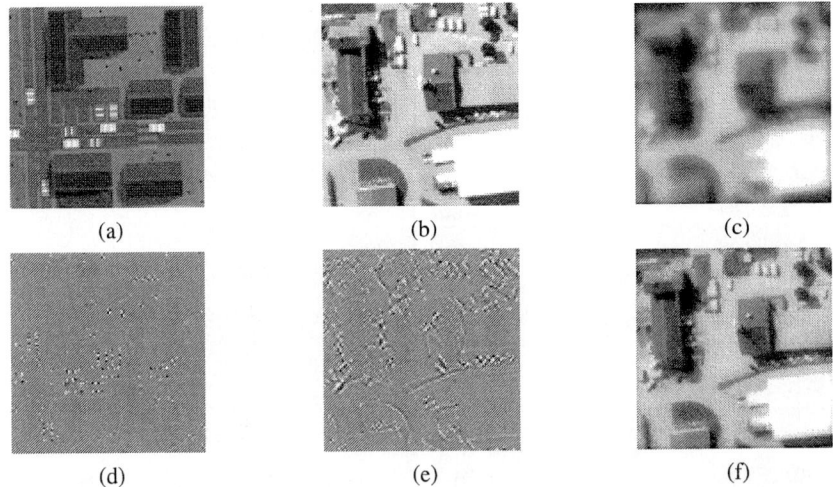

Fig. 1. (a) training synthetic image, (b) source aerial image S, (c) blurred image $S * A$, (d) sparsified training image, (e) sparsified source, (f) restored image.

Table 1. SIR, SIR_∞ and PSIR of the restored images.

Source		SIR [dB]	SIR_∞ [dB]	PSIR [dB]
S_1	Susy	17.7994	22.2092	22.6132
S_2	Aerial	17.0368	23.5482	9.6673
S_3	Gabby	19.3249	23.8109	29.8316
S_4	Hubble	14.5152	17.1552	19.8083

3×3. The sparsifying kernel obtained by training was very close to a corner detector. The signal-to-interference ratio in the deconvolution result was $SIR = 20.1561$ dB, $SIR_\infty = 25.7228$ dB.

In the second test, four natural source images were used: S_1 (Susy), S_2 (Aerial), S_3 (Gabby) and S_4 (Hubble) (Figure 2, top). Nearly-stable Lorenzian-shaped kernels were used to model the convolution system. This type of kernels characterizes scattering media, such as biological fluids and aerosols found in the atmosphere [15]. The observed images are depicted in Figure 2 (middle). Fast relative Newton step with kernel size set to 3×3 was used in this experiement. The smoothing parameter was set to $\lambda = 10^{-2}$. Corner detector was used as the sparsifying kernel. Optimization was terminated when the gradient norm reached 10^{-10}. Convergence was achieved in $10-20$ iterations (about 10 sec). The restored images are depicted in Figure 2 (bottom). Restoration quality results in terms of SIR, SIR_∞ and PSIR are presented in Table 1.

6 Conclusion

The QML framework, recently presented in the context of 1D deconvolution [4] is also attractive for BD of images. We presented an extension of the relative optimization

Susy Aerial Gabby Hubble

Fig. 2. Top: source images used in the simulations; middle: blurred images (observations); bottom: restored images.

approach to QML BD in the 2D case and studied the relative Newton method as its special case. Similarly to previous works addressing deconvolution in other spaces (e.g. [16]) and our studies of using sparse representation in the context of BBS, in BD the sparse prior appears very efficient as well. We showed a training approach for finding optimal sparse representations, yielding a general-purpose BD method. A particular class of LSI sparsifying transformations generalizes some previous results such as the total variation prior [12–14]. We also showed how optimal sparsifying transformations can be found by training.

Simulation results demonstrated the efficiency of the proposed methods. Although we have limited our attention to noiseless BD, it is important to emphasize that the sparsification framework is applicable to the noisy case as well. Sparsifying kernels are typically high-pass filters, since by their very nature sparse signals have high-frequency components. Such kernels have the property of amplifying noise – thus in case when the signal is contaminated by additive noise, using such kernels in undesired. To cope with the problem of noise, the signal should be smoothed with a low-pass filter F and afterwards the sparsifying kernel T should be applied. Due to commutativity of the convolution, it is equivalent to carrying out the sparsification with a smoothed kernel $T * F$.

Potential applications of our approach are in optics, remote sensing, microscopy and biomedical imaging, especially where the SNR is moderate. This approach is especially accurate and efficient in problems involving slowly-decaying (e.g. Lorenzian-shaped) kernels, which can be approximately inverted using a kernel with small support. Such kernels are typical of imaging through scattering media.

References

1. Chan, T.F., Wong, C.K.: Total variation blind deconvolution. (IEEE. Trans. Image Proc.) To appear.
2. Kaftory, R., Sochen, N.A., Zeevi, Y.Y.: Color image denoising and blind deconvolusion using the beltramy operator. In: Proc. 3rd Intl. Symposium on Image and Sig. Proc. and Anal. (2003) 1–4
3. Zibulevsky, M.: Sparse source separation with relative Newton method. In: Proc. ICA2003. (2003) 897–902
4. Bronstein, A.M., Bronstein, M., Zibulevsky, M.: Blind deconvolution with relative Newton method. Technical Report 444, Technion, Israel (2003)
5. Amari, S.I., Cichocki, A., Yang, H.H.: Novel online adaptive learning algorithms for blind deconvolution using the natural gradient approach. In: Proc. SYSID. (1997) 1057–1062
6. Pham, D., Garrat, P.: Blind separation of a mixture of independent sources through a quasi-maximum likelihood approach. IEEE Trans. Sig. Proc. **45** (1997) 1712–1725
7. Kisilev, P., Zibulevsky, M., Zeevi, Y.: Multiscale framework for blind source separation. JMLR (2003 (in press))
8. Zibulevsky, M., Pearlmutter, B.A.: Blind source separation by sparse decomposition. Neural Computation **13** (2001)
9. Zibulevsky, M., Kisilev, P., Zeevi, Y.Y., Pearlmutter, B.A.: Blind source separation via multinode sparse representation. In: Proc. NIPS. (2002)
10. Bronstein, A.M., Bronstein, M., Zeevi, Y.Y., Zibulevsky, M.: Quasi-maximum likelihood blind deconvolution of images using sparse representations. Technical report, Technion, Israel (2003)
11. Lewicki, M.S., Olshausen, B.A.: A probabilistic framework for the adaptation and comparison of image codes. J. Opt. Soc. Am. A **16** (1999) 1587–1601
12. Rudin, L.I., Osher, S., Fatemi, E.: Nonlinear total variation based noise removal algorithms. Physica D **60** (1992) 259–268
13. Blomgren, P., Chan, T.F., Mulet, P., Wong, C.: Total variation image restoration: numerical methods and extensons. In: Proc. IEEE ICIP. (1997)
14. Chan, T.F., Mulet, P.: Iterative methods for total variation image restoration. SIAM J. Num. Anal **36** (1999)
15. Moscoso, M., Keller, J.B., Papanicolaou, G.: Depolarization and blurring of optical images by biological tissue. J. Opt. Soc. Am. A **18** (2001) 948–960
16. Banham, M.R., Katsaggelos, A.K.: Spatially adaptive wavelet-based multiscale image restoration. IEEE Trans. Image Processing **5** (1996) 619–634

Separation of Convolutive Mixtures of Cyclostationary Sources: A Contrast Function Based Approach

Pierre Jallon[1], Antoine Chevreuil[1], Philippe Loubaton[1], and Pascal Chevalier[2]

[1] Université de Marne-la-Vallée UMR-CNRS 5141 5, boulevard Descartes
77454 Marne-La-Vallée Cedex 2
{jallon,chevreuil,loubaton}@univ-mlv.fr
[2] Thalés Communications EDS/SPM/SBP 160, bd Valmy 92700 Colombes
pascal.chevalier@fr.thalesgroup.com

Abstract. Fourth-order cumulants are quite popular in the field of blind separation of convolutive mixtures of stationary sources. Their use in the context of cyclo-stationary sources cannot be taken for granted because consistent estimation of the temporal mean of the fourth-order cumulants needs the knowledge of the cyclic frequencies of the received signal. In this paper, we introduce a cost function whose estimation does not need the knowledge of the cyclic frequencies. We show that under some reasonable sufficient conditions, its maximization allows to separate the sources.

1 Introduction

A convolutive mixture of K independent sources is a N-variate time series $(\mathbf{y}(n))_{n \in \mathbb{Z}}$ defined as the output of a multiple-input / multiple-output linear system with transfer function $\mathbf{H}(z)$ driven by a K-variate time series ($K \leq N$) $(\mathbf{s}(n))_{n \in \mathbb{Z}}$ whose components are statistically independent signals. Very often, the linear filter between $\mathbf{s}$ and $\mathbf{y}$ modelizes the effect of the propagation between the sources and the sensors array.

The blind separation of the convolutive mixture consists, for every source, in extracting on each sensor the signal which would be observed if this source was not corrupted by the others. In order to tackle this problem, several authors have proposed to use an iterative inverse filtering approach, also called a deflation approach (see [2], [3], [4], [12]). This approach consists in looking first for a N-input / 1-output filter, which driven by $\mathbf{y}(n)$, produces a filtered version of one of the source signal. It is then easy to substract the contribution of the extracted source to the observed signal $\mathbf{y}(n)$, thus producing a convolutive mixture of $K-1$ signals. The procedure is then iterated until all the sources are extracted.

In order to extract one of the source signals, a possible approach consists in looking for a N-input / 1-output filter with transfer function $\mathbf{g}(z)$ maximizing a contrast function (see [7], [11]). If we denote by $(r(n))_{n \in \mathbb{Z}}$ the signal $r(n) = [\mathbf{g}(z)]\mathbf{y}(n)$, the most popular contrast functions are constructed from the high-order cumulants of signal $(r(n))_{n \in \mathbb{Z}}$. In the following, if x is a complex valued random variable, we denote $c_4(x)$ the fourth-order cumulant defined by $c_4(x) = cum(x, x^*, x, x^*)$. It is well known that if each source signal is a non gaussian independent identically distributed (i.i.d.) sequence, then the local maxima of the function $\mathbf{g}(z) \rightarrow \left| \frac{c_4(r(n))}{(\mathbb{E}|r(n)|^2)^2} \right|$ correspond to filters which extract the sources. It has also been shown recently that this cost function is also relevant if the source signals are stationary sequences (see [5], [9], [10]). In particular, it is not required that the source signals coincide with filtered versions of i.i.d. sequences.

In certain kind of applications (e.g. digital communications), the source signals $(s_k)_{k=1,\ldots,K}$ are not stationary, but cyclostationary. Only a few previous works have addressed the source separation of cyclostationary signals. [1] considered the case of instantaneous mixtures, and proposed a second order statistics based approach. [6] proposed to adapt the concept of mutual information. [8] is also devoted to instantaneous mixtures, and studied the behavior of a non iterative fourth-order cumulant based method (the so-called Jade method). [8] pointed out that the contrast function to be maximized cannot in general be estimated consistently if the cyclic frequencies of the second order statistics of observation are unknown. However, [8] showed that if the second order statistics of the various source signals do not share the same cyclic frequencies, then their knowledge is not required.

In this paper, we consider the convolutive mixture case, and study the iterative inverse filtering approach using the fourth order kurtosis when the source signals are cyclostationary. If $r(n) = [\mathbf{g}(z)]\mathbf{y}(n)$, both $n \rightarrow c_4(r(n))$ and $n \rightarrow E|r(n)|^2$ are then almost periodic sequences, and we first establish in section 2 that the function $J(\mathbf{g})$ defined by

$$J(\mathbf{g}) = \left| \frac{<c_4(r(n))>}{(<E(|r(n)|^2)>)^2} \right| \quad (1)$$

is a contrast function. Here, if $(u(n))_{n \in \mathbb{Z}}$ is an almost periodic sequence, we define $<u(n)>$ as the temporal mean $<u_n> = \lim_{N \to \infty} \frac{1}{N} \sum_{n=0}^{N-1} u(n)$.

As explained below, this contrast is difficult to use in practice because the consistent estimation of $J(\mathbf{g})$ from a finite number of observations $\mathbf{y}(0), \ldots, \mathbf{y}(T-1)$ needs some knowledge on the cyclic frequencies of the second order statistics of $\mathbf{y}(n)$. In section 3, we introduce the cost function $\tilde{J}$ defined by

$$\tilde{J}(\mathbf{g}) = \left| \frac{<E(|r(n)|^4> - 2(<E(|r(n)|^2)>)^2 - |<E(r(n)^2)>|^2}{(<E(|r(n)|^2)>)^2} \right| \quad (2)$$

In contrast with J, $\tilde{J}(\mathbf{g})$ can be estimated consistently without any knowledge on the cyclic frequencies of $\mathbf{y}(n)$. We give two different conditions under which $\tilde{J}$ is a contrast. We first show that $\tilde{J}$ is a contrast if the source signals do not share the same second order cyclic frequencies. This result can be considered a generalization of [8] to the context of convolutive mixtures. Second, we introduce a technical sufficient condition having interesting implications in the particular context of digital communications. We finally present in section 4 some simulation experiments illustrating the above results.

2 A First Contrast Function

In this section, we first show that function J defined by (1) is a contrast. For this, we assume from now on that the cyclic spectrum at cyclic frequency 0 of each source signal is reduced to the constant 1 [1]. In this case, it is clear that if $\mathbf{f}(z)$ is a $1 \times K$ filter, then $< E|[\mathbf{f}(z)]\mathbf{s}(n)|^2 >= \|\mathbf{f}\|^2 = \sum_{k=1}^{K} \|f_k\|^2$ where $\|f_k\|$ is the $\mathbb{L}^2$–norm of $f_k(z) = \sum_{l \in \mathbb{Z}} f_k(l) z^{-l}$ defined by $\|f_k\|^2 = \sum_{l \in \mathbb{Z}} |f_k(l)|^2$.

In order to study the properties of function J, we put as usual $\mathbf{f}(z) = \mathbf{g}(z)\mathbf{H}(z)$, and remark that $r(n) = [\mathbf{g}(z)]\mathbf{y}(n)$ can be written as $r(n) = [\mathbf{f}(z)]\mathbf{s}(n)$. The following result holds.

Theorem 1 *Assume that:* $\forall k \in [1..K]$,

$$\sup_{\|f_k\|=1} |< c_4 (\lceil f_k(z) \rceil s_k(n)) >| \qquad (3)$$

if finite and reached for at least one filter $f_{k*}(z)$ such as $\|f_{k*}\| = 1$. Then the global maximum of J is finite, and reached by at least one filter $\mathbf{g}_*$ for which $r(n) = \lceil g_*(z) \rceil \mathbf{y}(n) = \lceil f_{k_0}(z) \rceil s_{k_0}(n)$ for some index k_0. Moreover, if $\mathbf{g}(z)$ is a local maximum of J, then $r(n) = [\mathbf{g}(z)]\mathbf{y}(n)$ coincides with a filtered version of one of the source signals.

Proof. The proof is similar to the proof of the main result of [5]. We however give a short overview in order to make this paper reasonnably self-contained. Let $\mathbf{g}(z)$ be a $1 \times N$ filter, and let $\mathbf{f}(z) = (f_1(z), \ldots, f_K(z))$ be the $1 \times K$ filter defined by $\mathbf{f}(z) = \mathbf{g}(z)\mathbf{H}(z)$. We put $\tilde{s}_k(n) = \lceil \frac{f_k(z)}{\|f_k\|} \rceil s_k(n)$ if $\|f_k\| \neq 0$ and $\tilde{s}_k(n) = 0$ if $\|f_k\| = 0$. Then, $J(\mathbf{g})$ can be written as

$$J(\mathbf{g}) = \frac{|< c_4(\sum_{k=1}^{K} \lceil f_k(z) \rceil s_k(n)) >|}{\left(\sum_{k=1}^{K} \|f_k\|^2\right)^2} = \frac{|\sum_{k=1}^{N} \|f_k\|^4 < c_4(\tilde{s}_k(n)) >|}{\left(\sum_{k=1}^{N} \|f_k\|^2\right)^2} \qquad (4)$$

As filter $\frac{f_k(z)}{\|f_k\|}$ is unit norm,

$$|< c_4(\tilde{s}_k(n)) >| \leq \sup_{\|e_k\|=1} |< c_4([e_k(z)]s_k(n)) >|$$

Hence, $J(\mathbf{g}) \leq \max_k \sup_{\|f_k\|=1} |< c_4([f_k(z)]s_k(n)) >|$ with equality and if and only if $f(z) = (0, \cdots, 0, f_{k_0}(z), 0, \cdots, 0)$ where $k_o = argmax\ sup_{\|f_k\|=1} |< c_4([f_k(z)]s_k(n)) >|$.

The claim related to the local maxima of J can be proved as in [5].

---
[1] if this is not the case, it is sufficient to replace each signal $s_k(n)$ by the signal obtained as the output of the filter $(S_{s_k}^{(0)}(e^{2i\pi f}))^{-1/2}$ driven by s_k, and to modify accordingly each column of $\mathbf{H}(z)$ to leave $\mathbf{y}(n)$ unchanged.

We now explain why contrast function J is difficult to use in practice. In effect,

$$J(\mathbf{g}) = \left| \frac{<\mathbb{E}(|r(n)|^4> -2 <(\mathbb{E}(|r(n)|^2)^2> - <|\mathbb{E}(r(n)^2)|^2>|}{(<\mathbb{E}(|r(n)|^2)>)^2} \right|$$

In practice, the second-order and fourth-order statistics of $\mathbf{y}(n)$ are unknown. Therefore, for each filter $\mathbf{g}(z)$, $J(\mathbf{g})$ and its gradient have to be consistently estimated from the available observations $\mathbf{y}(0),\ldots,\mathbf{y}(T-1)$ (i.e. the mean-square error has to converge to 0 when the sample size $T \to +\infty$). Under mild technical assumptions, the terms $<\mathbb{E}(|r(n)|)^4>$ and $<\mathbb{E}(|r(n)|^2)>$ can be consistently estimated by $\frac{1}{T}\sum_{n=0}^{T-1}|r(n)|^4$ and $\frac{1}{T}\sum_{n=0}^{T-1}|r(n)|^2$ respectively. However, it is not possible to estimate consistently $<(\mathbb{E}(|r(n)|^2)^2>$ and $<|\mathbb{E}(r(n)^2)|^2>$ without any knowledge of the second-order cyclic frequencies of $\mathbf{y}(n)$. We denote I is the set of frequencies α for which $<\mathbb{E}(\mathbf{y}(n+\tau)\mathbf{y}(n)^*)e^{2i\pi n\alpha}>\neq 0$ for some τ and I_c is the set of frequencies β for which $<\mathbb{E}(\mathbf{y}(n+\tau)\mathbf{y}(n))e^{2i\pi n\beta}>\neq 0$ for some τ. Then,

$$\mathbb{E}|r(n)|^2 = \sum_{\alpha_l \in I} R_r^{\alpha_l}(0)e^{-2i\pi\alpha_l n} \text{ and } \mathbb{E}(r(n))^2 = \sum_{\beta_l \in I_c} R_{c,r}^{\beta_l}(0)e^{-2i\pi\beta_l n} \quad (5)$$

where $R_r^{\alpha_l}(0) = <\mathbb{E}\left(|r(n)|^2\right)e^{+2i\pi\alpha_l n}>$ and $R_{c,r}^{\beta_l}(0) = <\mathbb{E}\left(r(n)^2\right)e^{+2i\pi\beta_l n}>$. Using the Parseval identity we get that , $<c_4(r(n))>$ can be written as

$$<c_4(r(n))> = <\mathbb{E}|r(n)|^4> -2\sum_{\alpha_l \in I}|R_r^{\alpha_l}(0)|^2 - \sum_{\beta_l \in I_c}\left|R_{c,r}^{\beta_l}(0)\right|^2 \quad (6)$$

All the terms in (6) can be estimated if the α_l and β_l are known. As in practice these cyclic frequencies are very often unknown, they have to be estimated from the available data.

3 A Modified Contrast

As the estimation of cyclic frequencies is not always an easy task, we propose in this section to replace J by the function $\tilde{J}$ defined by (2) and exhibit 2 sufficient conditions under which separation can be performed by maximizing $\tilde{J}$. In the following, we denote $I^* = I - \{0\}$ and $I_c^* = I_c - \{0\}$.

3.1 The Sources Do Not Share the Same Cyclic Frequencies

In this paragraph, we assume that two different sources do not have a common cyclic frequency, and state the following result.

Theorem 2 *Assume that:* $\forall k \in [1..K]$,

$$\sup_{\|f_k\|=1} |<E|[f_k(z)]s_k(n)|^4> -2(<E|[f_k(z)]s_k(n)|^2>)^2 - |<E([f_k(z)]s_k(n))^2>|^2| \quad (7)$$

if finite and reached for at least one filter $f_{k*}(z)$ such as $\|f_{k*}\| = 1$. Then, the global maximum of $\tilde{J}$ is finite and reached by a filter g_* such as $r(n) = \lceil g_*(z) \rceil y(n) = \lceil f_{k_0}(z) \rceil s_{k_0}(n)$ for some index k_0. Moreover, if $\mathbf{g}(z)$ is a local maximum of $\tilde{J}$, then $r(n) = [\mathbf{g}(z)]\mathbf{y}(n)$ coincides with a filtered version of one of the source signal.

Proof. Let $\mathbf{g}(z)$ be a $1 \times N$ filter and put as previously $\mathbf{f}(z) = \mathbf{g}(z)\mathbf{H}(z)$ and $\tilde{s}_k(n) = \lceil \frac{f_k(z)}{\|f_k\|} \rceil s_k(n)$ if $\|f_k\| \neq 0$ and $\tilde{s}_k(n) = 0$ if $\|f_k\| = 0$. Then, $r(n) = [\mathbf{f}(z)]\mathbf{s}(n)$ can be written as $r(n) = \sum_{k=1}^{K} \|f_k\| \tilde{s}_k(n)$. In order to express $\tilde{J}(\mathbf{g})$ in terms of $(\|f_k\|)_{k=1,...,K}$, we use the following expression of $\tilde{J}(\mathbf{g})$:

$$\tilde{J}(\mathbf{g}) = \frac{\left| <c_4(r(n))> + 2\sum_{\alpha_l \in I^*} |R_r^{(\alpha_l)}(0)|^2 + \sum_{\beta_l \in I_c^*} |R_{c,r}^{(\beta_l)}(0)|^2 \right|}{(<\mathbb{E}|r(n)|^2>)^2}$$

$<c_4(r(n))>$ is equal to $\sum_{k=1}^{K} \|f_k\|^4 <c_4(\tilde{s}_k(n))>$ and $(<\mathbb{E}|r(n)|^2>)^2 = (\sum_{k=1}^{K} \|f_k\|^2)^2$. We have now to express $|R_r^{(\alpha_l)}(0)|^2$ and $|R_{c,r}^{(\beta_l)}(0)|^2$. It is obvious that for each α_l, $R_r^{(\alpha_l)}(0) = \sum_{k=1}^{K} \|f_k\|^2 R_{\tilde{s}_k}^{(\alpha_l)}(0)$. Hence,

$$|R_r^{(\alpha_l)}(0)|^2 = \sum_{k=1}^{K} \|f_k\|^4 |R_{\tilde{s}_k}^{(\alpha_l)}(0)|^2 + 2 \sum_{k_1 < k_2} \|f_{k_1}\|^2 \|f_{k_2}\|^2 \Re\left\{ R_{\tilde{s}_{k_1}}^{\alpha_l}(0)(R_{\tilde{s}_{k_2}}^{\alpha_l}(0))^* \right\}$$

But, if $k_1 \neq k_2$, the term $R_{\tilde{s}_{k_1}}^{\alpha_l}(0)(R_{\tilde{s}_{k_2}}^{\alpha_l}(0))^*$ necessarily vanishes because α_l cannot be cyclic frequency of $\tilde{s}_{k_1}$ and $\tilde{s}_{k_2}$. Therefore, $|R_r^{(\alpha_l)}(0)|^2 = \sum_{k=1}^{K} \|f_k\|^4 |R_{\tilde{s}_k}^{(\alpha_l)}(0)|^2$, and similarly, $|R_{c,r}^{(\alpha_l)}(0)|^2 = \sum_{k=1}^{K} \|f_k\|^4 |R_{c,\tilde{s}_k}^{(\alpha_l)}(0)|^2$. Putting all the pieces together, we get that

$$\tilde{J}(\mathbf{g}) = \frac{\left| \sum_{k=1}^{K} \|f_k\|^4 \left(<\mathbb{E}|\tilde{s}_k(n)|^4> -2<\mathbb{E}|\tilde{s}_k(n)|^2>^2 -|<\mathbb{E}(\tilde{s}_k(n))^2>|^2 \right) \right|}{(\sum_{k=1}^{K} \|f_k\|^2)^2} \tag{8}$$

Reasoning as in the proof of Theorem (1), we get immediately that

$$\tilde{J}(\mathbf{g}) \leq \max_k \sup_{\|f_k\|=1} \left| <E|[f_k(z)]s_k(n)|^4> -2(<E|[f_k(z)]s_k(n)|^2>)^2 \right.$$

$$\left. -|<E([f_k(z)]s_k(n))^2>|^2 \right| \tag{9}$$

with equality if and only if $\mathbf{f}(z) = (0, \ldots, 0, f_{k_0}(z), 0, \ldots, 0)$ where k_0 maximizes the righthandside of (9). The claim related to the local maxima of $\tilde{J}$ can be proved as in [5].

3.2 The Case of Circular Sources

We now consider the case where the sources are circular, i.e. for each k, $E(s_k(\text{m}+\text{n})s_k(m)) = 0$ for each (m, n). In this context, $I_c = \{0\}$.

Theorem 3 *Assume that:* $\forall k \in [1..K]$,

$$\sup_{\|f_k\|=1} |<E|[f_k(z)]s_k(n)|^4>| \qquad (10)$$

if finite and reached for at least one filter $f_{k*}(z)$ such as $\|f_{k*}\| = 1$. Assume moreover that

$$\max_{i,j,i\neq j} \sup_{\sum_k \|f_k\|^2=1} \left|\Re\left\{\sum_{\alpha_k \in I, \alpha_k \neq 0} R_{\tilde{s}_i}^{\alpha_k}(0)(R_{\tilde{s}_j}^{\alpha_k}(0))^*\right\}\right| <$$

$$\max_k \sup_{\|f_k\|=1} |<E|[f_k(z)]s_k(n)|^4 > -2| \qquad (11)$$

where signals $(\tilde{s}_k)_{k=1,\ldots,K}$ are still defined by $\tilde{s}_k(n) = [\frac{f_k(z)}{\|f_k\|}]s_k(n)$. Then, the global maximum of $\tilde{J}$ is finite and reached by a filter g_* such as $r(n) = \lceil g_*(z) \rceil y(n) = \lceil f_{k_0}(z) \rceil s_{k_0}(n)$ for some index k_0.

Proof. Using (8), we obtain

$$\tilde{J}(\mathbf{g}) = \left|\frac{\sum_{k=1}^N \|f_k\|^4 (<E|\tilde{s}_k(n)|^4 > -2(<E|\tilde{s}_k(n)|^2>)^2) + 2\sum_{k_1<k_2}\|f_{k_1}\|^2\|f_{k_2}\|^2 \gamma_{k_1,k_2}}{\left(\sum_{k=1}^K \|f_k\|^2\right)^2}\right|$$

where $\gamma_{k_1,k_2} = \sum_{\alpha_l \in I^*} \Re\left\{R_{\tilde{s}_{k_1}}^{\alpha_l}(0)(R_{\tilde{s}_{k_2}}^{\alpha_l}(0))^*\right\}$. As filter $\frac{f_k(z)}{\|f_k\|}$ is unit norm, $<E|\tilde{s}_k(n)|^2>$ coincides with 1. We put $\delta_k = \sup_{\|f_k\|=1}|<E|[f_k(z)]s_k(n)|^4> -2|$, which, by (10), is finite. It is clear that,

$$\tilde{J}(\mathbf{g}) \leq \max_k \delta_k \frac{\sum_{k=1}^N \|f_k\|^4 + 2\sum_{k_1<k_2}\|f_{k_1}\|^2\|f_{k_2}\|^2|\gamma_{k_1,k_2}|/\max_k \delta_k}{\left(\sum_{k=1}^K \|f_k\|^2\right)^2} \qquad (12)$$

As $|\gamma_{k_1,k_2}|$ is strictly less than $\max_k \delta_k$, the equality in (12) implies that $\|f_{k_1}\|^2 \|f_{k_2}\|^2 = 0$ if $k_1 \neq k_2$, i.e. if $\mathbf{f}(z) = (0,\ldots,0,f_{k_0}(z),0,\ldots,0)$ where $k_0 = \text{argmax}_k \delta_k$.

The condition (9) is specially relevant in the context of digital communications. Assume that each source signal $s_k(n)$ is a sampled version $s_{a,k}(nT_e)$ of a continuous-time linearly modulated signal $s_{a,k}(t)$ that can be written as $s_{a,k}(t) = \sum_{j\in\mathbb{Z}} u_k(j)g_{a,k}(t-jT_k)$ (we assume that for each k, the sampling frequency T_e satisfies the Shannon condition). Here, the sequence $(u_k(j))_{j\in\mathbb{Z}}$ is a circular i.i.d. sequence representing the symbols sent by transmitter k, $\frac{1}{T_k}$ is the symbol rate of transmitter k, and function $g_{a,k}(t)$ is the so-called shaping filter. Signal $s_{a,k}(t)$ is cyclostationnary, and its non zero cyclic frequencies are the integer multiples of $\frac{1}{T_k}$. However, in practice, the support of the Fourier transform of $g_{a,k}(t)$ is equal to $[-\frac{1+\gamma_k}{2T_s}, \frac{1+\gamma_k}{2T_s}]$ where $0 \leq \gamma_k < 1$ is called the excess bandwith factor of the transmitter. In this context, $s_{a,k}(t)$ has only $\pm\frac{1}{T_k}$ as cyclic frequencies. Moreover, if $\gamma_k = 0$, then $s_{a,k}$ is stationary. Therefore, the smallest the excess bandwidth, the weakest the cyclic correlation coefficients at cyclic frequencies $\pm\frac{1}{T_k}$. It turns out that condition (11) is likely to hold if all the source signals have small excess bandwidth, a condition often met in practice.

4 Simulations

We first present the performance criterion that will be numerically evaluated. For this, we first precise that when a filtered version of a certain source signal has been extracted, it is easy to estimate its contribution on each sensor using a classical cancelling approach.

We define the separation rate $TSEP(k,l)$ of source k on sensor l by $TSEP(k, l) = \frac{\sum_{n=0}^{T-1} |\hat{c}_{k,l}(n) - c_{k,l}(n)|^2}{\sum_{n=0}^{T-1} |c_{k,l}(n)|^2}$ where $c_{k,l}(n)$ represents the contribution of source k on sensor l at time n (i.e. signal $[H_{l,k}(z)]s_k(n)$) and where $\hat{c}_{k,l}(n)$ denotes the estimate of $c_{k,l}$ provided by the source separation procedure. We also define the separation rate of source k as $TSEP(k) = \sum_{l=1}^{N} \alpha_{k,l} TSEP(k,l)$ where $\alpha_{k,l}$ denotes the ratio $\alpha_{k,l} = \frac{\sum_{n=0}^{T-1} |c_{k,l}(n)|^2}{\sum_{n=0}^{T-1} \sum_{j=1}^{N} |c_{k,l}(n)|^2}$. In each experiment, $K = 2$ and $N = 5$, and the estimated contrast functions are maximized using a gradient algorithm. In order to evaluate the performance of the various approaches, we plot for each source the empirical cumulative distribution functions of the separation rate (100 trials are generated).

The two source signals are linearly modulated by a circular symbol sequence randomly chosen at each trial among the following constellations : QPSK, 8 PSK, 16 PSK. The excess bandwidths and the symbol rates of the two source signals coincide, and are equal to 0.5 and $\frac{1}{4}$ respectively. In order to generate the mixing filter $\mathbf{H}(z)$, we have used a Rayleigh 3-path channel. The sample size corresponds to the observation of 3000 symbols.

The purpose of the experiment is to illustrate the fact that the sufficient condition (11) is verified in this case, and that contrast function $\tilde{J}$ provides better results than the use of contrast function J when the cyclic frequencies are known.

The obtained results are represented on figure 1. The cumulative distribution functions of the separation rates provided by the maximization of function J and of function $\tilde{J}$ are compared.

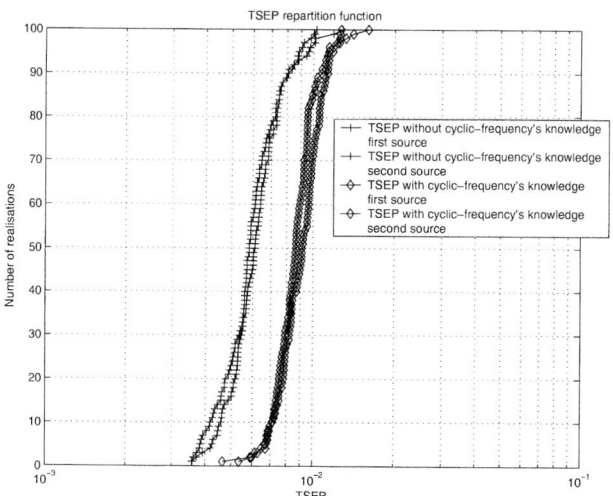

Fig. 1. Contrasts performances on over sampled circular sources using same baudrates.

In order to estimate function J, we assume the perfect knowledge of the cyclic frequencies, and $J(\mathbf{g})$ is estimated by $\hat{J}(\mathbf{g}) =$

$\frac{<|r(n)|^4>-2\sum_{\alpha_k}|\hat{R}_r^{\alpha_k}(0)|^2}{(<|r(n)|^2>)^2}$ where $< u(n) >$ stands for the temporal mean $< u(n) >= \frac{1}{T}\sum_{n=0}^{T-1} u(n)$ and where $\hat{R}^{\alpha_k}(0)$ is the estimated cyclic correlation coefficient at cyclic frequency α_k defined by $\hat{R}^{\alpha_k}(0) =< |r(n)|^2 e^{2i\pi n\alpha_k} >$. $\tilde{J}$ is estimated by $\hat{\tilde{J}}(\mathbf{g}) = \frac{<|r(n)|^4>-2|\hat{R}_r^0(0)|^2}{(<|r(n)|^2>)^2}$.

It can be seen that the results provided by function $\tilde{J}$ are better. This tends to indicate that in the present context, $\tilde{J}$ is a constrast, i.e. condition (11) holds, and that to estimate the cyclic frequencies of the received signal is not necessary to be able to use function J.

References

1. K. Abed-Meraim, Y. Xiang, J. Manton, Y. Hua, "Blind source separation using second-order cyclostationary statistics", *IEEE Trans. on Signal Processing*, vol. 49, no. 4, pp. 694-701, April 2001.
2. Ph. Loubaton and Ph. Regalia, "Blind deconvolution of multivariate signals: a deflation approach.", *Proceedings ICC*, pp. 1160-1164, Juin 1993.
3. N. Delfosse and Ph. Loubaton, "Adaptative blind separation of independent sources: a deflation approach.", *Signal Processing*, vol. 45, pp. 59-83, 1995.
4. Jitendra K. Tugnait, "Identification and deconvolution of multi-channel non gaussian processes using higher order statistics and inverse filter criteria.", *IEEE Transactions on Signal Processing*, vol. 45, pp. 658-672, 1997.
5. C. Simon, Ph Loubaton and C. Jutten, "Separation of a class of convolutive mixtures: a contrast function approach.", *Signal Processing*, vol. 81, pp. 883-887, 2001.
6. W. Wang, S. Sanei and J. A. Chambers, "Penalty function based joint diagonalization approach for convolutive blind separation of nonstationnary sources.", *IEEE Transactions on Signal Processing*, submitted in september 2003.
7. P. Comon, "Independant component analysis, an new concept ? ", *IEEE Transactions on Signal Processing*, vol 36, no. 3, pp 287-314, 1994.
8. A. Ferréol and P. Chevalier, "On the behavior of current second and higher order blind source separation methods for cyclostationnary sources.", *IEEE Transactions on Signal Processing*, vol 48, no. 6, pp 1712-1725, June 2000.
9. M. Kawamoto, Y. Inouye, "A deflation algorithm for the blind source-factor separation of MIMO-FIR channels driven by colored sources", *IEEE Signal Processing Letters*, vol. 10 , no. 11 , Nov. 2003, pp. 343 - 346.
10. M. Castella, J-C. Pesquet and A. Petropulu, "A family of frequency and time-domain contrasts for blind separation of convolutive mixtures of temporally dependant signals.", *IEEE Transactions on Signal Processing*, accepted for publication, 2003.
11. D. Donoho, "On minimum entropy deconvolution.", *Applied time series analysis II*, Academic press, New York, pp 565-608, 1981.
12. A. Hyvarinen, "A family of fixed-point algorithms for independant component analysis", *IEEE International Conference on Acoustics, Speech and Signal Processing*, pp 3917-3920, 1997.

A Continuous Time Balanced Parametrization Approach to Multichannel Blind Deconvolution

Liang Suo Ma and Ah Chung Tsoi

Office of Pro-Vice Chancellor (IT), University of Wollongong
Wollongong, NSW 2522, Australia
lsm01@uow.edu.au, ahchung.tsoi@arc.gov.au

Abstract. In this paper, we will apply a balanced parametrization approach to multichannel blind deconvolution problem with the mixer being modelled as a continuous time linear time invariant system. Such an approach has the advantages of (a) being a computationally robust method, compared with the controller canonical form representation or observer canonical form representation of the linear time invariant continuous time system, and (b) allowing the determination of the number of states required in the demixer. Our approach is validated through a computer simulation example using speech signals.

1 Introduction

The main objective of multichannel blind deconvolution problems is to recover the original latent source signals from a set of given observation data present at the sensor measurements. The source signals are mixed by a dynamical system, known as the mixer. The objective is to design a demixing algorithm which can recover the original latent source signals from the sensor measurements.

Dependent on the situation, one may assume that the mixer to be either a continuous time dynamical system, or a discrete time dynamical system. In this paper, we will assume that the mixer is modelled by a continuous time dynamical system. Accordingly the demixer is also modelled by a continuous time dynamical system.

In the literature, there are considerable amount of work performed on MBD problems with a discrete time mixer [3]. There is relatively less attention being paid to consider mixers which can be modelled by a continuous time dynamical system.

As the mixer is assumed to be a known linear time invariant (LTI) continuous time dynamical system, it is known that this class of systems is invariant under a coordinate transformation [6]. There are a number of possible canonical forms [6]. Some prominent canonical forms include: controller canonical form, observer canonical form, balanced realized canonical form.

To identify a canonical form from input output data, it is known that the observer canonical form or the controller canonical form are non-robust computationally [5], as sometimes the parameter estimation algorithm becomes unstable (often due to pole zero cancellations). It is generally acknowledged that the balanced realized canonical form offers a computationally more robust algorithm [2].

However, it is also acknowledged that if the system is unknown, it is quite difficult to recover a balanced realized canonical form [2]. Towards this end, there have been some work [7] in obtaining a balanced parametrization of an unknown linear system[1]. Such a balanced parametrization will produce a balanced realized canonical form from input output measurements.

As far as we are aware, there has not been any attempt in using a balanced parametrization to model the demixer in a MBD problem where the mixer is assumed to be a continuous time dynamical system. This will be the main aim of this paper: to introduce the possibility of modelling the continuous time demixer using a balanced parametrization.

This paper is organized as follows. In Section 2, we will review a general continuous time state space MBD algorithm given in [4]. In Section 3, we will introduce a balanced parametrization used in [2]. In Section 4, we will derive a continuous time balanced MBD algorithm based on the concepts introduced in Section 2 and Section 3. In Section 5, we will evaluate the proposed balanced parametrization algorithm through computer simulations. We will conclude in Section 6.

2 Continuous Time State Space Approach to MBD

For simplicity sake, we will focus on a noise-free MBD problem, and both the number of sources and the number of sensors are assumed to be n. This MBD problem was considered in [4]. Using a continuous time state space model, the mixer can be modelled as follows:

$$\dot{\overline{\mathbf{x}}}_t = \overline{A}\overline{\mathbf{x}}_t + \overline{B}\mathbf{s}_t \quad \mathbf{u}_t = \overline{C}\overline{\mathbf{x}}_t + \overline{D}\mathbf{s}_t \qquad (1)$$

where $\mathbf{s} \in \mathcal{R}^n$ is the source signal vector; $\mathbf{u} \in \mathcal{R}^n$ is the observation vector; $\overline{\mathbf{x}} \in \mathcal{R}^{\overline{N}}$ is the state vector; $\overline{N}$, generally unknown, is the number of states. The the observation data is given in the duration of $t \in [0,T]$.

The demixer can be modelled as:

$$\dot{\mathbf{x}}_t = A\mathbf{x}_t + B\mathbf{u}_t \quad \mathbf{y}_t = C\mathbf{x}_t + D\mathbf{u}_t \qquad (2)$$

where $\mathbf{u}$ serves as the inputs of the demixer; $\mathbf{y} \in \mathcal{R}^n$ is the vector of the recovered signals; $\mathbf{x} \in \mathcal{R}^N$ is the state vector, N is the number of states in the demixer. We will assume that $N \geq \overline{N}$. We will assume initially that N to be known. The parameter set of the demixer is defined as $\Omega \equiv \{A, B, C, D\}$.

In [4], a general continuous time state space MBD algorithm was derived through minimizing $l(\Omega) = -\log|\det(D)| - \sum_{k=1}^{n} \log P_k(y_k)$. The optimization

---

[1] In this paper, we note the difference between balanced realized canonical form, and a balanced parametrization. A balanced realized canonical form is obtained from a known linear system, while a balanced parametrization is to obtain a balanced realized canonical form from input output measurements.

problem was formulated as a constrained optimization problem, in which the state equation $\dot{\mathbf{x}} = A\mathbf{x} + B\mathbf{u}$ is treated as a constraint. Under this formulation, the MBD problem was solved using a Lagrange multipliers method.

The derivation of the general state space MBD algorithm [4] can be summarized as follows. Defined the performance index:

$$J(\Omega) \equiv \int_0^T dt\ L(t; \mathbf{x}, \mathbf{y}, \boldsymbol{\lambda}, \Omega) \tag{3}$$

where L is the Lagrangian and it is further defined as:

$$L(t; \mathbf{x}, \mathbf{y}, \boldsymbol{\lambda}, \Omega) \equiv l(\Omega) + \boldsymbol{\lambda}^T (A\mathbf{x} + B\mathbf{u} - \dot{\mathbf{x}}) \tag{4}$$

where the Lagrange multipliers $\boldsymbol{\lambda} \in \mathcal{R}^N$ are often called the adjoint states. Through these definitions, the constrained optimization problem is converted into an unconstrained optimization problem by minimizing the function (4).

To proceed, we need to specify $\mathbf{x}$ and $\boldsymbol{\lambda}$ by solving the following Euler-Lagrange variational equations:

$$\frac{\partial L}{\partial \boldsymbol{\lambda}} - \frac{d}{dt}\left(\frac{\partial L}{\partial \dot{\boldsymbol{\lambda}}}\right) = \mathbf{0} \quad \frac{\partial L}{\partial \mathbf{x}} - \frac{d}{dt}\left(\frac{\partial L}{\partial \dot{\mathbf{x}}}\right) = \mathbf{0} \tag{5}$$

with the boundary conditions $\mathbf{x}_0 = \mathbf{0}$ and $\boldsymbol{\lambda}_T = \mathbf{0}$, where $\mathbf{0}$ is an N-column null vector. As a result, we obtain:

$$\dot{\mathbf{x}} = A\mathbf{x} + B\mathbf{u} \quad \dot{\boldsymbol{\lambda}} = -A^T \boldsymbol{\lambda} - C^T \varphi(\mathbf{y}) \tag{6}$$

where $\varphi(\mathbf{y})$ is an n column vector of nonlinear activation functions, its k-th element is defined as $\varphi_k(y_k) \equiv -\frac{\partial \log P_k(y_k)}{\partial y_k}$.

The optimization problem unfortunately can not be solved analytically. A convenient method to solving this optimization problem is to use a gradient based method. This can be derived as follows:

$$\frac{\partial L}{\partial A} = -\boldsymbol{\lambda}\mathbf{x}^T \quad \frac{\partial L}{\partial B} = -\boldsymbol{\lambda}\mathbf{u}^T \quad \frac{\partial L}{\partial C} = -\varphi(\mathbf{y})\mathbf{x}^T \quad \frac{\partial L}{\partial D} = D^{-T} - \varphi(\mathbf{y})\mathbf{u}^T \tag{7}$$

Applying the rule of $\dot{X} = -\eta \frac{\partial L}{\partial X}$, we obtain the following parameter estimation algorithm:

$$\dot{A} = -\eta_A \boldsymbol{\lambda}\mathbf{x}^T \quad \dot{B} = -\eta_B \boldsymbol{\lambda}\mathbf{u}^T \tag{8}$$

$$\dot{C} = -\eta_C \varphi(\mathbf{y})\mathbf{x}^T \quad \dot{D} = \eta_D (I - \varphi(\mathbf{y})\mathbf{u}^T D^T) D \tag{9}$$

where I is an $n \times n$ identity matrix; η_X is a time-dependent learning rate with respect to X, X may be A, B, C or D.

We note the following points about this algorithm: (i) to improve the performance of the parameter estimation algorithm, the technique of natural gradient [1] has been employed in the parameter estimation algorithm of matrix D; (ii) it requires the number of states $\overline{N}$ known a prior; and (iii) the total number of parameters to be estimated is $(N + n)^2$.

3 Balanced Parametrization of Linear Systems

Apart from the general state space approach, the demixer can also be expressed in a balanced realized canonical form. Consider the dynamical system (2), we say that the system is balanced if there exist two $N \times N$ diagonal matrices P and Q, satisfying the following dual Lyapunov equations simultaneously:

$$AP + PA^T = -BB^T \qquad A^T Q + QA = -C^T C \qquad (10)$$

Matrices P, Q are known as controllability Grammian and observability Grammian respectively, and their diagonal elements $\{\sigma_i\}_{i=1}^N$ are known as Hankel singular values. If we assume that $\sigma_1 \geq \ldots \geq \sigma_n > 0$ and $\sigma_{r+1} \ll \sigma_r$ ($r \in [1, n-1]$) are satisfied, then it is possible to show that the order-reduced model of dimension r is asymptotically stable and minimal [5]. In addition, the behavior of the order-reduced model "approximates" that of the original LTI system [5].

The balanced realized canonical form can be obtained if A, B, C, and D are known. However, in our situation, we will need to find a set of parameters A, B, C and D from input output measurements such that the dual Lyapunov equations are satisfied. This is quite difficult. In the literature, there have been various attempts to devise a parametrization such that the parameters A, B, C, and D will result in a balanced realized canonical form [2]. In this paper, we will adopt the balanced canonical form used in [2], with which the linear system (2) can be balanced parameterized by the parameters $\Theta \equiv \{\Sigma, B, D, \Phi\}$, where

Σ: $\{\sigma_i\}_{i=1}^N$, satisfying $\sigma_1 > \cdots > \sigma_N > 0$;
Φ: $\{\phi_i\}_{i=1}^N$, $\phi_i \in \mathcal{R}^{(n-1)\times 1}$, $\phi_{pi} \in (-\frac{\pi}{2}, \frac{\pi}{2})$;
B: $\{B_i\}_{i=1}^N$, $B_i \in \mathcal{R}^{1\times n}$, $B_{i1} > 0$;
D: $n \times n$ real matrix.

Based on Θ, the elements of the parameters Ω can be obtained as follows:

B: $B^T = \begin{bmatrix} B_1^T & B_2^T & \cdots & B_N^T \end{bmatrix}$;
C: $C = \begin{bmatrix} C_1 & C_2 & \cdots & C_N \end{bmatrix}$;
A: $A = [A_{ij}]$ for $i, j = 1, 2, \cdots, N$;
D: $D = D$.

where the j-th column of C is constructed as:

$$C_j = V(\phi_j)\sqrt{B_j B_j^T} \qquad (11)$$

where $V(\phi_j) = \begin{bmatrix} v_{1j} & v_{2j} & \cdots & v_{nj} \end{bmatrix}^T$ and the elements of V are given as:

$$v_{pj} = \begin{cases} \cos\phi_{1j} \prod_{k=1}^{n-2} \cos\phi_{n-k,j}, & \text{for } p = 1 \\ \sin\phi_{p-1,j} \prod_{k=1}^{n-p} \cos\phi_{n-k,j}, & \text{for } 1 < p < n \\ \sin\phi_{n-1,j}, & \text{for } p = n \end{cases} \qquad (12)$$

The elements A_{ij} of the matrix A are given by:

$$A_{ij} = \begin{cases} -\frac{B_j B_j^T}{2\sigma_j}, & \text{for } i = j \\ \frac{\sigma_j B_i B_j^T - \sigma_i C_i^T C_j}{\sigma_i^2 - \sigma_j^2}, & \text{for } i \neq j \end{cases} \qquad (13)$$

Under the above parametrization, the continuous time LTI system with parameters Ω is balanced with $\mathrm{diag}(\sigma_1, \cdots, \sigma_N)$, where diag is a diagonal operator.

4 Balanced Parametrization Approach to MBD

Treating the parameters Ω as intermediate variables, it is possible to compute the gradients of the cost function (4) with respect to the parameters Θ. However, since the learning algorithm with respect to the parameters Ω have been given in section 2, if we can obtain the relationship between Θ and Ω, the balanced parameter estimation algorithm can be derived through combing these two parts.

4.1 The Relationship Between Θ and Ω

The matrices B and D are shared by the parameters Θ and Ω, thus we only need to consider ϕ_{pj} and σ_j, for $p = 1, 2, \cdots, n-1$ and $j = 1, 2, \cdots, N$.

Consider (11) and (12), if we define $R_{Bj} \equiv \sqrt{B_j B_j^T}$, we obtain:

$$\dot{C}_{p+1,j} = R_{Bj} \sum_{\ell=p}^{n-1} \xi_{kj}^{p+1} \dot{\phi}_{\ell j} + R_{Bj}^{-1} v_{p+1,j} B_j \dot{B}_j \qquad (14)$$

where we have used the definition $\xi_{kj}^{p+1} \equiv \frac{\partial C_{p+1,j}}{\partial \phi_{kj}}$, for $k = p, \cdots, n-1$, this gives:

$$\xi_{kj}^{p+1} = \begin{cases} -\sin\phi_{kj} \prod_{\ell=k+1}^{n-1} \cos\phi_{\ell j} \prod_{\ell=p}^{k-1} \cos\phi_{\ell j}, & k > p \\ \prod_{\ell=p}^{n-1} \cos\phi_{\ell j}, & k = p \end{cases}$$

From (14), we have the following relation:

$$\dot{C}_j^{n-1} = R_{Bj} \Delta_{(j)} \dot{\phi}_j + R_{Bj}^{-1} B_j \dot{B}_j \mathbf{v}_j^{n-1} \qquad (15)$$

where
$\dot{C}_j^{n-1} \equiv \begin{bmatrix} \dot{C}_{2j} & \dot{C}_{3j} & \cdots & \dot{C}_{nj} \end{bmatrix}^T$, $\mathbf{v}_j^{n-1} \equiv \begin{bmatrix} v_{2j} & \cdots & v_{nj} \end{bmatrix}^T$, $\dot{\phi}_j \equiv \begin{bmatrix} \dot{\phi}_{1j} & \cdots & \dot{\phi}_{n-1,j} \end{bmatrix}^T$
and $\Delta_{(j)} \equiv \begin{bmatrix} \xi_{1j}^2 & \cdots & \xi_{n-1,j}^2 \\ & \ddots & \vdots \\ & & \xi_{n-1,j}^n \end{bmatrix}$.

To express $\dot{\phi}_j$ with $\dot{C}_j^{n-1}$ and $\dot{B}_j$ from (15), we require $\det(\Delta_{(j)}) \neq 0$. Observe $\det(\Delta_{(j)}) = \prod_{k=1}^{n-1} \xi_{kj}^{k+1}$, hence we need to satisfy $\xi_{k-1,j}^k \neq 0$ for all k. From the definition of $\xi_{k-1,j}^k$, we know that this is equivalent to satisfying $\xi_{1j}^2 \neq 0$. In other words, we need $\prod_{k=1}^{n-1} \cos\phi_{kj} \neq 0$. In Section 3, we have defined $\phi_{pj} \in (-\frac{\pi}{2}, \frac{\pi}{2})$, this guarantees that the condition $\prod_{k=1}^{n-1} \cos\phi_{kj} \neq 0$ is satisfied.

From (15), we have the following relationship:

$$\dot{\phi}_j = R_{Bj}^{-1} \Delta_{(j)}^{-1} \left(\dot{C}_j^{n-1} - R_{Bj}^{-1} B_j \dot{B}_j \mathbf{v}_j^{n-1} \right) \qquad (16)$$

Note that in (16), we need to compute an inverse matrix $\Delta_{(j)}^{-1}$, however it is trivial because $\Delta_{(j)}$ is an upper-triangular matrix.

For the parameter set Σ, the derivation is relatively simple. Consider the diagonal elements of matrix A in (13), we can easily obtain $\frac{\partial A_{jj}}{\partial \sigma_j} = \frac{R_{Bj}^2}{2\sigma_j^2}$. Combining the chain rule $\dot{A}_{jj} = \frac{\partial A_{jj}}{\partial \sigma_j}\dot{\sigma}_j$, we obtain:

$$\dot{\sigma}_j = 2R_{Bj}^{-2}\sigma_j^2 \dot{A}_{jj} \tag{17}$$

4.2 Parameter Estimation Algorithm

Substitute $\dot{C}$ (in (9)) and $\dot{B}$ (in (8)) into (16), we obtain the following parameter estimation algorithm with respect to Φ:

$$\dot{\phi}_j = -\eta_\phi R_{Bj}^{-1} \Delta_{(j)}^{-1} \left[\varphi(\mathbf{y}_{n-1}) x_j^T + \eta_r R_{Bj}^{-1} B_j \lambda_j \mathbf{u}^T \mathbf{v}_j^{n-1} \right] \tag{18}$$

where η_r is the relative learning rate, which is used due to the fact that η_B and η_C are not necessarily identical. Similarly, substitute $\dot{A}$ (in (8)) into (17), we obtain the following parameter estimation algorithm with respect to Σ:

$$\dot{\sigma}_j = -\eta_\sigma R_{Bj}^{-2} \sigma_j^2 \lambda_j x_j^T \tag{19}$$

In summary, we have the following continuous time balanced MBD algorithm:

$$\dot{D} = \eta_D \left(I - \varphi(\mathbf{y})\mathbf{u}^T D^T \right) D \quad \dot{B} = -\eta_B \boldsymbol{\lambda}\mathbf{u}^T \quad \dot{\sigma}_j = -\eta_\sigma R_{Bj}^{-2} \sigma_j^2 \lambda_j x_j^T \tag{20}$$

$$\dot{\phi}_j = -\eta_\phi R_{Bj}^{-1} \Delta_{(j)}^{-1} \left[\varphi(\mathbf{y}_{n-1}) x_j^T + \eta_r R_{Bj}^{-1} B_j \lambda_j \mathbf{u}^T \mathbf{v}_j^{n-1} \right] \tag{21}$$

where η_X, X may be D, B, ϕ, or σ, is a time-dependent learning rate with respect to X.

Alternatively, the proposed balanced algorithm can be described using the following iterative updates:

$$D^{k+1} = D^k + \tau_D \dot{D}^k \quad B^{k+1} = B^k + \tau_B \dot{B}^k \tag{22}$$

$$\sigma_j^{k+1} = \sigma_j^k + \tau_\sigma \dot{\sigma}_j^k \quad \phi_j^{k+1} = \phi_j^k + \tau_\phi \dot{\phi}_j^k \tag{23}$$

where k is the iteration counter; τ_X is a suitably selected positive constant, which is related to time; $\dot{X}$ is given in (20) and (21).

Note, similar to the general state space algorithm [4], with respect to the implementation of the proposed balanced parametrization, to learn B, Φ and Σ, we need to specify the state $\mathbf{x}$ and the adjoint state $\boldsymbol{\lambda}$ by solving a two point boundary value problem with boundary conditions $\mathbf{x}_0 = \mathbf{0}$ and $\boldsymbol{\lambda}_T = \mathbf{0}$. Also note, in the proposed balanced parametrization algorithm, we need to estimate a total of $2nN + n^2$ parameters. This is a reduction of N^2 parameters compared with the general state space MBD algorithm [4], viz. $(N+n)^2$. In general, N may be large, thus, the saving may be significant.

Secondly, using the balanced parametrization approach, the number of states can be estimated by tracking the values of σ. Since the estimated values of σ are arranged in descending orders, if some of the estimated σ are small compared with others, then they can be neglected. Hence the number of states can be obtained accordingly.

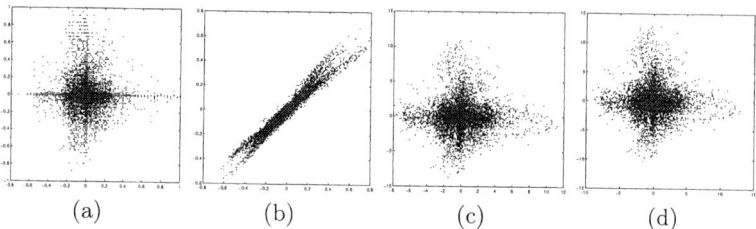

(a) (b) (c) (d)

Fig. 1. The scatter diagrams of (a) the sources; (b) the observations; (c), (d) the recovered signals obtained from the general algorithm [4] and the proposed algorithm respectively.

5 Computer Simulations

In this section, we will evaluate the proposed balanced parametrization algorithm through computer simulations. Our objectives are twofold: (i) to compare the performance of the proposed balanced parametrization algorithm with that of the general MBD algorithm [4]; and (ii) to examine the capacity of estimating the number of states N of the proposed balanced algorithm.

In these experiments, 10,000 samples of two-column observation data (Figure 1 (b)) are obtained by passing two speech sources (Figure 1 (a)) through a stable continuous time demixer (in the formulation of (1)) with 2 states, thus in this mixing environment, $n = 2$, $\overline{N} = 2$. The system matrices of the mixer is chosen such that (i) $\overline{D}^{-1}$ exists, (ii) the eigenvalues of the $\overline{A}$ are all on the left hand side of the complex plane, and (iii) $\overline{B}$ and $\overline{C}$ are randomly selected. We solve the continuous time dynamical system using a Runge-Kutta algorithm.

To run the general MBD algorithm [4] and the proposed balanced parametrization algorithm, we assume that $\overline{N}$ is unknown, but we believe that the current number of states ($N = 6$) of the demixer is greater than $\overline{N}$. The recovered signals obtained from both approaches are plotted in Figure 1 (c) and (d). The mean square error (MSE) and residual cross talk (X_{talk}) [2] obtained in both approaches are shown in Table 1. The evolution of the singular values obtained in the proposed balanced algorithm is plotted in Figure 2.

From Figure 1 and Table 1 we observe that, in the sense of mean square error and cross talk, the performance of the two algorithms are comparable for this simple MBD problem. From Figure 2, we observe that there is a big gap between the second largest singular value and the remaining singular values. This result shows that the proposed balanced parametrization algorithm can identify the correct number of states $\overline{N}$ quite well. This can be further verified by evaluating the relative weight of the first two principal singular values as a function of the total "amount of energy" (the sum of the singular values) in the system, which is 91%.

---

[2] Cross talk is defined as: $X_{talk} \equiv \frac{1}{T(n^2-n)} \sum_{i=1}^{n} \sum_{\substack{j=1 \\ j \neq i}}^{n} \sum_{t=1}^{T} \left[y_i(t) s_j(t) - s_i(t) s_j(t) \right]$.

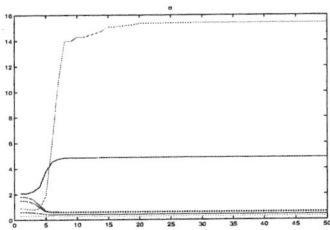

Fig. 2. Evolution of singular values in the proposed balanced parametrization algorithm.

Table 1. The mean square errors and cross talks (in the order of 10^{-3}) given by the general MBD algorithm [4] and the proposed balanced parametrization algorithm.

	General		Balanced	
Channel	I	II	I	II
MSE	1.0	1.6	0.9	1.6
X_{talk}	−0.3		−0.5	

6 Conclusion

In this paper we have considered a balanced parametrization approach to the MBD problem in the continuous time domain. Compared with the general state space MBD algorithm [4], the proposed balanced parametrization algorithm possesses the following advantages: (i) it can estimate the number of states of the mixer, this overcomes a practical problem in [4]; and (ii) it needs to estimate less number of parameters.

References

1. Amari, S: Natural gradient working efficiently in learning. Neural Computation. **10** (1998) 251–276
2. Chou, C.T., Maciejowski J.M.: System identification using balanced parametrizations. IEEE Trans. Auto. Contr. **42** (1997) 956–974
3. Cichocki, A., Amari, S.: Adaptive Blind Signal and Image Processing: Learning Algorithms and Applications. John Wiley & Sons. (2002)
4. Erten, G., Salam, F.: Voice extraction by on-line signal separation and recovery. IEEE Transactions on Circuits and Systems II: Analog and Digital Signal Processing. **46** (1998) 915–922
5. Glover, K.: All optimal Hankel norm approximations of linear multivariable systems and L^∞ error bounds. International Journal of Control. **39** (1984) 1115–1193
6. Kailath, T.: Linear Systems. Prentice Hall. (1980)
7. Tsoi, A.C., Ma, L.S.: Blind Deconvolution of Dynamical System Using a Balanced Parameterised State Space Approach. IEEE International Conference on Acoustics, Speech, and Signal Processing (ICASSP), Hongkong. **IV** (2003) 309–312

A Frequency-Domain Normalized Multichannel Blind Deconvolution Algorithm for Acoustical Signals

Seung H. Nam[1] and Seungkwon Beack[2]

[1] Dept. of Electronic Engineering, Paichai University, Taejon, Korea
shnam@pcu.ac.kr
[2] Multimedia Group, Information and Communications University, Taejon, Korea

Abstract. In this paper, a new frequency-domain normalized MBD algorithm is presented for separating convolutive mixtures of acoustical sources. The proposed algorithm uses unidirectional unmixing filters to avoid backward filtering in gradient terms. The gradient terms are then normalized in the frequency domain. As a result, separation and convergence performances are improved, while whitening effect is relieved greatly. Simulation results with real world recordings demonstrate superior performances of the proposed algorithm.

1 Introduction

Separation of acoustical mixtures in a room environment is very attractive for many practical applications such as robust speech recognition, echo cancellation, and object-based audio processing. The multichannel blind deconvolution (MBD) algorithm is one practical method for blind source separation. In the sequel, we describe the MBD algorithm with natural gradient and its shortcomings.

In convolutive mixing, the mixed signal at the sensor j is given by

$$x_j(k) = \sum_{i=1}^{n} \sum_{p=-\infty}^{\infty} a_{ji,p} s_i(k-p), \quad j = 1, 2, \cdots, m \quad (1)$$

where $s_i(k), i = 1, \cdots, n$, are source signals and $a_{ji,p}$ is the (j, i) element of the mixing system $\mathbf{A}(z) = \sum_{p=-\infty}^{\infty} \mathbf{A}_p z^{-p}$ at lag p. Similarly, the ith unmixing signal is given by

$$u_i(k) = \sum_{j=1}^{m} \sum_{p=-\infty}^{\infty} w_{ij,p}(k) x_j(k-p) \quad (2)$$

where $w_{ij,p}(k)$ is the (i, j) element at lag p of the unmixing system at time k, ie. $\mathbf{W}(z, k) = \sum_{p=-\infty}^{\infty} \mathbf{W}_p(k) z^{-p}$. The number of sensors m is assumed to be equal to or greater than the number of sources n in general.

In [1], the MBD algorithm with the natural gradient (NGMBD) is presented as

$$\triangle \mathbf{W}_p(k) = \mathbf{W}_p(k) - \mathbf{y}(k-L)\mathbf{v}^T(k-p) \quad (3)$$

where $\mathbf{y}(k) = f(\mathbf{u}(k))$ for some monotonic nonlinear function $f(\cdot)$ and

$$\mathbf{v}(k) = \sum_{q=0}^{L-1} \mathbf{W}_{L-q}^T \mathbf{u}(k-q) \tag{4}$$

where L is the length of truncated bidirectional unmixing filters. In (3), L sample delay is introduced in $\mathbf{y}(k)$ to accommodate anticausal parts of the unmixing filters. Step size may be normalized in the time domain for robustness as follows:

$$\mu_i(k) = \frac{\mu_0}{\beta + \sum_p y_i(k-p) u_i(k-p)} \tag{5}$$

where β is a small constant.

The NGMBD algorithm works very well if each source $s_i(k)$ is uncorrelated in time. Its performance is degraded severely, however, if it is applied to highly correlated nonstationary sources such as speech and audio. Performance degradation of the NGMBD algorithm for acoustical mixtures is twofold. One is slow convergence due to large eigenvalue spread of the cross-correlation matrix between $\mathbf{y}(k-L)$ and $\mathbf{u}(k-p)$. The other is whitening of unmixed sources since the the NGMBD algorithm (3) has equilibrium points

$$E\{y_i(k) u_j(k-l)\} = \delta_{ij} \delta_l. \tag{6}$$

Consequently quality of the unmixed acoustic signal is generally poor although it is still intelligible. This whitening problem has been treated by some researchers – post processing [2], a nonholonomic algorithm [3], and a linear predictive method [4].

In this paper, we propose a simple and efficient solution in the frequency domain to these difficulties. It is well known that eigenvalue spread of nonstationary sources can be lowered by implementing the algorithm in the frequency domain and employing a normalized step size for each frequency bin [5, 6]. Frequency domain implementations have been employed for various BSS algorithms for its computational efficiency [2, 4], and detail implementations of the NGMBD algorithm (3) have been discussed [7, 8]. However, normalization of the term $\mathbf{y}(k-L) \mathbf{v}^T(k-p)$ in (3) in the frequency domain has never been discussed.

For this purpose, we first examine the problem of the NGMBD algorithm in a single channel case. A modification to the NGMBD algorithm is discussed for normalization of gradient terms in the frequency domain. Then a new frequency-domain normalized MBD algorithm is proposed as a solution. Simulation results with speech mixtures recorded in real room environments are presented to demonstrate its superior performances.

2 An Alternative Form of the MBD Algorithm for Normalization in the Frequency Domain

2.1 Review of a Single Channel Case

Consider a single channel Bussgang deconvolution algorithm with natural gradient [1, 9]. In a single channel Bussgang deconvolution algorithm, the filter $w(k)$ is assumed to be finite and the output is expressed as

$$u(k) = \sum_{p=0}^{L-1} w_p(k) x(k-p). \qquad (7)$$

The Bussgang algorithm with natural gradient is then obtained by applying $w(z^{-1},k)w(z,k)$ to the standard gradient as

$$\triangle w_p(k) = y(k) \sum_{q=0}^{L-1}\sum_{r=0}^{L-1} x(k-p+q-r) w_q(k) w_r(k) \qquad (8)$$

where $y(k) = f(u(k))$ is the output of Bussgang nonlinearity to $u(k)$. If we assume that $w_r(k-p+q) \approx w_r(k)$ for $0 \le p \le L-1$ and $0 \le q \le L-1$, (8) can be rewritten approximately as

$$\begin{bmatrix} \triangle w_0(k) \\ \vdots \\ \triangle w_{L-1}(k) \end{bmatrix} = \begin{bmatrix} y(k)u(k) & \cdots & y(k)u(k+L-1) \\ \vdots & \ddots & \vdots \\ y(k)u(k-L+1) & \cdots & y(k)u(k) \end{bmatrix} \begin{bmatrix} w_0(k) \\ \vdots \\ w_{L-1}(k) \end{bmatrix}. \qquad (9)$$

Since future samples of $u(k)$ are involved in (9), $u(k)$ and $y(k)$ are delayed by L samples as in (3). The Bussgang algorithm is usually initialized with $w_p(0) = \delta_{p-q}$ for $0 \le q \le L-1$. Then the position of the leading tap w_q affects convergence of the algorithm. If $0 < q < L-1$, the converged filter $\{w_p(\infty)\}_{p=0}^{L-1}$ would be a delayed version of a bidirectional nonminimum phase filter.

The cross-correlation matrix between $y(k)$ and $u(k)$ is diagonal if the source signals are nearly white which is true in general for telecommunication signals. If the source signals are correlated in time, however, the cross-correlation would not be diagonal and eigenvalue spread would be large. Although natural gradient provides faster convergence than standard gradient, large eigenvalue spread would be very harmful to convergence of the algorithm.

To demonstrate this adverse effect, we examined the trajectories of equalizers with both standard and natural gradient. We used the same experimental setup as in [9] except that speech sources as well as noise sources are used. Figure 1 shows trajectories of algorithms that start from six different initial values to the optimal point $\mathbf{w}_{opt} = [1 \ 0.95]^T$. For white noise sources, trajectories 1(c) of natural gradient are direct than trajectories 1(a) of standard gradient as demonstrated in [9]. For a speech source, however, trajectories 1(d) of natural gradient become similar to trajectories 1(b) of standard gradient.

2.2 A Causal MBD Algorithm

It is noted that direct normalization of the gradient term in the frequency domain may not always provide satisfactory results because of backward filtering (4). To avoid backward filtering, future samples in (3) are simply ignored rather than being delayed. Then the single channel algorithm (9) becomes

$$\begin{bmatrix} \triangle w_0(k) \\ \vdots \\ \triangle w_{L-1}(k) \end{bmatrix} = \begin{bmatrix} y(k)u(k) & \cdots & 0 \\ \vdots & \ddots & \vdots \\ y(k)u(k-L+1) & \cdots & y(k)u(k) \end{bmatrix} \begin{bmatrix} w_0(k) \\ \vdots \\ w_{L-1}(k) \end{bmatrix} \qquad (10)$$

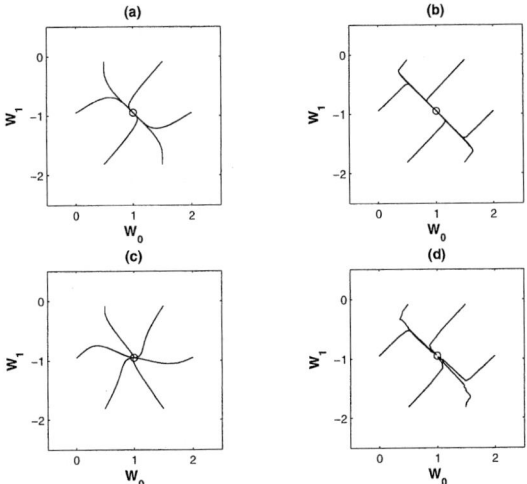

Fig. 1. Trajectories of single channel equalizers: (a) standard gradient-white source, (b) standard gradient-speech source, (c) natural gradient-white source, and (d) natural gradient-speech source.

or

$$\triangle w_p(k) = \sum_{q=0}^{p} y(k)u(k-p+q)w_q(k). \quad (11)$$

If we extending (11) into the multichannel case, the corresponding multichannel algorithm can be written as

$$\triangle \mathbf{W}_p(k) = \sum_{q=0}^{p} \left\{ \mathbf{I}\delta_{p-q} - \mathbf{y}(k)\mathbf{u}^T(k-p+q) \right\} \mathbf{W}_q(k). \quad (12)$$

Notice that (12) does not include backward filtering and unmixing filters are unidirectional and causal. The correlation term $\mathbf{y}(k)\mathbf{u}^T(k-p+q)$ can be easily normalized in the frequency domain. In fact, this is the same causal MBD algorithm that has been derived on the basis of the geometrical structures of the FIR manifolds [10]. The minimum phase algorithm (12) is known to have good convergence properties: equivariant property in the Lie group sense and nonsingularity of $\mathbf{W}_0$.

3 A Frequency-Domain Normalized MBD Algorithm

In this section, we propose a frequency-domain normalized form of the causal MBD (FNMBD) algorithm (12). Let

$$\bar{\mathbf{I}} = \begin{bmatrix} \bar{1} & \bar{0} \\ \bar{0} & \bar{1} \end{bmatrix}$$

where $\bar{1}$ and $\bar{0}$ denote vector of 1's and 0's of appropriate length, respectively. Then the frequency-domain normalized form of (12) is written as

$$\triangle \mathbf{W}(f,b) = \left\{ \bar{\mathbf{I}} - \mathbf{\Lambda}_\mathbf{y}^{-1}(f,b)\mathbf{Y}(f,b)\mathbf{U}^H(f,b)\mathbf{\Lambda}_\mathbf{u}^{-1}(f,b) \right\} \mathbf{W}(f,b) \qquad (13)$$

where $\mathbf{W}(f,b)$, $\mathbf{Y}(f,b)$, and $\mathbf{U}(f,b)$ are the Fourier transform of frame data $\mathbf{W}(b)$, $\mathbf{y}(b)$, and $\mathbf{u}(b)$, respectively, at block time b. Computation of (13) is performed in element-wise at each frequency f. Also, $\mathbf{\Lambda}_\mathbf{y}(f,b)$ and $\mathbf{\Lambda}_\mathbf{u}(f,b)$ are diagonal matrices with diagonal elements $\sqrt{\mathbf{P}_{y_i}(f,b)}$ and $\sqrt{\mathbf{P}_{u_i}(f,b)}$, respectively, that can be updated at each block time for each frequency f. The power spectra are updated for each frequency as follows:

$$\mathbf{P}_{y_i}(f,b) = (1-\gamma)\mathbf{P}_{y_i}(f,b-1) + \gamma |\mathbf{Y}_i(f,b)|^2 \qquad (14a)$$
$$\mathbf{P}_{u_j}(f,b) = (1-\gamma)\mathbf{P}_{u_j}(f,b-1) + \gamma |\mathbf{U}_j(f,b)|^2 \qquad (14b)$$

where $0 < \gamma < 1$.

It is clear that aliased parts in $\triangle \mathbf{W}(f,b)$ and $\mathbf{\Lambda}_\mathbf{y}^{-1}(f,b)\mathbf{Y}(f,b)\mathbf{U}^H(f,b)\mathbf{\Lambda}_\mathbf{u}^{-1}(f,b)$ should be discarded properly in the time domain. Furthermore, the unmixing filters are normalized to have unit power for proper scaling of unmixed signals.

At steady states, the update rule (13) has equilibrium points

$$\frac{E\left\{Y_i(f,b)U_j^*(f,b)\right\}}{\sqrt{E\left\{|Y_i(f,b)|^2\right\}E\left\{|U_j(f,b)|^2\right\}}} = \delta_{ij} \qquad (15)$$

whereas equilibrium points of (12) are given by

$$E\left\{y_i(k)u_j(k-l)\right\} = \delta_{ij}\delta_l. \qquad (16)$$

Clearly, equilibrium points (15) do not impose any compulsory constraints on the spectrum of unmixed signals whereas (16) forces whitening of unmixed sources. This shows that normalization of gradient terms in the frequency domain solves both convergence and whitening problems effectively.

4 Simulations

To demonstrate the performance of the proposed algorithm, we performed two sets of experiments using real world recordings available in web-sites. In the first experiment, we used the English-Spanish mixture in [2] to compare the proposed algorithm with two existing MBD algorithms: a frequency-domain implementation of the NGMBD algorithm [8] and a nonholonomic version of the causal MBD algorithm (12) [11]. Time-domain normalization is used for these existing MBD algorithms. Block length $M = 128$, frame length $N = 512$, filter length $L = 128$, and step size $\mu = 0.01$ are used for all algorithms. For the FNMBD algorithm, $\gamma = 0.5$ is used in (14) to update power spectrum. Figure 2 shows mixed and unmixed speech signals *just after* 1 iteration. Unmixed

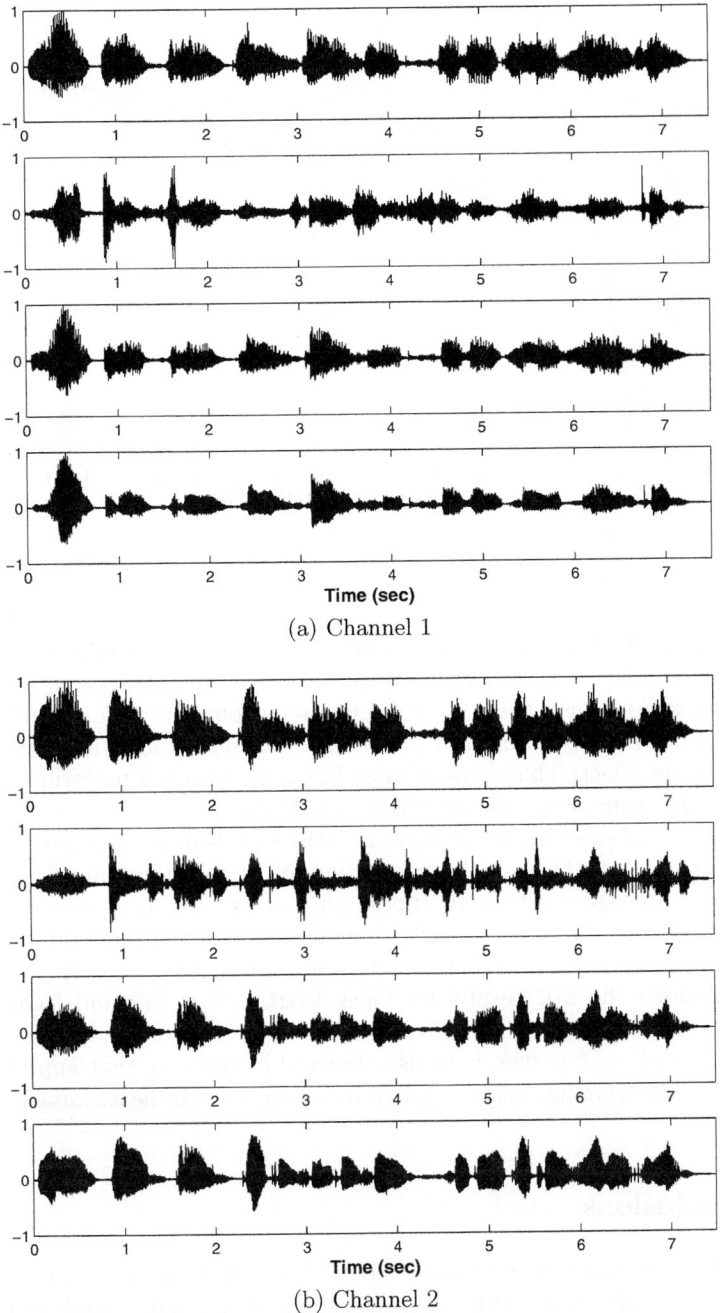

Fig. 2. Comparison of MBD algorithms using real world speech mixtures: mixed speech and unmixed speech by the NGMBD, the nonholonomic MBD, and the FNMBD algorithm from top to bottom for each channel.

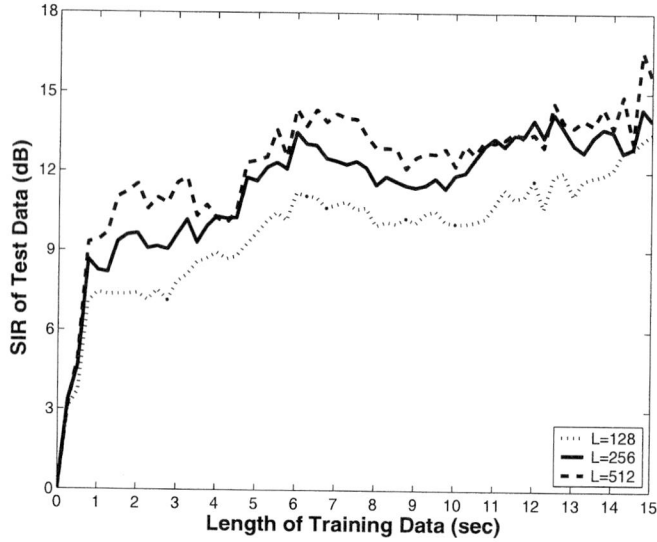

Fig. 3. SIR performance of the FNMBD algorithm ($\mu = 0.03$ and $\gamma = 0.5$).

outputs from the NGMBD algorithm are clearly whitened whereas those from the nonholonomic MBD algorithm are quite reverberant. Furthermore, unmixed outputs from both algorithms are not separated enough. On the other hand, the proposed FNMBD algorithm provides nearly clean separation without whitening or reverberant effect. This demonstrates fast and improved performance of the proposed algorithm over existing MBD algorithms.

In the second experiment, we measured the signal-to-interference ratio (SIR) for various filter lengths using the alternating speech recordings in [12]. Unmixing filters are trained iteratively 10 times using on increasing length of learning speech data. After training at each interval, the resulting unmixing filters are used to separate sources and SIR values are measured (see [12] for details). Figure 3 shows the SIR results for block length $M = 512$ and frame length $N = 2048$ for various unmixing filter lengths. It is observed that over 12 dB of SIR is achieved for $L = 256$. It is also observed in this case that approximately 5 seconds of the training data is required to converge when the learning is iterated 10 times.

5 Conclusions

A new frequency-domain normalized MBD algorithm is presented. It is derived from the MBD algorithm with natural gradient by assuming unidirectional unmixing filters and normalizing gradient terms in the frequency domain. The resulting algorithm provides faster convergence and improved separation for acoustic mixtures while whitening effects is greatly relieved. Simulations using real world recordings demonstrate superior performances of the proposed algorithm.

Acknowledgement

This work was supported in part by grant No. R05-2004-000-10290-0 from Ministry of Science & Technology of Korea.

References

1. Amari, S.I., Douglas, S.C., Cichocki, A., Yang, H.H.: Novel on-line adaptive learning algorithms for blind deconvolution using the natural gradient approach. In: Proc. IEEE 11th IFAC Symposium on System Identification, SYSID-97, Kitakyushu, Japan (1997) 1057–1062
2. Lee, T.W., Bell, A., Orglmeister, R.: Blind source separation of real world signals. In: Proc. IEEE Int. Conf. Neural Networks, Houston (June 1997) 2129–2135
3. Amari, S.I., Chen, T.P., Cichocki, A.: Nonholonomic orthogonal learning algorithms for blind source separation. Neural Computation **12** (2000) 1463–1484
4. Sun, X., Douglas, S.: A natural gradient convolutive blind source separation algorithm for speech mixtures. In: Proc. Int. Workshop on Independent Component Analysis and Signal Separation (ICA'01), San Diego, California (2001) 59–64
5. Ferrara, E.R.: Fast implmentation of LMS adaptive filters. IEEE Trans. on Acoustics Speech and Signal Processing **ASSP-28** (1980) 474–475
6. Haykin, S.: Adaptive Filter Theory. 4th edn. Prentice Hall (2002)
7. K. Na, S. Kang, K.L., Chae, S.: Frequency-domain implementation of block adaptive filters for ica-based multichannel blind deconvolution. In: Proc. IEEE Int. Conf. on Acoustics, Speech and Signal Processing (ICASSP'99). (1999)
8. Joho, M., Schniter, P.: Frequency domain realization of a multichannel blind deconvolution algorithm based on the natural gradient. In: Proc. Int. Workshop on Independent Component Analysis and Signal Separation (ICA'03). (2003)
9. S. Douglas, A.C., Amari, S.: Quasi-newton filtered-regressor algorithms for adaptive equalization and deconvolution. In: Proc. IEEE Workshop on Signal Processing Advances in Wireless Communications, Paris, France (1997) 109–112
10. Zhang, L., Cichocki, A., Amari, S.: Geometrical structures of FIR manifold and their application to multichannel blind deconvoluution. In: Proc. IEEE Workshop on Neural Networks for Signal Processing (NNSP'99), Madison, Wisconsin (1999) 303–312
11. Cichocki, A., Amari, S.I.: Adaptive Blind Signal and Image Processing: Learning algorithms and applications. Wiley (2002)
12. Fancourt, C., Parra, L.: Coherence function as a criterion for blind source separation. In: IEEE International Workshop on Neural Networks and Signal Processing 2001. (2001) 303–312

A Novel Hybrid Approach to the Permutation Problem of Frequency Domain Blind Source Separation

Wenwu Wang[1], Jonathon A. Chambers[1], and Saeid Sanei[2]

[1] Communications and Information Technologies Research Group
Cardiff School of Engineering, Cardiff University, Cardiff, CF24 0YF, UK
wenwu.wang@ieee.org, chambersj@cf.ac.uk
[2] Centre for Digital Signal Processing Research
King's College London, Strand London, WC2R 2LS, UK
saeid.sanei@kcl.ac.uk

Abstract. We explore the permutation problem of frequency domain blind source separation (BSS). Based on performance analysis of three approaches: exploiting spectral continuity, exploiting time envelope structure and beamforming alignment; we present a new hybrid method which incorporates a psychoacoustic filtering process for the misaligned permutations unable to be delt with by these approaches. We use a subspace based method (MUSIC) rather than conventional beamforming for the accurate estimation of the direction of arrivals (DOAs) of the source components, and a frequency dependent distance for the correlation of time envelopes. The proposed methods are compared with other approaches by signal to interference ratio (SIR) evaluation, and the new hybrid approach is shown to have the best performance.

1 Introduction

Convolutive BSS has recently received extensive interest within the signal processing community due to its potential applications in communications, speech processing, and medical imaging. An effective method of addressing this problem is to transform it into the frequency domain so that a series of complex-valued instantaneous BSS problems is solved separately using a conventional instantaneous mixing independent component analysis (ICA) framework. A crucial limitation associated with such a transformation is the permutation indeterminacy which is induced inherently by the general ICA approach. That is, the reconstructed source signals in the time domain will remain distorted if the permutations of the recovered frequency domain source components are not consistent with each other.

To address this problem, several approaches have been developed, which can be approximately classified as: (1) exploiting the continuity of the spectra of the recovered signals or the separation matrix [1] [2]; (2) Exploiting the time structure of the source components [3]; (3) applying beamforming techniques to

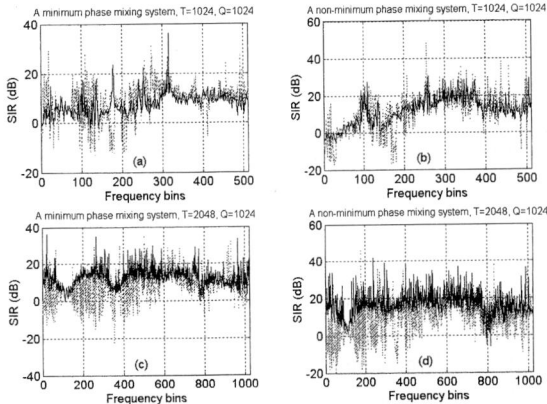

Fig. 1. SIR improvement across frequency axis before (dotted line) and after permutation alignment (solid line) using two methods: the separation matrices coupling over neighboring frequency bins (a) (b) and filter length constraint (c) (d).

the permutation alignment [4] [5]. These approaches may work well for carefully defined situations, but not necessarily for others. A recent work in [6] suggests that it is possible to combine the different properties of these approaches for developing a more robust and precise solution. In this paper, building upon this idea, we aim at developing a new hybrid approach, which is expected to benefit from some established results but have better performance. Additionally, we introduce some results of psychoacoustic research for reducing the permutation effect.

The remainder of the paper is organized as follows. Frequency domain BSS (FDBSS) together with its associated permutation problem is briefly described in Section 2. The various solutions are investigated in Section 3, which includes the introduction of the psychoacoutic filtering technique for the permutation problem. Section 4 summarizes the new hybrid approach and evaluates its performance. Finally, Section 5 concludes the paper.

2 Frequency Domain BSS and Permutation Problem

Assume that N source signals are recorded by M microphones (here we are particularly interested in acoustic applications), where $M \geq N$. The output of the j-th microphone is modeled as a weighted sum of convolutions of the source signals corrupted by additive noise, that is, $x_j(n) = \sum_{i=1}^{N} \sum_{p=0}^{P-1} h_{jip} s_i(n-p) + v_j(n)$, where h_{jip} is the p-th element of the P-point impulse response from source i to microphone j ($j = 1, \cdots, M$), s_i is the signal from source i, x_j is the signal received by microphone j, v_j is the additive noise, and n is the discrete time index. All signals are assumed zero mean. Using a discrete Fourier transformation (DFT), a frequency domain implementation of the mixing system is denoted as $\mathbf{X}(\omega, t) = \mathbf{H}(\omega)\mathbf{S}(\omega, t) + \mathbf{V}(\omega, t)$, where $\mathbf{S}(\omega, t)$ and $\mathbf{X}(\omega, t)$ are the time-frequency representations of the source vector and the mixture vector

Table 1. Overall SIR improvement before and after (B/A) applying the methods of filter constraint (FC) and separation matrices coupling (MC) respectively.

Systems/Methods	1/MC	1/FC	2/MC	2/FC	3/MC	3/FC
SIR in dB (B/A)	3.99/5.30	1.85/9.50	1.74/0.87	0.82/8.76	-1.31/-0.57	-0.10/10.50

respectively. Using the conventional ICA framework, $\mathbf{X}(\omega,t)$ can be separated at each frequency bin as $\mathbf{Y}(\omega,k) = \mathbf{W}(\omega)\mathbf{X}(\omega,k)$, where $\mathbf{Y}(\omega,k)$ is the time-frequency representation of the estimated source vector (assumed to be mutually independent), and k is the discrete time block index. Due to the inherent permutation ambiguity at each frequency bin, the recovered source components may have different permutations along the frequency axis so that the reconstructed source signals are still distorted in the time domain if the permutations are not correctly aligned. In the following discussion, we will use the penalty function based FDBSS algorithm developed in [9] for the separation of mixtures $\mathbf{X}(\omega,t)$, which exploits second order statistics (SOS) of nonstationary signals. We choose the penalty function to be in the form of a non-unitary constraint. The cost function is minimized by the gradient adaptation. Due to the limited space in this paper, we omit the implementation details which can be seen in [9].

3 Solutions to Permutation Problem

In this section, we will investigate some approaches briefly described in Section 1 and show some new results. We will use the SIR [2] as the performance index for the following evaluation, i.e.

$$SIR = 10\log\{(\sum_{\omega}\sum_{i}|H_{ii}(\omega)|^2 \langle|s_i(\omega)|^2\rangle)/(\sum_{\omega}\sum_{i\neq j}|H_{ij}(\omega)|^2 \langle|s_j(\omega)|^2\rangle)\}.$$

3.1 Exploiting Spectral Continuity

For this approach, either the recovered source components or separation matrices are assumed to have spectral similarities between neighboring frequency bins [1] [2]. In [1], an adaptive scheme was presented to apply frequency coupling for the unmixing matrices between neighboring frequency bins, that is $\Delta W_f \leftarrow \Delta W_f + k\Delta W_{f-1}$, where $0 < k < 1$. This intuitive scheme implicitly assumes that the permutations have been slightly changed during mixing, however it has limited performance for many cases, such as in Fig. 1 (a) and (b), where we can only identify a small SIR improvement along the frequency axis. In [2], a smoothness constraint was imposed on the unmixing filters in the time domain, that is, $Q < T$, and hence forced the solutions to be continuous in the frequency domain. As shown in Fig. 1 (c) and (d), compared with [1], this approach has a superior average performance along the frequency axis which is nevertheless, not consistent at every frequency, especially for some low frequencies. From Table 1, we find that the filter constraint approach is more robust with respect to the

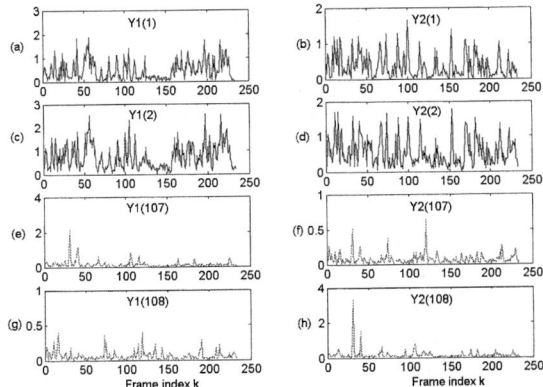

Fig. 2. The time envelopes of two separated source components at four different frequency bins; the upper four plots (a, b, c, d) represent two adjacent lower frequency bins, the lower four plots (e, f, g, h) represent two adjacent higher frequency bins.

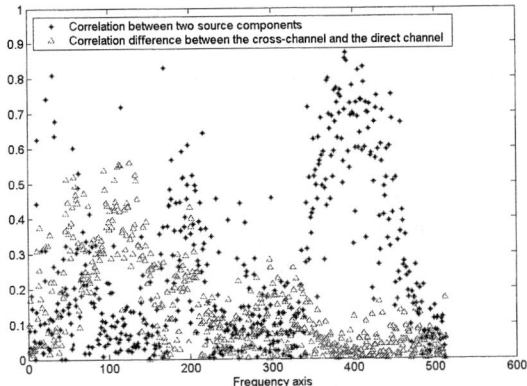

Fig. 3. Correlation value distribution along frequency axis.

mixing systems as compared with [1]. However, it is observed in [5] that the filter constraint may not be appropriate for a reverberant environment where a long filter may otherwise have a better performance. A merit of exploiting spectral continuity is that uniformity of the spectrum of the source signals has been preserved, which may not be shared by other approaches e.g. [3], where the frequencies have been processed separately. The identified drawbacks can be compensated by the approaches discussed in the following sections.

3.2 Exploiting Time Envelope Structure

This method was motivated by the time structure of speech signals [3] [6]. It is known that the source components at different frequency bins belonging to the same source signal should have similar shape in amplitude if they are modulated in a similar way. As a result, by measuring the correlation between the recovered source components at each frequency bin, we can determine the right order of the components in order to group them to the corresponding source. Mathematically, we define the time envelope of each extracted source component as $\mathcal{Y}_i(\omega, k) = |Y_i(\omega, k)|$, $i = 1, ..., N$. Fig. 2 shows an example of the time envelopes of the source components separated by the algorithms described in Section 2. From Fig. 2, we see that: 1) the envelopes from the same source signal at adjacent frequency bins are more similar to each other, such as (a) and (c), (e) and (h); 2) there exists the permutation problem since (e) corresponds to (h) but not (g). Therefore, by testing the correlations between the envelopes, we can determine the permutation for each frequency bin. A crucial problem in implementing this approach is, however, the selection of the frequency distance Δd_ω for the envelope correlation. In [3], the sum of the aligned frequencies is taken as the reference for the decision of the unpermuted frequencies, which unfortunately suffers from the fact that the envelopes with longer frequency distance do not necessarily have similar shapes (see Fig. 2 (a) and (g)). As a result, the permutation of the higher frequencies would not be accurately aligned since the correlation difference is small in this case (see Fig. 3). An alternative method for reducing this effect is to consider the correlation between the envelopes at neighboring frequency bins [6], however, it is sensitive to any misaligned frequency bins. To overcome this shortcoming, we propose to use the sum of the correlations as an approximate reference and conduct the correlations between neighboring frequency bins. Fig. 3 indicates that a fixed frequency distance is not appropriate for the envelope correlation. Therefore, we start the process from the frequency with the smallest correlation between the source components and adjust the distances to the correlation value at the current frequency between the source components.

3.3 Beamforming Alignment

Beamforming techniques have shown to be another promising approach for solving the permutation problem [4] [5], which is essentially motivated by the similarities between convolutive BSS and array signal processing. Comparatively, the model of convolutive BSS can be denoted by a phase and amplitude response, i.e., $\mathbf{y}(k) = e^{j\omega k}\mathbf{r}(\omega, \boldsymbol{\theta})$, where $\mathbf{r}(\omega, \boldsymbol{\theta}) = \mathbf{W}^H(\omega)\mathbf{D}(\omega, \boldsymbol{\theta})$, $\mathbf{D}(\omega, \boldsymbol{\theta}) = [\mathbf{d}(\omega, \theta_1), \cdots, \mathbf{d}(\omega, \theta_M)]$, $\mathbf{d}(\omega, \theta_j) = [e^{j\omega \tau_i(\theta_j)}]^H$ are steering vectors, and τ_i, $i = 1, \cdots, N$ denote propagation delays. The separation matrices for each frequency bin ω are analogously regarded as beamformers. Therefore, the DOAs of source components can be observed from every row of $\mathbf{W}(\omega)$ by plotting the directivity pattern, i.e. $F_i(\omega, \theta) = \sum_{k=1}^{M} W_{ik}(\omega) e^{j\omega(k-1)\tau_{ki}}$, where $\tau_{ki} = d_k \sin\theta_i/c$ is the time delay with respect to the ith source signal from the direction of θ_i, observed at the kth microphone with distance d_k, and c is the velocity of the sound. By estimating the DOAs at each frequency bin, the permutations can be determined in a straightforward way, sweeping or keeping the rows in $\mathbf{W}(\omega)$.

It has been suggested in [5] to use a low frequency range [1 $c/2d$) for the estimation of the DOAs of the sources (null directions) since their accurate estimates can not be guaranteed due to the existence of grating lobes at higher frequencies. However, it is also shown in [6] that for very low frequencies, null directions

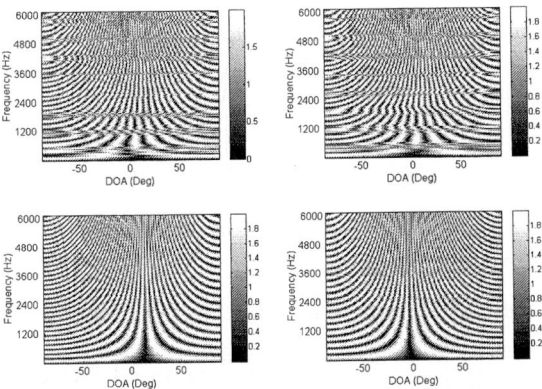

Fig. 4. The directivity pattern as a function of frequency before (upper two) and after (lower two) alignment by the MUSIC approach.

cannot be accurately estimated, due to the *flatness* of the directivity patterns. Another downside is that, unlike BSS which does not suffer from the prior information about the source location, it requires the two sources to be located up to a desired power resolution [7]. To give a more accurate estimate of the DOAs, we suggest to resort to subspace-based methods such as MUSIC [7]. To this end, we define the following MUSIC operator, $\bar{F}_i(\omega, \theta) = 1 / \left\| \tilde{\mathbf{P}}(\omega, \theta) \hat{\mathbf{a}}(\omega, \theta_i) \right\|_2$, where $\tilde{\mathbf{P}}(\omega, \theta)$ is the noise subspace formed by the estimate $\hat{\mathbf{A}}(\omega) = \mathbf{W}^{-1}(\omega)$, and $\hat{\mathbf{a}}(\omega, \theta_i)$ is the ith column of $\hat{\mathbf{A}}(\omega)$. Fig. 4 shows an example of the beam pattern of $\mathbf{W}(\omega)$ using $\bar{F}_i(\omega, \theta)$.

3.4 Psychoacoustic Post-filtering

To compensate for misaligned bins, a potential method is to exploit human perception for acoustic signals. Psychoacoutic studies reveal that, although human hearing ranges from about 20Hz to 20KHz, most of the energy of speech lies in the lower frequency band (with bandwidth normally less than 5KHz) [8]. The just-audible thresholds and critical bandwidths are not constant but non-uniform, non-linear across all frequencies and dependent on different sounds. This means that the average human does not have the same perception at all frequencies. This fact suggests that some frequencies can be cut due to the limitation of the human auditory system and the masking effect, however without loss of necessary information contained in speech. Based on this point, we propose to use a psychoacoustic model as a post-filter after the permutations initially aligned by the aforementioned approaches. This model exploits two properties of the human auditory system: absolute threshold of hearing (ATH) (also known as threshold of quiet) and auditory masking (AM). The tone masker and noise masker are calculated respectively and the maskers that are weaker than another masker within one critical bandwidth are attenuated, and the ATH is used as a reference for determining the global threshold. An experiment result by apply-

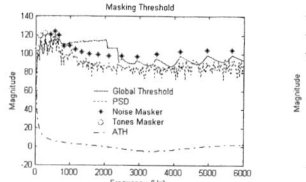 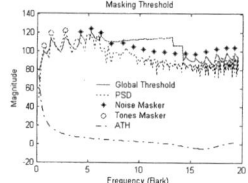

Fig. 5. Psychoacoutic post-filtering of one reconstructed speech signal from FDBSS output using threshold masking.

Table 2. SIR improvement of the various approaches

Methods	No alignment	[2]	[5]	[3]	[6]	proposed DOA	proposed hybrid
SIR_{av} (dB)	−0.33	9.59	10.04	6.23	11.35	10.89	14.12

ing this model to separate speech components is shown in Fig. 5, which clearly shows that there exists enough redundant information (including noise, see the masker above the global threshold) in the recovered source components that can be removed.

4 Approach Summary and Numerical Experiment

Based on the discussions of the above sections, our proposed hybrid approach for solving the permutations of $\mathbf{W}(\omega)$ is summarized as: 1) *performing filter constraint*; 2) *performing DOA alignment and detecting confidence*; 3) *retaining the frequency bins with high confidence, performing envelope correlation for the remaining frequencies, detecting confidence again*; 4) *performing psychoacoustic filtering for all the remaining frequency bins*. It should be noted that the procedure of confidence detection is to ensure a sufficiently high confidence for the permuted frequencies, which can be conducted in the same way as in [6].

We perform an experiment to evaluate the overall averaged performance of the proposed approach for three mixing systems which are identical to those used in Table 1. The result is compared with the method in [2] (using spectral continuity), [5] (using conventional beamforming), [3] (using time envelope), and [6] (using a combined approach). We artificially mix two speech signals (sampled at 12kHz with length of 9 seconds). $Q = 1024$, $T = 1024$ (for [2], $T = 2048$). For [6], $\triangle d_\omega = 3\triangle\omega$, where $\triangle\omega$ is the frequency resolution. The penalty function parameter is $\kappa = 0.1$ and the number of intervals used to estimate each cross-power-matrix is 7 (see [9]). The distance between two sensors is $1m$, the directions of the sources are respectively $19.68°$ and $-5.35°$. For the proposed method (step 3), $\triangle d_\omega$ decreases with a linear regulation from $10\triangle\omega$ to $\triangle\omega$ as frequency increases. From Table 2, we know that: 1) MUSIC has a superior performance over conventional beamforming (such as [5]) for the permutation alignment; 2) calculating the correlation over the whole frequency does not give an accurate alignment (see [5]) as compared with neighboring frequency coupling in [6]; 3) The proposed hybrid approach has a significantly improved performance due to the introduction of the psychoacoustic perception together with a more accurate DOA estimation and a dynamic frequency distance for envelope correlation.

5 Conclusion

A hybrid approach for solving the permutation problem of FDBSS has been presented. A psychoacoustic filtering technique has been effectively introduced to incorporate the human perception of sound in order to reduce the permutation effect at some frequency bins which are not accurately aligned. The subspace based MUSIC method has also been introduced to provide more accurate beam patterns along frequency bins. By varying the frequency intervals for envelope correlation, the nonstationarity of speech signals is nicely exploited. More extensive evaluations for the proposed approach including subjective tests using the mean opinion score (MOS) are currently under consideration.

References

1. P. Smaragdis, "Blind separation of convolved mixtures in the frequency domain," *Neurocomputing*, vol.22, pp. 21–34, 1998.
2. L. Parra and C. Spence, "Convolutive blind source separation of nonstationary sources," *IEEE Trans. on Speech Audio Proces.*, pp. 320–327, May 2000.
3. N. Murata, S. Ikeda, and A. Ziehe, "An approach to blind source separation based on temporal structure of speech signals," *Neurocomputing*, vol. 41, no. 1-4, pp. 1-24, Oct. 2001.
4. S. Kurita, H. Saruwatari, S. Kajita, K. Takeda, and F. Itakura, "Evaluation of blind signal separation method using directivity pattern under reverberant conditions," *Proc. ICASSP*, pp.3140-3143, 2000.
5. M. Z. Ikram and D. R. Morgan, "A beamforming approach to permutation alignment for multichannel frequency-domain blind speech separation," *Proc. ICASSP*, pp. 881-884, May 2002.
6. H. Sawada, R. Mukai, S. Araki, and S. Makino, "A robust and precise method for solving the permutation problem of frequency-domain blind source separation," *Proc. ICA*, Nara, Japan, Apr. 1-4, 2003.
7. H. Krim and M. Viberg, "Two decades of array signal processing research: the parametric approach," *IEEE SP Mag.*, pp. 67-94, Jul. 1996.
8. E. Zwicker and H. Fastl, "Psychoacoustics: facts and models," Springer, 2nd Ed.,1999.
9. W. Wang, J. A. Chambers, and S. Sanei, "Penalty function approach for constrained convolutive blind source separation," *Proc. ICA*, Granada, Spain, Sept. 22-24, 2004 (accepted).

Application of Geometric Dependency Analysis to the Separation of Convolved Mixtures

Samer Abdallah and Mark D. Plumbley

Centre for Digital Music, Queen Mary, University of London
{samer.abdallah,mark.plumbley}@elec.qmul.ac.uk
http://www.elec.qmul.ac.uk/digitalmusic

Abstract. We investigate a generalisation of the structure of frequency domain ICA as applied to the separation of convolved mixtures, and show how a geometric representation of residual dependency can be used both as an aid to visualisation and intuition, and as tool for clustering components into independent subspaces, thus providing a solution to the source separation problem.

1 Introduction

Geometric dependency analysis (GDA) was introduced in [1, ch. 8] as way to represent geometrically the residual dependencies in a distributed representation such as those generated by ICA, both as an aid to visualisation and as a basis for further processing. In this paper, we investigate how ICA and GDA, when applied to two-channel audio data, can yield a solution to the problem of separating convolutively mixed sources. The approach is conceptually quite simple in that it involves very few assumptions about the problem domain: the fact that there are two microphones is not explicitly modelled; neither is the assumption that the sources are mixed convolutively in the time domain. Instead, after training, the ICA weight matrix implicitly represents these aspects of the system. The final separation of the sources is based on clustering of components in a low-dimensional geometric space, which could in principle be done in an unsupervised manner, though in the present system it was done manually.

The description below will be in terms of a 2-by-2 (2 microphones, 2 sources) system, but can be generalised naturally to an m-by-m system.

2 An Overview of Frequency Domain Source Separation

We begin with an overview of a typical frequency-domain approach to the separation of convolved mixtures, (see, e.g., [2] for more details) emphasing how the entire system can be understood as a composition of constrained sparse matrices followed by a partition of the resulting components into two subspaces.

First, the signals from the two microphones are buffered into frames of length L samples, which we will denote by the vectors $\mathbf{x}_1$ and $\mathbf{x}_2$, both in $\mathbb{R}^L$. The next step, motivated by the duality between convolution in the time domain and multiplication in the frequency domain, is to compute the discrete Fourier transform to each frame. This is usually done using a complex-valued fast Fourier transform (FFT); however, each complex Fourier coefficient then represents a 2-dimensional subspace of $\mathbb{R}^L$. Since we aim to understand the overall process in terms of an analysis of subspaces, we choose to work instead with a real-valued Fourier transform: each coefficient then represents a 1-dimensional subspace, and higher dimensional subspaces must be formed explicitly by grouping components. In this case, the Fourier transform can be represented as an $L \times L$ orthogonal matrix $\mathbf{F}$, where the rows of $\mathbf{F}$ form an orthonormal basis of $\mathbb{R}^L$. Assuming L is even, these basis vectors are sinusoids covering $L/2 + 1$ different frequencies: the zero and Nyquist frequencies are represented by one basis vector each, while the other frequencies each inhabit 2-D subspaces spanned by two basis vectors in quadrature phase.

To do frequency domain ICA, the corresponding per-frequency subspaces from both microphones are brought together to form $L/2 + 1$ low-dimensional ICA problems, each of which is solved independently. These two steps can be represented as the product of a permutation matrix (to interleave the Fourier coefficients from the two channels), and a constrained *block diagonal* ICA weight matrix $\mathbf{V}$, where the first and last blocks are 2×2 and the rest are 4×4. The entire process so far can therefore be written as

$$\mathbf{s} = \mathbf{VP} \begin{pmatrix} \mathbf{F}^T & \mathbf{0} \\ \mathbf{0} & \mathbf{F}^T \end{pmatrix} \begin{pmatrix} \mathbf{x}_1 \\ \mathbf{x}_2 \end{pmatrix}, \qquad (1)$$

where $\mathbf{P}$ represents the permutation $[1, 2, \ldots, 2L] \mapsto [1, L, 2, L+1, \ldots, 2L]$, and $\mathbf{V}$ is of the form

$$\mathbf{W} = \begin{pmatrix} \mathbf{V}_0^{(2 \times 2)} & \mathbf{0} & \cdots & \mathbf{0} \\ \mathbf{0} & \mathbf{V}_1^{(4 \times 4)} & \cdots & \mathbf{0} \\ \vdots & \vdots & \ddots & \vdots \\ \mathbf{0} & \mathbf{0} & \cdots & \mathbf{V}_L^{(2 \times 2)} \end{pmatrix} \qquad (2)$$

Note also that the Fourier matrix can also be written as the product of $O(\log L)$ sparse matrices (hence the FFT algorithm). Indeed, it is the proliferation of sparse matrices that makes the computation rather tractable even for large frames.

Finally, the $2L$ components of $\mathbf{s}$ are partitioned into two groups (one for each source) containing representatives from each of the $L/2 + 1$ ICA sub-problems. The partition defines two orthogonal subspaces; we consider the problem solved if each subspace represents activity from only one source, in which case either source can be reconstructed in signal domain by setting the components in the other subspace to zero and inverting the system.

Our aim in this paper is to investigate what happens if the three matrices in (1) are replaced by a single unconstrained ICA weight matrix, and how an analysis of residual dependency can be used to partition the resulting components into two subspaces. Although this clearly requires more computation and much more training data to fit the larger number of parameters, the system is conceptually simpler, and shows how sensible processing strategies for dealing with stereo signals can emerge in an unsupervised way.

3 Unconstrained ICA of Buffered Stereo Data

The data for the unconstrained ICA system consists of the same packed stereo frames $\mathbf{x} \equiv (\mathbf{x}_1, \mathbf{x}_2) \in \mathbb{R}^N$ (with $N = 2L$) as described in the previous section. For these experiments, we used recordings from two microphones placed in a normally reverberant room along with two loudspeakers playing (38 s) short extracts from two different radio programmes. A female presenter was speaking in one programme, a male in the other. The signals were sampled at 16 kHz.

A natural gradient maximum-likelihood ICA algorithm [3] was used to estimate an $N \times N$ weight matrix $\mathbf{W}$, yielding the estimated independent components $\mathbf{s} = \mathbf{W}\mathbf{x}$ for each frame. The weight updates were of the form

$$\mathbf{W} \mapsto \mathbf{W} + \eta \langle \mathbf{I} - \varphi(\mathbf{s})\mathbf{s}^T \rangle \mathbf{W}, \qquad (3)$$

where $\langle \cdot \rangle$ denotes an average taken over the training data (or a smaller batch randomly sampled from the whole), η is the learning rate, and the function $\varphi : \mathbb{R}^N \to \mathbb{R}^N$ is the gradient of the negative log-prior on the components: $\varphi(\mathbf{s}) = -\nabla_\mathbf{s} \log p(\mathbf{s})$. A generalised exponential factorial prior was used:

$$p(\mathbf{s}) = \prod_{i=1}^N p(s_i), \qquad p(s_i) = \frac{\exp -|s_i|^{\alpha_i}}{2\Gamma(1 + \alpha_i^{-1})}. \qquad (4)$$

During training, the exponents α_i were periodically re-estimated from the data using a maximum-likelihood gradient method [4]. Some of the resulting stereo basis vectors (the columns of $\mathbf{A} = \mathbf{W}^{-1}$) are illustrated in fig. 1(a). If we take each component as an 'atom' of sound (in the case of perfect separation, it will come from just one of the sources), its stereo basis vector encodes how that atomic sound is received at the two microphones. Thus, the basis matrix contains information about the relative path delay and frequency-dependent response between each sources and microphone. Comparing the unconstrained ICA basis with the equivalent basis vectors for the frequency domain ICA system – some of which are illustrated in fig. 1(b) – it is clear that both systems exploit phase and amplitude differences to distinguish the sources, but only the unconstrained ICA system is able to exploit *time delays*, since, unlike the frequency domain system, in includes temporally localised basis vectors.

The goodness of fit of both ICA models was assessed in terms of the average log-probability of a frame:

$$L = \langle \log p(\mathbf{x}) \rangle = \langle \log \det \mathbf{W} + \log p_\mathbf{s}(\mathbf{W}\mathbf{x}) \rangle. \qquad (5)$$

This is related to the average coding cost-per-vector: the higher the score the lower the cost. The unconstrained ICA system achieved a score 1516.2, whereas the frequency domain ICA achieved 1347.9. In terms of coding cost, this is a difference of 168.3 nats (242.9 bits) per frame.

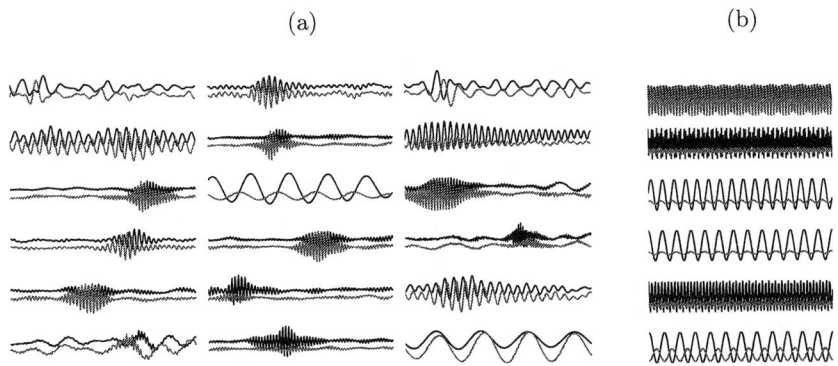

Fig. 1. (a) Some of the stereo basis vectors obtained by ICA of data recorded from two microphones with two sources present in the room. (The right-channel signal is offset below the left-channel.) The frame size for each channel was 256, so the ICA system is 512-dimensional. (b) Equivalent basis vectors implicit in frequency domain ICA of the same system, obtained by inverting the three matrices in (1).

4 Geometric Dependency Analysis

The aim of GDA is to represent each element of a distributed representation as a point in a metric space such the the distance between each pair is inversely related to the mutual information between the corresponding components. Truly independent components are pushed infinitely far apart, while dependent components form clusters or manifolds. As argued in [5], such residual dependencies can be useful in interpreting the representation and organising the next stage of processing. The method described in [1, ch. 8] involves estimating the mutual information between each pair of components s_i, s_j in terms of a nonlinear correlation coefficient in the range $[-1, 1]$,

$$\rho_f(S_i, S_j) = \mathrm{corr}[f_1(S_i), f_2(S_j)] = \frac{\mathrm{cov}[f_1(S_i), f_2(S_j)]}{\sqrt{\mathrm{var}\ f_1(S_i)\,\mathrm{var}\ f_2(S_j)}}, \quad (6)$$

where S_i and S_j denote the random variables whose realisations are s_i and s_j, f_1 and f_2 are rectifying (that is, even) nonlinear functions, and $f(S)$ is shorthand for the random variable obtained by applying the function f to realisations of S. From this, a matrix of pair-wise distances is defined:

$$D_{ij} = \sqrt{-2\log|\rho_f(S_i, S_j)|}. \quad (7)$$

Finally, multidimensional scaling [6] is used generate a spatial configuration of N points r_i in an E-dimensional metric space $\mathcal{M}$ such that their pair-wise distances according to a predetermined metric $d : \mathcal{M} \times \mathcal{M} \to \mathbb{R}^+$ approximate the correlation distances, that is $d(r_i, r_j) \approx D_{ij}$ for all pairs i, j.

Before presenting the results of GDA on the ICA systems described in previous sections, we describe a refinement to the nonlinear correlation that should in principle give a more accurate measure of dependence.

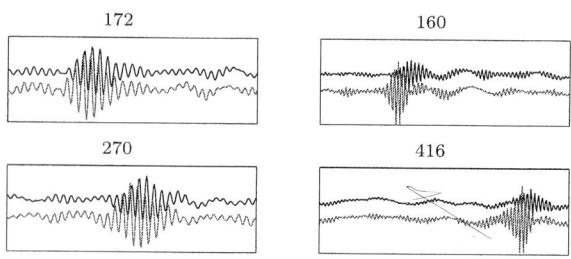

Fig. 2. Two pairs of basis vectors whose $\mathcal{F}$-correlation is increased by inclusion of lags. The original stereo data was buffered using a hop size of 32, and the function space $\mathcal{F}$ included lags up to 4, i.e. only the coefficients a_0 to a_4 in (10) were allowed to vary from zero. The $\mathcal{F}$-correlation of the pair on the left (172 and 270) increased from 0.42 to 0.83, while the correlation of the pair (160,416) increased from 0.51 to 0.87.

4.1 Estimating Residual Dependency Using the $\mathcal{F}$-Correlation

The $\mathcal{F}$-correlation [7] can be thought of as a generalisation nonlinear correlation: instead of two fixed functions f_1 and f_2, we allow f_1 and f_2 to range freely over a function space $\mathcal{F}$ and define $\rho_{\mathcal{F}}$ as the maximal correlation so obtained:

$$\rho_{\mathcal{F}} = \sup_{f_1, f_2 \in \mathcal{F}} \operatorname{corr}[f_1(S_1), f_2(S_2)]. \tag{8}$$

If $\mathcal{F}$ is a linear space, then the computation of the $\mathcal{F}$-correlation is equivalent to canonical correlation analysis (CCA) and can be solved as a generalised eigenvalue problem. The spectrum of canonical correlations can then be used to compute the $\mathcal{F}$-correlation and the so-called *generalised variance*, both of which can be used as measures of statistical dependence.

In this application, we used a function space spanned by a basis of *lagging* functions l_τ: if S represents the sequence of values of one component as successive frames are processed, then the lagged component $l_\tau(S)$ is the same sequence delayed by τ frames. In addition, we used as our rectification nonlinearity a form of *generalised energy* derived from the generalised exponential prior (4):

$$\mathcal{E}(s_i) = |s_i|^{\alpha_i}. \tag{9}$$

A typical element of this function space is

$$f : S_i \mapsto f(S_i), \quad f(S_i) = a_0 l_0(\mathcal{E}(S_i)) + a_1 l_1(\mathcal{E}(S_i)) + a_2 l_2(\mathcal{E}(S_i)) \ldots \tag{10}$$

where the a_τ are weighting coefficients. Note that, strictly speaking, because of the dependence of the generalised energy on the index of the component i, the functions f_1 and f_2 lie in two of N *separate* spaces $\mathcal{F}_i$ depending on the indices of the components S_i and S_j.

The motivation behind using a space of lagging functions is to capture temporal dependencies where activity in one component implies activity in another a certain time later. Fig. 2 shows two pairs of basis vectors whose $\mathcal{F}$-correlation is significantly increased by the inclusion of lags, which is unsurprising since they appear to be shifted versions of the same stereo waveform.

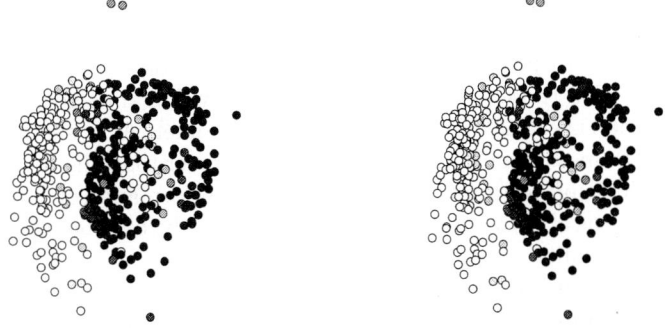

Fig. 3. MDS results obtained with a hop size of 32, lags 0 to 6, $q = 2$, using Kruskal's stress function (see [6]) in a 3-D Euclidean space. The shading indicates which side of a manually chosen separating plane each point lies. The points form two crescent-shaped clusters lying side-by-side with a gap in between them. (The two images are a stereo pair with the left-eye image on the left.)

4.2 Multidimensional Scaling Results

Given a particular weight matrix **W**, there are many variables involved in determining the final spatial configuration discovered by MDS: the hop size between frames and the lags used to compute the $\mathcal{F}$-correlation, the dimension E of the MDS embedding space, and stress function used in the MDS algorithm. Due to lack of space we can only illustrate one example here. The points in fig. 3 form two crescent-shaped clusters lying side by side, with, roughly, high frequency basis vectors are at one end and low frequencies are at the other. An inspection of the basis vectors in each cluster shows that those basis vectors which are readily interpretable as belonging to one or other source (such as those in fig. 2) are consistently segregated. In the next section we assess how well the original sources can be reconstructed on the basis of this partition.

5 Source Reconstruction and Evaluation

The reconstruction process involves setting some of the components s_i to zero and inverting the ICA system. Letting $\mathbf{y} \equiv (\mathbf{y}_1, \mathbf{y}_2)$ denote a reconstructed stereo frame, this can be expressed as $\mathbf{y} = \mathbf{W}^{-1}\mathbf{H}\mathbf{s} = \mathbf{W}^{-1}\mathbf{H}\mathbf{W}\mathbf{x}$, where **H** is a diagonal matrix. The process can be thought of as 'ICA domain filtering' in direct analogy with frequency domain filtering, the difference being that we use ICA instead of a Fourier transform before applying a diagonal operator and transforming back. In this application, we aim to place ones and zeros on the diagonal of **H** in order to select only those components which belong to a particular source. Given a partitioning of the components into two groups, we can therefore define two complementary ICA domain filters $\mathbf{H}^{(1)}$ and $\mathbf{H}^{(2)}$ to reconstruct the sources.

The partition was determined by manually positioning a separating plane between the two clusters of points found by MDS. Each source was reconstructed as a stereo signal (using a Hanning window to recombine overlapping frames) and compared with those obtained from a frequency domain algorithm [2] using the same frame size of 256 samples from each channel. At a sampling rate of 16 kHz, these frames are only 16 ms long – much shorter than the room's impulse response – which places limits on the potential performance of both systems. Hence, neither was able to separate the sources perfectly, but the reconstructions were of similar quality.

To make a quantitative evaluation, a different data set was used: the two sources were recorded in situ as before, but separately rather than simultaneously. The two (fully reverberant) stereo recordings were mixed artificially and analysed using previously fitted ICA model. The match between the separately recorded sources and their reconstructions was then measured by a matrix of correlation coefficients:

$$J_{kl} = \frac{\left[\sum_{ij} X_{ij}^k Y_{ij}^l\right]^2}{\sum_{ij} \left[X_{ij}^k\right]^2 \sum_{i'j'} \left[Y_{i'j'}^l\right]^2}, \qquad (11)$$

where X_{ij}^k is the jth sample from the ith channel (left or right) of the kth source, and Y_{ij}^l is similarly defined for the lth reconstruction. Perfect separation and reconstruction is achieved when $J_{kl} = \delta_{kl}$. The correlation matrices obtained for the two systems are tabulated below:

	recon. 1	recon. 2
source 1	0.357	0.218
source 2	0.089	0.421

(a) Frequency domain ICA

	recon. 1	recon. 2
source 1	0.400	0.189
source 2	0.230	0.295

(b) Unconstrained ICA

The unconstrained system does not quite achieve the contrast between sources in each reconstruction that the frequency domain method does; this may be because of sub-optimal convergence of the ICA model, either because of local minima, or because of over-fitting due to insufficient training data, though this is yet to be confirmed experimentally. Note that the frequency domain system was trained using a fixed-point algorithm, which generally gives better convergence than the natural gradient optimiser used in the unconstrained system.

6 Discussion and Conclusions

In this paper, we have shown that GDA is capable of generating meaningful, readily interpretable structures when applied stereo audio data, in this case, providing, on the basis of very few assumptions, a solution to the speaker separation problem comparable to that produced by a more specialised algorithm. The approach essentially boils down to finding independent subpaces in high dimensional data, which in this case, happens to be segments of a stereo signal. Hence, it can be compared with other systems that analyse the clustering of dependent components [5, 8].

One benefit of GDA is that it may reveal, as complex geometric forms, dependency structures that do not emerge clearly in other subspace analysis methods. A disadvantage of our approach in this particular application is that the unconstrained ICA phase of the process does not scale to long frames, because of the amount of training data required to adequately adapt the N^2 elements of $\mathbf{W}$. The weight matrix contains specific information about the physical configuration of the sources and microphones, as well general information about the statistical structure of speech. Thus, one approach might be to factorise the system along the lines of (1), but using a general speech-trained ICA transform [1, ch. 5] on each channel instead of a Fourier transform, followed by sparse matrix ICA in adaptive local 'modules' defined by residual dependency, as described in [9].

Acknowledgements

The authors would like to thank Nikolaos Mitianoudis for helpful discussions, for the test data used in these experiments, and for the frequency domain ICA results quoted in section 5. This work was supported by EPSRC grant GR/R54620.

References

1. Abdallah, S.A.: Towards Music Perception by Redundancy Reduction and Unsupervised Learning in Probabilistic Models. PhD thesis, Department of Electronic Engineering, King's College London (2002)
2. Mitianoudis, N., Davies, M.E.: New fixed-point solutions for convolved mixtures. In: 3rd Intl. Conf. on Independent Component Analysis and Source Separation (ICA2001), San Diego, California (2001)
3. Cardoso, J.F., Laheld, B.: Equivariant adaptive source separation. IEEE Trans. on Signal Processing **44** (1996) 3017–30
4. Everson, R., Roberts, S.: Independent component analysis: A flexible nonlinearity and decorrelating manifold approach. Neural Computation **11** (1999) 1957–1983
5. Hyvärinen, A., Hoyer, P., Inki, M.: Topographic independent component analysis. Neural Computation **13** (2001) 1527–1558
6. Cox, T., Cox, M.A.A.: Multidimensional Scaling. Chapman Hall/CRC, London (2001)
7. Bach, F.R., Jordan, M.I.: Kernel independent component analysis. Technical Report UCB/CSD-01-1166, Division of Computer Science, University of California, Berkely (2001)
8. Bach, F.R., Jordan, M.I.: Finding clusters in independent component analysis. In: 4th Intl. Symp. on Independent Component Analysis and Signal Separation (ICA2003), Nara, Japan (2003)
9. Matsuda, Y., Yamaguchi, K.: Linear multilayer ICA integrating small local modules. In: 4th Intl. Symp. on Independent Component Analysis and Signal Separation (ICA2003), Nara, Japan (2003)

Blind Deconvolution of SISO Systems with Binary Source Based on Recursive Channel Shortening

Konstantinos I. Diamantaras[1] and Theophilos Papadimitriou[2]

[1] Department of Informatics, TEI of Thessaloniki, Sindos 54101, Greece
kdiamant@it.teithe.gr
[2] Department of Int. Economic Relat. and Devel., Democritus University of Thrace
Komotini 69100, Greece
papadimi@ierd.duth.gr

Abstract. We treat the problem of Blind Deconvolution of Single Input - Single Output (SISO) systems with real or complex binary sources. We explicate the basic mathematical idea by focusing on the noiseless case. Our approach leads to a recursive channel shortening algorithm based on simple data gouping. The channel shortening process eventually results in an instantaneous binary system with trivial solution. The method is both deterministic and very fast. It does not involve any iterative optimization or stochastic approximation procedure. It does however, require sufficiently large datasets in order to meet the source richness condition.

1 Introduction

Binary signals received a lot of attention in the last decades, due to their application in digital/wireless communications. A typical problem in this area is the blind separation of multiple signals arriving at an antenna array when the sources are BPSK [1]. In [2], van der Veen investigated the problem of instantaneous Blind Source Separation (BSS) with binary sources, when the mixing operator is complex. The proposed solution is based on generalized eigenvalue decomposition and it is non-iterative. Diamantaras et al. in [3] found an analytical method to decompose mixtures of binary sources using only one observation. Similarly, Li et al. proposed a similar method, [4], where solvability of the problem was presented in details.

The Blind Deconvolution of a $n_1 \times n_2$ MIMO system consists on the estimation of the n_1 input signals given only the n_2 output signals, while the convolutive system is unknown. The solution was given through the eigen-decomposition of the output correlation matrix in [5]. In [6] the MIMO problem was reduced into multiple SIMO problems by multiplying each observation by its conjugate cyclic frequency. Ma et al. [7] proposed a method using the generalized eigenvalue decomposition of a matrix pencil formed by output auto-correlation matrices at different time-lags. A class of algorithms exploiting the statistical independence of the sources was presented in [8], by Yellin et al.. In [9], Tugnait proposed a cumulant maximization-based approach that decomposes the mixtures into sets of independent signal components at each sensor.

In this paper we investigate two similar problems: (a) Blind Deconvolution of real SISO systems with a real binary source, and (b) Blind Deconvolution of complex SISO systems with a complex source having binary real and imaginary parts. Both problems are studied in their ideal, noiseless case, because ths approach best exposes the underlying mathematical concepts. The introduction of noise would require a straightforward modification of the proposed algorithm. Necessary assumptions for the problem solvability are presented. In general, these assumptions require large datasets so that the source data set is rich enough in binary combinations. The proposed scheme recursively eliminates the channel parameters, i.e. it performs channel shortening. Once this iterative process concludes, the resulting system is a memoryless one. From that, the binary source estimation is straightforward. The proposed scheme is further developed in order to be applied in the case of complex source with binary parts and complex mixing operators. The paper concludes with results in both cases.

2 Blind Deconvolution of a SISO System: Real Binary Source

Let us observe the output of a real, noiseless SISO system

$$x(k) = \sum_{l=0}^{L} a(l) s(k-l), \quad k = 1, \cdots, K \tag{1}$$

or

$$x(k) = \mathbf{a}^T \mathbf{s}(k), \quad k = 1, \cdots, K \tag{2}$$

where $\mathbf{a}$ is the real, $(L+1)$-dimensional vector, x is the system output, and

$$\mathbf{s}(k) = [s(k), s(k-1), ..., s(k-L)]^T,$$

is the binary source $s(k) \in \{-1, +1\}$. Since the binary vector $\mathbf{s}$ has length $L+1$, it can take 2^{L+1} distinct values. Consequently, $x(k) \in \mathcal{X}$ can take at most 2^{L+1} values, i.e. $|\mathcal{X}| \leq 2^{L+1}$. We make the following assumption:

Assumption 1 *Every possible value of the output $x(k)$ corresponds to a unique source vector $\mathbf{s}(k)$, therefore the cardinality of $\mathcal{X}$ is exactly 2^{L+1}.*

This simply means that no pair of distinct sources will produce the same output. For the similar problem of blindly separating multiple binary sources $s_i(k)$, from a single linear mixture $x(k) = \sum_i a(i) s_i(k)$, Li et al. in [4] concluded that the following condition ensures solvability:

$$c_0 a(i_0) + ... + c_N a(i_N) \neq 0, \tag{3}$$

for any coefficients $c_k \in \{-1, +1\}$, and any subset $\{i_0, ..., i_N\}$ ($N < L$) of $\{0, 1, ..., L\}$.

Take any time instant k_0, and let $\mathbf{s}(k_0) = [c_0, c_1, ..., c_L]$, $c_i \in \{-1, +1\}$, be the source vector yielding the output $x(k_0) = r$. The successor observation at time instant $k_0 + 1$, can assume only two possible values, $r^s_{(1)}$ or $r^s_{(2)}$, depending on the corresponding source vector which can take one of the following two values $\mathbf{s}_{(1)}(k_0 + 1) = [+1, c_0, c_1, ..., c_{L-1}]$ or $\mathbf{s}_{(2)}(k_0 + 1) = [-1, c_0, c_1, ..., c_L]$. In our method it is essential that both pairs of consecutive values $[r, r^s_{(1)}]$ and $[r, r^s_{(2)}]$, will appear, at least once, in the output sequence $x(k)$, $k = 1, \cdots, K$. This is stated in the following assumption:

Assumption 2 *For any $r \in \mathcal{X}$, there are at least two indices $k_0, k_1 \in \{1, 2, \cdots, K\}$ such that $x(k_0) = r$, $x(k_0 + 1) = r^s_{(1)}$ and $x(k_1) = r$, $x(k_1 + 1) = r^s_{(2)}$.*

This assumption, of course, requires that the dataset is large enough. The successors $r^s_{(1)}$, $r^s_{(2)}$, of r can be found by simple observation of the output data set. Once this is done, it is straightforward to estimate $|a(0)|$ as follows:

$$\left| r^s_{(1)} - r^s_{(2)} \right| = \left| \mathbf{a}^T \mathbf{s}_{(1)}(k_0 + 1) - \mathbf{a}^T \mathbf{s}_{(2)}(k_0 + 1) \right|$$
$$= |a(0)(+1 - (-1))|$$
$$= 2|a(0)| \quad (4)$$

Moreover, the sum $\rho(r)$ of the successors of r is

$$\rho(r) = r^s_{(1)} + r^s_{(2)} = \mathbf{a}^T \mathbf{s}_{(1)}(k_0 + 1) + \mathbf{a}^T \mathbf{s}_{(2)}(k_0 + 1)$$
$$= a(0)(+1 + (-1)) + 2\sum_{i=1}^{L} a(i)c_i$$
$$= 2\sum_{i=1}^{L} a(i)c_i \quad (5)$$

Estimating $\rho(r)$ for every $r \in \mathcal{X}$ can lead to a new SISO system with a shortened channel. Indeed, substituting every observation $x(k) = r$ with $\rho(r)/2$, we obtain:

$$x^{(2)}(k) = \rho(r)/2 = \sum_{l=1}^{L} a(l)s(k-l) \quad (6)$$

It is clear that the new SISO system in Eq. 6 has the same taps as the original one in Eq. 2 except for the lack of $a(0)$. Of course the length of the new system is L, i.e. one less than the initial length $L + 1$. The above transformation can be recursively applied L times until the system is reduced into:

$$x^{(L+1)}(k) = a(L)s(k-L) = \pm a(L) \quad (7)$$

Since, system (7) is non-convolutive, the source estimation is a straightforward process. Notice that, at any time instant k, the output $x^{(L+1)}(k)$ will assume one of two values $+a(L)$ or $-a(L)$. So we can easily estimate the absolute value of the last filter tap as $\hat{a}(L) = |x^{(L+1)}(k)| = |a(L)|$, (any k) and from that we can estimate the source by:

$$\hat{s}(k-L) = x^{(L+1)}(k)/\hat{a}(L) = \sigma s(k-L) \tag{8}$$

where $\sigma = \pm 1$. In this process, of course, we lose the sign information but it is well known that the source sign is unobservable.

3 SISO Blind Deconvolution: Complex Binary Source

In the sequel we shall adopt the following notation convention: for any complex number c, c_R and c_I will denote the real and the imaginary parts of c respectively.

Now let us reconsider the SISO system of Eq. (2) where **a** is a complex, $(L+1)$-tap vector. The source $s(k)$ is also complex but with binary real and imaginary components: $s_R(k), s_I(k) \in \{1, -1\}$, i.e. $s(k) \in \mathcal{B} = \{1+j, 1-j, -1+j, -1-j\}$. As in the real case, the complex source vector $\mathbf{s}(k)$ has length $L+1$. Clearly, $\mathbf{s}(k)$ can take 4^{L+1} distinct values, and $x(k) \in \mathcal{X}_c$ can take at the most 4^{L+1} distinct values. Similarly to section 2 we assume that

Assumption 3 *Every possible value of the output $x(k)$ corresponds to a unique source vector $\mathbf{s}(k)$, therefore $|\mathcal{X}_c| = 4^{L+1}$.*

As in the real case, for any time instant k_0, the output $x(k_0) = r$, comes from a unique source vector $\mathbf{s}(k_0) = [c_0, c_1, ..., c_{L-1}, c_L]^T$, $c_i \in \mathcal{B}$. The successor vector, at time $k_0 + 1$, can now take four possible values, $r^s_{(1)}, r^s_{(2)}, r^s_{(3)}$, and $r^s_{(4)}$ as follows:

$$\mathbf{s}_{(1)}(k_0+1) = \begin{bmatrix} 1+j, c_1, ..., c_{L-1}, c_L \end{bmatrix}^T,$$
$$\mathbf{s}_{(2)}(k_0+1) = \begin{bmatrix} 1-j, c_1, ..., c_{L-1}, c_L \end{bmatrix}^T,$$
$$\mathbf{s}_{(3)}(k_0+1) = \begin{bmatrix} -1+j, c_1, ..., c_{L-1}, c_L \end{bmatrix}^T,$$
$$\mathbf{s}_{(4)}(k_0+1) = \begin{bmatrix} -1-j, c_1, ..., c_{L-1}, c_L \end{bmatrix}^T. \tag{9}$$

Again, it is essential that all the pairs of output values $[r, r^s_{(i)}], i = 1, 2, 3, 4$, will appear, at least once, in the output sequence $x(k)$, $k = 1, \cdots, K$:

Assumption 4 *For any $r \in \mathcal{X}_c$, there are at least four indices k_0, k_1, k_2, k_3 $\in \{1, 2, \cdots, K\}$ such that $x(k_0) = r$, $x(k_0+1) = r^s_{(1)}$, $x(k_1) = r$, $x(k_1+1) = r^s_{(2)}$, $x(k_2) = r$, $x(k_2+1) = r^s_{(3)}$, and $x(k_3) = r$, $x(k_3+1) = r^s_{(4)}$.*

Once we find the successors $r^s_{(i)}$, of a specific observation value $x(k) = r$, we compute $\rho(r)$:

$$\rho(r) = \sum_{i=1}^{4} r^s_{(i)} = \sum_{i=1}^{4} \mathbf{a}^T \mathbf{s}_{(i)}(k+1)$$
$$= [(1+j) + (1-j) + (-1+j) + (-1-j)]a(0) + 4\sum_{l=1}^{L} a(l)s(k-l)$$

Table 1. The true filter coefficients and their estimated absolute values.

	True a	Estimated $|a|$		True a	Estimated $|a|$
1	0.9235	0.9235	6	-0.1067	0.1067
2	-0.9398	0.9398	7	-0.6507	0.6507
3	0.9075	0.9075	8	0.6705	0.6705
4	0.4288	0.4288	9	0.9402	0.9402
5	0.2931	0.2931	10	-0.7301	0.7301

$$= 4 \sum_{l=1}^{L} a(l)s(k-l) \tag{10}$$

Thus substituting $x(k) = r$ by $\rho(r)/4$, $\forall k$, we obtain a shortened system:

$$x^{(2)}(k) = \rho(r)/4 = \sum_{l=1}^{L} a(l)s(k-l). \tag{11}$$

As in the real case, L repetitions lead to a memoryless system:

$$x^{(L+1)}(k) = a(L)s(k-L) \tag{12}$$

Now the source can be estimated only up to a multiplier $\lambda \in \{\pm 1, \pm j\}$. Indeed, λ is unobservable since

$$\begin{aligned} x^{(L+1)}(k) &= a(L)s(k-L) \\ &= (-1 \cdot a(L))(-1 \cdot s(k-L)) \\ &= (-j \cdot a(L))(j \cdot s(k-L)) \\ &= (j \cdot a(L))(-j \cdot s(k-L)) \end{aligned} \tag{13}$$

Using the memoryless system (12) we can estimate the source taking again a two step approach: First, we introduce the estimate $\hat{a}(L)$ and we randomly select any time instant k_0, assuming that $x(k_0) = a(L)s(k_0 - L) = \hat{a}(L)(1+j)$. We call $\lambda^{-1} = s(k_0 - L)/(1+j)$ and we note that $\lambda^{-1} \in \{\pm 1, \pm j\}$. So $\hat{a}(L) = x^{(L+1)}(k_0)/(1+j) = \lambda^{-1} a(L)$. Second, we estimate $s(k-L)$ by

$$\hat{s}(k-L) = x^{(L+1)}(k)/\hat{a}(L) = \lambda s(k-L). \tag{14}$$

4 Examples

Example 1. In this experiment we created a source dataset of 10,000 random binary numbers. This was convolved with a real valued filter of length 10 with coefficients randomly chosen in the interval $[-1, 1]$. Table 1 presents the true filter coefficients and their estimated absolute values. Furtermore, in this example, $\hat{s}$ is a perfect estimate of the true source, except for the sign: $\hat{s}(k) = -s(k)$, $\forall k$.

Example 2. In this example we used a source dataset of 20,000 complex binary samples. The following randomly generated complex filter of length three was used: $[a(0), a(1), a(2)]^T = [0.9003-0.0280j, -0.5377+0.7826j, 0.2137-0.5242j]^T$. Our estimated source was a perfect estimate of the true one except for the multiplier $\lambda = -j$.

5 Discussion and Conclusion

A novel blind method for deconvolving SISO systems with binary real or complex sources was presented in this paper. The method is based on the recursive shortening of the channel leading eventually in a linear memoryless system with trivial solution. In this work we only study the noiseless situation because we want to emphasize on the mathematical development of our approach. In this context, both examples presented above are simple verifications of the method. However, noise can be handled, as well, by a simple modification of the algorithm. For example, in a real SISO system with noise, we observe $y(k) = x(k) + e(k)$ instead of $x(k)$, where $e(k)$ is the noise component. In this case, the set $\mathcal{X}$ which contains the possible values of $x(k)$ has to be estimated using some clustering technique. One also needs a classification rule which will group the observations $y(k)$ into the proper values $x(k) = r$ of $\mathcal{X}$. Once this is achieved, the method can proceed as presented. The effects of noise in the algorithm performance will be studied in another contribution.

References

1. Anand, K., Mathew, G., Reddy, V.U.: Blind separation of multiple co-channel bpsk signals arriving at an antenna array. IEEE Signal Processing Letters **2** (1995) 176–178
2. van der Veen, A.J.: Analytical method for blind binary signal separation. IEEE Trans. on Signal Processing **45** (1997) 1078–1082
3. Diamantaras, K.I., Chassioti, E.: Blind separation of n binary sources from on observation: A deterministic approach. In: Proc. Second Int. Workshop on ICA and BSS, Helsinki, Finland (2000) 93–98
4. Li, Y., Cichocki, A., Zhang, L.: Blind separation and extraction of binary sources. IEICI Trans. Fundamentals **E86-A** (2003) 580–589
5. Gurelli, M.I., Nikias, C.L.: Evam: An eigenvector-based algorithm for multichannel blind deconvolution of input colored signals. IEEE Trans. Signal Processing **43** (1995) 134–149
6. Chevreuil, A., Loubaton, P.: Mimo blind second-order equalization method and conjugate cyclostationarity. IEEE Trans. Signal Processing **47** (1999) 572–578
7. Ma, C.T., Ding, Z., Yau, S.F.: A two-stage algorithm for mimo blind deconvolution of nonstationary colored signals. IEEE Trans. Signal Processing **48** (2000) 1187–1192
8. Yellin, D., Weinstein, E.: Criteria for multichannel signal separation. IEEE Trans. on Signal Processing **42** (1994) 2158–2168
9. Tugnait, J.K.: On blind separation of convolutive mixtures of independent linear signals in unknown additive noise. IEEE Trans. Signal Processeing **46** (1998) 3117–3123

Blind Deconvolution Using the Relative Newton Method

Alexander M. Bronstein, Michael M. Bronstein, and Michael Zibulevsky

Technion – Israel Institute of Technology, Department of Electrical Engineering
32000 Haifa, Israel
{alexbron,bronstein}@ieee.org,mzib@ee.technion.ac.il

Abstract. We propose a relative optimization framework for quasi maximum likelihood blind deconvolution and the relative Newton method as its particular instance. Special Hessian structure allows its fast approximate construction and inversion with complexity comparable to that of gradient methods. The use of rational IIR restoration kernels provides a richer family of filters than the traditionally used FIR kernels. Smoothed absolute value and the smoothed deadzone functions allow accurate and robust deconvolution of super- and sub-Gaussian sources, respectively. Simulation results demonstrate the efficiency of the proposed methods.

1 Introduction

Blind deconvolution problem appears in various applications related to acoustics, optics, geophysics, communications, control, etc. In the general setup of the single-channel blind deconvolution, the observed sensor signal x is created from the *source signal s* passing through a causal convolutive system

$$x_n = \sum_{k=0}^{\infty} a_k \, s_{n-k} + u_n, \tag{1}$$

with impulse response a and additive sensor noise u. The setup is termed *blind* if only x is accessible, whereas no knowledge on a, s and u is available. The problem of blind deconvolution aims to find such a deconvolution (or restoration) kernel w, that produces a possibly delayed waveform-preserving source estimate $\hat{s}_n = (w * x)_n \approx c \cdot s_{n-\Delta}$, where c is a scaling factor and Δ is an integer shift. Equivalently, the *global system response* $g = a * w$ should be approximately a Kroenecker delta, up to scale factor and shift. A commonly used assumption is that s is non-Gaussian.

Many blind deconvolution methods described in literature focus on estimating the impulse response of the convolution system $A(z)$ from the observed signal x using a causal finite length (FIR) model and then determining the source signals from this estimate [1–5]. Many of these methods use batch mode calculations and usually suffer from high computational complexity. Conversely, a wide class of the so-called *Bussgang-type* algorithms estimate directly the inverse kernel $W(z) = A^{-1}(z)$ by minimizing some functional using gradient descent iterations. These methods usually operate in the time domain and the gradient is usually derived by applying some non-linearity to the correlation of the observed signal and the estimated source. One of the most popular algorithms in this class is the constant modulus algorithm (CMA) proposed by Godard [6]. A review of these algorithms can be found in [7].

In their fundamental work, Amari *et al.* [8] introduced an iterative time-domain blind deconvolution algorithm based on the natural gradient learning, which was originally used in context of blind source separation [9–11] and became very attractive due to the so-called *uniform performance property* [11]. The natural gradient algorithm estimates directly the restoration kernel and allows real-time processing. Efficient frequency-domain implementation was presented in [12].

Natural gradient demonstrates significantly higher performance compared to gradient descent. In this work, we present a blind deconvolution algorithm based on the relative Newton method, which brings further acceleration. The relative Newton algorithm was originally proposed in the context of sparse blind source separation in [13, 14]. We utilize special Hessian structure to derive a fast version of the algorithm with complexity comparable to that of gradient methods. We focus our attention on a batch mode single-channel blind deconvolution algorithm with FIR restoration kernel and outline the use of IIR kernels. We use the smoothed absolute value for deconvolution of super-Gaussian sources, and propose the smoothed deadzone linear function for sub-Gaussian sources.

2 QML Blind Deconvolution

Under the assumption that the restoration kernel $W(z)$ is strictly stable, and the source signal is real and i.i.d., the normalized minus-log-likelihood function of the observed signal x in the noise-free case is [8]

$$\ell(x;w) = -\frac{1}{2\pi}\int_{-\pi}^{\pi} \log\left|W(e^{i\theta})\right| d\theta + \frac{1}{T}\sum_{n=0}^{T-1} \varphi(y_n), \qquad (2)$$

where $y = w * x$ is a source estimate; $\varphi(s) = -\log p(s)$, where $p(s)$ is the probability density function (PDF) of the source s. We assume that w is an FIR kernel supported on $n = -N, ..., N$, and denote its length by $K = 2N+1$. We will also assume without loss of generality that s is zero-mean. Cost function (2) can be also derived using negative joint entropy and information maximization considerations. In practice, the first term of $\ell(x;w)$ containing the integral is difficult to evaluate; however, it can be approximated to any desired accuracy using the FFT.

Consistent estimator can be obtained by minimizing $\ell(x;w)$ even when $\varphi(s)$ is not exactly equal to $-\log p(s)$. Such *quasi ML* estimation has been shown to be practical in instantaneous blind source separation when the source PDF is unknown or not well-suited for optimization [13]. The choice of $\varphi(s)$ and the consistency conditions of the QML estimator are discussed in Section 5.

The gradient of $\ell(x;w)$ w.r.t. w_i is given by

$$g_i = -q_{-i} + \frac{1}{T}\sum_{n=0}^{T-1} \varphi'(y_n)\, x_{n-i}, \qquad (3)$$

where q_n is the inverse DFT of W_k^{-1}. The Hessian of $\ell(x; w)$ is given by

$$H_{ij} = r_{-(i+j)} + \frac{1}{T} \sum_{n=0}^{T-1} \varphi''(y_n) \, x_{n-i} x_{n-j}, \qquad (4)$$

where r_n is the inverse DFT of W_k^{-2} (for derivation see [15]). Both the gradient and the Hessian can be evaluated efficiently using FFT.

3 Relative Optimization

Here we introduce a relative optimization framework for blind deconvolution. The main idea of relative optimization is to iteratively produce source signal estimate and use it as the observed signal at the next iteration. Similar approach was explored in [14] in the context of blind source separation.

Relative optimization algorithm

1. Start with initial estimates of the restoration kernel $w^{(0)}$ and the source $x^{(0)} = w^{(0)} * x$.
2. For $k = 0, 1, 2, \ldots$, until convergence
    3. Start with $w^{(k+1)} = \delta$.
    4. Using an unconstrained optimization method, find $w^{(k+1)}$ such that $\ell(x^{(k)}; w^{(k+1)}) < \ell(x^{(k)}; \delta)$.
    5. Update source estimate: $x^{(k+1)} = w^{(k+1)} * x^{(k)}$.
6. End

The restoration kernel estimate at k-th iteration is $\hat{w} = w^{(0)} * \ldots * w^{(k)}$, and the source estimate is $\hat{s} = x^{(k)}$. This method allows to construct large restoration kernels growing at each iteration, using a set of relatively low-order factors. In real application, it might be necessary to limit the filter length to some maximum order, which can be done by cropping w after each update. The relative optimization algorithm has uniform performance, i.e. its step at iteration k depends only on $g^{(k-1)} = a * w^{(0)} * \ldots * w^{(k-1)}$, since the update in Step 5 does not depend explicitly on a, but on the currents global system response only. When the input signal is very long, it is reasonable to partition the input into blocks and estimate the restoration kernel for the current block using the data of the previous block and the previous restoration kernel estimate.

3.1 Fast Relative Newton Step

A Newton iteration can be used in Step 4 of the relative optimization algorithm, yielding very fast convergence. However, its practical use is limited to small values of N and T, due to the complexity of Hessian construction, and solution of the Newton system. This complexity can be significantly reduced if special Hessian structure is exploited. Near the solution point, $x^{(k)} \approx cs$, hence $\nabla^2 \ell(x; \delta)$ evaluated at each relative Newton iteration becomes approximately $\nabla^2 \ell(cs; \delta)$. For a sufficiently large sample size (in practice, $T > 10^2$), the following approximation holds:

Proposition 1. *The Hessian $\ell(cs; \delta)$ has an approximate diagonal-anti-diagonal structure, with ones on the anti-diagonal.*

Proof. Substituting $w = \delta$, $x = cs$ and $y = \delta * x = cs$ into $\ell(x; w)$ in (4), one obtains

$$H_{ij} = \delta_{i+j} + \frac{1}{T} \sum_{n=0}^{T-1} \varphi''(cs_n) \, cs_{n-i} \, cs_{n-j}.$$

For a large sample size T, the sum approaches the corresponding expectation value. Invoking the assumption that s is zero-mean i.i.d., the off-diagonal and off-anti-diagonal elements of H vanish. □

Typical Hessian structure is depicted in Figure 1 (left). Under this approximation, the Newton system separates to K systems of linear equations of size 2×2

$$\begin{pmatrix} H_{-k,-k} & 1 \\ 1 & H_{kk} \end{pmatrix} \begin{pmatrix} d_{-k} \\ d_k \end{pmatrix} = - \begin{pmatrix} g_{-k} \\ g_k \end{pmatrix} \quad (5)$$

for $k = 1, ..., K$, and an additional equation

$$H_{00} \, d_0 = -g_0. \quad (6)$$

In order to guarantee decent direction and avoid saddle points, we force positive definiteness of the Hessian by inverting the sign of negative eigenvalues in system (5) and forcing small eigenvalues to be above some positive threshold. Computation of the Hessian approximation involves evaluation of its main diagonal only, which is of the same order as gradient computation. Approximate solution of the Newton system requires $\mathcal{O}(N)$ operations.

4 IIR Restoration Kernels

When the convolution system $A(z)$ has zeros close to the unit circle, the restoration kernel $W(z)$ has to be long in order to achieve good restoration quality. Therefore, when $W(z)$ is parameterized by the set of FIR coefficients $w_{-N}, ..., w_N$, the number of parameters to be estimated is large. Under such circumstances, it might be advantageous to use a rational IIR restoration kernel of the form

$$W(z) = \frac{h_{-N}z^N + ... + h_N z^{-N}}{(1 + b_1 z^{-1} + ... + b_M z^{-M})(1 + c_1 z + ... + c_L z^L)},$$

parameterized by $h_{-N}, ..., h_N$, $b_1, ..., b_M$ and $c_1, ..., c_L$. The asymptotic Hessian of $\ell(x; h, b, c)$ with respect to these coefficients, evaluated at $w = \delta$ (i.e., all the coefficients, except $h_0 = 1$ are set to zero) and $x = cs$ has the sparse structure depicted in Figure 1 (right) [16]. Approximate Newton system solution can be carried out using an analytical expression for the regularized inverse of the structured Hessian. Another possibility is to consider techniques for solution of sparse symmetric systems. In both cases, approximate Hessian evaluation and Newton system solution have the complexity of a gradient descent iteration.

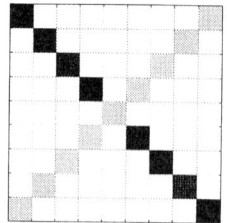 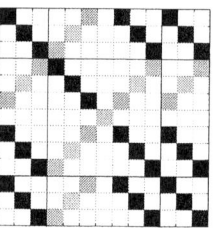

Fig. 1. Hessian structure at the solution point for FIR restoration kernel with $N = 3$ (left) and IIR restoration kernel with $N = M = L = 3$ (right). White represents near-zero elements.

5 The Choice of $\varphi(s)$

The choice of $\varphi(s)$ is limited first of all by the QML estimator consistency (or asymptotical stability) conditions, which guarantee that $w = a^{-1}$ is a stable minimum of $\ell(x; w)$ in the limit $T \to \infty$ [16].

When the source is super-Gaussian, e.g. sparse (sources common in seismology), or sparsely representable, a smooth approximation of the absolute value function usually obeys the asymptotic stability conditions [17, 18]. We use the following function [14]

$$\varphi_\lambda^{\mathrm{ABS}}(s) = |s| - \lambda \log\left(1 + \frac{|s|}{\lambda}\right), \quad (7)$$

which in the limit $\lambda \to 0^+$ yields an asymptoticall stable QML estimator if $\mathbb{E}|s| < 2\sigma^2 p(0)$, where $\sigma^2 = \mathbb{E}s^2$ [16]. In the particular case of strictly *sparse* sources, i.e. such sources that take the value of zero with some non-zero probability, *super-efficiency* is achieved in the limit $\lambda \to 0^+$ and in absence of noise [16].

In case of sub-Gaussian sources, common in digital communications, the family of power functions

$$\varphi_\mu^{\mathrm{PWR}}(s) = |s|^\mu \quad (8)$$

with the parameter $\mu > 2$ is usually a good choice for $\varphi(s)$. This function yields an asymptotically stable estimator for $\mathbb{E}|s|^{\mu+2} < (\mu+1)\sigma^2\,\mathbb{E}|s|^\mu$, which for the particular choice of $\mu = 4$ corresponds to negative kurtosis excess [16]. An increase of μ usually yields better performance. However, it is obvious that large values of μ imply high sensitivity to outliers due to the high powers. As a remedy, we propose to replace the power function with the *deadzone linear* function of the form

$$\varphi_\mu^{\mathrm{DZ}}(s) = \mu \cdot \max\{|s| - 1, 0\}, \quad (9)$$

which is often used for regression, data fitting and estimation [19]. This function has linear increase with controllable slope μ, and is known to have low sensitivity to outliers compared to the power function. Up to an additive constant, the deadzone linear function can be smoothly approximated by

$$\varphi_{\lambda,\mu}^{\mathrm{DZ}}(s) = \frac{\mu}{2}\left(\varphi_\lambda^{\mathrm{ABS}}(s-1) + \varphi_\lambda^{\mathrm{ABS}}(s+1)\right), \quad (10)$$

where the parameter λ controls the smoothness.

When the source PDF is compactly supported (e.g. digital communication signals), both the power function and the smoothed deadzone linear function yield super-efficient estimators in the limit $\mu \to \infty$. When in addition the source signal takes the values at the extremal points of the interval, s_{ext}, with some non-zero probability ρ, the use of the smoothed deadzone linear function achieves super-efficiency with $\lambda \to 0^+$ and *finite* μ. In the latter case, the estimator is asymptotically stable if $\mu\rho > 1$ and $2\sigma^2 \max\{(\mu\rho - 1)^2, 1\} > s_{\text{ext}}^2 \lambda \mu \rho$ [16].

6 Numerical Results

The convolution system was modelled by the empirically measured digital microwave channel impulse response from [20]. Two 10^4 samples long 2-level PAM and sparse normal i.i.d. processes were used as inputs. Input SNRs from 10 to 100 dB were tested. FIR restoration kernel with 33 coefficients was adapted in a block-wise manner, using blocks of length 33. The block fast relative Newton algorithm was compared to Joho's FDBD algorithm [12]. In both the power function with $\mu = 4$ was used for the PAM signal, whereas for the sparse source the smoothed absolute value with $\lambda = 10^{-2}$ was used in the relative Newton algorithm and the exact absolute value was used in the FDBD algorithm. In case of the PAM signal, performance was also compared to CMA with $p = 2$. Figure 2 (left) presents the restoration SIR averaged over 10 independent Monte-Carlo runs, as a function of the input SNR (95% confidence intervals are indicated on the plot). For SNR higher than 20 dB, the block relative Newton algorithm demonstrates an average improvement of about 4 dB compared to other methods for the PAM sources and about 7 dB for the sparse sources. Good restoration quality is obtained for SNR starting from 10 dB. Figure 2 (right) depicts the convergence of the compared algorithms, averaged over 10 independent runs with input SNR set to 20 dB.

Figure 3 (left) shows the SIR for the PAM source, averaged over 20 independent Monte-Carlo runs, wherein $\varphi(s)$ is chosen as the power function and the smoothed deadzone linear function. The comparison was performed both in the absence of noise, and in the presence of shot noise (sparse normal noise with 0.1% density, which introduced outliers into the signal). Unlike the power function, the proposed smoothed deadzone linear function appears to yield higher performance and demonstrates negligible sensitivity to outliers.

Advantages of an IIR restoration kernel can be seen in Figure 3 (right), which depicts the SIR for the sparse source, averaged over 10 Monte-Carlo runs, as a function of the number of optimization variables for different assignments of the degrees of freedom to restoration kernel numerator and denominator. A practically ideal SIR was achieved by the all-pole IIR kernel starting from 8 degrees of freedom. Additional simulation results can be found in [15, 18].

7 Conclusion

We have presented a relative optimization framework for QML single channel blind deconvolution and studied the relative Newton method as its particular instance. Diagonal-anti-diagonal structure of the Hessian in the proximity of the solution allowed to derive

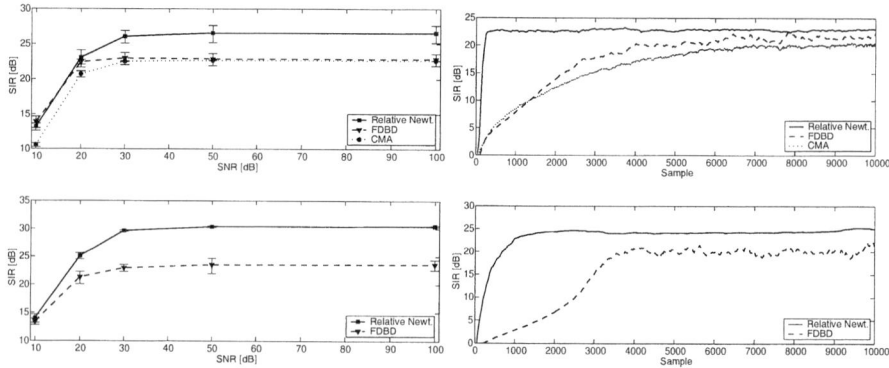

Fig. 2. Left: average SIR as a function of input SNR; right: average convergence in terms of SIR for input SNR of 20 dB. Top: 2-level PAM source; bottom: sparse source.

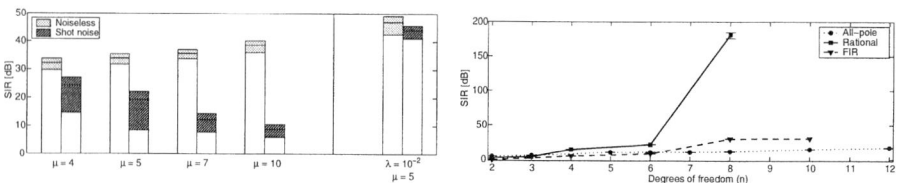

Fig. 3. Left: Average restoration SIR for the power function (left), and the smoothed deadzone linear function (two rightmost bars), with and without the presence of shot noise. Right: SIR as a function of degrees of freedom for different restoration kernel configurations.

a fast version of the relative Newton algorithm, with iteration complexity comparable to that of gradient methods. Additionally, we introduced rational restoration kernels, which often allow to reduce the optimization problem size. We also propose the use of the deadzone linear function for sub-Gaussian sources, which is significantly less sensitive to outliers than the commonly used non-linearities, and achieves super-efficient estimation in the absence of noise.

In simulation studies with super- and sub-Gaussian sources, the proposed methods exhibited very fast convergence and higher accuracy compared to the state-of-the-art approaches such as CMA and natural gradient-based QML algorithms. We are currently working on extending the presented approach to the multichannel and complex cases.

Acknowledgment

This research has been supported by the HASSIP Research Network Program HPRN-CT-2002-00285, sponsored by the European Commission, and by the Ollendorff Minerva Center.

References

1. Tong, L., Xu, G., Kailath, T.: Blind identification and equalization based on second-order statistics: A time domain approach. IEEE Trans. Inform. Theory **40** (1994) 340–349
2. Gurelli, M., Nikias, C.: EVAM: An eigenvector-based algorithm for multichannel bind deconvolution of input colored signals. IEEE Trans. Signal Processing **43** (1995) 134–149
3. Xu, G., Liu, H., Tong, L., Kailath, T.: Least squares approach to blind channel identification. IEEE Trans. Sig. Proc. **43** (1995) 2982–2993
4. Hua, Y.: Fast maximum likelihood for blind identification of multiple FIR channels. IEEE Trans. Sig. Proc. **44** (1996) 661–672
5. Gorokhov, A., Loubaton, P., Moulines, E.: Second order blind equalization in multiple input multiple output FIR systems: A weighted least squares approach. In: Proc. ICASSP. Volume 5. (1996) 2415–2418
6. Godard, D.N.: Self-recovering equalization and carrier tracking in two-dimensional data communication systems. IEEE Trans. Commun. **28** (1980) 1867–1875
7. Chi, C.Y., Chen, C.Y., Chen, C.H., Feng, C.C.: Batch processing algorithms for blind equalization using higher-order statistics. IEEE Sig. Proc. Magazine (2003) 25–49
8. Amari, S.I., Cichocki, A., Yang, H.H.: Novel online adaptive learning algorithms for blind deconvolution using the natural gradient approach. In: Proc. SYSID. (1997) 1057–1062
9. Cichocki, A., Unbehauen, R., Rummert, E.: Robust learning algorithm for blind separation of signals. Electronics Letters **30** (1994) 1386–1387
10. Amari, S.I., Douglas, S.C., Cichocki, A., Yang, H.H.: A new learning algorithm for blind signal separation. Advances in Neural Information Processing Systems **8** (1996) 757–763
11. Cardoso, J.F., Laheld, B.: Equivariant adaptive source separation. IEEE Trans. Sig. Proc. **44** (1996) 3017–3030
12. Joho, M., Mathis, H., Moschytz, G.S.: On frequency-domain implementations of filtered-gradient blind deconvolution algorithms. In: Proc. Asilomar Conf. Signals, Syst., Comput. (2002)
13. Pham, D., Garrat, P.: Blind separation of a mixture of independent sources through a quasi-maximum likelihood approach. IEEE Trans. Sig. Proc. **45** (1997) 1712–1725
14. Zibulevsky, M.: Sparse source separation with relative Newton method. In: Proc. ICA2003. (2003) 897–902
15. Bronstein, A.M., Bronstein, M., Zibulevsky, M.: Blind deconvolution with relative Newton method. Technical Report 444, Technion, Israel (2003)
16. Bronstein, A.M., Bronstein, M.M., Zibulevsky, M.: Relative optimization for blind deconvolution. IEEE Sig. Proc. (2004) Submitted. [Online] http://visl.technion.ac.il/bron/alex.
17. Zibulevsky, M., Pearlmutter, B.A., Bofill, P., Kisilev, P.: Blind source separation by sparse decomposition. In Roberts, S.J., Everson, R.M., eds.: Independent Components Analysis: Princeiples and Practice. Cambridge University Press (2001)
18. Bronstein, A.M., Bronstein, M.M., Zibulevsky, M., Zeevi, Y.Y.: Quasi maximum likelihood blind deconvolution of images using optimal sparse representations. Technical Report 455, Technion, Israel (2003)
19. Boyd, S., Vandenberghe, L.: Convex Optimization. Cambridge University Press (2003)
20. Giannakis, G.B., Halford, S.D.: Blind fractionally spaced equalization of noisy FIR channels: Direct and adaptive solutions. IEEE Trans. Sig. Proc. (**45**)

Blind Equalization Using Direct Channel Estimation

Hyung-Min Park[1], Sang-Hoon Oh[2], and Soo-Young Lee[1]

[1] Department of BioSystems, and Brain Science Research Center
Korea Advanced Institute of Science and Technology
Daejeon, 305-701, Republic of Korea
{hmpark,sylee}@kaist.ac.kr
[2] Department of Information Communication Engineering
Mokwon University
Daejeon, 302-729, Republic of Korea
shoh@mokwon.ac.kr

Abstract. In performing blind equalization, we propose a direct channel estimation method based on entropy-maximization of input signal with its known probability density function. That is, the proposed method estimates filter coefficients of the channel instead of equalizing filter coefficients which most of equalization methods try to estimate. Because the channel usually has a much shorter length than the equalizing filter, this method requires much smaller parameters to be estimated, and the channel can be equalized with much less computational demands. In addition, simulation results show that the proposed method can recover signals with a much smaller error than conventional methods.

1 Introduction

Blind equalization has become an important research problem in digital signal processing because of its desirable features and the challenge it poses to researchers in the field. If a training sequence is available, an adaptive equalizer can be easily adapted using the standard least-mean-squares (LMS) algorithm. However, there are many cases such as high data rate, bandlimited digital communication systems where the transmission of a training sequence is impractical or very costly. Therefore, blind adaptive equalization algorithms that do not rely on training signals need to be developed.

Let us consider a single-input-single-output (SISO) discrete-time linear system, in which the relationship between the input and the output signal is given by

$$x(n) = \sum_{k=0}^{L_m-1} h(k)s(n-k) + v(n). \tag{1}$$

The goal of blind equalization is to recover the input signal $s(n)$ from the output $x(n)$ without the assistance of a training sequence when the channel $h(k)$ is unknown. Typically, the input signal $s(n)$ is i.i.d., and the noise sequence $v(n)$ is modeled by a zero-mean white Gaussian noise process.

Many researchers have studied on the problem and proposed a number of blind equalization algorithms [1, 2]. In most of blind equalization methods, a causal finite-impulse-response (FIR) filter as a linear equalizer is used to recover the input signal $s(n)$. Hence, the equalizer model can be formulated by

$$u(n) = \sum_{k=0}^{L_a-1} w(k)x(n-k), \qquad (2)$$

where $w(k)$ is a filter coefficient of the equalizer. Since the blind equalization methods does not have a training sequence, adaptation of $w(k)$ usually makes use of some *a priori* statistical knowledge of the input signal $s(n)$.

In situations where the amplitude characteristics of $s(n)$ are roughly known, the class of Godard algorithms can be used [3]. Among the Godard algorithms, especially, the Sato algorithm and the constant modulus algorithm are very popular because of its simplicity [2]. If the probability density function of $s(n)$ is approximately known, an entropy-maximization algorithm for blind equalization can be derived by exploiting the higher order statistics (HOS) implicitly. It provides the same algorithm as the maximum likelihood estimation gives. Moreover, the signal is equalized with an improved convergence speed by applying the natural gradient to the blind equalization algorithm [4–6]. The entropy-maximization algorithm with the natural gradient is as follows [4]:

$$\Delta w(k) \propto w(k) - \varphi(u(n - L_a + 1))r(n - k), \qquad (3)$$

where

$$r(n) = \sum_{l=0}^{L_a-1} w(L_a - 1 - l)u(n - l). \qquad (4)$$

However, the equalizer requires a much longer filter length than the channel $h(k)$ since the equalizing filter approximates the inverse of the channel. Therefore, the number of estimated parameters is also very large, and it requires somewhat heavy computational loads to update filter coefficients of the equalizer. In addition, a large number of parameters degrade the recovered signal after convergence.

In this paper, we propose an equalizing method based on estimating a channel directly instead of the equalizing filter. This method requires much smaller parameters to be estimated. Therefore, we can equalize the channel with much less computational complexity and provide the recovered signal with a much smaller error than the conventional equalizing methods which estimate the inverse of the channel.

2 The Proposed Blind Equalization Algorithm

For simple derivation, the SISO linear system (1) can be represented in z-domain as

$$x(n) = H(z)s(n), \qquad (5)$$

where
$$H(z) = \sum_{k=0}^{L_m-1} h(k)z^{-k}. \tag{6}$$

In order to derive a new blind equalization algorithm, let us consider the input and the output signal of (5) over a N sample block, defined by the following vectors:
$$S = [s(0),\ s(1),\ \cdots,\ s(N-1)]^T,$$
$$X = [x(0),\ x(1),\ \cdots,\ x(N-1)]^T. \tag{7}$$

Both the input and the output signal, $s(n)$ and $x(n)$ are zeros for $n < 0$.

Then, we can write the output signal vector X as
$$X = \begin{bmatrix} h(0) & 0 & \cdots & 0 \\ h(1) & h(0) & \cdots & 0 \\ \vdots & \vdots & \ddots & \vdots \\ h(N-1) & h(N-2) & \cdots & h(0) \end{bmatrix} S. \tag{8}$$

Here, $h(L_m+1) = h(L_m+2) = \cdots = h(N-1) = 0$ by assuming that the length of the channel, L_m is much smaller than N.

The joint probability density of the output signal vector X can be given by
$$p(X) = \frac{p(S)}{|h(0)^N|}, \tag{9}$$

and $p(S) = p^N(s(n))$ for an i.i.d. input signal. Therefore, the log-likelihood of (9) is
$$L(H(z)) = -N\log|h(0)| + N\log p(s(n)). \tag{10}$$

An infinitesimal increment of the log-likelihood for an increment $dH(z)$ is
$$dL(H(z)) = L(H(z) + dH(z)) - L(H(z)). \tag{11}$$

With the score function defined by
$$\varphi(s(n)) = -\frac{d}{ds(n)}\log p(s(n)), \tag{12}$$

we have
$$d\log p(s(n)) = -\varphi(s(n))ds(n), \tag{13}$$

where $ds(n)$ is given in terms of $dH(z)$ as
$$ds(n) = -H^{-1}(z)dH(z)s(n). \tag{14}$$

Define a modified differential $dM(z)$ as
$$dM(z) = \sum_{k=-\infty}^{\infty} dm(k)z^{-k} = H^{-1}(z)dH(z). \tag{15}$$

Therefore,
$$d\log p(s(n)) = \varphi(s(n))dM(z)s(n). \tag{16}$$
In the similar way, we can show that
$$d\log|h(0)| = dm(0). \tag{17}$$
Thus, substituting (16) and (17) into (10) and (11) gives
$$dL(H(z)) = -Ndm(0) + N\varphi(s(n))dM(z)s(n). \tag{18}$$
Maximizing the log-likelihood in terms of $dM(z)$ provides the following learning algorithm,
$$\Delta M(z) \propto \frac{dL(H(z))}{dM(z)}. \tag{19}$$
Using (15), the natural gradient algorithm for updating $h(k)$ is given by
$$\begin{aligned}\Delta h(k) &\propto H(z)\frac{dL(H(z))}{dm(k)} \\ &\propto -H(z)\delta(k) + H(z)\varphi(s(n))s(n-k) \\ &= -h(k) + \varphi(s(n))q_k(n),\end{aligned} \tag{20}$$
where
$$q_k(n) = \sum_{l=0}^{L_m-1} h(l)s(n-k+l). \tag{21}$$

Note that the update of $h(k)$ depends on future values $s(n-k+l), k-l < 0$. In addition, it involves very intensive computation to compute all $q_k(n)$, $k = 0, \cdots, L_m - 1$, at each time step. Practically, the algorithm is modified by introducing an $L_m - 1$ sample delay to remove the non-causal terms and reusing past results assuming that $h(k)$ is not much changed over about $2L_m - 1$ time steps and $q_k(n) \approx q_0(n-k)$. Moreover, it is necessary to deal with complex-valued data for communication applications. With these considerations, the algorithm is modified as
$$\Delta h(k) \propto -h(k) + \varphi(s(n - L_m + 1))q^*(n - k), \tag{22}$$
where
$$q(n) = \sum_{l=0}^{L_m-1} h^*(L_m - 1 - l)s(n - l). \tag{23}$$

Taking the additive white Gaussian noise into consideration, the recovered signal follows the Pearson mixture model which is a mixture of the normal distributions [5,7]. Therefore, $s(n) - \tanh(s(n))$ can be used for the score function $\varphi(s(n))$. To deal with complex-valued data in communication systems, the score function becomes
$$\varphi(s(n)) = \Re\{s(n)\} - \tanh(\Re\{s(n)\}) + j\left[\Im\{s(n)\} - \tanh(\Im\{s(n)\})\right] \tag{24}$$

since one can consider that the real part of the signal $s(n)$ is independent of the imaginary part.

After estimating the channel $h(k)$ with the proposed method, we can recover the input signal $s(n)$ as

$$\hat{s}(n) = \sum_{k=0}^{L_a-1} \hat{h}^{-1}(k)x(n-k), \qquad (25)$$

where $\hat{s}(n)$ and $\hat{h}(k)$ denote estimations for $s(n)$ and $h(k)$, respectively. In order to compute $\hat{h}^{-1}(k)$, one can use lots of methods including [8]. In this paper, we adopt a method which makes use of reciprocal values of the results from the Fourier transform because of its simplicity.

3 Computational Complexity

In order to compare the computational loads of the proposed method and the conventional entropy-maximization method (3), let us consider the number of multiplications for complex-valued data. Assuming that the equalizing filter $w(n)$ has L_a taps, the conventional method requires approximately $4L_a$ multiplications for a sample. (Note that computational demands on the score function are negligible.) On the other hand, let us assume that the proposed method has L_m taps for the estimated channel $\hat{h}(n)$ and L_a taps for its inverse $\hat{h}^{-1}(n)$. In this case, the number of multiplications is approximately $3L_m + 2L_a + L_a \log_2 L_a$ using the fast Fourier transform.

When one performs adaptive learning, accumulating the update amounts over a sample block and updating the accumulation may often provide more stable convergence than 'one-by-one' updating. With the block updating, the conventional method requires about $3L_a$ multiplications for a sample whereas about L_a multiplications are required for a block. For the proposed method, the loads to compute $\hat{h}^{-1}(n)$, at most $L_a + L_a \log_2 L_a$ multiplications, are imposed for each block. Therefore, about $2L_m + L_a$ multiplications are required for a sample while the number of multiplications for a block is approximately $L_m + L_a + L_a \log_2 L_a$. Because a block usually contains a great many samples, the approximate ratio of the number of multiplications for the two methods, R is

$$R = \frac{2L_m + L_a}{3L_a}. \qquad (26)$$

It is worthy of note that L_a is much larger than L_m since a typical FIR filter usually requires much larger taps for an approximated inverse filter. Therefore, the computational complexity can be considerably reduced by the proposed algorithm.

4 Simulation Results

We illustrate the performance of the proposed blind equalization algorithm via simulations. We have tested the algorithm with quadrature-amplitude-modulated (QAM) signals whose amplitudes are $\sqrt{2}$. Experimental results were compared in terms of the intersymbol interference (ISI). It can be computed as

$$\text{ISI}(dB) = 10 \log \left(\frac{\sum_k |t(k)|^2 - \max_k |t(k)|^2}{\max_k |t(k)|^2} \right), \qquad (27)$$

where $t(k) = \hat{h}^{-1}(k) * h(k)$ or $t(k) = w(k) * h(k)$. We have chosen the length of $\hat{h}^{-1}(k)$ to be the same as the length of $w(k)$.

The output signal $x(n)$ of the SISO linear system (1) was generated using an 8 tap non-minimum phase channel shown in Fig. 1 [9]. In order to equalize the channel, we have employed a 17 tap filter for $\hat{h}(n)$ with tap-centering initialization. $\hat{h}^{-1}(n)$ has been computed for 64 taps. The step size was 0.001 with 100 samples for a block. Fig. 2 shows the ISI for the proposed method without noise. For comparison, the simulation on the conventional method (3) has been performed, and the result was included. In this simulation, a 64 tap filter was used as the equalizing filter $w(n)$, and other parameters had the same values as in the proposed method. In addition, we also compared it with the well-known method proposed in [10]. In order to consider the effect of noise, we repeated the simulations for the corrupted signals, and Fig. 3 shows the result for the signal whose SNR was 10dB. From these figures, it can be easily seen that the proposed method has a much smaller error than the conventional methods.

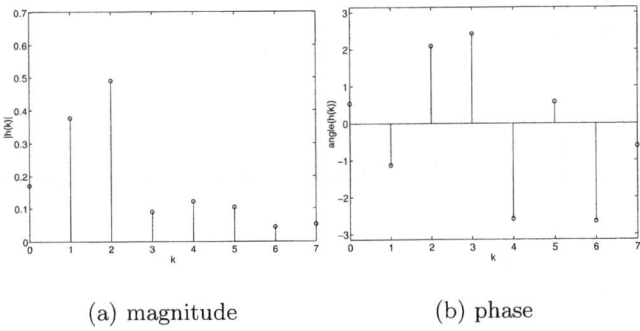

(a) magnitude (b) phase

Fig. 1. A minimum phase channel.

5 Conclusions and Further Works

In this paper, we proposed a blind equalization method using direct channel estimation. Under the assumption that the probability density function of the input signal is known, we derived the channel estimation algorithm by maximizing the entropy of the signal. By estimating filter coefficients of the channel directly, the method has much smaller parameters to be estimated than the conventional method which estimates the equalizing filters. Therefore, the proposed method can equalize the channel with much less computational complexity. Moreover,

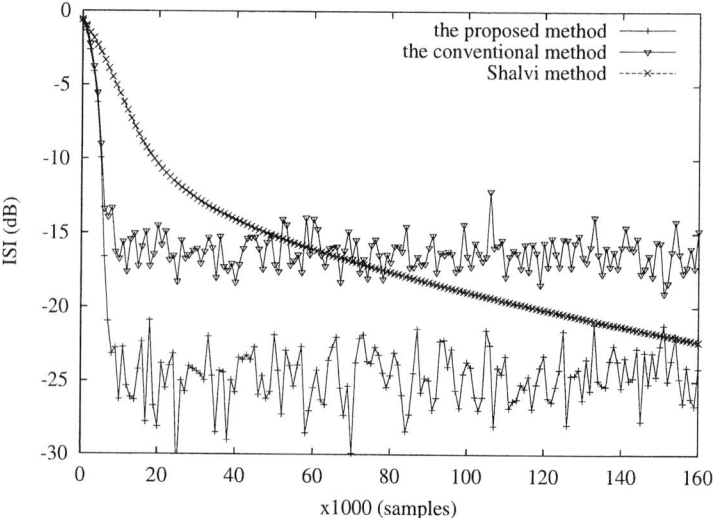

Fig. 2. ISI of the recovered signal without noise.

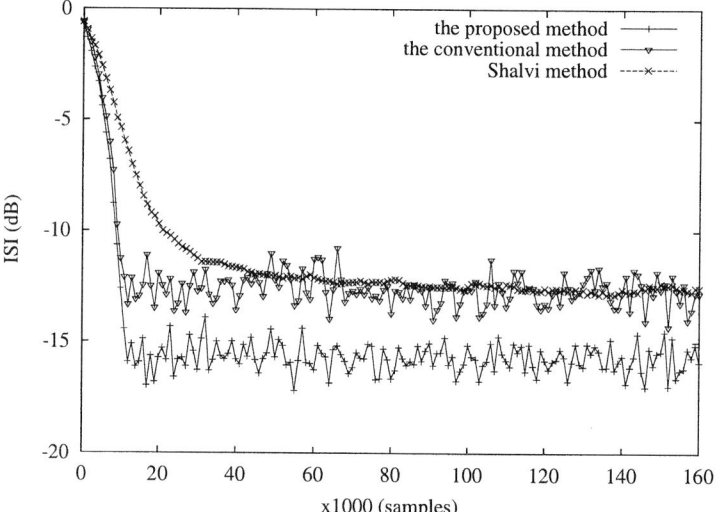

Fig. 3. ISI of the recovered signal with noise (10dB).

the simulations indicated that a much smaller error was contained in the recovered signal with the proposed method than the conventional method. As some further works, now we try to apply the method to other applications and extend it to multi-channel deconvolution.

Acknowledgment

This work was supported by the Brain Neuroinformatics Research Program sponsored by Korean Ministry of Science and Technology.

References

1. Haykin, S. (ed.): Blind Deconvolution. Prentice Hall, Englewood, (1994)
2. Ding, Z., Li, Y. (eds.): Blind Equalization and Identification. Marcel Dekker, New York, (2001)
3. Godard, D.N.: Self-recovering Equalization and Carrier Tracking in Two Dimensional Data Communication Systems. IEEE Trans. Comm., **28** (1980) 1867–1875
4. Amari, S., Douglas, S.C., Cichocki, A., Yang, H.H.: Multichannel Blind Deconvolution and Equalization Using the Natural Gradient. Proc. IEEE Workshop on Signal Processing Advances in Wireless Comm., Paris, France, (1997) 109–112
5. Lee, T.-W.: Independent Component Analysis. Kluwer Academic Publishers, Boston, (1998)
6. Bell, A.J., Sejnowski, T.J.: An Information-Maximization Approach to Blind Separation and Blind Deconvolution. Neural Computation, **7** (1995) 1129–1159
7. Lee, T.-W., Girolami, M., Sejnowski, T.J.: Independent Component Analysis using an Extended Infomax Algorithm for Mixed Sub-Gaussian and Super-Gaussian Sources. Neural Computation, **11** (1999) 417–441
8. Radlović, B.D., Kennedy, R.A.: Nonminimum-Phase Equalization and Its Subjective Importance in Room Acoustics. IEEE Trans. Speech and Audio Processing, **8** (2000) 728–737
9. IEEE Std 802.11a, Part11: Wireless LAN Medium Access Control (MAC) and Physical Layer (PHY) Specifications: High-Speed Physical Layer in the 5 GHz Band, (1999)
10. Shalvi, O., Weinstein, E.: New Criteria for Blind Deconvolution of Nonminimum Phase Systems (Channels). IEEE Trans. Information Theory, **36** (1990) 312–321

Blind MIMO Identification Using the Second Characteristic Function

Eran Eidinger* and Arie Yeredor

School of Electrical Engineering
Tel-Aviv University, Tel Aviv, 69978, Israel
{erane,arie}@eng.tau.ac.il

Abstract. We propose a novel algorithm for the identification of a Multi-Input-Multi-Output (MIMO) system. Instead of using "classical" high-order statistics, the mixing system is estimated directly from the empirical Hessian matrices of the second generalized characteristic function (GCF) at several preselected "processing points". An approximate joint-diagonalization scheme is applied to the transformed set of matrices in the frequency domain. This yields a set of estimated frequency response matrices, which are transformed back into the time domain after resolving frequency-dependent phase and permutation ambiguities. The algorithm's performance depends on the choice of processing points, yet compares favorably with other algorithms, especially at moderate SNR conditions.

1 Introduction

We address the following blind Multi-Input-Multi-Output (MIMO) Finite Impulse Response (FIR) model of order q:

$$\mathbf{x}[n] = \sum_{\ell=0}^{q} \mathbf{H}[\ell]\mathbf{s}[n-\ell] + \mathbf{v}[n] \quad (1)$$

where $\mathbf{s}[n], \mathbf{x}[n], \mathbf{v}[n] \in \mathbb{R}^N$ are the sources, observations and noise vectors, respectively. $\mathbf{H}[\ell]$ is the mixing matrix at lag ℓ, so that $h_{ij}[\ell]$ can be viewed as the impulse response of length q from source j to sensor i. The sources are assumed to be i.i.d. in time, and mutually independent in space. The noise is assumed to be spatially and spectrally white Gaussian with a known correlation matrix $\sigma_v^2 \mathbf{I}$. The goal is to estimate $\mathbf{H}[\ell], \ell = 0, 1 \ldots, q$ from T observations $\mathbf{x}[n], n = 0, 1, \ldots, T-1$.

Many of the existing approaches to this problem are based on high-order cumulants or Poly-Spectra (e.g., [1],[2] and [3]), which often (but not always) exhibit a relatively large estimation variance. Cumulants are well-known to be the high-order derivatives (at the origin) of the second Generalized Characteristic Function (GCF) of the observations' joint distribution (see definition below). An interesting alternative to the use of high-order derivatives at the origin is the

---

* This work has been supported by the Weinstein Institute for Signal Processing - Tel Aviv University, Israel. Its support is gratefully acknowledged.

use of second-order derivatives at selected off-origin points (termed "processing points"), as proposed by Yeredor in [4], [5]. These derivatives admit straightforward consistent estimates, in the form of specially-weighted empirical covariance matrices. Thus, in this paper we extend the GCF-based SISO identification algorithm of [5] into a MIMO algorithm named "CHAracteristic function MIMO Blind Identication" (CHAMBI). CHAMBI uses only the empirical Hessian matrices of the second GCF and attains the required diversity of these raw statistics by choosing the "processing points" from a continuous set rather than choosing discrete orders of derivatives (as higher-order cummulants). With an educated choice of processing points, CHAMBI performs better than Poly-Spectra algorithms (PSA), especially at moderate SNR conditions.

2 The Second Generalized Characteristic Function

The GCF of a vector of observations $\mathbf{x} \in \mathbb{R}^N$ at a "processing point" vector $\boldsymbol{\tau} \in \mathbb{C}^N$ is defined as $\phi_{\mathbf{x}}(\boldsymbol{\tau}) \triangleq E\left[e^{\boldsymbol{\tau}^T \mathbf{x}}\right]$ whereas the second GCF is defined as the natural logarithm $\psi_{\mathbf{x}}(\boldsymbol{\tau}) \triangleq \ln \phi_{\mathbf{x}}(\boldsymbol{\tau})$. In this paper we only consider $\boldsymbol{\tau} \in \mathbb{R}^N$. We assume that while we do not know the channel length q, we know at least an upper bound $L \geq q$. Let us define an extended version of $\mathbf{x}[n]$, and an associated "processing point" $\tilde{\boldsymbol{\tau}}$, both of length $(2L+1)N$:

$$\tilde{\mathbf{x}}[n] = [\mathbf{x}^T[n-L], \cdots, \mathbf{x}^T[n], \cdots, \mathbf{x}^T[n+L]]^T \; ; \; \tilde{\boldsymbol{\tau}} = [\boldsymbol{\tau}^T[L], \cdots, \boldsymbol{\tau}^T[0], \cdots, \boldsymbol{\tau}^T[-L]]^T \quad (2)$$

The processing points vector can thus be broken into $2L+1$ concatenated vectors $\boldsymbol{\tau}[\ell]$, each associated with $\mathbf{x}[n-\ell], -L \leq \ell \leq L$.

Consider the noiseless model. Using, for convenience, an infinite sum for the system's output (under the convention that $\mathbf{H}[\ell] = 0$ for $\ell \notin [0:q]$), we have

$$\tilde{\boldsymbol{\tau}}^T \tilde{\mathbf{x}}[n] = \sum_{k=-L}^{L} \boldsymbol{\tau}^T[k] \sum_{l=-\infty}^{\infty} \mathbf{H}[\ell]\mathbf{s}[n-k-\ell] \quad (3)$$

So that using $m = k + \ell$ we obtain

$$\tilde{\boldsymbol{\tau}}^T \tilde{\mathbf{x}}[n] = \sum_{m=-\infty}^{\infty} \sum_{\ell=m-L}^{m+L} \boldsymbol{\tau}^T[m-\ell]\mathbf{H}[\ell]\mathbf{s}[n-m] \triangleq \sum_{m=-\infty}^{\infty} \mathbf{a}^T[m]\mathbf{s}[n-m] \quad (4)$$

where $\mathbf{a}[m]$ can be interpreted as $\boldsymbol{\tau}[\ell]$ mixed by the MIMO system $\mathbf{H}[\ell]$

$$\mathbf{a}[m]^T \triangleq \sum_{\ell=m-L}^{m+L} \boldsymbol{\tau}^T[m-\ell]\mathbf{H}[\ell]. \quad (5)$$

We therefore obtain, due to the sources' i.i.d. time-structure (dropping the time index n due to stationarity) the first and second extended GCF-s:

$$\phi_{\tilde{\mathbf{x}}}(\tilde{\boldsymbol{\tau}}) = E\left[e^{\tilde{\boldsymbol{\tau}}^T \tilde{\mathbf{x}}}\right] = \prod_{m=-\infty}^{\infty} \phi_{\mathbf{s}}(\mathbf{a}[m]) \; ; \; \psi_{\tilde{\mathbf{x}}}(\tilde{\boldsymbol{\tau}}) = \sum_{m=-\infty}^{\infty} \psi_{\mathbf{s}}(\mathbf{a}[m]). \quad (6)$$

Differentiating $\psi_{\tilde{\mathbf{x}}}(\tilde{\boldsymbol{\tau}})$ once w.r.t. $\boldsymbol{\tau}[k]$, and again w.r.t. $\boldsymbol{\tau}[\ell]$ yields:

$$\mathbf{b}_{\tilde{\boldsymbol{\tau}}}[k] \triangleq \frac{\partial \psi_{\tilde{\mathbf{x}}}(\tilde{\boldsymbol{\tau}})}{\partial \boldsymbol{\tau}[k]} = \sum_{m=-\infty}^{\infty} \boldsymbol{\psi}_{\mathbf{s}}^{T}(\mathbf{a}[m]) \mathbf{H}[m-k] \tag{7}$$

$$\mathbf{C}_{\tilde{\boldsymbol{\tau}}}[k,\ell] \triangleq \frac{\partial^2 \psi_{\tilde{\mathbf{x}}}(\tilde{\boldsymbol{\tau}})}{\partial \boldsymbol{\tau}[k] \partial \boldsymbol{\tau}[\ell]} = \sum_{m=-\infty}^{\infty} \mathbf{H}[m-k] \boldsymbol{\Psi}_{\mathbf{s}}(\mathbf{a}[m]) \mathbf{H}^T[m-\ell], \tag{8}$$

where $\boldsymbol{\psi}_{\mathbf{s}}(\mathbf{a}[m]), \boldsymbol{\Psi}_{\mathbf{s}}(\mathbf{a}[m])$ denote the first and second derivatives (respectively) of $\psi_{\mathbf{s}}(\mathbf{a}[m])$ w.r.t. $\mathbf{a}[m]$.

An educated choice of $\tilde{\boldsymbol{\tau}}$ leads to first and second derivatives that for most values of m are equal to the sources' mean (**0**) and correlation ($\boldsymbol{\Sigma}_{\mathbf{s}}$), respectively. Specifically, if $\boldsymbol{\tau}[k] = \delta[k]\boldsymbol{\nu}$ so that $\tilde{\boldsymbol{\tau}} = [\mathbf{0}^T, \cdots, \boldsymbol{\nu}^T, \cdots, \mathbf{0}^T]^T$, then, by (5), we have

$$\boldsymbol{\psi}_{\mathbf{s}}(\mathbf{a}[m]) = \begin{cases} \boldsymbol{\psi}_{\mathbf{s}}(\mathbf{H}[m]^T \boldsymbol{\nu}) & 0 \le m \le q \\ \boldsymbol{\psi}_{\mathbf{s}}(\mathbf{0}) = E[\mathbf{s}] = \mathbf{0} & \text{o.w.} \end{cases}$$

$$\boldsymbol{\Psi}_{\mathbf{s}}(\mathbf{a}[m]) = \begin{cases} \boldsymbol{\Psi}_{\mathbf{s}}(\mathbf{H}[m]^T \boldsymbol{\nu}) & 0 \le m \le q \\ \boldsymbol{\Psi}_{\mathbf{s}}(\mathbf{0}) = E[\mathbf{s}\mathbf{s}^T] \triangleq \boldsymbol{\Sigma}_{\mathbf{s}} & \text{o.w.} \end{cases} \tag{9}$$

Since $\mathbf{C}_{\tilde{\boldsymbol{\tau}}}[k,\ell]$ depends on $\boldsymbol{\nu}$ (for our choice of $\tilde{\boldsymbol{\tau}}$), we shall replace the subscript $\tilde{\boldsymbol{\tau}}$ with $\boldsymbol{\nu}$. Exploiting the relation $\mathbf{R}_x[k-\ell] = \sum_{m=-\infty}^{\infty} \mathbf{H}[m-k] \boldsymbol{\Sigma}_s \mathbf{H}^T[m-\ell]$ we get:

$$\mathbf{C}_{\boldsymbol{\nu}}[k,\ell] = \mathbf{R}_x[k-\ell] + \sum_{m=0}^{q} \mathbf{H}[m-k] \mathbf{D}_{m,\boldsymbol{\nu}} \mathbf{H}^T[m-\ell], \tag{10}$$

where $\mathbf{D}_{m,\boldsymbol{\nu}} \triangleq \boldsymbol{\Psi}_{\mathbf{s}}(\mathbf{H}[m]^T \boldsymbol{\nu}) - \boldsymbol{\Sigma}_s$. Note that $\mathbf{D}_{m,\boldsymbol{\nu}}$ is diagonal due to the independence between sources [4]. Defining $\overline{\mathbf{C}}_{\boldsymbol{\nu}}[k,\ell] \triangleq \mathbf{C}_{\boldsymbol{\nu}}[k,\ell] - \mathbf{R}_x[k-\ell]$, we have

$$\overline{\mathbf{C}}_{\boldsymbol{\nu}}[k,\ell] = \sum_{m=0}^{q} \mathbf{H}[m-k] \mathbf{D}_{m,\boldsymbol{\nu}} \mathbf{H}^T[m-\ell] \tag{11}$$

Luckily, a straightforward consistent estimate of $\mathbf{C}_{\boldsymbol{\nu}}[k,\ell]$ at $\tilde{\boldsymbol{\tau}} = [\mathbf{0}^T, \cdots, \boldsymbol{\nu}^T, \cdots, \mathbf{0}^T]^T$ can be shown ([5], [4]) to be given by:

$$\widehat{\mathbf{C}}_{\boldsymbol{\nu}}[k,\ell] = \frac{1}{\sum_{n=q}^{T-q-1} w_n} \sum_{n=q}^{T-q-1} w_n (\mathbf{x}[n-k] - \bar{\mathbf{x}}_k)(\mathbf{x}[n-\ell] - \bar{\mathbf{x}}_\ell)^T \tag{12}$$

where $w_n = e^{\boldsymbol{\nu}^T \mathbf{x}[n]}$ and $\bar{\mathbf{x}}_k = (\sum_{n=q}^{T-q-1} w_n \mathbf{x}[n-k])/(\sum_{n=q}^{T-q-1} w_n)$.

3 Frequency Domain Separation

3.1 Transforming into the Frequency Domain

Joint diagonalization (JD) is the problem of, given a set of K matrices $\{\mathbf{M}_k\}$, finding a single matrix $\mathbf{A}_{JD}$ and a set of K diagonal matrices $\{\mathbf{D}_k\}$, such that:

$$\mathbf{M}_k = \mathbf{A}_{JD} \cdot \mathbf{D}_k \cdot \mathbf{A}_{JD}^H \,, k = 0, 1, \ldots, K-1 \tag{13}$$

Of course, there is no guarantee that such a set $\{\mathbf{D}_k\}$ and such a matrix $\mathbf{A}_{JD}$ exist. Often, while a set $\{\mathbf{M}_k\}$ is known to have such a structure, only an estimate of this set is accessible, so an approximate joint diagonalizer $\hat{\mathbf{A}}_{JD}$ of the set is sought, such that (13) holds "as closely as possible".

While (11) has a structure similar to the JD problem, the multiplication is replaced with a 2D convolution. This may be resolved by applying a $\tilde{L} \triangleq 2L + 1$ point 2D-DFT to $\overline{\mathbf{C}}_\nu[k, \ell]$ for $-L \le k, \ell \le L$, (equal to the DTFT sampled at the appropriate Fourier frequencies $\omega_k = \frac{2\pi k}{2L+1}, \omega_\ell = \frac{2\pi \ell}{2L+1}$). We use $\tilde{\mathbf{C}}_\nu$ as frequency domain notation to distinguish from $\overline{\mathbf{C}}_\nu$ in time.

$$\tilde{\mathbf{C}}_\nu(e^{j\omega_k}, e^{j\omega_\ell}) = \mathbf{H}(e^{-j\omega_k})\mathbf{D}_\nu(e^{j(\omega_k+\omega_\ell)})\mathbf{H}^T(e^{-j\omega_\ell}). \tag{14}$$

3.2 Formulation as a JD Problem

Essentially, (14) is already in the form of a JD problem if we choose $k = \ell$, but to avoid frequency-dependent phase and permutation ambiguities, some further manipulation is needed. Since only the Fourier frequencies are of interest, from here on the argument $(e^{j\omega_k})$ is dropped and replaced with a discrete index $[k]$.

Define $\mathbf{V}[k] (\triangleq \mathbf{V}(e^{j\omega_k}))$ such that

$$\mathbf{V}[k]\left(\mathbf{S}_{\mathbf{xx}}[k] - \sigma_v^2 \mathbf{I}\right)\mathbf{V}^H[k] = \mathbf{I}, \tag{15}$$

where $\mathbf{S}_{\mathbf{xx}}[k]$ is the spectrum matrix of $\mathbf{x}[n]$ at DFT index k. $\mathbf{V}[k]$ is thus a whitening matrix and $\mathbf{W}[k] \triangleq \mathbf{V}[k]\mathbf{H}[k]$ is unitary. It then follows that:

$$\mathbf{Y}_\nu[k, \ell] \triangleq \mathbf{V}[-k]\tilde{\mathbf{C}}_\nu[k, \ell]\mathbf{V}^H[-\ell] = \mathbf{W}[-k]\mathbf{D}_\nu[k + \ell]\mathbf{W}^H[-\ell]. \tag{16}$$

Choosing β as some integer we define a set of matrices:

$$\mathbf{M}^k_{\nu, \beta} \triangleq \mathbf{Y}_\nu[k, \beta - k]\mathbf{Y}^H_\nu[k, \beta - k] = \mathbf{W}[-k]\mathbf{D}_\nu[\beta]\mathbf{D}^H_\nu[\beta]\mathbf{W}^H[-k] \tag{17}$$

Note that $\mathbf{D}_\nu[\beta]\mathbf{D}^H_\nu[\beta]$ is a diagonal real-valued matrix.

We now have, at each frequency ω_k, a set of matrices which are jointly diagonalizable by a unitary matrix, $\mathbf{W}[-k]$. This is a variant of the JD scheme, called a unitary JD problem. A computationally simple algorithm, based on Jacobi rotation angles, has been introduced by Cardoso and Souloumiac [6] for this case. Note that for an approximate JD problem, A_{JD} can be estimated consistently under regularity conditions that are almost always satisfied (see [3]).

In theory, by finding the joint diagonalizer we can reconstruct the frequency response at each frequency and thus the overall system response. However, note that if $\mathbf{W}_1[-k]$ is a unitary joint diagonalizer, due to the structure of (13), so is:

$$\mathbf{W}_2[-k] = \mathbf{W}_1[-k] \cdot \mathbf{P}[-k] \cdot \mathbf{e}^{\Lambda[-k]} \tag{18}$$

with $\mathbf{P}[-k]$ a permutation matrix, $\Lambda[-k]$ a diagonal phase matrix and $e^{\Lambda[-k]}$ a diagonal matrix whose diagonal elements are the exponents of the diagonal of $\Lambda[-k]$. There usually also exists a scaling ambiguity that is inherent in the problem of BSS, which is resolved here implicitly by the pre-whitening (i.e. assuming the sources to be of unit variance). This means, that for the set of $\mathbf{M}_{\nu,\beta}^{-k}$ at frequency $-\omega_k$, the estimated joint diagonalizer $\widehat{\mathbf{W}}[k]$ can be used to estimate the original system response matrix up to frequency-dependent permutation and phase ambiguities as follows (the scaling ambiguity still exists but is not frequency-dependent):

$$\widehat{\mathbf{H}}_{\Lambda}[k] \triangleq \mathbf{V}^{-1}[k]\widehat{\mathbf{W}}[k] \approx \mathbf{H}[k]\mathbf{P}[k]e^{\Lambda[k]} \qquad (19)$$

3.3 Resolving the Permutation and Phase Ambiguities

The permutation ambiguity is easily resolved because of the structure of the diagonal matrices $\mathbf{D}_\nu[\beta]$ in (17), which evidently do not depend on ω_k, but only on the choice of β (which is identical for all frequencies). Thus, by ordering the diagonals in the same hierarchy, e.g., in increasing order, one can impose the same permutation matrix at all frequencies, and $\mathbf{P}[k]$ becomes simply $\mathbf{P}$. Such a frequency-independent permutation is acceptable because it merely implies a reordering of the sources in the time domain. The ordering method we chose will be explained in the simulations section.

Once the permutation ambiguity is resolved, (19) can be rewritten as $\widehat{\mathbf{H}}_\Lambda[k] \approx \mathbf{H}[k]\mathbf{P}e^{\Lambda[k]}$, and the phase ambiguity can be resolved, once again due to the fact that the diagonal matrices are not frequency dependent. For any integer α, consider the following diagonal matrix:

$$\mathbf{Q}_\nu[k] \triangleq \left(\mathbf{H}_\Lambda^{-1}[k]\right)^* \widetilde{\mathbf{C}}_\nu[k, \alpha - k]\mathbf{H}_\Lambda^{-1}[\alpha - k] = e^{j\Lambda[k]}\mathbf{P}^T\mathbf{D}_\nu[\alpha]\mathbf{P}e^{-j\Lambda[k-\alpha]} \qquad (20)$$

Bearing in mind that $\Lambda[k]$ is a diagonal phase matrix, we define phase matrices for the other diagonal elements in (20)

$$\mathbf{\Gamma}^\nu[k] \triangleq \arg\{\mathbf{Q}_\nu[k]\} \quad ; \quad \mathbf{\Theta}^\nu[\alpha] \triangleq \arg\{\mathbf{P}^T\mathbf{D}_\nu[\alpha]\,\mathbf{P}\}, \qquad (21)$$

where $\arg(z)$ denotes the phase of z. The phase of (20) can be rewritten as:

$$\mathbf{\Gamma}^\nu[k] = \mathbf{\Lambda}[k] + \mathbf{\Theta}^\nu[\alpha] - \mathbf{\Lambda}[k - \alpha]. \qquad (22)$$

Summing (22) over $k = 0, 1, \ldots, \widetilde{L} - 1$, the sums over $\Lambda[k - \alpha]$ and $\Lambda[k]$ cancel each other, and we have:

$$\mathbf{\Theta}^\nu[\alpha] = \frac{1}{\widetilde{L}} \sum_{k=0}^{\widetilde{L}-1} \mathbf{\Gamma}^\nu[k] \qquad (23)$$

Since we have access to $\mathbf{Q}_\nu[k]$, this means that $\mathbf{\Theta}^\nu[\alpha]$ can be computed from the observations.

Defining, for each diagonal element, $i = 1, 2, \ldots, N$, a vector of all frequencies excluding DC, as follows:

$$\boldsymbol{\lambda}_i \triangleq \left[\Lambda_{i,i}[1], \ldots, \Lambda_{i,i}[\widetilde{L} - 1]\right]^T \quad ; \quad \boldsymbol{\gamma}_i \triangleq \left[\Gamma_{i,i}[1], \ldots, \Gamma_{i,i}[\widetilde{L} - 1]\right]^T, \qquad (24)$$

the following proposition can help resolve the phase ambiguity:

Proposition 1. Let λ_i and γ_i be defined as above and let (22) hold. If α and $\widetilde{L}$ are co-prime, then the following set of equations holds:

$$\lambda_i = \mathbf{A}_{\widetilde{L},\alpha}^{-1} \gamma_i - \Theta_{i,i}[\alpha] \cdot \mathbf{A}_{\widetilde{L},\alpha}^{-1} \cdot \mathbf{1}_{[\widetilde{L}\times 1]} + \Lambda_{i,i}[0] \cdot \mathbf{1}_{[\widetilde{L}\times 1]}, \quad (25)$$

where $\mathbf{1}_{[\widetilde{L}\times 1]}$ denotes a vector of $\widetilde{L}$ ones and $\mathbf{A}_{\widetilde{L},\alpha}$ is an $\widetilde{L} \times \widetilde{L}$ matrix such that the diagonal elements are equal to one and the ith row contains a "-1" at column $(i - \alpha) \bmod \widetilde{L}$ for $i \neq \alpha$.

Proof. Equation (22) can be rewritten as

$$\mathbf{A}_{\widetilde{L},\alpha} \lambda_i = \gamma_i - \Theta_{i,i}[\alpha] \mathbf{1}_{[\widetilde{L}\times 1]} + \Lambda_{i,i}[0] \mathbf{e}_\alpha \quad (26)$$

where $\mathbf{e}_\alpha$ is a basis vector, all elements of which are zero, except for the α-th element, equal to 1. Since $\mathbf{A}_{\widetilde{L},\alpha}^{-1}$ exists if and only if α and $\widetilde{L}$ are co-prime, we multiply both sides by $\mathbf{A}_{\widetilde{L},\alpha}^{-1}$ and, noting that $\mathbf{A}_{\widetilde{L},\alpha}^{-1} \mathbf{1}_{[\widetilde{L}\times 1]} = \mathbf{e}_\alpha$, we get (25).

According to Proposition 1, by choosing α such that it is co-prime with $\widetilde{L}$, the phase matrix $\mathbf{\Lambda}[k]$ can be reconstructed up to some phase linear in k and a constant phase $\mathbf{\Lambda}[0]$. So, if $\widehat{\mathbf{\Lambda}}[k]$ is the diagonal matrix resulting from applying Proposition 1 to the estimates of $\mathbf{Q}_\nu[k]$ and $\mathbf{D}_\nu[\alpha]$, the system's final reconstruction is

$$\widehat{\mathbf{H}}[k] = \widehat{\mathbf{H}}_\Lambda[k] e^{-j\widehat{\mathbf{\Lambda}}[k]} \approx \mathbf{H}[k]\mathbf{P} e^{-j\mathbf{\Lambda}[0]-jk\mathbf{U}}, \quad (27)$$

where $\mathbf{U}$ is a diagonal matrix of integers. When an inverse DFT is applied the original system estimate is obtained up to an overall permutation ambiguity, an integer circular time-shift and a constant phase-shift for each channel.

4 The CHAMBI Algorithm

We present a closed-form algorithm, based on estimates (see (12)) of $\mathbf{C}_{\widetilde{\mathcal{T}}}[k,\ell]$:

1. Obtain a consistent estimate $\widehat{\mathbf{S}}_{\mathbf{xx}}[k]$ (e.g. using a (2L+1)-windowed correlogram) of the spectrum, and compute $\widehat{\mathbf{V}}[k]$, $k = 0, \ldots, \widetilde{L} - 1$, the frequency-dependent spatial whitening matrix.
2. Choose a set of P processing points $\{\nu_p\}_1^P$. Then, for each p, estimate $\widehat{\mathbf{C}}_{\nu_p}[k,\ell]$ (and thus $\overline{\widehat{\mathbf{C}}}_{\nu_p}[k,\ell]$ by subtracting $\widehat{\mathbf{R}}_x[k,\ell]$) for $-L \leq k,\ell \leq L$.
3. Apply an $\widetilde{L}$ size 2D DFT for every p to $\overline{\widehat{\mathbf{C}}}_{\nu_p}[k,\ell]$, and obtain $\widetilde{\widehat{\mathbf{C}}}_{\nu_p}[k,\ell]$. Using the appropriate $\widehat{\mathbf{Y}}_{\nu_p}[k,\ell]$, apply SVD to obtain estimates of the diagonals $\widehat{\mathbf{D}}_{\nu_p}[k+\ell]$ for all p and $[k+\ell]$.
4. For each pair β and ν_p grade the matrices $\widehat{\mathbf{D}}_{\nu_p}[\beta]$ based on the distinctness of the diagonals. For example, one can grade them by the maximal minimum absolute distance between the elements on the diagonal. Choose a subset of $\widetilde{P} \leq \widetilde{L}P$ "best graded" pairs of $(\nu,\beta)_{\tilde{p}}, \tilde{p} = 1, \ldots, \widetilde{P}$.

5. At each frequency ω_k, define a set for JD. For every pair $(\boldsymbol{\nu},\beta)_{\tilde{p}}$ find $\widehat{\mathbf{M}}_{\tilde{p}}^k \triangleq \widehat{\mathbf{M}}_{\boldsymbol{\nu}_{\tilde{p}},\beta_{\tilde{p}}}^k$ by computing $\mathbf{Y}_{\boldsymbol{\nu}_{\tilde{p}}}[k,\beta_{\tilde{p}}-k]\mathbf{Y}_{\boldsymbol{\nu}_{\tilde{p}}}^H[k,\beta_{\tilde{p}}-k]$. Apply a unitary JD scheme to the set $\left\{\widehat{\mathbf{M}}_{\tilde{p}}^k\right\}_{\tilde{p}=1}^{\tilde{P}}$ and thus obtain $\widehat{\mathbf{H}}_\Lambda[k]$, using the above mentioned method for resolving the permutation ambiguity.
6. Choose some $\boldsymbol{\nu} \in \{\boldsymbol{\nu}_p\}_1^P$, calculate $\widehat{\mathbf{Q}}_{\boldsymbol{\nu}}[k] = \widehat{\mathbf{H}}_\Lambda^*[k]^{-1}\widehat{\widetilde{\mathbf{C}}}_{\boldsymbol{\nu}}[k,\alpha-k]\widehat{\mathbf{H}}_\Lambda^{-1}[\alpha-k]$ and use it to resolve the phase ambiguity and get $\widehat{\mathbf{H}}(e^{j\omega_k})$.
7. Apply an inverse DFT to finally obtain $\widehat{\mathbf{H}}[\ell]$.

5 Simulation Results

We ran 200 Monte-Carlo trials, identifying a 2x2 nonminimum-phase system:

$$\mathbf{H}(z) = \begin{bmatrix} 1\text{ -}1.5537z^{-1}\text{ -}0.0363z^{-2}\text{+}0.5847z^{-3}\text{+}0.5093z^{-4} & 1\text{+}2.2149z^{-1}\text{+}1.0828z^{-2}\text{-}1.1731z^{-3}\text{-}0.8069z^{-4} \\ 1\text{+}0.9295z^{-1}\text{+}0.2453z^{-2}\text{-}0.7510z^{-3}\text{+}0.3717z^{-4} & 1\text{-}0.7137z^{-1}\text{-}1.5079z^{-2}\text{+}1.6471z^{-3}\text{-}1.2443z^{-4} \end{bmatrix} \tag{28}$$

We used zero-mean, unit variance sources with one-sided exponential distributions. Additive white Gaussian noise was applied to each sensor. Results are compared to those of the poly-spectra slices algorithm (PSA) suggested in [3] (using the same setup). SNR is measured by the ratio between the average sensor power and the noise variance.

We chose $L = 10$ (so that $\tilde{L} = 21$) and $P = 8$ processing points, spread over half a circle such that $\boldsymbol{\nu}_p = r[\cos(\pi p/P) \ \sin(\pi p/P)]^T$ with $r = 0.7$.

Out of $\tilde{L} \cdot P$ possible matrices at each frequency we chose the $\tilde{L} \cdot P/6$ best graded[1] matrices for the JD matrix set.

For the correction of the phase ambiguity, we chose $\alpha = 5$ (co-prime with $\tilde{L} = 21$). Assuming an unknown channel length, for performance analysis we used an estimated channel length of $L_e = L > q$, thus truncating the impulse response of length $\tilde{L}$ (obtained from the inverse DFT) to length L_e. The filters were artificially aligned before truncation. The performance measure per channel $h_{(i,j)}[\ell]$ is the Normalized Mean Square Error (NMSE), defined as the total square estimation error over all (truncated) taps, normalized by the total taps' energy. Results are presented in terms of the Overall NMSE (ONMSE), which is the NMSE averaged over all N^2 channels (and all Monte-Carlo trials).

Figure 1. shows the performance both in frequency and in time. Table 1. compares the performance of CHAMBI to that of PSA which was chosen for comparison because it is a relatively widely-used algorithm. CHAMBI is seen to outperform PSA and suffers less degradation than the PolySpectra algorithm at the lower SNR. It should also be noted that while the third-order PSA ([3]) cannot deal with symmetric distributions, CHAMBI has no such restrictions. Thus, for symmetric sources, PSA will be forced to resort to higher-order Poly-Spectra, which generally (but not always) admit less accurate estimates.

---
[1] The grade was computed as $grade(\mathbf{D}) = \min_{i \neq j}|\ln(D_{ii}/Djj)|$.

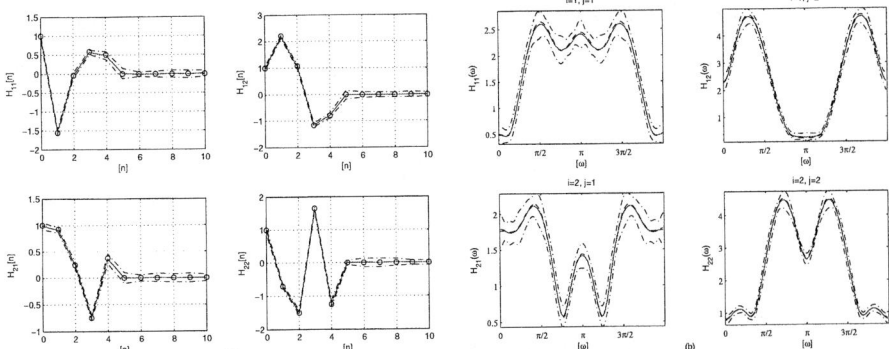

Fig. 1. Estimation of the 2×2 system in (28) (the same system used in [3], for comparison purposes) with SNR=$30dB$, using $T = 8192$ observations. (a) Truncated impulse response. True: circles, mean: solid, standard-deviation: upper and lower dash-dotted. (b) Magnitude of the frequency response. True: dashed, mean: solid, standard-deviation: upper and lower dash-dotted. The true magnitude is mostly hidden by the mean.

Table 1. ONMSE of CHAMBI and PSA for T=4096 and T=8192 observations.

Algorithm \ SNR	10dB	20dB	30dB
CHAMBI (4096)	0.0411	0.0397	0.0376
PSA (4096)	0.0502	0.0435	0.0407

Algorithm \ SNR	10dB	20dB	30dB
CHAMBI (8192)	0.0211	0.0183	0.0169
PSA (8192)	0.0405	0.0372	0.0227

6 Conclusions

We proposed a novel algorithm for blind MIMO identification / BSS of convolutive mixtures, based on the Hessian of the second GCF. Performance naturally depends on the choice of "processing points", and with proper choice the algorithm was shown to outperform a Poly-Spectra algorithm, especially at moderate SNR conditions. Further research would address data-adaptive optimization of the "processing points" selection.

References

1. Yellin, D., and Weinstein, E. Criteria for multichannel signal separation. *IEEE Transactions on Signal Processing*, 24(8):2156–2167, 1994.
2. Yellin, D., and Weinstein, E. Multichannel signal separation: methods and analysis. *IEEE Transactions on Signal Processing*, 44(1):106–118, 1996.
3. Chen, B., and Petropulu, A.P. Frequency domain blind mimo system identification based on second- and higher order statistics. *IEEE Trans. on Signal Processing*, 49(8):1677–1688, 2001.
4. Yeredor, A. Blind source separation via the second characteristic function. *Signal Processing*, 80(5):2000, 897-902.
5. Yeredor, A. Blind channel estimation using first and second derivatives of the characteristic function. *IEEE Signal Processing Letters*, 9(3):100–103, 2002.
6. Cardoso, J.-F., and Souloumiac, A. Jacobi angles for simultaneous diagonalization. *SIAM Journal on Matrix Analysis and Applications*, 17:161–164, 1996.

Blind Signal Separation of Convolutive Mixtures: A Time-Domain Joint-Diagonalization Approach

Marcel Joho

Phonak Hearing Systems, Champaign, IL, USA
joho@ieee.org

Abstract. We address the blind source separation (BSS) problem for the convolutive mixing case. Second-order statistical methods are employed assuming the source signals are non-stationary and possibly also non-white. The proposed algorithm is based on a joint-diagonalization approach, where we search for a single polynomial matrix that jointly diagonalizes a set of measured spatiotemporal correlation matrices. In contrast to most other algorithms based on similar concepts, we define the underlying cost function entirely in the time-domain. Furthermore, we present an efficient implementation of the proposed algorithm which is based on fast convolution techniques.

1 Introduction

1.1 Problem Formulation

Signal Mixing. The system setup is described as follows: M_s unknown mutually uncorrelated *source signals* s_m are filtered and mixed by an unknown time-invariant finite-length causal *convolutive mixing system* $\mathbf{A}^{M_x \times M_s} \triangleq \{\mathbf{A}_n\}_{n=0}^{N_a}$ resulting in M_x measurable sensor signals x_m. The source- and sensor-signals are stacked in vectors, $\mathbf{s}$ and $\mathbf{x}$, respectively. For simplicity we neglect any additive noise components. Hence, the convolutive mixing process is described as $\mathbf{x} = \mathbf{A} * \mathbf{s}$ where

$$\mathbf{x}(t) = (\mathbf{A} * \mathbf{s})(t) = \sum_{n=0}^{N_a} \mathbf{A}_n\, \mathbf{s}(t-n) \tag{1}$$

or, written in the z-domain,

$$\mathbf{x}(z) \triangleq \sum_t \mathbf{x}(t)\, z^{-t} = \mathbf{A}(z)\, \mathbf{s}(z). \tag{2}$$

Signal Separation. The M_x sensor signals x_m are mixed and filtered with a finite-length non-causal convolutive separation system $\mathbf{W}^{M_u \times M_x} \triangleq \{\mathbf{W}_n\}_{n=-N_w}^{N_w}$ resulting in M_u output signals u_m. The separation process is described as $\mathbf{u} = \mathbf{W} * \mathbf{x} = \mathbf{W} * \mathbf{A} * \mathbf{s}$, or written in the z-domain

$$\mathbf{u}(z) = \mathbf{W}(z)\, \mathbf{x}(z) = \mathbf{W}(z)\mathbf{A}(z)\, \mathbf{s}(z). \tag{3}$$

The objective of the blind-source-separation problem for the convolutive mixing case is to find a $\mathbf{W}(z)$ such that the global system can be written as

$$\mathbf{G}(z) = \mathbf{W}(z)\,\mathbf{A}(z) = \mathbf{P}\,\mathbf{D}(z) \tag{4}$$

where $\mathbf{D}(z)$ is a diagonal polynomial matrix and $\mathbf{P}$ is a permutation matrix. In the following we assume that $M_u = M_s \leq M_x$, and that the source signals and mixing system can be complex valued. Depending on $\mathbf{A}(z)$, M_x, and M_s perfect separation is possible for a finite N_w.

1.2 Mathematical Preliminaries

Basic Notation. The notation used throughout this paper is the following: Vectors are written in lower case, matrices in upper case. Matrix and vector transpose, complex conjugation and Hermitian transpose are denoted by $(.)^T$, $(.)^*$, and $(.)^H \triangleq ((.)^*)^T$, respectively. The sample index is denoted by t. The identity matrix is denoted by $\mathbf{I}$, a vector or a matrix containing only zeros by $\mathbf{0}$. $E\{.\}$ denotes the expectation operator. The Frobenius norm and the trace of a matrix are denoted by $\|.\|_F$ and $\mathrm{tr}\{.\}$, respectively. $\mathrm{diag}(\mathbf{A})$ zeros the off-diagonal elements of $\mathbf{A}$ and

$$\mathrm{off}(\mathbf{A}) \triangleq \mathbf{A} - \mathrm{diag}(\mathbf{A}) \tag{5}$$

zeros the diagonal elements of $\mathbf{A}$. The extension for polynomial matrices is defined straightforwardly as $\mathrm{off}(\mathbf{A}(z)) \triangleq \mathbf{A}(z) - \mathrm{diag}(\mathbf{A}(z)) = \sum_n \mathrm{off}(\mathbf{A}_n) z^{-n}$. Linear convolution between two sequences is denoted by $*$. Furthermore, we define

$$\mathbf{A}^\dagger(z) \triangleq \mathbf{A}^H(1/z^*) = \sum_n \mathbf{A}_n^H z^{+n}. \tag{6}$$

Signals. We use the following notation: $x_m(t)$ denotes the value of the signal x_m at discrete time t and $x_m \triangleq \{x_m(t)\}$ denotes the time series of signal x_m. Furthermore, we define $\mathbf{x}(t) \triangleq (x_1(1), \ldots, x_M(t))^T$ and $\mathbf{x} \triangleq (x_1, \ldots, x_M)^T = \{\mathbf{x}(t)\}$. The spatiotemporal correlation matrix between two signal vectors $\mathbf{u}$ and $\mathbf{x}$, and the corresponding z-transform of the correlation sequence are defined as

$$\mathbf{R}_{\mathbf{ux}}(\tau; t) \triangleq E\{\mathbf{u}(t)\,\mathbf{x}^H(t-\tau)\} \tag{7}$$

$$\mathbf{R}_{\mathbf{ux}}(z; t) \triangleq \sum_{\tau=-\infty}^{\infty} \mathbf{R}_{\mathbf{ux}}(\tau; t) z^{-\tau}, \tag{8}$$

respectively. For stationary signals we have $\mathbf{R}_{\mathbf{ux}}(\tau; t) = \mathbf{R}_{\mathbf{ux}}(\tau)$ and, hence, $\mathbf{R}_{\mathbf{ux}}(z; t) = \mathbf{R}_{\mathbf{ux}}(z)$.

Frobenius Norm. In the following we, will make use of some concepts from functional analysis [1]: Let $\mathbb{M}$ be the inner product space of complex matrixes. Given two matrices $\mathbf{A}$ and $\mathbf{B}$ with $\mathbf{A}, \mathbf{B} \in \mathbb{M}$, we define the *scalar product* of two matrices as $\langle \mathbf{A}, \mathbf{B} \rangle \triangleq \mathrm{tr}\{\mathbf{A}\mathbf{B}^H\}$. The induced norm is equivalent to the *Frobenius norm*, i.e. $\|\mathbf{A}\|_F \triangleq \sqrt{\langle \mathbf{A}, \mathbf{A} \rangle}$. Norms provide a convenient way to measure a distance between two matrices, as they induce a *metric* defined as $d(\mathbf{A}, \mathbf{B}) \triangleq \|\mathbf{A} - \mathbf{B}\|_F$.

Frobenius Norm for Polynomial Matrices. We can extend the definition of the Frobenius norm to polynomial matrices. Let $\mathbb{P}$ be the inner product space of complex polynomial matrixes. Let $\mathbf{A}(z) \triangleq \sum_n \mathbf{A}_n z^{-n}$ and $\mathbf{B}(z) \triangleq \sum_n \mathbf{B}_n z^{-n}$ be two matrix polynomials or Laurent series, i.e., their coefficients are complex matrices. If $\mathbf{A}(z)$ or $\mathbf{B}(z)$ have *finite energy*, i.e., $\sum_n \|\mathbf{A}_n\|_F^2 < \infty$ or $\sum_n \|\mathbf{B}_n\|_F^2 < \infty$, we can define the following *inner product*

$$\langle \mathbf{A}(z), \mathbf{B}(z) \rangle_{\mathcal{F}} \triangleq \sum_n \langle \mathbf{A}_n, \mathbf{B}_n \rangle = \sum_n \mathrm{tr}\{\mathbf{A}_n \mathbf{B}_n^H\}. \tag{9}$$

The inner product $\langle .,. \rangle_{\mathcal{F}}$ defines an induced norm on $\mathbb{P}$ given by

$$\|\mathbf{A}(z)\|_{\mathcal{F}} \triangleq \sqrt{\langle \mathbf{A}(z), \mathbf{A}(z) \rangle_{\mathcal{F}}} = \sqrt{\sum_n \|\mathbf{A}_n\|_F^2} \tag{10}$$

and a *metric* on $\mathbb{P}$ induced by the norm $d(\mathbf{A}(z), \mathbf{B}(z)) \triangleq \|\mathbf{A}(z) - \mathbf{B}(z)\|_{\mathcal{F}}$. It is not very difficult to show that the definitions of $\langle .,. \rangle_{\mathcal{F}}$ and $\|.\|_{\mathcal{F}}$ fulfill the properties of scalar products and norms [1], respectively. The induced metric $d(\mathbf{A}(z), \mathbf{B}(z)) \triangleq \|\mathbf{A}(z) - \mathbf{B}(z)\|_{\mathcal{F}}$ allows us to measure the "distance" between two polynomial matrices $\mathbf{A}(z)$ and $\mathbf{B}(z)$. In our case, we will use $d(.,.)$ to measure the distance between two spatiotemporal correlation matrices.

2 A Joint-Diagonalization Approach

2.1 Correlation Matrices

Stationary Source Signals. Assuming that the source signals s_m are stationary, the input spatiotemporal correlation matrix $\mathbf{R}_{\mathbf{xx}}(z)$ of the mixing process (2) is

$$\mathbf{R}_{\mathbf{xx}}(z) = \mathbf{A}(z)\mathbf{R}_{\mathbf{ss}}(z)\mathbf{A}^\dagger(z). \tag{11}$$

The output correlation matrix $\mathbf{R}_{\mathbf{uu}}(z)$ of the separation process (3) is

$$\mathbf{R}_{\mathbf{uu}}(z) = \mathbf{W}(z)\mathbf{R}_{\mathbf{xx}}(z)\mathbf{W}^\dagger(z) = \mathbf{W}(z)\mathbf{A}(z)\mathbf{R}_{\mathbf{ss}}(z)\mathbf{A}^\dagger(z)\mathbf{W}^\dagger(z). \tag{12}$$

Extension to Block-Wise Stationary Source Signals. If we relax the stationary assumption and assume that the source signals are non-stationary, but block-wise stationary, then Eq. (11) changes to

$$\mathbf{R}_{\mathbf{xx}}(z; t_p) = \mathbf{A}(z)\mathbf{R}_{\mathbf{ss}}(z; t_p)\mathbf{A}^\dagger(z) \tag{13}$$

where t_p denotes the center of the pth *snapshot* of $\mathbf{R}_{\mathbf{ss}}(z; t_p)$ and $\mathbf{R}_{\mathbf{xx}}(z; t_p)$. Since the output correlation matrix depends now also on t_p, (12) becomes

$$\mathbf{R}_{\mathbf{uu}}(z; t_p) = \mathbf{W}(z)\mathbf{R}_{\mathbf{xx}}(z; t_p)\mathbf{W}^\dagger(z) = \mathbf{W}(z)\mathbf{A}(z)\mathbf{R}_{\mathbf{ss}}(z; t_p)\mathbf{A}^\dagger(z)\mathbf{W}^\dagger(z) \tag{14}$$

assuming $\mathbf{A}(z)$ and $\mathbf{W}(z)$ are time-invariant. Since we assume that the source signals s_m are mutually uncorrelated for all t, $\mathbf{R}_{\mathbf{ss}}(z; t_p)$ has a diagonal structure for every snapshot. In the special case where all source signals are also white, then $\mathbf{R}_{\mathbf{ss}}(z; t_p) = \mathbf{R}_{\mathbf{ss}}(0)$. However, we do not require that the source signals need to be white.

2.2 Cost Function

Non-blind Cost Function. In the blind source separation setup, the source signals s_m are unknown. Let us assume for the moment that $\mathbf{R_{ss}}(z;t_p)$ is known for P *snapshots* at t_p ($p=1..P$). In this case a possible cost function for the (non-blind) source separation task is (recall that $\|\mathbf{R_{uu}} - \mathbf{R_{ss}}\|_{\mathcal{F}} = d(\mathbf{R_{uu}}, \mathbf{R_{ss}})$)

$$\mathcal{J}_0(\mathbf{W}(z)) \triangleq \sum_{p=1}^{P} \mathcal{J}_0'(t_p) = \sum_{p=1}^{P} \| \mathbf{R_{uu}}(z;t_p) - \mathbf{R_{ss}}(z;t_p) \|_{\mathcal{F}}^2 \quad (15)$$

$$= \sum_{p} \| \mathbf{W}(z) \mathbf{R_{xx}}(z;t_p) \mathbf{W}^{\dagger}(z) - \mathbf{R_{ss}}(z;t_p) \|_{\mathcal{F}}^2. \quad (16)$$

which obviously has a global minimum for $\mathbf{W}(z) = \mathbf{A}^{-1}(z)$.

Blind Cost Function. In the *blind signal separation* (BSS) problem we do not know the true source correlation matrices $\mathbf{R_{ss}}(z;t_p)$. Hence, we need to replace them by some estimates $\hat{\mathbf{R}}_{\mathbf{ss}}(z;t_p)$ in order to still use the cost function (15). Since we assume that the source signals s_m are mutually uncorrelated, we also assume that $\mathbf{R_{ss}}(z;t_p)$ has a diagonal structure. Therefore a possible choice is

$$\hat{\mathbf{R}}_{\mathbf{ss}}(z;t_p) = \operatorname{diag}\left(\mathbf{R_{uu}}(z;t_p)\right). \quad (17)$$

With this choice, we pretend that the diagonal entries of $\mathbf{R_{uu}}(z;t_p)$ coincide with those of $\mathbf{R_{ss}}(z;t_p)$ and simply ignore the nonzero off-diagonal elements of $\mathbf{R_{uu}}(z;t_p)$. The estimate (17) is consistent with the assumption of $\mathbf{R_{ss}}(z;t_p)$ having a diagonal structure. Inserting (17) into (15) yields the *blind* cost function

$$\mathcal{J}_1(\mathbf{W}(z)) \triangleq \sum_{p=1}^{P} \mathcal{J}_1'(t_p) \triangleq \sum_{p=1}^{P} \| \operatorname{off}\left(\mathbf{R_{uu}}(z;t_p)\right) \|_{\mathcal{F}}^2 \quad (18)$$

$$= \sum_{p} \| \operatorname{off}\left(\mathbf{W}(z) \mathbf{R_{xx}}(z;t_p) \mathbf{W}^{\dagger}(z)\right) \|_{\mathcal{F}}^2. \quad (19)$$

The cost function (19) attains its global minimum for a polynomial matrix $\mathbf{W}(z)$ which jointly diagonalizes all input correlation matrices $\mathbf{R_{xx}}(z;t_p)$. Because of our assumptions, the global minimum of (18) is, in fact, zero: Inserting (4) into (14) gives $\mathbf{R_{uu}}(z;t_p) = \mathbf{P}\mathbf{D}(z)\mathbf{R_{ss}}(z;t_p)\mathbf{D}^{\dagger}(z)\mathbf{P}^T$ which has a diagonal structure. In order to prevent the trivial solution $\mathbf{W}(z) \equiv \mathbf{0}$, which obviously minimizes (18) as well, we need to impose some additional constraints on $\mathbf{W}(z)$.

The optimization problem defined in (18) subject to some constraints, is referred to as a *joint-diagonalization problem*. In our case, we wish to find a polynomial matrix $\mathbf{W}(z)$ that jointly diagonalizes all products $\mathbf{W}(z)\mathbf{R_{xx}}(z;t_p)\mathbf{W}^{\dagger}(z)$. In fact, the cost function (18) can be seen, as the straightforward polynomial extention of a cost function commonly used in blind source separation for the instantaneous mixing case, see [2,3]. On the other hand, by setting $z = e^{j\omega}$ and evaluating ω at discrete frequency bins ω_i, (18) turns into a the cost function used in [4,5]. There the assumption has been made that the cost function in each frequency bin can be decoupled from the other frequency bins and therefore treated separately. Algorithms which treat each bin separately seem to have decent bin-wise convergence properties. Unfortunately, as it has been reported in the literature, the bin-wise decoupling of the adaptation also leads to a bin-wise permutation ambiguity of the separated source signals, which is commonly known in the context of blind source separation as *the permutation problem*.

2.3 Iterative Algorithm

Derivation of the Gradient. In order to minimize the cost function $\mathcal{J}_1$ we will use a steepest-descent algorithm. To this end, we need to derive the gradient of $\mathcal{J}_1$ with respect to the filter coefficients $\mathbf{W}_r$. We reformulate (18) in a similar way as carried out in [3]:

$$\mathcal{J}_1 = \sum_p \left\| \mathbf{R_{uu}}(z;t_p) - \text{diag}(\mathbf{R_{uu}}(z;t_p)) \right\|_{\mathcal{F}}^2 \tag{20}$$

$$= \sum_p \left\| \mathbf{R_{uu}}(z;t_p) \right\|_{\mathcal{F}}^2 - \sum_p \left\| \text{diag}(\mathbf{R_{uu}}(z)) \right\|_{\mathcal{F}}^2 \triangleq \mathcal{J}_1^{(a)} - \mathcal{J}_1^{(b)}. \tag{21}$$

Hereby we exploited $\langle \mathbf{R_{uu}}(z;t_p), \text{diag}(\mathbf{R_{uu}}(z;t_p)) \rangle_{\mathcal{F}} = \| \text{diag}(\mathbf{R_{uu}}(z;t_p)) \|_{\mathcal{F}}^2$. We derive the gradient for the two terms in (21) separately. By using the definition (10), we obtain after a few steps the two gradients

$$\nabla_{\mathbf{W}_r} \mathcal{J}_1^{(a)} = 4 \sum_p \sum_\tau \mathbf{R_{uu}}(\tau;t_p) \mathbf{R_{ux}}(r-\tau;t_p) \tag{22}$$

$$\nabla_{\mathbf{W}_r} \mathcal{J}_1^{(b)} = 4 \sum_p \sum_\tau \text{diag}(\mathbf{R_{uu}}(\tau;t_p)) \mathbf{R_{ux}}(r-\tau;t_p). \tag{23}$$

By combining (22) and (23), we obtain the overall gradient

$$\nabla_{\mathbf{W}_r} \mathcal{J}_1 = \nabla_{\mathbf{W}_r} \mathcal{J}_1^{(a)} - \nabla_{\mathbf{W}_r} \mathcal{J}_1^{(b)}$$

$$= 4 \sum_p \sum_\tau \text{off}(\mathbf{R_{uu}}(\tau;t_p)) \mathbf{R_{ux}}(r-\tau;t_p). \tag{24}$$

Update Equation. In the following, we consider only a finite interval, $\tau \in [-\tau_e, \tau_e]$, of $\mathbf{R_{uu}}(\tau;t_p)$ in the cost function (18). The slightly modified gradient (24) is then used to obtain the following time-domain update equation

$$\mathbf{W}_r[k+1] = \mathbf{W}_r[k] - 4\mu \sum_{p=1}^P \sum_{\tau=-\tau_e}^{\tau_e} \text{off}(\mathbf{R_{uu}}(\tau;t_p)[k]) \mathbf{R_{ux}}(r-\tau;t_p)[k] \tag{25}$$

where $[k]$ denotes the kth iteration and

$$\mathbf{R_{ux}}(\tau;t_p)[k] = \sum_m \mathbf{W}_m[k] \mathbf{R_{xx}}(\tau-m;t_p) \tag{26}$$

$$\mathbf{R_{uu}}(\tau;t_p)[k] = \sum_m \sum_n \mathbf{W}_m[k] \mathbf{R_{xx}}(\tau-m+n;t_p) \mathbf{W}_n^H[k]. \tag{27}$$

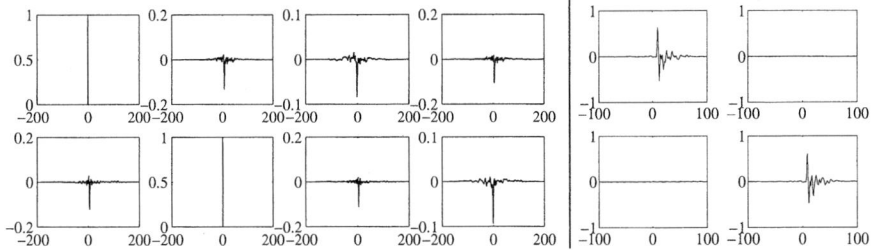

Fig. 1. Impulse responses of the 2×4 demixing filter $\mathbf{W}(z)$ (left) and the 2×2 global system $\mathbf{G}(z) = \mathbf{W}(z)\mathbf{A}(z)$ (right) after convergence.

Constraints. In order to prevent the algorithm of converging to the trivial solution $\mathbf{W}(z) \equiv \mathbf{0}$, additional constraints need to be imposed on $\{\mathbf{W}_r\}$ during the adaptation. The most common ones are to constrain $\|\mathbf{W}(z)\|_{\mathcal{F}} \equiv 1$ or $\mathrm{diag}(\mathbf{W}(z)) \equiv \mathbf{I}$ (sometimes referred to as the *minimum distortion principle* [6]).

3 Efficient Implementation in the Frequency Domain

In Fig. 2 we present an efficient implementation of the proposed joint-diagonalization algorithm for the convolutive mixing case. The algorithm works also for complex source signals and complex filter coefficients. Since (25), (26), and (27) are multichannel convolutional sums, we can compute them efficiently in the frequency domain by applying fast convolution techniques. Note that this procedure does not change the underlying cost function or the time-domain update equation if applied properly. The derivation and notation of the vectors are based on the same concepts as described in [7, Chapter 3 & Appendix F]. Even though the proposed algorithm is a major contribution of this paper, a detailed derivation is not possible at this point, due to lack of space.

4 Simulation Example

To verify the performance of the proposed algorithm, we setup an artificial mixing system with $M_s = 2$ source signals and $M_x = 4$ sensors. The mixing system $\mathbf{A}(z)$ is extracted from real measured HRTFs (head-related transfer functions) and have length $N_a = 100$. The correlation matrices $\mathbf{R}_{\mathbf{ss}}(z; t_p)$ of $P = 3$ snapshots are generated artificially to be diagonal matrices with diagonal elements $(\mathbf{R}_{\mathbf{ss}}(z; t_p))_{m,m} = b(z; t_p)\, b^\dagger(z; t_p)$ where $b(z; t_p)$ are randomly chosen filters of length 20. This setup simulates the case where the source signals are non-stationary *and* non-white. The input correlation matrices are computed as $\mathbf{R}_{\mathbf{xx}}(z; t_p) = \mathbf{A}(z)\mathbf{R}_{\mathbf{ss}}(z; t_p)\mathbf{A}^\dagger(z)$. This artificial generation of $\{\mathbf{R}_{\mathbf{ss}}(z; t_p)\}$ guarantees that the global minimum of $\mathcal{J}_1$ is, in fact, zero. The demixing system $\mathbf{W}(z)$ is a 2×4 matrix where each filter has length 199, ($N_w = 99$). The impulse responses of $\mathbf{W}(z)$ and $\mathbf{G}(z)$ after convergence are shown in Fig. 1. From

FCONVBSS-JD

Definitions:

$$\tilde{\mathbf{P}}_N \triangleq \begin{bmatrix} \mathbf{I}_{N+1} & \mathbf{0} & \mathbf{0} \\ \mathbf{0} & \mathbf{0}_{C-2N-1} & \mathbf{0} \\ \mathbf{0} & \mathbf{0} & \mathbf{I}_N \end{bmatrix}$$

$$\tilde{\mathbf{w}}_{mn}[k] \triangleq (w_{mn,0}[k], \ldots, w_{mn,N_w}[k], 0, \ldots, 0, w_{mn,-N_w}[k], \ldots, w_{mn,-1}[k])^T$$

Initialization ($\forall m, n, p$):

$$\tilde{\mathbf{w}}_{mn}[0] := \begin{cases} (1, 0, \ldots, 0)^T & \text{for } m = n \\ (0, 0, \ldots, 0)^T & \text{for } m \neq n \end{cases}$$

$$\bar{\mathbf{w}}_{mn}[0] := \text{FFT}(\tilde{\mathbf{w}}_{mn}[0])$$

$$r^{(p)}_{x_m x_n}(\tau) := E\{x_m(t_p)\, x_n^*(t_p - \tau)\} \quad \text{for } \tau \in \{-\tau_{\text{xx}}, \ldots, \tau_{\text{xx}}\}$$

$$\tilde{\mathbf{r}}^{(p)}_{x_m x_n} := (r^{(p)}_{x_m x_n}(0), \ldots, r^{(p)}_{x_m x_n}(\tau_{\text{xx}}), 0, \ldots, 0, r^{(p)}_{x_m x_n}(-\tau_{\text{xx}}), \ldots, r^{(p)}_{x_m x_n}(-1))^T$$

$$\bar{\mathbf{r}}^{(p)}_{x_m x_n} := \text{FFT}(\tilde{\mathbf{r}}^{(p)}_{x_m x_n})$$

For each loop k do ($\forall m, n, p$):

$$\bar{\mathbf{r}}^{(p)}_{u_m x_n}[k] := \sum_{l=1}^{M_x} \bar{\mathbf{w}}_{ml}[k] \odot \bar{\mathbf{r}}^{(p)}_{x_l x_n}$$

$$\bar{\mathbf{r}}^{(p)}_{u_m u_n}[k] := \sum_{l=1}^{M_x} \bar{\mathbf{w}}^*_{nl}[k] \odot \bar{\mathbf{r}}^{(p)}_{u_m x_l}[k]$$

$$\tilde{\mathbf{r}}^{(p)}_{u_m u_n}[k] := \text{IFFT}(\bar{\mathbf{r}}^{(p)}_{u_m u_n}[k])$$

$$\tilde{\mathbf{e}}^{(p)}_{mn}[k] := \begin{cases} \mathbf{0} & \text{for } m = n \\ \tilde{\mathbf{P}}_{\tau_e} \tilde{\mathbf{r}}^{(p)}_{u_m u_n}[k] & \text{for } m \neq n \end{cases}$$

$$\bar{\mathbf{e}}^{(p)}_{mn}[k] := \text{FFT}(\tilde{\mathbf{e}}^{(p)}_{mn}[k])$$

$$\Delta \bar{\mathbf{w}}_{mn}[k] := \sum_{p=1}^{P} \sum_{l=1}^{M_u} \bar{\mathbf{e}}^{(p)}_{ml}[k] \odot \bar{\mathbf{r}}^{(p)}_{u_l x_n}[k]$$

$$\bar{\mathbf{w}}'_{mn}[k+1] := \begin{cases} \bar{\mathbf{w}}_{mn}[k] & \text{for } m = n \\ \bar{\mathbf{w}}_{mn}[k] - \mu \cdot \Delta \bar{\mathbf{w}}_{mn}[k] & \text{for } m \neq n \end{cases}$$

$$\bar{\mathbf{w}}_{mn}[k+1] := \text{FFT}\left(\tilde{\mathbf{P}}_{N_w} \text{IFFT}\left(\bar{\mathbf{w}}'_{mn}[k+1]\right)\right)$$

Fig. 2. FCONVBSS-JD : Frequency-domain implementation of CONVBSS-JD All vectors have length C, which is also the FFT size. Since the linear convolutions are embedded in cyclic convolutions, the projection matrices $\tilde{\mathbf{P}}$ are necessary to extract only the linear-convolution part. In order to prevent circular wrap-around effects affecting the updates, the FFT size needs to be chosen large enough. The notation and concept behind the arrangement of the vector elements are taken from [7, Chapter 3 & Appendix F].

the vanishing off-diagonal impulse responses of $\mathbf{G}(z)$ it is clearly seen, that the proposed algorithm can perform almost perfect signal separation. Fig. 1 also indicates that the proposed algorithm does not suffer any permutation problem.

5 Conclusions

Many BSS algorithms for the convolutive mixing case are straightforward extensions of an instantaneous-mixing-case algorithm in the sense that the chosen cost function and corresponding update rule are applied independently in every frequency bin. This approach usually causes a so-called permutation problem. Our approach differs insofar that we define a single global cost function which penalizes all cross-correlations over all time-lags. Even though the update equation for the demixing system is derived in the time domain, most of the computation is carried out in the frequency domain. We would like to point out, that our main motivation to go into frequency domain was because of computational efficiency, similar to [8], and not to decouple the update equations, as done in [4,5]. Consequently, the proposed algorithm does not suffer from a so-called permutation problem, likewise to related pure time-domain algorithms described in [6,9,10].

References

1. Kreyszig, E.: Introductory Functional Analysis with Applications. John Wiley & Sons (1978)
2. Belouchrani, A., Abed-Meraim, K., Cardoso, J.F., Moulines, E.: A blind source separation technique using second-order statistics. IEEE Trans. Signal Processing **45** (1997) 434–444
3. Joho, M., Rahbar, K.: Joint diagonalization of correlation matrices by using Newton methods with application to blind signal separation. In: Proc. SAM, Rosslyn, VA (2002) 403–407
4. Parra, L., Spence, C.: Convolutive blind separation of non-stationary sources. IEEE Trans. Speech and Audio Processing **8** (2000) 320–327
5. Rahbar, K., Reilly, J.P.: Blind source separation algorithm for MIMO convolutive mixtures. In: Proc. ICA, San Diego, CA (2001) 224–229
6. Matsuoka, K., Nakashima, S.: Minimal distortion principle for blind source separation. In: Proc. ICA, San Diego, CA (2001) 927–932
7. Joho, M.: A Systematic Approach to Adaptive Algorithms for Multichannel System Identification, Inverse Modeling, and Blind Identification. PhD thesis, ETH Zürich (2000)
8. Joho, M., Schniter, P.: Frequency-domain realization of a multichannel blind deconvolution algorithm based on the natural gradient. In: Proc. ICA, Nara, Japan (2003) 543–548
9. Schobben, D.W.E.: Efficient Adaptive Multi-channel Concepts in Acoustics: Blind Signal Separation and Echo Cancellation. PhD thesis, Technical University Eindhoven (1999)
10. Krongold, B.S., Jones, D.L.: Blind source separation of nonstationary convolutively mixed signals. In: Proc. SSAP, Pocono Manor, PA (2000) 53–57

Characterization of the Sources in Convolutive Mixtures: A Cumulant-Based Approach

Susana Hornillo-Mellado[1], Carlos G. Puntonet[2], Rubén Martín-Clemente[1], Manuel Rodríguez-Álvarez[2], and Juan Manuel Górriz[3]

[1] Área de Teoría de la Señal y Comunicaciones, Universidad de Sevilla, Spain
{susanah,ruben}@us.es
[2] Departamento de Arquitectura y Tecnología de Computadores
Universidad de Granada, Spain
{carlos,mrodriguez}@atc.ugr.es
[3] Dpto. Ing. Sist. y Auto., Tecn. Electrónica y Electrónica, Univ. de Cádiz, Spain
juanmanuel.gorriz@uca.es

Abstract. This paper addresses the characterization of independent and non-Gaussian sources in a linear mixture. We present an eigensystem based approach to determine the number of independent components in the signal received by a single sensor. The temporal structure of the sources is also characterized using fourth-order statistics.

1 Introduction

In many situations, we observe the superposition of an unknown number of signals and noise when studying a physical phenomenon of interest. In mathematical form, the observed signal $x(n)$ can be written as

$$x(n) = \sum_{i=1}^{N} s_i(n) + r(n) \tag{1}$$

where $s_i(n)$ denotes the signal emitted by the i-th source and $r(n)$ stands for additive noise. Single-Sensor Source Separation is the problem of estimating the source signals $s_i(n)$ from $x(n)$; it is a challenging, still unsolved, problem (excepting the cases in which the source signals have non-overlapping spectra).

The aim of this research was to characterize the source signals $s_i(n)$ on the basis of the properties of higher-order statistics. The use of higher-order cumulants offers two main advantages: first of all, they are not affected by additive Gaussian noise. Secondly, cumulants are linear in the addition of independent variables. The latter property is very useful when considering mixtures like (1). Other results could be used to complement BSS of convolutive [3, 5, 8] or single-channel mixtures [1, 2, 4, 9].

This paper is organized as follows. In Section 2, we state some relevant hypothesis and fix notation. Section 3 presents a new cumulant matrix which collects useful information on the temporal structure of the source signals $s_i(t)$. Section 4 discusses some applications of the main theoretical results. Section 5 presents numerical experiments. Finally, Section 6 is devoted to the conclusions.

2 Model Assumptions and Notation

To begin with, we suppose the following hypotheses:

H1. Each source can be modeled as a moving-average (MA) process of order L, i.e.,

$$s_i(n) = \sum_{k=0}^{L} h_i(k)\, w_i(n-k) \qquad (2)$$

for $i = 1, \ldots, N$, where the excitation sequences $\{w_i(n)\}_{i=1}^{N}$ are non-Gaussian, zero-mean, i.i.d. processes with variance σ_i^2 and kurtosis κ_i.

H2. The source signals $\{s_i(n)\}_{i=1}^{N}$ are statistically independent among themselves.

H3. The additive noise $r(n)$ is stationary, normally distributed and independent from the sources.

H4. We assume that $N \leq L + 1$.

Hypotheses **H1**–**H3** can be usually assumed in practice. We will need hypothesis **H4** later on.

For purposes of notation, given any process $\{z(n)\}$ we define its covariance as

$$c_2^z(l) \stackrel{def}{=} cum(z(n), z(n+l)) \qquad (3)$$

and the fourth-order cumulant [10] of $\{z(n)\}$ as

$$c_4^z(l_1, l_2, l_3) \stackrel{def}{=} cum(z(n), z(n+l_1), z(n+l_2), z(n+l_3)) \qquad (4)$$

Note that, thanks to **H1**–**H3**, the cumulants (3) and (4) of $\{x(n)\}$ and the source signals $\{s_i(n)\}$ are well-defined.

3 Cumulant Matrix

Let $\mathbf{M}$ be the $(L+1) \times (L+1)$ symmetric cumulant matrix whose (i,j)-entry is given by

$$< \mathbf{M} >_{ij} = m_x(i-1, j-1) \qquad (5)$$

where we have defined

$$\boxed{m_x(p,q) = \sum_{k=-L}^{L-p} c_4^x(k, k+p, q)} \qquad (6)$$

This matrix has a very particular structure: it is shown in **Appendix A** that

$$\mathbf{M} = \sum_{i=1}^{N} \frac{\kappa_i}{\sigma_i^4} \mathbf{c}_2^{s_i}\, \mathbf{c}_2^{s_i\, T} = \mathbf{C}\,\mathbf{D}\,\mathbf{C}^T \qquad (7)$$

where $\mathbf{c}_2^{s_i}$ is the $(L+1) \times 1$ vector whose k-th entry is $c_2^{s_i}(k-1)$, $\mathbf{C}$ is the matrix whose columns are the covariance vectors $\mathbf{c}_2^{s_i}$, i.e.,

$$\mathbf{C} = (\mathbf{c}_2^{s_1} | \ldots | \mathbf{c}_2^{s_N})$$

and $\mathbf{D}$ is the $N \times N$ diagonal matrix whose entries are the fourth-order normalized cumulants κ_i/σ_i^4. It is supposed from now on that:

H5. The covariance vectors $\mathbf{c}_2^{s_1}, \ldots, \mathbf{c}_2^{s_N}$ are linearly independent (i.e., matrix $\mathbf{C}$ is full column rank).

Hypothesis **H5** is reasonable when the sources have different physical origins (In particular, sources with the same power spectra[1] are excluded). It follows that:

Property 1. The rank of matrix $\mathbf{M}$ is N [6].

This is interesting in the sense that *property 1* can be used to estimate the number N of sources.

The following property characterizes the covariances:

Property 2. Vectors $\mathbf{c}_2^{s_i}$ are a linear combination of the eigenvectors of $\mathbf{M}$ associated with nonzero eigenvalues.

Proof is given in **Appendix B**. Let $\mathbf{v}_1, \ldots, \mathbf{v}_N$ be those eigenvectors of $\mathbf{M}$ associated with nonzero eigenvalues. We cannot infer from (7) that $\mathbf{v}_1, \ldots, \mathbf{v}_N$ equal the covariances $\mathbf{c}_2^{s_1}, \ldots, \mathbf{c}_2^{s_N}$: as a matter of fact, the eigenvectors $\mathbf{v}_1, \ldots, \mathbf{v}_N$ are orthogonal (since $\mathbf{M}$ is symmetric) whereas $\mathbf{c}_2^{s_1}, \ldots, \mathbf{c}_2^{s_N}$ are usually not orthogonal. Mathematically, *property 2* only implies that there must exists an invertible $N \times N$ matrix $\mathbf{P}$ that relates $\mathbf{v}_1, \ldots, \mathbf{v}_N$ and $\mathbf{c}_2^{s_1}, \ldots, \mathbf{c}_2^{s_N}$ as follows:

$$(\mathbf{v}_1|\ldots|\mathbf{v}_N)\,\mathbf{P} = (\mathbf{c}_2^{s_1}|\ldots|\mathbf{c}_2^{s_N}) \tag{8}$$

Unfortunately, matrix $\mathbf{P}$ is completely unknown *a priori* and cannot be found in practice.

Finally, it can be stated that:

Property 3. Vectors $\mathbf{c}_2^{s_i}$ are orthogonal to all eigenvectors of $\mathbf{M}$ associated with zero eigenvalues.

Property 3 is an immediate consequence of *property 2*.

4 Applications

The generative model (1) appears in convolutive BSS and in the single-sensor BSS problem. In both cases:

A1. *Property 1* can be used to estimate the number N of sources.
A2. The covariance $\mathbf{c}_2^{s_i}$ of each estimated source $s_i(n)$ must satisfy *properties 2 and 3*. Hence, constraints can be derived from both properties that may be used to prevent BSS algorithms from converging to spurious solutions. For example, *property 3* implies that:

$$\mathbf{c}_2^{s_i\,T}\,\mathbf{U} = \mathbf{0}$$

---
[1] Power spectra is the Fourier transform of the covariance.

for $i = 1, \ldots, N$, where

$$\mathbf{U} = (\mathbf{u}_1 | \ldots | \mathbf{u}_{L+1-N})$$

and $\mathbf{u}_1, \ldots, \mathbf{u}_{L+1-N}$ are the eigenvectors of $\mathbf{M}$ associated with zero eigenvalues.

A3. If $N = 1$, it follows from (7) that the eigenvector associated with the unique nonzero eigenvalue of $\mathbf{M}$ is equal to the covariance vector $\mathbf{c}_2^{s_1}$ of the unique source $s_1(n)$ (up to a multiplicative constant). Thus, after estimating $\mathbf{c}_2^{s_1}$, we can filter the data $x(n)$ with a Wiener filter to enhance the source signal $s_1(n)$ from the noise $r(n)$ [7].

Note that, due to the presence of the noise $r(n)$, A1–A3 would not be feasible if only second-order statistics were used. By contrasts, the fourth-order cumulant function (6) is not affected by additive Gaussian noise.

5 Example

Let us consider a mixture of four signals plus noise

$$x(n) = \sum_{i=1}^{4} s_i(n) + r(n),$$

where $s_i(n) = h_i(n) * w_i(n)$ ('*' denotes 'convolution'), $\{w_i(n)\}_{i=1}^{4}$ are not-gaussian leptokurtic (i.e. with positive kurtosis) signals, each of which is obtained by raising to the third power a different i.i.d. Gaussian process and the noise $\{r(n)\}$ is i.i.d. and Gaussian. All of them are normalized to have unity variance. The impulse responses of the MA filters are randomly chosen as:

$h_1(n) = \delta(n) + 0.34\, \delta(n-1) + 0.12\, \delta(n-2) - 0.41\, \delta(n-3) + 0.65\, \delta(n-4) + 0.75\, \delta(n-5),$

$h_2(n) = \delta(n) - 0.35\, \delta(n-1) - 0.29\, \delta(n-2) - 0.28\, \delta(n-3) - 0.28\, \delta(n-4) + 0.66\, \delta(n-5),$

$h_3(n) = \delta(n) + 0.65\, \delta(n-1) + 0.52\, \delta(n-2) - 0.41\, \delta(n-3) + 0.55\, \delta(n-4) + 0.14\, \delta(n-5),$

and

$h_4(n) = \delta(n) - 0.50\, \delta(n-1) - 0.31\, \delta(n-2) - 0.81\, \delta(n-3) - 0.60\, \delta(n-4) + 0.24\, \delta(n-5),$

Figure 1 shows the power spectral density of each source, calculated using Welch's method with a Hanning window.

We used 7000 samples of $x(n)$ to estimate matrix $\mathbf{M}$ with $L = 6$. After each experiment, to normalize and facilitate the comparison, eigenvalues were divided by their maximum value.

The second column of table 1 shows the mean of the normalized eigenvalues, averaged over 1000 independent experiments (the quantity between brackets is the standard deviation). The third column of table 1 shows the 'true' normalized eigenvalues, i.e., the eigenvalues that would be obtained if there were no errors in the estimation of the cumulant matrix $\mathbf{M}$. It is observed in Table 1 that there

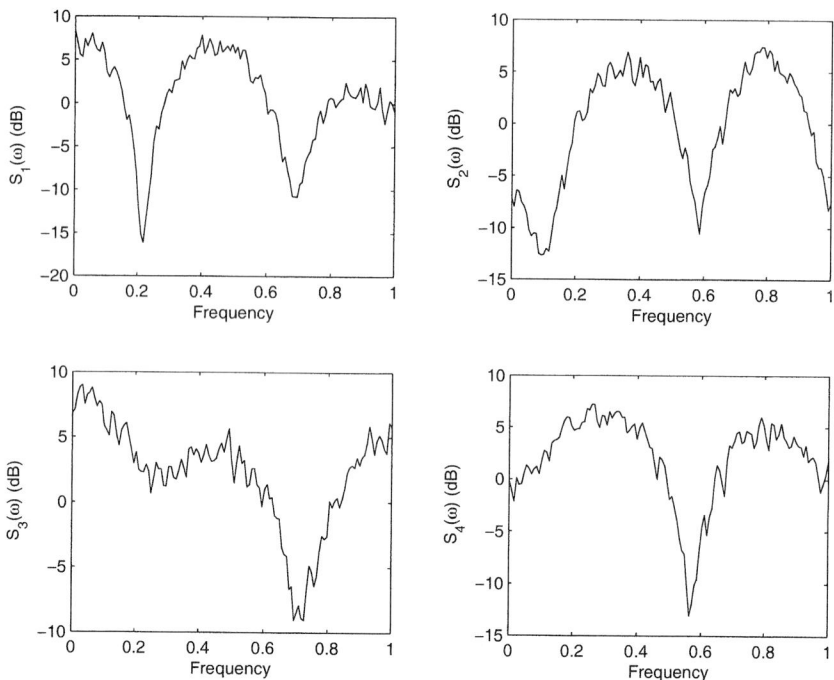

Fig. 1. Power Spectral Density of the four sources $s_1(n), s_2(n), s_3(n), s_4(n)$.

Table 1. Statistics of the eigenvalues of **M**.

Eigenvalue Number	Mean (Standard Deviation)	True Normalized Eigenvalue
1	1.0000 (0.0000)	1.0000
2	0.1282 (0.0303)	0.1234
3	0.0501 (0.0173)	0.0448
4	0.0186 (0.0113)	0.0155
5	0.0020 (0.0074)	0.0000
6	-0.0107 (0.0074)	0.0000
7	-0.0265 (0.0113)	0.0000

are three clearly-nonzero eigenvalues, indicating that the mixture is composed of at least $N = 3$ non-gaussian sources. We may need an additional criterion to decide whether a fourth-source is present or not (i.e., to decide whether the fourth eigenvalue is zero or not).

Let p_{ij} be the scalar product between the covariance $\mathbf{c}_2^{s_i}$ and the eigenvector of **M** that corresponds to the j-th normalized eigenvalue. Table 2 shows the mean value and the standard deviation of p_{ij} for $1 \leq i \leq 4$ and $1 \leq j \leq 7$.

In view of Table 2, p_{ij} seeems to be zero-mean for all i, j. However, the key is the standard deviation: observe that the standard deviations are large for $j = 1, 2, 3$, whereas they are small for $j = 5, 6, 7$. It is inferred that the covariances belong to the subspace spanned by the first, second and third eigenvectors,

Table 2. Statistics of the scalar product between $\mathbf{c}_2^{s_i}$ and the eigenvectors of $\mathbf{M}$.

p_{ij}	$i=1$	$i=2$	$i=3$	$i=4$
$j=1$	-0.1288 (2.45)	-0.1110 (1.7416)	-0.1191 (2.2495)	-0.1379 (2.3996)
$j=2$	-0.0124 (0.78)	0.0154 (0.8566)	-0.0441 (0.6651)	-0.0322 (0.7814)
$j=3$	0.0524 (0.42)	0.0233 (0.5616)	-0.0085 (0.3563)	-0.0576 (0.4778)
$j=4$	-0.0040(0.28)	0.0136 (0.3223)	0.0047 (0.2816)	-0.0370 (0.3133)
$j=5$	-0.0149 (0.21)	-0.0026 (0.2043)	0.0052 (0.2080)	0.0050 (0.2132)
$j=6$	-0.0052 (0.20)	-0.0122 (0.1742)	0.0006 (0.1988)	0.0036 (0.1928)
$j=7$	0.0051 (0.21)	-0.0028 (0.1674)	-0.0010 (0.2523)	0.0053 (0.2036)

whereas they are orthogonal to the subspace spanned by the fifth, sixth and seventh eigenvectors. Both conclusions agree with *properties 2* and *3*, respectively. Again, we may need an additional criterion to decide whether the covariances are orthogonal to the fourth eigenvector or not.

6 Conclusions

We have proposed a method that can be used to estimate the number of sources in a linear mixture and characterize their temporal structure. The method is robust against gaussian noise, since it is based on higher-order statistics.

Appendix A

Since $x(n) = \sum_{i=1}^{N} s_i(n) + r(n)$, it holds that $c_4^x(l_1, l_2, l_3) = \sum_{i=1}^{N} c_4^{s_i}(l_1, l_2, l_3)$. As a consequence

$$m_x(p,q) = \sum_{k=-L}^{L-p} c_4^x(k, k+p, q) = \sum_{i=1}^{N} \sum_{k=-L}^{L-p} c_4^{s_i}(k, k+p, q) \qquad (9)$$

Thanks to the multi-linearity property of the cumulants, the covariance of the source $s_i(n)$ can be written as

$$c_{s_i}^2(p) = \sum_{n=0}^{L} h_i(n) h_i(n+p) \sigma_i^2 \qquad (10)$$

and the fourth-order cumulant equals

$$c_4^{s_i}(p, q, r) = \sum_{n=0}^{L} h_i(n) h_i(n+p) h_i(n+q) h_i(n+r) \kappa_i \qquad (11)$$

Both cumulants can be easily related, as follows: from (10), we get

$$c_2^{s_i}(p) \, c_2^{s_i}(q) = \sum_{n=0}^{L} \sum_{k=0}^{L} h_i(n) h_i(n+p) \sigma_i^2 \, h_i(k) h_i(k+q) \sigma_i^2 \qquad (12)$$

Using in (12) the change of variables $k' = k - n$ and $n' = n$, we obtain:

$$c_2^{s_i}(p)\,c_2^{s_i}(q) = \sum_{n'=0}^{L} \sum_{k'=-n'}^{L-n'} h_i(n')h_i(n'+k')\,h_i(n'+k'+p)\,h_i(n'+q)\sigma_i^4 \quad (13)$$

Now, it is supposed that $p, q > 0$. Using that the MA coefficients $h_i(n) = 0$ if $n \notin \{0, \ldots, L\}$, it is readily obtained from (13) that

$$c_2^{s_i}(p)\,c_2^{s_i}(q) = \sum_{n'=0}^{L} \sum_{k'=-L}^{L-p} h_i(n')h_i(n'+k')\,h_i(n'+k'+p)\,h_i(n'+q)\sigma_i^4 \quad (14)$$

Then, comparing (11) with (14), it is deduced the following relation between cumulants:

$$c_2^{s_i}(p)c_2^{s_i}(q) = \frac{\sigma_i^4}{\kappa_i} m_{s_i}(p, q), \quad (15)$$

where we have defined

$$m_{s_i}(p, q) = \sum_{k=-L}^{L-p} c_4^{s_i}(k, k+p, q) \quad (16)$$

Then, inserting (16) in (9) and taking into account (15), we finally have:

$$m_x(p, q) = \sum_{i=1}^{N} \frac{\kappa_i}{\sigma_i^4} c_2^{s_i}(p)c_2^{s_i}(q) \quad (17)$$

which completes the proof □.

Appendix B

Let $\lambda_1 \geq \lambda_2 \geq \ldots \geq \lambda_{L+1}$ be the eigenvalues of $\mathbf{M}$. Furthermore, let $\mathbf{v}_1, \mathbf{v}_2, \ldots, \mathbf{v}_N$ be the unit-norm eigenvectors associated with $\lambda_1, \ldots, \lambda_N$ and $\mathbf{u}_1, \mathbf{u}_2, \ldots, \mathbf{u}_{L+1-N}$ be the eigenvectors corresponding to $\lambda_{N+1}, \ldots, \lambda_{L+1}$. Observe that $\lambda_{N+1}, \ldots, \lambda_{L+1}$ are all equal to zero since $rank(\mathbf{M}) = N$.

Using the definition of the eigenvalues yields:

$$\mathbf{M}\,\mathbf{v}_i = \lambda_i\,\mathbf{v}_i \quad (18)$$

Substituting (7) into (18) gives:

$$\sum_{i=1}^{N} \frac{\kappa_i}{\sigma_i^4} \mathbf{c}_2^{s_i} \left(\mathbf{c}_2^{s_i\,T} \mathbf{v}_i \right) = \lambda_i\,\mathbf{v}_i \quad (19)$$

or, equivalently,

$$\mathbf{v}_i = \sum_{i=1}^{N} \frac{\kappa_i}{\sigma_i^4 \lambda_i} \left(\mathbf{c}_2^{s_i\,T} \mathbf{v}_i \right) \mathbf{c}_2^{s_i} \quad (20)$$

which means that each eigenvector $\mathbf{v}_i$ can be expressed as a linear combination of the covariances $\mathbf{c}_2^{s_i}$.

Note that vectors $\mathbf{c}_2^{s_i}$ ($1 \leq i \leq N$) form a basis since they are linearly independent and, hence, span an N dimensional subspace. From this point of view, the preceding identity (20) just means that all eigenvectors $\mathbf{v}_i$ belong to this subspace. But these eigenvectors are orthogonal and therefore also form a basis for the subspace. Consequently, each covariance vector $\mathbf{c}_2^{s_i}$ can be represented as a linear combination of $\mathbf{v}_1, \ldots, \mathbf{v}_N$ as well. This completes the proof □.

References

1. S. Araki, S. Makino, A. Blin, R. Mukai and H. Sawada, "Blind Separation of More Speech than Sensors with Less Distortion by Combining Sparseness and ICA", *Proc. of IWAENC2003*, Kyoto, Japan, 2003.
2. L. Benaroya and F. Bimbot, "Wiener based source separation with HMM/GMM using a single sensor", in *Proc. of 4th International Conference on Independent Component Analysis and Blind Signal Separation (ICA 2003)*, Nara, Japan, 2003.
3. A. Cichocki and S. I. Amari, "Adaptive Blind Signal and Image Processing", *John Wiley and Sons*, 2002.
4. N. Doukas, T. Stathaki and P. Naylor "A single sensor souce separation approach to noise reduction", *Proc. of the Second World Congress of Nonlinear Analysis*, Athens, Greece, 1996.
5. C. Fevotte and C. Doncarli, "A Unified Presentation of Blind Separation Methods for Convolutive Mixtures using Block-Diagonalization", in *Proc. of 4th International Conference on Independent Component Analysis and Blind Signal Separation (ICA 2003)*, Nara, Japan, 2003.
6. G. Golub and C. van Loan "Matrix Computations", *The John Hopkins University Press*, 1996.
7. S. Haykin, "Adaptive Filter Theory", *Prentice Hall*,1991.
8. A. Hyvärinen, J. Karhunen and E. Oja "Independent Component Analysis", *John Wiley and Sons*, 2001.
9. T.-W. Lee, M.S. Lewicki, M. Girolami and T.J. Sejnowski "Blind Source Separation of More Sources than Mixtures Using Overcomplete Representations", *IEEE Signal Processing Letters*, Vol. 4, No. 4, 1999.
10. C. Nikias and A. Petropulu "Higher-order spectra analysis", *Prentice-Hall*, 1993.

CICAAR: Convolutive ICA with an Auto-regressive Inverse Model

Mads Dyrholm and Lars Kai Hansen

Informatics and Mathematical Modelling
Technical University of Denmark
2800 Kgs. Lyngby, Denmark

Abstract. We invoke an auto-regressive IIR inverse model for convolutive ICA and derive expressions for the likelihood and its gradient. We argue that optimization will give a stable inverse. When there are more sensors than sources the mixing model parameters are estimated in a second step by least squares estimation. We demonstrate the method on synthetic data and finally separate speech and music in a real room recording.

1 Introduction

Independent component analysis (ICA) of convolutive mixtures is a key problem in signal processing, the problem is important in speech processing and numerous other applications including medical, visual, and industrial signal processing, see, e.g., [1–5]. Convolutive ICA in its basic form concerns reconstruction of the $L+1$ mixing matrices A_τ and the N source signal vectors s_t of dimension K, from a D-dimensional convolutive mixture,

$$x_t = \sum_\tau A_\tau s_{t-\tau}. \tag{1}$$

We will assume L so large that all correlations in the process x can be 'explained' by the mixing process, and the source signal vectors are assumed temporally independent: $p(\{s_t\}) = \prod_{t=1}^{N} p(s_t)$. This is motivated by the observation that source signal auto-correlations can not be identified without additional a priori information [1]. This is most apparent in the frequency domain $A_\omega s_\omega$. A non-zero 'filter' $h(\omega)$ can be multiplied on a given source if $1/h(\omega)$ is applied to the corresponding column of the set of Fourier transformed mixing matrices A_ω.

Statistically motivated maximum likelihood schemes have been proposed, see e.g. [1,6–8]. The likelihood approach is attractive for a number of reasons. First, it forces a declaration of the statistical assumptions – in particular the a priori distribution of the source signals, secondly, the maximum likelihood solution is asymptotically optimal given the assumed observation model and the prior choices for the 'hidden' variables.

IIR representations of an inverse model have been proposed in e.g. [9,10]. In this paper we will invoke an auto-regressive IIR inverse model. This involves a linear recursive filter for estimation of the source signal and a non-linear recursive filter for maximum likelihood estimation of the mixing matrices. Our derivation formally allows the number of sensors to be greater than the number of sources.

2 Estimating the Sources Through a Stable Inverse

Let us define x, A, and s such that $x = As$ is a matrix product abbreviation of the convolutive mixture

$$\begin{bmatrix} x_N \\ x_{N-1} \\ \vdots \\ x_1 \end{bmatrix} = \begin{bmatrix} A_0 & A_1 & \cdots & A_L & & \\ & A_0 & A_1 & \cdots & A_L & \\ & & \ddots & & & \\ & & & & & A_0 \end{bmatrix} \begin{bmatrix} s_N \\ s_{N-1} \\ \vdots \\ s_1 \end{bmatrix} \quad (2)$$

which allows the likelihood to be written $p(x|\{A_\tau\}) = \int \delta(x - As) p(s) ds$.

2.1 Square Case Likelihood

In the square case, $D = K$, the likelihood integral evaluates to

$$p(x|\{A_\tau\}) = |\det A|^{-1} p(A^{-1}x). \quad (3)$$

Since A is upper block triangular we obtain $p(x|\{A_\tau\}) = |\det A_0|^{-N} p(A^{-1}x)$, furthermore, assuming i.i.d. source signals we finally get

$$p(\{x_t\}|\{A_\tau\}) = |\det A_0|^{-N} \prod_{t=1}^{N} p((A^{-1}x)_t). \quad (4)$$

The inverse operation $A^{-1}x$ is the multivariate AR(L) process

$$\tilde{s}_t = A_0^{-1} x_t - A_0^{-1} \sum_{\tau=1}^{L} A_\tau \tilde{s}_{t-\tau} \quad (5)$$

which follows simply by eliminating s_t in (1). In terms of (5) we now rewrite the negative log likelihood

$$\mathcal{L}(\{A_\tau\}) = N \log |\det A_0| - \sum_{t=1}^{N} \log p(\tilde{s}_t) \ , \ K = D. \quad (6)$$

2.2 Overdetermined Case Likelihood

When $D > K$ there are many inverse operations $A^{-1} : \mathbb{R}^D \mapsto \mathbb{R}^K$ which satisfy $A^{-1}A = I$. In this work we base the source estimates $\hat{s}_t$ on a particular choice of inverse operation, i.e. we define $\hat{s} = A^{-1}x$ by the multivariate AR(L) process

$$\hat{s}_t = A_0^{\#} x_t - A_0^{\#} \sum_{\tau=1}^{L} A_\tau \hat{s}_{t-\tau}, \quad (7)$$

where $A_0^\#$ denotes Moore-Penrose generalized inverse. The process (7) is inverse in the sense $A^{-1}A = I$ which means that when it is configured with the true mixing matrices it allows perfect reconstruction of the sources. Evoking (7) the likelihood integral can be evaluated to

$$\mathcal{L}(\{A_\tau\}) = \frac{N}{2}\log|\det A_0^T A_0| - \sum_{t=1}^{N} \log p(\hat{s}_t) \ , \ K \leq D. \tag{8}$$

The derivation of (8) is deferred to Sec. A for aesthetic reason, but note that (8) is based on our particular choice of inverse (7). For $K = D$ we note that (7) and (8) are identical to (5) and (6) respectively.

2.3 Optimization Yields a Stable Inverse

In praxis, convolution system matrices such as A are often found to be poorly conditioned and hence the inverse problem $\hat{s} = A^{-1}x$ sensitive to noise, see e.g. [11]. The extreme case for the inverse is it being *unstable* and sensitive to machine precision rounding errors. Fortunately, the maximum likelihood approach has a built-in regularization against this problem. This is seen from the likelihood noting that an ill-conditioned estimator $\{\hat{A}_\tau\}$ will lead to a divergent source estimate $\hat{s}_t$; but such large amplitude signals are exponentially penalized under the source pdf's typically used in ICA ($p(s) = \text{sech}(s)/\pi$). Therefore, our proposition is that it is 'safe' to use an iterative learning scheme for optimizing (8) because once it has been initialized with a well-conditioned convolution matrix A a learning decrease in (8) will lead to further refinements $\{\hat{A}_\tau\}$ which are stable in the context of equation (7). If no exact stable inverse exists the Maximum-Likelihood approach will give us a regularized estimator.

We propose here to use a gradient optimization technique. The gradient of the negative log likelihood w.r.t. $A_0^\#$ is given by

$$\frac{\partial \mathcal{L}(\{A\})}{\partial (A_0^\#)_{ij}} = -N(A_0^T)_{ij} - \sum_{t=1}^{N} \psi^T(\hat{s}_t) \frac{\partial \hat{s}_t}{\partial (A_0^\#)_{ij}} \tag{9}$$

where

$$\frac{\partial (\hat{s}_t)_k}{\partial (A_0^\#)_{ij}} = \delta(i-k)\left(x_t - \sum_{\tau=1}^{L} A_\tau \hat{s}_{t-\tau}\right)_j - \left(A_0^\# \sum_{\tau=1}^{L} A_\tau \frac{\partial \hat{s}_{t-\tau}}{\partial (A_0^\#)_{ij}}\right)_k \tag{10}$$

and $(\psi(\hat{s}_t))_k = p'((\hat{s}_t)_k)/p((s_t)_k)$. The gradient w.r.t. to the other mixing matrices is given by

$$\frac{\partial \mathcal{L}(\{A\})}{\partial (A_\tau)_{ij}} = -\sum_{t=1}^{N} \psi^T(\hat{s}_t) \frac{\partial \hat{s}_t}{\partial (A_\tau)_{ij}} \tag{11}$$

where

$$\frac{\partial (\hat{s}_t)_k}{\partial (A_\tau)_{ij}} = -(A_0^\#)_{ki}(\hat{s}_{t-\tau})_j - \left(A_0^\# \sum_{\tau'=1}^{L} A_{\tau'} \frac{\partial \hat{s}_{t-\tau'}}{\partial (A_\tau)_{ij}}\right)_k \tag{12}$$

These expressions allow for general gradient optimization schemes. A starting point for the algorithm is A_0 being random numbers and $A_\tau = 0$ for $\tau \neq 0$ – a stable initialization according to (7).

2.4 Re-estimating the Mixing Filters

When the dimension of x_t is strictly greater than the number of sources, $D > K$, the mixing matrices which figure as parameters for the learning process can not be taken as mixing filter estimates because $AA^{-1} \neq I \Rightarrow A\hat{s} \neq x$. Instead we here propose to estimate the mixing filters by least-squares. Multiplying (1) with $s_{t-\lambda}^T$ from right and taking the expectation we obtain the following normal equations

$$< x_t s_{t-\lambda}^T > = \sum_\tau A_\tau < s_{t-\tau} s_{t-\lambda}^T > \quad (13)$$

which is solved for A_τ by regular matrix inversion using the estimated sources and $< \cdot > = \frac{1}{N} \sum_{1=1}^{N}$. This system is unlikely to be ill conditioned because the sources are typically uncorrelated mutually and temporally.

2.5 Dimensionality Reduction

For lowering the training complexity we here propose to use a K-dimensional subspace representation of the data $y_t = U_K^T x_t$ where $U_K \in \mathbb{R}^{D \times K}$ is a projection. We can write a regular convolutive mixture where the number of sensors is now equal to K,

$$y_t = \sum_{\tau=0}^{L} B_\tau s_{t-\tau} \ , \ B_\tau = U_K^T A_\tau, \quad (14)$$

and note that the sources are unaltered by the projection. This means that we should be able to recover the sources from the projection using the square case of our algorithm. Once the sources have been estimated the D-by-K mixing matrices $\{A_\tau\}$ are estimated c.f. Sec 2.4.

3 Experiments

3.1 Simulation Data

We now illustrate the algorithm on a three-dimensional convolutive mixture of two sources, i.e. $D = 3$, $K = 2$. The true mixing filters are shown in the left panel of Fig.1 and set to decay within 30 lags, i.e. $L = 30$. The source signals, $N = 30000$, are both drawn from a Laplace distribution. 5000 consecutive samples is zeroed out from one of the sources, say 'Source-1'. Results are then evaluated from the estimated Source-1 by measuring the interference power P_i in the period where the true Source-1 is silent. We here define the Signal to Interference Ratio (SIR) P_s/P_i, where P_s is the signal power which is estimated in a period where both sources are active.

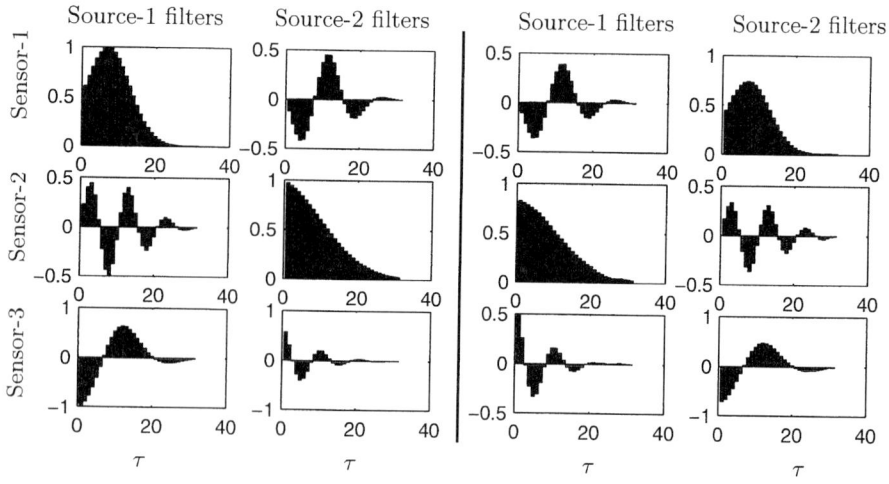

Fig. 1. (left) true mixing filters, (right) estimated mixing filters.

The data is projected onto the two major principal components and the sources $\hat{s}_t$ are estimated c.f. Sec. 2.5. The optimization scheme is Newton steps, i.e. updating $\{\hat{A}_\tau\}$ by $-H^{-1}g$ where g is the gradient vector and H^{-1} is the inverse Hessian which is estimated using the outer product approximation update per sample (see e.g. [12, page 153]). Convergence detected in 124 iterations. Obtained SIR = 19.3dB. The corresponding mixing filters estimated by (13) are then used as a starting guess for the general overdetermined algorithm using the original three-dimensional data as input. Convergence detected in 20 iterations. Obtained SIR = 34.2dB. Then we use (13) to estimate the corresponding mixing filters and the result is displayed in the right pane of Fig. 1.

3.2 Real Audio Recording

We now apply the proposed method to a 16kHz signal which was recorded indoor by two microphones and produced by a male speaker counting one-ten and a loud music source respectively. The microphones and the sources were located in the corners of a square. The signal is kindly provided by Dr. T-W. Lee, and is identical to the one used in [13]. We choose the number of mixing matrices $L = 50$. This time we use a BFGS Quasi-Newton optimization scheme (see e.g. [12, page 288]) convergence is reached in 490 iterations.

As noted, the source signals can only be recovered up to an arbitrary filter and we experience indeed a whitening effect on the sources. In [13] a low-pass filter was applied to overcome the whitening effect, hence, to make the sources 'sound more real'. In our presentation, because we have the forward model parameters, we reconstruct the microphone signals separately as they would sound if the other source was shut. This is simply achieved by propagating the given source signal through the estimated mixing model. Fig. 2 shows the recorded mixture along with the results of separation. For listening test and further analysis we

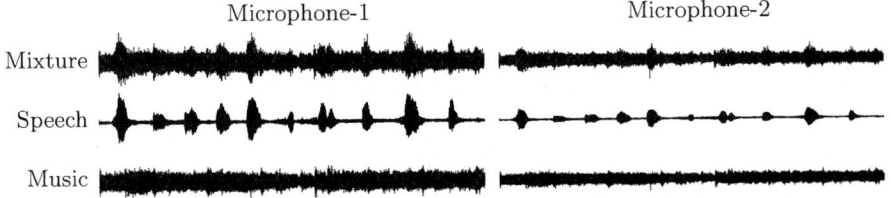

Fig. 2. Separation of real world sound signals. (Top row) The recorded mixture of speech and music. (Middle row) Separated speech reconstructed in the sensor domain. (Bottom row) Separated music reconstructed in the sensor domain.

have placed the resulting audio files at URL http://www.imm.dtu.dk/~mad/cicaar/sound.html. Again we evaluate the result by SIR; the interference power P_i as the mean power in ten manually segmented intervals in which the speaker is silent, and the signal power P_s is similarly estimated as the mean power in ten manually intervals where the speaker is clearly audible (and subtracting off the interference power). The SIR of the proposed algorithm and using the parameters described is SIR = 12.42 dB. The algorithm proposed by Parra and Spence [2] represents a state-of-the-art alternative for evaluation of performance. In the following table we give SIR's for the Parra-Spence algorithm using the implementation kindly provided by Stefan Harmeling[1] based on window lengths (N) and for three different numbers of un-mixing matrices (Q):

SIR (dB)	$Q = 50$	$Q = 100$	$Q = 200$
$N = 512$	11.9	11.8	12.3
$N = 1024$	12.0	12.2	12.5
$N = 2048$	11.9	12.0	12.3

The table indicates that in order to obtain a separation performance similar to that of the proposed algorithm the Parra-Spence inverse filter Q needs to be somewhat larger than the length of the IIR filter $L = 50$ we have used. Future quantitative studies are needed to substantiate this finding invoking a wider variety of signals and interferences.

4 Conclusion

We have proposed a maximum-likelihood approach to convolutive ICA in which an auto-regressive inverse model is put in terms of the forward model parameters. The algorithm leads to a stable (possibly regularized) inverse and formally allows the number of sensors to be greater than the number of sources. Our experiment shows good performance in a real world situation. In general, for *perfect* separation a stable un-regularized inverse must exist. An initial delay, e.g., is not minimum phase and no causal inverse exist. On the other hand, in

---

[1] http://ida.first.gmd.de/~harmeli/download/download_convbss.html

that case, the source can simply be delayed and thus remove the initial delay in the filter – exploiting the filter ambiguity. Such manoeuvre will in some cases make a real room impulse response minimum phase [14].

A Derivation of the Likelihood in the Overdetermined Case

We shall make use of the following definition: $\hat{s}_t(s_{t-1}, s_{t-2}, \ldots, s_{t-L}) \equiv A_0^\# x_t - A_0^\# \sum_{\tau=1}^{L} A_\tau s_{t-\tau}$. We can write the likelihood

$$p(X|\{A_\tau\}) = \int_{s_1} \int_{s_2} \cdots \left(\int_{s_N} p(s_N)\delta(f_N) ds_N \right) \prod_{t=1}^{N-1} p(s_t)\delta(f_t) ds_1 \ldots ds_{N-1}. \quad (15)$$

where $f_t \equiv x_t - \sum_{\tau=0}^{L} A_\tau s_{t-\tau}$. The first step in this derivation is to marginalize out s_N, using

$$\int_{s_N} p(s_N)\delta(f_N) ds_N = |A_0^T A_0|^{-1/2} p(\hat{s}_N^{(1)}) \quad (16)$$

where $\hat{s}_N^{(1)} = \hat{s}_N(s_{N-1}, \ldots, s_{N-L})$. Then we can rewrite the likelihood with one integral evaluated, i.e.

$$p(X|\{A_\tau\}) = |A_0^T A_0|^{-1/2} \int_{s_1} \int_{s_2} \cdots \int_{s_{N-1}} p(\hat{s}_N^{(1)}) \prod_{t=1}^{N-1} p(s_t)\delta(f_t) ds_1 \ldots ds_{N-1}. \quad (17)$$

Following the same idea to marginalize out s_{N-1} now using

$$\int_{s_{N-1}} p(\hat{s}_N^{(1)}) p(s_{N-1}) \delta(f_{N-1}) ds_{N-1} = |A_0^T A_0|^{-1/2} p(\hat{s}_N^{(2)}) p(\hat{s}_{N-1}^{(1)}) \quad (18)$$

where $\begin{cases} \hat{s}_{N-1}^{(1)} = \hat{s}_{N-1}(s_{N-2}, s_{N-3}, \ldots, s_{N-1-L}) \\ \hat{s}_N^{(2)} = \hat{s}_N(\hat{s}_{N-1}^{(1)}, s_{N-2}, \ldots, s_{N-L}) \end{cases}$. Then we can write the likelihood with two integrals evaluated

$$p(X|\{A_\tau\}) = |A_0^T A_0|^{-2/2} \int_{s_1} \int_{s_2} \cdots \int_{s_{N-2}} p(\hat{s}_N^{(2)}) p(\hat{s}_{N-1}^{(1)}) \prod_{t=1}^{N-2} p(s_t)\delta(f_t) ds_1 \ldots ds_{N-2}. \quad (19)$$

By repeating this procedure to evaluate all integrals we eventually get

$$p(X|\{A_\tau\}) = |A_0^T A_0|^{-N/2} \prod_{t=1}^{N} p(\hat{s}_t^{(t)}) \;,\; \begin{cases} \hat{s}_1^{(1)} = \hat{s}_1(s_0, s_{-1}, \ldots, s_{1-L}) \\ \hat{s}_2^{(2)} = \hat{s}_2(\hat{s}_1^{(1)}, s_0, \ldots, s_{2-L}) \\ \hat{s}_3^{(3)} = \hat{s}_3(\hat{s}_2^{(2)}, \hat{s}_1^{(1)}, \ldots, s_{3-L}) \\ \vdots \\ \hat{s}_t^{(t)} = \hat{s}_t(\hat{s}_{t-1}^{(t-1)}, \hat{s}_{t-2}^{(t-2)}, \ldots, \hat{s}_{t-L}^{(t-L)}) \end{cases} \quad (20)$$

Assuming s_t zero for $t \leq 0$ we finally get

$$p(X|\{A_\tau\}) = |A_0^T A_0|^{-N/2} \prod_{t=1}^{N} p(\hat{s}_t) \ , \ \hat{s}_t = \hat{s}_t(\hat{s}_{t-1}, \hat{s}_{t-2}, \ldots, \hat{s}_{t-L}). \quad (21)$$

References

1. Hagai Attias and C. E. Schreiner, "Blind source separation and deconvolution: the dynamic component analysis algorithm," *Neural Computation*, vol. 10, no. 6, pp. 1373–1424, 1998.
2. L. Parra, C. Spence, and B. De Vries, "Convolutive blind source separation based on multiple decorrelation," in *IEEE Workshop on Neural Networks and Signal Processing, Cambridge, UK, September 1998*, 1998, pp. 23–32.
3. Kamran Rahbar, James P. Reilly, and Jonathan H. Manton, "A frequency domain approach to blind identification of mimo fir systems driven by quasi-stationary signals," in *2002 IEEE International Conference on Acoustics, Speech, and Signal Processing*, 2002, pp. 1717–1720.
4. Jörn Anemüller and Birger Kollmeier, "Adaptive separation of acoustic sources for anechoic conditions: A constrained frequency domain approach," *IEEE transactions on Speech and Audio processing*, vol. 39, no. 1-2, pp. 79–95, 2003.
5. Mitianoudis N. and Davies M., "Audio source separation of convolutive mixtures," *IEEE transactions on Speech and Audio processing*, vol. 11:5, pp. 489–497, 2003.
6. Eric Moulines, Jean-Francois Cardoso, and Elizabeth Gassiat, "Maximum likelihood for blind separation and deconvolution of noisy signals using mixture models," in *Proc. ICASSP'97 Munich*, 1997, pp. 3617–3620.
7. Sabine Deligne and Ramesh Gopinath, "An em algorithm for convolutive independent component analysis," *Neurocomputing*, vol. 49, pp. 187–211, 2002.
8. Seungjin Choi, Sun ichi Amari, Andrezej Cichocki, and Ruey wen Liu, "Natural gradient learning with a nonholonomic constraint for blind deconvolution of multiple channels," in *International Workshop on Independent Component Analysis and Blind Signal Separation (ICA'99), Aussois, France, January 11–15 1999*, pp. 371–376.
9. K. Torkkola, "Blind separation of convolved sources based on information maximization," in *IEEE Workshop on Neural Networks for Signal Processing, Kyoto, Japan*, September 4-6 1996, pp. 423–432.
10. S. Choi and A. Cichocki, "Blind signal deconvolution by spatio-temporal decorrelation and demixing," in *Neural Networks for Signal Processing, Proc. of the 1997 IEEE Workshop (NNSP-97), IEEE Press, N.Y. 1997*, 1997, pp. 426–435.
11. Per Christian Hansen, "Deconvolution and regularization with toeplitz matrices," *Numerical Algorithms*, vol. 29, pp. 323–378, 2002.
12. Christopher M. Bishop, *Neural Networks for Pattern Recognition*, Oxford University Press, Inc., 1995.
13. Te-Won Lee, Anthony J. Bell, and Russell H. Lambert, "Blind separation of delayed and convolved sources," in *Advances in Neural Information Processing Systems*, Michael C. Mozer, Michael I. Jordan, and Thomas Petsche, Eds. 1997, vol. 9, p. 758, The MIT Press.
14. Stephen T. Neely and Jont B. Allen, "Invertibility of a room impulse response," *Journal of the Acoustical Society of America*, vol. 66, no. 1, pp. 165–169, July 1979.

Detection by SNR Maximization: Application to the Blind Source Separation Problem

Bernard Xerri and Bruno Borloz

Laboratory SIS/GESSY - ISITV
Université de Toulon et du Var
Av. Georges Pompidou, BP 56
83162 La Valette du Var Cedex, France
{xerri,borloz}@univ-tln.fr

Abstract. In this paper, we propose a method for the detection of stochastic signals embedded in additive noise applied to the blind source separation problem in the particular case of delayed speech sources.
The method proposed leads to a linear filter we call "constrained stochastic matched filter" (CSMF), which is optimal in the sense. that it maximizes the output signal-to-noise ratio (SNR) in a subspace whose dimension is fixed *a priori*.
We show that the second-order statistics of sources can be unknown.

1 Introduction

Detection arises in signal processing problems whenever a decision is to be made among two hypotheses concerning an observed waveform. Signal detection algorithms decide whether the waveform consists of 'noise alone' or 'signal corrupted by noise'. The objective of signal detection theory is to specify strategies for designing algorithms which minimize the average number of decision errors.

The problem of detecting a signal against a noise background is an old one [1] and numerous solutions have been proposed according to the kind of signal to detect: deterministic, known except for a few parameters, purely stochastic...
Theoretical considerations show that the key quantity to compute is the Likelihood Ratio (LR). When the probability density function (PDF) of noise is known, a LR test (LRT) is employed. However this method cannot be employed when PDF are unknown: then the output SNR maximization seems natural and can be a convenient way for detection. We restrict our topic to linear filters. The calculation of the output SNR requires the knowledge of second-order statistics of both signal and noise: in this paper, these statistics will be supposed to be known, through the the covariance matrices (of sources for the practical application in the BSS problem: we will see however that this assumption is too strong and can be not verfied).

In this scope, several approaches can be considered. The method proposed in this paper is derived from the matched filter in the sense that it shares the same philosophical framework: the maximization of a SNR. Derived methods have been proposed to face this problem when the signal to detect is no more perfectly known (channel nonlinearities, timing jitter, nonstationarities, modeling uncertainties ...). Approaches have been proposed consisting in modeling the available partial knowledge through uncertainty sets (minimax robust matched-filter for example).

The stochastic matched filter (SMF) was introduced to take into consideration the problem of detecting a stochastic signal whose second-order statistics are known [4]. It extends the aforesaid approaches and gives an optimal filter in the sense that it makes maximum the output SNR expressed in the form of a Rayleigh's quotient (written with covariance matrices of signal and noise); filtering consists in a projection onto an optimal direction.

A natural extension seems to take into account not only a direction but a subspace. Actually, it has been shown [4] that this way can statistically contribute to improve ROC curves.

The method proposed in this paper is a natural extension of the SMF method; it aim is to make maximum the output SNR in an aptly chosen subspace. This optimal filter is named 'Constrained Stochastic Matched Filter' (CSMF). Thus, we propose an extension of the notion of SNR maximization used to detect known deterministic signals (MF) or stochastic signals (SMF) corrupted by noise. It takes its place in the class of the subspace projection methods.

It can advantageously be applied to the BSS problem (with two sources) in the case where sources are broken by silent areas and delayed: this is a case of convolutive mixture. In this particular case, regions of interest can be detected, delays and gains estimated and the mixing model is invertible in the frequency domain. It can be shown that in fact the second-order statistics of observations are sufficient to solve this problem.

2 The Constrained Stochastic Matched Filter

2.1 Expression of the Power in a Subspace

The signal will be understood to be zero-mean and discrete, represented by a vector of E_N. s can always be expressed as follows: $\mathbf{s} = \sum_{i=1}^{N} \alpha_i \mathbf{v}_i = \mathbf{V}\mathbf{a}$ where the α_i are random variables, the $\{\mathbf{v}_i\}$ linearly independent unit vectors, $\mathbf{V} = [\mathbf{v}_1...\mathbf{v}_N]$. Noting $\mathbf{\Gamma}_s$ its covariance matrix and $P_s = tr\mathbf{\Gamma}_s$ its power, P_s can be calculated as follows:

$$P_s \triangleq tr\mathbf{\Gamma}_s = \sum_{i=1}^{N}\sum_{j=1}^{N} \mathsf{E}(\alpha_i \alpha_j)\mathbf{v}_i^\top \mathbf{v}_j.$$

This expression can be reduced to a unique sum in case where $\{\mathbf{v}_i\}$ is an orthonormal basis. Then $P_s = \sum_{i=1}^{N} \mathbf{v}_i^\top \mathbf{\Gamma}_s \mathbf{v}_i = tr\left(\mathbf{V}^\top \mathbf{\Gamma}_s \mathbf{V}\right)$.

Let us consider an integer p in $[1; N]$ and note E_p the linear vector subspace spanned by $\{\mathbf{v}_1, ..., \mathbf{v}_p\}$. As we try to find a subspace of dimension p and because the output SNR does not depend on the basis but only on the subspace itself, without loss of generality, we can consider that the $\{\mathbf{v}_i\}$ form an orthonormal basis. Noting $\mathbf{V}_p = [\mathbf{v}_1, ..., \mathbf{v}_p]$, the power of $\mathbf{s}$ in E_p is:

$$P_p = tr\left(\mathbf{V}_p^\top \mathbf{\Gamma}_s \mathbf{V}_p\right) = \sum_{i=1}^{p} \mathbf{v}_i^\top \mathbf{\Gamma}_s \mathbf{v}_i \qquad (1)$$

2.2 The Stochastic Matched Filter

This approach was developed [4] to take into consideration the case where $\mathbf{s}$ is a stochastic signal uncorrelated with noise. The output SNR can be written like a Rayleigh quotient

$$SNR = \frac{\mathbf{h}^\top \mathbf{\Gamma}_s \mathbf{h}}{\mathbf{h}^\top \mathbf{\Gamma}_n \mathbf{h}} \qquad (2)$$

where $\mathbf{h}$ is the linear filter to be found, i.e. an N-dimensional vector, and $\mathbf{\Gamma}_s$ and $\mathbf{\Gamma}_n$ the covariance matrices of respectively $\mathbf{s}$ and $\mathbf{n}$. If we consider normalized covariance matrices $\widetilde{\mathbf{\Gamma}}_s = \frac{\mathbf{\Gamma}_s}{tr\mathbf{\Gamma}_s}$ and $\widetilde{\mathbf{\Gamma}}_n = \frac{\mathbf{\Gamma}_n}{tr\mathbf{\Gamma}_n}$, then $SNR = \frac{tr\mathbf{\Gamma}_s}{tr\mathbf{\Gamma}_n} \cdot \frac{\mathbf{h}^\top \widetilde{\mathbf{\Gamma}}_s \mathbf{h}}{\mathbf{h}^\top \widetilde{\mathbf{\Gamma}}_n \mathbf{h}}$. Hence, $\frac{tr\mathbf{\Gamma}_s}{tr\mathbf{\Gamma}_n}$ is the initial SNR (before filtering), and maximizing (2) amounts to maximizing $\rho = \frac{\mathbf{h}^\top \widetilde{\mathbf{\Gamma}}_s \mathbf{h}}{\mathbf{h}^\top \widetilde{\mathbf{\Gamma}}_n \mathbf{h}}$ which must be interpreted like a gain on the SNR.

The maximal output SNR ρ_{max} is obtained for the filter $\mathbf{h} = \mathbf{u}_1$, say the eigenvector associated with the largest eigenvalue λ_1 of the matrix $\mathbf{C} = \mathbf{\Gamma}_n^{-1} \mathbf{\Gamma}_s$; then $\rho_{max} = \lambda_1$ and, in the sense of this maximization, the filter is said to be optimal: this filtering comes down to projecting the signal onto a 1D subspace. It can be proved that $\rho_{max} > 1$.

There exists a solution to this problem provided that both $\mathbf{\Gamma}_s$ and $\mathbf{\Gamma}_n$ are symmetric (real) and that $\mathbf{\Gamma}_n$ is positive definite. These conditions are verified here.

Conclusion. The initial objective consists in maximizing an output-SNR in a one-dimensional subspace. Calculations lead naturally to the eigenvectors of $\mathbf{\Gamma}_n^{-1} \mathbf{\Gamma}_s$, one of them being optimal. But, from intuitive arguments, experiments have shown that the use of more than one eigenvector (ESMF or extended SMF) could improve the detection of the signal of interest. Note that even if this basis maximizes the output-SNR in a one-dimensional subspace, there is no reason why it should do it also in a p-dimensional subspace ($p > 1$). This is why the approach of the constrained stochastic matched filter has been developed.

2.3 Expression of the Output SNR in a p-Dimensional Subspace

We propose an extension of the SMF; its objective is to maximize the output SNR in a subspace whose dimension is fixed *a priori*, say p. As the SNR does depend on the subspace in which it is calculated, and not on the basis used to describe this subspace, the expression (derived from (1)) of the output SNR in a subspace of dimension p takes the form:

$$SNR_p = \frac{\sum_{i=1}^{p} \mathbf{v}_i^\top \mathbf{\Gamma}_s \mathbf{v}_i}{\sum_{i=1}^{p} \mathbf{v}_i^\top \mathbf{\Gamma}_n \mathbf{v}_i} = \frac{tr\left(\mathbf{V}_p^\top \mathbf{\Gamma}_s \mathbf{V}_p\right)}{tr\left(\mathbf{V}_p^\top \mathbf{\Gamma}_n \mathbf{V}_p\right)} \quad (3)$$

where $\{\mathbf{v}_i\}$ is a set of p unknown N-dimensional unit orthogonal vectors. To find them, an algorithm needs to be developed (see section 2.5).

Our objective is to maximize this ratio. That means, p beeing chosen *a priori*, find the p-dimensional subspace E_p in which this value is maximum: this optimal subspace will be noted E_p^*.

Let consider the normalized covariance matrices $\widetilde{\mathbf{\Gamma}}_s = \frac{\mathbf{\Gamma}_s}{tr\mathbf{\Gamma}_s}$ and $\widetilde{\mathbf{\Gamma}}_n = \frac{\mathbf{\Gamma}_n}{tr\mathbf{\Gamma}_n}$. Then (3) can be written

$$SNR_p = \frac{tr\mathbf{\Gamma}_s}{tr\mathbf{\Gamma}_n} \cdot \frac{tr\left(\mathbf{V}_p^\top \widetilde{\mathbf{\Gamma}}_s \mathbf{V}_p\right)}{tr\left(\mathbf{V}_p^\top \widetilde{\mathbf{\Gamma}}_n \mathbf{V}_p\right)} \quad (4)$$

where the first term $\frac{tr\mathbf{\Gamma}_s}{tr\mathbf{\Gamma}_n}$ is the initial SNR (before projection onto E_p). The second term is noted ρ ; it appears as a gain on the SNR and it can be proved to be necessarily lower than the largest eigenvalue of $\widetilde{\mathbf{\Gamma}}_n^{-1}\widetilde{\mathbf{\Gamma}}_s$ [4].

Throughout the upcoming sections, we will focus our attention on this expression of the gain ρ and try to find E_p^*.

2.4 Properties of the Optimal Subspace E_p^*

Introduction. The optimal p-dimensional subspace E_p^* is spanned by a set of orthonormal vectors $\{\mathbf{x}_i\}$. In this subspace, the expression of the gain ρ described previously is maximum and is equal to:

$$\rho = \frac{tr\left(\mathbf{X}^\top \mathbf{A} \mathbf{X}\right)}{tr\left(\mathbf{X}^\top \mathbf{B} \mathbf{X}\right)}, \quad (5)$$

where $\mathbf{A} = \widetilde{\mathbf{\Gamma}}_s$, $\mathbf{B} = \widetilde{\mathbf{\Gamma}}_n$ (then $tr\mathbf{A} = tr\mathbf{B} = 1$) and $\mathbf{X} = [\mathbf{x}_1...\mathbf{x}_p]$.

The constraints can be expressed by $\mathbf{x}_i^\top \mathbf{x}_j = \delta_{ij}$. Clearly, p is given and the unknowns of our problem are the SNR ρ and the orthonormal $\mathbf{x}_i$'s, which must be calculated so as to maximize ρ. We are faced to an optimization problem with constraints which is usually solved by using a method of Lagrange multipliers. Let's define the following function:

$$L(\mathbf{X}, \mathbf{\Omega}) = \rho + tr\left(\mathbf{\Omega}\left(\mathbf{X}^\top \mathbf{X} - \mathbf{I}_p\right)\right), \quad (6)$$

where $\mathbf{\Omega} \equiv [\omega_{ij}]$ is a $p \times p$ symmetric matrix. This value is maximum when $\frac{\partial L}{\partial \mathbf{X}} = 0$, which leads to

$$(\mathbf{A} - \rho \mathbf{B})\mathbf{T} = \mathbf{T}\Delta_\mu \quad (7)$$

where $\mathbf{T}$ is a unitary eigenvectors matrix and Δ_μ a real diagonal matrix of entries μ_i, depending both on ρ. The optimal subspace E_p^* is spanned by p vectors $\mathbf{t}_i$ among the N satisfying (7). Simple considerations show that the $\{\mathbf{t}_i\}$'s to choose must be those for which $\sum_{i \in I} \mu_i = 0$ (I of $\{1,...N\}$ being a subset satisfying $card(I) = p$).

Conclusion. For the particular cases $p = 1$ and $N - 1$, the solution is directly calculable. For intermediate values of p, equations to solve are not linear, and we propose an algorithm to reach the solution.

From simple examples, it is possible to see that the optimal subspace E_p^* is not necessarily spanned by the eigenvectors of $\mathbf{B}^{-1}\mathbf{A}$, and even though this is the case, the eigenvectors are not necessarily the ones associated with its largest eigenvalues; then it is no use thinking of a recursive formulation on p to find E_p^*.

2.5 Algorithm to Determine the Optimal Subspace E_p^*

Considering an initial value ρ_0 of ρ (a reasonable one is the largest eigenvalue of $\mathbf{B}^{-1}\mathbf{A}$), we obtain the symmetric matrix $\mathbf{M}(\rho_0)$. One can calculate N the eigenvectors of $\mathbf{M}(\rho_0)$: $\mathbf{t}_i(\rho_0)$, and then choose among them the p ones $\{\mathbf{t}_i(\rho_0), i \in I_0 / card(I_0) = p\}$ for which

$$\rho_1 = \frac{\sum_{i \in I_0} \mathbf{t}_i^\top(\rho_0)\mathbf{A}\mathbf{t}_i(\rho_0)}{\sum_{i \in I_0} \mathbf{t}_i^\top(\rho_0)\mathbf{B}\mathbf{t}_i(\rho_0)} \tag{8}$$

is maximum. These p vectors generate a subspace $E_p^{(0)}$.

At last, one can calculate $\mathbf{M}(\rho_1)$, I_1 and the new subspace $E_p^{(1)}$ and iterate the process until $\Delta\rho_n = \rho_{n+1} - \rho_n < \varepsilon$. It remains to prove that this algorithm converges to the good solution ρ_{max}.

Study of Convergence. We will note $\mu_{i,n} = \mu_i(\rho_n)$, $\mathbf{t}_{i,n} = \mathbf{t}_i(\rho_n)$ and for the optimal values $\mu_i^* = \mu_i(\rho_{max})$ and $\mathbf{t}_i^* = \mathbf{t}_i(\rho_{max})$. We have proved that for ρ_{max}, there exists a subset $I \subset \{1,...N\}$ such as $card(I) = p$ and $\sum_{i \in I} \mu_i^* = 0$. At step n, it can be shown that the variation of ρ is

$$\Delta\rho_n \triangleq \rho_{n+1} - \rho_n = \frac{\sum_{i \in I_n} \mu_{i,n}}{\sum_{i \in I_n} \mathbf{t}_{i,n}^\top \mathbf{B} \mathbf{t}_{i,n}} = \frac{\sum_{i \in I_n} \mu_{i,n}}{-\sum_{i \in I_n} \frac{\partial \mu_{i,n}}{\partial \rho}}. \tag{9}$$

Of course, if $\sum_{i \in I_n} \mu_{i,n} = \sum_{i \in I_n} \mu_i^* = 0$, then $\Delta\rho_n = 0$. ρ_{max} is an attractive fixed-point of the algorithm. $\Delta\rho_n$ can be written

$$\sum_{i \in I_n} \mu_{i,n} + \Delta\rho_n \sum_{i \in I_n} \frac{\partial \mu_{i,n}}{\partial \rho} = \sum_{i \in I_n} \left(\mu_{i,n} + \Delta\rho_n \frac{\partial \mu_{i,n}}{\partial \rho}\right) = 0.$$

In the neighborhood of the solution, $\Delta \rho_n$ is small and this expression is a Taylor's series expansion of $\mu_{i,n+1}$. This equation becomes $\sum_{i \in I_n} \mu_{i,n+1} = 0$. Then by iterations, the algorithm converges to ρ_{max}.

Practical Remarks. In particular cases, it may be possible to find several subspaces of dimension p for which the SNR ρ is maximal; it is in fact not a problem, in such a case we can take an interest in finding a subspace of higher dimension than p with the same SNR, or we can add a new criterion to choose among those subspaces.

Practically, whatever the initial value taken for ρ_0, this algorithm converges to the solution.

3 Experimental Results

An experimental application of the CSMF to images is provided above, concerning the detection of a texture embedded in another one. ROC curves are given to illustrate the improvement obtained by projecting data on the optimal subspace of dimension $p > 1$: here E_5^* (see Fig.1).

This example shows the relevance of projecting the observation onto a subspace of dimension greater than one, even if the output SNR in this subspace is smaller than in a subspace of dimension one.

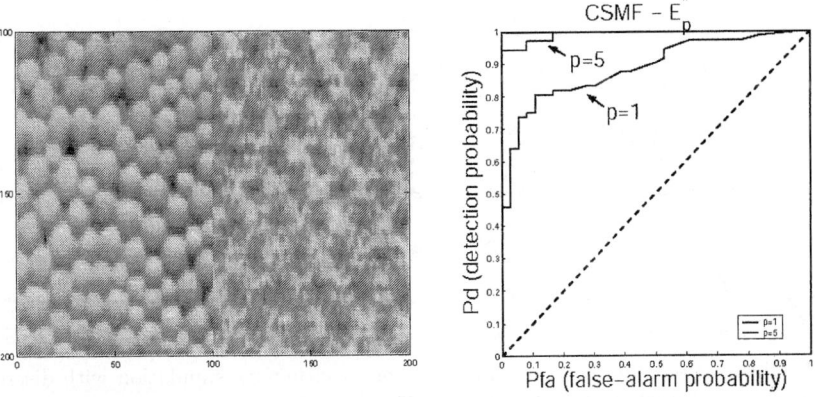

Fig. 1. Textures and ROC curves obtained for $p = 1$ and 5 filters.

4 Application to the BSS Problem

We consider two sources $x_i(t)$ (with different second-order statistics) in which there are silent areas: speech sources can be a good example of such signals. The mixing model leads to the following observations:

$$y_1(t) = x_1(t) + x_2(t) \\ y_2(t) = g_1 x_1(t - \tau_1) + g_2 x_2(t - \tau_2) \quad (10)$$

with the reasonable assumption that delays τ_i are quite smaller compared to the the duration of silence areas and $g_1 \neq g_2$; anyway delays greater than the coherence duration of signals are of no interest as they can be estimated easily in a different way.

Let us note Γ_{x_i} the covariance matrix of x_i ($i = 1, 2$). By opposing this both covariance matrices in the same way we maximized the SNR in a p-dimension subspace, we can try to highlight areas where only one source is present. For instance, when we try to detect x_1, considering x_2 as a noise, we use the subspace spanned by the p unit vectors $\mathbf{X}$ defined by the maximal value of the SNR $\rho = \frac{tr(\mathbf{X}^T \Gamma_{x_1} \mathbf{X})}{tr(\mathbf{X}^T \Gamma_{x_2} \mathbf{X})}$. The maximal and minimal values of ρ are respectively the greatest and smallest eigenvalues of $\Gamma_{x_2}^{-1} \Gamma_{x_1}$. The same can obviously be done for x_2 (then x_1 is the noise).

Each observation is filtered with two sets of filters, the first one to detect x_1, the second one to detect x_2. By applying a threshold to the power of observations after projection onto the optimal p-dimension subspaces, we highlight areas where only one signal is present (see Fig.2).

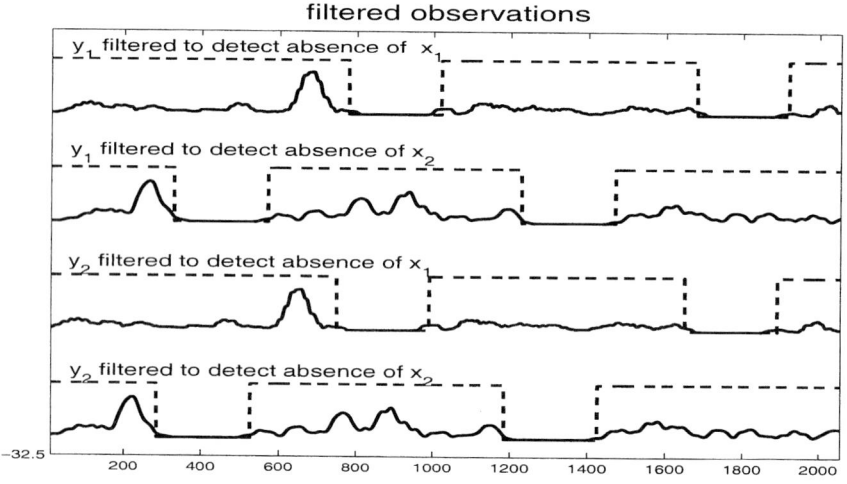

Fig. 2. Observations filtered to highlight either x_1 either x_2: simulation with discrete signals, $\tau_1 = 30$ (number of samples), $\tau_2 = 45$, $g_1 = 2.5$ and $g_2 = 3$.

In the case where the Γ_{x_i} are unknown, it is possible to use the covariance matrices of observations Γ_{y_i}. Simple calculation shows that it is equivalent to use the eigenvector associated with the largest eigenvalue of $\Gamma_{y_2}^{-1} \Gamma_{y_1}$ than those obtained from $\Gamma_{x_2}^{-1} \Gamma_{x_1}$. The ratio to maximize begins $\frac{tr(\mathbf{X}^T \Gamma_{x_1} \mathbf{X}) + tr(\mathbf{X}^T \Gamma_{x_2} \mathbf{X})}{\alpha tr(\mathbf{X}^T \Gamma_{x_1} \mathbf{X}) + \beta tr(\mathbf{X}^T \Gamma_{x_2} \mathbf{X})}$ which can be easily written in terms of ρ : $\frac{1+\rho}{\alpha+\beta\rho}$. It is easy to verify that this expression reaches its bounds when ρ is minimum or maximum. This is an important remark since in such a case the problem is absolutely blind.

Delays can be estimated with classical correlation and coherence methods. The gains g_i can be approximated either by a simple ratio in these areas (which should be a constant), either by a ratio of the power (for stationary signals). Knowing estimated values of τ_i and g_i, the reconstruction of sources x_i is performed through the frequency domain. The edges of reconstructed sources are all the worse that delays are great.

Experiments show that results are very satisfying: the mean-square error is about $4.7 \ 10^{-3}$ for both sources.

5 Conclusion

For the particular case of a delayed mixture of two sources with silent areas and different second-order statistics, the CSMF can advantageously be used in the separation problem. Those statistics can as well be unknown: then those, available, of observations are sufficient to perform separation. In fact, separation is realized in the frequency domain with estimated delays and gains. Experimental results are very convincing. In fact, silent areas are not necessary for stationary sources provided that covariance matrices of sources are discriminant enough. They just allow easier estimation of delays. Simulations have been performed successfully with stationary sources with different density spectral power and without silent areas, taking care of delays estimation.

References

1. A. Hero, *Signal Detection and Classification*, in 'The Digital Signal Processing Handbook' by Vijay K. Madisetti. CRC Press LLC, Series: Electrical Engineering Handbook, 1997
2. J. Barrere, G.Chabriel *A compact array for blind separation of sources*, IEEE Circuits and Systems, Part I, dec. 2001.
3. G. Chabriel, B. Xerri and J.F. Cavassilas, *Second Order Blind Identification of Slightly Delayed Mixtures*, ICA'99 First International Workshop on Independent Component Analysis and Signal Separation, pp. 75-79, Jan. 1999.
4. J.F. Cavassilas, B.Xerri, *Extension de la notion de filtre adapté. Contribution à la détection de signaux courts en présence de termes perturbateurs*, Revue Traitement du Signal Volume 10 n° 3, 1992, pp.215-221.
5. B.Xerri, B.Borloz, *An iterative method using conditional second order statistics applied to the blind separation problem*, IEEE trans. on Signal Processing, Feb. 2003.
6. J.F. Cardoso, P. Comon, *Independent component analysis, a survey of some algebraic methods*, Proc. ISCAS'96, vol.2, pp.93-96, 1996.
7. P. Comon, *Independent component analysis, a new concept ?*, Signal Processing, Elsevier, vol.36, n° 3, pp.287-314, Apr. 1994. Special issue on High-order Statistics.

Estimating the Number of Sources for Frequency-Domain Blind Source Separation

Hiroshi Sawada, Stefan Winter*, Ryo Mukai, Shoko Araki, and Shoji Makino

NTT Communication Science Laboratories, NTT Corporation
2-4 Hikaridai, Seika-cho, Soraku-gun, Kyoto 619-0237, Japan
{sawada,wifan,ryo,shoko,maki}@cslab.kecl.ntt.co.jp

Abstract. Blind source separation (BSS) for convolutive mixtures can be performed efficiently in the frequency domain, where independent component analysis (ICA) is applied separately in each frequency bin. To solve the permutation problem of frequency-domain BSS robustly, information regarding the number of sources is very important. This paper presents a method for estimating the number of sources from convolutive mixtures of sources. The new method estimates the power of each source or noise component by using ICA and a scaling technique to distinguish sources and noises. Also, a reverberant component can be identified by calculating the correlation of component envelopes. Experimental results for up to three sources show that the proposed method worked well in a reverberant condition whose reverberation time was 200 ms.

1 Introduction

Blind source separation (BSS) [1] is a technique for estimating original source signals solely from their mixtures at sensors. In some applications, such as audio acoustics, signals are mixed in a convolutive manner with reverberations. This makes the BSS problem more difficult to solve than an instantaneous mixture problem. Let us formulate the convolutive BSS problem. Suppose that N source signals $s_k(t)$ are mixed and observed at M sensors

$$x_j(t) = \sum_{k=1}^{N} \sum_l h_{jk}(l) s_k(t-l) + n_j(t), \tag{1}$$

where $h_{jk}(l)$ represents the impulse response from source k to sensor j and $n_j(t)$ is an additive Gaussian noise for each sensor. The goal is to obtain N output signals $y_i(t)$, each of which is a filtered version of a source $s_k(t)$. If we have enough sensors ($M \geq N$), a set of FIR filters $w_{ij}(l)$ of length L is typically used to produce separated signals

$$y_i(t) = \sum_{j=1}^{M} \sum_{l=0}^{L-1} w_{ij}(l) x_j(t-l) \tag{2}$$

at the outputs, and independent component analysis (ICA) [2] is generally used to obtain the FIR filters $w_{ij}(l)$. If the number of sensors is insufficient ($M < N$),

---

* The author is on leave from the Chair of Multimedia Communications and Signal Processing, University Erlangen-Nuremberg.

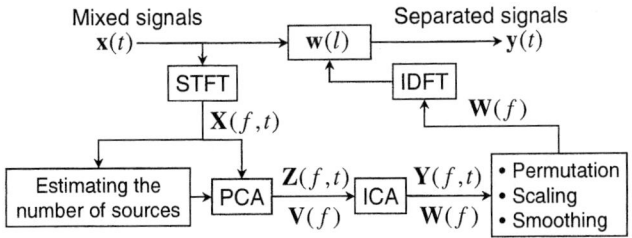

Fig. 1. Flow of frequency-domain BSS.

we need to rely on the sparseness of source signals, and approaches for separation become totally different [3]. Therefore, this paper focuses on cases where we have enough sensors ($M \geq N$).

There are two major approaches to the convolutive BSS problem. The first is time-domain BSS, where ICA is applied directly to the convolutive mixture model [4, 5]. The other approach is frequency-domain BSS, where complex-valued ICA for instantaneous mixtures is applied separately in each frequency bin [6–10]. The merit of frequency-domain BSS is that ICA for instantaneous mixtures is simpler and computationally more efficient than ICA for convolutive mixtures. We have implemented a frequency-domain BSS system that can separate three sources in real-time [10]. The price we must pay for this computational efficiency includes several additional problems that need to be solved for integrating the ICA solutions obtained separately in each frequency bin. The permutation problem is the best known. The permutation ambiguity of ICA should be aligned so that a separated signal in the time-domain contains the frequency components of the same source signal. We have proposed a method for solving the permutation problem [8], which performs well even if the number of sources is large [9, 10]. However, this method requires knowledge of the number of sources, and we assumed that the number was known apriori in these papers.

In this paper, we propose a method for estimating the number of sources N in the context of frequency-domain BSS. It is well known that the number of dominant eigenvalues of the spatial correlation matrix corresponds to the number of sources [11, 12]. However, it is difficult to decide whether an eigenvalue is dominant or not for sensor observations mixed in a reverberant condition as shown in Sec. 3. This difficulty has already been pointed out in [12], where they propose the use of support vector machines (SVM) to classify eigenvalue distributions and decide the number of sources. However, the SVM needs to be trained beforehand and experimental results were provided only for 1- or 2-source cases. Our proposed method is based on an analysis of ICA solutions obtained in the frequency domain as shown in Sec. 4. Experimental results for up to three sources show that the method worked well in a real reverberant condition.

2 Frequency-Domain BSS

This section describes frequency-domain BSS whose flow is shown in Fig. 1. First, time-domain signals $x_j(t)$ at sensors are converted into frequency-domain time-series signals $X_j(f,t)$ by short-time Fourier transform (STFT), where t is now down-sampled with the distance of the frame shift. Then, the number of sources N should be estimated from $\mathbf{X}(f,t) = [X_1(f,t), \ldots, X_M(f,t)]^T$. This part is the main topic of this paper, and will be discussed in Secs. 3 and 4.

After estimating the number of sources N, the dimension M of sensor observations $\mathbf{X}(f,t)$ is reduced to N typically by principal component analysis (PCA), $\mathbf{Z}(f,t) = \mathbf{V}(f)\mathbf{X}(f,t)$, where $\mathbf{V}(f)$ is an $N \times M$ matrix whose row vectors generate N principal components [13]. Even if $N = M$, PCA is useful as preprocessing. Then, complex-valued ICA $\mathbf{Y}(f,t) = \mathbf{B}(f)\mathbf{Z}(f,t)$ is applied, where $\mathbf{B}(f)$ is an N dimensional square matrix. Through these operations, the sensor observations $\mathbf{X}(f,t)$ are separated into independent components $\mathbf{Y}(f,t) = [Y_1(f,t), \ldots, Y_N(f,t)]^T$ by $\mathbf{Y}(f,t) = \mathbf{W}(f)\mathbf{X}(f,t)$, where $\mathbf{W}(f) = \mathbf{B}(f)\mathbf{V}(f)$. Note that $\mathbf{W}(f)$ is invertible if $\mathbf{V}(f)$ is full rank and $\mathbf{B}(f)$ is made unitary (by e.g. FastICA [2]).

The ICA solution $\mathbf{W}(f)$ in each frequency bin has permutation and scaling ambiguity: even if we permute the rows of $\mathbf{W}(f)$ or multiply a row by a constant, it is still an ICA solution. In matrix notation, $\mathbf{\Lambda}(f)\mathbf{P}(f)\mathbf{W}(f)$ is also an ICA solution for any permutation $\mathbf{P}(f)$ and diagonal $\mathbf{\Lambda}(f)$ matrix. The permutation ambiguity $\mathbf{P}(f)$ should be solved so that $Y_i(f,t)$ at all frequencies corresponds to the same source $s_i(t)$. We use the method described in [8]. The scaling ambiguity $\mathbf{\Lambda}(f)$ can be solved by making $Y_i(f,t)$ as close to a part of the sensor observation $\mathbf{X}(f,t)$ as possible. The minimal distortion principle (MDP) [4] makes $y_i(t)$ as close to $\sum_l h_{ii}(l)s_i(t-l)$, a part of $x_i(t)$, as possible. In the frequency domain, it is realized by $\mathbf{\Lambda}(f) = \text{diag}[\mathbf{W}^{-1}(f)]$ [7]. If $N < M$, the Moore-Penrose pseudoinverse $\mathbf{W}^+(f)$ is used instead of $\mathbf{W}^{-1}(f)$. Also, the scaling (3) that will be discussed in Sec. 4 can be used.

The aligned matrices $\mathbf{W}(f) \leftarrow \mathbf{\Lambda}(f)\mathbf{P}(f)\mathbf{W}(f)$ are the frequency responses of separation filters $\mathbf{w}(l)$. However, we need to be concerned about the circularity effect of discrete frequency representation. We perform spectral smoothing [14] for $[\mathbf{W}(f)]_{ij}$ to mitigate the circularity effect. Finally, time-domain filters $w_{ij}(l)$ are obtained by applying inverse DFT to the smoothed elements $[\mathbf{W}(f)]_{ij}$.

3 Conventional Eigenvalue-Based Method

This section describes a conventional eigenvalue-based method for estimating the number of sources in each frequency bin [11]. It performs eigenvalue decomposition for the spatial correlation matrix $\mathbf{R}(f) = \langle \mathbf{X}(f,t)\mathbf{X}(f,t)^H \rangle_t$ of sensor observations, where $\langle \cdot \rangle_t$ denotes the averaging operator and $\cdot^H$ denotes the conjugate transpose. Let $\lambda_1 \geq \cdots \geq \lambda_N \geq \cdots \geq \lambda_M$ be the sorted eigenvalues of $\mathbf{R}(f)$. If there is no reverberation, the number of dominant eigenvalues is equal to the number of sources N, and the remaining $M - N$ smallest eigenvalues are the same as the noise power: $\lambda_{N+1} = \cdots = \lambda_M = \sigma_n^2$. However, there are two problems in a real reverberant condition.

Reverberation. The number of dominant eigenvalues might be more than the number of source signals, if the reverberation of a mixing system is long and

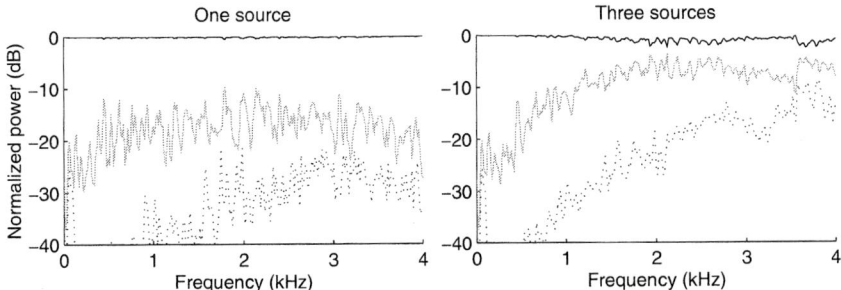

Fig. 2. Component powers estimated by the eigenvalue-based method.

strong. This is because the reverberation of a mixing system, i.e. the non-zero part of $h_{jk}(l)$, is usually longer than the STFT frame, and the reverberation component could be counted as a signal.

Unrecovered power. The number of dominant eigenvalues might be less than the number of source signals, if some of the column vectors of the mixing matrix are similar. In this case, the first few eigenvalues represent almost all powers. A typical situation can be seen in low frequencies, where the phase differences among sensors are very small.

Because of these two problems, the eigenvalue-based method does not work well in a real reverberant condition. Figure 2 shows component powers estimated by the eigenvalue-based method in an environment whose conditions are summarized in Fig. 3. The left hand plot shows a one-source case. Because of reverberations, the normalized power of the second principal components were around -20 dB. To distinguish the source and noises (including reverberations), a threshold of around -15 dB is good for the one-source case. However, if such a threshold is used for the three-source case shown in the right hand plot, the number of sources is estimated at two in most frequency bins. Therefore, it is hard to find a threshold that works well for both cases.

4 Proposed ICA-Based Method

In this section, we propose a new method for estimating the number of sources that solves the two problems mentioned above.

To solve the problem of unrecovered power, the proposed method recovers the power of each signal measured at sensors by using ICA and a scaling technique. It first applies ICA for $\mathbf{X}(f,t)$ without performing dimension reduction, i.e. assuming the number of sources N is equal to the number of sensors M. Because of the scaling ambiguity of ICA, the power of each component of the ICA solution $\mathbf{Y}(f,t) = \mathbf{W}(f)\mathbf{X}(f,t)$ is different from the power of each source or noise. If the real number of sources is less than M, $M-N$ noise components are generally enhanced.

To recover the power of each component measured at sensors, we use a scaling

$$\mathbf{\Lambda}(f) = \mathrm{sqrt}(\mathrm{diag}[\mathbf{W}^{-H}(f)\mathbf{W}^{-1}(f)]), \tag{3}$$

for ICA solution $\mathbf{Y}(f,t) = \mathbf{\Lambda}(f)\mathbf{W}(f)\mathbf{X}(f,t)$. Note again that $\mathbf{W}(f)$ is invertible if the smallest eigenvalue of the spatial correlation matrix $\mathbf{R}(f)$ is not zero. We call (3) *power-recovery scaling* since it recovers the power of the sensor observations as follows. Firstly, the total power of sensor observations is recovered at the outputs:

$$||\mathbf{Y}(f,t)||^2 = ||\mathbf{X}(f,t)||^2, \quad (4)$$

if the components of $\mathbf{Y}(f,t)$ are uncorrelated. Moreover, if ICA is properly solved and Y_i's are made mutually independent, the power of each source measured at sensors is recovered at each output:

$$|Y_i(f,t)|^2 = ||\mathbf{H}_{\Pi(i)}(f)S_{\Pi(i)}(f,t)||^2, \quad (5)$$

where Π is a permutation, S_k is the k-th source and $\mathbf{H}_k$ is the mixing vector of S_k. This equation (5) can be seen as decomposition of equation (4). We have proved both equations. However, the proofs are omitted here for space limit.

In this way, the power of each component $Y_i(f,t)$ of the ICA solution $\mathbf{Y}(f,t) = \mathbf{\Lambda}(f)\mathbf{W}(f)\mathbf{X}(f,t)$ approaches the real power of each source or noise measured at sensors $\mathbf{H}_{\Pi(i)}(f)S_{\Pi(i)}(f,t)$. Therefore, the power

$$\sigma_i^2 = \langle |Y_i(f,t)|^2 \rangle_t \quad (6)$$

can be used as a criterion for distinguishing sources and noises (including reverberations). Although the MDP explained in Sec. 2 can also be used for power estimation, the power recovered by the MDP contains only the power of a selected sensor $x_i(t)$, and is sensitive to the sensor selection. The power recovered by the power-recovery scaling (3) contains the power of all sensors, and is therefore more robust for power estimation.

The problem of reverberation discussed in Sec. 3 still needs to be solved. We observed that the envelope of a reverberant component has a strong correlation with the envelope of a source component. The correlation of two envelopes $|Y_{i1}(f,t)|$ and $|Y_{i2}(f,t)|$, $i1, i2 \in \{1,\ldots,M\}$, is defined as

$$\frac{\langle v_{i1}(t) \cdot v_{i2}(t) \rangle_t}{\sqrt{\langle v_{i1}^2(t) \rangle_t} \cdot \sqrt{\langle v_{i2}^2(t) \rangle_t}}, \text{ where } v_i(t) = |Y_i(f,t)| - \langle |Y_i(f,t)| \rangle_t. \quad (7)$$

When $Y_{i1}(f,t)$ is a source component and $Y_{i2}(f,t)$ is not a source component but includes the reverberation of source $i1$, the correlation of $|Y_{i1}(f,t-\Delta t)|$ and $|Y_{i2}(f,t)|$ with an appropriate time delay $-\Delta t$ tends to be large. Therefore, the correlation can be used as a measure to distinguish sources and reverberations.

The overall procedure of the proposed method is as follows.

1. Calculate independent components $Y_i(f,t)$ by using ICA and scaling (3).
2. If the normalized power $\sigma_i^2 / \sum_{k=1}^{M} \sigma_k^2$ of i-th component is smaller than a threshold, e.g. 0.01 (-20 dB), consider it to be a noise component.
3. If the normalized power $\sigma_i^2 / \sum_{k=1}^{M} \sigma_k^2$ is smaller than a threshold, e.g. 0.2, and one of the correlations (7) among other components is larger than a threshold, e.g. 0.5, consider it to be a reverberant component.
4. Otherwise, consider the i-th component to be a signal.

These thresholds can be determined beforehand by the power levels of background noise and reverberations.

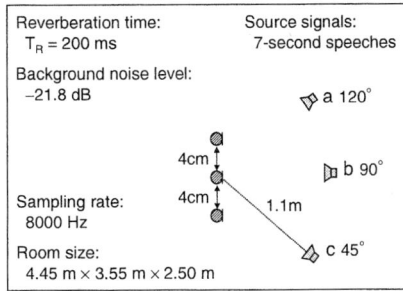 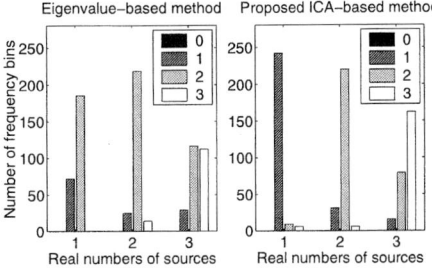

Fig. 3. Experimental conditions.   **Fig. 4.** Estimated numbers of sources.

Table 1. BSS results obtained with different estimation methods: the conventional eigenvalue-based method (Eig.) and the proposed ICA-based method (Prop.).

#sources (real)	1 (c)		2 (a,c)		3 (a,b,c)		
#sources (est.)	2 (Eig.)	1 (Prop.)	2 (Both)		2 (Eig.)	3 (Prop.)	
SIR (dB)	∞ ∞	∞	17.1 17.1		2.0 −1.2	13.6 15.4 13.3	
SDR (dB)	10.1 −4.4	∞	13.3 14.2		0.7 −3.5	9.4 10.5 10.2	

5 Experimental Results

We performed experiments to estimate the number of sources from sensor observations and to separate them into source signals. Sensor observations were generated by convolving source signals with impulse responses and then adding background noise. The impulse responses and the background noise were measured in the conditions summarized in Fig. 3. We tested cases of one, two and three sources, while the number of sensors was three for all cases. Figure 4 shows the numbers of sources estimated by using the conventional eigenvalue-based method and the proposed ICA-based method. The vertical axis shows the number of frequency bins for each estimated number of sources. The STFT frame size was 512, and thus the number of total frequency bins to cover 0–4000 Hz was 257. By taking the maximum vote, the ICA-based method successfully estimated the number of sources in all cases, whereas the eigenvalue-based method estimated the number of sources at 2 in all cases.

Table 1 shows the BSS results obtained with these estimations for the number of sources. The results were measured in terms of the signal-to-interference ratio (SIR) and signal-to-distortion ratio (SDR) of each output. To calculate the SIR of $y_i(t)$, it is decomposed as $y_i(t) = tar_i(t) + int_i(t)$, where $tar_i(t)$ is a filtered component of a target signal $s_{\Pi(i)}(t)$ and $int_i(t)$ is the remaining interference component. The SIR is the power ratio of $tar_i(t)$ and $int_i(t)$. The mapping Π was selected to maximize the SIR. To calculate the SDR of $y_i(t)$, the filtered component of the target signal is further decomposed as $tar_i(t) = \alpha_i \cdot ref_i(t) + e_i(t)$, where $ref_i(t)$ is a reference signal and α_i is a scalar that minimizes the error power of $e_i(t)$. We used $ref_i(t) = \sum_l h_{ii}(l) s_i(t-l)$ following the MDP. The SDR is the power ratio of $\alpha_i \cdot ref_i(t)$ and $e_i(t)$. The BSS performance was degraded if

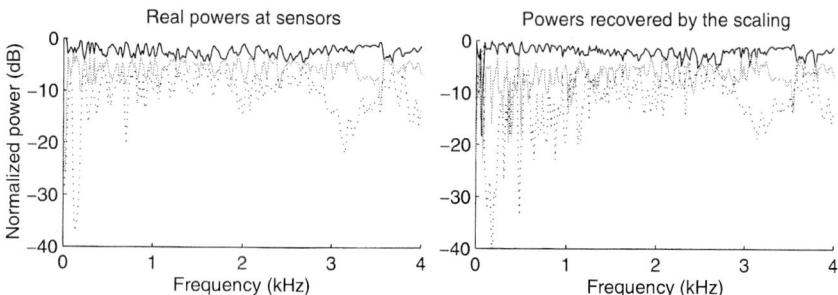

Fig. 5. Power recovery by scaling formula (3) when there are three sources.

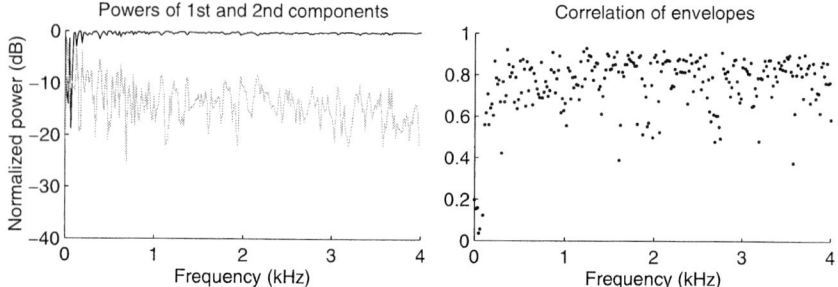

Fig. 6. Identifying reverberant components when there is one source.

the number of sources was incorrectly estimated. In the one-source case with the eigenvalue-based method, the number of sources was overestimated. Thus the source was decomposed into two outputs, and the SDRs were poor. In the three-source case, again with the eigenvalue-based method, the number of sources was underestimated. In this case, the output signals were still mixed, and thus the SIRs as well as SDRs were poor.

Figure 5 shows how well the powers of sources were recovered by ICA and the proposed scaling technique. The left hand plot shows the normalized powers of the three sources measured at sensors, and the right hand plot shows those estimated by ICA and the scaling formula (3). The powers were sufficiently well recovered to estimate the number of sources. Compared with the result obtained with the eigenvalue-based method (the right hand plot in Fig. 2), the advantage of the proposed method becomes clear.

Figure 6 shows how the proposed method copes with the reverberation problem. This case had only one source. The left hand plot shows the normalized power of the first and second largest components of the scaled ICA outputs. It was hard to decide solely from these normalized powers whether the second component was a signal or a noise because the powers of the second components were not sufficiently small in many frequency bins. However, by calculating the correlation of the envelopes between the first and second components, it became clear that the second component was a reverberation, i.e. a noise. The right hand plot shows the correlations, which were large enough (around 0.8) in many frequency bins.

6 Conclusion

We have proposed a method for estimating the number of sources in each frequency bin. Our method provides a solution for the two problems with the conventional eigenvalue-based method discussed in Sec. 3, and provides a good estimation even in a reverberant condition of $T_R = 200$ ms. With the proposed method, frequency-domain BSS can be practically applied without apriori knowledge of the number of sources.

References

1. Haykin, S., ed.: Unsupervised adaptive filtering (Volume I: Blind source separation). John Wiley & Sons (2000)
2. Hyvärinen, A., Karhunen, J., Oja, E.: Independent component analysis. John Wiley & Sons (2001)
3. Rickard, S., Balan, R., Rosca, J.: Blind source separation based on space-time-frequency diversity. In: Proc. ICA2003. (2003) 493–498
4. Matsuoka, K., Nakashima, S.: Minimal distortion principle for blind source separation. In: Proc. ICA 2001. (2001) 722–727
5. Douglas, S.C., Sun, X.: Convolutive blind separation of speech mixtures using the natural gradient. Speech Communication **39** (2003) 65–78
6. Smaragdis, P.: Blind separation of convolved mixtures in the frequency domain. Neurocomputing **22** (1998) 21–34
7. Murata, N., Ikeda, S., Ziehe, A.: An approach to blind source separation based on temporal structure of speech signals. Neurocomputing **41** (2001) 1–24
8. Sawada, H., Mukai, R., Araki, S., Makino, S.: A robust and precise method for solving the permutation problem of frequency-domain blind source separation. IEEE Trans. Speech and Audio Processing **12** (2004)
9. Mukai, R., Sawada, H., de la Kethulle, S., Araki, S., Makino, S.: Array geometry arrangement for frequency domain blind source separation. In: Proc. IWAENC2003. (2003) 219–222
10. Sawada, H., Mukai, R., Araki, S., Makino, S.: Convolutive blind source separation for more than two sources in the frequency domain. In: Proc. ICASSP 2004. (2004)
11. Wax, M., Kailath, T.: Detection of signals by information theoretic criteria. IEEE Trans. Acoustics, Speech, and Signal Processing **33** (1985) 387–392
12. Yamamoto, K., Asano, F., van Rooijen, W., Ling, E., Yamada, T., Kitawaki, N.: Estimation of the number of sound sources using support vector machines and its application to sound source separation. In: Proc. ICASSP 2003. (2003) 485–488
13. Winter, S., Sawada, H., Makino, S.: Geometrical interpretation of the PCA subspace method for overdetermined blind source separation. In: Proc. ICA2003. (2003) 775–780
14. Sawada, H., Mukai, R., de la Kethulle, S., Araki, S., Makino, S.: Spectral smoothing for frequency-domain blind source separation. In: Proc. IWAENC2003. (2003) 311–314

Estimating the Number of Sources in a Noisy Convolutive Mixture Using BIC

Rasmus Kongsgaard Olsson and Lars Kai Hansen

Technical University of Denmark, Informatics and Mathematical Modelling, B321
DK-2800 Lyngby, Denmark
rko@isp.imm.dtu.dk,lkh@imm.dtu.dk

Abstract. The number of source signals in a noisy convolutive mixture is determined based on the exact log-likelihoods of the candidate models. In (Olsson and Hansen, 2004), a novel probabilistic blind source separator was introduced that is based solely on the time-varying second-order statistics of the sources. The algorithm, known as 'KaBSS', employs a Gaussian linear model for the mixture, i.e. AR models for the sources, linear mixing filters and a white Gaussian noise model. Using an EM algorithm, which invokes the Kalman smoother in the E-step, all model parameters are estimated and the exact posterior probability of the sources conditioned on the observations is obtained. The log-likelihood of the parameters is computed exactly in the process, which allows for model evidence comparison assisted by the BIC approximation. This is used to determine the activity pattern of two speakers in a convolutive mixture of speech signals.

1 Introduction

We are pursuing a research program in which we aim to understand the properties of mixtures of independent source signals within a generative statistical framework. We consider *convolutive* mixtures, i.e.,

$$\mathbf{x}_t = \sum_{k=0}^{L-1} \mathbf{A}_k \mathbf{s}_{t-k} + \mathbf{n}_t, \quad (1)$$

where the elements of the source signal vector, $\mathbf{s}_t$, i.e., the d_s statistically independent source signals, are convolved with the corresponding elements of the filter matrix, $\mathbf{A}_k$. The multichannel sensor signal, $\mathbf{x}_t$, are furthermore degraded by additive Gaussian white noise.

It is well-known that separation of the source signals based on second order statistics is infeasible in general. Consider the second order statistic

$$\langle \mathbf{x}_t \mathbf{x}_{t'}^\top \rangle = \sum_{k,k'=0}^{L-1} \mathbf{A}_k \langle \mathbf{s}_{t-k} \mathbf{s}_{t'-k'}^\top \rangle \mathbf{A}_{k'}^\top + \mathbf{R},$$

where $\mathbf{R}$ is the (diagonal) noise covariance matrix. If the sources are white noise stationary, the source covariance matrix can be assumed proportional to the unit matrix without loss of generality, and we see that the statistic is symmetric to a common rotation of all mixing matrices $\mathbf{A}_k \to \mathbf{A}_k \mathbf{U}$. This rotational invariance means that the statistic is not informative enough to identify the mixing matrix, hence, the source time series.

However, if we consider stationary sources with *known*, non-trivial, autocorrelations $\langle \mathbf{s}_t \mathbf{s}_{t'}^\top \rangle = \mathbf{C}(t - t')$, and we are given access to measurements involving multiple values of $\mathbf{C}(t - t')$, the rotational degrees of freedom are constrained and we will be able to recover the mixing matrices up to a choice of sign and scale of each source time series. Extending this argument by the observation that the mixing model (1) is invariant to filtering of a given column of the convolutive filter provided that the inverse filter is applied to corresponding source signal, we see that it is infeasible to identify the mixing matrices if these arbitrary inverse filters can be chosen to that they 'whiten' the sources.

For non-stationary sources, on the other hand, the autocorrelation functions vary through time and it is not possible to choose a single common whitening filter for each source. This means that the mixing matrices may be identifiable from multiple estimates of the second order correlation statistic (2) for non-stationary sources. Parra and Spence [1] provide analysis in terms of the number of free parameters vs. the number of linear conditions.

Also in [1], the constraining effect of source non-stationarity was exploited by simultaneously diagonalizing multiple estimates of the source power spectrum. In [2] we formulated a generative probabilistic model of this process and proved that it could estimate sources and mixing matrices in noisy mixtures. A state-space model -a Kalman filter- was specialized and augmented by a stacking procedure to model a noisy convolutive mixture of non-stationary colored noise sources, and a forward-backward EM approach was used to estimate the source statistics, mixing coefficients and the diagonal noise covariance matrix. The EM algorithm furthermore provides an exact calculation of the likelihood as it is possible to average over all possible source configurations. Other approaches based on EM schemes for source inference are [3], [4] and [5]. In [6], a non-linear state-space model is proposed.

In this presentation we elaborate on the generative model and its applications. In particular, we use the exact likelihood calculation to make inference about the dimensionality of the model, i.e. the number of sources. Choosing the incorrect model order can lead to either a too simple, biased model or a too complex model. We use the so-called Bayes Information Criterion (BIC) [7] to approximate the Bayes factor for competing hypotheses.

The model is stated in section 2, while the learning in the particular model described in section 3. Model order selection using BIC is treated in section 4. Experiments for speech mixtures are shown in section 5.

2 The Model

As indicated above, the sources must be assumed non-stationary in order to uniquely retrieve the parameters and sources, since the estimation is based on second-order statistics. In line with [1], this is obtained by *segmenting* the signals into frames, in which the wide-sense stationarity of the sources is assumed. A separate source model is assumed for each segment. The channel filters and observation noise covariance are assumed stationary across segments in the entire observed signal.

The colored noise sources are modelled by AR(p) random processes. In segment n, source i is represented by:

$$s_{i,t}^n = f_{i,1}^n s_{i,t-1}^n + f_{i,2}^n s_{i,t-2}^n + \ldots + f_{i,p}^n s_{i,t-p}^n + v_{i,t}^n \qquad (2)$$

where $n \in \{1, 2, .., N\}$ and $i \in \{1, 2, .., d_s\}$. The innovation noise, $v_{i,t}$, is white Gaussian. In order to make use of well-established estimation theory, the above recursion is fitted into the framework of Gaussian linear models, for which a review is found in e.g. [8]. The Kalman filter model is an instance of this model that particularly treats continuous Gaussian linear models used widely in e.g. control and speech enhancement applications. The general Kalman filter with no control inputs is defined:

$$\mathbf{s}_t = \mathbf{F}\mathbf{s}_{t-1} + \mathbf{v}_t \qquad (3)$$
$$\mathbf{x}_t = \mathbf{A}\mathbf{s}_t + \mathbf{n}_t$$

where $\mathbf{v}_t$ and $\mathbf{n}_t$ are white Gaussian noise signals that drive the processes.

In order to incorporate the colored noise sources, equation (2), into the Kalman filter model, the well-known principle of *stacking* must be applied, see e.g [9]. At any time, the stacked source vector, $\bar{\mathbf{s}}_t^n$, contains the last p samples of all d_s sources:

$$\bar{\mathbf{s}}_t^n = \left[(\mathbf{s}_{1,t}^n)^\top \ (\mathbf{s}_{2,t}^n)^\top \ \ldots \ (\mathbf{s}_{d_s,t}^n)^\top \right]^\top$$

The component vectors, $\mathbf{s}_{i,t}^n$, contain the p most recent samples of the individual sources:

$$\mathbf{s}_{i,t}^n = \left[s_{i,t}^n \ s_{i,t-1}^n \ \ldots \ s_{i,t-p+1}^n \right]^\top$$

In order to maintain the statistical independency of the sources, a constrained format must be imposed on the parameters:

$$\bar{\mathbf{F}}^n = \begin{bmatrix} \bar{\mathbf{F}}_1^n & 0 & \cdots & 0 \\ 0 & \bar{\mathbf{F}}_2^n & \cdots & 0 \\ \vdots & \vdots & \ddots & \vdots \\ 0 & 0 & \cdots & \bar{\mathbf{F}}_{d_s}^n \end{bmatrix}, \ \bar{\mathbf{F}}_i^n = \begin{bmatrix} f_{i,1}^n & f_{i,2}^n & \cdots & f_{i,p-1}^n & f_{i,p}^n \\ 1 & 0 & \cdots & 0 & 0 \\ 0 & 1 & \cdots & 0 & 0 \\ \vdots & \vdots & \ddots & \vdots & \vdots \\ 0 & 0 & \cdots & 1 & 0 \end{bmatrix}$$

$$\bar{\mathbf{Q}}^n = \begin{bmatrix} \bar{\mathbf{Q}}_1^n & 0 & \cdots & 0 \\ 0 & \bar{\mathbf{Q}}_2^n & \cdots & 0 \\ \vdots & \vdots & \ddots & \vdots \\ 0 & 0 & \cdots & \bar{\mathbf{Q}}_{d_s}^n \end{bmatrix}, \ (\bar{\mathbf{Q}}_i^n)_{jj'} = \begin{cases} q_i^n & j = j' = 1 \\ 0 & j \neq 1 \lor j' \neq 1 \end{cases}$$

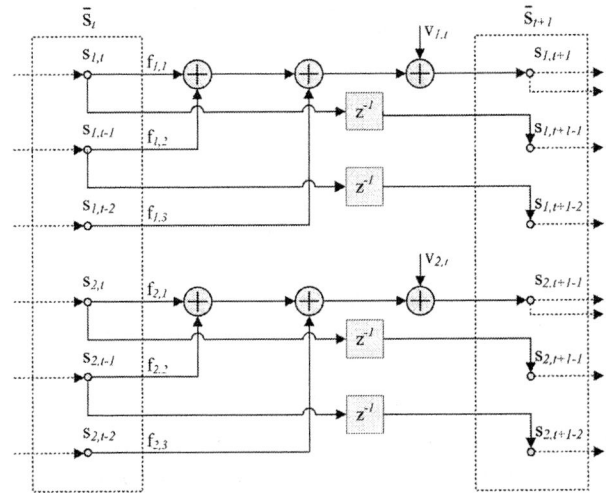

Fig. 1. The multiplication of $\bar{\mathbf{F}}$ on $\bar{\mathbf{s}}_t$ and the addition of innovation noise, $\mathbf{v}_t$, shown for an example involving two AR(3) sources. The special contrained format of $\bar{\mathbf{F}}$ simultaneously ensures the storage of past samples.

The matrix $\mathbf{A}$ of (3) is left unconstrained but its dimensions must be expanded to $d_x \times (p \times d_s)$ to reflect the stacking of the sources. Conveniently, its elements can be interpreted as the impulse responses of the channel filters of (1):

$$\bar{\mathbf{A}} = \begin{bmatrix} \mathbf{a}_{11}^\top & \mathbf{a}_{12}^\top & .. & \mathbf{a}_{1d_s}^\top \\ \mathbf{a}_{21}^\top & \mathbf{a}_{22}^\top & .. & \mathbf{a}_{2d_s}^\top \\ \mathbf{a}_{d_x 1}^\top & \mathbf{a}_{d_x 2}^\top & .. & \mathbf{a}_{d_x d_s}^\top \end{bmatrix}$$

where $\mathbf{a}_{ij} = [a_{ij,1}, a_{ij,2}, .., a_{ij,L}]^\top$ is the filter between source i and sensor j. Having defined the stacked sources and the constrained parameter matrices, the total model is:

$$\bar{\mathbf{s}}_t^n = \bar{\mathbf{F}}^n \bar{\mathbf{s}}_{t-1}^n + \bar{\mathbf{v}}_t^n$$
$$\mathbf{x}_t^n = \bar{\mathbf{A}} \bar{\mathbf{s}}_t^n + \mathbf{n}_t^n$$

where $\bar{\mathbf{v}}_t^n \sim (\mathbf{0}, \bar{\mathbf{Q}}^n)$ and $\mathbf{n}_t^n \sim (\mathbf{0}, \bar{\mathbf{F}}^n)$. Figures 1 and 2 illustrate the updating of the stacked source vector, $\bar{\mathbf{s}}_t$ and the effect of multiplication by $\bar{\mathbf{A}}$, respectively.

3 Learning

Having described the convolutive mixing problem in the general framework of linear Gaussian models, more specifically the Kalman filter model, optimal inference of the sources is obtained by the Kalman smoother. However, since the problem at hand is effectively *blind*, the parameters are estimated. Along the lines of, e.g. [8], an EM algorithm will be used for this purpose, i.e.

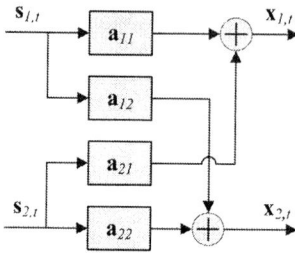

Fig. 2. The effect of the matrix multiplication $\bar{\mathbf{A}}$ on $\bar{\mathbf{s}}_t$ is shown in the system diagram. The source signals are filtered (convolved) with the impulse responses of the channel filters. Observation noise and the segment index, n, are omitted for brevity.

$\mathcal{L}(\theta) \geq \mathcal{F}(\theta, \hat{p}) \equiv \mathcal{J}(\theta, \hat{p}) - \mathcal{R}(\hat{p})$, where $\mathcal{J}(\theta, \hat{p}) \equiv \int d\mathbf{S}\hat{p}(\mathbf{S}) \log p(\mathbf{X}, \mathbf{S}|\theta)$ and $\mathcal{R}(\hat{p}) \equiv \int d\mathbf{S}\hat{p}(\mathbf{S}) \log \hat{p}(\mathbf{S})$ were introduced. In accordance with standard EM theory, $\mathcal{J}(\theta, \hat{p})$ is optimized wrt. θ in the M-step. The E-step infers the model posterior, $\hat{p} = p(\mathbf{S}|\mathbf{X}, \theta)$. The combined E and M steps are guaranteed not to decrease $\mathcal{L}(\theta)$.

3.1 E-Step

The forward-backward recursions which comprise the Kalman smoother is employed in the E-step to infer the source posterior, $p(\mathbf{S}|\mathbf{X}, \theta)$, i.e. the joint posterior of the sources conditioned on all observations. The relevant second-order statistics of this distribution in segment n is the posterior mean, $\hat{\mathbf{s}}_t^n \equiv \langle \bar{\mathbf{s}}_t^n \rangle$, and autocorrelation, $\mathbf{M}_{i,t}^n \equiv \langle \mathbf{s}_{i,t}^n (\mathbf{s}_{i,t}^n)^\top \rangle \equiv [\mathbf{m}_{i,1,t}^n \; \mathbf{m}_{i,2,t}^n \; .. \; \mathbf{m}_{i,L,t}^n]^\top$, along with the time-lagged covariance, $\mathbf{M}_{i,t}^{1,n} \equiv \langle \mathbf{s}_{i,t}^n (\mathbf{s}_{i,t-1}^n)^\top \rangle \equiv [\mathbf{m}_{i,1,t}^{1,n} \; \mathbf{m}_{i,2,t}^{1,n} \; .. \; \mathbf{m}_{i,L,t}^{1,n}]^\top$. In particular, $m_{i,t}^n$ is the first element of $\mathbf{m}_{i,1,t}^n$. All averages are performed over $p(\mathbf{S}|\mathbf{X}, \theta)$. The forward recursion also yields the likelihood $\mathcal{L}(\theta)$.

3.2 M-Step

The estimators are derived by straightforward optimization of $\mathcal{J}(\theta, \hat{p})$ wrt. the parameters. It is used that the data model, $p(\mathbf{X}, \mathbf{S}|\theta)$, factorizes. See, e.g., [8] for background, or [2] for details. The estimators for source i in segment n are:

$$\mu_{i,\text{new}}^n = \hat{s}_{i,1}^n$$

$$\Sigma_{i,\text{new}}^n = \mathbf{M}_{i,1}^n - \mu_{i,\text{new}}^n (\mu_{i,\text{new}}^n)^\top$$

$$(\mathbf{f}_{i,\text{new}}^n)^\top = \Big[\sum_{t=2}^\tau (\mathbf{m}_{i,t}^{1,n})^\top\Big] \Big[\sum_{t=1}^\tau \mathbf{M}_{i,t-1}^n\Big]^{-1}$$

$$q_{i,\text{new}}^n = \frac{1}{\tau - 1} \Big[\sum_{t=2}^\tau m_{i,t}^n - (\mathbf{f}_{i,\text{new}}^n)^\top \mathbf{m}_{i,t}^{1,n}\Big]$$

The stacked estimators, $\bar{\mu}_{\text{new}}^n, \bar{\Sigma}_{\text{new}}^n, \bar{\mathbf{F}}_{\text{new}}^n$ and $\bar{\mathbf{Q}}_{\text{new}}^n$ are reconstructed from the above as defined in section 2. The constraints on the parameters cause the above estimators to differ from those of the general Kalman model, which is not the case for $\bar{\mathbf{A}}_{\text{new}}$ and $\mathbf{R}_{\text{new}}$:

$$\bar{\mathbf{A}}_{\text{new}} = \Big[\sum_{n=1}^{N}\sum_{t=1}^{\tau} \mathbf{x}_t^n (\hat{\mathbf{s}}_t^n)^\top \Big]\Big[\sum_{n=1}^{N}\sum_{t=1}^{\tau} \bar{\mathbf{M}}_t^n\Big]^{-1}$$

$$\mathbf{R}_{\text{new}} = \frac{1}{N\tau}\sum_{n=1}^{N}\sum_{t=1}^{\tau} \text{diag}[\mathbf{x}_t^n(\mathbf{x}_t^n)^\top - \bar{\mathbf{A}}_{\text{new}}\hat{\mathbf{s}}_t^n(\mathbf{x}_t^n)^\top]$$

4 Estimating the Number of Sources Using BIC

In the following is described a scheme for determining d_s based on the likelihood of the parameters. A similar approach was taken in previous work, see [10]. Model control in a strictly Bayesian sense amounts to selecting the most probable hypothesis, based on the posterior probability of the model conditioned on the data:

$$p(d_s|\mathbf{X}) = \frac{p(\mathbf{X}|d_s)p(d_s)}{\sum_{d_s} p(\mathbf{X}, d_s)} \qquad (4)$$

In cases where all models, a priori, are to be considered equally likely, (4) reduces to $p(d_s|\mathbf{X}) \propto p(\mathbf{X}|d_s)$. The Bayes factor, $p(\mathbf{X}|d_s)$, is defined:

$$p(\mathbf{X}|d_s) = \int d\theta \, p(\mathbf{X}|\theta, d_s) p(\theta|d_s) \qquad (5)$$

Bayes information criterion (BIC), see [7], is an approximation of (5) to be applied in cases where the marginalization of θ is intractable:

$$p(\mathbf{X}|d_s) \approx p(\mathbf{X}|\theta_{ML}, d_s)\tau^{-\frac{|\theta|}{2}} \qquad (6)$$

The underlying assumptions are that (5) can be evaluated by Laplace integration, i.e. $\log p(\mathbf{X}|\theta, d_s)$ is well approximated by a quadratic function for large amounts of data ($\tau \to \infty$), and that the parameter prior $p(\theta|d_s)$ can be assumed constant under the integral.

5 Experiments

In order to demonstrate the applicability of the model control setup, a convolutive mixture of speech signals was generated and added with observation noise. The four models/hypotheses that we investigate in each time frame are that only one of two speakers are active, **1** and **2**, respectively, that both of them are active, **1+2**, or that none of them are active, **0**.

Recordings of male speech[1], which were also used in [11], were filtered through the $(2 \times 2 = 4)$ known channel filters:

$$\bar{\mathbf{A}} = \begin{bmatrix} 1.00 & 0.35 & -0.20 & 0.00 & 0.00, & 0.00 & 0.00 & -0.50 & -0.30 & 0.20 \\ 0.00 & 0.00 & 0.70 & -0.20 & 0.15, & 1.30 & 0.60 & 0.30 & 0.00 & 0.00 \end{bmatrix}$$

---

[1] Available at http://www.ipds.uni-kiel.de/pub_exx/bp1999_1/Proto.html.

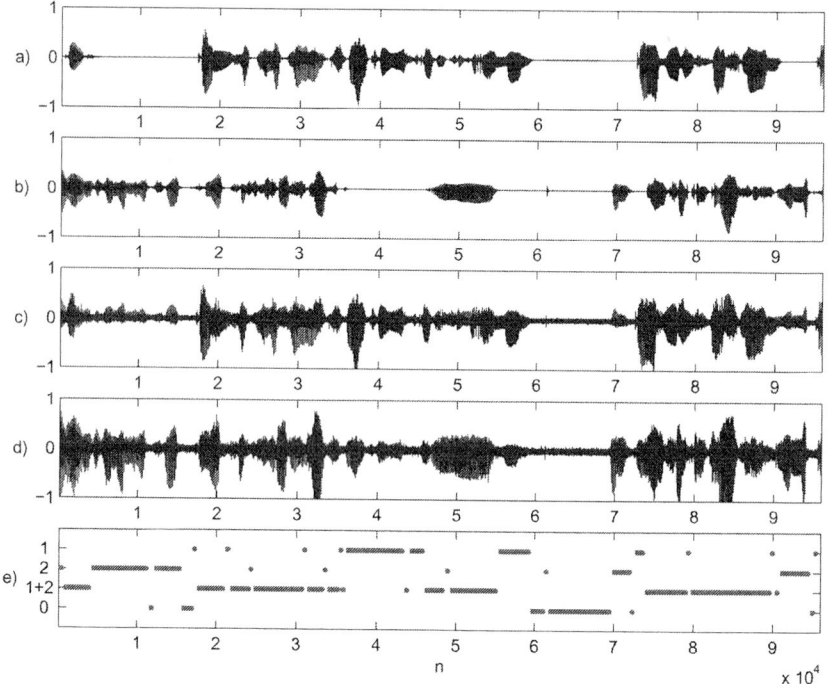

Fig. 3. From top to bottom, **a** & **b**) the original speech signals, **c** & **d**) the noisy mixtures and **e**) the most likely model in each segment. The four models are, **1**: first speaker exclusively active, **2**: second speaker exclusively active, **1+2**: both speakers simultaneously active and **0**: no speaker activity. A segment of 6 seconds of speech, sampled at $F_s = 16\text{kHz}$, is shown.

Observation noise was added to simulate SNR=15dB in the two sensor signals. KaBSS was then invoked in order to separate the signals and estimate $\bar{\mathbf{A}}$ and $\mathbf{R}$, as shown in [2]. The signals were segmented into frames of $\tau = 160$ samples. The obtained estimates of $\bar{\mathbf{A}}$ and $\mathbf{R}$ were treated as known true parameters in the following. In each segment and for each model-configuration, KaBSS was separately reinvoked to estimate the source model parameters, $\bar{\mathbf{F}}^n$, $\bar{\mathbf{Q}}^n$, and obtain the log-likelihood, $\mathcal{L}(\theta)$, of the various models. The four resulting $\mathcal{L}(\theta)$'s were then processed in the BIC model control scheme described in section 4. The number of samples in (6) were set to τ although the sensor signals are not i.i.d. This approximation is, however, acceptable due to the noisy character of speech. Figure 3 displays the source signals, the mixtures and the most likely hypothesis in each time frame. Convincingly, the MAP speech activity detector selects the correct model.

6 Conclusion

An EM algorithm, 'KaBSS', which builds on probabilistic inference in a generative linear convolutive mixture model with Gaussian sources was introduced in [2]. This contribution expands the model and its utility by showing that the exact computation of the log-likelihood, which is readily available as an output of the forward-backward recursion, can be exploited in a BIC-based model selection scheme. The result is an exploratory tool capable of determining the correct number of sources in a convolutive mixture. In particular, it was shown that the activity pattern of two speech sources in a convolutive mixture can be well estimated. Potential applications include the ability to select the correct model in speech enhancement and communication algorithms, hopefully resulting in more robust estimation.

References

1. Parra, L., Spence C., Convolutive blind separation of non-stationary sources. IEEE Transactions, Speech and Audio Processing (5), 320-7, 2000.
2. Olsson, R. K., Hansen L. K., Probabilistic blind deconvolution of non-stationary source. Proc. EUSIPCO, 2004, *submitted*.
3. Moulines E., Cardoso J. F., Gassiat E., Maximum likelihood for blind separation and deconvolution of noisy signals using mixture models, Proc. ICASSP (5), 3617-20, 1997.
4. Attias H., New EM algorithms for source separation and deconvolution with a microphone array. Proc. ICASSP (5), 297-300, 2003.
5. Todorovic-Zarkula S., Todorovic B., Stankovic M., Moraga C., Blind separation and deconvolution of nonstationary signals using extended Kalman filter. South-Eastern European workshop on comp. intelligence and IT, 2003.
6. Valpola H., Karhunen J, An unsupervised ensemble learning method for nonlinear dynamic state-space models. Neural Computation 14 (11), MIT Press, 2647-2692, 2002.
7. Schwartz G., Estimating the dimension of a model. Annals of Statistics (6), 461-464, 1978.
8. Roweis S., Ghahramani Z., Spence C., A unifying review of linear Gaussian models. Neural Computation (11), 305-345, 1999.
9. Doblinger G., An adaptive Kalman filter for the enhancement of noisy AR signals. IEEE Int. Symp. on Circuits and Systems (5), 305-308, 1998.
10. Højen-Sørensen P. A. d. F. R., Winther O., Hansen L. K., Analysis of functional neuroimages using ICA with adaptive binary sources. Neurocomputing (49), 213-225, 2002.
11. Peters B., Prototypische Intonationsmuster in deutscher Lese- und Spontansprache. AIPUK (34), 1-177, 1999.

Evaluation of Multistage SIMO-Model-Based Blind Source Separation Combining Frequency-Domain ICA and Time-Domain ICA

Satoshi Ukai[1], Hiroshi Saruwatari[1], Tomoya Takatani[1], Kiyohiro Shikano[1], Ryo Mukai[2], and Hiroshi Sawada[2]

[1] Nara Institute of Science and Technology
8916-5 Takayama-cho, Ikoma, Nara, 630-0192, Japan
sato-uk@is.aist-nara.ac.jp
[2] NTT Communication Science Laboratories
2-4, Hikaridai, Seika-cho, Soraku-gun, Kyoto, 619-0237, Japan

Abstract. In this paper, single-input multiple-output (SIMO)-model-based blind source separation (BSS) is addressed, where unknown mixed source signals are detected at the microphones, and these signals can be separated, not into monaural source signals but into SIMO-model-based signals from independent sources as they are at the microphones. This technique is highly applicable to high-fidelity signal processing such as binaural signal processing. First, we provide an experimental comparison between two kinds of the SIMO-model-based BSS methods, namely, traditional frequency-domain ICA with projection-back processing (FDICA-PB), and SIMO-ICA recently proposed by the authors. Secondly, we propose a new combination technique of the FDICA-PB and SIMO-ICA, which can achieve a more higher separation performance in comparison to two methods. The experimental results reveal that the accuracy of the separated SIMO signals in the simple SIMO-ICA is inferior to that of FDICA-PB, but the proposed combination technique can outperform both simple FDICA-PB and SIMO-ICA.

1 Introduction

Blind source separation (BSS) is the approach taken to estimate original source signals using only the information of the mixed signals observed in each input channel. In recent BSS works based on independent component analysis (ICA), various methods [1–4] have been proposed to deal with the separation of convolutive acoustical-sound mixtures, but these approaches only output each of the independent sound sources as a *monaural* signal. Accordingly, the separated sounds cannot maintain any spatial qualities of each sound source, e.g., directivity, localization, etc. This prevents any traditional BSS methods from being applied to binaural signal processing [5], or any high-fidelity acoustic signal processing.

In order to solve the problem, we should adopt a new blind separation framework in which Single-Input Multiple-Output (SIMO)-model-based BSS is considered. Here the term "SIMO" represents the specific transmission system in which the input is a single source signal and the outputs are its transmitted signals observed at multiple sensors. In the SIMO-model-based separation scenario, unknown multiple source signals which are mixed through unknown acoustical transmission channels are detected at the microphones, and these signals can be separated, not into monaural source signals but into SIMO-model-based signals from independent sources as they are at the microphones. Thus, the SIMO-model-based separated signals can maintain the spatial qualities of each sound source. Obviously the attractive feature is highly applicable to high-fidelity acoustic signal processing.

As an early contribution for SIMO-model-based BSS, Murata et al. have proposed frequency-domain ICA (FDICA) with projection-back processing [1] (hereafter we call it *FDICA-PB*). Also, we have proposed *SIMO-ICA* which consists of multiple time-domain ICAs (TDICAs) [6]. Following these methods, inspired by Nishikawa's multistage ICA approach [3], we are now studying and proposing a combination technique [7] of the FDICA-PB and SIMO-ICA, which can achieve a more higher separation performance with the low computational complexity. Our previous report [7], however, only showed limited and slightly unreliable experimental results in that the reverberation is too short, the number of data sets is little, and there is a lack of consistency on initial value in ICA. To improve them and provide more reliable evidences, this paper mainly describes an experimental evaluation of the proposed combination technique under more realistic conditions, and adds a discussion on the importance of the combination order. The experiments results explicitly reveal that the superiority of the proposed combination technique over the FDICA-PB or SIMO-ICA.

2 Mixing Process

In this study, the number of microphones is K and the number of multiple sound sources is L. The observed signals in which multiple source signals are mixed linearly are expressed as

$$\boldsymbol{x}(t) = \sum_{n=0}^{N-1} \boldsymbol{a}(n)\boldsymbol{s}(t-n), \qquad (1)$$

where $\boldsymbol{s}(t) = [s_1(t), \cdots, s_L(t)]^{\mathrm{T}}$ is the source signal vector, and $\boldsymbol{x}(t) = [x_1(t), \cdots, x_K(t)]^{\mathrm{T}}$ is the observed signal vector. Also, $\boldsymbol{a}(n) = [a_{kl}(n)]_{kl}$ is the mixing filter matrix with the length of N, $a_{kl}(n)$ is the impulse response between the k-th microphone and the l-th sound source, and $[X]_{ij}$ denotes the matrix which includes the element X in the i-th row and the j-th column. Hereafter, we only deal with the case of $K = L$ in this paper.

3 SIMO-Model-Based BSS 1: Conventional FDICA-PB

In the conventional FDICA-PB, first, the short-time analysis of observed signals is conducted by frame-by-frame discrete Fourier transform (DFT). By plotting the spectral values in a frequency bin of each microphone input frame by frame, we consider them as a time series. Hereafter, we designate the time series as $\boldsymbol{X}(f,t) = [X_1(f,t), \cdots, X_K(f,t)]^{\mathrm{T}}$.

Next, we perform signal separation using the complex-valued unmixing matrix, $\boldsymbol{W}(f) = [W_{lk}(f)]_{lk}$, so that the L time-series output $\boldsymbol{Y}(f,t)=[Y_1(f,t),\cdots,Y_L(f,t)]^{\mathrm{T}}$ becomes mutually independent; this procedure can be given as

$$\boldsymbol{Y}(f,t) = \boldsymbol{W}(f)\boldsymbol{X}(f,t). \qquad (2)$$

We perform this procedure with respect to all frequency bins. The optimal $\boldsymbol{W}(f)$ is obtained by, e.g., the following iterative updating:

$$\boldsymbol{W}^{[i+1]}(f) = \eta \Big[\boldsymbol{I} - \big\langle \boldsymbol{\Phi}(\boldsymbol{Y}(f,t))\boldsymbol{Y}^{\mathrm{H}}(f,t)\big\rangle_t\Big]\boldsymbol{W}^{[i]}(f) + \boldsymbol{W}^{[i]}(f), \qquad (3)$$

where $\langle\cdot\rangle_t$ denotes the time-averaging operator, $[i]$ is used to express the value of the i th step in the iterations, and η is the step-size parameter. In our research, we define the nonlinear vector function $\boldsymbol{\Phi}(\cdot)$ as $[e^{j\cdot\arg(Y_1(f,t))},\cdots,e^{j\cdot\arg(Y_L(f,t))}]^{\mathrm{T}}$, where $\arg(\cdot)$ represents an operation to take the argument of the complex value [4]. After the iterations, the permutation problem, i.e., indeterminacy in ordering sources, can be solved by [8].

Finally, in order to obtain the SIMO components, the separated signals are projected back onto the microphones by using the inverse of $\boldsymbol{W}(f)$ [1]. In this method, the following operation is performed.

$$Y_k^{(l)}(f,t) = \Big\{\boldsymbol{W}(f)^{-1}[\overbrace{0,\cdots,0}^{l-1}, Y_l(f,t), \overbrace{0,\cdots,0}^{L-l}]^{\mathrm{T}}\Big\}_k, \qquad (4)$$

where $Y_k^{(l)}(f,t)$ represents the l-th resultant separated source signal which is projected back onto the k-th microphone, and $\{\cdot\}_k$ denotes the k-th element of the argument.

The FDICA-PB has the advantage that **(F1)** this method is very fast and nonsensitive to the initial value in the iterative updating because the calculation of FDICA given by (3) and the projection-back processing given by (4) are simple. There exists, however, the disadvantages that **(F2)** the inversion of $\boldsymbol{W}(f)$ often fails and yields harmful results because the invertibility of every $\boldsymbol{W}(f)$ cannot be guaranteed, and **(F3)** the circular convolution effect inherent in FDICA is likely to cause the deterioration of the separation performance.

4 SIMO-Model-Based BSS 2: SIMO-ICA

The SIMO-ICA [6] consists of $(L-1)$ TDICA parts and a *fidelity controller*, and each ICA runs in parallel under the fidelity control of the entire separation system. The separated signals of the l-th ICA ($l = 1, \cdots L-1$) in SIMO-ICA are defined by

$$\boldsymbol{y}_{(\mathrm{ICA}l)}(t) = [y_k^{(\mathrm{ICA}l)}(t)]_{k1} = \sum_{n=0}^{D-1} \boldsymbol{w}_{(\mathrm{ICA}l)}(n)\boldsymbol{x}(t-n), \qquad (5)$$

where $\boldsymbol{w}_{(\mathrm{ICA}l)}(n) = [w_{ij}^{(\mathrm{ICA}l)}(n)]_{ij}$ is the separation filter matrix in the l-th ICA, and D is the filter length.

Regarding the fidelity controller, we calculate the following signal vector, in which the all elements are to be mutually independent,

$$\boldsymbol{y}_{(\mathrm{ICA}L)}(t) = \boldsymbol{x}(t - D/2) - \sum_{l=1}^{L-1} \boldsymbol{y}_{(\mathrm{ICA}l)}(t). \tag{6}$$

Hereafter, we regard $\boldsymbol{y}_{(\mathrm{ICA}L)}(t)$ as an output of a *virtual* "L-th" ICA. To explicitly show the meaning of the fidelity controller, we rewrite (6) as $\sum_{l=1}^{L} \boldsymbol{y}_{(\mathrm{ICA}l)}(t) - \boldsymbol{x}(t - D/2) = 0$. This equation means a constraint to force the sum of all ICAs' output vectors $\sum_{l=1}^{L} \boldsymbol{y}_{(\mathrm{ICA}l)}(t)$ to be the sum of all SIMO components $[\sum_{l=1}^{L} \sum_{n=0}^{N-1} a_{kl}(n) s_l(n - t + D/2)]_{k1} (= \boldsymbol{x}(t - D/2))$. Here the delay of $D/2$ is used as to deal with nonminimum phase systems. Using (5) and (6), we can obtain the appropriate separated signals and maintain their spatial qualities as follows.

Theorem: If the independent sound sources are separated by (5), and simultaneously the signals obtained by (6) are also mutually independent, then the output signals converge on unique solutions, up to the permutation, as

$$\boldsymbol{y}_{(\mathrm{ICA}l)}(t) = \sum_{n=0}^{L-1} \mathrm{diag}\left[\boldsymbol{a}(n)\boldsymbol{P}_l^{\mathrm{T}}\right] \boldsymbol{P}_l \boldsymbol{s}(n - t + D/2), \tag{7}$$

where $\boldsymbol{P}_l$ ($l = 1, \cdots, L$) are exclusively-selected permutation matrices which satisfy

$$\sum_{l=1}^{L} \boldsymbol{P}_l = [1]_{ij}. \tag{8}$$

Regarding a proof of the theorem, see [6]. Obviously the solutions given by (7) provide necessary and sufficient SIMO components, $\sum_{n=0}^{L-1} a_{kl}(n) s_l(n-t+D/2)$, for each l-th source.

In order to obtain (7), the natural gradient of Kullback-Leibler divergence of (6) with respect to $\boldsymbol{w}_{(\mathrm{ICA}l)}(n)$ should be added to the existing TDICA-based iterative learning rule [2] of the separation filter in the l-th ICA ($l = 1, \cdots, L-1$). The new iterative algorithm of the l-th ICA part ($l = 1, \cdots, L-1$) in SIMO-ICA is given as

$$\boldsymbol{w}_{(\mathrm{ICA}l)}^{[i+1]}(n) = \boldsymbol{w}_{(\mathrm{ICA}l)}^{[i]}(n)$$
$$- \alpha \sum_{d=0}^{D-1} \left[\left\{\mathrm{off\text{-}diag}\left\langle \boldsymbol{\varphi}(\boldsymbol{y}_{(\mathrm{ICA}l)}^{[i]}(t))\boldsymbol{y}_{(\mathrm{ICA}l)}^{[i]}(t - n + d)^{\mathrm{T}}\right\rangle_t\right\}\right.$$
$$\cdot \boldsymbol{w}_{(\mathrm{ICA}l)}^{[i]}(d) - \left\{\mathrm{off\text{-}diag}\left\langle \boldsymbol{\varphi}(\boldsymbol{x}(t - \frac{D}{2}) - \sum_{l=1}^{L-1} \boldsymbol{y}_{(\mathrm{ICA}l)}^{[i]}(t))\right.\right.$$
$$\left.\left.\cdot \left(\boldsymbol{x}(t - n + d - \frac{D}{2}) - \sum_{l=1}^{L-1} \boldsymbol{y}_{(\mathrm{ICA}l)}^{[i]}(t - n + d)^{\mathrm{T}}\right)\right\rangle_t\right\}\right]$$

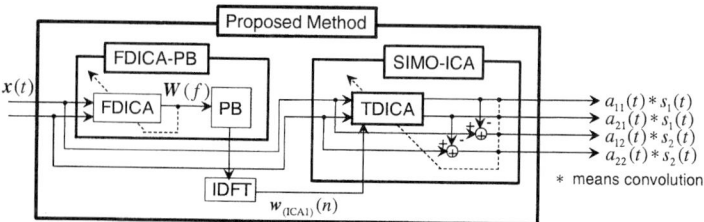

Fig. 1. Example of input and output relations in proposed method in the case of 2 sources with 2 microphones.

$$\cdot \left(\boldsymbol{I}\delta(d - \frac{D}{2}) - \sum_{l=1}^{L-1} \boldsymbol{w}_{(\mathrm{ICA}l)}^{[i]}(d) \right) \bigg], \tag{9}$$

where α is the step-size parameter, $\delta(n)$ is a delta function, i.e., $\delta(0) = 1$ and $\delta(n) = 0$ $(n \neq 0)$, and $\varphi(\cdot)$ is the nonlinear vector function, e.g., the l-th element is $y_l(t)/|y_l(t)|$. Also, the initial values of $\boldsymbol{w}_{(\mathrm{ICA}l)}(n)$ for all l should be different.

The SIMO-ICA has the following advantage and disadvantage. **(T1)** This method is free from both the circular convolution effect and the invertibility of the separation filter matrix. **(T2)** Since the SIMO-ICA is based on TDICA which involves more complex calculations than FDICA, the convergence of the SIMO-ICA is very slow, and the sensitivity to the initial settings of separation filter matrices is very high.

5 Proposed Combination Technique of FDICA-PB and SIMO-ICA

As described above, two kinds of SIMO-model-based BSS methods have some disadvantages. However, note that the advantages and disadvantages of FDICA-PB and SIMO-ICA are mutually complementary, i.e., (F2) and (F3) can be resolved by (T1), and (T2) can be resolved by (F1). Therefore, we propose a new multistage technique combining FDICA-PB and SIMO-ICA.

The proposed multistage technique is conducted with the following steps (see Fig. 1). In the first step, we perform FDICA to separate the source signals to some extent with the fast- and robust-convergence advantage (F1). After the FDICA, we generate a specific initial value $\boldsymbol{w}_{(\mathrm{ICA}l)}^{[0]}(n)$ for SIMO-ICA performed in the next step by using $\boldsymbol{W}(f)$ obtained from FDICA. This procedure is given by

$$\boldsymbol{w}_{(\mathrm{ICA}l)}^{[0]}(n) = \mathrm{IFFT}\left[\mathrm{diag}\left[\boldsymbol{W}(f)^{-1}\boldsymbol{P}_l^\mathrm{T} \right] \boldsymbol{P}_l \boldsymbol{W}(f) \right], \tag{10}$$

where $\boldsymbol{P}_l$ are set to be, e.g., (8), and IFFT[·] represents an inverse DFT with the time shift of $D/2$ samples. In the final step, we perform SIMO-ICA (9) to obtain resultant SIMO components with the advantage (T1).

Compared with the simple SIMO-ICA, this combination algorithm is not so sensitive to the initial value of the separation filter because FDICA is used for estimating the good initial value. Also, this technique has the possibility to provide a more accurate separation result over the simple FDICA because the resultant quality of the output signal is determined by the separation ability of the SIMO-ICA starting from the good initial state.

6 Experiments and Results

6.1 Conditions for Experiment

A two-element array with an interelement spacing of 4 cm is assumed. The speech signals are assumed to arrive from two directions, $-30°$ and $40°$. The distance between the microphone array and the loudspeakers is 1.15 m. Two kinds of sentences spoken by two male and two female speakers are used as the source speech samples. Using these sentences, we obtain 12 combinations. The sampling frequency is 8 kHz and the length of speech is limited to 7.5 seconds. To simulate the convolutive mixtures, the source signals are convolved with two kinds of impulse responses recorded in the experimental room which has a reverberation time (RT) of 150 ms or 300 ms. The length of the separation filter is set to be 2048 in both FDICA-PB and SIMO-ICA. The initial value in both the methods is null-beamformer whose directional null is steered to $\pm 45°$.

As an objective evaluation score, *SIMO-model accuracy* (SA) [9] is used to indicate a degree of similarity (mean-squared-error) between the SIMO-model-based BSSs' outputs and the original SIMO-model-based signals ($\sum_{n=0}^{N-1} a_{kl}(n)$ $s_l(n - t + D/2)$).

6.2 Comparison of Conventional Method and Proposed Method

Figure 2 and 3 show the results of SAs for FDICA-PB, SIMO-ICA, and the proposed combination technique in all speaker combinations, for each of reverberation conditions. In the results of the proposed combination technique, there exists a consistent improvement of SA compared with FDICA-PB as well as the simple SIMO-ICA. In RT = 150 ms, the average score of the improvement is 8.3 dB over SIMO-ICA, and is 2.9 dB over FDICA-PB. Also, in RT = 300 ms, the average score of the improvement is 5.1 dB over SIMO-ICA, and is 2.9 dB over FDICA-PB. From these results, we can conclude that the proposed combination technique can assist the SIMO-ICA in improving the separation performance, and successfully achieve the SIMO-model-based BSS under reverberant conditions.

6.3 Discussion on Combination Order

As described in the previous section, the combination of FDICA-PB and SIMO-ICA can contribute to the improvement of separation. In this combination, the advantage of FDICA-PB is useful in the initial step of separation procedure and thd advantage of SIMO-ICA is also useful in the later step. Therefore we use FDICA-PB as the first-stage BSS and SIMO-ICA as the second-stage BSS. In order to confirm the availability of this combination order, we compare the proposed combination with another combination in which SIMO-ICA is used in

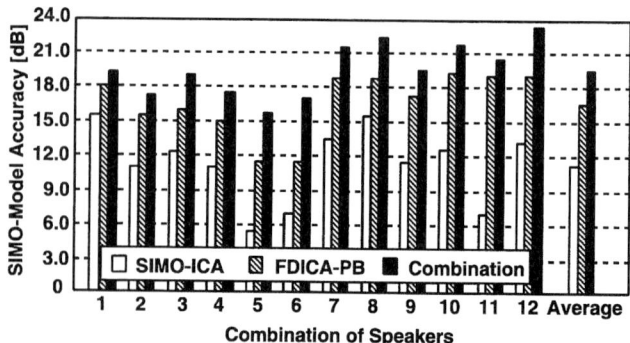

Fig. 2. Comparison of SIMO-model accuracy among conventional FDICA-PB, proposed SIMO-ICA, proposed combination technique (RT is 150 ms).

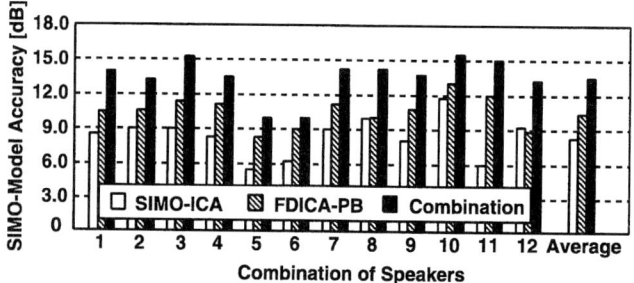

Fig. 3. Comparison of SIMO-model accuracy among conventional FDICA-PB, proposed SIMO-ICA, proposed combination technique (RT is 300 ms).

the first stage and FDICA-PB is used in the second stage (hereafter we designate this combination as "Swapped Combination").

The experiment of Swapped Combination was carried out in the following manner. Regarding SIMO-ICA part in Swapped Combination, parameters of SIMO-ICA part are the same as those of the simple SIMO-ICA in Sect. 6.2. Leaning of SIMO-ICA part is stopped at the peak of SA. As for FDICA-PB part in Swapped Combination, the analysis conditions and the parameter of FDPCA-PB are the same as those of the simple FDICA-PB given in Sect. 6.2.

We show the result of comparison of the simple SIMO-ICA, simple FDICA-PB, proposed combination, and Swapped Combination in Table 1. The result of Table 1 is the average of 12 experiments with different combinations of speakers. The average SA of 11.2 dB is obtained in Swapped Combination, and this performance is still better than that of simple SIMO-ICA and almost the same as that of simple FDICA-PB, but it is poorer than that of the proposed combination. In Swapped Combination, the SA is still improved by using FDICA-PB in the second stage, however, the separation performance is saturated because of the disadvantages (F2) and (F3) of FDICA-PB. This fact indicates that the proposed combination order (FDICA-PB in the first stage and SIMO-ICA in the second stage) is essential and the best.

Table 1. Comparison of SIMO-model accuracy among FDICA-PB, SIMO-ICA, proposed combination (Proposed), and Swapped Combination (Swapped) (unit is dB).

	SIMO-ICA	FDICA-PB	Proposed	Swapped
RT = 150 ms	11.3	16.7	**19.6**	17.1
RT = 300 ms	8.4	10.6	**13.4**	11.2

7 Conclusion

In this paper, first, the conventional FDICA-PB and the proposed SIMO-ICA were compared under a reverberant condition to evaluate the feasibility of SIMO-model-based BSS. Secondly, we proposed a new combination technique of FDICA-PB and SIMO-ICA to achieve the more higher separation performance compared with each of two methods. The experimental results revealed that the accuracy of the separated SIMO signals in the simple SIMO-ICA is inferior to that of FDICA-PB under low-quality initial value conditions, but the proposed combination technique of FDICA-PB and SIMO-ICA can outperform both simple FDICA-PB and SIMO-ICA. The average of the improvement was 8.3 dB over SIMO-ICA, and was 2.9 dB over FDICA-PB in RT = 150 ms, and 5.1 dB over SIMO-ICA, and was 2.9 dB over FDICA-PB in RT = 300 ms.

Acknowledgement. This work was partly supported by CREST Program "Advanced Media Technology for Everyday Living" of JST in Japan.

References

1. Murata, N., Ikeda, S.: An on-line algorithm for blind source separation on speech signals. In: Proc. NOLTA '98. Volume 3. (1998) 923–926
2. Choi, S., Amari, S., Cichocki, A., Liu, R.: Natural gradient learning with a nonholonomic constraint for blind deconvolution of multiple channels. In: Proc. Int. Workshop on ICA and BSS (ICA'99). (1999) 371–376
3. Nishikawa, T., Saruwatari, H., Kiyohiro, S.: Blind source separation of acoustic signal based on multistage ICA combining frequency-domain ICA and time-domain ICA. IEICE Trans. Fundam. **E86-A** (2003) 846–858
4. Sawada, H., Mukai, R., Araki, S., Makino, S.: Polar coordinate based nonlinear function for frequency domain blind source separation. IEICE Trans. Fundam. **E86-A** (2003) 590–596
5. Blauert, J.: Spatial Hearing. revised edn. Cambridge, MA: The MIT Press (1997)
6. Takatani, T., Nishikawa, T., Saruwatari, H., Shikano, K.: High-fidelity blind separation of acoustic signals using SIMO-model-based ICA with information-geometric learning. In: Proc. IWAENC2003. (2003) 251–254
7. Ukai, S., Saruwatari, H., Takatani, T., Mukai, R., Sawada, H.: Multistage SIMO-model-based blind source separtion combining frequency-domain ICA and time-domain ICA. In: ICASSP2004 (accepted). (2004)
8. Sawada, H., Mukai, R., Araki, S., Makino, S.: A robust and precise method for solving the permutation problem of frequency-domain blind source separation. In: Proc. Int. Sympo. on ICA and BSS. (2003) 505–510
9. Yamajo, H., Saruwatari, H., Takatani, T., Nishikawa, T., Shikano, K.: Evaluation of blind separation and deconvolution for convolutive speech mixture using SIMO-model-based ICA. In: Proc. IWAENC2003. (2003) 299–302

On Coefficient Delay in Natural Gradient Blind Deconvolution and Source Separation Algorithms

Scott C. Douglas[1], Hiroshi Sawada[2], and Shoji Makino[2]

[1] Department of Electrical Engineering
Southern Methodist University
Dallas, Texas 75275 USA

[2] NTT Communication Science Laboratories
NTT Corporation
Kyoto 619-0237 Japan

Abstract. In this paper, we study the performance effects caused by coefficient delays in natural gradient blind deconvolution and source separation algorithms. We present a statistical analysis of the effect of coefficient delays within such algorithms, quantifying the relative loss in performance caused by such coefficient delays with respect to delayless algorithm updates. We then propose a simple change to one such algorithm to improve its convergence performance.

1 Introduction

The related problems of blind source separation and multichannel deconvolution of convolutive signal mixtures have received much attention recently in the signal processing literature [1]–[4]. Interest in such tasks has been largely driven by the development of useful and powerful algorithms for separating and deconvolving such signal mixtures without specific knowledge of the source signals or the mixing conditions. Several of these techniques employ a natural gradient modification in which the estimated inverse model is applied to the parameter updates to improve convergence performance [5]. Since the system being estimated is a multichannel filter, such algorithms invariably employ filtered-gradient updates that have strong ties to classic procedures in model-reference adaptive control [6].

In developing solutions to blind deconvolution and source separation tasks, system designers have many design choices to make. It is often convenient to choose an algorithm structure that re-uses existing computed quantities to minimize the overall complexity of the coefficient updates. Another important design tool is the approximation of two-sided infinite-impulse response (IIR) systems by truncated finite-impulse response (FIR) models. Moreover, when a derived procedure requires signals that are non-causally related to the system's operation at any given time, signal and coefficient delays are often introduced within the algorithm updates to maintain causal operation. A combination of these design choices is often required to achieve a practical algorithm from its theoretical derivation. The effects of these design choices on overall system performance, however, is unclear. A careful study of any of their effects in a specific context would help system designers understand the tradeoffs involved in building blind deconvolution and source separation algorithms that are efficient, useful, and practical.

In this paper, we study the performance effects caused by coefficient delays in one well-known natural gradient multichannel blind deconvolution and source separation procedure [1]. Using a simplified adaptation model, we demonstrate through both analysis and simulation that, for a given convergence rate, algorithms that have coefficient delays within their updates exhibit worse performance than those without such delays. Simulations and analysis also show that recomputing the delayed equalizer output with the most-recent equalizer coefficients within the algorithm nonlinearity can improve this procedure's performance.

2 Coefficient Delay in Natural Gradient Adaptation

As described in [5], natural gradient adaptation is a modified gradient search in which the gradient search direction is modified by the Riemannian metric tensor for the associated parameter space. In [1], a simple but powerful algorithm for multichannel blind deconvolution and source separation is derived using the Kullback-Leibler divergence measure as the optimization criterion. This algorithm has been derived assuming a particular set of coefficient delays for the various signal quantities within the updates to minimize the number of arithmetic operations needed for its implementation. In order to account for the coefficient delays in this algorithm, we present a generalized version of this procedure for which the use of coefficient delays is carefully delineated. For notational simplicity, we shall focus on the single-channel blind deconvolution task in this paper, although our discussions could be easily extended to the multichannel case with minor effort.

Let $s(k)$ denote a sequence of i.i.d. random variables. We observe a filtered version of this sequence given by

$$x(k) = \sum_{i=0}^{\infty} a_i s(k-i), \qquad (1)$$

where a_i is the impulse response of an unknown mixing filter. We desire a linear filter $W(z)$ that extracts a scaled, time-delayed version of $s(k)$ from $x(k)$. A single-channel generalized version of the natural gradient algorithm derived in [1] computes an estimated source sequence as

$$y_n(k) = \sum_{l=0}^{L} w_l(n) x(k-l), \qquad (2)$$

where n denotes the time index of the equalizer filter coefficients $\{w_l(n)\}$, $0 \leq l \leq L$, and k the time-shift of the input signal. We compute a set of filtered output signals, given by

$$u_{n_1,n_2,\ldots n_{2L+2}}(k) = \sum_{q=0}^{L} w^*_{L-q}(n_{q+L+2}) y_{n_{q+1}}(k-q), \tag{3}$$

where $\{n_1, n_2, \ldots, n_{2L+2}\}$ denote time indices for the filter coefficients used in this calculation. Then, the $(L+1)$ coefficients $\{w_l(k)\}$ are updated as

$$w_l(k+1) = (1+\mu)w_l(k) - \mu f(y_{n_0^{(l)}}(k-L)) u^*_{n_1^{(l)}, n_2^{(l)}, \ldots n_{2L+2}^{(l)}}(k-l), \tag{4}$$

where $\{n_i^{(l)}\}$ denote the time indices of the coefficients used to update the lth equalizer tap. The above description employs $(2L+3)(L+1)$ different time indices for the coefficient updates, and a practical algorithm requires careful choice of the values of $\{n_i^{(l)}\}$ to allow both a computationally-efficient and statistically-effective algorithm. The only constraint imposed on the values of $\{n_i^{(l)}\}$ is that $n_i^{(l)} \leq k$ to maintain causality of the overall system.

The algorithm in [1] employs the following choices for the coefficient delays in (2)–(4): $n_0^{(l)} = k - L$ for all $0 \leq l \leq L$, $n_{q+1}^{(l)} = k - q - l$ and $n_{q+L+2}^{(l)} = k - l$ for $0 \leq q \leq L$. With these choices, both $y.(k)$ and $u.(k)$ become one-dimensional signals, such that delayed versions of $y.(k)$ and $u.(k)$ are all that are needed to implement the algorithm. The resulting procedure requires about four multiply/adds per filter tap to implement, not counting the nonlinearity computation $f(y_{k-L}(k-L))$. These choices, however, are not the best from the standpoint of system performance. As is well-known in adaptive control [6], algorithms that have the least adaptation delay within the coefficient updates usually perform the best, implying that $n_i^{(l)} = k$ should be chosen for the above procedure. The computational penalty paid for such an update is severe–the algorithm would require $(3L+4)$ multiply/adds per filter coefficient to implement. Clearly, a trade-off between algorithm complexity and algorithm performance must be made for practical reasons, especially when L is large. But how do coefficient delays affect overall performance of the system in a continuously-adapting scenario?

3 A Simplified Adaptation Model and Its Analysis

To better understand the performance effects caused by coefficient delays in natural gradient algorithms, we propose to study the following four single-coefficient adaptive systems operating on the i.i.d. sequence $x(k)$:

$$w(k+1) = (1+\mu)w(k) - \mu f(w(k)x(k))x(k)|w(k)|^2 \tag{5}$$

$$w(k+1) = (1+\mu)w(k) - \mu f(w(k-D)x(k))x(k)|w(k)|^2 \tag{6}$$

$$w(k+1) = (1+\mu)w(k) - \mu f(w(k)x(k))x(k)|w(k-D)|^2 \tag{7}$$

$$w(k+1) = (1+\mu)w(k) - \mu f(w(k-D)x(k))x(k)|w(k-D)|^2 \tag{8}$$

In these algorithms, D is an integer parameter that sets the coefficient delays within the updates. Eqn. (5) is similar in design to (2)–(4) when $n_i^{(l)} = k$ for all i and l. Eqn. (8) is similar in design to the procedure in [1]. The two algorithms in (6) and (7) are similar to versions of (2)–(4) in which coefficient delays appear in the cost function and in the natural gradient update modification, respectively.

By studying these variants, we can determine through analysis whether algorithms with significant coefficient delay within the updates, represented by $D \gg 1$, cause significant degradation in overall system performance. To make the analysis tractable, we shall make some additional assumptions regarding $x(k)$ and the form of $f(y)$. Specifically,

- $x(k) \sim \text{Unif}(-\sqrt{3}, \sqrt{3})$ is an i.i.d. uniformly-distributed sequence with unit variance, $m_4 = E\{x^4(k)\} = 1.8$, and $m^8 = E\{x^8(k)\} = 9$, and
- $f(y) = y^3$ is a cubic nonlinearity, such that the above procedures are locally-stable for negative-kurtosis $x(k)$.

Specific statistical assumptions and nonlinear update forms are often chosen to perform convergence analyses of linear adaptive filtering algorithms [7]. With these assumptions, we can determine the initial convergence behavior of the mean value of $w(k)$ over time as well as the steady-state value of the variance of $w(k)$ at convergence.

With these choices, the relation in (8) becomes

$$w(k+1) = (1+\mu)w(k) - \mu w^5(k-D)x^4(k) \qquad (9)$$
$$= (1+\mu)w(k) - \mu m_4 w^5(k-D) + \mu\left(x^4(k) - m_4\right)w^5(k-D) \quad (10)$$
$$= (1+\mu)w(k) - \mu m_4 w^5(k-D) + \mu\nu(k) \qquad (11)$$

where we have defined $\nu(k) = \left(x^4(k) - m_4\right)w^5(k-D)$ as a coefficient-dependent noise-like term that drives the deterministic nonlinear system given by

$$\overline{w}(k+1) = (1+\mu)\overline{w}(k) - \mu m_4 \overline{w}^5(k-D). \qquad (12)$$

Clearly, the initial convergence behavior of $w(k)$ is dominated on average by the dynamics of the corresponding deterministic system in (12), and the influence of the zero-mean signal $\nu(k)$ is significant only near convergence. Thus, we can simulate the behavior of the deterministic system in (12) with $\overline{w}(0) = w(0)$ to understand how $w(k)$ converges to its optimum value. It is straightforward to show that the stationary point of (12) occurs when $\overline{w}(k) = m_4 \overline{w}^5(k)$, or

$$\overline{w}_{ss} = (m_4)^{-1/4} = \pm 0.86334\ldots \qquad (13)$$

for uniformly-distributed unit-variance input signals.

To determine the variance of $w(k)$ at convergence, we can used a linearized analysis similar to that employed in [8]. Let $w(k) = \overline{w}_{ss} + \Delta(k)$, where $|\Delta(k)| \ll |\overline{w}_{ss}|$. Then, we can represent (8) as

$$\overline{w}_{ss} + \Delta(k+1) = (1+\mu)(\overline{w}_{ss} + \Delta(k)) - \mu m_4(\overline{w}_{ss}^5 + 5\overline{w}_{ss}^4 \Delta(k-D))$$
$$+ O(\mu\Delta^2(k-D)) + \mu\nu(k). \qquad (14)$$

Table 1. Analysis Results for Single-Coefficient Models

Eqn.	$\overline{w}(k+1)$ Update Relation	$H(z)$
(5)	$(1+\mu)\overline{w}(k) - \mu m_4 \overline{w}^5(k)$	$\dfrac{\mu z^{-1}}{1-(1-4\mu)z^{-1}}$
(6)	$(1+\mu)\overline{w}(k) - \mu m_4 \overline{w}^2(k)\overline{w}^3(k-D)$	$\dfrac{\mu z^{-1}}{1-(1-\mu)z^{-1}+3\mu z^{-D}}$
(7)	$(1+\mu)\overline{w}(k) - \mu m_4 \overline{w}^3(k)\overline{w}^2(k-D)$	$\dfrac{\mu z^{-1}}{1-(1-2\mu)z^{-1}+2\mu z^{-D}}$
(8)	$(1+\mu)\overline{w}(k) - \mu m_4 \overline{w}^5(k-D)$	$\dfrac{\mu z^{-1}}{1-(1+\mu)z^{-1}+5\mu z^{-D}}$

Using the relationship for $\overline{w}_{ss}$ in (13), (14) simplifies to

$$\Delta(k+1) - (1+\mu)\Delta(k) + \mu 5 m_4 \overline{w}_s^4 \Delta(k-D) = \mu\nu(k). \tag{15}$$

Taking z-transforms of both sides, we can relate $D(z)$, the z-transform of $\Delta(k)$, to $N(z)$, the z-transform of $\nu(k)$, as

$$D(z) = H(z)N(z), \tag{16}$$

where the transfer function $H(z)$ is given by

$$H(z) = \frac{\mu z^{-1}}{1-(1+\mu)z^{-1}+5\mu z^{-D}}. \tag{17}$$

Assuming that each $\nu(k)$ is i.i.d., the power of $\Delta(k)$ in steady state is given by

$$E\{\Delta^2(k)\}_{ss} = E\{\nu^2(k)\}_{ss} \sum_{l=0}^{\infty} h^2(l), \tag{18}$$

where $h(l)$ is the inverse z-transform of $H(z)$ and

$$E\{\nu^2(k)\}_{ss} = E\{(x^4(k)-m_4)^2\}w_{ss}^{10} \tag{19}$$
$$= (m_8 - m_4^2)w_{ss}^{10} = 1.325077\ldots \tag{20}$$

for uniformly-distributed input signals.

Similar analyses can be carried out for the algorithms in (5), (6), and (7), respectively. Table 1 shows the forms of the update relations for $\overline{w}(k)$ and the corresponding $H(z)$ in each case. The specific derivations are omitted for brevity.

These results allow us to fairly and accurately compare the performances of the different single-coefficient procedures in (5)–(8) by carefully maintaining certain convergence relationships between them. For example, we can choose different step size values μ for each procedure to maintain an identical convergence rate from a specific initial $w(0)$ and determine analytically the values of $E\{\Delta^2(k)\}_{ss}$ in steady-state. These results can also be compared with simulations of both the single-coefficient algorithms and their related blind deconvolution procedures in (2)–(4) for specific forms of coefficient delay.

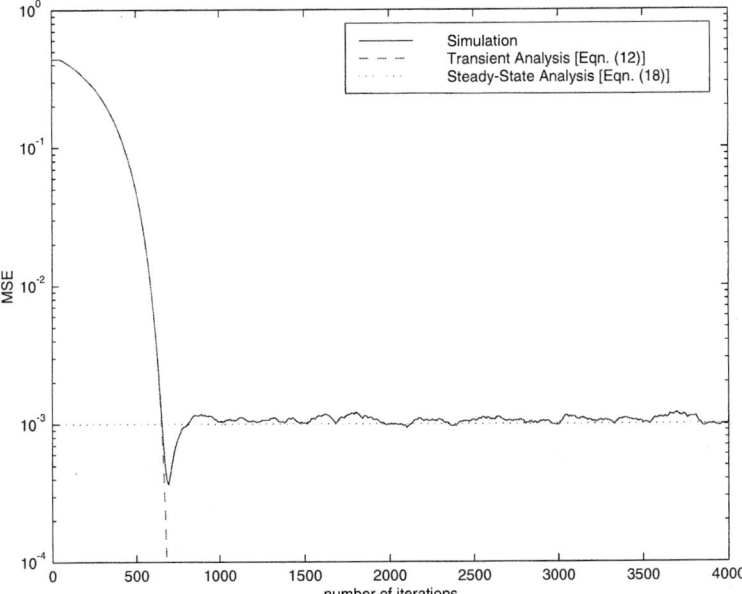

Fig. 1. Comparison of analysis and simulation results for the single-coefficient update in Eqn. (11).

4 Simulations

We now verify the analytical results presented in the previous section and compare these with simulation results from single-channel blind deconvolution tasks. From these results, we can gauge what performance degradations are caused by coefficient delays within the algorithm updates.

Our first set of simulations is designed to verify the analytical results for the single-coefficient systems in (5)–(8). For these simulations, $x(k)$ is a unit-variance Unif$[-\sqrt{3}, \sqrt{3}]$ random sequence, and we have arbitrarily chosen $D = 50$, $w(0) = 0.2$ and $\mu = 2.7062 \times 10^{-3}$. With these choices, our analyses predict that the procedure in (8) will converge to a steady-state variance of $E\{\Delta^2(k)\}_{ss} = 0.001$ in 586 iterations. Shown in Fig. 1 are (a) the evolution of $(\overline{w}(k) - |m_4^{-1/4}|)^2$, (b) the value of $E\{\Delta^2(k)\}_{ss}$ predicted from the analysis, and (c) the evolution of the coefficient MSE $E\{(w(k) - |m_4^{-1/4}|)^2\}$ as computed from ensemble averages of 1000 simulation runs. As can be seen, both $(\overline{w}(k) - |m_4^{-1/4}|)^2$ and $E\{\Delta^2(k)\}_{ss}$ are accurate predictors of the coefficient MSE during the transient and steady-state phases of adaptation, respectively.

Shown in Fig. 2 are the evolutions of the coefficient MSEs for the four systems in (5)–(8) as well as their predicted steady-state MSEs from (18). Here, we have chosen step sizes for each algorithm such that all of them converge from $w(0) = 0.2$ to an MSE of 0.001 in 563 iterations; thus, we can accurately compare the steady-state MSEs of each approach. As can be seen, the algorithm with

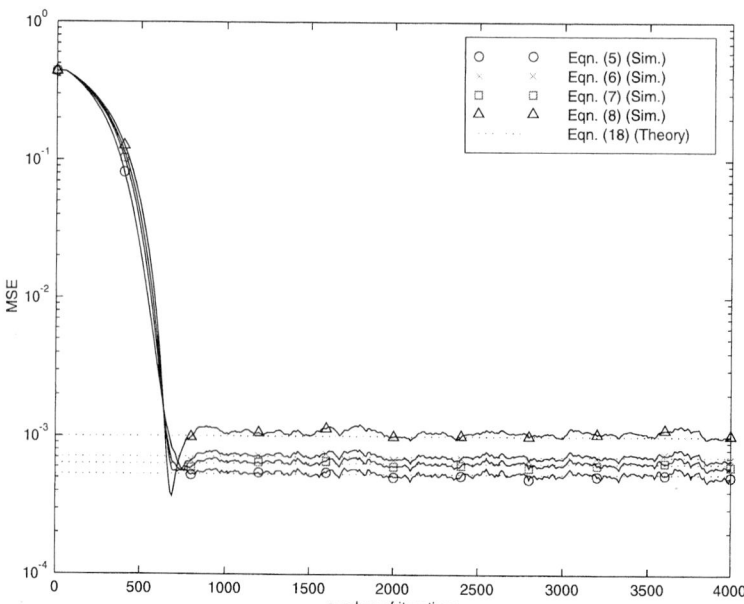

Fig. 2. Comparison of steady-state MSE analysis and simulation results for the single-coefficient update in Eqns. (5)–(8).

no adaptation delay performs the best, and the simulated algorithms in (6), (7), and (8) have steady-state MSEs that are 1.31, 0.78, and 2.99 dB greater, respectively, than that of (5). The steady-state MSE analysis determined from (18) and Table 1 predicts similar performance degradations of 1.26, 0.76, and 2.74 dB, respectively.

An adaptive filtering analysis is only useful if it is an accurate predictor of performance relationships for a given system and scenario. We now explore the simulated performance of the single-channel blind deconvolution procedure in (2)–(4) for various choices of adaptation delay within a particular blind deconvolution task in which $x(k)$ is generated from an i.i.d. uniformly-distributed sequence $s(k)$ as

$$x(k) = 0.7x(k-1) + \sum_{i=0}^{10}(0.4)^{10-i}s(k-i) \qquad (21)$$

This non-minimum-phase system cannot be equalized using simple linear prediction. The algorithms studied in this case are:

- **Case 1: All $n_i^{(l)} = k$.** This procedure most closely resembles (5).
- **Case 2: $n_0^{(l)} = k$ for all $0 \leq l \leq L$; $n_{q+1}^{(l)} = k - q - l$ and $n_{q+L+2}^{(l)} = k - l$ for $0 \leq q \leq L$.** This procedure most closely resembles (7) with $D = L$.
- **Case 3: The original procedure in [1].** This procedure most closely resembles (8) with $D = L$.

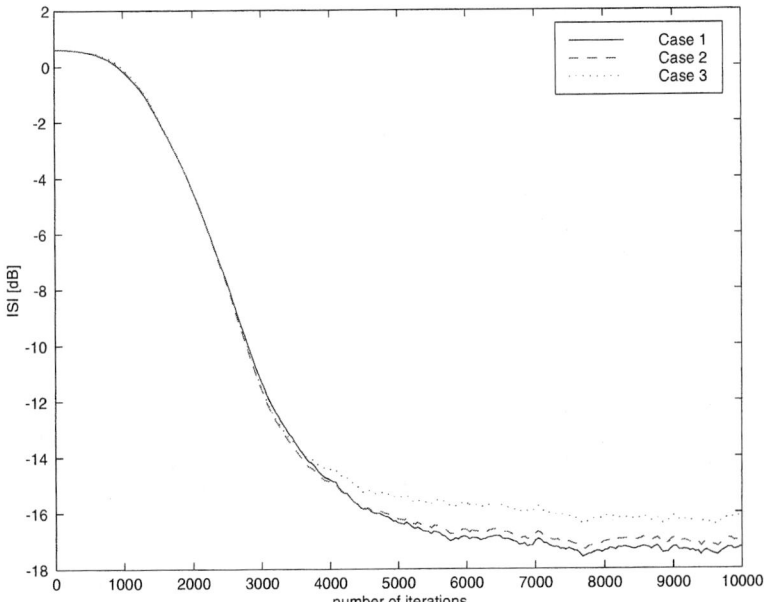

Fig. 3. Algorithm performance for a single-channel blind equalization task.

In each case, we have chosen $L = 50$. Step sizes of $\mu = 0.001, 0.00101,$ and 0.00102 were chosen to provide similar convergence rates of the averaged inter-symbol interferences $ISI(k) = (\sum_{i=0}^{1000} c_i^2(k))/(\max_{0 \le j \le 1000} c_j^2(k)) - 1$, where $c_i^2(k)$ is the combined channel-plus-equalizer impulse response. Shown in Fig. 3 are the results. The steady-state ISIs of the equalizers for Case 2 and Case 3 are 0.31 and 1.09 dB worse than that for Case 1, respectively. The performance of the Case 2 equalizer is especially noteworthy given its complexity; it only requires five multiply/adds per filter coefficient, a 25% increase over the algorithm in [1].

5 Conclusions

In this paper, we have studied the performance effects caused by coefficient delays in one well-known algorithm for blind deconvolution and source separation tasks. Through a simple analytical model, we show that algorithms with coefficient delays within their updates have worse adaptation performance than those without such delays. We also suggest a simple modification to improve this algorithm's adaptation performance. Simulations have been used to verify the accuracy of the analyses.

References

1. S. Amari, S. Douglas, A. Cichocki, H. Yang, "Multichannel blind deconvolution and equalization using the natural gradient," *Proc. 1st IEEE Workshop Signal Processing Adv. Wireless Commun.*, Paris, France, pp. 101–104, Apr. 1997.

2. R.H. Lambert and A.J. Bell, "Blind separation of multiple speakers in a multipath environment," *Proc. IEEE Int. Conf. Acoust., Speech, Signal Processing*, Munich, Germany, vol. 1, pp. 423-426, Apr. 1997.
3. L. Parra and C. Spence, "Convolutive blind separation of non-stationary sources," *IEEE Trans. Speech Audio Processing*, vol. 8, pp. 320-327, May 2000.
4. K. Matsuoka and S. Nakashima, "Minimal distortion principle for blind source separation," *Proc. 3rd Int. Workshop Indep. Compon. Anal. Signal Separation*, San Diego, CA, pp. 722–727, Dec. 2001.
5. S.C. Douglas and S. Amari, "Natural gradient adaptation," in *Unsupervised Adaptive Filtering, Vol. I: Blind Signal Separation*, S. Haykin, ed., (New York: Wiley, 2000), pp. 13-61.
6. B. Widrow and E. Walach, *Adaptive Inverse Control* (Upper Saddle River, NJ: Prentice-Hall, 1996).
7. S.C. Douglas and T.H.-Y. Meng, "Stochastic gradient adaptation under general error criteria," *IEEE Trans. Signal Processing*, vol. 42, pp. 1335-1351, June 1994.
8. S.C. Douglas, "Self-stabilized gradient algorithms for blind source separation with orthogonality constraints," *IEEE Trans. Neural Networks*, vol. 11, pp. 1490-1497, Nov. 2000.

On the FIR Inversion of an Acoustical Convolutive Mixing System: Properties and Limitations

Markus Hofbauer

Swiss Federal Institute of Technology, Zürich, Switzerland
hofbauer@isi.ee.ethz.ch

Abstract. In this paper we address the problem of Least-Squares (LS) optimal FIR inverse-filtering of an convolutive mixing system, given by a set of acoustic impulse responses (AIRs). The optimal filter is given by the LS-solution of a block-Toeplitz matrix equation, or equivalently by the time-domain Multi-Channel Wiener Filter. A condition for the minimum FIR filter length can be derived, depending on the number of sensors and sources and the AIR length, such that an exact FIR inverse exists, which perfectly separates and deconvolves all sources. In the general case, where an exact FIR solution does not exist, we discuss how SDR, SIR and SNR gains can be traded against each other. Results are shown for a set of AIRs, measured in an typical office room. Furthermore we present a method, which allows a time-domain shaping of the envelope of the global transfer function, reducing pre-echoes and reverberation.

1 Introduction

Blind Source-Separation methods aim at inverting a convolutive mixing system, using linear de-mixing filters. Examining the non-blind case, where the convolutive system of AIRs is perfectly known and determining the optimal FIR inverse (with respect to a quadratic cost function) demonstrates in principle the degree of achievable separation and deconvolution. The optimal FIR inverse is given by a block Toeplitz matrix equation. We derive a condition for the minimum FIR filter length, such that an exact FIR inverse exists, and also conditions for obtaining exact separation only, or exact deconvolution only. By appropriate weighting of the sources, the space of possible solutions, favoring the SIR, the SDR or SNR gain, can be sampled. Results are shown for a 4×4 set of AIRs, measured in an typical office room. In order to reduce reverberation, it is desirable to have an influence on the shape of the time-domain envelope of the global transfer function: we present a method using a weighting function.

1.1 Problem Formulation

In a reverberant and noisy environment with M mutually uncorrelated point sources s_m the ith sensor of an array with K sensors receives the signal $x_i[k]$

$$x_i[k] = \sum_{m=1}^{M} (h_{im} * s_m)\Big|_k + v_i[k] \qquad i = 1...K, \qquad (1)$$

where h_{im} is the acoustic impulse response (AIR) from source s_m, having n_h coefficients. The spatially incoherent sources and sensor noise appear as spatially uncorrelated components $v_i[k]$ at the ith sensor and are refered to as noise in the following.

We aim at finding a set of K FIR filters w_{mi}, $i = \{1, ..., K\}$ with n_w coefficients, such that output $y_m[k]$ is an estimate $\hat{s}_m[k-d]$ of the delayed source s_m.

$$y_m[k] = \hat{s}_m[k-d] = \sum_{i=1}^{K} (w_{mi} * x_i)\Big|_k. \quad (2)$$

Each output y_m has its own set of filters w_{mi}, extracting one of the sources s_m, and thus can be addressed separately. In the following we consider only one of the outputs, extracting the target source $s_{\tilde{m}}$ ($m = \tilde{m}$). Perfect deconvolution of the target source $s_{\tilde{m}}$ and separation from the other jammer sources s_m is obtained, if

$$\sum_{i=1}^{K} (h_{im} * w_i)\Big|_k = t_m[k] \qquad m = 1...M, \; k = 1...n_h + n_w - 1 \quad (3)$$

where $t_m[k]$ is the total (global) response from the source s_m to output y and is chosen as $t_{\tilde{m}}[k] = \delta[k-d]$ for the target source $s_{\tilde{m}}$, and $t_m[k] = 0$ for the other sources s_m. The minimum possible total delay $d = d_{\min}$ corresponds to the propagation time from the source $s_{\tilde{m}}$ to the sensors.

Since we are looking for a signal independent inverse of the AIR set, which is not optimized on the individual source spectras, we assume all sources to be white with equal power. The source powers are however weighted properly according to the desired inversion task, e.g. according to their actual powers $\sigma_{s_m}^2$.

2 Optimal FIR Inverse Filters

2.1 Exact FIR Inverses

The equation system (3), defining the optimal FIR inverse filters w_i is illustrated in Fig. 1. Each source/sensor adds a row/column of Toeplitz blocks to the block Toeplitz matrix $\bar{\mathbf{H}}$, which has dimension $(M(n_w + n_h - 1) \times Kn_w)$.

A plausible and common assumption [1],[2] for the AIR system is, that the filters h_{im} mutually do not share common zeros in the frequency domain and that the corresponding Matrix $\bar{\mathbf{H}}$ has full rank. Then there exists a set of FIR filters w_i, which fulfills (3) for the desired total transfer functions t_m, if Matrix $\bar{\mathbf{H}}$ is square or if $\dim(\bar{\mathbf{H}}) = (D_1 \times D_2)$, $D_1 \leq D_2$. This case applies if

$$M(n_w + n_h - 1) \leq Kn_w. \quad (4)$$

Solving (4) for n_w gives a condition for the minimum number of filter coefficients of w_i,

$$n_w \geq \left\lceil \frac{(n_h - 1)M}{K - M} \right\rceil \qquad K > M. \quad (5)$$

Thus, if there are less sources than sensors, there exist FIR filters w_i which achieve perfect separation and deconvolution of the sources. A proof for a similar condition for the case of only one source ($M = 1$) can be found in [3].

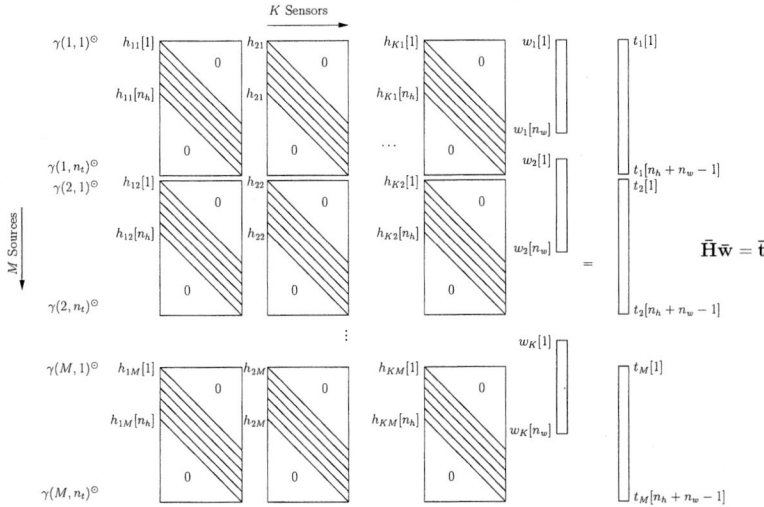

Fig. 1. Equation system (3) defining the FIR inverse filters w_i of an $K \times M$ AIR system (h_{im}) for the given desired total transfer functions t_m ($\odot$ denotes a scaling of an equation by $\gamma(m,k)$); n_w and n_h: filter lengths; length of t_m: $n_t = n_h + n_w - 1$.

If for instance there is one more sensor than sources ($K = M+1$), then choosing $n_w \geq M(n_h - 1)$ will allow perfect separation and deconvolution. If $K = aM$, $a \in \mathbb{Z}^+$, then $n_w \geq (n_h - 1)/(a-1)$ coefficients are required, i.e. filters w_i may be even shorter than the AIRs in this case.

2.2 Least Squares Optimal FIR Inverses

If the number of sources is equal or larger than the number of sensors then the optimal FIR inverse filters are given by the Least Squares solution of (3). This case ($K \leq M$) applies at any rate, if also the noise sources at the sensors are included in the LS-optimization since noise at the ith sensor can be regarded as a source 'sitting' at the sensor (with AIRs $h_{ii} = 1$ and $h_{ij} = 0$). Thus, when considering noise, K additional noise sources have to be placed in (3).

In the LS case, maximal separation (SIR), deconvolution (SDR) and a maximal SNR gain (see 2.4) can not be achieved simultaneously. However the SIR, SDR and SNR gains can be traded against each other. By weighting the sources with appropriate factors $\gamma(m,k)$

$$\gamma(m,k) = \gamma_m^\circ \gamma_m(k) \qquad k = 1...n_t = n_h + n_w - 1 \quad m = 1...M, \qquad (6)$$

the space of possible solutions, favoring the SIR, SDR or SNR gain, can be sampled. As shown in Fig. 1, both sides of equation (3) are scaled with $\gamma(m,k)$. While γ_m° is constant for each source (e.g. the source power $\sigma_{s_m}^2$), $\gamma_m(k)$ allows for a individual weighting of each equation. Choosing larger weights γ_m° for the noise sources will result in a solution favoring the SNR gain, larger weights for the target source will increase the SDR, and larger weights for the jammer sources will increase the SIR.

Maximal Separation (Best SIR). Maximal separation is achieved, if the jammer sources receive dominant weights γ_m°, and by choosing $t_m[k] = 0$ for all $m \neq \tilde{m}$. For the target source $s_{\tilde{m}}$ a total response $t_{\tilde{m}}[k]$ will result, which will introduce some additional distortion. A constraint, e.g. $t_{\tilde{m}}[d] = 1$ is required to avoid the all zero solution $w_i = 0$. In the *noiseless* case **perfect separation** (however *with distortion*) is possible, if

$$n_w \geq \left\lceil \frac{n_h(M-1) - M + 2}{K - M + 1} \right\rceil \qquad K \geq M, \tag{7}$$

since $\dim(\bar{\mathbf{H}}) = (D_1 \times D_2)$, $D_1 \leq D_2$ applies, if (7) holds. (Compared to (4) there are $n_w + n_h - 2$ fewer equations, since here $\gamma_{\tilde{m}}^\circ / \gamma_m^\circ \approx 0$ and $t_{\tilde{m}}[d] = 1$). If e.g. $K = aM$ (over-determined case), perfect separation is achieved with filters w_i shorter than the AIRs. In the square case ($K = M$) perfect separation is achievable, if $n_w \geq n_h(M-1) - M + 2$. With the definition of Matrix $\widehat{\mathbf{H}}^{K \times K} = [h_{im}]$ and since $\mathrm{adj}(\widehat{\mathbf{H}})\widehat{\mathbf{H}} = \det(\widehat{\mathbf{H}})\mathbf{I}$ [4], it follows, that $\mathrm{adj}(\widehat{\mathbf{H}})$ is the perfect separating solution for the $K = M$ case. Filters w_{mi} are then given by $w_{mi} = \widehat{\mathbf{H}}_{im}$, with $\widehat{\mathbf{H}}_{im}$ being the cofactors of $\widehat{\mathbf{H}}$. Thus w_{mi} will be a sum of different combinations of $K - 1$ convolutions of h_{im} (e.g. $h_{12} * h_{23} * ...$). The corresponding total target transfer function is causal if desired and is given by $t_{\tilde{m}} = \det(\widehat{\mathbf{H}})$, which is a sum of combinations of K convolutions of h_{im}. Since each convolution with an AIR introduces additional reverberation, the resulting distortion can be significant.

Perfect separation is not possible in the under-determined case $K < M$. Also, when including the noise sources in (3), the LS-solution will reduce the best possible SIR gain, depending on the weighting of jammer and noise sources.

Maximal Deconvolution (Best SDR). If a maximal deconvolution (minimal distortion, high SDR) of the target source is desired, all sources, except for the target, have to receive small or even zero weighting γ_m°. From (4) follows, that when neglecting all unwanted sources ($M = 1$), **perfect deconvolution** is in principal achievable, if at least two sensors are available $K \geq 2$ (with $n_w \geq n_h - 1$). Complete deconvolution may however result in a large SNR and SIR loss, especially if the delay d is chosen small [5]. Multi-channel deconvolution is caused by two mechanisms: single-channel type *inversion* (in general IIR) and *elimination by addition* of different sensor observations (possible with FIR). The latter is mainly responsible for possible SNR and SIR losses.

A common criterion used by BSS source separation algorithms is the **Minimal Distortion Principle** [6], where separation of the sources is aimed for, with the constraint, that no additional distortion (on top of the distortion/reverberation caused by the AIRs) should be introduced: the target source at the output is ought to be identical or close to its observation in one of the sensors, i.e. $y_{s_{\tilde{m}}} \approx s_{\tilde{m}} * h_{i\tilde{m}}$. The corresponding solution is obtained by setting $t_{\tilde{m}} = h_{i\tilde{m}}$ as desired total response in (3). Note that the source is not completely deconvolved in this case.

The **Minimum Variance Distortionless Response** Beamformer (MVDR-BF) [7] aims at complete deconvolution of the target source ($t_{\tilde{m}}[k] = \delta[k-d]$), while minimizing the variance of all undesired sources. Generalized from a model with simple propagation delays to general AIRs, the MVDR-BF filters w_i are calculated in the frequency domain [5] and are typically IIR. The MVDR-BF solution is obtained from (3) (approximately) by using a large n_w and dominant weights $\gamma_{\tilde{m}}^\circ$ for the target source.

Maximal SNR Gain (Best SNR). The solution leading to a maximal SNR gain is obtained by setting the weights γ_m° of all sources, except for the noise sources to small values or zero. A constraint, e.g. $t_{\tilde{m}}[d] = 1$ is required to avoid the all zero solution $w_i = 0$. The maximal SNR achieving solution is given by $w_i[.] = h_{i\tilde{m}}[-.]$, which follows from the Wiener solution (9) and (13), setting $\mathbf{R_{xx}} = \mathbf{I}$. If the AIRs are simple delays, $w_i[.]$ will reduce to the well known Delay-and-Sum Beamformer, which is the BF producing the highest white noise SNR gain [7].

2.3 Time-Domain Multi-channel Wiener Filter (MCWF)

If all sources are white and are given the weighting $\gamma(m,k) = \gamma_m^\circ$, the least squares solution of (3) coincides with the time-domain FIR Multi-Channel Wiener Filter MCWF, which is the MMSE solution. Defining the stacked data and filter vectors

$$\mathbf{x}_i[k] = [x_i[k]\, x_i[k-1] \cdots x_i[k-n_w+1]]^T \qquad (8)$$
$$\mathbf{x}[k] = [\mathbf{x}_1^T\, \mathbf{x}_2^T \cdots \mathbf{x}_K^T]^T$$
$$\mathbf{w}_i = [w_i[0]\, w_i[1] \cdots w_i[n_w-1]]^T$$
$$\mathbf{w} = [\mathbf{w}_1^T\, \mathbf{w}_2^T \cdots \mathbf{w}_K^T]^T,$$

the Multi-Channel Wiener Filter (with K inputs and one output) is given by

$$\mathbf{w}_{\text{MCWF}} = \mathbf{R_{xx}}^{-1} \mathbf{r}_{\mathbf{x}s_{\tilde{m}}}, \qquad (9)$$

where $\mathbf{R_{xx}}^{(Kn_w \times Kn_w)}$ is the autocorrelation matrix of the sensor signals

$$\mathbf{R_{xx}} = \mathrm{E}\{\mathbf{x}[k]\mathbf{x}[k]^T\} = \sum_{m=1}^{M} \gamma_m^{\circ\,2} \mathbf{R}_{\mathbf{x}_{s_m}\mathbf{x}_{s_m}} + \gamma_v^{\circ\,2} \mathbf{R_{vv}}, \qquad (10)$$

and $\mathbf{r}_{\mathbf{x}s_{\tilde{m}}}^{(Kn_w \times 1)}$ is the cross-correlation vector of $\mathbf{x}$ and $s_{\tilde{m}}$

$$\mathbf{r}_{\mathbf{x}s_{\tilde{m}}} = \gamma_{\tilde{m}}^{\circ\,2} \mathrm{E}\{\mathbf{x}[k] s_{\tilde{m}}[k-d]\}. \qquad (11)$$

For a given set of AIRs h_{im}, element $(\mathbf{R}_{\mathbf{x}_i\mathbf{x}_j})_{a,b}$ amounts to

$$(\mathbf{R}_{\mathbf{x}_i\mathbf{x}_j})_{a,b} = r_{x_i x_j}[a-b] =$$
$$= \sum_{m=1}^{M} \gamma_m^{\circ\,2} (h_{im}[-.] * h_{jm}[.] * r_{s_m}[.])\Big|_{k'=a-b} + \gamma_v^{\circ\,2} (\mathbf{R}_{\mathbf{v}_i\mathbf{v}_j})_{a,b}, \qquad (12)$$

where $h_{im}[-.]$ denotes time reversion. Element $\mathbf{r}_{\mathbf{x}_i s_{\tilde{m}}}(a)$ is given by

$$\mathbf{r}_{\mathbf{x}_i s_{\tilde{m}}}(a) = r_{x_i s_{\tilde{m}}}[a-1-d] = \gamma_m^{\circ\,2}(h_{i\tilde{m}}[-.] * r_{s_{\tilde{m}}}[.])\Big|_{k'=a-1-d}. \qquad (13)$$

With (12) and (13), the MCWF (9) can be determined. $\mathbf{R_{xx}}^{(Kn_w \times Kn_w)}$ and $\bar{\mathbf{H}}$ are matrices of *Toeplitz* blocks of typically very large dimension (e.g. 40000 × 40000). An algorithm, utilizing the block-Toeplitz structure can be used to solve (9) or (3) efficiently (Schur alg.).

2.4 SDR Gain, SIR Gain and SNR Gain

The realized degree of deconvolution, separation and noise reduction is measured by the Signal-to-Distortion Ratio SDR, the Signal-to-Interference Ratio SIR and the SNR, respectively. Since we assume sources and the noise to be white, the SDR of source $s_{\tilde{m}}$ at sensor 1 is calculated from $h_{1\tilde{m}}$ as

$$\text{SDR}_{x_1} := 10\log_{10}\left(\frac{\max(|h_{1\tilde{m}}[k]|^2)}{\sum_k |h_{1\tilde{m}}[k]|^2 - \max(|h_{1\tilde{m}}[k]|^2)}\right), \qquad (14)$$

i.e. the SDR is the ratio of the power of the main peak to the reverberation part in the AIR $h_{1\tilde{m}}$. The SDR at the output y we obtain from the total response $t_{\tilde{m}}[k]$:

$$\text{SDR}_y := 10\log_{10}\left(\frac{\max(|t_{\tilde{m}}[k]|^2)}{\sum_k |t_{\tilde{m}}[k]|^2 - \max(|t_{\tilde{m}}[k]|^2)}\right). \qquad (15)$$

The reduction of distortion or reverberation is then $\text{SDR}_{\text{gain}} = \text{SDR}_y - \text{SDR}_{x_1}$. Similarly the $\text{SIR}_{\text{gain}} = \text{SIR}_y - \text{SIR}_{x_1}$ of the power of the target source $s_{\tilde{m}}$ to all jammer sources s_m is given by:

$$\text{SIR}_{\text{gain}} := 10\log_{10}\left(\frac{\sum_k |t_{\tilde{m}}[k]|^2 \cdot \sum_{k,m\neq\tilde{m}} |h_{1m}[k]|^2}{\sum_{k,m\neq\tilde{m}} |t_m[k]|^2 \cdot \sum_k |h_{1\tilde{m}}[k]|^2}\right). \qquad (16)$$

Finally, the white noise gain $\text{SNR}_{\text{gain}} = \text{SNR}_y - \text{SNR}_{x_1}$ is obtained by:

$$\text{SNR}_{\text{gain}} := 10\log_{10}\left(\frac{\sum_k |t_{\tilde{m}}[k]|^2}{\sum_{i,k} |w_i[k]|^2 \cdot \sum_k |h_{1\tilde{m}}[k]|^2}\right). \qquad (17)$$

3 Shaping the Envelope of the Total Response

The LS-solution of (3) will result in some signal distortion, appearing in $t_{\tilde{m}}$. AIRs and filters w_i typically have several thousands of coefficients and thus will produce a long, noncausal global response $t_{\tilde{m}}$ with its main power concentrated at $k = d$ and slow decaying tails. These tails cause audible undesired artifacts: pre-echoes and (late) reverberation. It is therefore desirable to shape the envelope of $t_{\tilde{m}}$, such that the tails decay faster. We propose a method to shape the envelope of the total response $t_{\tilde{m}}$ by incorporating an appropriate weighting function $\gamma_{\tilde{m}}(k)$ in (3), scaling the equations associated with the target source:

$$\gamma_{\tilde{m}}(k) = (1+\epsilon) - \exp(-\tau[k-d]^2) + \beta\delta[k-d] \qquad \epsilon, \tau \approx 0, \quad \beta \approx 1. \qquad (18)$$

This will cause the LS-solution to drive the tails of $t_{\tilde{m}}[.]$ to zero, while permitting a degree of freedom for 'k around d', which is favorable for the LS-optimization. The effect of the total response shaping is shown in Fig. 2 f).

4 SDR, SIR and SNR Gains for a Typical Office Size Room

We have measured a $K \times M = 4 \times 4$ set of AIRs ($n_h = 3600$) in an office room (5m × 3.5m × 2.5m, $T_{60} \approx 400$ ms, $f_S = 8$ kHz), with a speaker microphone distance range of [1m-3m] and a sensor array spacing of [4cm 14cm 4cm]. Results of the inversion are shown in Fig. 2. The total transfer function $t_{\tilde{m}}$ of the target source, and exemplarily t_m of one of the jammers – the other jammers are comparable – are depicted. By applying a different weights γ_m° to the sources, the LS-solution of (3) favors the SDR, SIR or SNR gain. The achievable gains are indicated. **Case b)**: $K \times M = 4 \times 3$: in accordance with (5), which here demands $n_w \geq 10797$, (nearly) perfect deconvolution and separation is accomplished, since $n_w = 12000$ was chosen. The sensor noise sources were neglected here. **Case c)-e)**: $K \times M = 4 \times 4$, and noise being also considered (i.e. M is in fact increased by $K = 4$ noise sources to a total of M=8): now the Least Squares optimization applies, and the SDR, SIR, and SNR can be traded against each other, with maximal values of the SDR/SIR/SNR gain$_{\max}$ = 37/30/10dB. Plot **f)** shows the effect of shaping $t_{\tilde{m}}$ by incorporating the weighting function $\gamma_{\tilde{m}}(k)$ given by (18). It can be observed, that the tails of $t_{\tilde{m}}[k]$ vanish, at the cost of increased values 'around $k = d$'. The weighting function $\gamma_{\tilde{m}}(k)$ drives the tails of $t_{\tilde{m}}[k]$ to zero, reducing late reverberation and also the artifacts due to the non-causal side tail.

5 Conclusion

AIRs are typically non-minimum phase, having non-causal and very long-tailed IIR inverses. Nevertheless, under certain conditions, there exist causal FIR inverses which perfectly invert the AIR mixing system. If $K > M$, perfect separation and deconvolution is achievable, while separation only requires $K \geq M$. Deconvolution only, demands $K \geq 2$. We derived conditions for the corresponding minimum FIR filter lengths n_w. For the most common case ($K \leq M$), the LS-optimization will give a tradeoff between the SDR, SIR and SNR gains, which can be controlled by appropriate weighting of the sources. Confirmative results are shown for a 4 × 4 set of AIRs, measured in a typical office room. Finding the LS-optimal filters for the non-blind case demonstrates what one can expect from a BSS-algorithm in a similar acoustical setup. In order to additionally reduce

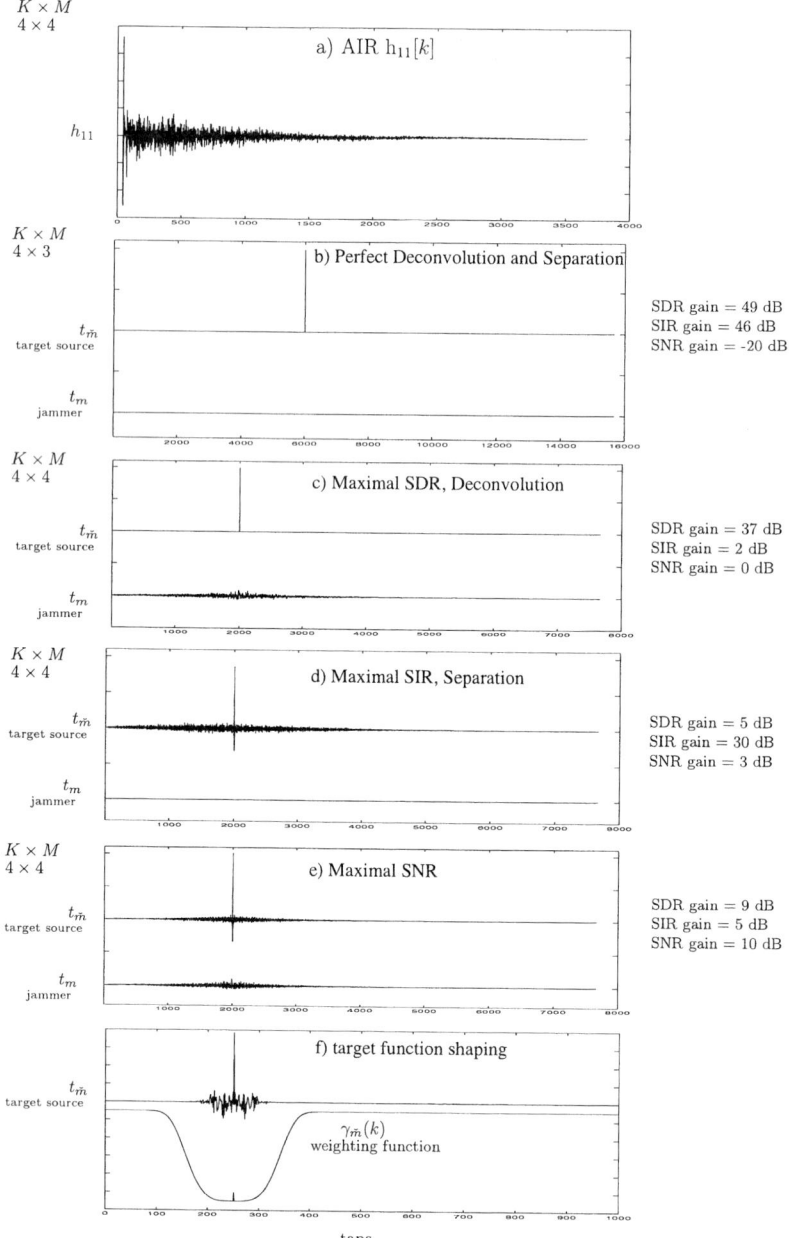

Fig. 2. FIR LS-inverse of an $K \times M = 4 \times 4$ AIR set ($n_h = 3600$, $T_{60} \approx 400$ms) of an office room: total transfer functions of target $t_{\tilde{m}}$, and t_m of one of the jammers; **a)** AIR h_{11}; **b)** perfect separation and deconvolution for the 4×3 case; **c)-e)** solutions favoring the SDR, SIR or SNR gain, obtained by different weighting γ_m° of the sources; **f)** effect of introducing the weighting function $\gamma_{\tilde{m}}(k)$: tails of $t_{\tilde{m}}$ are suppressed, reducing late reverberation (a simulated AIR set with $n_h = 500$ was used for case f)).

reverberation and artifacts we proposed a method using a weighting function $\gamma_m(k)$, which allows a time-domain shaping of the envelope of the global transfer function t_m, driving the undesired long tails to zero.

References

1. M. Miyoshi, Y. Kaneda: "Inverse filtering of room acoustics". IEEE T. Acoust. Speech and Signal Processing. (1988)
2. K. Rahbar, J.P. Reilly, J. H. Manton: "Blind identification of MIMO FIR systems driven by quasi-stationary sources using second order statistics": A frequency domain approach. IEEE T. Signal Processing. (2002)
3. G. Harikumar, Y. Bresler: "FIR Perfect Signal reconstruction from multiple convolutions: Minimum Deconvolver Orders". IEEE T. Signal Processing. (1998)
4. R. A. Horn, Ch. R. Johnson: "Matrix Analysis". Cambride University Press. (1999)
5. M. Hofbauer, H.-A. Loeliger: "Limitations of FIR Multi-Microphone Speech Derverberation in the Low-Delay Case". IWAENC. Kyoto. (2003)
6. K. Matsuoka: "Independent Component Analysis and Its Application to Sound Signal Separation". IWAENC. Kyoto. (2003)
7. M. Brandstein, D. Ward: "Microphone Arrays". Springer. (2001)

Overcomplete BSS for Convolutive Mixtures Based on Hierarchical Clustering

Stefan Winter*, Hiroshi Sawada, Shoko Araki, and Shoji Makino

NTT Communication Science Laboratories, NTT Corporation
2-4 Hikaridai, Seika-cho, Soraku-gun, Kyoto 619-0237, Japan
{wifan,sawada,shoko,maki}@cslab.kecl.ntt.co.jp

Abstract. In this paper we address the problem of overcomplete BSS for convolutive mixtures following a two-step approach. In the first step the mixing matrix is estimated, which is then used to separate the signals in the second step. For estimating the mixing matrix we propose an algorithm based on hierarchical clustering, assuming that the source signals are sufficiently sparse. It has the advantage of working directly on the complex valued sample data in the frequency-domain. It also shows better convergence than algorithms based on self-organizing maps. The results are improved by reducing the variance of direction of arrival. Experiments show accurate estimations of the mixing matrix and very low musical tone noise.

1 Introduction

High quality separation of speech sources is an important prerequisite for further processing like speech recognition. Often the underlying mixing process is unknown, which requires blind source separation (BSS). In general we can distinguish two cases depending on the number of sources N and the number of sensors M:

$N > M$	overcomplete BSS
$N \leq M$	(under-) complete BSS

Since undercomplete BSS ($N < M$) can be reduced to complete BSS ($N = M$) [1] we refer to both by complete BSS. Most approaches assume complete mixtures [2,3], but in reality often the contrary is true. While the area of overcomplete BSS has obtained more and more attention [4–12], it still remains a challenging task.

Several of the proposed algorithms are based on histograms and developed for only two sensors [4–6]. Some could, in principle, be enhanced for higher dimensions M. But since histograms are based on densities, the so called curse

---

* The author is on leave from the Chair of Multimedia Communications and Signal Processing, University Erlangen-Nuremberg.

of dimensionality [13] sets practical limits on the number of usable sensors. Another problem occurs with complex numbers, which cannot be handled straightforwardly by histograms, but are necessary if BSS is performed in the frequency-domain. Some methods approach complex numbers by applying real-valued algorithms to the real and imaginary part or amplitude and phase [7, 8], which is not always applicable. Some approaches extract features like the direction-of-arrival (DOA) or work on the amplitude relation between two sensor outputs [4, 5, 9, 10]. In both cases only two sensors can contribute, no matter how many sensors are available.

Other algorithms like GeoICA [12] or AICA [11] resemble self-organizing maps (SOM) and could more easily be applied to convolutive mixtures. However, their convergence depends on initial values.

In this paper we propose the use of hierarchical clustering embedded into a two-stage framework of overcomplete BSS to deal with convolutive mixtures in the frequency-domain. This method can work directly on the complex valued samples. While it does not limit the usable numbers of sensors, it also prevents the convergence problems which can occur with SOM based algorithms.

After estimating the mixing matrix in the first stage, a maximum a-posteriori (MAP) approach is applied to finally separate the mixtures, assuming statistical independence and Laplacian pdfs for the sources [14].

In Sec. 2 we first explain the general framework before we give details about the hierarchical clustering in Sec. 3 and the MAP based source separation in Sec. 4. After this, we present experimental results in Sec. 5 demonstrating the performance for convolutively mixed speech data in a real room with reverberation time $T_R = 130$ms.

2 General Framework

We will consider a convolutive mixing model with N sources $s_i(t)$ ($i = 1\ldots N$) and M ($M < N$) sensors that yield linearly mixed signals $x_j(t)$ ($j = 1\ldots M$). The mixing can be described by $x_j(t) = \sum_{i=1}^{N} \sum_{l=1}^{\infty} h_{ji}(l) s_i(t-l)$, where $h_{ji}(t)$ denotes the impulse response from source i to sensor j.

Instead of solving the problem in the time-domain, we choose a narrowband approach in the frequency domain by applying a short-time discrete Fourier transform (STDFT). Thus time-domain signals $\mathbf{s}(t) = [s_1(t),\ldots,s_N(t)]^T$ and $\mathbf{x}(t) = [x_1(t),\ldots,x_M(t)]^T$ are converted into frequency-domain time-series $\mathbf{S}(f,\tau) = [S_1(f,\tau),\ldots,S_N(f,\tau)]^T$ and $\mathbf{X}_\tau = \mathbf{X}(f,\tau) = [X_1(f,\tau),\ldots,X_M(f,\tau)]^T$ by an L-point STDFT, respectively. Thereby $f = 0, f_s/L, \ldots, f_s(L-1)/L$ (f_s: sampling frequency; τ: time dependence). Let us define $\mathbf{H}(f) \in \mathbb{C}^{M \times N}$ as a matrix whose elements are the transformed impulse responses. We call the column vectors $\mathbf{h}_i(f)$ ($i = 1,\ldots,N$) mixing vectors and approximate the mixing process by

$$\mathbf{X}(f,\tau) = \mathbf{H}(f)\mathbf{S}(f,\tau) \qquad (1)$$

This reduces the problem from convolutive to instantaneous mixtures in each frequency bin f. For simplicity we will omit the dependence on frequency and

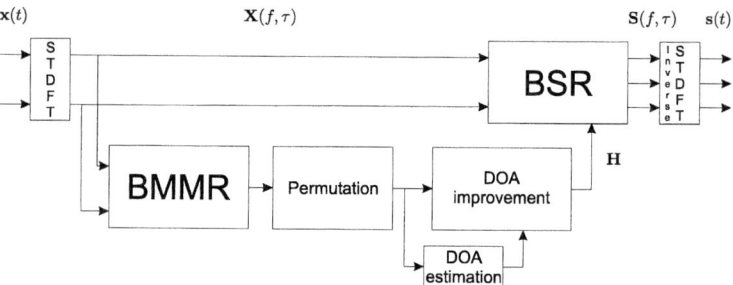

Fig. 1. Overall unmixing system.

time. Switching to the frequency domain has the additional advantage that the sparseness of the sources is increased [7]. This is very important, since the hierarchical clustering is based on the assumption of sparse sources.

The disadvantage of narrowband BSS in the frequency domain is the permutation problem, which results in wrong alignments of the frequency bins. In our framework we use a DOA based method to reduce the permutation problem [3]. We also apply the minimum-distortion-principle [2] to solve the scaling problem.

In complete BSS the mixing matrix **H** is square and (assuming full rank) invertible. Therefore the BSS problem can be solved by either inverting an estimate of the mixing matrix or directly estimating its inverse and solving (1) for **S**.

However, this approach does not work in overcomplete BSS where the mixing matrix is not invertible. Therefore we follow a two-stage approach as proposed in [7] consisting of blind mixing model recovery (BMMR) and blind source recovery (BSR). To estimate the mixing matrix in the BMMR step, we propose the use of hierarchical clustering as described in detail in Sec. 3. To eventually separate the signals in the BSR step, we utilize a MAP based approach. Finally the inverse STDFT is applied to obtain time-domain signals. The overall system is depicted in Fig. 1.

3 Blind Mixing Model Recovery

Several algorithms have been proposed so far for BMMR. They usually have in common that they assume a certain degree of sparseness of the original signals. In this paper we consider signals that are sparse in the time-frequency domain. That means that different signals are rarely active at the same time-frequency instant (f, τ). This assumption leads to the conclusion that the samples in the mixed vector space $\mathbf{X}(f, \tau)$ cluster around the true mixing vectors $\mathbf{h}_i(f)$. This becomes clear when we consider the most sparse case when only a single source is active. Let us rewrite (1) as

$$\mathbf{X}(f, \tau) = \sum_{i=1}^{N} \mathbf{h}_i(f) S_i(f, \tau) \qquad (2)$$

Assuming only one source active at (f,τ) means that the vector pointing to the resulting mixed sample $\mathbf{X}(f,\tau)$ is a scaled version of the corresponding mixing vector $\mathbf{h}_i(f)$. Depending on the actual sparseness of the source signals, the mixed signals will also have components of other signals and therefore be spread around the mixing vectors. In order to obtain a different cluster for each source signal S_i we assume a different mixing vector $\mathbf{h}_i(f)$ for each source signal.

3.1 Hierarchical Clustering

To avoid the problems discussed in Sec. 1, such as the curse of dimensionality or poor convergence, we propose the use of a hierarchical clustering algorithm following an agglomerative (bottom-up) strategy [13]. This means that at the beginning we consider each sample as a cluster that contains only one object. From there clusters are combined, so that the number of clusters decreases while the average number of objects per cluster increases. In the following we assume phase and amplitude normalized samples.

$$\mathbf{X} = \frac{\mathbf{X}}{|\mathbf{X}|_2} e^{-\varphi_{X_1}} \qquad (3)$$

where φ_{X_1} denotes the phase of the first component of $\mathbf{X}$.

The combination of clusters into new clusters is an iterative process and based on the distance between the current clusters. Starting from the normalized samples, the distance between each pair of clusters is calculated, resulting in a distance matrix. The two clusters with the least distance are combined and form a new binary cluster. This process is called linking and repeated until the final number of clusters has decreased to a predetermined number c, $N \leq c \leq P$ (P: total number of samples).

For measuring the distance between clusters, we have to distinguish between two different problems. First we need a distance measure $d(\mathbf{X}_{\tau_1}, \mathbf{X}_{\tau_2})$ that is applicable to M-dimensional complex vector spaces. While there are several possibilities, we currently use the Euclidean distance defined by

$$d(\mathbf{X}_{\tau_1}, \mathbf{X}_{\tau_2}) = \sqrt{<(\mathbf{X}_{\tau_1} - \mathbf{X}_{\tau_2}), (\mathbf{X}_{\tau_1} - \mathbf{X}_{\tau_2})^*>} \qquad (4)$$

where $<\cdot>$ stands for the inner product and $*$ for complex conjugation.

When a new cluster is formed, we need to enhance this distance measure to relate the new cluster to the other clusters. The method we employ here is called the nearest-neighbor technique. Let C_1 and C_2 denote two clusters as illustrated in Fig. 2. Then the distance $d(C_1, C_2)$ between these clusters is defined as the minimum distance between its samples by

$$d(C_1, C_2) = \min_{\mathbf{X}_{\tau_1} \in C_1, \mathbf{X}_{\tau_2} \in C_2} d(\mathbf{X}_{\tau_1}, \mathbf{X}_{\tau_2}) \qquad (5)$$

As mentioned earlier, most of the samples will cluster around the mixing vectors $\mathbf{h}_i$, depending on the sparseness of the original signals. Special attention must be paid to the remaining samples (outliers), which are randomly scattered

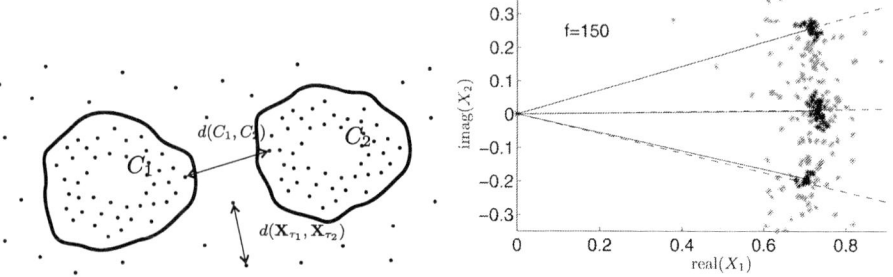

Fig. 2. Illustration of distances.   **Fig. 3.** Estimation of mixing vectors.

in the space between the mixing vectors. Usually they are far away from other samples and will be combined with other clusters only at higher levels of the clustering process (i.e. when only few clusters are left). This led us to the idea to set the final number of clusters at a high number

$$c \gg N \tag{6}$$

By doing so, we avoid linking these outliers with the clusters around the mixing vectors $\mathbf{h}_i$ and therefore distortions. This results in more robustness. More important, however, is the fact that we avoid combining desired clusters. Since the outliers are often far away from other clusters it might happen that desired clusters are closer to each other than to outliers. An example for the resulting clusters is shown in Fig. 3. Experimental details are given in Sec. 5.

3.2 Estimation of Mixing Matrix

Assuming that the clusters around the mixing vectors $\mathbf{h}_i$ have the highest densities and therefore the highest number of samples we finally choose the N largest clusters. Thereby the number of sources N must be known. To obtain the mixing vectors, we average over all samples of each cluster

$$\mathbf{h}_i = \frac{1}{|C_i|} \sum_{\mathbf{x} \in C_i} \mathbf{X}, \qquad 1 \leq i \leq N \tag{7}$$

where $|C_i|$ denotes the cardinality of cluster C_i. Thereby we assume that the influence of other sources has zero mean.

3.3 Advantages of Hierarchical Clustering

Among the most important advantages of the described hierarchical clustering algorithm is the fact that it works directly on the sample data in any vector space of arbitrary dimensions. The only requirement is the definition of a distance measure for the considered vector space. Therefore, it can easily be applied to complex valued data that occurs in frequency-domain convolutive BSS.

No initial values for the mixing vectors $\mathbf{h}_i$ are required. This means, in particular, that if the assumption of clusters with high densities around the mixing vectors is true, then the algorithm converges to those clusters.

Besides choosing a distance measure, there is only the single parameter c that determines the number of clusters. Experiments have shown that the choice for this parameter in the noiseless case is quite insensitive as long as it is above a certain limit that would combine desired clusters. Its choice is, in general, related to the sparseness of the sources. The sparser the signals are, the smaller the value of c can be chosen, because the number of outliers that must be avoided will be smaller.

While the considered signals must have some degree of sparseness, they do not have to be statistically independent at this point.

3.4 Reduction of DOA Variance

Experiments have shown that as long as there are clusters around the mixing vectors $\mathbf{h}_i$, the estimation results are of high quality. Even if the assumption of clear clusters is not true for all mixing vectors, the remaining ones are not influenced by poor estimation of others. In order to improve the wrongly estimated mixing vectors, we can utilize DOA information. While the mixing matrix is different for each frequency bin, the phase difference $\Delta\varphi_i$ between the components of a mixing vector $\mathbf{h}_i$ contains information about the relative physical position of its corresponding source. Assuming a linear sensor array in a far-field situation with plain wave fronts, the DOA θ_i is given by

$$\theta_i = cos^{-1}\left(\frac{\Delta\varphi_i v}{2\pi f d}\right) \qquad (8)$$

where v denotes the sound velocity, d the distance between the corresponding sensors. Since θ_i is theoretically constant for all frequency bins, we can consider the DOA of the i-th signal as a random variable (RV) θ_i with mean μ_i and variance σ_i^2. While even the DOA of the original mixing matrix has a variance larger than 0, the results for the estimated mixing matrix can be improved if the variance of its DOAs is reduced.

For this purpose we define a new RV $\widehat{\theta}_i$ with reduced variance by

$$\widehat{\theta}_i = \sqrt{\varepsilon}\theta_i + (1 - \sqrt{\varepsilon})\mu_i, \qquad 0 \le \varepsilon \le 1 \qquad (9)$$

While its mean is still μ_i, its variance $\widehat{\sigma}_i^{\,2}$ can be adjusted by ε and yields

$$\widehat{\sigma}_i^{\,2} = \varepsilon\sigma_i^2 \qquad (10)$$

We apply the new DOA by adjusting the phase of the mixing vectors $\mathbf{h}_i$. Since we do not need absolute DOA information, this improvement fully complies with the blind approach of BSS.

4 Blind Source Recovery

Since the mixing matrix cannot be inverted in overcomplete BSS, the unmixed signals cannot be directly obtained. Several approaches have been proposed to solve blind source recovery [14]. Among those we chose the shortest-path algorithm which is based on maximum a-posteriori (MAP) estimation, assuming statistical independence and Laplacian pdfs for the sources. Given the mixed signals $\mathbf{X}$ and the mixing matrix $\mathbf{H}$, the sources $\mathbf{S}$ are recovered by

$$\mathbf{S} = \arg \min_{\mathbf{X}=\mathbf{HS}} \sum_{i=1}^{N} |S_i| \quad (11)$$

This equation can be interpreted as finding the shortest-path decomposition, based on the mixing vectors $\mathbf{h}_i$ for each sample $\mathbf{X}_\tau$ separately. It means that each sample is assigned to exactly M signals. While (11) can, in general, be solved for real numbers by linear programming, we explicitly compute all $\binom{N}{M}$ possible decompositions and choose the one that minimizes $\sum_{i=1}^{N} |S_i|$. Taking a selection of M mixing vectors $\mathbf{h}_{i_1} \ldots \mathbf{h}_{i_M}$, the decomposition is calculated by

$$\mathbf{S} = [\mathbf{h}_{i_1} \ldots \mathbf{h}_{i_M}]^{-1} \mathbf{x} \qquad i_1, \ldots, i_M \in \{1, \ldots, N\} \quad (12)$$

5 Experimental Results

We performed experiments with the proposed algorithm using $N = 3$ speech signals and $M = 2$ sensors. The signals were taken from the Acoustical Society of Japan (ASJ) continuous speech corpus. The convolution was done with room impulse responses that were recorded at our laboratory. Further experimental conditions are given in Table 1. As performance measure, we used the signal-to-interference ratio $\text{SIR}_i = 10 \log \left(\frac{\sum_t y_i^s(t)^2}{\sum_t y_i^{if}(t)^2} \right)$ where $y_i^s(t)$ is the portion of $y_i(t)$ that comes from $s_i(t)$ and $y_i^{if}(t) = y_i(t) - y_i^s(t)$. We also evaluated the signal-to-distortion ratio (SDR) as described in [15].

Table 1. Experimental conditions.

Direction of sources	50°, 90°, 120°
Distance of sensors	40 mm
Length of source signals	7.4 seconds
Reverberation time T_R	130ms
Sampling rate	8 kHz
Window type	von Hann
Filter length	1024 points
Shifting interval	256 points
Cluster threshold c (const $\forall f$)	100
Variance factor ε	0.8

As an upper limit for the performance of the whole system, scenario 1 in Table 2 shows the separation results when the original mixing matrix is used. This means that the permutation problem does not occur and the BSR part is given the best possible input.

Table 2. Performance of different parts of the separation system.

$N = 3$, $M = 2$, $T_R = 130$ms		Source 1	Source 2	Source 3	Average
Scenario 1	SIR (dB)	14.8	13.9	11.7	13.50
	SDR (dB)	13.39	6.83	10.55	10.26
Scenario 2	SIR (dB)	10.5	6.4	9.3	8.73
	SDR (dB)	7.47	2.82	5.99	5.43
Scenario 3	SIR (dB)	11.1	9.9	8.9	9.95
	SDR (dB)	9.65	4.05	5.95	6.55

Scenario 2 gives the results if we use the estimated mixing matrix without reduction of DOA variance. The last scenario shows the results if the estimated mixing matrix is used together with reduction of DOA variance. Figure 3 gives an example for the clustering for $f = 1164$Hz. To visualize, the real part of the first component X_1 versus the imaginary part of the second component X_2 is plotted. The N largest clusters (black) around the original mixing vectors $\mathbf{h}_i$ (dashed) can be clearly seen and result in precise estimations (solid).

Subjective evaluation of the separated sources showed very low musical tone noise.

6 Conclusion

We proposed the application of hierarchical clustering embedded into a two-stage framework of overcomplete BSS for convolutive speech mixtures. This method can work directly on the complex mixture samples. It also prevents the convergence problems which can occur with SOM based methods like GeoICA. Experimental results confirmed that the assumption of sparseness and, therefore, clusters around the mixing vectors is sufficiently fulfilled for convolutively mixed speech signals in the frequency domain.

References

1. Winter, S., Sawada, H., Makino, S.: Geometrical interpretation of the PCA subspace method for overdetermined blind source separation. In: Proc. ICA 2003. (2003) 775–780
2. Matsuoka, K.: Independent component analysis and its applications to sound signal separation. In: Proc. IWAENC 2003, Kyoto (2003) 15–18
3. Sawada, H., Mukai, R., Araki, S., Makino, S.: A robust and precise method for solving the permutation problem of frequency-domain blind source separation. In: Proc. ICA 2003. (2003) 505–510
4. Yilmaz, O., Rickard, S.: Blind separation of speech mixtures via time-frequency masking. IEEE Transactions on Signal Processing (2004) (to appear).

5. Rickard, S., Yilmaz, O.: On the approximate W-disjoint orthogonality of speech. In: Proc. ICASSP 2002. (2002) 529–532
6. Vielva, L., Santamaria, I., Pantaleon, C., Ibanez, J., Erdogmus, D.: Estimation of the mixing matrix for underdetermined blind source separation using spectral estimation techniques. In: Proc. EUSIPCO 2002. Volume 1. (2002) 557–560
7. Bofill, P., Zibulevsky, M.: Blind separation of more sources than mixtures using sparsity of their short-time fourier transform. In: Proc. ICA 2000. (2000) 87–92
8. Bofill, P.: Underdetermined blind separation of delayed sound sources in the frequency domain. Neurocomputing **55** (2003) 627–641
9. Araki, S., Makino, S., Blin, A., Mukai, R., Sawada, H.: Blind separation of more speech than sensors with less distortion by combining spareseness and ica. In: Proc. IWAENC 2003. (2003) 271–274
10. Blin, A., Araki, S., Makino, S.: Blind source separation when speech signals outnumber sensors using a sparseness - mixing matrix estimation (SMME). In: Proc. IWAENC 2003. (2003) 211–214
11. Waheed, K., Salem, F.M.: Algebraic overcomplete independent component analysis. In: Proc. ICA 2003. (2003) 1077–1082
12. Theis, F.: Mathematics in independent component analysis. PhD thesis, University of Regensburg (2002)
13. Hastie, T., Tibshirani, R., Friedman, J.: The elements of statistical learning: data mining, inference, and prediction. Springer Series in Statistics. Springer-Verlag (2002)
14. Vielva, L., Erdogmus, D., Principe, J.C.: Underdetermined blind source separation using a probabilistic source sparsity model. In: Proc. ICA 2001. (2001) 675–679
15. Sawada, H., Mukai, R., de la Kethulle de Ryhove, S., Araki, S., Makino, S.: Spectral smoothing for frequency-domain blind source separation. In: Proc. IWAENC 2003. (2003) 311–314

Penalty Function Approach for Constrained Convolutive Blind Source Separation

Wenwu Wang[1], Jonathon A. Chambers[1], and Saeid Sanei[2]

[1] Cardiff School of Engineering, Cardiff University
Queen's Building, Cardiff, CF24 0YF, UK
wenwu.wang@ieee.org, chambersj@cf.ac.uk
[2] Division of Engineering, King's College London
Strand, London, WC2R 2LS, UK
saeid.sanei@kcl.ac.uk

Abstract. A new approach for convolutive blind source separation (BSS) using penalty functions is proposed in this paper. Motivated by nonlinear programming techniques for the constrained optimization problem, it converts the convolutive BSS into a joint diagonalization problem with unconstrained optimization. Theoretical analyses together with numerical evaluations reveal that the proposed method not only improves the separation performance by significantly reducing the effect of large errors within the elements of covariance matrices at low frequency bins and removes the degenerate solution induced by a null unmixing matrix, but also provides an unified framework to constrained BSS.

1 Introduction

Among open issues in BSS, recovering the independent unknown sources from their linear convolutive mixtures remains a challenging problem. To address this problem, we focus on the operation in the frequency domain [2]-[5] rather than the approaches developed in the time domain (see [1] for example), due to its simpler implementation and better convergence performance. Using a discrete Fourier transformation (DFT), a time-domain linear convolutive BSS model can be transformed into the frequency domain [2], i.e., $\mathbf{X}(\omega, k) = \mathbf{H}(\omega)\mathbf{S}(\omega, k) + \mathbf{V}(\omega, k)$, where $\mathbf{S}(\omega, k)$ and $\mathbf{X}(\omega, t)$ are the time-frequency vectors of the N source signals and the M observed signals respectively ($M \geq N$), k is the discrete time index. The objective of BSS is to find $\mathbf{W}(\omega)$ which is a weighted pseudo-inverse of $\mathbf{H}(\omega)$, so that the elements of estimated sources $\mathbf{Y}(\omega, k)$ are mutually independent, where $\mathbf{Y}(\omega, k) = \mathbf{W}(\omega)\mathbf{X}(\omega, k)$. To this end, we exploit the statistical nonstationarity of signals by using the following criterion [4]

$$\mathcal{J}(\mathbf{W}(\omega)) = \arg\min_{\mathbf{W}} \sum_{\omega=1}^{T} \sum_{k=1}^{K} \mathcal{F}(\mathbf{W})(\omega, k), \qquad (1)$$

where $\mathcal{F}(\mathbf{W}) = \|\mathbf{R}_Y(\omega, k) - diag[\mathbf{R}_Y(\omega, k)]\|_F^2$, $\|\cdot\|_F^2$ is the squared Frobenius norm, $diag(\cdot)$ is an operator which zeros the off-diagonal elements of a matrix, and $\mathbf{R}_Y(\omega, k)$ is the cross-power spectrum of the output signals at multiple times, i.e., $\mathbf{R}_Y(\omega, k) = \mathbf{W}(\omega)[\mathbf{R}_X(\omega, k) - \mathbf{R}_V(\omega, k)]\mathbf{W}^H(\omega)$, where $\mathbf{R}_X(\omega, k)$ and $\mathbf{R}_V(\omega, k)$ are respectively the covariance matrices of $\mathbf{X}(\omega, k)$ and $\mathbf{V}(\omega, k)$, and $(\cdot)^H$ denotes the Hermitian transpose operator. Minimization of this criterion is equivalent to joint diagonalization of $\mathbf{R}_Y(\omega, k)$ for all time blocks $k, k = 1, \ldots, K$, that is, $\mathbf{R}_Y(\omega, k)$ will become a diagonal matrix $\mathbf{\Lambda}_C(\omega, k)$ due to the independence assumption [4].

However, there exists degenerate effect at low frequency bins induced by the large errors within the elements of covariance matrices (see more details in Section 4). Moreover, a null unmixing matrix $\mathbf{W}(\omega)$ also minimizes the criterion and potentially leads to a degenerate solution. In this paper, we propose a new approach based upon penalty functions, which is motivated by nonlinear programming techniques for constrained optimization. Essentially, we reformulate of the constrained BSS discussed in Section 2 as an unconstrained optimization problem using penalty functions. We will show that this approach provides an effective way of overcoming the aforementioned problems and a framework of unifying the joint diagonalization with unitary and non-unitary constraint. The remainder of this paper is organized as follows. Constrained BSS problem is briefly discussed in Section 2. The penalty function approach is introduced in Section 3, which includes its mathematical formulation, convergence behavior, numerical stability, and algorithm summary. The experimental results and the conclusion are respectively given in Section 4 and Section 5.

2 Constrained Blind Source Separation

Although BSS employs the least possible information pertaining to the sources and the mixing system, there exists useful information in practice for developing various effective algorithms to separate the mixtures, such as the geometrically constrained parameter space with $\|\mathbf{w}\| = 1$ exploited in [12], orthonormal constraint on $\mathbf{W}(k)$ i.e., $\mathbf{W}(k)\mathbf{W}^T(k) = \mathbf{I}$ used in [11] and [13], a non-holonomic constraint on $\mathbf{W}(k)$ maintained by a natural gradient procedure in [15], the source geometric information constraint exploited in [17] and a non-negative constraint in [16]. The orthonormal constraint has also been addressed as the optimization problem on the *Stiefel manifold* or *Grassman manifold* in [14] [13]. A recent contribution in [10] justifies that imposing an appropriate constraint on the separation matrix $\mathbf{W}(k)$ or the estimated source signals with special structure provides meaningful information to develop a more effective BSS solution for practical applications.

3 Penalty Function Approach

Effectively, a constrained BSS problem can be reformulated as the following equality constrained optimization problem,

$$P_1: \quad \min \mathcal{J}(\mathbf{W}(\omega)) \quad s.t. \quad \mathbf{g}(\mathbf{W}) = \mathbf{0} \qquad (2)$$

where $\mathbf{g}(\mathbf{W}) = [g_1(\mathbf{W}), g_2(\mathbf{W}), \cdots, g_r(\mathbf{W})]^T \colon \mathbb{C}^{N \times M} \to \mathbb{R}^r$ denotes the possible constraints, $\mathcal{J} \colon \mathbb{C}^{N \times M} \to \mathbb{R}^1$, and $r \geq 1$ indicates there may exist more than one constraint. In the BSS context, $\mathcal{J}(\mathbf{W}(\omega))$ denotes the various joint diagonalization criteria (1), and $\mathbf{g}(\mathbf{W})$ represents various constraints such as unitary constraint $\mathbf{W}\mathbf{W}^H = \mathbf{I}$ or non-unitary constraint $\mathbf{W}\mathbf{W}^H \neq \mathbf{I}$. To convert (2) into an unconstrained optimization problem, we have to define suitable penalty functions since it is unlikely to find a generic penalty function optimal for all constrained optimization problems. Regarding the equality constraint, we introduce a class of exterior penalty functions given as follows.

Definition 1: Let $\mathcal{W}$ be a closed subset of $\mathbb{C}^{N \times M}$. A sequence of continuous functions $\mathcal{U}_q(\mathbf{W}) : \mathbb{C}^{N \times M} \to \mathbb{R}^1$, $q \in \mathbb{N}$, is a sequence of exterior penalty functions for the set $\mathcal{Z}$ if the following three conditions are satisfied: (i) $\mathcal{U}_q(\mathbf{W}) = 0$, $\forall \mathbf{W} \in \mathcal{W}$, $q \in \mathbb{N}$; (ii) $0 < \mathcal{U}_q(\mathbf{W}) < \mathcal{U}_{q+1}(\mathbf{W})$, $\forall \mathbf{W} \notin \mathcal{W}$, $q \in \mathbb{N}$; (iii) $\mathcal{U}_q(\mathbf{W}) \to \infty$, as $q \to \infty$, $\forall \mathbf{W} \notin \mathcal{W}$.

Fig. 1 shows a typical example of such a function. According to Definition 1, it is straightforward to show that a function $\mathcal{U}_q(\mathbf{W}) : \mathbb{C}^{N \times M} \to \mathbb{R}$ defined as follows forms a sequence of exterior penalty functions for the set $\mathcal{W}$,

$$\mathcal{U}_q(\mathbf{W}) \triangleq \zeta_q \|\mathbf{g}(\mathbf{W})\|_b^\gamma \tag{3}$$

where $q \in \mathbb{N}$, $\gamma \geq 1$, $\zeta_{q+1} > \zeta_q > 0$, $\zeta_q \to \infty$, as $q \to \infty$, where $b = 1, 2$, or ∞.

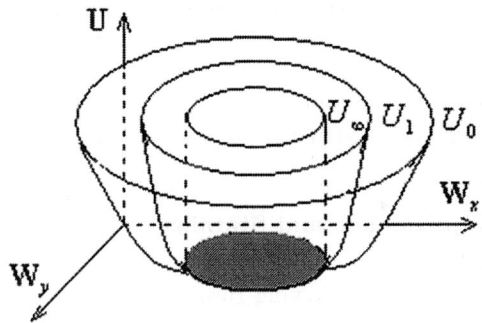

Fig. 1. $\mathcal{U}_i(\mathbf{W})$ ($i = 0, 1, \cdots, \infty$) are typical exterior penalty functions, where $\mathcal{U}_0(\mathbf{W}) < \mathcal{U}_1(\mathbf{W}) < \cdots < \mathcal{U}_\infty(\mathbf{W})$ and the shadow area denotes the subset $\mathcal{W}$.

After incorporating penalty functions, the new cost function becomes,

$$P_2: \quad \mathfrak{J}(\mathbf{W}(\omega)) = \mathcal{J}(\mathbf{W}(\omega)) + \boldsymbol{\kappa}^T \mathbf{U}(\mathbf{W}(\omega)), \tag{4}$$

where $\mathbf{U}(\mathbf{W}(\omega)) = [\mathcal{U}_1(\mathbf{W}(\omega)), \cdots, \mathcal{U}_r(\mathbf{W}(\omega))]^T$, whose elements take the form (3) which can be designed properly so that $\mathbf{W}(\omega) \neq 0$, $\mathcal{J}(\mathbf{W}(\omega))$ can be the form of (1), and $\boldsymbol{\kappa} = [\kappa_1, \cdots, \kappa_r]^T$ ($\kappa_i \geq 0$) are the weighted factors.

3.1 Convergence Behavior

The separation problem is thereby converted into an unconstrained optimization problem using joint diagonalization, i.e., $\min \mathfrak{J}(\mathbf{W}(\omega))$. The equivalence between (4) and (2), together with their critical points obey the following theorems.

Theorem 1: Let the set $\mathcal{W}$ be a closed subset of $\mathbb{C}^{N \times M}$ which satisfies $\mathbf{g}(\mathbf{W}) = \mathbf{0}$, the set $\mathbf{B}(\hat{\mathbf{W}}, \rho)$ be denoted by $\{\mathbf{W} \in \mathbb{C}^{N \times M} | \left\|\mathbf{W} - \hat{\mathbf{W}}\right\|_F \leq \rho\}$, where $\rho > 0$. Suppose that the following assumptions are satisfied: (a) If there exists a point $\mathbf{W}^*$ such that the level set $\{\mathbf{W} \in \mathbb{C}^{N \times M} | f(\mathbf{W}) \leq f(\mathbf{W}^*)\}$ is compact, and (b) there exists an optimal solution $\hat{\mathbf{W}}$ for problem (2) such that for $\forall \rho > 0$, $\mathbf{B}(\hat{\mathbf{W}}, \rho) \cap \mathcal{W}$ is not empty. Then: (i) For any given $i \in \mathbb{N}$, let $\mathbf{W}_i$ be an optimal solution to problem P_2 in (4) at the ith trial. Then any accumulation point $\hat{\mathbf{W}}$ of $\mathbf{W}_i (i = 0 \to \infty)$, is an optimal solution to problem P_1 in (2). (ii) For every $i \in \mathbb{N}$, let $\mathbf{W}_i$ be a strict local minimizer for problem P_2 in (4) at the ith trial, so that for some $\rho_i > 0$, $f_i(\mathbf{W}_i) < f_i(\mathbf{W})$ for all $\mathbf{W} \in \mathbf{B}(\mathbf{W}_i, \rho_i) = \{\mathbf{W} \in \mathbb{C}^{N \times M} | \left\|\mathbf{W} - \mathbf{W}_i\right\|_F \leq \rho\}$. If $\hat{\mathbf{W}}$ is an accumulation point of $\mathbf{W}_i (i = 0 \to \infty)$, and there exists a $\rho > 0$, such that $\rho_i \geq \rho$, for all $i \in \mathbb{N}$, then $\hat{\mathbf{W}}$ is a local minimizer for the problem P_1 in (2).

The proof of this theorem is omitted due to the limited space. It is worth noting that the assumption (a) in Theorem 1 is to ensure that problem (2) has a solution and the assumption (b) is to ensure that the closure of the set $\mathbf{B}(\hat{\mathbf{W}}, \rho) \cap \mathcal{W}$ contains an optimal solution to problem (2). The theorem implies that only given large enough penalty parameters, the new criterion (4) holds the same global and local properties as that without the penalty term. In practical situations, however, this means that the choice of the initial values of the penalty parameters has an important effect on the overall optimization accuracy and efficiency. Too small values will violate major constraints, and too large values may create an ill-conditioned computation problem [9]. This fact can also be observed from the eigenvalue structure of its Hessian matrix demonstrated in Section 3.2.

3.2 Numerical Equivalence and Stability

Assuming that $\mathfrak{J}(\mathbf{W})$ is twice-differentiable and calculating the perturbation matrix $\mathbf{\Delta}$ of $\mathbf{W}$, we have the following Hessian matrix

$$\nabla^2 \mathfrak{J}(\mathbf{W}) \stackrel{\Delta}{=} \nabla^2 \mathcal{F}(\mathbf{W}) + \kappa \frac{\partial \mathcal{U}(\mathbf{W})}{\partial \mathbf{W}^*} \nabla^2 g_i(\mathbf{W}) + \kappa \frac{\partial^2 \mathcal{U}(\mathbf{W})}{\partial \mathbf{W}^*} \nabla g_i(\mathbf{W}) \nabla g_i(\mathbf{W})^T \quad (5)$$

The conditions of *Theorem 1* indicate that as $\kappa \to \infty$, $\mathbf{W}$ will approach the optimum $\hat{\mathbf{W}}$. If $\hat{\mathbf{W}}$ is a regular solution to the constrained problem, then there exists unique Lagrangian multipliers $\bar{\lambda}_i$ such that $\frac{\partial \mathcal{U}(\hat{\mathbf{W}})}{\partial \mathbf{W}^*} + \sum \bar{\lambda}_i \nabla g_i(\mathbf{W}) = \mathbf{0}$ [7]. This means $\kappa \frac{\partial \mathcal{U}(\mathbf{W})}{\partial \mathbf{W}^*} \to \bar{\lambda}_i$ as $\mathbf{W} \to \hat{\mathbf{W}}$. The first two terms in (5) approach the Hessian of $\mathcal{F}(\mathbf{W}) + \sum \bar{\lambda}_i g_i(\mathbf{W})$. Considering the last term in (5), it can be shown that as $\kappa \to \infty$, $\nabla^2 \mathfrak{J}(\mathbf{W})$ has some eigenvalues approaching ∞, and others approach finite value. The infinite eigenvalues will lead to an ill-conditioned computation problem. Let ϵ be the step size in the adaptation, then in the presence of nonlinear equality constraints, the direction $\mathbf{\Delta}$ may cause any reduction of $\mathcal{F}(\mathbf{W} + \epsilon \mathbf{\Delta})$ to be shifted by $\kappa \mathcal{U}(\mathbf{W} + \epsilon \mathbf{\Delta})$. This requires the step size to be small to prevent the ill-conditioned computation problem induced by large eigenvalues with a trade-off of having a lower convergence rate. Such a theoretical analysis is verified in section 4.

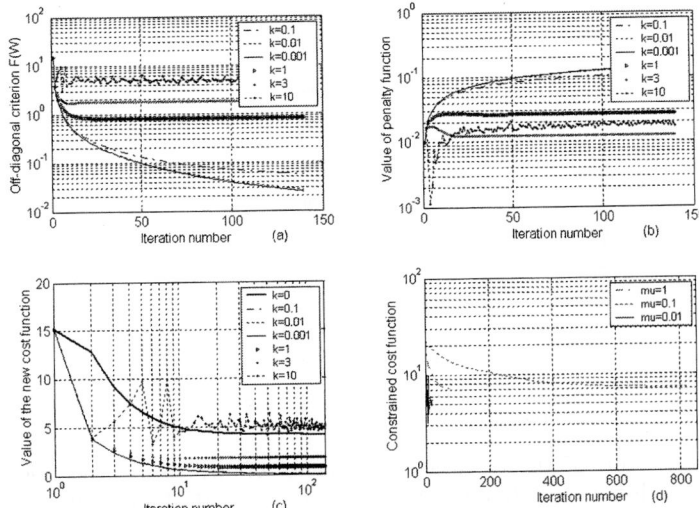

Fig. 2. Convergence behavior of the penalty function approach.

3.3 Approach Summary

Based on the discussions given in the above sections, the proposed algorithm by incorporating penalty functions is conducted as following steps (using the steepest descent gradient adaptation):

1). Initialize parameters N, M, D, T, K, $\mathbf{W}_0$, α, ξ, ς, IRN, $\mathbf{W}_0(\omega)$;
2). Convert the input mixtures $\mathbf{x}(n)$ to $\mathbf{X}(\omega, n)$; calculate the cross-power spectrum matrix $\hat{\mathbf{R}}_X(\omega, k) = \frac{1}{D}\sum_{m=0}^{D-1} \mathbf{X}(\omega, Dk+m)\mathbf{X}^H(\omega, Dk+m)$;
3). Calculate the cost function and update gradient:

   – for $i = 1$ to IRN
   * Update $\mu_{J_M}(\omega) = \alpha/(\sum_{k=1}^{K} \|\mathbf{R}_X(\omega, k)\|_F^2)$, and $\mu_{J_C}(\omega) = \xi/(\varsigma + \sum_{k=1}^{K} \left\|\frac{\partial J_C(\mathbf{W})(\omega,k)}{\partial \mathbf{W}^*(\omega)}\right\|_F)$ respectively;
   * Update $\mathbf{W}(\omega) \leftarrow \mathbf{W}(\omega) + \mu(\mu_{J_M}\frac{\partial \mathcal{J}}{\partial \mathbf{W}^*(\omega)} + \mu_{J_C}\frac{\partial \mathcal{U}}{\partial \mathbf{W}^*(\omega)})$;
   * Update $\mathfrak{J}_i(\mathbf{W}(\omega))$ using (4);
   * if $(\mathfrak{J}_i(\mathbf{W}(\omega)) > \mathfrak{J}_{i-1}(\mathbf{W}(\omega)))$ break;
   – end

4). Solve permutation problem $\mathbf{W}_{new}(\omega) \leftarrow \mathcal{P}(\mathbf{W}(\omega))$, where $\mathcal{P}$ is a function dealing with permutation operation (refer to [4]);
5). Calculate $\mathbf{Y}(\omega, k) = \mathbf{W}(\omega)\mathbf{X}(\omega, k)$ and reconstruct the time domain signals $\mathbf{y}(n) = IDFT(\mathbf{Y}(\omega, k))$;
6). Calculate the performance index, e.g., signal to interference ratio (SIR) [4].
7). End.

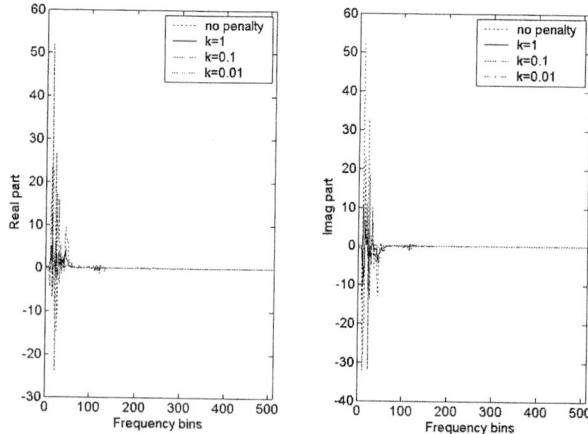

Fig. 3. Comparison of the off-diagonal elements of the cross-correlation matrices $\mathbf{R}_Y(\omega, k)$ at each frequency bin between the proposed method and that in [4] ($\kappa = 0$).

4 Numerical Examples

To examine the proposed method, we use an exterior penalty function with the form of $\|diag[\mathbf{W}(\omega) - \mathbf{I}]\|_F^2$ [6], and a variant of gradient adaptation $\kappa diag[\mathbf{W}(\omega) - \mathbf{I}]\mathbf{W}(\omega)$. A system with two inputs and two outputs (TITO) is considered for simplicity, that is, $N = M = 2$. Two real speech signals are used in the following experiments, which are available from [19]. In the first experiment, we artificially mix the two sources by a non-minimum phase system with $H_{11}(z) = 1 + 1.0z^{-1} - 0.75z^{-2}$, $H_{12}(z) = 0.5z^{-5} + 0.3z^{-6} + 0.2z^{-7}$, $H_{21}(z) = -0.7z^{-5} - 0.3z^{-6} - 0.2z^{-7}$, and $H_{22}(z) = 0.8 - 0.1z^{-1}$ [18]. Other parameters are set to be $T = 1024$, $K = 5$, $D = 7$, $\alpha = 1$, $\varsigma = 0.05$, $\xi = 0.2$, $\mathbf{W}_0(\omega) = \mathbf{I}$, and $\mu = 1$. We applied the short term FFT to the separation matrix and the cross-correlation of the input data. Fig. 2 show the convergence behavior by incorporating penalty functions. Fig 2 (a)-(c) indicate that, when increasing the penalty coefficient κ, not only the constraint is approached more quickly, but also the cost function converges faster. However, it is also observed that a large penalty κ (e.g. $\kappa = 10$) introduces the ill-conditional problem under a common step size. Such effect can be properly removed by reducing the step size, see Fig. 2 (d), where κ is fixed to be 10, but μ is changing (The adaptation stops when a threshold is satisfied). Theoretically, due to the independence assumption, the cross-correlation of the output signals should approach zero. Fig. 3 demonstrates that it is true at most frequency bins, however with exception for the low frequency bins. From Fig. 3, we see that this effect can be significantly reduced by using penalty functions.

In the second experiment, the proposed joint diagonalization method is compared with other two joint diagonalization criteria [4] [8]. The mixtures are obtained from the simulated room environment, which was implemented by a *room-mix* function available from [20]. The room is assumed to be a $10m \times 10m \times 10m$ cube with wall reflections computed up to the fifth order. The position matrices

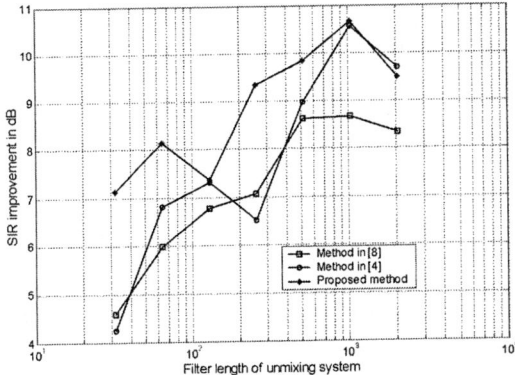

Fig. 4. SIR measurement for a simulated room environment with high reverberance.

of two sources and two sensors are respectively [2 2 5;8 2 5] and [3 8 5;7 8 5]. The SIR plot in Fig. 4 shows that: (i) incorporating a suitable penalty function can increase the SIR which indicates a better separation performance; (ii) the separation quality increases with the increasing filter length of the separation system; (iii) exploiting spectral continuity of the separation matrix (the proposed method and that in [4]) may have superior performance to the method (e.g., [8]) which considers the separation at each frequency bin independently.

5 Conclusion

The penalty function based joint diagonalization approach for frequency domain BSS has been presented. Its convergence behavior and numerical stability have also been discussed. Experimental evaluation indicates that the proposed approach improves the convergence performance as compared with cross-power spectrum based method, and significantly reduces the degenerate effect existing on the lower frequency bins therefore improves its separation performance. This approach also provides a unifying view to constrained BSS which is useful to develop suitable BSS algorithms using optimization techniques.

References

1. A. Cichocki and S. Amari, *Adaptive Blind Signal and Image Processing: Learning Algorithms and Applications*. John Wiley, Chichester, Apr. 2002.
2. P. Smaragdis, "Blind separation of convolved mixtures in the frequency domain," *Neurocomputing*, vol.22, pp. 21–34, 1998.
3. K. Rahbar and J. Reilly, "Blind source separation of convolved sources by joint approximate diagonalization of cross-spectral density matrices," *Proc. ICASSP*, May, 2001.
4. L. Parra and C. Spence, "Convolutive blind source separation of nonstationary sources," *IEEE Trans. on Speech and Audio Proc.*, pp. 320–327, May 2000.

5. W. Wang, J. A. Chambers, and S. Sanei, "A joint diagonalization method for convolutive blind separation of nonstationary sources in the frequency domain," *Proc. ICA*, Nara, Japan, Apr. 1-4, 2003.
6. M. Joho and H. Mathis, "Joint diagonalization of correlation matrices by using gradient methods with application to blind signal separation," *Proc. SAM*, Rosslyn, VA, 4-6, Aug. 2002.
7. M. S. Bazaraa, H. D. Sherali, and C. M. Shetty, *Nonlinear Programming Theory and Algorithms*, 2nd ed. John Wiley & Sons Inc., 1993.
8. N. Murata, S. Ikeda, and A. Ziehe. "An approach to blind source separation based on temporal structure of speech signals." *Neurocomputing*, vol. 41, pp. 1-24, 2001.
9. A. Cichocki and R. Unbehauen, *Neural Networks for Optimization and Signal Processing*, Wiley, 1993.
10. A. Cichocki and P. Georgiev, "Blind source separation algorithms with matrix constraints", *IEICE Trans. on Fundamentals of Elect. Comm. and Computer Science*, vol. E86-A, pp. 522-531, Mar. 2003.
11. J.-F. Cardoso and B. Laheld, "Equivariant adaptive source separation," *IEEE Trans. Signal Processing*, vol. 44, pp. 3017-3030, Dec. 1996.
12. S. C. Douglas, S. Amari, and S.-Y. Kung, "On gradient adaptation with unit norm constraints," *IEEE Trans. Signal Processing*, vol. 48, no. 6, pp. 1843-1847, June 2000.
13. S. C. Douglas, "Self-stabilized gradient algorithms for blind source separation with orthogonality constraints," *IEEE Trans. on Neural Networks*, vol. 11 no. 6, pp. 1490-1497, June 2000.
14. J. H. Manton, "Optimisation algorithms exploiting unitary constraints," *IEEE Trans. Signal Processing*, vol. 50, pp. 635–650, Mar. 2002.
15. S. Amari, T. P. Chen and A. Cichocki, "Nonholonomic orthogonal learning algorithms for blind source separation," *Neural Computation*, vol. 12, pp. 1463-1484, 2000.
16. M. D. Plumbley, "Algorithms for non-negative independent component analysis," *IEEE Transactions on Neural Networks*, vol. 14 no. 3, pp. 534- 543, May 2003.
17. L. Parra and C. Alvino, "Geometric Source Separation: Merging convolutive source separation with geometric beamforming", *IEEE Trans. on Speech and Audio Processing*, vol. 10, no. 6, pp. 352-362, Sept. 2002.
18. T. W. Lee, A. J. Bell, and R. Lambert, "Blind separation of delayed and convolved sources", *Advances in neural information processing systems 9*, MIT Press, Cambridge MA, pp. 758—764, 1997.
19. J. Anemüller, http://medi.uni-oldenburg.de/members/ane.
20. Westner, http://www.media.mit.edu/~westner.

Permutation Alignment for Frequency Domain ICA Using Subspace Beamforming Methods

Nikolaos Mitianoudis[1] and Mike Davies[2]

[1] Imperial College London, Electrical and Electronic Engineering, Exhibition Road
SW7 2AZ London, UK
n.mitianoudis@imperial.ac.uk
[2] Queen Mary London, Centre for Digital Music, Mile End Road
E1 4NS London, UK
michael.davies@elec.qmul.ac.uk

Abstract. In this paper, the authors address the *permutation ambiguity* that exists in *frequency domain Independent Component Analysis* of convolutive mixtures. Many methods have been proposed to solve this ambiguity. Recently, a couple of beamforming approaches have been proposed to address this ambiguity. The authors explore the use of *subspace methods* for permutation alignment, in the case of equal number of sources and sensors.

1 Introduction

Assume an array of M sensors $\underline{x}(n) = [x_1(n)\ x_2(n)\ \ldots\ x_M(n)]^T$ placed in a real room, capturing an auditory scene. Assume there are N sources in the auditory scene $\underline{s}(n) = [s_1(n)\ s_2(n)\ \ldots\ s_N(n)]^T$. To model the recording environment, one could use *FIR convolutive mixtures*.

$$x_i(n) = \sum_{j=1}^{N} \underline{a}_{ij} * s_j(n) \qquad i = 1, \ldots, M \qquad (1)$$

where $\underline{a}_{ij}$ represents an FIR filter modelling the transfer function between the ith sensor and the jth source. For the rest of the analysis, we will consider only the case of equal number of sensors and sources.

The convolutive mixtures problem can be addressed in the *time domain*, by estimating unmixing FIR filters $\underline{w}_{ij}$, assuming that the sources are *statistically independent*. The filters are adaptively estimated in the time domain, using the general framework of *Independent Component Analysis* (ICA).

$$u_i(n) = \sum_{j=1}^{N} \underline{w}_{ij} * x_j(n) \qquad i = 1, \ldots, N \qquad (2)$$

A more robust approach is to transfer the problem in the *frequency domain*. Consequently, the convolutive mixtures problem is transformed into several instantaneous mixtures problems. Many frequency domain ICA (FD-ICA) methods were proposed in literature. In [4], a fast FD-ICA framework was proposed with fast and robust results, compared to gradient-based methods. In frequency-domain methods, we encounter two interdeterminancies: the *scale* and the *permutation ambiguity*. The *scale ambiguity* (arbitrary source scaling) is rectified by mapping the separated sources to the observation space [3]. The *permutation ambiguity* (inherent ordering ambiguity of the instantaneous ICA model) produces an arbitrary ordering of sources along frequency. To tackle this problem, one should apply some mechanism to couple the sources along frequency. Some *source modelling* solutions exploit the coherence and the information between the frequency bands to align the permutations. There also exist some *channel modelling* solutions, assuming smooth filters, as a constraint to the unmixing algorithm.

In fact, the blind source separation systems can be considered array signal processing systems. A set of sensors arranged randomly in a room to separate the sources present is effectively a beamformer. Some methods [2,5,6] were proposed to solve the permutation problem using beamforming. In this paper, we investigate the idea of using subspace methods for permutation alignment in FD-ICA. Subspace methods produce more accurate alignment compared to the previously proposed methods using directivity patterns. We show that subspace methods even work in the case of equal number of sources and sensors.

2 Beamforming and Frequency-Domain ICA

A narrowband linear array of M sensors $\underline{x}(n)$, is defined as follows:

$$\underline{x}(n) = \sum_{i=1}^{N} \underline{a}(\theta_i) s_i(n) = [\underline{a}(\theta_1) \ \underline{a}(\theta_2) \ \ldots \ \underline{a}(\theta_N)] \underline{s}(n) \qquad (3)$$

where $\underline{a}(\theta_i) = [1 \ \alpha e^{-j2\pi f T_i} \ldots \alpha e^{-j2\pi f(M-1)T_i}]^T$, $T_i = d\sin\theta_i/c$, θ_i are the DOA, d is the intra-sensor distance and $c = 340 m/sec$. The array model is similar to the general Blind Source Separation model. The main objective is to estimate a filter $\underline{w}_i(f)$ to separate each source i. The *directivity pattern* (gain pattern) of the beamformer $\underline{w}_i(f) = [w_{i1} \ldots w_{iN}]$, can be expressed as follows:

$$F_i(f, \theta) = \sum_{k=1}^{N} w_{ik}^{ph}(f) e^{j2\pi f(k-1)d\sin\theta/c} \qquad (4)$$

In the context of FD-ICA, at a given frequency bin, the unmixing matrix can be interpreted as a *null-steering beamformer* that uses a *blind algorithm* (ICA) to place nulls on the interfering sources. The source separation framework does not use any information concerning the geometry of the auditory scene, but only the sources statistical profile. Inclusion of this additional information can help in aligning the permutations. Although, we are dealing with real room recordings, we assume that there is a consistent DOA along frequency for each source, belonging to the direct path signal signal. This is equivalent of approximating the room's transfer function with a single delay. The permutations of the unmixing matrices are flipped so that the directivity pattern of each beamformer is approximately "aligned". More specifically, having estimated the unmixing matrix $W(f)$ using FD-ICA, we permute the rows of $W(f)$, in order to align the permutations along the frequency axis. We form the directivity pattern (2), where $w_{ik}^{ph}(f) = W_{ik}(f)/|W_{ik}(f)|$ is the phase of the unmixing filter coefficient between the kth sensor and the ith source at frequency f. This approach can be considered a *channel modelling* technique.

However, in audio source separation, the sensors capture more than a single delay. The room's reflections tend to shift the "actual" DOA by a small arbitrary amount at each frequency. However, the average shift of DOA along frequency is not so significant and usually we can spot a main DOA. This implies that we can align the permutations in FD-ICA, using the DOA.

The reason why we are using beamforming for permutation alignment and not for separation is the poor estimate for DOA along frequency. The ICA algorithm can give very accurate separation. Instead, the slightly "shifted" DOA can help us in identifying the correct permutation of separated sources.

Next, we will address some ambiguities in DOA estimation and permutation alignment using directivity patterns, plus a novel mechanism to apply subspace techniques for permutation alignment.

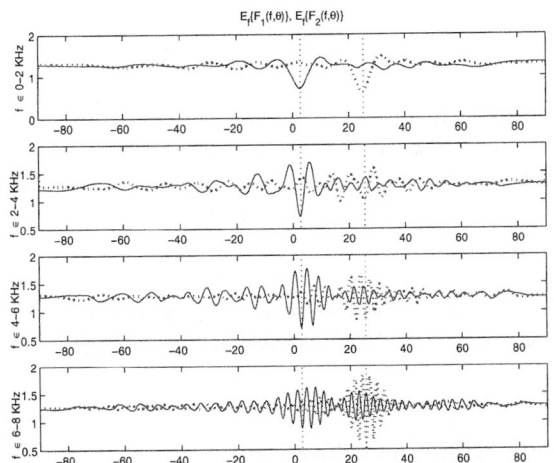

Fig. 1. Average Beampatterns along certain frequency bands for both sources.

2.1 DOA Estimation Ambiguity

Saruwatari et al [6] estimated the DOA by taking the statistics with respect to the direction of the nulls in all frequency bins and then tried to align the permutations by grouping the nulls that exist in the same DOA neighbourhood. On the other hand, Ikram and Morgan [2] proposed to estimate the sources' DOA in the lower frequencies, as it is less noisy than in higher frequencies. Parra and Alvino [5] used more sensors than sources along with *known* source locations and added this information as a geometric constraint to their unmixing algorithm.

In figure 1, we plot the average beampatterns along a certain frequency range $\mathcal{F}$, assuming a two sensor setup in a real room, where $d = 1m$. More specifically, we plot the average beampatterns between $0 - 2kHz$, $2 - 4kHz$, $4 - 6kHz$ and $6 - 8kHz$. We can see that in the lower frequencies, we get clear peaks denoting the directions of arrival. However, in higher frequencies, we get peaks at the same angle, but also multiple peaks around the main DOA. Observing the higher frequencies, we can not really define which of the peaks is the actual DOA. As a result, we may want to use only the lower subband $(0 - 2kHz)$ for DOA estimation.

It is simple to show that averaging beampatterns over a lower frequency band $\mathcal{F}$ will emphasize the position of the two DOAs. Hence, the following mechanism can be used for DOA estimation, without sorting the permutations along frequency.

1. Unmix the sources using an FD-ICA algorithm
2. For each frequency bin f and source i estimate the beamforming pattern $F_i(f,\theta)$.
3. Form the following expression for $\mathcal{F} = [0 - 2kHz]$

$$P(\theta) = \sum_{f \in \mathcal{F}} \sum_{i=1}^{N} |F_i(f,\theta)|^2 \qquad (5)$$

The minima of $P(\theta)$ will give an accurate estimate of the Directions of Arrival. The exact low-frequency range $\mathcal{F}$ we can use for DOA estimation is mainly dependent on the microphone spacing d. If we choose a small microphone spacing ($\sim$ cm), the ripples will start to appear at higher frequencies, as $f_{ripple} \sim c/2d$. However, as the microphones will be closer, the signals that will be captured will be more similar. Thus, the source separation SNR will decrease considerably, as our setup will degenerate to the less sensors than sources case. Therefore, the choice of sensor spacing is a tradeoff between *separation quality* and *beamforming pattern clarity*.

2.2 Permutation Alignment Ambiguity

Once we have estimated the DOA, we want to align the permutations along the frequency axis to solve the permutation problem in frequency domain ICA. There is a slight problem with that. Basically, all nulls, as explained in an earlier section, are slightly drifted due to reverberation. As a result, the classification of the permutations cannot be accurate.

One solution can be to look for nulls in a "neighbourhood" of the DOA. Then, we can do some classification, however, the definition of the neighbourhood is arbitrary. Hu and Kobatake [1] observed that for a room impulse response around $300ms$, the drift from the real DOA maybe $1 - 3$ degrees on average (this may be generally different at various frequencies). As a result, we can define the neighbourhood as 3 degrees around the DOA. However, in mid-higher frequencies there might be more than one null, making the classification even more difficult.

3 Permutation Alignment Using the MuSIC Algorithm

Another idea is to introduce *subspace methods*, as they tend to produce more "spiky" directivity patterns. The multiple nulls ambiguity still exists, however, the DOAs are more distinct and the permutation alignment should be more efficient. Although, in theory, we need to have more sensors than sources, it is possible to apply subspace methods in the case of equal number of sources and sensors. In our case, we will look at the MuSIC algorithm [7]. According to the MuSIC algorithm, one gets very localised estimates for the DOA by plotting the following function $M(\theta)$:

$$M(\theta) = \frac{1}{|P^\perp \underline{a}(\theta)|^2} \qquad \forall \ \theta \in [-\pi/2, \pi/2] \qquad (6)$$

where $P^\perp = (I - E_s E_s^H) = E_n E_n^H$, where $E_s = [\underline{e}_1, \underline{e}_2, \ldots, \underline{e}_N]$ contains the eigenvectors of $C_x = \mathcal{E}\{\underline{xx}^H\}$ that correspond to the desired source and $E_n = [\underline{e}_{N+1}, \ldots, \underline{e}_M]$ contains the eigenvectors of C_x that correspond to noise. The N peaks of the function $M(\theta)$ will denote the DOA of the N sources.

In [4], we proposed to rectify the *scale ambiguity* by mapping the separated sources back to the microphones' domain. Therefore, we have an observation of each source at each sensor, i.e. a more sensors than sources scenario. If we do not take any steps for the permutation problem, the ICA algorithm will unmix the sources at each frequency bin, however, the permutations will not be aligned along frequency. It is simple to demonstrate that mapping back to the observation space is not influenced by the permutation ambiguity [3]. Hence, after mapping we will have observations of each source at each microphone, however, the order of sources will not be the same along frequency. Using the observations of all microphones for each source, we can use MuSIC to find a more accurate estimation for the DOAs, using (6).

We can form "MuSIC directivity patterns" using $M(\theta)$ (6), instead of the original directivity patterns. To find more accurate DOA estimates, we can form $P(\theta)$ as expressed in (5), using $M(\theta)$ instead of the original directivity pattern. Finally, we can use the DOAs to align the "sharper" "MuSIC directivity patterns". The proposed algorithm can be summarised as follows:

1. Unmix the sources using the FD-ICA framework.
2. Map the sources back to the observation space, i.e. observe each source at each microphone.
3. Having observations of each source at each microphone, we apply the MuSIC algorithm to have more accurate DOA estimates along frequency.
4. Align permutations now, according to the DOAs estimated by MuSIC.

4 Experiments

In this section, we perform two experiments to verify the ideas analysed so far in this paper. The Fast FD-ICA algorithm [4] is used to unmix the data, without the Likelihood Ratio solution.

4.1 Experiment 1 – Single Delay

In the first experiment, two speech signals are mixed artificially using single delays between $5-6$ msecs at $16kHz$. We test the performance of the proposed solutions for the permutation problem, in terms of beamforming. In figure 2 (left), we can see a plot of $P(\theta)$ (5) for this case of a single delay. We averaged the directivity patterns over the lower frequency band $(0-2kHz)$ and as a result we can see two Directions of Arrival. The estimated DOAs will be used to align the permutations. Since we are modeling a single delay, we will not allow any deviations from the estimated DOAs. In figure 3 (left), we can see the general performance of this scheme for one of the sources. We can spot some mistakes in the mid-higher frequencies, verifying that it might be difficult to align the permutations there. In figure 2 (right), we can see a plot of $P(\theta)$ (5) using the MuSIC algorithm. We averaged the MuSIC directivity patterns over the lower frequency band $(0-2kHz)$. Now the peaks indicating the Directions of Arrival are now a lot more distinct and "spiky". In figure 3 (right), we can see that the permutations are correctly aligned using the more accurate MuSIC directivity plots.

4.2 Experiment 2 – Real Room Recording

Next, we perform a real world experiment. We used a university lecture room $\sim 7.5 \times 6m^2$ to record a 2 sources - 2 sensors experiment. We investigate the nature of real room directivity patterns as well as explore the performance of the proposed schemes for permutation alignment. In figure 4 (left), we can see a plot of $P(\theta)$ (5) for this case of real room recording. Averaging over the lower $2kHz$, we seem to get a very clear image of the main DOAs, giving us an accurate measure for this estimation task. We try to align to the permutations around the estimated DOAs allowing $\pm 3°$ deviation. In figure 5 (left), we see the results for one of the sources. We can spot that generally this scheme can perform robust permutation alignment in the lower frequencies, but considerable confusion exists in higher frequencies, as expected from our theoretical analysis. In figure 4 (right), we can see a plot of $P(\theta)$ (5), averaging the MuSIC directivity patterns over the lower frequency band $(0-2kHz)$. The two Directions of Arrival are more clearly identified from this graph. In figure 5 (right), we can see that most of the permutations are correctly aligned using the more accurate MuSIC directivity plots.

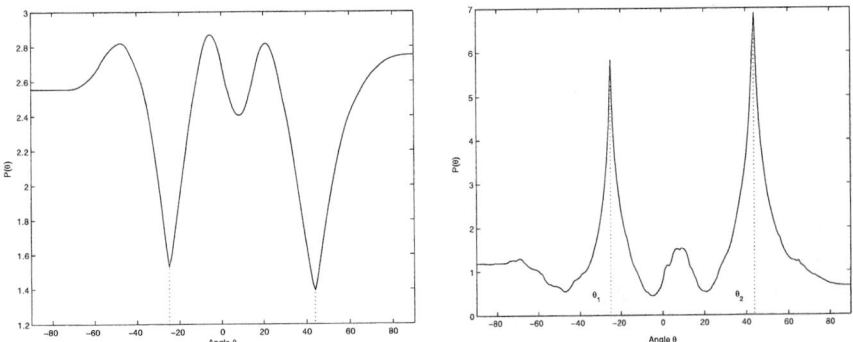

Fig. 2. Plotting $P(\theta)$ (eq. 5) using directivity patterns (left) and MuSIC directivity patterns (right) for the first 2kHz for the single delay case. Two distinct DOAs are visible.

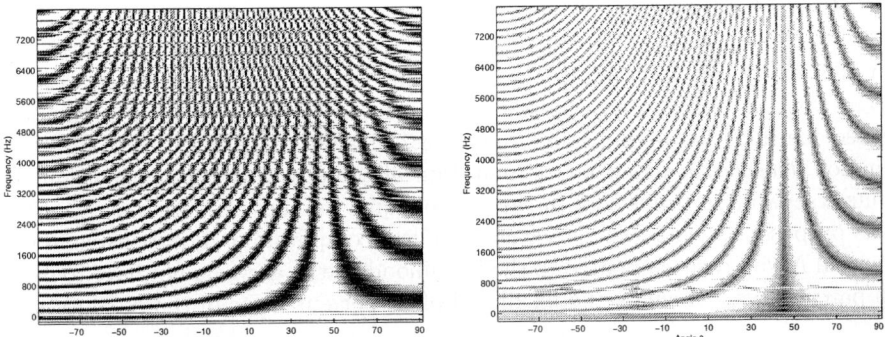

Fig. 3. Permutations aligned using the directivity patterns (left) and the MuSIC directivity patterns (right) in the single delay case.

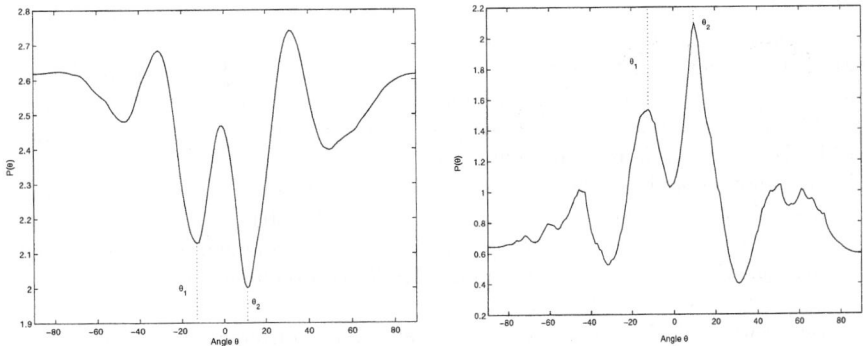

Fig. 4. Plotting $P(\theta)$ (eq. 5) using directivity patterns (left) and MuSIC directivity patterns (right) for the first 2kHz in the real room case. MuSIC enhances the positions of the DOAs.

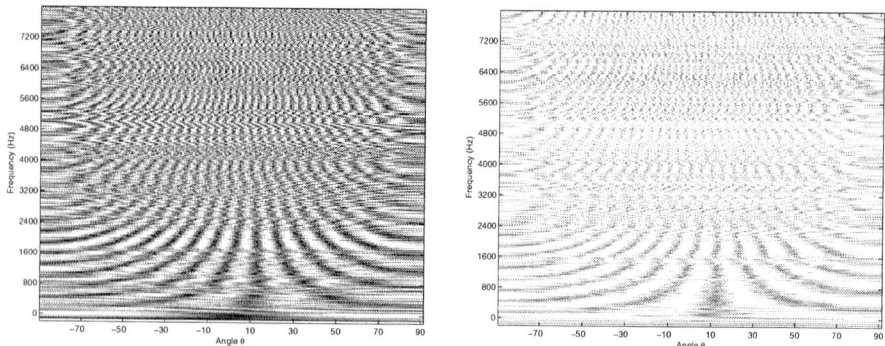

Fig. 5. Permutations aligned using the Directivity Patterns (left) and the MuSIC directivity patterns (right) in the real room case.

5 Conclusion

In this paper, we interpreted the Frequency-Domain audio source separation framework, as a Frequency-Domain beamformer. We reviewed some of the proposed methods for permutation alignment. In addition, a novel mechanism to employ *subspace methods* for permutation alignment in the frequency domain source separation framework in the case of equal number of sources and sensors was proposed. Such a scheme seems to be less computationally expensive in the general $N \times N$ case, compared to the Likelihood Ratio, as we do not have to work in pairs or even calculate the likelihood of all permutations of the N sources.

References

1. X. Hu and H. Kobatake. Blind source separation using ica and beamforming. In *Proc. Int. Workshop on Independent Component Analysis and Blind Signal Separation (ICA2003)*, pages 597–602, Nara, Japan, 2003.
2. M.Z. Ikram and D.R. Morgan. A beamforming approach to permutation alignment for multichannel frequency-domain blind speech separation. In *ICASSP*, 2002.
3. N. Mitianoudis. *Audio Source Separation using Independent Component Analysis*. PhD thesis, Queen Mary, University of London, 2004.
4. N. Mitianoudis and M. Davies. Audio source separation of convolutive mixtures. *Trans. Audio and Speech Processing*, 11(5):489–497, 2003.
5. L. Parra and C. Alvino. Geometric source separation: Merging convolutive source separation with geometric beamforming. *IEEE Transactions on Speech and Audio Processing*, 10(6):352–362, 2002.
6. H. Saruwatari, T. Kawamura, and K. Shikano. Fast-convergence algorithm for ica-based blind source separation using array signal processing. In *Proc. Int. IEEE WASPAA*, pages 91–94, New Paltz, New York, 2001.
7. R.O. Schmidt. Multiple emitter location and signal parameter estimation. *IEEE Trans. on Antennas and propagation*, AP-34:276–280, 1986.

QML Blind Deconvolution: Asymptotic Analysis

Alexander M. Bronstein, Michael M. Bronstein,
Michael Zibulevsky, and Yehoshua Y. Zeevi*

Technion - Israel Institute of Technology, Department of Electrical Engineering
32000 Haifa, Israel
{alexbron,bronstein}@ieee.org, {mzib,zeevi}@ee.technion.ac.il

Abstract. Blind deconvolution is considered as a problem of quasi maximum likelihood (QML) estimation of the restoration kernel. Simple closed-form expressions for the asymptotic estimation error are derived. The asymptotic performance bounds coincide with the Cramér-Rao bounds, when the true ML estimator is used. Conditions for asymptotic stability of the QML estimator are derived. Special cases when the estimator is super-efficient are discussed.

1 Introduction

Blind deconvolution arises in various applications related to acoustics, optics, medical imaging, geophysics, communications, control, etc. In the noiseless setup of single-channel blind deconvolution, the observed sensor signal x is created from the *source signal* s passing through a convolutive system with impulse response a, $x = a * s$. The setup is termed *blind* if only x is accessible, whereas no knowledge on w and s is available. Blind deconvolution attempts to find such a deconvolution (restoration) kernel w, that produces a possibly delayed waveform-preserving source estimate $\hat{s}_n = (w * x)_n \approx c \cdot s_{n-\Delta}$, where c is a scaling factor and Δ is an integer shift. Equivalently, the *global system response* $g = a * w$ should be approximately a Kroenecker delta, up to scale factor and shift. A commonly used assumption is that s is non-Gaussian.

Asymptotic performance of maximum-likelihood parameter estimation in blind system identification and deconvolution problems was addressed in many previous studies (see, for example, [1–4]). In all these studies, the Cramér-Rao lower bound (CRLB) for the system parameters are found, and lower bounds on signal reconstruction quality are derived. However, sometimes the true source distribution is either unknown, or not suitable for optimization, which makes the use of ML estimation impractical. In these cases, a common solution is to replace the true source PDF by some other function, leading to a *quasi ML* estimator. Such an estimator generally does not achieve the CRLB and a more delicate performance analysis is required. In [5,6], asymptotic performance analysis of QML estimators for blind source separation was presented.

In this study, we derive asymptotic performance bounds for a QML estimator of the restoration kernel in the single-channel blind deconvolution problem, and state the asymptotic stability conditions. We show that in the particular case when the true ML procedure is used, our bounds coincide with the CRLB, previously reported in literature.

---
* This research has been supported by the HASSIP Research Network Program HPRN-CT-2002-00285, sponsored by the European Commission, and by the Ollendorff Minerva Center.

2 QML Blind Deconvolution

Under the assumption that the restoration kernel w has no zeros on the unit circle, and the source signal is real and i.i.d., the normalized minus-log-likelihood function of the observed signal x in the noise-free case is [7]

$$\ell(x;w) = -\frac{1}{2\pi}\int_{-\pi}^{\pi} \log\left|W(e^{i\theta})\right| d\theta + \frac{1}{T}\sum_{n=0}^{T-1} \varphi(y_n), \quad (1)$$

where $W(e^{i\theta})$ stands for the discrete Fourier transform of w, $y = x * w$ is a source estimate, $\varphi(s) = -\log p(s)$ and $p(s)$ is the probability density function (PDF) of the source s_n. We will henceforth assume that the restoration kernel w_n has a finite impulse response, supported on $n = -N, ..., N$. We also assume without loss of generality that $\mathbb{E}s_n = 0$.

Consistent estimator can be obtained by minimizing $\ell(x;w)$ even when $\varphi(s)$ is not exactly equal to $-\log p(s)$. Such QML estimation has been shown to be practical in instantaneous blind source separation [5,8] and blind deconvolution [9,10] when the source PDF is unknown or not well-suited for optimization. For example, when the source is super-Gaussian (e.g. it is sparse or sparsely representable), a smooth approximation of the absolute value function is a good choice for $\varphi(s)$ [8]. It is convenient to use a family of convex smooth functions, e.g.

$$\varphi_\lambda(s) = |s| - \lambda \log\left(1 + \frac{|s|}{\lambda}\right) \quad (2)$$

with λ being a positive smoothing parameter, to approximate the absolute value [8]. $\varphi_\lambda(s) \to |s|$ as $\lambda \to 0^+$.

In case of sub-Gaussian sources, the family of functions

$$\varphi_\mu(s) = |s|^\mu \quad (3)$$

with the parameter $\mu > 2$ is usually a good choice for $\varphi(s)$ [9, 10].

2.1 Equivariance

A remarkable property of the QML estimator $\hat{w}(x)$ of a restoration kernel w given the observation x, obtained by minimization of $\ell(x;w)$ in (1), is its *equivariance*, stated in the following proposition:

Proposition 1. *The estimator $\hat{w}(x)$ obtained by minimization of $\ell(x;w)$ is equivariant, i.e., for every invertible h, $\hat{w}(h * x) = h^{-1} * \hat{w}(x)$, where h^{-1} stands for the impulse response of the inverse of h.*

Proof. Observe that for an invertible h,

$$\ell(h * x; h^{-1} * w) = -\frac{1}{2\pi}\int_{-\pi}^{\pi} \log\left|\frac{W(e^{i\theta})}{H(e^{i\theta})}\right| d\theta + \frac{1}{T}\sum_{n=0}^{T-1} \varphi((x*w)_n)$$

$$= \ell(x;w) + \frac{1}{2\pi}\int_{-\pi}^{\pi} \log\left|H(e^{i\theta})\right| d\theta.$$

Let $w = \operatorname{argmin} \ell(x; w)$. Then $\ell(h * x; h^{-1} * w) = \ell(x; w) + \text{const}$, hence w is a minimizer of $\ell(h * x; h^{-1} * w)$ as well. Consequently, $\hat{w}(h * x) = h^{-1} * \hat{w}(x)$. □

Equivariance implies that the parameters to be estimated (in our case, the coefficients, w_n, specifying the restoration kernel) form a group. This is indeed the case for invertible kernels with the convolution operation. In view of equivariance, we may analyze the properties of $\ell(w * x; \delta_n)$ instead of $\ell(x; w)$.

2.2 The Gradient and the Hessian of $\ell(x; w)$

The gradient and the Hessian of $\ell(x; w)$ in (1) are given by

$$\frac{\partial \ell(x; w)}{\partial w_k} = \frac{\partial}{\partial w_k} \left(-\frac{1}{2\pi} \int_{-\pi}^{\pi} \log |W(e^{i\theta})| \, d\theta + \frac{1}{T} \sum_{n=0}^{T-1} \varphi(y_n) \right) =$$

$$= -\frac{1}{4\pi} \int_{-\pi}^{\pi} \left(\frac{e^{-i\theta k}}{W(e^{i\theta})} + \left(\frac{e^{-i\theta k}}{W(e^{i\theta})} \right)^* \right) d\theta + \frac{1}{T} \sum_{n=0}^{T-1} \varphi'(y_n) \frac{\partial y_n}{\partial w_k}$$

$$= -w_{-k}^{-1} + \frac{1}{T} \sum_{n=0}^{T-1} \varphi'((x * w)_n) \, x_{n-k}, \qquad (4)$$

and

$$\frac{\partial^2 \ell(x; w)}{\partial w_k \partial w_l} = \frac{\partial}{\partial w_l} \left(-\frac{1}{2\pi} \int_{-\pi}^{\pi} \frac{e^{-i\theta k}}{W(e^{i\theta})} \, d\theta + \frac{1}{T} \sum_{n=0}^{T-1} \varphi'((x * w)_n) \, x_{n-k} \right)$$

$$= w_{-(k+l)}^{-2} + \frac{1}{T} \sum_{n=0}^{T-1} \varphi''((x * w)_n) \, x_{n-k} x_{n-l}, \qquad (5)$$

where w^{-1} denotes the impulse response of the inverse of w, and $w^{-2} = w^{-1} * w^{-1}$. At the solution point, where $w = ca^{-1}$, it holds that $x * w = cs$. Consequently the Hessian of $\ell(cs; \delta_n)$ is

$$\left(\nabla^2 \ell \right)_{kl} = \delta_{k+l} + \frac{c^2}{T} \sum_{n=0}^{T-1} \varphi''(cs_n) \, s_{n-k} s_{n-l}.$$

For a large sample size T, the average $\frac{1}{T} \sum_{n=0}^{T-1} \varphi''(cs_n) s_{n-k} s_{n-l}$ approaches the expected value $\mathbb{E}\varphi''(cs_n) s_{n-k} s_{n-l}$. Since s_n is assumed to be zero-mean i.i.d., the following structure of the Hessian at the solution point is obtained asymptotically:

$$\nabla^2 \ell(cs; \delta_n) \approx \begin{pmatrix} \ddots & & & & \reflectbox{$\ddots$} \\ & \gamma \sigma'^2 & & 1 & \\ & & \alpha c^2 + 1 & & \\ & 1 & & \gamma \sigma'^2 & \\ \reflectbox{$\ddots$} & & & & \ddots \end{pmatrix}, \qquad (6)$$

where $\sigma^2 = \mathbb{E}s^2$, $\sigma'^2 = (c\sigma)^2$, $\alpha = \mathbb{E}\varphi''(cs)s^2$, and $\gamma = \mathbb{E}\varphi''(cs)$.

3 Asymptotic Error Covariance Matrix

Let the restoration kernel, w, be estimated by minimizing the minus log likelihood function $\ell(x; w)$ defined in (1), where the true $-\log p(s)$ of the source is replaced by some other function $\varphi(s)$. We assume that w has sufficient degrees of freedom to accurately approximate the inverse of a. For analytic tractability, we assume that $\mathbb{E}\varphi''(cs)$, $\mathbb{E}s^2$, $\mathbb{E}\varphi''(cs)s^2$, $\mathbb{E}\varphi'^2(cs)$, $\mathbb{E}\varphi'(cs)s$ and $\mathbb{E}\varphi'^2(cs)s^2$ exist and are bounded. Note that the expected values are computed with respect to the true PDF of s.

Let $w^* = ca^{-1}$ be the exact restoration kernel (up to a scaling factor). It can be shown that w^* satisfies [11] $w^* = \operatorname{argmin}_w \mathbb{E}_x \ell(x; w)$. Let $\hat{w}$ be the estimate of the exact restoration kernel w^*, based on the finite realization of the data x, $\hat{w} = \operatorname{argmin}_w \ell(x; w)$. Note that $\nabla \ell(x; \hat{w}) = 0$, whereas $\nabla \ell(x; w^*) \neq 0$; yet $\mathbb{E}\nabla\ell(x; w^*) = 0$. Denote the estimation error as $\Delta w = w^* - \hat{w}$. Then, assuming $\|\Delta w\|$ is small, second-order Taylor expansion yields

$$\nabla \ell(x; w^*) \approx \nabla^2 \ell(x; w^*) \cdot (w^* - \hat{w}) = \nabla^2 \ell(x; w)\big|_{w=a^{-1}} \cdot \Delta w.$$

Due to the equivariance property, the former relation can be rewritten as

$$\nabla \ell(w^* * x; \delta_n) \approx \nabla^2 \ell(w^* * x; \delta_n) \cdot \Delta w.$$

Since $w^* = ca^{-1}$, we can substitute $w^* * x = cs$, and obtain $\nabla \ell(cs; \delta_n) \approx \nabla^2 \ell(cs; \delta_n) \cdot \Delta w$, or, alternatively, $\Delta w \approx \nabla^2 \ell(cs; \delta_n)^{-1} \cdot \nabla \ell(cs; \delta_n)$. For convenience, we will denote $\nabla \ell(cs; \delta_n)$ and $\nabla^2 \ell(cs; \delta_n)$ as g and $\nabla^2 \ell$, respectively. The covariance matrix of Δw is therefore given by

$$\Sigma_{\Delta w} = \mathbb{E}\Delta w \Delta w^T \approx \left(\nabla^2 \ell\right)^{-1} \cdot \mathbb{E}\nabla\ell\nabla\ell^T \cdot \left(\nabla^2 \ell\right)^{-T} = \left(\nabla^2 \ell\right)^{-1} \cdot \Sigma_{\nabla \ell} \cdot \left(\nabla^2 \ell\right)^{-1}.$$

For a large sample size, the asymptotic Hessian structure (6) can be used, allowing to split the asymptotic covariance matrix, $\Sigma_{\Delta w}$, into a set of 2×2 symmetric matrices of the form

$$\Sigma_{\Delta w}^{(k)} = \begin{pmatrix} \mathbb{E}(\Delta w_{-k})^2 & \mathbb{E}\Delta w_k \Delta w_{-k} \\ \mathbb{E}\Delta w_k \Delta w_{-k} & \mathbb{E}(\Delta w_k)^2 \end{pmatrix} \approx \begin{pmatrix} \gamma \sigma'^2 & 1 \\ 1 & \gamma \sigma'^2 \end{pmatrix}^{-1} \Sigma_{\nabla \ell}^{(k)} \begin{pmatrix} \gamma \sigma'^2 & 1 \\ 1 & \gamma \sigma'^2 \end{pmatrix}^{-1} \quad (7)$$

for $k \neq 0$, where $\Sigma_{\nabla \ell}^{(k)}$ is the covariance matric of g_{-k}, g_k, and an additional 1×1 element

$$\Sigma_{\Delta w}^{(0)} = \frac{\mathbb{E}g_0^2}{(\alpha c^2 + 1)^2}. \quad (8)$$

That is, the asymptotic error covariance matrix has a digaonal-anti-diagonal form. This implies that $\operatorname{cov}\Delta w_k \Delta w_{k'}$, for $k \neq k'$, $k \neq -k'$, decreases in the order of $1/T^2$ as $T \to \infty$. Taking the expectation of the gradient g_k, one obtains $\mathbb{E}g_k = -\delta_k + \mathbb{E}\varphi'(cs_n) cs_{n-k}$. Demanding $\mathbb{E}g_k = 0$, we obtain the following condition:

$$\mathbb{E}\varphi'(cs)cs = 1, \quad (9)$$

from where the scaling factor c can be found. Let us now evaluate the 2×2 gradient covariance matrix, $\Sigma_{\nabla \ell}^{(k)}$, for $k \neq 0$. Substituting $w = \delta_n$, $x = cs$ into (4) yields

$$g_k = \frac{\partial \ell(cs; \delta_n)}{\partial w_k} = -\delta_k + \frac{c}{T} \sum_n \varphi'(cs_n) s_{n-k}, \qquad (10)$$

which for $k \neq 0$ reduces to $g_k = \frac{1}{T} \sum_n \varphi'(cs_n) cs_{n-k}$. Taking the expectation w.r.t. s, and neglecting second-order terms, we obtain

$$\mathbb{E} g_k^2 = \frac{c^2}{T^2} \sum_{n,n'} \mathbb{E}\{\varphi'(cs_n)\varphi'(cs_{n'})\, s_{n-k} s_{n'-k}\} \approx \frac{c^2}{T} \mathbb{E}\varphi'^2(cs) \mathbb{E}s^2 = \frac{1}{T}\beta\sigma'^2$$

$$\mathbb{E} g_{-k}\, g_k = \frac{c^2}{T^2} \sum_{n,n'} \mathbb{E}\{\varphi'(cs_n)\varphi'(cs_{n'})\, s_{n+k} s_{n'-k}\} \approx \frac{1}{T} \mathbb{E}^2 \varphi'(cs) cs = \frac{1}{T},$$

that is,

$$\Sigma_{\nabla \ell}^{(k)} \approx \frac{1}{T} \cdot \begin{pmatrix} \beta\sigma'^2 & 1 \\ 1 & \beta\sigma'^2 \end{pmatrix},$$

where $\beta = \mathbb{E}\varphi'^2(cs)$. Substituting the former result to (7) yields after some algebraic manipulations

$$\operatorname{var} \Delta w_k \approx \frac{\beta\sigma'^2 \left(\gamma^2 \sigma'^4 + 1\right) - 2\gamma\sigma'^2}{T\left(\gamma^2 \sigma'^4 - 1\right)^2} \qquad (11)$$

$$\operatorname{cov} \Delta w_{-k} \Delta w_k \approx \frac{\gamma\sigma'^2 \left(\gamma\sigma'^2 - 2\beta\sigma'^2\right) + 1}{T\left(\gamma^2 \sigma'^4 - 1\right)^2} \qquad (12)$$

for $k \neq 0$. Note that the asymptotic variance depends on the sample size T and on parameters β, γ, c and σ'^2, which depend on the source distribution and on $\varphi(s)$ only.

Let us now address the case of $k = 0$. Neglecting second-order terms, the second moment of g_0 is given by

$$\mathbb{E} g_0^2 \approx -1 - 2\mathbb{E}\varphi'(cs)cs + \mathbb{E}^2 \varphi'(cs)cs + \frac{1}{T}\left(\mathbb{E}\varphi'^2(cs)(cs)^2 - \mathbb{E}^2 \varphi'(cs)cs\right) = \frac{c^2 \vartheta - 1}{T},$$

where $\vartheta = \mathbb{E}\varphi'^2(cs)s^2$. Hence, $\Sigma_{\nabla \ell}^{(0)} \approx (c^2 \vartheta - 1)/T$. Substituting $\Sigma_{\nabla \ell}^{(0)}$ into (8) yields

$$\operatorname{var} \Delta w_0 \approx \frac{c^2 \vartheta - 1}{T(\alpha c^2 + 1)^2}. \qquad (13)$$

Using $\operatorname{var}\Delta w_k$, an asymptotical estimate of restoration quality in terms of signal-to-interference ratio (SIR) can be expressed as

$$\operatorname{SIR} = \frac{\mathbb{E}\|cs\|_2^2}{\mathbb{E}\|w*x - cs\|_2^2} = \frac{|w_0^*|^2}{\mathbb{E}\|\Delta w\|_2^2} \approx \frac{T\left(\gamma^2 \sigma'^4 - 1\right)^2}{2N\left(\beta\sigma'^2\left(\gamma^2 \sigma'^4 + 1\right) - 2\gamma\sigma'^2\right)}. \qquad (14)$$

3.1 Cramér-Rao Lower Bounds

We now show that the asymptotic variance of the estimation error in (11), (13) matches the CRLB on the asymptotic variance of $\hat{w}_k$, when the true MLE procedure is used, i.e., when $\varphi(s) = -\log p(s)$. In this case, $c = 1$, $\sigma'^2 = \sigma^2$, and under the assumption that $\lim_{s \to \pm\infty} p(s) = 0$, it can be shown [12] that $\gamma = \beta$. Substituting c, σ'^2, γ into (11), we obtain for $k \neq 0$

$$\operatorname{var} \Delta w_k \approx \frac{\beta \sigma^2}{T(\beta^2 \sigma^4 - 1)} = \frac{1}{T} \cdot \frac{\mathcal{L}}{\mathcal{L}^2 - 1},$$

where $\mathcal{L} = \sigma^2 \cdot \mathbb{E}\varphi'^2(s)$ is known as Fisher's information for location parameter [4]. This result coincides with the CRLB on w_k developed in [4]. Similarly, under the assumption that $\lim_{s \to \pm\infty} p(s)s = 0$, it can be shown that $\theta = \alpha + 2$ [12]. Substituting $c = 1$ and the latter result into (13) yields

$$\operatorname{var} \Delta w_0 \approx \frac{1}{T} \cdot \frac{\vartheta - 1}{(\alpha + 1)^2} = \frac{1}{T} \cdot \frac{1}{\alpha + 1} = \frac{1}{TS},$$

where $\mathcal{S} = \operatorname{cum}\{\varphi'(s), \varphi'(s), s, s\} + \mathcal{L} + 1$ is the Fisher information for the scale parameter [4]. This result coincides with the CRLB on w_0 in [4]. Substituting the obtained β and γ into (14), yields

$$\operatorname{SIR} \approx \frac{T(\mathcal{L}^2 - 1)}{2N \cdot \mathcal{L}} \leq \frac{T\mathcal{L}}{2N}.$$

This result coincides with the asymptotic performance bound derived in [4].

3.2 Super-efficiency

Let us now consider the particular case of *sparse* sources, such sources that take the value of zero with some non-zero probability $\rho > 0$. An example of such distribution is the Gauss-Bernoully (sparse normal) distribution [12]. When $\varphi(s)$ is chosen according to (2), $\varphi'_\lambda(s) \to \operatorname{sign}(s)$ and $\varphi''_\lambda(s) \to 2\delta(s)$ as $\lambda \to 0^+$. Hence, for a sufficiently small λ,

$$\gamma = \mathbb{E}\varphi''(cs) \approx \frac{1}{\lambda} \int_{-\lambda/c}^{+\lambda/c} p(s)\, ds \approx \frac{\rho}{\lambda},$$

whereas β and c are bounded. Consequently, for $k \neq 0$

$$\operatorname*{plim}_{T \to \infty} T \cdot \operatorname{var} \Delta w_k \leq \frac{\beta}{\gamma^2 \sigma'^2} \leq \operatorname{const} \cdot \lambda^2, \tag{15}$$

where plim denotes the probability limit. Observe that this probability limit vanishes for $\lambda \to 0^+$, which means that the estimator $\hat{w}_k$ of w_k is *super-efficient*. Similarly, the sub-Gaussian QML estimator with $\varphi_\mu(s)$ defined in (3) is super-efficient for sources with compactly supported PDF.

4 Asymptotic Stability

A QML estimator $\hat{w}(x)$ of w^*, obtained by minimization of $\ell(x; w)$, is said to be *asymptotically stable* if $w = w^*$ is a local minimizer of $\ell(x; w)$ for infinitely large sample size. Asymptotic error analysis, presented in Section 3, is valid only when the QML estimator is asymptotically stable.

Proposition 2. *Let $\hat{w}(x)$ be the QML estimator of w. $\hat{w}(x)$ is asymptotically stable if the following conditions hold:*

$$\gamma > 0, \tag{16}$$
$$\gamma^2 \sigma'^4 > 1, \tag{17}$$
$$\alpha c^2 > -1. \tag{18}$$

Proof. The QML estimator is asymptotically stable if in the limit $T \to \infty$, $w = w^*$ is a local minimizer of $\ell(x; w)$, or due to equivariance, $w = \delta_n$ is a local minimizer of $\ell(cs; w)$. The first- and the second-order Karush-Kuhn-Tucker conditions

$$\plim_{T \to \infty} \nabla \ell(cs; \delta_n) = 0 \tag{19}$$
$$\plim_{T \to \infty} \nabla^2 \ell(cs; \delta_n) \succ 0 \tag{20}$$

are the necessary and the sufficient conditions, respectively, for existence of the local minimum. The necessary condition (19) requires that $\nabla \ell = 0$ as the sample size approaches infinity. For $k \neq 0$ we obtain from (10) that $\plim_{T \to \infty} g_k = \mathbb{E} \varphi'(cs) \cdot \mathbb{E} \, cs = 0$, and for $k = 0$, by choice of c, $\plim_{T \to \infty} g_0 = \mathbb{E} \varphi'(cs) cs - 1 = 0$. The sufficient condition (20) requires that $\nabla^2 \ell \succ 0$ as the sample size approaches infinity. Using the asymptotic Hessian given in (6), this condition can be rewritten as

$$\begin{pmatrix} \gamma \sigma'^2 & 1 \\ 1 & \gamma \sigma'^2 \end{pmatrix} \succ 0, \qquad \alpha c^2 + 1 \succ 0.$$

The latter holds if and only if $\gamma > 0$, $\gamma^2 \sigma'^4 > 1$ and $\alpha c^2 > -1$. □

It is observed that when $\varphi(s)$ is chosen to be proportional to $-\log p(s)$, $\hat{w}(x)$ is never asymptotically unstable. When $\varphi(s)$ is chosen according to (3), it can be shown that $c = (\mu \cdot \mathbb{E}|s|^\mu)^{-1/\mu}$, $\alpha = \mu(\mu-1)c^\mu \cdot \mathbb{E}|s|^\mu$, and $\gamma = \mu(\mu-1)c^{\mu-2} \cdot \mathbb{E}|s|^{\mu-2}$. For $\mu > 2$, it can be easily checked that conditions (16), (18) hold, hence, the asymptotic stability condition is $\mathbb{E}|s|^\mu < (\mu-1)\mathbb{E}s^2 \, \mathbb{E}|s|^{\mu-2}$. In the particular case when $\mu = 4$, the latter condition becomes $\kappa < 0$, where κ is the *kurtosis excess*, meaning that the estimator is asymptotically stable for sub-Gaussian sources.

When $\varphi(s)$ is chosen according to (2), there exists no analytic expression for the asymptotic stability conditions, except the case when $\lambda \to 0^+$. In the latter case, $\varphi' = \text{sign}(s)$ and $\varphi''(s) = 2\delta(s)$, from where $c = 1/\mathbb{E}|s|$, $\alpha = 2\mathbb{E}\delta(s)(cs)^2 = 0$, and $\gamma = 2\mathbb{E}\delta(s) = 2p(0)$. Observe that conditions (16), (18) hold again, hence the estimator is asymptotically stable if $\mathbb{E}|s| < 2p(0)\sigma^2$.

5 Conclusion

In order to be in a position to utilize the QML estimator of the restoration kernel in blind deconvolution, and to gain insight into the effect of the source distribution and the choice of $\varphi(s)$, it is important to quantify the asymptotic performance and establish stability conditions. For this purpose we derived simple closed-form expressions for the asymptotic estimation error, and showed that its covariance matrix has a diagonal-anti-diagonal form. An asymptotic estimate of the restoration quality in terms of SIR was also presented. The main conclusion from the performance analysis is that the asymptotic performance depends on the choice of $\varphi(s)$ essentially through the ratio $\mathbb{E}\varphi'^2(s)/\mathbb{E}^2\varphi''(s)$ of non-linear moments of the source. We demonstrated that for the true ML estimator, our asymptotic performance bounds coincide with the CRLB. Asymptotic stability conditions for the QML estimator have been shown as well. Extension to the MIMO case is presented in [13]. Particular cases wherein the families of functions $\varphi_\lambda(s)$ and $\varphi_\mu(s)$ yield super-efficient estimators were highlighted. More delicate analysis is required to determine whether zero variance can be achieved on a *finite* sample, and what is its minimum size. Such a result is important from both theoretical and practical viewpoints.

References

1. Bellini, S., Rocca, F.: Near optimal blind deconvolution. In: Proc. of IEEE Conf. Acoust., Speech, Sig. Proc. (1988)
2. Cardoso, J.F., Laheld, B.: Equivariant adaptive source separation. IEEE Trans. Sig. Proc. **44** (1996) 3017–3030
3. Shalvi, O., Weinstein, E.: Maximum likelihood and lower bounds in system identification with non-Gaussian inputs. IEEE Trans. Information Theory **40** (1994) 328–339
4. Yellin, D., Friedlander, B.: Multichannel system identification and deconvolution: performance bounds. IEEE Trans. Sig. Proc. **47** (1999) 1410–1414
5. Pham, D., Garrat, P.: Blind separation of a mixture of independent sources through a quasi-maximum likelihood approach. IEEE Trans. Sig. Proc. **45** (1997) 1712–1725
6. Cardoso, J.F.: Blind signal separation statistical principles. Proc. IEEE. Special issue on blind source separation **9** (1998) 2009–2025
7. Amari, S.I., Cichocki, A., Yang, H.H.: Novel online adaptive learning algorithms for blind deconvolution using the natural gradient approach. In: Proc. SYSID. (1997) 1057–1062
8. Zibulevsky, M.: Sparse source separation with relative Newton method. In: Proc. ICA2003. (2003) 897–902
9. Amari, S.I., Douglas, S.C., Cichocki, A., Yang, H.H.: Multichannel blind deconvolution and equalization using the natural gradient. In: Proc. SPAWC. (1997) 101–104
10. Bronstein, A.M., Bronstein, M.M., Zibulevsky, M.: Blind deconvolution with relative Newton method. Technical Report 444, Technion, Israel (2003)
11. Kisilev, P., Zibulevsky, M., Zeevi, Y.: Multiscale framework for blind source separation. JMLR (2003) In press.
12. Bronstein, A.M., Bronstein, M.M., Zibulevsky, M., Zeevi, Y.Y.: Quasi maximum likelihood blind deconvolution: Asymptotic performance analysis. IEEE Info. Theory (2004) Submitted. [Online] http://visl.technion.ac.il/bron/alex.
13. Bronstein, A.M., Bronstein, M.M., Zibulevsky, M., Zeevi, Y.Y.: Asymptotic performance analysis of MIMO blind deconvolution. Technical report, Technion, Israel (2004)

Super-exponential Methods Incorporated with Higher-Order Correlations for Deflationary Blind Equalization of MIMO Linear Systems

Kiyotaka Kohno[1], Yujiro Inouye[1], and Mitsuru Kawamoto[2]

[1] Department of Electronic and Control Systems Engineering, Shimane University
1060 Nishikawatsu, Matsue, Shimane 690-8504, Japan
kohno@yonago-k.ac.jp, inouye@riko.shimane-u.ac.jp
[2] Department of Electronic and Control Systems Engineering, Shimane University
1060 Nishikawatsu, Matsue, Shimane 690-8504, Japan
Bio-Mimetic Control Research Center, RIKEN, Moriyama, Nagoya 463-003, Japan
kawa@ecs.shimane-u.ac.jp

Abstract. The multichannel blind deconvolution of finite-impulse response (FIR) or infinite-impulse response (IIR) systems is investigated using the multichannel super-exponential deflation methods. In the conventional multichannel super-exponential deflation method [4], the so-called "second-order correlation method" is incorporated in order to estimate the contributions of an extracted source signal to the channel outputs. We propose a new multichannel super-exponential deflation method using higher-order correlations instead of second-order correlations to reduce the computational complexity in terms of multiplications and to accelerate the performance of equalization. By computer simulations, it is shown that the method of using fourth-order correlations is better than the method of using second-order correlations in a noiseless case or a noisy case.

1 Introduction

Multichannel blind deconvolution has recently received attention in such fields as digital communications, image processing and neural information processing [1],[2].

Recently, Shalvi and Weinstein proposed an attractive approach to single-channel blind deconvolution called the *super-exponential method* (SEM) [3]. Extensions of their idea to multichannel deconvolution were presented by Inouye and Tanebe [4], Martone [5], [6], and Yeung and Yau [7]. In particular, Inouye and Tanebe [4] proposed the multichannel super-exponential deflation method (MSEDM) using second-order correlations. Martone [6], and Kawamoto, Kohno and Inouye [8] proposed MSEDM's using higher-order correlations for instantaneous mixtures or constant channel systems. Adaptive versions of multichannel super-exponential algorithms are presented in [9].

In the present paper, we propose a new MSEDM using higher-order correlations for convolutive mixtures or dynamical channel systems, and show the effectiveness of the proposed method by computer simulations. Adaptive versions of proposed method will appear in a forthcoming paper.

The present paper uses the following notation: Let Z denote the set of all integers. Let $\boldsymbol{C}^{m\times n}$ denote the set of all $m \times n$ matrices with complex components. The superscripts T, $*$, H and $\dagger$ denote, respectively, the transpose, the complex conjugate, the complex conjugate transpose (Hermitian) and the (Moore-Penrose) pseudoinverse operations of a matrix. Let $i = \overline{1,n}$ stand for $i = 1, 2, \cdots, n$.

2 Assumptions and Preliminaries

We consider an MIMO channel system with n inputs and m outputs as described by

$$\boldsymbol{y}(t) = \sum_{k=-\infty}^{\infty} \boldsymbol{H}^{(k)} \boldsymbol{s}(t-k) + \boldsymbol{n}(t), \quad t \in Z, \qquad (1)$$

where
- $\boldsymbol{s}(t)$ n-column vector of input (or source) signals,
- $\boldsymbol{y}(t)$ m-column vector of channel outputs,
- $\boldsymbol{n}(t)$ m-column vector of Gaussian noises,
- $\boldsymbol{H}^{(k)}$ $m \times n$ matrix of impulse responses.

The transfer function of the channel system is defined by

$$\boldsymbol{H}(z) = \sum_{k=-\infty}^{\infty} \boldsymbol{H}^{(k)} z^k, \quad z \in C. \qquad (2)$$

For the time being, it is assumed for theoretical analysis that the noise term $\boldsymbol{n}(t)$ in (1) is absent.

To recover the source signals, we process the output signals by an $n \times m$ equalizer (or deconvolver) $\boldsymbol{W}(z)$ described by

$$\boldsymbol{z}(t) = \sum_{k=-\infty}^{\infty} \boldsymbol{W}^{(k)} \boldsymbol{y}(t-k), \quad t \in Z. \qquad (3)$$

The objective of multichannel blind deconvolution is to construct an equalizer that recovers the original source signals only from the measurements of the corresponding outputs.

We put the following assumptions on the systems and the source signals.

A1) The transfer function $\boldsymbol{H}(z)$ is stable and has full column rank on the unit circle $|z| = 1$ [this implies that the unknown system has less inputs than outputs, i.e., $n \leq m$, and there exists a left stable inverse of the unknown system].

A2) The input sequence $\{\boldsymbol{s}(t)\}$ is a complex, zero-mean, non-Gaussian random vector process with element processes $\{s_i(t)\}$, $i = \overline{1,n}$ being mutually independent. Moreover, each element process $\{s_i(t)\}$ is an i.i.d. process with a nonzero variance σ_i^2 and a nonzero fourth-order cumulant γ_i. The variances σ_i^2's and the fourth-order cumulants γ_i's are unknown.

A3) The equalizer $W(z)$ is an FIR system of sufficient length L so that the truncation effect can be ignored.

Remark 1: As to A1), if the channel system $H(z)$ is FIR, then a condition of the existence of an FIR equalizer is rank $H(z) = n$ for all nonzero $z \in C$ [10]. Moreover, if $H(z)$ is irreducible, then there exists an equalizer $W(z)$ of length $L \leq n(K-1)$, where K is the length of the channel system [10]. Besides, it is shown that there exists generically (or except for pathological cases) an equalizer $W(z)$ of length $L \leq \lceil \frac{n(K-1)}{m-n} \rceil$, where $\lceil x \rceil$ stands for the smallest integer that is greater than equal to x.

Let us consider an FIR equalizer with the transfer function $W(z)$ given by

$$W(z) = \sum_{k=L_1}^{L_2} W^{(k)} z^k, \tag{4}$$

where the length $L := L_2 - L_1 + 1$ is taken to be sufficiently large. Let $\tilde{w}_i$ be the Lm-column vector consisting of the tap coefficients (corresponding to the ith output) of the equalizer defined by

$$\tilde{w}_i := \left[w_{i,1}^T, w_{i,2}^T, \cdots, w_{i,m}^T \right]^T \in C^{mL}, \tag{5}$$

$$w_{i,j} = \left[w_{i,j}^{(L_1)}, w_{i,j}^{(L_1+1)}, \cdots, w_{i,j}^{(L_2)} \right]^T \in C^L, \tag{6}$$

where $w_{i,j}^{(k)}$ is the (i,j)th element of matrix $W^{(k)}$.

Inouye and Tanebe [4] proposed the *multichannel super-exponential algorithm* for finding the tap coefficient vectors $\tilde{w}_i$'s of the equalizer $W(z)$, of which each iteration consists of the following two steps:

$$\tilde{w}_i^{[1]} = \tilde{R}_L^\dagger \tilde{d}_i \qquad \text{for } i = \overline{1, n}, \tag{7}$$

$$\tilde{w}_i^{[2]} = \frac{\tilde{w}_i^{[1]}}{\sqrt{\tilde{w}_i^{[1]H} \tilde{R}_L \tilde{w}_i^{[1]}}} \quad \text{for } i = \overline{1, n}, \tag{8}$$

where $(\cdot)^{[1]}$ and $(\cdot)^{[2]}$ stand respectively for the result of the first step and the result of the second step. Let $\tilde{y}(t)$ be the Lm-column vector consisting of the L consecutive inputs of the equalizer defined by

$$\tilde{y}(t) := \left[\bar{y}_1(t)^T, \bar{y}_2(t)^T, \cdots, \bar{y}_m(t)^T \right]^T \in C^{mL}, \tag{9}$$

$$\bar{y}_i(t) := [y_i(t-L_1), y_i(t-L_1-1), \cdots, y_i(t-L_2)]^T \in C^L, \tag{10}$$

where $y_i(t)$ is the ith element of the output vector $y(t)$ of the channel system in (1). Then the correlation matrix $\tilde{R}_L$ is represented as

$$\tilde{R}_L = E\left[\tilde{y}^*(t) \tilde{y}^T(t) \right] \in C^{mL \times mL}, \tag{11}$$

and the fourth-order cumulant vector $\tilde{d}_i$ is represented as

$$\tilde{d}_i = E\left[|z_i(t)|^2 z_i(t)\tilde{y}^*(t)\right] - 2E\left[|z_i(t)|^2\right] E\left[z_i(t)\tilde{y}^*(t)\right]$$
$$- E\left[z_i^2(t)\right] E\left[z_i^*(t)\tilde{y}^*(t)\right] \in C^{mL}, \qquad (12)$$

where $E[x]$ denotes the expectation of a random variable x. We note that the last term can be ignored in case of $E[s_i^2(t)]=0$ for all $i = \overline{1,n}$, in which case $E[z_i^2(t)]=0$ for all $i = \overline{1,n}$.

3 A Super-exponential Deflation Method Incorporated with Higher-Order Correlations

The MSEDM proposed by Inouye and Tanebe [4] uses second-order correlations to estimate the contributions of an extracted source signal to the channel outputs. We utilize higher-order correlations instead of second-order correlations in order to estimate the contributions of an extracted source signal to the channel outputs. For notational simplicity, we confine ourselves to fourth-order correlations although our results are expandable to higher-order correlations.

For the details of the method of second-order correlations, see the equations from (55) through (58) in [4]. According to the discussions from (55) through (58) in [4], we calculate the higher-order cross-correlations of the equalizer outputs $z_i(t)$'s with the channel outputs $y_k(t)$'s, and define a possibly scaled and time-shifted estimate of the channel element $\hat{h}_{k,j_i}(\tau)$ as

$$\hat{h}_{k,j_i}(\tau) := \mathrm{cum}(z_i^*(t-\tau), z_i^*(t-\tau), z_i(t-\tau), y_k(t)). \qquad (13)$$

Then let us consider the contribution of $z_i(t)$ to the channel output $y_k(t)$ which is defined by

$$\hat{y}_{k,j_i}(\tau) := \frac{\sigma_{j_i}^2}{\gamma_{j_i}} \sum_\tau \hat{h}_{k,j_i}(\tau) z_i(t-\tau), \qquad (14)$$

where $\frac{\sigma_{j_i}^2}{\gamma_{j_i}}$ is introduced to adjust the difference between the scale of the contribution of $z_i(t)$ and the scale of the contribution of source $s_{j_i}(t)$ to the channel output $y_k(t)$. Subtract the above contribution from the channel output $\boldsymbol{y}(t)$ as

$$y_k^{(i)}(t) := y_k(t) - \hat{y}_{k,j_i}(t), \ k = \overline{1,m}. \qquad (15)$$

Let us analyze the above method. After the first cycle of the iteration, the first equalizer output $z_1(t)$ is a possibly scaled and time-shifted version of one of the channel input, that is,

$$z_1(t) = ds_{j_1}(t-k_1), \ j_1 \in \{1,2,\cdots,n\}, \qquad (16)$$

where $|d| = 1/\sigma_{j_1}$ and k_1 represents a delay-time, which may belong to the interval $[K+L_1, L_2]$ (see (28) for the derivation of the above relation), i.e.,

$$k_1 \in [K+L_1, L_2], \qquad (17)$$

in the case when the channel $\boldsymbol{H}$ is an FIR system with $\{\boldsymbol{H}(k)\}_0^{K-1}$, and the equalizer $\boldsymbol{W}$ is an FIR system with $\{\boldsymbol{W}(k)\}_{k=L_1}^{L_2}$.
Since

$$\begin{aligned}&\operatorname{cum}(z_i^*(t-\tau), z_i^*(t-\tau), z_i(t-\tau), y_k(t))\\ &= \operatorname{cum}(z_i^*(t), z_i^*(t), z_i(t), y_k(t+\tau)).\end{aligned} \quad (18)$$

it follows from (13)(with $i = 1$) and (16)

$$\begin{aligned}\hat{h}_{k,j_1}(\tau) &= \operatorname{cum}(z_1^*(t), z_1^*(t), z_1(t), y_k(t+\tau))\\ &= \sum_{j=1}^{n}\sum_{l} h_{k,j}(l)\operatorname{cum}(z_1^*(t), z_1^*(t), z_1(t), s_j(t+i-l))\\ &= d^{*2}d \sum_{j=1}^{n}\sum_{l} h_{k,j}(l)\\ &\quad \times \operatorname{cum}(s_{j_1}^*(t-k_1), s_{j_1}^*(t-k_1), s_{j_1}(t-k_1), s_j(t+\tau-l))\\ &= d^{*2}d h_{k,j_1}(k_1+\tau)\gamma_{j_1},\end{aligned}$$

which means

$$\hat{h}_{k,j_1}(\tau) = |d|^2 d^* \gamma_{j_1} h_{k,j_1}(k_1+\tau). \quad (19)$$

Substituting (16) and (19) into (14)(with $i = 1$) gives

$$\begin{aligned}\hat{y}_{k,j_1}(t) &= \frac{\sigma_{j_1}^2}{\gamma_{j_1}} \sum_{\tau} |d|^4 d^* \gamma_{j_1} h_{k,j_1}(k_1+\tau) s_{j_1}(t-\tau-k_1)\\ &= \sum_{\tau} h_{k,j_1}(k_1+\tau) s_{j_1}(t-\tau-k_1)\\ &= \sum_{\tau} h_{k,j_1}(\tau) s_{j_1}(t-\tau).\end{aligned} \quad (20)$$

Thus, we obtain from (15)

$$\begin{aligned}y_k^{(1)}(t) &= y_k(t) - \hat{y}_{k,j_1}(t)\\ &= \sum_{j=1}^{n}\sum_{\tau} h_{k,j}(\tau) s_j(t-\tau) - \sum_{\tau} h_{k,j_1}(\tau) s_{j_1}(t-\tau)\\ &= \sum_{j=1, j\neq j_1} h_{k,j}(\tau) s_j(t-\tau),\end{aligned} \quad (21)$$

which shows that $y_k^{(1)}(t)$ does not contain the contribution of the source $s_{j_1}(t)$. As one of the advantages of the above method, we can reduce the computational loads for calculating (13) and (14) as follows: Using the definitions of $\hat{h}_{k,j_i}$ and $[d_{i,j}]_\tau$ (see (13) and (44) in [4]), we have

$$\hat{h}_{k,j_i}(\tau) = [d_{i,k}]_{-\tau}^*, \quad (22)$$

Therefore, (14) becomes

$$\hat{y}_{k,j_i}(t) = \frac{\sigma_{j_i}^2}{\gamma_{j_i}} \sum_\tau \hat{h}_{k,j_i} z_i(t-\tau)$$

$$= \frac{\sigma_{j_i}^2}{\gamma_{j_i}} \sum_\tau [d_{i,k}]_\tau^* z_i(t+\tau)$$

$$= \frac{\sigma_{j_i}^2}{\gamma_{j_i}} \sum_{\tau=L_1}^{L_2} [d_{i,k}]_\tau^* z_i(t+\tau). \qquad (23)$$

Thus, (15) becomes

$$y_k^{(i)}(t) = y_k(t) - \frac{\sigma_{j_i}^2}{\gamma_{j_i}} \sum_{\tau=L_1}^{L_2} [d_{i,k}]_\tau^* z_i(t+\tau), \ k = \overline{1,m}, \qquad (24)$$

where the coefficients $[d_{i,k}]_\tau$'s are available at the first step (7) of the two-step iteration.

Some remarks are given below on conditions for the indices L_1 and L_2 of the equalizer. In the following discussion, we assume that delay-time k_1 in (16) is known or estimated ahead. If follows from (19) and (22) that

$$h_{k,j_1}(\tau+k_1) = \frac{1}{\alpha}\hat{h}_{k,j_1}(\tau) = \frac{1}{\alpha}[d_{i,j}]_{-\tau}^*, \qquad (25)$$

where $\alpha = |d^2|d^*\gamma_{j_1}$. When the channel $\boldsymbol{H}$ is an FIR system with $\{\boldsymbol{H}(k)\}_0^{K-1}$ and the equalizer $\boldsymbol{W}$ is an FIR system with $\{\boldsymbol{W}(k)\}_{k=L_1}^{L_2}$, the support of the function $h_{k,j_1}(\tau)$ is the interval $[0,K]$ and the support of the function $[d_{i,j}]_\tau$ is the interval $[L_1, L_2]$. Here, given a function $h(\tau)$ defined on Z, the subset $\{\tau \in Z : h(\tau) \neq 0\}$ is called the *support* of the function h. Therefore, in order for the sequence $[d_{i,j}]_{-\tau}^*$ to determine the values of the sequence $h_{k,j_1}(\tau+k_1)$ based on (25), the support of $h_{k,j_1}(\tau+k_1)$ should be included in the support of $[d_{i,j}]_{-\tau}^*$, that is,

$$[-k_1, -k_1+K] \subset [-L_2, -L_1], \qquad (26)$$

which implies

$$L_1 \leq k_1 - K, \ L_2 \geq k_1, \qquad (27)$$

or

$$k_1 \in [L_1 + K, L_2]. \qquad (28)$$

Thus the first tap index L_1 and the last tap index L_2 of the equalizer are chosen to satisfy the conditions

$$L_1 \leq k_1 - K, \qquad (29)$$

$$L_2 \geq k_1. \qquad (30)$$

4 Simulations

To demonstrate the effectiveness of the proposed method, many computer simulations were conducted. We considered an MIMO channel system with two inputs and three outputs, and assumed that the length of the channel is three ($K = 3$), that is, $\boldsymbol{H}^{(k)}$ in (1) was set to be

$$\boldsymbol{H}(z) = \begin{bmatrix} 1.00 + 0.60z + 0.30z^2 & 0.60 + 0.50z - 0.20z^2 \\ 0.50 - 0.10z + 0.20z^2 & 0.30 + 0.40z + 0.50z^2 \\ 0.70 + 0.10z + 0.40z^2 & 0.10 + 0.20z + 0.10z^2 \end{bmatrix}. \tag{31}$$

The length of the equalizer was chosen to be twelve ($L = 12$). We set the values of the tap coefficients to be zero except for $w_{12}(6) = w_{22}(6) = 1$. Two source signals were the 4-PSK and the 8-PSK signals, respectively. Three independent Gaussian noises (with identical variance σ_w^2) were added to the three outputs $y_i(t)$'s at various SNR levels. The SNR is defined as SNR:=10 $\log_{10}(\sigma_i^2/\sigma_w^2)$, where σ_i^2's are the variances of $s_i(t)$'s and are equal to one. As a measure of performance, we use the multichannel intersymbol interference ($\mathrm{M_{ISI}}$) defined in the logarithmic (dB) scale by

$$\mathrm{M_{ISI}} := 10 \log_{10} \left[\sum_{i=1}^{n} \frac{|\sum_{j=1}^{n} \sum_{t=-\infty}^{\infty} |g_{i,j}(t)|^2 - |g_{i,\cdot}|^2_{\max}|}{|g_{i,\cdot}|^2_{\max}} \right.$$

$$\left. + \sum_{j=1}^{n} \frac{|\sum_{i=1}^{n} \sum_{t=-\infty}^{\infty} |g_{i,j}(t)|^2 - |g_{\cdot,j}|^2_{\max}|}{|g_{\cdot,j}|^2_{\max}} \right], \tag{32}$$

where $|g_{i,\cdot}|^2_{\max}$ and $|g_{\cdot,j}|^2_{\max}$ are respectively defined by

$$|g_{i,\cdot}|^2_{\max} := \max_{j=1,2} |g_{i,j}|^2, \quad |g_{\cdot,j}|^2_{\max} := \max_{i=1,2} |g_{i,j}|^2. \tag{33}$$

Fig. 1 shows the averages of performance results over 10 independent Monte Carlo runs for the proposed method and the conventional method [4] in the noiseless case when the time-duration is from zero to 30 (1000 iterations were carried out in each time-duration). In each Monte Carlo run, $\tilde{\boldsymbol{R}}_L$ was estimated using 1,000 data samples. In each iteration of two steps (7) and (8), $\tilde{\boldsymbol{d}}_i$ was estimated using 1,000 data samples. It can be seen from Fig. 1 that our proposed method is better than the conventional method about 7dB when the second source signal is deconvolved.

Fig. 2 shows the averages of the performance results over 5 independent Monte Carlo runs when the SNR level was taken to be 0[dB], 5[dB], 10[dB], 15[dB], 20[dB] and ∞[dB], respectively. It can be seen from Fig. 2 that our proposed method is superior to the conventional method in the noiseless case, and our proposed method works as well as the conventional method even when the power of additive noise increases.

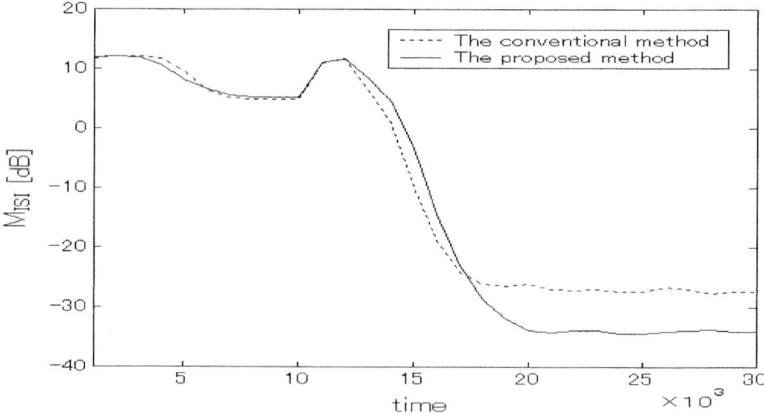

Fig. 1. Comparison between the proposed method and the conventional method (Noiseless case)

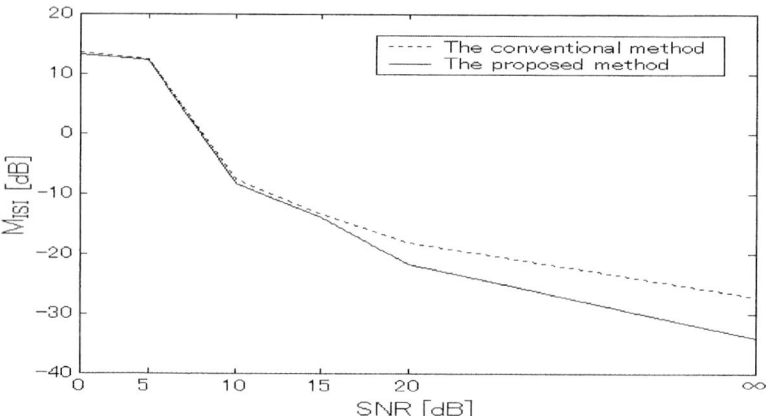

Fig. 2. Comparison between the proposed method and the conventional method (Noisy case)

5 Conclusions

We have proposed a new multichannel super-exponential deflation method using higher-order correlations instead of second-order correlations for estimating the contributions of an extracted source signal to the channel outputs in order to reduce the computational complexity and to accelerate the performance of equalization. By computer simulations, it is shown that the method of using fourth-order correlations is superior to the method of using second-order correlations in a noiseless case or a noisy case.

References

1. *Special issue on blind system identification and estimation*, Proc. *IEEE*, vol. 86, no. 10, pp. 1907-2089, Oct. 1998.
2. *Special section on blind signal processing: Independent component analysis and signal separation*, *IEICE Trans. on Fundamentals of Electronics*, vol. EA86-A, no. 3, pp. 522-642, Mar. 2003.
3. O. Shalvi and E. Weinstein, "Super-exponential methods for blind deconvolution," *IEEE Trans. Information Theory*, vol. 39, no. 2, pp. 504-519, Mar. 1993.
4. Y. Inouye and K. Tanebe, "Super-exponential algorithms for multichannel blind deconvolution," *IEEE Trans. Signal Processing*, vol. 48, no. 3, pp. 881-888, Mar. 2000.
5. M. Martone, "An adaptive algorithm for antenna array low-rank processing in cellular TDMA base stations," *IEEE Trans. Communications*, vol. 46, no. 5, pp. 627-643, May 1998.
6. M, Martone, "Fast adaptive super-exponential multistage beamforming for cellular base-station transceivers with antenna arrays," *IEEE Trans. Vehicular Tech.*, vol. 48, no. 4, Jul. 1999.
7. K. L. Yeung and S. F. Yau, "A cumulant-based super-exponential algorithm for blind deconvolution of multi-input multi-output systems," *Signal Process.*, vol. 67, pp. 141-162, 1998.
8. M. Kawamoto, K. Kohno and Y. Inouye, "Robust Super-Exponential Methods for Deflationary Blind Equalization of Instantaneous Mixtures," submitted to *IEEE Trans. Signal Processing*.
9. K. Kohno, Y. Inouye, M. Kawamoto and Tetsuya Okamoto, "Adaptive Super-Exponential Algorithms for Blind Deconvolution of MIMO Systems," accepted for presentation in ISCAS 2004.
10. Y. Inouye and R-W. Liu, "A system-theoretic foundation for blind equalization of an FIR MIMO channel system," *IEEE Trans. Circuits and Systems – 1, Fundam. Theory Appl.*, vol. 49, no. 4, pp. 425-436, Apr. 2002.

Blind Maximum Likelihood Separation of a Linear-Quadratic Mixture

Shahram Hosseini and Yannick Deville

Université Paul Sabatier, Laboratoire d'Acoustique, Métrologie, Instrumentation
Bat. 3R1B2, 118 route de Narbonne, 31062 Toulouse Cedex, France
{hosseini,ydeville}@cict.fr

Abstract. We proposed recently a new method for separating linear-quadratic mixtures of independent real sources, based on parametric identification of a recurrent separating structure using an *ad hoc* algorithm. In this paper, we develop a maximum likelihood approach providing an asymptotically efficient estimation of the model parameters. A major advantage of this method is that the explicit form of the inverse of the mixing model is not required to be known. Thus, the method can be easily generalized to more complicated polynomial mixtures.

1 Introduction

Little work has been dedicated to the BSS problem in nonlinear mixtures [1]-[8]. It is well known [1], [2] that the independence hypothesis is not sufficient for separating general nonlinear mixtures because of the very large indeterminacies which make the nonlinear BSS problem ill-posed. A natural idea for reducing the indeterminacies is to constrain the structure of mixing and separating models to belong to a certain set of transformations. This supplementary constraint can be viewed as a regularization of the initially ill-posed problem [5], [8].

In a recent paper [7], we studied a linear-quadratic mixture model which may be considered as the simplest (nonlinear) version of a general polynomial model. Our main aim is to develop an approach which can be easily extended to higher-order polynomial models. Hence, in [7] we proposed a recurrent separating structure whose realization does not require the knowledge of the explicit form of the inverse of the mixing model. A drawback of the proposed approach was that somewhat heuristic criteria had been chosen to identify the model parameters. In the present paper, we develop a rigorous method to identify the parameters of the separating structure in a maximum likelihood framework. Once more, the algorithm is developed so that the inverse of the mixing structure is not required to be known. Thus, it can be extended to more general polynomial mixtures.

2 Mixing and Separating Models

Suppose u_1 and u_2 are two independent random signals. Given the following nonlinear instantaneous mixture model

$$x_i = a_{i1}u_1 + a_{i2}u_2 + b_i u_1 u_2 \quad i=1,2 \quad (1)$$

we would like to estimate u_1 and u_2 up to a permutation and a scaling factor (and possibly an additive constant). For simplicity, let's denote $s_1 = a_{11}u_1$ and $s_2 = a_{22}u_2$. s_1 and s_2 will be referred to as the *sources* in the following. (1) can be rewritten as

$$x_1 = s_1 - l_1 s_2 - q_1 s_1 s_2$$
$$x_2 = s_2 - l_2 s_1 - q_2 s_1 s_2 \qquad (2)$$

in which $l_1 = -a_{12}/a_{22}$ and $l_2 = -a_{21}/a_{11}$ represent the linear contributions of the sources in the mixture, and $q_1 = -b_1/(a_{11}a_{22})$ and $q_2 = -b_2/(a_{11}a_{22})$ represent the quadratic contributions. The negative signs are chosen for simplifying the notations of the separating structure.

A more general form of the model (2), containing the additional terms s_1^2 and s_2^2, has been studied by a few authors [9], [10], for the special case of *circular* complex sources, when at least 5 mixtures are available. In the current work, however, we suppose that: 1) the sources are arbitrary real signals, and 2) only two mixtures are available.

The invertibility of the model (2) was briefly discussed in [7]. Here, we present a more rigorous analysis of this subject. Solving the model (2) for s_1 and s_2 leads to the following two pairs of solutions [7], which may be considered as two direct separating structures:

$$(f_1, f_2)_1 = ((-b_1 + \sqrt{\Delta_1})/2a_1, (-b_2 + \sqrt{\Delta_2})/2a_2)$$
$$(f_1, f_2)_2 = ((-b_1 - \sqrt{\Delta_1})/2a_1, (-b_2 - \sqrt{\Delta_2})/2a_2) \qquad (3)$$

where $\Delta_i = b_i^2 - 4a_i c_i$, $a_1 = q_2 + l_2 q_1$, $a_2 = q_1 + l_1 q_2$, $b_1 = q_1 x_2 - q_2 x_1 + l_1 l_2 - 1$, $b_2 = q_2 x_1 - q_1 x_2 + l_1 l_2 - 1$, $c_1 = x_1 + l_1 x_2$ and $c_2 = x_2 + l_2 x_1$. It can be easily verified that $\Delta_1 = \Delta_2 = J^2$, where J is the Jacobian of the mixing model (2) and reads

$$J = 1 - l_1 l_2 - (q_2 + l_2 q_1) s_1 - (q_1 + l_1 q_2) s_2 \qquad (4)$$

According to the variation domain of the two sources, three different cases may be considered:

1) $J < 0$ for all the values of s_1 and s_2. In this case (3) becomes:

$$(f_1, f_2)_1 = (s_1, s_2) \qquad (5)$$

$$(f_1, f_2)_2 = (-\frac{q_1 + l_1 q_2}{q_2 + l_2 q_1} s_2 - \frac{l_1 l_2 - 1}{q_2 + l_2 q_1}, -\frac{q_2 + l_2 q_1}{q_1 + l_1 q_2} s_1 - \frac{l_1 l_2 - 1}{q_1 + l_1 q_2}) \qquad (6)$$

Thus, the first direct separating structure in (3) leads to the actual sources and the second direct separating structure leads to another solution, equivalent to the first one up to a permutation, a scaling factor, and an additive constant.

2) $J > 0$ for all the values of s_1 and s_2. In this case, the first structure leads to the permuting solution, defined by (6), and the second structure to the actual sources (s_1, s_2). An example is shown in Fig. 1 for the numerical values $l_1 = -0.2$, $l_2 = 0.2$, $q_1 = -0.8$, $q_2 = 0.8$ and $s_i \in [-0.5, 0.5]$.

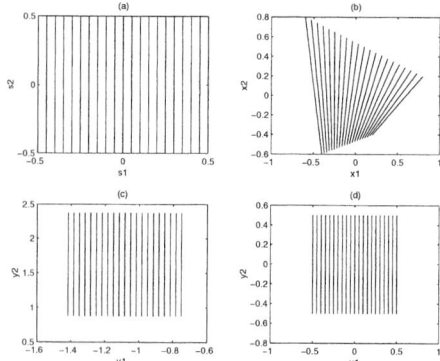

Fig. 1. Case when $J > 0$ for all the source values. Distribution of (a) sources, (b) mixtures, (c) output of the first direct separating structure, (d) output of the second direct separating structure.

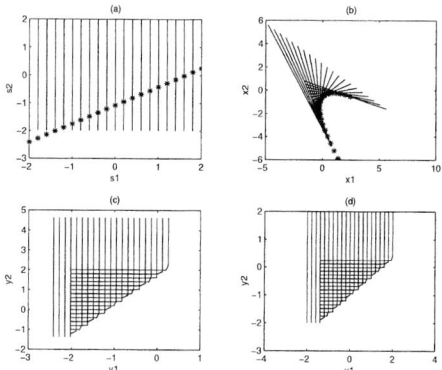

Fig. 2. Case when $J > 0$ for some values of the sources and $J < 0$ for the other values. Distribution of (a) sources, (b) mixtures, (c) output of the first direct separating structure, (d) output of the second direct separating structure.

3) $J > 0$ for some values of the sources and $J < 0$ for the other values. In this case, each structure leads to the non-permuted sources (5) for some values of the observations and to the permuted sources (6) for the other values. An example is shown in Fig. 2 (with the same coefficients as in the second case, but for $s_i \in [-2, 2]$). The permutation effect is clearly visible in the figure. One may also remark that the straight line $J = 0$ in the source plane is mapped to a conic section in the observation plane (shown by asterisks).

Thus, it is clear that the direct structures may be used for separating the sources if the Jacobian of the mixing model is always negative or always positive, *i.e.* for all the source values. Otherwise, although the sources are separated *sample by sample*, each retrieved signal contains samples of the two sources. This problem arises because the mixing model (2) is not bijective. This theoretically insoluble problem should not discourage us. In fact, our final objective is to extend the idea developed in the current study to more general polynomial models which will be used to approximate the nonlinear mixtures encountered in the real world. If these real-world nonlinear models are bijective, we can logically suppose that the coefficients of their polynomial approximations take values which make them bijective on the variation domains of the sources. Thus, in the following, we suppose that the sources and the mixture coefficients have numerical values ensuring that the Jacobian J of the mixing model has a constant sign.

The natural idea to separate the sources is to form a direct separating structure using any of the equations in (3), and to identify the parameters l_1, l_2, q_1 and q_2 by optimizing an independence measuring criterion. Although this approach may be used for our special mixing model (2), as soon as a more complicated polynomial model is considered, the solutions (f_1, f_2) can no longer be determined so that the generalization of the method to arbitrary polynomial models seems impossible. To avoid this limitation, we propose a recurrent structure. Such structures have been considered since the early work of Hérault and Jutten [11] and then in more complex configurations [12], [13]. We here extend them to linear-quadratic mixtures by introducing the structure shown in Fig. 3. Note that, for $q_1 = q_2 = 0$, this structure is reduced to the basic Hérault-Jutten network. It may be checked easily that, for fixed observations defined by (2), $y_1 = s_1$ and $y_2 = s_2$ corresponds to a steady state for the structure in Figure 3.

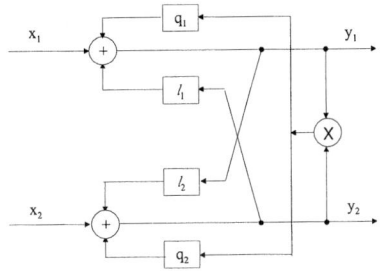

Fig. 3. Recurrent separating structure.

The use of this recurrent structure is more promising because it can be easily generalized to arbitrary polynomial models. However, the main problem with this structure is its stability. In fact, even if the mixing model coefficients are exactly known, the computation of the structure outputs requires the realization of the following recurrent iterative model

$$y_1(n+1) = x_1 + l_1 y_2(n) + q_1 y_1(n) y_2(n)$$
$$y_2(n+1) = x_2 + l_2 y_1(n) + q_2 y_1(n) y_2(n) \quad (7)$$

where a loop on n is performed for each couple of observations (x_1, x_2) until convergence is achieved.

In [7], we have studied the local stability of the model (7) and shown that this model is locally stable at the separating point $(y_1, y_2) = (s_1, s_2)$, if and only if the absolute values of the two eigenvalues of the Jacobian matrix of (7) are smaller than one. In the following, we suppose that this condition is satisfied.

3 Maximum Likelihood Estimation of the Model Parameters

Let $f_{S_1,S_2}(s_1, s_2)$ be the joint pdf of the sources, and assume that the mixing model is bijective so that the Jacobian of the mixing model has a constant sign on the variation domain of the sources. The joint pdf of the observations can be written as

$$f_{X_1,X_2}(x_1, x_2) = \frac{f_{S_1,S_2}(s_1, s_2)}{|J(s_1, s_2)|} \quad (8)$$

Taking the logarithm of (8), and considering the independence of the sources, we can write:

$$\log f_{X_1,X_2}(x_1, x_2) = \log f_{S_1}(s_1) + \log f_{S_2}(s_2) - \log |J(s_1, s_2)| \quad (9)$$

Given N samples of the mixtures X_1 and X_2, we want to find the maximum likelihood estimator for the mixture parameters $\mathbf{w} = [l_1, l_2, q_1, q_2]$. This estimator is obtained by maximizing the joint pdf of all the observations (supposing that the parameters in $\mathbf{w}$ are constant), which is equal to

$$E = f_{X_1,X_2}(x_1(1), x_2(1), \cdots, x_1(N), x_2(N)) \quad (10)$$

If $s_1(t)$ and $s_2(t)$ are two i.i.d. sequences, $x_1(t)$ and $x_2(t)$ are also i.i.d. so that $E = \prod_{i=1}^{N} f_{X_1,X_2}(x_1(i), x_2(i))$ and $\log E = \sum_{i=1}^{N} \log f_{X_1,X_2}(x_1(i), x_2(i))$. The cost function to be maximized can be defined as $L = \frac{1}{N} \log E$, which will be denoted using the temporal averaging operator $E_t[.]$ as

$$L = E_t[\log f_{X_1,X_2}(x_1(t), x_2(t))] \quad (11)$$

Using (9):

$$L = E_t[\log f_{S_1}(s_1(t))] + E_t[\log f_{S_2}(s_2(t))] - E_t[\log |J(s_1(t), s_2(t))|] \quad (12)$$

Maximizing this cost function requires that its gradient with respect to the parameter vector $\mathbf{w}$, i.e. $\frac{\partial L}{\partial \mathbf{w}}$, vanishes. Defining the score functions of the two sources as

$$\psi_i(u) = -\frac{\partial \log f_{S_i}(u)}{\partial u} \quad i = 1, 2 \quad (13)$$

and considering that $\frac{\partial \log |J|}{\partial \mathbf{w}} = \frac{1}{J} \frac{\partial J}{\partial \mathbf{w}}$, we can write

$$\frac{\partial L}{\partial \mathbf{w}} = -E_t[\psi_1(s_1) \frac{\partial s_1}{\partial \mathbf{w}}] - E_t[\psi_2(s_2) \frac{\partial s_2}{\partial \mathbf{w}}] - E_t[\frac{1}{J} \frac{\partial J}{\partial \mathbf{w}}] \quad (14)$$

Rewriting (2) in the vector form $\mathbf{x} = \mathbf{f}(\mathbf{s}, \mathbf{w})$ and considering $\mathbf{w}$ as the independent variable and $\mathbf{s}$ as the dependent variable, we can write, using implicit differentiation

$$\mathbf{0} = \frac{\partial \mathbf{f}}{\partial \mathbf{s}} \frac{\partial \mathbf{s}}{\partial \mathbf{w}} + \frac{\partial \mathbf{f}}{\partial \mathbf{w}} \tag{15}$$

which yields

$$\frac{\partial \mathbf{s}}{\partial \mathbf{w}} = -(\frac{\partial \mathbf{f}}{\partial \mathbf{s}})^{-1} \frac{\partial \mathbf{f}}{\partial \mathbf{w}} \tag{16}$$

Note that $\frac{\partial \mathbf{f}}{\partial \mathbf{s}}$ is the Jacobian matrix of the mixing model. Using (14) and (16), the gradient of the cost function L with respect to the parameter vector $\mathbf{w}$ is equal to (see the appendix for the computation details)

$$\begin{aligned}\frac{\partial L}{\partial \mathbf{w}} = -E_t\bigg[&\Big(\psi_1(s_1)(1-q_2 s_1)s_2 + \psi_2(s_2)(l_2+q_2 s_2)s_2 - (l_2+q_2 s_2)\Big)/J,\\&\Big(\psi_1(s_1)(l_1+q_1 s_1)s_1 + \psi_2(s_2)(1-q_1 s_2)s_1 - (l_1+q_1 s_1)\Big)/J,\\&\Big(\psi_1(s_1)(1-q_2 s_1)s_1 s_2 + \psi_2(s_2)(l_2+q_2 s_2)s_1 s_2 - (l_2 s_1+s_2)\Big)/J,\\&\Big(\psi_1(s_1)(l_1+q_1 s_1)s_1 s_2 + \psi_2(s_2)(1-q_1 s_2)s_1 s_2 - (s_1+l_1 s_2)\Big)/J\bigg] \end{aligned}\tag{17}$$

In practice, the actual sources and their density functions are unknown and will be replaced by the reconstructed sources, *i.e.* by the outputs of the separating structure of Fig 3, y_i, in an iterative algorithm. The score functions of the reconstructed sources can be estimated by any of the existing parametric or non-parametric methods. In our work, we used the kernel estimator proposed in [14] based on third-order cardinal splines. Using (17), the cost function (12) can be maximized by a gradient ascent algorithm which updates the parameters by the rule $\mathbf{w}(n+1) = \mathbf{w}(n) + \mu \frac{\partial L}{\partial \mathbf{w}}$. The learning rate parameter μ must be chosen carefully to avoid the divergence of the algorithm. Note that the algorithm does not require the knowledge of the explicit inverse of the mixing model (direct separating structures (3)). Hence, it can be easily extended to more general polynomial mixing models.

4 Simulation Results

The algorithm was tested using different combinations of subgaussian and supergaussian sources, where the subgaussien sources were uniformly distributed on $[-0.5, 0.5]$ and the supergaussian sources were laplacian with pdf $f_S(s) = 5exp(-10|s|)$. The distribution of the mixtures for two uniform sources is like that presented in Fig. 1.b. The distribution of the estimated sources y_1 and y_2 applying our algorithm is shown in Fig. 4. The rectangular shape of this distribution indicates that the independent components are retrieved. Table 1 represents the output Signal to Noise Ratio, defined as $SNR = 0.5 \sum_{i=1}^{2} 10 \log_{10} \frac{E[s_i^2]}{E[(y_i - s_i)^2]}$ achieved by our algorithm for 3 different combinations of the sources. In each case, the experiment was repeated 100 times, corresponding to different seed values of the random variable generator, using 1000 samples of the sources. The results confirm the good performance of the algorithm.

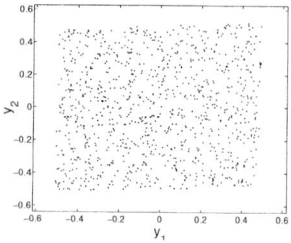

Fig. 4. Distribution of the estimated sources.

Table 1. Mean and Standard Deviation of output SNR (in dB) for different combinations of the sources.

	Mean(SNR)	STD(SNR)
s_1 and s_2 uniform	28.0	4.2
s_1 uniform, s_2 laplacian	27.8	3.8
s_1 and s_2 laplacian	26,8	3,1

5 Conclusion

Nonlinear blind source separation is a difficult, little studied problem. In this work, we investigated one of the simplest structured nonlinear models, *i.e.* the linear-quadratic model. As we aim at generalizing the idea developed in this study to more complicated polynomial models, we proposed a separating structure and an estimation method which do not make use of our knowledge on the explicit form of the inverse of the mixing model. The maximum likelihood approach, developed in this paper, provides an asymptotically efficient estimation of the model parameters and works very well in practice. Some of our objectives for completing this work are: a more precise stability analysis of the recurrent separating network, development of an equivariant estimating method using natural gradient, study of the separability problem, and generalizing the method to more complicated polynomial models and more sources.

Appendix: Details of Gradient Computation

Considering (2), we can write

$$\frac{\partial \mathbf{f}}{\partial \mathbf{s}} = \begin{pmatrix} 1 - q_1 s_2 & -l_1 - q_1 s_1 \\ -l_2 - q_2 s_2 & 1 - q_2 s_1 \end{pmatrix} \text{ and } \frac{\partial \mathbf{f}}{\partial \mathbf{w}} = \begin{pmatrix} -s_2 & 0 & -s_1 s_2 & 0 \\ 0 & -s_1 & 0 & -s_1 s_2 \end{pmatrix},$$

which implies, from (16)

$$\frac{\partial \mathbf{s}}{\partial \mathbf{w}} = \frac{-1}{J} \begin{pmatrix} 1 - q_2 s_1 & l_1 + q_1 s_1 \\ l_2 + q_2 s_2 & 1 - q_1 s_2 \end{pmatrix} \cdot \begin{pmatrix} -s_2 & 0 & -s_1 s_2 & 0 \\ 0 & -s_1 & 0 & -s_1 s_2 \end{pmatrix}$$

which yields

$$\frac{\partial s_1}{\partial \mathbf{w}} = \frac{1}{J}\Big[(1-q_2s_1)s_2\ ,\ (l_1+q_1s_1)s_1\ ,(1-q_2s_1)s_1s_2\ ,\ (l_1+q_1s_1)s_1s_2\Big]$$

$$\frac{\partial s_2}{\partial \mathbf{w}} = \frac{1}{J}\Big[(l_2+q_2s_2)s_2\ ,\ (1-q_1s_2)s_1\ ,(l_2+q_2s_2)s_1s_2\ ,\ (1-q_1s_2)s_1s_2\Big] \quad (18)$$

Considering (4)

$$\frac{\partial J}{\partial \mathbf{w}} = -\Big[l_2+q_2s_2, l_1+q_1s_1, l_2s_1+s_2, s_1+l_1s_2\Big] \quad (19)$$

(17) follows directly from (14), (18) and (19).

References

1. A. Hyvarinen and P. Pajunen, Nonlinear independent component analysis: Existence and uniqueness results, *Neural Networks*, 12(3), pp. 429-439, 1999.
2. A. Taleb and C. Jutten, Source separation in post-nonlinear mixtures, *IEEE Trans. on Signal Processing*, 47(10), pp. 2807-2820, 1999.
3. L. Almeida, Linear and nonlinear ICA based on mutual information, In *Proc. IEEE 2000 Adaptive Systems for Signal Processing, Communications, and Control Symposium (AS-SPCC)*, pp. 117-122, Lake Louise, Canada, October 2000.
4. J. Eriksson and V. Koivunen, Blind identifiability of class of nonlinear instantaneous ICA models. In *Proc. of the XI European Signal Proc. Conf. (EUSIPCO 2002)*, volume 2, pp. 7-10, Toulouse, France, September 2002.
5. A. Taleb, A generic framework for blind source separation in structured nonlinear models, *IEEE Trans. on Signal Processing*, 50(8), pp. 1819-1830, August 2002.
6. S. Hosseini and C. Jutten, On the separability of nonlinear mixtures of temporally correlated sources, *IEEE Signal Processing Letters*, 10(2), pp. 43-46, February 2003.
7. S. Hosseini, Y. Deville, Blind separation of linear-quadratic mixtures of real sources using a recurrent structure, in *Proc. IWANN*, vol.2, pp. 241-248, Mao, Menorca, Spain, June 2003.
8. C. Jutten, B. Babaie-Zadeh, S. Hosseini, Three easy ways for separating nonlinear mixtures?, *Signal Processing*, 84(2), pp. 217-229, February 2004.
9. M. Krob and M. Benidir, Blind identification of a linear-quadratic model using higher-order statistics, In *Proc. ICASSP*, vol. 4, pp. 440-443, 1993.
10. K. Abed-Meraim, A. Belouchrani, and Y. Hua, Blind identification of a linear-quadratic mixture of independent components based on joint diagonalization procedure, In *Proc. ICASSP*, pp. 2718-2721, Atlanta, USA, May 1996.
11. C. Jutten and J. Hérault, Blind separation of sources, part I: An adaptive algorithm based on neuromimetic architecture, *Signal Processing*, 24:1-10, 1991.
12. N. Charkani and Y. Deville, Self-adaptive separation of convolutively mixed signals with a recursive structure. Part I: Stability analysis and optimization of asymptotic behaviour, *Signal Processing*, 73(3)3, pp. 225-254, 1999.
13. N. Charkani and Y. Deville, Self-adaptive separation of convolutively mixed signals with a recursive structure. Part II: Theoretical extensions and application to synthetic and real signals, *Signal Processing*, 75(2), pp. 117-140, 1999.
14. D.-T. Pham, Fast algorithm for estimating mutual information, entropies and score functions, in *Proc. ICA*, pp. 17-22, Nara, Japan, April 2003.

Markovian Source Separation in Post-nonlinear Mixtures

Anthony Larue[1], Christian Jutten[1], and Shahram Hosseini[2,*]

[1] Institut National Polytechnique de Grenoble
Laboratoire des Images et des signaux (CNRS, UMR 5083)
F-38031 Grenoble Cedex, France
[2] Universite Paul Sabatier de Toulouse
Laboratoire d'Acoustique, Metrologie, Instrumentation
F-31062 Toulouse, France

Abstract. In linear mixtures, priors, like temporal coloration of the sources, can be used for designing simpler and better algorithms. Especially, modeling sources by Markov models is very efficient, and Markov source separation can be achieved by minimizing the conditional mutual information [1, 2]. This model allows to separate temporally colored Gaussian sources. In this paper, we extend this result for post-nonlinear mixtures (PNL) [3], and show that algorithms based on a Markov model of colored sources leads to better separation results than without prior, *i.e.* assuming iid sources. The paper contains theoretical developments, and experiments with auto-regressive (AR) source mixtures. PNL algorithms for Markov sources point out a performance improvement of about 7dB with respect to PNL algorithms for iid sources.

1 Introduction

First blind source separation methods, based on statistical independence of random variables and using higher (than 2) order statistics, does not take into account the temporal relation between successive source samples. However, early works [4–7] show that it is possible to exploit source temporal correlation by considering simultaneously a few variance-covariance matrices, with various delays. In recent works [1, 2], for linear mixtures, we proposed Markov models of the sources for taking into account the temporal relation between samples. In this paper, we generalize the method to post-nonlinear mixtures (PNL). The paper is organized as follows: Section 2 provides the main theoretical foundations, Section 3 details two practical issues of the algorithm, Section 4 reports the experiments, before the conclusions in Section 5.

---

* This work has been partly funded by the European project Blind Source Separation and applications (BLISS, IST-1999-14190).

2 Theoretical Assessments

2.1 Mixing and Separating Models

Post-nonlinear (PNL) mixtures of n sources, represented by the figure 1, are characterized by a linear instantaneous mixtures, associated to a mixing matrix $\mathbf{A}$, followed by component-wise nonlinear distortions f_i. Considering a suited separating structure (Fig. 1, right side), it can be shown [3] that, under mild conditions[1], output independence leads to source separation, with the same indeterminacy than linear mixtures. The vectorial notation $\mathbf{s}(t) = [s_1(t), \ldots, s_n(t)]^T$, also applied for e, x, z and y.

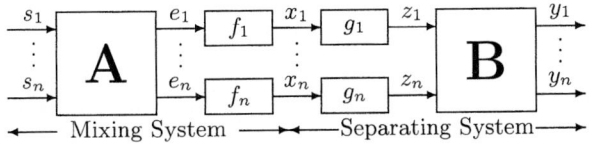

Fig. 1. The mixing-separating system for PNL mixtures.

Each source $s_i(t)$, $i = 1, \ldots, n$ is assumed to be temporally correlated (colored). It is modeled by a q-order Markov model, $i.e.$:

$$p_{s_i}(s_i(t)|s_i(t-1), \cdots, s_i(1)) = p_{s_i}(s_i(t)|s_i(t-1), \cdots, s_i(t-q)) \quad (1)$$

where p_{s_i} denotes the pdf of the random variable s_i.

2.2 Independence Criteria

Since output independence leads to source separation, a possible approach for separating source is to consider a criterion measuring the independence of the output $\mathbf{y}$. Following [8,2], one can use the conditional mutual information of $\mathbf{y}$, denoted by I:

$$I = \int p_{\mathbf{y}}(\mathbf{y}(t)|\mathbf{y}(t-1), \cdots, \mathbf{y}(t-q))$$
$$\times \log \frac{p_{\mathbf{y}}(\mathbf{y}(t)|\mathbf{y}(t-1), \cdots, \mathbf{y}(t-q))}{\prod_{i=1}^{n} p_{y_i}(y_i(t)|y_i(t-1), \cdots, y_i(t-q))} d\mathbf{y} \quad (2)$$

which is always nonnegative, and zero if and only if the variables $w_i(t) = y_i(t)|y_i(t-1), \cdots, y_i(t-q)$ are statistically independent for $i = 1, \cdots, n$, i.e. the signals $y_i(t)$, $i = 1, \cdots, n$ are independent Markovian process. Using the expectation operator $E[.]$, we can write:

$$I = E[\log p_{\mathbf{y}}(\mathbf{y}(t)|\mathbf{y}(t-1), \cdots, \mathbf{y}(t-q))]$$
$$- \sum_{i=1}^{n} E[\log p_{y_i}(y_i(t)|y_i(t-1), \cdots, y_i(t-q))] \quad (3)$$

---

[1] $\mathbf{A}$ is regular or full rank, with at least two non zero entries per row or per column, and f_i are invertible.

Considering the separation structure (Fig. 1), where $\mathbf{y}(t) = \mathbf{B}\mathbf{z}(t)$ and $z_i(t) = g_i(\theta_i, x_i(t))$ [2], Eq. (3) becomes:

$$I = E[\log p_{\mathbf{x}}(\mathbf{x}(t)|\mathbf{x}(t-1),\cdots,\mathbf{x}(t-q))]$$
$$- E\left[\log \left|\prod_{i=1}^{n} \frac{\partial g_i(\theta_i, x_i(t))}{\partial x_i(t)}\right|\right] - \log|\det(\mathbf{B})|$$
$$- \sum_{i=1}^{n} E[\log p_{y_i}(y_i(t)|y_i(t-1),\cdots,y_i(t-q))] \quad (4)$$

The first term being independent of $\mathbf{B}$ and $\Theta = [\theta_1,\ldots,\theta_n]$, the separation structure can be estimated by minimizing:

$$J(\mathbf{B}, \Theta) = -\sum_{i=1}^{n} E\left[\log \left|\frac{\partial g_i(\theta_i, x_i(t))}{\partial x_i(t)}\right|\right] - \log|\det(\mathbf{B})|$$
$$- \sum_{i=1}^{n} E[\log p_{y_i}(y_i(t)|y_i(t-1),\cdots,y_i(t-q))] \quad (5)$$

In practice, under the ergodicity conditions, the mathematical expectation (5) can be estimated by a time averaging, denoted $\widehat{J}(\mathbf{B}, \Theta)$, which requires the estimation of the conditional densities of the estimated sources. Asymptotically, extending the results for linear mixtures of Markovian sources [2], the equivalence of the mutual information minimization method with the Maximum Likelihood method still holds for PNL mixtures of Markovian sources.

2.3 Estimating Equation

Estimation of $\mathbf{B}$ and Θ can be done by minimizing $J(\mathbf{B}, \Theta)$. Using a gradient method, one obtain two sets of estimating equations, which are the gradients of $J(\mathbf{B}, \Theta)$ with respect to $\mathbf{B}$ and with respect to θ_i, $i = 1\ldots n$, i.e.:

$$\frac{\partial J(\mathbf{B}, \Theta)}{\partial \mathbf{B}} = -\mathbf{B}^{-T} + E\left[\sum_{l=0}^{q} \psi_{\mathbf{y}}^{(l)}(\mathbf{y}(t)|\mathbf{y}(t-1),\ldots,\mathbf{y}(t-q))\mathbf{z}^T(t-l)\right] \quad (6)$$

$$\frac{\partial J(\mathbf{B}, \Theta)}{\partial \theta_i} = -E\left[\frac{\partial^2 g_i(\theta_i, x_i(t))}{\partial x_i(t)\partial \theta_i}\left(\frac{\partial g_i(\theta_i, x_i(t))}{\partial x_i(t)}\right)^{-1}\right] + \quad (7)$$

$$E\left[\sum_{j=1}^{n} b_{ji} \sum_{l=0}^{q} \psi_{y_j}^{(l)}(y_j(t)|y_j(t-1),\ldots,y_j(t-q))\frac{\partial g_i(\theta_i, x_i(t-l))}{\partial \theta_i}\right]$$

where we define $q+1$ conditional score functions of a random variable w as $\psi_w^{(l)}(w_0|w_1,\ldots,w_q) = -\frac{\partial}{\partial w_l}\log p_w(w_0|w_1,\ldots,w_q)$, $l = 0,\ldots,q$, and we denote

---
[2] $g_i(\theta_i, x_i(t))$ is a parametric model of $g_i(.)$, where θ_i can represent a set of parameters.

$\psi_{\mathbf{y}}^{(l)}(\mathbf{y}(t)|\mathbf{y}(t-1),\ldots,\mathbf{y}(t-q))$ the n-th dimension vector whose i-th component is $\psi_{y_j}^{(l)}(y_j(t)|y_j(t-1),\ldots,y_j(t-q))$. One can remark that the gradients of the mutual information require first-order and second-order derivatives of the nonlinear mappings g_i's.

3 Algorithm

In this section, we focus on two points for practically implementing the algorithm. The first one concerns the estimation of conditional score functions. The second one is a trick for computing a good initialization point of the algorithm, which leads to enhanced speed of convergence. The algorithm is as follows:

1. initialization of the separating matrix $\mathbf{B}$ and the nonlinear parameters Θ
2. estimation of the conditional score functions
3. computation of the gradients (6) and (7)
4. updating of $\mathbf{B}$ and Θ according to a gradient descent
5. computation of the linearized observations z_i and the estimated sources y_i
6. normalization step

We iterate from 2 to 6 until convergence. The normalization step is required for taking into account scale indeterminacies in $\mathbf{B}$ and in g_i's estimations.

3.1 Estimating the Conditional Score Functions

For estimating the conditional score functions, we can firstly estimate the conditional densities and compute then the conditional score functions by computing the gradient of their logarithms. For a q-order Markovian source, the estimation of the conditional densities may be done using the estimation of the joint pdf of $q+1$ successive samples of each source by a kernel method, which is very time consuming and requires a lot of data. It must be also noticed that the distribution of the data in $(q+1)$-th dimensional space is sparse (curse of dimensionality) and not symmetric because of the temporal correlation between the samples. Thus, one should either use non symmetrical kernels or apply a pre-whitening transformation on data.

Recently, Pham [8] has proposed another algorithm for computing the conditional score functions. The method starts with a pre-whitening stage for obtaining non correlated temporal data. Pham suggests also that the time pre-whitening can allow to reduce the dimension of the used kernels because a great part of the dependence between the variables is cancelled. The influence of the pre-whitening on the estimation of the score functions is computed and will be later compensated using an additive term. Afterwards, the joint entropies of whitened data are estimated using a discrete Riemann sum and the third order cardinal spline kernels. The conditional entropies, defined as

$$H(y_i(t)|y_i(t-1),\cdots,y_i(t-q)) = -E[\log p_{y_i}(y_i(t)|y_i(t-1),\cdots,y_i(t-q))] \quad (8)$$

are computed by estimating the joint entropies:

$$H(y_i(t)|y_i(t-1),\cdots,y_i(t-q)) = H(y_i(t),y_i(t-1),\cdots,y_i(t-q))$$
$$-H(y_i(t-1),\cdots,y_i(t-q)) \qquad (9)$$

The estimator $\hat{H}(y_i(t)|y_i(t-1),\cdots,y_i(t-q))$ is a function of the observations $y_i(1),\cdots,y_i(N)$, where N is the sample number. The l-th component of the conditional score function in a sample point $y_i(n)$ is computed as:

$$\hat{\psi}_{y_i}^{(l)}(y_i(t)|y_i(t-1),\cdots,y_i(t-q))|_{t=n} = N \frac{\partial \hat{H}(y_i(t)|y_i(t-1),\cdots,y_i(t-q))}{\partial y_i(n-l+1)} \qquad (10)$$

The method is very powerful and provides a quite good estimation of the conditional score functions.

3.2 Initializing the Nonlinear Function

The convergence speed can been enhanced by choosing a relevant starting point, especially for the parameters of the functions g_i. As presented in [9, 10], the idea is based on two remarks: (i) each mixture of sources, e_i, is a random variable close to Gaussian, and (ii) due to the nonlinear distortions, the random variable, $x_i = f_i(e_i)$ is farther to the Gaussian than e_i. Consequently, the nonlinear transform $\hat{g}_i = \Phi^{-1} \circ F_{x_i}$, where Φ is the cumulative density function of the Gaussian and F_{x_i} is the cumulative density function of x_i, transforms x_i to a Gaussian random variable z_i. If x_i is exactly a Gaussian random variable, then $\hat{g}_i = f_i^{-1}$; if it is approximately Gaussian, it is a rough estimation of f_i^{-1}. Thus, we estimate the initial parameter θ_i by minimization of the mean square error between $\hat{g}_i$ and $g_i(\theta_i, .)$ Since the Gaussian assumption of x_i is not completely fulfilled, we used the above idea for computing a good starting point of the algorithm.

4 Experiments

The aim of this section is to check if a Markov model of the sources is able to improve the performance of the algorithm. We will consider two kinds of colored sources, modeled both by first order auto-regressive (AR) filters, whose input is an iid random signal with either a Gaussian or a uniform distribution. We restrict the study to post-nonlinear mixtures of 2 sources.

We compared two algorithms, the first one with 1-st order Markov model and the second one with order 0, i.e. without modeling source temporal correlation.

Each experiment is repeated about 16 times, with random choice of the AR coefficients, of the mixing matrix and of the nonlinear parameters:

- AR coefficients, ρ_i, $i = 1, 2$, are chosen so that $0.2 <| \rho_i |< 0.9$ and $| \rho_1 - \rho_2 |> 0.2$, since the source spectra must be different.

- The main diagonal entries a_{ii} of mixing matrix $\mathbf{A}$ are enforced to 1, while the other are chosen in the range $0.2 <| a_{ij} |< 1$. This choice allows to avoid mixing matrices close to diagonal matrices, which provide post-nonlinear observations x_i which would be still independent.

- Each nonlinear function f_i, $i = 1, 2$, defined by the relation (11) (see below) with parameter β_i, is chosen so that $0.1 < \beta_i < 5$.

4.1 Simple Nonlinear Functions

In this first set of experiments, we use three nonlinear invertible functions, f_i, depending on one parameter β, and their inverses, g_i, too. Since we can compute the theoretical parameter of g_i, this experiment allows to measure the parametric error in the estimation of the nonlinear function.

Example 1. The main advantage of this nonlinear function is to have a very simple inverse, expression of which is linear with respect to the parameter θ:

$$f(\beta,e) = \frac{sign(e)}{2\beta}(-1 + \sqrt{1+4\beta|e|}) \quad \Rightarrow \quad g(\theta,x) = x + \theta x|x| \qquad (11)$$

Examples 2 and 3. We can defined two others saturating non linear functions:

$$f(\beta,e) = sign(e) \times (\tanh(|e|))^{(1/\beta)} \Rightarrow g(\theta,x) = sign(x) \times \tanh^{-1}(|x|^\theta) \quad (12)$$

$$f(\beta,e) = \frac{\beta e}{\beta + |e|} \quad \Rightarrow \quad g(\theta,x) = \frac{\theta x}{\theta - |x|} \qquad (13)$$

The concavity of the function (12) can be varied with the parameter, but, contrary to the function (13) the magnitude of the saturation is not adjustable.

The figure 2 shows the shapes of the three functions defined by (11),(12) and (13) for three values of β: 0.1 in solid line, 1 in dashed line and 5 in dotted line.

Eq. (7) depends on the parametric models $g_i(\theta_i, x)$. In the experiments, we will use either specific models or polynomials. However, the derivatives with respect to parameters θ_i are evident and will not be given in this paper.

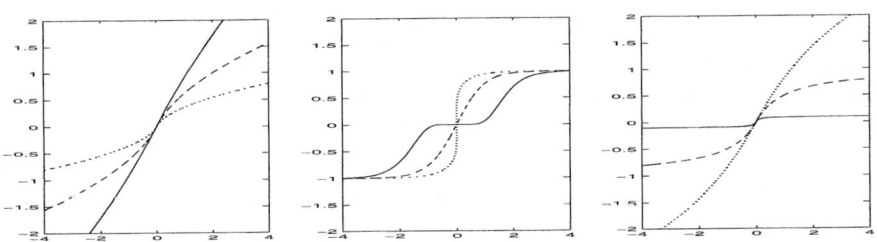

Fig. 2. Examples of simple nonlinear mappings. Left: non-linearity (11); middle: non-linearity (12); right: non-linearity (13).

4.2 Polynomial Approximation of Inverse Function

We also used a more general and more flexible model for estimating the nonlinear functions $g_i(\theta_i,.)$, based on a polynomial expression of $g_i(\theta_i,x) = \sum_{m=0}^{P} \theta_{im} x^m$. This expression being linear with respect to the parameter θ_{ik}, the gradient (7) becomes:

$$\frac{\partial J(\mathbf{B},\Theta)}{\partial \theta_{ik}} = -E\left[\frac{kx_i(t)^{k-1}}{\sum_{m=0}^{P} m\theta_{im}x_i(t)^{m-1}}\right] + \qquad (14)$$

$$E\left[\sum_{j=1}^{n} b_{ji} \sum_{l=0}^{q} \psi_{y_j}^{(l)}(y_j(t)|y_j(t-1),\ldots,y_j(t-q))x_i(t-l)^k\right]$$

The polynomial model is well suited to the inversion of saturating nonlinear transformation. For more general mappings, we could extend the model by adding rational powers in the polynomial.

4.3 Results

The following tables give the mean residual cross-talk, as well as max and min between brackets, expressed in dB for 16 random configurations. Since s_i and y_i are unit power signals, the residual cross-talk is $E[(y_i - s_i)^2]$.

With Simple Functions. We compare two algorithms based on minimization of the mutual information (MIM):

- the PNL algorithm developed in this paper, denoted MIM Markov 1, for $q = 1$ (order 1, Markov source),
- the PNL algorithm, denoted MIM iid, which does not take into account the source time structure (in fact, it correspond to MIM Markov 0, and is a special case of the algorithm developed in this paper).

Sources	MIM Markov 1	MIM iid
AR Gaussian	(-20.7) -16.2 (-14.1)	(-18.4) -8.9 (-5.0)
AR uniform	(-23.1) -16.1 (-13.6)	(-23.7) -14.5 (-12.3)

We remark that the Markov model of the source improves the performance: about 7 dB, on the average for Gaussian input, and 1.6 dB for uniform input. The improvement is then very sensitive, but much less important than for linear mixtures [2] (-34.5 dB for Markov model, and -24.2 dB for iid algorithm for uniformly distributed innovation process). This is mainly due to the nonlinear part of PNL: a small error in the nonlinear parameter estimation can imply a poor estimation of the separating matrix $\mathbf{B}$. We also remark that, like in linear mixtures, time correlation modeling (here with Markov models) allows to separate Gaussian sources.

With Polynomials. The functions g_i's are now modeled by 7-degree polynomials. We again compare the two algorithms, MIM Markov 1 and MIM iid, according to the notations of the previous paragraph.

Sources	MIM Markov 1	MIM iid
AR Gaussian	(-16.7) -14.0 (-11.8)	(-12.2) -7.8 (-5.1)
AR uniform	(-17.0) -14.3 (-11.9)	(-15.3) -11.1 (-5.4)

5 Conclusions

In this paper, we presented an algorithm modeling the temporal relation between successive source samples with a Markovian model, in post nonlinear mixtures. For various parametric model of the nonlinear mappings g_i's, Markov model of the sources provides a performance improvement for separating first order autoregressive sources. The computing time increases as 3^{q+1}, where q is the Markov model order: it is mainly due to the estimation of conditional score functions. Further works include (i) the comparaison of our algorithm with TDSEP [7] (using second order statistics) (ii) the relevance of suitability between the Markov order, q, and AR source order.

References

1. Hosseini, Sh., Jutten, C. and Pham, D. T.: Blind Separation of Temporally Correlated Sources Using A Quasi-Maximum Likelihood Approach. Proceedings ICA'01, San Diego (CA, USA) (2001) 586-590
2. Hosseini, Sh., Jutten, C. and Pham, D. T.: Markovian Source Separation. IEEE Trans. on Signal Processing **51** (2003) 3009-3019
3. Taleb, A., Jutten, C.: Source separation in post nonlinear mixtures. IEEE Trans. on Signal Processing **47** (1999) 2807-2820
4. Tong, L., Soon, V., Huang, Y. and Liu, R.:AMUSE: a new blind identification algorithm. Proceedings ISCAS'90, New Orleans (USA), 1990
5. Molgedey, L., Schuster, H. G.: Separation of a mixture of independent signals using time delayed correlation. Physical Review Letters **72** (1994) 3634-3636
6. Belouchrani, A., Abed Meraim, K., Cardoso, J.-F. and Moulines, E.: A blind source separation technique based on second order statistics. IEEE Trans. on Signal Processing **45** (1997) 434-444
7. Ziehe, A. and Müller, K.-R.: TDSEP: an efficient algorithm for blind separation using time structure. Proceedings of ICANN'98, Skövde (Sweden) (1998) 675-680
8. Pham, D. T.: Fast algorithm for estimating mutual information, entropies and score functions. Proceedings ICA'03, Nara (Japan) (2003) 17-22
9. Ziehe, A., Kawanabe, M., Harmeling, S., Müller, K.-R.: Blind separation of post-nonlinear mixtures using gaussianizing transformations and temporal decorrelation. Proceedings ICA'03, Nara (Japan) (2003) 269-274
10. Solé, J., Babaie-Zadeh, M., Jutten, C., Pham, D. T.: Improving algorithm speed in PNL mixture separation and Wiener system inversion. Proceedings ICA'03, Nara (Japan) (2003) 639-644

Non-linear ICA by Using Isometric Dimensionality Reduction

John A. Lee[1], Christian Jutten[2], and Michel Verleysen[1]

[1] Université catholique de Louvain (DICE)
Place de Levant, 3, B-1348 Louvain-la-Neuve, Belgium
{lee,verleysen}@dice.ucl.ac.be
[2] Institut National Polytechnique de Grenoble (LIS)
Avenue Félix Viallet, 46, 38031 Grenoble Cedex, France
Christian.Jutten@inpg.fr

Abstract. In usual ICA methods, sources are typically estimated by maximizing a measure of their statistical independence. This paper explains how to perform non-linear ICA by preprocessing the mixtures with recent non-linear dimensionality reduction techniques. These techniques are intended to produce a low-dimensional representation of the data (the mixtures), which is isometric to their initial high-dimensional distribution. A detailed study of the mixture model that makes the separation possible precedes a practical example.

1 Introduction

Independent Component Analysis [2,3] (ICA) aims at recovering a vector of unknown latent variables $\mathbf{x}$ starting from a vector of observed variables $\mathbf{y}$. Usually, the variables in $\mathbf{y}$ are assumed to be (noiseless) linear mixtures of $\mathbf{x}$, according to the generative model:

$$\mathbf{y} = \mathbf{A}\mathbf{x} , \tag{1}$$

where $\mathbf{A}$ is a full-rank $D \times P$ 'mixing' matrix, with $D \geq P$. In order to retrieve $\mathbf{x}$, the ICA model also assumes that all components of $\mathbf{x}$ have zero mean and are statistically independent from each other. Therefore, the goal of ICA is to identify $\mathbf{A}$ by determining the 'separating' matrix $\mathbf{B}$ in the reversed model

$$\mathbf{x} \approx \hat{\mathbf{x}} = \mathbf{B}\mathbf{y} . \tag{2}$$

Practically, ICA proceeds by defining a measure of independence E_{ICA} on $\hat{\mathbf{x}}$ and by maximizing it:

$$\hat{\mathbf{x}} = \arg\max E_{\text{ICA}}(\mathbf{B}\mathbf{y}) . \tag{3}$$

Thanks to the independence hypothesis on $\mathbf{x}$ and because the model is linear, it can be proved that E_{ICA} reaches its maximum when $\mathbf{B}\mathbf{A} = \mathbf{\Delta}\mathbf{\Pi}$ where $\mathbf{\Delta}$ and $\mathbf{\Pi}$ are respectively a diagonal and a permutation matrices [2,3].

When providing ICA with a non-linear model, the same statement should be true too. Unfortunately, it is not difficult to show that non-linear transformations of any set of variables allows building an infinity of variables that are independent from each other. Hence the maximization of E_{ICA} does not lead to the desired solution anymore. As a matter of fact, this has considerably slown down the investigation of non-linear ICA.

However, although ICA proves incompatible with a non-linear model in its full generality [6], it has been shown by several authors that non-linear ICA is still feasible in some specific cases. In particular, some work [8, 6] has been devoted to so-called post-non-linear mixtures (PNL), defined as follows:

$$\mathbf{y} = \mathbf{f}(\mathbf{A}\mathbf{x}) \; , \tag{4}$$

where $\mathbf{f}$ is a vector of D invertible and differentiable functions from $\mathbb{R}$ to $\mathbb{R}$. Under mild conditions on $\mathbf{A}$, the latent variables $\mathbf{x}$ can be retrieved using the same principle as in linear ICA. Indeed, by maximizing the independence of $\hat{\mathbf{x}} = \mathbf{g}(\mathbf{B}\mathbf{y})$, it is possible to identify $\mathbf{f}$ (as the inverse of $\mathbf{g}$) and $\mathbf{A}$ ($\mathbf{A}\mathbf{B} = \mathbf{\Delta}\mathbf{\Pi}$).

This paper explores another way to perform non-linear ICA. In PNL mixtures, the inversion of the non-linear functions is achieved by maximizing the independence. A slightly different and more complex model is proposed, consisting of two parts — a linear one and a non-linear one —, as in PNL mixtures. The main difference holds in the fact that the non-linear part of the model is identified by optimizing a criterion that does not relate to statistical independence. Actually, the non-linear part is inverted by computing an isometric transformation of the available data. Section 2 explains how isometric transformations can be integrated in an ICA model in a very natural way. In particular, Subsection 2.1 describes the particular metric that is used in the isometry. Section 3 gives some experimental results and Section 4 comments them. Finally, Section 5 draws the conclusions.

2 Mixture Model

The following generative model is considered:

$$\mathbf{y} = \mathbf{f}(\mathbf{z}) = \mathbf{f}(\mathbf{A}\mathbf{x}) \; , \tag{5}$$

where $\mathbf{f}$ is a smooth ($\mathcal{C}^\infty$) function from $\mathbb{R}^P$ to $\mathbb{R}^D$, $\mathbf{A}$ is a square $P \times P$ matrix and the vector $\mathbf{y}$ is assumed to be isometric to the vector $\mathbf{z}$. By 'isometric' it is meant that the distance measured between two realizations of $\mathbf{y}$ equals the distance measured between the corresponding realizations of $\mathbf{z}$.

Of course, if the Euclidean distance is used for both $\mathbf{y}$ and $\mathbf{z}$, then the mixing function $\mathbf{f}$ can only be a rotation matrix. More precisely, $\mathbf{f}(\mathbf{z}) = \mathbf{Q}\mathbf{z}$, where $\mathbf{Q}$ is a $D \times P$ matrix, resulting from the concatenation of P unit-norm vectors. This obviously raises little interest. Fortunately enough, the use of the Euclidean distance proves not mandatory at all. A couple of recent works [7, 4, 5] suggest using different metrics to measures distances on $\mathbf{y}$ and $\mathbf{z}$. In particular, the use of the so-called geodesic distance for $\mathbf{y}$ is advised, while keeping the Euclidean distance for $\mathbf{z}$.

2.1 Geodesic Distances

Geodesic distances are used in the fields of manifold learning and non-linear dimensionality reduction (NLDR) by distance preservation [7,4,5]. Given a P-dimensional smooth ($\mathcal{C}^\infty$) manifold $\mathcal{M}$ in a D-dimensional space, the geodesic distance between two points $\mathbf{y}_i$ and $\mathbf{y}_j$ of the manifold is measured along the manifold, unlike the Euclidean distance, which is measured along the line segment connecting the two points. Actually, the geodesic distance $\delta(\mathbf{y}_i, \mathbf{y}_j)$ is computed as the minimum arc length between the two points, an arc γ being a smooth one-dimensional submanifold. Hence,

$$\delta(\mathbf{y}_i, \mathbf{y}_j) = \min_{\gamma(\zeta)} \int_{\zeta_i}^{\zeta_j} \|\mathbf{J}_\zeta \mathbf{f}(\gamma(\zeta))\|_2 d\zeta \ , \tag{6}$$

where $\mathbf{y}_i = \gamma(\zeta_i)$ and $\mathbf{y}_j = \gamma(\zeta_i)$ are in $\mathcal{M}$, the function $\mathbf{f}(\mathbf{z})$ designates the parametric equations of $\mathcal{M}$ and $\mathbf{J}_\zeta$ is the Jacobian matrix with respect to ζ. It is easy to see that geodesic distances are equivalent to Euclidean ones if the manifold is linear (planar).

Figure 1 illustrates the purpose of geodesic distances in NLDR: such a metric allows measuring distances that are (almost) independent of the manifold embedding. Contrarily to Euclidean distances, geodesic ones do not change if the 'C'-shaped manifold in Fig. 1 is unrolled or unfolded. This shows how NDLR by distance preservation works: a low-dimensional embedding of the manifold is computed as the result of a (nearly) isometric transformation from $\mathbb{R}^D$ to $\mathbb{R}^P$.

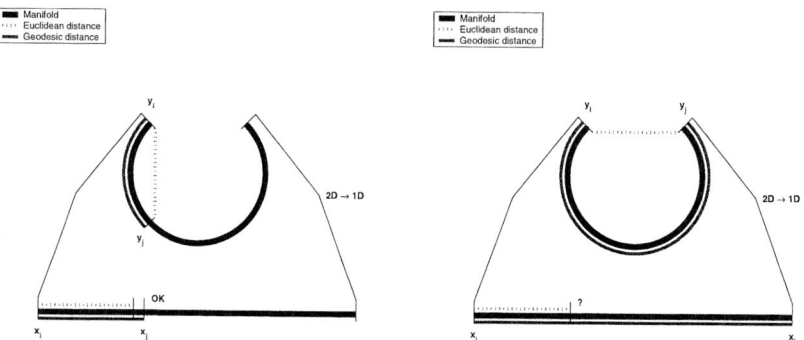

Fig. 1. Geodesic distances for dimensionality reduction in the case of a 'c'-shaped curve: for short (left) as well as long (right) distances, the geodesic distances makes possible the isometry between the manifold and its low-dimensional embedding.

In practice, computing geodesic distances from a finite-size sample $\mathbf{Y} = [\ldots, \mathbf{y}_i, \ldots, \mathbf{y}_j, \ldots]_{1 \leq i,j \leq N}$ is difficult. Fortunately, geodesic distances can be approximated by so-called graph distances, as illustrated in Fig. 2. The quality of that approximation is assessed in [1] (theoretical point of view) and [4,5] (practical issues).

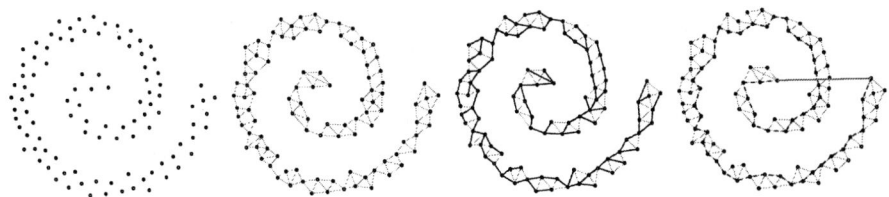

Fig. 2. Procedure to compute graph distances: (1st plot) a few manifold points are available, (2nd plot) each point becomes a graph vertex and is connected with its closest neighbors in order to obtain a graph, (3rd plot) after labeling the graph edges with their length, Dijkstra's algorithm is run on the graph, with the central point of the spiral as source vertex, (4th plot) the Euclidean and graph distances between the same two points.

2.2 Isometry

Back to the generative model in Eq. 5, it may be assumed as in the previous section that the vector $\mathbf{z}$ and the function $\mathbf{f}$ respectively contain the parameters of a manifold and its parametric equations. The hypothesis of isometry amounts to state that $\delta(\mathbf{y}_i, \mathbf{y}_j) = \|\mathbf{z}_i - \mathbf{z}_j\|_2$ for any corresponding pairs of realizations of $\mathbf{y}$ and $\mathbf{z}$. A manifold which satisfies that hypothesis is said to be Euclidean and has nice properties. For example, the minimization involved in the computation of the geodesic distance $\delta(\mathbf{y}_i, \mathbf{y}_j)$ may be dropped in Eq. 6. Indeed, by virtue of the isometry, it comes that

$$\begin{aligned}\delta(\mathbf{y}_i, \mathbf{y}_j) &= \|\mathbf{z}_i - \mathbf{z}_j\|_2 \\ &= \|\mathbf{z}_i - (\mathbf{z}_i + \alpha(\mathbf{z}_j - \mathbf{z}_i))\|_2 + \|(\mathbf{z}_i + \alpha(\mathbf{z}_j - \mathbf{z}_i)) - \mathbf{z}_j\|_2 \\ &= \delta(\mathbf{f}(\mathbf{z}_i), \mathbf{f}(\mathbf{z}_i + \alpha(\mathbf{z}_j - \mathbf{z}_i))) + \delta(\mathbf{f}(\mathbf{z}_i + \alpha(\mathbf{z}_j - \mathbf{z}_i)), \mathbf{f}(\mathbf{z}_j)) \quad , \end{aligned} \qquad (7)$$

where α is a real number between 0 and 1. These equalities simply demonstrate that the shortest geodesic arc between $\mathbf{y}_i$ and $\mathbf{y}_j$ is the image by $\mathbf{f}$ of the line segment going from $\mathbf{z}_i$ to $\mathbf{z}_j$: all points $\mathbf{f}(\mathbf{z}_i + \alpha(\mathbf{z}_j - \mathbf{z}_i))$ on that segment must also lie on the shortest path. Therefore, in the case of a Euclidean manifold, the arc $\boldsymbol{\gamma}(\zeta)$ in Eq. 6 can be written as

$$\zeta : [0,1] \subset \mathbb{R} \to \mathbb{R}^P, \zeta \mapsto \mathbf{z} = \boldsymbol{\gamma}(\zeta) = \mathbf{z}_i + \zeta(\mathbf{z}_j - \mathbf{z}_i) \qquad (8)$$

and the minimization in Eq. 6 becomes useless. Using the last result and knowing that geodesic distances equal Euclidean ones in a vector space, it comes that

$$\|\mathbf{z}_i - \mathbf{z}_j\|_2 = \delta(\mathbf{z}_i, \mathbf{z}_j) = \int_0^1 \|\mathbf{J}_\zeta \boldsymbol{\gamma}(\zeta)\|_2 d\zeta \quad \text{and} \qquad (9)$$

$$\delta(\mathbf{y}_i, \mathbf{y}_j) = \int_0^1 \|\mathbf{J}_\zeta \mathbf{f}(\boldsymbol{\gamma}(\zeta))\|_2 d\zeta = \int_0^1 \|\mathbf{J}_{\boldsymbol{\gamma}(z)} \mathbf{f}(\boldsymbol{\gamma}(\zeta)) \, \mathbf{J}_\zeta \boldsymbol{\gamma}(\zeta)\|_2 d\zeta \quad . \qquad (10)$$

As $\|\mathbf{z}_i - \mathbf{z}_j\|_2 = \delta(\mathbf{y}_i, \mathbf{y}_j)$, the equality $\|\mathbf{J}_\zeta \boldsymbol{\gamma}(\zeta)\|_2 = \|\mathbf{J}_{\boldsymbol{\gamma}(\zeta)} \mathbf{f}(\boldsymbol{\gamma}(\zeta)) \, \mathbf{J}_\zeta \boldsymbol{\gamma}(\zeta)\|_2$ must hold. This means that the Jacobian of a Euclidean manifold must be a D-by-P matrix whose columns are orthogonal vectors with unit norm. This leaves the norm of $\mathbf{J}_\zeta \gamma(\zeta)$ unchanged after left multiplication by $\mathbf{J}_{\gamma(z)}\mathbf{f}(\gamma(\zeta))$. More precisely, the Jacobian matrix can be written in a generic way as $\mathbf{J}_\mathbf{z}\mathbf{f}(\mathbf{z}) = \mathbf{Q}\mathbf{V}(\mathbf{z})$, where $\mathbf{Q}$ is a constant orthonormal matrix (a rotation matrix in the D-dimensional space) and $\mathbf{V}(\mathbf{z})$ a D-by-P matrix with unit-norm columns and only one non-zero entry per row. The last requirement ensures that the columns of $\mathbf{V}(\mathbf{z})$ are always orthogonal, independently from the value of $\mathbf{z}$.

Because of the particular form of its Jacobian matrix, a Euclidean P-manifold embedded in a D-dimensional space can always be written with the following 'canonical' parametric equations:

$$\mathbf{y} = \mathbf{Q}\mathbf{f}(\mathbf{z}) = \mathbf{Q}\left[f_1(z_{1\leq p\leq P}), \ldots, f_D(z_{1\leq p\leq P})\right]^T , \qquad (11)$$

where $\mathbf{Q}$ is the same as above, $\mathbf{J}_\mathbf{z}\mathbf{f}(\mathbf{z}) = \mathbf{V}(\mathbf{z})$ and $f_1, \ldots, f_D$ are constant, linear or non-linear continuous functions from $\mathbb{R}$ to $\mathbb{R}$. Hence, if $\mathbf{Q}$ is omitted, the parametric equation of each coordinate in the D-dimensional space of a Euclidean manifold depends on at most a single latent variable z_p.

Visually, in a three-dimensional space, a manifold is Euclidean if it looks like a curved sheet of paper.

2.3 Isometric Dimensionality Reduction

Using the above-mentioned ideas, it may be stated that if a manifold is Euclidean, its latent variables can be retrieved. More formally, knowing a sufficiently large set $\mathbf{Y} = [\ldots, \mathbf{y}_i, \ldots, \mathbf{y}_j, \ldots]_{1\leq i,j\leq N}$ of points drawn from a Euclidean P-dimensional manifold, it is possible to determine the corresponding values of the latent variables, up to a translation and a rotation, by finding an isometric P-dimensional representation $\mathbf{Z}$ of $\mathbf{Y}$.

From a practical point of view, an estimation $\hat{\mathbf{Z}}$ of $\mathbf{Z}$ can be computed using NLDR methods [7,4,5] that work by distance preservation. These methods precisely attempt to find a low-dimensional representation of high-dimensional points that is 'as isometric as possible'. If these methods use geodesic distances in the D-dimensional space and Euclidean distances in the P-dimensional space and if the manifold is Euclidean, then a prefect isometry is possible. This means that starting from an observation $\mathbf{y}_i$ the corresponding value $\mathbf{z}_i$ of the latent variables can be recovered or, in other words, that the function $\mathbf{y} = \mathbf{f}(\mathbf{z})$ can be perfectly inverted, up to the above-mentioned undeterminacies (translation and rotation). Within the framework of ICA, this also means that the non-linear part of the generative model in Eq. 5 can be inverted to find $\hat{\mathbf{z}} \approx \mathbf{z} = \mathbf{A}\mathbf{x}$; next, $\hat{\mathbf{x}}$ can be recovered by using a classical linear ICA method.

3 Experimental Results

In order to illustrate how isometric ICA works, the two following sources are proposed:

$$\mathbf{x} = \begin{bmatrix} \arccos(\cos(0.034\pi t)) \\ \sin(0.006\pi t) \end{bmatrix} . \qquad (12)$$

Thousand observations of **x** are available in the interval $[0.000 \leq t \leq 0.999]$ and are shown on the left of Fig. 3. Sources are artificially mixed as follows:

$$\mathbf{y} = 0.5 \begin{bmatrix} +\sqrt{2} & +1 & +1 \\ -\sqrt{2} & +1 & +1 \\ 0 & -\sqrt{2} & +\sqrt{2} \end{bmatrix} \begin{bmatrix} \cos(\pi z_1) \\ \sin(\pi z_1) \\ \pi z_2 \end{bmatrix} \ , \text{ where } \quad \mathbf{z} = \begin{bmatrix} 0.1 & 0.9 \\ 0.8 & 0.2 \end{bmatrix} \mathbf{x} \ . \quad (13)$$

It is easy to see that the above-stated conditions to apply isometric ICA are fulfilled. The mixtures are shown on the right of Fig. 3.

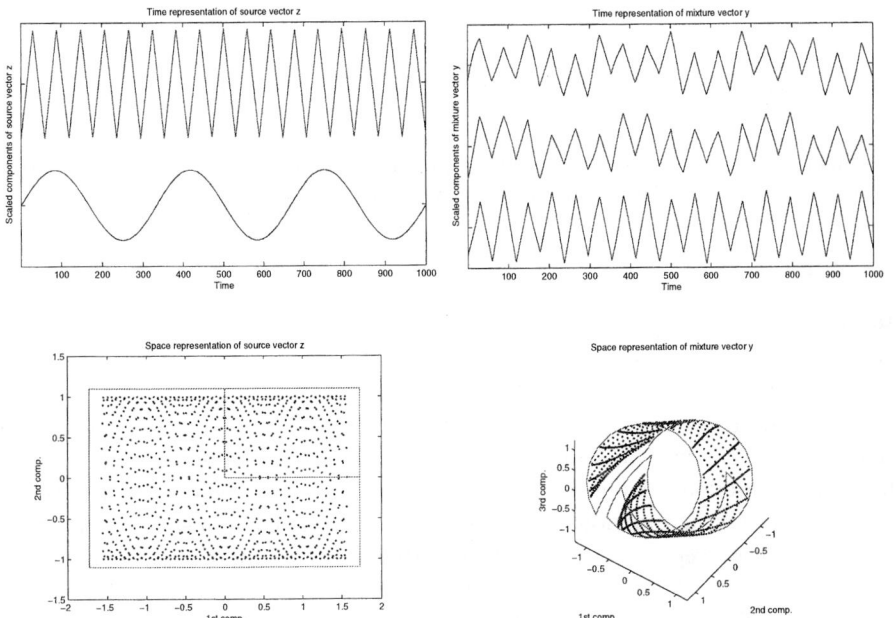

Fig. 3. Sources (left) and mixtures (right); time (top) and space (bottom) representations are given.

Starting from the mixtures, a first attempt to separate the sources consists in running a classical (linear) ICA method. For example, FastICA (deflation, tanh non-linear function) yields the result shown on the left of Fig. 4. During the whitening step, PCA indicates that *three* components are needed to explain 95% of the variance. Linear ICA succeeds rather well in recovering the serrated source (already clearly visible in the mixtures) but fails in the case of the sine.

Isometric ICA yields the result shown on the right of Fig. 4. To obtain that result, an isometric representation of the thousand available observations is computed with the method described in [4, 5]. This dimensionality reduction method works by gradient descent, contrarily to other methods which are purely algebraical [7]. The method indicates that an almost isometric two-dimensional rep-

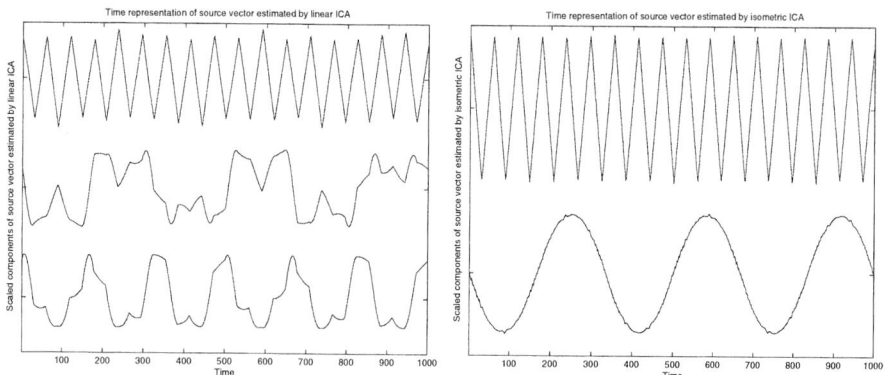

Fig. 4. Results computed by FastICA (left) and isometric ICA (right).

resentation of the observations is possible and computes it. Next, FastICA is run on the two remaining linear mixtures, leading to the result on the right of Fig. 4.

4 Discussion

Because of space constraints, other experiments cannot be included in this paper. However, here are some comments about the advantages and shortcomings of isometric ICA.

The main drawback of isometric ICA holds in the very restrictive conditions that must be satisfied to apply it. Exactly as for post-non-linear mixtures, the non-linear functions involved in the mixture process must be one-to-one. Moreover, all non-linear functions depending on the same component of $\mathbf{z}$ must be 'coupled', otherwise the norm of the corresponding column of the Jacobian matrix cannot be constant. Contrarily to PNL mixtures, isometric mixtures may be further multiplied by any rotation matrix $\mathbf{Q}$.

In practice, it has been shown experimentally that the isometry does not need to be absolutely perfect. Actually, the norms of the Jacobian columns may vary a little and slightly differ from each other. Similarly, the matrix $\mathbf{Q}$ does not need to be perfectly unitary. Even in those cases, the non-linear part of the model can be more or less well inverted, and better results are obtained than when using a simple linear ICA method.

Other practical issues of isometric ICA regard the quality of the isometric representation computed from the observations. Even if all above-mentioned conditions are satisfied, it must be ensured that the methods described in [7, 4, 5] work correctly. It has been shown in [4, 5] that the good approximation of the geodesic distances by the graph distances is very important. If the number of observations is low, if their distribution is very sparse in some regions, or simply if some parameters values are wrong, then the approximation becomes very rough. This may jeopardize the computation of the isometric representation and therefore the inversion of the non-linear function $\mathbf{f}$. As a direct consequence, the subsequent ICA step does not run in an optimal setting.

5 Conclusion

Although true non-linear ICA is impossible, several constrained models have been successfully proposed, especially post-non-linear mixtures. In this paper, a different model is proposed, in which the inversion of the non-linear part is based on geometrical considerations. More precisely, the proposed model assumes that variables at the input and output of its non-linear part are isometric, i.e. distances measured between two corresponding pairs of observations are equal. Because the isometry involves other distances than Euclidean ones, the associated transformation may be non-linear. In the case of the geodesic distance, which has become popular in the field of dimensionality reduction, conditions that makes a perfect isometry possible are studied in details. If those conditions are fulfilled, the non-linear part of the mixture model can be fully inverted and a linear ICA method may be run afterwards. Even if those conditions are rather restrictive, isometric ICA can tackle problems that other linear or non-linear ICA methods cannot solve. A simple example illustrates this fact.

Future work aims at comparing isometric ICA to post-non-linear ICA, from different points of view (computational costs, robustness, etc.). A further study of how isometric ICA behaves when the conditions of its model are not perfectly met is also planned.

References

1. M. Bernstein, V. de Silva, J. C. Langford, and J. B. Tenenbaum. Graph approximations to geodesics on embedded manifolds. Technical report, Stanford University, Stanford, December 2000.
2. P. Comon. Independent Component Analysis – A new concept? *Signal Processing*, 36:287–314, 1994.
3. A. Hyvärinen, J. Karhunen, and E. Oja. *Independent Component Analysis*. Wiley-Interscience, 2001.
4. J. A. Lee, A. Lendasse, and M. Verleysen. Curvilinear Distances Analysis versus Isomap. In M. Verleysen, editor, *Proceedings of ESANN'2002*, pages 13–20. D-Facto public., Bruges (Belgium), 2002.
5. J. A. Lee and M. Verleysen. Curvilinear distance analysis versus isomap. *Neurocomputing*, 2004. Accepted.
6. A. Taleb and C. Jutten. Source separation in postnonlinear mixtures. *IEEE Transactions on Signal Processing*, 47(10):2807–2820, 1999.
7. J. B. Tenenbaum, V. de Silva, and J. C. Langford. A global geometric framework for nonlinear dimensionality reduction. *Science*, 290(5500):2319–2323, December 2000.
8. H. H. Yang, Amari S., and A. Cichocki. Information-theoretic approach to blind separation of sources in non-linear mixtures. *Signal Processing*, 64(3):291–300, 1998.

Postnonlinear Overcomplete Blind Source Separation Using Sparse Sources

Fabian J. Theis[1,2] and Shun-ichi Amari[1]

[1] Brain Science Institute, RIKEN
2-1, Hirosawa, Wako-shi, Saitama, 351-0198, Japan
fabian@theis.name, amari@brain.riken.go.jp
[2] Institute of Biophysics, University of Regensburg
D-93040 Regensburg, Germany

Abstract. We present an approach for blindly decomposing an observed random vector $\mathbf{x}$ into $\mathbf{f}(\mathbf{As})$ where $\mathbf{f}$ is a diagonal function i.e. $\mathbf{f} = f_1 \times \ldots \times f_m$ with one-dimensional functions f_i and $\mathbf{A}$ an $m \times n$ matrix. This postnonlinear model is allowed to be overcomplete, which means that less observations than sources ($m < n$) are given. In contrast to Independent Component Analysis (ICA) we do not assume the sources $\mathbf{s}$ to be independent but to be sparse in the sense that at each time instant they have at most $m - 1$ non-zero components (Sparse Component Analysis or SCA). Identifiability of the model is shown, and an algorithm for model and source recovery is proposed. It first detects the postnonlinearities in each component, and then identifies the now linearized model using previous results.

Blind source separation (BSS) based on ICA is a rapidly growing field (see for instance [1,2] and references therein), but most algorithms deal only with the case of at least as many observations as sources. However, there is an increasing interest in (linear) overcomplete ICA [3–5], where matrix identifiability is known [6], but source identifiability does not hold. In order to approximatively detect the sources [7], additional requirements have to be made, usually sparsity of the sources.

Recently, we have proposed a model based *only* upon the sparsity assumption (summarized in section 1) [8]. In this case identifiability of both matrix and sources given sufficiently high sparsity can be shown. Here, we extend these results to postnonlinear mixtures (section 2); they describe a model often occurring in real situations, when the mixture is in principle linear, but the sensors introduce an additional nonlinearity during the recording [9]. Section 3 presents an algorithm for identifying such models, and section 4 finishes with an illustrative simulation.

1 Linear Overcomplete SCA

Definition 1. *A vector* $\mathbf{v} \in \mathbb{R}^n$ *is said to be* k-sparse *if* $\mathbf{v}$ *has at most* k *non-zero entries.*

If an n-dimensional vector is $(n-1)$-sparse, that is it includes at least one zero component, it is simply said to be *sparse*. The goal of *Sparse Component Analysis* of level k (k-SCA) is to decompose a given m-dimensional random vector $\mathbf{x}$ into

$$\mathbf{x} = \mathbf{A}\mathbf{s} \tag{1}$$

with a real $m \times n$-matrix $\mathbf{A}$ and an n-dimensional k-sparse random vector $\mathbf{s}$. $\mathbf{s}$ is called the *source vector*, $\mathbf{x}$ the *mixtures* and $\mathbf{A}$ the *mixing matrix*. We speak of *complete*, *overcomplete* or *undercomplete* k-SCA if $m = n$, $m < n$ or $m > n$ respectively. In the following without loss of generality we will assume $m \leq n$ because the undercomplete case can be easily reduced to the complete case by projection of $\mathbf{x}$.

Theorem 1 (Matrix identifiability). *Consider the k-SCA problem from equation 1 for $k := m - 1$ and assume that every $m \times m$-submatrix of $\mathbf{A}$ is invertible. Furthermore let $\mathbf{s}$ be sufficiently rich represented in the sense that for any index set of $n - m + 1$ elements $I \subset \{1, ..., n\}$ there exist at least m samples of $\mathbf{s}$ such that each of them has zero elements in places with indexes in I and each $m - 1$ of them are linearly independent. Then $\mathbf{A}$ is uniquely determined by $\mathbf{x}$ except for left-multiplication with permutation and scaling matrices.*

Theorem 2 (Source identifiablity). *Let $\mathcal{H}$ be the set of all $\mathbf{x} \in \mathbb{R}^m$ such that the linear system $\mathbf{A}\mathbf{s} = \mathbf{x}$ has an $(m-1)$-sparse solution $\mathbf{s}$. If $\mathbf{A}$ fulfills the condition from theorem 1, then there exists a subset $\mathcal{H}_0 \subset \mathcal{H}$ with measure zero with respect to $\mathcal{H}$, such that for every $\mathbf{x} \in \mathcal{H} \setminus \mathcal{H}_0$ this system has no other solution with this property.*

The above two theorems show that in the case of overcomplete BSS using $(m-1)$-SCA, both the mixing matrix and the sources can uniquely be recovered from $\mathbf{x}$ except for the omnipresent permutation and scaling indeterminacy. We refer to [8] for proofs of these theorems and algorithms based upon them. We also want to note that the present source recovery algorithm is quite different from the usual sparse source recovery using l_1-norm minimization [7] and linear programming. In the case of sources with sparsity as above, the latter will not be able to detect the sources.

2 Postnonlinear Overcomplete SCA

2.1 Model

Consider n-dimensional k-sparse sources $\mathbf{s}$ with $k < m$. The *postnonlinear mixing model* [9] is defined to be

$$\mathbf{x} = \mathbf{f}(\mathbf{A}\mathbf{s}) \tag{2}$$

with a diagonal invertible function $\mathbf{f}$ with $\mathbf{f}(0) = 0$ and a real $m \times n$-matrix $\mathbf{A}$. Here a function $\mathbf{f}$ is said to be *diagonal* if each component f_i only depends on x_i. In abuse of notation we will in this case interpret the components f_i of $\mathbf{f}$ as functions with domain $\mathbb{R}$ and write $\mathbf{f} = f_1 \times \ldots \times f_m$. The goal of *overcomplete postnonlinear k-SCA* is to determine the mixing functions $\mathbf{f}$ and $\mathbf{A}$ and the sources $\mathbf{s}$ given only $\mathbf{x}$.

Without loss of generality consider only the complete and the overcomplete case (i.e. $m \leq n$). In the following we will assume that the sources are sparse of level $k := m - 1$ and that the components f_i of $\mathbf{f}$ are continuously differentiable with $f_i'(t) \neq 0$. This is equivalent to saying that the f_i are continuously differentiable with continuously differentiable inverse functions (diffeomorphisms).

2.2 Identifiability

Definition 2. *Let $\mathbf{A}$ be an $m \times n$ matrix. Then $\mathbf{A}$ is said to be* mixing *if $\mathbf{A}$ has at least two nonzero entries in each row. And $\mathbf{A} = (a_{ij})_{i=1\ldots m, j=1\ldots n}$ is said to be* absolutely degenerate *if there are two columns $k \neq l$ such that $a_{ik}^2 = \lambda a_{il}^2$ for all i and fixed $\lambda \neq 0$ i.e. the normalized columns differ only by the sign of the entries.*

Postnonlinear overcomplete SCA is a generalization of linear overcomplete SCA, so the indeterminacies of postnonlinear SCA contain at least the indeterminacies of linear overcomplete SCA: $\mathbf{A}$ can only be reconstructed up to scaling and permutation. Also, if $\mathbf{L}$ is an invertible scaling matrix, then

$$\mathbf{f}(\mathbf{A}\mathbf{s}) = (\mathbf{f} \circ \mathbf{L})\big((\mathbf{L}^{-1}\mathbf{A})\mathbf{s}\big),$$

so $\mathbf{f}$ and $\mathbf{A}$ can interchange scaling factors in each component.

Two further indeterminacies occur if $\mathbf{A}$ is either not mixing or absolutely degenerate. In the first case, this means that f_i cannot be identified if the i-th row of $\mathbf{A}$ contains only one non-zero element. In the case of an absolutely degenerate mixing matrix, sparseness alone cannot detect the nonlinearity as the counterexample $\mathbf{A} = \begin{pmatrix} 1 & 1 \\ 1 & -1 \end{pmatrix}$ and arbitrary $f_1 \equiv f_2$ shows.

If $\mathbf{s}$ is an n-dimensional random vector, its image (or the support of its density) is denoted as $\operatorname{im}\mathbf{s} := \{\mathbf{s}(t)\}$.

Theorem 3 (Identifiability). *Let $\mathbf{s}$ be an n-dimensional k-sparse random vector ($k < m$), and $\mathbf{x}$ an m-dimensional random vector constructed from $\mathbf{s}$ as in equation 2. Furthermore assume that*

(i) *$\mathbf{s}$ is fully k-sparse in the sense that $\operatorname{im}\mathbf{s}$ equals the union of all k-dimensional coordinate spaces (in which it is contained by the sparsity assumption),*
(ii) *$\mathbf{A}$ is mixing and not absolutely degenerate,*
(iii) *every $m \times m$-submatrix of $\mathbf{A}$ is invertible.*

If $\mathbf{x} = \hat{\mathbf{f}}(\hat{\mathbf{A}}\hat{\mathbf{s}})$ is another representation of $\mathbf{x}$ as in equation 2 with $\hat{\mathbf{s}}$ satisfying the same conditions as $\mathbf{s}$, then there exists an invertible scaling $\mathbf{L}$ with $\mathbf{f} = \hat{\mathbf{f}} \circ \mathbf{L}$, and invertible scaling and permutation matrices $\mathbf{L}', \mathbf{P}'$ with $\mathbf{A} = \mathbf{L}\hat{\mathbf{A}}\mathbf{L}'\mathbf{P}'$.

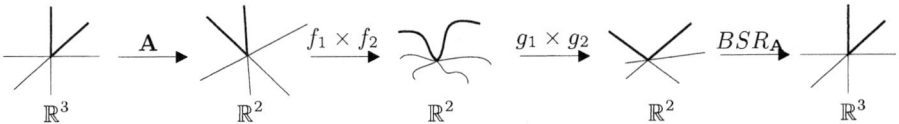

Fig. 1. Illustration of the proof of theorem 3 in the case $n = 3, m = 2$. The 3-dimensional 1-sparse sources (leftmost figure) are first linearly mapped onto $\mathbb{R}^2$ by **A** and then postnonlinearly distorted by $\mathbf{f} := f_1 \times f_2$ (middle figure). Separation is performed by first estimating the separating postnonlinearities $\mathbf{g} := g_1 \times g_2$ and then performing overcomplete source recovery (right figure) according to the algorithms from [8]. The idea of the proof now is that two lines spanned by coordinate vectors (thick lines, leftmost figure) are mapped onto two lines spanned by two columns of **A**. If the composition $\mathbf{g} \circ \mathbf{f}$ maps these lines onto some different lines (as sets), then we show that (given 'general position' of the two lines) the components of $\mathbf{g} \circ \mathbf{f}$ satisfy the conditions from lemma 1 and hence are already linear.

The proof relies on the fact that when **s** is fully k-sparse as formulated in 3(i), it includes all the k-dimensional coordinate subspaces and hence intersections of k such subspaces, which give the n coordinate axes. They are transformed into n curves in the **x**-space, passing through the origin. By identification of these curves, we show that each nonlinearity is homogeneous and hence linear according to the previous section. The proof is omitted due to lack of space. Figure 1 gives an illustration of the proof in the case $n = 3$ and $m = 2$. It uses the following lemma (a generalization of the analytic case presented in [10]).

Lemma 1. *Let $a, b \in \mathbb{R} \setminus \{-1, 0, 1\}, a > 0$ and $f : [0, \varepsilon) \to \mathbb{R}$ differentiable such that $f(ax) = bf(x)$ for all $x \in [0, \varepsilon)$ with $ax \in [0, \varepsilon)$. If $\lim_{t \to 0+} f'(t)$ exists and does not vanish, then f is linear.*

Theorem 3 shows that **f** and **A** are uniquely determined by **x** except for scaling and permutation ambiguities. Note that then obviously also **s** is identifiable by applying theorem 2 to the linearized mixtures $\mathbf{y} = \mathbf{f}^{-1}\mathbf{x} = \mathbf{As}$ given the additional assumptions to **s** from the theorem.

For brevity, the theorem assumes in (i) that im **s** is the whole union of the k-dimensional coordinate spaces — this condition can be relaxed (the proof is local in nature) but then the nonlinearities can only be found on intervals where the corresponding marginal densities of **As** are non-zero (however in addition, the proof needs that locally at 0 they are nonzero). Furthermore in practice the assumption about the image of **s** will have to be replaced by assuming the same with non-zero probability. Also note that almost any $\mathbf{A} \in \mathbb{R}^{mn}$ in the measure sense fulfills the conditions (ii) and (iii).

3 Algorithm for Postnonlinear (Over)Complete SCA

The separation is done in a two-stage procedure: In the first step, after geometrical preprocessing the postnonlinearities are estimated using an idea similar to the one used in the identifiability proof of theorem 3, also see figure 1. In the second stage, the mixing matrix $\mathbf{A}$ and then the sources $\mathbf{s}$ are reconstructed by applying the linear algorithms from [8], section 1, to the linearized mixtures $\mathbf{f}^{-1}\mathbf{x}$. So in the following it is enough to reconstruct $\mathbf{f}$.

3.1 Geometrical Preprocessing

Let $\mathbf{x}(1), \ldots, \mathbf{x}(T) \in \mathbb{R}^m$ be i.i.d.-samples of the random vector $\mathbf{x}$. The goal of *geometrical preprocessing* is to construct vectors $\mathbf{y}(1), \ldots, \mathbf{y}(T)$ and $\mathbf{z}(1), \ldots, \mathbf{z}(T)$ $\in \mathbb{R}^m$ using clustering or interpolation on the samples $\mathbf{x}(t)$ such that $\mathbf{f}^{-1}(\mathbf{y}(t))$ and $\mathbf{f}^{-1}(\mathbf{z}(t))$ lie in two linearly independent lines of $\mathbb{R}^m$. In figure 1 they are to span the two thick lines which already determine the postnonlinearities.

Algorithmically, $\mathbf{y}$ and $\mathbf{z}$ can be constructed in the case $m = 2$ by first choosing far away samples (on different 'non-opposite' curves) as initial starting point and then advancing to the known data set center by always choosing the closest samples of $\mathbf{x}$ with smaller modulus. Such an algorithm can also be implemented for larger m but only for sources with at most one non-zero coefficient at each time instant, but it can be generalized to sources of sparseness $m-1$ using more elaborate clustering.

3.2 Postnonlinearity Estimation

Given the subspace vectors $\mathbf{y}(t)$ and $\mathbf{z}(t)$ from the previous section, the goal is to find $\mathcal{C}^1$-diffeomorphisms $g_i : \mathbb{R} \to \mathbb{R}$ such that $g_1 \times \ldots \times g_m$ maps the vectors $\mathbf{y}(t)$ and $\mathbf{z}(t)$ onto two different *linear* subspaces.

In abuse of notation, we now assume that two *curves* (injective infinitely differentiable mappings) $\mathbf{y}, \mathbf{z} : (-1, 1) \to \mathbb{R}^m$ are given with $\mathbf{y}(0) = \mathbf{z}(0) = 0$. These can for example be constructed from the discrete sample points $\mathbf{y}(t)$ and $\mathbf{z}(t)$ from the previous section by polynomial or spline interpolation. If the two curves are mapped onto lines by $g_1 \times \ldots \times g_m$ (and if these are in sufficiently general position) then $g_i = \lambda_i f_i^{-1}$ for some $\lambda \neq 0$ according to theorem 3. By requiring this condition only for the discrete sample points from the previous section we get an approximation of the unmixing nonlinearities g_i. Let $i \neq j$ be fixed. It is then easy to see that by projecting $\mathbf{x}, \mathbf{y}$ and $\mathbf{z}$ onto the i-th and j-th coordinate, the problem of finding the nonlinearities can be reduced to the case $m = 2$, and g_2 is to be reconstructed, which we will assume in the following.

$\mathbf{A}$ is chosen to be mixing, so we can assume that the indices i, j were chosen such that the two lines $\mathbf{f}^{-1} \circ \mathbf{y}, \mathbf{f}^{-1} \circ \mathbf{z} : (-1, 1) \to \mathbb{R}^2$ do not coincide with the coordinate axes. Reparametrization ($\bar{\mathbf{y}} := \mathbf{y} \circ \mathbf{y}_1^{-1}$) of the curves lets us further assume that $\mathbf{y}_1 = \mathbf{z}_1 = \mathrm{id}$. Then after some algebraic manipulation, the condition that the separating nonlinearities $\mathbf{g} = g_1 \times g_2$ must map $\mathbf{y}$ and $\mathbf{z}$ onto lines can be written as $g_2 \circ y_2 = a g_1 = \frac{a}{b} g_2 \circ z_2$ with constants $a, b \in \mathbb{R} \setminus \{0\}$, $a \neq \pm b$.

So the goal of *geometrical postnonlinearity detection* is to find a $\mathcal{C}^1$-diffeomorphism g on subsets of $\mathbb{R}$ with

$$g \circ y = cg \circ z \qquad (3)$$

for an unknown constant $c \neq 0, \pm 1$ and given curves $y, z : (-1, 1) \to \mathbb{R}$ with $y(0) = z(0) = 0$. By theorem 3, g (and also c) are uniquely determined by y and z except for scaling. Indeed by taking derivatives in equation 3, we get $c = y'(0)/z'(0)$, so c can be directly calculated from the known curves y and z.

In the following section, we propose to solve this problem numerically, given samples $y(t_1), z(t_1), \ldots, y(t_T), z(t_T)$ of the curves. Note that here it is assumed that the samples of the curves y and z are given at the *same* time instants $t_i \in (-1, 1)$. In practice, this is usually not the case, so values of z at the sample points of y and vice versa will first have to be estimated, for example by using spline interpolation.

3.3 MLP-Based Postnonlinearity Approximation

We want to find an approximation $\tilde{g}$ (in some parametrization) of g with $\tilde{g}(y(t_i)) = c\tilde{g}(z(t_i))$ for $i = 1, \ldots, T$, so in the most general sense we want to find

$$\tilde{g} = \operatorname{argmin}_g E(g) := \operatorname{argmin}_g \frac{1}{2T} \sum_{i=1}^{T} (g(y(t_i)) - cg(z(t_i)))^2. \quad (4)$$

In order to minimize this energy function $E(g)$, a single-input single-output *multilayered neural network (MLP)* is used to parametrize the nonlinearity g. Here we choose one hidden layer of size d. This means that the approximated $\tilde{g}$ can be written as

$$\tilde{g}(t) = \mathbf{w}^{(2)\top} \bar{\sigma}\left(\mathbf{w}^{(1)} t + \mathbf{b}^{(1)}\right) + b^{(2)}$$

with weigh vectors $\mathbf{w}^{(1)}, \mathbf{w}^{(2)} \in \mathbb{R}^d$ and bias $\mathbf{b}^{(1)} \in \mathbb{R}^d$, $b^{(2)} \in \mathbb{R}$. Here σ denotes an activation function, usually the logistic sigmoid $\sigma(t) := (1 + e^{-t})^{-1}$ and we set $\bar{\sigma} := \sigma \times \ldots \times \sigma$, d times. The MLP weights are restricted in the sense that $\tilde{g}(0) = 0$ and $\tilde{g}'(0) = 1$. This implies $b^{(2)} = -\mathbf{w}^{(2)\top} \bar{\sigma}(\mathbf{b}^{(1)})$ and $\sum_{i=1}^{d} w_i^{(1)} w_i^{(2)} \sigma'(b_1^{(1)}) = 1$.

Especially the second normalization is very important for the learning step, otherwise the weights could all converge to the (valid) zero solution. So the outer bias is not trained by the network; we could fix a second weight in order to guarantee the second condition — this however would result in an unstable quotient calculation. Instead it is preferable to perform network training on a submanifold in the weight space given by the second weight restriction. This results in an additional Lagrange term in the energy function from equation 4

$$\bar{E}(\tilde{g}) := \frac{1}{2T} \sum_{j=1}^{T} (\tilde{g}(y(t_j)) - c\tilde{g}(z(t_j)))^2 + \lambda \left(\sum_{i=1}^{d} w_i^{(1)} w_i^{(2)} \sigma'(b_1^{(1)}) - 1 \right)^2 \quad (5)$$

with suitably chosen $\lambda > 0$.

Learning of the weights is performed via backpropagation on this energy function. The gradient of $\bar{E}(\tilde{g})$ with respect to the weight matrix can be easily

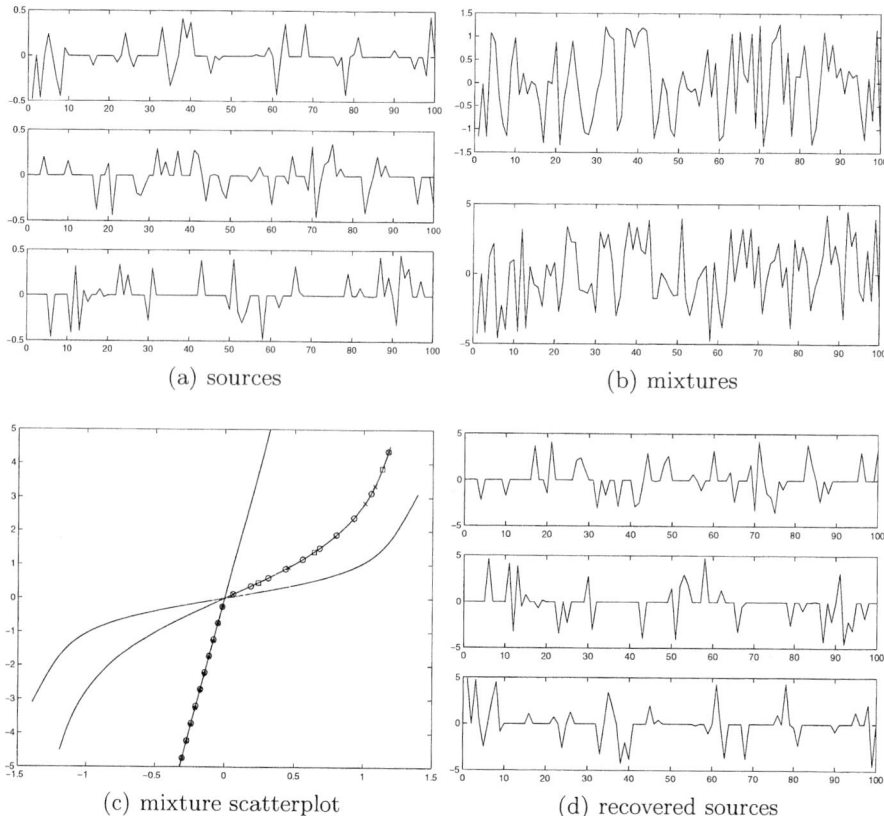

Fig. 2. Example: (a) shows the 1-sparse source signals, and (b) the postnonlinear overcomplete mixtures. The original source directions can be clearly seen in the structure of the mixture scatterplot (c). The crosses and stars indicate the found interpolation points used for approximating the separating nonlinearities, generated by geometrical preprocessing. Now, according to theorem 3, the sources can be recovered uniquely, figure (d), except for permutation and scaling.

calculated from the Euclidean gradient of g. For the learning process, we further note that all weights $w_i^{(j)}$ should be kept nonnegative in order to ensure invertibility of $\tilde{g}$.

In order to increase convergence speed, the Euclidean gradient of g should be replaced by the *natural gradient* [11], which in experiments enhances the algorithm performance in terms of speed by a factor of roughly 10.

4 Experiment

The postnonlinear mixture of three sources to two mixtures is considered. 10^5 samples of artificially generated sources with one non-zero coefficient (drawn uniformly from $[-0.5, 0.5]$) are used. We refer to figure 2 for a plot of the sources, mixtures and recoveries. The sources were mixed using the postnonlinear mixing model $\mathbf{x} = f_1 \times f_2(\mathbf{As})$ with mixing matrix $\mathbf{A} = \begin{pmatrix} 4.3 & 7.8 & 0.59 \\ 9 & 6.2 & 10 \end{pmatrix}$ and postnonlinearities $f_1(x) = \tanh(x) + 0.1x$ and $f_2(x) = x$. For easier algorithm visualization and evaluation we chose f_2 to be linear and did not add any noise.

MLP based postnonlinearity detection algorithm from section 3.3 with natural gradient-descent learning, 9 hidden neurons, a learning rate of $\eta = 0.01$ and 10^5 iterations gives a good approximation of the unmixing nonlinearities g_i. Linear overcomplete SCA is then applied to $g_1 \times g_2(\mathbf{x})$: for practical reasons (due to approximation errors, the data is not fully linearized) instead of the matrix recovery algorithm from [8] we use a modification of the geometric ICA algorithm [4], which is known to work well in the very sparse one-dimensional case to get the recovered mixing matrix $\hat{\mathbf{A}} = \begin{pmatrix} -0.46 & -0.81 & -0.069 \\ -0.89 & -0.58 & -1.0 \end{pmatrix}$, which except for scaling and permutation coincides well with $\mathbf{A}$. Source recovery then gives a (normalized) signal-to-noise ratios (SNRs) of these with the original sources are high with 26, 71 and 46 dB respectively.

References

1. Cichocki, A., Amari, S.: Adaptive blind signal and image processing. John Wiley & Sons (2002)
2. Hyvärinen, A., Karhunen, J., Oja, E.: Independent component analysis. John Wiley & Sons (2001)
3. Lee, T., Lewicki, M., Girolami, M., Sejnowski, T.: Blind source separation of more sources than mixtures using overcomplete representations. IEEE Signal Processing Letters **6** (1999) 87–90
4. Theis, F., Lang, E., Puntonet, C.: A geometric algorithm for overcomplete linear ICA. Neurocomputing **56** (2004) 381–398
5. Zibulevsky, M., Pearlmutter, B.: Blind source separation by sparse decomposition in a signal dictionary. Neural Computation **13** (2001) 863–882
6. Eriksson, J., Koivunen, V.: Identifiability and separability of linear ICA models revisited. In: Proc. of ICA 2003. (2003) 23–27
7. Chen, S., Donoho, D., Saunders, M.: Atomic decomposition by basis pursuit. SIAM J. Sci. Comput. **20** (1998) 33–61
8. Georgiev, P., Theis, F., Cichocki, A.: Blind source separation and sparse component analysis of overcomplete mixtures. In: Proc. of ICASSP 2004, Montreal, Canada (2004)
9. Taleb, A., Jutten, C.: Indeterminacy and identifiability of blind identification. IEEE Transactions on Signal Processing **47** (1999) 2807–2820
10. Babaie-Zadeh, M., Jutten, C., Nayebi, K.: A geometric approach for separating post non-linear mixtures. In: Proc. of EUSIPCO '02. Volume II., Toulouse, France (2002) 11–14
11. Amari, S., Park, H., Fukumizu, K.: Adaptive method of realizing gradient learning for multilayer perceptrons. Neural Computation **12** (2000) 1399–1409

Second-Order Blind Source Separation Based on Multi-dimensional Autocovariances

Fabian J. Theis[1,2], Anke Meyer-Bäse[2], and Elmar W. Lang[1]

[1] Institute of Biophysics
University of Regensburg, D-93040 Regensburg, Germany
[2] Department of Electrical and Computer Engineering
Florida State University, Tallahassee, FL 32310-6046, USA
fabian@theis.name

Abstract. SOBI is a blind source separation algorithm based on time decorrelation. It uses multiple time autocovariance matrices, and performs joint diagonalization thus being more robust than previous time decorrelation algorithms such as AMUSE. We propose an extension called mdSOBI by using multidimensional autocovariances, which can be calculated for data sets with multidimensional parameterizations such as images or fMRI scans. mdSOBI has the advantage of using the spatial data in all directions, whereas SOBI only uses a single direction. These findings are confirmed by simulations and an application to fMRI analysis, where mdSOBI outperforms SOBI considerably.

Blind source separation (BSS) describes the task of recovering the unknown mixing process and the underlying sources of an observed data set. Currently, many BSS algorithm assume independence of the sources (ICA), see for instance [1,2] and references therein. In this work, we consider BSS algorithms based on time-decorrelation. Such algorithms include AMUSE [3] and extensions such as SOBI [4] and the similar TDSEP [5]. These algorithms rely on the fact that the data sets have non-trivial autocorrelations. We give an extension thereof to data sets, which have more than one direction in the parametrization, such as images, by replacing one-dimensional autocovariances by multi-dimensional autocovariances.

The paper is organized as follows: In section 1 we introduce the linear mixture model; the next section 2 recalls results on time decorrelation BSS algorithms. We then define multidimensional autocovariances and use them to propose mdSOBI in section 3. The paper finished with both artificial and real-world results in section 4.

1 Linear BSS

We consider the following *blind source separation* (BSS) problem: Let $\mathbf{x}(t)$ be an (observed) stationary m-dimensional real stochastical process (with not necessarily discrete time t) and $\mathbf{A}$ an invertible real matrix such that

$$\mathbf{x}(t) = \mathbf{A}\mathbf{s}(t) + \mathbf{n}(t) \tag{1}$$

where the source signals $\mathbf{s}(t)$ have diagonal *autocovariances*

$$\mathbf{R}_\mathbf{s}(\tau) := \mathbf{E}\left((\mathbf{s}(t+\tau) - \mathbf{E}(\mathbf{s}(t)))(\mathbf{s}(t) - \mathbf{E}(\mathbf{s}(t)))^\top\right)$$

for all τ, and the additive noise $\mathbf{n}(t)$ is modelled by a stationary, temporally and spatially white zero-mean process with variance σ^2. $\mathbf{x}(t)$ is observed, and the goal is to recover $\mathbf{A}$ and $\mathbf{s}(t)$. Having found $\mathbf{A}$, $\mathbf{s}(t)$ can be estimated by $\mathbf{A}^{-1}\mathbf{x}(t)$, which is optimal in the maximum-likelihood sense (if the density of $\mathbf{n}(t)$ is maximal at 0, which is the case for usual noise models such as Gaussian or Laplacian noise). So the BSS task reduces to the estimation of the mixing matrix $\mathbf{A}$. Extensions of the above model include for example the complex case [4] or the allowance of different dimensions for $\mathbf{s}(t)$ and $\mathbf{x}(t)$, where the case of larger mixing dimension can be easily reduced to the presented complete case by dimension reduction resulting in a lower noise level [6].

By centering the processes, we can assume that $\mathbf{x}(t)$ and hence $\mathbf{s}(t)$ have zero mean. The autocovariances then have the following structure

$$\mathbf{R}_\mathbf{x}(\tau) = \mathbf{E}\left(\mathbf{x}(t+\tau)\mathbf{x}(t)^\top\right) = \begin{cases} \mathbf{A}\mathbf{R}_\mathbf{s}(0)\mathbf{A}^\top + \sigma^2\mathbf{I} & \tau = 0 \\ \mathbf{A}\mathbf{R}_\mathbf{s}(\tau)\mathbf{A}^\top & \tau \neq 0 \end{cases} \quad (2)$$

Clearly, $\mathbf{A}$ (and hence $\mathbf{s}(t)$) can be determined by equation 1 only up to permutation and scaling of columns. Since we assume existing variances of $\mathbf{x}(t)$ and hence $\mathbf{s}(t)$, the scaling indeterminacy can be eliminated by the convention $\mathbf{R}_\mathbf{s}(0) = \mathbf{I}$. In order to guarantee identifiability of $\mathbf{A}$ except for permutation from the above model, we have to additionally assume that there exists a delay τ such that $\mathbf{R}_\mathbf{s}(\tau)$ has pairwise different eigenvalues (for a generalization see [4], theorem 2). Then using the spectral theorem it is easy to see from equation 2 that $\mathbf{A}$ is determined uniquely by $\mathbf{x}(t)$ except for permutation.

2 AMUSE and SOBI

Equation 2 also gives an indication of how to perform BSS i.e. how to recover $\mathbf{A}$ from $\mathbf{x}(t)$. The usual first step consists of whitening the no-noise term $\tilde{\mathbf{x}}(t) := \mathbf{A}\mathbf{s}(t)$ of the observed mixtures $\mathbf{x}(t)$ using an invertible matrix $\mathbf{V}$ such that $\mathbf{V}\tilde{\mathbf{x}}(t)$ has unit covariance. $\mathbf{V}$ can simply be estimated from $\mathbf{x}(t)$ by diagonalization of the symmetric matrix $\mathbf{R}_{\tilde{\mathbf{x}}}(0) = \mathbf{R}_\mathbf{x}(0) - \sigma^2\mathbf{I}$, provided that the noise variance σ^2 is known. If more signals than sources are observed, dimension reduction can be performed in this step, and the noise level can be reduced [6].

In the following without loss of generality, we will therefore assume that $\tilde{\mathbf{x}}(t) = \mathbf{A}\mathbf{s}(t)$ has unit covariance for each t. By assumption, $\mathbf{s}(t)$ also has unit covariance, hence $\mathbf{I} = \mathbf{E}\left(\mathbf{A}\mathbf{s}(t)\mathbf{s}(t)^\top\mathbf{A}^\top\right) = \mathbf{A}\mathbf{R}_\mathbf{s}(0)\mathbf{A}^\top = \mathbf{A}\mathbf{A}^\top$ so $\mathbf{A}$ is orthogonal. Now define the *symmetrized autocovariance* of $\mathbf{x}(t)$ as $\bar{\mathbf{R}}_\mathbf{x}(\tau) := \frac{1}{2}\left(\mathbf{R}_\mathbf{x}(\tau) + (\mathbf{R}_\mathbf{x}(\tau))^\top\right)$. Equation 2 shows that also the symmetrized autocovariance $\mathbf{x}(t)$ factors, and we get

$$\bar{\mathbf{R}}_\mathbf{x}(\tau) = \mathbf{A}\bar{\mathbf{R}}_\mathbf{s}(\tau)\mathbf{A}^\top \quad (3)$$

for $\tau \neq 0$. By assumption $\bar{\mathbf{R}}_\mathbf{s}(\tau)$ is diagonal, so equation 3 is an eigenvalue decomposition of the symmetric matrix $\bar{\mathbf{R}}_\mathbf{x}(\tau)$. If we furthermore assume that $\bar{\mathbf{R}}_\mathbf{x}(\tau)$ or equivalently $\bar{\mathbf{R}}_\mathbf{s}(\tau)$ has n different eigenvalues, then the above decomposition i.e. $\mathbf{A}$ is uniquely determined by $\bar{\mathbf{R}}_\mathbf{x}(\tau)$ except for orthogonal transformation in each eigenspace and permutation; since the eigenspaces are one-dimensional this means $\mathbf{A}$ is uniquely determined by equation 3 except for permutation. In addition to this separability result, $\mathbf{A}$ can be recovered algorithmically by simply calculating the eigenvalue decomposition of $\bar{\mathbf{R}}_\mathbf{x}(\tau)$ (AMUSE, [3]).

In practice, if the eigenvalue decomposition is problematic, a different choice of τ often resolves this problem. Nontheless, there are sources in which some components have equal autocovariances. Also, due to the fact that the autocovariance matrices are only estimated by a finite amount of samples, and due to possible colored noise, the autocovariance at τ could be badly estimated. A more general BSS algorithm called SOBI (second-order blind identification) based on time decorrelation was therefore proposed by Belouchrani et al. [4]. In addition to only diagonalizing a single autocovariance matrix, it takes a whole set of autocovariance matrices of $\mathbf{x}(t)$ with varying time lags τ and jointly diagonalizes the whole set. It has been shown that increasing the size of this set improves SOBI performance in noisy settings [1].

Algorithms for performing joint diagonalization of a set of symmetric commuting matrices include gradient descent on the sum of the off-diagonal terms, iterative construction of $\mathbf{A}$ by Givens rotation in two coordinates [7] (used in the simulations in section 4), an iterative two-step recovery of $\mathbf{A}$ [8] or more recently a linear least-squares algorithm for diagonalization [9], where the latter two algorithms can also search for non-orthogonal matrices $\mathbf{A}$. Joint diagonalization has been used in BSS using cumulant matrices [10] or time autocovariances [4,5].

3 Multidimensional SOBI

The goal of this work is to improve SOBI performance for random processes with a higher dimensional parametrization i.e. for data sets where the random processes $\mathbf{s}$ and $\mathbf{x}$ do not depend on a single variable t, but on multiple variables $(z_1, \ldots, z_M)$. A typical example is a source data set, in which each component s_i represents an image of size $h \times w$. Then $M = 2$ and samples of $\mathbf{s}$ are given at $z_1 = 1, \ldots, h$, $z_2 = 1, \ldots, w$. Classically, $\mathbf{s}(z_1, z_2)$ is transformed to $\mathbf{s}(t)$ by fixing a mapping from the two-dimensional parameter set to the one-dimensional time parametrization of $\mathbf{s}(t)$, for example by concatenating columns or rows in the case of a finite number of samples. If the time structure of $\mathbf{s}(t)$ is not used, as in all classical ICA algorithms in which i.i.d. samples are assumed, this choice does not influence the result. However, in time-structure based algorithms such as AMUSE and SOBI results can vary greatly depending on the choice of this mapping, see figure 2.

Without loss of generality we again assume centered random vectors. Then define the *multidimensional covariance* to be

$$\mathbf{R}_\mathbf{s}(\tau_1, \ldots, \tau_M) := \mathbf{E}\left(\mathbf{s}(z_1 + \tau_1, \ldots, z_M + \tau_M)\mathbf{s}(z_1, \ldots, z_M)^\top\right)$$

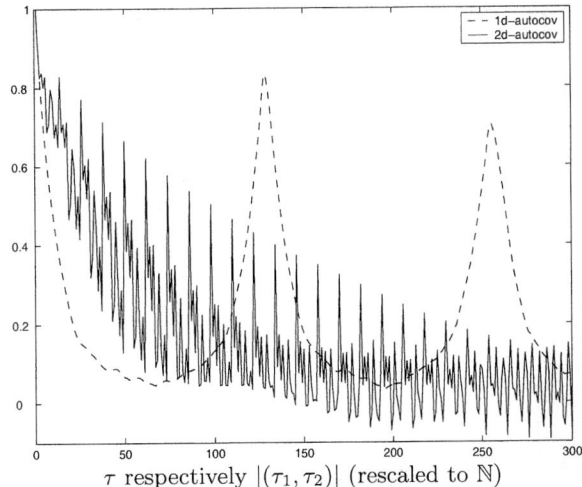

τ respectively $|(\tau_1, \tau_2)|$ (rescaled to $\mathbb{N}$)

Fig. 1. Example of one- and two-dimensional autocovariance coefficient of the grayscale 128×128 Lena image after normalization to variance 1.

where the expectation is taken over $(z_1, \ldots, z_M)$. $\mathbf{R_s}(\tau_1, \ldots, \tau_M)$ can be estimated given equidistant samples by replacing random variables by sample values and expectations by sums as usual.

The advantage of using multidimensional autocovariances lies in the fact that now the multidimensional structure of the data set can be used more explicitly. For example, if row concatenation is used to construct $\mathbf{s}(t)$ from the images, horizontal lines in the image will only give trivial contributions to the autocovariance (see examples in figure 2 and section 4). Figure 1 shows the one- and two-dimensional autocovariance of the Lena image for varying τ respectively (τ_1, τ_2) after normalization of the image to variance 1. Clearly, the two-dimensional autocovariance does not decay as quickly with increasing radius as the one-dimensional covariance. Only at multiples of the image height, the one-dimensional autocovariance is significantly high i.e. captures image structure.

Our contribution consists of using multidimensional autocovariances for joint diagonalization. We replace the BSS assumption of diagonal one-dimensional autocovariances by diagonal multi-dimensional autocovariances of the sources. Note that also the multidimensional covariance satisfies the equation 2. Again we assume whitened $\mathbf{x}(z_1, \ldots, z_K)$. Given a autocovariance matrix $\bar{\mathbf{R}}_{\mathbf{x}} \left(\tau_1^{(1)}, \ldots, \tau_M^{(1)} \right)$ with n different eigenvalues, multidimensional AMUSE ($mdAMUSE$) detects the orthogonal unmixing mapping $\mathbf{W}$ by diagonalization of this matrix.

In section 2, we discussed the advantages of using SOBI over AMUSE. This of course also holds in this generalized case. Hence, the multidimensional SOBI algorithm ($mdSOBI$) consists of the joint diagonalization of a set of symmetrized multidimensional autocovariances

$$\left\{ \bar{\mathbf{R}}_{\mathbf{x}} \left(\tau_1^{(1)}, \ldots, \tau_M^{(1)} \right), \ldots, \bar{\mathbf{R}}_{\mathbf{x}} \left(\tau_1^{(K)}, \ldots, \tau_M^{(K)} \right) \right\}$$

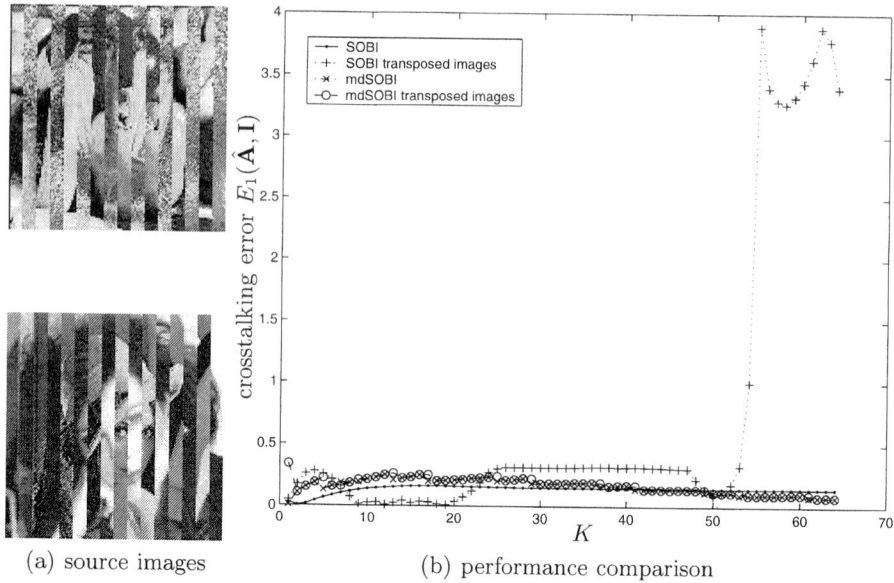

(a) source images (b) performance comparison

Fig. 2. Comparison of SOBI and mdSOBI when applied to (unmixed) images from (a). The plot (b) plots the number K of time lags versus the crosstalking error E_1 of the recovered matrix $\hat{\mathbf{A}}$ and the unit matrix $\mathbf{I}$; here $\hat{\mathbf{A}}$ has been recovered by bot SOBI and mdSOBI given the images from (a) respectively the transposed images.

after whitening of $\mathbf{x}(z_1, \ldots, z_K)$. The joint diagonalizer then equals $\mathbf{A}$ except for permutation, given the generalized identifiability conditions from [4], theorem 2. Therefore, also the identifiability result does not change, see [4]. In practice, we choose the $(\tau_1^{(k)}, \ldots, \tau_M^{(k)})$ with increasing modulus for increasing k, but with the restriction $\tau_1^{(k)} > 0$ in order to avoid using the same autocovariances on the diagonal of the matrix twice.

Often, data sets do not have any substantial long-distance autocorrelations, but quite high multi-dimensional close-distance correlations (see figure 1). When performing joint diagonalization, SOBI weighs each matrix equally strong, which can deteriorate the performance for large K, see simulation in section 4.

Figure 2(a) shows an example, in which the images have considerable vertical structure, but rather random horizontal structure. Each of the two images consists of a concatenation of stripes of two images. For visual purposes, we chose the width of the stripes to be rather large with 16 pixels. According to the previous discussion we expect one-dimensional algorithms such as AMUSE and SOBI to perform well on the images, but badly (for number of time lags $\gg 16$) on the transposed images. If we apply AMUSE with $\tau = 20$ to the images, we get excellent performance with a low crosstalking error with the unit matrix of 0.084; if we however apply AMUSE to the transposed images, the error is high with 1.1. This result is further confirmed by the comparison plot in figure 2(b); mdSOBI performs equally well on the images and the transposed

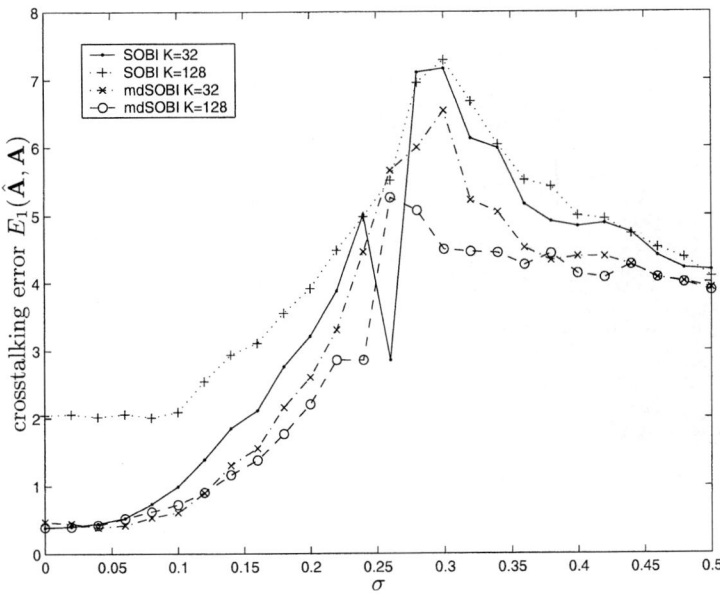

Fig. 3. SOBI and mdSOBI performance dependence on noise level σ. Plotted is the crosstalking error E_1 of the recovered matrix $\hat{\mathbf{A}}$ with the real mixing matrix $\mathbf{A}$. See text for more details.

images, whereas performance of SOBI strongly depends on whether column or row concatenation was used to construct a one-dimensional random process out of each image. The SOBI breakpoint of around $K = 52$ can be decreased by choosing smaller stripes. In future works we want to provide an analytical discussion of performance increase when comparing SOBI and mdSOBI similar to the performance evaluation in [4].

4 Results

Artificial Mixtures. We consider the linear mixture of three images (baboon, black-haired lady and Lena) with a randomly chosen 3×3 matrix $\mathbf{A}$. Figure 3 shows how SOBI and mdSOBI perform depending on the noise level σ. For small K, both SOBI and mdSOBI perform equally well in the low noise case, but mdSOBI performs better in the case of stronger noise. For larger K mdSOBI substantially outperforms SOBI, which is due to the fact that natural images do not have any substantial long-distance autocorrelations (see figure 1), whereas mdSOBI uses the non-trivial two-dimensional autocorrelations.

fMRI Analysis. We analyze the performance of mdSOBI when applied to fMRI measurements. fMRI data were recorded from six subjects (3 female, 3 male, age 20–37) performing a visual task. In five subjects, five slices with 100

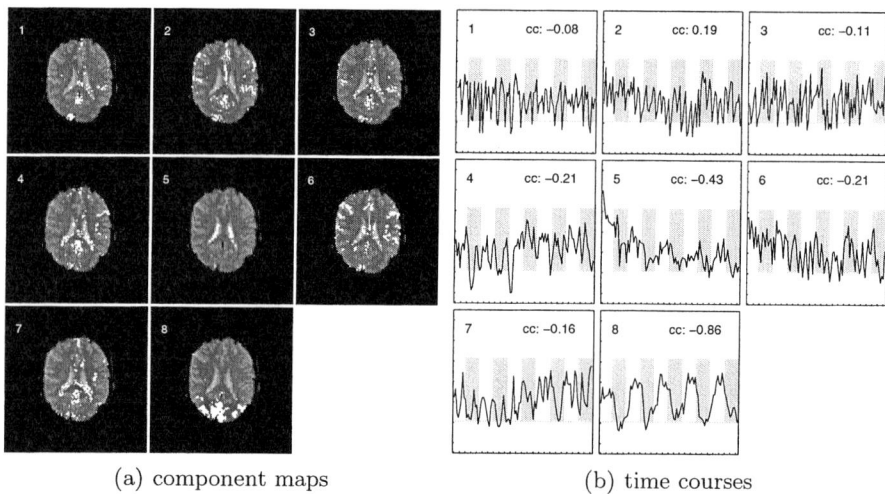

(a) component maps   (b) time courses

Fig. 4. mdSOBI fMRI analysis. The data was reduced to the first 8 principal components. (a) shows the recovered component maps (white points indicate values stronger than 3 standard deviations), and (b) their time courses. mdSOBI was performed with $K = 32$. Component 5 represents inner ventricles, component 6 the frontal eye fields. Component 8 is the desired stimulus component, which is mainly active in the visual cortex; its time-course closely follows the on-off stimulus (indicated by the gray boxes) – their crosscorrelation lies at $cc = -0.86$ – with a delay of roughly 2 seconds induced by the BOLD effect.

images (TR/TE = 3000/60 msec) were acquired with five periods of rest and five photic simulation periods with rest. Simulation and rest periods comprised 10 repetitions each, i.e. 30s. Resolution was $3 \times 3 \times 4$ mm. The slices were oriented parallel to the calcarine fissure. Photic stimulation was performed using an 8 Hz alternating checkerboard stimulus with a central fixation point and a dark background with a central fixation point during the control periods. The first scans were discarded for remaining saturation effects. Motion artifacts were compensated by automatic image alignment (AIR, [11]).

BSS, mainly based on ICA, nowadays is a quite common tool in fMRI analysis (see for example [12]). Here, we analyze the fMRI data set using spatial decorrelation as separation criterion. Figure 4 shows the performance of mdSOBI; see figure text for interpretation. Using only the first 8 principal components, mdSOBI could recover the stimulus component as well as detect additional components. When applying SOBI to the data set, it could not properly detect the stimulus component but found two components with crosscorrelations $cc = -0.81$ and -0.84 with the stimulus time course.

5 Conclusion

We have proposed an extension called mdSOBI of SOBI for data sets with multidimensional parametrizations, such as images. Our main contribution lies in replacing the one-dimensional autocovariances by multi-dimensional autocovariances. In both simulations and real-world applications mdSOBI outperforms SOBI for these multidimensional structures.

In future work, we will show how to perform spatiotemporal BSS by jointly diagonalizing both spatial and time autocovariance matrices. We plan on applying these results to fMRI analysis, where we also want to use three-dimensional autocovariances for 3d-scans of the whole brain.

Acknowledgements

The authors would like to thank Dr. Dorothee Auer from the Max Planck Institute of Psychiatry in Munich, Germany, for providing the fMRI data, and Oliver Lange from the Department of Clinical Radiology, Ludwig-Maximilian University, Munich, Germany, for data preprocessing and visualization. FT and EL acknowledge partial financial support by the BMBF in the project 'ModKog'.

References

1. Cichocki, A., Amari, S.: Adaptive blind signal and image processing. John Wiley & Sons (2002)
2. Hyvärinen, A., Karhunen, J., Oja, E.: Independent component analysis. John Wiley & Sons (2001)
3. Tong, L., Liu, R.W., Soon, V., Huang, Y.F.: Indeterminacy and identifiability of blind identification. IEEE Transactions on Circuits and Systems **38** (1991) 499–509
4. Belouchrani, A., Meraim, K.A., Cardoso, J.F., Moulines, E.: A blind source separation technique based on second order statistics. IEEE Transactions on Signal Processing **45** (1997) 434–444
5. Ziehe, A., Mueller, K.R.: TDSEP – an efficient algorithm for blind separation using time structure. In Niklasson, L., Bodén, M., Ziemke, T., eds.: Proc. of ICANN'98, Skövde, Sweden, Springer Verlag, Berlin (1998) 675–680
6. Joho, M., Mathis, H., Lamber, R.: Overdetermined blind source separation: using more sensors than source signals in a noisy mixture. In: Proc. of ICA 2000, Helsinki, Finland (2000) 81–86
7. Cardoso, J.F., Souloumiac, A.: Jacobi angles for simultaneous diagonalization. SIAM J. Mat. Anal. Appl. **17** (1995) 161–164
8. Yeredor, A.: Non-orthogonal joint diagonalization in the leastsquares sense with application in blind source separation. IEEE Trans. Signal Processing **50** (2002) 1545–1553
9. Ziehe, A., Laskov, P., Mueller, K.R., Nolte, G.: A linear least-squares algorithm for joint diagonalization. In: Proc. of ICA 2003, Nara, Japan (2003) 469–474
10. Cardoso, J.F., Souloumiac, A.: Blind beamforming for non gaussian signals. IEE Proceedings - F **140** (1993) 362–370
11. Woods, R., Cherry, S., Mazziotta, J.: Rapid automated algorithm for aligning and reslicing pet images. Journal of Computer Assisted Tomography **16** (1992) 620–633
12. McKeown, M., Jung, T., Makeig, S., Brown, G., Kindermann, S., Bell, A., Sejnowksi, T.: Analysis of fMRI data by blind separation into independent spatial components. Human Brain Mapping **6** (1998) 160–188

Separating a Real-Life Nonlinear Mixture of Images

Luís B. Almeida and Miguel Faria

INESC ID and IST, R. Alves Redol, 9, 1000-029 Lisboa, Portugal
luis.almeida@inesc-id.pt
http://neural.inesc-id.pt/~lba

Abstract. This manuscript presents results obtained using an ICA technique in a real-life nonlinear image separation problem: the separation of the images of the two pages of a paper document when the image from the back page shows through, superimposed on the image of the front page. For this manuscript, two images were printed on opposite sides of a sheet of onion skin paper, and then both sides of the sheet were scanned. The scanned images contained a markedly nonlinear mixture of the original images. Nonlinear ICA, using the MISEP technique, was used to recover the original images. It showed to be able to achieve a reasonable, but not perfect separation. The best results were obtained with a separating system which was somewhat customized, based on prior knowledge about the mixture process, and which used explicit regularization.

1 Introduction

When scanning or photographing a paper document, the image of the back page sometimes shows through. This is normally due to partial transparency of the paper or to bleeding of the ink through the paper. In either case, the image that is acquired consists of a mixture of the original images contained in each of the pages, and it would be of interest to be able to eliminate the superposition and recover the original images. Since it is possible to acquire both sides of the document, two different mixtures of the original images can be obtained, and therefore ICA is a natural candidate for source separation. Often, however, the mixture is substantially nonlinear, and linear ICA techniques are not adequate. This constitutes, therefore, an interesting test case for nonlinear ICA methods.

We decided to implement a difficult version of this problem, using a relatively transparent paper ("onion skin"), resulting in a mixture that is both strong and significantly nonlinear. We show separation results obtained with (1) linear ICA, (2) nonlinear ICA (MISEP method) with the basic separating structure, and (3) nonlinear ICA (MISEP) using a customized separating structure that incorporates some knowledge about the mixing process. The latter results correspond to the best separation, which is still not perfect, leaving room for improvement.

There are still very few published results of source separation on nonlinear mixtures of real-life data. An example is [1]. Other applications to real-life data, e.g. [2,3], do not provide any means to assess whether the extracted components correspond to actual sources.

Fig. 1. Photographs used for the second test problem. The right hand photograph has been horizontally flipped, to correspond with the position in which it appears in the mixture and separation images.

2 Source Images, Printing, Acquisition and Preprocessing

We present separation results on two test problems. For each problem we printed two gray scale images on opposite pages of onion skin paper. The images of the first problem were artificially generated, each of them consisting of parallel bars with randomly chosen gray levels. In one of the images the bars were oriented horizontally and in the other they were oriented vertically. We don't present those images here to save space, but the results shown ahead clarify the images' contents. The second pair of images consisted of the photos shown in Fig. 1.

We used a monochrome laser printer at a resolution of 1200 dpi, with the printer's default halftoning system. Both sides of the onion skin paper were scanned in monochrome mode with a resolution of 100 dpi. A low resolution was purposely selected for scanning, so that the printer's halftoning grid would not be strongly noticeable in the scanned images. The inner face of the scanner's cover (facing the back page of the onion skin paper) was white, originating a strong mixture of the contents of both pages in the acquired images.

After acquisition one of the images was horizontally flipped, to make the orientations of both images match. The two images were then coarsely aligned by hand, using alignment marks printed together with the images. It was found that even a careful alignment based just on those marks could not properly align all parts of the images, probably due to some slight geometrical distortions introduced by the scanner. Therefore, after the coarse manual alignment an automatic alignment procedure was run. For this purpose the images were first increased in resolution by a factor of four in each direction (using bicubic interpolation) so that the alignment could be made with a precision of $1/4$ pixel. The alignment procedure operated on 100×100 pixel squares (corresponding to 25×25 pixel squares in the scanned images), and was based on finding the maximum of the local correlation between both images. After the automatic alignment the images were brought back to their original resolution. The preprocessing was completed

Fig. 2. Mixture components (preprocessed acquired images) in the 'bars' and 'photos' problems.

by scaling the intensity range of each image to the interval [0,1]. Figure 2 shows the mixture components after preprocessing.

3 Outline of the Separation Method

All separation tests, both linear and nonlinear, were based on the MISEP method. A detailed description of the method is given in [4]. A brief outline is given here, to clarify the main concepts and the nomenclature. The method is an extension of the well known INFOMAX method [5], and is based on the minimization of the mutual information of the extracted components. The structure of the system that is used for separation is shown in Fig. 3. The observations o_i enter the **F** block, which performs the separation, yielding the separated components y_i. The ψ_i blocks are used only during the training phase, and yield the auxiliary outputs z_i. In INFOMAX the ψ_i blocks implement nonlinearities which are fixed a priori. In MISEP these blocks are adaptive, learning nonlinearities suited to the components' statistical distributions. In the linear mode, MISEP uses a linear **F** block, and corresponds to INFOMAX with adaptive output nonlinearities. In the nonlinear mode, the **F** block is a nonlinear parameterized system (a multilayer perceptron – MLP – in our case). The whole system is trained by maximizing the joint entropy of the auxiliary outputs z_i. This results both in

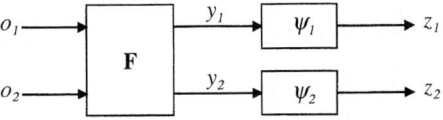

Fig. 3. Structure of the system used for separation.

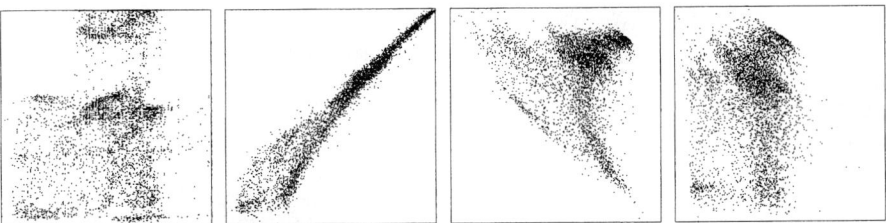

Fig. 4. Scatter plots (left to right): original photos, before printing; mixture; linear separation; nonlinear separation through the improved method.

the adaptation of the ψ_i blocks to the distributions of the extracted components and in the minimization of the mutual information of those components.

4 Results

All tests used a training set of 5000 points randomly selected from the preprocessed images. The ψ_i blocks were implemented with MLPs with one hidden layer of 10 sigmoidal units each.

Figure 4 shows several scatter plots corresponding to the photos problem. The leftmost plot corresponds to the two original images, before printing, and shows that the images were not completely independent. The next plot corresponds to the preprocessed acquired images, and shows that the mixture was nonlinear: a linear mixture would correspond to a 'parallelogram' distortion of the original distribution. This scatter plot also shows that the mixture was almost singular in the lighter parts of the images. The remaining plots correspond to separation results, and are discussed ahead.

4.1 Linear Separation

Linear separation was used as a standard against which to compare the results of nonlinear separation. The linear separation system used the MISEP method as described in [4]. The **F** block was linear, performing just a product by the separation matrix. The separation results are shown in Fig. 5[1]. We can see that

---

[1] All separated images were subject to a normalization of the intensity histogram before printing, to compensate for the nonlinearities that are sometimes introduced by nonlinear ICA [4]. This facilitates the comparison of separation results.

Fig. 5. Linear separation results.

only a partial separation was achieved, as expected. The third plot from the left, in Fig. 4, corresponds to the linearly separated components. It shows, again, that the mixture was nonlinear. A linear method can't separate it completely.

4.2 Basic Nonlinear Separation

The first set of nonlinear separation tests used a separating system similar to those used in [4]. Block **F** consisted of an MLP with one hidden layer of sigmoidal units, with linear output units and with direct "shortcut" connections between inputs and outputs. The hidden layer had 20 units, 10 of which were connected to each of the output units. The separations shown both in this and in the next section were obtained with 1000 training epochs, corresponding to about 15 minutes in a 1.6 GHz Centrino processor programmed in Matlab.

The results that were obtained had a relatively large variability, sometimes being better than those of linear separation, and sometimes worse. Figure 6 shows two "extremes" of the range os results that were obtained, for the component that had the largest variability in the photos problem. Somewhat infrequently (in about 10% of the tests) the system yielded results much outside this range. Normally the source images were then strongly mixed, in the extracted components. This variability is probably related to the ill-posedness of nonlinear ICA [6]. Attempts to make the outputs more stable by means of explicit regular-

Fig. 6. Two "extreme" results of nonlinear separation with the basic method. The same extracted source is shown in both images, the difference being only in the random initialization of network weights.

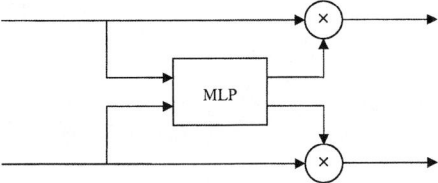

Fig. 7. Structure of the **F** block in the improved nonlinear method.

ization through weight decay didn't show much success: a weak regularization would not constrain the system enough, and a stronger regularization would make it essentially equivalent to a linear separator.

4.3 Improved Nonlinear Separation

The results of the basic nonlinear separation tests led us to try to incorporate more prior information in the separation system. The shape of the mixture distribution, together with a qualitative knowledge of the physical mixing process, led us to hypothesize that an **F** block with the structure shown in Fig. 7 would yield a more stable separation. In this structure each output is obtained by multiplying the corresponding input by a variable gain. The gains are computed by the MLP. These gains should be rather smooth functions, which we expected to be able to adequately constrain through regularization. The MLP that we used had a hidden layer of 20 sigmoidal units, linear output units and no direct connections from inputs to outputs. Ten of the hidden layer's units were connected to each of the output units.

The separation results produced by this network, with adequate regularization through weight decay, were significantly more stable than those form the basic nonlinear method (although, as with the basic method, the system pro-

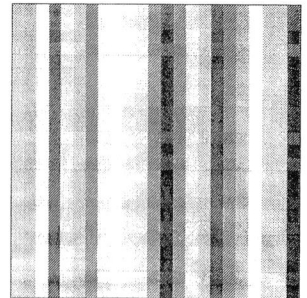
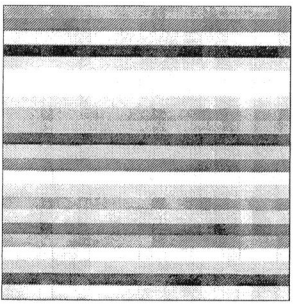

Fig. 8. Nonlinear separation of the bars images with the improved nonlinear method.

Fig. 9. Nonlinear separation of the photographs with the improved nonlinear method. The upper images correspond to "best" results, and the lower ones to "worst" ones.

duced rather wrong results in about 10% of the tests). Figures 8 and 9 show results for the two problems, obtained by training the system with the same set of parameters. Figure 9 also gives an idea of the degree of variability of the results on the photos problem. These results were somewhat better than those obtained with linear separation. The rightmost plot of Fig. 4 confirms that a better degree of independence was achieved, than with linear ICA. This plot also suggests, however, that the non-independence of the original images had somewhat a negative impact on the separation: in an attempt to "fill" the upper left corner of the distribution, the system tilted the upper part of the distribution somewhat, resulting in some amount of mixing in the extracted components.

Both in the linear and nonlinear methods, the separation results show an amount of noise that is significantly higher than that of the mixture components. This may not be easily noticeable in the printed version of this paper, but should be visible to readers of the electronic version, by zooming in on the pictures. This noise probably comes mostly from a residual effect of the printer's halftoning process and from slight inhomogeneities of the onion skin paper. The noise was then amplified by the inversion of the quasi-singular mixture.

5 Conclusion

We have shown results of separation of a real-life nonlinear mixture of images. The results show that nonlinear ICA can outperform linear ICA in this problem, but they also shows that the ill-posedness of nonlinear ICA makes stabilization of the separation somewhat difficult.

Two main directions for improvement are envisaged, based on these results. On the one hand, it is desirable to develop better methods to stabilize the ICA results, possibly by making use of more prior information. On the other hand, it is of interest to develop separation criteria that are better suited to this specific problem than independence. These may yield better results, both in terms of quality and of stability, and may make the separation less affected by the possible statistical dependence of the original images.

References

1. Haritopoulos, M., Yin, H., Allinson, N.: Image denoising using SOM-based nonlinear independent component analysis. Neural Networks **15** (2002) 1085–1098
2. Lappalainen, H., Honkela, A.: Bayesian nonlinear independent component analysis by multi-layer perceptrons. In Girolami, M., ed.: Advances in Independent Component Analysis, Springer-Verlag (2000) 93–121
3. Lee, S.I., Batzoglou, S.: Application of independent component analysis to microarrays. Genome Biology **4** (2003) R76 http://genomebiology.com/2003/4/11/R76.
4. Almeida, L.B.: MISEP – Linear and nonlinear ICA based on mutual information. Journal of Machine Learning Research **4** (2003) 1297–1318 http://www.jmlr.org/papers/volume4/almeida03a/almeida03a.pdf.
5. Bell, A., Sejnowski, T.: An information-maximization approach to blind separation and blind deconvolution. Neural Computation **7** (1995) 1129–1159
6. Hyvarinen, A., Pajunen, P.: Nonlinear independent component analysis: Existence and uniqueness results. Neural Networks **12** (1999) 429–439

Independent Slow Feature Analysis and Nonlinear Blind Source Separation*

Tobias Blaschke and Laurenz Wiskott

Institute for Theoretical Biology, Humboldt University Berlin
Invalidenstraße 43, D-10115 Berlin, Germany
{t.blaschke,l.wiskott}@biologie.hu-berlin.de
http://itb.biologie.hu-berlin.de/{~blaschke,~wiskott}

Abstract. We present independent slow feature analysis as a new method for nonlinear blind source separation. It circumvents the indeterminacy of nonlinear independent component analysis by combining the objectives of statistical independence and temporal slowness. The principle of temporal slowness is adopted from slow feature analysis, an unsupervised method to extract slowly varying features from a given observed vectorial signal. The performance of the algorithm is demonstrated on nonlinearly mixed speech data.

1 Introduction

Unlike in the linear case the nonlinear Blind Source Separation (BSS) problem can not be solved solely based on the principle of statistical independence [Hyvärinen and Pajunen, 1999; Jutten and Karhunen, 2003]. Performing nonlinear BSS with Independent Component Analysis (ICA) requires additional information about the underlying sources or to regularize the nonlinearities. Since source signal components are usually more slowly varying than any nonlinear mixture of them we consider to require the estimated sources to be as slowly varying as possible. This can be achieved by incorporating ideas from Slow Feature Analysis (SFA) [Wiskott and Sejnowski, 2002] into ICA.

After a short introduction to linear BSS, nonlinear BSS and SFA we will show a way how to combine SFA and ICA to obtain an algorithm that solves the nonlinear BSS problem.

2 Linear Blind Source Separation

Let $\mathbf{x}(t) = [x_1(t), \ldots, x_N(t)]^T$ be a linear mixture of a source signal $\mathbf{s}(t) = [s_1(t), \ldots, s_N(t)]^T$ and defined by

$$\mathbf{x}(t) = \mathbf{As}(t), \qquad (1)$$

with an invertible $N \times N$ mixing matrix $\mathbf{A}$. Finding a mapping

$$\mathbf{u}(t) = \mathbf{QWx}(t), \qquad (2)$$

---

* This work has been supported by the Volkswagen Foundation through a grant to LW for a junior research group.

such that the components of **u** are mutually statistically independent is called Independent Component Analysis (ICA). The mapping is often divided into a whitening mapping **W**, resulting in uncorrelated signal components y_i with unit variance, and a successive orthogonal transformation **Q**, because one can show [Comon, 1994] that after whitening an orthogonal transformation is sufficient to obtain independence. It is well known that ICA solves the linear BSS problem [Comon, 1994]. There exists a variety of algorithms performing ICA and therefore BSS (see e.g. [Cardoso and Souloumiac, 1993; Lee et al., 1999; Hyvärinen, 1999]). Here we focus on a method using only second-order statistics introduced by Molgedey and Schuster [1994]. The method consists of optimizing an objective function, subject to minimization, which can be written as

$$\Psi_{ICA}(\mathbf{Q}) = \sum_{\substack{\alpha,\beta=1\\\alpha\neq\beta}}^{N} \left(C_{\alpha\beta}^{(\mathbf{u})}(\tau)\right)^2 = \sum_{\substack{\alpha,\beta=1\\\alpha\neq\beta}}^{N} \left(\sum_{\gamma,\delta=1}^{N} Q_{\alpha\gamma} Q_{\beta\delta} C_{\gamma\delta}^{(\mathbf{y})}(\tau)\right)^2, \qquad (3)$$

operating on the already whitened signal **y**. $C_{\gamma\delta}^{(\mathbf{y})}(\tau)$ is an entry of a symmetrized time delayed covariance matrix defined by

$$\mathbf{C}^{(\mathbf{y})}(\tau) = \left\langle \mathbf{y}(t)\mathbf{y}(t+\tau)^T + \mathbf{y}(t+\tau)\mathbf{y}(t)^T \right\rangle, \qquad (4)$$

and $\mathbf{C}^{(\mathbf{u})}(\tau)$ is defined correspondingly. $Q_{\alpha\beta}$ denotes an entry of **Q**. Minimization of Ψ_{ICA} can be understood intuitively as finding an orthogonal matrix **Q** that diagonalizes the covariance matrix with time delay τ. Since, because of the whitening, the instantaneous covariance matrix is already diagonal this results in signal components that are decorrelated instantaneously and at a given time delay τ. This can be sufficient to achieve statistical independence [Tong et al., 1991].

2.1 Nonlinear BSS and ICA

An obvious extension to the linear mixing model (1) has the form

$$\mathbf{x}(t) = F(\mathbf{s}(t)), \qquad (5)$$

with a function $F(\cdot) \; \mathbb{R}^N \to \mathbb{R}^M$, that maps N-dimensional source vectors **s** onto M-dimensional signal vectors **x**. The components x_i of the observable are a nonlinear mixture of the sources and like in the linear case source signal components s_i are assumed to be mutually statistically independent. Unmixing is in general only possible if $F(\cdot)$ is an invertible function, which we will assume from now on.

The equivalence of BSS and ICA in the linear case does in general not hold for a nonlinear function $F(\cdot)$ [Hyvärinen and Pajunen, 1999; Jutten and Karhunen, 2003]. To solve the nonlinear BSS problem additional constraints on the mixture or the estimated signals are needed to bridge the gap between ICA and BSS. Here we propose a new way to achieve this by adding a slowness objective to the independence objective of pure ICA. Assume for example a sinusoidal signal component $x_i = \sin(2\pi t)$ and a second component that is the square of the first $x_j = x_i^2 = 0.5(1 - \cos(4\pi t))$ is given. The second component is more quickly varying due to the frequency doubling induced by the squaring. Typically nonlinear mixtures of signal components are more quickly varying than the original components. To extract the right source components one should therefore prefer the slowly varying ones. The concept of slowness is used in our approach to nonlinear BSS by combining an ICA part that provides the independence of the estimated source signal components with a part that prefers slowly varying signals over more quickly varying ones. In the next section we will give a short introduction to Slow Feature Analysis building the basis of the second part of our method.

3 Slow Feature Analysis

Assume a vectorial input signal $\mathbf{x}(t) = [x_1(t), \ldots, x_M(t)]^T$ is given. The objective of SFA is to find an in general nonlinear input-output function $\mathbf{u}(t) = \mathbf{g}(\mathbf{x}(t))$ with $\mathbf{g}(\mathbf{x}(t)) = [g_1(\mathbf{x}(t)), \ldots, g_R(\mathbf{x}(t))]^T$ such that the $u_i(t)$ are varying as slowly as possible. This can be achieved by successively minimizing the objective function

$$\Delta(u_i) := \langle \dot{u}_i^2 \rangle, \qquad (6)$$

for each u_i under the constraints

$$\langle u_i \rangle = 0 \quad \text{(zero mean)}, \qquad (7)$$
$$\langle u_i^2 \rangle = 1 \quad \text{(unit variance)}, \qquad (8)$$
$$\langle u_i u_j \rangle = 0 \ \forall j < i \quad \text{(decorrelation and order)}. \qquad (9)$$

Constraints (7) and (8) ensure that the solution will not be the trivial solution $u_i = \text{const}$. Constraint (9) provides uncorrelated output signal components and thus guarantees that different components carry different information. Intuitively we are searching for signal components u_i that have on average a small slope.

Interestingly Slow Feature Analysis (SFA) can be reformulated with an objective function similar to second-order ICA, subject to maximization [Blaschke et al., 2004],

$$\Psi_{\text{SFA}}(\mathbf{Q}) = \sum_{\alpha=1}^{M} \left(C_{\alpha\alpha}^{(\mathbf{u})}(\tau) \right)^2 = \sum_{\alpha=1}^{M} \left(\sum_{\beta,\gamma=1}^{M} Q_{\alpha\beta} Q_{\alpha\gamma} C_{\beta\gamma}^{(\mathbf{y})}(\tau) \right)^2. \qquad (10)$$

To understand (10) intuitively we notice that slowly varying signal components are easier to predict, and should therefore have strong auto correlations in time. Thus, maximizing the time delayed variances produces slowly varying signal components.

4 Independent Slow Feature Analysis

If we combine ICA and SFA we obtain a method, we refer to as Independent Slow Feature Analysis (ISFA), that recovers independent components out of a nonlinear mixture using a combination of SFA and second-order ICA. As already explained, second-order ICA tends to make the output components independent and SFA tends to make them slow. Since we are dealing with a nonlinear mixture we first compute a nonlinearly expanded signal $\mathbf{z} = \mathbf{h}(\mathbf{x})$ with $\mathbf{h}(\cdot)$ $\mathbb{R}^M \to \mathbb{R}^L$ being typically monomials up to a given degree, e.g. an expansion with monomials up to second degree can be written as

$$\mathbf{h}(\mathbf{x}(t)) = [x_1, \ldots, x_N, x_1 x_1, x_1 x_2, \ldots, x_M x_M]^T - \mathbf{h}_0^T, \qquad (11)$$

when given an M-dimensional signal $\mathbf{x}$. The constant vector $\mathbf{h}_0^T$ is used to make the expanded signal mean free. In a second step $\mathbf{z}$ is whitened to obtain $\mathbf{y} = \mathbf{W}\mathbf{z}$. Thirdly we apply linear ICA combined with linear SFA on $\mathbf{y}$ in order to find the estimated source signal $\mathbf{u}$. Because of the whitening we know that ISFA, like ICA and SFA, is solved by finding an orthogonal $L \times L$ matrix $\mathbf{Q}$. We write the estimated source signal $\mathbf{u}$ as

$$\mathbf{v} = \begin{pmatrix} \mathbf{u} \\ \tilde{\mathbf{u}} \end{pmatrix} = \mathbf{Q}\mathbf{y} = \mathbf{Q}\mathbf{W}\mathbf{z} = \mathbf{Q}\mathbf{W}\mathbf{h}(\mathbf{x}), \qquad (12)$$

where we introduced $\tilde{\mathbf{u}}$ since R, the dimension of the estimated source signal $\mathbf{u}$, is usually much smaller than L, the dimension of the expanded signal. While the u_i are statistically independent and slowly varying the components $\tilde{u}_i$ are more quickly varying and may be statistically dependent on each other as well as on the selected components.

To summarize, we have an M dimensional input $\mathbf{x}$ an L dimensional nonlinearly expanded and whitened $\mathbf{y}$ and an R dimensional estimated source signal $\mathbf{u}$. ISFA searches an R dimensional subspace such that the u_i are independent and slowly varying. This is achieved at the expense of all $\tilde{u}_i$.

4.1 Objective function

To recover R source signal components u_i $i = 1, \ldots, R$ out of an L-dimensional expanded and whitened signal $\mathbf{y}$ the objective reads

$$\Psi_{\text{ISFA}}(u_1, \ldots, u_R; \tau) = b_{\text{ICA}} \sum_{\substack{\alpha, \beta = 1, \\ \alpha \neq \beta}}^{R} \left(C_{\alpha\beta}^{(\mathbf{u})}(\tau)\right)^2 - b_{\text{SFA}} \sum_{\alpha = 1}^{R} \left(C_{\alpha\alpha}^{(\mathbf{u})}(\tau)\right)^2, \qquad (13)$$

where we simply combine the ICA objective (3) and SFA objective (10) weighted by the factors b_{ICA} and b_{SFA}, respectively. Note that the ICA objective is usually applied to the linear case to unmix the linear whitened mixture $\mathbf{y}$ whereas here it is used on the nonlinearly expanded whitened signal $\mathbf{y} = \mathbf{W}\mathbf{z}$. ISFA tries to minimize Ψ_{ISFA} which is the reason why the SFA part has a negative sign.

4.2 Optimization Procedure

From (12) we know that $\mathbf{C}^{(\mathbf{u})}(\tau)$ in (13) depends on the orthogonal matrix $\mathbf{Q}$. There are several ways to find the orthogonal matrix that minimizes the objective function. Here we apply successive Givens rotations to obtain $\mathbf{Q}$. A Givens rotation $\mathbf{Q}^{\mu\nu}$ is a rotation around the origin within the plane of two selected components μ and ν and has the matrix form

$$Q_{\alpha\beta}^{\mu\nu} := \begin{cases} \cos(\phi) & \text{for } (\alpha, \beta) \in \{(\mu, \mu), (\nu, \nu)\} \\ -\sin(\phi) & \text{for } (\alpha, \beta) \in \{(\mu, \nu)\} \\ \sin(\phi) & \text{for } (\alpha, \beta) \in \{(\nu, \mu)\} \\ \delta_{\alpha\beta} & \text{otherwise} \end{cases} \qquad (14)$$

with Kronecker symbol $\delta_{\alpha\beta}$ and rotation angle ϕ. Any orthogonal $L \times L$ matrix such as $\mathbf{Q}$ can be written as a product of $\frac{L(L-1)}{2}$ (or more) Givens rotation matrices $\mathbf{Q}^{\mu\nu}$ (for the rotation part) and a diagonal matrix with elements ± 1 (for the reflection part). Since reflections do not matter in our case we only consider the Givens rotations as is often used in second-order ICA algorithms (see e.g. [Cardoso and Souloumiac, 1996]).

We can therefore write the objective as a function of a Givens rotation $\mathbf{Q}^{\mu\nu}$ as

$$\Psi_{\text{ISFA}}(\mathbf{Q}^{\mu\nu}) = b_{\text{ICA}} \sum_{\substack{\alpha,\beta=1 \\ \alpha \neq \beta}}^{R} \left(\sum_{\gamma,\delta=1}^{L} Q_{\alpha\gamma}^{\mu\nu} Q_{\beta\delta}^{\mu\nu} C_{\gamma\delta}^{(y)}(\tau) \right)^2 - b_{\text{SFA}} \sum_{\alpha=1}^{R} \left(\sum_{\beta,\gamma=1}^{L} Q_{\alpha\beta}^{\mu\nu} Q_{\alpha\gamma}^{\mu\nu} C_{\beta\gamma}^{(y)}(\tau) \right)^2 , \tag{15}$$

Assume we want to minimize Ψ_{ISFA} for a given R, where R denotes the number of signal components we want to unmix. Applying a Givens rotation $\mathbf{Q}^{\mu\nu}$ we have to distinguish three cases

- **Case 1** Both axes u_μ and u_ν lie inside the subspace spanned by the first R axes ($\mu, \nu \leq R$): The sum over all squared cross correlations of all signal components that lie outside the subspace is constant as well as those of all signal components inside the subspace. There is no interaction between inside and outside, in fact the objective function is exactly the objective for an ICA algorithm based on second-order statistics e.g. TDSEP or SOBI [Ziehe and Müller, 1998; Belouchrani et al., 1997]. In [Blaschke et al., 2004] it has been shown that this is equivalent to SFA in the case of a single time delay.
- **Case 2** Only one axis, w.l.o.g. u_μ, lies inside the subspace, the other, u_ν, outside ($\mu \leq R < \nu$): Since one axis of the rotation plane lies outside the subspace, u_μ in the objective function can be optimized at the expense of $\tilde{u}_\nu$ outside the subspace. A rotation of $\pi/2$, for instance would simply exchange components u_μ and u_ν. This gives the possibility to find the slowest and most independent components in the whole space spanned by all u_i and $\tilde{u}_j$ ($i = 1, \ldots, R$, $j = R+1, \ldots, L$) in contrast to Case 1 where the minimum is searched within the subspace spanned by the R components in the objective function.
- **Case 3** Both axes lie outside the subspace ($R < \mu, \nu$): A Givens rotation with the two rotation axes outside the relevant subspace does not affect the objective function, and can therefore be disregarded.

It can be shown that like in [Blaschke and Wiskott, 2004] the objective function (15) as a function of ϕ can always be written in the form

$$\Psi_{\text{ISFA}}^{\mu\nu}(\phi) = A_0 + A_2 \cos(2\phi + \phi_2) + A_4 \cos(4\phi + \phi_4), \tag{16}$$

where the second term on the right hand side vanishes for Case 1. There exists a single minimum (if w.l.o.g. $\phi \in \left[-\frac{\pi}{2}, \frac{\pi}{2}\right]$) that can easily be calculated (see e.g.[Blaschke and Wiskott, 2004]). The derivation of (16) involves various trigonometric identities and, because of its length, is documented elsewhere[1].

---
[1] http://itb.biologie.hu-berlin.de/~blaschke

It is important to notice that the rotation planes of the Givens rotations are selected from the whole L-dimensional space whereas the objective function only uses information of correlations among the first R signal components u_i.

Successive application of Givens rotations $\mathbf{Q}^{\mu\nu}$ leads to the final rotation matrix $\mathbf{Q}$ which is in the ideal case such that $\mathbf{Q}^T \mathbf{C}^{(y)}(\tau) \mathbf{Q} = \mathbf{C}^{(v)}(\tau)$ has a diagonal $R \times R$ submatrix $\mathbf{C}^{(u)}(\tau)$, but it is not clear if the final minimum is also the global one. However, in various simulations no local minima have been found.

4.3 Incremental Extracting of Independent Components

It is possible to find the number of independent source signal components R by successively increasing the number of components to be extracted. In each step the objective function (13) is optimized for a fixed R. First a single signal component is extracted ($R = 1$) and than an additional one ($R = 2$) etc. The algorithm is stopped when no additional signal component can be extracted. As a stopping criterion every suitable measure of independence can be applied; we used the sum over squared cross-cumulants of fourth order. In our artificial examples, this value is typically small for independent components, and increases by two orders of magnitudes if the number of components to be extracted is greater than the number of original source signal components.

5 Simulation

Here we show a simple example, with two nonlinearly mixed signal components as shown in Figure 1. For comparison we chose a mixture from [Harmeling et al., 2003] defined by

$$x_1(t) = (s_2(t) + 3s_1(t) + 6)\cos(1.5\pi s_1(t)),$$
$$x_2(t) = (s_2(t) + 3s_1(t) + 6)\sin(1.5\pi s_1(t)). \quad (17)$$

We used the ISFA algorithm with different nonlinearities (see Tab. 1). Already a nonlinear expansion with monomials up to degree three was sufficient to give good unmixing results. In all cases ISFA did find exactly two independent signal components. Using all monomials up to degree five led to results that showed virtually no difference between estimated and true source signal (see Fig. 1). A linear BSS method failed completely to find a good unmixing matrix.

6 Conclusion

We have shown that connecting the ideas of slow feature analysis and independent component analysis into ISFA is a possible way to solve the nonlinear blind source separation problem. SFA enforces the independent components of ICA to be slowly varying which seems to be a good way to discriminate between the original and nonlinearly distorted source signal components. A simple simulation showed that ISFA is able to extract the original source signal out of a nonlinear mixture. Furthermore ISFA can predict the number of source signal components via an incremental optimization scheme.

Table 1. Correlation coefficients of extracted (u_1 and u_2) and original (s_1 and s_2) source signal components

	linear		degree 2		degree 3		degree 4		degree 5		kTDSEP	
	u_1	u_2	u_1	u_2	u_1	u_2	u_1	u_2	u_1	u_2	u_1	u_2
s_1	-0.890	0.215	0.936	0.013	0.001	0.988	0.002	-0.996	0.998	-0.000	0.990	-
s_2	-0.011	-0.065	-0.027	0.149	-0.977	0.006	0.983	-0.000	-0.000	0.994	-	0.947

Correlation coefficients of extracted (u_1 and u_2) and original (s_1 and s_2) source signal components for linear ICA (first column) and ISFA with different nonlinearities (monomials up to degree 2,3,4, and 5). Note, that the source signal can only be estimated up to permutation and scaling, resulting in different signs and permutations of u_1 u_2. The correlation coefficients for kTDSEP were taken from [Harmeling et al., 2003] with same mixture but different source signal.

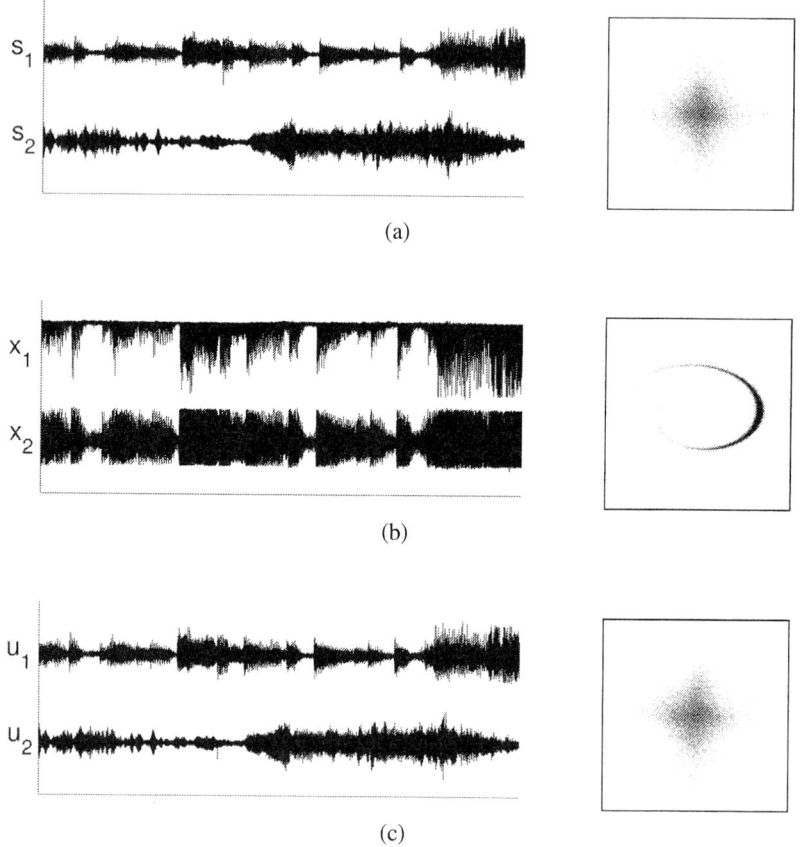

Fig. 1. Waveforms and Scatter-plots of **(a)** the original source signal components s_i, **(b)** the nonlinear mixture, and **(c)** recovered components with nonlinear ISFA (u_i). As a nonlinearity we used all monomials up to degree 5.

References

Belouchrani, A., Abed Meraim, K., Cardoso, J.-F., and Éric Moulines (1997). A blind source separation technique based on second order statistics. *IEEE Transactions on Signal Processing*, 45(2):434–44.

Blaschke, T. and Wiskott, L. (2004). CuBICA: Independent component analysis by simultaneous third- and fourth-order cumulant diagonalization. *IEEE Transactions on Signal Processing*, 52(5).

Blaschke, T., Wiskott, L., and Berkes, P. (2004). What is the relation between independent component analysis and slow feature analysis? *in preparation*.

Cardoso, J.-F. and Souloumiac, A. (1993). Blind beamforming for non Gaussian signals. *IEE Proceedings-F*, 140:362–370.

Cardoso, J.-F. and Souloumiac, A. (1996). Jacobi angles for simultaneous diagonalization. *SIAM J. Mat. Anal. Appl.*, 17(1):161–164.

Comon, P. (1994). Independent component analysis, a new concept? *Signal Processing, Elsevier*, 36(3):287–314. Special issue on Higher-Order Statistics.

Harmeling, S., Ziehe, A., Kawanabe, M., and Müller, K.-R. (2003). Kernel-based nonlinear blind source separation. *Neural Computation*, 15:1089–1124.

Hyvärinen, A. (1999). Fast and robust fixed-point algorithms for independent component analysis. *IEEE Transactions on Neural Networks*, 10(3):626–634.

Hyvärinen, A. and Pajunen, P. (1999). Nonlinear independent component analysis: existence and uniqueness results. *Neural Networks*, 12(3):429–439.

Jutten, C. and Karhunen, J. (2003). Advances in nonlinear blind source separation. In *Proc. of the 4th Int. Symp. on Independent Component Analysis and Blind Signal Separation (ICA2003)*, pages 245–256, Nara, Japan.

Lee, T.-W., Girolami, M., and Sejnowski, T. (1999). Independent component analysis using an extended Infomax algorithm for mixed sub-Gaussian and super-Gaussian sources. *Neural Computation*, 11(2):409–433.

Molgedey, L. and Schuster, G. (1994). Separation of a mixture of independent signals using time delayed correlations. *Physical Review Letters*, 72(23):3634–3637.

Tong, L., Liu, R., Soon, V. C., and Huang, Y.-F. (1991). Indeterminacy and identifiability of blind identification. *IEEE Transactions on Circuits and Systems*, 38(5).

Wiskott, L. and Sejnowski, T. (2002). Slow feature analysis: Unsupervised learning of invariances. *Neural Computation*, 14(4):715–770.

Ziehe, A. and Müller, K.-R. (1998). TDSEP – an efficient algorithm for blind separation using time structure. In *8th International Conference on Artificial Neural Networks (ICANN'98)*, pages 675 – 680, Berlin. Springer Verlag.

Nonlinear PCA/ICA for the Structure from Motion Problem

Jun Fujiki[1], Shotaro Akaho[1], and Noboru Murata[2]

[1] National Institute of Advanced Industrial Science and Technology
Umezono, Tsukuba-shi, Ibaraki 305-0035, Japan
jun-fujiki@aist.go.jp
[2] Waseda University, Okubo, Sinjuku-ku, Tokyo 169-8555, Japan

Abstract. Recovering both camera motion and object shape from multiple images, called structure from motion problem, is an important and essential problem in computer vision. Generally, the result of the structure from motion problem has an ambiguity represented by a three-dimensional rotation matrix. We present two kinds of specific criteria such as independence of parameters to fix the ambiguity by choosing an appropriate rotation matrix in the sense of computer vision. Once some criterion is defined, the fixing of the ambiguity is reduced to a nonlinear extension of the PCA/ICA. We examine the efficiency through synthetic experiments.

1 Introduction

Recovering both camera motion and object shape from multiple images is an important and essential problem in the field of computer vision, and the results are used for various fields such as man-machine interface, virtual reality system and auto-control robot system. The problem is called *structure from motion problem*. A perspective projection is suitable for representing a pin-hole camera theoretically. However, recovering from perspective images is a non-linear inverse problem, which is sensitive and unstable in numerical computation. Therefore, the affine approximations of the perspective projection have been proposed. These approximations can be resolved into the orthographic projection[1]. Then without loss of generality, we only consider to the orthographic projection. To solve the structure from motion problem under orthographic projection, many methods have been presented, and the factorization method [4] is known as an excellent method to solve the problem.

As a matter of fact, an image of an object (without background) is determined only by the relative position between a camera and the object. This means that there is an ambiguity of choosing the Euclidean coordinate system to describe the result of recovering, that is, the reconstruction of the motion and the shape. The ambiguity is represented by a three-dimensional rotation matrix. However, in computer vision, no attention has been paid to the ambiguity because it has no influence on the recovered shape of the object explicitly, and fixing the ambiguity by setting the camera coordinate system of the first image without consideration.

However, there might exist a special coordinate system which is suitable for getting images. To fix the ambiguity, that is, to choose an appropriate Euclidean coordinate system, we introduce two criteria. One criterion is defining the most statistically reliable view, and the other is extracting the most informative components of camera motion. As described in later sections, the problem to choose the appropriate rotation matrix can be treated as a nonlinear extension of PCA (principal component analysis) and ICA (independent component analysis) [3].

The framework of nonlinear PCA/ICA has three distinctive points compared with the original PCA/ICA. Firstly, each sample point is a rotation matrix and not a point belonging to Euclidean space. Secondly, an objective function such as variance and kurtosis is not defined in the sample space but defined in another space that is nonlinearly mapped from the sample space. Lastly, the freedom of transformation is strictly restricted to rotation compared with the original ICA, otherwise the sample points do not belong to the space of rotation matrix any more. These three properties are also different from the existing nonlinear extensions of ICA[5].

In the following sections, we first review the factorization method and explain the existence of ambiguity of rotation. We next explain the geometrical representation of the rotation matrix. Then we give two kinds of criteria to choose the rotation matrix by formulating it as a nonlinear extension of PCA/ICA. We also give a simple experiment by using synthetic data.

2 Factorization Method for Orthographic Projection

The factorization method for orthographic projection[4] is an excellent method for recovering both motion and shape simultaneously only from multiple orthographic images without knowing any information on physical positions of camera. The method is known as providing high stability in numerical computations and relatively high quality of reconstruction.

In the context of the factorization method, we can set the object is stable and only by the camera is moving without loss of generality because the images are determined only the relative position between camera and object.

Let $\{i_f, j_f\}$ be the orthonormal basis on the f-th image plane, k_f be the unit vector along optical axis, $C_f = (i_f, j_f, k_f)^T$ be the camera basis matrix which forms the f-th camera coordinate system, and s_p be the world coordinate of the p-th feature point. We also define the camera coordinate of p-th point on the f-th image as $X_{fp} = (X_{fp}, Y_{fp}, Z_{fp})^T$, and the image coordinate of p-th point on the f-th image as $x_{fp} = (x_{fp}, y_{fp})^T$. When considering the orthographic projection, using the relative coordinate from some feature point named $*$-th feature point (or center-of-mass of the object) is convenient. By using the relative coordinate $s_p^* = s_p - s_*$, $X_{fp}^* = X_{fp} - X_{f*}$ and $x_{fp}^* = x_{fp} - x_{f*}$, there holds

$$X_{fp}^* = C_f s_p^*$$

(see figure 1), and the representation of the orthographic projection is given by

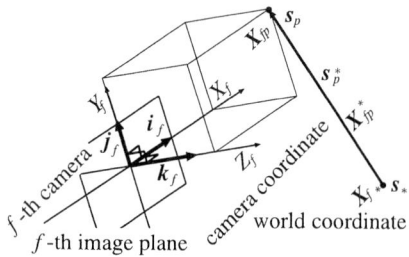

Fig. 1. Camera coordinate and world coordinate.

$$x^*_{fp} = \begin{pmatrix} 1 & 0 & 0 \\ 0 & 1 & 0 \end{pmatrix} X^*_{fp} \iff x^*_{fp} = \begin{pmatrix} i^T_f \\ j^T_f \end{pmatrix} s^*_p.$$

Let a measurement matrix W^*, a motion matrix M and a shape matrix S^* be defined as

$$W^* = \begin{pmatrix} x^*_{11} & \cdots & x^*_{1P} \\ \vdots & \ddots & \vdots \\ x^*_{F1} & \cdots & x^*_{FP} \end{pmatrix}, \quad M = \begin{pmatrix} M_1 \\ \vdots \\ M_F \end{pmatrix}$$

$$M_f = \begin{pmatrix} i^T_f \\ j^T_f \end{pmatrix}, \quad S^* = (s^*_1, \ldots, s^*_P),$$

there holds

$$W^* = \underset{(2F \times 3)}{M} \underset{(3 \times P)}{S^*}.$$

Note that rank $W^* \leq 3$.

We can easily compute $C_f = (i_f, j_f, i_f \times j_f)^T$ from $M_f = (i_f, j_f)^T$, then the decomposition of W^* into MS^* attains the reconstruction of the camera motion and the object shape. However, the decomposition of W^* into MS^* is not unique because the decomposition of $W^* = \widehat{M}_{(2F \times 3)} \widehat{S}^*_{(3 \times P)}$ derives another decomposition $(\widehat{M}A)(A^{-1}\widehat{S}^*)$ where A is arbitrary 3×3 invertible matrix. Hence, $\widehat{M}$, $\widehat{S}^*$ are only the affine reconstruction. To upgrade the affine reconstruction to Euclidean reconstruction, the matrix A should be computed to satisfy the conditions named metric constraints which comes from the orthogonality of the basis of image coordinate as $i^T_f i_f = j^T_f j_f = 1$ and $i^T_f j_f = 0$:

$$M_f M_f^T = \widehat{M}_f Q \widehat{M}_f^T = \begin{pmatrix} 1 & 0 \\ 0 & 1 \end{pmatrix} \quad (f = 1, \ldots, F) \tag{1}$$

where $Q = AA^T$ and $\widehat{M}_f = \begin{pmatrix} \widehat{M}_1^T & \cdots & \widehat{M}_F^T \end{pmatrix}^T$.

Once the matrix $Q = AA^T$ is computed, A is easily obtained by the Cholesky decomposition. Let the Cholesky decomposition of Q be LL^T where L is a low triangle matrix, the general solution of A is $A = LU$ where U is any orthogonal matrix. After computing A, the Euclidean reconstruction of the motion $\widehat{M}A$ and the shape $A^{-1}\widehat{S}^*$ are derived. This is the procedure of the factorization method.

Note that corresponding to $\det U > 0$ or $\det U < 0$, a pair of reconstructions are derived from orthographic images under point correspondences, and the pair is mutually reflection called Necker reversal. It is well-known that we cannot choose one of the pair as a true reconstruction only from point correspondences[2].

The most important thing here is that the Euclidean reconstruction is not unique. Let C_f be the Euclidean reconstruction of motion obtained by the above algorithm, another Euclidean reconstruction is easily obtained by multiplying any rotation matrix R to all $\{C_f\}_{f=1}^F$.

3 Representation of the Rotation Matrix

We use the Euler angle representation of a rotation matrix to understand the camera basis matrix C_f. As is in the left of figure 2, each image plane is represented by a point on the unit sphere and its unit tangent vector correspond to x-axis ($\boldsymbol{i}_f$). The representation of many images is as in the middle of figure 2.

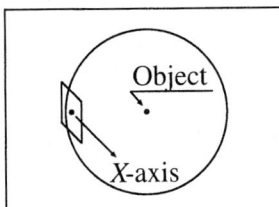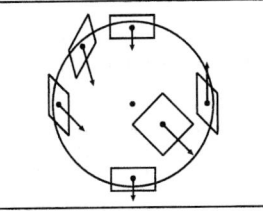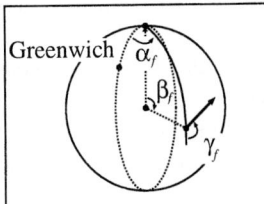

Fig. 2. Camera direction and the unit sphere.

Let us define an Euler angle representation of C_f as

$$C_f = C(\alpha_f, \beta_f, \gamma_f) = \begin{pmatrix} \cos\alpha_f & -\sin\alpha_f & 0 \\ \sin\alpha_f & \cos\alpha_f & 0 \\ 0 & 0 & 1 \end{pmatrix} \begin{pmatrix} \cos\beta_f & 0 & \sin\beta_f \\ 0 & 1 & 0 \\ -\sin\beta_f & 0 & \cos\beta_f \end{pmatrix} \begin{pmatrix} \cos\gamma_f & -\sin\gamma_f & 0 \\ \sin\gamma_f & \cos\gamma_f & 0 \\ 0 & 0 & 1 \end{pmatrix},$$

the position of f-th image from $(0,0,1)^{\mathrm{T}}$ (North pole) is given as in the right of figure 2. The geometrical meaning of each Euler angle is as follows:

- α_f determines a longitude of the f-th image,
- β_f determines a latitude of the f-th image,
- γ_f determines a horizon on the f-th image.

4 Fixing the Rotation Ambiguity

In order to determine the rotation matrix, we apply two kinds of criteria described in the following two subsections. In 4.3, we give a general form of these problems.

4.1 PCA on the Sphere

The first one is based on the confidence of the reconstruction. The distribution of the directions of projection affects the degree of confidence of the reconstruction, because frequently observed directions give higher confidence. Therefore, there exists the most reliable direction, and we can set the coordinate system to the view from that direction.

In order to find the most reliable direction, we apply a PCA-like method. Each data is correspond to the rotation matrix named camera basis matrix and we use the camera direction (optical axis) represented as a point on the sphere shown in the previous section. The sphere is a nonlinear two dimensional space therefore we have to extend the framework of PCA to the data on the sphere. Once we define mean and variance on the sphere, the problem is to find an axis on the sphere starting from the mean such that the vertical axis has the least variance (equivalently the most concentrated axis). We can fix the rotation ambiguity by putting the mean as origin and fixing the axis.

Now let us formulate the above framework. The f-th camera direction is given by a three dimensional vector $\boldsymbol{k}_f = \boldsymbol{i}_f \times \boldsymbol{j}_f$ which is on the sphere because $|\boldsymbol{k}_f| = 1$. Therefore, it is not appropriate to treat $\boldsymbol{k}_f$ as a point on the three-dimensional Euclidean space but a point on the sphere as the two dimensional space.

We use a notation $\boldsymbol{k}_f^R$ as rotating $\boldsymbol{k}_f$ by a rotation matrix R, and let us determine the objective function for $\boldsymbol{k}_f^R$ as follows.

First, we determine the origin of the sphere by $\overline{\boldsymbol{k}^R} = (0,0,1)^T$ or $(0,0,-1)^T$, which is chosen so as to minimize the sum of squared distance along the geodesic on the sphere from samples. Then we map each sample point to the two dimensional Euclidean space by the following function:

$$(p_f^R, q_f^R) = (l_f^R \cos \theta_f^R, l_f^R \sin \theta_f^R), \quad (2)$$

where l_f^R is the distance along the geodesic from the origin $\overline{\boldsymbol{k}^R}$ to $\boldsymbol{k}_f^R$, and θ_f^R is the angle of the longitude from the direction of $(1,0,0)$. This map is equivalent to azimuthal equidistant projection of the sphere.

By this map, we define the mean and the variance for each axis by

$$(\mu_p^R, \mu_q^R) = E_f[(p_f^R, q_f^R)], \quad (3)$$

$$((\sigma_p^R)^2, (\sigma_q^R)^2) = E_f[(p_f^R - \mu_p^R)^2, (q_f^R - \mu_q^R)^2]. \quad (4)$$

The problem of PCA on the sphere is to find the rotation matrix R such that it minimizes the variance σ_p^2 (or equivalently maximizes the variance σ_q^2) with keeping the mean μ_p and μ_q be zero, that is,

$$R_{\mathrm{opt}} = \arg \max_R (\sigma_q^R)^2 - \lambda((\mu_p^R)^2 + (\mu_q^R)^2), \quad (5)$$

where λ is a Lagrange multiplier.

In the practical implementation, we set λ by an appropriate constant (e.g. $\lambda = 1$). Note that the additional term of the mean μ_p and μ_q does not appear in the original PCA, because the original PCA is a method in a linear space and the shift does not affect the result.

4.2 ICA on the Angle Space

The second criterion to determine the rotation is to extract the most informative (equivalently independent) components of the camera positions. If we use the coordinate system whose axis corresponds to the physical freedom of camera, it would be useful to analyze or deal with the physical properties due to the camera motion. The description based on the most informative components is often robust against noise. In that sense, this criterion represents another aspect of confidence of the reconstruction. Here, each data point is a camera position which is represented by the point of three dimensional orthogonal matrix, which is a nonlinear function of the physical angles. To solve the problem, we present an extension of the independent component analysis (ICA) incorporating the nonlinearity.

In this case, we assume that the physical freedoms are given by the Euler angle representation. The problem is to find the independent components in the angle space (α, β, γ), where the transformation is given by

$$C(\alpha_f, \beta_f, \gamma_f) R = C(\alpha_f^R, \beta_f^R, \gamma_f^R). \tag{6}$$

Note that the map from a three-dimensional vector $(\alpha_f, \beta_f, \gamma_f)$ to another three-dimensional vector $(\alpha_f^R, \beta_f^R, \gamma_f^R)$ is the nonlinear map.

It is difficult to define a cost function for nonlinear transformation of ICA in general. In this paper, we assume that the nonlinear should be well-approximated by linear function at least locally, and apply the same objective function as the original ICA (we use kurtosis in the experiments).

4.3 General Formulation

The two methods described in the above two sections can be summarized in a general way as follows.

1. Samples C_f are from the space of the orthogonal matrix, C_f should be rotated to C_f^R by a rotation matrix R.
2. A nonlinear map $g(C_f^R)$ is given. $((p^R, q^R)$ representation in PCA on the sphere, and $(\alpha_f^R, \beta_f^R, \gamma_f^R)$ representation in ICA on the angle space.
3. A cost function $L(g(C_f^R))$ is defined (variance or kurtosis)
4. The problem is to find R that minimizes $L(g(C_f^R))$.

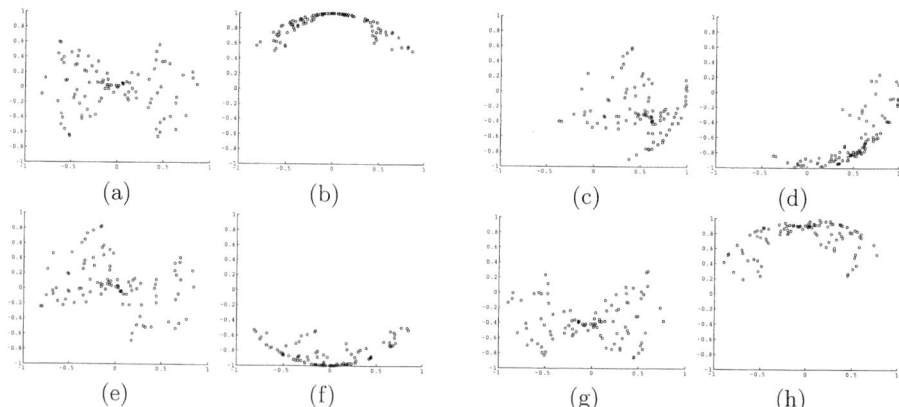

Fig. 3. Plots of $k_f = i_f \times j_f$. (a) and (b): The original C_f^0. (c) and (d): rotated C_f^1. (e) and (f): Recovered by the PCA on the sphere. (g) and (h): Recovered by the ICA on the angle space. In each pair, the left figures ((a),(c),(e),(g)) of pairs show the first and the second elements plot, and the right figures ((b),(d),(f),(h)) show the first and the third elements plot.

4.4 Optimization Method

Because of the higher nonlinearity than the original PCA/ICA, it is difficult to optimize the cost function in a efficient and stable way. Since the dimensionality three of the optimization is rather small, we apply the alternating optimization in which the optimization is carried out for each component angle. In the optimization for each angle, we applied the hill-climbing method in which we move the parameter by comparing with the neighborhood points of the current parameter. However, it may be possible to develop more efficient algorithm and it is left as a future work.

5 Experiment

First, we generated 100 samples of $\alpha_f^0, \beta_f^0, \gamma_f^0$ independently from the uniform distribution on $[-\pi/3, \pi/3)$. Since $\alpha_f^0, \beta_f^0, \gamma_f^0$ are independent, it is an unknown target of the ICA on the angle space. In the space of k_f, the origin and the mean of samples are close.

Next we calculated the camera position C_f^0 by the Euler angle representation.

Then, we prepared a 3×3 orthogonal matrix R^1 randomly, and multiplied it to C_f^0 and got a training samples of the camera position C_f^1. We applied the two proposed algorithms to $\{C_f^1\}$.

In figure 3, the points in the k_f space are plotted. Figure 3(a) and (b) show the generated data, and (c) and (d) show the training samples.

The result of the PCA on the surface in given in (e) and (f). The variance defined in the sphere of vertical axis has smaller value in (e) than in (a).

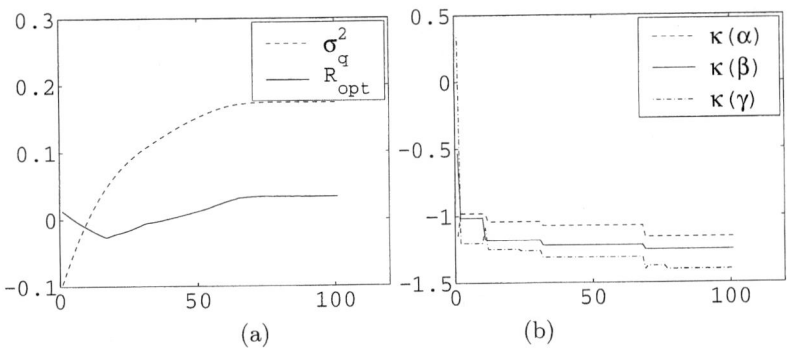

Fig. 4. Plot of the objective functions. (a): Cost of PCA on the sphere, (b): Kurtosis in ICA on the angle space.

The result of the ICA on the angle is shown in (g) and (h). In this case, the independent directions are successfully found and directions of axes are similar to the original one. The origin is shifted because we only used the kurtosis that is less sensitive against the shift (but not completely independent because of the nonlinearity).

The behavior of objective function through learning is plotted in figure 4. In spite of the primitive algorithms, the objective function is minimized/maximized quickly.

References

1. J. Fujiki, T. Kurata and M. Tanaka, "Iterative factorization method for object recognition," Proc. SPIE98, San Diego, Vision Geometry VI, vol.3454-18, pp.192–201, July 1998.
2. J.Fujiki, "Motion and shape from sequences of images under feature correspondences," Trans. of IEICE on Information and Systems, Vol.E82-D., No.3, pp.548-557, Mar 1999.
3. A.Hyvärinen, J.Karhunen and E.Oja, *"Independent Component Analysis,"* John Wiley & Sons, 2001.
4. C. Tomasi and T. Kanade. Shape and motion from image streams under orthography: a factorization method. *IJCV*, 9(2):137–154, 1992.
5. H. Valpola, X.Giannakopoulos, A.Honkela and J.Karhunen, Nonlinear independent component analysis using ensemble learning: Experiments and discussion, *Proc. ICA2000*: 351–356, 2000.

Plugging an Histogram-Based Contrast Function on a Genetic Algorithm for Solving PostNonLinear-BSS

Fernando Rojas Ruiz[1], Carlos G. Puntonet[1], Ignacio Rojas Ruiz[1], Manuel Rodríguez-Álvarez[1], and Juan Manuel Górriz[2]

[1] Dpto. Arquitectura. y Tecnología de Computadores, University of Granada, Spain
{frojas,carlos,irojas,mrodriguez}@atc.ugr.es
[2] Area de Electrónica, University of Cádiz, Spain
juanmanuel.gorriz@uca.es

Abstract. This paper proposes a novel Independent Component Analysis algorithm based on the use of a genetic algorithm intended for its application to the problem of blind source separation on post-nonlinear mixtures. We present a simple though effective contrast function which evaluates individuals of each population (candidate solutions) based on estimating the probability densities of the outputs through histogram approximation. Although more sophisticate methods for probability density function approximation exist, such as kernel-based methods or k-nearest-neighbor estimation, the histogram presents the advantage of its simplicity and easy calculation if an appropriate number of samples is available.

1 Introduction

The guiding principle for ICA is statistical independence, meaning that the value of any of the components gives no information on the values of the other components. This method differs from other statistical approaches such as principal component analysis (PCA) and factor analysis precisely in the fact that is not a correlation-based transformation, but also reduces higher-order statistical dependencies. The extensive use of ICA as the statistical technique for solving blind source separation (BSS), may have lead in some situations to the erroneous utilization of both concepts as equivalent. In any case, ICA is just the *technique* which in certain situations can be sufficient to solve a given *problem*, that of blind source separation. In fact, statistical independence insures separation of sources in linear mixtures, up to the known indeterminacies of scale and permutation. However, generalizing to the situation in which mixtures are the result of an unknown transformation (linear or not) of the sources, independence alone is not a sufficient condition in order to accomplish blind source separation successfully. Indeed, in [5] it is formally demonstrated how for nonlinear mixtures, an infinity of mutually independent solutions can be found that have nothing to do with the unknown sources. Thus, in order to successfully separate The observed signals into a wave-preserving estimation of the sources, we need additional information about either the sources or the mixing process.

This paper is structured as follows: Section 2 introduces the post-nonlinear model as an alternative to the unconstrained pure nonlinear model. Afterwards, in Section 3, the basis of the genetic algorithm is described: independence measure, probability density function estimation and evolutionary method depiction. Some experiments are shown in Section 4, using speech and synthetic signals. Finally, a few conclusion remarks and future lines of research terminate this paper.

2 Nonlinear Independent Component Analysis

2.1 Post-non-linear Model

The linear assumption is an approximation of nonlinear phenomena in many real world situations. Thus, the linear assumption may lead to incorrect solutions. Hence, researchers in BSS have started addressing the nonlinear mixing models, however a fundamental difficulty in nonlinear ICA is that it is highly nonunique without some extra constraints, therefore finding independent components does not lead us necessarily to the original sources [5].

Blind source separation in the nonlinear case is, in general, impossible. Taleb and Jutten [11] added some extra constraints to the nonlinear mixture so that the nonlinearities are independently applied in each channel after a linear mixture (see Fig.1). In this way, the indeterminacies are the same as for the basic linear instantaneous mixing model: invertible scaling and permutation.

The mixture model can be described by the following equation:

$$\mathbf{x}(t) = \mathbf{F}\left(\mathbf{A} \cdot \mathbf{s}(t)\right) \quad (1)$$

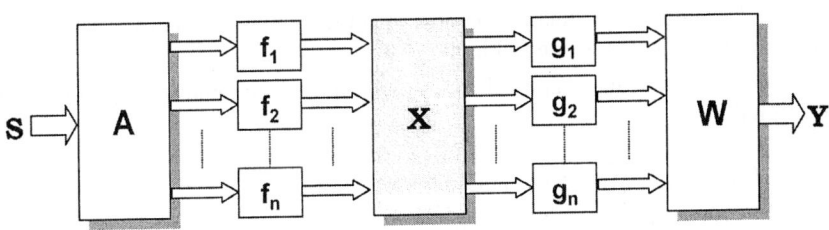

Fig. 1. Post-nonlinear model.

The unmixing stage, which will be performed by the algorithm here proposed is expressed by Equation (2):

$$\mathbf{y}(t) = \mathbf{W} \cdot \mathbf{G}\left(\mathbf{x}(t)\right) \quad (2)$$

The post-nonlinearity assumption is reasonable in many signal processing applications where the nonlinearities are introduced by sensors and preamplifiers, as usually happens in speech processing. In this case, the nonlinearity is assumed to be introduced by the signal acquisition system.

3 Genetic Algorithm for Source Separation

3.1 Mutual Information Approximation

The proposed algorithm will be based on the estimation of mutual information, value which cancels out when the signals involved are independent. Mutual information I between the elements of a multidimensional variable $\mathbf{y}$ is defined as:

$$I(y_1, y_2, ..., y_n) = \sum_{i=1}^{n} H(y_i) - H(y_1, y_2, ..., y_n).\quad(3)$$

where *H(x)* is the entropy measure of the random variable or variable set *x*.

For Eq. 5, in the case that all components $y_1 ..... y_n$ are independent, the joint entropy is equal to the sum of the marginal entropies. Therefore, mutual information will be zero. In the rest of the cases (not independent components), the sum of marginal entropies will be higher than the joint entropy, leading thus to a positive value of mutual information.

In order to exactly compute mutual information, we need also to calculate entropies, which likewise require knowing the analytical expression of the probability density function (PDF) which is generally not available in practical applications of speech processing. Thus, we propose to approximate densities through the discretization of the estimated signals building histograms and then calculate their joint and marginal entropies. In this way, we define a number of bins *m* that covers the selected estimation space and then we calculate how many points of the signal fall in each of the bins (B_i $i = 1, ..., m$). Finally, we easily approximate marginal entropies using the following formula:

$$H(y) = -\sum_{i=1}^{n} p(y_i) \log_2 p(y_i) \approx -\sum_{j=1}^{m} \frac{Card(B_j(y))}{n} \log_2 \frac{Card(B_j(y))}{n}.\quad(4)$$

where *Card(B)* denotes cardinality of set B, *n* is the number of points of estimation *y*, and B_j is the set of points which fall in the jth bin.

The same method can be applied for computing the joint entropies of all the estimated signals:

$$H(y_1, ..., y_p) = \sum_{i=1}^{p} H(y_i | y_{i-1}, ..., y_1)$$

$$\approx -\sum_{i_1=1}^{m} \sum_{i_2=1}^{m} ... \sum_{i_n=1}^{m} \frac{Card(B_{i_1 i_2 ... i_p}(y))}{n} \log_2 \frac{Card(B_{i_1 i_2 ... i_p}(y))}{n}.\quad(5)$$

where *p* is the number of components which need to be approximated and *m* is the number of bins in each dimension.

Therefore, substituting entropies in Eq.5 by approximations of Eqs.6 and 7, we obtain an approximation of mutual information (Eq. 8) which will reach its minimum value when the estimations are independent:

$$Est(I(\mathbf{y})) = \sum_{i=1}^{p} Est(H(y_i)) - Est(H(\mathbf{y})) =$$

$$= -\sum_{i=1}^{p} \left[\sum_{j=1}^{m} \frac{Card(B_j(y_i))}{n} \log_2 \frac{Card(B_j(y_i))}{n} \right] + ...$$

$$... + \sum_{i_1=1}^{m} \sum_{i_2=1}^{m} ... \sum_{i_n=1}^{m} \frac{Card(B_{i_1 i_2 ... i_p}(y))}{n} \log_2 \frac{Card(B_{i_1 i_2 ... i_p}(y))}{n}.\quad(6)$$

where *Est(X)* stands for "estimation of x".

Next section describes an evolution based algorithm that minimizes the contrast function defined in Eq. 8, escaping from local minima.

3.2 Proposed Genetic Algorithm

A genetic algorithm (GA) evaluates a population of possible solutions and generates a new one iteratively, with each successive population referred to as a generation. Given the current generation at iteration t, G(t), the GA generates a new generation, G(t+1), based on the previous generation, applying a set of genetic operations. Aside from other aspects regarding genetic algorithms, the key features that characterize a genetic algorithm are the encoding scheme and the evaluation or fitness function.

First of all, it should be recalled that the proposed algorithm needs to estimate two different mixtures (see Eq. 4): a family of nonlinearities g which approximates the inverse of the nonlinear mixtures f and a linear unmixing matrix W which approximates the inverse of the linear mixture A [9,10]. This linear demixing stage will be performed by the well-known FastICA algorithm by Hyvärinen and Oja [4]. To be precise, FastICA will be embedded into the genetic algorithm in order to approximate the linear mixture.

Therefore, the encoding scheme for the chromosome in the post-nonlinear mixture will be the coefficients of the odd polynomials which approximate the family of nonlinearities g. Fig. 2 shows an example of polynomial approximation and encoding of the inverse non-linearities.

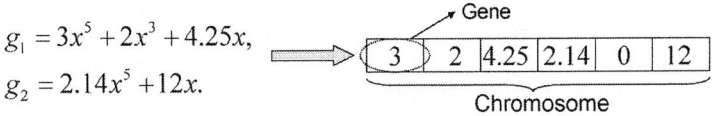

$$g_1 = 3x^5 + 2x^3 + 4.25x,$$
$$g_2 = 2.14x^5 + 12x.$$

Fig. 2. Encoding example for $p=2$ signals and polynomials up to grade 5.

The fitness function is easily derived from Eq. 8 which is precisely the inverse of the approximation of mutual information, so that the genetic algorithm maximizes the fitness function, which is more usual in evolution programs literature.

$$Fitness(\mathbf{y}) = \frac{1}{Est(I(\mathbf{y}))}. \tag{7}$$

Expression (7) obeys to the desired properties of a contrast function [2], that is, a mapping ψ from the set of probability densities $\{p_x, x \in \mathbb{R}^N\}$ to $\mathbb{R}$ satisfying the following requirements:

i. $\psi(p_x)$ does not change if the components of x_i are permuted.
ii. $\psi(p_x)$ is invariant to invertible scaling.
iii. If x has independent components, then $\psi(p_{Ax}) \leq \psi(p_x)$, $\forall A$ invertible.

Regarding other aspects of the genetic algorithm, the population (i.e. set of chromosomes) was initialized randomly within a known interval of search for the polynomial coefficients. The genetic operators involved were "Simple One-point Crossover" and "Non-Uniform Mutation" [8]. Selection strategy is elitist, keeping the best individual of a generation for the next one.

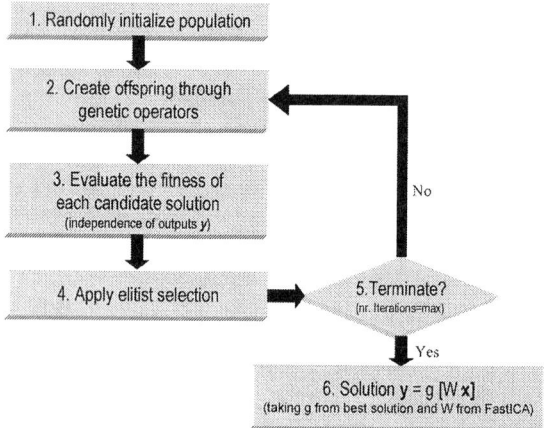

Fig. 3. Genetic algorithm scheme for post-nonlinear blind separation of sources.

4 Simulations

This section illustrates the validity of the genetic algorithm here proposed and investigates the accuracy of the method. We combined voice signals and noise nonlinearly and then try to recover the original sources. In order to measure the accuracy of the algorithm, we evaluate it using the Mean Square Error (MSE) and the Crosstalk in decibels (Ct):

$$MSE_i = \frac{\sum_{t=1}^{N}(s_i(t)-y_i(t))^2}{N} \qquad Ct_i = 10\log\left(\frac{\sum_{t=1}^{N}(s_i(t)-y_i(t))^2}{\sum_{t=1}^{N}(s_i(t))^2}\right) \qquad (8)$$

4.1 Two Voice Signals

This experiment corresponds to a "Cocktail Problem" [] situation, that is, separating one voice from another. Two voice signals corresponding to two persons saying the numbers from one to ten in English and Spanish were non-linearly according to the following matrix and functions:

$$\mathbf{A} = \begin{bmatrix} 1 & 0.87 \\ -0.9 & 0.14 \end{bmatrix}, \mathbf{F} = [f_1(x) = f_2(x) = \tanh(x)]. \qquad (9)$$

Then the genetic algorithm was applied (population size=40, number of iterations=60). Polynomials of fifth order were used as the approximators for $g=f^{-1}$. Performance results and a plot of the original and estimated signals are briefly depicted below (Figure 5.b, right).

MSE(y_1, s_1) = 0.0012   Crosstalk(y_1, s_1) (dB) = -17.32 dB,
MSE(y_2, s_2) = 0.0009   Crosstalk(y_2, s_2) (dB) = -19.33 dB.

As can be seen, estimations (**y**) are approximately equivalent to the original sources (**s**) up to invertible scalings and permutations. E.g. estimation y_1 is scaled and inverted in relation to s_1.

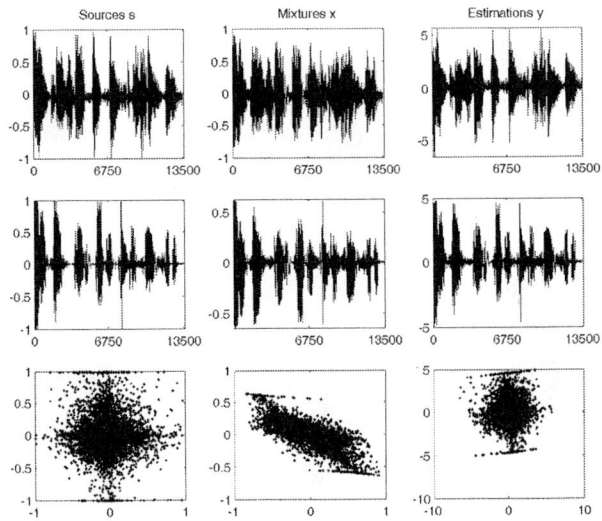

Fig. 4. Sources, mixtures and estimations (along time and scatter plots in the bottom line) for two voice signals.

4.2 Three Image Signals

In this experiment, source signals correspond to three image signals ("Lena", "Cameraman", "University of Granada emblem"). They were mixed according to the PNL scheme (Fig. 1) with the following values:

$$A = \begin{bmatrix} 0.9 & 0.3 & 0.6 \\ 0.6 & -0.7 & 0.1 \\ -0.2 & 0.9 & 0.7 \end{bmatrix}, f = \begin{bmatrix} f_1(x) = \tanh(x), \\ f_2(x) = \tanh(0.8 \cdot x), \\ f_3(x) = \tanh(0.5 \cdot x) \end{bmatrix}. \quad (10)$$

In this case, the simulation results draw a slightly worse performance than the former case, due to the increase of the dimensionality from to two to three sources:

MSE(y_1, s_2) = 0.0006   Crosstalk(y_1, s_2) (dB) = -16.32 dB,
MSE(y_2, s_1) = 0.0010   Crosstalk(y_2, s_1) (dB) = -12.12 dB.
MSE(y_3, s_3) = 0.0011   Crosstalk(y_3, s_3) (dB) = -11.58 dB.

Original images can be clearly distinguished through the estimations, although some remains of the other images interfere. Also note that an inversion in the signal results obviously in the negative estimation of the source (e.g. the cameraman).

5 Concluding Remarks

In this work, an specific case of nonlinear source separation problem has been tackled by an ICA algorithm based on the use of genetic algorithms. As the separation of sources through the independence basis only is impossible in nonlinear mixtures, we assumed a linear mixture followed by a nonlinear distortion in each channel (Post-Non-Linear model) which constraints the solution space. Experimental results showed

Fig. 5. Original images (s), PNL-mixtures (x) and estimations (y) applying the genetic algorithm.

promising results, although future research will focus on the adaptation of the algorithm for higher dimensionality and stronger nonlinearities.

Acknowledgement

This work has been supported by the CICYT Spanish Project TIC2001-2845.

References

1. G.Burel, Blind separation of sources: A nonlinear neural algorithm, Neural Networks, vol.5, pp.937-947, 1992.
2. P. Comon, Independent component analysis, a new concept?, Signal Processing, vol. 36, no. 3, pp. 287--314, 1994.
3. D.E. Goldberg, Genetic Algorithms in Search, Optimization and Machine Learning,AddisonWesley, Reading, MA, 1989.
4. A.Hyvärinen and E.Oja, A fast fixed-point algorithm for independent component analysis. Neural Computation, 9 (7), pp.1483-1492, 1997.
5. A. Hyvärinen and P. Pajunen. Nonlinear Independent Component Analysis: Existence and Uniqueness results. Neural Networks 12(3): pp. 429-439, 1999.
6. C. Jutten, J. Karhunen, Advances in Nonlinear Source Separation, Proceedings of the 3rd International Conference On Independent Component Analysis and Signal Separation (ICA2003). pp. 245-256, April 1-4, Nara (Japan), 2003.

7. T-W.Lee, B.Koehler, R.Orglmeister, Blind separation of nonlinear mixing models, In IEEE NNSP, pp.406-415, Florida, USA, 1997.
8. Z. Michalewicz, Genetic Algorithms + Data Structures = Evolution Programs, Springer-Verlag, New York USA, Third Edition, 1999.
9. F. Rojas, I. Rojas, R.M. Clemente, C.G. Puntonet. Nonlinear Blind Source Separation using Genetic Algorithms, in Proceedings of the 3rd International Conference On Independent Component Analysis and Signal Separation (ICA2001). pp. 400-405, December 9-13, San Diego, CA, (USA), 2001.
10. F. Rojas, C.G., Puntonet, M. Rodríguez-Álvarez, I. Rojas, Evolutionary Algorithm Using Mutual Information for Independent Component Analysis. Lecture Notes in Computer Science, LNCS, ISSN 0302-9743, Vol. 2687, pp. 233-240, 2003.
11. A.Taleb, C.Jutten, Source Separation in Post-Nonlinear Mixtures, IEEE Transactions on Signal Processing, vol.47 no.10, pp.2807-2820, 1999.
12. A. Ziehe, M. Kawanabe, S. Harmeling, K.R. Müller, Blind Separation of Post-nonlinear Mixtures using Linearizing Transformations and Temporal Decorrelation, J. of Machine Learning Research Special Issue on Independent Components Analysis, pp.1319-1338, 2003.

Post-nonlinear Independent Component Analysis by Variational Bayesian Learning

Alexander Ilin and Antti Honkela

Helsinki University of Technology, Neural Networks Research Centre
P.O. Box 5400, FI-02015 HUT, Espoo, Finland
{alexander.ilin,antti.honkela}@hut.fi
http://www.cis.hut.fi/projects/bayes/

Abstract. Post-nonlinear (PNL) independent component analysis (ICA) is a generalisation of ICA where the observations are assumed to have been generated from independent sources by linear mixing followed by component-wise scalar nonlinearities. Most previous PNL ICA algorithms require the post-nonlinearities to be invertible functions. In this paper, we present a variational Bayesian approach to PNL ICA that also works for non-invertible post-nonlinearities. The method is based on a generative model with multi-layer perceptron (MLP) networks to model the post-nonlinearities. Preliminary results with a difficult artificial example are encouraging.

1 Introduction

The problem of ICA have been studied by many authors in recent years. The general goal of ICA is to estimate some unknown signals (or sources) from a set of their mixtures by exploiting only the assumption that the mixed signals are statistically independent. The linear ICA model is well understood (see e.g. [1] for review) while the general nonlinear ICA and related nonlinear blind source separation (BSS) are more difficult problems from both theoretical and practical points of view [2, 1]. In fact, the general nonlinear ICA problem is ill-posed and most approaches to it are better classified as nonlinear BSS, where the goal is to estimate the specific sources that have generated the observed mixtures.

Post-nonlinear mixtures are a special case of the nonlinear mixing model studied first by Taleb and Jutten [3]. They are interesting for their separability properties and plausibility in many real world situations. In the PNL model, the nonlinear mixture has the following specific form:

$$x_i(t) = f_i \left[\sum_{j=1}^{M} a_{ij} s_j(t) \right] \quad i = 1, \ldots, N \quad (1)$$

where $x_i(t)$ are the N observations, $s_j(t)$ are the M independent sources, a_{ij} denotes the elements of the unknown mixing matrix A and $f_i : \mathbb{R} \to \mathbb{R}$ are a set of scalar to scalar functions sometimes also called post-nonlinear distortions.

Most of the existing ICA methods for PNL mixtures assume that the source vectors $\mathbf{s}(t)$ and the observations $\mathbf{x}(t)$ are of the same dimensionality (i.e. $N = M$) and that all post-nonlinear distortions f_i are invertible. In this case, under certain conditions on the distributions of the sources (at most one Gaussian source) and the mixing structure ($\mathbf{A}$ has at least 2 nonzero entries on each row or column), PNL mixtures are separable with the same well-known indeterminacies as in the linear mixtures [4, 3].

However, as was shown in [5], overdetermined PNL mixtures (when there are more observations x_i than sources s_j, i.e. $N > M$) can be separable even when some of the distortions f_i are non-invertible functions. In [5], the general nonlinear factor analysis (NFA) model [6]

$$\mathbf{x}(t) = \mathbf{f}\left(\mathbf{s}(t), \boldsymbol{\theta}_f\right) + \mathbf{n}(t) \qquad (2)$$

followed by the linear FastICA post-processing [1] was successfully applied to recover the independent sources from this kind of PNL mixtures.

In the present paper, we restrict the general NFA model of Eq. (2) to the special case of PNL mixtures of Eq. (1) and derive a learning algorithm based on variational Bayesian learning. In the resulting model, which we call post-nonlinear factor analysis (PNFA), the sources $s_j(t)$ are assumed to be Gaussian and therefore the nonlinear ICA problem can be solved by first learning the roughly Gaussian sources and then rotating them using any linear ICA algorithm to recover the independent components [6, 7].

The rest of the paper is structured as follows. First, the PNFA model is introduced in Sec. 2. The learning algorithm used to estimate the model is presented in Sec. 3 and the results of an experiment with a difficult artificial example in Sec. 4. The paper concludes with discussion in Sec. 5.

2 Post-nonlinear Factor Analysis Model

Most PNL ICA methods [3, 8] separate sources by inverting the mixing model (1) and therefore by estimating the following separating structure

$$s_j(t) = \sum_{i=1}^{N} b_{ji} g_i(x_i(t), \boldsymbol{\theta}_i) \qquad j = 1, \ldots, M. \qquad (3)$$

This approach implicitly assumes the existence of the inverse of the component-wise nonlinearities $g_i = f_i^{-1}$, and therefore fails in separable PNL mixtures with non-invertible distortions f_i [5].

To overcome this problem, we present the Bayesian PNFA algorithm which instead learns the generative model (1) in the following form (see Fig. 1):

$$x_i(t) = f_i\left[y_i(t), \boldsymbol{W}_i\right] + n_i(t) = f_i\left[\sum_{j=1}^{M} a_{ij} s_j(t), \boldsymbol{W}_i\right] + n_i(t) \qquad (4)$$

where $y_i(t) = \sum_{j=1}^{M} a_{ij} s_j(t)$ and $n_i(t)$ is the observation noise. The post-nonlinear component-wise distortions f_i are modelled by multi-layer preceptron (MLP) networks with one hidden layer:

$$f_i(y, \boldsymbol{W}_i) = \mathbf{D}_i \phi(\mathbf{C}_i y + \mathbf{c}_i) + d_i. \qquad (5)$$

Here the parameters $\boldsymbol{W}_i$ of the MLPs include the column vectors $\mathbf{C}_i$, $\mathbf{c}_i$, row vector $\mathbf{D}_i$ and scalar d_i. A sigmoidal activation function ϕ that operates component-wise on its inputs is used.

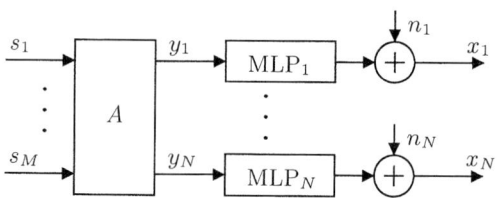

Fig. 1. The model structure of PNFA.

Implementing the Bayesian approach, we express all the model assumptions in the form of the joint distribution of the observations $\boldsymbol{X} = \{\mathbf{x}(t)|t\}$, the sources $\boldsymbol{S} = \{\mathbf{s}(t)|t\}$ and other model parameters $\boldsymbol{\theta} = \{\theta_i|i\}$.

Assuming independent Gaussian noise $n_i(t)$ yields the likelihood

$$p(\boldsymbol{X} \mid \boldsymbol{S}, \boldsymbol{\theta}) = \prod_{i,t} N\left(x_i(t);\, f_i\left[y_i(t), \boldsymbol{W}_i\right],\, e^{2v_{n,i}}\right) \qquad (6)$$

where $N(x;\, \mu,\, \sigma^2)$ denotes a Gaussian density for variable x having mean μ and variance σ^2, and the variance parameter has lognormal hierarchical prior. The sources $s_j(t)$ are assumed to be Gaussian and have the prior

$$p(\boldsymbol{S} \mid \boldsymbol{\theta}) = \prod_{j,t} N\left(s_j(t);\, 0,\, e^{2v_{s,j}}\right). \qquad (7)$$

The parameters of the prior distributions (such as the variance parameters $v_{n,i}$, $v_{s,j}$) as well as the other model parameters (such as the parameters $\boldsymbol{W}_i$ of the component-wise MLPs) are further assigned Gaussian priors making the prior $p(\boldsymbol{\theta})$ of the parameters hierarchical. For example, the noise parameters $v_{n,i}$ of different components of the data share a common prior:

$$p(v_{n,i} \mid \boldsymbol{\theta} \setminus v_{n,i}) = N\left(v_{n,i};\, m_{v_n},\, e^{2v_{v_n}}\right) \qquad (8)$$

and the hyperparameters m_{v_n}, v_{v_n} have very flat Gaussian priors.

3 Learning

In this section, the variational Bayesian learning algorithm used to learn the model, is introduced.

3.1 Variational Bayesian Learning

The PNFA model is learned using variational Bayesian method called ensemble learning [9–11]. It has recently become very popular in linear ICA [12–15] but it has been applied to nonlinear BSS [6, 16, 7] as well. Reasons for the popularity of ensemble learning include the ability to easily compare different models and its resistance to overfitting, which is especially important in applications with nonlinear models.

As a variational Bayesian method, ensemble learning is based on approximating the posterior distribution of the sources and model parameters $p(\boldsymbol{S}, \boldsymbol{\theta}|\boldsymbol{X})$ with another, simpler distribution $q(\boldsymbol{S}, \boldsymbol{\theta})$. The approximation is fitted by minimising the cost function

$$\mathcal{C} = \left\langle \log \frac{q(\boldsymbol{S}, \boldsymbol{\theta})}{p(\boldsymbol{S}, \boldsymbol{\theta}, \boldsymbol{X})} \right\rangle = D_{KL}(q(\boldsymbol{S}, \boldsymbol{\theta})||p(\boldsymbol{S}, \boldsymbol{\theta}|\boldsymbol{X})) - \log p(\boldsymbol{X}) \qquad (9)$$

where $\langle \cdot \rangle$ denotes expectation over the distribution $q(\boldsymbol{S}, \boldsymbol{\theta})$ and $D_{KL}(q||p)$ is the Kullback-Leibler divergence between the distributions q and p. The approximation is restricted to be of fixed simple form, such as a multivariate Gaussian with a diagonal covariance used in PNFA.

3.2 Learning the Model

Most terms of the cost function in Eq. (9) are simple expectations over Gaussian variables that can be evaluated analytically. The only difficulties arise from the likelihood term

$$\mathcal{C}_x = \langle -\log p(\boldsymbol{X}|\boldsymbol{S}, \boldsymbol{\theta}) \rangle \qquad (10)$$

that has to be approximated somehow.

With the Gaussian noise model, the likelihood term can be written as

$$\mathcal{C}_x = \sum_{t,i} \left\langle -\log N(x_i(t); f_{i,t}, \sigma_n^2) \right\rangle$$
$$= \sum_{t,i} \left[\frac{1}{2} \left\langle \log \sqrt{2\pi\sigma_n^2} \right\rangle + \left\langle \frac{1}{2\sigma_n^2} \right\rangle \left([x_i(t) - \langle f_{i,t} \rangle]^2 + \text{Var}[f_{i,t}] \right) \right] \qquad (11)$$

where $f_{i,t} = f_i[y_i(t), \boldsymbol{W}_i]$ and $\text{Var}[\cdot]$ denotes variance under $q(\boldsymbol{S}, \boldsymbol{\theta})$. This can be thus evaluated if the mean and variance of the outputs of the MLP networks are known. Once the cost function can be computed, it can be minimised numerically. The minimisation is performed by a gradient based algorithm similar to one used in [6].

3.3 Evaluation of the Statistics of MLP Outputs

To simplify the notation, subindices i will be dropped in this section. The mean and variance of the inputs $y(t)$ of the MLP networks can be computed exactly. Assuming these are Gaussian, the mean and variance of the MLPs $f(y(t), \boldsymbol{W})$

can easily be evaluated using e.g. Gauss-Hermite quadrature, which in this scalar case for y using three points is equivalent to unscented transform.

The above discussion ignores the variance of the network weights $\boldsymbol{W}$. Their effect could be included by performing the unscented transform on full combined input of $y(t)$ and $\boldsymbol{W}$, but that would increase the computational burden too much. As the variances of the weights are usually small, their effects are represented sufficiently well by using first-order Taylor approximation of the network with respect to them [17]. Thus the mean of the output is approximated as

$$\langle f_t \rangle = \sum_j w_j f(\hat{y}_j(t), \overline{\boldsymbol{W}}) \tag{12}$$

where w_j are the weights and $\hat{y}_j(t) = \langle y(t) \rangle + t_j \operatorname{Var}[y(t)]^{1/2}$ are the basis points of the Gauss-Hermite quadrature corresponding to the abscissas t_j, and $\overline{\boldsymbol{W}}$ denotes the mean of the weights $\boldsymbol{W}$.

Correspondingly, the variance is approximated by a combined Gauss-Hermite and Taylor approximation

$$\operatorname{Var}[f_t] = \sum_j w_j \left[\left(f(\hat{y}_j(t), \overline{\boldsymbol{W}}) - \langle f_t \rangle \right)^2 + \nabla_{\boldsymbol{W}} f(\hat{y}_j(t), \overline{\boldsymbol{W}}) \operatorname{Cov}[\boldsymbol{W}] \nabla_{\boldsymbol{W}} f(\hat{y}_j(t), \overline{\boldsymbol{W}})^T \right]. \tag{13}$$

4 Experiments

The proposed PNFA algorithm was tested on a three-dimensional PNL mixture of two independent sources. The sources were a sine wave and uniformly distributed white noise. The PNL transformation used for generating the data contained two non-invertible post-nonlinear distortions:

$$\mathbf{y} = \begin{bmatrix} 1.2 & 0.2 \\ 1 & 0.7 \\ 0.2 & 0.8 \end{bmatrix} \mathbf{s} \qquad \mathbf{x} = \begin{bmatrix} (y_1 - 0.5)^2 \\ (y_2 + 0.4)^2 \\ \tanh(2y_3) \end{bmatrix}. \tag{14}$$

The observations were centered and normalised to unit variance and observation noise with variance 0.01 was added. The number of samples was 400.

The PNFA model was trained by trying different model structures, i.e. different numbers of hidden neurons in the PNL MLPs (5), and several random initialisations of the parameters to be optimised. The source initialisation was done by the principal component analysis of the observations. The best PNFA model[1] had 5 neurons in the hidden layers of all MLPs.

The PNL distortions learned by the best model after 10000 iterations is presented in Fig. 2: The post-nonlinearities f_i are estimated quite well except for

---

[1] The best model has the smallest value of the cost function (9) which corresponds to the maximum lower bound of the model evidence $p(\boldsymbol{X}|model)$.

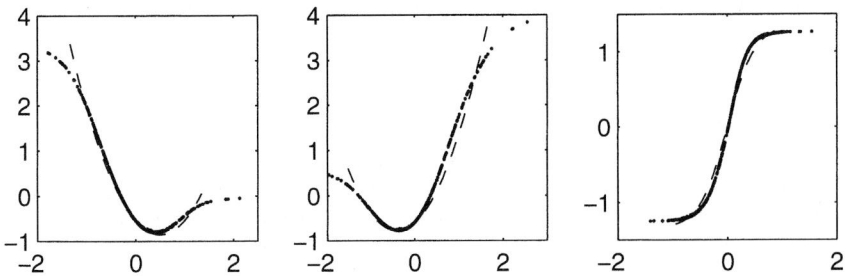

Fig. 2. The estimated post-nonlinear distortions f_i against the functions used for generating the data (the dashed line). Each point in the figure corresponds to a single observation.

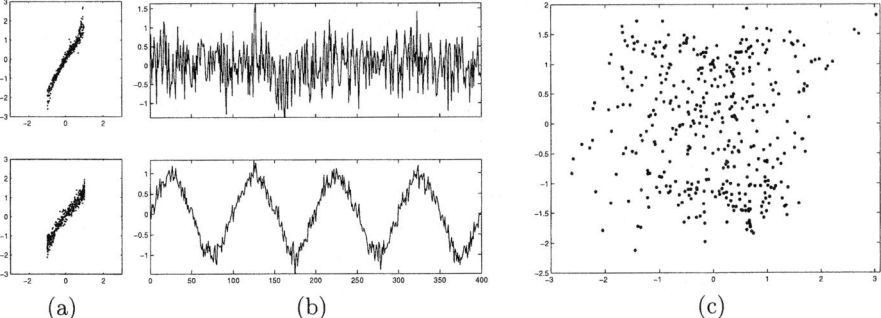

Fig. 3. The sources found by the PNFA and further rotated with the FastICA algorithm. (a) – the scatter plots; (b) – the estimated time series; (c) – the distribution of the sources. The signal-to-noise ratio is 12.95 dB.

some points at the edges. The difficulties mostly affect the two quadratic functions which are difficult to model with such small MLP networks and relatively few observations, especially at those edges.

The sources found by PNFA were further rotated by the FastICA algorithm to obtain independent signals (see Fig. 3). The scatter plots in Fig. 3a show how well the original sources were reconstructed. Each point corresponds to one source $s_i(t)$. The abscissa of a point is the original source which was used for generating the data and the ordinate is the estimated source. The optimal result would be a straight line which would mean that the estimated values of the sources coincide with the true values. Again, the sources were estimated quite well except for some points at the edges.

This result is somewhat natural due to the great difficulty of the test problem: There are only two bounded sub-Gaussian sources in the mixture and their linear combinations are quite far from Gaussianity assumed by PNFA. Another difficulty is the complex PNL mapping with a small number of observations and several non-invertible post-nonlinear distortions. Removing any of the observations from the mixture would make the mixing process non-injective and the separation problem unsolvable.

5 Discussion

In this paper, we presented a new Bayesian algorithm for learning the post-nonlinear mixing structure. The algorithm which we call post-nonlinear factor analysis is based on modelling the component-wise post-nonlinear distortions by MLP networks and using variational Bayesian learning.

An important feature of the proposed technique is that it learns the generative model of the observations while most existing PNL methods estimate the complementary separating structure. This makes the algorithm applicable to some post-nonlinear ICA problems unsolvable for the alternative methods.

We tested PNFA on a very challenging ICA problem and the obtained experimental results are very promising. The PNFA algorithm complemented by a linear ICA method was able to recover original sources from a globally invertible PNL mixture with non-invertible post-nonlinear distortions. This cannot be achieved by existing alternative methods [5].

The presented results are still preliminary and further investigations of the algorithm are needed. For example, the problem with local minima appears more severe for PNL mixtures with non-invertible distortions. Another interesting question is whether PNFA can improve the source restoration quality compared to the general NFA method applied to PNL problems.

An important issue is how the proposed PNL ICA technique works in higher-dimensional problems: Due to the Gaussianity assumption for the sources, the performance of the algorithm may be better for a greater number of mixed sources. Also, we are planning to implement a mixture-of-Gaussians model for the sources like in [12, 6] in order to improve the source estimation quality.

Acknowledgements

This work was partially done in the Lab. des Images et des Signaux at Institut National Polytechnique de Grenoble (INPG) in France. The authors would like to thank Sophie Achard, Christian Jutten and Harri Valpola for the fruitful discussions and help. This research has been partially funded by the European Commission project BLISS.

References

1. A. Hyvärinen, J. Karhunen, and E. Oja, *Independent Component Analysis*. J. Wiley, 2001.
2. A. Hyvärinen and P. Pajunen, "Nonlinear independent component analysis: Existence and uniqueness results," *Neural Networks*, vol. 12, no. 3, pp. 429–439, 1999.
3. A. Taleb and C. Jutten, "Source separation in post-nonlinear mixtures," *IEEE Trans. on Signal Processing*, vol. 47, no. 10, pp. 2807–2820, 1999.

4. C. Jutten and J. Karhunen, "Advances in nonlinear blind source separation," in *Proc. of the 4th Int. Symp. on Independent Component Analysis and Blind Signal Separation (ICA2003)*, pp. 245–256, 2003. Invited paper in the special session on nonlinear ICA and BSS.
5. A. Ilin, S. Achard, and C. Jutten, "Bayesian versus constrained structure approaches for source separation in post-nonlinear mixtures," in *Proc. International Joint Conference on Neural Networks (IJCNN 2004)*, 2004. To appear.
6. H. Lappalainen and A. Honkela, "Bayesian nonlinear independent component analysis by multi-layer perceptrons," in *Advances in Independent Component Analysis* (M. Girolami, ed.), pp. 93–121, Berlin: Springer-Verlag, 2000.
7. H. Valpola, E. Oja, A. Ilin, A. Honkela, and J. Karhunen, "Nonlinear blind source separation by variational Bayesian learning," *IEICE Transactions on Fundamentals of Electronics, Communications and Computer Sciences*, vol. E86-A, no. 3, pp. 532–541, 2003.
8. A. Taleb and C. Jutten, "Batch algorithm for source separation in postnonlinear mixtures," in *Proc. Int. Workshop on Independent Component Analysis and Signal Separation (ICA'99)*, (Aussois, France), pp. 155–160, 1999.
9. G. E. Hinton and D. van Camp, "Keeping neural networks simple by minimizing the description length of the weights," in *Proc. of the 6th Ann. ACM Conf. on Computational Learning Theory*, (Santa Cruz, CA, USA), pp. 5–13, 1993.
10. D. J. C. MacKay, "Developments in probabilistic modelling with neural networks – ensemble learning," in *Neural Networks: Artificial Intelligence and Industrial Applications. Proc. of the 3rd Annual Symposium on Neural Networks*, pp. 191–198, 1995.
11. H. Lappalainen and J. Miskin, "Ensemble learning," in *Advances in Independent Component Analysis* (M. Girolami, ed.), pp. 75–92, Berlin: Springer-Verlag, 2000.
12. H. Attias, "Independent factor analysis," *Neural Computation*, vol. 11, no. 4, pp. 803–851, 1999.
13. H. Lappalainen, "Ensemble learning for independent component analysis," in *Proc. Int. Workshop on Independent Component Analysis and Signal Separation (ICA'99)*, (Aussois, France), pp. 7–12, 1999.
14. J. Miskin and D. J. C. MacKay, "Ensemble learning for blind source separation," in *Independent Component Analysis: Principles and Practice* (S. Roberts and R. Everson, eds.), pp. 209–233, Cambridge University Press, 2001.
15. W. Penny, R. Everson, and S. Roberts, "ICA: model order selection and dynamic source models," in *Independent Component Analysis: Principles and Practice* (S. Roberts and R. Everson, eds.), pp. 299–314, Cambridge University Press, 2001.
16. H. Valpola and J. Karhunen, "An unsupervised ensemble learning method for nonlinear dynamic state-space models," *Neural Computation*, vol. 14, no. 11, pp. 2647–2692, 2002.
17. A. Honkela, "Approximating nonlinear transformations of probability distributions for nonlinear independent component analysis," in *Proc. International Joint Conference on Neural Networks (IJCNN 2004)*, 2004. To appear.

Temporal Decorrelation as Preprocessing for Linear and Post-nonlinear ICA

Juha Karvanen and Toshihisa Tanaka

Laboratory for Advanced Brain Signal Processing
Brain Science Institute, RIKEN
2-1 Hirosawa, Wako-shi, Saitama 351-0198, Japan
juha.karvanen@hut.fi, t.tanaka@ieee.org

Abstract. We present a straightforward way to use temporal decorrelation as preprocessing in linear and post-nonlinear independent component analysis (ICA) with higher order statistics (HOS). Contrary to the separation methods using second order statistics (SOS), the proposed method can be applied when the sources have similar temporal structure. The main idea is that componentwise decorrelation increases non-Gaussianity and therefore makes it easier to separate sources with HOS ICA. Conceptually, the non-Gaussianizing filtering matches very well with the Gaussianization used to cancel the post-nonlinear distortions. Examples demonstrating the consistent improvement in the separation quality are provided for the both linear and post-linear cases.

1 Introduction

In independent component analysis (ICA), the goal is to present the observed signals as linear (or nonlinear) combinations of statistically independent components (source signals). Most of methods for ICA fall in either of the following categories

1. Higher order statistics (HOS) ICA. The sources are required to be non-Gaussian. The possible time structures of the signals are not utilized. Higher order statistics (e.g. fourth order cumulants) are optimized in the algorithms.
2. Second order statistics (SOS) ICA. The non-Gaussianity assumption is not needed. The separation is based on different time structures of the sources. Second order statistics (covariances with different time delays) are optimized.

In addition, some methods using nonstationary or time-frequency distributions are proposed.

We consider the case where the mixing is instantaneous (linear or post-nonlinear) and the sources are non-Gaussian, stationary and mutually independent. In addition, the sources are assumed to have similar temporal structure (i.e. same spectra). Note that this assumption is the opposite of the assumption of different temporal structures that is needed in the SOS methods. Since the sources are non-Gaussian and the mixing is instantaneous, the HOS methods are applicable. The ordinary HOS methods, however, do not employ temporal structures. In this paper, we propose a straightforward way to use temporal decorrelation preprocessing (TDP) in HOS ICA.

Our work is related to many recent papers on ICA and blind source separation (BSS). An overview on different ICA and BSS methods and related approaches is given in [1]. In the framework of convolutive mixing, linear prediction and temporal decorrelation are considered e.g. in [2–4]. Temporal predictability is used as separation criterion in [5, 6]. The concept of subband decomposition is considered in [7, 5, 8]. An interesting attempt to use temporal structures is also presented in [9].

The idea that ICA can be applied to the innovation processes was presented in [10]. Unlike in the approaches mentioned above, the temporal properties are used here only in the preprocessing for the instantaneous ICA. In this paper, we study how the preprocessing can be carried out with decorrelating filters and extend the idea of preprocessing for the post-nonlinear ICA. The key ideas are presented first for the linear ICA in Section 2 and then the concept is applied to the post-nonlinear ICA in Section 3. Examples are provided in Section 4. Section 5 concludes the paper.

2 Temporal Decorrelation in Linear ICA

We consider the linear instantaneous mixing model

$$\mathbf{x}(t) = \mathbf{A}\mathbf{s}(t), \tag{1}$$

where the sources $\mathbf{s}(t) = [s_1(t), s_2(t), \ldots, s_m(t)]^\mathsf{T}$ are mutually independent and at mostly one of the sources is Gaussian. In addition, we assume that the sources have similar temporal structure (same spectra). In ICA, our goal is to estimate $\mathbf{y}(t) = \mathbf{W}\mathbf{x}(t)$ such that $\mathbf{y}(t)$ is a permutated and scaled estimate of the sources $\mathbf{s}(t)$.

Now assume that the filter $\mathbf{h} = [h(0), h(1), h(2), \ldots, h(L)]^\mathsf{T}$ is applied to each component of $\mathbf{x}(t)$ and filtered signals $\tilde{\mathbf{x}}(t)$ are obtained. The filtering may be written as

$$\tilde{\mathbf{x}}(t) = \mathbf{X}(t)\mathbf{h} = \bigl(\mathbf{A}\mathbf{S}(t)\bigr)\mathbf{h} = \mathbf{A}\bigl(\mathbf{S}(t)\mathbf{h}\bigr) = \mathbf{A}\tilde{\mathbf{s}}(t), \tag{2}$$

where

$$\mathbf{X}(t) = \begin{pmatrix} x_1(t) & x_1(t-1) & \ldots & x_1(t-L) \\ x_2(t) & x_2(t-1) & \ldots & x_2(t-L) \\ \vdots & \vdots & \ddots & \vdots \\ x_m(t) & x_m(t-1) & \ldots & x_m(t-L) \end{pmatrix} \tag{3}$$

$$\mathbf{S}(t) = \begin{pmatrix} s_1(t) & s_1(t-1) & \ldots & s_1(t-L) \\ s_2(t) & s_2(t-1) & \ldots & s_2(t-L) \\ \vdots & \vdots & \ddots & \vdots \\ s_m(t) & s_m(t-1) & \ldots & s_m(t-L) \end{pmatrix}. \tag{4}$$

The equation (2) shows that we can apply ICA to the filtered signals $\tilde{\mathbf{x}}(t)$ instead of $\mathbf{x}(t)$ and still obtain the same separating matrix $\mathbf{W}$. Theoretically, this result applies for any nonzero filter $\mathbf{h}$, but in practice the filter should be carefully chosen. It is important to notice that although the separating matrix $\tilde{\mathbf{W}}$ is estimated from $\tilde{\mathbf{x}}(t)$, the estimated independent components are $\mathbf{y}(t) = \tilde{\mathbf{W}}\mathbf{x}(t)$, not $\tilde{\mathbf{y}}(t) = \tilde{\mathbf{W}}\tilde{\mathbf{x}}(t)$.

We propose that the preprocessing filter $\mathbf{h}$ should result in temporal decorrelation. This corresponds to computing the residuals of the original signal and a linear predictor. Temporal decorrelation is expected to increase non-Gaussianity and thus make it easier to estimate the separating matrix. Here the implicit assumption is that the sources are time series with independent, non-Gaussian innovation processes. When these innovations are summed, Gaussianity increases. The opposite operation, linear prediction, reduces Gaussianity.

In practice, temporal decorrelation can be applied if the temporal structure of the signals is approximately similar. However, equation (2) holds only if the same filter is used for all signals. Therefore, the filter applied can be, for instance, an average of the componentwise linear predictors. The key property is that the filtering reduces autocorrelation for all signals.

3 Temporal Decorrelation in Post-nonlinear ICA

In the post-nonlinear ICA model, invertible unknown nonlinear distortions f_i are applied componentwise

$$x_i(t) = f_i \left(\sum_{j=1}^{m} a_{ij} s_j(t) \right). \tag{5}$$

The algorithms for the post-nonlinear mixtures consist of the cancellation of the nonlinear distortions and linear ICA. Methods for post-nonlinear ICA are reviewed in [11]. A natural way to cancel the nonlinear distortions is Gaussianization proposed by Ziehe et al. [12]. The motivation is that the linear mixtures $\sum_{j=1}^{m} a_{ij} s_j(t)$ are nearly Gaussian and the nonlinear transformations make them less Gaussian. Gaussianization may be used to approximate the signals before the nonlinear transformations. Gaussianization is performed employing the result that any continuous distribution can be transformed to any other continuous distribution [13]. More specifically, a random variable x with a cumulative distribution function (cdf) F_x can be transformed to random variable y with cdf F_y defining

$$y = F_y^{-1}(F_x(x)). \tag{6}$$

In the case of Gaussianization, this leads to formula

$$v_i = \Phi^{-1}(F(x_i)) \approx \Phi^{-1}\left(\frac{\mathrm{rank}(x_i)}{T+1}\right), \tag{7}$$

where Φ is the Gaussian cdf, $\mathrm{rank}(x_i)$ denotes rankings in ordered data and $+1$ in denominator is needed to avoid infinite values.

The obvious problem of Gaussianization is that theoretically it makes the signals exactly Gaussian and thus non-separable by the HOS methods. To overcome this problem, the SOS methods were recommended in [12]. The SOS methods are, however, useful only if all sources have different temporal structure. We propose that the HOS methods with the TDP can solve the problem when all sources have similar temporal structure. The proposed algorithm for post-nonlinear mixtures consist of the following components:

1. Gaussianization
2. Temporal decorrelation
3. HOS ICA

After Gaussianization the procedure is similar to the linear case.

4 Examples

In this section, we present simulation examples that demonstrate the performance gain due to the TDP. We generate mutually independent sources that have similar time structure. The sources are instantaneously mixed and post-nonlinear the separation results with and without the temporal decorrelation preprocessing are compared. Two alternative ARMA(1,1) (autoregressive moving average) models are considered as time structures of the sources:

– Strong autocorrelation model

$$s_i(t) = -0.4 s_i(t-1) + u_i(t) + 0.5 u_i(t-1) \tag{8}$$

– Weak autocorrelation model

$$s_i(t) = -0.1 s_i(t-1) + u_i(t) + 0.1 u_i(t-1). \tag{9}$$

The innovation processes u_i are mutually independent and have uniform or Laplacian distribution. For temporal decorrelation we use linear prediction with autoregressive order 3. The predictor coefficients are estimated by the autocorrelation method [14]. Pearson-ICA [15] and JADE [16] are chosen as the ICA algorithms.

In the first example, we studied the linear model. 12 sources were mixed: 6 sources had uniform innovation process and 6 sources had Laplacian innovation process. Sample size (length of signals) was 5000. A full-rank mixing matrix **A** was randomly generated.

The results from 1000 experiments are summarized in Tables 1 and 2. Signal to interference ratio (SIR(dB)=$-10\log_{10}$(MSE), MSE stands for mean square error) between the sources and their sign and the scale adjusted estimates are used to measure the quality of separation.

In the case of strong autocorrelation (Table 1), the TDP clearly improved separation: Pearson-ICA alone achieved the median SIR 17.16 dB, whereas Pearson-ICA with temporal decorrelation gave the median SIR 26.56 dB. The improvement is consistent: the results with the TDP were better in all 1000 experiments.

The conclusions are essentially similar for the JADE algorithm: the overall performance was slightly worse than with Pearson-ICA, but JADE with temporal decorrelation gave consistently better results than JADE without temporal decorrelation. SOBI [17] (a SOS algorithm) failed because the assumption of different time structures is violated.

In the case of weak autocorrelation (Table 2), the TDP also improved separation. Although the difference between the results with and without temporal decorrelation was not large, the results with temporal decorrelation were better in almost all experiments.

Table 1. Separation of linear mixtures of 12 strongly autocorrelated sources. The median SIR and a nonparametric 95% confidence interval from 1000 experiments are reported. The column 'Best' indicates the percentage of the experiments where the method gave the best SIR value.

Method	Median	95% interval	Best
TDP & Pearson-ICA	26.56 dB	(25.35, 27.85)	99.9%
Pearson-ICA	17.16 dB	(15.71, 18.85)	0.0%
TDP & JADE	21.11 dB	(19.32, 22.91)	0.1%
JADE	15.16 dB	(13.43, 16.90)	0.0%
SOBI	-0.16 dB	(-0.55, 0.32)	0.0%

Table 2. Separation of linear mixtures of 12 weakly autocorrelated sources.

Method	Median	95% interval	Best
TDP & Pearson-ICA	26.78 dB	(25.53, 27.98)	97.6%
Pearson-ICA	26.01 dB	(24.79, 27.22)	2.4%
TDP & JADE	21.18 dB	(19.37, 22.79)	0.0%
JADE	20.80 dB	(18.96, 22.42)	0.0%
SOBI	-0.16 dB	(-0.53, 0.33)	0.0%

Next, the same experimental settings were applied in the post-nonlinear case. As post-nonlinear distortions we used the following nonlinearities

$$f_i(x_i) = \tanh(2x_i), \qquad i \leq m/2 \qquad (10)$$
$$f_i(x_i) = x_i^3, \qquad i > m/2. \qquad (11)$$

In separation, Gaussianization (7) was first performed. Then Pearson-ICA and JADE were used either with or without temporal decorrelation.

The results are shown in Tables 3 and 4. It is immediately seen that the separation of post-nonlinear mixtures is a more difficult problem than the separation of linear mixtures. The SIRs varied lot and in some cases the performance was relatively poor. The performance gain due to temporal decorrelation is, however, visible also in the post-nonlinear case. In the case of strong autocorrelation, the TDP with Pearson-ICA gave the best separation in 943 experiments out of 1000. In the case of weak autocorrelation, the TDP with Pearson-ICA was the best in 870 experiments but the differences were almost negligible.

Table 3. Separation of post-linear mixtures of 12 strongly autocorrelated sources.

Method	Median	95% interval	Best
TDP & Pearson-ICA	9.36 dB	(5.43, 20.71)	93.5%
Pearson-ICA	7.89 dB	(4.29, 15.84)	5.7%
TDP & JADE	7.00 dB	(3.83, 18.21)	0.8%
JADE	4.83 dB	(1.78, 14.12)	0.0%
SOBI	-0.17 dB	(-0.55, 0.30)	0.0%

Table 4. Separation of post-linear mixtures of 12 weakly autocorrelated sources.

Method	Median	95% interval	Best
TDP & Pearson-ICA	9.87 dB	(5.28, 18.85)	87.7%
Pearson-ICA	9.73 dB	(5.20, 18.65)	12.1%
TDP & JADE	6.46 dB	(3.14, 16.73)	0.2%
JADE	6.35 dB	(2.97, 16.58)	0.0%
SOBI	-0.17 dB	(-0.56, 0.28)	0.0%

We also studied the effect of increasing the number of sources in the post-nonlinear case. In our simulation, the number of sources varied from 2 to 20. The strong autocorrelation model (8) was used for the sources. Half of the sources had uniform innovations and half of the sources had Laplacian innovations. For each number of the sources we generated 101 realizations using full-rank random mixing matrices and post-nonlinear distortions (10).

In Figure 1 the median SIRs from 101 experiment are presented. Again, the TDP clearly improved the separation. An interesting phenomenon was seen in the medians when temporal decorrelation was not applied: the median SIR for 6 sources is higher than the median SIR for 2 or 4 sources. No such peak occurred when the TDP was applied and the median SIRs for 2 and 4 sources were over 20 dB. The same example was also repeated using sub-Gaussian sources (uniform innovations) only or super-Gaussian sources (Laplacian innovations) only. The results were effectively similar to Figure 1.

5 Conclusion

In this paper, we present a straightforward way to use temporal decorrelation as preprocessing in HOS ICA. The proposed method can be applied when the sources have similar temporal structure, i.e. when the SOS methods fail. The source signals are assumed to be time series with independent, non-Gaussian innovation processes. Our examples demonstrate consistent improvement in separation due to the TDP both in linear and post-nonlinear case. The performance gain is related to the strength of autocorrelation.

In future, we will consider the different estimation methods for temporal decorrelation. Another interesting open problem is ICA in the case where some sources have the same temporal structure and some sources have different temporal structure. The effect of Gaussianization in the post-nonlinear ICA is also worth of further studies.

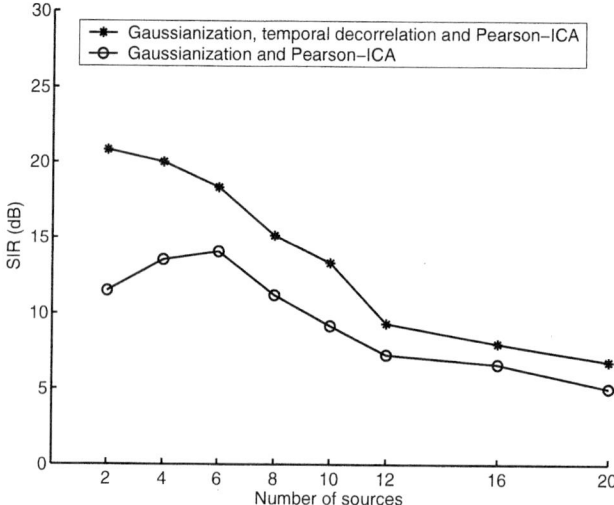

Fig. 1. Post-nonlinear ICA (Gaussianization + Pearson-ICA) with and without the temporal decorrelation preprocessing. The medians of the overall SIRs from 101 experiments are reported as a function of the number of the sources.

References

1. Cichocki, A., Amari, S.I.: Adaptive Blind Signal and Image Processing: Learning Algorithms and Applications. Wiley (2002)
2. Lee, T.W., Ziehe, A., Orglmeister, R., Sejnowski, T.: Combining time-delayed decorrelation and ICA: towards solving the cocktail party problem. In: Proc. ICASSP98. Volume 2. (1998) 1249–1252
3. Kokkinakis, K., Zarzoso, V., Nandi, A.K.: Blind separation of acoustic mixtures based on linear prediction analysis. In: Proc. Fourth International Symposium on Independent Component Analysis and Blind Signal Separation (ICA2003). (2003) 343–348
4. Nishikawa, T., Sarauwatari, H., Shikano, K.: Stable learning algorithm for blind separation of temporally correlated signals combining multistage ICA and linear prediction. In: Proc. Fourth International Symposium on Independent Component Analysis and Blind Signal Separation (ICA2003). (2003) 337–342
5. Cichocki, A., Rutkowski, T., Siwek, K.: Blind signal extraction of signals with specified frequency band. In: Proc. International Workshop on Neural Networks for Signal Processing. (2002)
6. Mandic, D.P., Cichocki, A.: An online algorithm for blind extraction of sources with different dynamical structures. In: Proc. Fourth International Symposium on Independent Component Analysis and Blind Signal Separation (ICA2003). (2003) 645–650
7. Tanaka, T., Cichocki, A.: Subband decomposition independent component analysis and new performance criteria. In: Proc. of 2004 IEEE International Conference on Acoustics, Speech, and Signal Processing (ICASSP 2004). (2004)
8. A. Cichocki, S. Amari, K. Siwek, T. Tanaka et al.: ICALAB Toolboxes, http://www.bsp.brain.riken.jp/ICALAB. (2002)

9. Jung, A., Kaiser, A.: Considering temporal structures in independent component analysis. In: Proc. Fourth International Symposium on Independent Component Analysis and Blind Signal Separation (ICA2003). (2003) 95–100
10. Hyvärinen, A.: Independent component analysis for time-dependent stochastic processes. In: Proc. Int. Conf. on Artificial Neural Networks (ICANN'98). (1998) 541–546
11. Jutten, C., Karhunen, J.: Advances in nonlinear blind source separation. In: Proc. Fourth International Symposium on Independent Component Analysis and Blind Signal Separation (ICA2003). (2003) 245–256
12. Ziehe, A., Kawanabe, M., Harmeling, S., Müller, K.R.: Blind separation of post-nonlinear mixtures using Gaussianizing transformations and temporal decorrelation. In: Proc. Fourth International Symposium on Independent Component Analysis and Blind Signal Separation (ICA2003). (2003) 269–274
13. Stuart, A., Ord, J.K.: Kendall's Advanced Theory of Statistics: Distribution Theory. Sixth edn. Volume 1. Edward Arnold (1994)
14. Jackson, L.: Digital Filters and Signal Processing. Second edn. Kluwer Academic Publishers (1989)
15. Karvanen, J., Koivunen, V.: Blind separation methods based on Pearson system its extensions. Signal Processing **82** (2002) 663–673
16. Cardoso, J., Souloumiac, A.: Blind beamforming for non Gaussian signals. IEE-Proceedings-F **140** (1993) 362–370
17. Belouchrani, A., Meraim, K.A., Cardoso, J.F., Moulines, E.: A blind source separation technique based on second order statistics. IEEE Transactions on Signal Processing **45** (1997) 434–444

Tree-Dependent and Topographic Independent Component Analysis for fMRI Analysis

Anke Meyer-Bäse[1], Fabian J. Theis[1,2],
Oliver Lange[1,3], and Carlos G. Puntonet[4]

[1] Department of Electrical and Computer Engineering
Florida State University, Tallahassee, Florida, 32310-6046, USA
[2] Institute of Biophysics, University of Regensburg
D-93040 Regensburg, Germany
[3] Department of Clinical Radiology, Ludwig–Maximilians University
Munich 80336, Germany
[4] Department of Architecture and Computer Technology
E-18071 University of Granada, Spain

Abstract. Recently, a new paradigm in ICA emerged, that of finding "clusters" of dependent components. This striking philosophy found its implementation in two new ICA algorithms: tree–dependent and topographic ICA. Applied to fMRI, this leads to the unifying paradigm of combining two powerful exploratory data analysis methods, ICA and unsupervised clustering techniques. For the fMRI data, a comparative quantitative evaluation between the two methods, tree–dependent and topographic ICA was performed. The comparative results were evaluated based on (1) correlation and associated time–courses and (2) ROC study. It can be seen that topographic ICA outperforms all other ICA methods including tree–dependent ICA for 8 and 9 ICs. However, for 16 ICs topographic ICA is outperformed by both FastICA and tree–dependent ICA (KGV) using as an approximation of the mutual information the kernel generalized variance.

1 Introduction

Functional magnetic resonance imaging with high temporal and spatial resolution represents a powerful technique for visualizing rapid and fine activation patterns of the human brain. Among the data–driven techniques, ICA has been shown to provide a powerful method for the exploratory analysis of fMRI data [1, 2]. ICA is an information theoretic approach which enables to recover underlying signals, or independent components (ICs) from linear data mixtures. Therefore, it is an excellent method to be applied for the spatial localization and temporal characterization of sources of BOLD activation. ICA can be applied to fMRI both temporal [3] or spatial [2]. Spatial ICA has dominated so far in fMRI applications because the spatial dimension is much larger than the temporal dimension in fMRI. However, recent literature results have suggested that temporal and spatial ICA yield similar results for experiments where two predictable task–related components are present.

In this paper, we perform a detailed comparative study for fMRI among the tree–dependent and topographic ICA with standard ICA techniques. The employed ICA algorithms are the TDSEP [4], JADE [5], the FastICA [6], the tree–dependent ICA [7], and topographic ICA which combines topographic mapping with ICA [8].

In a systematic manner, we will compare and evaluate the results obtained based on each technique and present the benefits associated with each paradigm.

2 Models of Spatial ICA in fMRI

According to the principle of functional organization of the brain, it was suggested for the first time in [2] that the multifocal brain areas activated by performance of a visual task should be unrelated to the brain areas whose signals are affected by artifacts of physiological nature, head movements, or scanner noise related to fMRI experiments. Every single above mentioned process can be described by one or more spatially–independent components, each associated with a single time course of a voxel and a component map. It is assumed that the component maps, each described by a spatial distribution of fixed values, is representing overlapping, multifocal brain area of statistically dependent fMRI signals. This aspect is visualized in Figure 1. In addition, it is considered that the distributions of the component maps are spatially independent, and in this sense uniquely specified. It was shown in [2] that these maps are independent if the active voxels in the maps are sparse and mostly nonoverlapping. Additionally it is assumed that the observed fMRI signals are the superposition of the individual component processes at each voxel. Based on these assumptions, ICA can be applied to fMRI time–series to spatially localize and temporally characterize the sources of BOLD activation.

In the following we will assume that $\mathbf{X}$ is a $T \times M$ matrix of observed voxel time courses (fMRI signal data matrix), $\mathbf{C}$ is the $N \times M$ random matrix of component map values, and A is a $T \times N$ mixing matrix containing in its columns the associated time–courses of the N components. Furthermore, T corresponds to the number of scans, and M is the number of voxels included in the analysis.

The spatial ICA (sICA) problem is given by the following linear combination model for the data:

$$\mathbf{X} = \mathbf{AC} \qquad (1)$$

where no assumptions are made about the mixing matrix $\mathbf{A}$ and the rows $\mathbf{C_i}$ being mutually statistically independent.

Then the ICA decomposition of $\mathbf{X}$ can be defined as an invertible transformation:

$$\mathbf{C} = \mathbf{WX} \qquad (2)$$

where $\mathbf{W}$ is an unmixing matrix providing a linear decomposition of data. $\mathbf{A}$ is the pseudoinverse of $\mathbf{W}$.

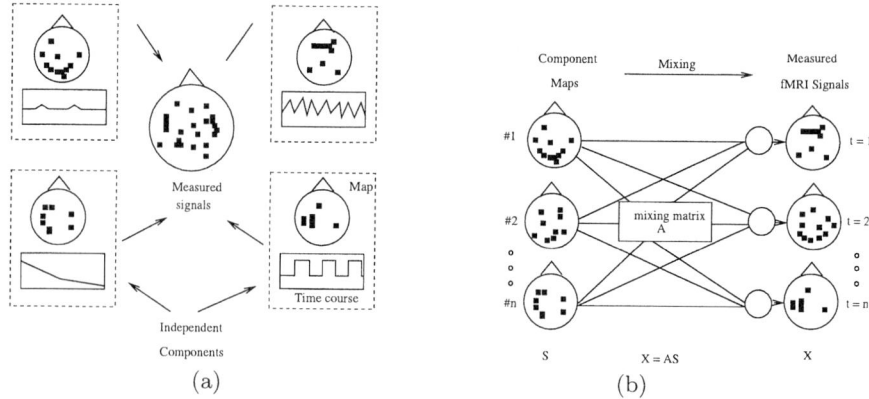

Fig. 1. Visualization of ICA applied to fMRI data. (a) Scheme of fMRI data decomposed into independent components, and (b) fMRI data as a mixture of independent components where the mixing matrix **M** specifies the relative contribution of each component at each time point [2].

3 Tree-Dependent Component Analysis

The paradigm of TCA is derived from the theory of tree–structured graphical models. In [9] was shown a strategy to approximate optimally an n–dimensional discrete probability distribution by a product of second–order distributions, or the distribution of the first–order tree dependence. A tree is an undirected graph with at most a single edge between two nodes. This tree concept can be easily interpreted with respect to ICA. A graph with no edges means that the random variables are mutually independent and this pertains to ICA. The connected components of the graphical model can be viewed as "clusters" of dependent components, and thus the decomposition of the source variables yields to dependent components within a cluster and independent outside a cluster.

The idea of approximating discrete probability distributions with dependence trees can be easily translated to ICA [7]. In classic ICA, we want to minimize the mutual information of the estimated components $\mathbf{s} = \mathbf{Wx}$. Thus, the result derived in [9], can be easily extended and becomes the tree–dependent ICA.

The objective function for TCA is given by $J(\mathbf{x}, \mathbf{W}, t)$ and includes the demixing matrix $\mathbf{W}$. Thus, the mutual information for TCA becomes

$$J(\mathbf{x}, \mathbf{W}, t) = I^t(\mathbf{s}) = I(s_1, \cdots, s_m) - \sum_{(u,v) \in t} I(s_u, s_v) \qquad (3)$$

$\mathbf{s}$ factorizes in a tree t.

In TCA as in ICA, the density $p(\mathbf{x})$ is not known and the estimation criteria have to be substituted by empirical contrast functions. As described in [7], we will employ three types of contrast functions: (i) approximation of the entropies being part of equation (3) via kernel density estimation (KDE), (ii) approximation of the mutual information based on kernel generalized variance (KGV), and (iii) approximation based on cumulants using Gram–Charlier expansions (CUM).

4 Topographical Independent Component Analysis

The paradigm of topographic ICA has its roots in [10] where a combination of invariant feature subspaces [11] and independent subspaces [12] is proposed.

To introduce a topographic representation in the ICA model, it is necessary to relax the assumption of independence among neighboring components s_i. This makes it necessary to adopt an idea from self-organized neural networks, that of a lattice. It was shown in [8] that a representation which models topographic correlation of energies is an adequate approach for introducing dependencies between neighboring components.

In other words, the variances corresponding to neighboring components are positively correlated while the other variances are in a broad sense independent. The architecture of this new approach is shown in Figure 2.

This idea leads to the following representation of the source signals:

$$s_i = \sigma_i z_i \qquad (4)$$

where z_i is a random variable having the same distribution as s_i, and the variance σ_i is fixed to unity.

The variance σ_i is further modeled by a nonlinearity:

$$\sigma_i = \phi\left(\sum_{k=1}^{n} h(i,k) u_k\right) \qquad (5)$$

where u_i are the higher order independent components used to generate the variances, while ϕ describes some nonlinearity. The neighborhood function $h(i,j)$ can either be a two-dimensional grid or have a ring-like structure. Further u_i and z_i are all mutually independent. The classic ICA results from the topographic ICA by setting $h(i,j) = \delta_{ij}$.

5 Results and Discussion

FMRI data were recorded from five subjects performing a visual task. For each subject, five slices with 100 images (TR/TE=3000/60msec) were acquired with five periods of rest and five photic simulation periods with rest. Simulation and rest periods comprised 10 repetitions each, i.e. 30s. Resolution was $3 \times 3 \times 4$ mm. The slices were oriented parallel to the calcarine fissure. Photic stimulation was performed using an 8 Hz alternating checkerboard stimulus with a central fixation point and a dark background with a central fixation point during the control periods [13]. The first scans were discarded for remaining saturation effects. Motion artifacts were compensated by automatic image alignment (AIR, [14]).

The clustering results were evaluated by (1) artifactual– and task–related activation maps, (2) associated time–courses and (3) ROC curves.

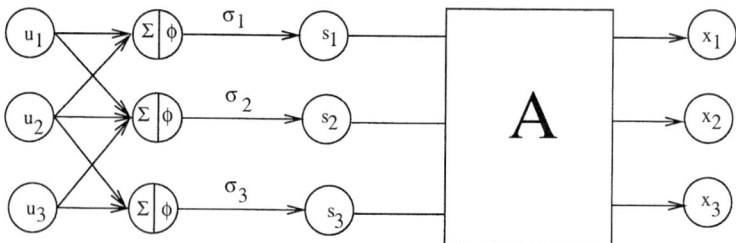

Fig. 2. Topographic ICA model [8]. The variance generated variables u_i are randomly generated and mixed linearly inside their topographic neighborhoods. This forms the input to nonlinearity ϕ, thus giving the local variance σ_i. Components s_i are generated with variances σ_i. The observed variables x_i are obtained as with standard ICA from the linear mixture of the components s_i.

5.1 Estimation of the ICA Model

To decide to what extent spatial ICA of fMRI time–series depends on the employed algorithm, we have first to look at the optimal number of principal components selected by PCA and used in the ICA decomposition. ICA is a generalization of PCA. In case no ICA is performed, then the number of independent components equals zero, and this means there is no PCA decomposition performed.

In the following we will give the set parameters. For PCA, no parameters had to be set. For FastICA we choose: (1) $\epsilon = 10^{-6}$, (2) 10^5 as the maximal number of iterations, and (3) the nonlinearity $g(u) = \tanh u$. And last, for topographic ICA we set: (1) stop criterium is fulfilled if the synaptic weights difference between two consecutive iterations is less than $10^{-5} \times$ Number of IC, (2) the function $g(u) = u$, and (3) 10^4 is the maximal number of iterations.

It is significant to find a fixed number of ICs that can theoretically predict new observations in same conditions, assuming the basic ICA model actually holds. To do so, we compared the six proposed algorithms for 8, 9, and 16 components in terms of Receiver Operating Characteristic (ROC) analysis using correlation map with a chosen threshold of 0.4. The obtained results are plotted in Figure 3. It can be seen that topographic ICA outperforms all other ICA methods for 8 and 9 ICs. However, for 16 ICs topographic ICA is outperformed by both FastICA and tree–dependent ICA (KGV) using as an approximation of the mutual information the kernel generalized variance.

5.2 Characterization of Task-Related Effects

For all subjects, and runs, unique task–related activation maps and associated time–courses were obtained by the tree–dependent and topographic ICA techniques. The correlation of the component time course most closely associated with the visual task for the these two techniques is shown in Table 1 for IC=8,9, and 16. From the Table, we see for the tree–dependent ICA a continuous increase

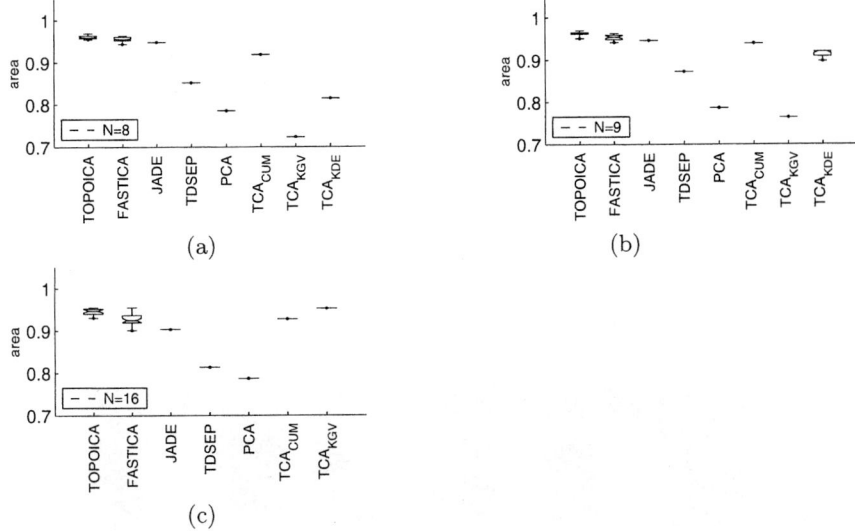

Fig. 3. Results of the comparison between tree–dependent ICA, topographic ICA, Jade, FastICA, TDSEP, and PCA on fMRI data. Spatial accuracy of ICA maps is assessed by ROC analysis using correlation map with a chosen threshold of 0.4. The number of chosen independent components for all techniques is in (a): IC=8, (b): IC=9, and (c): IC=16.

Table 1. Comparison of the correlations of the component time course most closely associated with the visual task for tree–dependent and topographic ICA for IC=8,9, and 16.

	Tree–dependent ICA	Topographic ICA
IC=8	0.78	0.85
IC=9	0.91	0.87
IC=16	0.92	0.86

for the correlation coefficient while for the topographic ICA this correlation coefficient decreases for IC=16.

An interesting aspect can be observed if we compare the computed reference func tions at the maximum correlation for topographic ICA with tree–dependent ICA. Figure 4 visualizes the computed reference functions for the two model-free methods. We see that the reference function for the tree–depedent ICA approximates better the shape of the stimulus function. Figure 5 shows task- and artifactual–related activation maps for tree–dependent ICA.

6 Conclusions

In the present paper, we have experimentally compared four standard ICA algorithms already adopted in the fMRI literature with two new algorithms, the tree– dependent and topographic ICA. The goal of the paper was to determine

788 Anke Meyer-Bäse et al.

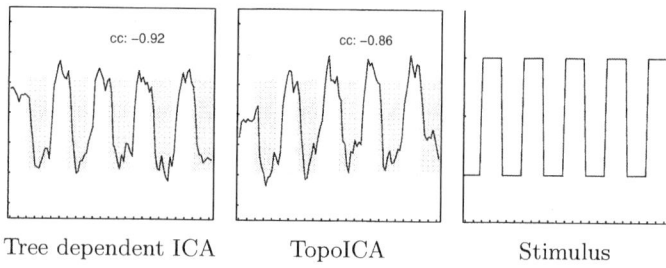

Tree dependent ICA   TopoICA   Stimulus

Fig. 4. Computed reference functions for the two techniques tree–dependent and topographic ICA (IC=16).

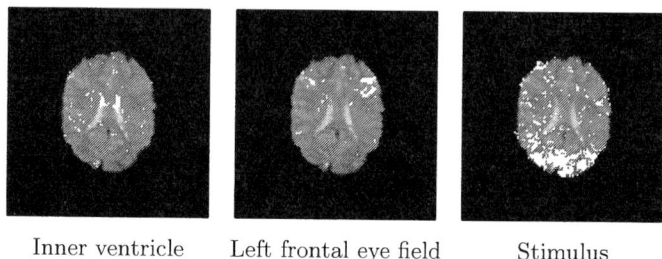

Inner ventricle   Left frontal eye field   Stimulus

Fig. 5. Computed activation maps (artifactual and task–related) for tree–dependent ICA (IC=16).

the robustness and reliability of extracting task–related activation maps and time–courses from fMRI data sets. The success of ICA methods is based on the condition that the spatial distribution of brain areas activated by task performance must be spatially independent of the distributions of areas affected by artifacts. It can be seen that topographic ICA outperforms all other ICA methods for 8 and 9 ICs. However, for 16 ICs topographic ICA is outperformed by both FastICA and tree–dependent ICA (KGV) using as an approximation of the mutual information the kernel generalized variance. The applicability of the new algorithms is demonstrated on experimental data.

Acknowledgement

The authors would like to thank Dr. Dorothee Auer from the Max Planck Institute of Psychiatry in Munich, Germany, for providing the fMRI data.

References

1. Arfanakis, K., Cordes, D., Haughton, V., Moritz, C., Quigley, M., Meyerand, M.: Combining independent component analysis and correlation analysis to probe interregional connectivity in fmri task activation datasets. Magnetic Resonance Imaging **18** (2000) 921–930

2. McKeown, M., Jung, T., Makeig, S., Brown, G., Jung, T., Kindermann, S., Bell, A., Sejnowski, T.: Analysis of fmri data by blind separation into independent spatial components. Human Brain Mapping **6** (1998) 160–188
3. Biswal, B., Ulmer, J.: Blind source separation of multiple signal sources of fmri data sets using independent component analysis. Journal of Computer Assisted Tomography **23** (1999) 265–271
4. Ziehe, A., Müller, K.: Tdsep - an efficient algorithm for blind separation using time structure. Proc. ICANN **2** (1998) 675–680
5. Cardoso, J.F., Souloumiac, A.: Blind beamforming for non gausssian signals. IEE Proceedings-F **140** (1993) 362–370
6. Hyvarinen, A., Oja, E.: Independent component analysis: algorithms and applications. Neural Networks **13** (2000) 411–430
7. Bach, F.R., Jordan, M.I.: Beyond independent components: Trees and clusters. Journal of Machine Learning Research **4** (2003) 1205–1233
8. Hyvarinen, A., Hoyer, P.: Topographic independent component analysis. Neural Computation **13** (2001) 1527–1558
9. Chow, C.K., Liu, C.N.: Approximating discrete probability distributions with dependence trees. IEEE Transaction on Information Theory **14** (1968) 462–467
10. Hyvarinen, A., Hoyer, P.: Emergence of phase- and shift-invariant features by decomposition of natural images into independent feature subspaces. Neural Computation **12** (2000) 1705–1720
11. Kohonen, T.: Emergence of invariant-feature detectors in the adaptive-subspace self-organizing map. Biological Cybernetics **75** (1996) 281–291
12. Cardoso, J.F.: Multidimensional independent component analysis. Proc. IEEE ICASSP, Seattle **4** (1998) 1941–1944
13. Wismüller, A., Lange, O., Dersch, D., Leinsinger, G., Hahn, K., Pütz, B., Auer, D.: Cluster analysis of biomedical image time–series. International Journal on Computer Vision **46** (2002) 102–128
14. Woods, R., Cherry, S., Mazziotta, J.: Rapid automated algorithm for aligning and reslicing pet images. Journal of Computer Assisted Tomography **16** (1992) 620–633

Using Kernel PCA for Initialisation of Variational Bayesian Nonlinear Blind Source Separation Method

Antti Honkela[1], Stefan Harmeling[2], Leo Lundqvist[1], and Harri Valpola[1]

[1] Helsinki University of Technology, Neural Networks Research Centre
P.O. Box 5400, FI-02015 HUT, Espoo, Finland
{antti.honkela,leo.lundqvist,harri.valpola}@hut.fi
[2] Fraunhofer FIRST.IDA, Kekuléstr. 7, 12489 Berlin, Germany
harmeli@first.fhg.de

Abstract. The variational Bayesian nonlinear blind source separation method introduced by Lappalainen and Honkela in 2000 is initialised with linear principal component analysis (PCA). Because of the multilayer perceptron (MLP) network used to model the nonlinearity, the method is susceptible to local minima and therefore sensitive to the initialisation used. As the method is used for nonlinear separation, the linear initialisation may in some cases lead it astray. In this paper we study the use of kernel PCA (KPCA) in the initialisation. KPCA is a rather straightforward generalisation of linear PCA and it is much faster to compute than the variational Bayesian method. The experiments show that it can produce significantly better initialisations than linear PCA. Additionally, the model comparison methods provided by the variational Bayesian framework can be easily applied to compare different kernels.

1 Introduction

Nonlinear blind source separation (BSS) and related nonlinear independent component analysis (ICA) are difficult problems. Several different methods have been proposed to solve them in a variety of different settings [1, 2]. In this work, we attempt to combine two different methodologies used for solving the general nonlinear BSS problem, the kernel based approach [3, 4] and the variational Bayesian (VB) approach [5, 6]. This is done by using sources recovered by kernel PCA as initialisation for the sources in the variational Bayesian nonlinear BSS method.

Kernel PCA (KPCA) [3] is a nonlinear generalisation of linear principal component analysis (PCA). It works by mapping the original data space nonlinearly to a high dimensional feature space and performing PCA in that space. With the kernel approach this can be done in a computationally efficient manner. One of the drawbacks of KPCA in general is the difficulty of mapping the extracted components back to the data space, but in the case of source initialisation, such mapping is not needed.

The variational Bayesian nonlinear BSS method presented in [5] is based on finding a generative model from a set of sources through a nonlinear mapping to the data. The sources and the model are found by using an iterative EM-like algorithm. Because of the flexible multilayer perceptron (MLP) network used to model the nonlinearity and general ill-posed nature of the problem, the method requires a reasonable initialisation to provide good results. In the original implementation, the initialisation was handled by computing a desired number of first linear principal components of the data and fixing the sources to those values for some time while the MLP network was adapted. The linear initialisation is robust and seems to work well in general, but a nonlinear initialisation provided by KPCA should lead to better results and faster learning.

In the next section, kernel PCA and variational Bayesian nonlinear BSS methods will be presented in more detail. Experimental results of using KPCA initialisation for VB approach are presented in Section 3. The paper concludes with discussion and conclusions in Sections 4 and 5.

2 The Methods

In this section, kernel PCA and the variational Bayesian nonlinear BSS method will be introduced briefly. For more details, see the referenced papers.

2.1 Kernel PCA

Kernel principal component analysis (kernel PCA) was introduced in [3] as a nonlinear generalisation of principal component analysis. The idea is to map given data points from their input space $\mathbb{R}^n$ to some high-dimensional (possibly infinite-dimensional) feature space $\mathcal{F}$,

$$\Phi : \mathbb{R}^n \to \mathcal{F}, \tag{1}$$

and to perform PCA in $\mathcal{F}$. The space $\mathcal{F}$ and therewith also the mapping Φ might be very complicated. However, employing the so-called kernel trick, kernel PCA avoids to use Φ explicitly: PCA in $\mathcal{F}$ is formulated in such a way that only the inner product in $\mathcal{F}$ is needed (for details see [3]). This inner product can be seen as some nonlinear function, called *kernel function*,

$$\begin{aligned} \mathbb{R}^n \times \mathbb{R}^n &\to \mathbb{R} \\ (\mathbf{x}, \mathbf{y}) &\mapsto k(\mathbf{x}, \mathbf{y}), \end{aligned} \tag{2}$$

which calculates a real number for each pair of vectors from the input space. Deciding on the form of the kernel function, defines implicitly the feature space $\mathcal{F}$ (and the mapping Φ). The kernel functions used in this paper are shown in Table 1. These functions are not proper Mercer kernels and the "covariance matrix" evaluated in feature space is not positive semidefinite. Most eigenvalues are nevertheless positive and the corresponding components are meaningful, so the negative eigenvalues can be simply ignored.

2.2 Variational Bayesian Nonlinear BSS

Denoting the observed data by $\boldsymbol{X} = \{\mathbf{x}(t)|t\}$ and the sources by $\boldsymbol{S} = \{\mathbf{s}(t)|t\}$, the generative model for the VB nonlinear BSS method can be written as

$$\mathbf{x}(t) = \mathbf{f}(\mathbf{s}(t), \boldsymbol{\theta}_\mathbf{f}) + \mathbf{n}(t), \tag{3}$$

Table 1. Summary of the kernels used in the experiments

Function	Values of parameter κ used
$\tanh(\kappa(\mathbf{x} \cdot \mathbf{y}))$	$10^{-3}, 10^{-2.5}, 10^{-2}, \ldots, 10^{1.5}, 10^2$
$\mathrm{arsinh}(\kappa(\mathbf{x} \cdot \mathbf{y}))$	$10^{-3}, 10^{-2.5}, 10^{-2}, \ldots, 10^{1.5}, 10^2$

where $\mathbf{f}$ is the unknown nonlinear (mixing) mapping modelled by a multilayer perceptron (MLP) network with weights and parameters $\boldsymbol{\theta_f}$, and $\mathbf{n}(t)$ is Gaussian noise. The sources $\boldsymbol{S}$ are usually assumed to have a Gaussian prior, which leads to a PCA like nonlinear factor analysis (NFA) model. This can be extended to a full nonlinear BSS method by either using a mixture-of-Gaussians source prior or using standard linear ICA as post-processing for the sources recovered by NFA. As the latter method is significantly easier and produces almost as good results, it is more commonly used [5,6].

The NFA model is learned by a variational Bayesian learning method called ensemble learning. As a variational Bayesian method, ensemble learning is based on finding a simpler approximation to the true posterior distribution $p(\boldsymbol{S}, \boldsymbol{\theta}|\boldsymbol{X})$ of the sources and model parameters $\boldsymbol{\theta}$. The approximation $q(\boldsymbol{S}, \boldsymbol{\theta})$ is fitted by minimising the cost function

$$\mathcal{C} = E_q \left[\log \frac{q(\boldsymbol{S}, \boldsymbol{\theta})}{p(\boldsymbol{S}, \boldsymbol{\theta}, \boldsymbol{X})} \right] = D_{KL}(q(\boldsymbol{S}, \boldsymbol{\theta}) \| p(\boldsymbol{S}, \boldsymbol{\theta}|\boldsymbol{X})) - \log p(\boldsymbol{X}), \quad (4)$$

where $D_{KL}(q\|p)$ denotes the Kullback-Leibler divergence between the distributions q and p. The remaining evidence term is a constant with respect to the parameters of the model so the cost is minimised when the Kullback-Leibler divergence is minimised. Because the Kullback-Leibler divergence is always non-negative, the cost function yields an upper bound for $-\log p(\boldsymbol{X})$ and consequently a lower bound for model evidence $p(\boldsymbol{X})$. The values of the cost function can be thus used for model comparison with smaller values indicating larger lower bounds on model evidence [7,8]. In our case, the approximating distribution $q(\boldsymbol{S}, \boldsymbol{\theta})$ is restricted to be a multivariate Gaussian with a diagonal covariance.

2.3 Learning and Initialisation of the VB Method

The variational Bayesian learning algorithm of the NFA model is based on iterative updates of the parameters of the approximating distribution. The means and diagonal elements of the covariance correspond to estimated values and variances of the different sources and weights. The sources and MLP network weights are updated by minimising the cost in Eq. (4) with a gradient based algorithm. The optimal values of other model parameters such as noise variances and parameters of the hierarchical priors can be solved exactly if the other parameters are assumed to be fixed.

Because of the iterative nature of the update algorithms and especially because the MLP network is very prone to local optima, the method needs a good initialisation to produce good results. Earlier, a given number of first linear PCA components has been used as initialisation of the posterior means of the sources while the means of the weights have been initialised randomly. The variances of all parameters are initialised to small constant values. The means of the sources are then kept fixed for the first 50 iterations while the network adapts to model the mapping from the PCA sources to the observations [5].

In this work, the principal components extracted with the linear algorithm are replaced with components extracted with the nonlinear kernel PCA algorithm. Otherwise the learning proceeds in the same way as before. The flow of information in the method is illustrated in Fig. 1.

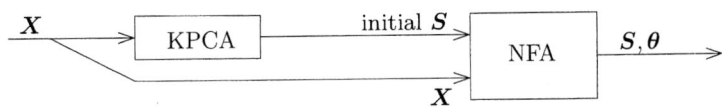

Fig. 1. A block diagram of the learning method

3 Experiments

The experiments were conducted using the same artificial data set that was used in [9]. The data was generated by mapping 4 super-Gaussian and 4 sub-Gaussian sources with a random MLP to a 20 dimensional space and adding some noise. The number of samples used was 1000. The NFA model used an MLP network with 10 inputs (sources), 30 hidden neurons and 20 outputs. The model can prune unneeded sources so using too many causes no problems[1].

In order to get the initialisations for the sources, kernel PCA was applied to the data. A number of different types of kernels and parameters were used as listed in Table 1. These were then all used for brief simulations with the NFA algorithm to see which provided the best results. The results in terms of cost function value attained after 1000 iterations are illustrated in Fig. 2. The figure shows that larger parameter values tend to produce better results although variations between neighbouring values can be large.

The results of the experiments were evaluated based on both the attained values of the cost function in Eq. (4) and the signal-to-noise ratios (SNRs) of the optimal linear reconstruction from the estimated source subspace to the true sources. The two statistics are strongly correlated, as illustrated in Fig. 3. This shows that the ensemble learning cost function is a very good measure of the quality of the found solution. This is in agreement with the results reported in [9] for a hierarchical nonlinear model.

Based on the results shown in Fig. 2, the parameter value $\kappa = 10^{1.5} \approx 31.6$ was chosen as best candidate for the tanh kernel and the value $\kappa = 100$ for the arsinh kernel. The simulations for these kernels and linear initialisation were

---

[1] Matlab code for KPCA and NFA methods used in the experiments is available at http://www.lis.inpg.fr/pages_perso/bliss/deliverables/d20.html.

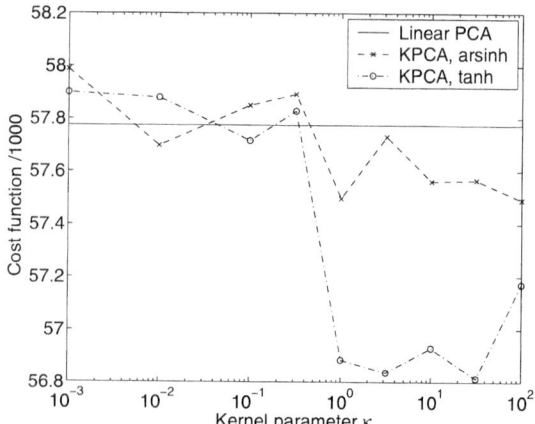

Fig. 2. Comparison of cost function values attained with different kernels and their parameter values after 1000 iterations of the NFA algorithm. The lines show the mean result of 10 simulations with different random MLP initialisations for kernel PCA with tanh and arsinh kernels and linear PCA

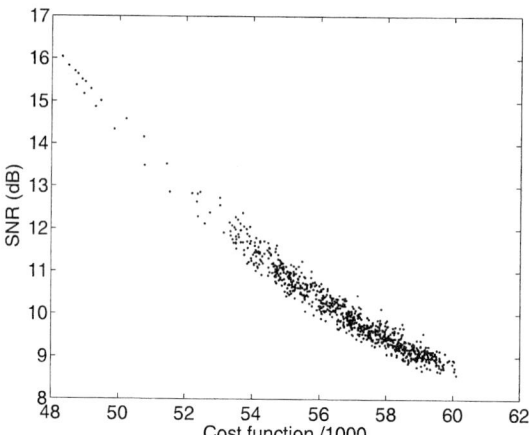

Fig. 3. Signal-to-noise ratio of the optimal linear reconstruction of the true sources from the estimated source subspace as a function of the cost function value attained in different stages of different simulations, some of which were run for up to 50000 iterations

then continued for 4000 more iterations. The SNRs attained at different stages of learning on average in 10 simulations with these initialisations are illustrated in Fig. 4. The results show that kernel PCA is able to provide a consistent improvement of about 1 dB in signal-to-noise ratio to the results attained in equal time with linear PCA initialisation.

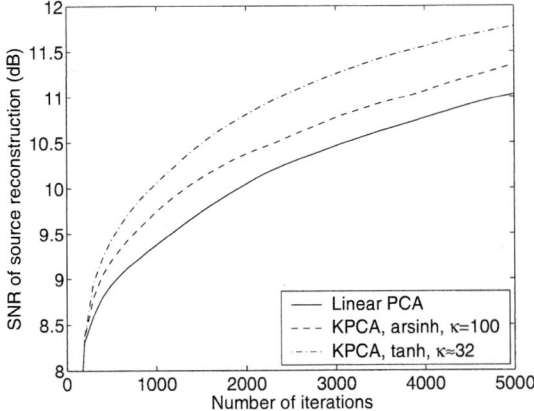

Fig. 4. Comparison of signal-to-noise ratios attained with linear PCA and kernel PCA initialisations. The results shown here are the mean of 10 simulations with different random MLP initialisations

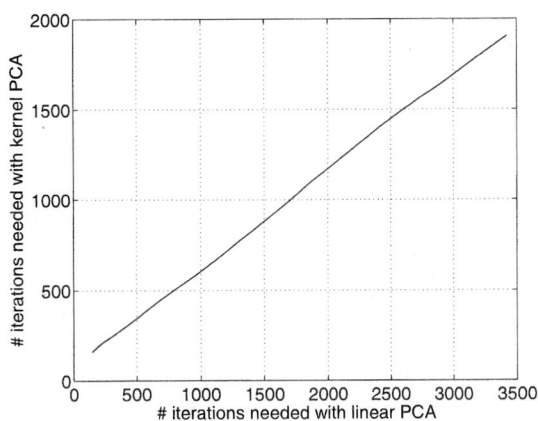

Fig. 5. The number of iterations needed on average to attain the same level of cost function value with linear PCA initialisation as a function of number of iterations needed with kernel PCA initialisation

Looking at the same result from time perspective, the kernel PCA initialisation can speed up learning significantly. This can be seen from Fig. 5, which shows a comparison of numbers of iterations needed with different initialisations on average in 10 simulations with tanh kernel to reach a given level of cost function value. The figure shows that as good results can be attained with kernel PCA initialisation using only slightly more than half of the time needed with linear PCA initialisation.

4 Discussion

The signal-to-noise ratios reported in the experiments were evaluated for optimal linear reconstruction from the estimated source subspace to the true sources. As noted in [9], these optimal results are presumably about 1 dB higher than completely blind application of linear ICA would produce. The optimal reconstruction was selected for comparison because it needed to be evaluated often and was more efficient to evaluate than running linear ICA every time and avoided a possible source of error.

In order to find out which kernels were the best ones, the signal-to-noise ratios were also evaluated for the components extracted with linear PCA and kernel PCA with different kernels. Surprisingly these SNRs had little correlation with how well NFA worked with different initialisations. The best SNR among the initialisations was attained by linear PCA followed by the kernels that were closest to linear. These were however not the ones that produced the best overall results. Fortunately the best kernels could be identified rather quickly from the cost function values attained during learning.

5 Conclusions

The experiments show that kernel PCA can provide significantly better initialisation for nonlinear factor analysis than linear PCA. The lower bound of model evidence provided by the cost function correlates strongly with the quality of the results as measured by the signal-to-noise ratio of optimal linear reconstruction of true sources from the estimated sources, thus allowing easy evaluation of results. The cost function can also be evaluated in more realistic situations, whereas the SNR cannot.

From variational Bayesian perspective, the kernel PCA initialisations are good complement to the nonlinear BSS method. Considering the significant computational demands of the basic method, the computation time required for kernel PCA and even kernel selection is more or less negligible. From kernel point of view, the variational Bayesian NFA is an interesting complement to KPCA as it allows relatively easy comparison of different kernels and parameter values.

Acknowledgements

This work was supported in part by the IST Programme of the European Community, under the project BLISS, IST-1999-14190, and under the PASCAL Network of Excellence, IST-2002-506778. This publication only reflects the authors' views.

References

1. A. Hyvärinen, J. Karhunen, and E. Oja, *Independent Component Analysis*. J. Wiley, 2001.

2. C. Jutten and J. Karhunen, "Advances in nonlinear blind source separation," in *Proc. of the 4th Int. Symp. on Independent Component Analysis and Blind Signal Separation (ICA2003)*, pp. 245–256, 2003. Invited paper in the special session on nonlinear ICA and BSS.
3. B. Schölkopf, A. Smola, and K.-R. Müller, "Nonlinear component analysis as a kernel eigenvalue problem," *Neural Computation*, vol. 10, no. 5, pp. 1299–1319, 1998.
4. S. Harmeling, A. Ziehe, M. Kawanabe, and K.-R. Müller, "Kernel-based nonlinear blind source separation," *Neural Computation*, vol. 15, no. 5, pp. 1089–1124, 2003.
5. H. Lappalainen and A. Honkela, "Bayesian nonlinear independent component analysis by multi-layer perceptrons," in *Advances in Independent Component Analysis* (M. Girolami, ed.), pp. 93–121, Berlin: Springer-Verlag, 2000.
6. H. Valpola, E. Oja, A. Ilin, A. Honkela, and J. Karhunen, "Nonlinear blind source separation by variational Bayesian learning," *IEICE Transactions on Fundamentals of Electronics, Communications and Computer Sciences*, vol. E86-A, no. 3, pp. 532–541, 2003.
7. G. E. Hinton and D. van Camp, "Keeping neural networks simple by minimizing the description length of the weights," in *Proc. of the 6th Ann. ACM Conf. on Computational Learning Theory*, (Santa Cruz, CA, USA), pp. 5–13, 1993.
8. D. J. C. MacKay, "Developments in probabilistic modelling with neural networks – ensemble learning," in *Neural Networks: Artificial Intelligence and Industrial Applications. Proc. of the 3rd Annual Symposium on Neural Networks*, pp. 191–198, 1995.
9. H. Valpola, T. Östman, and J. Karhunen, "Nonlinear independent factor analysis by hierarchical models," in *Proc. 4th Int. Symp. on Independent Component Analysis and Blind Signal Separation (ICA2003)*, (Nara, Japan), pp. 257–262, 2003.

A Geometric Approach for Separating Several Speech Signals*

Massoud Babaie-Zadeh[1,2], Ali Mansour[3],
Christian Jutten[4], and Farrokh Marvasti[1,2]

[1] Multimedia Lab, Iran Telecom Research Center (ITRC), Tehran, Iran
mbzadeh@yahoo.com, marvasti@itrc.ac.ir
[2] Electrical Engineering Department, Sharif University of Technology, Tehran, Iran
[3] E3I2, ENSIETA, Brest, France
mansour@ieee.org
[4] Institut National Polytechnique de Grenoble (INPG), Laboratoire des Images et des Signaux (LIS), Grenoble, France
Christian.Jutten@inpg.fr

Abstract. In this paper a new geometrical approach for separating speech signals is presented. This approach can be directly applied to separate more than two speech signals. It is based on clustering the observation points, and then fitting a line (hyper-plane) onto each cluster. The algorithm quality is shown to be improved by using DCT coefficients of speech signals, as opposed to using speech samples.

1 Introduction

Blind Source Separation (BSS) or Independent Component Analysis (ICA) consists in retrieving unknown statistically independent signals from their observed mixtures, assuming there is no information about the original source signals, or about the mixing system (hence the term *Blind*).

For linear instantaneous mixtures $\mathbf{x}(t) = \mathbf{A}\mathbf{s}(t)$, where the sources $\mathbf{s}(t) \triangleq (s_1(t), \ldots, s_N(t))^T$ are (unknown) statistically independent signals, the observation signals are denoted $\mathbf{x}(t) \triangleq (x_1(t), \ldots, x_N(t))^T$, and $\mathbf{A}$ is the $N \times N$ (unknown) mixing matrix. In this paper, the number of observations and sources are assumed to be equal. The problem is then to estimate the source vector $\mathbf{s}(t)$ only by knowing the observation vector $\mathbf{x}(t)$.

One approach to solve the problem is to determine a *separating matrix* $\mathbf{B}$ such that the outputs $\mathbf{y}(t) \triangleq \mathbf{B}\mathbf{x}(t)$ become statistically independent. This independence insures the estimation of the sources, up to a scale and a permutation indeterminacy [1].

Another approach is the geometric source separation algorithm, which has been first introduced in [2]. In this approach (for 2-dimensional case), it is

---

* This work has been partially funded by the European project Blind Source Separation and applications (BLISS, IST 1999-14190), by Iran Telecom Research Center (ITRC) and by Sharif university of technology.

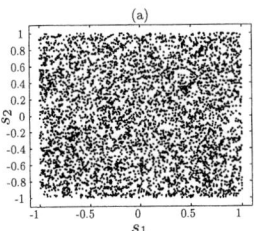 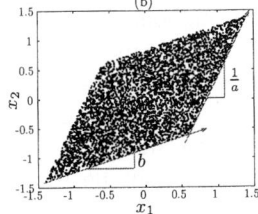

Fig. 1. Distribution of a) source samples, and b) observation samples.

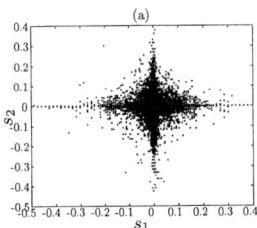 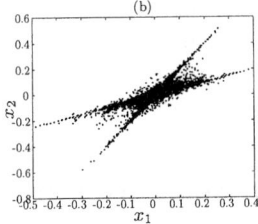

Fig. 2. Distribution of a) two speech samples, and (b) their mixtures.

first noted that because of the independence of source signals, $p_{s_1 s_2}(s_1, s_2) = p_{s_1}(s_1) p_{s_2}(s_2)$, where p stands for the Probability Density Function (PDF). Consequently, for bounded sources, the points (s_1, s_2) will be distributed in a rectangular region (Fig. 1-a). Now, because of the scale indeterminacy, the mixing matrix is assumed to be of the form (normalized with respect to diagonal elements):

$$\mathbf{A} = \begin{pmatrix} 1 & a \\ b & 1 \end{pmatrix} \quad (1)$$

Under the transformation $\mathbf{x} = \mathbf{A}\mathbf{s}$, the rectangular region of the s-plane will be transformed into a parallelogram (Fig. 1-b), and the slopes of the borders of this parallelogram are $1/a$ and b. In other words, for estimating the mixing matrix, it is sufficient to determine the slopes of the borders of the distribution of the observation samples.

Although this approach is not easily generalized to higher dimensions, it is successful in separating two sources, provided that their distributions allow a good estimation of the borders of the parallelogram (*e.g.* uniform and sinusoidal sources). However, this technique cannot be used in separating speech signals because the PDF of a speech is mostly concentrated about zero. This comes from the fact that in a speech signal, there are many low energy (silence or unvoiced) sections. Consequently, as it can be seen in Fig. 2, it is practically impossible to find the borders of the parallelogram when the sources are speech signals. This is explained by a probabilistic manner in [3]: the probability of having a point in the borders of the parallelogram is very low.

Although for speech signals the borders of the parallelogram are not visible in Fig. 2, there are two visible "axes", corresponding to lines $s_1 = 0$ and $s_2 = 0$ in the s-plane (throughout the paper, it is assumed that the sources and hence the observations have zero-means). The slopes of these axes, too, determine a and b in (1). In other words, for speech signals, instead of finding the borders, we try to find these axes. This idea is used in [3] for separating speech signals by utilizing an "angular" histogram for estimating these axes. In this method, the resolution of the histogram cannot be too fine (requires more data points), and cannot be too coarse (bad estimation of the mixing matrix). Moreover, this approach cannot be easily generalized to mixtures of more than two speech signals.

In this paper, we propose another approach for estimating these "axes" based on line (or hyper-plane) fitting. The main idea is to fit two lines on the scatter plot of observations, which will be the required axes. This approach does not suffer from the problem of the resolution of a histogram. Moreover, we will see that this approach can be directly used in higher dimensions.

2 Speech Separation by Line Fitting

2.1 Two Dimensional Case

As it is explained in the previous section, the main idea of our method is to estimate the slopes of two axes of the scatter plot of observations (Fig. 2-b). These axes corresponds to the lines $s_1 = 0$ and $s_2 = 0$ in the scatter plot of sources. The existence of these lines is a result of many low-energy sections of a speech signal. For example, the points with small s_1 and different values for s_2 will be concentrated about the axis $s_1 = 0$.

However, we do not use (1) as a model for mixing matrix, because it has two restrictions. Firstly, in this model, it is implicitly assumed that the diagonal elements of the actual mixing matrix are not zero, otherwise infinite values for a and b may be encountered (this situation corresponds to vertical axes in the x-plane). Secondly, this approach is not easy to be generalized to higher dimensions.

Instead of model (1), let us consider a general "separating matrix" $\mathbf{B} = [b_{ij}]_{2\times 2}$. Under the transformation $\mathbf{y} = \mathbf{B}\mathbf{x}$, one of the axes must be transformed to $y_1 = 0$, and the other to $y_2 = 0$. In other words, for every (x_1, x_2) on the first axis:

$$\begin{pmatrix} 0 \\ y_2 \end{pmatrix} = \begin{pmatrix} b_{11} & b_{12} \\ b_{21} & b_{22} \end{pmatrix} \begin{pmatrix} x_1 \\ x_2 \end{pmatrix} \Rightarrow b_{11}x_1 + b_{12}x_2 = 0 \qquad (2)$$

That is, the equation of the first axis is $b_{11}x_1 + b_{12}x_2 = 0$. In a similar manner, the second axis will be $b_{21}x_1 + b_{22}x_2 = 0$. Consequently, for estimating the separating matrix, the equations of the two axes must be found in the form of $\alpha_1 x_1 + \alpha_2 x_2 = 0$, and then each row of the separating matrix is composed of the coefficients of one of the axes. For finding the axes we suggest is to "fit" two straight lines on the scatter plot of the observations.

It is seen that by this approach, we are not restricted to non-vertical axes (non-zero diagonal elements of the mixing matrix). More interestingly, this approach can be directly used in higher dimensions, as stated below.

2.2 Higher Dimensions

The approach stated above can be directly generalized to higher dimensions. For example, for 3 speech signals and 3 sources, the low-energy (silence and unvoiced) values of s_1 with different values of s_2 and s_3 will form the plane $s_1 = 0$ in the 3-dimensional scatter plot of sources. Hence, in this 3-dimensional scatter plot, there are 3 visible planes: $s_1 = 0$, $s_2 = 0$ and $s_3 = 0$. These planes will be transformed to three main planes in the scatter plot of observations. With calculations similar as (2), it is seen that each row of the separating matrix is composed of the coefficients of one of these main planes in the form of $\alpha_1 x_1 + \alpha_2 x_2 + \alpha_3 x_3 = 0$.

Consequently, for N-dimensional case, N (hyper-)planes in the form of $\alpha_1 x_1 + \cdots + \alpha_N x_N = 0$ must be first "fitted" onto the scatter plot of observations. Then, each row of the separating matrix is the coefficients $(\alpha_1, \ldots, \alpha_N)$ of one of these (hyper-)planes.

3 Line Fitting

To use the idea of the previous section, we need a method for fitting two lines (or N hyper-planes) onto the scatter plot of observations.

3.1 Fitting a Straight Line onto a Set of Points

First of all, consider the problem of fitting a line onto K data points $(x_i, y_i)^T$, $i = 1 \ldots K$. In the traditional least squares method, this is done by finding the line $y = mx + h$ which minimizes $\sum_{i=1}^{K}(y - y_i)^2 = \sum_{i=1}^{K}(mx_i + h - y_i)^2$. This is equivalent to minimizing the "vertical" distances between the line and the data points, as shown in Fig. 3-a. This technique is mainly used in linear regression analysis where there are errors in y_i's, but not in x_i's.

However, in our application of fitting a line onto a set of points, a better measure is minimizing the sum of "orthogonal distances" between the points and the line, as shown in Fig. 3-b. Moreover, as discussed in the previous sections, we are seeking a line in the form $ax + by = 0$. Consequently, the best fitted line is determined by minimizing $\sum_{i=1}^{K} d_i^2$, where d_i is the orthogonal distance between the i-th point and the line:

$$d_i = \frac{|ax_i + by_i|}{\sqrt{a^2 + b^2}} \tag{3}$$

However, $ax + by = 0$ is not uniquely determined by a pair (a, b), because (ka, kb) represents the same line. To obtain a unique solution, the coefficients are normalized such that $a^2 + b^2 = 1$. To summarize, the best fitted line $ax + by = 0$ is obtained by minimizing $\sum_{i=1}^{K}(ax_i + by_i)^2$ under the constraint $a^2 + b^2 = 1$.

N-Dimensional Case. In a similar manner, an N-dimensional hyper-plane $\alpha_1 x_1 + \alpha_2 x_2 + \cdots + \alpha_N x_N = 0$ is fitted onto a set of K data points $\mathbf{x}_i = (x_1^{(i)}, x_2^{(i)}, \ldots, x_N^{(i)})^T$, $i = 1, \ldots, K$ by minimizing the cost function:

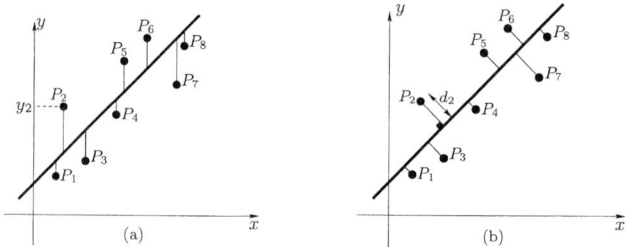

Fig. 3. a) Least squares line fitting, b) Orthogonal line fitting.

$$\mathcal{C}(\alpha_1,\ldots,\alpha_N) = \sum_{i=1}^{K} \left(\alpha_1 x_1^{(i)} + \cdots + \alpha_N x_N^{(i)}\right)^2 \quad (4)$$

under the constraint $g(\alpha_1,\ldots,\alpha_N) \equiv \alpha_1^2 + \cdots + \alpha_N^2 - 1 = 0$.

Solution. Using Lagrange multipliers, the optimum values for $\alpha_1,\ldots,\alpha_N$ satisfy $\nabla \mathcal{C} = \lambda \nabla g$. After a few algebraic calculations, this equation is written in the matrix form:

$$\mathbf{R_x}\boldsymbol{\alpha} = \frac{\lambda}{K}\boldsymbol{\alpha} \quad (5)$$

where $\boldsymbol{\alpha} \triangleq (\alpha_1,\ldots,\alpha_N)^T$ and $\mathbf{R_x} \triangleq \frac{1}{K}\sum_{i=1}^{K}\mathbf{x}_i \mathbf{x}_i^T$ is the correlation matrix of data points. Equation (5) shows that λ/K and $\boldsymbol{\alpha}$ are eigen value and eigen vector of the correlation matrix $\mathbf{R_x}$, respectively. Moreover:

$$\mathcal{C} = \sum_{i=1}^{K} \left(\boldsymbol{\alpha}^T \mathbf{x}_i\right)^2 = \sum_{i=1}^{K} \boldsymbol{\alpha}^T \mathbf{x}_i \mathbf{x}_i^T \boldsymbol{\alpha} = K \boldsymbol{\alpha}^T \mathbf{R_x} \boldsymbol{\alpha} = \lambda \boldsymbol{\alpha}^T \boldsymbol{\alpha} = \lambda$$

and hence for minimizing the cost function, λ must be minimum. Consequently, the solution of the hyper-plane fitting problem is given by the eigen vector of the correlation matrix which corresponds to its minimum eigen value.

Discussion. It is interesting to think about the conjunction of the above approach to Principal Component Analysis (PCA). Note that $\boldsymbol{\alpha}$ is the vector perpendicular to the plane $\alpha_1 x_1 + \cdots + \alpha_N x_N = 0$, and the above theorem states that this vector must be chosen in the direction with the minimum spread of data points, which is compatible with our heuristic interpretations of plane (line) fitting. This method has old foundations in mathematics [4], and somewhat called Principal Component Regression (PCR) [5].

3.2 Fitting 2 Straight Lines (N Hyper-planes)

However, as stated in Section 2, for 2 mixtures of 2 sources our problem is to fit 2 lines onto the observation points, not just 1 line. In other words, as it is seen in Fig. 2, we need to divide the data points into 2 clusters, and then to fit a line onto the points of each cluster. The extension to N mixtures of N sources

> - Initially distribute the points into clusters $S_1, \ldots, S_N$ (*e.g.* random initialization).
> - Loop:
>   1. Fit a line (hyper-plane) onto each set of points S_i (we call it l_i).
>   2. Recalculate the clusters: Let S_i be the set of all points which are closer to line (hyper-plane) l_i than other lines (hyper-planes), that is:
> $$S_i = \{\mathbf{x} \mid d(\mathbf{x}, l_i) < d(\mathbf{x}, l_j), \forall j \neq i\}$$
> - Repeat until convergence.

Fig. 4. Algorithm of fitting two lines (N hyper-planes) onto a set of points.

is straightforward: we need to divide the data into N clusters, and then to fit a hyper-plane onto the points of each cluster. Mathematically, this is equivalent to minimizing the following cost function (for the N-dimensional case):

$$C = \sum_{\mathbf{x}_i \in S_1} d^2(\mathbf{x}_i, l_1) + \sum_{\mathbf{x}_i \in S_2} d^2(\mathbf{x}_i, l_2) + \cdots + \sum_{\mathbf{x}_i \in S_N} d^2(\mathbf{x}_i, l_N) \qquad (6)$$

where S_j is the j-th cluster of points and $d^2(\mathbf{x}_i, l_j)$ denotes the perpendicular distance of the i-th point from the j-th plane.

Having divided the points into clusters $S_1, \ldots, S_N$, the previous section gives us the best line fitted onto the points of each cluster. For clustering the data points, we use the algorithm stated in Fig. 4, which is inspired from the k-means (or Lloyd) algorithm for data clustering [6]. Its difference with k-means is that in k-means, each cluster is mapped onto a point (point $\rightarrow$ point), but in our algorithm each cluster is mapped onto a line or hyper-plane (point $\rightarrow$ line).

The following theorem is similar to a corresponding theorem for the k-means algorithm [6].

Theorem 1. *The algorithm of Fig. 4 converges in a finite number of iterations.*

Proof. At each iteration, the cost function (6) cannot be increased. This is because in the first step (fitting hyper-planes onto the clusters) the cost function is either decreased or does not change. In the second step, too, the redistribution of the points in the clusters is done such that it decreases the cost function or does not change it. Moreover, there is a finite number of possible clustering of finite number of points. Consequently, the algorithm must converge in a finite number of iterations. □

Initialization. The fact that the cost-function is non-increasing in the algorithm, shows that the algorithm may get trapped in a local minimum. This is one of major problems of the k-means algorithm, too. It depends on the initialization of the algorithm, and become more severe when the dimensionality increases. In k-means, one approach is to run the algorithm with several randomly chosen initializations, and then to take the result which produces the minimum cost-function.

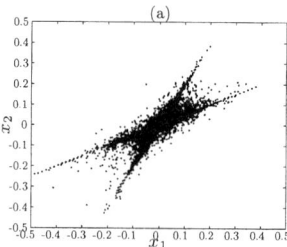 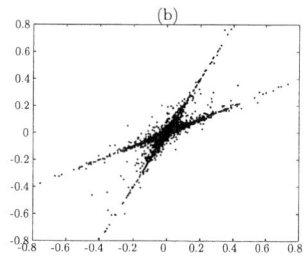

Fig. 5. Distribution of a) the observations, and b) their DCT coefficients (right).

4 Final Algorithm, and Its Improvement by Using DCT

The final separation algorithm is now evident. First, run the algorithm of Fig. 4. After convergence, there are N lines (hyper-planes) $l_i : \alpha_{i1}x_1 + \cdots + \alpha_{iN}x_N = 0$, $i = 1 \ldots N$. Then, the i-th row of the separating matrix is $(\alpha_{i1}, \ldots, \alpha_{iN})$.

However, the separation quality of the algorithm can be improved, with a simple trick. Recall that the success of the algorithm is because of the existence of two visible "axes" in Fig. 2. These axes were formed because of the small-valued (low-energy) parts of one speech and other parts of the second one. Now, recall that the Discrete Cosine Transform (DCT) coefficients of a speech frame (10-20 msec) contain a lot of nearly zero values. Moreover, DCT is a linear transformation, and hence, the DCT coefficients of the observations are a mixture of the DCT coefficients of the original speeches with the same mixing matrix. Therefore, it seems that it is a good idea to apply the algorithm on the DCT coefficients of observations instead of themselves. Figure 5 shows an example of the scatter plot of observations, and that of their DCT coefficients. It is seen visually that the "axes" are more visible in the scatter plot of DCT coefficients. Consequently, one expects to get better results by applying the algorithm on the DCT coefficients of the observations, as is confirmed by our experiments, too.

5 Experimental Results

Many simulations have been conducted to separate 2, 3 or 4 sources. In all these simulations, typically less than 30 iterations are needed to achieve separation. The experimental study shows that local minima depends on the initialization phase of the algorithm and on the number of sources (local minima have been never encountered in separating two sources).

Here, the simulation results of 4 typical speech signals (sampled with 8KHz sampling rate) are presented. In all the experiments, the diagonal elements of the mixing matrix are 1, while all other elements are 0.5. For each simulation, 10 random initializations are used, and then the matrix which creates minimum cost-function is taken as the answer.

To measure the performance of the algorithm, let $\mathbf{C} \triangleq \mathbf{BA}$ be the global mixing-separating matrix. Then, we define the Signal to Noise Ratio by (assuming no permutation) SNR_i(in dB) $\triangleq 10 \log_{10} \frac{c_{ii}^2}{\sum_{j \neq i} c_{ij}^2}$. This criterion shows

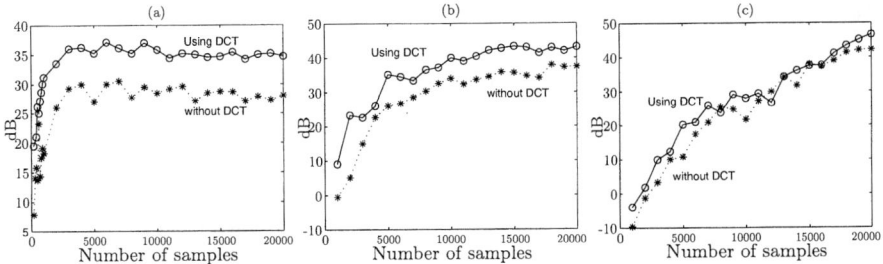

Fig. 6. Separation result in separating N speech signals, a) $N = 2$, b) $N = 3$, c) $N = 4$.

how much the global matrix $\mathbf{C}$ is close to the identity matrix. As a performance criterion of the algorithm, we take the average of the SNR's of all outputs: SNR $= \frac{1}{N}\sum_i \text{SNR}_i$. To virtually create different source signals, each speech signals is shifted randomly in time (more precisely, each speech signal is shifted $128k$ samples, where k is a randomly chosen integer). This results in a completely different source scatter plot, and virtually creates a new set of source signals. Then, for each experiment, the algorithm is run 50 times (with 50 different random shifts), and the averaged SNR is calculated.

Figure 6 shows this averaged SNR's with respect to number of samples, for separating 2, 3 and 4 speech signals. The figure clearly shows the ability of the algorithm for speech separation, and the advantage obtained by using DCT coefficients. Moreover, it is seen that when the number of sources increases, more data samples are required to reach a given separation quality. This was expected, because the algorithm is based on the "sparsity" of the speech signals. In other words, for forming the planes, it is required that one speech signal is low-energy (silence/unvoiced), and the others are not. If p is the probability of being in a low energy state, the probability of sparsity is $p(1-p)^{(N-1)}$, which decreases exponentially with N. Consequently, it is expected that the required number of data samples grows exponentially with N.

6 Conclusion

In this paper, a geometrical approach for separating several speech signals has been presented. It has been shown that for speech signals (or other sources whose PDF's are concentrated about zero), the ICA can be accomplished by a clustering of observation samples and then applying a PCA on each cluster and taking the smallest principal component. Although this approach was based on geometric interpretations, its final algorithm is completely algebraic.

Initialization is the main problem of this algorithm. Finding better initialization approaches is currently under study.

References

1. P. Comon, "Independent component analysis, a new concept?," *Signal Processing*, vol. 36, no. 3, pp. 287–314, 1994.
2. C. Puntonet, A. Mansour, and C. Jutten, "A geometrical algorithm for blind separation of sources," in *Actes du XVème Colloque GRETSI 95*, Juan-Les-Pins, France, Septembre 1995, pp. 273–276.
3. A. Prieto, B. Prieto, C. G. Puntonet, A. Cañas, and P. Martín-Smith, "Geometric separation of linear mixtures of sources: Application to speech signals," in *ICA 99*, Aussois, France, January 1999, pp. 295–300.
4. K. Pearson, "On lines and planes of closest fit to systems of points in space," *The London, Edinburgh and Dublin Philosophical Magazine and Journal of Science*, vol. 2, pp. 559–572, 1901.
5. W. F. Massy, "Principal component regression in exploratory statistical research," *Journal of American Statistical Association*, vol. 60, pp. 234–256, March 1965.
6. A. Gersho and R. M. Gray, *Vector Quantization and signal compression*, Kluwer Academic Publishers, 1992.

A Novel Method for Permutation Correction in Frequency-Domain in Blind Separation of Speech Mixtures

Christine Serviere[1] and Dinh-Tuan Pham[2]

[1] Laboratoire des Images et des signaux, BP 46, 38402 St Martin d'Hère Cedex, France
Christine.serviere@inpg.fr
[2] Laboratoire de Modélisation et Calcul, BP 53, 38041 Grenoble Cedex, France
Dinh-Tuan.Pham@imag.fr

Abstract. This paper presents a method for blind separation of convolutive mixtures of speech signals, based on the joint diagonalization of the time varying spectral matrices of the observation records and a novel technique to handle the problem of permutation ambiguity in the frequency domain. Simulations show that our method works well even for rather realistic mixtures in which the mixing filter has a quite long impulse response and strong echoes.

1 Introduction

There has been many works on blind separation of convolutive audio signal [3,8,9], but successful application in realistic setting is still elusive [2], due mainly to the long impulse response of the mixing filter. Time domain approach would be too computational heavy, not to mention the difficulty of convergence, since it requires the adjustment of too many parameters. Therefore frequency domain approach is often adopted, which has the advantage that it reduces the problem to a set of independent problems of separation of instantaneous mixtures in each frequency bin. But the finite Fourier transform tends to produce nearly Gaussian variables and it is well known that blind separation of instantaneous mixtures requires non Gaussianity. Fortunately, speech signals are highly non stationary so one can exploit this nonstationarity to separate their mixture and use only their second order statistics [5], which leads to a joint diagonalization problem. This approach has been developed in two earlier papers of the authors [6,7]. Actually the idea of exploiting nonstationarity has been introduced even earlier by Para and Spence [3], but these authors used an ad-hoc criterion, while we use a criterion based on the Gaussian mutual information and related to the maximum likelihood. Such criterion has in fact been considered in [9], but without using the nonstationarity idea.

Although the methods in [6,7] work reasonably well, the main problem in a frequency domain approach, namely the permutation ambiguity, is still not satisfactory solved. This is the biggest challenge in blind separation of audio signal. In this paper, we present a novel technique to solve this problem, which provides much better results than those in [6,7]. Actually our method starts from a solution of [6] and improves it, based on the consideration of the time variation of the signal energy in each frequency bin. Such consideration has also appeared in [1,7], but is exploited here in a quite different way.

2 Model and Methods

The problem considered corresponds theoretically to the blind separation of convolutive mixtures: the observed sequences $\{x_1(t)\}, \ldots, \{x_K(t)\}$ are related to the source sequences $\{s_1(t)\}, \ldots, \{s_K(t)\}$ through a mixing filter with impulse response matrix $\{\mathbf{H}(n)\}$, of general element $\{H_{kj}(n)\}$, as

$$x_k(t) = \sum_{n=-\infty}^{\infty} \sum_{j=1}^{K} H_{kj}(n) s_j(t-n), \quad 1 \leq k \leq K, \tag{1}$$

The goal is to recover the sources through another filtering operation $\mathbf{y}(t) = \sum_{n=-\infty}^{\infty} \mathbf{G}(n)\mathbf{x}(t-n)$ where $\mathbf{x}(t) = [x_1(t) \cdots x_K(t)]^\mathbf{T}$ ($\mathbf{T}$ denoting the transpose), $\{\mathbf{G}(l)\}$ is the impulse response matrix of the separation filter and $\mathbf{y}(t) = [y_1(t) \cdots y_K(t)]^\mathbf{T}$ is the recovered source vector. In the blind context, the idea is to adjust the filter $\{\mathbf{G}(n)\}$ such that the reconstructed sources $\{y_k(t)\}$ are as mutually independent as it is possible. By adopting a second order approach, we are in fact focused on the inter-spectra between the reconstructed sources at every frequency. But since we are dealing with nonstationary signals, we shall consider the time varying spectra, that is the localized spectra around each given time point. It is precisely the time evolution of these spectra which helps us to separate the sources. From (1), the time varying spectrum of the vector observation sequence $\{\mathbf{x}(t)\}$ is $S_\mathbf{x}(t,f) = \mathbf{H}(f)S_\mathbf{s}(t,f)\mathbf{H}^*(f)$ where $\mathbf{H}(f) = \sum_{n=-\infty}^{\infty} e^{inf2\pi}\mathbf{H}(n)$ denotes the frequency response of the mixing filter at frequency f, $S_\mathbf{s}(t,f)$ is the diagonal matrix with diagonal elements being the time varying spectra of the sources and $*$ denotes the transpose conjugated. The spectrum of the reconstructed source vector $\mathbf{G}(f)S_\mathbf{x}(t,f)\mathbf{G}^*(f)$ should be diagonal and as in [6,7], the following diagonalization criterion (up to a constant term) is used

$$\sum_t \left\{ \frac{1}{2} \log \det \operatorname{diag}[\mathbf{G}(f) S_\mathbf{x}(t,f) \mathbf{G}^*(f)] - \log \det |\mathbf{G}(f)| \right\} \tag{2}$$

where $\operatorname{diag}(\cdot)$ denotes the operator which builds a diagonal matrix from its argument and the summation is over the time points of interest. This criterion is to be minimized with respect to $\mathbf{G}(f)$ and a simple and fast algorithm [4] is available for this purpose.

In practice, the spectrum $S_\mathbf{x}(t,f)$ is estimated over a (high resolution) grid of frequencies. It is important to have good estimator, since the final separation would depend on it. This paper introduces a different and better estimation method than in [6,7]. We form the short term periodogram using a *Hanning taper window*

$$P_\mathbf{x}(\tau, f) = \frac{2}{3N} \left[\sum_t H_N(t-\tau)\mathbf{x}(t)e^{2\pi i f t} \right] \left[\sum_t H_N(t-\tau)\mathbf{x}(t)e^{2\pi i f t} \right]^*.$$

where H_N is the Hanning taper windows of length N: $H_N(t) = 1 - \cos(2\pi t/N + \pi/N)$ for $0 \leq t < N$, 0 otherwise. The above periodogram will be averaged over m consecutive equispaced points $\tau_1, \ldots, \tau_m$ yielding the estimated spectrum at time $(\tau_1 + \tau_m + N - 1)/2$:

$$\hat{S}_\mathbf{x}\left(\frac{\tau_1 + \tau_m + N - 1}{2}, f\right) = \frac{1}{m} \sum_{k=1}^{m} P_x(\tau_k, f)$$

The frequencies are taken to be of the form $f = n/N, n = 0, \ldots, N/2$, with N being chosen to be a power of 2, to take advantage of the Fast Fourier Transform. The frequency resolution is thus determined by the taper window length and the time resolution by $m\delta$ where $\delta = \tau_i - \tau_{i-1}$ is the spacing between the τ_i. Using $\delta > 1$ helps to reduce the computational cost but slightly degrades the estimator: actually δ can be a small fraction of N without a significant degradation. Of course a compromise between time and frequency resolution has to be made to get a reasonably low variance of the estimator. Our method is more flexible for adjusting these resolutions than that of [6,7] and further its use of tapering helps to reduce the bias.

3 The Permutation Ambiguity Problem

The advantage of the frequency domain approach, as explained in the introduction, comes however with a price. The joint diagonalization only provides the matrices $\mathbf{G}(f)$ up to a scale change and a permutation: if $\mathbf{G}(f)$ is a solution then so is $\mathbf{\Pi}(f)\mathbf{D}(f)\mathbf{G}(f)$ for any diagonal matrix $\mathbf{D}(f)$ and any permutation matrix $\mathbf{\Pi}(f)$. The scale ambiguity is however intrinsic to the blind separation of convolutive mixtures and cannot be lifted.

In [6] we have proposed a method to solve the permutation ambiguity problem based on the continuity of the frequency response of the separation filter, which is more or less equivalent to constrain the separating filter to have short support in the time domain [8,9]. Although it can detect most of frequency permutation jumps, its weakness is that even a single wrong detection can cause wrong permutations over a large block of frequency. To avoid this problem, a complementary method based on an idea similar to that in [1], which introduces some frequency coupling [8] is proposed in [7]. The glottis is the main energy for speech production and emits a broadband sound with spectral peaks at the harmonics of the speaker's pitch frequency. Then the vocal tract is filtering this broadband sound and the result speech signal can be seen as an amplitude modulation due to the succession of phonemes which constitutes speech. Based on this observation, the main idea is that, for a speech signal, the energy over different frequency bins appears to vary in time in a similar way, up to a gain factor. For example, one would expect that its energy would be nearly zero in all frequency bins in a period of pause and be maximum in all frequency bins for speech period. To check this similarity, [1] proposes to recover the permutation ambiguity by exploiting correlations on amplitude spectrograms, i. e. the module of the time varying spectra. But this is awkward and very time consuming as there are $K^2L(L-1)/2$ correlations to be computed, L denoting the number of frequency bins. The method proposed in [7] avoids this problem by associating each frequency bin with a profile (of relative variation of the spectral energy) and compares them with a reference profile. More specifically, after the step of joint diagonalization, the spectra of the reconstructed sources $\hat{S}_y(t, f)$, can be computed as the k-th diagonal element of $\mathbf{G}(f)\hat{S}_x(t,f)\mathbf{G}^*(f)$. As each spectra is recovered up to a gain factor, we consider the "profiles" $E(f,\cdot;j)$, defined as the logarithm of the j-th diagonal element of $\mathbf{G}(f)\hat{S}_x(\cdot,f)\mathbf{G}^*(f)$. Thus, they are defined up to an additive constant, hence by centering all profiles by subtracting its time average this additive constant is eliminated and the notation E' will be used for centered profiles. In [7], these profiles are compared with reference profiles associated with each sources

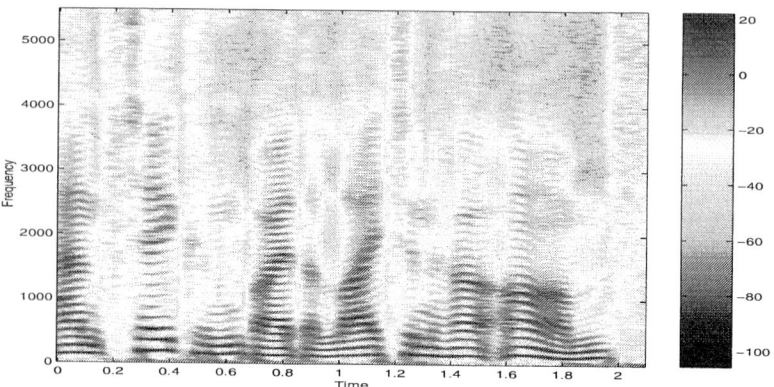

Fig. 1. Time-frequency representation of a speech signal in dB

(but not depending on frequency) to determine which sources they come from. The reference profiles, in turn, are constructed iteratively by averaging profiles previously identified as coming from the same sources.

The method in [7] assumes that profiles from the same sources, but at different frequencies are still more similar than those from other sources. This may not be true as profiles in fact can vary considerably across frequency (see figure 1). In this paper we abandon this assumption and only assume that profiles vary *smoothly* with frequency. Thus we work with profiles averaged on a bandwidth $[f - M : f + M]$:

$$F_\mathbf{y}(f, k; \cdot) = \frac{1}{2M+1} \sum_{l=f-M}^{f+M} E'(l, k; \cdot).$$

These averaged profiles are used to detect the block permutation errors arisen in application of the method in [6]. Consider for simplicity the case of two sources and two sensors, we consider the difference between the profiles of the two reconstructed sources at the first step of the separating system:

$$D_1(f, k) = F_\mathbf{y}(f, k; 1) - F_\mathbf{y}(f, k; 2).$$

Suppose there is a permutation of the separation filter $\mathbf{G}(f)$ at frequency bin f_0. Between $f_0 - M$ and $f_0 + M$, the two outputs correspond to two different sources and the profiles are also permuted.

$$D_1(f_0 - M, k) = F_\mathbf{S}(f_0 - M, k; 1) - F_\mathbf{S}(f_0 - M, k; 2).$$
$$D_1(f_0 + M, k) = F_\mathbf{S}(f_0 + M, k; 2) - F_\mathbf{S}(f_0 + M, k; 1).$$

If we assume that the averaged profiles are changing slowly enough, the difference $D_1(f_0 - M, k)$ and $D_1(f_0 + M, k)$ will be of opposite sign, whatever the time index k. Consequently, a sign change exists in $D_1(f, k)$ for *all time index k* at some frequency bin near f_0. This is illustrated in figure 2, in which for each time index k, the curve

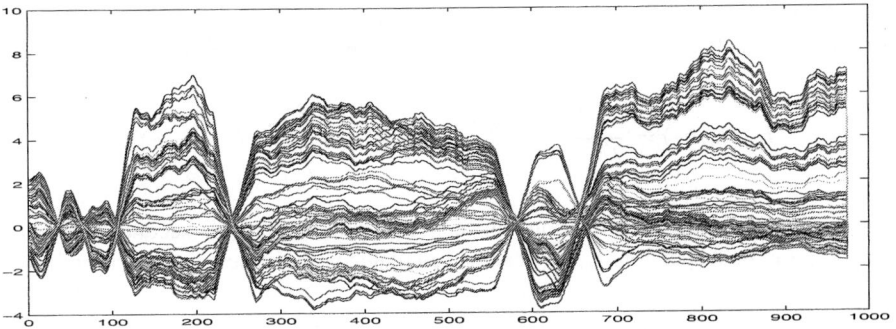

Fig. 2. Differences $D_1(f,k)$ between averaged profiles in function of frequency bins

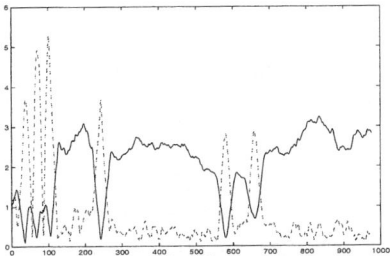

Fig. 3. $\sigma^2_{D_1}$ (solid) and $\sigma^2_{D_2}$ (dashed) before permutation correction

Fig. 4. $\sigma^2_{D_1}$ (solid) and $\sigma^2_{D_2}$ (dashed) after permutation correction

$D_1(f,k)$ is plotted as a function of f. These curves change sign correctly at six frequencies where the sources must be permuted. If we examine the same curves after elimination of the permutations, we remark that all the sign changes have disappeared. It can be deduced from this, that at each frequency bin f_0 where the sources are permuted, the dispersion of the values $D_1(f_0, k)$ will be minimum. The minima can then detect the beginning and the end of a frequency block to permute. Suppose that the time frequency representation is computed on K time blocks. As the profiles are centered by construction, the mean value of $D_1(f_0, k), k = 1, \ldots, K$ is zero and its dispersion is:

$$\sigma^2_{D_1(f_0)} = \sum_{k=1}^{K} \|D1(f_0, k)\|^2 \qquad (3)$$

The dispersion $\sigma^2_{D_1(f_0)}$ of the data $D_1(f,k)$, shown in figure 2, is plotted in solid line on figures 3 and 4, before and after performing permutations correction. In figure 3, the six minima are actually permutation frequencies. They occur correctly at the six sign changes (see figure 2). After permutation correction, these minima disappear, as can be seen in figure 4.

In order to detect a possible permutation at any frequency bin f, we introduce a second function difference $D_2(f, k)$, based on new profiles $H_\mathbf{y}(f, k; \cdot)$ of $\mathbf{y}(t)$. Similar to $F_\mathbf{y}(f, k; \cdot)$, they are constructed by averaging on the band $[f - M : f + M]$ but we impose a permutation on the second part of the band $[f+1 : f+M]$. The outputs are permuted on the band $[f+1 : f+M]$ versus the outputs on the band $[f-M : f]$:

$$H_\mathbf{y}(f,k;\cdot) = \frac{1}{2M+1}\left(\sum_{l=f-M}^{f} E'(l,k;\cdot) + \sum_{l=f+1}^{f+M} E'(l,k;\pi)\right).$$

where π denotes the permutation between the two outputs. A second difference $D_2(f,k)$ and its dispersion $\sigma^2_{D_2(f_0)}$ can be calculated with the new averaged profiles:

$$D_2(f,k) = H_\mathbf{y}(f,k;1) - H_\mathbf{y}(f,k;2)$$

$$\sigma^2_{D_2(f_0)} = \sum_{k=1}^{K} \|D_2(f_0,k)\|^2$$

The dispersion $\sigma^2_{D_2(f_0)}$ is plotted in dotted line before (figure 3) and after (figure 4) elimination of the permutation. If f_0 is a permutation frequency, $H_\mathbf{y}(f_0,k;\cdot)$ will be the profiles of the corrected sources and the dispersion $\sigma^2_{D_2(f_0)}$ will be bigger than $\sigma^2_{D_1(f_0)}$ as there will be no sign change in the difference of profiles $H_\mathbf{y}(f_0,k;\cdot)$. The two curves $\sigma^2_{D_1(f_0)}$ and $\sigma^2_{D_2(f_0)}$ are crossing when permutation must be detected. On the contrary, when a frequency band is correctly permuted, the profiles $F_\mathbf{y}(f,k;\cdot)$ are good and the dispersion $\sigma^2_{D_1(f)}$ is maximum for this band and bigger than $\sigma^2_{D_2(f)}$. The curves are no more crossing for this band. When all permutation are corrected, the profiles $H_\mathbf{y}(f,k;\cdot)$ only add false permutation and impose sign change in the function $D_2(f,k)$. The dispersion $\sigma^2_{D_2(f)}$ is then always smaller than $\sigma^2_{D_1(f)}$.

The permutation detection can be done in an iterative way. We compute $\sigma^2_{D_1(f)}$ and $\sigma^2_{D_2(f)}$ and detect the global minimun of $\sigma^2_{D_1(f)}$, which occurs at f_0, say. Then we permute the two outputs for all frequency higher than f_0. We re-compute the new profiles $F_\mathbf{y}(f,k;\cdot)$ and $H_\mathbf{y}(f,k;\cdot)$ and the new functions $\sigma^2_{D_1(f)}$ and $\sigma^2_{D_2(f)}$ and re-detect the new global minimun of $\sigma^2_{D_1(f)}$ and so on until $\sigma^2_{D_1(f)} > \sigma^2_{D_2(f)}$ for all f. This method is easy to implant and shows very good results even for short signals.

4   Design and Simulation Results

We considered two mixtures of two real sound sources from pre-measured room impulse responses. These responses are obtained from Alex Westner (found in http://sound.media.mit.edu/ica-bench), which uses a library of impulse responses measured off a real 3.5m × 7m × 3m conference room. The sources are speech signals of 2s sampled at 11 kHz (24000 samples). These responses are quite long, up to 8192 lags, but become quite small at high lags so that we can truncate them to 256 lags and still retaining all echoes. The four impulse responses are shown in figure 5.

First, the joint diagonalization is processed on spectral matrices. The block length is $N = 2048$ with an overlap of 75% (yielding 41 time blocks). The spectral matrices are estimated as detailed in section 2 and the averaged profiles $F_\mathbf{y}(f,k;\cdot)$ are then

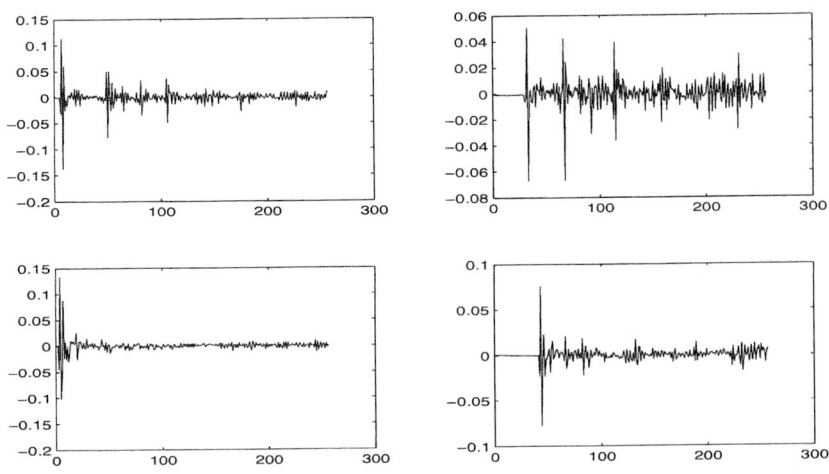

Fig. 5. Impulse responses of the mixing filter

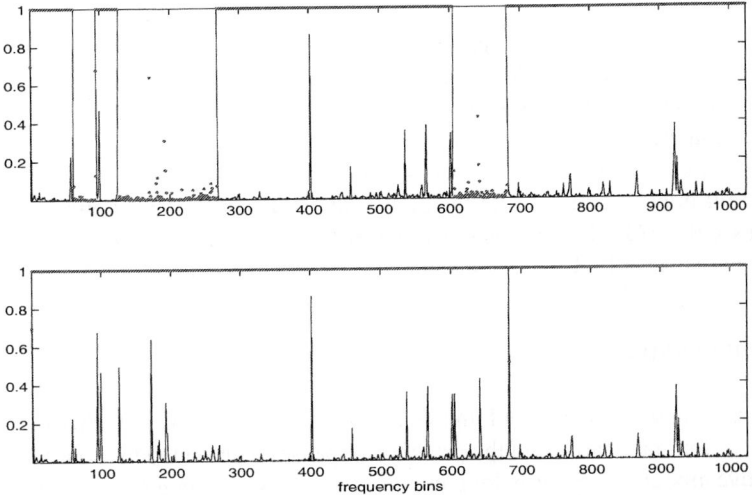

Fig. 6. Separation index (red dots) and its inverse (solid blue) truncated at 1, before (upper panel) and after (lower panel) applying the new permutation correction

constructed by averaging on 50 frequency bins ($M = 25$). As in [6], we consider the performance index

$$r(f) = |(\mathbf{GH})_{12}(f)(\mathbf{GH})_{21}(f)/[(\mathbf{GH})_{11}(f)(\mathbf{GH})_{22}(f)]|^{1/2}$$

where $(\mathbf{GH})_{ij}(f)$ is the ij element of the matrix $\mathbf{G}(f)\mathbf{H}(f)$. For a good separation, this index should be close to 0 or infinity (in this case the estimated sources are permuted). When r crosses the value 1, this means that a permutation has occurred. Figure 6 plots $\min(r, 1)$ and $\min(1/r, 1)$ versus frequency (in Hz), before and after apply-

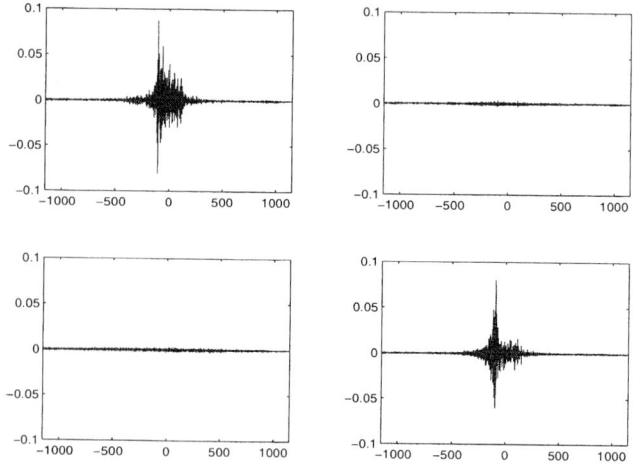

Fig. 7. Impulse responses of the global filter $(\mathbf{G} * \mathbf{H})(n)$

ing the new method of frequency permutation correction (but always with a preliminary correction by the method in [6]). One can see that the new method eliminates all permutation errors (relative to a global permutation) except two errors occuring in isolated frequency channels, which are not visible in the plot. But the method in [7] (results not shown for lack of space) still leaves a large bloc of permutation errors.

The four impulse responses of the global filter $(\mathbf{G} * \mathbf{H})(n)$ are shown in figure 7. One can see that $(\mathbf{G} * \mathbf{H})_{11}(n)$ is much bigger than $(\mathbf{G} * \mathbf{H})_{12}(n)$ and $(\mathbf{G} * \mathbf{H})_{22}(n)$ is also bigger than $(\mathbf{G} * \mathbf{H})_{21}(n)$, meaning that the sources are well separated.

5 Conclusion

We have introduced a method for blind separation of speech signals, which exploits their specificity: non stationarity and the presence of pauses. Our method is able to separate convolutive mixtures with fairly long impulse responses containing strong echoes.

References

1. J. Anemüler and B. Kollmeier. Amplitude modulation decorrelation for convolutive blind source separation. In *Proceeding of ICA 2000 Conference*, pages 215–220, Helsinki, Finland, June 2000.
2. R. Mukai, S. Araki, and S. Makino. Separation and dereverberation performance of frequency domain blind source separation. In *Proceeding of ICA 2001 Conference*, pages 230–235, San-Diego, USA, December 2001.
3. L. Parra and C. Spence. Convolutive blind source separation of non-stationary sources. *IEEE Trans. on Speech and Audio Processing*, 8(3):320–327, May 2000.
4. D. T. Pham. Joint approximate diagonalization of positive definite matrices. *SIAM J. on Matrix Anal. and Appl.*, 22(4):1136–1152, 2001.

5. D. T. Pham and J.-F. Cardoso. Blind separation of instantaneous mixtures of non stationary sources. *IEEE Trans. Signal Processing*, 49(9):1837–1848, 2001.
6. D. T. Pham, Ch. Servière, and H. Boumaraf. Blind separation of convolutive audio mixtures using nonstationarity. In *Proceeding of ICA 2003 Conference*, Nara, Japan, April 2003.
7. D. T. Pham, Ch. Servière, and H. Boumaraf. Blind separation of speech mixtures based on nonstationarity. In *Proceeding of the ISSPA 2003 Conference*, Paris, France, July 2003.
8. P. Smaragdis. Blind separation of convolved mixtures in the frequency domain. In *International Workshop on Independence & Artificial Neural Networks*, University of La Laguna, Tenerife, Spain, February 1998.
9. H.-C. Wu and J. C. Principe. Simultaneous diagonalization in the frequency domain (SDIF) for source separation. In *Proceeding of ICA 1999 Conference*, pages 245–250, Aussois, France, January 1999.

Convolutive Acoustic Mixtures Approximation to an Instantaneous Model Using a Stereo Boundary Microphone Configuration

Juan Manuel Sanchis, Francisco Castells, and José Joaquín Rieta

Universidad Politécnica de Valencia
46730 Gandia Spain
{jmsanch,jjrieta,fcastells}@eln.upv.es

Abstract. In this work it is demonstrated that, taking into account the conditions of a convolutive mixture of acoustic signals, it is possible to configure a mixture system whose separation model accomplishes the conditions of an a instantaneous mixture. This system is achievable by using stereo boundary microphones; this type of coincident microphones can be used to uniform delays of the propagation channels and to reduce the number of reflections that characterize the impulse response of the system. By means of coincident boundary microphones techniques, instantaneous BSS algorithms are applicable, thus providing optimal results with less computational cost. This system is validated in both anechoic and reverberant chambers.

1 Introduction

The main problem that presents the evaluation of the algorithms of blind separation of sources (BSS) lays on the non availability of the original signal sources and the mixture system that originate the observed available signals. This implicit problem that arises when dealing with real mixtures can be solved using synthetic signals, where the original sources are known [1], [2]. In fact, the validation of BSS algorithms with synthetic recordings may represent an initial start point towards its further application to real mixtures. Certainly, by analyzing the separation quality in the case of synthetic mixtures, the main characteristics of the algorithm can be evaluated in some basic situations, thus permitting the study of its feasibility to achieve the main objective of separating the independent sources as accurately as possible. Hence, different scenarios can be created depending on the statistical properties of the sources and on the mixture process, and the effect that these factors may introduce in the separation performance of the algorithm can be evaluated [3].

Some works have focused their study on the influence of the mixture system on the results obtained by BSS algorithms. In this sense, the effect of the acoustic environment has been analyzed [4], since the impulse response that models the mixture system is directly related to the reverberation and the density of reflections that characterize the acoustic chamber. Those studies put into evidence that chambers with long impulse responses and high reflections densities imply a decrease in the separation performance.

However, there are very few studies regarding the characteristics and configuration of the transducers employed for the reception of the observations that serve as inputs for the BSS algorithm [5], [6]. At the moment the available bibliography always establishes a placement of the microphone array following the technique of separated microphones (Fig.1a). The microphones are placed according to certain criteria, normally at equispaced points [5].

In opposition to this technique, we propose the use of the coincident microphones technique, jointly with a disposition that minimizes the number of captured reflections. Using boundary stereophonic microphones will be shown that the quality of the estimated sources is improved.

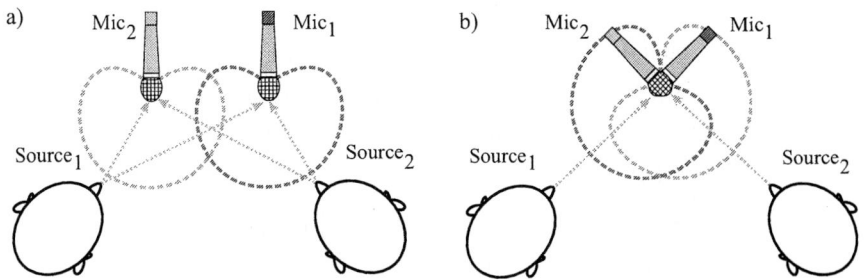

Fig. 1. Configuration for the reception of two sound sources using (a) separate microphones in front of (b) coincident microphones.

2 Approximation

In the configuration of coincident microphones (Fig. 1b), two directive microphones are perpendicularly placed such that both transducers are located at the same point. Following this configuration, the acoustic signal will ideally reach both microphones at the same time instant. Due to the directional characteristic of the microphones, the sources will be captured with different level. In the example of Fig.1b, the microphone 1 will capture more signal from source 2 than from source 1, and on the contrary, the microphone 2 will capture more signal from source 1. Consequently, two observations or mixed signals are obtained in a different manner, such that there are no time differences between the sources that are present at each mixture. That is to say, ideally we are being able to obtain an instantaneous mixture system for acoustic signals.

Analyzing the proposed mixture system for the case of two sources and two observations, and being $\mathbf{s}[n] = [s_1[n]\ s_2[n]\ ]^T$ the sources vector, $\mathbf{x}[n] = [x_1[n]\ x_2[n]\ ]^T$ the observations vector and $h_{ji}[n]$ the impulse response of a LTI system that connects the ith observation with the jth source ($i, j = 1,2$), we can express the mixture model as

$$\begin{bmatrix} x_1[n] \\ x_2[n] \end{bmatrix} = \begin{bmatrix} h_{11}[n] & h_{12}[n] \\ h_{21}[n] & h_{22}[n] \end{bmatrix} * \begin{bmatrix} s_1[n] \\ s_2[n] \end{bmatrix} \quad (1)$$

Supposing that the directive property of the microphone i^{th} towards the source j^{th}, d_{ji} is frequency independent, the impulse response can be decomposed in a term that reflects the characteristics of the acoustic environment $h'_{ji}[n]$ multiplied by a directivity factor d_{ji}

$$h_{ji}[n] = d_{ji} \, h'_{12}[n], \quad i,j = 1,2 \tag{2}$$

In addition, when two microphones are summoned in the same point, the impulse response from a certain source towards each one of the microphones can be supposed to be approximately the same:

$$\begin{aligned} h'_{11}[n] &= h'_{12}[n] = h'_{1}[n] \\ h'_{22}[n] &= h'_{21}[n] = h'_{2}[n] \end{aligned} \tag{3}$$

Taking into account these considerations, we can rewrite (1) as:

$$\begin{aligned} \begin{bmatrix} x_1[n] \\ x_2[n] \end{bmatrix} &= \begin{bmatrix} d_{11} \, h'_{11}[n] & d_{21} \, h'_{21}[n] \\ d_{12} \, h'_{12}[n] & d_{22} \, h'_{22}[n] \end{bmatrix} * \begin{bmatrix} s_1[n] \\ s_2[n] \end{bmatrix} \\ \begin{bmatrix} x_1[n] \\ x_2[n] \end{bmatrix} &= \begin{bmatrix} d_{11} \, h'_1[n] & d_{21} \, h'_2[n] \\ d_{12} \, h'_1[n] & d_{22} \, h'_2[n] \end{bmatrix} * \begin{bmatrix} s_1[n] \\ s_2[n] \end{bmatrix} \\ \begin{bmatrix} x_1[n] \\ x_2[n] \end{bmatrix} &= \begin{bmatrix} d_{11} & d_{21} \\ d_{12} & d_{22} \end{bmatrix} \cdot \begin{bmatrix} s_1[n] * h'_1[n] \\ s_2[n] * h'_2[n] \end{bmatrix} \end{aligned} \tag{4}$$

These results lead us to consider that the observed signals constitute an instantaneous mixture of source signals modified by the effect of a concrete acoustic environment. The problem becomes the separation of instantaneous mixtures, where the independent components represent the signal that would have been recorded in the case that just the corresponding source had been active. If the main objective is a spatial separation of cue source and the minimisation of the interferences introduced by other sources, this perspective of the problem would be optimal.

In a practical application it will be more complicated to accomplish the requirements established in the proposed model. Firstly, a small positioning error will exist in the transducers that conform the array of coincident microphones and secondly, the reception response of the microphone, i.e. the directivity, does not remain constant with the frequency. However, in spite of these factors, with this configuration we can force that the convolutive mixture is generated according to an instantaneous model, achieving higher separation degree.

3 Real Room Experiment

To verify the previous ideas, some experiments using both configurations of microphones, coincident and separated, have been carried out. These experiments consist of a simple scenario of two sources and two microphones (2x2) summoned in two different acoustic environments: in an anechoic chamber and in a recording studio. These configurations are illustrated in Fig. 2.

For each one of the rooms the following steps were carried out:
1. Different acoustic signals are emitted by each speaker (intermittent voice + guitar), firstly simultaneously and next, separately. Hence, both mixed signals and separated signals are captured by the microphones.
2. Convolutive [7], [8] and instantaneous [9] BSS algorithms are applied to the observations. In the case of convolutive algorithms, the longitude (taps) of the separation filter was set from 1 (instantaneous mixing) until the maximum permitted by the algorithm.

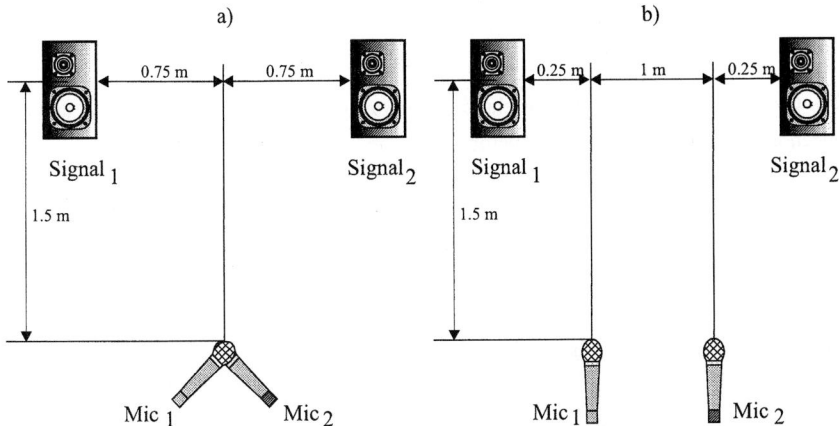

Fig. 2. Configuration of sources and microphones inside the acoustic chamber used for evaluating the influence of the microphones, according to coincident (a) and separated schemes (b).

The measurement of the separation degree is carried out by computing the parameter Signal to Interference Ratio (*SIR*), defined in [10].

The sources signals that had been registered independently (only one speaker emitting), can be used to verify that the impulse response h_{i1} is very similar to h_{i2} in the case of coincident microphones, as it had been previously hypothesized. The cross-correlation (h_{11}/h_{12}) and the autocorrelation (h_{11}/h_{11}) were computed for each configuration. The cross-correlation and autocorrelation functions corresponding to the impulse responses h_{11} and h_{12} recorded in an anechoic chamber are presented in fig. 3 and fig. 4 in the case of separated and coincident microphones, respectively. In addition, the cross-correlation and autocorrelation functions corresponding to the impulse responses h_{11} and h_{12} recorded in a recording studio are presented in fig. 5 and fig. 6 as well.

In these figures we can observe the following:
- In the configuration of coincident microphones the autocorrelation and the cross-correlation functions resemble each other, in both anechoic and reverberant chambers, with no delay differences on the registered signals.
- In the configuration of separated microphones, there exists some similarity between the correlation functions as well, although it is observed a temporal shift due to different arrival times.

- Taking into consideration the autocorrelation function as the main signal, the cross-correlation of the measures in the reverberant room presents a higher degree of distortion, being more significant in the case of separated microphones.

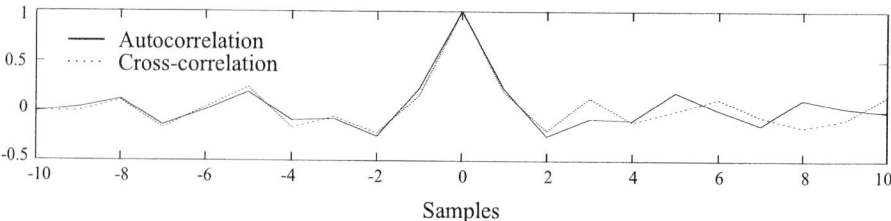

Fig. 3. Cross-correlation and autocorrelation functions of the impulse responses h_{11} and h_{12} in an anechoic chamber for coincident microphones (sample frequency = 44100 Hz).

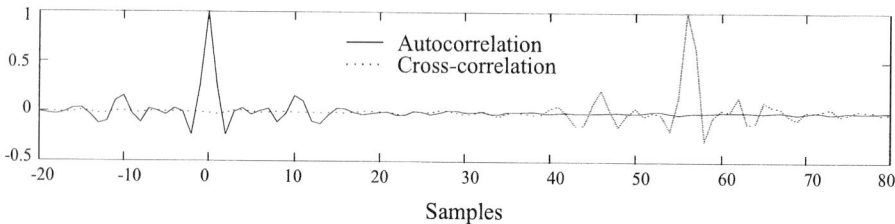

Fig. 4. Cross-correlation and autocorrelation functions of the impulse responses h_{11} and h_{12} in an anechoic chamber for separated microphones (sample frequency = 44100 Hz).

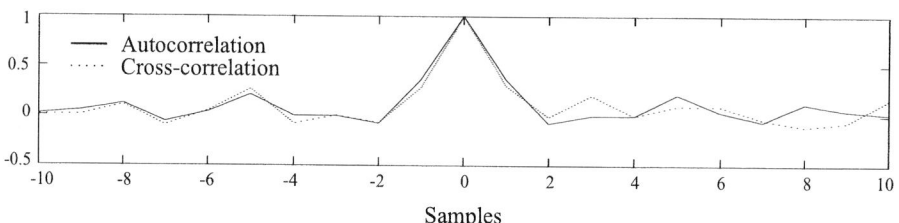

Fig. 5. Cross-correlation and autocorrelation functions of the impulse responses h_{11} and h_{12} in a recording studio for coincident microphones (sample frequency = 44100 Hz).

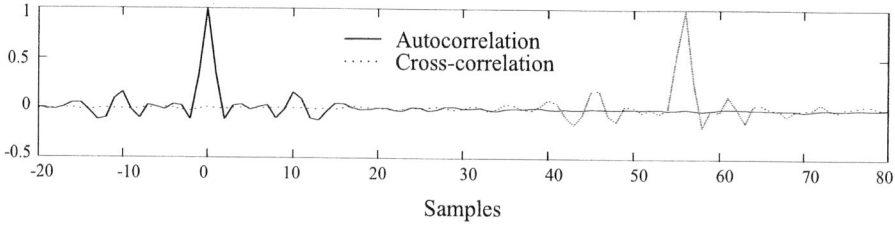

Fig. 6. Cross-correlation and autocorrelation functions of the impulse responses h_{11} and h_{12} in a recording studio for separated microphones (sample frequency = 44100 Hz).

The previous analysis regarding the correlation functions confirms that the observations obtained by coincident microphones in low-reverberant chambers can be approximated to instantaneous mixtures.

In practice, these observations do not correspond to an ideal instantaneous mixture, but at least, the number of samples of the separation filter is minimised. Indeed, this is an advantage, since convolutive algorithms provide better results with shorter filter lengths. The distortion that appears in the cross-correlation for the configuration of coincident microphones is due to the reflections in a real acoustic environment. Furthermore, any temporal delay between the cross-correlation function in reverberant and anechoic chambers is due to these reflections. However, this difference of time is much less than the existent using a configuration of separated microphones. Therefore, the separation filter length will be necessarily shorter, thus reducing the computational cost of the separation algorithm.

4 Convolutive BSS

By applying the convolutive BSS algorithms [7], [8] and varying the filter tap number, the independent sources are estimated, and the performance is measured according to the *SIR* parameter. The performance obtained with coincident and separated microphones are compared in figs. 7 and 8 for the anechoic chamber and the recording studio respectively.

From the experimental results, it can be assessed that:

- The configuration with coincident microphones permitted the best approximations, independent of the acoustic chamber. The resulting filter lengths were actually short (ideally it should be one unique sample). In particular, the algorithm contributes with a separation that is comparable with both configurations of microphones when high longitudes of the separation filter are employed.
- The results obtained with coincident microphones seem to be independent of the acoustic characteristics of the enclosure.

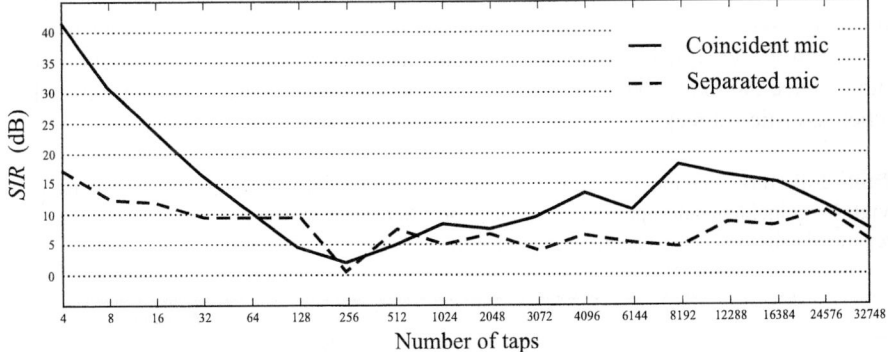

Fig. 7. *SIR* obtained with convolutive BSS algorithm for coincident and separated microphones in a anechoic chamber.

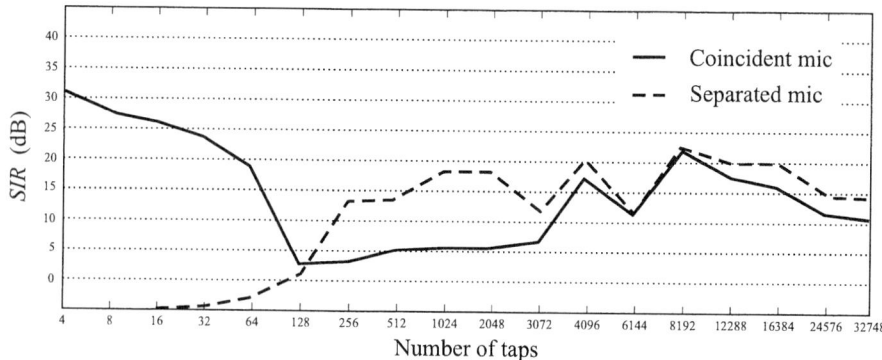

Fig. 8. *SIR* obtained with convolutive BSS algorithm for coincident and separated microphones in a recording studio.

- The convolutive BSS algorithm obtained better results for short filter lengths. Due to convergence limitations of the algorithm, it was not possible to test filter lengths close to one sample. However, extrapolating the tendency of the results, it is reasonable to think that the algorithm would have provided good results if an instantaneous mixture model had been applied.

5 Instantaneous BSS

In order to validate whether the mixtures registered by coincident microphones can be approximated to instantaneous mixtures, an instantaneous BSS algorithm has been applied to the observations [9]. The results in terms of *SIR* are reflected in Table 1. As it can be observed, the separation degree of the estimated sources is very satisfactory. Hence, the theoretical approximation to instantaneous mixtures has been corroborated empirically.

The main difference of the estimated sources via instantaneous BSS with respect to convolutive algorithms is that in the first case, the estimated sources conserve the acoustic effect due to the chamber, whereas in the second case this effect is minimized. However, it does not generally constitute a limitation of the separation algorithm, since the main objective is usually to minimize the interferences introduced by other acoustic sources.

Table 1. Performance measurement in terms of *SIR* using a configuration of two sources registered by coincident microphones in an anechoic chamber and a radio studio. The sources are separated using an instantaneous BSS approach.

	Voice	Guitar	Average
Anechoic chamber	40.9 dB	20.1 dB	30.5 dB
Recording studio	24.7 dB	18.8 dB	21.7 dB

6 Conclusion

With this work it has been demonstrated that in simple configurations of two sources and two microphones, the separation results using convolutive BSS algorithms are optimal taking into consideration a lower taps number. It is even possible to apply instantaneous BSS algorithms obtaining satisfactory results. With the approximation to the instantaneous case, it is possible to recover the source signals convolved with the impulse response of the chamber with a considerable decrease in the computational cost. In opposition to these methods, convolutive BSS algorithms aim to recover the sources exempt of the acoustic effect of the chamber.

References

1. H. Sahlin and H. Broman. "Signal separation applied to real world signals," *Proceedings of Int. Workshop on Acoustic Echo and Noise Control*, London, UK, September, 1997.
2. D. Schobben, K. Torkkola, and P. Smaragdis, "Evaluation of blind signal separation methods," *1st International Conference on Independent Component Analysis and Blind Signal Separation (ICA'99)*, Aussois, France, pp. 261-266, January 1999.
3. J. M. Sanchis, *"Evaluation of mixture conditions in convolutive blind source separation for audio applications,"* Ph. D. Thesis, Universidad Politecnica de Valencia, December 2003.
4. R. Mukai, S. Araki and S. Makino, "Separation and dereverberation performance of frequency domain blind source separation," *3rd International Conference on Independent Component Analysis and Blind Signal Separation*, San Diego, California, USA, pp. 230-235, December 2001.
5. D. V. Rabinkin et al., "Optimun microphone placement for array sound capture," *Proceedings of 133rd Meeting of the Acoustical Society of America*, State College, Pennsylvania, USA, pp. 227-239, June 1997.
6. J. R. Hopgood, P. J. W. Rayner and P. W. T. Yuen, "The effect of sensor placement in blind source separation," *IEEE Workshop on Applications of Signal Processing to Audio and Acoustics*, New Paltz, New York, USA, October 2001.
7. D. Schobben, *Real-time Adaptative Concepts in Acoustics*, KluwerAcademic Publishers, 2001.
8. R. H. Lambert, *Multichannel Blind Deconvolution: FIR Matrix Algebra and Separation of Multipath Mixtures*, Ph. D. Thesis, University of Southern California, USA, 1996.
9. V. Zarzoso. *Closed-form higher-order estimators for blind separation of independent source signals in instantaneous linear mixtures*. PH. D. Thesis, University of Liverpool, UK, October 1999.
10. S. Araki, R. Mukai, S. Makino, T. Nishikawa, H. Saruwatari, "The fundamental limitation of frequency domain blind source separation for convolutive mixture of speech," *IEEE Transactions on Speech and Audio Processing*, vol. 11, no 2, pp. 109-116, March 2003.

DOA Detection from HOS by FOD Beamforming and Joint-Process Estimation

Pedro Gómez Vilda, R. Martínez, Agustín Álvarez Marquina, Victor Nieto Lluis, María Victoria Rodellar Biarge, F. Díaz, and F. Rodríguez

Departamento de Arquitectura y Tecnología de Sistemas Informáticos
Universidad Politécnica de Madrid
Campus de Montegancedo, s/n, 28660 Boadilla del Monte, Madrid, Spain

Abstract. Array Beamforming is a powerful technique in Speech Enhancement, Noise Reduction, Source Separation, etc. for which powerful techniques have been developed [8]. Nevertheless, large Arrays present several inconveniences, as are sensor equalization, complex DOA algorithms, high costs, large computational requirements, etc. This lead to exploring other possible structures based on paired sensors, as First-Order Differential Beamformers (FODB) [2]. These structures may be steered to aim their sharp notch to the desired source, which may be removed from the output, and complementarily reconstructed using several methods, as direct or spectral subtraction [1], or joint-process estimation [5]. The main problem that these systems present is DOA estimation in the presence of reverberation. Through this paper it is shown that the use of Higher Order Statistics may help in detecting DOA's. Results for simulated source separation, and DOA detection in a real room are given and discussed.

1 Introduction

An FODB is a structure using two microphones and a combination of signals as given in Fig. 1, where the parameter β is the *steering factor* of the FODB.

This structure may be used as in Fig. 2, where the source separation capabilities of the FODB are exploited to obtain an estimation of a source by complementation. The angular transfer function of the FODB will be a sharp notch aiming at a certain angle φ as a function of β:

$$F(\varphi = \varphi_i) = 1 - \delta(\varphi_i); \quad -\pi \le \varphi_i \le \pi \quad (1)$$

φ_i being the angular DOA where source s_i is located, and δ Dirac's delta function (see Fig.30).

Therefore, the output of the FODB will contain information coming from any DOA except from φ_i. The output of the filter will be defined in general as:

$$y = xF(\varphi_i) \quad (2)$$

x being the equivalent input to the FODB, which may be evaluated from the signals arriving to each sensor.

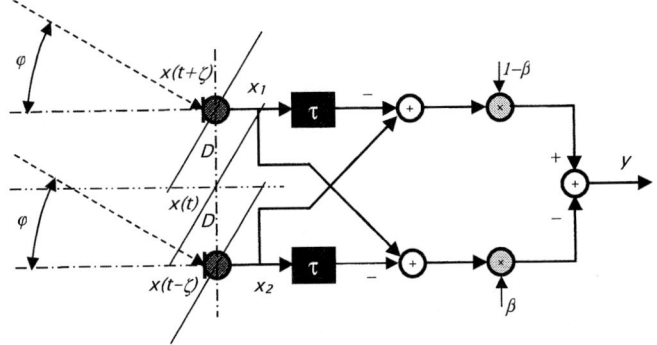

Fig. 1. First-Order Differential Beamformer, x_1 and x_2 being the inputs to both microphones y the beamformer output, and β the *steering factor*.

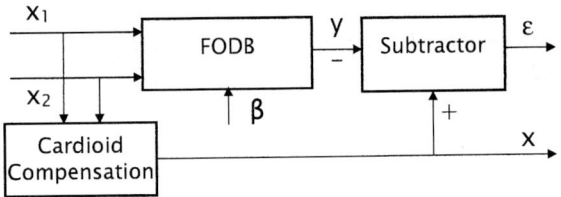

Fig. 2. Simplified structure of the Source Separator, x_1 and x_2 being the outputs of both microphones m_1 and m_2, x the output of the *equivalent cardioid microphone*, y the beamformer output, and ε the *estimation of the detected source*.

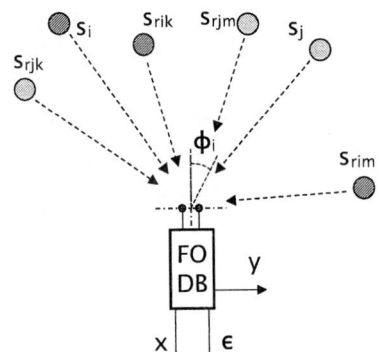

Fig. 3. Source Composition model, s_i and s_j are two (real) sources, s_{rik} and s_{rjm} being the respective multiple-path arrivals (apparent sources) corresponding to each real source.

The following assumptions are to be made regarding the signals arriving to the FODB:
- Sources are mutually independent.
- Reverberations are dependent to their corresponding sources within a given time-lag.
- Reverberations corresponding to one source are independent from those corresponding to another.

Having in mind these assumptions the following definitions will be introduced:

Let S be the set of all sources (real or apparent) inducing signal on both microphones m_1 and m_2, defined by the pair (φ_j, s_j): $s_j \in S : \varphi_j \in [-\pi, \pi]$.

Let S_{id} be the set of sources (real or apparent) dependent to the given source s_i:

$$S_{id} = \{s_j \in S : E\{s_{i,n} s_{j,n+k}\} \neq 0; \forall k \in \mathbf{Z}\} \quad (3)$$

Let S_{io} be the set of sources (real or apparent) independent to the given source s_i:

$$S_{io} = \{s_j \in S : E\{s_{i,n} s_{j,n+k}\} = 0; \forall k \in \mathbf{Z}\} \quad (4)$$

Let x_i be the component of x contributed by the source being aimed to, s_i.

Let x_{id} be the component of x contributed by S_{id}, or *dependent component*:

$$x_{id} = \sum_{\forall s_j \in S_{id}} h(s_j) \quad (5)$$

Let x_{io} be the component of x contributed by S_{io} or *independent component*:

$$x_{io} = \sum_{\forall s_j \in S_{io}} h(s_j) \quad (6)$$

where $h(s_j)$ explains the influence of the propagation media, sensor transfer function, and pre-processing stages on the incoming sound. It will assumed that this function shows a linear behavior. As a consequence of the above, it will be assumed that the following properties hold:

$$E\{x_{id,n} x_{io,n+k}\} = 0; \forall k \in \mathbf{Z} \Rightarrow x_{id} \perp x_{io} \quad (7)$$

The consequence of the above is that the input signal of the FODB may be split into two parts, mutually independent to each other within a time span, these being $x_i + x_{id}$ and x_{io}.

It will be implied that $x_i + x_{id}$ is identified with the contributions associated to source s_i, whilst x_{io} is the signal induced by other sources (independent from s_i) and their respective reverberation paths. The relations among the different signals are expressed in vector form in Fig. 4.

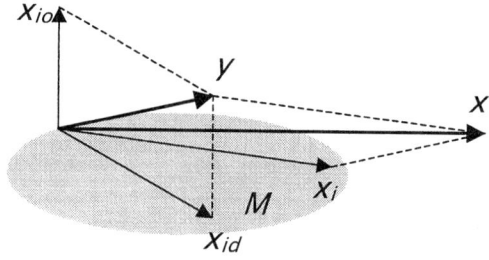

Fig. 4. Geometrical relations among the different signals: x_i and x_{id} define a plane M, x_{io} being orthogonal to it.

2 Signal Separation

The operation of the *subtractor* in Fig. 2 is carried out using JPE (Joint-Process Estimation) due to the presence of channel inequalities and delays in both x_1, x_2, rendering it impossible to subtract simply one trace from another. Instead the structure in Fig. 5.a is used, for which the relationships holding among the different signals are the following:

$$\hat{s} = \Im\{r\} \qquad (8)$$
$$e = s - \hat{s} = s - \Im\{r\} \qquad (9)$$

where $\Im\{*\}$ is the linear operator representing the operation performed by the *JPE* adaptive filtering.

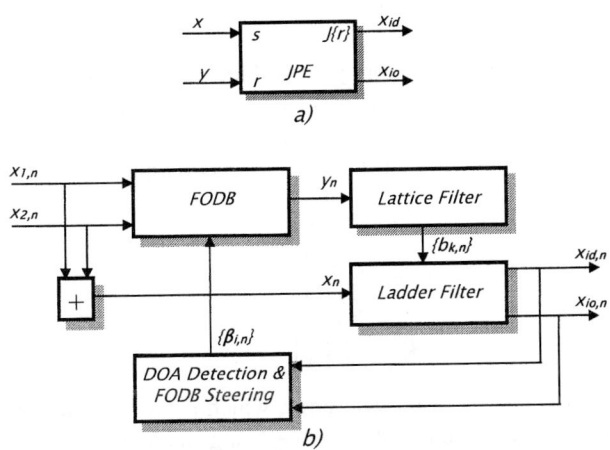

Fig. 5. a) *JPE* used. b) Whole structure implementing source separation.

It is well known that when the operator $\Im\{*\}$ has been optimally adapted the norm of the estimation error will be minimum in a least squares sense [5]:

$$\|s - \Im_o\{r\}\| = min\|s - \Im\{r\}\| \qquad (10)$$

In what follows, it will be assumed that the process of *JPE* has been carried to this condition, under which the following orthogonalization properties hold:

$$E\{e_n \hat{s}_{n+k}\} = 0; 0 \le k \le K \Rightarrow e \perp \hat{s} \qquad (11)$$
$$E\{e_n r_{n+k}\} = 0; 0 \le k \le K \Rightarrow e \perp r \qquad (12)$$

where K is the order of the adaptive filter, which will be used to extract x_i subtracting y from x. With this in mind the *JPE* will recover the common components between the reference and the estimated signals (direct path), and produce an error which will be the uncorrelated (or complementary) part between x and y, for which the following associations are established:

$$s = x_n; \quad r = y_n$$

$$\hat{s} = \Im\{y_n\}$$

$$e = \begin{cases} \hat{x}_{in}; \varphi = \varphi_{si} \\ \to 0 \text{ other angles} \end{cases}$$

This set of relationships is implemented by the structure given in Fig. 5.b. A lattice-ladder filter [7] algorithm supporting the structure given above will be used.

3 DOA Determination

One of the most important problems yet to be solved at this point, is detecting the presence of a source from the analysis of the statistics of the resulting signals. Using second order statistics at this point is not sensitive enough, as the behavior of second order statistics with the sweeping angle shows bowl-like curves with a single minimum pointing to a certain *center of gravity* corresponding to an average of sources and intensities as shown in [4]. A more powerful technique for DOA determination is based on Independent Component Analysis [6], as it may be shown that Higher-Order Statistics are more sharply affected by the directions of arrival than Second-Order ones. The criteria used in the present study will be based on the following estimators:

$$E_{e_n} = E\{e_n^2\} \tag{13}$$

$$\eta = \frac{E\{e_n^2\}}{E\{\hat{s}_n^2\}} \tag{14}$$

$$K_{e_n} = E\{e_n^4\} \tag{15}$$

The first one is the energy of the JPE error, which is an estimator of the contribution of the source s_i, the presence of a source being marked by a maximum of this function. The second one is a relationship between the energy of the contribution of the source and its complementary part. It should be expected a maximum in the numerator and a minimum in the denominator when a source is present. The third one is the Kurtosis of the JPE error, and should be a sharper estimator than second-order ones, presenting a higher sensitivity to changes in the independence of the signals involved.

4 Results and Discussion

To check the performance of the methods described, two sets of experiments have been carried out. The first one used pure sinusoidal tones at different frequencies and positions, simulated to calibrate the processing system. The results for three sources of *500 Hz (-12.25º), 1 kHz (0º)* and *2kHz (+12.25º)* with same amplitudes may be seen in Fig. 6. On top the energy of the independent component e_n estimated from (13) is presented. It may be seen that the positions of the sources are correctly signaled by the maxima of this estimator. Comparing the results given in the middle trace from (14) and the one at the bottom from (15) it can be deducted that the capability of the three estimators to signal the presence of a source is equivalent, but the angular selectivity of (15) is higher than the other two. Having set this important fact, a second set of experiments was carried out with real recordings using the framework given in Fig. 7, where a couple of loudspeakers s_1 and s_2 are placed in the same plane as a pair of microphones m_1 and m_2 at a distance of *150 cm.* within a sound-proof chamber. The separation between microphones used in the experiments described is *2D=5 cm*. The angular position of both sources relative to the array are $-22.5°$ and $22.5°$. The experiments consisted in the detection of a sinusoidal tone of *1 kHz* played on the right loudspeaker (with respect to the array axis, at+*22.5°*) while white noise was played on the left loudspeaker.

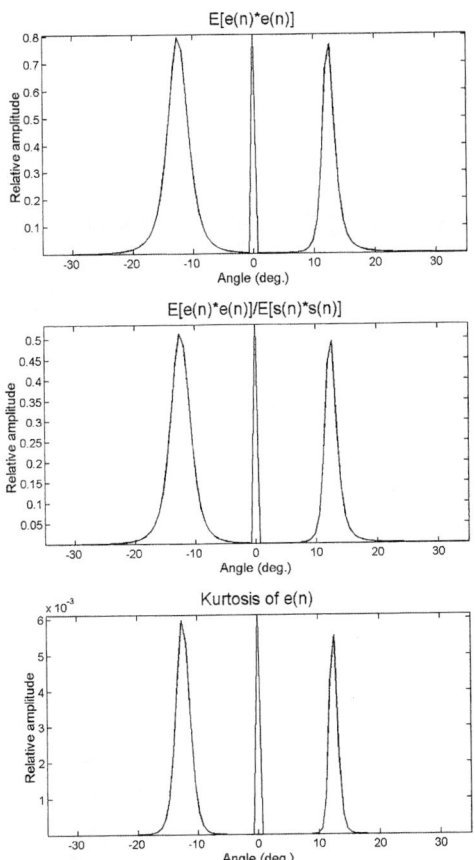

Fig. 6. Results from simulations with three sources of *0.5 kHz*, *2kHz* and *1kHz* at *–12.25°*, *0°* and *+12.25°*, respectively. Top: Energy of e_n. Middle: Ratio between the energies of e_n and $\hat{s}_n$. Bottom: Coefficient of Kurtosis of e_n.

The SNR of the sinusoidal tone with reference to noise was around *–5dB*. A frame of sound *1 sec.* long was recorded at a sample rate of *11,025 Hz*, dividing it into frames *128 samples* long, on a sliding window with *50%* overlapping. The input power spectrum may be seen in Fig. 8 (top).

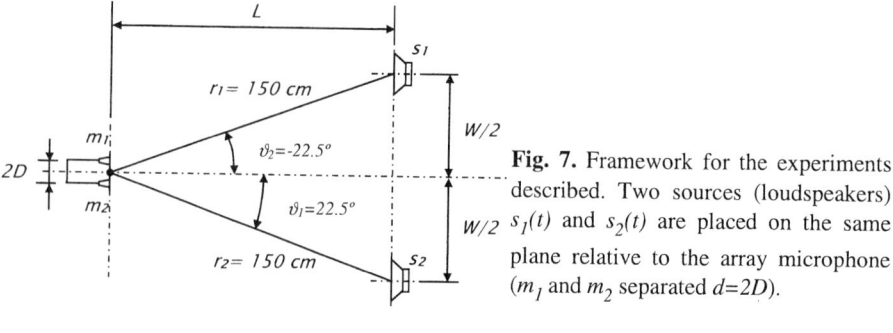

Fig. 7. Framework for the experiments described. Two sources (loudspeakers) $s_1(t)$ and $s_2(t)$ are placed on the same plane relative to the array microphone (m_1 and m_2 separated $d=2D$).

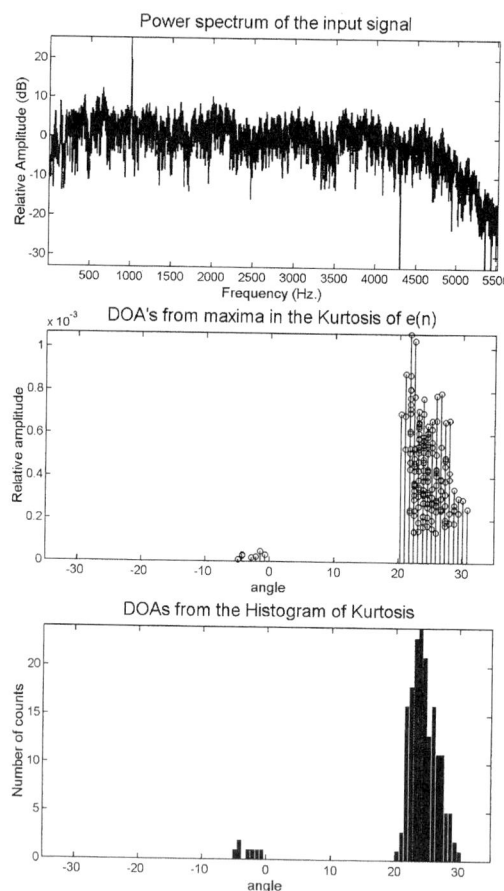

Fig. 8. Top: Power spectra of the input signal x_n. Middle: Estimation of the maxima in the Kurtosis of the independent component e_n for the case described. Bottom: Histogram giving the number of detections for each DOA from the Kurtosis of e_n.

For each frame, the best *DOA* was detected accordingly to the maxima in the Kurtosis measured on the source estimator e_n as shown in Fig. 8 (middle). The number of detections per angle of arrival was counted and accumulated on a histogram, which is given in Fig. 8 (bottom). From these results several conclusions may be drawn out:

- The capability of the *FODBs* to establish an unbalance between the contribution due to the source aimed to, and the background contributions, is enough to justify the use of *FODB* with other separation methods, as spectral subtraction [3] or joint-process estimation.
- The separation introduced by joint-process estimation is large enough to allow DOA detection using the energy of the independent component of the input signal.
- Other HOS-based detection methods may show a better capability for DOA detection, providing better angular selectivity.

In experiments with real signals the capability of detecting DOAs from Kurtosis have been proven.

Deviations in the results obtained in the real framework with respect to the ideal model may (as the deviation in the bell-shape of DOA histogram) may be due to multiple-path fusion of signals, non point-like sources, and near-field effect, these constituting future lines of research.

Acknowledgments

This research is being carried out under grants TIC99-0960, TIC2002-02273 and TIC2003-08756 from the Programa Nacional de las Tecnologías de la Información y las Comunicaciones (Spain).

References

1. Álvarez, A., Gómez, P., Nieto, V., Martínez, R., Rodellar, V., "Speech Enhancement and Source Separation supported by Negative Beamforming Filtering", Proc. of the 6th ICSP, Beijing, China, August 26-29, 2002, pp. 342-345.
2. Elko, G. W., "Microphone array systems for hands-free telecommunication", Speech Communication, Vol. 20, No. 3-4, 1996, pp. 229-240.
3. Gómez, P., Álvarez, A., Martínez, R., Nieto, V., Rodellar, V., "Optimal Steering of a Differential Beamformer for Speech Enhancement", Proc. of EUSIPCO'02, Vol. III, Toulouse, France, 3-6 September, 2002, pp. 233-236.
4. Gómez, P., Álvarez, A., Martínez, R., Nieto, V., Rodellar, V., "Time-Domain Steering of a Differential Beamformer for Speech Enhancement and Source Separation", Proc. of the 6th ICSP, Beijing, China, August 26-29, 2002, pp. 338-341.
5. Haykin, S., Adaptive Filter Theory, Prentice-Hall, Englewood Cliffs, N. J., 1996.
6. Hyvärinen, A., Karhunen, J., Oja, E., Independent Component Analisis, John Wiley & Sons, New York, 2001.
7. Proakis, J. G., Digital Communications, Mc Graw-Hill, 1989.
8. Van Trees, H. L., Optimum Array Processing, John Wiley, N. Y. 2002.

Nonlinear Postprocessing for Blind Speech Separation

Dorothea Kolossa and Reinhold Orglmeister

TU Berlin, Berlin, Germany
D.Kolossa@ee.tu-berlin.de
http://ntife.ee.tu-berlin.de/personen/kolossa/home.html

Abstract. Frequency domain ICA has been used successfully to separate the utterances of interfering speakers in convolutive environments, see e.g. [6],[7]. Improved separation results can be obtained by applying a time frequency mask to the ICA outputs. After using the direction of arrival information for permutation correction, the time frequency mask is obtained with little computational effort. The proposed postprocessing is applied in conjunction with two frequency domain ICA methods and a beamforming algorithm, which increases separation performance for reverberant, as well as for in-car speech recordings, by an average 3.8dB. By combined ICA and time frequency masking, SNR-improvements up to 15dB are obtained in the car environment. Due to its robustness to the environment and regarding the employed ICA algorithm, time frequency masking appears to be a good choice for enhancing the output of convolutive ICA algorithms at a marginal computational cost.

1 Introduction

Frequency domain blind source separation can be employed to obtain estimates of clean speech signals in reverberant environments. One successful approach uses independent component analysis to obtain an estimate of the mixing system (i.e. the room transfer function) and subsequently inverts it. Applying this unmixing system to the signals yields estimates of the short time spectra of the speech signals $\hat{S}_{1..n}(k, \Omega)$.

This ICA-based estimate can be further enhanced taking advantage of the approximate disjoint orthogonality of speech signals. Two signals $s_1(t)$ and $s_2(t)$ are called W-disjoint orthogonal, when the support of their windowed Fourier transforms do not overlap, i.e. when

$$S_1(k, \Omega) S_2(k, \Omega) = 0 \quad \forall k, \Omega, \tag{1}$$

for the window function $W(t)$, where k refers to the frame number and Ω to the frequency bin. This condition does not hold exactly for interfering speech signals, however, it is true approximately for an appropriate choice of time frequency representation, as shown in [10].

Thus, a postprocessing scheme is proposed as follows: in each frequency bin Ω and at each frame k, the magnitudes of the ICA outputs are compared. Based on the assumption of disjoint orthogonality, only one of the outputs should have a non-zero value at any given frame and bin. Therefore, only the frequency bin with the greatest magnitude is retained, the other frequency bins are set to zero. An overview of the entire system is given in Figure 1. While the approach was first tested on a frequency domain implementation of JADE [5], it is also successful as postprocessing for other ICA and beamforming algorithms.

The remainder of this paper is organized as follows. Section 2 gives an overview of the entire signal processing system and describes the ICA and beamforming algorithms which were used to arrive at an initial speech signal estimate $\hat{\mathbf{S}}(k, \Omega)$. Subsequently, Section 3 deals with the nonlinear postprocessing stage. The algorithm was evaluated on three data sets: real-room recordings made in a reverberant office environment, the ICA99 evaluation data sets, and in-car speech data, which was recorded in cooperation with DaimlerChrysler[1]. Details of the evaluation data and methods are given in Section 4. Finally, in Section 5, the results are collected and conclusions are drawn.

2 Algorithms

The block diagram of the algorithm is shown in Figure 1 for the case of two signals. While the algorithm is applicable for demixing an arbitrary number of sources, provided that they meet the requirement of approximate disjoint orthogonality, it was tested here only for the case of two sources and sensors.

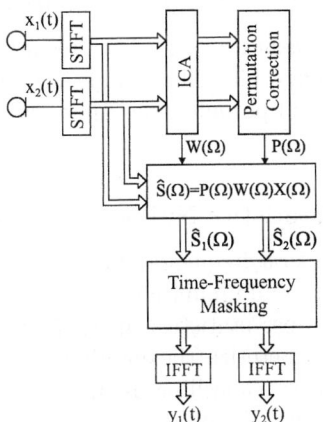

Fig. 1. Overview of the algorithm

First, the microphone signals, sampled at 16kHz, are transformed into the time frequency domain via STFT using a Hamming window of 512 samples, i.e. 32ms duration, and a frame shift of 8ms. In the ICA stage, the unmixing filters

---
[1] The authors wish to thank DaimlerChrysler for the cooperation and support.

$\mathbf{W}(\Omega)$ are determined for each frequency bin. This can be accomplished with any ICA algorithm, provided it operates on complex data. For this work, two different ICA approaches were tested, and were also compared to a fixed direction nullbeamformer. The unmixing filters, determined by ICA, are applied to the microphone signals to obtain initial speech estimates $\hat{S}(k,\Omega)$. The permutation problem is solved by beampattern analysis, which is done assuming that the incoming signal obeys the farfield beamforming model, i.e. all incoming sound waves are planar. In this case, the directivity patterns of a demixing filter $\mathbf{W}(\Omega)$ can be calculated as a function of the angular frequency $\omega = \frac{F_s}{N}\Omega$ and the angle of incidence of the signal relative to broadside, φ, via

$$F_l(\Omega,\varphi) = \sum_{k=1}^{2} W_{lk}(\Omega) exp(j\frac{\Omega F_s d \sin\varphi}{N \cdot c}). \qquad (2)$$

Here, N is the number of frequency bins and F_s the sample rate. The permutation matrix $\mathbf{P}(\Omega)$ is determined by aligning the minima of directivity patterns between frequency bins, as described in [6].

This result of this procedure is, on each channel, a linear, filtered combination of the input signals. Since speech signals are sparse in the chosen time frequency representation, subsequent time frequency masking (TF masking) can be used to further suppress noise and interference in those frames and bins, where the desired signal is dominated by interference. Finally, the unmixed signals $\mathbf{Y}(\Omega,k)$ are transformed back into the time domain using the overlap-add method.

2.1 Complex JADE with Beampattern Correction

A frequency domain implementation of JADE results in a set of unmixing matrices, one for each frequency bin. The scaling problem is avoided by using a normalized mixing model and permutations are corrected by beampattern analysis as described above.

2.2 Minimum Cross Statistics Nullbeamforming

The second algorithm is also a frequency domain convolutive approach, which is based on searching for the minimum cross cumulant nullbeamformer in each frequency bin. Here, the cross statistics up to fourth order are used, similar to [2]. The idea is to parameterize the unmixing system in such a way that it becomes a nullbeamformer, cancelling as many directional interferers as the number of microphones allows. When the microphones are sufficiently close and well adjusted so that no damping occurs, and when the sources obey the farfield model, the mixing matrix can be written as

$$\mathbf{X}(j\omega) = \mathbf{A}_{ph}(j\omega) \cdot \mathbf{S}(j\omega) \qquad (3)$$

with the phase shift mixing matrix

$$\mathbf{A}_{ph}(j\omega) = \begin{bmatrix} 1 & 1 \\ e^{-j\omega\frac{d}{c}\sin(\varphi_1(\omega))} & e^{-j\omega\frac{d}{c}\sin(\varphi_2(\omega))} \end{bmatrix} \qquad (4)$$

which depends on the angular frequency ω, the speed of sound c and the distance d between microphones. To cancel one of the signals, the inverse of the mixing model

$$\mathbf{W}(j\omega) = \frac{|e_1 - e_2|}{e_1 - e_2} \begin{bmatrix} -e_2 & 1 \\ e_1 & -1 \end{bmatrix} \quad (5)$$

is used, with

$$e_1 = e^{-j\omega \frac{d}{c} \sin(\varphi_1(\omega))} \quad \text{and} \quad e_2 = e^{-j\omega \frac{d}{c} \sin(\varphi_2(\omega))}. \quad (6)$$

This nullbeamformer is optimized for each frequency bin separately so that it is possible to compensate phase distortions introduced by the impulse response. The optimization is carried out by stochastic gradient descent for the cost function

$$J(\hat{S}_1, \hat{S}_2) = E(|\hat{S}_1 \cdot \hat{S}_2|) + |Cum(\hat{S}_1, \hat{S}_2)|, \quad (7)$$

where $Cum(\hat{S}_1, \hat{S}_2)$ refers to the fourth order cross-cumulant of $\hat{S}_1$ and $\hat{S}_2$.

2.3 Why Parameterize Each Bin?

Both ICA algorithms find an unmixing system separately in each frequency bin, and subsequently use only those time frequency points, in which one ICA output dominates the others by a set margin. This approach is strongly reminiscent of a family of algorithms described by Yilmaz and Rickard ([10]), where the following mixing model was used in the windowed Fourier transform domain:

$$\begin{bmatrix} X_1(\omega, \tau) \\ X_2(\omega, \tau) \end{bmatrix} = \begin{bmatrix} 1 & \ldots & 1 \\ a_1 e^{-j\omega \delta_1} & \ldots & a_N e^{-j\omega \delta_N} \end{bmatrix} \cdot \begin{bmatrix} S_1(\omega, \tau) \\ \vdots \\ S_N(\omega, \tau) \end{bmatrix} \quad (8)$$

The main difference of this mixing model is that the delay δ is not adjusted independently in different frequency bins. In anechoic environments, in which the far-field beamforming assumption is valid, it is sufficient to use one angle of incidence estimate φ, corresponding to one delay estimate δ, for all frequencies. In this case, source separation perfomance does not profit notably from introduction of frequency variant nulldirections as shown by [1]. However, when reverberation or noise is present in the signal, phase shift varies strongly over frequency. Thus it becomes difficult to estimate one best direction of arrival (DOA) for each source, and demixing performance suffers from localization errors. To assess the improvements gained from the extra computational effort of an ICA stage, we compared the separation performance of the two above algorithms to that of a constant DOA nullbeamformer, which was pointed to the directions giving minimum cross statistics of the outputs. This beamformer was used in the same structure as the ICA algorithms.

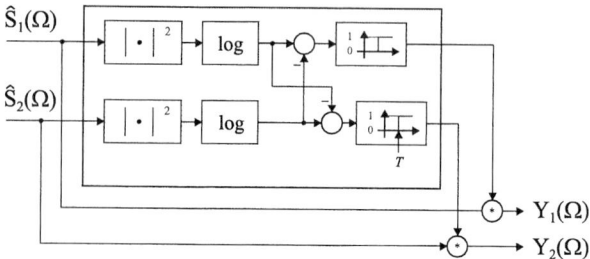

Fig. 2. Postprocessing for the 2x2 case

3 Nonlinear Postprocessing

In the postprocessing stage, a time frequency mask is applied to the ICA or beamformer outputs, as shown in Figure 2 for the special case of two signals. The time-frequency mask is determined from the ratio of demixed signal energies, which provides an estimate of the local SNR. The masking function

$$M_i = \Psi\left(log(|\hat{S}_i(\Omega)|^2) - \max_{\forall j \neq i} log(|\hat{S}_j(\Omega)|^2) - \frac{T}{10}\right) \qquad (9)$$

is obtained by comparing this SNR-estimate to an acceptance threshold T, with Ψ defined by

$$\Psi(x) = \begin{cases} 0 \text{ for } -\infty \leq x \leq 0, \\ 1 \text{ for } 0 < x < \infty. \end{cases} \qquad (10)$$

The threshold T was varied between -3dB and 5dB, with higher thresholds leading to better SNR gains but in some test cases to musical noise.

4 Evaluation

To test the proposed postprocessing method, three datasets were used on which separation was carried out with and without nonlinear postprocessing.

4.1 Datasets

ICA1999 Evaluation Data (Real Room). The tracks, which were suggested for evaluating ICA performance for the 1999 ICA Workshop [4], are sampled at 16KHz and are 10 seconds long (160000 samples). A male and a female speaker are speaking simultaneously and there is some background noise.

Reverberant Room Recording. Recordings were made in an office room with dimensions of about 10m × 15m × 3.5m. The distance between the loudspeakers and the two microphones (Behringer ECM 8000) was set to one meter. At this distance, the reverberation time was measured to be 300ms. Speech signals from the TIDigits database [9] were played back and recorded in two different setups of loudspeakers, with the angles of incidence, relative to broadside, as shown in Table 1.

Table 1. Recording configurations

config	θ_1	θ_2	recordings
A	45°	-25°	speaker 1, speaker 2, both speakers
B	10°	-25°	speaker 1, speaker 2, both speakers

In-Car Speech Data. In the final dataset, recordings were made inside a Mercedes S 320 at standstill and at 80 and 100km/h. Speech from the TIDigits database was reproduced with artificial heads and recorded simultaneously with four cardioid microphones, an eight channel microphone array mounted in the center of the ceiling near the rearview mirror, and two reference signals on a 16 channel-recorder. For evaluation, two recordings were used, one of a male and a female speaker and one of two male speakers. The impulse response of the car was measured, and the reverberation time was determined to lie between 60 and 150ms, depending on the position of the artificial head relative to the microphone.

4.2 Results

Evaluation of Separation Performance. To measure separation quality, the SNR improvement between the mixed and the demixed signal is used. For this purpose, two SNRs are calculated: the SNR at the input of the ICA stage and the output SNR. The output SNR is proposed as a measure of separation performance in [8] and it is calculated for channel j via:

$$SNR_{out,j} = 10 \log_{10} \frac{E(y_{j,j}^2)}{E(\sum_{i \neq j} y_{j,i}^2)} \quad (11)$$

Here, the term $y_{j,i}$ stands for the jth separation output, which is calculated with the microphone signals recorded using only source i active. The input SNR is calculated in a similar way, so that the SNR improvement is obtained by:

$$SNRI_j = 10 \log_{10} \frac{E(y_{j,j}^2)}{E(\sum_{i \neq j} y_{j,i}^2)} - 10 \log_{10} \frac{E(x_{j,j}^2)}{E(\sum_{i \neq j} x_{j,i}^2)} \quad (12)$$

with $x_{j,i}$ denoting the jth microphone signal when only source i is active.

To determine the influence of time frequency masking on the performance of ICA algorithms, the SNR improvement was calculated with and without nonlinear postprocessing for the three datasets of actual recordings. Table 2 shows the comparison.

The best values are marked in bold. As can be seen, nonlinear postprocessing adds between 1 and 6dB, on average 3.8dB, to the output SNR. Also, it is interesting to see that ICA performance in the noisy recordings (ICA99 and in-car data) is significantly higher than that of the constant DOA beamformer. When the threshold for the local SNR is increased, the SNR can be improved

Table 2. Average SNR improvements for real room recordings

	JADE without TF Mask	JADE with TF Mask	MCC Null-beamformer without TF Mask	MCC Null-beamformer with TF Mask	Fixed DOA NBF without TF Mask	Fixed DOA NBF with TF Mask
Reverberant Room (A)	5.2dB	5.5dB	7.3dB	9.3dB	6.8dB	**9.7dB**
Reverberant Room (B)	5.8dB	7.1dB	6.4dB	**10.1dB**	5.3dB	9.4dB
ICA 99 Dataset	2.9dB	**8.3dB**	2.5dB	4.4dB	0.7dB	3.0dB
Car Data standstill	13.8dB	**15.4dB**	8.8dB	12.0dB	4.6dB	10.9dB
Car Data 100kmh	6.3dB	**12.3dB**	5.4dB	10.4dB	3.1dB	8.5dB

Table 3. Average SNR improvements for all configurations

	JADE	MCC Null-beamformer	Fixed DOA Estimate
no TF-Mask	7.9dB	6.3dB	4.2dB
-3dB	9.2dB (+1.3dB)	7.6dB (+1.3dB)	6.0dB (+1.8dB)
0dB	8.9dB (+1.0dB)	8.3dB (+2.0dB)	7.3dB (+3.1dB)
3dB	10.0dB (+2.1dB)	9.8dB(+3.5dB)	8.1dB (+3.9dB)
5dB	10.4dB (+2.5dB)	10.2dB (+3.9dB)	9.2dB (+5.0dB)

further, on the other hand, listening quality can profit from lower thresholds. The average SNR improvement for different thresholds is shown in Table 3, where the average was taken over all datasets.

The values in parentheses are the SNR gains due to TF masking.

5 Conclusions

A combination of ICA and time frequency masking has been applied to in car speech recordings as well as to reverberant room recordings and artificial speech mixtures. In the car environment, SNR improvements of 15dB and more can be obtained with this combination, and SNR improvements due to time frequency masking alone in the range of 3dB and more are noted for most test cases. In the scenarios considered, using an frequency variant look direction improved separation performance by an average 1.9dB. However, in the noisy test cases, the output SNR of the ICA processor was greater than that of frequency invariant processing by a margin of 4.5dB.

Generally speaking, time frequency masking as a postprocessing step for frequency domain ICA algorithms can improve signal separation significantly. In the simplest form, where a signal in a frequency bin is retained only if its magnitude exceeds that of all other signals, the extra computational effort is negligible, and additional SNR gains of 5dB and more can be obtained. The postprocessing has been tested in conjunction with two ICA algorithms and one beamformer, and it can be expected to yield similar improvements on other frequency domain source separation algorithms.

References

1. Balan R.; Rosca J. and Rickard S.: Robustness of Parametric Source Demixing in Echoic Environments. Proc. Int. Workshop on Independent Component Analysis and Blind Signal Separation, San Diego, California (2001) 144–149
2. Baumann, W.; Kolossa, D. and Orglmeister, R.: Beamforming-based convolutive source separation. Proceedings ICASSP '03 **5** (2003) 357–360
3. Baumann, W.; Kolossa, D. and Orglmeister, R.: Maximum Likelihood Permutation Correction for Convolutive Source Separation. Proc. Int. Workshop on Independent Component Analysis and Blind Signal Separation, Nara, Japan (2003) 373–378
4. Available at URL: *http://www2.ele.tue.nl/ica99/*
5. Cardoso J.-F., High order contrasts for independent component analysis, Neural Computation **11** (1999) 157–192
6. Kurita, S.; Saruwatari, H.; Kajita, S.; Takeda, K. and Itakura, F.: Evaluation of blind signal separation method using directivity pattern under reverberant conditions, Proceedings ICASSP '00 **5** (2000) 3140 – 3143
7. Parra L. and Alvino C.: Geometric Source Separation: Merging convolutive source separation with geometric beamforming. IEEE Trans. on Speech and Audio Processing **10:6** (2002) 352–362
8. Schobben, D.; Torkkola, K. and Smaragdis, P.: Evaluation of Blind Signal Separation. Proc. Int. Workshop on Independent Component Analysis and Blind Signal Separation, Aussois, France (1999)
9. TIDigits Speech Database: Studio Quality Speaker-Independent Connected-Digit Corpus. Readme file on CD-ROM.
See also at URL: *http://morph.ldc.upenn.edu/Catalog/LDC93S10.html*
10. Yilmaz, Ö. and Rickard, S.: Blind Separation of Speech Mixtures via Time-Frequency Masking. Submitted to IEEE Transactions on Signal Processing (2003)

Real-Time Convolutive Blind Source Separation Based on a Broadband Approach*

Robert Aichner, Herbert Buchner, Fei Yan, and Walter Kellermann

Multimedia Communications and Signal Processing
University of Erlangen-Nuremberg
Cauerstr. 7, D-91058 Erlangen, Germany
{aichner,buchner,wk}@LNT.de

Abstract. In this paper we present an efficient real-time implementation of a broadband algorithm for blind source separation (BSS) of convolutive mixtures. A recently introduced matrix formulation allows straightforward simultaneous exploitation of nonwhiteness and nonstationarity of the source signals using second-order statistics. We examine the efficient implementation of the resulting algorithm and introduce a block-on-line update method for the demixing filters. Experimental results for moving speakers in a reverberant room show that the proposed method ensures high separation performance. Our method is implemented on a standard laptop computer and works in realtime.

1 Introduction

The problem of separating convolutive mixtures of unknown time series arises in several application domains, a prominent example being the so-called cocktail party problem, where individual speech signals should be extracted from mixtures of multiple speakers in a usually reverberant acoustic environment. Due to the reverberation, the original source signals $s_q(n)$, $q = 1, \ldots, Q$ of our separation problem are filtered by a linear multiple input and multiple output (MIMO) system before they are picked up by the sensors. BSS is solely based on the fundamental assumption of mutual statistical independence of the different source signals. In the following, we further assume that the number Q of source signals $s_q(n)$ equals the number of sensor signals $x_p(n)$, $p = 1, \ldots, P$. An M-tap mixing system is thus described by $x_p(n) = \sum_{q=1}^{P} \sum_{\kappa=0}^{M-1} h_{qp,\kappa} s_q(n-\kappa)$, where $h_{qp,\kappa}$, $\kappa = 0, \ldots, M-1$ denote the coefficients of the filter from the q-th source to the p-th sensor.

In BSS, we are interested in finding a corresponding demixing system, where the output signals are described by $y_q(n) = \sum_{p=1}^{P} \sum_{\kappa=0}^{L-1} w_{pq,\kappa} x_p(n-\kappa)$ with $q = 1, \ldots, P$. The separation is achieved by forcing the output signals y_q to be mutually statistically decoupled up to joint moments of a certain order. For convolutive mixtures, frequency-domain BSS is very popular since all techniques

---

* This work was partly supported by the ANITA project funded by the European Commission under contract IST-2001-34327.

originally developed for instantaneous BSS may be applied independently in each frequency bin. This bin-wise processing, implying a narrowband signal model is denoted here as *narrowband approach* and is described, e.g., in [6]. In the context of instantaneous BSS and narrowband approaches for convolutive BSS it is known that on real-world signals with some time-structure *second-order statistics* generates enough constraints to solve the BSS problem in principle by utilizing nonstationarity or nonwhiteness [6]. Unfortunately, this traditional narrowband approach exhibits several limitations as, e.g., circular convolution effects may arise, and the permutation problem, which is inherent in BSS, may then also appear independently in each frequency bin so that extra repair measures become necessary. In [2, 3] a class of *broadband* algorithms was derived, for both the time domain and frequency domain, i.e., the frequency bins are no longer considered to be independent for unrestricted time-domain signals. These algorithms are based on second-order statistics exploiting simultaneously nonwhiteness and nonstationarity and inherently avoid the above-mentioned problems. In this paper we present an efficient realization of one of these broadband algorithms which has led to a robust real-time implementation.

2 Generic Block Time-Domain BSS Algorithm

2.1 Matrix Formulation

To obtain a block processing broadband algorithm simultaneously exploiting nonwhiteness and nonstationarity of the source signals, it was shown in [2] that we need to introduce a block output signal matrix

$$\mathbf{Y}_q(m) = \begin{bmatrix} y_q(mL) & \cdots & y_q(mL-L+1) \\ y_q(mL+1) & \ddots & y_q(mL-L+2) \\ \vdots & \ddots & \vdots \\ y_q(mL+N-1) & \cdots & y_q(mL-L+N) \end{bmatrix}, \quad (1)$$

and reformulate the convolution as

$$\mathbf{Y}_q(m) = \sum_{p=1}^{P} \mathbf{X}_p(m) \mathbf{W}_{pq}, \quad (2)$$

with m being the block time index and N denoting the block length. The $N \times L$ matrix $\mathbf{Y}_q(m)$ incorporates L *time-lags* in the correlation matrices into the cost function defined in Sect. 2.2, which is necessary for the exploitation of the nonwhiteness property. To ensure linear convolutions for all elements of $\mathbf{Y}_q(m)$, the $N \times 2L$ matrices $\mathbf{X}_p(m)$ and $2L \times L$ matrices $\mathbf{W}_{pq}$ are given as

$$\mathbf{X}_p(m) = \begin{bmatrix} x_p(mL) & \cdots & x_p(mL-2L+1) \\ x_p(mL+1) & \ddots & x_p(mL-2L+2) \\ \vdots & \ddots & \vdots \\ x_p(mL+N-1) & \cdots & x_p(mL-2L+N) \end{bmatrix}, \quad (3)$$

$$\mathbf{W}_{pq} = \begin{bmatrix} w_{pq,0} & 0 & \cdots & 0 \\ w_{pq,1} & w_{pq,0} & \ddots & \vdots \\ \vdots & w_{pq,1} & \ddots & 0 \\ w_{pq,L-1} & \vdots & \ddots & w_{pq,0} \\ 0 & w_{pq,L-1} & \ddots & w_{pq,1} \\ \vdots & & \ddots & \vdots \\ 0 & \cdots & 0 & w_{pq,L-1} \\ 0 & \cdots & 0 & 0 \end{bmatrix}, \quad (4)$$

where the matrices $\mathbf{X}_p(m)$, $p = 1, \ldots, P$ in (2) are Toeplitz matrices due to the shift of subsequent rows by one sample each. The matrices $\mathbf{W}_{pq}$ exhibit a Sylvester structure, where each column is shifted by one sample containing the current weights $\mathbf{w}_{pq} = [w_{pq,0}, w_{pq,1}, \ldots, w_{pq,L-1}]^T$ of the MIMO filter of length L from the p-th sensor channel to the q-th output channel.

To allow a convenient notation of the algorithm combining all channels, we write (2) compactly as

$$\mathbf{Y}(m) = \mathbf{X}(m)\mathbf{W}, \quad (5)$$

with the matrices

$$\mathbf{Y}(m) = [\mathbf{Y}_1(m), \cdots, \mathbf{Y}_P(m)], \quad (6)$$
$$\mathbf{X}(m) = [\mathbf{X}_1(m), \cdots, \mathbf{X}_P(m)], \quad (7)$$
$$\mathbf{W} = \begin{bmatrix} \mathbf{W}_{11} & \cdots & \mathbf{W}_{1P} \\ \vdots & \ddots & \vdots \\ \mathbf{W}_{P1} & \cdots & \mathbf{W}_{PP} \end{bmatrix}. \quad (8)$$

2.2 Cost Function and Generic Broadband Algorithm

Based on (5) we use a cost function first introduced in [2] as a generalization of [8]:

$$\mathcal{J}(m) = \sum_{i=0}^{m} \beta(i,m) \left\{ \log \det \operatorname{bdiag} \mathbf{Y}^H(i)\mathbf{Y}(i) - \log \det \mathbf{Y}^H(i)\mathbf{Y}(i) \right\}, \quad (9)$$

where β is a weighting function with finite support that is normalized according to $\sum_{i=0}^{m} \beta(i,m) = 1$ allowing on-line or block-on-line realizations of the algorithm. For a properly chosen $\beta(i,m)$ (see Sect. 2.3) the nonstationarity of the signals is utilized for the separation. Since we use the matrix formulation (5) for calculating the short-time correlation matrices $\mathbf{Y}^H(m)\mathbf{Y}(m)$, the cost function inherently includes all L time-lags of all auto-correlations and cross-correlations of the BSS output signals. The bdiag operation on a partitioned block matrix consisting of several submatrices sets all submatrices on the off-diagonals to zero. In our case, the block matrices refer to the different signal channels and are of

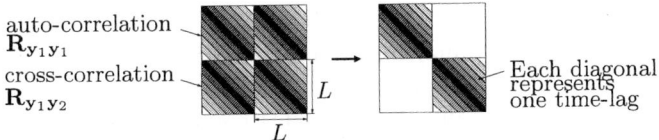

Fig. 1. Illustration of (9) for the 2 × 2 case

size $L \times L$. The cost function becomes zero if and only if all block-offdiagonal elements of $\mathbf{Y}^H \mathbf{Y}$, i.e., the *output cross-correlations over all time-lags*, become zero (see Fig. 1). Therefore, in addition to the nonstationarity (9) explicitly exploits the nonwhiteness property of the output signals.

In [2,3] it was shown that the natural gradient derivation of (9) with respect to $\mathbf{W}$ leads to an iterative algorithm with the following coefficient update:

$$\nabla_{\mathbf{W}}^{\text{NG}} \mathcal{J}(m) = 2 \sum_{i=0}^{m} \beta(i,m) \mathcal{Q}(i), \qquad (10)$$

$$\mathcal{Q}(i) = \mathbf{W}(i) \left\{ \mathbf{R}_{\mathbf{yy}}(i) - \text{bdiag}\, \mathbf{R}_{\mathbf{yy}}(i) \right\} \text{bdiag}^{-1} \mathbf{R}_{\mathbf{yy}}(i), \qquad (11)$$

where the $PL \times PL$ short-time correlation matrices $\mathbf{R}_{\mathbf{yy}}$ are consisting of the channel-wise $L \times L$ submatrices $\mathbf{R}_{\mathbf{y}_p \mathbf{y}_q}(m) = \mathbf{Y}_p^H(m) \mathbf{Y}_q(m)$.

2.3 Approximated Version and Efficient Implementation

Starting from the update equation (10) we first address implementation details concerning the update term $\mathcal{Q}(i)$ of the i-th block which are applicable regardless of the choice of the weighting function $\beta(i,m)$. In the last paragraph we specify $\beta(i,m)$ to obtain a block-on-line update rule.

Step 1: Estimation of the Correlation Matrices Using the Correlation Method.
In principle, there are two basic methods to estimate the output correlation matrices $\mathbf{R}_{\mathbf{y}_p \mathbf{y}_q}(m)$ for nonstationary signals: the so-called correlation method, and the covariance method as they are known from linear prediction problems [7]. We consider here the correlation method which leads to a lower computational complexity and follows as a special case of the more accurate covariance method if we assume stationarity within each block. This leads to a Toeplitz structure of $\mathbf{R}_{\mathbf{y}_p \mathbf{y}_q}(m)$ which can be expressed as

$$\mathbf{R}_{\mathbf{y}_p \mathbf{y}_q}(m) = \left[r_{y_p y_q}(m, v-u) \right]_{L \times L} \qquad (12)$$

$$r_{y_p y_q}(m, v-u) = \begin{cases} \sum_{n=mL}^{mL+N-v+u-1} y_p(n+v-u) y_q(n) & \text{for } v-u \geq 0 \\ \sum_{n=mL+|v-u|}^{mL+N-1} y_p(n+v-u) y_q(n) & \text{for } v-u < 0 \end{cases} \qquad (13)$$

Step 2: Approximation of the Normalization.
A straightforward implementation of (11) together with (12), (13) leads to a complexity of $\mathcal{O}(L^2)$ due to the inversion of P auto-correlation Toeplitz matrices $\mathbf{R}_{\mathbf{y}_q \mathbf{y}_q}$ of size $L \times L$ which are normalizing the update (as also known from

the recursive least-squares (RLS) algorithm in supervised adaptive filtering [5]). Thus, for an efficient implementation suitable for reverberant environments requiring a large filter length L we use an approximated version of (11) which was first heuristically introduced in [1, 10] and theoretically derived in [2]. The efficient version is obtained by approximating the auto-correlation submatrices in the normalization term by the output signal powers, i.e.,

$$\tilde{\mathbf{R}}_{\mathbf{y}_q \mathbf{y}_q}(m) = \left(\sum_{n=mL}^{mL+N-1} y_q^2(n) \right) \mathbf{I} = \sigma_{y_q}^2(m) \mathbf{I} \qquad (14)$$

for $q = 1, \ldots, P$. Thus, the matrix inversion is replaced by an element-wise division. This is comparable to the normalization in the well-known normalized least mean squares (NLMS) algorithm in supervised adaptive filtering approximating the RLS algorithm [5].

Step 3: Efficient Implementation of the Matrix-Matrix Multiplication.
In the remaining channel-wise matrix product of $\mathbf{W}_{pt}(m)$ and the Toeplitz matrices $\frac{\mathbf{R}_{\mathbf{y}_t \mathbf{y}_q}(m)}{\sigma_{y_q}^2(m)}$, $p, q, t = 1, \ldots, P$ in (11) we can exploit the Sylvester structure of $\mathbf{W}_{pt}(m)$ for an efficient implementation. Firstly, it has to be ensured that the update $\mathcal{Q}(i)$ exhibits again a channel-wise Sylvester structure in the form of (4). A simple way to impose this constraint is to calculate only the first L elements of the first column of the matrix product which contain the filter weights update $\Delta \mathbf{w}_{pq}(m)$ (see Fig. 2a). Secondly, it can be shown that this matrix product denotes a linear convolution of the filter weights $\mathbf{w}_{pt}$ with each column of $\frac{\mathbf{R}_{\mathbf{y}_t \mathbf{y}_q}(m)}{\sigma_{y_q}^2(m)}$ due to the Sylvester structure of $\mathbf{W}_{pt}$. By implementing this operation as a fast convolution using fast Fourier transforms (FFTs) the computational complexity can be reduced to $\mathcal{O}(\log L)$.

Step 4: Update Using a Block-on-Line Weighting Function
The weighting function $\beta(i, m)$ allows for different realizations of the algorithm, e.g., off-line or on-line [3]. Similar to the approach in [9] we are combining the on-line and off-line approach in a so-called block-on-line method (Fig. 2b). In Table 1 a pseudo-code for the block-on-line implementation of this efficient algorithm is given exemplarily for the filter update $\Delta \mathbf{w}_{11}$ for $P = 2$. In the block-on-line approach a block of $KL + N$ input signal samples is acquired denoted by the on-line block index m' (see Fig. 2b). K denotes the number of blocks within the offline part and thus the data is segmented into K blocks of length N with off-line block index m ($m = m' \cdot K$) and is processed by an off-line algorithm with $j_{\max}$ iterations. By simultaneously processing K blocks we exploit the nonstationarity of the signals. The implementation of the off-line part is shown in Steps 3-9 of Table 1, where j denotes the iteration number and μ_{off} is the stepsize of the off-line part.

Concerning the initialization of $\mathbf{w}_{pq}(m')$ in Step 3 for $m' = 1$ and $j = 1$, it can be shown using (4), (11) that the first coefficients of the filters $\mathbf{w}_{pp}(m')$ must be unequal to zero. Thus we use unit impulses for the first filter tap in each $\mathbf{w}_{pp}(m')$. The filters $\mathbf{w}_{pq}(m')$, $p \neq q$ are set to zero.

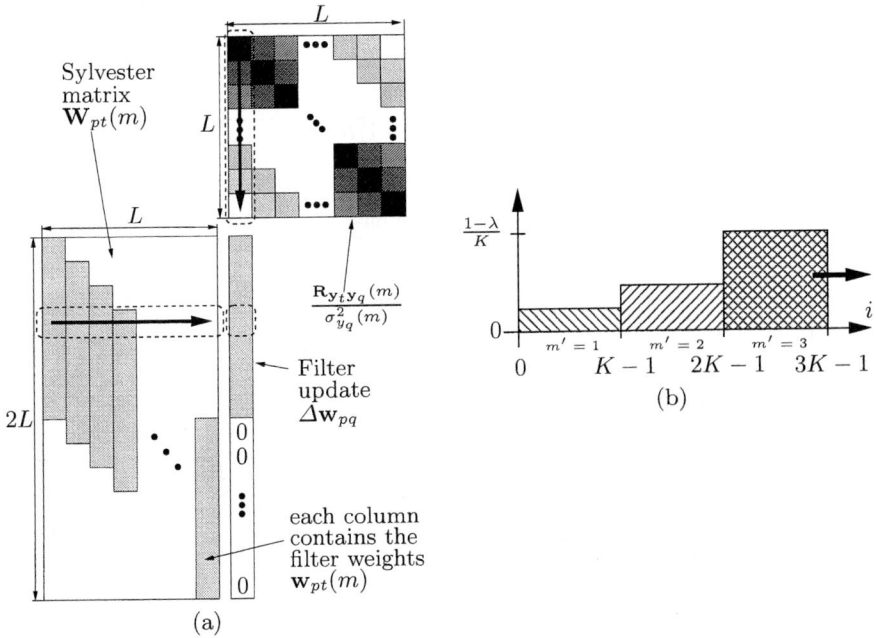

Fig. 2. (a) Illustration of the channel-wise matrix-matrix product. (b) Weighting function $\beta(i, m)$ for block-on-line implementation

The update $\Delta \tilde{\mathbf{w}}_{pq}^{j_{\max}}(m')$ (Step 9) is then used as input of the on-line part of the block-on-line algorithm. The recursive update equations of the on-line part yield the final filter weights $\mathbf{w}_{pq}$ used for separation (Step 10). Here λ denotes the exponential forgetting factor ($0 \leq \lambda < 1$) and μ_{on} is the stepsize of the on-line part. The demixing filter weights $\mathbf{w}_{pq}(m')$ of the current block m' are then used as initial values for the off-line algorithm of the next block (Step 11).

Analogously to supervised block-based adaptive filtering, the approach followed here can also be carried out with overlapping data blocks in both, the on-line and off-line part to increase the convergence rate and to reduce the signal delay. Overlapping is done by simply replacing the time index mL and $m'KL$ in the equations by $m\frac{L}{\alpha_{\text{off}}}$ and $m'\frac{KL}{\alpha_{\text{on}}}$, respectively. The overlap factors $1 \leq \alpha_{\text{off}}, \alpha_{\text{on}} \leq L$ should be chosen suitably to obtain integer values for the time index. For clarity, however, the overlap factors are omitted in Table 1.

3 Experiments and Real-Time Implementation

The experiments have been conducted using speech data convolved with the impulse responses of a real office room (580cm × 590cm × 310cm), with a reverberation time $T_{60} = 200$ ms and a sampling frequency of $f_s = 16$ kHz. A two-element microphone array with an inter-element spacing of 20 cm was used

Table 1. Pseudo code of efficient broadband algorithm implementation exemplarily shown for the update $\Delta \mathbf{w}_{11}$ in the 2×2 case

On-line part:
1. Acquire $KL + N$ new samples $x_p((m'-1)KL), \ldots, x_p(m'KL + N - 1)$ of the sensors x_p, $p = 1, 2$ and on-line block index $m' = 1, 2, \ldots$
2. Generate K blocks $x_p(mL), \ldots, x_p(mL + N - 1)$ with off-line block index $m = (m'-1)K, \ldots, m'K - 1$ to enable off-line iterations
Off-line part:
Compute for each iteration $j = 1, \ldots, j_{\max}$:
Compute for each block $m = (m'-1)K, \ldots, m'K - 1$:
3. Compute output signals $y_q(mL), \ldots, y_q(mL + N - 1)$, $q = 1, 2$ by convolving x_p with filter weights $\mathbf{w}_{pq}^{j-1}(m')$ from previous iteration
4. Calculate the signal energy of each block m $r_{y_1 y_1}(m, 0) = \sum_{n=mL}^{mL+N-1} y_1^2(n)$
5. Compute 1st column of cross-correlation matrix $\mathbf{R}_{\mathbf{y}_2\mathbf{y}_1}(m)$ by $r_{y_2 y_1}(m, v - u)$ for $v - u = -L + 1, \ldots, 0$ according to (13)
6. Normalization by elementwise division $r_{y_2 y_1}(m, v - u)/r_{y_1 y_1}(m, 0)$ for $v - u = -L + 1, \ldots, 0$
7. Compute the matrix product $\mathbf{W}_{12}(m) \frac{\mathbf{R}_{\mathbf{y}_2\mathbf{y}_1}(m)}{\sigma_{y_1}^2(m)}$ as a convolution according to Fig. 2a. Each filter weight update $\Delta w_{11,\kappa}^j$, $\kappa = 0, \ldots, L - 1$ is therefore calculated as: $\Delta w_{11,\kappa}^j(m') = \frac{1}{K} \sum_m \sum_{n=0}^{L-1} w_{12,n}(m) r_{y_2 y_1}(m, n - \kappa)/r_{y_1 y_1}(m, 0)$
8. Update equation for the off-line part: $\mathbf{w}_{11}^j(m') = \mathbf{w}_{11}^{j-1}(m') - \mu_{\text{off}} \Delta \mathbf{w}_{11}^j(m')$
9. Repeat Steps 3-8 for $j_{\max}$ iterations and calculate the overall update for the current m' as: $\Delta \mathbf{w}_{11}^{j_{\max}}(m') = \sum_{j=1}^{j_{\max}} \Delta \mathbf{w}_{11}^j(m')$
On-line part:
10. Compute the recursive update of the on-line part yielding the demixing filter $\mathbf{w}_{11}(m')$ used for separation: $\Delta \mathbf{w}_{11}(m') = \lambda \Delta \mathbf{w}_{11}(m' - 1) + (1 - \lambda) \Delta \mathbf{w}_{11}^{j_{\max}}(m')$ $\mathbf{w}_{11}(m') = \mathbf{w}_{11}(m' - 1) - \mu_{\text{on}} \Delta \mathbf{w}_{11}(m')$
11. Compute Steps 4-10 similarily for the other channels and use the demixing filter $\mathbf{w}_{pq}(m')$ as the initial filter for the offline part $\mathbf{w}_{pq}^0(m' + 1) = \mathbf{w}_{pq}(m')$

for the recording. The speech signals arrived from two different directions, $-45°$ and $45°$. After 10 seconds one speaker position was changed from $-45°$ to $0°$. Sentences spoken by two male speakers from the TIMIT speech corpus [4] were selected as source signals. To evaluate the performance, the signal-to-interference ratio (SIR) averaged over both channels was calculated in each block which is defined as the ratio of the signal power of the target signal to the signal power from the jammer signal. Simulation results for the algorithm implemented in the real-time system are given in Fig. 3. The parameters were chosen as $L = 1024$, $N = 2048$, $K = 4$, $\alpha_{\text{on}} = 4$ resulting in a latency of 2048 samples (128 msec).

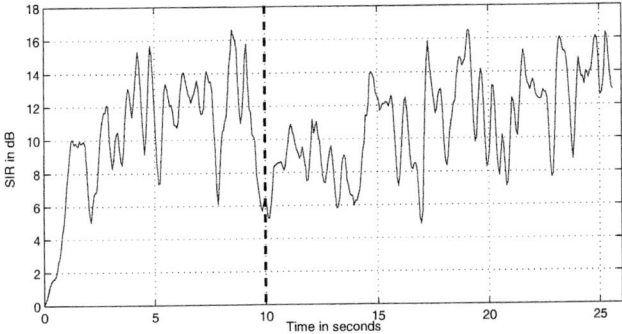

Fig. 3. Experimental results for the efficient block-on-line algorithm with an instantaneous speaker position change at 10 seconds.

The offline-part was calculated for $j_{\max} = 10$ iterations and the stepsizes for on-line and off-line part were chosen as $\mu_{\text{on}} = \mu_{\text{off}} = 0.002$ with $\lambda = 0.2$. It can be seen in Fig. 3 that the algorithm is robust against speaker movements and converges quickly due to the block-on-line structure.

Our scalable real-time system is implemented on a regular laptop using C++ in combination with the efficient Intel Integrated Performance Primitives (IPP) library. The demonstrator is applicable to $P \times P$ scenarios ($P = 2, 3, \ldots$) and works both under Linux and Windows operating systems. The computational load on an 1.6 GHz Intel Pentium 4 Processor for the above-mentioned parameter settings is approximately 70%. A video showing the capability of the system in reverberant rooms can be found at www.LNT.de/~aichner/bss_video.html

4 Conclusions

In this paper we presented a real-time implementation of an efficient BSS algorithm based on a general class of broadband algorithms. The system is robust to speaker movements and exhibits a low latency, showing the applicability of this method to real-world scenarios.

References

1. R. Aichner, S. Araki, S. Makino, T. Nishikawa, and H. Saruwatari. Time-domain blind source separation of non-stationary convolved signals by utilizing geometric beamforming. In *Proc. Neural Networks for Signal Processing*, pp 445–454, 2002.
2. H. Buchner, R. Aichner, and W. Kellermann. A generalization of a class of blind source separation algorithms for convolutive mixtures. In *Proc. Int. Symp. on Independent Comp. Analysis and Blind Signal Separation (ICA)*, pp 945–950, 2003.
3. H. Buchner, R. Aichner, and W. Kellermann. Blind source separation for convolutive mixtures: A unified treatment. In J. Benesty and Y. Huang, editors, *Audio Signal Processing for Next-Generation Multimedia Communication Systems*. Kluwer Academic Publishers, Boston, Feb. 2004.

4. J.S. Garofolo et al. TIMIT acoustic-phonetic continuous speech corpus, 1993.
5. S. Haykin. *Adaptive Filter Theory*. Prentice Hall Inc., Englewood Cliffs, NJ, 4th edition, 2002.
6. A. Hyvaerinen, J. Karhunen, and E. Oja. *Independent Component Analysis*. John Wiley & Sons, 2001.
7. J.D. Markel and A.H. Gray. *Linear Prediction of Speech*. Springer, Berlin, 1976.
8. K. Matsuoka, M. Ohya, and M. Kawamoto. Neural net for blind separation of nonstationary signals. *IEEE Trans. Neural Networks*, 8(3):411–419, 1995.
9. R. Mukai, H. Sawada, S. Araki, and S. Makino. Robust real-time blind source separation for moving speakers using blockwise ICA and residual crosstalk subtraction. In *Proc. ICA*, pages 975–980, 2003.
10. T. Nishikawa, H. Saruwatari, and K. Shikano. Comparison of time-domain ICA, frequency-domain ICA and multistage ICA for blind source separation. In *Proc. European Signal Processing Conference*, volume 2, pages 15–18, Sep. 2002.

A New Approach to the Permutation Problem in Frequency Domain Blind Source Separation

Koutaro Kamata, Xuebin Hu, and Hidefumi Kobatake

Graduate School of Bio-Applications and Systems Engineering
Tokyo University of Agriculture & Technology
2-24-16 Naka-cho, Koganei-shi, Tokyo, 184-8588 Japan
{kamakoht,huxb,kobatake}@cc.tuat.ac.jp

Abstract. Frequency domain blind source separation has the great advantage that the complicated convolution in time domain becomes multiple efficient multiplications in frequency domain. However, the inherent ambiguity of permutation of ICA becomes an important problem that the separated signals at different frequencies may be permuted in order. Mapping the separated signal at each frequency to a target source remains to be a difficult problem. In this paper, we first discuss the inter-frequency correlation based method [1], and propose a new method using the continuity in power between adjacent frequency components of same source. The proposed method also implicitly utilizes the information of inter-frequency correlation, as such has better performance than the previous method.

1 Introduction

Blind source separation (BSS) has received extensive attentions in signal and speech processing, machine intelligence, and neuroscience communities. The goal of BSS is to recover the unobserved original sources without any prior information given only the sensor observations that are unknown linear mixtures of the independent source signals. If the mixture is instantaneous, we can directly employ independent component analysis (ICA) to achieve the task. In real environment, due to multi-path propagation and reverberation, the signals impinged on an array of microphones are convolutive mixtures of sources. BSS may be implemented in time domain by learning the time-domain coefficients of the unmixing filter. However, the filter may need to be thousands of taps long to properly invert the mixing. Computationally, it is lighter to move to the frequency domain as convolution with long filter in the time domain becomes efficient multiplications in the frequency domain under certain conditions [1, 2]. This has the great advantage that ICA still could be directly used to achieve the separation.

Frequency domain BSS brings out the problem that standard ICA indeterminacy of scaling and permutation appears at each output frequency bin. The scaling problem could be easily solved by putting the separated frequency components back to the microphones with the inverse matrices. However, permutation remains to be a difficult problem. We need to map a separated component at each frequency to a target source signal so as to properly reconstruct the separated signal in the time domain. Various proposals have been reported using different continuity criteria to overcome the permutation problem [2]. Nevertheless, it is still open to satisfying and rigorous solutions.

There are some inherent limitations on various proposals. For example, one method makes use of the coherency of separating matrices at neighbor frequencies. However, the coherency only exists in very simple environment, it does not hold in most case. Another approach is based on direction of arrival (DOA) estimation in array signal processing. By analyzing the directivity patterns formed by the separating matrix, source directions can be estimated and therefore permutation can be solved. The inherent limitation of this approach is that the sources must be spaced apart away. Otherwise it could not work well due to the variation of DOA at different frequencies [3] and the unavoidable error in DOA estimation. Ikeda et al. proposed an approach employing the inter-frequency correlation of signal envelopes to align permutation because source signals are speech [1]. It seems a sound solution as inter-frequency correlation does exist at adjacent frequencies for speech signals. This approach is related with our method and is to be discussed in section 3. A recently proposed method [4] combines the DOA estimation with inter-frequency correlation, and also incorporates the harmonic structure of sound. It seems to achieve the best performance up to now, but has the disadvantage that the location of microphones should be known in advance. In other word, it is not completely *blind*.

This paper proposes a new method based on a similar but different assumption from Ikeda's method. We assume that there exists continuity in power (amplitude) between the waveforms of adjacent frequency components of same source. Based on the assumption, we propose to use the distance between the signals at adjacent frequencies to align the separated signals. As the information from distance implicitly includes the information of inter-frequency correlation, the proposed method does not conflict with the Ikeda's method but includes more helpful information. Consequently, it has a better performance than the previous method.

Section 2 briefly describes the frequency domain BSS system. In section 3, we review the inter-frequency correlation based method and present the proposed method. Section 4 gives the comparison test results, and followed by the conclusion at the last.

2 Frequency Domain BSS

The BSS system employed in this paper is summarized as follows. Source signals are assumed to be independent with each other, zero mean, and are denoted by a vector $s(t) = (s_1(t), \cdots, s_N(t))^T$. In real environment, by ignoring the noise, the observations can be approximated with convolutive mixtures of source signals,

$$x(t) = A * s(t) = \left(\sum_i a_{ik} * s_i(t) \right), \quad (1)$$

where A is an unknown polynomial matrix, a_{ik} is the impulse response from source i to microphone k, and the asterisk symbol * refers to convolution operation. The observed mixtures are decomposed into frequency domain by performing short-time discrete Fourier transform. Then the convolutive mixing problem becomes multiple instantaneous mixing problems.

$$X(\omega, t) = A(\omega) S(\omega, t). \quad (2)$$

The unmixing filter $W(\omega)$ is derived using the Infomax algorithm [5]. The learning rule is defined as follows,

$$W_{i+1}(\omega) = W_i(\omega) + \eta\left[I - \varphi(Y(\omega,t))Y^H(\omega,t)\right]W_i(\omega)$$
$$\varphi(Y) = 2\tanh(\text{Re}(Y)) + 2j\tanh(\text{Im}(Y))$$
(3)

where η is a factor that determine the convergence speed, $\varphi(\cdot)$ is a nonlinear score function. The scaling problem is solved by filtering the individual output of the unmixing filter using the inverse matrices separately. The unmixing filter becomes,

$$W'_{i,\omega} = W_\omega^{-1}\delta(i,i)W_\omega,$$ (4)

where W_ω denotes the derived unmixing filter, $W'_{i,\omega}$ denotes the unmixing filter that outputs the i-th source signal, and $\delta(i,i)$ denotes a "delta matrix" of which only the (i,i) element equals to one and all the remaining elements are zeros.

The permutation problem is the main topic of this paper and is to be described in section 3. After solving the ambiguity of scaling and permutation, the derived unmixing filters are then transformed back to time domain through inverse Fourier transform. The time domain unmixing filter is derived as follows,

$$W_{i,t} = F^{-1}\left[W'_{i,\omega}(\omega_k)H(\omega_k)\right] \cdot ham(t),$$ (5)

where F^{-1} denotes inverse Fourier transformation, $ham(t)$ denotes the Hamming window, and $H(\omega_k)$ is a circular time shift operator. When the window length is N, a time shift of $N/2$ is experimentally good, i.e., $H(\omega_k) = e^{i\pi k}$ [6].

3 Continuity Based Approach

In this section, we first review the inter-frequency correlation based method proposed by Ikeda et al. In [1], it is assumed that if the split band-passed signals originate from the same source signal, they are under the influence of a similar modulation in amplitude. In other word, there is correlation exist between the envelopes of Fourier components of same source. The operator to take the envelope, ε is defined as

$$\varepsilon\hat{s}_\omega(t_s;i) = \frac{1}{2M}\sum_{t_s'=t_s-M}^{t_s+M}\sum_{k=1}^{K}\left|\hat{s}_{k,\omega}(t_s';i)\right|,$$ (6)

where $\hat{s}_\omega(t_s;i)$ denotes the frequency component of the i-th source, and $\hat{s}_{k,\omega}(t_s';i)$ denotes the input of the i-th source component into the k-th ($k=1,\cdots,K$) sensor. M is the number of time steps for taking the moving average, and t_s refers to the sequence number of windows.

The permutation is solved using the correlation between the envelopes of separated signals. First, the sequence of frequency ω to solve the permutation is determined by sorting the similarity between the separated components in an increasing order. The similarity is defined as follows,

$$sim(\varepsilon Y_i(\omega,t), \varepsilon Y_j(\omega,t)) = \frac{\varepsilon Y_i(\omega,t) \cdot \varepsilon Y_j(\omega,t)}{\|\varepsilon Y_i(\omega,t)\|\|\varepsilon Y_j(\omega,t)\|},$$ (7)

where, "·" denotes inner product, and ‖ ‖ denotes the norm. For ω_1, assign the order as it is. For ω_k, find the alignment that maximizes the correlation between the envelope with the aggregated envelope from ω_1 through ω_{k-1} of the aligned source.

In Ikeda's method, because permutation is solved in increasing order of similarity, it is implemented in a random frequency sequence. This implies that the aligned frequencies may be apart away from the frequency to be decided. Fig. 1 shows an example. The envelopes of same source have high correlation at adjacent frequencies (see the top and middle rows of Fig.1). However, correlation does not hold when frequencies are apart away (see the bottom row of Fig.1). Consequently, using the sum of envelopes of decided frequencies can not ensure a good job. The inter-frequency correlation should only be used within an adjacent frequency band.

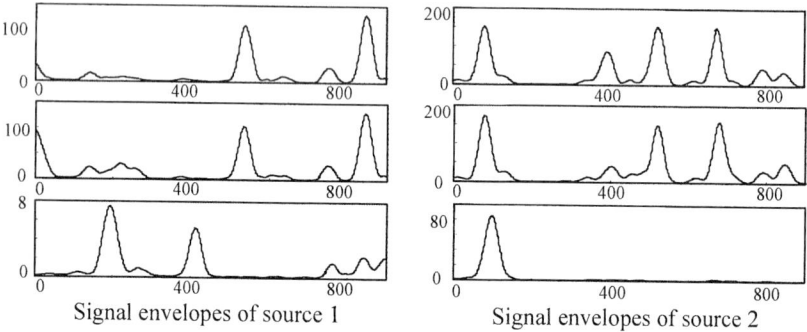

Fig. 1. Example of envelopes at different frequencies. From up to down, they are the envelopes of source 1 and 2 at 390.6, 398.4, 1953.1 Hz, respectively.

The inter-frequency correlation based method only uses the correlation between signal envelopes. From Fig.1, we see that it is also reasonable to assume that there exists continuity in power between adjacent frequency components. In other word, the power will not change dramatically between neighbor frequencies. Based on the assumption, we proposed to use the distance between the signal vectors at adjacent frequencies to align the separated signals. The new assumption does not conflict with the correlation assumption but includes more helpful information. The distance criterion implicitly utilizes the information of inter-frequency correlation. As such, it should have a better performance than the previous method. For example, it is possible that the Fourier components of different sources may be relatively correlated. In such case, the previous method might be difficult to deal with. However, if the powers of the highly correlated envelopes of different sources are quite different, the continuity based method will be competent.

We use the continuity of power within a neighboring frequency band. This has the advantage that allows a separation failure at adjacent frequencies, which is sometimes unavoidable due to reasons like the low-independence between original source components [7]. Additionally, solving the permutation within a short band instead of only the immediate neighboring frequency eliminates the risk to transfer a misalignment to all the subsequent frequencies.

The distance between two signal vectors $s_i(\omega_k,t)$, $s_j(\omega_r,t)$, is defined as,

$$d_{i,j}(\omega_k,\omega_r) = \left(\sum_t |v_i(\omega_k,t) - v_j(\omega_r,t)|^p\right)^{1/p} \quad (8)$$

$$v_i(\omega,t) = \ln|s_i(\omega,t)|^2 \quad (9)$$

where, ω_k denotes the frequency at which permutation is to be decided, and ω_r denotes the frequency to be used as reference. p is a constant. If p equals to one, $d_{i,j}$ is the sum of the absolute difference, and when p equals to two, it is the Euclidean distance. Before calculating the distance, natural logarithm is taken for reducing the effect from the variation of amplitude. The proposed approach consists of the following steps:

- Using equation (7) to find ω_1 at which the separated signals are the most uncorrelated. The order of ω_1 is set as it is.
- For ω_k, we first find the most reliable reference frequency ω_r in the band of $[\omega_{k-L},\cdots\omega_{k-1}]$ when spread to increasing direction, or in $[\omega_{k+1},\cdots,\omega_{k+L}]$ when spread to decreasing direction. L is the band width.

$$\omega_r = \arg\max_{\omega_r}(F) \quad (10)$$

$$F = |d_{i,i}(\omega_k,\omega_r) - d_{i,j}(\omega_k,\omega_r)| \quad (11)$$

where, F denotes a relative distance between one source to another. It is a measure of reliability of a permutation decision. A higher F means the signals at ω_k and ω_r of same source have stronger continuity in power, and the signals of different sources are more apart away. In other words, the decision made with ω_r as the reference is more reliable.

- Assigning $s_i(\omega_k,t)$ to j-th source if $d_{i,i}(\omega_k,\omega_r) > d_{i,j}(\omega_k,\omega_r)$, else, assigning it as it is.

Figure 2 illustrates the effectiveness of the proposed method. We use two sound signals, and compare the proposed method with Ikeda's method with regard to the reliability index F. For the proposed method, we set p equals 2, bandwidth L equals to 5, and evaluate $d_{i,i}$ and $d_{i,j}$ for each ω using equation (8). $c_{i,i}$ and $c_{i,j}$ are the correlation factors used in the previous method. $d_{i,i}$ appears much more stable than $c_{i,i}$, which implies the defined distance in (8) is a better measure than the correlation in making the decision whether two components belong to the same source or not. In Fig. 2-left, $c_{i,i}$ and $c_{i,j}$ are very close to each other at some frequencies, where this is not the case between $d_{i,i}$ and $d_{i,j}$. With the proposed approach, a more reliable and better performance could be expected.

4 Experimental Results

We evaluate the proposed method in a 2-source 2-microphone system. Twenty pairs of different sound signals were used to simulate the observed mixtures using the impulse response from RWCP Sound Scene Database in Real Acoustic Environment. The reverberation time is 300 msec. The performance was evaluated under the re-

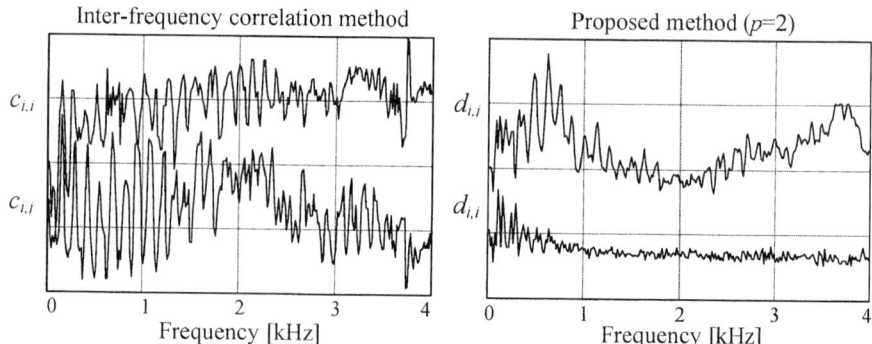

Fig. 2. Comparison of the proposed method with Ikeda's method on the difference between $(c_{i,i}, c_{i,j})$, $(d_{i,i}, d_{i,j})$. The difference reflects the reliability of alignment. In the proposed method, the band width L is set to 5.

verberation time changes from 20 to 100 msec (Using the beginning 20 to 100 msec of the real response. Note: more than 99 percent of energy of the impulse response are within the first 100 msec). The sampling rate was 8 kHz, DFT length was 64 msec, Hamming window was used, and window shift was 2 msec. The proposed method is tested at p set to 1 and 2, respectively. For investigating the up-limit of permutation solution, we use the original source signal as the reference for solving the permutation. The separating performance is evaluated using the following defined signal-to-noise ratio,

$$SNR_{ij} = 10\log_{10}\frac{\sum_t signal_i(t;j)^2}{\sum_t error_i(t;j)^2} \quad (12)$$

$$signal_i(t;j) = a_{ij}s_j(t) \quad (13)$$

$$error_i(t;j) = y_i(t;j) - signal_i(t;j) \quad (14)$$

Fig. 3 shows the separation performance at different reverberation times. The result is the average of twenty trials. Perm1, 2, 3, 4 refer to the previous method, the proposed method when p equals to 1 and 2, and using the real signal as the reference, respectively. At the reverberation time of 20 ms, compare with the previous method, the proposed method achieved about 6.8 dB improvement when p equals to 1, and 7.4 dB improvement when p equals to 2, and closed to the up-limit. When the reverberation time gets longer, because the separation performance itself decreased (SNR of perm4 decreased with reverberation time), the overall performance decreased gradually. Nevertheless, the proposed method still achieved better performance. At the reverberation time of 100 ms, about 2.0 and 1.6 dB higher performances were achieved when p equals to 1 or 2, respectively.

Fig.4 shows the error rates of previous method and the proposed method at different reverberation times. Fig.5 shows the error distributions of Ikeda's method and the new method when reverberation time equal to 20 or 100 msec, respectively. The vertical coordinate is the sum of the errors in the twenty trails. These figures demonstrated the improvement achieved by the proposed method, especially at the low frequencies.

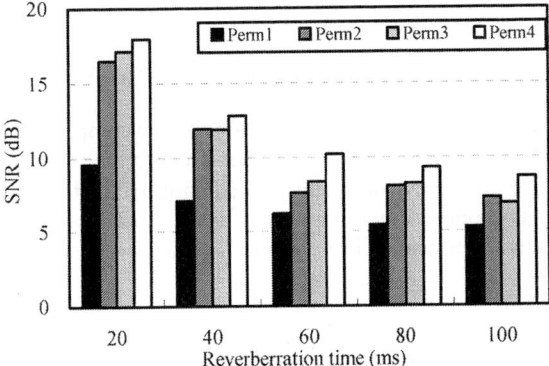

Fig. 3. Simulation test results. Perm1: Ikeda's method; Perm2 and 3: the proposed method when p equals to 1 or 2, respectively; Perm4: solving permutation by source signals.

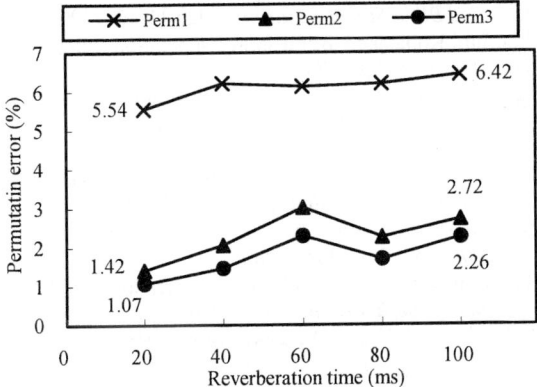

Fig. 4. Error rates of the previous method and the new method at various reverberation time.

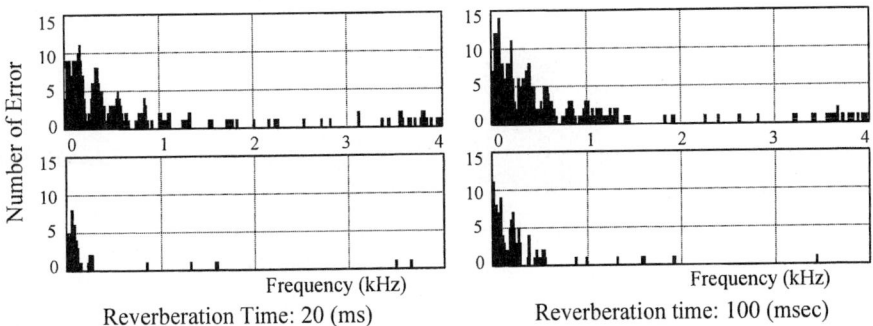

Fig. 5. Error distributions of the previous method (upper row) and the new method (lower row) when reverberation time equals to 20 or 100 msec. The *number of error* refers to the total times of errors happened in the 20 trials at each frequency.

5 Conclusion

This paper proposed a new permutation solution for frequency domain blind source separation. It is proposed based on a sound assumption that there exists continuity in amplitude between the waveforms of adjacent frequency components of same source.

The proposed method has the advantage that it keeps the useful information from correlation and introduces new favorable information from the continuity of amplitude. This makes the proposed method to do a better job than the previous method in aligning the separated components in frequency domain blind source separation.

References

1. S. Ikeda, N. Murata, "A method of blind separation based on temporal structure of signals," In *Proceedings of The Fifth International Conference on Neural Information Processing (ICONIP'98 Kitakyushu)*, pp. 737-742, 1998.
2. Kari Torkkola, "Blind separation for audio signals – are we there yet?" *Proc. Workshop on Independent Analysis and Blind Signal Separation*, Jan 11-16, 1999.
3. Xuebin Hu and Hidefumi Kobatake, "Blind source separation using ICA and beamforming", *Proc. of ICA2003*, pp. 597-602, April 2003.
4. H. Sawada, R. Mukai, S. Araki, S. Makino, "A Robust and Precise Method for Solving the Permutation Problem of Frequency-Domain Blind Source Separation," *Proc. of ICA2003*, pp. 505-510, Apr. 2003.
5. A. Bell, and T. Sejnowski, "An information maximization approach to blind separation and blind deconvolution," *Neural Computation*, 7: 1129-1159, 1995.
6. F. Asano, Y. Motomura, H. Asoh and T. Matsui, "Effect of PCA filter in blind source separation," In *Proc. ICA2000*, pp. 57-62, June 2000.
7. Xuebin Hu and Hidefumi Kobatake, "Blind speech separation - the low-independence problem and solution", *Proc. of the International Conference on Acoustics, Speech, and Signal Processing (ICASSP2003)*, vol. V, pp. 281-284, April, 2003.

Adaptive Cross-Channel Interference Cancellation on Blind Source Separation Outputs

Changkyu Choi, Gil-Jin Jang, Yongbeom Lee, and Sang Ryong Kim

Human Computer Interaction Laboratory
Samsung Advanced Institute of Technology
Mt. 14-1, Nongseo-Ri, Giheung-Eup, Yongin-Si, Gyeonggi-Do 449-712, Korea
{changkyu_choi,giljin.jang,leey,srkim}@samsung.com
http://myhome.naver.com/flyers/

Abstract. Despite an abundance of research outcomes of blind source separation (BSS) in many types of simulated environments, their performances are still not satisfiable to apply to the real environments. The major obstacle may seem the finite filter length of the assumed mixing model and the nonlinear sensor noises. This paper presents a two-step speech enhancement method with stereo microphone inputs. The first step performs a frequency-domain BSS algorithm with no prior knowledge of the mixed source signals and generates stereo outputs. The second step further removes the remaining cross-channel interference by a spectral cancellation approach using a probabilistic source absence/presence detection technique. The desired primary source is detected every frame of the signal, and the secondary source is estimated in the powerspectral domain using the other BSS output as a reference interference source. Then the secondary source is subtracted to remove the cross-channel interference. Our experimental results show good separation enhancement performances on the real recordings of speech and music signals compared to the conventional BSS methods.

1 Introduction

Separation of multiple signals from their superposition recorded at several sensors is an important problem that shows up in a variety of applications such as communications, biomedical and speech processing. The class of separation methods that require no source signal information except the number of mixed sources are often referred to blind source separation (BSS) [1]. In real recording situations with multiple microphones, each source signal spreads in all directions and reaches each microphone through "direct paths" and "reverberant paths." The observed signal by the jth microphone is expressed as

$$x_j(t) = \sum_{i=1}^{N} \sum_{\tau=0}^{\infty} h_{ji}(\tau) s_i(t-\tau) + n_j(t) = \sum_{i=1}^{N} h_{ji}(t) * s_i(t) + n_j(t), \quad (1)$$

where $s_i(t)$ is the ith source signal, N is the number of sources, $x_j(t)$ is the observed signal, and $h_{ji}(t)$ is the transfer function from source i to sensor j. The noise term $n_j(t)$ refers to the nonlinear distortions due to the characteristics of the recording devices. The assumption that the sources never move often fails due to the dynamic nature of the acoustic objects [2]. Moreover the practical systems should set a limit on the length of an impulse response, and the limited length is often a major performance bottleneck in realistic situations [3].

This paper proposes a post-processing technique for eliminating the remaining cross-channel interference at the BSS output. Our method is motivated by adaptive noise cancellation (ANC) [4]. The proposed method considers one BSS output as noisy signal and the other as reference noise source, and performs cancellation in the powerspectral domain as the conventional spectral subtraction methods do [5]. The advantage of the powerspectral subtraction is the effective absorption of small amount of mismatch between the actual filter and the estimated one, and the generation of cleanly denoised signals. The disadvantage is the introduction of the musical noises due to the below-zero spectral components as a result of the subtraction. With the help of source absence/presence detection prior to the subtraction, we reduce the error of the cancellation factor estimation and hence minimize the musical noises. Experimental results show that our proposed method has a superior performance to the frequency-domain BSS method in realistic conditions.

2 Frequency-Domain Blind Source Separation

The frequency domain blind source separation algorithm for the convolutive mixture cases is to transform the original time-domain filtering architecture into an instantaneous BSS problem in the frequency domain [6]. For simplicity, we consider stereo input, stereo output convolutive cases only. Using short time Fourier transform, Equation 1 is rewritten as

$$\mathbf{X}(\omega, n) = \mathbf{H}(\omega)\mathbf{S}(\omega, n) + \mathbf{N}(\omega, n), \qquad (2)$$

where ω is a frequency index, $\mathbf{H}(\omega)$ is the 2×2 square mixing matrix, $\mathbf{X}(\omega, n) = [X_1(\omega, n) \; X_2(\omega, n)]^T$ and $X_j(\omega, n) = \sum_{\tau=0}^{T-1} e^{-i2\pi\omega\tau/T} x_j(t_n + \tau)$, representing the DFT of the frame of size T with shift length $\lfloor \frac{T}{2} \rfloor$ starting at $t_n = \lfloor \frac{T}{2} \rfloor (n-1) + 1$ where $\lfloor \cdot \rfloor$ is a flooring operator, and corresponding expressions apply for $\mathbf{S}(\omega, n)$ and $\mathbf{N}(\omega, n)$ [1]. The unmixing process can be formulated in a frequency bin ω:

$$\mathbf{Y}(\omega, n) = \mathbf{W}(\omega)\mathbf{X}(\omega, n), \qquad (3)$$

where 2×1 vector $\mathbf{Y}(\omega, n)$ is an estimate of the original source $\mathbf{S}(\omega, n)$ disregarding the effect of the noise $\mathbf{N}(\omega, n)$. The convolution operation in the time

---

[1] In our manuscript, we denote lowercase letters with argument t for the time-series, and capital letters with argument ω and n for the Fourier transform at frequency ω for the nth frame. When the letters are boldfaced, they are column vectors whose components are accompanying the same arguments.

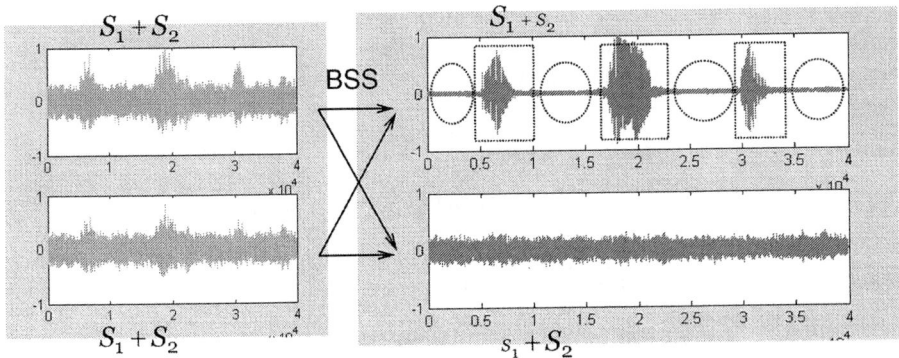

Fig. 1. The separability of the ordinary BSS algorithm. Left two signals are sensor inputs, and right two signals are BSS outputs. The original sources are rock music and speech signals [10]. There exists no speech signal in the ellipse-marked parts but still remains a small amount of rock music signal.

domain corresponds to the element-wise complex multiplication in the frequency domain. The instantaneous ICA algorithm we use is the non-holonomic information maximization [7] that guarantees an orthogonal solution:

$$\Delta \mathbf{W} \propto \left[\varphi(\mathbf{Y})\mathbf{Y}^H - \text{diag}\left(\varphi(\mathbf{Y})\mathbf{Y}^H \right) \right], \quad (4)$$

where ^{H} is the Hermitian transpose, and the polar nonlinear function $\varphi(\cdot)$ is defined by $\varphi(\mathbf{Y}) = [Y_1/|Y_1| \; Y_2/|Y_2|]^T$ [8]. A disadvantage of this decomposition is that there arise the permutation problem in each independent frequency bin [2]. The problem is solved by the time-domain spectral smoothing [9].

3 Adaptive Cross-Channel Interference Cancellation

3.1 Cross-Channel Interference Detection

Figure 1 illustrates the input and the output of the ordinary BSS system. The output signals still contain cross-channel interference that is audible and identifiable by human listeners. However, in the first output, if we assume that the speech signal is present only in the region enclosed by rectangles (call them active blocks), apparently the region enclose by ellipses (inactive blocks) contains the music signal only. The existence of the cross-channel interference can be described by the presence of the primary source. When the primary source is present, it often coexists with the secondary source, and the interference occurs. Therefore we define the interference probability by the presence probability of the primary source, which is modeled by complex Gaussian distributions [11]. The probabilities are used to properly estimate the interference cancellation factors regarding the cross-channel output as a reference noise source.

For each frame of the ith BSS output, we denote a set of all the frequency components for a frame by $Y_i(n) = \{Y_i(\omega, n) | \omega = 1, \ldots, T\}$, and two hypotheses $H_{i,0}$ and $H_{i,1}$ are given which respectively indicate the absence and presence of the primary source:

$$H_{i,0} : Y_i(n) = \tilde{S}_j(n)$$
$$H_{i,1} : Y_i(n) = \tilde{S}_i(n) + \tilde{S}_j(n), \ i \neq j, \quad (5)$$

where the $\tilde{S}_i$ is a filtered version of S_i. Conditioned on $Y_i(n)$, the source absence/presence probabilities are given by

$$p(H_{i,m} | Y_i(n)) = \frac{p(Y_i(n) | H_{i,m}) p(H_{i,m})}{p(Y_i(n) | H_{i,0}) p(H_{i,0}) + p(Y_i(n) | H_{i,1}) p(H_{i,1})}, \quad (6)$$

where $p(H_{i,0})$ is *a priori* probability for source i absence, and $p(H_{i,1}) = 1 - p(H_{i,0})$ is that of the cross-channel interference. Assuming the probabilistic independence among the frequency components,

$$p(Y_i(n) | H_{i,m}) = \prod_\omega p(Y_i(\omega, n) | H_{i,m}). \quad (7)$$

Then the source absence probability becomes

$$p(H_{i,0} | Y_i(n)) = \left[1 + \frac{P(H_{i,1})}{P(H_{i,0})} \prod_\omega^T \frac{p(Y_i(\omega, n) | H_{i,1})}{p(Y_i(\omega, n) | H_{i,0})} \right]^{-1}. \quad (8)$$

The posterior probability of $H_{i,1}$ is simply $p(H_{i,1} | Y_i(n)) = 1 - p(H_{i,0} | Y_i(n))$, which indicates the amount of cross-channel interference at the ith BSS output. In the following sections, we explain the cancellation of the co-channel interference and the statistical models for the component densities $p(Y_i(\omega, n) | H_{i,m})$.

3.2 Cross-Channel Interference Cancellation

Because the assumed mixing model of ANC is a linear FIR filter architecture, directly applying ANC may not model the linear filter's mismatch to the realistic conditions — nonlinearities due to the sensor noise and the infinite filter length. Therefore we add a nonlinear feature adopted in spectral subtraction [5]:

$$|U_i(\omega, n)| = f(|Y_i(\omega, n)| - \alpha_i b_{ij}(\omega) |Y_j(\omega, n)|),$$
$$\angle U_i(\omega, n) = \angle Y_i(\omega, n), \ i \neq j, \quad (9)$$

where α_i is the over-subtraction factor, $Y_i(\omega, n)$ is the ith component of the BSS output $\mathbf{Y}(\omega, n)$, $b_{ij}(\omega)$ is the cross-channel interference cancellation factor for frequency ω from channel j to i, and the bounding function $f(\cdot)$ is defined by

$$f(a) = \begin{cases} a & \text{if } a \geq \varepsilon \\ \varepsilon & \text{if } a < \varepsilon \end{cases}, \quad (10)$$

where the positive constant ε sets a lowerbound on the spectrum value. The nonlinear operator $f(\cdot)$ suppresses the remaining errors of the BSS, but may introduce musical noises as most spectral subtraction techniques suffer.

3.3 Probability Model and Cancellation Factor Update

If the subtraction in Equation 9 successfully removes the cross-channel interference, the spectral magnitude $|U_i(\omega,n)|$ would be zero in inactive frames. We evaluate the posterior probability of $Y_i(\omega,n)$ given each hypothesis by the complex Gaussian distributions of $|U_i(\omega,n)|$:

$$p(Y_i(\omega,n)|H_{i,m}) \simeq p(U_i(\omega,n)|H_{i,m}) \propto \exp\left[-\frac{|U_i(\omega,n)|^2}{\lambda_{i,m}(\omega)}\right], \quad (11)$$

where $\lambda_{i,m}$ is the variance of the subtracted frames. When $m=1$ it is the variance of the primary source, and when $m=0$ it is of the secondary source. The variance $\lambda_{i,m}$ can be updated at every frame by the following probabilistic averaging formula:

$$\lambda_{i,m} \Leftarrow \{1 - \eta_\lambda p(H_{i,m}|Y_i(n))\}\lambda_{i,m} + \eta_\lambda p(H_{i,m}|Y_i(n))|U_i(\omega,n)|^2, \quad (12)$$

where the positive constant η_λ defines the adaptation frame rate. The primary source signal is expected to be at least "emphasized" by BSS. Hence we assume that the amplitude of the primary source should be greater than that of the interfering source, which is primary in the other BSS output channel. While updating the model parameters, it might happen that the variance of the enhanced source, $\lambda_{i,1}$, becomes smaller than $\lambda_{i,0}$. Since such cases are undesirable, we explicitly change two models when

$$\sum_\omega \lambda_{i,0}(\omega) > \sum_\omega \lambda_{i,1}(\omega). \quad (13)$$

The next step is updating the interference cancellation factors. First we compute the difference between the spectral magnitude of Y_i and Y_j at frequency ω and frame n:

$$\delta_i(\omega,n) = |Y_i(\omega,n)| - b_{ij}(\omega)|Y_j(\omega,n)|. \quad (14)$$

We define the cost function J by ν-norm of the difference multiplied by the frame probability:

$$J(\omega,n) = p(H_{i,0}|Y_i(n)) \cdot |\delta_i(\omega,n)|^\nu. \quad (15)$$

The gradient-descent learning rules for b_{ij} at frame n is

$$\Delta b_{ij}(\omega) \propto -\frac{\partial J(\omega,n)}{\partial b_{ij}(\omega)} = p(H_{i,0}|Y_i(n)) \cdot |\delta_i(\omega,n)|^{\nu-1} Y_j(\omega,n). \quad (16)$$

According to the earlier findings about natural sound distributions, ν is set to be less than 1 for highly kurtotic speech signals [12], greater than 1 for music signals [13], and 2 for pure Gaussian random noises. In the case of the speech signal mixtures, we assign $\nu = 0.8$ for $p(H_{i,1}|Y_i(n))$, and $\nu = 1.5$ for $p(H_{i,0}|Y_i(n))$ to fit to the distribution of the musical noises that are frequently observed in the inactive frames by the result of spectral subtraction.

4 Evaluation

We conducted experiments designed to demonstrate the performance of the proposed method. The test data are recorded in a normal office room. Two loudspeakers play the different sources and two omnidirectional microphones simultaneously record the mixtures at a sampling rate of 16 kHz. The left speaker plays one of male and female speech signals, and the right speaker plays one of 5 different sounds at a time. The speech signals are a series of full sentence utterances, and the music signals are a pop song, a rock with vocal sounds, and a soft instrumental music. The distance between the sensors is 50cm, between the speakers is 50cm, and between the sensor and the speaker is 100cm. The length of the frame for the frequency domain BSS was 512 samples, and the same length is used for the cross-channel interference cancellation algorithm.

The separation results are compared by signal to interference ratio (SIR), which we define the logarithm of the primary source power to the secondary source power ratio in a channel:

$$\text{SIR}(u_i) \,[\text{dB}] = 10\log_{10}\left[\frac{E_1(u_i)}{E_2(u_i)}\right] \simeq 10\log_{10}\left[\frac{E_{1+2}(u_i) - E_2(u_i)}{E_2(u_i)}\right],$$

where $E_1(u_i)$ and $E_2(u_i)$ are the average power of primary and secondary source in signal u_i, and $E_{1+2}(u_i)$ is the average power when cross-interference occurs. When the two sources are uncorrelated, we can approximate $E_1 \simeq E_{1+2} - E_2$. Because the exact signal is unable to obtain, we exploit the interference probabilities to evaluate the source powers:

$$E_2(u_i) = \frac{\sum_n P(H_{i,0}|Y_i(n))\left\langle u_i(t)^2\right\rangle_n}{\sum_n P(H_{i,0}|Y_i(n))}, \quad E_{1+2}(u_i) = \frac{\sum_n P(H_{i,1}|Y_i(n))\left\langle u_i(t)^2\right\rangle_n}{\sum_n P(H_{i,1}|Y_i(n))},$$

where $\left\langle u_i(t)^2\right\rangle_n$ is the average sample energy of frame n.

Table 1 reports the SIR improvements of the proposed method. By applying the frequency-domain BSS, there was 4 dB SIR enhancement on the average. But by the proposed post-processing method on the BSS output, there was

Table 1. Computed SIRs of the input signals, the BSS outputs, and the interference-canceled results with the proposed method. 'mixture' columns are types of sources mixed in the stereo input. 'f1' and 'f2' are female speeches, 'm1' and 'm2' are male speeches, and 'g1' to 'g3' are three different music signals. All the values are in dB.

mixture	Input	BSS only	Proposed	mixture	Input	BSS only	Proposed
f1-g1	6.37	7.13	11.04	m1-g1	7.91	10.37	16.15
f1-g2	3.84	8.75	16.57	m1-g2	4.19	8.81	16.36
f1-g3	1.89	5.74	11.11	m1-g3	0.87	4.84	10.97
f1-f2	3.08	6.45	10.9	m1-f2	2.54	9.42	15.74
f1-m2	7.23	10.92	16.82	m1-m2	6.74	11.72	17.46
average	4.48	7.80	**13.29**	average	4.45	9.03	**15.34**
increase	-	+3.32	**+5.49**	increase	-	+4.58	**+6.30**

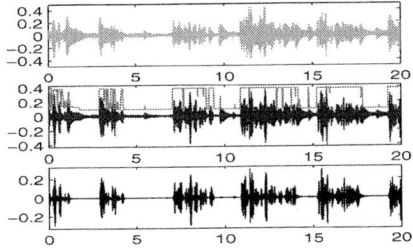
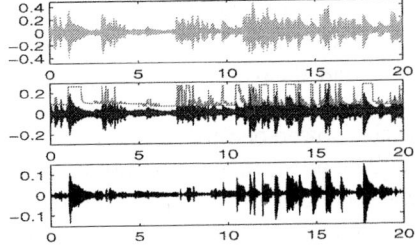

Fig. 2. Step-by-step enhancement result of the developed stereo-input separation method. The first row is the input mixture signal, and the second row is the separation result of the frequency-domain blind source separation (BSS) algorithm. The cross-channel interference probability is computed and represented by the red lines on the waveforms. Using the interference probabilities, the leftover cross-channel interferences are removed and represented in the third row.

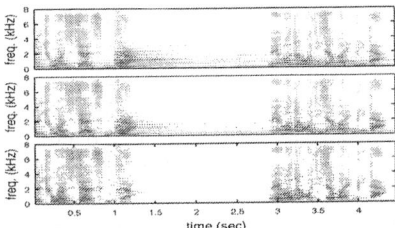
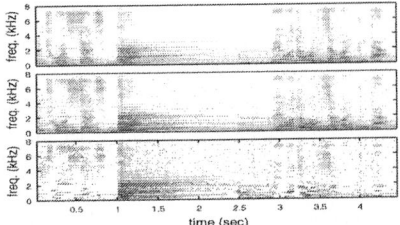

Fig. 3. Spectral view of the separation results. The orders are the same as Figure 2 and the first 4.5 seconds are selected and shown. As the steps proceed, the mixed components are removed and the primary sources are emphasized. However, it is observed that musical noises are newly introduced especially in the music source. Listening to the final results one can hear the mumbling-like musical tones that were not observed in the original mixtures. This is the classical problem of the ordinary spectral subtraction and can be reduced by careful tuning of the system parameters. All the audio files are available at http://myhome.naver.com/flyers/.

additional 6 dB average enhancement. Figure 2 plots the stepwise processing of the proposed method on the mixture f1-g2. The cross-channel interference in the BSS outputs is significantly removed in the final results.

5 Conclusions

The ordinary BSS algorithms have inherent separation errors due to the mismatch between the assumed linear model and the real transfer functions. We proposed a post-processing technique that is applicable to such realistic environments. It has been a similar effort to compensate the separation errors for stationary noise sources [14]. In the proposed method, we deal with nonstationary natural noise sounds on the assumption that the number of sources and the number of sensors are strictly two, and each of the blind source separation system outputs has a primary source and a secondary source signal identified by their relative power. The proposed algorithm considers one BSS output as noisy signal and the other output as reference noise source, and the cancellation is done in the powerspectral domain as the conventional spectral subtraction methods do. The advantage of the powerspectral subtraction is that it effectively absorbs the small amount of mismatch between the actual filter and the estimated one, and generates cleanly denoised signals. The disadvantage is the introduction of the musical noises due to the half-wave rectification. With the help of source absence/presence detection prior to the subtraction, we reduce the error of the cancellation filter estimation and hence minimize the musical noises.

References

1. Bell, A.J., Sejnowski, T.J.: An information-maximization approach to blind separation and blind deconvolution. Neural Computation **7** (1995) 1004–1034
2. Torkkola, K.: Blind signal separation for audio signals - are we there yet? In: Proc. ICA99, Aussois, France (1999) 261–266
3. Araki, S., Makino, S., Aichner, R., Nishikawa, T., Saruwatari, H.: Subband based blind source separation with appropriate processing for each frequency band. In: Proc. ICA2003, Nara, Japan (2003) 499–504
4. Widrow, B., Glover, J.R., McCool, J.M., Kaunitz, J., Williams, C.S., Hearn, R.H., Zeidler, J.R., Dong, E., Goodlin, R.C.: Adaptive noise cancelling: principles and applications. Proceedings of the IEEE **63** (1975) 1692–1716
5. Boll, S.F.: Suppression of acoustic noise in speech using spectral subtraction. IEEE Trans. Acous., Speech and Signal Processing, ASSP **27** (1979) 113–120
6. Smaragdis, P.: Blind separation of convolved mixtures in the frequency domain. Neurocomputing **22** (1998) 21–34
7. Choi, S., Amari, S., Cichocki, A., wen LIU, R.: Natural gradient learning with a nonholonomic constraint for blind deconvolution of multiple channels. In: Proc. ICA99, Aussois, France (1999) 371–376
8. Sawada, H., Mukai, R., Araki, S., Makino, S.: Polar coordinate based nonlinear function for frequency-domain blind source separation. In: Proc. ICASSP, Orlando, Florida (2002)
9. Parra, L., Spence, C.: Convolutive blind separation of non-stationary sources. IEEE Trans. Speech and Audio Processing **8** (2000) 320–327
10. Lee, T.W., Bell, A.J., Orglmeister, R.: Blind source separation of real world signals. In: Proc. ICNN, Houston, USA (1997) 2129–2135
11. Kim, N.S., Chang, J.H.: Spectral enhancement based on global soft decision. IEEE Signal Processing Letters **7** (2000) 108–110
12. Jang, G.J., Lee, T.W., Oh, Y.H.: Learning statistically efficient features for speaker recognition. In: Proc. ICASSP, Salt Lake City, Utah (2001)
13. Bell, A.J., Sejnowski, T.J.: Learning the higher-order structures of a natural sound. Network: Computation in Neural Systems **7** (1996) 261–266
14. Visser, E., Otsuka, M., Lee, T.W.: A spatio-temporal speech enhancement scheme for robust speech recognition in noisy environments. Speech Communication **41** (2003) 393–407

Application of the Mutual Information Minimization to Speaker Recognition / Verification Improvement

Jordi Solé-Casals[1] and Marcos Faúndez-Zanuy[2]

[1] Signal Processing Group, University of Vic (Catalonia, Spain)
jordi.sole@uvic.es
http://www.uvic.es/eps/recerca/ca/processament/inici.html
[2] Escola Universitària Politècnica de Mataró, UPC (Catalonia, Spain)
faundez@eupmt.es

Abstract. In this paper we propose the inversion of nonlinear distortions in order to improve the recognition rates of a speaker recognizer system. We study the effect of saturations on the test signals, trying to take into account real situations where the training material has been recorded in a controlled situation but the testing signals present some mismatch with the input signal level (saturations). The experimental results for speaker recognition shows that a combination of several strategies can improve the recognition rates with saturated test sentences from 80% to 89.39%, while the results with clean speech (without saturation) is 87.76% for one microphone, and for speaker identification can reduce the minimum detection cost function with saturated test sentences from 6.42% to 4.15%, while the results with clean speech (without saturation) is 5.74% for one microphone and 7.02% for the other one.

1 Introduction

This paper proposes a non-linear channel distortion estimation and compensation in order to improve the recognition rates of a speaker recognizer. Mainly it is studied the effect of a saturation on the test signals and the compensation of this non-linear perturbation. This paper is organized as follows. Section 2 describes the Wiener model, its parameterization, and obtains the cost function based on statistical independence. Section 3 summarizes the speaker recognition/verification application. Finally, section 4 deals the experiments using the blind inversion in conjunction with the speaker recognition/verification application.

2 Non-parametric Approach to Blind Deconvolution of Nonlinear Channels

When linear models fail, nonlinear models appear to be powerful tools for modeling practical situations. Many researches have been done in the identification and/or the inversion of nonlinear systems. These works assume that both the input and the output of the distortion are available [1]; they are based on higher-order input/output cross-correlation [2], bispectrum estimation [3, 4] or on the application of the Bussgang and Prices theorems [5, 6] for nonlinear systems with Gaussian inputs. However, in a real world situations, one often does not have access to the distortion input. In this case, blind identification of the nonlinearity becomes the only way to solve the problem. This paper is concerned by a particular class of nonlinear systems, composed by a linear filter followed by a memoryless nonlinear distortion (figure 1, top). This class of nonlinear systems, also known as a Wiener system, is a nice and mathematically attracting model, but also a realistic model used in various areas [7]. We use a fully blind inversion method inspired on recent advances in source separation of nonlinear mixtures. Although deconvolution can be viewed as a single input/single output (SISO) source separation problem in convolutive mixtures (which are consequently not cited in this paper), the current approach is actually very different. It is mainly based on equivalence between instantaneous postnonlinear mixtures and Wiener systems, provided a well-suited parameterization.

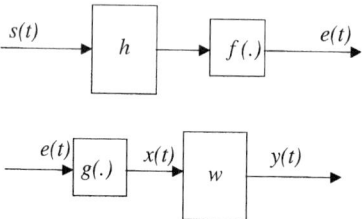

Fig. 1. The unknown nonlinear convolution system (top) and the proposed inversion structure (bottom).

2.1 Model and Assumptions

We suppose that the input of the system $S=\{s(t)\}$ is an unknown non-Gaussian independent and identically distributed (i.i.d.) process, and that subsystems h, f are a linear filter and a memoryless nonlinear function, respectively, both unknown and invertible. We would like to estimate $s(t)$ by only observing the system output. This implies the blind estimation of the inverse structure (figure 1, bottom), composed of similar subsystems: a memoryless nonlinear function g followed by a linear filter w. Such a system is known as a Hammerstein system. Let **s** and **e** be the vectors of infinite dimension, whose t-th entries are $s(t)$ or $e(t)$, respectively. The unknown input-output transfer can be written as:

$$e = f(\mathbf{Hs}) \quad (1)$$

where:

$$\mathbf{H} = \begin{pmatrix} \cdots & \cdots & \cdots & \cdots & \cdots \\ \cdots & h(t+1) & h(t) & h(t-1) & \cdots \\ \cdots & h(t+2) & h(t+1) & h(t) & \cdots \\ \cdots & \cdots & \cdots & \cdots & \cdots \end{pmatrix} \quad (2)$$

is an infinite dimension Toeplitz matrix which represents the action of the filter h to the signal s(t). The matrix **H** is non-singular provided that the filter h is invertible, i.e. satisfies $h^{-1}(t)*h(t) = h(t)*h^{-1}(t) = \delta(t)$, where δ(t) is the Dirac impulse. The infinite dimension of vectors and matrix is due to the lack of assumption on the filter order. If the filter h is a finite impulse response (FIR) filter of order N_h, the matrix dimension can be reduced to the size N_h. In practice, because infinite-dimension equations are not tractable, we have to choose a pertinent (finite) value for N_h. Equation (1) corresponds to a post-nonlinear (pnl) model [8]. This model has been recently studied in nonlinear source separation, but only for a finite dimensional case. In fact, with the above parameterization, the i.i.d. nature of s(t) implies the spatial independence of the components of the infinite vector **s**. Similarly, the output of the inversion structure can be written $y = \mathbf{W}x$ with $x(t) = g(e(t))$. Following [8, 9] the inverse system (g, w) can be estimated by minimizing the output mutual information, i.e. spatial independence of **y** which is equivalent to the i.i.d. nature of y(t).

2.2 Cost Function

The mutual information of a random vector of dimension n, defined by

$$I(Z) = \sum_{i=1}^{n} H(z_i) - H(z_1, z_2, \ldots, z_n) \qquad (3)$$

can be extended to a vector of infinite dimension, using the notion of *entropy rates* of stationary stochastic processes [10]:

$$I(Z) = \lim_{T \to \infty} \frac{1}{2T+1} \left\{ \sum_{t=-T}^{T} H(z(t)) - H(z(-T), \ldots, z(T)) \right\} = H(z(\tau)) - H(Z) \qquad (4)$$

where τ is arbitrary due to the stationarity assumption. We can notice that $I(Z)$ is always positive and vanishes iff z(t) is i.i.d. Since S is stationary, and h and w are time-invariant filters, then Y is stationary too, and $I(Y)$ is defined by:

$$I(Y) = H(y(\tau)) - H(Y) \qquad (5)$$

Using the Lemma 1 of [9], the last right term of equation (5) becomes:

$$H(Y) = H(X) + \frac{1}{2\pi} \int_{0}^{2\pi} \log \left| \sum_{t=-\infty}^{+\infty} w(t) e^{-jt\theta} \right| d\theta \qquad (6)$$

Moreover, using $x(t) = g(e(t))$ and the stationarity of $E = \{e(t)\}$:

$$H(X) = \lim_{T \to \infty} \frac{1}{2T+1} \left\{ H(e(-T), \ldots, e(T)) + \sum_{t=-T}^{T} E[\log g'(e(t))] \right\} = H[E] + E[\log g'(e(\tau))] \qquad (7)$$

Combining (6) and (7) in (5) leads finally to:

$$I(Y) = H(y(\tau)) - \frac{1}{2\pi} \int_{0}^{2\pi} \log \left| \sum_{t=-\infty}^{+\infty} w(t) e^{-jt\theta} \right| d\theta - E[\log g'(e(\tau))] - H[E] \qquad (8)$$

3 Speaker Recognition/Verification

One of the main sources of degradation in speaker recognition is the mismatch between training and testing conditions. For instance, in [11] we evaluated the relevance of different training and testing languages, and in [12] we also studied other mismatch, such as the use of different microphones. In this paper, we study a different source of degradation: different input level signals in training and testing. Mainly we consider the effect of a saturation. We try to emulate a real scenario where a person speaks too close to the microphone or to loud, producing a saturated signal. Taking into account that the perturbations are more damaging when they are present just during training or testing but not in both situations, we have used a clean database and artificially produced a saturation in the test signals. Although it would be desirable to use a "real" saturated database, we don't have this kind of database, and the simulation give us more control about "how the algorithm is performing". Anyway, we have used a real saturated speech sentence in order to estimate the nonlinear distortion using the algorithm described in section 2 and the results have been successful. Figure 2 shows a real saturated speech frame and the corresponding estimate of the NL perturbation.

3.1 Database

For our experiments we have used a subcorpora of the Gaudi database, that follows the design of [13]. It consists on 49 speakers acquired with a simultaneous stereo recording with two different microphones (AKG C-420 and SONY ECM66B). The speech is in wav format at fs=16 kHz, 16 bit/sample and the bandwidth is 8 kHz. We have applied the potsband routine that can be downloaded from: http://www.ee.ic.ac.uk/hp/staff/dmb/voicebox/voicebox.html in order to obtain narrow-band signals. This function meets the specifications of G.151 for any sampling frequency. The speech signals are pre-emphasized by a first order filter whose transfer function is $H(z)=1-0.95z^{-1}$. A 30 ms Hamming window is used, and the overlapping between adjacent frames is 2/3. One minute of read text is used for training, and 5 sentences for testing (each sentence is about two seconds long).

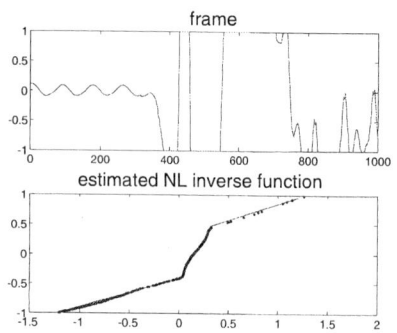

Fig. 2. Saturated frame and the estimated channel function.

3.2 Speaker Recognition / Verification Algorithm

We have chosen a second-order based measure for the recognition of a speaker. In the training phase, we compute for each speaker empirical covariance matrices (CM) based on feature vectors extracted from overlapped short time segments of the speech signals, i.e., $C_j = \hat{E}[x_n x_n^T]$, where $\hat{E}$ denotes estimate of the mean and x_n represents the features vector for frame n. As features representing short time spectra we use mel-frequency cepstral coefficients. In the speaker-recognition system, the trained covariance matrices (CM) for each speaker are compared to an estimate of the covariance matrix obtained from a test sequence from a speaker. An arithmetic-harmonic sphericity measure is used in order to compare the matrices [14]: $d = \log(\text{tr}(C_{test} C_j^{-1}) \text{tr}(C_j C_{test}^{-1})) - 2\log(l)$, where $\text{tr}(\cdot)$ denotes the trace operator, l is the dimension of the feature vector, C_{test} and C_j is the covariance estimate from the test speaker and speaker model j, respectively. In the speaker-verification system, the algorithm is basically the previous one, were have applied the following equation in order to convert the distance measure d into a probability measure p: $p = e^{-0.5d}$, and the system has been evaluated using the DET curves [15], with the following detection cost function (DCF): $DCF = C_{miss} \times P_{miss} \times P_{true} + C_{fa} \times P_{fa} \times P_{false}$ where C_{miss} is the cost of a miss, C_{fa} is the cost of a false alarm, P_{true} is the a priori probability of the target, and $P_{false} = 1 - P_{true}$. The optimal value is indicated in each plot with a "o" mark. We have used $C_{miss} = C_{fa} = 1$.

4 Experiments and Conclusions

Using the database described in section 3, we have artificially generated a test signal database, using the following procedure:
- All the test signals are normalized to achieve unitary maximum amplitude.
- A saturated database has been artificially created using the following equation:
- $x' = tanh(kx)$, where k is a positive constant.

The training set remains the same, so no saturation is added. In order to show the improvement due to the compensation method, figure 3 shows one frame that has been artificially saturated with a dramatic value ($k=10$), the original, and the recovered frame applying the blind inversion of the distortion.

Using the original (clean) and artificially generated database (saturated) we have evaluated the identification rates and the minimum DCF. For the saturated test sentences scenario, we have estimated one different channel model for each test sentence, applying the method described in section 2. This is a way to manage real situations where the possible amount of saturation is not known in advance and must be estimated for each particular test sentence.

In order to improve the results an opinion fusion is done, using the scheme shown in figure 4. Thus, we present the results in three different combination scenarios for speaker recognition:

- Just one opinion (1 or 2 or 3 or 4).
- To use the fusion of two opinions (1&2 or 2&3).
- The combination of the four available opinions.

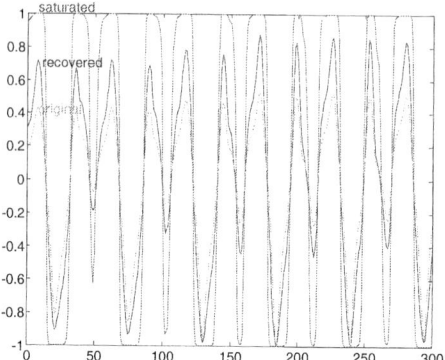

Fig. 3. Example of original, saturated, and recovered frame using the proposed procedure.

Table 1, for speaker recognition experiments, and Table 2, for speaker verification experiments, show the results for $k=2$ in all this possible scenarios using two different combinations [16] rules (arithmetic and geometric mean, [17]), with a previous distance normalization [18].

Main conclusions are:

- The use of the NL compensation improves the obtained results with the same conditions than without this compensation block.
- The combination between different classifiers improves the results. These results can be even more improved using a weighted sum instead a mean. Anyway, we have preferred a fixed combination rule than a trained rule.
- We think that using a more suitable parameterization, the improvements would be higher.

Table 1. Results for several classifiers, shown in figure 4.

Combination		Recognition rate
1 (AKG+NL compensation)		83.67 %
2 (AKG)		82.04 %
3 (SONY+NL compensation)		80.82 %
4 (SONY)		80 %
1&2	Arithmetic	84.9 %
	Geometric	84.9 %
1&3	Arithmetic	89.39%
	Geometric	87.35%
2&4	Arithmetic	88.16%
	Geometric	86.53 %
1&2&3&4	Arithmetic	88.16 %
	Geometric	87.76 %

Table 2. Minimum Detect Cost Function for several classifiers, shown in figure 4.

Combination		Minimum DCF
1 (AKG+NL compensation)		6.42 %
2 (AKG)		5.74 %
3 (SONY+NL compensation)		6.59 %
4 (SONY)		7.02 %
1&2	Arithmetic	5.95 %
	Geometric	5.95 %
1&3	Arithmetic	4.15 %
	Geometric	4.89 %
3&4	Arithmetic	6.99 %
	Geometric	6.21 %
2&4	Arithmetic	4.61 %
	Geometric	5.53 %
1&2&3&4	Arithmetic	4.43 %
	Geometric	5 %

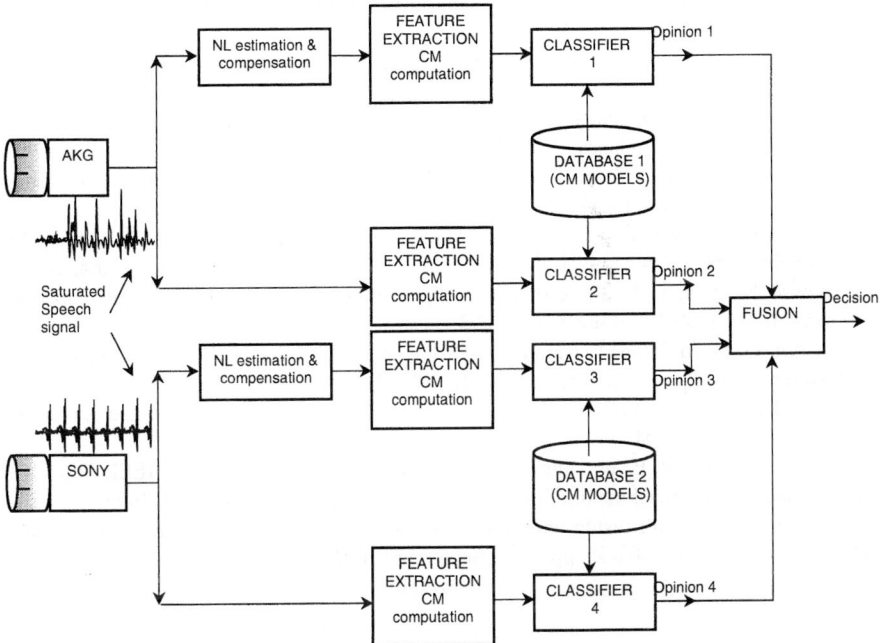

Fig. 4. General Scheme of the recognition system.

Acknowledgments

This work has been supported by COST action 277, University of Vic under the grant R0912, FEDER & CICYT TIC-2003-08382-C05-02.

References

1. S. Prakriya, D. Hatzinakos. Blind identification of LTI-ZMNL-LTI nonlinear channel models. Biol. Cybern., 55 pp. 135-144 (1985).
2. S.A. Bellings, S.Y. Fakhouri. Identification of a class of nonlinear systems using correlation analysis. Proc. IEEE, 66 pp. 691-697 (1978).
3. C.L. Nikias, A.P. Petropulu. Higher-Order Spectra Analysis – A Nonlinear Signal processing Framework. Englewood Cliffs, NJ: Prentice-Hall (1993).
4. C.L. Nikias, M.R.Raghuveer. Bispectrum estimation: A digital signal processing framework. Proc. IEEE, 75 pp. 869-890 (1987)
5. E.D. Boer. Cross-correlation function of a bandpass nonlinear network. Proc. IEEE, 64 pp. 1443-1444 (1976)
6. G. Jacoviti, A. Neri, R. Cusani. Methods for estimating the autocorrelation function of complex stationary process. IEEE Trans. ASSP, 35, pp. 1126-1138 (1987)
7. J. Solé, C. Jutten, A. Taleb "Parametric approach to blind deconvolution of nonlinear channels". Ed. Elsevier, Neurocomputing 48 pp.339-355, 2002
8. A. Taleb, C. Jutten. Source separation in postnonlinear mixtures. IEEE Trans. on S.P., Vol. 47, n°10, pp.2807-20 (1999).
9. Taleb, J. Solé, C. Jutten. Quasy-Nonparametric Blind Inversion of Wiener Systems. IEEE Trans. on S.P., Vol. 49, n°5, pp.917-924 (2001).
10. T.M. Cover, J.A. Thomas. Elements of Information Theory. Wiley Series in Telecommunications (1991)
11. A. Satué, M. Faúndez-Zanuy "On the relevance of language in speaker recognition" EUROSPEECH 1999 Budapest, Vol. 3 pp.1231-1234
12. C. Alonso, M. Faúndez-Zanuy, "Speaker identification in mismatch training and testing conditions".Vol. II, pp. 1181-1184, IEEE ICASSP'2000, Istanbul
13. J. Ortega, J. Gonzalez & V. Marrero, "Ahumada: a large speech corpus in spanish for speaker characterization and identification", Speech Communication 31, pp.255-264, 2000.
14. F. Bimbot, L. Mathan "Text-free speaker recognition using an arithmetic-harmonic sphericity measure." pp.169-172, Eurospeech 1993.
15. A. Martin, G. Doddington, T. Kamm, M. Ordowski, and M. Przybocki, "The DET curve in assessment of detection performance", V. 4, pp.1895-1898, Eurospeech 1997
16. M. Faundez-Zanuy "Data fusion in biometrics". In press, IEEE Aerospace and Electronic Systems Magazine, 2004
17. J. Kittler, M. Hatef, R. P. W. Duin & J. Matas "On combining classifiers". IEEE Trans. On pattern analysis and machine intelligence, Vol. 20, N° 3, pp. 226-239, march 1998.
18. Sanderson "Information fusion and person verification using speech & face information". IDIAP Research Report 02-33, pp. 1-37. September 2002

Single Channel Speech Enhancement: MAP Estimation Using GGD Prior Under Blind Setup

Rajkishore Prasad, Hiroshi Saruwatari, and Kiyohiro Shikano

Graduate School of Information Science, Nara Institute of Science and Technology,
Nara, Japan
{kishor-p,sawatari,shikano}@is.aist-nara.ac.jp

Abstract. This paper presents a statistical algorithm using *Maximum A Posteriori* (MAP) estimation for the enhancement of single channel speech, contaminated by the additive noise, under the blind framework. The algorithm uses Generalized Gaussian Distribution (GGD) function as a prior probability to model magnitude of the Spectral Components (SC) of the speech and noise in the frequency domain. An estimation rule has been derived for the estimation of the SC of the clean speech signal under the presence of additive noise signal. Since the parsimony of the GGD distribution depends on its shape parameter, it provides flexible statistical model for the data with different distribution, e.g. impulsive, Laplacian, Gaussian, etc. The enhancement result for Laplacian noise have been presented and compared with that of the conventional Wiener filtering, which assumes Gaussian distribution for SCs of both the speech and noise.

1 Introduction

There have been developments of different algorithms for the enhancement of speech signal in the Discrete Fourier Transform (DFT) domain assuming the Probability Distribution Function (PDF) of the DFT coefficients of speech and noise as Gaussian. However, such an assumption about the distribution of the DFT coefficient of a quasi-stationary segment of speech is not true. Different researchers have used different PDF models, e.g., Potter and Boll in [1] proposed and used Gamma distribution or Laplacian Distribution (LD) while Ephraim and Malah in [2] have used Gaussian Distribution (GD) to derive an enhancement algorithm in the DFT domain. Recently, in [3] LD model has been used to derive an enhancement algorithm. The statistical distribution of the speech spectral component, obtained by the Short-Time Fourier Transform (STFT) analysis, depends on the signal content of each quasi-stationary segment. Thus statistical distribution of the SCs of a signal is not exactly same in each frequency bin. The use of LD or GD with fixed parameters for the SCs in each frequency bin fails to model inherent variation in the signal. Similar, mismatch between actual and used statistical model for the noise signal also arises. In the most of the proposed speech enhancement algorithms, Gaussian noise is frequently considered. However, many real world noise signals such as chair crack, clapping, object dropping etc. are neither Gaussian nor exactly Laplacian [4]. The PDF of one of such noises is shown in Fig.1. The parsimony of PDF of spectral components of such noises is also different in different frequency bins. Obviously, it cannot be accurately modeled with Gaussian or Laplacian distributions with the fixed parameters for each frequency bin.

In this paper we present an algorithm under the blind setup, using MAP estimation, for the speech enhancement in the DFT domain using GGD function as a flexible prior probability model for the DFT coefficients of both the noise and speech [5]. The proposed method is blind in the sense that its functioning relies only on the information fetched from the noisy speech signal.

2 Signal Model in Additive Noise

For the single channel signal capture, the observed speech $y(n)$ in the presence of additive noise $d(n)$ is given by

$$y(n) = x(n) + d(n) \qquad (1)$$

where $x(n)$ represents clean speech signal, n is the time-index, and random noise $d(n)$ is uncorrelated with the clean speech signal. The aim of the enhancement technique is to estimate clean signal $\hat{x}(n)$ from the observed noisy signal $y(n)$. As we said earlier that our aim is to make estimation in the DFT domain, where DFT coefficients of the clean speech are estimated. The observed speech signal is subjected to STFT analysis to produce time-frequency series of the speech [5]. Thus by taking STFT the signal model in Eq.(1), it can be represented as follows in the frequency domain

$$Y(f) = X(f) + D(f). \qquad (2)$$

Speech enhancement algorithm, thus in frequency domain makes modification in $Y(f)$ by $G(f)$ to estimate the spectral component $\hat{X}(f)$ of the clean speech .i.e.

$$X(f) = G(f).Y(f). \qquad (3)$$

The modification function $G(f)$ is called gain function.

3 MAP Estimation Under GGD Prior

We propose here to model speech spectral components by GGD [5]. The PDF of the GGD is parameterized by the mean μ, scale factor α, and shape parameter β. The GGD PDF for zero mean Random Variable (RV) z is given by

$$f_{GG}(z;\mu,\alpha,\beta) = \frac{\beta}{2\alpha\Gamma(1/\beta)} \exp\left(-[|z-\mu|/\alpha]^\beta\right) = A\exp(-[b|(z)|]^\beta)$$

where $A = \frac{b\beta}{2\Gamma(1/\beta)}$, $b = \frac{1}{\alpha} = \frac{1}{\sigma}\sqrt{\frac{\Gamma(3/\beta)}{\Gamma(1/\beta)}}$; $\Gamma(x) = \int_0^\infty e^{-t}t^{x-1}dt$=Gamma PDF; $\qquad (4)$

-∞<z<∞; α>0; β>0; σ=stdv.

The shape parameter β determines the shape of the distribution. For β=1 and β=2 distributions are Laplacian and Gaussian respectively and distribution tends to become uniform as β→∞. The shapes of distribution for the different values of β are shown in Fig.3. Thus use of GGD as a statistical model for the spectral components of speech and noise can provide flexibility as well as adaptation in the algorithm. Under the proposed framework, samples of $X(f)$ and $D(f)$ will be represented by GGD with parameters estimated from the data.

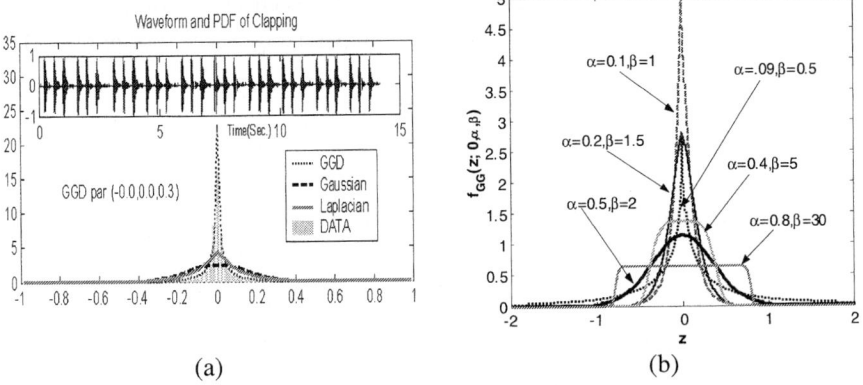

Fig. 1. (a) Plot of the clapping noise. Noise waveform is also shown inside the figure panel. The fittings of GD, LD, and GGD function are also shown in the histogram of the noise. GGD par (mean, scale, shape) are also shown. Actual value of scale par. is .0016. (b) GGD family for different values of scale (α) and shape (β) parameters

Let in the *kth* frequency bin $Y_k = R_k e^{iv_k}$ represents noisy signal and $X_k = a_k e^{i\alpha_k}$ represents spectral components of clean signal in the polar form. Thus the problem of estimating clean signal can be formulated in terms of estimation of magnitude a_k and phase α_k. The joint MAP estimate of a_k and α_k is given by the mode of the posterior distribution $p(a_k, \alpha_k | Y_k)$. Here we will use GGD to represent joint and marginal PDFs of the magnitude of the DFT coefficients of the signals as follows

$$p(a_k) = A_x e^{-\left[\frac{|a_k|}{b_x}\right]^{\beta_x}}, \textit{ for } 0 \le a_k \le \infty. \quad (5)$$

The scale parameter is represented by b. The PDF of phase α_k, follows uniform distribution as the analysis window position, used in STFT analysis is arbitrary. Thus

$$p(\alpha_k) = \textit{Uniform distribution} = \begin{cases} \dfrac{1}{2\pi} & \textit{for } -\pi \le \alpha_k \le \pi \\ 0 & \textit{else} \end{cases}, \quad (6)$$

$$p(a_k, \alpha_k) = \frac{A_x}{2\pi} e^{-\left[\frac{|a_k|}{b_x}\right]^{\beta_x}}, \textit{ for } 0 \le a_k \le \infty, \quad (7)$$

$$p(Y_k | X_k) = p(Y_k | a_k, \alpha_k) = \textit{Noise PDF} = \frac{A_N}{2\pi} e^{-\left[\frac{|Y_k - X_k|}{b_n}\right]^{\beta_n}}, \textit{ for } 0 \le |Y_k - X_k| \le \infty, \quad (8)$$

$$A_x = \frac{\beta_x}{2 b_x \Gamma(1/\beta_x)} = \frac{\beta_x}{2 \Gamma(1/\beta_x)} \frac{1}{\sigma_x} \sqrt{\frac{\Gamma(3/\beta_x)}{\Gamma(1/\beta_x)}}; \quad \sigma_x = \sqrt{E\{a_k^2\}} \text{ ,and} \quad (9)$$

$$A_N = \frac{\beta_n}{2 b_n \Gamma(1/\beta_n)} = \frac{\beta_n}{2 \Gamma(1/\beta_n)} \frac{1}{\sigma_n} \sqrt{\frac{\Gamma(3/\beta_n)}{\Gamma(1/\beta_n)}}; \quad \sigma_n = \sqrt{E\{|D_k^2|\}}, \quad (10)$$

where β_x, β_n represent shape parameters for clean speech and noise respectively. Now MAP estimation is obtained by maximizing the posterior PDF $p(a_k, \alpha_k | Y_k)$ given by

$$p(a_k, \alpha_k | Y_k) = p(X_k | Y_k) = \frac{p(Y_k | a_k, \alpha_k) p(a_k, \alpha_k)}{p(Y_k)}. \quad (11)$$

Since $p(Y_k)$ is constant, only numerator of the above is of interest in maximizing the posterior density. Thus dropping denominator for the optimization we get

$$p(a_k, \alpha_k | Y_k) = p(Y_k | a_k, \alpha_k) p(a_k, \alpha_k) = \frac{A_N A_x}{4\pi^2} e^{-\left[\frac{|Y_k - X_k|^{\beta_n}}{b_n^{\beta_n}}\right] - \left[\frac{|a_k|^{\beta_x}}{b_x^{\beta_x}}\right]}. \quad (12)$$

Its natural logarithmic function is optimized which is given by

$$J = \ln[p(Y_k | a_k, \alpha_k) p(a_k, \alpha_k)] = -\frac{|R_k e^{iv_k} - a_k e^{i\alpha_k}|^{\beta_n}}{b_n^{\beta_n}} - \frac{|a_k|^{\beta_x}}{b_x^{\beta_x}} + \ln \frac{A_N A_x}{4\pi^2}. \quad (13)$$

Now for the highest posterior PDF, differentiating Eq.(13) w.r.t. α_k and a_k and equating derivatives to zero result in

$$\frac{dJ}{d\alpha_k} = 0 \Rightarrow [R_k^2 + a_k^2 - 2R_k a_k \cos(v_k - \alpha_k)]^{0.5\beta_n - 1}[R_k a_k \sin(v_k - \alpha_k)] = 0,$$
$$\Rightarrow \sin(v_k - \hat{\alpha}_k) = 0 \Rightarrow v_k = \hat{\alpha}_k, \quad (14)$$

which shows that phase of SCs of the noisy speech and clean speech are same. Similar treatment of Eq.(13) w.r.t a_k gives

$$\frac{dJ}{da_k} = 2B[R_k^2 + a_k^2 - 2R_k a_k \cos(v_k - \alpha_k)]^{\beta}[-R_k + a_k] - \frac{\beta_x a_k^{\beta_x - 1}}{b_x^{\beta_x}}[sign(a_k)]^{\beta_x} = 0, \quad (15)$$

where $B = 0.5\beta_n / b_n^{\beta_n}$; $\beta = 0.5\beta_n - 1$. Further simplification results in the following radical (power) equation

$$\beta_x a_k^{\beta_x - 1} b_x^{-\beta_x} sign(a_k)^{\beta_x} = 2B(R_k - a_k)^{2\beta + 1} \Rightarrow a_k^{\beta_x - 1} = P(R_k - a_k)^{\beta_n - 1}, \quad (16)$$

where $P = b_x^{\beta_x} \beta_n / b_n^{\beta_n} \beta_x$. Solution of this equation gives MAP estimation for the polar magnitude of the spectral component of the clean speech signal. However, it may be cumbersome to find analytical solution of Eq.(16) that will be valid for all the values of parameters of the GGD used to model noise and speech. Using Newton-Rapshon method, the numerical solution is given as

$$a_k(new) = a_k(old) - \frac{a_k^{\beta_x - 1} - P(R_k - a_k)^{\beta_n - 1}}{(\beta_x - 1)a_k^{\beta_x - 2} + P(\beta_n - 1)(R_k - a_k)^{\beta_n - 2}}. \quad (17)$$

This solution is sensitive to the initial values. The good initial values can be obtained as the special case solution of the Eq.(16). For $\beta_x = \beta_n = 2$, both the noise and speech signal have Gaussian (assumption working under the conventional Wiener filtering) PDF and solution of the Eq.(16) is given by

$$\hat{a}_k = \frac{b_x^2}{b_x^2 + b_n^2} R_k = \frac{\sigma_x^2}{\sigma_x^2 + \sigma_n^2} R_k, \quad (18)$$

which is Wiener filter and can be used as the initial value for the iterative solution in Eq.(17). In Eq.(17), the scale and shape parameters of clean speech and noise data are required which can be estimated from the noisy data only. The scale and shape parameters of GGD for noise and clean speech are estimated using Maximum Likelihood (ML) estimation technique [6]. The noise-only frames and noisy speech frames are first labeled using Voice Activity Detection (VAD). For VAD we have used negentropy as a measure [7], as it is independent of SNR level of the signal. The basic idea is that during the speech part signal is less chaotic than during the noise only frames and thus entropy-based measure can discriminate noise-only and noisy speech frames. The negentropy of each frame in DFT domain is obtained in terms of Differential Entropy (DE) of GGD function, which depends on the scale and shape parameters. The negentropy $H(\beta)$ is computed as the difference of DE of Gaussian RV (with same variance as of SC of speech) and DE of SCs of speech modeled by the GGD $f_{GG}(0, \alpha, \beta)$. Accordingly, negentropy $H(\beta)$ is given by

$$H(\beta) = DE(\alpha, \beta = 2) - DE(\alpha, \beta) = \log\left[\frac{\beta}{2}\sqrt{\frac{6.29\Gamma\left(3/\beta\right)}{\Gamma\left(1/\beta\right)^3}}\right] + \left(0.5 - \frac{1}{\beta}\right). \quad (19)$$

A threshold value based on the global statistics of the data is used the separate the noise frames and noisy speech frames. The threshold value of the negentropy is estimated under assumption that noise-only frame and noisy speech frames have same probability p at the threshold. Accordingly, the threshold H_{TH} is given by

$$H_{TH} = \left[\frac{-\alpha_e^{\beta_e} \log\left(2p\Gamma(1/\beta_e)\right)}{\beta_e}\right]^{1/\beta_e}, \quad (20)$$

where subscript e indicates negentropy. The GGD parameters for noise are estimated from the noise-only frames using ML technique. The shape parameter of GGD model for the clean speech can be obtained from the kurtosis K_x

$$K_x = \Gamma(1/\beta_x)\Gamma(5/\beta_x)\left[\Gamma(3/\beta_x)\right]^{-2} \Rightarrow \beta_x = F^{-1}(K_x) \quad (21)$$

The kurtosis K_x of the clean speech is estimated from

$$K_x = (\sigma_y^2 - \sigma_n^2)^{-2}\left[K_y\sigma_y^4 - 4S_x\sigma_x^3\mu_n - 4S_n\sigma_n^3\mu_x - 6\sigma_y^2\sigma_n^2 - (K_n - 6)\sigma_n^4\right] \quad (22)$$

$$S_x = Skewness = \sigma_x^{-3}\left[S_y\sigma_y^3 - 3(\sigma_y^2 - \sigma_n^2)\mu_n - 3(\mu_y - \mu_n)\sigma_n^2 - S_n\sigma_n^3\right] \quad (23)$$

$$\sigma_y^2 = \sigma_x^2 + \sigma_n^2; \quad \mu_y = \mu_x + \mu_n$$
$$\hat{\sigma}_x^2 = \max\left(\sigma_y^2 - \sigma_n^2, 0\right) \quad (24)$$

in which the subscripts n denotes noise and y denotes noisy speech signal, S, σ and μ denotes skewness, standard deviation and mean of the signal indicated by the subscripts. The scale parameter of the GGD for clean speech is then obtained using σ_x and β_x in Eq.(14). The whole process of the speech enhancement in the DFT domain is shown in Fig.2.

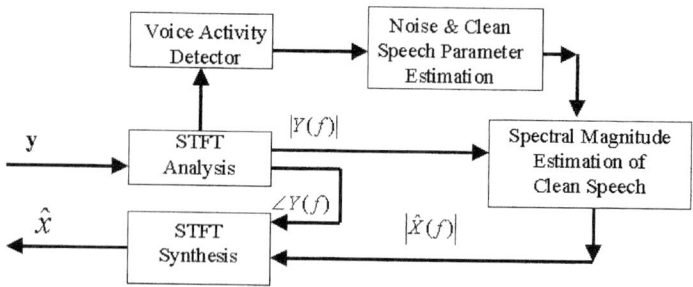

Fig. 2. Speech enhancement scheme used to estimate spectral components of clean speech in the DFT domain. The phase of noisy data is used to reconstruct the original signal.

4 Experiments and Results

For the experimental purpose we have used speech data from ASJ continuous speech corpus for research [5] the and noise data from the NOISEX-92 database available at http://mi.eng.cam.ac.uk/comp.speech/Section1/Data/noisex.html. The result of negentropy based VAD for the White Gaussian Noise (WGN) and clapping are shown in Fig.4. In that figures, patterns in spectrograms revel chaos of noisy speech and noise-only frames, which is also tracked by the negentropy curve. The negentropy of each frame is also plotted over noisy and clean speech waveforms. The characteristic of noise suppression rule is shown in the Fig.4(a) and (b). These gain curves were obtained using Eq.(17) for the 5000 samples of RV generated for given GGD parameters. The β parameters of GGD for clean speech was held constant at 1.2.It was done so, as the average value β for $|X(f)|$ was found around 1[5], but at exactly 1 Eq.(16) vanishes. The value of β parameter estimated for the clean speech and noise from the noisy data (clapping noise) are shown in Fig.5. The true estimation is difficult e.g. it can be seen in Fig.5 how the actual and estimated β differs in different frequency bins. However, the denoising capacity of the algorithm highly depends on these estimated parameters and accurate estimation of these parameters for clean speech and noise is another research issue. The experiment was also performed to denoise speech signal corrupted by babble and clapping noise. The result is placed in the Table 1. If he blind estimation of GGD parameters for the noise and speech spectral components are not very close to true value, denoising performance have been found to be poorer.

Table 1. Denoising performance with using blindly estimated parameters of GGD.

Noise↓	SNR(dB) after MAP ESTIMATION					SNR (dB) after Wiener Filtering				
Input SNR→	-5dB	0dB	5dB	10dB	15dB	-5dB	0dB	5dB	10dB	15dB
Clapping	1.65	3.55	8.37	12.00	16.27	0.86	3.10	5.59	10.97	15.57
Babble	2.03	5.79	9.927	13.34	17.27	0.56	3.62	8.46	12.74	16.77

Single Channel Speech Enhancement 879

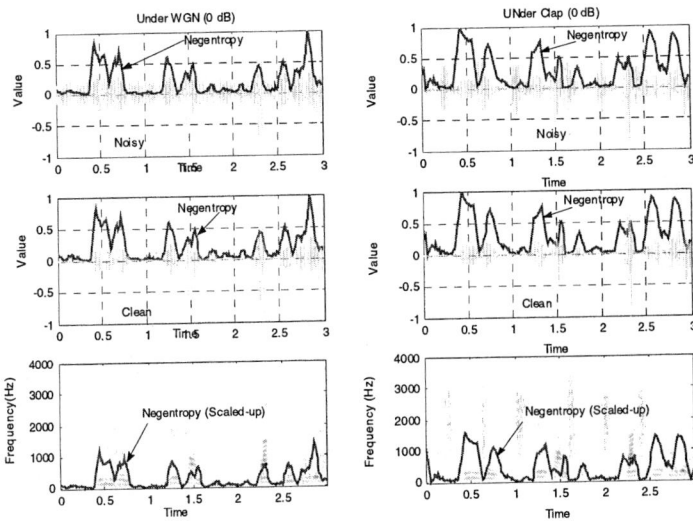

Fig. 3. VAD under WGN (Left column) and Clap noise (right coloumn). SNR=0 dB.

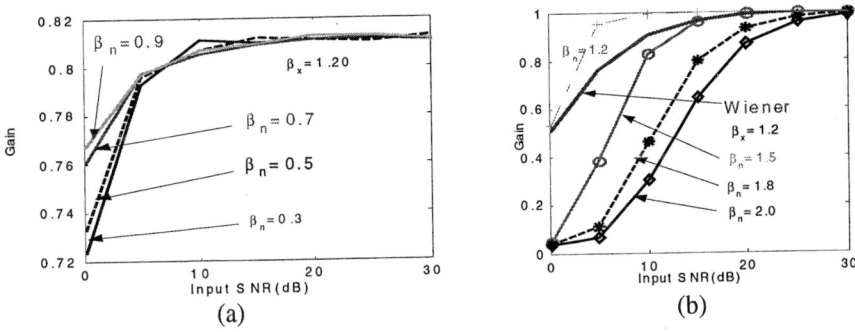

Fig. 4. The nature of MAP estimation rule for different values of GGD parameters of noise and speech under the varying SNR of speech.

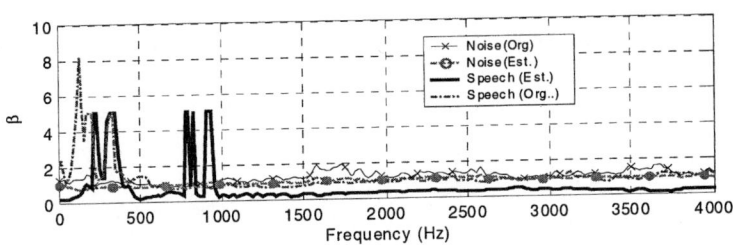

Fig. 5. Estimated β of clean speech and noise. True β and estimated β differs a lot.

References

1. Porter J., Boll S.: Optimal Estimators for Spectral Restoration of Noisy Speech. Proc. IEEE ICASSP (1984) 18A.2.1-18.A.2.4
2. Ephraim A., Malah D.: Speech Enhancement Using a Minimum Mean Square Error Short-Term Spectral Amplitude Estimator. IEEE Trans. Acoust., Speech, Signal Processing, Vol. ASSP-32, no. 6 (Dec.1984) 1109-21
3. Martin R., Colin B.: Speech Enhancement in the DFT Domain Using Laplacian Priors. Proc. IWAENC, Kyoto, Japan (2003) 87-90
4. Panayotis G.G., Panagiotis T., and Chris K.: Alpha-Stable Modeling of Noise and Robust Time Delay Estimation in the Presence of Impulsive Noise. IEEE Trans. on Multimedia, Vol.1, No.3 (1999) 291-301
5. Prasad R.K., Saruwatari H., Shikano K.: Probability Distribution of Time-Series of Speech Spectral Components. IEICE Trans. Fundamental of Elec., Com., CS., Vol.E87-A, No.3, (2004) 584-597
6. Varanasi M.K. and Aazhang B.: Parametric Generalized Gaussian Density Estimation J. of Acost. Society of America, Vol. 86(4), (1989) 1404-1414
7. Hyvarinen A., Karhunen J., Oja E.: Independent Component Analysis. John Wiley & Sons, Inc., New York (2001)

Stable and Low-Distortion Algorithm Based on Overdetermined Blind Separation for Convolutive Mixtures of Speech

Tsuyoki Nishikawa[1], Hiroshi Saruwatari[1],
Kiyohiro Shikano[1], and Atsunobu Kaminuma[2]

[1] Graduate School of Information Science, Nara Institute of Science and Technology
8916-5 Takayama-cho, Ikoma-shi, Nara, 630-0192, Japan
{tsuyo-ni,sawatari,shikano}@is.aist-nara.ac.jp
[2] Nissan Research Center, NISSAN MOTOR CO., LTD.
1 Natsushima-cho, Yokosuka-shi, Kanagawa 237-8523, Japan

Abstract. We propose a new algorithm with a stable learning and low-distortion based on overdetermined blind separation for the convolutive mixture of the speech. To improve the separation performance, we have proposed multistage ICA, in which frequency-domain ICA and time-domain ICA (TDICA) are cascaded. For temporally correlated signals, we must use TDICA with a nonholonomic constraint to avoid the decorrelation effect. However, the stability cannot be guaranteed in the nonholonomic case. Also, in the holonomic case, the sound quality of the separated signal is distorted by the decorrelation effect. To solve the problem of the stability, we perform TDICA with the holonomic constraint. To avoid the distortions, we estimate the distortion components by TDICA with the holonomic constraint and we compensate the sound qualities by using the estimated components. The stability of the proposed algorithm can be guaranteed by the holonomic constraint, and the proposed compensation work prevents the distortion. The experiments in a reverberant room reveal that the algorithm results in higher stability and higher separation performance.

1 Introduction

Blind source separation (BSS) is an approach for estimating original source signals only from the information of the mixed signals observed in each input channel. This technique is applicable to high-quality hand-free speech recognition systems. Many BSS methods based on independent component analysis (ICA) [1] have been proposed [2,3] for the acoustic signal separation. However, the performances of these methods degrade seriously, especially under heavily reverberant conditions. In order to improve the separation performance, we have proposed multistage ICA (MSICA) involving subarray processing [4], in which frequency-domain ICA (FDICA) [3,5] and time-domain ICA (TDICA) [2,6] are cascaded (see Fig. 1). In this method, first, we divide the observed signals in a microphone array into the observed signals in the subarrays. In every subarray,

FDICA can find an approximate solution to separate the sources to a certain extent, and finally TDICA can remove the residual crosstalk components from FDICA. Therefore, the improvement of TDICA is a primary issue because the quality of resultant separated signals is determined by TDICA.

In this paper, we discuss a stability issue of the TDICA algorithm, and newly propose a stable and low-distortion algorithm based on MSICA using subarray processing for temporally correlated signals, e.g., speech signals. First, the following points are explicitly noted: (1) The stability of learning in the conventional TDICA with a holonomic constraint [2] is highly acceptable. However, the method cannot work well for speech signals due to the deconvolution property; i.e., the separated speech is harmfully distorted by the whitening process. (2) To decrease the whitening effect, TDICA with a nonholonomic constraint has been proposed [6]. This method, however, includes an inherent drawback that the stability of learning cannot be guaranteed. We have already solved these problem in the specific mixing model, where the number of microphones is equal to that of sources [7]. However this method cannot be applied to the model, where the number of microphones is larger than that of sources, because the compensation of the sound qualities of separated signals becomes difficult.

In order to solve the problems, we propose a novel approach in that we perform TDICA with the holonomic constraint for solving the problem of the stability. Also, for avoiding the distortions, we estimate the distortion components by TDICA with the holonomic constraint and we compensate the sound qualities by using the estimated components. The stability of the proposed algorithm can be guaranteed by the holonomic constraint, and the proposed compensation method prevents the ICA from performing the decorrelation. Also, the proposed method can apply some types of models, where the number of microphones is not equal to that of sources. The experimental results under a reverberant condition reveal that the proposed algorithm provides the higher stability and the higher separation performance than the conventional MSICA.

2 Conventional MSICA and Problems

2.1 Sound Mixing Model of Microphone Array

In general, the observed signals $\boldsymbol{x}(t) = [x_1(t), \cdots, x_K(t)]^\mathrm{T}$ in which multiple source signals $\boldsymbol{s}(t) = [s_1(t), \cdots, s_L(t)]^\mathrm{T}$ are convolved with room impulse responses (see Fig. 1) are obtained as $\boldsymbol{x}(t) = \sum_{\tau=0}^{P-1} \boldsymbol{a}(\tau) \boldsymbol{s}(t - \tau)$, where K is the number of array elements (microphones) and L is the number of multiple sound sources. Here, $\boldsymbol{a}(\tau) = [a_{ij}(\tau)]_{ij}$ ($[\cdot]_{ij}$ denotes the matrix in which the ij-th element is $[\cdot]$) is the mixing filter matrix, and P is the length of the impulse response.

2.2 BSS Algorithm Based on MSICA Using Subarray Processing [8]

Figure 1 shows the procedure of the MSICA using subarray processing. In the MSICA, we regard the K observed signals as the combinations of the L ($< K$)

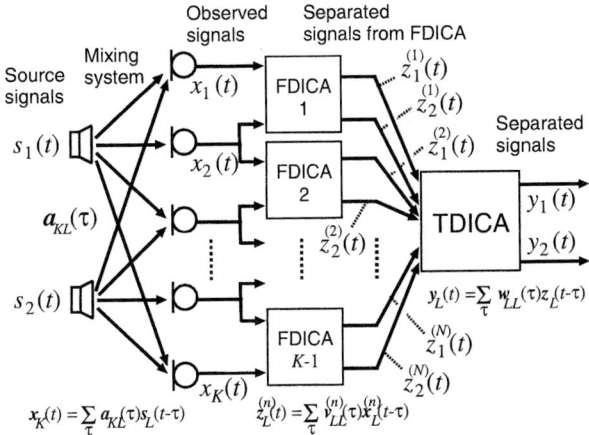

Fig. 1. Blind source separation procedure performed in original MSICA using subarray processing which has been previously proposed by the authors [8].

observed signals, and we regard this combination as a subarray. First, we divide the whole inputs into $K-1$ subarrays, and we perform FDICA in every subarray. The output signals $\boldsymbol{z}^{(n)}(t) = [z_1^{(n)}(t), \cdots, z_L^{(n)}(t)]^{\mathrm{T}}$ from FDICA in the n-th subarray can be given as $\boldsymbol{z}^{(n)}(t) = \sum_{\tau=0}^{Q-1} \boldsymbol{v}^{(n)}(\tau)\, \boldsymbol{x}^{(n)}(t-\tau)$, where $\boldsymbol{v}^{(n)}(\tau) = [v_{ij}^{(n)}(\tau)]_{ij}$ is the separation filter matrix of FDICA in the n-th subarray and Q is the length of the separation filter of FDICA. In FDICA part, we optimize $\boldsymbol{v}(\tau)$ so that the narrow-band output signals are mutually independent in every frequency. Also, $\boldsymbol{x}^{(n)}(t) = [x_n(t), x_{n+1}(t), \cdots, x_{n+L-1}(t)]^{\mathrm{T}}$. Second, we regard all output signals from FDICA in $K-1$ subarrays as the input signals for TDICA, and we remove the residual crosstalk components from FDICAs. The resultant separated signals $\boldsymbol{y}^{(n)}(t)$ can be given as $\boldsymbol{y}(t) = \sum_{\tau=0}^{R-1} \boldsymbol{w}(\tau)\, \boldsymbol{z}(t-\tau)$, where $\boldsymbol{w}(\tau)$ is the separation filter matrix and R is the length of the separation filter of TDICA. Also, $\boldsymbol{z}(t) = [z_1^{(1)}(t), \cdots, z_1^{(K-1)}(t), z_2^{(1)}(t), \cdots, z_2^{(K-1)}(t), \cdots z_L^{(1)}(t), \cdots, z_L^{(K-1)}(t)]^{\mathrm{T}}$. In TDICA part, we optimize $\boldsymbol{w}(\tau)$ so that the separated signals are mutually independent.

The selection of TDICA is an important issue because the quality of resultant separated signals is determined by TDICA. We have two choices for TDICA algorithms with a holonomic constraint [2] and a nonholonomic constraint [6]. In the next section, detailed explanations for each algorithm and their problems are described.

2.3 Conventional Holonomic TDICA (H-TDICA)

Amari proposed the TDICA algorithm which optimizes the separation filter by minimizing the Kullback-Leibler divergence (KLD) between the joint probability density function and the marginal probability density function of the separated signals [2]. The iterative equation of the separation filter $\boldsymbol{w}^{(\mathrm{H})}(\tau)$ to minimize the KLD is given as

$$\boldsymbol{w}_{i+1}^{(\mathrm{H})}(\tau) = \boldsymbol{w}_i^{(\mathrm{H})}(\tau) + \alpha \sum_{d=0}^{Q-1} \left\{ \boldsymbol{I}\delta(\tau - d) - \langle \boldsymbol{\phi}(\boldsymbol{y}(t))\boldsymbol{y}(t - \tau + d)^{\mathrm{T}} \rangle_t \right\} \boldsymbol{w}_i^{(\mathrm{H})}(d), \quad (1)$$

where $\langle \cdot \rangle_t$ denotes the time-averaging operator, i is used to express the value of the i-th step in the iterations, α is the step-size parameter and $\boldsymbol{I}$ is the identity matrix. $\delta(\tau)$ is Dirac delta function, where $\delta(0) = 1$ and $\delta(n) = 0$ ($n \neq 0$). Also, we define the nonlinear vector function $\boldsymbol{\phi}(\boldsymbol{y}(t)) \equiv \tanh(y_1(t)), \cdots, \tanh(y_L(t))]^{\mathrm{T}}$.

2.4 Conventional Nonholonomic TDICA (NH-TDICA)

The H-TDICA forces the separated signals to have the characteristic that their higher-order autocorrelation is $\delta(\tau)$, i.e., the signals are temporally decorrelated. This performance might have a negative influence on the source separation. In order to solve the problem, Choi proposed a modified TDICA algorithm with a nonholonomic constraint [6]. In this algorithm, the constraint for the diagonal component of $\{\cdot\}$ part in Eq. (1), i.e., the higher-order autocorrelation of separated signals, is set to be arbitrary. The iterative equation of the separation filter $\boldsymbol{w}^{(\mathrm{NH})}(\tau)$ is given as

$$\boldsymbol{w}_{i+1}^{(\mathrm{NH})}(\tau) = \boldsymbol{w}_i^{(\mathrm{NH})}(\tau) - \alpha \sum_{d=0}^{Q-1} \left\{ \text{off-diag}\big(\langle \boldsymbol{\phi}(\boldsymbol{y}(t))\boldsymbol{y}(t - \tau + d)^{\mathrm{T}} \rangle_t \big) \right\} \boldsymbol{w}_i^{(\mathrm{NH})}(d), \quad (2)$$

where off-diag($\cdot$) is the operation for setting every diagonal element of matrix as zero. We have also introduced Eq. (2) in the original MSICA using subarray processing [8] to separate the mixed speech which corresponds to the temporally correlated signal by utilizing the flexibility of the nonholonomic constraint.

2.5 Problems in Conventional TDICAs

The advantage and disadvantage of conventional TDICAs can be summarized as follows. (1) The stability of learning in H-TDICA is satisfactory. However, the method cannot work well for speech signals due to the deconvolution property; i.e., the separated speech is harmfully distorted by the whitening process. (2) On the other hand, NH-TDICA possibly performs no deconvolution, i.e., NH-TDICA is applicable to speech signals. This method, however, includes the inherent drawback that the stability of learning cannot be guaranteed as described in Sect. 4.2 and [7]. (3) To solve the problems (1) and (2), we have already proposed the algorithm combining MSICA and linear prediction [7]. However this method does not apply to the model, where the number of microphones larger than that of sources because the compensation to the sound qualities of separated signals become difficult. Thus, we should propose the new algorithm, in which the source-separation of temporally correlated signals such as speech without the distortion by the decorrelation effect in the model where the number of microphones larger than that of sources.

3 Proposed Stable and Low-Distortion MSICA

This section describes a new algorithm with stable and low-distortion property based on MSICA using subarray processing. In the iterative learning in the proposed algorithm, we perform H-TDICA. However the separated signals are distorted by the decorrelation effect as described in Sect. 2.3. Therefore we estimate the the components which contribute to the distortion and we compensate the sound qualities of the separated signals.

First, we explicate the mechanism of H-TDICA algorithm. The iterative learning of H-TDICA Eq. (1) can be decomposed into the components which contribute to the separation and decorrelation given; these are given as

$$w_{i+1}^{(H)}(\tau) = w_0(\tau) + \alpha \sum_{j=0}^{i} w_j^{(D)}(\tau) + \alpha \sum_{j=0}^{i} w_j^{(S)}(\tau), \tag{3}$$

$$w_j^{(D)}(\tau) = \sum_{d=0}^{Q-1} \Big\{ I\delta(\tau - d) - \text{diag}(\langle \phi(y(t))y(t - \tau + d)^T \rangle_t) \Big\} w_j^{(H)}(d), \tag{4}$$

$$w_j^{(S)}(\tau) = - \sum_{d=0}^{Q-1} \Big\{ \text{off-diag}(\langle \phi(y(t))y(t - \tau + d)^T \rangle_t) \Big\} w_j^{(H)}(d), \tag{5}$$

where $w_0(\tau)$ is the initial filter for the iterative learning, $w_j^{(D)}(\tau)$ is the component which contributes to the decorrelation, and $w_j^{(S)}(\tau)$ is the component which contributes to the separation. The iterative learning of NH-TDICA is the algorithm in which $w_j^{(D)}(\tau) = \mathbf{0}$. This modification yields the source separation without the decorrelation effect [6, 7], but, the stability in the iterative learning degrades. We note that the only source separation is performed by the nonholonomic constraint; indeed it is experimentally proved that $w_j^{(S)}(\tau)$ is the component which contribute to the separation. Also, we can understand that $w_j^{(D)}(\tau)$ is the component which contribute to the decorrelation. Therefore we estimate the distortion components by using $w_j^{(D)}(\tau)$ and we compensate the sound quality of the separated signal. The desired optimized separation filter $w^{(S)}(\tau)$ is obtained by subtracting the component to the distortion from $w_N^{(H)}(\tau)$:

$$w^{(S)}(\tau) = w_N^{(H)}(\tau) - \alpha \sum_{i=0}^{N-1} w_i^{(D)}(\tau), \tag{6}$$

where N is the number of iterations.

The compensation in the proposed method is achieved by recovering the higher-order autocorrelation of the separated signal to be that of the input signals for TDICA. If we apply the conventional simple TDICA, the higher-order autocorrelation of the separated signal approximates to that of the mixed signal. The distortion occurs because the autocorrelation of the mixed signal and that of the desired source signal at the microphone point are completely different. Therefore the application of this algorithm to the conventional simple TDICA is not effective. On the other hand, in the application of this algorithm to MSICA, the input signals for TDICA is the output signals from FDICA. The typical separation performance in FDICA is about 8 to 10 dB under the condition that the reverberation time is 300 ms [4,5] and the mel cepstral distortion between the observed signal with the single source component at the microphone and the output signals from FDICA is about 2 to 3 dB. From this, we can judge that the autocorrelation of the output signal from FDICA which corresponds to the input signal for TDICA part in MSICA and that of the desired source signal at the microphone point are similar. Therefore it is possible that the autocorrelation of the separated signal approximates to that of the output signal from FDICA, and we can compensate the distortion effectively in the proposed method.

4 Experiments and Results

4.1 Experimental Setup

A six-element array with the interelement spacing of 2.83 cm is assumed. The speech signals are assumed to arrive from two directions, $-40°$ and $20°$. The distance between the microphone array and the loudspeakers is 2.0 m. Two kinds of sentences spoken by two male and two female speakers are used as the original speech samples and the sampling frequency is 8 kHz. Using these sentences, we obtain 12 combinations with respect to speakers and source directions. In these experiments, we use the following signals as the source signals: the original speech convolved with the impulse responses specified by the reverberation times of 300 ms. We use the impulse responses recorded in a real room selected from Real World Computing Partnership (RWCP) sound scene database [9]. In order to evaluate the separation performance, we used the *noise reduction rate* (NRR) [4], which is defined as the output signal-to-noise ratio (SNR) in dB minus input SNR in dB. Also, to evaluate the sound quality, we used the mel cepstral distortion (MelCD) between the single source component at the microphone and the output signals from TDICAs. The filter length in FDICA is 1024 taps and the filter length in TDICA is 2048 taps.

4.2 Experimental Results and Discussion

In this study, we compare the following MSICAs: **Conventional MSICA1:** FDICA is followed by NH-TDICA, **Conventional MSICA2:** FDICA is followed by H-TDICA, and **Proposed MSICA:** FDICA is followed by the proposed TDICA algorithm, Figures 2 (a) and (b) show the NRR results of the conventional MSICA1, MSICA2, and the proposed MSICA for different iteration points. Figures 3 (a) and (b) show the mel cepstral distortion between the observed signal with the single source component at the microphone and the output signals from (a) the conventional MSICA1 and the proposed MSICA or (b) the conventional MSICA2.

These values were averages of all of the combinations with respect to speakers and source directions. From these results, the following are revealed. First, in the conventional MSICA1 in which the NH-TDICA is used, the behavior of the NRR is not monotonic (see Fig. 2 (a)) and there is remarkably consistent deterioration. The MelCD of the conventional MSICA1 in the initial step of the iterative learning is superior, but, the MelCD degrades as the number of iterations is increased

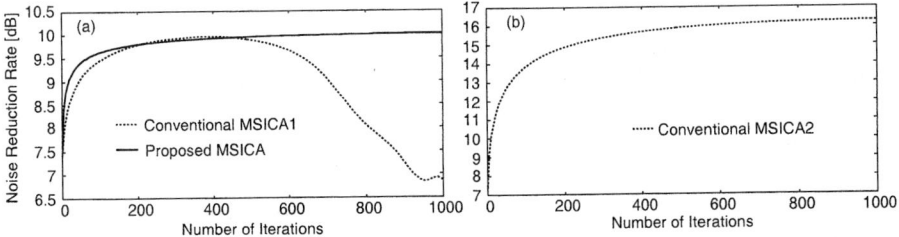

Fig. 2. Comparison of the noise reduction rates in (a) conventional MSICA1 and proposed MSICA, and (b) conventional MSICA2.

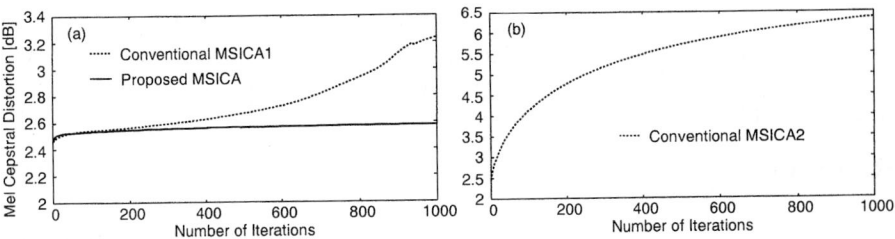

Fig. 3. Comparison of the mel cepstral distortion in (a) conventional MSICA1 and proposed MSICA, and (b) conventional MSICA2.

(see Fig. 3 (a)). Secondly, regarding the separation performance of the conventional MSICA2 in which the H-TDICA is used, the separation performance of the conventional MSICA2 is obviously superior to that of the proposed MSICA (see Fig. 2 (b)). However, the MelCD degrades as the number of iterations is increased (see Fig. 3 (b)). We speculate that the *specious* performance in MSICA2 is due to the exceeding emphasis of high-frequency components by the whitening effect of H-TDICA. In general, the separation in the high-frequency region is easier than that in low-frequency region [10] because the reverberation is shorter as the frequency increases. Thus, MSICA2 gains the improvement of the NRR only in the high-frequency region, and consequently we can conclude that MSICA2 is useless for separating the speech signals from the practical viewpoint.

On the other hand, in the proposed MSICA, there is no deterioration of NRR (see Fig. 3 (a)). Therefore, the separation performances are almost completely retained during all of the iterations and the proposed MSICA is effective for the stability in the iterative learning. Also, there are almost no degradation of the MelCD in the proposed MSICA than that of the conventional MSICA2. From these results, we can conclude that the proposed algorithm is effective for improving the stability of the learning.

5 Conclusion

We proposed a new algorithm with a stability and low-distortion for overdetermined BSS based on MSICA using subarray processing. In the proposed algorithm, to solve the problem of the stability, we perform TDICA with the holonomic constraint. Also, to avoid the distortions, we estimate the distortion components by TDICA with the holonomic constraint and we compensate the sound qualities by using the estimated components. The stability of the proposed algorithm can be guaranteed by the holonomic constraint, and the proposed compensation method prevents the distortion. The experimental results under a reverberant condition revealed that the proposed algorithm provides the higher stability and the higher separation performance, compared with the conventional MSICA including H-TDICA or NH-TDICA.

Acknowledgement

This work was partly supported by NISSAN MOTOR CO., LTD. in Japan and Core Research for Evolutional Science and Technology (CREST) Program "Advanced Media Technology for Everyday Living" of Japan Science and Technology Agency (JST).

References

1. Comon, P.: Independent component analysis, a new concept? In: Signal Processing. Volume 36. (1998) 287–314
2. Amari, S., Douglas, S., Cichocki, A., Yang, H.: Multichannel blind deconvolution and equalization using the natural gradient In: Proc. SPAWC97. (1997) 101–104
3. Murata, N., Ikeda, S.: An on-line algorithm for blind source separation on speech signals In: Proc. International Symposium on Nonlinear Theory and Its Application. (1998) 923–926
4. Nishikawa, T., Saruwatari, H., Shikano, K.: Blind source separation of acoustic signals based on multistage ICA combining frequency-domain ICA and time-domain ICA In: IEICE Trans. Fundamentals. **E86-A** (2003) 846–858
5. Araki, S., Mukai, R., Makino, S., Nishikawa, T., Saruwatari, H.: The fundamental limitation of frequency domain blind source separation for convolutive mixtures of speech In: IEEE Trans. Speech and Audio Processing. **11**, (2003) 109–116
6. Choi, S., Amari, S., Cichocki, A., Liu, R.: Natural gradient learning with a nonholonomic constraint for blind deconvolution of multiple channels. In: Proc. International Workshop on ICA and BSS. (1999) 371–376
7. Nishikawa, T., Saruwatari, H., Shikano, K.: Stable learning algorithm for blind separation of temporally correlated acoustic signals combining multistage ICA and Linear Prediction In: IEICE Trans. Fundamentals. **E86-A** (2003) 2028–2036
8. Nishikawa, T., Abe, H., Saruwatari, H., Shikano, K.: Overdetermined Blind Separation for Convolutive Mixtures of Speech Based on Multistage ICA using Subarray Processing In: Proc. ICASSP2004. (accepted). (2004)
9. Nakamura, S., Hiyane, K., Asano, F., Nishiura, T., Yamada, T.: Acoustical sound database in real environments for sound scene understanding and hands-free speech recognition In: Proc. International Conference on Language Resources and Evaluation. (2000) 965–968
10. Aichner, R., Araki, S., Makino, S., Nishikawa, T., Saruwatari, H.: Time domain ICA blind source separation of non-stationary convolved signals by utilizing geometric beamforming In: Proc. IEEE International Workshop on Neural Networks for Signal Processing (2002) 445–454

Two Channel, Block Adaptive Audio Separation Using the Cross Correlation of Time Frequency Information

Daniel Smith, Jason Lukasiak, and Ian Burnett

Whisper Laboratories, SECTE, University of Wollongong
dsmith@titr.uow.edu.au, {jasonl,i.burnett}@elec.uow.edu.au

Abstract. TIFCORR is a Blind Signal Separation technique that is well suited to separating audio signals, requiring each signal to be sparse in only a local time-frequency region of their representation [1]. TIFCORR can suffer from inconsistencies in mixing system estimation, thus we present a modified algorithm incorporating k-means clustering [2] to improve estimation robustness. To improve the data efficiency of TIFCORR, we also include an adaptive weighting function for mixing column estimates. These modifications transform our algorithm into a block adaptive algorithm with the ability to track time-varying mixtures.

1 Introduction

Blind Signal Separation (BSS) techniques attempt to separate unknown signals s, from observations x, which is the result of the signals s being mixed by system A. In addition to the observations x, conventional BSS techniques use Independent Component Analysis (ICA), requiring signals in the mixture to be statistically independent, non-Gaussian and stationary in order to separate them successfully [3]. In some cases, these assumptions are not satisfied by audio signals, thus BSS techniques that use an alternate mechanism have been considered. These are based upon the sparsity [1,4–6] or equivalently disjoint orthogonality [7,8] of the signal's time-frequency (T-F) representation. (i.e. signals that do not overlap in the T-F domain of the mixture). The techniques in [5–8] however, require a high degree of sparsity across the complete representation of all input signals to achieve robust separation. This requirement can be shown to be satisfied for speech signals in some approximate sense [9], but is not reliable across all mixtures (i.e. when formants of speech signal's have significant overlap) and weakens considerably as the number of signals in the mixture increase [9].

TIFCORR and TIFROM, introduced by [1] and [4], reduce such strong assumptions of sparsity, requiring input signals to be sparse in only localised T-F regions of the mixture's representation to achieve robust separation. TIFCORR and TIFROM utilise cross correlation and variance across a series of successive T-F windows respectively, to identify sparse T-F regions that correspond to mixing column estimates. Despite using different statistics to identify mixing columns, TIFCORR and TIFROM have a similar estimation structure that suffers from inconsistency where mixing column identification is biased towards signals that have T-F series that are highly sparse. As a consequence, mixing columns of signals with T-F series that are less sparse are often overlooked during estimation.

We recently addressed this estimation problem in TIFROM [10], modifying the framework to improve estimation quality and data efficiency. In this paper we show that similar modifications can also be successfully incorporated into TIFCORR. K-means clustering [2] is incorporated to improve the quality of mixing system estimation, while a weighted estimate improves TIFCORR's data efficiency, enhancing its ability to track time-varying mixing systems.

The paper is organised as follows. In Section 2, we introduce the TIFCORR algorithm and discuss in greater detail the specific limitations associated with its mixing system estimation and ways to remedy them with k-means clustering. In Section 3, k-means clustering modifications are detailed and experiments are conducted to compare the performance of our modified algorithm and TIFCORR in terms of mixing system estimation quality. We incorporate a block adaptive weighting into the modified algorithm and compare it's performance to TIFCORR with respect to both stationary and time-varying mixtures in Section 4. The conclusions of our work are presented in Section 5.

2 TIFCORR Approach to Separation [1]

Consider two linear instantaneous mixed observations of two real signals $s_j(n)$:

$$x_1(n) = a_{11}s_1(n) + a_{12}s_2(n) \tag{1}$$
$$x_2(n) = a_{21}s_1(n) + a_{22}s_2(n)$$

where a_{ij} are the mixing coefficients of the mixing system A. TIFCORR employs a simple approach to separation in this mixed system, by estimating each signal's mixing columns $C_1 = \frac{a11}{a21}$ and $C_2 = \frac{a12}{a22}$, under the following assumptions:

1. For each signal s_j, there exists at least a series (Υ_u, k) of time-adjacent T-F windows of mixed observations where either s_j occurs alone or where $s_j \gg s_i$. This will be referred to as TIFCORR's sparsity assumption. In addition, every T-F series (Υ_u, k) must always possess at least one signal.
2. In each series (Υ_u, k) of time-adjacent T-F windows the signals are uncorrelated.

The short-time Fourier transform (STFT) of the mixed observations $X_j(m, k)$, centered on short time window m and frequency k are computed. TIFCORR then computes the cross correlation across a series Υ_u of time-adjacent, centered T-F windows of the mixed observations $(X_{jc}(\Upsilon_u, k))$ obtaining the normalised cross correlation coefficient $cc(\Upsilon_u, k)$:

$$cc(\Upsilon_u, k) = \frac{E(X_{1c}(\Upsilon_u, k)X_{2c}(\Upsilon_u, k))}{\sqrt{E(X_{1c}(\Upsilon_u, k)X_{1c}(\Upsilon_u, k))E(X_{2c}(\Upsilon_u, k)X_{2c}(\Upsilon_u, k))}} \tag{2}$$

It is proven in [1] that under assumptions (1-2), when only a single signal is present in a T-F series, $cc(\Upsilon_u, k)=1$, and when both signal's are present $cc(\Upsilon_u, k) < 1$. The correlation coefficient associated with T-F series where $cc(\Upsilon_u, k)=1$ is:

$$ce(\Upsilon_u, k) = \frac{E(X_{1c}(\Upsilon_u, k)X_{1c}(\Upsilon_u, k))}{E(X_{1c}(\Upsilon_u, k)X_{2c}(\Upsilon_u, k))} = \frac{a_{1j}}{a_{2j}} \quad (3)$$

corresponding to the mixing column of the signal present in the series. Thus the TIFCORR identification process involves searching all T-F series, for the $cc(\Upsilon_u, k)$ closest to unity. The $ce(\Upsilon_u, k)$ of the series closest to unity becomes the first mixing column estimate C_{je}. The second mixing column estimate C_{ie} is found as the $ce(\Upsilon_u, k)$ of the series in the set $Q\epsilon|ce(\Upsilon_u, k) - C_{je}| > T$ where $cc(\Upsilon_u, k)$ is closest to unity. T is a threshold set to determine the minimum difference between the mixing columns. Both mixing columns are used to estimate the separation matrix A^{-1} as in [1].

2.1 Limitations of the TIFCORR Approach

Even under the condition that signals comply with TIFCORR's assumptions, inconsistency in mixing column estimation can be experienced. This is primarily due to the bias in identifying mixing column estimates, created by the different degrees to which signals comply with the sparsity assumption. The T-F series of C_{je} is sparser than the T-F series C_{ie}, and thus the normalised cross correlation measure $(cc(\Upsilon_u, k))$ of the C_{je} series will be closer to unity than $cc(\Upsilon_u, k)$ of the C_{ie} series. Thus, C_i is estimated incorrectly if the T-F series differ in their compliance with TIFCORR's sparsity assumption to the extent that the $cc(\Upsilon_u, k)$ of the C_{ie} estimate is further from unity than $cc(\Upsilon_u, k)$ of weak C_{je} estimates i.e. in the vicinity of $C_j \pm T$.

Estimation of the second mixing column therefore relies upon choosing a series that approaches unity $cc(\Upsilon_u, k)$, but more importantly, a series with a $ce(\Upsilon_u, k)$ that is distinct from the first mixing column estimate. K-means clustering allows series to be partitioned into distinct clusters in $ce(\Upsilon_u, k)$ space. Although clustering of mixing columns estimates was previously mentioned as a footnote in [1], our clustering approach differs as it does not derive each mixing column estimate "as an average of all identified occurences". Our approach employs a single $ce(\Upsilon_u, k)$ series from each cluster to estimate the corresponding mixing column. Implementation details of the k-means clustering will be discussed in Section 3.

3 New TIFCORR Architecture

To resolve the estimation problem discussed in Section 2.1, we modify the TIFCORR algorithm. The new architecture (TIFCmod) consists of the original TIFCORR algorithm, but includes the following modifications after computation of $ce(\Upsilon_u, k)$ in (3):

1. K-means clustering [2] is conducted on the correlation coefficients $ce(\Upsilon_u, k)$ and $cc(\Upsilon_u, k)$ belonging to the set of series S, which are partitioned into two clusters $P_o \in (P_1, P_2)$ such that:

$$(d_{av}(P_o, S))_{min} \tag{4}$$

where S is an $n \times 2$ matrix of the correlation coefficients for the n series where $|1 - cc(\Upsilon_u, k)| < c_{max}$ and d_{av} is the average distortion. The threshold c_{max} is an upper limit for determining which series have the level of sparsity required to be considered for mixing column estimation. In the case where k-means clustering attempts to perform outside its bound $n \leq 1$, or perform outside TIFCORR estimation bounds $|ce(P_1) - ce(P_2)| < T$, the threshold is increased $c_{max} = c_{max} + \triangle c_{max}$ until $n \geq 2$ and $|ce(P_1) - ce(P_2)| > T$.
2. The normalised correlation centroid $cc(P_o)$ of both clusters are examined. If $|1 - cc(P_o)| < c_{min}$, $C_{je} = ce(P_o)$, as c_{min} is a threshold that identifies series with a high level of sparsity so that estimation accuracy is ensured. Otherwise C_{je} is estimated as in the TIFCORR algorithm, but under the condition that if $C_{je} = ce(\Upsilon_u, k) \epsilon P_1$ then C_{ie} is estimated from the series belonging to the other cluster, that is $C_{ie} = ce(\Upsilon_u, k) \epsilon P_2$.

3.1 TIFCmod Results

To verify that TIFCmod displays improvement over TIFCORR for mixing matrix estimation, we apply the algorithms to 6 audio mixtures that are 2.5s in length and sampled at 8000 Hz. Each pair of audio signals were mixed by 24 different stationary mixing models. All mixtures were passed to the algorithms in data blocks sized:

$$\begin{aligned}blocksize &= overlap * framesize * (fps + 1) \\ &+ overlap * framesize * (seriesnum - 1)\end{aligned} \tag{5}$$

where the $framesize$ = 20ms, $overlap$ = 0.5 of a frame, number of adjacent frames per series (fps) = 6, 8 and number of series in each block of data $(seriesnum)$ = 1....180. Mixing ratio estimation and data block update are performed every 40ms. The threshold (T) is not given in [1, 4], but our empirical results indicate that a suitable value for T is 15% of the first ratio. For TIFCmod, the c_{min}, $\triangle c_{max}$ and c_{max} heuristics were obtained from an extensive empirical study of the cross correlation of mixing column estimates, with c_{min} = 0.000695, $\triangle c_{max}$ = 0.002 and c_{max} = 0.00466.

To measure the quality of each algorithm's mixing ratio estimation we used the Interference Measurement (IM) as a criteria:

$$IM = \frac{1}{2} \sum_{j=1}^{2} (p_j^T * p_j - max(p_j)^2)^{\frac{1}{2}} \tag{6}$$

where p is the product of the separation and mixing matrix, and p_j is a column of p. IM is a measure of mixing system identification, measuring p's average

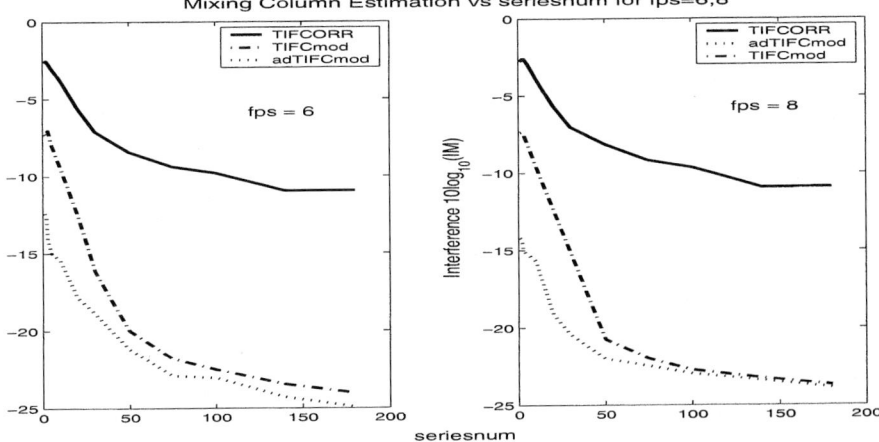

Fig. 1. The average ($10*log_{10}IM$) interference vs *seriesnum* for TIFCORR, TIFCmod and adTIFCmod, across 6 audio mixtures and 24 stationary mixing matrix.

distance from a scaled, permuted diagonal matrix corresponding to perfect estimation of the mixing channel. It is related to the measure used in [11].

Figure 1 shows TIFCORR and TIFCmod average log distortion ($10*log_{10}IM$) across six pairs of audio mixtures and 24 stationary mixtures with respect to *seriesnum* for $fps = 6, 8$. Although TIFCmod outperforms TIFCORR for mixing column estimation across all *seriesnum*, it is evident that TIFCmod increases it's estimation advantage over TIFCORR as *seriesnum* increases. This is because k-means clustering of TIFCmod only produces significant estimation improvement if it possesses series that represent all mixing columns, providing an accurate basis for clustering. In general, a data block with a larger number of *seriesnum* will possess a greater number of T-F series that are highly sparse. This provides TIFCmod with a better representation of the mixing column space, further improving upon TIFCORR's mixing system estimation.

4 Adaptive Block Based TIFCORR

Figure 1 indicates TIFCmod's estimation performance decreases, approaching TIFCORR's performance as the number of *serienum* decrease. With smaller block sizes, the number of T-F series that comply with TIFCORR's sparsity assumption are scarce and thus clustering is conducted across a more erroneous representation of the mixing column space. To reduce the influence of poor mixing system estimates, we propose a weighting (running average) function that uses the confidence of the estimate to determine the update weight for the mixing columns. As we can measure our confidence in the accuracy of the $C_{je}(t)$ estimate from its cross correlation (cc_{je}), the weighting function we utilise is:

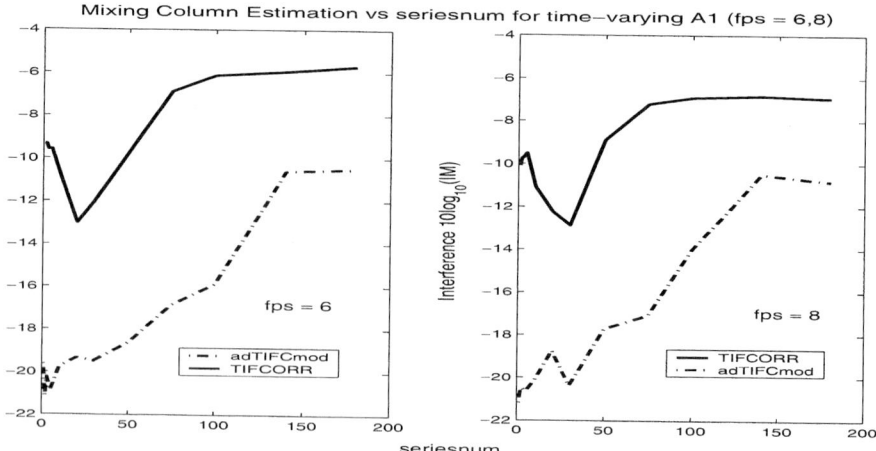

Fig. 2. The average ($10 * log_{10} IM$) interference vs *seriesnum* for TIFCORR and adTIFCmod, across 6 audio mixtures and the time varying mixture $A1$.

$$C_{jwe}(t) = C_{je}(t) \quad if \ cc_{je} \leq c_{min} \quad (7)$$
$$C_{jwe}(t) = (1 - cw) * C_{jwe}(t-1) + cw * C_{je}(t)$$
$$if \ c_{min} < cc_{je} < c_{max}$$
$$C_{jwe}(t) = C_{jwe}(t-1) \quad if \ cc_{je} \geq c_{max}$$

where $cw = \frac{c_{max} - cc_{je}}{c_{max} - c_{min}}$. Poor estimates of $C_{je}(t)$ are thus penalised or excluded in $C_{jwe}(t)$. The k-means clustering (Section 3) also uses these weighted estimates as initial conditions for the next block estimate. The block adaptive algorithm (adTIFCmod) combines weighted estimates and k-means clustering[1].

4.1 adTIFCmod Results

The experiment from Section (3.1) was repeated for the adaptive algorithm (adTIFCmod). Figure 1 shows that adTIFCmod achieves improvement upon TIFCmod for mixing column estimation across all *seriesnum*, and in particular, much better estimation quality for smaller data blocks, as adTIFCmod's average IM advantage (for $fps = 6,8$) of 0.45dB for *seriesnum* > 50 increases to 6.1dB for *seriesnum* < 10. Therefore the adTIFCmod algorithm offers far superior performance to TIFCORR across all *seriesnum*, but most importantly, an improved data efficiency that highlights the potential adTIFCmod has to estimate time varying mixtures.

A second experiment was conducted to demonstrate that the adTIFCmod algorithm has the ability to track a time-varying mixture and offer superior performance to TIFCORR. Both algorithms were applied to the same six pairs of

---

[1] TIFROM estimation also suffers as a result of signals having weak, differing levels of compliance to the sparsity assumption, thus the modifications of adTIFCmod can be applied in an equivalent way.

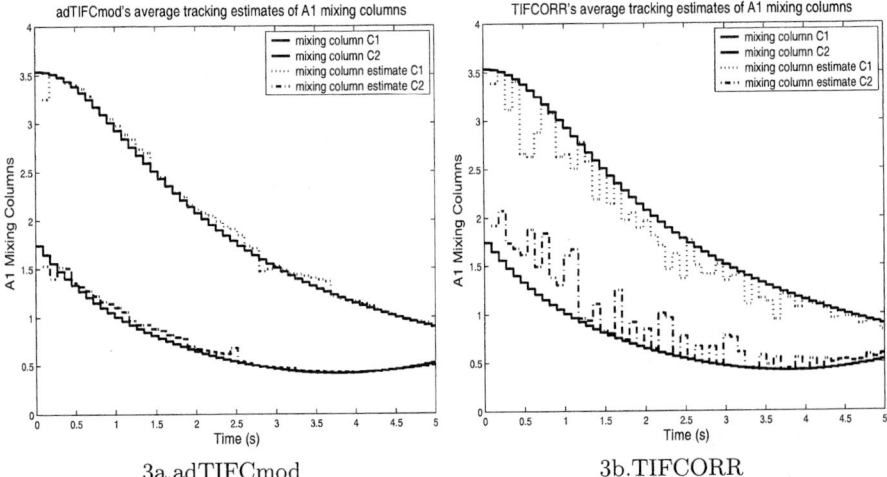

Fig. 3. adTIFCmod and TIFCORR's average tracking estimates of the $A1$ mixture across 6 audio mixtures for $fps = 8$ and $seriesnum = 2$. Estimates of $A1$ are updated every 90ms.

audio signals from the experiment in Section (3.1), however signals were extended to 5s in length. The $A1$ mixing system was generated at 90ms intervals under the assumption that $A1$ varies inversely with the distance of each sensor to source. Both sources move in a circular path, at a constant velocity of 4ms clockwise and 2ms anticlockwise, around two sensors situated 2m to the left and right of the circle center. The actual mixing columns of $A1$ are shown as the solid lines of Figure 3a and 3b. The T, c_{max}, Δc_{max} and c_{min} heuristics, $framesize$, $overlap$, $seriesnum$ range and fps were the same as in the previous experiment. Mixing column estimates and data blocks were updated every 90ms.

Figure 2 shows TIFCORR's and adTIFCmod's average log distortion ($10 * log_{10} IM$) across six audio mixtures for the mixing system $A1$ in relation to $seriesnum$ and $fps = 6, 8$. Figure 2 indicates that in tracking $A1$, adTIFCmod outperforms TIFCORR across the range of $seriesnum$, increasing it's average IM advantage (for $fps = 6, 8$) from 3.5dB to 10.5dB as $seriesnum$ decreases. We attribute this to:

1. AdTIFCmod's superior data efficiency to TIFCORR. TIFCORR estimation is poor for smaller data blocks ($1 \leq seriesnum \leq 10$) as there are few or no T-F series corresponding to mixing columns. The weighted estimates of adTIFCmod however, reduce the influence of poor estimates allowing the $A1$ mixing system to be successfully traced with a minimal number of series ($1 \leq seriesnum \leq 10$).
2. The decrease in adTIFCmod's estimation advantage over TIFCORR, as data blocks increase in size from 90ms i.e. $seriesnum > 3$ for $fps = 6$ and $seriesnum > 1$ for $fps = 8$. As the data blocks become larger than 90ms, they span at least one mixing matrix change, making TIFCORR and adTIFCmod estimation of the current mixture inaccurate.

The adTIFCmod and TIFCORR algorithm's average tracking estimates of the $A1$ columns (for the 6 audio pairs) are shown as the dotted lines of Figure 3a and Figure 3b respectively. Figure 3 illustrates adTIFCmod's superior performance for tracking the time-varying system $A1$ with a small number of series ($seriesnum = 2$ and $fps = 8$). The average adTIFCmod estimates trace the $A1$ mixing columns with greater accuracy than the average TIFCORR estimates that are oscillatory across their tracking path. In addition, TIFCORR mixing columns of some audio pairs were estimated as different versions of the same mixing column, due to the estimation inconsistency that was discussed in Section 2.1. The adTIFCmod algorithm overcomes this problem, through k-means clustering, or in the presence of a poor estimate, a weighting function that reduces its influence in the current block.

5 Conclusion

The TIFCORR framework was modified to resolve inconsistencies regarding mixing column estimation. As a consequence our algorithm adTIFCmod, was shown to offer significant improvements in estimation performance compared to the original algorithm, with an average IM improvement of 12.4dB across all fps and $seriesnum$ for stationary mixing systems. In addition, the improved data efficiency of adTIFCmod enabled us to demonstrate that our architecture could operate in real time, tracking a time-varying instantaneous mixture.

References

1. Y.Deville: Temporal and time frequency correlation based blind source separation methods. In: Proc.4th International Symposium on Independent Component Analysis and Blind Signal Separation (ICA 2003). (2003) 1059–1064
2. A.Gersho, R.Gray: Vector Quantization and Signal Compression. Kluwer Academic Publishers (1992)
3. A.Hyvarinen, J.Karhunen, E.Oja: Independent Component Analysis. John Wiley & Sons (2001)
4. F.Abrard, Y.Deville, P.White: From blind source separation to blind source cancellation in the undetermined case: A new approach based on time-frequency analysis. In: Proc.3rd International Conference on Independent Component Analysis and Blind Source Separation (ICA2001). (2001) 734–739
5. C.Choi: Real time binaural blind source separation. In: Proc.4th International Symposium on Independent Component Analysis and Blind Signal Separation (ICA 2003). (2003) 567–572
6. P.Bofill, M.Zibulevsky: Undetermined blind source separation using sparse representations. Signal Processing **81** (2001) 2353–2362
7. R.Balan, J.Rosca, S.Rickard: Scalable non-square blind source separation in the presence of noise. In: Proc.IEEE International Conference on Acoustics, Speech and Signal Signal Processing (ICASSP2003). Volume 5. (2003) 293–296
8. A.Jourjine, S.Rickard, O.Yilmaz: Blind separation of disjoint orthogonal signals: Demixing n sources from 2 mixtures. In: Proc.IEEE Conference on Acoustics, Speech, and Signal Processing (ICASSP2000). Volume 5. (2000) 2985–2988

9. S.Rickard, O.Yilmaz: On the w-disjoint orthogonality of speech. In: Proc.IEEE International Conference on Acoustics, Speech, and Signal Processing (ICASSP2002). Volume 1. (2002) 529–532
10. D.Smith, J.Lukasiak, I.Burnett: A block-adaptive audio separation technique based upon time-frequency information. In: Submitted to EUSIPCO-2004. (2004)
11. A.Cichoki, S.Amari: Adaptive Blind Signal and Image Processing: Learning Algorithms and Applications. John Wiley & Sons (2002)

Underdetermined Blind Separation of Convolutive Mixtures of Speech with Directivity Pattern Based Mask and ICA

Shoko Araki, Shoji Makino, Hiroshi Sawada, and Ryo Mukai

NTT Communication Science Laboratories, NTT Corporation
2-4 Hikaridai, Seika-cho, Soraku-gun, Kyoto 619-0237, Japan
{shoko,maki,sawada,ryo}@cslab.kecl.ntt.co.jp

Abstract. We propose a method for separating N speech signals with M sensors where $N > M$. Some existing methods employ binary masks to extract the signals, and therefore, the extracted signals contain loud musical noise. To overcome this problem, we propose using a directivity pattern based continuous mask, which masks $N - M$ sources in the observations, and independent component analysis (ICA) to separate the remaining mixtures. We conducted experiments for $N = 3$ with $M = 2$ and $N = 4$ with $M = 2$, and obtained separated signals with little distortion.

1 Introduction

In this paper, we consider the blind source separation (BSS) of speech signals observed in a real environment, i.e., the BSS of convolutive mixtures of speech. Recently, many methods have been proposed to solve the BSS problem of convolutive mixtures [1]. However, most of these methods consider the determined or overdetermined case. In contrast, we focus on the underdetermined BSS problem where the N source signals outnumber M sensors.

Several methods have been proposed for underdetermined BSS [2–5]. There are two approaches, and both approaches rely on the sparseness of the source signals. One extracts each signals with time-frequency binary masks [2], and the other is based on ML estimation, where the sources are estimated after mixing matrix estimation [3–5]. In [2], the authors employ a time-frequency binary mask (BM) to extract each signal, and they have applied it to real speech mixtures. However, the use of binary masks causes too much discontinuous zero-padding to the extracted signals, and they contain loud musical noise.

To overcome this, we have proposed combining binary masks and ICA (BMICA) to solve the underdetermined BSS problem [6] especially for $N = 3$ and $M = 2$. This method consists of two stages: (1) one source removal with a binary mask and (2) separation of the remaining mixtures with ICA (for details see Sec. 3.3). As this one source removal extracts more time-frequency points than the BM method, it causes less zero-padding than the BM method, and therefore, we have been able to separate signals with less musical noise. However, the BMICA still employs a binary mask.

Therefore we have also proposed to utilize a directivity pattern based continuous mask (DCmask) instead of a binary mask at the source removal stage (DCmask and ICA: DCICA) [7]. The DCmask has a small gain for the DOAs of sources to be masked, and has a large gain for other directions. Because the DCmask is a non-binary mask, we can avoid the zero-padding. However, in [7], as we masked at most $M-1$ sources, we applied the DCICA only for $N \leq (M-1)+M$.

In this paper, to release this limit, we propose a method for masking $N - M$ sources for an arbitrary number of sources N. Our proposal is to utilize the directivity pattern of a null beamformer (NBF), which makes nulls towards given $N - M$ directions, formed by $V = N - M + 1$ *virtual* microphones. We conducted experiments for $N = 3$ with $M = 2$ and $N = 4$ with $M = 2$, and the experimental results show that our method can separate signals with little distortion.

2 Problem Description

In real environments, N source signals s_i observed by M sensors are modeled as convolutive mixtures $x_j(n) = \sum_{i=1}^{N} \sum_{l=1}^{L} h_{ji}(l) \, s_i(n-l+1)$ $(j = 1, \cdots, M)$, where h_{ji} is the L-taps impulse response from a source i to a sensor j. Our goal is to obtain separated signals $y_k(n)$ $(k = 1, \cdots, N)$ using only the information provided by observations $x_j(n)$. Here, we consider the case of $N > M$.

This paper employs a time-frequency domain approach because speech signals are more sparse in the time-frequency domain than in the time-domain [5] and convolutive mixture problems can be converted into instantaneous mixture problems at each frequency. In the time-frequency domain, mixtures are modeled as $\mathbf{X}(\omega, m) = \mathbf{H}(\omega)\mathbf{S}(\omega, m)$, where $\mathbf{H}(\omega)$ is an $M \times N$ mixing matrix whose ji component is a transfer function from a source i to a sensor j, $\mathbf{S}(\omega, m) = [S_1(\omega, m), \cdots, S_N(\omega, m)]^T$ and $\mathbf{X}(\omega, m) = [X_1(\omega, m), \cdots, X_M(\omega, m)]^T$ denote short-time Fourier transformed sources and observed signals, respectively. ω is the frequency and m is the time-dependence of the short-time Fourier transformation (STFT). We assume that sources are mutually independent and that each source has a sparse distribution in a time-frequency domain. These assumptions are approximately true for speech signals. Moreover, $\mathbf{Y}(\omega, m) = [Y_1(\omega, m), \cdots, Y_N(\omega, m)]^T$ denotes the STFT of separated signals.

3 Conventional Methods

3.1 Classification of Time-Frequency Points with Sparseness

Several methods have been proposed [2–8] for solving the underdetermined BSS problem, and they all utilize source sparseness. When signals are sufficiently sparse, it can be assumed that sources do not overlap very often. Therefore, a histogram of ($\frac{|X_i(\omega,m)|}{|X_j(\omega,m)|}, \angle \frac{X_i(\omega,m)}{X_j(\omega,m)}$) ($i \neq j$) for example, contains N peaks. Furthermore, we can classify the observation sample points $X_j(\omega, m)$ into N classes according to the histogram, which is what the BM method does (see Sec. 3.2).

In this paper, we utilize omnidirectional microphones, therefore we use the phase difference $\varphi(\omega, m) = \angle \frac{X_i(\omega,m)}{X_j(\omega,m)}$ ($i \neq j$) between two observations. A histogram of the direction of arrival (DOA) $\theta(\omega, m) = \cos^{-1} \frac{\varphi(\omega,m)c}{\omega d}$ (d: the microphone space, c: the speed of sound) has N peaks (Fig. 1). Each peak corresponds to each source. Let these peaks be $\tilde{\theta}_1, \tilde{\theta}_2, \cdots, \tilde{\theta}_N$ where $\tilde{\theta}_1 \leq \tilde{\theta}_2 \leq \cdots \leq \tilde{\theta}_N$ (Fig. 1), and the signal from $\tilde{\theta}_\xi$ be $\tilde{S}_\xi$ ($\xi = 1, \cdots, N$).

3.2 Conventional Method 1: With Only Binary Masks (BM)

As alluded to in Sec. 3.1, we can extract each signal using time-frequency binary masks (e.g., [2]). We can extract each signal with a binary mask

$$[\text{BM}] \quad M_{\text{BM}}^\xi(\omega, m) = \begin{cases} 1 & \tilde{\theta}_\xi - \Delta \leq \theta(\omega, m) \leq \tilde{\theta}_\xi + \Delta \\ 0 & \text{otherwise} \end{cases} \quad (1)$$

by calculating $Y_\xi(\omega, m) = M_{\text{BM}}^\xi(\omega, m) X_j(\omega, m)$ where Δ is an extraction range parameter.

Although we can obtain separated signals with binary masks (1), the signals are discontinuously zero-padded by binary masks, and therefore, we hear musical noise in the outputs. Moreover, the performance depends on the parameter, Δ.

3.3 Conventional Method 2: With Binary Mask and ICA (BMICA)

To overcome the musical noise problem, we have proposed using both a binary mask and ICA (BMICA) [6]. The BMICA has two stages. At the first stage, using the sparseness assumption, we *remove* the $N - M$ sources from the observations with a binary mask. Then in the second stage, we apply ICA to the remaining mixtures to obtain M separated signals.

Let $\Theta_S = \{\tilde{\theta}_{s(1)}, \cdots, \tilde{\theta}_{s(M)}\}$ be the set of DOAs of M signals to be separated and $\Theta_R = \{\tilde{\theta}_{r(1)}, \cdots, \tilde{\theta}_{r(N-M)}\}$ be the set of DOAs of $N - M$ signals to be removed (Fig. 1). To define the masks, let $\mathcal{I}_S = \{\mathsf{s}(1), \cdots, \mathsf{s}(M)\}$ be the set of indexes of Θ_S and $\mathcal{I}_R = \{\mathsf{r}(1), \cdots, \mathsf{r}(N - M)\}$ be the set of indexes of Θ_R.

For an index set $\mathcal{I}$, we define an area $\mathbb{A}$ by the following procedure:

1. $\mathbb{A} \leftarrow \emptyset$
2. if $1 \in \mathcal{I}$, $\mathbb{A} \leftarrow \mathbb{A} \cup [0°, \tilde{\theta}_1]$
3. if $N \in \mathcal{I}$, $\mathbb{A} \leftarrow \mathbb{A} \cup [\tilde{\theta}_N, 180°]$
4. for every index i such that $i \in \mathcal{I}$ and $i + 1 \in \mathcal{I}$, $\mathbb{A} \leftarrow \mathbb{A} \cup [\tilde{\theta}_i, \tilde{\theta}_{i+1}]$

We define the separation area $\mathbb{A}_S$ by using $\mathcal{I}_S$, and the removal area $\mathbb{A}_R$ by using $\mathcal{I}_R$. We also define the transition area $\mathbb{A}_T = \bar{\mathbb{A}}_S \cap \bar{\mathbb{A}}_R$ (Fig. 1).

In the first stage, unlike the BM method where each source is extracted, we attempt to *remove* $N - M$ sources from Θ_R using a binary mask

$$[\text{BMICA}] \quad M_{\text{BMICA}}(\omega, m) = \begin{cases} 1 & \theta(\omega, m) \in \mathbb{A}'_S \\ 0 & \text{otherwise} \end{cases} \quad (2)$$

by calculating $\hat{\mathbf{X}}(\omega, m) = M_{\text{BMICA}}(\omega, m) \mathbf{X}(\omega, m)$, where $\mathbb{A}'_S = \mathbb{A}' \cup \mathbb{A}_S$, $\mathbb{A}' = \bigcup_{1 \leq i \leq M} [\tilde{\theta}_{\mathsf{s}(i)} - \Delta, \tilde{\theta}_{\mathsf{s}(i)} + \Delta]$ and Δ is an extraction range parameter. Here,

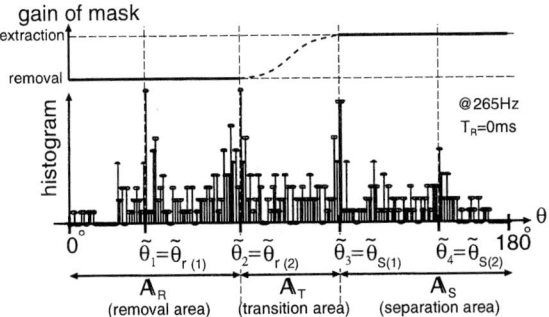

Fig. 1. Example histogram. ($N = 4$. Two male and two female combination with STFT frame size $T = 512$. $T_R = 0$ ms). An example of the area definition is also drawn for $N=4$, $M=2$. Here $\Theta_S = \{\tilde{\theta}_3, \tilde{\theta}_4\}$ and $\Theta_R = \{\tilde{\theta}_1, \tilde{\theta}_2\}$. Signals from $\tilde{\theta}_1$ and $\tilde{\theta}_2$ are masked in the 1st stage, and signals from $\tilde{\theta}_3$ and $\tilde{\theta}_4$ are separated in the 2nd stage

$\hat{\mathbf{X}}(\omega, m)$ are expected to be mixtures of M signals from Θ_S. Therefore, in the second stage, we apply a standard ICA to these remaining mixtures.

We expect the zero-padding of the separated signals to cause less trouble because we extract more time-frequency points at the 1st stage than with the BM method. However, as BMICA still employed a binary mask for the source removal, the zero-padding to the separated signals still remained. Moreover, we have to find a reasonable Δ. This is not an easy problem and we relied on manual setting.

4 Proposed Method: Directivity Pattern Based Continuous Mask and ICA (DCICA)

Although the basic scheme (Fig. 2) of our proposed method is the same as that of BMICA, here we utilize non-binary masks at the 1st stage.

[1st Stage] $N - M$ Source Removal with New DC Mask: Here, we utilize a directivity pattern based continuous mask (DCmask) instead of a binary mask M_{BMICA}. When we have M microphones, we can utilize $M \times M$ ICA at the 2nd stage if we can mask $N - M$ signals. This can be realized by applying a mask that has $N - M$ nulls towards the DOAs Θ_R of the signals to be removed.

One way to obtain such a mask is to utilize the directivity pattern of a null beamformer (NBF), which makes nulls towards given $N - M$ directions Θ_R, formed by $V = N - M + 1$ (virtual) microphones. Here, V is not necessarily equal to M because *a mask is determined only by the number of signals to be removed at the 1st stage*: remember that we do not need information on the microphone number M when designing the masks for BM and BMICA methods.

Here, we assume that the number of sources N is known or estimated beforehand, e.g., from a histogram such as that shown in Fig. 1. First we form a $(V \times V)$ matrix $\mathbf{H}_{\text{NBF}}(\omega)$ whose ji element $H_{\text{NBF}ji}(\omega) = \exp(j\omega\tau_{ji})$, where

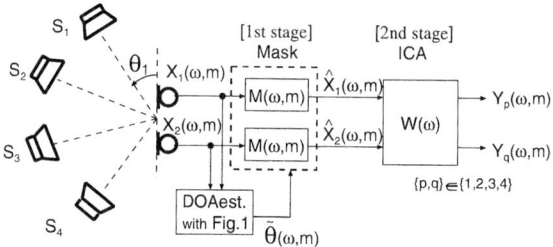

Fig. 2. Block diagram of proposed method. $N=4$ and $M=2$ case is drawn for example

$\tau_{ji} = \frac{d_j}{c}\cos\tilde{\theta}_i$, d_j is the position of the j-th virtual microphone, c is the speed of sound, $\{\tilde{\theta}_i\ (i=2,\cdots,V)\} = \Theta_R$, and $\tilde{\theta}_1 = \theta_c \notin \mathbb{A}_R$ from which the signal's gain and phase are constrained at a constant value. By making a $(V \times V)$ matrix $\mathbf{H}_{\mathrm{NBF}}(\omega)$, we can remove $N-M$ signals even if $N >$ (the number of nulls formed by M sensors) + (the number of outputs of a standard ICA)= $(M-1)+M$.

Then one of the directivity patterns of the NBF, $\mathbf{W}(\omega) = \mathbf{H}_{\mathrm{NBF}}^{-1}(\omega)$, is

$$F(\omega,\theta) = \sum_{k=1}^{V} W_{1k}(\omega)\exp\left(j\omega d_k \cos\theta/c\right). \tag{3}$$

In this paper, we use the directivity pattern of the NBF as our mask,

[DCICA 1] $M_{\mathrm{DC1}}(\omega,m) = F(\omega,\theta(\omega,m))$. (4)

This is our new mask, the DCmask. Figure 3 shows an example of the gain pattern of a DCmask.

We can also use a modified directivity pattern, for example,

[DCICA 2] $M_{\mathrm{DC2}}(\omega,m) = \begin{cases} c_s & \theta(\omega,m) \in \mathbb{A}_S \\ F(\omega,\theta(\omega,m)) & \theta(\omega,m) \in \mathbb{A}_T \\ c_r & \theta(\omega,m) \in \mathbb{A}_R \end{cases}$ (5)

where c_s is a constant (e.g., $\min_{\tilde{\theta}_i \in \Theta_s}|F(\omega,\tilde{\theta}_i)|$) and c_r is a small constant (e.g., the minimum value of the directivity pattern). By the mask M_{DC2}, the constant gain c_s is given to the M signals in the area $\mathbb{A}_S$. Moreover, this M_{DC2} changes smoothly in the transition area $\mathbb{A}_T$.

The source removal is achieved by $\hat{\mathbf{X}}(\omega,m) = M_{\mathrm{DC}k}(\omega,m)\mathbf{X}(\omega,m)$ ($k=1$ or 2). It should be noted that the DCmask is applied to all channels (Fig. 2), because ICA in the 2nd stage needs M inputs that maintain the mixing matrix information.

Because M_{DC1} and M_{DC2} are spatially smooth in the transition area $\mathbb{A}_T$, it is expected that the discontinuity of the extracted signals by these DCmasks is less serious than that by a mask M_{BM} in the BMICA.

[2nd Stage] Separation of Remaining Sources by ICA: Because the remaining signals $\hat{\mathbf{X}}$ are expected to be mixtures of M signals, we separate the signals using $M \times M$ ICA. The separation process is formulated as

$$\mathbf{Y}(\omega,m) = \mathbf{W}(\omega)\hat{\mathbf{X}}(\omega,m), \tag{6}$$

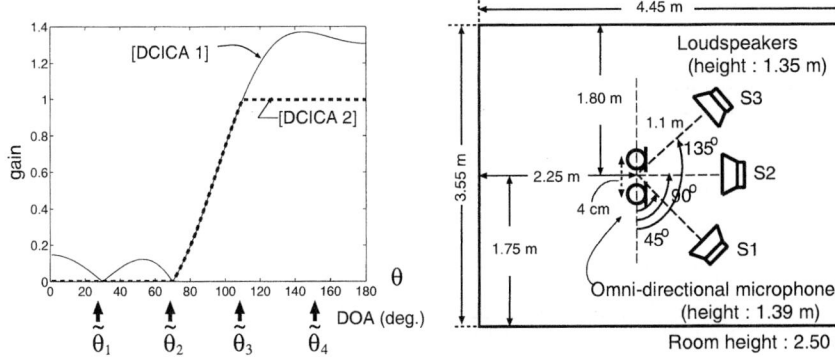

Fig. 3. Example mask pattern

Fig. 4. Room for reverberant tests. $T_R = 130$ ms

where $\hat{\mathbf{X}}$ is the masked observed signal, $\mathbf{Y}(\omega, m) = [Y_1(\omega, m), \cdots, Y_M(\omega, m)]^T$ is the separated output signal, and $\mathbf{W}(\omega)$ represents an $(M \times M)$ separation matrix. $\mathbf{W}(\omega)$ is determined so that the output signals become mutually independent.

Note that we need several masks with nulls towards different directions to obtain all N separated signals because our system has only M outputs.

5 Experiments

5.1 Experimental Conditions

We conducted anechoic tests and reverberant tests. For the anechoic tests ($T_R = 0$ ms), we mixed speech signals using the mixing matrix $H_{ji}(\omega) = \exp(j\omega\tau_{ji})$, where $\tau_{ji} = \frac{d_j}{c}\cos\theta_i$, d_j is the position of the j-th microphone, and θ_i is the direction of the i-th source. The source directions were 45°, 90° and 135° ($N=3$), and 30°, 70°, 90° and 150° ($N=4$). For the reverberant tests, the speech data was convolved with impulse responses recorded in a real room (Fig. 4) whose reverberation time was $T_R = 130$ ms. As the original speech, we used Japanese sentences spoken by male and female speakers. We investigated three combinations of speakers.

The STFT frame size T was 512 and the frame shift was 256 at a sampling rate of 8 kHz. The Δ value for the conventional methods was 15° in DOA ($N = 3$) 10° in DOA ($N = 4$).

The adaptation rule of ICA we used was $\mathbf{W}_{i+1}(\omega) = \mathbf{W}_i(\omega) + \eta \left[\mathbf{I} - \langle \Phi(\mathbf{Y}) \mathbf{Y}^H \rangle \right] \cdot \mathbf{W}_i(\omega)$, where $\Phi(\mathbf{y}) = \phi(|\mathbf{y}|) \cdot e^{j \cdot \angle(\mathbf{y})}$, $\phi(x) = \text{sign}(x)$. To solve the permutation problem of frequency domain ICA, we employed the DOA and correlation approach [9], and to solve the scaling problem of frequency domain ICA, we used the minimum distortion principle [10].

5.2 Performance Measures

We used the signal to interference ratio (SIR) and the signal to distortion ratio (SDR) as measures of separation performance and sound quality, respectively:

Table 1. Results of $N = 3$, $M = 2$ simulations. (a) T_R=0ms, (b) T_R=130ms

(a)

pq	SIR1	SIR2	SIR3	SDR1	SDR2	SDR3
BM	18.0	8.9	18.4	7.9	11.5	8.3
BMICA 12	12.6	5.9		18.1	15.2	
BMICA 23		6.1	13.0		13.6	17.4
BMICA 13	16.9		16.4	11.7		11.7
DCICA1 12	16.2	4.9		15.2	13.1	
DCICA1 23		4.6	16.3		13.2	15.6
DCICA1 13	18.2		18.7	11.3		11.9
DCICA2 12	12.7	5.8		19.0	16.3	
DCICA2 23		5.6	13.0		15.9	18.0

pq: $\Theta_s = \{\tilde{\theta}_p, \tilde{\theta}_q\}$ [dB]

(b)

pq	SIR1	SIR2	SIR3	SDR1	SDR2	SDR3
BM	12.3	6.3	11.0	5.0	13.9	5.8
BMICA 12	9.8	5.5		7.8	15.9	
BMICA 23		5.5	9.2		14.5	9.3
BMICA 13	11.9		12.5	6.9		7.2
DCICA1 12	13.6	4.1		7.0	11.2	
DCICA1 23		3.9	11.7		14.4	8.6
DCICA1 13	10.0		11.3	5.6		8.0
DCICA2 12	10.9	5.1		8.3	13.9	
DCICA2 23		4.5	8.7		16.3	9.2

pq: $\Theta_s = \{\tilde{\theta}_p, \tilde{\theta}_q\}$ [dB]

$$\text{SIR}_i = 10 \log \frac{\sum_n y_{is_i}^2(n)}{\sum_n (\sum_{i \neq j} y_{is_j}(n))^2} \text{ and } \text{SDR}_i = 10 \log \frac{\sum_n x_{ks_i}^2(n)}{\sum_n (x_{ks_i}(n) - \alpha y_{is_i}(n-D))^2},$$

where y_i is the estimation of s_i, and y_{is_j} is the output of the whole separating system at y_i when only s_j is active, and $x_{ks_i} = h_{ki} * s_i$ ($*$ is a convolution operator). α and D are parameters to compensate for the amplitude and phase difference between x_{ks_i} and y_{is_i}.

The SIR and SDR values were averaged over three speaker combinations.

5.3 Experimental Results

Applicability of ICA at the 2nd Stage Before trying to separate signals with our method, we investigated the masking performance. The percentage of each signal power extracted by M_{DC1} was $S_1:S_2:S_3:S_4 = 78:20:1:1$, and by M_{DC2} was 50:47:2:1 (N=4 (all female), M=2, T_R= 0ms, $\Theta_S = \{\tilde{\theta}_1, \tilde{\theta}_2\}$), for example. Two signals are dominant and other two signals are small. Therefore, we can use (2×2) ICA at the 2nd stage.

Separation results Table 1 (a) shows the experimental results for $T_R = 0$ ms and $N = 3, M = 2$. With BM method, the SDR values were unsatisfactory, and a large musical noise was heard. In contrast, with our proposed method (DCICA), we were able to obtain high SDR values without any serious deterioration in the separation performance SIR. Although the SDR values were slightly degraded compared with those by BMICA, we heard no musical noise with DCICA. Some sound samples can be found at our web site [11].

In DCICA1, SIR2 was degraded. This is because the gain for $\tilde{\theta}_2$ was less than the gain for $\tilde{\theta}_1$ or $\tilde{\theta}_3$. In DCICA2, which had constant gains for $\tilde{\theta}_2$ and $\tilde{\theta}_1$ or $\tilde{\theta}_3$, the SIR2 was improved and we obtained high SDR values.

Tables 2 shows the results for $N = 4$ and $M = 2$. We can apply our method for $N = 4$.

Table 1 (b) shows the results of reverberant tests for $T_R = 130$ ms ($N = 3, M = 2$). In the reverberant case, due to the decline of sparseness, the performance with all methods was worse than when $T_R = 0$ ms. However, we were

Table 2. Results of $N = 4$, $M = 2$ simulations. T_R=0ms

pq	SIR1	SIR2	SIR3	SIR4	SDR1	SDR2	SDR3	SDR4
BM	16.7	9.6	7.7	16.7	4.4	7.1	7.5	4.7
BMICA 12	11.3	6.7			8.9	9.2		
BMICA 34			5.4	10.5			9.3	10.1
DCICA1 12	14.1	3.4			9.2	7.7		
DCICA1 34			3.6	14.3			8.8	9.7
DCICA2 12	10.9	5.4			10.7	11.3		
DCICA2 34			4.4	9.8			11.2	12.2

pq: $\Theta_s = \{\tilde{\theta}_p, \tilde{\theta}_q\}$ [dB]

able to obtain higher SDR values with DCICA than with the BM method even in a reverberant environment without musical noise.

It should be noted that it remains difficult to separate signals at the center position with any method.

6 Conclusion

We proposed utilizing a directivity pattern based continuous mask and ICA for BSS when speech signals outnumber sensors. Our method avoids discontinuous zero-padding, and therefore, can separate the signals with no musical noise.

References

1. Haykin, S.: Unsupervised adaptive filtering. John Wiley & Sons (2000)
2. Rickard, S., Yilmaz, O.: On the W-disjoint orthogonality of speech. In: Proc. ICASSP2002. (2002) 529–532
3. Theis, F.J., Puntonet, C.G., Lang, E.W.: A histogram-based overcomplete ICA algorithm. In: Proc. ICA2003. (2003) 1071–1076
4. Vielva, L., Erdogmus, D., Pantaleon, C., Santamaria, I., Pereda, J., Principe, J.C.: Underdetermined blind source separation in a time-varying environment. In: Proc. ICASSP2002. (2002) 3049–3052
5. Bofill, P., Zibulevsky, M.: Blind separation of more sources than mixtures using sparsity of their short-time Fourier transform. In: Proc. ICA2000. (2000) 87–92
6. Araki, S., Makino, S., Blin, A., Mukai, R., Sawada, H.: Blind separation of more speech than sensors with less distortion by combining sparseness and ICA. In: Proc. IWAENC2003. (2003) 271–274
7. Araki, S., Makino, S., Sawada, H., Mukai, R.: Underdetermined blind speech separation with directivity pattern based continuous mask and ICA. In: EUSIPCO2004. (2004)
8. Blin, A., Araki, S., Makino, S.: Blind source separation when speech signals outnumber sensors using a sparseness-mixing matrix combination. In: Proc. IWAENC2003. (2003) 211–214
9. Sawada, H., Mukai, R., Araki, S., Makino, S.: Convolutive blind source separation for more than two sources in the frequency domain. In: Proc. ICASSP2004. (2004)
10. Matsuoka, K., Nakashima, S.: A robust algorithm for blind separation of convolutive mixture of sources. In: Proc. ICA2003. (2003) 927–932
11. http://www.kecl.ntt.co.jp/icl/signal/araki/dcica.html

A Digital Watermarking Technique Based on ICA Image Features

Wei Lu, Jian Zhang, Xiaobing Sun, and Kanzo Okada

Singapore Research Laboratory
Sony Electronics (S) Pte Ltd., Singapore
{wei.lu,jian.zhang,xiaobing.sun,kanzo.okada}@ap.sony.com

Abstract. A novel digital watermarking technique based on ICA image features is proposed in this paper. This new watermarking technique is provided for both high-quality visual imperceptibility and robust & effective watermark detection. An adaptive-transform approach is employed in this technique, which is different from the conventional DCT or Wavelet transformations. The learned image-adaptive ICA features with localized, oriented and band-pass characters represent similar properties exhibited by the primary and secondary visual cortexes in human vision system (HVS). It enables a powerful masking effect to hide extra information into images with very little visual changes to human eyes. The embedding and detection of watermarks on ICA coefficients whose distribution is super-Gaussian in nature are found to be effective and robust even when only a classical spread-spectrum method is used. Additionally, the adaptive watermarking on suitable images and image regions is achieved implicitly owing to the merit that the ICA bases are automatically learnt from images. The experiments of the blind image watermarking system demonstrate its advantages on good image quality and robustness under various attacks such as image compression, geometric distortion and noises, in comparison with some conventional methods.

1 Introduction

Independent Component Analysis (ICA) has become a popular and promising method to solve many signal processing problems. It has been used not only as an adaptive blind signal separation (BSS) tool but also as a statistical model for audio and video data analyses, like feature extraction. The fundamentals of independence and sparseness lead to the representations and interpretation which exhibit remarkable similarity to human perception in different media, e.g. image [1]. Therefore, we think that ICA models can perform well in image processing tasks acquiring features related to human perceptions; one of many such applications is the digital watermarking.

Digital watermarking is a technology to encode additional information, i.e. watermarks, into host data. The encoding is done in such a way to keep the modifications as imperceptible as possible, while the correct watermarks that is detectable in its decoding process can serve the purposes as copyright protection, document authentication, data transfers, etc [2]. Such an invisible watermarking is well suited to hiding necessary data and keeping contents in a perceptually original form.

Over the last decade, there have been many watermarking algorithms developed in the image space, Fourier, DCT, Mellin-Fourier transforms and wavelet domains [2][3]. One of the major limitations is that they are seldom based on human vision system (HVS) that may be regarded as the final judge for a successful watermarking technique. There have also been several ICA-based watermarking methods proposed in the last few years. Noel and Szu are the first to point out that the de-mixing algorithm based on ICA seeking statistically factorized probability density can yield a watermarking technique using unsupervised neural networks [4]. Yu, et. al. [5] and Shen, et. al. [6] employed similar ideas to realize the watermarking system using BSS methods. Also, Liu, et. al. [7] has attempted to use the ICA separation on DWT coefficients to achieve better detection results than classical methods.

In this paper, we propose a new ICA-based watermarking method in which embedding and detection are performed on HVS-alike image features that are represented by ICA basis/filter functions. Gonzalez-Serrano, et. al. had a simple trial algorithm based on the similar idea [8]. Nevertheless, the technique is extended much further in this paper as not only developing an ICA-based watermarking system but also focusing on the advantages of excellent image quality (imperceptibility) and robust watermark detection/extraction by watermarking on ICA image domain.

2 A Watermarking System Based on ICA Image Features

2.1 ICA Model

In ICA representation of images, the linear image synthesis model is given by

$$\mathbf{x} = \mathbf{Ac} \tag{1}$$

where each image patch is represented by vector $\mathbf{x}$, the ICA basis functions form the columns of the matrix $\mathbf{A}$, the coefficients or weights of the bases are given by the vector $\mathbf{c}$. In contrast, a patch can be transformed into the coefficient vector $\mathbf{c}$ by $\mathbf{W}$:

$$\mathbf{c} = \mathbf{Wx} \tag{2}$$

where ICA filter functions form the rows of matrix $\mathbf{W}$ and $\mathbf{W} = \mathbf{A}^{-1}$.

2.2 Watermarking System

The whole ICA-based watermarking system includes three parts: ICA feature learning, watermark embedding and watermark detection, as shown in Fig. 1.

(1) ICA feature learning
It is unique in this watermarking technique to have a learning process so as to obtain ICA transforms, $\mathbf{A}$ and $\mathbf{W}$, adaptively from particular images, instead of using any mathematically pre-defined functions in conventional watermarking systems. The procedures are briefly described as follows:
1) randomly select image patches and form data matrix $\mathbf{X}$ with zero means.
2) remove the row-wise correlation by the PCA method and obtain a whitening matrix, $\mathbf{V}$, with eigenvector at each row.
3) apply a particular ICA learning algorithm, e.g. fastICA [12] for its fixed-point algorithm and fast learning process, to learn the orthogonal demixing matrix, $\mathbf{dW}$.
4) form the matrix $\mathbf{W} = \mathbf{dW} \times \mathbf{V}$, in which rows represent ICA filters. The columns of its inverse matrix $\mathbf{A} = \mathbf{W}^{-1}$ represent the corresponding basis functions.

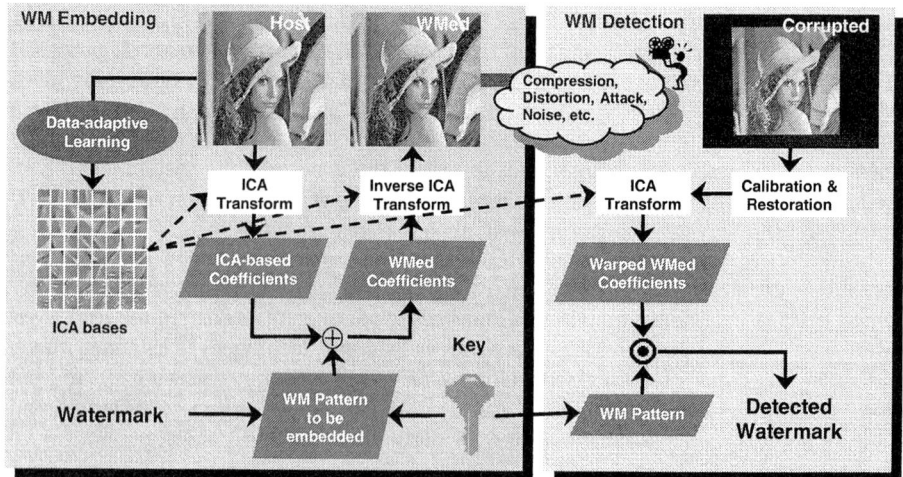

Fig. 1. The flowchart of the whole ICA-based watermarking scheme.

(2) Watermark embedding
The embedding procedure includes three stages:
1) decompose each 8 × 8 pixels, in its vector form **x**, adjacently connected in an image into 64 ICA coefficients **c** for each block by applying the filter matrix **W**.
2) select suitable coefficients based on some schemes which can ensure the image imperceptibility and robustness (described in next section), encode watermark information into these coefficients by a spread spectrum approach as follows: Watermark patterns are generated as two sequences of random numbers ϖ_i^0 and ϖ_i^1 ($i = 1 \ldots M$) by transforming the narrow-band watermark bits (0/1) with private keys to orthogonal and wide-spectrum random series, where M denotes the number of watermark bits. A linear additive embedding is used to perform a simple and effective modulation on the selected ICA coefficients $f(n)$, $n = 1 \ldots N$, by using the formula:

$$f'(n) = f(n)(1 + \alpha \cdot \varpi_i(n)) \qquad (3)$$

Where α is a strength factor to control the image quality and robustness, $f'(n)$ is the modified ICA coefficients carrying the watermark information, and N is the number of coefficients used for embedding one WM bit, possibly cross multiple blocks.
3) reconstruct an image from all coefficients by using the basis matrix **A**.

(3) Watermark detection:
The detection procedure includes two stages as follows:
1) decompose images into ICA coefficients, same as the first stage of embedding.
2) detect/extract watermark bits without using the host image, correlate the watermark pattern, for example, ϖ_i^0, directly with the selected ICA coefficients $f'(n)$ by adopting the central limit theorem,

$$\delta_i^0 = \frac{\sum^N f'(n) \cdot \varpi_i^0(n)}{N} \qquad (4)$$

and then compare the correlation value δ_i^0 with detection threshold T_i^0 which is the average value between the autocorrelation of ϖ_i^0 and the correlation of ϖ_i^0 and ϖ_i^1; vice versa for detecting/extracting the watermark bit ϖ_i^1.

3 Analysis on ICA Transforms and Coefficients

In this section, we discuss two particular achievements in the present technique: good watermark imperceptibility and robust watermark detection.

3.1 Imperceptibility

Human Vision System (HVS)
In conventional methods, watermarks are embedded in image's spatial domain or frequency domain through DCT or Wavelet. The methods using fixed basis functions usually weaken the watermark imperceptibility due to their artificial patterns. More desired approach is expected to emerge image-adaptive schemes by combining with HVS [9].

It is now admitted that the HVS splits the visual stimuli from the retina of the eye into many different components through different tuned channels. The characteristics of a component in the visual field are location, orientation, spatial frequency. Signals that have similar characteristics use the same channels from the eye to the cortex [10]. It appears that such signals interact and subject to non-linear effects. Masking that is one of such effects occurs when a signal cannot be seen because of another signal with close characteristics but at a higher energy level [11].

ICA Transform
As researchers have observed [1], the ICA basis/filter functions have the similar properties to the primary and secondary cells of human visual cortex: localized and oriented in image space, band-passed in frequency domain, as shown in Fig. 2(a). This is very different from the fixed mathematic basis functions in DCT and Wavelet, arranged horizontally and vertically in frequency, Fig. 2(b) and (c).

Matching to HVS can improve the imperceptibility by embedding watermarks into these features because the mid-frequency edge details are the important factors affect human eyes and they have the perceptual masking effect on the changes applied to human eyes. The adaptive watermarking on suitable images and image regions is achieved implicitly because of the merit of ICA bases learnt automatically.

ICA Coefficients
The coefficients obtained through one ICA transform will be statistically very independent from others. Thus, the amplitudes of ICA coefficients usually have a sparse probability distribution, also known as super-Gaussian distribution, as the histogram shown in the lower-right corner of Fig. 3(a).

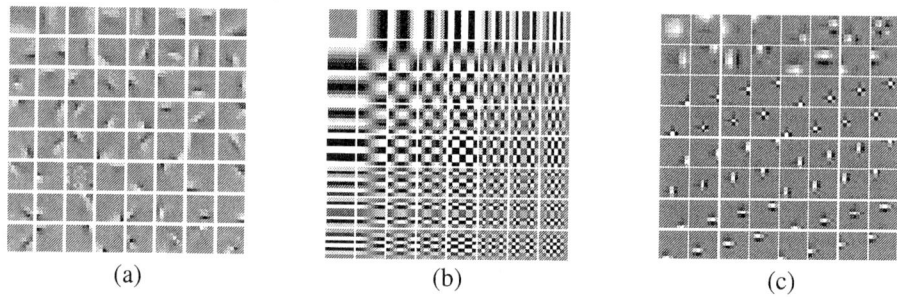

Fig. 2. Illustration of basis functions of (a) ICA image features, (b) DCT and (c) Wavelet.

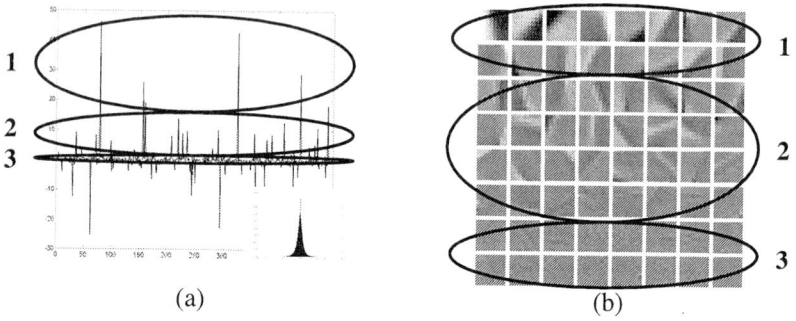

Fig. 3. Classification of ICA coefficients and ordering of ICA basis functions.

As illustrated in Fig. 3(a), ICA coefficients can be classified into three types in terms of their magnitudes: a small number of coefficients marked by 1 have big magnitudes; the coefficients marked by 2 have middle range magnitudes with enough numbers to represent significant features; the coefficients pointed by 3 have small magnitudes, close to zero. Therefore, the order of the ICA basis/filter functions can be sorted according to the average energy of their corresponding coefficients, which may result in the sequence of class 1, 2 and 3. The sorted basis functions shown in Fig. 3(b) are also ordered from low to high frequency.

From experimental investigation, it is suitable to classify the 1st − 16th bases into class 1 and the 17th − 48th bases into class 2, but a threshold may be needed to exclude the exceptional coefficients. The bases in class 3 associated to small energies are usually high-frequency features. In the spread spectrum method, we choose coefficients in class 2 for watermarking. The watermarks added onto class 2 coefficients are perceptually masked by the similar characteristics of image features in class 1 and the large difference of coefficient magnitudes between class 1 and 2, which results in the excellent imperceptibility. We should avoid modulating watermarks in class 3 as they are the least robust components in image processing.

3.2 Robustness

High Tolerance & Significant Features
Because of the high energies on the class 1's coefficients and the particular primary-cortex channels corresponding to them, the watermarks may be added with a relatively large strength ratio, α in Eq. (3), but they still remain invisible. The high ratio increases the watermark robustness under image processing and distortions.

Attacks by using image processing techniques usually try to weaken or remove the watermarks but still keeping the image contents rarely affected, like image compression. As our embedding affects the edge details that are the major elements in images, the attacks may unexpectedly worsen the image quality dramatically. In addition, most imaging tools remove high-frequency components due to its insignificancy to our eyes, so watermarks are less affected as band-passed in the mid-band frequency.

Small Magnitudes
As an oblivious extraction indicated in Eqs. (3) and (4), the host coefficients $f(n)$ have to be small in order to minimize the bias effect in detection. Therefore, class 1 is only good at masking effect, whereas class 2 is ideal for the blind detection as their magnitudes are relatively small. It is very accurate when detection is performed without distortions. Under various attacks, we can use more coefficients across blocks for one-bit watermarking to increase the extraction accuracy significantly.

Restoration of Geometric Distortion
Geometric distortion is common in A-D/D-A conversion and image acquisition while transmitting and receiving watermarked images. We use an image restoration method based on image boundary and significant feature points in the system to recover back the original image geometry. Due to the limited manuscript size, the details will be described in our later publications.

4 Experiments

4.1 Imperceptibility

To compare the image quality of watermarked images, we conducted experiment of embedding same watermark (30 bits) into the 512 × 512 gray Lena image by using four watermarking algorithms based on ICA, image space, DCT and Wavelet domains. The strength of watermarks was kept same for all algorithms as the PSNR between the original image and the watermarked image was maintained at 38dB. The Lena's face regions of the original image and four watermarked images are illustrated in Fig. 4. There is no visible difference between the original, Fig. 4(a), and the ICA's, Fig. 4(b). However, we can easily notice the rigid and repeated patterns on the space's, Fig. 4(c), the Japanese Tatami-Mat like patterns on DCT's, Fig. 4(d), and the horizontal and vertical grain textured patterns on Wavelet's watermarked images, Fig. 4(e). This experiment verified the excellent imperceptibility that the present technique can achieve as the watermarks are embedded on significant image features in natural images and the masking effect also reduces the visual impact effectively.

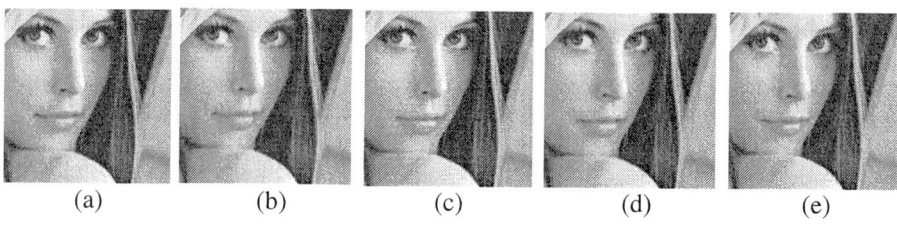

(a) (b) (c) (d) (e)

Fig. 4. Comparison of visual quality on the Lena image (face region) among (a) its original, (b) ICA-based, (c) space-based, (d) DCT-based, and (e) Wavelet-based watermarked images.

4.2 Detection Robustness for Watermarked Images

Several standard images, like Lena, Boat, Baboon, Airplane and Peppers, have been used to test the robustness of watermark detection using the present technique. Three types of distortions have been applied onto the watermarked images with PSNR at 38dB, such as (1) cross-compression by JPEG at quality 65 and JPEG2000 at ratio 0.1, (2) 2D geometric distortion of scaling, rotation and skewing and 3D geometric projection in image acquisition, (3) digital camera capturing in real environment, which involves A-D/D-A conversion, luminance changes, etc. Three types of watermarks have been used to simulate small, medium and large watermark load, such as (1) digit number '1234567890' (31 bits), (2) characters 'Sony Singapore' (98 bits), (3) Sony logo images (around 400 bits).

Table 1. Bit error rate (BER) of our ICA-based watermarking technique under robustness test of cross-compression, 2D/3D geometric distortion and camera capturing in real environment for watermarks in digit, character and image' forms.

WM Robust (BER)	Digit (31 bits)	Char. (98 bits)	Logo (400 bits)
Cross - Compression	0 %	~ 0 %	5-15 %
Geometric Distortion	0 %	0 %	0 %
Camera Capture	~ 0 %	3 %	5-10 %

Table 1 shows the bit error rate (BER) of watermark detection in each type of distortion versus three kinds of watermarks. The present technique is rather robust for cross image compression at low and medium watermark loads. But the BER is increased when the number of watermark bits is large as the length of watermark pattern, N in Eq. (4), is reduced. Our system has very robust watermark detections for the images restored back from geometric distortion, even for the large watermark load. In the situation of capturing images using digital camera in real environment, the results show that the present technique is stably robust in such a complex situation involving many kinds of distortions when the watermark has around 50~70 bits in average. But, it is still acceptable that the BER is around 10% when the images with hundreds of bits are used as watermarks.

5 Conclusion

For digital watermarking techniques, one good source of inspiration has always been the human perception system. Most of the image features obtained by ICA transform represent properties exhibited by the primary and secondary cells of human visual cortexes. The present paper has shown that the ICA-based digital watermarking is indeed quite promising in real applications, possessing notable advantages of better image quality and more robust detection over the conventional methods.

References

1. A. J. Bell and T. J. Sejnowski, "The 'Independent Components' of Natural Scenes are Edge Filters", *Vision Research*, 37(23):3327-3338, 1997.
2. I. J. Cox, J. Kilian, T. Leighton and T. Shamoon, "Secure Spread Spectrum Watermarking for Multimedia", *IEEE Transactions on Image Processing*, vol. 6, no. 12, pp.1673-1687, Dec, 1997.
3. X.-G. Xia, C. G. Boncelet, and G. R. Arce, "A Multiresolution Watermark for Digital Images", *Proceedings of ICIP'97*, Santa Barbara, CA, USA, October 26-29, 1997, Vol I, pp. 548-551.
4. S. Noel and H. Szu, "Multimedia authenticity with independent-component watermarks," in *14th Annual International Symposium on Aerospace/Defense Sensing Simulation, and Controls*, Orlando, Florida, April 2000.
5. D. Yu, F. Sattar and K.-K. Ma, "Watermark Detection and Extraction using Independent Component Analysis Method," *EURASIP Journal on Applied Signal Processing*, vol. 2002, no. 1, pp. 92-104, 2002.
6. M.-F. Shen, X.-J. Zhang, L.-S. Sun, P. J. Beadle, and F. H. Y. Chan, "A Method for Digital Image Watermarking Using ICA," *4th International Symposium on Independent Component Analysis and Blind Signal Separation (ICA2003)*, Nara, Japan, pp. 209-214, April 2003.
7. J. Liu, X.-G. Zhang, J.-D. Sun and M. A. Lagunas, "A Digital Watermarking Scheme based on ICA Detection," *4th International Symposium on Independent Component Analysis and Blind Signal Separation (ICA2003)*, Nara, Japan, pp. 215-220, April 2003.
8. F.J. González-Serrano, H.Y. Molina-Bulla, J.J. Murillo-Fuentes "Independent Component Analysis applied to Digital Image Watermarking," *Proc. IEEE International Conference on Acoustics, Speech and Signal Processing (ICASSP'2001)*, Salt Lake City, USA. pp. 1997-2000. IEEE Press. 2001.
9. C. Podilchuk and W. Zeng, "Image-adaptive watermarking using visual models", *IEEE Journal on Selected Areas in Communications*, vol. 16, no. 4, pp.525-539, 1998.
10. John Wiley, L. A. Olzak, and J. P. Thomas, "Handbook of Perception and Human Performance. Volume 1: Sensory Processes and Perception. Chapter 7: Seeing Spatial Patterns." University of California, Los Angeles, California, 1986.
11. G. E. Legge, "Spatial Frequency Masking in Human Vision: Binocular Interactions", *Journal of Optical Society in America*, 69(6): 838-847, June 1979.
12. A. Hyvarinen, "Fast and Robust Fixed-point Algorithms for Independent Component Analysis", *IEEE Transactions on Neural Networks*, 10(3): 626-634, May 1999.

A Model for Analyzing Dependencies Between Two ICA Features in Natural Images

Mika Inki

Neural Networks Research Centre
Helsinki University of Technology
P.O. Box 5400, FI-02015 HUT, Finland

Abstract. In this paper we examine how the activation of one independent component analysis (ICA) feature changes first and second order statistics of other independent components in image patches. Essential for observing these dependencies is normalizing patch statistics, and selecting patches according to activation. We then estimate a model predicting the conditional statistics of a component using the properties of the corresponding feature as well as those of the conditioning feature.

1 Introduction

Independent component analysis has been used successfully in analyzing image data, even though the model is fundamentally insufficient for describing images. In ICA the observed data is expressed as a linear transformation of latent variables that are nongaussian and independent. We can express the model as

$$\mathbf{x} = \mathbf{As} = \sum_i \mathbf{a}_i s_i, \quad (1)$$

where $\mathbf{x} = (x_1, x_2, \ldots, x_m)$ is the vector of observed random variables, $\mathbf{s} = (s_1, s_2, \ldots, s_n)$ is the vector of latent variables called the independent components (ICs) or source signals, and $\mathbf{A}$ is an unknown constant matrix, called the mixing matrix. The columns of $\mathbf{A}$ are often called features or basis vectors. Exact conditions for the identifiability of the model were given in [2], and several methods for estimation of the classic ICA model have been proposed in the literature, see [5] for a review.

The assumption of independence is fundamental in ICA. Most types of natural data (e.g. image data) do not, however, have such independent (linear) features that ICA attempts to find. It is important to know about the data structures not captured by the ICA model. It is possible to use these structures to extend and improve the ICA model of image data. The usefulness of this approach can be motivated by the link between ICA features and cortical simple cell receptive fields, see [7]. Our approach here has similarities to some analyses made with Gabor functions [1], or with higher-order ICA models [4] and with our earlier paper [6], but the approach of using a parametric model to predict the changes in the statistics of one IC due to the activity of another has not, to our knowledge, been used earlier. We chose our model so that its properties are easily analyzable, yet the model is able to capture most of the structures.

2 Analysis of Conditional Dependencies

We will investigate how the statistics of image patches (image windows, data samples) change, when we know one specific IC is highly active. By activity we mean that the absolute value of the estimated IC ($|\mathbf{y}_i|$, $\mathbf{y}_i = \mathbf{w}_i^T \mathbf{x}$) exceeds some threshold α. In these cases the component can be considered to describe something essential appearing in the patch, i.e. part of an edge or line. We denote the indexes for which IC i exceeds α by

$$I_{\alpha,i} = \{t \mid |y_i(t)| > \alpha\}, \tag{2}$$

and the subset of the whole data associated with $I_{\alpha,i}$ by $\mathbf{X}_{\alpha,i}$.

There are two essential steps of preprocessing we do here that are important for observing the statistics. The first is normalizing variances of individual patches or patch norms (these differ by an irrelevant scaling factor). With this, and with the reduction of the mean and whitening of the data, contrast variations between the patches are mostly eliminated. After this normalization, a certain level of activation means that the feature contributes a specified portion of the content (variance) of the patch.

Note that we demand that the patch variances equal unity and that the data is white simultaneously. The requirement for whitening can be written as

$$\text{cov}(\mathbf{x}) = \mathbf{I} - \frac{1}{n}\mathbf{1}, \tag{3}$$

where the latter term on the right side results from eliminating the patch means. For fixed t (i.e. for each patch) we require for the mean, variance and norm:

$$\text{mean}_i(x_i(t)) = 0, \quad \text{var}_i(x_i(t)) = 1 \iff \|\mathbf{x}(t)\| = \sqrt{n}. \tag{4}$$

We will later discuss how we enforce all these requirements simultaneously.

The second step of preprocessing before examining the dependencies is normalization by the sign of the active component. There is no inherent sign attached to an image patch as components can be positive or negative mostly regardless of each other. But, for example two collinear edge detectors may very well exhibit consistently the same signs when either one is highly active as this corresponds to having an edge in the patches to which both react. We will denote the sign-normalized data associated with independent component i and threshold α as $\mathbf{Z}_{\alpha,i}$:

$$\mathbf{Z}_{\alpha,i} = \{\mathbf{z}(t) \mid \mathbf{z}(t) = \mathbf{x}(t)\text{sign}(y_i(t)), \quad t \in I_{\alpha,i}\}. \tag{5}$$

We will denote the vectors in $\mathbf{Z}_{\alpha,i}$ as $\mathbf{z}_{\alpha,i}$.

3 Data Selection and Preprocessing

As data we used 24 images of landscapes, plants and animals. The images were taken with a digital camera (Canon Ixus 400), converted to grayscale, and block averaged in four by four pixel blocks (and downscaled by the same factor). An area of 512 by 384 pixels was then selected of each of the downscaled images. The downscaling should pretty much negate artifacts brought by color interpolation, noise reduction and even compression. The images were saved in a 16-bit grayscale TIFF-format (after the histograms were stretched to cover the 16-bit range). The original images, and the 16-bit TIFFs can be found at the web address http://www.cis.hut.fi/inki/images/.

Next we sampled 200000 12 by 12 pixel patches from these images. The mean was subtracted from each patch. Sometimes, due to the fact that usually the upper parts of the images are the brightest (e.g. parts of the sky are visible), even though the mean value has been subtracted from each patch, the pixels still do not have zero mean. Therefore we randomly assigned a new sign to each of the patches. The patches were then stacked into 144-dimensional vectors.

We then whitened the data and normalized patch variances. As patch variance normalization affects the covariance of the data, whitening and patch normalization were repeated (alternately) a total of ten times. After this, the largest (nonzero) eigenvalue of the covariance matrix was less than a millionth larger than the smallest. Note that this whole process can be described as whitening and patch normalization, as it corresponds to multiplying the data with a matrix (product of all the whitening matrices) and then normalizing the patches.

FastICA [3] in symmetric mode using the hyperbolic tangent nonlinearity was then used to perform ICA on the data. The basis vectors we found can be seen on the left side of Figure 1. These are presented here in the original, not whitened space.

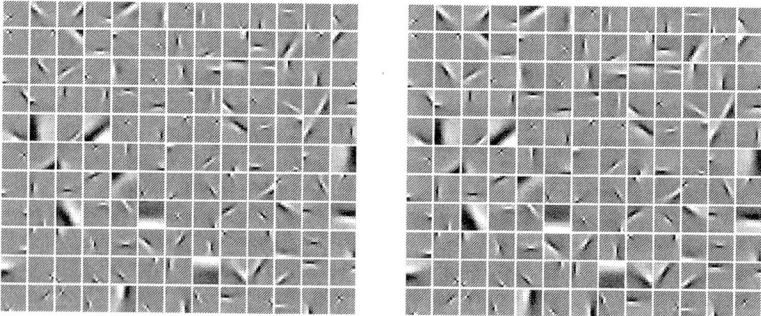

Fig. 1. Left: ICA basis found in normalized data. Right: Mean values of the patches when normalizing by the sign of the active component and using threshold $\alpha = 3$.

We will have to limit our analysis of the statistics to a single value of α. As a compromise between the number of samples, and size of the dependencies, we selected $\alpha = 3$ as the baseline value. In this case 87.7 percent of all patches are included in at least one of the active sets $I_{3,i}$ (Eq. 2), average size of which is 2650 patches, and the active feature contributes at least 6.3 % of patch variance.

4 Dependencies in the ICA Basis

In order to analyze the qualities of the dependencies, we decided to estimate a model where the properties of the features predict the change in a statistic for a given value of α. So, the inputs are calculated from the properties of the conditioning and conditioned feature, and the output should be the statistic, e.g. variance of the conditioned component.

4.1 A Model for Analyzing the Dependencies

In order to estimate the properties of the features, we fitted Gabor functions to the ICA features on the left side of Figure 1. As Gabor functions we used real-valued two-dimensional functions:

$$g(\mathbf{r}) \propto \exp(-\sum_{i=1}^{2} \frac{r_i^2}{2b_i^2}) \cos(2\pi\omega r_1 + \theta). \qquad (6)$$

Here $\mathbf{r}$ is the two dimensional position vector, b_i:s are the widths in corresponding dimensions of $\mathbf{r}$, θ is the phase and ω the frequency. Any Gabor function can now be obtained with a rotation, translation, and scaling of $g(\mathbf{r})$. Let us denote the angle of this rotation as β.

Our model was of the type

$$R'_{i,j} = f_l(\prod_{k=1}^{l-1} f_k(G_{i,j}(k))), \qquad (7)$$

where $R'_{i,j}$ is the estimate of the dependency $R_{i,j}$ between the conditioning IC i and the conditioned IC j, $G_{i,j}(k)$ is the k:th value measured from the features corresponding to ICs i and j. As f_k, $k < l$ we used functions consisting of evenly spaced five points (the smallest of which is at the smallest value of corresponding G_i, the largest at the largest value) that were interpolated with piecewise cubic Hermite interpolation as implemented in Matlab version 6.5.

The function f_l consisted of eleven unevenly spaced points. The first point is at the smallest value of $\prod_{j=1}^{l-1} f_i(G_i)$, i.e. the zero percent mark, the second at the five percent mark (where five percent of $\prod_{j=1}^{l-1} f_i(G_i)$ are smaller), third at the ten percent mark, then 20%, 35%, 50%, 65%, 80%, 90% 95% and the final one at the 100% mark. Additionally, f_l was required to be monotonically increasing and positive. We used $l = 5$ here, and the model had therefore a total of 31 free parameters. Of these 31 parameters, four are actually redundant, as the scaling in f_5 can offset the scaling in any (and all) of the other functions.

As $R_{i,j}$ we used the variances of the conditioned components, as well as the absolute values of the mean values of the conditioned components. Variances highlight large dependencies better than standard deviations, which is partly why we chose to model them. Additionally, we achieved best fits with these choices. We fitted Gabor functions to the ICA features and, in order to have sensible results, excluded the smallest features that cannot be so well described as Gabor functions from the analysis. We picked 98 of the best fitting features, so there were a total of 9506 ($= 98^2 - 98$) examples of $R_{i,j}$ for further analysis.

We used four variables as $G_{i,j}(k)$:s. The first was $G_{i,j}(1) = \log(b_i/b_j)$, where b_i and b_j are the widths of the fitted Gabors for the conditioning and conditioned components respectively. The second was a measure of the difference between the orientations of the components, $G_{i,j}(2) = |\sin(\beta_i - \beta_j)|$. The third was a collinearity measure: new features are obtained by ignoring the attenuation of the Gabors along the edge, i.e. along r_2. These new features are normalized (w.r.t. inner product with themselves), and $G_{i,j}(3)$ is the absolute value of the inner product of these new features for the conditioning and conditioned component. The final variable $G_{i,j}(4)$ is obtained by discarding the cosine part of the Gabors, normalizing, and taking the inner product of these new features, i.e. it was an overlap measure of the functions, which depends on distance and size difference.

Of these, only the first variable G_1 can capture nonsymmetric properties of the dependency, i.e. if the places of the conditioning and conditioned component are exchanged, the dependency can change.

4.2 Results

We fitted the model in equation (7) to the statistics for $\alpha = 3$. The minimum of $\sum_{i,j}(R_{i,j} - R'_{i,j})^2$ was searched by Matlab's fminsearch -function which uses the Nelder-Mead method not requiring derivatives. The error we ended up with was 0.2656 of the variance of $R_{i,j}$ for second order statistics, and 0.4787 of the variance of $R_{i,j}$ for first order statistics.

We have plots of the individual f_k:s, $k \leq 4$, in Figure 2. The functions for second order statistics have been plotted with solid lines. The first function f_1 shows a maximum at zero, i.e. when the functions are of the same size, and slightly unsymmetric behaviour (so the model is slightly unsymmetric). The second variable has a maximum at zero, i.e. when the orientations of the components are identical (or differ by π). The third component shows a maximum when the modified features overlap the most, i.e. are in a sense collinear. The fourth shows a maximum when the overlap of the features is the greatest. These functions are plotted in (natural) logarithmic scale, and their precise values do not matter, for which reason their maximum values have been normalized. The differences between their ranges does matter, and one can see that this is the biggest in the case of the fourth function, and smallest in the case of the first function. This would strongly suggest that the fourth component is the most important for the fit, and the first is the least important.

We have also plotted the individual f_k:s for first order statistics with dashed lines in Figure 2. As one can see, the f_k:s have similar shapes to the corresponding functions for second order statistics, yet there are differences. Again, judging by the ranges of the functions, the fourth variable (overlap) appears to be the most important for the fit, but now the importance of the second variable (difference of orientation) seems to be smaller, and the third variable (collinearity) appears to be more important.

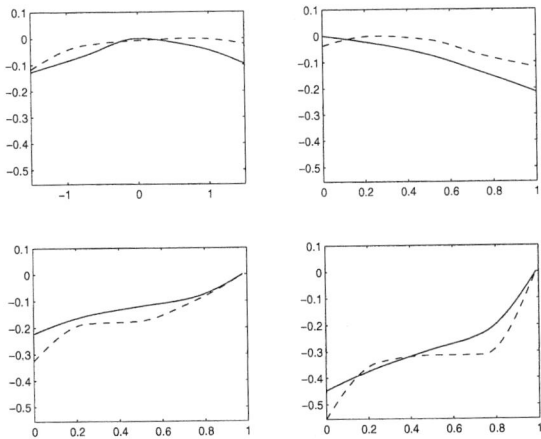

Fig. 2. Logarithms of the f_k:s, $k \leq 4$. Top left: Logarithmic difference in feature width. Top right: Difference in angle. Bottom left: Collinearity measure. Bottom right: Overlap measure. First and second order statistics with dashed and solid lines, respectively.

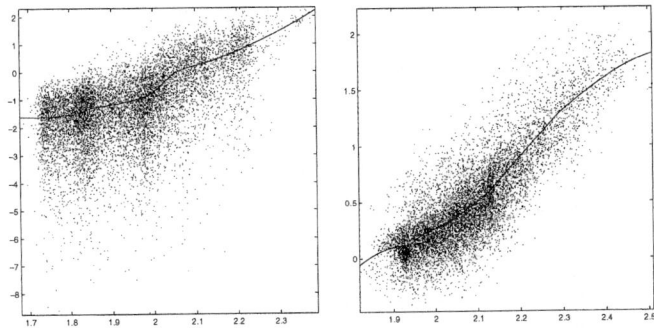

Fig. 3. Logarithm of the observed statistics on the y-axis, $\log(\prod_{k=1}^{l-1} f_k(G_{i,j}(k)))$ on the x-axis. Left side: First order statistics. Right side: Second order statistics. Also plotted in both figures is the correspoding f_l.

We have scatterplots in Figure 3, where on the x-axis are the values obtained by multiplying the $f_k(G_{i,j}(k))$:s, $k \leq 4$, and on the y-axis the observed values $R_{i,j}$. Both axes are in logarithmic scale. On the left side we have the scatterplot for the first order statistics, and on the right side the scatterplot for second order statistics. Also plotted in the figure (with a solid line) is the function f_l. The interpolation of function f_l was done on logarithmic scale.

It appears that for second order statistics, the function f_l has a somewhat sigmoidal shape. This is sensible, as very low values indicate that all the functions f_k give a low value, but already for example insignificant overlap of the features is enough to make the them virtually independent (and the significance of the other factors should be reduced). Similar argument can be made of very high values. It is harder to say anything of the shape of f_l for first order statistics.

Another way of exploring how important the different variables are for the fit is by excluding one variable from the analysis, and fitting the model again. For the second order statistics, excluding the first variable produced an error of 0.2869, excluding the second 0.3668, excluding the third 0.3084, and excluding the fourth 0.4775. This supports our earlier conjecture that the fourth variable is the most important, and the first the least important in the fit.

For the first order statistics, the errors were 0.4923, 0.5128, 0.5846, and 0.6844, for excluding the first, second, third, and fourth variable respectively. This supports our earlier reasoning that the second variable is not so important for the fit as the third, which makes sense as collinearity of two features means they basically describe the same edge at different positions. As can be seen in Figure 1, the mean values for high activation essentially express how the feature on average continues (extends to orthogonal dimensions). The mean value features are longer than the original feature and extend further from the zero crossing. Orientation is not so important for first order statistics because similar orientation without collinearity does not produce a consistent edge.

So, one can say that the most important factor for the size of the dependency (first or second order) is the overlap of the features. Note that the way in which we measure overlap depends on distance between the features and on size difference. However, one can't say that overlap is the only factor to be considered. We also attempted to use additional parameters ($G_{i,j}(k)$, $k > 4$) in the model, but could not achieve essentially better fits.

4.3 Assessing the Validity of the Model

In order to test the validity of our model, we also fitted a multilayer perceptron (MLP) network to the same variables. An MLP should be able to fit into the dependencies between the parameters, whereas in our model the parameters are essentially independent w.r.t. their contribution to the dependency, barring for the effect of f_l. We used Matlab's Neural Network Toolbox for creating and training the MLP. The input and target variables were the same as earlier.

For second order statistics, with five hidden layer neurons (and as many parameters as in our model), we obtained a very similar error measure: 0.2625. But with an MLP it is harder to interpret the properties of the fitted model. With fifty hidden layer neurons, i.e. a total of 301 parameters, we obtained an error of 0.200. For first order statistics, with five hidden layer neurons, we obtained an error measure of 0.4580. With fifty hidden layer neurons, the error was 0.3492. These values are sufficiently close to the errors we obtained with our model with less free parameters (and less chance of overfitting) for us to say that most of the information available in the four variables is captured by our model.

Note also that we can estimate a lower bound for the error in the fit (without overfitting). We made a new version of the data, where by construction the value of the conditioning (active) IC does not affect other components. We call this $\mathbf{U}_{\alpha,i}$. For each component i and every patch $\mathbf{z}_{\alpha,i}(t)$ (Equation 5), we keep the active component, and select randomly a patch $\mathbf{x}(t_2)$ from which we take the other components. We multiply these other components so that the variance of the new patch is normalized. That is:

$$\mathbf{u}_{\alpha,i}(t) \leftarrow \mathbf{P}_i \mathbf{z}_{\alpha,i}(t) + (\mathbf{I} - \mathbf{P}_i)\mathbf{x}(t_2) \frac{\sqrt{n - \|\mathbf{P}_i \mathbf{z}_{\alpha,i}\|^2}}{\|(\mathbf{I} - \mathbf{P}_i)\mathbf{x}(t_2)\|}, \quad (8)$$

where $\mathbf{P}_i$ projects the data into a subspace spanned by component i. We calculated the variance of $R_{i,j}$ from this control data, and it was as low as 0.0061 for second order statistics, even though our best fit with MLP was only 0.200. When fitting our model and the MLP network, we have the added difficulty of Gabor parameter estimation and choosing Gabor parameters for further use. This is significant, especially as the ICA features are not perfectly Gabor functions. We can assume that this is for a large part responsible for the difference between the best error and our noise estimate. For first order statistics this lower bound for the fit was 0.0468, which is still significantly lower than our best fits.

5 Conclusions

Here we studied residual dependencies in the ICA model for image data by examining what effect the activation of one feature has for first and second order statistics of other features. Changes in these statistics tell if the features are usually present simultaneously, and if they have similar signs. We showed how the changes can be largely explained by a model with a few basic properties of the features as parameters, including their overlap, collinearity and orientation.

The results obtained here can offer some useful information for image analysis, or processing, or may offer a small insight into the workings of biological visual systems.

References

1. R.W. Buccigrossi and E.P. Simoncelli. Image compression via joint statistical characterization in the wavelet domain. *IEEE Transactions on Image Processing*, 8(12):1688–1701, 1999.
2. P. Comon. Independent component analysis—a new concept? *Signal Processing*, 36:287–314, 1994.
3. A. Hyvärinen. Fast and robust fixed-point algorithms for independent component analysis. *IEEE Trans. on Neural Networks*, 10(3):626–634, 1999.
4. A. Hyvärinen, P. O. Hoyer, and M. Inki. Topographic independent component analysis. *Neural Computation*, 13(7), 2001.
5. A. Hyvärinen, J. Karhunen, and E. Oja. *Independent Component Analysis*. Wiley Interscience, 2001.
6. M. Inki. Examining the dependencies between ICA features of image data. In *Proc. of ICANN/ICONIP 2003*, Istanbul, Turkey, 2003.
7. B. A. Olshausen and D. J. Field. Emergence of simple-cell receptive field properties by learning a sparse code for natural images. *Nature*, 381:607–609, 1996.

An Iterative Blind Source Separation Method for Convolutive Mixtures of Images

Marc Castella and Jean-Christophe Pesquet

Université de Marne-la-Vallée / UMR-CNRS 8049
5 bd Descartes, Champs-sur-Marne
77454 Marne-la-Vallée CEDEX 2, France
{castellm,pesquet}@univ-mlv.fr

Abstract. The paper deals with blind source separation of images. The model which is adopted here is a convolutive multi-dimensional one. Recent results about polynomial matrices in several indeterminates are used to prove the invertibility of the mixing process. We then extend an iterative blind source separation method to the multi-dimensional case and show that it still applies if the source spectra vanish on an interval. Based on experimental observations we then discuss problems arising when we want to separate natural images: the sources are non i.i.d. and have a band limited spectrum; a scalar filtering indeterminacy thus remains after separation.

1 Introduction

Due to its numerous applications such as passive sonar, seismic exploration, speech processing and multi-user wireless communications, blind source separation (BSS) has been an attractive and fruitful research topic for the last few years. Independent Component Analysis (ICA) has also found interesting applications in image processing, but in this context, the original framework of instantaneous mixture has been mainly considered (e.g. [1, 2]).

However mixtures may be more complicated in practice, and the spread of each source over several pixels may require to address the general model of convolutive mixtures [3]. The single-channel blind deconvolution of images has been extensively studied and solutions have been proposed which usually involve some regularization and use some prior information [4]. Multichannel acquisition, which provides several different blurred version of a single image, also allows to improve the image restoration quality. This Single Input/Multiple Output (SIMO) case has been extensively studied (see e.g. [5, 6]) and is not addressed here. However, little attention has been paid to the general Multiple Input/Multiple Output (MIMO) case, where independent sources are mixed on different sensors.

This paper considers MIMO convolutive mixtures of independent multi-dimensional signals: the problem is described in Section 1. Two main difficulties arise in the 2D case: finding invertibility conditions for the 2D mixing process and ability to deal with non i.i.d. sources which may have band-limited (or rapidly decaying) spectra. The former problem is discussed in Section 3. The latter one is addressed in Section 4 which presents the separation method. Finally, simulation results in Section 5 show the validity of the proposed method and outline the specificities of convolutive source separation for images.

2 Problem Statement

We consider $N \in \mathbb{N}^*$ two-dimensional signals which, for $i \in \{1,\ldots,N\}$ are denoted by $(s_i(\boldsymbol{n}))_{\boldsymbol{n} \in \mathbb{Z}^2}$. Though our theoretical results apply to the general p-dimensional case, we will be more particularly interested in images. For the sake of readability, we shall equivalently use either a two-dimensional notation (n_1, n_2) or a boldface character $\boldsymbol{n}$. The N former signals are referred to as *source* signals, which generate $Q \in \mathbb{N}^*$ *observation* signals according to the following 2D convolutive mixture model:

$$\boldsymbol{x}(n_1, n_2) = \sum_{(k_1, k_2) \in \mathbb{Z}^2} \boldsymbol{M}(k_1, k_2) \boldsymbol{s}(n - k_1, n - k_2) = \sum_{\boldsymbol{k}} \boldsymbol{M}(\boldsymbol{k}) \boldsymbol{s}(\boldsymbol{n} - \boldsymbol{k}). \quad (1)$$

We use here vector notations where $\boldsymbol{s}(\boldsymbol{n}) = (s_1(\boldsymbol{n}), \ldots, s_N(\boldsymbol{n}))^T$ and $\boldsymbol{x}(\boldsymbol{n}) := (x_1(\boldsymbol{n}), \ldots, x_Q(\boldsymbol{n}))^T$ are respectively the source and observation vectors, and $(\boldsymbol{M}(\boldsymbol{k}))_{\boldsymbol{k} \in \mathbb{Z}^2}$ is a set of $Q \times N$ matrices which corresponds to the impulse response of the mixing system. BSS aims at inverting the above described process, with no precise knowledge about the mixing process or the sources. The *separating* system is modeled as a linear convolutive structure and reads:

$$\boldsymbol{y}(n_1, n_2) = \sum_{(k_1, k_2) \in \mathbb{Z}^2} \boldsymbol{W}(k_1, k_2) \boldsymbol{x}(n - k_1, n - k_2) = \sum_{\boldsymbol{k}} \boldsymbol{W}(\boldsymbol{k}) \boldsymbol{x}(\boldsymbol{n} - \boldsymbol{k}) \quad (2)$$

where $(\boldsymbol{W}(\boldsymbol{k}))_{\boldsymbol{k} \in \mathbb{Z}^2}$ is the impulse response of the separating filter of size $N \times Q$ and $\boldsymbol{y}(\boldsymbol{n}) = (y_1(\boldsymbol{n}), \ldots, y_N(\boldsymbol{n}))^T$ is the separation result. Ideally, $\boldsymbol{y}(\boldsymbol{n})$ reduces to the original source vector, up to a permutation and a scalar filtering indeterminacy. Some assumptions have to be made in addition to the aforementioned convolutive model for the source separation task to be achievable:

A.1 The source processes $(s_i(\boldsymbol{n}))_{\boldsymbol{n} \in \mathbb{Z}^2}, i \in \{1, \ldots, N\}$ are statistically mutually independent and stationary.
A.2 The mixing system is stable (i.e. its impulse response is summable) and admits a summable inverse.

Assumption A.1 is a key assumption in BSS and ICA, whereas A.2 is necessary to be able to separate the sources. Invertibility conditions for multivariate systems are discussed in detail in Section 3.

Let us further emphasize that we do not require the sources to be i.i.d. and, contrary to other separation methods in the non i.i.d. context, we do not exploit the spectral diversity of the sources to realize source separation. Indeed, images may exhibit similar spectral characteristics. In most of the works dealing with convolutive mixtures in the same context, sources are generally supposed to be i.i.d.[3]. From the fact that the sources are non i.i.d., it follows that each source can only be recovered up to a scalar filtering, in addition to the well-known permutation ambiguity. The scalar filtering issue does not appear in instantaneous mixtures, as it reduces to a scaling factor ambiguity.

3 Invertibility Conditions

The considered separation method is valid for all kind of filters, both with infinite impulse response (IIR) and with finite impulse response (FIR). However, considering FIR filters allows us to provide simple conditions for the invertibility of the mixing process.

3.1 Finite Impulse Response Assumption

We assume in the following:

A.3 The mixing filter is FIR and $M(k_1, k_2) = 0$ if $k_1 \notin \{0, \ldots, L_1 - 1\}$ or $k_2 \notin \{0, \ldots, L_2 - 1\}$.

In the 1D case, it is well known that, under primeness conditions, the mixing system admits an inverse (see [7] and references therein). This result is based on results concerning polynomial matrices and extends to the multivariate case. Let us first define the following z-transform of the mixing system:

$$M[z_1, z_2] := \sum_{(k_1,k_2)\in\mathbb{Z}^2} M(k_1,k_2) z_1^{-k_1} z_2^{-k_2} = M[z] = \sum_k M(k) z^{-k}. \quad (3)$$

The z-transform of the separating system is defined in the same way and is denoted by $W[z]$. Equations (1) and (2) can then be formally written:

$$x(n) = M[z]s(n) \quad \text{and} \quad y(n) = W[z]x(n). \quad (4)$$

The goal of BSS consists in finding $W[z]$ such that the global transfer function $G[z] := W[z]M[z]$ is diagonal up to a permutation. Conditions for the existence of such an inverse are discussed in the next section.

3.2 Primeness Properties and Invertibility

Invertibility properties of the mixing system rely on primeness properties of the polynomial matrix $M[z]$. Although some theoretical results may be found in the literature [8, 9] for multi-dimensional signals, some of them are not easily accessible. The ring of Laurent polynomials in indeterminates $z = (z_1, \ldots, z_p)$ and with coefficients in $\mathbb{C}$ is denoted by $\mathbb{C}[z]$. Primeness properties in $\mathbb{C}[z]$ are somewhat more complicated than in the case of polynomials in one indeterminate as there exist four distinct notions of primeness. We will be particularly interested in the following definition:

Definition 1. *A polynomial matrix $M[z] \in \mathbb{C}[z]^{Q \times N}$ is said to be* right zero prime *if $Q \geq N$ and the ideal generated by its maximal order minors is the ring $\mathbb{C}[z]$ itself.*

An equivalent definition of right-zero coprimeness can be obtained after slight modifications of known results:

Property 1. *A polynomial matrix $M[z] \in \mathbb{C}[z]^{Q \times N}$ is right zero prime if and only if its maximal order minors have no common zero in $(\mathbb{C}^*)^p$.*

The invertibility of the mixing system is ensured by the following property [8]:

Property 2. *A polynomial matrix $M[z] \in \mathbb{C}[z]^{Q \times N}$ is right zero prime if and only if it has a polynomial left inverse, or equivalently if and only if there exists $W[z] \in \mathbb{C}[z]^{Q \times N}$ such that $W[z]M[z] = I_N$*

The above property provides a necessary and sufficient condition for the mixing system to be invertible. Since however the mixing system is supposed to be unknown, an interesting point would be to know if a randomly generated polynomial matrix is likely to have a polynomial left inverse or not. The answer was partially given in [9]:

Property 3. *If the $Q \times N$ polynomial matrix $M[z]$ has coefficients drawn from a continuous density function and if $\binom{Q}{N} := \frac{Q!}{N!(Q-N)!} > p$, then $M[z]$ is almost surely invertible.*

In particular, for images $p = 2$ and one can see that a mixing system with coefficients driven from a continuous density function is almost surely invertible as soon as there are more sensors than sources ($Q > N$). Finally, bounds on the order of the separating filter have been given [9]: although they are quite large, they give a maximum order for a possible separating system.

4 Separation Method

In this section, we will see how a 1D iterative separating method can be used for the separation of multi-dimensional sources. Among the possible approaches, iterative and deflation-like methods appear especially appealing as they allow the separation of non i.i.d. sources. In addition, they do not present spurious local maxima, unlike many global MIMO approaches.

4.1 An Iterative Approach

Contrast Function for the Extraction of One Source. We first consider the extraction of one source and denote by $w[z]$ one row of the separating system $W[z]$. Let $g[z] := w[z]M[z]$ denote the corresponding row of the global system. Contrast functions are a practical tool to tackle BSS as they reduce it to an optimization problem: by definition, a contrast function is maximum if and only if separation is achieved. Since we consider here an iterative approach and since the global filter $g[z]$ is a Multiple Input/Single Output (MISO) one, a contrast is maximum if and only if the global scalar output is a scalar filtered version of one source. Consider the following function:

$$J(w) := |\text{Cum}^4[y(n)]| \tag{5}$$

where $y(n) := g[z]s(n)$ is the global scalar output corresponding to $w[z]$ and where $\text{Cum}^4[.]$ denotes the fourth-order auto-cumulant. It has been proved that the function J (which depends on w or equivalently on $y(n)$) constitutes a contrast for both i.i.d. [10] and non i.i.d. sources [11], if it is maximized under the constraint:

C.1 $\text{E}\{|y(n)|^2\} = 1$.

The method has been used in the 1D case [11,12]. In the 2D case, it can be implemented as follows: define the vectors $\mathcal{W}$ and $\mathcal{X}(n)$ which are composed of the terms $w_j(k)$ and $x_j(n-k)$, respectively, when k varies in $\{0,\ldots,L_1-1\} \times \{0,\ldots,L_2-1\}$ and j varies in $\{1,\ldots,Q\}$. One can then write: $y(n) = \mathcal{W}\mathcal{X}(n)$. The optimization of (5) is then be carried out with a batch, iterative algorithm, where constraint C.1 is imposed at each iteration by a re-normalization step. The optimization procedure can be written in a such way that $\mathcal{X}(n)$ and $\mathcal{W}$ are the only required inputs. This means that 1D separation procedures can be used under the appropriate modifications of the definition of $\mathcal{W}$ and $\mathcal{X}(n)$.

Extraction of the Remaining Sources. After having separated one source, deflation approaches subtract its contribution from the observations by a least square approach. The former procedure is then applied again on a new observation vector. If P sources $(P < N)$ have been extracted and $y_1(n),\ldots,y_P(n)$ denote the obtained outputs, we alternatively suggest to carry out the optimization of J under the constraint:

C.2 $\forall i \in \{1,\ldots,P\}, \forall k \quad \text{E}\{y(n)y_i^*(n-k)\} = 0$.

It can be proved that constraint C.2 prevents from separating twice the same source. Furthermore, C.2 is a linear constraint on $\mathcal{W}$, which can hence easily be taken into account.

4.2 Validity of the Method for Sources with Non-positive Definite Auto-correlation

Natural images are highly correlated and their spectrum is mostly concentrated on low frequencies. We consider the limit case when the source spectrum is positive on a set Ω and vanishes on its complementary set $\overline{\Omega}$. Let us see the consequences when $\overline{\Omega}$ is with non zero measure. Writing $g[z] = (g_1[z],\ldots,g_N[z])$, let fix $i \in \{1,\ldots,N\}$ and define:

$$\|g_i\|_i = \Big(\sum_{k,l} g_i(k)g_i^*(l)\gamma_i(l-k)\Big)^{\frac{1}{2}} \tag{6}$$

Fig. 1. Original images used as sources

where $\gamma_i(\boldsymbol{k})$ is the autocorrelation of the i-th source. If $\overline{\Omega} = \emptyset$ the sequence $\gamma_i(\boldsymbol{k})$ is definite positive and $\|.\|_i$ is a norm. On the contrary, if $\overline{\Omega}$ is with non zero measure, $\|.\|_i$ is a semi-norm only and the proof in [11] no longer applies, since it is based on the possibility to write $g_i[z] = \|g_i\|_i \frac{g_i[z]}{\|g_i\|_i}$ for any non-zero filter .

Fortunately, one can consider working over the subset of filters which are identically zero on the frequency band $\overline{\Omega}$, and the proof in [11] can then be easily adapted. However, one can see that after separation, the part of the global filter which operates on the band $\overline{\Omega}$ is left free. This part has indeed no influence neither on the separator outputs nor on the contrast J. This may however lead to numerical difficulties with sources with limited band spectrum.

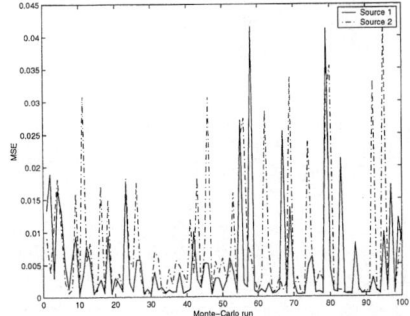

Fig. 2. MSE after separation of two uniform i.i.d. 2D sources

5 Effectiveness of the Procedure

The previous results have been tested on convolutive mixtures of images. In our experiments, there were 2 source images and 3 sensors, so that invertibility is almost surely guaranteed as soon as the coefficients of the mixing system are drawn from a continuous probability density function.

5.1 Simulation Results with i.i.d. Sources

We first verified the validity of our assertions with i.i.d. sources. The study was carried out on a set of 100 Monte-Carlo runs. The coefficients of the mixing systems were drawn randomly from a Gaussian zero-mean unit-variance distribution and the sources were i.i.d. uniform, unit-variance and zero-mean. The length of the mixing system was set to $L_1 = L_2 = 2$ whereas the length of the separator was set to $D_1 = D_2 = 5$. The images were of size 256×256. Results are plotted in Figure 2 and show the mean square reconstruction error (MSE)

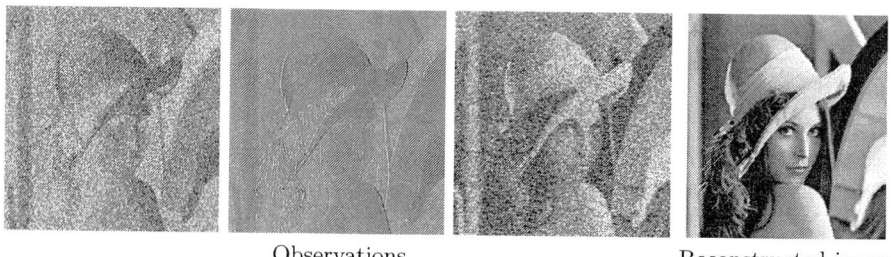

Observations   Reconstructed image

Fig. 3. Separation of a natural image and a noise source

Observations

Output of the separation algorithm   Least square reconstruction

Fig. 4. Separation of two images

for each source an Monte-Carlo run. As the sources are i.i.d., the scalar filtering ambiguity is known to reduce to a simple delay. All MSE were below 0.043 and the mean value over all realization was 5.5×10^{-3}. Naturally, the invertibility of the mixing filter must be ensured in order to obtain good results. Hence the separator should be long enough and a shorter separator led in our experiments to degraded performances. These experimental results prove both the validity of Properties 2, 3 and the ability of the method to separate sources.

5.2 Natural Images

The separation of a natural image and a noise image was tested on a filter of length $L_1 = 3, L_2 = 1$ and a separator of length $D_1 = 4, D_2 = 1$. The sources are the first two ones in Figure 1 and the mixtures are given in Figure 3. Another interesting example involving now two natural sources is given by the mixture of the last two images given in Figure 1 with a filter of length $L_1 = 2, L_2 = 1$

and a separator of length $D_1 = 2, D_2 = 1$. The observations are represented in Figure 4. As previously said, there is no guarantee to recover the original unfiltered sources at the output of the method: this is particularly well illustrated in Figure 4, where we roughly recognize high-pass filtered versions of the original sources. It is interesting to note however that in a certain number of experiments, other filtered versions of the original sources can be obtained as well, including filters which can be close to the identity. On the contrary, the noise in Figure 3 being i.i.d., it can be recovered up to a delay and scaling factor only.

Suppose that each source has a non convolutive contribution on one sensor. Then, it is possible to recover the sources by subtracting the output of the algorithm from the sensors by a least square approach. The results are given in Figure 4, and one can see that the original sources are well recovered. In other cases, one may resort to other image processing techniques in order to solve the remaining blind SISO deconvolution problem, when the sources have been separated.

Acknowledgment

The authors are grateful to Professor Maria Elena Valcher from University of Padova for fruitful discussions.

References

1. Cardoso, J.F.: Independent component analysis of the cosmic microwave background. In: Proc. of ICA'03, Nara, Japan (2003) 1111–1116
2. Hyvärinen, A., Hoyer, P.O., Hurri, J.: Extensions of ica as models of natural images and visual processing. In: Proc. of ICA'03, Nara, Japan (2003) 963–974
3. Comon, P.: Contrasts for multichannel blind deconvolution. IEEE Signal Processing Letters **3** (1996) 209–211
4. Kundur, D., Hatzinakos, D.: Blind image deconvolution. IEEE Signal Processing Mag. **13** (1996) 43–64
5. Giannakis, G.B., Heath, R.W.: Blind identification of multichannel FIR blurs and perfect image restoration. IEEE Trans. on Image Processing **9** (2000) 1877–1896
6. Šroubek, F., Flusser, J.: Multichannel blind iterative image restoration. IEEE Trans. on Image Processing **12** (2003) 1094–1106
7. Gorokhov, A., Loubaton, P.: Subspace based techniques for blind separation of convolutive mixtures with temporally correlated sources. IEEE Trans. Circuits and Systems I **44** (1997) 813–820
8. Fornasini, E., Valcher, M.E.: nD polynomial matrices with applications to multidimensional signal analysis. Multidimensional Systems and Signal Processing **8** (1997) 387–408
9. Rajagopal, R., Potter, L.C.: Multivariate MIMO FIR inverses. IEEE Trans. on Image Processing **12** (2003) 458–465
10. Tugnait, J.K.: Identification and deconvolution of multichannel linear non-gaussian processes using higher order statistics and inverse filter criteria. IEEE Trans. Signal Processing **45** (1997) 658–672
11. Simon, C., Loubaton, P., Jutten, C.: Separation of a class of convolutive mixtures: a contrast function approach. Signal Processing (2001) 883–887
12. Tugnait, J.K.: Adaptive blind separation of convolutive mixtures of independent linear signals. Signal Processing **73** (1999) 139–152

Astrophysical Source Separation Using Particle Filters

Mauro Costagli[1], Ercan E. Kuruoğlu[1], and Alijah Ahmed[2]

[1] Istituto di Scienza e Tecnologie dell'Informazione Alessandro Faedo
Area della Ricerca CNR, Via Moruzzi 1, 56124 Pisa, Italy
{mauro.costagli,ercan.kuruoglu}@isti.cnr.it
[2] EADS Astrium, Gunnels Wood Road, Stevenage SG1 2AS, UK

Abstract. In this work, we will confront the problem of source separation in the field of astrophysics, where the contributions of various Galactic and extra-Galactic components need to be separated from a set of observed noisy mixtures. Most of the previous work on the problem perform blind source separation, assume noiseless models, and in the few cases when noise is taken into account assume Gaussianity and space-invariance. However, in the real scenario both the sources and the noise are space-varying. In this work, we present a novel technique, namely *particle filtering*, for the non-blind (Bayesian) solution of the source separation problem, in case of non-stationary sources and noise, by exploiting available a-priori information.

1 Introduction

Blind signal separation has been applied with a certain degree of success in applications ranging from speech processing to fMRI and from financial time series analysis to telecommunications. In these applications, generally a classical blind source separation technique, namely ICA, has been employed. In most of the applications, ICA has been employed in a simple form assuming equal number of sources and observations, stationary mixing, and either noise free mixtures or in presence of noise, adopting a Gaussian stationary model. In most such work, the time (or space) structure in the signal (or image) is completely ignored, turning the technique into an emsemble data analysis technique rather than a time-series analysis one. Moreover, ICA is blind and does not consider any prior information.

However, in some applications these assumptions are highly questionable. In most real life environments noise is present, sometimes is non-Gaussian and even non-stationary. The mixing matrix might change in time or space as well as the sources which might be highly non-stationary. There may exist a wealth of prior information regarding the time (or space) structure of the data and its statistical distribution as well as the mixing matrix. In this work, we will present a relatively novel technique, namely particle filtering, which can potentially account for all these different features and exploit most of its potentials in a novel and real application, namely the separation of independent components in astrophysical radiation maps.

2 Astrophysical Source Separation

This new application has been motivated by the need to analyse vast amounts of astrophysical radiation maps that will be available with launch of the Planck satellite in 2007 by the European Space Agency (ESA) [10] which will provide measurements in nine different frequency channels ranging from 30 GHz to 857 GHz, with a spectral and spatial resolution much higher than the previous NASA missions COBE and WMAP. This data is the superposition of various independent astrophysical sources, among which the most important one is the *Cosmic Microwave Background* (CMB) for various reasons: it is the relic radiation remaining from the first light radiation in the universe released at the Big Bang. Therefore, CMB provides a picture of the universe shortly after it has started. Secondly, it houses vital information to determine the values of certain cosmological parameters, the high-sensitivity calculation of which in turn would help us decide between competing theories for the evolution of the universe. The signal measured in CMB experiments is however contaminated not only by the intrinsic noises due to the satellite microwave detectors but also by astrophysical contaminants (the so-called *foregrounds*). The most relevant foregrounds are the dust emission, synchrotron (caused by the interaction of the electrons with the magnetic field of the galaxy) and the free-free radiation (due to the the interaction of hot electrons with the interstellar gas), while other contaminations come from extragalactic microwave sources and from the so-called Sunyaev-Ze'ldovich effect [10]. Before achieving cosmological information from the statistical analysis of the CMB anisotropies, all these components must be separated from the intrinsic CMB signal. The problem, therefore, is conveniently formulated as the source separation from linear instantaneous mixtures:

$$\mathbf{y}_{1:n,t} = \mathbf{H}_t \boldsymbol{\alpha}_{1:m,t} + \mathbf{w}_{1:n,t} \ , \tag{1}$$

where $\mathbf{y}_{1:n,t}$, $\boldsymbol{\alpha}_{1:m,t}$ and $\mathbf{w}_{1:n,t}$ are column vectors, representing the n observations, the m sources and the n additive noise samples at time t respectively. $\mathbf{H}_t$ is the $n \times m$ real valued mixing matrix, and is allowed to vary in t.

The problem has been dealt with using other methods by several researchers in the literature including Baccigalupi et al. [3] and Maino et al. [14] who implemented the FastICA algorithm and its noisy version which had limited success in the presence of significant noise. A source model was introduced by Kuruoglu et al. in [12] implementing the Independent Factor Analysis (IFA) technique which also included the noise in the mixing model. Despite this added flexibility, IFA uses a fixed source model which lacks freedom in modelling source model parameters and moreover could not deal with non-stationary noise which is the case in our problem. Snoussi et al. [15] utilise an EM algorithm in the spectral domain, making use of some generic priors for the sources. Cardoso et al. [4] perform blind source separation via spectral matching. Both of these works assume stationary noise and signals, which is not the case for the astrophysical image separation problem, and they both suffer from common drawbacks of the EM algorithm, i.e. local optimality and computational complexity. All of these approaches are blind or semi-blind techniques which do not exploit a wealth of information about the sources the astrophysics theory provides us with. To be able to incorporate these prior information a full Bayesian formulation was derived in [13] which utilises MCMC techniques but unfortunately it does not address non-stationarity and does not consider the auto-correlation structure in images.

In this work a different approach, named *Particle Filtering*, which avoids all of these problems is proposed. It deals with the nonstationarity of the noise, allows very flexible modelling of the sources and conveniently enables the utilisation of available prior information including dependence structure in the images. Particle filtering is an extension of Kalman filtering which can model nonlinear systems and non-Gaussian signals.

3 Particle Filtering

The general filtering problem for non-stationary sources can be expressed with

$$\boldsymbol{\alpha}_t = f_t(\boldsymbol{\alpha}_{t-1}, \mathbf{v}_t) \qquad \mathbf{y}_t = h_t(\boldsymbol{\alpha}_t, \mathbf{w}_t) \ . \qquad (2)$$

The state equation describes the evolution of the state $\boldsymbol{\alpha}_t$ over t, f_t is a possibly nonlinear function, $\boldsymbol{\alpha}_{t-1}$ is the state at the previous step, and $\mathbf{v}_t$ is called dynamic noise process. The observation equation describes the evolution of the data $\mathbf{y}_t$ at step t through a possibly nonlinear function h_t given the current state $\boldsymbol{\alpha}_t$ and the observation noise realization $\mathbf{w}_t$ at time step t. In the classical case of linear f_t and h_t, and Gaussian distributed $\mathbf{v}_t$ and $\mathbf{w}_t$, the filtering problem reduces to Kalman filtering. Particle filtering is a relatively novel technique that provides a solution to the general nonlinear, non-Gaussian filtering problem using numerical Bayesian (sequential Monte Carlo) techniques. Although known since late 60's the technique has received interest only recently finding successful applications especially in tracking problems (see [7] for examples). Very recently it has also been applied to solve source separation problems ([1, 2, 8]). In particular, Everson and Roberts [8] considered a linear instantaneous mixing in which the sources and the noise are stationary while the mixing matrix is non-stationary. They assumed generalised Gaussian models for the sources which are fixed but unknown. Andrieu and Godsill [2] considered the problem of convolutional mixing instead and adopted a parametric model (time-varying AR) for the sources which were assumed to be Gaussian. The mixing was also assumed to be evolving according to a time-varying AR Gaussian process. In our problem, non-stationarity is in the sources and the noise rather than the mixing and we consider the source model parameters as random to fully exploit the potentials of Bayesian modelling. Moreover, our model requires an instantaneous mixing and the astrophysical sources need to be modelled with Gaussian mixtures due to their multi-modality as shown in [12] rather than by Gaussian or generalised Gaussian densities. Therefore, in this work we follow a formulation very similar to that in [1].

The basis of the particle filtering is the representation of continuous pdfs with discrete points (*particles*), as in $p_N(d\boldsymbol{\alpha}_{0:t}|\mathbf{y}_{1:t}) = \frac{1}{N}\sum_{i=1}^{N}\delta_{\boldsymbol{\alpha}_{0:t}^{(i)}}(d\boldsymbol{\alpha}_{0:t})$, where $\delta_{\boldsymbol{\alpha}_{0:t}^{(i)}}$ denotes the delta-Dirac mass and N is the number of points where the continuous pdf is discretised. In this case, a MMSE estimate of a function of interest $I(f_t)$ can be obtained as:

$$I_{\text{MMSE}_N}(f_t) = \int f_t(\boldsymbol{\alpha}_{0:t})p_N(d\boldsymbol{\alpha}_{0:t}|\mathbf{y}_{1:t}) = \sum_{i=1}^{N} f_t\left(\boldsymbol{\alpha}_{0:t}^{(i)}\right). \tag{3}$$

Unfortunately, it is usually impossible to sample from the posterior distribution since it is, in general, multivariate, non-standard, and only known up to a proportionality constant. A classical solution is to use the *importance sampling* method, in which the true posterior distribution is replaced by an *importance function* $\pi(\boldsymbol{\alpha}_{0:t}|\mathbf{y}_{1:t})$ which is easier to sample from. Provided that the support of $\pi(\boldsymbol{\alpha}_{0:t}|\mathbf{y}_{1:t})$ includes the support of $p(\boldsymbol{\alpha}_{0:t}|\mathbf{y}_{1:t})$, we get the identity

$$I(f_t) = \frac{\int f_t(\boldsymbol{\alpha}_{0:t})w(\boldsymbol{\alpha}_{0:t})\pi(\boldsymbol{\alpha}_{0:t}|\mathbf{y}_{1:t})d\boldsymbol{\alpha}_{0:t}}{\int w(\boldsymbol{\alpha}_{0:t})\pi(\boldsymbol{\alpha}_{0:t}|\mathbf{y}_{1:t})d\boldsymbol{\alpha}_{0:t}}, \tag{4}$$

where $w(\boldsymbol{\alpha}_{0:t}) = \frac{p(\boldsymbol{\alpha}_{0:t}|\mathbf{y}_{1:t})}{\pi(\boldsymbol{\alpha}_{0:t}|\mathbf{y}_{1:t})}$ is known as the *importance weight*. Consequently, it is possible to obtain a Monte Carlo estimate of $I(f_t)$ using N particles $\{\boldsymbol{\alpha}_{0:t}^{(i)}; i = 1,\cdots,N\}$ sampled from $\pi(\boldsymbol{\alpha}_{0:t}|\mathbf{y}_{1:t})$:

$$\bar{I}_N(f_t) = \frac{\frac{1}{N}\sum_{i=1}^{N} f_t\left(\boldsymbol{\alpha}_{0:t}^{(i)}\right)w\left(\boldsymbol{\alpha}_{0:t}^{(i)}\right)}{\frac{1}{N}\sum_{j=1}^{N} w\left(\boldsymbol{\alpha}_{0:t}^{(i)}\right)} = \sum_{i=1}^{N} f_t\left(\boldsymbol{\alpha}_{0:t}^{(i)}\right)\tilde{w}_t^{(i)}, \tag{5}$$

where the *normalised importance weights* $\tilde{w}_t^{(i)}$ are given by: $\tilde{w}_t^{(i)} = \frac{w\left(\boldsymbol{\alpha}_{0:t}^{(i)}\right)}{\sum_{j=1}^{N} w\left(\boldsymbol{\alpha}_{0:t}^{(i)}\right)}$. This integration method can be interpreted as a sampling method, where the posterior distribution is approximated by:

$$\bar{p}_N(d\boldsymbol{\alpha}_{0:t}|\mathbf{y}_{1:t}) = \sum_{i=1}^{N} \tilde{w}_t^{(i)}\delta_{\boldsymbol{\alpha}_{0:t}^{(i)}}(d\boldsymbol{\alpha}_{0:t}). \tag{6}$$

When the importance function is restricted to be of the general form:

$$\pi(\boldsymbol{\alpha}_{0:t}|\mathbf{y}_{1:t}) = \pi(\boldsymbol{\alpha}_{0:t-1}|\mathbf{y}_{1:t-1})\pi(\boldsymbol{\alpha}_t|\boldsymbol{\alpha}_{0:t-1},\mathbf{y}_{1:t}) = \pi(\boldsymbol{\alpha}_0)\prod_{k=1}^{t}\pi(\boldsymbol{\alpha}_k|\boldsymbol{\alpha}_{0:k-1},\mathbf{y}_{1:k})$$

the importance weights and hence the posterior can be evaluated recursively. We model each source by a finite mixture of Gaussians, so:

$$p(\alpha_{i,t}) = \sum_{j=1}^{q_i}\rho_{i,j}\mathcal{N}(\alpha_{i,t};\mu_{i,j,t},\sigma_{i,j,t}^2); \sum_{j=1}^{q_i}\rho_{i,j} = 1, \tag{7}$$

where $\rho_{i,j}$ is the weight of the jth Gaussian component of the ith source, q_i is the number of Gaussian components for the ith source, while $\mu_{i,j,t}$ and $\sigma_{i,j,t}$ are the parameters which describe each Gaussian component. We define an index variable z_i which takes on a finite set of values $Z_i = \{1, \cdots, q_i\}$ and determines the active Gaussian component in the mixture at time t, so that $p(\alpha_{i,t}|z_{i,t} = j) = \mathcal{N}(\alpha_{i,t}; \mu_{i,j}, \sigma_{i,j}^2)$ and $p(z_{i,t} = j) = \rho_{i,j}$.

At time t let $\mathbf{z}_{1:m,t} \triangleq [z_{1,t} \cdots z_{m,t}]^T$. It is possible to describe the discrete probability distribution of $\mathbf{z}_{1:m,t}$ using the i.i.d. model: in this case, the indicators of the states $z_{i,t}$ have identical and independent distributions. If we want to introduce temporal correlation beween the samples of a particular source, we have to consider the first-order Markov model case, where the vector of the states evolves as a homogeneous Markov chain for $t > 1$:

$$p(\mathbf{z}_{1:m,t} = \mathbf{z}_l | \mathbf{z}_{1:m,t-1} = \mathbf{z}_j) = \prod_{i=1}^{m} p(z_{i,t} = [\mathbf{z}_l]_i | z_{i,t-1} = [\mathbf{z}_j]_i) = \prod_{i=1}^{m} \tau_{j,l}^{(i)}, \quad (8)$$

where $\tau_{j,l}^{(i)}$ is an element of the $q_i \times q_i$ real valued *transition matrix* for the states of the ith source, denoted by $\boldsymbol{\tau}^{(i)}$. The state transition can be thus parametrised by a set of m transition matrices $\boldsymbol{\tau}^{(i)}$, $i \in \{1, \cdots, m\}$.

Given the observations $\mathbf{y}_t$ (assuming that the number of sources m, the number of Gaussian components q_i for the ith source, and the number of sensors n are known), we would like to estimate all the following unknown parameters of interest, grouped together:

$$\boldsymbol{\theta}_{0,t} = [\boldsymbol{\alpha}_{1:m,0:t}, \mathbf{z}_{1:m,0:t}, \{\mu_{i,j,0:t}\}, \{\sigma_{i,j,0:t}^2\}, \{\boldsymbol{\tau}_{0:t}^{(i)}\}] \quad (9)$$

where we recall that $\boldsymbol{\alpha}_{1:m,0:t}$ are the sources, $\mathbf{z}_{1:m,0:t}$ is the matrix of the indicator variables which determines which Gaussian component is active at a particular time for each source, $\{\mu_{i,j,0:t}\}$ and $\{\sigma_{i,j,0:t}^2\}$ are the means and the variances of the jth Gaussian component of the ith source and $\{\boldsymbol{\tau}_{0:t}^{(i)}\}$ is the transition matrix for the evolution of $z_{i,0:t}$.

In order to reduce the size of the parameter set to be estimated, we will find the values of the mixing matrix subsequently, by means of the *Rao-Blackwellisation* technique [5]. As it was introduced before, it is not easy to sample directly from the optimal importance distribution: this is the reason why a sub-optimal method will be employed throughout taking the importance distribution at step t to be the *prior distribution* of the sources to be estimated. The mixture of Gaussians model allows for an easy factorization of the prior distribution into several easy-to-sample distributions related to the parameters which describe the model itself. Detailed information about the algorithm used here can be found in [6].

4 Numerical Experiments

The algorithm has been tested on two 64×64 mixtures of CMB and synchrotron radiation, at 100 GHz and 30 GHz. Synthetic but realistic maps of the sources have been provided by the Planck Technical Working Group [10]: in particular, the CMB map is generated synthetically to follow a Gaussian distribution as it is implied by the cold dark matter model widely accepted for CMB. The synchrotron template was obtained by extrapolating the 408 MHz radio map of Haslam et al. [9] to Planck frequency channels and resolution. The real antenna noise RMS maps are used to generate the additive space-varying noise samples. The average SNR is 10 dB.

We use three Gaussian components to approximate the synchrotron posterior distribution, and 1000 particles are generated at each step, for each parameter of interest. It is obvious that better approximations could be obtained by increasing the number of Gaussian components for each source, albeit an increase in computational cost. For the index distribution for the Gaussian component in the mixture a Dirichlet prior is adopted, while means and variances are drawn from Gaussian distributions centered at the value of the previous particles and with variance determined by drift parameters. The antenna RMS maps are completely known in our algorithm, therefore noise variances at each pixel are fixed in our implementation.

Since we are not aware of any other work which considers separation of non-stationary sources under non-stationary noise environment, we will compare our results with those obtained by the FastICA algorithm, which is one of the most widespread methods in source separation. As seen in Fig. 1, FastICA fails to recover CMB and for the synchrotron gives an estimate with high interference while the particle filter succeeds in recovering the original maps albeit some noise. Signal-to-interference results in Table 1 quantify this performance.

Table 1. Signal to Interference Ratio (SIR) values for FastICA and Particle Filtering.

SIR	FastICA	Particle Filtering
CMB	1.14 dB	10.46 dB
Synchrotron	1.62 dB	19.16 dB

5 Conclusions

In this work, we have presented a relatively novel technique, namely particle filtering, for the separation of the independent components in astrophysical images. In contrast with the other work in literature, this method provides a very flexible framework which can successfully account for the non-stationarity in the receiver noise and the sources, as well as the prior knowledge about the sources and the mixing matrix. The technique, in addition to providing point estimates, gives us the posteriors for the sources and the mixing matrix out of which inference can be made on various statistical measures. We demonstrated on realistic data that the particle filter provides significantly better results in comparison with one of the most widespread algorithms for source separation (FastICA), especially in the case of low SNR.

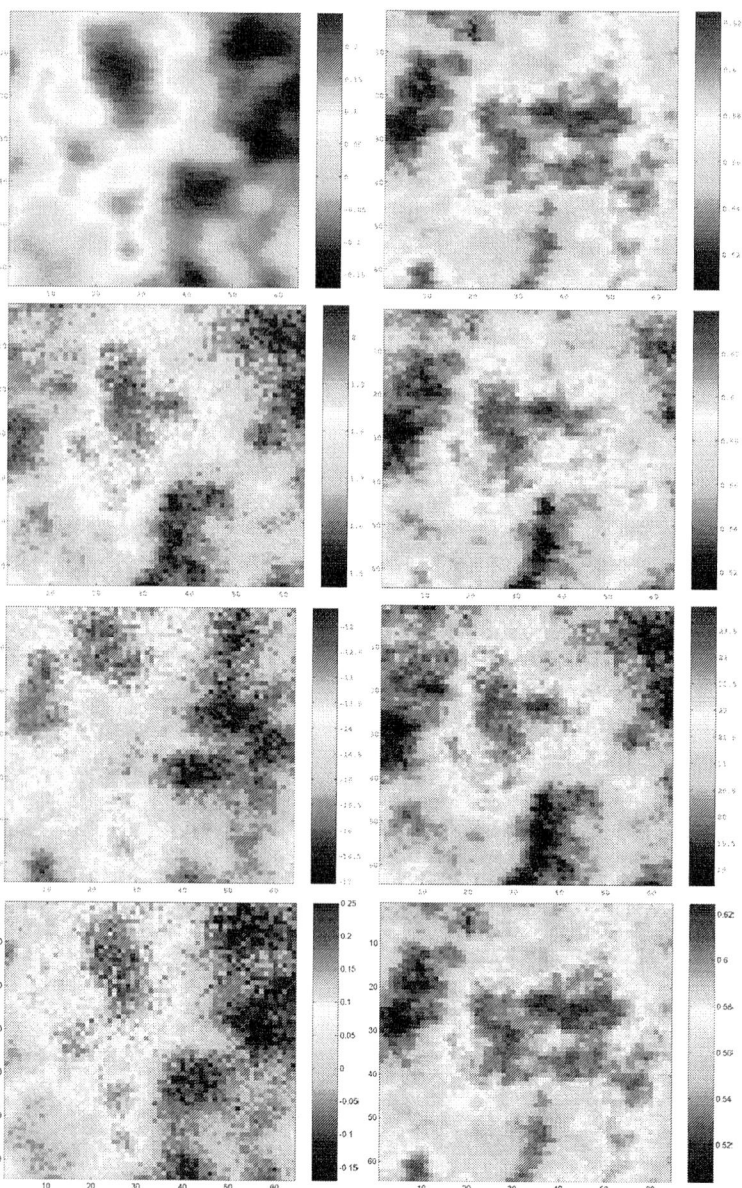

Fig. 1. From top to bottom: original CMB and Synchrotron signals; mixtures at 100 GHz and 30 GHz; FastICA estimates; particle filter estimates.

A fundamental step in the implementation of the particle filter algorithm is the choice of the importance function: in this work we have used the prior distribution as the importance function due to its analytical ease. Unfortunately this choice is far from being optimal since it does not allow us to exploit any information about the observed data. Currently, we are testing other importance functions that consider information from observations.

References

1. Ahmed A., Andrieu C., Doucet A., Rayner P. J. W.: On-line Non-stationary ICA Using Mixture Models. Proc. IEEE ICASSP, (2000) Vol. **5**, 3148–3151.
2. Andrieu C., Godsill S.J.: A Particle Filter for Model Based Audio Source Separation. Int. Work. on ICA and Blind Signal Separation, ICA 2000, Helsinki, Finland.
3. Baccigalupi C, Bedini L., Burigana C., De Zotti G., Farusi A., Maino D., Maris M., Perrotta F., Salerno E.: Neural Networks and the Separation of Cosmic Microwave Background and Astrophysical Signals in Sky Maps. Monthly Notices of the Royal Astronomical Society, **318** (2000), 769–780.
4. Cardoso J.-F., Snoussi H., Delabrouille J., Patanchon G.: Blind Separation of Noisy Gaussian Stationary Sources. Application to Cosmic Microwave Background Imaging. Proc. EUSIPCO, **1** (2002), 561–564.
5. Casella G., Robert C. P.: Monte Carlo Statistical Methods. Springer, (1999).
6. Costagli M., Kuruoğlu E. E., Ahmed A.: Source Separation of Astrophysical Images Using Particle Filters. ISTI-CNR Pisa, Italy – Technical Report 2003-TR-54.
7. Doucet A., De Freitas J. F. G., Gordon N. J.: Sequential Monte Carlo Methods in Practice. Springer-Verlag (2001).
8. Everson R. M., Roberts S. J.: Particle Filters for Non-stationary ICA. Advances in Independent Components Analysis, M. Girolami (Ed.) 23–41, Springer (2000).
9. Haslam C. G. T., Salter C. J., Stoffel H., Wilson W. E.: A 408 MHz All-Sky Continuum Survey. II - The Atlas of Contour Maps. Astronomy & Astrophysics, **47** (1982), 1.
10. http://astro.estec.esa.nl/planck/: The Home Page of Planck.
11. Hyvärinen A., Oja E.: A Fast Fixed-point Algorithm for Independent Component Analysis. Neural Computation, **9 (7)** (1997), 1483–1492.
12. Kuruoğlu E. E., Bedini L., Paratore M. T., Salerno E., Tonazzini A.: Source Separation in Astrophysical Maps Using Independent Factor Analysis. Neural Networks, **16** (2003), 479–491.
13. Kuruoglu, E. E., Comparetti, P. M.: Bayesian Source Separation of Astrophysical Images Using Markov Chain Monte Carlo. Proc. PHYSTAT (Statistical Problems in Particle Physics, Astrophysics and Cosmology), September 2003.
14. Maino D., Farusi A., Baccigalupi C., Perrotta F., Banday A. J., Bedini L., Burigana C., De Zotti G., Grski K. M., Salerno E.: All-Sky Astrophysical Component Separation with Fast Independent Component Analysis (FastICA). Monthly Notices of the Royal Astronomical Society, **334** (2002), 53–68.
15. Snoussi H., Patanchon G., Macias-Perez J., Mohammad-Djafari A., Delabrouille J.: Bayesian blind component separation for cosmic microwave background observation, AIP Proceedings of MaxEnt, (2001), 125–140.

Independent Component Analysis in the Watermarking of Digital Images

Juan José Murillo-Fuentes

Area de teoria de la Señal y Comunicaciones, Universidad de Sevilla
Paseo de los Descubrimientos sn, Sevilla 41092, Spain
murillo@esi.us.es
http://viento.us.es/~murillo

Abstract. The author proposes a new solution to the robust watermarking of digital images. This approach uses Independent Component Analysis (ICA) to project the image into a basis with its components as statistically independent as possible. The watermark is then introduced in this representation of the space. Thus, the change of basis is the key of the steganography problem. The method is improved with standard techniques such as spread-spreading mark generation, perceptual masking, and holographic properties. Some results are included to illustrate that the performance of the ICA watermarking is close to that of well-known methods.

1 Introduction

The ease of copying and editing digital images facilitates their unauthorized manipulation and misappropriation. Watermarking is one of the most common solutions proposed to protect owners rights. It consists of embedding another signal or mark into the to be protected host image. Watermark may be visible or invisible. This last watermark is designed to be transparent to the observer or user and to be detected by the owner. Invisible marks use the properties of the human visual system to minimize the perceptual distortion introduced. We have two main categories in the transparent watermarking. On the one hand we aim the watermark to be detected after severe attacks to enforce copyright ownership. This is the robust watermarking problem [1]. On the other hand, we may be interested in the detection of any change, i.e, in image authentication. In this case we face the fragile watermarking (FW) [2] of the host image.

Regarding RW and FW systems, some of them embed the mark in the spatial domain [3]. On the other hand we have methods working in any transform domain, either in the DCT or the DWT [1], [4], [5], [6]. Recently, independent component analysis (ICA) has been proposed in RW as watermarking detection algorithms [7], [8] or as a transform domain where to embed the mark [9]. In this paper we focus on this late approach. We will review the ICA watermarking method in [9] to greatly improve its performance. These results may be easily applied to the fragile watermark by including the ideas in [3], [5] or [6]. The paper is organized as follows. In Section 2 we develop a new embedding and

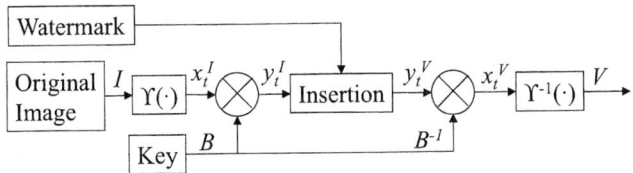

Fig. 1. Embedding algorithm.

detection method based on ICA. In Section 3 we will include some experiments to illustrate its performance. We shall end with conclusions.

2 Watermarking with ICA

The problem of independent component analysis (ICA) [10], [11] consists of obtaining from a set of components another set as statistically independent as possible. In the instantaneous linear model we just consider a $l \times 1$ ergodic vector x_t projected into a space of l independent components y_t. In matrix form this change of basis is represented by a $l \times l$ matrix B.

$$y_t = Bx_t \qquad (1)$$

where $t = 1, 2, \ldots$ are a set of samples. Very much literature have been devoted to ICA algorithms. We will use here the SICA algorithm [12].

If we divide an image, I in patches[1] of size $(k \times k)$ and then reshape them into column vectors, x_t, we have the k^2 components for that patch [13], [14]. We will denote this transformation as $x_t^I = \Upsilon(I, k)$. We then may apply ICA to project them into its independent components (IC) to which we perform any image processing [15]. The ICA watermarking method developed in this paper is based on this approach, since we embed the watermark in these IC as follows.

2.1 Embedding

In Fig. 1 we include a general basic architecture for a ICA based embedding algorithm. We assume than no other information than a key is needed in the detection algorithm in Fig. 2 to estimate the watermark. The embedding algorithm in Fig. 1 may be decomposed into the steps bellow.

Algorithm 1: Embedding.

1. *Image components*. Compute the components x_t^I of the $n \times m$ cover image I using $k \times k$ blocks.
2. *ICA components*. Compute its IC, $y_t^I = Bx_t^I$, using an ICA projection B, the key of the insertion method.

---

[1] We will assume no overlapping of these blocks.

3 *ICA watermarked image components.* Compute the IC of the marked image, y_t^V, by updating y_t^I with the watermark, W.
4 *Restoring the watermarked image.* Restore the watermarked image V from components $x_t^V = B^{-1} y_t^V$.

One of the advantages of the algorithm rest on the IC projection matrix B. An analysis of the image IC components for several images shows [15] that IC of images with similar features may be restored from a common set of basis (rows of the separating matrix B). Hence, it is possible to use ICA to define a set of basis functions to encode a group of images such as natural images or text scans. We may also conclude [15], [16] that the ICA projection B computed for one image may be successfully applied in the processing of another one. Particularly, if they are the same class (text images, natural scenes,...). We will exploit this feature by using a separating matrix B for a group of images. It is interesting to notice that this projection is a key of the method, as it is needed in the complete removal of the watermark.

The main function to define in Fig. 1 is the block *insertion*, i.e., step 3 of the embedding algorithm. There are several possibilities depending on the watermark generation. In [9] the authors proposed another image as watermark. In this paper we embed a spread spectrum mark, i.e., a message "modulated" by means of spread spectrum techniques (SS) [1], [4]. We pay special attention to methods hiding every bit of the message over the entire image ("holographic" property [17]) as the whole message may be recovered after cropping based attacks and they present better synchronization properties [18]. We propose the watermark to have the size of one component, $n/k \times m/k$. This watermark is computed as the circular convolution of a key-dependent pseudorandom image P and an image containing the bits of the message Q

$$W = P \otimes Q \qquad (2)$$

Let's M be a $p \times p$ matrix whose pixels are the bits of the message. We define matrix Q as follows

$$Q(i,j) = \sum_{rs} M(r,s) \delta(i - r \cdot n_r/2, j - s \cdot n_c/2) \qquad (3)$$

where $n_r = n/(k \cdot p)$ and $n_c = m/(k \cdot p)$. Hence matrix Q is a zero valued matrix except for the bits of the message, located at the center of each $n_r \times n_c$ block. Note that the convolution in (2) may be easily carried out by means of the two-dimensional Fourier transform. Once we have the watermark we perform a perceptual masking [19] to improve the invisibility of the watermark. Now we have the watermark ready to be embedded.

In the insertion block in Fig. 1 we have as inputs the IC components $y_t(i)$, $i = 1, ..., k^2$ arranged in descending order of magnitude, i.e., variance. This gives us the low frequency component first, then IC with the edges of the image and the very high-frequency, in this ordering. Hence, and similarly to other frequency transform watermarking algorithms [1], we propose to place the watermark into

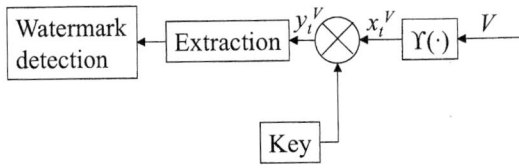

Fig. 2. Detection algorithm.

the r highest magnitude independent components, i.e., we embed the mark in the first components. We reshape the watermark into a row vector, y_t^W, to insert it in the cover image IC, $\boldsymbol{y}_t^V$, as follows

$$y_t^V(h) = y_t^I(h) + \alpha_h y_t^W \quad h = 1, \ldots, k^2 \tag{4}$$

where α_h is a scaling factor to control the perception of the watermark. We can view α_h as a relative measure of how much we must alter component h to alter the perceptual quality of the document. In the embedding of the watermark we have proposed to add the watermark component to the image. Other techniques such as the multiplicative or exponential approaches are possible [1]. In [9] we proposed a replacement of high-frequency components instead. This leads us to a blind approach but with worse performance.

2.2 Detection

The aim of this subsection is the detection of the watermark $\boldsymbol{W}$ from the watermarked image $\boldsymbol{V}$. We go back on the steps of the embedding Algorithm 1, as in Fig. 2. We first compute the components $\boldsymbol{x}_t^V$ and then the IC $\boldsymbol{y}_t^V$ by using the key $\boldsymbol{B}$. Finally, we estimate and detect the watermark. The watermark detection yields

Algorithm 2: Detection.

1. *Watermarked image components.* Compute the components $\boldsymbol{x}_t^V$ of the watermarked image $\boldsymbol{V}$ by dividing it in $k \times k$ patches.
2. *Watermarked image ICA components.* Compute the independent components $\boldsymbol{y}_t^V$ of the image as $\boldsymbol{y}_t^V = \boldsymbol{B}\boldsymbol{x}_t^V$.
3. *Watermark ICA components.* Extract the watermark from $\boldsymbol{y}_t^V$.
4. *Detection.* Estimate the message and the probability of watermark detection.

We have embedded a message modulated using spread spectrum techniques. This approach has multiple advantages. The main one is we achieve that detection and synchronization by simple correlation, i.e., a matched filter is a good detector. The watermark may be estimated, similarly to the detection in [1], as follows. We first subtract the cover image to the watermarked one, $\boldsymbol{J} = \boldsymbol{V} - \boldsymbol{I}$, and compute its IC, $\boldsymbol{y}_t^J = \boldsymbol{B}\boldsymbol{x}_t^J$. Then we average all components $h : \alpha_h \neq 0$, improving the signal (watermark) to noise (image+attacks) ratio,

$$\hat{\boldsymbol{y}}_t^W = \sum_{h:\alpha_h \neq 0} \boldsymbol{y}_t^J(h) \tag{5}$$

and reshape the resulting vector into matrix $\hat{\boldsymbol{W}}$. Then we estimate matrix $\boldsymbol{Q}$ in (3) by simple correlation as

$$\hat{\boldsymbol{Q}} = \boldsymbol{P} \otimes \hat{\boldsymbol{W}} \tag{6}$$

In order to compute the probability of detection we can locate the peaks of the correlation result $\hat{\boldsymbol{Q}}$ and compare their heights to the mean plus standard deviation of the rest of pixels. As this matrix is divided into $n_r \times n_c$ blocks we may sum every block

$$\mathcal{Q}(i,j) = \sum_{p=1}^{n/(kn_r)} \sum_{q=1}^{m/(kn_c)} \hat{\boldsymbol{Q}}(i + (p-1)n_r, j + (q-1)n_c) \tag{7}$$

and then compare the maximum point

$$c_{mx} = \max_{ij} \mathcal{Q}(i,j) \tag{8}$$

to the rest of values. In the following, we will denote by (i_{mx}, j_{mx}) those indexes satisfying $c_{mx} = \mathcal{Q}(i_{mx}, j_{mx})$. In order to get an statistic measure of this comparison we assume each entry $z = \{\mathcal{Q}(i_{mx}, j_{mx}) : (i,j) \neq (i_{mx}, j_{mx})\}$ to be distributed as a Gaussian random variable. We first estimate its mean $\bar{z}$ and variance σ_z^2. Then we compute the probability of detection as the probability of every other point different from c_{mx} to be lower than c_{mx} as

$$p_d = F_z(c_{mx}, \bar{z}, \sigma_z^2)^{n_r n_c - 1} \tag{9}$$

where $F_z(c, \bar{z}, \sigma_z^2) = Pb(z \leq c)$ is the cumulative distribution for z. Notice that we do not assume synchronization as we do not use $c_{mx} = \mathcal{Q}(n_r/2, n_c/2)$ but the maximum value for all (i,j). Finally, the bit error rate may be computed by estimating the message bits as follows,

$$\hat{M}(p,q) = \hat{\boldsymbol{Q}}(i_{max} + p \cdot n_r, j_{max} + q \cdot n_c) \tag{10}$$

3 Experimental Results

We next include an example of robust watermarking applied to the 512×512 intensity image, in the range (0,1), of Lena in Fig.3.a. We first computed $\boldsymbol{x}_t^I$ with $k = 3$ and then the IC of the image as $\boldsymbol{y}_t^I = \boldsymbol{B}\boldsymbol{x}_t^I$, where matrix $\boldsymbol{B}$ was the one obtained for another image. The watermark was generated as the spread version of a 2-dimensional message of 8×8 bits. The watermark was first multiplied by a perceptual mask of the image. Then it was added to the IC of the image number $h = 1$. Hence, we embed the watermark into the 11% of the components or transform domain coefficients. The final peak signal-to-noise ratio (PSNR) was 41 dB. The watermarked image is included in 3.b. We performed the following attacks and obtain the probability of detection and number of erroneous bits included in Tab.1. We first added white Gaussian noise with standard deviation $\sigma = 0.1$, see Fig.3.c. In Fig.3.d we requantized the image to 2^2 levels. Then we

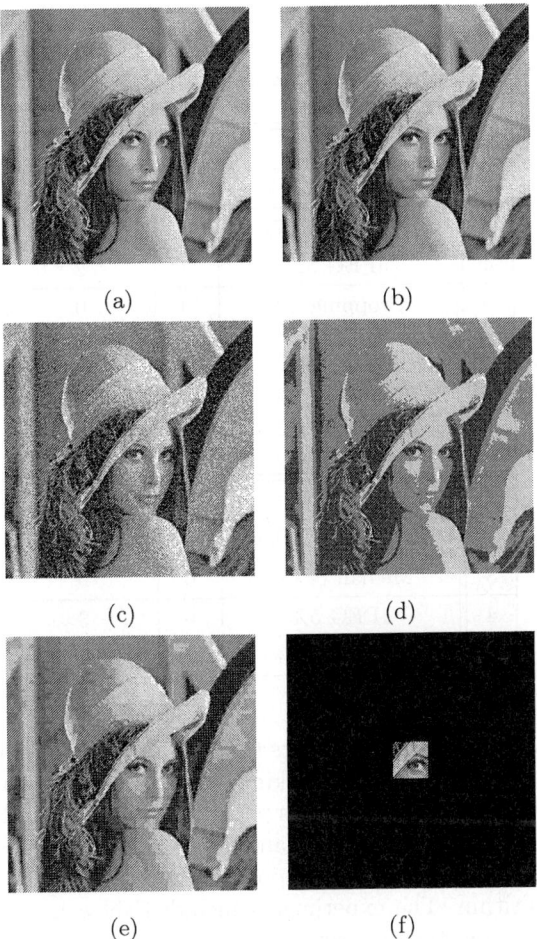

Fig. 3. Illustration of some attacks on the watermarked Image of Lena: original (a) and watermarked (b) images along with AWGN (c), quantizing (d), 7×7 median filtering (e), JPEG compression (f) and cropping (g) attacks.

applied 7 × 7 median filter. In Fig.3.e the image was JPEG compressed to 5% of its original size. Finally, we cropped the 98% of the image, see Fig. 3.f. The same features obtained for the DCT algorithm may be observed in Tab.2. We used the same watermarking generation to embed the watermark in the 128 × 128 most significant DCT transform coefficients, i.e., in the 7% of the transform coefficients. We may conclude that the ICA and DCT approaches have quite a similar performance.

4 Conclusions

In this paper we present a new approach to image watermarking based on independent component analysis. The starting point is the ICA based image pro-

Table 1. Probability of detection, p_d, and number of erroneous bits for different attacks performed on the ICA watermarked image of LENA PSNR=41 dB.

N^o	Attack	$1-p_d$	N^o Err Bits
1	AWGN	0	1
2	Quantization 2^2 levels	0	1
3	Median (7×7)	0	2
4	JPEG 5%	0	2
5	Cropping 2%	0	0

Table 2. Probability of detection, p_d, and number of erroneous bits for different attacks performed on the DCT watermarked image of LENA PSNR=41 dB.

N^o	Attack	$1-p_d$	N^o Err Bits
1	AWGN	0	0
2	Quantization 2^2 levels	0	2
3	Median (7×7)	0	0
4	JPEG 5%	0	3
5	Cropping 2%	1	2

cessing in [15]. We apply these concepts to write a new watermarking algorithm. The problem of robust blind watermarking is addressed. The keys of the steganographic method are the change of basis performed by applying ICA to the image and the seed to generate the pseudorandom sequence in the SS watermark. A perceptual mask and a holographic approach have been introduced to greatly improve the algorithm. The experiments included show how this new method successes in extracting the watermark even when the image has been severely attacked. The results were quite close to that of the DCT method if we embed the image into the first IC, i.e, in the low frequencies. We may conclude that ICA may be successfully applied to the digital watermarking of images.

References

1. Cox, I., Kilian, J., Leighton, T., Shamoon, T.: Secure spread spectrum watermarking for multimedia. IEEE Trans. on Image Processing **6** (1997) 1673–1687
2. Lin, E., Delp, E.: A review of fragile image watermarks. In: Multimedia and Security Workshop (ACM Multimedia '99) Multimedia Contents, Orlando, Florida (1999) 25–29
3. Wolfgang, R., Delp, E.J.: Fragile watermarking using the vw2d watermark. In: SPIE/IS&T International Conference on Security and Watermarking of Multimedia Contents. Volume 3657., San José, CA (1999) 204–213
4. Hernández, J.R., Pérez-González, F.: The impact of channel coding on the performance of spatial watermarking for copyright protection. In: Proc. ICASSP'98. Volume V., Seattle, USA (1998) 2973–2976

5. Wu, M., Liu, B.: Watermarking for image authentification. In: IEEE International Conference on Image Processing. Volume 2., Chicago, Illinois (1998) 437–441
6. Kundur, D., Hatzinakos, D.: Towards a telltale watermarking tehnique for tamper-proofing. In: IEEE International Conference on Image Processing. Volume 2., Chicago, Illinois (1998) 409–413
7. Sattar, D.Y.F., Ma, K.K.: Watermark detection and extraction using independent component analysis method. In: IEEE-EURASIP NSIP'01 Workshop, Baltimore, USA (2001)
8. Liu, J., Zhang, X., Sun, J., Lagunas, M.A.: A digital watermarking scheme based on ICA detection. In: Proc. ICA2003, Nara, Japan (2003) 215–220
9. Murillo-Fuentes, J., Molina-Bulla, H., González-Serrano, F.: Independent component analysis applied to digital image watermarking. In: Proc. ICASSP'01. Volume III., Salt Lake City, USA (2001) 1997–2000
10. Comon, P.: Independent component analysis, a new concept? Signal Processing **36** (1994) 287–314
11. Caamaño-Fernandez, A., Boloix-Tortosa, R., Ramos, J., Murillo-Fuentes, J.J.: High order statistics in multiuser detection. IEEE Trans. on Man and Cybernetics A. Accepted for publication (2004)
12. Murillo-Fuentes, J.J., González-Serrano, F.J.: A sinusoidal contrast function for the blind separation of statistically independent sources. IEEE Trans. on Signal Processing. Accepted for publication (2004)
13. Bell, A.J., Sejnowski, T.J.: Edges are the independent components of natural scenes. In Mozer, M.C., Jordan, M.I., Petsche, T., eds.: Advances in Neural Information Processing Systems. Volume 9., The MIT Press (1997) 831
14. Hyvärinen, A., Karhunen, J., Oja, E.: Independent component analysis. John Willey and Sons (2001)
15. Lee, T., Lewicki, M., Sejnowski, T.: Unsupervised classification with non-gaussian mixture models using ICA. In: Advances in Neural Information Processing Systems. Volume 11., Cambridge, MA, The MIT Press (1999) 58–64
16. Bugallo, M.F., Dapena, A., Castedo, L.: Image compression via independent component analysis. In: Learning, Leganés (2000)
17. Bruckstein, A., Richardson, T.: A holographic transform domain image watermarking method. Circuits, Systems, and Signal Processing **17** (1998) 361–389
18. Mora-Jimenez, I., Navia-Vazquez, A.: A new spread spectrum watermarking method with self-synchronization capabilities. In: Proc. ICIP2000, Vancouver, BC, Canada (2000)
19. Wolfgang, R., Podilchuk, C., Delp, E.J.: Perceptual watermarks for digital images and video. Proc. of the IEEE **87** (1999) 1108–1126

Spatio-chromatic ICA of a Mosaiced Color Image

David Alleysson[1] and Sabine Süsstrunk[2]

[1] Laboratory for Psychology and NeuroCognition, CNRS UMR 5105
Université Pierre-Mendès France, Grenoble, France
David.Alleysson@upmf-grenoble.fr
[2] Audiovisual Communications Laboratory
École Polytechnique Fédérale de Lausanne, Switzerland
Sabine.Susstrunk@epfl.ch

Abstract. We analyze whether Independant Component Analysis (ICA) is an appropriate tool for estimating spatial information in spatio-chromatic mosaiced color images. In previous studies, ICA analysis of natural color scenes (Hoyer et al. 2000; Tailor et al., 2000; Wachtler et al., 2001; Lee et al. 2002) have shown the emergence of achromatic patterns that can be used for luminance estimation. However, these analysis are based on fully defined spatio-chromatic images, i.e. three or more chromatic values per pixel. In case of a reduced spatio-chromatic set with a single chromatic measure per pixel, such as present in the retina or in CFA images, we found that ICA is not an appropriate tool for estimating spatial information. By extension, we discuss that the relationship between natural image statistics and the visual system does not remain valid if we take into account the spatio-chromatic sampling by cone photoreceptors.

1 Introduction

The statistical analysis of natural scenes, as viewed by human observers, has given new insight into the processing and functionality of the human visual system. Pioneer work by Field (1987) and Barlow (1989) has established the relation between redundancy reduction in natural scenes and the visual system's receptive fields. Using gray-scale natural scene imagery, Olshausen & Field (1996) show that representing images with sparse (less redundant) code leads to spatial basis functions that are oriented, localized, and band-pass, and resemble the receptive field structures of the primary cortex cells. Bell & Sejnowski (1997) found that sparseness could be appropriately formalized using Independent Component Analysis (ICA), and show that independent components of natural scenes act as edge filters.

For the case of color, Buchsbaum & Gottschalk (1983) use Principal Component Analysis (PCA) of L, M, and S cone signals to derive post-receptoral mechanisms: luminance and opponent chromatic channels (blue minus yellow, and red minus green). They propose that this de-correlated coding reduces the information transmitted to the optical nerve. Later, Attick & Redlich (1992) formalized the relation between natural color scenes and retinal functions. They show that a retinal filter is consistent with a whitening process of the natural scene structure when noise is taken into account.

Finally, the use of hyperspectral images to simulate cone responses has allowed to precisely analyze the spatio-chromatic structure of natural scenes and confirmed previous studies (Wachtler et al., 2001). Ruderman et al. (1998) show that the principal components of natural color images, as sampled by cones, are consistent with post-receptoral receptive fields and provide reduced signals. Using ICA, Hoyer et al. (2000), Tailor et al. (2000) and Lee et al. (2002) found that natural color image statistics could account for simple and complex color opponent receptive fields in the primary cortex.

From these studies, it seems that the post-receptoral mechanisms of the human visual system correspond to a statistical analysis of natural scenes and provide a redundancy reduction. But none of these studies take into account that cone sampling already results in a reduced spatio-chromatic signal. In the retina, the three types of cones (L, M and S) form a mosaic such that only a single chromatic sensitivity is sampled at each spatial location. Thus, the spatio-chromatic signals are already reduced by a factor of three compared to fully defined spatio-chromatic signals of a natural scene (or color image). Doi et al. (2003) did propose a study where the cone mosaic is taken into account. They used a local arrangement of cones (127), from which they sampled LMS responses to construct vectors and perform ICA analysis. Although this method gives interesting results, it is still not realistic for simulating cone sampling since only a small part of the entire mosaic is used. Their study actually corresponds to analyzing the signal of a part of the retina scanning a natural scene.

In a previous paper (Alleysson & Süsstrunk, 2004a), we studied whether we can find a similar correspondence with the processing of the human visual system by statistical analysis of natural color images sampled with a spatio-chromatic mosaic. In that preliminary study, we have restricted our analysis to Principal Component Analysis (PCA), a second order statistical analysis that performs a simple de-correlation of a signal. We used RGB color images instead of LMS images constructed from hyperspectral data, and we assumed a regular arrangement of RGB samples instead of a random arrangement, such as given by the cone distribution in the retina. Actually, this experimental set-up coincides with many digital camera sensors, since most are single-chip and use a Color Filter Array (CFA) to provide color responses. Such systems sample a single chromatic sensitivity per pixel and need to interpolate the missing information to render color images (Alleysson et al., 2002). We then investigated if a spatio-chromatic analysis using PCA is able to help the reconstruction of the spatial information.

The conclusion of the previous study is that spatial information cannot be recovered with a second order statistical analysis. In this paper, we extend our approach and study if a higher order statistical analysis, given by Independant Component Analysis, provides a method to separate spatial information from chromatic information in a spatio-chromatic mosaiced color image. If such an approch is successfull, it would provide insights into the human visual processing and help design better demosaicing algorithms for CFA images.

2 Spatio-chromatic ICA of Color Images

Independent Component Analysis is a high order statistical analysis, which supposes that the signals measured by a sensor (observations) are an unknown linear mixing of unknown independent sources. Rather than exploiting the correlation between observations, as is done with Principal Component Analysis, high order statistics are used in ICA.

Suppose N measures of T random variables x_{ij} ($i \in [1..N], j \in [1..T]$). ICA presumes that these observations come from an instantaneous linear mixture (i.e. verified for all i), given by $\mathbf{A}$ of T independent sources $\mathbf{s}_j$, as follows:

$$\mathbf{x}_j = \mathbf{s}_j \mathbf{A} \qquad (1)$$

Here we consider only the case where the number of sources is equal to the number of observations. ICA analysis will try to find T independent vectors $\mathbf{u}_j$, which are representatives of the T vectors $\mathbf{x}_j$ through a linear transformation $\mathbf{B}$: $\mathbf{u}_j = \mathbf{x}_j \mathbf{B}$ for all i. The estimated vectors $\mathbf{u}_j$ are independent when their probability distributions $f_\mathbf{u}$ factorize in $f_\mathbf{u}(\mathbf{u}) = \Pi_j f_{\mathbf{u}_j}(\mathbf{u}_j)$ or, equivalently, when there is zero mutual information between them: $I(\mathbf{u}_j, \mathbf{u}_k) = 0$ (Bell & Sejnowski, 1997).

In this paper, we have used a freeware Matlab Toolbox called *FastICA* (Hyvärinen, 2004) for the ICA analysis, which uses an approximation of the negative normalized entropy (negentropy) as a criterion for independance, and is equivalent to a fixed point infomax or maximum likelihood approach (Hyvärinen & Oja, 2000).

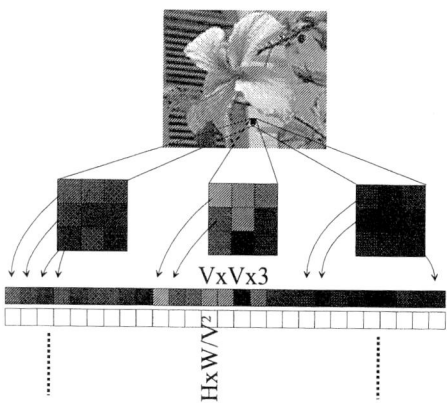

Fig. 1. Decomposition of the original image into a spatio-chromatic vector

Given an image $I_{i,j,c}$, defined by a three-dimensional matrix of size $H \times W \times 3$, we can construct a two-dimensional matrix X that contains for each row a vector x_j composed of spatial neighbors of size V of a pixel for all three color layers (see Figure 1). In our previous paper (Alleysson & Süsstrunk, 2004a), we used the neigborhood of each pixel. In this paper, we reduced the matrix size by using only the neigborhood of a pixel separated by the interval of V. The estimation of the unmixing matrix B gives the same result in both conditions, certainly because of the stationarity of spatial variables in images. Thus, the size of X is $(HW/V^2) \times (3V^2)$. This matrix, on which we can apply an ICA, can be interpreted as containing on each row a representation of the spatio-chromatic random variables of a color image.

Once the matrix X is constructed from the image, the ICA analysis is performed. The resulting matrices U_e, A_e, B_e are the estimated sources $[u_j]$ for all i, mixing, and unmixing matrices, respectively. The repesentation of estimated sources is not very usefull. Sources are estimated through a permutation and scaling indetermination (Hyrävinen & Oja, 2000), which could modify the role and gain of color, horizontal, and vertical variables in the spatio-chromatic estimated sources. But the column a_i of A_e represent the basis functions that are applied to the sources to form the observations. In Figure 2, these functions are represented for a 3×3 neigborhood in the original image. They are ordered in decreasing order, according to $\sum_i |a_{ij}|$. Each function is recast on a $3 \times 3 \times 3$ arrangement and is represented as a color patch rescaled between $[0, 1]$.

Fig. 2. Basis functions of the ICA of the color image ordered according to $\sum_i |a_{ij}|$

It is possible to partially reconstruct the image using only a few basis functions. This can be achieved by exchanging the entries of the undesired column vectors in A_e with zero. Call A_1 the matrix A_e in which some column vectors have been replaced with zeros. The partially reconstructed two-dimensional image matrix is then given by $X_1 = U_e A_1$ from which we can reconstruct an image. As an example, Figure 3 shows reconstruction examples with the first, second, third, fourth, and seventh basis functions, as well as the reconstruction with first through seventh components.

Fig. 3. Partial reconstruction of the original image (Figure 2) using only a few basis functions. (Top) Using the first, the second, and the third basis function. (Bottom) The reconstruction using the fifth, the seventh, and using first through seventh basis functions

The results of Figures 2 and 3 are compatible with pevious results in spatio-chromatic ICA (Hoyer et al. 2000, Wachtler et al., 2001; Tailor et al. 2000), where it was found that basis functions are composed of achromatic band-pass oriented functions and red/green and blue/yellow basis functions. We can see in Figure 2 that several basis functions are achromatic and do not carry chromatic information, but a part of the spatial information. This is even better illustrated in the partial image reconstruction. For example, the reconstruction with only the first component (top-left of Figure 3) is mainly achromatic with coarse luminance. Also, the reconstruction with only the third and fourth basis functions contain horizontal, respectively vertical achromatic information of the original image. Incidentally, all achromatic basis functions in Figure 2 carry part of spatial information of the original image, and the chromatic basis functions are composed of red/green and blue/yellow components.

3 Spatio-chromatic ICA of a Mosaiced Images

Retinal images and images captured with single-chip digital cameras with a Color Filter Array (CFA) have only a single chromatic value per spatial location. Figure 4 illustrates a mosaiced image according to the Bayer CFA. It has been shown that such image can be decomposed into a sum of an achromatic image with full spatial resolution plus a subsampled chromatic image (Alleysson et al. 2002). The sub-sampling of chromatic information at the sensor level does not affect luminance but only chrominance. Moreover, luminance and chrominance have localized energies in the Fourier domain, allowing their estimation by frequency selection. A full color image is then obtain by interpolating chrominance and add it to the estimated luminance. In many cases, however, luminance and chrominance alias in the Fourier domain when their representations occupy too large of a region in the frequency spectrum and overlap (Alleysson & Süsstrunk, 2004b). In that case, the frequency selection algorithm cannot separate the two. Thus, one can investigate if a statistical method allows separating luminance and chrominance in a mosaiced image. Moereover, a linear decomposition method should work because the luminance and chrominance are composed linearly in a mosaiced image.

Since there are achromatic basis functions in an ICA decomposition of a color image, we can hope that there are also achromatic functions in an ICA decompostion of a mosaiced image. In that case, it should be possible to use only these achromatic functions to reconstruct the luminance part of a mosaiced image, and to improve existing demosaicing algorithms.

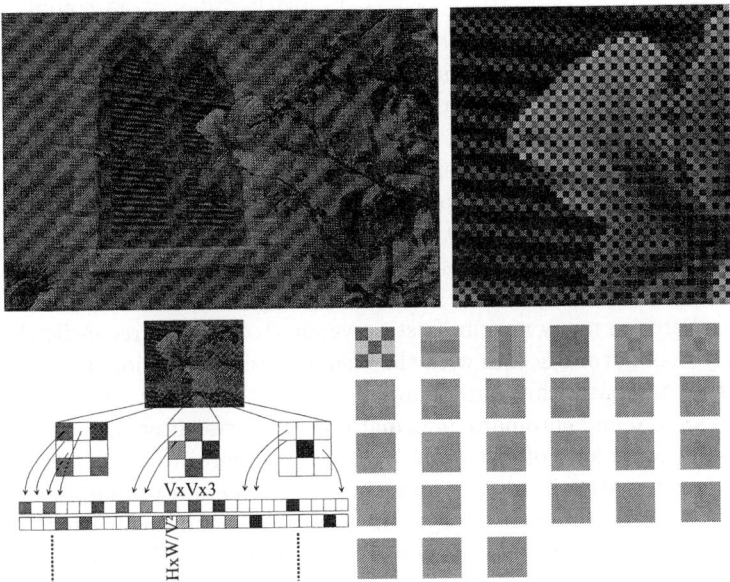

Fig. 4. (Top) A mosaiced image according to the Bayer CFA. The image appears green as there are twice as many green pixels as red and blue pixels. (Bottom) The de-composition of a mosaiced image into a two-dimensional matrix and the resulting basis functions of the ICA decomposition

It could be seen in Figure 4 that none of the basis function have an achromatic characteristic. This mean that, like the PCA, ICA is not a good tool for estimating the achromatic information in a mosaiced color image.

4 Conclusion

Principal and Independent Component Analysis are not adequate methods to estimate statistically the luminance information in a mosaiced color image. Possible reasons for this failure could be the difference in spatial resolution of the luminance and chromatic signal in a mosaiced image that is not taken into account by using a spatio-chromatic decomposition based on a fixed neigborhood size. Also, even if the luminance and chrominance additively compose a mosaiced image, it is possible that in case of aliasing between them, a non-linear method should be used. Finally, *FastICA* provides a way to obtain independent signals from mixed ones, but there may be other ICA methods that do converge to an independence between achromatic and chromatic signals.

With regards to the visual system, the relationship between statistical analysis of natural scenes and the visual system does not remain clear when considering the first step of neural activity in the visual system, i.e. the sampling of visual information by cones. It was quite intuitive that the decorrelation process of the PCA described in the previous paper (Alleysson & Süsstrunk, 2004a) would not be able to separate spatial and chromatic information in spatio-chromatic samples. For example, in the schema proposed by Bell & Sejnowski (1997), decorrelation is the first step of visual processing and is part of retinal processing. There is a correspondance between retinal receptive fields and a decorrelation process. Atick & Redlich 1992 also favor the hypothesis of decorrelation by retinal functions. However, it is known that the separation of luminance and chominance does not completly arise in the retina, because there are pathways at the output of the retina that contain multiplexed spatial and chromatic signals. But it is less intuitive that an independant decomposition is still not able to perform such a separation, as this analysis is supposedly happening in the brain, at least at the primary cortex level.

Thus, there are many new interesting venues to study to reconcile the statistical analysis of natural scenes with the visual system's functionality, taking into account the real spatio-chromatic sampling by the cones. One is to find if the separation of spatial and chromatic information could arise after the primary cortex level by some other kind of statistical analysis. Another is to find if the processing of the reduced spatio-chromatic signals in the retina and primary cortex could be equivalent to a statistical analysis of a complete, fully populated chromatic signal. In that case, the visual system could give insights for statistical analysis of natural scenes based on reduced information. For example, a mosaiced image contains already a kind of decorrelated information because of subsampling (if neighboring pixels have high correlation, their correlation decreases after subsampling). It is therefore plausible that the retino-cortical projection provides an independent transformation, without having recourse to a complicated and iterative process.

References

Alleysson, D., Süsstrunk, S., Hérault, J. Color demosaicing by estimating luminance and opponent chromatic signals in Fourier domain. Proc. IS&T/SID 10th Color Imaging Conference, Scottsdale, 2002, 331-336.

Alleysson, D., Süsstrunk, S., Spatio-chromatic PCA analysis of a mosaiced image. Proc. IS&T 2nd European Conf. on Color in Graphics, Image and Vision (CGIV'04). Aachen, Germany, 2004a, 311-314.

Alleysson, D., Süsstrunk, S., Aliasing in digital cameras. SPIE Electronic Imaging Newsletter, Special Issue on Smart Image Acquisition and Processing, **14**(1), 2004b, 1.

Atick J.J., Redlich A.N., What does the retina know about natural scenes? Neural Computation **4**, 1992, 196-210.

Barlow H.B., Unsupervised learning. Neural Computation **1**, 1989, 295-311.

Bauchsbaum G., Gottschalk A., Trichromacy, opponent colours coding and optimum colour information in the retina. Proc. R. Soc. Lond. **B220**, 1983, 89-113.

Bell A.J., Sejnowski T.J., The independent components of natural scenes are edge filters. Vis. Res., 1997, 3327-3338.

Doi E., Inui T., Lee T.W., Wachtler T., Sejnowski T.J., Spatio-chromatic receptive field properties derived from information-theoretic analyses of cone mosaic response to natural scenes. Neural Comp. **15**, 2003, 397-417.

Field D.J., Relations between the statistics of natural images and the response properties of cortical cells. J. Opt. Soc. Am. A **4**, 1987, 2379-2394.

Hoyer, P.O. and Hyvärinen A. Independent Component Analysis Applied to Feature Extraction from Colour and Stereo Images. Network: Computation in Neural Systems, **11**, 2000, (3):191-210.

Hyvärinen, A., Oja, E. Independent Component Analysis: Algorithms and Applications. Neural Networks, **13**, 2000, 411-430.

Hyvärinen, A. The FastICA package for Matlab.
http://www.cis.hut.fi/projects/ica/fastica/, 2004.

Lee T., Wachtler T. and Sejnowski T.J., Color opponency is an efficient representation of spectral properties in natural scenes. Vis. Res. **42**, 2002, 2095-2103.

Olshausen B.A., Field D.J., Emergence of simple-cell receptive field properties by learning a sparse code for natural images. Nature **381**, 1996, 607-609.

Ruderman D.L., Cronin T.W., Chiao C.C., Statistics of cone responses to natural images: implication for visual coding. J. Opt. Soc. Am., **15**, 1998, 2036-2045.

Tailor D.R, Finkel L.F., Buchsbaum G., Color-opponent receptive fields derived from independent component analysis of natural images. Vis. Res, **40**, 2000, 2071-2076.

Wachtler T., Lee T.W., Sejnowski T.J., Chromatic structure of natural scenes. J. Opt. Soc. Am. A **18**, 2001, 65-77.

An Extended Maximum Likelihood Approach for the Robust Blind Separation of Autocorrelated Images from Noisy Mixtures

Ivan Gerace[1], Francesco Cricco[1], and Anna Tonazzini[2]

[1] Dipartimento di Matematica e Informatica
Università degli Studi di Perugia
Via Vanvitelli, 1, I-06123 Perugia, Italy
{gerace,cricco}@dipmat.unipg.it
[2] Istituto di Scienza e Tecnologie dell'Informazione
Consiglio Nazionale delle Ricerche
Via G. Moruzzi 1, I-56124 Pisa, Italy
{anna.tonazzini}@isti.cnr.it

Abstract. In this paper we consider the problem of separating autocorrelated source images from linear mixtures with unknown coefficients, in presence of even significant noise. Assuming the statistical independence of the sources, we formulate the problem in a Bayesian estimation framework, and describe local correlation within the individual source images through the use of suitable Gibbs priors, accounting also for well-behaved edges in the images. Based on an extension of the Maximum Likelihood approach to ICA, we derive an algorithm for recovering the mixing matrix that makes the estimated sources fit the known properties of the original sources. The preliminary experimental results on synthetic mixtures showed that a significant robustness against noise, both stationary and non-stationary, can be achieved even by using generic autocorrelation models.

1 Introduction

Formerly developed for signal processing problems, such as the "cocktail party" problem in audio and speech processing, Blind Source Separation (BSS) and Independent Component Analysis (ICA) techniques have recently shown a great potentiality for solving important image processing and computer vision problems [1]. Indeed, in many imaging fields, we have to cope with observations or maps that are linear mixtures of images with unknown coefficients. These maps cannot be properly interpreted, unless some strategy is adopted for separately extracting the various component images. The linear assumption is often physically grounded, as for instance in the case of radiation sky maps in astrophysics [2]. In other cases, it represents a reasonable approximation of more complex combination phenomena, as for overlapped texts in ancient or degraded documents and in palimpsests [3].

Earlier ICA methods for linear BSS were only based on the assumption of the mutual statistical independence of the sources, enforced in different manners [4][5], and were designed for noiseless mixtures [6]. Since in most real-world applications dealing with noisy data is an unavoidable need, several methods have then been proposed for the separation of noisy mixtures [7][8][9]. Nevertheless, though providing satisfactory estimates of the mixing matrix, these methods still produce noisy source estimates, due to the typical ill-conditioning of the mixing matrix. On the other hand, autocorrelation constraints have been proved to be effective for achieving stable solutions in many inverse problems, and especially in those dealing with images, where these constraints correspond to natural features of real physical maps and scenes. Even for the highly underdetermined BSS problem, a way to jointly obtain robust estimates for both the mixing matrix and the sources could be to incorporate into the problem available information about autocorrelation properties of the single sources [10].

The Bayesian estimation setup offers a natural and flexible way to account for prior knowledge we may have about a problem, and permits to formulate the BSS problem as the joint Maximum A Posteriori (MAP) estimation of the mixing matrix and the sources [11]. In this paper, we apply Bayesian estimation to regularize the blind separation of noisy mixtures of images by means of local autocorrelation constraints. To keep generality, we incorporate generic local smoothness properties for the individual sources through Markov Random Fields (MRF) models, that allow also for retaining the independence assumption of the ICA approach. According to the most general assessment of BSS, we assume that no prior information is available on the mixing matrix, but extend the MRF model to account for regularity constraints on the image edges as well. This is an important issue, since edges constitute essential features to be preserved in an image, for analysis and understanding purposes. Furthermore, instead of the usual joint MAP estimation of both the mixing and the sources, we propose a novel estimation strategy, based on the point of view that the best mixing matrix is the one that makes the related estimated sources fit the known properties of the original sources. Thus, we reformulate the problem as the estimation of the mixing alone, based on the source priors, while the sources are kept clamped to their MAP estimate, for any status of the mixing. This can be viewed as an extension to noisy data of the Maximum Likelihood ICA approach for noiseless data. From the theoretical scheme, reasonable approximations are derived which allow for reducing the computational complexity, and finding a remedy to other drawbacks, such as the unavailability of analytical formulas for the sources viewed as functions of the mixing, and non-convexity of the priors. These will make the method computational efficient and still effective.

The paper is organized as follows. In Section 2, the principles of Bayesian ICA will be revised, and the joint estimation approach will be reformulated in terms of estimation of the mixing alone, based on the source priors. In Section 3, our choice of the edge-preserving priors will be described, and the estimation algorithm will be derived. Finally, Section 4 will be devoted to preliminary experimental results and concluding remarks.

2 Bayesian ICA and Estimation Strategy

The data generation model for a linear and instantaneous BSS problem is:

$$\mathbf{x}(t) = A\mathbf{s}(t) + \mathbf{n}(t) \qquad t = 1, 2, ..., T \qquad (1)$$

where $\mathbf{x}(t)$ is the column vector of the measurements, $\mathbf{s}(t)$ is the vector of the unknown sources, and $\mathbf{n}(t)$ is the noise or measurement error vector, at location t, and A is the unknown mixing matrix. Of course, in imaging location t stands for the couple of pixel indices. Although not necessary, for simplicity sake we assume the same number N of measured and source signals, so that A is an $N \times N$ matrix. Vectors $\mathbf{s}_i = (s_i(1), s_i(2), ..., s_i(T))^T$, $i = 1, 2, ..., N$, represent the lexicographically ordered notation of the various sources, and $\mathbf{s} = (\mathbf{s}(1), ..., \mathbf{s}(T))$, is the matrix whose t-th column contains the N sources at location t and whose i-th row is the source $\mathbf{s}_i$. These definitions extend to data and noise as well.

Obviously, solving the system in eq. (1) with respect to both A and $\mathbf{s}$ would be an undetermined problem, unless more information is exploited. The kind of information used in the ICA approach is independence and non-Gaussianity of the sources. Assuming to know the prior distribution for each source, the joint prior distribution for $\mathbf{s}$ is thus given by:

$$P(\mathbf{s}(t)) = \prod_{i=1}^{N} P_i(s_i(t)) \quad \forall t. \qquad (2)$$

In the noiseless case, the separation problem can be formulated as the maximization of eq. (2), subject to the constraint $\mathbf{x} = A\mathbf{s}$. This is equivalent to the search for a matrix W, $W = (\mathbf{w}_1, \mathbf{w}_2, ..., \mathbf{w}_N)^T$, such that, when applied to the data $\mathbf{x}$, it produces the set of vectors $\mathbf{w}_i^T \mathbf{x}$ that are maximally independent, and whose distributions are given by the P_i. By taking the logarithm of eq. (2), ICA algorithms solve the following problem:

$$\hat{W} = arg \max_{W} \sum_{t} \sum_{i} log P_i(\mathbf{w}_i^T \mathbf{x}(t)) + T log |det(W)|. \qquad (3)$$

which corresponds to a Maximum Likelihood (ML) estimation of W, in that no a priori information on W are exploited. Matrix $\hat{W}$ is an estimate of A^{-1}, up to arbitrary scale factors and permutations of the columns. Hence, each vector $\hat{\mathbf{s}}_i = \hat{\mathbf{w}}_i^T \mathbf{x}$ is one of the original source vectors up to a scale factor. To enforce non-Gaussianity, generic super-Gaussian or sub-Gaussian distributions can be used as priors for the sources. These have proven to give very good estimates for the mixing matrix and for the sources as well, no matter of the true source distributions, which, on the other hand, are usually unknown [6].

When the data are noisy and/or information is available on the unknowns, the problem of estimating the mixing matrix A and the source samples $\mathbf{s}$ could be stated as the joint Maximum A Posteriori (MAP) estimation problem:

$$(\hat{\mathbf{s}}, \hat{A}) = arg \max_{\mathbf{s}, A} P(\mathbf{s}, A, |\mathbf{x}) = arg \max_{\mathbf{s}, A} P(\mathbf{x}|\mathbf{s}, A) P(\mathbf{s}) P(A) \qquad (4)$$

where, from the independence assumption, $P(\mathbf{s})$ is given as in eq. (2). This problem is usually approached by means of alternating componentwise maximization with respect to the two sets of variables in turn. In [12], we proposed an implementation scheme which ensures convergence and has limited computational complexity. This is based on an overall simulated annealing for the estimation of A, interrupted at the end of each Metropolis cycle, to perform an update of the sources $\mathbf{s}$. Such a scheme can cope with very general assumptions about the involved distributions. In a first application, we adopted edge-preserving convex priors, assumed white, Gaussian noise and no prior for the mixing, so that the problem resulted globally convex, and the scheme was ensured to converge to the global maximum by means of a gradient ascent updating for $\mathbf{s}$ and an analytic updating formula for A. Satisfactory results were obtained for the separation of images from noisy mixtures, also in the case of non-stationary noise, and some robustness against cross-correlated sources was observed as well.

In this paper, still for non-informative $P(A)$, we propose a novel formulation of the estimation process, and a more exhaustive autocorrelation model, which includes regularity of both intensity and edges in the images. This model will be presented in the next section. The estimation process is formulated as follows:

$$\hat{A} = arg \max_A P(\hat{\mathbf{s}}(A)) \tag{5}$$

$$\hat{\mathbf{s}}(A) = arg \max_\mathbf{s} P(\mathbf{x}|\mathbf{s}, A) P(\mathbf{s}). \tag{6}$$

The rationale for this new formulation is the looking for the mixing matrix that makes the estimated sources fit the a priori knowledge we possess about the original sources. Thus, the original joint MAP estimation is reformulated as the ML estimation of the mixing alone, based on the source priors, while the sources are kept clamped to their MAP estimate, for any status of the mixing. The dependence of the mixing on the data is indirectly retained through the sources.

The scheme in eqs. (5)-(6) has been successfully proposed for the blind restoration of a single blurred and noisy image, when no knowledge is available about the blur operator [13]. When used in BSS, it can be viewed as a direct extension to the noisy case of the ML ICA for noiseless data described above. In that case, eq. (5) is directly equivalent to eq. (3), where matrix W is intended as the inverse of A, while eq. (6) simply amounts to $\mathbf{s} = W\mathbf{x}$, being the data noiseless. In our case, since the data are noisy, the dependence of the sources on the mixing matrix and the data cannot be a simple linear relationship, which would amplify noise. Thus we established this dependence in the usual form of a regularized estimate, based both on the data and on the priors.

3 The MRF Model and the Algorithm

In this paper, we adopt generic, local smoothness MRF models for the sources, augmented to account for information about the features of realistic edge maps. In the Gibbs/MRF formalism our priors are given by:

$$P_i(\mathbf{s}_i) = \frac{1}{Z_i} \exp\left\{-U_i(\mathbf{s}_i)\right\} \tag{7}$$

where Z_i is the normalizing constant and $U_i(\mathbf{s}_i)$ is the prior energy in the form of a sum of potential functions, or stabilizers, over the set of cliques of interacting locations. The number of different cliques, as well as their shape, is related to the extent of correlation among the pixels, while the functional form of the potentials determines the correlation strength, and various features of the image edges. In our case we express the regularity of edges by penalizing parallel, adjacent edges, and chose $U_i(\mathbf{s}_i)$ as:

$$U_i(\mathbf{s}_i) = \sum_t \sum_{(r,z) \in N_t} \psi_i\left((s_i(t) - s_i(r)), (s_i(r) - s_i(z))\right) \qquad (8)$$

where N_t is the set of the two couples of adjacent locations (r, z), $z < r$, that, in the 2D grid of pixels, precede location t in horizontal and in vertical. Note that extra edge regularity constraints, such as continuation, could be easily included as well. As stabilizers ψ_i, all having same functional form but possibly different hyperparameters, in order to graduate the constraint strength in dependence of the source considered, we chose the following functions [14]:

$$\psi_i(\xi_1, \xi_2) = \begin{cases} \begin{cases} \lambda_i \xi_1^2 & \text{if } |\xi_1| < \theta \\ \alpha_i & \text{if } |\xi_1| \geq \theta \end{cases} & \text{if } |\xi_2| < \theta \\ \begin{cases} \lambda_i \xi_1^2 & \text{if } |\xi_1| < \bar{\theta} \\ \alpha_i + \varepsilon_i & \text{if } |\xi_1| \geq \bar{\theta} \end{cases} & \text{if } |\xi_2| \geq \theta. \end{cases} \qquad (9)$$

In eq. (9), λ_i is a positive weight, the so-called regularization parameter, the quantity $\theta = \sqrt{\alpha_i/\lambda_i}$ has the meaning of a *threshold* on the gradient above which a discontinuity is expected, while $\bar{\theta} = \sqrt{(\alpha_i + \varepsilon_i)/\lambda_i}$ is a *suprathreshold*, higher than the threshold, to lower the expectation of an edge when a parallel, close edge is likely to be present.

The solution of the problem in eqs. (5)-(6), in view of the adopted priors of eqs. (7), (8) and (9), presents some computational difficulties. Indeed, in general it is not possible to derive analytical formulas for the sources viewed as functions of A, and the priors are not convex. Thus, a simulated annealing (SA) algorithm has to be adopted for the updating of A and, for each proposal of a new status for it, the sources must be computed through numerical estimation. Nevertheless, some reasonable approximations can be adopted to reduce the complexity of the original problem, while keeping the effectiveness of the approach. First of all, due to the usual small number of mixing coefficients, SA is not particularly cumbersome in this case. On the other hand, based on the feasible assumption that small changes in A do not affect too much the sources, these can be updated only after significant modifications, e.g. at the end of a complete visitation of all the mixing coefficients. Furthermore, though the posterior is non-convex as well, the image models we adopted allow for performing the MAP source estimation through efficient deterministic non-convex optimization algorithms, such as the

Graduated Non-Convexity (GNC) algorithm. A GNC-like algorithm for the specific stabilizer in eq. (9) was derived in [14], in the case of image denoising. In [15], the same algorithm has been extended to account for images degraded by a linear operator. In this form, the algorithm is suitable to be applied to our present separation problem. The whole blind separation algorithm reduces thus to an alternating scheme governed by an external simulated annealing for the estimation of A, according to eq. (5), interrupted at the end of each Metropolis cycle, to perform an update of the sources **s**, according to eq. (6).

4 Experimental Results and Concluding Remarks

In this section we show preliminary results of the application of our algorithm to the blind separation of images reflecting the local autocorrelation assumptions adopted. In a synthetic setting, we were able to quantitatively evaluating the performance of our approach against that of the FastICA algorithm [6].

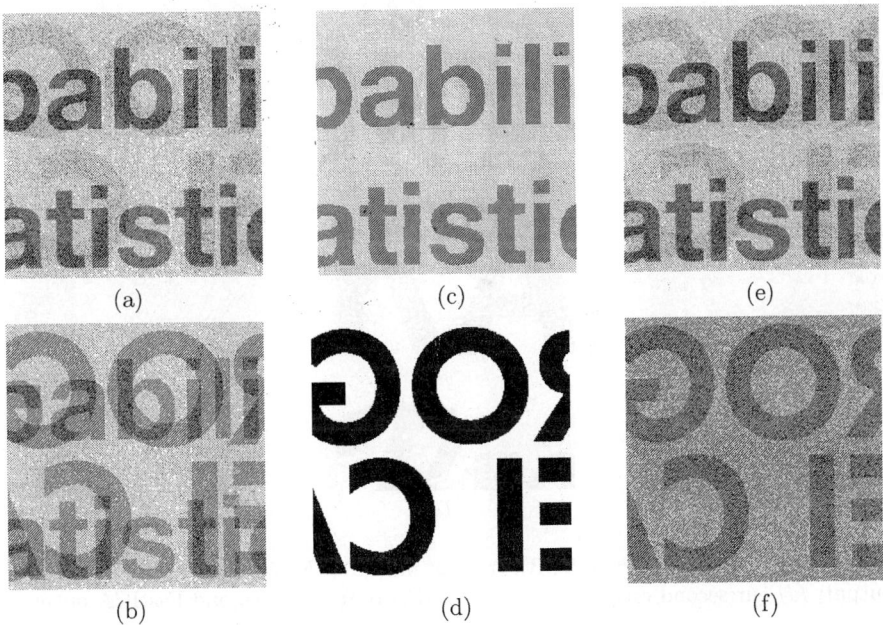

Fig. 1. Text images (SNR=10.5 dB): *(a)* first mixture; *(b)* second mixture; *(c)* our first output; *(d)* our second output; *(e)* first FastICA output; *(f)* second FastICA output.

Figures 1(a)-(b) show two scans of real texts, exhibiting a cross-correlation of 0.11%, mixed with a randomly generated matrix and then added with a white Gaussian noise (SNR=10.5 dB). We assumed different hyperparameters for stabilizer of eq. (9), to account for the different scale of the characters in the two sources. This allowed us to predetermine the order of the outputs, thus avoiding the permutation indeterminacy of BSS. We obtained the results shown in Figures 1(c)-(d), while the FastICA outputs are shown in Figures 1(e)-(f) for comparison. The better performance of our method in separating and denoising is apparent. The original A, and the estimated mixing matrices (after column rescaling), obtained with our method and with FastICA, were respectively:

$$A = \begin{bmatrix} 0.7605 & 0.1884 \\ 0.4753 & 0.4710 \end{bmatrix} \quad \hat{A}_1 = \begin{bmatrix} 0.7605 & 0.1884 \\ 0.4656 & 0.4308 \end{bmatrix} \quad \hat{A}_2 = \begin{bmatrix} 0.7605 & 0.1884 \\ 0.2866 & 0.3248 \end{bmatrix}$$

With several randomly selected mixing matrices and several realizations of both stationary and mildly non-stationary noise, we always obtained similar results. Note, however, that while FastICA estimates even very different matrices for different noise realizations, our method is instead very stable. Figure 2 shows

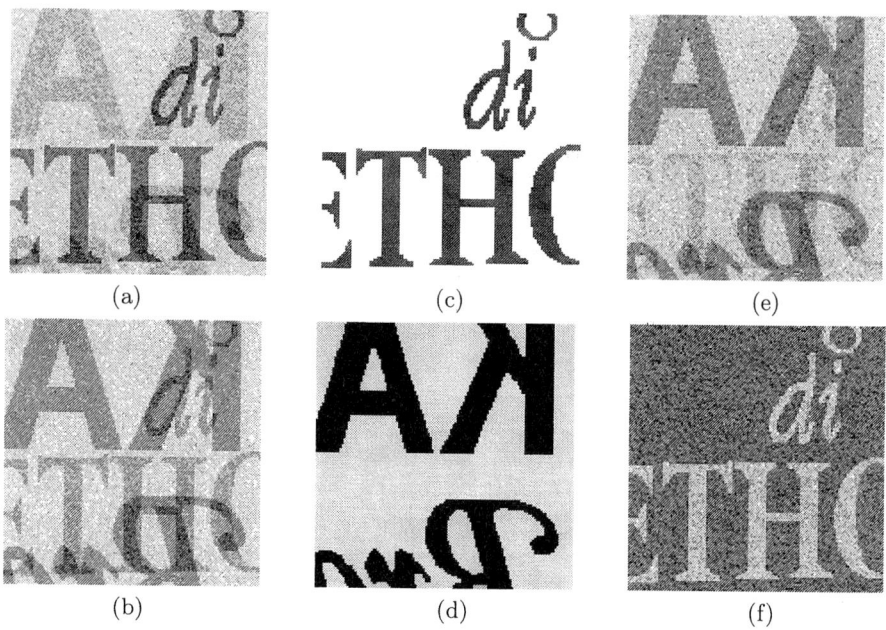

Fig. 2. Text images (SNR=10.5 dB): *(a)* first mixture; *(b)* second mixture; *(c)* our first output; *(d)* our second output; *(e)* first FastICA output; *(f)* second FastICA output.

the results of another experiment still on noisy (SNR=10.5 dB), mixed texts. Note the accurate reconstruction of the italics, thin characters obtained with our method. The original and estimated mixing matrices (after column rescaling and permutation) were:

$$A = \begin{bmatrix} 0.6987 & 0.2985 \\ 0.4013 & 0.5897 \end{bmatrix} \quad \hat{A}_1 = \begin{bmatrix} 0.6987 & 0.2985 \\ 0.4088 & 0.5802 \end{bmatrix} \quad \hat{A}_2 = \begin{bmatrix} -0.6987 & 0.2985 \\ -0.2467 & 0.6248 \end{bmatrix}$$

It is to be noted that the hyperparameter selection is performed, at present, in a heuristic way, based on the visual inspection of the mixtures. However, our estimation scheme could be easily added with an extra step where the hyperparameters are jointly estimated. This is especially important to manage nonstationary images or non-stationary noise showing a large variability, where it is expected that also the hyperparameters should be considered space-varying, and thus cumbersome to select by trial-and-error. Extensions of the method in these directions are planned, as well as further investigations for testing the robustness of the method against sensibly cross-correlated sources. Finally, experiments with real mixtures are currently on course, with application to the separation of overlapped texts in palimpsests. In this case, the mixtures are given by the Red, Green and Blue channels of the color document, or by others multispectral views, e.g. in the non-visible range.

References

1. Cichocki, A, Amari, S.: Adaptive Blind Signal and Image Processing. Wiley, New York (2002).
2. Kuruoglu, E., Bedini, L., Paratore, M.T., Salerno, E., Tonazzini, A.: Source separation in astrophysical maps using independent factor analysis. Neural Networks **16** (2003) 479–491.
3. Tonazzini, A., Bedini, L. Salerno, E.: Independent Component Analysis for document restoration. Int. J. on Document Analysis and Recognition, to appear, 2004.
4. Comon, P.: Independent Component Analysis, a new concept? Signal Processing, **36** (1994) 287-314.
5. Hyvärinen, A., Karhunen, J., Oja, E.: Independent Component Analysis. John Wiley, New York (2001).
6. Hyvärinen, A.: Fast and Robust Fixed-Point Algorithms for Independent Component Analysis. IEEE Trans. NN **10** (1999) 626-634.
7. Hyvärinen, A.: Gaussian moments for noisy independent component analysis. IEEE Signal Proc. Letters **6** (1999) 145-147.
8. Moulines, E., Cardoso, J.-F., Gassiat, E.: Maximum likelihood for blind separation and deconvolution of noisy signals using mixture models. Proc. ICASSP'97 **5** (1997) 3617-3620.
9. Attias, H.: Independent Factor Analysis. Neural Computation **11** (1999) 803-851.
10. Tong L., Liu R.W., Soon V.C., Huang Y.-F.: Indeterminacy and identifiability of blind identification. IEEE Trans. CS **38** (1991) 499–509.
11. Knuth, K.: Bayesian source separation and localization. Proc. SPIE'98 Bayesian Inference for Inverse Problems (1998) 147-158.
12. Tonazzini, A. Bedini, L., Kuruoglu, E., Salerno, E.: Blind separation of autocorrelated images from noisy mixtures using MRF models. Proc. ICA'03 (2003) 675–680.
13. Gerace, I., Pandolfi, R., Pucci, P.: A new estimation of blur in the blind restoration problems", Proc. ICIP'03 (2003) 4.
14. Bedini, L., Gerace, I., Tonazzini, A.: A deterministic algorithm for reconstructing images with interacting discontinuities. Graph. Models Image Proc. **56** (1994) 109–123.
15. Gerace, I., Pucci, P., Boccuto, A., Discepoli, M., Pandolfi, R.: A New Technique for Restoring Blurred Images with Convex First Approximation. Submitted (2003).

# Blind Separation
of Spatio-temporal Data Sources

Hilit Unger and Yehoshua Y. Zeevi

Department of Electrical Engineering, Technion – Israel Institute of Technology
Haifa 32000, Israel
hilitg@tx.technion.ac.il, zeevi@ee.technion.ac.il

Abstract. ICA and similar techniques have been previously applied to either one-dimensional signals or still images. We consider the problem of blind separation of dynamic sources, i.e. functions of both time and two spatial variables. We extend the Sparse ICA (SPICA) approach and apply it to a sliding data cube, defined by the two dimensions of the visual scene and the extent in time over which the mixing problem can be considered to be stationary and linear. This framework and formalism are applied to two special problems encountered in two different fields: The first deals with separation of dynamic reflections from a desired moving visual scene, without having any *a priori* knowledge on the structure of the images and/or their statistics. The second problem concerns blind separation of 'neural cliques' from the background firing activity of a neural network. The approach is generic in that it is applicable to any linearly mixed dynamic sources.

1 Introduction

Most of the research devoted to the problem of Blind Source Separation (BSS) has been concerned with either one-dimensional functions of time or static images (for references, see [1]). Yet, many physical systems generate linear mixtures of dynamic data sets. In biomedical applications, for example, those encountered in functional MRI, one is interested in the dynamic activity of specific loci of the brain. Another application concerns video sequences acquired through a semireflective medium and thereby contaminated by superimposed reflections. The video captures the dynamics of events. Since most real-world scenarios are dynamic, it is desirable to extend the BSS techniques to functions of both time and space.

Our first application deals with separation of dynamic images, such as video signals. In such applications it is desirable to eliminate reflections superimposed on a dynamic scene recorded through a glass windshield of a moving vehicle, or eliminate the reflections of the sun superimposed on the image of the visual environment observed through the cockpit of an airplane. The video sequence acquired in such cases can be represented as a three-dimensional (volumetric) cube, in which spatial images are stacked along a third axis (Fig. 1).

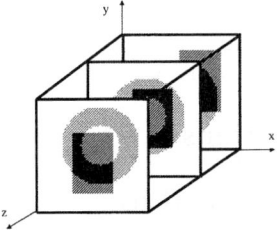

Fig. 1. Video sequence considered as a volumetric (cubic) date set. Shown is a data set comprised of three consecutive frames obtained from the sequence. Note the relative movement of the objects.

The second application presented here is concerned with the recording and analysis of biological neural networks, where there is a concerted effort to decipher the simultaneous messages signaled by the spatio-temporal firing patterns typical of the firing activity of massively connected neural networks. This application motivated our current study.

2 Sparse ICA (SPICA)

In Blind Source Separation an N-channel sensor signal x_i is generated by M unknown scalar source signals s_i, linearly mixed together by an unknown constant $N \times M$ mixing matrix A. In matrix notation, the N-dimensional vector of mixtures, X, is equal to the product of the $N \times M$ mixing matrix by the M-dimensional sources vector, S:

$$X = A \cdot S. \qquad (1)$$

Under the assumption that the sources are statistically independent, the BSS method yields an estimate of $\tilde{A}$, the unknown mixing matrix, without prior knowledge of the sources and/or the mixing process. The sources are recovered (up to permutation and scale) by using an inverse of the estimated mixing matrix, provided it exists:

$$\tilde{S} = \tilde{A}^{-1} \cdot X. \qquad (2)$$

It has been shown that when sources are sparse, they can be easily recovered from their linear mixtures using simple geometrical methods [2],[5]. This is based on the observation that whenever sources are sparse, there is a high probability that each data point in each mixture will result from the contribution of only one source. If we plot the N-dimensional scatter plot wherein each axis represents one of the mixtures, a co-linear cluster emerges with a specific orientation for each source. It can be shown that the coordinates of the vectors representing the centroids of these clusters correspond to the columns of the mixing matrix A. The simplest way to estimate the mixing matrix is to calculate the orientations of the clusters and select the optimal M angles from the histogram of angles. Another algorithm projects the data points onto a hemisphere, then uses clustering (such as Fuzzy C-means) in order to recover the orientations. Another related maximum-likelihood-based approach is the well-known Infomax [3].

3 Sparse Decompositions

3.1 Overcomplete Representations

Natural images and image sequences are not typically sparse. In order to exploit the methods previously described, we have to apply a transformation that yields a sparse representation of the signals. It has been shown that for a wide range of natural images, smoothed derivative operators yield a good sparsification results [4]. However, an overcomplete representation obtained, for example, by the Wavelet Packet transform (WPT, proposed in [5]) matches better the specific structure of a given set of images and thereby yields better sparsification which, in turn, facilitates and improves the estimation of the mixing matrix.

3.2 WP Transform

According to the formalism of the Wavelet Packet transform, a signal is recursively decomposed into its approximation (L) and detail (H) subspaces.

In the case of 2D signals, using separable wavelets, the signal is decomposed into its approximation and vertical, horizontal and diagonal details sub-images. For 3-dimensional data cube, the signal is decomposed into 8 sub-volumes (Fig. 2).

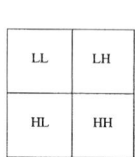
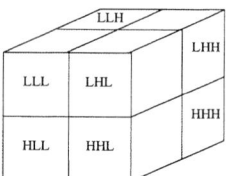

Fig. 2. WP decomposition. Left: 2D decomposition. Right: 3D decomposition.

We chose to use a separable transformation, for the sake of simplicity, by transforming rows first, then columns and then time (depth) axis. Nonseparable wavelets offer certain advantages, but are much more complex to deal with [6]. Their application in the context of sparsification is beyond the scope of this study.

3.3 Source Separation Using the WPT

After the mixture signals are decomposed into WP tree nodes using the WPT [5], a quality criterion is calculated for each node. The quality criteria should assign high values for sparse nodes and lower values for less sparse nodes. Common choices for quality criteria are entropy or global distortion The best node (or the top few nodes) is chosen and used as input data for the BSS algorithm.

Using the WPT has another advantage: because of downsampling in the process of the transform, the number of data points in each node is significantly smaller than the number of data points in the mixture signals, which speeds up the separation process.

4 BSS of Dynamic Reflections

Fig. 3 depicts an example of a typical scenario wherein a virtual (reflected) image is superimposed on a visual scene.

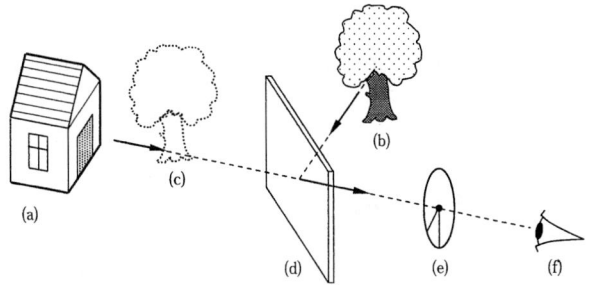

Fig. 3. A typical optical setup including a semireflecting windshield: (a) - object 1, (b) - object 2, (c) - virtual image of object 2, (d) - a semireflective lens, (e) - polarizer, (f) - camera. (adopted from [4]).

In the context of separation of reflections, the BSS problem usually reduces to the case of M=2 sources. The observed mixture is then given by

$$x(\xi 1, \xi 2, t) = a_{11} s1(\xi 1, \xi 2, t) + a_{12} s2(\xi 1, \xi 2, t), \qquad (3)$$

where x, s1 and s2 are dynamic images, usually acquired as video sequences. It is assumed here that the dynamics of the image and of the superimposed reflections are limited to planar translation of rigid bodies. The more difficult problem of non-planar motion and rotation as well as non-rigid distortion are beyond the scope of this paper, and will be dealt with elsewhere. Likewise, the coefficients a_{11} and a_{12} are assumed to be constant, approximating spatial invariance and linear mixing [4].

Since the reflected light is polarized, by using a linear polarizer, the relative weights of the two mixed video sequences can be varied to yield N different mixtures of the form:

$$x_n(\xi 1, \xi 2, t) = a_{n1} s1(\xi 1, \xi 2, t) + a_{n2} s2(\xi 1, \xi 2, t) : n = 1, \ldots, N. \qquad (4)$$

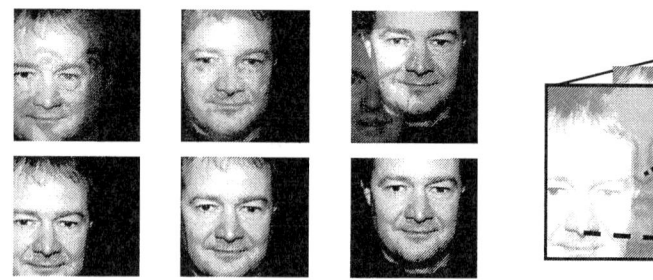

Fig. 4. Left frame of 6 images, simulation of blind separation of dynamic (moving) image from a superimposed reflection: frames from one mixture (up), and frames from one sequence of a recovered source (bottom). Right: data cube of one mixture. The arrows trace the trajectories of movements of the image and reflection, relatively to a stationary background.

Thus, we can use two or more video sequences obtained with different polarizations and separate objects and reflections. Simulation results are shown in Fig. 4.

5 BSS of Neural Cliques

In recent years, new optical [7] and electrical [8] imaging techniques for simultaneous recording of activity of populations of neurons in the brain tissue were developed. Whereas traditional methods for detection of action potentials in neurons were limited to a small number of neurons, it is now possible to record massive neural activity with spatial resolution of a single cell and a temporal resolution of a single action potential. It is therefore important to develop new techniques for the analysis of such activity.

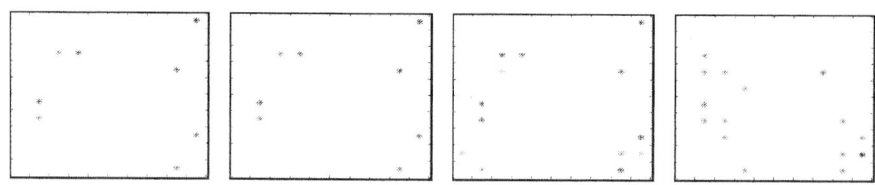

Fig. 5. Four states characterizing activity of an artificial neural network. The firing patterns depict functional phenomenon of localization.

The study of large populations of neurons enables to identify and analyze neural phenomena such as Synfire chains [9]: waves of synchronous neural activity that propagate over different areas of the biological neural network. It is believed that such separated activities represent processes related to higher level brain functions, e.g. percepts. Examining such spatio-temporal patterns of firing neurons, or 'neural cliques', led us to the assumption that there are underlying sources that are mixed together into each observed firing pattern. To understand the concept of cliques in the context of spatio-temporal neural network activity, recall the representation of spatio-temporal data as a cubical data set (Fig. 1). Here each frame corresponds to a slice along time axis of duration Δt. A clique then corresponds to correlated pattern of activity of two or more such slices of duration $T > \Delta t$.

To provide some intuitive insight into the analysis of neural cliques by means of Blind Source Separation technique, we generate data using CSIM circuit-tool; a simulator for neural networks [11]. The network connectivity is randomized, and one input neuron excites a random subset of the network. The output discrete spiking activity is then converted into continuous analog signal which, in turn, is quantized for further computation. It is interesting to observe that such a random network that is not endowed with any spatial localization structure, exhibits functional localization such as depicted in Fig. 5.

We do not have prior knowledge of the number of sources, therefore we need to estimate it by using the PCA technique [10] or geometrical version of an ICA-type approach, that permits separation of a larger number of sources than the given number of mixtures [5].

We assume that each neural clique has a finite (yet unknown) duration and that the neural activity is quasi-stationary over time, i.e. mixing coefficients remain constant over the duration of the clique, but may vary over longer periods of time. The separation problem is still endowed with a large number of degrees of freedom: the duration of the examined mixtures, the starting frame of each observation and the number of observations considered. Choosing those parameters carefully is crucial in order to achieve meaningful results. The optimal parameters for this problem are yet to be studied.

Using our BSS approach, we then project the data deduced from slices onto a scatter plot wherein each axis represents activity in one mixture slice. Each point then represents the activity of a neuron at a specific time in the slice.

Investigating the mutual activity of two slices, one often observes that two slices that are selected within the duration of co-activation do not necessarily exhibit coincidence of spike activity. In fact, the spatio-temporal activity may be almost exclusively restricted to only one slice. Under these circumstances, the distribution of activity over the scatter plot will form either a vertical or a horizontal cluster (Fig. 6 left). To compare with, when the second slice is partially co-active in space and time, more clusters emerge over the scatter plot (Fig. 6 middle). These clusters should provide some insight into the structure of the network, and functionally are indicative of clique-type activity. The full meaning of such embodiment of co-activation has yet to be further studied. It should be observed though, that unlike the previous application of video data, here we face a non-linear phenomenon that limits the power of ICA-type techniques. Nevertheless, the formalism and approach of projecting the data onto a scatter plot is powerful in gaining some insight into the structure of non-linearly interacting sources (or cliques).

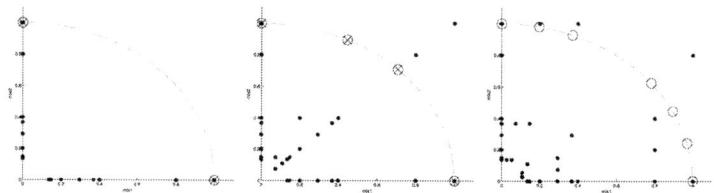

Fig. 6. Left, two slices with no co-activations. Middle, two slices with 2 emergent cliques. Cluster centers are marked with ×. Right, uncorrelated activity.

6 Conclusions

The extension of the Sparse ICA approach to three-dimensional problems broadens the range of ill-posed BSS problems that can be dealt with efficiently by providing relatively simple solutions to complex problems. Yet, the underlying assumptions of stationarity and linearity are not always met. More powerful results have yet to await the extension of these sparse ICA technique to the non-linear and non-stationary regime. This ambitious approach is under investigation.

The example of removal of reflections from a video sequence demonstrates that the sparse ICA approach is easily extended into the three-dimensional space and provides good results in the case of dynamic reflections.

Finding sources of neural activity in neural networks is a much more demanding and challenging task. Unlike the physics of separation of superimposed reflections, which can, to a good approximation, be considered linear, the neural cliques separation is necessarily non-linear, and most likely non-stationary. Nevertheless, as we have demonstrated here, the novel approach of using BSS techniques in isolation of the fingerprints of coherent neural activity from a neural network, can be instrumental in highlighting the functions of biological neural networks. It may be also instrumental in studies attempting to reverse engineer the structures of linear skeletons of such networks using spatio-temporal spiking activity.

Acknowledgement

Research supported in part by the Ollendorff Minerva center, by the HASSIP Research Network Program HPRN-CT-2002-00285, sponsored by the European Commission and by the Fund for Promotion of Research at the Technion.

References

1. Special issue on Independent Components Analysis. In: J. Machine Learning Research. Volume 4. (2003)
2. Zibulevsky, M., Pearlmutter, B.A.: Blind source separation by sparse decomposition in a signal dictionary. Neural Comp. **13** (2001) 863–882

3. Cardoso, J.: Infomax and maximum likelihood for blind source separation. IEEE Signal Processing Letters **4** (1997) 112–114
4. Bronstein, A., Bronstein, M., Zibulevsky, M., Zeevi, Y.Y.: Separation of reflections via sparse ICA. In: ICIP03. (2003) 313–316
5. Kisilev, P., Zibulevsky, M., Zeevi, Y.Y.: A multiscale framework for blind separation of linearly mixed signals. J. Mach. Learn. Res. **4** (2003) 1339–1363
6. Stanhill, D., Zeevi, Y.Y.: Two-dimensional orthogonal wavelets with vanishing moments. IEEE Transactions on Signal Processing **44** (1996) 2579–2590
7. Smetters, D., Majewska, A., Yuste, R.: Detecting action potentials in neuronal populations with calcium imaging. Methods **18** (1999) 215–221
8. Shahaf, G., Marom, S.: Learning in networks of cortical neurons. J. of Neuroscience **21** (2001) 8782–8788
9. Abeles, M.: Corticonics, neural circuits of the cerebral cortex. Cambridge University Press (1991)
10. Duda, R.O., Hart, P.E., Stork, D.G.: Pattern Classification. Wiley (2000)
11. http://www.lsm.tugraz.at/csim/index.html.

Data Hiding in Independent Components of Video

Jiande Sun, Ju Liu, and Huibo Hu

School of Information Science and Engineering, Shandong University
Jinan 250100, Shandong, China
{jd_sun,juliu,huibo_hu}@sdu.edu.cn

Abstract. Independent component analysis (ICA) is a recently developed statistical technique which often characterizes the data in a natural way. Digital watermarking is the main technique for copyright protection of multimedia digital products. In this paper, a novel blind video watermarking scheme is proposed, in which ICA is applied to extract video independent components (ICs), and a watermark is embedded into the ICA domain by using a 4-neighboring-mean-based method. The simulation shows that the scheme is feasible. And without degrading the video quality, it is robust to MPEG-2 compression and able to temporally synchronize.

1 Introduction

After the invention of digital video, its copyright protection issues have become important as it is possible to make unlimited copies of digital video without quality loss. Video watermark is a proposed method of video copyright protection. Watermarking of digital video has taken increasingly more significance lately. Emergence of the video technologies such as DVD, consumer-grade DV authoring and editing tools, video streaming, QuickTime TV initiative and video on demand have all been contributing factors. To be useful, video watermark should be perceptually invisible, blind detection, temporal resynchronization, robust to MPEG compression, etc. Many video watermark algorithms have been proposed, most of which have embedded watermark into the extracted frame feature [1], [2] or block-based motion feature [3], [4], [5]. However, these features are all based on frames, i.e. they are not the real features of video. What is more, embedding into such features does not consider the nature independent components of video and the results are weak in terms of robustness.

Independent Components Analysis (ICA) is a novel signal processing and data analysis method developed in the research of blind signals separation. Using ICA, even without any information of the source signals and the coefficients of transmission channel, people can recover or extract the source signals only from the observations according to the stochastic property of the input signals. It has been one of the most important methods of blind source separation and received increasing attentions in pattern recognition, data compression, image analyzing and so on [6], [7], for the ICA process derives features that best present the data via a set of components that are as statistically independent as possible and characterizes the data in a natural way.

In this paper, the FastICA algorithm [8] is used to extract the independent components (ICs) of video. The video ICs are watermarked by modifying their wavelet coefficients according to the 4-neighboring-mean-based algorithm [9]. Simulations show that the watermark is imperceptive and can be detected blindly. In addition, the scheme is robust to MPEG-2 compression and can re-synchronizing temporally. ICA and video ICs extraction are presented in section 2. Section 3 describes the proposed scheme. And simulations in Section 4 show its robustness to MPEG compression and the ability to temporal resynchronization. Finally, there is a conclusion.

2 Independent Component Analysis

Recently, the research on ICA (Independent Component Analysis) and its application has been the focus in the field of signal processing. To do ICA is to find a certain linear transform which can decompose the objective vectors and make the components of it as independent as possible. Though ICA comes from blind source separation or blind signal separation, it is used widely in many other fields, such as feature extraction, data compression and image analysis, etc.

2.1 Problem Formulation

Assume that there are m sensors and n source signals in the mixture system. The relationship between sources and observations is:

$$x = As \qquad (1)$$

where $x = [x_1, x_2, \cdots, x_m]^T$ are m mixtures, $s = [s_1, s_2, \cdots, s_n]^T$ are n mutually independent unknown sources and A is a mixing matrix. ICA is to estimate the source signal s or a de-mixing matrix C only from the observation signal x according to the statistical characteristic of s.

Many ICA algorithms have been proposed [8], [10], [11]. Now we give a simple description of the FastICA algorithm developed by Hyvainen and Oja [8] which we used to extract the video ICs in this paper.

2.2 FastICA Algorithm

The FastICA algorithm used in this paper is a fixed-point algorithm for independent component analysis (ICA), which provides good decomposition results efficiently. It pre-whitens the observation by performing Principal Component Analysis (PCA). The observed signal x is transformed to $v = Tx$, whose components are mutually uncorrelated and all have unit variance.

The objective function of FastICA by kurtosis is:

$$kurt(D^T v) = E\{(D^T v)^4\} - 3[E\{(D^T v)^2\}]^2 = E\{(D^T v)^4\} - 3\|D\|^4 \qquad (2)$$

The de-mixing matrix learning algorithm is:

$$D(k) = E\{v(D(k-1)v)^3\} - 3D(k-1) \qquad (3)$$

where k is the number of iteration. The de-mixing matrix is $C = D^T T$. So the estimation of source signal $\hat{s}$ is gotten through $\hat{s} = Cx$.

2.3 Video ICs

In the literature [12], Hateren and Ruderman show their results of performing independent component analysis on video sequences of natural scenes, which are qualitatively similar to spatiotemporal properties of simple cells in primary visual cortex. Fig. 1 describes this process.

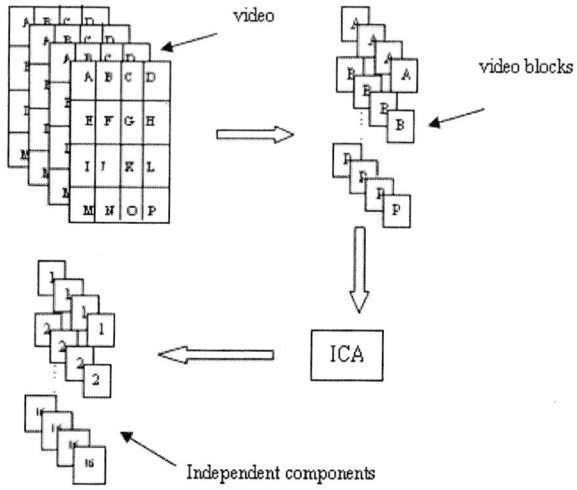

Fig. 1. The description of video block separation by ICA. The video is divided into video blocks, which are taken as the observation of ICA. Through ICA, the independent video components are got, which are called ICs in the following section.

3 Watermark Embedding and Detection

3.1 Selection of Video Independent Components

Before watermark embedding, where the data will be hidden should be determined. The extracted video ICs are considered as independent videos, whose frames are the slices just like the sub-images shown in Fig.2. The slices of the same order in their respective ICs are collected, among which the one that has the maximum variance is selected to be embedded watermark. The slice with larger variance is considered that it embodies more information including texture information, motion, etc.

3.2 4-Neighboring-Mean-Based Embedding Algorithm

Pseudo-random 0,1 sequences are used as watermarks W_n embedded into the wavelet domain of the selected slices. If the coefficient $p_{i,j}$ will be watermarked, the mean of

 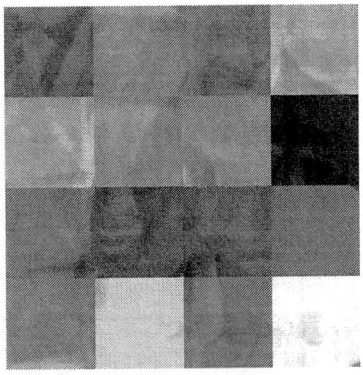

Fig. 2. The first frame of the experimental video is on the left. And on the right, there are the slices of the 16 extracted video ICs, which are the same order in ICs. These ICs can be regarded as some independent video, whose frames are this kind of slices.

its four neighboring coefficients is $mp_{i,j} = \frac{1}{4}(p_{i,j-1} + p_{i-1,j} + p_{i,j+1} + p_{i+1,j})$. And these four coefficients are not watermarked. The embedding algorithm is following.

$$\begin{cases} p'_{i,j} = p_{i,j} + \alpha(p_{i,j} - mp_{i,j}), C_n(m) = 1 & p_{i,j} > mp_{i,j} \\ p'_{i,j} = p_{i,j} - \alpha(mp_{i,j} - p_{i,j}), C_n(m) = 0 & p_{i,j} < mp_{i,j} \end{cases} \quad (4)$$

where, $m = 1, 2, \cdots, r$, r is the length of watermark sequence, n is the slice order in the video ICs, α is a weight. The r-long sequence C_n records the relationship between the coefficient and its 4-neighboring mean. So the symbol F_n is gotten, depending on which the existence of watermark is determined.

$$F_n(m) = XOR[C_n(m), W_n(m)] \quad (5)$$

XOR denotes exclusive or.

3.3 Watermark Detecting Algorithm

Before detecting watermark, the received video should be resynchronized temporally. After synchronization, the relation recording sequences C'_n are gotten from the received video just as do in the embedding procedure. Then pseudo-random sequences are selected to obtain corresponding extracted symbol $F'_{n,rand}$. The similarity between the original symbol and the extracted one determines whether there is the watermark or not. It is defined as following.

$$Sim(F_n, F'_{n,rand}) = 1 - \frac{\sum_{m=1}^{r} XOR[F_n(m), F'_{n,rand}(m)]}{r} \quad (6)$$

where the subscript *rand* is the state of pseudo-random sequence. If there is an outstanding peak at the state *rand*, the randth sequence is the watermark.

4 Simulations

The experimental video consists of 16 256×256 frames. And the whole video is divided into 16 video blocks of the same size. We select 16 numbers from 100 to 250 as the seed to generate pseudo- random 0,1 sequences, which will be used as the watermark. Every to-be-watermarked slice is watermarked with different sequence.

Table 1. The PSNR of every frame is listed. In video, the motion of active objects is the most important factor affecting the video quality. So though the mean of PSNR is 26.3506 dB, the video quality is still subjectively good.

frame order	1	2	3	4	5	6	7	8
PSNR(dB)	25.113	26.27	26.283	25.404	24.958	21.392	23.879	24.027
frame order	9	10	11	12	13	14	15	16
PSNR(dB)	27.446	27.889	27.634	28.489	28.178	28.31	28.742	27.596

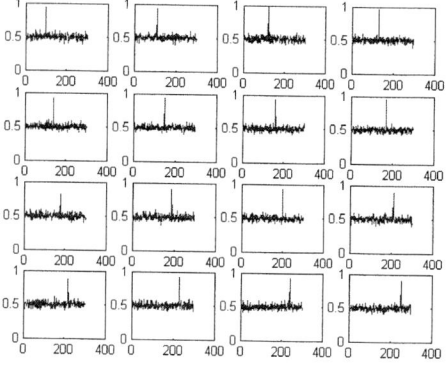 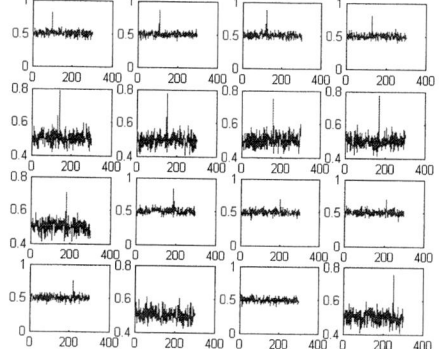

Fig. 3. The similarity between the original symbol and the extracted one of each watermarked slice is listed. Here the watermarked video does not receive any processing. X-axis denotes the state of pseudo-random sequences, while the y-axis is the similarity. The detection peaks almost all appear at the very random states.

Fig. 4. After the video is MPEG-2 compressed and de-compressed, the similarity is listed again. Obviously the similarities are affected by the compression, but the detection peaks are still outstanding at the very states. X-axis denotes the state of pseudo-random sequences, while the y-axis is the similarity.

5 Conclusion

In this paper, a novel blind video watermarking scheme is presented, which is based on the video independent component extraction by FastICA. The watermark is embedded into the slices of the obtained video ICs according to the 4-neighboring-mean algorithm. The simulation shows the feasibility of this scheme. The results demonstrate that the watermark does not degrade the video quality and can be blindly detected. Its robustness to MPEG-2 compression and the ability to temporal resynchronization are good. The robustness to other attacks, such as collusion, and the tradeoff between the robustness and invisibility are the next research focus. And new application of ICA in video analysis is also interesting.

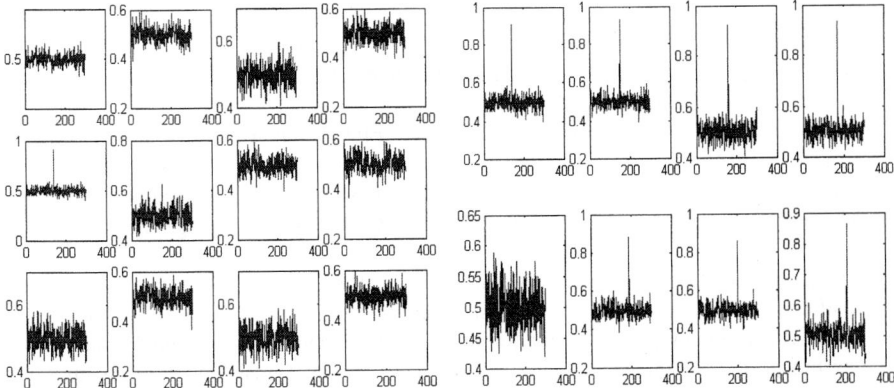

Fig. 5. The similarities listed here are the detection results of resynchronization experiment. The first 12 frames are used as a group to be ICA decomposed and watermarked. The other 4 frames are in another group. But at the receive terminal, only the last 12 fames are received to detect watermark. The detection results without resynchronization are listed on the left. Given the single peak in the 5th sub-image, the first frame of the received video is determined to be the 5th frame of the original video and the temporal synchronization is achieved. The detection results of the common 8 frames after temporal synchronization are listed on the right. X-axis denotes the state of pseudo-random sequences, while the y-axis is the similarity.

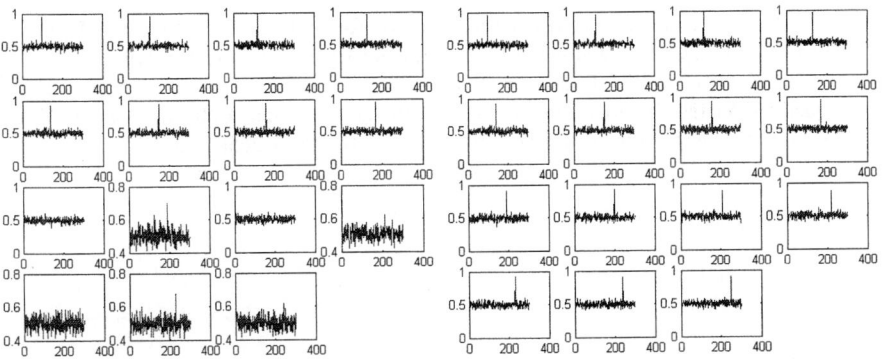

Fig. 6. The similarities listed here are the detection results of frame dropping experiment. The 9th frames are dropped at the receive terminal. The detection results without resynchronization are list on the left, in which the detection results of the first 8 frames are correct, and the detection results are damaged from the 9th frame. That shows that the first dropped frame is the 9th frame. Maybe there are many dropped frames, but the 9th is the first one is confirmed. The detection results after temporal synchronization are shown on the right. If there are many frames dropped, the resynchronization can be achieved after finding all the dropped frames. X-axis denotes the state of pseudo-random sequences, while the y-axis is the similarity.

References

1. Bhardwaji, A., Pandey, T.P. and Gupta, S.: Joint Indexing and Watermarking of Video Using Color Information, IEEE 4th Workshop on Multimedia Signal Processing, (2001)333 – 338
2. Hongmei Liu, Nuo Chen, Jiwu Huang, Xialing Huang and Shi, Y.Q.: A Robust DWT-Based Video Watermarking Algorithm, IEEE International Symposium on Circuits and Systems, Vol. 3, (2002)631 – 634
3. Jun Zhang, Maitre, H., Jiegu Li and Ling Zhang: Embedding Watermark in MPEG Video Sequence, IEEE 4th Workshop on Multimedia Signal Processing, (2001)535 – 540
4. Bodo, Y., Laurent and N., Dugelay, J.: Watermarking Video, Hierarchical Embedding in Motion Vectors, Proceedings of International Conference on Image Processing,Vol. 2,(2003) 739 – 742
5. Zhongjie Zhu, Gangyi Jiang, Mei Yu and Xunwei Wu: New Algorithm for Video Watermarking, 6th International Conference on Signal Processing, Vol.1(2002)760 - 763
6. Jarmo Hurri, Aapo Hyvarinen, Juha Karhunen and Erkki Oja: Image Feature Extraction Using Independent Component Analysis, Proc. IEEE Nordic Signal Processing Symposium, Espoo Finland (1996)
7. Jan Larsen, Lars Kai Hansen, Thomas Kolenda and Finn Arup Nielsen: Independent Component Analysis in Multimedia Modeling, 4th international symposium on independent component analysis and blind signal separation (ICA2003), Japan (2003) 687-695
8. Hyvrinen A., Oja.E: A Fast Fixed-Point Algorithm for Independent Component Analysis [J], Neural Computation, vol 9, no 7 (1997) 1483-1492
9. Ikpyo Hong, Intaek Kim, Seung-Soo Han: A Blind Watermarking Technique Using Wavelet Transform, Proceedings of IEEE International Symposium on Industrial Electronics, vol. 3, (2001) 1946 – 1950
10. A. Hyvärinen: Survey on Independent Component Analysis, Neural Computing Surveys, 2 (1999) 94-128
11. J. Liu, K. B. Nie, and Z. He: Blind Separation by Redundancy Reduction in A Recurrent Neural Network, Chinese Journal of Electronics, vol.10, no. 3 (2001) 415-419
12. J.H. van Hateren and D.L.Ruderman: Independent Component Analysis of Natural Image Sequences Yields Spatio-Temporal Filters Similar to Simple Cells in Primary Visual Cortex, proceedings of the Royal Society of London B (1998), 265(1412)2315-2320

3D Spatial Analysis of fMRI Data on a Word Perception Task

Ingo R. Keck[1], Fabian J. Theis[1], Peter Gruber[1], Elmar W. Lang[1],
Karsten Specht[2], and Carlos G. Puntonet[3]

[1] Institute of Biophysics, Neuro- and Bioinformatics Group
University of Regensburg, D-93040 Regensburg, Germany
{Ingo.Keck,elmar.lang}@biologie.uni-regensburg.de
[2] Institute of Medicine, Research Center Jülich, D-52425 Jülich, Germany
k.specht@fz-juelich.de
[3] Departamento de Arquitectura y Tecnologia de Computadores
Universidad de Granada/ESII, E-1807 Granada, Spain
carlos@atc.ugr.es

Abstract. We discuss a 3D spatial analysis of fMRI data taken during a combined word perception and motor task. The event - based experiment was part of a study to investigate the network of neurons involved in the perception of speech and the decoding of auditory speech stimuli. We show that a classical general linear model analysis using SPM does not yield reasonable results. With blind source separation (BSS) techniques using the FastICA algorithm it is possible to identify different independent components (IC) in the auditory cortex corresponding to four different stimuli. Most interesting, we could detect an IC representing a network of simultaneously active areas in the inferior frontal gyrus responsible for word perception.

1 Introduction

Since the early 90s [1,2], functional magnetic resonance imaging (fMRI) based on the blood oxygen level dependent contrast (BOLD) developed into one of the main technologies in human brain research. Its high spatial and temporal resolution combined with its non-invasive nature makes it to an important tool to discover functional areas in the human brain work and their interactions.

However, its low signal to noise ratio (SNR) and the high number of activities in the passive brain require sophisticated analysis methods which can be divided into two classes:

- model based approaches like the general linear model which require prior knowledge of the time course of the activations,
- model free approaches like blind source separation (BSS) which try to separate the recorded activation into different classes according to statistical specifications without prior knowledge of the activation.

In this text we compare these analysis techniques in a study of an auditory task. We show an example where traditional model based methods do not yield reasonable results. Rather blind source separation techniques have to be used to get meaningful and interesting results concerning the networks of activations related to a combined word recognition and motor task.

1.1 Model Based Approach: General Linear Model

The general linear model as a kind of regression analysis has been the classic way to analyze fMRI data in the past [3]. Basically it uses second order statistics to find the voxels whose activations correlate best to given time courses. The measured signal for each voxel in time $y = (y(t_1), ..., y(t_n))^T$ is written as a linear combination of independent variables $y = Xb + e$, with the vector b of regression coefficients and the matrix X of the independent variables which in case of an fMRI-analysis consist of the assumed time courses in the data and additional filters to account for the serial correlation of fMRI data. The residual error e ought to be minimized. The normal equation $X^T X b = X^T y$ of the problem is solved by $b = (X^T X)^{-1} X^T y$ and has a unique solution if XX^T has full rank. Finally a significance test using e is applied to estimate the statistical significance of the found correlation.

As the model X must be known in advance to calculate b, this method is called "model-based". It can be used to test the accuracy of a given model, but cannot by itself find a better suited model even if one exists.

1.2 Model Free Approach: BSS Using Independent Component Analysis

In case of fMRI data blind source separation refers to the problem of separating a given sensor signal, i.e. the fMRI data at the time t

$$x(t) = A[s(t) + s_{noise}(t)] = \sum_{i=1}^{n} a_i s_i(t) + \sum_{i=1}^{n} a_i s_{noise,i}(t)$$

into its underlying n source signals s with $a_i(t)$ being its contribution to the sensor signal, hence its mixing coefficient. A and s are unique except for permutation and scaling. The functional segregation of the brain [3] closely matches the requirement of spatially independent sources as assumed in spatial ICA. The term $s_{noise}(t)$ is the time dependent noise. Unfortunately, in fMRI the noise level is of the same order of magnitude as the signal, so it has to be taken into account. As the noise term will depend on time, it can be included as additional components into the problem. This problem is called "under-determined" or "over-complete" as the number of independent sources will always exceed the number of measured sensor signals $x(t)$.

Various algorithms utilizing higher order statistics have been proposed to solve the BSS problem. In fMRI analysis, mostly the extended Infomax (based on entropy maximisation [4,5]) and FastICA (based on negentropy using fix-point iteration [6]) algorithm have been used so far. While the extended Infomax algorithm is expected to perform slightly better on real data due to its adaptive nature, FastICA does not depend on educated guesses about the probability density distribution of the unknown source signals. In this paper we choose to utilize FastICA because of its low demands on computational power.

2 Results

First, we will present the implementation of the algorithm we used. Then we will discuss an example of an event-designed experiment and its BSS based analysis where we were able to identify a network of brain areas which could not be detected using classic regression methods.

2.1 Method

To implement spatial ICA for fMRI data, every three-dimensional fMRI image is considered as a single mixture of underlying independent components. The rows of every image matrix have to be concatenated to a single row-vector and with these image-vectors the mixture matrix X is constructed.

For FastICA the second order correlation in the data has to be eliminated by a "whitening" preprocessing. This is done using a principal component analysis (PCA) step prior to the FastICA algorithm. In this step a data reduction can be applied by omitting principal components (PC) with a low variance in the signal reconstruction process. However, this should be handled with care as valuable high order statistical information can be contained in these low variance PCs. The maximal variations in the timetrends of the supposed word-detection ICs in our example account only for 0.7 % of the measured fMRI Signal.

The FastICA algorithm calculates the de-mixing matrix $W = A^{-1}$. Then the underlying sources S can be reconstructed as well as the original mixing matrix A. The columns of A represent the time-courses of the underlying sources which are contained in the rows of S. To display the ICs the rows of S have to be converted back to three-dimensional image matrixes.

As noted before because of the high noise present in fMRI data the ICA problem will always be under-determined or over-complete. As FastICA cannot separate more components than the number of mixtures available, the resulting IC will always be composed of a noise part and the "real" IC superimposed on that noise. This can be compensated by individually de-noising the IC. As a rule of thumb we decided that to be considered a noise signal the value has to be below 10 times the mean variance in the IC which corresponds to a standard deviation of about 3.

2.2 Example: Analysis of an Event-Based Experiment

This experiment was part of a study to investigate the network involved in the perception of speech and the decoding of auditory speech stimuli. Therefor one- and two-syllable words were divided into several frequency-bands and then rearranged randomly to obtain a set of auditory stimuli. The set consisted of four different types of stimuli, containing 1, 2, 3 or 4 frequency bands (FB1-FB4) respectively. Only FB4 was perceivable as words.

During the functional imaging session these stimuli were presented pseudo-randomized to 5 subjects, according to the rules of a stochastic event-related paradigm. The task of the subjects was to press a button as soon as they were sure that they had just recognized a word in the sound presented. It was expected that in case of FB4 these four types of stimuli activate different areas of the auditory system as well as the superior temporal sulcus in the left hemisphere [8].

Prior to the statistical analysis the fMRI data were pre-processed with the SPM2 toolbox [9]. A slice-timing procedure was performed, movements corrected, the resulting images were normalized into a stereotactical standard space (defined by a template from the Montreal Neurological Institute) and smoothed with a gaussian kernel to increase the signal-to-noise ratio.

Classical Fixed-Effect Analysis. First, a classic regression analysis with SPM2 was applied. No substantial differences in the activation of the auditory cortex apart from an overall increase of activity with ascending number of frequency bands was found in three subjects. One subject showed no correlated activity at all, two only had marginal activity located in the auditory cortex (figure 1 (c)). Only one subject showed obvious differences between FB1 and FB4: an activation of the left supplementary motor area, the cingulate gyrus and an increased size of active area in the left auditory cortex for FB4 (figure 1 (a),(b)).

Spatial ICA with FastICA. For the sICA with FastICA [6] up to 351 three-dimensional images of the fMRI sessions were interpreted as separate mixtures of the unknown spatial independent activity signals. Because of the high computational demand each subject was analyzed individually instead of a whole group ICA as proposed in [10]. A principal component analysis (PCA) was applied to whiten the data. 340 components of this PCA were retained that correspond to more than 99.999% of the original signals. This is still 100 times greater than the share of ICs like that shown in figure 3 on the fMRI signal. In one case only 317 fMRI images were measured and all resulting 317 PCA components were retained.

Then the stabilized version of the FastICA algorithm was applied using tanh as non-linearity. The resulting 340 (resp. 317) spatially independent components (IC) were sorted into different classes depending on their structural localization within the brain. Various ICs in the region of the auditory cortex could be identified in all subjects, figure 2 showing one example. Note that all brain images in this article are flipped, i.e. the left hemisphere appears on the right side of the picture. To calculate the contribution of the displayed ICs to the observed fMRI data the value of its voxels has to be multiplied with the time course of its activation for each scan (lower subplot to the right of each IC plot). Also

3D Spatial Analysis of fMRI Data on a Word Perception Task

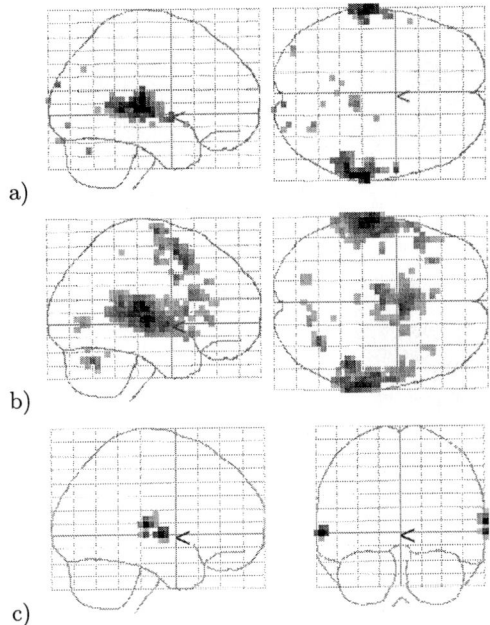

Fig. 1. Fixed-effect analysis of the experimental data. No substantial differences between the activation in the auditory cortex correlated to (a) FB1 and (b) FB4 can be seen. (c) shows the analysis for FB4 of a different subject.

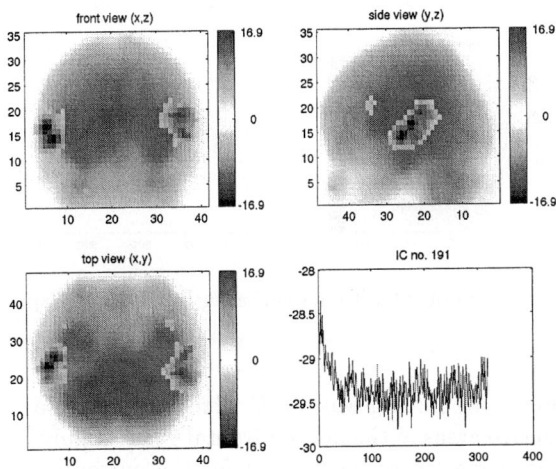

Fig. 2. Independent component located in the auditory cortex and its time course.

a component located at the position of the supplementary motor area (SMA) could be found in all subjects.

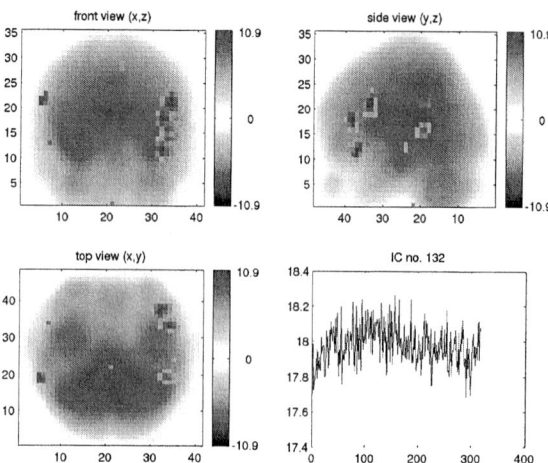

Fig. 3. Independent component which correspond to a proposed subsystem for word detection.

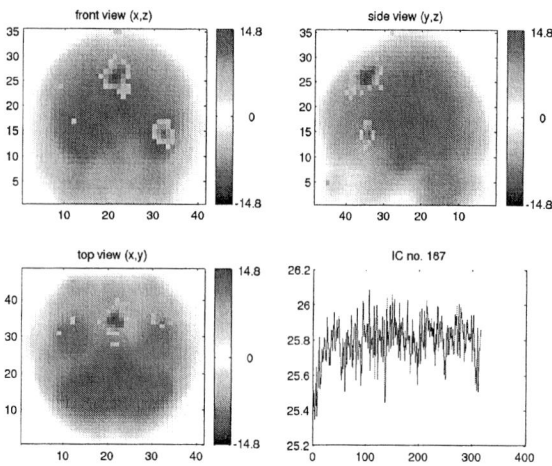

Fig. 4. Independent component with activation in Broca's area (speech motor area).

The most interesting finding was an IC which represents a network of three simultaneously active areas in the inferior frontal gyrus (figure 3) in one subject. This network was suggested to be a center for the perception of speech in [8]. Figure 4 shows an IC (of the same subject) that we assume to be a network for the decision to press the button. All other subjects except one had ICs that correspond to these networks, although often separated into different components. The time course of both components matches visually very well (figure 5) while their correlation coefficient remains rather low ($k_{corr} = 0.36$), apparently due to temporary time- and baseline-shifts.

Comparison of the Regression Analysis Versus ICA. To compare the results of the fixed-effect analysis with the results of the ICA the correlation coefficients between the expected time-trends of the fixed-effect analysis and the time-trends of the ICs were calculated. No substantial correlation was found: 87 % of all these coefficients were in the range of -0.1 to 0.1, the highest coefficient found being 0.36 for an IC within the auditory cortex (figure 2). The correlation coefficients for the proposed word detection network (figure 3) were 0.14, 0.08, 0.19 and 0.18 for FB1–FB4. Therefor it is quite obvious that this network of areas in the inferior frontal gyrus cannot be detected with a classic fixed-effect regression analysis.

While the reasons for the differences between the activation-trends of the ICs and the assumed time-trends are still subject to on-going research, it can be expected that the results of this ICA will help to gain further information about the work flow of the brain concerning the task of word detection.

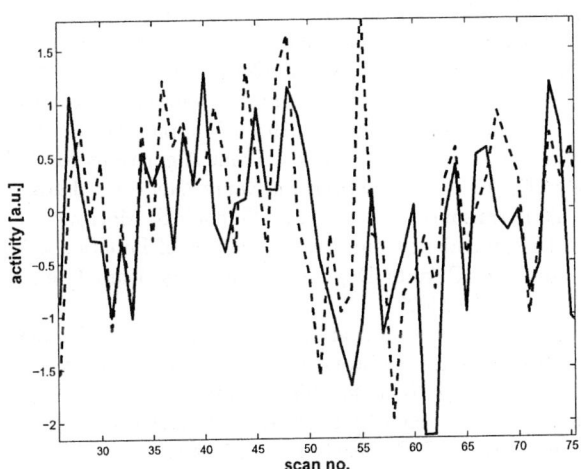

Fig. 5. The activation of the ICs shown in figure 3 (dotted) and 4 (solid), plotted for scan no. 25–75. While these time-trends obviously appear to be correlated, their correlation coefficient remains very low due to temporary baseline- and time-shifts in the trends.

3 Conclusions

We have shown that ICA can be a valuable tool to detect hidden or suspected links and activity in the brain that cannot be found using the classical approach of a model-based analysis like the general linear model. While clearly ICA cannot be used to validate a model (being in itself model-free), it can give useful hints to understand the internal organization of the brain and help to develop new models and study designs which then can be validated using a classic regression analysis.

Acknowledgment

This work was supported by the BMBF (project ModKog).

References

1. K. K. Kwong, J. W. Belliveau, D. A. Chester, I. E. Goldberg, R. M. Weisskoff, B. P. Poncelet, D. N. Kennedy, B. E. Hoppel, M. S. Cohen, R. Turner, H-M. Cheng, T. J. Brady, B. R. Rosen, "Dynamic magnetic resonance imaging of human brain activity during primary sensory stimulation", Proc. Natl. Acad. USA **89**, 5675–5679 (1992).
2. S. Ogawa, T. M. Lee, A. R. Kay, D. W. Tank, "Brain magnetic-resonance-imaging with contrast dependent on blood oxygenation", Proc. Natl Acad. Sci. USA **87**, 9868–9872 (1990).
3. R. S. J. Frackowiak, K. J. Friston, Ch. D. Frith, R. J. Dolan, J. C. Mazziotta, "Human Brain Function", Academic Press, San Diego, USA, 1997.
4. A.J. Bell, T.J. Sejnowski, "An information-maximisation approach to blind separation and blind deconvolution", Neural Computation, **7**(6), 1129–1159 (1995).
5. M.J. McKeown, T.J. Sejnowski, "Independent Component Analysis of FMRI Data: Examining the Assumptions", Human Brain Mapping **6**, 368–372 (1998).
6. A. Hyvärinen, "Fast and Robust Fixed-Point Algorithms for Independent Component Analysis", IEEE Transactions on Neural Networks **10**(3), 626–634 (1999).
7. F. Esposito, E. Formisano, E. Seifritz, R. Goebel, R. Morrone, G. Tedeschi, F. Di Salle, "Spatial Independent Component Analysis of Functional MRI Time-Series: To What Extent Do Results Depend on the Algorithm Used?", Human Brain Mapping **16**, 146–157 (2002).
8. K. Specht, J. Reul, "Function segregation of the temporal lobes into highly differentiated subsystems for auditory perception: an auditory rapid event-related fMRI-task", NeuroImage **20**, 1944–1954 (2003).
9. SPM2: http://www.fil.ion.ulc.ac.uk/spm/spm2.html, July 2003.
10. V.D. Calhoun, T. Adali, G.D. Pearlson, J.J. Pekar, "A Method for Making Group Inferences from Functional MRI Data Using Independent Component Analysis", Human Brain Mapping **14**, 140–151 (2001).

Decomposition of Synthetic Multi-channel Surface-Electromyogram Using Independent Component Analysis

Gonzalo A. García, Kazuya Maekawa, and Kenzo Akazawa

Dep. of Bioinformatic Eng., Osaka Univ., Osaka, Japan
{gonzalo,kazuya,akazawa}@ist.osaka-u.ac.jp

Abstract. Independent Component Analysis (ICA) can be used as a signal preprocessing tool to decompose electrode-array surface-electromyogram (s-EMG) signals into their constitutive motor-unit action potentials [García et al., IEEE EMB Mag., vol. 23(5) (2004)]. In the present study, we have established the effectiveness and the limitations of ICA for s-EMG decomposition using a set of synthetic signals. In addition, we have selected the best-suited algorithm to perform s-EMG decomposition by comparing the effectiveness of two of the most popular standard ICA algorithms.

Introduction

The central nervous system (CNS) sends commands to the muscles by trains of electric impulses (firing) via alpha-motoneurons (α-MNs), whose bodies (*somas*) are located in the spinal cord. The terminal axons of an α-MN innervate a group of muscle fibres. A motor unit (MU) consists of an α-MN and the muscle fibres that it innervates (see Fig. 1A). The electric activity of a firing MU (motor unit action potential – MUAP) can be detected by intramuscular or surface electrodes, and the signal obtained is defined electromyogram (EMG) [1].

The CNS regulates the force exerted by the muscle using two different mechanisms; recruitment of MUs and modulation of active MUs' firing rate [2]. The study of the firing pattern of α-MNs and some features of MUAP waveforms, such as shape and amplitude, is important for neurophysiological studies (e.g., CNS motor-control strategies) as well as for the diagnosis of motor neuron diseases [3]. These studies are generally carried out analyzing EMG detected by intramuscular recordings, obtained by placing needle electrodes inside the muscle; MUAP waveforms are in fact sharp enough to identify single MUs. This technique is however uncomfortable for patients and is time consuming for physicians.

A non-invasive, painless alternative is to analyse surface EMG (s-EMG) gathered by electrodes placed on the skin above the target muscle [4, 5]. The acquisition and processing of s-EMG signals are well-established techniques; nevertheless, s-EMG signals are a temporal and spatial summation of several MUAPs that result in a complex interference pattern of difficult interpretation (see Fig. 1B). For this reason, the application of classical techniques used for the analysis of intramuscular recordings [6] cannot guarantee an effective decomposition of s-EMG signals. Moreover, s-EMGs have a low signal-to-noise ratio (SNR) in comparison with intramuscular recordings due to the filtering effect of the tissues existing between electrodes and muscle fibres [7, 8] and the noise originated in the electrodes [9].

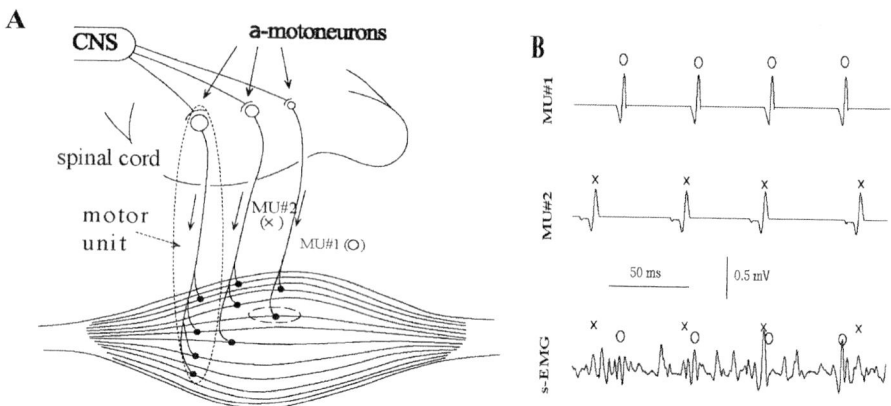

Fig. 1. (A) Outline of the neuromuscular system. (B) Example of signal obtained from a surface electrode showing the superimposed activity of several Mus.

A number of studies investigating s-EMG signals decomposition have already been carried out [10-12]; but they can only be applied to s-EMGs recorded at low force levels. Studies estimating the MUs general firing activity at higher contraction levels have also been carried out [13-16]. However, their purpose is not the full decomposition of s-EMG.

We have recently developed an algorithm able to decompose eight-channel s-EMG signals into their constitutive MUAP trains (MUAPTs) at high contraction levels (up to 30 and even 60% of the subjects' maximum voluntary contraction – MVC) [17]. This algorithm is based on the application of signal-conditioning filters, Independent Component Analysis (ICA) [18-20], and a template-matching technique [21]. The objective of the present study is to prove the effectiveness and understand the limitations of the ICA step by applying it to a set of synthetic data. We also carried out a comparative study on two of the most popular standard ICA algorithms (JADE and FastICA) to establish which is the best suited for decomposing s-EMG signals.

Methods

Synthetic Signals Generation

We developed a multi-channel s-EMG generator based on Disselhorst-Kulg *et al.*'s model [22], employing Andreassen and Rosenfalck's conductivity parameters [23], and the firing rate statistical characteristics described by Clamann [24]. Using this program, we produced four eight-channel synthetic s-EMG signals of 1.5 s length. These signals were composed of different numbers of motor units (namely, 3, 5, 8, and 10) distributed randomly, and firing at 20 firings/second. Each MU was composed by 50 muscle fibres distributed uniformly in an area of 1 mm radius.

ICA Algorithms

We applied two different ICA algorithms to the same set of synthetic data, namely Cardoso and Souloumiac's joint approximate digitalization of eigen-matrices (JADE) [25] and Hyvarinen and Oja's fast fixed-point algorithm for ICA (FastICA) [26]. To each pair of independent components (ICs) obtained, we applied the following independence (mutual information) measures: Principe's quadratic mutual information (QMI) [27], Kulback-Leibler information distance (K-LD) [28], Renyi's entropy measure [28], mutual information measure (MuIn) [27], Rosenblatt's squared distance functional (RSD) [28], Skaug and Tjøstheim's weighted difference (STW) [28], cross-correlation (Xcor), and Amari's separation performance index (AI) [29]. All the measures were normalized with respect to the maximum value obtained when applied to each IC with itself (maximum mutual information).

In order to measure the MUAP waveforms enhancement of a target MUAPT compared to the other MUAPTs obtained by applying ICA, we designed and applied to all the data obtained an additional measure; the peak-to-peak signal-to-interference ratio (SIR_{pp}) [17]. This measure was defined as the mean value of a MUAPT N firings peak in its closest channel (where it appeared the strongest) divided by the mean of the N strongest peaks belonging to the other MUAPTs present in the same channel. The signal improvement was calculated as the quotient of the SIR_{pp} before and after applying the ICA algorithms.

Results and Discussion

To establish the optimal conditions for s-EMG separation, we applied ICA to the generated set of synthetic multi-channel s-EMG signals.

The ICA model presents some requirements [18-20] that we fulfilled as follows; (1) The mixing matrix must be linear: in our case, the tissue crossed by the signals acts as a low-pass filter [7], and for this reason the signal components suffer a different attenuation depending on their frequency. However, the source signals have a very narrow bandwidth (the 70% of the power of their spectrum is normally condensed in approximately only 50 Hz, being the sampling frequency 10 KHz); therefore, we can assume a linear mixing process. (2) The mixing process and the source signals must be stationary: since the signals are recorded at isometric, constant force tension, we can assume the mixing process and the source signals to be stationary. (3) The source signals must be non-Gaussian: MUAPTs are obviously non-Gaussian as they are highly leptokurtic (*e.g.*, in the three-MU signal, the MUAPTs' normalized kurtosis mean was 13.2, s.d. 1.0).

Only two ICA requirements were not totally fulfilled by the MUAPTs; (4) The source signals must be independent: to check the independence between the different MUAPTs, we applied a χ^2 test (at $\alpha=0.05$) over the contingency table formed by pairs of source signals [30]. The percentage of MUAPTs pair-wise independent was 33, 30, 25 and 42% for the s-EMGs formed by 3, 5, 8, and 10 MUs, respectively (mean: 32.6, s.d.: 7.2). However, JADE was able to enhance also the MUAP waveforms belonging to non pair-wise independent MUAPTs. (5) The mixing process must be instantaneous: the medium is anisotropic [7], producing different delays in different directions, so we obtain a convolution of the original signals rather than an instantaneous mixing.

This might be the reason why when using real s-EMG signals we do not get results as accurate as those obtained when using synthetic s-EMG [17].

For some of the independence measures it was necessary to estimate both joint and marginal probability density function (p.d.f.) of the signals. Parametric p.d.f. estimation methods require previous knowledge of the underlying densities and, moreover, it is necessary to find a known p.d.f. fitting the considered data. The most commonly non-parametric method used is the Parzen window [31, 32]. However, this method needs a smoothing kernel; its selection and width are critical for ensuring the convergence of the estimate [32, 33]. In our case, the p.d.f.'s has three characteristic peaks, and in spite the Parzen window method is able to accurately reveal those features, this can be achieved only when the window width is selected properly. Furthermore, it might happen that a parameter appropriate for one region is entirely unsuitable for a different region [34].

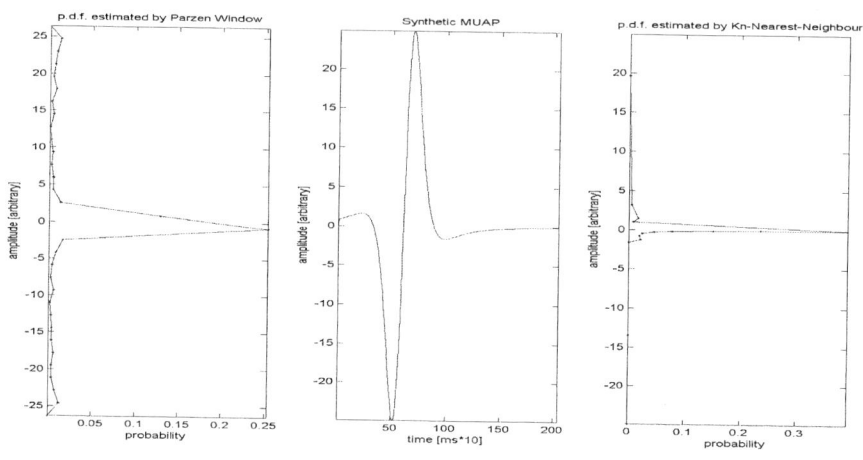

Fig. 2. Comparison of p.d.f. estimation methods.

For the above-mentioned reasons, we decided to use a data-driven method; K_n-nearest-neighbour (K_nNN) [34]. An illustrative example is given in Fig. 2, where the estimation of a MUAP waveform's p.d.f. by both methods with the same number of points is shown. We can observe how the K_nNN method performs a "zoom" in the areas where there is higher number of observations. A MUAPT's p.d.f. is very similar to the one shown in Fig. 2, except that the peak at zero is sharper.

We applied two different standard ICA algorithms (namely, JADE and FastICA) to the set of four synthetic signals. In order to compare the performance of both methods, we calculated the aforementioned measures for each source signal, and then subtracted the mean of the normalized measures corresponding to JADE from the ones corresponding to FastICA. In Fig. 3 the results obtained for the s-EMG composed of eight MUs are shown. The results for the other synthetic signals were very similar. The dissimilarity measure is the inverse (1-) of the normalized cross-correlation calculated for each MUAPT-IC pair. Higher value of a measure indicates worse performance. As all the differences are above zero, it is evident that JADE is more suitable than FastICA for the examined signals. The fluctuations in the performance of

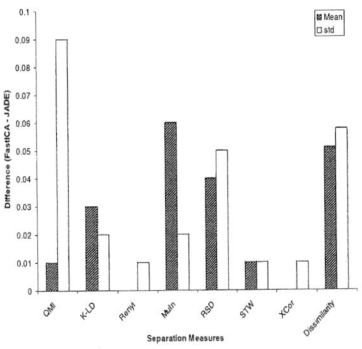 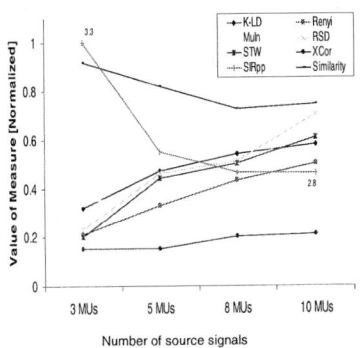

Fig. 3. Performance comparison between JADE and FastICA.

Fig. 4. JADE performance for different numbers of source signals.

FastICA (the s.d.) for each source signal were also bigger. The AI for the separating matrix of JADE was four times better than the one for FastICA in the eight-MU signal. In addition, some of the ICs given by FastICA were inverted, which could become a problem when trying to make an automatic decomposition algorithm. No sign-inversion was detected in any JADE output.

One essential requirement for ICA algorithms is that the number of source signals cannot be bigger than the number of available recorded channels. To study the sensitivity of the number of MUAPTs (source signals) on the JADE performance, we applied this algorithm to the four generated s-EMG signals and calculated for each number of MUs the aforementioned measures (Fig. 4). We observed how the increasing number of MUAPTs composing each s-EMG signal produces an increase in the mutual information between ICs pairs, and at the same time a steep decrease of the similarity (normalized to 1 to fit into the graph) and the SIR_{pp} improvement. This shows that the performance of JADE is highly affected by the number of source signals. However, a high improvement (above 2.5) respect the original signals could be observed even in the case of 10 MUAPTs (more source signals than channels).

Effect of Inter-channel Delay on JADE Performance

As mentioned above, another ICA requirement is that the mixing process must be instantaneous. In our real s-EMG signals, we have detected a total inter-channel delay not greater than 4 samples (0.4 ms), which was solved applying a minimum-square error technique [17]. This is a necessary step because, as shown in Fig. 5, the performance steeply decreases as the inter-channel delay increases.

We added progressive delays in the eight-MU signal. As shown in Fig. 5, from inter-channels delays of 3 samples, the SIR_{pp} improvement and the ICs cross-correlation (compared with the one of the original s-EMG channels) equal 1. This indicates that no improvement has been achieved in comparison to the original s-EMG signal. In addition, the AI increases steeply (its maximum value is 7 [35]). The reason of this JADE performance decrease could be that JADE identifies a signal in a channel and the same – but delayed – signal in a different channel as corresponding to different source signals.

Effect of Noise Addition on JADE Performance

To study the performance of ICA in the presence of noise, we used the real signals obtained in our previous work [17] recorded at 0% of subjects' MVC. The noise signal was first normalized and then added at increased amplitudes so that we could examine different signal-to-noise ratios (SNRs).

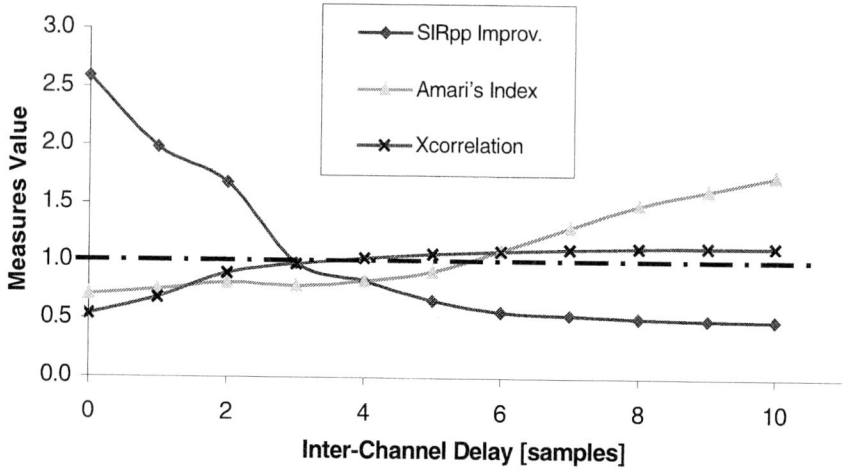

Fig. 5. JADE performance for different inter-channel delays.

The pre-processing step of our algorithm was able to decrease the power of the noise, therefore improving the performance of the JADE algorithm. However, our algorithm was not able to decompose s-EMG signals when the SNR was below 1.5.

Conclusion

In this work, we have shown how ICA can separate in different ICs the MUAPTs composing a set of synthetic s-EMG signals. We demonstrated this by measuring the decrease of mutual information between ICs respect the original s-EMG channels, and by comparing the original source signals with their respective ICs. The enhancement obtained is sufficient to allow the proper functioning of the template-matching step of our decomposition algorithm [17]. The JADE algorithm resulted more effective than FastICA in separating the s-EMG signals.

Added noise did not affect strongly JADE performance. The main limitation found was related to the inter-channel delay of the signals. When using real s-EMG signals, we have to face an additional problem; the convolution suffered by the original signals after crossing the forearm tissues. To overcome this problem we are currently investigating the applicability of convolutive ICA algorithms on s-EMG signals.

Acknowledgements

The authors would like to thank Dr. S. Rainieri for helpful discussion. This work was partly supported by the Ministry of Education, Culture, Sports, Science, and Technology of Japan (Grant-in-Aid for Scientific Research). G.A.G. is supported by a grant from the same Ministry (*Monbukagakusho*).

References

1. J.V. Basmajian and C.J. De Luca: Muscle Alive. Their Functions Revealed by Electromyography. 5th edn., Williams & Wilkins, Baltimore (1985)
2. E. Henneman, G. Somjen, and D.O. Carpenter: Functional significance of cell size in spinal motoneurons. J. Neurophysiol., vol.28, pp. 560-580 (1965)
3. A.M. Halliday, S.R. Butler, and R. Paul (Eds): A Textbook of Clinical Neurophysiology, John Wiley & Sons, New York (1987)
4. R. Merletti, D. Farina, and A. Granata: Non-Invasive Assessment of Motor Unit Properties with Linear Electrode Arrays, In: Clinical Neurophysiology: From Receptors to Perception (*EEG Suppl. 50*). Editors: G. Comi, C.H. Lücking, J. Kimura, and P.M. Rossini, pp. 293-300 (1999)
5. B. Freriks and H.J. Hermens: Roessingh Research and Development b.v., European Recommendations for Surface ElectroMyoGraphy, results of the SENIAM project (1999), ISBN: 90-75452-14-4 (CD-ROM)
6. D. Stashuk: EMG signal decomposition: how can it be accomplished and used?, J. Electromyogr. Kines., vol. 11, pp. 151-173 (2001)
7. E.J. De la Barrera and T.E. Milner: The effect of skinfold thickness on the selectivity of surface EMG, Electroencephalography and Clinical Neurophysiology, vol. 93, pp. 91-99 (1993)
8. D. Farina, C. Cescon, and R. Merletti: Influence of anatomical, physical, and detection-system parameters on surface EMG, Biol. Cybern., vol. 86, pp. 445-456 (2002)
9. E. Huigen, A. Peper, and C.A. Grimbergen: Investigation into the origin of the noise of surface electrodes, Med. Biol. Eng. Comput., vol. 40, pp. 332-338 (2002)
10. Z. Xu, S. Xiao, and Z. Chi: ART2 neural network for surface EMG decomposition, Neural Computing & Applications, vol. 10, pp. 29-38 (2001)
11. T.-Y. Sun, T.-S. Lin, and J.-J. Chen: Multielectrode surface EMG for noninvasive estimation of motor unit size, Muscle & Nerve, vol. 22, pp. 1063-1070 (1999)
12. P. Bonato, Z. Erim, and J.A. Gonzalez-Cueto: Decomposition of superimposed waveforms using the cross time frequency transform, *In*: Proc. of the 23rd Ann. Intl. Conf. of the IEEE EMBS, Istanbul, pp. 1066-1069 (2001)
13. D. Stashuk and Y. Qu: Robust method for estimating motor unit firing-pattern statistics, Medical & Biological Engineering & Computing, vol. 34, pp. 50-57 (1996).
14. S. Karlsson, J. Yu, and M. Akay: Time-frequency analysis of myoelectric signals during dynamic contractions: a comparative study, IEEE Transactions on Biomedical Engineering, vol. 47 (2), pp. 228-238 (2000)
15. P. Zhou and W.Z. Rymer: Estimation of the number of motor unit action potentials in the surface electromyogram, *In:* Proc. of the 1st International IEEE EMBS Conference on Neural Engineering, Capri Island, Italy, pp. 372-375 (2003)
16. J.-Y. Hogrel: Use of surface EMG for studying motor unit recruitment during isometric linear force ramp, Journal of Electromyography and Kinesiology, vol. 13(5), pp. 417-23 (2003)
17. G.A. García, R. Okuno, and K. Akazawa: Decomposition Algorithm for Surface Electrode-Array Electromyogram in Voluntary Isometric Contraction, IEEE BME Magazine, vol. 23(5) (2004) [*Accepted*]

18. P. Comon: Independent component analysis, a new concept?, Signal Processing, vol. 36, pp. 287-314 (1994)
19. J.-F. Cardoso: Blind signal separation: statistical principles, Proceedings of the IEEE, vol. 86(10), pp. 2009-2025 (1998)
20. Hyvärinen and E. Oja: Independent component analysis: algorithms and applications, Neural Networks, vol. 13, pp. 411-430 (2000)
21. B. Mambrito and C.J. De Luca: A technique for the detection, decomposition and analysis of the EMG signal, Electroencephalogra. Clin. Neurophysiol., vol.58, pp.175-188 (1984)
22. C. Disselhorst-Kulg, J. Silny, and G. Rau: Estimation of the Relationship Between the Noninvasively Detected Activity of Single Motor Units and Their Characteristic Pathological Changes by Modelling, J. of Electromyography and Kinesiology, vol. 8, pp. 323-335 (1998)
23. S. Andreassen and A. Rosenfalck: Relationship of Intracellular and Extracellular Action Potentials of Skeletal Muscle Fibers, CRC Crit. Rev. Bioeng., vol. 6(4), pp. 267-306 (1981)
24. H.P. Clamann: Statistical Analysis of Motor Unit Firing Patterns in a Human Skeletal Muscle, Biophysical Journal, vol. 9, pp. 1233-1251 (1969)
25. J.F. Cardoso and A. Souloumiac: Blind Beamforming for Non-Gaussian Signals, IEE Proceedings-F, vol. 140(6) (1993)
26. A. Hyvärinen and E. Oja: A Fast Fixed-Point Algorithm for Independent Component Analysis, Neural Computation, vol. 9, MIT Press, Cambridge, pp. 1483-1492 (1997)
27. D. Xu, J.C. Principe, J. Fisher III, and H.-C. Wu: A Novel Measure for Independent Component Analysis (ICA), ICASSP '98, Seattle, vol.2, pp. 1161-1164 (1998)
28. D. Tjøstheim: Measures of Dependence and Tests of Independence, Statistics, vol. 28, pp. 249-282 (1996)
29. S. Amari, A. Cichocki, and H.H. Yang: A New Learning Algorithm for Blind Signal Separation, Advances in Neural Information Processing Systems, vol. 8, Touretzky, Mozer and Hasselmo Ed., MIT Press (1996)
30. A. Nortes-Checa: Estadística Teórica y Aplicada, 3rd ed., Ediciones Sol, Madrid (1987), pp. 541-543
31. E. Parzen: On Estimation of a Probability Density Function and Mode, Annals of Mathematical Statistics, vol. 33, pp. 1065-1076 (1962)
32. M.P. Wand and M.C. Jones: Kernel Smoothing, Chapman & Hall, London (1995)
33. D. Erdogmus, L. Vielva, and J.C. Principe: Nonparametric Estimation and Tracking of the Mixing Matrix for Underdetermined Blind Source Separation, Proc. of ICA'01, San Diego, California, pp. 189-194 (2001)
34. R.O. Duda, P.E. Hart, and David G. Stork: Pattern classification, 2nd ed., Wiley, New York (2001)
35. F.R. Bach and M.I. Jordan: Kernel Independent Component Analysis, Journal of Machine Learning Research, vol. 3 (2002), pp. 1-48

Denoising Using Local ICA and a Generalized Eigendecomposition with Time-Delayed Signals

Peter Gruber[1], Kurt Stadlthanner[1], Ana Maria Tomé[2], Ana R. Teixeira[2], Fabian J. Theis[1], Carlos G. Puntonet[3], and Elmar W. Lang[1]

[1] Institute of Biophysics, University of Regensburg, 93040 Regensburg, Germany
elmar.lang@biologie.uni-regensburg.de
[2] Dept. de Electrónica e Telecomunicações/IEETA
Universidade de Aveiro, 3810 Aveiro, Portugal ana@ieeta.pt
[3] Dep. Arquitectura y Tecnologia de Computadores
Universidad de Granada, 18071 Granada, Spain
carlos@atc.ugr.es

Abstract. We present denoising algorithms based on either local independent component analysis (ICA) and a minimum description length (MDL) estimator or a generalized eigenvalue decomposition (GEVD) using a matrix pencil of time-delayed signals. Both methods are applied to signals embedded in delayed coordinates in a high-dim feature space Ω and denoising is achieved by projecting onto a lower dimensional signal subspace. We discuss the algorithms and provide applications to the analysis of 2D NOESY protein NMR spectra.

1 Introduction

Blind source separation (BSS) techniques have been shown to solve the problem of removing the prominent water artifact in 2D NOESY protein NMR spectra [9]. An algebraic algorithm [13], [14] based on a GEVD of a matrix pencil (GEVD-MP) has proven especially efficient in this respect. The results indicated, however, that the statistical separation process introduces unwanted additional noise into the reconstructed protein spectra (see Fig. 2). Hence denoising as a postprocessing step appeared necessary. Many denoising algorithms have been proposed [1], [4], [7] and [16] including algorithms based on local linear projective noise reduction. Noise is generally assumed to be additive Gaussian white noise whereas the signal comes from a deterministic source usually. This implies, using basic differential geometry, that the signal embedded in a high-dimensional feature space of delayed coordinates resides within a sub-manifold of the space of delayed coordinates. The task is to detect this signal manifold.

We propose denoising algorithms based either on local ICA using k-means clustering of the embedded signals and an MDL estimator of the dimension of the signal subspace or on a GEVD using a matrix pencil of time-delayed signals (GEVD-dMP). All experiments refer to the polypeptide P11 [10] to which both algorithms, local ICA and GEVD-dMP, have been applied.

2 Denoising Using Local ICA

The algorithm we present is a local projective denoising algorithm. The idea is to embed the noisy signal into a high dimensional feature space of delayed signals. The denoising is then achieved by locally projecting the embedded signal onto a lower dimensional subspace which contains the characteristics of the noise free signal. The algorithm is based on local ICA using an minimum description length (MDL) criterion for parameter selection. To perform ICA we will use the popular FastICA algorithm by Hyvärinen and Oja [3], which performs ICA by maximizing the non-Gaussianity of the signal components.

Consider a signal $x(t)$ at discrete time steps $t = t_1, \ldots, t_n$, $\Delta t = t_n - t_{n-1}$ of which only its noise corrupted version $x_N(t) = x(t) + N(t)$ is measured. $N(t)$ are samples of a random variable with Gaussian distribution. First the noisy signal is transformed into a high-dimensional signal $\tilde{x}$ in the m-dimensional space of delayed coordinates according to $\tilde{x}_N(t) := (x_N(t), \ldots, x_N(t + (m-1)\tau \mod n))^T$, $\tau = c \cdot \Delta t$, $c \in \mathcal{N}$. Then the problem is localized by selecting k clusters of the delayed time series $\{\tilde{x}_N(t) \mid t = t_1, \ldots, t_n\}$ using a k-means cluster algorithm [5]. Now we can analyze these k m-dimensional signals using FastICA. We used an MDL criterion [6] to estimate the dimension p_{MDL} of the subspace onto which we project after using ICA:

$$p_{MDL} = \underset{p=1,\ldots,m}{\operatorname{argmin}} \left\{ -\ln\left(\frac{\Pi_{j=p+1}^{m} \lambda_j^{\frac{1}{m-p}}}{\frac{1}{m-p}\sum_{j=p+1}^{m}\lambda_j}\right)^{(m-p)n} + \left(pm - \frac{p^2}{2} + \frac{p}{2} + 1\right) \right.$$
$$\left. \cdot \left(\frac{1}{2} + \ln\gamma\right) - \frac{pm - \frac{p^2}{2} + \frac{p}{2} + 1}{p} \sum_{j=1}^{p}\ln\left(\lambda_j\sqrt{\frac{2}{n}}\right) \right\} \quad (1)$$

Here λ_j represents the ordered eigenvalues of the covariance matrix of the signal and γ represents a parameter of the MDL estimator. The MDL criterion is a maximum likelihood estimator of the number of signal components for data with additional white Gaussian noise. Using ICA we extract $p_{MDL} + 1$ independent components (ICs) of the signal (one additional component for the noise). Like in all MDL based algorithms noise reduction is achieved by projection of the signal onto a p_{MDL}-dimensional subspace. For PCA one applicable method is to select the largest components in terms of signal variance. For ICA applied to data with a non-Gaussian distribution we select the noise component as the component with the smallest value of the kurtosis. For non-stationary data with stationary noise we identify the noise by the least variance of its autocorrelation.

To reconstruct the noise reduced signal we reverse the clustering process to obtain a signal $\tilde{x}_e : \{1, \ldots, n\} \to \mathbb{R}^m$ and then average over the candidates in the delayed data.

$$x_e(t) := \frac{1}{m} \sum_{i=0}^{m-1} [\tilde{x}_e(t - i \cdot \tau \mod n)]_i \quad (2)$$

The selection of optimal parameters m and k can again be based on an MDL criterion for the detected noise $e := x - x_e$. Accordingly we project these signals e for different m and k in a high dimensional space of delayed coordinates and choose the parameters m and k such that the MDL criterion with respect to the eigenvalues of the correlation matrix of e is minimal.

3 Denoising Using GEVD-dMP

We present an algorithm similar to the recently proposed algorithm dAMUSE [12], [11] which can be used to solve BSS problems and simultaneously denoise the estimated source signals. Consider sensor signals x_i embedded in a high-dim feature space Ω of delayed signals. The trajectory matrix [8] of the sensor signals $x_i(t_0)$ and their M delayed versions $x_i(t_0 + m\Delta t)$, $m = 0, ..., M - 1$ computed for a set of L samples is given by ($t_0 = 0$ for simplicity)

$$\mathbf{X}_i = \begin{bmatrix} x_i((M-1)\Delta t) & x_i(\tau + (M-1)\Delta t) & \cdots & x_i(L\tau) \\ x_i((M-2)\Delta t) & x_i(\tau + (M-2)\Delta t) & \cdots & x_i(L\tau - \Delta t) \\ \vdots & \vdots & \cdots & \vdots \\ x_i(0) & x_i(\tau) & \cdots & x_i(L\tau - (M-1)\Delta t) \end{bmatrix} \quad (3)$$

where τ^{-1} is the sampling rate. Considering a group of N L-dim sensor signals, $\mathbf{x}_i$, $i = 1 \ldots N$, the trajectory matrix of the set will be a concatenation of the component trajectory matrices computed for each sensor. Assuming that each sensor signal is a linear combination of N underlying but unknown source signals ($\mathbf{s}_i$), a matrix $\mathbf{S}$ can be written in analogy to eqn(3). Then the sensor signals can be expressed as $\mathbf{X} = \mathbf{AS}$, where the mixing matrix $\mathbf{A} = \mathbf{a} \otimes \mathbf{I}$ is a block matrix with a diagonal matrix $a_{ij}\mathbf{I}_{M \times M}$ in each block. The matrix $\mathbf{I}_{MxM}$ is the identity matrix and the mixing coefficient a_{ij} relates the sensor signal i with the source signal j.

Considering NMR spectra it seems natural to deal with data in the frequency domain. Hence a data matrix $\hat{\mathbf{X}}$ is constructed by Fourier transforming every row of $\mathbf{X}$ to the frequency domain. Additionally a filtered version $\hat{\mathbf{X}}_f$ of $\hat{\mathbf{X}}$ is generated by computing the Hadamard product between the rows of $\hat{\mathbf{X}}$ and the frequency response function of an appropriate filter. Then a matrix pencil $(\mathbf{R}_{x,f}, \mathbf{R}_x)$ is formed where $\mathbf{R}_x$ is the correlation matrix of the unfiltered and $\mathbf{R}_{x,f}$ is the correlation matrix of the filtered signals. According to the linear mixing model the correlation matrix $\mathbf{R}_x$ can then be related to a corresponding matrix in the source signal domain via:

$$\mathbf{R}_x = \hat{\mathbf{X}}\hat{\mathbf{X}}^H = \mathbf{A}\mathbf{R}_s\mathbf{A}^H = \mathbf{A}\hat{\mathbf{S}}\hat{\mathbf{S}}^H\mathbf{A}^H \quad (4)$$

Analogously the correlation matrix of the filtered signals $\mathbf{R}_{x,f}$ is related to the correlation matrix $\mathbf{R}_{s,f}$ of the filtered sources.

Then the two pairs of matrices $(\mathbf{R}_{x,f}, \mathbf{R}_x)$ and $(\mathbf{R}_{s,f}, \mathbf{R}_s)$ represent a congruent pencil [15] with identical eigenvalues, i.e. $\mathbf{D}_x = \mathbf{D}_s$ and corresponding eigenvectors which are related by $\mathbf{E}_s = \mathbf{A}^H \mathbf{E}_x$ in case of non-degenerate eigenvalues.

Assuming that all sources are uncorrelated, the matrices $\mathbf{R}_s$ and $\mathbf{R}_{s,f}$ are block diagonal with block matrices along the diagonal given by $\mathbf{R}_{mm} = \langle \hat{\mathbf{s}}_m (\hat{\mathbf{s}}_m)^H \rangle$ and $\mathbf{R}_{mm,f} = \langle \hat{\mathbf{s}}_{m,f} (\hat{\mathbf{s}}_{m,f})^H \rangle$. The eigenvector matrix of the GEVD of the pencil $(\mathbf{R}_{s,f}, \mathbf{R}_s)$ is also block-diagonal with the block matrix $\mathbf{E}_{mm}$ on the diagonal being the $M \times M$ eigenvector matrix of the GEVD of the pencil $(\mathbf{R}_{mm}, \mathbf{R}_{mm,f})$. The independent components can be estimated from linearly transformed sensor signals via

$$\mathbf{Y} = \mathbf{E}_x^H \mathbf{X} = \mathbf{E}_x^H \mathbf{A} \mathbf{S} = \mathbf{E}_s^H \mathbf{S} \tag{5}$$

and turn out to be filtered versions of the underlying source signals.

To simultaneously perform BSS and denoising the GEVD of $(\mathbf{R}_{x,f}, \mathbf{R}_x)$ is determined in a two step procedure. First, the EVD of $\mathbf{R}_x = \mathbf{U}\mathbf{V}\mathbf{U}^H$ is calculated but only its l largest eigenvalues and corresponding eigenvectors are considered assuming that small eigenvalues are related to noise only. Following this dimension reduction the $l \times NM$ matrix $\mathbf{Q} = \mathbf{V}^{-1/2} \mathbf{U}^H$ is defined and the EVD problem of the matrix $\mathbf{C} = \mathbf{Q} \mathbf{R}_{x,f} \mathbf{Q}^H$ is solved. Then the eigenvector matrix $\mathbf{E}_x$ of the matrix pencil is

$$\mathbf{E}_x = \mathbf{Q}^H \mathbf{E}_C \tag{6}$$

where $\mathbf{E}_C$ is the eigenvector matrix of $\mathbf{C}$.

4 Denoising of Reconstructed NMR Spectra

Both algorithms have been applied recently to artificially generated signals and random noise to test their performance and evaluate their properties [2], [11]. These results suggest that a local ICA approach is more effective when the signal is infested with a large amount of noise whereas local PCA seems to be better suited for signals with high SNRs. Comparable investigations have been performed with the dAMUSE algorithm also with similar results [11].

In the following the local ICA based denoising algorithm uses the component kurtosis for noise selection and the MDL criterion to determine the number of independent signal components. The MDL criterion is used a second time to optimize the dimension m of the embedding space and the number k of neighborhoods to be used in local ICA. We apply the local ICA denoising algorithm as well as a local PCA and a Kernel-PCA based denoising algorithm to 2D NOESY NMR spectra of the polypeptide P11 [10].

The local ICA algorithm has been applied only to those independent components (IC), obtained with a GEVD-MP algorithm [10], related with the water peak. These ICs have been considered the signal now and were embedded in an m-dim feature space of delayed coordinates. There a k-means cluster algorithm was used to detect clusters of similar feature vectors. An MDL criterion has been used to estimate optimal values for the feature space dimension m and the number k of nearest neighbors. On these clusters an ICA has been performed locally

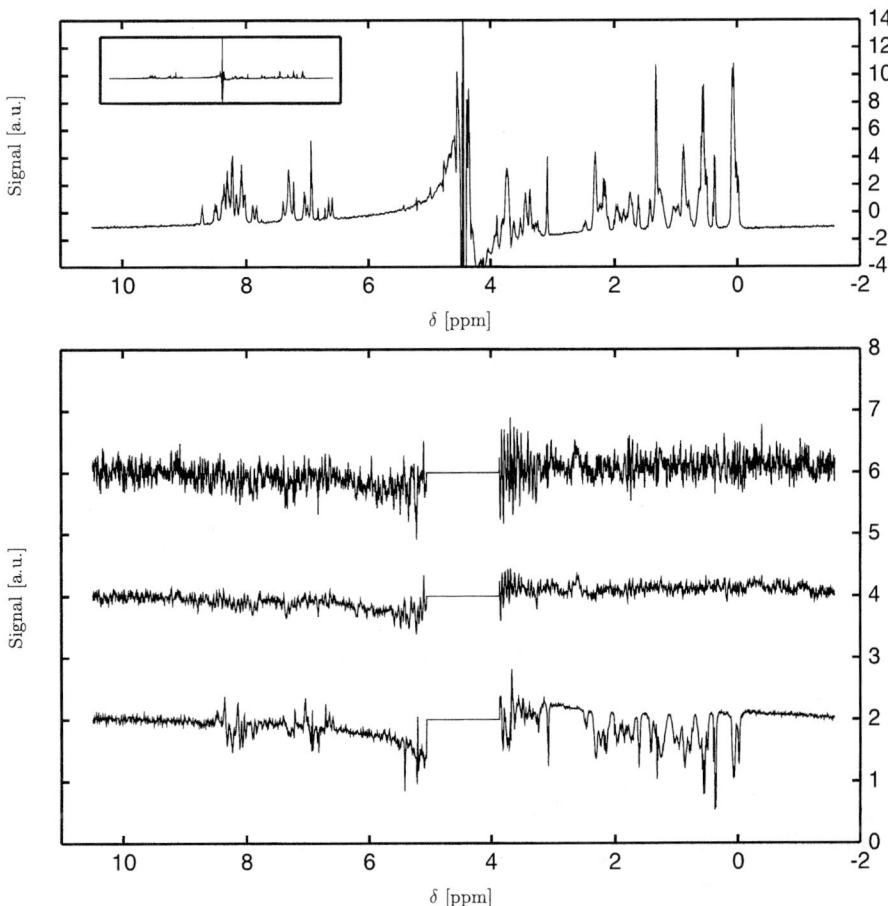

Fig. 1. The graph uncovers the differences of local ICA and Kernel-PCA denoising. The original spectrum is displayed on top with an insert showing the full water peak. The three curves represent the difference of the original and reconstructed spectra using *top* GEVD-MP, *middle* local ICA denoising, and *bottom* Kernel-PCA denoising. Note that the graphs are vertically translated by 2, 4 and 6 *a.u.*, respectively. Also note that both graphs have an identical scale.

using the fastICA algorithm. Again an MDL criterion was used to estimate an optimal number of signal components to be used to reconstruct the noise-reduced signals. These now represent "noise-free" versions of the ICs obtained with the GEVD-MP algorithm. Calculating the difference of the ICs obtained from the GEVD-MP algorithm directly and those reconstructed after the local ICA denoising has been applied, yields the noise contribution to these components. Now noise-reduced artifact-free protein spectra can be reconstructed using those ICs from the GEVD-MP algorithm not assigned to the water peak and the noise components obtained after removing the noise-free water components. Adding

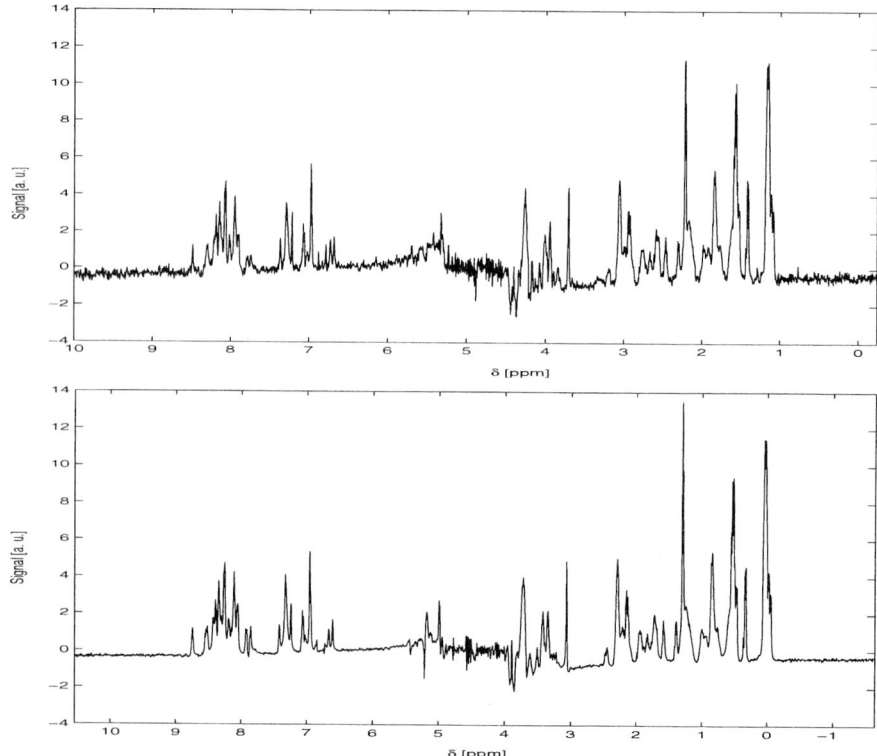

Fig. 2. Reconstructed spectrum of the polypeptide P11 using top: GEVD-MP [10] bottom: GEVD-dMP.

the noise components during the reconstruction process is essential as the noise which the ICs of the GEVD-MP algorithm convey only results from the statistical nature of the separation process. Hence all noise contributions to all ICs must be added up for them to compensate and result in the low noise content of the experimental spectra as can clearly be seen from figure 1. On the part of the spectrum away from the water artifact, we could estimate the increase of the SNR defined by

$$SNR(\boldsymbol{x}, \boldsymbol{x}_N)[dB] := 20 \log_{10} \frac{||\boldsymbol{x}||}{||\boldsymbol{x} - \boldsymbol{x}_N||} \qquad (7)$$

with the original spectrum as reference. We calculated a SNR of 17.3 dB of the noisy spectrum and a SNR of 21.6 dB after applying local ICA denoising.

We compare the reconstructed artifact-free protein spectrum of the local ICA denoising algorithm to the result of a Kernel-PCA based denoising algorithm [7] using a gaussian kernel in figure 1. The figure depicts the differences between the denoised reconstructed spectra and the original spectrum in regions away from the water peak. Local ICA denoising reduces the noise without changing the intensity of the protein signals, whereas Kernel-PCA denoising [7] clearly distorts the peak amplitudes of the protein resonances as well. This is detrimental to any spatial structure determination of the protein and is not acceptable to NMR spectroscopists.

The GEVD-dMP algorithm can be used elegantly to *simultaneously* separate the water artifact from the protein spectrum and perform denoising of the reconstructed spectrum. The starting point for the application of GEVD-dMP to P11 were 128 time domain signals taken from a 2D NOESY NMR experiment where each of the 128 signals consisted of 2048 data points. For every signal the component trajectory matrix $\mathbf{X}_i$, $i = 1, ..., 128$, (cf. eqn(3)) was formed by using $M = 1$ time delays of size $\Delta t = 2 \cdot \tau$. Thus the resulting trajectory matrix $\mathbf{X}$ was of size 256×2046. The matrix $\hat{\mathbf{X}}$ was determined by Fourier transforming each row of $\mathbf{X}$ to the frequency domain. After calculating the correlation matrix $\mathbf{R}_x = \hat{\mathbf{X}}\hat{\mathbf{X}}^H$ and a gaussian shaped filter of width $\sigma = 1$ was applied to every row of $\hat{\mathbf{X}}$ leading to the matrix $\hat{\mathbf{X}}_f$. Its correlation matrix was determined by $\mathbf{R}_{x,f} = \hat{\mathbf{X}}_f \hat{\mathbf{X}}_f^H$.

In the two step GEVD procedure, after the first EVD only the $l = 95$ largest eigenvalues of the 256×256 correlation matrix $\mathbf{R}$ were considered in order to reduce noise. Then the 95×2046 matrix $\mathbf{Q}$ was computed and the EVD of the matrix $\mathbf{C} = \mathbf{Q}\mathbf{R}_{x,f}\mathbf{Q}^H$ was performed which eventually lead to the eigenvector matrix $\mathbf{E}_x$ of the matrix pencil (eqn 6). Finally, those estimated components of $\mathbf{Y}$ (eqn. 5) which showed a high spectral density at the resonance frequency of the water protons were set to zero to reconstruct the artifact-free protein spectra. Note that with less than the 95 largest eigenvalues the separation of the water and the protein signals failed whereas considering more than 100 of the largest eigenvalues lead to a drastic increase in noise. Fig. 2 compares the results obtained by the standard GEVD-MP and the GEVD-dMP algorithms corresponding to SNRs of 17.3 dB and 22.43 dB, respectively.

5 Conclusions

Water artifact separation from 2D NOESY NMR protein spectra with statistical techniques like ICA introduce unwanted noise into the independent components obtained. We present noise reduction techniques using local ICA and an MDL based selection of the signal subspace as well as a GEVD-dMP algorithm. Both algorithms are based on local projective methods imposed on signals embedded in a high-dim feature space of delayed coordinates. The proposed methods are very effective in reducing the noise and show better results than a Kernel-based PCA method. Whereas local ICA denoising needs another GEVD or ICA preprocessing step to effect the artifact separation, GEVD-dMP provides both the artifact removal and the denoising in one stroke and is computationally very efficient with comparable results.

References

1. A. Effern, K. Lehnertz, T. Schreiber, P. David T. Grunwald, and C.E. Elger. Nonlinear denoising of transient signals with application to event-related potentials. *Physica D*, 140:257–266, 2000.
2. Peter Gruber, Fabian J. Theis, Ana Maria Tomé, and Elmar W. Lang. Automatic denoising using local independent component analysis. In *Proc. International ICSC Conference on Engineering of Intelligent Systems, ESI'2004*, 2004.
3. A. Hyvärinen and E. Oja. A fast fixed-point algorithm for independent component analysis. *Neural Computation*, 9:1483–1492, 1997.
4. Aapo Hyvärinen, Patrik Hoyer, and Erkki Oja. *Intelligent Signal Processing*, chapter Image Denoising by Sparse Code Shrinkage. IEEE Press, 2001.
5. A.K. Jain and R.C. Dubes. *Algorithms for Clustering Data*. Prentice Hall: New Jersey, 1988.
6. A.P. Liavas and P.A. Regalia. On the behavior of information theoretic criteria for model order selection. *IEEE Transactions on Signal Processing*, 49:1689–1695, 2001.
7. S. Mika, B. Schölkopf, A. Smola, K. Müller, M. Scholz, and G. Rätsch. Kernel PCA and denoising in feature spaces. *Adv. Neural Information Processing Systems, NIPS11*, 11, 1998.
8. V. Moskvina and K. M. Schmidt. Approximate projectors in singular spectrum analysis. *SIAM Journal Mat. Anal. Appl.*, 24(4):932–942, 2003.
9. K. Stadlthanner, A. M. Tomé, F. J. Theis, W. Gronwald, K. R. Kalbitzer, and E. W. Lang. Blind source separation of water artifacts in NMR spectra using a matrix pencil. In *Fourth International Symposium On Independent Component Analysis and Blind Source Separation, ICA'2003*, pages 167–172, Nara, Japan, 2003.
10. K. Stadlthanner, A. M. Tomé, F. J. Theis, W. Gronwald, K. R. Kalbitzer, and E. W. Lang. On the use of independent component analysis to remove water artifacts of 2D NMR protein spectra. In *7th Portuguese Conference on Biomedical Engineering, BIOENG'2003*, Lisbon, Portugal, 2003.
11. A. R. Teixeira, A. M. Tomé, E. W. Lang, and K. Stadlthanner. dAMUSE - A Tool for BSS and Denoising. *ICA2004, submitted*, 2004.
12. Ana R. Teixeira, A. P. Rocha, R. Almeida, and A. M. Tomé. The analysis of heart rate variability using independent component signals. In *2nd Intern. Conf. on Biomedical Engineering, BIOMED'2004*, pages 240–243, Innsbruck, Austria, 2004. IASTED.
13. Ana Maria Tomé. Blind source separation using a matrix pencil. In *Int. Joint Conf. on Neural Networks, IJCNN'2000*, Como, Italy, 2000.
14. Ana Maria Tomé. An iterative eigendecomposition approach to blind source separation. In *3rd Intern. Conf. on Independent Component Analysis and Signal Separation, ICA'2003*, pages 424–428, San Diego, USA, 2001.
15. Ana Maria Tomé and Nuno Ferreira. On-line source separation of temporally correlated signals. In *European Signal Processing Conference, EUSIPCO2002*, Toulouse, France, 2002.
16. Rolf Vetter, J.M. Vesin, Patrick Celka, and Jens Krauss Philippe Renevey. Automatic nonlinear noise reduction using local principal component analysis and MDL parameter selection. *Proceedings of the IASTED International Conference on Signal Processing Pattern Recognition and Applications (SPPRA 02) Crete*, pages 290–294, 2002.

MEG/EEG Source Localization
Using Spatio-temporal Sparse Representations

Alexey Polonsky and Michael Zibulevsky[*]

Technion - Israel Institute of Technology, Department of Electrical Engineering,
32000 Haifa, Israel
alexeyp@tx.technion.ac.il, mzib@ee.technion.ac.il

Abstract. Inverse MEG/EEG problem is known to be ill-posed and no single solution can be found without utilizing some prior knowledge about the nature of signal sources, the way the signals are propagating and finally collected by the sensors. The signals are assumed to have a sparse representation in appropriate domain, e.g. wavelet transform, and spatial locality of sources is assumed, the fact that MEG/EEG data comes from physiological source justifies such assumption. Spatial information is utilized through MEG/EEG forward model, which is used when looking for an inverse solution. Finally, we formulate an optimization problem that incorporates both the sparsity and the locality assumptions, and physical considerations about the model. The optimization problem is solved using an augmented Lagrangian framework with truncated Newton method for the inner iteration.

1 Introduction

1.1 MEG/EEG Inverse Problem

Neural activities of the brain are accompanied with ionic currents that produce weak magnetic and electric fields. Those fields are non invasively measured by highly sensitive sensors. The data collected on the sensors reflects some distribution of brain activity. The final goal is to localize those activities and possibly recover their time courses. In MEG/EEG inverse problem the brain is modeled as a mesh of voxels, each voxel represents a current dipole source [1]. The inverse problem should be solved given the forward model [3], [4] and the response of the sensors.

1.2 Spatial-temporal Approach

The voxels activity can be described by an $N:T$ matrix S where N is the number of voxels and T is a number of coefficients that represent the activity of each voxel. Assuming that number of active voxels is relatively small, even if we stay in time domain (the coefficients are time samples), the solution would have a sparse structure. The left plot on Fig. 1 depicts time samples of synthetic voxels activity $S^{(t)}$ ($N = 200$ voxels, $T^{(t)} = 60$ time samples, only 10 voxels are active).

---

[*] This research has been supported by the HASSIP Research Network Program HPRN-CT-2002-00285, sponsored by the European Commission, and by the Ollendorff Minerva Center.

It is possible to achieve a sparser representation by applying appropriate transform to time courses of each voxel. A short time Fourier transform and a wavelet transform are known to produce good results when compressing signals from various natural sources [8]. By applying a wavelet transform separately to each voxel in $S^{(t)}$, a much sparser structure is obtained. The right plot on Fig. 1 depicts a resulting matrix $S^{(WT)}$.

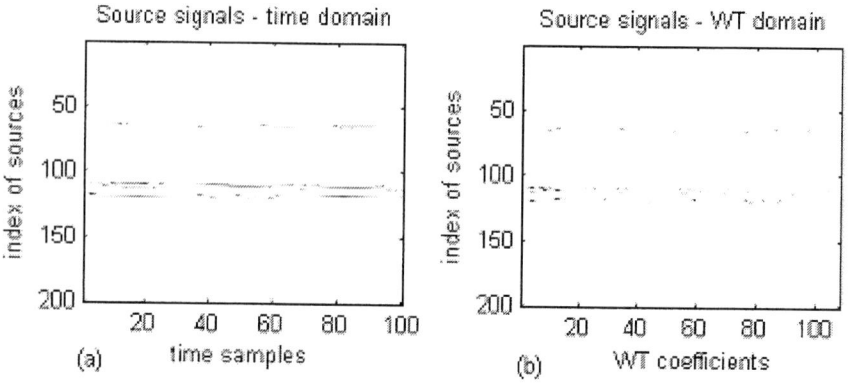

Fig. 1. (a) time course of voxels activity; (b) WT coefficients of voxels activity.

Notice, that this view on voxels activity utilizes both temporal and spatial information. Temporal information is utilized once all time courses (in a raw or transformed form) are used to form the matrix S. Spatial information can be utilized by considering a known forward model for MEG/EEG (i.e. the propagation rule that tells what is the response of sensors given voxels activity). Finally, sparsity of the solution is assumed, thus we should be able to achieve a good separation of the sources.

2 Problem Formulation

Let's denote:

$s = vec(S)$ - matrix S stacked column wise

$S = mat(s)$ - the reverse of vec

s_{ij} - i'th row-j'th column element of S

S_i - i'th row of S reshaped as a column vector

A - block diagonal $MT \times NT$ matrix consisting of T diagonal blocks, where each diagonal block equals A (size $M \times N$).

The MEG/EEG inverse problem can be stated in terms of optimization problem $\min_S f_{obj}(S)$ with an objective function of the following general form:

$$f_{obj} = \frac{1}{2} \cdot \|AS - X\|_F^2 + w_1 \cdot \Psi_1(S) + \ldots + w_k \cdot \Psi_k(S) \tag{1}$$

where $\|\circ\|_F^2$ is the Frobenius matrix norm, Ψ_k are some convex scalar functions with minimum at 0 and w_k are positive scalar weights. While $\frac{1}{2} \cdot \|AS - X\|_F^2$ forces the forward model response AS to be close to the sensors readings X, functions Ψ_k reflect a priori knowledge and assumptions about S.

In this work we used two such functions. The first one enforces sparsity of the solution:

$$\Psi_1(S) = \|s\|_1 = \sum_{i,j} |s_{ij}|, \qquad (2)$$

We further assume locality of sources. Active sources can be represented by several non-zero coefficients while non-active sources ideally have all-zero coefficients. Hence, we want to give a high penalty for stand-alone coefficients in random locations of matrix S and low penalty for coefficients that represent the same voxel, i.e. belong to the same row of matrix S. One possible choice for such penalty function is a sum of row-wise L2 norms $\sum_{i=1}^{N} \|S_i\|_2$, where S_i is an i'th row of S. This expression is not differentiable at 0. In order to make it smooth, we use the following technique:

$$\Psi_2(S) = \sum_{i=1}^{N} \sqrt{\|S_i\|_2^2 + \varepsilon}, \qquad (3)$$

The resulting objective function is

$$f_{obj} = \frac{1}{2} \cdot \|AS - X\|_F^2 + w_1 \cdot \|s\|_1 + w_2 \cdot \sum_{i=1}^{N} \sqrt{\|S_i\|_2^2 + \varepsilon}, \qquad (4)$$

where A is an $M:N$ gain matrix of forward model, S is an $N:T$ matrix of sources coefficients, X is an $M:T$ matrix of sensors coefficients and w_1, w_2, ε are empiric values that are tuned during simulations.

Let $C = S\Phi$ be a row-wise wavelet or other transform coefficients of S, which we expect to be sparse for true solution. Denote $Y = X\Phi$ the coefficients of the sensor signal X. Our objective can be then reformulated as

$$f_{obj} = \frac{1}{2} \cdot \|AC\Phi^{-1} - Y\|_F^2 + w_1 \cdot \|c\|_1 + w_2 \cdot \sum_{i=1}^{N} \sqrt{\|C_i\|_2^2 + \varepsilon}, \qquad (5)$$

where Φ^{-1} is the corresponding inverse transform. In the case of the orthonormal operator Φ^{-1}, we can write equivalently

$$f_{obj} = \frac{1}{2} \cdot \|AC - Y\|_F^2 + w_1 \cdot \|c\|_1 + w_2 \cdot \sum_{i=1}^{N} \sqrt{\|C_i\|_2^2 + \varepsilon} \qquad (6)$$

Note, that both optimization problems with respect to S and with respect to C have identical objective functions (6) and (4). For convenience, in the further text S will denote a variable of a generic optimization problem, i.e. S will stand for S or C and X will stand for X or Y.

Typically, in an MEG/EEG problem there are thousands of voxels, hundreds of sensors and hundreds of time course samples/coefficients, which results in $10^5 - 10^6$ optimization variables (matrix S or C). This is a large-scale optimization problem, and special optimization techniques should be used.

3 Solution

3.1 Moving from Unconstrained Non-smooth Objective Function to Smooth but Constrained Objective

The objective function (4) or (6) is non-differentiable because of the non-smooth second term (the sum of the absolute values). A common way to make it smooth is to express S (or C) as a difference of two non-negative terms $S = S_+ - S_-$ where $S_+ \geq 0$ and $S_- \geq 0$. The resulting optimization problem is:

$$\min\left(\frac{1}{2} \cdot \|A(S_+ - S_-) - X\|_F^2 + w_1 \cdot \|vec(S_+ - S_-)\|_1 + w_2 \cdot \sum_{i=1}^{N} \sqrt{\|(S_+ - S_-)_i\|_2^2 + \varepsilon} \right) \quad (7)$$

s.t. $S_+ \geq 0$, $S_- \geq 0$

Or in a more compact formulation:

$$\min\left(\frac{1}{2} \cdot \|\tilde{A}\tilde{S} - X\|_F^2 + w_1 \cdot \|vec(\tilde{S})\|_1 + w_2 \cdot \sum_{i=1}^{N} \sqrt{\|\tilde{S}_i\|_2^2 + \|\tilde{S}_{i+N}\|_2^2 + \varepsilon} \right), \quad (8)$$

s.t. $\tilde{S} \geq 0$

where

$$\tilde{A}_{M \times 2N} \equiv [A_{M \times N} \quad -A_{M \times N}]$$

$$\tilde{S}_{2N \times T} \equiv \begin{bmatrix} S_+ \\ S_- \end{bmatrix}$$

We choose to solve this constrained problem (8) in a framework of augmented Lagrangian [5], [6]. For convenience, in the further text S will stand for $\tilde{S}$ and A will stand for $\tilde{A}$.

3.2 Augmented Lagrangian Method and Truncated Newton Method

Consider an optimization problem with inequality constraints:

$$\min_s f(s), \text{ s.t. } g_j(s) \leq 0, \ 1 \leq j \leq r \quad (9)$$

The Lagrangian of this problem is:

$$L(s, \mu) = f(s) + \sum_{j=1}^{r} \mu_j \cdot g_j(s) \quad (10)$$

According to Lagrange multipliers theory, under certain conditions, the solution of the **constrained** optimization problem above is identical to the solution of an **uncon-**

strained optimization problem $\min_s L(s, \mu^*)$, where μ^* is a vector of optimal Lagrange multipliers.

Augmented Lagrangian algorithm is a numeric implementation of the above idea. It iteratively searches both for s^* and μ^*, the saddle point of $L(s,\mu)$, by solving an unconstrained optimization problem. The objective function of this problem is called the aggregate function of augmented Lagrangian algorithm and is given by:

$$F_p(s,\mu) = f(s) + \sum_{j=1}^{r} \varphi_p(g_j(s), \mu_j), \tag{11}$$

where $\varphi_p(g_j(s), \mu_j)$ is a penalty function for inequality constraints. The penalty function is chosen as described in [6].

In our case, this is a large-scale optimization problem. Thus data storage beyond $O(N)$ becomes prohibitively high (N is a number of variables) and computational load needed to find a search direction per iteration should also be proportional to N. Though the second order derivative information is theoretically available, the explicit calculation is not feasible.

Truncated Newton method [2] overcomes these obstacles and features linear storage and linear computational load per iteration, while exploiting second order information. It makes truncated Newton an appealing choice for large scale problems.

At each iteration, a search direction d is found by approximately solving a set of Newton equations $\nabla^2 f \cdot d \approx -\nabla f$, which is a linear system with respect to d. An approximate solution is most effectively found by conjugate gradients method. Global convergence is ensured by a back tracking line search [2] (Armijio rule) along direction d. The optimization is "truncated" after a certain (fixed) number of iterations, hence the name of the method. There is no need to calculate the Hessian explicitly, only a Hessian-vector product of the form $\nabla^2 f \cdot v$ needs to be calculated for an arbitrary vector v. This product is calculated at the cost similar to gradient calculation.

3.3 Gradient and Hessian-Vector Product of Aggregate Function

A thorough development for gradient, Hessian and Hessian-vector product of objective function (8) and of the aggregate function can be found in our research report [9].

The gradient and the Hessian of the constraints vector $g(s) = -s$ are given by

$$\nabla g(s) = [\nabla g_1(s) \ \nabla g_2(s) \ \ldots \ \nabla g_{2NT}(s)] = -I$$

$$H_{g_j}(s) \equiv \nabla^2 g_j(s) = 0, \quad 1 \le j \le 2NT$$

After substitution we obtain the gradient and the Hessian-vector product of the aggregate function:

$$\nabla F_p(s,\mu) = \nabla f_{obj}(s) - \begin{bmatrix} \varphi'_p(g_1(s), \mu_1) \\ \varphi'_p(g_2(s), \mu_2) \\ \ldots \\ \varphi'_p(g_{2NT}(s), \mu_{2NT}) \end{bmatrix} \tag{12}$$

$$\nabla^2 F_p(s,\mu) \cdot v = \nabla^2 f_{obj}(s) \cdot v + \begin{bmatrix} \varphi_p''(g_1(s),\mu_1) \cdot v_1 \\ \varphi_p''(g_2(s),\mu_2) \cdot v_2 \\ \ldots \\ \varphi_p''(g_{2NT}(s),\mu_{2NT}) \cdot v_{2NT} \end{bmatrix} \quad (13)$$

The gradient and the Hessian-vector product of the objective function (8) are given by

$$\nabla f_{obj}(s) = A^T A \cdot s + w_1 \cdot \underline{1} - A^T \cdot x + w_2 \cdot \begin{bmatrix} p_1 \cdot S_1 \\ \ldots \\ p_N \cdot S_N \\ p_1 \cdot S_{N+1} \\ \ldots \\ p_N \cdot S_{2N} \end{bmatrix} \quad (12)$$

$$\nabla^2 f_{obj}(s) \cdot v = A^T A \cdot v + w_2 \cdot \begin{bmatrix} p_1 \cdot V_1 - p_1^3 S_1 (S_1^T V_1) \\ \ldots \\ p_N \cdot V_N - p_N^3 S_N (S_N^T V_N) \\ p_1 \cdot V_{N+1} - p_1^3 S_{N+1} (S_{N+1}^T V_{N+1}) \\ \ldots \\ p_N \cdot V_{2N} - p_N^3 S_{2N} (S_{2N}^T V_{2N}) \end{bmatrix}_{2NT \times 1} \quad (135)$$

where $\underline{1}$ is an all-ones $2NT \times 1$ vector, V is an arbitrary $2N \times T$ matrix, $v = vec(V)$, v_i is an i'th element of v, $p_i \equiv (S_i^T S_i + S_{i+N}^T S_{i+N} + \varepsilon)^{-1/2}$.

Recall that A denotes block diagonal matrix consisting of T diagonal blocks, where each diagonal block equals A.

4 Simulations

We generated data for 20 sensors, 200 dipole sources. The generated signals were 50 samples long. Active sources were chosen at random locations and non-zero coefficients were spread with 10% sparsity. A temporally and spatially white Gaussian noise, $\sigma = 0.01 \cdot \frac{\|x\|_1}{MT}$, was added to the sensors signals. The coordinates of voxels and forward model were taken from results of experiments with a human skull phantom [4], [7].

Generated and recovered sources signals are depicted on Fig. 2 and Fig. 3 respectively. The signals are recovered and localized almost correctly. Only one signal (signal 3 on Fig. 2) is missing. This is an expected behaviour. This signal has the least number of non-zero coefficients and is rejected, because of the sources locality assumption, which is incorporated into the optimization problem through the third term of objective function.

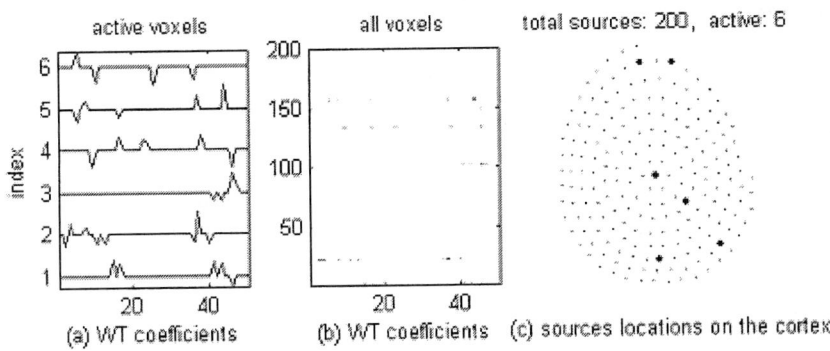

Fig. 2. Generated sources signals, (a) WT coefficients of active sources; (b) WT coefficients of all sources; (c) locations of sources on the cortex (active sources are marked with bold dots).

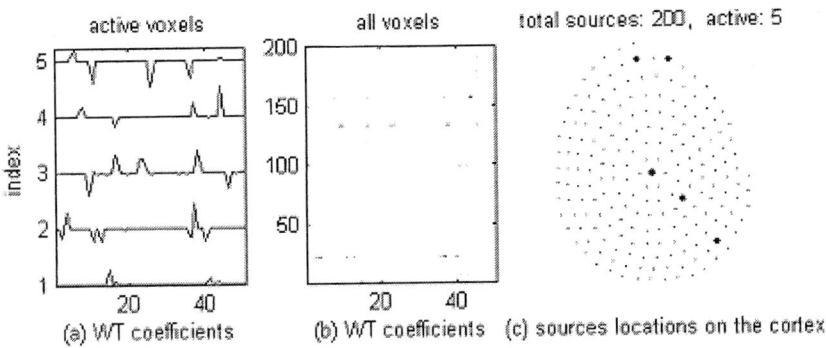

Fig. 3. Recovered sources signals, (a) WT coefficients of active sources; (b) WT coefficients of all sources; (c) locations of sources on the cortex (active sources are marked with bold dots).

5 Conclusions

We have explored a spatial-temporal approach for a solution of an inverse MEG/EEG problem. The solution was based on two physiological assumptions: sparsity of wavelet coefficients of the signals and spatial locality of sources. We have formulated a corresponding optimization problem that incorporates the physiological assumptions and forces the forward model response to be close to the observed data.

The simulation results confirm that the framework of optimization problem allows to efficiently utilize the available temporal and spatial information. The simulations showed that the solution of optimization problem is close to generated sources signals even when sensors readings are spoiled with noise. Sparsity assumption proved to be a driving force that takes an optimization to the right solution. The simulations also showed that as a result of sources locality assumption, signals with small number of coefficients might be suppressed. Recall, that this assumption is incorporated into optimization problem through the third term of objective function.

A clear advantage of spatial-temporal approach is utilizing as much of the available information as possible. And a framework of optimization problem allows incorporating of physiological and physical assumptions and considerations into a single model.

References

1. M. Hamalainen, R. Hari, R. J. Ilmoniemi, J. Knuutila, and O. V. Lounasmaa, "Magnetoencephalography – theory, instrumentation, and applications of noninvasive studies of the working human brain", Rev.Mod. Phys., vol. 65, pp. 413–497, Mar. 1993.
2. Stephen G. Nash and Ariela Sofer, "Linear and Nonlinear Programming", McGraw-Hill, New York, 1996.
3. J.P. Ary, S.A. Klein, and D.H. Fender, "Location of sources of evoked scalp potentials: corrections for skull and scalp thickness", IEEE Trans. Biomed. Eng. 28:447-452, 1981.
4. Brain Storm group, public data of MEG and EEG experiments with a human skull phantom http://neuroimage.usc.edu/.
5. Dimitri P. Bertsekas, "Nonlinear Programming", 2'nd edition, Athena Scientific, 1999.
6. A. Ben-Tal, M. Zibulevsky, "Penalty/Barrier Multiplier Methods for Convex Programming Problems", SIAM Journal on Optimization v. 7 # 2, pp. 347-366, 1997.
7. R. M. Leahy, J. C. Mosher, M. E. Spencer, M. X. Huang, and J. D.Lewine, "A study of dipole localization accuracy for MEG and EEG using a human skull phantom", Electroencephalogr. Clin. Neurophysiol.,vol. 107, pp. 159–173, Aug. 1998.
8. Stephane Mallat, "A Wavelet Tour of Signal Processing", 2'nd edition, Academic Press, 1999.
9. A. Polonsky, M. Zibulevsky, "MEG/EEG Source Localization Using Spatio-Temporal Sparse Representations", research report currently in preparation.

Reliable Measurement of Cortical Flow Patterns Using Complex Independent Component Analysis of Electroencephalographic Signals

Jörn Anemüller[1,2], Terrence J. Sejnowski[1,2], and Scott Makeig[1,2]

[1] Swartz Center for Computational Neuroscience, Institute for Neural Computation,
University of California San Diego, La Jolla, California
[2] Computational Neurobiology Laboratory,
The Salk Institute for Biological Studies, La Jolla, California

Abstract. Complex independent component analysis (ICA) of frequency-domain electroencephalographic (EEG) data [1] is a generalization of real time-domain ICA to the frequency-domain. Complex ICA aims to model functionally independent sources as representing patterns of spatio-temporal dynamics. Applied to EEG data, it may allow non-invasive measurement of flow trajectories of cortical potentials. As complex ICA has a higher complexity and number of parameters than time-domain ICA, it is important to determine the extent to which complex ICA applied to brain signals is stable across decompositions. This question is investigated for the complex ICA method applied to the 5-Hz frequency band of data from a selective attention EEG experiment[1].

1 Complex ICA of Frequency-Domain EEG Signals

The goal of complex ICA for frequency-domain EEG signals [1] is to replace the standard static source model with a more dynamic one that allows modeling each source as having a spatio-temporally varying activation pattern. The spatio-temporal dynamics of each source may be the result of, e.g., the spatial propagation of neural activity across the cortex, as observed in invasive animal recordings [2]. In contrast, instantaneous time-domain ICA would at best be able to approximate spatio-temporal source dynamics by one or more static ICA sources.

Taking into account spatio-temporal dynamics of sources leads to a convolutive model of source signal superposition, expressed in the frequency-domain as instantaneous mixing with complex-valued superposition coefficients that vary across frequencies. The frequency-domain approach to EEG signal analysis allows different dominant functional sources in different frequency bands – an observation supported by the functionally distinct frequency bands observed in human EEG [3].

Here, we give a brief overview of the processing stages of the complex ICA algorithm for EEG signals, cf. Fig. 1. For a detailed description, the reader is referred to [1]. The measured EEG data $\mathbf{x}(t) = [x_1(t), \ldots, x_M(t)]^T$ are first transformed into the frequency-domain using the standard techniques of short-time Fourier or wavelet decomposition. This yields the spectral data $\mathbf{x}(T,f) = [x_1(T,f), \ldots, x_M(T,f)]^T$, where

---

[1] Supported by the German Research Council DFG (J. A.), and by the Swartz Foundation.

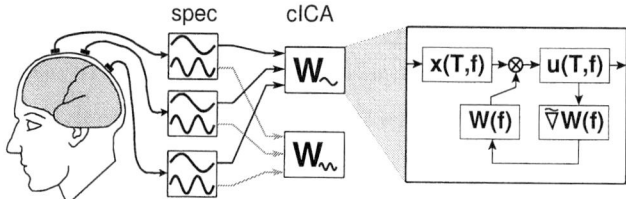

Fig. 1. Schematic representation of the processing stages of the complex frequency-domain ICA algorithm. Left ('spec'): the recorded electrode signals are decomposed into different spectral bands. Center ('cICA'): Complex ICA decomposition is performed within each spectral band. Right: Iteration steps performed by complex ICA for estimation of each separating matrix $\mathbf{W}(f)$.

f denotes frequency, and T denotes temporal position of the analysis window center. The goal of the complex ICA decomposition is to find for each frequency f a complex matrix $\mathbf{W}(f)$ that decomposes the measured signals $\mathbf{x}(T, f)$ into signals

$$\mathbf{u}(T, f) = \mathbf{W}(T) \mathbf{x}(T, f) \qquad (1)$$

so that the components of $\mathbf{u}(T,f) = [u_1(T,f), \ldots, u_M(T,f)]^T$ are statistically independent. In practice, full independence usually cannot be achieved under the linear separation model (1). Rather, the decomposition makes the signals $\mathbf{u}(T, f)$ as independent as possible. The matrix $\mathbf{W}(f)$ may be estimated using a complex generalization of the infomax ICA algorithm. For details, see [1]. We have found that, applied to data from a visual selective attention experiment, complex ICA separates physiologically plausible components while achieving a higher degree of independence between signal components in each frequency band.

The improved quality of signal separation is largely due to the higher number of degrees of freedom in the complex algorithm, allowing one complex matrix per frequency band, as opposed to standard time-domain ICA which estimates a single real-valued matrix for the entire data. The question of reliability may be particularly important for the complex ICA algorithm since the increased number of parameters to be estimated might lead to instability of the obtained solutions under perturbations of the data or internal algorithm parameters.

In the remainder of the paper, we drop the frequency index f (i.e., $\mathbf{u}(T) \equiv \mathbf{u}(T, f)$ etc.), since reliability is evaluated separately for each frequency band.

2 Reliability Analysis of Complex Independent Components

Before introducing the approach for the reliability analysis of complex ICA, we briefly review previous work on the reliability analysis of real-valued ICA decompositions. Meinecke et al. [4] used resampling methods to study the stability of blind source separation (BSS) algorithms. This approach may be characterized as a 'local' approach: The original data were first decomposed into separated component signals. Then, bootstrap data sets were generated from the *separated* signals and again decomposed. The obtained separating matrices were characterized by 'small deviations from the identity matrix in every Bootstrap sample' [4]. Meinecke et al. analyzed 8-channel electrocardiographic (ECG) data and 49-channel magnetoencephalographic (MEG) data that were projected to the first 23 and 25 principal component subspaces, respectively, during preprocessing. Only a few reliable one-dimensional components were obtained (2 and 3, respectively, as mentioned in [4]). Signal subspaces with multidimensional components were also found, a case not investigated in the present paper.

Recently, Duann et al. [5] studied the reliability of real infomax ICA [6] using functional magnetic resonance imaging (fMRI) data. Variability was induced by different random shuffling of the training data order in ten repeated decompositions, so that the algorithm's gradient update steps were applied to the data in different orders in different decompositions. This approach may be characterized as testing 'global' reliability: The ICA algorithm was applied to the mixed data, resulting in convergence to solutions in which the same ICA components could occur in different orders in different decompositions. Matching components between pairs of ICA decompositions, Duann et al. obtained most consistent results using a correlation criterion. The 600 dimensional recorded data were projected onto their first 100 principal components during preprocessing. All 100 separated components were reliably reproduced in each run of the algorithm.

To analyze the reliability of complex ICA [1], we generalize the 'global' approach of [5] to complex data, repeating decompositions on mixed data and finding best-matching components in different decompositions after training.

We investigate three sources of variability as test conditions:

Bootstrap data selection. We generate R bootstrap data sets $\mathbf{x}^{(r)}(T)$, $r = 1, \ldots, R$, from the original data set $\mathbf{x}^{(o)}(T)$ by drawing (with replacement) each sample of $\mathbf{x}^{(r)}(T)$ at random from the original data set. The bootstrap sets have same size as the original set. Complex ICA is performed on the original and on each of the bootstrap data sets, yielding independent components $\mathbf{u}^{(o)}(T), \mathbf{u}^{(1)}(T), \ldots, \mathbf{u}^{(R)}(T)$. The initial estimate of the separating matrix is the identity matrix for each decomposition. This condition tests the stability of the algorithm with respect to small variations in the data, i.e., it allows us to assess how stable the algorithm is with respect to data generated from the same (empirical) distribution.

Training data order selection. Our implementation of the complex ICA algorithm performs optimization in a semi-online fashion on small blocks of data points, in the same way as the standard implementation of the infomax ICA algorithm [5,7]. The order in which data points are used for gradient evaluation is chosen at random prior to each sweep through the whole data, and could affect optimization and convergence. To test variability with respect to this training data shuffling, we perform complex ICA decompositions of the original data set $\mathbf{x}^{(o)}(T)$ for R different seed settings of the random number generator that generates the shuffling. Hence, the training data order is different for each of the R decompositions, resulting in independent components $\mathbf{u}^{(1)}(T), \ldots, \mathbf{u}^{(R)}(T)$. Again, the initial estimate of the separating matrix is chosen as the identity matrix for each decomposition. The training data order condition tests the stability of the implementation's gradient optimization procedure.

Weight initialization selection. The complex ICA algorithm uses by default the identity matrix as the initial estimate of the separating matrix. To test variability with respect

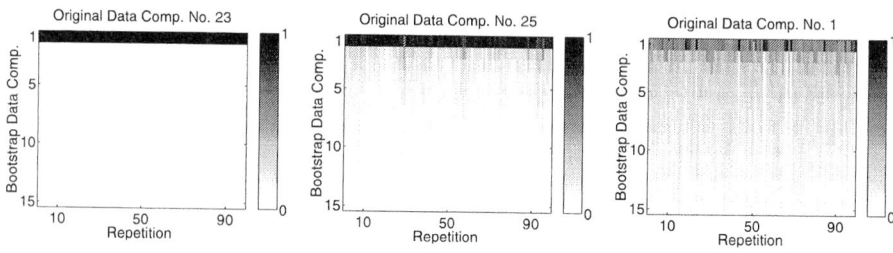

Fig. 2. Effectiveness of the method employed for matching components in different repetitions of the ICA decomposition. For three components obtained from the original data set (from left to right: 23, 25, 1), and for each of the 99 bootstrap data sets (abscissa), the graphs display the correlation coefficients for the 15 (out of 31) bootstrap data set components (ordinate) that had highest correlations with the original data set component. Each column was sorted from highest (top) to lowest (bottom). In each repetition, the bootstrap component with the highest correlation (row 1), was defined as best-matching the original component. Original components 23 (left) and 25 (center) had median correlation coefficients of 0.9996 and 0.9112, respectively, with their best-matching bootstrap components. These values correspond to the highest and lowest correlation values among components found as reliable (i.e., having a median correlation coefficient of 0.9 or higher). Original component 1 (right) had median correlation coefficient of 0.5761 with its best-matching bootstrap components, corresponding to the smallest correlation value of all (unreliable) components. Since except for outliers the best-matching bootstrap components (top row) for the reliable components (left and center panels) show significantly higher correlation than their next-best matching bootstrap components (rows 2 to 15), the employed scheme for matching components is appropriate.

to this initial condition, we generate R initialization matrices $\mathbf{W}_{init}^{(r)}$, $r = 1, \ldots, R$, with coefficients drawn randomly from a Gaussian distribution with zero mean and unit variance, resulting after decompositions in independent components $\mathbf{u}^{(1)}(T), \ldots, \mathbf{u}^{(R)}(T)$. The original data set $\mathbf{x}^{(o)}(T)$ is used as input, and the same training data order is used for all R decompositions. The initialization condition tests the ability of the gradient-based optimization to escape local minima and find global minima.

Complex independent components obtained, in any of the three conditions, with repeated decompositions may occur in different orders in different decompositions due to the permutation invariance of the ICA solution. Therefore, pairs of best-matching components in the original decomposition $\mathbf{u}^{(o)}(T)$ and each repeated decomposition $\mathbf{u}^{(r)}(T)$ must be found. Because of the scaling invariance of the ICA problem, corresponding components may differ by an unknown scaling factor, which in the complex case includes an arbitrary phase shift (i.e., multiplication by a unit-norm complex number). To find best-matching components, we employ the complex correlation coefficient of their activity time courses, similar to the procedure that has been determined in [5] as optimal for real-valued components. Define as $\mathbf{u}^{(o)}(T)$ the components obtained from the original data $\mathbf{x}^{(o)}(T)$ with original training data order and identity matrix initialization $\mathbf{W}_{init}^{(o)} = \mathbf{I}$. Denote by $\mathbf{u}^{(r)}$ the components obtained from the r-th repetition in one of the three investigated conditions. We compute for all repetitions r the correlation matrix $\mathbf{C}^{(r)}$ with (i, j)–entry

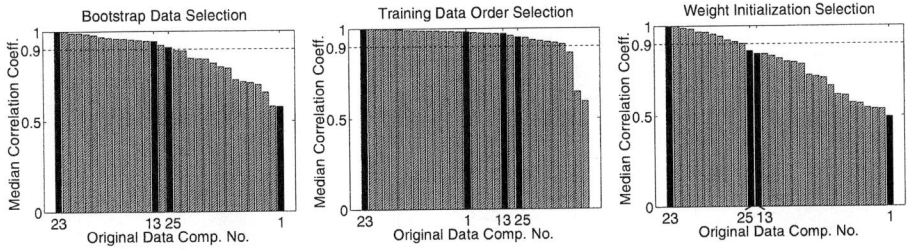

Fig. 3. Reliability of the obtained components. The median correlation coefficient (ordinate) of each original data set component's (abscissa) time course with its best-matching bootstrap data set components is shown for bootstrap data (left), training data order (center), and weight initialization selection (right), respectively. Components were sorted from highest correlation (left) to lowest (right). Components used as examples in this manuscript are marked by black bars, with their component number in the original data set decomposition indicated on the abscissa. Using a threshold of a minimum 0.90 median correlation coefficient (dashed line) for reliability, the training data order condition (center) yielded the highest number of reliable components (28 out of 31), bootstrap data (left) produced 16 reliable components, and weight initialization 11 (right).

$$\left[\mathbf{C}^{(r)}\right]_{ij} = \rho_{ij}^{(r)} = \left| \left\langle u_i^{(o)}(T) \left(u_j^{(r)}(T)\right)^* \right\rangle_T - \mu_i^{(o)} \left(\mu_j^{(r)}\right)^* \right| / \left[\sigma_i^{(o)} \sigma_j^{(r)}\right] \quad (2)$$

corresponding to the magnitude correlation coefficient of component i in the original decomposition and component j in the r-th repeated decomposition. Here, $\langle \cdot \rangle_T$ denotes expectation computed as time average; $*$ complex conjugation; $|\cdot|$ magnitude; and $\mu_i^{(o)}$, $\sigma_i^{(o)}$ and $\mu_j^{(r)}$, $\sigma_j^{(r)}$ denote mean and standard deviation of the i-th original component and of the j-th component of the r-th repetition, respectively. For each repetition r and each original data component i, we assign component $\tilde{j}$,

$$\tilde{j}(i,r) = \underset{j}{\operatorname{argmax}} \left(\rho_{ij}^{(r)}\right), \quad (3)$$

that has maximum correlation coefficient as the best-matching component for original component i. Component $u_i^{(o)}(T)$ is considered 'reliable' if the median (across repetitions) correlation coefficient with its matching components,

$$\tilde{\rho}_i = \underset{r}{\operatorname{median}} \left(\rho_{i,\tilde{j}(i,r)}^{(r)}\right) \quad (4)$$

reaches or exceeds a chosen threshold value $\tilde{\rho}_{\text{thresh}}$.

3 Experiment Paradigm

We employed complex ICA [1] to analyze data from a visual spatial selective attention experiment in which the subject was asked to respond by a button press as quickly as possible each time a target stimulus appeared in an attended location [7]. Analysis included 582 trials, each 1 s long, time locked to target stimulus presentations to

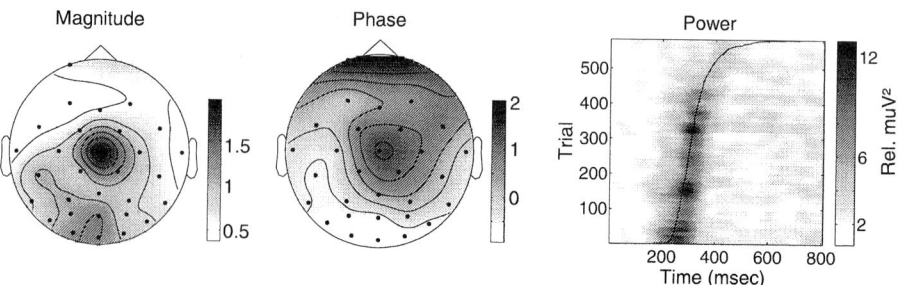

Fig. 4. Component 13 of the 5-Hz band represents a component whose activity was tightly linked to subject behavior (button press). It proved reliable in the bootstrap data and training data order conditions, and was just below the reliability threshold in the weight initialization condition. The graphs show (from left to right) the magnitude of the associated complex scalp map, the scalp map's phase, and the ERP-image plot of 5-Hz component signal power in different experimental trials (ordinate) and times after stimulus onset (abscissa). Subject response time in each trial is overlaid on the ERP image (black trace). For better visualization, the ERP image has been smoothed across trials with a 30-trials wide rectangular window. The impression of temporal smearing is induced by the length (200 ms) of the spectral decomposition window. The ERP image shows that component energy is maximum at the subject response time. The best-matching component obtained by real time-domain ICA did not show such a tight relationship to behavior (data not shown here). The component map exhibits a phase gradient around the central focus of activation, that reflects spatio-temporal dynamics of the underlying cortical activation.

one subject. The data were recorded from 31 EEG electrodes at a sampling rate of 256 Hz. Spectral decomposition was performed with a hanning-windowed Fourier basis of length 50 samples, with a window shift of 1 sample between successive analysis windows. Reliability analysis was confined to data in the 5-Hz band, due to processing limitations. For each of the three conditions (bootstrap data, training data order, and weight initialization selection), $R = 99$ decompositions were performed, each resulting in 31 complex independent components. In the bootstrap data condition, $R = 99$ bootstrap data sets were generated from the original 5-Hz band data after spectral decomposition. The 5-Hz original and bootstrap data, respectively, were sphered during preprocessing. The adaptation rate of the complex ICA algorithm was lowered successively, and optimization was halted when the total weight-change induced by one sweep through the data was smaller than 10^{-6} relative to the Frobenius norm of the weight-matrix. Convergence was attained in each decomposition within 222 iterations or less. To define component reliability, a minimum median correlation coefficient $\tilde{\rho}_{thresh} = 0.90$ was chosen, as suggested by the results presented in Figs. 2 and 3.

4 Results

The main results are presented in Figs. 2 to 5. Assigning best-matching complex components in different decompositions by means of the correlation coefficient between component activities is an appropriate method of identifying reliable components. The correlation of a reliable component with its best-matching component in different de-

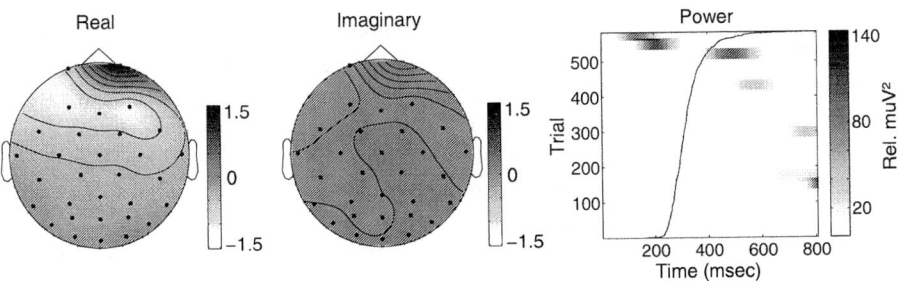

Fig. 5. Component 23 of the 5-Hz band was the most reliable in all three test conditions. The graphs display (from left to right) the real and imaginary parts of the component's complex scalp map, and its 5-Hz power ERP image, respectively (similar to Fig. 4). ERP image and scalp maps indicate that this component can be linked to a right eye artifact (possibly a right eyelid twitch) with a very sparse activity pattern. The small imaginary part of the complex scalp map indicates a low degree of spatio-temporal dynamics for this component.

compositions was always significantly higher than with its second-best matching component. This is illustrated in Fig. 2 for three example components obtained in the bootstrap data condition.

Reliable complex components were found under each of the three test conditions. However, the number of reliable components (out of 31 components in total) varied across conditions, with the bootstrap condition producing 16, the shuffling condition 28, and the initialization condition 11; see Fig. 3 for details. The reliable components constituted nested subsets: All components that were reliable in the weight initialization condition were also reliable in the bootstrap data condition, and all components reliable in the bootstrap data condition were also reliable in the training data order condition.

Among the reliable components, we found physiologically plausible components that were closely linked to behavior, e.g., the motor-response related component in Fig. 4, and 'artifactual' components like the right eye blink related component in Fig. 5. Some reliable components showed a clear phase shift across electrode positions with highest signal energy in the associated component scalp maps (Fig. 4), indicating an activity pattern with a strong spatio-temporal dynamics. Other reliable components did not exhibit such a phase shift (Fig. 5), showing that complex ICA may also produce near-static scalp maps, if appropriate for components of the data.

5 Discussion and Conclusion

Complex ICA [1] applied to EEG recordings produces physiologically plausible and behaviorally relevant components. We have demonstrated here the degree of reliability of such components in the 5-Hz band. These results are complementary to the physiological and behavioral evidence that complex ICA may be used to faithfully model brain processes. Some of the reliable components showed phase shifts across electrode positions in the associated complex scalp maps that could not have been obtained using standard time-domain ICA methods. This finding suggests that complex ICA for EEG signals can non-invasively measure the spatio-temporal flow patterns of cortical activation. It remains to be shown that these results extend also to other frequency bands.

We have shown that matching complex components across decompositions by means of their activity time course correlations is an effective way to investigate the effects of several variabilities on the results produced by the complex ICA algorithm. Variability in the data set, as implemented by resampling techniques, may be regarded as the variation of highest interest, since it tests reliability by generating data that might have been measured instead of the original data. It is reassuring that under this test about half of the 31 separated complex ICA components proved reliable. It remains an open question why we found a higher fraction of (one-dimensional) independent components than reported in [4]. This could, e.g., be due to the different algorithms or data sets used, or to our use of a 'global' instead of a 'local' approach. It appears unlikely that we have mistaken some multidimensional components for one-dimensional components, since the reliability threshold was chosen to be fairly high. Varying the training data order resulted in even smaller variations in the ICA results. Duann et al. [5] obtained even higher reliability, which might be attributed to differences in algorithm, data set, or training parameters. Varying the initial conditions of the algorithm resulted in a much larger variability in the outcome, which might not be surprising since global convergence has not been proven for ICA algorithms (including complex ICA). Nevertheless, it is remarkable that a significant number of components remained highly reliable even under this perturbation.

The methods investigated here provide complex ICA brain signal analysis with a quantitative indication of the numerical stability of components. Further studies are needed to confirm whether the quantitatively reliable components always coincide with components having meaningful physiological interpretations and covarying with subject behavior.

References

1. J. Anemüller, T. J. Sejnowski, and S. Makeig. Complex independent component analysis of frequency-domain electroencephalographic data. *Neural Networks*, 16:1311–1323, 2003.
2. A. Arieli, A. Sterkin, A. Grinvald, and A. Aertsen. Dynamics of ongoing activity: Explanation of the large variability in evoked cortical responses. *Science*, 273:1868–1871, 1996.
3. H. Berger. Über das Elektroencephalogramm des Menschen (On the electroencephalogram of man). *Archiv für Psychiatrie und Nervenkrankheiten*, 87:527–570, 1929.
4. F. Meinecke, A. Ziehe, M. Kawanabe, and K.-R. Müller. A resampling approach to estimate the stability of one-dimensional or multidimensional independent components. *IEEE Transactions on Biomedical Signal Processing*, 49(12):1514–1525, December 2002.
5. J.-R. Duann, T.-P. Jung, S. Makeig, and T. J. Sejnowski. Consistency of infomax ICA decomposition of functional brain imaging data. In *Proceedings of the fourth international workshop on independent component analysis and blind signal separation*, pages 289–294, Nara, Japan, April 2003.
6. A. J. Bell and T. J. Sejnowski. An information maximization approach to blind separation and blind deconvolution. *Neural Computation*, 7:1129–1159, 1995.
7. S. Makeig, M. Westerfield, T.-P. Jung, S. Enghoff, J. Townsend, E. Courchesne, and T. J. Sejnowski. Dynamic brain sources of visual evoked responses. *Science*, 295:690–694, 2002.

Sensor Array and Electrode Selection for Non-invasive Fetal Electrocardiogram Extraction by Independent Component Analysis

Frédéric Vrins[1], Christian Jutten[2], and Michel Verleysen[1]

[1] Université catholique de Louvain, Machine Learning Group
Place du Levant, 3, 1380 Louvain-la-Neuve, Belgium
{vrins,verleysen}@dice.ucl.ac.be
[2] Institut National Polytechnique de Grenoble, Images and Signals Laboratory,
Avenue Félix Viallet 46, 38031 Grenoble, France
jutten@lis.inpg.fr

Abstract. Recently, non-invasive techniques to measure the fetal electrocardiogram (FECG) signal have given very promising results. However, the important question of the number and the location of the external sensors has been often discarded. In this paper, an electrode-array approach is proposed; it is combined with a sensor selection algorithm using a mutual information criterion. The sensor selection algorithm is run in parallel to an independent component analysis of the selected signals. The aim of this method is to make a real time extraction of the FECG possible. The results are shown on simulated biomedical signals.

1 Introduction

In order to improve the accuracy of their diagnosis, detect the fetal distress and avoid unnecessary caeserian deliveries, the obstetricians are interested in completing the information given by the fetal heart rate variability (FHRV) by a waveform analysis of the fetal electrocardiogram (FECG) signal. Moreover, FECG could be a very efficient way for *in utero* fetal heart monitoring and pathology detection during the pregnancy.

Today, this signal can be catched during the labour through a sensor located on the scalp of the fetus, and its diagnostic reliability is confirmed. Obviously, this method can only be applied when the fetal membranes are broken, *i.e.* during the delivery.

To make earlier FECG-based fetal monitoring possible, it can be interesting to develop a non-invasive method to extract this signal. In addition to the possibility of an earlier analysis, a non-invasive method to measure the FECG signal has other advantages. For instance, such method is less stressful for the fetus, because there is no contact between its body and the measurement instrumentation. Furthermore, as the sensors are located on the pregnant woman's abdomen, sanitary precautions are less crucial.

Recently, some authors have shown that this problem can fit into the blind source separation (BSS) framework, where the 'mixtures' are the signals recorded by external sensors, and the original sources are signals emitted by maternal and fetal muscles. Previous works based on this method for the FECG extraction have given promising results [1–3]. In practice, even if the results of the extraction are satisfactory, some assumptions on the model can be slightly violated (e.g. linearity and instantaneity of the mixture), and the location and the number N of the external sensors $\mathcal{S} = \{S_1, \ldots, S_N\}$ (recording signals $\mathcal{X} = \{X_1, \ldots, X_N\}$ respectively) is a question still under debate.

In this paper, a hundred-electrodes belt (N=100) [4], located around the pregnant woman's abdomen, is used. In order to be able to extract m sources in real-time, a subset $\mathcal{X}^\star$ of $n < N$ signals $X_i^\star$ recorded by *selected sensors* ($S_i^\star$, respectively with $1 \leq i \leq n$) will be processed by a BSS algorithm (discarding all other electrodes). It will be shown that choosing an appropriate criterion for the selection of the n signals gives interesting results: the extraction can be performed on few sensor signals. Furthermore, it seems that in some cases, an optimal number $n \geq m$ of selected signals appears: the quality of the FECG extraction – possibly after projection by principal component analysis (PCA) – using only $\mathcal{X}^\star$ is improved by comparison to the performances reached if the whole set $\mathcal{X}$ of signals is used in the extraction process ($n = 100$).

In the following of this paper, we will first stress the importance of the FECG signal for the obstetricians. In the next section, we will discuss on a non-invasive (parallel) process to extract this signal. In section 3, the sensor selection algorithm, based on the mutual information, is detailed. Finally, simulation results are presented, before concluding.

2 FECG Measurement Process

2.1 Non-invasive Measurement

A non-invasive method to extract the FECG signal seems thus very attractive. Unfortunately, sensor signals record mixtures of electrical components, due to the electrical activity of several physical sources: the fetal and maternal hearts, the diaphragm and the uterus, among others. The fetal contributions (due to the fetal heart muscular activity) are minor by comparison with these electrical sources, and classical signal processing (like de-noising, filtering, ...) does not allow us to recover the FECG.

One of the most recently investigated methods in order to recover the FECG is BSS. Indeed, the sensor signals actually record a mixture of the electrical signals emitted by the original sources. If the sources are mutually independent and if their mixture is linear, instantaneous and noise-free (or of negligible power), the well-known method of independent component analysis (ICA) is able to recover the original sources, up to a scale factor and permutation [1–3]. In the FECG case, these indeterminations do not matter, because the analysis focuses only on waveforms. Note that the identification of the FECG signal among all the estimated sources signals is a quite easy task, but extracting a complete PQRST complex requires much more effort, especially due to the residual noise.

2.2 Optimal Number and Location of External Sensors

In non-invasive methods, the sensors must be obviously external. In the ideal case of source separation (linear, instantaneous and noise-free mixtures of independent sources), a necessary condition to perfectly recover the original sources is that the number of external sensors must be greater or equal to the number of original sources. As a consequence, the location of sensors does not seem important. Nevertheless, an additional condition exists on the sensors: they must record 'different mixtures'. Indeed, if the number of sensors is equal to the number of sources but two sensors record exactly the same signal, the system is overcomplete and the inversion of the mixing system becomes impossible (null determinant). Similarly, all sources must be involved in the recordings with a non-zero variance. These considerations and the very low power of the FECG signal (by comparison to the electrical environment) explain why the location of the electrodes is an important problem. It is reasonable to think that relevant locations of the electrodes can improve the extraction of the FECG signal, while others can deteriorate it [5]. Moreover, as the fetus moves, it is clear that it does not exist an optimal location for the sensors, constant in time. Furthermore, some electrodes may record irrelevant signals (for example because of a poor contact between the mother's skin and the sensor itself). For all these reasons, it seems to be careful to place a lot of electrodes on the mother's body, possibly further discarding some of them by some selection algorithm.

2.3 Fetal ECG Extraction

The previous discussion justifies the sensor-array approach for the FECG application. In this section, a belt of hundred electrodes (located around the abdomen of the mother) is first presented. Next, the processing of the signals recorded by these sensors (in order to allow a real-time extraction of the FECG) is briefly explained.

A Hundred Electrodes Belt. Consider an array of ten rows and ten columns[1] of electrodes [4], located around the pregnant woman's abdomen (each row i and column j are labelled from 1 to 10). The sensor located at the intersection of the ith row and j^{th} column is noted S_{ID}, with $ID(i,j) = (i-1) \times 10 + j$. The associated recorded signal is X_{ID}.

Parallel Processing. Recall that most of the ICA algorithms separate up to as many sources as sensors. If the number of sources is lower than the number of electrodes, convergence problems (switching problems due to the permutation indeterminacy) may appear. In order to avoid this problem, a dimension reduction by PCA is usually first applied. Nevertheless the projection of the

---

[1] Of course, other grid geometry of electrodes can be proposed.

100-dimensional data[2] on the subspace of the original sources ($m \ll 100$) requires the eigenvalue decomposition of the mixture covariance matrix, which can be critical from a computational complexity point of view. In order to allow a real-time extraction, the discarding of some electrodes must be investigated: this will also reduce the computation time of the FECG separation.

The first parallel processing is the electrode selection algorithm. It consists in selecting/discarding several electrodes from the initial set $\mathcal{S} = \{S_1, \ldots, S_{100}\}$ in order to select 'interesting' sensors to reduce the computational time of the extraction process, keeping satisfactory performances of the FECG recovery. Only a subset of $n < 100$ electrodes ($\mathcal{S}^\star = \{S_1^\star, \ldots, S_n^\star\}$) will be analyzed by the source extraction algorithm. The duration of the selection algorithm may be greater than the separation process. An ICA algorithm will then process the signals recorded by the selected sensors $S_i^\star$'s, possibly after PCA if $n > m$. The selection algorithm is run continuously in parallel to the separation one, to select new sets of n electrodes. After each new subset $\mathcal{S}^\star$ is built, the PCA/ICA is run on $\mathcal{S}^\star$, in order to process *in real-time* the associated signals; then the selection algorithm is restarted (with $\mathcal{S} = \{S_1, \ldots, S_{100}\}$).

3 Sensor Selection Algorithm

In this section, an unsupervised criterion for the selection of electrodes is presented. Its interpretation is emphasized in the context of the FECG extraction.

3.1 An Information-Based Criterion

In order to extract correctly the FECG signal even with a low number of sensors, it seems natural to look for signals that drive different electrical components due to the fetal heart's activity. Note that such task may not be accomplished by a classical spectral analysis, because in pathological cases, the frequencies of the maternal and fetal hearts may be very close. By contrast, the probability density functions (pdf) of the sensor signals may contain interesting information. The pdf are estimated here by the Parzen estimator with isotropic Gaussian kernels of standard deviation equal to 0.02 (see [6] for more details)

Consider a signal X_{Ref} (which will be called in the following the 'reference'), which is the closest one from the pure maternal ECG signal ($X_{Ref} \simeq \text{MECG}$). In the simulations below, X_{Ref} is recorded by sensor S_{50}. By contrast to this signal, the temporal structure of other ones may contain electrical components due to the muscular activity of the fetal myocardium. One can observe on Fig. 1(a) that when X_{Ref} takes values equal to zero (horizontal dashed line), X_i (with $i = \{6, 36\}$) can take values different from zero due to i) the centered noise on X_i and ii) the fetal R-waves (located on the vertical dotted lines). These

---

[2] Using one electrode as a reference, a 100-electrode array provides 99 signals, which are the voltage difference between the reference and the 99 remaining electrodes. Changing the reference electrode can provide up to $100 \times 99/2 = 4950$ different signals.

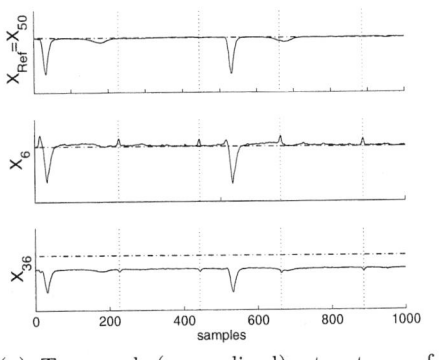
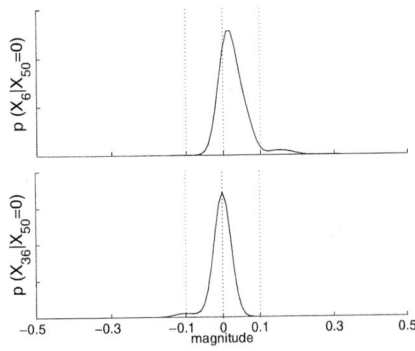

(a) Temporal (normalized) structure of three recorded signals.

(b) Conditional density functions of X_6 and X_{36} w.r.t. $X_{Ref} = 0$.

Fig. 1. Temporal and (conditional) statistical structure of three recorded signals.

considerations explain why the conditional density (cpdf) $p(X_i|X_{Ref} = 0)$ does not reduce to a symmetric Gaussian (because of the kernel estimator) function (see Fig. 1(b)); symmetric result would be obtained for $p(X_j|X_{Ref} = 0)$ if the difference between X_j and X_{Ref} was only due to symmetric effects (a.o. noise). The R-wave is approximatively triangular (i.e. with uniform distribution) and is not centered, contrarily to the noise. Consequently, the asymmetry of the cpdf $p(X_i|X_{Ref} = 0)$ is mainly due to the fetal R-waves: it is thus an interesting way to identify signals that drive important fetal contribution.

Nevertheless, a simple measure of the cpdf asymmetry is not robust. Indeed, this function corresponds (up to a scale factor ensuring a unitary area) to a particular 'slice' ($X_{Ref} = 0$) of the joint pdf between the reference and the recorded signals. This slice becomes irrelevant if an offset appears on X_{Ref}. In order to circumvent this problem, it is preferable to consider the specificity of the 'shape' of the whole pdf. For this reason, another criterion for selection is preferred, based on the mutual information (MI), noted I (see [7]). Of course, the aim of the preprocessing detailed in the previous section is to reduce the number of electrodes to be processed by the separation algorithm; its role is thus mainly to reduce the dimensionality of the 'effective' inputs. The first selected electrode is S_{Ref}. In the algorithm (detailed in Fig. 2), while $k < n$, the sensor $S_k^* \doteq S_i$ ($S_i \in \mathcal{S}$) is selected if X_i minimizes the sum of the MI's with the previously selected signals (line 6 in Fig. 2). Next, this sensor is removed from $\mathcal{S}$.

3.2 Interpretation of the SenSelec Algorithm

According to the meaning of I (see [7]), the selected signals will be quite independent (because of the minimization of a MI-based criterion). Therefore, the SenSelec algorithm constitutes a good preprocessing for ICA (that consists in finding the rotation of signals that rends them as independent as possible). It must be stressed that in the case of the FECG extraction, the selection of the

SenSelec $(\mathcal{S}, Ref, n)$

1 $S_1^\star \leftarrow S_{Ref}$ // reference electrode
2 $\mathcal{S} \leftarrow \mathcal{S}/\{S_{Ref}\}$
3 $\mathcal{S}^\star \leftarrow \{S_1^\star\}$
4 **for** $k \leftarrow 2$ **to** n **do**
5   **for** $i \leftarrow 1$ **to** $100 - (k-1)$ **do**
6        $\mathcal{C}(i) \leftarrow \sum_{j=1}^{k-1} I(X_j^\star, X_i | S_i \in \mathcal{S})$ // cost function
7     $j \leftarrow argmin_i(\mathcal{C})$ // ID of winner sensor
8     $S_k^\star \leftarrow S_j$ // winner sensor
9     $\mathcal{S} \leftarrow \mathcal{S}/\{S_j\}$ // removing winner sensor
10    $\mathcal{S}^\star \leftarrow \mathcal{S}^\star \bigcup \{S_k^\star\}$ // update selected subset
11 **Return** $\mathcal{S}^\star$; // set of selected sensors

Fig. 2. Electrode selection algorithm. The cost function $\mathcal{C}$ is based on the mutual information between the selected and unselected electrodes.

electrodes is actually done according to the fetal contributions in the signals [6]. For instance, $X_2^\star$ minimizes the MI with $X_{Ref} \simeq$ MECG (i.e. which is the most independent from it); here, $X_2^\star$ drives an important fetal contribution (in the simulations below, $X_2^\star = X_6$, see Fig. 1(a)). The shape of $I(X_i, X_{Ref})$ is given in Fig. 3. We can observe that the 'distance' between X_i and X_{Ref} mainly varies along the columns.

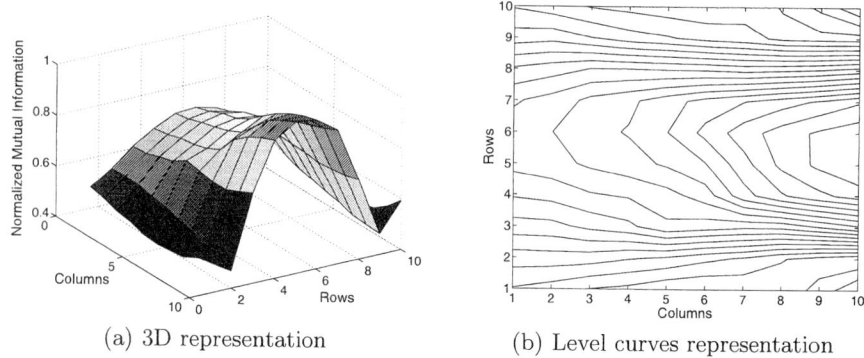

(a) 3D representation   (b) Level curves representation

Fig. 3. Mutual information between each recorded signal and the reference vs the location of the sensor.

4 Performances of the Fetal ECG Extraction

In this work, simulated signals have been used. This is useful in order to be able to use correlation-based criterions between the estimated FECG and the true one. The simulator used here is a realistic model of the electrical interaction in the maternal body. It was shown that in such 'real-world' mixtures, it is difficult to find a reliable blind criterion to estimate the quality of the extraction

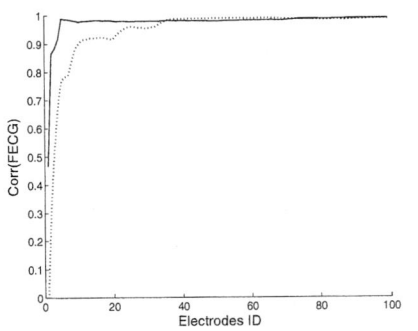 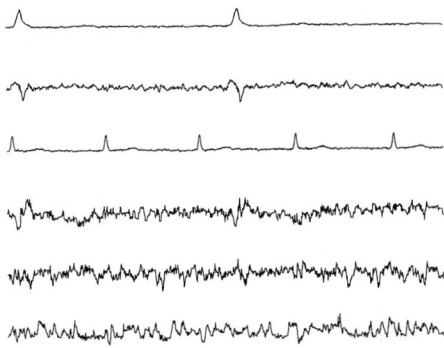

(a) Correlation between the true and the estimated FECG's vs n, using i) SensSelec($\mathcal{S}, 50, n$) (solid line) and ii) sets of electrodes taken in numerical order (dotted line).

(b) Separated sources if the six first signals are selected with SensSelec($\mathcal{S}, 50, 6$). The optimal extraction of the FECG is reached for $n = n^\star = 6$, third separated signal.

Fig. 4. Extraction performances vs n (left) and separated sources for $n = n^\star$ (right).

of each source, one by one [8]. Furthermore, the hardware instrumentation is quite expensive, and the simulation is a useful step before material realization. The model used to simulate the signals includes both real measurement and simulated data for the sources and the mixing environment. More details about this model can be found in [9]. The selection algorithm is applied here on these simulated sensor signals.

If the number n of selected signals is greater than the – supposed – number m of sources, then the signals are projected by PCA on the sources subspace (m has been taken equal to 6). In order to test the validity of the algorithm, we have plotted in Fig. 4(a) the correlation curve between the original FECG and the estimated one (the extraction was done using the JADE algorithm [10]). We can observe that if we project the six first selected signals, the correlation between the original FECG and the estimated one is even greater than if we had projected directly the hundred signals on the source subspace. This optimal value of n is denoted $n^\star$ (here, $n^\star = 6$). The associated estimated sources are given in Fig. 4(b).

5 Discussion and Conclusion

It was explained why many sensors (say N) should be involved in the measurement process to non-invasively extract the fetal ECG signal. But difficult problems to extract the sources in real-time occur because of this high dimensionality; for example, the computational cost related to the projection can be high. In order to circumvent this problem, a sensor selection method was derived, using an unsupervised criterion based on the mutual information. The first 'selected' signal ($X_1^\star$) must be chosen by other means. The MI criterion was shown to be linked to the fetal content if $X_1^\star$ is close from the pure maternal ECG. The selection algorithm builds a subset of the original sensors; the associated signals will be processed (possibly after a PCA to guarantee a square mixing system) by a BSS algorithm. Selecting a subset of signals has three major advantages. First, it reduces the computational cost of the extraction process, and allows us to separate the sources in real time. Secondly, if the number of selected electrodes is 'well chosen' ($n = n^*$), it can be possible to obtain better extraction performances than if all the sensors were involved in the extraction process ($n = N$). In real situations however, the original source is unknown and a 'blind' criterion must be used to measure the 'FECG extraction quality', instead of the correlation; this task may reveal difficult. Note that in practice, taking $n > n^*$ does not seems to be very thorny from the FECG separation performances point of view. Third, the selection process is able to choose an optimal electrode set, despite the fetal motion.

Further investigations will include extension of the selection process in noisy mixtures, the test on actual FECG data and the determination of n^*.

References

1. De Lathauwer, L., De Moor, B., Vandewalle, J.: Fetal electrocardiogram extraction by blind source subspace separation. IEEE Trans. Biomed. Eng. 47 (2000) 567–572
2. Vigneron, V., Paraschiv-Ionescu, A., Azancot, A., Jutten, C., Sibony, O.: Fetal electrocardiogram extraction based on non-stationary ica and wavelet denoising. In: 7th Symposium on Signal Processing and Appl., Paris, France (2003) 69–72
3. Mareossero, D.E., al.: Independent component analysis for fetal electrocardiogram extraction: A case for the data efficient mermaid algorithm. Neural Networks for Signal Processing (2003, Toulouse, France) 399–408
4. Vigneron, V., Azancot, A., Sibony, O., Herail, C., Jutten, C.: Dispositif materiel et logiciel d'extraction d'ecg foetal. enveloppe Soleau, Institut National de la Protection Industrielle (France) (2002)
5. Vrins, F., Lee, J., Verleysen, M., Vigneron, V., Jutten, C.: Improving independent component analysis performances by variable selection. In: 13th IEEE workshop on Neural Networks for Signal Processing (NNSP 2003), Toulouse (France) (2003) 359–368
6. Vrins, F., Vigneron, V., Jutten, C., Verleysen, M.: On the extraction of the snore acoustic signal by independent component analysis. In: The second IASTED conf. on Biomedical Engineering (BioMED04), Innsbruck (Austria) (2004)
7. Cover, T.M., Thomas, J.A.: Elements of information theory. Wiley and sons (1991)
8. Vrins, F., Archambeau, C., Verleysen, M.: Towards a local separation performances estimator using common ica contrast functions ? In Verleysen, M., ed.: proceedings of ESANN'04, the 12th European Symposium on Artificial Neural Networks, Bruges (Belgium), d-side publications (2004) 211–216
9. Schmidt, M.: Sensor array for fetal ecg signals. simulation, sensor selection and source separation. Master of science, INPG, Laboratoire des Images et des Signaux, Grenoble, France (2003) supervised by V. Vigneron and C. Jutten.
10. Cardoso, J.F.: Source separation using higher order moments. Proceedings of the IEEE, Int. Conf. on Acoustics, Speech and Signal Processing (ICASSP'89) (1989, Glasgow, England) 2109–2112

A Comparison of Time Structure and Statistically Based BSS Methods in the Context of Long-Term Epileptiform EEG Recordings

Christopher J. James and Christian W. Hesse

Signal Processing and Control Group, ISVR, University of Southampton,
Southampton SO17 1BJ, UK
C.James@soton.ac.uk

Abstract. Blind source separation (BSS) techniques are increasingly being applied to the analysis of biomedical signals in general and electroencephalographic (EEG) signals in particular. The analysis of the long-term monitored epileptiform EEG presents characteristic problems for the implementation of BSS techniques because the ongoing EEG has time varying frequency content which can be both slowly varying and yet also include short bursts of neurophysiologically meaningful activity. Since statistically based BSS methods rely on sample-estimates, which generally require larger window sizes, these methods may extract neurophysiologically uninformative components over short data segments. Here we show that BSS techniques using signal time structure succeed in extracting neurophysiologically meaningful components where their statistical counterparts fail. To this end we use an algorithm that extracts linear mixtures of nonstationary sources, without pre-whitening, through joint diagonalisation of a number of windowed, lagged cross-covariance matrices. We show that this is extremely useful in tracking seizure onset in the epileptiform EEG.

1 Introduction

Multi-channel electroencephalographic (EEG) recordings capture ongoing brain activity which is useful in the diagnosis of many brain disorders. This activity can be interpreted as a number of brain-sources whose outputs vary over time. Under certain conditions, some sources dominate and specific types of brain activity are then associated with specific brain states – such is the case with rhythmic activity in the epileptiform EEG. Artifacts also appear consistently in the recorded EEG and, although not brain sources, can also be treated as distinct sources of information from the many ongoing brain sources. It would be particularly useful to be able to automatically isolate, visualise and track multiple neurophysiologically meaningful sources underlying the ongoing EEG recordings. However EEG signals are typically nonstationary and have spatial and temporal correlations, all of which generally complicate automated EEG analysis techniques

A variety of methods for Blind Source Separation (BSS) and Independent Component Analysis (ICA) [1,2] have been applied to EEG for extracting artifacts and neurophysiologically meaningful components [3]. Generally, for these methods to work, the EEG is assumed to be a linear mixture of statistically independent, possibly nonstationary, sources which may be decomposed using either statistical and information theoretic signal properties, or signal time structure. In the first of these approaches the signals are treated as samples of random variables where temporal ordering is irrelevant, and works by factorising marginal distributions using higher order moments or cumulants, or by minimising mutual information [4]. The popular method of Fast ICA is an example of the latter approach and will be used for the sake of comparison in this work. Whilst statistical approaches are frequently used in the analysis of EEG [3,5], it is more intuitive, and indeed it has been argued [1], that using time structure for the decomposition may be more appropriate for such time-series data.

Source decomposition on the basis of signal time structure may be achieved through temporal decorrelation (TD). For sources with stationary waveforms and unique power spectra, the time structure is adequately captured by temporal cross-covariances [6,7]. However, the EEG is generally considered nonstationary over longer durations, and a TD approach would not be expected to yield useful results. However, the stationarity constraint of TD based methods may be relaxed by using a series of short-time windowed TDs as seen in [8]. The decorrelation operation in time structure BSS methods involves the joint diagonalisation of a set of symmetric matrices which reflect the spatio-temporal covariance structure of the source mixture. Furthermore, algorithms have recently been developed for non-orthogonal joint diagonalisation that process signal covariances directly with no need for pre-whitening (note that whitening *is* required for Fast ICA), one such algorithm is given by [9] which we will term $LSDIAG_{TD}$ here.

In this work we perform a set of experiments on real and synthetically generated seizure EEG using both statistically based and temporal decorrelation methods. The aim is to assess the ability of both to extract neurophysiologically meaningful information under normal (clinical) recording conditions.

2 Methods

In the standard, noise free, formulation of the BSS problem, the observed signals $\mathbf{x}(t)$ are assumed to be a linear mixture of an equal number of unknown but statistically independent source signals $\mathbf{s}(t)$, i.e., $\mathbf{x}(t)=\mathbf{A}\mathbf{s}(t)$, where the square mixing matrix $\mathbf{A}$ is also unknown but invertible. The problem is solvable up to a permutation, and sign and power indeterminacy of the sources, by finding an appropriate de-mixing matrix $\mathbf{W}=\mathbf{A}^{-1}$ which allows estimation of the source waveforms by $\mathbf{s}(t) = \mathbf{W}\mathbf{x}(t)$.

2.1 Fast ICA

The statistically based BSS method we use in this study is Fast IC, which is a fast, fixed-point iterative algorithm that undertakes to find projections that maximise the non-Gaussianity of components by their kurtosis or negentropy. A common preprocessing step for Fast ICA is whitening, whereby the observed time-series are linearly decorrelated and scaled to unit variance. With this algorithm the problem is posed as an optimisation problem with the independent components as its solution which can be extracted one at a time. A more flexible and reliable approximation of negentropy was introduced such that

$$J(\mathbf{y}) \approx \rho [E\{G(\mathbf{y})\} - E\{G(v)\}]^2 , \qquad (1)$$

where ρ is a positive constant, v is a zero mean, unit variance Gaussian and $G(.)$ can be any non-quadratic function (a number of which are suggested by [4]).

2.2 BSS Through Temporal Decorrelation

The TD approaches such as TDSEP [6] or SOBI [10] exploit the fact that, due to statistical independence, the source covariance matrix $\mathbf{C}_\tau^s = \mathbf{W}\mathbf{C}_\tau^x\mathbf{W}^T$ is diagonal for all time lags $\tau = 1,2,3, \ldots$, where T denotes matrix transpose and $\mathbf{C}_\tau^x$ is the signal covariance matrix.

In essence, $\mathbf{W}$ transforms the signal covariances into the source covariances, which are diagonal due to assumed independence. Estimation of $\mathbf{W}$ reduces to the well-researched problem of joint (approximate) diagonalisation of the stack of matrices given by $\mathbf{W}\mathbf{C}_\tau^x\mathbf{W}^T$, for which a fast and efficient new algorithm was recently proposed (LSDIAG$_{TD}$) [9].

To counteract the possible issues the stationarity constraint of TD based methods may raise when applied to the EEG, [8] suggested a method that jointly exploits the nonstationarity and temporal structure of sources. The method uses multiple time-delayed correlation matrices of the observed data, each of which is evaluated within a different time window of the data in order to estimate the overall demixing matrix.

2.3 The Data

Both the Fast ICA and the LSDIAG$_{TD}$ method were in turn applied to a synthetically generated multichannel seizure EEG and a real 20 s segment of seizure EEG. The EEG was digitally recorded in a long-term epilepsy monitoring unit (EMU), at a rate of 200 samples/sec with a low-pass filter of 65 Hz and with 12 bits resolution. Twenty-five electrodes were used for recording the scalp EEG, placed according to the modified 10-20 electrode placement system. The data was recorded with a reference at position FCz and the data matrices in each case were then mean corrected in the columns (i.e., an average referential montage was assumed) and in the rows.

3 Results

3.1 Synthetic Seizure EEG

The synthetic seizure EEG consists of a real 20s segment of multichannel, ongoing EEG onto which a rhythmic signal was superimposed. The synthetic seizure was generated through a 6 Hz sine wave exponentially weighted in the amplitude so that the amplitude of the signal rises above zero at the 5 s mark and continues to increase in amplitude exponentially. The real ongoing 20 s segment of EEG used was chosen such that it showed no visible ictal activity of any kind. The synthetic seizure was superimposed on each measurement channel as though it was emanating from a dipolar source normal to the cortex somewhere in the right parietal region. Fig. 1 (a) depicts the resulting synthetic seizure EEG epoch with the topography of the seizure shown in the inset and Sz showing the underlying ictal activity.

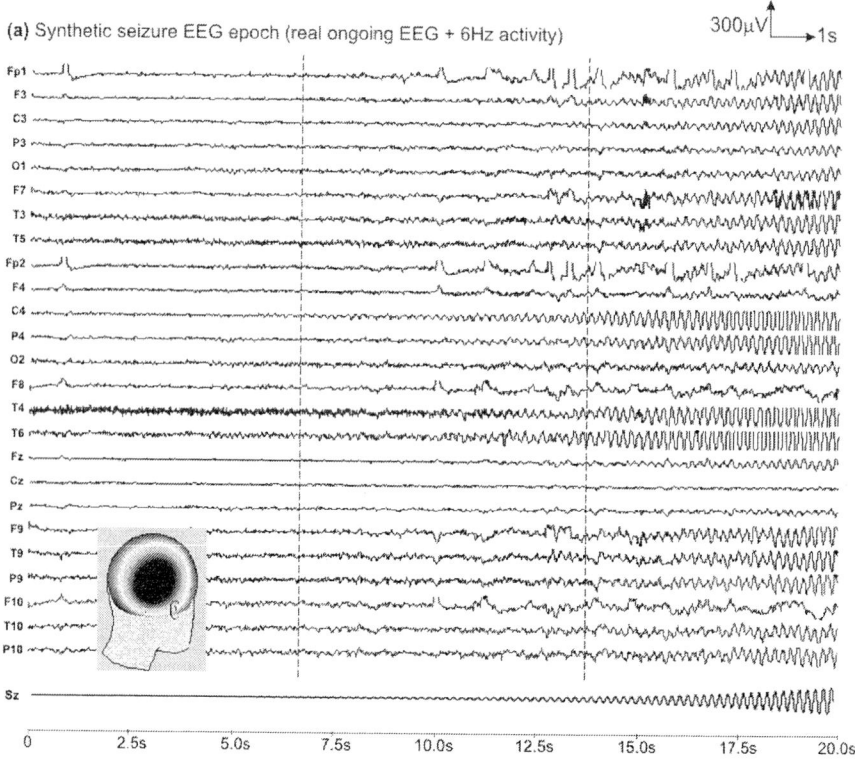

Fig. 1. 20 s epoch of real scalp EEG recorded with 10-20 electrode system (ref FCz) with a synthetic seizure onset (6 Hz sine-wave). Sz shows the rhythmic 6 Hz sine-wave seizure onset (~6 s into recording).

Figure 2(a) shows the result of applying Fast ICA to the synthetic segment of Fig. 1 and selected components of interest are depicted. IC1, IC2 and IC3 depict quite noisy components that exhibit the later stages of the rhythmic 6 Hz of the synthetic seizure. In each case the seizure is only really visible when the SNR of the seizure is quite large and IC2 in particular is heavily contaminated with (mainly) ocular artifact. The method also extracted ocular artifact (IC4) and electrode artifact as seen in IC5 – amongst others.

Figure 2(b) shows the equivalent results after applying $LSDIAG_{TD}$ as explained in the previous section. Three non-overlapping short-time windows, each of about 6.7 s (boundaries indicated in Fig. 1) were used to generate separate sets of temporal correlations of up to 50 lags. The three sets of matrices were then stacked and the $LSDIAG_{TD}$ algorithm applied to jointly diagonalise the cross-covariance matrices. IC1 depicts a clear (single) synthetic seizure component with a strong right parietal focus, of importance is that there is virtually no artifact present in this component. IC2 and IC3 depict artifacts as extracted by Fast ICA (ocular artifact and electrode artifact), whereas IC4 and IC5 depict two other components which were not present in the Fast ICA derived components. IC4 depicts a further slow, rhythmic ocular artifact and IC5 depicts a burst of 10 Hz activity at about 12.5 s, although this is contaminated by ocular artifact earlier on.

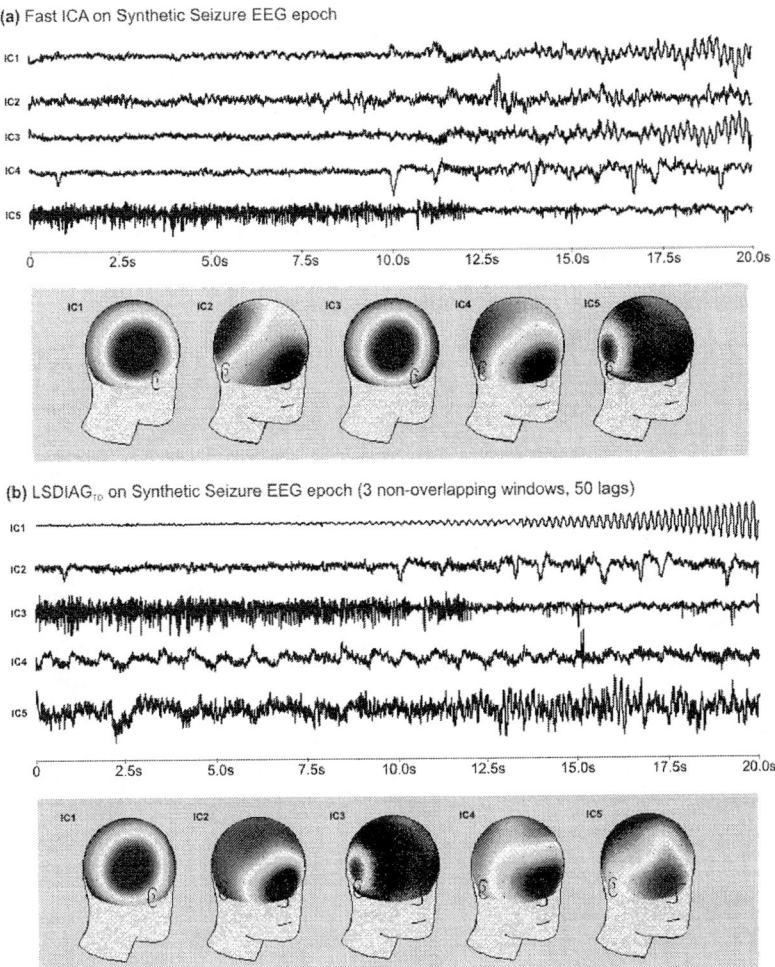

Fig. 2. Result of applying (a) Fast ICA and (b) LSDIAG$_{TD}$ to the 20 s epoch of scalp EEG depicted in Fig. 1. (a) Selected components: IC1, 2 & 3 depict noisy seizure (including ocular artifact) IC4 depicts ocular artifact and IC5 electrode artifact (T4). (b) Selected components: IC1 the extracted seizure component, IC2 ocular artifact, IC3 electrode artifact (T4), IC4 ocular artifact and IC5 ocular artifact contaminated rhythmic activity at around 12.5 s mark.

3.2 Real Seizure EEG

Figure 3 depicts a 20 s segment of real EEG recorded from the long-term EMU. It contains a seizure which has a left-temporal focus consisting of rhythmic 5 Hz activity with a visible onset at around the 7.5 s mark. Both the Fast ICA and the LSDIAG$_{TD}$ algorithm where repeatedly applied to a series of consecutive overlapping short, 3 s windows of multichannel EEG starting from the 1st sample and skipping 250 ms (50 samples) each time. Highlighted in Fig. 3 (a) are the multichannel segments labeled I, II and the greyed out segment just after seizure onset is visible. For the most part it was very difficult to (subjectively) identify even single components of seizure activity at any point in the process when using Fast ICA. Fig. 3(b) shows two Fast ICA components from section I and II which *may* be considered epileptiform (this is based mostly on their topographies). Fig 3(c) shows two components from sections I and II as extracted by LSDIAG$_{TD}$ which show much clearer 5 Hz rhythmic activity and have strong left temporal foci. Furthermore other neurophysiologically relevant components were also extracted such as the strong 10 Hz activity shown in Fig. 3(d). The inset of Fig 3(a) depicts components derived by LSDIAG$_{TD}$ just after the seizure onset is visible – no similar components could be identified through Fast ICA.

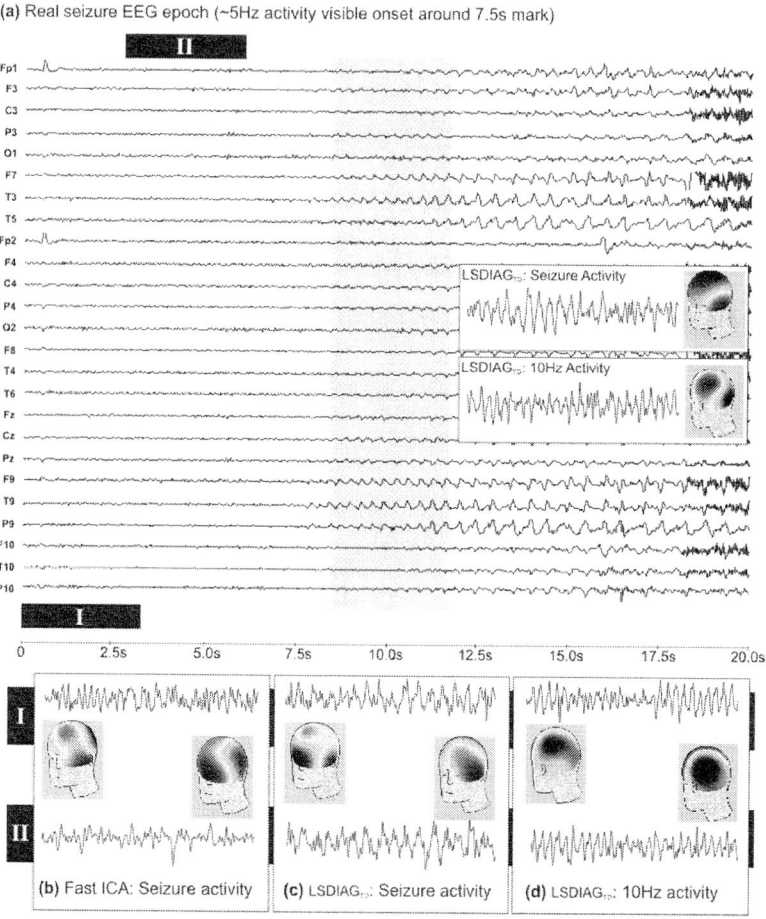

Fig. 3. (a) 20 s epoch of real scalp EEG recorded with modified 10-20 electrode system (ref FCz). Inset shows output of LSDIAG$_{TD}$ on short (3 s) segment highlighted at seizure onset. Fast ICA failed to extract any recognisable neurophysiologically meaningful components. (b) – (d) depict the application of Fast ICA and LSDIAG$_{TD}$ to two consecutive 3 s windows starting 8 s prior to visible seizure onset. (b) Fast ICA derived seizure components; (c) LSDIAG$_{TD}$ extracts strong seizure components with appropriate spatial distribution and (d) LSDIAG$_{TD}$ also extracts other meaningful activity (10 Hz activity).

4 Discussion

The previous section shows that both Fast ICA and LSDIAG$_{TD}$ can produce neurophysiologically meaningful results when applied to the ongoing EEG. However, the analysis of ongoing EEG can provide obstacles to the successful implementation of popular BSS methods such as Fast ICA. Here we present two situations where such methods do not perform as well as other TD based methods do.

In the first case applying Fast ICA to a long segment of seizure EEG did not result in the successful extraction of a clear (single) seizure component. Worryingly (from a neurophysiological perspective) the method depicted up to three components with poor temporal representation with albeit stronger (and similar) topographies and the components also failed to show the full extent of the reach of the ictal activity. With LSDIAG$_{TD}$ a much clearer picture of the underlying activity was obtained. Primarily only a single clear and well focused seizure component was extracted, and the temporal extent of this component was quite evident. Furthermore, other neurophysiologically meaningful components were extracted which Fast ICA did not. Due to the nature of the method relying on the temporal statistics of the underlying sources then applying LSDIAG$_{TD}$ to longer data segments carries with it the application to nonstationary sources. However, due to the assumption that the mixing matrices do not change over time, it is then possible to take advantage of this with LSDIAG$_{TD}$ through a set of stacked cross-covariance matrices. The success of this method was clearly shown with the synthetic ictal EEG.

5 Conclusions

Methods for BSS that use signal time structure have been shown to be immensely useful in extracting neurophysiologically meaningful components underlying EEG measurements. It has been shown that short segments of EEG along with wider, nonstationary segments, can be handled equally successfully by this technique. The technique implemented in this study works through the joint diagonalisation of the cross-covariance matrices at different lags of the multichannel EEG data. A logical progression to this work is the use of *time-frequency* based approaches as these are well suited for analysing, filtering and de-noising non-stationary time-series. We are now exploring the use of *wavelet BSS* using the same joint diagonalisation scheme as for LSDIAG$_{TD}$ which offers an integrated, versatile and efficient framework for analysing nonstationary multichannel signals in general, with promising results when applied to multichannel epileptiform EEG data in particular.

Acknowledgments

This work is funded by EPSRC Grant #GR/S13132/01.

References

1. Hyvärinen A., Karhunen J. and Oja E.: Independent component analysis, J. Wiley and Sons, New York (2001)
2. Roberts S. and Everson R.: Independent component analysis: principles and practice, Cambridge University Press, Cambridge (2001)

3. James C.J. and Lowe D.: ICA in Electromagnetic Brain Signal Analysis, in Proc. Int. Conf. on Neural Networks and Expert Systems in Medicine and Healthcare (NNESMED 2001), Milos Island, Greece, pp. 197-202 (2001)
4. Hyvärinen A. and Oja E.: A fast fixed-point algorithm for independent component analysis, Neural Computation, **9**, pp. 1483-1492 (1997)
5. James C.J. and Gibson O.J.: Temporally Constrained ICA: an Application to Artifact Rejection in Electromagnetic Brain Signal Analysis, IEEE Transactions on Biomedical Engineering, **50**(9), pp. 1108-1116 (2003)
6. Ziehe A. and Müller K.-R.: TDSEP - an efficient algorithm for blind separation using time structure, in Proc. Int. Conf. on Artificial Neural Networks (ICANN'98), Skovde, Sweden. pp. 675-680 (1998)
7. Belouchrani A. and Amin M.G.: Blind source separation based on time-frequency signal representations, IEEE Trans. Signal Processing, **46**(11), pp. 2888-2897 (1998)
8. Choi S., Cichocki A. and Belouchrani A.: Second order nonstationary source separation, Journal of VLSI Signal Processing, **32**, pp. 93-104 (2002)
9. Ziehe A., Laskov P., Müller K.-R. and Nolte G.: A linear least-squares algorithm for joint diagonalization, in Proc. Int. Conf. on Independent Component Analysis and Blind Signal Separation (ICA2003), Nara, Japan. pp. 469-474 (2003)
10. Belouchrani A., Abed-Meraim K., Cardoso J.-F. and Moulines E.: A blind source separation technique using second order statistics, IEEE Trans. Signal Processing, **45**(2), pp. 434-444 (1997)

A Framework for Evaluating ICA Methods of Artifact Removal from Multichannel EEG

Kevin A. Glass[1], Gwen A. Frishkoff[2], Robert M. Frank[1], Colin Davey[3], Joseph Dien[4], Allen D. Malony[1], and Don M. Tucker[1,2]

[1] NeuroInformatics Center, 5219 University of Oregon, Eugene, OR 97403
{kglass,rmf,malony}@cs.uoregon.edu
[2] Department of Psychology, University of Oregon, Eugene, OR 97403
[3] Electrical Geodesics Inc.,1600 Millrace Dr. Suite 307, Eugene, OR 97403
[4] Department of Psychology, University of Kansas, Lawrence, KS 66045

Abstract. We present a method for evaluating ICA separation of artifacts from EEG (electroencephalographic) data. Two algorithms, Infomax and FastICA, were applied to "synthetic data," created by superimposing simulated blinks on a blink-free EEG. To examine sensitivity to different data characteristics, multiple datasets were constructed by varying properties of the simulated blinks. ICA was used to decompose the data, and each source was cross-correlated with a blink template. Different thresholds for correlation were used to assess stability of the algorithms. When a match between the blink-template and a component was obtained, the contribution of the source was subtracted from the EEG. Since the original data were known a priori to be blink-free, it was possible to compute the correlation between these "baseline" data and the results of different decompositions. By averaging the filtered data, time-locked to the simulated blinks, we illustrate effects of different outcomes for EEG waveform and topographic analysis.

1 Introduction

Accurate assessment of signal decomposition methods such as ICA should account for multiple parameters that affect the decomposition, including characteristics of the input data (properties of the signal and noise activity) and properties of different ICA algorithms and implementations (e.g., contrast functions, tolerance levels). The theoretical underpinning of ICA and its various algorithms have been extensively discussed in the literature [1,2,3], and experiments have been designed to demonstrate the effectiveness of the procedure (for example, see [4]). However, there are few empirical studies measuring the effectiveness of ICA algorithms, and even fewer discussing these measures in the context of specific applications. One reason for the lack of empirical studies is the lack of empirical measures of effectiveness [5].

To this end, the present paper describes a new method for evaluation of ICA decompositions and applies this method to the problem of artifact extraction from multichannel EEG (electroencephalographic) data. The goal of this application was to compare the efficacy of two ICA algorithms, FastICA [6] and Infomax [3], in removing blinks from EEG signals. However, the procedure can be generalized to other problems and algorithms. Our technique, described below, is similar to Harmeling, et al. [5] and Zibulevski and Zeevi [7] except that our approach uses realistic data, thus giving the user a familiar basis for qualitative comparisons.

The results of our tests demonstrate the quantitative and qualitative utility of measures in evaluating ICA decomposition. With this method, it is possible to characterize the sensitivity of different ICA methods to multiple variables and perhaps, in future applications, to determine the appropriateness of different ICA methods for particular data analysis goals. Further, in addition to quantitative measures, we evaluated the effects of different ICA results on EEG waveforms and topographies. This allowed us to visualize the results and to examine the practical implications of different statistical outcomes.

2 Methods

EEG Acquisition and Preprocessing. EEG data were acquired from 256 scalp electrodes EEG net (Electrical Geodesics, Inc) referenced to Cz in a language task described elsewhere. Data contaminated by blinks were manually marked and removed, providing a blink-free EEG ("baseline") for evaluating the success of the blink removal. The EEG was downsampled to 34 channels, making it feasible to examine spatial and temporal properties of all 34 extracted components.

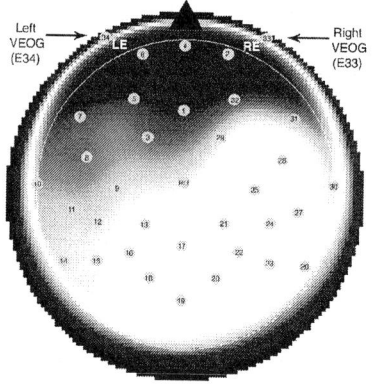

Fig. 1. Blink topography. Red, positive. Blue, negative. LE = left eye. RE = right eye.

Creating the Blink Template. Thirty-two segments of data with representative blinks were segmented from the continuous data. The segments were aligned to the peak of each blink and averaged to derive a blink template (Fig. 1).

Construction of Synthesized Datasets. To construct the synthesized data, the raw EEG data were inspected for ocular artifacts, and all trials contaminated with blink activity were removed from the recording, resulting in a "blink-free" EEG, to which a stream of blinks with known spatial and temporal characteristics was added (Fig 2).

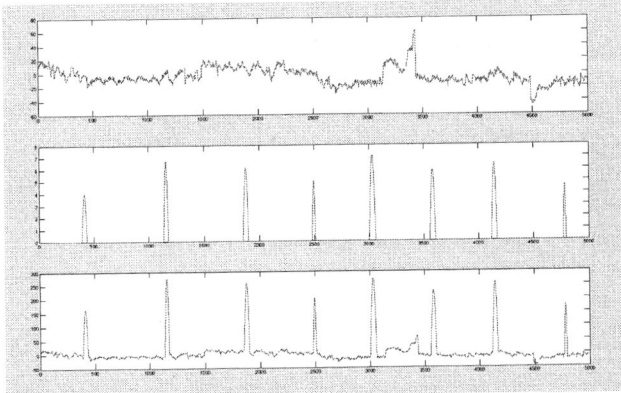

Fig. 2. Construction of "synthetic" data. Top panel, original data (~10 sec). Center panel, simulated blinks (Dataset #7). Bottom panel, original data plus simulated blinks.

To assess the robustness of the two algorithms and their sensitivity to data parameters, seven such datasets were constructed. The datasets differed with respect to blink amplitude, blink duration, and inter-blink interval. Datasets 1-5 contained blink activations of constant duration with inter-blink spacing of 400 milliseconds and 5000 milliseconds, respectively. Intensity of the blink activations ranged from 25% (Set #1) to 400% (Set #5) the intensity of the largest non-blink activity. Datasets 6 and 7 contained blinks of variable duration, spacing and intensity (Table 1).

Table 1. Test data set characteristics.

Data Set	Blink Strength	Inter-blink Spacing (ms)	Blink Duration (ms)
1	25%	5000	400
2	50%	5000	400
3	100%	5000	400
4	200%	5000	400
5	400%	5000	400
6	50%-200%	635-2500	312-5000
7	255-400%	312-5000	25-400

ICA Algorithms and Blink Removal Procedures. Both ICA algorithms were implemented in Matlab. The Infomax code [8] is an enhanced version of the Infomax algorithm of Bell and Sejnowski [2]; the FastICA code [9] uses a fixed-point algorithm. To remove blinks, we used a modified version of the ICABlinkToolbox [10,11].

The FastICA decomposition was performed using two contrast functions, the cubic (default) contrast function and a hyperbolic tangent (tanh) function. In the initial tests, the tanh function outperformed the cubic function. Therefore, in subsequent analyses, we used the tanh contrast function only. The Infomax decomposition used the developer's default settings. The projections of the components onto the EEG detector array ("spatial correlates" for short) were correlated with the blink template. Then contribution of the highest correlated component was removed from the dataset and the cleaned and original datasets were compared to measure the quality of the ICA algorithm's decomposition.

Metrics. The covariance between corresponding channels of the ICA-filtered EEG data and the original EEG data was computed for each dataset. To provide qualitative metrics for comparison of the different algorithms, we averaged the original and ICA-filtered data, time-locking the averages to the peak of the simulated blinks. The resulting averages should therefore accentuate residual blink activity after data cleaning. This procedure provides a visual reference for the significance of the correlation values.

3 Results

The overall (grand average) correlation between the original and cleaned data, for both ICA algorithms was 0.95 or better for FastICA and 0.969 for Infomax. When broken down for the separate electrodes, the lowest correlations occurred for channels 2, 4, and 6: depending on the particular dataset, and the threshold for blink identification, correlations at these channels ranged from about 0.55 to about 0.70. This is not unexpected, since these channels are located just above the eyes (Fig. 1).

A more detailed comparison of the results for FastICA and Infomax revealed several important differences. The most salient difference is that Infomax decompositions varied little across the datasets, whereas the FastICA decompositions showed considerable variation (Fig. 3). This suggests that changes in the properties of the blink data may affect factor extraction, allocation of variance across the factors, or both. As mentioned previously, FastICA implemented with the default (cubic) contrast function fared considerably worse than the implementation with the tanh contrast function. Therefore, subsequent analyses focused on the comparison of Infomax and FastICA using the tanh contrast function. Figure 4 demonstrates that the periorbital channels show the worst correlations. In addition, the largest differences between Infomax and FastICA are observed over these same channels, where blink activity is most pronounced.

Infomax was similarly robust to changes in tolerance (threshold for correlation with blink template), whereas FastICA on average showed worse accuracy at lower tolerances (data not shown here). In general, Infomax was more stable and more robust to changes in properties of the data and ICA implementation.

Further inspection of the ICA decompositions revealed that where FastICA was less successful, more than one spatial projector correlated strongly with the blink template, a strong correlation being any correlation above the experimentally determined threshold of 0.90. For example, as illustrated in Figure 5 above, FastICA-1, one of the least successful decompositions performed for this report, contained 6 projectors that matched the blink template > 0.90 as compared to InfoMax and FastICA-1, which each contained only one.

To illustrate the effects of successful and less successful ICA decompositions, we examined the ICA-cleaned data for different FastICA and Infomax runs (Dataset 5) after removing the source that was perfectly correlated with the blink template. Because FastICA gave more variable results across runs, we selected one example of a successful FastICA run (run 2) and one example of a less successful run (run 1). Although "the same" source was removed from the data in each case, the effects were very different, reflecting misallocation of variance when additional sources showed a close (but less than perfect) match to the blink template, as illustrated in Figure 6.

The failure of FastICA (run 1) that is evident in the averaged waveforms is also visible in the topographic distribution of the filtered data (Fig. 7). Note the resemblance of the topography for FastICA (run 1) to the blink template (Fig. 1). This outcome appears to reflect misallocation of variance to additional components in the decomposition [5].

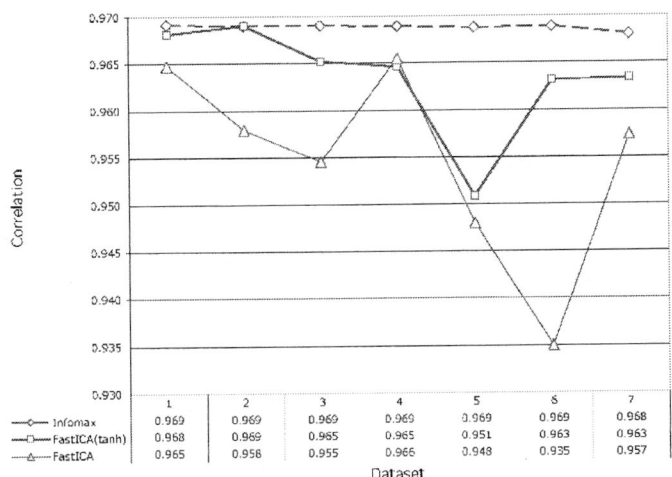

Fig. 3. Graph of correlations between original and ICA-filtered data across the seven datasets. Thin line, FastICA with cubic constrast function. Thick line, FastICA with tanh contrast function. Dotted line, Infomax.

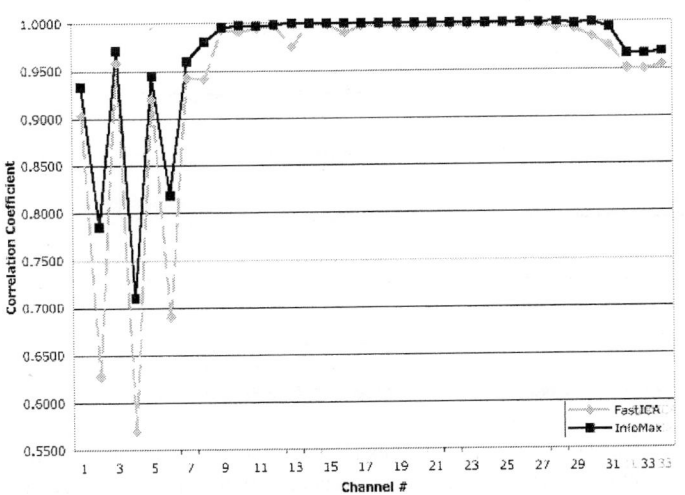

Fig. 4. Correlation between original & ICA-filtered data across the 34 electrodes.

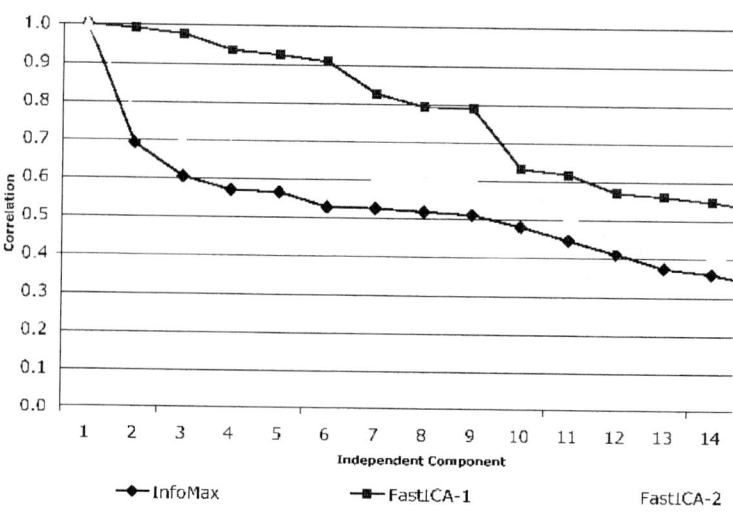

Fig. 5. Correlation between the spatial projectors of the independent component activations and the synthetic blink activitity template. The figure shows the 14 components with the strongest correlations.

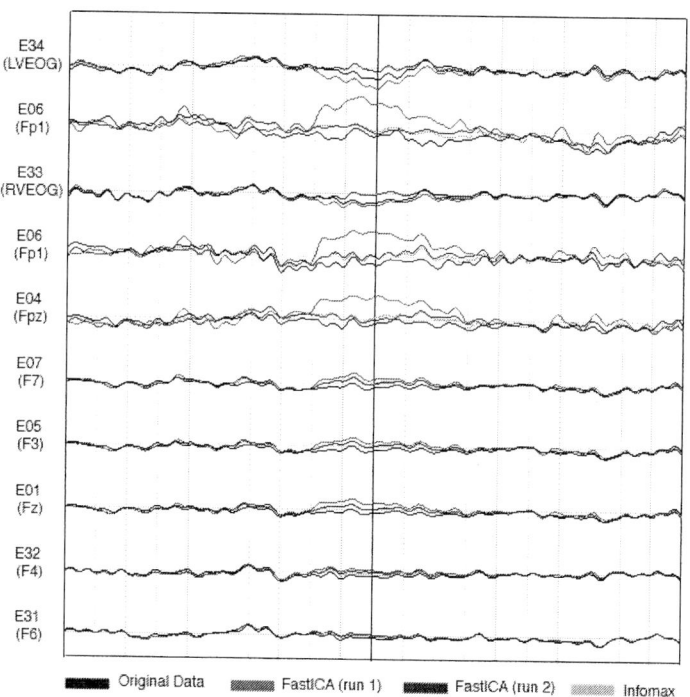

Fig. 6. EEG waveforms, averaged to the peak of the blink activity. Note residual blinks in run 1 for FastICA, where more than one source was strongly correlated with the blink template, and the source activations revealed misallocation of variance (cf. Fig 5).

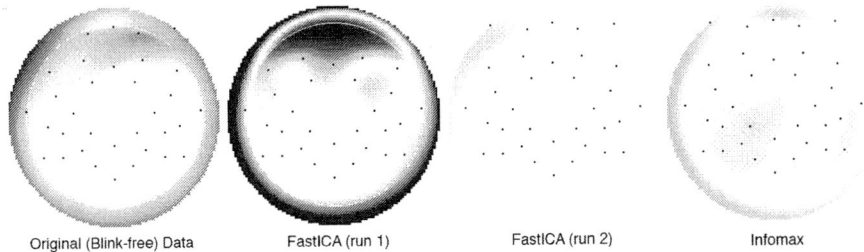

Original (Blink-free) Data   FastICA (run 1)   FastICA (run 2)   Infomax

Fig. 7. Topography of blink-averaged data, centered at peak of blink activity. Red, positive voltage. Blue, negative voltage. FastICA run1 is the less successful decomposition. Note the remaining blink activity at this time point.

4 Discussion

In this report we have demonstrated a new method for evaluation of ICA for removal of blink activity from multichannel EEG. The grand average correlation suggest that Infomax and FastICA were highly accurate in their ability to separate out the simulated blinks from the EEG. In every ICA run, exactly one of the extracted components showed a perfect correlation with the blink topography used to construct the simulated blinks. On the other hand, the activations corresponding to this source differed across runs and across ICA algorithms and implementations. In every case, the source activations were less than perfectly correlated with the time series for the simulated blinks. Infomax showed the closest correspondence, while FastICA was more variable, showing excellent correspondence on some runs, and misallocation of variance on other runs. Future studies will examine causes of misallocation of variance, extend this method to account for other data parameters, and compare results for Infomax and FastICA with other ICA algorithms and implementations.

References

1. Cardoso, J.-F. and P. Comon (1996). "Independent Component Analysis, A Survey of Some Algebraic Methods." Proc. ISCAS'96, vol.2, pp. 93-96.
2. Hyvarinen, A. and E. Oja. (1999). "Independent Component Analysis: A Tutorial." Downloaded from http://www.cs.helsinki.fi/u/ahyvarin/papers/index.html, March 12, 2004.
3. Bell, A. J. and T. J. Sejnowski (1995). An Information Maximization Approach to Blind Separation and Blind Deconvolution. San Diego, Institute for Neural Computation, UCSD, San Diego CA: 1-38.
4. Aichner, R., H. Buchner, S. Araki, S. Makino (2003). On-line Time-domain Blind Source Separation of Nonstationary Convolved Signals." 4rd Intl. Conf. on Indep. Comp. Ana. and Blind Sig. Sep., Nara, Japan.
5. Harmeling, S., F. Mieinecke, et al. (2003). "Analysing ICA Components by Injecting Noise." 4rd Intl. Conf. on Indep. Comp. Ana. and Blind Sig. Sep., Nara, Japan.
6. Hyvarinen, A. (1999). "Fast and Robust Fixed-Point Algorithms for Independent Component Analysis." IEEE Trans. on Neural Nets. 10(3): 626-634.
7. Zibulevski, M. and Y. Y. Zeevi (2001). Source Extraction Using Sparse Representation. 3rd Intl. Conf. on Indep. Comp. Ana. and Blind Sig. Sep., San Diego, CA.

8. EEGLab download site. http://www.sccn.ucsd.edu/eeglab/downloadtoolbox.html. March 12, 2004.
9. FastICA download site. http://www.cis.hut.fi/projects/ica/fastica/. March 12, 2004.
10. ICAToolBox download site. http://people.ku.edu/~jdien/. March 12, 2004.
11. Dien, J. (1998). Issues in the application of the average reference: Review, critiques, and recommendations. Behav. Res. Methods, Instruments, and Computers, 30(1), 34-43.

A New Method for Eliminating Stimulus Artifact in Transient Evoked Otoacoustic Emission Using ICA

Ju Liu, Yu Du, Jing Li, and Kaibao Nie

School of Information Science and Engineering, Shandong University,
Jinan, 250100, Shandong, China
juliu@sdu.edu.cn

Abstract. How to eliminate the stimulus artifact from the TEOAE measurement is a key question in TEOAE test. In this paper, a new method to eliminate stimulus artifact by ICA is proposed. First, four linear increasing stimulating sounds are used, and the waveforms recorded are the mix of TEOAEs and stimulus artifacts. Because stimulus artifact and TEOAE are independent statistically, and stimulus artifacts are line-increasing with stimulus while TEOAEs are nonlinear increasing with the trend to saturation gradually, their mix coefficients in mixed signals are different. The independent components and mix matrix can be estimated using ICA algorithm, and stimulus artifact is one of the independent components. Then we get rid of the artifact and remixed these independent components in order to separate stimulus artifacts from TEOAEs. Finally, compared with traditional DNLR algorithm, it is proved that the method we proposed is right and more effective.

Keywords: TEOAE, Stimulus artifact, ICA

1 Introduction

Otoacoustic Emission(OAE) is very low level sound produced from cochlea and can be measured by a sensitive microphone placed in outer ear canal[1]. Transient evoked otoacoustic emission (TEOAE) is the OAE response of the ear transiently evoked by a stimulating sound, such as a click. TEOAE can be detected from 100% of the normal hearing ears. The absence of TEOAE usually means the damages of outer hear cells (OHC) and the abnormity of audition. So TEOAE can serve as an objective and noninvasive means to assess cochlear status and have been widely used in physiology research and clinical applications.

Artifact is the reflection of stimulating sound in outer ear canal during the early 5-6 ms after stimulus. It is mixed with TEOAE in both time domain and frequency domain. Stimulus artifact decreases the signal noise ratio of TEOAE. TEOAE is very difficult to be distinguished from stimulus artifact because the energy of stimulus artifact is usually much larger than TEOAE's in the early time during measurement period. So, how to eliminate stimulus artifact is a key question in TEOAE measurement. In present the most prevailing method to get rid of stimulus artifact is derived nonlinear response (DNLR), which relies on the principle that stimulus artifacts are line-increasing with stimulus while TEOAEs are nonlinear increasing and turn to saturate gradually[2]. But TEOAEs maybe not saturate if the stimulating sound is weak, so using DNLR must lead to the loss of part of useful signal and decrease of SNR in that condition. And DNLR is not suitable if we only review the waveform of TEOAE in a particular sound pressure stimulus.

Independent component analysis (ICA) is a new signal processing algorithm in recent years, which can separate statistical independent signals from the mixed signals of those even when the mixed coefficients is unknown. Because stimulus artifact is the direct reflection of stimulus while TEOAE is active sound produced by cochlea, their waveforms should be statistical independent. And stimulus artifacts are linear increasing with stimulus while TEOAEs are nonlinear increasing, so waveforms got from different sound pressure stimulus can compose a mixed signal matrix. Basing on the model, we analyze mixed signals by ICA and separate stimulus artifacts and TEOAEs successfully. The experiment results prove that the mathematics model we proposed is proper . By clearing the mixed coefficient of stimulus artifact we manage to eliminate the stimulus artifacts completely.

2 ICA Approach

ICA is also called blind sources separation, first being proposed by Herault and Jutten in 1980s[3]. Its definition is described as below:

Giving m random variables' observed values $\{x_1(t), x_2(t), \cdots x_m(t)\}$, where t is the sample time. Assuming those values are linearly mixed results by n independent components $\{s_1(t), s_2(t), \cdots s_n(t)\}$:

$$\begin{bmatrix} x_1(t) \\ x_2(t) \\ \vdots \\ x_m(t) \end{bmatrix} = \begin{bmatrix} a_{11} a_{12} \cdots a_{1n} \\ a_{21} a_{22} \cdots a_{2n} \\ \vdots \\ a_{m1} a_{m2} \cdots a_{mn} \end{bmatrix} \begin{bmatrix} s_1(t) \\ s_2(t) \\ \vdots \\ s_n(t) \end{bmatrix} \quad (1)$$

Equation (1) can be written as matrix-vectors:

$$\mathbf{x}(t) = \mathbf{A}\mathbf{s}(t) \quad (2)$$

Where mixing matrix **A** is unknown. The aim of ICA algorithm is to work out separating matrix **W** and get source signal **s**(t) from observed signal **x**(t) using **W**. Assuming **y**(t) is the estimated vector of **s**(t), then separated results can be written as:

$$\mathbf{y}(t) = \mathbf{W}\mathbf{x}(t) \quad (3)$$

Two restrictions are needed to guarantee the solvability of the ICA model. First, because the linear addition of Gaussian signals remains Gaussian signal, which can't be separated, so there must be no more than one Gaussain siganl of the source signals. Second, it requires $m \geq n$ and mixing matrix **A** must be full-order[4].

It can be seen from equation (1) that **x**(t) are the mixed results by **s**(t), so they are not independent. The main idea of ICA is to eliminate the two order and more than two order correlation among components by maximizing a certain function related to their independency, so as to make them as independent as they could and estimate source signals $s(t)$. The common functions include Kurtosis, Information Maximization (Infomax), Maximum Likelihood Estimation (MLE), Minimum Mutual Information (MMI), and so on [5,6].

3 ICA Model for TEOAE

3.1 ICA Model Without Noise

First, if we don't consider the effect of noise, there are stimulus artifacts and TEOAEs in the sounds we recorded. For the same ear, if we only change the sound pressure of stimulus while don't change the measurement condition, such as instruments and the probe's place in ear, not the waveforms but the amplitudes of the stimulus artifacts and TEOAEs' are changed. The sounds recorded by different sound pressure of stimulus can be seen as the mix of stimulus artifacts and TEOAEs with different coefficients. Assuming stimulus artifact's waveform is $s_1(t)$, TEOAE's waveform is $s_2(t)$, linearly increasing stimulus's intensity is written as $A_0, 2A_0 \ldots mA_0$, and the sounds recorded $x_i(t)$ are the mix of stimulus artifacts and TEOAEs:

$$\begin{cases} x_1(t) = a_{11}s_1(t) + a_{12}s_2(t) \\ x_2(t) = a_{21}s_1(t) + a_{22}s_2(t) \\ \cdots \cdots \\ x_m(t) = a_{m1}s_1(t) + a_{m2}s_2(t) \end{cases} \quad (4)$$

Where $a_{11}, a_{21}, \cdots, a_{m1}$ are the coefficients of stimulus artifacts linearly increasing with stimulus, with $a_{11} : a_{21} : \cdots : a_{m1} = 1 : 2 : \cdots : m$; $a_{12}, a_{22}, \cdots, a_{m2}$ are the coefficients of TEOAEs nonlinearly increasing with stimulus. Generally speaking, the TEOAEs' increasing rate is lower than stimulus artifacts' and the former tend to be saturated if the stimulus is strong enough. Because TEOAEs are the cochlear active sounds while stimulus artifacts are the stimulus's direct reflection in outer ear canal, it is reasonable to assume they are independent. And because they are all nonstationary signals, independent components in ICA model can be separated if only $m \geq 2$.

3.2 ICA Model with Noise

Noise's effect is inevitable in TEOAEs' measurement, including non-Gaussian noise such as impulse interference and SOAEs and Gaussian noise such as white noise. Assuming the i_{th} time measurement includes non-Gaussian noise $n_i(t)$ and Gaussian noise $N_i(t)$, and equation (4) can be rewritten as:

$$\begin{cases} x_1(t) = a_{11}s_1(t) + a_{12}s_2(t) + n_1(t) + N_1(t) \\ x_2(t) = a_{21}s_1(t) + a_{22}s_2(t) + n_2(t) + N_2(t) \\ \cdots \cdots \\ x_m(t) = a_{m1}s_1(t) + a_{m2}s_2(t) + n_m(t) + N_m(t) \end{cases} \quad (5)$$

The analyses for mixing coefficients of the first and the second column are the same as section 2.1. Where m Gaussian noises don't satisfy the first separation constrained condition of ICA, but after enough time coherence average the σ of white trends to zero and have little obvious effect to the separated results of ICA model. As far as non-Gaussian noises $n_i(t)$ is concerned, the m signals got in measurement is independent with $s_1(t)$ and $s_2(t)$. So if we don't consider the effect of white noise, there are $m+2$ independent components, meaning $n = m+2$. So $m < n$ don't satisfy the second constrain condition of ICA and the all $m+2$ components can't be separated. Our aim isn't to separate all components but to eliminate the stimulus artifact. Assuming the SNR is high enough, at first we can whiten the signals, then arrange all components' eigenvalues in energy order, last separate the independent components by ICA algorithm. By this means the signals $s_1(t)$ and $s_2(t)$ with bigger energy can be separated, while the rest m smaller signals remain in the $m-2$ mixed signals.

3.3 Eliminating Stimulus Artifact by ICA

According to the analyses of section 2.2, we can conclude that m should be more than 3 to guarantee that $s_1(t)$ and $s_2(t)$ can be separated properly. First, preprocess mixed signals by centralizing and whitening, and then separate them by ICA, so we can get $s_1(t)$, $s_2(t)$ and the rest $m-2$ components. In the same time we can calculate separating matrix $\mathbf{W}$, and the inverse matrix of $\mathbf{W}$ is mixing matrix $\mathbf{A}$.

It is not difficult to pick out $S_1(t)$ and $S_2(t)$ by observing the waveform character of each component and the corresponding mixing coefficients. For example, stimulus artifacts' waveforms decrease gradually, and the corresponding mixing coefficients $a_{11}, a_{21}, \cdots, a_{m1}$ linearly increase with stimulus. While TEOAEs' frequency is increasing gradually and the lasting time is longer than stimulus artifacts', and the corresponding mixing $a_{12}, a_{22}, \cdots, a_{m2}$ are nonlinearly increasing with stimulus.

In order to eliminate stimulus artifacts, we can set stimulus artifacts' column vector in $\mathbf{A}$ to zero, and remix according to equation (1). The m signals include no stimulus artifact. Because each TEOAE's increasing rate may be different, the rest $m-2$ components maybe include some useful signals of TEOAEs besides useless noises, so the $m-2$ components should be reserved.

4 Results and Analysis

We used the OAE measurement system developed by our laboratory, which based on computer sound card. And we control I/O of sound card by using API of Windows programming. Peripheral equipment is analog band pass filter: 300~8000Hz. Then the recorded signal was amplified and inputted into computer and was digital filtered bandpassed from 600 to 6000Hz. At last the signals were synchronous averaged and time domain windowed. The advantage of this system is that it can be adjusted swiftly as need.

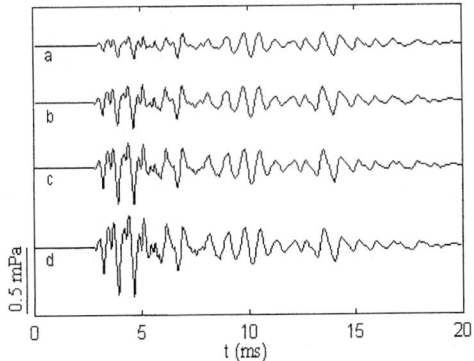

Fig. 1. This shows TEOAE waveforms stimulated by 15mPa, 30mPa, 45mPa and 60mPa click sounds, which including stimulus artifacts. They were numbered a, b, c, d in order.

4.1 Separating by ICA

The four signals obtained can be regarded as the mix of several independent components, $\mathbf{x}(t)$ in equation (1). We calculated $\mathbf{s}(t)$ and $\mathbf{A}$ using Hyvärinen's FastICA fixed-point algorithm[7]. The results are showed in fig.2 and table 1.

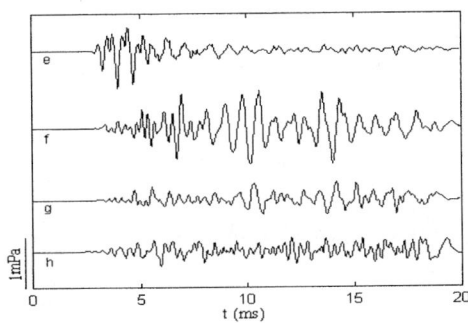

Fig. 2. Separated results by FastICA, numbered e, f, g, h in order.

Table 1. $\mathbf{A}$ is the mixed matrix of TEOAE's ICA model.

0.1244	0.1128	0.0498	0.0244
0.2456	0.1441	-0.0163	0.0266
0.3664	0.1571	-0.0515	-0.0076
0.4848	0.1609	-0.0765	0.0147

According to equation (1), the waveforms in figure.1 could be got through multiplying the corresponding row coefficients in **A** by the four waveforms in fig.2. For example, a=0.1244×e+0.1128×f+0.0498×g+0.0244×h, and so on.

Although there is uncertainty in ICA separated results, we can distinguish them by observing the waveforms' character and the corresponding coefficients in **A** . For example, in fig.2, it is obvious that the waveform e' characters looks like stimulus artifact most, and this conclusion would be confirmed by observing the linearly increasing coefficients in **A** . For the same method, waveform f looks like TEOAEs most and the corresponding coefficients, the second column of **A** , are nonlinearly increasing with stimulus, trending to saturation gradually. We calculated the proportion of each component a, b, c, d, and the results were that e was 46.4%, f was 43.8%, g was 7.4% and h was 2.4%. It is obvious that the energy of g and h is very small in total energy, and their mixing coefficients were irregular. So they could be the mixing of some small components such as noise and SOAEs.

4.2 Eliminating Stimulus Artifacts

We can judge e was stimulus artifact waveform based on the former analysis. So we should set the first column of **A** zero, or set the first row of s(t) zero, and remix **x**(t) by equation (1), then the stimulus artifacts would be eliminated completely. The TEOAEs' waveforms excluding stimulus artifacts was showed as fig.3. And the results can be proved effective to compare with DNLR method. To multiply waveform a by 3 and subtract c, then divide it by 2, and DNLR's result is gotten. Showed in fig.4.

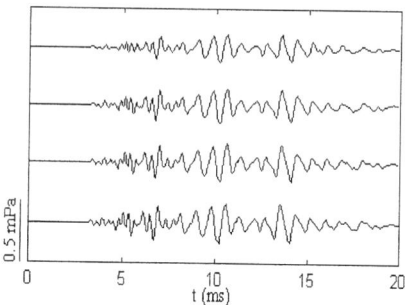

Fig. 3. TEOAEs waveforms Excluding stimulus artifacts. Comparing fig.3 with fig.1, it can be found that the stimulus artifacts existed in former period had been eliminated while the useful signals exited in middle and latter period had been reserved.

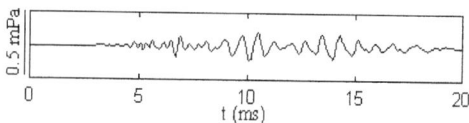

Fig. 4. DNLR result, which shows the same waveform to the first waveform in fig2, and proves ICA method's separated result is right.

Seen from the coefficients, stimulus artifacts linearly increased with stimulus, therefore the effect to eliminate stimulus artifacts for this two method is the same. But when the stimulus is weak, TEOAEs would not saturate. In that condition using DNLR method would loss some useful signals, while ICA method hasn't the disadvantage.

5 Conclusion

In this paper we first tried to apply ICA in the analyses of TEOAE and obtained good results. Seen from experiment results, ICA method could eliminate stimulus artifacts effectively, and solve some disadvantages of traditional DNLR method, such as SNR decrease, loss of useful signal energy and being not able to measure TEOAE in particular sound pressure. On the other hand, ICA algorithm is more complex than DNLR and needs more computing time, so it is only fit for out-line analysis. Whether the separation can succeed depends on whether the signals have enough high SNR, so the recording environments should be very quiet and the recording time should be prolonged a longer time. And if the probe place in ear canal changes during measurement, stimulus artifacts waveforms in former period wouldn't coincide with the latter, which would also leads to the failure of separating.

Acknowledgements

This paper is supported by the National Natural Foundation of China(NO.30000041) and the Natural foundation of Shandong Provence(NO.Y2000G13).

References

1. Whitehead ML, Stagner BB, Lonsbury-Martin BL, et al. Measurement of otoacoustic emissions for hearing assessment[J]. IEEE Engineering in Medicine and Biology .1994, 13(2):210-226
2. P.Ravazzani, F.Grandori. Evoked otoacoustic emissions: nonlinearities and response interpretation[J]. IEEE Trans on Biomedical Engineering .1993,40(5):500-504
3. Jutten C, Herault J. Blind separation of sources, Part I: An adaptive algorithm based on neuromimetic[J]. Signal Processing. 1991,24(1):1-10
4. Tong L, Liu R, Soon VC, et al. Indeterminacy and identifiability of blind identification[J]. IEEE Trans on Circuits and Systems. 1991,38(5): 499-506
5. Hyvärinen A, Oja E. Independent component analysis: algorithms and applications[J]. Neural Networks. 2000,13(4-5): 411-430
6. Liu Ju, He Zhenya. A Survey of Blind Source Separation and Blind Deconvolution[J]. Chinese Journal of Electronics.(In Chinese)
7. Hyvärinen A, Oja EA. A fast fixed-point algorithm for independent component analysis[J]. Neural Computation. 1997,9(7):1483-1492

An Efficient Time-Frequency Approach to Blind Source Separation Based on Wavelets

Christian W. Hesse and Christopher J. James

Signal Processing and Control Group, ISVR, University of Southampton,
Highfield, Southampton, SO17 1BJ, UK
c.w.hesse@soton.ac.uk

Abstract. Time-frequency representations based on wavelets, such as the discrete wavelet (DWT) and wavelet packet (WPT) transforms, offer an efficient means of analysing, de-noising and filtering non-stationary signals. They furthermore provide a rich description of time-varying frequency content of a signal, that is useful for the problem of blind source separation (BSS). We present and explore a multispectral decorrelation approach, whereby linear mixtures of sources with unique time-frequency signatures are separated, without pre-whitening, through joint diagonalisation of wavelet sub-band covariance matrices. Compared with BSS algorithms using temporal decorrelation only, wavelet BSS works well for stationary and non-stationary synthetic mixtures, with stable performance as the number of sources increases. Combined with conventional wavelet analysis and filtering techniques, wavelet BSS offers an integrated, versatile and efficient framework for analysing non-stationary multichannel signals in general, with promising results when applied to multichannel electroencephalographic (EEG) data.

1 Introduction

Multichannel biomedical signals are typically non-stationary and have spatial and temporal correlations. For example, the electroencephalogram (EEG) measures electrophysiological brain activity using recording electrodes positioned on the scalp. The EEG reflects spatial and temporal brain dynamics which are of neurophysiological interest in both clinical and experimental contexts. Non-stationarity and spatial temporal correlations complicate EEG analysis, with efforts being further hampered by contamination of the signals due to artifacts of physiological and non-physiological origin.

Time-frequency approaches are well suited for analysing, filtering and de-noising non-stationary time-series, and are increasingly used in EEG [1]. The discrete wavelet (DWT) and wavelet packet (WPT) transforms [2] decompose a signal using a set of *wavelet* basis functions, which are localised in frequency and time. Thus, wavelet transform coefficients reflect the correlation between the signal and the wavelet basis at different time and frequency scales, providing a rich description of signal time-frequency structure, which is sparse, and can be statistically optimal for the WPT. Filtering and de-noising can be achieved by zeroing selected coefficients prior to transform inversion.

A variety of methods for independent component analysis (ICA) and blind source separation (BSS) [3, 4] have also been applied to EEG for extracting artifacts and nerophysiologically meaningful components [5]. The observed signals are assumed to be a linear mixture of statistically independent source waveforms, which may be separated by their statistical properties or their time structure. The former approach treats signals as samples of random variables where temporal ordering is irrelevant, and factorises marginal distributions using higher order moments [6] or cumulants [7], or by minimising mutual information [8]. Although the statistical approach is common in EEG, it has been argued [3] that methods using time structure may be more appropriate when analysing time-series data, such as biomedical signals.

BSS on the basis of signal time structure is generally achieved through temporal or multispectral decorrelation. If time structure is represented by temporal cross-correlations or cross-covariances [9–11], the sources are assumed to have stationary waveforms with unique power spectra, and time shifted but otherwise identical waveforms cannot be distinguished. A more plausible assumption for non-stationary biomedical signals, such as EEG, is that the sources have unique time-frequency signatures. In such instances, the performance of temporal decorrelation algorithms can be improved by considering sets of temporal correlations over short time windows [12]. An alternative approach is multispectral decorrelation based on cross-correlations of band-pass filtered versions of the signals [13] or spatial time-frequency distributions [14].

The decorrelation operation in time structure BSS methods involves joint (approximate) diagonalisation of a set of symmetric matrices which reflect the spatial temporal, or the spatial time-frequency covariance structure of the source mixture. The diagonalising matrix, an estimate of the de-mixing matrix (i.e. the inverse of the unknown mixing matrix), transforms the observed signal covariance matrices into source covariance matrices, which are (approximately) diagonal due to statistical independence. The accuracy and stability of methods which constrain the de-mixing matrix to be orthogonal [9, 15, 16] are sensitive to errors introduced by pre-whitening, and close spacing of the eigenvalues of the covariance matrices. Recent algorithms for non-orthogonal joint diagonalisation [17, 18] estimate the de-mixing matrix from the signal covariance directly, and without the need for pre-whitening.

Wavelets are not routinely applied in conjunction with ICA/BSS. When they are, their use seems motivated primarily by the favorable statistical properties of sparse representations [19]. Representation of signal time structure by DWT for multispectral decorrelation has been advocated previously [20]. However, implementation of *Wavelet ICA* has been limited to only two sub-bands by the decorrelation methods available at the time [9]. Wavelet ICA/BSS is a desirable alternative to existing temporal and multispectral decorrleation methods, not least because efficient wavelet based time-frequency representations reduce the computational cost of covariance estimation, and are free from the cross-term issues associated with STFDs. Here, we extend wavelet ICA into a full multispectral decorrelation method using non-orthogonal joint diagonalisation.

2 Temporal and Multispectral BSS Methods

Wavelets and Wavelet Packets. Computation of the DWT involves repeated filtering, downsampling and partitioning of the signal into orthogonal detail (high pass) and approximation (low pass) components. Successive transformations of the approximation part halve the length each time. While the maximum decomposition level is limited by the number of signal samples and filter coefficients, it can be increased arbitrarily by zero-padding the signal. The final approximation and detail parts from each level form the DWT sub-bands, so that at decomposition level w the transform has $w + 1$ frequency bands.

The WPT extends the DWT by repeated filtering, downsampling and partitioning of both the approximation and the detail components. Each step doubles the number of sub-bands, so that at decomposition level w the WPT has up to 2^w sub-bands, thereby increasing increases the frequency resolution. A consequence of repeated filtering of approximation and detail components is a change in wavelet shape, so that the signal is effectively transformed by a set of wavelet bases, called wavelet packets.

BSS Through Temporal Decorrelation. In the standard formulation of the BSS problem, n observed signals $\mathbf{x}(t)$ are assumed to be a linear mixture of m unknown but (statistically) independent source signals $\mathbf{s}(t)$ where the mixing matrix $\mathbf{A}$ is unknown but invertible and the number of sources is the same as the number of observed signals

$$\mathbf{x}(t) = \mathbf{A}\mathbf{s}(t). \tag{1}$$

The problem is solvable up to a permuation, and sign and power indeterminacy of the sources, by finding an appropriate matrix $\mathbf{W} = \mathbf{A}^{-1}$ which allows estimation of the source waveforms through inversion

$$\mathbf{s}(t) = \mathbf{W}\mathbf{x}(t). \tag{2}$$

Due to statistical independence, the source covariance matrix $\mathbf{C}^s_\tau$ is diagonal for all time lags $\tau = 0, 1, 2, \ldots$, and related to the corresponding signal covariance matrix $\mathbf{C}^x_\tau$ through

$$\mathbf{C}^s_\tau = \mathbf{W}\mathbf{C}^x_\tau\mathbf{W}^T, \tag{3}$$

where T denotes matrix transposition. Estimation of $\mathbf{W}$ is achieved through joint diagonalisation of the set of matrices $\{\mathbf{W}\mathbf{C}^x_\tau\mathbf{W}^T\}$, defined by the lags τ.

BSS Through Multispectral Decorrelation. Analogous to temporal decorrelation, mutlispectral decorrelation in wavelet BSS involves joint diagonalisation of a set of matrices $\mathbf{C}^x_\omega$ which describe the signal covariance structure across different DWT or WPT sub-bands, indexed by ω. If the source waveforms are independent, the source sub-band covariances $\mathbf{C}^s_\omega$ are diagonal

$$\mathbf{C}^s_\omega = \mathbf{W}\mathbf{C}^x_\omega\mathbf{W}^T. \tag{4}$$

Joint diagonalisation of DWT or WPT sub-bands is related to the wavelet Karhunen-Loève transform [21], which is uses (orthogonal) decorrelation of the wavelet spectrum or of individual wavelet sub-bands for de-noising.

Joint Diagonalisation. The aim of joint diagonalisation is to determine the matrix $\mathbf{W}$, so that each member of a set of K square matrices $\{\mathbf{C}_k\}$ becomes as diagonal as possible by the similarity transformation $\mathbf{F}_k = \mathbf{W}\mathbf{C}_k\mathbf{W}^T$. One measure of the diagonality of a square matrix $\mathbf{F}_k$ is the sum of the squared off-diagonal elements, or "off" criterion. Thus, $\mathbf{W}$ is the joint diagonaliser of the set of matrices $\{\mathbf{C}_k\}$ if it minimises the off criterion

$$\mathbf{W} = \operatorname{argmin} \sum_{k=1}^{K} \sum_{i \neq j} \| \left(\mathbf{W}\mathbf{C}_k\mathbf{W}^T\right)_{ij} \|^2. \tag{5}$$

Our implementation of temporal decorrelation and wavelet BSS use non-orthogonal joint diagonalisation based on the linear least-squares algorithm by [18] to determine $\mathbf{W}$ without pre-whitening of the data.

3 Simulations

We compared the performance of the temporal decorrelation method with wavelet BSS (DWT and WPT) in separating different numbers (2 to 25) of linearly mixed (random, non-singular $\mathbf{A}$ with unit-norm columns) synthetic source signals with random stationary or non-stationary waveforms (8192 samples). We manipulated the residual dependencies among the sources (none or random), since these may obtain in real data and affect the accuracy of the mixing matrix estimate. Performance was based on the average estimation error over 20 runs.

Performance Measure. Given a known mixing matrix $\mathbf{A}$ and an estimate $\hat{\mathbf{W}}$ of its inverse $\hat{\mathbf{W}} \simeq \mathbf{W} = \mathbf{A}^{-1}$, the model estimation error was quantified in terms of the distance of the matrix $\mathbf{G} = \hat{\mathbf{W}}\mathbf{A}$ from a permutation matrix using a formula adapted from equation (15) of [12], which computes an average of the absolute values of the rows and columns of $\mathbf{G}$ normalised by their respective row or column absolute maxima.

Stationary Signals. The synthetic stationary random signals were generated using autoregressive processes (zero-mean, weighted moving averages) with 50 lag coefficients (random, positive, unit-norm and unique for each source), according to

$$s(t) = \sum_{i} a_i s(t - \tau_i) + \phi(t), \tag{6}$$

where a_i is the coefficient associated with lag τ_i and $\phi(t) \sim \mathcal{N}(0,1)$ is a standard normal random deviate. Random residual temporal correlations were retained, or eliminated by temporal decorrelation BSS, before scaling the waveforms to random variance and re-mixing them with the specified mixing matrix $\mathbf{A}$.

Non-stationary Random Waveforms. The time-frequency structure of the synthetic non-stationary random signals was specifically tailored to wavelet analysis. Waveforms were generated by applying the inverse DWT or WPT to cubed, unit-variance, normal random deviates, subsequently rescaled to have random variance. Residual source dependency was either retained, or eliminated by decorrelating individual sub-bands before inversion. All wavelet transforms were based on the fourth order Daubechies wavelet and decomposition level 5. Temporal decorrelation was based on time lags $\tau \in [0, 1, 2, 3, ..., 50]$.

Results. Figure 1 shows the model estimation errors for temporal decorrelation and wavelet BSS for stationary and non-stationary source mixtures. As one would expect, the most accurate results obtain when the data models of the source waveforms and the BSS methods match. Nevertheless, all methods achieve satisfactory performance. Interestingly, for exact source models wavelet BSS is relatively more accurate on stationary data, than temporal decorrelation is on non-stationary data. In the presence of random residual source dependency there is no appreciable performance difference for stationary sources, but WPT BSS performs best on non-stationary sources. On the whole, the wavelet BSS methods successfully separate both stationary and non-stationary source waveforms, with performance comparable to or better than the temporal decorrelation method.

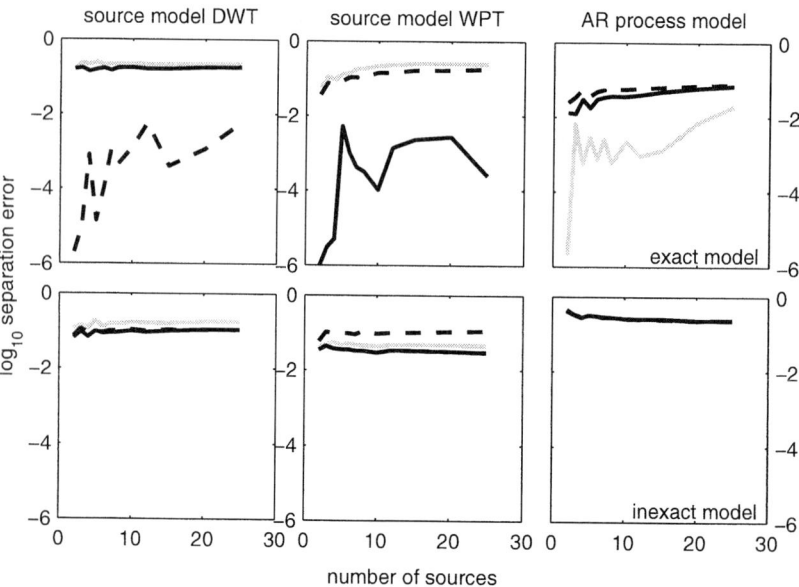

Fig. 1. Model estimation errors for temporal decorreltation BSS (solid grey) and wavelet BSS based on WPT (solid black) and DWT (dashed black), as a function of the number of sources. Columns reflect stationary (right) and non-stationary (left, middle) sources. Rows reflect the effects of absence (top) and presence (bottom) of random residual dependencies among the sources.

4 Application to EEG Data

We applied wavelet BSS to a 20 second segment of 25-channel ictal EEG (sampled at 200 Hz), with an epileptic seizure developing over the right temporal lobe about 6 seconds into the recording. To adequately cover the 1 to 70 Hz range of physiologically relevant EEG activity, the signals were appropriately zero-padded and decomposed to level 8. Covariances from sub-bands above 70 Hz were excluded during diagonalisation to reduce noise effects. Figure 2 shows the EEG traces and the sources extracted using wavelet (WPT) BSS, sorted by source strength. The three strongest sources (S1, S2, S3) represent the the onset and propagation of the seizure. Other components such as eye blinks, eye movements and muscle activity are clearly separated (S4, S9, S6, S8, S16, S17).

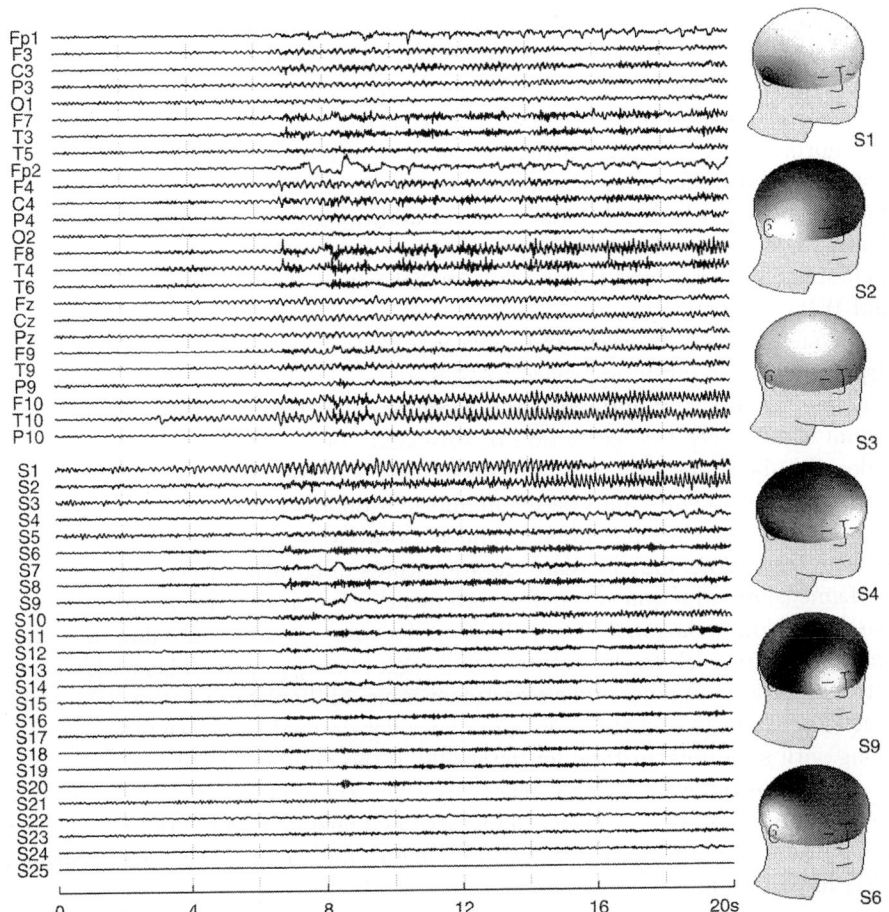

Fig. 2. 25-channel ictal EEG (top), source waveforms extracted with WPT based BSS (bottom) sorted by variance (power), and selected source scalp topographies (right).

5 Discussion

Wavelet transforms offer a computationally efficient representation of signal time structure that can be exploited for blind source separation by means of multispectral decorrelation. Through application of non-orthogonal diagonalisation [18] to wavelet sub-band covariance matrices, we extend earlier work on wavelet ICA [20] and obtain an efficient multispectral approach to BSS that does not require pre-whitening of the data. Simulations showed that wavelet BSS successfully separates linear mixtures of stationary and non-stationary random synthetic signals, with stable performance as the number of sources increases, and accuracy that is comparable to or better than BSS by temporal decorrelation. Furthermore, wavelet BSS could extract physiologically meaningful sources from a segment of non-stationary ictal EEG.

Compared with temporal decorrelation BSS, wavelet BSS is better suited for analysing non-stationary signals, allows estimation of a frequency-tuned mixing matrix, and is computationally more efficient. The cost of calculating all wavelet sub-band covariance matrices is the same as calculating lag 0 signal covariance. Extension to several time lags quickly exceeds the initial overhead of the wavelet transform, and additional matrices greatly increase the workload of diagonalisation algorithms. Moreover, the wavelet transform may be used for signal filtering and de-noising, which otherwise requires additional, separate processing steps. Thus, wavelet BSS shows much promise, and further performance comparisons including STFD BSS and ICA methods which can exploit the sparseness of DWT and WPT are underway.

Some aspects of wavelets and wavelet packets may further enhance the multispectral decorrelation approach to BSS, but require careful examination. For example, statistically optimal WPT decomposition trees are desirable for individual signals, yet the extraction of sub-band covariances from channels with different sub-band structure is a problem. One solution might be to stop further decomposition of sub-bands based on information loss estimated across all channels, rather than for individual channels. Moreover, the use of shift invariant wavelet transforms may increase the robustness of covariance estimates.

Dimension reduction is usually conveniently combined with pre-whitening of the data. Non-orthogonal diagonalisation does not require whitening, however, and algorithms such as [18] assume a full rank mixing matrix. This makes accurate estimation of model order a pertinent issue, and appropriate constraints for joint diagonalisation in the over- and under-determined cases need to be found, along with statistical criteria for determining the number of sources.

In conclusion, wavelet BSS, especially when combined with wavelet filtering and de-noising, offers an efficient, integrated and flexible framework for analysing non-stationary multichannel signals in general, and biomedical signals such as EEG in particular.

Acknowledgments

This work is funded by EPSRC Grant #GR/S13132/01.

References

1. Bradley, A.P., Wilson, W.J.: On wavelet analysis of auditory evoked potentials. Clinical Neurophysiology **115**(5) (2004) 1114–1128
2. Mallat, S.: A wavelet tour of signal processing. 2nd edn. Academic Press, San Diego, CA (1999)
3. Hyvärinen, A., Karhunen, J., Oja, E.: Independent component analysis. John Wiley and Sons, New York (2001)
4. Roberts, S., Everson, R.: Independent component analysis: principles and practice. Cambridge University Press, Cambridge (2001)
5. James, C.J., Lowe, D.: ICA in electromagnetic brain signal analysis. in Proc. Int. Conf. on Neural Networks and Expert Systems in Medicine and Healthcare (NNESMED 2001), Milos Island, Greece, (2001) 197–202
6. Hyvärinen, A., Oja, E.: A fast fixed-point algorithm for independent component analysis. Neural Computation **9** (1997) 1483–1492
7. Cardoso, J.-F., Souloumiac, A.: Blind beamforming for non Gaussian signals. IEE Proceedings-F **140**(6) (1993) 362–370
8. Bell, A.J., Sejnowski, T.J.: An information-maximization approach to blind separation and blind deconvolution. Neural Computation **7** (1995) 1483–1492
9. Molgedey, L., Schuster, H.G.: Separation of a mixture of independent signals using time delayed correlations. Physical Review Letters **72**(23) (1994) 3634–3636
10. Belouchrani, A., Abed-Meraim, K., Cardoso, J.-F., Moulines, E.: A blind source separation technique using second order statistics. IEEE Trans. Signal Processing **45**(2) (1997) 434–444
11. Ziehe, A., Müller, K.-R.: TDSEP - an efficient algorithm for blind separation using time structure. in Proc. Int. Conf. on Artificial Neural Networks (ICANN'98), Skövde, Sweden (1998) 675–680
12. Choi, S., Cichocki, A., Belouchrani, A.: Second order nonstationary source separation. Journal of VLSI Signal Processing **32** (2002) 93–104
13. Cichocki, A., Belouchrani, A.: Source separation of temporally correlated source using bank of band pass filters. Proc. Int. Conf. on Independent Component Analysis and Blind Signal Separation (ICA2001), San Diego, USA (2001) 173–178
14. Belouchrani, A., Amin, M.G.: Blind source separation based on time-frequency signal representations. IEEE Trans. Signal Processing **46**(11) (1998) 2888–2897
15. Cardoso, J.-F., Souloumiac, A.: Jacobi angles for simultaneous diagonalization. SIAM J. Matrix Anal. Applicat. **17**(1) (1996) 161–164
16. Pham, D.-T.: Joint approximate diagonalization of positive definite matrices. SIAM J. Matrix Anal. Applicat. **22**(4) (2001) 1136–1152
17. Yeredor, A.: Non-orthogonal joint diagonalization in the least-squares sense with application in blind source separation. IEEE Trans. Signal Processing **50**(7) (2002) 1545–1553
18. Ziehe, A., Laskov, P., Müller, K.-R., Nolte, G.: A linear least-squares algorithm for joint diagonalization. in Proc. Int. Conf. on Independent Component Analysis and Blind Signal Separation (ICA2003), Nara, Japan (2003) 469–474
19. Roberts, S., Roussos, E., Choudrey, R.: Hierarchy, priors and wavelets: structure and signal modelling using ICA. Signal Processing **84**(2) (2004) 283–297
20. Koehler, B.-U., Orglmeister, R.: Independent component analysis of electroencephalographic data using wavelet decomposition. Proc. Mediter. Conf. on Medical and Biological Engineering and Computing, Lemesos, Cyprus (1998)
21. Starck, J.-L., Querre, P.: Multispectral data restoration by the wavelet Karhunen-Loève transform. Signal Processing **81** (2001) 2449–2459

Blind Deconvolution of Close-to-Orthogonal Pulse Sources Applied to Surface Electromyograms

Ales Holobar and Damjan Zazula

University of Maribor, Faculty of Electrical Engineering and Computer Science,
Smetanova 17, 2000 Maribor, Slovenia
{ales.holobar,zazula}@uni-mb.si
http://www.storm.uni-mb.si

Abstract. Surface electromyogram (SEMG) decomposition technique suitable for identification of complete motor unit (MU) firing patterns during low level isometric voluntary muscle contractions is introduced. The approach is based on joint-diagonalization of whitened correlation matrices of SEMG recordings. It supposes constant and finite system impulse responses and more measurements than sources. Preliminary tests on synthetic signals prove 95% accuracy in detection of source pulses down to the signal-to-noise ratio of 10 dB. In the case of real SEMG, recorded with an array of 61 electrodes during low level contraction of biceps brachii muscle of three subjects 2.5 MUs active with the mean firing rate of 11.8 Hz were identified on average.

1 Introduction

Surface electromyography (SEMG) has become a rather developed and matured measuring technique. Its recent advancements open the possibilities of extensive field SEMG acquisition with multi-electrode pick-ups placed on practically arbitrary muscles or muscle groups. Such signal recordings now contain enough reliable information for a major step forward in obtaining diagnostically relevant parameters. One of the most challenging issues having been tackled since a long time but still lacking confidence and general robustness certainly remains the EMG decomposition to its constituent components, i.e. to the motor-unit action potentials (MUAPs) and to the innervation pulse trains (IPTs).

The existing computer-aided EMG decomposition methods have been mainly focused on the intra-muscular EMG signals. Being based on the pattern recognition and clustering in time domain, on spatial filters, and on time-scale analysis most of the methods fail when MUAPs become superimposed. While the SEMG measurements may be considered compound signals which are generated by statistically pretty independent signal sources, i.e. motor units (MUs), at least in low contraction force conditions, a variety of the blind source separation (BSS) methods seem applicable [5].

In [4] a novel technique suitable for decomposition of convolutive mixtures of close-to-orthogonal pulse sources with constant and finite unit sample responses was introduced. This paper discusses its application to surface EMG. In Sections 2 and 3 the assumed data model and the decomposition approach are quickly enlightened from the SEMG viewpoint. Simulation and experimental protocols with EMG signals are discussed in Section 4, while Section 5 presents the decomposition results. The paper is concluded with discussion in Section 6.

2 Data Model

Under the assumption of isometric muscle contractions at constant contraction forces the sampled multi-channel surface EMG can be modelled as a discrete linear time-invariant (LTI) multiple-input-multiple-output (MIMO) system [6]. Each channel in such a system is considered a MU with its response in the form of MUAP as captured by a surface electrode, while the channel inputs correspond to the innervation pulse train:

$$x_i(n) = \sum_{j=1}^{N} \sum_{l=0}^{L-1} h_{ij}(l) s_j(n-l) + v_i(n) ; \quad i=1,...,M, \qquad (1)$$

where, $\mathbf{x}(n) = [x_1(n),....,x_M(n)]^T$ stands for the transposed vector of M SEMG recordings (measurements), $\mathbf{s}(n) = [s_1(n),....,s_N(n)]^T$ denotes the vector of N MU innervation pulse trains (sources), and $\mathbf{v}(n) = [v_1(n),....,v_M(n)]^T$ is the noise vector. $h_{ij}(l)$ stands for the unit sample response (MUAP) of the i-th source as detected in the j-th SEMG recoding. For the simplicity reasons we will suppose the length of all impulse responses (MUAPs) equal to L. We further suppose the number of measurements greater than the number of sources $M>N$.

The additive noise $v_i(n)$ is modeled as stationary, temporally and spatially white zero-mean Gaussian random process, being independent from the sources

$$E[\mathbf{v}(n+\tau)\mathbf{v}^*(n)] = \sigma^2 \delta(\tau)\mathbf{I}, \qquad (2)$$

where $E[\cdot]$ stands for mathematical expectation, $\delta(.)$ for the Dirac impulse (delta function), σ^2 for the noise variance, and $\mathbf{I}$ denotes the identity matrix.

To extend relation-ship (1) to multiplicative MIMO vector form, the vector $\mathbf{x}(n)$ has to be augmented by K delayed repetitions of each measurement [1]:

$$\overline{\mathbf{x}}(n) = [x_1(n),....,x_1(n-K+1),....,x_M(n),....,x_M(n-K+1)]^T, \qquad (3)$$

where K is an arbitrary large integer which satisfies $KM > N(L+K)$. Extending the noise vector in the same manner, (1) can be rewritten as

$$\overline{\mathbf{x}}(n) = \mathbf{A}\overline{\mathbf{s}}(n) + \overline{\mathbf{v}}(n). \qquad (4)$$

where $\mathbf{A}$ stands for the so called mixing matrix of size $KM \times N(L+K)$ which contains the unit sample responses $h_{ij}(l)$:

$$\mathbf{A} = \begin{bmatrix} \mathbf{H}_{11} & \cdots & \mathbf{H}_{1N} \\ \vdots & \ddots & \vdots \\ \mathbf{H}_{M1} & \cdots & \mathbf{H}_{MN} \end{bmatrix} \text{ with } \mathbf{H}_{ij} = \begin{bmatrix} h_{ij}(0) & \cdots & h_{ij}(L) & \cdots & 0 \\ \vdots & \ddots & \ddots & \ddots & \vdots \\ 0 & \cdots & h_{ij}(0) & \cdots & h_{ij}(L) \end{bmatrix}, \qquad (5)$$

while the extended vector of sources $\bar{\mathbf{s}}(n)$ takes the following form:

$$\bar{\mathbf{s}}(n) = [s_1(n),....,s_1(n-L-K+1),....,s_N(n),....,s_N(n-L-K+1)]^T. \quad (6)$$

Following the above assumptions the correlation matrix of extended measurements can be expressed as:

$$\mathbf{R}_{\bar{\mathbf{x}}}(\tau) = \lim_{T \to \infty} \frac{1}{T} \sum_{n=1}^{T} \bar{\mathbf{x}}(n)\bar{\mathbf{x}}^*(n+\tau) = \mathbf{A}\mathbf{R}_{\bar{\mathbf{s}}}(\tau)\mathbf{A}^T + \delta(\tau)\sigma^2 \mathbf{I}, \quad (7)$$

where $\mathbf{R}_{\bar{\mathbf{s}}}(\tau)$ denotes the correlation matrix of sources and $\bar{\mathbf{x}}^*(n)$ stands for the conjugate transpose of $\bar{\mathbf{x}}(n)$. Taking into account implicite BSS indeterminancy and MU refractory period we can suppose the variance of all extended sources equal to 1 ($r_{ii} = \lim_{T \to \infty} \frac{1}{T} \sum_{n=1}^{T} \bar{s}_i(n)\bar{s}_i^*(n) = 1$). Hence, the correlation matrix of the extended sources at zero lag $\tau = 0$ can be set equal to the identity matrix:

$$\mathbf{R}_{\bar{\mathbf{s}}}(0) = \mathbf{I}. \quad (8)$$

3 Decomposition Method

Following the decomposition approach in [4] the mixing matrix $\hat{\mathbf{A}}$ is estimated in two steps. Firstly the measurements are whitened (second-order decorrelated) by so called whitening matrix $\mathbf{W}$ satisfying

$$\mathbf{W}\mathbf{A}\mathbf{R}_{\bar{\mathbf{s}}}(0)\mathbf{A}^H \mathbf{W}^H = \mathbf{W}\mathbf{A}\mathbf{A}^H \mathbf{W}^H = \mathbf{I}. \quad (9)$$

According to (8) and (9) the whitening matrix $\mathbf{W}$ can be obtained as an inverse square root of the observation correlation matrix $\mathbf{R}_{\bar{\mathbf{x}}}(0)$ [1]. The mixing matrix $\mathbf{A}$ is now transformed to unknown $N(L+K) \times N(L+K)$ unitary matrix $\mathbf{U}$ [1]:

$$\mathbf{W}\mathbf{A} = \mathbf{U}. \quad (10)$$

In the second step the matrix $\mathbf{U}$ is identified by exploiting the cross-correlations of the augmented sources. Some algebra upon (6), (7) and (10) produces [4]:

$$\mathbf{Q}_{\bar{\mathbf{x}}}(-\tau,\tau) = \mathbf{W}\mathbf{R}_{\bar{\mathbf{x}}}(-\tau)\mathbf{W}^H \mathbf{W}\mathbf{R}_{\bar{\mathbf{x}}}(\tau)\mathbf{W}^H = \mathbf{U}\mathbf{R}_{\bar{\mathbf{s}}}(-\tau)\mathbf{R}_{\bar{\mathbf{s}}}(\tau)\mathbf{U}^H + \delta(\tau)\sigma^2 \mathbf{I} =$$
$$= \mathbf{U}(\mathbf{C}_{\bar{\mathbf{s}}}(\tau) + \delta(\tau)\sigma^2 \mathbf{I})\mathbf{U}^H \quad (11)$$

where

$$\mathbf{C}_{\bar{\mathbf{s}}}(\tau) = \mathbf{R}_{\bar{\mathbf{s}}}(-\tau)\mathbf{R}_{\bar{\mathbf{s}}}(\tau) = \begin{bmatrix} \mathbf{D}_1(\tau) & 0 & \cdots & 0 \\ 0 & \mathbf{D}_2(\tau) & \ddots & \vdots \\ \vdots & \ddots & \ddots & 0 \\ 0 & \cdots & 0 & \mathbf{D}_N(\tau) \end{bmatrix} \quad (12)$$

with $(L+K) \times (L+K)$ $\mathbf{D}_i(\tau)$ matrices defined as

$$\mathbf{D}_i(\tau) = r_{ii}^2 \begin{bmatrix} \sum_{j=-L-K+1}^{0} \delta(\tau-j) & 0 & \cdots & 0 \\ 0 & \sum_{j=-L-K+2}^{1} \delta(\tau-j) & \ddots & \vdots \\ \vdots & \ddots & \ddots & 0 \\ 0 & \cdots & 0 & \sum_{j=0}^{L+K-1} \delta(\tau-j) \end{bmatrix} \qquad (13)$$

The matrices $\mathbf{Q}_{\bar{x}}(-\tau,\tau)$ effectively suppress the noise and are all diagonal in the basis of the columns of the matrix $\mathbf{U}$. Hence, the missing matrix $\mathbf{U}$ can be obtained as a joint-diagonalizing matrix [3] of the $\mathbf{Q}_{\bar{x}}(-\tau,\tau)$ matrices. To guarantee the uniqueness of the unitary matrix which simultaneously diagonalizes the set of $\mathbf{Q}_{\bar{x}}(-\tau,\tau)$ matrices the condition $\forall i, \forall j \quad i \neq j: r_{ii} \neq r_{jj}$ must be met [2]. However, due to the whitening step all the sources have unit variance: $\forall i, r_{ii} = 1$. Processing the pulse sources with time varying firing frequencies (due to the fatigue effect the MUs' firing frequencies are expected to decrease in time) this problem can easily be avoided by limiting the calculation of the zero-lagged correlation matrix in (9) and nonzero-lagged correlation matrices in (11) to two different time subintervals. This prevents the equalization of the r_{ii} factors in (13), and consequently, guarantees the uniqueness of $\mathbf{U}$ [4]. In order to increase the numerical robustness, several sets of $\mathbf{Q}_{\bar{x}}(-\tau,\tau)$ matrices (each set calculated at different time interval) should be joint-diagonalized. More strict treatment of this problem can be found in [4].

Once the mixing matrix $\hat{\mathbf{A}}$ is reconstructed, the original sources are identified as

$$\bar{\mathbf{s}}(n) = \mathbf{A}^{\#} \bar{\mathbf{x}}(n), \qquad (14)$$

where $\mathbf{A}^{\#}$ denotes a pseudo-inverse of the matrix $\mathbf{A}$.

4 Simulations and Experiments with Real SEMG Signals

To test its performance the described decomposition approach was applied to both synthetic and real surface EMG signals.

4.1 Synthetic Signals

Synthetic surface EMG signals were generated using the advanced EMG simulator [6]. The number of active MUs was set to 5, 10 and 20, respectively, while SNR ranged from 5 dB to 20 dB, in steps of 5 dB. 5 simulations were performed for each number of active MUs. In each simulation runs the depth of MUs in the anisotropic muscle layer (uniformly distributed over [3,10] mm), the number of fibres (uniformly distributed over [50,300]), and conduction velocity (normally distributed with mean of 4 m/s and standard deviation of 1 m/s) were randomly selected. MU territories were supposed circular with 20 fibres/mm². The innervation zones with spread of 5 mm were placed in the middle of the fibers with the semi-fiber length of 70 *mm*. The MUs' firing rate was normally distributed with mean of 15 Hz and standard deviation of 4Hz. Fatigue induced decrease of MU firing rate was limited to 1 Hz per 10 s of the simulated signals. The inter-pulse interval (IPI) variability was modelled zero-mean Gaussian with the variance equal to 10 % of the IPI mean. Signals from each simulation run were corrupted by additive noise (10 realisations of noise for each SNR).

Detection system consisted of rectangular 1×1 mm electrodes arranged in 10 lines and 5 columns with the inter-electrode distance of 5 mm. The array of electrodes was centred over innervation zone (columns aligned with the direction of fibres). Single differential recordings in duration of 30 s were sampled at 1024 Hz.

4.2 Experimental Protocol with Real SEMG Signals

Real SEMG signals were recorded in Laboratorio di Ingegneria del Sistema Neuromuscolare (LISiN), Centro di Bioingegneria, Politecnico di Torino, Italy. Three healthy male subjects (age 27.3 ± 3.2 years, height 179 ± 3 cm, and weight of 65.5 ± 2.6 kg) participated in our study. Firstly the dominant arm of the subjects was placed into the isometric brace at 120°. Skin was slightly abraded with abrasive paste and moistened to improve the electrode-skin contact. The array of 61 electrodes (arranged in 13 lines and 5 columns without the four corner electrodes) was placed over the biceps with its third electrode row centred over the innervation zone (columns aligned with the muscle fibres). 30 s long SEMG signals were recorded at isometric voluntary contractions sustained at 5 % and 10 % of maximum voluntary contraction (MVC). The contraction force was measured by the torque sensor and displayed on the oscilloscope to provide the visual feedback to the subjects.

The detected signals were amplified (gain set to 10000) by a 64-channel EMG amplifier (LISiN; Prima Biomedical & Sport, Treviso, Italy), band-pass filtered (-3 dB bandwidth, 10 Hz – 500 Hz), and sampled at 1024 *Hz* by 12-bit A/D converter. Longitudinal single-differential recording technique was applied with the adjacent electrode pairs along the columns in the electrode array, what resulted in 59 SEMG recordings.

5 Decomposition Results

The 30 s long synthetic signals were first divided into three successive 10 s long subintervals. The zero-lagged correlation matrix in (9) was calculated from the second interval, while the two sets of $Q_{\bar{x}}(-\tau,\tau)$ matrices, one from the first and the other from the last subinterval were joint-diagonalized. The length of MUAPs was estimated to 20 samples, while the number of active MUs was set to N=5, N=10 and N=20, respectively. Before comparing to the original sources the estimations of each firing pulse train were normalized, classified, aligned according to the pulse triggering times and finally summed together. The decomposition results are outlined in Tables 1 and 2. Representatives of reconstructed innervation pulse trains are depicted in Fig. 1.

Blind Deconvolution of Close-to-Orthogonal Pulse Sources 1061

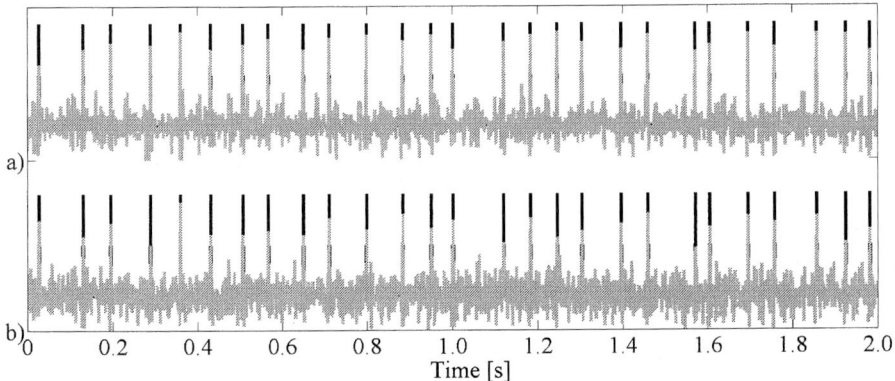

Fig. 1. Original synthetic MU innervation train (*black*) of the first MU (164 fibers, depth of 5.3 mm and firing rate of 13.75 Hz) and the decomposed innervation trains (*grey*), reconstructed in the case of 10 active MUs at a) SNR=15 dB, and b) SNR=10 dB

Table 1. The number of identified MUs (mean ± std. dev.) in dependence of the number of active MUs and the signal-to-noise ratio. Results were obtained on the synthetic SEMG signals

SNR [dB]	5	10	15	20
5 active MUs	1.93 ± 1.03	2.83 ± 0.95	2.87 ± 0.82	3.8 ± 1.11
10 active MUs	1.21 ± 0.91	2.40 ± 0.74	4.16 ± 0.85	5.33 ± 1.13
20 active MUs	1.33 ± 1.12	2.27 ± 0.71	3.47 ± 0.52	5.40 ± 1.24

Table 2. Percentage (mean ± std. dev.) of correctly detected pulses in identified MU innervation pulse trains. The numbers of detected pulses are normalized by the number of pulses in corresponding original synthetic pulse trains and averaged over all identified MUs

SNR [dB]	5	10	15	20
5 active MUs	91 % ± 9 %	93 % ± 8 %	95 % ± 8 %	98 % ± 5 %
10 active MUs	88 % ± 8 %	90 % ± 6 %	93 % ± 5 %	95 % ± 6 %
20 active MUs	81 % ± 9 %	86 % ± 8 %	89 % ± 6 %	91 % ± 6 %

Exactly the same decomposition procedure was applied to the real surface EMG. The 30 s long signals were first divided into three successive 10 s long intervals. The whitening matrix was calculated from the second interval, and the $\mathbf{Q}_{\tilde{\mathbf{x}}}(-\tau,\tau)$ matrices from the first and last interval. The length of MUAPs was approximated to 20 samples. The number of MUs active in the detection volume was estimated to 5 with 5 % MVC contractions, and 10 with 10 % MVC contractions. The estimations of each firing pulse train were first normalized, classified, aligned and finally summed together. The results are depicted in Table 3 and Figs 2 and 3.

Table 3. The number of identified MUs (*No. of MUs*) and their average firing rates (mean ± standard deviation) calculated from two 5 s long intervals, one at the beginning (F_1) and the other at the end (F_2) of 30 s long signal. SEMG signals were recorded during an isometric 5 % and 10 % MVC measurements of the dominant biceps brachii of three healthy male subjects

	5 % MVC			10 % MVC		
Subject	1	2	3	1	2	3
No. of MUs	3	2	3	2	2	3
F_1 [Hz]	11.0±0.1	10.5±0.8	15.7±1.7	12.0±0.4	11.0±0.5	12.0±2.2
F_2 [Hz]	10.6±0.2	9.9±0.1	14.6±1.0	10.7±0.3	10.4±0.6	11.1±2.5

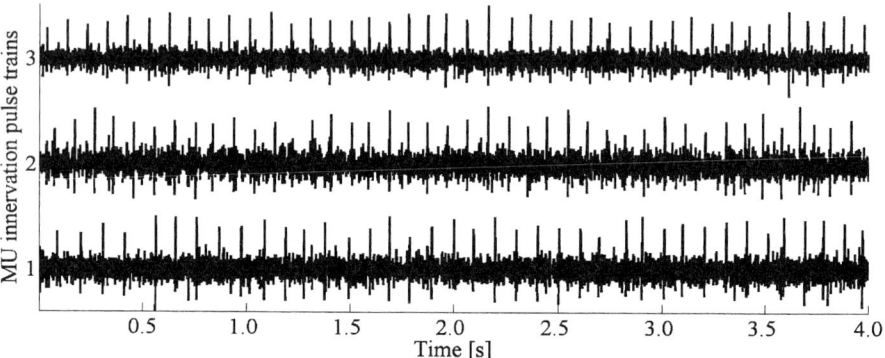

Fig. 2. MU innervation trains reconstructed from 30 s long real SEMG signal recorded during an isometric 5 % MVC measurement of the dominant biceps brachii muscle of subject 1 (age 26 years, height 176 cm, weight 68 kg)

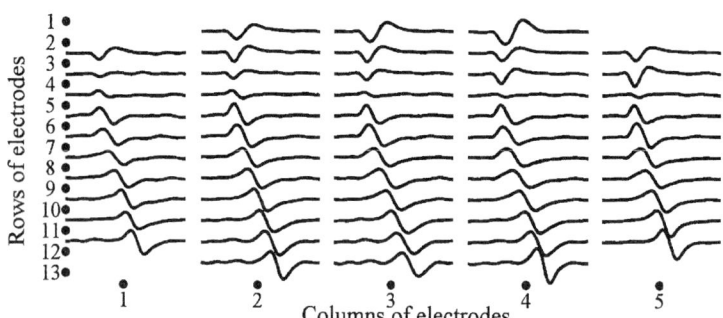

Fig. 3. The MUAPs corresponding to the first MU (Fig. 2) reconstructed by spike triggered sliding window averaging technique. Each MUAP is depicted between the two adjacent electrodes (*black circles*) constituting the corresponding single-differential electrode pair. SEMG was recorded during an isometric 5 % MVC measurement of the dominant biceps brachii of subject 1 (age 26 years, height 176 cm, weight 68 kg)

6 Discussion

As demonstrated by the results in Section 5, the decomposition method successfully suppresses the influence of the additive white noise while sufficiently resolving the superimpositions of MUAPs. When processing the synthetic signals almost all pulse trains were reconstructed down to the SNR of 10 dB (Table 2). The performance drops with lower SNRs. At SNR of 5 dB only 88 % of original firing pulses in 1.5 MUs were identified on average. On the other hand, no significant influence of the number of active MUs was disclosed. In the case of 10 and 20 active MUs only the strongest MUs (superficial MUs with large number of muscle fibres) were identified, whereas the MUs located deep in the muscle were treated as a background noise.

Experimental results coincide with those from simulations. On average 2.5 MUs were reconstructed from both, 5 % and 10 % MVC measurements. With the real signals no direct evaluation of the decomposition method is possible. However, the regularity of the inter-pulse intervals, decreasing firing rates, and finally, the MUAP shapes reconstructed by spike triggered sliding window averaging technique (Fig. 3) provide strong indirect evidences that the reconstructed pulse sequences truly correspond to the MU innervation trains. The presented method thus contributes a new insight to the non-invasive analysis of single MU properties.

References

1. Abed-Meraim, K., Belouchrani, A., Leyman, A.R.: Blind Source Separation Using Time-Frequency Distributions. In: Boashash, B. (ed.): Time frequency Signal Processing & Applications, Elsevier (2003)
2. Belouchrani, A., Abed-Meraim, K.: Blind source separation based on time-frequency signal representation. IEEE Trans. On Signal Processing, Vol. 46, No. 11, (1998) 2888-2898
3. Cardoso, J.F., Souloumiac, A.: Jacobi angles for simultaneous diagonalization. SIAM J. Mat. Anal. Appl., Vol. 17, No. 1, (1996) 161-164
4. Holobar, A., Zazula, D.: A novel approach to convolutive blind separation of close-to-orthogonal pulse sources using second-order statistics. EUSIPCO 2004
5. Hyvarinen, A., Karhunen, J., Oja, E.: Independent Component Analysis, John Wiley & sons, Inc. New York (2001)
6. Farina, D., Merletti, R.: A novel approach for precise simulation of the EMG signal detected by surface electrodes. IEEE Trans. Biomed. Eng., Vol. 48, (2001) 637-646

Denoising Mammographic Images Using ICA

P. Mayo[1], Francisco Rodenas Escriba[2], and Gumersindo Verdú Martín[1]

[1] Chemical and Nuclear Engineering Department, Polytechnic University of Valencia, Camino de Vera s/n 46022, Valencia, Spain
{pmayo,gverdu}@iqn.upv.es
[2] Applied Mathematics Department, Polytechnic University of Valencia,
Camino de Vera s/n 46022, Valencia, Spain
frodenas@mat.upv.es

Abstract. Digital mammographic image processing often requires a previous application of filters to reduce the noise level of the image while preserving important details. This may improve the quality of digital mammographic images and contribute to an accurate diagnosis. Denoising methods based on linear filters cannot preserve image structures such as edges in the same way that methods based on nonlinear filters can do it. Recently, a nonlinear denoising method based on ICA has been introduced [1,2] for natural and artificial images. The functioning of the ICA denoising method depends on the statistics of the images. In this paper, we show that mammograms have statistics appropriate for ICA denoising and we demonstrate experimentally that ICA denoising is a suitable method to remove the noise of digitised mammographys.

1 Introduction

Nowadays the mammography is the most effective technique for detecting breast occult tumours. The low contrast of the small tumours to the background, which is sometimes close to the noise, makes that small breast cancer lesions can hardly be seen in the mammography [3]. In this sense an image preprocessing to reduce the noise level of the image preserving the mammography structures, is an important item to improve the detection of mammographic features.

Classically, denoising methods have been based on apply linear filters as the Wiener filter to the image, however linear methods tend to blur the edge structure of the image. Several denoising methods based on nonlinear filters have been introduced to avoid this problem [4, 5]. For this reason, we will check here whether a nonlinear denoising technique based on independent component analysis is a suitable method to denoise mammographic images.

Independent component analysis (ICA) is a method to represent a set of multidimensional data vectors in a basis where the components are as independent as possible [6].

ICA denoising methods rely on the fact that the transformed components have sparse (supergaussian) distributions, so that the denoising techniques attempt to reduce gaussian noise by shrinkage (soft thresholding) of these sparse components. The choice of a shrinkage function depends on the statistical distribution of each sparse component [7]. The paper presents our first results on application of ICA to denoise mammographic images, showing that statistical distributions of the independent components of these images are appropriated to apply the *Sparse Code Shrinkage* [1].

2 Independent Component Analysis

In ICA an observed random vector is expressed as a linear transformation of another variables that are nongaussian and statistically independent. Denote by x the n-dimensional data vector, in our case the vector contains the pixel gray levels of an image window. The basic ICA model may be expressed as (see [6,8] for a survey):

$$x = As \qquad (1)$$

where $x=[x_1,...,x_n]^T$ is the vector of observed data, $s=[s_1,...,s_m]^T$ is the vector of independent components, called source signals, and A is a constant full-rank $n \times m$ matrix, named the mixing matrix. The column vectors a_i $i=1,...,m$, of A are called the basis vectors of ICA. Independent components and mixing matrix are determined by requiring that the coeficients s_i, are mutually independent or as independent as possible.

ICA can be viewed as an extension of standard principal component analysis (PCA) where the coefficients of the expansion must be mutually independent instead of uncorrelated as in PCA case. ICA basis vectors a_i are generally not mutually orthogonal, in contrast with standard PCA where the transformartion is orthogonal.

The ICA basis functions are data dependent in the sense that they are obtained from a set of training data. Training data are considered realizations of random vector x with similar statistical properties. The ICA basis vectors can be considered as image building blocks, they describe spatial frequency, capturing the inherent features of the training data [9]. The independent components are estimated by determining an $m \times n$ separating matrix W, so that the components s_i of the linear transformed vector s have maximally non-gaussian distributions and are mutually uncorrelated,

$$s = Wx. \qquad (2)$$

The separating matrix W is determined using an algorithm that optimizes iteratively statistical independence of the components of s. Then W is estimated presenting a set of training data to the ICA algorithm. The algorithm performing ICA that we have used is Hyvärinen's fixed-point algorithm [10], often called *FastICA* algorithm.

If the number of independent components m is equal to n, the mixing matrix A is the inverse of W, $W=A^{-1}$. It is useful to choose $m<n$, i.e. the number of independent components is less than the number of components of the original data x. In this case, A is the pseudoinverse of W. This dimension reduction can be accomplished by PCA, since PCA transformation optimally compresses the data vectors.

In order to apply the ICA algorithm, the original data x must be preprocessed. First, data are centered, i.e, we subtract the data mean. The second step, called whitening, is to remove the second order statistical dependence in the data. Whitened data have unit variance and are uncorrelated. Whitening can be done using standard PCA, which simultaneously may be used to reduce the data dimension. The m-dimensional whitened vector is obtained by the linear transformation

$$y = Vx \qquad (3)$$

where the PCA $m \times n$ whitening matrix is of the form $V=D^{-1/2}E^T$. Using the eigenvalue decomposition of the covariance matrix of x, the diagonal matrix D contains the m greatest eigenvalues of the covariance matrix and the columns of the orthogonal matrix E contain the corresponding eigenvectors.

After whitening, we seek an orthogonal matrix B which maximizes some given measure of supergaussianity of the componentes of the vector $s=By$, this may be done by using FastICA algorithm [10]. The separating equation is then

$$s = By = BVx = BD^{-1/2}E^Tx = Wx. \quad (4)$$

In [2,7] an orthogonal separating matrix is introduced by

$$W_{ortho} = W(W^TW)^{-1/2}. \quad (5)$$

This matrix retains the sparseness of the original ICA transformed components and produces acceptable denoising results [7].

3 ICA Denoising Method

If the ICA model holds, the independent components are sparse which means that each component has a supergaussian distribution. This is fundamental to apply the approach of [1] to eliminate the gaussian noise from a nongaussian random variable.

Denote by f the noisy observed random variable, the model for f can be expressed:

$$f = c + n \quad (6)$$

where c is a non-gaussian random variable corrupted by an additive gaussian noise $n \sim N(0, \sigma^2)$. The method introduced in [1] to denoise the observed data f proposes an estimate of c given by

$$\hat{c} = g(f) \quad (7)$$

where g is a function depending on the probability density distribution of c. The function g is a shrinkage function that can be considered a soft thresholding operator applied to the values of f.

In [2] two parametrizations of density models that are suitable for the densities encountered in image denoising are developed. Both parametrizations depend on two parameters and they model different degrees of non-gaussianity, in particular, the mildly and strongly sparse densities found for the natural and artificial images. The method of denoising a random vector consists in applying the method described above for a scalar variable to each component separately. In general, there is no guarantee that the vector components are sparse. To solve this problem, the vector is linearly transformed in such a way that the resulting components are as nongaussian as possible. This is the ICA transformation. Therefore, the denoising method for a vector variable consists in applying on the ICA independent components a component-wise denoising using the appropriate shrinkage fuction g for each component.

4 Applying ICA Denoising to Mammograms

The idea of the ICA denoising [2,7] procedure that we wish to test with mammographic images is: the data are transformed into a representation with statistical suitable properties, a shrinkage function is applied to the transformed components and finally, the transformation is inverted. The denoising procedure is based on the property that ICA transformed components of the image have supergaussian distributions, then the analysis of the density functions of the independent components of mammographic images in the transform domain is a crucial point to test if the denoising method is suitable for denoising these images. In summary, the main steps of *Sparse Code Shrinkage* [7] applied to mammographic images are:

1. Estimating the transformation and the shrinkage functions from the training data:
   - Estimate the ICA separating matrix W using a representative noise-free set of mammograhic images z. This training data is a set of noise-free mammographic images obtained from the MiniMammographic database with the same statistical properties as the images we wish to denoise.
   - For every independent component of the transformed vector $s=Wz$, estimate a density model for the probability density function using the parametrizations introduced in [2]. The probability density parametrization of s_i gives the corresponding shrinkage function g that will be used to denoise s_i.
2. Denoising an image:
   - Arrange the noisy image as a matrix X whose columns are the vector data containing the pixel gray level of an image window. Transform each vector into the sparse basis by using the separating ICA matrix, $S=WX$.
   - Shrinkage each sparse component by applying the corresponding nonlinearity g. Thus, the estimated components are $ŝi=g(wiTX)$.
   - Finally, the denoised version of the original image is obtained by using the pseudo inverse transformation W-1, $\hat{X}=W-1\hat{S}$, and reconstructing the image from the columns of $\hat{X}$.

5 Experiments

5.1 Training Data

The training images to estimate the ICA transform are selected from the MIAS MiniMammographic Database, provided by the Mammographic Image Analysis Society. The mammograms are digitized at 200 micron pixel edge, resulting images with 1024x1024 pixel resolution. The criteria for image selection has been that the training set must be representative of the mammographic images, so we have chosen ten images containing all kinds of typical mammogram regions as spiculated masses, distortion in breast architecture, asymmetries and clustered microcalcifications (Fig. 1).

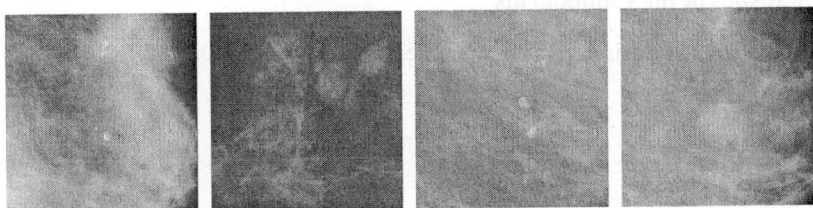

Fig. 1. Some typical regions of training mammographic images

5.2 Sampling

Each image of the training data set is linearly normalized, so that pixels have zero mean and unit variance. The data vectors x are obtained from 20000 image patches of size 16x16 that are taken at random from the training images, so these subimages are vectorized into 256-dimensional vectors which are used as the mixed data of the ICA model. From each column of X of 256 components the local mean is subtracted to discard low frequency components of the images and to obtain greater generality of the data.

5.3 Whitening and Dimension Reduction

The zero-mean vectors are linearly transformed into uncorrelated and unit variance variables using the whitening matrix $V=D^{-1/2}E^T$. Simultaneously, PCA transformation is used to reduce the dimension of the original data. The dimension reduction is performed retaining the 98% of the initial variance.

5.4 ICA Algorithm

After whitened, FastICA algorithm [10] is applied to find the independent components s_i, $i=1...m$. and the ICA separating matrix W. The nonlinear function used by the algorithm is the typical choise $g(u) = tanh(u)$, related with the normalized kurtosis local maxima criterion.

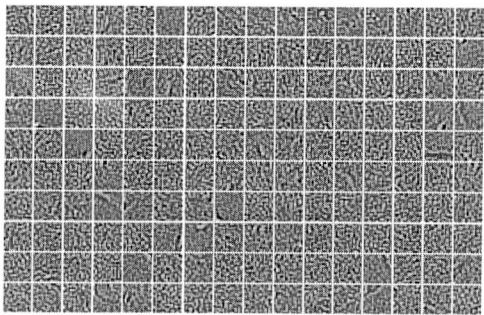

Fig. 2. ICA separating filters for the mammographic training images

5.5 Statistics of the Components

The normalized kurtosis k is a measurement of the non gaussianity of the distributions. It is defined as:

$$k(s) = \frac{E[s^4]}{(E[s^2])^2} - 3 \qquad (8)$$

In order to estimate the sparseness of the independent components, we have sampled 40000 image patches at random locations from the same dataset that is used for estimation of the transform. Then we transformed these samples using the estimated ICA transform and we have calculated the kurtosis for each one of the components.

The normalized kurtosis values encountered for every transformed component are greater than zero (Fig 3), so ICA transform finds sparse representations of the original vectors. As a result, the statistics of the independent components of mammographic images are suitable for the ICA-based denoising procedure.

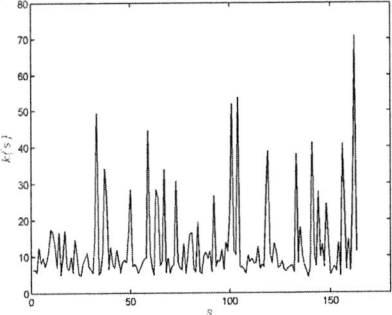

Fig. 3. Normalized kurtosis for the independent components in the transformed domain

We have checked that all component densities encountered in mammographic images fit the parametrization given in [2] corresponding to mildly sparse densities. This parametrization depends on two parameters that were calculated from the probability density function estimated by using a non parametric histogram technique.

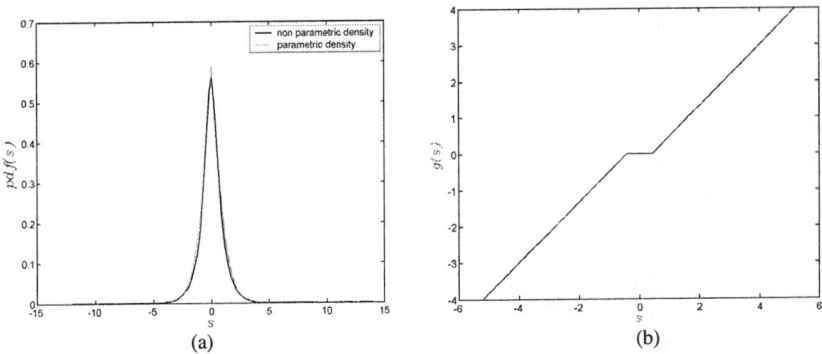

Fig. 4. (a) Parametrization of the density function for an independent component (b) Corresponding shrinkage nonlinearity g given by the parametric density model

The shrinkage functions are, according to [2], which correspond to the mildly sparse density parametrization. Using the parameter values for each density functions, the corresponding shrinkage functions g are obtained.

5.6 Denoising a Mammographic Image

For a mammographic image of the database, we have applied the denoising algorithm taking a sliding window approach of the image [7], so we have applied the algorithm to all vectors obtained from every possible 16x16 window of the image. Then, each 256-dimensional vector is preprocessed by whitening and its dimension is reduced, resulting vectors are transformed into the sparse basis and the estimated nonlinearity shrinkage functions are applied to every component of each vector. After that, we have inverted the transformation to obtain estimates of the denoised vectors. Finally, it is necessary the reconstruction of the denoised image from the denoised vectors.

Since we have considered the sliding window approach, then each pixel has 256 different suggested values and we have computed the final result as the mean of these values. This is a type of local filtering, considering all the possible 16x16 neighbourhood around each pixel.

6 Results

A 256x256 pixel subimage of a random image from the Mias MiniMammographic Database has been chosen for denoising experiments. Gaussian noise of standard deviation 0.3 was added. The results of the denoising procedure are shown in Fig.5. The denoising results obtained by using the standard ICA matrix are compared in Fig.5 with the results using the orthogonal ICA transformation (5) and with the results of the *wavelet shrinkage* denoising method proposed by Donoho [4].

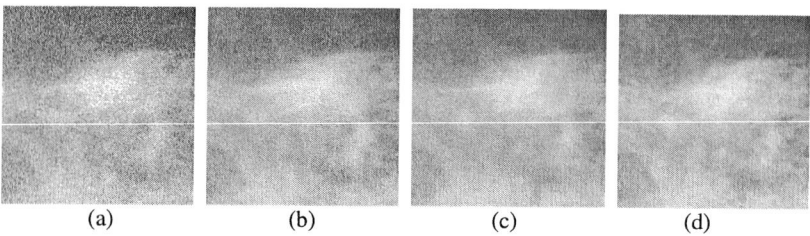

(a)   (b)   (c)   (d)

Fig. 5. (a) Mammographic image from the database with noise added. (b), (c) Images after denoising method with standard and orthogonal ICA transformation, respectively. (d) Image after wavelet denoising with coifflet10 family level 1

The reasons because of the comparison with wavelet denoising method has been included are: first, denoising methods using wavelet decomposition have been successfully apply to mammographic images [11] and second, the authors in [9,7] have shown that ICA filters of natural images are closely related with wavelet bases.

7 Conclusions

We have applied to mammographic images a denoising method based on applying a shrinkage function to the independent components of the original image (*Sparse Code Shrinkage* [2]). Theoretically, this method is valid if the components have sparse probability density functions. We have shown experimentally that the densities of the independent components for mammographic images are supergaussian and we have checked that this ICA-based denoising method gives a good noise reduction when it is applied to mammographic images while the features in the image are retained.

References

1. Hyvärinen, A.: Sparse code shrinkage: denoising of nongaussian data by maximum likelihood estimation. Neural Computation 11 7 (1999) 1739–1768.
2. Hyvärinen, A., Hoyer, P., Oja, E.: Image Denoising by Sparse Code Shrinkage.In: Hykin, S., Kosko, B. (eds): Intelligent Signal Processing, IEEE Press, (2001).

3. Dengler, J., Behrens, S., Desaga, J.F.: Segmentation of microcalcifications in mammograms, IEEE Transactions on Medical Imaging 12 (1993), 634-642.
4. Donoho, D., Johnstone, I., Kerkyacharian, G., Picard, D.: Wavelet Shrinkage: Asymptopia? Journal of the Royal Statistical Society B57 (1995), 301-369.
5. Catté, F., Lions, P., Morel J., Coll T.: Image Selective Smoothing and Edge Detection by Nonlinear Diffusion. SIAM Numerical Analysis 29 (1992), 182-193.
6. Hyvärinen A., Oja E.: Independent Component Analysis: algorithms and applications. Neural Networks 13 (2000), 411-430.
7. Hoyer, P.: Independent component analysis of image denoising, Master's Thesis, Helsinki University of Technology, 1999.
8. Comon, P., Independent component analysis–a new concept?. Signal Processing 36 (1994), 287–314.
9. Hurri, J., Hyvärinen, A., Oja E.: Wavelets and natural image statistics, in: M. Frydrych, J. Parkinnen, A. Visa (Eds.), Scandinavian Conference on Image Analysis, Finland, 1997.
10. Hyvärinen, A.: Fast and robust fixed-point algorithms for independent component analysis. IEEE Trans. Neural Networks 10 3 (1999), 626–634.
11. Sakellaropoulos, P., Costaridou, L., Pabayiotakis G.: A wavelet-based spatially adaptive method for mammographic contrast enhancement. Physics in Medicine and Biology 48 6 (2003), 787-803.

Independent Component Analysis of Pulse Oximetry Signals Based on Derivative Skew

Paul F. Stetson

Nellcor / Tyco Healthcare, Technology Development
Pleasanton CA USA 94588
stetson@isl.stanford.edu

Abstract. This paper describes the application of ICA to signals from a pulse oximeter, a device for measuring arterial blood oxygen content. Since the arterial pulse signal resembles a sawtooth wave, taking the time derivative enhances the skew of this component relative to that of the interference. Third-order ICA techniques can take advantage of this to aid the separation. This paper shows an example of ICA used on pulse oximetry signals and then presents simulations to demonstrate that, for pulse oximetry signals, ICA based on third-order statistics of the derivative is indeed superior to fourth-order ICA.

1 Introduction

Some of the earliest applications of ICA have successfully used fourth-order (4°) cumulant methods [1] and the infomax method [2], [3] on biomedical signals, yet so far ICA has not been applied to pulse oximetry. It turns out that pulse oximeter signals lend themselves especially well to ICA due to the nature of their time-derivative.

Pulse oximetry is a standard technique for monitoring the oxygen content of arterial blood in the operating room, intensive care unit, and other critical care settings [4], [5]. It measures the absorption of light passing through the body, which varies as blood pulses into the tissue with each beat of the heart. The pulse rate can be found from the variations in the amount of transmitted light, and the ratio of the pulse amplitudes from two wavelengths can be accurately calibrated to give the oxygen saturation of the pulsatile blood [6], [7].

Obtaining reliable pulse rates and saturations in the presence of large interference sources is a major challenge in pulse oximetry. Adaptive comb filtering [6] has been used to pass only the portion of the signal at the pulse rate and its harmonics. Blind source separation techniques [8] present a new approach in that they can separate out the arterial pulse component in a single iteration, based solely on the input statistics.

While the general use of ICA as a means of blindly separating independent signal sources is fairly well known, its implementation in pulse oximetry presents some unique circumstances. For instance, most ICA techniques are based on fourth-order cumulants, as the signals and noise commonly encountered in communications have zero third-order cumulant (skew), and cumulants of higher than fourth order are difficult to estimate accurately. However, because the arterial component of an oximetry signal has somewhat of a sawtooth nature, the third-order statistics can be enhanced by taking the time derivative. Since the skew of the time-derivative of the arterial signal is generally much greater in magnitude than that of the interference [9], the performance of ICA may be enhanced by using unmixing coefficients that are derived from the 3° cumulants of the derivatives of the signals.

Applying ICA to pulse oximetry, the mixture signals correspond to signals obtained at multiple wavelengths, which include both the desired signal and the undesired noise components. Source components refer to the arterial signal as well as interference data, which may be caused by motion, light interference, respiratory artifacts, and instrumental and environmental noise.

2 Methods

2.1 ICA Algorithm

For the set of signals, we assume

$$\mathbf{x}(k) = \mathbf{A}\,\mathbf{s}(k)$$

where the matrix $\mathbf{A}$ mixes the signal and noise sources $s(k)$, without including past values. We seek to recover those sources by finding the unmixing matrix $\mathbf{B}$ which is ideally the inverse of $\mathbf{A}$:

$$\mathbf{z}(k) = \mathbf{B}\mathbf{x}(k) = \mathbf{B}\mathbf{A}\mathbf{s}(k)$$

ICA methods which give closed solutions for blocks of data [10],[11] were chosen in order to avoid convergence issues that may arise with adaptive techniques when the probability distribution of the inputs takes on unexpected forms [2]. After the Principal Component Analysis of the data, the higher-order decorrelation is performed in three ways: by using fourth-order method described in [11], by using its third-order counterpart, or by calculating the unmixing coefficients $\mathbf{B}$ based on the 3rd-order statistics of the time-differentiated signals and applying $\mathbf{B}$ to the undifferentiated signals. Since

$$\mathbf{x}(k) - \mathbf{x}(k-1) = \mathbf{A}\,[\mathbf{s}(k) - \mathbf{s}(k-1)]$$

the mixing coefficients for the time-differentiated signals are the same as those of the undifferentiated signals. However, since the statistics for those differentiated signals are not the same, the ICA results will differ.

2.2 Simulation of Noisy Signals

In order to precisely assess the performance of different ICA methods on pulse oximetry signals, 15-second segments of clean signal and heavy motion artifact were synthetically mixed. Forty-eight pairs of clean and noisy segments of data were identified from ten subjects and a variety of oximeters. The segments were first normalized to zero mean and unit variance. Although any underlying pulse component was filtered out of the noise segments, some incidental correlation remained between the signal and noise segments, giving sources which are not completely independent.

Signals were then mixed to give the inputs to the ICA:

$$\mathbf{x}(k) = \begin{bmatrix} SNR_{in} & 1 \\ SNR_{in}m_A & m_V \end{bmatrix} \begin{bmatrix} s(k) \\ n(k) \end{bmatrix} = \mathbf{A}\,\mathbf{s}(k)$$

where SNR_{in} is the input SNR, $s(k)$ is the clean segment, $n(k)$ is the noisy segment, and m_A and m_V reflect the oxygen saturation of the arterial and venous blood. This assumes that the pulsatile signal is pure arterial blood and that the interference appears as the motion of venous blood.

The outputs are checked to see if a sign change is appropriate. The output corresponding to the clean signal is identified by its higher correlation with the clean segment. The outputs **z** and unmixing matrix **B** are rearranged so that the first Independent Component corresponds to the clean signal.

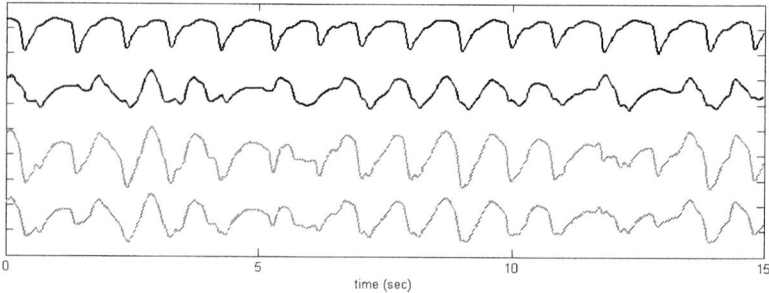

Fig. 1. Clean arterial pulse source (top trace), noise source (second trace), and mixed signal and noise sources, simulating absorption measurements at two wavelengths (bottom two traces). $SNR_{in} = 0$dB.

2.3 Evaluating Results

The separation was evaluated by estimating the change in Mutual Information [12] relative to the original sources. Since

$$I(\mathbf{z}) = \sum_i H(z_i) - H(\mathbf{s}) - \log(|\det(\mathbf{BA})|)$$

and

$$I(\mathbf{s}) = \sum_i H(s_i) - H(\mathbf{s})$$

then

$$\Delta I = I(\mathbf{z}) - I(\mathbf{s})$$
$$= \sum_i [H(z_i) - H(s_i)] - \log(|\det(\mathbf{BA})|)$$

Entropies $H(z_i)$ and $H(s_i)$ were estimated from histograms of the data, using a rectangle approximation for the probability distribution.

Another criterion for evaluating separation was the calculation of the SNR of the first Independent Component. This was determined from the net mixing and unmixing matrix

$$\mathbf{BA} = \begin{bmatrix} b_{11}SNR_{in} + b_{12}SNR_{in}m_A & b_{11} + b_{12}m_V \\ b_{21}SNR_{in} + b_{22}SNR_{in}m_A & b_{21} + b_{22}m_V \end{bmatrix}$$

thus giving the SNR of the Independent Component corresponding to the clean signal:

$$SNR_1 = SNR_{in} \frac{b_{11} + b_{12}m_A}{b_{11} + b_{12}m_V}.$$

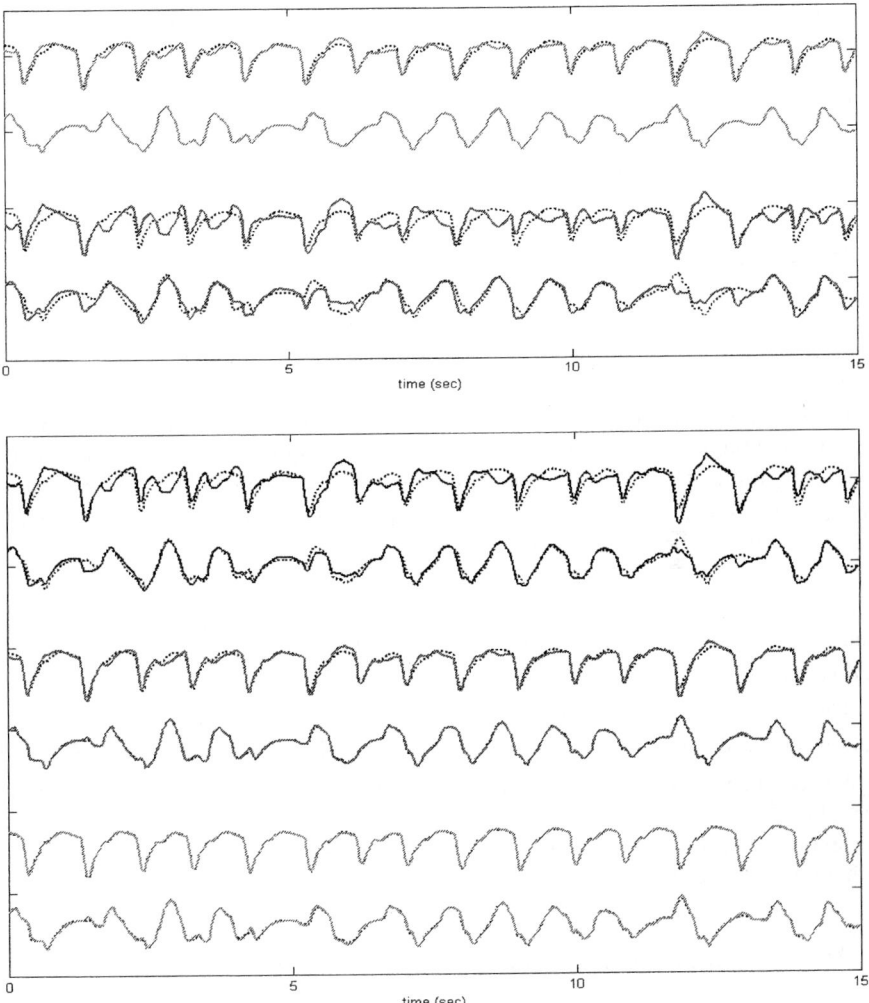

Fig. 2. a) Results from SVD PCA at SNR_{in} = -40dB (top two traces) and SNR_{in} = 0dB (bottom two traces). Signal and noise sources are shown in dashed lines. b) Results from 4° ICA (top two traces), 3° ICA (middle two traces), and ICA based on the 3rd-order statistics of the derivative (bottom two traces). ICA results are independent of SNR_{in}.

3 Results

3.1 An Example of ICA of Pulse Oximetry Signals

Fig. 1 shows clean and noisy signal segments and the mixed signal and noise segments, simulating oximetry signals from two wavelengths. In Fig. 2(a), we see an example of the dependence of the SVD result on SNR_{in}. In Fig. 2(b), ICA results are presented for this example. For these simulations, ICA results are independent of SNR_{in}.

3.2 Overall Performance

Fig. 3 displays the estimated change in Mutual Information ΔI for the original, unmixed segments (dashed line) and for the various techniques assessed here. The mean ΔI over all 48 pairs of signal and noise segments is shown for a range of values of SNR_{in}.

Fig. 4 shows the mean SNR of the component selected as the arterial signal (SNR_1). Again, the mean is taken over the 48 pairs of signal and noise segments. The dashed line indicates $SNR_1 = SNR_{in}$, below which the separation techniques fail to improve SNR over what is available at the input $x_1(t)$.

Below about −20 dB, SVD may slightly outperform ICA on the undifferentiated signals, although this difference was not found to be highly significant ($p>0.01$). The low-SNR_{in} limit of SNR_1 was found to be due to the incidental 2nd-order correlation between the signal and noise sources. When this correlation is regressed out prior to mixing, SNR_1 can be observed to increase as $1/SNR_{in}$ at low SNR_{in}. For actual signals and interference, though, the assumption of independent sources is not completely correct over a finite time window, so such correlation should be present in the simulation. The location of the minimum SNR_1 for SVD is also dependent upon the arterial and venous oxygen saturations.

The ΔI and SNR_1 for 3° and 4° ICA showed no significant difference when the signals were not differentiated. When based on the derivative, though, the SNR_1 for 3° ICA was much higher than for the other techniques ($p<0.001$). ΔI was lower than for the other ICA methods, but less significantly ($p<0.02$). Below about −20dB, there was no significant difference between ΔI for SVD and 3° ICA based on the derivative, although SNR_1 was significantly higher ($p<0.01$).

4 Discussion and Conclusion

The ability of SVD, 3° ICA, 3° ICA of the derivative, and 4° ICA to separate mixed source signals was evaluated using two criteria: ΔI, the estimated change in Mutual Information relative to the sources, and SNR_1, the SNR of the component that corresponds to the clean data source. Note that SNR_1 does not reflect the ability of the algorithm to correctly reproduce the noise source, but that identifying the noise source is important to achieving a low Mutual Information at the outputs.

SVD showed similar results from the two criteria, both strongly dependent on SNR_{in}. As SNR_{in} decreases below about 0 dB, ΔI and SNR_1 actually improve as the noise source dominates one Principal Component and the other component is identified as the signal source. For the ICA techniques, the initial rotation of the PCA result is immaterial, and the final ICA results are independent of SNR_{in}. Still, SVD is comparable to or even slightly better than ICA on undifferentiated signals at very low SNR_{in}.

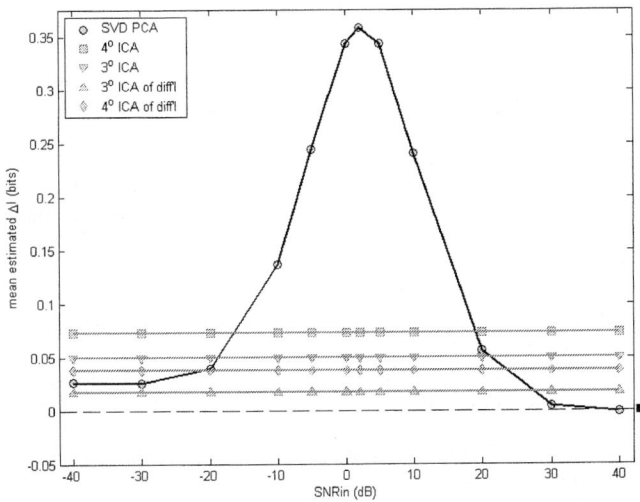

Fig. 3. Mean estimated increase in Mutual Information as a function of input SNR.

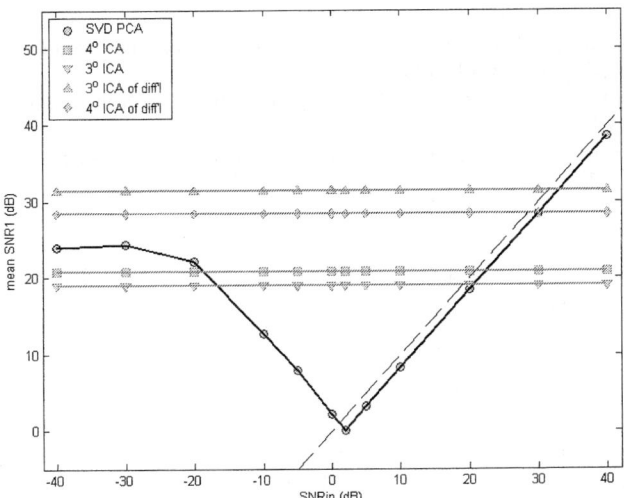

Fig. 4. Mean SNRs of the output corresponding to the clean arterial signal.

Furthermore, when the signals were not differentiated, 3° ICA offered no significant advantage over 4° ICA. However, when the time-differentiated signals were used to determine the unmixing coefficients, 3° ICA gave a much higher SNR_1.

The value of ICA in pulse oximetry has been demonstrated here by an illustrative example and by evaluating the separation of simulated signal-noise mixtures for the estimated change in Mutual Information of the outputs and the SNR of the arterial signal component. Because the signal of interest resembles a sawtooth wave, taking the derivative greatly enhances its skew. Because differentiation does not have the same effect on the skew of the interference sources, third-order ICA of the derivative signal has a clear advantage over other techniques for this application.

References

1. R. Vigário et al., "Independent Component Analysis for Identification of Artifacts in Magnetoencephalographic Recordings", in *Neural Information Processing Systems 10*, pp. 145-151, 1998.
2. A. Bell and T.J. Sejnowski, "An Information-Maximization Approach to Blind Separation and Blind Deconvolution", *Neural Computation 7*, no. 6, pp. 1004-1034, 1995.
3. S. Makeig et al., "Independent Component Analysis of Electroencephalographic Data", in *Advances in Neural Information Processing Systems 8*, pp. 145-151, 1996.
4. J.W. Severinghaus and J.F. Kelleher, "Recent Developments in Pulse Oximetry," *Anesthesiology*, vol. 76, no. 6, pp. 1018-1038, June 1992.
5. R.R. Kirby, R.W. Taylor, J.M. Civetta, *Handbook of Critical Care*, 2nd ed., Lippincott-Raven, Philadelphia, PA, USA: 1997, pp. 135-138.
6. C.R. Baker, Jr and T.J. Yorkey, "Method and Apparatus for Estimating Physiological Parameters Using Model-Based Adaptive Filtering", U.S. Patent No. 5,853,364, December 29, 1998.
7. J.E. Corenman et al., "Method and Apparatus for Detecting Optical Pulses", U.S. Patent No. 4,911,167, March 27, 1990.
8. P.F. Stetson, "Blind Source Separation of Pulse Oximetry Signals", U.S. Patent No. 6,701,170, March 2, 2004.
9. D.B. Swedlow and R.S. Potratz, "Oximeter with motion detection for alarm modification", U.S. Patent No. 5,368,026, November 29, 1994.
10. P. Comon, "Independent Component Analysis, a New Concept?", *Signal Processing*, vol. 36, no. 3, pp. 287-314, April 1994.
11. J.F. Cardoso and A. Souloumiac, "Blind Beamforming for Non-gaussian Signals", *IEE Proceedings F*, vol. 140, no. 6, pp. 362-370, Dec. 1993.
12. T.M. Cover and J.A. Thomas, *Elements of Information Theory*, Wiley-Interscience, 1991.

Mixing Matrix Pseudostationarity and ECG Preprocessing Impact on ICA-Based Atrial Fibrillation Analysis

José Joaquín Rieta, César Sánchez,
Juan Manuel Sanchis, Francisco Castells, and José Millet

Bioengineering, Electronics, Telemedicine and Medical Computer
Science Research Group. Valencia University of Technology
Carretera Nazaret–Oliva s/n, 46730, Gandía (Valencia) Spain
{jjrieta,jmsanch,fcastells,jmillet}@eln.upv.es

Abstract. In this work two relevant considerations in the ICA-based estimation of atrial activity (AA) in atrial fibrillation (AF) episodes from real electrocardiogram (ECG) recordings are presented. Firstly, the impact of low-pass filtering preprocessing on the extraction quality of AA is analyzed, showing an average improvement over 17% in spectral concentration (SC) when low-pass filtering is applied after ICA with respect to the application of the same filtering before ICA. Secondly, it is demonstrated that the ICA mixing matrix obtained from one AF segment can also be used to estimate the AA present in different segments of the same recording, thus proving the pseudostationarity of the mixing matrix. Results over 32 AF segments show a mean cross-correlation of $\overline{R}_{dp} = 81.5\%$ between the directly estimated AA and the estimated using presudostationarity. Changes in spectral concentration from one case to the other ($\overline{\Delta SC}_{dp} = 1.4\%$) are negligible.

1 Introduction

One of the most important research areas where independent component analysis (ICA) techniques have proved their success is in biomedical engineering [1], with a relevant increase of novel applications during the past years. Regarding the electrocardiogram (ECG), it is well known the extraction of the fetal ECG from maternal recordings [2], the separation of breathing artifacts and other disturbances [3], analysis of ST segments for ischemia detection [4], identification of humans using the ECG [5], ventricular arrhythmia detection and classification [6] and the study of atrial fibrillation (AF).

AF is the most common sustained arrhythmia encountered by clinicians and occurs in approximately 0.4% to 1.0% of the general population. Its prevalence increases with age, and up to 10% of the population older than 80 have been diagnosed with AF [7]. ICA and methods related to blind signal separation have also been applied to AF. In this sense, principal component analysis (PCA) has been used both to extract the atrial activity (AA) from the 12–lead surface ECG in patients with AF [8] and to measure the degree of local organization of this arrhythmia [9]. Regarding ICA, it has also been applied for the extraction of AA in AF episodes from the surface ECG [10, 11], the suppression of artifacts from internal epicardial recordings [12] and the discrimination among supraventricular arrhythmias [13]. This contribution presents the impact of traditional filtering steps, used for ECG preprocessing, that may decrease ICA performance in the estimation of AA in AF episodes. Next the paper shows the empirical demonstration of the instantaneous linear mixing model of an AF recording, through the corroboration of the ICA mixing matrix pseudostationarity (MMPS).

2 Methods

The suitability of ICA to extract the AA from the ECG in patients with AF has been already demonstrated [11] and the results compared to other AA estimation methods [14]. Nevertheless, the ICA-based optimal estimation of AA still remains as an open issue in continuous development [15]. The following sections deal with these considerations, starting with a short description of AF and its ICA model.

2.1 Atrial Fibrillation and ICA Suitability

The manifestation of AF is characterized by uncoordinated atrial activation with consequent deterioration of atrial mechanical function [7]. AF occurs when the electrical impulses in the atria degenerate into a chaotic pattern, resulting in an irregular and rapid heartbeat due to the unpredictable conduction of these impulses across the atrioventricular node [7]. On the ECG, AF is described by the replacement of P waves by fibrillatory waves that vary in size, shape, and timing, being the topic of intensive research because it is the most common sustained cardiac arrhythmia [7]. ICA can be applied to AF due to the fulfillment of these conditions [16]: independence of the sources, nongaussianity and nonorthogonal observations generated by instantaneous linear mixing of the sources [11]. These considerations can be proved through the study of the electrophysiological mechanisms regarding the generation of AF and the matrix-form solution of the forward problem of electrocardiography [17].

2.2 Impact of ECG Preprocessing

Preprocessing is used over ECG recordings to improve the later analysis or processing stages. The most widely used involves notch filtering, to cancel out mains interference, high pass filtering, to eliminate baseline wandering and low-pass filtering to reduce thermal and muscular noise [18]. Additionally, ICA works much better with low noise data, thus allowing for a better separation of the independent components [19]. On the other hand, linear filtering does not affect the fulfillment of the ICA model [19], hence, having a column vector of observations $\mathbf{x}(t)$ obtained by linearly mixing a column vector of sources $\mathbf{s}(t)$ with a mixing matrix $\mathbf{A}$, the ICA model is $\mathbf{x}(t) = \mathbf{A}\mathbf{s}(t)$ [19].

Filtering the set of observed signals to obtain a new observation vector $\mathbf{x}_f(t)$, will generate the ICA model $\mathbf{x}_f(t) = \mathbf{A}\mathbf{s}_f(t)$, where the mixing matrix $\mathbf{A}$ is the same as before and the source column vector $\mathbf{s}_f(t)$ corresponds to the set of sources $\mathbf{s}(t)$ filtered with the same filtering applied to $\mathbf{x}(t)$.

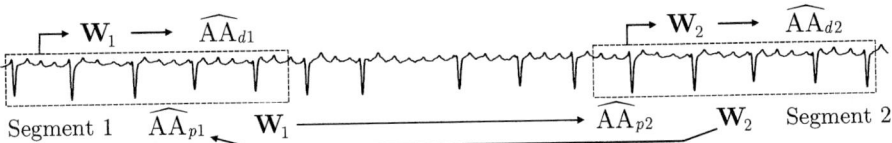

Fig. 1. Proposed methodology for the mixing matrix pseudostationarity analysis. In each AF segment the AA can be directly estimated ($\widehat{AA}_{d1}$ and $\widehat{AA}_{d2}$) or indirectly via the pseudostationarity property ($\widehat{AA}_{p1}$ and $\widehat{AA}_{p2}$).

A problem with low-pass filtering is that it reduces the information in the data, since high-frequency features of the data are lost. Hence, this information reduction may involve a reduction of independence [19]. In addition, low-pass filtering performs some kind of averaging over the data, and sums tend to increase Gaussianity [19], thus, decreasing ICA performance.

2.3 Mixing Matrix Pseudostationarity

The signal vector from the AF recording can be identified with the observations in ICA, the set of sources being composed of the independent atrial and ventricular activities and other nuisance signals. The mixing matrix entries will can be associated to the transfer coefficients relating the potentials from the heart towards the body surface [17]. Apart from the direct observation or the comparison to other techniques [14], one way to corroborate that an AF recording satisfies the ICA model, may consist of verifying the similarities of the mixing (or separation) matrices between two different segments. If the matrix is preserved then it will be reasonable to consider that AF fulfills the ICA model. Nonetheless, the study and proper electrophysiological interpretation of the variations across the 144 entries of a 12 × 12 matrix is not easy to perform.

As an alternative each AF recording can be divided into two non-overlapped segments (see Fig. 1). In the first segment we can define the directly estimated atrial activity, $\widehat{AA}_{d1}$, as the activity obtained after performing ICA over it. This will also give us the mixing and separation matrices for the first segment, $\mathbf{A}_1$ and $\mathbf{W}_1$, respectively, where $\mathbf{W}_1 = \mathbf{A}_1^{-1}$. With the same procedure, it is possible to obtain $\widehat{AA}_{d2}$ and matrices $\mathbf{A}_2$ and $\mathbf{W}_2$ for the second segment. Hence, the verification of the MMPS will consist of applying the separation matrices $\mathbf{W}_1$ and $\mathbf{W}_2$ over the second and first ECG segments, respectively, to obtain the activities $\widehat{AA}_{p2}$ and $\widehat{AA}_{p1}$ (see Fig. 1). Evaluating the similarity between $\widehat{AA}_{d1}$ and $\widehat{AA}_{p1}$, for segment 1, and $\widehat{AA}_{d2}$ and $\widehat{AA}_{p2}$, for segment 2, it will be possible to assess the preservation of the matrix. Considering the real world limitations, it is reasonable to define this behavior as MMPS.

2.4 Measurement of Atrial Activity Estimation Quality

The estimation of AA from real AF recordings represents an inverse problem where true sources are impossible to observe. Considering the spectral mor-

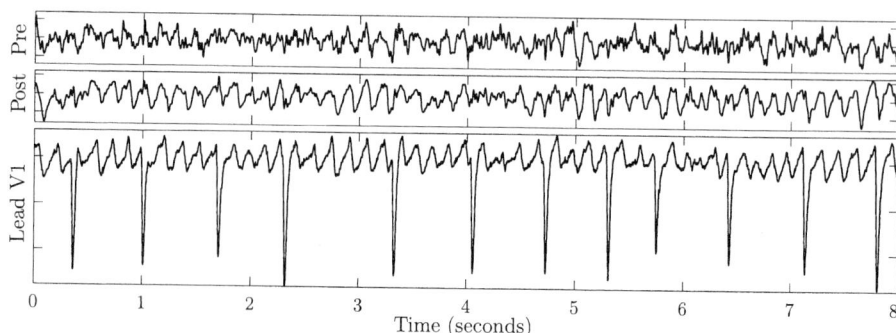

Fig. 2. Estimated AA waveforms from patient #10 for the pre-ICA and post-ICA low-pass filtering. Lead V1 from the same patient is included for comparison purposes.

phology of AA, with a very pronounced peak, no harmonics and insignificant amplitudes above 15Hz [8, 20], it is possible to define a performance extraction index to evaluate the AA extraction quality based on the spectral concentration, that can be defined as [15]

$$SC = \left(\int_{0.82f_p}^{1.17f_p} P_{AA}(f)df \right) \bigg/ \left(\int_{0}^{\frac{f_s}{2}} P_{AA}(f)df \right) \quad (1)$$

where f_p is the frequency of the AF main peak, P_{AA} is the power spectral density (PSD) of the AA and f_s is the sampling frequency. SC can evaluate the spectral variation due to the presence of other nuisance signals outside the main peak band. Hence, for a concrete AA, a higher SC will indicate a more efficient elimination of non-AA components. The PSD was obtained via the Welch–WOSA method, discarding the content above 20Hz due to its low contribution.

3 Results and Discussion

The low-pass filtering impact was by its application before and after ICA. The study was carried out over the authors' own database comprising 12-lead AF recordings from 16 patients sampled at 1kHz with segments of 8 seconds. This recording time is long enough because it includes several cardiac cycles and other human events, like breathing, that may affect the transfer coefficients. The low-pass filtering applied in this study was a linear phase Chebyshev type II digital filter, with no ripple in the pass-band and 40dB ripple in the stop-band, the cut-off frequency being 70Hz. Next, the FastICA algorithm was used due to its robust performance and fast convergence [19]. Fig. 2 plots the AA filtering before (pre) and after (post) the ICA stage. The same Fig. also shows lead V1 from the ECG, because it is considered as the lead with higher AA contribution [7]. As can be seen, the post-ICA AA is much more approximated to the AA contained in the ECG. Fig. 3 plots the PSD of the AA signals from Fig. 2. The main atrial frequency and spectral morphology is quite similar for both signals, but

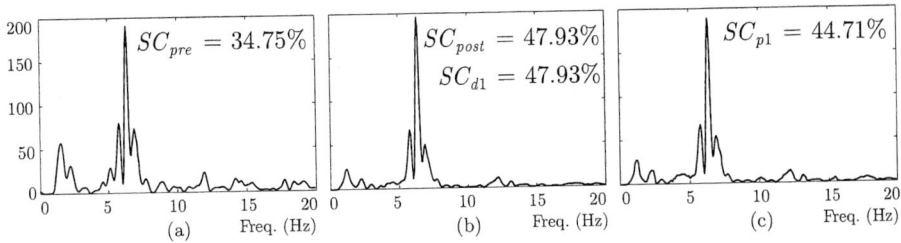

Fig. 3. PSD and SC for the AA in the first segment of patient #10 (Fig 2). (a) Pre-ICA low-pass filtering. (b) Post-ICA low-pass filtering and directly estimated activity $\widehat{AA}_{d1}$ using $\mathbf{W}_1$. (c) Indirectly estimated activity $\widehat{AA}_{p1}$ via $\mathbf{W}_2$ from the second segment.

Table 1. Percentages of SC for each patient and mean values for the whole database. SC_{pre} is for low-pass filtering previous to ICA and SC_{post} for the reverse methodology.

	Pat.01	Pat.02	Pat.03	Pat.04	Pat.05	Pat.06	Pat.07	Pat.08
SC_{pre}	30.4	37.9	41.9	45.8	49.9	34.2	31.1	21.6
SC_{post}	63.0	44.9	57.7	47.1	84.5	47.0	45.4	55.8
$\Delta_{post-pre}$	32.6	6.9	15.8	1.3	34.5	12.7	11.2	34.2

	Pat.09	Pat.10	Pat.11	Pat.12	Pat.13	Pat.14	Pat.15	Pat.16
SC_{pre}	36.6	34.7	38.2	40.5	43.0	28.1	33.3	36.7
SC_{post}	43.3	47.9	49.9	65.4	61.3	42.2	44.3	62.0
$\Delta_{post-pre}$	6.7	13.1	11.6	24.9	18.3	14.1	11.0	25.3

$$\overline{SC_{pre}} = 36.7 \quad \overline{SC_{post}} = 53.9 \quad \overline{\Delta_{post-pre}} = 17.1$$

the pre-ICA estimation (Fig. 3.a) shows larger spectral components below 5Hz and above 10Hz that are notably reduced with post-ICA (Fig. 3.b). Hence, the SC for this latter case is higher and the AA extraction performance is better.

Table 1 shows preprocessing (SC_{pre}) and postprocessing (SC_{post}) values for all the patients, $\Delta_{post-pre}$ being the percentage of increase in SC. Similar analysis have been performed regarding notch filtering and high-pass filtering. The results, though are not presented in this contribution, show an irrelevant impact over the ICA separation performance.

To study the MMPS a recording of 20 seconds in length was selected, then, FastICA was applied over the first 8 seconds (segment 1), giving us $\widehat{AA}_{d1}$, $\mathbf{A}_1$ and $\mathbf{W}_1$. Segment 2 was comprised of the last 8 seconds, obtaining $\widehat{AA}_{d2}$, $\mathbf{A}_2$ and $\mathbf{W}_2$. The post-ICA low-pass filtering strategy was selected and applied over both segments. Next, by using the separation matrices, the AA estimation was obtained indirectly to verify the MMPS: $\widehat{AA}_{p1}$ was extracted using $\mathbf{W}_2$ and $\widehat{AA}_{p2}$ with $\mathbf{W}_1$, respectively. Fig. 4 shows the result over the AF segment in Fig. 2. The similarity obtained is significant, thus corroborating the MMPS. Also for this AA pair, the cross-correlation percentage is 81.7%.

Regarding SC for the AA pairs, Fig. 3.b plots the PSD and SC for $\widehat{AA}_{d1}$ from Fig. 4 and the result for $\widehat{AA}_{p1}$ can be seen in Fig. 3.c. Comparing both Figs.

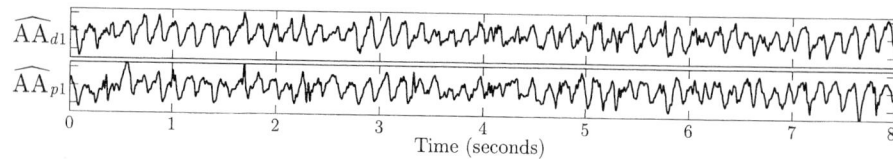

Fig. 4. Waveform of the directly estimated atrial activity $\widehat{AA}_{d1}$ from the first segment of patient #10 (see Fig. 2) and the same activity obtained via the MMSP $\widehat{AA}_{p1}$.

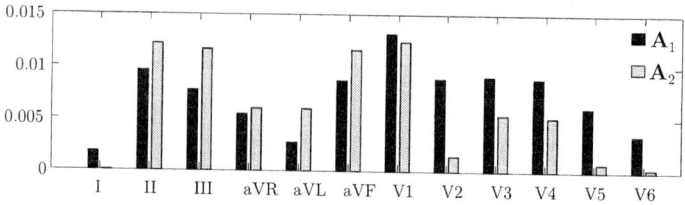

Fig. 5. Column comparison for the entries, in absolute value, of the mixing matrices $\mathbf{A}_1$ and $\mathbf{A}_2$ related to the AA source from the ECG recording of Fig. 2.

the differences in spectral morphology and main peak frequency are negligible. There is a variation in the main peak amplitude and the spectral concentration (SC_{d1} vs SC_{p1}) quite acceptable when dealing with real AF recordings.

Fig. 5 plots the columns of the mixing matrices $\mathbf{A}_1$ and $\mathbf{A}_2$ associated to the AA source. Observe the similarities in the leads with largest AA contribution. Finally Fig. 6 concentrates the MMPS results for the 16 patients in the database. Above each patient number there are six bars, the first three ones for the first AF segment and the others for the second segment. Within each bar group, the first bar indicates SC_{d1} in percentage, i.e., the spectral concentration of $\widehat{AA}_{d1}$. The second bar is SC_{p1} associated to $\widehat{AA}_{p1}$ and the third bar is cross-correlation percentage, R_{dp1}. The same is applicable for the other bar group on the right, but for the second segment. The differences between the SC_d and SC_p pairs are generally small, being the cross-correlation below 75% in only 6 of the 32 analyzed cases. The mean difference between SC_d and SC_p is $\overline{\Delta SC}_{dp} = 1.47\%$ in spectral concentration. Regarding the correlation value, its mean for the same 32 situations is $\overline{R}_{dp} = 81.56 \pm 10.74\%$, thus reinforcing the large similarity between $\widehat{AA}_d$ and $\widehat{AA}_p$ for all the cases analyzed.

Note that this application assumes the fulfillment of the ICA model, hence, it will only be possible to derive the spatial filters (mixing matrix) and the sources from the ECG, when the physical sources associated to heart's activity are spatially stationary [1]. Contraction of the atria during fibrillation, the ventricles in the cardiac cycle or any other relative movement from sources to observations, could violate the assumption of spatial stationarity. The authors consider that these variations do not affect significantly the ICA model in AF: firstly, results of the AA have demonstrated its validity [10,11], secondly, the main atrial frequency of AA obtained using ICA is in agreement with the results obtained via other accepted AA extraction techniques [20]. Finally, the MMPS corroboration gives the definitive support to say that the 12-lead ECG of an AF recording fulfills the instantaneous linear mixing ICA model.

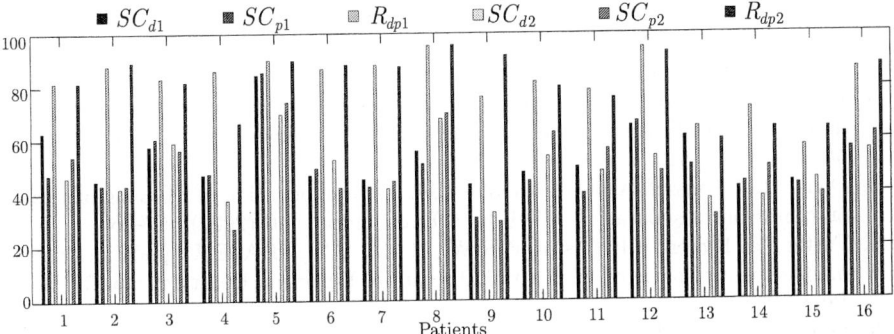

Fig. 6. MMPS analysis results for the whole patient database.

4 Conclusions

This contribution has demonstrated that preprocessing has a significant impact on the separation performance of ICA-based AA estimation. The study has proved that low-pass filtering, though itself is a linear operation, involves a data reduction that decreases the quality of the ICA-based AA estimation. To solve this problem, the ICA approach has to be applied before the low-pass filtering and then, any other post-ICA processing could be performed over the data. Besides, the impact of notch and baseline wander filtering is not relevant.

Moreover, it has been demonstrated that the AA from one segment can be recovered with the separation matrix from other segment. This observation, defined as the MMPS, gives the definitive support to the fulfillment of the ICA model for AF. In addition, this fact may imply the birth of other studies on the evolution and periodicity of the mixing matrix entries across the cardiac cycle or the patient's own breathing, not only in atrial fibrillation, but in other supraventricular arrhythmias and cardiac pathologies where atrial and ventricular activities can be regarded as decoupled or independent.

Acknowledgements. This work was partly funded by the research incentive program of the Valencia University of Technology and TIC2002-00957. The authors would like to thank the cardiologists from the Univeristary Clinical Hospital of Valencia (Spain), for their clinical advice and kind help in obtaining the signals.

References

1. Jung, T.P., Makeig, S., Lee, T.W., McKeown, M.J., Brown, G., Bell, A.J., Sejnowski, T.J.: Independent component analysis of biomedical signals. International Conference on Independent Component Analysis and Blind Signal Separation (ICA) **2** (2000) 633–644
2. Zarzoso, V., Nandi, A.K.: Noninvasive fetal ECG extraction: blind separation versus adaptive noise cancellation. IEEE Trans. Biomed. Eng **48** (2001) 12–18
3. Barros, A.K., Mansour, A., Ohnishi, N.: Adaptive blind elimination of artifacts in ECG signals. International Workshop on Independence & Artificial Neural Networks (I&ANN'98) (1998) 1380–1386

4. Stamkopoulos, T., Diamantaras, K., Maglaveras, N., Strintzis, M.: ECG analysis using nonlinear PCA neural networks for ischemia detection. IEEE Transactions on Signal Processing **46** (1998) 3058–3067
5. Biel, L., Pettersson, O., Philipson, L., Wide, P.: ECG analysis: A new approach in human identification. IEEE Transactions on Instrumentation and Measurement **50** (2001) 808–812
6. Owis, M.I., Youssef, A.B.M., Kadah, Y.M.: Characterisation of ECG signals based on blind source separation. Med. Biol. Eng Comput. **40** (2002) 557–564
7. Fuster, V., Ryden, L.E., Asinger, R.W., et al.: ACC/AHA/ESC guidelines for the management of patients with atrial fibrillation. European Heart Journal **22** (2001) 1852–1923
8. Langley, P., Bourke, J.P., Murray, A.: Frequency analysis of atrial fibrillation. IEEE Computers in Cardiology **27** (2000) 65–68, Boston, MA, USA
9. Faes, L., Nollo, G., Kirchner, M., Olivetti, E., et al.: Principal component analysis and cluster analysis for measuring the local organisation of human atrial fibrillation. Med. Biol. Eng Comput. **39** (2001) 656–663
10. Rieta, J.J., Zarzoso, V., Millet, J., Garcia, R., Ruiz, R.: Atrial activity extraction based on blind source separation as an alternative to QRST cancellation for atrial fibrillation analysis. IEEE Computers in Cardiology **27** (2000) 69–72, Boston.
11. Rieta, J.J., Castells, F., Sanchez, C., Igual, J.: ICA applied to atrial fibrillation analysis. International Conference on Independent Component Analysis and Blind Signal Separation (ICA) **4** (2003) 59–64, Nara, Japan
12. Liu, J.H., Kao, T.: Removing artifacts from atrial epicardial signals during atrial fibrillation. International Conference on Independent Component Analysis and Blind Signal Separation (ICA) **4** (2003) 179–183, Nara, Japan
13. Rieta, J.J., Millet, J., Zarzoso, V., Castells, F., Sanchez, C., Garcia, R., Morell, S.: Atrial fibrillation, atrial flutter and normal sinus rhythm discrimination by means of blind source separation and spectral parameters extraction. IEEE Computers in Cardiology **29** (2002) 25–28, Memphis, TN
14. Langley, P., Rieta, J.J., Stridh, M., Millet, J., Sornmo, L., Murray, A.: Reconstruction of atrial signals derived from the 12-lead ECG using atrial signal extraction techniques. IEEE Computers in Cardiology **30** (2003) 129–132, Greece.
15. Castells, F., Ruiz, R., Rieta, J.J., Millet, J.: An integral atrial wave identification approach based on spatiotemporal source separation: Clinical validation. IEEE Computers in Cardiology **30** (2003) 717–720, Thessaloniki, Greece.
16. Cardoso, J.F.: Blind signal separation: Statistical principles. Proceedings of the IEEE **86** (1998) 2009–2025
17. Rieta, J.J., Castells, F., Sanchez, C., Moratal, D., Millet, J.: Bioelectric model of atrial fibrillation: Applicability of blind source separation techniques for atrial activity estimation in atrial fibrillation episodes. IEEE Computers in Cardiology **30** (2003) 525–528, Thessaloniki, Greece
18. Tompkins, W.J.: Biomedical digital signal processing: C languaje examples and laboratory experiments for the IBM PC. Prentice Hall, New Jersey (1993)
19. Hyvarinen, A., Karhunen, J., Oja, E.: Independent Component Analysis. John Wiley & Sons, Inc. (2001)
20. Langley, P., Stridh, M., Rieta, J.J., Sörnmo, L., Millet, J., Murray, A.: Comparison of atrial rhythm extraction techniques for the detection of the main atrial frequency from the 12-lead ECG in atrial fibrillation. IEEE Computers in Cardiology **29** (2002) 29–32, Memphis, (TN)

'Signal Subspace' Blind Source Separation Applied to Fetal Magnetocardiographic Signals Extraction

Giulia Barbati[1], Camillo Porcaro[1], and Carlo Salustri[2]

[1] Center of Medical Statistics and Information Technology, AfaR, San Giovanni Calibita, Fatebenefratelli Hospital, Isola Tiberina, 00186 Rome, Italy
[2] Institute of Cognition Science and Technology (CNR) - MEG Unit, San Giovanni Calibita, Fatebenefratelli Hospital, Isola Tiberina, 00186 Rome, Italy

Abstract. In this paper we apply Independent Component Analysis to magnetocardiographic data recorded from the abdomen of pregnant women. In particular, we include a dimensionality reduction in the 'Cumulant Based Iterative Inversion' algorithm to achieve a 'signal subspace' subdivision, which enhances the algorithm's efficacy in resolving the signals of interest from the recorded traces. Our results show that the proposed two-step procedure is a powerful means for the extraction of the cardiac signals from the background noise and for a sharp separation of the baby's heart from the mother's.

1 Introduction

Rate and waveform of the fetal heart are fundamental indicators of how a pregnancy is proceeding. Unfortunately, the electric (ECG) and magnetic (MCG) signals produced by the baby's heart are always mixed up with those produced by the mother's heart and both are embedded in a background noise generated by breathing, muscle contractions, bowel movements, and other biological and electronic causes. Moreover, due to the variability of the fetus' position, the fetal heart wave is in some cases barely distinguishable from the mother's. For all these reasons, an efficient extraction of the heart signals from the background noise and a sharp separation of maternal and fetal signals are prerequisites for an assessment of the fetus' condition.

Independent Component Analysis (ICA) is a statistical technique, which resolves the unknown individual contributions to a signal from recordings of their mixtures. ICA has being successfully applied in diverse fields, as for example in neuroscience [1,2], and recently to explore the feasibility of analyzing fetal electrocardiograms [3,4,5]. In this paper we applied ICA to magnetocardiographic data recorded from the abdomen of pregnant women. The first step for noninvasive fetal ECG/MCG extraction is usually separating the signal "subspace" from the noise "subspace", i.e., separating maternal and fetal heartbeats from the noise; in a second step the cardiac signals are separated from each other. Thus, the first step implies the definition of a 'signal subspace' and of its dimensionality, on the basis of the number of recording channels. To date, in biomedical applications, not much attention seems to have been devoted to the selection procedure of the signal subspace dimension. In this paper we address this problem by presenting a practical procedure that integrates a semi-automatic 'dimensionality reduction' step in an already existing ICA algorithm.

In particular, we modify the 'Cumulant Based Iterative Inversion' algorithm[1] [6,7] to include a 'Signal Subspace' subdivision, which allows the user to control the number of components to be estimated in an 'interactive' fashion by rounding the eigenvalues distribution. Our results prove that this two-step procedure is a powerful means for the extraction of MCG signals from the background noise and for a sharp separation of the baby's heart from the mother's.

2 Blind Source Separation

If we distribute an array of sensors over the mother's abdomen, each sensor will record a weighted mixture of the magnetic fields generated by multiple simultaneous sources (both biological and artifactual), the weights depending on the sensors locations relative to the sources. The general problem of identifying and separating simultaneous sources of activity, about which we have no information, goes under the name of Blind Source Separation (BSS). If we assume that N sources generate N fields $s_1(t)$ ...... $s_N(t)$, whose M weighted mixtures $x_1(t)$ ...... $x_M(t)$ are recorded by $M \geq N$ distributed sensors, our recordings can be written in matrix notation: $\mathbf{x}(t) = \mathbf{A}\mathbf{s}(t)$.

The MxN matrix **A**, which is unknown to us, is called the *'mixing'* matrix since it mixes up the N sources **s**(t), which in turn cannot be measured directly. We can think of the vector **s**(t) as representing both 'interesting' sources (in our case maternal and fetal heartbeats) and 'uninteresting' interferences, as for example artefacts (fetal movements, uterus contractions, etc.) and system interferences; **n**(t) is an M-dimensional unknown vector representing additive Gaussian noise, both instrumental and environmental, and it is assumed to be independent of **s**(t). Our objective is of course to find the "interesting" components of the source vector **s**(t), and to eliminate or reduce the influence of the noise vector **n**(t).

2.1 Dimensionality Reduction

We started our procedure by 'whitening' the data. Whitening is a process that transforms the original data into a new set of data that are uncorrelated, have zero mean and unit variance. Generally, data are whitened by Singular Value Decomposition (*svd*) of the original data's covariance matrix $R_{xx} = E\{xx^T\}$:

$$svd(R_{xx}) \rightarrow \Lambda^{-1/2} V^T \ x = Qx = \tilde{x}. \qquad (1)$$

where the MxM matrix **V** contains the eigenvectors associated with the eigenvalues of $\Lambda = diag\{\lambda_1 \geq \lambda_2 \geq ... \geq \lambda_m\}$ in descending order and the whitening matrix **Q** is such that $R_{\tilde{x}\tilde{x}} = E\{\tilde{x}\tilde{x}^T\} = I_m$.

Instead of the standard whitening approach described by eq.(1), we used a 'quasi-whitening' algorithm proposed in [8], applicable to cases in which the noise covariance matrix can be modeled as: $R_{nn} = \sigma_n^2 I_m$ and the noise variance is relatively small

---
[1] Implemented in the package ICALAB of Cichocki and Amari (version 2.0, October 2003, freely downloadable at: www.bsp.brain.riken.jp) under the name ERICA: Equivariant Robust ICA-based on cumulants.

(i.e. the SNR is high above some threshold). We judged that our data matched these conditions since both the mother's and the fetus' heartbeats generally appeared quite visible in our traces, despite the noise. We assumed that the first k of the m eigenvalues in Λ form a k-dimensional signal subspace, whereas the remaining m-k define a (m-k)-dimensional noise subspace; k is of course to be estimated. The noise variance σ_n^2 is estimated as the mean value of the m-k 'minor' eigenvalues, and the KxM 'quasi-whitening' matrix is defined as:

$$\tilde{Q} = \hat{\Lambda}^{-\frac{1}{2}} V_S^T = (\Lambda_S - \sigma_n^2 I_k)^{-1/2} V_S^T . \qquad (2)$$

where the MxK matrix V_S contains the eigenvectors associated with the eigenvalues of the KxK matrix $\Lambda_S = diag\{\lambda_1 > \lambda_2 > ... > \lambda_k\}$ of the signal subspace, and the k-dimensional data vector can then be expressed as: $\tilde{x} = \tilde{Q}x$. The matrix $\tilde{Q}$ is such that $R_{\tilde{x}\tilde{x}} = E\{\tilde{x}\tilde{x}^T\} \cong I_k$, where the subscript k indicates the order of I. Note that the matrix $R_{\tilde{x}\tilde{x}}$ is not exactly the identity matrix in this quasi-whitening procedure, but this is not a problem since the ICA algorithm we used does not require an orthogonal mixing matrix[2] [7]. Several ways have been proposed to estimate k in the signal processing literature; in [8] it is suggested to find empirically a gap in the eigenvalues distribution of the data covariance matrix; in alternative, the same authors also consider the use of one of two Information Theory criteria, namely Akaike's Information Criterion (AIC) and the Minimum Description Length criterion (MDL). Nevertheless, beside the fact that both these procedures present relevant computational problems, they have been proven to provide estimates of the number of sources that are rough, very sensitive to SNR variations and especially to the number of data samples available [10]. Recently, a 'Rank Detection' (RD) criterion has been proposed to evaluate the above mentioned gap by means of numerical analysis procedures and it has been indicated to give a 'maximally stable' decomposition of a data covariance matrix into signal and noise subspaces [10]. The authors show that RD is insensitive to variations in the SNR and the number of data samples; moreover, they show that RD may indicate whether a classification into signal and noise subspaces is stable and well-conditioned or unstable and ill-conditioned, and they verify this property in microwave radio channels; in the present work we computed AIC, MDL and RD following formulas reported in [9,10]. In our proposal, that we will call 'Eigen-Round' (ER) procedure, we applied the following scheme to estimate k, taking into account the particular characteristics of the data available and including a 'control cycle' on subsequent independent components:

(a) After the M eigenvalues of the data covariance matrix are delivered in decreasing order $\lambda_1 \geq \lambda_2 \geq ... \geq \lambda_M$, where M corresponds to the number of channels, each eigenvalue λ_j is divided by the order of magnitude 10^i of the greatest eigenvalue

---

[2] Note that our objective in this work was not to compare the algorithm chosen by us with others: once k is determined, the separation phase performs in a similar way using other common ICA algorithms, for example FastICA or JADE. Moreover, beside the fact that a comparative measure of performance is difficult to define in real cases, these algorithms require an orthogonal mixing matrix.

λ_1 (for example: if $\lambda_1 = 98765.4321$, then every λ_j is divided by 10^5 i.e. $i=5$). In this way we obtain a new set of eigenvalues $\widetilde{\lambda}_j = \lambda_j / 10^i$.

(b) In this way, each $\widetilde{\lambda}_j$ is rounded up to the *i-th* decimal digit and we check whether and which two eigenvalues are identical:

(b.1) If $\widetilde{\lambda}_j = \widetilde{\lambda}_{j+1}$, then the signal subspace dimension is *k=j-1* and the round up parameter remains *i*. Quasi-whitened data are *k*-dimensional, and we proceed to the ICA phase:

(b.1.1) if the algorithm converges, then we proceed to judge, by visual inspection, the *'quality of the separation'* (quality is defined satisfactory when maternal and fetal heartbeats result positively represented by separated components):

(b.1.1.1) (b.1.1.1) if the quality is satisfactory, the process ends;

(b.1.1.2) (b.1.1.2) if the quality is not satisfactory, then we set *i=i-1* and go back to step (b);

(b.1.2) if the algorithm does not converge, then we set i=i-1 and go back to step (b).

(b.2) If there are no identical eigenvalues, we set i=i-1 and go back to step (b).

Of course this *Eigen*values *Round*ing procedure (based on the selected rounding parameter *i*) differs from other criteria mainly in the fact that it contains an empirical *'interactive'* control cycle (the user moves through the described steps on the basis of his/her judgment of quality and performance) making it a semi-automatic method; attempts to develop mechanisms of *automatic* identification are currently in progress.

3 Results and Conclusions

3.1 Application to Fetal Data

Magnetocardiographic data were recorded from five pregnant women participating in our hospital's pregnancy assistance program, using a 28-channel system used also for magnetoencephalography, placed over the mother's abdomen. The system features 16 first-order axial gradiometers and 9 magnetometers, plus 3 balancing magnetometers for noise cancellation, all coupled to low noise dc-SQUIDs with an overall sensitivity of about 5-7 fT/Hz$^{1/2}$. Gradiometers and magnetometers are characterized by a noise of 5-6 fT/ Hz$^{1/2}$ and 7-9 fT/Hz$^{1/2}$ respectively. The 25 measuring sites were uniformly distributed over the mother's abdomen covering an area of about 180 cm^2. Mothers sat on a comfortable plastic chair inside a magnetically shielded room (Vacuumschmelze GMBH) where the system is located. There was no need for special preparations and all recordings took the minimal technical time. Data were acquired continuously for 4 minutes at 250 Hz sampling rate with a 0.48-64 Hz band pass prefiltering.

3.2 Results

In all five recordings the proposed two-step procedure extracted independent components that could be clearly associated to maternal and fetal cardiac sources. Maternal cardiac signals generally dominated the recorded MCGs, but the fetus/mother amplitude ratios turned out very different from case to case. The number of data samples was large (more than 150000 time points) and, in the absence of any dimensionality reduction, AIC and MDL criteria did not perform useful reduction. RD appeared too 'restrictive', producing signal subspace dimensions from one to four components and not demixing mother and fetus heartbeat (see figures 4 and 6). It must be noted that we deal with two separate subspaces, the signal and the noise ones; one first of all looks for a gap between the signal and the noise but, in doing so, one comes across another gap inside the signal subspace due to the difference in energy between maternal and fetal signals that could be significantly large too (this happened in the vast majority of our cases); consequently, we noted that in applying RD we ran the risk of either seeing the fetal component mixed together with the noise one, or mixed with one of the mother's, which is precisely the opposite of what we wanted to do. In Table 1 we report values of k for the four criteria and the five MCG cases: as mentioned above, AIC and MDL did not perform dimensionality reduction and therefore resulted useless to eliminate noise; for RD and our ER procedure we estimated the corresponding k components and verify the quality of separation. Statistical evaluations on bigger MEG databases are currently under development.

Table 1. Signal subspace dimensions selected by AIC, MDL, RD and ER for the five cases.

	Case A	Case B	Case C	Case D	Case E
AIC	25	25	25	25	25
MDL	25	25	25	25	25
RD	4	3	1	1	2
ER	9	11	8	15	10

We show figures of two representative cases: case A, in which the fetus' cardiac amplitude is comparable to the mother's, and case B, in which the fetus' signal is barely visible in the recorded traces.

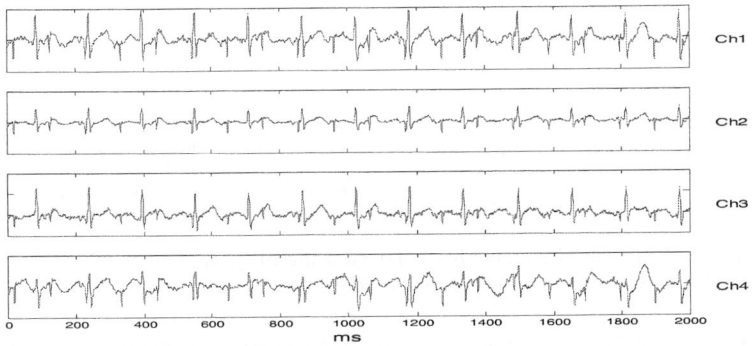

Fig. 1. Case A: a time excerpt of the first four recording channels in which maternal and fetal heartbeats are clearly visible.

In case A the high-amplitude QRS complexes of both mother and fetus allowed an easy eye identification of the presence of both cardiac sources and Fig.1 shows a time excerpt of the first four recording channels.

1092   Giulia Barbati, Camillo Porcaro, and Carlo Salustri

In case B, instead, the fetal heartbeat was barely visible as shown in Fig.2.

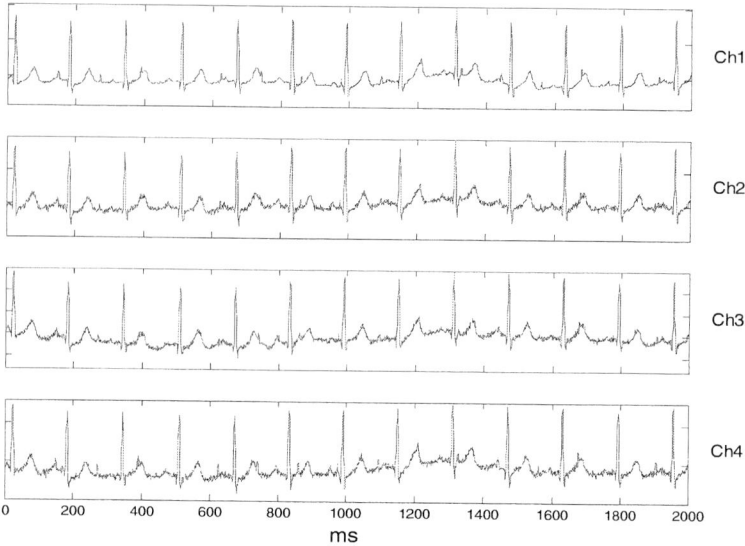

Fig. 2. Case B: a time excerpt of the first four channels in which the maternal heartbeat is clearly visible whereas the fetal heartbeat is hardly visible.

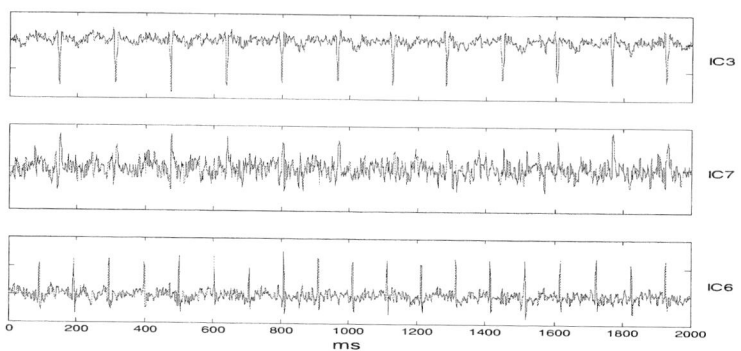

Fig. 3. Case A: Three out of the nine ICs obtained using ER procedure, representing maternal (IC3 and IC7) and fetal (IC6) heartbeats respectively.

In Fig.3 we selected three out of the nine Independent Components (ICs) obtained with k determined by ER procedure for case A; Fig.4 shows ICs obtained with k determined by the RD criterion.

Note that in Fig.3 maternal and fetal heartbeats result clearly separated in three different ICs; in Fig.4, IC3 contains instead both cardiac sources.

Fig.5, for case B, shows five out of the eleven ICs obtained with k determined by ER; Fig.6 shows the ICs obtained with k determined by RD.

'Signal Subspace' Blind Source Separation 1093

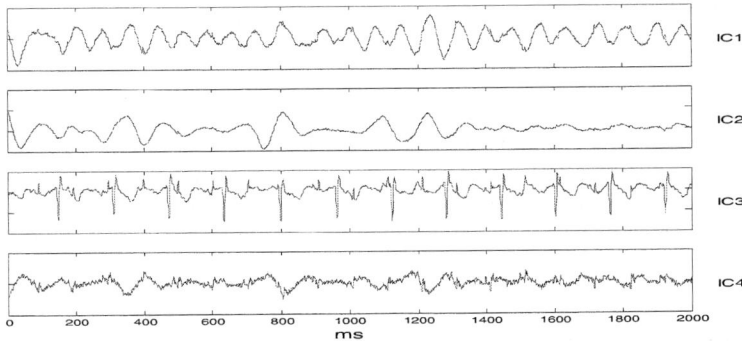

Fig. 4. Case A: The four ICs obtained using RD and representing both maternal and fetal heartbeats (IC3, IC4).

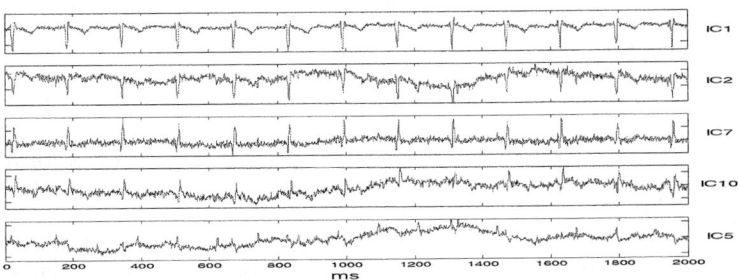

Fig. 5. Case B: Five out of eleven ICs obtained using ER procedure, representing maternal (IC1, IC2, IC7, IC10) and fetal (IC5) heartbeats.

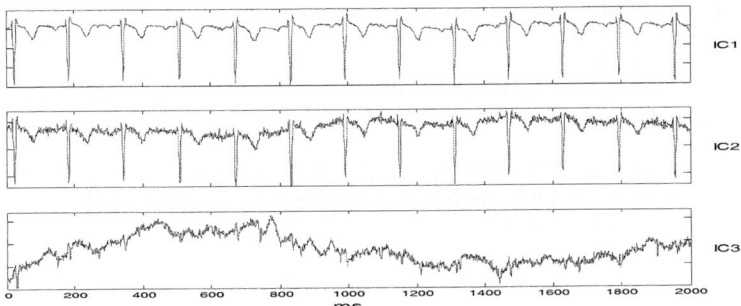

Fig. 6. Case B: The three ICs obtained using RD, representing both maternal and fetal heartbeats: IC1 and IC2 contain both cardiac sources with strong dominance of the mother; IC3 contain both cardiac sources with relevant presence of noise.

Note that in Fig.5 four ICs clearly isolated the maternal heartbeat and one IC (IC5) contained the fetal cardiac wave; in Fig.6, when estimating only three ICs, the fetal heartbeat remained mixed with the maternal heartbeat in all components.

3.3 Conclusions

In this paper we have assumed the fetus' and the mother's hearts as statistically independent sources of magnetic fields: under this assumption we have proposed, from a practical point of view, the use of a two-step BSS procedure which, by defining two signal subspaces (signal and noise) and identifying a number of independent components, extracted the cardiac signals from the background noise and effectively separated the fetus' from the mother's traces. We have shown the effectiveness of this method even in cases where the fetal signal is almost invisible to the naked eye in the MCG trace.

References

1. Jung, T.P., Makeig, S., McKeown, M.J., Bell, A.J., Lee, T.W., Sejnowski, T.J. "Imaging Brain Dynamics Using Independent Component Analysis", in Proceedings of the IEEE, 2001, vol. 89(7), pp. 1107-1122
2. Ziehe, A., Nolte, G., Sander, T., Muller, K.R., and Curio, G., "A comparison of ICA-based artifact reduction methods for MEG", in Proc. 12th Int. Conf. Biomagnetism, Espoo, Finland, 2001, pp. 895-898
3. De Lathauwer, L., De Moor, B., and Vandewalle, J., "Fetal Electrocardiogram extraction by blind source subspace separation", IEEE Trans. Biomed. Eng., vol. 47(5), pp. 567-572, May 2000
4. Zarzoso, V., Nandi, A. K., and Bacharakis, E., "Maternal and Foetal ECG Separation using Blind Source Separation Methods", IMA Journal of Mathematics Applied in Medicine & Biology, 14, 207-225 (1997)
5. Zarzoso, V., and Nandi, A. K., "Noninvasive Fetal Electrocardiogram Extraction: Blind Separation vs Adaptive Noise Cancellation", IEEE Trans. Biomed. Engineering 48(1), 12-18 (2001)
6. Cruces-Alvarez, S., Cichocki, A., and Castedo-Ribas, L., "An Iterative Inversion Approach to Blind Source Separation", IEEE Trans. on Neural Networks, vol. 11, pp. 1423-1437, 2000
7. Cruces-Alvarez, S., Castedo-Ribas, L., and Cichocki, A., "Robust blind source separation algorithms using cumulants", Neurocomputing, vol. 49, pp. 87-118, 2002
8. Cichocki, A. and Amari, S., Adaptive Blind Signal and Image Processing, John Wiley & Sons, Chichester, 2002
9. Karhunen, J., Cichocki, A., Kasprzak, W., Pajunen P., "On neural blind separation with noise suppression and redundancy reduction", Int. Journal of Neural Systems, 8(2): 219-237, 1997
10. Liavas, A.P., Regaglia, P.A., Delmas, J-P., "Blind channel approximation: effective channel order determination", IEEE Trans. Signal Processing, 47: 3336-3344, 1999

Suppression of Ventricular Activity in the Surface Electrocardiogram of Atrial Fibrillation

Mathieu Lemay[1], Jean-Marc Vesin[1], Zenichi Ihara[1], and Lukas Kappenberger[2]

[1] Signal Processing Institute, EPFL, Lausanne, Switzerland
http://itswww.epfl.ch/
[2] Division of cardiology, CHUV, Lausanne, Switzerland

Abstract. The analysis of the surface electrocardiogram is potentially useful for the study of atrial fibrillation. Since the ventricular activity is much stronger than the atrial activity, one has to suppress it. To this end, we applied two ICA algorithms to a data set of surface electrocardiogram signals recorded in clinical conditions. We also propose a procedure to judge the quality of the suppression of ventricular activity and the extraction of atrial activity. We apply this procedure to our extracted activities and discuss our results.

1 Introduction

Atrial fibrillation (AF) is the most common type of human arrhythmia. Its clinical description is that of "absolute arrhythmia". The diagnosis of AF has been assessed for years by visual inspection of the surface electrocardiogram (ECG). Its underlying mechanism involving self-sustained multiple reentrant waves has been discovered several decades ago. Although there are many substrate abnormalities that may cause AF, they are generally considered as the same arrhythmia problem. However, at a closer look, AF is not a uniform phenomenon. Rather it is a collection of more or less sustained atrial disorders, whose underlying mechanisms and substrates are not clearly defined and have never been linked to structural entities.

In physiologic conditions, the pacemaker function of the heart is ensured by the sino-atrial node. The node cells initiate regular waves of depolarization through atria and ventricles, at a rate of 60-100 times per minute at rest. These cells may fire as fast as 180-200 times per minute at peak exercise. During AF, the sino-atrial node becomes the secondary pacemaker, being superseded by the atrial myocardium which may fire at rates of 350-600 times per minute.

Due to the much higher amplitude of the electrical ventricular activity (VA) on the surface ECG, isolation of the atrial activity (AA) component is crucial for the study of AF. Various techniques may be used to obtain an ECG signal without any ventricular component. One of these techniques is the segmentation of the AA in the surface ECG. One identifies the location of the QRST complexes (various QRS detection algorithms are available [1]) and extracts the segments without any ventricular component by time-oriented operations. This technique is robust, but the information contained in the rejected parts (i.e with VA) is lost. Another popular technique is QRS cancellation. The average beat subtraction approach (ABS) and the use of adaptive recurrent filtering are the principal methods proposed to this end [2]. Slight changes in the QRS morphology affect these methods and, more important, they require some interpolation. This may affect the efficiency of QRS suppression.

As noted in [3] and [4], there are two characteristics defining the electrical activity of the AF in surface ECG that make possible the use of independent component analysis (ICA) methods. First, due to the bioelectrical independence of the atrial and ventricular regions, the ECG signals can be viewed as a weighted sum of atrial and ventricular electrical sources, noise and artifacts. Second, the distributions of the atrial and ventricular activities are nongaussian (subgaussian and supergaussian, respectively). Because of these two characteristics, one of the independent components obtained from the surface ECG lead signals should correspond to atrial activity only. In [3], an ICA approach based on whitening (decorrelation and normalization) followed by maximum-likelihood estimation of the Givens rotation matrix [5] was used. In [4], an additional second-order blind source separation step was applied. In both works, validation was performed on simulated AF signals. In [6], a validation of BSS-based AA extraction using spectral parameters has been proposed. It focuses also on the spectral differences between AF and atrial flutter (a "degenerate" form of atrial fibrillation).

In this article we apply two different ICA algorithms to real surface ECGs. The first algorithm [7] is an extended version of the algorithm in [5]. The second one [8] is based on an adaptive estimation of the marginal probability density functions of the sources. We assess the performance of these algorithms in terms of VA suppression and AA extraction directly on the clinical data. Of course it is not possible to quantify accurately these performances since the true AA and VA are unknown, but we define several spectral and temporal criteria, some qualitative and some quantitative, to judge the success of VA cancellation. We also briefly discuss the physiological relevance of the demixing coefficients.

2 Clinical Importance of AF

AF is an uncoordinated electrical activation of the atria, classified as supraventricular arrhythmia. The anarchic depolarization of the atria leads to an inefficient atrial contraction and thus to a compromised function of the whole heart. On the surface ECG, the P-wave is replaced by fibrillation waves (F-waves) varying in shape, size, and timing, with rates over 350 cycles per minute. These features are associated with an irregular ventricular response, which is often rapid in case of physiologic atrio-ventricular conduction.

AF is the most common arrhythmia in clinical practice and it is responsible for about one third of hospitalizations for arrhythmia problems. The prevalence of AF is 0.4% of the general population [9]. AF is more frequent in elderly, as its prevalence doubles with each decade of age, from 0.5% at ages between 50-59 years to almost 9% at ages between 80-89 years [10],[11]. With the increase of life expectancy the prevalence is expected to double in the next fifty years. The yearly incidence of AF, also related to advancing age, is less than 0.1% among people of age under 40 years and over 1.5% among those above 80 years [12],[13].

AF is an important clinical entity because of the increased risk of morbidity and mortality (1.5 to 1.9 fold in Framingham study [14]). The most frequent consequences are hemodynamic function impairment (loss of atrial synchronized contraction, irregular and inadequately rapid ventricular rate), atriogenic thromboembolic [15] events and tachycardia induced atrial and ventricular cardiomyopathy [16].

3 Algorithms

The standard context of ICA for linear instantaneous mixtures as presented here uses the following notations. If $\mathbf{s}(t) = [s_1(t), s_2(t), ..., s_m(t)]^T$ denotes the source vector at time t, one assumes that the observation $\mathbf{x}(t) = [x_1(t), x_2(t), ..., x_n(t)]^T$ is given by:

$$\mathbf{x}(t) = M\mathbf{s}(t), \quad (1)$$

with M the mixing matrix. Under the assumption of mutual independence of the sources, the goal is to estimate a demixing matrix W such that the output vector

$$\mathbf{u}(t) = W\mathbf{x}(t) \quad (2)$$

recovers the sources $\mathbf{s}(t)$ up to permutation and scaling.

3.1 Extended Maximum-Likelihood

In the case of two sources, $\mathbf{s}(t) = [s_1(t), s_2(t)]^T$, one performs first a decorrelation and normalization of observations with a whitening matrix B, i. e. $\mathbf{z}(t) = B\mathbf{x}(t)$, so that the demixing is now expressed by $\mathbf{u}(t) = W\mathbf{z}(t)$. Then only a Givens rotation matrix Q defined by:

$$Q = \begin{bmatrix} \cos\theta & -\sin\theta \\ \sin\theta & \cos\theta \end{bmatrix} \quad (3)$$

has to be estimated to perform the demixing of the two sources. The angle θ is estimated through a maximum-likelihood procedure (see [5] for details). To separate more than two sources, an algorithm based on the iterative application of the former procedure to pairwise combinations of observations has been proposed [7]. We refer to this algorithm as EML in the remaining of this paper.

3.2 Hybrid Independent Component Analysis by Adaptive Look-Up Table Activation Functions

This ICA method, described in [8], is based on the natural-gradient version of the stochastic minimal mutual information (MMI) learning rule [17]. This learning rule expresses the update $\triangle W$ of the demixing matrix W by:

$$\triangle W = \eta \left[\mathbf{I} - \Phi(\mathbf{u})\mathbf{u}^T \right] W, \quad (4)$$

where η is a positive learning step size and the vector Φ is defined by:

$$\Phi(\mathbf{u}) := \left(-\frac{r'_1(u_1)}{r_1(u_1)}, -\frac{r'_2(u_2)}{r_2(u_2)}, \ldots, -\frac{r'_n(u_n)}{r_n(u_n)}\right)^T, \quad (5)$$

In this equation, the functions $r_i(u_i)$ stand, in the optimal case for the marginal probabilities of the vector output components $u_i(t)$, which are unknown. In this ICA method, those marginal densities and their derivatives are iteratively estimated with time-varying histograms. We refer to this algorithm as adaptive LUT (for histogram look-up table) in the remaining of this paper.

4 Data Collection and Settings

We used the surface 12-lead ECGs from 16 patients with AF (various substrates) recorded in clinical conditions. The sampling frequency was 500 Hz, and the record duration was 80 seconds. We used all the leads for the suppression of VA. The preprocessing consisted in baseline modulation removal using a high-pass filter with cutoff frequency 0.5 Hz.

In the LUT, a pre-whitening was also applied to the surface ECG signals, and 64-bin histograms were used. The algorithm was run once on the signals to ensure convergence and the extraction results were obtained during the second run.

5 Experimental Results

5.1 Typical Example of AA Extraction

Figure 1 displays the ECG signal from the V1 lead (top) and the recovered sources most probably corresponding to the AA activity (middle and bottom) extracted with EML and adaptive LUT. Since these recovered sources do not correspond to an AA recorded on one of the leads, but rather to a weighted sum of lead AA activities, we refer to them as the synthesized AA (sAA). The white

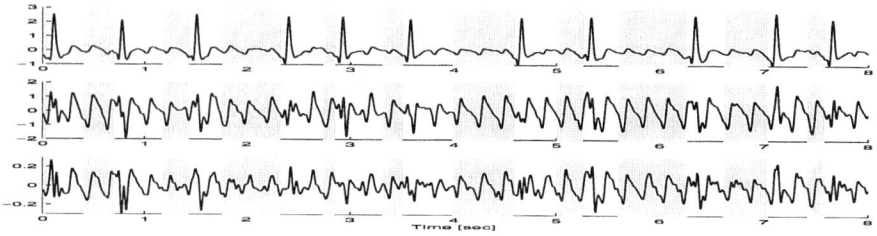

Fig. 1. Top: ECG V1 lead signal. Middle: sAA signal obtained by the EML method. Bottom: sAA signal obtained by the adaptive LUT method. White areas correspond to the location of QRST complexes.

regions (grey regions) correspond to the location of the QRST waves, i. e. to VA + AA (respectively AA only). Visually, the sAA presents the characteristic aspect of AF and VA seems to be adequately suppressed, although some remnant of VA is discernible in the 2th, 5th, 8th and 10th QRST wave locations.

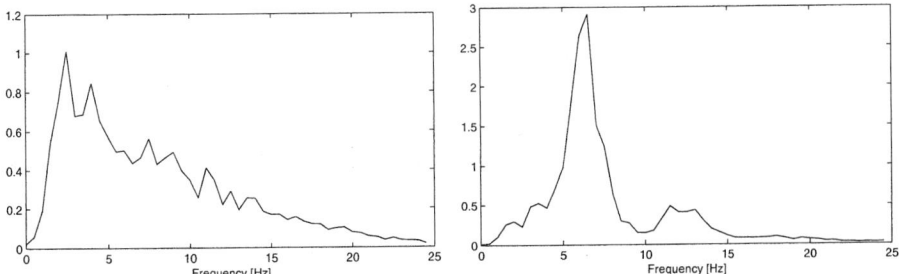

Fig. 2. Left: PSD of V1 lead signal. Bottom: PSD of sAA signal obtained by the adaptive LUT method.

In Figure 2, we have plotted the estimates of the power spectral densities (psd) of the V1 lead ans the sAA obtained by the adaptive LUT method. One observes that the psd of sAA (right) is characterized by a sharp peak around 6.5 Hz while the low- and high-frequency components of the raw ECG signal corresponding to VA (left) are mostly absent.

5.2 Assessment of VA Suppression and AA Extraction

Based on the clinical data, it is impossible to quantify accurately the quality of AA extraction and VA suppression, since those true activities were unknown. Nonetheless, if sAA corresponds indeed to AA only, the following observations must hold:

- The sAA must be strongly correlated with the raw ECG signal in the time intervals where only AA is present (i. e. outside the QRST complexes) and a lot less in the time intervals with VA. Comparison between this two cross-correlations should obviously be performed using estimated normalized cross-covariances, due to the indeterminacy in sAA variance and the amplitude differences between various leads.
- Note however that some correlation in the latter case is inevitable since AA is present during those complexes. Even more, this correlation should be due only to this AA, because sAA and VA should be uncorrelated. During time intervals where both VA and AA are present, one should have $cov[sAA \cdot (VA + AA)] = cov[sAA \cdot AA]$. That is, the cross-covariance between sAA and an ECG lead signal should be close, be it estimated during time intervals where only AA is present or during those where both VA and AA are present.

Segments of 20 seconds were considered in the test. We have applied the following criteria to our extraction results:

1. Comparison between the normalized cross-covariances estimated on all the AA-only and VA + AA segments. The ratio of these two measures had to be larger than 1.2.
2. Visual validation of the estimated psd of sAA. We checked if the decrease of the psd in the frequency bands characteristic of VA (1-2.5 Hz and 15-60 Hz) was significant enough, and if the spectral power was concentrated in the 5-12 Hz frequency band.
3. Comparison between estimated non-normalized cross-covariances estimated on all the AA-only and VA + AA segments. The ratio of these two measures had to be larger than 0.5 and smaller than 2.

Successive selection using these three criteria gave the following results with LUT:

1. On the 16 recordings, the test on the normalized cross-covariance ratio gave a suitable result (minimum value 1.43 and maximum value 4.70) in 15 recordings.
2. On these 15 recordings, the visual validation of psd gave a suitable result in 13 recordings.
3. On these 13 recordings, the test on the non-normalized cross-covariance ratio gave a suitable result (minimum value 0.91 and maximum value 1.97) in 11 recordings.

Successive selection using these three criteria gave the following results with EML:

1. On the 16 recordings, the test on the normalized cross-covariance ratio gave a suitable result (minimum value 1.41 and maximum value 3.14) in 13 recordings.
2. On these 13 recordings, the visual validation of psd gave a suitable result in 12 recordings.
3. On these 12 recordings, the test on the non-normalized cross-covariance ratio gave a suitable result (minimum value 0.95 and maximum value 1.83) in 9 recordings.

6 Discussion

At first glance, the results are good: 11 excellent sAA on 16 recordings for the adaptive LUT algorithm and 9 excellent sAA for the EML. However, these results was better with a strong preprocessing step; high-pass filtering at 3 Hz. Our clinical data contain a baseline in the same low frequency band that characterized ventricular rythm (1-2.5 Hz). Our initial purpose was to preprocess our data minimally. In this light, our results are lot more positive and give a clear indication that ICA is indeed a promising tool for AF analysis.

Our results indicate also that adaptive LUT performs better than EML, with respect to the specific data set we used. However, this difference may be due to the fact we did not implement the adaptive version of EML [4]. It may be that the specificities of the source mixture in ECG, i. e. the fact that VA is not always present, makes adaptive algorithms more suitable than batch ones.

Of interest also is the way the various synthesized atrial activities are built in the successful trials, that is, how the various ECG signals are linearly combined to obtain them. Table 1 shows the normalized weights (appropriate row of the $W \cdot B$) for successful trials with V1, V2, V5 and V6 leads. With more than 4 leads, the preprocessing step combine different leads that are correlated. In this case, it is difficult to determine the contribution of each leads. Table 1 shows that the weights (contributions) for the V1 lead are much larger: this seems intuitively appealing, since this lead is located on the upper right part of the thorax, close of the atria and far from the ventricles. The weights across the recordings are also quite similar, which indicates that the extraction dynamic is stable. The suppression of VA works well with 4 different leads or more when one of them is the V1 lead.

Table 1. Normalized weights of the five successful trials.

Successful trails	Normalized weights			
	V1	V2	V5	V6
sAA #1	**0.5695**	-0.2118	-0.0302	0.1887
sAA #2	**0.5356**	-0.2752	0.0404	0.1488
sAA #3	**0.5371**	-0.2277	-0.0179	0.2173
sAA #4	**0.3865**	-0.0860	-0.1322	0.3953
sAA #5	**0.6284**	-0.1578	-0.0183	0.1956

7 Conclusion

We have presented in this paper a preliminary investigation on the potential of ICA for atrial activity extraction in surface ECG atrial fibrillation signals. We have proposed some criteria to assess the quality of this extraction and we have applied them to the extracted atrial activities we obtained with two ICA algorithms published in the literature. The results are good and encouraging if one takes into account the limited quality of our data set.

References

1. Köhler,B.-U., Hennig,C.,Orglmeister,R.: The Principles of Software QRS Detection. IEEE Engineering in medecine and biology (2002) 42–57
2. Stridh,M., Sörnmo,L.: Spatiotemporal QRST cancellation techniques for atrial fibrillation analysis in the surface ECG. Signal Processing Report **SPRN-44** (1998) 1–30

3. Rieta,J.J., Zarzoso,V., Millet-Roig,J., García-Civera,R., Ruiz-Granell,R.: Atrial Activity Extraction Based on Blind Source Separation as an Alternative to QRST Cancellation for Atrial Fibrillation Analysis. IEEE Computers in Cardiology **27** Boston, MA (2000) 69–72
4. Castells, F., Igual,J., Rieta,J.-J., Sanchez,C., Millet,J.: Atrial Fibrillation Analysis Based on ICA Including Statistical and Temporal Source Information. ICASSP 2003 Hong Kong (2003) 59–64
5. Zarzoso,V., Nandi,A.K.: Blind Separation of Independent Sources for Virtually Any Source Probability Density Function. IEEE Trans. Signal processing **47-9** (1999) 2419–2432
6. Millet-Roig,J., Rieta,J.J., Zarzoso,V., Cebrián,A., Castells,F., Sánchez,C., García-Civera,R.: Surface-ECG Atrial Activity Extraction via Blind Source Separation: Spectral Validation. IEEE Computers in Cardiology **29** Memphis, TN (2002) 605–608
7. Zarzoso,V., Nandi,A.K.: Adaptive Blind Source Separation for Virtually Any Source Probability Density Function. IEEE Trans. on signal processing **48-2** (2000) 477–488
8. Fiori,S.: Hybrid independent component analysis by adaptive LUT activation function neurons. Neural Networks **15** (2002) 85–94
9. Stewart,S., Hart,C.L., Hole,D.J., McMurray,J.J.: Population prevalence, incidence, and predictors od atrial fibrillation in the Renfrew/Paisley study. Heart **86** (2001) 516–521
10. Kannel,W.B., Wolf,P.A., Benjamin,E.J., Levy,D.: Prevalence, incidence, prognosis, and predisposing conditions for atrial fibrillation: population-based estimates. Am. J. Cardiol. **82** (1998) 2N–9N
11. Furberg,C.D. et al.: Prevalence of atrial fibrillation in elderly subjects (the Cardiovascular Health Study). Am. J. Cardiol. **74** (1994) 236–241
12. Psaty,B.M. et al.: Incidence of and risk factors for atrial fibrillation in older adults. Circulation **96** (1997) 2455–2461
13. Wolf,P.A., Abbott,R.D., Kannel,W.B.: Atrial fibrillation: a major contributor to stroke in the elderly: the Framingham Study. Arch. Intern. Med. **147** (1978) 1561–1564
14. Benjamin,E.J.: Impact of atrial fibrillation on the risk of death: the Framingham Heart Study. Circulation **98** (1991) 946–952
15. Wolf,P.A., Abbott,R.D., Kannel,W.B.: Atrial fibrillation as an independent risk factor for stroke: the Framingham Study. Stroke **22** (1991) 983–988
16. Shinbane,J.S. et al.: Tachycardia-induced cardiomyopathy: a review of animal models and clinical studies. J. Am. Coll. Cardio. **29** (1997) 709–715
17. Haykin,S.: Unsupervised adaptive filtering. J. Wiley & Sons, N.Y. **Volume 1** (2000)

Unraveling Spatio-temporal Dynamics in fMRI Recordings Using Complex ICA

Jörn Anemüller[1,2], Jeng-Ren Duann[1,2], Terrence J. Sejnowski[1,2], and Scott Makeig[1,2]

[1] Swartz Center for Computational Neuroscience, Institute for Neural Computation, University of California San Diego, La Jolla, California
[2] Computational Neurobiology Laboratory, The Salk Institute for Biological Studies, La Jolla, California, and Howard Hughes Medical Institute

Abstract. Independent component analysis (ICA) of functional magnetic resonance imaging (fMRI) data is commonly carried out under the assumption that each source may be represented as a spatially fixed pattern of activation, which leads to the instantaneous mixing model. To allow modeling patterns of spatio-temporal dynamics, in particular, the flow of oxygenated blood, we have developed a convolutive ICA approach: spatial complex ICA applied to frequency-domain fMRI data. In several frequency-bands, we identify components pertaining to activity in primary visual cortex (V1) and blood supply vessels. One such component, obtained in the 0.10-Hz band, is analyzed in detail and found to likely reflect flow of oxygenated blood in V1[1].

1 Introduction

The blood oxygenation level dependent (BOLD) contrast measured by fMRI recordings depends on the change in level of oxygenated blood with neural activity. ICA has been successful at finding independent spatial components that vary in time [1], but there may also be spatio-temporally dynamic patterns in fMRI recordings of brain activity.

Convolutive models are a way to account for dynamic flow patterns. In convolutive models, each source process is characterized by the spatio-temporal pattern it elicits and by the time-course of activation of this pattern. The signal accounted for by each source process is obtained by convolving the spatio-temporal source pattern with its time-course of activation. The mixed (measured) data are obtained by summing over the contributions of all source processes. Separation of mixed activity generated by several of such processes is not possible for instantaneous ICA algorithms since the convolutive mixing is beyond the scope of their instantaneous mixing assumption.

The convolutive separation problem can be solved by performing all computations in the frequency-domain since the convolution in the time-domain factorizes into a multiplication in the frequency-domain. Separation is performed by applying a complex ICA algorithm to the complex-valued data in each frequency-band.

The use of this procedure for the analysis of electroencephalographic (EEG) data has recently been presented elsewhere [2]. Here, we present the application of the

---
[1] Supported by the German Research Council DFG (J. A.), and by the Swartz Foundation.

method to fMRI data. Compared to EEG data, fMRI data are characterized by their high spatial resolution at a low temporal sampling rate. fMRI data are commonly analyzed by spatial ICA decomposition, where time-points correspond to input dimensions and voxels to samples. This is in contrast to temporal ICA used for EEG, where sensors constitute input dimensions and time-points samples. To apply complex ICA to fMRI signals, we analogously apply spatial complex ICA to frequency-domain fMRI data.

2 Methods

Consider measured signals x_{ti}, where t denotes time and i denotes voxels. Their spectral time-frequency representations $x_{Ti}(f)$ are computed using the short-term Fourier transformation

$$x_{Ti}(f) = \sum_{\tau} x_i(T+\tau) h(\tau) e^{-i2\pi f \tau/2K} \quad (1)$$

where f denotes center frequency, and $h(\tau)$ is a Hanning window. Hence, data of size [times t × voxels i] are transformed into data of size [times T × voxels i × frequencies f].

For each frequency band f, the signals are modeled to be generated from independent sources $s_{Ti}(f)$ by multiplication with frequency-specific mixing coefficients $a_{TT'}(f)$,

$$x_{Ti}(f) = \sum_{T'} a_{TT'}(f) s_{T'i}(f), \quad (2)$$

which in matrix notation reads

$$\mathbf{X}(f) = \mathbf{A}(f)\mathbf{S}(f). \quad (3)$$

Complex ICA is used to obtain the independent components using the linear projection

$$\mathbf{U}(f) = \mathbf{W}(f)\mathbf{X}(f), \quad (4)$$

where $u_{Ti}(f)$ and $w_{TT'}(f)$ represent complex spatial component patterns and the separating matrix, respectively.

Hence, as in real-valued ICA for fMRI signals, a spatial ICA decomposition is performed, where time-points correspond to input dimensions and voxels to samples. For each frequency-band, we obtain a set of complex-valued independent components, each characterized by its associated complex time-course $\mathbf{a}_{T'}(f)$ and complex-valued spatial pattern $\mathbf{s}_{T'}(f)$, with T' denoting component number.

The complex infomax ICA algorithm [2,3] is a generalization of the real-valued infomax ICA algorithm [4] and models the sources as complex random variables with a circular symmetric, super-Gaussian probability density function. Analysis of the statistics of frequency-domain fMRI data strongly indicates that these assumptions are fulfilled. Matrices $\mathbf{W}(f)$ are found using natural gradient optimization [5]

$$\tilde{\nabla}\mathbf{W}(f) = \nabla\mathbf{W}(f)\mathbf{W}(f)^H \mathbf{W}(f) = \left(\mathbf{I} - \mathbf{V}(f)\mathbf{U}(f)^H\right)\mathbf{W}(f), \quad (5)$$

where $\mathbf{V}(f)$ is a non-linear function of the source estimates $\mathbf{U}(f)$:

$$v_{Ti}(f) = \text{sign}(u_{Ti}(f)) g(|u_{Ti}(f)|), \quad (6)$$

$$\text{sign}(z) = \begin{cases} 0 & \text{if } z = 0, \\ z/|z| & \text{if } z \neq 0. \end{cases} \quad (7)$$

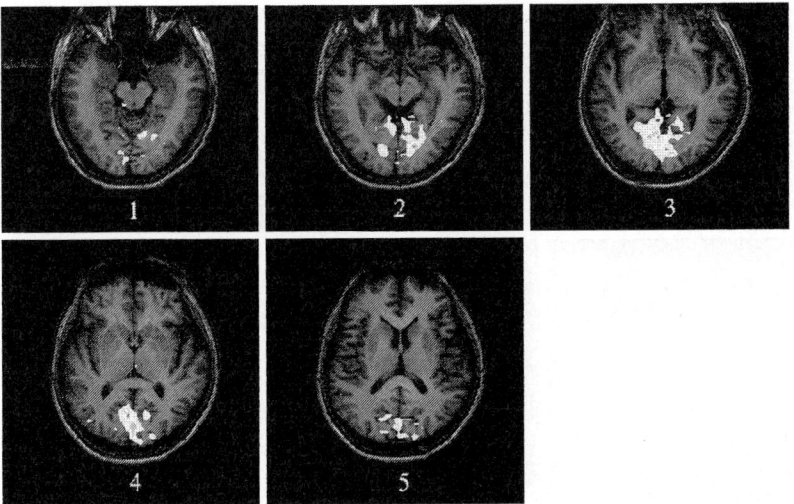

Fig. 1. Magnitude map of the component region of activity (ROA) for complex component IC2 obtained by complex ICA in the 0.1-Hz frequency-band. The ROA extends over visual area V1 and blood supply vessels. Light yellow colors indicate component magnitude in ROA. Structural image of the recorded areas is plotted in darker gray tones. The component ROA is interpolated to the higher resolution of the structural scan for better visualization. (Note: The electronic version of this document contains color figures for better visualization and can be obtained from the first author).

I denotes the identity matrix and the function $g(\cdot) : \mathbb{R} \rightarrow \mathbb{R}$ is a real-valued non-linearity, chosen as $g(x) = \tanh(x)$.

3 Results

The experimental data were from a 250 s experimental session consisting of ten epochs with stimulus onset asynchrony (SOA) of 25 s. An 8-Hz flickering checkerboard stimulus was presented to one subject for 3.0 s at the beginning of each epoch. 500 time-points of data were recorded at a sampling rate of 2 Hz (TR=0.5) with resolution $64 \times 64 \times 5$ voxels, field-of-view 250×250 mm^2, slice thickness 7 mm. Standard preprocessing included removal of off-brain and low-intensity voxels, reducing the data to 2863 voxels. For more experiment details refer to [6]. The data of this experiment are freely available as part of the FMRLAB toolbox for fMRI ICA analysis [7].

Spectral decomposition was performed using the windowed discrete Fourier transformation (1) with a Hanning window of length 40 samples, a window shift of 1 sample, and frequency-bands $0.05, 0.10, \ldots, 1.00$ Hz. This resulted in data split into 20 bands, each with 461 time-points and 2863 voxels.

Spatial complex ICA decomposition was performed within each frequency-band. In a preprocessing step, input dimensionality in each band was reduced from 461 to 50 by retaining only the subspace spanned by the (complex) eigenvectors corresponding

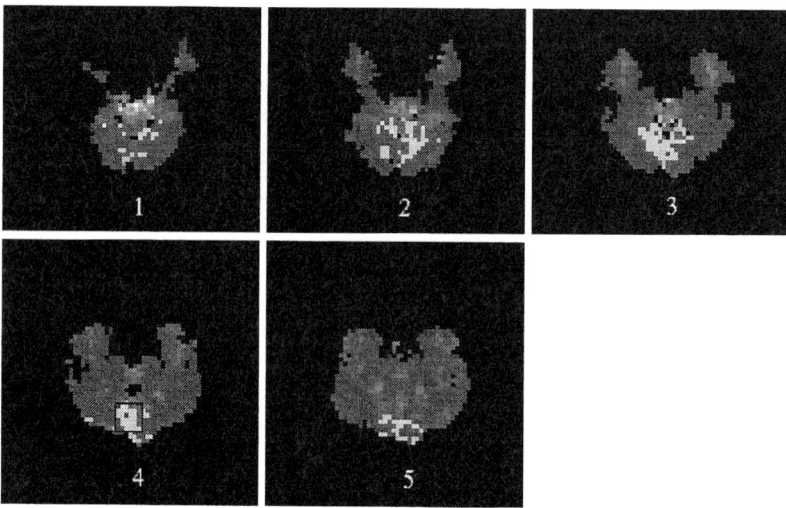

Fig. 2. Phase map of the component ROA for complex component IC2. Light green colors indicate component phase in region of activity. Note that in contrast to Fig. 1 the information is displayed at the lower spatial resolution of the functional recordings. The voxels marked by a blue square in slice 4 are investigated further in Fig. 4.

to the 50 largest eigenvalues of the data matrix $\mathbf{X}(f)$. Complex ICA decomposed this subspace into 50 complex independent components per band.

Motivated by previous results of real-valued infomax ICA on the same data [6], we were interested in components with a region of activity (ROA) near primary visual cortex V1. One such component was found in several frequency-bands, with a time-course of activation that closely reflected the SOA of the visual stimulus. Time-locking of component activity to stimulus presentation was particularly reliable for component number 2 (IC2) in the 0.10-Hz band. IC2 was the single clearly V1-related component in this band. The following analysis is restricted to this particular component.

Fig. 1 displays the magnitude of the complex spatial component map of IC2 in the ROA of the five recording slices. The ROA was determined from z-scores of the component map by transforming each component map to zero mean and unit variance, and setting a heuristic threshold of 1.5. The extent of IC2 from the centrally located main blood vessels to primary visual cortex is clearly visible, in particular in slices 3 and 4. The complex component's phase in the ROA is displayed in Fig. 2. Slices 3 and 4 display a phase shift from the upper left border of the component ROA image towards the lower right border. The phase shift indicates a time lag in the activation of the component voxels when transformed back into the time-domain which will be further investigated below. Fig. 3 shows power (squared magnitude) and phase of the component time-course of activation. Component power clearly reflects the pattern of visual stimulation with an SOA of 25 s, with peaks in power that follow stimulation with a time lag of about 9 s, and a high dynamic range between component activity and inactivity. The component phase regularly advances and appears to be time locked to

Unraveling Spatio-temporal Dynamics in fMRI Recordings Using Complex ICA 1107

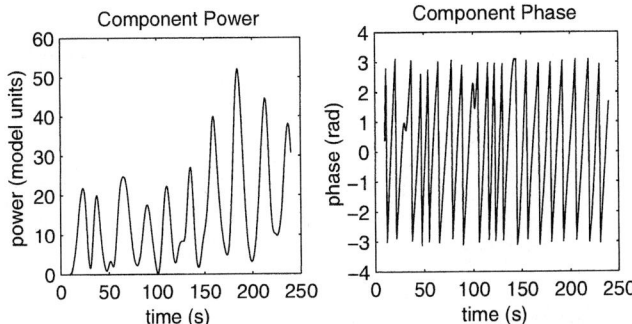

Fig. 3. Time-course of power (left) and phase (right) of complex component IC2 in the 0.1-Hz frequency-band. Note the time-locking of amplitude and phase to stimulus presentation in 25 seconds intervals. The first and last 10 seconds of the experiment are not shown because computation of the spectral components was stopped when the analysis window (length 20 s) reached the edges of the recording. The time-interval from 179.5 s to 187.0 s around the largest component power peak is investigated further in figure 4.

stimulus presentation, possibly to a lower degree during the periods from 0 s to 50 s and from 100 s to 130 s, which could be due to the subject's level of attention to the stimulus.

Complex voxel activity induced by the component may be obtained by backprojecting the complex time-course to the complex spatial map, i.e., by forming the product $\mathbf{a}_{T'}(f)\mathbf{s}_{T'}(f)$, where T' denotes component number, $\mathbf{a}_{T'}(f)$ the corresponding column of the mixing matrix $\mathbf{A}(f)$, and $\mathbf{s}_{T'}(f)$ the corresponding row of the source matrix $\mathbf{S}(f)$. Transforming the complex frequency-domain voxel activity to the real time-domain reduces – in the case of a window-shift of one sample and a single frequency-band – to taking the real-part. We performed these steps to analyze time-domain voxel activity induced by the component near the largest component power peak between 179.5 s and 187.0 s of the experiment. Fig. 4 displays the activity within a patch of 24 voxels located in recording slice 4, marked by a blue square in Fig. 2. Following stimulus presentation at 175.0 s, activity in the patch started to increase with a time lag of about 4.5 s, first in the voxels most centrally located in the brain (top row of voxels in each plot of Fig. 4), and propagating within about 1 s to the posterior voxels of to primary visual cortex (bottom row in each plot of Fig. 4). Analogously, voxel activity decreased first in the top row of voxels before decreasing in the bottom rows.

To investigate whether similar time lag effects can be found *without* ICA processing, we also computed the 0.10-Hz band activity of the recorded data at the 24 voxels that have been investigated in Fig. 4, using the same spectral decomposition that has been used for the complex ICA decomposition. Activity accounted for by recorded data and by IC2 was separately averaged within each voxel row, starting with row 1 for the most centrally located voxels, and up to row 6 for the voxels in the posterior position. The resulting averages are plotted in Fig. 5 for recorded data and for component induced activity. Since the signals are band-limited, we obtain oscillatory activity with positive and negative swings. The analysis of relative time lags and amplitudes near the peak of

1108 Jörn Anemüller et al.

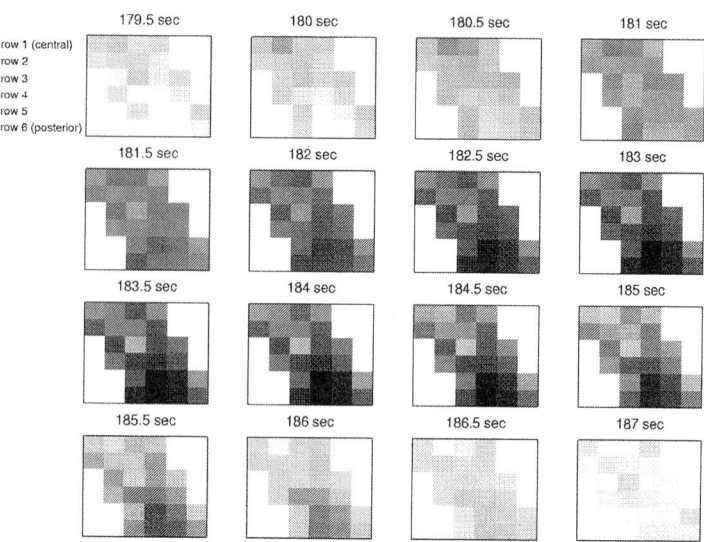

Fig. 4. Backprojected component activity from complex component IC2. Complex component time-course was backprojected to corresponding activity at the voxels and transformed to the time-domain. Shown is the activity of 24 voxels in visual area V1, the position of which is marked by a blue box in slice 4 of Fig. 2. The flickering-checkerboard stimulus was presented for 3.0 s at experiment time 175.0 s (not shown). Activation started to increase with a time lag of about 4.5 s, with first increase occuring at the centrally-located voxels (top rows) and propagated to the posterior voxels (bottom rows) within approximately 1 s. This is compatible with over-supplied oxygenated blood propagating in the posterior direction and being washed out through the drainage vein from area V1.

component power (at 184.5 s) is not influenced by this fact. In the component induced activity, the more centrally located voxels are activated between 0.5 s and 1.0 s prior to the posterior voxels. The time lag increases monotonously with more posterior voxel position. This gradient of posterior voxels being activated later than the central voxels is also reflected in the activity of the recorded voxels signals. However, the voxels in row 3 form an exception since their extremal activation occurs even after the posterior voxels are activated. The analysis of activation amplitudes in Fig. 5 gives similar results: The component induced amplitude increases monotonously towards more posterior voxel position. Overall, this tendency is also found in the recorded signals, but some exceptions occur, e.g., amplitude in row 2 is smaller than in row 1.

4 Discussion and Conclusion

We analyzed fMRI signals using a convolutive ICA approach which enabled us to model patterns of spatio-temporal dynamics. Parameters for this model were efficiently estimated in the frequency-domain where the convolution factorizes into a product. Our method consists of three processing stages: 1) Computing time-frequency representations of the recorded signals, using short-term Fourier transformation. 2) Separation

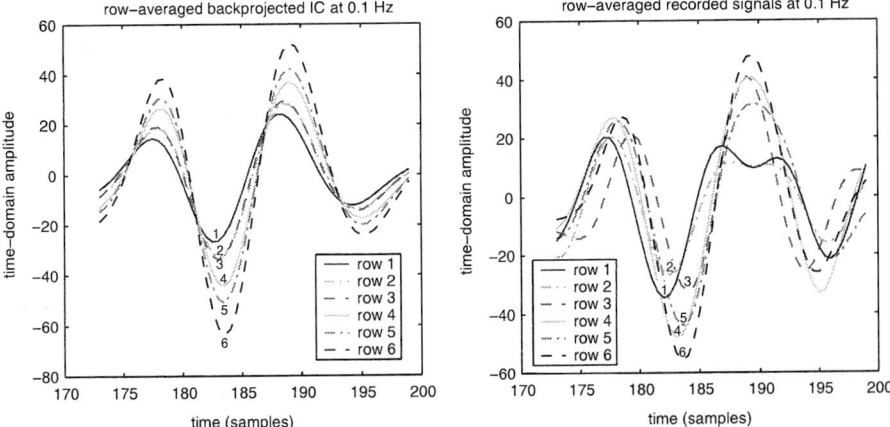

Fig. 5. Left: Average time-courses near largest component power peak (at 184.5 s) for each row of 0.1-Hz band time-domain backprojected component activations displayed in Fig. 4. Row 1 corresponds to the most centrally located voxels, row 6 to the posterior ones. Right: Corresponding average time-courses computed from the recorded activations in the 0.1-Hz band of the same voxels. For the average IC activation (left), the voxel-rows are activated in the order $1 - (2,3) - (4,5,6)$ with row 6 being activated with a time lag of about 1 second with respect to row 1. This lag is compatible with blood supply propagating across the patch in the posterior direction. In the average recorded activations (right), the voxel-rows are activated in the order $1 - 2 - 4 - (5,6) - 3$. With the exception of row 3, this also indicates a posterior direction of propagation. The most posterior voxel-row of backprojected component IC2 shows strongest activation which is plausible since it is closest to the drainage vein. The same tendency is found in the recorded signals, but ordering of amplitude of voxel-rows is not as monotonous as for IC2. Backprojected IC activations may represent a cleaner picture of the stimulus related process with respect to phase- and amplitude-gradient, because activity of other ongoing brain processes is canceled out.

of the measured signals into independent components using spatial complex infomax ICA in each frequency-band. 3) Computing the corresponding dynamic voxel activation pattern induced by each independent component in the time-domain.

From data of a visual stimulation fMRI experiment we obtained a complex component in the 0.1-Hz band with a component map ROA extending across primary visual cortex and its blood supply vessels. By reconstructing the spatio-temporal activation pattern accounted for by this component, we identified a time lag of about 1 s between activation of central and posterior voxels. A related time lag, but distributed less regularly, could be observed in the 0.1-Hz frequency-band of the measured signals. The amplitude of component-induced voxels activations increased in the posterior direction. Also this trend could be seen in the recorded signals, but it was less systematic than for the ICA processed signals.

Both observations are compatible with the physiology underlying generation of the fMRI signal. The posterior voxels in the component ROA are the ones closest to the posterior drainage vein. The convergence of over-supplied oxygenated blood towards the drainage vein may therefore result in the large amplitudes for these voxels. The temporal delay between activation of central and posterior voxels is consistent with the propagation of over-supplied oxygenated blood from the centrally located arteries to the posterior drainage vein. Similar temporal delays have been observed from optical recordings of intrinsic signals, related to blood oxygenation, in monkey visual cortex [8].

These results may indicate that frequency-domain complex infomax ICA can capture patterns of spatio-temporal dynamics in the data. It is reassuring that similar dynamics could also be observed in the recorded (mixed) signals, making the possibility of the complex ICA results being mere processing artifacts implausible. On the other hand, the spatio-temporal dynamics emerged with a higher degree of regularity and physiological plausibility from the complex ICA results than from the measured data. Separation of the stimulus evoked activity from interfering, ongoing brain activity by the complex ICA method appears as the natural explanation for this observation.

Here, we have focused the analysis on a single frequency-band. Taking into account information from other frequency-bands in which components have been found near V1 should allow us to reconstruct the full time-domain spatio-temporal dynamics associated with visual stimulation.

In conjunction with previous results reported on modeling the spatio-temporal dynamics in EEG signals with complex ICA [2], the results presented here are a further indication that convolutive models may be useful for analyzing a wide range of data.

References

1. M. J. McKeown, T.-P. Jung, S. Makeig, G. Brown, S. S. Kindermann, T. W. Lee, and T. J. Sejnowski. Spatially independent activity patterns in functional MRI data during the Stroop color-naming task. *Proc. Natl. Acad. Sci. U. S. A.*, 95:803–810, 1998.
2. J. Anemüller, T. J. Sejnowski, and S. Makeig. Complex independent component analysis of frequency-domain electroencephalographic data. *Neural Networks*, 16:1311–1323, 2003.
3. J. Anemüller and B. Kollmeier. Adaptive separation of acoustic sources for anechoic conditions: A constrained frequency domain approach. *Speech Communication*, 39(1-2):79–95, Jan 2003.
4. A. J. Bell and T. J. Sejnowski. An information maximization approach to blind separation and blind deconvolution. *Neural Computation*, 7:1129–1159, 1995.
5. S. Amari, A. Cichocki, and H. H. Yang. A new learning algorithm for blind signal separation. In D. Touretzky, M. Mozer, and M. Hasselmo, editors, *Advances in Neural Information Processing Systems 8*, pages 757–763, Cambridge, MA, 1996. MIT Press.
6. Jeng-Ren Duann, Tzyy-Ping Jung, Wen-Jui Kuo, Tzu-Chen Yeh, Scott Makeig, Jen-Chuen Hsieh, and Terrence J. Sejnowski. Single-trial variability in event-related BOLD signals. *NeuroImage*, 15:823–835, 2002.
7. J.-R. Duann, T.-P. Jung, and S. Makeig. FMRLAB: Matlab software for independent component analysis of fMRI data. World Wide Web publication, 2003. http://sccn.ucsd.edu/fmrlab.
8. R. M. Siegel, J. R. Duann, T. P. Jung, and T. J. Sejnowski. Independent component analysis of intrinsic optical signals for gain fields in inferior parietal cortex of behaving monkey. In *Society for Neuroscience Abstracts 28*, 2002.

Wavelet Domain Blind Signal Separation to Analyze Supraventricular Arrhythmias from Holter Registers

César Sánchez[1], José Joaquín Rieta[2], Francisco Castells[2],
Raúl Alcaraz[1], and José Millet[2]

[1] E.U.P.Cuenca, University of Castilla-La Mancha,
Campus Universitario s/n, 16071, Cuenca, Spain
cesar.sanchez@uclm.es
[2] BeT. DIEo, Valencia University of Technology,
Carretera Nazaret-Oliva s/n, 46730, Gandía (Valencia), Spain
{jjrieta,fcastells,jmillet}@eln.upv.es

Abstract. Detection of atrial activity (AA) is quite important in the study and monitoring of supraventricular arrhythmias. This study shows the possibility of AA extraction from atrial fibrillation (AF) episodes in Holter registers using only two leads with a new technique, the Wavelet Domain in Blind Source Separation (WDBSS). Our principal aim is to join a processing stage with Blind Source Separation (BSS) with methodologies based on wavelet transform. A first stage with Discrete Wavelet Transform (DWT) increases the spectral information, decomposing the considered signal in a set of coefficients with different temporal and spectral features. A second stage with BSS uses this information to extract the AA. The obtained improvements are the increase of spectral concentration (in the band of 5-8 Hz) and the lack of residual complexes. In WDBSS, the use of several leads from the ECG is needless, which could be applied for to the detection of different arrhythmias in Holter registers, where the number of leads is reduced, like the paroxysmal atrial fibrillation.

1 Introduction

About 2-4% of people above 60 years suffers from AF. These numbers rise to 12% in people above 75 [1]. This arrhythmia (AF) is one of the most common and can be classified in three types: paroxysmal, permanent and persistent. The paroxysmal AF appears in episodes with lengths below 48 hours and in most cases have to be detected in Holter registers where the number of leads is reduced.

The isolated study of the registered atrial activity (AA) in the electrocardiogram (ECG) is necessary for the detection and characterization of AF in these cases. This study requires a previous extraction or cancellation of the ventricular activity (VA) which is spectrally overlapped and has larger amplitude level than AA.

Nowadays, there are several techniques that can extract the AA with a good performance- Blind Source Separation [2], Spatio-Temporal Cancellation [3,4]- but poor results are obtained when the number of used reference signals (leads) is less than three, or when the duration of these signals is reduced. On the other hand, classic techniques- Average Beat Subtraction (AVBS) [5]- have developed AA extraction from only one lead, but these systems are very sensitive to the presence of ectopic complexes.

In previous works [6], the possibilities of Discrete Packet Wavelet Transform (DPWT) have been presented as a possible cancellation technique of VA in registers with synthesized AF and reduced number of leads. The obtained results showed an AA with spectral and temporal behaviour very similar to the expected AA. However, the presence of residual QRS complexes and the low performance in the case of real AF episodes required a second decomposition with Discrete Wavelet Transform (DWT). This second process obtained signals without QRS complexes, but the wave form of the extracted AA was distorted and it could not be identified as an AF by cardiologist.

In this paper, the WDBSS is presented as an improvement of the methodologies based on the DPWT and DWT for the AA extraction. The WDBSS consists of an analysis of several decomposition levels obtained with the wavelet transform using Blind Source Separation algorithms. The spectral concentration levels in the typical band of AF and the completed elimination of QRS complex justify the development of this new method.

2 Theory

The DPWT is a generalization of the wavelet transform that joins spectral and temporal analysis. The signal is decomposed in basic blocks corresponding to different frequency bands. Local and global parameters of the original signal can be identified using certain characteristics of these basic blocks.

The wavelet transform of a signal $f(t)$ can be expressed as follows, in its most general form:

$$C(a,b) = \int_R f(t)\psi_{a,b}(t)dt$$
$$\psi_{a,b}(t) = \frac{1}{\sqrt{a}}\psi\left(\frac{t-b}{a}\right) \tag{1}$$

The function $\psi_{a,b}$ is a dilated and displaced version of the "mother wavelet" ψ, where the parameters a and b indicate scale and translation respectively. The possibility of reconstruction of the original signal from some of the obtained basic blocks has been used for noise and interferences elimination in ECG, abnormal pattern recognition, complex detection, etc.

On the other hand, BSS, as a processing tool, is able to recover signals from a linear combination of these same signals. The simplest model of BSS takes on the presence of n statically independent signals and n observed linear and instantaneous mixtures. In this work, the independence and nongaussianity of the atria and ventricle as signal sources is taken on. Recent works have studied the propagation mechanisms and uncoordinated atrial activation of the AF to demonstrate this assumption [7].

The independent component analysis based on higher order statistics is the support for the different methodologies that solve the problem of BSS. The BSS model in its more compact form is given by

$$x(t) = \sum_{j=1}^{n} a_{ij} \cdot s_j(t)$$
$$x(t) = A \cdot s(t) \tag{2}$$

where, s(t) is a vector of n V_l columns which contains the estimated sources, x(t) is the vector of the mixtures and A is the square mixing matrix. As BSS tries to recover s(t) from the observations, x(t) is necessary to estimate the matrix A. If the ICA methods can estimate the separation matrix, the independent sources can be expressed as follows

$$y(t) = W \cdot s(t) \quad (3)$$

where y(t) is the estimated sources and W is the inverse matrix of A.

3 Database

The signals used are created from recordings of an own database of ECG, with signals obtained at the Cardiac Electrophysiology Laboratory of the University Clinical Hospital in Valencia and diagnosed by cardiologists. All the registers have been pre-processed and normalized to remove possible fluctuations of the base line, interferences, noises, etc. Leads V1 and V5 have been used from 12-lead ECG and from Holter system. The final configuration of the database is shown in Table 1.

Table 1. Registers in database.

	12-leads ECG.	Holter System
Number of registers	29	15

4 Method

The principal aim of this study is to provide enough useful information to the BSS implemented system, only from the V1 and V5 leads to achieve efficiently the AA extraction. This rise of useful information is obtained by increasing the observed mixtures of the signal from the decomposition of each lead into six transformed signals using a wavelet transform with the corresponding levels. This idea has been studied in several recent works that probe the increase of the quality blind source separation if the sparse representability of the sources is exploited [8].

This methodology can be expressed by

$$V1 \Rightarrow [C\_V1_1, C\_V1_2, ..., C\_V1_6]$$
$$V5 \Rightarrow [C\_V5_1, C\_V5_2, ..., C\_V5_6] \quad (4)$$
$$x(t) = [C\_V1_i; C\_V5_i] = A \cdot s_{V1,V5}(t)$$

where the coefficients $C\_V i_j$ represent the six obtained signals from the original leads V1 and V5 yielding several spectral and temporal representations for each lead, that can be considered as different mixtures. This wavelet stage uses denoising techniques that intensify the AA in the sources.

The function *"symlet7"* and six levels of decomposition are the wavelet family and configuration that offers the best performance. Several BSS algorithms have been tested, but the *FastIca* is the most efficient for the BSS stage and offers the lowest computational load.

Spectral analyses have been used to identify the AA between the obtained signals, s_1 and s_2. The signal with a principal frequency peak in the band of 5-8 Hz and higher spectral concentration in this range –as it is usual in an AF episode- is identified as AA.

5 Results

The leads V1 and V5 of an example ECG from the used database and the final extraction using only DPWT and WDBSS are shown in Figure 1. In the results with DPWT, there are residual QRS complexes which are eliminated using the second stage with BSS. The characteristic "*f*" waves hold the form observed in the original lead perfectly. In the case of DPWT, the comparison of the signal fragments corresponding to the AA shows high values of correlation too, but the presence of the mentioned VA reduces the final performance.

The spectral concentrations of the estimated AA in the different methodologies are shown in figures 2 and 3. The dispersion of the spectral distribution is lower in the results obtained with WDBSS and a mean peak of frequency can be identified in the range between 5 and 8 Hz clearly. This fact makes it easier to identify this AA as an atrial fibrillation episode, from a spectral point of view.

In Figure 4, we can appreciate the differences between the wave form of the original leads and the form of the main extracted sources (identified as possible AA) using the WDBSS method and the extracted signals using only BSS algorithms (JADE, FastICA). As it can be observed, the stage implemented without the previous wavelet decomposition doesn´t work properly in any case, comparing with the WDBSS method. The obtained sources don´t show the typical and expected "f" waves and the spectral study can´t identify these signals as an atrial activity.

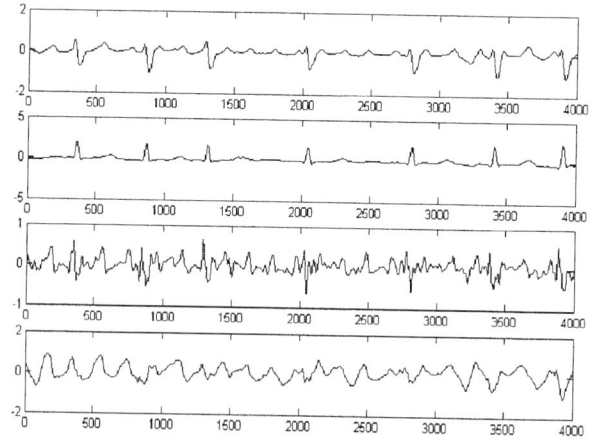

Fig. 1. Original leads V1 and V5 (upper) and extracted AA using DPWT (middle) and WDBSS (low).

The previous methodologies have been applied to the registers of Table 1. The obtained results of principal frequency peak and spectral concentration in the typical band of AF, with mean values and standard deviation, are shown in Table 2.

Table 2. Spectral parameters obtained with DPWT and WDBSS.

	WDBSS	DPWT
Principal Peak (Hz)	5.62 ± 1.02	5.84 ± 1.98
Spect.Concentration	0.43 ± 0.04	0.14 ± 0.05

The spectral concentration in the band 5-8 Hz is much lower in the case of DPWT (<15%). If the WDBSS is used, these values increase more than 200%. The analysis with higher order statistics allows the processing of signals with abnormal beats, and the high values of standard deviation of the principal peak show the lower reliability of the DPWT system in the case of real AF episodes.

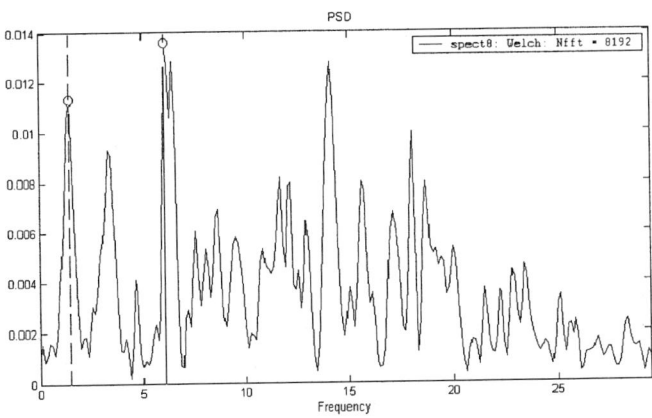

Fig. 2. AA spectral distribution estimated with DPWT.

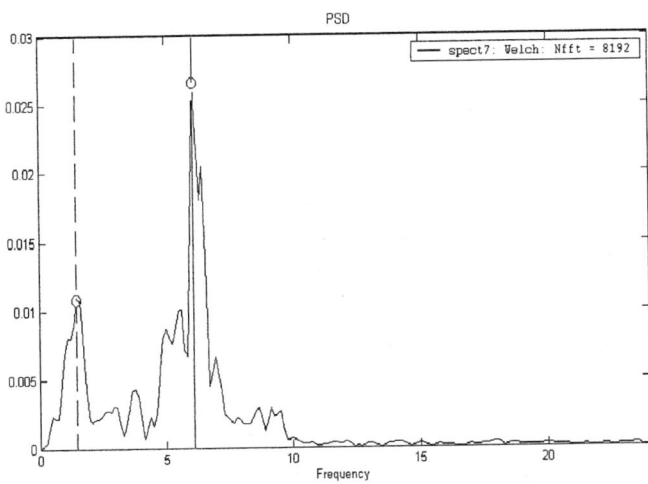

Fig. 3. AA spectral distribution estimated with WDBSS.

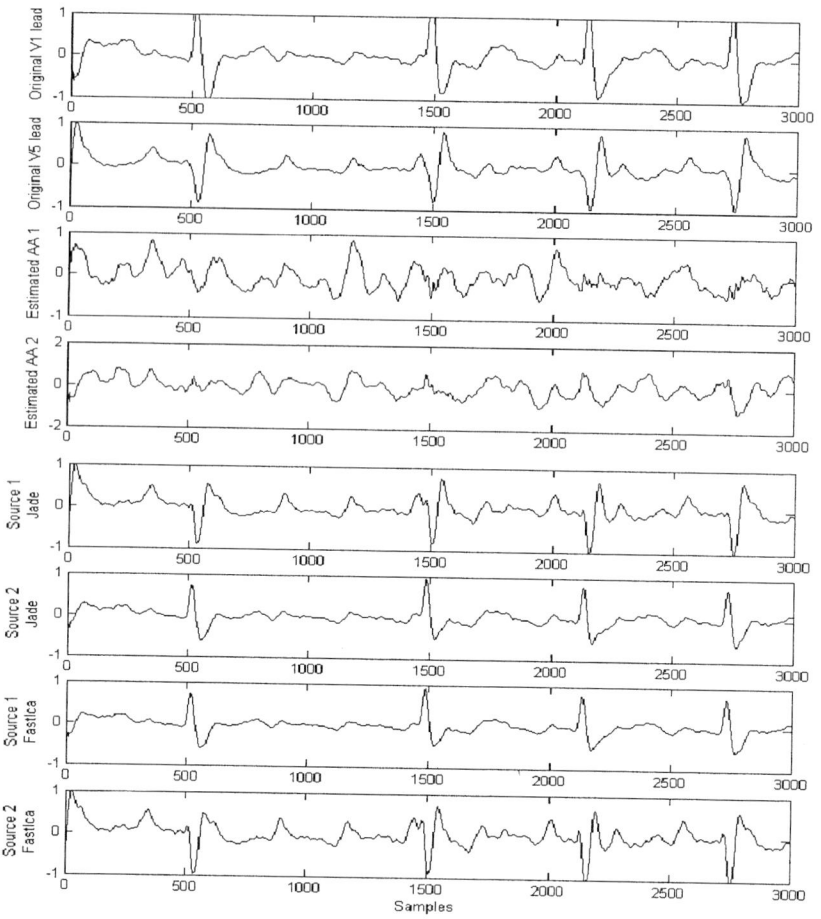

Fig. 4. Wave form comparisons between original V1 and V5 leads, main extracted sources with WDBSS and extracted sources using JADE and FastIca methods without wavelet decomposition.

6 Conclusions

Throughout this paper, the possibilities, as an AA extraction technique, of the system that implement the DPWT and BSS jointly have been shown. The initial hypothesis of statistical independence atria-ventricle and the increase of useful information obtaining from the wavelet decomposition have been demonstrated.

The reliability of this method in the case of real AF episodes has been probed, in contrast with the results of the DPWT method or the use of BSS methodologies without a previous decomposition.

This is an important step to find an atrial extraction method applicable in short duration registers with a reduced number of leads. In this sense, this process should be applicable in arrhythmia detection and analysis, like paroxysmal atrial fibrillation, which have to be usually detected from Holter systems.

Acknowledgements

This work was partly funded by several research grants from the D.I.E.E.A. in the University of Castilla-La Mancha and the TIC2002–00957. The authors would like to thank cardiologists R. Ruiz, S. Morell, F.J. Chorro and R. Garcia, for their clinical advices and kind help in obtaining the signals and Dr. J. I. Albentosa from the University of Castilla-La Mancha for his corrections.

References

1. Falk RH. Medical progress: Atrial fibrillation. New England Journal of Medicine 2001; 344(14):1067-1078.
2. Rieta JJ, Castells F, Sanchez C, Igual J. ICA applied to atrial fibrillation analysis. International Conference on Independent Component Analysis and Blind Signal Separation (ICA) 2003; 4:59-64.
3. Stridh M, Sörnmo L. Spatiotemporal QRST cancellation techniques for analysis of atrial fibrillation. IEEE Trans Biomed Eng 2001;48(1):105-111.
4. Langley P, Stridh M, Rieta JJ, Sörnmo L, Millet-Roig J, Murray A. Comparison of Atrial Rhythms Extraction Techniques for the Estimation of the Main Atrial Frequency from the 12-lead Electrocardiogram in Atrial Fibrillation . IEEE Computers in Cardiology 2002;29:29-32.
5. Shkurovich S, Sahakian AV, Swiryn S. Detection of atrial activity from high-voltage leads of implantable ventricular defibrillators using a cancellation technique. IEEE Trans Biomed Eng 1998;45(2):229-234.
6. Sánchez C, Millet J, Rieta JJ, Ródenas J, Castells F. Packet Wavelet Decomposition: An Approach to Atrial Activity Extraction. IEEE Computers in Cardiology 2002;29: 33-36.
7. Rieta J.J., Millet J., Zarzoso V., Castells F., Sánchez C., Garcia R., Morell S. Atrial Fibrillation, Atrial Flutter And Normal Sinus Rhythm Discrimination By Means of Blind Source Separation And Spectral Parameters Extraction. IEEE Computers in Cardiology 2002;29:25-28.
8. Zibulesvsky M., Pearlmutter B.A., Blind Source Separation by Sparse Decomposition. Neural Computations 2001;13(4).

A New Auditory-Based Index to Evaluate the Blind Separation Performance of Acoustic Mixtures

Juan Manuel Sanchis[1], José Joaquín Rieta[1], Francisco Castells[1], and José Millet[2]

[1]Universidad Politécnica de Valencia,
46730 Gandia, Spain
{jmsanch,jjrieta,fcastells}@eln.upv.es
[2]Universidad Politécnica de Valencia,
46020 Valencia, Spain
jmillet@eln.upv.es

Abstract. A new method to evaluate the performance of convolutive blind signal separation (BSS) algorithms in acoustic mixtures is presented. The method is able to compute the spectral level enhancement, in a frequency-band based fashion, between the estimated and the residual sources, thus combining two previously defined parameters in time and frequency domain: the signal to interference ratio and the spectral preservation index. The new index is able to compute the quality of separation performing the spectral computations over a set of logarithmically spaced frequency bands in a similar way as the human auditory system. Obtained results clearly verify that this methodology is a more realistic approach to evaluate the convolutive BSS results of acoustic mixtures.

1 Introduction

In the blind source separation (BSS) problem the objective is to separate multiple sources, mixed through an unknown mixing system (or channel), using only the system output data (observed signals) and in particular with the absence (or least amount) of information about the sources or the mixing system. Today, it is well known the successful application of BSS techniques in a wide variety of fields including speech processing, data communication, biomedical signal processing, etc.

An extensive part of the available literature has been directed toward the simpler case of instantaneous mixtures; i. e., when the observed signals are generated through a linear combination of the sources and no time-delays are involved in the mixing model [1-3]. A more challenging case arises when dealing with convolutive mixing systems; i.e., where the sources are mixed through a linear filtering operation and the observed signals are linear combinations of the sources and their corresponding scaled and delayed versions [4-6]. A difficult practical example, and possibly one of the most extended convolutive BSS problem, regards the separation of audio signals mixed in a reverberant environment.

To evaluate the results of BSS algorithms in acoustic mixtures two different methods are available in the literature. The first one is based on the impulse response associated to the mixing channels and the separation filters. The quality of separation is calculated by convolving the mixing and separation systems and measuring the approximation degree of the result to the unitary system [7], [8]. The second one establishes a comparison between the nuisance sources and the desired source for each separated signal [9], [10]. To perform the comparison, the aforementioned methods need the availability of the original sources and the mixing system, which is a problem for real world situations. Additionally, other method mainly used in real situations, evaluates the signals' subjective quality of separation, because the original sources and the mixing system are unavailable for comparison purposes. Another relevant aspect is to evaluate the spectral preservation of the recovered signals but few methods in the bibliography consider this problem [8].

In the present work, a new index to evaluate the separation performance is presented based on two parameters, previously defined in the bibliography, for measuring the quality of separation from two different points of view: signal to interference ratio (*SIR*) in the computed time interval [10] and the spectral preservation index (*SPI*) [8]. These indexes will be used together for the definition of the new spectral enhancement index (*SEI*).

2 Signal to Interference Ratio

The estimation of *SIR* can be performed both in synthetic and real world situations. In the first case, the original source signals s_j ($j = 1,..., N$), the mixing matrix $\mathbf{H}^{MxN}$, the observations x_i ($i = 1,...,M$), the estimated separation matrix $\mathbf{W}^{NxM}$, and the estimated source signals y_j ($j = 1,...,N$) are available. For real world situations, the only available data are the microphone signals, the estimated signals and the separation matrix.

In a synthetic case, the *SIR* for a source signal s_j can be defined as the difference between the *SIR* of s_j in the estimation y_j, and the *SIR* of s_j in the observation x_i. Hence, the difference between the *SIR* of the estimated (SIR_j^e) and observed (SIR_j^o) signals is the parameter where s_j presents the largest power contribution

$$SIR_j = SIR_j^e - SIR_j^o, \quad j = 1,...,N \quad (1)$$

where

$$SIR_j^o = 10\log\frac{E\{(h_{jj} * s_j)^2\}}{E\{(\sum_{\substack{k=1\\k\neq j}}^{N} h_{jk} * s_k)^2\}}, \quad j = 1,...,N \quad (2)$$

being $E\{\cdot\}$ the mathematical expectation operator and h_{ji} the mixing matrix entries. Taking into account that the global transfer function of the mixing-separation system can be defined as $\mathbf{G} = \mathbf{W} * \mathbf{H}$, one can formulate the *SIR* of the estimated signal as

$$SIR_j^e = 10\log\frac{E\{(g_{jj} * s_j)^2\}}{E\{(\sum_{\substack{k=1\\k\neq j}}^{N} g_{jk} * s_k)^2\}}, \quad j = 1,...,N \quad (3)$$

where g_{jk} are the entries of global matrix.

When an ideal separation is not achievable ($\mathbf{G} \neq \mathbf{I}$), Eq. (3) performs the comparison between a modified source signal and the residual sources present in the estimation.

Regarding real world situations, where we have no access to the original signals s_j and the mixing matrix $\mathbf{H}$, the measurement of the *SIR* index given in Eq. (1) would not be applicable, unless just one of the sources is active during a certain time interval [11]. In this case, the observed signal by the microphone x_i can be defined as

$$x_i = \sum_{j=1}^{N} h_{ji} * s_j = \sum_{j=1}^{N} x_{ji}, \qquad j = 1,...,N \qquad (4)$$

Therefore, we can define the observed *SIR*, with respect to the rest of the signals present in that observation, as

$$SIR_j^o = 10 \log \frac{E\{(x_{ji})^2\}}{E\{(\sum_{\substack{k=1 \\ k \neq j}}^{N} x_{ki})^2\}}, \qquad j = 1,...,N \qquad (5)$$

and the estimated *SIR*, given the separation matrix $\mathbf{W}$, as the quality of separation of source s_j, with respect to the rest of the signals present in that estimation, as

$$SIR_j^e = 10 \log \frac{E\{(\sum_{i=1}^{M} w_{ij} * x_{ji})^2\}}{E\{(\sum_{i=1}^{M} w_{ij} * \sum_{\substack{k=1 \\ k \neq j}}^{N} x_{ki})^2\}}, \qquad j = 1,...,N \qquad (6)$$

3 Spectral Preservation

A complementary way to estimate the performance of a BSS algorithm lies in the analysis of the preservation capability, with respect to the spectral components of the signal, by comparing the spectral content of the estimated sources to the original ones. This kind of measure has special importance when the main goal is focused on recovering the signal with the highest fidelity, in contrast with the situations where the relevance of fidelity is shaded by intelligibility. Hence, the spectral preservation index (*SPI*) can be as [8]

$$SPIj = E\{|P_{s_j}(f) - P_{y_j}(f)|^2\}, \qquad j = 1,...,N \qquad (7)$$

where $P_{sj}(f)$ is the power spectral density (PSD) of the jth original source and $P_{yj}(f)$ the PSD of the jth estimated signal.

Note that the aforementioned defined parameters, *SIR* and *SPI*, present some deficiencies when compared to the human auditory system. Regarding the computation of *SIR*, the mean value of the total signal power is obtained over a concrete time interval, with no consideration of the spectral content of the source signals. Hence, this index gives no information about the real quality of separation because the spectral overlapping degree of the sources is not considered at all. In the case of *SPI*, the spectral content of the sources is analyzed but the computation is performed linearly over the frequency axis, hence, the perceptual behavior of the human auditory system is not considered.

4 Spectral Enhancement Index

In order to define an auditory-based index, capable of measuring the quality of separation in a similar way as the human auditory system, the spectral computations should be performed over a logarithmic frequency axis, more specifically, in critical bands. Considering this fact, is it possible to define the spectral enhancement index *SEI* that can be obtained following this steps: firstly, the contribution of each independent source s_j is obtained separately on the observation point. This will give us the contribution of the jth source into the ith observation in the same way as has been defined in Eq. (4). The second step consists of obtaining the difference between the PSD of the source with largest power contribution into the ith observation (x_{ji}) with respect to the PSD of the rest of the sources in that observation, which can be defined as

$$SEI^o_{ji}(f) = P_{x_{ji}}(f) - \sum_{\substack{k=1 \\ k \neq 1}}^{N} P_{x_{ki}}(f), \quad \begin{array}{l} j=1,\dots,N \\ i=1,\dots,M \end{array} \quad (8)$$

The next step deals with the application of the BSS algorithm and the evaluation of the quality of separation over each estimated source. Then, the difference between the PSD of the jth source with largest power contribution into the lth estimation (y_{jl}) with respect to the PSD of the rest of the sources in that estimation, can then be computed as

$$SEI^e_{jl}(f) = P_{y_{jl}}(f) - \sum_{\substack{k=1 \\ k \neq j}}^{N} P_{y_{kl}}(f), \quad j,l=1,\dots,N \quad (9)$$

Finally, the *SEI* for the jth source can be defined as

$$SEIj = E\{SEI^e_{jl}(f)|_{1/3} - SEI^o_{ji}(f)|_{1/3}\}, \quad j=1,\dots,N \quad (10)$$

where $SEI^e_{jl}(f)|_{1/3}$ and $SEI^o_{ji}(f)|_{1/3}$ are the spectral enhancement indexes for the estimated and observed sources evaluated in 1/3 octave bands, respectively.

5 Results

The proposed methodology has been applied over real recordings corresponding to an acoustic guitar (s_1) and a male speaker (s_2) in the recording studio of Gandia Higher School of Technology. The resulting impulse response between each source and two microphones are plotted in Fig. 1.

The signals were 4 seconds in length and sampled at 44.1 kHz. Fig. 2 plots the PSD of source s_1 and s_2 in the observation x_1. After the application of the estimated separation matrix **W** (Fig. 3) limited to 16384 taps, it is possible to obtain a set of two estimated sources. Fig. 4 plots the PSD of the estimated source s_1 and the interfering source s_2 in the estimation y_1.

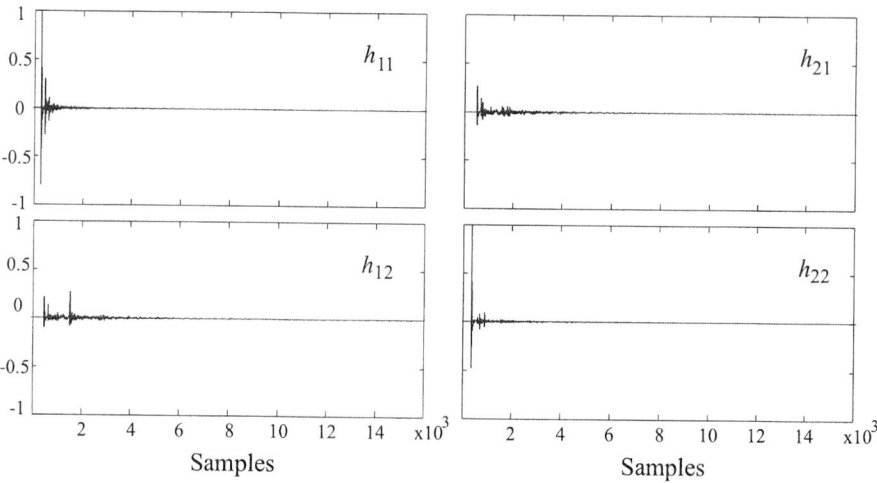

Fig. 1. Mixing matrix **H**, composed by the impulse response among the position of two sources and two microphones located in a recording study.

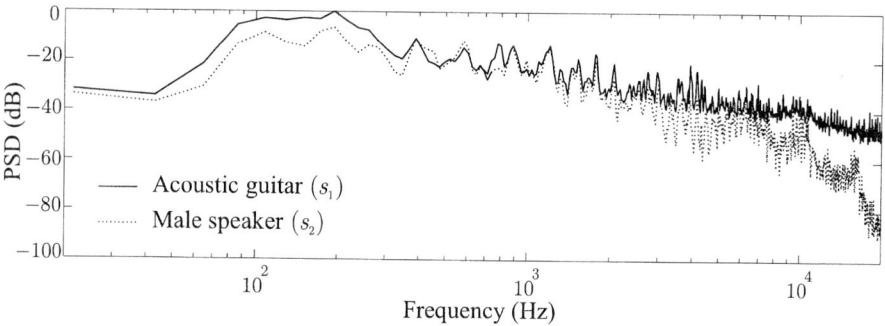

Fig. 2. PSD of the sources s_1 and s_2 in the observation x_1.

Finally, the difference between $SEI_{11}^e(f)$ (Eq. (9)) and $SEI_{11}^o(f)$ (Eq. (8)), in 1/3 octave bands, can be observed in Fig. 5. As shown in this Fig. the quality of the separation decays substantially in the mid-range frequencies, indicating a poor separation performance in this frequency band that would be masked with the use of the previously mentioned indexes *SIR* and *SPI*, thus corroborating that *SEI* is a much more realistic approach to compute the BSS separation.

When a global separation index is required, it is possible to obtain the SEI_1 index as the mean value of the partial band indexes across the whole bandwidth shown in Fig. 5.

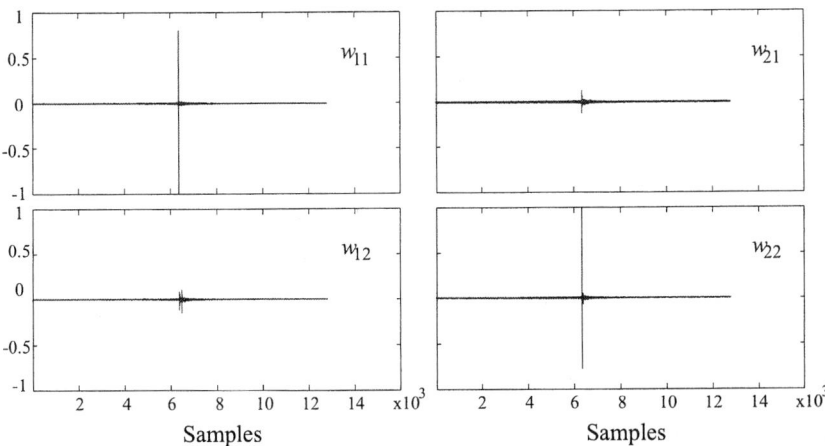

Fig. 3. Separation matrix **W**, obtained from the 16384 taps Hamming window of the mixing matrix inverse ($\mathbf{H^{-1}}$).

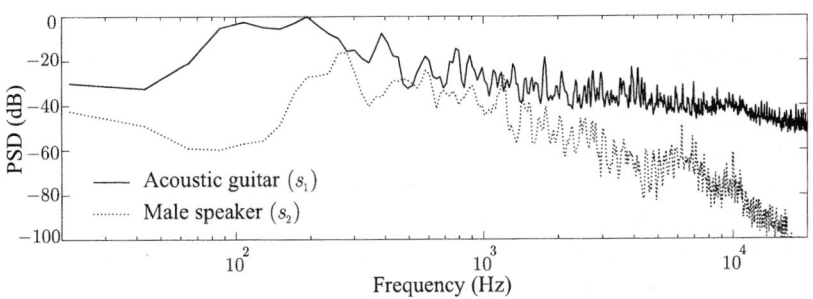

Fig. 4. PSD of the sources s_1 and s_2 in the estimation y_1.

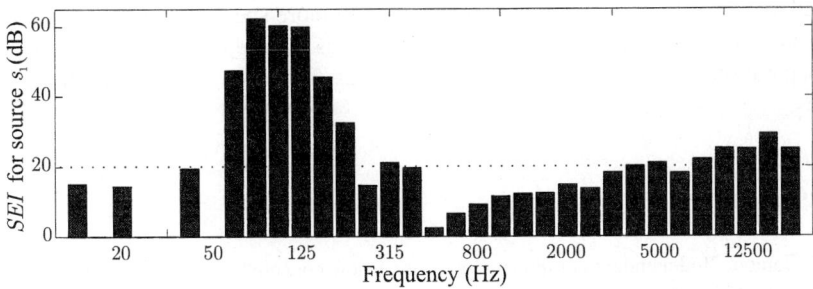

Fig. 5. SEI_1 evaluated in 1/3 octave bands in estimated signal y_1.

Table 1. *SIR* and *SEI* values obtained in the you are considered signal y_1 by lineal average and by average of the value in 1/3 octave bands, respectively.

SIR_1	SEI_1
30.1 dB	21.1 dB

SIR values at different frequencies are linearly averaged in order to obtain the global *SIR*. Using this performance measurement, the *SIR* values at high frequencies are considerably overweighted in comparison to the *SIR* values at low and medium frequencies, and have a significant impact on the global value. Therefore, the performance index provided by *SIR* is far from the performance index that would be obtained if the logarithmic characteristic of the human auditory system had been considered.

This divergence between the values given by *SIR* and *SEI* is even more significant in the cases where the involved audio signals present limited bandwidth. In those situations, both *SIR* and *SPI* will quantify with much lower accuracy the separation degree of the estimated source in contrast to the real perception of the human auditory system.

6 Conclusion

A new index for measuring the separation performance of convolutive BSS algorithms for acoustic mixtures has been defined. The method considers the quality of separation between the estimated source and the secondary sources as a function of frequency . Moreover, the measurement is performed over a set of frequency bands logarithmically spaced similar to the human auditory system.

The most widely used methods to evaluate the quality of separation have been compared to the new proposed methodology in order to show that the results obtained with the latter are much more approximated to reality. This observation can be corroborated through the introduction of the logarithmically spaced frequency-band based computations. Therefore, this methodology establishes a new measurement procedure closer to human perception and, though more spectral and statistic parameters could be defined, this new index can be used as a more realistic basis to evaluate the separation performance in acoustic mixtures of present and future BSS algorithms.

References

1. P. Comon, "Independent component analysis, A new concept?," *Signal Processing*, vol. 36, pp. 287-314, 1994.
2. A. Bell and T. J. Sejnowski, "An information maximization approach to blind separation and blind deconvolution," *Neural Computation*, no. 7, pp. 1129-1159, 1995.
3. J. Cardoso and B. Laheld, "Equivariant adaptative source separation," *IEEE Transaction on Signal Processing*, vol. 44, pp. 3017-3030, Dec. 1996.

4. H. Sahlin and H. Broman, "MIMO signal separation for FIR channels: a criterion and performance analysis," *IEEE Transaction on Signal Processing*, vol. 48, pp. 642-649, March. 2000.
5. E. Weinstein, M. Feder, and A. V. Oppenheim, Multi-channel signal separation by decorrelation", *IEEE Transaction on Signal Processing*, vol. 1, pp. 404-413, Oct. 1993.
6. J. K. Tugnait, "Adaptative blind separation of convoluitve mixtures of independent signals," *Signal Processing*, vol. 73, pp. 139-152, 1999.
7. S. Cruces. *An unified view of Blind Source Separation Algoritmhs*. Ph. D. Thesis, University of Vigo, 1999.
8. K. Kokkinakis, V. Zarzoso and A. Nandi, "Blind separation of acoustic mixtures based on linear prediction analysis," *International Conference on Independent Component Analysis and Blind Signal Separation (ICA)*, vol. 4, pp. 59-64, Nara, Japan, April 2003.
9. S. Ikeda and N. Murata, "A method of blind separation based on temporal structure of signals," *Proceedings of the International Conference on Neural Information Processing, (ICONIP'98)*, pp.737-742, Kitakyushu, Japan, October 1998.
10. S. Araki, R. Mukai, S. Makino, T. Nishikawa, H. Saruwatari, "The fundamental limitation of frequency domain blind source separation for convolutive mixture of speech," *IEEE Transactions on Speech and Audio Processing*, vol. 11, no 2, pp. 109-116, March 2003.
11. D. Schobben, K. Torkkola, and P. Smaragdis, "Evaluation of blind signal separation methods," *International Conference on Independent Component Analysis and Blind Signal Separation (ICA)*, vol. 1, pp. 261-266, Aussois, France, January 1999.

An Application of ICA to Identify Vibratory Low-Level Signals Generated by Termites

Juan Jose G. de la Rosa[1], Carlos G. Puntonet[2],
Juan Manuel Górriz, and Isidro Lloret

[1] University of Cádiz, EPS-Electronics Instrumentation Group,
Av. Ramón Puyol S/N, 11202 Algeciras-Cádiz, Spain
juanjose.delarosa@uca.es
http://www2.uca.es/grup-invest/instrument_electro/ppjjgdr/jjgdr.htm
[2] University of Granada, Department of Architecture and Computers Technology,
ESII, C/Periodista Daniel Saucedo, 18071 Granada, Spain

Abstract. An extended robust independent components analysis algorithm based on cumulants is applied to identify vibrational alarm signals generated by soldier termites (*reticulitermes grassei*) from background noise. A seismic accelerometer is employed to characterize acoustic emissions. To support the proposed technique, vibrational signals from a low cost microphone were masked by white uniform noise. Results confirm the validity of the method, taken as the basis for the development of a low cost, non-invasive, termite detection system.

1 Introduction

Termites damage structures world-wide irreparably. The costs of this harm could be significantly reduced through earlier detection. Detection is also important because environmental laws are becoming more restrictive with termiticides due to their health threats. Besides, only about 25 percent of the building structure is accessible, and the conclusions depend very much on subjectiveness [1]. Thus, new techniques have been developed to gain accessibility. But at best they are considered useful only as supplements. Acoustic methods have emerged as an alternative.

When wood fibers are broken by termites they produce acoustic signals which can be monitored using *ad hoc* resonant acoustic emission (AE) piezoelectric sensors which include microphones and accelerometers, targeting subterranean infestations by means of spectral and temporal analysis. The drawback is the relative high cost and their practical limitations (biophysical factors).

Modern signal processing techniques can be used to distinguish insect sounds from background noise with good reliability in soil, because sound insulating properties of soil help reduce interference. Besides, such techniques have been successfully used in relatively noisy urban environments [2], [3].

The particular contribution of this study is to show that a robust ICA cumulant-based algorithm is capable of separating termite alarm signals, generated in wood and recorded using a low cost microphone, from background noise. This could be the basis of separating low-level termite activity signals from background urban noise using cheap equipment with non-invasive sensors. A seismic accelerometer was used to characterize the frequency contents. Data were acquired in the "Costa del Sol" (Malaga, Spain), in subterranean wood structures and roots.

The paper is structured as follows: Section 2 summarizes the methods for acoustic detection of termites; Section 3 defines the ICA model and outlines the characteristics of emissions in wood; Section 4 describes the experiments carried out. Conclusions are drawn in Section 5.

2 Acoustic Detection of Termites: Characteristics and Devices

Acoustic emission (AE) is the elastic energy that is spontaneously released by materials undergoing deformation. This energy travels through the material as a stress and can be detected using a piezoelectric transducer.

Termites use a sophisticated system of vibratory long distance alarm. When disturbed in their extended galleries, soldiers produce vibratory signals by drumming their heads against the substratum [4]. The signals consist of pulse trains which propagate through the substrate with pulse repetition rates (beats) in the range of 10-25 Hz, with burst rates around 500-1000 ms, depending on the species [3]. Workers perceive the vibrations, become alert and tend to escape. Figure 1 shows a typical drumming signal produced by a soldier by taping its jaws against a chip of wood. It comprises two four-impulse bursts. Each of the pulses arises from a single, brief tap of the jaw.

Signals' amplitudes were highly variable and depend on the wood and strength of the taps. Power spectrum of a single impulse shows that significant drumming responses are produced over the range 200 Hz-10 kHz and the carrier frequency is around 2600 Hz. The spectrum is not flat as a function of frequency as one

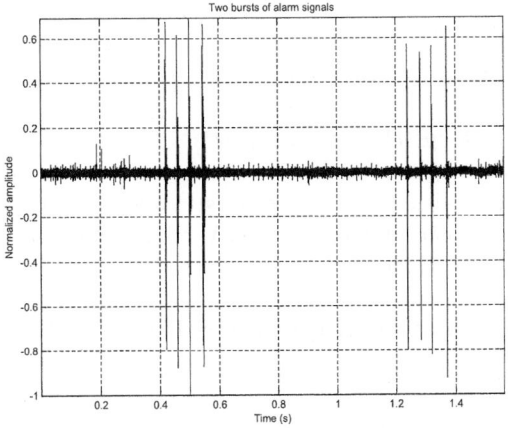

Fig. 1. Two bursts of a typical AE alarm signal produced by a soldier.

would expect for a pulse-like event. This is due to the frequency response of the microphone, and also to the frequency-dependent attenuation coefficient of the wood.

AE sensors have been used primarily for detection of termites in wood [5], but there is also the need of detecting termites in trees and soil surrounding building perimeters. Soil and wood have a much longer coefficient of sound attenuation and distortion than air($\sim$600 dB m^{-1}, compared with 0.008 dB m^{-1} in the air), and the coefficient increases with frequency [2].This attenuation reduces the detection range of the emission to 2-5 cm in soil and 2-3 m in wood, as long as the sensor is in the same piece of material [5].

3 The ICA Model and Its Properties

3.1 Outline of ICA

Blind source separation (BSS) by ICA is receiving attention because of its applications in many fields such as speech recognition, medicine and telecommunications [6],[7],[8]. Statistical methods in BSS are based in the probability distributions and the cumulants of the mixtures. The recovered signals (the source estimators) have to satisfy a condition which is modelled by a contrast function. The underlying assumptions are the mutual independence among sources and the non-singularity of the mixing matrix [6],[9],[10].

Let $\mathbf{s}(t) = [s_1(t), s_2(t), \ldots, s_m(t)]^T$ be the vector of unknown sources (statistically independent), where the superscript represents transpose. Independence means one source provides no further information about any other [11]. The mixture of the sources is modelled by

$$\mathbf{x}(t) = \mathbf{A} \cdot \mathbf{s}(t) \quad (1)$$

where $\mathbf{x}(t) = [x_1(t), x_2(t), \ldots, x_m(t)]^T$ is the available vector of observations and $\mathbf{A} = [a_{ij}] \in \Re^{m \times n}$ is the unknown mixing matrix, modelling the environment in which signals are mixed, transmitted and measured [12]. We assume that $\mathbf{A}$ is a non-singular n×n square matrix. The goal of ICA is to find a non-singular n×m separating matrix $\mathbf{B}$ such that extracts sources via

$$\hat{\mathbf{s}}(t) = \mathbf{y}(t) = \mathbf{B} \cdot \mathbf{x}(t) = \mathbf{B} \cdot \mathbf{A} \cdot \mathbf{s}(t) \quad (2)$$

where vector $\mathbf{y}(t) = [y_1(t), y_2(t), \ldots, y_m(t)]^T$ is an estimator of the sources [13],[14]. The separating matrix has a scaling freedom on each of its rows because the relative amplitudes of sources in $\mathbf{s}(t)$ and columns of $\mathbf{A}$ are unknown [6],[10],[14]. The transfer matrix $\mathbf{G} \equiv \mathbf{BA}$ relates the vector of independent original signals to its estimators [15].

3.2 The Implementation of the Algorithm

High order statistics, known as cumulants, are used to infer new properties about the data of non-Gaussian processes [16],[17]. Before cumulants, such processes had to be treated as if they were Gaussian. Cumulants and polyspectra reveal information about amplitude and phase, whereas second order statistics are phase-blind [18],[19]. The relationship among the cumulant of r stochastic signals and their moments of order $p, p \leq r$, can be calculated by using the *Leonov-Shiryayev* formula [17],[18]

$$Cum(x_1,...,x_r) = \sum (-1)^k \cdot (k-1)! \cdot E\{\prod_{i \in v_1} x_i\}$$
$$\cdot E\{\prod_{j \in v_2} x_j\} \cdots E\{\prod_{k \in v_p} x_k\} \quad (3)$$

where the addition operator is extended over all the set of v_i ($1 \leq i \leq p \leq r$) and v_i compose a partition of $1,...,r$.

It has been proved that a set of random variables are statistically independent if their cross-cumulants are zero [14]. This property can be used to define a contrast function. A criteria to obtain this function is to minimize the distance between the cumulants of the sources $\mathbf{s}(t)$ and the outputs $\mathbf{y}(t)$. But in a real situation sources are unknown, so it is necessary to involve the observed signals. Separation of the sources can be developed using the following contrast function based on the entropy of the outputs [9],[14]

$$H(\mathbf{z}) = H(\mathbf{s}) + log[det(\mathbf{G})] - \sum \frac{\mathbf{C}_{1+\beta, y_i}}{1+\beta} \quad (4)$$

where $\mathbf{C}_{1+\beta, y_i}$ is the $1+\beta$th-order cumulant of the ith output, $\mathbf{z}$ is a non-linear function of the outputs y_i, $\mathbf{s}$ is the source vector, $\mathbf{G}$ is the global transfer matrix of the ICA model and $\beta > 1$ is an integer verifying that $\beta + 1$-order cumulants are non-zero.

Using this contrast function it has been demonstrated [14] that the separating matrix can be obtained by means of the following recurrent equation

$$\mathbf{B}^{(h+1)} = [\mathbf{I} + \mu^{(h)}(\mathbf{C}_{y,y}^{1,\beta}\mathbf{S}_y^\beta - I)]\mathbf{B}^{(h)} \quad (5)$$

where $\mathbf{S}_y^\beta$ is the matrix of the signs of the output cumulants. Equation (5) can be interpreted as a quasi-Newton algorithm of the cumulant matrix $\mathbf{C}_{y,y}^{1,\beta}$. The learning rate parameters $\mu^{(h)}$ and η are related by

$$\mu^{(h)} = \min(\frac{2\eta}{1+\eta\beta}, \frac{\eta}{1+\eta\|\mathbf{C}_{y,y}^{1,\beta}\|_p}) \quad (6)$$

with $\eta < 1$ to avoid $\mathbf{B}^{(h+1)}$ being singular; $\|.\|_p$ denotes de p-norm of a matrix. The adaptative equation (5) converges, if the matrix $\mathbf{C}_{y,y}^{1,\beta}\mathbf{S}_y^\beta$ tends to the identity.

4 Results and Discussions

Data acquisition took place in a basement, using a low-cost microphone, *Ariston* CME6 model, with a sensibility of 62±3 (dB) and a bandwidth of 100 Hz-8 kHz, connected to the sound card of a portable computer (96000 Hz, sample frequency).

High-pass filtering suppresses non-relevant low-frequency coupling from the sensor and the environment, obtaining two zero-mean normalized bursts (sources 1 and 2). Normalized kurtosis are 212.93, and 211.09, respectively; which shows that ICA is expected to work. The third and forth sources consist of two uniform distributed noise signals with enough amplitude to mask the burst. The mixing matrix is a 4×4 matrix whose elements are chosen from uniformly distributed random numbers within 0 and 1. No pre-whitened was applied in order to manipulate four mixtures.

In order to compare this method with traditional ones, based on power spectrum comparisons, we compared the power spectra of the separated signals to the original sources of *reticulitermes grassei*.

AE methods work under the hypothesis of considering the vibratory signals as pulse trains. Characterization was developed using a seismic accelerometer (KB12V, MMF). Figure 2 shows a comparison between the impulse response (upper graph) of the accelerometer and the spectrum of the drumming signals, which let us conclude the 2600 Hz peak corresponding to the carrier [1],[3].

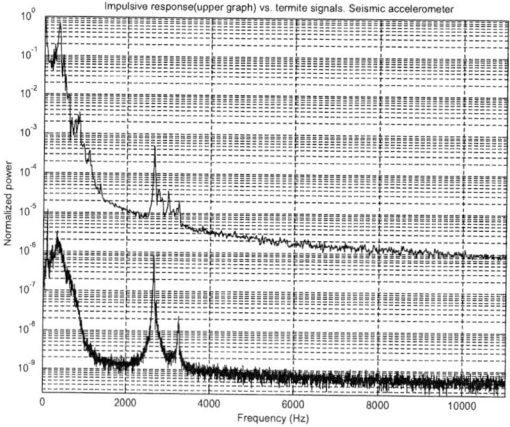

Fig. 2. Comparison between impulsive response and spectrum of vibratory alarm signals.

Figure 3 shows the original filtered sources and the mixtures, which give very little information about the original sources. Comparing the separated results, in figure 4, with the source signals in figure 3, a number of differences are found. First, the amplitudes are amplified to some extent due to the changes in the demixing matrix, implying that original amplitude information has lost. Second, there are time shifts between the original sources and the recovered signals. Three, the sequences are arranged as the same way as the original, although this can be changed.

Figure 5 show the normalized power spectrum of the second output. The spectra of the separated signals $y_1(t)$ and $y_2(t)$ show the same carrier frequency,

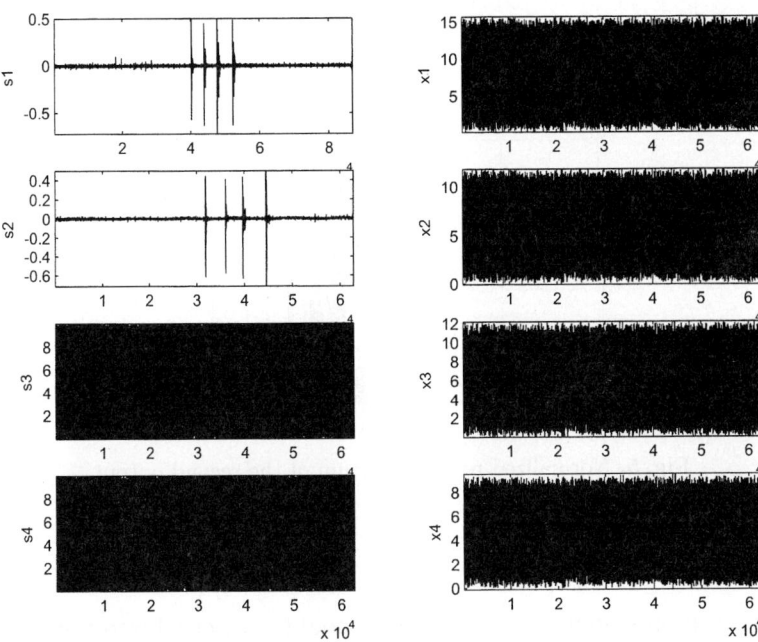

Fig. 3. The sources and their mixtures. Horizontal units: 1/96000 (s).

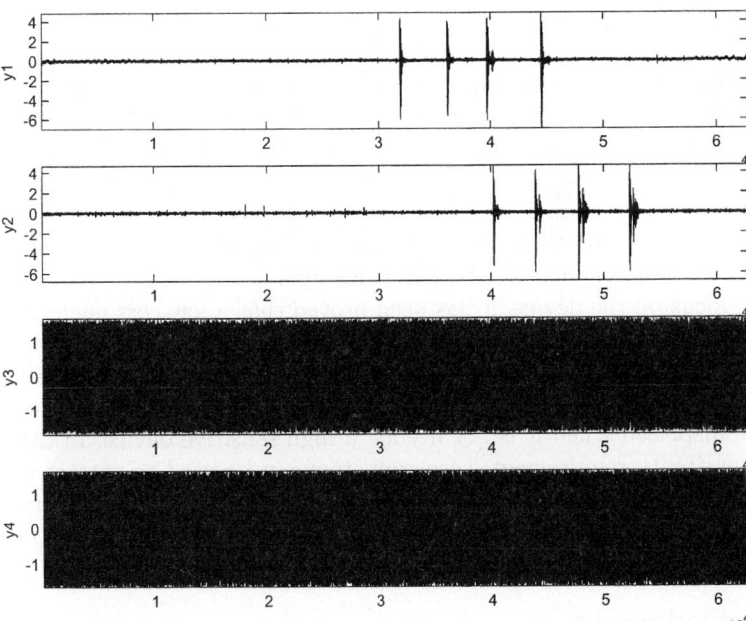

Fig. 4. The separation results by the ICA algorithm. Horizontal units: 1/96000 (s).

confirming the validity of the proposed method based on the traditional spectra-based method.

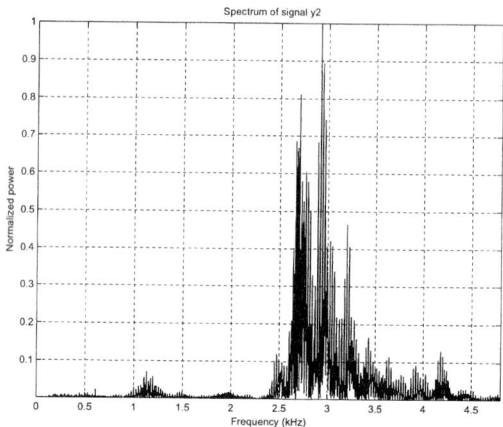

Fig. 5. Normalized power spectrum of the second output.

5 Conclusions

ICA has been presented as a novel method used to detect vibratory signals from termite activity. This method is far different from traditional ones, as power spectrum, which obtain an energy diagram of the different frequency components, with the risk that low-level sounds could be masked.

This experience shows that the algorithm is able to separate the sources with small energy levels in comparison to the background noise. This is explained away by statistical independence basis of ICA, regardless of the energy associated to each frequency component. Results of the spectra let us conclude that the separation has been performed correctly, because the same spectral shape as the accelerometer response is outlined. In this stage we have proved the validity of ICA over a pre-processed set of signals. No frequency-domain comparison is made; a time-domain characterization is enough.

If we focus on the device, it has been proved that a low-cost microphone can be used for insect-detection purposes. This is so because in case of high-level background noise, even if it is white, as it has been proved, ICA is capable of extracting the burst of impulses. This means that accelerometers-based equipment could be displaced when it is not needed a high sensitive device. In the case of a high sensibility requirement, accelerometers can be used to extract distorted information which would be processed by ICA to extract the vibratory signals produced by insects.

Acknowledgment

The authors would like to thank the *Spanish Ministry of Science and Technology* for funding the project DPI2003-00878, and the *Andalusian Autonomous Government Division* for funding the research with *Contraplagas Ambiental S.L.*

References

1. Robbins, W., Mueller, R., Schaal, T., Ebeling, T.: Characteristics of acoustic emission signals generated by termite activity in wood. In: Proceedings of the IEEE Ultrasonic Symposium. (1991) 1047–1051
2. Mankin, R., Fisher, J.: Current and potential uses of acoustic systems for detection of soil insects infestations. In: Proceedings of the Fourth Symposium on Agroacoustic. (2002) 152–158
3. Connétable, S., Robert, A., Bouffault, F., Bordereau, C.: Vibratory alarm signals in two sympatric higher termite species: Pseudacantotermes spiniger and p. militaris (termitidae, macrotermitinae). Journal of Insect Behaviour **12** (1999) 90–101
4. Röhrig, A., Kirchner, W., Leuthold, R.: Vibrational alarm communication in the african fungus-growing termite genus macrotermes (isoptera, termitidae). Insectes Sociaux **46** (1999) 71–77
5. Mankin, R., Osbrink, W., Oi, F., Anderson, J.: Acoustic detection of termite infestations in urban trees. Journal of Economic Entomology **95** (2002) 981–988
6. Puntonet, C.: New Algorithms of Source Separation in Linear Media. PhD thesis, University of Granada, Department of Architecture and Technology of Computers, Spain (1994)
7. Mansour, A., Barros, A., Onishi, N.: Comparison among three estimators for higher-order statistics. In: The Fifth International Conference on Neural Information Processing, Kitakyushu, Japan (1998)
8. A. Mansour, N. Ohnishi, C.P.: Blind multiuser separation of instantaneous mixture algorithm based on geometrical concepts. Signal Processing **82** (2002) 1155–1175
9. Puntonet, C., Mansour, A.: Blind separation of sources using density estimation and simulated annealing. IEICE Transactions on Fundamental of Electronics Communications and Computer Sciences **E84-A** (2001)
10. Hyvärinen, A., Oja, E.: Independent Components Analysis: A Tutorial. Helsinki University of Technology, Laboratory of Computer and Information Science (1999)
11. Lee, T., Girolami, M., Bell, A.: A unifying information-theoretic framework for independent component analysis. Computers and Mathematics with Applications **39** (2000) 1–21
12. Zhu, J., Cao, X.R., Ding, Z.: An algebraic principle for blind source separation of white non-gaussian sources. Signal Processing **79** (1999) 105–115
13. Cardoso, J.: Blind signal separation: statistical principles. Proceedings of the IEEE **9** (1988) 2009–2025
14. Górriz, J.: Hybrid Algorithms for Time-Series Modelling Using AR-ICA Tecnniques. PhD thesis, University of Cádiz, Department of Systems' Engineering and Electronics, Spain (2003)
15. Ham, F., Faour, N.: Infrasound Signal Separation using Independent Component Analysis. Sponsored by the Boeing Company, Contract No. 7M210007 (2002)
16. Hinich, M.: Detecting a transient signal by biespectral analysis. IEEE Trans. Acoustics **38** (1990) 1277–1283
17. Nykias, C., Mendel, J.: Signal processing with higher-order spectra. IEEE Signal Processing Magazine (1993) 10–37
18. Mendel, J.: Tutorial on higher-order statistics (spectra) in signal processing and system theory: Theoretical results and some applications. Proceedings of the IEEE **79** (1991) 278–305
19. Swami, A., Mendel, J., Nikias, C.: Higher-Order Spectral Analysis Toolbox User's Guide. (2001)

Application of Blind Source Separation to a Novel Passive Location

Gaoming Huang[1,2], Luxi Yang[1], and Zhenya He[1]

[1] Dept. of Radio Engineering, Southeast University, Nanjing, 210096, China
{lxyang,zyhe}@seu.edu.cn
[2] Naval University of Engineering, Wuhan, 430033, China
redforce@sohu.com

Abstract. The location of emitter by passive sensor arrays has received considerable attention in recent years. It is important in Electronic Warfare (EW) system. To solve the problems of a passive location method based on broadcast & television signals, this paper proposes a novel locating approach mainly based on Blind Source Separation (BSS). It realizes the signal separation, echo extraction, and estimation of time difference in strong direct wave environments. By the combination of the Direction of Arrival (DOA) estimation, the location issue can be transferred into the time difference & direction location, and therefore a high precision can be obtained. The implementation of this passive location method is simple and computationally efficient. The simulations show that the novel passive location approach based on the BSS method is efficient and feasible.

1 Introduction

Precise attack is an important task in modern warfare especially in information warfare. To fulfill it, it is crucial to obtain the position of the target, which requires location techniques. Generally attacking target needs the assistant of radar, which may be attacked by Anti-Radiation Missile (ARM). This is the fatal weakness of radar and the working distance of radar is relatively limited. While the passive location can obtain the position of emitter only by the receiving signals. Traditional passive location methods are limited by the measuring precision. A novel passive location technology is based on broadcast & television signals [1-4], The representative equipment is "Silent Sentry"[1]. The uniqueness of Silent Sentry is its innovative Passive Coherent Location (PCL) technology developed by Lockheed Martin Mission Systems, which uses everyday broadcast signals, such as those for television and radio, to illuminate, detect and track objects. It enables a large range and real-time detecting as well as precise tracking. By utilizing broadcast transmitters and signals available throughout the world, Silent Sentry casts a "wider net" than current radar systems and provides new levels of early detection, to reveal tangible proof of intruders and can enable rapid, defensive reaction to threats. However, the direct wave and reflected wave are inevitably existed simultaneously in real world environments. The reflected wave is often weak compared with direct wave. The practicability of passive location based on broadcast signals still exists some problems such as how to extract these weak signals, how to cancel scatter cluster and interference, how to separate receiving mixture signals and how to improve the performance of the parameter estimation etc. This paper proposes a novel location method based on BSS to solve the aforementioned problems.

In recent years, particularly after Herault and Jutten's work [5], the research of BSS has become a hotspot in signal processing. A lot of BSS algorithms have been proposed [6-11]. BSS can effectively separate or extract source signals only by the mixing measurements, which is just the processing that passive location requires. Hereby we proposed applying BSS to passive location based on broadcast signals.

The organization of this paper is as follows: In section 2, we formulate the issue of passive location and describe the problem to be solved. In section 3, we present our location algorithm. Simulations conducted in section 4 show the effectiveness of the algorithm, and finally is the conclusion.

2 Problem Description

2.1 Location System

Target position can be determined mainly by the parameters of DOA, time difference and the Doppler frequency shift. There are many methods for estimating DOA and Doppler frequency shift. The center problem of this paper is to extract reflected wave signal in strong direct wave background and estimate the time difference. Then the location problem can be solved based on estimation of time difference and DOA. Here we assume that there exist a transmitting station including a television station and a FM broadcast station, and a receiving station, as shown in figure 1.

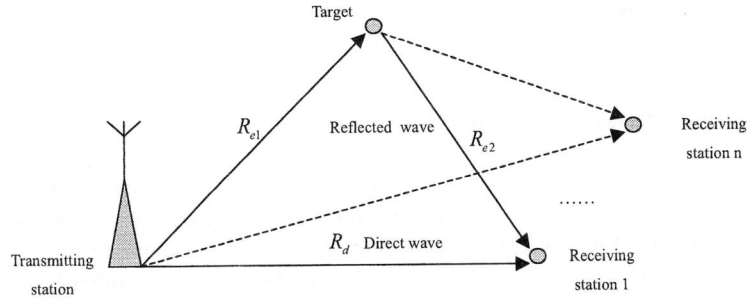

Fig. 1. The location system.

In order to implement coherence processing, the receiving station utilizes two antennas with different orientations to receive signals from space-selective channels. One is for direct wave, which aims at the position of transmitting station and receives broadcast & television signals. The other is for reflected wave, which aims at the position of target and receives the echo of the target.

2.2 Signal Model

Television signal is a strong AM carrier wave and is suitable for high precision estimation of time difference, but it is degraded in the case of moving target. Broadcast signal is a kind of FM signal, its property is just contrary to AM signal. Simultaneously utilizing FM broadcast signal and AM television signal, an excellent detecting and estimating performance can be acquired. Here AM and FM signals can be describe as:

$$\begin{cases} s_{AM}(t) = \left[A_0 + s_m(t) \right] \cos(2\pi f_0 t) = A_0 \left[1 + k_{AM} s_m(t) \right] \cos(2\pi f_0 t) \\ s_{FM}(t) = \cos(2\pi f_0 t + \phi(t)) \end{cases} \quad (1)$$

where A_0, k_{AM} are constants, f_0 is the carrier frequency, $s_m(t)$ is the modulating signal, $\phi(t)$ is a modulating component.

Assuming each receiving station has two receiving antennas, the received direct wave signal $x_d(t)$ and the target reflecting signal $x_r(t)$ can be described as:

$$\begin{cases} \mathbf{X}_d = \mathbf{H}_d \cdot \mathbf{R}_d \cdot \mathbf{S} + \mathbf{n} = \mathbf{A} \cdot \mathbf{S} + \mathbf{n} \\ \mathbf{X}_r = \mathbf{H}_r \cdot \mathbf{R}_r \cdot \mathbf{S}' + \mathbf{n}' = \mathbf{B} \cdot \mathbf{S}' + \mathbf{n}' \end{cases} \quad (2)$$

where $\mathbf{A}, \mathbf{B}$ are $m \times n$ mixing matrices of direct wave channels and reflected wave channels respectively, which are the product of receiving antenna responding functions $\mathbf{R} = \left[\mathbf{r}(\theta_1), \cdots, \mathbf{r}(\theta_n) \right]$ and the mixing matrices $\mathbf{H}$ during the signals transmitting process. $\mathbf{X}_d = [x_{d1}, x_{d2}]^T$, $\mathbf{X}_r = [x_{r1}, x_{r2}]^T$, $\mathbf{S} = [s_{AM}, s_{FM}]^T$, $\mathbf{S}' = [s'_{AM}, s'_{FM}]^T$, $\mathbf{n}, \mathbf{n}'$ are the additive noises. $s'_{AM}(t) = s_{AM}(t - \Delta \tau)$ and $s'_{FM}(t) = s_{FM}(t - \Delta \tau)$, $\Delta \tau$ is the delay of reflected wave relatively to direct wave. Then the problem transfers into two BSS problem in noisy case. Here signals and noise are all assumed mutual independent.

2.3 Algorithm Requirement Analysis

According to figure 1 and Eq.(2), the location problem can be divided into several cases, that is, transmitting signal and transmitting position is either known or unknown, the transmitting position is either in the visual range scope or not. They can be reduced to two problems: one is time difference estimation, the other is the detection of weak signals. On the other hand, noise exists all the time.

3 Location Algorithm

The algorithm of this passive location technology can be divided into four steps: normalization, blind separation, correlation to time difference, location computation.
The processing of location is shown in figure 2.

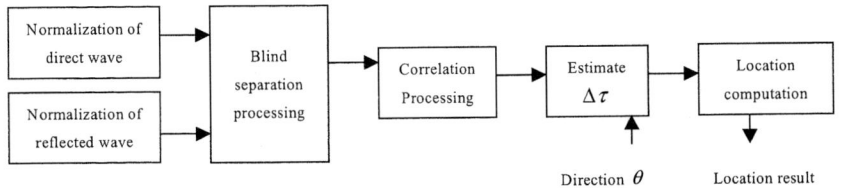

Fig. 2. The processing of location system.

3.1 Normalization Processing

The power ratio of direct wave and reflected wave is about 100 dB, The power ratio of interference and target echo is about -40dB[4]. Normalization processing is a key to the success of this location algorithm. We can possible avoid the effect of direct wave to reflected wave and obtain a preferably separate and correlating results. The normalization processing of $\mathbf{X}$ and to $\mathbf{S}$ is as:

$$\mathbf{X}_s = \frac{\mathbf{X}}{\|\mathbf{X}\|+b} \quad \mathbf{S}_s = \frac{\mathbf{S}}{\|\mathbf{S}\|+b} \quad (3)$$

where b is a normalization departure parameter, usually chosen as a little value: $b = 1\times 10^{-10}$, $\mathbf{X}_s$ is the normalization result of receiving mixture signals $\mathbf{X}$, $\mathbf{S}_s = (s_{AMs}, s_{FMs}, s'_{AMs}, s'_{FMs})^T$ is the normalization result of $\mathbf{S} = (s_{AM}, s_{FM}, s'_{AM}, s'_{FM})^T$, which are the blind separation signals.

3.2 Blind Separation Processing

Signals may be interfered by the influence of outside interference and inner noise during the processing of transmission. Gaussian noisy signal is the important object that location processing should face. We can begin with measure the non-Gaussianity of signals by the technique of BSS. There are many methods for measuring of non-Gaussianity such as kurtosis, differential entropy, negentropy and mutual information etc [6][7]. A.Hyvärinen proposed a noisy ICA fixed-point algorithm based on Gaussian moments [12][13] in 1999, which is fit for the BSS request of this paper.

According to the definition and property of Gaussian moments [13], the noisy ICA model can be realized by a contrast function for maximizing the quasi-whitened data $\tilde{\mathbf{x}}$ as follows:

$$\max_{\|\mathbf{w}\|=1} \left| E\{\varphi_{d(\mathbf{w})}^{(k)}(\mathbf{w}^T\tilde{\mathbf{x}})\} - E\{\varphi_c^{(k)}(v)\} \right|^p \quad (4)$$

with $d(\mathbf{w}) = \sqrt{c^2 - \mathbf{w}^T \tilde{\mathbf{R}}_{nn} \mathbf{w}}$, where c^2 is the variance. Performing the optimization in Eq. (4), the preliminary form of the fixed-point iteration for quasi-whitened data may be obtained as:

$$\mathbf{w}^* = E\{\tilde{\mathbf{x}}\varphi_{d(\mathbf{w})}^{(k+1)}(\mathbf{w}^T\tilde{\mathbf{x}})\} - (\mathbf{I} + \tilde{R}_{nn})\mathbf{w}E\{\varphi_{d(\mathbf{w})}^{(k+2)}(\mathbf{w}^T\tilde{\mathbf{x}})\} \qquad (5)$$

where $\mathbf{w}^*$, the new value of $\mathbf{w}$, is normalized to unit norm after every iteration. The fixed-point algorithm in Eq. (5) can be considerably simplified by adapting the value of c after every iteration, which gives finally the following algorithm with bias removal for quasi-whitened data:

$$\mathbf{w}^* = E\{\tilde{\mathbf{x}}g(\mathbf{w}^T\tilde{\mathbf{x}})\} - (\mathbf{I} + \tilde{R}_{nn})\mathbf{w}E\{g'(\mathbf{w}^T\tilde{\mathbf{x}})\} \qquad (6)$$

Only one independent component can be obtained after every iteration. Using the decorrelation method as noisy-free case, all of the independent components can be obtained at last.

3.3 Correlation Processing

After completing normalization and blind separation processing, the following work is to compute the delay $\Delta\tau$ of reflected wave relatively to direct wave. Here we utilize correlation processing. To direct wave s and reflected wave s', as the different of distance, the appear time of reflected signal is also different. In this case, we must consider the resemble property of the two signals while the time varying. Delaying $\Delta\tau$ to s'_n and change it to $s'_{n-\tau}$, then the coefficient $r_{ss'}$ can be written as:

$$r_{ss'}(\tau) = \sum_{n=-\infty}^{\infty} s_n s'_{n-\tau} \qquad (7)$$

When τ varying from $-\infty$ to $+\infty$, $r_{ss'}(\tau)$ is a function of τ. We call $r_{ss'}(\tau)$ as the coefficient of s_n and s'_n, τ is the time delay of s'_n. When $|r_{ss'}(\tau)|$ reaches the maximal value at τ_0, then τ_0 is the time difference $\Delta\tau$ of the two signals.

During the correlation processing, we take the separated direct wave as the tracking wave and conduct a second tracking iterative separation for reflected wave. According to the principle of blind separate, we can take the direct wave signal and reflected signal as the mixture of source signals and noisy signals after obtain the time difference $\Delta\tau$. An optimal estimation of source signals can be acquired by tracking processing. Then a more exact time difference value can be obtained.

3.4 Location Computation

When the time difference $\Delta\tau$ is obtained and combined with the direction value θ obtained by optimal estimating method, the location problem is transferred into a simple direction finding and time difference measuring location. Practically the target may be regarded as a transmitting station. Taking the location baseline as the receiving and transmitting station, $\Delta\tau$ is the transmitting time, the location issue is shown as figure 3.

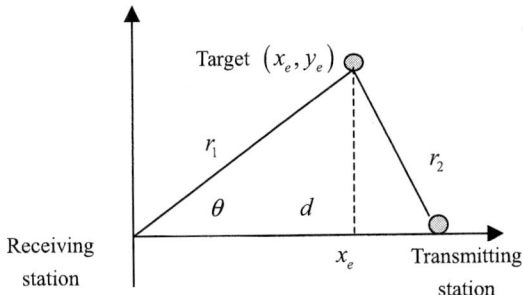

Fig. 3. The location principle, a coordinate can be established as follow: receiving station as the origin, the direction of receiving station to transmitting station as x axis. Assuming the location of target is (x_e, y_e) in this coordinate, the distance between the receiving station and transmitting is d, the distance between receiving station and target is r_1, the distance between transmitting station and target is r_2.

Then the equations can be established as follows:

$$\begin{cases} x_e = r_1 \cos\theta \quad y_e = r_1 \sin\theta \\ \Delta\tau \cdot c = r_1 + r_2 - d \end{cases} \quad (8)$$

where $c = 3 \times 10^8 \, m/s$, Utilizing Pythagorean theorem as follows:

$$(d - x_e)^2 + r_1^2 \sin^2\theta = r_2^2 \quad (9)$$

Combining with Eq. (8), r_1 can be obtained as:

$$r_1 = \frac{c^2 \Delta\tau^2 + 2dc\Delta\tau}{2(c\Delta\tau + d) - 2d\cos\theta} \quad (10)$$

Substituting equation Eq. (10) to Eq. (8), then (x_e, y_e) can be obtained as:

$$x_e = \frac{c^2 \Delta\tau^2 + 2dc\Delta\tau}{2(c\Delta\tau + d) - 2d\cos\theta} \cos\theta \quad y_e = \frac{c^2 \Delta\tau^2 + 2dc\Delta\tau}{2(c\Delta\tau + d) - 2d\cos\theta} \sin\theta \quad (11)$$

4 Simulations

According to the algorithm analysis in the section 3, some corresponding simulations were conducted as: Normalization - Blind separation - Time difference estimation by correlation- Location computation. Here we assume the distance between the transmitting station and the receiving station is 90km, the relative direction of target to transmitting station and the receiving station is 30°, the relative time difference is $110 \mu s$, the broadcast signal signals model like Eq. (1). The sampling frequency is $6 \times 10^5 \, Hz$, correspond sampling interval is $1.67 \times 10^{-6} \, s$, relative delay is 66 sampling intervals. The simulation steps are taken as follows:

(1) The first step is normalization processing. This step normalizes the receiving mixtures, which can avoid different power standard affecting later blind separating processing.
(2) The second step is blind source separation. The noisy mixture signals scilicet are the signals from the received channels shown as figure 4. The separation results are shown in figure 5, the first and the second signals are the separate results of direct waves, the third and the fourth signals are the separate results of reflected waves. The separate signals all have some distortion especially the separation of reflected signals.

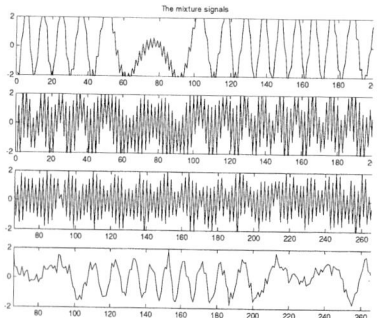

Fig. 4. The receiving mixtures. Fig. 5. The blind separation results.

(3) The third step is correlation processing. The correlation results can be obtained as shown in figure 6. From the correlation coefficient of FM signal in figure 6, we can know that the delay is about 66 sampling intervals. So the time difference is $66 \times 1.67 \times 10^{-6} s = 110 \mu s$, which is consistent to the experiment condition set in advance. The best reflected signals can be obtained by iteration BSS and tracking processing as shown in figure 7.

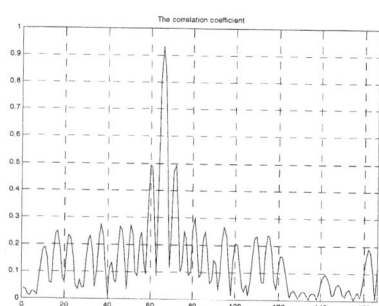 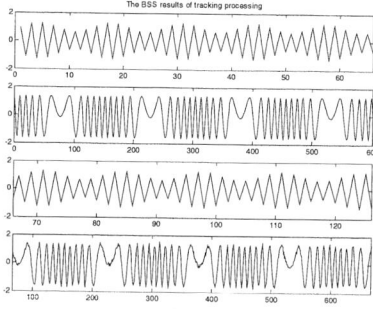

Fig. 6. The correlation results. Fig. 7. The tracking BSS processing results.

(4) The fourth step is location computation. Substituting the aforementioned results value and known value to Eq. (10), we can obtain $r_1 = 77.94 Km$, then substitute r_1 to Eq. (11), the position of target can be obtained as $(67.50, 38.97)$ (Km), which is accurate to the experiment conditions.

5 Conclusion

This paper proposes an approach applying blind source separation to a novel passive location method based on broadcast signals. It provides an algorithm for the engineering implementation of this location method and solves a major problem in passive location based on broadcast signals. The real world signal environment is very complicated. So the future work will aim at the separation of convolutive and nonlinear mixtures, and the problem for time varying channels. After the solution of these problems, this new passive location technology will play an important role in military and civilian affairs.

References

1. Nordwall B.D.: "Silent Sentry"-a new type of radar [J], Aviation Week & Space Technology, 30, (1998) 25-26
2. Griffiths H.D., Long N.R.W.: Television-based bistatic radar [J] IEE Proc. Pt.F, 133(7), (1986) 649-657
3. Sahr, J.D., Lind, F.D.: The manastash Ridge Radar: A passive bistatic radar for upper atmospheric radio science, Radio Sci., 32,(1997) 2345-2358
4. Howland, P.E.: Target tracking using television-based bistatic radar, IEE Proceedings, Radar, Sonar, Navigation, Vil.146, No.3, June, (1999) 166-174
5. Jutten, C., Herault, J.: Blind separation of sources, Part I:An adaptive algorithm based on neuromimetic,[J].Signal Processing, 24(1), (1991) 1-10
6. Hyvärinen, A., Karhunen,J. and Oja,E.: Independent Component Analysis. J. Wiley (2001)
7. V. David Sánchez A.: Frontiers of research in BSS/ICA, Neurocomputing 49, (2002) 7-23
8. Mansour, A., Barros, A. K. and Ohnishi, N.: Blind Separation of Sources: Methods, Assumptions and Application, In Special Issue on Digital Signal Processing in IEICE Transactions on Fundamentals of Electronics, Communications and Computer Sciences. Vol. E83-A (8), (2000) 1498-1512
9. Cardoso, J.: Blind Signal Separation: Statistical Principle, Proceedings of the IEEE, Vol.86, NO.10, October (1998) 2009-2025
10. Smaragdis, P.: Blind Separation of Convolved Mixtures in the Frequency Domain, Neurocomputing 22 (1998) 21-34.
11. Dapena, A., Serviere, C. and Castedo, L.: Separation of convolutive mixtures in the frequency-domain using only two frequency bins, Acoustics, Speech, and Signal Processing, 2002 IEEE International Conference on, Volume: 2, 12-17 May (2002) 1633 –1636.
12. Hyvärinen, A.: Fast ICA for noisy data using Gaussian moments, ISCAS '99, Proceedings of the 1999 IEEE International Symposium on, Vol.5 (1999) 57 -61
13. Hyvärinen, A.: Gaussian moments for noisy independent component analysis, IEEE Signal Processing Letters, Vol. 6, Issue: 6 (1999) 145-147

Blind Source Separation in the Adaptive Reduction of Inter-channel Interference for OFDM*

Rafael Boloix-Tortosa and Juan José Murillo-Fuentes

Area de Teoria de la Señal y Comunicaciones, Universidad de Sevilla,
Paseo de los Descubrimientos sn, Sevilla 41092, Spain
rboloix@us.es
http://viento.us.es/~rboloix

Abstract. In this paper, we propose a new application of blind source separation (BSS) in OFDM systems with zero padding (ZP). We first focus on the model of ZP-OFDM to reduce it to an instantaneous linear mixing matrix model. Then we apply blind separation source (BSS) techniques based on the natural gradient (NG) and particularly EASI and M-EASI algorithms. Hence, we endow the method with adaptive properties and extend the use of BSS to complex symbols. This novel method blindly cancels the inter-channel interference (ICI) introduced by the channel with no use of cyclic prefix, meaning a power efficiency improvement and avoiding the recovery difficulties when some of the subcarriers are hit by a channel frequency response null. We include some experiments to illustrate the excellent performance of the proposed application even when ill-conditioned channels.

1 Introduction

Orthogonal Frequency Division Multiplexing (OFDM) [1] has attracted the attention of the digital communications community as it has been proposed as transmission system for many of the last generation systems such as the digital audio broadcast (DAB), the WiFi IEEE 802.11a or the Asynchronous Digital Subscriber Line (ADSL). In OFDM, blocks of symbols are transmitted in parallel over several narrowband subchannels (subcarriers) which experience almost flat fading. If the channel is just additive White Gaussian Noise (AWGN) the orthogonality of the subcarriers ensures perfect reception. However, as the signal passes through a time-dispersive channel both inter-symbol interference (ISI) between blocks and inter-channel interference (ICI) between subcarriers is induced. We can avoid the first one by means of Zero Padding (ZP) or Cyclic Perfix (CP) techniques. In this paper we will focus on ZP as it allows saving transmitting power and symbol recovery even when some of the subcarriers are hit by channel frequency response nulls. Since ZP does not avoid ICI, we propose a new blind adaptive solution based on Blind Source Separation (BSS) to cancel it. We

---
* Thanks to Spanish goverment for funding TIC-2003-03781.

will first introduce BSS algorithms based on the statistical independence of the outputs, paying attention to cope with complex constellations. Then we shall describe how the transmitting plus the receiving model of ZP-OFDM yields a linear instantaneous mixing matrix model of the data symbols. Finally, we will relate both adaptive BSS and ZP-OFDM to find an adaptive blind ICI removing method. Finally our simulations will prove the good performance of that detector even for ill-conditioned channels.

2 Blind Source Separation

Consider a set of M measured signals

$$\boldsymbol{x}(t) = [x_1(t), x_2(t), \ldots, x_M(t)]^T \tag{1}$$

being a linear instantaneous mixture of N mutually independent unknown signals or *sources*

$$\boldsymbol{s}(t) = [s_1(t), s_2(t), \ldots, s_N(t)]^T. \tag{2}$$

In matrix form we could express this mixing as

$$\boldsymbol{x}(t) = \boldsymbol{A}\boldsymbol{s}(t). \tag{3}$$

The problem of Blind Source Separation (BSS) is the reconstruction of those unknown sources from the set of mixtures. Where *blind* stands for the lack of knowledge about the sources or the mixing matrix $\boldsymbol{A}$. In some methods, the spatial independence of the sources is the key to separate them. If the mixing matrix $\boldsymbol{A}$ is non-singular, $\boldsymbol{x}(t)$ is a stationary ergodic random sequence, and no more than one Gaussian distributed sources is present in the mixture, forcing the statistical independence of the outputs yields the sources as follows

$$\boldsymbol{y}(t) = \boldsymbol{B}\boldsymbol{x}(t) = \boldsymbol{B}\boldsymbol{A}\boldsymbol{s}(t) = \boldsymbol{C}\boldsymbol{s}(t), \tag{4}$$

where ideally matrix $\boldsymbol{C}$ should be the identity matrix. However, the original scaling and arrangement cannot be estimated from the mere independence assumption. In this sense, $\boldsymbol{C}$ is regarded as a non-mixing matrix if it has one and only one nonzero element in each row and column.

The separating matrix $\boldsymbol{B}$ can be decomposed into the product of a whitening $\boldsymbol{W}$ and a rotation $\boldsymbol{V}$ matrix. The whole process yields

$$\boldsymbol{y}(t) = \boldsymbol{B}\boldsymbol{x}(t) = \boldsymbol{V}\boldsymbol{W}\boldsymbol{A}\boldsymbol{s}(t) = \boldsymbol{V}\boldsymbol{z}(t) \tag{5}$$

Notice that if the outputs are decorrelated and normalized to unit-variance, the problem reduces to the computation of unitary matrix $\boldsymbol{V}$.

Considering the sources corrupted by additive white Gaussian noise, the model of the problem yields

$$\boldsymbol{x}(t) = \boldsymbol{A}\boldsymbol{s}(t) + \boldsymbol{n}(t), \tag{6}$$

and the separation process yields

$$\boldsymbol{y}(t) = \boldsymbol{B}\boldsymbol{x}(t) = \boldsymbol{C}\boldsymbol{s}(t) + \hat{\boldsymbol{n}}(t) \tag{7}$$

Several algorithms have been proposed to solve the BSS problem based on the statistical assumption (see [2] and references therein). These algorithms differ in the way the source independence assumption is exploited. Some batch techniques are based on the minimization of second and fourth-order cumulant-based contrast functions [3], [4]. Some other techniques are based on a geometric approach to ICA [5]. Adaptive or online algorithms may also be based on optimizing a loss function using the relative [6] or natural gradient [7], as learning law. If we face the noisy case we may use some alternatives [8], [9], [10]. As the MIMO model in (6) corresponds to several structures in digital and radio communications, BSS has been successfully applied in the beamforming of antenna arrays [7], multiuser detection (MUD) in CDMA [11] or polarization discrimination in radiolinks. In this paper we focus on OFDM receivers. Let's first introduce the BSS algorithms used.

2.1 Adaptive BSS: The Natural Gradient

The steepest descent method updates matrix B according to the direction of the gradient of a loss function $L(B)$. The natural or relative gradient proposes to use $\tilde{\nabla} L(B) = \tilde{\nabla} E[l(B)] = \nabla L(B) B^T B$. The stochastic version uses the instantaneous value $\tilde{\nabla} l(B)$. The learning law yields

$$B \longleftarrow B - \lambda \tilde{\nabla} l(B) B^T B. \tag{8}$$

Under the independence assumption, the maximum likelihood (ML) approach is widely used to derive the loss function [2]. It can be shown that we can achieve independence at the output cancelling the following estimating equation: $H_{ML}(y) = E[\nabla l(B) B^T] = E[\varphi(y) y^T - I]$. The stochastic learning law now yields

$$B \longleftarrow B - \lambda H_{ML}(y) B, \tag{9}$$

and normalizing

$$B \longleftarrow B - \lambda \frac{\varphi(y) y^T - I}{1 + \lambda |\varphi(y)^T y|} B, \tag{10}$$

where in the ML approach the activation or score function is $\varphi_i(y_i) = -q'_i(y_i)/q_i(y_i)$, being $q_i(\cdot)$ the probability density function (pdf) of the i-th source. Since sources are suppose to be unknown, $q_i(\cdot)$ are also unknown and each author proposes his own activation function. A family of them for sources with negative or positive kurtoses may be found in [6]. In [7] authors developed a relative gradient based adaptive method known as EASI (Equivariant Adaptive Separation via Independence). Its learning laws may be written as follows

$$B \longleftarrow B - \lambda [\frac{y y^T - I}{1 + \lambda y^T y} + \frac{\varphi(y) y^T - y \varphi(y)^T}{1 + \lambda |y^T \varphi(y)|}] B. \tag{11}$$

These learning laws show, at low noise, a good behaviour independent of the mixing matrix (uniform performance). Since sources in digital communications are zero-mean sources with symmetric pdf, the authors in [8] introduced the

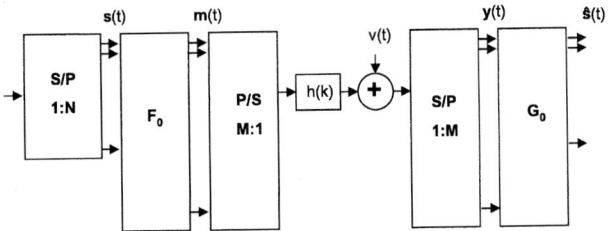

Fig. 1. Discrete-time baseband equivalent model for the block-by-block transceiver.

signum function to reduce the bias caused by noise. The estimation function yields

$$\boldsymbol{H}_{MEASI} = \frac{\boldsymbol{y}sgn(\boldsymbol{y})^H - \boldsymbol{I}}{1 + \lambda \cdot sgn(\boldsymbol{y})^H \boldsymbol{y}} + \frac{1}{\alpha} \frac{\varphi(\boldsymbol{y})\boldsymbol{y}^H - sgn(\boldsymbol{y})\varphi(\boldsymbol{y})^H}{1 + \lambda |\boldsymbol{y}^H \varphi(\boldsymbol{y})|}, \qquad (12)$$

where $sgn(\boldsymbol{y}) = sgn(\Re(\boldsymbol{y})) + jsgn(\Im(\boldsymbol{y})), \boldsymbol{y} \in \mathbb{C}$. This learning law improves the stability of the algorithm and endow the method with phase recovery properties and more robustness against noise.

3 Application to Zero-Padding OFDM

3.1 System Description

Orthogonal Frequency Division Multiplexing (OFDM) [1] is an important example of a block based multicarrier (MC) communication system. OFDM, and many other modern communication systems, including discrete multitone modulation (DMT), could be described by the generalized block-by-block transceiver model [12] shown in Fig. 1. In the S/P block, the source symbol sequence is divided in blocks of N symbols. Each symbol of a block would be transmitted in parallel through a different subchannel, modulating a different subcarrier. In order to avoid ISI, or interblock interference (IBI), the tth block of N data symbols is extended giving a block of $M > N$ transmitted symbols. Therefore, we assume the channel response, $h(k)$ to be shorter than $M - N$. Usually this redundancy is made by the addition of a cyclic-prefix or zero-padding of the original block of symbols. This could be written in matrix form [12] as

$$\boldsymbol{m}(t) = \boldsymbol{F}_0 \boldsymbol{s}(t), \qquad (13)$$

where $\boldsymbol{s}(t)$ is a $N \times 1$ vector that contains the block of data symbols and $\boldsymbol{F}_0$ is a $M \times N$ matrix that introduces the redundancy and, in standard OFDM systems, performs the modulation of each subcarrier by its corresponding symbol. Hence, $\boldsymbol{m}(t)$ is a vector that contains the block of the M transmitted symbols. Notice that each entry of $\boldsymbol{m}(t)$ depends on every entry of $\boldsymbol{s}(t)$. We convert $\boldsymbol{m}$ to serial, transmit them over the channel, and then convert them to parallel. The received tth block of symbols, vector $\boldsymbol{y}(t)$, is the result of the convolution of the transmitted block with the channel impulse response $h(k)$ and additive noise. This can be written [12] as

$$y(t) = \sum_{k=-\infty}^{\infty} H_k m(t-k) + v(t), \quad (14)$$

where $M \times M$ matrix H_k is computed as

$$H_k = \begin{pmatrix} h(kM) & h(kM-1) & \cdots & h(kM-M+1) \\ h(kM+1) & h(kM) & \cdots & h(kM-M+2) \\ \vdots & \vdots & \ddots & \vdots \\ h(kM+m-1) & h(kM+P-2) & \cdots & h(kM) \end{pmatrix} \quad (15)$$

A receiver G_o could be developed to estimate the symbols as

$$\hat{s}(t) = G_0 \sum_{k=-\infty}^{\infty} H_k m(t-k) + G_0 v(t). \quad (16)$$

3.2 Zero-Padded OFDM Transmission

If we assume that the channel impulse response $h(k) = 0$, for $k < 0$ and $k > L$, and select $M \geq N + L$ and $M > 2L$ then (16) simplifies to

$$\hat{s}(t) = G_0 H_0 F_0 s(t) + G_0 H_1 F_0 s(t-1) + G_0 v(t), \quad (17)$$

where H_1 has nonzero elements only in its $L \times L$ top right submatrix. In this case we could eliminate interblock interference (IBI) selecting G_0 and F_0 so that $G_0 H_1 F_0 = 0$. In [12], and for zero-padded transmission, authors propose $G_0 = GR$ and $F_0 = TF$, where G and F will be defined later, and

$$T = \begin{pmatrix} I_{(M-L)} \\ 0_{L \times (M-L)} \end{pmatrix}$$
$$R = I_M \quad (18)$$

The vector of symbols $\hat{s}(t)$ yields

$$\hat{s}(t) = GHFs(t) + n(t), \quad (19)$$

where $H = RH_0T$, i.e.,

$$H = \begin{pmatrix} h(0) & 0 & \cdots & 0 \\ \vdots & \ddots & \ddots & \vdots \\ h(L) & \ddots & \ddots & 0 \\ 0 & \ddots & \ddots & h(0) \\ \vdots & \ddots & \ddots & \vdots \\ 0 & \cdots & 0 & h(L) \end{pmatrix} \quad (20)$$

In simple standard OFDM systems, and when $M = N + L$, matrix $\boldsymbol{F} = \boldsymbol{D}^H \boldsymbol{\Delta}_T$ and matrix $\boldsymbol{G} = \boldsymbol{\Delta}_R[\boldsymbol{D}\ \boldsymbol{0}_{N \times L}]$, where $\boldsymbol{D}$ is the $N \times N$ normalized DFT matrix:

$$[\boldsymbol{D}]_{k,n} = (N)^{-1/2} \exp(-j2\pi kn/N), \tag{21}$$

for $0 \leq k, n \leq N-1$, and $\boldsymbol{\Delta}_T$ and $\boldsymbol{\Delta}_R$ are power determining diagonal matrices.

3.3 Adaptive Blind Receiver

Zero-padding of the data symbols eliminates the IBI. However, as symbols are transmitted across the channel, we have inter-channel interference (ICI). We may consider that each of the N data symbols of a block corresponds to one of N independent sources. Hence, symbols in (19) are the result of mixing those sources in a noisy environment using a mixing matrix $\boldsymbol{A} = \boldsymbol{GHF}$. In this case, we can use any of the well known BSS techniques to solve this problem and recover $\boldsymbol{s}(t)$ from the measured mixture signal $\hat{\boldsymbol{s}}(t)$, eliminating the ICI, i.e.,

$$\boldsymbol{y}(t) = \boldsymbol{B}_{OFDM} \hat{\boldsymbol{s}}(t) = \boldsymbol{B}_{OFDM} \boldsymbol{GHF} \boldsymbol{s}(t) + \hat{\boldsymbol{n}}(t) = \boldsymbol{C}\boldsymbol{s}(t) + \hat{\boldsymbol{n}}(t) \tag{22}$$

In this work we propose to use the natural gradient to compute $\boldsymbol{B}_{OFDM}$ and exploit the benefits of the EASI or M-EASI algorithms; superefficiency and equivariance.

4 Performance Evaluation

We consider here the performance of the adaptive blind OFDM receiver. Transmitted symbols are equiprobable and independent 4-QAM symbols. The data blocks are of length $N = 32$, and the channel has impulse response $h(k) = \{0.6121, -0.5331 - 0.4481j, 0.369j, 0.0513 - 0.0388j\}$, i.e, $L = 3$. This channel is selected to obtain a significant ICI. The magnitude of the DFT of the impulse response of the channel is shown in Fig. 2.a. The length of the block of transmitted symbols is set to $M = N + L = 35$, and this block is a zero-padded version of the data block. The total number of blocks transmitted is 25000, i.e., $t = 0, 1, \ldots, 25000$. At reception, a direct detection is performed. Also, EASI and M-EASI algorithms are used as adaptive natural gradient based blind source separation techniques in order to recover original data blocks. The adaptation coefficient for both algorithms was set to $\lambda = 1/t + 10^{-6}$. Fig. 2.b shows the estimated bit error rate (BER) performance of the system with and without using the blind source separation techniques at SNR-per-bit between -3 a 26 dB. SNR-per-bit is obtained as the ratio of the average energy of bit after the receiver $\boldsymbol{G}$ to N_0, being $N_0/2$ the two-sided power spectral density of the additive white Gaussian noise (AWGN) after the receiver. It is interesting to notice that in the ZP-OFDM with no ICI cancellation, BER does not decreases as SNR $\to \infty$. With the BSS technique, ICI is removed and BER could be greatly reduced with an increase of only a few dB in SNR. Fig. 2.b also shows

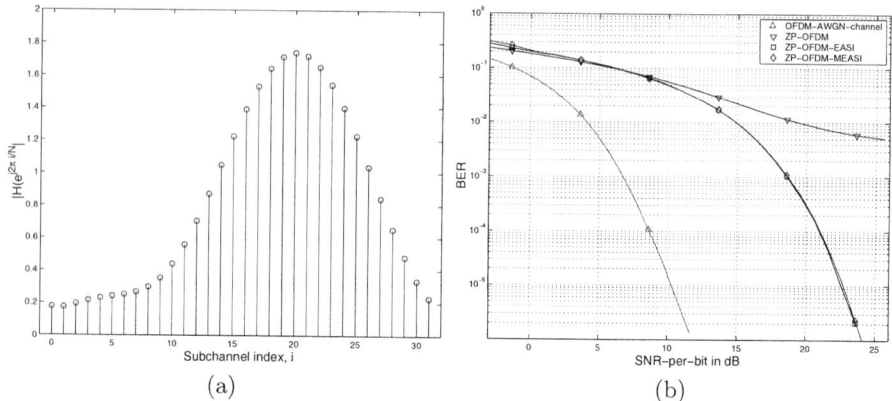

Fig. 2. (a) Frequency response of the channel. (b) BER performance.

the calculated BER without using the blind source separation techniques when the channel impulse response is $h[n] = 1$, i.e., in an AWGN channel.

To show an example of how ICI is removed with BSS, the received symbols (constellation) sent through subchannel $i = 2$ are shown in Fig.3.a., along with those symbols after blind separation with EASI (Fig.3.b.) and M-EASI (Fig.3.c.) for a value of $SNR - per - bit = 28dB$. Notice that the dispersion of the points is reduced and, therefore, the probability of error.

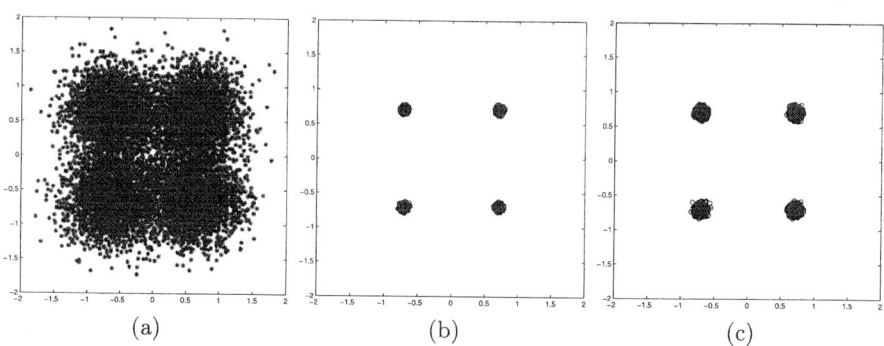

Fig. 3. Symbols received through subchannel 2 for $SNR-per-bit = 28dB$: (a) without applying BSS techniques, (b) using EASI and (c) using M-EASI.

5 Conclusions

In this paper, we have faced the ZP-OFDM problem as a new application of BSS tecniques. We have shown that this new solution blindly reduce the inter-channel interference (ICI) due to the propagation channel. We first write the model for ZP-OFDM to reduce it to an instantaneous linear mixing matrix model, i.e, to a blind separation source (BSS) problem. We recall some adaptive BSS techniques for complex sources such as those based in the natural gradient (NG), and particularly EASI and M-EASI algorithms. This novel method blindly cancels the ICI with no use of cyclic prefix, meaning a power efficiency improvement and avoiding the recovery difficulties when some of the subcarriers are hit by channel frequency response null. Some included experiments illustrate the excellent performance of the proposed application even when the channel is ill-conditioned. The results could be easily extended to any other constellations.

References

1. Edfors, O., Sandell, M., van de Beek, J.J., Landström, D., Sjöberg, F.: An introduction to orthogonal frequency-division multiplexing. Tutorial (1996) http://courses.ece.uiuc.edu/ece459/spring02/ofdmtutorial.pdf
2. Cardoso, J.F.: Blind signal separation: Statistical principles. Proc. of the IEEE **86** (1998) 2009–2025
3. Comon, P.: Independent component analysis, a new concept? Signal Processing **36** (1994) 287–314
4. Murillo-Fuentes, J.J., González-Serrano, F.J.: A sinusoidal contrast function for the blind separation of statistically independent sources. IEEE Tras. on Signal Processing. Acepted for publication (2004)
5. Theis, F., Jung, A., Puntonet, C., Lang, E.: Linear geometric ICA: Fundamentals and algorithms. Neural Computation **15** (2002) 1–21
6. Amari, S.I.: Natural gradient works efficiently in learning. Neural Computation **10** (1998) 251–276
7. Cardoso, J.F., Laheld, B.H.: Equivariant adaptive source separation. IEEE Trans. on Signal Processing **44** (1996) 3017–3030
8. Murillo-Fuentes, J., González-Serrano, F.: Median equivariant adaptive separation via independence: application to communications. Neurocomputing **49** (2002) 389–409
9. Cichocki, A., Douglas, S., Amari, S.: Robust techniques for independent component analysis (ICA) with noisy data. Neurocomputing (1998) 113–129
10. Douglas, S., Cichocki, A., Amari, S.: Self-whitening algorithms for adaptive equalization and deconvolution. IEEE Trans. on Signal Processing **47** (1999) 1161–1165
11. Caamaño-Fernández, A., Boloix-Tortosa, R., Ramos, J., Murillo-Fuentes, J.J.: High order statistics in multiuser detection. IEEE Tras. on Man and Cybernetics A. Acepted for publication (2004)
12. Ding, Y., Davidson, T.N., Lou, Z.Q., Wong, K.M.: Minimum ber block precoders for zero-forcing equalitation. IEEE Trans. on Signal Processing **51** (2003) 2410–24–23

BSS for Series of Electron Energy Loss Spectra

Danielle Nuzillard[1] and Noël Bonnet[2,3]

[1] LAM, UFR Sciences, Moulin de la Housse, B.P. 1035, 51687 Reims Cedex 2, France
Danielle.nuzillard@univ-reims.fr
[2] UMRS Inserm 514 (IFR 53); Hopital Maison Blanche;
45 rue Cognacq Jay, 51092 Reims Cedex, France
[3] LERI, Rue des Crayères, B.P. 1035, 51687 Reims Cedex 2, France
Noel.bonnet@univ-reims.fr

Abstract. The analysis of series of electron energy-loss spectra recorded with the spectrum-line technique is usually performed with two methods : the spatial difference approach and multivariate statistical analysis. Such spectra are strongly correlated, thus we associate Blind Source Separation techniques to specific pre and post processing. The principle is presented and illustrations are given through a simulation example and an experimental one.

1 Introduction

Improvements of Electron Energy Loss Spectroscopy (EELS) techniques have made possible the development of many experimental studies in the field of core-loss spectroscopy, including the study and use of near-edge structures. Among these techniques, the spectrum-image technique (or its spectrum-line variant) plays an important role for performing spatially resolved EELS.

Besides experimental advances, such as the increase of spatial and spectral resolutions, data analysis techniques also play their role in the improvement of the technique. The spatial-difference technique can be cited as one of the data analysis technique that helped to get useful information from series of spectra recorded across interfaces [1], [2]. Another set of methods, dedicated to larger sets of spectra, comes from Multivariate Statistical Analysis (MSA) [3], [4], [5]. The last developments in this category concern the oblique analysis method [6]. These two groups of methods proved already useful. But they are also prone to several drawbacks that make that more sophisticated methods for analysing series of spectra were anticipated.

In the work described in this paper, we explore the ability of Blind Source Separation (BSS), or Independent Component Analysis (ICA) to extract pertinent spectra and also the variation of the composition across the interface. The raw data do not fulfill the conditions of independence, thus the BSS is associated with pre and post processing. The paper is organized as follows. In the next section, we briefly summarize the methods already in use for analysing series of spectra. In the following section, we present the interest of complementary processing associated to ICA/BSS dealing with correlated signals that do not fulfill the required independence of the sources. Then, we illustrate the application in the field of EELS. Finally, we draw some conclusions and offer some tracks for future work.

2 Usual Data Analysis

We concentrate on data analysis methods aiming at discovering unknown spectra within series of recorded spectra. This problem arises for specimens where the EEL spectrum is dominated by the matrix component and an unknown component is present locally at low concentration. The problem may also be to quantify the variation of composition of (known and unknown) components across an interface. Two groups of methods are already in use for trying to cope with these problems.

2.1 The Spatial Difference Approach

This approach deals with two spectra, one of them S_M being recorded in the matrix region and the other S_I being recorded where a variation of the chemical composition is expected. Two variants of the method can be used:

– The difference between the two spectra is computed, after some sort of normalization was performed on one of the spectra [1].

$$S_D = S_I - \alpha S_M \qquad (1)$$

– The spectrum in the region of unknown composition is modelled as a linear combination of the matrix spectrum and the unknown spectrum [2]:

$$S_I = \alpha S_M + \beta S_U \qquad (2)$$

The difficulty is to estimate the coefficient α. This is often done through a trial-and-error approach. The drawbacks of the spatial difference approach have been identified for a long time: with the first variant, care must be taken to avoid confusing the new spectrum with artifacts due to any energy drift, for instance [7]. With the second variant, the approach remains subjective with respect to the scaling factor, even if general guidelines have been devised for choosing it properly [2], [8], [9].

2.2 Multivariate Statistical Analysis

With the development of the spectrum-line approach [10], or even more with the spectrum-image approach, it becomes worth processing a series of spectra at once rather than pairs of spectra. Multivariate techniques are able to do so and to extract the different sources of information contained in such complete series [3], [4], [5]. More specifically, a series of spectra can be decomposed into an average spectrum $\bar{S}$ and a set of orthogonal components ψ_k such as any spectrum S_i can be represented as in equation (3).

$$S_i = \bar{S} + \sum_{k=1} \varphi_{ik}\psi_k \quad \text{with} \psi_k.\psi_{k'} = \delta_{kk'} \tag{3}$$

where δ is the Kronecker symbol. This decomposition (through Principal Components Analysis, Correspondence Analysis, Karhunen-Loeve Analysis, etc) may be useful for the qualitative interpretation of the data set (number of real components, besides noise, for instance) or for performing a multivariate improvement of the signal-to-noise ratio. However, it does not help very much in the discovery of a *new* chemical component, through its (unknown) spectrum. The main reason for that is that the decomposition performed by MSA is made into orthogonal components, also called abstract components. Thus, the principal components cannot be identified with the real components, *i.e.* the spectra of the different elements that enter into the composition of the studied area [5]. One way to go further is to make an oblique analysis after the orthogonal analysis. This means rotating the axes of representation in such a way that they possess a more physical meaning [6]. Although several automatic approaches have been suggested (in other fields of application) for performing the rotation of the axes automatically, the only way we found to perform it with EEL spectra is again through a trial-and-error procedure. This is not completely satisfactory and justify to try other approaches, such as BSS.

3 BSS and ICA

The general problem to solve is that of mixture of signals x_k, the signals originating from n sources s_j being mixed and recorded by p sensors:

$$x_k = \sum_{i=1}^{n} a_{ik}s_i + n_k \quad k = 1, ..., p \quad \text{or} \quad X = AS + N \tag{4}$$

where a_{ij} are the proportions, n_k is the noise of the sensors.

Provided the sources can be considered as statistically independent, solutions can be found to recover them, hence the name Blind Source Separation (BSS). These solutions trace back to the work of Jutten and Herault [11]. They often consist in finding an inverse filter that provides outputs as much independent as possible, hence the name Independent Component Analysis (ICA).

In the context of EEL spectroscopy, our purpose is not to search for the optimal solution, but to demonstrate that BSS could be a useful alternative technique. The behaviour of EEL spectra at the interface of Si-SiO2 is a linear process without lag propagation. It is described by the instantaneous model in equation (4) where $p < n$. Data Spectra are corrupted with a faint noise, which is mainly a photon-noise slightly correlated with the sources, but not with itself. The searched spectra refer to different materials and are, in that sense, physically independent. However they can be mathematically correlated (for instance their scalar product is not null). They do not fit the required property of pairwise independence, but their derivatives fulfill these conditions. The process of deriving the data can be iterated as in equation (6). High order derivation is equivalent to introduce high order pass filter.

$$dx_i(t) = \sum_{j=1}^{N} a_{ij} ds_j(t) + dn_j(t) \quad \text{or} \quad \mathcal{D}X = A.\mathcal{D}S + \mathcal{D}N \qquad (5)$$

$$\mathcal{D}^k X = A.\mathcal{D}^k S + \mathcal{D}^k N \qquad (6)$$

Before calculating the derivatives[1], the introduction of a smoothing filter is compulsory to rub the noise. Its length is obtained by trials and errors.

We selected the SOBI algorithm [12], with which we obtained good results in the real space [13], [15] and in the reciprocal space [14], [18]. This algorithm is robust to the noise because it looks for a unitary matrix which diagonalizes a set of cross-covariance matrices by minimizing the off-diagonal terms [16]. This matrix is parameterised with a complex Given rotation and the criterion is minimized by successive rotations. The algorithm provides derivative sources spectra $\hat{\mathcal{D}S}$.

$$\hat{\mathcal{D}S} = \hat{A}^H R_x^{-1} \mathcal{D}X. \qquad (7)$$

Separated derivative spectra are either integrated or the separating matrix is applied on the data spectra since the filtering is linear.

Finally the spectra are improved by applying a positivity constraint on both spectra and proportions as in [17].

4 Results

4.1 A Simulated Three-Component Mixture

Here we consider the problem raised in [19] of a Si-SiO2 interface where the two spectra far from the interface are known and a third unknown component is suspected close to the interface. The aim of the study is to infer the shape of the unknown spectrum and to deduce the variation of composition across the interface, *i.e.* the weights of the different spectra in the mixture. As in [6], we start with a simulation of the problem, inserting an hypothetical spectrum as the third component, in order to check to what extent we are able to cope with the problem.

Dealing with a three-component mixture and a number of spectra across the interface greater than 3, we have two possibilities. We can process several triplets of spectra, including two pure spectra and one composite spectrum. This will give us several estimations of the unknown component. Alternatively, we can process all the spectra at once. This will give us one estimation of the unknown spectra.

---

[1] An alternative way to compute the derivative is to perform a frequential filtering. We tried many filters in the real space and the reciprocal space.

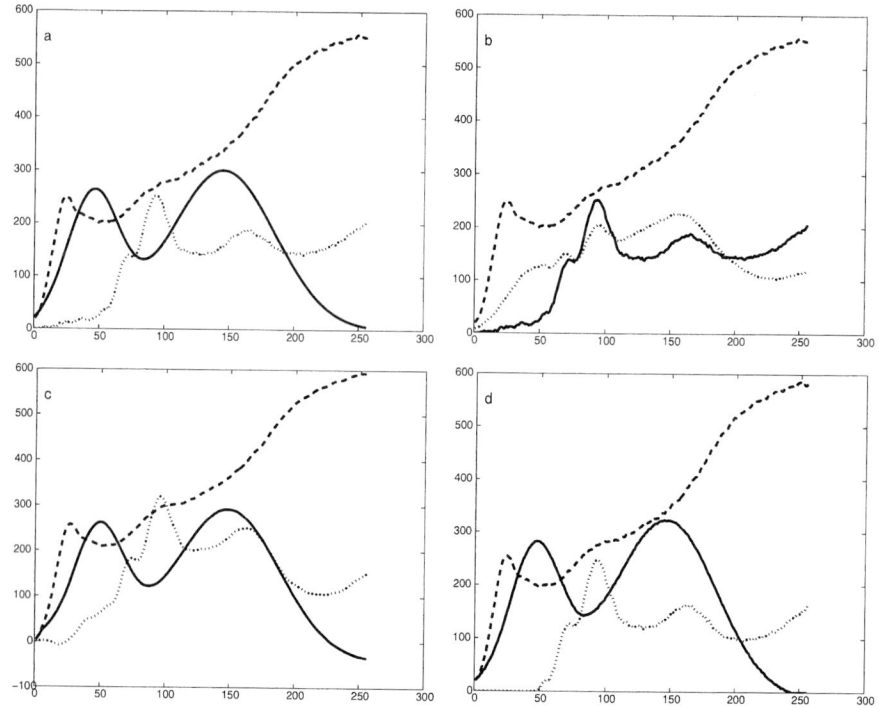

Fig. 1. Recovering of an unknown spectrum from 3 mixtures with two known spectra; a: the three initial spectra; b: 3 mixtures of 3 spectra; c: 3 retrieved spectra; d: 3 retrieved spectra with positivity constraint.

Fig. 1 displays the results obtained through the first approach, with two known spectra and one unknown spectrum, composed of two Gaussian bumps. The mixture matrix used to produce the simulation is A, and the estimated mixture matrix is $\hat{A}$.

$$A = \begin{pmatrix} 1 & 0.0 & 0.0 \\ 0.58 & 0.42 & 0.0 \\ 0.0 & 0.0 & 1.0 \end{pmatrix} \quad \hat{A} = \begin{pmatrix} 0.90 & 0.0 & 0.10 \\ 0.58 & 0.34 & 0.08 \\ 0.0 & 0.04 & 0.96 \end{pmatrix} \quad (8)$$

Fig. 1c shows the result obtained without imposing any constraint on the solution. Fig. 1d displays the result obtained when a positivity constraint is imposed to the spectra and the weight of spectra in the mixture.

Fig. 2 displays the results obtained through the second approach, where 21 spectra are processed all together. The 21 mixtures of spectra, representing what could be recorded as a spectrum-line across an interface, are displayed in Fig. 2a. The three spectra recovered without the positivity constraint are shown in Fig. 2b while the spectra recovered when using the positivity constraint are shown in Fig. 2c. Fig. 2d shows the mixture matrix (*i.e.* the weight of the three spectra in the mixtures) across the interface. These values are very close to the

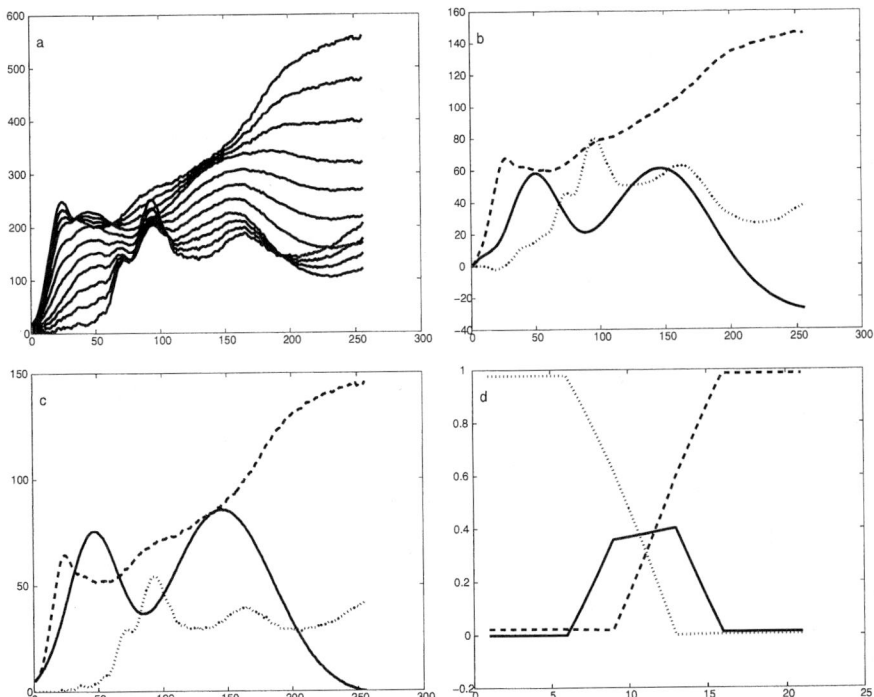

Fig. 2. Recovering of an unknown spectrum from a series of mixtures involving three spectra; a. series of 21 spectra across the interface; b. 3 retrieved spectra; c. 3 retrieved spectra with a positivity constraint; d. proportion of the components across the interface.

values introduced during the simulation of the mixtures. This approach with a number of mixture spectra larger than the number of components in the mixtures provides better results than the previous approach, with as many mixture spectra than components.

4.2 An Experimental Three-Component Mixture

Then, we process an experimental set of 13 electron energy-loss spectra, recorded across a Si-SiO2 interface for an energy loss between 99 and 124 eV [6][2,3].

Fig. 3a contains the recorded spectra while Fig. 3b shows the three estimated components and Fig. 3c displays the weight of these three components across the interface. As expected, the third component is mainly located close to the interface, but diffuses on both sides of the bulk.

---

[2] The authors thank N. Brun and C. Colliex, Laboratoire de Physique des Solides, Orsay for courtesy.

[3] More spectra than sources are necessary because one aim of the study is to determine the variation of composition across the interface.

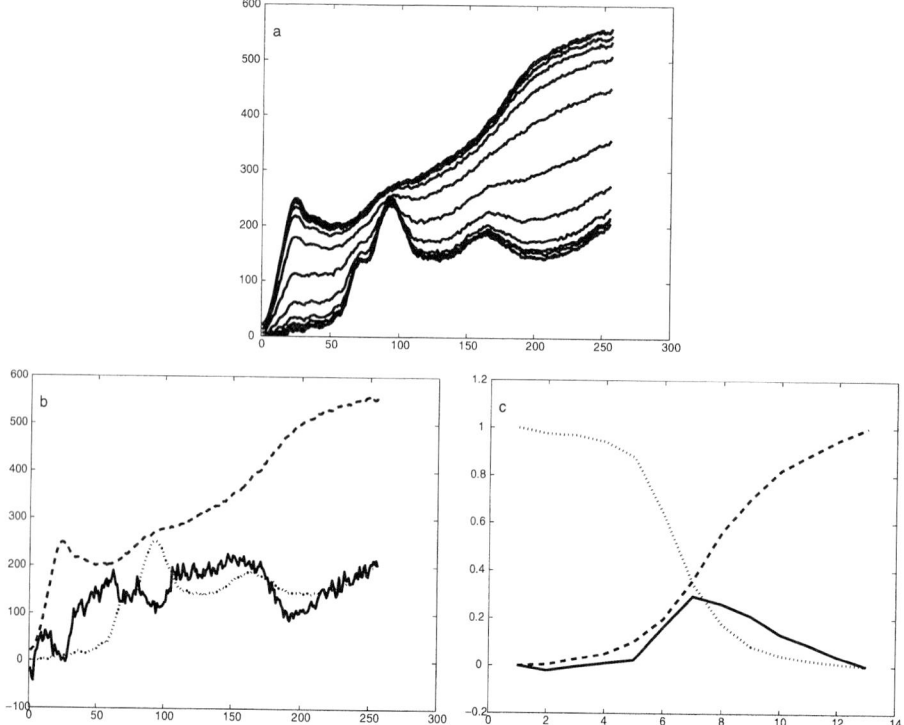

Fig. 3. Real example; a. series of 21 spectra across the interface; b. 3 retrieved spectra with a positivity constraint; c. proportion of pure components across the interface.

5 Discussion and Future Work

Throughout this work, we investigated a BSS method for analysing series of EEL spectra, to complement the two existing methods: the spatial difference method and multivariate statistical analysis. This new approach is very efficient, provided the unknown sources are really independent. For raw EEL spectra, it appears that this is not the case. However, this appears to be approximately the case for first derivative spectra. Thus, we could demonstrate that ICA-BSS can be a good alternative to the spatial difference method and to MSA. We do not claim that this method will systematically outperform previously existing methods. But we claim that it may be a useful complementary tool to explore the content of spectrum-lines or spectrum-images.

The work described here is only an introduction to this new method. More work remain to be done before the technique can be made really efficient:

- the different variants of ICA-BSS must be tried in order to check whether one of them is more appropriate for EEL spectra,
- different variants of spectrum derivatives in the real space (first or second order derivatives, more or less smoothing) as well as in the Fourier space

have also to be checked. At the present time, this choice of parameters is made by trials and errors. A significant improvement would be made if an automatic choice could be made, based on the spectral content of spectra and noise.

References

1. H. Müllejans, J. Bruley, Ultramicroscopy 53 (1994) 351.
2. H. Müllejans, J. Bruley, J. Microsc. 180 (1995) 12.
3. N. Bonnet, E. Simova, X. Thomas, Microsc. Microanal. Microstruct. 2 (1991) 129.
4. N. Bonnet, E. Simova, S. Lebonvallet, H. Kaplan, Ultramicroscopy 40 (1992) 1.
5. N. Bonnet, 14th International Congress on Electron Microscopy. Cancun (Mexique). Sept. 1998. Proceedings p. 133-134.
6. N. Bonnet, N. Brun, C. Colliex, Ultramicroscopy 77 (1999) 97-112.
7. D. A. Muller, Ultramicroscopy 78 (1999) 163.
8. H. Gu, M. Ceh, S. Stemmer, H. Müllejans, M. Rûhle, Ultramicroscopy 59 (1995) 215.
9. H. Gu, Ultramicroscopy 76 (1999) 159.
10. C. Colliex, M. Tencé, E. Lefèvre, C. Mory, H. Gu, D. Bouchet and C. Jeanguillaume, Mikrochimica Acta 114/115 (1994)71.
11. C. Jutten, J. Hérault, Signal Proc. 24 (1991) 1.
12. A. Belouchrani, K. Abed-Meraim, J.-F. Cardoso, and E. Moulines. A blind source separation technique using second-order statistics, IEEE Trans. Signal Proc. 45 (1997) 434-444.
13. D. Nuzillard, J.-M. Nuzillard, Application of Blind Source Separation to 1-D and 2-D NMR Spectroscopy, IEEE SPL 5 (1998) 209-211.
14. D. Nuzillard, A. Bijaoui, Blind source separation and analysis of multi-spectral astronomical images, Astronomy & Astrophysics, Suppl. Ser. 147(2000) 129-138.
15. D. Nuzillard, S. Bourg, J-M. Nuzillard, Model-free Analysis of Mixtures by MNR, Journal of Magnetic Resonance, 133(1998) 358-363.
16. J.-F. Cardoso and A. Souloumiac, IEE Proc.-F 40(1993)362.
17. D. Nuzillard, J.-M. Nuzillard, BSS applied to non-orthogonal signals, ICA'99, Aussois, France, 25-30(1999).
18. D. Nuzillard, J.M. Nuzillard, Second Order Blind Source Separation, Signal Processing,(2003).
19. N. Brun, C. Colliex, J. Rivory and K. Yu-Zhang, Microsc. Microanal. Microstruct. 7 (1996) 161.

HOS Based Distinctive Features for Preliminary Signal Classification

Maciej Pędzisz and Ali Mansour

École Nationale Supérieure des Ingénieurs
des Études et Techniques d'Armement (ENSIETA),
Laboratoire "Extraction et Exploitation de l'Information
en Environnements Incertains" (E^3I^2), Brest, France
{pedzisma,mansour}@ensieta.fr

Abstract. We consider the problem of preliminary classification of digitally modulated signals. The goal is to simplify further signal analysis (synchronization, signal separation, modulation identification and parameters estimation) by making initial separation among the most known classes of signals. Proposed methodology is mainly based on Higher Order Statistics (HOS) of the distributions of instantaneous amplitude and frequency. The experimental results emphasize the performance of the proposed set of features.

1 Introduction

In Communication Intelligence (COMINT), knowledge of signal's frequency structure is essential to recognize underlying modulation type and measure its parameters. Up to now, all frequency synchronization algorithms consider only one signal, they need a big number of symbols and a long time to converge. Thus, making a preliminary signal classification based on frequency invariant features, will much simplify further processing, allowing applications of signal-specific synchronization, source separation and modulation classification techniques.

In [1], authors presented empirical results in Blind Source Separation (BSS) using overcomplete Independent Component Analysis (ICA) representations. They demonstrated fidelity of their algorithm in the case of 2 mixtures of 3 speech signals. Separation of 2 audio sources from a single sensor is the subject covered in [2]. Proposed method generalizes the Wiener filtering with Gaussian Mixture distributions and Hidden Markov Models. A time-frequency filtering based on the Pseudo Wigner-Ville distribution is considered in [3]. Performance of the presented algorithm was validated using a mixture of 2 voice recordings. In [4], sparse factorization approach with K-means clustering algorithm applied to BSS problem is discussed. Provided results reveal the performance of the algorithm in case of 10 face images (6 mixtures), as well as 8 speech signals (5 mixtures). Authors of [5], derive algebraic means for ICA in the case of undetermined mixtures. Their results are based on the structure of the fourth-order cumulant tensor. Sixth-order statistics and the virtual array concept are addressed in [6]. It was shown that their algorithm can be used to increase the effective aperture of an antenna array, and so to identify the mixture of more sources than sensors. The case of binary source separation is covered in [7] and [8]. Their algorithm uses the structure of the probability distributions of the observed data. Simulations showed that the method can successfully separate at least up to 10 binary sources at different noise levels.

On the other hand, modulation recognition algorithms ([9], [10], [11], [12]) deal with the cases where some a priori information is available (carrier frequency, symbol timing, ...) and there is only one signal in additive noise. In this contribution, we try to fill the gap between synchronization & modulation recognition methods, and source (signal) separation algorithms based on one observation (undetermined problem). Using the proposed set of features, we are able to distinguish among the most common known signal types, and so, choose the appropriate methodology for further signal processing.

2 Signal Models

2.1 Mono-component Signal

Let's assume working in the conditions where signal's carrier frequency is not known. The received complex baseband signal (after imperfect demodulation) can be expressed as a sum of two uncorrelated components:

$$s(t) = A_c(t)e^{j(\omega_r t + \Theta_r)} + n(t) \qquad (1)$$

where $A_c(t)$ is a signal complex envelope, ω_r is a residual frequency, Θ_r is a phase of the residual frequency, and $n(t)$ corresponds to a zero-mean, additive white gaussian complex noise.

Using the concept of the complex envelope, we can express any linearly modulated signal as:

$$A_c(t) = A \sum_k d_k h(t - kT - \tau), \qquad k \in \{1, 2, \ldots, K\} \qquad (2)$$

where A is a constant amplitude, d_k describe signal constellation, $h(t)$ is a pulse shaping function, T is a symbol duration, τ is an out-of-synchronization error (due to imperfect demodulation), and K is a number of available symbols. For the most known M-ary linear modulations (MASK – M-ary Amplitude Shift Keying, MQAM – M-ary Quadrature Amplitude Modulation, MPSK – M-ary Phase Shift Keying), we have:

$$d_k^{\text{MASK}} = a_k, \qquad a_k \in \{\pm(2m-1) : m = 1, 2, \ldots, M/2\} \qquad (3)$$

$$d_k^{\text{MQAM}} = a_k + jb_k, \qquad a_k, b_k \in \{\pm(2m-1) : m = 1, 2, \ldots, \log_2(M) - 2\} \qquad (4)$$

$$d_k^{\text{MPSK}} = e^{j\varphi_k}, \qquad \varphi_k \in \{\tfrac{2\pi}{M}(m-1) : m = 1, 2, \ldots, M\}. \qquad (5)$$

In the nonlinear case (MFSK – M-ary Frequency Shift Keying), we can write:

$$A_c(t) = Ae^{j \sum_k d_k \Delta_\omega (t - kT - \tau) h(t - kT - \tau)}, \qquad k \in \{1, 2, \ldots, K\} \qquad (6)$$

where Δ_ω is a frequency deviation, and d_k can be expressed as:

$$d_k^{\text{MFSK}} \in \{\pm(2m-1) : m = 1, 2, \ldots, M/2\}. \tag{7}$$

It is assumed that variables a_k, b_k and φ_k in equations (3), (4) and (5), as well as d_k in (7) are independent and identically distributed (i.i.d. processes). It is assumed also that all modulation states are equiprobable (which is always accomplished when source coding is applied) and the pulse shaping function $h(t)$ is rectangular.

2.2 Multi-component Signal

Taking into consideration the mono-component model of a linear modulation ((1) and (2)), we can write a general formula for a multi-component signal as:

$$\begin{aligned} S(t) &= \sum_{i=1}^{L} A_{c_i}(t) e^{j(\omega_{r_i} t + \Theta_{r_i})} + n_i(t) \\ &= \sum_{i=1}^{L} A_i \sum_k d_{k_i} h_i(t - kT_i - \tau_i) e^{j(\omega_{r_i} t + \Theta_{r_i})} + n(t) \end{aligned} \tag{8}$$

where L is a number of mono-component signals and $n(t)$ is a term which absorbed all noise contributions $n_i(t)$.

In COMINT applications, it is often sufficient to consider: there are two signals in the mixture ($L = 2$), and applied modulation types are MPSK. Additionally, we assume that signal amplitudes are identical ($A_1 = A_2$)[1]. We are not considering prior knowledge about:

- residual frequencies and phases (ω_{r_i} and Θ_{r_i});
- symbol durations (T_i);
- synchronization errors (τ_i).

3 Distinctive Features

3.1 Preliminary Results

Based on the signal models (1) and (8), we can rewrite the received signal as:

$$s_r(t) = p(t) + jq(t) = A_i(t) e^{j\phi_i(t)} \tag{9}$$

where $p(t)$ and $q(t)$ are in-phase and quadrature components, $A_i(t)$ is an instantaneous amplitude and $\phi_i(t)$ is an instantaneous phase. Then, we can define:

$$A_i(t) = |s_r(t)|, \quad \phi_i(t) = \arg\{s_r(t)\}, \quad \omega_i(t) = \frac{d\phi_i(t)}{dt} = \frac{p(t)\frac{dq(t)}{dt} - q(t)\frac{dp(t)}{dt}}{p^2(t) + q^2(t)} \tag{10}$$

where $\omega_i(t)$ is an instantaneous frequency. In general, $\omega_i(t)$ is defined using the concept of the analytic signal [13].

---

[1] General case $A_1 \neq A_2$ will be addressed elsewhere.

It is well known that the probability density function (PDF) of A_i of any MPSK/MFSK signal can be expressed in terms of its constant amplitude A (Eq. (2)) and noise variance σ_n^2 (Eq. (1)) by means of Rice distribution [14]:

$$f_{A_i}(A_i;\, A, \sigma_n^2) = \frac{A_i}{\sigma_n^2} e^{-\frac{A_i^2+A^2}{2\sigma_n^2}} I_0\left(\frac{A_i A}{\sigma_n^2}\right), \quad A_i \geq 0 \tag{11}$$

where $I_0(x)$ is the modified Bessel function of order 0.

If $A = 0$ (NOISE), then PDF of A_i becomes Rayleigh. For the MQAM class of signals, we can write the corresponding PDF as a mean of $f_{A_i}(A_i;\, A_l, \sigma_n^2)$ over all distinctive amplitudes A_l.

The second distribution which can be considered as distinctive in signal classification is the PDF of ω_i [15]. For a single-carrier modulation (MPSK, MQAM), we have:

$$f_{\omega_i}(\omega_i;\, A, \sigma_n^2) = \vartheta^{-1} v_i^{-\frac{3}{2}} e^{-\frac{A^2}{2\sigma_n^2}} {}_1F_1\left(\frac{3}{2}, 1;\, \frac{A^2}{2\sigma_n^2 v_i}\right) \tag{12}$$

where $v_i = 1 + \omega_i^2/\vartheta^2$, $\vartheta^2 = \int_{-\infty}^{+\infty} \omega_i^2 \gamma(\omega)\,d\omega / \int_{-\infty}^{+\infty} \gamma(\omega)\,d\omega$, $\gamma(\omega)$ is a power spectral density (PSD) of noise, and ${}_1F_1(\alpha, \beta;\, x)$ is a confluent hypergeometric function defined as:

$$_1F_1(\alpha, \beta;\, x) = \sum_{k=0}^{+\infty} \frac{\Gamma(\alpha+k)\Gamma(\beta) x^k}{\Gamma(\alpha)\Gamma(\beta+k) k!}, \quad \beta \neq 0, -1, -2, \ldots \tag{13}$$

It is obvious that in the multi-carrier case (MFSK), the PDF of ω_i can be expressed as a mean over all distinctive (carrier) frequencies.

Finally, when $A \gg \sigma_n$, we can approximate both distributions by the Gaussians [13], [15], [16]:

$$f_{A_i}(A_i;\, A, \sigma_n^2) \approx \mathcal{N}(A_i;\, A, \sigma_n^2), \quad f_{\omega_i}(\omega_i;\, A, \sigma_n^2) \approx \mathcal{N}(\omega_i;\, 0, \tfrac{B\sigma_n}{2\sqrt{3}A}) \tag{14}$$

where $\mathcal{N}(x;\, \mu, \sigma^2) \triangleq \frac{1}{\sigma\sqrt{2\pi}} \exp\left[-\frac{(x-\mu)^2}{2\sigma^2}\right]$, and B is a noise effective bandwidth.

3.2 Features Extraction

The main objective in preliminary signal classification is to find a set of characteristics which allows distinction among different classes of signals. Based on distributions of A_i and ω_i, we can extract normalized cumulants [17] of order 3 γ_3 (skewness) and 4 γ_4 (kurtosis) as:

$$\gamma_3 = \frac{\kappa_3}{\kappa_2^{3/2}}, \quad \gamma_4 = \frac{\kappa_4}{\kappa_2^2} \tag{15}$$

where cumulants κ_r and corresponding moments m_r are defined by:

$$\kappa_2 = m_2 - m_1^2 \tag{16}$$
$$\kappa_3 = m_3 - 3m_2 m_1 + 2m_1^3 \tag{17}$$
$$\kappa_4 = m_4 - 4m_3 m_1 - 3m_2^2 + 12 m_2 m_1^2 - 6 m_1^4 \tag{18}$$
$$m_r = \int_{-\infty}^{+\infty} x^r f(x)\,dx\,. \tag{19}$$

Other sets of characteristics can be obtained by using Renyi's quadratic entropy [18]:

$$H_2 = -\log\left[\int_{-\infty}^{+\infty} f^2(x)\,dx\right] \quad (20)$$

and by solving a polynomial regression on the logarithm of a PDF:

$$\log(f(x)) \approx \sum_k a_k x^k. \quad (21)$$

3.3 Features Selection and Dimensionality Reduction

It is obvious that limiting the number of features will make learning and testing faster and demanding less memory. Aside from this, feature space of a lower dimension may enable more accurate classifiers for a finite learning set.

Based on the characteristics presented in the previous section, experiments have been conducted to choose the most discriminative set of features:

- features based on A_i: γ_3^A, γ_4^A, H_2^A, a_3^A, a_2^A, a_1^A, a_0^A (3-rd degree polynomial is sufficient to describe asymmetry and flatness of considered distributions);
- features based on ω_i: γ_4^ω, H_2^ω, a_4^ω, a_2^ω, a_0^ω (PDF of ω_i is symmetrical about the mean, so all the features based on asymmetry were eliminated).

Once they have been selected, one can apply the Linear Discriminant Analysis to verify the importance of chosen features. Using the Fisher's criterion [19]:

$$J_F = \mathrm{tr}\{\mathbf{T}\} = \mathrm{tr}\{\mathbf{S}_w^{-1}\mathbf{S}_b\} \quad (22)$$

where $\mathbf{S}_w$ is the within-class covariance matrix (the sum of covariance matrices computed for each class separately), and $\mathbf{S}_b$ is the between-class covariance matrix (the covariance matrix of class means), we found:

- all selected features are of equal importance – among differents combinations of features, the whole set is the most discriminative;
- features from A_i are best to separate between classes of signals with symmetric A_i PDF (MPSK, MFSK) and asymmetric (NOISE, MQAM and MIXTURE);
- features from ω_i are best to separate between classes of signals with unimodal ω_i PDF (MPSK, MQAM and MIXTURE) and multimodal (MFSK).

It should be noted, that using eigenvectors of matrix $\mathbf{T}$, it is possible to reduce dimensionality of the feature vector

$$\mathbf{x} = [\gamma_3^A, \gamma_4^A, H_2^A, a_3^A, a_2^A, a_1^A, a_0^A, \gamma_4^\omega, H_2^\omega, a_4^\omega, a_2^\omega, a_0^\omega]^T \quad (23)$$

by means of linear transformation:

$$\mathbf{y} = \mathbf{W}\mathbf{x} \quad (24)$$

where eigenvectors corresponding to largest eigenvalues of $\mathbf{T}$ form the rows of the transformation matrix $\mathbf{W}$.

4 Simulations

To evaluate the performance of the proposed set of features, extensive simulations were conducted on the signals: NOISE, MPSK (2, 4 and 8), MFSK (2 and 4), MQAM (16 and 32) and MIXTURE (2xBPSK, 2xQPSK and BPSK & QPSK). All signals were composed of 512 samples, 5 samples per symbol, 1000 different realizations. Signal to Noise Ratio (SNR) was varying from 0 dB up to 30 dB. The residual frequencies ω_{r_i}, the corresponding phases Θ_{r_i}, as well as the symbol timings T_i, were chosen randomly according to Nyquist sampling theorem. Corresponding results (SNR = 5 dB) are shown in Fig. 1.

5 Conclusion

It is evident that selected set of features is very efficient even for low SNR. Perfect classification can be obtained for the classes NOISE, MPSK and MFSK for SNR > 5 dB, however distinction between MQAM and MIXTURE is far from being "sufficient enough".

Although classification in a 2D space was used for visualization purposes, one should not limit himself during constructing a final classifier. Adding another set of characteristics (based for example on Time-Frequency Distributions (TFD)), may be more attractive in more than 2 dimensions. Also, making classifier hierarchical or using some nonlinear mappings (MMI [20], NPCA [21]), may increase separability of the classes. These topics will be covered in a future work.

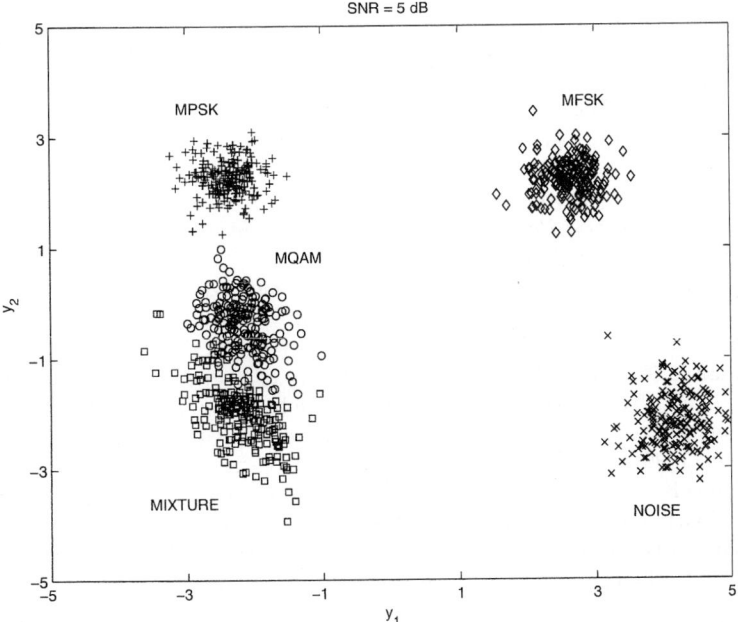

Fig. 1. Signals in a 2D space after dimensionality reduction (SNR = 5 dB).

References

1. Lee T.-W., Lewicki M.S., Girolami M., Sejnowski T.J.: *Blind Source Separation of More Sources Than Mixtures Using Overcomplete Representations*. IEEE Sig. Proc. Let., Vol. 6, No. 4, April 1999.
2. Benaroya L., Bimbot F.: *Wiener Based Source Separation with HMM/GMM Using a Single Sensor*. Journée AS Séparation de Sources et GdR ISIS, Paris, 12 June 2003.
3. Mansour A., Kawamoto M., Puntonet C.: *A Time-Frequency Approach to Blind Separation of Under-Determined Mixture of Sources*. Proc. of the IASTED International Conference on Applied Simulation and Modelling, Marbella, Spain, 3-5 September 2003.
4. Li Y., Cichocki A., Amari S.: *Sparse Component Analysis for Blind Source Separation with Less Sensors than Sources*. ICA 2003, Nara, Japan, April 2003.
5. De Lathauwer L., De Moor B., Vandewalle J., Cardoso J.-F.: *Independent Component Analysis of Largery Undetermined Mixtures*. ICA 2003, Nara, Japan, April 2003.
6. Albera L., Ferréol A., Comon P., Chevalier P.: *Sixth Order Blind Identification of Undetermined Mixtures (BIRTH) of Sources*. ICA 2003, Nara, Japan, April 2003.
7. Diamantaras K.I.: *Blind Separation of Multiply Binary Sources using a Single Linear Mixture*. ICASSP 2000, Istanbul, Turkey, June 2000.
8. Diamantaras K.I., Chassioti E.: *Blind Separation of n Binary Sources from one Observation: A Deterministic Approach*. ICA 2000, Helsinki, Finland, 19-22 June 2000.
9. Azzouz E.E., Nandi A.K.: *Automatic Modulation Recognition of Communication Signals*. Kluwer Academic Publishers, 1996.
10. Soliman S.S., Hsue S.-Z.: *Signal Classification Using Statistical Moments*. IEEE Transactions on Communications, Vol. 40, No. 5, May 1992.
11. Hero A.O., Hadinejad-Mehram H.: *Digital Modulation Classification Using Power Moment Matrices*. ICASSP 1998, Seattle, USA, May 1998.
12. Hipp J.E.: *Modulation Classification Based on Statistical Moments*. MILCOM 1986, Monterey, USA, October 1986.
13. Proakis J.G.: *Digital Communications*. McGraw-Hill, 4-th edition, 2001.
14. Haykin S.: *Communication Systems*. John Wiley & Sons, Inc., 3-rd edition, 1994.
15. Levin B.R.: *Theoric Bases of Statistical Technics in Radio*. Radio and Communications (in russian), 3-rd edition, 1989.
16. Blachman N.M.: *Gaussian Noise–Part II: Distribution of Phase Change of Narrow-Band Noise Plus Sinusoid*. IEEE Trans. on Inf. Theory, Vol. 34, No. 6, November 1988.
17. Kendall M.G., Stuart A.: *The Advanced Theory of Statistics*. Charles Griffin, 1958.
18. Haykin S.: *Unsupervised Adaptive Filtering*. John Wiley & Sons, Inc., 2000.
19. Fukunaga K.: *Introduction to Statistical Pattern Recognition*. Academic Press, 2-nd edition, 1990.
20. Torkkola K.: *Feature Extraction by Non-Parametric Mutual Information Maximization*. Journal of Machine Learning Research 3, p. 1415-1438, 2003.
21. Chalmond B., Girard S.: *Nonlinear Data Representation for Visual Learning*. Rapport de Recherche, No 3550, November 1998.

ICA as a Preprocessing Technique for Classification*

V. Sanchez-Poblador[1], Enric Monte-Moreno[1], and Jordi Solé-Casals[2]

[1] TALP Research Center
Universitat Politècnica de Catalunya, Catalonia, Spain
enric@gps.tsc.upc.es
http://gps-tsc.upc.es/veu/personal/enric/enric.html
[2] Signal Processing Group, University of Vic, Catalonia, Spain
jordi.sole@uvic.es
http://www.uvic.es/eps/recerca/ca/processament/inici.html

Abstract. In this paper we propose the use of the independent component analysis (ICA) [1] technique for improving the classification rate of decision trees and multilayer perceptrons [2], [3]. The use of an ICA for the preprocessing stage, makes the structure of both classifiers simpler, and therefore improves the generalization properties. The hypothesis behind the proposed preprocessing is that an ICA analysis will transform the feature space into a space where the components are independent, and aligned to the axes and therefore will be more adapted to the way that a decision tree is constructed. Also the inference of the weights of a multilayer perceptron will be much easier because the gradient search in the weight space will follow independent trajectories. The result is that classifiers are less complex and on some databases the error rate is lower. This idea is also applicable to regression

1 Introduction

The problem of classification consists on deciding a class membership of an observation vector [2]. Usually this observation vector consists of features that are related. The classification algorithm has to take a decision after the analysis of several features even though they can be mutually related in difficult ways. The dependencies between the features have an influence on the learned classifier. In the case of a decision tree, each node analyses a single feature, or a lineal combination of features, and the selection of a feature for a given level of the tree, is made in a greedy way [3]. As the decisions in the tree are made on a feature or subset of features basis, it would help if one could transform the features in such a way that instead of having the discriminative information spread through all the features, on could make each feature independent of the others and consequently simplify the decision process. This simplification of the decision process gives trees with a lower number of nodes. It is well known that there is a relationship between the complexity of a classifier and the generalization error [4]. Generally this complexity is dealt by pruning the tree. We pro-

---

* This work has been partially supported by the Spanish CICyT project ALIADO, by the EU integrated project CHIL and by the University of Vic under the grant R0912

pose transforming the input so that the resulting vector has the property that each component is independent of the others. We shall do this by means of ICA In the case of classifying by means of a decision tree, as there is no statistical dependency between features the number of decisions (i.e. nodes) is lower, as it is confirmed by our experimental results. In the case of training a multilayer perceptron, the inference of the weights is made by a gradient search, which is known to be very inefficient if the features are highly correlated [2]. It is also known that the incorrelation preprocessing of the inputs of a multilayer perceptron improves the convergence of the algorithm because near a minimum the form of the error function can be approximated locally by a hyper-parabola. This explains the improvement that can be achieved by the use of algorithms such as the conjugate gradient or the Levenberg-Marquardt. Notice that the characteristics of these algorithms are adapted to the fact that the data can have correlated features. So a process of whitening the data or using these improvements of the gradient algorithms means that we are making a strong hypothesis about the data. We propose to preprocess the data in such a way that the features will be mutually independent, and therefore the gradient descent will follow a smooth surface, even if high order moments between features are present in the original pattern. In the past, ICA has been used for training decision trees [6], where the authors searched for an unsupervised method for training decision trees. This method uses the fact that by means of ICA feature space is transformed in such a way that data is aligned to the axis, therefore, they suppose that the components that go to a given node correspond somehow to a given class. As we will see in section 2, this is true depending on the distribution of the data. In our case, we use explicitly the fact that data is labeled, and the classifiers are constructed in a supervised way.

2 Independent Component Analysis as Preprocessing Tool

Independent component analysis supposes that the observation vector was generated by a linear mixture of a set of independent random variables. This hypothesis might not be true for all the classification problems, as the generation of the observations do not always come from a linear mixture. For instance, in the classification task for the echocardiogram database, we have a set of variables, which although are related, are not generated by a linear mixture. For instance in the case of the echocardiogram database, the Left ventricular end-diastolic dimension (LVDD), and the age of the patient affected by the infarction/heart attack, are related, i.e. sick hearts are related to high values of the LVDD, and with older patients. But the age cannot be interpreted as a linear mixture of two different causes, which also generate the LVDD. As can be seen in the scatter plot of figure 1, the two classes do not follow different directions in the feature space, therefore an ICA preprocessing will not find rotations or transforms that separate classes. This effect will be seen in the results, where the preprocessing with ICA does not improve results. Nevertheless, in the case of the crabs database (see figure 2), the classes are spread along different directions, therefore, the effect of the ICA preprocessing will be the alignment of the classes with the axis, which will increase the mean distance between the instances of different classes, and at the same time sets the orientation of the class boundaries in parallel with the axes. And also, as shown in the figure, decision boundaries are simpler. In this case the whitening preprocessing yields the most simple border. In the case of vowel2 database, as can be seen in figure 3, either whitening or transforming with an ICA processing does not change significantly the relative position between the points in the feature space. In this case, the complexity of the decision boundary remains more or less the same, and therefore the recognition rate is practically the same in the three cases. In this case, there is no improvement because of the geometry of the problem. The classes are distributed in such a way that the orientation of the samples of each class is different.

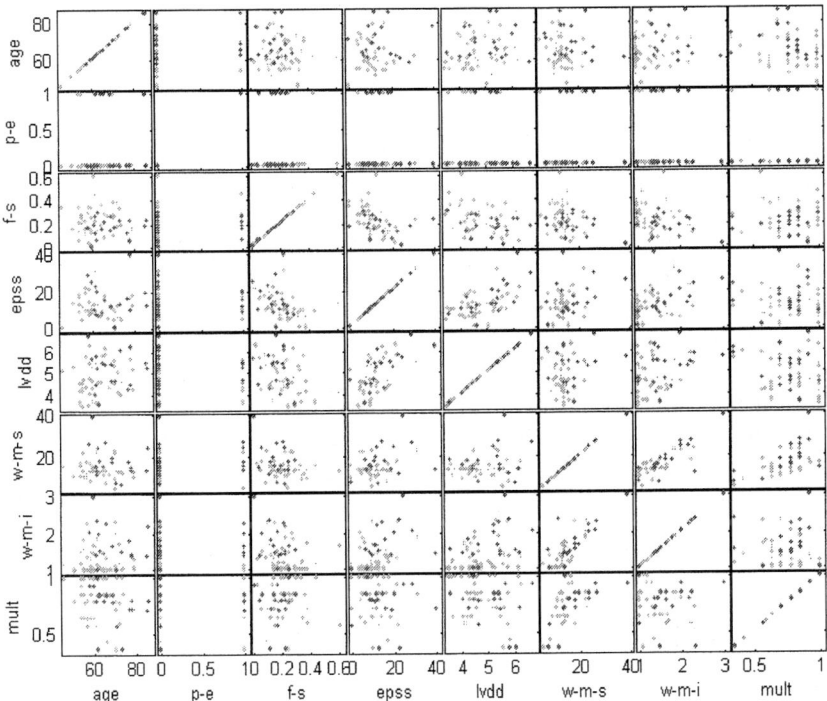

Fig. 1. Scatter plot of the echocardiogram database, before preprocessing.

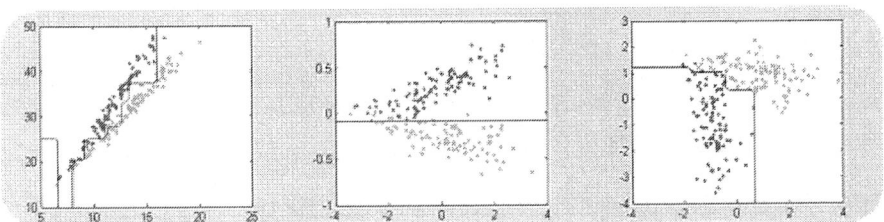

Fig. 2. Scatter plot for two features of the crabs database (length, width of carapace), the scatter plot after whitening and after ICA. The decision boundaries correspond to a decision tree.

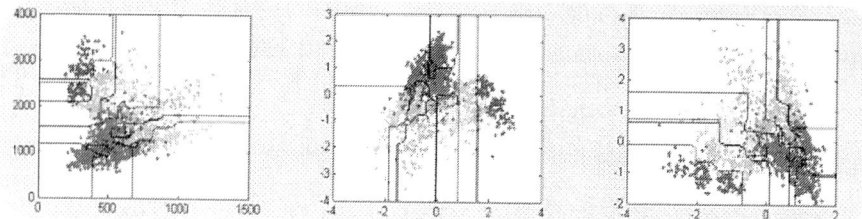

Fig. 3. Scatter plot for two features (first two formants) of the vowel2 database, after whitening and after ICA. The decision boundaries correspond to the decision tree.

Another example where the recognition rate does not change due to the preprocessing is the case of breast cancer. Figure 4, shows the scatter plot for two combinations of features, with the decision boundaries made either by a multilayer perceptron or by a decision tree. As can be seen, the distribution of samples is similar (except for a rotation), in all cases, and therefore the decision boundaries have more or less the same complexity. The error rate is more or less similar for all the cases, perhaps slightly lower for the case of a preprocessing with ICA. But in the case of the echocardiogram database, the results are different. In figure 5, we show the scatter plot for different components. It can be seen that the distribution without preprocessing or with a PCA, give distribution of points where classes are not aligned with the axes, which does not happen after the ICA processing. The result is that the class borders are smoother in the case of the multilayer perceptron, and in the case of decision trees, the borders are aligned with features. Therefore, we can expect that the classification results will improve after ICA in the cases were the feature space has a structure where the data is aligned with certain directions.

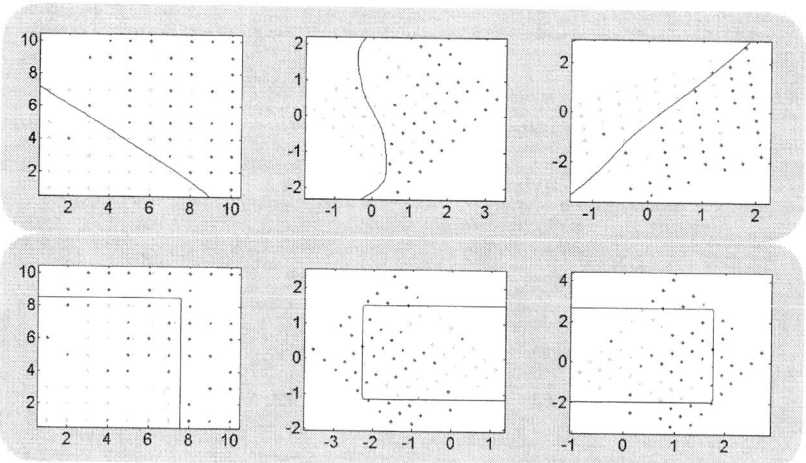

Fig. 4. Scatter plot for two features of the breast cancer database. Upper figures, features (1, 7), lower figure features (7, 8). Scatter plot without processing, after whitening and after ICA. The depicted decision boundaries correspond in the upper figures to a multilayer perceptron and in the lower to a decision tree.

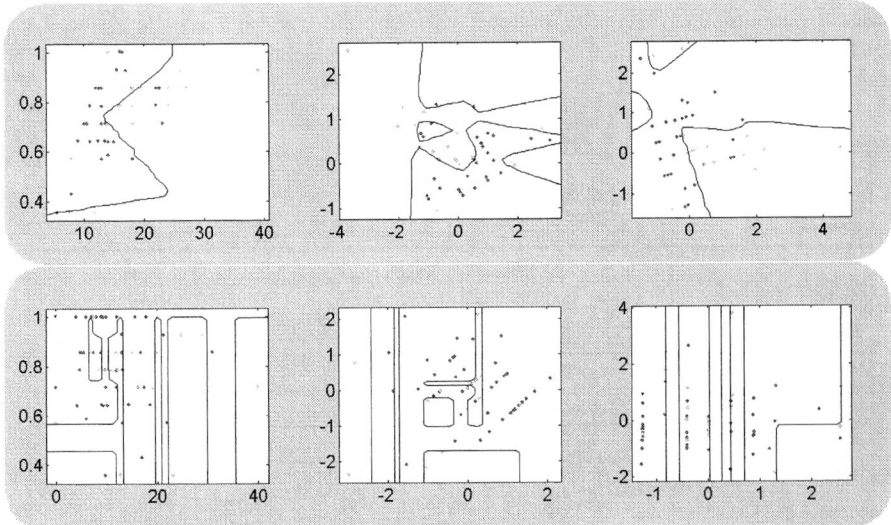

Fig. 5. Scatter plot for two features of the echocardiogram database. Upper figures, features (6, 8), lower figure features (4, 8). Scatter plot without processing, after whitening and after ICA. The depicted decision boundaries correspond in the upper figures to a multilayer perceptron and in the lower to a decision tree.

3 Experimental Frameworks

In order to test the effect of an ICA preprocessing and compare with a whitening preprocessing, we did an experiment on 9 databases [5], which are summarized in table 1. Both preprocessing were done without a dimensionality reduction. The databases were divided into train/test. In the case of the multilayer perceptron this partition was done in order to use the validation sub-database for stopping the training phase. In order to compute the test results, we did 50 bootstrap samples of the data base. For each bootstrap replication, the bootstrap data was used for training (in the case of the multilayer perceptron for training and validation), and for test the samples that were not selected, therefore the relation train/test was about 65%/35 % depending on the realization. As it is well known that the CART algorithm is unstable [7], and neural nets fall into local minima, the results that we present in the tables are the mean of the 10 best results,. i.e. the mean of the best error rates and complexity of the classifiers. The pruning criterion of the decision trees and the number of hidden layer units were selected by cross validation. The ICA transformation was done by means of the Jade algorithm [8]. We used two different classifiers; decision trees and multilayer perceptrons. For the decision trees we used the CART algorithm [3], and the multilayer perceptron was trained by means of the Levenberg-Marquart algorithm. The results presented in tables (2,3,4) are somehow lower than the ones found in literature, because the objective of the paper was to assert the effect of the preprocessing, which meant dealing with the instability of the classifiers (i.e. local minima of the multilayer perceptron, different tree structure with slightly different database). In order to smooth the variability of the classifiers, the train and test databases were selected by bootstrap, which meant that the training database was poorer than the standard experiments, which assign about 90% the data, instead of a 65%.

Table 1. Data Set Summary.

Data Set	Size	Inputs	Classes
biomed	209	4	2
breast cancer	683	9	2
crabs	200	6	2
echo	62	8	2
sonar	208	60	2
titanic	2201	3	2
vowel	990	10	11
wine	178	13	3
vowel2	1520	4	10

4 Results

In Tables 2, 3 and 4 we present the classification errors and complexity of two different classifiers; decision trees and multilayer perceptrons. We compare a baseline version without data preprocessing (table 2), with the results obtained by the whitening preprocessing (table 3), and ICA preprocessing (table 4). The first conclusion that can be drawn from the results is that some databases benefit from the ICA processing, while in others the benefit is so small that is within the confidence margins, so we cannot assert that there is a real benefit or even in some cases there is a degradation. Improvements are consistent with the classification algorithms, that is, the preprocessing improves the results in both the trees and the multilayer perceptron, although the performance between classifiers can be different. The explanation of this different behavior is related to the distribution of the data on the feature space and is explained in section 2. In cases where the distribution of the classes in the feature space has regularity such that classes can be aligned in certain directions there is a clear improvement. A sign of it is that the databases that improve the error rate with ICA, also improve when we whitened the data. The improvement gotten by the whitening is in most cases lower than the one obtained by the use of ICA. The whitening process improved in 6 of the 9 databases in the case of decision trees, and 4 of the 9 in the case of the multilayer perceptron. The ICA preprocessing yielded improvements in 5 out of 9 cases for the case of the decision trees and 6 cases out of 9 for the multilayer perceptron. On the other hand, the complexity of the decision trees changes depending on the preprocessing, and is related to the improvement/degradation of the recognition rate. In the case of decision trees an improvement on the recognition rate is associated with a simpler structure (3 best improvements), while in the cases when the recognition rate does not change or degrades the number of nodes of the decision tree increases. The best results with a multilayer perceptron were not associated to smaller networks, the benefit of the ICA

Table 2. Test set Errors (%) without processing.

Data Set	Decision Tree	Multilayer perceptron	Complexity	
			Tree/ Mean Number of nodes	MLP/ Nodes hidden layer
biomed	12.77	09.30	24	6
breast cancer	3.49	2.55	42	6
crabs	9.04	03.79	28	6
echocardio	21.96	21.41	4	8
sonar	25.53	15.97	29	10
titanic	29.13	19.97	1	6
vowel	29.80	24.13	558	20
wine	6.42	0.74	34	10
vowel2	21.17	13.44	522	20

Table 3. Test set Errors (%) whitening pre processing.

Data Set	Decision Tree	Multilayer perceptron	Complexity	
			Tree/ Mean Number of nodes	MLP/ Nodes hidden layer
biomed	13.15	08.36	48	10
breast cancer	3.36	02.20	27	6
crabs	2.28	02.34	12	6
echocardiogram	20.74	22.50	2	2
sonar	19.08	17.59	39	6
titanic	28.03	20.11	4	8
vowel	37.56	21.05	648	20
wine	04.92	01.64	22	8
vowel2	25.95	14.24	577	20

smaller networks, the benefit of the ICA was due to the smoother error surface that gave a lower number of local minima.

5 Conclusions

We have shown that the use of ICA as a preprocessing tool can improve the classification results when the feature space has a certain structure. This improvement does not happen always and is related to the fact that the classes distribution in the feature spaces is such, that after the ICA preprocessing the samples of classes are better aligned with the dimensions. In these cases, ICA gives better results than PCA.

Table 4. Test set Errors (%) ICA preprocessing.

Data Set	Decision Tree	Multilayer perceptron	Complexity	
			Tree/ Mean Number of nodes	MLP/ Nodes hidden layer
biomed	14.58	09.17	41	10
breast cancer	04.79	02.63	50	8
crabs	06.85	02.47	25	10
echocardiogram	20.05	19.40	2	10
sonar	24.57	15.46	44	10
titanic	27.73	20.14	3	8
vowel	48.70	20.77	799	20
wine	20.86	01.47	53	10
vowel2	20.15	13.33	349	20

References

1. Hyvärinen, J. Karhunen, E. Oja: Independent Component Analysis, Wiley, 2001
2. Richard O. Duda, Peter E. Hart, David G. Stork: Pattern Classification (2nd Edition) Wiley Interscience, 2000
3. Breiman, Friedman, Olshen, Stone: Classification And Regression Trees (CART) Chapman & Hall, 1984
4. Machine Learning Tom M. Mitchell McGraw-Hill, 1997
5. Blake, C.L. & Merz, C.J.: UCI Repository of machine learning databases [http://www.ics.uci.edu/~mlearn/MLRepository.html]. Irvine, CA: University of California, Department of Information and Computer Science.(1998).
6. Petteri Pajunen, Mark Girolami: Implementing decisions in binary Decision Trees using ICA, 2nd International Workshop on ICA and BSS, 483–487, Helsinki, 2000.
7. L. Breiman: Bagging predictors. Machine Learning, 24(2): 123-140, 1996.
8. Jean-François Cardoso and Antoine Souloumiac: Jacobi angles for simultaneous diagonalization, (SIAM) J. Mat. Anal. Appl. jan, 1996.

Joint Delay Tracking and Interference Cancellation in DS-CDMA Systems Using Successive ICA for Oversaturated Data

Tapani Ristaniemi and Toni Huovinen

Institute of Communications Engineering,
Tampere University of Technology
P.O.Box 553, FIN-33101, Tampere, Finland
{Tapani.Ristaniemi,Toni.Huovinen}@tut.fi

Abstract. Recent studies have found successive interference cancellation (SIC) schemes employing blind source separation (BSS) methods to be promising competitors for conventional interference cancellation schemes in DS-CDMA systems. One reason for this is the inherent ability of BSS to mitigate many kinds of interference sources e.g. multi-access, multi-path and out-of-cell interferences. In addition, "more sources than observations"-problem can be somewhat circumvented by combining the SIC-ideology to BSS. Hence, BSS-SIC -type receivers can be used also in highly loaded systems where both conventional ICA and conventional interference cancellation usually fails. Recently proposed BSS-SIC receivers have needed accurate estimates for the user delays. This is mandatory for any interference canceller, since subtraction of a inaccurately estimated source from the original data actually enhances interference. In this paper we propose a scheme in which the ICA solution is used to tracking of users' delays, too, which helps to avoid the performance losses due to the inaccuracies in conventional delay tracking circuitry.

1 Introduction

The capacity of a Direct-Sequence Code Division Multiple Access (DS-CDMA) system is in practice limited by the multiple access interference (MAI), which arises due to the non-ideal crosscorrelations between the spreading sequences. Conventional detection considers this interference as background noise and thus becomes inadequate in highly loaded systems. The other extreme from computational and performance points of view is individually optimum detection [1], which has been followed by numerous sub-optimum solutions with much lower complexity.

Interference rejection/cancellation has been considered one of the most attractive class of suboptimal solutions, and has been studied extensively in the past [2]–[5]. Parallel and successive interference cancellation (PIC and SIC, respectively) are the main categories within this class, describing the procedure by which the interference is subtracted from the original data, after regenerating the interference from the tentatively estimated data. This procedure can, naturally, be repeated many times resulting in multi-stage interference cancellation schemes. Needless to say, any interference subtractive receiver performs the better the more accurately tentative decisions are being made. This is because the interference level is then reduced the most. Otherwise, from a DS-CDMA signal point of view, two different situations will occur:

(a) The particular user signal is not completely taken out from the original data or it is even strengthened; or in addition to that,
(b) a fictitious multipath component is generated and added to the original data.

The former happens e.g. if all the other parameters except the (complex-valued) amplitude is correctly estimated. The latter happens if the phase of the spreading code is inaccurately estimated. To mild the consequences of the situation (a) one can e.g. estimate the reliability of tentative decision from the soft decisions. Doing this way one actually chooses in favor of accepting only partial mitigation (which happens almost surely) rather than accepting occasional interference enhancement. In this paper we consider the avoidance of the situation (b), for which the most natural solution is more accurate delay estimation.

One relatively new idea is to employ blind source separation (BSS) techniques [6] in interference subtractive receivers. In [7] it was shown how the parametric form of the mixing matrix can efficiently be used to refine the ICA solution, and hence avoid interference enhancement while subtracting that source from the original data. The BSS-SIC -type receiver was further developed in [8] for highly loaded systems. Recall that a major drawback for standard BSS is the case where the number of source signals (to be blindly extracted from the received data) is greater than the number of observations made. This is a commonplace situation in communications applications in which cases standard BSS model doesn't hold anymore. The key finding in [8] was the ability of a BSS-SIC -type receiver structure to somewhat circumvent the "more sources than observations"-problem in the sense that adequate performance (in terms of bit-error probability) is still achievable even in extremely highly loaded system, whereas conventional parallel and successive interference cancellation only remain at a moderate level.

In this paper we further develop the BSS-SIC -type receiver (at the expense of negligible increase in computation) to cope with erroneous timing estimates. Recall that the timing information, that is, the phase of the band-spreading code, should be known accurately to avoid interference enhancement during the subtraction phase. Roughly speaking, the ICA solution implicitly includes the timing information, thus giving a possibility to estimate it due to the fact that the parametric form of the mixing matrix is known. Numerical examples are given to show how the joint delay tracking and interference cancellation based on BSS-SIC -type receiver tolerates quite inaccurate initial parameter estimates, unlike conventional parallel and successive interference cancellation schemes as well as advanced LMMSE-PIC [9] receiver. The latter receiver estimates each user with an optimal linear MMSE detector, after which subtraction follows.

2 Signal Model

Consider a DS-CDMA uplink channel with additive white gaussian noise (AWGN) [10]. The well-known form for the single-path data, $r(t)$, is assumed:

$$r(t) = \sum_{m=1}^{M} \sum_{k=1}^{K} b_{km} a_k s_k(t - mT - \tau_k \frac{T_c}{T}). \quad (1)$$

Here b_{km} is the mth symbol sent by kth user. The complex coefficient of the kth user's channel is denoted by a_k, which is assumed to remain the same during the data block of, say, M symbols. $s_k(\cdot)$ is kth user's binary chip sequence, supported by $[0, T)$, where T is the symbol duration. T_c is the chip duration. The user delay is denoted $\tau_k = d_k + \delta_k$, where $d_k \in \{0, \ldots, C-1\}$, C is number of chips in the spreading code and $\delta \in (-\frac{1}{2}, \frac{1}{2})$. The delays are assumed to remain constant during the block of M data symbols. $n(t)$ denotes noise.

The received continuous-time signal is assumed to be sampled by chip-matched filtering, and using processing window size of two symbols. The sampled data has the form [11], [12]

$$\mathbf{r}_m \overset{\text{def}}{=} \sum_{k=1}^{K} a_k (b_{k,m-1} \underline{\mathbf{g}}_k + b_{km} \mathbf{g}_k + b_{k,m+1} \overline{\mathbf{g}}_k) + \mathbf{n}_m \quad (2)$$

Here $\mathbf{n}_m$ denotes noise vector and the code vectors of length $2C$ are defined as

$$\begin{aligned}
\underline{\mathbf{g}}_k &\overset{\text{def}}{=} \underline{\mathbf{g}}_k(\delta_k, d_k) = (1 - |\delta_k|)\, \underline{\mathbf{c}}_k(d_k) + |\delta_k|\, \underline{\mathbf{c}}_k(d_k + \text{sign}(\delta_k)) \\
\mathbf{g}_k &\overset{\text{def}}{=} \mathbf{g}_k(\delta_k, d_k) = (1 - |\delta_k|)\, \mathbf{c}_k(d_k) + |\delta_k|\, \mathbf{c}_k(d_k + \text{sign}(\delta_k)) \\
\overline{\mathbf{g}}_k &\overset{\text{def}}{=} \overline{\mathbf{g}}_k(\delta_k, d_k) = (1 - |\delta_k|)\, \overline{\mathbf{c}}_k(d_k) + |\delta_k|\, \overline{\mathbf{c}}_k(d_k + \text{sign}(\delta_k))
\end{aligned} \quad (3)$$

where

$$\begin{aligned}
\underline{\mathbf{c}}_k(d_k) &\overset{\text{def}}{=} [s_k[C - d_k + 1] \ldots s_k[C]\ \mathbf{0}_{2C - d_k}^T]^T \\
\mathbf{c}_k(d_k) &\overset{\text{def}}{=} [\mathbf{0}_{d_k}^T\ s_k[1] \ldots s_k[C]\ \mathbf{0}_{C - d_k}^T]^T \\
\overline{\mathbf{c}}_k(d_k) &\overset{\text{def}}{=} [\mathbf{0}_{C + d_k}^T\ s_k[1] \ldots s_k[C - d_k]]^T.
\end{aligned} \quad (4)$$

With a simple manipulation, we can get a compact representation for the data,

$$\mathbf{r}_m = \mathbf{G}\mathbf{b}_m + \mathbf{n}_m. \quad (5)$$

The $2C \times 3K$ dimensional code matrix $\mathbf{G}$ contains the code vectors and path strengths, while the $3K$-vector $\mathbf{b}_m$ contains the symbols:

$$\begin{aligned}
\mathbf{G} &\overset{\text{def}}{=} [\cdots a_k \underline{\mathbf{g}}_k\ a_k \mathbf{g}_k\ a_k \overline{\mathbf{g}}_k \cdots] \\
\mathbf{b}_m &\overset{\text{def}}{=} [\cdots b_{k,m-1} b_{km} b_{k,m+1} \cdots]^T.
\end{aligned} \quad (6)$$

The DS-CDMA signal model (5) is readily a linear noisy ICA model with $m = 2C$ observations of $n = 3K$ source components.

Without loosing any generality we assume that the delay of a user k is pre-estimated to be equal to d_k, that is, $\hat{\tau}_k = d_k$, even though $d_k + \delta_k$ would be the correct one. Consequently, δ_k is considered as an residual delay estimation error after delay tracking circuitry. It is thus typical to consider this error to obey zero mean Gaussian distribution.

Recall that conventional single user detection makes a decision according to

$$\hat{b}_k = \mathrm{sgn}(a_k^* \mathbf{r}_m^H \mathbf{c}_k(\hat{\tau}_k)) \tag{7}$$

from which we see that there is a timing misalignment equal to δ_k between the local replica of the spreading code and that of received data.

3 BSS/ICA Based Successive Interference Cancellation

By an interference subtractive receiver we loosely speaking mean an iterative multi-user receiver, where the estimated interference is subtracted from the received signal prior to the estimation of a particular user. The principle of this kind of receiver [7, 8] utilizing BSS/ICA is quite straightforward and is here only shortly revisited. Namely, the received signal is first separated by ICA. After separation, users are identified by their spreading codes. Hence, the receiver is semi-blind. Strictly speaking, for the user identification a correlation

$$\rho(k, k') = \frac{\mathbf{c}_k(\hat{\tau}_k) \mathbf{w}_{k'}^H}{\|\mathbf{c}_k(\hat{\tau}_k)\| \|\mathbf{w}_{k'}\|} \tag{8}$$

is computed for each k, k', where $\mathbf{w}_{k'}$ corresponds to the ICA basis vector of k':th source. Next, a threshold, ρ_1, is set to indicate proper detection. This is needed because subtraction of erroneously detected signal would actually enhance interference. Thus, only the signal corresponding to the user k with $|\rho(k, k')| > \rho_1$ for some k' is subtracted from the receiver signal. After subtraction of all such users' signals, the next ICA separation is performed for the interference-subtracted signal. The order of ICA model decreases in subtraction, which help ICA to recover the remaining users. The procedure is repeated successively until all users are detected. Furthermore, one can easily show, that phase of complex number $\rho(k, k')$ in (8) equals to phase shift produced by ICA. This makes it possible to correct ICA's phase ambiguity. The successive ICA receiver is illustrated in Fig. 1.

It is of primary importance to see that what is actually subtracted from the original data is a tentative decision of the form $a_k \mathbf{c}_k(\hat{\tau}_k) \hat{b}_k$ rather than $\mathbf{w}_k \hat{b}_k$. Hence the ICA solution is first refined (according to the knowledge of the parametric form of the mixing matrix) before subtraction is performed.

4 Joint Delay Tracking and Interference Cancellation Using BSS/ICA

The receiver structure described above is now developed to cope with inaccuracies in delay estimation. Recall that ICA performs purely in blind manner.

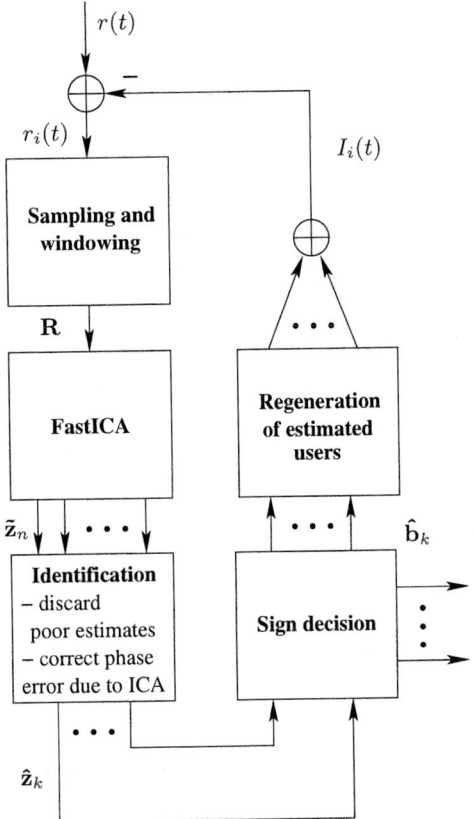

Fig. 1. Group-wise successive ICA receiver structure. $r_i(t)$ is received data after the $(i-1)$th interference subtraction round $(r_1(t) = r(t))$. $\mathbf{R} = [\mathbf{r}_1 \ldots \mathbf{r}_M]$ is matrix containing sampled data. Vectors $\tilde{\mathbf{z}}_n$ are source components by FastICA separation. Number of these vectors decreases in each round. Soft and hard decisions of users' symbols are denoted by $\hat{\mathbf{z}}_k$ and $\hat{\mathbf{b}}_k$, respectively. $I_i(t)$ denotes re-generated interference due to the users detected in ith round.

The first occasion where some a prior knowledge of the users is needed is the user identification phase (8). Naturally, the timing information can also be used to generate a good initial value for ICA iterations, and hence speed up the separation [11]. Anyway, what lousy timing estimate ultimately does is that it worsens the user identification. More importantly, given that a user is nevertheless identified having an erroneous timing estimate, the subtraction of that user enhance interference the more the bigger was the timing inaccuracy. To avoid that situation the delay of each identified user could first be refined according to

$$\hat{\delta}_k \leftarrow \arg\max_\delta \frac{\mathbf{g}_k(\delta, \hat{\tau}_k)\mathbf{w}_{k'}^H}{\|\mathbf{g}_k(\delta, \hat{\tau}_k)\|\|\mathbf{w}_{k'}\|} \tag{9}$$

where the maximization of the correlation is performed in a close neighborhood of $\hat{\tau}_k$, e.g. $\delta \in (-1/2, 1/2)$. The signal corresponding to the user k is re-built after delay refinement, only after which it is beneficial to go to the subtraction phase:

$$\mathbf{r}_m \leftarrow \mathbf{r}_m - a_k(b_{k,m-1}\underline{\mathbf{g}}_k(\hat{\delta}_k, \hat{\tau}_k) + b_{km}\mathbf{g}_k(\hat{\delta}_k, \hat{\tau}_k) + b_{k,m+1}\overline{\mathbf{g}}_k(\hat{\delta}_k, \hat{\tau}_k)) \qquad (10)$$

5 Numerical Experiments

The performance of the proposed receiver is studied with numerical experiments. Each of K users is assumed to transmit data blocks of $M = 5000$ QPSK symbols. Symbols are spread using Gold codes of length $C = 31$. Two service classes are assumed, which is modelled as a power difference of 10 dB between the two user groups. Inside both groups all the users are assigned the same power. All the results are based on average bit error rates (BER) over 1000 independent repetition. BER values are computed with respect of all K users.

The length of the all the receivers is $2C$ and hence both the LMMSE-PIC and LMMSE detectors are truncated to that length. Recall that the optimal length for asynchronous data would be MC (the whole block of symbols) which is not a sensible choice for the receiver length in practise [1].

As an "Basic ICA" (one of the reference methods) we used the version of FastICA algorithm, which is able to operate also in oversaturated systems [13]. This was used also in successive ICA receivers.

The results demonstrate that the proposed receiver is clearly more immune to delay estimation errors than the reference ones. This can been seen in Fig. 2. If the delays are known perfectly, LMMSE-PIC-receiver outperforms successive ICA-receiver (Fig. 2 (a)). However, the proposed receiver is superior to LMMSE-PIC, if there is even a slight error in delay estimation (Fig. 2 (b)). Fig. 3 depicts the performance of successive ICA-receiver as a function of the number of FastICA iterations performed. The figure shows that both successive ICA-receivers overtake e.g. LMMSE-PIC-receiver very fast when system have delay estimation errors.

Finally, we also notice from the experiments that the FastICA with successive interference cancellation performs clearly better than FastICA alone, given that the delays are tracked, too.

6 Conclusions

In this paper we considered successive interference cancellation (SIC) schemes employing blind source separation (BSS) methods in DS-CDMA systems. Especially, a BSS-SIC -type receiver structure where the users' delays are simultaneously tracked was proposed and evaluated via numerical examples. The main finding was that SIC-ideology combined with ICA for oversaturated systems is quite beneficial given that certain key parameters like users' delays are tracked simultaneously. The example cases assumed $m = 2C = 62$ observations from the mixture of $n = 3K = 66 - 90$ sources.

Joint Delay Tracking and Interference Cancellation in DS-CDMA Systems 1179

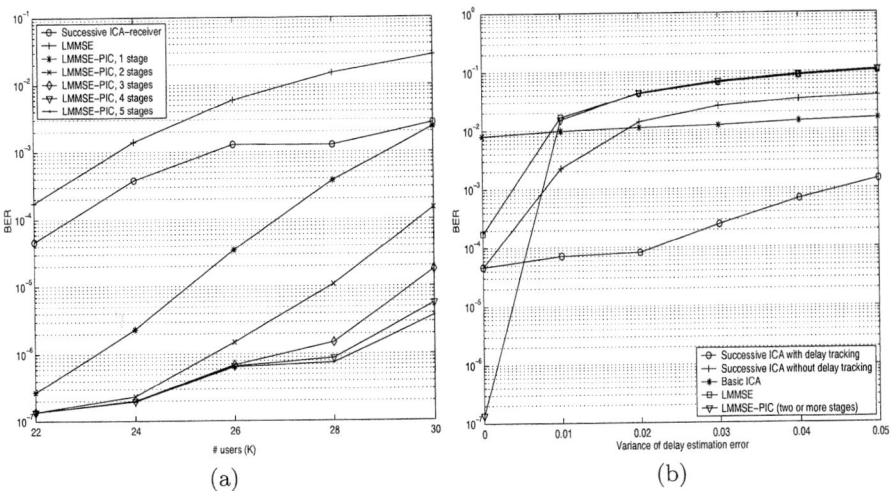

Fig. 2. (a) Bit-error-rates as a function of number of users in case of perfect delay estimation. (b) Bit-error-rates as a function of variance of delay estimation error in a system of $K = 22$ users. In both figures, users are split in two power groups and SNR is fixed to 20 dB wrt. the weakest users. In addition, 110 FastICA iterations are used in successive ICA receiver.

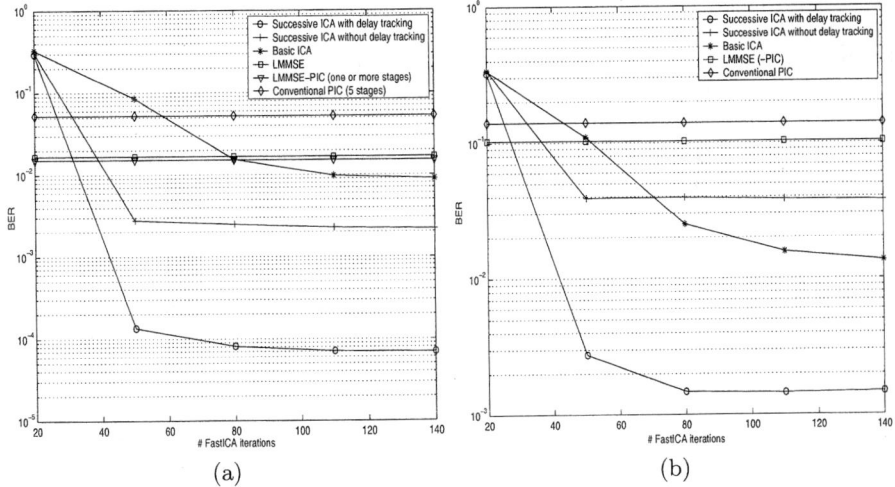

Fig. 3. Bit-error-rates as a function of number of FastICA iterations. The $K = 22$ users are split in two power groups. SNR is fixed to 20 dB wrt. the weakest users and users' estimated delays are assumed to have an error δ_k with variances $\sigma_\delta^2 = 0.01$ (a) and $\sigma_\delta^2 = 0.05$ (b) (unit of error corresponds to one chip time).

References

1. S. Verdú, *Multiuser detection*, Cambridge University Press, 1998.
2. L. Milstein, "Interference rejection techniques in spread spectrum communications", *Proc. of the IEEE*, vol. 66, June 1988, pp. 657–671.
3. J. M. Holtzman, "DS/CDMA successive interference cancellation", *Proc. IEEE ISSSTA '94*, vol. 1, 1994, pp. 69–78.
4. M. K. Varanasi, B. Aazhang, "Multistage detection in asynchronous code-division multiple-access communication", *IEEE Transactions on communication*, vol. 38(4), 1990, pp. 509–519.
5. M. Latva-aho, J. Lilleberg, "Parallel interference cancellation in multiuser detection", *Proc. IEEE 4th ISSSTA '96*, vol. 3, 1996, pp. 1151–1155.
6. A. Hyvärinen, J. Karhunen, E. Oja, *Independent Component Analysis*, Wiley, New York, 2001.
7. T. Huovinen, T. Ristaniemi, "Blind source separation based successive interference cancellation in the DS-CDMA uplink", *Proc. IEEE ISCCSP2004*, March 2004, to appear.
8. T. Huovinen, T. Ristaniemi, "DS-CDMA Capacity Enhancement Using Blind Source Separation Based Group-Wise Successive Interference Cancellation", *Proc. IEEE 5th SPAWC2004*, July 2004, subm.
9. M. Latva-aho, M. Juntti, K. Kasanen, "Residual interference suppression in parallel interference cancellation receivers", *Proc. IEEE ICC'99*, vol. 2, 1999, pp. 927–931.
10. A. Viterbi, *CDMA: Principles of Spread Spectrum Communications*, Addison-Wesley, 1995.
11. T. Ristaniemi and J. Joutsensalo, "Advanced ICA-based receivers for block fading DS-CDMA channels", *Signal Processing*, vol. 82, 2002, pp. 417–431.
12. S. Bensley, B. Aazhang, "Subspace-based channel estimations for code division multiple access communication systems", *IEEE Trans. Commun.*, vol. 42, pp. 1009–1020, Aug. 1996.
13. A. Hyvärinen, R. Cristescu, E. Oja, "A fast algorithm for estimating overcomplete ICA basis for image windows", *Proc. International Joint Conference on Neural Networks*, pp. 894–899, 1999.

Layered Space Frequency Equalisation for MIMO-MC-CDMA Systems in Frequency Selective Fading Channels

Sonu Punnoose*, Xu Zhu, and Asoke K. Nandi

Signal Processing and Communications Group,
Department of Electrical Engineering and Electronics,
The University of Liverpool, Brownlow Hill, Liverpool, L69 3GJ, UK
{sppadic,xuzhu,a.nandi}@liverpool.ac.uk

Abstract. The capacity gain offered by MIMO (Multiple Input Multiple Output) MC-CDMA (Multi-Carrier Code Division Multiple Access) systems, which exploits the transmit and receive diversity inherent in multipath propagation, along with the robustness provided by MC-CDMA scheme utilising OFDM (Orthogonal Frequency Division Multiplexing) in fading channels is potentially significant. Past papers have considered the Single-Carrier MIMO model as well as the single user MISO and SISO models. In this paper, a Layered Space Frequency Equalisation (LSFE) approach for the Multi-User MIMO MC-CDMA system is presented and is shown to yield significantly better performance compared to the conventional MMSE (Minimum Mean Square Error) equaliser method. It is shown that LSFE can provide a viable solution to the problem of multi-user detection in MIMO MC-CDMA systems.

1 Introduction

The concept of multiple antennas at the transmitter and receiver has been gaining momentum and various papers [6], [7] have already dealt with the extension of MIMO technology to basic OFDM transmission scheme. It aims to combine the advantages of OFDM [1], [2] namely the simple equalisation with the increased data-rates of MIMO technology, brought about as a direct result of the multi-path phenomenon, common to wireless communications systems.

OFDM [1] converts the frequency selective channel into many parallel flat-fading channels thereby simplifying the equalisation process. MC-CDMA combines the OFDM transmission scheme with the CDMA system [3], [4], where the spreading is done in the frequency domain. Different equalisation techniques are available for MC-CDMA systems such as EGC (equal gain combining), MRC (maximum ratio combining) and MMSE being the more common ones. The MIMO model can also be applied to MC-CDMA (Multi-Carrier Code Division

---

* This work is supported by the Overseas Research Studentship (ORS) Awards Commitee U. K. and the University of Liverpool.

Multiple-Access) systems increasing data rate and spectrum efficiency by using spatial multiplexing gains as well as receive diversity inherent in MIMO systems.

The BLAST technique addresses the Layered Equalisation in time domain [10] while LSFE does the same in frequency domain [5]. Both techniques involve a multi-stage process, where at each stage, ordering of the different transmitted streams is done based on a MSE (Mean Square Error) criterion. The interference caused by the detected stream is subsequently removed from the received signal to leave behind the undetected streams.

In this paper we extend the LSFE method to MIMO-MC-CDMA system in frequency selective channels, showing the significant performance gain achieved over the MMSE detector. Our work is different in that we apply the BLAST philosophy in the frequency domain namely the LSFE. In MIMO-OFDM systems, the use of LSFE method yielded an improved performance as compared to the normal MMSE method [5] for a single user. The MMSE coefficients were calculated on a per-carrier basis before the SIC (Successive Interference Cancellation) stage.

The following notations are used throughout the paper - $[.]^T$, $[.]^H$ denote the transpose and conjugate transpose respectively, vector are represented in lower-case, while matrices are in boldface capital and $\mathbf{I}$ represents the identity matrix.

2 MIMO-MC-CDMA System Model

This section will briefly describe the MIMO MC-CDMA system with N_t transmit antenna and N_r receive antenna. The block diagram of the system is given below. User information represented as $d(k)$, where k is the sample index, is first spread in the frequency domain before being modulated using the IFFT as in OFDM scheme. The transmitter consists of a copier, which makes N copies of the each user's data (N is the number of sub-carriers). Each of the N copies are then spread with a single spreading code bit for that carrier. The Walsh-Hadamard spreading code of size N was used to spread each users data. Thus each user's data is spread in frequency domain where all carriers carry the same data [3]. After summing all the users spread bits, the spread signal is then transmitted using the OFDM scheme (i.e. using IFFT for modulation) with the addition of suitable guard band to counter the ISI (Inter-Symbol Interference) caused by multipath propagation.

The fading channel between a transmit-receive antenna pair is assumed to be frequency selective quasi-static fading [6] and is described by the channel impulse response vector of length L as [8] $h_{\mu\nu} = [h_{\mu\nu}(0), ..., h_{\mu\nu}(L-1)]^T$. The received signal at the νth receive antenna after guard band removal and FFT processing can be expressed as [6], [7]

$$y^\nu(n) = \sum_{\mu=1}^{N_t} \mathbf{H}_{\mu\nu}(n) x^\mu(n) + w^\nu(n) \qquad (1)$$

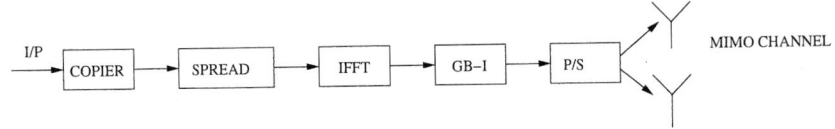
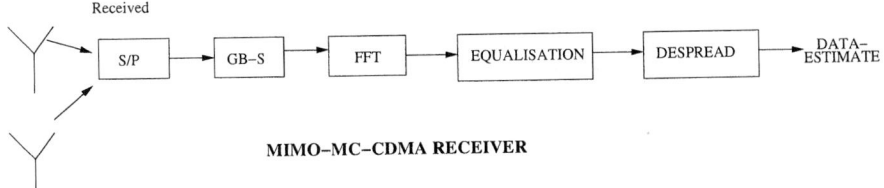

Fig. 1. MIMO-MC-CDMA.

where $\mathbf{H}_{\mu\nu}(n)$ as shown below, is the subchannel gain from the μth transmit antenna to the νth receive antenna evaluated on the nth subcarrier. $x^\mu(n)$ denotes the spread symbol of all users on the nth carrier transmitted on the μth transmit antenna, while $y^\nu(n)$ represents the received symbol on the nth subcarrier obtained at the νth receive antenna. The additive noise is $w^\nu(n)$, which represents a zero-mean complex Gaussian process with variance N_0.

$$\mathbf{H}_{\mu\nu}(n) = \sum_{l=0}^{L} h_{\mu\nu}(l) e^{-j(2\pi/N)ln} \tag{2}$$

The received signal is equalised depending on the technique used, before despreading the signals to obtain each user's information. The following sections explain the two equalisation techniques compared in this paper with emphasis on the LSFE approach that we have proposed for this system.

3 MMSE Equalisation

The MMSE equaliser is a standard equalisation technique used often in communication systems and is briefly described as below. The concept behind the MMSE equaliser/detector is the minimisation of the cost function 'J'(i.e.Mean Square Error (MSE)) as shown here [1], [4], [9] $J = E[|d_q - w_q^H y|^2]$, where y denotes the received signal, while d_q represents the data transmitted on stream q. The weight coefficients w are obtained using $w^{\text{mmse}} = \mathbf{R}_y^{-1}\mathbf{H}$. As the input symbols are independent of the noise and are uncorrelated, the above is simplified [1], [7] as $\mathbf{R}_y = \mathbf{HH}^H + N_0\mathbf{I}$, where w^{mmse} is the MMSE coefficient vector, y is the received signal, $\mathbf{R}_y$ is the autocorrelation of the recieved signal and $\mathbf{I}$ is the identity matrix. The channel matrix $\mathbf{H}$ denotes the entire channel frequency response matrix of size $N_r * N \times N_t * N$ and N_0 is the noise power. After despreading the detected signals are then obtained as $d = w^{\text{mmse}} y$.

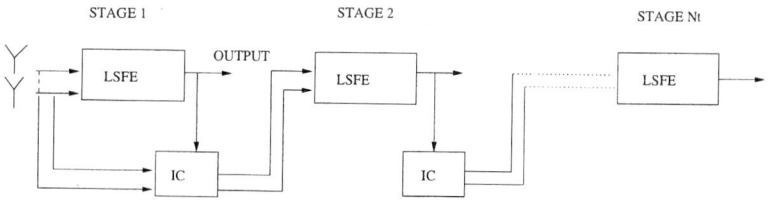

IC– INTERFERENCE CANCELLATION

Fig. 2. LSFE Method.

4 Layered Space Frequency Equalisation

4.1 System Model

This technique is the equivalent of the BLAST (Bell Labs Layered Space Time equalisation) method [10] in frequency domain, which combines FDE (Frequency Domain Equalisation) with interference cancellation to perform detection, which consists of N_t stages, where N_t is the number of transmit antenna [5]. The FFT demodulated received signal is equalised based on stream ordering per *carrier*.

This ordering (selection) of the streams is done based on the MMSE criteron [5], i.e. the stream with the least MSE is selected and detected. In the MIMO-MC-CDMA scheme, the LSFE technique described above is modified such that the ordering of the streams is based on the sum of all MSE values. The remaining streams are treated as interference. The influence of the detected stream is then cancelled from the received signal. This method is represented in the Fig. 2.

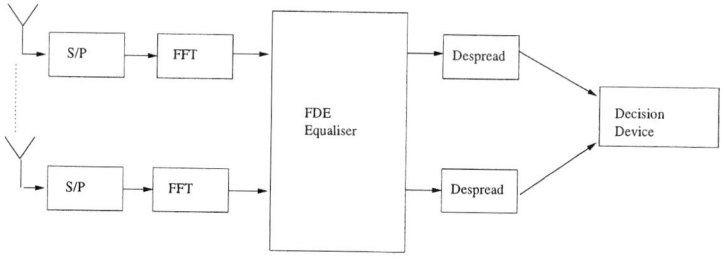

Fig. 3. Particular LSFE Stage.

The LSFE technique at one stage is illustrated in Fig. 3 [5], where the FDE block performs equalisation as well as interference cancellation, before the hard decision is made by the decision device. At the end of each stage, the interference from the already detected streams is cancelled from the received signals yielding modified received signals with less interference. Thus all streams are detected and their corresponding interference are cancelled in a successive manner.

4.2 LSFE Algorithm

The received signal after FFT processing at a particular stage can be written as [5], $\mathbf{Y}[n] = \sum_q \widetilde{\mathbf{H}_q}[n]\mathbf{X_q}[n]$, where q denotes the stream, $\widetilde{\mathbf{H}_q}$ represents the channel frequency response corresponding to the stream q, $\mathbf{X}_q$ represents the signal transmitted on stream q and n is the sub-carrier index. The Frequency Domain Equaliser coefficients are calculated based on the MMSE criterion. Assuming perfect channel estimation (CSI) at the receiver, the soft estimate of the detected stream can be expressed as

$$\widetilde{d}(k) = \sum_{i=0}^{N-1} \mathbf{W}[i]\mathbf{Y}[i] \tag{3}$$

where $\mathbf{Y}[i]$ is the FFT of the received signal vector y and $\mathbf{W}$ represents the FDE weight vectors expressed as

$$\mathbf{W}[n] = \mathbf{R}^{-1}[n]\mathbf{F}[n] \tag{4}$$

where $\mathbf{R}[n]$ is the autocorrelation matrix of the channel response for each carrier n expressed as $\mathbf{R}[n] = \sum_\mu \widetilde{\mathbf{H}}_\mu[n]\widetilde{\mathbf{H}}_\mu^H[n] + N_0 I$, where $\widetilde{\mathbf{H}}_\mu[n]$ is the channel frequency response for carrier n on the μth transmit antenna, N_0 is the noise power and $F[n]$ is the channel sub-matrix corresponding to the stream that is being detected. $\mathbf{F}[n] = \{\widetilde{\mathbf{H}}_1[n].....\widetilde{\mathbf{H}}_N[n]\}$. The MSE for each carrier is then calculated using the formula [5]

$$\mathbf{MSE}_n = 1 - \frac{1}{N}\sum_{n=0}^{N-1} \widetilde{\mathbf{H}}_{nt}^H[n] R^{-1}[n]\widetilde{\mathbf{H}}_{nt}[n] \tag{5}$$

The decision variable which determines the ordering of the streams is given by the sum of all the MSE values on all carriers indexed by n as shown below $\mathbf{MSE}_{decision} = \sum_n \mathbf{MSE}_n$. The selected stream is then detected, despread and its interference cancelled from the received signal to leave only the remaining undetected streams in the received signal. The despread signal is then passed onto the decision device to yield the hard estimate of the detected stream.

5 Simulation Results

The MIMO-MC-CDMA system was implemented for the 2×2 and 4×4 cases in frequency selective fading channels with channel order of 4. The BER vs SNR is plotted for each of the cases specified with the SNR (Signal to Noise ratio) defined as the ratio of the trace of the auto-correlation of the transmitted signal to the noise-power N_0 defined earlier. The average power at each receive antenna is equal to the total transmitted power. The system was simulated with a symbol rate of 32Kbps using BPSK modulation with the number of sub-carriers set to 32 and number of users equal to 4. The BER vs SNR plots for the 2×2 MIMO-MC-CDMA system is shown in Fig. 4 and that of the 4×4 case is presented

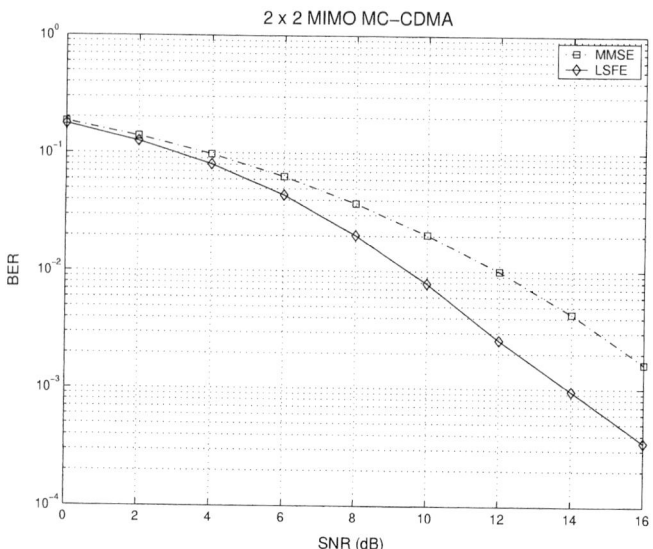

Fig. 4. Performance in the case of 2×2 MIMO-MC-CDMA system.

in Figure 5. Each point in these Figures summarises 1000 realisations In both the plots shown. The solid line denotes the performance of the LSFE technique, while the dash-dotted line indicates the performance of the MMSE method. The plots show the significant improvement in performance of the LSFE technique as compared to the standard MMSE technique for both systems. This is to due to the better interference cancellation at each stage of the LSFE technique, resulting in removal of the detected stream, leaving behind only the undetected components in the received signal.

Fig. 6 compares the performance of LSFE technique for different number of receive antennae. The 2×8 system yields the best performance compared to the 2×4 and 2×2 systems. This highlights the improvement in performance that can be obtained in MIMO systems using more receive antenna (i.e. the advantages of receive diversity). The improved performance is obtained by utilising the multi-path propagation phenomenon whereby each stream is transmitted over a number of paths from the transmitter to the receiver. Therefore once the effects of the channel are equalised, multiple copies of the transmitted streams are obtained at the receive antennas resulting in spatial diversity gain and hence the vastly improved performances observed with larger number of receive antenna.

6 Conclusions

In this paper we have implemented the LSFE equalisation technique for multi-user MIMO MC-CDMA systems over frequency selective channels, which is compared with the MMSE technique that equalises in one stage. The multi-stage

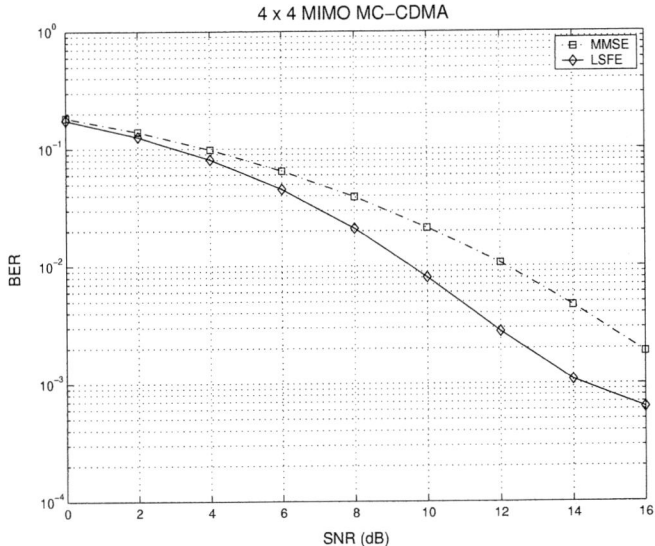

Fig. 5. Performance in the case of 4x4 MIMO-MC-CDMA system.

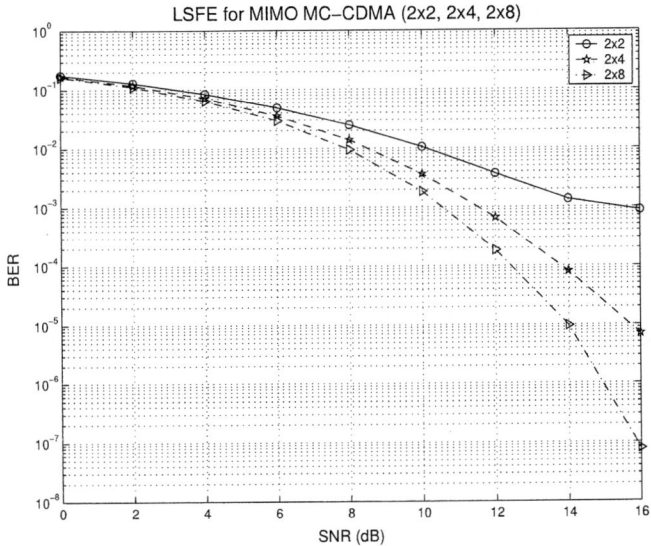

Fig. 6. Performances of LSFE for 2x2, 2x4 and 2x8 MIMO-MC-CDMA systems.

LSFE approach performs significantly better, as a result of the successive interference cancellation of the appropriately selected stream, from the received signal at each stage. This leaves only the undetected data streams and their interference for the successive stages thereby resulting in an improved BER performance as indicated in the plots. Although the cases considered here were for 4 users, this can easily be extended for more users. The combination of receive diversity (i.e. more receive antennas) with multi-stage equalisation and interference cancellation of the LSFE, provide a viable means of meeting the high-data rate and spectral efficiency requirements of next generation mobile technology.

References

1. Z. Wang, G. B. Giannakis,"Wireless Multi–Carrier Communications –Where Fourier Meets Shannon", IEEE Signal Processing Magazine, pp. 29-48, May 2000.
2. W. Y. Zou, Y. Wu, "Coded OFDM–An Overview", IEEE Transactions on Broadcasting, Vol. 41, No. 1, pp. 1-8, March 1995.
3. S. Hara, R. Prasad, "Design and Performance of Multi-carrier CDMA System in Frequency-Selective Rayleigh Fading Channels", IEEE Transactions on Vehicular Technology, Vol. 48, No. 5, pp. 1584-1595, September 1999.
4. S. L. Miller, B. J. Rainbolt,"MMSE Detection of Multicarrier CDMA", IEEE Journal on Selected Areas in Communications, Vol. 18, No. 11, pp. 2356-2362, November 2000.
5. X. Zhu, R. D. Murch, "Layered Space–Frequency Equalization for Uncoded and Space-Time Block Coded MIMO Systems", in IEEE Global Telecommunications Conference (GLOBECOM), Vol. 1, pp. 1002-1006, November 2002.
6. Z. Liu, Y. Xin and G. B. Giannakis, "Space Time Frequency Coded OFDM Systems over Frequency-Selective Fading Channels", IEEE Transactions on Signal Processing, Vol. 50, No. 10, pp. 2465-2476, October 2002.
7. Z. Liu and G. B. Giannakis,"Space–Time Block Coded Multiple Access Through Frequency-Selective Fading Channels", IEEE Transactions on Communications, Vol. 49, No. 6, pp. 1033-1044, June 2001.
8. T. S. Rappaport, *Wireless Communications–Principles And Practice*.
9. D. Gesbert,"Robust Linear MIMO Receivers: A Minimum Error–Rate Approach", IEEE Transactions on Signal Processing, Vol. 51, No. 11, pp. 2863-2871, November 2003.
10. G. J. Foschini,"Layered Space Time Architecture For Wireless Communication in a Fading Environment When Using Mulitple Antennas", *Bell Labs Tech. Journal* Vol.1, pp. 41-59, 1996.

Multiuser Detection and Channel Estimation in MIMO OFDM Systems via Blind Source Separation

Luciano Sarperi*, Asoke K. Nandi, and Xu Zhu

Signal Processing and Communications Group,
Department of Electrical Engineering and Electronics,
The University of Liverpool, Brownlow Hill, Liverpool, L69 3GJ, UK
{lsarperi,a.nandi,xuzhu}@liverpool.ac.uk

Abstract. This paper proposes a blind multiuser detection and channel estimation method for multiple-input multiple-output (MIMO) orthogonal frequency-division multiplexing (OFDM) systems based on blind source separation (BSS). Using multiple antennas for transmission and reception the multiuser detection problem can be cast as a BSS problem of linear instantaneous mixtures. The proposed method uses BSS in each subcarrier to detect the user signals. The source order and scaling indeterminacies inherent to BSS methods are overcome by exploiting the cross correlation between the subcarriers introduced by convolutional encoding. Additionally, the method provides an estimate of the channels. It is shown that this method provides significant performance enhancement over previous work.

1 Introduction

OFDM systems have recently attracted great interest. Some of the reasons for this are low complexity of implementation, the fact that the orthogonality of the subcarriers is maintained when transmitted through a linear multipath channel and ease of equalisation. OFDM has been adopted in digital audio/video broadcasting standards in Europe and has been proposed for digital cable television systems and wireless networks such as IEEE802.11a [2]. In a multiple-input multiple-output (MIMO) OFDM system, multiple transmit antennas and multiple receive antennas are employed and the same subcarriers are used by all transmitters. The use of MIMO systems is motivated by the significant capacity gains over single-input single-output (SISO) systems [5], [6], [7], [9].

BSS methods allow the recovery of source signals from the observation of the mixtures only, with unknown source signals and mixing transformation. However, some assumptions must be fulfilled by the source signals and the mixing transformation [8]. In a MIMO OFDM system these assumptions are met if the source signals are mutually statistically independent and at least as many receive antennas as transmit antennas are employed.

---
* This work is supported by the Overseas Research Studentship (ORS) Awards Committee UK and the University of Liverpool

In [1] a BSS based OFDM MIMO multiuser detection method was proposed, which employs BSS only in one subcarrier and unmixes the remaining subcarriers using a minimum mean square error (MMSE) approach. The advantage of this BSS-MMSE method is the low computational cost, however, errors tend to propagate from subcarrier to subcarrier since the previous subcarrier is used as a reference to unmix the current subcarrier.

Our work is different in that we use BSS in all subcarriers to perform multiuser detection for MIMO OFDM systems. The separation of the multiple users is carried out in the same fashion as the separation of convolved mixtures in the frequency domain, see for example [3]. It is shown that the proposed method can obtain significant performance enhancement compared to the method in [1].

This paper is structured as follows. In section 2 the system model is introduced, section 3 proposes the BSS based multiuser detection method, simulation results are given in section 4 and conclusions are drawn in section 5.

2 System Model

In OFDM systems data are transmitted in blocks using inverse FFT (IFFT) at the transmitter and FFT at the receiver. By adding a guard interval between each block, inter-block interference (IBI) can be eliminated. Usually, during the guard interval a cyclic prefix (CP), consisting of redundant symbols, is transmitted at the beginning of each block [2].

At the transmitter, binary source data $d_u(n)$ with identical independent distribution (i.i.d.) and unit variance is encoded using a convolutional encoder with real valued impulse response $c(n)$ and length F to obtain the encoded signal for transmit antenna u

$$s_u(n) = \sum_{l=0}^{F-1} c(l) d_u(n-l). \tag{1}$$

The encoded signal is then transmitted in blocks $\mathbf{s}_u(i) = [s_u(iN), s_u(iN+1), \cdots, s_u(iN+N-1)]^T$ where i is the block index and the block length N corresponds to the number of subcarriers. Using a CP of sufficient length, the frequency-selective channel is transformed into a flat-fading channel for each subcarrier [2].

Considering N_t transmit antennas and N_r receive antennas (see Fig. 1) we can build blocks of received signals per subcarrier k for $k = 0, 1, \cdots, (N-1)$: $\mathbf{r}(k) = [r_1(iN+k), r_2(iN+k), \cdots, r_{N_r}(iN+k)]^T$ where $r_v(\cdot)$ is the signal from receive antenna v. The total received signal per subcarrier k becomes now

$$\mathbf{r}(k) = \mathbf{H}(k)\mathbf{s}(k) + \mathbf{n}(k) \tag{2}$$

with

$$\mathbf{H}(k) = \begin{pmatrix} H_{11}(k) & \cdots & H_{1N_t}(k) \\ H_{21}(k) & \cdots & H_{2N_t}(k) \\ \vdots & \ddots & \vdots \\ H_{N_r 1}(k) & \cdots & H_{N_r N_t}(k) \end{pmatrix} \tag{3}$$

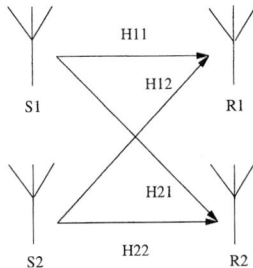

Fig. 1. MIMO OFDM system with $N_t = N_r = 2$.

where $H_{vu}(k)$ is the k-th DFT coefficient of the channel between transmit antenna u and receive antenna v. The signal transmitted per subcarrier k is $\mathbf{s}(k) = [s_1(iN + k), s_2(iN + k), \cdots, s_{N_t}(iN + k)]^T$ where $s_u(\cdot)$ is the signal transmitted by transmit antenna u and $\mathbf{n}(k) = [n_1(iN + k), n_2(iN + k), \cdots, n_{N_r}(iN + k)]^T$ is the additive white gaussian noise (AWGN) vector with zero mean and variance N_0. In [1] it was recognised that (2) corresponds to a BSS problem of linear instantaneous mixtures with the complex valued mixing matrix $\mathbf{M}(k) = \mathbf{H}(k)$, sources $\mathbf{s}(k)$ and mixtures $\mathbf{r}(k)$.

3 BSS Based Multiuser Detection

In the proposed BSS-only multiuser detection and channel estimation method BSS is used in each subcarrier to obtain an estimate of the sources $\hat{\mathbf{s}}(k)$ and the mixing matrix $\hat{\mathbf{M}}(k)$ from the received mixtures $\mathbf{r}(k)$. The order and scaling indeterminacies inherent in BSS methods necessitate a post-BSS processing to order the estimated sources and mixing matrices in the same way for all subcarriers $k = 0, 1, \cdots, (N-1)$.

The ordering and scaling of the estimated sources and mixing matrices rely on the correlation introduced by the convolutional encoder with impulse response $c(n)$ in (1). As presented in [1], the cross correlation between the transmitted signals from transmit antenna p in one subcarrier and transmit antenna q in a neighbouring subcarrier is

$$\gamma_{pq} = \mathrm{E}[s_p(iN + k)s_q(iN + k \pm 1)^*] = \begin{cases} \sum_{p=0}^{F-2} c(p)c(p+1) & \text{when } p = q \\ 0 & \text{when } p \neq q \end{cases} \quad (4)$$

where $\mathrm{E}[\cdot]$ is the expectation with respect to i. The following method provides reordering and scaling of the estimated sources $\hat{\mathbf{s}}(k) = [\hat{s}_A(iN+k)\hat{s}_B(iN+k)]^T$ and mixing matrix $\hat{\mathbf{M}}(k)$ for an $N_t = N_r = 2$ MIMO system, which can be extended to more transmit and receive antennas. First, the sources in subcarrier $k = 0$ are separated. Next, to reorder and scale the remaining subcarriers $k = 1, 2, \cdots, (N-1)$ the following cross correlations

$$\rho_1 = \mathrm{E}[\hat{s}_A(iN+k)\hat{s}_1(iN+k-1)^*] \tag{5}$$
$$\rho_2 = \mathrm{E}[\hat{s}_B(iN+k)\hat{s}_2(iN+k-1)^*] \tag{6}$$
$$\rho_3 = \mathrm{E}[\hat{s}_A(iN+k)\hat{s}_2(iN+k-1)^*] \tag{7}$$
$$\rho_4 = \mathrm{E}[\hat{s}_B(iN+k)\hat{s}_1(iN+k-1)^*] \tag{8}$$

of the estimated sources are obtained (see Fig. 2), where $\mathrm{E}[\cdot]$ is the expectation with respect to i. Then, the following metrics based on the difference between the cross correlations of the estimated sources and the true cross correlations in (4) are obtained

$$\delta_1 = (|\rho_1| - \gamma_{11})^2 + (|\rho_2| - \gamma_{22})^2 \tag{9}$$
$$\delta_2 = (|\rho_4| - \gamma_{11})^2 + (|\rho_3| - \gamma_{22})^2 \tag{10}$$

assuming that the encoded signals $s_u(\cdot)$ are real valued and $\mathrm{var}[\hat{s}_A(\cdot)] = \mathrm{var}[\hat{s}_B(\cdot)] = \mathrm{var}[\hat{s}_1(\cdot)] = \mathrm{var}[\hat{s}_2(\cdot)] = \mathrm{var}[s_u(\cdot)]$, where $\mathrm{var}[\cdot]$ is the variance. The estimated sources $\hat{s}_A(iN+k)$ and $\hat{s}_B(iN+k)$ are now reordered and scaled as follows to obtain the same order and scaling as in the previous subcarrier $(k-1)$:

$$\hat{\mathbf{s}}(k) = \begin{cases} \begin{bmatrix} \hat{s}_A(iN+k)\frac{|\rho_1|}{\rho_1} \\ \hat{s}_B(iN+k)\frac{|\rho_2|}{\rho_2} \end{bmatrix} & \text{when } \delta_1 < \delta_2 \\ \begin{bmatrix} \hat{s}_B(iN+k)\frac{|\rho_4|}{\rho_4} \\ \hat{s}_A(iN+k)\frac{|\rho_3|}{\rho_3} \end{bmatrix} & \text{when } \delta_1 \geq \delta_2 \end{cases} \tag{11}$$

The estimate of the channel $\hat{\mathbf{H}}(k)$ can be obtained by reordering and scaling the estimated mixing matrix $\hat{\mathbf{M}}(k)$ as follows

$$\hat{\mathbf{H}}(k) = \begin{cases} \mathbf{M}(k)\,\mathrm{diag}(\frac{\rho_1}{|\rho_1|}, \frac{\rho_2}{|\rho_2|}) & \text{when } \delta_1 < \delta_2 \\ \mathbf{M}(k)\,\mathbf{J}\,\mathrm{diag}(\frac{\rho_4}{|\rho_4|}, \frac{\rho_3}{|\rho_3|}) & \text{when } \delta_1 \geq \delta_2 \end{cases} \tag{12}$$

where $\mathrm{diag}(d_1, d_2)$ is a diagonal matrix with diagonal elements d_1, d_2 and $\mathbf{J}$ is the 2x2 exchange matrix with ones on the antidiagonal and zeros elsewhere.

4 Simulations

Simulations were carried out for a $N_t = N_r = 2$ MIMO system. Both our BSS-only method and the BSS-MMSE method in [1] employed the JADE BSS algorithm in [4]. BPSK source data d_u was used. The encoded and modulated signal was transmitted through channels with the following impulse responses:

$$\begin{aligned} h_{11}(n) &= [\,1.00;\,0.30;\,-0.10\,] \\ h_{12}(n) &= [\,1.00;\,-0.60;\,0.08\,] \\ h_{21}(n) &= [\,1.00;\,0.00;\,-0.25\,] \\ h_{22}(n) &= [\,1.00;\,0.30;\,-0.28\,] \end{aligned} \tag{13}$$

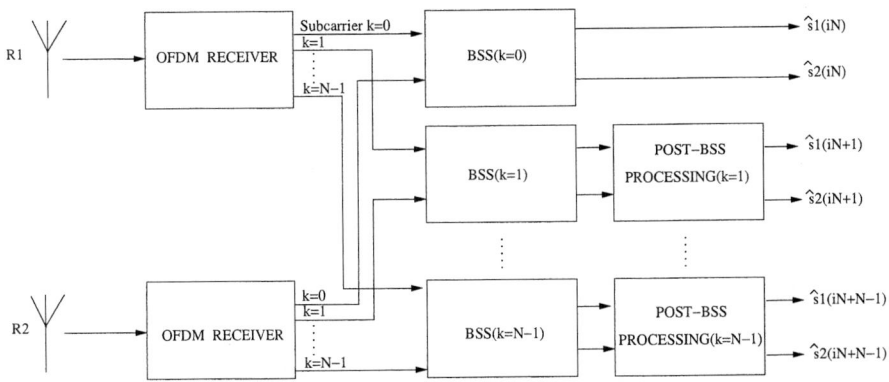

Fig. 2. BSS-only OFDM receiver for $N_t = N_r = 2$.

The impulse response of the convolutional encoder was $c(n) = [1; 0.5; -0.25]$ and the number of subcarriers was $N = 20$ as in [1]. AWG noise was added at the receiver to obtain the desired SNR level. Both methods used the noisy received signal $\mathbf{r}(k)$ only.

For perfect separation of the sources in the noiseless case the global matrix $\mathbf{G} = \hat{\mathbf{H}}(k)^{-1}\mathbf{H}(k)$ is of the form $\text{diag}(\alpha_1, \alpha_2)$, where α_p is a scalar and assuming that the sources were recovered in the original order. The quality of the source separation was measured using the mean square error (MSE) between a scaled version of $\mathbf{G}$ and the identity matrix $\mathbf{I}$.

$$\mathbf{Q} = \begin{pmatrix} \frac{g_{11}}{g_{11}} & \frac{g_{12}}{g_{11}} \\ \frac{g_{21}}{g_{22}} & \frac{g_{22}}{g_{22}} \end{pmatrix} - \mathbf{I} \qquad (14)$$

Using (14) the MSE was obtained by averaging 100 runs:

$$\text{MSE} = 10 \log_{10} \left\{ \mathrm{E} \left[\sum_{p=1}^{2} \sum_{q=1}^{2} |q_{pq}|^2 \right] \right\} \qquad (15)$$

Simulations were carried out for Ns = 1000, 5000 and 10000 symbols per subcarrier and at SNR levels of 20 dB and 10 dB. Figs. 3 and 4 were obtained with Ns = 1000 and SNR = 20 dB and 10 dB respectively. They show that both methods obtain better performance at higher SNR levels, but while the MSE of the BSS-MMSE method increases progressively for higher subcarrier numbers k, the MSE of the BSS-only method remains at a low level for all subcarriers. This is because the BSS-MMSE method uses the sources separated in the previous subcarrier as a reference to unmix the current subcarrier while in the BSS-only method the quality of the source separation does not depend on the previous subcarrier. The performance in the first subcarrier $k = 0$ is similar for both methods, as both use BSS in the first subcarrier.

For Ns = 5000 in Fig. 5 the performance of both methods is better than with Ns = 1000 in Fig. 3 and Fig. 4. A further increase to Ns = 10000 in Fig. 6 does for both methods not lead to a significant improvement.

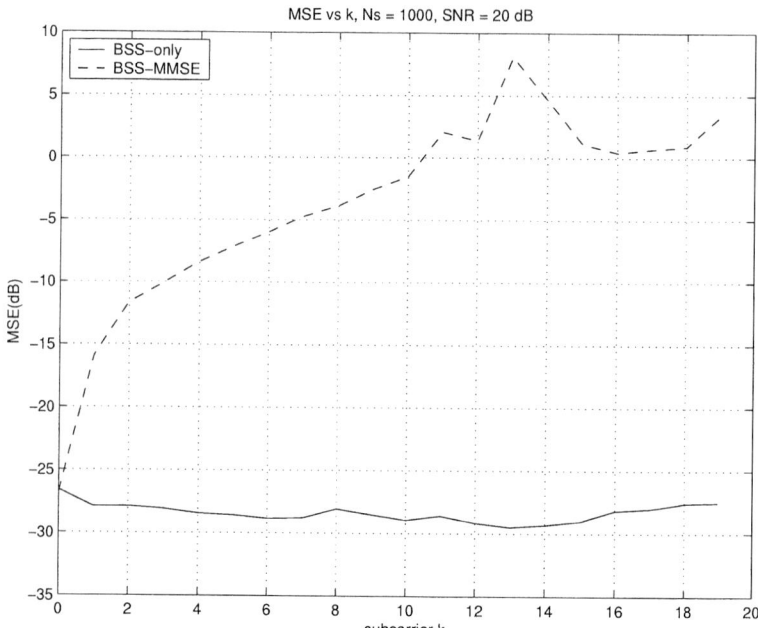

Fig. 3. MSE (dB) vs subcarrier k at SNR = 20 dB for Ns = 1000 symbols per subcarrier.

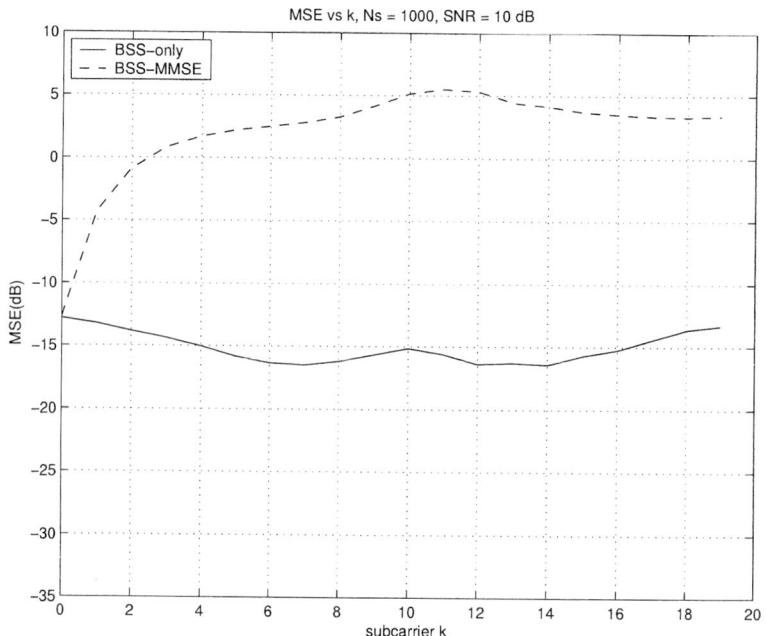

Fig. 4. MSE (dB) vs subcarrier k at SNR = 10 dB for Ns = 1000 symbols per subcarrier.

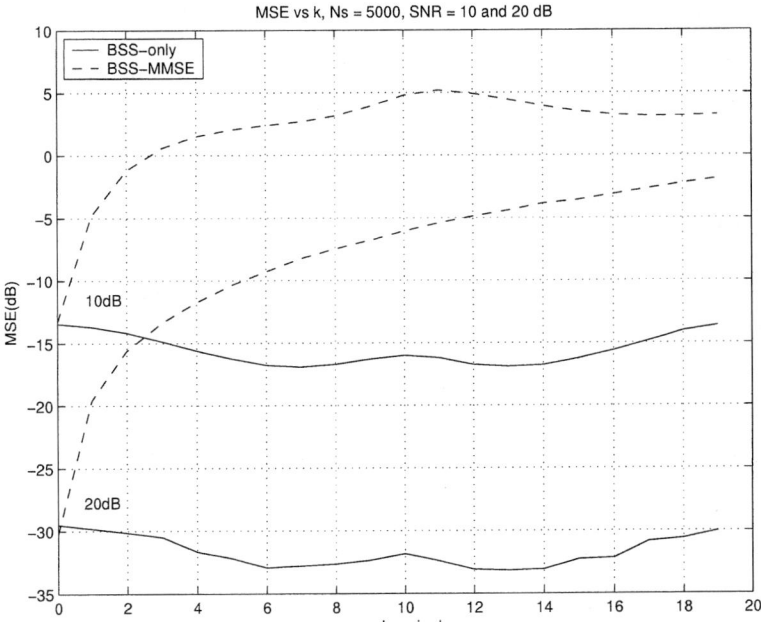

Fig. 5. MSE (dB) vs subcarrier k at SNR = 10 and 20 dB for Ns = 5000 symbols per subcarrier.

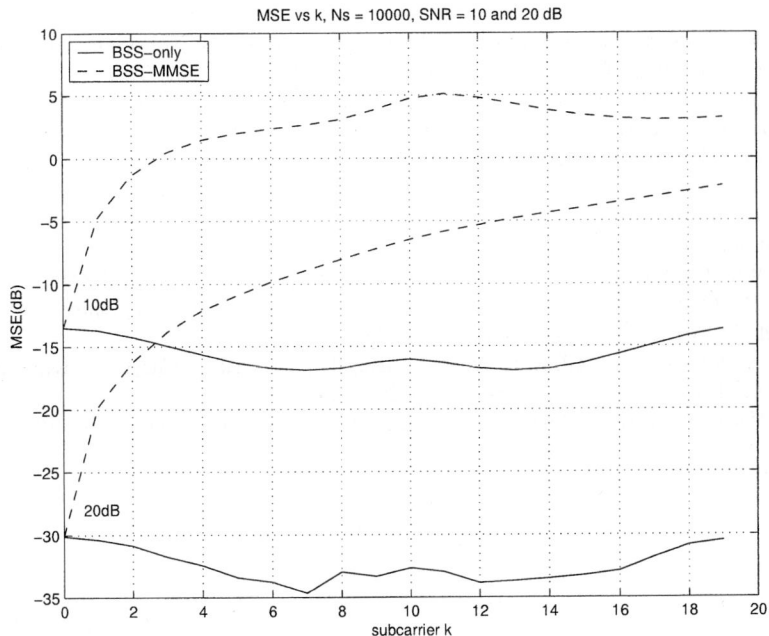

Fig. 6. MSE (dB) vs subcarrier k at SNR = 10 and 20 dB for Ns = 10000 symbols per subcarrier.

The simulations show that the performance improvement of the BSS-only method is significant, however, the computational requirements are higher.

Depending on the source signals and the channels, the choice of a particular BSS method may play an important role for the successful separation in both approaches.

5 Conclusions

The BSS-only multiuser detection and channel estimation method proposed in this paper has been found to clearly outperform the BSS-MMSE method in [1]. The BSS-only method obtains a nearly constant performance for all subcarriers while the performance of the BSS-MMSE method degrades progressively for increasing subcarrier numbers. Future work will focus on reducing the complexity of our BSS based receiver.

References

1. D. Iglesia, A. Dapena, C. J. Escudero, "Multiuser Detection in MIMO OFDM Systems Using Blind Source Separation", in *Proc. Sixth Baiona Workshop on Signal Processing in Communications (WSPC03)*, Baiona, Spain, 2003, pp. 41–46.
2. Z .Wang, G. B. Giannakis, "Wireless Multicarrier Communications", *IEEE Signal Processing Magazine*, Vol. 17, No. 3, May 2000, pp. 29–48.
3. P. Smaragdis, "Blind Separation of Convolved Mixtures in the Frequency Domain", *Neurocomputing*, Vol. 22, No. 1–3, November 1998, pp. 21–34.
4. J. F. Cardoso, A. Souloumiac, "Blind Beamforming for Non-Gaussian Signals", *IEE Proceedings F*, Vol. 140, No. 6, December 1993, pp. 362–370.
5. H. Bolcskei, D. Gesbert, A. J. Paulraj, "On the Capacity of OFDM-Based Multi-Antenna Systems", in *Proc. IEEE Int. Conf. Acoust., Speech, Sig. (ICASSP)*, Istanbul, Turkey, June 2000, pp. 2569-2572.
6. Z. Liu, G. B. Giannakis, "Space-Time Block Coded Multiple Access Through Frequency-Selective Fading Channels", *IEEE Transactions on Communications*, Vol. 49, No. 6, June 2001, pp. 1033-1044.
7. B. Lu, G. Yue, X. Wang, "Performance Analysis and Design Optimization of LDPC-Coded MIMO OFDM Systems", *IEEE Transactions on Signal Processing*, Vol. 52, No. 2, February 2004, pp. 348–361.
8. A. K. Nandi (ed.), "Blind Estimation using Higher-Order Statistics", Kluwer Academic Publishers, Dordrecht, The Netherlands, 1999.
9. S. Verdu, "Multiuser Detection", Cambridge Press, Cambridge, U.K., 1998.

Music Transcription with ISA and HMM

Emmanuel Vincent and Xavier Rodet

IRCAM, Analysis-Synthesis Group
1, place Igor Stravinsky
F-75004 PARIS
emmanuel.vincent@ircam.fr

Abstract. We propose a new generative model for polyphonic music based on nonlinear Independent Subspace Analysis (ISA) and factorial Hidden Markov Models (HMM). ISA represents chord spectra as sums of note power spectra and note spectra as sums of instrument-dependent log-power spectra. HMM models note duration. Instrument-dependent parameters are learnt on solo excerpts and used to transcribe musical recordings as collections of notes with time-varying power and other descriptive parameters such as *vibrato*. We prove the relevance of our modeling assumptions by comparing them with true data distributions and by giving satisfying transcriptions of two duo recordings.

1 Introduction

In this article we consider the problem of polyphonic music transcription. A musical excerpt can be considered as a time-varying mixture of notes from several musical instruments, where the sound of a given note evolves across time and is described with a set of descriptors (instantaneous power, instantaneous frequency, timbre, *etc*). Given a single-channel musical excerpt and knowing which instruments are playing, we aim at inferring the notes played by each instrument and their descriptors. This can be considered as a semi-blind source separation problem where "meaningful parameters" are extracted instead of waveforms [1]. The main difficulty is that sounds from different instruments are not disjoint in the time-frequency plane and that information about quiet sounds may be masked by louder sounds. Usual approaches are reviewed in [2].

Independent Subspace Analysis (ISA) is a well-suited model for music transcription. Linear ISA describes the short-time power spectrum ($\mathbf{x}_t$) of a musical excerpt as a sum of typical power spectra (or components) ($\mathbf{\Phi}_h$) with time-varying weights (e_{ht}). This is expressed as $\mathbf{x}_t = \sum_{h=1}^{H} e_{ht}\mathbf{\Phi}_h + \epsilon_t$ where the modeling error (ϵ_t) is a Gaussian noise [3]. Each note from each instrument is represented by a subspace containing a few components. ISA has been applied to transcription of MIDI-synthesized solo harpsichord [3] and drum tracks [4]. However its robustness to real recording conditions and its ability to discriminate musical instruments have not been studied yet.

The results of existing methods show that linear ISA has three limitations reagarding its possible application to the transcription of real musical recordings.

The first limitation is that the modeling error is badly represented as an additive noise term since the absolute value of ϵ_t is usually correlated with $\mathbf{x}_t$. The modeling error may rather be considered as multiplicative noise (or as additive noise in the log-power domain) [3]. This is confirmed by instrument identification experiments, which use cepstral coefficients (or equivalently log-power spectra) as timbre features instead of power spectra [5, 2, 6]. The second limitation is that summation of power spectra is not an efficient way of representing the time evolution of note spectra. Many components are needed to represent small fundamental frequency (f0) variations in *vibrato*, wide-band noise during attacks or energy rise of higher harmonics in *forte*. It can easily be seen that summation of log-power spectra is more efficient. The third limitation is that ISA results are not often directly interpretable since many estimated notes with short duration or low power need to be removed before obtaining a readable musical score.

To solve these limitations, we derive here a new nonlinear ISA model considering both summation of power spectra and of log-power spectra and we also study the use of factorial Hidden Markov Models (HMM) as note duration priors.

The structure of the article is as follows. In Section 2 we propose a generative model for polyphonic music combining ISA and HMM. In Section 3 we explain how to learn the model parameters on solo excerpts and how to perform transcriptions. In Section 4 we discuss the relevance of our assumptions and we show two transcription examples. We conclude on possible improvements of the generative model.

2 Generative Model for Polyphonic Music

2.1 A Three-Layer Generative Model

Let $(\mathbf{x}_t)$ be the short-time log-power spectra of a given polyphonic musical excerpt. As a general notation in the following we use bold letters for vectors, regular letters for scalars and parentheses for sequences. Transcribing $(\mathbf{x}_t)$ consists in retrieving for each time frame t, each instrument j and each note h from the semitone scale both a discrete state $E_{jht} \in \{0, 1\}$ denoting presence/absence of the note and a vector of continuous note descriptors $\mathbf{p}_{jht} \in \mathbb{R}^{K+1}$. We assume a three-layer probabilistic generative model, where high-level states (E_{jht}) generate middle-level descriptors $(\mathbf{p}_{jht})$ which in turn generate low-level spectra $(\mathbf{x}_t)$. These three layers are termed respectively state layer, descriptor layer and spectral layer. In this Section we describe these layers successively.

2.2 Spectral Layer with Nonlinear ISA

Let us denote $\mathbf{m}_{jt}$ the power spectrum of instrument j at time t and $\mathbf{\Phi}'_{jht}$ the log-power spectrum of note h from instrument j at time t. We write the note descriptors as $\mathbf{p}_{jht} = [e_{jht}, v^1_{jht}, \ldots, v^K_{jht}]$, where e_{jht} is the log-energy of note h from instrument j at time t and (v^k_{jht}) are "variation variables" describing the differences between the mean spectrum of this note and its spectrum at time t.

Following the discussion of linear ISA limitations in Section 1, we assume

$$\mathbf{x}_t = \log\left[\sum_{j=1}^{n} \mathbf{m}_{jt} + \mathbf{n}\right] + \epsilon_t, \tag{1}$$

$$\mathbf{m}_{jt} = \sum_{h=1}^{H_j} \exp(\mathbf{\Phi}'_{jht}) \exp(e_{jht}), \tag{2}$$

$$\mathbf{\Phi}'_{jht} = \mathbf{\Phi}_{jh} + \sum_{k=1}^{K} v_{jht}^k \mathbf{U}_{jh}^k, \tag{3}$$

where $\exp(.)$ and $\log(.)$ are the exponential and logarithm functions applied to each coordinate. The vector $\mathbf{\Phi}_{jh}$ is the total-power-normalized mean log-power spectrum of note h from instrument j and the $\mathbf{U}_{jh}^k$ are L_2-normalized "variation spectra" related to the "variation variables". The vector $\mathbf{n}$ is the power spectrum of the background noise. The modeling error ϵ_t is supposed to be a Gaussian white noise with variance $\sigma_\epsilon^2 \mathbf{I}$.

Equations (1-2) can be approximated by a simpler nonlinear model using the maximum over each coordinate [7].

2.3 Descriptor Layer

We assume that note descriptors $\mathbf{p}_{jht}$ are conditionally independent given the note state E_{jht}. We set the parametric conditional priors

$$P(\mathbf{p}_{jht}|E_{jht}=1) = \mathcal{N}_{\mu e_{jh}, \sigma e_{jh}}(e_{jht}) \prod_{k=1}^{K} \mathcal{N}_{\mu v_{jhk}, \sigma v_{jhk}}(v_{jht}^k), \tag{4}$$

$$P(\mathbf{p}_{jht}|E_{jht}=0) = \delta_{-\infty}(e_{jht}) \prod_{k=1}^{K} \delta_0(v_{jht}^k), \tag{5}$$

where $\mathcal{N}_{\mu,\sigma}(.)$ is the Gaussian distribution of mean μ and variance σ^2 and $\delta_\mu(.)$ the Dirac distribution centered in μ. For most instruments the parameters μe_{jh}, σe_{jh}, μv_{jhk} and σv_{jhk} can be shared for all notes h.

2.4 State Layer with Factorial Markov Chains

Finally, we suppose that the states E_{jht} are independent for different (j,h). This results in independence of the descriptors $\mathbf{p}_{jht}$ for different (j,h), which is the usual ISA assumption. We consider two state models: a product of Bernoulli priors and a factorial Markov chain [8], whose equations are respectively

$$P(E_{jh,1},\ldots,E_{jh,T}) = \prod_{t=1}^{T} (P_Z)^{1-E_{jht}} (1-P_Z)^{E_{jht}}, \tag{6}$$

$$P(E_{jh,1},\ldots,E_{jh,T}) = P(E_{jh,1}) \prod_{t=2}^{T} P(E_{jht}|E_{jh,t-1}), \tag{7}$$

where the initial and transition probabilities $P(E_{jh,1})$ and $P(E_{jht}|E_{jh,t-1})$ are themselves Bernoulli priors that can be shared for all notes h.

The silence probability P_Z is a sparsity factor: the higher it is, the less notes are set as present in a transcription. P_Z can also be expressed with Markov transition probabilities as $P_Z = (1 + P(1|0)P(0|1)^{-1})^{-1}$. So both models can be used to model the exact sparsity of the transcriptions.

The difference is that the Markov prior adds some temporal structure and favors transcriptions where E_{jht} is the same on long time segments. HMM have also been used with ISA for source separation in biomedical applications, but in the simpler case of noiseless invertible mixtures [9].

3 Learning and Transcribing

Now that we have described each layer of the generative model, we explain in this Section how to learn its parameters on solo excerpts and use them to transcribe polyphonic excerpts. We define the instrument model $\mathcal{M}_j$ as the collection of the fixed parameters related to instrument j: the spectra Φ_{jh} and $\mathbf{U}_{jh}^k$, the means and variances μe_{jh}, σe_{jh}^2, μv_{jhk} and σv_{jhk}^2, and the sparsity factor P_Z.

3.1 Weighted Bayes

The probability of a transcription $(E_{jht}, \mathbf{p}_{jht})$ is given by the weighted Bayes law

$$P_{\text{trans}} = P((E_{jht}, \mathbf{p}_{jht})|(\mathbf{x}_t),(\mathcal{M}_j)) \propto (P_{\text{spec}})^{w_{\text{spec}}} (P_{\text{desc}})^{w_{\text{desc}}} P_{\text{state}}, \qquad (8)$$

involving probability terms $P_{\text{spec}} = \prod_t P(\epsilon_t)$, $P_{\text{desc}} = \prod_{jht} P(\mathbf{p}_{jht}|E_{jht},\mathcal{M}_j)$ and $P_{\text{state}} = \prod_{jh} P(E_{jh,1},\ldots,E_{jh,T}|\mathcal{M}_j)$ and correcting exponents w_{spec} and w_{desc}. Weighting by w_{spec} with $0 < w_{\text{spec}} < 1$ improves the quality of the Gaussian white noise model for ϵ_t. This mimics the existence of dependencies between values of ϵ_t at adjacent time-frequency points and makes the noise distribution less "peaky" [10].

3.2 Learning and Transcription Algorithms

Transcribing an excerpt $(\mathbf{x}_t)$ with instrument models $(\mathcal{M}_j)$ means maximizing the posterior P_{trans} over $(E_{jht}, \mathbf{p}_{jht})$.

Transcription with Bernoulli state priors is carried out by reestimating iteratively the note states with a jump procedure. At start all states E_{jht} are set to 1, then at each iteration at most one note is added or subtracted at each time t to improve the posterior probability value. Inference with Markov state priors is done with Viterbi decoding. The factorial state space is reduced approximately beforehand by performing inference with a Bernoulli state prior and a low sparsity factor P_Z to rule out very unprobable notes. In both cases the note descriptors are estimated with an approximate second order Newton method.

The modeling error variance σ_ϵ, the correcting exponents w_{spec} and w_{desc} and the mean note duration are set by hand to achieve a good compromise between insertion and deletion errors, whereas the power spectrum of the background noise $\mathbf{n}$ is estimated from the data in order to maximize the posterior.

It is interesting to note that the Newton method updates involve the quantity $\pi_{jhtf} = \exp(\Phi'_{jhtf})\exp(e_{jht})[\sum_{h'=1}^{H_j}\exp(\Phi'_{jh'tf})\exp(e_{jh't})]^{-1}$ that is the power proportion of note h into the model spectrum at time-frequency point (t, f). When note h is masked by other notes, $\pi_{jhtf} \approx 0$ and the value of the observed spectrum x_{tf} is not taken into account to update e_{jht} and v^k_{jht}. This proves that nonlinear ISA performs "missing data" inference [2] in a natural way.

Learning the parameters of a model $\mathcal{M}_j$ on a solo excerpt $(\mathbf{x}_t)$ from instrument j is done by maximizing the posterior P_{trans} iteratively over $(E_{jht}, \mathbf{p}_{jht})$ and over $\mathcal{M}_j$. Models could also be learnt directly on polyphonic excerpts [8], but learning on solo excerpts is more accurate because notes are less likely to be masked.

Re-estimation is done again with a Newton method. Note that when $K > 1$ the spectra $\mathbf{U}^k_{jh}$ are identifiable only up to a rotation. The size of the model and the initial parameters are fixed by hand. Experimentally we noticed that when the mean spectra Φ_{jh} are initialized as spectral peaks at harmonic frequencies they keep this structure during the learning procedure (only the power and the width of the peaks vary). So the learnt spectra really represent "notes" and there is no need of a supervised learning method for the transcriptions to make sense.

3.3 Computation of the Short-Time Log-Power Spectra

The performance of the generative model can be improved by choosing a good time-frequency distribution for $\mathbf{x}_t$. In the field of music transcription, nonlinear frequency scales giving more importance to low frequencies are preferable to the linear Fourier scale. Indeed the harmonics of low frequency notes are often masked by harmonics from high frequency notes, so that information in low frequencies is more reliable. Moreover, the modeling of *vibrato* with (3) is relevant only if small f0 variations induce small spectral variations. In the following we use a bank of filters linearly spaced on the auditory-motivated ERB frequency scale $f_{ERB} = 9.26\log(0.00437 f_{Hz} + 1)$ and we compute log-powers on 11 ms frames (a lower threshold is set to avoid dropdown to $-\infty$ in silent zones).

4 Experiments

Let us now present a few experiments to justify our modeling assumptions and evaluate the performance of the model. We transcribe two real duo recordings: an excerpt from Pachelbel's canon in D arranged for flute and cello and an excerpt from Ravel's sonata for violin and cello. We learn instrument models (with $K = 1$) on one-minute solo excerpts taken from other CDs.

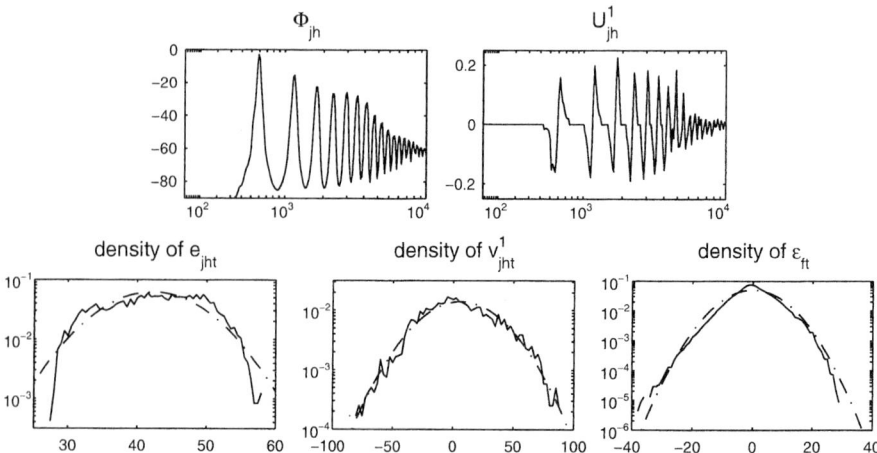

Fig. 1. Top: learnt violin spectra Φ_{jh} and $\mathbf{U}_{jh}^1$ with $h =$ MIDI 74. Horizontal axis is frequency in Hertz, vertical axis is in Decibels for Φ_{jh} and has no unity for $\mathbf{U}_{jh}^1$. Bottom: densities of e_{jht}, v_{jht}^1 and ϵ_{ft} shared for all h and for 70 Hz $\leq f \leq$ 5000 Hz (plain line) compared with Gaussians (dash-dotted line). Horizontal axis is in Decibels.

Fig. 1 shows the violin spectra for $h =$ MIDI 74. The mean spectrum Φ_{jh} contains peaks at harmonic frequencies. And the "variation spectrum" $\mathbf{U}_{jh}^1$ looks like the derivative versus frequency of Φ_{jh} which represents the first order linear approximation of small f0 variations. Fig. 1 also shows the densities of the variables e_{jht} and v_{jht}^1 measured on the violin learning excerpt (after transcription with ISA+Bernoulli) and the density of the modeling error ϵ_{ft} measured on the first duo (after transcription with ISA+Markov). These densities are all close to Gaussians. We conclude from this that the model actually captured the main spectral characteristics of these musical sounds.

Fig. 2 shows the transcription of the two duos.

The first duo is a difficult example since notes played by the flute are nearly always harmonics of the notes played by the cello, and also belong to the playing range of the cello. The results show that our model is both able to identify most of the notes and to associate them with the right instrument. Results with Markov state model are also a bit better than results with Bernoulli state model, since some spurious short duration notes are removed. There remains 2 note deletions and 10 notes insertions. The inserted notes all have very short durations and could be removed with more complex state models involving rhythm or forcing instruments to play one note at a time (plus reverberation of previous notes).

With the second duo as input, the model is able to identify the notes, but completely fails in associating them with the right instrument. This is not surprising since violin and cello have very close timbral properties, and more complex state models should be used to separate the two note streams. Considering only note transcription and not instrument identification, there are 11 note insertions that have again very short durations. Note that both instruments play *vibrato*, and that this can be seen in the oscillating values of $v_{2,ht}^1$.

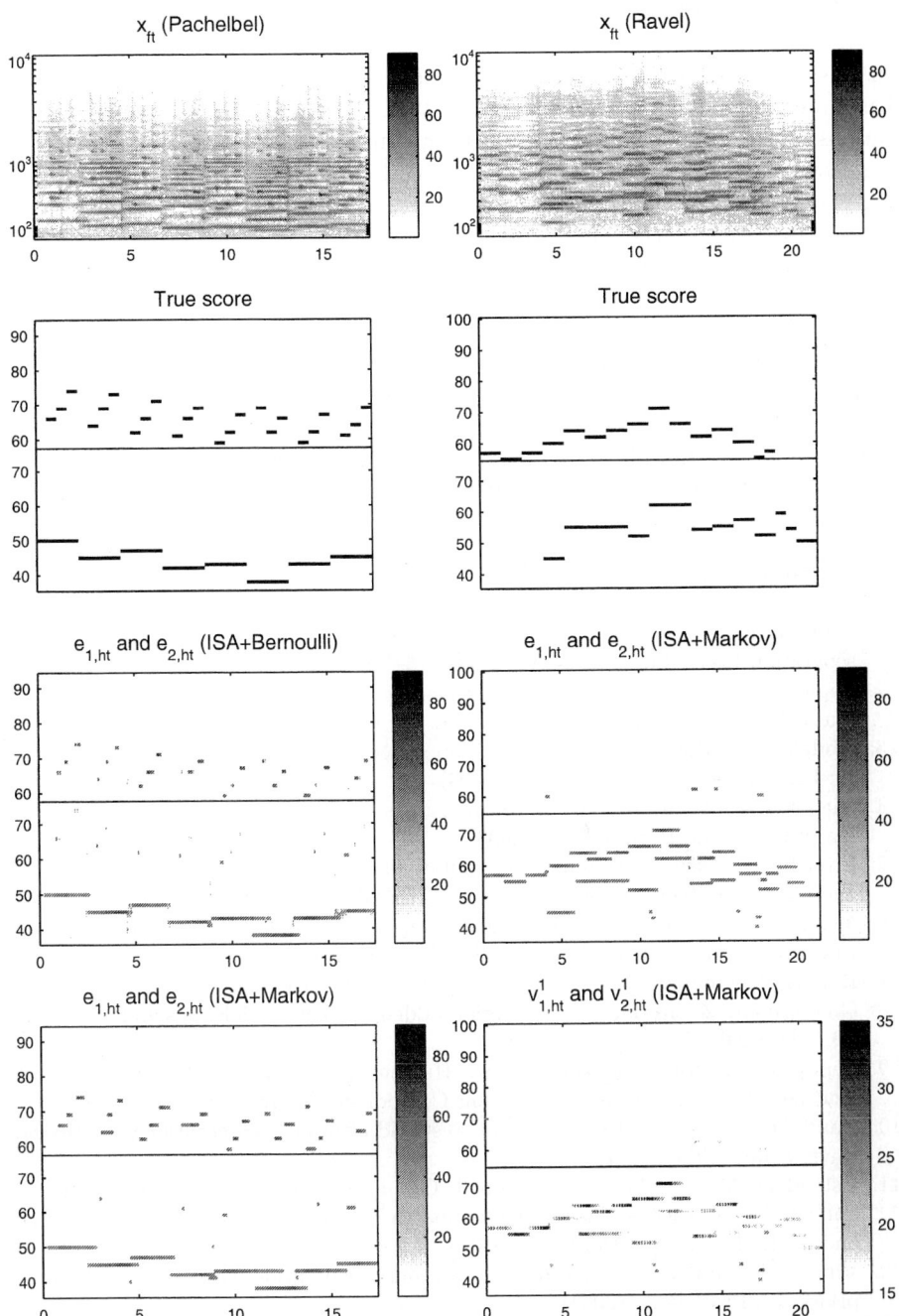

Fig. 2. Transcription of two duo recordings. Top: spectrograms of the recordings. Middle: true scores of the recordings. Below: some estimated note descriptors using various state priors. Horizontal axis is time in seconds, vertical axis is frequency in Hertz (top) or note pitch on the MIDI scale (middle and below). The color range is in Decibels.

5 Conclusion

In this article we proposed a new method for polyphonic music transcription based on nonlinear ISA and factorial HMM. This model overcomes the limitations of usual linear ISA by using both summation of power spectra and of log-power spectra and by considering the modeling error as additive noise in the log-power domain. These modeling assumptions were verified on learning data. We also performed satisfying transcriptions of two rather difficult duo recordings.

As noted above, the quality of the transcriptions could be improved by using more complex state models. We are currently considering three directions: constraining instruments to play only one note at a time (plus reverberation of previous notes), using segment models [11] instead of HMM for a better modeling of note durations, and setting temporal continuity priors [4] on the variables e_{jht} and v_{jht}^k. We are also investigating the use of instrument models as structured source priors for semi-blind source separation. We present some results in stereo underdetermined mixtures in a companion article [12].

References

1. Vincent, E., Févotte, C., Gribonval, R.: A tentative typology of audio source separation tasks. In: Proc. ICA. (2003) 715–720
2. Eggink, J., Brown, G.: Application of missing feature theory to the recognition of musical instruments in polyphonic audio. In: Proc. ISMIR. (2003) 125–131
3. Abdallah, S., Plumbley, M.: An ICA approach to automatic music transcription. In: Proc. 114th AES Convention. (2003)
4. Virtanen, T.: Sound source separation using sparse coding with temporal continuity objective. In: Proc. ICMC. (2003)
5. Eronen, A.: Musical instrument recognition using ICA-based transform of features and discriminatively trained HMMs. In: Proc. ISSPA. (2003)
6. Mitianoudis, N., Davies, M.: Intelligent audio source separation using Independent Component Analysis. In: Proc. 112th AES Convention. (2002)
7. Roweis, S.: One microphone source separation. In: Proc. NIPS. (2000) 793–799
8. Ghahramani, Z., Jordan, M.: Factorial hidden Markov models. Machine Learning **29** (1997) 245–273
9. Penny, W., Everson, R., Roberts, S.: Hidden Markov Independent Components Analysis. In: Advances in Independent Component Analysis. Springer (2000)
10. Hand, D., Yu, K.: Idiot's bayes - not so stupid after all ? International Statistical Review **69** (2001) 385–398
11. Ostendorf, M., Digalakis, V., Kimball, O.: From HMMs to segment models: a unified view of stochastic modeling for speech recognition. IEEE Trans. on Speech and Audio Processing **4** (1996) 360–378
12. Vincent, E., Rodet, X.: Underdetermined source separation with structured source priors. In: Proc. ICA. (2004)

On Shift-Invariant Sparse Coding

Thomas Blumensath* and Mike Davies

Queen Mary, University of London
Department of Electronic Engineering
Mile End Road
London, E1 4NS, UK

Abstract. The goals of this paper are: 1) the introduction of a shift-invariant sparse coding model together with learning rules for this model; 2) the comparison of this model to the traditional sparse coding model; and 3) the analysis of some limitations of the newly proposed approach. To evaluate the model we will show that it can learn features from a toy problem as well as note-like features from a polyphonic piano recording. We further show that the shift-invariant model can help in overcoming some of the limitations of the traditional model which occur when learning less functions than are present in the true generative model. We finally show a limitation of the proposed model for problems in which mixtures of continuously shifted functions are used.

1 Introduction

First let us introduce the traditional sparse coding model as:

$$x = As + \epsilon, \quad (1)$$

where $x \in \mathbb{R}^M$, $s \in \mathbb{R}^N$, $A \in \mathbb{R}^{M \times N}$, $\epsilon \in \mathbb{R}^M$ and $N > M$. The likelihood is defined as:

$$p(x|A, s) \sim \mathcal{N}(As, \Sigma_\epsilon) \quad (2)$$

and the prior $p(s)$ is a sparse distribution.

Learning in this model is generally achieved by an iterative algorithm. The problem of learning the matrix A can be formulated as finding the maximum likelihood estimate of:

$$p(x|A) = \int p(x|A, s)p(s) \, ds . \quad (3)$$

The gradient of the logarithm of this likelihood can be rewritten as [1]:

$$\frac{\partial \log p(x|A)}{\partial A_{mn}} = \left\langle \frac{\partial}{\partial A_{mn}} \log p(x|A, s) \right\rangle_{p(s|A,x)}, \quad (4)$$

---

* This work was partly supported by EPSRC Grant GR/R54620.

where $\langle \cdot \rangle$ denotes expectations. For the sparse coding model, (4) can be written as:
$$\frac{\partial \log p(\boldsymbol{x}|\boldsymbol{A})}{\partial \boldsymbol{A}_{mn}} = \left\langle \boldsymbol{\Sigma}_\epsilon^{-1}(\boldsymbol{x} - \boldsymbol{A}\boldsymbol{s})\boldsymbol{s}^T \right\rangle_{p(\boldsymbol{s}|\boldsymbol{A},\boldsymbol{x})} . \qquad (5)$$
The full posterior $p(\boldsymbol{s}|\boldsymbol{A},\boldsymbol{x})$ can not be evaluated analytically. In [2, 3] Gibbs sampling was used to sample from $p(\boldsymbol{s}|\boldsymbol{A},\boldsymbol{x})$. In [4] a delta approximation of the density $p(\boldsymbol{s}|\boldsymbol{A},\boldsymbol{x})$ at the MAP estimate of $\boldsymbol{s}$ was proposed and in [5] the use of a Gaussian approximation of $p(\boldsymbol{x}|\boldsymbol{A})$ around $\boldsymbol{s}_{MAP}$ was suggested. To find the MAP estimate different constrained optimisation routines can be used [4–9].

2 Shift-Invariant Sparse Coding

If $\boldsymbol{x}$ is a block taken from a discrete time series $x[t]$ the above model can be seen to be non-invariant to shifts $\boldsymbol{x} = \{x[t+l]\}$ in general. We therefor introduce a shift-invariant model formulation as follows:
$$\boldsymbol{x} = \sum_{k \in K, l \in L} \boldsymbol{a}_{kl} s_{kl} + \boldsymbol{\epsilon} = \boldsymbol{A}\boldsymbol{s} + \boldsymbol{\epsilon} , \qquad (6)$$
where now the functions (i.e. the columns of $\boldsymbol{A}$) can occur at any location t. The indices k and l denote the values corresponding to the kth function and the lth shift.

To derive the learning rules for the above model, the gradient in (4) has to be evaluated with respect to the pth component of the function $\boldsymbol{a}_k$ which we will write as a_{kp}. In order to keep the notation simple, we assume the noise to be i.i.d. Gaussian with variance σ_ϵ. We will also use s_{kl} to denote the coefficient related to function $\boldsymbol{a}_k$ with a time shift of l where we let l be zero to denote no time shift; i.e. we use $\boldsymbol{a}_{k0} = \boldsymbol{a}_k = [a_{k1}, a_{k2}, \cdots, a_{kL}]^T$ and $\boldsymbol{a}_{k4} = [0,0,0,0,a_{k1},a_{k2},\cdots,a_{k(L-4)}]^T$ and so forth.

We can calculate the derivative with respect to a_{kp} as:
$$\frac{\partial \log p(\boldsymbol{x}|\boldsymbol{A},\boldsymbol{s})}{\partial a_{kp}} = -\frac{1}{\sigma_\epsilon} \sum_m \left[\left(x_m - \sum_{\bar{k} \in K, l \in L} a_{\bar{k},m+l} s_{\bar{k}l} \right) s_{k,p-m} \right] . \qquad (7)$$

If $\boldsymbol{x}$ and $\boldsymbol{a}_k$ are both in $\mathbb{R}^M$, we can write this expression as a convolution and derive a gradient update rule for the set of functions $\{\boldsymbol{a}_k\}$ as:
$$\Delta\{\boldsymbol{a}_k\} \propto \sigma_\epsilon^{-1} \langle \boldsymbol{\epsilon} \star \{s_k\} \rangle_{p(\boldsymbol{s}|\boldsymbol{A},\boldsymbol{x})} , \qquad (8)$$
where $\star$ is the convolution operator and $\boldsymbol{\epsilon} = x_m - \sum_{\bar{k} \in K, l \in L} a_{\bar{k},m+l} s_{\bar{k}l}$. This expression can then be used to find approximations to the learning rule similar to those in [4, 5].

This formulation offers three advantages over the traditional model: 1) many of the operations become convolutions which can be implemented efficiently; 2) the number of parameters to be learned is much smaller which leads to a more efficient use of data and faster convergence; and 3) a certain robustness against the effects of using a model with less functions than contribute to the signal is found (see section 4).

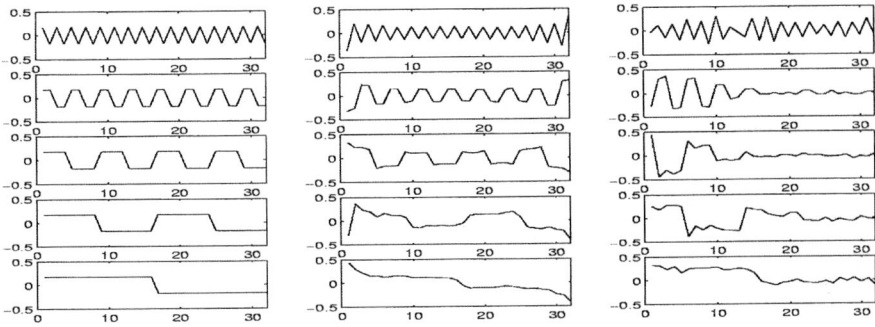

Fig. 1. The five original functions (left), the functions learned with the shift-invariant model (centre) and some functions learned with the non-shift-invariant model (right).

3 Results

To evaluate the proposed model we will first use a toy problem and then compare the learned functions with those learned with the traditional model. We generate data using the generative model of (6) with five different functions. These functions are shown in the left panel of figure 1. The coefficients s were sampled from a mixture density, being zero with probability 0.99 or Gaussian distribution otherwise.

When training the proposed algorithm with realisations of this model, the five functions in the middle panel of figure 1 were learned. Here we used the sampling strategy described in [3] to sample from the posterior $p(s|A, x)$ during learning.

We repeated the experiment using the same training data with the update rule in (5) and using a matrix A large enough to include all possible functions and their shifts. Five of the learned functions are shown in the right panel of figure 1. It was found that the matrix A contained five different types of functions that resembled the underlying functions at different shifts. Examples of functions learned at shifted positions are shown in the second to fourth row in the right panel of figure 1.

To further compare the performance of our proposed model with the traditional model, we reduced the size of matrix A to include only 64 functions (a two times over-complete set). The 64 functions learned without the shift-invariance constraint are shown in the left panel of figure 2. It can be seen that some of the learned functions are windowed versions of the original features. When learning only 1 function with the shift-invariant model (which also gives us a two times over-complete set of functions), the function in the right panel of figure 2 emerged. This function resembles the fifth function in panel 1 of figure 1 and is also similar to the fifth learned function in panel 2 of figure 1. This shows that by enforcing shift-invariance, features can still be identified even when learning less functions than are present in the signal. This is not true for the traditional model where in this case filtered features are found.

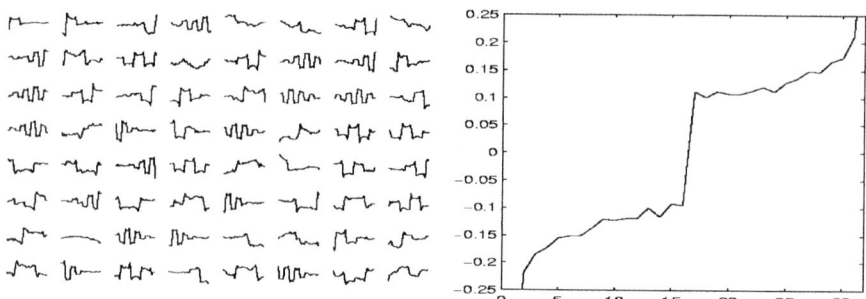

Fig. 2. The functions learned with the non-shift-invariant model using too few functions (left) and the one function learned with the shift-invariant model (right).

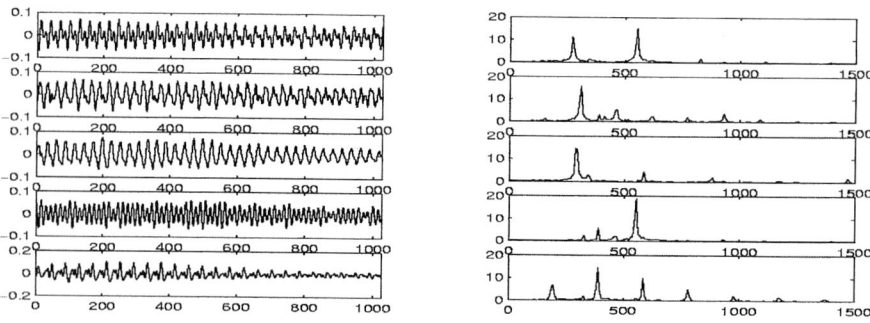

Fig. 3. A selection of 5 harmonic functions learned from the piano signal (left). The spectrum of the 5 functions showing frequencies below 1500Hz (right).

To test the algorithm on a real world signal and to present a possible application of the model, we used the proposed algorithm to extract individual notes from a polyphonic piano recording [10]. We used a recording of L. von Beethoven's Sonata for Piano No. 12, in A flat, Scherzo (Allegro molto) which was summed to mono and resampled at 8 kHz. We then learned 50 functions with a function length of 1024 samples. To reduce the computational complexity, we used an inference process in which we first selected a small number of functions at certain shifts depending on the correlation between those functions and the training data. We also used the additional constraint that a function could not overlap with a shifted version of itself by more than 50%. We then used an Iterative Re-weighted Least Squares algorithm to find s_{MAP} using only the selected functions. See [10] for details.

After 100,000 iterations, the algorithm had learned a set of 50 features, 10 of which had not been updated much. For the other 40 functions, a clear harmonic structure emerged. 35 of these functions were found to have different fundamental frequencies corresponding to notes of the western equally tempered 12 tone scale in a range from C#2 to A5.

In the left panel of figure 3 we show a random selection of the learned notes with their magnitude spectrum shown in the right panel of figure 3. We found that the learned functions did not contain much high frequency energy. Frequencies above 1500 Hz were insignificant and are therefore not shown. A possible reason for this will be investigated in section 5.

In order to evaluate whether learned functions represent individual notes in the original recording we repeated the above experiment with a recording of L. von Beethoven's Bagatelle No. 1 op. 33. This recording was made using a MIDI controlled acoustic piano so that the exact MIDI representation was available together with the audio file. We found that 61% of the coefficients exactly identified the notes present (7% of the coefficients identified notes in a different octave) which shows the similarity between notes and learned features.

4 Sensitivity of the Model Size

In this section we study the influence of the number of functions used in the traditional sparse coding model. We will assume here that the observed signal follows the model of (6). $\hat{a}_{\check{k}o}$ and $\hat{s}_{\check{k}}$ will be used to denote the estimated function and coefficient respectively. $a_{k,p+l}$ and s_{kl} will refer to the true underlying process generating the observation. We now use the indices k and l to denote the functions and the associated shifts of the underlying process, while $\check{k}$ indexes the learned functions. (7) can be written as:

$$\Delta \hat{a}_{\check{k}p} = \mu \Sigma_\epsilon^{-1} \int \left(\sum_{k,l} a_{k,p+l} s_{kl} - \sum_{\check{k}} \hat{a}_{\check{k}p} \hat{s}_{\check{k}} \right) \hat{s}_{\check{k}} \; p(\hat{s}|\boldsymbol{x}, \hat{\boldsymbol{A}}) \; d\hat{s} \; .$$

This leads to an update rule of:

$$\hat{a}_{\check{k}p}^{[r+1]} = \left(\hat{a}_{\check{k}p}^{[r]} - \mu \Sigma_\epsilon^{-1} \int \hat{\mathbf{x}} \hat{s}_{\check{k}} p(\hat{s}|\boldsymbol{x}, \hat{\boldsymbol{A}}) \; d\hat{s} \right) \qquad (9)$$

$$+ \mu \Sigma_\epsilon^{-1} \int \sum_{k,l \in \widetilde{K}} a_{kp} s_k \hat{s}_{\check{k}} p(\hat{s}|\boldsymbol{x}, \hat{\boldsymbol{A}}) \; d\hat{s} \qquad (10)$$

$$+ \mu \Sigma_\epsilon^{-1} \int \sum_{k,l \in K} a_{kp} s_k \hat{s}_{\check{k}} p(\hat{s}|\boldsymbol{x}, \hat{\boldsymbol{A}}) \; d\hat{s} \; , \qquad (11)$$

where the last term has been split into those indices k and l for which s_{kl} and $\hat{s}_{\check{k}}$ are conditionally independent (the set $\widetilde{K}$) and dependent (the set K) given the current model and observation.

The first term in which we use $\hat{\mathbf{x}} = \sum_{\check{k}} \hat{a}_{\check{k}p} \hat{s}_{\check{k}}$ is a normalising term preventing unlimited function growths and does not depend on the $\boldsymbol{a}_k$.

The second term will be zero due to the assumed conditional independence between s_{kl} and $\hat{s}_{\check{k}}$ for $k \in \widetilde{K}$.

The third term can be written as $\mu \Sigma_\epsilon^{-1} \sum_{k,l \in K} a_{k,p+l} s_{kl} \bar{\hat{s}}_{\check{k}}$. Here $\bar{\hat{s}}_{\check{k}}$ is used to denote the conditional mean of $\hat{s}_{\check{k}}$. For many updates, this equation is a weighted average of the functions $a_{k,p+l}$ weighted by $s_{kl}\hat{\bar{s}}_{\hat{k}}$. As the coefficients s_{kl} are i.i.d., the weights are $\hat{\bar{s}}_{\hat{k}}$, which depend on the estimation procedure used. If the set K only includes shifted versions of the same function (i.e. only one index k, but several indices l), then this averaging is equal to filtering of one function a_k.

If the number of functions used to model a signal is less than the number of functions in the signal at all locations, then dependencies between $\hat{s}_{\hat{k}}$ and several s_{kl} have to occur. Dependencies can also occur as a result of the inference process or the approximations to the learning rule used.

To analyse the possible dependencies which can occur due to the incorrect model size, we will assume that all learned functions have converged to some of the true functions. The dependency between $\hat{s}_{\hat{k}}$ and s_{kl} (and therefore the exact form of the averaging process described above) then depends on which of the functions a_{kl} are modelled by each function $\hat{a}_{\hat{k}}$. The function chosen to model a function which has not been learned, will depend on the decrease in reconstruction error when using this function. The highest decrease in this error will be achieved by modelling a function in the signal with the same function at the exact location. If this function is not available at this location, a function at a different location or a different function has to be used.

In the following list three forms of dependencies which can occur are given together with the influence they have on the learned functions:

- A function can be modelled with a slightly shifted version of itself. If several slightly shifted functions are modelled by a single function, then the average update of this function will be a low-pass filtered version of the true function.
- A windowed periodic function can be modelled with a version of itself which is shifted by multiples of the period. A weighted averaging over such function shifts will lead to a windowing of the learned function.
- A missing function can also be modelled with a different function. The chosen function is likely to share a strong frequency component and will be at a location at which both functions have the same phase for this component. Averaging will then increase this frequency component but might decrease other frequency components, as the phase for those other components might not match.

This seems to suggest that if the number of functions to be learned is less than the number of functions in the signal, windowed and filtered functions will emerge. However, the above derivation used the traditional sparse coding formulation. If shift-invariance is explicitly enforced and if the inference process is working correctly, then the first two effects (i.e. the filtering and the windowing) will not occur. This observation was made in the previous section, when the number of functions learned was less than the number of functions in the signal. The functions learned with the traditional model were often windowed or filtered versions of the functions underlying the observations. In the shift-invariant model this observation was not made.

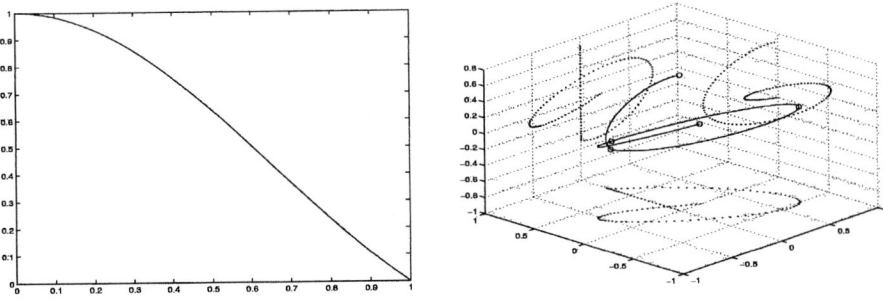

Fig. 4. Filtering due to sampling (left) and path of a shifted function (right).

5 Sampling and Shift-Invariant Sparse Coding

We have found that the functions learned from real data did not contain much high frequency energy. This seems to be the result of the proposed model which does not account for shifts by fractional samples. In this section we will investigate the influence of this approximation when learning functions from a sampled continuous signal, in which functions can be located at any position.

A continuously shifted function occupies a 1-dimensional manifold in the sampled space. This path is shown in the right panel of figure 4 for a 3-dimensional signal. (The circles show the location of the sampled signal and its shifted versions. The doted lines are the projections onto the xy, xz and yz planes.)

By assuming that during learning, functions are averaged over a window of one sample centred at each sample, the frequency response of the averaging process can be found as the Fourier transform of a rectangular window. The amplitude response is $\frac{sin(\pi f)}{\pi f}$ where f is the normalised frequency. This sync function is plotted in the left panel of figure 4. It can be seen that the effect of continuously shifted functions in the original signal leads to learning of low-pass filtered functions. This effect can not easily be overcome without substantially increasing computational complexity.

6 Summary and Discussion

We have developed a sparse coding model, which is invariant to discrete shifts. We investigated the performance of this model using a toy problem as well as real world data. We also compared the proposed model with the model which does not take shifts into account. A theoretical analysis of the influence of the number of functions learned in sparse coding showed that, in some situations, the features learned can be averaged versions of the features present in the observation. This is the result of not using enough functions to represent the signal. In the standard sparse coding model three different effects can occur: low-pass filtering, windowing and averaging of different functions. The model proposed here is not influenced by the first two effects and will only average over different functions.

We also showed experimentally that it is possible to learn a shift-invariant representation with both models in the case in which the number of functions was known. However, for most time-series or images, the required number of functions is not known and it could be shown that in this case, enforcing shift-invariance can help in extracting the underlying structure. The enforcement of shift-invariance has two further advantages. It can lead to a faster algorithm, as many operations become convolutions, which can be implemented efficiently and it does offer faster convergence, as less parameters have to be adapted.

Finally, the influence of sampling signals composed of continuously shifted features was analysed. Here filtering of the features occurred as an inherent property of the model. This is a possible explanation for the missing high frequency energy in the results obtained from the experiments with real world data. This effect is present in both, the original as well as in the shift-invariant model as both approaches can only model functions at discrete shifts. Overcoming this problem with a model formulation as used here, seems not feasible without a sharp increase in computational complexity.

References

1. S. Abdallah, *Towards Music Perception by Redundancy Reduction and Unsupervised Learning in Probabilistic Models*. PhD thesis, King's College London, February 2003.
2. B. A. Olshausen and K. Millman, "Learning sparse codes with a mixture-of-Gaussians prior," *Advances in Neural Information Processing Systems*, 2000.
3. P. Sallee and B. A. Olshausen, "Learning sparse multiscale image representations," *Advances in Neural Information Processing Systems*, 2003.
4. B. A. Olshausen and D. J. Field, "Emergence of simple-cell receptive field properties by learning a sparse code for natural images," *Nature*, no. 381, pp. 607–609, 1995.
5. M. S. Lewicki and T. J. Sejnowski, "Learning overcomplete representations," *Neural Computation*, no. 12, pp. 337–365, 2000.
6. K. Kreutz-Delgado, J. F. Murray, B. D. Rao, K. Engan, T.-W. Lee, and T. J. Sejnowski, "Dictionary learning algorithms for sparse representation," *Neural Computation*, vol. 15, pp. 349–396, 2003.
7. M. A. T. Figueiredo and A. K. Jain, "Bayesian learning of sparse classifiers," in *IEEE Computer Society Conference on Computer Vision and Pattern Recognition - CVPR'2001*, (Hawaii), pp. 35–41 Vol. 1, IEEE, December 2001.
8. S. S. Chen, D. L. Donoho, and M. A. Saunders, "Atomic decomposition by basis pursuit," *SIAM Journal of Scientific Computing*, vol. 20, no. 1, pp. 33–61, 1998.
9. Y. Li, A. Cichocki, and S. Amari, "Sparse component analysis for blind source separation with less sensors than sources," *ICA03*, pp. 89–94, 2003.
10. T. Blumensath and M. Davies, "Unsupervised learning of sparse and shift-invariant decompositions of polyphonic music," in *IEEE International Conference on Acoustics, Speech, and Signal Processing*, 2004.

Reliability in ICA-Based Text Classification

Xavier Sevillano, Francesc Alías, and Joan Claudi Socoró

Communications and Signal Theory Department
Enginyeria i Arquitectura La Salle. Universitat Ramon Llull
Pg. Bonanova 8, 08022 - Barcelona, Spain
{xavis,falias,jclaudi}@salleURL.edu

Abstract. This paper introduces a novel approach for improving the reliability of ICA-based text classifiers, attempting to make the most of the independent components of the text data. In this framework, two issues are adressed: firstly, a relative relevance measure for category assignment is presented. And secondly, a reliability control process is included in the classifier, avoiding the classification of documents belonging to none of the categories defined during the training stage. The experiments have been conducted on a journalistic-style text corpus in Catalan, achieving encouraging results in terms of rejection accuracy. However, similar results are obtained when comparing the proposed relevance measure to the classic magnitude-based technique for category assignment.

1 Introduction

In the last decade, text classification (TC) has become a focus of interest for the information retrieval research community, due to the need to organize the rapidly growing amount of digital documents available worldwide. The main approaches for TC belong to the machine learning paradigm (see [1] for an extensive review).

Generally, the TC process is divided in two phases: training and test. Training consists of building a classifier from a collection of documents (*training set*). Testing is the process of classifying a new group of documents (*test set*) according to the structure learned during the training phase.

In the context of TC, the application of Independent Component Analysis (ICA) is based on the assumption that a document collection (or corpus) is generated by a combination of several thematic topics [2–4]. Thus, the independent components (IC) obtained by the ICA algorithms define statistically independent *clusters* of documents, allowing their thematic classification.

Mainly, Independent Component Analysis has been used for text retrieval [2, 5], text classification [3] and text clustering [4, 6]. Another interesting application of ICA is chat topic spotting [7, 8], which takes into account the temporal dimension of text data. ICA has also been employed as a feature extraction technique in order to improve the performance of Naïve Bayes [9] and Support Vector Machines [10] classifiers. Moreover, in the context of multi-domain text-to-speech synthesis [11], an ICA-based hierarchical text classifier was presented in [12].

This paper presents a first step towards making the most of the information provided by ICA-based text classifiers in order to increase the reliability of the classifier. In this sense, we introduce i) a relative relevance measure for category assignment, and ii) a novel reliability evaluation process to inhibit the ICA-based text classifier from categorizing documents belonging to none of the categories defined during the training stage.

This paper is structured as follows: section 2 reviews the fundamentals of the application of ICA in text classification. In section 3, the proposals for improving the reliability in ICA-based text classification are presented. Section 4 describes the conducted experiments and, finally, the conclusions of our work are discussed in section 5.

2 Fundamentals of ICA for Text Classification

Text classification is defined as the task of assigning a collection of documents $D = \{d_1, d_2, \ldots, d_{|\mathcal{D}|}\}$ to a set of *predefined* categories $C = \{c_1, c_2, \ldots, c_{|\mathcal{C}|}\}$ [1].

Many TC methods are based on the vector space model (VSM) representation [1]. As a result, each document (d_j) is defined as a vector of weights $(\mathbf{w}_j)$ related to the terms composing the text. Thus, a document corpus containing $|\mathcal{D}|$ documents and $|\mathcal{T}|$ terms is represented by means of a *term by document matrix* $\mathbf{X}$:

$$\mathbf{X} = \begin{pmatrix} & d_1 & d_2 & d_3 & \cdots & d_{|\mathcal{D}|} \\ t_1 & w_{11} & w_{12} & w_{13} & \cdots & w_{1|\mathcal{D}|} \\ t_2 & w_{21} & w_{22} & & & \vdots \\ t_3 & \vdots & & \ddots & \ddots & \vdots \\ \vdots & \vdots & & & \ddots & \vdots \\ t_{|\mathcal{T}|} & w_{|\mathcal{T}|1} & \cdots & \cdots & \cdots & w_{|\mathcal{T}||\mathcal{D}|} \end{pmatrix} \quad (1)$$

In order to reduce the dimensionality of $\mathbf{X}$ and avoid overfitting, term space reduction techniques are applied: i) topic-neutral words such as prepositions, conjuctions, etc. are removed (*stoplisting*), ii) different morphological inflections of a word are merged into a same form (*stemming*) and iii) terms that occur in few documents are eliminated (*document frequency thresholding*) [1].

Using ICA for TC is related to the application of Latent Semantic Analysis (LSA) [13]. This term extraction technique projects the data onto an orthogonal *document* space[1] of reduced dimensionality by means of Singular Value Decomposition (SVD), extracting the K principal components from the data. This procedure is equivalent to the whitening and dimension reduction preprocessing steps that are usually applied to simplify the ICA problem [14].

---

[1] Data can be also be projected onto a *term* space by using LSA, which allows keyword identification (see [3,4]), but this is beyond the scope of this paper.

Applying an ICA algorithm on the LSA data yields a matrix $\mathbf{S}$ containing K independent components (IC):

$$\mathbf{S} = (\mathbf{s}_1\ \mathbf{s}_2\ \ldots\ \mathbf{s}_K)^T = \begin{pmatrix} s_{11} & s_{12} & \cdots & s_{1|\mathcal{D}|} \\ s_{21} & s_{22} & \cdots & s_{2|\mathcal{D}|} \\ \vdots & \vdots & \ddots & \\ s_{K1} & s_{K2} & \cdots & s_{K|\mathcal{D}|} \end{pmatrix} \quad (2)$$

Each independent component ($\mathbf{s}_k$ for $k = \{1, \ldots, K\}$) defines a cluster of documents with a common thematic category. Consequently, documents are classified under K categories.

The ICA-based text classifier comprises the separating matrix $\mathbf{W}$ and the IC of the documents of the training set, obtained by the ICA algorithm. Hence, new unseen documents (*queries*, q_j) composing the test set are classified by comparing their IC with the training IC, after projecting them onto the ICA space by means of matrix $\mathbf{W}$.

Moreover, a hierarchical organization of documents can be derived by performing a sweep of values of K (space dimensionality) in the neighbourhood of the number of thematic categories contained in the corpus ($|\mathcal{C}|$) [12].

3 Towards a Reliable ICA-Based Text Classifier

The performance of a text classifier is challenged when classifying unseen documents (i.e. during the test stage). Its reliability can be improved if *i*) the misclassification of queries that should be categorized under one of the categories contained in the training set is minimized, and *ii*) the rejection of queries belonging to none of the training categories (*out-of-domain*, OOD) is maximized. The following sections describe our proposals for addressing both issues.

3.1 A Relative Measure of Relevance

Selecting the most suitable category to a given document depends on the degree of belonging (*relevance*) of that document to such category. In the classic TC literature [1], the concept of relevance is defined by means of a function $CSV_k(d_j)$ (categorization status value) that measures the appropriateness of classifying document d_j under category c_k:

- in *hard* (fully-automated) text classification, document d_j is assigned to the category that attains a maximum $CSV_k(d_j)$.
- in *soft* (semi-automated) text categorization, a ranking of the categories —according to their CSV_k— is presented to the user.

In ICA-based TC, the categorization status value has been commonly defined as the magnitude of the independent components of each document [6] (see equation 3), often normalized by a *softmax* transformation [3,4].

$$CSV_k(d_j) = s_{kj} \quad (3)$$

Nevertheless, Kabán and Girolami [4] suggest avoiding direct comparison of IC values, due to their possible different scaling, which can produce misclassifications. Taking this idea into account, we propose a relative $CSV_k(d_j)$, which is defined in equation 4.

$$CSV_k(d_j) = F_k(s_{kj}) \qquad (4)$$

where F_k denotes the cumulative distribution function (*cdf*) of the k-th independent component. This method is insensitive to the distribution of the IC values, as it measures the *relative* relevance of document d_j to category c_k.

3.2 Evaluating the Reliability of the Classification Decisions

A reliable classifier should be able to measure the appropriateness of its decisions, notifying the user about the confidence level of each classification (in an interactive TC system) or rejecting the query (in a fully automatic TC framework). In this section, we propose a method for evaluating the reliability of the classification decisions, based on the definition of a *relevance index* vector containing the CSV_k values corresponding to document d_j (equation 5).

$$\mathbf{RI}(d_j) = [\, CSV_1(d_j) \quad CSV_2(d_j) \quad \ldots \quad CSV_K(d_j)\,]^T \qquad (5)$$

In order to reject OOD queries, it is necessary to define a *confidence region* for each category during the training phase. The confidence region of category c_k (CR_k) is defined in terms of two parameters:

- the mean value of $\mathbf{RI}(d_j)$, $\mu\,(\mathrm{CR}_k)$
- the standard deviation of $\mathbf{RI}(d_j)$, $\sigma\,(\mathrm{CR}_k)$

both computed over the space of training documents correctly classified under the k-th category.

During the test phase, each query q_j is classified under the k-th category if $\mathbf{RI}(q_j)$ is fully contained within the confidence region CR_k, i.e. a *conditional* classification is performed.

4 Experiments

The following experiments have been conducted on a collection of articles extracted from the Catalan newspaper AVUI, compiled during three periods of time in 2000, 2003 and 2004. This journalistic-style corpus is composed of 180 documents (3400 terms) divided into four thematic domains (fields): $D = \{politics$ (POL: 60 documents), *music* (MUS: 40 documents), *theatre* (TEA: 40 documents), *economy* (ECO: 40 documents)$\}$. Documents are represented in the VSM, weighted by their normalized tf×idf [1] and the document frequency threshold is experimentally fixed to 4.

The classifier is trained with the 90% of the documents of the corpus. All experiments are conducted following a 10-fold cross-validation scheme, where the training and test sets are chosen randomly. Moreover, the classifier is obtained applying the version of FastICA [14] that maximizes the skewness of the IC [4].

4.1 Comparison of Relevance Measures

This experiment compares the classic method for CSV_k computation (magnitude-based) to the proposed cdf-based technique. Table 1 presents the classification accuracy (A_c) [1] attained by both methods during the training and test phases. Both measures offer similar classification accuracies, but depending on the distribution of the IC values, misclassifications are differently scattered. The proposed cdf-based relevance measure offers a modest increase of the averaged classification accuracy during the test stage, while attaining similar results during the training phase.

Table 1. Classification accuracies obtained by the ICA-based classifier comparing the magnitude-based to the cdf-based CSV_k function during the training and test phases.

Relevance measure	A_c			
TRAINING	POL	MUS	TEA	ECO
Magnitude-based	.787	1	1	.986
cdf-based	.802	.951	1	.989
TEST	POL	MUS	TEA	ECO
Magnitude-based	.817	.900	1	.975
cdf-based	.800	1	1	.975

4.2 Performance of the Classification Reliability Control

The following experiments evaluate the performance of the proposed conditional categorization method presented in section 3. In order to test the ability of the classifier to reject OOD queries, the test set described at the beginning of section 4 was enlarged by adding documents concerning *society* (SOC: 60 documents) and *sports* (SPO: 40 documents).

Firstly, the optimal definition of the confidence regions CR_k is studied. Table 2 presents the *microaveraged* rejection accuracy[2] (A_r^μ) obtained by the proposed reliability control process for three definitions of CR_k. The best results are obtained when the upper and lower bounds of CR_k are set to $\mu(CR_k) \pm 2\sigma(CR_k)$, respectively.

Figure 1 illustrates the suitability of the optimal confidence region definition for rejecting OOD queries, as none of them (see the third and fourth rows of figure 1) is fully contained in the confidence regions of the training categories, in contrast to the rest of the queries (shown on the first and second rows of figure 1). The depicted data correspond to one arbitrarily chosen experiment.

After performing the 10-fold cross-validation experiments, the rejection accuracies for each category (using the optimal definition of the confidence regions) are presented in table 3.

---

[2] A_r^μ is defined as the ratio between the number of rejected OOD documents (*true positives*) plus the number of non-rejected classifiable documents (*true negatives*) to the total number of test documents.

Table 2. Microaveraged rejection accuracies obtained by the reliability control process for three confidence region definitions during the test process (μ and σ stand for μ (CR$_k$) and σ (CR$_k$), respectively).

CR$_k$	A_r^μ
$\mu \pm \sigma$	.463
$\mu \pm 2\sigma$	.812
$\mu \pm 3\sigma$	.645

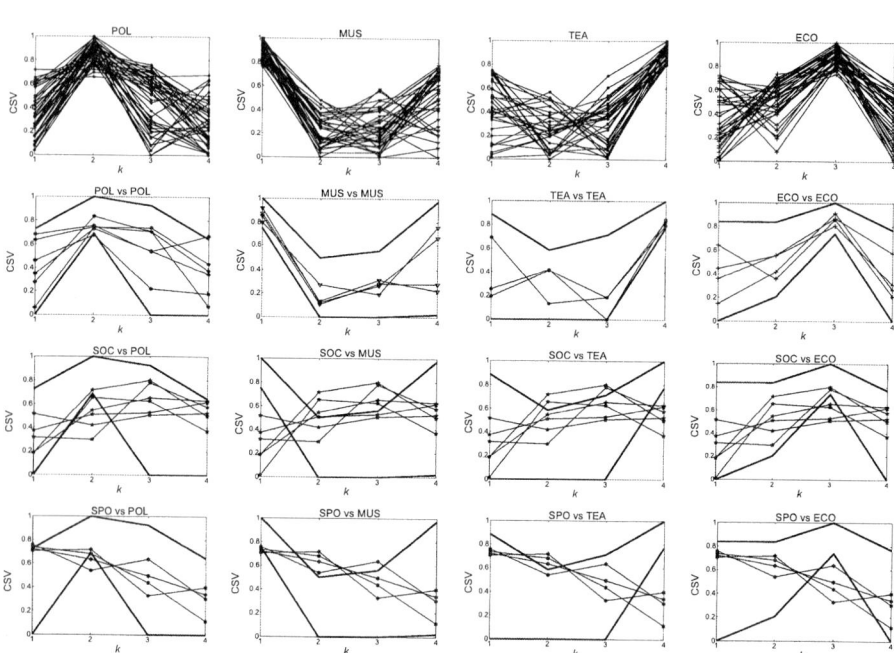

Fig. 1. The top row shows the components of the relevance index vectors $\mathbf{RI}(d_j)$ of the training documents (from left to right: *politics*, *music*, *theatre* and *economy*). The second row presents the components of $\mathbf{RI}(d_j)$ of the test documents (thin line) -from left to right: *politics*, *music*, *theatre* and *economy*- compared to the upper and lower bounds of the confidence regions of the corresponding training category (thick line). The third row presents the components of $\mathbf{RI}(d_j)$ of the *society* test documents (thin line) compared to the upper and lower bounds of the confidence regions of each training category (thick line). And finally, the bottom row presents the components of $\mathbf{RI}(d_j)$ of the *sports* test documents (thin line) compared to the upper and lower bounds of the confidence regions of each training category (thick line).

Note that the OOD queries are notably rejected (see the fifth and sixth columns of table 3). In average, only the 15% of the test documents corresponding to the training categories are rejected. Nevertheless, approximately the 50% of these rejected documents would end up in misclassifications if no reliability control was employed. Furthermore, only a 4% of the non-rejected test documents

Table 3. Rejection accuracies (A_r) for each category using optimal CR_k during the test process.

Category	A_r	Category	A_r	Category	A_r
POL	.300	TEA	.075	SOC	.850
MUS	.150	ECO	.100	SPO	.825

are misclassified. As a conclusion, a more reliable ICA-based text classifier is obtained, without a significant loss of overall classification accuracy.

5 Conclusions

In this paper we have presented a first step towards reliable ICA-based text classification. In this context, a relative CSV function has been evaluated in contrast to the magnitude-based relevance measure. Although the proposed measure is insensitive to the IC distributions, little impact on the classification accuracy is observed. In addition, a method for controlling the reliability of the classification decisions has been presented. The experiments demonstrate good rejection accuracies of out-of-domain queries, without a significant loss of performance.

Further studies will be focused on evaluating the performance of the proposed relevance measure on other text corpora and improving the reliability control process. Moreover, the presented reliability postprocessing can provide useful information in the context of hierarchical text classifiers (e.g. for choosing the most appropriate level of hierarchy) and dynamic text classification systems (e.g. for discovering new categories).

References

1. Sebastiani, F.: Machine Learning in Automated Text Categorisation. ACM Computing Surveys, Vol. 34, Nr. 1 (2002) 1–47
2. Isbell, C.-L. and Viola, P.: Restructuring Sparse High Dimensional Data for Effective Retrieval. Adv. in Neural Information Processing Systems, 11 (1999) 480–486
3. Kolenda, T., Hansen, L.K. and Sigurdsson, S.: Independent Components in Text. In: Girolami, M. (ed.): Advances in Independent Component Analysis. Springer-Verlag, Berlin Heidelberg New York (2000) 241–262
4. Kabán, A. and Girolami, M.: Unsupervised Topic Separation and Keyword Identification in Document Collections: a Projection Approach. Dept. of Computing and Information Systems, University of Paisley. Technical Report Nr. 10 (2000)
5. Kim, Y.-H. and Zhang, B.-T.: Document Indexing Using Independent Topic Extraction. Proc. of the 3rd International Conference on Independent Component Analysis and Blind Signal Separation (ICA'2001). San Diego, USA (2001) 557–562
6. Kolenda, T.: Clustering text using Independent Component Analysis. Inst. of Informatics and Mathematical Modelling, Tech. University of Denmark. T.R.(2002)
7. Kolenda, T. and Hansen, L. K.: Dynamical Components of Chat. Inst. of Informatics and Mathematical Modelling, Tech. University of Denmark. T.R.(2000)

8. Bingham, E.: Topic Identification in Dynamical Text by Extracting Minimum Complexity Time Components. Proc. of the 3rd International Conference on Independent Component Analysis and Blind Signal Separation (ICA'2001). San Diego, USA (2001) 546–551
9. Bressan, M. and Vitrià, J.: Improving Naive Bayes Using Class Conditional ICA. In Garijo, F., Riquelme, J. and Toro, M. (Eds.), Advances in Artificial Intelligence-Iberamia, Springer Verlag Series. (2002) LNAI 2527:1–10
10. Takamura, H. and Matsumoto, Y.: Feature Space Restructuring for SVMs with Application to Text Categorization. Proc. of the Conference on Empirical Methods in Natural Language Processing (EMNLP'2001). Pittsburgh, USA (2001) 51–57
11. Alías, F., Iriondo, I. and Barnola, P.: Multi-domain Text Classification for Unit Selection Text-to-Speech Synthesis. Proc. of the 15th International Congress of Phonetic Sciences (ICPhS'2003). Barcelona, Spain (2003) 2341–2344
12. Sevillano, X., Alías, F. and Socoró, J.C.: ICA-Based Hierarchical Text Classification for Multi-domain Text-to-Speech Synthesis. Proc. of the 29th International Conference on Acoustics, Speech and Signal Processing (ICASSP'2004). Montreal, Quebec, Canada (2004) (to appear)
13. Deerwester, S., Dumais, S.-T., Furnas, G.-W., Landauer, T.-K. and Harshman, R.: Indexing by Latent Semantic Analysis. Journal American Society Information Science, Vol. 6, Nr. 41 (1990) 391–407
14. Hyvärinen, A., Karhunen, J. and Oja, E.: Independent Component Analysis. John Wiley and Sons, New York (2001)

Source Separation on Astrophysical Data Sets from the WMAP Satellite

Guillaume Patanchon[1], Jacques Delabrouille[2], and Jean-François Cardoso[3]

[1] University of British Columbia, Canada
[2] CNRS/PCC, UMR 7553, Paris, France and APC, Paris France
[3] CNRS/LTCI, UMR 5141, Paris, France
http://tsi.enst.fr/~cardoso

Abstract. This paper presents and discusses the application of blind source separation to astrophysical data obtained with the WMAP satellite. Blind separation permits to identify and isolate a component compatible with the Cosmic Microwave Background, and to measure both its spatial power spectrum and its emission law. Both are found to be compatible with the present concordance cosmological model. This application confirms the usefulness of ICA in cosmological applications.

1 Introduction

The detection of Cosmic Microwave Background (CMB) fluctuations and the measurement of their spatial power spectrum has been, over the past three decades, subject of intense activity in the cosmology community.

The CMB, discovered in 1965 by Penzias and Wilson, is a relic radiation emitted some 13 billion years ago, when the Universe was about 370.000 years old. This spectrum of this radiation fits extremely well the spectrum of a blackbody at a temperature of 2.726 Kelvin. However, minute spatial fluctuations can be detected at the level of a few tens of micro Kelvin. These fluctuations are understood as tracing the primordial homogeneities which gave rise to present large scale structures as galaxies and clusters of galaxies.

The importance of measuring these fluctuations to constrain cosmological scenarios describing the history and properties of our Universe is now well established. The measurement of their statistical properties (and in particular of their spatial power spectrum) permits to drastically constrain the cosmological parameters describing the matter content, the geometry, and the evolution of our Universe [Jungman et al. 1996]. The accuracy required for precision tests of the cosmological scenarios, however, is such that it is necessary to disentangle in the data the contribution of several distinct astrophysical sources, all of which emit radiation in the frequency range used for CMB observations [Bouchet & Gispert 1999]. In addition, all measurements being noisy, it is necessary to identify, characterize, and remove as well as possible contributions due to instrumental noise.

CMB emission peaks in the millimeter-wave domain of the electromagnetic spectrum. In this wavelength range however, several astrophysical sources of radiation contribute to the total emission. Some of them are quite well known (or modeled), and others not so well. It is of utmost importance then, for the interpretation of the observations, to separate the various emissions. This can be achieved (to some extent) by the joint processing of various observations performed at various wavelengths.

To first order, the observation at a given wavelength λ can be modeled as a linear superposition of a number of sources: if J frequency channels are used to form J sky maps, the intensity $X_j(\theta, \phi)$ of the radiation at wavelength λ_j coming from direction (θ, ϕ) is used to form a $J \times 1$ vector $\{\mathbf{X}(\theta, \phi)\}$ which is modeled as an instantaneous mixture of K independent sources contaminated by independent noise $\mathbf{N}(\theta, \phi)$:

$$\mathbf{X}(\theta, \phi) = \mathbf{AS}(\theta, \phi) + \mathbf{N}(\theta, \phi) \qquad (1)$$

where $\mathbf{A}$ is an *a priori* unknown $J \times K$ matrix.

However, a strong prediction of the standard Big Bang model of cosmology is that the spectral emission law of CMB anisotropies is given, to first order, by the derivative (with respect to temperature) of the blackbody law, taken at the temperature of the CMB. This assumption completely determines the column of $\mathbf{A}$ corresponding to the CMB component *provided the instrument is perfectly calibrated*. Blind component separation, which permits to estimate the emission law of sky components, provides a unique tool for checking this prediction/assumption (blackbody law and perfect calibration).

In this paper, recently acquired WMAP data (see next section) are processed using a method for the blind separation of noisy mixtures.

2 Data from the Wilkinson-MAP Space Probe

The WMAP space probe, launched by NASA in 2001, is a large telescope for full sky imaging at 5 different wavelengths with a diffraction-limited resolution ranging from about 0.2 degrees to 0.9 degrees.[Bennett et al. 2003].

After a one-year proprietary period, WMAP data now are freely available to the scientific community[1]. In its most usable format, the data set is a collection of 10 maps: four maps at λ =3.2 mm (94 GHz), two maps at λ =4.9 mm (61 GHz), and one at each of λ =7.3 mm, 9.1 mm, and at 13.0 mm (41, 33 and 23 GHz).

Astrophysical Emissions. In the millimeter wave band, the total sky emission can be considered as the superposition of two main classes of sources: diffuse emissions and compact (point-like) sources. Diffuse emissions are due to extended processes on the sky. For all of these processes, the observed photons are emitted by large scale distributions of emitters – gases of particles. Among these, one expects emission from the hot plasma in the early Universe (the CMB), greybody emission from cold dust particles in the galaxy (dust emission), emission from

[1] http://lambda.gsfc.nasa.gov/

relativistic electrons spiraling in the galactic magnetic field (synchrotron), free–free (Bremsstrahlung) emission from ionized gas. Compact source emission is due to distant objects (galaxies, quasars, clusters of galaxies), the angular size of which is smaller than the resolution of the instrument.

A Simple Linear Model. For each physical source of emission, the total observable intensity $I(\lambda, \theta, \phi)$ at wavelength λ, in direction (θ, ϕ), depends on the physical properties of the source (clumpiness, emission process, temperature, etc...). For most diffuse emissions, the intensity approximately factorizes as $I(\lambda, \theta, \phi) \simeq a(\lambda) s(\theta, \phi)$. This is the approximation which justifies using the instantaneous mixture model (1). It is expected to be excellent (better than 0.1%) for some components (e.g. CMB) and not so good (at the 10 % level) for other, such as dust emission.

Spherical Harmonics. Because WMAP images the whole sky, harmonic analysis becomes *spherical harmonic analysis*. One use the *spherical harmonics*: a doubly indexed set $Y_{\ell,m}(\theta, \phi)$ of functions which form an orthonormal basis on the sphere. A scalar function on the sphere $F(\theta, \phi)$ can be represented by its harmonic coefficients in a doubly indexed expansion

$$F(\theta, \phi) = \sum_{\ell=0}^{\infty} \sum_{m=-\ell}^{m=+\ell} F(\ell, m) Y_{\ell,m}(\theta, \phi) \qquad (2)$$

where, in a slight abuse of notation, the same symbol is used for the function $F(\theta, \phi)$ and for its set of harmonic coefficients $F(\ell, m)$. In the spherical harmonic domain, model (1) reads

$$\mathbf{X}(\ell, m) = \mathbf{A}\mathbf{S}(\ell, m) + \mathbf{N}(\ell, m) \qquad (3)$$

for $\ell \geq 0$ and $-\ell \leq m \leq \ell$.

The Harmonic Spectrum. For a stationary isotropic random Gaussian field $F(\theta, \phi)$ on the sphere, the $F(\ell, m)$ coefficients are uncorrelated and

$$\mathrm{E}(F(\ell, m) F(\ell', m')) = C_\ell \delta_{\ell\ell'} \delta_{mm'}.$$

The sequence $\{C_\ell\}_{l \geq 0}$ is called the "harmonic spectrum" or spatial power spectrum (of the stationary isotropic process). It is the main quantity of interest for cosmology.

3 Separation of Instantaneous Mixtures in Colored Noise

3.1 Blind Parameter Estimation via Spectral Matching

Following the method described in [Delabrouille et al. 2003], we compute sample spectral covariance matrices:

$$\widehat{\mathbf{R}}_{\mathbf{X}}(q) = \frac{1}{n_q} \sum_{(\ell, m) \in \mathcal{D}_q} \mathbf{X}(\ell, m) \mathbf{X}(\ell, m)^\dagger \quad (q = 1, \ldots, Q) \qquad (4)$$

where q labels a domain $\mathcal{D}_q$ of (ℓ, m) space and $n_q = \text{card}(\mathcal{D}_q)$. For our application, it is customary to estimate band averages of the power spectra in domains $\ell_1 < \ell \leq \ell_2$, with m taking all available values for each ℓ, i.e. $-\ell \leq m \leq \ell$.

If the observations $\mathbf{X}$ match the multi-component model of equation 3, then $\widehat{\mathbf{R}}_\mathbf{X}(q)$ is an estimator of $\mathbf{R}_\mathbf{X}(q)$:

$$\mathbf{R}_\mathbf{X}(q) = \mathbf{A}\mathbf{R}_\mathbf{S}(q)\mathbf{A}^\dagger + \mathbf{R}_\mathbf{N}(q) \quad (q = 1, \ldots, Q) \tag{5}$$

with $\mathbf{R}_\mathbf{S}(q)$ and $\mathbf{R}_\mathbf{N}(q)$ defined similarly to $\mathbf{R}_\mathbf{X}(q)$. The parameters of interest (mixing matrix and band-averaged spatial power spectra of components) are precisely $\mathbf{A}$ and $\mathbf{R}_\mathbf{S}(q)$, the latter being diagonal for uncorrelated components. Noise spatial power spectra, being of no particular use for astrophysics, are nuisance parameters which must be estimated, so that the full set of parameters of the data is $\theta = \{\mathbf{A}, \mathbf{R}_\mathbf{S}, \mathbf{R}_\mathbf{N}\}$.

These parameters are estimated as by maximizing the (Whittle approximation to the) likelihood. This boils down to minimizing

$$\phi(\theta) = \sum_{q=1}^{Q} n_q\, D\left(\widehat{\mathbf{R}}_\mathbf{X}(q), \mathbf{R}_\mathbf{X}(q; \theta)\right) \tag{6}$$

where $D(\cdot, \cdot)$, the Kullback-Leibler divergence, is a measure of divergence between two positive $n \times n$ matrices, defined by

$$D(\mathbf{R}_1, \mathbf{R}_2) = \frac{1}{2}\left[\text{tr}\left(\mathbf{R}_1\mathbf{R}_2^{-1}\right) - \log\det(\mathbf{R}_1\mathbf{R}_2^{-1}) - n\right] \tag{7}$$

Optimization is achieved with dedicated algorithms based on the EM method, followed by few steps of a descent method (BFGS algorithm) to speed-up the convergence [Cardoso & Pham].

3.2 Estimation of Component Maps

Once the parameters have been estimated, component maps are reconstructed using Wiener filter in the spherical harmonic domain:

$$\widehat{\mathbf{S}}(\ell, m) = [\widehat{\mathbf{A}}^t \widehat{\mathbf{R}}_\mathbf{N}(q)^{-1}\widehat{\mathbf{A}} + \widehat{\mathbf{R}}_\mathbf{S}(q)^{-1}]^{-1}\widehat{\mathbf{A}}^t\widehat{\mathbf{R}}_\mathbf{N}(q)^{-1}\mathbf{X}(\ell, m) \tag{8}$$

where $\widehat{\mathbf{R}}_\mathbf{S}(q)$ and $\widehat{\mathbf{R}}_\mathbf{S}(q)$ are the (diagonal) matrices from the parameter vector $\widehat{\theta}$ resulting from the minimization of (6). It is understood that the value of q used for computing $\widehat{\mathbf{S}}(\ell, m)$ in (8) is such that $(\ell, m) \in \mathcal{D}_q$.

3.3 Dealing with Real Data

In this ICA analysis of CMB, we focus on checking whether WMAP data contains a component which has both a CMB power spectrum and the expected emission law for temperature fluctuations of the cosmological blackbody. This is not as easy as it may sound since actual astrophysical emissions have a more complex structure than the simple model of equation 3 can capture:

- Known *galactic* emissions (dust, synchrotron, free-free) are positive and concentrated towards the galactic plane;
- In addition, galactic component emissions may be correlated for physical reasons (thus violating the independence assumption);
- The emission laws (the elements of the mixing matrix) of the galactic components are known to depend slightly on position and on scale;
- Point sources are not modeled satisfactorily by a linear mixture, and their distribution is poorly represented by the Whittle approximation;
- Instrumental noise is not stationary over the sky;
- Detectors have different spatial resolutions (depending on λ, to start with).

For the present analysis, we deal with these complications in the simplest possible way.

The impact of the first four items in the list is drastically reduced by cutting out (masking) from the analysis those regions of the sky which are most contaminated by galactic emission. Also, we cut out from the analysis small regions around bright point sources at known locations. This mask permits to isolate a region (the major fraction of the sky) where the emission is thought to be dominated by CMB fluctuations.

We ignore noise non stationarity. Irregularities are smooth enough on the sky to induce only small (and small range) correlations in harmonic space.

Detector-dependent spatial resolution can be easily accounted for (to first order) in Fourier space by including a multiplicative term B_ℓ which describes the spectral shape of the point spread function.

We restrict the analysis to multipole ℓ values between 10 and 400, which encompasses all of the first acoustic peak observed by WMAP using the highest frequency channels only. The upper limit permits to minimize the impact of a background of residual unresolved point sources and of uncertainties in the modeling of the beam of the instrument.

Using an incomplete sky, as any windowing, does affect spectral estimation. The spectra estimated on the cut sky have to be corrected to take into account the impact of this "window function". In our results, the Master approach [Hivon et al. 2002] of Hivon *et al.* is applied to correct spectral estimates from partial sky coverage effects.

4 Results

The application of SMICA to WMAP data gives the following results:

- Two components are clearly identified: one is the CMB, and the second emission from the high latitude galaxy.
- A third component is marginally detected, which may correspond to large scale variations of galactic emission law,
- The identified CMB component has an emission law compatible with the expected derivative of a blackbody, to excellent accuracy. Its power spectrum displays a peak around $\ell = 200$ (first acoustic peak) compatible with the measurement announced by the WMAP team (figure 1).

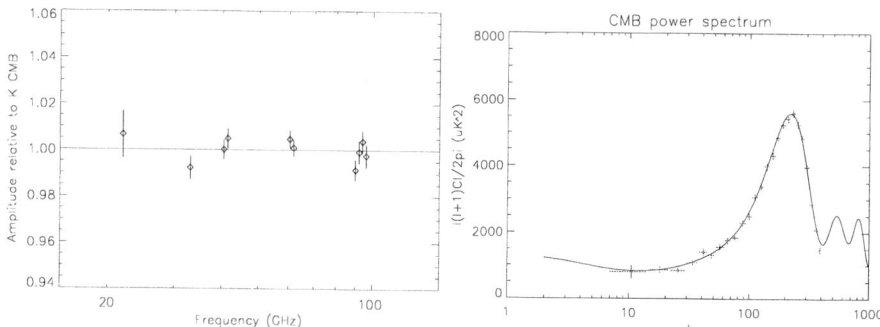

Fig. 1. Left: Measured CMB fluctuation mixing matrix, compared to theoretical expectations. Data points in the same frequency channel have been slightly offset in frequency for readability. This figure shows that the first component identified blindly in the WMAP data, and seen consistently in all WMAP channels, has a CMB fluctuation emission law to within less than one per cent (error bars here exclude WMAP calibration errors). Right: Measured CMB spatial power spectrum, compared to the theoretical model matching best WMAP team measurement (solid black line). This figure shows that the first component identified blindly in the WMAP data, and seen consistently in all WMAP channels, has a spatial power spectrum consistent with what has been measured by WMAP using only the two highest frequency channels.

- The galactic component has an emission law compatible with synchrotron emission in all WMAP channels (proportional to $\nu^{-2.7}$) except those at 90 GHz, for which a significant excess is seen. This excess is interpreted as due to galactic dust emission correlated with the synchrotron (either for physical reasons or – more likely – by happenstance because of concentration of galactic emissions at low galactic latitude).
- Component maps are reconstructed using a Wiener filter, and are displayed in figure 2.

Error bars for all quantities are computed from the Hessian at the point of convergence, with a correction factor $1/\sqrt{f_{sky}}$ accounting for part–sky coverage.

5 Conclusion

This work is a follow up to preliminary presentation of SMICA at the 2003 ICA workshop. Improvements relative to previous work include: full-sky processing (use of spherical harmonics), estimation of noise of arbitrary spectral shape and beam correction.

The SMICA blind component separation has been applied successfully to WMAP data. The analysis is performed on ℓ varying from about 10 to 400, which encompasses all of the first acoustic peak of the CMB. We find that all 10 WMAP maps comprise a common astrophysical component identified here as CMB. The emission law of this first component is compatible with CMB

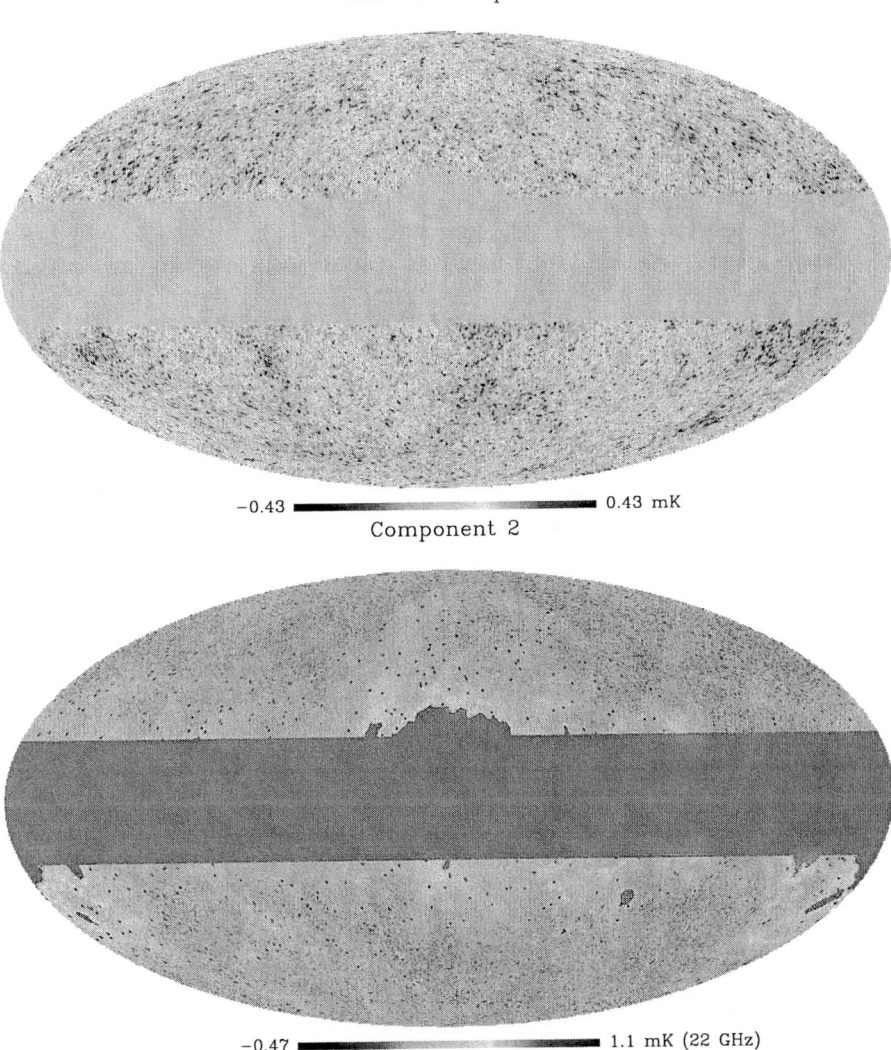

Fig. 2. Top: The map of CMB fluctuations obtained after Wiener filtering of the data using parameters estimated with SMICA. Bottom: The map of the second component (galactic emission) obtained in the same way.

fluctuations to within about one per cent, and its spatial power spectrum in the range $10 \leq \ell \leq 400$ compatible with the measurement obtained by the WMAP team. The second component has an emission law proportional to $\nu^{-2.7}$ except in the W (90 GHz) channel for which a significant excess is observed. This excess is thought to be due to dust emission contribution to the component. These results demonstrate the usefulness of blind component separation for astrophysical data analysis.

References

[Jungman et al. 1996] Jungman, G., et al., 1996: Cosmological-parameter determination with microwave background maps. Phys. Rev. D **54** (1996) 1332

[Bouchet & Gispert 1999] Bouchet, F.R., Gispert, R.: Foregrounds and CMB experiments I. Semi-analytical estimates of contamination. New Astronomy **4** (1999) 443

[Delabrouille et al. 2003] Delabrouille, J., Cardoso, J.-F., Patanchon, G.: Multi-Detector Multi-Component spectral matching and applications for CMB data analysis. Monthly Notices of the RAS **346** (2003) 1089

[Cardoso & Pham] Cardoso, J.-F, Pham, D.T.: Optimisation issues in noisy gaussian ICA. Proceedings of ICA 2004.

[Hivon et al. 2002] Hivon, E. et al. MASTER of the Cosmic Microwave Background Anisotropy Power Spectrum: A Fast Method for Statistical Analysis of Large and Complex Cosmic Microwave Background Data Sets. The Astrophysical Journal **567** (2002) 2–17

[Bennett et al. 2003] Bennett, C. L. et al. First-Year Wilkinson Microwave Anisotropy Probe (WMAP) Observations: Preliminary Maps and Basic Results. The Astrophysical Journal Suppl. **148** (2003) 1–27

Multidimensional ICA for the Separation of Atrial and Ventricular Activities from Single Lead ECGs in Paroxysmal Atrial Fibrillation Episodes

Francisco Castells[1], Cibeles Mora[2], José Millet[2], José Joaquín Rieta[1], César Sánchez[3], and Juan Manuel Sanchis[1]

[1] Universidad Politécnica de Valencia
46730 Gandia, Spain
{fcastells,jjrieta,jmsanch}@eln.upv.es

[2] Universidad Politécnica de Valencia
46022 Valencia, Spain
jmillet@eln.upv.es, cimomo@doctor.upv.es

[3] Universidad de Castilla la Mancha, Cuenca, Spain
cesar.sanchez@uclm.es

Abstract. The analysis of paroxysmal atrial fibrillation requires the previous estimation of the atrial activity (AA) from one lead ECG. Considering the statistical properties of the cardiac electrical activities, it follows that both AA and ventricular activity (VA) present a high redundancy degree at different time intervals, whereas AA keeps independent from VA. This contribution adopts a multidimensional independent component analysis (MICA) formulation in order to find a set of components that minimises the mutual information existing in the ECG signal at different intervals. The independent components can be grouped in VA and AA subspaces, what enables the reconstruction of the AA at each observation point from the AA subspace. The proposed approach is validated with a significant database composed of simulated and real AF recordings.

1 Introduction

Signal processing techniques have been widely employed in biomedical applications. In particular, the analysis of electrocardiograms (ECG) has provided important advances in the understanding, characterisation and diagnosis of cardiac arrhythmias. One of them is atrial fibrillation (AF), which consists of a malfunction of the atrium characterised by a modification of the normal atrial activity (AA) pattern on the ECG [1].

The proper characterisation of AF from non-invasive techniques (e.g. the ECG) requires the analysis of the atrial fibrillatory signal. However, the signal recorded at the surface skin level is a mixture of the ventricular activity (VA) and AA, and a previous step that cancels the VA, i.e. the QRS complex and the T wave, is essential [2]. The problem of the AA estimation from the multilead ECG has been recently modelled as a blind source separation problem (BSS), which is able to extract separately VA and AA [3].

Multichannel signal processing methods are applicable in this context to persistent AF [3][4], where the signals are usually recorded at an electrophysiology lab using the 12-lead standard ECG. However, in the case of early stages of AF, i.e. paroxysmal AF, where the arrhythmia starts and terminates spontaneously, these techniques are no longer valid, since the signals are recorded by means of an ambulatory system (i.e. holter) which usually stores the biological signals provided by no more than two or three electrodes. Such a reduced number of available leads is insufficient for independent component analysis (ICA) techniques to achieve the separation of VA and AA satisfactorily, since the ECG also contain additional components due to muscular movements, thermal noise, mains interference and other nuisances.

Motivated by the observation that VA and AA present specific patterns which are highly correlated in different time intervals, we propose to utilize this temporal redundancy to separate VA and AA from a single lead. Following this consideration, different segments of the ECG signal can be regarded as several observations with a high degree of mutual information. The main goal of this contribution is to model the separation of VA and AA from a single lead as a BSS problem. The proposed methodology will be validated with a significant database composed of simulated AF ECGs and paroxysmal AF episodes obtained from holter recordings.

2 Physiological Considerations

Atrial fibrillation is a cardiac arrhythmia in which normal atrial electrical activation is substituted by continuous activation, with multiple wavelets depolarising the atria simultaneously. On the ECG, normal atrial activity (P wave) is no longer visible, being substituted by rapid oscillations or fibrillatory waves characterised by a certain cycle length within the range 120-250ms depending on the patient [5]. Fig. 1 shows an example of normal sinus rhythm (NSR) and AF.

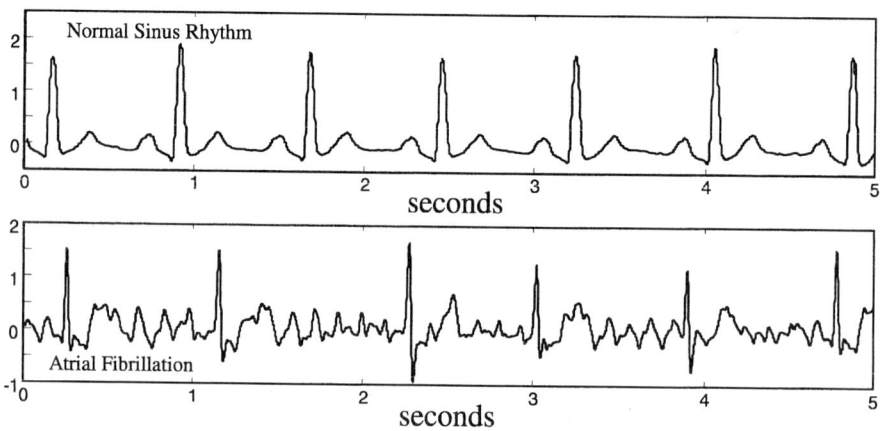

Fig. 1. Examples of normal sinus rhythm (NSR) and atrial fibrillation (AF) episodes.

VA is characterised by the QRS complex and T wave, which are due to the ventricular depolarisation and repolarisation respectively. Successive QRST waves are consequence of the same bioelectrical activity, and hence the corresponding waveforms present a high degree of redundancy [2][4][5]. The ventricular response in AF depends on electrophysiological properties of the atrioventricular node, and the R-R interval becomes more irregular than in normal sinus rhythm (NSR). Fig. 2 shows an example of VA and its beat waveform.

AA consists of continuous wavelets whose spectrum presents a main frequency peak (typically around 6Hz) [2][6]. Therefore, its autocorrelation function exhibits significant values at non-zero lags, being equal to null only at the following time lags:

$$\tau_n = \frac{1+2k}{4}T,$$

being T the cycle length and k an integer index. Fig. 2 also illustrates an example of AA and its corresponding spectrum.

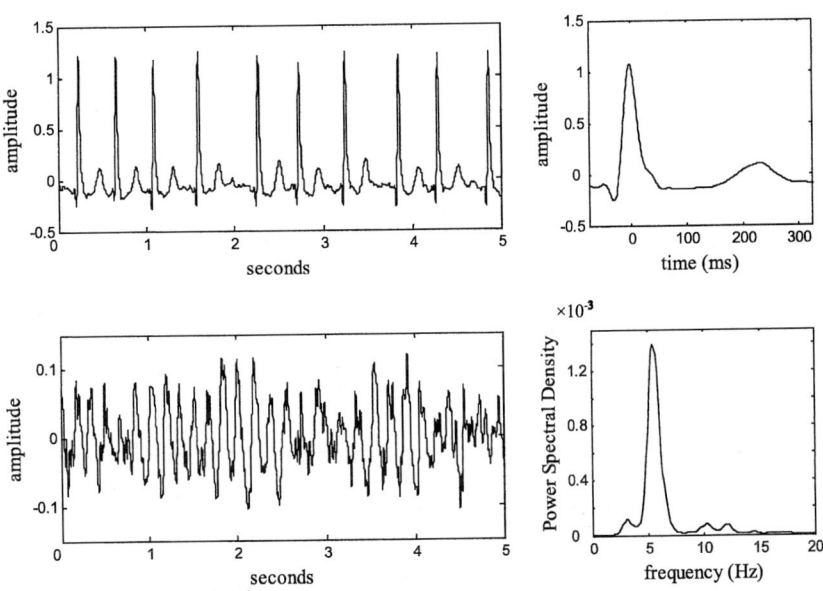

Fig. 2. Example of VA and its corresponding beat waveform and example of AA and its corresponding spectrum, with a frequency peak at 5.5Hz.

3 Methods

Motivated by the observation that both VA and AA present certain time dependence, whereas both activities remain uncoupled, we propose to extract different components corresponding to different biological activities by minimising the mutual information contained in the ECG at different cardiac beats.

The QRST waveforms can be obtained from the ECG by using an R-detector that identifies the position of the cardiac beats [7] and employing an n-length window that covers the whole Q-T interval. Considering a total number of m cardiac beats, the observations can be rewritten as an m-length vector $\mathbf{x}(t)$ which is indeed a combination of the independent components $\mathbf{s}(t)$:

$$\mathbf{x}(t) = \mathbf{A}\mathbf{s}(t),$$

where $\mathbf{A}$ is the mixing matrix. The problem of recovering the independent components follows the basis of a BSS model [8][9]. These components mainly consist of an important ventricular component related to the QRST wave $s_{VA}(t)$, several components related to the AA subspace $\mathbf{s}_{AA}(t)$ and other nuisance sources that conform the noise subspace $\mathbf{s}_n(t)$. In those cases where there is more than one unique shape for the QRST wave, those waveforms are considered as new independent components related to the VA subspace $\mathbf{s}_{VA}(t)$. The mixing matrix can be decomposed in three matrices of sizes $m \times m_{VA}, m \times m_{AA}$ and $m \times m_n$, such that $\mathbf{A}=[\mathbf{A}_{VA}, \mathbf{A}_{AA}, \mathbf{A}_n]$ is full column rank.

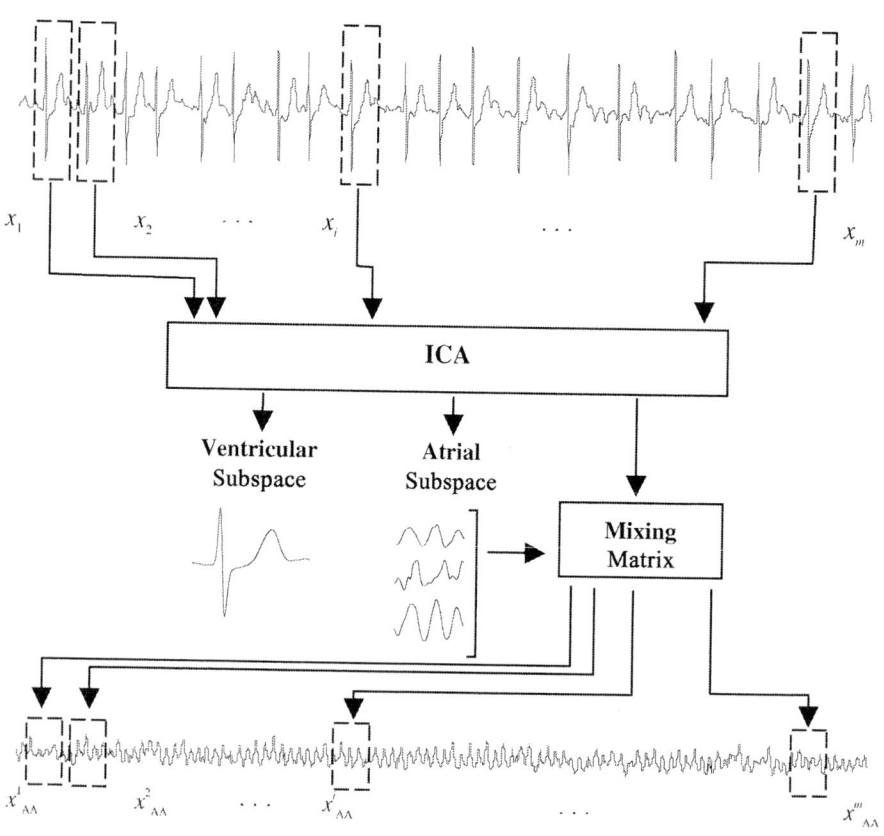

Fig. 3. Illustration of the multidimensional ICA approach for the estimation of the AA source.

Hence, by identifying the AA subspace from the whole set of independent components, the AA at each observation interval $x_{AA}^i(t)$ can be recomposed using the mixing submatrix $\mathbf{A}_{AA}$. This conception is closely related to the more generalised notion of multidimensional independent component analysis (MICA) [10] that aims to reconstruct the contribution of each independent subspace into the observation signals. The estimated AA wave in the ECG can be finally obtained by mapping back the AA content in the observations $\mathbf{x}_{AA}(t)$ to the corresponding time intervals. Fig. 3 illustrates the methodology employed for the estimation of the AA source.

The identification of the ventricular and atrial subspaces can be carried out automatically since the ventricular components are those that contribute with more power to the ECG signal. Therefore, the independent components can be reordered in a similar way as principal component analysis (PCA) [11], which establishes an arbitrary order according to the variance degree of the uncorrelated components.

4 Database

The fact that the AA is unknown in real recordings hinders an in-depth performance evaluation of the proposed methodology. Hence, suitable simulated AF ECGs with known AA content must be designed, which allows us to compare the estimated and the original AA. Ultimately the method is to be applied over actual AF episodes, and thus a database of such recordings is also employed to demonstrate the suitability of the algorithm in real scenarios.

The first database is composed of 10 simulated AF ECGs, which have been generated by adding VA and AA extracted from real AF patients. All recordings are 30 seconds in length and digitised at a sampling rate of 1 KHz and an amplitude resolution of 16 bits. The second database consists of 10 paroxysmal AF recordings obtained from Holter systems, and hence, digitised at a sampling rate of 250Hz and an amplitude resolution of 12 bits. All recordings were resampled at 1 KHz.

5 Results

Firstly, the proposed methodology was applied to the database of simulated AF recordings. In all cases, it was possible to remove the QRS complex and the T-wave. The estimated AA was compared with the original AA in terms of Pearson correlation indices, being of 0.774±0.106 in average. The performance measurement obtained for each patient is detailed in Table 1.

Table 1. Correlation values of the estimated AA and original AA for the simulated AF ECGs.

P1	P2	P3	P4	P5	P6	P7	P8	P9	P10
0.717	0.744	0.699	0.854	0.888	0.762	0.906	0.861	0.879	0.697

Taking into account that the AA presents much less amplitude than the VA, these results indicate that the proposed approach is able to estimate the AA component free from QRST residua. To illustrate the quality of the AA extraction, the estimated AA corresponding to patient 6 is compared with the original AA in fig. 4.

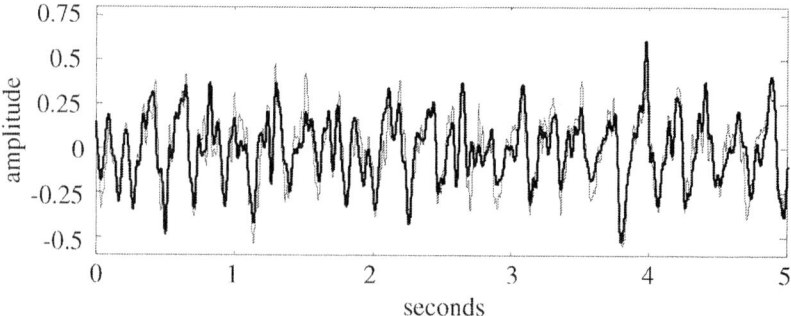

Fig. 4. Comparison of the estimated AA (solid line) with the real AA (dashed line) in a typical case (patient 6). Notice that the QRST complex has been successfully cancelled.

Secondly, the same methodology was applied to the database of recordings that have been obtained from patients that suffer from paroxysmal AF, which is the final aim of the method. The number of beats for each ECG varied from 36 to 49 depending on heart rate of the patient. The number of beats available determines the number of observations. Therefore, the length of the ECG signal has a direct impact in the performance of the AA extraction. From our experiments, we suggest that the ECG length should be between 30 and 60 seconds. In all patients the QRST complex was successfully cancelled. In 6 out of 10 patients the VA subspace was conformed just by one component due to the regularity of the QRST waveform. In the remaining cases two or three VA components were identified. The number of components that corresponded to the AA subspace varied from 4 to 10. The rest of components scarcely contributed to the ECG signal and could be considered as the nuisance subspace. The fact that so few components appear in VA and AA subspaces confirms the initial assumption that VA and AA were highly dependent at different intervals.

The parameter of the AA signal that has showed major clinical importance has been the detection of the main frequency peak f_p, and the presence or absence of harmonics, which includes some information regarding the organization of the electrical depolarization/repolarization of the atria. In all cases, the main frequency could be detected, which is specified in Table 2 for each patient.

Table 2. Main frequency of the AA signal for each AF patient.

	P1	P2	P3	P4	P5	P6	P7	P8	P9	P10
f_p (Hz)	5.7	4.8	4.7	3.2	6.9	6.8	3.8	4.6	5.0	5.4

The ECG signal of patient 1 and the corresponding AA estimation is represented in fig. 5. Notice that the VA has been completely cancelled, whereas the AA has been preserved.

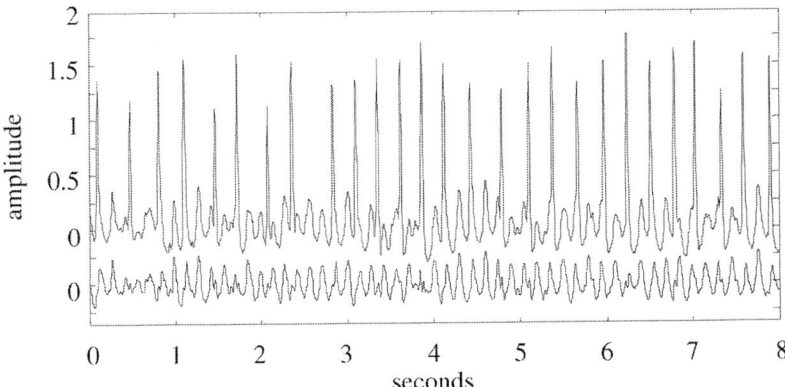

Fig. 5. AA estimation from a patient that suffered from paroxysmal AF (below). The corresponding ECG signal has been plotted above in order to facilitate visual comparison.

6 Conclusions

The estimation of AA in paroxysmal AF episodes requires the implementation of QRST cancellation techniques for only one lead. Therefore, existing AA estimation techniques for multilead ECGs are suitable for persistent AF arrhythmias, but are no longer valid in the case of paroxysmal AF. However, the fact that electrical cardiac activities show a high degree of dependency at different time intervals allows us to take profit from this information.

ICA techniques have the capability of minimising the mutual information in a given set of observations. By means of a multidimensional ICA approach, it is possible to transform the initial problem of separating VA and AA components from one single lead into a new formulation that consists of separating independent components from multiple observations. In addition, the number of available observations does not depend on the number of sensors or electrodes, but on the temporal duration of the signal. This property allows the technical simplification of portable devices, since it is required just two electrodes (one electrode is always employed as a reference). In addition, size, weight and battery consumption can be optimised.

With this solution, it is given one step ahead in the characterisation and treatment of the most frequent cardiac arrhythmia in its earlier stages. The correct estimation of the AA is highly important in clinical practice, and has a direct impact on the decision of the most suitable treatment strategy.

Acknowledgements

The authors would like to acknowledge the helpful support received from Servicio de Hemodinámica of the Hospital Clínico Universitario the Valencia, and specially from Ricardo Ruiz, Salvador Morell and Roberto García Civera, for providing signals and for the high quality of their clinical advice. This study has been partly funded by TIC2002-00957 and the Universidad Politécnica de Valencia (UPV).

References

1. V.Fuster, L. Ryden, et al., ACC/AHA/ESC guidelines for the management of patients with atrial fibrillation, *Journal of the American College of Cardiology 38* (2001) 1231-1265.
2. A. Bollmann, K. Sonne, H.D. Esperer, I. Toepffer, J.J. Langberg, H.U. Klein. Non-invasive assessment of fibrillatory activity in patients with paroxysmal and persistent atrial fibrillation using the Holter ECG. Cardiovasc Res. 1999;44:60-6.
3. J.J. Rieta, F. Castells, C. Sanchez, V. Zarzoso, J. Millet, "Atrial activity extraction for atrial fibrillation analysis using blind source separation", *IEEE Trans. Biomed. Eng.*, Vol. 51, 2004, pp. 1176-86.
4. M. Stridh, L. Sörnmo, Spatiotemporal QRST Cancellation Techniques for Analysis of Atrial Fibrillation, *IEEE Trans. Biomed. Eng.* 48 (2001), 105-111.
5. M. Allessie, K.Konings, M.Wijffels, *Atrial Arrhythmias – State of the Art: Electrophysiological Mechanism of Atrial Fibrillation*, J.P. DiMarco and E.N. Prystowsky, Eds. Arrnonk, NY: Futura Publ. Co., 1995.
6. M. Stridh, Signal Characterization of Atrial Arrhythmias using the Surface ECG. PhD. dissertation, Lund University (Sweden), 2003
7. J. Pan, WJ. Tompkins. "A real-time QRS detection algorithm". *IEEE Trans. Biomed. Eng.* Vol. 32, pp. 230-236, 1985
8. P.Comon, Independent Component Analysis – a new concept?, *Signal Processing 36* (1994) 287-314.
9. A.Hyvärinen, J.Karhunen, E.Oja., *Independent Component Analysis*, John Willey & Sons, Inc., Ed. 2001.
10. J.F. Cardoso (1998). Multidimensional independent componentanalysis Proc. ICASSP '98. Seattle.
11. I.T. Joliffe, *Principal Component Analysis*, Springer Verlag, 2002.

Music Indexing Using Independent Component Analysis with Pseudo-generated Sources

E.S. Gopi, R. Lakshmi, N. Ramya, and S.M. Shereen Farzana

Sri Venkateswara College of Engineering, Pennalur, Sriperumbudur - 602 105
esgopi@svce.ac.in

Abstract. In this paper we present a new approach towards Singing Voice/ Music segmentation using Independent Component Analysis. If the singing voice and the background music are assumed to be two independent signals mixed to form the song, Independent Component Analysis can be used to separate them. ICA requires at least two sources in order to separate two mixed signals, whereas in this case only a single source, i.e. the recording of the song, is available. Another pseudo source is generated from the single source using Discrete Wavelet Transform and the discrimination between singing voice and music is done using a Feed Forward Back Propagation Neural Network.

1 Introduction

Methods to accurately separate the segments of a song containing singing voice from those containing only background music are in great demand as preprocessing tools to Automatic Lyric Transcription and Automatic Singer Identification. Lyric Transcription and Singer Identification need be applied only to the singing voice and hence removal of music segments will greatly improve the performance of both.

Currently, most methods used for singing voice/ music discrimination make use of speech recognizers, exploiting the resemblance between singing voice and speech. But speech recognizers are language dependent and hence there is a need for a method which is language independent and hence universally applicable.

A solution to this problem can be obtained by utilizing Independent Component Analysis to separate the singing voice from the music. The singing voice and background instrumental music are considered as two independent signals mixed in an unknown proportion in the song. For all songs only one source of the mixture is available, i.e. the studio recording is available. ICA requires n sources to separate n mixed signals and hence another source of the song must be obtained before ICA can be applied.

As the frequency range of singing voice and background music is non-overlapping for most frequencies, Wavelet Transform can be used to decompose the song into its various frequency components. By giving a different gain to different frequency components and then reconstructing the signal, several signals can be obtained, which can be used as pseudo sources.

The output of the ICA is then processed and fed to a Feed Forward Back Propagation Neural Network which makes the final decision as to whether the segment contains singing voice or not. The block diagram shown in the figure below gives the flow of our method.

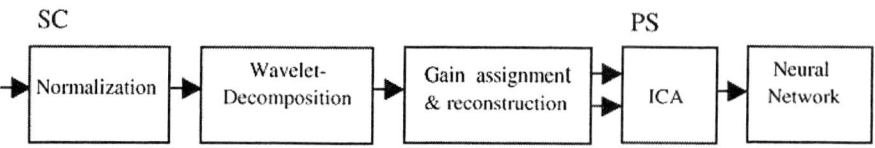

Fig. 1. Block diagram showing flow of project, SC – song clippings, PS – Pseudo sources.

2 Preprocessing Tasks

Initially the data is discretized by sampling at a rate of 44100 samples/second. The distribution of values of the samples of the song is very wide. In order to achieve uniformity in the processing of the data, the distribution of the values must be made uniform. Hence preprocessing must be done on all the data.

2.1 Mean Variance Normalization

In order to obtain a more or less uniform distribution of values, the segments are Mean Variance Normalized. The normalized segments have uniform mean and variance. The normalization of the data is done as follows,

$$\Delta = [\sigma_d *(X(i) - \mu_d) / \sigma], \quad (1)$$
$$X_n(i) = \mu_d + \Delta \quad \text{if } X(i) > \mu , \quad (2)$$
$$X_n(i) = \mu_d - \Delta \quad \text{if } X(i) < \mu . \quad (3)$$

In the above equations σ_d and μ_d are the desired standard deviation and desired mean respectively. σ and μ are the standard deviation and mean of the segment being normalized. $X(i)$ and $X_n(i)$ are the original song segment and the normalized song segment.

3 Generation of Pseudo Sources

After preprocessing the song data, pseudo sources must be generated before ICA is applied. The generation of pseudo sources is achieved through the Discrete Wavelet Transform.

3.1 The Discrete Wavelet Transform (DWT)

Wavelet transform is capable of providing the time and frequency information simultaneously, hence giving a time-frequency representation of the signal. The DWT analyzes the signal at different frequency bands with different resolutions by decomposing the signal into a coarse approximation and detail information. DWT employs two sets of functions, called scaling functions and wavelet functions, which are associated with low pass and high pass filters, respectively. The decomposition of the signal into different frequency bands is simply obtained by successive high pass and low pass filtering of the time domain signal. The original signal x[n] is first passed through a half band high pass filter g[n] and a low pass filter h[n]. A single stage of decomposition is mathematically represented as follows,

$$y_{high}[k] = \Sigma_n x[n] \cdot g[2k-n], \quad (4)$$

$$y_{low}[k] = \Sigma_n x[n] \cdot h[2k-n]. \quad (5)$$

Here $y_{high}[k]$ and $y_{low}[k]$ are the outputs of the high pass and low pass filters. The above procedure can be repeated for further decomposition.

The DWT of the original signal is then obtained by concatenating all coefficients starting from the last level of decomposition.

The frequencies that are most prominent in the original signal will appear as high amplitudes and those that are not so prominent will appear as low amplitudes in that region of the DWT signal that includes those particular frequencies. Thus by scaling the DWT coefficients the prominence of the frequencies in the reconstructed signal can be altered.

For the reconstruction, the signals at every level are upsampled by two, passed through the synthesis filters g'[n], and h'[n] (high pass and low pass, respectively), and then added. As the analysis and synthesis filters are identical to each other, except for a time reversal, the reconstruction formula is

$$x[n] = \Sigma_k (y_{high}[k] \cdot g[-n +2k]) + (y_{low}[k] \cdot h[-n +2k]). \quad (6)$$

Hence by scaling the DWT coefficients in an appropriate manner, different frequencies can be suppressed or enhanced in the reconstructed signal.

The given song data is wavelet decomposed to ten levels giving ten detail coefficients (D1, D2, D3, D4, D5, D6, D7, D8, D9, D10) with D1 having high frequency and D10 the lowest frequency. There is one approximation coefficient A10.

3.2 Gain Assignment

As signals with different frequency components can be obtained using DWT on the original signal, the assignment of gains or scaling factors to the DWT coefficients used for reconstructing these signals is very important. Three methods have been designed for gain assignment:

- Constant gain assignment
- Dynamic gain assignment
- Gain assignment based on harmonicity

Constant Gain Assignment. In this method a constant set of gains is assigned to the DWT coefficients. For one pseudo source the gains are assigned such that the highest frequency coefficients have highest gain and the lowest frequency coefficients have the lowest gains. For the other pseudo source the gains are assigned in reverse order such that the highest frequency coefficients have lowest gain and lowest frequency coefficients have highest gain.

Dynamic Gain Assignment. In this method the gain assignment is varied based on the frequency content of the signals in the song.

A two second segment of the song is first wavelet decomposed into ten levels. The energy content of each of the ten detail coefficients and the approximation coefficient was determined. The coefficients were then sorted in the ascending order of energy. The original song was then wavelet decomposed into ten levels and the gains were assigned in the sorted order obtained earlier. For the first pseudo source the coefficient with the highest energy was assigned the highest gain and the coefficient with the lowest energy was assigned the lowest gain. For the second pseudo source the gains were assigned in the reverse order.

Table 1. Constant gain assignment for two pseudo sources.

Pseudo source 1		Pseudo source 2	
Coefficient	Gain	Coefficient	Gain
D1	0.0	D1	1.0
D2	0.1	D2	0.9
D3	0.2	D3	0.8
D4	0.3	D4	0.7
D5	0.4	D5	0.6
D6	0.5	D6	0.5
D7	0.6	D7	0.4
D8	0.7	D8	0.3
D9	0.8	D9	0.2
D10	0.9	D10	0.1
A10	1.0	A10	0.0

As the order of gain assignment is based on the frequency content of a sample of the original song, the gain assignment varies from song to song.

Gain Assignment Based on Harmonicity Content. Singing voice has higher harmonicity than instrumental background music. Assuming that singing voice lies in the frequency range of 200Hz to 2000Hz, the data is first passed through a Chebyshev band pass IIR filter, which allows the vocal range to pass through while attenuating all other frequencies. This filtered signal is then passed through an inverse comb filter bank, which detects high amounts of harmonicity, by maximally attenuating the harmonic signals. A measure of harmonicity H is given by,

$$H = E_{(original)} / \min [E_{(filtered)}] . \tag{7}$$

Here $E_{(original)}$ is the energy of the original data and $E_{(filtered)}$ is the energy of the filtered data. The value of H will be high for non-harmonic signals and low for signals with high harmonicity.

The original song is segmented into two-second clippings and the value of H for each is determined. Then the data clippings are arranged in the ascending order of H. Twenty clippings with the least value of H and twenty clippings with the highest value of H are wavelet decomposed and the coefficients arranged in ascending order of energy content. Then the coefficients that occur maximally in a particular position in the ascending order are chosen to form a sequence for the assignment of gains. For one pseudo source the coefficients are assigned gains such that the coefficient occurring first in the sequence gets the highest gain and the coefficient occurring last has the least gain. For the other pseudo source the coefficients are assigned gains in the reverse order.

4 Independent Component Analysis

After the assignment of appropriate gains the data is then reconstructed to form two pseudo sources. This data is then processed using ICA. The algorithm used for Independent Component Analysis is the FastICA algorithm, which makes use of the fixed-point iteration scheme. The decorrelation approach used is 'symmetric' i.e. all the independent components are estimated in parallel. The non-linearity used is tanh, which is given mathematically as,

$$g(u) = \tanh(a1*u). \tag{8}$$

Here $a1$ is a constant equal to 1.

The outputs of ICA are then downsampled by replacing every 5000 samples of the data by their variance. The downsampled data is then made uncorrelated and orthogonal using the Hotelling Transform.

5 Feed Forward Back Propagation Neural Network

This is a feed forward neural network trained using the back propagation algorithm. The training is supervised with a finite number of inputs and a desired output. The errors in the output are propagated backwards and the weights of the connections between the layers are adjusted accordingly. The neural network has 9 input neurons and 1 output neuron. The learning rate is 0.02. The output of the neural network was biased with 0.9 in favor of segments with singing voice and 0.1 in favor of segments with only music.

The input to this neural network is a vector containing the variances corresponding to each output of the ICA. The output of the neural network is 0 if there is music only and 1 singing voice is present along with instrumental background music.

6 Evaluation Experiments

6.1 Audio Data Collection

The audio data collected were movie soundtracks in the language Tamil. The audio data contained singing voices of both male and female singers and a wide variety of instruments in the background music. The data was segmented into 2-second segments before processing. The sampling rate for all the data was 44100 samples / second.

6.2 Evaluation

The system was evaluated using all the three methods of gain selection and the success percentage for the discrimination of segments with and without singing voice determined. The data used for evaluation was similar to the data used for training the neural network.

The following sets of figures show the plots of the various signals generated during processing for each method of gain assignment.

1242   E.S. Gopi et al.

Constant Gain Assignment

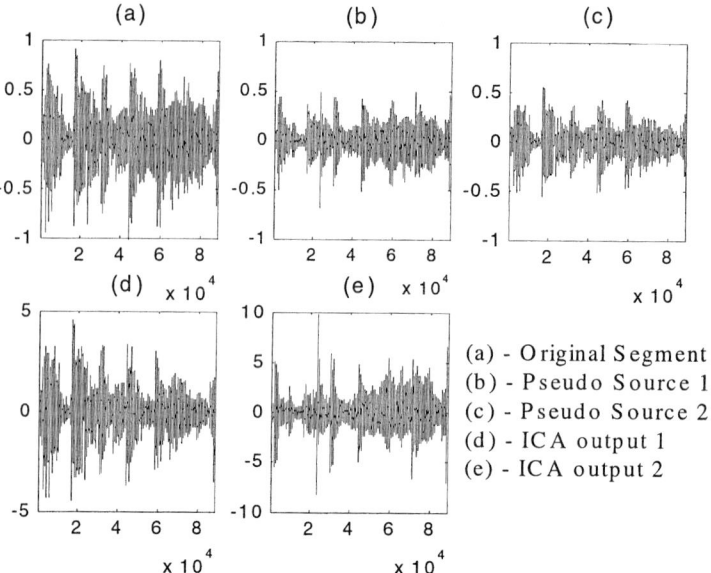

Fig. 2. The signals generated at various stages of processing. The values along the x-axis refer to the No. of samples and those along the y-axis is the amplitude of the signal.

Dynamic Gain Assignment

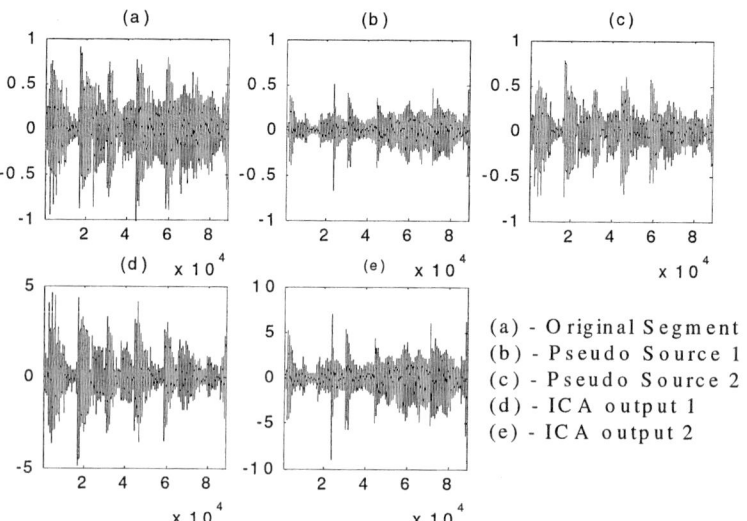

Fig. 3. The signals generated at various stages of processing .The values along the x-axis refer to the No. of samples and those along the y-axis is the amplitude of the signal.

Gain Assignment Based on Harmonicity Content

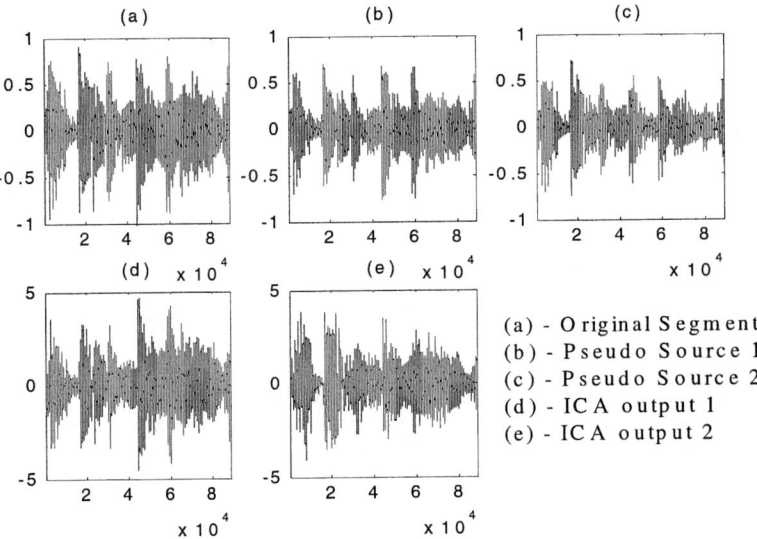

Fig. 4. The signals generated at various stages of processing. The values along the x-axis refer to the No. of samples and those along the y-axis is the amplitude of the signal.

6.3 Results

The results obtained using each method of gain selection are listed below in Table 2. The percentage success in differentiating segments with singing voice and those without, the overall percentage success and the Sum Squared Error (SSE) obtained on training the neural network are tabulated

Table 2. Experimental results for various gain assignment methods.

Gain Assignment Method	Sum Squared Error (SSE)	Percentage Success		
		Singing Voice	Music	Total
Constant	3.781	46.0	56.0	51.0
Dynamic	2.119	58.0	64.0	61.0
Harmonicity Based	2.4207	56.0	68.0	62.0

7 Conclusions

As seen from the results above stated, the method of gain assignment using harmonicity is the most efficient method of generating pseudo sources for ICA and hence discrimination of singing voice/ music. This method is quite successful in the cases wherein the data segments contained either only singing voice with background music or only instrumental music. Segments in which there is part music and part singing voice along with music (transition segments) pose a problem in the discrimination of singing voice/ music.

8 Future Work

The system described in our paper was modified to accommodate the transition segments. The evaluation of the modified system has provided satisfactory results and research is being done to design the complete system.

References

1. Adam L. Berenzweig and Daniel P.W. Ellis: Locating Singing Voice Segments within Music Signals, IEEE Workshop on Applications of Signal Processing to Audio and Acoustics, (2001).
2. A. Hyvarinen and E. Oja. Independent Component Analysis: Algorithms and Applications. Neural Networks, 13(4-5): 411-430, (2000).
3. Robi Polikar, The Wavelet Tutorial.
   (http://users.rowan.edu/~polikar/WAVELETS/WTtutorial.html)
4. Youngmoo E. Kim and Brian Whitman, Singer Identification in Popular Music Recordings using Voice Coding Features, Proc. 2002 International Symposium on Music Information Retrieval, Paris, France,(Oct. 2002).

Lie Group Methods for Optimization with Orthogonality Constraints

Mark D. Plumbley

Department of Electronic Engineering, Queen Mary University of London,
Mile End Road, London E1 4NS, UK
mark.plumbley@elec.qmul.ac.uk

Abstract. Optimization of a cost function $J(\mathbf{W})$ under an orthogonality constraint $\mathbf{WW}^T = \mathbf{I}$ is a common requirement for ICA methods. In this paper, we will review the use of *Lie group* methods to perform this constrained optimization. Instead of searching in the space of $n \times n$ matrices $\mathbf{W}$, we will introduce the concept of the Lie group $SO(n)$ of orthogonal matrices, and the corresponding *Lie algebra* $\mathfrak{so}(n)$. Using $\mathfrak{so}(n)$ for our coordinates, we can multiplicatively update $\mathbf{W}$ by a rotation matrix $\mathbf{R}$ so that $\mathbf{W}' = \mathbf{RW}$ always remains orthogonal. Steepest descent and conjugate gradient algorithms can be used in this framework.

1 Introduction

The independent component analysis problem has a natural 2-step solution: *whitening* followed by *orthogonal rotation* [1]. Given pre-whitened observation vectors $\mathbf{z}$, and a linear transformation $\mathbf{y} = \mathbf{Wz}$, the latter 'rotation' step requires optimization of some function $J = J(\mathbf{W})$ subject to an orthogonality constraint $\mathbf{WW}^T = \mathbf{I}$ on the solution. For standard ICA, J is typically a kurtosis or negentropy measure. For non-negative ICA, we can use a mean squared negativity measure $J = \frac{1}{2}E(|\mathbf{y}_-|^2)$ where $[\mathbf{y}_-]_i = \min(y_i, 0)$ is a negative-rectified version of the output $\mathbf{y}$ [2]. A simple approach to function minimization would be to perform steepest-descent search, $\mathbf{W}_{k+1} = \mathbf{W}_k - \eta \nabla_{\mathbf{W}} J$ where η is a small constant and $[\nabla_{\mathbf{W}} J]_{ij} = \partial J / \partial w_{ij}$ is the gradient of J in $\mathbf{W}$-space. For example, for non-negative ICA, we have $\nabla_{\mathbf{W}} J = E(\mathbf{y}_- \mathbf{z}^T)$. However, this ignores the constraint $\mathbf{WW}^T = \mathbf{I}$: how do we find a minimum of J, subject to the constraint $\mathbf{WW}^T = \mathbf{I}$?

One approach is to modify J with the addition of a penalty term which has a minimum when the constraint is satisfied. Another is to re-impose the constraint (e.g. through Gram-Schmidt orthogonalization) after each update of $\mathbf{W}$. We can also restrict changes to $\mathbf{W}$ so that components in any direction that would change the quantity $\mathbf{WW}^T - \mathbf{I}$ are eliminated, and self-stabilized algorithms can be constructed that tend to reduce deviations away from $\mathbf{WW}^T \approx \mathbf{I}$ (see e.g. [3]). However, these methods do not constrain $\mathbf{W}$ to be orthogonal at all times: we are continually having to fight against the tendency of $\mathbf{W}$ to "drift away" from the constraint surface.

In this paper we will briefly review an approach to this problem that has gained some interest recently: the *Lie group* method [4, 5]. In this approach we represent the possible movements of **W** using a set of coordinates which only allows **W** to take values on a *manifold* that satisfies the constraints. The coordinates **B**, forming a *Lie algebra*, identify a matrix **W** in our Lie group and manifold using the exponential map $\mathbf{W} = \exp \mathbf{B}$. We will see that this elegant property allows us to construct methods for ICA type problems which always maintain the orthogonality constraints we require.

2 Illustration: The Lie Group of Unit-Length Complex Numbers

Consider the complex numbers $z = x + iy$. If we want to 'move about' in the space of complex numbers, addition is a simple way to do it. However, suppose we were only interested in the unit-length complex numbers, those for which $|z| = 1$. These are no longer "closed" under addition: e.g. while 1 and i are each unit-length, $1 + i$ is not.

However, the unit-length complex numbers are closed under *multiplication*. To see this, it is convenient to use the 'length and angle' notation $z = re^{i\theta}$ (Fig. 1(a)). We would know that $r = 1$ for any unit-length complex number,

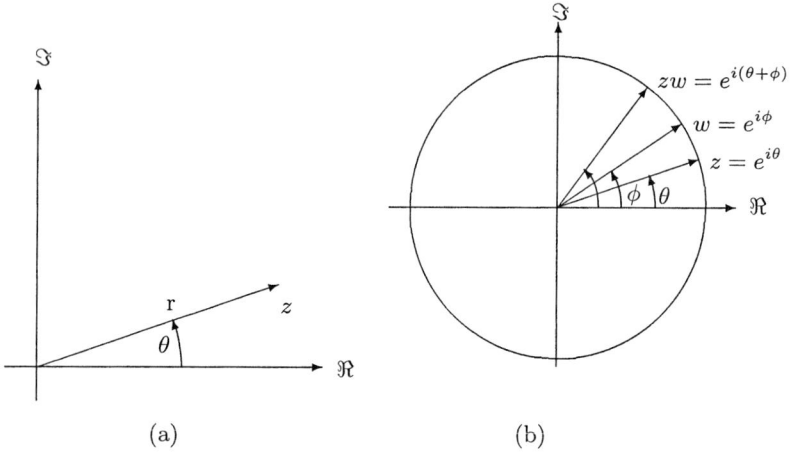

Fig. 1. Complex value (a) z in r-θ notation, and (b) product of unit-length complex numbers z and w.

so we could simply write $z = e^{i\theta}$. We can now see that multiplication of any two unit-length complex numbers z and $w = e^{i\phi}$ will give us a third unit-length complex number $zw = e^{i(\theta+\phi)}$. In fact, the unit-length complex numbers form a *group* G. It is easy to check that they satisfy the group axioms:

1. Closed under the operation: if $z, w \in G$, then $zw = y \in G$;
2. Associativity: $z(wy) = (zw)y$ for $z, w, y \in G$;
3. Identity element: $I \in G$, such that $Iz = zI = z$;
4. Each element has an inverse: z^{-1} such that $z^{-1}z = zz^{-1} = I$;

Our unit-length complex numbers are also commutative, i.e. $zw = wz$, so this is a commutative, or *Abelian* group.

While addition of unit-length complex numbers is of no use to us here, we notice that multiplication of complex numbers has a direct correspondence to addition of angles θ. So another way to 'move about' in the space of unit-length complex numbers is to add *angles* to get a desired angle θ', then convert this to a unit-length complex number using the exponential function $z' = e^{i\theta'}$.

This also makes clear another property of our group: it is "smooth". A local region looks like the real line $\mathbb{R}$, and we can use real-valued coordinates (e.g. the angles θ) to describe a path over this local region. Every local region can be given a coordinate system $\mathbb{R}$, and any overlaps between regions can be 'stitched together' smoothly, so our group with this set of coordinate systems forms a *manifold* [6]. The fact that each local coordinate system is one-dimensional ($\mathbb{R}$) means that we have a one-dimensional manifold. The smoothness in the group means that we have what is called a *Lie group*. The theory of Lie groups is particularly important in physics, such as for general relativity, and has more recently been used in robotics and computer vision. The key to how this helps with our optimization is how we form *derivatives* over this group.

Let $z(t) = e^{i\theta(t)}$ be a unit-length complex number changing with time. Then the time derivative of z is

$$dz/dt = \frac{d}{dt}e^{i\theta(t)} = i(d\theta/dt)e^{i\theta(t)} = i\omega z \qquad (1)$$

where $\omega = d\theta/dt$. In other words, the derivative of z is proportional to iz, which is at right angles to z (Fig. 2). This means, in particular, that the derivative at z

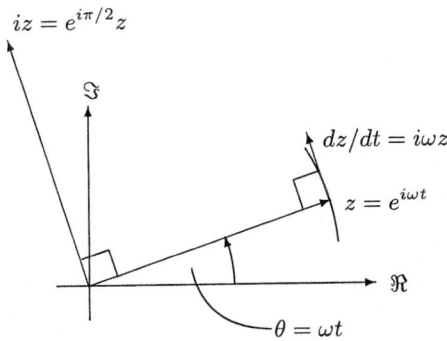

Fig. 2. Derivative of $z = \exp(i\omega t)$.

is *tangent* to the group, pointing 'along' the direction that z changes. The space of all possible derivatives at z is called the *tangent space* T_z at z. We have to be careful to say 'at z' since the tangent spaces T_w and T_z at different unit-length complex numbers w and z will be different.

How does this help with our optimization over orthogonal matrices $\mathbf{WW}^T = \mathbf{I}$? It turns out that the set of orthogonal 2×2 matrices also forms a manifold and Lie group which is very similar to the unit-length complex numbers, so these ideas carry over directly. More generally, orthogonal $n \times n$ matrices can be decomposed into a *block diagonal* form of 2×2 matrices, the *Jordan canonical form* [6,7], so knowing how these unit-length complex numbers behave (and hence the 2×2 matrices) is key to visualizing the general case.

3 The Lie Group of (Special) Orthogonal Matrices

For any integer $n \geq 1$, the set of $n \times n$ orthogonal matrices $\mathbf{W}$ with real entries, i.e. those real matrices that satisfy $\mathbf{WW}^T = \mathbf{I}$, form a group under matrix multiplication, given the symbol $O(n)$. We can check the group axioms if we wish: for example, if $\mathbf{W}$ and $\mathbf{Z}$ are orthogonal, then $\mathbf{V} = \mathbf{WZ}$ is also orthogonal since $\mathbf{VV}^T = \mathbf{WZ}(\mathbf{WZ})^T = \mathbf{WZZ}^T\mathbf{W}^T = \mathbf{WW}^T = \mathbf{I}$.

However, further investigation reveals that the Lie group $O(n)$ actually consists of two disconnected parts: those matrices with determinant $+1$, and those with determinant -1. We cannot get smoothly from one half of $O(n)$ to the other: every time we multiply by a matrix with determinant -1 we 'flip' from one half to the other. We therefore restrict ourself to the so-called 'special' orthogonal matrices $SO(n)$, meaning those matrices in $O(n)$ with determinant 1.

Consider for the moment the special case $n = 2$. The matrices in $SO(2)$ can all be written in the following form:

$$\mathbf{W} = \begin{pmatrix} \cos\theta & \sin\theta \\ -\sin\theta & \cos\theta \end{pmatrix} \quad (2)$$

for some $0 \leq \theta < 2\pi$, and multiplication of these matrices satisfies

$$\begin{pmatrix} \cos\theta & \sin\theta \\ -\sin\theta & \cos\theta \end{pmatrix} \begin{pmatrix} \cos\phi & \sin\phi \\ -\sin\phi & \cos\phi \end{pmatrix} = \begin{pmatrix} \cos(\theta+\phi) & \sin(\theta+\phi) \\ -\sin(\theta+\phi) & \cos(\theta+\phi) \end{pmatrix} \quad (3)$$

so that multiplication of matrices $\mathbf{W} \in SO(2)$ corresponds to addition of angles θ. We can therefore see that the group $SO(2)$ with matrix multiplication behaves in exactly the same way as the unit-length complex numbers with complex number multiplication. Both can be specified completely by an angle $0 \leq \theta < 2\pi$, with the operation of addition modulo 2π. (Groups that "act the same" like this are said to be *isomorphic*.) So given some $\mathbf{W} = \begin{pmatrix} c & s \\ -s & c \end{pmatrix} \in SO(2)$, to move to a new matrix we just need to find the angle $\theta = \arctan(s,c)$, move to a new angle θ', then transform to the new matrix $\mathbf{W}' = \begin{pmatrix} \cos\theta' & \sin\theta' \\ -\sin\theta' & \cos\theta' \end{pmatrix}$. The constraint $\mathbf{WW}^T = \mathbf{I}$ is maintained automatically.

Now, to apply a gradient-based search method, we will need to calculate derivatives. For the special case of $n = 2$, if we let $\theta = t\phi$ and differentiate

$\mathbf{W} = \begin{pmatrix} \cos\theta & \sin\theta \\ -\sin\theta & \cos\theta \end{pmatrix}$ with respect to t, we get

$$\frac{d}{dt}\mathbf{W} = \begin{pmatrix} -\sin\theta & \cos\theta \\ -\cos\theta & -\sin\theta \end{pmatrix} \cdot \phi = \begin{pmatrix} 0 & \phi \\ -\phi & 0 \end{pmatrix}\mathbf{W}. \tag{4}$$

Now we know that for scalars, if $dz/dt = bz$ then $z = e^{tb}$. We can check that this also works for matrices. The matrix exponent of $t\mathbf{B}$ is, by definition

$$\exp(t\mathbf{B}) = \mathbf{I} + t\mathbf{B} + \frac{t^2\mathbf{B}^2}{2!} + \cdots + \frac{t^k\mathbf{B}^k}{k!} + \cdots \tag{5}$$

from which is is straightforward to verify that

$$\frac{d}{dt}\exp(t\mathbf{B}) = 0 + \mathbf{B} + \mathbf{B}\frac{t\mathbf{B}}{1!} + \cdots + \mathbf{B}\frac{t^{k-1}\mathbf{B}^{k-1}}{(k-1)!} + \cdots = \mathbf{B}\exp(t\mathbf{B}) \tag{6}$$

and therefore (since $\theta = t\phi$) we can write

$$\mathbf{W} = \begin{pmatrix} \cos\theta & \sin\theta \\ -\sin\theta & \cos\theta \end{pmatrix} = \exp\Theta \quad \text{where} \quad \Theta = \begin{pmatrix} 0 & \theta \\ -\theta & 0 \end{pmatrix}. \tag{7}$$

(Note that we must be rather careful with this matrix exponential: since matrix multiplication is not commutative, $\exp(\mathbf{A} + \mathbf{B}) \neq \exp(\mathbf{A})\exp(\mathbf{B})$ in general.) In fact, given any skew-symmetric $\Phi \neq \mathbf{0}$, all of the elements in SO(2) can be specified by a single real parameter t, as $\mathbf{W}(t) = \exp(t\Phi)$.

For optimization over our 2×2 orthogonal matrices SO(2), we now have a clear way to proceed: instead of searching over the space of matrices $\mathbf{W}$, we can search over the space of angles θ, or alternatively, the space of skew-symmetric matrices Θ. To get back to an orthogonal matrix, we apply the exponential map $\mathbf{W} = \exp\Theta$.

4 Searching over SO(n) Using the Lie Algebra $\mathfrak{so}(n)$

The ideas that we have seen for SO(2) generalize to SO(n). However, we have to be a little careful for $n > 2$, due to the non-commutation of matrices that we mentioned earlier, meaning that $\exp(\mathbf{A})\exp(\mathbf{B}) \neq \exp(\mathbf{B})\exp(\mathbf{A})$. This is true even for very small $\mathbf{A}$ and $\mathbf{B}$: for small scalar ϵ we find

$$\exp(\epsilon\mathbf{A})\exp(\epsilon\mathbf{B}) - \exp(\epsilon\mathbf{B})\exp(\epsilon\mathbf{A}) = \epsilon^2[\mathbf{A},\mathbf{B}] + O(\epsilon^3) \tag{8}$$

where the matrix commutator, or *bracket*, $[\cdot,\cdot]$ is defined by $[\mathbf{A},\mathbf{B}] = \mathbf{AB} - \mathbf{BA}$.

Interestingly, the commutator of two skew-symmetric matrices $\mathbf{A} = -\mathbf{A}^T$ and $\mathbf{B} = -\mathbf{B}^T$ is itself skew-symmetric: $[\mathbf{A},\mathbf{B}]^T = -[\mathbf{A},\mathbf{B}]$. This set of $n \times n$ skew-symmetric matrices, closed under addition, multiplication by scalars, and the bracket operation, is an example of a *Lie algebra*, and is denoted $\mathfrak{so}(n)$. We can therefore map from an element $\mathbf{W}$ in the Lie group SO(n) to an element $\mathbf{B}$ in the Lie algebra $\mathfrak{so}(n)$ using the matrix logarithm operator (the inverse of

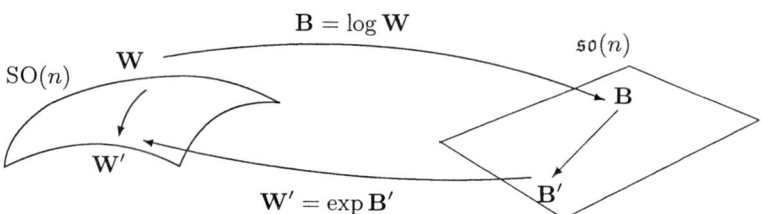

Fig. 3. Motion from **W** on the Lie group SO(n) by mapping **B** to the Lie algebra $\mathfrak{so}(n)$, moving to $\mathbf{B}' \in \mathfrak{so}(n)$, and mapping back to $\mathbf{W}' \in \text{SO}(n)$.

$\exp(\cdot)$), move to a new element $\mathbf{B}' \in \mathfrak{so}(n)$, and then map from $\mathbf{B}'$ to $\mathbf{W}' = \exp(\mathbf{B}') \in \text{SO}(n)$ using the matrix exponential (Fig. 3). Since $\mathfrak{so}(n)$ is a vector space, it is easy to stay in $\mathfrak{so}(n)$, because it is closed under addition of elements and multiplication by scalars. It also has dimension $n(n-1)/2$, the number of independent entries in an $n \times n$ skew-symmetric matrix, and hence is a "smaller" space to search over than the original n^2-dimensional space of matrices **W**. In this way we can make sure that our matrices **W** always stay on SO(n).

In fact we can use a slightly modified form of this method, to avoid calculating the initial logarithm. Due to the group properties of SO(n), we know that $\mathbf{W}' = \mathbf{RW}$ for some "rotation" matrix $\mathbf{R} \in \text{SO}(n)$. Moving from $\mathbf{W} = \mathbf{IW}$ to $\mathbf{W}' = \mathbf{RW}$ is equivalent to moving from **I** to **R**, and we already know that $\log \mathbf{I} = \mathbf{0}$. Our modified method is therefore

1. Start at $\mathbf{0} \in \mathfrak{so}(n)$, equivalent to $\mathbf{I} \in \text{SO}(n) = \exp(\mathbf{0})$
2. Move about in $\mathfrak{so}(n)$ from **0** to $\mathbf{B} \in \mathfrak{so}(n)$
3. Use $\exp(\cdot)$ to map back into SO(n), giving $\mathbf{R} = \exp(\mathbf{B})$
4. Calculate $\mathbf{W}' = \mathbf{RW} = \exp(\mathbf{B})\mathbf{W} \in \text{SO}(n)$.

Of course we must start from a **W** that is orthogonal: we typically choose the identity matrix $\mathbf{W}_0 = \mathbf{I}$.

5 Gradient in $\mathfrak{so}(n)$ and Geodesic Flow

To choose the search direction in $\mathfrak{so}(n)$ (B-space), we need to calculate the gradient of J in B-space, $\nabla_\mathbf{B} J$. Calculating this, which must be skew-symmetric since the matrices **B** must remain skew-symmetric, we get [8]

$$\nabla_\mathbf{B} J = (\nabla_\mathbf{W} J)\mathbf{W}^T - \mathbf{W}(\nabla_\mathbf{W} J)^T. \tag{9}$$

This gives us the search direction for a steepest-descent search in B-space, instead of using our original steepest-descent search in W-space. For example, for non-negative ICA, we get $\nabla_\mathbf{B} J = E(\mathbf{y}_- \mathbf{y}^T - \mathbf{y}\mathbf{y}_-^T)$ which is clearly equivariant, since it only depends on **y** [9].

Using this approach, for the steepest-descent algorithm with small constant update factor η, we start at $\mathbf{B} = \mathbf{0} \in \mathfrak{so}(n)$, move to $\mathbf{B}' = -\eta \nabla_\mathbf{B} J$, map to $\mathbf{R} =$

$\exp(\mathbf{B}') \in \mathrm{SO}(n)$, and finally perform a multiplicative update $\mathbf{W}_{k+1} = \mathbf{R}\mathbf{W}_k$. Putting this all into one equation, we get

$$\mathbf{W}_{k+1} = \exp(-\eta \nabla_\mathbf{B} J|_{\mathbf{B}=0})\mathbf{W}_k \tag{10}$$

which is the *geodesic flow* method introduced to ICA by Nishimori [7]. In the non-negative ICA case this gives $\mathbf{W}_{k+1} = \exp(-\eta E(\mathbf{y}_-\mathbf{y}^T - \mathbf{y}\mathbf{y}_-^T))\mathbf{W}_k$. Since $\mathbf{W}$ always remains in SO(n), the constraint $\mathbf{W}\mathbf{W}^T = \mathbf{I}$ is maintained without the use of penalty functions or any constraint re-imposition [5].

6 Related Methods

Now we have the basic approach of working in $\mathfrak{so}(n)$ instead of SO(n), we can also implement faster search methods equivalent to line search or conjugate gradients. For example, for a repeated line search (a *geodesic search* [2]), we can proceed as follows. We choose a search direction in the direction of steepest descent, i.e. $\mathbf{H} = -\nabla_\mathbf{B} J/|\nabla_\mathbf{B} J|$, make large steps along $\mathbf{B}(t) = t\mathbf{H}$ to get close to a minimum of J at t^*, update $\mathbf{W}_{k+1} = \exp(t^*\mathbf{H})\mathbf{W}_k$, and repeat with a new line search direction until the gradient or error is as small as we wish.

If we can calculate second derivative information, we can also use Newton updates in our line search to find the minimum. Also, due to the rotational structure of the group SO(n), we can also use a Fourier expansion in some situations [10].

Edelman, Arias and Smith [11] constructed conjugate gradient algorithms on Stiefel manifolds (a generalization of SO(n)), and this approach was used by Martin-Clemente et al [12] for ICA. For the conjugate gradients method on manifolds such a SO(n) the basic idea is the same as in the more usual Euclidean space, but it needs a little care to ensure that the various gradients that are used are all 'transported' to the same point in an appropriate way before they are used [11].

In passing, we mention that the exponential map is not the only map from the skew-symmetric matrices $\mathfrak{so}(n)$ to the orthogonal matrices SO(n). For example, the Cayley transform $\mathbf{W} = (\mathbf{I} + \mathbf{B})/(\mathbf{I} - \mathbf{B})^{-1}$, can also be used for SO(n) and other related manifolds [13].

7 Conclusions

We have briefly reviewed the use of Lie group methods to optimize a cost function $J(\mathbf{W})$ under an orthogonality constraint $\mathbf{W}\mathbf{W}^T = \mathbf{I}$, a common requirement for ICA methods (including non-negative ICA). We have seen that we can search in the Lie algebra $\mathfrak{so}(n)$ of skew-symmetric matrices, instead of the original space of matrices, and multiplicatively update $\mathbf{W}$ by a rotation matrix $\mathbf{R}$ so that $\mathbf{W}' = \mathbf{R}\mathbf{W}$ always remains orthogonal. The simplest case of steepest descent over $\mathfrak{so}(n)$ corresponds to Nishimori's *geodesic flow*, and this approach can also be generalised to line search and conjugate gradient methods.

Acknowledgements

This work was partially supported by EPSRC grant GR/R54620, and by EU-FP6-IST-507142 project SIMAC (Semantic Interaction with Music Audio Contents: www.semanticaudio.org). An extended discussion of this subject will be presented in [8].

References

1. Comon, P.: Independent component analysis - a new concept? Signal Processing **36** (1994) 287–314
2. Plumbley, M.D.: Algorithms for nonnegative independent component analysis. IEEE Transactions on Neural Networks **14** (2003) 534–543
3. Douglas, S.C.: Self-stabilized gradient algorithms for blind source separation with orthogonality constraints. IEEE Transactions on Neural Networks **11** (2000) 1490–1497
4. Iserles, A.: Brief introduction to Lie-group methods. In Estep, D., Tavener, S., eds.: Collected Lectures on the Preservation of Stability Under Discretization (Proceedings in Applied Mathematics Series). SIAM (2002)
5. Fiori, S.: A theory for learning by weight flow on Stiefel-Grassman manifold. Neural Computation **13** (2001) 1625–1647
6. Schutz, B.: Geometrical Methods of Mathematical Physics. Cambridge University Press, Cambridge, UK (1980)
7. Nishimori, Y.: Learning algorithm for ICA by geodesic flows on orthogonal group. In: Proceedings of the International Joint Conference on Neural Networks (IJCNN'99). Volume 2., Washington, DC (1999) 933–938
8. Plumbley, M.D.: Geometric methods for non-negative ICA: Manifolds, Lie groups and toral subalgebras (2004) Submitted to *Neurocomputing*.
9. Cardoso, J.F., Laheld, B.H.: Equivariant adaptive source separation. IEEE Transactions on Signal Processing **44** (1996) 3017–3030
10. Plumbley, M.D.: Optimzation using Fourier expansion over a geodesic for non-negative ICA (2004) To appear in Proceedings of the International Conference on Independent Component Analysis and Blind Signal Separation, ICA2004.
11. Edelman, A., Arias, T.A., Smith, S.T.: The geometry of algorithms with orthogonality constraints. SIAM J. Matrix Anal. Appl. **20** (1998) 303–353
12. Martin-Clemente, R., Puntonet, C.G., Acha, J.I.: Blind signal separation based on the derivatives of the output cumulants and a conjugate gradient algorithm. In Lee, T.W., Jung, T.P., Makeig, S., Sejnowski, T.J., eds.: Proceedings of the International Conference on Independent Component Analysis and Signal Separation (ICA2001), San Diego, California. (2001) 390–393
13. Yamada, I., Ezaki, T.: An orthogonal matrix optimzation by dual Cayley parametrization technique. In: Proc. 4th Intl. Symp. On Independent Component Analysis and Blind Signal Separation (ICA2003), Nara, Japan. (2003) 35–40

A Hierarchical ICA Method for Unsupervised Learning of Nonlinear Dependencies in Natural Images

Hyun-Jin Park and Te-Won Lee

Institute for Neural Computation (INC), University of California San Diego (UCSD),
9500 Gilman Drive, La Jolla, CA 92093-0523
{hjinpark,tewon}@ucsd.edu

Abstract. Capturing dependencies in images in an unsupervised manner is important for many image-processing applications and understanding the structure of natural image signals. Linear generative models such as independent component analysis (ICA) have shown to capture low level features such as oriented edges in images. However ICA only captures linear dependency due to its linear model constraints and its modeling capability is limited. We propose a new method for capturing nonlinear dependencies in natural images. It is an extension of the linear ICA method and builds on a hierarchical representation. It makes use of lower level linear ICA representation and a subsequent mixture of Laplacian distribution for learning the nonlinear dependencies. The model is learned via the EM algorithm and it can capture variance correlation and high order structures in a consistent manner. We visualize the learned variance structure and demonstrate applications to image segmentation and denoising.

1 Introduction

Unsupervised learning algorithms that capture sophisticated representations can provide a better understanding of neural information processing and better algorithms for signal processing applications. Recently, adaptive techniques have gained popularity due to many potential applications. For example, independent component analysis (ICA) has been effective at learning representations for natural images that have similar properties than receptive fields in the visual cortex. Also those learned features has been applied for feature extraction, denoising and image segmentation [1],[2]. In ICA model, statistics of natural image patches are modeled by Eq.(1) [3], where X is the measured data, A is a linear transformation matrix, and u is a source signal vector whose distribution is a product of sparse distributions such as generalized Laplacian distributions[2].

$$X = Au, \qquad P(u) = \prod_i P(u_i). \qquad (1)$$

ICA is effective at learning linear dependency between pixels [3], but its modeling capability is limited up to linear dependency. The representations learned by ICA algorithm are relatively low-level, but in biological systems, there are more high-level representations that are correlated with contours, textures, and objects, which are not represented by the linear ICA. In previous approaches, it is shown that by capturing nonlinear dependencies beyond linear, one can learn higher-level representations similar to biological systems [4],[5],[6]. Studies beyond linear ICA model are focused on variance correlation of source signal after ICA. In linear ICA, source signals are assumed to be independent. But in real image signals, the variances of ICA source signals tend to be correlated [4],[5],[6]. In ICA model, a source signal is not linearly predictable from others, but a source signal is still 'predictable' in a nonlinear manner given variance dependency. This higher-order dependency cannot be captured by linear ICA model.

Hyvarinen suggested to use a special distribution to model variance correlation[4],[5]. His model uses distribution of Eq.(2) to model two dependent sources. This distribution shows variance dependencies similar to those found in natural image signals. His model can learn grouping of dependent sources (Subspace ICA) or topographic arrangements of variance correlated sources (Topographic ICA) [4],[5]. But the learned grouping or topography does not provide higher order structures explicitly nor an analytic description of the variance correlation. It is not clear how to compute the higher order signal encodings and its applicability to signal processing is limited.

$$P(u_1, u_2) = c \exp\left(-\sqrt{u_1^2 + u_2^2}\right). \qquad (2)$$

Recently, Lewicki proposed a hierarchical 2-stage model where the 1st stage is an ICA model and the 2nd stage is a generative model for variance of sources [6]. In the 2nd stage, variances for the ICA source distribution is linearly generated as Eq.(3). In Eq. (3), u, v and λ are vectors and B is a matrix.

$$P(u \mid \lambda) = c \exp\left(-\left|\frac{u}{\lambda}\right|^q\right), \qquad \log[\lambda] = Bv. \qquad (3)$$

Important contribution of his model is that it explicitly learns variance correlations in the ICA source signals. It can learn higher order structures encoded in 'variance basis'. Those variance bases can be interpreted as textures or structures in natural images. But treating the variance as another random variable introduces a high complexity and rough approximations, and, its application to signal processing is limited.

In this paper, we propose a parametric mixture model that can learn nearly independent variance correlated source signals after ICA. Our model allows direct representation of variance correlation structure and reveals high order structure in natural image signals. It provides a simple parametric PDF for ICA source signals and can be used for better signal processing such as denoising. We derive the parametric mixture model and its learning algorithm in Section 2. Then we visualize learned variance structure for each mixture in Section 3, which shows high-level structures. Finally, we demonstrate the application of the learned model to image segmentation and Bayesian denoising.

2 Model of Nonlinear Dependency

We propose a hierarchical 2-stage model where the 1st stage is an ICA model and the 2nd stage is a mixture model that captures variance correlated source priors (figure 1).

The correlation of variance in natural images reflects different types of regularities in real world. Such regularities can be called as 'contexts'. A basic assumption is that different contexts are caused by different variances. To model such context dependent variances, we use a mixture model where each mixture is a Laplacian distribution with same 0-mean but different variances. The ICA matrix A is learned in prior independently of the 2nd stage. The 2nd stage is trained given the first stage ICA source signals.

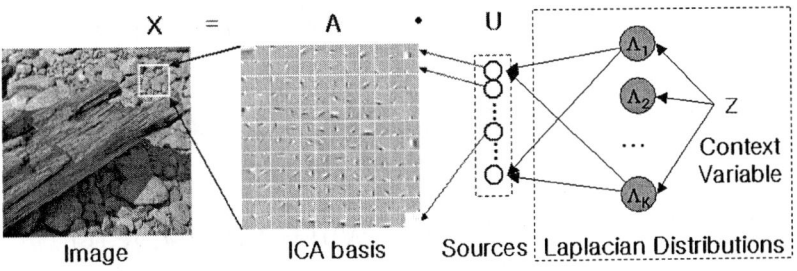

Fig. 1. Two stage model for learning nonlinear dependency. In the 1st stage, image patches are encoded by ICA model. In the 2nd stage, ICA source signal U is modeled by a mixture model. Z is a discrete random vector that represents which mixture is responsible for the given patch.

In figure 1, the context variable Z 'selects' which Laplacian will represent ICA source signal u for each image patch. Each Laplacian distribution is a factorial multi-dimensional distribution. The advantage of the Laplacian distribution for modeling context is that we can model a sparse distribution using a single distribution. In addition, the conventional ICA is approximated well as a special case of our model.

2.1 Mixture of Laplacian Distribution Model

The first stage of our model is the conventional ICA. For the second stage, we define a PDF for mixture of M-dimensional Laplacian distribution as Eq.(4), where M is the dimension of input data, N is the number of data, and K is the number of mixtures.

$$P(U|\Lambda,\Pi) = \prod_n^N P(\vec{u}_n|\Lambda,\Pi) = \prod_n^N \sum_k^K \pi_k P(\vec{u}_n|\vec{\lambda}_k) = \prod_n^N \sum_k^K \pi_k \prod_m^M \frac{1}{(2\lambda_{k,m})} \exp\left(-\frac{|u_{n,m}|}{\lambda_{k,m}}\right). \quad (4)$$

Where $u_n = (u_{n,1},...,u_{n,m},...,u_{n,M})$ is the n-th data same, $\lambda_k = (\lambda_{k,1},...,\lambda_{k,m},...,\lambda_{k,M})$ is the variance k-th Laplacian distribution, π_k is the probability of mixture k ($\sum_k \pi_k = 1$), $U = (\vec{u}_1,\vec{u}_2,...,\vec{u}_i,...,\vec{u}_N)$, $\Lambda = (\vec{\lambda}_1,\vec{\lambda}_2,...,\vec{\lambda}_k,...,\vec{\lambda}_K)$, and $\Pi = (\pi_1,...,\pi_K)$.

It is not easy to maximize Eq.(4) directly, and we use EM (expectation maximization) algorithm. We can rewrite the likelihood by introducing a hidden variable Z.

$$P(U,Z|\Lambda,\Pi) = \prod_n^N P(\vec{u}_n,Z|\Lambda,\Pi) = \prod_n^N \left[\prod_k^K \left[(\pi_k)^{z_k^n} \prod_m^M \left(\left(\frac{1}{2\lambda_{k,m}}\right)^{z_k^n} \cdot \exp\left(-z_k^n \frac{|u_{n,m}|}{\lambda_{k,m}}\right)\right)\right]\right]. \quad (5)$$

Where z_k^n is a binary random variable, which is 1 if mixture k generated n-th data sample, 0 otherwise. $Z = (z_k^n)$ is a matrix of z_k^n, and $\sum_k z_k^n = 1$ for all n = 1...N.

EM algorithm works by maximizing log of the likelihood in Eq.(5), averaged over the hidden variable Z. The expectation of log likelihood can be written as Eq. (6). The expectation $E\{z_k^n | U, \Lambda, \Pi\}$ can be evaluated using Eq. (7), given the data U and parameters Λ and Π. We can use temporary estimations Λ' and Π' for Λ and Π in EM method. The normalization constant c_n can be computed by normalizing Eq. (7) [11].

$$E\{\log P(U,Z | \Lambda, \Pi)\} = \sum_{n,k} E\{z_k^n | U, \Lambda, \Pi\} \left[\log(\pi_k) + \sum_m \left(\log(\frac{1}{2\lambda_{k,m}}) - \frac{|u_{n,m}|}{\lambda_{k,m}} \right) \right]. \quad (6)$$

$$E\{z_k^n\} \equiv E\{z_k^n | U, \Lambda', \Pi'\} = \sum_{z_k^n=0}^{1} z_k^n P(z_k^n | u_n, \Lambda', \Pi') = P(z_k^n = 1 | u_n, \Lambda', \Pi')$$
$$= P(u_n | z_k^n = 1, \Lambda', \Pi') P(z_k^n = 1 | \Lambda', \Pi') / P(u_n | \Lambda', \Pi') \quad (7)$$
$$= \left(\prod_m^M \frac{1}{2\lambda_{k,m}'} \exp(-\frac{|u_{n,m}|}{\lambda_{k,m}'}) \cdot \pi_k' \right) / P(u_n | \Lambda', \Pi') = \frac{1}{c_n} \prod_m^M \frac{\pi_k'}{2\lambda_{k,m}'} \exp(-\frac{|u_{n,m}|}{\lambda_{k,m}'}).$$

The EM algorithm works by maximizing equation (6) given $E\{z_k^n\}$ in Eq. (7). $E\{z_k^n\}$ can be computed given the data U and parameters Λ' and Π' estimated in previous iteration of EM algorithm. Then, we need to maximize Eq. (6) over Λ and Π. The maximization problem can be solved analytically and the details can be found in [11]. The maximization solution and the final EM algorithm are summarized in figure 2.

---

1. Initialize $\pi_k = \frac{1}{K}$, $\lambda_{k,m} = E\{|u_m|\} + e$ (e is a small random noise)
2. Calculate the Expectation by

$$E\{z_k^n\} \equiv E\{z_k^n | U, \Lambda', \Pi'\} = \frac{1}{c_n} \prod_m^M \frac{\pi_k'}{2\lambda_{k,m}'} \exp(-\frac{|u_{n,m}|}{\lambda_{k,m}'})$$

3. Maximize the log likelihood given the Expectation

$$\lambda_{k,m} \leftarrow \left(\sum_n E\{z_k^n\} \cdot |u_{n,m}| \right) / \left(\sum_n E\{z_k^n\} \right), \quad \pi_k \leftarrow \left(\sum_n E\{z_k^n\} \right) / \left(\sum_k \sum_n E\{z_k^n\} \right)$$

4. If (converged) stop, otherwise repeat from step 2.

Fig. 2. Outline of EM algorithm for Learning the Mixture Model.

3 Experimental Results

Here, we show the model learning procedure and visualization of learned variances. Then, we provide application examples for image segmentation and denoising. The parametric PDF provided by our model is an important advantage for designing adaptive signal processing applications. Utilizing this, we 'identify' the hidden context variable Z for given image patch in image segmentation, detecting high-level structures of natural images. In denoising, our model is used as a better source prior for Bayesian MAP (maximum a posteriori) inference.

3.1 Learning Nonlinear Dependencies in Natural Images

As shown in figure 1, the first stage of our model is linear ICA model. The ICA matrix A and $W(=A^{-1})$ are learned by the FastICA algorithm [7]. We sampled $10^5(=N)$ data of 16x16 patches from images of natural scenes and use them for both the 1st and 2nd stages. The ICA source signals are computed and used for learning the second stage. ICA input dimension is 256, and source dimension is set to 160(=M). The number of mixtures is set to 16, 64, or 256(=K). Training by the EM algorithm is fast and takes about 100 iterations for convergence (takes 0.5 hour in a P4 1.7GHz PC).

For the visualization of the learned variance, we adapted the visualization method from [6]. For a Laplacian distribution, each component of variance vector corresponds to an image basis. For each image basis has a wavelet like shape and we can compute its center positions in image and frequency space. Then we can map a variance value to a color at the corresponding centers in image and frequency space as figure 3.

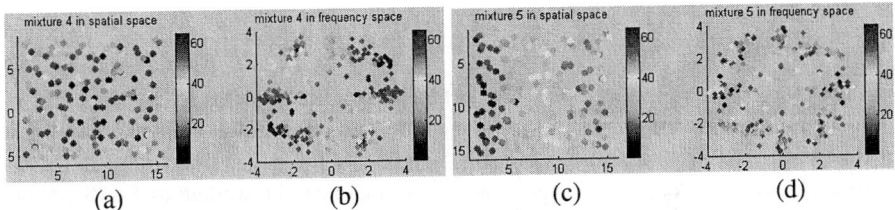

Fig. 3. Visualization of learned variances – two examples. Subfigure (a) and (b) visualizes a variance vector for mixture #4. (c) and (d) visualizes a variance vector for mixture #5. High variance values are mapped to red color and small values are mapped to blue color. Variance of mixture #4 shows strong localization of high variance values in frequency space but not in image space. This variance represents a texture consists of oriented edges spread over the image. Mixture #5 shows localization in image space but not in frequency space. It represents image patches whose left side is filled with textures while the right side is relatively clean.

In figure 3, high and low valued variances are smoothly clustered in image or frequency space. The model does not know spatial or frequency centers of the ICA basis and such 'regularities' are learned from the remaining dependencies in the source signals. They contain regular textural structures such as sky, grass, and woods. In other words, it is such regular structures that caused nonlinear dependency and variance correlation. We verify this relationship in segmentation experiments.

3.2 Application: Unsupervised Image Segmentation

The idea behind our model is that images can be modeled as mixture of 'contexts' with different variance correlation. We can show what caused such variance correlation by unsupervised image segmentation. As our model assumes a hidden variable Z, we can compute the posterior probability of Z given the image patch using Eq. (4). By selecting a mixture with highest probability we can 'segment' contexts in image. Figure 4 shows an example image and its segmentation results.

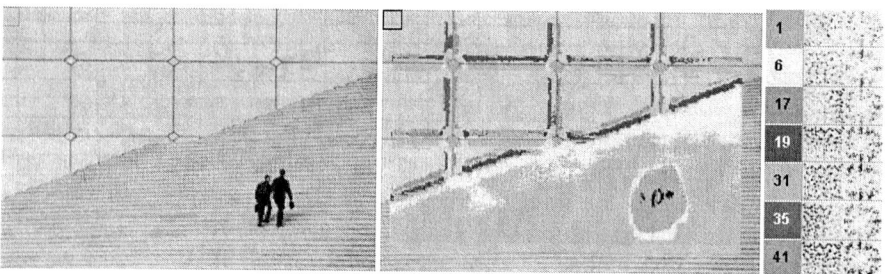

Fig. 4. Unsupervised image segmentation example. Test image (left) and segmented image (middle). The right side shows color plates with corresponding mixture variance structures. We showed a subset of color labels for clear visualization. Also we used a model with 16 mixtures. A box at left-top of the original image denotes the size of patches used.

3.3 Application: Image Denoising

The proposed mixture model provides a better parametric model for ICA source prior and hence an improved model of the image statistics. We can take advantage of this in Bayesian MAP (maximum a posteriori) estimation for image denoising. If we assume a Gaussian noise n, the image generation model can be written as Eq.(8). Now, we can obtain the MAP estimation of ICA source signal u using Eq.(15) and we can reconstruct the estimated original image through $\hat{X} = A\hat{u}$.

$$X = Au + n \quad , \quad n \sim N(0, \sigma_n^2) \tag{8}$$

$$\hat{u} = \underset{u}{argmax} \log P(u \mid X, A) = \underset{u}{argmax} \left[\log P(X \mid u, A) + \log P(u) \right] \tag{9}$$

In Eq. (9), P(X| u,A) is a Gaussian distribution and P(u) can be varied depending on the source model [1],[2],[9]. For ICA with Laplacian prior, we can rewrite Eq. (9) as Eq. (10). Let's call this ICA MAP with Laplacian prior [2]. Also, we can use a refined prior using our mixture model. The mixture prior can be approximated by single most explaining Laplacian distribution k. For this case, we can reuse Eq. (10) by replacing λ_m with $\lambda_{k,m}$. This method can be called as ICA MAP with Mixture prior.

$$\hat{u} = \underset{u}{argmax} \left(-\frac{|X - Au|^2}{2\sigma_n^2} - \sum_m \frac{|u_m|}{\lambda_m} \right) \tag{10}$$

In addition to two ICA-based methods, we also evaluated three other methods including BayesCore [8], BayesJoint [9], and Wiener filter. BayesCore is similar to ICA MAP method except that it uses wavelet transformation and Bayes marginal estimator instead of MAP estimator [8]. BayesJoint is also a Bayes marginal estimator with wavelet transformation and Gaussian source prior where the variances of sources are linearly correlated with each other [9].

General settings for the experiment are like this. For the maximization of Eq. (10), we can use orthogonalized ICA transform ($W = A^T$) where an efficient deterministic solution is possible [2],[9]. For all denoising methods, we assume that the noise variance is known and denoising tests are performed with images different from training images. For both wavelet-based methods, we used QMF (quadratic mirror filter) type wavelet decomposition. For measuring denoising performance, we used signal to noise ratio (SNR) and structural similarity measure (SSIM) which measures perceptual similarity between two images [10]. Figure 5 and 6 summarize the denoising performance of 5 algorithms on an image 'mountain'. Denoising by our model outperforms others. Especially our denoising method is much better in terms of perceptual quality.

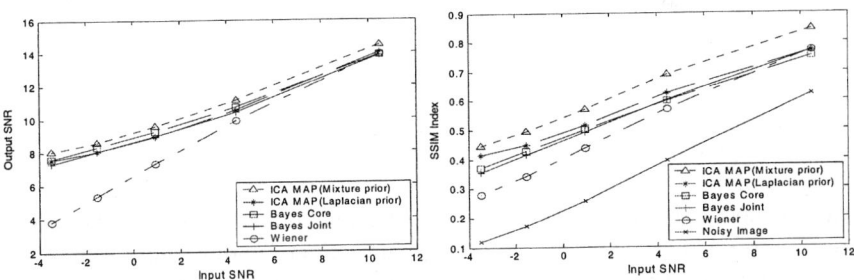

Fig. 5. SNR and SSIM for 'mountain' image over different noise variances. Image var. = 1.0, SSIM index is 1 if the denoised image is same as the original image, and $0 \leq$ SSIM index ≤ 1

4 Discussions

We proposed a model to learn nonlinear dependencies in natural images. The proposed mixture of Laplacian distribution is a generalization of the conventional independent source priors and can model variance dependencies given natural image signals. Experimental results show that the proposed model can learn variance correlated signal groups as different mixtures and learn high-level structures highly correlated with underlying physical properties. Our model provides an analytic prior of nearly independent and variance-correlated signals, which is not viable in previous models.

The learned variances of the mixture model show structured localization in image and frequency space, which reveals regularity in natural images. We showed two application examples. In the image segmentation application, we show how the learned model can be used to discover hidden contexts. In the image denoising application, we show the advantage of our parametric model in Bayesian denoising scheme.

Fig. 6. Denoising results of 'mountain' image. Original image variance = 1.0, input noise standard variance = 0.9. NSV denotes noise standard variance, and SM denotes structural similarity index (SSIM). Subfigures (d~h) show results from 5 different denoising methods. (c) is a segmentation result that is used for denoising by ICA MAP with mixture prior

For further investigation, there are several ways to extend our model. First, we can exploit the regularity of the image segmentation result to learn even higher-level structures building on additional hierarchies over the current model. Second, we can extend our model to fully adaptive model by adapting both the first stage (linear generative matrix) and second stage (source mixture prior). Thirdly, more applications should be investigated. Application to image coding seems promising since our model provides a better analytic prior. Application to image segmentation should be exploited further.

References

1. G. E. Hinton & T. J. Sejnowski, *Unsupervised Learning: Foundations of Neural Computation* (edited), MIT Press, Cambridge, Massachusetts, 1999.
2. A. Hyvarinen, P. O. Hoyer, and E. Oja, Image Denoising by Sparse Code Shrinkage, In S. Haykin and B. Kosko (eds), *Intelligent Signal Processing*, IEEE Press, 2001.
3. A.J. Bell and T.J. Sejnowski, An information-maximization approach to blind separation and blind deconvolution, *Neural Computation* 7:1129-1159, 1995.
4. A. Hyvarinen, P. O. Hoyer, Emergence of phase and shift invariant features by decomposition of natural images into independent feature subspaces, *Neural Computation*, 12(7), 2000.
5. A. Hyvarinen, P. O. Hoyer, Topographic Independent Component Analysis, *Neural Computation*, 13(7):1525:1558, 2001.
6. M. S. Lewicki and Y. Karklin, Learning higher-order structures in natural images, *Network: Comput. Neural Syst.* 14 (August 2003) 483-499.
7. A. Hyvaerinen and E. Oja, A fast fixed-point algorithm for independent component analysis, *Neural Computation*, vol. 9, 1997, pp. 1483–1492.

8. E. P. Simoncelli and E. H. Anderson, Noise removal via Bayesian wavelet coring, Proceedings of the 3rd IEEE International Conf. on Image Processing, vol. 1, 1996, pp. 379-382.
9. E. P. Simoncelli, Bayesian Denoising of Visual Images in the Wavelet Domain, in *Bayesian Inference in Wavelet Based Models*, Springer-Verlag, 1999, pp 291-308.
10. Z. Wang, A. C. Bovik, H. R. Sheikh and E. P. Simoncelli, Image quality assessment: From error visibility to structural similarity, IEEE Trans. on Image Proc., vol. 13, no. 4, Apr. 2004.
11. H.J.Park, and T.W.Lee, Unsupervised learning of nonlinear dependencies in natural images, Intl. Journal of Image Science and Technology, to be published, 2004.

Author Index

Abdallah, Samer 540
Abed-Meraim, Karim 113
Acha, José I. 33
Afsari, Bijan 437
Ahmed, Alijah 930
Aichner, Robert 840
Akaho, Shotaro 750
Akazawa, Kenzo 985
Alcaraz, Raúl 1111
Alías, Francesc 1213
Alleysson, David 946
Almeida, Luís B. 734
Álvarez Marquina, Agustín 824
Amari, Shun-ichi 718
Andina, Diego 160
Andrzejak, Ralph G. 209
Anemüller, Jörn 1009, 1103
Araki, Shoko 461, 610, 652, 898
Artés-Rodríguez, Antonio 271

Babaie-Zadeh, Massoud 9, 798
Barbati, Giulia 1087
Barma, Tanusree Deb 97
Barrón-Adame, J. Miguel 160
Barros, Allan Kardec 129
Beack, Seungkwon 524
Benidir, Messaoud 113
Bickel, Peter J. 225
Bijaoui, Albert 97
Blanco, D. 248
Blanco, Yolanda 73
Blaschke, Tobias 742
Blumensath, Thomas 1205
Boloix-Tortosa, Rafael 1142
Bonnet, Noël 1150
Borloz, Bruno 602
Bronstein, Alexander M. 406, 500, 554, 677
Bronstein, Michael M. 406, 500, 554, 677
Buchner, Herbert 840
Buendía Buendía, Fulgencio S. 160
Burnett, Ian 889

Cardoso, Jean-François 41, 1221
Carrion, M.C. 248
Castella, Marc 922

Castells, Francisco 18, 816, 1079, 1111, 1118, 1229
Chambers, Jonathon A. 177, 532, 661
Chen, Aiyou 225
Chevalier, Pascal 508
Chevreuil, Antoine 508
Choi, Changkyu 857
Comon, Pierre 105
Costagli, Mauro 930
Cricco, Francesco 954
Cruces, Sergio 57

Dapena, Adriana 358
Davey, Colin 1033
Davies, Mike 152, 669, 1205
Delabrouille, Jacques 1221
de la Rosa, Juan Jose G. 1126
De Lathauwer, Lieven 295, 335
Deville, Yannick 279, 694
Diamantaras, Konstantinos I. 548
Díaz, F. 824
Dien, Joseph 1033
Doron, Eran 390
Douglas, Scott C. 634
Du, Yu 1041
Duann, Jeng-Ren 1103
Durán, Iván 57
Dyrholm, Mads 594

Eidinger, Eran 570
Erdogmus, Deniz 26, 185, 311

Fadaili, El Mostafa 366
Faria, Miguel 734
Farzana, S.M. Shereen 1237
Faúndez-Zanuy, Marcos 865
Fernández, Juan Charneco 233
Févotte, Cédric 398
Frank, Robert M. 1033
Frishkoff, Gwen A. 1033
Fujiki, Jun 750

García, Gonzalo A. 985
Georgiev, Pando 121
Gerace, Ivan 954

Glass, Kevin A.   1033
Godsill, Simon J.   398
Gopi, E.S.   1237
Górriz, Juan Manuel   256, 414, 586, 758, 1126
Gosálbez, Jorge   470
Grassberger, Peter   209
Gribonval, Rémi   201
Gruber, Peter   977, 993

Hansen, Lars Kai   594, 618
Harmeling, Stefan   217, 790
He, Zhenya   1134
Hesse, Christian W.   1025, 1048
Hoegaerts, Luc   335
Hofbauer, Markus   643
Holobar, Ales   1056
Honkela, Antti   766, 790
Hori, Gen   144
Hornillo-Mellado, Susana   33, 256, 586
Hosseini, Shahram   279, 694, 702
Hu, Huibo   970
Hu, Xuebin   849
Huang, Gaoming   1134
Huovinen, Toni   1173

Ibáñez, Jesús   185
Igual, Jorge   18, 470
Ihara, Zenichi   1095
Ilin, Alexander   766
Inki, Mika   914
Inouye, Yujiro   374, 685
Ito, Daisuke   343
Ito, Masanori   129, 374

Jallon, Pierre   508
James, Christopher J.   1025, 1048
Jang, Gil-Jin   857
Jaulin, Luc   81
Joho, Marcel   578
Jutten, Christian   9, 81, 168, 263, 702, 710, 798, 1017

Kamata, Koutaro   849
Kaminuma, Atsunobu   881
Kappenberger, Lukas   1095
Karvanen, Juha   774
Katayama, Yusuke   129
Kawamoto, Mitsuru   374, 685

Kawanabe, Motoaki   136
Keck, Ingo R.   977
Kellermann, Walter   840
Kim, Sang Gyun   445
Kim, Sang Ryong   857
Kobatake, Hidefumi   849
Kohno, Kiyotaka   685
Kokkinakis, Kostas   486
Kolossa, Dorothea   832
Kopriva, Ivica   240
Kraskov, Alexander   209
Krishnaprasad, Perinkulam S.   437
Kudo, Hiroaki   129
Kudo, Mineichi   193
Kuruoğlu, Ercan E.   930

Lagrange, Sebastien   81
Lakshmi, R.   1237
Lang, Elmar W.   287, 726, 977, 993
Lange, Oliver   782
Larue, Anthony   702
Lee, John A.   710
Lee, Soo-Young   562
Lee, Te-Won   1253
Lee, Yongbeom   857
Leiva-Murillo, José M.   271
Lemay, Mathieu   1095
Li, Jing   1041
Liu, Ju   970, 1041
Lloret, Isidro   1126
Lluis, Victor Nieto   824
Loubaton, Philippe   508
Lu, Wei   906
Lukasiak, Jason   889
Lundqvist, Leo   790

Ma, Liang Suo   516
Maekawa, Kazuya   985
Makeig, Scott   1009, 1103
Makino, Shoji   461, 610, 634, 652, 898
Malony, Allen D.   1033
Mansour, Ali   453, 798, 1158
Martín-Clemente, Rubén   33, 350, 586
Martínez, R.   824
Marvasti, Farrokh   798
Matsuda, Yoshitatsu   303
Matsumoto, Tetsuya   129
Mayo, P.   1064
McLaughlin, S.   248
Meinecke, Frank C.   217

Meyer-Bäse, Anke 726, 782
Millet, José 18, 1079, 1111, 1118, 1229
Miralles, Ramón 470
Mitianoudis, Nikolaos 669
Model, Dmitri 319
Monte-Moreno, Enric 1165
Mora, Cibeles 1229
Moreau, Eric 366
Mukai, Ryo 461, 610, 626, 898
Mukai, Toshiharu 129, 374
Mulgrew, B. 248
Müller, Klaus-Robert 89, 136, 217
Murata, Noboru 343, 750
Murillo-Fuentes, Juan José 938, 1142

Nakamura, Atsuyoshi 193
Nam, Seung H. 524
Nandi, Asoke K. 486, 1181, 1189
Nayebi, Kambiz 9
Nie, Kaibao 1041
Nielsen, Morten 201
Nishikawa, Tsuyoki 881
Nuzillard, Danielle 97, 1150

O'Grady, Paul D. 430
Oh, Sang-Hoon 562
Ohata, Masashi 374
Ohnishi, Noboru 129, 374
Oja, Erkki 1
Okada, Kanzo 906
Olsson, Rasmus Kongsgaard 618
Orglmeister, Reinhold 832

Papadimitriou, Theophilos 548
Paraschiv-Ionescu, Anisoara 263
Park, Hyun-Jin 1253
Park, Hyung-Min 562
Patanchon, Guillaume 1221
Pearlmutter, Barak A. 430, 478
Pędzisz, Maciej 1158
Pesquet, Jean-Christophe 922
Pham, Dinh-Tuan 41, 807
Plumbley, Mark D. 49, 540, 1245
Polonsky, Alexey 1001
Porcaro, Camillo 1087
Prasad, Rajkishore 873
Príncipe, José Carlos 26, 185, 311
Punnoose, Sonu 1181
Puntonet, Carlos G. 33, 256, 350, 414, 586, 758, 782, 977, 993, 1126

Rajih, Myriam 105
Ramya, N. 1237
Rao, Yadunandana N. 26, 311
Rieta, José Joaquín 18, 816, 1079, 1111, 1118, 1229
Ristaniemi, Tapani 1173
Rivet, Bertrand 263
Rodellar Biarge, María Victoria 824
Rodenas Escriba, Francisco 1064
Rodet, Xavier 327, 1197
Rodríguez, F. 824
Rodríguez-Álvarez, Manuel 350, 586, 758
Rojas Ruiz, Fernando 350, 414, 758
Rojas Ruiz, Ignacio 350, 758
Ruiz, D.P. 248

Sahmoudi, Mohamed 113
Salazar, Addison 470
Salmerón, Moisés 256, 414
Salustri, Carlo 1087
Samadi, Samareh 9
Sánchez, César 1079, 1111, 1229
Sanchez-Poblador, V. 1165
Sanchis, Juan Manuel 816, 1079, 1118, 1229
Sanei, Saeid 177, 532, 661
Santamaría, Ignacio 185
Särelä, Jaakko 65
Sarperi, Luciano 1189
Saruwatari, Hiroshi 626, 873, 881
Sawada, Hiroshi 461, 610, 626, 634, 652, 898
Schechner, Yoav Y. 422
Sejnowski, Terrence J. 1009, 1103
Serviere, Christine 807
Sevillano, Xavier 1213
Shikano, Kiyohiro 626, 873, 881
Shwartz, Sarit 422
Smaragdis, Paris 494
Smith, Daniel 889
Socoró, Joan Claudi 1213
Solé-Casals, Jordi 865, 1165
Specht, Karsten 977
Spyrou, Loukianos 177
Stadlthanner, Kurt 287, 993
Stetson, Paul F. 1072
Stögbauer, Harald 209
Sun, Jiande 970
Sun, Xiaobing 906

Süsstrunk, Sabine 946
Szu, Harold 240

Tabrikian, Joseph 382
Takatani, Tomoya 626
Takeuchi, Yoshinori 129
Takigawa, Ichigaku 193
Tanaka, Toshihisa 774
Teixeira, Ana R. 287, 993
Theis, Fabian J. 121, 718, 726, 782, 977, 993
Thirion-Moreau, Nadège 366
Todros, Koby 382
Tomé, Ana Maria 287, 993
Tonazzini, Anna 954
Toyama, Jun 193
Tsoi, Ah Chung 516
Tucker, Don M. 1033

Ukai, Satoshi 626
Unger, Hilit 962

Valpola, Harri 65, 790
Vandewalle, Joos 295, 335
Vega-Corona, Antonio 160
Verdú Martín, Gumersindo 1064
Vergara, Luis 470
Verleysen, Michel 710, 1017
Vesin, Jean-Marc 1095
Vielva, Luis 185
Vigneron, Vincent 81, 168, 263

Vilda, Pedro Gómez 824
Vincent, Emmanuel 327, 1197
Vrins, Frédéric 1017

Wang, Wenwu 177, 532, 661
Winter, Stefan 610, 652
Wiskott, Laurenz 742
Wolfe, Patrick J. 398

Xerri, Bernard 602
Xu, Jian-Wu 311

Yamaguchi, Kazunori 303
Yan, Fei 840
Yang, Luxi 1134
Yeredor, Arie 89, 390, 570
Yoo, Chang D. 445
Yuan, Zhijian 1

Zador, Anthony M. 478
Zarzoso, Vicente 18
Zazo, Santiago 73
Zazula, Damjan 1056
Zeevi, Yehoshua Y. 500, 677, 962
Zhang, Jian 906
Zhu, Xu 1181, 1189
Zibulevsky, Michael 319, 406, 422, 500, 554, 677, 1001
Ziehe, Andreas 89

Lecture Notes in Computer Science

For information about Vols. 1–3139

please contact your bookseller or Springer

Vol. 3263: M. Weske, P. Liggesmeyer (Eds.), Object-Oriented and Internet-Based Technologies. XII, 239 pages. 2004.

Vol. 3260: I. Niemegeers, S.H. de Groot (Eds.), Personal Wireless Communications. XIV, 478 pages. 2004.

Vol. 3258: M. Wallace (Ed.), Principles and Practice of Constraint Programming – CP 2004. XVII, 822 pages. 2004.

Vol. 3256: H. Ehrig, G. Engels, F. Parisi-Presicce (Eds.), Graph Transformations. XII, 451 pages. 2004.

Vol. 3255: A. Benczúr, J. Demetrovics, G. Gottlob (Eds.), Advances in Databases and Information Systems. XI, 423 pages. 2004.

Vol. 3254: E. Macii, V. Paliouras, O. Koufopavlou (Eds.), Integrated Circuit and System Design. XVI, 910 pages. 2004.

Vol. 3253: Y. Lakhnech, S. Yovine (Eds.), Formal Techniques in Timed, Real-Time, and Fault-Tolerant Systems. X, 397 pages. 2004.

Vol. 3250: L.-J. (LJ) Zhang, M. Jeckle (Eds.), Web Services. X, 300 pages. 2004.

Vol. 3249: B. Buchberger, J.A. Campbell (Eds.), Artificial Intelligence and Symbolic Computation. X, 285 pages. 2004. (Subseries LNAI).

Vol. 3246: A. Apostolico, M. Melucci (Eds.), String Processing and Information Retrieval. XIV, 316 pages. 2004.

Vol. 3242: X. Yao, E. Burke, J.A. Lozano, J. Smith, J.J. Merelo-Guervós, J.A. Bullinaria, J. Rowe, P. Tiño, A. Kabán, H.-P. Schwefel (Eds.), Parallel Problem Solving from Nature - PPSN VIII. XX, 1185 pages. 2004.

Vol. 3241: D. Kranzlmüller, P. Kacsuk, J.J. Dongarra (Eds.), Recent Advances in Parallel Virtual Machine and Message Passing Interface. XIII, 452 pages. 2004.

Vol. 3240: I. Jonassen, J. Kim (Eds.), Algorithms in Bioinformatics. IX, 476 pages. 2004. (Subseries LNBI).

Vol. 3239: G. Nicosia, V. Cutello, P.J. Bentley, J. Timmis (Eds.), Artificial Immune Systems. XII, 444 pages. 2004.

Vol. 3238: S. Biundo, T. Frühwirth, G. Palm (Eds.), KI 2004: Advances in Artificial Intelligence. XI, 467 pages. 2004. (Subseries LNAI).

Vol. 3232: R. Heery, L. Lyon (Eds.), Research and Advanced Technology for Digital Libraries. XV, 528 pages. 2004.

Vol. 3229: J.J. Alferes, J. Leite (Eds.), Logics in Artificial Intelligence. XIV, 744 pages. 2004. (Subseries LNAI).

Vol. 3225: K. Zhang, Y. Zheng (Eds.), Information Security. XII, 442 pages. 2004.

Vol. 3224: E. Jonsson, A. Valdes, M. Almgren (Eds.), Recent Advances in Intrusion Detection. XII, 315 pages. 2004.

Vol. 3223: K. Slind, A. Bunker, G. Gopalakrishnan (Eds.), Theorem Proving in Higher Order Logics. VIII, 337 pages. 2004.

Vol. 3221: S. Albers, T. Radzik (Eds.), Algorithms – ESA 2004. XVIII, 836 pages. 2004.

Vol. 3220: J.C. Lester, R.M. Vicari, F. Paraguaçu (Eds.), Intelligent Tutoring Systems. XXI, 920 pages. 2004.

Vol. 3219: M. Heisel, P. Liggesmeyer, S. Wittmann (Eds.), Computer Safety, Reliability, and Security. XI, 339 pages. 2004.

Vol. 3217: C. Barillot, D.R. Haynor, P. Hellier (Eds.), Medical Image Computing and Computer-Assisted Intervention – MICCAI 2004. XXXVIII, 1114 pages. 2004.

Vol. 3216: C. Barillot, D.R. Haynor, P. Hellier (Eds.), Medical Image Computing and Computer-Assisted Intervention – MICCAI 2004. XXXVIII, 930 pages. 2004.

Vol. 3212: A. Campilho, M. Kamel (Eds.), Image Analysis and Recognition. XXIX, 862 pages. 2004.

Vol. 3211: A. Campilho, M. Kamel (Eds.), Image Analysis and Recognition. XXIX, 880 pages. 2004.

Vol. 3210: J. Marcinkowski, A. Tarlecki (Eds.), Computer Science Logic. XI, 520 pages. 2004.

Vol. 3208: H.J. Ohlbach, S. Schaffert (Eds.), Principles and Practice of Semantic Web Reasoning. VII, 165 pages. 2004.

Vol. 3207: L.T. Yang, M. Guo, G.R. Gao, N.K. Jha (Eds.), Embedded and Ubiquitous Computing. XX, 1116 pages. 2004.

Vol. 3206: P. Sojka, I. Kopecek, K. Pala (Eds.), Text, Speech and Dialogue. XIII, 667 pages. 2004. (Subseries LNAI).

Vol. 3205: N. Davies, E. Mynatt, I. Siio (Eds.), UbiComp 2004: Ubiquitous Computing. XVI, 452 pages. 2004.

Vol. 3203: J. Becker, M. Platzner, S. Vernalde (Eds.), Field Programmable Logic and Application. XXX, 1198 pages. 2004.

Vol. 3202: J.-F. Boulicaut, F. Esposito, F. Giannotti, D. Pedreschi (Eds.), Knowledge Discovery in Databases: PKDD 2004. XIX, 560 pages. 2004. (Subseries LNAI).

Vol. 3201: J.-F. Boulicaut, F. Esposito, F. Giannotti, D. Pedreschi (Eds.), Machine Learning: ECML 2004. XVIII, 580 pages. 2004. (Subseries LNAI).

Vol. 3199: H. Schepers (Ed.), Software and Compilers for Embedded Systems. X, 259 pages. 2004.

Vol. 3198: G.-J. de Vreede, L.A. Guerrero, G. Marín Raventós (Eds.), Groupware: Design, Implementation and Use. XI, 378 pages. 2004.

Vol. 3195: C.G. Puntonet, A. Prieto (Eds.), Independent Component Analysis and Blind Signal Separation. XXIII, 1266 pages. 2004.

Vol. 3194: R. Camacho, R. King, A. Srinivasan (Eds.), Inductive Logic Programming. XI, 361 pages. 2004. (Subseries LNAI).

Vol. 3193: P. Samarati, P. Ryan, D. Gollmann, R. Molva (Eds.), Computer Security – ESORICS 2004. X, 457 pages. 2004.

Vol. 3192: C. Bussler, D. Fensel (Eds.), Artificial Intelligence: Methodology, Systems, and Applications. XIII, 522 pages. 2004. (Subseries LNAI).

Vol. 3191: M. Klusch, S. Ossowski, V. Kashyap, R. Unland (Eds.), Cooperative Information Agents VIII. XI, 303 pages. 2004. (Subseries LNAI).

Vol. 3190: Y. Luo (Ed.), Cooperative Design, Visualization, and Engineering. IX, 248 pages. 2004.

Vol. 3189: P.-C. Yew, J. Xue (Eds.), Advances in Computer Systems Architecture. XVII, 598 pages. 2004.

Vol. 3187: G. Lindemann, J. Denzinger, I.J. Timm, R. Unland (Eds.), Multiagent System Technologies. XIII, 341 pages. 2004. (Subseries LNAI).

Vol. 3186: Z. Bellahsène, T. Milo, M. Rys, D. Suciu, R. Unland (Eds.), Database and XML Technologies. X, 235 pages. 2004.

Vol. 3185: M. Bernardo, F. Corradini (Eds.), Formal Methods for the Design of Real-Time Systems. VII, 295 pages. 2004.

Vol. 3184: S. Katsikas, J. Lopez, G. Pernul (Eds.), Trust and Privacy in Digital Business. XI, 299 pages. 2004.

Vol. 3183: R. Traunmüller (Ed.), Electronic Government. XIX, 583 pages. 2004.

Vol. 3182: K. Bauknecht, M. Bichler, B. Pröll (Eds.), E-Commerce and Web Technologies. XI, 370 pages. 2004.

Vol. 3181: Y. Kambayashi, M. Mohania, W. Wöß (Eds.), Data Warehousing and Knowledge Discovery. XIV, 412 pages. 2004.

Vol. 3180: F. Galindo, M. Takizawa, R. Traunmüller (Eds.), Database and Expert Systems Applications. XXI, 972 pages. 2004.

Vol. 3179: F.J. Perales, B.A. Draper (Eds.), Articulated Motion and Deformable Objects. XI, 270 pages. 2004.

Vol. 3178: W. Jonker, M. Petkovic (Eds.), Secure Data Management. VIII, 219 pages. 2004.

Vol. 3177: Z.R. Yang, H. Yin, R. Everson (Eds.), Intelligent Data Engineering and Automated Learning – IDEAL 2004. XVIII, 852 pages. 2004.

Vol. 3176: O. Bousquet, U. von Luxburg, G. Rätsch (Eds.), Advanced Lectures on Machine Learning. IX, 241 pages. 2004. (Subseries LNAI).

Vol. 3175: C.E. Rasmussen, H.H. Bülthoff, B. Schölkopf, M.A. Giese (Eds.), Pattern Recognition. XVIII, 581 pages. 2004.

Vol. 3174: F. Yin, J. Wang, C. Guo (Eds.), Advances in Neural Networks - ISNN 2004. XXXV, 1021 pages. 2004.

Vol. 3173: F. Yin, J. Wang, C. Guo (Eds.), Advances in Neural Networks – ISNN 2004. XXXV, 1041 pages. 2004.

Vol. 3172: M. Dorigo, M. Birattari, C. Blum, L. M. Gambardella, F. Mondada, T. Stützle (Eds.), Ant Colony, Optimization and Swarm Intelligence. XII, 434 pages. 2004.

Vol. 3171: A.L.C. Bazzan, S. Labidi (Eds.), Advances in Artificial Intelligence – SBIA 2004. XVII, 548 pages. 2004. (Subseries LNAI).

Vol. 3170: P. Gardner, N. Yoshida (Eds.), CONCUR 2004 - Concurrency Theory. XIII, 529 pages. 2004.

Vol. 3166: M. Rauterberg (Ed.), Entertainment Computing – ICEC 2004. XXIII, 617 pages. 2004.

Vol. 3163: S. Marinai, A. Dengel (Eds.), Document Analysis Systems VI. XI, 564 pages. 2004.

Vol. 3162: R. Downey, M. Fellows, F. Dehne (Eds.), Parameterized and Exact Computation. X, 293 pages. 2004.

Vol. 3160: S. Brewster, M. Dunlop (Eds.), Mobile Human-Computer Interaction – MobileHCI 2004. XVII, 541 pages. 2004.

Vol. 3159: U. Visser, Intelligent Information Integration for the Semantic Web. XIV, 150 pages. 2004. (Subseries LNAI).

Vol. 3158: I. Nikolaidis, M. Barbeau, E. Kranakis (Eds.), Ad-Hoc, Mobile, and Wireless Networks. IX, 344 pages. 2004.

Vol. 3157: C. Zhang, H. W. Guesgen, W.K. Yeap (Eds.), PRICAI 2004: Trends in Artificial Intelligence. XX, 1023 pages. 2004. (Subseries LNAI).

Vol. 3156: M. Joye, J.-J. Quisquater (Eds.), Cryptographic Hardware and Embedded Systems - CHES 2004. XIII, 455 pages. 2004.

Vol. 3155: P. Funk, P.A. González Calero (Eds.), Advances in Case-Based Reasoning. XIII, 822 pages. 2004. (Subseries LNAI).

Vol. 3154: R.L. Nord (Ed.), Software Product Lines. XIV, 334 pages. 2004.

Vol. 3153: J. Fiala, V. Koubek, J. Kratochvíl (Eds.), Mathematical Foundations of Computer Science 2004. XIV, 902 pages. 2004.

Vol. 3152: M. Franklin (Ed.), Advances in Cryptology – CRYPTO 2004. XI, 579 pages. 2004.

Vol. 3150: G.-Z. Yang, T. Jiang (Eds.), Medical Imaging and Augmented Reality. XII, 378 pages. 2004.

Vol. 3149: M. Danelutto, M. Vanneschi, D. Laforenza (Eds.), Euro-Par 2004 Parallel Processing. XXXIV, 1081 pages. 2004.

Vol. 3148: R. Giacobazzi (Ed.), Static Analysis. XI, 393 pages. 2004.

Vol. 3147: H. Ehrig, W. Damm, J. Desel, M. Große-Rhode, W. Reif, E. Schnieder, E. Westkämper (Eds.), Integration of Software Specification Techniques for Applications in Engineering. X, 628 pages. 2004.

Vol. 3146: P. Érdi, A. Esposito, M. Marinaro, S. Scarpetta (Eds.), Computational Neuroscience: Cortical Dynamics. XI, 161 pages. 2004.

Vol. 3144: M. Papatriantafilou, P. Hunel (Eds.), Principles of Distributed Systems. XI, 246 pages. 2004.

Vol. 3143: W. Liu, Y. Shi, Q. Li (Eds.), Advances in Web-Based Learning – ICWL 2004. XIV, 459 pages. 2004.

Vol. 3142: J. Diaz, J. Karhumäki, A. Lepistö, D. Sannella (Eds.), Automata, Languages and Programming. XIX, 1253 pages. 2004.

Vol. 3140: N. Koch, P. Fraternali, M. Wirsing (Eds.), Web Engineering. XXI, 623 pages. 2004.